Understanding Light Microscopy

Current and future titles in the Royal Microscopical Society – John Wiley Series

Published

Principles and Practice of Variable Pressure/Environmental Scanning Electron Microscopy (VP-ESEM)
Debbie Stokes

Aberration-Corrected Analytical Electron Microscopy
Edited by Rik Brydson

Diagnostic Electron Microscopy – A Practical Guide to Interpretation and Technique
Edited by John W. Stirling, Alan Curry & Brian Eyden

Low Voltage Electron Microscopy – Principles and Applications
Edited by David C. Bell & Natasha Erdman

Standard and Super-Resolution Bioimaging Data Analysis: A Primer
Edited by Ann Wheeler and Ricardo Henriques

Electron Beam-Specimen Interactions and Applications in Microscopy
Budhika Mendis

Biological Field Emission Scanning Electron Microscopy 2V Set
Edited by Roland Fleck and Bruno Humbel

Understanding Light Microscopy
Jeremy Sanderson

Forthcoming

Correlative Microscopy in the Biomedical Sciences
Edited by Paul Verkade and Lucy Collinson

The Preparation of Geomaterials for Microscopical Study: A Laboratory Manual
Owen Green and Jonathan Wells

Electron Energy Loss Spectroscopy
Edited by Rik Brydson and Ian MacLaren

Understanding Light Microscopy

Jeremy Sanderson
Mammalian Genetics Unit, MRC Harwell Institute, Oxford, United Kingdom

RMS

WILEY

This edition first published 2019

Registered Offices
John Wiley & Sons, Inc., 111 River Street, Hoboken, NJ 07030, USA
John Wiley & Sons Ltd, The Atrium, Southern Gate, Chichester, West Sussex, PO19 8SQ, UK

Editorial Office
The Atrium, Southern Gate, Chichester, West Sussex, PO19 8SQ, UK

For details of our global editorial offices, customer services, and more information about Wiley products visit us at www.wiley.com.

Wiley also publishes its books in a variety of electronic formats and by print-on-demand. Some content that appears in standard print versions of this book may not be available in other formats.

Library of Congress Cataloging-in-Publication Data

Names: Sanderson, J. B. (Jeremy B.), author.
Title: Understanding light microscopy / by Jeremy Sanderson.
Description: First edition. | Hoboken, NJ : John Wiley & Sons, 2018. | Series: Royal Microscopical Society/John Wiley Series | Includes bibliographical references and index. |
Identifiers: LCCN 2018004188 (print) | LCCN 2018009681 (ebook) | ISBN 9781118696743 (pdf) | ISBN 9781118696811 (epub) | ISBN 9780470973752 (cloth)
Subjects: LCSH: Microscopy.
Classification: LCC QH205.2 (ebook) | LCC QH205.2 .S32 2018 (print) | DDC 570.28/2–dc23
LC record available at https://lccn.loc.gov/2018004188

Cover Design: Wiley
Cover Images: © Cochlear hair cells courtesy of Doctor Mike Bowl. All other cover photos courtesy of Jeremy Sanderson.

Set in 9/11pt OptimaLTStd by Aptara Inc., New Delhi, India

Dedicated to the memory of:

Savile Bradbury
6th February 1931 – 29th November 2001

Teacher, mentor, friend

Contents

About the Author

Jeremy Sanderson is the Bio-imaging Facility Manager at MRC Harwell Institute, and a Chartered Scientist. He has been working in core imaging facilities for 16 years, and as a light and electron microscopist for over three decades. He joined the Royal Microscopical Society in 1987, has taught regularly on its instruction courses; has also served on the RMS Council, and assists with RMS teaching qualifications and educational outreach. Starting as an NHS histologist, Mr Sanderson then worked at Oxford University before going on to the Max Planck Institute of Molecular Cell Biology and Genetics in Dresden. From there he returned to Sheffield University to set up and manage a light microscopy imaging facility before joining the Medical Research Council in 2009.

Acknowledgements

My thanks go to my family and many friends and colleagues who have supported me and helped me to write this book. Please forgive me if space dictates brevity, and I do not name you all. I have derived much support from my loving family. To my wife, Yvonne, who has been a pillar of encouragement throughout: I could not have any better by my side; she has given to me more than any man could, or should, ask for. To my family: Alastair, Heather, Richard, Vicky, David, Val, Fiona, Derek and Hilary, who love me (but profess not to understand what I do). I am grateful to our three children for so considerately becoming economically self-sufficient and leaving home during the gestation of this book, thus enabling me to devote my time to it. I particularly wish to thank my dear brother David for his lifelong support and inspiration. He is a consummate professional in his chosen field and without parallel.

I could not have got this far without the support of Dr Peter Evennett and Dr Christopher Hammond. They have given to me richly and have taught me all I know. To them I owe a profound debt. Few people can successfully cover the fields of biological and materials microscopy, and I am no exception. I am therefore particularly grateful to Chris Hammond for permission to use his draft material, for his support and encouragement throughout and for his reading of my work.

I owed a considerable debt to Savile Bradbury who, *inter alia*, helped me as editor of my first book. Our joint friend Dr Brian Bracegirdle very kindly made time to read the entire first draft; I am indebted to his memory. I am grateful for the help that I have received from Drs Graham Brown, David Crook, Peter O'Toole, Kurt Anderson, Alex Sossick, Nicola Lawrence, Jason Swedlow, John Chapman, Steve Metcalfe, Paul Sheppard, Paul Southey, Aubrey Lambert, Markus Neumann, Simon Ameer-Beg, Emmanuel Reynaud, Andy Hill, Aman Sood, Chris Power, Rosy Manser, Ingela Parmryd, Andy Parker, Susan Cox, Liz Darley, Jeremy Graham, Martin Thomas, Rüdi Bader, Carsten Hoyer, Sue Wilson, Ben Ewins, Dan Phillips, Nic Sergent, Lorraine Berry, Ami Jackson, Alastair Cropp, Rene Hessling, Oliver Clarke, Michael Zölffel, Ashley Crane, Serena Kerrigan, Brad Amos, Gail McConnell, Spike Walker and Owen Green, who have very kindly read draft chapters for me, helped with advice, or loaned objectives and equipment. I also wish to record my gratitude to the following friends at work who have supported me: Drs Andy Greenfield, Charlotte Dean, Paul Potter, Dominic Norris, Dani Sagar, Samantha Laber, Jenny Keynton and Becca Walker, who have given generously of their time and advice. I am also grateful to Kevin Glover, formerly of the MRC MGU Imaging Service, for his very considerable expertise with photographing equipment for the illustrations. I am particularly grateful for the support of the MRC Unit Director, Professor Steve Brown, and the Chief Operating Officer, Dr Nanda Rodrigues, in the genesis of this book. Two further wonderful friends I must surely thank for their wholehearted support are Chris Delger, my lifelong school friend, and also Philip Garner an inestimable wingman, scion of England and of holy church, who has kept me on the straight and narrow.

The world of the professional and amateur microscopist is a select one; more importantly it is filled with friendly and helpful folk whom I count fortunate to have as friends and colleagues. The microscope is a wonderful instrument, which I hope you will realise from the pages of this book. To the very competent and highly professional staff in the Royal Microscopical Society office, who have helped me considerably over the years, as well as friends and colleagues in the wider microscopical *Gemeinschaft*: thank you all. A good friend, thank you Clive, has suggested – tongue in cheek – that I ought also to explain in this book what a microscope *cannot* do. Whilst the light microscope is an extremely sophisticated and versatile instrument capable of many things, it cannot unfortunately solve political or relationship issues, nor the deep spiritual or philosophical questions of humankind. That said, it will reveal to you a wondrous world of superb order of the infinitesimally small.

Many thanks too, and not least, to my publisher at Wiley: Jenny Cossham, who has faith and perseverance in abundance like no other. She has supported this book nobly from start to finish. It has also been a considerable pleasure to work with Sarah Keegan, my project editor, Aldo Rosas, my copy editor, Katherine Wong, my production editor, Janine Maer, Emma Strickland and Beth Dufour to bring this book to publication. I am very fortunate indeed to have had the precise and extremely thorough support of Kat throughout the proofreading and correction stage. Indeed, every author needs a Kat: my fulsome and humble thanks. Working with these very talented people, I have the highest regard for their professionalism, precision and thoroughness. I am truly grateful for all they have done for me.

Of course, any errors of commission or omission are mine alone, not the responsibility of the above-named. I welcome comments and criticism for improvement of this book, and possibly encouragement (which we all thrive on) should you wish to give it.

Look-Up Guide to Feature Boxes by Theme

Autobiographical Boxes

Information Boxes

Exercise Boxes

Glossary and Definitions

Aberration
An optical defect inherent in the design of a lens causing an optical system not to form a perfect image.

Absorption
The transformation of radiant energy (e.g. light) to a different form of energy (typically heat) by interaction with matter. This gives rise to colour when absorption is wavelength dependent. At the atomic level, quanta of energy are transferred from the radiant energy field (e.g. light beam) to electrons within atoms and molecules, causing attenuation of the radiant energy.

Acceptance angle
The angle between the extreme rays of light collected by an objective lens (see angular aperture).

Accommodation
The act of adjusting the refractive power of the eye to allow sharp imaging of an object located at a particular distance.

Achromatic
Free of colour. Absence of chromatic aberration.

Achromatic lens
A lens, or microscope objective, comprising two or more elements (usually of crown and flint glass) corrected for chromatic aberration, i.e. to have the same focal length with respect to two or more selected colours (e.g. blue and red). Such a lens is also corrected for spherical aberration and other aperture-dependent defects with respect to one (e.g. green) or more colours (see also apochromatic lens and semi-apochromatic lens).

Acousto-optical tunable filter (AOTF)
Used in microscopy as a tunable bandpass filter. An AOTF comprises a tellurium dioxide or quartz anisotropic crystal to which a piezoelectric transducer is bonded. Radio signals from 40–70 MHz produce a controllable acoustic wave through the crystal in the visible spectrum causing it to act as a diffraction grating with a bandpass characteristic.

Aerial image
A real image existing in a plane in space, not normally visible to the eye or a camera or other detector.

Airy disc and Airy pattern
Diffraction of a point source of light at a circular lens aperture forms a central bright disc surrounded by a sequence of concentric alternating dark and light rings, the entirety of which is called an Airy pattern. The diameter of the pattern depends upon the wavelength of the illuminating light and the diameter of the circular aperture. The central part of this pattern is called the Airy disc. It is the smallest point to which a beam of light can be focused. A point source will always form an Airy pattern.

Amplitude
The strength of a wave field at a particular location and instant, referring here to the electric field of light.

Angular aperture
Half the maximum plane angle subtended by a lens at the centre of either an object field or image field by two opposite marginal rays when the lens is used in its correct working position.

NB. The *RMS Dictionary of Light Microscopy* notes: sometimes the full plane angle is taken as the angular aperture, but this causes certain inconveniences in geometrical optical or technical lighting calculations. The acceptance angle (see above) defines the full plane angle subtended by a lens.

Aperture
A fixed or variable opening through which light may pass. Aperture may also, in context, refer as a shortened name for numerical aperture. See also Note 2 in the Notes section.

Aperture diaphragm
A diaphragm in the plane of an aperture of an optical system which delimits its pupil, or cross section of all rays bundles of light. This term usually refers specifically to the iris diaphragm situated in the illumination set of conjugate planes of a microscope and is usually located beneath the microscope condenser in its front focal plane, where it controls the angle of convergence of the solid cone of light emerging from the condenser.

Aplanatic
Corrected for spherical aberration and coma by complying with the sine condition (see sine condition).

Apochromatic lens
A lens, or microscope objective, in which axial chromatic aberration is minimised for at least three wavelengths, the sine condition fulfilled for two wavelengths and spherical aberration and other aperture-dependent defects minimised with respect to one (e.g. green) or more colours. The most highly corrected of the three classes of objective (see also achromatic lens and semi-apochromatic lens).

Back focal plane
The second, or rear, focal plane of a lens. It usually refers specifically to the objective where all rays diffracted by interaction of light with the object pass. The back focal plane of the objective is one of the illumination, or aperture, set of conjugate focal planes, which also include the lamp filament; the aperture diaphragm in the front focal plane of the condenser and the exit pupil of the eyepiece.

Bandwidth
The difference between the upper and lower limits of a frequency band. The amount of data transmitted at a fixed rate, usually expressed in hertz. May also be used to cite the part of the spectrum passed by a filter, in which case the units are nanometres.

Baud rate
Number of bits per second at which a digital signal is transmitted between computers.

Beam
A beam is a group of parallel, or approximately unidirectional, rays (see also ray).

Bertrand lens
A supplementary lens system that may be introduced into the microscope body-tube above the objective to transfer an image of the back focal plane of the objective into the primary image plane of the microscope. Together with the eyepiece the Bertrand lens forms a viewing telescope for inspecting the back focal plane of the objective. Sometimes called an Amici-Bertrand lens. Used for conoscopic observation in polarised light microscopy and adjustment of the illumination system, especially in conjunction with phase-contrast microscopy. Named after Émile Bertrand, the 19th century French mineralogist.

Bioluminescence
The production and emission of 'cold' light by a living organism, as a result of the enzymatic catalysis of the oxidation of luciferin. The enzyme is either luciferase or a photoprotein (e.g. GFP).

Birefringence
The property of possessing different refractive indexes with respect to light polarised in different azimuths set orthogonally to one another. For optically uniaxial media, the difference between the extraordinary principal refractive index and the ordinary one, viz: $\Delta n = n_e - n_o$. The quantitative expression of the maximum difference in refractive index due to double refraction.

Bit
A contraction of the words 'binary digit'. A bit is the smallest unit of data: it is either on or off, black or white, a 'one' or a 'zero'. Eight bits are arranged into one byte of information. A bitmap is a spatially mapped array of bits comprising a raster image (see also raster image).

Bit depth
The number of bits used to represent intensity values of an analogue stream of photons converted to a digitised value at each sampling point as a discrete pixel, or 'picture element' which forms a raster image.

Bright-field
Microscopy or illumination system in which the direct (i.e. zero order) light passes through the objective aperture and illuminates the background against which the image is seen. It is the most usual form of transmitted-light microscopy.

Brightness
The visual perception of a source which appears to be radiating or reflecting light, elicited by the luminance of a visual target. Brightness is not necessarily proportional to luminance. The electric vector of light is the parameter which causes the sensation of brightness. Brightness should not be used to describe illuminance or radiance (see also light intensity and Chapter 1, endnote 1).

Cardinal distances
In geometrical optics the distances between the cardinal points of a lens. These are: the focal lengths *f* on the object and *f′* on the image side; the object distance, u, and image distance, v; the intersection distances (s and s′) between the vertex and the object/image points respectively; the hiatus between the principal points. The vertex is the intersection of the surface of the lens with the optical axis. The hiatus is the distance along the optical axis between the principal planes h and h′ of a lens.

Cardinal points
The reference points on the axis of a lens. These are: the focal points, F and F′; the principal points, H and H′; the nodal points, N and N′; the object point, O, and the image point, I or O′.

Chromatic aberration
An inherent optical defect of a lens, due to dispersion by the material from which it is made, which causes different colours to be brought to a focus at different distances from the lens. It degrades image quality and may be seen as colour fringes surrounding edges of features in the image. This aberration is corrected, or minimised, by combining lens elements from glass, or other materials, of different dispersive power.

Coherence
The correspondence, or equivalence, of different wave trains of light of relative phase emanating from a common source, which permit the waves to interfere. Coherent radiation can be produced only by a very small (point) source, by a very small part of a larger source – effectively a point source – or by a laser (see also incoherence).

Collimation
The process of aligning two or more optical axes with respect to one another, or to a reference axis.
The act of making rays originating from one point parallel to one another. This is usually done with a converging lens or a concave mirror.

Compensator
A retardation plate of variable optical path difference, itself used to measure optical path-length differences within a birefringent (i.e. doubly-refracting) specimen.

Condenser
A single converging (positive) lens, or a group of such lenses, used in the illuminating path to collect light and concentrate it by convergence into the specimen plane and evenly illuminate the object being examined.

Condenser diaphragm
The iris diaphragm usually located beneath the microscope condenser in its front focal plane, situated in the illumination set of conjugate planes of a microscope, where it controls the angle of convergence of the solid cone of light emerging from the condenser. Also sometimes referred to as the illuminating aperture iris (IAI) or illuminating aperture diaphragm (IAD).

Conjugate plane(s)
Planes in an optical system that are equivalent such that an object placed in one plane will be imaged sharply at each subsequent plane in that series. Two separate sets of conjugate planes – the Aperture (illumination) set and the Field (imaging) set exist in the microscope. These planes are perpendicular to the optical axis at the conjugate points of an optical system; the conjugate points being those points in both object space and image space which are imaged upon one another.

Conoscopic image
The interference pattern and diffraction image seen at the back aperture of the objective in the rear focal plane.

Contrast
The difference in tonal intensity, brightness or colour from point to point within the object or image, and variations in the intensity ratio between the object and its background as transferred into the image.

Contrast transfer function (CTF)
As regards the modulation transfer function (MTF) when the brightness of the periodic object varies as square wave, the MTF becomes the CTF (see also MTF). A graphical measurement of the ability of an imaging system to transfer contrast from the object to the image.

Convolution
The mathematical operation on two functions to produce a third resultant function. In microscopy the resulting image of a complex object is the linear shift-invariant convolution of the true object and the point spread function through a dioptric lens (see also deconvolution and point spread function, PSF).

Correction collar
A graduated scale fitted to the barrel of a high numerical aperture dry objective, or sometimes to a long working distance or immersion objective, that acts to move internal lens elements to correct or minimise spherical aberration introduced by the thickness of the coverslip and/or refractive index mismatch between the specimen preparation and immersion medium.

Cut-off frequency
Diffraction at the image plane causes all imaging systems to act as low pass filters with a finite ability to resolve detail. This is the cut-off frequency, k_{max}, which is equal to $2NA/\lambda$ shown on an OTF plot as the Abbe limit: the smallest spatial frequency in the object that can be transferred into, and discriminated in, the image. For fluorescence microscopy, λ_{em} is used (see also optical transfer function).

Dark-field
Microscopy or illumination system in which the direct light is prevented from passing through the objective aperture. The image is formed from light scattered and diffracted by features in the object, the detail appearing bright against a dark background (see also bright-field).

Deconvolution
Inverse of the convolution operation. In fluorescence microscopy, using a calculated or experimentally-derived point spread function to computationally recover an image degraded by the process of convolution.

Depth of field
The depth of sharpness in object space; the axial depth of the space on both sides of the object plane within which the object can be moved without detectable loss of sharpness in the image and within which features of the object appear acceptably sharp in the image while the position of the image plane is maintained.

Depth of focus
The depth of sharpness in image space; the axial depth of the space on both sides of the image plane within which the image appears acceptably sharp while the positions of the object plane and the objective are maintained.

Dichromatic mirror
Also known as a dichroic mirror, the essential component of an epi-fluorescence microscope, an interference filter which selectively reflects shorter-wavelength excitation radiation whilst simultaneously transmitting the longer-wavelength emission fluorescent signal.

Diffraction
The spreading of light after passing the edge of an object, or after passing through a small aperture, particularly features that act as such within the object from which an image is formed.

Dioptre
Unit of refractive power expressed as the reciprocal of the focal length of a lens in metres. See also Note 4 in the Notes section.

Direct light
Light that enters the objective which undergoes no change in propagation direction after passing through the object field or being specularly reflected off the surface of the object. It gives rise to the zero-order of the diffraction image formed in the back focal plane of the objective.

Dispersion
The change in phase velocity of a wave train as a function of wavelength when passing from one medium to another, causing the separation of white light or several colours into individual component spectral colours. The phenomenon whereby the refractive index of optical glass varies with wavelength.

Dynamic range
The number of values in a signal that may be individually distinguished, expressed as a ratio or logarithmic value (e.g. decibels) of the difference between the smallest and largest signal values.

Emission
Release of radiant energy. Conversion of the energy of an excited atom or molecule into a photon.

Empty magnification
High magnification that increases the size of an image without any concomitant increase in resolution, due to a limitation upon the resolving power of the optical system.

Entrance pupil
The apparent minimum cross section of all ray bundles on the object side of a lens.

Excitation
The input (sometimes by absorption) of energy into matter leading to the emission of radiation.

Exit pupil
Point at which bundles of refracted rays emerging from a lens on the image side cross over (i.e. at its back focal plane). In the case of the microscope eyepiece, the point where the observer's eye should be located.

Extinction
The condition whereby an optically anisotropic specimen appears dark when viewed between crossed polars. The extinction factor is a measure of the perfection of extinction with crossed polars, expressed as a ratio of intensities: $X_{\parallel}/X_{\perp}$.

Eye relief
The distance between the eye lens of the microscope eyepiece to the eye point, or exit pupil of the microscope.

Field diaphragm
A diaphragm limiting the field of view, usually fitted just after the lamp collector lens or at the primary image plane in the eyepiece. The term normally refers to the illuminated field diaphragm (IFD) situated in the base of a modern microscope, which is a member of the field, or imaging, set of conjugate planes, which also include the specimen plane; the primary image plane and the retina or camera detector.

Field of view
The visible area seen through the microscope when the specimen is in focus. The greater the magnification, the smaller the field of view, and vice versa.

Fluorescence
The phenomenon of selective absorption of quanta of short-wavelength radiation and re-emission of signal at a longer wavelength. A form of photoluminescence.

Fluorite objective
A semi-apochromatic lens objective intermediate in its correction between an achromat and apochromat. So-called because fluorite, having low dispersion, was once used in the construction of these and apochromatic objectives to minimise and thus correct for chromatic aberration. Fluorite objectives are often colloquially referred to by brand names, such as 'Neofluar' or 'Fluotar' (see also semi-apochromatic lens).

Focal length
The distance, f or f', measured along the optical axis from the principal plane of the lens, objective, or group of lenses, to the focal point.

Focal plane
The plane at right angles to the optical axis at the focal point.

Focal point
The point where light rays originating from a distant source set at infinity come to a focus and form an image after passing through a lens. If the source is set at the focal point, these rays emerge parallel to each other after passing through the lens.

Gain
The numerical factor by which an amplifier increases signal strength. In a light detector the number of electrons counted per detected photon. In a CCD or sCMOS camera, this factor is unity. In a photomultiplier tube (PMT) the gain ranges from 1 to $\approx 10^6$.

Gamma
The function relating digital intensity to display brightness. A nonlinear operation used to encode and decode luminance values since these are not generally a linear function of the input signal.

Glare
Non image-forming stray light, which diminishes contrast. Such scattered stray light generally arises from the improper adjustment or use of field or aperture diaphragms (see also stray light).

Graticule
A pattern such as a scale or a grid, together with its support, which is placed in an object or image plane within the microscope or optical system. It is used for measurement, alignment, location reference or stereological analysis. Most commonly situated at the primary image plane in an eyepiece. Also sometimes referred to as a reticle.

Illuminated field diaphragm (IFD)
An adjustable iris diaphragm limiting the field of view situated in the base of a modern microscope, which is a member of the field, or imaging, set of conjugate planes, which also include the specimen plane; the primary image plane and the retina or camera detector.

Illuminating aperture diaphragm (IAD)
An adjustable iris diaphragm situated in the illumination set of conjugate planes of a microscope located in the front focal plane of the condenser, where it controls the angle of convergence of the solid cone of light emerging from the condenser. Sometimes called the IAI, or illuminating aperture iris.

Image
A representation, necessarily less than perfect, of an object made by a lens or mirror, which is located in another plane, formed by means of light (photons) or other form of radiation (e.g. electrons).

Incoherence
An adjective expressing the property of light waves not satisfying the condition of coherence and thus being incapable of interference (see also coherence). The lack of a fixed phase relationship between two, or more, waves.

Interference
Alteration of the total intensity of light and the amplitude sum produced by the interaction of two coherent light waves, either constructively or destructively. The phenomenon is used to convert optical path-length differences in the object into intensity variations in the image, thus providing contrast.

Köhler illumination
A system devised by August Köhler for producing even illumination of the specimen using small uneven light sources, often possessing a definite filament structure, which would otherwise interfere with observation and/or analysis of the image.

Light Intensity
Luminous intensity is the perceived power per unit solid angle, measured in candelas (cd = lumens/steradian), whereas is luminous flux is the total perceived power emitted in *all* directions; it is measured in lumens and is proportional to the square of the amplitude of the light wave. Illuminance is the luminous flux incident upon a surface and luminous excitance the luminous flux emitted from a surface, both being measured in lux (lumens/m^2). Luminance, measured in cd/m^2, is luminous flux per unit solid angle per unit projected source area. Light output is expressed in lumens, whereas light intensity is measured in lux (see also brightness and Chapter 1, endnote 1).

Magnifying power
The ability of an optical system to produce a magnified image, usually expressed as the lateral or angular magnification of the image (see also Note 1 in the Notes section). When applied to the objective, the term refers to the lateral magnification of the primary real image by the objective alone when corrected for a finite image distance, or by an infinity-corrected objective and its associated tube-lens together. When quoted for an infinity-corrected objective without the tube-lens, the angular magnification is given (e.g. 40x).

Mechanical tubelength
The distance between the shoulder of the objective where it screws into the nosepiece and the top of the microscope body-tube into which the eyepiece fits: the length of the body-tube in its simplest form without any intermediate lenses. This distance is commonly 160 mm for finite tubelength objectives, whereas for infinity-corrected objectives the tubelength is hypothetically considered to be infinite.

Modulation transfer function (MTF)
Mathematical function describing the ability of an optical or electronic device to transfer a signal as a function of the spatial or temporal frequency of that signal (see also CTF, OTF).

Monochromator
Any device used for isolating portions of the spectral output of a complex radiant energy (e.g. light) source.

Noise
Random fluctuations in the signal intensity recorded in an image that are not due to the object being imaged. Unwanted interference in a signal or display. Particularly prevalent in low-light conditions; shot noise is random noise due to quantum uncertainties when small numbers of photons are recorded (see also signal-to-noise ratio).

Numerical aperture (NA)
A measure n.sin α of the light-gathering power of the objective or condenser, devised by Ernst Abbe, where n is the refractive index of the medium between the front lens of the objective and the specimen preparation, and α is the half-angle (see acceptance angle) of the cone of rays entering the objective. This NA value is usually found engraved on objectives, and sometimes the condenser, and is a factor in the mathematical formula expressing the resolving power of the microscope. The greater the numerical aperture value, the greater the resolving power and also the light flux gathered by the objective.

Objective
The principal part of the microscope, consisting of a lens, or a series of lens elements, its mount and any associated parts. The objective (together with a separate tube-lens in an infinity-corrected microscope) forms the primary image. The NA of the objective is a fundamental parameter (see also numerical aperture).

Optical axis
A straight line joining the centres of curvature of lens surfaces.

Optical index
A term devised by Edward Nelson, being the ratio of the numerical aperture to the magnifying power of an objective. The OI describes mathematically how closely information is packed into the primary image. It can be used to ensure that the analogue image is sampled adequately by a digital camera. As the optical index increases, the sampling requirements become more stringent.

Optical indicatrix
The visual representation of refractive index by an imaginary spatial figure in the shape of an ellipsoid of revolution, or triaxial ellipsoid, representing the refractive indexes of a doubly refracting material as distances between the centre of the ellipsoid and points on its surface and the vibration of plane-polarised light related to the refractive indexes by the directions of the said distances. Not to be confused with a ray velocity surface representation.

Optical transfer function (OTF)
The optical transfer function plot illustrates graphically how any imaging system transfers resolvable object information into the image. It is a combination of the modular transfer function (MTF) and the phase transfer function (PTF). The MTF describes how the imaging system handles both resolution and contrast; the PTF describes the inherent aberrations. Conventionally, the OTF is normalised to unity for zero spatial frequency. Strictly, the Fourier transform of the PSF is equivalent to the OTF, providing the phase component of the optical transfer function is negligible (see also CTF, cut-off frequency, MTF).

Optical tubelength
Distance between the back focal plane of the objective and the primary image plane (the front focal plane of the eyepiece). Relevant only to microscopes fitted with objectives corrected for a *finite* primary image distance.

Overcorrection
Error in the correction of spherical aberration leading to an unsharp image lacking in contrast. Caused by using a thicker coverslip or an optical tubelength longer than intended by the designer when computing the objective used. May be corrected using a correction collar, if fitted (see also undercorrection).

Parcentral
The quality applied to objectives grouped on the nosepiece of the microscope such that a particular specimen detail in the centre of the field of view with one objective remains centrally placed when another objective of different magnifying power is rotated into position on the optical axis.

Parfocal
The quality applied to objectives such that practically no change of focus has to be made when an objective of one magnifying power and depth of field is substituted for another.

Photon
Massless quantum particle of radiant energy of the electromagnetic field, whose value, E, is equal to hν.

Pixel
A portmanteau of the words 'picture element'. A small two dimensional geometrically-square non-overlapping Cartesian subunit, the smallest addressable sampling point forming a raster image. A voxel is the three dimensional equivalent.

Point spread function (PSF)
The impulse response of a focused imaging system to a point source or point object. The PSF is a spatial description of the optical transfer function of the imaging system: i.e. the signature of the microscope. The intensity distribution in the image from a convolved sub-diffraction point source. In incoherent imaging systems (e.g. fluorescence microscopes) the image formation process is linear.

Primary image
The enlarged real image of the specimen formed by the objective lens. The primary image plane (PIP) being the one of the imaging (field) set of conjugate planes, and where the primary image is formed.

Principal focus
The focal point for a beam of light rays parallel to the optical axis when refracted by a lens or reflected by a mirror.

Principal plane
For a thin lens, the plane perpendicular to the optical axis passing through the centre of the lens and from which the focal lengths are measured. A thick lens has two principal planes separated by the hiatus. The principal point(s) occur where the principal plane(s) intersect the optical axis.

Pupil
Cross section of all rays bundles of light.

Ramsden disc
Older term used for the exit pupil of the eyepiece.

Raster
The pattern of lines or points traced, or systematically applied to an area by rectilinear scanning, most commonly in microscopy by a laser beam regularly and repeatedly sampling the specimen.

Raster image
An image composed of pixels.

Ray
A geometrical abstraction; a straight line describing the direction of propagation of radiation, which in an isotropic medium is normal to the propagated wavefront. Ray bundles are all those rays bounded by a diaphragm. A beam is a group of parallel, or approximately unidirectional, rays.

Real image
An image which can be formed on a screen.

Rear focal plane
The second, or back, focal plane of a lens. It usually refers specifically to the objective where all rays diffracted by interaction of light with the object pass. The rear focal plane of the objective is one of the illumination, or aperture, set of conjugate focal planes, which also include the lamp filament; the aperture diaphragm in the front focal plane of the condenser and the exit pupil of the eyepiece.

Refraction
The change in velocity of light (with or without a change in direction) upon crossing the boundary, or interface, between media of different refractive indexes. A change in the direction of propagation of radiation determined by the change in velocity of propagation upon passing through an optically non-homogeneous medium or upon passing from one medium to another in a direction other than normal to the interface.

Refractive index
The ratio (n) of the speed of light in a vacuum (c) to that given in a given medium $n = c/v$.

Resolution
The smallest distance between two points in an object that can just be distinguished as separate points in the image.

Resolving power
The ability of a microscope, or of an individual objective, to resolve or make visible fine detail present within the object in the image.

Reticle
A pattern such as a scale or a grid, together with its support, that is placed in an object or image plane within the microscope or optical system (see also graticule).

Sampling
The process of transforming a continuous analogue signal into a finite set of discrete digital values; each sampling point commonly being represented by an individual pixel or voxel. An 8-bit pixel, the most common type, allows for 256 (2^8) intensity tones of black, grey and white values.

Semi-apochromatic lens
A lens, or microscope objective, intermediate in its correction and complexity of construction between an achromat and apochromat (see also fluorite objective, achromatic lens and apochromatic lens).

Signal-to-background ratio (SBR)
A measure of contrast in the image. The ratio of image signal versus the average background level. Optical sectioning removes the background contribution to the image, giving higher SBR values.

Signal-to-noise ratio (SNR)
The ratio of intensity of the object represented in the image as signal versus the average noise level, which consists of unwanted, random, variation in intensity (i.e. noise) per sampling point not directly generated by the object.

Sine condition
A mathematical formulation derived by Ernst Abbe (and sometimes called the Abbe sine condition) of a condition that must be fulfilled by an aplanatic lens or other optical system to produce sharp images together of objects that lie off-axis as well as those that lie on-axis. Such an aplanatic system is free from both spherical aberration and coma.

Spectrum
The orderly arrangement of the dispersion of radiant energy derived from atomic or molecular vibration arranged by wavelength.

Spherical aberration
An inherent optical failure of a lens to form a sharp image due to the curvature, or 'figure' of the lens surfaces, which refract the rays of light such that peripheral and axial rays are brought to a focus at different distances from the lens. Commonly the peripheral rays are brought to a focus closer to the lens than the axial rays. Like chromatic aberration, spherical aberration degrades image quality; these two aberrations being the most serious the lens designer has to contend with.

Stop
A term used loosely for a diaphragm of fixed size (see also Note 2 in the Notes section).

Stray light
Light that does not contribute to image formation but arises from the scattering or reflection of light by the object, in lenses, or by obstacles in the light path and which reduces contrast in the image (see also glare).

Tubelength
See mechanical tubelength and optical tubelength.

Tube-lens
An intermediate lens designed as the last element of an infinity-corrected objective, located in either the body of the microscope tube or the viewing tube. The tube-lens is remote from the objective to allow any extra equipment conveniently to be placed in the parallel light path whilst maintaining the primary image at its proper location.

Undercorrection
Error in the correction of spherical aberration leading to an unsharp image lacking in contrast. Caused by using a thinner coverslip or mechanical tubelength shorter than intended by the designer when computing the objective used. May be corrected using a correction collar, if fitted (see also overcorrection).

Virtual image
An image that cannot be formed on a surface but which may be converted to a real image by the eye, or other converging lens system.

Visual acuity
Refers to the sharpness and clarity of vision and our ability to resolve, or distinguish, fine detail with the naked eye.

Voxel
A 3D sampling element (see also pixel).

Wave train
A sequence of light waves of equal phase, amplitude, wavelength and polarisation state travelling in the same direction with equal velocity.

Widefield
Microscopy whereby the entire field of view is imaged at once, as opposed to confocal fluorescence microscopy where the image is acquired by raster scanning of point illumination. The term is almost always used when referring to fluorescence microscopy.

Working distance
The free distance between the objective front lens and the upper surface of the coverslip, or the specimen if uncovered, when the latter is in focus. Generally, the greater the magnifying power of the objective, the smaller the working distance.

These definitions have been compiled from several sources, but many have been taken verbatim from the RMS Dictionary of Light Microscopy (1989) RMS Handbook No. 15, compiled by the Nomenclature Committee of the RMS, namely: Drs Bradbury, Evennett, Haselmann & Piller.

Notes

1. Magnification

The magnifying powers of objectives are cited in this book as 40x throughout for consistency. This is the correct notation for infinity-corrected objectives and also accessory optical components, such as the magnifying glass, magnification changers and eyepieces. In these cases, magnification is an angular measure. With a magnifying glass or eyepiece, our eyes act as the 'tube-lens' to form a real image from a lens which would otherwise form a virtual image. Because an infinity-corrected lens does not form a real image, it must be compared with something, and the focal length of the tube-lens (e.g. 180 mm) is chosen as the reference distance with which to compare its focal length. 40x thus describes the magnification of a 4.5 mm focal length objective (i.e. 180/4.5). However, for the older finite tubelength microscopes, with agreed standard tubelengths of 160 mm (sometimes 170 mm) and 250 mm, for biological and metallurgical stands respectively, magnification is a ratio of the image distance from the centre of the lens to the object distance from the same lens centre, and the proper notation should be 40:1. See Chapter 3 for more about geometrical optics.

2. Aperture

Apertures are associated with angles; they are not merely holes but the area of a lens that passes light. Therefore, for a builder to refer to an aperture in a wall is wrong: it is a hole. Informally 'aperture' is used to describe numerical aperture (usually of the objective), when the meaning is clear. A diaphragm may (i.e. in front focal plane of the condenser), or may not (i.e. at the illuminated field diaphragm) be an aperture diaphragm. A stop usually refers (loosely) to a diaphragm of fixed aperture; strictly it is an opaque area concentric with the optical axis. As such, a stop is used in dark-field illumination.

3. Spelling

This book is written by an Englishman who eschews Windows spell-checkers. The letter s is therefore used in place of z, as appropriate. Shakespeare detested z, describing it as unnecessary (*King Lear*, Act 2, scene 2, line 65). Many of my friends are American, and so I ask the forbearance and understanding of my American readers. A mere slip of Coleridge's pen must have spelt artefact when he coined the term in 1834. The stem of the third declension noun *ars* in compound words is *arti-*. Some books put a diacritic acute *accent aigu* on Abbe's name, thus: Abbé. He may have been a key figure in – arguably – the finest optical company ever and consequently a great High Priest of imaging, but Abbe was never a French abbot. Authors are fallible and make mistakes, but this one is fundamental. Any microscopy book that commits this sin should be shunned.

4. Terminology

The Royal Microscopical Society's *Dictionary of Light Microscopy*, although over 25 years old at the time of publication, is a very useful and little-known resource, which has been followed herein. Language both alters and evolves; words and meanings change subtly over time. For example, in the dictionary the term 'darkground' is favoured over dark-field (as opposed to bright-field). Some hold that the term should properly be darkground, since if the entire field were dark there would be nothing to see: it is merely the back*ground* that is dark. This is very logical reasoning, but the term bright-ground is never used. For consistency, and to help understanding, the term dark-field is used in this book, since it is congruent with bright-field. Incident illumination is often used to mean reflected light or top lighting, but all illumination is incident upon the specimen, either upon the top or on the bottom of the specimen. Epi-illumination is the term for reflected-light illumination. Sometimes dia-illumination is used as a term for transmitted-light illumination (e.g. in Nikon product brochures). The somewhat imprecise term 'dioptre adjustment control' has been retained for convenience. The dioptre is a unit of refractive power expressed in reciprocal metres, whereas the adjustment control compensates for the refractive power of the eye to enable relaxed viewing through the microscope.

5. Mathematical formulae

Formulae are merely precise ways of encapsulating an idea that would otherwise require a long sentence. It may help to ask yourself what is the idea behind the formula. As far as possible, mathematical explanations and formulae have been kept to a minimum, and where possible have been 'quarantined' in their own boxes within the text.

6. Information boxes

Microscopy is a dynamic science, that has evolved through human enquiry and endeavour; the work of a few important individuals has been highlighted to show this and to inform the reader. The box of information on August Köhler was the first written, there being to date very few biographical details in English.

7. References and Internet Resources

Where references are cited, they are given so that you, the reader, can pursue a topic in more depth and/or gain access to further technical aspects of the subject that cannot be included here for reasons of space. They are deliberately fulsome in those chapters that cover fast-evolving subjects, such as light-sheet and super-resolution microscopy. The Key Reading suggestions are papers that, in the author's opinion, are significant and/or contribute extra interest and understanding to the subject of the chapter concerned. The Further Reading is included so that those who wish may broaden their vista of the subject. The decision to include Internet Resources has been made because these are conveniently and immediately available for download. The difficulty with material from the Internet lies in the fact that it is not peer-reviewed or in some cases does not respect copyright and can contain both subtle and gross errors of omission or commission. Furthermore, web pages may become dated. This is not as readily obvious as with printed matter and, of course, URL links may become broken or taken down over time. All links were checked for connectivity during the book production period of June–October 2018.

8. Illustrations

I have tried, as far as possible, throughout this book to include illustrations of microscopes, objectives and other components equally from between the manufacturers of imaging equipment. I am grateful to all those in the microscope trade who have assisted me so willingly. Where company names and brands are given, this information is intended to help scientists who are new to microscopy and who would value initial guidance. Although I may have my own preferences, I have tried to be even-handed, not least because I have friends employed in all the major manufacturing companies, and seek not to cause upset or offence. Where companies and products have been inadvertently omitted, I can only apologise.

Introduction

Presumably you are reading this book because you are interested in light microscopy and are either self-taught, attending a course, responsible for using and maintaining a microscope at work – or all of these things. The author, and those who have supported writing this book, have taught about light microscopes for many years, are passionate about microscopy and wish to help you. That is why this book has been written: to help you understand, set up and use a research light microscope properly.

The aim in writing this book is to *educate* and not merely to *train* you in the use of the light microscope. The distinction is important. Education involves learning about the individual parts of the microscope and how they function, a process inseparable from *understanding* how and why they work as they do, particularly appreciating their limitations – hence the title of this book *Understanding Light Microscopy*. Training, on the other hand, consists only of knowing how to operate the various microscope controls, from switching on the lamp to pressing the digital camera button: procedures set out in the manufacturers' instruction manuals. Education is a journey from instruction to understanding. In one sense it is never complete, although you may stop where you wish.

An important aspect of education is that it is a far more enjoyable and rewarding process than training; being educated in the use of the microscope you will truly become in control of the instrument and empowered in several ways: knowing the 'envelope' of proper operation you will be able to adjust the microscope properly, acquire and record far better images – which is the primary goal. You will also be able recognise and correct faults where they may occur and be able to devise special techniques to suit your own specimens.

These potential goals may be daunting, but each step taken is small. You do not need to know very much physics or mathematics[1] to understand microscopy or to work your way through this book. Essentially, the hard work was done by several geniuses from the late 17th century to the late 19th century, one of these being Ernst Carl Abbe, of whom more later. The author is an image facility manager who started out as a biological sciences technician; he is not a physicist or optician. So, please do not worry; mathematical formulae have been kept to an absolute minimum. Each physical concept and mathematical formula or procedure is explained clearly as they arise.

When we want to see something in more detail, we bring it closer to our eyes. At some point no further detail can be seen or we cannot get sufficiently close to the object. In both cases we use an optical aid to help us distinguish, or resolve, further detail. If we cannot get sufficiently close, we might use a telescope or a pair of binoculars. Where we can get close up to the object, we use a magnifying glass or microscope.

It is commonly thought that the function of the microscope is to magnify, and of course they do magnify. This is the layman's view:

> 'Yes, I have a pair of eyes,' replied Sam, 'and that's just it. If they wos a pair o' patent double million magnifyin' gas microscopes of hextra power, p'raps I might be able to see through a flight o' stairs and a deal door; but bein' only eyes, you see my wision's limited.
>
> [Sam Weller to Sergeant Buzfuz in: *The Posthumous Papers of the Pickwick Club*, chapter 34, page 240; Charles Dickens (1837)]

However, magnifying power is not the principal function of the microscope, which is resolving power – the ability to perceive, or discriminate, fine detail in the object and transfer that to the image. A second function is to provide contrast in the image, which is of particular importance for transparent specimens that would otherwise be invisible. Only when the microscope can provide adequate resolution and contrast will it be useful to magnify the image for comfortable viewing. There is little that the microscope user can do to improve resolution of the image – this property is determined by the objective lens and has to be paid for, though it is commonly made worse by misuse and misadjustment of the microscope. Likewise, magnification is usually fixed by design of the objective and eyepieces (and any intermediate lens system) by the manufacturer. Contrast, on the other hand, can often be improved in simple do-it-yourself ways, which can be devised by the ingenious and informed user.

A microscope is any instrument designed to provide information about the fine details of an object. The term thus includes the single-lens magnifying glass (termed the *simple microscope*) the familiar *compound microscope* which uses an objective lens and an eyepiece, electron microscopes, and also instruments that form their images from a mechanical or electrical interaction between a

---

[1] It is entirely possible to understand the underlying principles of microscopy without a formal qualification in physics or mathematics. In his study Einstein kept three portraits: Newton, Maxwell and Faraday. He considered them all to have profoundly underpinned progress in science, physics and technology. However, Faraday was noted for his mathematical inability, and used no complex equations in any of his 450 or so publications.

stylus and the surface of the specimen. In this book we shall confine ourselves to microscopes whose images are formed by the visible part of the electromagnetic spectrum, which we call *light*.

There are many types of light microscope, which differ because of the kinds of specimen they are designed to examine and the techniques used. The traditional microscope is the one generally used for biological work and configured for transmitted light (light that passes through a specimen, rather like light illuminating a stained-glass window in a church). Most explanations in this book will centre on this design, partly because it is so common but also because with the optical components arranged one after the other, it is the easiest to explain and also to understand.

Microscopes designed for use in materials science and also for fluorescence microscopy are configured to examine opaque or fluorescently-stained transparent specimens by reflected light, which illuminates through the objective lens. The objective therefore also acts as its own condenser. These microscopes are generally rather easier to use than the transmitted-light design, because whenever the objective lens is changed, the condenser is also changed appropriately and no major readjustment or centring are required.

Transmitted- and reflected-light microscopes are made in both the conventional upright design in which the specimen is examined from above, and also the inverted form where it is examined from below. Inverted microscopes are particularly useful (and widespread) in biological research laboratories, where they are used to view specimens in water or culture medium. An example would be cell monolayers or tissue explants adherent on a coverslip or grown on the bottom surface of a culture vessel, such as a Petri dish with a coverslip bottom.

In the last thirty years, light microscopy has benefited tremendously from advances in computing, fluorescent markers and laser technology. Fluorescence microscopy has become a ubiquitous technique in biological and medical science in recent years, and most biological microscopes can be adapted to illuminate the specimen with short-wavelength radiation and to observe the emission of longer-wavelength light from fluorescent compounds and proteins stained, or inherent within, the sample.

Most microscopes are designed to operate at high resolution, but a consequence of this is that they have extremely shallow depth of field where the full extent of the object is in focus in the image. Stereomicroscopes adopt a different approach, sacrificing ultimate resolution for increased depth of field. Stereomicroscopes are useful for examining solid objects by reflected light (including fluorescence), although they can be used with transmitted light also. The design essentially uses two microscopes, one for each eye but offset in slightly different directions to produce a three-dimensional effect. In this respect, stereomicroscopes are more like our eyes than the compound microscopes described above and so are particularly suited for educating school children.

These technological advances that heralded the renaissance of the light microscope have come at a price. Unlike other scientific instruments, which will not work unless they are adjusted properly, the light microscope will give an image – however poor – at the flick of a switch, however badly it may be set up. For this reason superficially, practicing microscopy appears simple, but scientific journals are replete with examples of poor quality images. Imaging is more than just 'the bit tacked on the end' of the experiment.

> To correctly capture images using a modern microscope, researchers must have a good grasp of optics, an awareness of the microscope's complexity…such skills can take months or even years to master [which] owing to inexperience or the rush to publish, are all too often squeezed into hours or days. … Inept microscopy, and subsequent analysis, can easily generate results that are misleading or wrong.
>
> [Pearson, H. (2007). 'The Good, the Bad and the Ugly'. *Nature* vol 447 (7141): 138–140, p. 138.]

Modern research microscopes are both very expensive and complex, because viewing anything but the thinnest fluorescent cells and tissues requires optical sectioning to discriminate fine detail within the sample. In order to provide a range of different approaches to detecting and visualising the precise location of fluorescent markers by optical sectioning and kindred techniques such as multi-photon microscopy, several different fluorescence microscopes may routinely be used. Therefore, research institutions and university departments increasingly pool their resources to fund and operate these instruments within centralised facilities.

Image facilities vary enormously and are usually very busy places. Some incorporate both light and electron microscopes together with flow cytometers and high-throughput plate readers; others house light and electron microscopes separately. By their very nature, light microscope facilities have many clients: typical numbers are between 150 and 300 users in any one university or institute. Generally, people who manage imaging facilities have been drawn into the role via a research background as a scientific investigator. Some facilities have several staff to advise about, maintain and operate equipment; others have a sole manager/operator.

If you are reading this book prior to using an imaging facility, it may help to bear in mind that staff, although knowledgeable, may not have as much time as they would wish to help you. Also, there may be greater emphasis placed on image acquisition than subsequent analysis – not least because there is an 'activation energy' associated with the learning curve for new software and your particular application. However, good facility managers should not only be familiar with their equipment but should be also approachable towards their client base.

Image acquisition and analysis is time consuming. The *raison d'être* of this book is to enable you to set up your microscope and collect good images competently and efficiently, unaided.

Treating the microscope like a powerful holiday-slide viewer ignores the fact that with the microscope, we are using light for 'something it wasn't designed for' – studying little things whose dimensions are close to the wavelength of the illuminating light itself. When we use a microscope we do not look at the object directly, but at an *image* of it. Therefore – like a painting or a sculpture – the microscopical image is only a *representation* of the object, and that image can be altered by the quality of the microscope, its objective lens and how light interacts with the specimen.

The prime purpose of this book is to guide you in avoiding these pitfalls whether you are beginning in microscopy or have been a microscope user for some time and wish to learn more. Lack of instruction is nothing new. Henry Baker's *The Microscope made Easy* (1742) had this advice in chapter 15, entitled 'Cautions in Viewing Objects', which is still pertinent. Baker warns us thus:

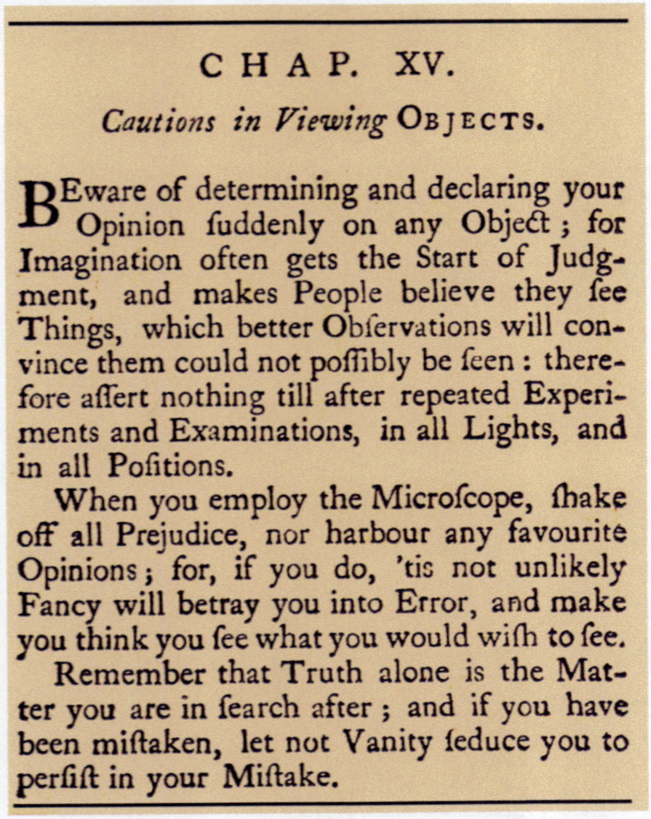

CHAP. XV.

Cautions in Viewing OBJECTS.

BEware of determining and declaring your Opinion ſuddenly on any Object; for Imagination often gets the Start of Judgment, and makes People believe they ſee Things, which better Obſervations will convince them could not poſſibly be ſeen: therefore aſſert nothing till after repeated Experiments and Examinations, in all Lights, and in all Poſitions.

When you employ the Microſcope, ſhake off all Prejudice, nor harbour any favourite Opinions; for, if you do, 'tis not unlikely Fancy will betray you into Error, and make you think you ſee what you would wiſh to ſee.

Remember that Truth alone is the Matter you are in ſearch after; and if you have been miſtaken, let not Vanity ſeduce you to perſiſt in your Miſtake.

Figure I.1 *Source: Baker, H. (1742). 'The Microscope made Easy', Chapter 15, pp. 62–63.*

Most books and courses start teaching microscopy by explaining some optics prior to evaluating the instrument itself. The layout of this book is broadly similar, but it is important not to forget the specimen, the object of study in the first place, whose fine detail we wish to resolve, to study and to understand. It is the interaction of light with the specimen that gives rise to any detectable signal. If this specimen is living, is marked with fluorescent probes and is being studied over time, then the constraints on collecting suitable images are different – and intrinsically more challenging – than for fixed immobile samples.

Chapters 1–3 in this book cover the need for microscopes, some basic optical principles and a treatment of ray optics. Chapters 4 and 5 explain how different microscopes differ in design according to their function and how to apply ergonomics to the safe use of the microscope. Chapters 6–8 cover the objective and other components of the microscope, as well as a treatment of optical aberrations. Chapter 9 explains how to adjust and set up a microscope properly. Chapter 10 tells you about diffraction and wave optics: how images are formed. Chapters 11 and 12 explain how to generate contrast with both transmitted and reflected light. Chapters 13 and 14 explain the theoretical fundamentals of polarised light and how to apply these in a practical manner. Chapters 15–26 cover all aspects of the theoretical and practical application of fluorescence microscopy, including optical sectioning, colocalisation and super-resolution imaging. Chapter 27 advises on which fluorescence technique to use, as well as giving advice on using an imaging facility. Chapters 28 and 29 discuss sample preparation for biological and materials specimens respectively, while the final two chapters, 30 and 31, cover how best to record the image.

As stated at the outset, the aim is not merely to instruct but to help you understand. Where applicable, explanations are underpinned by a few simple experiments that you can try yourself to aid your understanding. Illustrated text boxes are used to emphasise key facts or convey points of interest. The didactic approach comes from experience teaching students, those attending the Royal Microscopical Society's Summer Schools in Light Microscopy, and also from managing light microscope imaging facilities over several years. I and my colleagues acknowledge the debt from students for their questions and feedback about our teaching. We know that many students come to us with misconceptions, which they are honest and humble enough to admit to, and so ask for help. We admire them.

Jeremy Sanderson
Oxford, January 2018

1

Our Eyes and the Microscope

Outline

Understanding Light Microscopy, First Edition. Jeremy Sanderson.

1.1 Introduction

'Ah, I see!' demonstrates how dependent we are on our sight, and how it is linked inextricably with understanding. We are a visually oriented species, and our two eyes are our 'window on the world'. It is easy to forget that not all animals depend upon sight to the extent that humans do. Bats use echolocating sonar to navigate, and bees exploit ultraviolet (UV) light to discriminate foliage. At the other end of the visible spectrum, vipers sense infrared radiation to detect warm blooded prey. For those creatures whose habitat is underground (e.g. moles), vision necessarily cedes priority to other senses such as touch and smell. Other animals, chiefly birds, are more dependent upon their sight than humans, with a much greater proportion of their head devoted to large eyes.

Via the optic nerve, the eyes are a direct extension of the brain. The human eye is a wonderful device. If it were a camera, it might boast autofocus, wide-angle lens, auto-exposure, high sensitivity to light, automatic colour balancing, one-hundred megapixel resolution (each retina contains around 100 million sensitive cells) and 3D imaging (when used in pairs). We can see in bright daylight or discriminate a single light in the dark several miles away. The eye has naturally evolved to suit its principal function: helping its owner navigate the world.

Many modern microscopes are designed to be operated from a computer screen, rather than be used by viewing the image directly down the eyepieces. Nevertheless, all images must eventually be seen by our eyes, so in this chapter we shall discuss how our eyes function. We shall consider how well our eyes are adapted for their purpose but also the limitations and aberrations that they suffer from. These limitations are relevant to microscopy, affecting how we acquire and interpret scientific images (i.e. data) using the light microscope. We shall see how a magnifying glass works and why the microscope was developed to assist our vision.

1.2 How Our Eyes Work

Any image must ultimately be seen by the eye. Light is electromagnetic radiation that stimulates the eye. It is merely a fraction of the entire electromagnetic spectrum (Figure 1.1). Only a small proportion of solar radiation reaches the earth's surface. We depend on the ozone layer for protection; atmospheric dust, smoke, air molecules and water vapour also absorb a significant proportion of insolation, or incident solar radiation. Our eyes evolved from aquatic animals and contain a significant amount of water. Human sensitivity to electromagnetic radiation (Figure 1.2) corresponds closely to the wavelengths of minimum water absorbance, located away from harmful UV radiation and towards the infrared end of the spectrum.

Our eyes respond to the visible part of the electromagnetic spectrum from near UV at a wavelength of 380 nm to deep red at 710 nm. This stimulation depends on both the energy (frequency, expressed as wavelength) and the quantity (number of photons) of light. The wavelength of light is perceived as colour, and the quantity of light (expressed as the amplitude of the light wave) is seen as intensity[1] (see also Appendix A1.1.3 and Table A1.1, pages 45–46 in: Tilley, 2010). Suppose we have four LEDs: blue emitting at 490 nm; green at 555 nm; far red at 670 nm and infrared at 940 nm (commonly used for remote controls) all emitting the same radiant flux of 5 mW absolute power, measured in radiometric units. If we measure the respective light output of each of these LEDs in photometric units, the green LED will be the brightest (3.4 lm); the blue will be the second one (0.75 lm), and the far red will be the third (0.1 lm). The infrared LED will have a recorded emission of zero lumens (this example is taken from Tilley, 2010).

Further details about the electromagnetic spectrum and the nature of light are discussed in Chapter 2, Section 2.2 onwards. Light itself has no inherent colour; our perception of different hues is fundamentally a complex judgment experienced as a sensation by our brains. We discriminate colour very well, although brightness less so. This is why we choose paint in a range of colours, rather than different intensities.

The retina is the photo-sensitive tissue of the eye (Figure 1.3), within which there are two types of receptors. These are called rods and cones. Rods are more numerous (about 100 million per eye) and are better suited to night vision because they are about 100 times more sensitive to light than cones. Rods are thus adapted to low-intensity light, whereas cones (about 5 million per eye) give us our

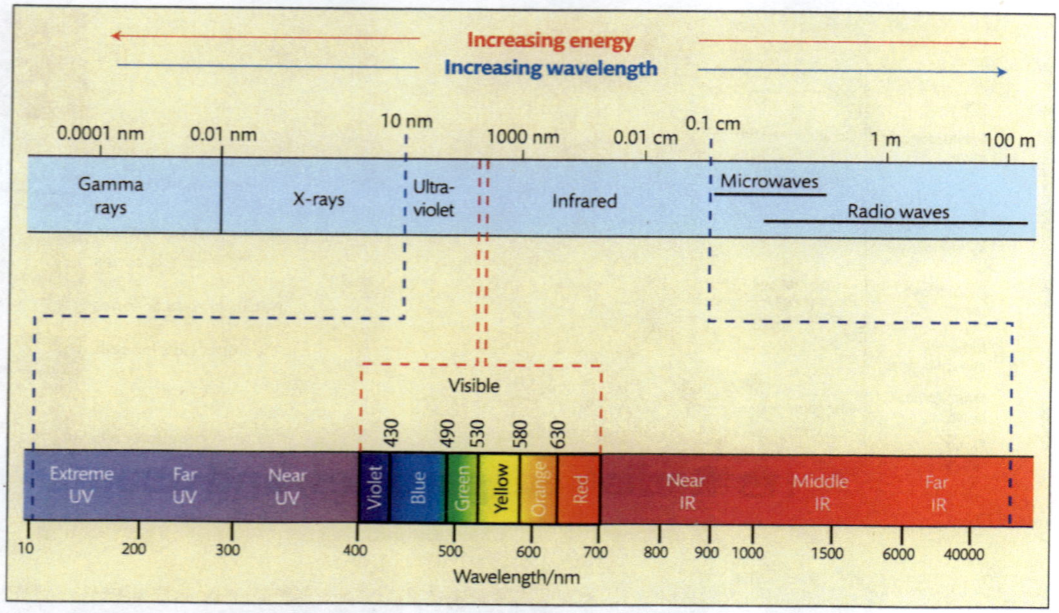

Figure 1.1 ***The Electromagnetic spectrum*** *The electromagnetic spectrum extends from beyond low frequencies used for modern radio to very high-frequency gamma radiation at the short-wavelength end of the spectrum, encompassing wavelengths from thousands of kilometres down to a fraction of a nanometre. Humans are able to discriminate the visible wavelengths from long wavelength, low energy, red at around 710 nm to short wavelength, high-energy violet radiation at around 380 nm. A very small number of people can see into the infrared, and people suffering from cataracts who have their lens removed (i.e. who are aphakic) can see into the UV, even if they can't focus well. Source: Glencross et al. 2011. Reproduced with permission from Oxford University Press.*

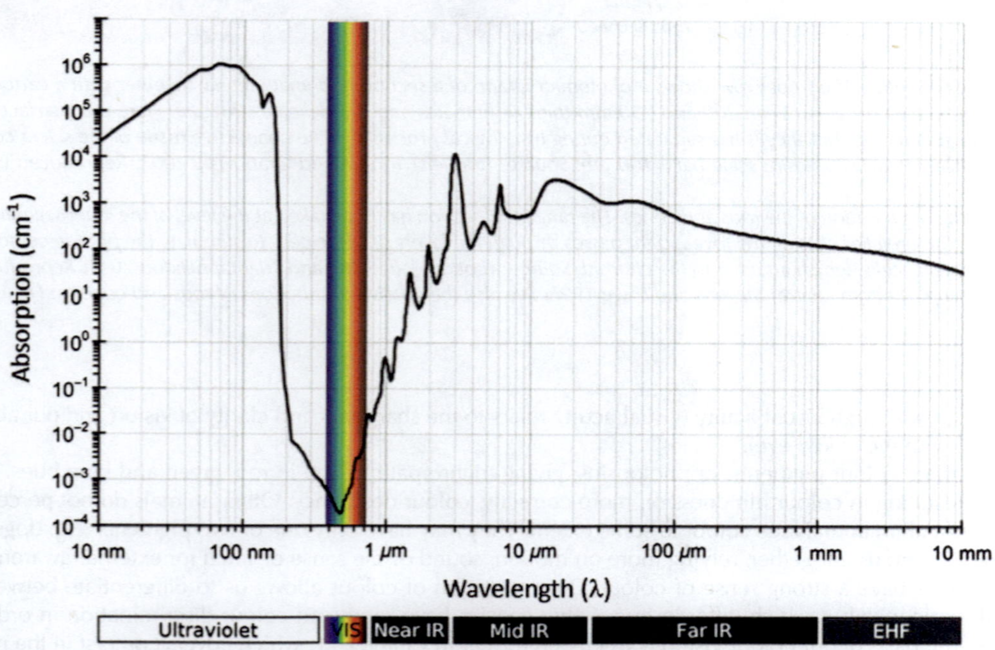

Figure 1.2 ***Human sensitivity to light*** *Visible light corresponds closely to the wavelengths of minimum water absorbance on the electromagnetic spectrum. See also Chapter 2, Section 2.5. Since our eyes are composed largely of water, it is advisable to use a heat filter (e.g. Schott KG5) in the light path to cut out any absorption of infrared light. Source: Adapted from Kebes at English Wikipedia under the terms of the Creative Commons Attribution licence, CC BY-SA 3.0 (http://creativecommons.org/licenses/by-sa/3.0).*

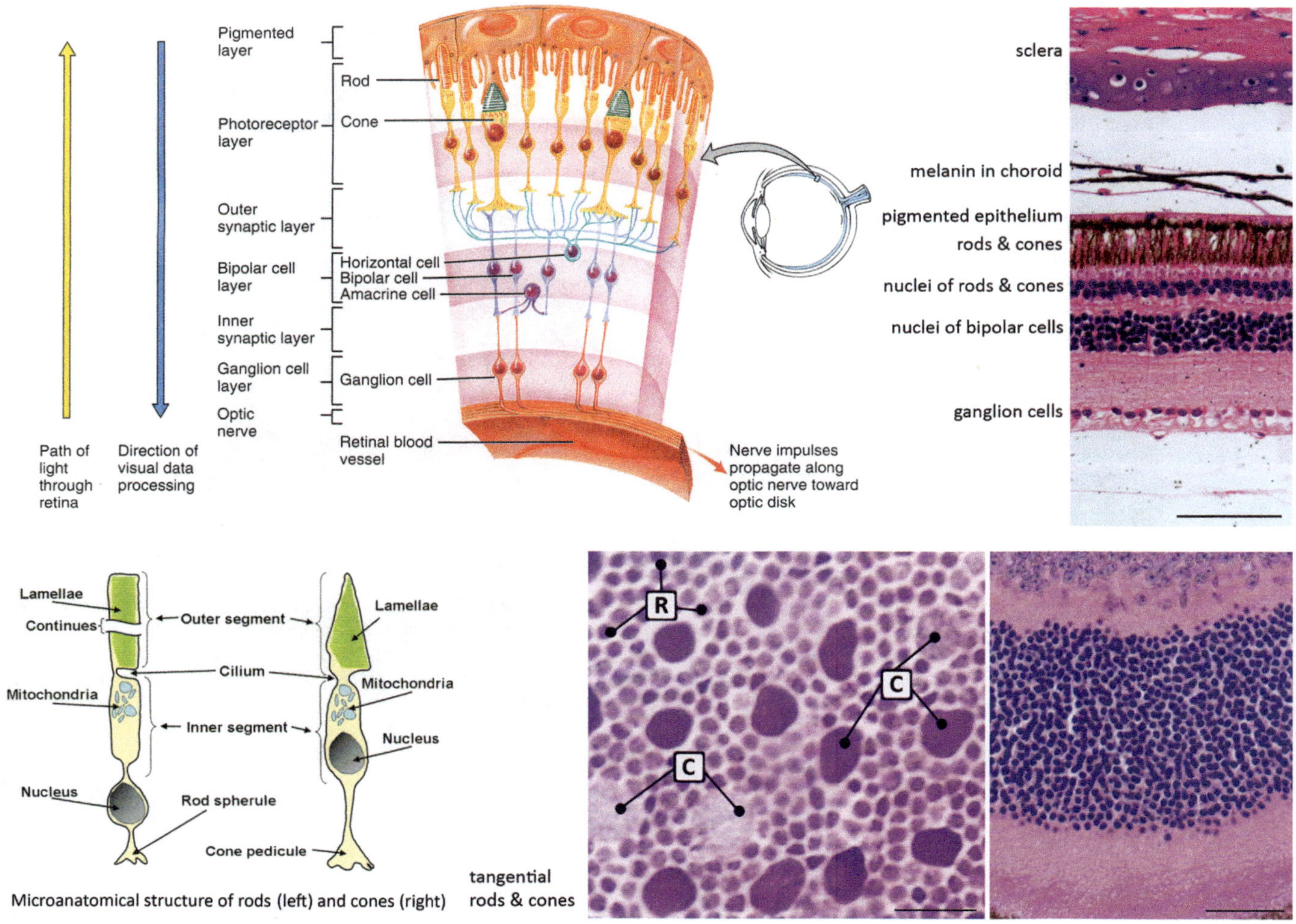

Figure 1.3 **Structure of the retina** *The upper row shows a photomicrograph of a section of primate retina, together with a cartoon diagram showing the organisation of the retina, which is between 200 and 250 mm thick in primates and is a delicate structure. Here the interface between the dense black choroid and the pigmented epithelial cells has separated during histological processing. The connective tissue of the sclera coat is seen at the top of the section. Section: haematoxylin & eosin; scale bar = 100 μm. Source: (top left) Tortora and Grabowski 2003. Reproduced with permission from John Wiley & Sons.*

The lower row shows the organisation of the rods and cones. The tangential section has been taken at the level of the inner segment of the photoreceptors. R = rods, C = cones; this area lies outside the fovea. Light passes through the layers of nerve cells to stimulate the photoreceptors. The retina lies on a layer of pigmented epithelial cells, which attach it to the choroid. Source: (bottom left) Allen and Triantaphillidou 2011. Reproduced with permission from Taylor & Francis Group. (bottom centre) Stevens and Lowe 1996. See also the relative distributions of rods and cones in Figure 1.12.

daylight vision in colour with high visual acuity (visual acuity refers to the sharpness and clarity of vision and our ability to resolve, or distinguish, fine detail with the naked eye).

Most humans have three colour pigments, or photopsins, giving trichromatic vision in red, green and blue hues. Some people lack one of the pigments, resulting in colour blindness or, more correctly, colour-deficiency. Other animals do not perceive colour as well as we do, though almost all mammals see colour to some extent. They may have only one or two pigments (e.g. dogs and cats), or may have different pigments from us altogether, relying more on motion, sound or the sense of smell for external awareness. Tree-dwelling and fruit-eating mammals have a strong sense of colour. Discrimination of colour allows us to differentiate between objects whose surfaces have equal luminance but which differ in hue. Other species have sacrificed colour discrimination in order to see better in low-light conditions. The basis of our colour vision is discussed further in Chapter 11, which covers contrast in the microscope image, and also in Chapter 31, which discusses how best to display the recorded image for the benefit of the colour-blind.

Our eyes are very sensitive to brightness: rods can signal the absorption of a single photon. From full sunlight to starlight represents a luminance ratio of 10 million to 1. We don't see equal increments of luminance as equal increments of brightness; rather[2] we see *logarithmic* increments of luminance as equal brightness steps (Figure 1.4). Therefore, as luminance increases, we require larger changes to discriminate a noticeable difference. As the illuminance (i.e. intensity) of light around us falls, we switch from daylight, or photopic,

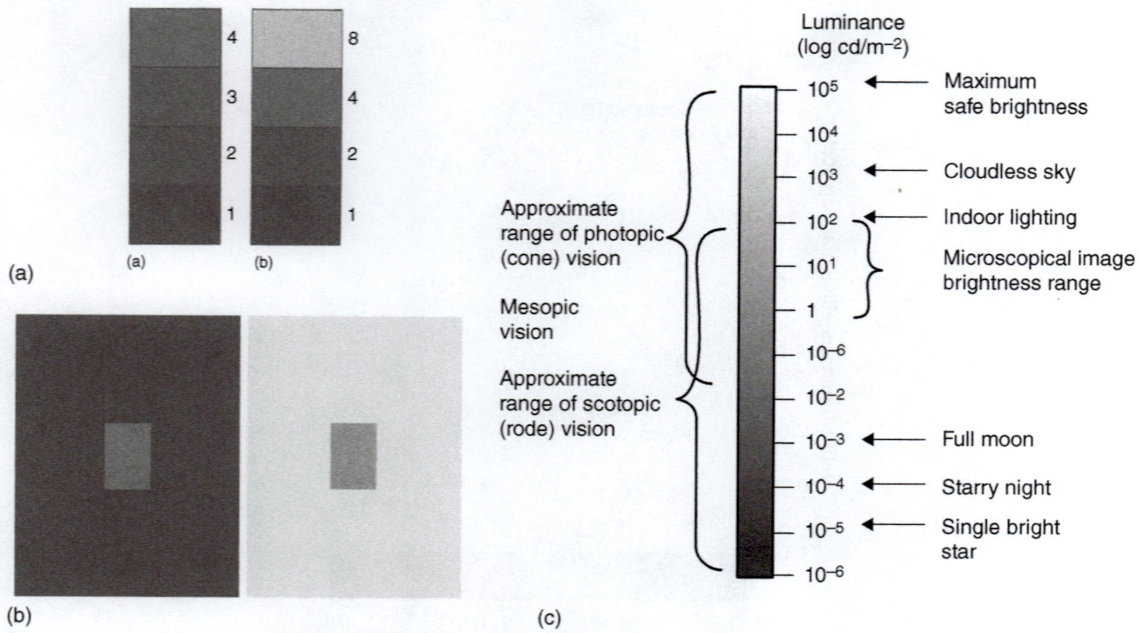

Figure 1.4 **Response of the eye to luminance** *Each greyscale luminance step of brightness on the left-hand scale differs by an equal arithmetic ratio, whereas those on the right vary logarithmically by the power of two. In the absence of a background, the eye sees the logarithmic scale as equal brightness steps, not so the arithmetic one. This is known as the Weber-Fechner law: the relationship between a stimulus and its perception is logarithmic. This is unlike a semiconductor CCD sensor, which discriminates brightness in a linear fashion. The eye responds logarithmically so that familiar objects tend to maintain their perceived reflectivity in changing light conditions, as our eyes adapt to the change. Otherwise (like exposure values using a camera) we would continually notice changing brightness and colours as the sun slipped behind, and emerged from, cloud. This also explains why we don't see stars in daytime, because the ratio of intensity of the stars to the sky is very small, whereas it changes at night as the intensity coming from the sky is small.*

However, if the backgrounds between areas of the same grey intensity are very different, this can alter our perception. In (b), the central grey rectangles are printed at the same intensity but are perceived differently, because of the contrasting adjacent backgrounds.

(c) shows the eleven orders of magnitude that the eye is able to perceive, although not all at once. At any given time, the eye is adapted to only 2–3 orders of magnitude on the scale shown.

Source: Falk et al. 1986. Reproduced with permission from John Wiley & Sons.

vision to night, or scotopic, vision. This occurs at luminance levels around 10^{-2} to 10^{-6} cd/m^2; below this light level we cannot read even large print text or recognise very small details, because the fovea[3] is inactive.

Since rhodopsin, contained in the rod cells, bleaches rapidly in bright light, this causes temporary blindness of the sensitive night vision. The practical consequence is that, although partial recovery may take 10 minutes, full recovery can take four times longer and affect visual acuity for viewing dimly-lit specimens (e.g. fluorescent samples). Daylight vision by the cone cells (Figure 1.5) adapts much more rapidly to changing light levels, adjusting well to a change such as coming indoors out of sunlight in just a few seconds. Rods are insensitive to light of wavelengths longer than about 640 nm. Red illumination in a darkened microscope room is useful[4] because red light bleaches rhodopsin inefficiently and thus maintains dark adaptation for high-quality microscopy.

Although rods can detect a photon as a single quantum of light, we are quite bad at seeing in the dark. This has implications when observing fluorescent specimens against a relatively dark background. Figure 1.6 explains why this is so. There is a statistical uncertainty associated with the detection of light by any photoreceptor (this is explained with respect to digital cameras in Chapter 30, Section 30.10). Although a rod can detect a single photon, a threshold exists to avoid spontaneous activation (and depletion) of rhodopsin: a minimum of six rods must be simultaneously stimulated with $\approx$10 photons in order to register a response at the threshold of vision. In Figure 1.6 there is a central dark area in the image. However, even at the visual threshold, the contrast is insufficient to discriminate this central area from the background. Only at 1000 times threshold is the central dark area surrounded by enough light areas to be visible with sufficient contrast. The converse is also true: small self-luminous fluorescent objects in the sample have to be sufficiently large to stimulate a group of rods in order to be seen against a large dark background.

We can see – at best – 60 shades of grey, particularly if there is an abrupt (i.e. straight edge) boundary between each shade and the next. The computer can record many more grey levels, shades, or intensities, irrespective of boundaries. This facet of digital imaging, and its consequences, is explained further in Chapter 18 explaining the operation of the confocal microscope in Box 18.1, and in Chapter 30, in Section 30.6, on recording the digital image. Suffice to say here that (a) computers can be used to enhance contrast so that details, and data, which we would otherwise miss, are made visible and (b) we need to use saturation look-up tables (LUTs) to ensure that all the dynamic range of electronic detectors (such as CCD and sCMOS cameras for widefield microscopy and photomultiplier (PMT) tubes for point-scanning confocal microscopes) is used effectively.

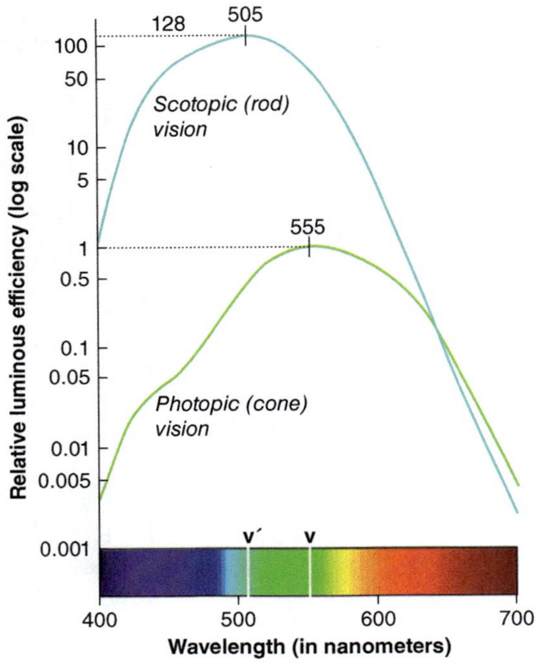

Figure 1.5 **Photopic and scotopic vision** *Daylight vision in bright light is called photopic vision. It occurs above 10 cd/m^2 when the rod photoreceptors become saturated, as illuminance rises. As the light levels around us fall, mesopic vision occurs when both rods and cones are active. At luminance less than 10^{-2} cd/m^2, photopic vision occurs. Sensitivity is plotted relative to peak photopic sensitivity on a log vertical scale.*

At light levels where the cones function, the eye is more sensitive to yellowish-green light (555 nm) than other colours because this stimulates the two most common (M and L) of the three kinds of cones almost equally. At lower light levels, where only the rods function, sensitivity is greatest at 505 nm, a blueish-green wavelength. Source: Handprint.com 2009. Reproduced with permission from B. MacEvoy.

1.3 The Anatomy of the Eye

The eyeball is a sphere approximately 24 mm in diameter with the cornea bulging slightly outwards. Figure 1.7 shows the features of the human eye that are important for our purposes. Essentially three transparent structures are held in position by three coats of tissue. The sclera is the white fibrous, protective, outer layer of the eye; the cornea is fused as an extension to this coat. A small part of the sclera is seen from close distance around the iris; hence the term 'the white of the eye'. The choroid is a pigmented black membrane that has two functions: it prevents unfocused light from shining through the sides of the eye and prevents light that enters the pupil from reflecting inside the eye. The third coat, the retina, lies directly on top of the choroid. The three transparent structures within the eye are the aqueous humour, the lens and the vitreous humour.

Any optical instrument – and the eye is no exception – must have its image-forming surfaces (e.g. the cornea and lenses) kept in a steady position with respect to each other and the detector (screen, CCD/sCMOS face-plate, PMT tube, film plane or retina) upon which the image is formed. The flexible tissues of the eyeball and the two surfaces of different radii, or 'figures', of the lens are kept firm by the pressure of the internal fluids of the eye. The aqueous humour lies between the cornea and front, or anterior, surface of the lens. It is continually drained and replaced by fresh fluid. The more gelatinous vitreous humour fills the larger space of the eyeball between the rear, or posterior, surface of the lens and the retina. The vitreous humour is not replaced like the aqueous humour and tends to liquefy very slightly as we age.

When the vitreous humour becomes less viscous, cellular debris or clumped strands from the vitreous humour can float freely. Being less transparent, these *floaters* cast fleeting shadows on the retina, following the motion of the eye whilst drifting through the vitreous humour. They are more prevalent in short-sighted eyes, and become increasingly annoying as we age (see also Chapter 9, Section 9.13). Floaters are most likely to be seen when using objectives and eyepieces of high magnification, especially if the objective is of relatively low aperture and the microscope is set up with the adjustment of the illumination set too low such that it does not fill the objective with light. The proper method of adjusting the microscope is explained in Chapter 9, Section 9.11. If a floater appears, moving the eye from side to side or up and down can create an internal current that will help move the floater away from the line of sight.

The lens of the eye is composed of transparent cells held in place by suspensory ligaments connected to the ciliary body, which focuses the lens. The majority of the focusing (approximately 60%) to form the image is done by the fixed cornea[5], while the bi-convex lens adjusts to allow fine focusing to maintain a clear image. This focusing is called accommodation and is discussed further below. The focal length of the relaxed lens is 17.1 mm and 14.2 mm when fully accommodated. The iris, which determines the pigmentation

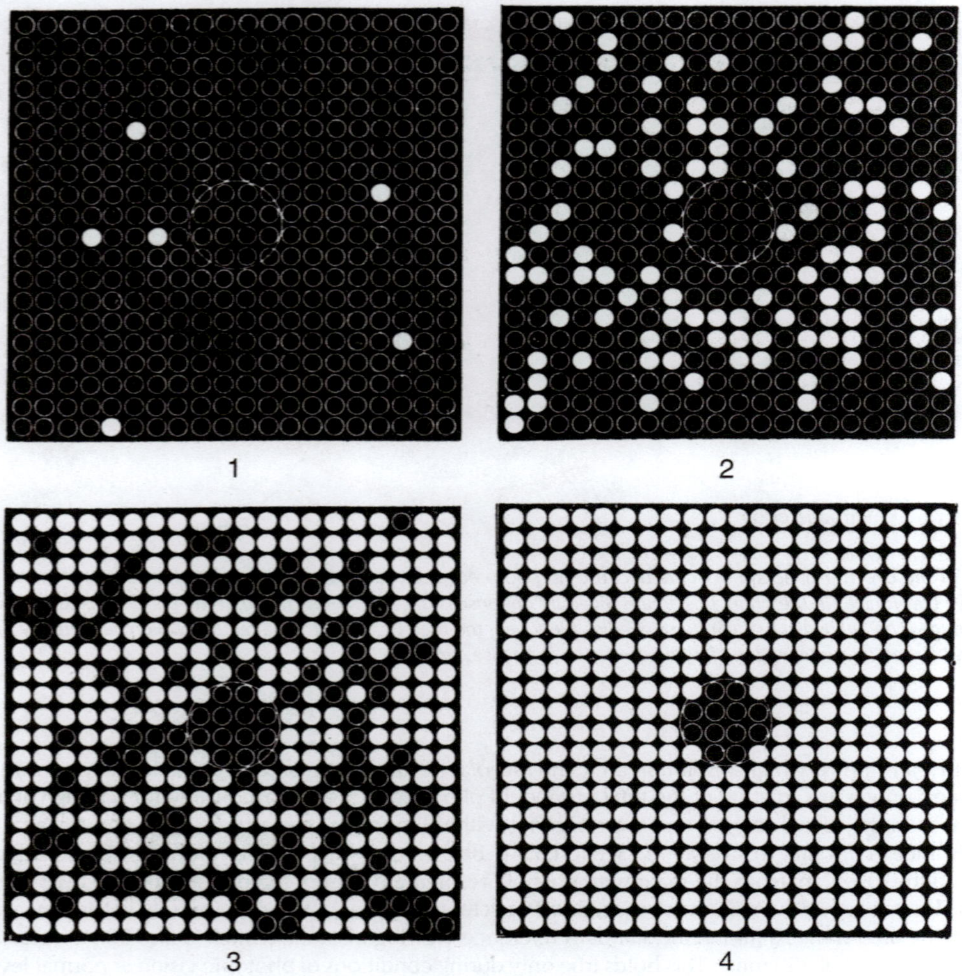

Figure 1.6 ***Sensitivity of scotopic vision*** *All four panels of the figure show the image on the retina of the same (bright) field containing a central dark patch. In panel 1 the illuminance is so low that only 6 of the 100 receptors receive a photon – at or just below the absolute threshold of vision. In panel 2 the illuminance is ten times higher, but the central dark patch is hidden within the noise of the random background photons. In panel 3, at 100 times threshold, the central dark patch is only just discernible but not convincingly so, and it is only in panel 4, at 1000 times threshold, that the central dark area is readily visible against the background with sufficient contrast. Source: Land and Nilsson 2001. Reproduced with permission from Oxford University Press.*

of our eye, dilates automatically in response to differing light intensities (irradiance), to regulate the diameter of the central opening or aperture – the *pupil* – of the eye[6]. In very bright light, the iris contracts to give a pupil of approximately 2 mm or *f*/8.3. This diameter increases in young adults to about 8 mm (or *f*/2.1) or in the elderly to about 5 mm (or *f*/3.3) at low-light intensities.

Within the retina, the sizes of the rods and cones differ slightly. Cones are typically 40–50 μm long. A rod photoreceptor is longer (about 60 μm) and narrower than a cone: about 2 μm in diameter, whilst a cone is 2–8 μm in diameter, depending upon its location in the retina. With rods we see in black and white, or monochrome. Overall, there are roughly 1.5×10^6 ganglion nerve cells per eye, with as many as 150 rods connected to a single nerve cell via bipolar cells. This summation, or convergence, increases sensitivity, which is useful when the light intensity is low. However, summation reduces definition – things look less sharp. A single rod is only slightly more sensitive than a single cone. However, cones are innervated as few as 2–6 per nerve cell, making them approximately 30 times less sensitive to light. Cones are essential for resolving fine detail sharply and for colour vision.

The area of the retina directly behind the lens is called the macula. It is densely innervated and contains the *fovea centralis*, which has the highest concentration of singly-innervated cones, giving excellent visual acuity in our central vision. The entire fovea is about 1.5 mm in diameter, comprising approximately 5° of the visual field. The rod-free area of the fovea is 0.5 mm in diameter, subtending 1.7°. Whilst the majority of cones (about 94%) are located in the peripheral retina, the 60 000 cones of the fovea are much slimmer. Here they measure 1–2.5 μm in diameter and are packed together in a hexagonal pattern. At the fovea, the layers overlying the photoreceptor layer are much thinner, giving the appearance of a pit (Figure 1.8) to the cross section of the whole retina at this point. Within the central

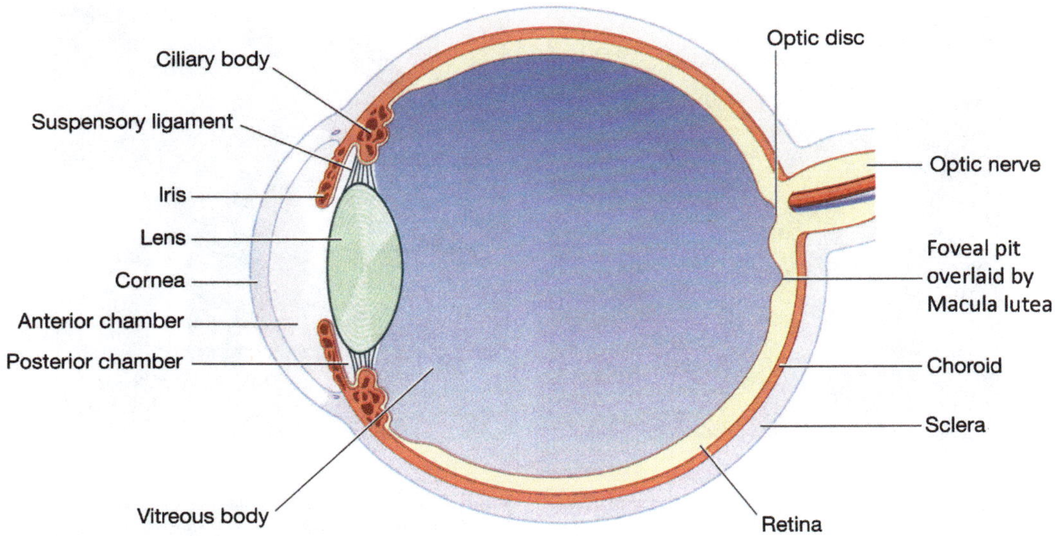

Figure 1.7 **Anatomy of the eye** *The blind spot occurs because no photoreceptors can exist where the optic nerve and retinal blood supply join the eyeball. Because the eye tiles images in the brain, this area is 'filled in' unconsciously, despite the fact that the blind spot extends to about 6° in diameter and is therefore quite large. The yellowish macula lutea overlies the fovea, the area of highest visual acuity, which lies directly on the visual axis of the eye. Source: Waugh and Grant 2001. Reproduced with permission from Elsevier.*

0.35 mm (1°) of the fovea is the very high-resolution area, the *foveola*, located at the bottom of the foveal pit. Here in the foveola there are about 12 000 cones; only a tiny fraction of the total number of photoreceptors are responsible for our high acuity vision.

The size of the cones in the foveal region, and their packing within the retina, has a direct bearing on visual acuity. The average cone-cone centre distance is 3.8 µm. In Chapter 3, Section 3.11, Box 3.4 is an explanation of how the packing of the light-sensitive cones, which make up the fovea, matches the maximum possible resolving power of the eye. At very best we cannot discriminate, or resolve, detail that is closer together than 72 arcseconds, (i.e. 36 cycles per degree, CPD; visual acuity is often measured in CPD), which is equivalent to resolving 0.09 mm (90 µm) at the standard reference viewing distance (which is also sometimes called the nearest, or least, distance of distinct vision) of 250 mm[7]. This holds true only during conditions of photopic vision at normal levels of light intensity; when we are dependent upon our rods alone, in scotopic conditions, our visual acuity falls off.

Outside of the fovea, there is summation of both rod and cone cells. Thus the bandwidth of visual processing is highest for visual fields directly in front of us. The fovea comprises less than 1% of the area of the retina, but its importance to our vision is demonstrated

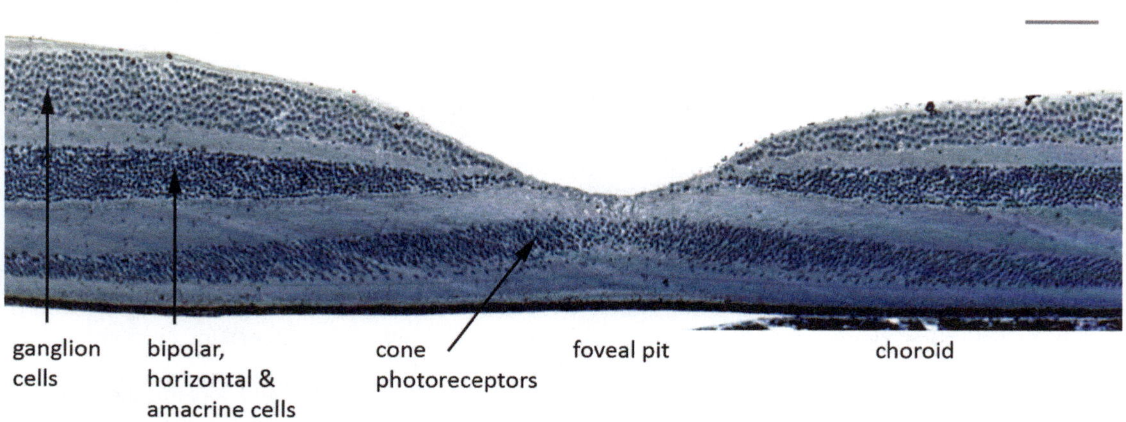

Figure 1.8 **The fovea and foveal pit** *This area of highest visual acuity contains slimmer cones and no rod photoreceptors. The organisation of each layer within the retina is also shown in Figure 1.3. Section: monkey, toluidine blue; scale bar = 100 µm. For further details, refer to Kolb, H. (2007); https://www.ncbi.nlm.nih.gov/books/NBK11556/.*

by the fact that over 50% of the visual cortex is devoted to processing the information derived from this small region. Since the fovea contains no rods, it is insensitive to low intensity light signals. Astronomers know this: in order to observe a dim star, they use averted vision.

Broadly speaking, the density of receptors and the degree of summation determine the degree of visual acuity: the greater the number of singly-innervated receptors, the greater the ability of the eye to distinguish individual objects at a distance. We have a mean value of about 200 000 cones per mm^2, which is good for a mammal but is quite low compared to a large raptor, which has about 400 000 cones per mm^2 as well as more ganglion cell connections between the receptors and the brain, since it requires a high visual acuity for hunting[8]. We may have limited visual acuity compared with a falcon, but it is much better than that of a honeybee, which has only 1% of our visual acuity and would not be able to see if there were any books on the shelf across the room, let alone read the titles. However, a honeybee does not need to read but does need to navigate. It can respond to the polarisation[9] vector of light and uses this to navigate to food sources.

Approximately two thirds of the population is right-eye dominant and one third, left-eye dominant, preferring visual input from one eye over the other. This has consequences for adjusting and using the microscope, which is covered in Chapter 9, Section 9.11. The diameter of the *exit pupil* of the beam of light leaving the eyepiece is determined by the diameter of the pupil of the iris, so that no information about the image is lost. This is explained further in Chapter 8, Section 8.6.

1.4 Aberrations of the Eye

Because our eyes have a single lens, they suffer to some extent from both spherical and chromatic aberration. Our eyes may also suffer from aberrations such as astigmatism, for which we require corrective spectacle lenses. The subject of optical aberrations and how they affect the performance of the microscope in forming an image is discussed later in Chapter 6. In the microscope these lens aberrations can be corrected by cementing single lenses together into doublet or triplet combinations. These lens groups require combining convex (positive) and concave (negative) lenses; animal eyes are only bulging, or biconvex – they have no negative configuration. Doublet or triplet elements have a longer focal length than is feasible in a small eye. The eye must therefore rely on natural methods to correct optical aberrations.

Spherical aberration manifests itself as unsharpness in the image (Figure 1.9). Spherical aberration in the eye is minimised by the different curvature (known as the 'figure' of the lens by opticians) of each of the two surfaces of the elliptical biconvex lens. Also, under daylight illumination, our pupils are smaller; this restricts light paraxially, close to the optical axis, and thus minimises the effects of both spherical and chromatic aberration. Additionally, a smaller pupil also ensures that light falls only on the central fovea for sharp vision. Whilst a larger pupil lets in more light to the eye and too small a pupil suffers from image blurring due to the effects of diffraction, there are no advantages in the eye having a pupil too large (Figure 1.10); otherwise, aberrations limit image quality. The direct effect of chromatic aberration in the eye manifests itself as different colours appearing slightly out of focus with respect to one another. Figure 1.11 demonstrates this.

The entire macula is covered with a yellow carotenoid pigment, very similar in hue to the yolk of an egg, called the macula lutea. Together the cornea, lens and macula filter out much of the harmful UV light. The pigmented macula absorbs blue light also. Since blue light refracts more than red light, absorption of these shorter wavelengths helps minimise the effects of chromatic aberration and glare from scattered light. We have approximately 64% of red (L) cones and 32% of green (M) cones but only 4% of blue (S) cones (Figure 1.12), and these are largely absent from the fovea, where visual acuity is highest. Additionally, our brains respond to light falling on-axis upon the fovea preferentially to other parts of the retina (the Stiles-Crawford effect). This reduces perception of any scattered blue wavelengths, helping further to minimise chromatic aberration.

Eyes (and lenses) suffering from astigmatism[10] have an unequal curvature of the cornea, the lens or both. This inequality causes the light to be refracted, or bent, to a different degree at different meridians of the cornea or lens. As Figure 1.13 shows, rays that propagate along two perpendicular planes, for example, will be brought to a focus by an astigmatic lens at different distances along the optical axis from each other. Figure 1.14 shows a test for diagnosing astigmatism. When using the microscope, it is necessary to determine whether your spectacles are corrected for astigmatism (see Box 9.5, Chapter 9). This dictates whether you must retain your spectacles for observation down the microscope, or whether you can remove them, if desired. Many people have a mild degree of astigmatism, often without realising it. The condition is severe enough to necessitate ophthalmic correction by spectacles or contact lenses in one third of the population.

Our eyes are sometimes likened to a photographic camera, but unlike the exposure of film in an analogue camera or the capture of photons and conversion to an electronic signal in a digital camera, our vision is not instantaneous but rather a continuous process. We build up a panoramic view by extremely rapid scanning microsaccadic movements, which our brains then tile into a sharp image. We do this for three reasons: scanning allows the fovea – the region of highest cone density and the greatest sensitivity – continually to image a scene. The fovea sees only the central 1.7° of the visual field, roughly equivalent to the width, at arm's length, of both thumbnails placed side by side. Without microsaccades, our vision would be severely limited: only our central vision would be at all sharp, with the rest an indistinct blur (Figure 1.15). Secondly, scanning allows the brain to tile images and thus fill in over the blind spot, the area without photoreceptors where the optic nerve connects to the brain. Thirdly, staring fixedly at a scene without microsaccadic scanning would also cause the rods and cones temporarily to cease operating, since they must constantly regenerate their pigment, which is bleached by light.

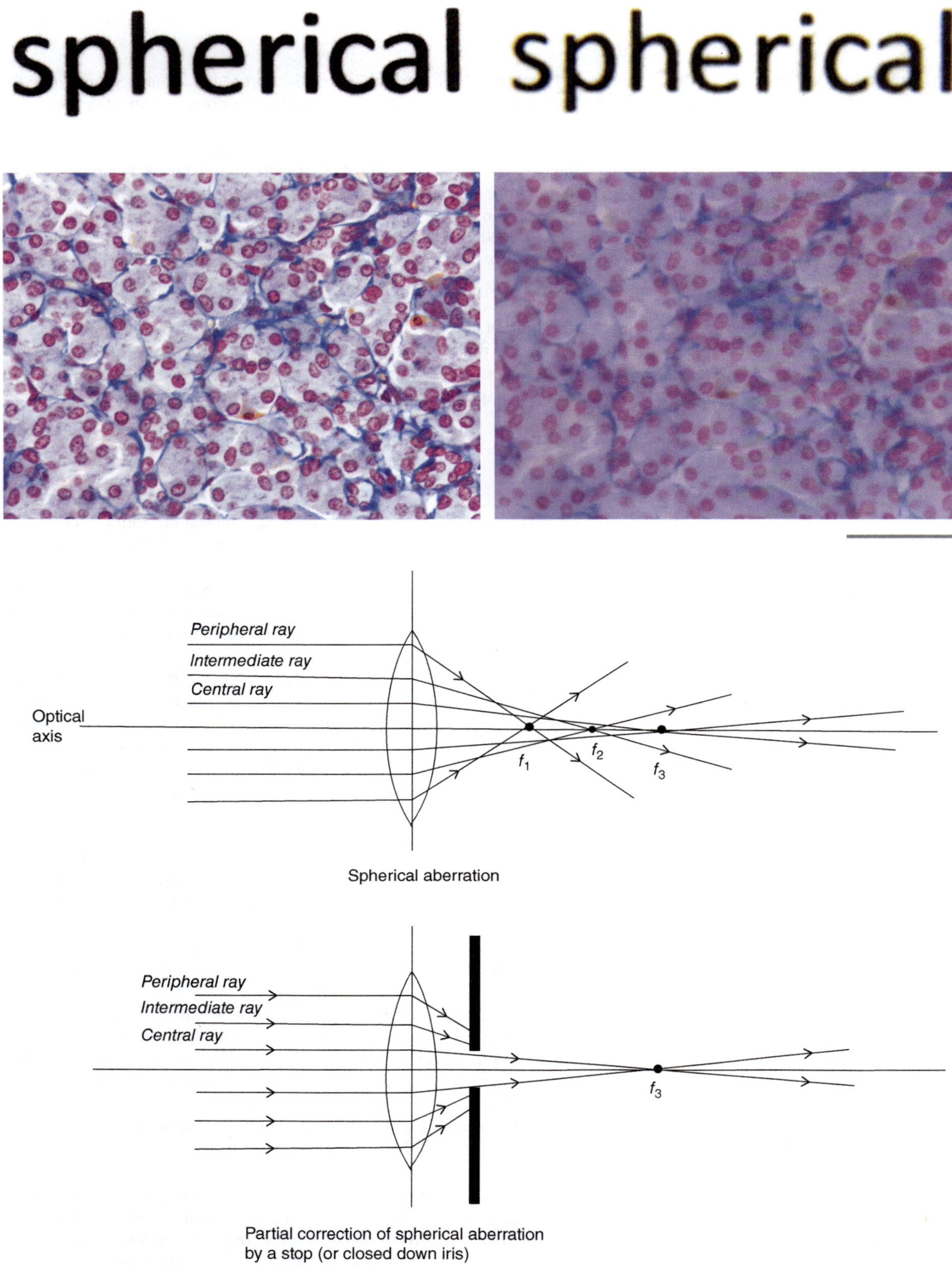

Figure 1.9 **Spherical aberration** *The top two illustrations were taken with a highly corrected camera lens (left) and a single lens that had failed in the quality control process (right), of the word 'spherical'. Spherical aberration is most noticeable at the edges of the image; notice how the letter R is smeared and unsharp. A smaller pupil ensures that only paraxial light rays fall on the central fovea for sharp vision.*

The bottom two illustrations show spherical aberration in the microscopical image. The image on the left is taken with a corrected lens. The right-hand image of the same field of view shows spherical aberration. The image is blurred and unsharp; image contrast is also severely degraded. Section: monkey pancreas, acid fuchsin & azan; scale bar = 50 μm. Source: (top) Courtesy of Mr K. Glover. (bottom) Bradbury 1967. Reproduced with permission from Elsevier.

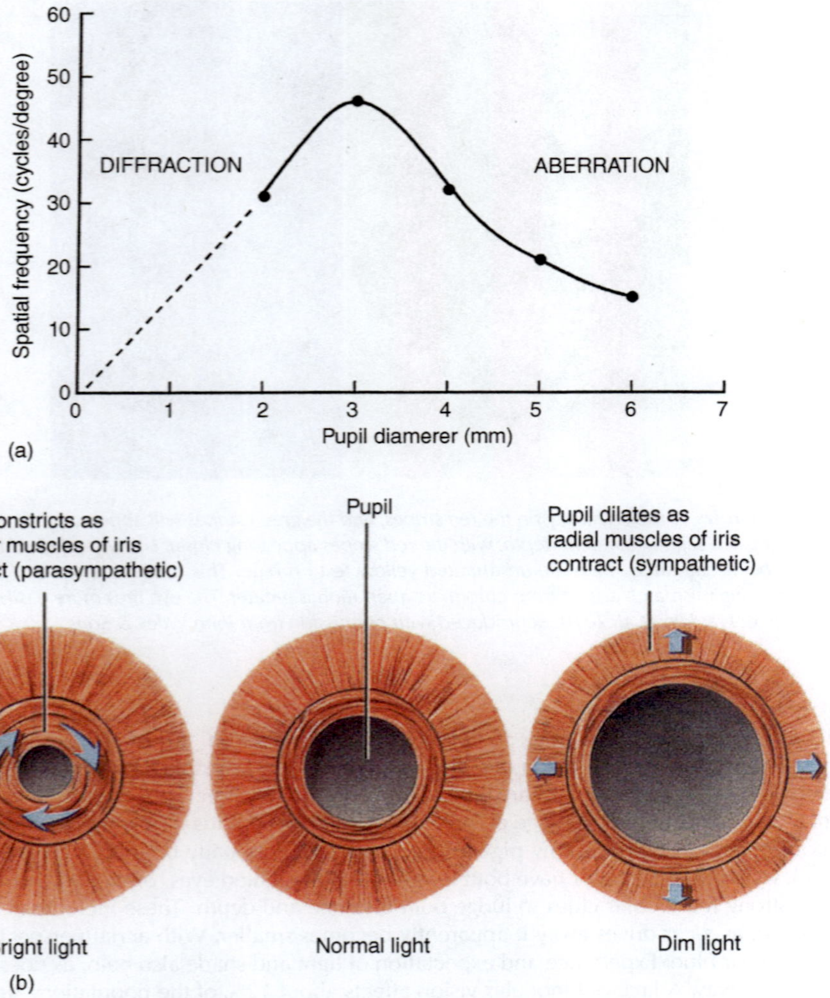

Figure 1.10 ***Optimum pupil diameter*** *In bright light, under photopic conditions, a pupil of 3 mm is optimal between the effects of diffraction and spherical aberration and (to a lesser extent) chromatic aberration. As illuminance falls, visual acuity is limited by photon noise rather than optical quality. Source: (a) Land and Nilsson 2001. Reproduced with permission from Oxford University Press. (b) Tortora and Grabowski 2003. Reproduced with permission from John Wiley & Sons.*

1.5 Binocular and Stereoscopic Vision

Despite the limitations of a single lens and consequent aberrations, our eyes are suitably adapted for their function. We can see into the distance and also close up to a near distance of distinct vision of about 250 mm. We can detect peripheral movement with ease, although with eyes at the sides of their heads, many animals are able to see very large visual fields, in some cases behind themselves, without moving their heads. This is clearly useful for detecting moving predators; the trade-off is having very limited 3D vision, which only extends in a narrow sector to their front for these animals. Predators, on the other hand, tend to have their eyes positioned on the front of their heads. We have forward-facing eyes on the front of our faces separated by the inter-pupilliary, or interocular, distance of between 50 and 75 mm. With each eye offset by about 5° from the optical axis through the centre of our head, this disparity gives us binocular vision. Each eye presents different views simultaneously to the brain. Complex neural circuitry matches each set of points from one view with the equivalent set from the other eye. We exploit binocular disparity to fuse in the brain the image from each eye to provide visual depth and three-dimensionality, allowing us successfully to navigate the world around us.

Parallax is the apparent displacement of an object due to a change in the position of the observer. For example, the needle on an analogue speedometer, because it is raised away from the face of the instrument dial, will indicate a different speed to a passenger in the car than the driver. The passenger sees the needle from an angle and observes an apparent speed; the driver observes the needle face on and notes the true speed of the car.

Figure 1.11 **Chromatic aberration** *Try to focus sharply on the red stripes, and the green stripes will appear slightly unsharp and out of focus. The brain will interpret each colour as being at a slightly different depth, with the red stripes appearing closer. Likewise, the red text on a blue background is harder to read (despite these colours being saturated) than the unsaturated yellow text on blue. This is because red and blue occur at opposite ends of the visible spectrum, and so the focusing mismatch arising from chromatic aberration is greater. The eye tires more easily trying to accommodate and focus on each colour separately. Source: Franklin et al. 2010. Reproduced with permission from John Wiley & Sons.*

The process by which the brain exploits parallax to compute depth perception and to estimate distances is known as stereopsis. The etymology of the word is from the Greek *stereo* meaning 'solid', and *opsis* meaning 'power of sight'. Birds, such as chickens and pigeons, who have eyes on the sides of their heads, are unable to view a single object stereoscopically with both eyes. They cannot use parallax to judge distances, as we do, and must judge distance by moving their heads – viewing an object with each eye independently. This is referred to as motion parallax[11] and is why pigeons can be seen continually bobbing their heads up and down. People who are stereo-blind, who lack vision in one eye, or have both long- and short-sighted eyes, often have poor depth perception, but it is not entirely absent. We learn strong monocular clues to judge both distance and depth. These include a change in the size of the retinal image and linear perspective: as a car drives away it apparently becomes smaller. With aerial perspective a change in colour occurs, that is, distant mountains appear blue. Experience and expectation of light and shade also help, as does interposition: an object which is overlapped appears further away. A lack of binocular vision affects about 12% of the population. Stereo-blindness will not prevent you from adjusting and using a microscope normally.

Béla Julesz, a Hungarian neuroscientist, developed the computer-generated random-dot stereogram that, with a stereoscope, allowed the brain to see 3D shapes from a two 2D flat images. Within the random dots a pattern is horizontally displaced in each image. Observers with binocular vision can process this disparate information to see a 3D image in depth above the random background. This proved that depth perception is a neurological phenomenon, rather than a purely visual process. Christopher Tyler, a student of Julesz, together with Maureen Clarke further developed the autostereogram (employing a repeating pattern) so that the 3D effect could be seen without the aid of special glasses or a stereoscope. These were later to give rise to the popular *Magic Eye* books.

1.6 Why We Need Optical Aids

When we want to see something in more detail, we bring it closer to our eyes. It is difficult or impossible, for example, to read a newspaper at a distance of several metres. As the newspaper is brought closer to the eye, the image is spread further over the retina (Figure 1.16), and the print becomes easier to read until a certain distance is reached, when it is no longer possible to see it clearly in focus. Alternatively, the object we are viewing may be so distant that it is not possible to get sufficiently close to bring it closer to the eye and see the detail that we would like.

In both cases we can use an optical aid to help us distinguish, or resolve, further detail. If we cannot get sufficiently close, we might use a telescope or a pair of binoculars. Where we can get close up to the object, we use a magnifying glass or microscope.

These examples highlight two of the limitations of the eye. When the newsprint is far from the eye, only a small image of it is formed on the retina. Several details may fall on any one sensitive photoreceptor so that individual fine details are not discriminated, or resolved. As the paper is moved closer, the size of the image increases and falls on more sensitive cells in the retina, enabling more detail to be

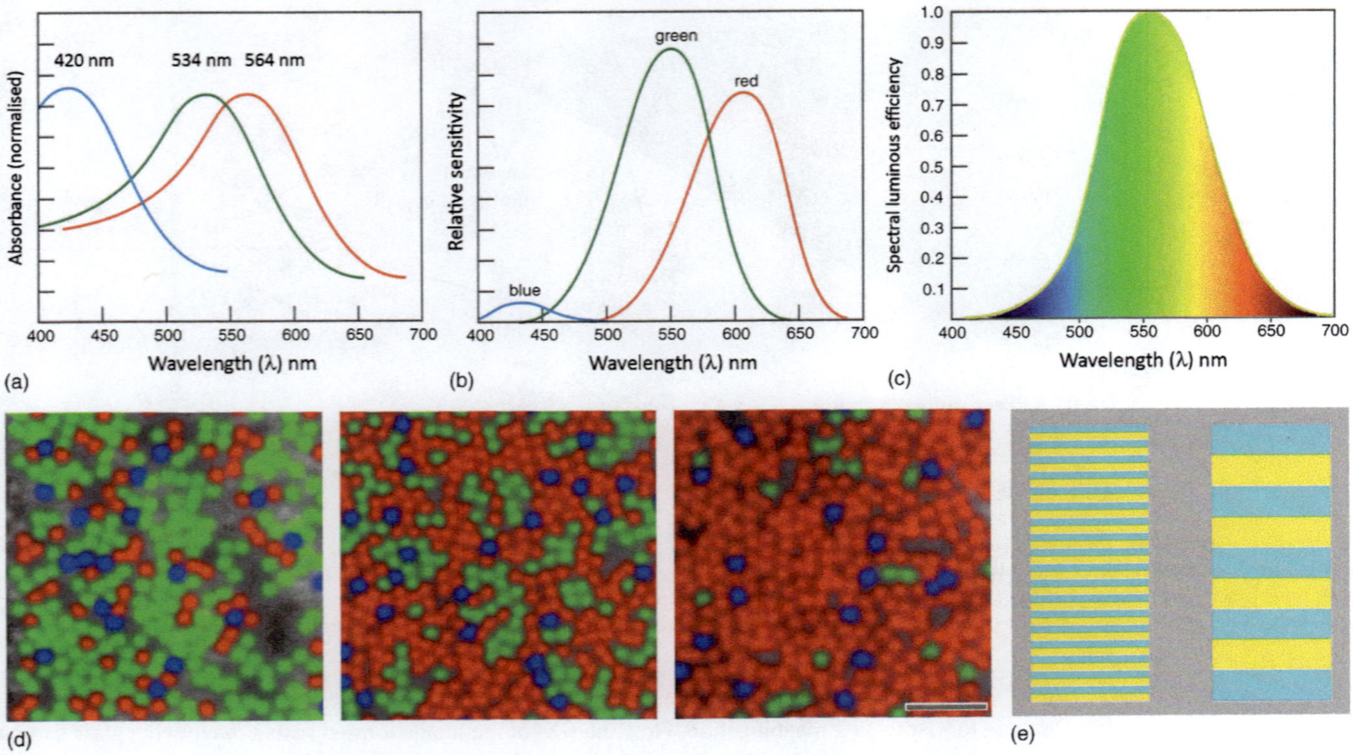

Figure 1.12 **Distribution of cones in the retina** *(a) shows the sensitivity of each of the three cones to the visible spectrum as a function of wavelength. (b) shows the relative sensitivity of red, green and blue cones to one another, and (c) shows the overall photopic sensitivity of the eye.*

We have very approximately 60% of red (L; long) cones and 30% of green (M; medium) cones but only 5% of blue (S; short) cones. These three types have peak wavelengths near 564 nm, 534 nm and 420 nm, respectively. The images are false coloured so that red, green and blue are used to represent the L, M and S cones respectively. (The true colours of these cones are bluish-purple, purple and yellow). The proportion of S cones is relatively constant across eyes, ranging from 3.9 to 6.6% of the total population of cone photoreceptors. However, the proportions of M and L vary widely. The mosaics in (d) illustrate the enormous variability in the L:M cone ratio – the left-hand panel has an L:M ratio of 1:3; the middle panel, 2:1 and the right-hand panel, an extreme 16:1. The scale bar represents 5 arcminutes, which is approximately equivalent to 25 μm.

It is because there are only a few blue (S) cones is that blue and yellow is seen with slightly poorer spatial resolution than other colours. This is demonstrated in (e). When held extended, at a large distance from the eye, the thicker blue and yellow stripes appear more saturated, and the thinner set of stripes appear less saturated. This is because the thin stripes are closer to the spatial resolution limit of the S cone mosaic. The consequence of this for microscopical fluorescence images (e.g. for CFP or YFP) is to pseudo-colour the label of interest white (for high contrast) or green (for high sensitivity) rather than cyan or yellow when discriminating fine detail in the image.

Source: (a,b) Tilley 2011. Reproduced with permission from John Wiley & Sons. (c) Skatebiker at English Wikipedia, via Wikimedia Commons. (d) Professor H. Hofer. Courtesy of Professor Hofer. (e) Adapted from Professor D. Heeger. http://www.cns.nyu.edu/~david/courses/perception/lecturenotes/color/color.html

discerned. The important factor here is the *viewing angle*, subtended at the eye by a feature in the object, and which increases as the object is moved closer to the eye.

The other limitation is exhibited when the lens of the eye reaches the end of its close-focusing adjustment (Figure 1.17, its *accommodation*), usually with the object at a distance of about 250 mm for a normal adult eye. Moving the object closer will now no longer help because, although the viewing angle is increased, the lens cannot bend sufficiently to maintain the image in focus. This *near point* defines the nearest distance of distinct vision, sometimes called the least distance of distinct vision. It is the point at which the lens of the eye is maximally accommodated.

Young people generally have a near point closer than 250 mm, but as we age, our power of accommodation decreases. The lens and ciliary muscle both become less flexible, and we gradually lose the power to accommodate. Our near point recedes, and we must perforce hold a newspaper (or this book!) further from our eyes[12] in order to read the print. This condition is called presbyopia (Figure 1.18) and is the reason why older people need spectacles for close vision.

Magnification can be expressed as a ratio between the tangent of the viewing angle when an object is observed with a magnifying system (such as a magnifying glass) and the tangent of the viewing angle subtended by the same object when observed by the naked eye at an agreed distance. A viewing distance of 250 mm, between the object and the vertex of the cornea, is taken as the International

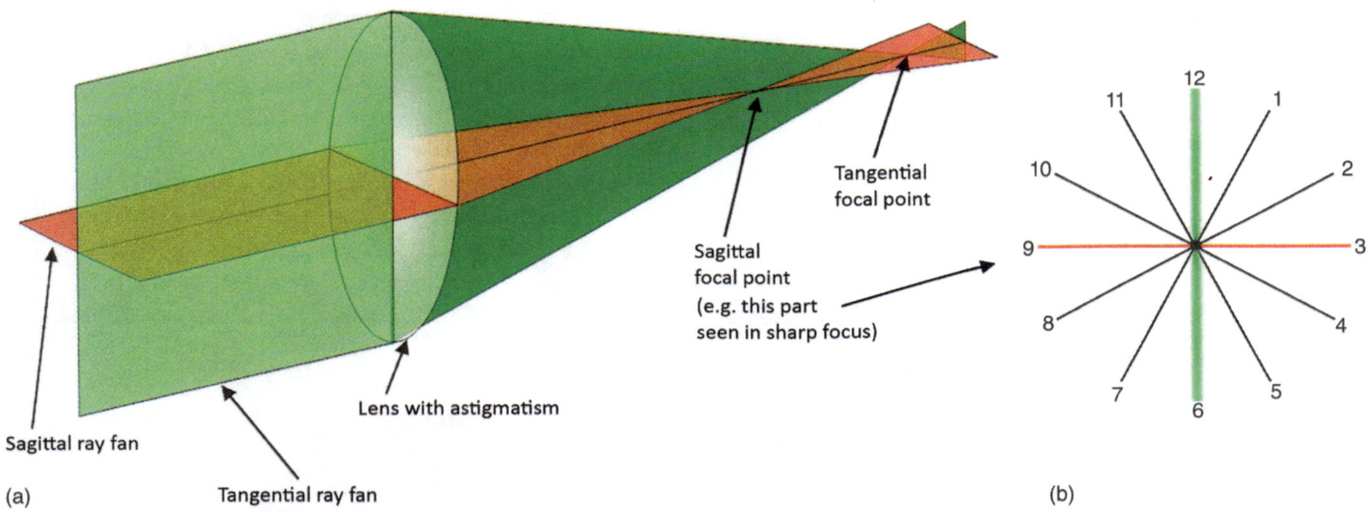

Figure 1.13 **Astigmatism** *The vertical and horizontal planes are identified as* tangential *and* sagittal *meridians, respectively. When imaged with an astigmatic lens, part of an object (e.g. a vertical bar) in the tangential plane (shown here in green) will be focused at a different point to a separate part of the same object (e.g. a horizontal bar) in the sagittal plane (shown here in red). This means that the entire object can never be wholly seen in sharp focus without the aid of corrective lenses. Source: Allen and Triantaphillidou 2011. Reproduced with permission from Taylor & Francis Group.*

Standard *Reference Viewing Distance* for expressing magnification. How the size of the image formed with a magnifying glass can be calculated is shown in Chapter 3.

With a magnifying glass or eyepiece, our eyes form a real image from a lens, which would otherwise form a virtual image. A magnifying lens in front of the eye enables the eye to focus an image from an object lying within the nearest distance of distinct vision. The magnifying glass produces an upright virtual image. This image is the same side of the lens as the object, apparently at infinity. The lens of the eye can thus relax to view the object instead of accommodating. The image is perceived by the eye as if it were at a viewing distance of 250 mm, the reference viewing distance. The eye can form a real image (Figure 1.19), yet the object lies only an inch or so away from the observer. A microscope eyepiece can conveniently be used as a simple magnifier: for example, for inspecting whether an objective requires cleaning. Invert the eyepiece and look through it in the 'reverse' direction. Since it has a short focal length (25 mm), the front of the eyepiece will have to be held very close to the surface of the objective, as explained in Chapter 3, Section 3.8.

In ophthalmology, and when using single lenses as simple magnifiers, the overall magnification is calculated as an *angular* measure. The angle θ subtended by the object at the nodal point of the eye determines the retinal image size. Sometimes it is advantageous not to have to calculate linear magnification from the ratio of image size to object size. Either it may not be possible to approach the object to measure its size or the image may be a virtual one so that its size cannot be directly measured.

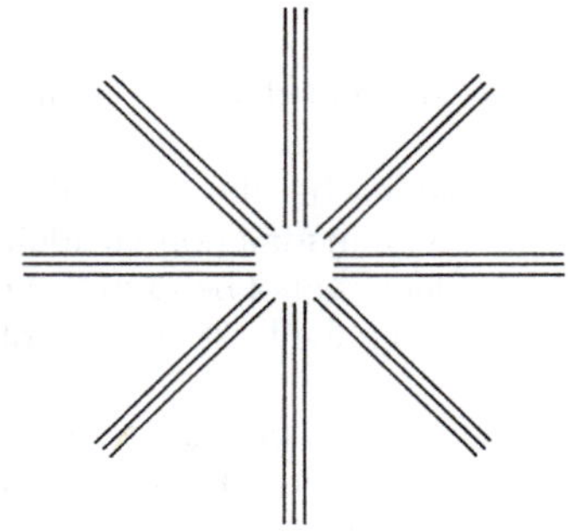

Figure 1.14 **Test for astigmatism** *Close one eye, and view this figure without glasses or contact lenses. Hold the figure sufficiently close so that all lines look blurred. Steadily move the figure away until all or only one set of lines is in sharp focus. If all lines are in sharp focus, you don't have astigmatism. If one set is in focus, rotate the figure to see if the lines become blurred (another set may become sharper). Re-orientate the figure as before; continue to move it until lines perpendicular to the first set are sharply focussed. Repeat the procedure with your glasses or contact lenses to see if your astigmatism is corrected. Source: Falk et al. 1986. Reproduced with permission from John Wiley & Sons.*

Figure 1.15 **The visual field** *If our eyes were to operate in a fixed fashion without employing microsaccadic movements to allow our brains to tile together a wider and sharper field of view, this is what we would see – assuming that the photopsins of the cones are not bleached. Our vision would suffer from spherical aberration, being particularly noticeable at the edges of our field of view.*

1.7 Using Lenses to Correct Eye Defects

The far point is the furthest point at which an object can be placed so that a clear image is formed on the retina. For people with normal vision, the far point lies at infinity (Figure 1.17). Emmetropia (from Greek *emmetros*, 'well-proportioned') describes the state of the normal eye when an object placed at infinity is in focus with the lens relaxed. For all practical purposes, infinity is considered to be greater than 30–40 times the focal length of the lens. For our eyes, this is greater than 6 metres or 20 feet away (6 metres is roughly 350 times 17 mm, the approximate focal length of the normal eye). The wavefront of light rays coming from an object at infinity are essentially parallel, and the rays are focused on the retina without any effort of accommodation.

Myopia is a condition, often called near-sightedness, whereby images that are far away from the eye appear blurred. This is either due to axial myopia or refractive myopia (Figure 1.20). In the case of axial myopia, the globe of the eyeball is too long, so the surface of the retina does not lie at the rear focal plane of the lens. Consequently a clear image cannot be formed. If the eyeball is of the correct dimensions but excessive curvature exists either in the cornea or the lens, then refractive myopia occurs.

Hyperopia (or hypermetropia; far-sightedness) exists when an unsharp retinal image is again formed (Figure 1.21), but in this case behind the retina, rather than in front as for myopia. Either the globe is squashed and too short, which is axial hyperopia, or the cornea

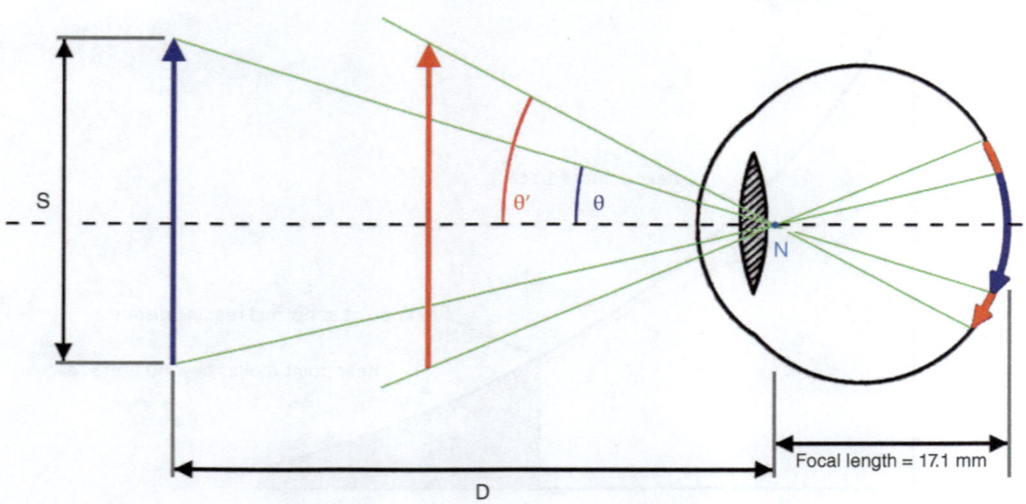

Figure 1.16 **Ray diagram of an image being formed on the retina** *The diagram shows the ray construction of a distant arrow in blue with the central ray passing through the nodal point of the lens of the eye. The nodal point, N, lies between the front and back focal points of the relaxed eye, which is 17.1 mm; it is effectively the centre of the lens. The visual angle θ is the angle subtended by the blue arrow at the eye, usually stated in degrees of arc. As the arrow is brought closer to the eye, the image is spread further over the retina, is detected by more and more cones and becomes easier to see. The visual angle, θ, in degrees, is given by $\theta = 2 \times \arctan(S/2D)$. Source: Adapted from Abbott 1984 with permission from Pearson.*

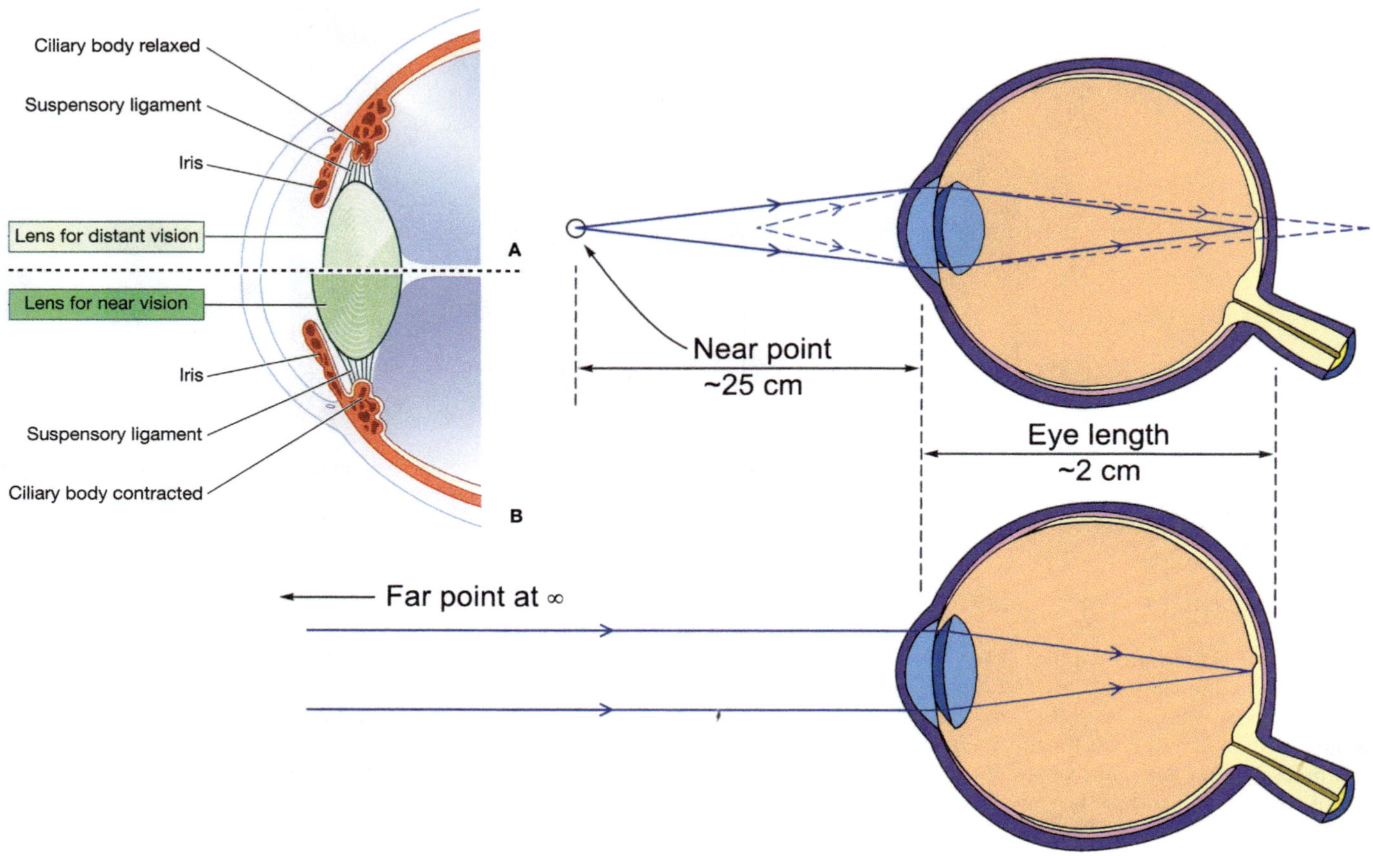

Figure 1.17 **Shape of the lens during accommodation** *In the relaxed eye the lens is about 3.6 mm thick at the centre; in the accommodated eye, it thickens to about 4.5 mm. The focal length changes from about 17 mm to about 14 mm, respectively. Source: (left) Waugh and Grant 2001. Reproduced with permission from Elsevier. (right) Adapted from Franklin et al. 2010 with permission from John Wiley & Sons.*

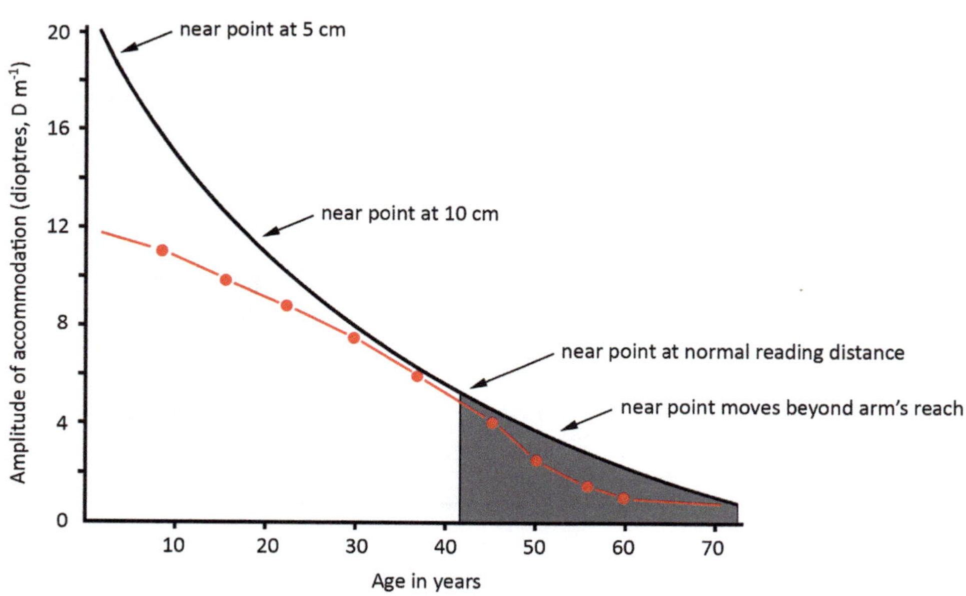

Figure 1.18 **Presbyopia** *As we age, our ability to accommodate is gradually lost as the lens and ciliary muscle become less flexible. The inverse hyperbolic curve shows how the near point recedes with age. The red line shows how the dioptric power of our eye decreases with age, from around 10 dioptres in mid-teens to 1 dioptre at 60-plus years. The dioptre measures the refracting power of a lens and is the reciprocal of the equivalent focal length in metres. The greater the dioptre value, the shorter the lens focal length. Source: Dr P. A. Howarth. Adapted with permission from Dr Howarth.*

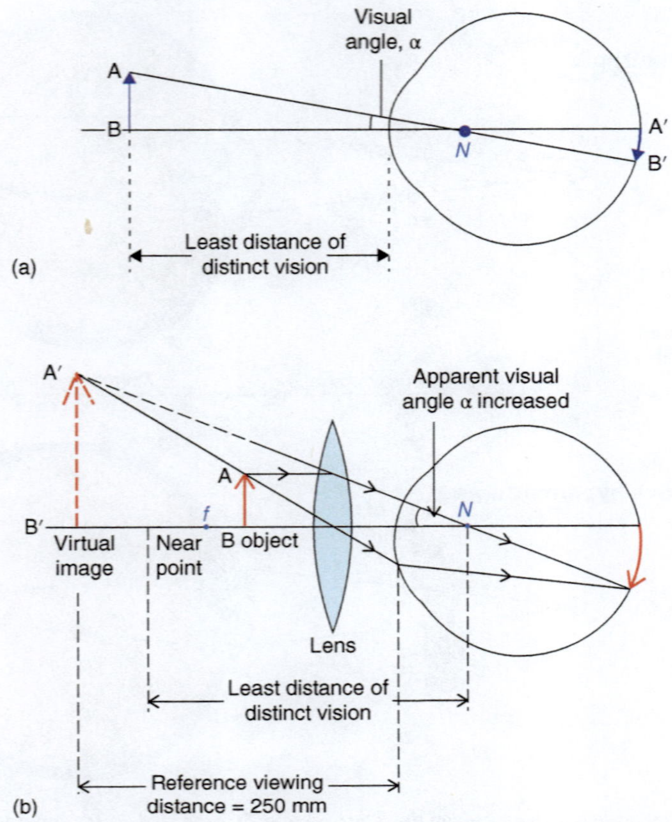

Figure 1.19 ***Action of a single lens to increase visual acuity*** *A positive (biconvex) lens acting as a magnifying glass or simple microscope. The object is located within the near point of the eye and within the focal length of the lens. The virtual image so formed is erect, laterally correct and magnified, and the apparent visual angle subtended at the eye is increased. Source: Bradbury 1967. Reproduced with permission from Elsevier.*

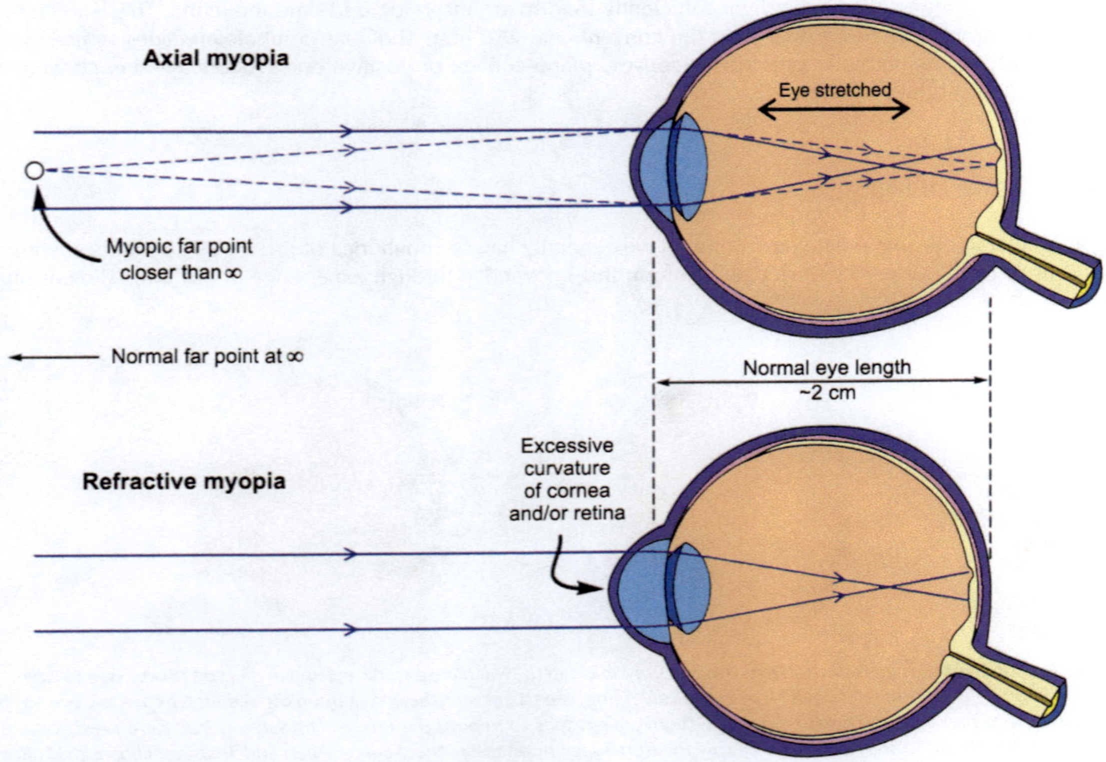

Figure 1.20 ***Myopia*** *In myopia the image is focused in front of the retina. Source: Franklin et al. 2010. Reproduced with permission from John Wiley & Sons. Compare this figure to Figure 1.17.*

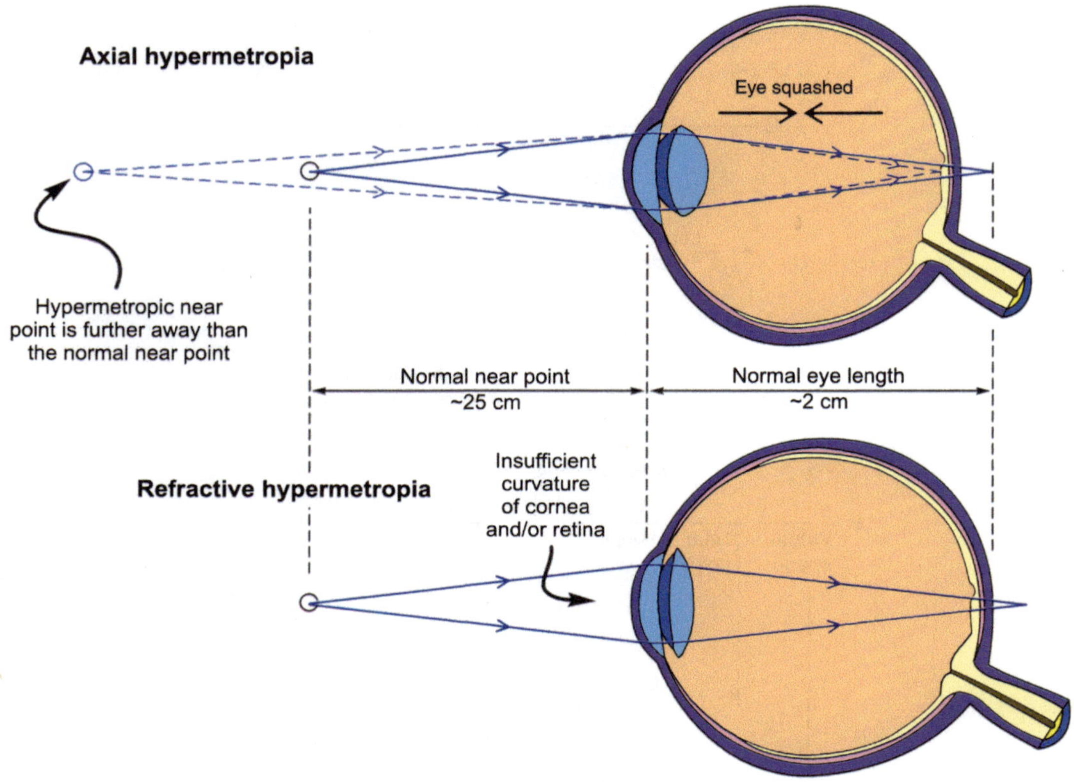

Figure 1.21 **Hyperopia** *In hyperopia the image is focused behind the retina. Source: Franklin et al. 2010. Reproduced with permission from John Wiley & Sons.*

and/or lens cannot bend or refract the rays of light sufficiently to form an image focused onto the retina. This is refractive hyperopia. The near point of a hyperopic eye lies further than the conventional 250 mm. The near point also recedes as presbyopia gradually occurs (Figure 1.18). Opticians normally prescribe biconvex, plano-convex or positive convex-concave (i.e. convergent) meniscus lenses (Figure 1.22) to correct these defects.

1.8 Seeing the Scientific Image

When viewing an object in our immediate environment, we generally have a rough idea of its size by comparison with other objects or people around us or from our own knowledge. We form this knowledge through experience of our three-dimensional world and

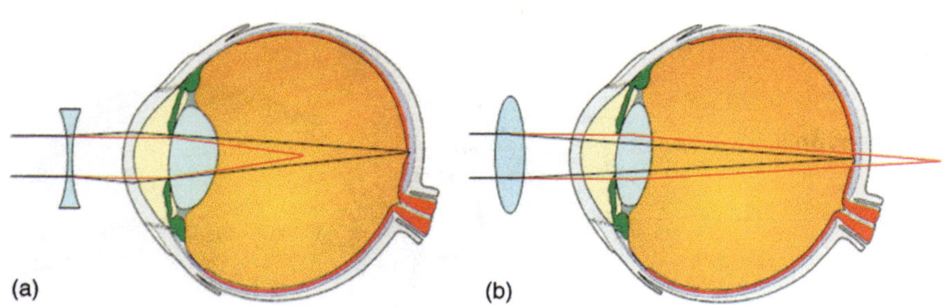

Figure 1.22 **Correction of myopia, hyperopia and presbyopia** *A myopic eye (a) will have excessive positive dioptric power (see Chapter 3, Section 3.4, for an explanation of dioptres) and is corrected with a negative (diverging) lens to refocus the image upon the retina. A hyperopic eye (b) has insufficient positive dioptric power, requiring correction with a positive (converging) lens to refocus the image. The same is true for a presbyopic eye, which has lost the power to accommodate sufficiently to form a sharply focused image upon the retina. Source: Allen and Triantaphillidou 2011. Reproduced with permission from Taylor & Francis Group.*

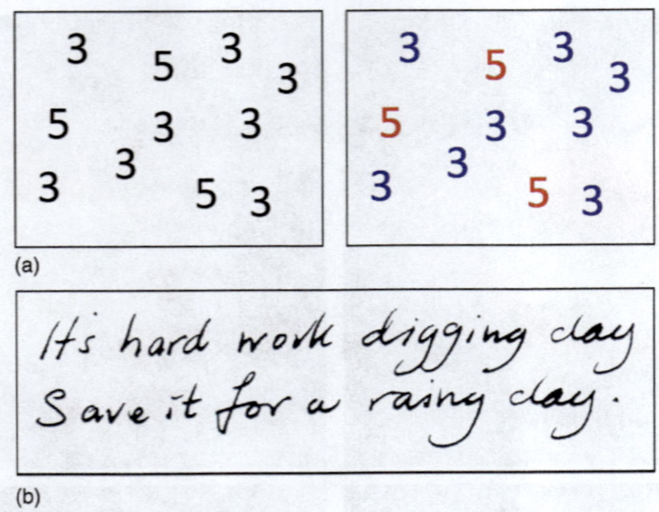

Figure 1.23 **Using colour to discriminate patterns and detail** *The eye is good at recognising patterns, colours and shapes: colour is useful for perceiving related groups quickly; pattern recognition helps to understand an otherwise ambiguous message, arising from use of the identically written words 'clay' and 'day'. Source: Russ 2004. Reproduced with permission from John Wiley & Sons.*

its perspective. The eye is very good at colour and pattern recognition and separating shapes from each other. The primary function of human vision is recognition, whether searching for food, avoiding danger or seeking a mate. We do this by comparison, based on experience. This hardwired ability is excellent in enabling us to cross the road and meet a friend, for example, but it is not suited to analysing scientific images, for which we have no requirement in evolutionary terms.

Because we scan our visual field rapidly and our mind 'fills in' details from previous experience, we naturally invest only the briefest amount of time assessing our environment for food, danger or mates. We can recognise a familiar scene with the barest minimum of visual clues because of the human ability to discriminate patterns, colours and shapes. This is what cartoonists rely upon to convey their message. Figure 1.23 illustrates our ability to employ colour to lift specific details from amongst a pattern or to use both context and redundancy to correctly understand a message when the image is wrong or ambiguous. The drawback is that we may miss subtle changes in our images: alterations in contrast, grey levels, colour or dynamic changes over time, if these changes are not abrupt. We certainly cannot hold all the details of an image in our mind's eye; our mental images are transient, comprising scant detail at low resolution. We are also susceptible to change blindness, and may fail to notice subtle changes details in successive images.

Computers, however, are very good at processing large amounts of data, at analysing and measuring, which we cannot do effectively (e.g. particle tracking through a Z-stack data set: we track particles by toggling back and forth between consecutive frames to decide on correspondences). Many microscopical samples are sections of tissues or flat surfaces of opaque materials. In these cases the images are two-dimensional representations of part of a 3D object. Without specific experience of the object being studied, it is very difficult for most people to place images of plane sections into a proper context. We are more used to seeing the external surfaces of solid objects. For most people it is challenging to imagine the appearance and shape of a 2D section cut through a familiar object (Figure 1.24). Without prior experience, it is more difficult to perform the exercise in reverse and to reconstruct a 3D object from a set of plane images (Figure 1.25). Yet this is what microscopists have to do.

While computers excel at handling data sets in ways that we find challenging or impossible, it is, however, very difficult to write effective algorithms and programmes to enable computers to fully 'discriminate' and analyse patterns and shapes in the way that humans do. Facial recognition is a case in point: we perceive a face as very different if only a few key features are altered, but it is very difficult to programme a computer to do the same with equal sophistication. When we recognise someone or something, we unconsciously endow the familiar person or scene with a full set of characteristics, according to our expectation and preconceptions. In short we have 'seen and interpreted' not '*observed* and interpreted'. This makes us prone to bias. When publishing scientific images, we can easily be biased; journal editors are aware that our 'typical' cell may really be 'the nicest picture that we have'.

Images are not merely pretty pictures, they are scientific data: a microscopical image is essentially a sampled 'photon intensity map'. The context of the image is important for interpreting data correctly. This point is made in the introduction to this book, in advice given by Henry Baker from one of the earliest instruction manuals on microscopy, when instruments collected far less complex data sets than they do nowadays.

1.9 Sizes of Objects

The eye-brain combination is very poor at estimating – let alone quantifying – size, length, area or volume. While we have a very well-developed sense of recognising familiar objects and patterns, we find it harder to describe and may fail to detect those that are

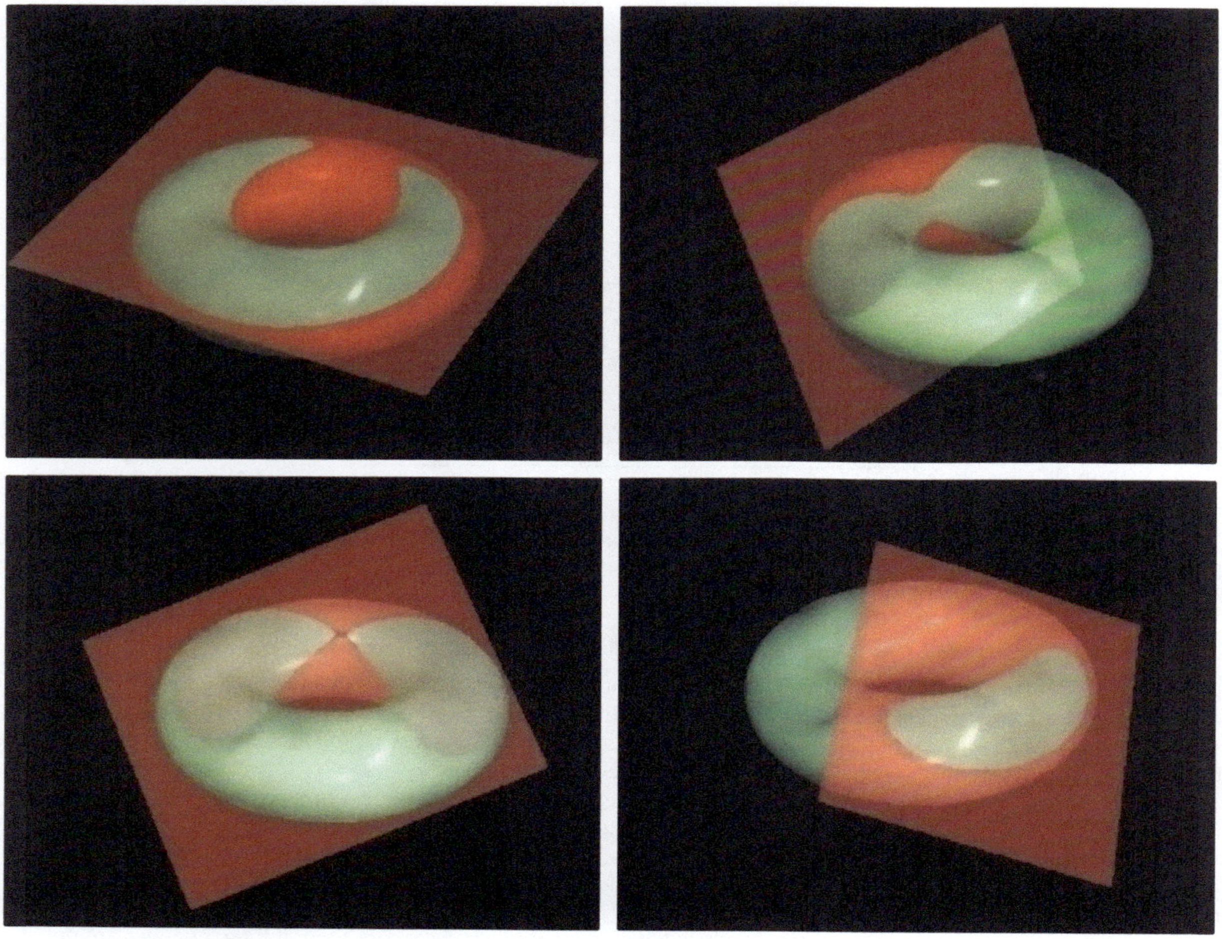

Figure 1.24 **Calculating the appearance and shape of a 2D section** *A bagel is a simple torus, which we recognise easily. It is more difficult to predict the shape of various surfaces as the bagel is cut through with a knife. Source: Russ 2004. Reproduced with permission from John Wiley & Sons.*

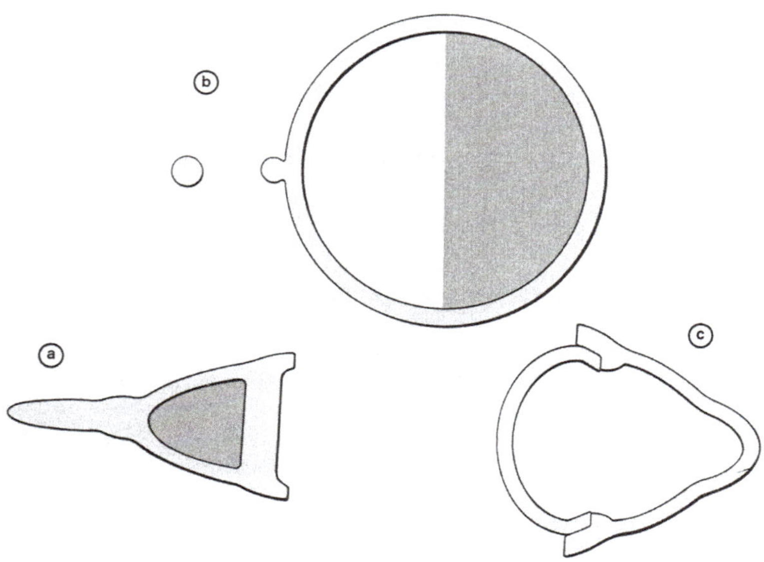

Figure 1.25 **Reconstructing a 3D object from a set of plane images** *Here the reverse problem to that presented in Figure 1.24 is illustrated. Three 2D sections derived from cutting through a familiar object are shown. It is difficult to visualise the original 3D object in our minds. Once presented with the answer (which may be found at the end of the chapter) the solution appears very simple! Source: Rogers 1983. Reproduced with permission from Elsevier.*

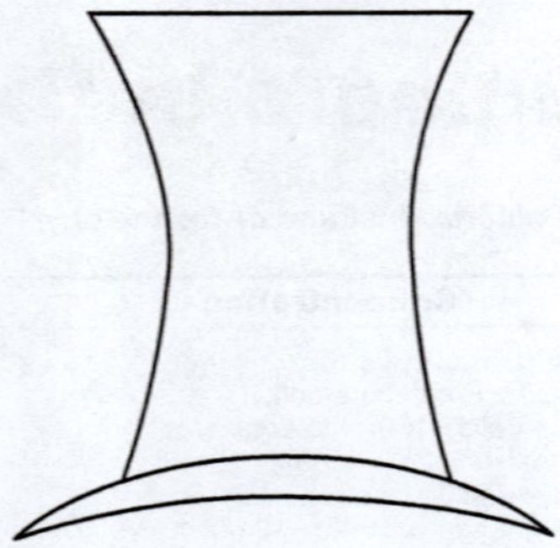

Figure 1.26 ***Top hat illusion*** *Our brains generally consider vertical dimensions to be more important than horizontal ones. Both dimensions are actually equal in length. Source: Russ 2004. Reproduced with permission from John Wiley & Sons.*

unfamiliar. Optical illusions depend upon these two weaknesses in order to work. It is also why eyewitness accounts of accidents or crime scenes vary so much and are often unreliable. Figure 1.26 shows a simple example: our brains generally consider vertical dimensions to be more important than horizontal ones. The top hat is, in fact, drawn within a square, and both dimensions are equal in length. Nevertheless, measurement is an essential part of the description and understanding of any object and/or its image – to put it into a proper perspective.

In microscopy, unless we are very familiar with the two-dimensional image presented to us, we have no idea of scale. The problem is compounded when studying a small field of view (at high magnification) with few or no features of reference that we can recognise. Is the image taken with a 10x or a 100x objective? Has it been further enlarged? Unless such features whose size we know are present in the image, we have no way of knowing the overall size of our image or the size of smaller components within it. We need to include an absolute linear scale with our images (for a description of how to do this, see Chapter 31, Section 31.8), or else we need to know the size limits of the objects we are studying. An appreciation of the scale and size of the images that we take with the microscope is important (although the importance of this is often overlooked) so that we can communicate this to our audience. Quite apart from being scientifically rigorous, giving our audience a scale with our image(s) imparts confidence and trust in what we have to say.

This book is only concerned with the light microscope, which will magnify from approximately 10 times[13] at the lower extreme to about 1500 times as an upper limit and resolve detail down to about 0.2 μm. The light microscope uses photons to illuminate the object and so form an image, but much greater detail may be resolved and seen if electrons are used. An electron microscope can resolve detail down to 0.2 nm under favourable circumstances.

A useful way to remember these values is 'the rule of twos'. The limit of resolution of the human eye is very approximately 0.2 mm at 250 mm distance; the limit of resolution of the light microscope 0.2 μm; the limit of the electron microscope, approximately 0.2 nm (depending on the sample).

From ants to atoms, there is a widely diverse range of both living and nonliving biological objects studied using the microscope. They range in size from metres to nanometres (Figure 1.27). Representative sizes of cells and organelles are given in Box 1.1. Until recently there was a dearth of published information on the sizes of cells and organelles, but the very useful BioNumbers online database (for the URL Internet link, refer to the end of the chapter) has redressed this issue. Entries have supporting references and a unique ID number; hence an audit trail underpins all data entered. Although the life we see around us appears very diverse, it is remarkably similar at the molecular level in the size range from 10 μm to 0.1 nm. The light microscope is used to study the upper part of this range: from 10 μm to 0.2 μm. The rapidly developing field of so-called super-resolution microscopy (see Chapter 26) – better called diffraction unlimited fluorescence microscopy – is driving the lower limit of resolution down towards 0.1 μm and beyond.

Box 1.1 The Size of Biological Structures

Many microscope users who come to use core facility microscopes often come to acquire images of labelled proteins and cell structures within tissues with little, if any, idea regarding the size of the structures they wish to study. So much thought and preparation has gone into the experimental design and study of the biological relevance of their proteins of interest that they have neglected to think sufficiently about the size, thickness and volumes of the samples they have prepared. A proper understanding of the sizes of the cells and tissues you look at under the microscope will allow you to decide whether to acquire an image at the limit of resolution or not, to choose the correct objective(s) and make any necessary adjustment to the pixel size of the PMT or camera detector for successful image acquisition. The associated database can be found at http://Bionumbers.org.

SnapShot: Key Numbers in Biology

Uri Moran,[1] Rob Phillips,[2] and Ron Milo[1]
[1]Weizmann Institute of Science, Rehovot, Israel; [2]California Institute of Technology, Pasadena, CA, USA

Cell

Cell size

Bacteria (*E. coli*): ≈0.7-1.4 μm diameter, ≈2-4 μm length, ≈0.5-5 μm³ in volume; 10^8-10^9 cell/ml for culture with OD_{600}≈1

Yeast (*S. cerevisiae*): ≈3-6 μm diameter ≈20-160 μm³ in volume

Mammalian cell volume: 100-10,000 μm³; HeLa cell: 500-5000 μm³ (adhering to slide ≈15-30 μm diameter)

Length scales inside cells

Nucleus volume: ≈10% of cell volume

Cell membrane thickness: ≈4-10 nm

"Average" protein diameter: ≈3-6 nm

Base pair: 2 nm (D) x 0.34 nm (H)

Water molecule diameter: ≈0.3 nm

Energetics

Membrane potential ≈70-200 mV → 2-6 k_BT per electron (k_BT ≡ thermal energy)

Free energy (ΔG) of ATP hydrolysis under physiological conditions
≈40-60 kJ/mol → ≈20 k_BT/molecule ATP; ATP molecules required during an *E. coli* cell cycle ≈10-50 × 10^9

ΔG° resulting in order of magnitude ratio between product and reactant concentrations:
≈6 kJ/mol ≈60 meV ≈2 k_BT

Concentration

Concentration of 1 nM:
in *E. coli* ≈1 molecule/cell;
in HeLa cells ≈1000 molecules/cell

Characteristic concentration for a signaling protein: ≈10 nM-1 μM

Water content: ≈70% by mass; general elemental composition (dry weight) of *E. coli*: ≈$C_4H_7O_2N_1$; Yeast: ≈$C_6H_{10}O_3N_1$

Composition of *E. coli* (dry weight): ≈55% protein, 20% RNA, 10% lipids, 15% others

Protein concentration: ≈100 mg/ml = 3 mM. 10^6-10^7 per *E. coli* (depending on growth rate); Total metabolites (MW < 1 kDa) ≈300 mM

Division, replication, transcription, translation, and degradation rates

at 37°C with a temperature dependence (Q10) of ≈2-3

Cell cycle time (exponential growth in rich media): *E. coli* ≈20-40 min; budding yeast 70-140 min; HeLa human cell line: 15-30 hr

Rate of replication by DNA polymerase:
E. coli ≈200-1000 bases/s;
human ≈40 bases/s. Transcription by RNA polymerase 10-100 bases/s

Translation rate by ribosome: 10-20 aa/s

Degradation rates (proliferating cells):
mRNA half life < cell cycle time;
protein half life ≳ cell cycle time

Diffusion and catalysis rate

Diffusion coefficient for an "average" protein: in cytoplasm D≈5-15 μm²/s → ≈10 ms to traverse an *E. coli* → ≈10 s to traverse a mammalian HeLa cell; small metabolite in water D≈500 μm²/s

Diffusion-limited on-rate for a protein:
≈10^8-10^9 $s^{-1}M^{-1}$ → for a protein substrate of concentration ≈1 μM the diffusion-limited on-rate is ≈100-1000 s^{-1} thus limiting the catalytic rate k_{cat}

Genome sizes and error rates

Genome size:
E. coli ≈5 Mbp
S. cerevisiae (yeast) ≈12 Mbp
C. elegans (nematode) ≈100 Mbp
D. melanogaster (fruit fly) ≈120 Mbp
A. thaliana (plant) ≈120 Mbp
M. musculus (mouse) ≈2.6 Gbp
H. sapiens (human) ≈3.2 Gbp
T. aestivum (wheat) ≈16 Gbp

Number of protein-coding genes:
E. coli = 4000; *S. cerevisiae* = 6000;
C. elegans, A. thaliana, M. musculus, H. sapiens = 20,000

Mutation rate in DNA replication:
≈10^{-8}-10^{-10} per bp

Misincorporation rate:
transcription ≈10^{-4}-10^{-5} per nucleotide
translation ≈10^{-3}-10^{-4} per amino acid

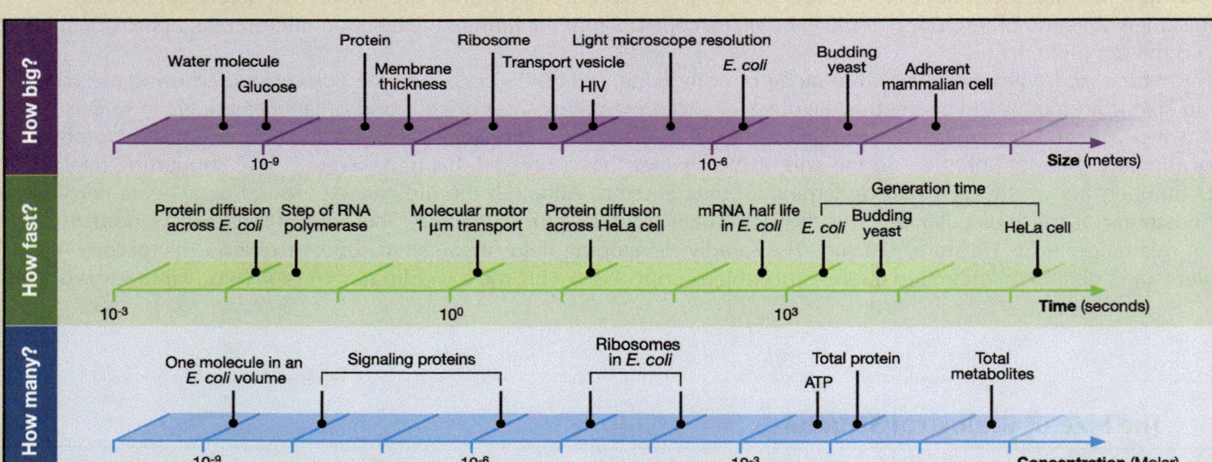

Useful biological numbers extracted from the literature. Numbers and ranges should only serve as "rule of thumb" values. References are in the online annotated version at www.BioNumbers.org. See the website and original references to learn about the details of the system under study including growth conditions, method of measurement, etc.

1 Cell *141*, June 25, 2010 ©2010 Elsevier Inc. DOI 10.1016/j.cell.2010.06.019

See online version for legend and references.

Source: Moran et al. 2010. Reproduced with permission from Elsevier.

These values are approximate. Prokaryotes can range between 1 and 10 μm; eukaryote animal cells can be from 10 to 30 μm and higher plant cells generally have a wider distribution of overall size from 10 to 100 μm.

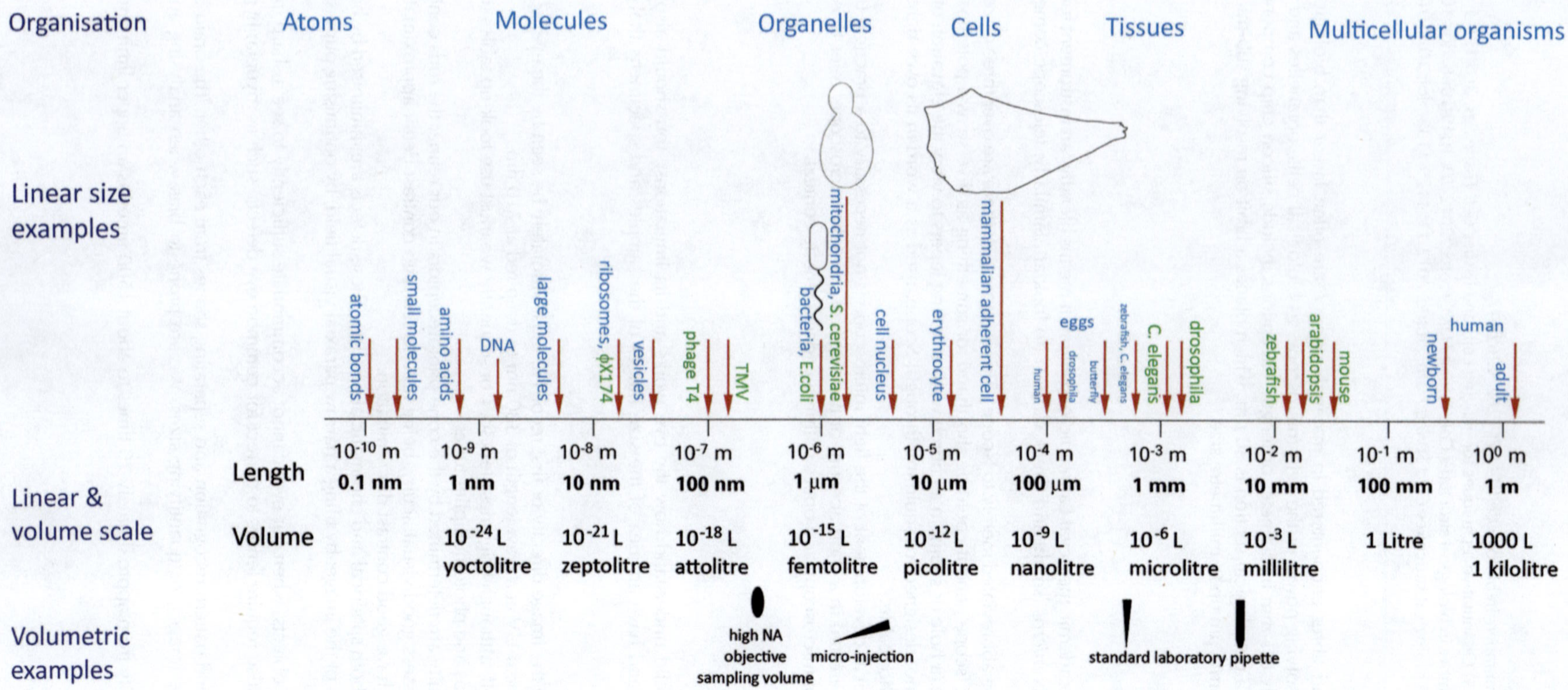

Figure 1.27 ***The size range of specimens studied with microscopes*** *The human eye will resolve details down to approximately 0.2 mm; the light microscope will resolve detail down to 0.2 μm; the electron microscope will resolve detail as fine as 0.2 nm. These are best-case values in each case, but it is possible to study fine detail throughout nine orders of magnitude. Source: Adapted from Davey and Lord 2003 with permission from Oxford University Press.*

References

Moran, U et al (2010) Snapshot: Key numbers in biology *Cell 141/7*: 1262–1262.e1
Von Bartheld, CS & Wouters, FS (2015) Quantitative techniques for imaging cells and tissues *Cell Tissue Res. 360/*1: 1-4
West, MJ (2012) Estimating object number in biological structures *Cold Spring Harb. Protoc.* 2012 *10*: 1049-1066. doi: 10.1101/pdb.top071423
West, MJ (2012) Estimating volume in biological structures *Cold Spring Harb. Protoc.* 2012 *11*: 1129-1139. doi: 10.1101/pdb.top071787

The sizes of particles, crystals and fibres encountered in materials science are often larger than biological specimens. They do not usually approach the limit of resolving power of the light microscope, as biological cells, organelles and macromolecules do. In metallurgy, fine crystals are generally 0.1 mm in diameter or length; exceptions include silicon chip components and sintered metal products. The standard thickness of a petrological section is 30 μm, which places a limit on resolving sub-micrometre detail, and the structures observed are generally from 1 μm up to millimetre size.

1.10 Nomenclature and History

The term 'microscope' comes from academic medieval Latin, *microscopium*. It means literally 'an instrument for viewing what is small', derived in turn from the Greek prefix *micros* (small) and Greek *skopein* (to look at). Similarly, telescope comes from the Greek prefix *tele* (afar).

The term microscope should not be abbreviated merely to 'scope'. This is imprecise: there are over three dozen scientific or engineering instruments which use the suffix 'scope', and it could be shorthand for something as diverse as a polemoscope or 'jealousy glass' for use at the opera, or a hagioscope (a hole or squint in a church wall) enabling lepers to view the elevation of the host from outside a medieval church – a microscope is not designed for squinting through! Scope, used as a word in its own right, means opportunity, or else refers to the slack left in an anchor cable.

Knowing the evolution and historical development of the light microscope is not necessary to a practical understanding of how to operate it successfully. For those interested in a short account of the origin of the light microscope, refer to Appendix 6. Appendix 7 lists significant dates in the history of microscopy and optics within a wider scientific context.

1.11 Chapter Summary

At the end of this chapter you should understand how the eye works and its limitations. You should also know why we need a microscope to study small objects and have an idea of the size ranges of the samples and specimens that are examined with the microscope.

1. Any image, whether we observe the image directly or first record it, must ultimately be seen by the eye.
2. Our eyes respond to light from near UV at a wavelength of 380 nm to deep red at 710 nm.
3. We discriminate colour very well, although brightness less so. Consequently we must use look-up tables (LUTs) to set the limits of dynamic range for digital cameras and photomultiplier tubes.
4. Our colour vision and visual acuity are determined by the cone photoreceptors in our retina; the rods control the sensitivity of our dark-adapted sight. Humans possess good visual acuity, being able to see high contrast detail approximately 0.1 mm in size. We possess stereoscopic vision and have good contrast discrimination.
5. The single-lens eye suffers from both spherical and chromatic aberration. These defects are minimised by both physiological design and psychological interpretation of the image: by tiling a narrow paraxial visual field, by optimising pupil size and by reducing our response to blue light.
6. We require lenses to view small objects, when our eyes cannot accommodate sufficiently to see a sharp image at sufficient size in order to resolve fine detail. We also require lenses to correct for common eye defects, such as myopia, hyperopia, presbyopia and astigmatism.
7. The eye is very good at colour and pattern recognition and separating shapes from each other. The brain is poor at reconstructing a 3D object from a set of plane images, or quantifying sizes. We are prone to bias when analysing and interpreting scientific images.
8. The light microscope will magnify from approximately 10 times to about 1500 times as an upper limit and resolve detail down to 0.2 μm.

Answer to Figure 1.25

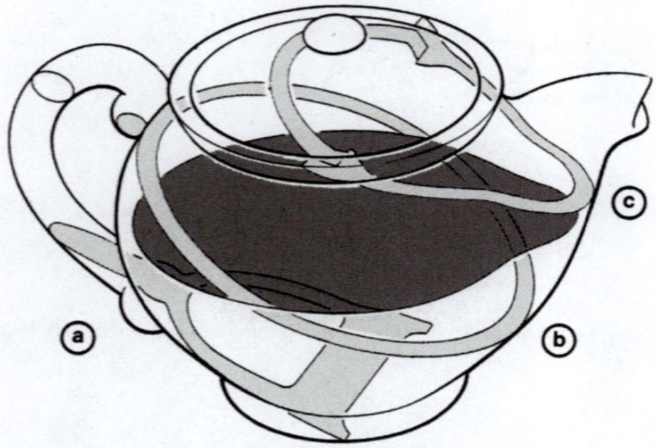

Source: Rogers 1983. Reproduced with permission from Elsevier.

Key Reading

Gross, H., et al. (2008). *The eye*. In: Gross, H., Blechinger, F., and Achtner, B., (ed.) *Handbook of Optical Systems* Vol. 4, Survey of Optical Instruments. Weinheim: Wiley-VCH, pp. 1–87.

- A comprehensive chapter, with all the facts and figures about the eye well explained and easily accessible. At the time of writing, it is available as a free download from the Wiley website.

References

Abbott, A.F. (1984). 'Lenses and optical instruments'. In: *Ordinary Level Physics*, 4th Ed. Heinemann Educational Books.
Bradbury, S. (1967). *The Evolution of the Microscope*. Oxford: Pergamon Press.
Falk, D., Brill, D., and Stork, D. (1986). *Seeing the light*. New York: John Wiley & Sons.
Franklin, K. et al. (2010). *Introduction to Biological Physics for the Health & Life Sciences*. New York: John Wiley & Sons.
Harvey, L. et al. (2008). *Molecular Cell Biology*, 6th Ed. WH Freeman & Co.
Heeger, D. (2006). 'Perception Lecture Notes: Color'. Department of Psychology, New York University.
Hofer, H. (2005). 'Different sensations from cones with the same photopigment'. *Journal of Vision* May 19 5/5: 444–454.
Jenkin, R., (2010). 'The human visual system'. In: Allen, E. and Triantaphillidou, S. (ed.) *The Manual of Photography*, 10th Ed., Focal Press, Elsevier.
Kolb, H. 'Facts and Figures Concerning the Human Retina'. (2005). In: Kolb H., Fernandez, E., Nelson, R., (ed.). Webvision: The Organization of the Retina and Visual System [Internet]. Salt Lake City (UT): University of Utah Health Sciences Center; 1995-.
Land, M.F. and Nilsson, D.E. (2001). *Animal Eyes*, 1st Ed. Oxford: OUP.
Lodish, H. et al. (2008) *Molecular Cell Biology*, 6th Ed. WH Freeman & Co.
MacEvoy, M. (2009). 'Photopic & scoptopic sensitivity functions'.
Rogers, A.W. (1983). *Cells and Tissues: An introduction to Histology and Cell Biology*. Academic Press.
Rostgaard, J. and Qvortrup, K. (1999). 'A note about retinal structure and visual acuity'. *Acta Opthalmologica Scandinavica 77/1*: 45–49.
Russ, John C. (2004). 'Seeing the Scientific Image'. Proceedings of the Royal Microscopical Society *Dec. 39/4*: 267–281.
Stevens, Alan and Lowe, James. (1996). Chapter 19, 'Special Senses'. In: *Human Histology*, 2nd Rev. Ed. (1996). Mosby, Elsevier Health Sciences.
Tilley, R.J. (2010). *Colour and the optical properties of materials* 2nd Ed. Chichester: Wiley.
Tortora, S. (2003). *Principles of Anatomy & Physiology*, 10th Ed. New York: John Wiley & Sons.
Wang, Q. et al. (2011). 'Spectroscopy'. In: Glencross, H., Ahmed, N., and Wang, Q. (ed.) *Biomedical Science Practice* (2011) Fundamentals of Biomedical Science series, Oxford: Oxford University Press.
Waugh, A. and Grant, A. (2001). 'The special senses'. In: *Anatomy and Physiology in Health and Illness*, 9th Ed. Churchill, Livingstone: Harcourt Health Sciences.
Zimmerman, T., Pepperkok, R., Stevens, D., Girod, A., and Reitdorf, J. (2003). In: Davey, J. and Lord, M. *Essential Cell Biology: a Practical Approach*, Vol. 1: Cell Structure. Oxford University Press.

Further Reading

Land, M.F. and Nilsson, D.E. (2012). *Animal Eyes*. 2nd Ed. Oxford: OUP.
Mason, P. (1982). *The light fantastic*. London: Penguin.
McIlwain, J.T. (1996). *An Introduction to the Biology of Vision*. Cambridge: Cambridge University Press.

Originally published as three articles in the *Proceedings of the Royal Microscopical Society*, the article is available on-line from John Russ' website: http://www.drjohnruss.com/downloads/seeing.pdf (accessed 1 June 2018)

Part 1: *Proceedings of the Royal Microscopical Society* June 39/2: 97–114.
Part 2: *Proceedings of the Royal Microscopical Society* Sept 39/3: 179–193.
Part 3: *Proceedings of the Royal Microscopical Society* Dec 39/4: 267–281.

Internet Resources

BioNumbers database. URLs:
1. http://bionumbers.hms.harvard.edu/
2. http://en.wikipedia.org/wiki/Bionumbers

Optical radiation terminology and units (Newport). URL: http://assets.newport.com/webDocuments-EN/images/Optical_Radiation_Terms_And_Units.PDF

Notes

1. The familiar terms 'intensity' and 'brightness' are subjective attributes, having no measurable units. The term 'irradiance' or 'radiant flux density', which measures the radiative flux of electromagnetic radiation incident on a surface in watts per square metre (W/m^2) should be used instead of 'intensity'. However 'intensity' is widely understood as a popular term for the qualitative measure of the amount of light present, or incident, upon a surface. Two parallel systems are used to measure light. Radiometric quantities measure absolute electromagnetic power; photometric quantities characterise light as perceived by the human eye. The eye has different sensitivities to light of different wavelengths within the visible spectrum. The photometric unit equivalent to irradiance (E_e) is illuminance (E_v). Standards organizations recommend that photometric quantities be denoted with a suffix 'v' (for visual) to avoid confusion with radiometric quantities.

   Luminous flux (Φ_v) is the measure [in lumens; lm = cd·sr] of the perceived power of light, the rate of flow of energy – over the visible spectrum – emitted by a source in all directions. When adapted for bright conditions (photopic vision), the eye is most sensitive to green-yellow light at 555 nm. Luminous flux (also called luminous power) is the photometrically-weighted radiant flux, or power. Light with the same radiant flux (Φ_e; watts) at other wavelengths, say red, has a lower luminous flux. A 1000 watt heater gives off a large radiant flux, but as a light source, it radiates very few lumens. Most of the energy is in the infrared, leaving only a dim glow visible to the eye. Luminous flux differs from radiant flux, which is the measure of the *total* power of light emitted in all directions and measured in watts (W), because luminous flux is adjusted to reflect the varying sensitivity of the human eye to different wavelengths of light. For monochromatic light of 555 nm, 1 watt is equal to 683 lumens. The solid angle in every direction all around a point is equal to 4π sr (steradians). The total luminous flux of a one-candela point source is 4π lumens.

   Luminous intensity (I_v) is slightly different from luminous flux. Luminous intensity is the perceived power *per unit solid angle* projected in a given direction, a measure [in candelas; cd = lm/sr] i.e. of luminous flux per unit solid angle. Radiant intensity (I_e), the corresponding radiometric unit, is measured in watts per steradian (W/sr).

   Illuminance is a measure [lux = lm/m^2] of how much luminous intensity from a source is incident on a surface, spread over a given area; it is a measure of the *intensity* of the incident light. Illuminance is inversely proportional to area. Luminance (L_v) is illuminance ($lm/m^2/sr$) per unit solid angle, whilst the radiometric equivalent is radiance (L_e), which is irradiance ($W/m^2/sr$) per unit solid angle per unit projected area: the flux per unit area received on a surface, for example, by a detector.

   Luminance is a measure [cd/m^2] of the 'density' of the luminous intensity per unit of light travelling in a given direction, and of how much luminous flux will be detected by an eye looking at an illuminated surface. It describes the amount of light that passes through or is emitted from a particular area, and falls within a given solid angle. Luminance most resembles the perception of brightness. Luminance is often used to characterise emission or reflection from flat, diffuse surfaces. Lightness is the perceptual correlate of diffuse reflectance. For example, a well-lit room has a combined high luminous flux; the light is divergent, so it fills the room. A laser pointer has very low luminous flux (it cannot illuminate a room) but a dangerously high luminous intensity (and luminance) off a shiny surface. Luminance is sometimes incorrectly referred to by the term luminosity.

   Luminous efficacy (η) is a measure [lumen per watt = lm/W] of how well a light source produces visible light. It is expressed as the ratio of luminous flux to radiant flux or luminous flux to power consumption, depending on context.
2. For those interested, this discrimination of luminance is discussed further in Box 11.1. It is because of the logarithmic response of the eye to luminance that gamma correction is required in digital displays.
3. The fovea is a tiny depression in the retinal tissue (see Figure 1.8) within the centre of the macula lutea of the retina which contains only cones. In the light-adapted state, the fovea not only has the highest visual acuity, but also especially high sensitivity for luminance and colour discrimination. Approximately half of the nerve fibres in the optic nerve carry information from the fovea, whilst the remainder serve the rest of the retina.
4. A darkroom should not be painted black, but have white walls and controllable lighting. The white walls reflect rather than absorb the light, so allowing management of the intensity level and colour of the illumination. EM Nelson, a famous 19th century microscopist (Box 31.2) took great care to ensure his night vision was well adapted for all his high-resolution microscopy.
5. When we swim under water without a face mask, because the refractive index of the cornea (n = 1.376) is very close to that of water (n = 1.333) we are reminded of the role of the cornea in focusing the image onto the retina. Whilst swimming we can only see an image which is very blurred indeed, if at all. This is because hardly any focusing occurs at the submerged cornea; we lose 40 dioptres of focusing power. For an explanation of how dioptres are calculated, see Chapter 3.
6. See the definition and meaning of the term 'aperture' in Note 2, in the Notes section.
7. This minimum resolved distance is a rule of thumb, since the eye, like any device, has an MTF (modulation transfer function) response (see Chapter 10). Absolute visual acuity depends upon the nature of the object, but for adjacent high-contrast black-white bars, the calculation is as follows: If the eye can differentiate either a black bar or a white bar of a high-contrast pattern in 72 arcsec then, since one minute of arc (60

arcsec) = 1/60 degree, a single line pair (1 cycle) must be twice the width of each line, so there are 30 cycles in a degree, and (30 x 1.2) = 36 cycles per degree in 72 arcsec. For individuals with 6/4.5 Snellen visual acuity (see Rostgaard and Qvortrup, 1999) the maximal resolving power is 0.8 arcminute (24 CPD) or 3.8 μm at the retina, which is the approximate diameter of cones within the fovea. See also Box 3.4, Chapter 3, Section 3.11.

8. The size of the retinal area, and hence visual acuity, is determined by the size of the eyeball. Provided that the centre-centre spacing between the photoreceptors remains constant, a larger posterior focal length improves visual acuity. A focusing lens allows for close vision and spreading a larger image over the retina. A large eye size also improves sensitivity, since a large pupil reduces diffraction and lets in more light. A larger pupil relative to eye length gives a smaller *f*-number and increased image brightness; for example, the giant squid in the deep ocean have the largest eyes of the animal kingdom: a 27 cm eyeball with a pupil of 90 mm and aperture approximately *f*/1. Visual acuity is also dependent upon the irradiance and contrast of the object. Humans have better contrast discrimination than most birds and better visual acuity.
9. Polarised light propagates as a wave within a certain field. That is, the vibrations are partially or completely supressed in certain directions, whereas natural light contains multiple wave trains of different amplitudes, wavelengths and polarisation states all travelling, or propagating, in the same direction. Plane-polarised light (see Chapter 13) has the electric field vector of its wave trains all oriented parallel to one another, such that there exists only one vibration direction. With circularly polarised light, the orientation of the field changes, and includes all planes of orientation.
10. On-axis astigmatism is described here; third-order, or off-axis, astigmatism is described in Chapter 6. It is a misconception to think that laser beams are naturally highly collimated: laser diodes have highly astigmatic beams that require optical correction in order to be useful.
11. Our eyes lock onto the fixation point in order to keep a specific 3D location at the centre of our vision. The locus of fixation points that produce a single image at corresponding retinal points of the two eyes, when the eyes are fixating binocularly, is called the horopter. With motion parallax, objects nearer than the point of fixation will move in a direction opposite to the direction of head movement, and those objects farther away will move in the same direction. Employing motion parallax in this way can be useful when focusing onto the film plane or CCD faceplate with a focusing graticule on a photo-microscope.
12. As the near point recedes with age, this causes a corresponding decrease in visual acuity and resolving power of the eye, which occurs without the quality of the retina being compromised, or involved.
13. This is only true for 'high power microscopes' and not stereomicroscopes, which can reduce the size of the image in relation to the object. Super-resolution microscopes can circumvent the classical resolving power limit.

2

Light

Outline

Understanding Light Microscopy, First Edition. Jeremy Sanderson.

2.1 Introduction

Having emphasised in Chapter 1 that all images are seen by the eye, this chapter considers the start of the imaging process: light. We shall discuss what light is and how it interacts with matter. We shall consider how the velocity of light is altered by materials denser than air and how rays of light are bent or refracted.

Light is so central to our perception that for much of the time we are unaware of it and take it for granted, except perhaps when viewing a gorgeous sunset or glimpsing crepuscular rays behind clouds. Although light surrounds us and is essential for life, it is anything but ordinary. As one popular science writer has put it: 'Light, the universal metaphor for understanding and revelation, is astonishingly opaque' (Brooks, 2012). The Greeks were aware of the importance of light, referring to dying as 'losing the light'. Yet it is only in the last 200 years that humankind has begun to understand light's unique properties. Indeed, Benjamin Franklin remarked in a letter discussing science to a colleague, written in April 1752, 'I must own that I am much in the *Dark* about *Light*...'

2.2 The Nature of Light

Ancient and medieval scholars made some progress in understanding the nature of light, being aware that it travelled in straight lines (we call this the rectilinear propagation of light; it is the basis of geometrical optics). For our purposes it is convenient to start in the mid-17th century with Robert Hooke, Christiaan Huygens and Sir Isaac Newton. Around 1666 Newton admitted a ray of light though a prism (Figure 2.1) and explained the decomposition of white light into spectral colours by refraction. Although Newton investigated some aspects of the wave nature of light, such as the formation of Newton's rings (for an explanation of this phenomenon see Chapter 14, Section 14.11), his writings emphasised the particulate ('corpuscular') model. Both Robert Hooke and Christiaan Huygens, contemporaries of Newton, came to very different conclusions about the nature of light. Huygens was convinced that regarding light as a series of wavefronts better explained refraction, but he was unable to extend this to explain the phenomenon of diffraction (Chapter 10). The corpuscular model failed to explain refraction satisfactorily, but because of Newton's stature, his view was to dominate science until the mid-19th century. Not only that, Newton's disagreement with both Huygens and Hook led to permanent antagonism[1].

It was the work of the polymath Thomas Young from 1803 onwards, who demonstrated the effects of the interference of coherent light rays, that truly marked the beginning of humankind's significant understanding of the nature of light. Building upon Young's investigation of diffraction and also on the earlier work of Christiaan Huygens, Augustin-Jean Fresnel showed that light propagated entirely as a transverse wave, with no longitudinal vibration whatsoever. He also explained both the rectilinear propagation of light and diffraction effects. Later in the mid-19th century, from the data of simple electrical experiments, James Clerk Maxwell formulated his equations to describe electromagnetic waves that could travel at approximately the known speed of light. Maxwell considered this fact more than a coincidence. In a key paper published in 1865 he wrote:

> 'The ... results seem to show that light and magnetism are affections of the same substance, and that light is an electromagnetic disturbance propagated ... according to electromagnetic laws.'

Towards the end of the 19th century, in a successful attempt to prove Maxwell's equations, Heinrich Hertz demonstrated the existence of radio waves. He showed that these belonged to the electromagnetic spectrum, of which light was part.

2.3 The Nature of Waves

Energy is carried from one place to another by two means. Either it is transferred directly by the movement of matter or is carried by a wave. In the former case, two boats may ram one another, causing a transport, or displacement, of mass. Conversely, a boat may

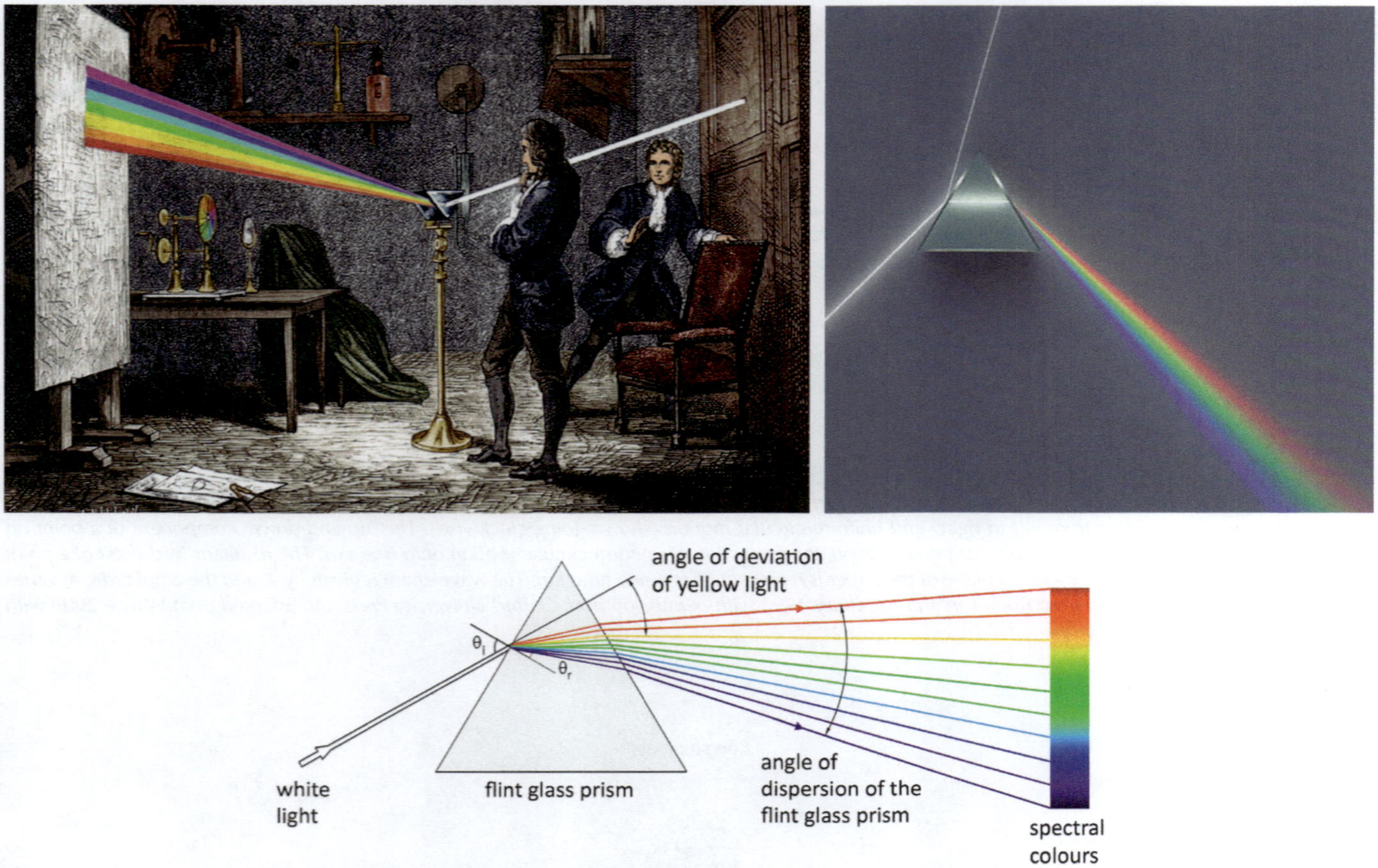

Figure 2.1 **Newton's prism experiment** *In 1666 Isaac Newton decomposed sunlight into visible spectral colours. An animated gif can be found on Wikipedia (see the Internet Resources listed at the end of the chapter). By using a lens to recombine spectral colours into white light, and also by passing isolated single colours through a second prism (his* experimentum crucis*), Newton proved that a prism does not 'add' colour to white light* per se *but that dispersion is a result of refraction.*

Some books show a second prism recombining the spectrum with a gap between the two prisms, but this is incorrect; a lens is required, as Newton showed. Using three prisms, this can be done, but two prisms can only recombine spectral colours to form white light if the second is inverted, and placed very *close to the first so that the colours exit the second prism parallel and recombine to form white light.*

Source: (top left) Apic / Contributor. Reproduced with permission from Getty Images. (top right) Spigget (Own work) Reproduced under the terms of the Creative Commons Attribution Share-Alike licence, CC BY-SA 3.0 (http://creativecommons.org/licenses/by-sa/3.0).

be rocked by an approaching wave. It will bob up and down (or oscillate) in unison with that wave, but not necessarily be displaced. Waves can therefore transfer energy *without* transferring mass. We can also use waves to transmit information.

Mechanical waves are formed when there is a disturbance in a medium from a resting state. Wave motions have the property of periodicity: the disturbance repeats regularly in space and time. This is shown in Figure 2.2. For example, sound waves exist in gases or liquids as longitudinal mechanical pressure waves, also called compression waves. This means that the particles of the supporting medium (air or liquid) travel back and forth, as successive compression and rarefaction towards and away from the source, *along* the axis of propagation. The motion of the wave can be different from the motion of the medium on which the wave moves. Thus, a wave pulse moves through a medium, while the carrier medium itself does not travel.

In a longitudinal wave (e.g. a mechanical sound wave), the medium vibrates in the same direction as the propagation of the wave (Figure 2.3). Where the motion of the medium is perpendicular to the motion of the wave, the wave is called a transverse wave (e.g. an electromagnetic wave). When travelling through solids, sound waves can exist as both longitudinal and transverse waves. Polarisation effects (Chapter 13) are associated with transverse waves but not with longitudinal ones.

Electromagnetic waves (including light) exist wholly as transverse waves, in which the particles travel up and down, in a plane of vibration perpendicular to the direction of propagation (Figure 2.4). Both water and light waves are examples of transverse waves. When you drop a stone into a pond, the waves propagate radially away from the point of origin, but the actual water droplets travel up in crests and down in troughs while the wave propagates along the surface of the water. A wave has seven characteristics (some of these are illustrated in Figure 2.5):

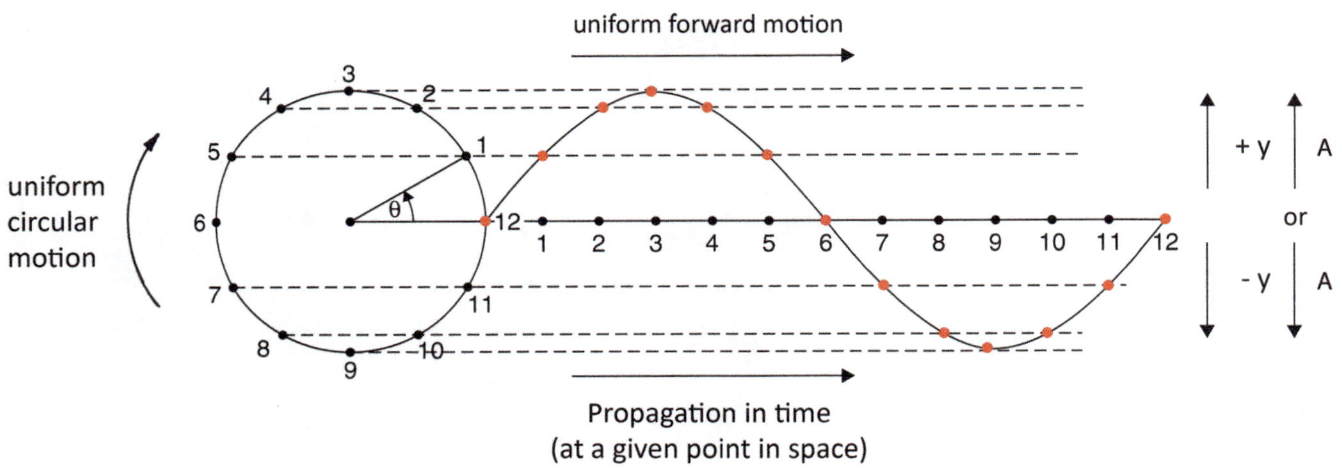

Figure 2.2 **Characteristics of a wave** *A wave repeats regularly in space and time. A wave may be regarded as the ray trace of a combination of uniform forward motion (shown going from left to right) and uniform circular motion (shown going clockwise). The 'up-and-down' component of a point on the circle moving forward is called simple harmonic motion, the projection of uniform circular motion onto one axis. The resultant (red dots) of a point propagating along a wave vibrating in the plane of the paper is represented as a sine function. The wavelength is given by λ, and the amplitude, A, varies from +y to −y. Source: Adapted from Robinson and Bradbury 1992 with permission from Oxford University Press and adapted from Wayne 2009 with permission from Elsevier.*

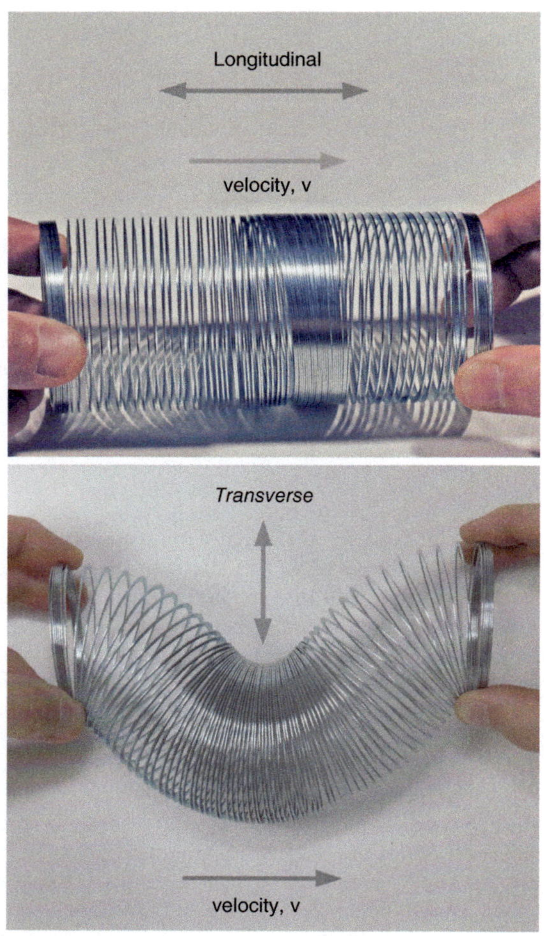

Figure 2.3 **Longitudinal and transverse waves** *Sound waves in air propagate as longitudinal waves. The gas molecules undergo a series of compressions and rarefactions, equivalent to the crests and troughs of the wave. This back-and-forth motion of compression and rarefaction is characteristic of longitudinal waves. Just as in the case of a transverse wave, the velocity, v, is equal to frequency times wavelength ($v = f\lambda$).*

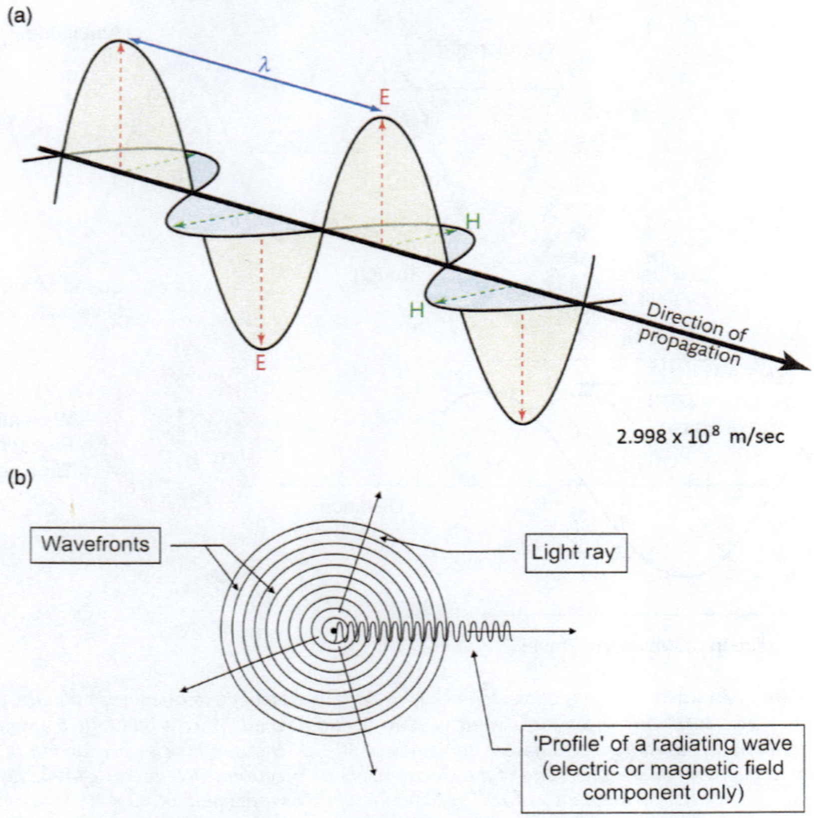

Figure 2.4 ***The wave nature of electromagnetic radiation*** *(a) An electromagnetic wave is one in which electrical and magnetic fields fluctuate together. This figure shows a wave stopped at a fixed moment in time. The electric field (E) and magnetic fields (here H; sometimes denoted B) are orientated at right angles to one another, and both are perpendicular to the direction of propagation, or travel, of the wave. To simplify matters, in this book we shall disregard the magnetic field and only consider the electric field. Source: Glencross et al. 2011. Reproduced with permission from Oxford University Press.*

(b) A wavefront can be defined as the propagation of a position of constant phase. Since the frequency of light >10^{14} Hz, measuring absolute phase is not useful. When two or more waves interact, precise measurement of relative phase (often denoted by φ) is possible, because the waves interfere to give overall changes in amplitude, perceived as a change in intensity and contrast. The wavefront is perpendicular to the direction of propagation, with separation between wavefronts equal to a single, or multiple wavelength(s). Source: Allen and Triantaphillidou 2011. Reproduced with permission from Taylor & Francis Group.

1. Phase – two points moving with the same velocity, which also have the same displacement from the undisturbed resting state, are said to be in phase. The phase, or relative position of a point on the wave, can be expressed as an angle. One cycle or wavelength corresponds to 2π radians or 360°. A phase difference occurs when one wave lags behind, or precedes another, in time and space; this difference is often denoted by the Greek letter phi (φ).
2. Wavelength – the shortest distance between two points that are moving in phase. This is often depicted as the distance between two adjacent crests. It is denoted by the Greek letter lambda (λ), with units in nanometres (nm).
3. Wavenumber – is the reciprocal of the wavelength (i.e. $1/\lambda$), representing the number of wavelengths per unit length. It is useful in spectroscopy, with units of cm^{-1}. The angular wavenumber, k, is equivalent to $2\pi/\lambda$.
4. The frequency of the wave – denoted by *f* or the Greek letter nu (ν) – is the number of waves (e.g. measured at their crests) that pass a fixed point. A wave whose crests pass a fixed point every second has a frequency of one Hertz (Hz), that is, one cycle per second, or simply one reciprocal second (s^{-1}).
5. The period of a wave denoted by tau (T or τ) is the time taken to complete one full cycle and is the reciprocal of the frequency; its unit is the second.
6. The velocity, denoted by v (not to be confused with ν for frequency), of the wave is the speed of the wave in a given direction. Velocity gives both the speed *and* direction of propagation. It is the product of frequency and wavelength: $v = f \times \lambda$ (or $v = \nu\lambda$) and has units of metres/second.
7. Amplitude is the height, or maximum displacement, of the wave crest above the undisturbed position. It is not the distance between the top and bottom of a wave cycle.

Since sound travels as a mechanical wave through a medium, a vacuum will halt the propagation of the sound wave. Electromagnetic waves (e.g. radio waves, microwaves and light), however, do not require the support of a medium, and so are capable of travelling

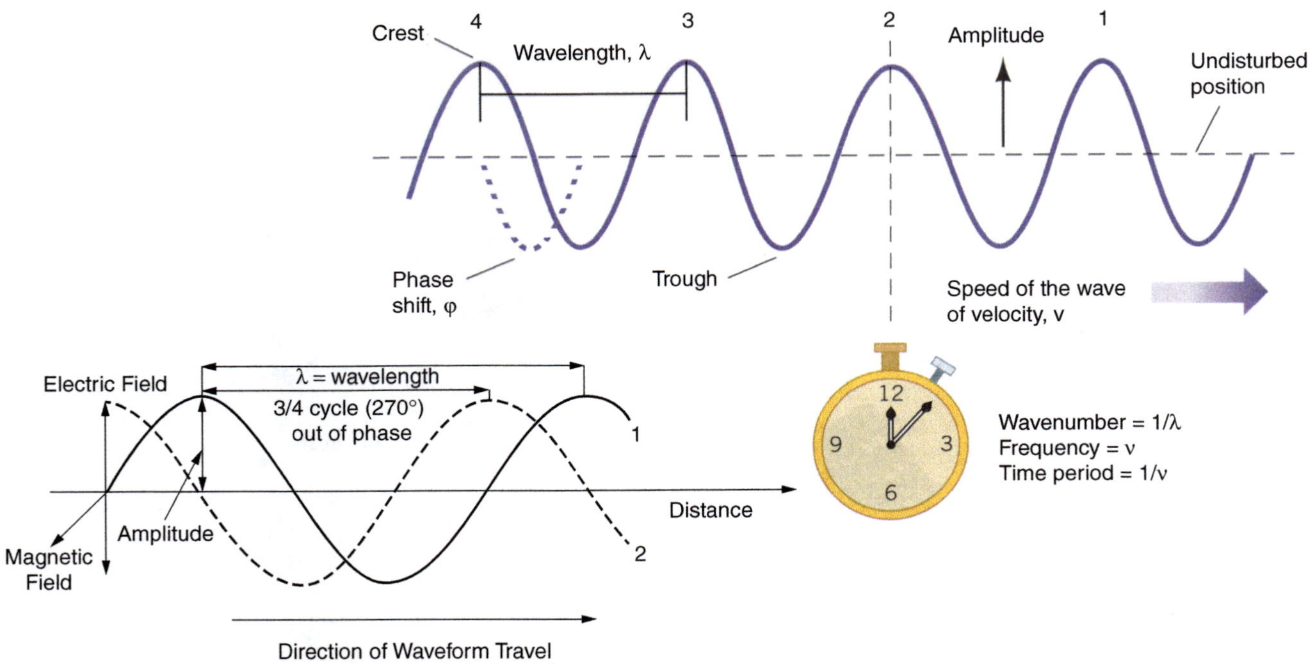

Figure 2.5 **Important characteristics of a wave** *A wave is characterised by velocity, wavelength, frequency, amplitude, phase and polarisation vectors. For all electromagnetic waves: $\lambda \times f = v \approx 3 \times 10^8$ m/sec. Since the velocity of an electromagnetic wave is constant in a vacuum, an electromagnetic wave can possess any wavelength or frequency. Thus radio waves have a frequency of 30–300 thousand cycles per second (kHz) and have correspondingly (long) wavelengths from 10–1 km respectively. At the other end of the electromagnetic spectrum, high-energy gamma rays have very short wavelengths of very high frequency. Gamma rays have a typical frequency of 10^{22} Hz and therefore a wavelength of $\approx 3 \times 10^{-14}$ m or 30 femtometres. Source: (top) Trefil and Hazen 2004. Reproduced with permission from John Wiley & Sons. (bottom) Bushberg 2014. Reproduced with permission from Wolters Kluwer.*

through a vacuum. The self-supporting nature of electromagnetic radiation means that the measured speed of the wave is constant regardless of whether the velocity of the emitter or detector used to measure it changes. The frequency of light is related to its speed and wavelength as follows: $v = c/\lambda$. When light passes into, or through, a transparent medium, the speed is reduced, but the frequency remains constant, and so the wavelength changes.

Electromagnetic waves, such as light, are self-propagating, and unlike other wave phenomena (e.g. transverse water waves), they are *not* a disturbance of a medium. For many years, light was considered to move through an invisible substance called the 'luminiferous æther'. This was finally proved not to exist by the negative outcome of the Michelson-Morley experiment.

2.4 Wave-Particle Duality

By the end of the 19th century, Maxwell's theoretical model of light as continuously oscillating electric and magnetic fields seemed complete. However, several experimental observations (principally black-body radiation and the photoelectric effect) could not be wholly explained by Maxwell's wave model of light. We need only understand the fundamental details of how the wave-particle duality of light arose. For those interested, a more detailed explanation can be found in chapter 4 of *Light* (1989) by Michael Sobel, listed under Further Reading.

Max Planck was a very conservative scientist, working in thermodynamics, trying to explain the 'irregular' behaviour of radiation emitted by an incandescent black body. It was with reluctance that he proposed the hypothesis that energy is radiated and absorbed as discrete quanta rather than as a continuous stream of energy (see Figure 2.6 and Box 2.1). Planck cautiously insisted that quantum energy transfer was merely an aspect of the processes of absorption and emission of radiation and had nothing to do with the physical reality of the radiation itself. We now accept the Planck postulate that electromagnetic energy is emitted in quantised form, this being a fundamental principle of quantum mechanics. Because Planck could not justify his apparently counterintuitive conclusion on the basis of Newtonian classical physics, he considered quantisation a mathematical trick and a heuristic approach that would eventually be superseded. Einstein is best known for developing the general theory of relativity, but it is the first of four papers that he published in 1905 on the photoelectric effect, and for which he was later awarded the Nobel Prize, that is important for our purposes. Einstein's contribution was to correct Planck's view and show that electromagnetic waves (including light) inherently transfer energy as discrete quanta, rather than energy flowing continuously along the wave. The only reason that the energy flow appears continuous is because the quanta are so small in magnitude.

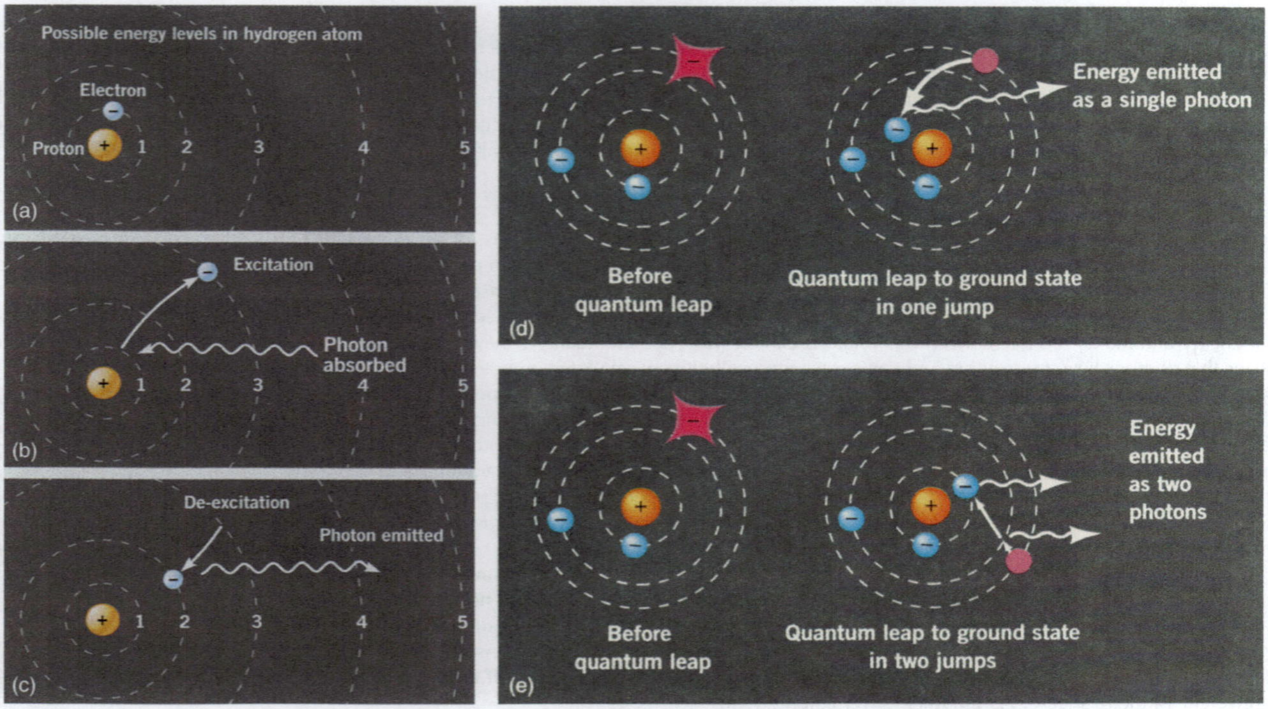

Figure 2.6 **Photons as energy quanta** *Just as we can stand upon individual steps in a flight of stairs, it is convenient to consider an atom (Bohr's model) with discrete energy levels, or orbitals. Since energy cannot be created or destroyed, but can only change its location or form, then if an electron changes its position within the atom, it will absorb or emit a photon as a packet of energy. Radiation is the transfer of energy as electromagnetic particles (photons) or waves. Electrons may jump between the energy levels shown in (a) from absorbing (b) or releasing (c) a quantum of energy in the form of a photon. When a photon of energy is absorbed by an atom or molecule, an electron will pass from a lower energy orbital to a higher state. The difference in energy is proportional to the frequency of the photon. An electron can move from a higher to lower energy orbital in a single step (d) or multiple quantum leaps (e). The energy (E) of any electromagnetic wave is given as $E = hc/\lambda = h\nu$, where h equals Planck's constant. The introduction of Planck's constant in the equation above means that a specific wavelength has a discrete quantum of energy associated with it. For a photon of light to be absorbed or emitted, its energy must exactly match the difference in energy between two energy levels in an atom (or molecule). Source: Trefil and Hazen 2004. Reproduced with permission from John Wiley & Sons.*

Box 2.1 Wave-Particle Duality: Planck, Einstein and de Broglie

According to Planck, each energy element, E, is proportional to its frequency, ν:

$$E = h\nu \quad \text{also} \quad E = hc/\lambda \quad \text{thus} \quad \lambda = hc/E$$

where h is Planck's constant: 6.626×10^{-34} joule.sec, λ is the wavelength, and c is the velocity of light in a vacuum: 2.998×10^{8} m/s.

Planck's constant can be expressed in several different units. It is 6.626×10^{-27} erg.sec, or 4.136×10^{-15} eV.s. The joule-second is a unit equal to a joule multiplied by a second, used to measure action (as defined in physics) and therefore has the dimensions of energy integrated over time. The joule (kg.m^2/s) is the SI unit, the erg the unit of the CGS system. The introduction of Planck's constant in the equations above means that a specific wavelength has a discrete quantum of energy associated with it. For a photon of light to be absorbed or emitted, its energy must exactly match the difference in energy between two energy levels in an atom (or molecule, see Figure 2.6). The fact that Planck's constant is vanishingly small reflects the fact that everyday objects are made of a *large* number of particles.

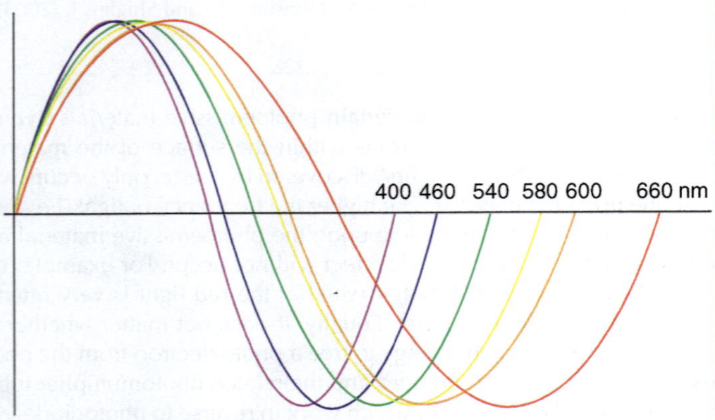

Source: Green et al. 2003. Reproduced with permission from GIA.

Einstein showed that an X-ray photon carries much more energy than a lower-frequency, longer-wavelength, photon of visible light. Also violet light ($\approx 8 \times 10^{14}$ Hz; 800 THz or $\approx$3 eV) has almost twice the energy of a photon of red light ($\approx 4 \times 10^{14}$ Hz; 400 THz or 1.5 eV) where 1 eV = 1.602×10^{-19} J. We are all continuously bathed in longer-wavelength, low-energy, radio waves without ill effect, but even short exposure to short-wavelength, high-energy, X-rays or γ-rays causes us considerable harm. Thus green light at 550 nm wavelength (from the equations above) has a frequency (v) of $c/\lambda = 545$ THz and an energy of 3.61×10^{-19} joules, or 0.361 attojoules.

While Einstein showed that electromagnetic waves are particulate, de Broglie, the French theoretical physicist, turned this concept upon its head. He proposed that since waves could have particle-like properties, then also matter could have wave-like properties: that streams of particles can behave like waves. Thus (as well as a photon in light microscopy) an electron can also act as a wave, with its wavelength given as:

$$\lambda = h/p \quad \text{(as a particle)} \qquad \text{also} \qquad \lambda = h/mv \quad \text{(as a wave)}$$

Hence, the momentum, p, of an electron moving with velocity v equals mv (m is the mass of the electron, 9.11×10^{-28} g at rest, or zero velocity). This formula of de Broglie is useful in electron microscopy where electron, rather than photon, waves are used for increased resolving power. Thus $p = mv$, and by analogy with the equation above, $mv = h/\lambda$ *and* $\lambda = h/mv$, so $p = h/\lambda$.

We know intuitively that waves exert pressure from observing the power of waves at sea. From Einstein's famous equation, $E = mc^2$, we know that energy and matter are interconvertible. Photons exhibit wave-particle duality – they have the properties of both waves and particles. This means that photons have momentum and can exert a force, albeit very small. For a photon with wavelength λ or energy E, this is $h/\lambda c$ or E/c^2.

$E = hc/\lambda$ relates, through the constant h, a particle like property (E) to a wave-like property (λ or v). Planck's equation may be combined with Einstein's equation, $E = mc^2$, i.e. $E = hc/\lambda$; hence, $mc = h/\lambda$. Now, mc (mass × velocity) represents a momentum, p. In short photons carry a momentum, and when a beam of light is absorbed by, or reflected from, an opaque object, it exerts a pressure called the *radiation pressure* of light – a phenomenon first predicted by Maxwell. For a further explanation, see https://en.wikipedia.org/wiki/Radiation_pressure and http://scienceworld.wolfram.com/physics/RadiationPressure.html

In his final novel, *The Last Theorem*, the late Arthur C. Clarke imagined solar yacht races with astronauts competing to reach the Moon and back by photon power. This has now become reality with the announcement in May 2010 by the Japanese Aerospace Exploration Agency of the launch of a satellite, named Ikaros, powered by a solar sail. JAXA plans to control the path of Ikaros by changing the angle at which sunlight particles bounce off the lightweight 14 metre square, 7.5 μm thick, silver-coloured sail.

Source: IKAROS 2009. Reproduced with permission from JAXA.

Ikaros is an acronym for Interplanetary Kite-craft Accelerated by Radiation of the Sun, and refers to the mythical character Icarus who flew too close to the sun and fell into the sea.

Thus all moving particles – not just photons or electrons – have wave-like properties, including pieces of chalk that (used to be) thrown at dozing or talkative students in lectures but, as a simple calculation will show, the corresponding wavelengths are very small. If a piece of chalk 20 grammes is thrown at a speed of 50 km/hr at a student, then the wavelength will be:

$\lambda = 4.13 \times 10^{-15} / (20 \times 50) = 4 \times 10^{-18}$ m. This is well beyond the wavelength of gamma rays.

Equally, the momentum of a photon with $\lambda = 500$ nm will be:

$p = h/\lambda$ or 6.63×10^{-34} J.sec/(500×10^{-9} m) $= 1.33 \times 10^{-27}$ kg.m/s. The radiation pressure, P, is given by $(1/c) \cdot \Phi_E$ where Φ_E is the energy or radiant flux, and so P is of the order of 6×10^{-6} Pa. Light of sufficiently high irradiance can move microscopical particles, operating as laser tweezers.

Reference

Green, B., Reinitz, I., Gilbertson, A., Johnson, M., and Shigley, J. (2003) 'Light Dispersion, Color Wavelengths and Fire'. Gemological Institute of America.

When light is shone upon certain photoemissive materials (typically visible light for alkali metals, near UV for other metals, and far UV for non-metals), electrons within the surface of the material absorb the energy and escape as free electrons (referred to as photoelectrons). This effect, first discovered by Hertz, only occurs when light possessing a quantum of energy above a certain threshold is shone upon the material. The higher the frequency of light (i.e. the shorter the wavelength), the more energy is available to cause the photoelectric effect. Depending upon the photoemissive material and the energy required to free an electron, there is a cut-off point below which the photoelectric effect will not occur. For example, the photoelectric effect may occur with UV, blue or green light but not red light. It does not matter whether the red light is very intense or not, no photoelectrons are emitted when red light is shone on the surface of the material. Equally, it does not matter whether the blue or green light is very dim; the *frequency* is such that the light possesses sufficient energy to free a photoelectron from the photoemissive material. The photoelectric effect has been put to good use in microscopy: photodiodes and the related photomultiplier tubes (PMTs) used for confocal microscopy use this technology, as do light-emitting diodes (LEDs), which work in reverse to photodiodes, giving off light instead of converting photons to an electrical signal.

Einstein realised that the photoelectric effect could be explained by extending the radiation/absorption hypothesis of Planck to describe light itself as being composed of *discrete* quanta rather than wholly being a continuous wave. The higher the frequency of light, the greater the energy of the photon, as evidenced by the photoelectric effect. Thus a photon of blue light has almost twice the energy (see Box 2.1) of a photon of red light. Equally cautious, like Planck, about overturning received dogma of the wave nature of light, Einstein used the word 'heuristic' in the title of his paper, insisting that his radical idea was a helpful conceptual aid and shouldn't be understood as a physical reality. Later experiments showed that these quanta of electromagnetic radiation possessed momentum, and could be considered as massless particles, ultimately to be named *photons*. Newton was (partially) back in vogue.

Table 2.1 Wave-Particle Duality: Explaining the Behaviour of Light

Phenomenon	Explained in terms of waves	Explained in terms of particles
Reflection	Yes	Yes
Refraction	Yes	Yes
Interference	Yes	No
Diffraction	Yes	No
Polarisation	Yes	No
Photoelectric effect	No	Yes

Therefore, to the question 'what is light' only a partial answer can be given: it may be described in terms of photon 'particles' of energy quanta or as waves, according to the phenomena that need to be described (Table 2.1). We thus refer to the wave-particle duality of light. The wave picture of light corresponds to classical physics; the photon picture is quantum mechanical. These concepts are not contradictory but rather complementary: physics has yet to describe a fully unified model for electromagnetic radiation. Light, and its interaction with matter, is best considered in terms of photons. At the simplest level, the statistical behaviour of a large number of photons is represented by an electromagnetic wave. Heisenberg's uncertainty principle limits our knowledge of quantum dynamic events, such as electromagnetic radiation. Thus light is not merely a smooth sine wave, but one whose amplitude and phase are 'smudged' by the travel of photon particles. As a generalisation, light behaves as discrete quanta or 'particles' when it is produced from, or transformed into, other types of energy, but is like a wave when traveling from one point to another.

All atoms are characterised by one or more electrons circulating the nucleus in well-defined orbitals. Light originates from the transitions of electrons that occur when atoms or molecules absorb energy by virtue of being at transiently elevated temperature, such as in filament lamps[2]. This process is called incandescence. When the excited electron returns to the ground state, the energy used to excite the electron is given off, either as non-radiant energy or light[3]. The light emitted from an incandescent source is the result of spontaneous emission. The photons have the same frequency, but there are no phase relationships between the individual waves: the light is said to be incoherent. Also the vibration direction of each wave is random: the light is unpolarised. On the other hand, coherent rays of light emanate from the *same* light source, have a constant phase relationship with respect to one another and thus can interfere with each other. Lasers (by virtue of stimulated emission) produce coherent monochromatic light. Because of their importance in microscopy, the principle of lasers is described in Box 2.2.

Box 2.2 Lasers

Together with the transistor, the manufacture of the laser has revolutionised our daily lives; the transistor led to the development of modern electronics, the laser to photonics. The name laser is an acronym for **L**ight **A**mplification by **S**timulated **E**mission of **R**adiation.

Light emitted spontaneously from 'ordinary' (i.e. incandescent) sources or semiconductor light-emitting diodes is incoherent. Einstein postulated two possible types of emission: either atoms in an excited state can move randomly from the resting 'ground' state and emit a photon (spontaneous emission) or else, stimulated emission can occur.

Einstein realised that an incident photon of a frequency equal to the energy separating the two orbitals causes the electron in the higher state to relax and thus emit a further photon of equivalent energy to the incident one. Significantly, these photons are coherent, having not only the same wavelength but also the same phase, direction and polarisation. Because the photons are in phase (both spatial and temporal coherence), the waves are amplified.

Laser light is largely monochromatic because of the laser material and the design of the resonant, or optical, cavity. A single pure frequency just perfectly fits between the two mirrors. By changing the spacing of the laser's mirrors, it is possible to change the frequency, and hence the colour, of the light.

Stimulated emission is impossible to achieve when matter is in thermodynamic equilibrium. For stimulated emission to dominate, the majority of atoms must be excited. This is known as *population inversion* and must be kick-started by energy input from an external source. The first laser used a flashlight as a starter; this external energy source is required continually to maintain population inversion (the process is called pumping). The energy may be supplied as an electrical current or as light at a different wavelength, either by a flash lamp or another laser.

A laser therefore essentially consists of a gain medium inside a highly-reflective resonant cavity with an external energy source as a pump. The gain medium (e.g. crystals doped with rare-earth or transition metal ions; gases; semiconductors or dyes) is a material whose properties permit light amplification by stimulated emission. In its simplest form, the resonant cavity consists of two mirrors arranged such that light bounces back and forth repeatedly through the gain medium. One of the two mirrors is partially transparent; the laser beam is emitted through this mirror.

Coherent stimulated emission was, originally, easier to achieve with longer-wavelength electromagnetic radiation – in fact radio waves can be similarly coherent. The forerunner of the laser was the 'maser' – analogous microwave amplification by stimulated emission of radiation. In 1954 James Gordon, Herbert Zeiger and Charles Townes built the first maser by passing a stream of ammonia molecules through an electric field in a microwave cavity. The excited molecules were focused into a beam, the others being defocused. This setup acted both as an amplifier and oscillator to emit coherent radiation. The maser concept was extended to the visible spectrum by Arthur Schawlow and Charles Townes. Independent work was carried out by Gordon Gould, who coined the term laser.

A prolonged legal battle arose over the patent rights, which was eventually settled in Gould's favour. Having previously pointed out that the amplification process is achieved by 'light oscillation' within the optical cavity, Schawlow suggested that the correct acronym ought to be 'loser'. Fortunately, the laser was named in analogous fashion to the earlier maser.

The first working laser was demonstrated by Theodore Maiman (1927–2007) on May 16th 1960. Due to a deluge of papers on the maser and laser concepts, the journal *Physical Review* turned down Maiman's paper, which was accepted by *Nature* instead. An artificial ruby crystal doped with chromium atoms provided the electronic energy levels for the stimulated emission cycle, producing laser light at 694 nm. Maiman faced initial difficulties from his employers who, according to his wife, wanted him to work on computers or 'do something useful'. He threatened to leave and build a laser in his garage.

At the outset, there was much hyperbole concerning lasers, which was not helped by their potential to cause harm (Maiman characterised the first laser as having a power of one 'Gillette' as it could burn through one Gillette razor blade). Early press reports talked of 'death rays' and unbelievable military power, but Maiman and colleagues emphasised the laser's healing possibilities, including the ability to focus within single cells. When lasers were first invented, they were referred to as 'a solution looking for a problem'. As microscopists, we use lasers for illumination, but they have a wide range of uses in diverse fields: from CO_2 gas lasers, which are used accurately to cut and weld metal, also for precision engraving, to lasers used for bar code readers, printers, audio players and data storage, automatic light switching, fibre optic transmission, navigation systems, and precision surgery.

Lasers, because they produce broadly collimated coherent, highly directional, light (i.e. a very narrow beam that diverges, or spreads out, very little with increasing distance from the source and thus forms only a small spot a hundred metres or more away) can potentially cause irreversible eye and skin damage over a long distance if these tissues are inadvertently exposed to powerful laser sources. The irradiance of the laser in W/m^2 is the power divided by the cross-sectional area of the beam.

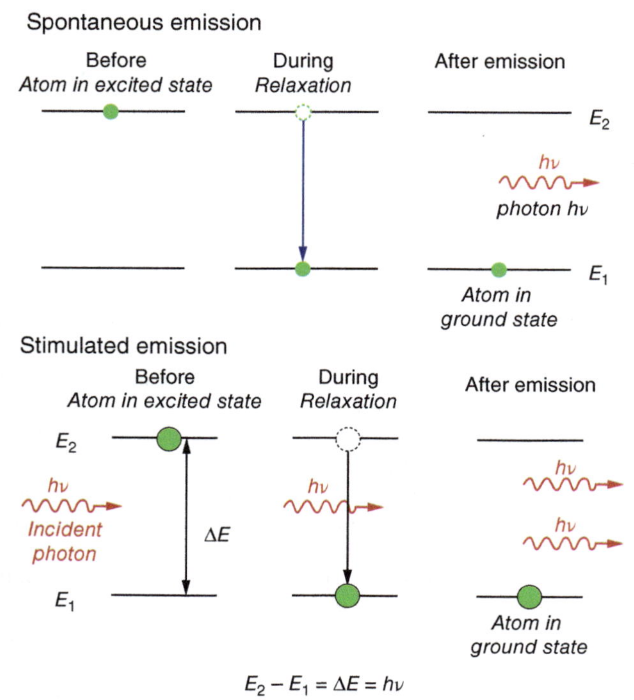

Source: (top) Ilmari Karonen. Reproduced under the terms of the Creative Commons Attribution Share-Alike licence, CC-BY-SA-3.0 (http://creativecommons.org/licenses/by-sa/3.0/), via Wikimedia Commons. (bottom) V1adis1av (Own work). Reproduced under the terms of the Creative Commons Attribution Share-Alike licence, CC BY-SA 4.0, via Wikimedia Commons.

At the time of writing, lasers are classified in the UK according to BS EN 60825-1:2014. In the USA, the equivalent standard is Federal Performance Standard for Laser Products, 21CFR1040.10. The British Standard sets out seven classes of laser: Class 1, Class 1M, Class 2, Class 2M, Class 3R, Class 3B and Class 4 defined according to the accessible emission limit (AEL), which in turn is based upon maximum permissible exposure (MPE) levels. Briefly, Class 1 should present no eye hazard during normal use; Class 1 M lasers produce either a highly divergent beam or a large diameter beam but can be harmful to the eye if the beam is viewed using optical magnifier. The Class 1 AEL is 0.39 mW. Class 2 lasers emit less than 1 mW of power. Protection is via the body's natural involuntary aversion response; exposure should not be longer than 0.25 seconds. Class 2 M lasers can be harmful if viewed with an optical magnifier. Class 3R and Class 3B lasers emit in the ranges 1–5 mW and 5–500 mW respectively; Class 3R lasers may be hazardous when viewed directly or by specular reflection, but diffusely reflected radiation from a Class 3R laser is not normally hazardous. Most continuous wave (CW) lasers used for microscopy are Class 3B. Direct and specularly reflected beams are hazardous: safety interlocks and/or protective eyewear are required. Class 4 lasers have an AEL of 500 mW and over. They present both eye and skin hazards from both direct and diffusely reflected radiation. Pulsed multi-photon lasers fall into this class.

ANSI and IEC laser classification	Class 1		Class 2		Class 3		Class 4	Notes
Sub-class	Class 1	Class 1M	Class 2	Class 2M	Class 3R	Class 3B	Class 4	
U.S. FDA laser classification	Class I	No special FDA class	Class II	No special FDA class	Class IIIa (definition is different but results are similar)	Class IIIb	Class IV	Newer ANSI/IEC number classes are now preferred over older FDA Roman numeral classes
Human-accessible laser power (for visible light)	For visible light, emits beam less than 0.39 milliwatts, or beam of any power is inside device and is not accessible during operation.		Emits visible beam of less than 1 milliwatt.		For visible light, emits beam between 1 and 4.99 milliwatts.	For visible light, emits beam between Class 3R limit (e.g. 5 milliwatts) and 4.99.9 milliwatts	For visible light, emits beam of 500 milliwatts (1/2 watt) or more	Non-visible lasers emitting infrared or ultraviolet are not included in this chart. Only visible lasers are discussed.
Caution/warming indication	No special caution/ warning indication		No special caution/ warning indication		CAUTION	WARNING	DANGER	
Label descriptive text		DO NOT VIEW DIRECTLY WITH OPTICAL INSTRUMENTS	DO NOT STARE INTO BEAM	DO NOT STARE INTO BEAM OR EXPOSE USERS OF TELESCOPIC OPTICS	AVOID DIRECT EYE EXPOSURE	AVOID EXPOSURE TO BEAM	AVOID EYE OR SKIN EXPOSURE TO DIRECT OR SCATTERED RADIATION	For visible-light lasers, the word 'light' can be used instead of 'radiation'. The latter is more accurate for lasers emitting infrared and ultraviolet radiation.

Source: Adapted from LaserSafetyFacts.com with permission from P. Murphy.

References

Hecht, J. (2010). *Beam: The Race to Make the Laser*. OUP.
Hecht, J (2008). *Understanding Lasers*, 3rd Ed. Wiley-Blackwell.
Laser Safety Facts at http://www.lasersafetyfacts.com/laserclasses.html (accessed 20 April 2018).
UK classification BS EN 60825-1:2014. URL: https://www.gov.uk/government/publications/laser-radiation-safety-advice/laser-radiation-safety-advice#fnref:1 (accessed 1 June 2018)
US classification 21CFR1040.10. URL: https://www.fda.gov/Radiation-EmittingProducts/ElectronicProductRadiationControlProgram/GettingaProducttoMarket/RegulatedProducts/default.htm and https://www.accessdata.fda.gov/scripts/cdrh/cfdocs/cfcfr/CFRSearch.cfm?FR=1040.10

Each atom emits an individual photon, which are packaged together as a wave packet, or wave train[4], (with a beginning and end point). Each packet carries just one quantum of energy. These photons overlap with each other and vibrate randomly, so light can be considered as a continuous wave, leading to the sum of all photons vibrating in all directions (since the plane of vibration of the E-vectors may be at any angle – not just up and down as with water waves) as unpolarised light transverse to the direction of propagation.

The speed of propagation for light is the well-known constant c with the exact[5] value of 299,792,458 m/s, and more convenient approximate value of 3×10^8 m/s. In imperial terms, this is 186,282.397 miles/second (or 670,616,629 million miles an hour). To put this magnitude of speed into context, a photon (in air) would take 134 milliseconds, or just over 1/10th of a second, to circumnavigate the earth, take 3.34 nanoseconds to travel one metre and take 1.2 seconds to reach earth when reflected from the surface of the moon. The wave of the photon packet has a definite length, which depends upon the speed of light and the period of emission. Thus a wave from an incandescent filament lamp has a period of 10^{-8} s and is approximately 3 metres long (2.998×10^8 m/s $\times 10^{-8}$ s).

Light also exhibits momentum – albeit extremely small; for those interested see Box 2.1. The momentum of light is equal to the radiant power of light (otherwise called the radiant flux) divided by the speed of light, c. In practical terms the pressure exerted by a photon on objects in its path is extremely small ($\approx 1 \times 10^{-27}$ N), because of the large magnitude of the value for c. The momentum of beams of light is used in applications such as laser tweezers (with values $\approx 2.5 \times 10^{-12}$ N) to manipulate viruses, bacteria and macromolecules such as DNA strands.

2.5 The Visible Light Spectrum

Newton first used the word *spectrum* when describing his optical experiments. The spectrum of visible light, with wavelengths from about 700 nm (7×10^{-7} m – red light) to about 400 nm (4×10^{-7} m – violet/blue light) constitutes a *very small* part of the whole electromagnetic spectrum, which extends from about 1000 km for very long radio waves to about 10 picometres and lower for γ-rays[6]: see Figure 1.1, in Chapter 1. Although the extent of the visible spectrum is small, of all the energy emitted by the sun and stars, the greatest amount is located within the range extending from the ultraviolet (ca. 10–400 nm) to the infra-red (ca. 700–1500 nm).

When Newton famously split white light into the colours of the spectrum in the 1660s, he initially recorded only five spectral colours (red, yellow, green, blue and violet). The colours indigo and orange were considered boundary colours between blue-violet and orange-red, respectively. When Newton finally published *Opticks* in 1704, he included seven colours: red, orange, yellow, green, blue, indigo and violet). These are remembered by the mnemonics 'Richard of York gave battle in vain' or the more general one I teach my students: 'run off you girls, boys in view'. Newton chose seven colours out of a belief that there was a connection between the colours and musical notes. He expanded the boundary colours orange and indigo into being equal to the others from an aesthetic point of view.

The division of the spectrum (which is continuous) into named colours (Table 2.2) is somewhat arbitrary and merely for convenience. Light possesses no actual colour *per se*; colour is a perception of our brains. Many people cannot distinguish the tone indigo from blue and violet. Also, some authorities include cyan between blue and green, even though it is slightly desaturated. Figure 2.7 shows the visible spectrum rendered as accurately as can be by printing. Pure spectral colours are formed by single wavelengths within the visible spectrum. They lie along the curved boundary (the spectral locus) of the CIE (1931) chromaticity diagram (see Chapter 11, Section 11.5). White and black are not spectral colours, neither are those desaturated colours obtained by mixing white, grey or black with spectral or other non-spectral colours. Colour theory includes purple as a non-spectral colour. Purple hues form the lower boundary of the chromaticity diagram.

Most harmful high-energy short-wavelength electromagnetic radiation is blocked by the earth's atmosphere. Some longer-wavelength infrared radiation is also absorbed, leaving an 'optical window' through which visible wavelengths pass largely unattenuated. Figure 1.2 in Chapter 1 shows how human sensitivity to electromagnetic radiation not only matches this window but corresponds to those wavelengths minimally absorbed by water.

2.6 Important Properties of Light

Throughout this book we will refer to different properties of light (Figure 2.8), which are important for light microscopy:

- **Monochromatic** – waves having the same wavelength (λ) or vibrational frequency (ν). They are therefore perceived to be of the same colour.
- **Coherent** – waves (of a stated wavelength) that maintain the same phase relationship whilst travelling through space and time.

Table 2.2 The Visible with the Ultraviolet and Infrared Part of the Electromagnetic Spectrum

			Frequency (ν, Hz)	
	Colour	Wavelength (λ, nm)	$\nu = c / \lambda$	mid-value (THz)
Ultraviolet	UV-C	100–280	3×10^{15}–1.1×10^{15}	2000 THz
	UV-B	280–315	1.1×10^{15}–9.5×10^{14}	980 THz
	UV-A	315–380	9.5×10^{14}–7.9×10^{14}	870 THz
Visible	violet	380–420	7.9×10^{14}–7.1×10^{14}	750 THz
	(indigo)	421–445	7.1×10^{14}–6.7×10^{14}	690 THz
	blue	446–485	6.7×10^{14}–6.3×10^{14}	650 THz
	(cyan)	486–495	6.3×10^{14}–6.1×10^{14}	620 THz
	green	496–570	6.1×10^{14}–5.3×10^{14}	570 THz
	yellow	571–590	5.3×10^{14}–5.1×10^{14}	520 THz
	orange	591–620	5.1×10^{14}–4.8×10^{14}	500 THz
	red	621–710	4.8×10^{14}–4.2×10^{14}	450 THz
	(far red)	711–740	4.2×10^{14}–4.1×10^{14}	415 THz
Infrared	Near IR(IR-A)	740–1400	4.1×10^{14}–2.1×10^{14}	300 THz
	Infrared (IR-B)	1.4×10^{3}–3×10^{3}	2.1×10^{14}–1×10^{14}	150 THz
	Far IR (IR-C)	3×10^{3}–1×10^{6}	1×10^{14}–3×10^{11}	50 THz

Table 2.2 gives the accepted wavelength range for each spectral colour combined from several sources, although the cut-off point of each division can never be absolute. The human eye is relatively insensitive to indigo's frequencies, and some otherwise well-sighted people cannot distinguish indigo from blue and violet.

Figure 2.7 **The visible spectrum** *The visible spectrum rendered into the sRGB colour space. Pure spectral colours cannot be reproduced exactly in print or by electronic display. The range of colours (colour space) that can be so reproduced with a given colour reproduction system is called the gamut. sRGB is a standard RGB colour space created cooperatively by Hewlett Packard and Microsoft. Source: Spigget (own work). Reproduced under the terms of the Creative Commons Attribution Share-Alike licence, CC BY-SA 3.0 (http://creativecommons.org/licenses/by-sa/3.0).*

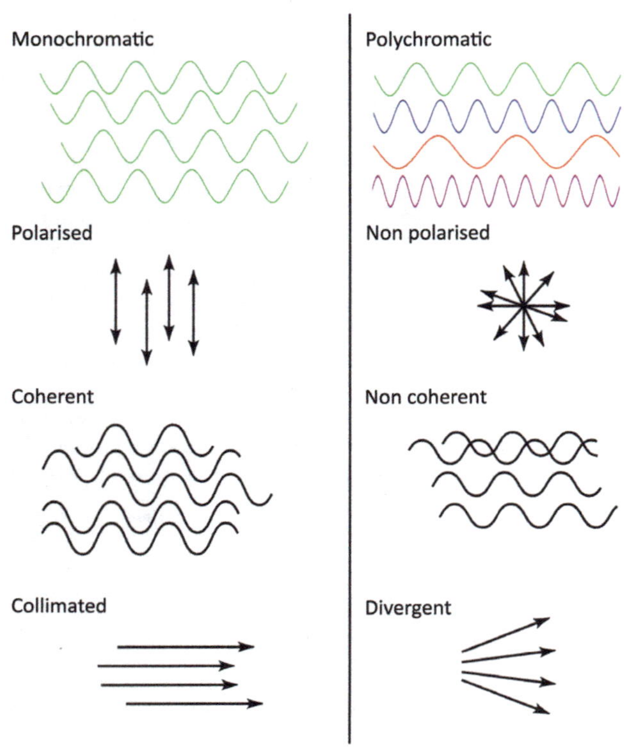

Figure 2.8 **Properties of light** *Eight waveforms depicting variations in the quality of light. Source: Adapted from Murphy 2001 with permission from John Wiley & Sons.*

- **Collimated** – waves whose paths of propagation are parallel without convergence or divergence but not necessarily with the same wavelength, coherence or polarisation.
- **Polarised** – waves whose E vectors vibrate in planes that are parallel to one another. Such polarisation may be linear, circular or elliptical.

Laser light possesses all these qualities: it is coherent, monochromatic, polarised and collimated. As such it is a powerful illumination source in confocal microscopy.

In studying geometrical optics, it is helpful to trace the path of a light wave with a single line trace perpendicular to the wavefront. This is called the ray approximation method, and a ray is an idealised narrow beam of light. For this reason, geometrical optics is sometimes called ray optics. Rays of light propagate in straight lines in homogeneous media; they reflect and refract at smooth, or even, interfaces between materials of different refractive indexes.

When the object is very large compared to the wavelength of light, such as when we use a camera or a single-lens magnifying glass, we can consider light as rays and use geometric optics (see Chapter 3) to understand how the image is formed. However, when the size of the object being studied is approximately the same size as the wavelength of light, then the effects of diffraction (which always occurs) become much more noticeable, and we must consider light as waves, rather than rays. This is because ray theory cannot describe phenomena such as interference and diffraction, which require wave theory (involving consideration of the phase of the wave). How light forms an image by diffraction is explained in Chapter 10, and the special case of polarised light, in Chapters 13 and 14.

2.7 Interaction of Light with Matter

Light interacts with the microscopical specimen in one of six ways:

- Light may pass straight through matter and be partially absorbed. This is how coloured stains work.
- Light may be reflected or refracted by matter.
- Light may be scattered or diffracted (i.e. a regular form of scattering) by matter.
- The light wave may undergo a phase change as a result of passing through matter. A phase shift may occur in the path of one wave with respect to another.
- Light may be polarised by passing through, or reflecting off, matter.
- Light may excite matter to cause the emission of fluorescence.

To form an image, light is diffracted by the matter of the specimen. Part of this diffracted light is then refracted by the objective lens and recombined to form the (primary) image. The controlled manipulation of the interaction of light with matter is the basis of contrast formation in the image. This is described in Chapter 11.

2.8 Reflection of Light

Reflections of waves (including light) occur when the medium through which they propagate changes abruptly from one type to another, causing a corresponding change in the velocity of the wave. In the case of an opaque medium, the wave is reflected off the surface. The reflection of light off surfaces is crucial for us to be able to see objects around us. Most people naturally think only of mirrors when considering reflection, but all non-luminous objects reflect light falling upon their surfaces. The law of reflection states that the angle of incidence is equal to the angle of reflection (Figure 2.9a). The two angles are always measured from the normal[7] to the surface, the normal being the line drawn perpendicular to the surface at the point at which the incident ray of light falls on the surface.

If the surface is very smooth and flat (e.g. highly polished), the rays of light forming the incident beam all reflect in the same direction. This is called specular reflection, from the Latin for mirror (Figure 2.9b and Figure 2.9c). If the two media have the same refractive index, there is no change in velocity, and so no reflection or refraction occurs.

For all intents and purposes, a flat surface giving rise to specular reflection is smooth on a scale comparable to one quarter of the wavelength of light (i.e. within 100–150 nm). Light incident upon a rough or matt surface (i.e. a sheet of matt, not glossy, paper) is diffusely reflected (Figure 2.10a). The individual rays appear to strike the surface at different angles of incidence. Diffuse reflections are far less noticeable than specular reflections but are responsible for subtle photographic lighting, as well as shading and colouring, as captured by artists. However, at the *microscopical* level, the law of reflection still holds true. Each individual ray is reflected such that the angles of incidence and reflection are equivalent, but normals to different points of the surface point in different directions to one another. The individual rays are, therefore, not reflected in an identical fashion in one direction (Figure 2.10b) but are scattered in different directions; the reflections are randomly oriented.

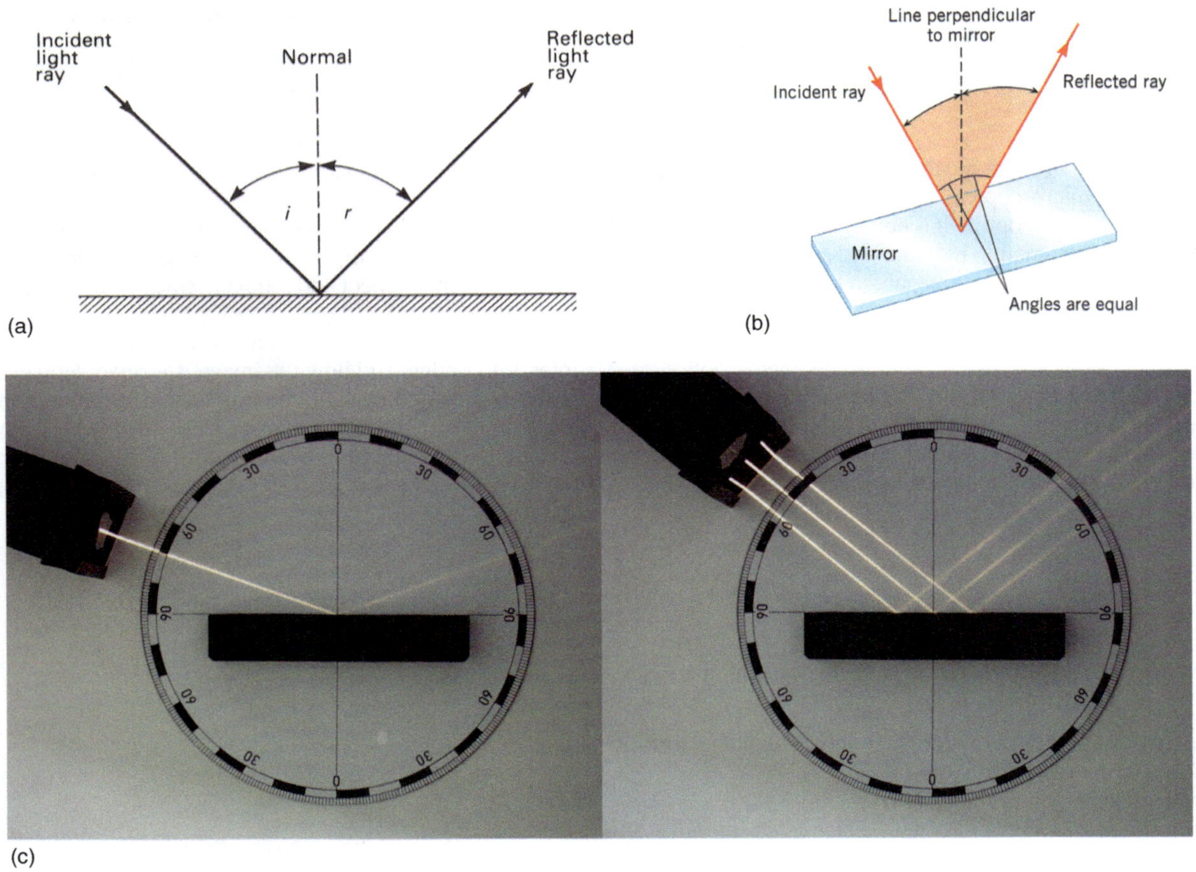

Figure 2.9 **Specular reflection of light** *The incident ray, normal and reflected ray are all coplanar: in three dimensions, they all lie within a single plane, as shown in (a) and (b). (c) shows a single specularly reflected ray of light compared with three reflected rays. In both cases, the angles of incidence and angles of reflection are all equal. Source: (a) Abbott 1984. Reproduced with permission from Pearson. (b) Trefil and Hazen 2004. Reproduced with permission from John Wiley & Sons. (c) Dr Z. Sándor 2005.*

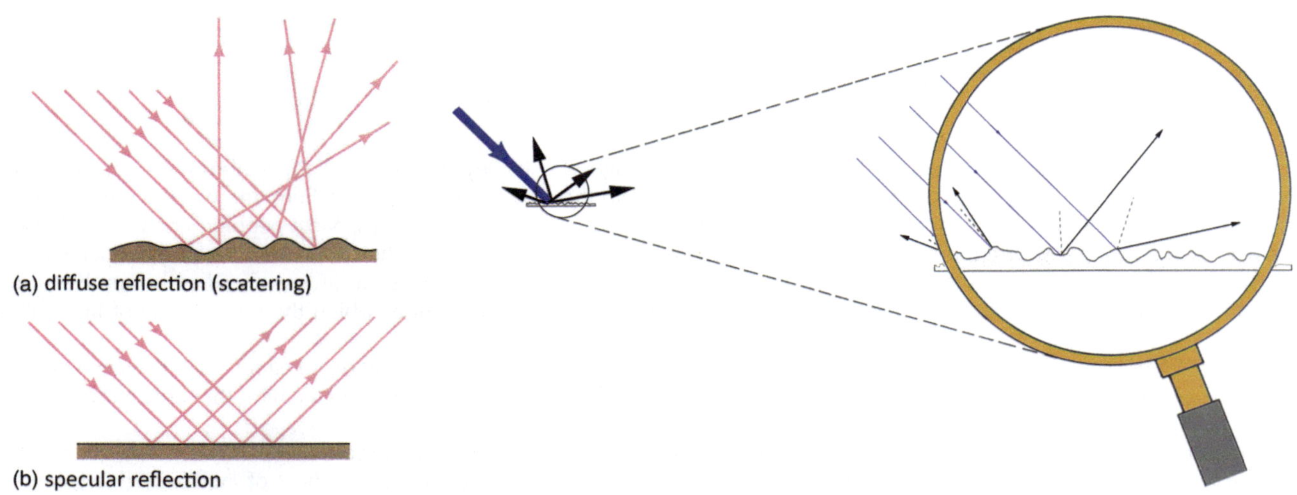

Figure 2.10 **Diffuse reflection of light** *At the microscopical level, the law of reflection holds true for diffuse reflection as for specular reflection. Although not all highly polished, all objects reflect light to a greater or lesser degree. A typical matt black surface will reflect at least 2.5% of incident light, while a high-gloss surface will reflect 90% or more light. Source: (left) Trefil and Hazen 2004. Reproduced with permission from John Wiley & Sons. (right) Falk et al. 1986. Reproduced with permission from John Wiley & Sons; Wayne 2009. Reproduced with permission from Elsevier.*

2.9 Refraction of Light

We now consider the propagation of light from a rarer (e.g. air) isotropic medium to a denser isotropic medium (e.g. water or glass). Light incident upon a transparent surface is partially reflected at the surface but is mostly transmitted. The key physical process that gives rise to the bending, or refraction[8], of the light beam is the slowing down of the light as it enters the denser medium.

For all practical purposes the speed of light in a vacuum (299,792,458 m/s) and in air may be considered equivalent. The speed of light in dry air at a standard temperature of 0°C and pressure of 1 atmosphere is *very slightly* less than in a vacuum: it is 2.99704764×10^8 m/s, so the refractive index, RI (see definition below), of air is 1.0002926.

The refractive index (denoted by the letter 'n'; sometimes the Greek letter eta, η, is erroneously used – and in this text abbreviated to RI) is simply defined as:

$$n = \frac{\text{speed of light in vacuum (or air)}}{\text{speed of light in the medium}} = \frac{c}{v}$$

As a rule of thumb, materials that have higher density have larger refractive indexes. Thus, when a ray of light passes into a transparent, denser medium, then the ray of light bends towards the normal of the interface, and the angle of refraction, θ_r, is less than the angle of incidence, θ_i (Figures 2.11a–c). Two points are important here:

- The incident and refracted rays are on opposite sides of the normal at the point of incidence and all three are in the same plane (i.e. coplanar).
- The law of refraction (Snell's law) states that when light passes from one medium to another, the sine of the angle of incidence bears a constant ratio to the sine of the angle of refraction (see Box 2.3).

Some light is also reflected at the interface between the two media (Figure 2.12); this is considered further in Chapter 12 on reflected-light microscopy and Chapter 13 on polarised light microscopy.

Hence, refraction can be understood in terms of a change in velocity. The figure in Box 2.3 shows this. A wavefront A-B travelling in air (n_1) falls upon a glass surface. At point A, the wavefront changes velocity and slows as it enters the glass, while the velocity of point B, in air, is unchanged. Although the time taken to travel distance A-C in glass is the same as the time taken to travel B-D in air, the distance A-C is shorter than B-D, and so the light is refracted.

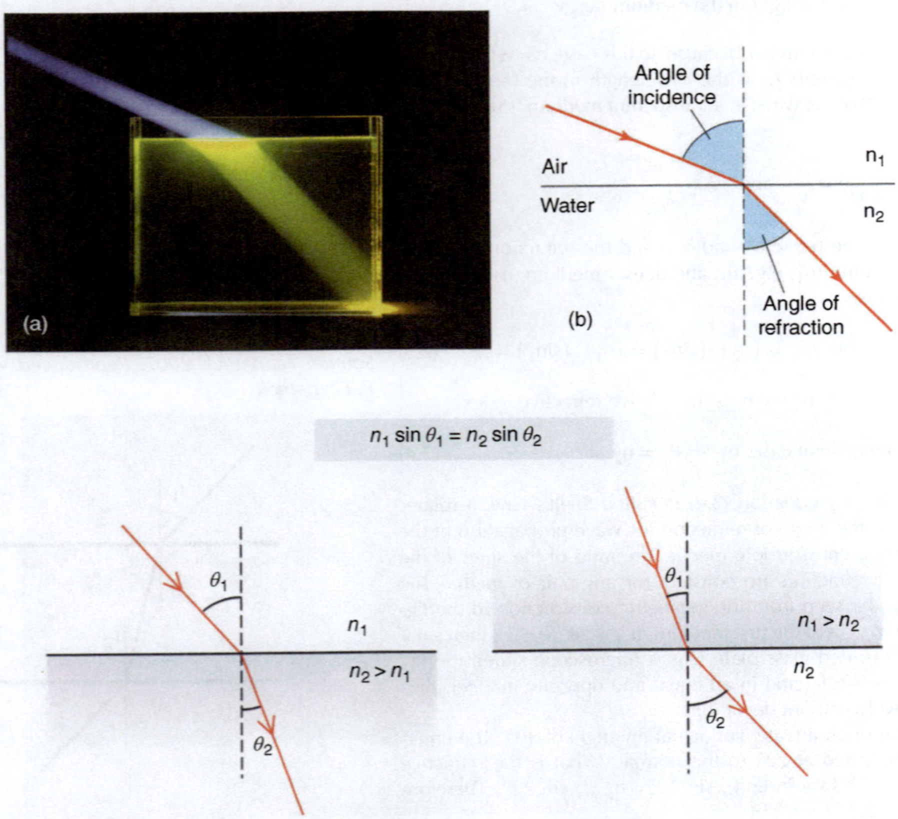

Figure 2.11 **The refraction of light** *The velocity of light is inversely proportional to the refractive index of the material through which it propagates. UV light shining through a bath of fluorescent liquid shows the effect of refraction. For any pair of transparent media (n_1, n_2) the ratio of the sine of the angle of incidence and the sine of the angle of refraction is constant. When plotted graphically in the form $y = mx + b$, the constant of proportionality, m in this equation, equals the refractive index for light travelling from air into any material. Source: (top) Trefil and Hazen 2004. Reproduced with permission from John Wiley & Sons. (bottom) Franklin et al. 2010. Reproduced with permission from John Wiley & Sons.*

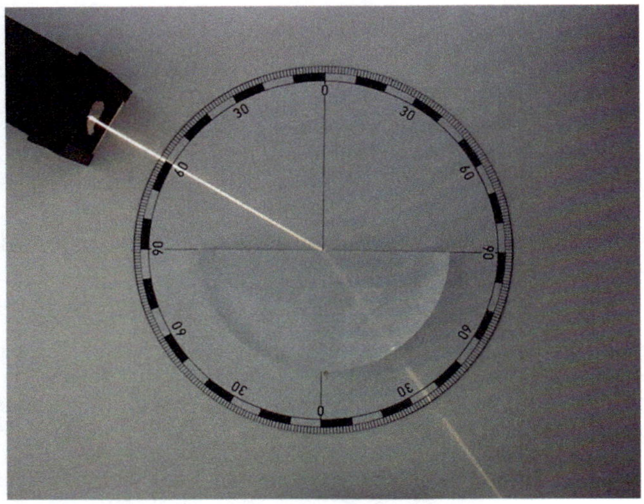

Figure 2.12 ***The refraction of light*** *When a beam of light is refracted, some light is also reflected at the interface between two transparent media. Source: Dr Z. Sándor 2005. These relationships are explained further in Figure 12.1.*

Box 2.3 Snell's Law and the Mathematics of Refraction

Because

$$n = \frac{\text{speed of light in vacuum (or air)}}{\text{speed of light in the medium}} = \frac{c}{v} \quad (2.1)$$

The refractive index, n, of a vacuum is 1 because, in this case, $c = v$. Now, since $c = \lambda\nu$ (or λf) and $v = \lambda_m \nu$ (where λ_m is the wavelength in the medium) and since the frequency (ν or f) of the wave in a transparent medium is unchanged, we have:

$$n = \frac{c}{v} \quad \text{and} \quad n = \frac{\lambda}{\lambda_m} \quad (2.2)$$

The relationship between the refractive index n and the refraction of a light beam passing between a rarer (n_1, e.g. air) and denser medium (n_2, e.g. glass or water) is usually expressed:

$$\sin i/\sin r = v_1/v_2 = n_2/n_1 = (n_2).(\sin_2) = (n_1).(\sin_1) = n_r \quad (2.3)$$

and $(n_1).(\sin i) = (n_2).(\sin r) = n_r$ where n_r is the relative refractive index.

$$\text{or in the general case: } n_1.\sin\theta_1 = n_2.\sin\theta_2 \quad (2.4)$$

The relationship described by equation (2.4) is called Snell's Law. It relates the angle of incidence to the angle of refraction for wave propagation at the boundary between transparent isotropic media. The ratio of the sines of the angles of incidence and refraction are constant for any pair of media. The proof of Snell's Law can be seen from the geometric construction in the figure. Here, $p/q = \sin i/\sin r$. Where the first medium, n_1, is air (RI = 1), then $\sin i/\sin r = n_2$. For a parallel-sided glass plate (e.g. a microscope slide), the ray passing through the glass is refracted in an equal and opposite manner, thus passing through, effectively without deviation.

A ray in air is incident upon a water surface at an angle of 30°. The emergent refracted ray is measured at 22° to the normal. What is the refractive index of water? From Snell's Law in (2.4), $\sin 30° = n_{water} \sin 22°$. Therefore $n_{water} = 0.5/\sin 22° = 0.5/0.375 = 1.333$.

Conversely, the angle of the refracted ray can be calculated knowing the refractive index. For glass of RI = 1.5172, $\theta_{glass} = \arcsin\ [n_{air}.\sin\theta_{air}/n_{glass}] = \arcsin\ [1.00 \times 0.5/1.5172] = 19.24°$.

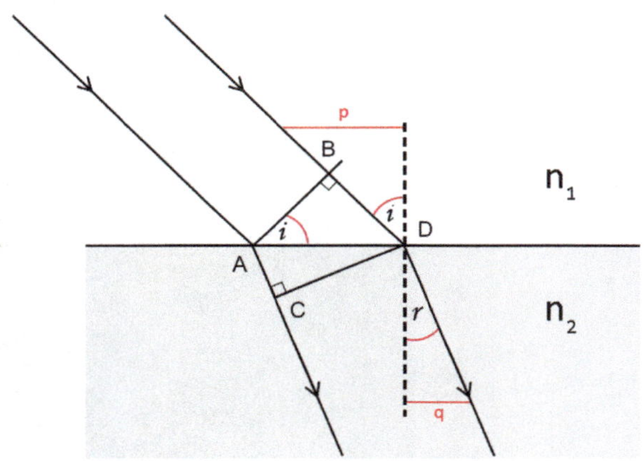

Source: Fitzpatrick 2007. Reproduced with permission from R. Fitzpatrick.

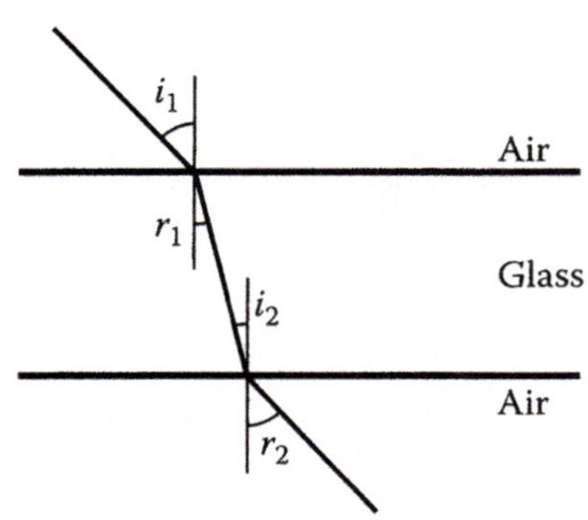

$r_1 = i_2$ and so $i_1 = r_2$

Source: Saxby 2011. Reproduced with permission from Taylor & Francis.

We now consider the refraction and reflection of light when the incident ray is within the denser material (e.g. water or glass, so $n_1 > n_2$). At the critical of incidence, where $\sin i = \theta_c$, the angle of refraction is 90° to the normal, and thus $\sin r = 1$. Therefore $n_2/n_1 = \sin\theta_c$ and $\theta_c = \arcsin(n_2/n_1)$. The critical angle for any medium can easily be calculated, because the sine of the critical angle is the inverse of the refractive index: $\sin\theta_c = 1/RI$. For glass, taking $n = 1.5172$, the critical angle is given by: $\sin\theta_c = 1/1.5172 = 0.6591$, and $\theta_c = 41.231°$. Other values are given in Table 2.3.

Thus, when light travelling through a medium of higher refractive index n_1 into a medium with lower refractive index, n_2 (so $n_1 > n_2$), where $\sin i\ (n_1)/n_2 > 1$, and no value of θr fulfills Snell's Law, the light is reflected back into the denser medium, and is totally internally reflected. It does not pass into the medium of lesser refractive index.

References

Fitzpatrick, R. (2007). 'Geometric Optics'. Lecture Notes. University of Texas at Austin.
Saxby, G. (2011). *The Science of Imaging*, 2nd Ed. Boca Raton: CRC Press. Taylor & Francis.

The refractive index is a ratio, a dimensionless parameter greater than unity, with no associated units. Refraction is a phenomenon associated with any wave that passes from a medium of one density to a medium of another density and thus changes speed at the boundary, or interface. In water the velocity of light is 2.24777449×10^8 m/s (RI = 1.33334 at n_D^{20}), and in glass the velocity will vary according to the type of glass, and the wavelength at which the measurement is taken (usually the mean of the yellow Fraunhofer sodium D lines at 589.29 nm at a specified temperature, hence the symbol n_D^{20}). Assuming a refractive index for pure crown glass of 1.5167 at 20°C, the speed of light through crown glass is approximately 1.976×10^8 m/sec. For both precision and accuracy, RI values should be cited at a particular standard wavelength and at a particular temperature.

Refraction is not a phenomenon exclusive to light but is most often associated with this form of radiation. Further, light will refract when it hits the surface of an opaque medium, such as lead. However, we do not notice any refraction, as we do with a transparent substance, because the path length of the light is extremely short before the photons are absorbed. Therefore, for all practical purposes, light is only refracted by, and passes through, transparent media.

The velocity of light is inversely proportional to the refractive index of the material through which it propagates. The slower the velocity of transmission in a medium, the higher the refractive index. In terms of velocity change we invoke Fermat's principle, which (simply expressed) states that a ray of light takes the shortest time to travel between two points (the phenomenon is sometimes called the principle of least time and is used to define a ray of light).

Consider two points, one in air (A) and the other in glass (D) (Figure 2.13a). Light travelling in a straight line A → B → D is slowed down as it passes through the glass from B to D. However, if it travels first to C the time spent in travelling the longer distance A to C is more than offset by the time in the shorter distance C to D. It is a fairly simple proof that the minimum overall time corresponds to the angles of incidence, θ_i, and refraction, θ_r, given in Snell's law[9]. Fermat's principle also applies to reflection: the equal angles of incidence and reflection are again a consequence of light taking the shortest optical path between two points. For the same reason light is reversible: it will follow exactly the same path if its direction of travel is reversed. We shall see in the next chapter how this is important when considering conjugate points.

The path taken by the light ray A → C → D, called the *optical path*, has an important corollary in image formation. It may appear that the axial ray passing through the centre of the lens (Figure 2.13b) takes a shorter time to arrive at the image point than the rays passing closer to the edges of the lens. But this is not so; the axial ray passes through – and is slowed down – by a greater thickness of glass. Provided the lens has been properly corrected for this to happen, all the rays take the *same* time to travel from object to image: they all have the *same* optical paths. Or, to put it another way, the conjugate points of a lens, or indeed any imaging system, are points for which the light rays between the two points have the same optical paths.

A pencil or a spoon inserted into a half-full beaker of water (Figure 2.14a) will show two things: (a) the submerged part of the pencil will appear to bend slightly towards the surface of the water, and (b) the curvature of the glass beaker and the water it contains will

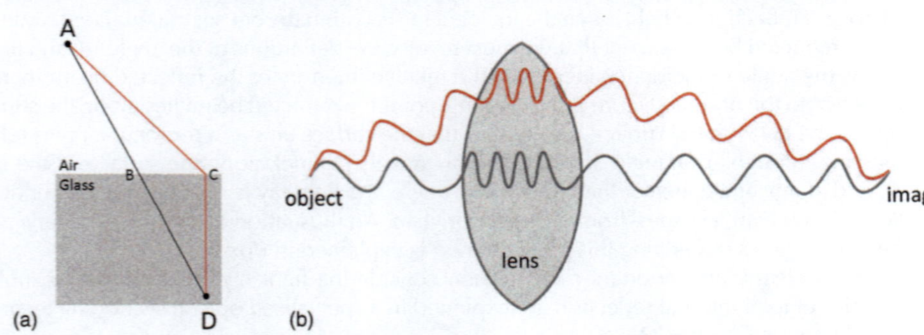

Figure 2.13 **Fermat's principle of least time** *(a) Light takes the shortest optical path between two points. See the text for explanation. Source: Adapted from Murphy 2001 with permission from John Wiley & Sons.*

(b) The number of vibrations, as well as the transit time, experienced by waves travelling between an object point and the conjugate image point are the same regardless of whether a given wave passes through the middle or periphery of a lens. A natural consequence of the principle of reversibility of light is that the object and image positions can be interchanged. The points corresponding to object and image are called conjugate points. Source: Adapted from Murphy 2001 with permission from John Wiley & Sons.

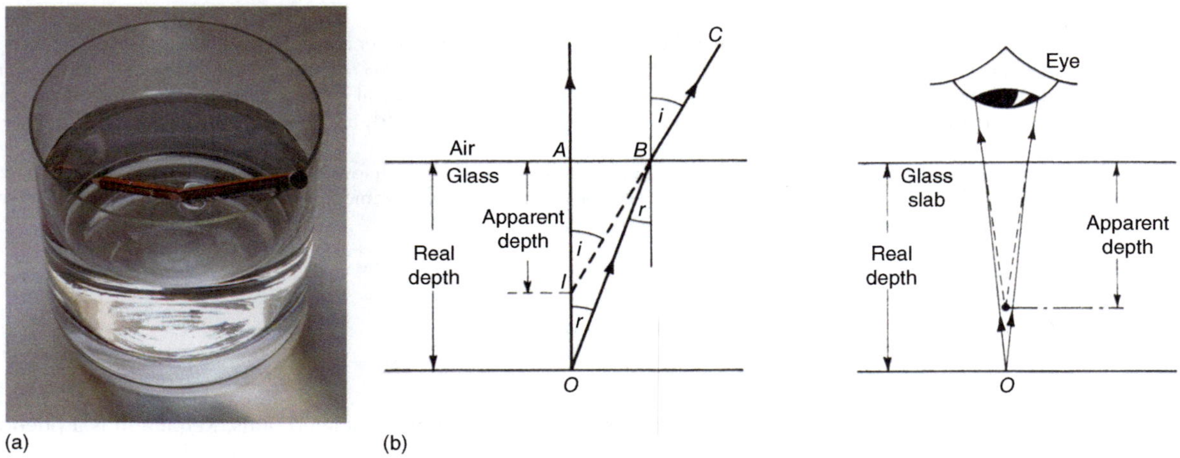

Figure 2.14 **Refraction and image formation** *Refraction of the image-forming rays causes the image to appear higher in the fluid (or embedded in a slide preparation) than it actually is. This holds true only when the observer is in the rarer medium of lower RI, and the object is in the denser medium of higher RI. If the observation is made from the denser medium of higher RI into the rarer medium, of lower RI, then the object appears further away. Source: (a) Tilley 2011. Reproduced with permission from John Wiley & Sons. (b) Abbott 1984. Reproduced with permission from Pearson.*

act as a magnifying lens to give an enlarged image of that part of the pencil that is under the surface of the water. The bending and discontinuity seen at the water-air interface are due to refraction.

Now take an opaque mug, or pan, and place a coin on the bottom. Angle your head so that the coin just disappears behind the side of the pan. Now, without moving your head carefully fill (or get an assistant to fill) the pan with water. The coin will come into view because of refraction of the light reflected off it. Furthermore, the coin will appear 'higher' through the water than it actually is, because of the refraction of the rays of light forming the image. If the ray trace of the refracted ray is extrapolated to the apparent position of the coin, it will appear higher in the water. The principle is illustrated in Figure 2.14b. For the same reason, fishermen must aim their spear behind and below the apparent position of the fish in order to be certain of spearing their quarry.

Most (80%) of the refraction of light rays to form an image on the retina occurs at the surface of the cornea, which has an RI of 1.376. The aqueous humour has an RI of 1.336 and the vitreous humour, 1.337. Therefore, when we swim underwater without goggles, image-forming rays are refracted much less and so cannot be completely focused by the lens alone. In short, we suffer temporarily from extreme long-sight, or hyperopia.

When light travels through a medium of lower refractive index, n_1 (i.e. air), into a medium with higher refractive index, n_2 (i.e. water, so $n_1 < n_2$), there will always be reflection and refraction for all angles of incidence. A ray of light which impinges upon a surface at 90° (i.e. normal, or orthogonal to the surface) will pass straight through without deviation, although it will be slowed down according to the refractive index of the substance. If a material suspended in a fluid possesses the same refractive index, then the system is homogeneous. The rays of light do not change in velocity; therefore, no refraction will occur. We shall meet this situation later when we consider homogeneous immersion for improving resolving power (Chapter 7, Section 7.5). The consequence of the material and surrounding fluid having identical, or very nearly the same, refractive indexes is near or total invisibility (Figure 2.15) and greatly reduced image contrast.

2.10 Total Internal Reflection

We now consider the refraction and reflection of light when the incident ray is within the denser material (e.g. water or glass, so $n_1 > n_2$). Again, Snell's law applies to the refracted beam (except that we must reverse the definitions of the angles θ_i and θ_r, (refer back to Figure 2.11c and see Figure 2.16a). As the angle of incidence increases, the relative intensity of the reflected beam increases up to the point where the angle of refraction is 90° to the normal (Figure 2.16b). At this point the refracted beam lies along the surface. At higher angles of incidence, where the angle of the ray with the surface is so obtuse that the surface acts as a mirror, the ray is reflected off the surface of the medium instead of passing through. The angle at which the ray travels parallel along the surface of the medium is called the *critical angle* (see Table 2.3), and at any angle greater than the critical angle, the light ray is reflected off the surface (Figure 2.16c). The light is *totally internally reflected* – no light 'escapes' from the water or glass. An illustration of total internal reflection, using laser light, is shown in Figure 2.17. The mathematics underlying this phenomenon is explained in Box 2.3.

Total internal reflection (TIR) is clearly an important matter when considering light rays passing at high oblique angles through glass/air interfaces. The properties of total internal reflection are exploited in a specialised optical sectioning technique in fluorescence microscopy, TIRF, which is explained in Chapter 23.

2.11 Dispersion

Although the frequency, v, of light does not change as it travels from one medium to another, the refractive index of a material will vary according to the wavelength of light passing through it. Thus 'blue bends best' and will have a slightly higher

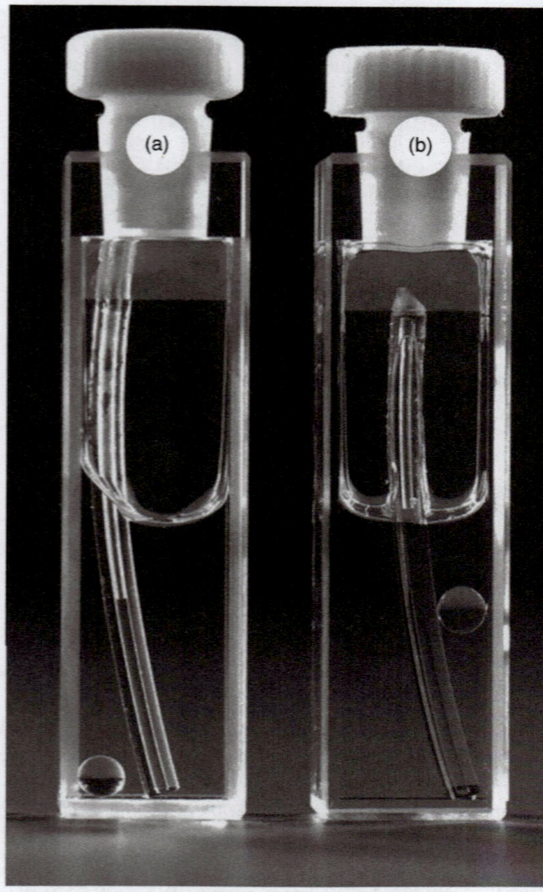

Figure 2.15 **Homogeneous immersion** *A refractive index 'mismatch' gives rise to contrast in the image, and the perception of boundaries between the object and the fluid it is immersed (or mounted) in. Here two cuvettes contain a piece of plastic tubing (n = 1.55) and a glass bead (n = 1.52). In water (n = 1.33), seen in (a), scattering is strong and the two objects are easily visible. In 100% glycerol (n = 1.47), seen in (b), scattering of light at the boundaries is reduced. Consequently, the objects are much less visible, and blend with the background. If immersed in a medium of RI = 1.55, the plastic tubing would 'disappear', being matched completely. The practical application of homogeneous immersion is covered in Chapter 18, Section 18.8 onwards. Source: Lakey 2009. Reproduced with permission from the Royal Society.*

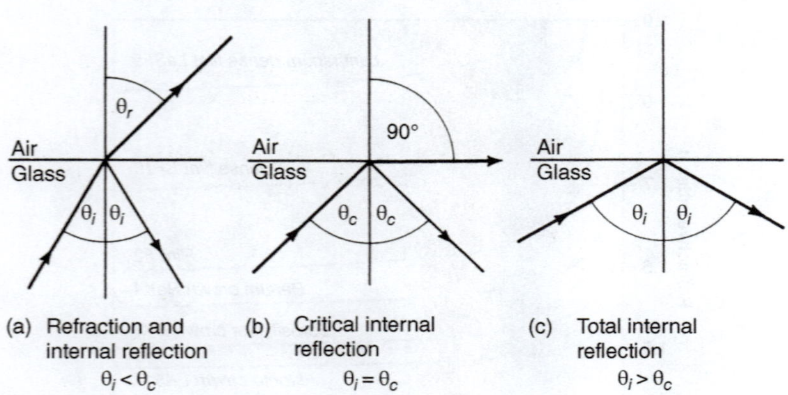

Figure 2.16 **Total internal reflection** *The relationship between the angle of incidence (θ_i) and the critical angle (θ_c) when light passes from a transparent medium having a high refractive index to one with a lower refractive index. When the angle of incidence, θ_i, is less than θ_c, refraction occurs in the usual manner. When $\theta_i = \theta_c$, the angle of refraction, θ_r, is 90°, and the refracted ray travels along the interface between the two media. When θ_i is greater than θ_c, the ray is totally internally reflected. Source: Abbott 1984. Reproduced with permission from Pearson.*

Table 2.3 Refractive Indexes and Critical Angles of Various Optical Glasses, Compared to Water and Diamond

Medium	Refractive index (n_D^{20})*	Approx. density @ 20°C (kg/m^3)	Critical angle (°)
Vacuum	1.0000	0	–
Air	1.0003	1.204	88.597° ≈ 90°
Borosilicate Crown BK7 glass	1.5167	2400–2800 (2500)	41.239°
Flint F2 glass	1.6184	2900–5900 (3700)	38.174°
Water	1.3334	1000	48.607°
Diamond	2.4173	3520	24.440°

*Refractive index measured at Fraunhofer D line (589.3 nm) from a sodium vapour lamp.

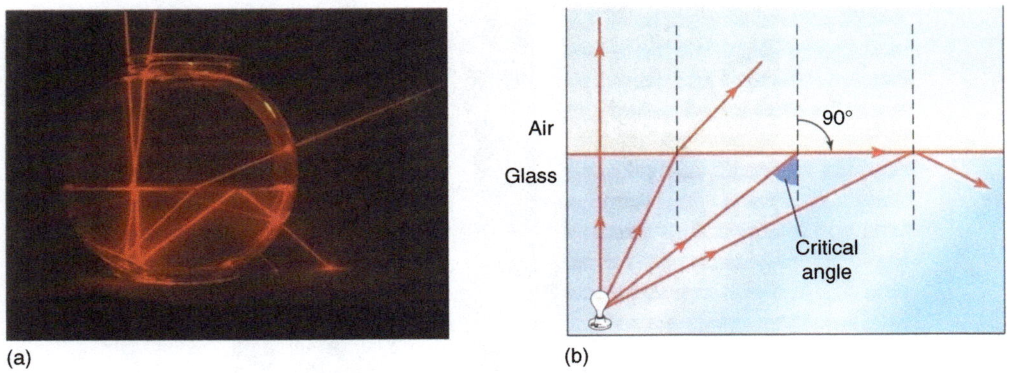

Figure 2.17 ***Total internal reflection by laser light*** *A laser beam shining through water in a half-filled fish bowl illustrates total internal reflection. The critical angle exists only when light passes from a medium of higher RI to one of lower RI. The magnitude of the critical angle depends on both media at the boundary. It is not a quantity that is determined solely by one medium. Source: Trefil and Hazen 2004. Reproduced with permission from John Wiley & Sons.*

refractive index (e.g. for borosilicate crown glass $RI_{blue} = 1.522$ compared to $RI_{red} = 1.513$) than red (Figure 2.1 and Figure 2.18), since blue light is slowed down more by the glass. This phenomenon is called *dispersion* and is best known from Newton's experiments[10] with a prism to split white light into its component spectral colours. Dispersion is the rate of variation of refractive index with wavelength. Its effect is manifested as chromatic aberration.

Scientists usually measure the coefficient of dispersion between the Fraunhofer C and F spectral lines. These lines have values of 486.1 nm and 656.3 nm respectively (see Table 2.4). The values of the C and F lines lie closer to what our eyes can distinguish than the B and G lines (see Box 2.4), which are otherwise used to measure dispersion, but lie in the deep red and indigo part of the spectrum and so may be hard to see for most humans. The third characteristic point[11] for which single refractive index values are normally cited is the mean of the sodium D lines at 589.29 nm.

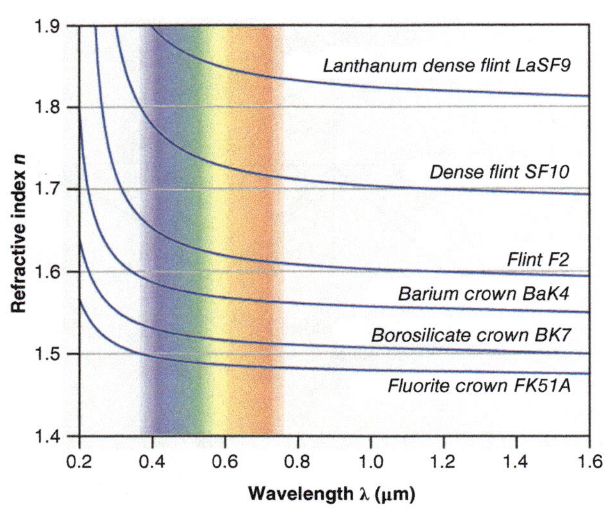

Figure 2.18 ***Dispersion*** *The refractive index plots of the optical glasses listed in Table 2.4, showing how the refractive index varies with wavelength over the UV, visible and IR components of the electromagnetic spectrum. Source: Adapted from DrBob under the terms of the Creative Commons Attribution Share-Alike licence CC-BY-SA-3.0 (http://creativecommons.org/licenses/by-sa/3.0/), via Wikimedia Commons. (spectral band) Franklin et al. 2010. Reproduced with permission from John Wiley & Sons.*

Table 2.4 *Refractive Indexes and Dispersion of Various Optical Glasses, Compared to Water and Diamond*

Glass type/material	Refractive index at 430.8 nm (G) indigo	486.1 nm (F) blue	589.3 nm (D) yellow	656.3 nm (C) red	686.7 nm (B) red	Dispersion (constringence)	Abbe No. (V_d)
Lanthanum dense Flint LaSF9	1.887	1.869	**1.850**	1.843	1.839	0.048	32.7
Dense Flint SF10	1.764	1.746	**1.728**	1.721	1.718	0.046	29.1
Flint F2	1.643	1.632	**1.618**	1.614	1.613	0.030	34.3
Barium Crown BaK4	1.582	1.576	**1.569**	1.566	1.565	0.017	56.9
Borosilicate Crown BK7	1.527	1.522	**1.517**	1.514	1.513	0.010	64.6
Fluorite Crown FK51A	1.494	1.491	**1.487**	1.485	1.484	0.010	81.2
Water	1.338	1.336	**1.333**	1.331	1.331	0.007	66.6
Diamond	2.452	2.436	**2.417**	2.410	2.407	0.044	54.5

Because of their well-defined wavelengths, Fraunhofer lines are often used to characterise the refractive index and dispersion of optical materials. The coefficient of dispersion, or constringence, of a transparent material is the difference between the refractive indexes of the G and B Fraunhofer lines. The Abbe number (V_d) is a commonly used measure of dispersion. The Abbe number = $(n_D - 1)/(n_F - n_C)$, where n_F, n_C and n_D are the refractive indexes for the Fraunhofer F, C and D lines (Fraunhofer labelled his lines starting with A in the far red; major lines have upper case letters and minor ones, lower case). The Abbe number (V_d) is calculated from the Fraunhofer spread, or mean dispersion, Δ, between the hydrogen C-line, at 656.3 nm, and the hydrogen F-line, at 486.1 nm. Sometimes the Abbe number is measured (V_e) with the green mercury e-line, blue cadmium C′ and red cadmium F′ lines, C′ and F′ (546.1, 480.0 and 643.8 nm respectively). Crown glass, with low dispersion, has a high value Abbe number; the converse is true for flint glass. The graphical plot of the refractive indexes of these optical glasses is shown in Figure 2.18. A material with a high Abbe number exhibits very little dispersion because different wavelengths will have nearly the same refractive index. Since a typical flint glass may have a dispersive power, Δ, of 1/30 (Abbe number = 30) and a crown glass only half the dispersive power (Δ = 1/60; Abbe number = 60), a positive achromatic doublet, with a net converging power, can be made by combining a strongly positive crown glass lens with a weaker negative flint glass lens.

Box 2.4 Diamonds, Dispersion and Refractive Index

The figure shows refractive index values (n) for diamond, two types of glass and water. Note, from the slope of the curves, that the refractive index varies with wavelength, being larger for blue/violet light than for red. This variation is called *dispersion*, and the amount of dispersion (the slope of the curves) also varies from glass to glass. When white light is passed through a prism and the visible light decomposed into its spectral colours, then dispersion gives rise to *chromatic aberration* in simple lenses. By constructing lenses from glasses of different dispersive power and refractive indexes, chromatic aberration can be partly, or nearly wholly, overcome (achromatic and apochromatic lenses).

In optics, the coefficient of dispersion value (e.g. 0.044 for diamond) is inversely proportional to the Abbe number, or V-number, which is used to classify glass and other optically transparent materials. Flint glass generally has an Abbe number of <50, and crown glass >50; diamond has an Abbe number of 54.5 and distilled water 66.6 (see Table 2.3).

The Abbe number is useful for classifying glass types, as well as for calculating the necessary focal lengths for manufacturing lenses. An Abbe diagram is a graphical plot of the Abbe number versus refractive index. This allows the rational selection of glass to form the appropriate lens figures for combining different lenses into aberration-minimised glass doublets. Abbe numbers are only a useful measure of dispersion for visible light; for other wavelengths (e.g. with multi-photon lasers) the group velocity dispersion, or GVD, is used.

Diamond has the highest refractive index across the whole wavelength range and also the highest hardness of all crystals. It is the pronounced refraction of light in diamond (and also its strong reflection from the internal surfaces) that gives diamond its prized sparkling quality. How many purchasers of diamond jewellery know that what they pay high prices for is simply hardness and refractive index?

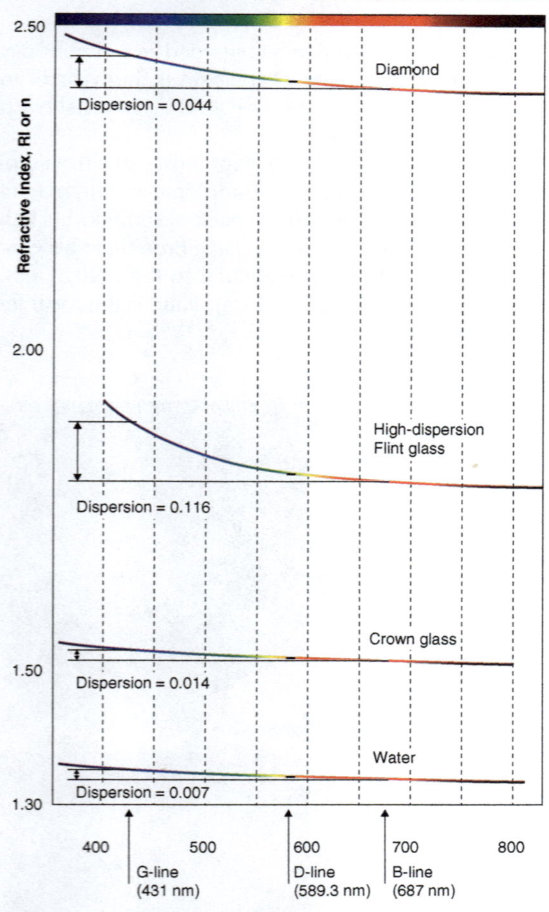

Source: Adapted from Green et al. 2003 with permission from GIA.

Reference

Green, B., Reinitz, I., Gilbertson, A., Johnson, M., and Shigley, J. (2003) 'Light Dispersion, Color Wavelengths and Fire'. Gemological Institute of America.

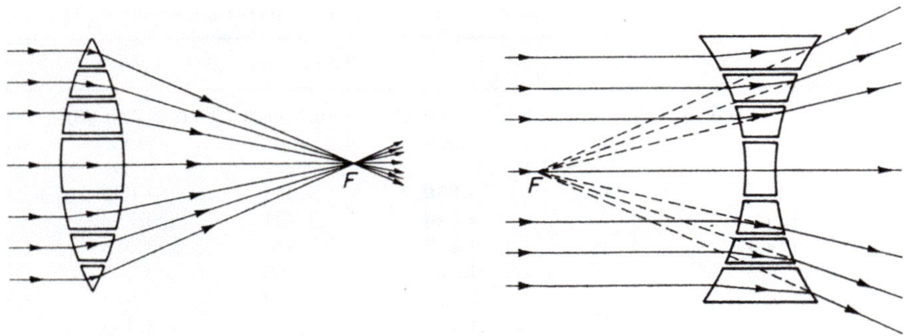

Figure 2.19 **Action of a lens** *A lens may be considered to be a continuous set of a large number of tiny prisms, with each individual face refracting light to bring it to a focus. Although the term 'lens' commonly refers to a shaped piece of glass or transparent plastic used with visible light, any object which refracts electromagnetic radiation is also a lens. For example, a microwave lens can be made of paraffin wax. Source: Abbott 1984. Reproduced with permission from Pearson.*

We noted at the beginning of this chapter that waves can be used to carry information. Electromagnetic waves are particularly good for this, as the use of radio (mobile phones), microwaves (satellites) and light (fibre optics) attest. However, dispersion can be a problem, particularly in fibre-optics communication. The sharp pulses required for digital optical communication are spread out, and information is either attenuated or lost. Modern materials are now used in fibre-optic cables to transmit information pulses with minimal dispersion. However, dispersion is not all bad: high dispersion is valued in gemstones, giving diamond[12] its characteristic 'fire', or flashes of colour.

2.12 Refraction and the Action of Lenses

In its simplest form, a lens can be regarded as a continuous series of prisms (Figure 2.19), with each face refracting the rays of light from the object to form an image. Dispersion is the basis of inevitable chromatic aberration in objectives and other optical components, as discussed in Chapter 6, Section 6.4. A graph showing the dispersion curve for borosilicate and other glass types is shown in Figure 2.18.

All of the fundamentally important lenses of the microscope are positive, or converging, lenses – that is, the glass is thicker in its centre than at its edges, like the traditional simple magnifying glass. Most of us have held a magnifying glass and used it to form an image of the sun – possibly on the back of somebody's hand, when it hurts, or on a piece of paper which might catch fire (Figure 2.20). Here, parallel light from the sun is being brought to a *focus*[13] at the principal focus. The (principal) focal plane of the lens passes through the focal point, and lies perpendicular to the optical axis. The distance between the principal axis, which passes through the centre of a thin[14] convex lens, and the focal plane is the focal length of the lens (Figure 2.21). The lens will produce an image of any distant object in its focal plane.

Figure 2.20 **Focusing a plano-convex lens** *See also endnote 13. What small boy isn't aware of the qualities of 'a burning glass'? Temperatures in a solar furnace can reach 3500°C (6330°F), sufficient to melt steel or rock.*

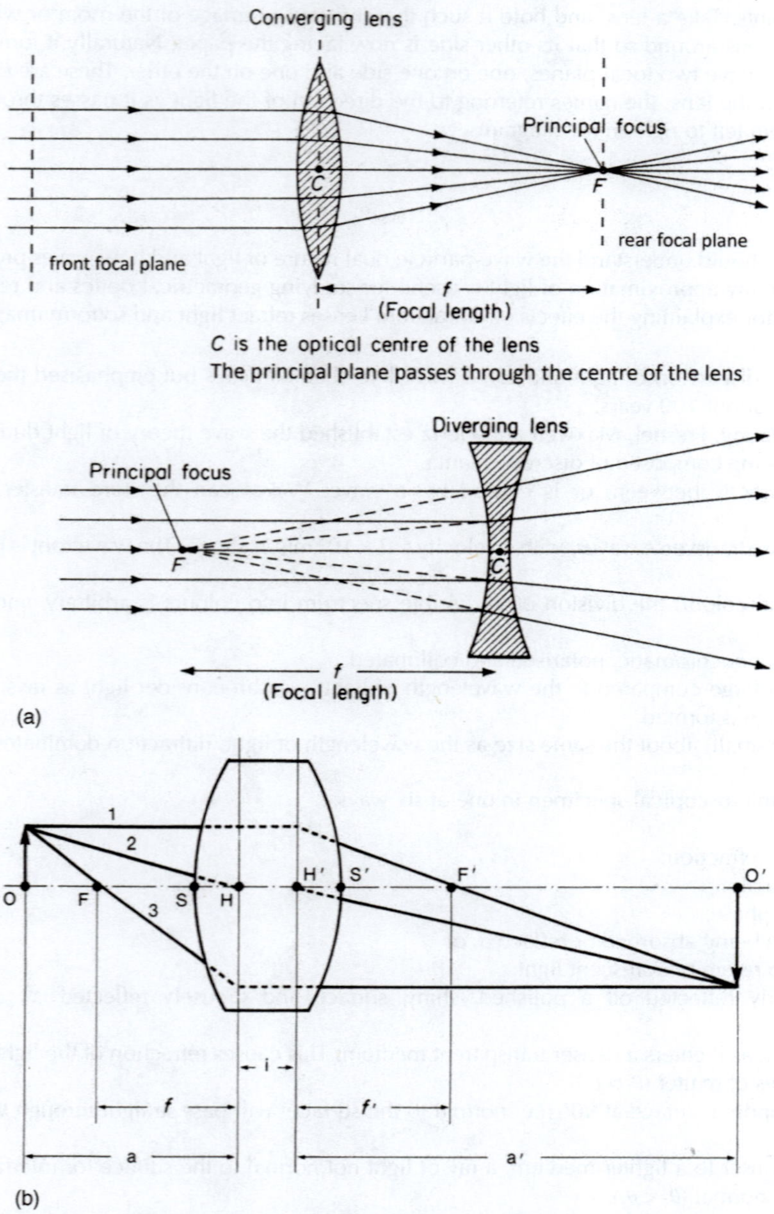

Figure 2.21 ***Cardinal points and planes of simple lenses*** *The principal axis, also called the optical axis (O-O′), passes through the centre of the lens, perpendicular to the mid-line of the lens, and intersects the lens surface at each vertex (pole) of the lens (S, S′). The vertices of the lens occur at each of the points where the principal axis meets each surface of the lens.*

For ease we regard lenses as thin, as shown here in (a), but in practice a lens is 'thick' composed of matched groups of individual lens elements, as shown in (b). Cardinal points provide a way to simplify the optical properties of corrected thick lenses and predict their behaviour using ray tracing and thin lenses. The three pairs of cardinal points all lie on the principal axis; they are the focal points (F, F′), the principal points (H, H′) and the nodal points. (b) shows the cardinal elements of a thick lens in air. The principal points (H, H′) are the points where the principal planes cross the optical axis, with distance i the hiatus. We can regard refraction occurring (as shown) at the principal planes, such that a and a′ are respectively the object distance and image distance from the object point, O, and image point, O′. The focal points (F, F′) define the focal lengths, f and f′. The front and rear nodal points occur where an incident oblique ray and the corresponding refracted ray cross the optical axis. The incident ray passes through the front nodal point; the refracted ray passes through the rear nodal point and exits the lens parallel to the incident ray (see Figure 3.5b, Section 3.3). If the medium on both sides of the lens is the same (e.g. air) then the front and rear nodal points coincide with the front and rear principal points, respectively. The equivalent focal length (EFL) is measured from the nodal points. The surface vertices (S, S′) are the points where each optical surface crosses the optical axis. The surface vertices of the lens of the eye are called the anterior and posterior poles of the lens. Source: (a) Abbott 1984. Reproduced with permission from Pearson. (b) Bradbury et al. 1989.

Each lens has two focal points. Take a lens, and hold it such that it forms an image of the room or window in front of you onto a piece of paper. Now turn the lens around so that its other side is now facing the paper. Naturally it forms an image in just the same manner. In other words, lenses have two focal planes, one on one side and one on the other. These are called the first and second, or front and back, focal planes of the lens, the names referring to the direction of the light as it passes through the lens. By convention, light is shown as travelling from left to right in ray diagrams.

2.13 Chapter Summary

At the end of this chapter you should understand the wave-particle dual nature of light and how waves propagate, carrying photons, or 'wave packets', of energy. The ray approximation of light is useful for studying geometrical optics and refraction, while consideration of light as a wave is essential for explaining the effects of diffraction. Lenses refract light and so form images.

1. Newton investigated the dispersion of light and some wave-like characteristics but emphasised the corpuscular nature of light. This view dominated for about 100 years.
2. The work of Huygens, Young, Fresnel, Maxwell and Hertz established the wave theory of light during the 19th century. Einstein described light as also being composed of discrete quanta.
3. Energy is transferred directly between, or is carried by, a wave. Waves can therefore transfer energy *without* transferring mass.
4. Electromagnetic waves are transverse waves, with a velocity $\approx 3 \times 10^8$ m/s *in vacuo*. The wavefront is perpendicular to the direction of propagation.
5. Light possesses no actual colour; the division of the visible spectrum into colours is arbitrary, and our perception of colour is psychophysical.
6. Laser light is coherent, monochromatic, polarised and collimated.
7. When the object is very large compared to the wavelength of light, we can consider light as rays, and use geometric optics to understand how the image is formed.
8. When the object is very small, about the same size as the wavelength of light, diffraction dominates, and we must consider light as waves.
9. Light interacts with the microscopical specimen in one of six ways:
   - by absorption;
   - through reflection or refraction;
   - by scattering or diffraction;
   - through a change in phase;
   - by polarisation while being absorbed or reflected; or
   - through excitation to re-emit fluorescent light.
10. A ray can be specularly reflected off a polished, shiny, surface and diffusely reflected off a non-homogeneous, rough, surface.
11. A ray of light slows down as it enters a denser transparent medium. This causes refraction of the light towards the normal with the interface of the two states of matter ($\theta_i > \theta_r$).
12. A ray of light incident upon a surface at 90° (i.e. normal to the surface) will pass straight through without deviation, although it will be slowed down.
13. When passing from a denser to a lighter medium, a ray of light not normal to the surface (or interface) of the two media will be refracted away from the normal ($\theta_i < \theta_r$).
14. At angles greater than the critical angle, light is totally internally reflected off the surface.
15. The speed of light is 299,792,458 m/s in a vacuum. A vacuum transmits all frequencies of light with the same velocity.
16. The speed of light is inversely proportional to the refractive index of the material through which it propagates.
17. Dispersion is the change in phase of a complex (e.g. white) light wave, when passing from one material to another, such that the monochromatic wavelengths are separated.
18. The wavelength of light vibrating at a particular frequency will become shorter when passing from one medium into another – of higher refractive index – which transmits light more slowly.
19. The important lenses of the microscope are positive or converging lenses. These refract light to form an image.
20. Rays entering a converging lens parallel to the optical axis come to a focus at the principal focal point.
21. Each lens has two focal points. These are the front and rear focal points.
22. Each lens has two corresponding focal planes: the front and rear (or back) focal plane.
23. The principal focal plane of the lens passes through the focal point and lies perpendicular to the optical axis. The lens will produce an image of any distant object in its focal plane.
24. The distance between the principal axis, which passes through the centre of a thin convex lens, and the focal plane is the focal length of the lens.

Key Reading

Brooks, M. (2012). 'What is light?' In: *Can we travel through time? The 20 Big Questions of Physics*. London: Quercus Publishing, pp. 201–213.

- A short and very readable summary of the duality of light, the photoelectric effect and the importance of lasers.

Johnsen, S. (2012). *The optics of life: a biologist's guide to light in nature*. Princeton and Oxford: Princeton University Press.

- A 300-page text explaining optics for biologists in non-mathematical fashion. A unified and sufficiently detailed account of the emission of light, absorption, scattering and interference is given together with chapters explaining fluorescence and polarisation. A further two chapters explain simply the units employed for measuring light and how to measure light flux. The final chapter introduces the concept of quantum mechanics.

References

Abbott, A.F. (1984). 'Lenses and optical instruments'. In: *Ordinary Level Physics*, 4th Ed. Heinemann Educational Books.
Allen, E. and Triantaphillidou, S. (2010) *The Manual of Photography*, 10th Ed., Focal Press, Elsevier.
Bradbury et al. (1989). *RMS Dictionary of Light Microscopy*, RMS Handbook No. 15 OUP.
Bushberg, J.T. (2002) *The essential physics of medical imaging*, 2nd Ed. Lippincott Williams & Wilkins.
Falk, D., Brill, D., and Stork, D. (1986). *Seeing the light*. New York: John Wiley & Sons.
Franklin, K. et al (2010). *Introduction to Biological Physics for the Health & Life Sciences*. New York: John Wiley & Sons.
Lakey, H.J. (2009). 'Neutrons for biologists: a beginner's guide, or why you should consider using neutrons'. *Jour. Roy. Soc. Interface*. PMC2843970.
Murphy D.B. (2001). *Fundamentals of Light microscopy & Electronic Imaging*. Wiley-Liss.
Robinson, P.C. and Bradbury, S. (1992). *Qualitative Polarised-Light Microscopy, Microscopy Handbook No. 9*. Oxford: Oxford Science Publications.
Tilley, R.J. (2011). *Colour and the Opical Properties of Materials*. New York: John Wiley & Sons.
Trefil, J. and Hazen, R.M. (2004). *Physics Matters*. New Jersey: John Wiley & Sons Inc.
Wang, Q. et al. (2011). 'Spectroscopy'. In: Glencross, H., Ahmed, N., and Wang, Q. (ed.) *Biomedical Science Practice* (2011) Fundamentals of Biomedical Science series, Oxford: Oxford University Press.
Wayne, R. (2009). *Light & Video Microscopy*, 1st Ed. Amsterdam: Academic Press, Elsevier.

Further Reading

Bahaa, E.A.S. and Teich, M.C. (2007). *Fundamentals of Photonics*, 2nd Ed., New York: Wiley.
Falk, D., Brill, D., and Stork, D. (1986). *Seeing the Light*. New York: John Wiley & Sons.
Fara, P. (2015). 'Newton shows the light: a commentary on Newton (1672) "A letter … containing his new theory about light and colours…"' *Philos Trans A Math Phys Eng Sci. 373*(2039): 20140213.
Nature Milestones, Photons, URL: https://www.nature.com/milestones/milephotons/index.html and https://www.nature.com/milestones/milephotons/timeline.html (accessed 1 June 2018)
Saxby, G. (2002). *The Science of Imaging: an introduction* 2nd Ed. Boca Raton: CRC Press.
Silfvast, W.T. (2005). *Lasers*, Module 5/10. In: Guenther, A, Pedrotti, L.S., and Roychoudhuri, C., Ed., *SPIE Fundamentals of Photonics*.
Sobel, M.I. (1989). *Light*. Chicago and London: The University of Chicago Press.

Internet Resources

Dispersion of light through a prism, animated gif. URL: https://en.wikipedia.org/wiki/File:Light_dispersion_conceptual_waves.gif.
Dispersion factsheet (Bellingham and Stanley). URL: https://www.thomasnet.com/pdf.php?prid=100269.
Demonstration of Newton's *Experimentum Crucis* by Dr Jonathan Hare (M4V file). URLs: http://www.creative-science.org.uk/prism.html and https://www.youtube.com/watch?v=uucYGK_Ymp0.
The Institute of Physics teaching webpages. URL: http://tap.iop.org/.
Simple harmonic motion describing a wave. URL: https://en.wikipedia.org/wiki/Simple_harmonic_motion.
Refractive Index database. URL: http://refractiveindex.info.
Proof of Snell's law using Fermat's principle. URL: http://hyperphysics.phy-astr.gsu.edu/hbase/phyopt/Fermat.html.
Lasers: (Phoenix Laser Safety LLC). URL: https://www.youtube.com/embed/EnKxKmaJvfY.

Notes

1. It is thought by some that Newton's famous remark, in a letter to Hooke dated 5th February 1675 damning him with faint praise, in which he concludes: '*If I have seen further it is by standing on ye sholders of Giants*' is a jibe to Hooke and Huygens, who were both of short stature.
2. All objects above absolute zero emit electromagnetic radiation. A spectroscope is chiefly used by astronomers and chemists to determine the nature of materials. When heated to incandescence, many materials emit light that is a characteristic signature of the elements that make up their atomic structure. The absorption spectra of materials can also be studied. Joseph Fraunhofer developed the spectroscope to test the refractive power of optical glasses and lenses. A spectroscope disperses light into a very wide spectrum of colours using a prism or (nowadays) a diffraction grating. As such, it is another instrument – apart from the microscope – used to study the interaction of light with matter.
3. The wavelength of light emitted can be determined by the formula $\lambda = hc/\Delta E$, where ΔE equals the discrete transitional energy difference between the excited and ground state electron orbitals in joules. Planck's constant relates the amount (or quanta, or 'packet') of energy a photon carries with the oscillation frequency of its electromagnetic wave (see also Box 2.1).
4. A wave packet, a term coined by Einstein in order to express the idea that light travels with discrete 'packets' of Energy, *E*, may generally be (but not wholly) identified with a photon. The term photon was widely adopted from the late 1920s.
5. Since 1983 the metre has been defined by international agreement as the length of the path travelled by light in a vacuum during a time interval of 1/299,792,458 of a second. This makes the speed of light *exactly* 299,792,458 km/s.
6. X-rays and γ-rays should, strictly, be called X-*waves* and γ-*waves*.
7. The term normal comes from the Latin *norma*, describing a carpenter's try square
8. The term refraction comes from the Latin *refringere*, meaning to break away. Whilst Newton used the term 'refraction', he also used the root word directly, referring to 'refrangible' rays.
9. The proof is given in optics textbooks; also see the URL links under Internet Resources. However, the significance of Fermat's principle may be better understood by a 'real life' analogy. Suppose that A represents a holiday-maker on a sandy beach (air), the glass a rocky region of foreshore and D the only exit point from otherwise unclimbable cliffs! The tide is coming in and the holiday-maker must arrive at D in the shortest possible time – but whereas he or she can walk quickly over the sand (air), progress across the rocky foreshore (glass) is much slower. Again, the path taken will (or should) be from A to C to D, the 'amount' of deviation from a straight line being a measure of the difference in speed of walking across the sand and rocks, i.e. the refractive index, air to glass.
10. See the Wikipedia animation given in the list of Internet Resources at the end of the chapter.
11. Usually RI values are given without any reference to wavelength, but since RI values are wavelength dependent, for the refractive index value to have any meaning, a particular wavelength should be cited. RI values can change with temperature also, so temperature values as well as wavelength should be cited when precision is required, hence the symbol n_D^{20}, for example, is used.
12. Because diamond has a higher refractive index than glass, in theory a homogeneous immersion diamond lens, with coverglass, immersion and mounting media all matching the RI of diamond would have a higher resolving power than a glass lens of the same aperture. Certainly, the diamond lens could be made much thinner than a glass one, ideal for spectacle lenses.
13. The word *focus*, coined by Kepler in the 17th century, is derived from the Latin word for hearth or fireplace and refers to the concentration of the sun's heat in a small spot; in German it is even called *Brennpunkt*, literally meaning 'burning point'. The THEMIS solar furnace research facility at Odeillo in France can reach temperatures up to 3500°C, sufficient to melt steel or rock.
14. The optical properties of thick lenses are complex. Even though most lenses used in microscopes are thick lenses, it is easier – and sufficient for our purposes – to trace ray paths with thin lenses.

3

Basic Microscope Optics

Outline

Understanding Light Microscopy, First Edition. Jeremy Sanderson.

3.1 Introduction

We have seen in Chapter 1 how light waves radiate from a point source and in Chapter 2 how a lens can be regarded as a continuous series of prisms that will refract light and bring it to a focus. We have also assumed, whilst discussing the refraction of light, that it travels in straight lines, a phenomenon known as the rectilinear propagation of light. These phenomena are illustrated in Figure 3.1. Later, in Chapter 10, we shall see how light is 'bent' by diffraction and indeed how image formation (at any scale) depends fundamentally upon diffraction by the object and the interference of light. However, where the size of the object is very much larger than the wavelength of

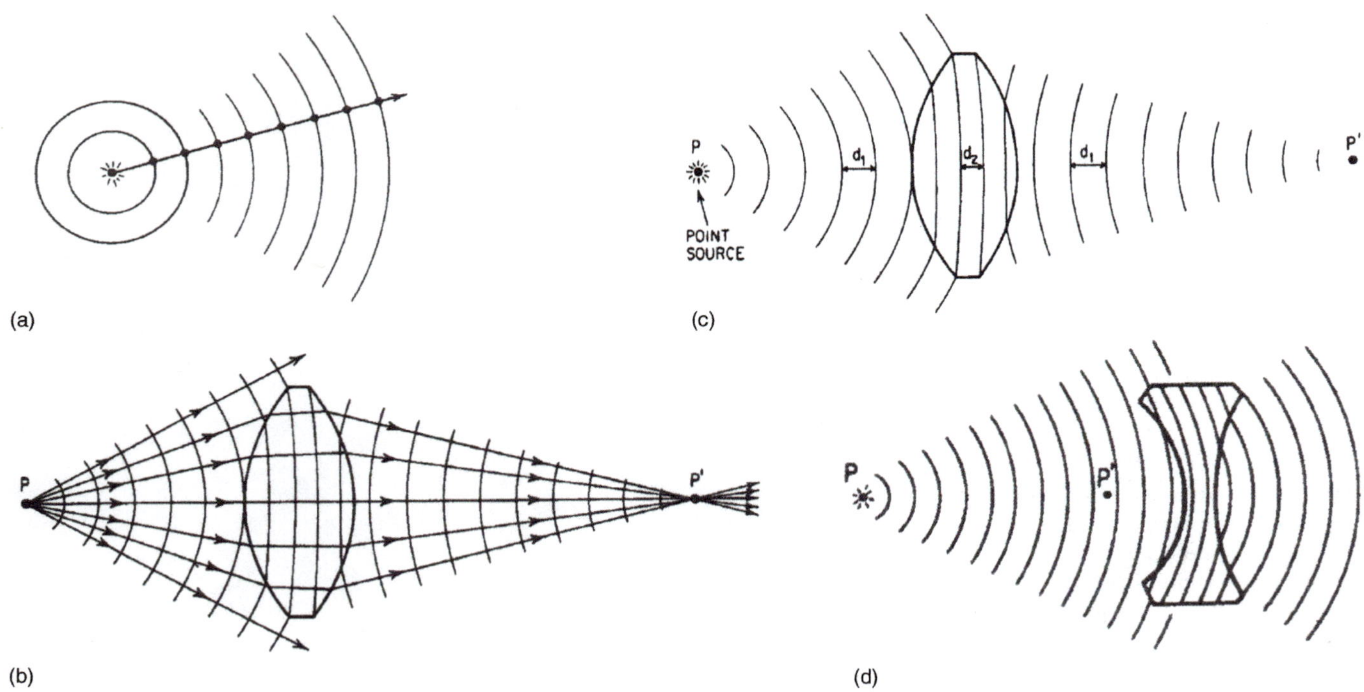

Figure 3.1 ***Interaction of light with lenses*** *The rectilinear propagation and diffraction of light by positive and negative lenses. (a) Shows the rectilinear propagation of a ray of light from a point source. (b) Shows several rays being brought to a focus (P′) to form an image of P by a positive, biconvex converging lens. The wavefronts are also shown. (c) Shows the wavefronts alone and how the wavelength d_1 decreases to d_2 in proportion to the refractive index of the lens before increasing again upon exiting the lens. (d) Shows the equivalent wavefronts for a negative, biconcave diverging lens; the virtual image point is located at P′. In this case, because the rays arise from P and are not parallel, P′ is not equivalent to the focal point, F′. The front focal plane, F, is sometimes referred to as F_1, and the back focal plane as F_2 instead of F′; both notations are used in the figures in this chapter. Source: Smith 2000. Reproduced with permission from SPIE.*

light illuminating it (e.g. when using cameras or telescopes), we can invoke Fermat's principle, which states that a ray of light takes the shortest time to travel between two points, and thus we can discount the effect of diffraction.

Geometrical optics together with ray-tracing can be used to describe simply the actions of lenses and predict the image formed. This is useful in order to understand how our microscope forms both erect and inverted images. In this chapter those optical principles relevant to the microscope are explained. Ray optics also allows us to appreciate, for example, why high-magnification objectives have short working distances and to exercise due care in their use.

When ray-tracing, analysing or designing an optical system, the usual convention is to show a ray of light as travelling from left to right. The Cartesian sign convention is adopted, whereby all distances are measured from the vertex of the lens surface or the optical centre of the lens on the principal plane(s). In the case of a so-called 'thin' lens, having zero thickness, the centre of the lens and the vertex are equivalent. The numerical value of a distance that is measured in the same direction as that in which light is travelling (left to right) is given a positive sign. Conversely, the numerical value of a distance that is measured in the opposite direction to that in which light is travelling is given a negative sign. All heights measured above the optical axis are positive, and those below are negative. The focal length of a converging lens is positive, and that of a diverging lens is negative.

3.2 Lenses

There are four basic optical components that we should consider. These are: the mirror, the prism, the lens and diffraction grating. A mirror, if present in the microscope stand, is located in the base to reflect the illumination through 90° and upwards along the optical axis of the microscope. Lenses of course are found in the bullseye lens of the illuminating optics – the condenser, the objective and the eyepiece or any projection lens in the microscope. Those microscopes that have spectral detectors will possess either a prism and/or a diffraction grating. Of these components, the lens is the most important. The chief properties of a lens are its focal length and its aperture. The term 'lens' is derived from the Latin for the vegetable lentil (*lens culinaris*), and the Romans' observation that a biconvex lens is lentil-shaped. Although 'lens' commonly refers to the action upon visible light of a shaped piece of glass – or transparent plastic – bounded by spherical surfaces, any object that refracts electromagnetic radiation is also a lens. For example, a microwave lens can be made of paraffin wax. Electrons as well as photons can be used to form images[1], the advantage of electrons being their much shorter wavelength (ca. 4 picometres for 80–100 kV electrons) about 100 000 times smaller than photons in the deep-blue region of the visible spectrum.

A ray of light is refracted towards the normal when it enters the lens, and it is then refracted away from the normal as it exits (see Box 2.3). Lenses are dioptric, that is, they use only refraction to form an image. On the other hand, a mirror is catoptric, using reflection to focus light and form an image. Catadioptric optical components employ both refraction and reflection; certain dark-field condensers (Chapter 11, Section 11.4.2) are catadioptric condensers. The ability of a lens to refract light is determined by the refractive index of the lens material as well as the curvature of the lens surface, both of which determine[2] the focal length of the lens. This is true whether the lens is a positive (e.g. converging, type with biconvex or plano-convex surfaces) or a negative, diverging, lens with concave surfaces.

The magnifying glass, and the one-lens simple microscope (which are optically identical) both use positive, or biconvex, lenses. The compound microscope is so-called because it uses two lenses – the objective and the eyepiece – to produce the final magnification of the image. Not all lenses can be used as magnifying lenses. Diverging lenses form a diminished erect virtual image on the object side of the lens.

A real image (Figure 3.2a) is one that is formed by the actual intersection of rays that can be seen on a screen or captured directly by a recording device, such as the eye or a CCD camera. A virtual image (Figure 3.2b) cannot be projected onto a screen. It can only be seen with the aid of another lens or system of lenses. A virtual image is formed by the *apparent* intersection of rays when back-projected from their direction of propagation. This may be the lens of the eye (which then projects the image onto the retina; Figure 3.2c) or it may be a lens system such as a centring telescope, or else a Bertrand lens, which is used to view an aerial image in the microscope.

A positive lens (Figure 3.2a) converges incident rays, has at least one convex surface and forms a magnified real image. A negative lens causes incident rays to diverge; they are thinner in the middle than the edges and so have at least one concave surface. Negative lenses (Figure 3.2b) do not form a real image but a virtual one, and when held in front of the eye, they will reduce, or demagnify, the image.

Most lenses have one or both of their two surfaces configured as parts of the surfaces of spheres which are perpendicular to the lens axis. Toric, or sphero-cylindrical, lenses have surfaces figured with two different radii of curvature, orthogonal to one another. These second types of lenses are deliberately manufactured astigmats, with different focal power in different meridians, in order to correct visual astigmatism of equal and opposite power. Non-spherical (aspheric) lens surfaces can be figured manually by the manufacturer, and these confer great advantage in having absolutely minimal spherical aberration.

For economic reasons (with a very few exceptions) the surfaces of lenses (Figure 3.3) are automatically mass-produced accurately and reproducibly as spherical or plane surfaces. Both convex lenses (having a positive focal length) and concave lenses (negative focal length) are used by microscope designers, often grouped together as a series of doublets to form a single optical unit. For our purposes we can understand how images are formed by ray-tracing if we can think of the microscope condenser, objective and eyepiece lenses in each case as a single, thin[3], convex lens (Figure 3.4).

3.3 Thin and Thick Lenses

Assuming lenses are thin very much simplifies matters when ray-tracing in our consideration of geometrical ray optics. These principles were first elucidated by Isaac Newton and so are sometimes called Newtonian optics. An explanation of the thin-lens equation and

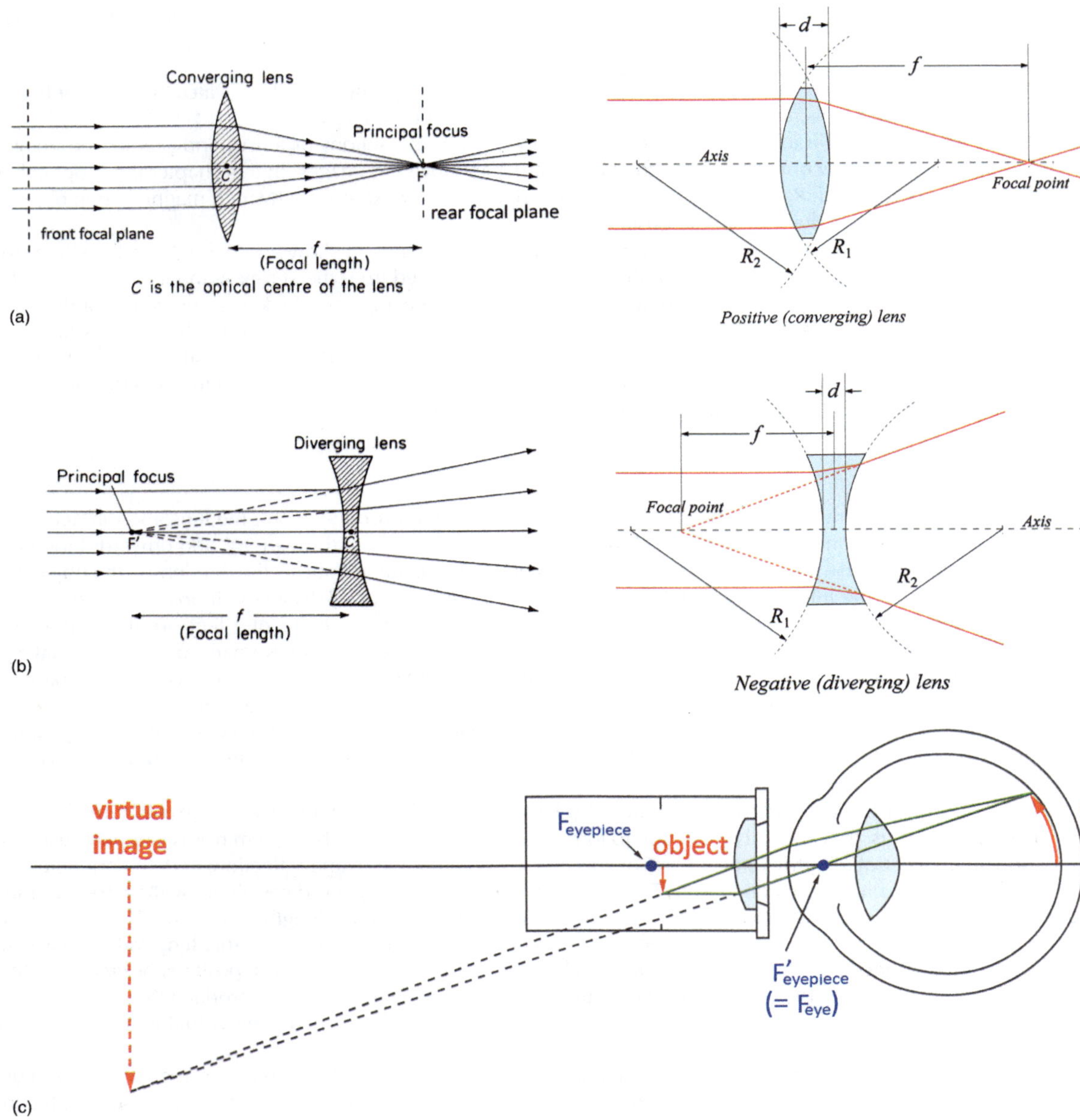

Figure 3.2 **Real and virtual images** *While negative diverging lenses always produce virtual images, converging lenses are capable of producing both real and virtual images. An inverted real image is produced when the object is located at a distance greater than one focal length from a positive converging lens.*

Converging lenses form virtual images if the object distance is shorter than the focal length (c). In this case, the focal point marked $F_{eyepiece}$ is that of the eye lens of the eyepiece. For clarity the field lens is not shown. When an object is located inside of the focal point of a positive converging lens, the image will be virtual, upright, larger than the object and located on the same side of the lens as the object: the lens acts as a magnifying glass. Depending upon the location of the object relative to the focal point of the lens, the image may be real or virtual and, therefore, magnified or demagnified.

Unlike converging lenses, a diverging lens always produces an image that is located on the object side of the lens – it is a virtual image; the image is upright and reduced in size. The location of the object – its distance from the lens – does not affect the characteristics of the image.

The lens-maker's formula need not concern us as microscopists. It is given in optics texts and is included here for completeness. The focal length in air can be calculated from $1/f = (n-1)\cdot(1/R_1 - 1/R_2)$ where n is the refractive index of the lens, and R_1 and R_2, the radii of the two spherical surfaces of the thin lens, being the radius of the front surface and rear surface, respectively, of the lens.

The Cartesian sign convention is used here. By this convention light travels from left to right when displayed on the computer screen, printed out or published in a book. In constructing a ray diagram, all distances are measured from a reference surface, such as optical centre of a lens or the refracting surface (the vertex or pole) of a lens. The distance from the lens to the object (designated 'u') is negative if it is in front of the lens, to the left – towards the origin of the ray of light. The distance from the lens to the image (designated 'v') is positive if it lies on the far side, behind the lens, to the right – away from the origin of the light ray. These distances of the object and image from the lens are called conjugate distances. Heights above the optical axis are positive, and those below are negative. Angles measured clockwise from the optical axis are negative.

If the lens equation yields a negative image distance, then the image is a virtual image on the same side of the lens as the object. If it yields a negative focal length, then the lens is a diverging lens. A positive lens thus has a positive focal length, and a negative lens, a negative focal length. The real-is-positive convention allows for u, v and f all to be positive values; it is not used here.

Source: (a,b) Abbott 1984. Reproduced with permission from Pearson. (c) Adapted from Wayne 2009 with permission from Elsevier. (diagram top right) Adapted from DrBob under the terms of the Creative Commons Attribution Share-Alike licence CC-BY-SA-3.0 (http://creativecommons.org/licenses/by-sa/3.0/), via Wikimedia Commons.

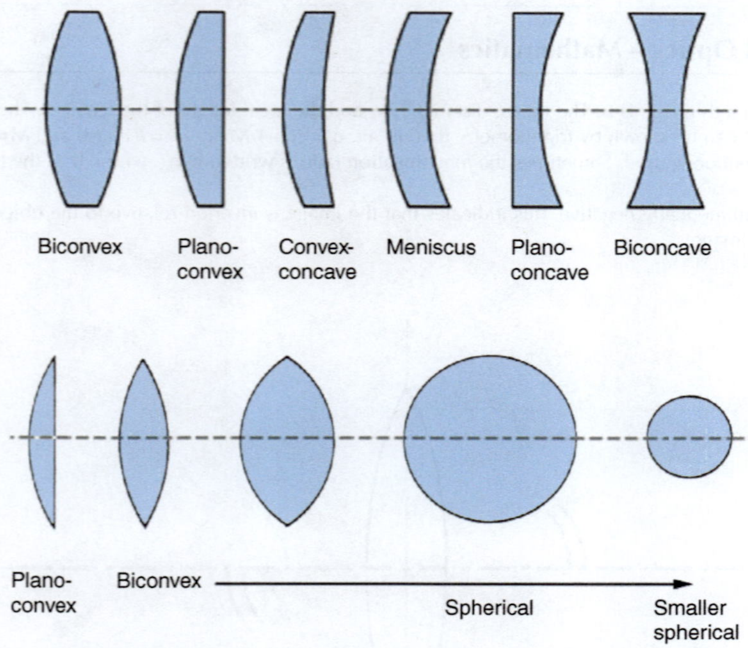

Figure 3.3 ***Positive and negative lenses*** *Different configurations and different possible powers of positive and negative lenses. Source: (top) Adapted from ElfQrin (Own Work) under the terms of the Creative Commons Attribution Share-Alike licence CC-BY-SA-3.0 (http://creativecommons.org/licenses/by-sa/3.0/), via Wikimedia Commons. (bottom) Adapted from Bradbury 1967 with permission from Elsevier.*

of magnification is given in Box 3.1. For reasons that we shall see later, microscope lenses are usually more complicated than simple, single pieces of glass, often consisting of many elements. It is the job of the microscope designer (using complex computer programmes) to manage the design and proper function of these thick lenses and groups of thick lenses, which form objectives and other microscope components. In order to understand how rays of light form images, we shall also make two other assumptions that do not hold true in practice: namely that the lens aperture is small compared to the focal length and that the refractive index of the lenses is the same for all wavelengths and there is no dispersion or other aberration present.

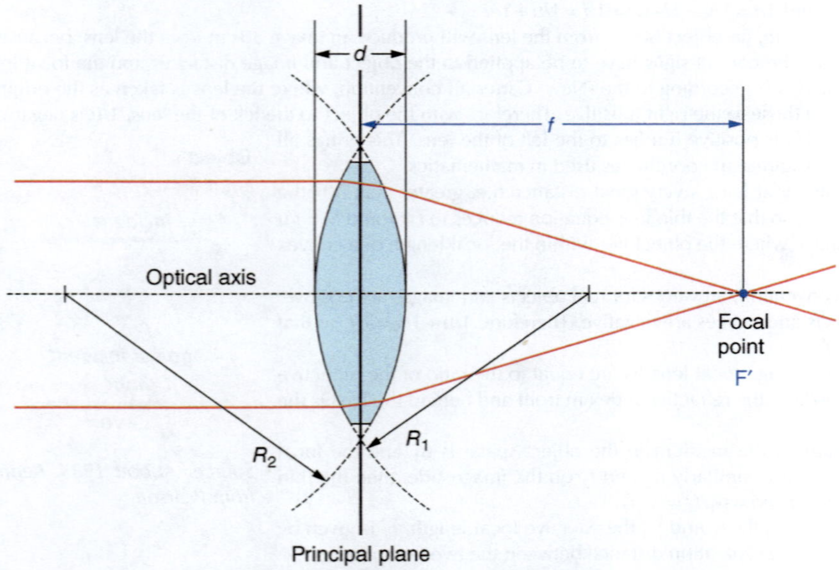

Figure 3.4 ***The thin lens*** *A lens can be considered as 'thin' when its thickness (often denoted as 'd', the distance between the surface vertices, or poles) is very much less than its focal length, the radii of curvature of the figure of the lens, or the distances (u) from the lens to the object and (v) from the lens to the image. Source: DrBob. Reproduced under the terms of the Creative Commons Attribution Share-Alike licence CC-BY-SA-3.0 (http://creativecommons.org/licenses/by-sa/3.0/), via Wikimedia Commons.*

Box 3.1 Geometrical Optics – Mathematics

The distance of the object from a thin lens is **u**. The image distance is **v**, and the focal length of the lens is ***f***. The magnification, M, of the lens is given by the ratio: $M = v/u$. It can be shown by trigonometry that, in air, $u = f(1 + 1/M)$ and $v = f(1 + M)$ and $M = (v - f)/f$. These formulae hold when the Real-is-Positive convention is used. Sometimes the magnification ratio is written h'/h_o, where h′ is the height of the image and h_o the height of the object.

Where v/u, h′ and M are arithmetically negative, this indicates that the image is *inverted* relative to the object (as in the case of a positive converging lens forming a real image).

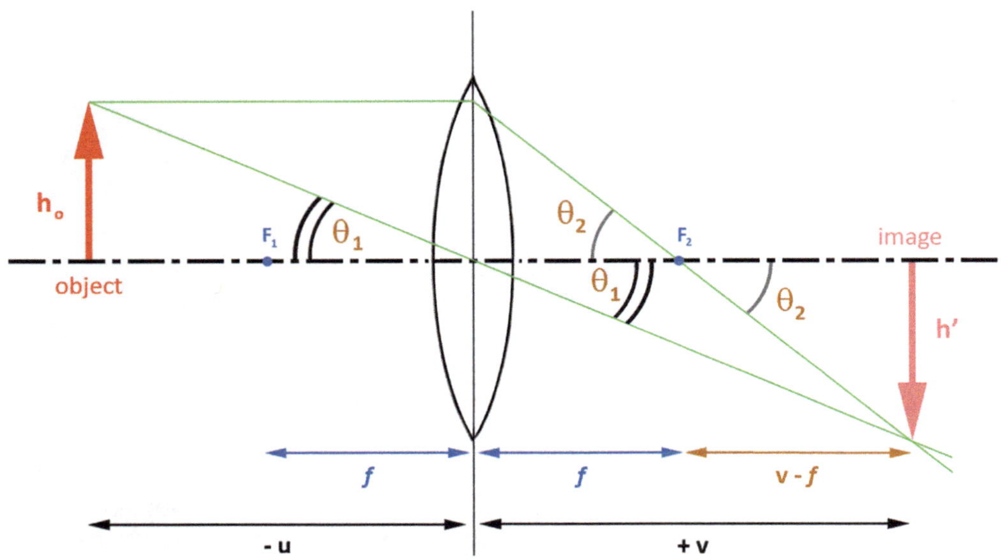

Source: Adapted from Meek 1976 with permission from John Wiley & Sons.

The thin lens equation is: $1/f = 1/u + 1/v$ This can be deduced from two expressions for the similar triangles in the figure above.

From θ_1: $h_o/h' = -u/v$ From θ_2: $h_o/h' = f/(v - f)$

Therefore, $-u/v = f/(v - f)$ and $1/f = 1/v - 1/{-u} \Rightarrow 1/f = 1/u + 1/v$

For a lens of focal length 20 cm, an object 60 cm from the lens will produce an image 30 cm from the lens, because $1/60 + 1/30 = 1/20$.

When using lens formulae, the correct signs have to be applied to the object and image distances and the focal length so that the formulae are mathematically correct. This is according to the (New) Cartesian convention, where the lens is taken as the origin so that so that distances from the left are negative and those to the right, positive. Therefore with the object to the left of the lens, 1/u is negative:

The numerical value of $-1/u$ is positive if u lies to the left of the lens. This brings all signs into agreement with the Cartesian coordinates used in mathematics.

If the object lies at infinity, or at least a very great distance (i.e. greater than 30x the focal length) 1/u tends to zero, so that the thin lens equation reduces to $f \approx v$ and $M \approx f/u$ and $Mu \approx f$. For a virtual image where the object lies within the focal length of a convex lens, $1/u - 1/v = 1/f$.

For the Real-is-Positive convention, all values for real objects and images are positive, whilst those of virtual objects and images are negative. Therefore: $1/u + 1/v = 1/f$, so that $u = vf(v - f)$ and $v = uf/(u - f)$.

The ratio of the object and image focal lengths are equal to the ratio of the refractive indexes, thus: $f/f' = -n/n'$. Where the refractive index in front and behind the lens is the same (e.g. air), then $f = -f'$.

Where the refractive index of the medium in the object space is n_1 and the focal length on the object side is f_1 and similarly n_2 and f_2 on the image side, then the thin lens formula becomes: $(n_1/u + n_2/v) = n_1/f_1 = n_2/f_2$

For two thin lenses of focal lengths f_1 and f_2, the effective focal length, *F*, is given by $F = f_1 f_2/(f_1 + f_2 - s)$, where s is the separation distance between the two components.

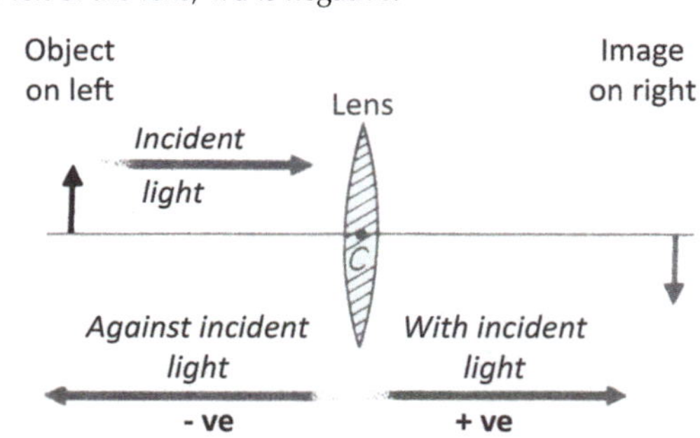

Source: Abbott 1984. Reproduced with permission from Pearson.

From the thin lens equation, it can be proved that for a real image to form the *minimum* distance between the object and the image is equal to four times the focal length: $u + v = 4f$, i.e. the object and the image are at twice the focal length and the magnification = 1.

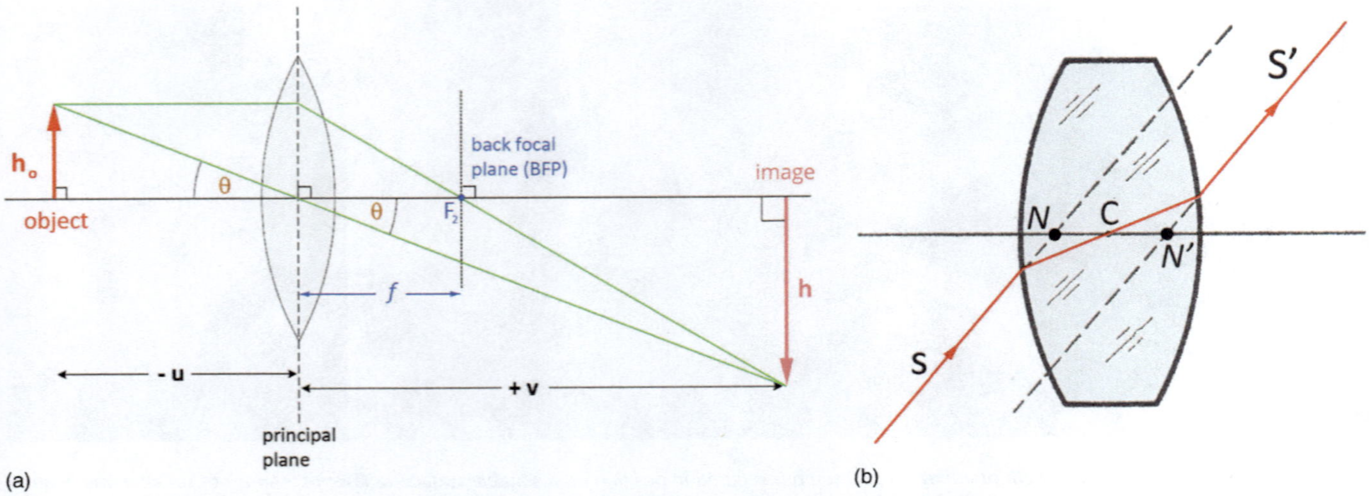

Figure 3.5 ***The thin and thick lens compared*** *(a) shows the relationship between the focal length, focal point, the object size and image size when considering a lens as 'thin'. (b) shows how the two nodal points (which coincide at the centre of a thin lens) are the effective 'centres' and the cardinal point from which the focal length is measured. A ray passing through the first nodal point will be refracted such that it appears to pass through the second nodal point at the other side of the thick, or compound, lens. See also Figure 2.21. Source: Pluta 1988.*

In the case of thick lenses, all we need note here is that the single principal plane existing in a thin lens becomes – in a thick lens – two principal planes (Figure 2.21) that for a thick lens in air are coincident with the first and second nodal planes, respectively. (These planes are not coincident when the refractive index of the medium surrounding the lens is different for object and image spaces, such as immersion objectives.) An oblique ray incident on a thin lens passes through the central point (C) of the principal plane undeviated (see Figure 3.5a). For the equivalent situation in a thick lens, the nodal plane N emerges undeviated on a parallel path at nodal plane N′ (see Figure 3.5b). For all practical purposes, microscope objectives and other optical components are formed from combinations of thick lenses, but consideration of thin lenses allows us to understand their behaviour in terms of geometrical optics.

3.4 The Focal Length of a Lens

The most important parameter of any lens is its focal length (usually designated by the letter *f*). Parallel rays from infinity are brought to a focus by a thin lens at the rear focal point, which lies in the rear focal plane of the lens, as shown in Figure 3.2a and Figure 3.4 (this is how an image is formed in a camera). Conversely, rays entering a lens that emanate from, or pass through, the focal point exit the lens on the opposite side as parallel rays. The distance from the midline (or principal plane) of the lens to this focal point is the focal length of the lens[4]. The focal length is given by the simple Gaussian form of the lens equation $1/u + 1/v = 1/f$ (see also Box 3.1).

We saw in Chapter 2 that each thin lens has two focal lengths (*f* on the object side and *f′* on the image side) and, as a consequence, two focal planes. In the case of the lenses in our eyeballs or cameras, it is the *second* focal plane that is of importance – the plane where the image is formed on the retina or CCD – and the first focal plane is of no interest. In a projector the transparency or image-forming device is situated in the *first* focal plane, and the second focal plane is of no interest.

When considering the lenses of the microscope, and the objective lens in particular, *both* focal planes are of interest. Of course an object and its resulting image are often not positioned precisely at infinity or in one of the focal planes – see Figure 3.8. When the object is placed at the first focal point, a real, inverted, image is 'formed' at infinity – parallel light leaves the lens, and the eye sees only a blurred patch of light. Conversely, the image of an object 'at infinity' is formed at the second focal point of the lens. These phenomena are exploited in the design of Köhler illumination.

The simplest way to determine the focal length of a thin lens is to aim it at a bright source of light that is 'infinitely distant'. This can be the sun[5] (as in Figure 2.20), or for all practical purposes, a bright light arranged to lie at least 30x the focal length of the lens away from the screen onto which an image is formed. Figure 3.6 shows one way in which this can be done. The focal lengths of microscope lenses range from about 50 mm for low-magnification (low power) objectives to about 2 mm for high-magnification objectives.

Another method to compute the focal length is to calculate *u*, the distance from the source to the lens, and *v*, the distance from the lens to the projection screen where the point source is brought to a focus by the lens to form an image, by forming an image on a screen from a point source (e.g. a card perforated with a fine hole, having a lamp as an extended source behind the card). Then the distances *u* and *v* are measured. This second method is useful where it is not possible to use an infinitely distant bright source of light. It is also possible to calculate the dioptric power, and hence the focal length of a lens, by projecting an image onto a screen placed 250 mm (the reference viewing distance) from the second focal plane of the lens.

The dioptre, *D*, is the unit of measurement of the refracting power of a lens or curved mirror. When the refractive index between the lens and the image is air, the dioptre is the reciprocal of the equivalent focal length in metres (i.e. power, *D*, equals 1/focal length)

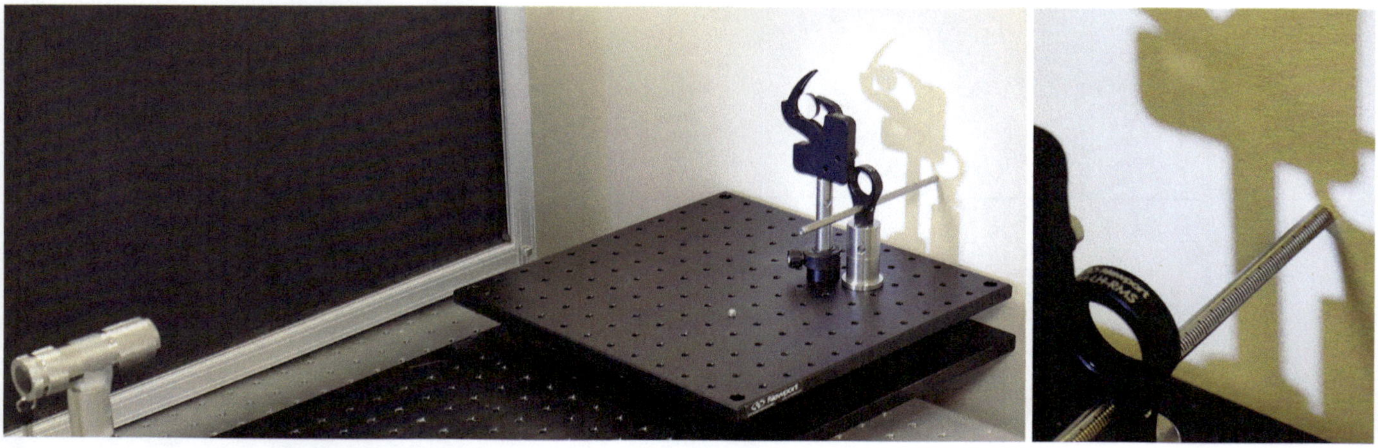

Figure 3.6 **Calculating the focal length of a lens** *An LED torch is secured in position on a suitable surface (a laser pointer is not suitable: the beam is much too tightly collimated). The lens is held in the light path, and the distance from the focal spot can be measured with a ruler. The same procedure can be carried out more accurately, as shown here, on a custom-designed optical bench. The lens is an achromat of 8.0 cm focal length; the scale rod is marked in millimetre divisions, ordered in groups of 5 mm. Source: Courtesy of Dr S. Kerrigan.*

expressed in m^{-1} (reciprocal metres). The higher the power in dioptres, the shorter the focal length and the greater the magnifying power of the lens.

Microscopes used to be classified in terms of focal length rather than magnifying power. Nowadays we evaluate objectives according to their numerical aperture[6] and magnification values. Optometrists use reciprocal focal lengths, or dioptre values. The advantage of expressing the power of a lens in dioptres rather than focal length is that when thin lenses are placed close together, their powers add. This is useful when prescribing corrective lenses. It is helpful to know what the dioptre values of each of your eyes and/or corrective lenses are. The eyepieces can then be adjusted for viewing free of eye strain. This topic is covered in Chapter 9, Section 9.11.

3.5 Ray-Tracing

Each lens has two focal points, front and rear, and the distance from the centre of the lens to the focal point is the focal length of that lens. The plane passing through the focal point normal to the axis is the focal plane of the lens. Three properties enable us to construct a ray diagram for a particular lens:

1. When parallel rays of light (i.e. from infinity) enter a lens, they come to a focus at the rear focal point of that lens.
2. Conversely, rays passing from the object through a focal point will be refracted by the lens and leave it parallel to the optical axis.
3. Rays that pass through the centre of the lens pass through undeviated.

Therefore, if you know the focal length of a lens and the position of the object, you can construct a ray trace using just two of these three conjugate ray pairs (compare Figure 3.5a and Figure 3.7) and determine the position of the image formed by a thin lens.

The object and image positions on either side of a lens can be interchanged. This is a consequence of the principle of reversibility of light rays. The object and image lie at conjugate points, or conjugate foci. The planes passing through these conjugate points and perpendicular to the optical axis are the conjugate planes. These are important in practical microscopy and are explained further in Section 3.6.

3.6 Lenses in the Microscope

As the object is moved closer to the lens, its image moves further away on the other side (see the Stanford Gaussian lens formula applet under Internet Resources at the end of the chapter). When the image distance is smaller than the object distance, the image will be smaller than the object by the ratio of these distances; this is usually the case with the camera's lens. Figure 3.8 shows how lenses form images with differing object and image distances. Different optical instruments (e.g. camera, projector, telescope) are similar to the microscope (see Section 3.12 at the end of this chapter) but use lenses of different refractive power and magnifying power placed at different object and image distances along the optical axis.

An object placed within the focal length of a convex lens forms an erect, magnified virtual image. In this case the object distance (u) < the focal length (f). No real image exists that can be projected upon a screen, but the eye can see a virtual image from the far side of the lens. This is how both the eyepiece and a magnifying glass, or one-stage (simple) microscope, function.

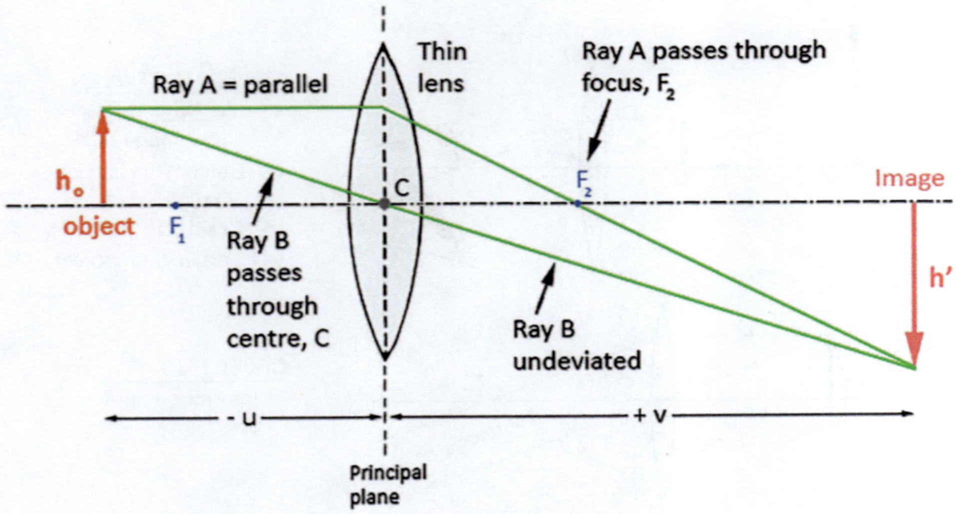

Figure 3.7 **Constructing the size and position of the image by ray-tracing** *Showing how the image size and position is calculated by ray-tracing through a thin lens (compare this figure with Figure 3.5a). Source: Adapted from Meek 1976 with permission from John Wiley & Sons.*

When the object is placed precisely in a focal plane of the lens, u equals f, and the image will be formed at infinity on the other side of the lens. As in the situation above, the image cannot be projected upon a screen. A magnifying glass may also work in this way. In reverse, parallel light from a distant object forms an image at the focal point. This is the way a telescope and a camera focused at infinity work; the image is smaller than the object.

When $u > f$, a real image is always formed. When the object is positioned between one and two focal lengths from the lens ($2f > u > f$), the image is formed beyond two focal lengths on the other side of the lens. A real magnified image is formed that is now larger than the object. This is how a slide projector works. As the image distance (v) increases, so does the magnification, since magnification is given by v/u. Also, the shorter the focal length of a lens, the greater the magnifying power.

Depending upon the tubelength of the microscope (see Chapter 7, Section 7.3 and Section 7.9) the objective works in one of these two ways to form an image. In the modern infinity-corrected objective design, an object is placed at the front focal plane of the objective, and an object formed at 'infinity' is then brought to a focus at the primary image plane by a tube-lens. This has advantages, which we shall discuss later in Chapter 7. In the older finite tubelength arrangement, the object is placed at a position between one and two focal lengths of the lens, and a magnified image is formed directly at the primary image plane.

At a special position, when the object is twice the focal length away from the lens ($u = 2f$, and so $v = 2f$ also), the image is formed at twice the focal length on the other side; because object and the image distances are now the same, the image will be the same size as the object. This situation occurs when using a macro lens on a camera for close-up work.

Conversely, an object placed beyond two focal lengths from the lens (when $u > 2f$) will form a real inverted image demagnified and reduced in size. This is how a camera works. Demagnification of the primary image is counter-intuitive in the microscope but is useful to increase the field of view where the chip of the CCD faceplate is smaller than the visual field of view. In this case, a 0.5x or 0.63x demagnifying C-mount adapter is used to connect the camera to the phototube of the microscope.

All of these configurations of lenses, focal planes, objects and images are to be found in the microscope. In the case of microscope lenses it may be important to consider *both* focal planes, and this is especially true of the objective lens. The specimen is placed in or just outside the first, or *front*, focal plane of the lens, and the second, or *back*, focal plane is found one focal length behind the objective, either outside or within the lens mount, depending on the focal length. The back focal plane of the objective is the place where parallel light entering the front of the objective is brought to a focus; it is where the film would be if the objective were a camera lens.

3.7 Conjugate Planes

Conjugate foci are any pair of points such that an object placed at one point will form a real image at the other point. The object point and image point of a lens system are conjugate, or 'linked together'. Since all the light paths from the object to the image are reversible, it follows that if the object were placed where the image is, an image would be formed at the original object position. The space inhabited by a conjugate focal point is extended to include a focal plane. In the microscope, planes in which an object in one forms an image in another are known as *conjugate* planes (this terminology is taken from the Latin *conjugare* meaning joined together, or linked). The most obvious of these conjugate planes are the object plane and the retina of the eye, but there are others. It is very important to understand the principle of conjugate planes in order to set up the microscope in a rational way (see also Section 9.10).

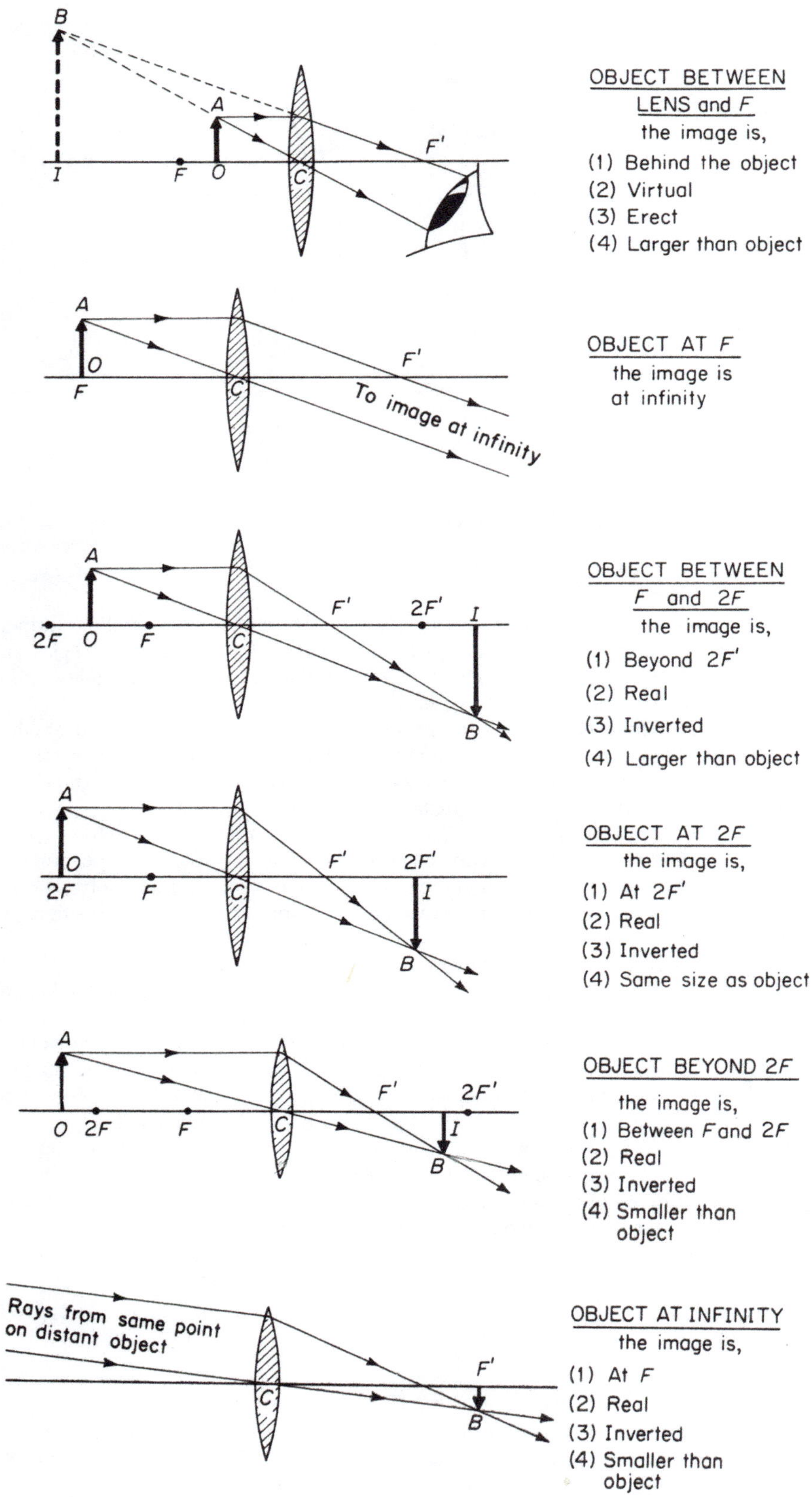

Figure 3.8 **Formation of images by lens with differing object and image distances** *(See text for explanation). Source: Abbott 1984. Reproduced with permission from Pearson.*

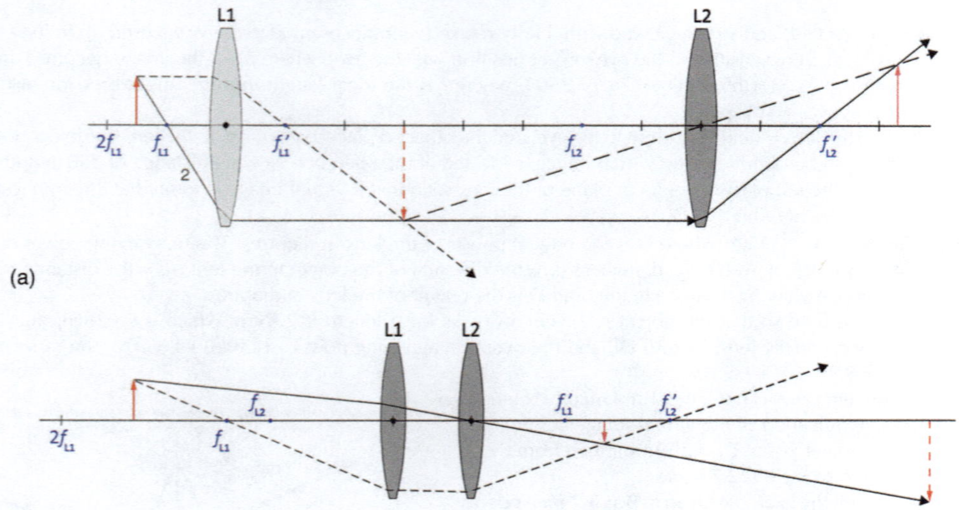

$$\text{Equivalent focal length, F} = \frac{f_1 f_2}{f_1 + f_2 - d} \quad \text{or} \quad \frac{1}{f} = \frac{1}{f_1} + \frac{1}{f_2} - \frac{d}{f_1 f_2}$$

(b)

Figure 3.9 ***Formation of images by a pair of convex lenses*** *Two convex lenses that are separated by a distance greater than the sum of their focal lengths (top) will together form a real erect (upright) image. When two thin lenses are separated by a distance smaller than either focal length, a real inverted demagnified image (bottom) is formed. For the derivation of the mathematical equations giving the equivalent focal length and also that of the dioptric power of the combined lenses, see Wayne (2013), pages 30–32. Source: Wayne 2009. Reproduced with permission from Elsevier.*

The importance of these conjugate planes in practical microscopy will become apparent when considering, and applying, Köhler's principle of illumination when adjusting and setting up the microscope. We will see, when setting up the microscope for Köhler illumination, that there are two sets of conjugate planes: (a) the *imaging*, or *field*, *set* of conjugate planes and (b) the *illuminating*, or *aperture*, *set* of conjugate planes. Furthermore, both these sets of conjugate planes are independent of one another in a correctly adjusted microscope. The effect of closing the illuminated field diaphragm (IFD) situated in the imaging set of conjugate planes, which controls the area of the specimen that is illuminated, should not be seen directly in the illuminating set of conjugate planes, in which the condenser diaphragm – controlling the angle of illumination – lies. The converse is also true.

Conjugate planes are also essential to an understanding of confocal microscopy – indeed, if terminology were to have developed a little differently, we might speak of *confocal* planes, confocal meaning 'in focus together'. Before returning to consider the arrangement of lenses in a compound microscope, we shall consider how lenses form a magnified image, when we need to use them and how the microscope evolved from a simple magnifying glass.

3.8 The Single Lens Magnifier, or Loupe

As explained in Chapter 1 (Section 1.6) the near point of individual eyes differs. When a single convex lens is held close to an object and used as a magnifying glass, such that distance between the lens and the object is shorter than the focal length of the lens, it is used as a magnifying glass, and an erect virtual image (Figure 3.10 and Box 3.2) is observed. The virtual image is usually represented as situated at 250 mm from the observer's eye, but it may actually be formed anywhere between this distance and infinity, depending on how close the object is held in front of the lens and the natural accommodation of the observer's eye. Therefore, a standard reference viewing distance of 250 mm has been agreed.

Box 3.2 Magnifying Glasses and Virtual Images

The magnification seen with a magnifying glass depends on how the glass and the object are held, relative to the eye. This is apparent if the head and object are both held still and the glass itself is moved. Try this with individual letters on the keys of a computer keyboard, for example.

If the lens is held such that its front focal point is focused on the object, then the distance from the object to the lens is equivalent to the focal length. The image appears to be located at infinity ($v = \infty$ and $1/f = 1/u$).

The image-forming rays are parallel, and so are viewed with a fully relaxed, non-accommodated, eye focused on infinity. A good image can be obtained easily: the focus is not very sensitive to the eye's exact position (v). However, where $u > f$, the image becomes unsharp.

The magnification is thus given as $M = (h/f)/(h/D_v) = D_v/f = 250/f$, where f is the focal length in mm. This expression also holds true for the determining the magnifying power of an eyepiece.

In the second case, if instead the lens is held very close to the eye, and the object is placed very close to the lens, the eye is not relaxed. However a larger angular magnification can be obtained. The virtual image is located at the reference viewing distance of 250 mm, then $M = (250/f) + 1$. The object is at a distance (u) to the left of the front focal plane of the lens such that a virtual image is formed at 250 mm to the left of the lens. The magnification is therefore: $M = (h/u)/(h/D_v) = D_v/u$

Another way to express this is $M = \frac{1}{4}D + 1$, where D is the optical power of the lens in dioptres. The magnifying power of a magnifying lens can also be found from: $M = 250/(d_e - d_o) - (D.\ d_e.\ d_o)$ where d_o = the distance of the object to the lens, d_e = the distance from the magnifying lens to the eye, 250 = the reference viewing distance in mm and D is the power of the lens in dioptres.

Suppose a magnifying glass is held so that the object is 2.8 cm away, its focal length is 2.8 cm, which is 0.028 m, thus the dioptric power is $1/0.028 = +35.7$ dioptres for a convex lens, ($\approx +36$ D), and the overall magnifying power is: $(36/4) + 1 = 10$x. The same result is found with $250/f + 1$. $f = 28$ mm, $250/28 = 8.93$; $8.9 + 1 \approx 10$x.

The difference between the two cases is not large. For a biconvex lens of focal length 25 mm, the magnification of the virtual image at infinity is: $250/25 = 10$x. If the virtual image lies at $D_v = 250$ mm, then from the thin lens equation, $u = 22.7$, and $M = 250/22.7 = 11$x.

Comparing a negative lens with the example given in Box 3.1 for a positive lens: the focal length is the same, at 20 cm; the object is again placed (u) at 15 cm from the lens. Using the thin lens equation, $1/v - 1/u = 1/f$, then: $1/60 - 1/15 = -1/20$. The image is then situated (v) at -60 cm. The negative value for v denotes a virtual image, and the negative value for f indicates that the image is to the left of the lens, on the side of the first focal point, F_1. Since $-v/-u = 4$, the image is four times the height of the object.

When a magnifying glass is held so that the object lies outside its focal length, it acts as any other convex lens (Figure 3.7). When the object lies between F_1 and $2F_1$, the image is real, inverted and larger than the object. When the object is at 2F, the image is real, inverted and the same size as the object. When the object lies at > 2F, the real image is inverted and is de-magnified (although counter-intuitive, try looking at a tree or building through a magnifying glass!).

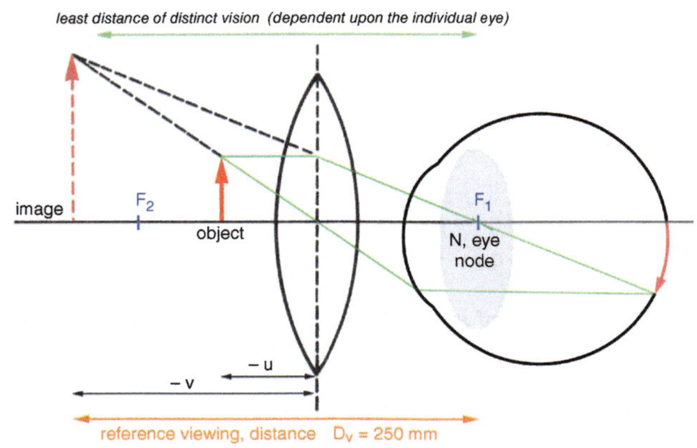

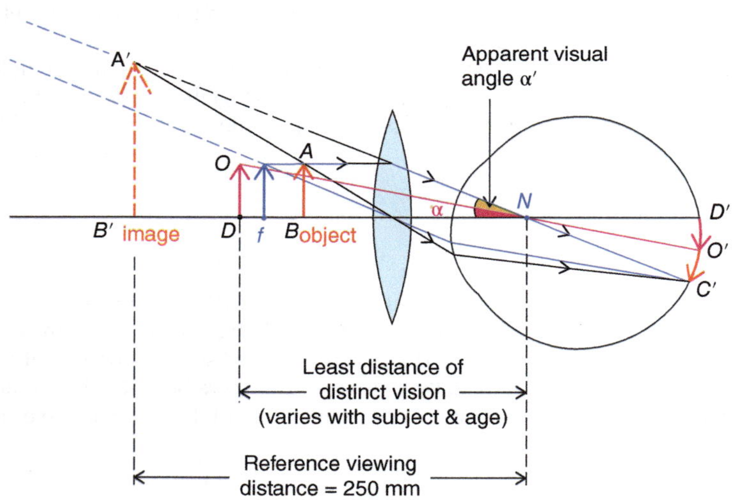

Figure 3.10 ***How a magnifying glass works*** *An object (D-O) placed at any distance up to the nearest distance of distinct vision will subtend a small angle (shown here as α) at the node of the eye to form an image D′-O′. Since the nearest distance of distinct vision varies between individuals due to many factors including visual acuity and age, a standardised reference viewing distance of 250 mm is taken as the standard by which angular magnification is calculated. If the same-sized object is now placed at the first focal point (f) of a magnifying lens, the angle α is effectively increased to α′, thus increasing the retinal image (shown now as D′-C′). An erect virtual image will appear to be formed at infinity. If the object is now moved closer to the eye than the first focal point of the lens (e.g. at A-B), an erect virtual image will be seen (at A′-B′). By convention, for a magnifying lens (or magnifying glass), the magnification is the ratio between the tangent of the viewing angle subtended by the object when viewed through the lens and the tangent of the viewing angle of the same object when viewed by the eye alone at the reference viewing distance (see also Figure 1.19 and Box 3.2). Source: Adapted from Bradbury 1967 with permission from Elsevier.*

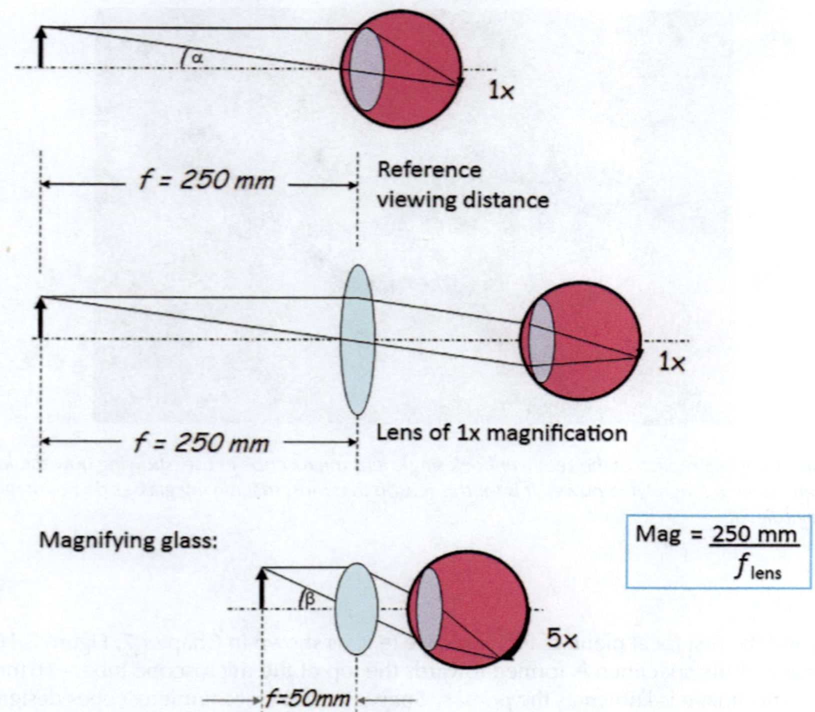

Figure 3.11 ***The reference viewing distance*** *How the reference viewing distance is used to calculate the magnification of a simple lens. See the text and also Box 3.2 for further explanation. Source: Reproduced with permission from Carl Zeiss Ltd.*

Figure 3.11 shows how an object placed at the reference viewing distance forms a 1x image on the retina. The same 1x image is formed if a lens with a focal length of 250 mm is used to view the object. If a lens of shorter focal length is used, a magnified image is formed.

For optical instruments with an eyepiece, or for a single lens forming a virtual image, the linear dimension of the image cannot be given since in these cases magnification is an angular measure and the image size is proportional to the tangent of the angle subtended by the object at the focal point of the lens (See the comments on magnification in Note 1, in the Notes section of the front matter). By convention, for magnifying glasses, telescopes and microscope eyepieces, where the size of the object is a linear dimension and the apparent size is an angle, the magnification is the ratio between the apparent (angular) size as seen in the eyepiece and the angular size of the object when placed at the reference viewing distance of 25 cm from the eye (as if observed by a 'normal' unaided eye). The formulae given in Box 3.2 show how to calculate the power of a magnifying glass.

Magnifiers are limited by having short working distances between the object and magnifying lens and also small depths of field. Loupes are used by jewellers, watchmakers, surgeons and dentists for intricate work. Simple hand loupes generally magnify a maximum of 10x.

3.9 The Compound Microscope

A compound microscope is one in which the image of the object is presented to the eye by a combination of objective and eyepiece. The advantage of a compound lens system, rather than a single simple lens magnifier, is that much higher magnifications can be achieved. To obtain high magnification with a single-lens magnifier, the lens itself must be placed close to the eye (Figure 3.12), to increase the angle subtended and thus the size of the image projected onto the retina. The observer's visual acuity must also be good, for it is otherwise impossible to place the lens very close to the eye, and at the very least it is uncomfortable.

The highest magnification achieved by Leeuwenhoek with a single (double-convex) lens magnifier, or simple microscope, was 275x. With a compound microscope, magnifications up to 1400x are possible; not only that, but two other consequences flow from using a compound microscope. Firstly the specimen can be illuminated much more efficiently, and secondly, the lens system in a compound microscope can be made with much higher numerical aperture. This, in turn, means better resolving power – the whole purpose of the microscope – as well as increased light-gathering capacity.

As we shall see, the objective is by far the most important lens of the microscope – indeed, we might even say that the objective *is* the microscope, and all the other parts are there to support its function. In the traditional microscope (with a finite tubelength), the

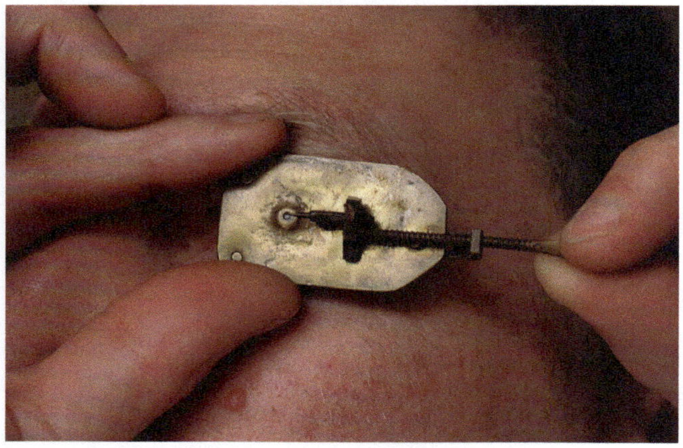

Figure 3.12 **The single lens microscope** *A replica of the Leeuwenhoek single lens microscope in use, showing how the lens has to be placed very close to the eye in order to benefit from the high magnifying power. It is for this reason that most magnifying glasses do not magnify more than about 10 times. Source: Courtesy of Dr B. Bracegirdle.*

specimen is located just beyond the first focal plane of the objective (e.g. as shown in Chapter 7, Figure 7.16 and also in Figure 3.14a), in such a position that an image of the specimen is formed towards the top of the microscope tube – 10 mm below the top according to the International Standard. This image is known as the *primary image*. In more recent microscopes designed for 'infinity correction', the object is placed precisely in the first focal plane of the objective (Figure 3.14b) so that the image passes up the tube in the form of parallel rays, which are then converged by a 'tube-lens' to form the primary image (Figure 3.14c); the end result is thus the same. A second lens, the eyepiece, is arranged so that its first focal plane coincides with the primary image plane (Chapter 8, Figure 8.12b). The imaging rays then leave the eyepiece parallel and enter the eye (Figure 3.13), where they are converged on to the retina. The eyepiece is really little more than a magnifying glass arranged to fit the tube of the microscope.

Indeed, the eyepiece can be useful as a magnifier for quickly inspecting the front surface of objectives. Take an eyepiece out of the eyepiece tube, and invert it. The standard eyepiece is 10x, so the focal length is 25 mm (250/10). You will therefore have to hold the eyepiece closer to the object than this distance (because the effective principal plane and second nodal point of the eye-lens doublet is contained within the body of the eyepiece). You will also need to be under, or nearby, good lighting in order for this method to work. Since the surface of the objective is placed within the front focal point of the eyepiece, the image rays enter divergent to form an erect virtual image. It does not matter very much how far your eye is placed from the eyepiece.

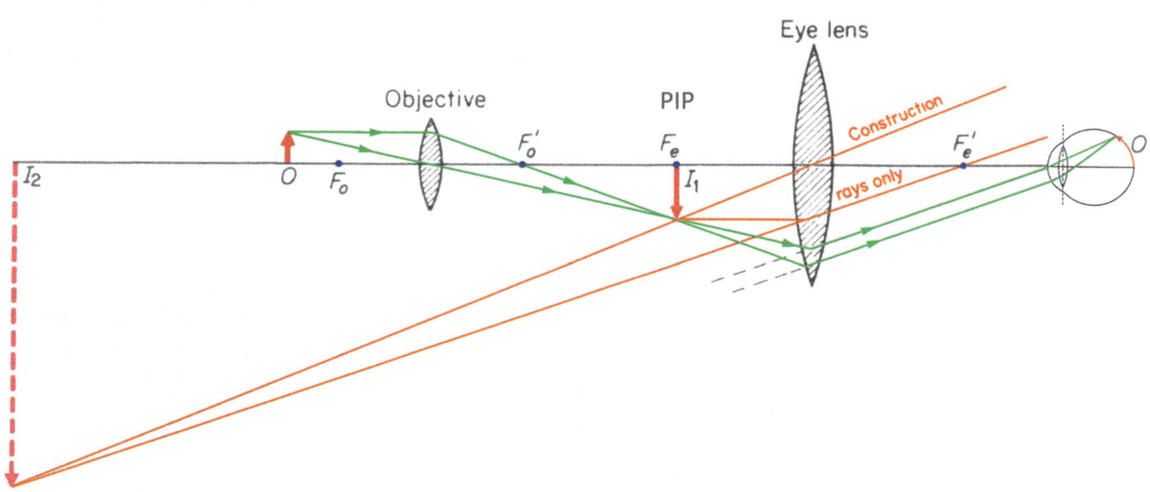

Figure 3.13 **The compound microscope** *Construction of the ray path of the compound microscope with the objective, eyepiece and eye acting together to form an image of much greater magnification than that possible with a single lens microscope. For clarity, and to simplify the ray-tracing, the objective and eyepiece are shown as single thin lenses. Because the front focal plane of the eyepiece lies at the primary image plane (PIP) where I_1 is formed, the rays leaving the eyepiece are parallel, therefore construction rays are required to locate the position of the virtual image, I_2. Source: Adapted from Abbott 1984 with permission from Pearson.*

3.10 Magnification in the Compound Microscope

Novice users of the microscope usually want to use the highest magnification objective possible from the outset. This is unwise; we find it difficult to judge size and scale accurately, and our visual clues come from our experience of our (three-dimensional) world around us. Using even a low magnification 10x objective with a 10x eyepiece gives a total magnification of 100x of a two-dimensional field of view. This requires a large conceptual jump to properly understand the relative size and nature of the image in front of us. High-magnification objectives necessarily give very small fields of view. Unless it is imperative to see a lot of image detail at the limit of the resolving power, it is better – regardless of whether you are an experienced microscopist or a beginner – to use low-magnification objectives to give sufficient detail in the largest possible field of view. Before using the microscope, consider using an inverted eyepiece to give a 10x overview of the slide preparation.

The objective forms an image in the primary image plane. The primary image is further magnified by the eyepiece. The primary image plane is located at a distance from the back focal plane of the objective, which is called the optical tubelength (also see Chapter 7, Section 7.9) and which is typically 160 mm in a finite tubelength microscope. The overall magnification of the objective and the eyepiece (or ocular) in the compound microscope is therefore given by:

$$M = Mag_{obj} \cdot Mag_{eye}.$$

This is also the case for an infinity-corrected microscope, as shown in Figure 3.14. Any additional magnification factor (M_c) due to the binocular head, tube-lens or a magnification changer must also be included; thus:

$$M_{Total} = (160/f_{obj}) \cdot (250/f_{eye}) \cdot M_c.$$

Because the lateral magnification of a finite tubelength objective is a ratio between the object to lens distance and the lens to image distance, the magnifying power is properly expressed as a ratio: e.g. 40:1. Since the object is placed at the front focal plane of the objective in an infinity-corrected system and the primary image is theoretically at infinity, the objective magnification is calculated in the same way as the eyepiece (which produces a virtual image). The objective magnification is calculated by dividing the reference focal length (the optical tubelength, a value between 160 and 200 mm) by the focal length of the objective. The magnification is expressed

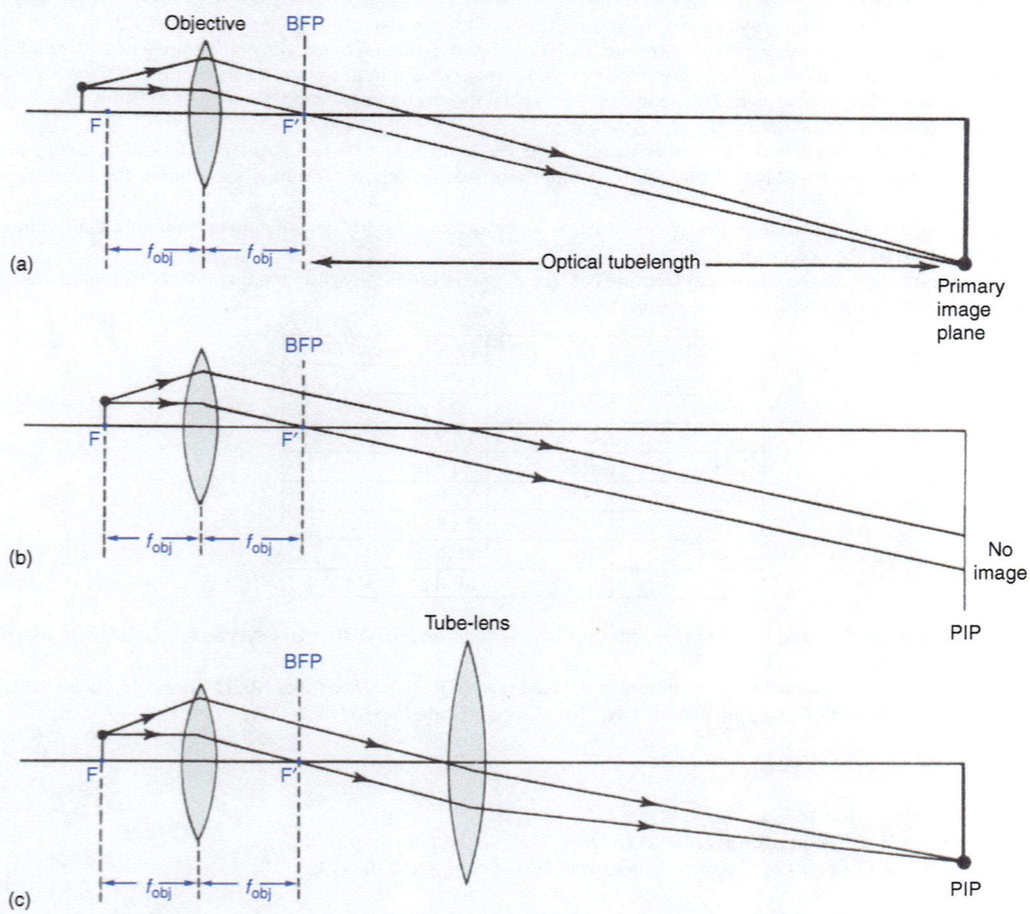

Figure 3.14 ***Image formation in the compound microscope*** *The objective forms an image in the primary image plane, which is further magnified by the eyepiece. The ray paths of both the older finite tubelength microscope (a) and the current infinity-corrected microscope (b) are shown, together with how the tube-lens in the infinity-corrected microscope (c) forms the primary image. See the text for further explanation. Source: Adapted from Bradbury 1984 with permission from Oxford University Press.*

as an angle rather than a ratio (i.e. 40x and not 40:1) because the viewing angle that the virtual image subtends is determined by the reference focal length.

Ideally, infinity-corrected objectives have to be parfocal. Therefore the reference focal length, as well as the focal length of the tube-lens, is fixed. This means that the focal length of each objective must change. An objective of shorter focal length will project a larger primary image than one of longer focal length. This does not matter, for all that is required in the infinity-corrected system is for the object to be placed at the front focal plane of each objective. Common objectives have focal lengths ranging from about 100 mm for the extremely low magnifications, down to about 1.5 mm for a 100x objective.

3.11 Useful Magnification

The primary function of the microscope is to resolve fine detail, not to magnify. Without sufficient resolving power, merely enlarging an image will not improve the ability to distinguish detail but will, in fact, degrade it. Prior to the early 1870s, when Ernst Abbe worked out a complete understanding of microscope image formation by diffraction, microscope manufacturers in the mid-19th century made objectives with excessive magnification (in some cases circa 150:1 with a numerical aperture of only 0.95 – the highest possible for a dry objective) hoping to see ever more detail. Such over-enlargement is called 'empty magnification'.

However, the converse is also true: the need to enlarge the fine detail resolved by the microscope objective (as determined by numerical aperture, or NA) sufficiently to match the minimum resolved distance seen by the detector, either the eye or a camera. Without sufficient magnification, the eye will not discriminate the resolved detail transferred into the image by the objective. Hence there are both upper and lower limits of magnification. The calculations are given in Box 3.3.

Box 3.3 Upper and Lower Limits of Magnification

The upper and lower limits of magnification are often given in microscopy texts as NA 500x and NA 1000x.

The calculations are as follows:

The minimum detail resolved by a young eye with very good visual acuity was given in Chapter 1 (Section 1.2) as 72 arcseconds, or 1.2 minutes of arc, which is equivalent to resolving detail 0.09 mm, or 90 μm, apart at the reference viewing distance of 250 mm. Usually, the limit of visual acuity is taken to be 1 minute of arc. Taking a conservative estimate of 2 minutes of arc (α in Figure 3.10) for normal visual acuity, this corresponds to a minimum distance resolved by the eye of 0.15 mm at the reference viewing distance of 250 mm. The minimum distance resolved by an objective[i], d, is: $d = 1.22\lambda/2.NA_{objective}$.

Dividing 0.15 mm by the equation for d, with λ expressed in mm (i.e. $0.15/[0.61\lambda/NA]$, or $0.15 \times NA/(0.61 \times 0.00055)$, where $\lambda = 0.00055$ mm), will give a minimum magnification of 447.NA. If the common blue laser wavelength of 488 nm is substituted, the expression calculates as NA 504x.

Hence the minimum magnification is usually cited at NA 500x. Below this value, the image will appear brilliantly sharp, but the eye will not perceive all the detail resolved by the objective.

For an objective of magnification M and numerical aperture NA, the diameter of the Airy disc (the diameter between the two first minima) in micrometres is: $D = 1.22\lambda.M/NA$. For a typical series of objectives,

Objective	NA	Ø Airy disc at PIP
5x	0.16	21.0 μm
10x	0.30	22.3 μm
10x	0.45	14.9 μm
20x	0.65	20.6 μm
20x	0.8	16.8 μm
40x	0.75	35.8 μm
40x	1.3	20.6 μm
60x	1.4	28.8 μm

a mean value for the image of the Airy disc at the primary image plane is circa 22 μm. With a 10x eyepiece this Airy disc will be magnified to 220 μm at the retina.

Rearranging the equation for the minimum resolved distance: $d = 0.61 \times \lambda/NA$ to give $NA = 1.22\lambda.M/2D$. Therefore, for an Airy disc of 220 μm, where $\lambda = 0.55$ μm, maximum theoretical magnification is NA 976x. This is rounded up to NA 1000x.

References

Danz, R. (2005). Carl Zeiss Innovation 15: 14.

Loveland, R.P. (1970). Chapter 2. In: *Photomicrography*, Vol 1. New York: John Wiley, p. 90.

Strong, J. (2004). Chapter 15. In: *Concepts of Classical Optics*. New York: Dover, pp. 336–338.

[i] This unit, the radius of the Airy disc, is sometimes referred to as a 'resel' or resolution element, in the same way that a pixel is a 'picture element'. For further details, see Chapter 10 on diffraction and image formation.

The overall magnification value of the microscope is obtained by multiplying the initial magnifying power of the objective by that of the eyepiece and any auxiliary apparatus, such as a magnification changer or any modular insert such as an epi-fluorescence unit carrying a magnification factor (i.e. 1.25x such as that on an older finite tubelength microscope). Working at the lower limit of useful magnification (i.e. NA 500x) can be uncomfortable, particularly for those without excellent visual acuity. It therefore pays to increase the magnification by two or three times this minimum value to make allowance not only for visual acuity but also objective quality and lighting conditions.

The upper limit of useful magnification is cited as NA 1000x. The absolute minimum resolved distance of the light microscope is 0.2 µm (i.e. with blue light and a maximum NA of 1.4). Since the unaided eye (in most cases) can comfortably resolve 0.2 mm, then a magnification factor of 1000x is required. Beyond this figure, empty magnification generally occurs. This upper limit of useful magnification of NA 1000x assumes good normal eyesight. Since not everyone has 6/6 vision or maximum visual acuity, or the observation conditions might be suboptimal, then this NA 1000x figure must sometimes be exceeded. Further information regarding the resolving power of the eye is given in Box 3.4.

Box 3.4 Optics of the Human Eye

The eye can function as either a telescope or camera or (if viewed in reverse) can be considered as a water-immersion objective with the retina as the 'object'. This is superbly and simply explained by Hammond (2015), and experimentally verified in an elegant study by Rostgaard and Qvortrup (1999). The resolving power of the telescope is normally given as an angular limit of resolution, whereas it is more usefully expressed as a spatial limit in the microscope. The resolving power, or visual acuity, of the eye is normally expressed as an angular limit[i] but can also be expressed as a spatial limit.

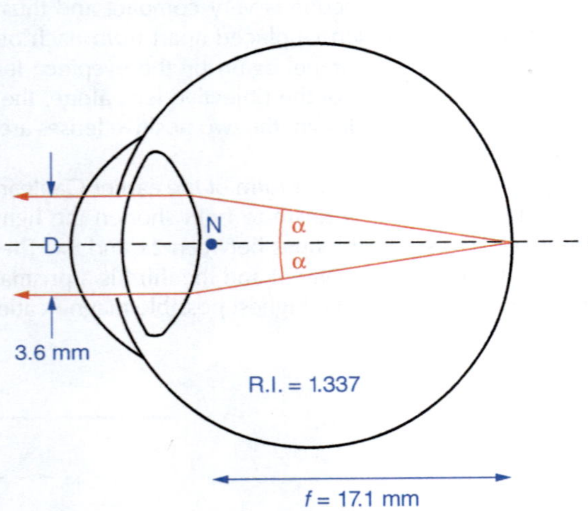

The focal length of the average relaxed eye is taken to be 17.1 mm, and the refractive index of the vitreous humour is 1.337. The pupil diameter, D, varies according to the intensity of light and will range from 2 mm for light-adapted to 8 mm for dark-adapted subjects. A median value for normal pupil diameter is 3.6 mm. From these values we get an *f*-No. of 17.1/3.6 = 4.8.

The half-acceptance angle of the eye is given by $\tan \alpha = \frac{1}{2} \times 3.6/17.1$ mm, or $\alpha = 6.0°$. For these small angles, $\sin \alpha \approx \tan \alpha = 0.105$. Since the refractive index of the vitreous humour is ≈1.34, from these values the numerical aperture of the eye ($n.\sin \alpha$) = 1.34 × 0.105 ≈ 0.14.

Depending on the pupil diameter the *f*-No. will vary from *f*/2.1 (with a dark-adapted pupil) to *f*/8.6 in very bright light. The equivalent numerical aperture range is NA 0.23–0.06.

Substituting these NA values into Rayleigh's formula for the minimum resolved distance ($d = 0.61\lambda/NA$), with a wavelength of 0.555 mm (maximum photopic sensitivity), gives a minimum resolved distance value for the human eye of 2.4 µm for a pupil diameter of 3.6 mm. Assuming appropriate Nyquist sampling, this agrees with the sizes of the cones in the *fovea centralis* from 1 µm to 2.5 µm. Taking the cone-cone centre distance as 3.8 µm from the work of Rostgaard & Qvortrup (see Figure 4), a single cone cell in the fovea subtends an angular limit of resolution of 0.8 minutes of arc ($\tan \alpha = 3.8/17.1 \times 10^{-3} \Rightarrow \alpha = 0.0128°$, or 0.77 ≈ 0.8 arcminutes).

This value of 0.8 minutes of arc as the minimum angle of resolution agrees with that found with the Snellen chart, used by ophthalmologists. A vision of 6/4.5 equals subtending an angle of 0.8 arcminute or 3.8 µm, close to the maximum possible. 6/6 vision equals subtending an angle of 1.0 arcminutes, or 4.8 µm. It is approximately equal to the minimum resolved distance of the naked eye of 90 µm at the least distance of distinct vision of 250 mm.

How many pixels are needed to match the resolution of the human eye? Each pixel must subtend an angle no larger than 0.4 arcminute, to comply with the minimum requirement for Nyquist sampling. Consider a print on a sheet of A4 paper (21 × 29 cm) viewed at 150 cm, which is close to the optimal viewing distance. Along the long edge, the print subtends an angle of 11°, and an angle of 8° along the shorter edge. Since there are 60 arcminutes in one degree, the pixel dimension must subtend an angle ≤ 0.4/60, or ≤ 0.007°. Thus (11 × 60/0.4) × (8 × 60/0.4) ≈ 2 megapixels as a minimum recording density to show detail at the limits of human visual acuity.

References

Hammond, C. (2015). Chapter 7.5. In: *The basics of crystallography and diffraction*, 4th ed. OUP, p. 189.
Parkin, A. (2016). Section 10.6 'Snellen acuity'. In: *Digital Imaging Primer.* Heidelberg: Springer, pp. 172–176.
Rostgaard, J. and Qvortrup, K. (1999). 'A note about retinal structure and visual acuity'. *Acta Opthalmologica Scandinavica* 77/1: 45–49.

[i] The microscope resolves specimen details which are close to the objective, and must take into account the refractive index of the medium between specimen and objective, so a spatial resolution limit is a more useful expression of resolving power than the angular aperture used for eyes and telescopes where the focal length of the objective is relatively much larger and the specimen located much further from the objective than when viewed with a microscope.

3.12 Similarities between Different Optical Instruments

We are concerned in this book with the light microscope, but other optical aids have been developed. The most common are spectacles to correct refractive errors and aberrations, and these may act also act as simple magnifiers. Closely related is the single lens magnifying glass, which – as when made with a lens of very high magnifying power, rather than the lower power lens used in a magnifier or loupe – was successfully used by Antoni van Leeuwenhoek as a simple microscope.

The camera and slide projector form an image in the same way as a simple single lens but with the object-image distances being different (see Figure 3.8). In the case of the camera, the object distance is larger than the image distance, whereas with the slide projector the object distance is smaller than the image distance (as in a microscope) with the object situated between $2f$ and f but with each instrument employing lenses of different focal length.

The telescope was developed shortly before the microscope at the start of the 17th century, the simplest design being the Galilean telescope (Figure 3.15; see also Appendix 6), which utilises a converging (biconvex or plano-convex) lens for the objective and a diverging (biconcave or plano-concave) lens as the eyepiece. The later, more sophisticated, design is the Keplerian telescope, which employs biconvex (or plano-convex) lenses for both objective and eyepiece.

The angular magnifying power of both designs of telescope is $-f_o/f_e$. In the case of the Galilean telescope the image is erect, whilst for the Keplerian design it is inverted (which does not matter when viewing stars). Although of limited magnifying power and field of view, the Galilean telescope is very compact and thus is suitable for use as a collimator to expand or reduce the diameter of a beam of light. With the lenses placed apart from each other at a distance equal to ($f_o + f_e$, e.g. 30 cm + −10 cm = 20 cm), parallel input rays are rendered parallel again by the eyepiece lens. The Galilean telescope is compact because its overall length is determined by the focal length of the objective lens alone, the two lenses being separated by the difference between their focal lengths, whereas in the Keplerian design, the two positive lenses are separated by the sum of their focal lengths, which dictates the length of the telescope.

The opera glass is a binocular form of the earlier Galilean telescope, whilst the modern binocular is a compact form of the Keplerian refracting telescope with a prism to both shorten the light path and form an erect image. Simple hand loupes generally magnify a maximum of 10x; surgical loupes between 2x and 5x. There are three types of loupe. The first is a simple single magnifying lens, the second is a Galilean arrangement and the third is a prismatic systems of Keplerian akin to telescopes/binoculars. These loupes reflect and focus the light to give the highest possible magnification. These latter may be built into spectacles for precise and exacting work.

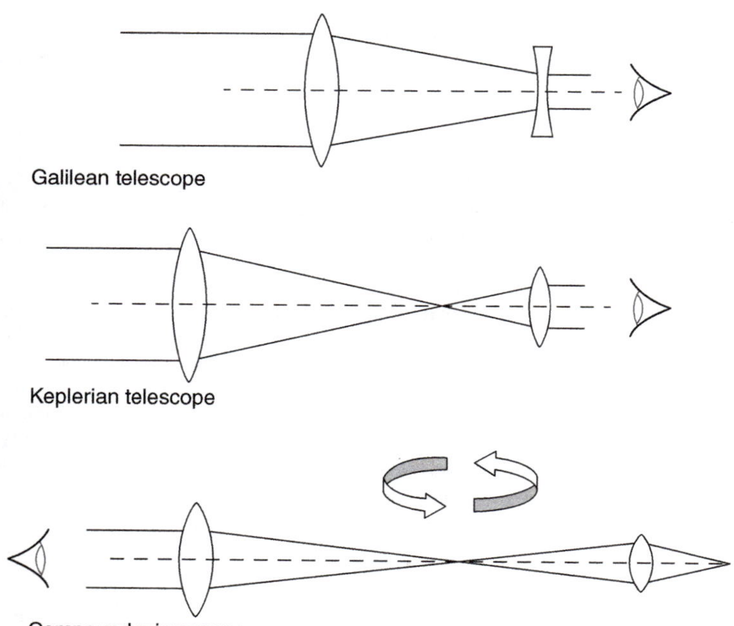

Figure 3.15 **The microscope and telescope compared** *Since the rays from the object glass in a Galilean telescope never cross at the back focal point, the image is upright, being a virtual image formed by the negative eye lens. This is the advantage of this design. However, when the Galilean telescope is pointed at any object not (as shown) on the optical axis, the ray path permits only a very small field of view. This is because the front focal plane of the concave eye lens, which is the exit pupil of the telescope, is situated between both lenses. The pupil of the eye therefore acts as an aperture stop, restricting the field of view.*

The Keplerian telescope does not suffer from the same restricted field of view because the ray path crosses the optical axis, and the exit pupil of the telescope lies beyond the eye lens, so the pupil of the eye can accept light rays with the same off-axis angle without vignetting and restriction of the field of view. The Keplerian telescope, therefore, is not limited in magnification to the same extent as the Galilean telescope. The compound microscope – in its simplest form – is merely an inversion of the Keplerian telescope design. Source: Adapted from Hartley 1993 with permission from Senecio Press.

Some optical systems employ telecentric lenses, particularly for low magnification and/or metrology and machine vision applications. Non-telecentric lenses give different magnifications at different object-to-lens distances; also, objects of different sizes look as if they have the same dimensions when they subtend the same viewing angle. A telecentric lens has its entrance pupil (object-space telecentric) or exit pupil (image-space telecentric) at infinity. This means that the magnification of the image is independent of the distance between the object and the lens, or its position in the field of view with respect to the optical axis. Telecentric lenses therefore correct for differences in magnification arising from parallax errors (which we naturally use to interpret our 3D world). Image space telecentricity also allows for extremely uniform image plane illumination.

3.13 Chapter Summary

At the end of this chapter you should understand how dioptric lenses form images and how the focal length of the lens, the position of the object and the position of the image are all important in the microscope. The image formed by one group of lenses can act as the object for a subsequent set, these being situated at conjugate foci. The primary function of the microscope is to resolve fine detail, but there needs to be sufficient magnification to perceive this resolved detail, although not so much that empty magnification occurs.

1. Fermat's principle states that a ray of light takes the shortest time to travel between two points.
2. Where the size of the object is very much larger than the wavelength of light, geometrical optics with ray-tracing can be used to describe the action of lenses and image formation.
3. A lens is any object that refracts electromagnetic radiation.
4. Lenses are dioptric – they refract light to form an image.
5. A real image is formed by the intersection of rays. This real image can be viewed on a screen or seen directly with the eye or a CCD camera.
6. A virtual image is formed by the *apparent* intersection of rays when back-projected from their direction of propagation. It cannot be projected onto a screen. A virtual image can only be seen with the aid of another lens (including the eye) or system of lenses.
7. The focal length, f, of a lens is the distance from the midline (or principal plane) of the lens to the focal point.
8. A positive lens converges incident rays, has at least one convex surface and forms a magnified real image.
9. A negative lens forms a demagnified virtual image.
10. Microscope lenses consist of many elements but can be approximated by thin lenses.
11. Parallel rays from infinity are brought to a focus by a thin lens at the back focal point, which lies in the back focal plane of the lens.
12. Rays entering a lens that emanate from, or pass through, the focal point exit the lens on the opposite side as parallel rays.
13. The dioptre, D, is a measure of the refracting power of a lens or curved mirror. It is expressed in reciprocal metres (m^{-1}). $1/f$ = power of the lens in dioptres. Convex lenses have +ve dioptre values; concave lens have -ve dioptre values.
14. Each lens has two focal points, front and rear, and therefore two focal planes.
15. The numerical aperture (NA) of a lens is a measure of the ability of the lens (or objective) to resolve fine detail and of the light-gathering power of the lens/objective.
16. As the object is moved closer to the lens, its image moves further away on the other side.
17. When the image distance is smaller than the object distance, the image will be smaller than the object by the ratio of these distances; e.g. as with a camera's lens.
18. An object placed within the focal length, f, of a convex lens forms an erect, magnified, virtual image.
19. When the object distance, u, is greater than f, a real image is always formed.
20. When the object is positioned between one and two focal lengths from the lens, a real magnified image is formed that is now larger than the object.
21. When the object is two focal lengths away from the lens, the image will be the same size as the object.
22. An object placed beyond two focal lengths from the lens will form a real, inverted image demagnified and reduced in size.
23. Conjugate foci are any pair of points such that at an object placed at one point will form a real image at the other point(s) in any set of conjugate planes. The object and image lie at conjugate points, or conjugate foci.
24. For a lens forming a virtual image (e.g. a magnifying glass), magnification is expressed as the ratio between the apparent (angular) size as seen with the lens and the angular size of the object when placed at the reference viewing distance of 25 cm from the eye.
25. The magnifying power of a finite tubelength objective is the ratio between the object-to-lens distance and the lens-to-image distance, e.g. 40:1, and should properly be cited thus, but for all purposes is usually written x40 or 40x.
26. The magnifying power of an infinity-corrected objective is an angular measure, i.e. 40x.
27. Useful magnification lies between 500x numerical aperture and 1000x numerical aperture.

Key Reading

Blackwell, J.M. and Thornton, S.W. (1996). *Mastering Optics*. Maidenhead, Berks: McGraw-Hill Europe.

- A practical introduction, advertised as 'non-academic', this book explains the essential concepts clearly and succinctly. It has a minimum of mathematical formulae and is very well illustrated.

References

Abbott, A.F. (1984). *Ordinary Level Physics*, 4th Ed. Heinemann Educational Books.
Bradbury, S. (1967). *The Evolution of the Microscope*. Pergamon Press.
Bradbury, S. (1984). *An Introduction to the Optical Microscope*, 1st Ed. *Royal Microscopical Society Handbook No. 1*. Oxford University Press.
Meek, G.A. (1976). *Practical Electron Microscopy for Biologists*, 2nd Ed. John Wiley & Sons.
Hartley, W.G. (1993). *The Light Microscope: Its Use and Development*. Senecio Press.
Pluta, M. (1988). Chapter 2. In: *Advanced Light Microscopy: Principles and Basic Properties*, Vol. 1. Amsterdam: Elsevier, pp. 129–286.
Smith, W.J. (2000). *Modern Optical Engineering: the Design of Optical Systems*, 3rd Ed. New York: McGraw-Hill.
Wayne, R. (2009). *Light and Video Microscopy*. Amsterdam/Oxford: Elsevier.
Wayne, R. (2013). *Light & Video Microscopy* 2nd Ed. Amsterdam: Academic Press, Elsevier.

Further Reading

Hecht, E. (2013). *Optics* 4th Ed. Pearson.
Lens and optical component suppliers often provide catalogues with useful technical data and helpful tutorials. A selection of these, available on-line at the time of writing, are given below.

Internet Resources

Basic optics – SPIE Fundamentals of Photonics, Module 1.3. URL: http://spie.org/x17229.xml.
PowerPoint on Optics, CBSE Physics. URL: http://physics.2lv.in/wp-content/uploads/2011/07/6m_OPTICS.pdf.
Thin lens exercises – Department of Physics & Astronomy University of Mississippi. URL: http://www.phy.olemiss.edu/~thomas/weblab/222%20Lab%20Manual/222_lab_man_spr2011_update/12_ThinLens_proce_Sp_2011.pdf.
Stanford Gaussian lens formula applet. URL: https://graphics.stanford.edu/courses/cs178/applets/gaussian.html
Edmund Optics. URL: http://www.edmundoptics.com/resources.
Thorlabs. URL: http://www.thorlabs.com.
Newport Corporation. URL: http://www.newport.com/ (see under Resources).
Melles Griot. URLs: http://mellesgriot.com and http://pdf.directindustry.com/
PhET Geometrical Calculator. URL: https://phet.colorado.edu/sims/geometric-optics/geometric-optics_en.html.

Notes

1. Electron microscope lenses are positive lenses. They use primarily electromagnetic, or sometimes electrostatic, fields to converge electron rays to a focus. The rays of electrons do not change in velocity (which is approximately half to two-thirds that of light for an 80–100 kV electron microscope as used in biological electron microscopy). Whereas refraction of the rays of photons in a glass or plastic lens occurs at a sharp interface between the lens and air or immersion medium with a consequent change in velocity, refraction in an electron lens is continuous because there is no sharp interface between the refracting magnetic field and the medium (i.e. vacuum) through which the ray is propagated. Because of limitations in the design of even the very best electron lenses, the theoretical resolving power of the EM has not been approached by a factor of 100: two orders of magnitude. If we were to have LM lenses equivalent to those for electron microscopy, it would be akin to using the bottoms of milk bottles to form images!
2. Optically, it is the *change* in refractive index between two media (i.e. air and the lens material) that is important. Similarly, the curvature, or 'figure', of the lens surface is equivalent to the curvature of the interface between these two media of differing refractive index.
3. The principal axis is the line joining the centres of curvature of the lens. The principal plane of the lens is perpendicular to the optic axis, from which the focal lengths are measured. The principal plane passes through the optical centre of the lens. The principal point lies where the principal plane intersects the optical axis. In a thin lens, there is only one principal point, and this can be regarded as the optical centre of the lens. The surface vertex (pole) each side of the lens is where the surface of the lens intersects the optical axis. The principal points, nodal points and focal points, together with the object and its image, all lie on the optical axis. These four sets of points are collectively known as the cardinal points.
4. The effective, or equivalent, focal length of a thick lens is the composite focal length of a number of separated glass elements, or groups of elements. In this case, the focal length is measured from the second nodal plane, N_2.
5. NEVER look at the sun directly or with an optical aid. To do so may well result in permanent damage to your eyes. Always use a properly designed solar filter, or view an image of the sun projected onto a matt surface (Figure 2.20).
6. The numerical aperture (NA) of a lens is a measure of the light-gathering power of the lens (or objective) and – more importantly – the ability of the lens or objective to resolve fine detail. Because the numerical aperture is a product of two ratios, it is dimensionless and the scale therefore has no units.

4
Microscope Anatomy and Design

Outline

4.1 Introduction

In Chapter 1, the basic principles of how our eyes work and how we perceive images have been explained. Chapters 2 and 3 have discussed the nature of light and how it behaves according to the laws of geometrical optics. In this chapter we turn to the design of the microscope, to understand how various lenses and other optical components are held in conjunction with one another on the optical axis to produce useful images.

The properties and limitations of the human eye dictate the design of the microscope and its lenses, even though nowadays the image may be presented to a camera or photomultiplier tube and displayed on a computer monitor. The image must be presented at infinity, so that the image-forming rays are parallel on entering the eye. When the image is viewed directly with a microscope, rather than through a camera/computer monitor interface, the exit pupils from each eyepiece must be 3–5 mm in diameter to match the pupils of the eyes and separated by about 65 mm to match the inter-ocular distance. Since the eye and cameras are sensitive only to variations in brightness

Understanding Light Microscopy, First Edition. Jeremy Sanderson.

and colour, information from the object must be manipulated so that the image is presented as brightness and/or colour differences. The common requirements of microscopes mean that, although different manufacturers may follow different design philosophies, there will always be the same basic controls to look out for and learn to use.

In this respect a light microscope is like a car: we take the engine, wheels and brakes and other fundamental parts for granted. We look for the means to switch on the engine, steer the car, brake and operate the various electrical components, which may be designed and placed differently according to the microscope function and the particular manufacturer. Personal preference, naturally, plays a part in the selection of microscope brands, just as it does in car choice. In the same way, the basic design and functional parts are similar between stands, and many of these may be hidden from, or out of the direct control of, the microscopist. What we are looking for are the operating controls common to all stands to enable us to 'drive' the microscope successfully.

4.2 The Function of the Light Microscope

The compound light microscope is so called because it uses two lenses, the objective and the eyepiece, to produce the final magnification of the image. There may also be intermediate magnifying components (usually for photomicrography) between the objective and eyepieces. Other intermediate accessories, for example an epi-fluorescence accessory, may also introduce a magnification factor (e.g. 1.25x). Thus:

$$\mathrm{Mag_{final}} = \mathrm{Mag_{obj}} \times \mathrm{Mag_{intermediate}} \times \mathrm{Mag_{eyepiece}}$$

The advantage of a compound lens system is that much higher magnifications can be achieved than with a simple microscope. To obtain high magnification with a single-lens magnifier, the lens itself must be placed close to the eye, as explained in Chapter 3, Section 3.9. The highest magnification achieved by Leeuwenhoek with his single-lens simple microscope was 275x. With a compound microscope, magnifications up to 1400x are possible. Not only that, but two other consequences flow from using a compound microscope. First, the specimen can be illuminated much more efficiently, and second, the lens system in a compound microscope can be made with much higher numerical aperture. This, in turn, means better resolving power – the whole purpose of the microscope – and increased light-gathering capacity.

In the traditional (i.e. finite tubelength) microscope, the specimen is located just beyond the first focal plane of the objective, in such a position that an image of the specimen is formed 10 mm below the top of the tube according to the International Standard. This image is known as the *primary image*. In more recent microscopes designed for 'infinity correction' (see Chapter 7, Section 7.10), the object is placed precisely in the first focal plane of the objective so that the image passes up the tube in the form of parallel rays, which are then converged by a 'tube-lens' to form the primary image; the end result is thus the same. A second lens, the eyepiece, is arranged so that its first focal plane coincides with the primary image plane. The imaging rays then leave the eyepiece parallel and enter the eye, where they converge onto the retina.

Of the various components, the objective is the primary lens system, and it is sometimes said that the objective *is* the microscope. The condenser is important, too, because it must focus sufficient light in a controlled manner onto the specimen. Both the objectives and condenser contain multiple lens elements and perform close to their theoretical limits. The three essentials for microscopy are:

(1) **resolving power**;
(2) **contrast** or visibility; and
(3) **magnification**, in that order.

As we noted in the introduction, the primary function of the microscope is not to magnify but resolve fine detail. Clearly the fine detail in an image cannot be made visible without sufficient contrast. The eyepieces are only necessary to present the primary image formed by the objective to the eye or final detector sufficiently magnified so that all the resolved detail can be seen. In some cases, with high-quality CCD cameras, an eyepiece is not required because all the resolved detail can be discriminated on the monitor.

Since microscopes are nowadays mainly operated through a computer interface, it is very important that due consideration is given to the design of the associated software package and its ease of use. A recent trend, initiated by Nikon and now followed by all four major manufacturers, is to use a common software platform applicable to both stereo and compound microscopes as well as widefield fluorescence and confocal stands. At the time of writing, Nikon offer NIS-Elements; Olympus offers cellSens software, which has replaced the Cell^M and related software; the Leica Application Suite (LAS) provides a common user interface for all Leica products, while Zen software (Zeiss Efficient Navigation) is currently available as Zen Blue for widefield microscopes and as Zen Black for confocal systems. Each manufacturer supplies stripped-down 'lite' versions for off-line viewing of image files in their individual proprietary image formats.

4.3 Inspection of the Microscope

The optical axis is arranged vertically, from the base of the microscope, through the condenser, objectives and eyepieces to the eye or other detector (Figure 4.1). On research microscope stands, certain components such as the stage, objective nosepiece, condenser and

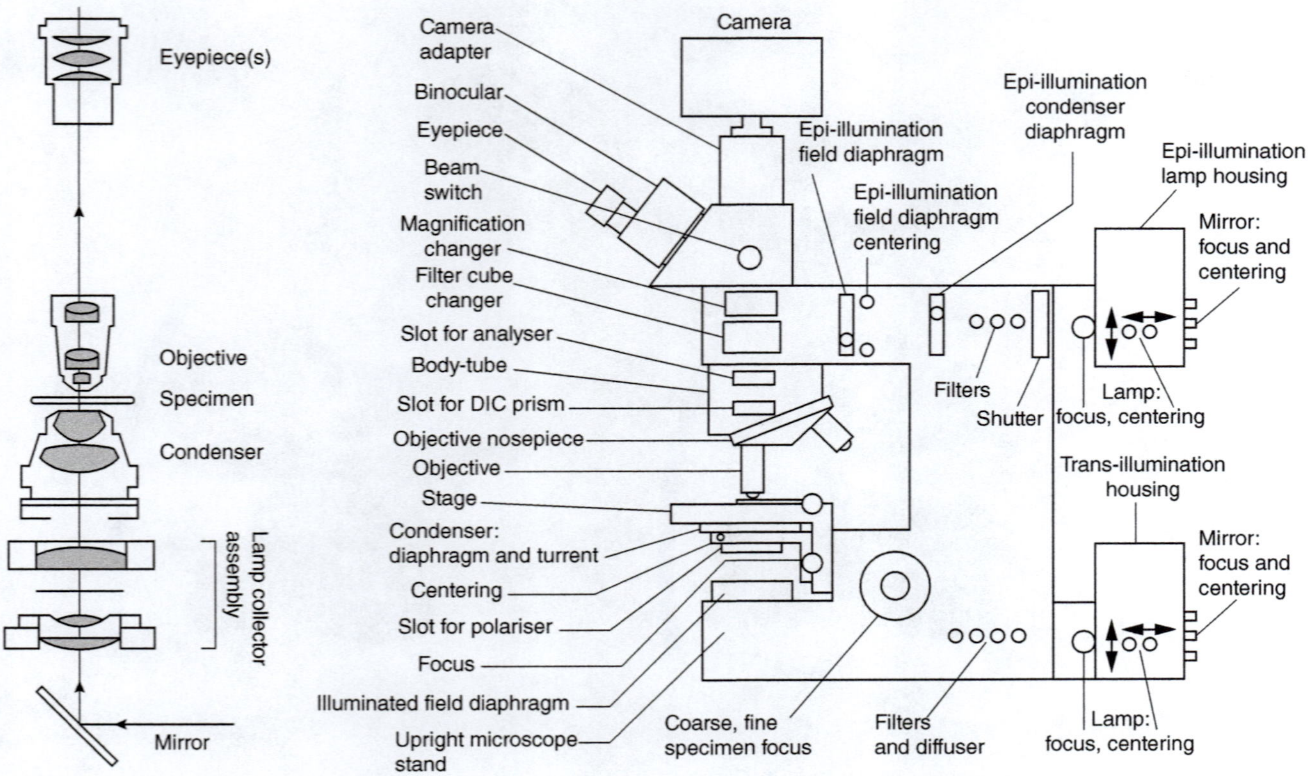

Figure 4.1 **Parts of a microscope** *Source: (left) Wayne 2009. Reproduced with permission from Elsevier. (right) Watkins and St. Croix 2013. Reproduced with permission from John Wiley & Sons.*

possible fluorescence filters, are often motorised. Look for the following components on your microscope and any other you may have access to:

- Microscope stand and light source
- Microscope base and power supply
- Illumination control(s)
- Coarse and fine focus controls (are these co-axial?)
- Illuminated field diaphragm – explained in Chapter 9
- Condenser substage
- Any filters inserted into the base or condenser substage – explained in Chapter 11
- Polariser and analyser filters for polarised or DIC microscopy, if any – explained in Chapter 11
- Condenser – whether simple bright-field type or universal design for contrast methods – explained in Chapter 8
- Illuminating aperture (condenser) diaphragm – explained in Chapter 9
- The stage – is this mechanical or automatic?
- Translational stage controls (are these co-axial?) and slide holder
- Objective(s) on the (rotating) nosepiece
- Any prisms inserted above the objective, or in the microscope body, for Differential Interference Contrast – explained in Chapter 11
- Any slots for accessories in the body-tube
- Magnification changer, if any, in the body-tube
- Epi-illuminator, if any, in the body-tube
- Fluorescence filters, if any
- Monocular, binocular or trinocular head
- Eyepieces – look for independent focusing ability if the eyepiece is designed for measurement
- Dioptre adjustment controls – either on the eyepiece holder of the binocular tube or the eyepiece
- Any camera attached
- Any focusing telescope
- Any separate automatic metering controls

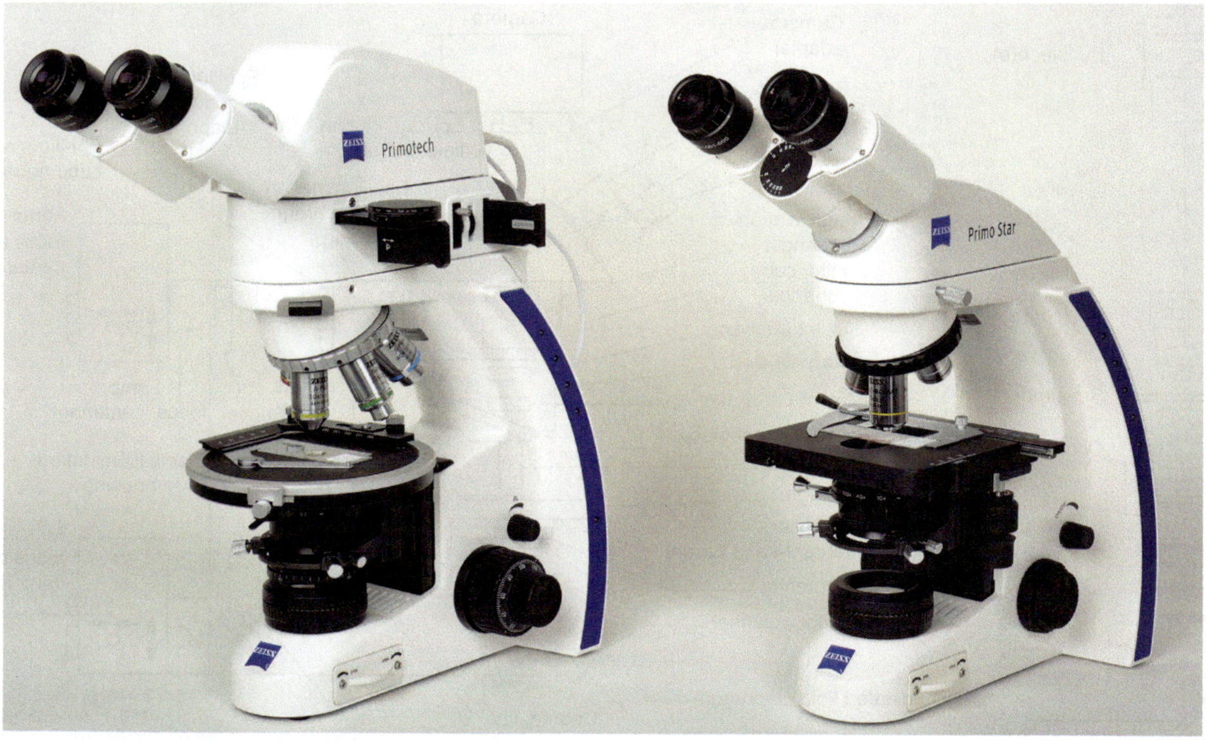

Figure 4.2 **Two upright microscopes** *Source: Courtesy of Mr K. Glover.*

Discover the common components and operating controls on different designs of microscope stand. The figures in this chapter show upright rather than inverted configurations for ease in identifying the various components on a modular microscope. Inverted stands are important, particularly for tissue culture and studies of living cells; they are discussed in Section 4.5.4.

4.4 Anatomy of the Microscope

Figure 4.1 shows the possible parts and components that may be found on a modern upright light microscope. Figure 4.2 shows two upright transmitted-light microscopes of different sophistication and complexity. Each of the four major manufacturers offer similar designs, both as entry level stands and as research stands. Modular designs (Figure 4.3), which allow a microscope to be built for a specific purpose from a range of different components, allow for easy upgrade and configuring of a stand to suit a particular application in an economical and cost-effective manner. Not every microscope will have all the following components listed, but most will be found on the microscope in your possession.

A Stand and Light Source

Trace the power supply from the mains supply to the light source or entry into the microscope stand. Identify the on-off switch and any rheostat controls. There may be a separate control to preset the lamp intensity (colour temperature) for photomicrography.

Identify the type of lamp(s) fitted to the microscope. Usually there is only one lamp, but on a large research stand there may be two separate light sources: one for transmitted light and one for (fluorescence) epi-illumination or reflected light work. Increasingly, LEDs are used now rather than incandescent lamps. Lights and illumination are discussed further in Chapter 9.

Determine whether the lamp has its own centring controls, or whether it is of the pre-centred type. If there is only one lamp, there may be blanking caps (or not, if lost!) that cover the incident light port. There may be a focus control for the lamp-collector (bullseye) lens that is crucial to Köhler illumination. Alternatively, it is more likely that the position of the lamp-collector lens cannot be located because it is hidden within the base of the microscope stand. Within the base of the microscope stand there may be built-in neutral-density and colour filters; ascertain the presence and operation of these.

Also on the stand there will be focusing controls. These usually move the stage, although on older or inverted designs of microscope, the focusing is likely to act on the nosepiece carrying the objectives. Modern focus controls are coaxial: the fine focus is contained

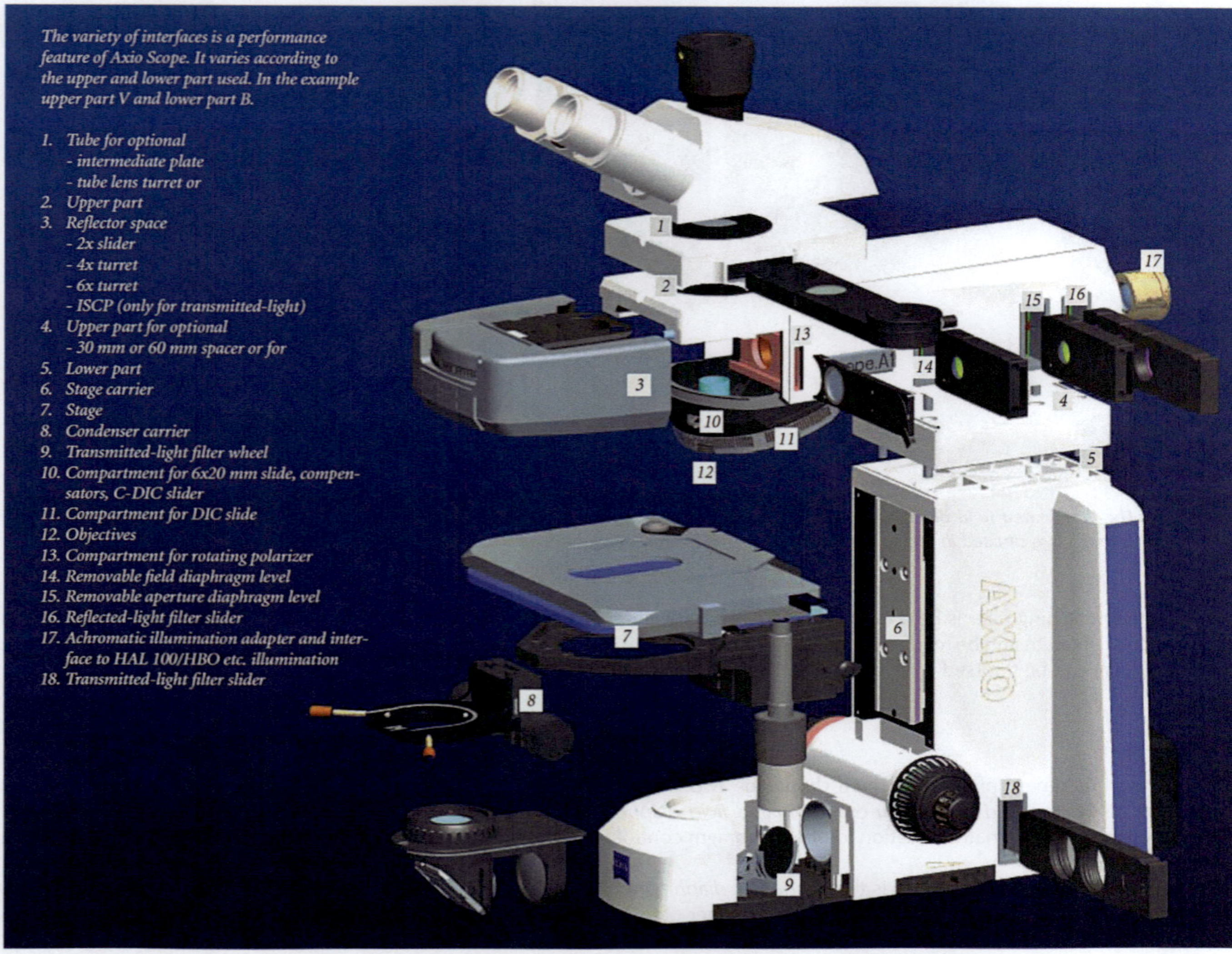

Figure 4.3 ***Exploded modular design*** *This figure, taken from a manufacturer's catalogue, shows the benefit of the modern modular design. Parts can be changed easily to modify the stand for different functions or to future-proof a purchase for later upgrades. Source: Reproduced with permission from Carl Zeiss Ltd.*

within the coarse focus knob. This fine focus along the z-axis may be graduated in 1 or 2 µm divisions. Generally, the focus controls are also under electronic control on modern research microscopes and so may not give any detectable mechanical resistance.

B Substage Apparatus

Determine how the condenser is carried on the substage holder underneath the stage. This holder will have a rack-and-pinion control to focus the condenser vertically along the optical axis. The condenser may be mounted on a dovetail fitting and slide into place on the substage holder; alternatively, it may be inserted within a retaining ring or else will clip into position against a spring. This latter solution does not have a retaining screw, but the dovetail and retaining ring designs will have these.

A well-designed microscope stand will have two centring control thumbscrews set at 120° to each other on the substage apparatus. These are used to centre the condenser onto the optical axis when adjusting the microscope for Köhler illumination. The condenser centring controls may be absent on basic microscope stands. This set of centring controls are not to be confused with a set of centring controls on so-called 'universal' condensers, which are for centring the individual positions of the carousel independently onto the optical axis of the microscope, for example when aligning phase contrast annuli. The function of the condenser is discussed further in Chapter 8.

If the microscope is fitted with a polar for polarised-light work or for Differential Interference Contrast (DIC), this may be a simple filter dropped into the position of the illuminated field diaphragm. However, this is not the ideal place[1] to locate any filter. It is more

Figure 4.4 ***The illuminated field diaphragm*** *All microscopists need to know where the controls for both the illuminated field diaphragm and the condenser diaphragm are situated in order to correctly adjust the microscope for Köhler illumination. Source: Courtesy of Mr K. Glover.*

likely that the polarising filter is held in the substage apparatus just below the condenser. The filter should have the means to slide in and out of the light path and be rotatable (possibly against a scale marked in degrees) so that the analyser, usually between the objective and eyepiece(s), can be 'crossed' with respect to the polariser. This is explained in more detail in Chapter 14, Section 14.2.

C Diaphragms

On every well-designed microscope there should be two diaphragms. The first is the one situated at the base of the instrument (Figure 4.4) called the *field diaphragm* or *illuminated field diaphragm* (IFD). This is usually controlled by an adjustable control on its circumference. An alternative location for the diaphragm control is on the side of the base of the microscope stand, controlled by a vertical thumbwheel.

The second diaphragm (Figure 4.5) is the *condenser diaphragm* or *illuminating aperture diaphragm* (IAD). This is usually built into the condenser. There will be a control to alter the diameter of this diaphragm. On some microscopes, this lever may be set against a graduated[2] scale, marked in units of numerical aperture.

D Condensers

The simplest type of condenser is the Abbe condenser, properly called the Abbe illuminator. This may or may not have a flip-out top lens, which is removed from the optical path for low magnification objectives. This type of condenser will have a lever control operating the top lens component and a lever or knurled ring for opening and closing the condenser diaphragm. Condensers will either be a dry or an

Figure 4.5 ***The condenser diaphragm*** *Both designs of condenser control on uprights stands (left and centre panels) are shown here graduated in units of numerical aperture. This helps adjust the correct setting of the diaphragm, but graduations are not always provided. The condenser on the inverted stand (right panel) is motorised, the control being shown on the condenser housing. Source: Courtesy of Mr K. Glover.*

oil-immersion design (see Figures 8.1 to 8.6). The numerical aperture of a dry condenser is always ≤ NA 0.9, whereas an oil-immersion condenser will have a value between NA 1.0 and NA 1.4.

There are 'universal' condensers as alternatives to the Abbe illuminator. These have a carousel with various types of contrast-enhancing annuli, prisms or stops. The carousel allows you to rotate the required apparatus into the optical path. The microscope may have phase-contrast annuli of different sizes; there may be different Wollaston prisms for Differential Interference Contrast, or there may be one or more dark-field patch stops. Each of these annuli, prisms or stops will be marked, and the direction of shear for the Wollaston prism should, ideally, also be marked. To see the contrast-enhancing inserts for phase-contrast, DIC and dark-field in the condenser, remove it – hold it securely – turn it upside down and rotate the carousel.

Fluorescence microscopes, almost without exception, use an epi-illuminator. This may have both the illuminated field diaphragm and condenser diaphragm controls, although sometimes only the field diaphragm control is fitted. Further details are given in Chapter 9, Section 9.12. This epi-condenser will have either a slider or rotating carousel to hold the fluorescence filter cubes in place. These should be exchangeable, and their function is explained in Chapter 15, Section 15.7 onwards. The epi-condenser should be fitted with a shutter, or else a blanking plate, if fitted, can be used for the same purpose to close off the light path and prevent photobleaching when the fluorescent signal is not being observed or photographed.

E Stage and Stage Controls

The stage can either be fixed or rotatable. Rotating stages are more commonly seen on polarising microscopes and those configured for DIC, petrological and material science work. Older microscopes used spring stage clips to hold the slide in place on the stage. This is now uncommon. On the top of the stage there should be a slide holder with some form of spring-loaded arm to keep the slide in place while the stage is moved laterally. There are usually scales on each lateral x- and y-axis, with vernier markers, to allow for measurement of the size of the specimen and for recording the position of any feature of interest on the slide.

Drop-down coaxial controls, to the right or left hand of the stage provide for ergonomic translational control of the stage. On modern research-grade stands, these functions may be motorised, particularly if the microscope is programmed to collect Z-stacks or configured for live-cell imaging.

F Objectives

The objective is the most important part of the microscope. There will usually be a series of 4, 5 or 6 objectives screwed into a rotating nosepiece. Note the markings on the objectives: there should at least be the **magnification** and **numerical aperture** of the objective marked. Much older objectives had their focal length (e.g. 1/12″) marked instead of magnification. The significance of this is explained in Chapter 7, Box 7.1.

An objective may have the magnification marked without units (e.g. merely 40) or, if it is an objective with a finite tubelength, as a ratio (e.g. 40:1). See the comments on magnification in Note 1, in the Notes section of the front matter for an explanation of why these two different tubelength systems have different magnification notations. Modern infinity-corrected objectives have their magnification marked, e.g. 40x. Modern objectives also have a coloured ring around them to indicate the magnification, and this is standardised according to the colour of the spectrum. Further details are given in Chapter 7, Section 7.3.

The tubelength (e.g. 160 or ∞) should be marked on the objective; 160 for finite tubelength objectives, or ∞ for infinity-corrected tubelength objectives, respectively. The 160 notation[3] stands for the distance in millimetres of the mechanical tubelength for objectives designed for finite tubelength microscopes. Modern microscopes are almost invariably infinity-corrected designs. This is to allow extra accessories to be added without compromising image quality. An explanation is given in Chapter 7, Section 7.10.

The class of objective (e.g. achromat, semi-apochromat or apochromat) should also be marked. Sometimes semi-apochromat objectives are called 'fluorite' or 'Neofluar' objectives because crystalline fluorspar was formerly used in their construction (see also Section 7.2). If the class of objective is not marked, it is probably an achromat. Most modern objectives are corrected for a flat field across most of the field of view, and these will be marked with the prefix *plan-* (e.g. plan-Neofluar).

For numerical aperture values greater than NA 0.4 or so, the thickness of the coverslip becomes important, and if this is the case for the objective in question, you should see the inscription '0.17' marked on the barrel of the objective, meaning that the standard thickness of coverslip, 0.17 mm, should be used. Dry objectives with a very high numerical aperture and also some immersion objectives will have a coverslip correction collar. Besides the physical rotating collar-grip around the objective, this may be indicated by the inscription 'Corr' or 'Korr'.

Non-immersion objectives are intended to be used dry, but most are *not* so marked. It is up to the user to deduce this. This important point is often overlooked. Objectives that require an immersion medium between the specimen or coverslip and their front lens are usually marked to indicate this fact. This may be by the prefix 'Imm' or a coloured ring around the objective. Oil and water are common immersion fluids, but it is possible to use glycerine (or silicone, introduced by Olympus) particularly on objectives designed for live-cell imaging. You should normally see a black ring around the front of the objective to indicate oil immersion and a white ring for water immersion. A glycerine-immersion objective will have an orange ring and a silicone objective, a red ring. A multi-immersion objective with a rotatable collar (to adjust the internal lens elements according to the immersion fluid selected) will also have a red ring around it. In each case, the correct immersion fluid for the objective *must* be used.

Those objectives designed for phase contrast or Differential Interference Contrast (DIC) will usually have this fact included in the objective inscription. Phase-contrast objectives will either have the abbreviation 'Ph' and/or their inscription in green lettering. Objectives for DIC or polarised-light work will either have 'DIC' or 'Pol' and/or their inscriptions in red lettering.

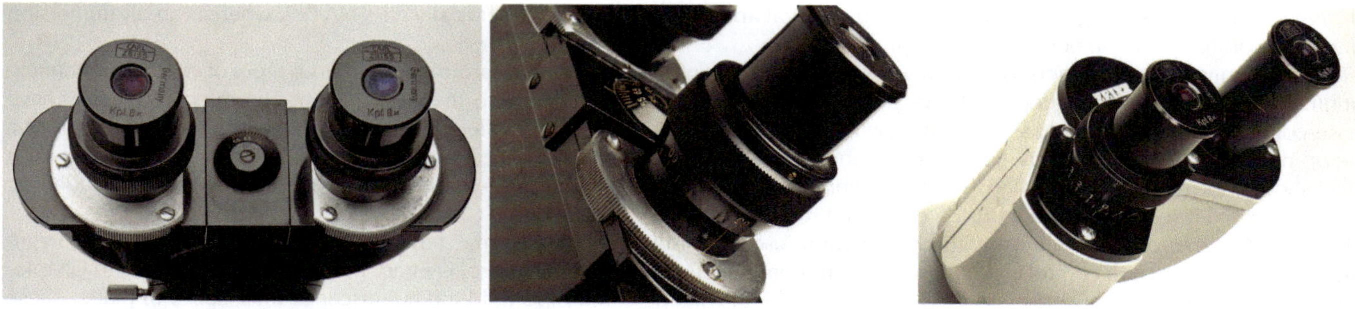

Figure 4.6 **Binocular head designs and dioptre adjustment controls** *When using the laterally sliding head (Jentsch) design, seen here on the left, first the inter-ocular distance must be set for the individual user, and then a compensation (middle panel) must be made by adjusting each eyepiece to match that distance; otherwise, the correction made by the designer for spherical aberration will be upset. Some Jentsch binocular head designs automatically lower the eyepiece tubes into the binocular head as they are separated for a greater inter-pupillary distance, so keeping the mechanical tubelength constant. This older Zeiss design does not, and so the tubelength compensation correction must be set on the eyepiece tube, as shown.*

The advantage of the Siedentopf design of binocular tube (right-hand panel) rotating about a fixed point is that the inter-pupillary distance can be adjusted between users without causing any alteration in the mechanical tubelength of the microscope. The scale on the eyepiece tube is for dioptre adjustment, according to the eyes of each user, not for optical tubelength adjustment. Source: Courtesy of Mr K. Glover.

In most cases the manufacturer's name is included and sometimes (on older objectives) a part number and unique serial number. One very useful piece of information that is often lacking (except for Nikon objectives which *do* include this information) is the working distance of the objective. The significance of this is explained in Chapter 7, Section 7.6.

- Never drop or otherwise strain objectives or other optical components of the microscope.
- Never force the focus controls of the objective and condenser, and always watch the lens surface as it approaches the specimen, to avoid damage.
- Never touch the anti-reflection coatings on the front of lenses or filters.
- *Do* keep the objectives clean!
- Clean objectives *only* with the correct soft fibre lens-cleaning tissues and correct solvent.

G Eyepieces

The eyepieces will have their magnification engraved on them. It will always be one number, e.g. x10 or 15x, never a ratio. The common values are 10x; sometimes 8x or 12x or 6.3x eyepieces are found.

The inscription 'WF' may be present to indicate a wide field of view. The field-of-view number (FN) may be given as a single figure, e.g. 18, 20 or 22. The field-of-view number is the diameter in millimetres of the field diaphragm. Using the formula in Chapter 8, Section 8.10, this number can be used to calculate the diameter and area of the object field in millimetres and so give an approximate indication of the size of the image.

Some eyepieces are made with high exit pupils (e.g. Figure 8.10), to enable spectacle wearers to use these comfortably. In this case a little spectacles symbol is usually seen on the rim. If the eyepieces are compensating types to be used with apochromats, this may be indicated by the abbreviation 'Comp' or 'Komp'.

If the eyepiece is designed as a measuring eyepiece, to hold a graticule at the primary image plane, it will have a helical or push-pull focusing capability, which will be independent of the dioptre adjustment control on the eyepiece holder on the binocular head.

H Binocular/Trinocular Head

There are very few monocular microscope stands in use these days, the eyepieces commonly being inserted as a pair into a binocular head. Alternatively, a trinocular pattern may be provided for photomicrography. If the microscope has a trinocular head, then there should be a slider, or other control, to split all or part of the light to the camera port and/or the eyepieces. This is a 50-50%, 80-20% or 100-0% split to the camera.

The binocular head should have either a Jentzsch pattern, i.e. a horizontal sliding adjustment, (Figure 4.6) to set the correct inter-ocular distance, or ideally a modern, i.e. Siedentopf pattern, adjustment rotating about a common centre. The latter is more common on modern research microscope stands and can be seen on both microscopes illustrated in Figure 4.2.

The binocular head should ideally have two adjustable dioptre controls for balancing the focus for each eye according to the strength of the eye in dioptres. This is to avoid eyestrain (see ergonomics in Chapter 5, and Section 8.7). Well-designed microscopes have two independently adjustable eyepiece dioptre controls. These are usually on the binocular head eyepiece tubes (Figure 4.6) but may sometimes be integrated into the eyepiece (Figure 4.7). Less well-designed stands (often for economic reasons) have only one adjustable dioptre control on the binocular head.

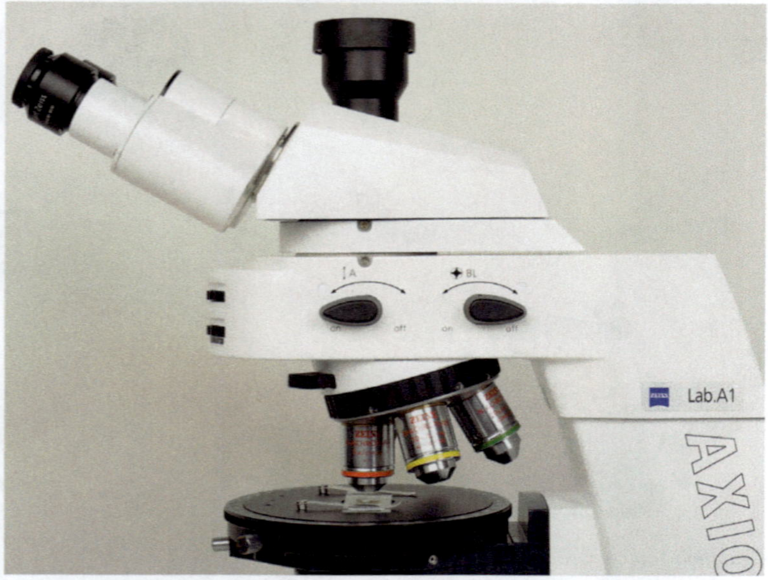

Figure 4.7 **Bertrand lens** *A Bertrand lens (the control for which is shown here denoted BL) is a simple converging lens that may be switched into the light path to project the back focal plane of the objective into the primary imaging plane, allowing conoscopic viewing of the interference figures of crystals and mineral grains in thin section without removing the eyepiece(s). It is sometimes called the Amici-Bertrand lens. When used in conjunction with the eyepiece, it forms a telescope, which may also be used to adjust the annuli in a phase-contrast condenser. Source: Courtesy of Mr K. Glover.*

Our eyes are most comfortable when viewing distant objects, from which the light arrives at the eyes essentially parallel (see Figures 9.19 and 9.20). A microscope must therefore present its image in the form of parallel rays, arriving as if from infinity. Secondly, because the pupils of our eyes are generally between 3 and 5 mm in diameter under normal lighting conditions, the bundles of rays carrying the image information from the eyepieces – the exit pupils – should also have a similar diameter.

If the imaging rays left the eyepiece in the form of a disc the size of a dinner plate, only a small part of the information would enter the eyes and be perceived. Conversely, if the exit pupils were tiny pinpoints of light, they would cast harsh shadows onto the retina of disturbing deposits within the vitreous part of the eyeball. These translucent features are *muscae volitantes*, often better known as 'floaters'. For more on how to minimise the effect of these floaters on the image, see Chapter 9, Section 9.13. The eyepiece is therefore designed such that the bundle of imaging rays leaving the microscope is about 3 mm in diameter – the usual size of a relaxed pupil.

I Magnification Changers and Other Controls

Any magnification changer is usually situated in between the upper body of the stand and the binocular/trinocular head. On an inverted microscope, this may be a push-pull rod below the objective nosepiece, or else is under electronic control. Check to see whether there is a camera adaptor or C-mount, and what its magnification factor is. Further details on how to match the most suitable camera adapter to your microscope is given in Chapter 31, Section 31.7.

A microscope designed for polarising work or materials science should have a built-in Bertrand lens either as part of the same wheel holding the magnification-changing lenses or situated within the body (Figure 4.7) of the microscope. This is used in conoscopic work in crystal analysis, which is briefly discussed in Chapter 14, Section 14.17. A Bertrand lens is also useful if fitted to a phase-contrast or DIC microscope for aligning the phase annuli and Wollaston prisms when setting up the microscope.

Fluorescence microscopes, almost without exception, all use an epi-illumination configuration (Figure 4.8), seen here on an entry-level fluorescence microscope with LEDs providing the short-wavelength illumination for fluorescence. The filter cubes may either be held in a slider or, more commonly, within a carousel as shown.

Refer to the diagram in Figure 4.1 of a generalised upright light microscope stand. This shows any other components that could possibly be present on your microscope. The stand may have filter control pushrods built into the base. Alternatively, there may be a filter holder, or swing-out tray, attached just below the condenser or substage holder. It may also be possible to swing a diffuser in and out of the illumination path.

On an older microscope fitted with a substage mirror, and designed for use with an external light source, identify the illuminated field diaphragm, the bullseye collector lens and the focus control for the bullseye collector lens on the stand of the external lamp.

J Co-Observation/Discussion and Comparison Optical Systems

Co-observation units are often found in pathology or histology laboratories, where several users may need to discuss, or learn from, a particular slide. Forensic laboratories use comparison microscopes – two instruments joined by a common binocular head for comparing two fields of view – in ballistics analysis and identification, for example. Figure 4.9 shows examples of both of these pieces of equipment.

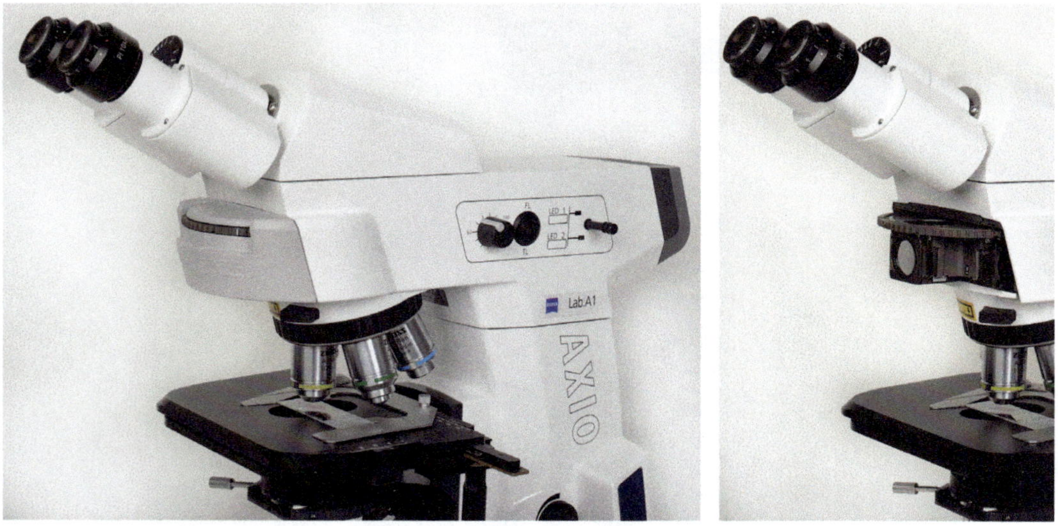

Figure 4.8 **Epi-illumination configuration** *Source: Courtesy of Mr K. Glover.*

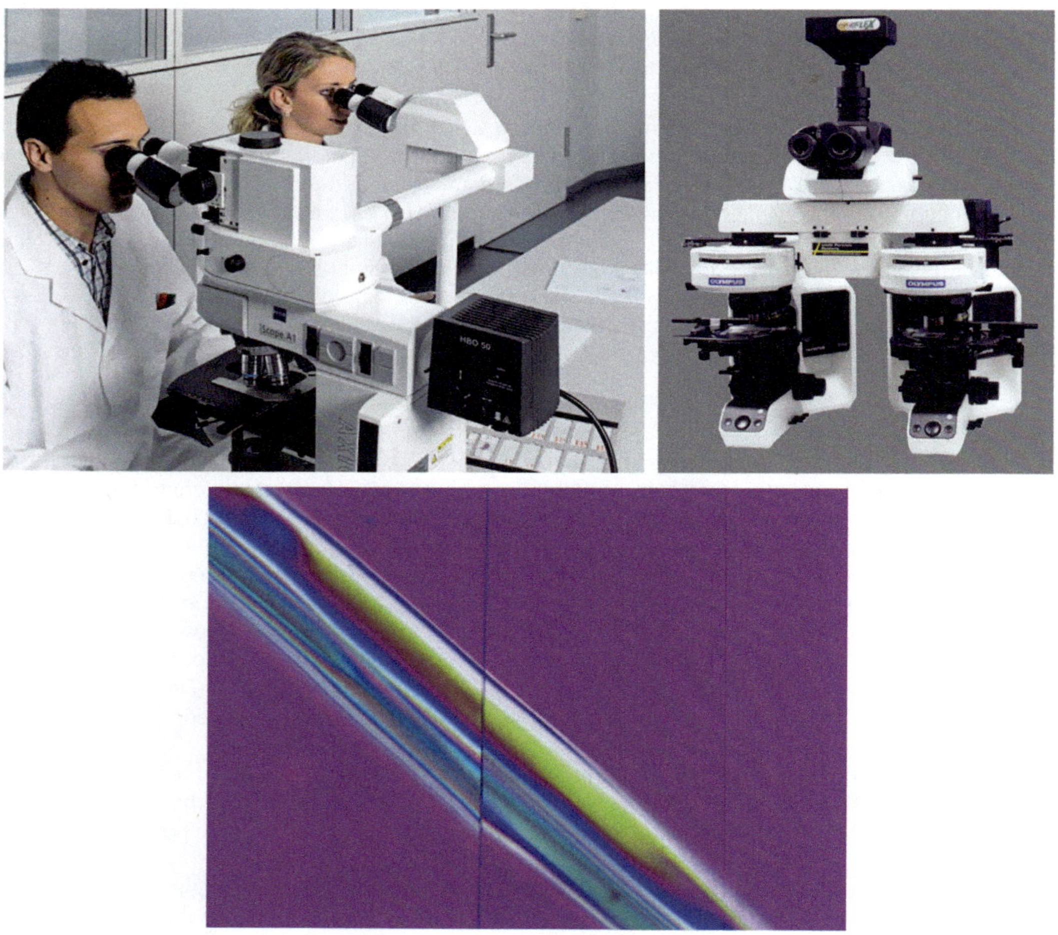

Figure 4.9 **Discussion and comparison bridges** *The discussion bridge shown here (top left) is manufactured by Zeiss, although very similar designs are available from all manufacturers. The comparison bridge shown here (top right) is manufactured by Leeds Forensic Systems, but Leica is also a market leader in forensic microscopes and macroscopes. The bottom panel shows a subject fibre being compared with a reference fibre. Source: (top left) Reproduced with permission from Carl Zeiss Ltd. (top right) Leeds Forensic Systems. Reproduced with permission from ForenteQ. (bottom) Reproduced with permission from ForenteQ.*

K Anti-Vibration Tables

High-quality research microscopes with sensitive cameras more often than not demand the use of an anti-vibration table or other means of dampening and preventing the periodic effect of mechanical vibrations from being seen in the recorded image (see Box 4.1 and Figure 18.11). Vibration and its isolation involve resonance and simple harmonic motion. Every object on a flexible support has a fundamental natural frequency at which it naturally vibrates once set into motion. A tuning fork is designed to vibrate at a known natural frequency, depending upon the mass and/or length of the tines. For example, a tuning fork for the musical note A_4, the A above middle C, used for orchestral tuning, vibrates at a frequency of 440 Hz. Dampening dissipates mechanical energy, attenuating vibrations more quickly than would otherwise naturally occur. Immerse the tuning fork's tines in water, and the vibrations are almost instantly attenuated.

There are three sources of vibration: ground motion, acoustic noise (see Box 4.1) and direct vibration from the equipment itself. Microscopy equipment has natural frequencies from 10 to 100 Hz. Fortunately, it is not difficult to dampen these vibrations.

Box 4.1 Sources of Vibration

	Typical frequency	Isolation solution
Swaying buildings	0.1–5 Hz	Move to ground floor or basement
Upper floor buildings	1–10 Hz	Air tables, passive rubber or elastomer mounts
Air compressors and pumps	4–20 Hz	Install on soft mount dampers
Building resonance	5–50 Hz	Air tables, passive rubber or elastomer mounts
Street traffic	5–100 Hz	Move to a quiet location
Machinery	10–200 Hz	Install on soft mount dampers
Motorised equipment	10–500 Hz	Use active damping piezoelectric actuators
Acoustic vibrations	above 20 Hz	Use both passive and active solutions

Common machinery problems include pump cavitation, unbalanced fans and flow excitation from air-handling systems. For air-conditioning vents, which may be placed above a microscope, or live-cell system, consider using air baffles to direct the air flow away so that draughts of air don't alter focus or image quality through temperature and/or vibration.

Air isolators are the most cost-effective method for isolating low-frequency vibrations from 1.5 to 3 Hz. Springs are effective from about 3 to 10 Hz; rubber, elastomer or cork pads from 3 to 40 Hz, depending upon size. Isolation of sub-hertz vibrations requires passive negative stiffness isolators, or active control methods.

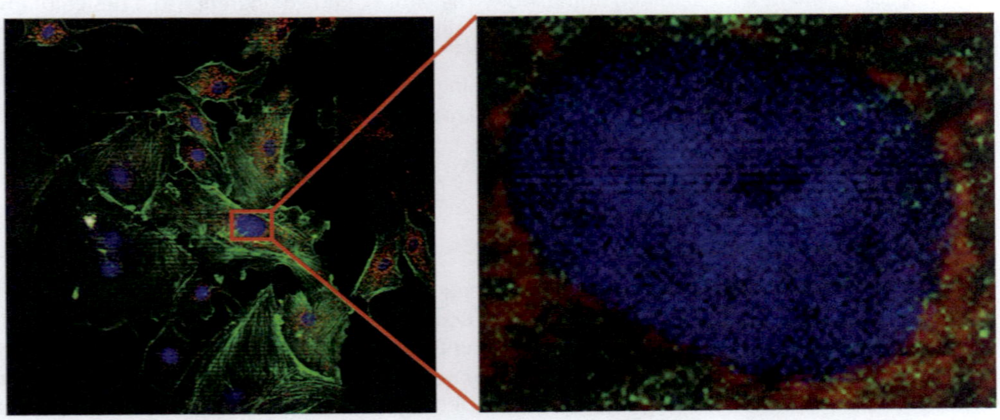

The effect of low-frequency vibration can easily be seen in the image. The field of view of endothelial cells (left) was taken with the anti-vibration air table inactive and lowered onto its legs without an air cushion. The high-magnification image of the single nucleus (right) was taken with the air table correctly pumped up, but the table has been deliberately tapped during the first part of the scan to collect the image.

The easiest way to dampen unwanted vibration is passively: using pneumatic air tables; rubber mats or blocks; cork or elastomer pads (e.g. Fabreeka; see the Internet Resources at the end of the chapter); a spring mechanism or tennis balls. These latter may be held in a ring to prevent them from rolling. Damping dissipates energy, which reduces vibration transmitted at the natural resonance frequency. The earliest anti-vibration tables used granite or slate, but these are not as effective as modern solutions. The key is to use a material that vibrates at a lower natural frequency than the equipment requiring isolation. The lower the natural frequency of the isolator, the better the protection against low-frequency vibrations – by far the most critical to instrument performance.

Air tables are usually built with an internal honeycomb structure sandwiched between two steel plates. This maximises stiffness and optimises rigidity with respect to weight. The high stiffness helps push resonance to higher, less detrimental, frequencies. Alternatively, active control using electronic sensors and control systems can be employed; these are more effective at sub-hertz frequencies, but it costs more to isolate these lower frequencies. Suitably quiet long-lived air compressors for microscope air tables are manufactured by Jun-air (indeed, they are so quietly efficient that it is all too easy to forget their presence and keep the motor topped-up with lubricating oil).

4.5 Different Types of Microscope Design

4.5.1 Transmitted-Light Microscopes

The traditional microscope is the one generally used for biological work and configured for transmitted light (light that passes through a specimen). Most explanations in this book will centre on this design, partly because it is so common, but also because the optical components are arranged one after the other; it is the easiest to explain and also to understand.

4.5.2 Reflected-Light Microscopes

Microscopes designed for use in materials science are configured to examine opaque specimens by reflected light, which illuminates the object through the objective lens. The objective therefore also acts as its own condenser. These microscopes are generally rather easier to use than the transmitted-light design, because whenever the objective lens is changed, the condenser is also changed appropriately, and no major readjustment or centring are required. Their use is discussed in Chapter 12.

4.5.3 Fluorescence Microscopes

In the last thirty years, light microscopy has benefited tremendously from advances in computing, fluorescent stains and laser technology. Fluorescence microscopy has become a ubiquitous technique in biological and medical science in recent years. Most biological microscopes can be adapted to illuminate the specimen with short-wavelength radiation and to observe the emission of longer-wavelength light, from fluorescent stains or fluorescent proteins, within the sample. Figure 4.8 shows an entry-level epi-fluorescence microscope with a carousel fitting for holding fluorescence filter cubes.

4.5.4 Upright and Inverted Microscopes

Transmitted- and reflected-light microscopes are made in both the conventional upright design, in which the specimen is examined from above, and also the inverted form, where it is examined from below. The earliest known reference to a design of inverted microscope known to the author was published in the mid-19th century by Smith, in 1852, although the upright design predominated until well into the second half of the 20th century.

Inverted microscopes are particularly useful (and widespread) in biological research laboratories, where they are used to view specimens in water or a culture medium. An example would be cell monolayers or tissue explants adherent on a coverslip or grown on the bottom surface of a culture vessel.

4.6 Stereomicroscopes and Macroscopes

Most microscopes are designed to operate at high resolution, but a consequence of this is that they have extremely shallow depth of field when the object is in focus in the image plane. These compound microscopes also have a small field of view at the high magnifications required for high resolution of detail. Stereomicroscopes adopt a different approach, sacrificing ultimate resolving power for increased depth of field. Two other advantages of eschewing ultimate resolving power are gains in (a) field of view and (b) working distance.

Stereomicroscopes are useful for examining solid objects by reflected light (including fluorescence), although they can be used with transmitted light also. Stereomicroscopes come into their own where three-dimensional observation and perception of depth and contrast is critical to the interpretation of specimen structure or where the sample is too large for examination with a compound light microscope (Figure 4.10). These instruments are also essential when a large working space is required for micromanipulation of the specimen.

The stereomicroscope essentially uses two microscopes, one for each eye, but offset in slightly different directions (by about 8–10°) to produce a laterally correct image with a three-dimensional effect. In this respect, stereomicroscopes are more like our eyes than the compound microscopes described above and so are particularly suited for educating schoolchildren.

There are two basic designs of stereomicroscope (Figure 4.11). The older, Greenough, system consists of two individual and separate microscopes converging with paired objective lenses. However, if one tube is used for photomicrography, the subject must be tilted so that it is *orthogonal* to that particular light path. The other is the CMO (common main objective) design, which is better adapted to photomicrography. Because the optical path of the common main objective stereomicroscope produces a parallel bundle of light rays between the body and microscope, the CMO design is especially suited to modern infinity optics and allows extra units, such as fluorescence illuminators, to be incorporated into the stereomicroscope. When using a CMO stereomicroscope, the centre portions of

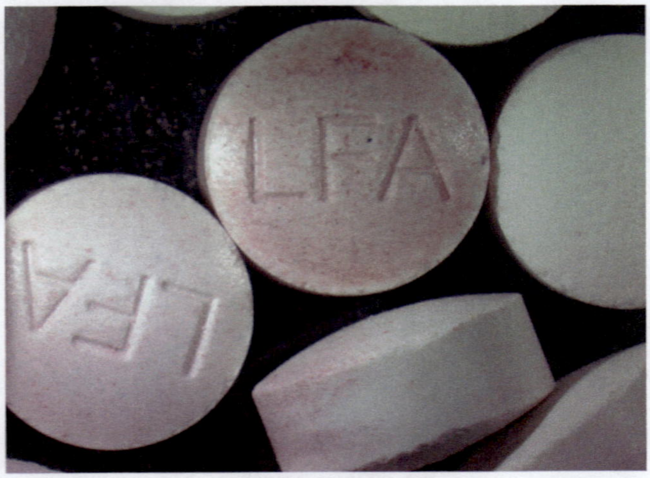

Figure 4.10 **Stereomicroscope specimen** *This pharmaceutical sample has been chosen to illustrate the advantages of a modern stereomicroscope. These tablets, which measure 1 cm in diameter, show the very large depth of field achievable. The image was taken using a Zeiss Stereo Discovery V.20 fitted with a plan-Apo S 0.63x objective with a free working distance of 81 mm.*

the image may appear slightly elevated so that a flat specimen appears convex. This is perspective distortion. It is virtually eliminated in high-quality CMO stereomicroscopes but may still be present in cheaper designs.

The Greenough stereomicroscope design suffers from an artifact called the *keystone* or *tombstone effect*, caused by the inclination of each microscope tube to the common optical axis. Only the central regions of each image are in simultaneous focus at identical magnifications, and the area on the left-hand side of the right-eye image thus appears slightly smaller than that on the right-hand side of the same image, and vice versa for the left eye. The brain usually compensates for this effect so that it is often not noticed; however, the keystone effect can be responsible for fatigue and eyestrain during prolonged use of a Greenough stereomicroscope.

Macroscopes are dedicated to photomacrography and are an alternative to stereomicroscopes. Like a compound microscope, the optical axis is arranged vertically, from the base of the microscope, through the objective and eyepieces to the detector. They do not exhibit parallax shift, as do stereomicroscopes. Older macroscope designs, such as the Leica M420 and the Zeiss Tessovar for example,

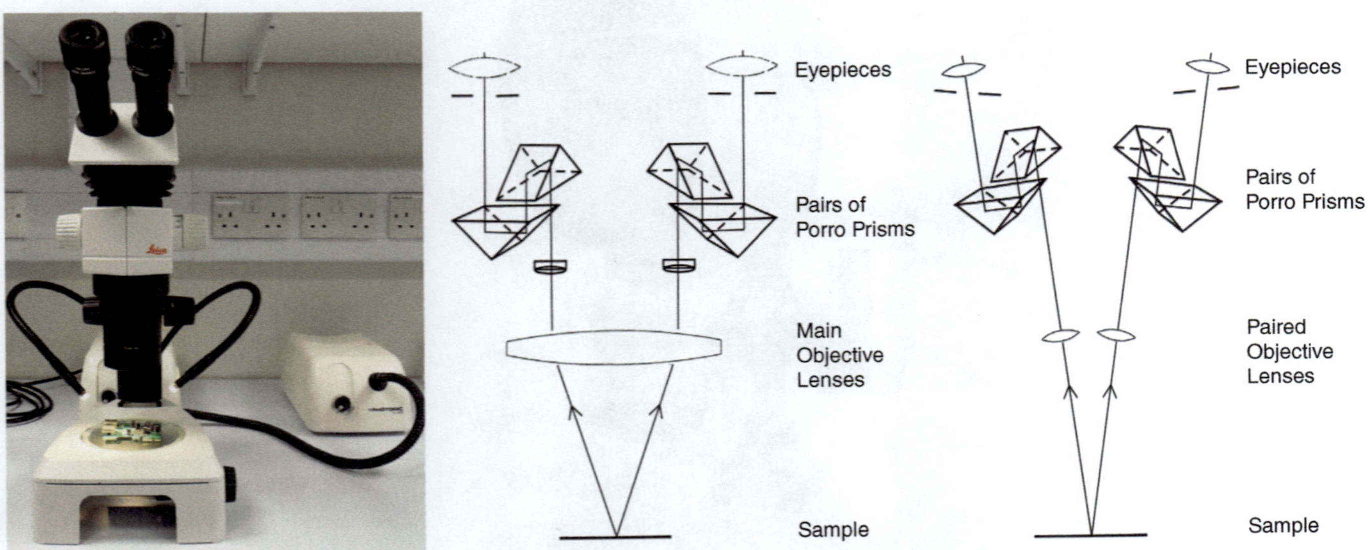

Figure 4.11 **Stereomicroscope designs and ray paths** *The two principal designs of stereomicroscope. The Greenough design was introduced by Zeiss in 1896, and the CMO (Common Main Objective) design was introduced in 1946. The older Greenough design (right) is still used with an inclined stereo angle between 11 and 16 degrees inclination from the optical axis for entry-level instruments and where a compact design is required. The CMO design (left and centre) is less fatiguing to use and allows for addition of modular components. Having one large objective, the CMO design has a greater light-gathering power and often higher numerical aperture than an equivalent Greenough stereomicroscope. Crucially, the objective axis is perpendicular to the object field, making the CMO design much better for measurement and inspection purposes. The stereo effect comes from the tilted beam path of the two telescope ray paths after exiting the objective. Because the objective is used only at the periphery, a high-quality lens must be used to avoid zonal aberrations. Source: (diagram) Bradbury and Bracegirdle 1998.*

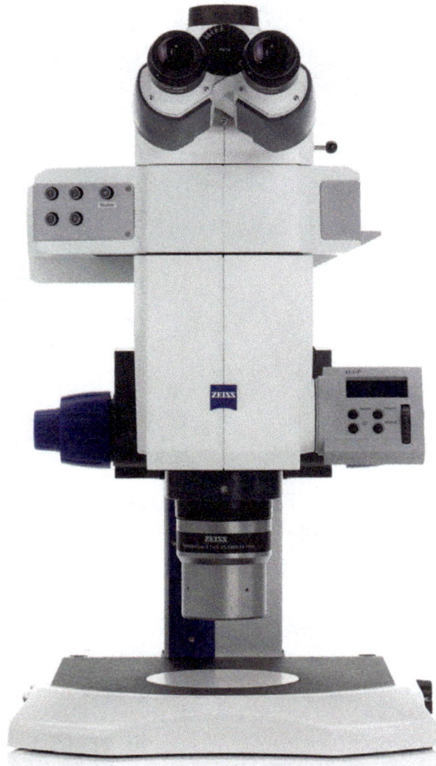

Figure 4.12 **The Axio Zoom V.16 macroscope** *Source: Reproduced with permission from Carl Zeiss Ltd.*

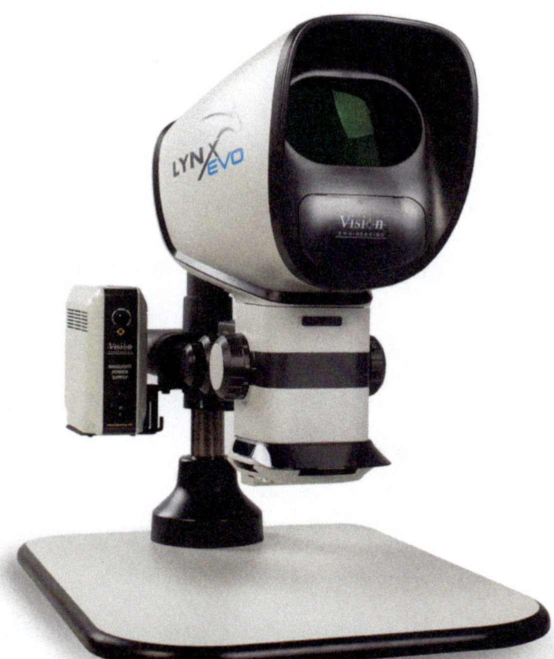

Figure 4.13 **The Lynx inspection stereomicroscope** *Source: Vision Engineering Ltd.*

have been superseded by the Leica Z16 APO A, the Zeiss SteREO Lumar V12 and the latest Axio-Zoom V16 (Figure 4.12), which also can profitably be used for fluorescence work. At the time of writing, Olympus manufactures the MVX10 macroscope and Nikon the Multizoom AZ100. Vision Engineering specialises in manufacturing ergonomic inspection stereomicroscopes and macroscopes such as the Lynx (Figure 4.13). These modern macroscopes combine the benefits of a stereomicroscope with an extensive zoom system, thus overlapping the magnification range between a stereomicroscope and compound microscope without sacrificing the large field of view or long working distance.

4.7 Chapter Summary

There are many designs of microscope, ranging from toys for children – some of which are of good design – to professional research microscopes fully equipped with all manner of accessories. The educational outreach programme of the Royal Microscopical Society recommends the Timstar Junior Microscope MS-2. It is even possible to buy an attachment to fit in front of the lens of a mobile phone to convert it into a microscope.

1. The properties and limitations of the eye dictate the design of the microscope.
2. The common requirements of microscopes mean that there will always be the same basic controls to look for.
3. The compound light microscope is so called because it uses two lenses, the objective and the eyepiece, to produce the final magnification of the image.
4. The three essentials for light microscopy are resolving power, contrast and magnification, in that order.
5. Transmitted- and reflected-light microscopes may be made in upright or inverted configurations according to purpose.
6. Research microscopes often require some form of vibration isolation.
7. Stereomicroscopes confer an advantage of field of view and working distance, at the expense of resolving power.
8. Nearly all modern stereomicroscopes are of the common main objective design, rather than the older Greenough design.

References

Bradbury, S. and Bracegirdle, B. (1998). *Introduction to Light Microscopy*, 2nd Ed. *Royal Microscopical Society Handbook No. 42*. Bios Scientific Publishers.
Watkins and St. Croix. (2013). *Current Protocols Select Imaging & Microscopy*. Hoboken, New Jersey: Wiley.
Wayne, R. (2009). *Light and Video Microscopy*. Amsterdam/Oxford: Elsevier.

Further Reading

Freere, R. (1996). 'Stereomicroscopes, macroscopes and low-magnification imaging: a review of the advantages and limitations of these and similar devices'. *Proceedings of the Royal Microscopical Society 31*/1: 9–15.
Gras, H. (2014). 'A multiaxis device for in-focus manipulation of objects under a dissecting microscope'. *Jour. Microscopy 256*/1: 1–5.
Greenberg, G. and Boyde, A. (1993). 'Novel method for stereo-imaging in light microscopy at high magnifications'. *Neuroimage 1*/2: 121–128.
See also the associated correspondence and authors' reply in *Neuroimage* (1995) *2*/1: 85–87.
Hammond, C. (1996). 'Principles of stereoscopy in the single objective binocular microscope'. *Quekett Journal of Microscopy 37*/7: 566–574.

Internet Resources

Tutorials on the different types of lamp available to microscopists. URL: http://zeiss-campus.magnet.fsu.edu/articles/lightsources/index.html.
Anti-vibration information. URLs:

1. https://bit.ly/2CYotl4
2. https://bit.ly/2R51eZO
3. https://www.newport.com/t/fundamentals-of-vibration.

The microscope mechanical drawtube. URL: http://www.microscopy-uk.org.uk/mag/indexmag.html?http://www.microscopy-uk.org.uk/mag/artnov04/pjtube.html.

Selected Company Information

Vistek (now Newport) selector guide. URLs: http://www.vistekinc.com/ and http://www.vistekinc.com/passive-vibration/microscope-platform.
Newport selector guide. URL: https://www.newport.com/c/vibration-isolators.
Spiers Robertson. URL: https://www.speirsrobertson.co.uk/17-microscope-platforms-and-tables.
Optophase. URL: http://www.optophase.com/Copie%20de%20Table-anti-vibration.html.
Fabreeka pads. URL: https://www.fabreeka.com/?lang=en-uk.
Solent scientific incubators. URL: https://www.solentsci.com/microscope_incubators.htm.
Piezo stages. URLs: http://www.piezoconcept.com and http://www.piezoconcept.com/all-products/z-insert-2/.
The stereomicroscope. URL: http://www.microscopy-uk.org.uk/mag/artdec11/rjk-stereo1.pdf.

Timstar Junior Microscope for schoolchildren. URL: https://www.timstar.co.uk/microscopes.
see also the RMS schools activity kits. URL: https://www.rms.org.uk/discover-engage/microscope-activity-kits.html.
KeepLoop mobile phone attachment. URL: http://www.keeploop.com.
How to build a microscope from a smartphone. URL: http://www.iflscience.com/technology/how-build-smartphone-microscope-stand-10/.

Manufacturer Lite Software Versions

Leica. URLs: https://www.leica-microsystems.com/products/microscope-software/details/product/leica-las-x-ls/#downloads and https://www.leica-microsystems.com/products/microscope-software/ PC only, not Mac OS.
Nikon. URL: https://www.nikoninstruments.com/en_GB/Products/Software/NIS-Elements-Advanced-Research/NIS-Elements-Viewer Both PC & Mac.
Olympus. URL: https://www.olympus-lifescience.com/en/software/cellsens/ PC only, not Mac OS.
Zeiss. URL: https://www.zeiss.com/microscopy/int/downloads/zen.html PC only, not Mac OS.

Notes

1. This is because anything situated at the illuminated field diaphragm will be imaged into the specimen plane. Any dirt or scratches on the filter will be seen in the final image. See also Section 9.10 on conjugate planes in the microscope.
2. Since numerical aperture is a product of two ratios, it is dimensionless. The scale therefore has no units.
3. Sometimes (e.g. on older Leitz microscopes) the mechanical tubelength will be 170 mm, and the inscription on the objective will be 170. Antique English microscopes were designed for a mechanical tubelength of 10″, or 250 mm, to correspond with the reference viewing distance, formerly known as 'the nearest distance of distinct vision'. Later reflected light stands for metallurgy were computed for an MTL of 200 mm. Monocular microscopes may also have a drawtube to alter the mechanical tubelength to correct for spherical aberration. See the internet URL link given in the Internet Resources.

5
Ergonomics

Outline

5.1 Introduction

It is common to find people sitting awkwardly in front of their microscope or sometimes resorting to propping up the stand on books to use it at a suitable height. Rather, adjust your seating to suit you. It is also very common to find that the eyepieces have not been properly adjusted to suit each individual user. On a properly-designed microscope it is possible to adjust the dioptre controls (Section 8.7) of each eyepiece to match the refractive power of each eye. For each individual user this will inevitably be different from other users of a shared microscope. Failure to check the microscope and carry out this simple procedure as necessary will very likely lead to eye strain, particularly during prolonged periods of use.

Understanding Light Microscopy, First Edition. Jeremy Sanderson.

Comfort at the microscope is particularly important if you plan to spend a session of any length using it (pathologists or cyto-screeners, for example). Microscopists are thus chiefly at risk from (a) eye fatigue and (b) torso and limb strain and musculoskeletal disorders (MSD). Repetitive strain injury (RSI) is most commonly related to the strain upon, or overuse and inflammation of, muscles, tendons and soft tissues of the upper part of the body, such as the wrists, forearms, shoulders, elbows, back or neck.

In the last few years, all the large microscope manufacturers have incorporated ergonomic principles into the design of their compound microscopes and stereomicroscopes to reduce both eye strain, MSD and RSI. Each manufacturer offers an extensive range of armrests, adjustable binocular tubes and other modular components to minimise or reduce fatigue whilst working at the microscope.

Employers have a duty of care to protect the health, safety and welfare of their employees as well as others affected by their business. Conversely, employees have a duty to take care of their own health and safety and that of others who may be affected by their actions at work. Workers must cooperate with employers and co-workers to help everyone meet their statutory requirements.

5.2 Ergonomics

Ergonomics is a science concerned with the 'fit' between people and their work. The online Free Dictionary defines ergonomics as: 'the applied science of equipment design, intended to maximise productivity by reducing operator fatigue and discomfort'. The aim of ergonomics, or human engineering, is to reduce the potential for accidents, injury and long-term ill-health, whilst also improving performance and productivity.

Box 5.1 Ergonomics

The illustration, dated 1890, shows Frank Crisp Honorary Secretary and also Treasurer of the RMS using a Victorian inclined microscope. This standard design of the day required the user to lean over to use it. An upright head weighs approximately 5–8 kg. Tilting the head forward by 15° increases the load to approximately 16 kg! Leaning forward like this puts even more stress onto the neck, shoulders and lower back. However, the ill effects of this posture had been noted as early as 1835. Sir David Brewster wrote:

'The worst of all positions is that in which we look downwards vertically'

(Rules for microscopic observations, No. 3, page 288, part IV, *Treatise on Optics* 1835)

Source: Science & Society Picture Library / Contributor. Reproduced with permission from Getty Images.

Headaches	60–80%
Eyestrain	20–50%
Neck	50–60%
Shoulders	65–70%
Lower arms	65–70%
Wrists	40–60%
Hands & fingers	40–50%
Back (total)	70–80%
Lower back	65–70%
Legs and feet	20–35%

A series of studies show that the neck, shoulders, arms and lower back are most prone to musculoskeletal disorders and repetitive strain injury, whilst using a microscope for extended periods. Use a chair that provides good back support and follow the recommendation given in the text.

Headaches and eyestrain are also significant indicators of bodily stress. It is important to adjust the eyepieces on your microscope properly.

Take breaks, at something like every 15–20 minutes. Close your eyes briefly (do not rub them); then focus on something in the distance. This helps relax your eyes. Every 30–60 minutes, get up to stretch and move. Stretching exercises are given in the text.

Surveys have shown that strain is most likely to occur in the neck, shoulders, wrists and back region when using the microscope. This is because the ideal distance between the stage and eyepieces should be between 45 and 55 cm with the eyepieces inclined at a maximum of 30° to the horizontal, whereas, in fact, most microscopes have much shorter distances of about 25–30 cm and eyepieces inclined at 60°. This forces the microscopist to bend both neck and back in order to keep the arms and wrists comfortable (Figure 5.1). Leaning forward can be very hard on your back, neck and shoulders, increasing the pressure on your spine by a factor of four or more. Your spinal column is shaped more or less like a very shallow 'S' curve. It is important to maintain the natural S-curve of the back, whether sitting or standing. The most important part of this curve is in the lower back. Good lumbar support from a well-designed chair is essential, to maintain the proper curve in the small of your back.

The neck bones are part of the spinal column and thus are subject to the same requirements of maintaining the natural S-curve of the back. Prolonged twisting and bending of the neck or holding the head forward can be as stressful for the neck as for the lower back. If, alternatively, the microscope is raised to maintain a straight neck and back, this leads to stress in the wrists and arms. In order to operate the focusing knobs and other controls on the stand, the wrists must be bent and arms raised off the bench. As a consequence, it is not uncommon for microscopes to be found perched in an unsafe manner upon a variety of books, telephone directories or piles of

Figure 5.1 **Ergonomic microscope design – 1** *This older design of microscope, whilst perfectly good, was manufactured in the 1960s when the importance of ergonomics was much less understood or recognised. The user is forced to hunch over the instrument, leading to discomfort during extended use. Source: Courtesy of Mr K. Glover.*

paperwork in order to create the least damaging 'ergonomic' solution for use. The neutral posture (Figure 5.2) for your arms is to keep your elbows at your sides and your shoulders relaxed but not rounded. Work should be done at about elbow height, with the hand in the same plane as the forearm. Some modern stands (Figure 5.3) allow the neck and back to be kept straight whilst also allowing a stress-free neutral posture to be adopted for the arms.

The risk factors associated with musculoskeletal disorders are forceful exertion, repetition and awkward posture. Microscopists are particularly prone to muscle tension from sitting with awkward postures. Sitting for hours on end, with both head and arms held forward, bent over a microscope, having rounded shoulders and curved spine is not an activity for which the body is well adapted. If the feet

Figure 5.2 **Ergonomic microscope design – 2** *This microscope has been designed to allow the microscopist to use it with a straight back and the shoulders, arms and wrists in a neutral posture. Source: Courtesy of Mr K. Glover.*

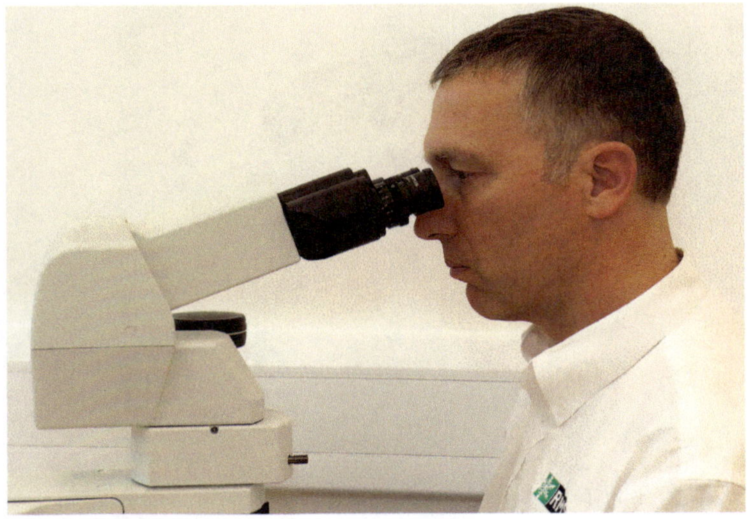

Figure 5.3 **Ergonomic microscope design – 3** *Another modern microscope design allowing extended use whilst maintaining the natural S-curve of the back and without requiring the neck to be bent forward – a stress-free neutral posture. Source: Courtesy of Mr K. Glover.*

are placed on footrests, the position is further exaggerated. Working for long periods with your back in a C-curve can place strain on your back. The design of older microscopes often forced microscopists to lean over them standing in an inverted 'V' posture (see Box 5.1), which is even more stressful to the spine than adopting a curved posture.

Pain that is likely to arise from MSDs can normally be identified as such if it tends to go away when changing tasks, is alleviated when moving from one position to another or if the pain and discomfort is reduced at weekends when not at work. If you use the microscope for more than an hour or so, take regular breaks – five minutes per hour – and exercise to loosen the relevant muscle groups. Remember, the human body is designed to move. Therefore, gently move and rotate or stretch joints and limbs. Also exercise the eyes by taking time to move from the workstation and focus on distant objects. Keep an upright back, and use an ergonomic chair to facilitate good posture. Here are six simple exercises (Figure 5.4) that can be undertaken to counter tension and fatigue:

1. Back extension
2. Back rotation. Both back extension and rotation, to improve flexibility and to counter loads on the spine. Use the back of the chair as a pivot, and as a surface to massage the back. Using your hand as an anchor, twist round to look behind. Repeat the rotation on each side.
3. Neck tuck, to stretch the neck.
4. Shoulder rolls, shrug your shoulders, and then roll them, trying to make the blades meet. Relax and repeat.
5. Back stretch. Start with the feet comfortably apart, swing the arms, and then arch your back. Repeat the movement rhythmically.
6. Side stretch. Start with the feet apart, side stretch with your arm over your head. Repeat on the other side.

Not everyone is in a position to buy a new ergonomically-designed microscope or workstation. Until relatively recently, microscopes were offered on the basis that 'one size fits all' and made no account for inevitable variability in the size and position adopted by users. Older microscopes were not designed with ergonomic operation as a priority. Therefore, the key to improved ergonomics is in finding a means to adapt the microscope to suit you, the user, rather than forcing you to adopt an awkward or uncomfortable posture or position.

Try to develop an awareness of posture, and be comfortable. Maintain the natural curve of the back, and use a lumbar support if needed; do not lean too far forward. Adopt a sitting position so that your feet are firmly on the floor, the pressure of the chair seat transmitted through the back of the thighs and your back is entirely supported by the chair back. If you do need to bend forward, do so from the hips, rather than curve your back for an unduly long time.

Ergonomically well-designed chairs should have at least five castors for stability; the height-adjustable seat should cushion the legs comfortably and firmly, with the thighs horizontal or angled very slightly downwards towards the knees. It is important that the seat back is also height-adjustable to support the lumbar region of your back and maintain the natural S-curve of the back. The back of the chair should be angled slightly about 5–20° from the vertical to allow comfortable lumbar support and to maintain a neutral back posture. Top-quality chairs have adjustment for seat depth to enable you to sit against the backrest whilst also fully supporting your legs. There should be even pressure on the underside of your thighs and 2–3 fingers gap between the front of your chair and the back of your knees when seated against the backrest.

Adjust the height of the chair or support the microscope securely on an extra base so that your neck and head are bent as little as possible. The neck should be bent forward as little as possible (with the head no more than 10–15° from the vertical) and with the eyes looking downward approximately 30–45°. Your eyes should not be straining to look downwards into the eyepieces. An ideal addition to a microscope that is not sufficiently high is to have a modern eye-level riser or adjustable binocular head assemblies (Figure 5.5). Your upper arms should be perpendicular to the bench and your forearms resting comfortably on the bench, with your wrists straight, able

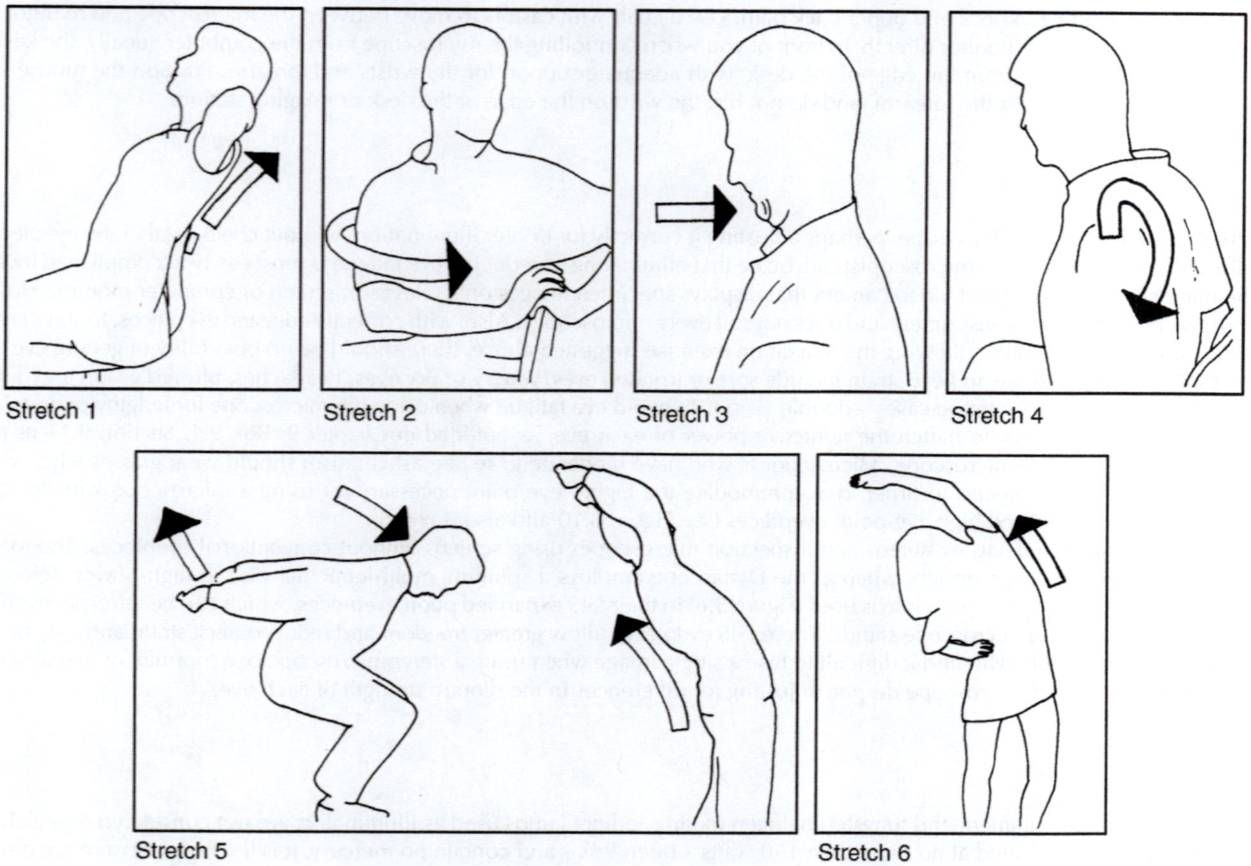

Figure 5.4 ***Exercises to avert posture pain and RSI injuries*** *See the text for further details. Source: Haines and McAtamney July 1993. Reproduced with permission from John Wiley & Sons.*

to reach focusing and operating controls on the microscope stand without unduly bending your wrists. Watch out for contact stress: leaning your forearms against the hard edge of a work table creates a pressure point.

Most advanced microscopes are now computer controlled, requiring the use of a monitor, mouse and keyboard. The monitor should be approximately an arm's length away (70–80 cm) from your eyes, which should be level with the top of the viewing area when sitting upright in a neutral posture. Adjust the text size or increase the size of the software window (or see an optician for glasses!) rather than be constantly leaning forward to use the image-acquisition software. The monitor should be adjustable for vertical height, swivel and tilt; some monitor designs may require a monitor riser to achieve the correct height. If the monitor is to one side, you may be constantly

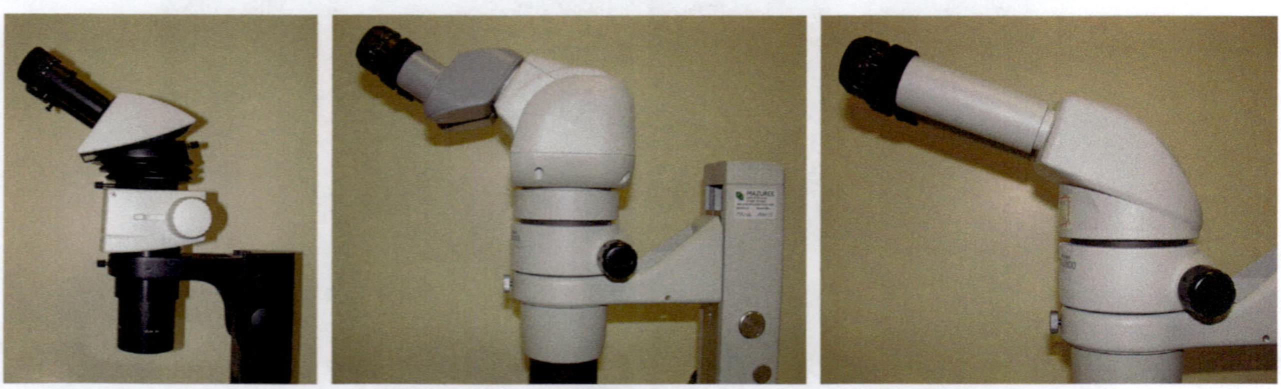

Figure 5.5 ***Adjustable binocular head assembly and eye-level risers*** *Two designs of adjustable binocular head fitted to stereomicroscopes are shown. In addition, an eye-level riser has been fitted to the Nikon stereomicroscope (centre panel) to allow the instrument to be used by individuals of different heights – an important consideration when equipment is shared and is in frequent use. The same stereomicroscope is shown in the right-hand panel with a fixed, non-adjustable binocular head and without the eye-riser spacer fitted. The modular design allows the riser to be quickly and easily exchanged, as required.*

turning your head, leading to neck and upper back pain. Use a chair with castors to move between the microscope and monitor when required and try to keep the monitor directly in front of you when controlling the microscope from the computer. Ideally, the keyboard should be placed 10–15 cm from the edge of the desk, with adequate support for the wrists and forearm. Position the mouse so that you have sufficient support for the forearm, and do not rest the wrist on the edge of the desk or imaging station.

5.3 Eye Fatigue

It is natural to begin to use a microscope, perhaps adjusting it correctly for Köhler illumination, without checking that the eyepieces are best adjusted for your use. Some microscopists advocate that eliminating or reducing eye fatigue is most easily accomplished by equipping the microscope with a digital video camera that displays specimen images on a television screen or computer monitor. However, this is not practicable for every instrument and does not suit every microscopist. Also, with correctly adjusted eyepieces, to suit the refractive power of your eyes, and by following the relaxation exercise suggested above, there should be no possibility of getting eye strain.

The symptoms of eye fatigue and eye strain include sore or irritated eyes, watery or dry eyes, headaches, blurred vision and difficulty focusing, dizziness and – in extreme cases – double vision. To avoid eye fatigue when using the microscope for lengthy periods, adjust the eyepiece dioptre controls, to match the refractive power of each eye, as outlined in Chapter 9, Box 9.5, Section 9.11 as part of setting-up and adjusting the microscope. Microscopists who have moderate to severe astigmatism should wear glasses when viewing specimens through the eyepieces. In order to accommodate the higher eye point necessary for using a microscope with spectacles, manufacturers offer specialised high eye-point eyepieces (see Figure 8.10 and also Box 9.5).

Vision Engineering manufactures stereo- and inspection microscopes using screens without conventional eyepieces. The Mantis is a low-power stereomicroscope design, whereas the Dynascope employs a spinning multi-lenticular disc in high-power stereomicroscopes. The same lenticular disc principle is used (Figure 5.6) in their ISIS expanded pupil eyepieces, which can be retrospectively fitted to conventional compound microscope stands. These ISIS eyepieces allow greater freedom and reduced neck strain and may be useful to the small group of people who find it difficult to fuse a single image when using a stereomicroscope or a 'normal' or classical design binocular head compound microscope despite adjusting for differences in the dioptre strength of each eye.

5.4 Safety and Lamps

The low voltage tungsten filament and tungsten-halogen (quartz-iodine) lamps used as illuminators are not considered especially hazardous. They are generally rated at no more than 150 watts – often less – and contain no mercury. It is the high-pressure arc discharge lamps and lasers used for fluorescence microscopy that pose a greater hazard in the workplace.

Tungsten-halogen lamps emit visible and infrared (IR) light (see Figure 9.2). Some workers consider that they can cause conjunctivitis if operated at full voltage, with too great a light flux over a lengthy period. One possibility is to fit a heat-blocking filter (e.g. Schott KG5) into the light path to prevent the IR component of the light reaching the eye. Such filters are available as 32 mm diameter discs that can either be fitted into a filter holder at the base of the microscope, into a filter tray underneath the condenser, or merely placed over the illuminated field diaphragm (see Chapter 4, Section 4.4C and Chapter 9, Section 9.10) at the base of the microscope.

Take care that the control rheostat is turned down before switching the lamp on. A sudden burst of current through the lamp will shorten its lifetime. Use a lamp of the correctly rated wattage for the lamp housing and the transformer. When replacing lamps, isolate

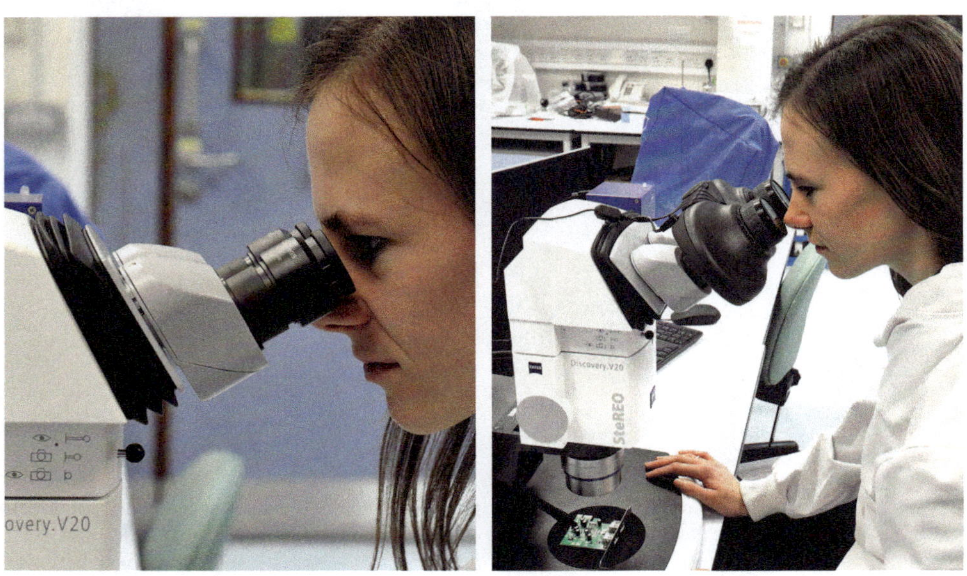

Figure 5.6 ***ISIS expanded pupil eyepieces (right-hand panel) compared to conventional eyepieces (left-hand panel)*** *Source: Courtesy of Mr K. Glover.*

the power supply – this is common sense. Otherwise, incandescent tungsten-halogen lamps are relatively safe, and even though they operate at high temperatures, they pose little hazard risk.

When changing lamps, do not handle the glass envelope of the bulb with your fingers. Natural grease will create a hot spot on the lamp surface and may weaken it. Most bulbs have pin electrodes that fit into a pre-centred holder; otherwise, follow the manufacturer's instructions that came with your microscope for guidance on replacing and centring the lamp. If you no longer have these instructions, either consult the company's local representative or refer to the generic instructions on changing lamps in Chapter 9, Section 9.7.

5.4.1 Fluorescence and Short-Arc Lamps

Replacing a pre-centred metal halide lamp for fluorescence is straightforward and very similar to changing a pre-centred tungsten-halogen lamp for bright-field illumination. Changing a short-arc fluorescence lamp is a little more involved and requires alignment after it has been replaced. Because both the metal-halide and short-arc discharge lamps and the lamp housing get hot, you should wait at least 30 minutes before attempting to change the lamp, in order to let everything cool sufficiently before opening the lamp unit. Most modern microscope lamp units have interlocks, but any discharge lamp should *never* be switched on whilst the cover is removed from the lamp housing! Severe damage to the eyes will result from looking at the fluorescent light source, and this must be guarded against, also, when aligning a new lamp on-axis after a lamp refit.

Most fluorescence lamp starter/transformer units have a bulb lifetime counter incorporated, which can be reset when the lamp is changed. Mercury short-arc lamps last about 300 hours with a significant fall-off in radiant flux output after 200 hours. If a timer is not present, get the users of the microscope to record the length of their operating sessions in a notebook kept with the microscope. This also means that people can see whether others will follow them once their study session is finished. The reason for this is that it is better to leave a discharge lamp on if there are only a few void hours between users during the day. This prolongs the lifetime of the lamp, which is otherwise shortened by repeated switching on and off. The notebook should also help ensure that the lamp is not inadvertently left on overnight. Old tungsten-halogen bulbs and short-arc discharge lamps (which contain a small amount of mercury) should be disposed of in accordance with local health and safety rules for your institution.

5.5 Dismantling Microscopes

Most components on modern modular microscopes can be changed and/or dismantled by the user. The standard tool for this purpose is a 3 mm hexagonal Allen key, and Olympus provide a slot in their stands to keep this with the microscope. Being able to dismantle the microscope provides great flexibility and future-proofs the possibility of upgrading. However, it also exposes an unwary user to safety risks, for example removing filter cubes or blanking plates in an epi-fluorescence unit. Zeiss have a policy of using a 3 mm Allen key for all those items that can be exchanged by the user (e.g. binocular head, cameras, complete epi-fluorescence units) but reserve other sizes (e.g. 2.5 mm Allen key) for releasing and changing filter sets in a rotating carousel, which it is not expected that a user would change without proper instruction and training.

Specific legislation relates to manually handling heavy objects at work and/or carrying out repetitive tasks. This may involve moving or transporting heavy microscope stands. In such cases, evaluate whether the lifting operation can be avoided or else performed using a mechanical aid, for example a trolley and/or a lifting hoist. Investigate whether the load can be split or subdivided. If not, then keep your back straight, bend your knees, keep the load close to the body and certainly ensure any lifting task is well within your physical capability.

5.6 Health and Safety Risk Assessments

Most countries have legal requirements pertaining to health and safety. In the United Kingdom, the Health and Safety Executive (HSE) is the government body charged with implementing health and safety legislation in the workplace. The Health and Safety at Work Act dates from 1974 and is the key legislation in the UK. Information about this and subsequent regulations is given in the Internet Resources at the end of the chapter. In the United States, the Occupational Safety & Health Administration (OSHA) is the federal agency charged with ensuring safe and healthy conditions at work. Online information about the agency is likewise given in the Internet Resources at the end of the chapter.

The HSE provides a guide to risk assessment (given at the end of the chapter). It suggests a five-step approach to evaluating and managing risks: identify the hazards, decide who is at risk, evaluate the risk and decide upon precautions, record and implement these precautions, review the risk assessment periodically. Figure 5.7 shows how the risk factors of a particular hazard may be evaluated for a risk assessment, and an example of these risk factors applied to specific microscopy applications is given. The Control of Substances Hazardous to Health (COSHH) regulations of 2002 is a statutory requirement which must also be taken into account when writing risk assessments. The equivalent in the USA is the OSHA's Hazard Communication Standard (HCS).

Apart from musculoskeletal and eye fatigue factors, the most likely risk to microscopists is from solvents used to clean immersion oil from objectives and slides or dishes. COSHH factors are associated with immersion oils (low risk) and cleaning fluids (higher risk). The vendor has a statutory duty to provide a material safety data sheet (MSDS) for these substances. This data sheet is not a risk assessment per se but should be used to help write your own risk assessment. The OSHA has very similar requirements for risk assessment and hazard management; web links are given in the Internet Resources at the end of the chapter. All users of equipment requiring a risk

Table 1 Task Risk Assessment	
Frequency Factor (FF): How likely is it that the identified hazard will occur while carrying out the task? 1 = Improbable - less than once per year 2 = Possible - once per year 3 = Occasional - twice or more per year 4 = Frequent - more than once every three months 5 = Regular - more than once per month	Severity Factor (SF): What are the potential consequences if the identified hazard does occur? 0 = No identified effect 1 = Trivial effect – first aid may be required, but no lost time (away from work) 2 = Minor effect – lost time – medical attention and rest of the day off work. 3 = Moderate effect – up to three days off of work. 4 = Major injury – more than three days off work Or HSE reportable. 5 = Death of one or more persons

Table 2 Risk Rating							
Frequency		Severity					
		None 0	Trivial 1	Minor 2	Moderate 3	Major 4	Death 5
Improbable	1	0	1	2	3	4	5
Possible	2	0	2	4	6	8	10
Occasional	3	0	3	6	9	12	15
Frequent	4	0	4	8	12	16	20
Regular	5	0	5	10	15	20	25

Table 3 Summary of the Risk Rating Classification			
Risk Rating	Classification	Action Required	Timescale For Control Application
0 - 3	LOW	Obvious and cheap solutions only	6 months
4 - 8	MEDIUM	Further consideration is required	3 months
9 - above	HIGH	Action must be taken	1 month

Figure 5.7 **Risk assessment evaluation** *Typical procedure for evaluating equipment and protocol risk for writing a risk assessment as used at the author's place of work. Source: Courtesy of MGU Harwell.*

assessment have to read and understand the assessment and sign to say they have done so. The HSE and OSHA require workplaces to keep these records and review risk assessments regularly, according to the level of risk.

5.7 Chapter Summary

This chapter explains how to sit at and work with a microscope safely. Modern microscopes are ergonomically designed, so it is easier to specify modular components to reduce the risk of musculoskeletal disorders when working with a new microscope. Advice is given on the potential risk of lamps and how to operate these safely.

1. It is important to sit at and work comfortably with the microscope.
2. Microscopists are prone to musculoskeletal disorders through adopting awkward posture.
3. Simple exercises are given to help relax during long study sessions with the microscope.
4. Advice is given on reducing eye fatigue and working safely at a microscope and imaging workstation.
5. Advice is given on operating incandescent lamps safely.
6. Advice is given on dismantling microscopes and handling heavy components safely.
7. A brief overview of the statutory requirements for health and safety at work in the UK and USA is given.

Reference

Haines, H. and McAtamney, L. (1993). 'Applying Ergonomics to Improve Microscopy Work'. *Microscopy & Analysis* July 1993, *1*: 15–17. Wiley.

Further Reading

Fritzsche, F.R. et al. (2012). 'Occupational health risks of pathologists – results from a nationwide online questionnaire in Switzerland'. *BMC Public Health* 2012, *12*: 1054.

George, E. (2010). 'Occupational hazard for pathologists: microscope use and musculoskeletal disorders'. *Am. Jour. Clinical Pathology 133*/4: 543–548.

Helander, M., Grossmith, E.J., and Prabhu, P. (1991). 'Planning and implementation of microscope work'. *Applied Ergonomics 22*/1: 36–42.

Sillanpää, J., Nyberg, M., and Laippala, P. (2003). 'A new table for work with a microscope, a solution to ergonomic problems'. *Applied Ergonomics 34*/6: 36–42.

Sundaragiri, K.S. et al. (2014). 'Ergonomics in an oral pathology laboratory: Back to basics in microscopy'. *Jour. Ora. Maxillofac. Pathol. 18*(Suppl 1): S103–S110. doi: 10.4103/0973-029X.141341.

Thompson, S.K., Mason, E., and Dukes, S. (2003). 'Ergonomics and cytotechnologists: reported musculoskeletal discomfort'. *Diagnostic Cytopathol. 29*/6: 364–367.

Internet Resources

Background introduction to ergonomics. URL: http://www.working-well.org/articles/archive.html#lab.

RSI management. URLs: https://web.eecs.umich.edu/~cscott/rsi.html and http://www.rsiaction.org.uk/rsi-conditions-and-prevention/what-people-with-rsi-need-from-employers/.

All four major manufacturers have information on ergonomics on their own websites or the web pages that they sponsor. Leica produce a good ergonomics handbook, which is now available online:

Leica ergonomics handbook. URL: https://www.leica-microsystems.com/fileadmin/downloads/Leica%20MS5/Application%20Notes/Handbuch_ERGO_englischTeil1.pdf.

Nikon. URL: https://www.microscopyu.com/microscopy-basics/basic-microscope-ergonomics.

Olympus. URL: https://www.olympus-lifescience.com/en/microscope-resource/.

Zeiss. URL: http://zeiss-campus.magnet.fsu.edu/referencelibrary/basics/ergonomics.html.

Most countries have legal requirements pertaining to health and safety. In the UK, the Health & Safety Executive publishes the following information:

General introduction. URL: http://www.hse.gov.uk/pubns/hsc13.pdf.

Musculoskeletal disorders (MSDs). URL: http://www.hse.gov.uk/msd/index.htm.

Risk assessments. URL: http://www.hse.gov.uk/pubns/indg163.pdf.

Control of Substances Hazardous to Health (COSHH). URLs: http://www.hse.gov.uk/coshh/ and http://www.hse.gov.uk/pubns/indg136.pdf.

In the USA, the Occupational Safety & Health Administration regulates and enforces occupational health & safety legislation.

OSHA (HCS). URL: https://www.osha.gov/dsg/hazcom/ghd053107.html#introduction.

Like their UK counterparts, the advice given on microscopes comes under preventing musculoskeletal disorders. URLs: https://www.osha.gov/SLTC/ergonomics/index.html and https://www.osha.gov/Publications/laboratory/OSHAfactsheet-laboratory-safety-ergonomics.pdf.

In addition, there is also advice on the OSHA webpage for handling asbestos samples for microscopical analysis. URLs: https://www.osha.gov/pls/oshaweb/owadisp.show_document?p_table=STANDARDS&p_id=10005 and https://www.osha.gov/pls/oshaweb/owadisp.show_document?p_table=STANDARDS&p_id=9997.

6

Optical Aberrations of the Microscope

Outline

Understanding Light Microscopy, First Edition. Jeremy Sanderson.

6.1 Introduction

The previous chapters show how lenses work, how a microscope of modern modular design is constructed and how to use it in a relaxed and stress-free manner. A microscope is a series of lenses arranged on-axis to give an image appropriate for viewing by eye or with a camera. We shall see in the next chapter how the objective lens – the primary component of the microscope – works. Although lenses manufactured nowadays are consistently of extremely high quality, they nevertheless inherently possess aberrations. The term 'aberration' comes from the Latin to mean 'to go astray', which describes what the light rays do.

Quite apart from the normal effect of diffraction, these aberrations impair the ability of a lens (principally the objective lens) to form an image that is an exact representation of the object. Optical aberrations arise from the dioptric nature of how lenses work. They can be minimised, but not entirely eliminated. Very often a solution to correct one aberration makes others worse. Therefore, an important part of microscope design is manufacturing lenses and their elements so that all these aberrations are kept within acceptable limits. The manufacturer must prioritise the various aberration corrections according to the intended function of the particular optical component (i.e. condenser, objective, eyepieces). This requires the skill and experience of the optical designer.

This chapter discusses the nature and effects of these optical aberrations. Most aberrations are taken care of by the manufacturer when designing the objective and other optical components. As such, the effect of spherical and chromatic aberration is rarely seen in the microscope image, since almost all microscope objectives are corrected for these basic aberrations to some extent. As discussed in Chapter 4, the components of modern microscopes of modular design are mechanically interchangeable. However, components from different manufacturers are generally *not* optically interchangeable. Providing that matched components from one manufacturer are used, most optical aberrations need not concern us. They are mentioned here for completeness.

Spherical aberration (Figure 6.1b) and chromatic aberration (Figure 6.1c) are the two most serious of the seven lens aberrations; the microscopist must be aware of these two, at least. Both spherical and chromatic aberration can be controlled and minimised (Figure 6.1d) by using combinations of glass lens elements with different dispersive powers. Three types[1] of objective are sold, depending upon the degree of correction: **achromats**, **semi-apochromats** (often colloquially referred to as 'Neofluar' or 'Fluar' objectives) and **apochromats**. Further information on these three classes of objective is given in Chapter 7, Section 7.2 and also Table 7.1. Spherical aberration is proportional to the fourth power of the diameter of the lens and inversely proportional to the third power of the focal length. It is pronounced in objectives of short focal length (i.e. high magnification) and of high numerical aperture. So-called 'high-dry' objectives are particularly susceptible to spherical aberration occurring in the image. There is one very simple way in which the effect of chromatic aberration can be annulled: a narrow band-pass filter (or monochromator) can be used to provide monochromatic illumination. All the other types of aberration still occur with monochromatic light.

6.2 Lens Aberrations

Aberrations are grouped (Table 6.1) as on-axis (paraxial) or off-axis types. The three on-axis aberrations are spherical aberration, astigmatism and (longitudinal) chromatic aberration. Off-axis aberrations – which have a greater potential effect in high numerical aperture objectives – are:

- coma,
- field curvature,
- astigmatism,
- distortion and
- lateral (transverse) chromatic aberration.

The monochromatic aberrations are sometimes referred to as Seidel aberrations, after Philipp von Seidel, who first wrote about them in 1857. The other collective term for them is 'third-order aberrations' from the type of mathematical equation that describes them.

Because mirrors do not exhibit aberration in the same way as refracting lenses do, catoptric (or reflecting) microscopes were constructed for a short period before about 1830. Quite apart from the inability to combine lenses to cancel out aberrations, in the

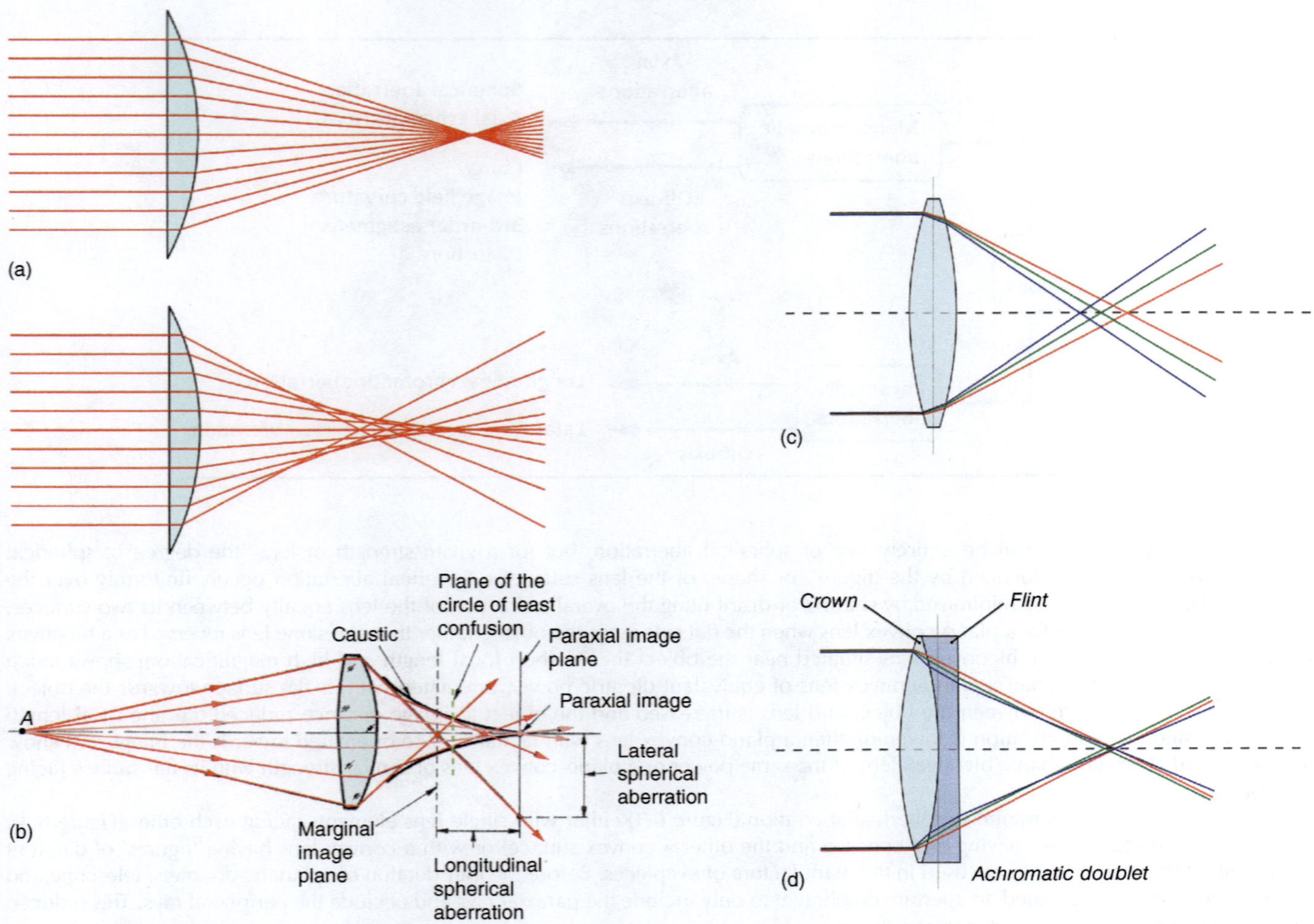

Figure 6.1 ***Spherical and chromatic aberration*** *These illustrations show (a) the ideal and spherical aberration (b) the ray diagram and caustic curve of spherical aberration in detail (c) how chromatic aberration arises from dispersion and (d) how lens elements of different dispersive power are combined to correct these aberrations. The illustration shows the correction of longitudinal chromatic aberration, but spherical aberration can also be corrected by using doublets of flint and crown glass (see Figures 6.4 and 6.5). The flint glass has a higher dispersive power (Table 2.4) than the crown glass, and since the crown glass element is convex and the flint glass concave, these individual dispersive 'errors' cancel out to give a corrected doublet.*

When the marginal rays come to a focus closer to the lens than the axial rays, as shown in (a), this is referred to as under-corrected spherical aberration. Conversely, when the marginal focus is located beyond the axial focus, the lens is said to suffer from over-corrected spherical aberration. The caustic curve is the envelope of the surface of revolution defined by spherically aberrated rays. The intersection of the caustic with the marginal rays defines the plane of the circle of least confusion, where the spherically aberrated image will be sharpest.

Source: (a) Adapted from Mglg, via Wikimedia Commons. (b) Pluta 1988. (c,d) Adapted from DrBob under the terms of the Creative Commons Attribution Share-Alike licence CC-BY-SA-3.0 (http://creativecommons.org/licenses/by-sa/3.0/), via Wikimedia Commons.

18th century, early optical glass was inhomogeneous and of such poor quality that microscopes employing mirrors gave superior images. With the advent of achromatic microscope objectives, these microscopes using mirrors fell out of use. Further information can be found about modern catoptric mirror objectives in Section 7.13.

6.3 Spherical Aberration

The most economical way to manufacture lenses is by grinding at least one of their surfaces to the shape, or 'figure', of a sphere. However, light rays are refracted differently through different parts of the lens because parts of the lens surface curve to differing extents. A ray of light close (i.e. paraxial) to the optical axis is refracted by a lens with a spherically-figured surface much less than a more divergent (peripheral) ray, which strikes the lens at a greater angle. Consequently, the peripheral ray is brought to a focus much nearer to the principal plane of the lens than is the paraxial ray. The result is best focus at a 'zone of confusion'. The image is not sharp, with blurring increasing towards the edge of the lens (Figure 6.2). The greater the aperture of the uncorrected lens, the more pronounced the spherical aberration will be.

Table 6.1

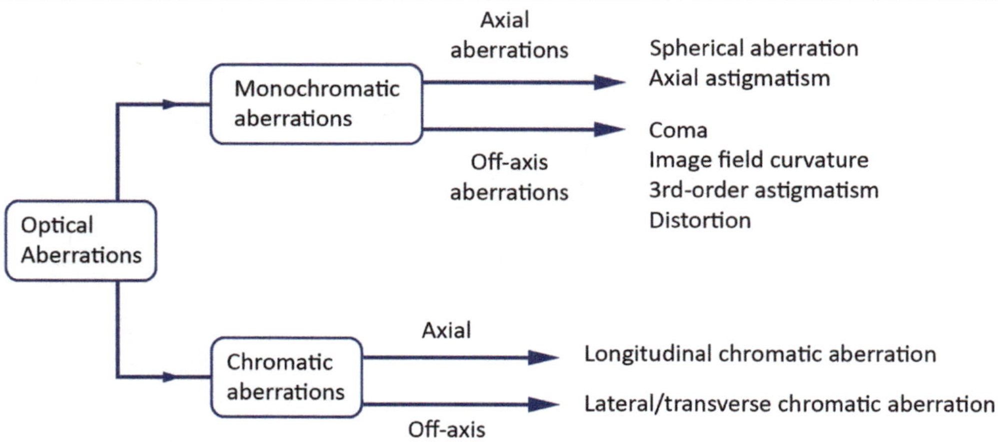

No single spherical lens can be entirely free of spherical aberration, but for a given strength of lens, the degree of spherical aberration is very much influenced by the 'figure', or shape, of the lens surface(s). Spherical aberration occurs uniformly over the whole field of view. It can be minimised by sharing or distributing the overall refraction of the lens equally between its two surfaces. Spherical aberration is less for a plano-convex lens when the flat side faces the object, rather than the same lens reversed or a biconvex lens. For example, a positive biconvex lens situated near the object (i.e. of short focal length and high magnification) shows much more spherical aberration than a plano-convex lens of equivalent dioptric power, orientated with its flat surface towards the object. Conversely, if the distance between the object and lens is increased and the object to image distance reduced (i.e. the focal length is increased and the magnification is lowered), then a plano-convex lens with its flat surface orientated towards the image will show less[2] spherical aberration than a biconvex lens of the same power or a plano-convex lens of similar strength with its flat surface facing the object.

It is therefore possible to minimise spherical aberration (Figure 6.3) either with single lens elements facing each other (Figure 6.4), or by using meniscus lenses (having one concave and the other a convex surface) or with a convex lens having 'figures' of different radii. Both of these approaches are used in the manufacture of eyepieces. Before the introduction of aplanatic doublets, telescope and microscope manufacturers used an aperture diaphragm to only include the paraxial rays and occlude the peripheral rays. This reduced the effective aperture and degraded image detail.

Fortunately, another solution is also available to lens designers. A positive convex lens can be combined with a negative concave element having an opposing degree of spherical aberration (Figure 6.5). Since this solution (Figure 6.1d) is also used to minimise chromatic aberration, doublets comprising positive and negative lens elements of different types of glass can be designed to minimise

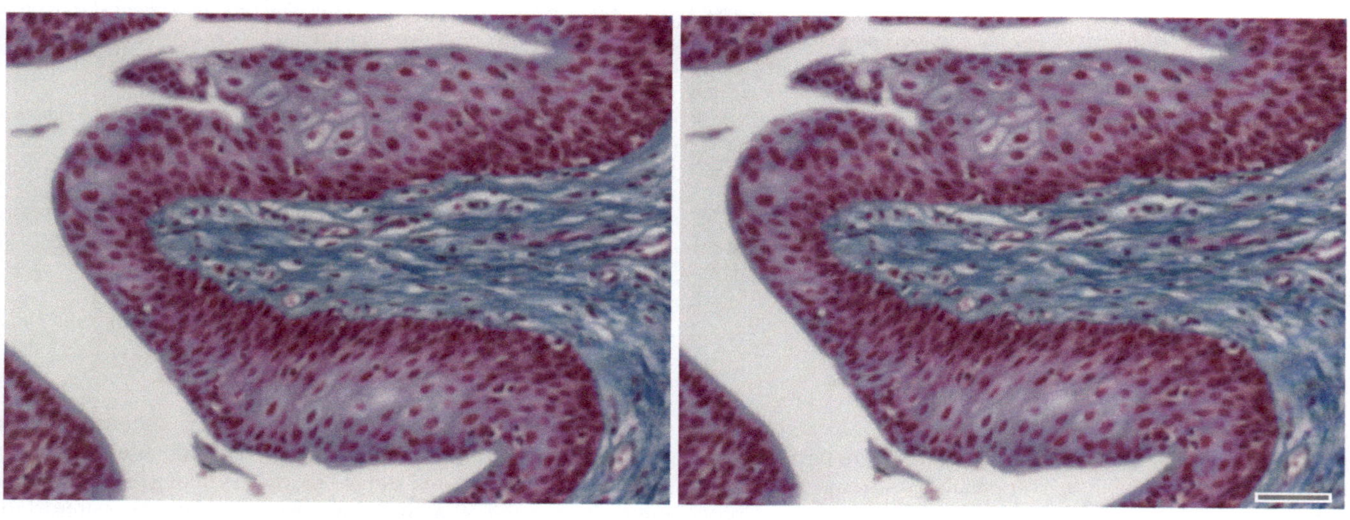

Figure 6.2 **Spherical aberration in the image** *An image of urethra with (left-hand panel) a high-quality image free of spherical aberration and on the right the same image suffering from spherical aberration. In this case, an extra coverslip was attached to the preparation deliberately to upset the calculated cardinal distances between object and image planes and artificially introduce spherical aberration. Scale bar = 50 μm.*

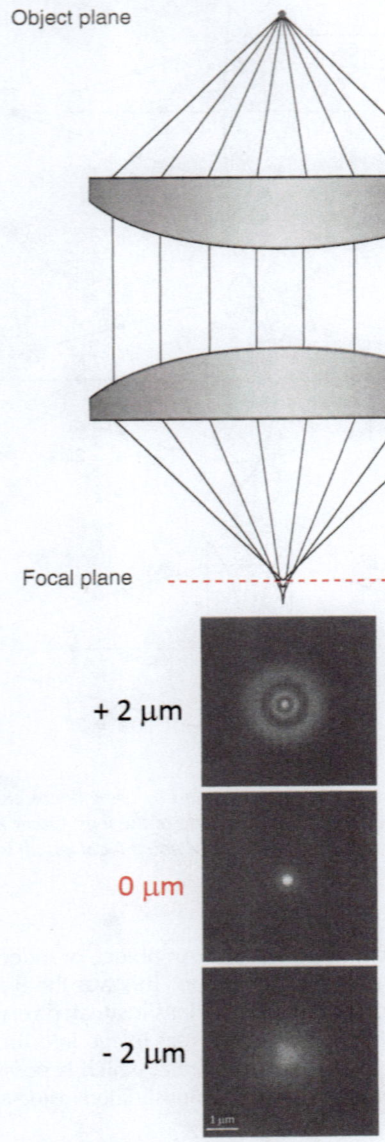

Figure 6.3 **The ray path of spherical aberration** *The ray path of under-corrected spherical aberration arising in an uncorrected objective, showing the image appearance of a point above and below the focal plane. Source: Goodwin 2007. Reproduced with permission from Elsevier.*

both aberrations at once. Nevertheless, spherical aberration is minimised by combining a strong positive convex crown glass lens with a weaker negative concave or plano-concave flint glass lens. Chromatic aberration is corrected by combining lens elements of differing dispersive power. Although the fundamental principles for correction in each case are different, is fortuitous that both aberrations can be corrected together. Chromatic and spherical aberration cannot be completely eliminated, but they can be reduced to within acceptable limits. All that need concern us here is to appreciate that combining lens elements in this way produces achromatic optical components that also have minimal spherical aberration.

If these two lens elements of an achromatic doublet are cemented together, reflection at the glass-air interfaces is considerably reduced. The doublet is also stable and durable. Nevertheless, this means that the optical designer now has only three radii available for adjustment to minimise spherical aberration. If, however, the two lens elements can be separated by a small air gap, this offers a fourth independent radius for design purposes. Furthermore, altering the distance of the lens elements with respect to one another also helps to minimise spherical aberration. This is the basis of the correction collar used in high-magnification objectives of high numerical aperture (i.e. high-dry objectives), which are prone to suffer from spherical aberration by reason of the design correction parameters being upset. The correction collar allows the microscopist to adjust the lens elements to once again minimise spherical aberration if a coverslip of incorrect thickness or a refractive index mismatch is introduced from using an incorrect or mismatched mountant or immersion medium.

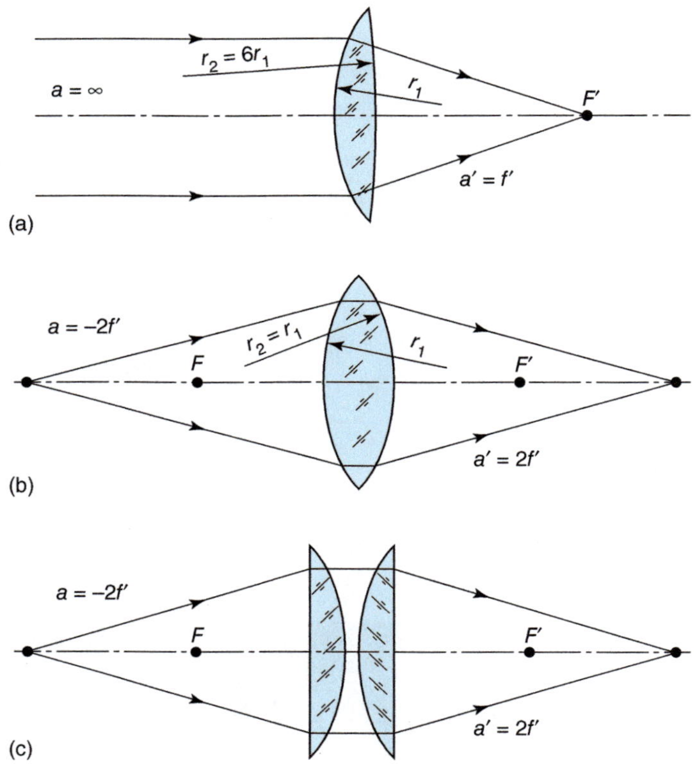

Figure 6.4 **Correcting spherical aberration** *Spherical aberration can be minimised by using a lens element with one figure six times the radius of the other or with object and image conjugate points placed equidistant at 2f either side of the lens. Other lens configurations and arrangements are shown in this figure. The Ramsden eyepiece has two plano-convex single lens elements of equal focal length with both convex surfaces facing each other, and a separation of 2f/3, where f is the common focal length. Source: Pluta 1988.*

An aplanatic combination of lenses has two pairs of aplanatic points. An object, or indeed an image, placed at one aplanatic point will be imaged at the other point more or less free of spherical aberration. This was the discovery made by Lister (Bracegirdle, 1987) whereby he constructed the first achromatic aplanatic objective. The first lens is situated very close to the object, within its focal length, so that it forms a virtual image, which the second lens re-images as a real image into the primary image plane of the microscope. A further improvement for correcting an objective (Box 6.1) is due to Amici which is possibly less well known than Lister's work on constructing an aplanatic achromatic objective. Light emerging from a point situated inside a sphere of glass will have a virtual image at

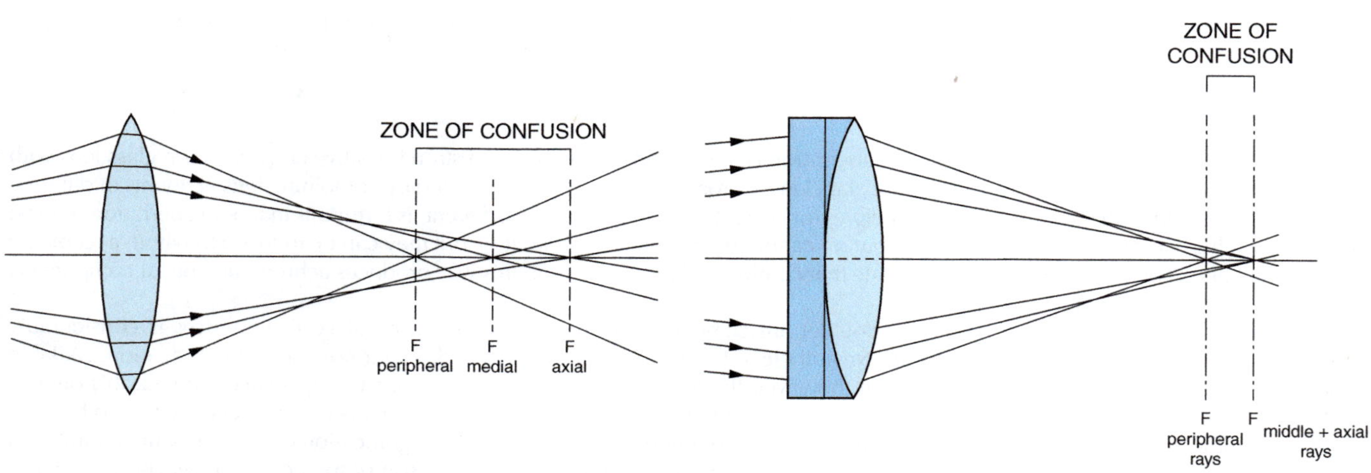

Figure 6.5 **Correcting spherical aberration** *The ray path of spherical aberration arising in an uncorrected lens element and how it may be corrected – and the zone of confusion significantly reduced – using a doublet combination of crown and flint glass. Source: Adapted from Bradbury and Bracegirdle 1998 with permission from Taylor & Francis Group.*

a second aplanatic point. If the sphere is truncated and replaced by a hyperhemispherical lens completed by a homogenous immersion medium, the front surface of which has the object at its centre, then an image virtually free from spherical aberration is formed, thus conforming with Abbe's sine condition; see also Box 6.2[3]. Amici's aplanatic hemispherical lens improved spherical aberration and consequently the light-gathering capacity of the objective.

Box 6.1 Correcting Spherical Aberration – Aplanatic Points

It is possible to minimise spherical aberration with a single lens figured to certan radii or with lens elements facing each other; this approach is used in the design of eyepieces and low power objectives and is illustrated in Figure 6.4. A different method of correcting spherical aberration is used for medium and high magnification objectives.

A spherical refracting lens (or a doublet) will give rise to a pair of conjugate aplanatic points where spherical aberration and coma are both zero. An object, or indeed an image, placed at one aplanatic point will be imaged at the other point free of spherical aberration and coma. This is the principle of aplanatic foci, and was the discovery made by Lister whereby he constructed the first achromatic aplanatic objective.

In the figure, below, for a sphere of refractive index n and radius r the aplanatic points lie, respectively, on circles in the plane of the paper of radius r/n and $r.n$. All rays emerging from A will, after refraction at the surface of the sphere, appear to originate from A′. If the object is arranged to lie on the sphere at r/n (e.g. at point A) a virtual image A′ formed at $r.n$ (at point A′) will be free of spherical aberration.

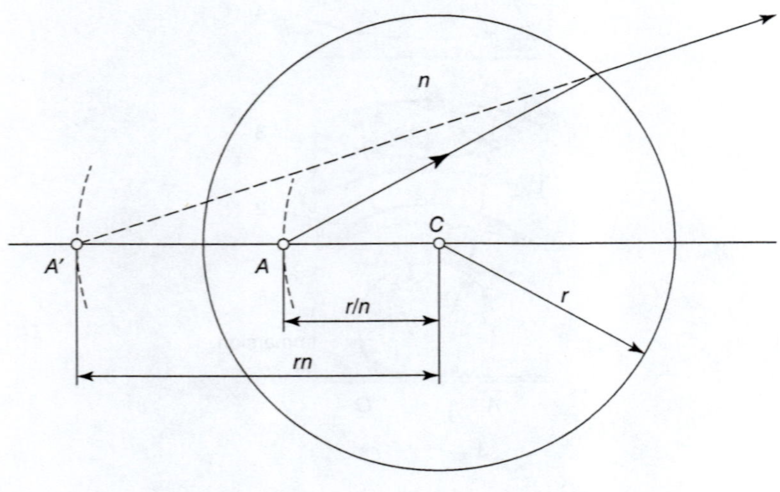

Source: Pluta 1988.

(Continued)

Since objectives are corrected for an object to lie at a calculated aplanatic point, and for the primary image to also lie in a specific plane, the spherical correction designed into the objective can very easily be upset inadvertently by the user:

- either by using a mountant of incorrect refractive index in making the preparation;
- by using the wrong thickness of coverslip;
- by placing the specimen such that it lies remotely away from the coverslip and not in contact with its surface (or – the same thing – imaging details lying too deep within a thick specimen);
- by using the wrong immersion fluid between the objective front lens and the preparation;
- by using an incorrect tubelength; or
- by damaging the objective lens.

It is relatively easy for the microscopist to select an appropriate thickness of coverslip. It is usually not a problem to arrange for the specimen to be situated at the under-surface of the coverslip (e.g. cells grown on coverslips are perfectly situated), for the correct thickness of mountant to be used and for the correct immersion medium to be selected. Section 6.10 describes two simple tests for evaluating the presence and extent of both spherical and chromatic aberration in the image.

Of all the optical aberrations, spherical aberration is potentially the most troublesome. Unless specimen preparations are made with due care (particularly with respect to coverslip thickness when making mounted and covered preparations), spherical aberration can cause image degradation. Although manufacturers design objectives such that spherical aberration is minimised, it is possible for these corrections to be upset by the microscopist and thus reintroduce spherical aberration into the image. The major causes of spherical aberration in this case being inappropriate coverslip thickness and refractive index mismatch. More is said about these topics in Chapter 7 on objectives, in Chapter 18 on operating the confocal microscope and also in Chapter 28 on biological specimen preparation.

Box 6.1 *(Continued)*

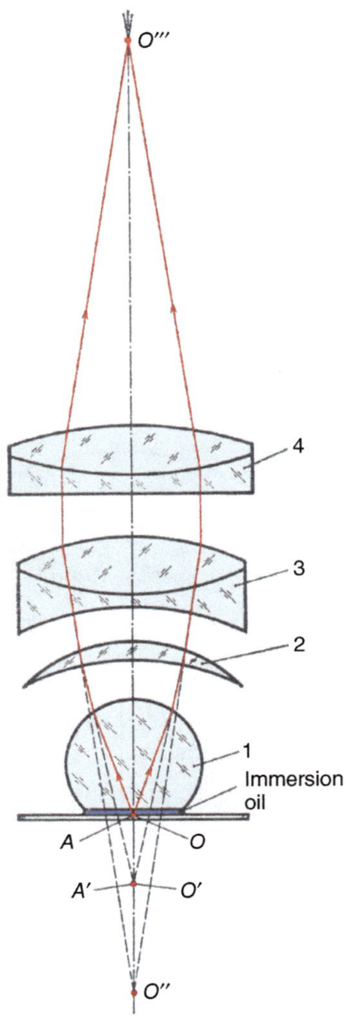

Source: Pluta 1988.

The front lens of an oil-immersion objective (1) can simply be a truncated sphere with its plane or concave front surface cut so that, with a homogeneous immersion fluid whose refractive index is equal to the refractive index of the glass, the object is located at aplanatic point A. Together with a second meniscus lens (2) whose first and second surfaces have centres of curvature located at A′ and A respectively, an image formed at O″ – also free of spherical aberration – can further form the primary image at O″′ via suitable doublets (3, 4) which receive light rays of small (i.e. nearly paraxial) divergence and so can easily be used to correct residual chromatic and other Siedel aberrations. For simplicity, a finite tubelength system is shown.

Notice that the distance with respect to one another of these two achromatic doublets is crucial. If these two lenses are separated by a small distance, the lens designer has four independent radii to work with and figure, whereas if they are glued together as a corrected doublet, there are only three separate radii to work with, and it is not possible to move the lens elements with respect to one another. Herein lies the reason why high-dry objectives of high magnification (Chapter 7, Sections 7.7 and 7.8) require correction collars to minimise spherical aberration: high power (high magnification) objectives are very prone to spherical aberration but, by altering the relative posiiton of these lens elements at will by using a correction collar operating through a helical mount, spherical aberration can be much reduced.

The microscopist should use higher magnification objectives with due regard for the correct coverslip thickness, correction collar setting and the correct immersion fluid, so that the conditions leading to aplanatism are not disturbed.

Reference

Pluta, M. (1988). *Advanced Light Microscopy*, vol. 1. Amsterdam, Oxford: Elsevier.

Box 6.2 Abbe's Sine Condition

The Abbe sine condition must be fulfilled by a lens in order for both paraxial and off-axis points to be imaged sharply. The condition applies to an optical system that is aplanatic, i.e., free from both spherical aberration and coma and where the lateral magnification is constant for all rays contributing to the image.

The formula for numerical aperture involves the sine of the half angle and not the tangent (as is the case when calculating the aperture and *f*-number of a camera lens) because image formation in the microscope, involving widely divergent rays of light (diffracted by the fine detail of the specimen) has to obey Abbe's sine condition, that is: the ratio of sines is constant for all angles (not just the extreme oblique angles). Therefore the ratio of the numerical aperture on the object side to the numerical aperture on the image side must equal the lateral magnification, namely:

$$\mathrm{Mag} = n.\sin U / n'.\sin U', \text{ or } n.h.\sin U = n'.h'.\sin U',$$

where n′ is the refractive index, h′ is the height of the image on the image side and U' is the half-angle of the marginal ray; thus n.sinU' is the NA on the image side.

Conversely, the *f*-No. is the ratio between the focal length and the diameter of the camera lens (*f*/D). The focal length is parallel to the optical axis, and so f/D = cotangent $\alpha/2$, rather than being a sine function (as is numerical aperture), which defines the marginal ray accepted by the angular aperture of the objective. The formula for Abbe's sine condition can be proved by trigonometry, and such a proof can be found in books on optics (e.g. Mansuripur, 2009, or Conrady, 1905).

Whereas the Abbe sine condition refers to points perpendicular to the optical axis, the corresponding Herschel condition refers to where image points lie along the optical axis, affecting longitudinal magnification. Since both the Abbe sine condition and the Herschel condition can only be satisfied simultaneously in the optically trivial case of image formation by a plane mirror, dioptric lenses cannot form an image of any voluminous object without aberration.

References

Conrady, A.E. (1905). 'The optical sine-condition'. *Monthly Notices of the Royal Astronomical Society 65*: 501–509.
Gross, H. (2007). *Handbook of Optical Systems*, vol. 3. Aberration Theory and Correction of Optical Systems. Weinheim: Wiley-VCH.
Mansuripur, M. (2009). 'Abbe's sine condition', Chapter 1. In: *Classical Optics and its Applications*, 2nd ed. Cambridge: CUP, pp. 9–22.

With an uncorrected condenser, spherical aberration is very evident with a small light source, because the lens elements are not corrected to the same degree as for objectives. Since Köhler illumination effectively uses a large light source, this reduces the effect of spherical aberration. If annular illumination is used (e.g. as for dark-field or phase contrast microscopy), the degree of spherical aberration present is further reduced.

6.4 Chromatic Aberration

Chromatic aberration arises because all materials from which lenses can be made have a different refractive index for each colour. In other words, chromatic aberration occurs because the refractive index of a transparent solid (or liquid) medium varies with wavelength[4]. This is why white light refracted by a prism is dispersed into a coloured spectrum. The refractive index is[5] generally higher for shorter (RI_{blue} = 1.522; 'blue bends best') than for longer (RI_{red} = 1.513) wavelengths. The effect of chromatic aberration is to focus different wavelengths of light at different focal points along the optical axis (Figure 6.1c; Figure 6.6). For this reason this particular form of aberration is also known as longitudinal, or axial, chromatic aberration to distinguish it from chromatic aberration occurring transversely to the optical axis, which is called lateral chromatic aberration. This second type of chromatic aberration is considered later (Section 6.9); for the time being, we shall discuss only longitudinal chromatic aberration.

As Figure 6.6 shows, an image suffering from longitudinal chromatic aberration can exhibit a colour tint throughout and not just at the edges of the object. It is well known that Newton discovered prismatic dispersion in 1666. Regrettably, on the basis of some rather unfortunate experiments, he concluded that dispersion is directly proportional to refraction, and it was thus impossible to correct chromatic aberration using lenses. We now know that this is not true. Dispersion varies with different types of glass, and it *is* possible to minimise chromatic aberration due to dispersion simply by combining lenses of different types of glass into doublets[6]. There are two principal types of glass: crown glass, with a lower refractive index and less dispersion (see Table 2.3 & Table 2.4), and flint glass, with a high refractive index and dispersion. By combining a positive convex crown glass lens with a negative concave flint glass lens (Figure 6.1d), each lens element cancels out the aberrations arising in the other.

Crown glass is the most common glass and accounts for about 95% of world production. It is a 'soda-lime' formulation, containing silicon dioxide (silica), sodium oxide (soda) and calcium oxide (lime). As it is spun to form window glass, it assumes the shape of a crown, from which it takes its name. Flint glass, developed later in the 17th century, once contained flint powder, but now owes its properties to lead dioxide additives to make the glass both heavy and clear. Barium glasses have refractive indexes comparable to the flints but have lower dispersion values. Other heavy elements are used to make flint glasses, such as lanthanum and the rare earths. Following a long search for suitable glasses, Ernst Abbe (Chapter 7, Section 7.2) eventually turned to using fluorite particularly for

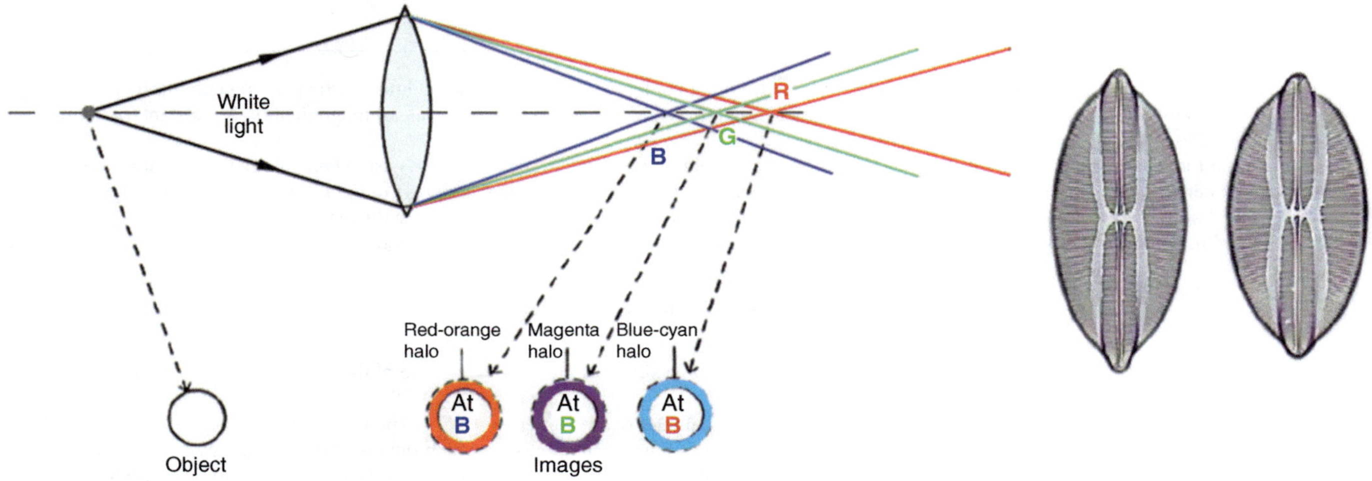

Figure 6.6 ***Chromatic aberration in the image*** *The colour tints possible with uncorrected lenses or poorly corrected objectives. Source: Adapted from White 1966, with permission from Elsevier.*

semi-apochromatic objectives, because it has particularly low dispersion across the spectrum. Unfortunately it is doubly-refractive and so is unsuitable for observations with polarised light. Fluorite has long since been replaced by a suitable choice of artificial glass[7].

Another practical solution for managing chromatic aberration is available to the microscopist. The chromatic aberration of both an uncorrected lens as well as a corrected achromatic objective can be eliminated by using a coloured filter to view or record the image with monochromatic light. This stratagem is often overlooked and is also useful even with well-corrected apochromat objectives.

6.5 Coma

This aberration manifests itself as streaking in the image; the name in fact comes from the Greek for hair. A practical example of this can be seen in Figure 2.20, where an uncorrected biconvex lens is being used to focus the sun's rays. Any tilt in the lens shows the sun's round image as a comet-shaped streak.

Coma arises from the wavefront being off-centre (Figure 6.7) and is thus an off-axis aberration. It is similar to spherical aberration, because the refraction of an off-axis image point is different for different parts of the curvature (or 'figure') of the lens surface. The presence of coma in an image generally indicates that lens elements within the objective have been – or have become – misaligned. The image streak is usually therefore blurred very much like a comet-shaped flare: lying with a wide, blurred tail away from the optical axis. Manufacturers take a great deal of trouble to design out this aberration by selecting optimum combinations of lens configurations. Coma is very rarely encountered in an objective unless, of course, it has been dropped at some point. An aplanatic lens (see Boxes 6.1 & 6.2) is one that is effectively free from both spherical aberration and coma. Good-quality condensers will be aplanats.

6.6 Astigmatism

Axial astigmatism occurs where rays that propagate in two perpendicular planes are refracted to differing extents and thus have different foci (Figure 6.8 and Figure 6.9). The image is smeared longitudinally at different focal points; the visual effect (Figure 1.13) being a blurring along one axis in the image. Axial astigmatism does not occur with a perfectly spherical lens. It is this form of astigmatism that can occur in the eye. It may also occur because lens elements in the objective are not in perfect alignment with respect to one another. This can happen if the objective is dropped, for example.

Astigmatism is often found associated with curvature of field. They are corrected together; a lens corrected thus is called an anastigmat. Astigmatism can also occur when forming an image from objects that are off-axis (Figure 6.10). This is called radial, oblique or third-order astigmatism and can be seen with an otherwise perfect lens. Light from an off-axis point is brought to two different foci. On one curved image surface, all the points appear as short lines radial from the optical axis, and in the plane orthogonal to the first, as tangential lines (see also Figure 1.13). Astigmatism mainly affects the periphery of the field of view and, like most aberrations, is more troublesome with high numerical aperture objectives.

Astigmatism is quite common: about 1 in 3 people have astigmatic eyes. It is important to determine whether you have astigmatism (see Box 9.5), or not. This dictates whether you must retain your spectacles for observing down the microscope or whether you may

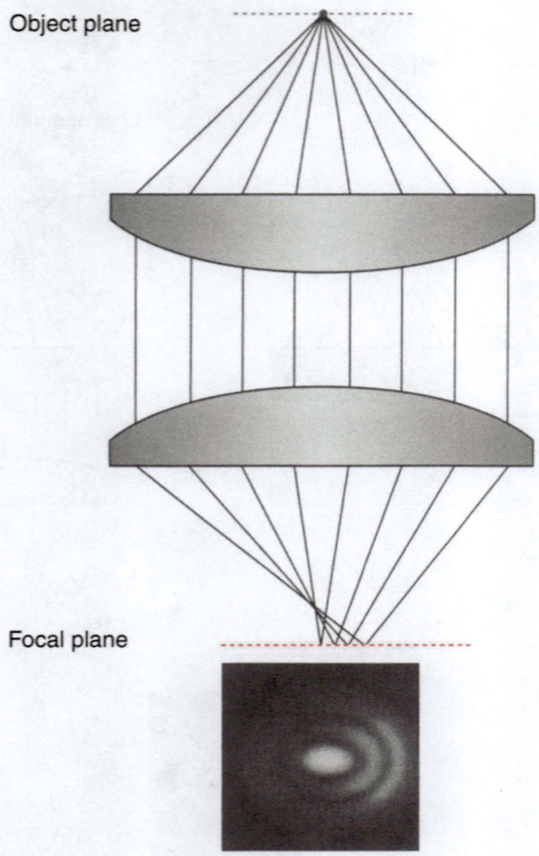

Figure 6.7 **The ray path of coma** *The ray path of an image point suffering from coma. Source: Goodwin 2007. Reproduced with permission from Elsevier.*

remove them, if desired. Astigmatism occurs in the eye because different meridians in the eye have different refractive power, either caused by irregularities in the cornea, the lens or both. This aberration requires toric lenses for correction. Astigmatism cannot simply be minimised by dioptre adjustment with a spherical lens, as is the case with myopia, hyperopia or presbyopia.

6.7 Distortion

Distortion is the result of differential magnification across the field of view. This can arise either because the magnifying power of the lens varies from the centre to the periphery of the field of view, or it arises when an image is formed off axis. This aberration often arises from the incorrect placing of a stop (a diaphragm of fixed aperture) in the optical train. In this case, those parts of the off-axis image so formed at the periphery will be magnified to a slightly greater extent than the part of the image situated near the optical axis.

The aberration is easily recognised when imaging an object with square edges. An electron microscope grid with a square mesh is ideal. Where positive distortion occurs, with an increase in magnification along the off-axis distance, images of straight lines near the periphery of the field of view will be bent inwards like a pin-cushion (Figure 6.11). Likewise, in the case of negative distortion, with a decrease in magnification along the off-axis distance, so-called barrel distortion occurs.

6.8 Image Field Curvature

The image of a thin flat object will, if uncorrected, not lie in a flat image plane but be curved with the ends of the image turned in towards the lens (Figure 6.12). This is because the off-axis marginal image-forming rays travel a slightly longer distance than the paraxial rays. Field curvature leads to a sharp image in the centre of the field of view but blurring at the periphery. It is not possible to secure an image that is both sharp in the paraxial region as well as at the margins of the field of view. As with other aberrations, field curvature

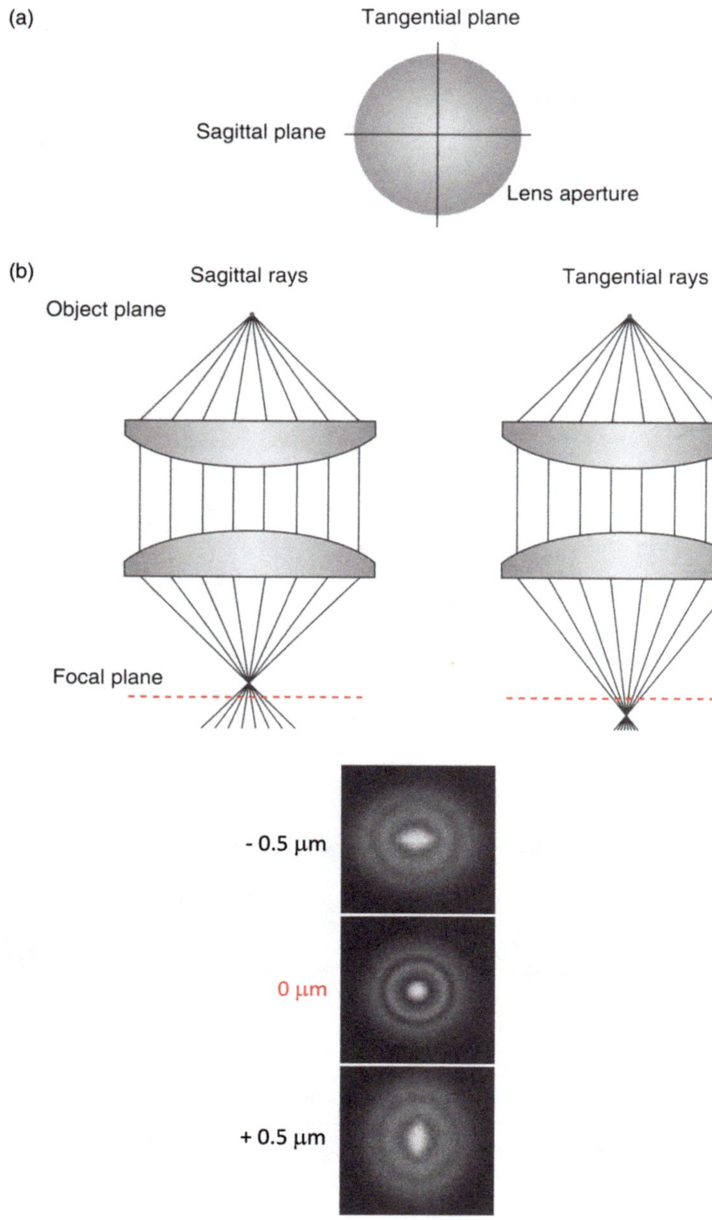

Figure 6.8 **The ray path of astigmatism** *The ray path and effect of imaging a point through an astigmatic lens. The overall effect is a smeared and elongated point spread function that rotates through 90° as the ray propagates from the point of focus of the sagittal plane to that of the tangential plane. Source: Goodwin, 2007. Reproduced with permission from Elsevier.*

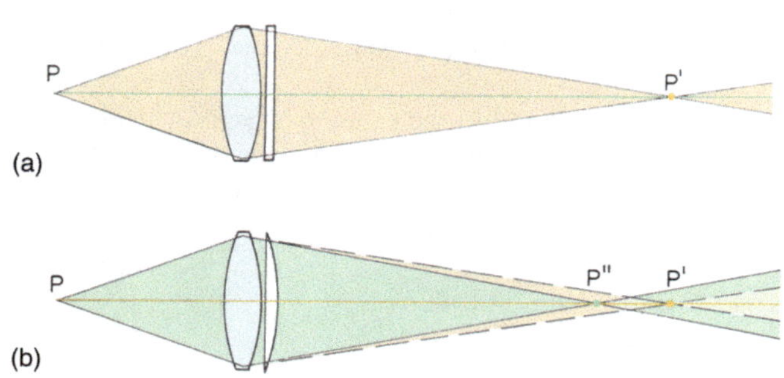

Figure 6.9 **On-axis astigmatism** *If a perfect lens has a plano-cylindrical lens placed after it such that the axis of the cylinder lies in the sagittal plane (coloured here orange; compare Figure 1.13) within the plane of the paper (i.e. looking down onto the lens from above), then no distortion will occur, and the point P′ remains unchanged for – in this respect – the plano-cylindrical lens behaves as a parallel-sided pane of glass. In (b), the view is now turned 90°; the tangential, or meridional, plane (coloured green) is now in the plane of the paper. The rays are refracted by the curved figure of the plano-cylindrical lens so that point P″ is imaged nearer to the lenses than point P′. Source: Adapted from Loveland 1970, with permission from John Wiley & Sons.*

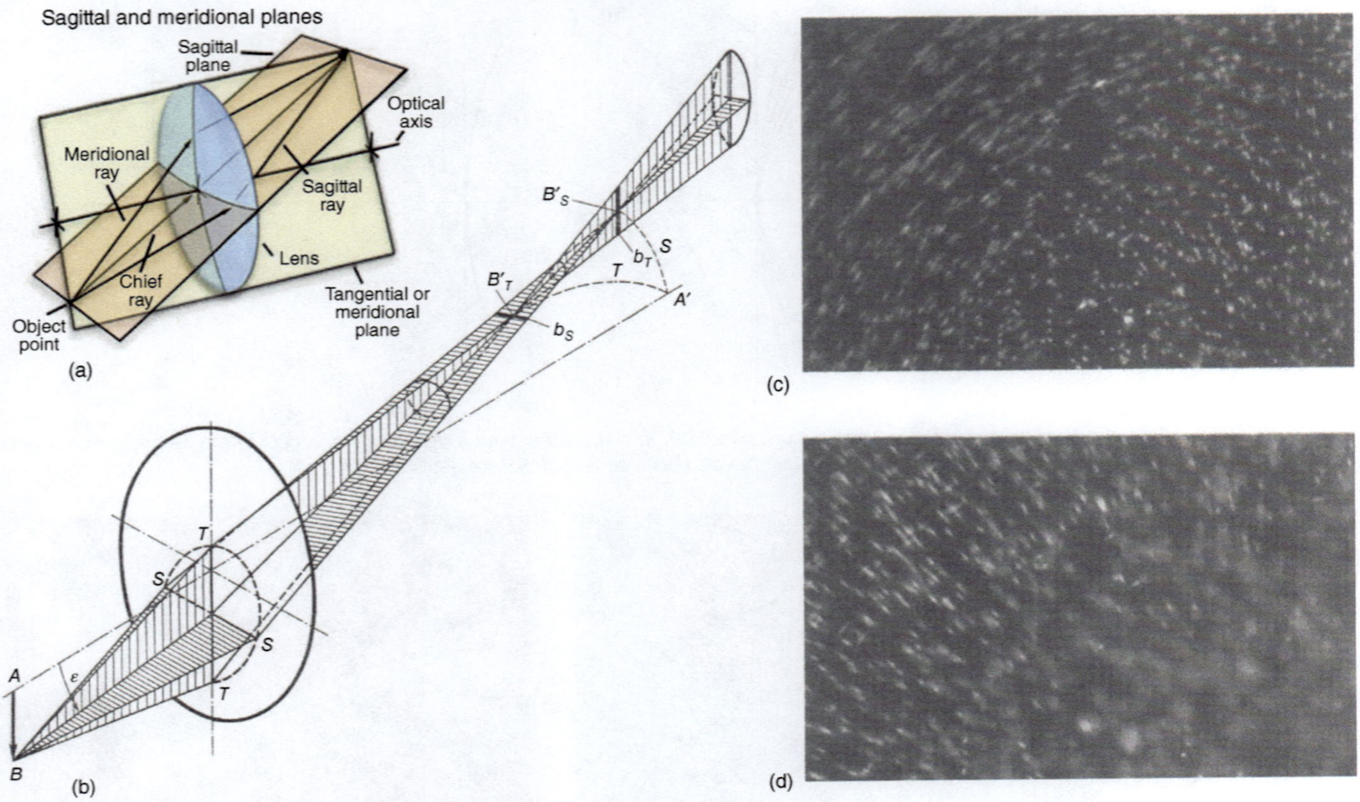

Figure 6.10 **Off-axis astigmatism** *In (a) the chief ray of an off-axis object point, passes through the centre of the lens. (b) The ray path of an off-axis point suffering from 3rd-order astigmatism. (c) An image of fluorescent beads showing the extreme edge of a field-of-view where the objective suffers from both off-axis astigmatism and field curvature. The centre of the field-of-view lies beyond the lower right-hand corner. At one focal point, the beads in the lower right of the picture nearer the optical axis are in focus whilst the peripheral beads appear as tangential streaks, or arcs. At the corresponding focus, whilst the central beads are still rounded, but out-of-focus, the peripheral beads now appear in the image as radial lines. Source: (a) Davidson, 1998–2015. (b) Pluta 1988. (c,d) Cox 2007. Reproduced with permission from Taylor & Francis.*

is particularly obvious with objectives of high initial magnifying power and large numerical aperture. This aberration is not so serious with visual observation, partly because of the natural accommodation of the eye, but also because it is easy to adjust the fine focus as required while scanning the field of view.

Because the plane of the film/CCD faceplate is flat, field curvature is particularly noticeable in photomicrography. The most obvious way to reduce field curvature is only to record the central part of the field of view. The other way is to use flat-field objectives corrected for image field curvature. This is done by distributing[8] the overall field curvature between the several individual lens elements that comprise the objective, or optical component. Old microscope objectives (Figure 6.13) will not be corrected for field curvature, for it was not until 1938 that Hans Boegehold, working for Carl Zeiss, solved this problem. Modern objectives are now routinely made as flat-field objectives, which usually have the prefix 'F' or 'Plan' in their name.

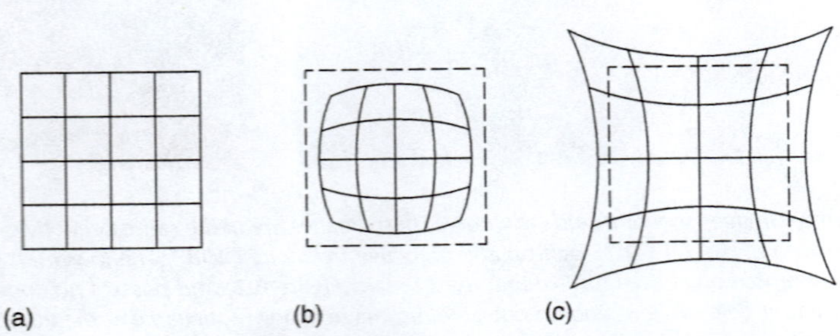

Figure 6.11 **Distortion** *Distortion: (a) shown the perfect non-distorted image; (b) an image suffering from negative distortion, giving rise to a barrel shape and (c) positive distortion which leads to so-called pincushion distortion. Source: Pluta 1988.*

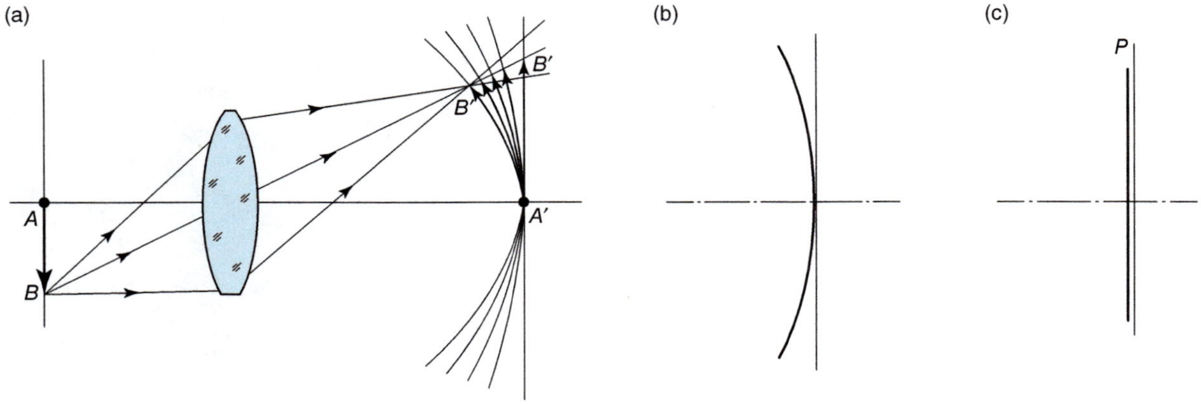

Figure 6.12 **Image field curvature** Field curvature often occurs, panel (a), in conjunction with astigmatism. In panel (b) the astigmatism has been corrected, leaving only field curvature. In panel (c) field curvature has also been corrected. Source: Pluta 1988.

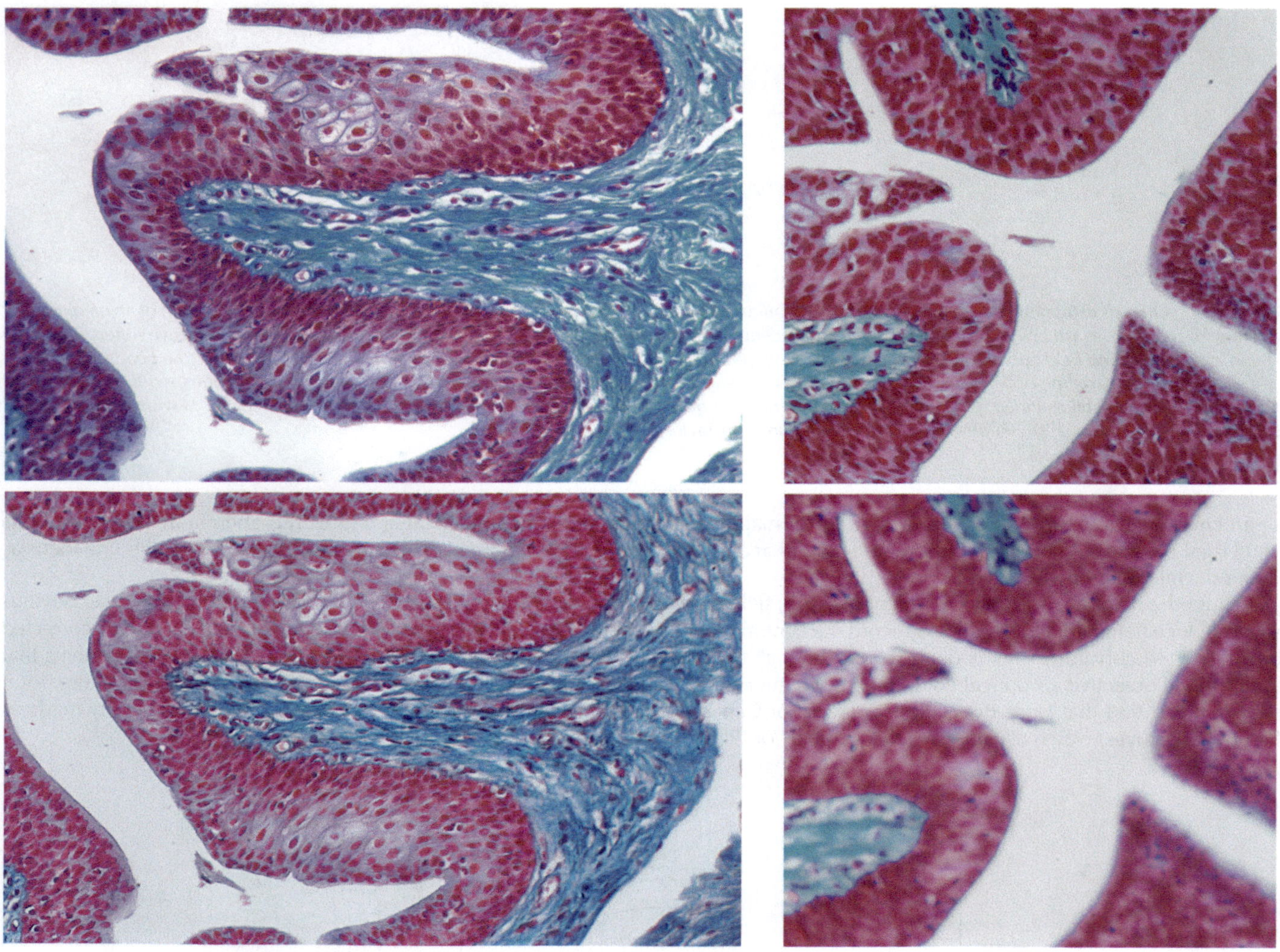

Figure 6.13 **Field curvature in the image** Images showing field curvature. These images are of the same field of view as for Figure 6.2 showing spherical aberration. The top-left panel is taken with a 10:1/NA 0.22 achromatic objective from circa 1980. The lower-left panel is taken with a 10:1/NA 0.30 achromat dating from 1909. Both are non-plan objectives (unlike that used to take Figure 6.2) and possess no correction for field curvature.

The modern objective has a slightly flatter field of view, and an out-of-focus image is only noticeable at the edge of the visual field. Since the camera records the central field of view, the effect of field curvature is less noticeable. The top-right panel shows the centre of the field of view taken with a 16:1 semi-apochromatic NA 0.40 Neofluar objective, again with no correction for field curvature, from circa 1970. The focus has to be adjusted by 26 μm to bring detail at the edge of the field of view into sharp focus. The lower-right panel shows the central field taken with this adjustment applied and now out of focus.

6.9 Lateral or Transverse Chromatic Aberration

The differing dispersive power of lenses means that the image is magnified by slightly different amounts according to the wavelength of the light (Figure 6.14). Thus if an image projected onto a screen transverse to the optical axis is viewed through coloured filters, it will show images of slightly different height depending upon which filter is used. This is chromatic difference of magnification. Under white light illumination, the result is colour fringes, seen particularly at the edges of features situated at the periphery of the image (Figure 6.15). Bright off-axis image points are spread into a short spectrum with the blue rays coming to a focus nearer the optical axis and the red focused more towards the periphery of the field of view. Therefore, the height of the red image is larger than that of the blue image with a positive, or converging, lens.

Chromatic difference of magnification used to be corrected by specially designated compensating eyepieces that were deliberately designed to have an equal and opposite error (i.e. an error of the same magnitude but opposite sign) in order to cancel out the aberration from the objective present in the image. Since a diverging lens produces an opposite sequence of colours, weak negative eyepieces could be constructed that cancelled out the chromatic difference of magnification in the objective. The objective and eyepiece(s) were now matched and functioned as a single optical unit.

These eyepieces have the designation 'C', 'Comp' or the German equivalent 'Komp' marked on them. They are particularly intended for use with apochromatic objectives, whose correction could not be fully achieved in the objective itself. A compensating eyepiece can be identified from the yellow-orange rim around the internal diaphragm. Non-compensating eyepieces have a faint blue-purple rim or else none at all. The rule was (and still should be) don't mix the objectives of one manufacturer with the compensating eyepieces of another.

In the late 1970s chrome-free, or CF, optics were designed by Zenji Wahimoto of Nikon and simultaneously by Horst Riesenberg of Zeiss. Modern glass formulations and optical computing means that lateral chromatic aberration can be entirely corrected in the objective, and this is done by the Japanese firms Nikon and Olympus. With infinity-corrected optics, the residual correction can still done in two stages but in the tube-lens rather than the eyepieces, allowing the user easily to exchange eyepieces of different magnifying power. This approach is adopted by the German firms, Leica and Zeiss. Once again, this means that – for best results – objectives should not be swapped onto the microscope stand of a different manufacturer.

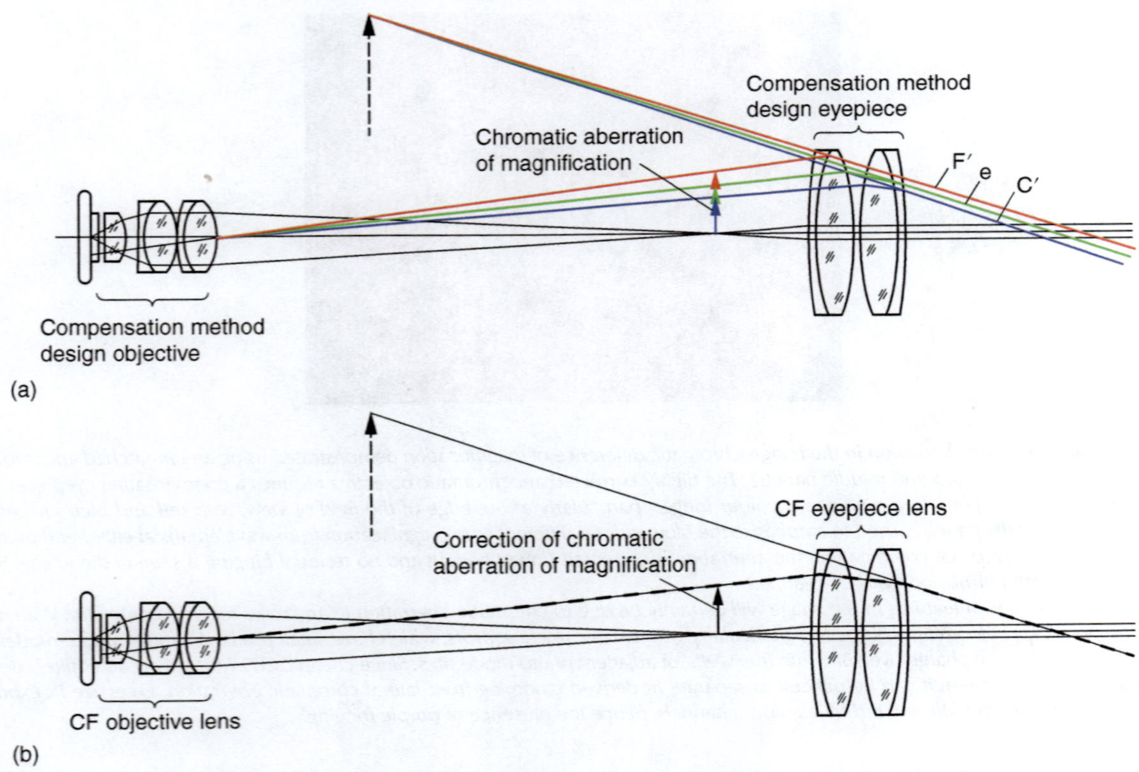

Figure 6.14 ***Lateral chromatic aberration and its correction*** *Chromatic difference of magnification in an older apochromat with (a) an unmatched and (b) a matched compensating eyepiece. Modern objectives are now 'chrome free' or have the residual errors corrected in the tube-lens of the infinity-corrected microscope. Source: Pluta 1988.*

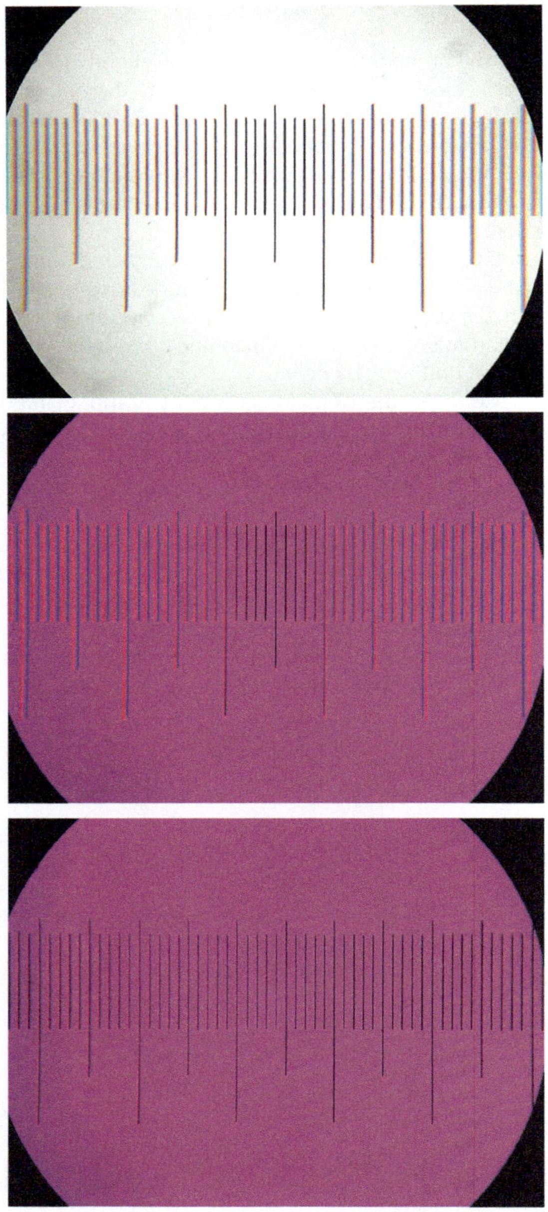

Figure 6.15 **Lateral chromatic aberration in the image** *Chromatic difference of magnification demonstrated using an unmatched apochromatic objective and eyepiece combination (top and middle panels). The highly corrected apochromatic objective requires a compensating eyepiece. The result of using an unmatched non-compensating eyepiece is colour fringes, particularly at the edge of the field of view, with red and blue images of different sizes. The purple filter (middle panel) is used to emphasise the blue and red differences in magnification from wavelengths at either end of the spectrum. With the correct objective-eyepiece combination, the aberration is corrected (lower panel), and no residual fringing is seen in the image. Source: Dr P. Evennett. Reproduced with permission from Dr Evennett.*

Purple fringing at the edges of features in the image will certainly be due to chromatic aberration in an image recorded onto film. With digital media there is a second possibility: blooming. This occurs particularly with small camera sensors, which have small pixels. When the photo-diodes receive too much signal, the photo-electron charge overspills into the 'wells' of adjacent photo-diode sites. Since cheap CCDs (not having an IR filter) detect infrared in addition to the visible spectrum, it can be difficult to separate IR-derived blooming from lateral chromatic aberration. Excessive IR exposure causes higher signal intensity values in both red and blue colour channels, hence the presence of purple fringing.

6.10 Testing for Chromatic and Spherical Aberration

Without complex optical equipment, lens evaluation and testing by the average user is a subjective affair, but even with a small number of special slides it is possible to make comparative assessments and form a valid judgement as to how good your objectives are and whether they are suitable for the job for which they are designed. As a minimum, all that is required is:

- a diatom test plate (for resolving power and subjective impression of image quality);
- a preparation of triple-stained fluorescent beads;
- a thin well-stained specimen (e.g. histological section or blood smear);
- a smoky carbon film; and
- a silvered coverslip (or silver tape, the sort that can be bought in gift and card shops) mounted on a slide as a star test.

The use of the diatom test plate for evaluating resolving power is described and illustrated in Chapter 10 (see Section 10.12). Because diatoms have well-defined edges yet are unstained, they can be useful (see Figure 6.6) in judging whether chromatic aberration is present to any degree in the image. Taken with the results of the tests described below, the diatom preparation can help evaluate overall image quality and performance of the objective.

Image shift (Figure 6.16 and Figure 27.7) will occur to some degree between colour channels in fluorescence microscopy as result of chromatic aberration. It might range from a fraction of a micrometre offset between channels to several micrometres, depending upon the quality of the objective. Chromatic shift can be assessed using triple-coloured fluorescently stained polystyrene beads, such as those sold by Thermo Fisher Scientific (Multi-Speck beads: M-7901; Tetraspeck beads: T-7284, or mounted on a slide: T-14792). A Z-stack taken through a single triple-stained bead will reveal the degree of chromatic aberration, which can be usually corrected with image analysis software. This particular test for fluorescence microscopy is dealt with further in Chapter 27, Section 27.6.

Figure 6.17 shows the effect of spherical aberration. The slide used is a very well-fixed 0.5 μm semi-thin section stained with toluidine blue for electron microscopy. This type of very thin well-stained slide is ideal for checking the presence of any softness and lack of contrast in the image, which very often arises from spherical aberration. A blood smear with well-stained leucocytes (white blood cells) can serve the same purpose: the essential requirement is that the preparation be thin and sufficiently contrasty to focus precisely at the sharpest point in the image. A well-corrected achromatic objective should not show any noticeable spherical aberration.

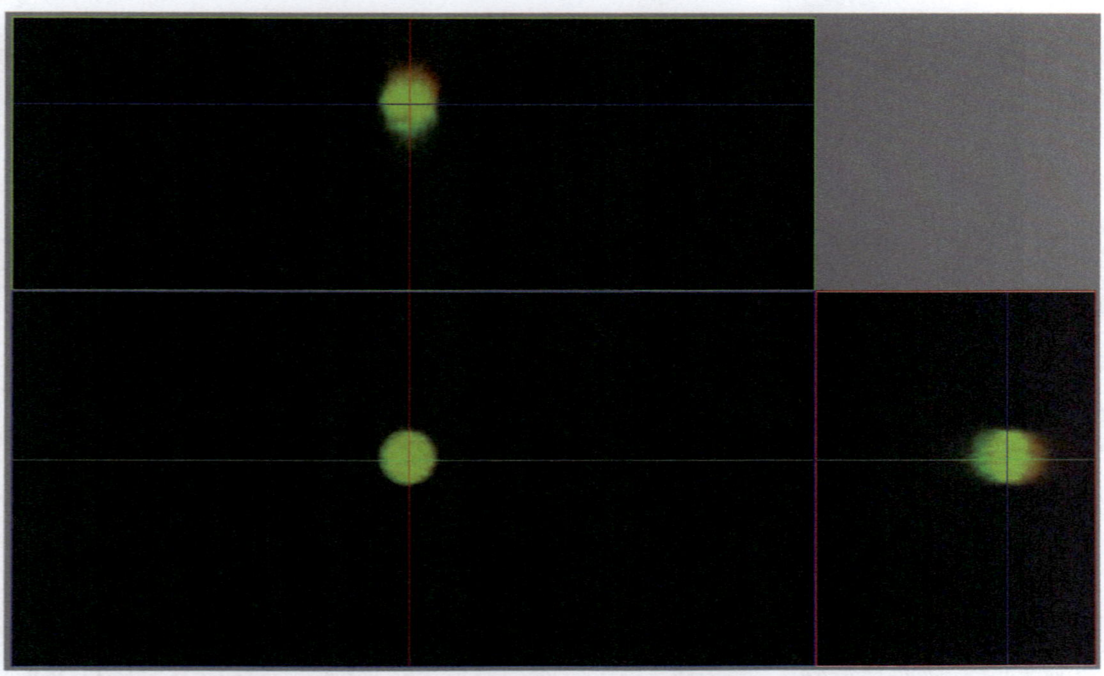

Figure 6.16 **Colour shift between channels in fluorescence microscopy** *Colour shift occurring between different fluorescent colour channels as a result of residual chromatic aberration. The fluorescent bead is stained throughout with both red and green fluorescent dyes. This is an orthogonal view showing x-y, x-z and y-z views simultaneously. A colour shift will occur even with highly corrected apochromat objectives. How great this appears depends not only on the perfectness of the correction but also on the overall magnification (i.e. including any zoom factor on a confocal microscope) of the final image.*

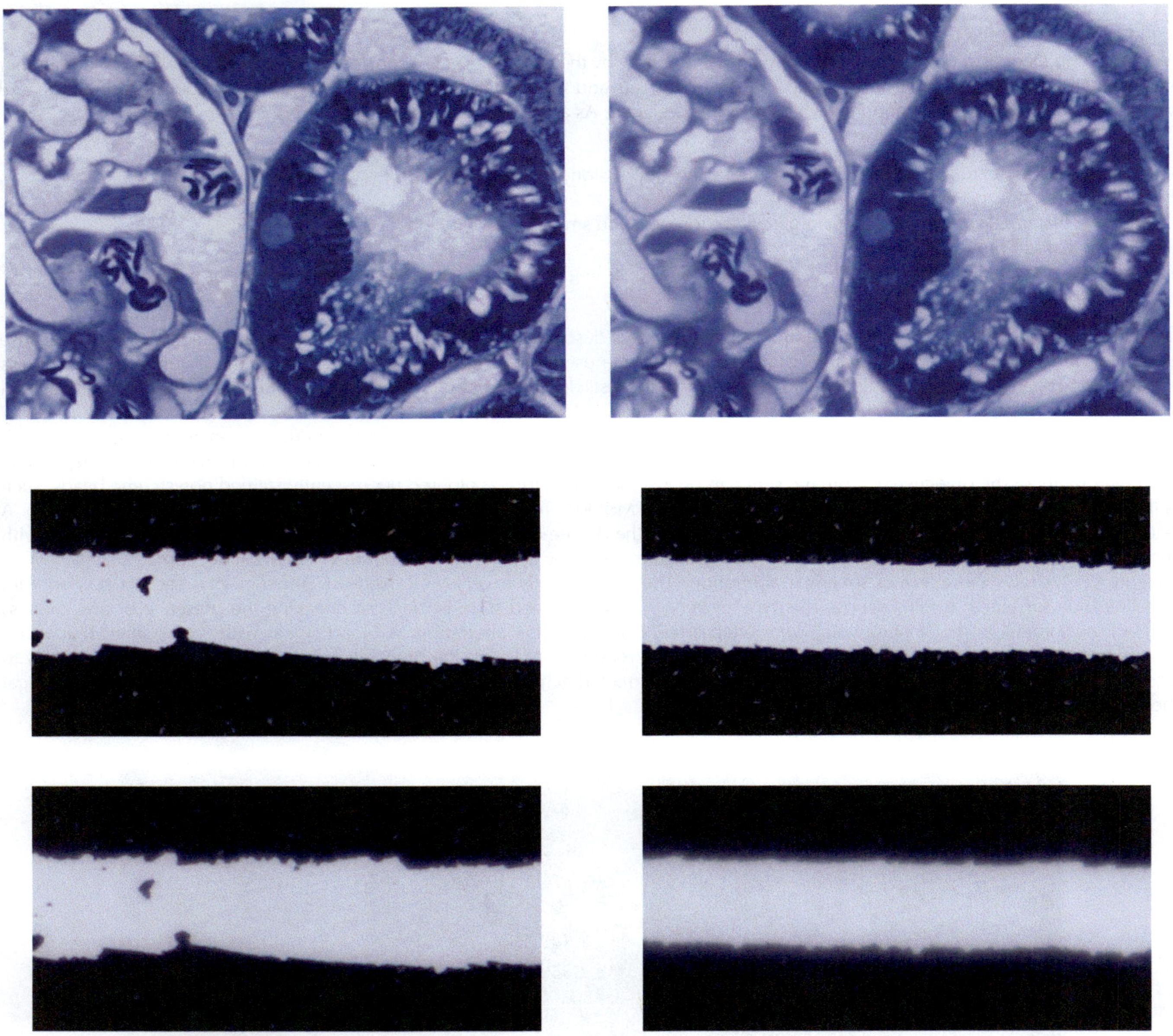

Figure 6.17 **Correcting spherical aberration in the image** *Thin (0.5 μm) perfusion-fixed kidney sample stained with toluidine blue. This specimen has been chosen as a test slide because it is very thin – well within the depth of field of most objectives – and has excellent preservation of detail. All images were taken with an Olympus plan-apochromat 60x/NA 0.90 high-dry objective with a 0.11–0.23 mm range correction collar.*

Top left: image taken with the correction collar adjusted to give the sharpest image with minimal spherical aberration. The collar was set at 0.175 mm. The coverslip thickness is unknown, but if the section were affixed to the slide and the coverslip were exactly 0.170 mm thick, then the thickness of the layer of mounting medium would be 5 μm – which falls within the average value of a maximum of 30 μm for the thickness of mountant applied on top of the section.

Top right: the collar was deliberately set to the limit of its range at 0.11 mm. Note the spherical aberration introduced into the image with consequent loss of contrast, image sharpness and blurring of fine detail leading to loss of resolution.

Middle: images of Abbe test plate taken with (left) 0.17 mm setting on the correction collar and at the 0.17 mm position under the coverslip of the test plate; (right) 0.23 collar setting at the 0.23 mm position on the test plate. Both images are sharp, showing no spherical aberration.

Bottom: mild spherical aberration (left) the 0.17 mm position under the coverslip of the test plate has been imaged with the correction collar set to 0.23 mm. Severe spherical aberration (right) the 0.23 mm position on the test plate has been imaged with the collar set to 0.11 mm. Note the loss of contrast, indistinctness and blurring of the image.

The smoky carbon film and the star test are both used to evaluate the presence of spherical aberration and also chromatic aberration in the image. Examine the edges of the smaller clumps of carbon or the edges of the scratched line. If the objective has a noticeable degree of chromatic aberration, you will see green and magenta fringes around small particles, particularly at the periphery of the field of view. This should be evident to some degree on achromat objectives, rather than on apochromats. Above the focal plane, yellow-green or green fringes of residual colour occur; below focus, magenta fringes are seen.

Protocol 6.1 The Carbon Test Slide

1. Take a No. $1^1/_2$ coverslip. Either use one sold by Zeiss as high tolerance (part No. 47-40-30-9000) 18 mm^2, thickness = 0.170 mm $\pm$ 0.005 mm: ISO 8255-1:2011 (Figure 6.18), or measure one from an ordinary box of No. $1^1/_2$ coverslips with a Vernier micrometer gauge. These coverslips are made by Marienfeld GmbH & Co. Kg. Their product number for these 18 x 18 mm coverslips is 0107032.

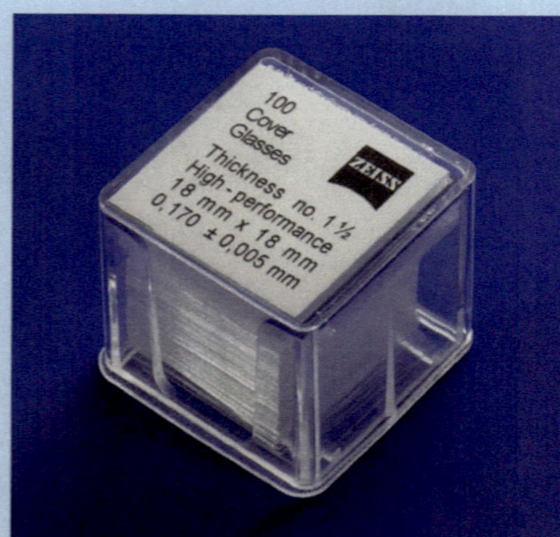

Figure 6.18 ***High-tolerance coverslips*** *High-tolerance coverslips for use with objectives of high numerical aperture, to help minimise spherical aberration in the image. The coverslips in this box are guaranteed to be within $\pm$ 5 μm of 0.17 mm. Source: Courtesy of Mr A. Ford.*

2. Single matches are insufficient to produce a thick enough carbon film. Use a candle or otherwise a resinous Maya tinder stick[9], used outdoors for laying fires. Take care that no smoke detectors are in the vicinity.
3. Scratch the surface of the carbon with a pin, or opened-out paper clip, to make a series of fine lines through the film.
4. Mount the coverslip with the carbon film downwards onto a microscope slide using a hard-setting mountant, such as DPX.

The presence of spherical aberration will, for an unfocused image of a point source object lying away from the focal plane, be seen either as a series of concentric rings or as a blurred point. This is shown in Figure 6.19 for both under-corrected and over-corrected spherical aberration and in Figure 6.20 when observing particles in the carbon test slide.

It is seldom that a modern microscope objective exhibits field curvature: most have a flat-field. However, an easy way to check this and evaluate the depth of field of the central area is to use a stage micrometer and read off the extent of that part of the central field that is in focus.

The degree to which achromats, semi-apochromats and apochromats are corrected can be represented visually by plotting the paraxial focal distance against wavelength, as shown in Figure 6.21. The image of a point in white light shows colour dispersion and is blurred. This uncorrected spectral distribution is called the primary spectrum, whereas the residual colour remaining after correction is the secondary spectrum.

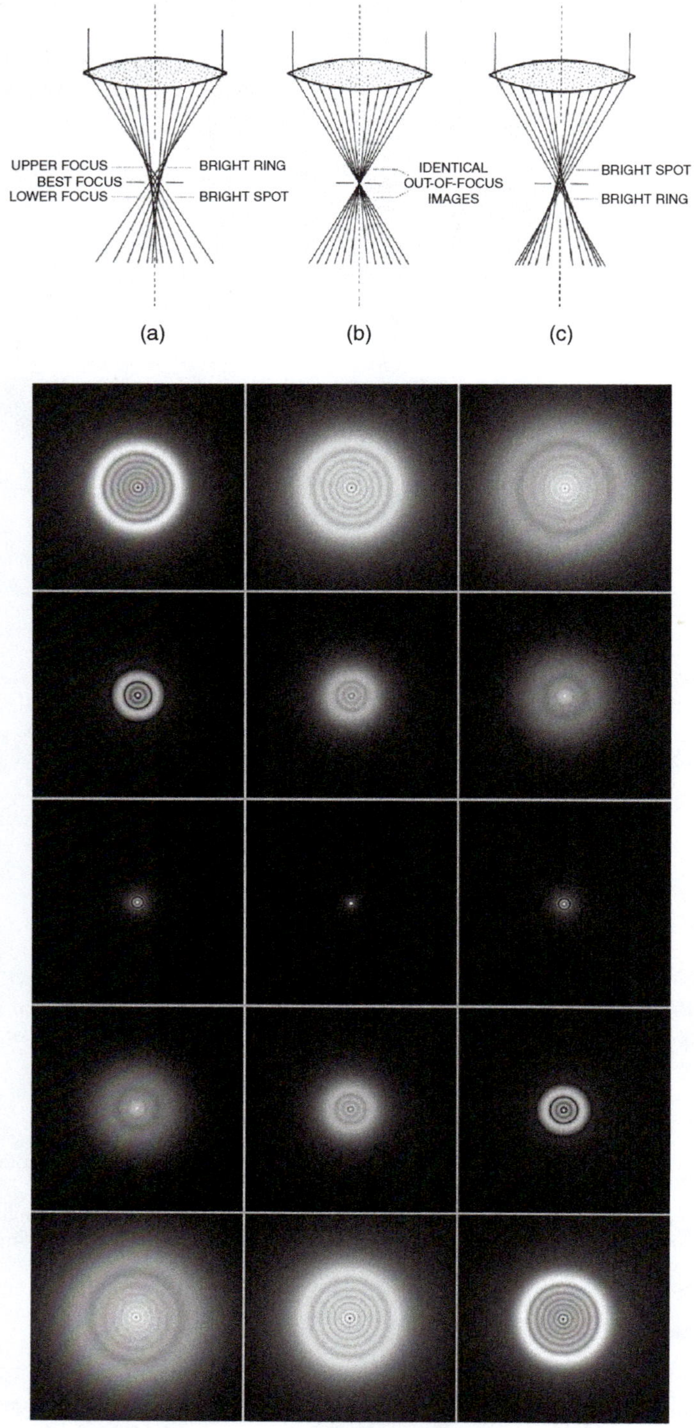

Figure 6.19 **Evaluating spherical aberration** *The effect of spherical aberration on the point spread function. In column (a) the image suffers from spherical under-correction, with a series of bright rings above the plane of focus and a bright spot with blurred rings below the plane of focus. The middle column, (b), shows the image from a theoretically perfect microscope system, and column (c), the effect of over-correction on the image, with a bright central spot and blurred rings above the plane of focus and bright rings below. Source: Oldfield 1994. Reproduced with permission from Elsevier. (image) Adapted from Mdf, via Wikimedia Commons.*

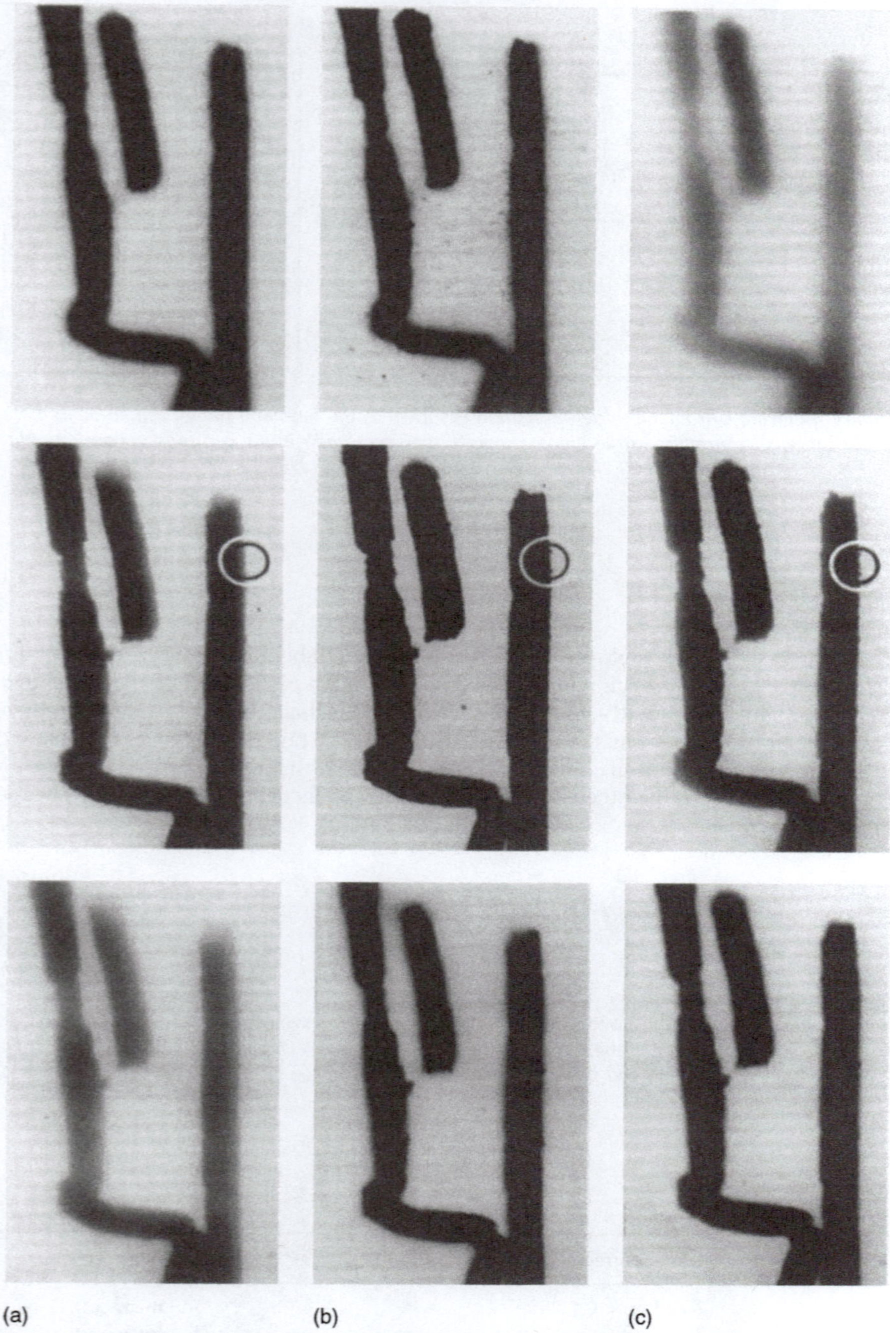

Figure 6.20 **Spherical aberration in the image** *Each column equates to those in Figure 6.19. The left-hand column (a) shows under-correction (a too-thin No. 1 grade coverslip has been used to make the preparation). The central column (b) exhibits a well-corrected image (the carbon particles have been deposited upon a measured No. $1^1/_2$ grade coverslip, which has been checked by measurement to be exactly 0.170 mm thick) and the right-hand column (c) shows the effect of over-correction, where two No. $1^1/_2$ coverslips have been used.*

From top to bottom of the figure, each row shows images taken above the plane of focus (top), at the plane of focus (middle) and below the plane of focus (bottom) in each case. Compare each image panel with the corresponding image of the PSF in Figure 6.19. The circle shows the portion of carbon film shown as the region of critical focus.

Source: Oldfield 1994. Reproduced with permission from Elsevier; Rost and Oldfield 2000. Reproduced with permission from Cambridge University Press.

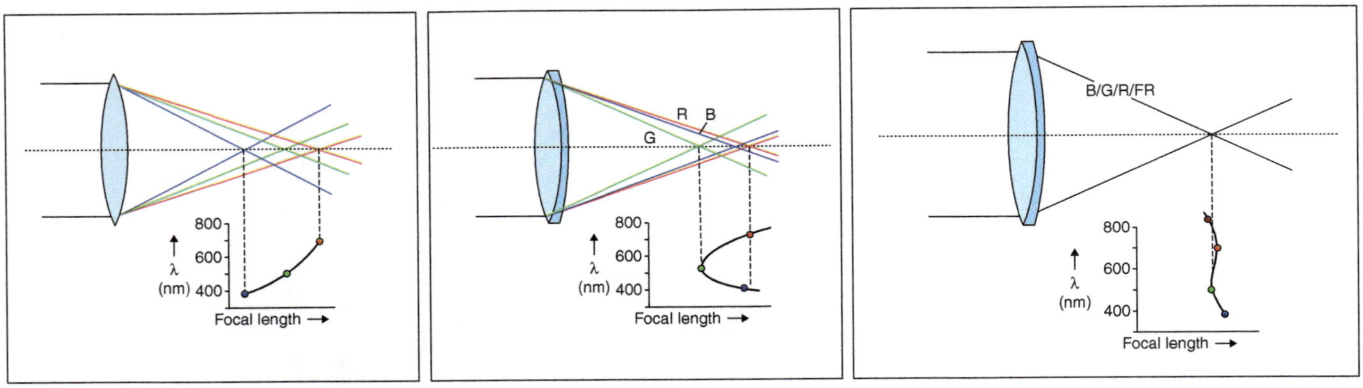

Figure 6.21 **Residual secondary spectrum** *Chromatic under-correction is shown in the left-hand panel; simple achromatic correction, here for two wavelengths – blue and red and spherical aberration in the green region of the spectrum – in the centre panel and apochromatic correction for four colours in the right-hand panel. Source: Adapted from Oldfield 1994, with permission from Elsevier.*

6.11 The Star Test

Although little-used now, the star test can be employed to recognise spherical aberration. An opaque metal film deposited onto the underside of a measured coverslip[10] 0.17 mm thick will have diffraction-limited pinholes in it. These will give an Airy disc from the point source and upon going through focus should – in the absence of spherical aberration – expand symmetrically on either side of focus (Figure 6.19). If spherical aberration is present, the image will be blurred on one side of focus, and action can be taken (Table 6.2) to correct this. Nevertheless, an incorrectly-adjusted correction collar will give substantially worse results than none at all. By extension, a badly-adjusted correction collar will also render the expensive apochromat effectively worse than the cheapest achromat, so care is needed in the use of any correction collar.

Protocol 6.2 The Star Test Slide

1. Take a measured No. $1^1/_2$ coverslip, as in Protocol 6.1.
2. Evaporate an opaque film of aluminium onto this coverslip, and mount it film face-down onto a microscope slide with a hard-setting mountant, such as DPX. Since the film is on the underside of the coverslip, the exact quantity and thickness of mountant is immaterial.
3. Ideally – to eliminate chromatic aberration – the test should be performed with a green filter in place.

Table 6.2 Causes and correction of spherical aberration

Cause	Under-corrected aberration	Over-corrected aberration
Coverslip	Too thin: < 0.17 mm	Too thick: > 0.17 mm
Optical tubelength	Too short	Too long
Drawtube*	Pull out: increase	Push in: decrease
Correction collar	Lens elements too close together	Lens elements too far apart
Immersion medium	Refractive index too low	Refractive index too high
Diagnostic effect seen in the image	Greater expansion of image on focusing up	Greater expansion of image on focusing down

* If your microscope is an older monocular model fitted with a drawtube, the optical tubelength can be very simply adjusted to correct for spherical aberration by pulling out or pushing in the drawtube. The disadvantage of using a drawtube to adjust for spherical aberration in this way is that it will introduce a greater change in the magnification of the primary image than adjustment using a correction collar.

If a hazy disc with a bright spot in the centre is seen on focusing upwards and a sharply defined annular ring seen on focusing downwards, the cover glass is too thin, the system is spherically under-corrected and the tubelength should be increased. The rule of thumb is 'ring down tube up'. Conversely, if a hazy disc of light with a bright central spot is seen on focusing downwards and a bright ring on focusing upwards, the cover glass is too thick, hence the system is over-corrected and the tubelength should be decreased. Do not use an alteration of the tubelength with the drawtube to give a slight change in magnifying power, since this obviously will introduce unwanted spherical aberration (With low–numerical aperture objectives you would gain more magnification change and less change in spherical correction than with high-NA objectives. This because the depth of focus in the primary image plane is much greater for low-NA objectives).

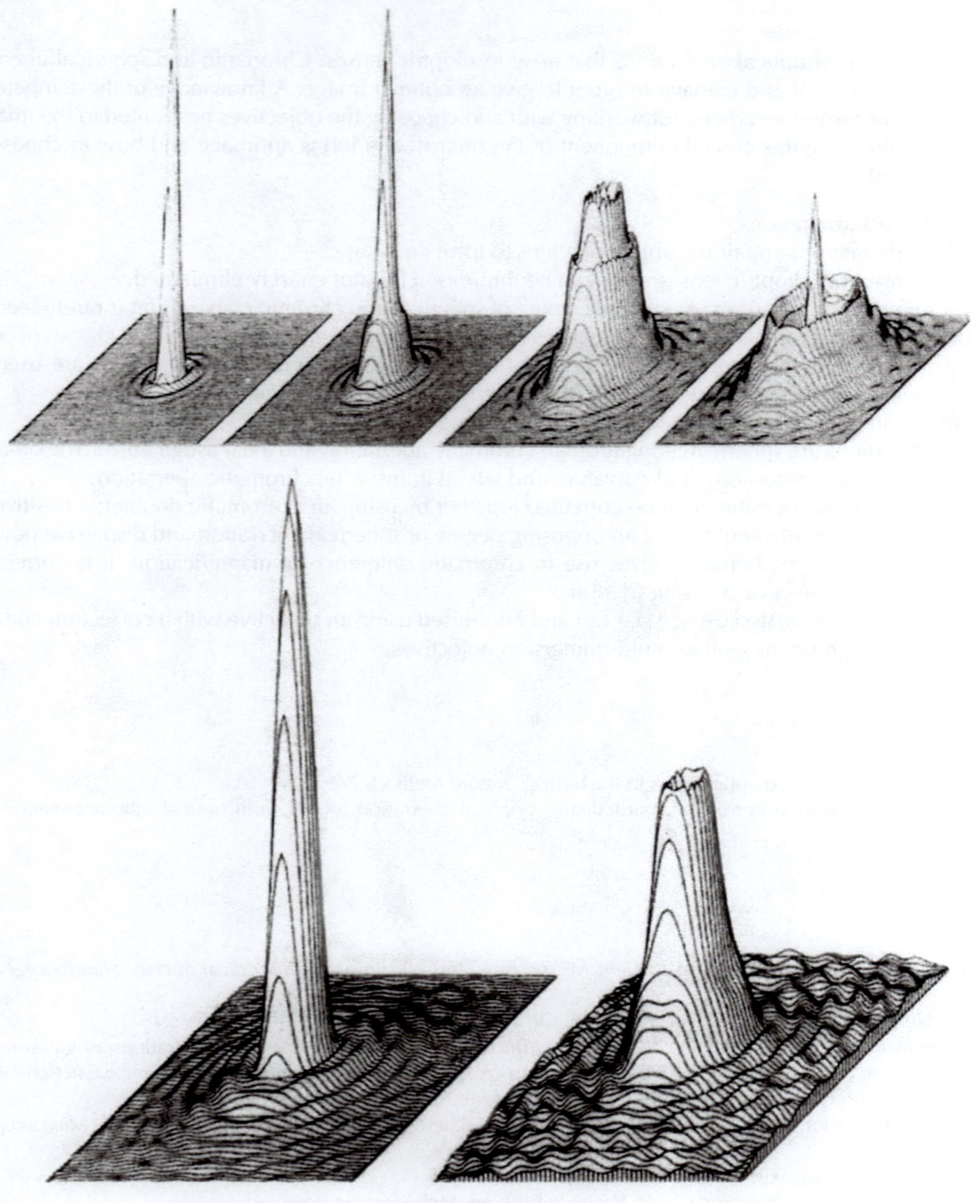

Figure 6.22 **Effect of defocusing and of spherical aberration on the PSF** *The top panel (a) shows the effect (going from left to right) on the intensity distribution of the point spread function as the microscope is gradually defocused from Δ = 0 on the left to fully defocussed on the right: defocusing changes both the size and intensity of the PSF. The bottom panel (b) shows the equivalent change in intensity above (left-hand figure) and below (right-hand figure) the point of best focus in a system suffering from spherical aberration. Source: Pawley 2006. Reproduced with permission from Springer.*

Examine the slide for a suitable pinhole. Theoretically, a diffraction-limited pinhole is ideal, but the images from these are of low intensity and can show interference rings, which hide any spherical aberration in the objective. In practice a pinhole from 0.5 to 3 μm in diameter is to be preferred, depending upon the resolving power and magnification of the objective.

The effect of spherical aberration on the point spread function, or PSF, causes both a reduction in intensity of the central portion (the Airy disc) and a broadening of the Airy pattern. The proportion of attenuation (Figure 6.22) of the Airy disc intensity, taking the non-aberrated PSF as 100%, is known as the Strehl ratio. Values above 0.8 are usually taken to mean that the objective is diffraction limited. Manufacturers usually select objectives with high Strehl ratios for super-resolution microscopy, and Zeiss supply these for Airyscan imaging.

6.12 Chapter Summary

This chapter explains the nature of optical aberrations that arise in dioptric lenses. Chromatic and spherical aberration are the most important of these to take account of and manage in order to give an optimal image. A knowledge of these inherent aberrations and how they are minimised helps formulate criteria for working with and choosing the objectives best suited to the imaging task. The next chapter on objectives explains how this crucial component of the microscope forms an image and how to choose the most suitable objective design for your work.

1. Lenses inherently possess aberrations.
2. Aberrations (aside of diffraction) impair the ability of a lens to form an image.
3. Optical aberrations arise from dioptric lenses; they can be minimised but not entirely eliminated.
4. The manufacturer minimises aberrations. As such, the effect of spherical and chromatic aberration is rarely seen in the microscope image. Achromats, semi-apochromats ('fluorite' objectives) and apochromats are the three principal classes of corrected objectives.
5. The microscopist must be aware of spherical and chromatic aberration. Well-corrected objectives are used when minimising chromatic and spherical aberration is important (e.g. fluorescence colocalisation studies).
6. Spherical aberration can be inadvertently introduced into the image.
7. The three on-axis aberrations are spherical, (longitudinal) chromatic aberration and axial astigmatism. The off-axis aberrations are: coma, 3rd-order astigmatism, distortion, field curvature and lateral (transverse) chromatic aberration.
8. Both chromatic and spherical aberration can be corrected together by using an achromatic doublet: a positive convex lens combined with a negative concave element having an opposing degree of spherical aberration and dispersive power.
9. Lateral, or transverse, chromatic aberration gives rise to chromatic difference of magnification. It is corrected either by using matched compensating objectives or by using CF optics.
10. Spherical aberration can be evaluated using a star test and minimised using an objective with a correction collar. These correction collars can be found on high-dry as well as multi-immersion objectives.

Key Reading

Heintzmann, R. (2010). 'Correcting distorted optics: back to the basics'. *Nature Methods 7*/2: 108–110.
- A readable introduction to aberration correction, particularly spherical aberration arising from optical inhomogeneities and refractive index mismatch.

References

Bradbury, S. and Bracegirdle, B. (1988). *Introduction to Light Microscopy*, 2nd ed. *Royal Microscopical Society Handbook No. 42*. Bios Scientific Publishers.
Cox, G. (2007). *Optical Imaging techniques in Cell Biology*, 1st ed. CRC Press, Taylor & Francis.
Davidson, M. (1998). *Molecular Expressions Microscopy Primer*. http://micro.magnet.fsu.edu/primer/java/aberrations/astigmatism/index.html.
Goodwin, P.C. (2007). 'Evaluating optical aberration using fluorescent microspheres: methods, analysis, and corrective actions'. In: Greenfield, S. and David, E.W. (eds.), *Digital Microscopy*, 3rd ed. Amsterdam, Boston, Oxford: Academic Press, Elsevier.
Keller, H.E. (2006). 'Objective lenses for confocal microscopy'. In: Pawley, J.B. (ed.), *Handbook of Biological Confocal Microscopy*, 3rd ed. New York: Springer.
Loveland, R.P. (1970). *Photomicrography*, vol. 1. New York, London: Wiley.
Oldfield, R. (1994). *Light Microscopy: An Illustrated Guide*. Wolfe Publishing, Moby imprint (Elsevier).
Pluta, M. (1988). *Advanced Light Microscopy*, vol. 1. Amsterdam, Oxford: Elsevier.
Rost, F. and Oldfield, R. (2000). *Photography with A Microscope*. Cambridge: Cambridge University Press.
White, G.W. (1966). *Introduction to Microscopy*. London: Butterworths.

Further Reading

Arimoto, R. and Murray, J.M. (2004). 'A common aberration with water-immersion objectives'. *Journal of Microscopy 216*/1: 49–51.
Bracegirdle, B. (1987). 'Famous microscopists: Joseph Jackson Lister, 1786–1869'. *Proceedings of the Royal Microscopical Society 22*/5: 273–297.
Kam, Z. et al. (1997). 'Three-dimensional microscopy in thick biological samples: a fresh approach for adjusting focus and correcting spherical aberration'. *Bioimaging 5*: 40–49.
Pluta, M. (1988). Chapter 2. In: *Advanced Light Microscopy*, vol 1: *Principles and Basic Properties*. Amsterdam: Elsevier, pp. 155–179.

Internet Resources

Spherical aberration: http://hyperphysics.phy-astr.gsu.edu/Hbase/geoopt/aber.html and http://www.telescope-optics.net/secondary_spectrum_spherochromatism.htm.
Nikon CF 160 Optics Overview: http://www.cyto.purdue.edu/cdroms/micro1/7_spon/nikon/id87.htm.

Notes

1. Two other classes of objectives are referred to in Pluta (1988). These are monochromats and CF objectives. Monochromats are a very rare type of objective, lacking chromatic correction across the visible spectrum and also lacking correction for field curvature. Monochromatic objectives are principally used for imaging at one wavelength (usually ± 1 nm) in the ultraviolet, since below 300 nm the range of suitable optical materials (e.g. quartz) for making objective lenses is limited. Below 350 nm, the refractive index is very strongly dependent upon wavelength. This makes it impossible to correct for chromatic aberration over the entire spectrum. CF objectives have chromatic aberration corrected independently of the eyepiece or tube-lens. They are discussed in Section 6.9, on lateral chromatic aberration.
2. This is because spherical aberration is proportional to the fourth power of the diameter of the lens and inversely proportional to the third power of the focal length and so is more pronounced in high magnification objectives (particularly dry ones) of short focal length.
3. Abbe's sine condition: where the ratio of the product of refractive indexes and angular apertures on object side to image side equal the magnifying power, such that this ratio is constant for both paraxial and peripheral rays forming the image. Abbe regarded coma as a consequence of difference in image height from one lens zone to another and for a spherically corrected lens to be free of coma, the magnification of the image formed by paraxial and marginal rays must be equal. See also Chapter 7, Section 7.4, on numerical aperture & resolving power.
4. The frequency, v, of light does not change as it travels from one medium to another.
5. These refractive index values are for borosilicate crown BK7 – see Table 2.4.
6. Neither crown nor flint glasses transmit well below 420 nm, so other materials are required to make doublets that work optimally with UV and the near-UV 405 nm laser line. Fused silica glass is often used for the positive element and either UV-grade calcium fluoride or lithium fluoride for the negative lens element.
7. The development of high-quality optical glass originated with Otto Schott in partnership with Carl Zeiss. The firm is still at the forefront of glass manufacture. When a suitable glass has been cast for use from calculations, it is made in a single melt. The finished product is quartered and stored at opposite corners of the factory basement, as an insurance policy.
8. The reciprocal value of the overall field curvature is called the Petzval sum. When this is zero, the objective (e.g. an anastigmat) is corrected for field curvature (and, usually, also for astigmatism). Plan-achromat and plan-apochromat objectives contain *thick* positive meniscus lens elements, which correspond to negative elements. These balance the field curvature of other positive lens elements in the objective.
9. Maya tinder sticks are available from outward-bound and survival shops for fire lighting.
10. See Section 7.7 as to why objectives requiring coverslips are corrected to work with coverslips 0.17 mm thick and why cells or tissues should be mounted directly on the underside of the coverslip.

7

The Microscope Objective

Outline

Understanding Light Microscopy, First Edition. Jeremy Sanderson.

7.1 Introduction

Having discussed the nature of light, how our eyes work, shown how individual lenses form images and then how a microscope of modern modular design is constructed and how to use the microscope in a relaxed and stress-free manner, this chapter puts this learning into practice and discusses how the microscope objective works. The objective is the most important lens of the microscope, all the other parts being designed to support its function. The properties of the objective lens determine the quality of the final image. The most important characteristic of an objective is its *aperture*, which sets the limit to the resolving power of the microscope but also determines the brightness of the image and the depth of field.

Up until the 1980s it was common to find significant variation between different examples of the same objective. Some objectives would meet the minimum specification; others failed to do so. If the company representative could be persuaded, it was prudent to examine as many lenses as possible before purchasing the one that performed best. Nowadays, not only are the specifications improved, especially in terms of longer working distances and larger spectral transmittance windows, but variability between different objectives of the same design has been virtually eliminated. Computer-aided design and quality control has largely been responsible for this progress.

7.2 Different Types of Objective

There are three different classes of objective, for it makes no economic sense to correct every objective for optical aberrations as perfectly as possible. When an objective is not required to resolve the smallest detail achievable (e.g. a basic microscope for routinely checking histological stains, for tissue culture, for checking thin sections or for teaching), then the best and most expensive objectives are clearly not needed. A less–well corrected objective has a lower numerical aperture and so confers the advantage of a longer working distance for ease of use. This facilitates handling and manipulation of the sample on the stage of the microscope. Another example may be in fluorescence microscopy where a simpler objective will mean less glass is used in its construction giving a better signal (e.g. UV fluorescent stains are best seen with mid-corrected objectives). All manufacturers therefore offer the following three types of objective:

- **achromats**;
- **semi-apochromats**; and
- **apochromats**.

Box 7.1 The Information Written on the Objective

Since the objective is the microscope – at least it is the most important part of the instrument – it is as well to know exactly what information is carried on the barrel of the objective, and how this can help you select the correct objective for the job.

The most important information is the magnification of the objective and its numerical aperture. These are the minimum found on any objective. If there is no indication of higher correction, assume that the objective is an achromat.

The magnification can easily be seen by a coloured ring around the barrel. Conventionally these colours approximately follow the order of the visible spectrum (low-magnification red to high-magnification blue).

Generally, a black line around the front of the barrel indicates oil immersion, and a white line water-immersion. These are the most common options. Sometimes, in addition to the bands, 'Oil', Oel, H-I or HI will indicate oil immersion and 'water' or W for water immersion. However, **'Dry' is never marked.** Check the numerical aperture to ascertain whether the objective is dry or not. In most cases an NA < 1 means the objective is to be used dry, although there are a few exceptions for highly corrected apochromats. Don't get oil onto dry objectives!

Nikon put the free working distance of their objectives on the barrel. This is useful, since you then don't have to refer to the manufacturer's web pages or literature for the information.

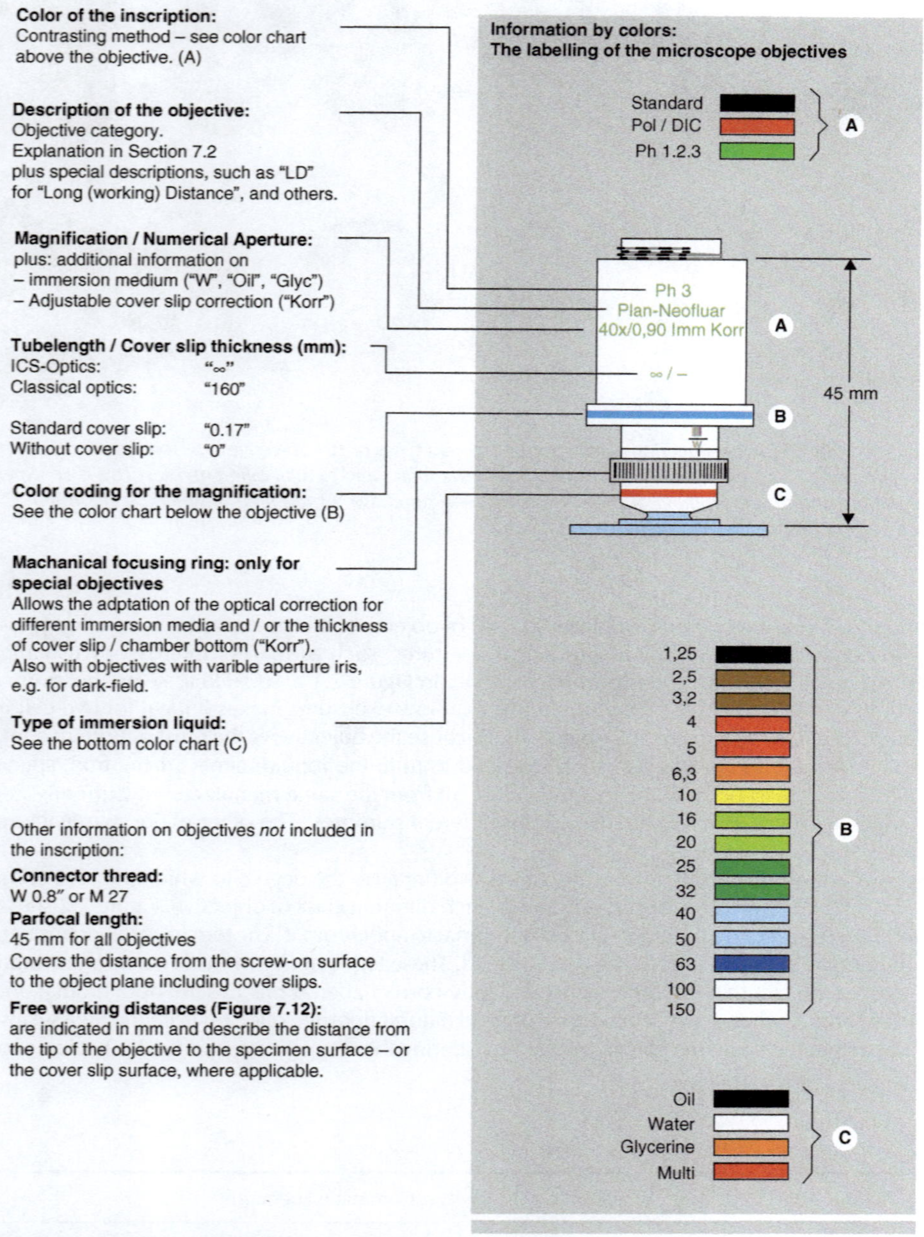

Source: Reproduced with permission from Carl Zeiss Ltd.

In previous years, the magnification tended not to be written on the objective but the focal length was given instead. By knowing the optical tubelength of the microscope (the distance between the back focal plane of the objective and the primary image plane), the magnification of the objective could be calculated, since M_{obj} = OTL/focal length. Thus the 6.0 mm focal length NA 0.95 objective shown would magnify 26.5x, which is very good indeed. A 1/12th inch gives (for the standard finite tubelength of 160 mm) an initial magnification at the objective, M_{obj} = 75:1 or 75x.

Figure 7.1 **10x Objective selection** *A selection of different types of 10x objective available from one leading manufacturer to show how different versions of the same magnification objective are designed and manufactured to fulfil different roles. This is an expensive process, but a range of objectives allows for versatility and future-proofing of the microscope you purchase. At the time of writing, Zeiss listed 37 10x objectives; Leica make 32; Nikon list 15 objectives and Olympus nine different 10x objectives.*

Flat-field versions are offered in all three classes of objective. It is nowadays very rare not to have 'plan' objectives; an exception would be in order to achieve *very* high[1] numerical apertures, such as in TIRF microscopy (Chapter 23, Section 23.3) or as the most basic achromat objective on an entry-level microscope. In Figure 7.1 a 10x 'Fluar' semi-apochromat objective with a very high NA of 0.5 is shown. Its ability to capture a very bright image at low exposures makes it ideal for live-cell imaging of cell monolayers.

Figure 7.1 shows a range of 10x objectives. Because the objective is the most important component of the microscope, manufacturers offer a range of different designs, each intended to fulfil the requirements arising from specific and differing applications. The figure shows 15 objectives of the same magnification, all from the same manufacturer. Currently Zeiss offer 37 different 10x objectives, with numerical apertures from NA 0.2 to 0.5, for different purposes. The other major manufacturers likewise offer a range of objectives to precisely suit your purposes.

Achromats, semi-achromats and apochromats differ as to the degree to which they are corrected for spherical and chromatic aberration. Table 7.1 shows the corrections found in each different class of objective. Figure 7.2 shows a plot of chromatic correction, showing the improved correction progressing from achromat to apochromat. The term semi-apochromat is sometimes little heard of, and the terms 'Neofluar' or 'Fluotar' or similar are used instead. These terms derive from manufacturers' brand names for semi-apochromat objectives and refer to the fact that fluorite was used to fully correct aberrations in semi-apochromat and apochromat lenses[2]. This harks back to the 19th century when Ernst Abbe (Box 7.2) used natural fluorite collected from the Swiss Alps to improve lens corrections[3], since there was not at that time suitable glass formulations offering the dispersive properties that calculations had shown were required. Nowadays,

Table 7.1 *Objective Classes and Their Corrections*

	Spherical aberration	Chromatic aberration	Cost
Achromat	Corrected for 1 colour: green at about 550 nm	Corrected for two colours: blue (typically 486 nm) and red (656 nm)	**1**x – base cost for comparison (about £ 460 for a 10x/NA 0.25) (about £ 735 for a 40x/NA 0.65)
Semi-apochromat ('Fluorite' etc.)	Corrected for 2 colours (blue and green)	Corrected for 2–3 colours (blue, green and red)	**2–3**x (about £ 1460 for a 10x/NA 0.30) (about £ 1980 for a 40x/NA 0.75)
Apochromat	Corrected for 2–3 colours (blue, green and red)	Corrected for 4–5 colours (deep blue/UV, blue, green and red and near-IR)	**5**x (about £ 2120 for a 10x/NA 0.45)
		Usually correction for 4 wavelengths, but in some special cases for near-IR also	**8–10**x (about £ 5920 for a 40x/NA 0.95 Korr) (about £ 6800 for a 40x/NA 1.3 oil imm)

All three classes of objective also have 'Plan' versions, which are corrected for field curvature.
Modern Zeiss objectives are fully colour-corrected no longer for only 3–4 spectral lines but for a full spectral range. This corresponds to a correction of up to 14 spectral lines.

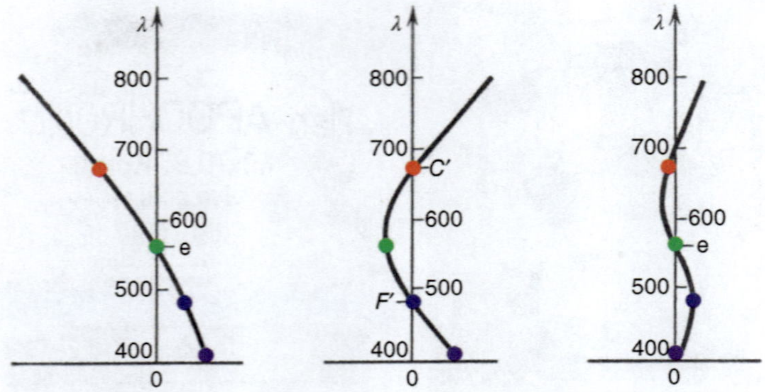

Figure 7.2 **Chromatic correction plots** *These three graphs show the focal position according to wavelength for an uncorrected (left), achromat (centre) and apochromat (right) objective. The deep blue (purple) point is 410 nm; the blue (Fraunhofer F′ line) at 486 nm; the green (e line) at 546 nm and the red (C′ line) at 656 nm. The focal shift from zero through the spectrum indicates the extent of chromatic correction designed in the objective. Modern curves are much flatter across the spectrum, but these show the principle. Source: Pluta 1988.*

synthetic lanthanum fluoride is used for its low dispersive properties in place of fluorite. Semi-apochromats have better transmittivity of light in the blue and near-UV part of the spectrum and so are favoured for fluorescence microscopy.

Achromats are only corrected for spherical aberration at one wavelength, usually 546 nm or 550 nm, in the green region of the spectrum. Particularly when viewing contrasty objects, a pale magenta fringe will be seen around the edge of the object (see Figure 6.6). This magenta fringe is the complementary colour to green and constitutes the residual, or secondary, spectrum of the objective. The image can be improved by the use of a green filter, and if a monochrome image is all that is necessary, there is every reason to use a green filter to collect a high-quality image.

By 1886 Abbe had successfully overcome the design limitations to produce the first apochromatic objectives (Figure 7.3), making corrections for chromatic and spherical aberration at different lens doublet surfaces, rather than at the same ones as had been proposed by Lister. Indeed, Abbe introduced[4] the term 'apochromatic' to emphasise the full chromatic correction inherent in this new design of objective. These new apochromats allowed for a higher numerical aperture as a direct consequence of the much improved aberration correction, for optical aberrations always become significantly worse at the periphery of the field of view. At the same time Abbe also introduced a compensating eyepiece to correct for residual lateral chromatic aberration (Chapter 6, Section 6.9) not fully corrected in the apochromatic objective.

Modern apochromats have much higher numerical apertures than was formerly the case. It is not uncommon to have a 10x/NA 0.45, 20x/NA 0.7 or NA 0.8 plan-apo objective or a 40x Plan-Neofluar objective of NA 1.30 or NA 1.40, whereas older objectives might only have half these numerical apertures. This is significant: it is much harder to design objectives of low magnifying power than it is to produce high-magnification objectives with concomitantly high numerical aperture. The trend for large NA objectives has been driven by the demand for extra brightness and resolving power to make visible (relatively) dim fluorescent images, where the numbers of photons forming the signal is limited (unlike bright-field microscopy, where it is a simple matter merely to turn up the rheostat on the lamp control). High NA objectives – from NA 1.49 to 1.70 – are also essential for TIRF microscopy (Chapter 23). Computer-controlled optical design software, rather than manual computation and ray-tracing, has revolutionised the design of new objectives. Modern objectives are better corrected over more wavelengths (in some cases for five, seven or even a dozen wavelengths over the 365–1000 nm range) rather than three, as is the case for older lenses. Modern objectives also have greater transparency at each end of the visible spectrum and beyond. The degree to which an objective can transmit a given wavelength of light allows the designer to compute its transmission profile. These graphs are usually released by the manufacturer and, if needed, can be found in the specification table of each objective in the manufacturer's objective database (for the URLs see Table 7.2).

Table 7.2 *Microscope Objective Information Online*

Leica:	https://www.leica-microsystems.com/products/microscope-objectives/ https://www.leica-microsystems.com/products/microscope-objectives/objective-classes/
Nikon:	https://www.nikoninstruments.com/Products/Optics https://www.nikoninstruments.com/Product-Selectors/Objective-Selector https://www.nikoninstruments.com/Products/Optics/Apochromat-Objectives
Olympus:	https://www.olympus-ims.com/en/microscope/uis2/ https://www.olympus-ims.com/en/microscope/slmpln/ http://www.olympusamerica.com/files/seg_bio/olympus_specialty_objectives.pdf
Zeiss:	https://www.zeiss.com/microscopy/us/products/microscope-components/objectives.html https://www.micro-shop.zeiss.com/?s=445046279979e2™l=en™p=uk™f=o

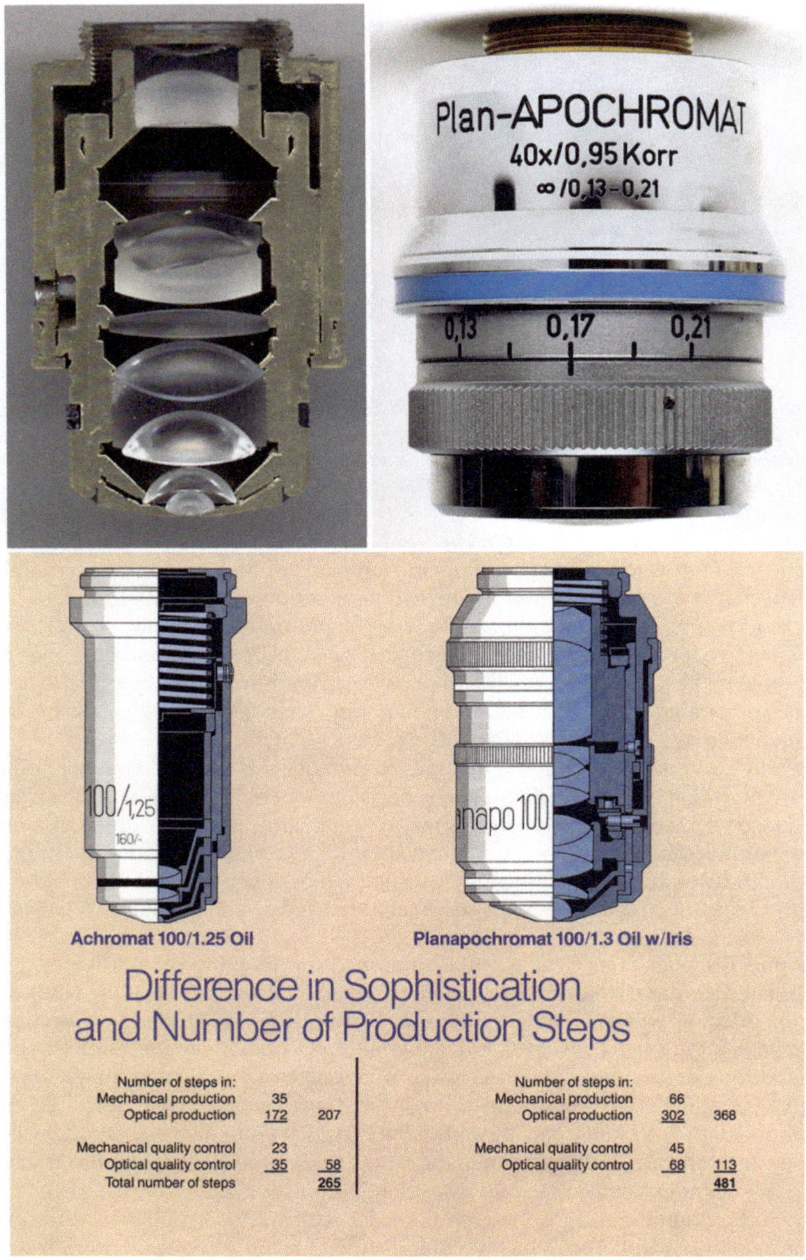

Figure 7.3 ***Apochromat objectives*** *A sectioned plan-apochromat objective together with marketing information from Zeiss, both of which show the numbers of lens elements in modern objectives and the steps required to manufacture these. Compensating eyepieces must be used with older designs of apochromat objectives in order to correct residual aberration present in the primary image. Such eyepieces are marked with 'C' a 'K' or else 'Comp'. Source: (top) Courtesy of Mr K. Glover. (bottom) Reproduced with permission from Carl Zeiss Ltd.*

In order to design higher NA values for given magnifications, the objectives themselves are now much larger with wider back apertures. Traditionally, the mounting screw threads on objectives conformed to an international standard laid down by the Royal Microscopical Society, hence called the 'RMS thread' (see ISO 8038 parts 1 & 2, 2011 and Table 2 in the *RMS Dictionary of Light Microscopy* (1989) on page 72). This standard is currently used in the production of infinity-corrected objectives by Olympus and also by Zeiss for some objectives. Leica, Nikon and Zeiss have adopted the larger thread sizes M25 (25-millimetre objective thread) M27 and M32 objective thread sizes. See the Internet Resources at the end of the chapter for further details of relevant International Standards.

7.3 Inscriptions on the Objective

Besides these three classes of objective – achromat, semi-apochromat and apochromat – there are many different types of objective within each class, each designed for a particular imaging application. Manufacturers engrave the barrel of the objective with information

Table 7.3 Objective Inscriptions

Abbreviation	Type
Achro, Achromat, (nothing)	Achromat correction
Fluar, NeoFluar, Fluotar, Fl, FL, Semi-Apo	Semi-apochromat correction
Apo, Apochromat	Apochromat correction
Plan, Pl, PL, Plano, NPL, UPLAN	Flat field (field curvature) correction
CF, CFI	Chrome-Free (chromatic aberration correction)
CFI, ICS or usually ∞	Infinity-corrected objective
160 (very occasionally 170)	Mechanical tubelength for finite tubelength objective
Met, M, NC, NCG	Objective for use with no coverslip or cover glass (e.g. a metallurgical objective)
0.17	Coverslip of correct thickness must be used
/-	Coverslip correction is not required; cover glass thickness is not important
EF	Extended field of view (field of view < plan)
L, LL, LD, LWD, ELWD, SLWD, ULWD	Long working distance (extra/super/ultra LWD)
Corr, Korr, W-Corr, CR	Objective with correction collar
I, Iris, W-Iris, NA range given (e.g. NA 0.6–NA 1.3)	Objective with adjustable iris in BFP (back focal plane of objective)
Oil, Oel, HI, black line on barrel nose	Oil-immersion objective
Water, W, white line on barrel nose	Water-immersion objective
Gly, Sil, orange or red line on barrel nose	Glycerol or silicone-immersion objective
Multi-imm	Multi-immersion objective (will have a correction collar)
Ph, Phaco, PC, Phase Ph#/No. (e.g. Ph1, Ph2, Ph3) (lettering may be in green)	Phase contrast
Pol, SF (lettering may be in red)	Strain-free objective for polarised light work
DIC, NIC (lettering may be in red)	Differential (Nomarski) Interference Contrast
U, UV, Universal	Transmitting down to 340 nm for UV work
D	For use with dark-field only
H-D (Hell-Dunkel)	For use with bright-field and dark-field
H	For use with a heated stage

(see Table 7.3 and Box 7.1) to inform you what the objective is designed to do, what its magnification is and its numerical aperture. The figure in Box 7.1 is taken from an educational brochure from one of the major manufacturers and shows how the information is laid out on the objective barrel. The basic information will be the magnification and the numerical aperture.

If the objective is designed for phase contrast, there should be the inscription 'Ph', and additionally the writing may be in green, to remind the user that phase-contrast microscopy benefits from the use of a green filter in the illumination path. Likewise, strain-free objectives suitable for polarised light or interference contrast microscopy may be engraved in red or have the word 'Pol' on the barrel. Objectives for standard bright-field or dark-field microscopy will be engraved in black lettering.

Objectives carry a colour-coded ring around the barrel. This was introduced by Dr Kurt Michel at Zeiss in 1953 and is useful for identifying quickly the magnifying power in use. The colours progress through the colours of the visible spectrum, starting with black for 1x or 1.25x, brown for 2x or 2.5x, red for 4x or 5x, orange for 6.3 or 8x, yellow for 10x, light green for 16x and 20x, dark green for 25x and 32x, light blue for 40x and 50x, dark blue for 60x or 63x and white for 80x or 100x and above.

Besides the colour code for the magnification, it is important to be aware of the rings just above the front lens of the objective that indicate whether the objective is an immersion one or not. Dry objectives are almost never marked 'dry'. They really ought to be, for some people make the mistake that all high-magnification objectives work with oil. Look for a black ring (it may not always be obvious, but it *will* be present if the objective is an oil-immersion design). Likewise a water-immersion objective has a white ring. There are two other types of immersion media that are much less used: glycerol-immersion objectives (Leica; Martini et al., 2002), which have an orange ring, and the Olympus silicone-immersion objectives, which have a red ring. A red ring also denotes that an objective is 'multi-immersion' (Figure 7.4) and can be used with oil, water or glycerol, depending upon which immersion medium has the closest refractive index match to the preparation. All that is required is to set the adjustment correction collar to the optimum position to give the sharpest, clearest image.

Most modern objectives are of the infinity tubelength design, and this will be indicated on the objective with an infinity symbol, thus: ∞. If the objective is of the older finite tubelength design, this will be indicated with '160' to show the mechanical tubelength (Figure 7.5). There will then be a forward slash (/) and information on whether a coverslip (also called a cover glass) is required or not. If the objective is designed to use a coverslip of the standardised thickness, the inscription '0.17' will be engraved on the barrel. If the numerical aperture is sufficiently low that coverslip thickness is unimportant, and does not introduce spherical aberration (i.e. generally NA less than 0.4), then a dash (-) is used. Those objectives that are designed to be used to observe the surface of the specimen directly without the use of a coverslip (sometimes colloquially referred to as 'uncovered' objectives) will have a zero (0) inscribed here.

These are the main items of information to look out for. Normally, the class of the objective is also abbreviated ('Apo' or 'Fluar') with 'Plan' or 'Pl' for flat-field correction for field curvature. On older objectives it is not common to state the aberration correction class if the objective is merely an achromat. Manufacturers tend to follow naming conventions, but they also have idiosyncrasies of style, and it is sometimes confusing as to what each abbreviation means. Table 7.3 shows some of the abbreviations in use. Figure 7.6 shows

Figure 7.4 **Water, oil and immersion objectives** *A selection of high numerical aperture immersion objectives. In the top left-hand panel is a 40x/NA 1.4 oil-immersion objective, one of the very best available in terms of resolving power and light-gathering capability. The objective next to it is an oil immersion 40x/NA 1.3. There is very little difference in ultimate resolving power, but this objective has an adjustable collar to control a variable diaphragm in the back focal plane of the objective allowing the numerical aperture to be varied from NA 0.6 to 1.3. This objective is ideal for dark-field microscopy. In the lower left-hand panel, on the 40x/NA 1.2 water-immersion objective, the very similar adjustment ring is a correction collar to correct for any introduction of spherical aberration in the image, according to the thickness of coverslip used, refractive index inhomogeneity and elevated temperature for live-cell imaging. The red fiducial mark is used when operating the microscope at 37°C. The objective with a green magnification ring is a 25x/NA 0.8 multi-immersion objective. It has a similar correction collar to correct for spherical aberration according to the mountant used and whether a coverslip is used or not. This is a very versatile objective for photomicrography and for live-cell imaging. In the two right-hand panels of this figure are four immersion objectives using silicone as an immersion medium. They are designed for live-cell imaging, for silicone immersion reduces the potential refractive index mismatch between the living specimen and the glass of the Petri dish or microscopical preparation.*

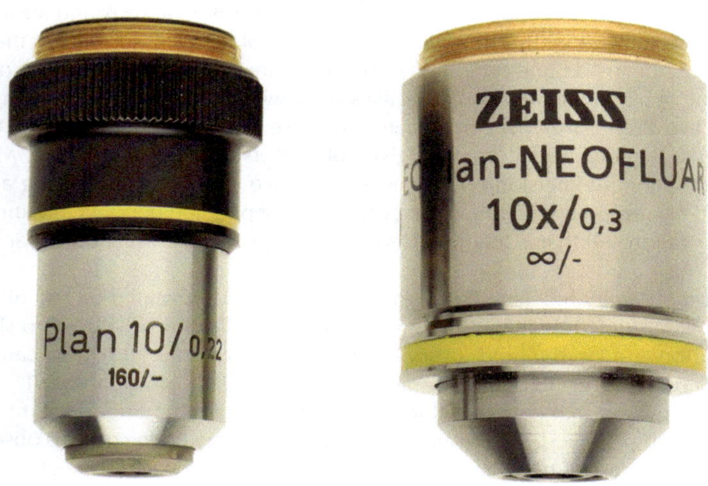

Figure 7.5 **Finite tubelength and infinity tubelength objectives compared** *A 10:1 objective (left) used with 160 mm finite mechanical tubelength, and a 10x objective (right) for use with infinity-corrected microscopes. Both have similar numerical apertures, these being sufficiently low that the thickness of coverslip used in the slide preparation is immaterial.*

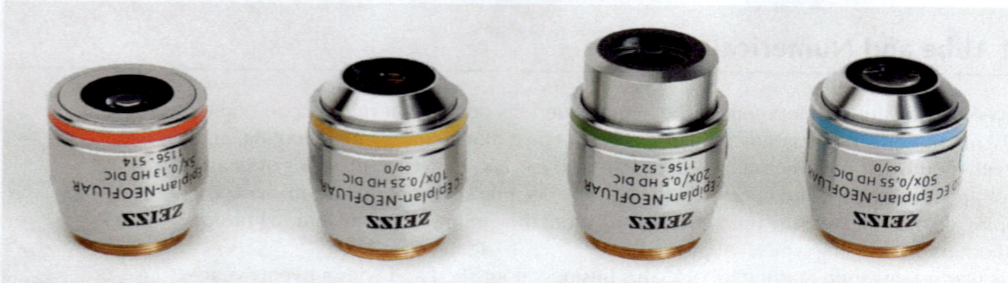

Figure 7.6 ***Epiplan objectives for viewing opaque specimens.***

epi-plan objectives. These objectives act as their own condenser, being used to study opaque specimens. These objectives are further illustrated and described in Chapter 12 on reflected-light microscopy.

7.4 Numerical Aperture and Resolving Power

Imagine that every ray of light leaving a point on the object is carrying some information about the object and that rays leave the object in all directions as shown in Figure 7.7. Clearly only some of these rays will enter the objective; some rays, and thus some information, will not be collected. The amount of information collected will depend on the aperture of the objective and the *angle* subtended by the objective at the object. The aperture of the objective is usually expressed[5] as a dimensionless number, the **numerical aperture** (NA). The numerical aperture value is derived from the angle between the axis of the microscope and the most oblique ray that is able to contribute to the image (denoted by α or sometimes u or θ) and the refractive index of the medium (properly *n* or sometimes incorrectly η) between the slide and the lens (Figure 7.8; also Box 7.2). This angle, α, is sometimes also called the half-angle of aperture. The well-known formula, derived by Ernst Abbe in his calculations[6] is:

$$\mathrm{NA} = \mathrm{n} \cdot \sin\alpha$$

Because the largest value that the sine of an angle may have is 1 (for 90°) and the refractive index of air is 1, the largest possible NA for a dry lens is 1. In practice, dry lenses have apertures no larger than about NA 0.95. This is because the specimen cannot be mounted directly onto the front lens of the objective! There must also be a minimum working distance to allow the objective to focus through the coverslip to the specimen in the microscope preparation. Thus the greatest angular aperture of a high-dry NA 0.95 objective is 143.6° (arcsin 0.95 = 71.8°). By the same token, the greatest angular aperture of an NA 1.4 oil-immersion objective is 134.6° (arcsin 1.4/1.518 = 67.3°; see Figure 7.8).

In the case of the common biological slide preparation with a coverslip, light is refracted as it passes from the coverslip into air. The more obliquely the rays of light leave the object, the greater is the angle of refraction. Depending upon the refractive index of the glass, light refracted at the glass-air interface at an angle greater than the critical angle (about 41°) is reflected back into the glass. This refraction at the coverslip-air interface is thus another factor influencing the ability of the objective lens to collect oblique rays from the object[7]. The use of immersion oil enables rays leaving the specimen more obliquely to be collected, by eliminating the refraction at the coverslip surface and at the front lens of the objective (Figure 7.8). This is because the oil used is chosen to have a refractive index approximately equal to that of glass, typically 1.518, and this permits objective numerical apertures up to about NA 1.4.

As will be explained later (Chapter 10, Section 10.7), the fineness of detail resolvable laterally at the specimen can be calculated from the equation:

$$r_{\mathrm{Airy}} = 1.22 \cdot \lambda/(\mathrm{NA}_{\mathrm{obj}} + \mathrm{NA}_{\mathrm{cond}})$$

where r_{Airy} is the minimum distance resolvable (usually the term 'd' or simply 'r' is used instead of r_{Airy}), and λ is the wavelength of the light (approximately 0.5 µm). Depending upon the wavelength of light and whether the illumination is axial or oblique, the minimum possible lateral resolving power (in the xy-plane) is 0.176 µm, and the minimum axial resolving power (in the z direction) is between 0.5 and 0.8 µm. The axial resolution ($z_{\mathrm{min}} = 2\lambda n/\mathrm{NA}^2$) of a widefield or confocal microscope is usually 3–4 times worse ($z_{\mathrm{min}}/r_{\mathrm{Airy}} = 3.28\ n/\mathrm{NA}$) in the z direction than laterally, in the x and y directions. This has a significant effect when working with small structures and collecting 3D Z-stacks near the resolution limit of the microscope.

Box 7.2 Ernst Abbe and Numerical Aperture

Source: Reproduced with permission from Carl Zeiss Ltd.

Following Joseph Lister's successful development of the achromatic lens doublet for microscope objectives in the late 1820s, microscope manufacture and design was put on a much more rational footing, and businesses consequently flourished both in continental Europe and the UK. However, the basis of image formation was still not understood fully, and microscope producers became skilled at making microscope objectives on an empirical basis, lacking any theoretical foundation: they followed manufacturing protocols that worked without understanding why.

Carl Friedrich Zeiss was no different. Starting in 1846, his business flourished and within twenty years had made 1000 microscopes. Zeiss insisted on high standards – he wanted his name to be synonymous with the highest quality and performance – and it was not unknown for him personally to take a hammer to those instruments that failed the test. However, the empirical method was time consuming and expensive, and success was not wholly guaranteed. Some competitors even said that their microscopes were *'not* like those made in Jena'!

In 1866, Zeiss asked Ernst Abbe, then 26 years old, to join him and place microscope manufacture on a rational footing. Abbe was a highly gifted and energetic personality with outstanding qualifications. His father was an extremely hard-working spinner, yet the family lived in poverty. In later life Abbe worked very hard with Zeiss to ensure good employment conditions for his workforce, and the Carl Zeiss Stiftung (Foundation) continues today.

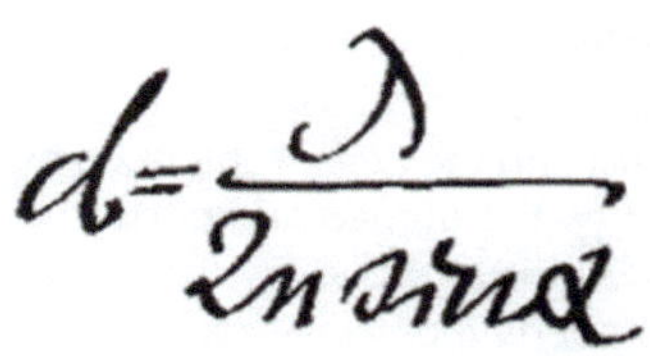

Source: Reproduced with permission from Carl Zeiss Ltd.

By 1871 Abbe had completed his calculations for lens design, with the elucidation of the Abbe sine condition and a complete understanding of microscope image formation by diffraction. Now competitors would claim that their microscopes 'followed the scientific principles established in Jena'. In Abbe's own words:

> '[objective manufacture] was based on a precise study of the materials used, the designs concerned are specified by computation to the last detail – every curvature, every thickness, every aperture of a lens, so that any groping around is excluded'.

Abbe realised that the resolving power was not unlimited but depended upon numerical aperture and wavelength. These terms were related in Abbe's famous formula: $d = \lambda/2n.\sin\alpha$. The term $n.\sin\alpha$ is the numerical aperture (NA), which is useful for comparing the resolving power of one objective with another. Abbe devised a simple means of measuring the NA of an objective using the Abbe Apertometer. A very high-refractive index glass block ($n_D^{20} = 1.625$) was used to project an image of a scale into the back focal plane of the objective, from which the numerical aperture could be read directly. It works with both dry and immersion objectives. This useful instrument is no longer made, but it is possible to make one yourself. A few years ago directions for doing so were published in the *Proceedings of the Royal Microscopical Society*.

When he died, in 1905, two memorials were erected in Jena to Ernst Abbe. The larger, more central one in Carl-Zeiß-Platz is an octagonal temple-like building. However, if you go north, via Johannisplatz and onto Fürstengraben, you will find the Abbe memorial opposite the Botanical Garden. The sphere appears to be made of two lenses – condenser and objective. Around the circumference, in between, you will see orders of diffracted rays and the resolving power formula*.

Source: KaurJmeb 2005. Reproduced under the terms of the Creative Commons Attribution Share-Alike licence, CC-BY-SA 3.0, via Wikimedia Commons.

References

Bower, D.I. (2016). 'Ernst Abbe: creator of the Zeiss Foundation'. Transl. of 1951 volume by N. Günther.

Haselmann, H. (1991). 'A mini-Apertometer'. *Proceedings of the Royal Microscopical Society 26/4*: 217–221.

*This is Jan Hinsch's suggestion; I am grateful to Peter Evennett for pointing it out to me.

The poorer resolving power along the optical axis arises because the lens does not capture all the rays of light emanating from an object but only a few. There is diminished coverage along the optical axis when focusing from above. The ideal would be to place the object at the centre of a spherical objective lens, clearly an impossible proposition. Imaging an object placed at the centre of this spherical lens would mean that the resolution laterally (in x-y) and axially (in z) would be equal and the pixels generated isotropic. As it is, with a real lens, light from the lateral xy-plane enters the lens with a greater semi-angle (higher NA) than those from the axial z-plane, and so the image point spread function (PSF) is not a circular spot but one elongated in the direction of the z-axis.

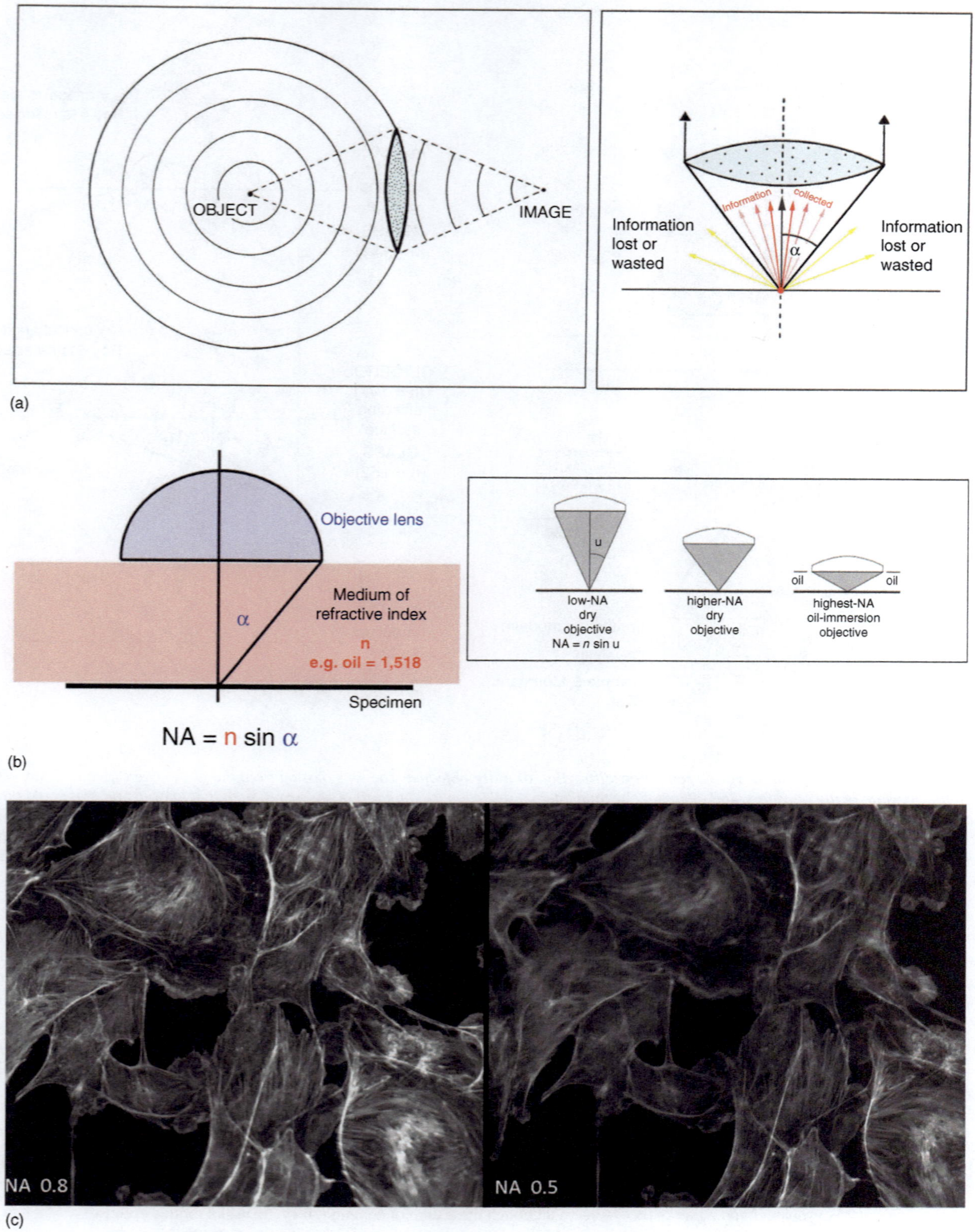

Figure 7.7 **Diffracted rays leaving the specimen** *Only a proportion of all rays diffracted by the object are collected to form the image (a). The numerical aperture value is derived from product of the half-angle of acceptance (α) by the objective and the refractive index (n) of the medium between the objective and the object being examined (b). Numerical aperture determines the resolving power of the objective and relative brightness of the image. The space between object and objective may be filled with air (n = 1.000), or an immersion medium, such as water or Immersol W (n = 1.334); glycerol (n = 1.456) or oil (e.g. n = 1.518 or similar value). Objectives of differing NA will exhibit different half-angles of aperture; the notation u can also be used instead of α. The lower two panels (c) in this figure show the cytoskeleton of flat adherent bovine pulmonary endothelial (BPAE) cells, both taken with 20x objectives. The image of the cell on the left was acquired with a plan-apo 20x/NA 0.8 objective with a resolving power limit at 488 nm of 0.37 μm, while that on the right was taken with a plan-achromat 20x/NA 0.5 objective with a resolving power at the same 488 nm wavelength of 0.60 μm. The image on the right is less bright, also. Source: (a,b) Adapted from Oldfield 1994 with permission from Elsevier.*

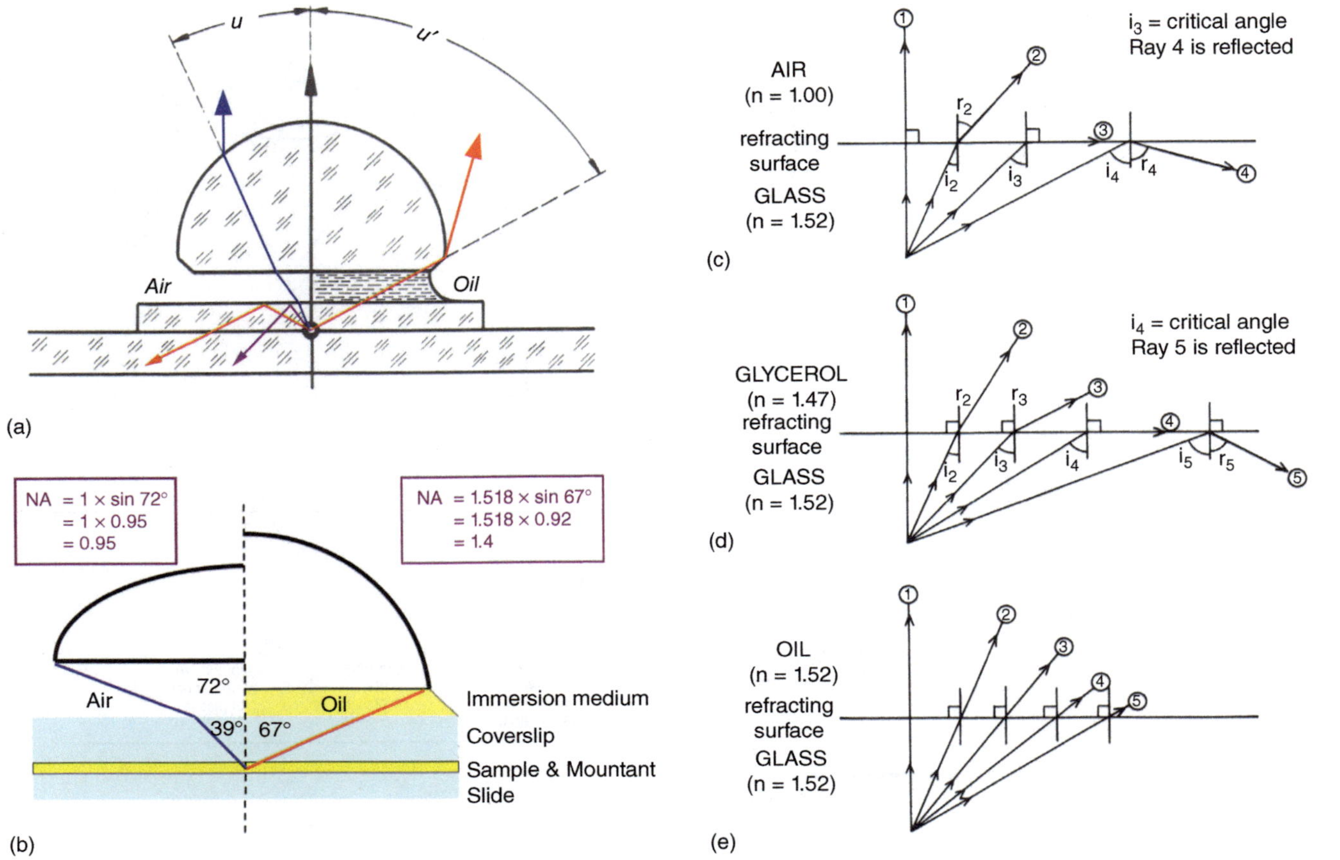

Figure 7.8 ***Function of immersion oil versus refraction/diffraction of a dry objective*** *The rays passing from the glass coverslip into air are refracted away from the aperture of the objective, and ultimately by total internal reflection for the most oblique rays, such that information that could otherwise contribute to resolved fine detail in the image is lost. With an appropriate immersion medium, homogeneous immersion is entirely, or very nearly, achieved, allowing the objective to capture much more oblique diffracted rays. The half-angle, here designated by u, is much smaller in air than the equivalent u′ is for oil immersion. Source: (a) Reproduced with permission from Carl Zeiss Ltd. (b) Adapted from Dr P. Evennett with permission from Dr Evennett. (c,d,e) Allman and Lawrence 1972.*

The magnification of an objective is very much secondary to its aperture, but it is nonetheless important to consider it. Objectives of larger aperture generally have higher magnification so that the eye can more easily see the finer image detail resolved; conversely, low magnification objectives cannot take advantage of the resolution of fine detail provided by a large aperture. There is thus a relationship between aperture and magnification, but it is only an approximate one. It is the aperture that you pay for – the magnification is free; indeed a 40x/1.4 objective will be more expensive than a 100x/1.4, most likely because the 40x/NA 1.4 objective has a greater NA:magnification ratio (optical index; Chapter 31, Section 31.7) and a longer working distance. Figure 7.9 shows two objectives of similar numerical aperture but with different magnifications. The 20x apochromatic objective is of higher quality than the 40x objective and resolves the same detail into a much better-corrected image having a larger field of view.

Exercise 7.1 Comparison of Magnification and Resolving Power

If you have a pair of 25x eyepieces and a pair of 6x eyepieces (although these are not common, they are useful for this exercise),

1. Set up two microscopes, each with a diatom test plate as the specimen. Set up one microscope with the 25x eyepieces and a 10x/NA 0.25 or NA 0.30 objective and the other microscope with the 6x eyepieces and a 40x/NA 0.65 or NA 0.7 objective;
2. Focus on the *Stauroneis phoenocenteron* or *Pleurosigma angulatum* diatoms (refer to Table 10.3 and Box 10.4). Record what you see in each case.

The images of each diatom will both be magnified to approximately the same extent: 240 times. However, more fine detail will be resolved in the image of the diatom viewed with the NA 0.65 objective than with the NA 0.25 objective. In the latter case, it should

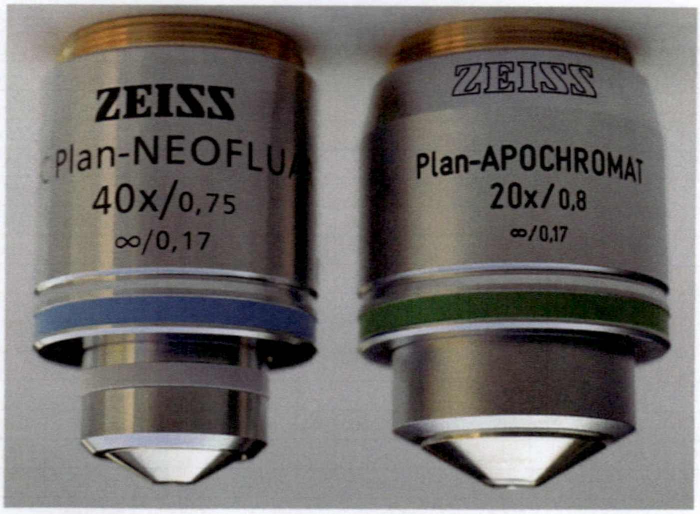

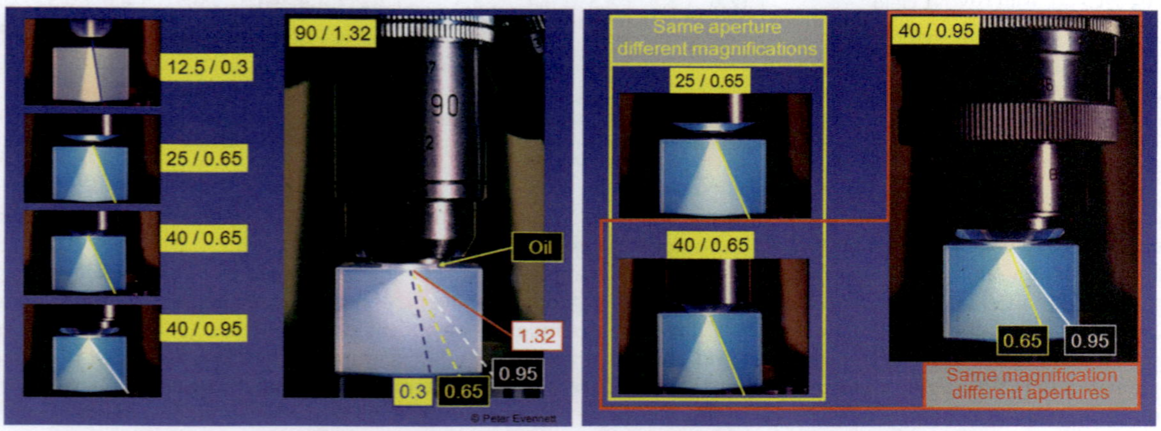

Figure 7.9 ***Objectives of similar numerical aperture and different magnification & correction*** *The 20x/NA 0.8 has the advantage of better correction and a larger field of view. This 20x design is a superb dry objective with a very high numerical aperture for its magnifying power. As a result it is an excellent workhorse on both widefield and confocal microscopes. It is, of course, nearly twice as expensive than the 40x/NA 0.75 objective. The 40x semi-apochromatic objective has fewer lens elements and slightly better transmittivity below 350 nm and above 800 nm. If you have the choice, use the objective that is best suited to your purposes. Source: (top) Courtesy of Mr K. Glover. (bottom) Dr P. Evennett. Reproduced with permission from Dr Evennett.*

be impossible to resolve the striae in the images of each diatom. This should underline the message that of the three essentials for microscopy – resolving power, contrast and magnification – resolving power is more important than magnification. The purpose of the objective is to resolve in the image the fine detail present that is in the object. Magnification is only required to make the image large enough for the eye, or camera, to see the resolved detail.

The standardised values of magnification (Table 7.4) are based on a decimal geometric series that increases in steps of 1.6x and sometimes 1.25x. It is good to have sufficient low magnification objectives on the microscope nosepiece. These give a greater field of view and consequently more information about the specimen. The most common eyepiece magnification is 10x, although a lower magnification eyepiece may be useful for the visual observation of fluorescent samples. High-quality objectives from the four main manufacturers can be relied upon to be within ±1% of the stated magnification value.

7.5 Homogeneous Immersion

Immersion fluid is used to increase the ability of the objective to collect very oblique diffracted rays, which would otherwise – in air – be refracted away at the glass-air interface from the objective. This is explained in Figure 7.8b. Oil is the most commonly used immersion medium, but water (RI 1.334) is also employed for aqueous preparations and/or live-cell imaging. Other immersion media include glycerol (RI 1.456), and Olympus now offer silicone as an immersion medium formulated for live-cell time-lapse work. Ideally, the oil should exactly match the specimen, the mounting medium, the coverslip and the immersion medium (i.e. RI 1.518, approximately equal to that of crown glass, RI 1.517). Under these circumstances, homogeneous immersion exists, and the light rays pass undeviated

Table 7.4 Standard Magnification Values

The standard magnifications are chosen so that they increase by a factor of 1.6 times.																				
Thus:	(a)			2.5x		4x		6.3x		10x		16x		25x		40x		63x		100x
	(b)	1.25x	2x		3.2x		5x		8x		12.5x		20x		32x		50x		80x	
The standard magnification series								Low power		1x	1.25x	1.6x	2x	2.5x	3.2x	4x	5x	6.3x	8x	
								High power		10x	12.5x	16x	20x	25x	32x	40x	50x	63x	80x	100x
Objective selection 1										2.5x	4x or 5x or 6.3x		10x	16x	20x	25x	40x	50x	63x	100x
Objective selection 2										2.5x			10x		20x		40x		63x	
Objective selection 3											4x or 5x or 6.3x		10x				40x			100x
Objective selection 4 (minimum)													10x				40x			100x
Magnification changers							0.5x	0.63x	0.8x	1x	–	1.6x	2x							
Eyepieces										3.2x	4x	5x	6.3x	8x	10x	12.5x	16x	20x	25x	32x
Eyepiece selection 1															10x					
Eyepiece selection 2													6.3x		10x				25x	
Diameter of the field of view (mm) for 10x eyepiece with FoV No. = 22 (assuming tube-lens = 1x)					Objective			1.25x	2.5x	4x	5x	6.3x	10x	16x	20x	25x	40x	63x	80x	100x
					Field diameter (mm)			17.6	8.8	5.5	4.4	3.5	2.2	1.4	1.1	0.9	0.6	0.3	0.3	0.2

If you have ever wondered why the major German manufacturers produce a 63x objective rather than round this to 60x (as the Japanese manufacturers do), it is to comply with an international standard geometrical series of magnifications for light microscopes. The Renard R10 series of preferred numbers makes each succeeding step in magnification a factor of approximately 1.25 (i.e. an increase of about 25%); the product of any two values is a standard value. ISO 8039 also notes that the magnification values 1.5, 15, 30, 60 and 150 are also in use.

from the object, in a straight line, to be collected by the objective, without refraction. For further explanation of this important topic, see also Section 18.8 in Chapter 18 on operating the confocal microscope.

The influence of temperature upon the refractive index of immersion oil is often overlooked. Some workers claim that it is possible to use immersion oil designed for use at room temperature at 37°C without any adverse effect. If the temperature at which the immersion objective is used differs greatly from the standard[8] 23°C, the refractive index of the immersion oil will change. This can make a significant difference when heating the microscope, objective and preparation for a live-cell experiment. The refractive index decreases with an increase in temperature (the Cargille website says that a difference of 1°C in room temperature from the stated value causes a change in the index of the immersion oil of approximately 0.0004). Therefore a graded immersion oil of a high refractive index (obtainable from Cargille) should be used to ensure as near homogeneous immersion as possible. This is particularly important for deconvolution work. The DeltaVision microscope has a calculation algorithm in its software to calculate the correct refractive index oil to use when setting up an imaging experiment.

Some objectives designed for live-cell work have a red fiducial mark on the correction collar (see Figure 7.4) to set when using the objective at the higher temperature of 37°C; any alteration to the correction collar to adjust for refractive index mismatch or difference in coverslip thickness can be made from this mark.

In practice, samples may be mounted in media of different refractive index from that of glass: living samples mounted in water or ionic media will have a much lower refractive index, much nearer to the refractive index of water (RI = 1.333), with an overall refractive index of 1.36, or so. There are four main consequences of refractive index mismatch:

- Where the mountant or the immersion medium and the sample have different refractive indexes (and are not homogeneous), the optical path length is not constant but changes during focusing. The location of the focal plane does not correspond to the distance traversed by the mechanical focus. As a consequence, the sample will appear elongated or squashed along the z-axis. Where the objective immersion medium RI > specimen/mountant RI, the sample appears elongated, and vice versa. Rescaling the z-axis with a factor equivalent to the ratio of the two refractive indexes will restore a 3D data set to its proper dimensions.
- Refractive index is dependent upon wavelength (see endnote 8). If the wrong media are used for the particular objective in question, chromatic aberration will be introduced into the image. For example, a multi-labelled fluorescent bead may have a different z-axis registration for each channel; this may have consequences for the colocalisation of cellular structures.
- A refractive index mismatch will introduce spherical aberration into the image, and if imaging living, watery, specimens with an oil-immersion objective, this will be unavoidable. The situation is made worse if the preparation is warmed to provide an amenable environment for living tissue.
- The consequence of spherical aberration is a squashed point spread function. This in turn has two consequences: a signal intensity drop when focussing into the sample and a loss of resolution.

Provided that the specimen is mounted as close to the under-surface of the coverslip as possible and an objective with a correction collar is used, it should be possible to minimise the blurring that will arise from spherical aberration. Potential refractive index mismatching

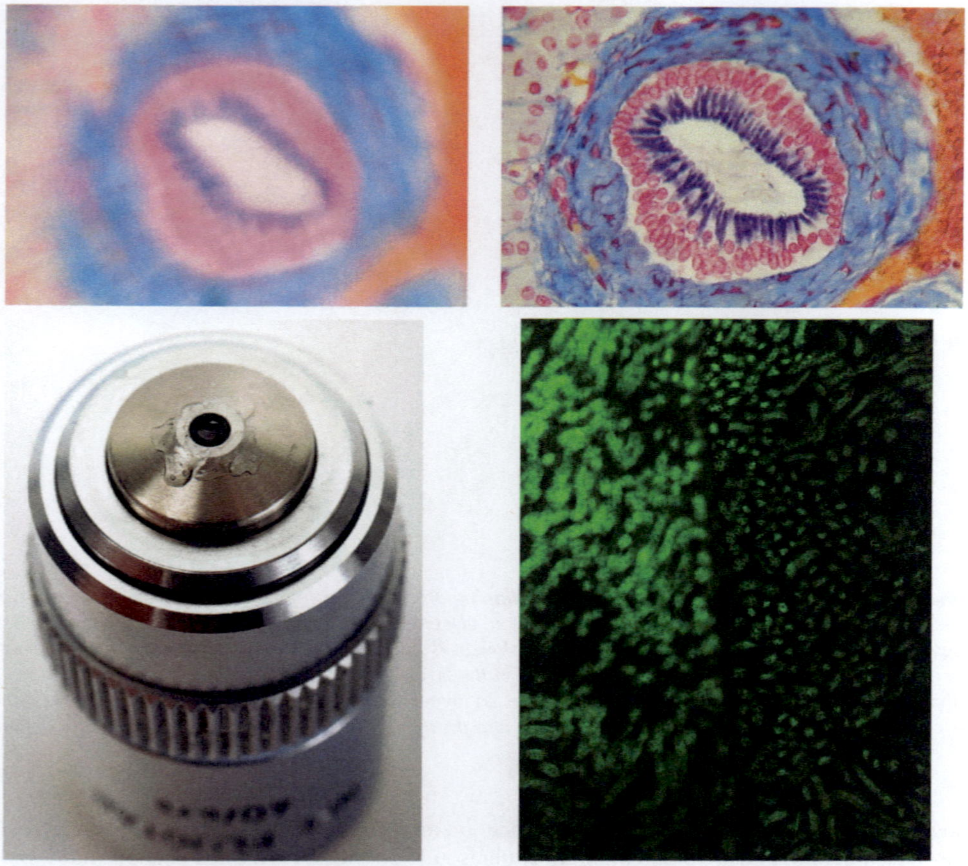

Figure 7.10 ***Immersion objective used without correct immersion medium*** *The two panels (top), show the image degradation arising from oil left to dry onto the front of an objective. In this case (bottom left), a dry objective with a short working distance has picked up oil from the bottom of a slide preparation. An immersion objective with dried-on oil will also give a poor image. In the lower right-hand panel this image of kidney with a broken coverslip half covering the preparation shows the spherical aberration introduced by using the incorrect coverslip, or none at all, when a coverslip of precise thickness is mandatory. Source: (bottom left) Courtesy of Mr K. Glover.*

between the preparation and the immersion medium can also be minimised if a water-, glycerol- or silicone-immersion objective is used instead.

Some immersion oils are designated with the suffix 'F' or 'N', thus: 518 N and 518F. Immersion oil with the suffix 'F' should be used for fluorescence imaging. It has a higher Abbe number ($V \approx 46$ compared to ≈ 42 for 'N' oil) and therefore exhibits less dispersion. This is important when imaging multicolour fluorescent specimens, where dispersion and chromatic aberration should be minimised.

An immersion objective should *not* be used without oil, water, glycerol or silicone, as appropriate to its design. An immersion objective used dry will give an image exhibiting *gross* spherical aberration (Figure 7.10) so much so that it will be impossible to focus sharply or appreciate the detail that should potentially be resolved. Clearly, a dry objective should also never be used with any oil, water or other immersion fluid. They are not generally marked 'dry', as previously explained, but will have a numerical aperture ≤ 0.95 and a curved front lens.

As explained in Section 7.2 and in Section 7.3 on objective inscriptions, oil-immersion objectives have a black ring on the barrel, and water-immersion objectives have a white ring; they will also have 'oil', 'oel' or 'W' for 'water' as part of their description and have a flat front lens. These are the two most commonly encountered immersion objectives. When using a water-immersion objective for any length of time, particularly when the objective and the preparation are to be kept warm at 37°C for live-cell work, there is the risk that the small volume of immersion fluid will evaporate, and this can be very difficult, if not impossible, to replace without stopping the time-lapse experiment prematurely. Zeiss manufacture Immersol™ W, which is immersion oil corrected for the refractive index of water. This does not evaporate in the same way[9]. Zeiss also manufacture thermally-insulated objectives (see Section 7.13) as well as an extremely versatile auto water-immersion system (Figure 7.15), which may be retrofitted to the inverted Axio-Observer stand, is applicable to a number of objectives and is ideal for multi-position time-lapse work.

If you keep a bottle of Immersol W next to the microscope with regular immersion oil, make sure that the two bottles are clearly marked: in my facility I put different coloured tape around each oil bottle or a blue sticker on top (Figure 7.11), to distinguish them at a glance. It has been known for the wrong immersion oil to be inadvertently applied by a user and much time wasted trying to focus the image. In this case, the spherical aberration will be so severe that focusing cannot be done. Very occasionally, it may be that it is

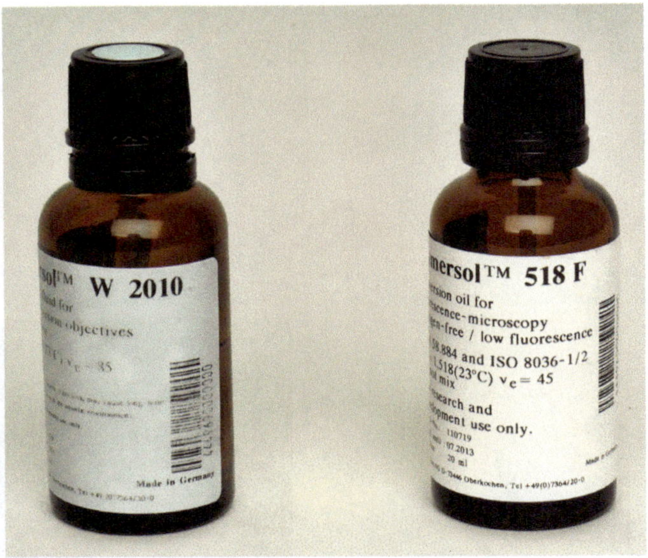

Figure 7.11 **Water-immersion objectives, multi-immersion objective dipping objective and Immersol W** *Homogeneous immersion is the theoretical ideal but is rarely achieved in practice. Water-immersion, multi-immersion, glycerol (Leica) and silicone (Olympus) objectives are all used to reduce refractive index inhomogeneities and improve image quality. Dipping objectives are useful where water immersion is required together with a long working distance. Immersol W (seen here on the left with a blue spot on the lid) is available from Zeiss and has the refractive index of water. It will not evaporate. The last thing you want is a small volume of water used as an immersion coupling to evaporate on an inverted microscope whilst in the middle of a time-lapse experiment. Any attempt to pipette fresh medium into the air gap will upset the focus and ruin the experiment. Source: Courtesy of Mr K. Glover.*

required to use dipping objectives (for their long working distance) on an inverted microscope rather than as normally on an upright microscope. In this case, it is possible to use a small quantity of ultrasound gel, which is a thick aqueous immersion medium – thicker than Immersol W, which will not work. Ideally, you should use a colourless brand of gel; those marketed in the UK tend to have a blue colourant in them. Water-immersion objectives, since they are designed to operate with a medium of different refractive index to glass, can be susceptible to any inadvertent tilt in the coverslip in a way that will not affect an (homogeneous) oil-immersion objective. The image can be aberrated in a manner akin to coma. This phenomenon has been investigated and reported by Arimoto and Murray (2004).

Box 7.3 Focusing an Immersion Objective

The depth of field of a high numerical-aperture, high-magnification objective is very shallow (about 0.4–0.7 μm). The correct focal plane can easily be missed when attempting to focus the specimen, particularly one with little or no contrast. Furthermore, the working distance of an immersion objective is so small (for a 60x/NA 1.4 plan-apochromat it is 0.12–0.19 mm, depending upon the manufacturer) that care is needed to ensure the objective front lens and the coverslip do not touch. The same holds true for specimens examined without a coverslip: the objective front lens must not come into contact with the top surface of the specimen.

For an upright microscope:

1. Locate the sample with a low-magnification dry objective and position the microscope slide so that the area of interest is squarely in the central portion of the stage opening. You can use the small circle of light projected by the condenser, and seen from sideways-on on the top surface of the slide, to locate the area of interest in the illuminated field of view.
2. Lower the stage, or raise the objective upwards, with the coarse focus control to give some clearance.
3. Place a drop of immersion fluid onto the coverslip. One way to do this is to swing the immersion objective partially into position where it will be the next objective rotated into the optical pathway directly above the specimen. Use the space created between this and the neighbouring objective to apply a small drop of the correct immersion fluid to the slide.
4. Look from side-on, and carefully raise the stage or lower the objective with the coarse focus until the front lens makes contact with the drop of fluid. As contact is made, there will sometimes be a bright 'flash' as light passes through the immersion medium.
5. Look down the eyepieces, and continue focusing *very* slowly with the fine focus control. The specimen should come into clear focus, but be aware that with such an objective having a very shallow depth of field, the focal plane will flash in and out of view very quickly. Be careful not to go past focus.
6. If you do overshoot the focal plane, you should feel the front of the objective touch the coverslip. The front of the objective should be spring-loaded, in case of this possibility. In this case, focus the other way with the fine focus only. At the correct focal plane, the image will come into sharp focus.

For an inverted microscope:

1. Place a drop of immersion fluid onto the coverslip, as before.
2. Lower the objective with the coarse focus to give some clearance.
3. Invert the slide, and insert it into the slide holder.
4. Watch the fluid drop from side-on and wind up the objective, with care; this can be done with the coarse focus control.
5. When the front of the objective breaks the meniscus of the fluid drop and it flattens out, continue focusing upwards with the fine focus control until the diameter of the fluid drop is about a little less than 1 cm in diameter.
6. Look down the eyepieces, and continue focusing *very* slowly with the fine focus control. The specimen should come into clear focus.
7. If you overshoot the focal plane, the objective front lens will probably push the slide upwards in the slide holder or push it off the stage, in which case you should feel that contact has been made.
8. In this case, focus the other way with the fine focus only. At the correct focal plane, the image will come into sharp focus.

Immersion objectives and those dry ones with short working distances usually have a sprung-loaded barrel that will retract a certain distance to prevent damage to the front lens and/or slide if the focal plane is missed and the preparation and objective allowed to come into contact with one another. This retractable front can be raised and twisted to lock it upwards out of the way when the objective is not in use. On an infinity-corrected objective, it is possible inadvertently to use the objective in this manner, but it will not be parfocal, and there is the very real risk of damage to the preparation or another objective if it is used in the way and an attempt is then made to swing another objective into position. Check that the immersion objective is properly extended before using it.

Avoid trapping air in the oil when applying it to the microscope. The bubbles, acting as lenses, destroy the image or reduce its clarity. The problem is usually avoided if a sufficiently large drop is used – but not too much. Wait a little, and allow a sufficiently large drop to form that it falls off the applicator by gravity, without touching the applicator to the slide. Dabbing increases the likelihood of bubbles.

As you scan the surface of an immersed microscope slide, periodically observe the position of the stage with respect to the objective to ensure that there is sufficient fluid between the objective and the coverslip. When scanning over a large area, it may be necessary to re-apply some immersion fluid if the layer breaks or becomes too thin. It will be obvious if this is the case, because you will see the fluid boundary move across the field of view as air is drawn in.

On an inverted microscope you can apply a small drop of immersion fluid to the slide before inverting it and still use a low-magnification objective (e.g. no higher than 10x) with a long working distance to safely scan the field of view. If you use the applicator to apply the oil or water directly to the inverted objective by gravity, instead of prior application to the slide before inversion (or bottom surface of a glass-bottomed Petri-dish), take care not to touch the front lens of the objective directly, but allow the drop of immersion medium to flow onto the front lens of the objective.

Immersion oil left on the front of objectives (Figure 7.10) will dry to form a layer that is difficult to remove, running the risk of scratching the anti-reflection coating whilst attempting to do so. Also, it is all too easy to get oil transferred onto the front of a dry objective with a short working distance. If this happens, the image will be indistinct. If this is suspected, remove the dry objective and examine it, either under a stereomicroscope (screw the objective into the lid of its pot to act as a base) or with an inverted eyepiece under good light. Otherwise, point the objective towards a strong, preferably homogeneous, ceiling light, and examine the objective through its rear. You will be able to see any oil contamination clearly on the front lens. Advice on cleaning objectives is given in Appendix 4. Old immersion oil is prone to separate, particularly if held below 18°C. Under the Internet Resources, at the end of the chapter, advice is given on how to redissolve these deposits[10]. Where a microscope is situated in a cold room, Cargille immersion oil has been reported to have a much lower 'cloud point' and so is suitable for this application.

The objectives on inverted microscopes are particularly prone to damage from immersion fluid (oil or water) that is left on the objective and not cleaned off after use or from the application of too much fluid. If excess oil is allowed to run down over the objective, it can seep between the joint of the retractable barrel and so contaminate the internal doublets of the objective. It is impossible to clean an objective that has suffered in this way. Even if it is returned to the manufacturer, they may judge it uneconomic to disassemble, clear and reassemble the objective. Instead you will be charged 50–70% of the value of a replacement internal unit within the barrel of your old objective. To avoid this occurring, there has been a lot of discussion on the confocal and microscopy list servers to share ideas about how to best protect objectives from oil seepage. Some people bind the objective with parafilm; in this case the retraction on the barrel probably won't work. A common solution is to use hair scrunchies, which have the advantage of being colour coded, although they will need to be changed frequently. Other light microscope facility managers use felt 'O' rings or pipe cleaners. Zeiss make an Aquastop for their inverted microscopes to prevent gross aqueous spillage from Petri dishes and perfusion chambers onto objectives, filter sets, analysers and other components. This equipment is not suitable for preventing oil seepage.

Summary

1. Never use an immersion lens without the proper fluid.
2. Never get oil or immersion fluid onto a dry objective.
3. Clean up all fluid, particularly oil, immediately when finished.

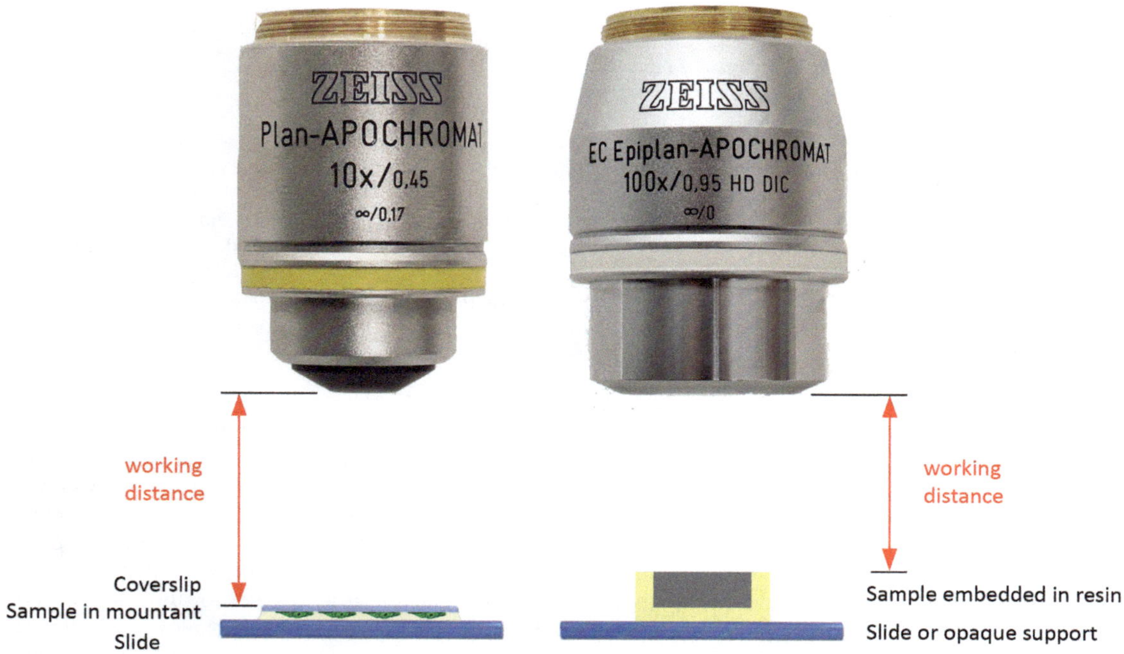

Figure 7.12 ***Illustration of free working distance*** *It is worth knowing the working distance of your objectives – although only Nikon inscribe these values on their objectives – because they set an absolute limit to how far into cells or tissues an objective may focus on a confocal microscope. Working distance can be varied by the manufacturer over a wide range but is shorter in more highly corrected objectives of higher numerical aperture.*

7.6 Working Distance

The working distance (strictly called the free working distance) is the distance between the front of the objective and the upper surface of the coverslip (also called a cover glass) *when the objective is in focus*. The objective should be sharply focused on the specimen properly mounted on the undersurface of a 0.17 mm thick coverslip. For uncovered specimens, the free working distance is measured from the surface of the sample to the front of the objective (Figure 7.12). Some microscopists think that the working distance tells them how deep into their sample the objective can focus without producing a spherically aberrated image. In most cases, this is not true. As a general rule (although there are exceptions), working distance values tend to decrease with increasing numerical aperture and also with increasing magnification.

Nikon mark their objectives with the working distance; otherwise construct your own table, and keep a copy by the microscope. Refer to the manufacturers' websites given in Table 7.2, where you can find the free working distances of your objectives. Table 7.5 shows some selected objectives and their respective working distances, together with other relevant objective parameters.

It is important to have some idea of the working distance of your objectives, particularly the higher magnification ones, where the working distance is usually very small. Firstly, it will help you avoid ramming the objective into the slide as you try and find a specimen and get it into focus. Secondly, because working distances can be so small, the coverslip (if required and if used) must face the objective, which is designed to form an image 'looking' through the coverslip and not the thickness of the slide. In any case, if the slide is put on the microscope stage the wrong way up, with the coverslip on the side furthest from the objective, it usually is not possible to focus the objective onto the specimen because the thickness of the slide exceeds the working distance of the objective.

It is possible to buy objectives designed with long working distances (LWD; see Figure 7.1, Table 7.3 and Figure 7.13). They are required when using inverted tissue culture microscopes. These LWD objectives (and the recent Olympus SLWD objectives), will inevitably have a lower numerical aperture than their routine counterparts. For example, a 20x plan-apochromat will have an NA of 0.7 or 0.8. NA, whereas an LWD 20x objective has a longer focal length, and so the numerical aperture will be less, usually NA 0.45. Long working distance objectives – particularly high-dry ones – will normally also have correction collars (Figure 7.4 and Figure 7.13) on them in order to minimise the spherical aberration introduced by tissue culture vessels being thicker than the standard coverslip.

7.7 The Importance of Coverslip Thickness

In Chapter 6 Section 6.3 it was mentioned that the objective is only corrected for spherical aberration provided that that the object lies at a calculated aplanatic point. This is a simple matter as far as uncovered preparations are concerned, where the surface of the object is observed directly. However, biologists generally prepare their specimens as a sandwich between the slide, which acts as a transparent support, and a thinner piece of glass, the coverslip or cover glass, which acts as a top seal. This allows for a certain amount of fluid, or

Table 7.5 Typical Depth of Field, Depth of Focus, Resolution and Working Distance Values

Selected Objective	Standard Magnification Colour	Refractive index (n) of immersion medium in front of objective	Lateral resolution (μm) $d = 0.61\lambda/NA$	Axial resolution (μm) $(z_{min}) = 2n\lambda/NA^2$	Visual depth of field, DoF (μm) object side = $(\lambda n/2NA_{obj}^2 + n.75/Mag_{TOT}.NA_{obj})$	Approximate depth of field, DoF (μm) $(=n\lambda/2NA^2)$	Approximate depth of focus (mm) image side = $([\lambda n/NA_{obj}^2] \times Mag_{TOT}^2)$	Brightness factor $(NA^4/M_{obj}^2) \times 10^4$	Optical Index $NA \times 1000/Mag_{obj}$	Example Working Distance (mm)
2.5x / NA 0.075 dry		n = 1.0 (air)	3.97	174	83	43	54	0.05	30	Plan-neofluar = 9.5 mm
5x / NA 0.13 dry		n = 1.0 (air)	2.29	57.8	25.5	14	72	0.11	26	Plan-achromat = 5.5 mm
6.3x / NA 0.16 dry		n = 1.0 (air)	1.86	38.13	17.4	10	76	0.17	25.4	Plan-achromat = 4.9 mm UVI for microdissection = 12.8 mm (LWD)
10x / NA 0.25 dry		n = 1.0 (air)	1.19	15.62	7.0	4	78	0.4	25	Plan-achromat 1 = 3.9 mm Plan-achromat 2 = 10 mm (LWD)
10x / NA 0.30 dry		n = 1.0 (air)	0.99	10.84	5.5	3	54	0.8	30	Plan-neofluar = 5.5 mm Plan-fluorite = 16 mm (LWD)
20x / NA 0.5 dry		n = 1.0 (air)	0.60	3.90	1.8	1	78	1.6	25	Plan-neofluar = 2.0 mm
20x / NA 0.75 dry		n = 1.0 (air)	0.40	1.74	0.9	0.4	35	7.9	37.5	Plan-apochromat 1 = 0.60 mm Plan-apochromat 2 = 0.36 mm
20x / NA 0.8 dry		n = 1.0 (air)	0.37	1.53	0.9	0.4	31	10.2	40	Plan-apochromat 1 = 0.55 mm
20x / NA 1.0 W water		n = 1.333 (water)	0.30	1.30	0.8	0.3	26	25.0	50	Plan-apochromat 1 = 2.3 mm Plan-apochromat 2 = 2.0 mm Plan-apochromat 3 = 1.95 mm
25x / NA 0.8 Multi-immersion		n = 1.333 (water) n = 1.456 (glycerol) n = 1.518 (oil)	0.37	2.03 2.22 2.31	0.7 1.1 1.2	0.5 0.6 0.6	64 69 72	6.6	32	Plan-neofluar = 0.21 mm Plan-apochromat = 0.57 mm (LWD)
30x / NA 1.05 silicone		n = 1.406 (silicone)	0.28	1.24	0.6	0.3	56	13.5	35	UPL-S apochromat = 0.80 mm
40x / NA 0.75 dry		n = 1.0 (air)	0.40	1.74	0.7	0.4	139	2.0	18.8	Plan-neofluar 1 = 0.63 mm Plan-neofluar 2 = 0.37 mm
40x / NA 1.2 W water		n = 1.333 (water)	0.25	0.90	0.4	0.2	72	13.0	30	Plan-apochromat 1 = 0.19 mm Plan-apochromat 2 = 0.2 – 0.16 mm Plan-apochromat 3 = 0.28 mm

(continued)

Table 7.5 (Continued)

Selected Objective	Standard Magnification Colour	Refractive index (n) of immersion medium in front of objective	Lateral resolution (μm) $d = 0.61\lambda/NA$	Axial resolution (μm) $(z_{min}) = 2n\lambda/NA^2$	Visual depth of field, DoF (μm) object side = $(\lambda n/2NA_{obj}^2 + n.75/Mag_{TOT}.NA_{obj})$	Approximate depth of field, DoF (μm) $(=n\lambda/2NA^2)$	Approximate depth of focus (mm) image side = $([\lambda n/NA_{obj}^2] \times Mag_{TOT}^2)$	Brightness factor $(NA^4/M_{obj}^2) \times 10^4$	Optical Index $NA \times 1000/Mag_{obj}$	Example Working Distance (mm)
40x / NA 1.30 oil		n = 1.518 (oil)	0.23	0.88	0.4	0.2	70	17.9	32.5	Plan-neofluar = 0.16 mm Plan-apochromat = 0.21 mm
40x / NA 1.40 oil		n = 1.518 (oil)	0.21	0.76	0.4	0.2	60	24.0	35	Plan-apochromat = 0.13 mm
63x / NA 1.2 W water		n = 1.333 (water)	0.25	0.90	0.3	0.2	179	5.2	19	UPL-S apochromat = 0.28 mm
60x / NA 1.30 silicone		n = 1.406 (silicone)	0.23	0.81	0.3	0.2	146	7.9	20.6	UPL-S apochromat = 0.30 mm
63x / NA 1.40 oil		n = 1.518 (oil)	0.21	0.76	0.3	0.2	150	9.7	22.2	Plan-apochromat 1 = 0.19 mm Plan-apochromat 2 = 0.14 mm
100x / NA 1.30 oil		n = 1.518 (oil)	0.23	0.88	0.3	0.2	438	2.9	13	Epiplan-neofluar = 0.31 mm Plan-apochromat = 0.20 mm
100x / NA 1.40 oil		n = 1.518 (oil)	0.21	0.76	0.3	0.2	378	3.8	14	Plan-apochromat 1 = 0.17 mm Plan-apochromat 2 = 0.13 mm Plan-apochromat 2 = 0.09 mm

The minimum distance, d, resolved by a given microscope objective in the lateral (x,y) planes is $d_{x,y} = 0.61.\lambda/NA$, where 0.61 is a constant, λ is the wavelength of the illuminating radiation and NA is the numerical aperture of the objective. The minimum distance (d_z, or z_{min}) resolved by a given microscope objective in the axial (z) plane is $2.n.\lambda/(NA)^2$, where n is the refractive index of the immersion medium between the objective and the sample. Lateral resolution is always 3–4 times better than axial resolution. All values are calculated for the wavelength 0.488 μm and the refractive index of the immersion fluids standardised to the Fraunhofer mecury e line (546.1 nm) at 23°C. The Mag_{TOT} and Mag_{TOT}^2 values assume a 10x eyepiece is used.

Ideally, *as a minimum requirement*, you should collect sections in the widefield microscope at sampling intervals equal to half the z resolution or in the confocal microscope: half the optical slice as determined by the pinhole diameter, otherwise, you lose resolution along the z-axis. For further explanation on the parameters for 2D and 3D sampling, obeying Nyquist's criterion, see Section 31.6 and Table 31.3.

For the visual depth of field calculation, a 10x eyepiece, with no extra tube factors, is assumed. For example, Mag_{TOTAL} for the 20x objectives is 200 times. The brightness factor illustrates why it is difficult to get a good fluorescence image with a large field of view at low initial magnifying powers. Either use a higher magnification (and hence higher NA) objective at a reduction zoom on a scanning microscope or tile the image, to give a larger field of view. Tiling can be done with a motorised stage and good software on either a confocal or widefield microscope.

The optical index is a means of comparing the efficiency of different objectives, in the same way that NA can be compared directly. The larger the optical index, the better the objective (see Chapter 31, Section 31.7).

Figure 7.13 **Objectives with correction collars** *These semi-apochromat and apochromat objectives require a correction collar to adjust the position of the internal lens elements with respect to one another to reduce any spherical aberration re-introduced into the image either by an incorrect coverslip thickness and/or refractive index mismatch. The objective shown in (a) is a high-dry objective, which is particularly sensitive to spherical aberration. The objective in (b) is a long working distance objective designed for use on a tissue culture microscope. The objective in (c) is a water-immersion objective, which, because it is an immersion objective, is not so sensitive to differences in coverslip thickness, but it still requires an adjustment collar to correct for spherical aberration introduced by refractive index mismatch. The objective in panel (d) is the glycerol-immersion apochromat objective manufactured by Leica for live-cell imaging, designed to minimise RI mismatch between the sample, the mountant and the glass coverslip of the preparation. Source: Courtesy of Mr K. Glover.*

mounting medium, to cover the specimen, while the coverslip acts to seal the preparation and protect it from damage or drying out. That is all very well, but introducing a plane-parallel piece of glass into the optical train between the specimen and the objective will upset the calculations that the manufacturer has made in order to correct for spherical aberration. Objective designers can, in turn, take account of this and once again minimise spherical aberration when viewing the sample through a coverslip[11]. However, this can normally only be done for a single thickness. All manufacturers therefore arrange to further correct for a *specific* thickness of coverslip, to ensure sharp focus. This is laid down by International Standard[12] to be 0.17 mm (170 μm) thick.

As light leaves the coverslip, it is refracted at the air-glass interface causing rays to be bent (Figure 7.8). If we trace these rays back to their apparent origin, from the point of view of the objective, the object appears to originate from different positions with respect to the coverslip according to the obliquity of the rays diffracted by the object as they are refracted at the air-glass interface of the coverslip surfaces (Figure 7.14). The effect is the same as viewing a stick inserted into water (Figure 2.14), which appears bent to the observer. The effect is particularly noticeable with dry objectives of high numerical aperture (high-dry objectives) and where some difference of refraction (i.e. a refractive index mismatch) exists between the coverslip, immersion medium, mounting medium and any constituent that may remain of the cells or tissue slice. Theoretically, in a truly homogeneous immersion system, rays are not refracted from their path, but no immersion system is so exact and, whilst the effect is much reduced, spherical aberration due to refractive index mismatch

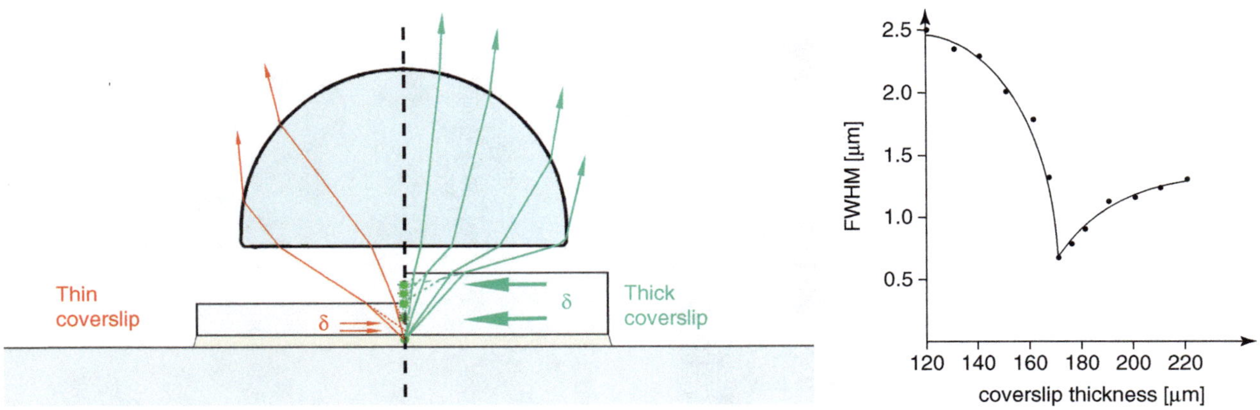

Figure 7.14 ***Illustration of ray path for spherical aberration and the coverslip and Graph showing image quality, spherical aberration and c/s thickness*** *The diagram shows how, from an object O mounted at the under-surface of the coverslip, diffracted rays of differing obliquity refract to differing extents at the glass-air interface at the upper surface of the coverslip, thus appearing to originate at different depths along the optical axis. The designer can correct this effect for a given thickness of coverslip (0.17 mm). Using a coverslip of incorrect thickness upsets this correction and will introduce spherical aberration (see also Figure 6.17) into the image. Spherical aberration affects the point spread function, as measured by the full width half maximum, shown in the graph (for a 63x/NA 1.2 WI objective). Image quality decreases to a greater extent when using a coverslip thinner than the optimum thickness of 170 µm. The notation δ shows the difference between the apparent positions of the object as seen by the objective. Source: Pawley 2006. Reproduced with permission from Springer.*

will still occur when coverslips of an incorrect thickness are used, or a water-immersion objective (RI = 1.33) is used to observe watery cells (say, RI = 1.38–1.42) through a glass coverslip (RI = 1.52).

For really exacting work of the highest standard, it is best to measure the coverslip before use with a vernier micrometer gauge. Square or oblong coverslips ought to be measured at centre and (if large) at the corners also, while circular coverslips, unless really large, need only be measured at the centre. Previous studies have shown that there is a considerable range of variation amongst the coverslips in any box from the nominal thickness quoted. The Royal Microscopical Society recommends (Table 7.6) using a tolerance of ±3 µm about the 170 mm standard, with a relaxed tolerance of ±6 µm. Unpublished work by the author found that as few as 51% of coverslips were within the relaxed RMS standard of ±6 µm (164–176 µm) about the standard thickness, and that the greatest variation occurred within, rather than between, boxes of 100 pieces. It is now possible to buy boxes of coverslips (see Figure 6.18) that fall within

Table 7.6 *Thickness Range of Coverslips*

Box grade No.	Thickness (mm)	
# 0	0.085–0.13	
# 1	0.13–0.16	
# 1½	0.16–0.19	
# 2	0.19–0.25	
# 3	0.25–0.35	
Numerical aperture (objective)	**Admissible deviation from 0.17 mm (mm)**	**Total allowed range (mm)**
0.08–0.30	0.15	0–0.3
0.31–0.45	0.07	0.10–0.24
0.46–0.55	0.05	0.12–0.22
0.56–0.65	0.03	0.14–0.20
0.66–0.75	0.02	0.15–0.19
0.76–0.85	0.01	0.16–0.18
0.86–0.95	0.005	0.165–0.175

Source: Data from Rawlins 1992 and Pluta 1988.

References

Pluta, M. (1988). Table 2.9. *Advanced Light Microscopy*, vol. 1. Amsterdam, Oxford: Elsevier, p. 233.
Rawlins, D.J. (1992). Table 6.3. *Light Microscopy*. Oxford: Bios Scientific Publishers.
Sanderson, J.B. (1994). *Biological Microtechnique*. Oxford: Bios Scientific Publishers.

this relaxed tolerance (e.g. Zeiss, part No. 47-40-30-9000) 18 mm^2, thickness = 0.17 ± 0.005 mm. See the Marienfeld website given under Internet Resources at the end of the chapter.

For the use of a measured coverslip to have any meaning, the specimen *must* be mounted directly, (and the specimen weighted, if needs be during drying time after mounting) or cells cultured directly, onto the surface of the coverslip. Not only must the specimen be mounted, or cultured, in contact with the underside of the coverslip, but the specimen must also be less than 6 µm thick for exacting work. The refractive index (RI) of the sample must lie between 1.523 and 1.525, and the temperature lie within 20 ± 2°C.

More often than not, these conditions are *not* adhered to. It is easier, and much quicker, in fact to mount the section or whole mount sample directly onto the slide, because this is less likely to break. One study (Aumonier and Setterington, 1967) found that if the sample (e.g. a tissue slice) is mounted onto the glass slide and mountant spread between the specimen and the underside of the coverslip (as with histological preparations), the thickness of mountant ranged from 10 to 76 µm. The mean thickness of mountant above the section was 32 µm. In effect the objective 'sees' a coverslip (i.e. including the transparent mountant) about 200 µm thick. If coverslips are *not* measured and sections mounted directly upon the slide, it is advisable to use No.1 thickness coverslips *with* a thin layer of mountant. If a thinner coverslip is used without this extra layer of mounting medium, image quality will decrease, as the graph in Figure 7.14 shows.

Table 7.6 shows the thickness range found in boxes of coverslips. Do not attempt to use a No. 0 coverslip to increase the working distance that the objective will penetrate into the sample, nor use a No.2 coverslip in the belief that it is thicker and therefore offers greater mechanical protection to the sample. For the reasons given, you are advised preferably to use coverslips from a box No. 1½ or, if mounting on the slide, thinner coverslips from a box No. 1. For all exacting high-quality work, use measured coverslips or those from a box of high-tolerance coverslips.

7.8 Objective Correction Collars

The effect of coverslip thickness is negligible for objectives of numerical aperture less than about NA 0.4. Objectives over NA 0.8 are so sensitive (Table 7.6) that it is imperative to correct for spherical aberration when using such an objective for demanding work. There are some multi-immersion objectives which have a correction collar (see Figures 7.4 & 7.13) similar to those found on high-dry objectives. The purpose is the same: to correct for spherical aberration introduced by refractive index mismatch or a coverslip that is not exactly 0.17 mm thick. Turning these collars (see Box 7.4) alters the spacing of those internal lens elements of the objective that are mounted on a helical screw. By this means spherical aberration is corrected, and it should be possible to focus the image sharply. Achieving good focus and a sharp image using an objective fitted with a correction collar does require some experience and a lot of patience, because as the collar is adjusted, so the focus of the image alters. An online Java tutorial has been published on the Nikon NicroscopyU website (this may require an additional Java plug-in to work). The URL is given in the Internet Resources section at the end of the chapter.

Box 7.4 Using an Objective Correction Collar

The thickness of the coverslip used in making the preparation becomes important when using an objective with a numerical aperture greater than about NA 0.4. The diffracted rays of light that form the image are refracted at the glass-air interface of the top surface of the coverslip. Viewed from the perspective of the objective, the object (sandwiched between coverslip and slide), will appear to originate from different depths according to the obliquity of the rays leaving the specimen.

The result is a hazy or unsharp image resulting from spherical aberration that cannot be focused sharply. This effect is particularly noticeable with a dry (non-immersion) objective and becomes more pronounced the higher the numerical aperture of the objective used. What the manufacturer can do is to compute the parameters of the lens doublets within the objective so that spherical aberration is minimised for one particular coverslip thickness. By international convention that thickness is 0.17 mm, or 170 µm.

Thus your specimen should be situated in direct contact with the under-surface of a coverslip that is 0.17 mm thick, with no air gap or mounting medium in between. If the coverslip is thicker or thinner or has mounting medium in between, then the *apparent* thickness of the coverslip (as seen through the objective) will no longer be 0.17 mm. Spherical aberration will therefore be present in the image, to a greater or lesser degree.

Source: Dr P. Evennett. Reproduced with permission from Dr Evennett.

So-called 'high-dry' objectives are fitted with a correction collar. This enables a lens doublet within the objective to move vertically on a helical screw and so minimise any residual spherical aberration. The effect of non-standard coverslip thickness is very much reduced with immersion objectives, because the refraction of light at the glass-oil or glass-water interface is very much reduced. However, some highly corrected NA 1.2 water-immersion apochromats for live-cell imaging are fitted with a correction collar to minimise refractive index mismatch (use the black fiducial line for room temperature and the red line when working at 37°C).

The procedure for setting the correction collar properly is as follows:

1. Ensure that the microscope has first been properly adjusted for Köhler illumination (Box 9.4, Chapter 9, Section 9.11).
2. Set the fiducial line on the collar to 0.17 mm.
3. Focus the specimen as best you can.
4. Move the collar slightly one way, noting the direction.
5. Refocus with the fine focus, and decide whether the image is better or worse.
6. If it is better, continue to move the collar in the direction taken, and repeat 5.
7. If it is worse, turn the collar the other way, and repeat 5.
8. Reiterate until the best image possible has been secured.

Be aware that, particularly on an inverted microscope with a live-cell heating chamber around the microscope, it may be very difficult to see the scales on the objective and its collar and equally difficult to grasp the collar and move it delicately without moving or disturbing the position of the objective and its focus. All this requires practice; consider whether an alternative objective without a correction collar can get the image with the same quality. If it can, use it!

It is worth remembering, particularly when using an inverted microscope within a live-cell heating incubator placed around the microscope, that it may be very difficult to see the scales on the objective and its collar and equally difficult to grasp the collar and move it delicately without moving or disturbing the position of the objective and its focus. All this requires practice; consider whether an alternative objective without a correction collar can get the image with the same quality.

The other option, for those who use the Leica SP8 confocal, or a Leica widefield microscope based on an inverted Leica DMI 6000 stand, is to use a plan-apo 40x/NA 1.10 water-immersion and/or a 63x/NA 1.20 W objective with a motorised correction collar that is fully software controlled. In the autumn of 2016, Zeiss also announced the introduction of a range of five similar objectives with automatically controlled correction collars (Figure 7.15) for confocal and with long working distance for live-cell imaging microscopy: the plan-apo 25x/NA 0.8; C-apo 40x/NA 1.2 W and the C-apo 63x/1.2 W, which fit their Axio-Observer stands.

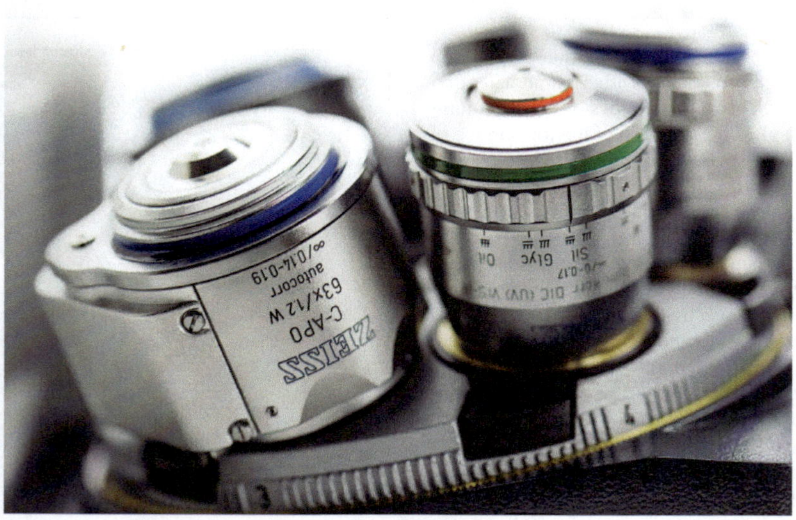

Figure 7.15 **Automatic correction collars** *When using an inverted microscope within a live-cell heating incubator placed around the microscope, it can be very difficult to see the scales on the objective and its collar and awkward or impossible to grasp the collar and rotate it without moving or disturbing the position of the objective and its focus. This is where automated correction collars come into their own for making the necessary delicate adjustment. Source: Reproduced with permission from Carl Zeiss Ltd.*

7.9 Parfocality and Tubelength

Most microscopes will have a rotating nosepiece to change objectives easily and quickly. For convenience, manufacturers compute their objectives to be parfocal. This means that the image will remain in focus when changing from one objective to another (if a small change in focus *is* required, this is due to accommodation in the eye of the observer). To mechanically ensure parfocality, objectives are constructed to be of the same length. This is 45 mm (Leica, Olympus, Zeiss) or 60 mm (Nikon) measured from the shoulder of the objective flange to the object plane.

Since the focal length of each objective of different magnification will change, the microscope must maintain focus when changing from one objective to another. Microscopes maintain focus by having a fixed distance between the object and primary image plane. This is equal to the objective parfocal distance plus the distance between the objective shoulder and the primary image plane (PIP). On older finite tubelength microscopes this distance is 195 mm; on infinity-corrected microscopes it is hypothetically infinite. The eyepiece is constructed so that its front focal plane, designed to be coincident with the PIP, is located 10 mm inside the microscope body-tube from the top of the eyepiece tube on the binocular head (or the body-tube on a monocular microscope). This 10 mm dimension is the parfocalising distance of the eyepiece.

To ensure that the image is sharp and that the diaphragm at the front focal plane of the eyepiece is precisely located where the objective will form the primary image, there is a standard distance, the mechanical tubelength (MTL), which is the distance from the objective thread collar to the top of the tube (the eyepiece locating surface) into which the eyepiece slips. On older microscopes the MTL equals 160 mm; this arrangement is called the finite tubelength system. On infinity-corrected microscopes this distance is hypothetically considered to be infinite[13]. Conventionally the English designers first used 10 inches (250 mm) for the mechanical tubelength. Towards the close of the 19th century the 'continental' standard of 160 mm began to be widely adopted, although Leitz adopted 170 mm and 250 mm for some metallurgical microscopes.

The optical tubelength (OTL) is different from the mechanical tubelength. The OTL is the distance between the back focal plane of the objective and the primary image plane. The OTL is *not* a standardised length because objectives of different magnifying power have different focal lengths: the higher the magnification of the objective, the shorter the focal length. Hence the back focal plane of the objective is located further within the lens elements of the objective. Only very simple achromats of low power will have a back focal plane located outside the barrel of the objective.

7.10 Infinity-Corrected Microscopes

In the finite tubelength system, the ray paths are necessarily convergent between the objective and primary image plane (Figure 7.16a). Therefore, when extra equipment is added to the modular microscope stand (such as a fluorescence epi-illuminator and filter sets, magnification changer, DIC prisms or polarising equipment), the once perfectly corrected optical system now has an effective tubelength greater than 160 mm, and the convergence of the rays towards the primary image plane is upset. To avoid introducing spherical aberration, the ray path has to be expanded with a negative lens system. It is complex for the designer to elongate the tubelength, but not only that, the extra glass elements lead to lower transmittivity of light and a consequent reduction in signal intensity. This is particularly unfavourable for fluorescence microscopy; therefore all the major firms have now turned[14] to manufacturing infinity-corrected optics.

In the case of an infinity-corrected objective, the object lies in the front focal plane of the objective so that *parallel* imaging rays emanate from the back of the objective (Figure 7.16b). This is sometimes referred to as 'infinity space'. No real image is formed, and (in theory) the microscope has an infinite optical tubelength[15]. With these objectives a Telan lens, or tube-lens, at the top of the microscope body, just in front of the binocular head system, converges the rays to a focus at the primary image plane. With parallel rays emanating from the objective, any extra equipment can conveniently be placed in the light path without having to incorporate many extra lens elements to maintain the primary image at its proper location. Furthermore, DIC components introduced into an infinity-corrected system do not alter, or shift, the location of the primary image.

Infinity-corrected objectives are marked '∞' on the barrel and *cannot* be exchanged for finite tubelength objectives (marked '160 mm'; Box 7.1 and Figure 7.5). The tube-lens can be regarded as the last part of the objective: integral to it but distinct and well separated from the main objective body. Parfocality is maintained since objectives with a higher magnification have a shorter focal length, and the parallel rays from the rear of the objective travel slightly further to the tube-lens than the rays from a lower magnification objective with a longer focal length. With infinity optics, therefore, the concept of optical tubelength no longer applies. The magnification produced by an infinity-corrected objective is calculated by dividing the focal length of the tube-lens (i.e. the reference focal length) by the focal length of the objective. The reference focal length ranges between 160 and 200 mm depending upon the manufacturer (see also the Notes in the front matter).

There is a limit to the infinity space behind the objective. Light originating from the periphery of the field of view travels at a diagonal angle (see Figure 7.16b) to the optical axis as it propagates towards the tube-lens. The diameter of the tube-lens therefore places a limit on the length of the infinity space.

In the early days of microscopy, objective lenses were characterised by their focal length, which was inscribed on the barrel (for examples, see Box 7.1 and Figure 7.19). The mechanical tubelength of a monocular microscope could be altered by using a drawtube. For objectives without correction collars, spherical aberration – arising from the use of coverslips of incorrect thickness – could be corrected by using the drawtube (White, 1974). Pushing the drawtube into the microscope body-tube shortened the mechanical tube-length from the standard value and compensated for a coverslip which was too thick (or a coverslip of correct thickness with excess

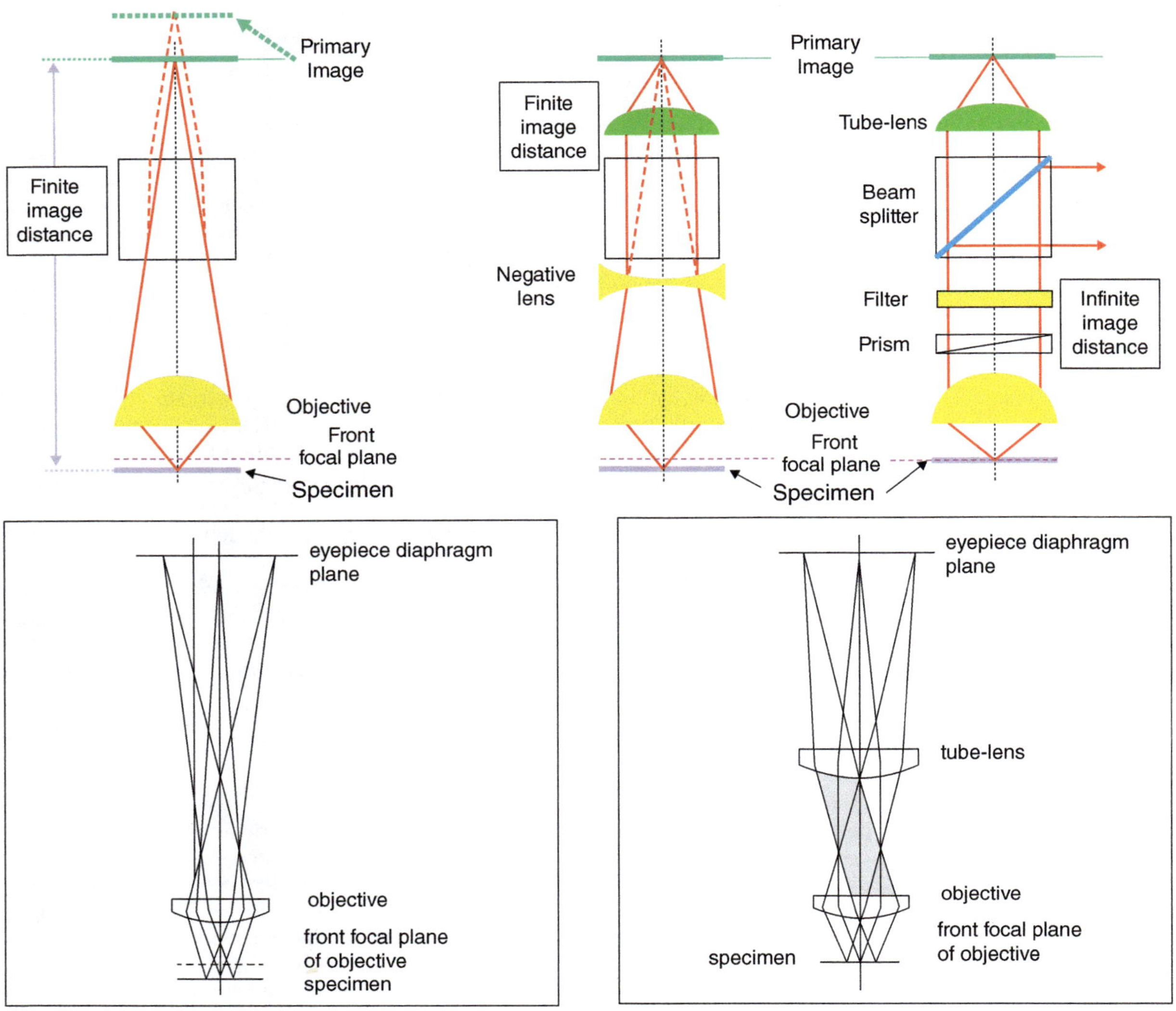

Figure 7.16 ***Finite and infinite tubelength ray paths/infinity space from an infinite tubelength system together with diagonal ray path of peripheral rays*** *So-called Telan optics are used with finite tubelength corrected microscopes to generate an infinity space into which extra components can be inserted, but the negative and positive lens elements that make up the Telan lens may introduce extra magnification and also have potential to reduce contrast and image quality through flare and stray light. Another advantage of infinity-corrected systems is that focusing can be performed by moving the objective, rather than having to move the stage. Source: (top) Dr P. Evennett. Reproduced with permission from Dr Evennett. (bottom) Watkins and St. Croix 2013. Reproduced with permission from John Wiley & Sons.*

mountant between coverslip and specimen). Likewise, pulling the drawtube out of the body lengthened the mechanical tubelength and so compensated for a thin coverslip.

7.11 Depth of Field and Depth of Focus

It is important to have an idea of how much of the thickness of the specimen will be focused sharply by the objective. This is particularly so when using optical sectioning techniques in fluorescence microscopy[16] (Chapter 17, Section 17.2 and following) so that the correct settings can be calculated and applied when collecting three-dimensional data sets. The terms 'depth of field' and 'depth of focus' are not only very similar sounding, but they also refer to distances along the optical axis, albeit one is located in front of the objective and the other behind it. To make matters more confusing, some books and papers use the terms erroneously, employing the term 'depth of focus' in relation to the sample when they actually mean depth of field!

Depth of field is the depth of sharpness in object space, defined as the axial depth of the space on both sides of the object plane within which the object can be moved without detectable loss of sharpness in the image and within which features of the object appear acceptably sharp in the image while the position of the image plane is maintained. Depth of focus is the axial depth of the space on both sides of the image plane within which the image appears acceptably sharp while the positions of the object plane and of the objective are maintained (see Box 7.5).

Box 7.5 Depth of Field and Depth of Focus

There is wide discrepancy in the literature in the formulae cited for calculating depth of field, partly because different assumptions are made regarding the limits to depth of field, the setting accuracy of the microscope, physiological and psychophysiological factors and the minimum angle of visual resolution.

The simplest formula for depth of field is:

$$\text{DoF} = \lambda \cdot n/2(\text{NA}_{obj})^2,$$

where λ is the wavelength of illumination, n is the refractive index of the immersion medium and NA_{obj} is the objective numerical aperture. The proof is given in Pluta (1988) pages 351–352.

The eye, however, can accommodate, and therefore *visual* depth of field is greater than that recorded photographically or by an electronic detector. From work done by Max Berek, chief designer for Leitz in 1927, an extra term must be added, which is $(n.250\,000.\omega/\text{Mag}_{TOTAL}.\text{NA}_{obj})$,

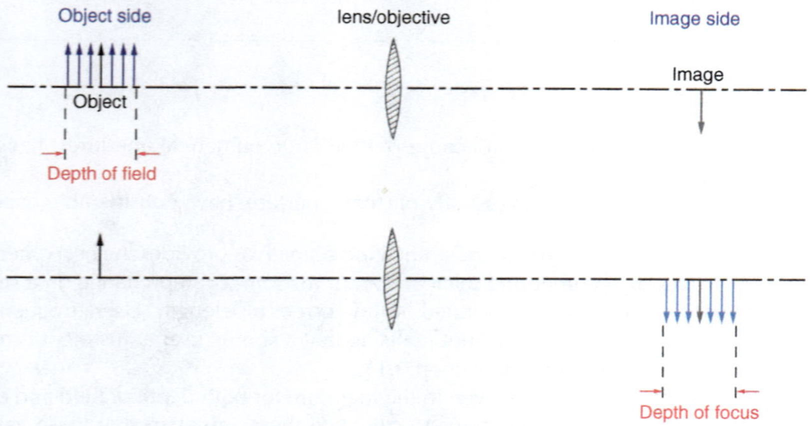

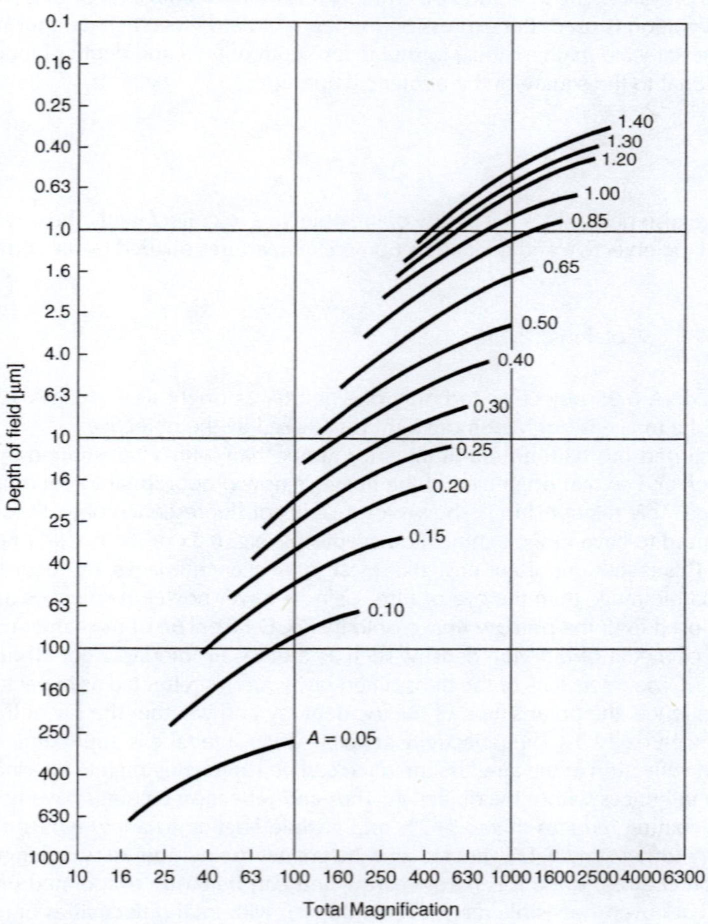

Source: Pluta 1988.

where 250 000 is the reference viewing distance in micrometres; ω = the tangent of the limit of visual acuity (usually taken to be 0.0003) and Mag_{TOTAL} is the overall magnification.

Therefore, for visual observation the total DoF = $\lambda.n/(2\text{NA}_{obj})^2 + n.75/(\text{Mag}_{TOTAL}.\text{NA}_{obj})$

The figure shows values for visual depth of field taken from Figure 3.44 in Pluta (1988) according to overall magnification for objectives of different numerical aperture (A) which agree broadly with the expression above.

These values for the depth of field apply to incoherently illuminated (i.e. emitting) point sources (i.e. $\text{NA}_{cond} \approx \text{NA}_{obj}$); the depth of field increases by up to a factor of 2 as the coherence of illumination increases (i.e. as $\text{NA}_{cond} \rightarrow 0$). The diffraction-limited 'wave-optical' depth of field term $(\lambda.n/(\text{NA}_{obj})^2)$ shrinks inversely with the square of the numerical aperture. The result is that axial resolution (see values in Table 7.5) and the thickness of optical sections in the image are affected by the numerical aperture much more than is the lateral resolution of the microscope.

The depth of focus, on the image side, is given by:

$$\left[\lambda n/\text{NA}^2_{obj}\right] \times \text{Mag}^2_{TOTAL}$$

References

Brattgård, S-O. (1953). 'Microscopical determinations of the thickness of histological sections'. *J. Microscopy 74/2*: 113–122.

Delly, J. (1988). *Photography through the Microscope*, 9th ed. Rochester: Kodak.

Pluta, M. (1988). *Advanced Light Microscopy*, vol. 1: 'Principles and basic properties', Section 3.7.1. Amsterdam: Elsevier, pp. 349–361.

Depth of ***field*** thus refers to ***object space*** outside the microscope and depth of ***focus*** to ***image space*** inside the microscope. A useful mnemonic is that the *field* of view is that part of the *object* that is being examined, and the *focus* is the point at which parallel rays converge *after passing through a lens*.

Two important practical points are:

- high-magnification objectives (because of their large numerical apertures) have extremely limited depth of field yet have relatively large depth of *focus*; and
- low-magnification objectives (usually of small aperture) have considerable depth of field but extremely shallow depth of *focus*.

The restricted depth of focus of low-magnification objectives provides the best criterion for correct setting of microscope eyepiece tubes. A set of objectives on a compound microscope, or a zoom- or step-change on a stereomicroscope, will maintain focus throughout the magnification range *only* when operated at the correct tubelength. Users frequently blame the microscope when it fails to maintain focus throughout its range, perhaps not realising that a simple user adjustment is required. A protocol for the proper adjustment of the microscope is given in Chapter 9, Section 9.11.

Several different equations are given in the literature for both depth of field and depth of focus. This is because different authors have used different criteria and assumptions to calculate these values so that these values can vary by as much as a factor of three! Max Berek, who was a designer for Leitz, first investigated the means of calculating depth of field and published his results in 1927. The values cited in this book are derived from Berek's formula.

The depth of field of a microscope objective, whether fitted to a stereomicroscope or compound microscope, is determined by both geometrical and diffraction (wave) optics, as well as aberrations present in the lens and also whether it is used for image recording with a camera, an electronic detector or by eye. If direct visual observation is used, the expression for depth of field must include a further factor to allow for accommodation by the eye. Box 7.5 gives the selected mathematical formulae for depth of field and depth of focus. To a first approximation, the depth of field is inversely proportional to the square of the numerical aperture.

7.12 Brightness of the Image

Besides determining the resolving power, the numerical aperture also determines the ability of the objective to collect light. The brightness of the image is proportional to the product of the squares of the objective and condenser numerical apertures divided by the square of the magnification:

$$[NA_{obj}^2 \times NA_{cond}^2/\text{Total mag}^2].$$

Thus a 40x/NA 0.65 objective will be half as bright as a 40x/NA 0.95 objective and only one quarter as bright as a 40x/NA 1.30 objective. The increased numerical aperture is the reason why the image is brighter: more light is gathered by the objective.

One of the main reasons that fluorescence microscopy is carried out with an epi-illuminator rather than with a transmitted-light configuration is because the objective acts as its own condenser, and so that brightness of the image is now proportional to the fourth power of the numerical aperture (NA^4), brightness $\propto NA^4/mag^2$. This relationship is shown for a series of fluorescence objectives in Figure 15.8, in Section 15.7. Older fluorescence microscopes used to have low-magnification eyepieces (e.g. 6.3x or 8x instead of the usual 10x) in order to keep the brightness of the image high. This is less important now that most fluorescence images are recorded with high-quality digital cameras that have a much higher dynamic range than the eye or film. Using a lower power eyepiece is also rendered obsolete by the fact that a demagnifying lens is often used to fit the primary image onto the CCD faceplate of the camera.

When light is refracted at the interfaces between two types of media of differing density, such as a glass-air interface, not all light will pass from the first into the second medium (see Figure 2.12). The intensities of the transmitted ray, which is refracted as expected, and the reflected ray will change according to the angle of incidence, the polarisation of the incident ray and whether the ray of light interferes with other coherent rays reflected off other surfaces (see Box 12.1). The reflectivity at an air-water interface is approximately 0.02 (2%) and at an air-glass interface is 0.04. However, this 4% reflection at the air-glass interface can be sufficiently high to attenuate contrast of the image through cumulative reflection at multiple interfaces within the objective. Thus anti-reflection coatings have been developed to improve visibility in the microscope[17]. A single coating reduces this to 1–2% and a triple coating to a few tenths of a percent. The greater the number of coatings, the smaller the amount of stray light reflected and the greater the number of wavelengths covered. Magnesium fluoride is often used as an anti-reflection coating, since it is hard-wearing and can be easily evaporated onto substrates under vacuum. Broadband anti-reflection coatings that cover the visible range (400–700 nm) with total reflectivities of less than 0.5% are commonly achievable.

The fact that reflection of light is markedly affected by interference of light reflected at nearby surfaces means that if a lens surface is coated with a thin transparent coating of dielectric material, then if the optical path difference is $1/4\lambda$, the light reflected off the second surface of the coating will be $1/2\lambda$ out of phase and so will interfere destructively with that reflected off the first surface. This effect can nowadays be controlled very finely by lens manufacturers but was first observed by Lord Rayleigh in 1886, when he found that old, slightly tarnished pieces of glass transmitted more light, and suffered less from annoying reflections, than new, clean pieces.

7.13 Specialised Objectives

The major manufacturers offer very much the same types of objectives – both for transmitted and reflected light – according to the three different classes of achromats, semi-apochromats and apochromats. For example, all manufacturers offer long-working distance objectives[18] for observing cells in culture flasks as well as strain-free objectives for work with polarised light. However, there are specialised objectives, as well as some objectives designed for specific purposes, that can only be obtained from certain manufacturers. Some of the more important designs are mentioned here.

All the manufacturers offer water-immersion and so-called 'dipping objectives' for examining cells and tissues that are either living or else still held in an aqueous medium. Dipping objectives (Figure 7.17) have long working distances and so should be used with upright microscopes. They are suitable for electro-physiological work involving patch clamping (Molleman, 2002) or other manipulative techniques. The ceramic casing around the objective acts not only as an electrical insulator but can also help minimise temperature changes, which can affect the specimen and potentially cause focus drift. Zeiss now manufacture the 25x/0.8 i LCI Plan-Neofluar objective and 63x/1.3 i LCI Plan-Neofluar objective, designed for thermal insulation. To overcome focus drift in the z-axis over time by allowing the imaging system to reach thermal equilibrium, both the incubator and the microscope should be switched on at least four hours – preferably overnight – before any time-lapse experiment. Further practical details regarding live-cell imaging is given in Chapter 28.

Whilst water-immersion objectives are best suited for imaging aqueous specimens, multi-immersion objectives (Figure 7.4) have a correction collar to help minimise spherical aberration when imaging specimens located some distance away from the undersurface of the coverslip. They are some of the most versatile objectives produced and are offered by all the manufacturers. Extra-long working-distance objectives have been developed for multi-photon microscopy, often in conjunction with specialised tissue-clearing techniques, such as Sca*l*e and Clarity. Olympus, Leica and Zeiss offer so-called Sca*l*e and Clarity objectives with working distances between 4 and 8 mm and numerical apertures of either NA 1.0 or NA 0.9. Where a large field of view is required, for example when screening dermatology samples, very low-magnification objectives (Figure 7.17) are available.

Reflecting microscopes were mainly used in the 18th and early 19th centuries, before the introduction of Lister's aplanatic foci, because it was easier to correct for spherical and chromatic aberrations in the image using mirrors rather than lenses. Quite apart from the inability at that time to combine lenses to cancel out aberrations, early optical glass was inhomogeneous and of such poor quality that mirror-based microscopes gave superior images. A change in magnification is effected by changing the eyepiece on a mirror-based reflecting microscope. As we have seen from Chapter 6 and the exercise in this chapter, it is better to change the objective so as to avoid empty magnification and loss of resolving power. A successful reflecting microscope of the early 19th century was designed by Amici yet after Lister's groundbreaking work on aplanatic foci for correcting the lenses of refracting microscopes Amici was to describe his own reflecting microscope as 'a historical souvenir'.

In the mid-20th century, C. H. Burch of Bristol introduced Schwarzschild mirror objectives. Unlike their forebears, these were intended to be complementary to refracting microscopes rather than direct competitors. For equivalent numerical apertures, catoptric reflecting objectives have much greater working distances. For example, a 10x/NA 0.22 mirror objective will have a working distance of 30 mm compared to an dioptric 10x/NA 0.22 objective with a working distance between 4 and 9 mm, depending on the design. Likewise, a 40x mirror objective of NA 0.65 has a working distance of 8 mm compared to 0.5 mm for an equivalent refracting objective using dioptric lenses. These allow microsurgery on living tissues or the examination of objects that would otherwise be inaccessible. Furthermore, mirror objectives allow the use of ultraviolet illumination for increased resolving power and examination of tissues by autofluorescence.

With a mirror objective (Figure 7.18) it is possible to focus using the visible part of the spectrum and then change to UV illumination without needing to refocus, as is the case when using high-transmission objectives with quartz lens elements in a refracting microscope.

Figure 7.17 **Very low magnification and dipping objectives** *Dipping objectives are useful for examining liquid specimens on an upright microscope. These specialist low-magnification objectives are used where a very large field of view must be examined, for example when screening dermatology samples following Mohs surgery. Source: (centre) Courtesy of Mr K. Glover. (right) Courtesy of Olympus Europa SE & Co. KG.*

Figure 7.18 **Mirror reflecting objective** *Long working distance mirror objectives are nowadays very rarely used but can be useful in some industrial inspection applications. Source: Courtesy of Mr K. Glover.*

A mirror objective also allows examination across the entire spectrum from UV to infrared, while quartz lenses can only be used at the one wavelength for which they were designed. For all their advantages, mirror objectives were only ever used for niche applications. It is time consuming and uneconomical to produce the aspherical mirror figures to eliminate spherical aberration. The mirror pairs must be kept accurately aligned, and their surfaces are prone to lose their reflectivity through mechanical damage or atmospheric corrosion.

Metrological and inspection microscopes are best fitted with telecentric objectives. Conventional lenses magnify an object to differing extents depending upon the object-to-lens distance (u). A telecentric objective maintains a constant image size and constant perspective. This allows accurate measurement of objects irrespective of their distance from the lens and/or position within the field of view. Otherwise, perspective distortion will adversely affect accuracy when measuring objects; for example, a circular feature will appear elliptical. Object-space telecentric lenses are the most common design. These maintain constant magnification within a range of distances from the lens and across the field of view. Image-space telecentric lenses are used with sensors that do not accommodate wide angles of incidence, to avoid 'vignetting'.

Since the objective is the microscope and high-quality objectives are expensive, all the major manufacturers offer guidance on choosing the most appropriate objective for the task, or series of tasks, that your microscope is called upon to perform. Table 7.7 gives some principles to consider when choosing objectives.

7.14 Delamination and Fungal Contamination

Cemented lens doublets separate, wholly or in part, when the adhesive joining them together fails. Separation can also occur in wide-angle eyepieces, Nomarski DIC Wollaston prisms and (on older equipment) the negative Telan lenses found in the rotating objective nosepiece. Very often the rear doublet of an objective is the element most likely to be affected. Delamination can be seen as a faint, usually grey, film spreading across the lens element and – in bad cases – Newton's rings will also be seen. There are two likely causes of

Table 7.7 *Selecting an Objective*

Your selection should be determined by both the sample and your application.

1. Define the magnification – due to specimen size.
2. Define the smallest detail you wish to see.
3. Calculate the resolution.
4. Choose the lowest magnification consistent with this resolving power for the largest field of view and increased image brightness.
5. Choose the objective best-suited for particular applications:
   (a) contrast method
   (b) live-cell or inverted microscope work, use LWD objectives if the sample is thick (do not use LWD objectives for high-resolution work: they have comparatively low numerical apertures)
   (c) fluorescence work
6. Use an apochromat for multicolour fluorescence work and image recording.
7. Use a lower-corrected objective, with fewer glass elements, for live-cell work, or for imaging into the UV region.
8. Match the immersion medium to the RI of the sample and the mountant.
9. Use dipping objectives for physiology or direct immersion into water/aqueous solution.
10. Use strain-free objectives for polarised-light work.
11. Use DIC-optimised objectives for thick (100–200 μm) samples; optically section with DIC.
12. If the unstained sample is very thin, use phase contrast, or dark-field illumination (phase contrast not recommended for magnification > 40x for BF & DIC. Absorption of light by the phase ring is in the order of 88% overall intensity).

delamination. Firstly, excessive heat from the illumination system. It is always wise to use a Scott KG5 filter (or similar) to cut down the infrared (IR) transmission through the microscope. Secondly, storing objectives for long periods in conditions where the ambient temperature fluctuates considerably can also predispose towards delamination. An extremely knowledgeable and well-respected applications specialist keeps their objectives at constant temperature in a wine cooler. A well-illustrated article by David Walker on Microscopy-UK (2006) shows images taken with objectives suffering from delamination. Loss of image contrast results from delamination; with care it is possible to minimise this using post-acquisition digital manipulation with corrections for tonal range and colour cast.

Fungal contamination is rare outside the tropics, and modern objectives have an anti-fungal agent incorporated into them, but this is not the case for older objectives. Keeping the room air dry is the best defence, and good air conditioning will help. Otherwise install a low-wattage infrared lamp above at least 150 cm (5 feet) above the microscope, in a cupboard, to keep the relative humidity below 60%. It is good practice to cover microscopes against dust ingress, but in the tropics this should not be done, to allow dry air to circulate and reduce the risk of fungal spores settling and germinating. Surface-growing fungi can be wiped off, but corrosion marks will remain due to etching by organic acid produced by the fungus.

7.15 Handling Objectives

The objective forms its image by having a relatively large lens-to-image distance (*v*) in relation to the object-to-lens distance (*u*). Because the image is a very much magnified representation of the object, in order to minimise aberrations adequately, the degree of correction must be very high and the tolerance latitude very low. Most objectives are corrected for errors to within $\lambda/20$, a very small distance,

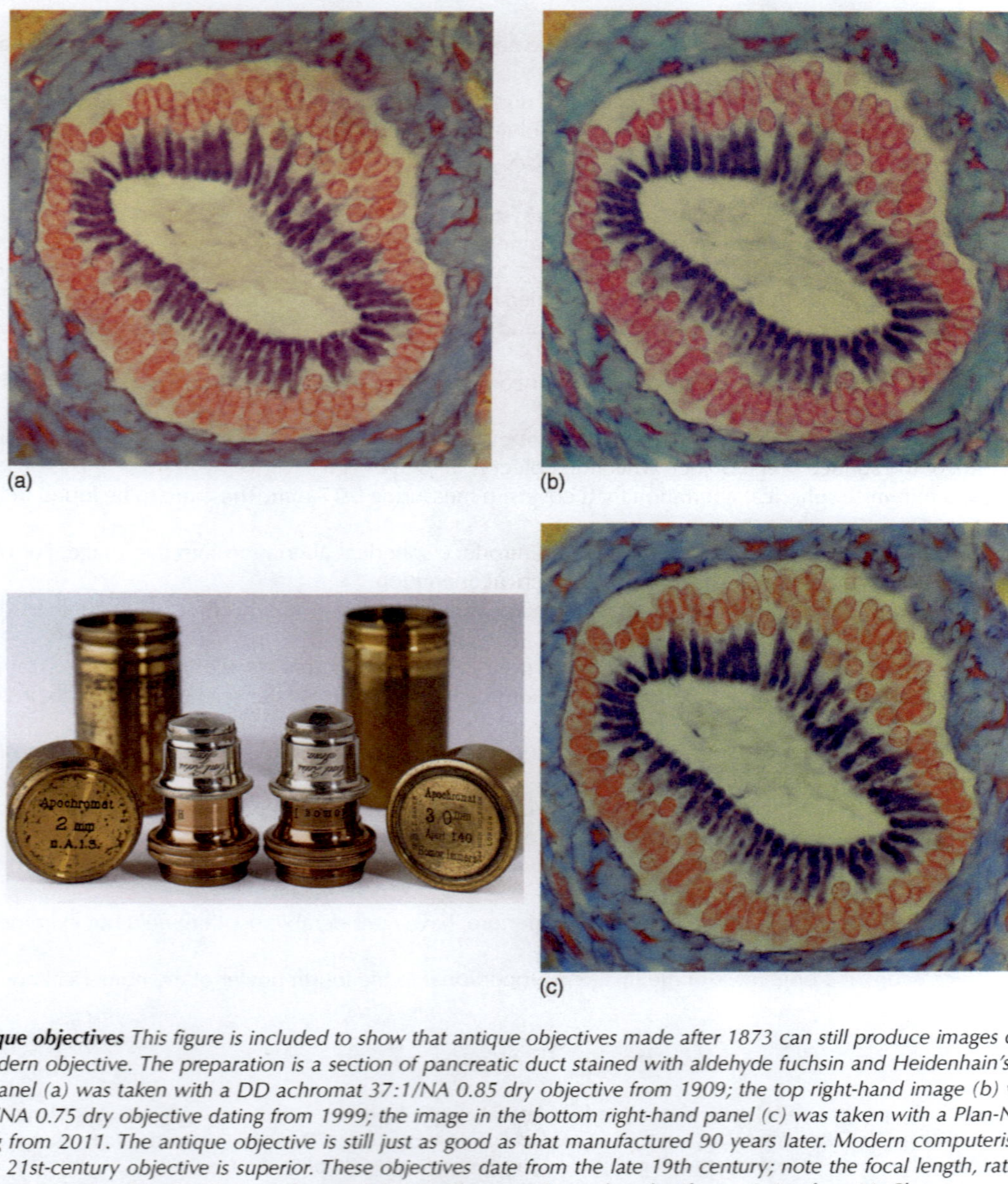

Figure 7.19 **Antique objectives** *This figure is included to show that antique objectives made after 1873 can still produce images comparable to those secured with a modern objective. The preparation is a section of pancreatic duct stained with aldehyde fuchsin and Heidenhain's azan. The image in the top left-hand panel (a) was taken with a DD achromat 37:1/NA 0.85 dry objective from 1909; the top right-hand image (b) was acquired with a Plan-Neofluar 40x/NA 0.75 dry objective dating from 1999; the image in the bottom right-hand panel (c) was taken with a Plan-Neofluar 40x/NA 1.3 oil objective dating from 2011. The antique objective is still just as good as that manufactured 90 years later. Modern computerised objective design methods mean the 21st-century objective is superior. These objectives date from the late 19th century; note the focal length, rather than magnifying power, used to characterise the objective. Source: (bottom left) J.D. Chapman. Reproduced with permission from Mr Chapman.*

about 20–30 nm. Objectives are constructed of multiple lens elements which are therefore very precisely aligned. Consequently, objectives can be thought of having an infinite Young's modulus of elasticity! In other words: they do not bounce. Dropping an objective of even simple construction, like a 10x achromat, through a few centimetres onto the laboratory bench carries a high risk of permanently damaging it.

Hold objectives with both hands. It is easy to have an objective inadvertently slip whilst removing it from the nosepiece of the microscope. The sudden drop as it winds free of the nosepiece thread can take you by surprise, particularly if space is tight and you try to remove the objective single-handed and the objective is a high-quality apochromat that is comparatively heavy. Do not try and carry more than one objective at a time: instead make separate trips, as needed – in case you trip!

Finally, it is worth remembering that from the 19th century, with the elucidation of the principles of image formation in the microscope by Ernst Abbe, formulated and elucidated by 1868 and published in 1873, the light microscope was essentially perfect. Image formation was limited only by the laws of physics. Many people at the time considered that microscope design could not be improved. Until the widespread adoption of infinity optics from the 1980s onwards, the objectives of yesteryear could just as easily be used on a modern finite tubelength microscope. Figure 7.19 shows a photomicrograph of a histological section taken with three objectives made a century apart. On close inspection of the original, there is a slight softening in the images formed by the antique objective, but otherwise the differences are very slight – testament to the very high quality of German manufacture. The image of starch grains under polarised light with a full wave compensation plate, in Chapter 14 (Exercise 14.17) has been taken with a 40:1/NA 1.25 apochromat water-immersion objective from 1898, equivalent in its day to a 21st-century 40x C-Apochromat objective.

7.16 Chapter Summary

1. The three main classes of objective, corrected for chromatic and spherical aberration to different extents, are: **achromats**, **semi-apochromats** and **apochromats**.
2. The inscriptions on the objective indicate its magnification, numerical aperture and function.
3. The higher the numerical aperture value, the greater the resolving power and the greater the brightness of the image.
4. The detail resolved in the image equals $1.22.\lambda/(NA_{obj} + NA_{cond})$ or $0.61.\lambda/(NA_{obj})$ in epi-illumination microscopy, where the objective acts as its own condenser.
5. Older objectives conform to the (160 mm) finite tubelength system. Modern objectives are designed to have infinite tubelengths. The two types of objective cannot be interchanged on the same microscope stand.
6. Immersion fluid is used to increase the ability of the objective to collect very oblique diffracted rays. Generally, objectives below NA 1 are used dry, whilst all objectives > NA 1 require immersion.
7. Oil-immersion objectives have a black ring on the barrel, and water-immersion objectives have a white ring. Multi-immersion objectives usually have a red ring on the barrel.
8. An immersion objective should *not* be used dry or with the incorrect immersion fluid. All immersion objectives should be cleaned carefully after use (see Appendix 4).
9. The free working distance is the distance between the front of the objective and the upper surface of the coverslip or (for an uncovered specimen) the surface of the object, when the object is in focus.
10. Objective designers minimise spherical aberration for a coverslip measuring 0.17 mm. These are to be found in boxes designated No. $1^1/_2$.
11. Below NA 0.4, a coverslip of incorrect thickness does not introduce spherical aberration into the image. For objectives of high NA, a correction collar is sometimes fitted to minimise spherical aberration.
12. Modern objectives are parfocal. The image will remain in focus when changing from one objective to another. They are generally also parcentral.
13. With an infinity-corrected objective, the object lies at the front focal plane so that *parallel* rays emanate from the back of the objective. This infinity space allows extra equipment (e.g. fluorescence illuminators/filter sets; DIC prisms) to be placed in the light path. Infinity-corrected objectives are marked '∞' on the barrel.
14. Depth of field is the axial depth of the space on both sides of the object plane within which the object can be moved without detectable loss of sharpness in the image.
15. Depth of focus is the axial depth of the space on both sides of the image plane within which the image appears acceptably sharp while the positions of the object plane and of the objective are maintained.
16. High-magnification objectives (because of their large numerical apertures) have extremely limited depth of *field* yet have relatively large depth of *focus*.
17. Low-magnification objectives (usually of small numerical aperture) have considerable depth of *field* but extremely shallow depth of *focus*.
18. In fluorescence microscopy the brightness of the image is proportional to the fourth power of the numerical aperture.

Key Reading

Benham, G.S. (2002). Practical aspects of objective lens selection for confocal and multiphoton digital imaging techniques Chapter 6. In: Matsumoto, B., ed. *Methods in Cell Biology 70 Cell biological applications of confocal microscopy* 2nd ed. Amsterdam: Academic Press, pp. 245–299.

- A very comprehensive and practical chapter on selecting objectives for confocal and multi-photon microscopy.

Piston, D.W. (1998). 'Choosing objective lenses: the importance of numerical aperture and magnification in digital optical microscopy'. *Biol. Bulletin 195*/1: 1–4.

- A short, concise account of the principles discussed in this chapter.

References

Allman, M. and Lawrence, D.F. (1972). *Geological Laboratory Techniques*. London: Blandford Press.

Arimoto, R. and Murray, J.M. (2004). 'A common aberration with water-immersion objective lenses'. *Jour. Microscopy 216*/1: 49–51.

Aumonier, F.J. and Setterington, R. (1967). 'Some notes on the mounting of histological sections'. *Proc. Roy. Microsc. Soc. 2*/4: 428–429.

Martini, N. et al. (2002). 'A new high-aperture glycerol immersion objective lens and its application to 3D-fluorescence microscopy'. *Jour. Microscopy 206*/2: 146–151.

Molleman, A. (2002). *Patch Clamping: An Introductory Guide to Patch Clamp Electrophysiology*. Chichester: Wiley.

Oldfield, R. (1994). *Light Microscopy: An Illustrated Guide*. Wolfe, an imprint of Mosby-Year Book Europe (Elsevier).

Pawley, J.B. (2006). *Handbook of Biological Confocal Microscopy*, 3rd ed. New York: Springer.

Pluta, M. (1988). *Advanced Light Microscopy*, vol. 1: 'Principles and basic properties'. Amsterdam: Elsevier, pp. 349–361.

Watkins, S. and Croix, C.S. (eds.) (2013). *Current Protocols Select: Methods and Applications in Microscopy and Imaging*. Wiley.

White, G.W. (1974). 'The correction of tubelength to compensate for coverglass thickness variations'. *Quekett Jour. Microscopy 32*/10: 411–420.

Further Reading

Abbe, E. (1881). 'On the estimation of aperture in the microscope'. *Jour. Roy. Microsc. Soc. 1*/3: 388–423.

Berek, M. (1927). 'Grundlagen der Tiefenwahrnehmung im Mikroskop'. *Marburger Sitzungsberichte 61*: 189–223.

Gerth, K. (2005). *Ernst Abbe: Scientist, Entrepreneur, Social Reformer*. Jena: Verlag Dr. Bussert & Stadeler.

Gunther, N. (1951). 'Ernst Abbe: Creator of the Carl Zeiss Foundation', translated by Bower, D.I. (2016). https://bit.ly/2DlyojV

Hartley, W.G. (2000). 'Numerical aperture and its determination'. *Quekett Jour. Microscopy 38*/6: 435–443.

Haselmann, H. (1986). 'One hundredth anniversary of Abbe's apochromats'. *Proc. Roy. Microsc. Soc. 21*/4: 197–202.

Internet Resources

Typical lens designing software WinLens. http://www.qioptiq-shop.com/Optik-Software/Winlens-Optical-Design-Software/.

ImageJ objective parameter calculator. https://imagej.nih.gov/ij/plugins/calc-objective/index.html.

Olympus Objective FAQ. https://www.olympus-lifescience.com/en/microscope-resource/.

Nikon MicroscopyU Correction collar tutorial. https://www.microscopyu.com/tutorials/adjustment-of-objective-correction-collars.

Infinity optics explanation. https://www.microscopyu.com/microscopy-basics/infinity-optical-systems and http://zeiss-campus.magnet.fsu.edu/tutorials/basics/objectivemagnification/indexflash.html.

Nikon CF 160 optics. http://www.cyto.purdue.edu/cdroms/micro1/7_spon/nikon/id87.htm.

Telecentric lenses. https://www.opto-engineering.com/index.php?/resources/telecentric-lenses-tutorial and https://www.edmundoptics.com/resources/application-notes/imaging/advantages-of-telecentricity/.

Delamination, David Walker (2006). http://www.microscopy-uk.org.uk/mag/artaug06/dw-delam.html.

Avoiding fungal contamination. http://micro.magnet.fsu.edu/primer/anatomy/cleaning.html.

Cargille immersion oil. http://www.cargille.com and http://www.cargille.com/immeroil.shtml.

Marienfeld website. http://www.marienfeld-superior.com.

Precision coverslips. https://www.marienfeld-superior.com/cover-glasses.html.

International Standards ISO/TC 172/SC 5 Microscopes and Endoscopes

See:

Microscopes>Cover glasses>Part 1: Dimensional tolerances, thickness and optical properties. ISO 8255-2:2011.

Optics and optical instruments>Microscopes>Cover glasses>Part 2: Quality of materials, standards of finish and mode of packaging. ISO 8255-1:2013.

Microscopes>Imaging distances related to mechanical reference planes>Part 1: Tubelength 160 mm. ISO 9345-1:2012.

Microscopes>Imaging distances related to mechanical reference planes>Part 2: Infinity-corrected optical systems. ISO 9345-2:2014.

Microscopes>Designation of microscope objectives>Part 1: Flatness of field/Plan ISO 19012-1:2013. Part 2: Chromatic correction ISO 19012-2:2013 Part 3: Spectral transmittance. ISO 19012-3:2015.

Microscopes>Immersion liquids for light microscopy. ISO 8036:2015.

Optics and optical instruments>Microscopes>Reference system of polarized light microscopy. ISO 8576:1996.

Notes

1. Fluar objectives are designed especially for qualitative and quantitative analyses of ion movements and for particularly critical fluorescence methods (e.g. chromosome studies in cytogenetics and where high brightness leads to low exposures in time-lapse microscopy). Fluar objectives are distinguished by extremely high numerical apertures and high transmission for wavelengths from 340 nm. Field flattening is sufficient in the centre of the field of view to allow the use of CCD cameras.
2. The fluorite objective was first designed by Abbe and Rudolf Winkel in 1884; the first apochromat was offered for sale in the 1886 Zeiss catalogue. The Neofluar was manufactured by Zeiss in 1953 using artificial fluoride from Schott Glass. Nowadays fluorite is still used in specialised objectives on account of its very low refractive index (RI = 1.4328) and anomal dispersion (RI decreases with wave frequency, ν). With normal dispersion, RI increases with frequency and decreases with wavelength.
3. So commercially sensitive was the use of fluorite, that Abbe marked an 'X' on the data sheet for the fluorite element, so as to keep it secret from the other optical companies. The 'water-clear' fluorite 'Flussspath' came from a scree slope of the Oltschikopf, near the village of Brienzwiler in the canton of Bern, on the eastern edge of the Bernese Oberland.
4. Abbe's definition of apochromatism was the following. Apochromat: an objective corrected to a common focus for three widely separated wavelengths and for spherical aberration/coma for two widely separated wavelengths.
5. Modern manufacturing processes are so reliably reproducible that the NA value inscribed on the objective barrel can be accepted. Most NA values are accurate to two decimal places. Formerly, especially amongst amateur microscopists, a great deal of time and effort was expended in measuring numerical aperture. For a reference on how to do this, see Box 7.2. Since numerical aperture is a product of two ratios, it is dimensionless. The scale therefore has no units.
6. In a letter dated 13th of January 1878, Abbe stated that the numerical aperture provides 'the exact theoretical measure of the resolving power of an objective'.
7. The practical advantage of immersion objectives was known from the middle of the 19th century. These objectives proved popular: Zeiss, for example, introduced water-immersion objectives in 1871; oil-immersion in 1877 and by 1904, in the space of just 27 years (when many fewer people used microscopes) they had sold their 10 000th objective for homogeneous oil-immersion; on average one a day.
8. Immersion oil from the major microscope manufacturers, or Cargile, is produced according to the German DIN (Deutsches Institut für Normung) specification 58-884 and ISO 8036-1/2. The refractive index of immersion oil is specified at one wavelength, generally the Fraunhofer mecury e line (546.1 nm) because this wavelength is near the middle of the visible spectrum where the eye has the greatest sensitivity (Other standard refractive index measurements are usually cited at the sodium D lines mean value of 589.29 nm; see Table 2.4 in Chapter 2). DIN/ISO specify that the refractive index of general-purpose immersion oil at the mercury e line at 23°C shall be 1.5180 ± 0.0004 (DIN), ± 0.0005 (ISO), with a dispersion (Abbe number, V_e) of 44 ± 5 at 23°C.
9. Zeiss recommend that Immersol W should only be used with a preparation sealed with a coverslip and not be used for open preparations (use 518N or 518F in this case). Immersol W will attack dyes and, if smeared accidently on the colour coding ring of the objective, attack that coloured material also.
10. Zeiss has since modified Immersion oil 518F so that this is far less of a problem, but crystals can still form below 18°C, particularly if, say, the sample slide has excess 518F oil on the coverslip and is stored in a fridge. However, Cargille oils FF and DF in comparison have a cloud point of −6°C [far lower than Zeiss' 518F oil] – hence the reason we had to use Cargille FF in the cold lab. If air conditioning gives rise to cloudiness problems, it is also worth using Cargille oils.
11. The reason that chromatic aberration does not need to be corrected further and that coverslip thickness only affects spherical aberration in the image is because *all* wavelengths are equally affected.
12. The details of the International Standard are given at the end of the chapter. On some older British and American objectives, the standard coverslip thickness was taken to be 0.18 mm. This is now obsolete: the standard is 0.17 mm.
13. For further details and definitions, see the International Standards cited at the end of this chapter and also in Appendix 1 of the *RMS Dictionary of Light Microscopy* Figure 2 on page 64 and Table 2, pages 72–73.
14. Introduced by the firm of Reichert in the 1930s; adopted by all manufacturers from the 1980s onwards.
15. The tubelength in infinity-corrected optical systems (also known as the tube-lens focal length) ranges from 160 to 200 mm. Meiji, Leica and Nikon all have a focal length of 200 mm, Olympus uses 180 mm, and Zeiss 165 mm. The objective parfocal distances are all 45 mm, except for Nikon CFI60 objectives, which is 60 mm.
16. As a rule of thumb, a 'thick' fluorescence specimen is one which is at least 10 times the depth of field of the objective.
17. Single-layer anti-reflection coatings were first used in the manufacture of objectives by Zeiss, as a result of the research done by A Smakula. Much earlier Fraunhofer and Lord Rayleigh had discovered that old lenses with weathered surfaces displayed higher transmission. Changes caused by corrosion reduced the refractive index in the upper layers of glass. This resulted in decreased reflection and therefore increased transmission. See Figure 7-26, page 159, in Pawley (2006).
18. Long working distance objectives inevitably have a lower numerical aperture than equivalent objectives of the same magnification having shorter working distances.

8

Condensers and Eyepieces

Outline

Understanding Light Microscopy, First Edition. Jeremy Sanderson.

8.1 Introduction

The objective, as we have seen in the last chapter, is the most important, best-corrected optical element of the microscope. All the other components – lamp, condenser, eyepiece(s) – are simply means either of providing light to the objective or of delivering the resolved detail of the specimen in the primary image (formed by the objective) to the observer's brain. This chapter discusses, firstly, the function of the condenser and, secondly, the eyepiece.

8.2 The Condenser

The condenser is so-called because its purpose is to concentrate and focus light onto the specimen, to illuminate it sufficiently so that the field of view is brightly and evenly illuminated and the entire back focal plane of the objective is filled with light, thus enabling a good-quality image to be formed. The condenser for transmitted-light microscopy fulfils three functions:

- Firstly, it provides an area of even illumination in the field of view at the specimen plane.
- Secondly, it illuminates the aperture of the objective uniformly with light of sufficient yet controllable angle.
- Thirdly, it provides a means of specialised illumination to regulate and control contrast-enhancement.

Modern condensers consist of a self-contained lens system that can be moved along the optical axis of the microscope to focus light onto the specimen. The condenser is supported on a substage mechanism to permit this focusing to occur. On well-designed research microscopes, the condenser is a modular component. It can be detached from the substage mechanism so that other condensers may be exchanged. This is done either to match the magnifying power of the objective in use or to enable a particular contrast-enhancement method to be used. Various annuli, stops or prisms placed in the condenser are part of these contrast-enhancement techniques. Ideally they should be located in the front focal plane of the condenser, to correspond with the back focal plane of the objective, as is explained in the next chapter (Section 9.10). The substage mechanism may additionally carry one or more swing-out holders into which coloured filters or a polarising filter can be inserted.

8.3 Condenser Designs

Condensers are normally classified according their function (e.g. bright-field, dark-field, phase contrast, DIC) and/or how they are corrected for optical aberrations. For example, they can be grouped according to whether they are uncorrected for chromatic and spherical aberration, are corrected for chromatic aberration only (achromatic) or both chromatic and spherical aberration (achromatic-aplanatic). Equally, condensers may be classified according to the numerical aperture of the maximum cone of light they can supply to the condenser.

In order illuminate large fields of view yet also fill the aperture of all the available objectives (with numerical apertures from around NA 0.14 to NA 1.4) with light, several different designs of condenser are made. Low-power condensers up to about NA 0.25 are available (Figure 8.1) for objectives up to 5x magnification. Above 5x, for use with dry objectives, condensers with a swing-out top lens (Figure 8.6) and a maximum numerical aperture of NA 0.9 are made.

Figure 8.1 **Low-power condensers** *Two low-power condensers for use with objectives of less than 5x magnification. Source: Courtesy of Mr K. Glover.*

Figure 8.2 ***Achromatic-aplanatic oil-immersion condensers*** *Three achromatic-aplanatic oil-immersion condensers of high numerical aperture. On the left are two bright-field condensers; on the right a multi-purpose universal condenser designed for polarising and DIC microscopy, with strain-free optical components marked in red. The S and T notations on this particular condenser are Zeiss-specific markings indicating how the Wollaston prisms in the modular universal condenser turret should be adjusted for different designs of microscope. Source: Courtesy of Mr K. Glover.*

For 10x objectives and above, the top lens – which has a short focal length – should be swung into the light path in order to provide a sufficient cone of illumination to fill the objective. For oil-immersion objectives, high-power oil-immersion condensers (Figure 8.2), with numerical apertures up to NA 1.4, are made. These condensers have to be oiled to the underside of the slide, which is a messy business but a necessary one, in order to resolve and discriminate detail at the limit of the resolving power of the objective. Whereas an objective must be used dry or with an immersion fluid, depending on the design, a condenser designed to work as an (oil) immersion condenser will work dry (Figure 8.3) but will obviously be limited to a working numerical aperture of less than NA 1. Only when it

(a) (b) (c)

Figure 8.3 ***An oil-immersion condenser limited to working dry*** *An oil-immersion condenser limited to working dry. A fluorescent block placed on top of the condenser shows the cone of light emanating from the condenser that can be accepted by the objective. The left-hand panel (a) shows the condenser stopped down to about NA 0.7. The middle panel (b) shows how the cone of light is limited to NA 1 when the condenser is working under dry conditions. The right-hand panel (c) shows the underside of the fluorescent block oiled to the top lens of the condenser and the condenser diaphragm opened fully. A cone of light of the highest aperture is achievable to completely match the numerical aperture of an oil-immersion objective. Source: Adapted from James 1976 with permission from Springer.*

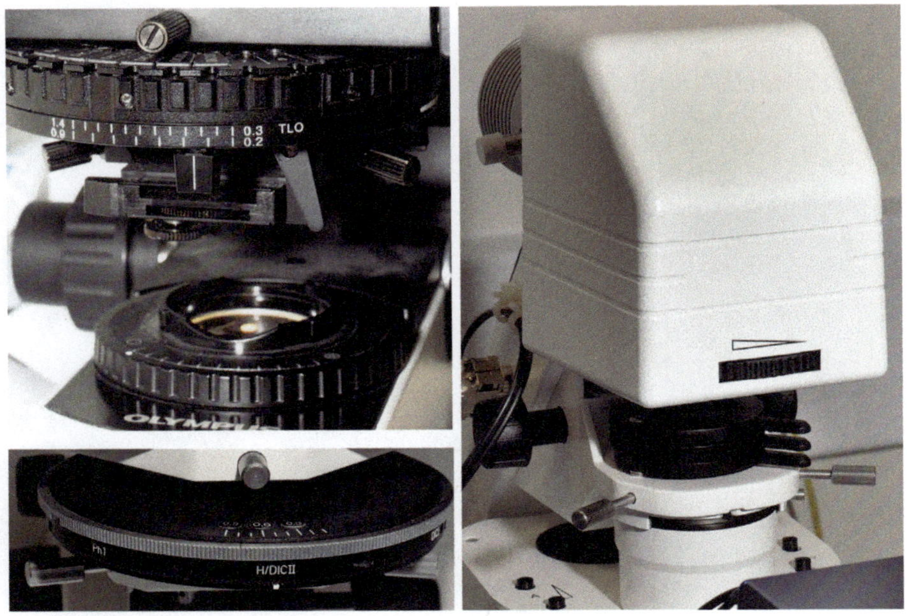

Figure 8.4 ***The centring controls on a condenser for Köhler illumination adjustment*** *The centring controls offset at 120° on a condenser are for Köhler illumination adjustment. If these are not built into the condenser, they are usually present on the substage condenser carrier, as shown for the inverted microscope in the right-hand panel. The top-left panel shows these two centring controls on the Olympus universal condenser. Compare this to Figure 8.5 of the same condenser highlighting the different carousel centring controls. Source: Courtesy of Mr K. Glover.*

is used immersed will it be possible to take advantage of the full resolving power of the condenser and completely fill the back focal plane of a high-NA immersion objective with light, as intended.

The most obvious effect of a defocused condenser in bright-field microscopy is a significant loss of resolving power giving, in turn, a 'rotten' image with diffraction haloes around each point in the image (see Step 16, Box 9.4). The same result occurs if the top (short-focus) lens is omitted or left flipped-out when a high-power objective is used, and the back focal plane of the objective is not completely filled with light. When phase contrast microscopy is attempted with a wrongly focused condenser, the annulus in the condenser often will not match the diameter of the phase ring in the back focal plane of the objective, and any contrast enhancement is lost.

Since the condenser is required to fill the rear aperture on the objective (in the back focal plane) with light, it has a direct influence of the overall resolving power of the microscope (Baker, 1952). This is discussed in Chapter 10 (Section 10.11). So that stray light, or glare, does not reduce the contrast in – and hence the visibility of – the image, the numerical aperture of the condenser must be set at slightly less than that that of the objective. This important point is reiterated in the next chapter (Section 9.11) on adjusting the microscope for Köhler illumination.

All good-quality condensers should have the means to centre the condenser on the optical axis when adjusting the microscope for Köhler illumination. These controls (Figure 8.4) are usually in the form of knurled screws on either side of the optical axis, underneath the condenser body, and offset with respect to one another by 120°. These centring controls are quite different from those that separately centre each position in the carousel of a universal condenser (see Section 8.4). The centring controls in this latter case (Figure 8.5) are usually recessed hexagonal screws, requiring the appropriate Allen key (usually of 3 mm diameter) to operate them. They are recessed so that they are not inadvertently knocked out of alignment, and they are either built into the front or the rear of the condenser body.

8.4 Different Types of Condenser

Because the condenser does not form the image, there is less of a requirement for it to be highly corrected and optical aberrations minimised. In the 19th century, whilst the English diatomists in the pursuit of ultimate resolution of the image used an inverted objective as a condenser, the Germans on the continent were well aware that medical progress could be made by equally good scientific microscopy using much less well-corrected condensers as illuminators. To this end, Abbe designed the illuminator named after him.

The two-lens Abbe illuminator (Figure 8.6) is the simplest design and is widely used for bright-field microscopy with transmitted light. It is uncorrected for chromatic and spherical aberrations, and as a consequence shows colour fringing around the condenser diaphragm leaves, which can be seen when adjusting the microscope for Köhler illumination. The Abbe illuminator gives good results with low-magnification objectives, but for objectives above numerical aperture about NA 0.6, a significant amount of spherical and chromatic aberration will be present, resulting in reduced image quality. Spherical aberration in the condenser will lead to uneven illumination of the field of view at high numerical aperture.

Figure 8.5 **The centring controls on a multi-purpose universal condenser** *Two differing designs of centring controls are shown. On the Olympus condenser (top) one of the offset controls for Köhler illumination adjustment can be seen to the right. The separate controls for adjusting the centration of the individual dark-field patch stops, phase annuli and DIC prisms are the pairs of recessed grub screws, visible on the rim of the condenser body, which are adjusted with a hexagonal Allen key. On the older design of Zeiss condenser (bottom) the centring controls for the phase annuli, etc. are the knurled handwheel in the centre of the foreground and the sliding lever with the knurled handgrip above it, to the left. Source: Courtesy of Mr K. Glover.*

Better-corrected achromatic-aplanatic[1] condensers (Figure 8.2) are normally found on research microscopes[2]. Normally these condensers are designed with numerical apertures greater than 1, to be immersed to the underside of the slide, but they can also be used dry. These corrected condensers give a sharper image of the border of the field diaphragm, more even illumination and better contrast in the image. The correction of an achromatic-aplanatic condenser is equivalent to that of an achromatic objective. Apochromatic corrected condensers confer no additional advantage. It is possible to use objectives as condensers in transmitted-light mode, but their fixed numerical aperture and short working distances mean that it is difficult to illuminate the specimen evenly.

So-called 'universal' condensers (Figure 8.4; Figure 8.7) are multi-functional. They consist of a revolving disc holding a selection of aperture diaphragms, filters, patch stops, phase plates or Wollaston prisms for differential interference contrast (DIC). This arrangement permits changing the illumination required from one contrast method to another conveniently and easily.

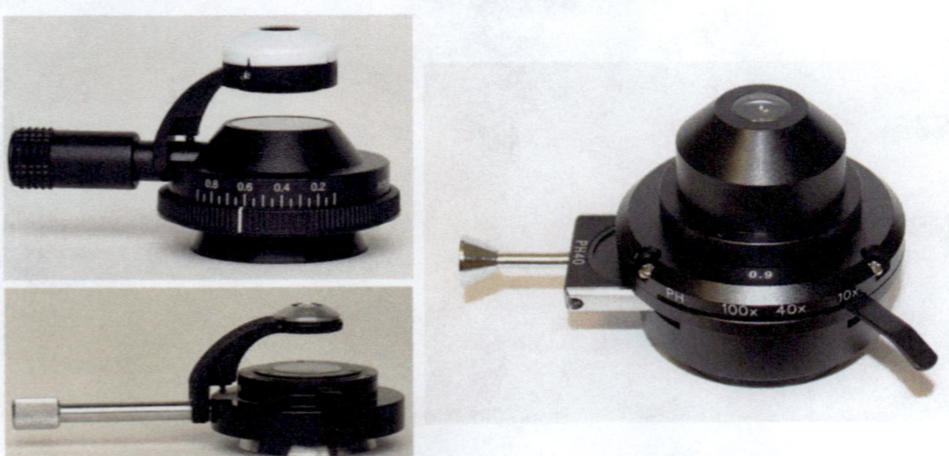

Figure 8.6 **The two-lens Abbe illuminator** *The two-lens Abbe illuminator design of condenser: simple, versatile and widely used. Three different designs are shown here from Olympus and Zeiss. On the right is one which includes a phase-contrast annulus on a slider. Source: Courtesy of Mr K. Glover.*

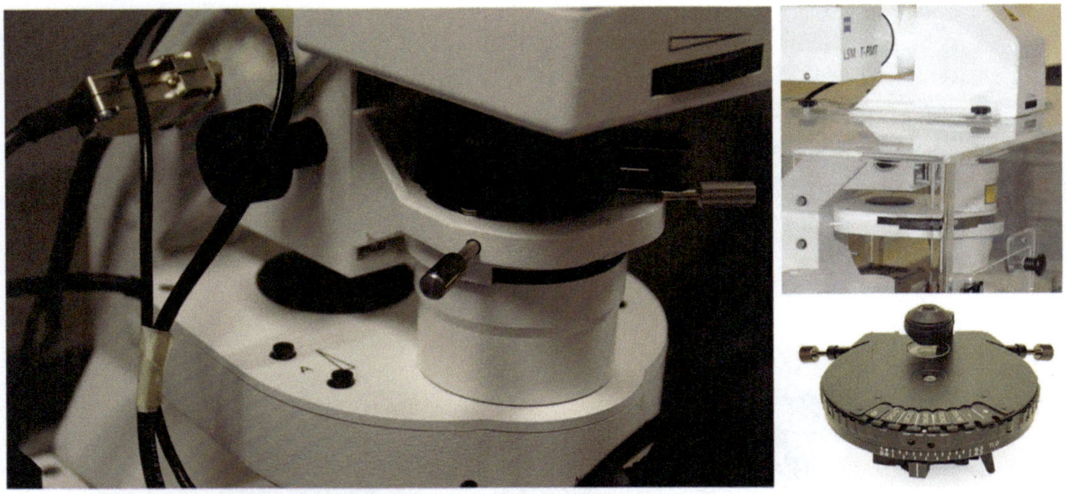

Figure 8.7 ***A universal condenser on an inverted microscope*** *The panel on the left shows a motorised universal condenser, also shown in Figure 8.4. This particular instrument is configured for phase contrast and DIC for observing unstained cells. The multi-functional universal condenser holds the means of changing the illumination for both these contrast-enhancement modes as well as for bright-field microscopy. Compare this to a manual universal condenser for inverted (top-right panel) and upright (bottom-right) microscopes. The Olympus condenser for transmitted-light microscopy can also be seen in place in Figure 8.4. Source: Courtesy of Mr K. Glover.*

8.5 The Substage Mirror

Microscopes from the 18th up to the mid-20th century were fitted with mirrors. Earlier designs of microscope, which can still be purchased second-hand, that do not have built-in illumination are fitted with a mirror for use with an external light source. This mirror either hangs from the substage mechanism or sits on the microscope base (Figure 8.8), mounted in a fork which allows it to be turned in all directions. The surfaces of these mirrors were concave on one side and flat on the other. Since the condenser requires parallel beams of light to focus onto the field of view (see Figure 8.3), the mirror should be used with the flat surface. The concave surface is for use with very low magnification (e.g. 1x or 2x) objectives to illuminate the field of view when no condenser is present or else has been removed.

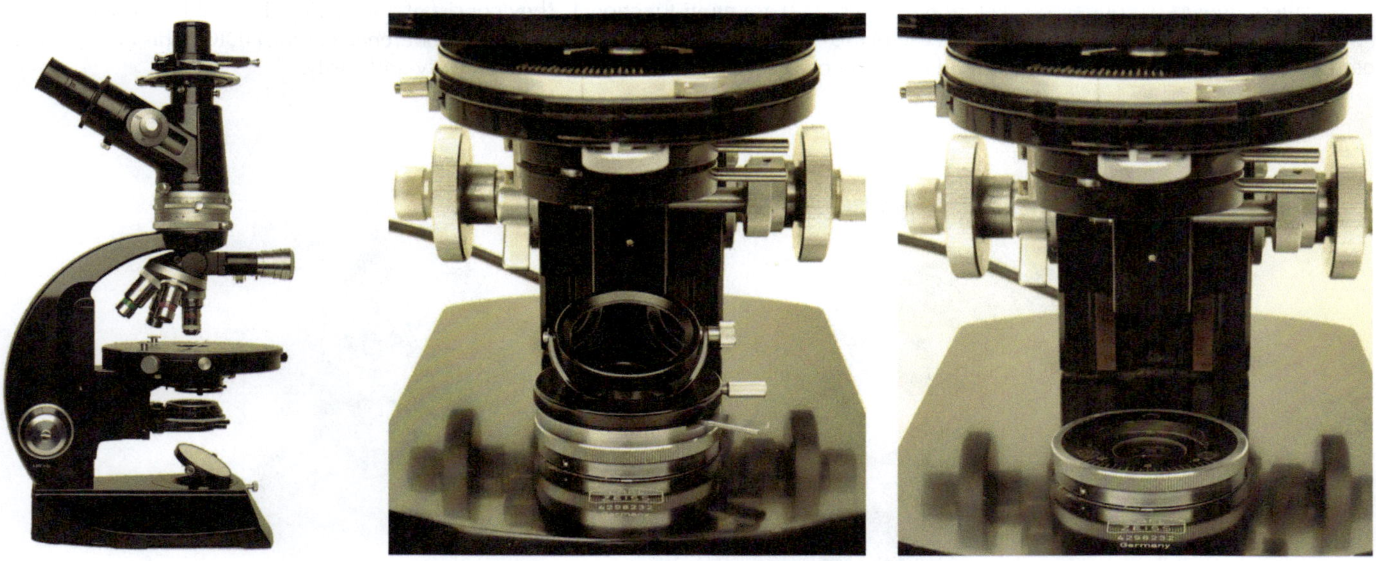

Figure 8.8 ***Two designs of mirror*** *Two designs of mirror, found on older microscope stands. The Vickers M74 stand (left) is configured for both reflected- and transmitted-light polarised-light microscopy. The design of this stand may be compared with Figure 12.9. The mirror on the Zeiss stand for use with an independent lamp is superbly designed (centre) and may be removed (right) so that the internal transmitted light source may be used. Source: (left) Hammond, 2002. Reproduced with permission from The Quekett Microscopical Club. (centre; right) Courtesy of Mr K. Glover.*

Modern microscopes have built-in illumination; they dispense with the external mirror and the condenser is an integral part of this system for lighting the specimen. A mirror placed at 45° in the base of the microscope, usually just underneath the illuminated field diaphragm, projects the light upwards onto the lower surface of the condenser. The systematic use of condensers from the mid-19th century led to an improvement in resolution. Significantly, Robert Koch discovered the tuberculosis bacillus in 1882, the same year that Abbe introduced his simple two-lens illuminator.

An auxiliary lens may be fitted beneath the condenser in some microscopes with built-in illumination. When used with the condenser, this lens serves the same purpose as the concave mirror. It enlarges the image of the lamp filament without altering the angular aperture of the illuminating cone of light.

8.6 The Eyepiece

The eyepiece (or ocular) in a compound microscope fulfils three purposes:

- Firstly, to convert the rays diverging from the primary image plane into parallel rays for delivery to the eye or (with a further intermediate lens) to the camera.
- Secondly, to sufficiently magnify the image detail formed by the objective into the primary image at the primary image plane (PIP) so that the eye or camera can see this resolved detail.
- Thirdly, to complete the correction of residual aberrations in the primary image not corrected by the objective.

The principal function of the eyepiece is to deliver parallel or near-parallel rays of light within the aperture of the eye (the eye can receive parallel rays of light from distant objects, or slightly divergent pencils of light from near objects).

The eye is best suited to seeing objects at infinity,[3] where, from objects located far off, parallel light rays enter the eye and are focused onto the retina. This is a throwback to the primitive lives of *Homo sapiens* as a hunter-gatherer. The eye can receive slightly divergent light rays, the degree of divergence being determined by the power of accommodation of the lens of the eye, until the near point, or least distance of distinct vision, is reached. The eye can also receive light from a wide angular range – about 150° approximately. In a microscope the cone of light entering our eyes consists of bundles or 'pencils' of parallel rays across the entire angular range that the eye can see. Each pencil of light arises from a single point in the object being brought to a focus on the retina. The microscope eyepiece is therefore designed to deliver parallel, or near-parallel, bundles of light rays to the lens within the aperture of the eye.

There are two means by which light rays may be made parallel to enter the eye (Figure 8.9). The first is by using a concave lens for the eyepiece, the Galilean method. The second is to use a convex lens, a design which is attributed to Kepler. The Galilean eyepiece is still used today, principally in opera glasses. It forms an erect image – an advantage – which is relatively aberration free. Its disadvantages are its small field of view, especially at higher magnifications, and the fact that it cannot be used with a graticule.

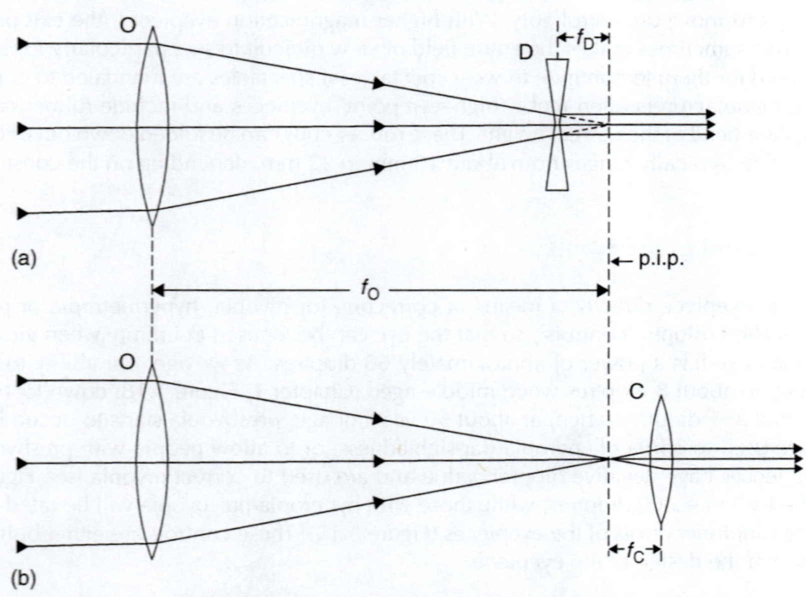

Figure 8.9 ***Two ways in which divergent light from the objective is made parallel*** *Two ways in which divergent light from the objective (whether the objective lens or the tube-lens) that forms the primary image, is made parallel for transmission into the eye. (a) shows the 'Galilean' method using a diverging lens (D) placed before the primary image whereas (b) shows the 'Keplerian' method using a convergent lens (C) placed after the primary image. In this figure 'p.i.p' indicates the primary image plane; f_C, f_D and f_O indicate focal lengths. Source: Hammond 2002. Reproduced with permission from The Quekett Microscopical Club.*

All eyepieces use the Keplerian method to deliver parallel rays of light to the eyes. These parallel rays of light proceed out of the eyepiece, first converging to the optical axis, crossing over and then diverging. The crossover point is called the eye-point, where the exit pupil (sometimes called by its older name, the Ramsden disc) of the microscope is formed. This eye-point is the place where the eye[4] is placed in order to see the whole field of view. If the eye is closer or further away, the more oblique pencils will not be intercepted and the field of view will be reduced. The diameter of the exit pupil should be equal to or smaller than the diameter of the (entrance) pupil of the eye. The diameter of the exit pupil is determined by the focal length, or power of the eye lens of the eyepiece.

Hold a piece of tracing paper or ground glass above the eyepiece of the microscope to locate the Ramsden disc (little spot of light of smallest diameter), to measure its distance above the eyepiece lens and to record the data for eyepieces of different powers and optical construction. The diameter of the pupil of the eye can change from 2 to 8 mm depending upon the light intensity, being smaller in brightly lit conditions, such as routine bright-field microscopy. The exit pupil of the microscope must therefore be about 3–4 mm and equal to or smaller than the pupil, for the eye to be able to capture all of the image formed at the primary image plane.

An eye lens of shorter focal length situated closer to the primary image plane gives a narrower set of pencils of parallel rays and a smaller Ramsden disc situated closer to the eyepiece lens itself. As with objectives, the higher the magnification of the eyepiece, the smaller is the field of view. The converse applies to eye lenses of longer focal length. A compromise is needed between the diameter of the Ramsden disc and its distance (usually expressed as 'eye relief') from the eye lens. In practice this compromise limits eyepiece powers in the range 4x (Ramsden disc ca. 4 mm diameter, eye relief ca. 15 mm) to 15x (Ramsden disc ca. 1.5 mm diameter, eye relief ca. 5 mm). In modern eyepieces consisting of several lens elements, there is of course a much greater optical flexibility, but the eyepiece powers used in practice (e.g. 10x) remain much the same.

It is possible to have any combination of objective and eyepiece, but it is more efficient to have a pair of eyepieces of fixed magnification (usually 10x) and to increase the overall magnification by rotating another objective of higher power (held as part of a group on a multi-position nosepiece) into place. Although it is possible to change the eyepieces for a pair of different magnification, there are several reasons for changing the objective and not the eyepieces (see also the exercise in Section 7.4). Firstly, the objective resolves fine detail in the image: the eyepiece merely magnifies the image formed by the objective; it does not provide any further resolving power. Secondly, the eye-point and the diameter of the exit pupil will change with differing magnification eyepieces, and this can be inconvenient. Thirdly, it is convenient to have parfocal objectives and to swing these rapidly into place.

Eyepieces were formerly available in various magnifications (Table 7.4), but the 10x eyepiece is the one most commonly used today – about 90% of all eyepieces sold are 10x magnification. Higher magnification eyepieces (e.g. 16x or 25x) are useful for stereomicroscopes where high-quality objectives, with a relatively high numerical aperture-to-magnification ratio, can use a high-magnification eyepiece as a secondary magnifier without introducing 'empty magnification' into the image. Increasing the overall magnification this way rather than using a shorter focal length objective to achieve a higher magnification means that a long working distance can be maintained. For this reason, higher magnification eyepieces (16x, 20x and 25x) are popular on stereomicroscopes.

The eyepiece magnification is cited as '8 times' or 'ten times' because it is responsible for the *angular* magnification of the final virtual image formed by the eyepiece at infinity from the primary image. The virtual image is converted into a real image by a convergent lens, either the microscopist's eye or another lens forming a real image onto film or a CCD chip or when projected onto a screen.

Beginners in microscopy often find difficulty in holding their head sufficiently steady at the correct height at the plane of the exit pupil for the field(s) of view not to move uncontrollably. With higher magnification eyepieces, the exit pupil is formed closer to the top eye lens of the eyepiece. This sometimes makes the entire field of view difficult to see, particularly for spectacle wearers. A certain distance, or *eye relief*, is required for them to continue to wear spectacles if spectacles are mandated to correct astigmatism in the eye (see also Box 9.5). Therefore, manufacturers often make 'high–eye point' eyepieces and include rubber cups around the eyepiece to assist the observer in holding their head at the correct height. These rubber cups can be folded down out of the way if desired. Eye relief (Figure 8.10) on modern objectives typically ranges from about 15 mm to 22 mm, depending on the construction of the eyepiece.

8.7 Dioptre Adjustment Control of the Eyepiece

While it is not a purpose of the eyepiece directly, a means of correcting for myopia, hypermetropia or presbyopia in our eyes may be found in the adjustable so-called 'dioptre controls', so that the eye can be focused at infinity when viewing the image through the microscope. The relaxed human eye has a power of approximately 60 dioptres. As we age, our ability to focus closely reduces from about 15 dioptres when young, to about 8 dioptres when middle-aged (Chapter 1, Figure 1.18) down to 1–2 dioptres in old age. Our focusing power reduces to about 3–5 dioptres when, at about 50 years of age, presbyopia starts to occur. Positive convex lenses have positive dioptric value and are used to correct hyperopia (far-sightedness) or to allow people with presbyopia to read at close range. Negative, concave, diverging lenses have negative dioptric value and are used to correct myopia (see Figure 1.22). Eyes having mild myopia will have a power of −1.00 to −3.00 dioptres, while those with hyperopia/presbyopia will be rated at +1.00 to +3.00 dioptres. These values can be set on the dioptre controls of the eyepieces (Figure 8.15). These controls are either built into the binocular head of the microscope or else are part of the design of the eyepiece.

8.8 Eyepiece Designs

Theoretically, only a single lens (the eye lens of a modern eyepiece) is required as a secondary magnifier, but it would, in consequence, have to be very large. (A ray trace makes this clear; see, for example, Figure 3.13.) The first lens of the eyepiece (the eyepiece field lens[5])

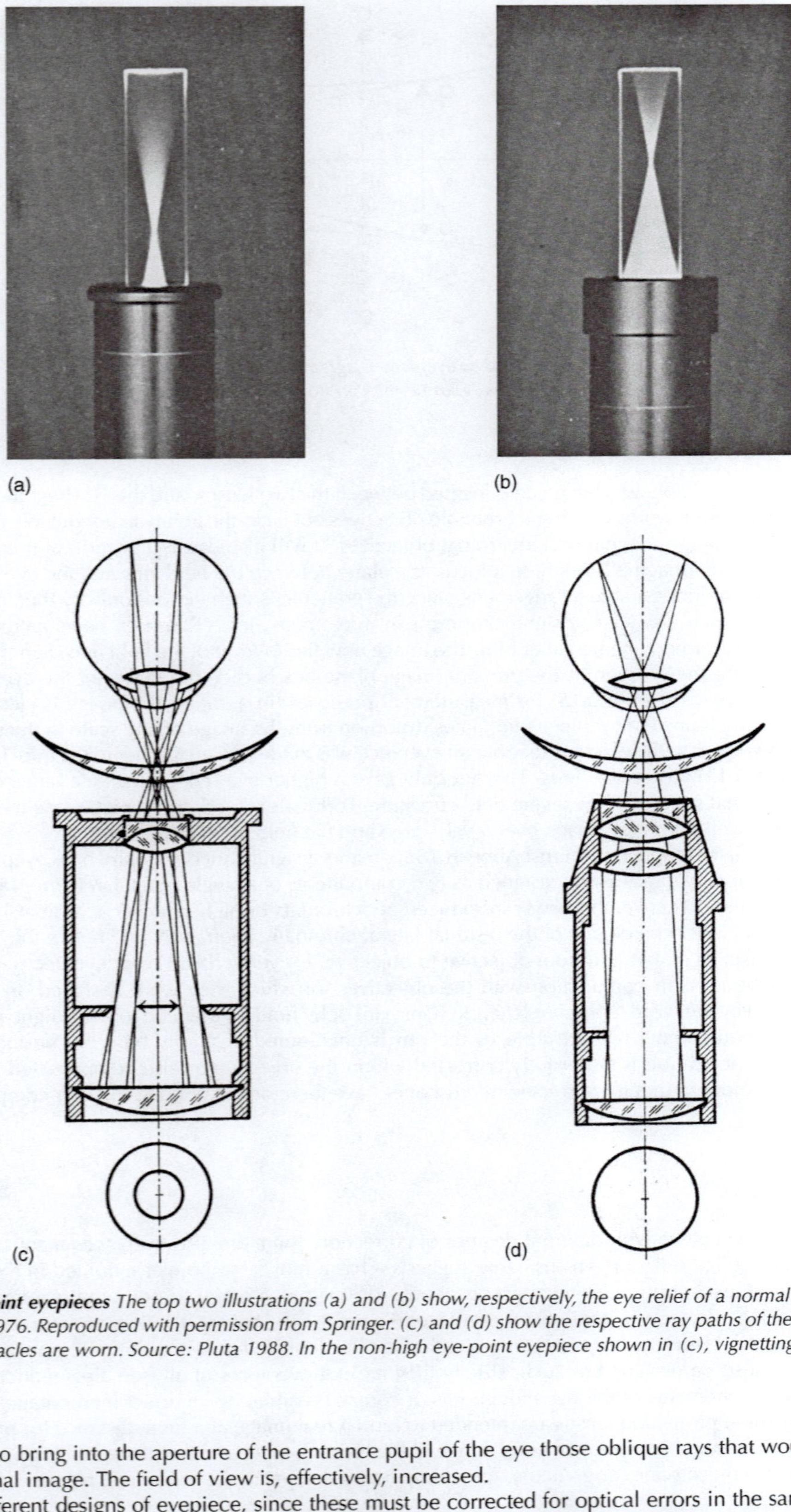

Figure 8.10 ***High eye-point eyepieces*** *The top two illustrations (a) and (b) show, respectively, the eye relief of a normal 10x and high eye-point 12.5x eyepiece. Source: James 1976. Reproduced with permission from Springer. (c) and (d) show the respective ray paths of the light bundles passing through the eyepieces when spectacles are worn. Source: Pluta 1988. In the non-high eye-point eyepiece shown in (c), vignetting of the primary image can be seen.*

is added (Figure 8.11) to bring into the aperture of the entrance pupil of the eye those oblique rays that would otherwise be vignetted and missed from the final image. The field of view is, effectively, increased.

There are several different designs of eyepiece, since these must be corrected for optical errors in the same way that objectives are corrected. The first practical design was the so-called Huygens eyepiece (a so-called negative eyepiece, nowadays termed an internal-diaphragm eyepiece) consisting of two plano-convex lenses with the plane sides towards the eye separated by an air gap (Figure 8.11 and Figure 8.12a). Arranged in this way, the spherical and chromatic errors of these two uncorrected lenses largely cancel each other

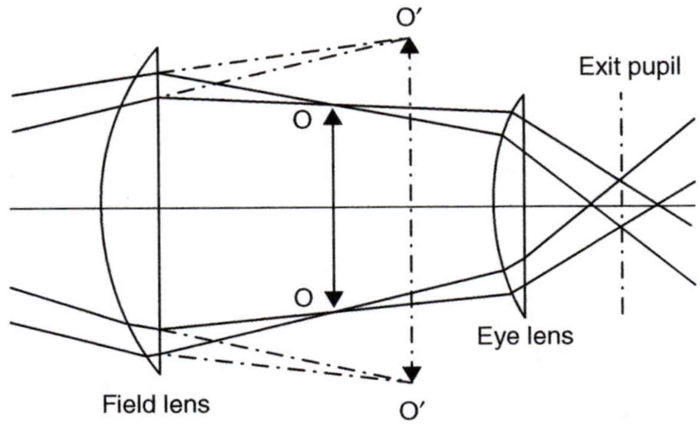

Figure 8.11 **Diagram explaining the function of the field lens of an eyepiece** *Diagram explaining the function of the field lens of an eyepiece. See the text for further explanation. Source: Hammond 2002. Reproduced with permission from The Quekett Microscopical Club; Bradbury and Bracegirdle 1998. Reproduced with permission from Taylor & Francis Group.*

out. The focal plane of the internal[6] diaphragm eyepiece is located between the two lenses, and this is where the primary image plane is arranged to lie. This very simple design works well for achromatic objectives but lacks the ability to adequately correct lateral chromatic difference of magnification of semi-apochromat or apochromat objectives. It will also cause distortion of image points off the optical axis. Furthermore, since the primary image is brought to a focus at a plane *between* the field lens and the eye lens, any graticule must be placed between them. This is an undesirable arrangement, since the graticule is then viewed *only* by the uncorrected eye lens.

To rectify these aberrations, which can lead to significant error in micrometry, Jesse Ramsden developed the external-diaphragm (positive) eyepiece (Figure 8.12b), formerly named after him. The image now lies *in front of* the field lens. Spherical aberration is further corrected in this design[7]. Because the position of the primary image plane lies outside the lenses of the eyepiece, it is much easier to insert a graticule into the eyepiece (Figure 8.13) for measurement purposes; in addition the image is clearer than that seen with the earlier Huygenian design. The Ramsden eyepiece eliminates distortion from the image of any scale in the eyepiece but is not fully achromatised. A further improvement to the external-diaphragm eyepiece was made by Carl Kellner in the mid-19th century by adopting achromatised doublets (Figure 8.13) for the eye lens. This not only gave a higher eye relief but also a larger field of view. However, these eyepieces gave an image that suffered from severe field curvature. Fortunately modern eyepieces use meniscus lenses to correct for this curvature and so give a well-corrected image over a very large and flat field of view.

The compensating eyepiece, first introduced by Ernst Abbe in 1886, marks a significant development in eyepiece design, for this was the first time that the objective and eyepiece were designed as two components of a single optical system. Abbe showed that residual lateral chromatic aberration in the objective (the newly introduced apochromats or higher power achromats) could be compensated for in the design of the eyepiece. Since the degree of the residual lateral chromatic aberration, and hence the compensation required, is not fixed but varies from maker to maker and from objective to objective, for very critical work eyepieces – compensating or non-compensating – should only be used in conjunction with the objectives for which they were designed. In practice compensating eyepieces may be distinguished by the word 'comp' or letter 'K' (German) or by holding them up to white light and observing an orange rim round the diaphragm. In non-compensating eyepieces the rim is uncoloured or faintly blue. In the modern infinity-corrected microscopes, lateral chromatic aberration is either fully corrected within the objective or else compensated for in the design of the tube-lens. Eyepieces for use in modern infinity-corrected microscopes have therefore reverted to the non-compensating type.

8.9 Different Types of Eyepiece

Besides the different designs of eyepiece with different degrees of correction, there are also eyepieces manufactured to fulfil specific functions. The pointer eyepiece (Figure 8.14a) – as its name suggests – has a moveable pointer mounted in the primary image plane. These tend not now to be used, except on older microscope stands, for it is much easier to point out features using the mouse on a computer monitor. For polarising microscopy, goniometer eyepieces (Figure 8.14b), having a 360° protractor scale, allow the interfacial angles of crystals to be measured. Likewise, this measuring function can be achieved post-acquisition of the image, using software.

Projection 'eyepieces' may also be designed as such; strictly they are not eyepieces at all (see also endnote 6) since they are not designed to fulfil the optical requirements of the eye though it is of course possible, by incorrect manipulation of the microscope, to see an image through them. These projection lenses are intended to form a real image at a finite distance for photomicrography.

Because the primary image is formed just below the top flange of the viewing tube, and the plane of this primary image is conjugate with the specimen plane, an eyepiece scale, or graticule, for taking measurements in the image can be conveniently inserted into the lower part of the eyepiece.

Eyepieces designed to hold a graticule (or reticule[8]) for measurement are often colloquially known as 'micrometer eyepieces'. The eye lens in this design of eyepiece has a focusing mechanism – usually a helical thread or possibly a sliding mount – which permits the graticule to be focused sharply. This adjustment is completely separate to the dioptre focusing control, which will normally be found on

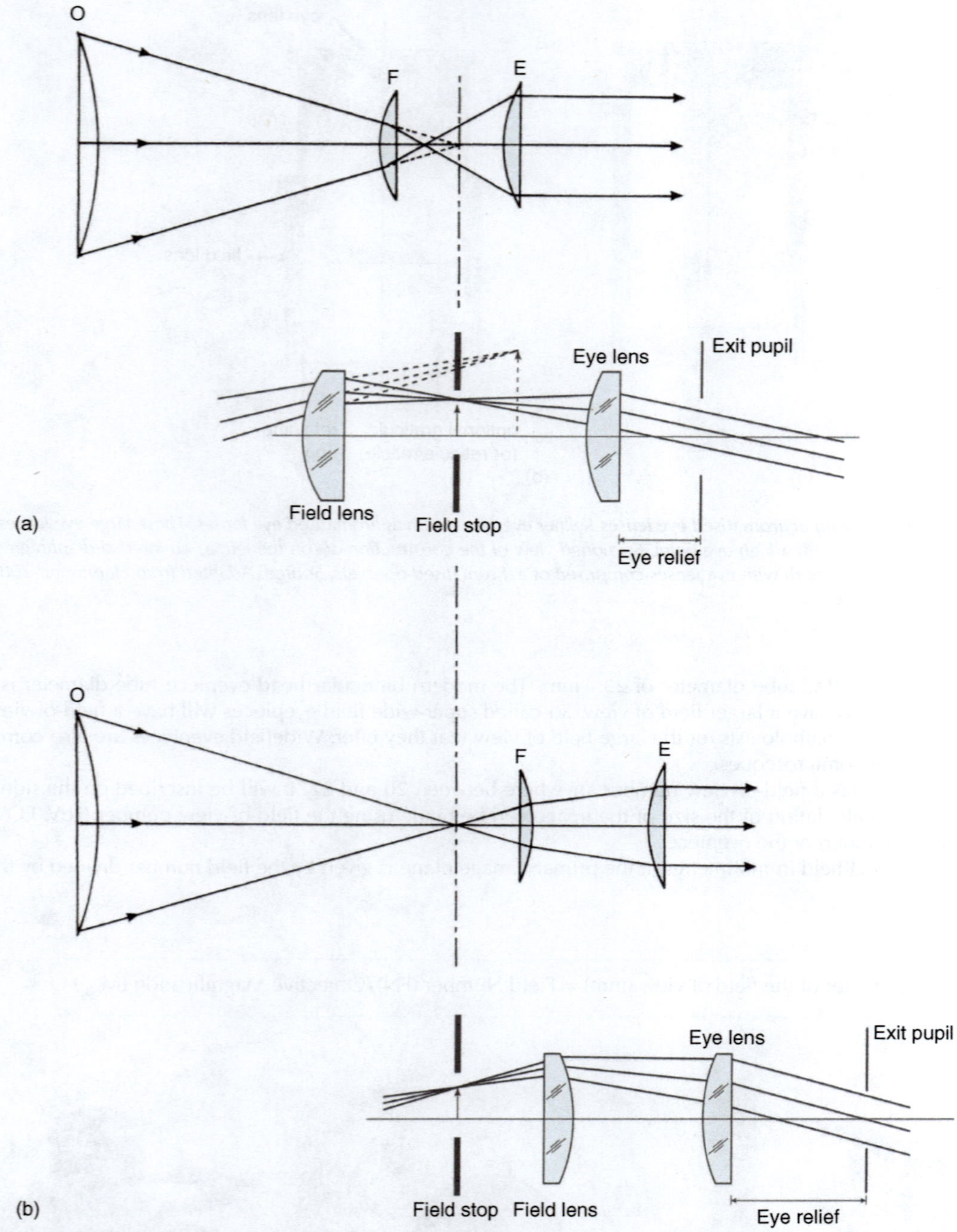

Figure 8.12 ***Eyepiece designs*** *The design of (a) the internal-diaphragm (Huygenian) and (b) the improved external-diaphragm (Ramsden) eyepiece. The focal length of the field lens of the internal-diaphragm design is twice that of the eye lens (f), and the spacing between the two lenses is 1.5f to minimise aberration in the image. In the external-diaphragm design, the focal length of each lens is f and the spacing is about 2/3 f. Source: Hammond 2002. Reproduced with permission from The Quekett Microscopical Club; Blackwell and Thornton 1996. Reproduced with permission from S. Thornton.*

the binocular head of the microscope. For assessing volume fractions and counting areas in statistical work, a net graticule, or reticle, is used. A net reticle comprises a series of squares (rather like a chessboard of transparent squares) covering the field of view.

8.10 Field-of-View Number

The field of view, or object field, is that part of the image field that is imaged onto the observer's retina. The diameter of the field of view can be calculated using the field-of-view number[9] (FN), which is physically limited by the diameter of the eyepiece tube to about

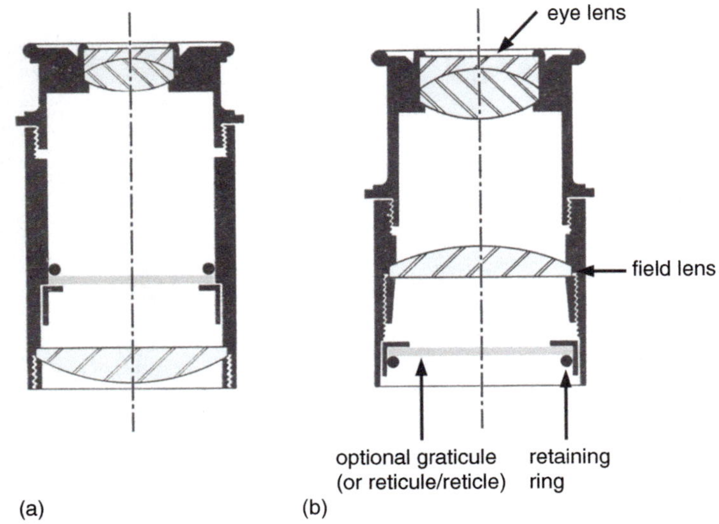

Figure 8.13 **Kellner eyepieces with achromatised eye lenses** *Kellner eyepieces with achromatised eye lenses. These large eye lenses have a large field of view and can also be designed with a high eye-point. Sectioned view of the construction of (on the left, a) an internal-diaphragm and (on the right, b) an external diaphragm eyepiece both with eye lenses comprised of achromatised doublets. Source: Adapted from Hammond 2002 with permission from The Quekett Microscopical Club.*

20 mm for the conventional RMS tube diameter of 23.4 mm. The modern binocular head eyepiece tube diameter is standardised at 30 mm, allowing eyepieces to have a larger field of view. So-called super-wide field eyepieces will have a field-of-view number of 25 or 26.5. They are favoured by pathologists for the large field of view that they offer. Widefield eyepieces are also commonly found on lower-power dissecting stereomicroscopes.

A modern 10x eyepiece has a field-of-view number anywhere between 20 and 22. It will be inscribed on the side of the eyepiece (Figure 8.15)[10]. An initial calculation of the size of the image can be made using the field-of-view number (FoV, FOV, FN, otherwise known simply the field number) of the eyepiece.

The diameter of the object field in millimetres at the primary image plane is given by the field number divided by the magnification of the objective:

Diameter of the field of view (mm) = Field Number (FN)/Objective Magnification (M_{obj}).Q

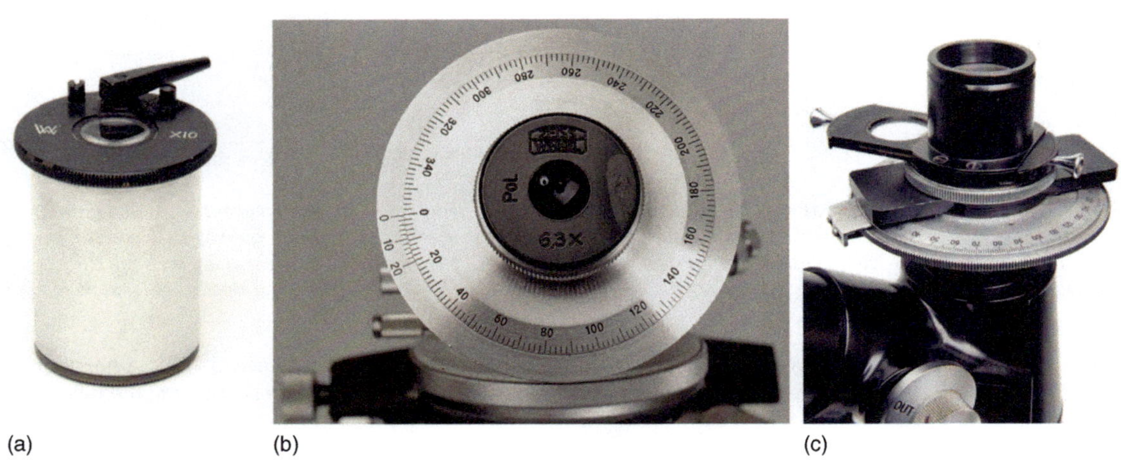

Figure 8.14 **Pointer and goniometer eyepieces** *The pointer (a) and (b) goniometer eyepieces. The Dick-Wright polarising microscope eyepiece (c). These are not often used for their function can be performed easily in software. This figure shows that the eyepiece, in its position at the primary image plane and exit pupil of the microscope, could evolve into sophisticated multi-functional designs. The Dick-Wright eyepiece enables a quartz wedge compensator to be inserted and uses an iris diaphragm to isolate conoscopic figures of crystal grains. Source: Courtesy of Mr K. Glover.*

Figure 8.15 **Location of the field-of-view number on an eyepiece** *The FoV number, here 22, is inscribed on the barrel of the eyepiece. Source: Courtesy of Mr K. Glover.*

where Q is any other magnification factor. If a magnification-changing device is fitted between the objective and eyepiece, its magnification factor (Q) should also be considered when calculating magnification at the primary image. Normally the tube-lens factor can be assumed as 1x, unless it is inscribed on the microscope; sometimes it will be 1.25x.

Therefore for a 20x objective, using an eyepiece with a FN = 26.5 and a tube-lens factor of 1.25x, the diameter of the field of view is: 26.5/(20 × 1.25) mm = 1.06 mm. Other values are given in Table 7.4. The area of the field of view can now be calculated from $A = \frac{\pi}{4} \times D^2$. In our example, the area of the field of view is 1.66 mm^2. If the object *fills* the diameter of the field of an eyepiece whose FOV is 20 mm at the primary image plane, with a 10x objective and no Q value, then the size of the object is 20/10 mm = 2 mm. An ImageJ field-of-view calculator is available; see the Internet Resources at the end of the chapter for the URL link.

8.11 Linear Measurement with the Microscope

In all types of eyepiece, a fixed platform, or diaphragm, is often found as a rim around the internal part of the tube of the eyepiece at the plane of the primary image so that a scale (usually called a graticule) can be inserted into the eyepiece and superimposed on the image of the specimen seen at the primary image plane. The Ramsden, or external-diaphragm eyepiece, presents the most conveniently placed location for the scale. When the graticule – which is marked out in arbitrary units – is calibrated properly, the size of the specimen can be calculated. A scale bar can then be added, for example, in any published image. A more accurate method of taking linear measurements is by using a filar micrometer eyepiece. In this case the graticule is arranged so that turning a graduated drum on the side of the eyepiece moves a fiducial line either side of a central, static zero line. An illustration of a filar micrometer eyepiece is shown in Figure 31.13. Instructions on how to use a micrometer eyepiece to calibrate the image, make a linear measurement and superimpose a scale bar are given in Chapter 31, Section 31.8. Some microscopists may regard using a stage micrometer to calibrate an eyepiece graticule as 'old school', preferring to use image-acquisition software to place a scale on their image, but the software has to be calibrated so it is worth knowing how to use a stage micrometer for both calibration and measurement.

8.12 Microscopes without Eyepieces

With increasingly sophisticated digital CCD and sCMOS cameras, eyepieces or projection lenses for photomicrography are increasingly not required. The camera is coupled directly to the microscope. Sometimes the digital sensor is small enough that a reduction lens must be included in the optical train to ensure no vignetting of the image recorded by the CCD faceplate, rather than any eyepiece providing the customary secondary magnification. The ultimate evolution of this trend is to have a microscope without eyepieces at all[11]. On some modern slide-scanning or high-throughput microscopes, selecting a suitable field of view and capturing an image is all performed via a monitor display linked to the CCD camera or other electronic detector. Incubator microscopes, for facilitating cell culture (Chapter 28, Section 28.6) also have no eyepieces.

8.13 Chapter Summary

This chapter explains the purpose of the condenser for providing even and controllable illumination of the microscope specimen. The different designs of condenser are described. The purpose of the eyepiece to convert rays diverging from the primary image plane into parallel rays for the eye is also discussed. The two principal designs of eyepiece, the Huygens (internal-diaphragm) and Ramsden (external-diaphragm) eyepieces, together with their better-corrected Kellner equivalents, are described and illustrated. Compensating eyepieces for the correcting lateral chromatic aberration, for use with apochromatic objectives, are described, as are so-called micrometer eyepieces for measurement with the microscope.

1. The function of the condenser is to provide an area of even illumination in the field of view at the specimen plane.
2. The condenser illuminates the aperture of the objective uniformly with light of sufficient yet controllable angle.
3. The condenser provides a means of specialised illumination to regulate contrast enhancement.
4. Several different designs of condenser are made to illuminate large fields of view and also fill the aperture of objectives of different magnification.
5. A condenser not correctly focused will cause a significant loss of resolving power.
6. The centring controls of a condenser for adjusting Köhler illumination are separate for those to centre phase-contrast annuli.
7. The condenser does not have to be as well corrected for aberrations as the objective. The achromatic-aplanatic condenser is corrected for spherical aberration and coma.
8. The function of the eyepiece is to provide parallel rays to the eye. It is also to magnify the image detail resolved by the objective so that it can be seen by the eye.
9. The eyepiece may also complete the correction of residual aberrations in the primary image. These types of eyepiece are called 'compensating' eyepieces and must be matched to the objective used.
10. The eyepiece magnification is cited as '8 times' or 'ten times' because the eyepiece is responsible for the *angular* magnification of the final virtual image formed by the eyepiece at infinity.
11. High eye-point eyepieces are designed to allow the continued use of spectacles whilst operating the microscope.
12. Two principal designs of eyepiece exist; these are the internal-diaphragm (Huygenian) and external-diaphragm (Ramsden) designs. Both are now made as improved Kellner achromatic eyepieces.
13. The primary image plane is located at the position of the diaphragm in the eyepiece. This allows transfer scales to be inserted for measurement and calibration of the object and field of view.

Key Reading

Bradbury, S. (1991). *Basic Measurement Techniques for Light Microscopy*. Royal Microscopical Society Handbook No. 23, Oxford University Press.
- This RMS Handbook is the most complete description of basic measurement techniques with the light microscope.

References

Baker, J.R. (1952). 'Remarks on the effect of the aperture of the condenser on resolution by the microscope'. *Quart. Jour. Microsc. Sci 93/4*: 375–377.
Blackwell, J. and Thornton, S. (1996). *Mastering Optics*. McGraw-Hill Book Company Europe.
Bradbury, S. and Bracegirdle, B. (1998). *Introduction to Light Microscopy*, 2nd ed. *RMS Microscopy Handbook No. 42*. Oxford: Bios Scientific Publishers.
Hammond, C.H. (2002). 'The evolution of the eyepiece'. *Quekett Journal of Microscopy 39*: 365–378.
James, J. (1976). *Light Microscopic Techniques in Biology and Medicine*. Martinus Nijhoff Medical Division.
Murphy, D.B. and Davidson, M.W. (2013). *Fundamentals of Light Microscopy and Electronic Imaging*, 2nd ed. New Jersey: Wiley Blackwell. John Wiley.
Pluta, M. (1988). *Advanced Light Microscopy, Vol. 1*. Amsterdam, Oxford: Elsevier.

Further Reading

Baker, J.R. (1966). 'Experiments on the function of the eye in light microscopy'. *Journal of the Royal Microscopical Society 85/2*: 231–254.
Hartley, W.G. (1951). 'The determination of the numerical aperture of a condenser'. *Journal of the Royal Microscopical Society 71/4*: 429–432.
Hopkins, H.H. and Barham, P.M. (1950). 'The influence of the condenser on microscopic resolution'. *The Proceedings of the Physical Society 63/10*: 737–744, which is discussed by John Baker (1952).

Internet Resources

Graticule/reticule diameters for common microscopes (Pyser SGI): https://www.pyseroptics.com/graticules/technical-guidance-notes/reticle-diameters-for-common-microscopes/.
ImageJ field-of-view calculator: https://imagej.nih.gov/ij/plugins/fov-calculator.html.

Notes

1. Aplanatic means corrected for spherical aberration and coma.
2. The converse does not always hold. Many research microscopes, especially inverted stands, will have condensers of quite low power and long working distance, for examining living cells and tissues. Inverted confocal microscopes may not have a high-power achromatic-aplanatic condenser, firstly so that it does not require oiling to the slide, secondly because any bright-field image formed by the transmitted-light detector PMT does not generally have to be of ultimate resolving power, unlike the fluorescence image, since it is usually acquired for morphological purposes so that the high-quality fluorescence image may be overlaid on it.
3. For practical purposes, infinity means a distance greater than about thirty lens focal lengths – about 5 metres in our case.
4. Strictly, the entrance pupil of the eye, although we do this naturally without thought.
5. A field lens also provides homogeneous illumination of the field to which it relates. It is placed in, or close to, a field plane (i.e. the object plane or any plane conjugate with it). The field plane most commonly refers to the primary image plane (PIP) and to the eyepiece but not necessarily so.
6. These eyepieces have been termed 'positive' and 'negative' respectively, an ambiguous, unsatisfactory and misleading terminology. The terms 'external-diaphragm' and 'internal-diaphragm' are precise and unambiguous. Strictly, a negative eyepiece is not a true eyepiece, but rather a diverging lens used as a projection lens. It cannot be used as an eyepiece because, by nature of its design, the exit pupil is located internally within the lens groups.
7. An explanation and proof of how spherical and chromatic aberration are minimised may be found in: Hammond, C. (2002) 'The evolution of the eyepiece'. This reference is given in the Further Reading section.
8. The term reticule, or sometimes reticle, is often used in North America to describe a graticule.
9. The value for field-of-view number, FN, is determined by the focal length (f_{eye}) and magnification of the eyepiece (Mag_{eye}) as well as the angle of view, or apparent viewing angle (A or α). This value A is shown in Figure 3.10 as the apparent visual angle. The formula is: $FN = 2f_{eye} \cdot \tan{}^1\!/_2 A = 500/Mag_{eye} \cdot \tan{}^1\!/_2 A$.
10. The FN is the actual diameter, in millimetres, of the field diaphragm of an external-diaphragm (positive, or Ramsden) eyepiece, and the apparent diameter of the virtual image of the field diaphragm in an internal-diaphragm (negative, or Huygenian) eyepiece.
11. Many practicing microscopists of long-standing find it 'unnatural' to use a 'microscope' without eyepieces that is operated remotely via a monitor. It is much easier to rapidly scan and locate suitable fields of view by eye, even if all the subsequent image acquisition is performed digitally via the computer monitor. Not having an eyepiece in the fluorescence emission path can help considerably to maximise the intensity of the signal collected. For example, the OMX structured illumination super-resolution microscope dispenses with eyepieces, but a second microscope, fitted with eyepieces for rapid visual operation and with both fields of view linked digitally in synchrony, provides the means to locate the right object for examination. The Olympus LV200 for recording very low intensity bioluminescent signals minimises the optical path, so dispenses entirely with eyepieces and further uses a 5x reduction lens to 'condense' the signal immediately onto the CCD faceplate. In this case, the fewer optical components and glass elements, the better.

9

Illumination in the Microscope

Outline

Understanding Light Microscopy, First Edition. Jeremy Sanderson.

9.1 Introduction

This chapter divides into two parts. In the first part we consider the light sources used to illuminate the specimen. Incandescent lamps and mercury arc lamps are increasingly being replaced by metal-halide lamps and LEDs, respectively. The second part of the chapter explains how these light sources are used to evenly illuminate the specimen in a controllable manner. The classical and ubiquitous method normally employed to do this is by adjusting the microscope according to the principles of Köhler illumination. Well-designed microscopes allow this to be done easily.

9.2 Incandescent Lamps

The tungsten lamp (Figure 9.1) has been used for many years as a light source for microscopy. They are usually rated between 6 and 12 volts and between 10 and 100 watts and so require transformers. Because of the heat generated, these transformers and lamp houses are usually external to the microscope stand. The spectrum (Figure 9.2) of a tungsten lamp ranges from about 300–1500 nm, rising into the infrared. Because of the red-shifted spectrum, these lamps have to be run with a relatively high voltage in order to maintain both sufficient brightness (the perception of luminance) and the correct colour temperature (Chapter 31, Section 31.4). As tungsten lamps age, their light reduces in intensity and also the lamp becomes blackened. This blackening occurs as tungsten evaporates off the filament (so making it increasingly thin until it fails) to be deposited upon the glass envelope of the bulb. To minimise the blackening, tungsten lamps (e.g. the older style domestic light bulb) tend to be rather large, to minimise the build-up of the tungsten film. For these reasons, most microscopes using incandescent small filament lamps now use tungsten-halogen (also called quartz-iodine, or Q-I) lamps (Figure 9.1), which do not blacken. Further information is given online: see the Internet Resources at the end of the chapter.

Tungsten-halogen lamp filaments run at a very high temperature (around 2500 to 3000°C) and there is a significant pressure (3 bar when cold; 7–8 bar when operational) within the lamp, so the envelope (which attains a lower operating temperature, around 700°C) is made of a silica-glass (fused-quartz) mixture or high-melting point aluminosilicate glass and not merely ordinary glass, which would otherwise approach melting point and soften. The high operational temperature of a tungsten-halogen lamp gives a very stable, very high luminance spectrum (Figure 9.2), which is not as red-shifted as with tungsten lamps. Furthermore, the lamp does not blacken, since the halides inside the envelope help redeposit the tungsten preferentially back onto the filament. This permits the use of smaller bulb envelopes and an increased lifetime for the tungsten-halogen lamp. These incandescent lamps are filled with an inert gas, usually

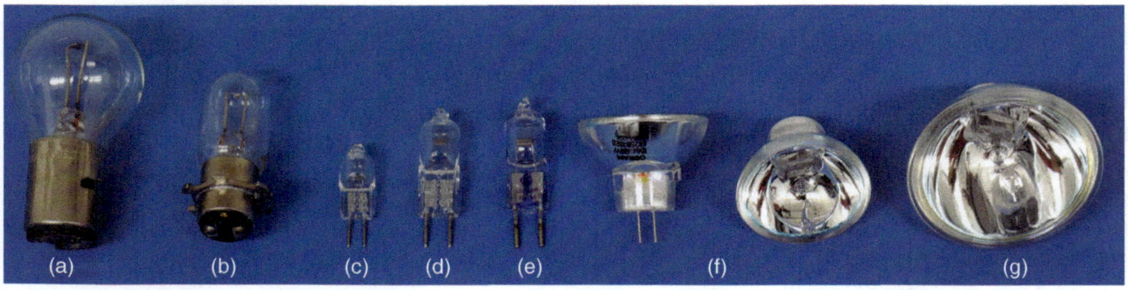

Figure 9.1 ***Incandescent lamps*** *Seven incandescent lamps commonly used in microscopes are shown. From left to right (a) 12 V 60 W bayonet tungsten lamp; (b) 6 V 15 W tungsten-halogen double contact bayonet with special flange, a prefocus design; (c) 6 V 30 W type G4 (miniature bi-pin) of 9 mm diameter and 30 mm long. A G4 base has two pins measuring 0.7 mm in diameter, located 4 mm apart; (d) 12 V 100 W type GY6.35 Osram xenophot, of 11.5 mm diameter and 44 mm long. A GY6.35 base has two pins measuring 1.25 mm in diameter, located 6.35 mm apart; (e) 12 V 100 W type GY6.35 of General Electric manufacture, rated the same as (d) and intended for the same purpose; the filament is slightly thicker; (f) two views, front and back, of a tungsten-halogen 8 V 20 W reflector. An MR11 design with a fixed focal length of 26 mm, a G4 base 32 mm long, with a reflector of 35 mm diameter; (g) a larger 12 V 100 W GZ6.35 reflector. A 16 design, focal length 32 mm, with a reflector of 51 mm diameter and length 42 mm. Source: Courtesy of Mr K. Glover.*

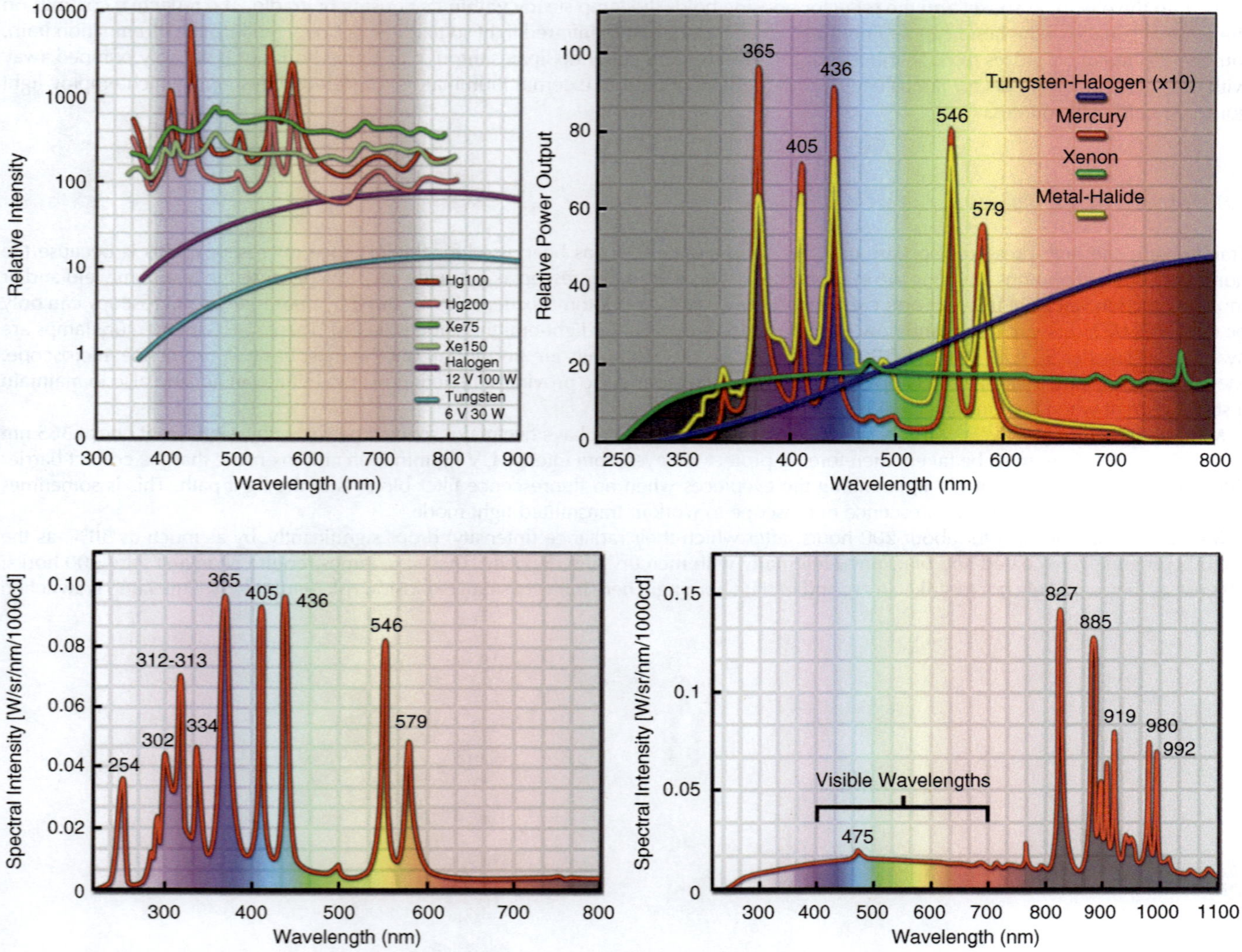

Figure 9.2 ***Spectra of various light sources*** *High-pressure short-arc mercury lamps, rated at 50 and 100 watts have been the common illuminants for fluorescence microscopy for many years. They have a lifetime between 200 and 300 hours and require careful alignment upon replacement. Mercury arc lamps have a number of spectral lines in the ultraviolet region between 250 and 350 nanometres and in the infrared above 1000 nm. Approximately half of the radiant flux usefully lies between 350 and 700 nm, in the visible region. There are major spectral lines at 365 nm (11%), 405 nm (4%), 436 nm (13%), 546 nm (7%) and 579 nm (8%).*

Because of the existence of these spectral peaks, xenon lamps have been used particularly for qualitative and quantitative measurements in fluorescence microscopy: they exhibit a more uniform spectral output. The top left-hand figure shows the relative intensity of various light sources. The 75 watt Xenon lamp has a higher intensity than the 100 watt short-arc mercury lamp because although the 100 watt lamps have greater overall output, the distance between the arcs is greater, giving a lower overall intensity output.

The top right-hand figure shows the comparison of power output over the visible spectrum for these illuminators, whist the figures in the lower panels show the individual spectra for a short-arc mercury and xenon arc lamp.

Source: Murphy et al. 2013. Reproduced with permission from John Wiley & Sons.

krypton; Osram uses xenon in microscopy lamps for improved performance, and these lamps are designated HLX (**H**alogen, **L**ow-voltage and **X**enon).

Manufacturers recommend that the fused-quartz silica glass envelope of tungsten-halogen lamps should not be handled directly. Natural oils from the skin of your fingertips can create a potential hotspot and damage the glass envelope. Hold the bulb by the porcelain base or connection pins. Better still, wear gloves or use a folded paper tissue or towel, or the package the lamp comes in, to handle the lamp when inserting it into the lamp unit. If the glass envelope is contaminated, it must be cleaned with alcohol and allowed to dry before use.

Some designs of tungsten-halogen lamps incorporate an integral reflector with an elliptic paraboloid surface. This is similar in design to a car headlamp and serves the same purpose: to gather light rays emitted in all directions by the lamp and direct them forwards into the illumination system. These reflector lamps are manufactured as sealed pre-centred units. They are simple to replace: the pins fitting into the electrical socket and the reflector housing holds the lamp steady within its housing or cradle. The reflective coatings on the internal surface of the lamp can be fine-tuned to be transparent to infrared light so that it is not directed into the illumination train. Infrared (IR) radiation passes through the rear of the reflector and either dissipates through the lamp house or is actively pumped away with an electrical fan. Reflector lamps (Figure 9.1) are widely used in external illuminators for stereomicroscopy, which employ light guides as cold light sources.

9.3 Metal Halide, Xenon and Mercury Lamps

Traditionally, the high-pressure short-arc mercury lamp (Figure 9.3) has been used for fluorescence microscopy. This is because the number of emitted photons reaching the eye, or other detector, from fluorescence is usually very low because the quantum yield and or molar extinction values of fluorophores can be quite low. The high photon-flux densities required by fluorescence microscopy can only be delivered by lasers, arc lamps and now – for many applications – by light-emitting diodes (LEDs). These short-arc mercury lamps are available as smaller 50 watt or larger 100 watt lamps: the 100 watt units are best for all but the most basic fluorescence microscope. Mercury[1], xenon and metal-halide lamps require control transformers to provide sufficient power to initiate an arc and also to maintain a stable current when the lamp is operating.

Mercury lamps do not give a constant spectral output (Figure 9.2) but have five major spectral peaks, with a significant one at 365 nm in the ultraviolet. Care must be taken, therefore, to protect the eyes from intense UV illumination and to ensure that the correct barrier filters are in place to prevent UV light exiting the eyepieces when no fluorescence filter block is in the light path. This is sometimes necessary when configuring a fluorescence microscope to work in transmitted-light mode.

Mercury arc lamps burn for about 200 hours, after which their radiance (intensity) drops significantly, by as much as 30%, as the lamp is progressively coated and blackened internally with mercury film deposits. These arc lamps require changing after 300 hours; otherwise, they are likely not to strike an arc and initiate burning. There is a very small explosion risk, particularly if the lamp is switched

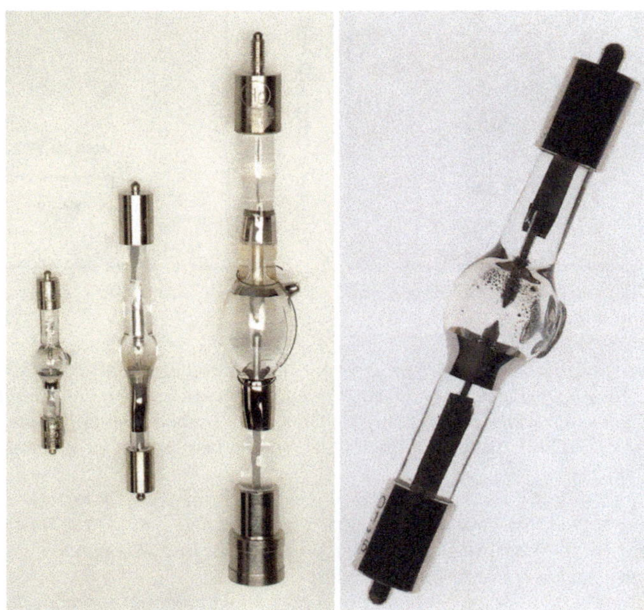

Figure 9.3 **Short-arc mercury and xenon arc lamps** *50 watt mercury (left), 100 watt mercury (middle) and 75 watt xenon (right) short-arc gas discharge lamps for comparison are shown in the left-hand panel. The electrodes are made of tungsten; current flows from the thinner cathode (the bottom electrode in this illustration) to the anode, which is larger to help radiate away excess heat. The outer coil, visible on the xenon lamp, is to help with arc formation. These lamps operate at great pressure, so the envelope – filled with mercury or xenon – is made of quartz glass.*

Arc lamps operate from a DC current via a transformer. When the current is switched on and the starter button pressed, an arc is initiated between the tips of the two electrodes. It helps to ensure optimum lifetime if when a new lamp is first burnt, it is initially left on for a few hours. The arc gradually burns a small pit in the anode, and this forms a preferential strike point when the lamp is next turned on and an arc initiated – exactly like lightning striking a conductor.

The diameters of the ends of the xenon and 100 watt mercury lamps are different, being DC lamps; the anode is 9 mm, slightly larger in diameter than the 7.5 mm cathode. This ensures correct orientation when fitting a new lamp. In the right-hand panel, it can be seen that the ends of the smaller 50 watt AC lamp are equal: the lamp must be inserted with the end having the silver coating on the bottom part of the bulb. Also notice (in the right-hand panel) the condensed drops of mercury within the bulb, which vaporise during operation to give the characteristic mercury spectral lines of the short-arc Hg lamp. Source: (left-hand panel) Courtesy of Mr K. Glover. (right-hand panel) Reproduced with permission from Carl Zeiss Ltd.

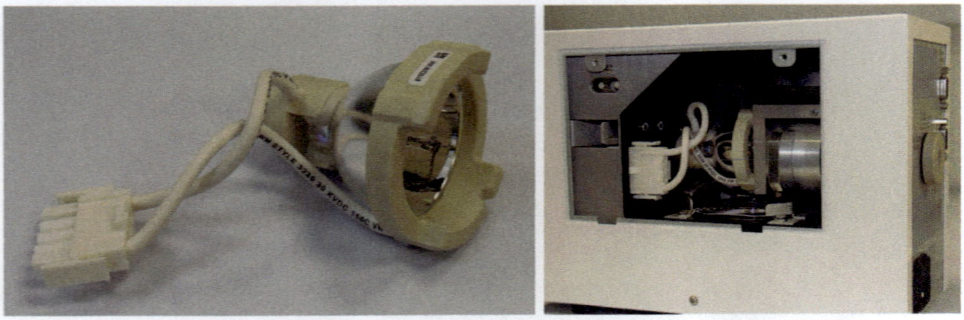

Figure 9.4 **Metal halide lamp** *Metal-halide lamps are cheaper to operate than mercury lamps. Whilst the individual units are more expensive, they have a lifetime of 2000 hours. From switch on, it takes around five minutes to achieve the full-colour spectrum and emission stability at the design operating temperature. Most units have an inbuilt timer preventing them from actuating if the lamp is still hot. Source: Courtesy of Mr K. Glover.*

on again before being allowed to cool properly. For that reason, high-pressure fluorescence lamps (this advice applies to both arc lamps and metal-halide designs) should be allowed half an hour or so to cool to ambient temperature before being switched on again. It is good practice to leave the lamp on for an extended session and to use a booking system so that the next user of the microscope is not inconvenienced.

Xenon arc lamps (Figure 9.3) are generally used in preference to mercury lamps for quantitative fluorescence microscopy, because they offer a more constant (i.e. flatter) illumination intensity across the visible spectrum. To achieve this, they are operated with DC current and use stabilising circuitry to maintain a specific current. Xenon lamps are rated at 75 or 150 watts. They cost about the same as mercury lamps to run, being more expensive to buy but running for 3–5 times as long. They should be changed after 1000 hours of burning time. The orientation of a xenon lamp is critical to its optimum operation and longevity; preferably they should be run in an upright position. In those lamps that are designed for vertical operation, the anode is positioned at the top and the cathode at the bottom of the lamp unit. This configuration is rotationally symmetrical and yields excellent arc performance. In contrast, lamps operated in a horizontal configuration produce arcs that require stabilisation in order to reduce premature and accelerated electrode wear.

Metal-halide lamps (Figure 9.4) are increasingly replacing the high-pressure short-arc mercury lamp. Like xenon lamps, they have a higher initial cost but last longer, burning for 2000 hours or more. Unlike short-arc lamps, metal-halide lamps give out a light flux at their rated intensity for about 70% of their lifetime and do not exhibit a quickly decreasing output intensity. They are manufactured as sealed units and are much simpler to exchange than either mercury or xenon lamps. Not only do metal-halide lamps run for much longer, they are also safer than either mercury or xenon lamps. The metal-halide units have an elliptical reflector to feed the maximum light flux into the epi-illuminator via a liquid light guide, which helps give an even illumination. Consequently, they do not require a collimator lens, as does a mercury lamp or xenon lamp. Another very important function of the light guide is to ensure that the illumination entering the microscope is much cooler than would otherwise be the case with a directly coupled incandescent lamp. These metal-halide lamps have the same spectral profile and the same spectral peaks as mercury lamps but give off a lower percentage of their radiation as heat[2]. All these lamps – mercury, xenon and metal-halide – contain mercury vapour, and so the relevant precautions should be taken when disposing of them.

9.4 Cold Light Sources

Most light sources used in microscopy are incandescent – that is, they emit light as a result of heating a filament by running an electrical current at high, though variable, resistance through the filament. These sources inevitably waste energy, which is emitted as heat, meaning that the design of the microscope and the lamp house must accommodate this production of heat and dissipate it without damage to the instrument or the specimen. To avoid these disadvantages, there are now two main types of cold light source available, which are increasingly being used in microscopy.

Firstly, there are the fibre-optic 'liquid light' guides (Figure 9.5). These illuminators provide even illumination through liquid light guides or swan-neck fibre optics, although they operate with incandescent lamps as a source. The reflector lamps are described in Section 9.2. The light travels along a transparent core by total internal reflection; it is constrained within the fibre by an external cladding of lower refractive index. The heat is dissipated out of the lamp house directly through the reflector (which is transparent to IR). Thus the illumination is effectively cold when it reaches the specimen. The second type (Figure 9.6) – a truly cold light source – is the semiconductor light-emitting diode, or LED.

Liquid light guides have a limited lifetime, and reportedly a limited shelf-life (3 years) regardless of use. Try to ensure the optical fibre is not bent: sharp bends, kinks or twists are a certain way to reduce both the lifetime and efficiency of any light guide. For many microscopes, the transformer for the metal-halide lamp is placed very close to a back wall: commonly a sharp right angle is introduced into the fibre-optic without consideration for the effect upon the liquid light guide for the duration of its lifetime. Also try, if possible, to use the shortest length of light guide between the light source and the microscope.

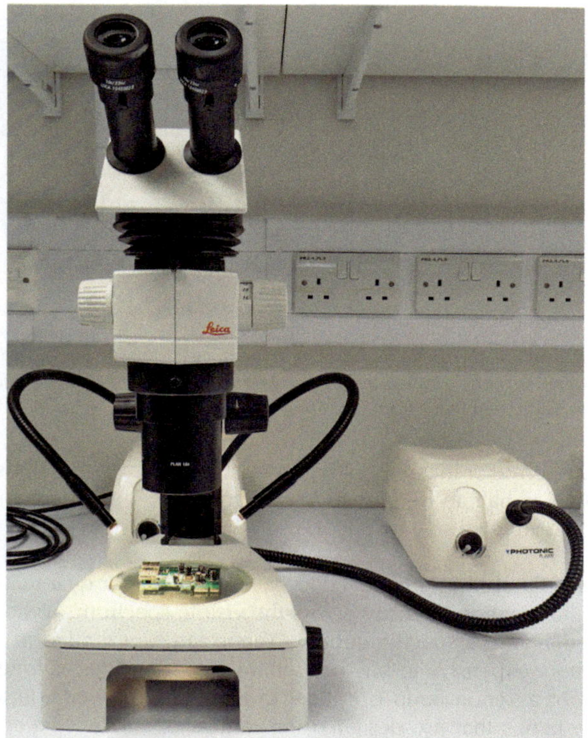

Figure 9.5 **A liquid light-guide illuminator** *This cold light source is particularly useful for illuminating specimens viewed with a stereomicroscope, because the direction of the light may be easily changed. Source: Courtesy of Mr K. Glover.*

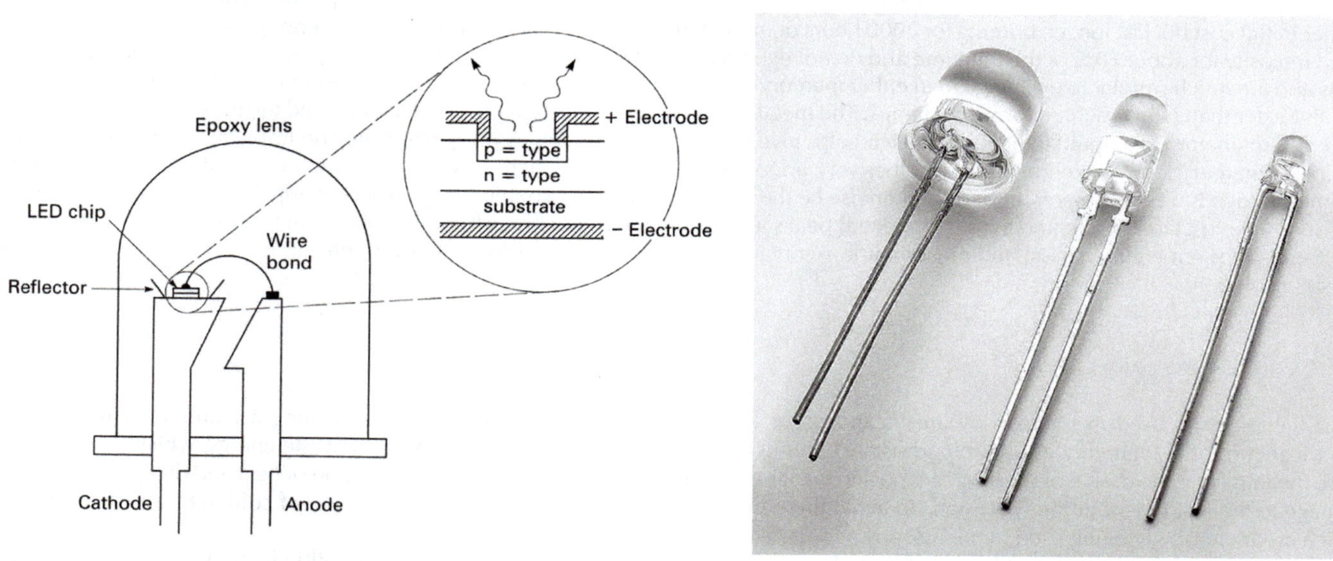

Figure 9.6 **Diagram of LED construction** *A diagram of the construction of a light-emitting diode and a selection of the individual components before they are built into a light source. Source: Freeman and Hull 2003. Reproduced with permission from Elsevier. (image) afrank99 (own work). Reproduced under the terms of the Creative Commons Attribution Share-Alike licence, CC BY-SA 2.5 (http://creativecommons.org/licenses/by-sa/2.5), via Wikimedia Commons.*

9.5 Light-Emitting Diodes

In 1962 Nick Holonyak[3] Jr., who worked for General Electric, developed the first visible-spectrum light-emitting diode in the red part of the spectrum. Light-emitting diodes are stable, efficient[4] and intense illumination sources, which are gradually replacing traditional tungsten-halogen lamps in both the domestic and scientific markets. They are illuminated solely by the movement of electrons in a semiconductor material, and they don't have a filament that gets hot or burns out. LEDs are now available in all colours from near UV

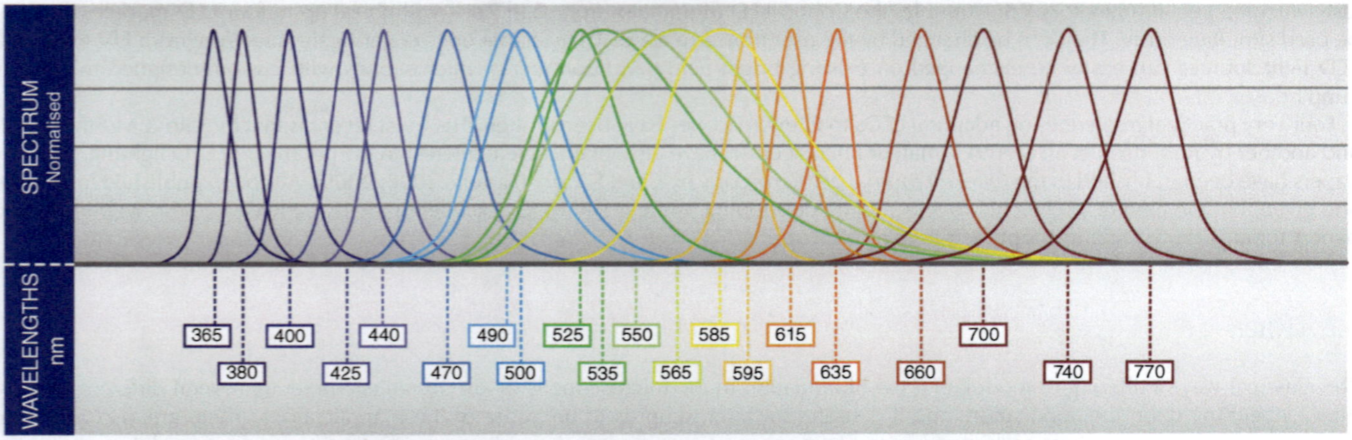

Figure 9.7 ***The wavelengths available with modern LEDs*** *Source: CoolLED. Reproduced with permission from CoolLED.*

to infrared (Figure 9.7) and have the advantage that heavy transformers are no longer required – the current, and not the voltage, passing through the device controls the output intensity. This allows microscope stands to be self-contained and much more portable. LEDs (Figure 9.6) are solid-state devices, so they are far more robust than filament or arc lamps. These solid-state lights, unlike incandescent sources, have an extremely long lifetime – in excess of 10 000 hours compared with a few hundreds of hours for short-arc lamps and a few thousand hours for metal-halide lamps.

Full-spectrum 'white' LED lamp houses are now available (Figure 9.8) and suitable for microscopy, although their peak output is around 400–480 nm. This is suitable for observing GFP and other green-emitting fluorophores. CoolLED are a specialist firm offering bespoke units to fit most fluorescence microscopes. They have a well-designed website (see Internet Resources at the end of this chapter) where you can check your model of microscope or contact the firm for technical help and advice. LEDs tend to have a narrower bandwidth than mercury, xenon or metal-halide lamps, thus minimising heating or photo-toxicity of samples. LEDs themselves switch on and off instantly, attaining full brightness in microseconds, thus eliminating high-speed shutters (and the associated vibration) for some applications. The shutter is one of the slowest components of the microscope, and its moving parts will induce vibration. Switching LEDs on and off only for the duration of the exposure and not having a shutter is very advantageous for live-cell microscopy. The total photon dose (which is photo-toxic) can be significantly reduced.

LEDs can be constructed so that both UV and IR emissions are avoided; they have low IR emission anyway. Furthermore, incandescent lamps have relatively broad spectral peaks that require isolating with filter sets and so require slow-moving filter wheels (e.g. in the order of 50–100 msec, or more) to switch from one specific wavelength to another. LEDs have relatively narrow excitation (a FWHM bandwidth[5] of 20–40 nm); a tilting-grating monochromator working with a white light LED can switch wavelengths in 1–2 msec. At the time of publishing, LEDs still only produce significant amounts of light at low power. For applications requiring greater power with high current loads, the luminous flux from LEDs is significantly reduced. Nevertheless, the major companies make

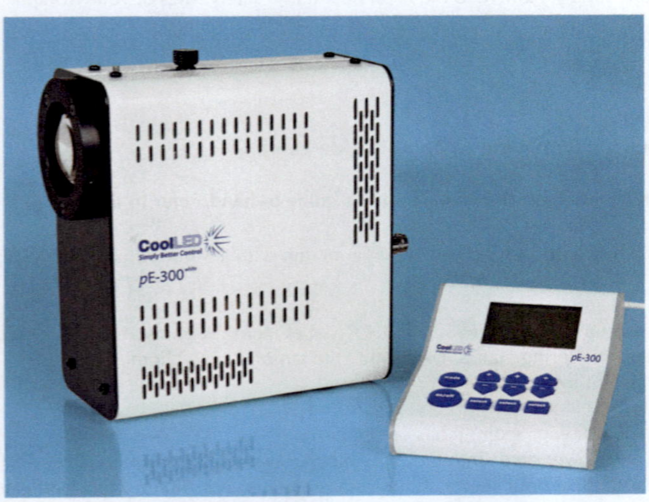

Figure 9.8 ***A full-spectrum LED light source*** *These units can be retrospectively fitted to existing microscopes and can be obtained from independent manufacturers, such as CoolLED. Source: CoolLED. Reproduced with permission from CoolLED.*

fast-switching LED illuminators. For example, the Colibri LED illuminator from Carl Zeiss will house up to four LED modules that can be used simultaneously. They can be changed by the user from a range of 10 available units covering the spectrum from UV to far red. LED light sources can easily be exchanged on existing bright-field and fluorescence microscopes with custom-designed third-party lamp houses.

Four very practical references on adapting LEDs to fit microscopes have been written. The most recent is that by Sato & Murthy (2012) and another by Robertson et al. (2009). Amateur microscopists have also taken a great interest in the potential of LED lighting. Two key papers have been published in the *Quekett Journal of Microscopy* by Carel Sartory (Sartory, 2008; 2009). Although published in 2008/9, and given that the field of LED technology is rapidly developing, there is still much excellent practical advice in these well-illustrated papers for anyone wishing to adapt their own microscope.

9.6 Filters

The classical way of altering or modulating the illumination in the microscope is by use of filters. There are several different types of filter, depending upon the application, and it is useful for microscopists to be aware of these applications. There are also two main filter designs: absorption filters and interference filters. Absorption filters, such as glass and plastic filters, are stained throughout the mass of the filter. They tend to be cheaper than interference filters; they have no surface coatings, which run the risk of being scratched off or damaged, and their placement in the illuminating beam is not critical. Interference filters have multiple very thin coats of metal interspersed with coats of equally thin dielectric (i.e. non-conducting) materials. These layers are prone to scratch damage, so they must be cleaned (Chapter 15, Section 15.8) very carefully, and they must be inserted into the illuminating beam with the reflective surface facing the light source. The advantage of interference filters is that they have very steep cut-in and cut-off transmission boundaries, much more so than coloured glass or plastic filters. An entire industry has therefore developed for designing and manufacturing interference filters for fluorescence microscopy.

Neutral-density filters are used to attenuate the intensity of light across the entire visible spectrum. Colour filters may be used with transmitted- and reflected-light microscopes to isolate specific regions of the spectrum, whereas blocking filters (e.g. Schott GG395 for UV and Schott KG1/KG5 or BG38 for near IR) are used to block harmful radiation.

9.7 Changing Lamps

General protocols for changing microscope lamps are given below. You are advised, wherever possible, to refer to the manufacturers' instructions provided with, or relevant to, your microscope. LEDs are built-in at the factory and so by definition are pre-centred and do not require changing. Most lamp houses have pre-centred fittings; certainly this is true for reflector lamps and for the sealed unit metal-halide lamps that are increasingly replacing the short-arc high-pressure mercury lamps for fluorescence microscopy. The metal-halide lamps often have a built-in 'inteli-sensor', which logs the number of hours the lamp has been switched on. Both the electrical connection for the lamp and this sensor have a design polarity; it is very easy to replace this type of pre-centred lamp.

> It should go without saying: ALWAYS isolate the lamp from the electricity supply before opening up the unit. Do not be tempted to take a short cut and leave it plugged into the mains supply. Adhere to the local health and safety rules of your workplace. Wear safety glasses, or goggles.

Protocol 9.1 Generic Instructions for Replacing a Metal-Halide Lamp

These are generalised instructions only. If you have the manufacturer's guide to hand, refer to this.

1. Disconnect the power supply and remove the outer cover of the lamp unit with a 3 mm Allen hex key. Open the bulb bracket arm and pull out the old bulb.
2. Unclip the two sets of wires.
3. With the lamp facing the rear of the unit, and the 'THIS SIDE OUT' label facing you, insert the new bulb into the groove of the lamp bracket. The middle of the bulb will snap into the spring clamp. Ensure that the lamp is seated home correctly.
4. Close the bulb bracket arm by pushing it in and to the left.
5. Attach the 4-pin Intelli-sensor into place. It will only go one way round.
6. Attach the 2-pin lamp white power leads plug; again, it has orientation polarity.
7. Tuck the Intelli-lamp lead harness and power cords into their retaining grooves in the lamp housing.
8. Refit the cover and turn on the lamp. The hour meter should read 0 hours, not 8888, which indicates a bulb that is over-hours and will not ignite.

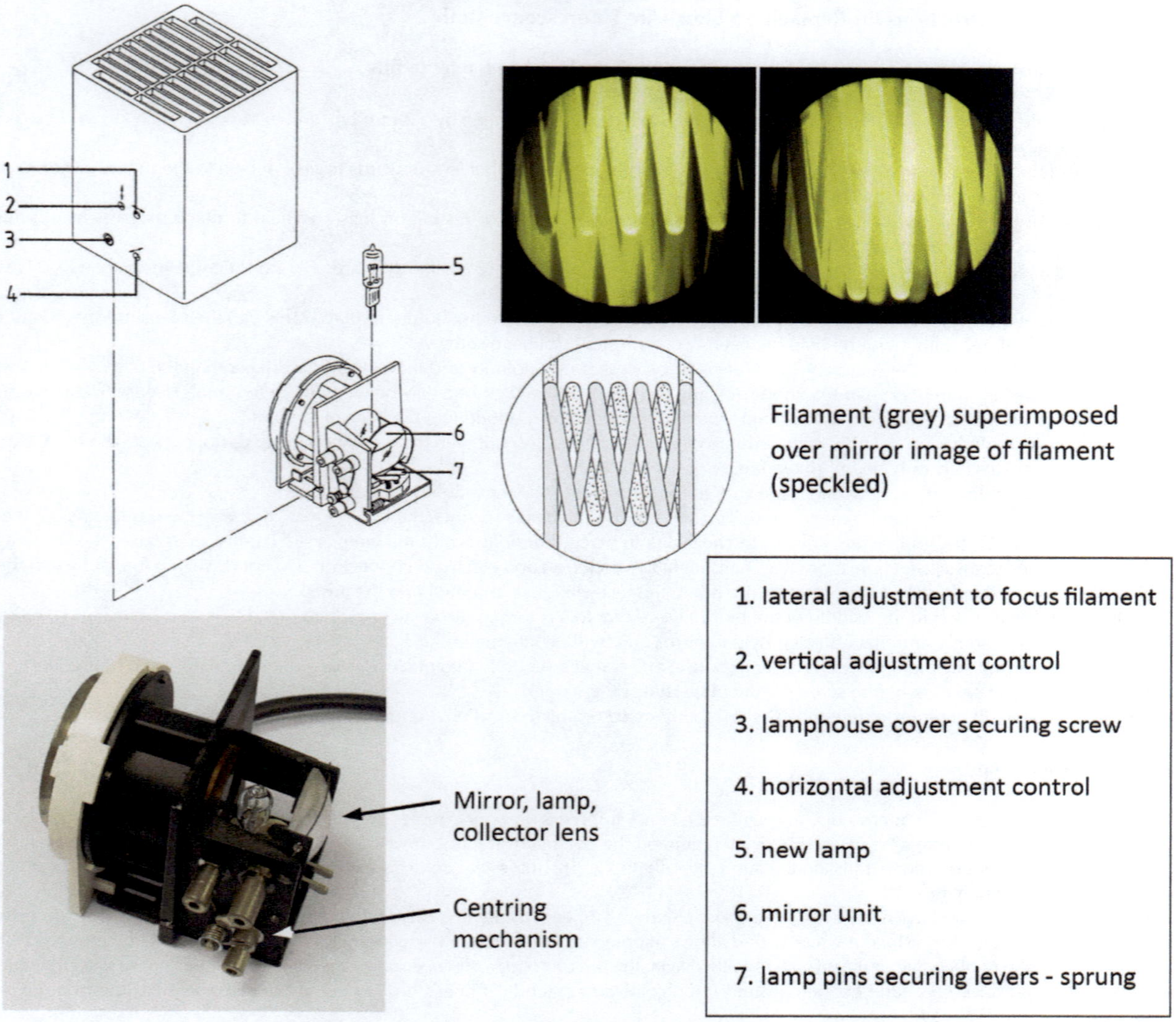

Figure 9.9 ***Changing a tungsten-halogen lamp and centring it*** *See the text for further generic details. Source: (top) Murphy et al. 2013. Reproduced with permission from John Wiley & Sons. (bottom left) Courtesy of Mr K. Glover.*

The power should always be isolated before removing the lamp house from the microscope, and/or the cover on the lamp house. Tungsten-halogen lamps should be handled with powder-free gloves or held between two pieces of lens-cleaning tissue. The design of the electrical contacts will be a bayonet fitting, a screw fitting or else pins. The lamp housing can then be replaced and the power turned back on.

For changing tungsten-halogen lamps, it is best to refer to the manufacturer's instructions, since the design of these bulbs and the lamp houses vary much more than for metal-halide and short-arc mercury lamps. Depending upon the design and sophistication of the lamp housing, there may be both filament and mirror centring controls present, with or without a separate lamp collimating lens (Figure 9.9). If the bulb requires centring, remove an eyepiece and inspect the image of the filament in the back focal plane of the objective. A focusing telescope (such as used for phase-contrast microscopy or else a Bertrand lens[6]) can be used to magnify this image, to allow for more precise adjustment. Focus the telescope so that the aperture delimiting the back focal plane of the objective is seen in sharp focus. Check that any diffuser present in the illumination train in front of the filament has been removed or swung out of the light path. Use the lamp collector lens – if present – to focus the image of the filament, and then centre this image in the field of view. Superimpose the mirror (reflection) image of the filament over that of the image of the filament itself, and interdigitate the two (Figure 9.9) so that the back focal plane of the objective is evenly illuminated. The diffuser can be replaced, if desired.

Replacing mercury and xenon short-arc lamps is very similar to the procedure described for incandescent lamps, above. Again, check that the lamp transformer has been isolated from the mains: the lamp housing cover of a short-arc lamp should *never* be removed whilst the electrical power remains connected.

Protocol 9.2 Generic Instructions for Replacing a Short-Arc Fluorescence Bulb

These are generalised instructions only. If you have the manufacturer's guide to hand, refer to this.

1. You may need to take the entire lamp unit off the microscope. It is normally attached by a 3 mm hex grub-screw. An Allen key will be required to undo this.
   The Olympus lamp house units exist as two designs. The bulb holder unit either lifts upwards to give access to the mercury bulb or else it pulls laterally out from the side of the lamp house.
2. The access panel of the lamp unit can then be removed. It is normally hinged or resting on lugs and held in place by cross-head screws or captive screws (which may require a 2.5 mm hexagonal Allen key).
   The older Leica design of lamp house has a captive fuse plug, which can be pulled forward. Do not attempt to dismantle this with a screwdriver.
3. Lay the unit down. If there is a bulb holder in place, undo the screws that secure the holder in place. This can then be slid forward out of the lamp housing. The bulb may otherwise be held between two springs by slot head screws.
4. Release the old bulb from its holder (Figure 9.11). Notice that it possesses a polarity and that one end is thicker than the other.
5. Normally the cathode (–) is thinner than the anode (+), and the lamp can only go one way between these terminals (Figure 9.11). The smaller diameter cathode inserts into the heat sink at the top, and the larger diameter anode into the base of the unit.
   A 50 watt short-arc bulb has no polarity. Both ends are of equal diameter. Insert it with the writing on the ends uppermost. There is **UP** marked on the top electrode. This may only be imprinted faintly on the bulb.
6. Remove the old bulb and insert the new one. Take care not to handle the glass envelope with your fingers.
   Release the bottom screw first, and remove the bulb. Then remove the heat sink once the lamp is free. This way there is less risk of the lamp breaking under leverage if the grub-screw holding the heat sink in place is undone whilst the lamp is still captive in its base.
7. Tighten the electrodes screws just a little more than hand-tight to provide a good electrical connection. Do not do this so forcefully that the bulb breaks! Replace the new bulb in reverse order: secure it into the heat sink first, and then into the base.
   The 50 watt bulb has a seal in the middle of the bulb. Ensure that this is turned 90°away from the optical axis (see Figure 9.11). Otherwise the photon flux from the bulb, and that reflected by the mirror unit, will be attenuated.
8. Slide the bulb holder unit back into place in the lamp house, and screw it securely into place. Replace the side of the lamp house, ensuring the two lugs are correctly in position before screwing the plate shut.
9. These bulbs contain small amounts of mercury. Dispose of them according to local environmental guidelines.

Aligning a fluorescence lamp

1. Replace the lamp unit onto the microscope and ignite the lamp. If there is an hour counter, set this to zero.
2. Identify the lamp-house controls (see Figure 9.13 for a guide). The collimator control moves a collimating lens axially. This focuses the bulb image. The two side controls move/translate the lamp laterally in x,y. The three rear controls focus the mirror image bulb axially in z and also separately move it laterally in x,y.
3. Some designs of microscope have a built-in centring tool (Figure 9.14), which can be pushed into the optical path. Some manufacturers provide a special 'telescope' with a translucent window that allows alignment of the lamp. If these are not available, proceed as follows:
4. Select the lowest power objective or remove an objective from the microscope nosepiece. Take care when doing this not to drop the objective. Use two hands, and have one hand cupped around the objective to catch it, if needs be. Do not get fingerprints onto the front lens of the objective.
5. Place a piece of white paper, or lens tissue, on the stage (Figure 9.12). Turn in a fluorescence filter cube (do not do this viewing only raw UV light). A green filter cube is easiest on the eyes. If the light is too bright, put in a neutral-density filter; otherwise, double up on the thickness of paper.
6. The first adjustment that usually needs to be checked is the real image vertical control. Adjust this control until both the real and mirror images are visible together in the field.
7. Adjust the condenser collimator lens control until the real image of the bulb burning between the two electrodes is at its smallest size and is in focus. Ideally, the tips of the two electrodes should be seen sharply in focus.
8. Adjust the real image position using the built-in controls (if any) or Allen keys, as required so that it is centred up and down, and is just to the side of the horizontal centre line.
9. Move the bulb centring/translation controls until an equal and opposite movement of the mirror image of the lamp can also be seen.
   Remember that although the real image controls move both the real and mirror images, the mirror image controls move only the mirror image.
10. Once you have located the mirror image of the lamp, focus this also upon the paper, using the separate control, to just the other side of the centre line.
11. Use both sets of controls, for the lamp and its mirror image, to bring both images to the centre of the field so that they are just touching one another.
12. Use the condenser collimator control to spread the fluorescence photon flux so that the field of view is evenly illuminated.
13. Replace the objective.
14. Replace the paper with an evenly stained slide (e.g. an H&E preparation of a liver or kidney section).
15. Focus the slide in the microscope field of view and move it up and down and from one side to the other to reveal any lack of even intensity of fluorescence across the field of view.
16. The arc between the two electrodes takes up to 8 hours to burn preferential initiation points between the two electrodes. Try to let a newly installed lamp burn for this length of time before first switching off.

Figure 9.10 **Design of a mercury-arc lamp house** *On the left, the lamp alignment window can be seen within the frame of the microscope stand. Source: Courtesy of Mr K. Glover.*

Some designs have a cut-out switch, which activates when the cover is removed. Arc lamps are under pressure, even when cold, so safety glasses, or goggles, should be worn when changing lamps. The modern designs of lamp house (Figure 9.10) for the short-arc mercury lamps have spring-loaded retention of the lamp, rather than a screw fitting, which carries the risk of snapping the lamp upon tightening. If you are working with the older screw fitting design, remove the lamp (with the heat sink still attached) from the lamp holder first. When re-assembling the new lamp, screw the upper cooling heat sink (Figure 9.11) onto the bulb first, and then screw tight the lower fitting. This way, the bulb is less likely to break by snapping when torque is applied to the screw fittings. Check that the bulb arc area is on-axis with the collimator lens and the reflector mirror. Check also that the electrical contacts are secure but not over-tightened. Replace the lamp housing cover and – if provided – re-zero the hour counter on the transformer. Reconnect the electrical supply, switch on the transformer and ignite the lamp with the arc initiation igniter control.

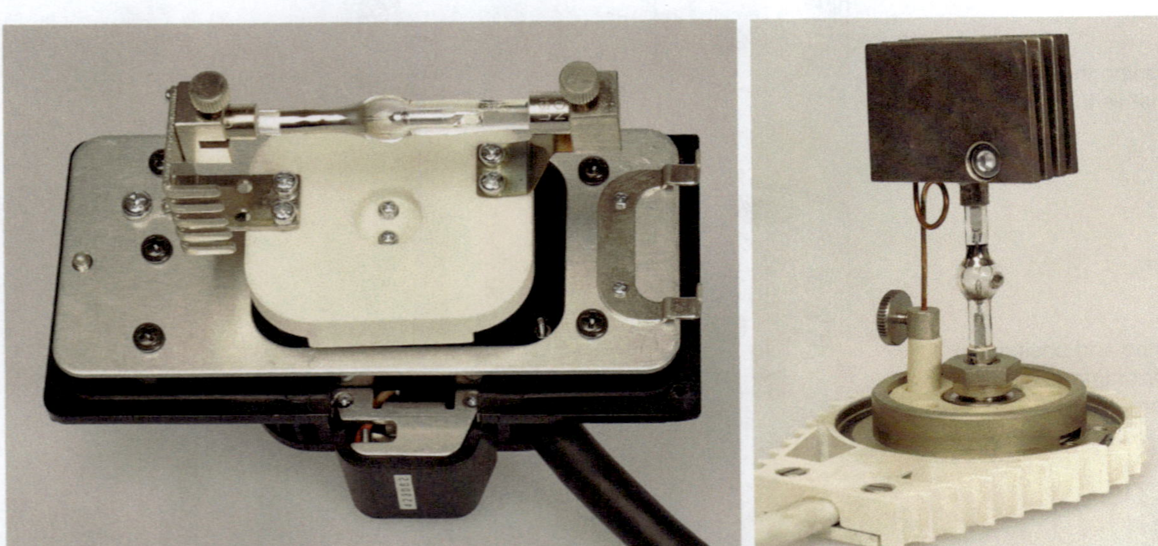

Figure 9.11 **Heat sinks on short-arc lamps** *Two designs are shown. The Nikon heat sink looks like fingers on the top (anode) of the mercury lamp; the Zeiss design on the 50 watt lamp is similar to that in the previous figure for the 100 watt lamp. Source: Courtesy of Mr K. Glover.*

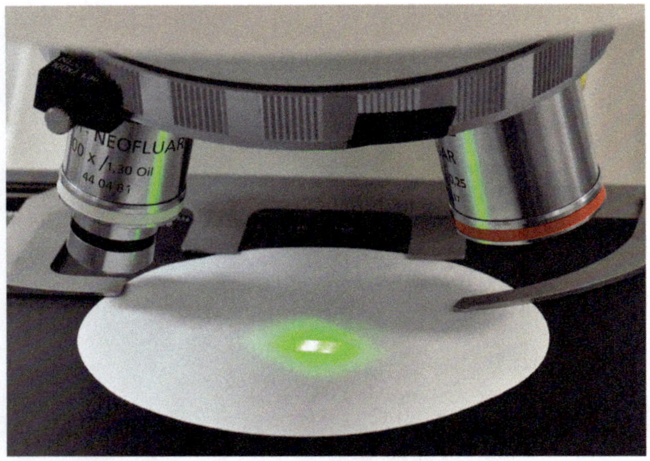

Figure 9.12 **Changing a fluorescence lamp and centring it** *The anode on 50 W lamps is the same size as the cathode. In this case, without a polarity to help replace the lamp in its correct orientation, the silver top of the bulb must be inserted downwards into the holder. Source: Courtesy of Mr K. Glover.*

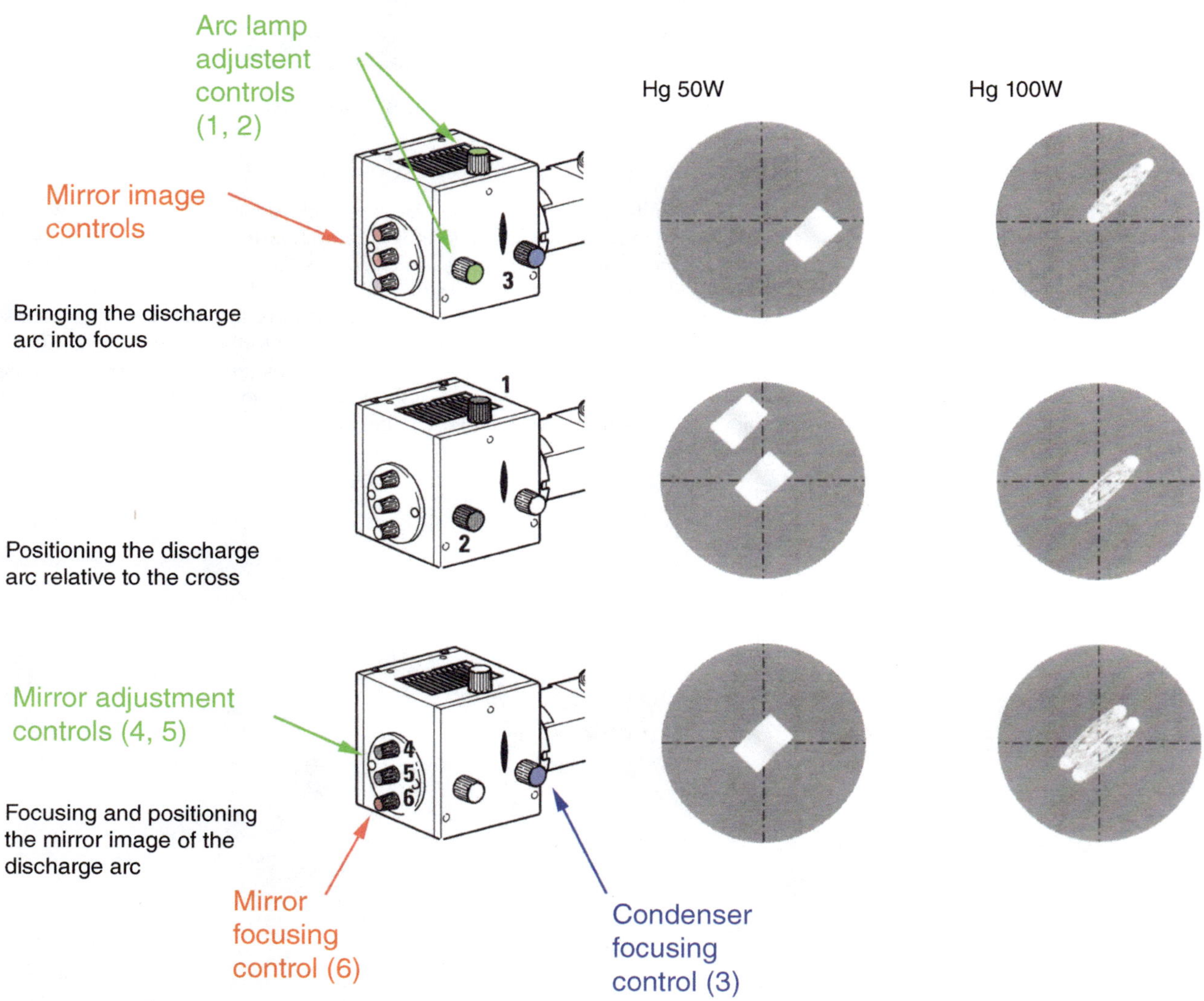

Figure 9.13 **Aligning a 100 watt and 50 watt short-arc mercury lamp** *Source: Adapted from Leica 2003 with permission from Leica.*

Figure 9.14 ***Sliding window to aid fluorescence changing*** *Source: Courtesy of Mr K. Glover.*

Some fluorescence lamp houses are also fitted with a captive sliding mirror and window assembly (Figure 9.14) to allow easier centration of the mercury arc and its mirror image – a pushrod on the side of the microscope enables the mirror to be slid into the optical path for centration. These features all make lamp exchange both quick and efficient and easier than with older designs of lamp house. Olympus fluorescence microscopes may use a liquid light guide to deliver fluorescence illumination to the microscope. The light guide is removed at the lamp house, and a special centring tool (similar to the captive mirror/window assembly) inserted to align the lamp arc and its mirror image. Alignment cannot be done without this tool, since the lamp house opening is protected with a spring-loaded safety shutter. Once aligned, check the evenness of illumination in the field of view with a histological section, as described above.

9.8 Requirements of the Illumination System

Although the numerical aperture of the objective determines the ultimate resolving power of the microscope, this will not be achieved unless the specimen is appropriately illuminated. The microscope can only perform optimally and exploit the full resolving power of the objective if the illumination is correctly adjusted and the entire back focal plane of the objective filled with light in a controlled manner. The illumination should be uniform and controllable so that stray light and glare (Box 9.1) – which do not contribute in any way to the image and merely reduce visibility or contrast – are excluded.

A well-designed illumination system should:

- provide uniform illumination at the specimen over an adjustable *area*, corresponding to that being observed; and
- fill the aperture of the objective with light over an adjustable *angle*, corresponding to the NA of the objective.

Box 9.1 Stray Light and Glare

Stray light can arise from being scattered or reflected by the object without contributing to image formation. It may also arise from obstacles in the light path, either being luminescent themselves or reflecting light into the field of view.

Glare is slightly different, although the terms 'glare' and 'stray light' are often used synonymously; put simply, glare is the result of stray light. Glare is the increased flux of light either disturbing the proper function of the eye (as with headlights 'blinding' a driver) within the field of view or is caused by very high illuminance (and reduced visibility) across the whole field of view.

Lenticular glare is caused by reflection of light from the glass-air interfaces of the lens elements making up the objective. This is nowadays very much reduced because of high-quality anti-reflection coatings on the lenses. This is the reason for careful cleaning of objectives: a scratched front lens will invariably have a damaged anti-reflection coating, which will admit stray light into the image.

The usual rule for bright-field microscopy using transmitted light is to close down the condenser diaphragm to correspond to 80% of the aperture of the objective. Otherwise, stray light is refracted off the edges of the lens mounts and is reflected off the inside of the microscope tube, reducing contrast. Since this 80% figure is only a guide – and very much specimen dependent – use your own judgement, but you may have to close the condenser diaphragm more than this in order to be able to see a useful image with sufficient contrast. However, if the illuminated field diaphragm is closed to well within the field of view so that only the object of interest in the centre of the field is illuminated, then the condenser diaphragm may usually be opened greater than 80% of the diameter of the objective aperture.

In the reflected-light microscope, the judicious use of a polariser can help reduce glare, in much the same way that a polariser in sunglasses blocks sunlight reflected specularly off water or road surfaces. In this case, the reflected sunlight is mostly polarised horizontally, parallel to the reflecting surface, and so the glare can be blocked by polarisers in the sunglasses with their permitted vibration direction set vertically. Not all stray light is eliminated, for some light that is reflected diffusely will be reflected or scattered several times, and so be unpolarised.

The use of homogeneous (e.g. oil) immersion will also help reduce glare. Stray light reflected off the underside of the glass slide and top surface of the coverslip is eliminated or, at least, reduced. It is possible that some glare can arise from reflections from the eyepiece, spectacles, or the external room illumination. If you see glare on a camera image for which you cannot identify the cause, there is always the possibility that it is entering the microscope via the eyepieces of a trinocular photo-tube from the room lights (for example, see Figure 12.14). Some microscopes dedicated to photomicrography have been designed with extra-wide body-tubes, to minimise the reflections reducing contrast in the image. It is also true that some high-end research microscopes have been sold with black velvet lining the inside of the body and photo-tube to minimise glare.

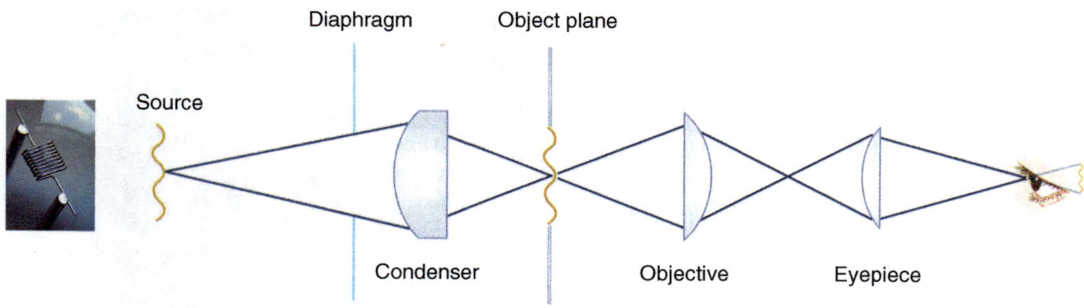

Figure 9.15 **Source-focused illumination** *Source: Bancroft and Gamble 2008. Reproduced with permission from Elsevier.*

9.9 Illumination in the Microscope: Source-Focused and Köhler Illumination

Nowadays we expect the specimen to be illuminated uniformly over the entire field of view. However, prior to the widespread use of electric light, this could be accomplished only by using a cloudy sky or the flame from an oil lamp as the light source. This was because the illumination was focused by the condenser directly into the plane of the specimen. This simple type of illumination (Figure 9.15) is called 'source-focused', 'critical' or 'Nelsonian' illumination[7] (after E. M. Nelson, who championed it in the 19th century).

Suitable daylight conditions were not always possible, and the artificial light from candles or oil lamps was not only of low intensity, but the differing intensity across the field of view also, from movement of the flame, made photomicrography difficult and frustrating. Gas lamps with woven mantles and subsequently electric filament lamps were introduced into microscopy towards the end of the 19th century. These provided clean, reliable, intense and immediate sources of illumination. However, both the gas mantle with its square weave and the electric lamp filament possess definite structure, which, if imaged onto the specimen by source-focused illumination, will interfere with interpretation of the image.

One way of circumventing the use of a small lamp filament and providing even illumination across the field of view is to use a sheet of white paper illuminated from behind to project even illumination onto the condenser. However, August Köhler (Box 9.2) considered the use of any form of diffuser wasteful of light. In 1893 he devised the method of illumination bearing his name to circumvent this problem of using a small filament without a diffuser.

If Köhler had placed his irregular light source at infinity, the filament structure would cease to be problematical; equally if he had placed the lamp at the front focal plane of the condenser it would throw parallel, and thus even, illumination onto the specimen plane. There are disadvantages with both approaches. In the first instance, with the light source at infinity, it is too small and too dim to be of any practical use. In the second instance, situated right underneath the condenser, the size of the lamp is restricted, and it will heat the specimen unnecessarily.

Box 9.2 August Köhler

Who was the man whose name has been given to the procedure by which the light microscope is set up and adjusted?

August Karl Johann Valentin Köhler was born on the 4th of March 1866 In the state of Hessen, in the city of Darmstadt. His father worked as a municipal official in Darmstadt, and after attending the local Ludwig-Georgs-Gymnasium (grammar school), the young August began his zoological studies when he was 18 years old at Darmstadt Technical University. He finished his studies at the universities of both Heidelberg and Gießen covering a wide range of subjects from biological sciences to chemistry, physics and mineralogy.

Following his graduation in 1888, Köhler returned to his old gymnasium, this time as a school teacher, before starting as an assistant to, and PhD student of, J.W. Spengel at Gießen University Zoological Institute. His project, one that depended heavily on photomicrography, involved the taxonomic classification of, amongst others, the air-breathing sea snail, or false limpet, *Syphonaria*. It was towards the end of his doctorate studies that Köhler published his 1893 paper: 'A new illumination method for photomicrographic purposes'. The ray path from the original paper is shown here.

Seven years later, Köhler's paper motivated Ernst Abbe to invite him to join the staff at Carl Zeiss, which he did on the 1st of October 1900 and stayed until his retirement, in 1945. In an attempt to improve the resolving power of the microscope, Köhler carried out much research into ultraviolet microscopy and was instrumental in the development of the fluorescence microscope. He was also responsible for introducing parfocal and also

Source: Zeiss. Reproduced under the terms of the Creative Commons Attribution Share-Alike licence, CC-BY-SA 2, via Flickr.

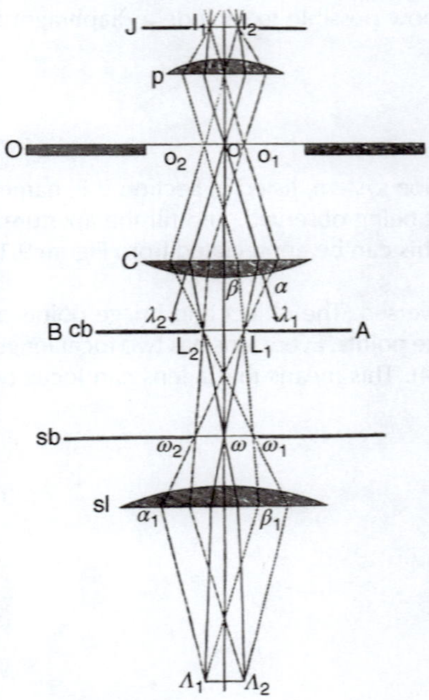

Source: Köhler 1893.

flat-field objectives (designed by Hans Boegehold) and for standardising the nomenclature of objectives and eyepieces. Köhler pushed forward the development of the commercial phase-contrast microscope in the 1940s. In the course of his work, he wrote 77 other scientific papers, besides the one for which he is famous, and contributed book chapters to edited volumes, as well as filing several patents both in Europe and the USA.

Köhler has been recorded as 'always kindly, helpful and modest' with a particular willingness to help his colleagues. Nevertheless, he was very conservative and, when shown the new all-black enamelled microscope without any obvious shiny brass, remarked: "Now just paint a silver palm frond on it, and the funeral will be ready". August Köhler never married, and for such a conservative man it is surprising that he reacted against convention by living with his long-term partner, Ida. They had no children, and she died in 1942.

In 1922 Köhler was appointed professor at the University of Jena, and he was awarded an honorary medical degree by the Faculty of Medicine. There is no record of his pursuing any hobbies or interests outside of microscopy. At the division of Germany in 1945, Köhler was already retired; he remained in Jena and towards the end of his life was not taken with the top workforce to the West when Jena was handed over by American troops to the Russians. He suffered in his last years from stomach complaints; the privations of post-war Germany under the occupying Soviets led to his dying of starvation. This was not uncommon at that time, but is a terrible conclusion to the life of such a talented, formerly well-to-do and energetic man. Köhler died in Jena, aged 82, on the 12th of March 1948.

Reference

Koehler, A. (1894). 'New method of illumination for phomicrographical purposes'. *Journal of the Royal Microscopical Society 14*: 261–262.

Köhler moved the light source away from the front focal plane of the condenser but introduced an extra lens into the illumination train, which we now call the bullseye condenser or (more properly) the lamp collector lens. This lamp collector lens projects an *image* of the light source[8] into the front focal plane of the condenser (Figure 9.16). A small and irregular light source, having a definite filament structure, can now provide a uniform and intense patch of light of considerable area. When imaged into the object plane of a

Figure 9.16 **Lamp collector image on the condenser front focal plane** *The lamp collector lens projects an image of the light source into the front focal plane of the condenser. Once correctly set, the lamp/collector lens combination should not normally require readjustment. The condenser will require frequent readjustment, principally due to objectives of differing magnification, but also because of variations in the thickness of slides.*

microscope, this patch of light serves as an ideal secondary source of illumination. This image of the lamp collector lens, bounded by the illuminated field diaphragm[9], provides a flawless field of uniform illumination. It is now possible to include a diaphragm for regulating the illuminating aperture at the front focal plane of the condenser.

9.10 Conjugate Planes for Köhler Illumination

A microscope correctly set up for Köhler illumination fulfils the requirements of the illumination system, listed in Section 9.8, namely, provide uniform illumination at the specimen over an adjustable *area*, corresponding to that being observed, and fill the aperture of the objective with light over an adjustable *angle*, corresponding to the NA of the objective. This can be appreciated from Figure 9.17, which shows the two sets of conjugate planes together in a transmitted-light microscope.

Because of the principle of the reversibility of light rays, the object and image may be reversed. The object and image points are called conjugate points. A conjugate plane occurs perpendicular to the optical axis at conjugate points. Every lens has two focal lengths and thus two focal planes (see Chapter 3, Section 3.10 and also Figure 2.21 and Figure 3.14). This means that a lens can focus two different sets of rays simultaneously.

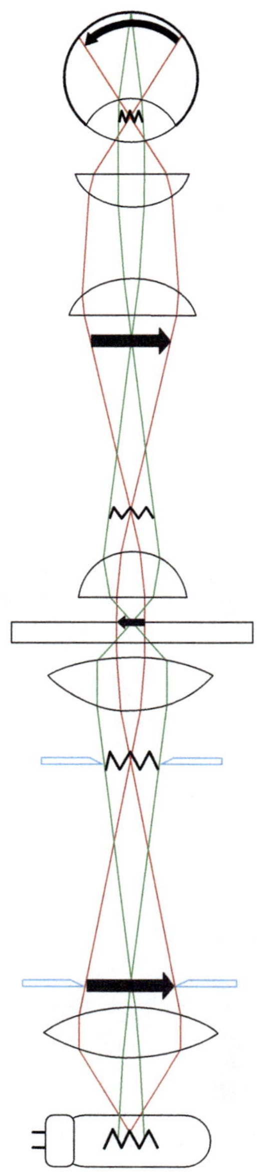

Figure 9.17 ***Two sets of conjugate planes*** *Ray paths of the finite tubelength transmitted-light microscope showing (a) the aperture set (red rays) and (b) the field set (green rays) of conjugate planes superimposed together on one another – as they are in the microscope. Sometimes (rarely) these are called conoscopic and orthoscopic sets of conjugate planes, particularly in books on polarised light microscopy. Source: Adapted from Wootton et al. 1995 with permission from Cambridge University Press. Courtesy of Mr S. Thomas.*

It is therefore possible for the two sets of conjugate planes found in Köhler illumination to exist together. The two conjugate sets interleave with each other, starting with the lamp filament in the illumination (or aperture) set of conjugate planes:

	Aperture (illumination) set	Field (imaging) set
1.	Lamp filament	
2.		Illuminated field diaphragm (IFD)
3.	Front focal plane of the condenser	
4.		Specimen plane
5.	Back focal plane of the objective (BFP)	
6.		Primary image plane (PIP)
7.	Exit pupil of the eyepiece (Ramsden disc)	
8.		Retina/camera

The lamp is deliberately placed in the *aperture* set of planes. It is imaged by the lamp collector lens into the front focal plane of the condenser. Thus, the structure of the illuminating lamp filament, in the aperture set of conjugate planes, cannot disturb the illumination of the object or its images, all of which lie in the field set of conjugate planes.

The illustration in Figure 9.18 shows the two sets of conjugate planes, this time as separate ray diagrams. In reality these conjugate planes occur together in the microscope optical train because each lens can focus two different sets of rays simultaneously. Hammond

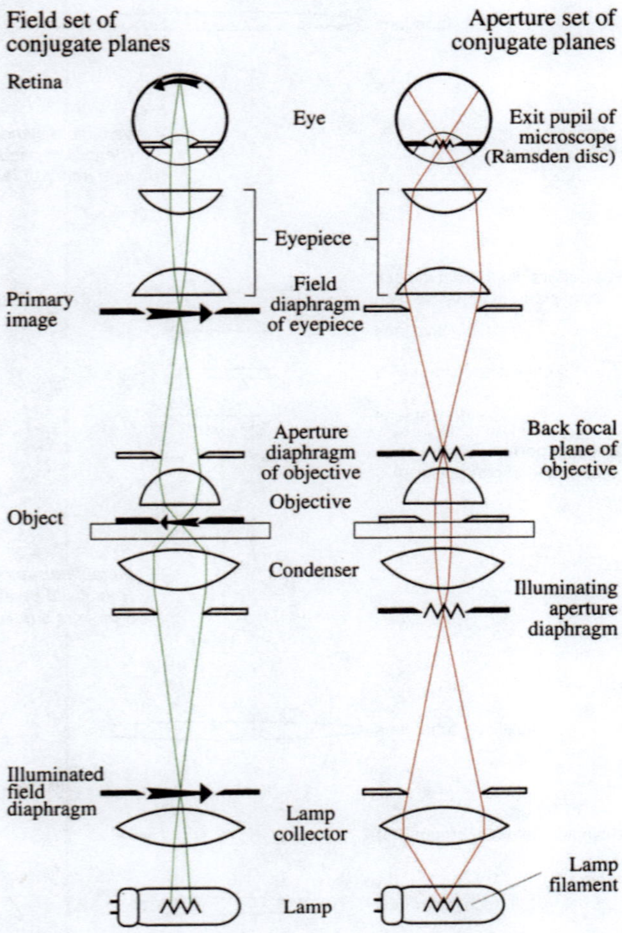

Figure 9.18 **Two sets of conjugate planes shown separately** *In the accompanying diagrams, only on-axis rays arising from the filament (right) or passing through the object (left) have been shown. Other similar diagrams could be drawn for non-axial rays, which would come to a focus in the same planes as the axial rays. The ray paths are shown separately for clarity. In reality, they are present together, as shown in the previous figure.*

The diagrams illustrate another elegant feature of Köhler illumination: that rays arising from each individual point on the filament pass through all parts of the object under observation, and that each individual point on the object receives light from all parts of the filament admitted by the condenser aperture. Source: Wootton et al. 1995. Reproduced with permission from Cambridge University Press.

(1998) discusses the construction of a ray diagram of the conjugate points and planes of the aperture set and the field set of rays for a finite tubelength microscope. This design of microscope is nowadays used much less. Figure 9.19 shows the corresponding ray diagram for a transmitted light infinity-corrected microscope, which employs an objective/tube-lens combination. Figure 9.20 shows the corresponding (simpler) ray diagrams – as both separate ray paths and the two superimposed upon one another – for a reflected-light microscope, where the objective acts as its own condenser. For further details, see Hammond & Heath (2006) and an updated and expanded version of this paper in Evennett et al. (2009).

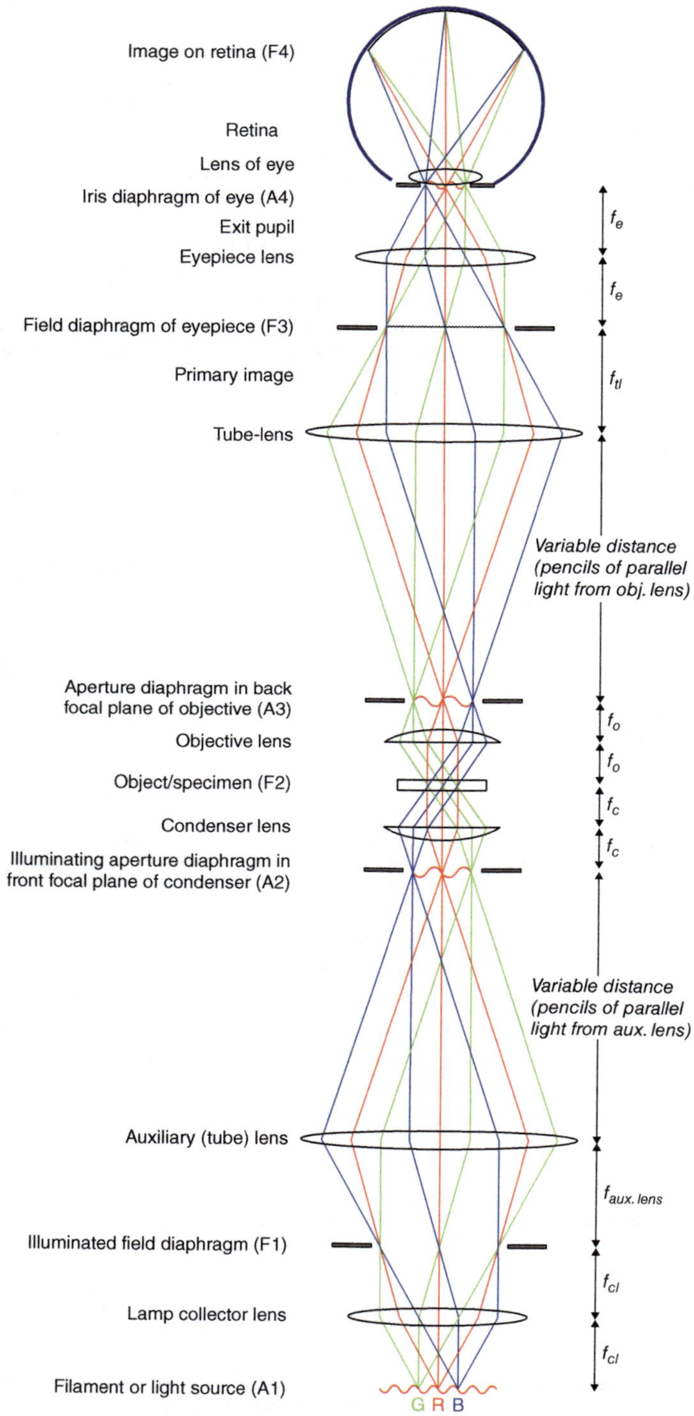

Figure 9.19 ***The two sets of conjugate planes for an infinity-corrected transmitted-light microscope*** *Most modern microscopes are designed as infinity-corrected microscopes, with a separate tube-lens. The two sets of ray paths for the aperture set and the field set of conjugate planes are shown together for this design of microscope. Source: Hammond 2006. Reproduced with permission from John Wiley & Sons.*

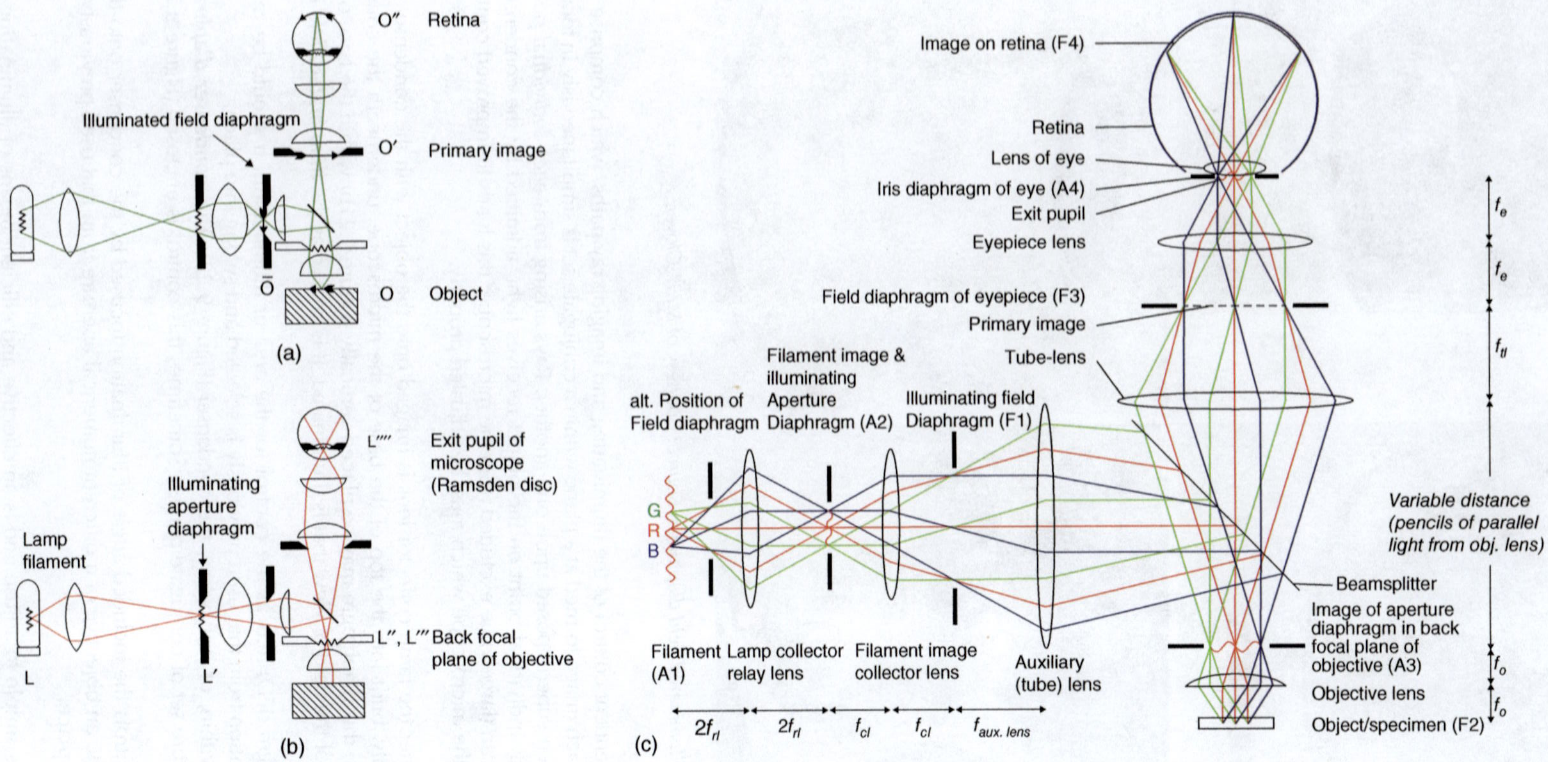

Figure 9.20 **Conjugate planes in a reflected-light microscope** *The reflected-light microscope is simpler to adjust for Köhler illumination. Here the rays paths for a finite tubelength design are shown separately (a,b), as well as the ray paths of both sets of conjugate planes (c) for an infinity-corrected reflected-light microscope. Source: (a) Adapted from Bradbury et al. 1989 with permission from Oxford University Press. (b) Bradbury et al. 1989. Reproduced with permission from Oxford University Press. (c) Hammond 2006. Reproduced with permission from John Wiley & Sons.*

Figure 9.21 **Possible locations of the illuminated field diaphragm** *Source: Courtesy of Mr K. Glover.*

Köhler illumination allows independent control of the illuminating and imaging ray paths, which comprise two different, and independent, sets of conjugate planes. Each image formed at a plane within a conjugate set is simultaneously in focus with the other images in that set. These images can be seen superimposed upon one another. Rays arising from *each individual point* on the filament pass through all parts of the object. Each individual point on the object receives light, admitted by the condenser, from all parts of the filament. Non-image-forming light can (largely) be excluded from the microscope; this leads to improved contrast in the image.

There are two iris diaphragms in the microscope which must be adjusted properly.

- An iris diaphragm placed just after the lamp collector lens is imaged onto the object with the condenser lens. This diaphragm is normally found lying horizontally, built into the top of the base of the microscope, underneath the condenser and its substage (Figure 9.21a). On some designs, this diaphragm may be placed vertically (Figure 9.21b) within the base of the microscope. Since this diaphragm limits the extent of the field of view that is illuminated, it is known as the ***illuminated field diaphragm***. It lies in the field set of conjugate planes.

  The illuminated field diaphragm (IFD) acts as the control for the *area* of illumination. It should be re-adjusted each time an objective of a different magnification from that used previously is selected and swung into place.
- A second diaphragm is found within, or just below, the condenser (Figure 9.22). This ***condenser diaphragm*** (illuminating aperture diaphragm) lies in the aperture set of conjugate planes. Sometimes this control is enclosed (Figure 9.23) in a motorised condenser unit.

  The condenser diaphragm controls the included *angle* of illumination focused by the condenser onto the object (Figure 9.24). It should be re-adjusted each time an objective of a different numerical aperture from that used previously (which is normally the case) is selected and swung into place.

Köhler's method of illumination is simple to adjust, and is an effective and efficient means of illuminating the specimen. A better method has not been devised or adopted. In October 1993, to commemorate the centenary of the publication of Köhler's paper, the *Proceedings* of the Royal Microscopical Society carried a reprint of Köhler's paper together with a number of commentaries and articles placing Köhler illumination into perspective.

Figure 9.22 **Location of the condenser diaphragm** *Source: Courtesy of Mr K. Glover.*

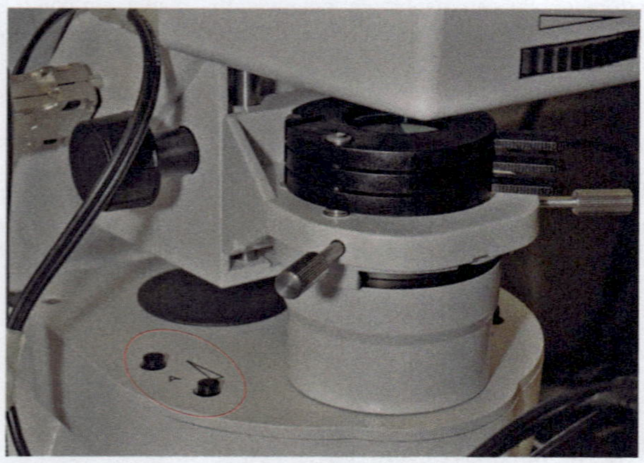

Figure 9.23 **Location of the condenser diaphragm in a motorised condenser** *This is an example of a motorised condenser on an inverted transmitted-light microscope. Source: Courtesy of Mr K. Glover.*

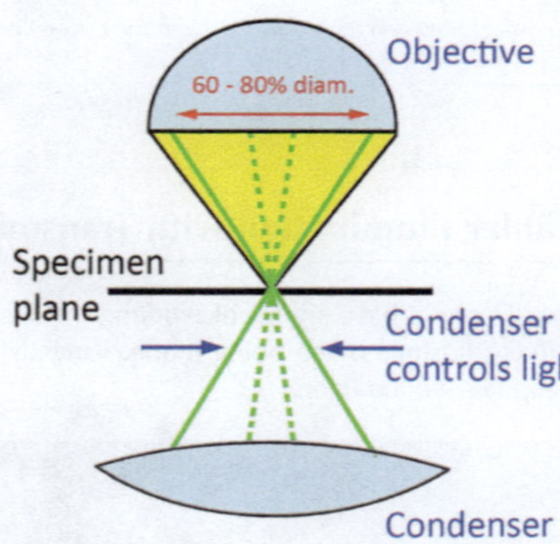

Figure 9.24 **Angle of illumination focused by the condenser**

9.11 Adjusting a Microscope for Köhler Illumination

Setting up Köhler illumination with a built-in light source (see Box 9.4) is the most commonly used method and should be employed each time the microscope is switched on. In fact, this protocol should be so familiar that adjustment of the microscope for optimum resolution and image formation before use becomes automatic. The use of an external lamp for the light source is very rare nowadays and largely the preserve of skilled amateurs who desire ultimate control over all aspects of the illumination and the microscope, for contrast enhancement, or who may choose to operate an older design of microscope having a mirror below the condenser and substage apparatus. A description of adjusting a microscope in this way has been included for completeness. A protocol for setting up the microscope with source-focused illumination is not given. It is very simple: the lamp and condenser are both focused simultaneously upon the specimen plane. Details of how to set up the microscope for source-focused illumination may be found in Barron (1965) or Bradbury & Bracegirdle (1998).

Box 9.3 Understanding the Effect of the Two Diaphragms Used to Set Up Köhler Illumination in the Microscope

This exercise demonstrates why the illuminated field diaphragm must be set according to the *magnification* of the objective used, because this determines the *area* observed, and why the illuminating aperture (condenser) diaphragm, which controls the angle of illumination, is set according to the objective *numerical aperture*.

1. To visualise these use a block of slightly turbid glass; on our teaching courses we use pieces of fluorescent coloured perspex with polished edges. Otherwise, make your own: take an empty transparent coverslip box. Fill it with a *very* dilute solution of milk in water. To 10 mL water add a couple of drops (e.g. 50 μL) of milk. Pour this into the cover glass box. Another alternative is a 20 mL clear glass scintillation vial[i].
2. Use a condenser designed for oil-immersion[ii]. Raise the condenser to the top of its travel. Place a drop of immersion oil on the condenser's top lens, and lower the turbid block or its substitute on to the microscope stage.
3. Open both the illuminated field diaphragm and the condenser illuminating aperture diaphragm. Switch on the lamp, and observe the rays as they pass between condenser and objective.
4. Open and close the illuminating aperture (condenser) diaphragm: note how this diaphragm controls the included *angle* of the rays. This is shown in panel (a) [condenser diaphragm set to a narrow angle: NA = 0.25] and panel (b) [condenser diaphragm control opened to a large angle: NA 0.65].
5. Leave the illuminating aperture diaphragm in the centre of its range and close the *illuminated field diaphragm*; note that the angle of illumination remains unchanged, but the *area* illuminated decreases; see panel (c). (This is represented by the diameter of the 'neck' at the apex of the inverted cone of rays leaving the object plane; compare panel (b) and panel (c). It is not as noticeable in this demonstration as when changing the condenser illuminating diaphragm but happens nonetheless).

(a) (b) (c)

[i] E.g. Perkin Elmer 20 mL Econo glass scintillation vial, Product No. 1210-138.
[ii] A dry condenser can be used, but immersion oil ensures optical contact between the condenser and the coverslip box, or vial. This gives a clearer result.

Box 9.4 Adjusting the Microscope for Köhler Illumination with Transmitted Light

1. Turn on the light source and swing in a 10x objective. This will have sufficient working distance (the clearance between the front of the objective and the slide) not to hit the slide while initial adjustments are being made. If needs be, hold a piece of paper in front of the illuminated field diaphragm at the base of the microscope to see the light.

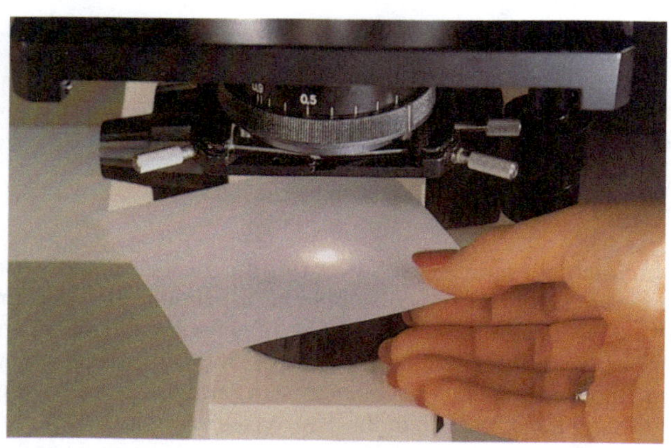

Source: Kapitza 1997. Reproduced with permission from ZEISS Microscopy.

2. Raise the condenser to the top of its travel. Open fully the illuminated field (lamp) diaphragm (IFD) and the illuminating aperture (condenser) diaphragm (IAD).

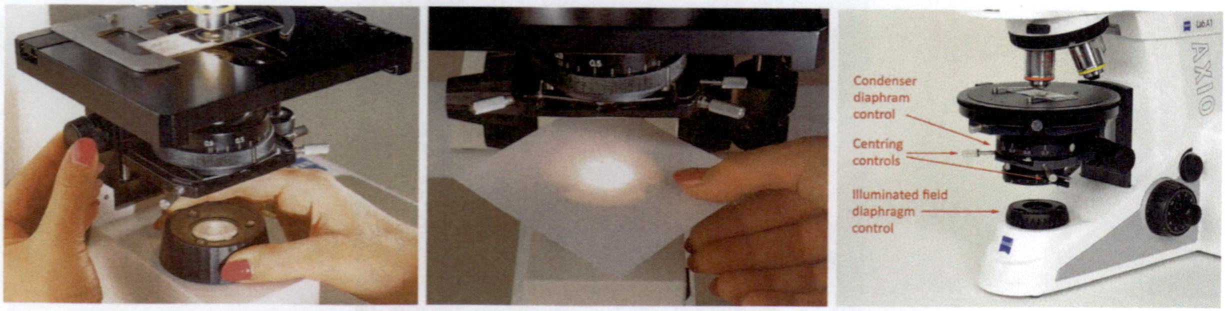

Source: Kapitza 1997. Reproduced with permission from ZEISS Microscopy.

3. If the lamp is one that is external to the microscope, the IFD is the diaphragm fitted to the lamp. Behind that is the control for the lamp collector lens, which is fixed and thus not available for adjustment on a microscope with a built-in light source. Align the external lamp about 150–180 mm in front of the microscope. The distance from the front focal plane of the condenser from substage to mirror and from mirror to lamp should be about 200–250 mm.
4. Light should be discernible in the eyepieces. If needs be, adjust the intensity of the lamp. This can be done by using the rheostat intensity control or by inserting neutral-density filters if you wish to maintain the correct colour temperature for photomicrography.
5. When using the microscope, your eyes should be relaxed. Before using the microscope, first look into the distance, i.e. at 'infinity', and then check that the inter-ocular distance between the two eyepieces via the folding (or sliding) bridge on the binocular head is set correctly for you to see one circle of light instead of two. The folding Siedentopf design is shown above, and the sliding Jentsch design below in this figure.

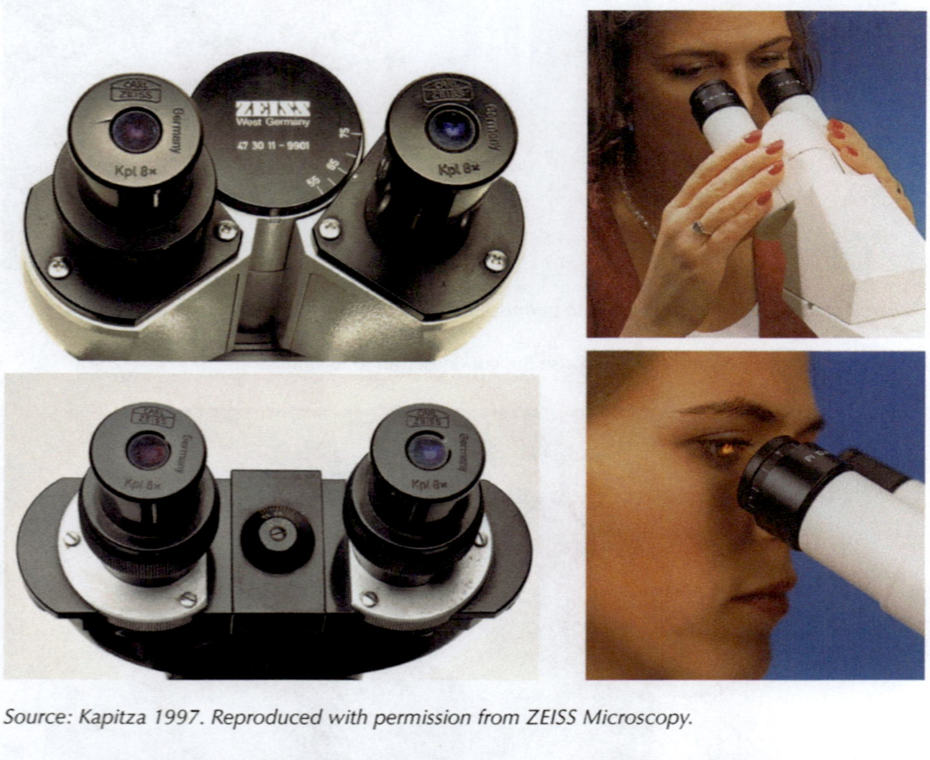

Source: Kapitza 1997. Reproduced with permission from ZEISS Microscopy.

6. Place a slide on the stage, checking that the coverslip is uppermost (or on an inverted microscope downwards) facing towards the objective – we all forget this at one time or another!
7. Use the coarse focus, to reduce the distance between slide and objective to a minimum, closer than the focal point of the objective. Look side-on at the gap between the objective and slide whilst doing this, to ensure that they do not collide. Remind yourself which way to turn the coarse focus in order to focus away from the objective. Looking down the eyepieces, you should see the image of the specimen very much out of focus.

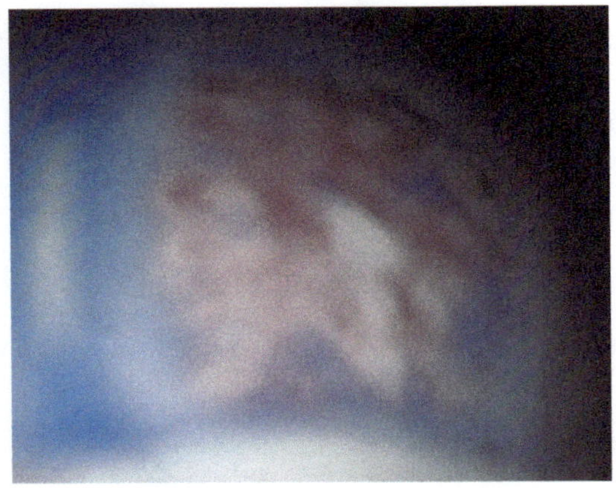

Source: Kapitza 1997. Reproduced with permission from ZEISS Microscopy.

8. Now, whilst looking into the eyepiece(s), increase the distance between slide and objective using the coarse focus control, and stop when the image of the specimen comes into focus. Adjust as necessary with the fine focus control. This sets the specimen in correct relation to the objective.

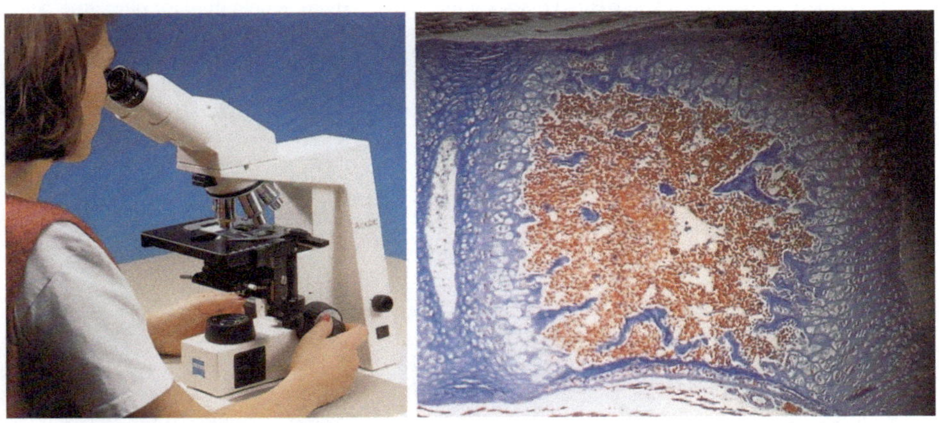

Source: Kapitza 1997. Reproduced with permission from ZEISS Microscopy.

9. The correct height of the condenser must now be set. Raise the condenser up to its limit.

Source: Kapitza 1997. Reproduced with permission from ZEISS Microscopy.

10. Close down the illuminated field diaphragm (IFD) almost to a pinhole, or at least to its smallest extent. This diaphragm is usually situated in the base of the microscope, underneath the condenser and substage assembly. If the lamp is an external one, the field diaphragm will be the iris built into the lamp at the front, after the lamp collector lens. Rack the condenser down slowly until the image of this diaphragm is sharply in focus at the specimen plane, superimposed upon the image of the object.

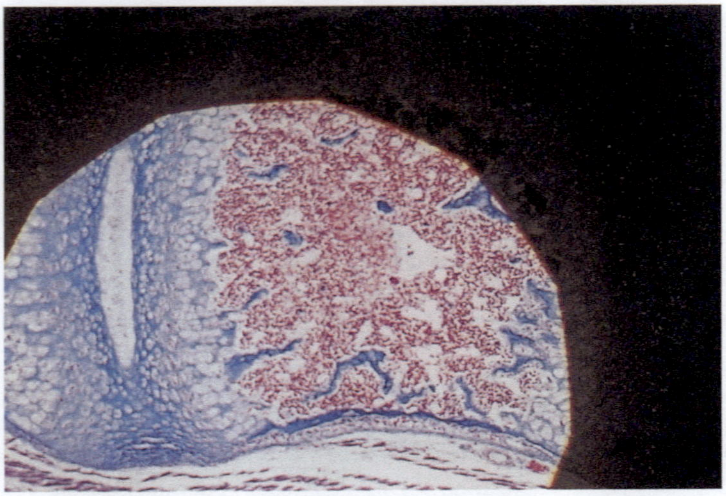

Source: Kapitza 1997. Reproduced with permission from ZEISS Microscopy.

11. Open up the field diaphragm until its image almost reaches the edge of the field of view. This way, it is easy to operate the two centring screws such that the narrow penumbral shadow just inside the field of view and surrounding the illuminated part of the image of the specimen can be made of parallel width. Centre the condenser with its adjusting screws, if provided – they usually are on well-designed research microscopes. With an external lamp, move the mirror and/or the lamp.

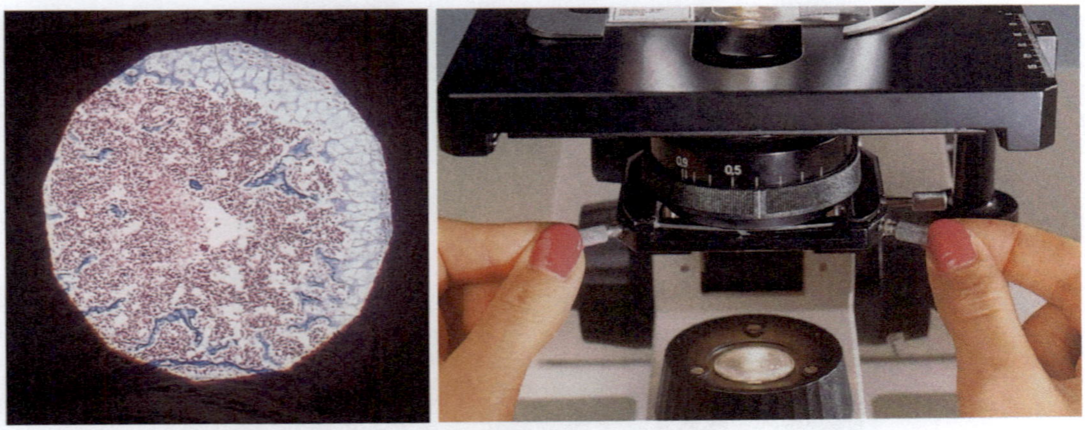

Source: Kapitza 1997. Reproduced with permission from ZEISS Microscopy.

12. Open the diaphragm further by a small amount until the image of the diaphragm lies just outside the field of view. Do not open it too much; otherwise, stray light (Box 9.1) will reduce contrast in the image. The IFD should confine the area illuminated to that under observation by a particular objective lens. (For contrast enhancement methods such as phase contrast and dark-field, which employ their own annular diaphragms in the condenser, the IFD must be left fully open.)

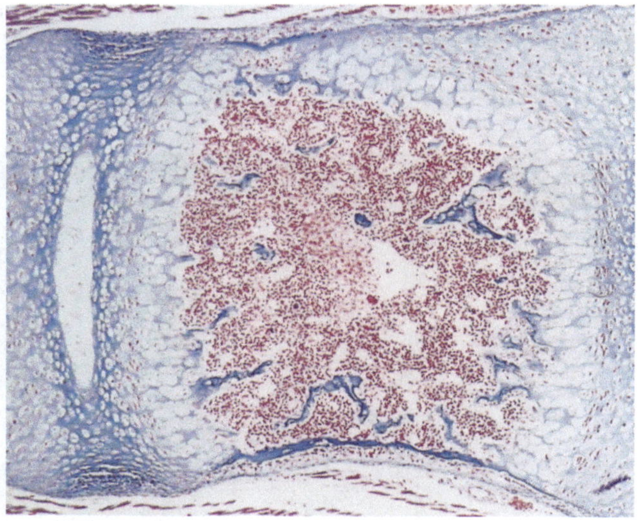

Source: Kapitza 1997. Reproduced with permission from ZEISS Microscopy.

13. The correct height of the condenser has now been set. The condenser iris can now be adjusted so that the aperture of the variable cone of light supplied by the condenser can be correctly matched to the (generally) fixed numerical aperture (NA) of the objective.
14. If using an external lamp, an image of the lamp filament should be focused onto the front focal plane of the condenser (see Figure 9.16) using the focus adjustment control for the lamp collector lens, which is behind the IFD control on the external lamp. Ensure that the image is central by small movements of the lamp; do not move the mirror, since this has already been adjusted to give axial illumination (NB a small dental mirror can help observe the underside of the condenser conveniently and comfortably). Adjust the position of the lamp slightly so that the image of the lamp filament just fills the aperture of the substage condenser. If this cannot be done without moving the lamp further away, the lamp collector lens should be replaced by one of shorter focal length, if possible.
15. Remove an eyepiece[i] and look into the microscope body-tube to inspect the back focal plane of the objective, which should be seen as a disc of light at the base of the tube. The rays of light are brought to a focus in the back focal plane of the objective, where a (usually fixed) diaphragm is located. This *aperture diaphragm of the objective* limits the effective numerical aperture of the objective.

Source: Kapitza 1997. Reproduced with permission from ZEISS Microscopy.

[i] In addition to removing an eyepiece, the back focal plane of the objective may be observed by other means. Insert a phase-contrast centring telescope, or screw a low-power objective to the bottom of the drawtube (if fitted). These methods provide a magnified image of the back focal plane of the objective for inspection. If a magnification changer is included, this often includes a supplementary lens system – a Bertrand lens – for the same purpose. A final method is to place an inverted eyepiece over the exit pupil of the eyepiece to view the exit pupil of the microscope.

16. Adjust the condenser aperture diaphragm until the image of this iris is just a little smaller than the diameter of the disc of light (about 80% diameter), which represents the full aperture of the objective (see also Figure 9.24). Replace the eyepiece.

In theory, the aperture of the condenser should equal that of the objective (see the image below). However, in this case stray light refracted from the extreme edges of the objective lens elements and reflected from the surface of the body-tube would cause an appreciable loss in contrast. It is worse, however, to close down the aperture diaphragm too far: this will cause serious degradation in image quality and loss of resolving power. This condenser diaphragm should *not* be used to control brightness in the image; rather, use the rheostat control on the lamp transformer, or (where the intensity of the lamp must remain constant, as for colour photomicrography for example) use neutral-density filters. Closing down the condenser aperture diaphragm from its optimum position will increase contrast and may sometimes be the only method of introducing sufficient contrast to visualise the image at all, but this contrast is gained at the expense of resolution. The rule that the condenser illuminating aperture diaphragm is set open at approximately 80% of the numerical aperture of the objective is for guidance and good practice; by all means break it if you know what you are doing!

However, closing the condenser diaphragm too far will cause the image gradually to become a less faithful representation of the object: outlines of all features will appear thickened, and adjacent features will merge. Decreasing the aperture of the condenser will also increase the depth of field and bring into focus dust and other contamination from the surfaces of the specimen preparation normally invisible in the properly adjusted microscope.

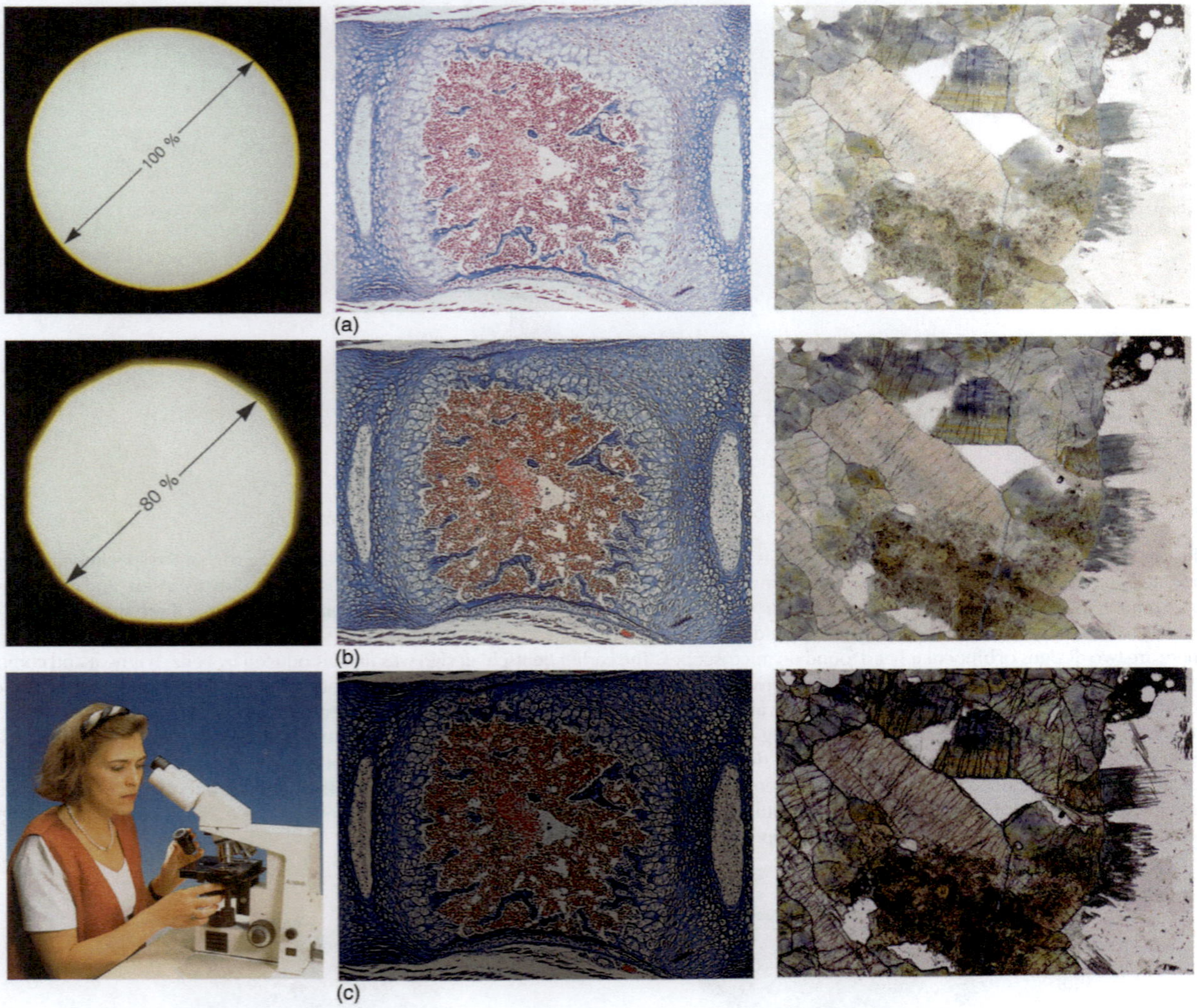

Source: Kapitza 1997. Reproduced with permission from ZEISS Microscopy.

This illustration of six images shows the effect of adjusting the condenser on a biological tissue section and a thin section of the mineral biotite. The figures of the two images in the top row (a) show the effect of opening the condenser diaphragm too widely: stray light reduces contrast in the image. The pair of images in the middle row (b) show the aperture of the condenser diaphragm is set to about 80% of the NA of the objective. This is not a hard-and-fast rule but depends upon the nature of the specimen as well as the quality and degree of correction of the objective (see also the comments in Box 9.1). An experienced microscopist can set the diameter of the condenser diaphragm by instinct: opening the diaphragm to its maximum, then gradually closing it whilst looking for a slight darkening as the stray light is cut out and for a concomitant increase in contrast of the image. The two figures in the bottom row (c) show the effect of closing down the condenser diaphragm too far. Whilst contrast may *appear* to improve further, diffraction lines and haloes serve to give rise to a 'rotten image', and the full resolving power of the objective is not utilised.

When changing to another objective of different magnification, this will have a different field of view (requiring a change in the diameter of the illuminated field) or numerical aperture (requiring a different illuminating aperture), and both diaphragms will need to be adjusted. With objectives below ten times, the field of view may not be fully illuminated even with the field diaphragm fully open. In this case, the top lens of the condenser should be either swung out or unscrewed from the condenser assembly. Do not defocus the condenser to enlarge the illuminated area, since uneven illumination and a loss of resolving power and image quality will result.

Sometimes, although rarely, a 63x or 100x objective will be 'locked up', that is, its front end – which is sprung to give way in case of being brought inadvertently into contact with the slide – may be retracted against the spring when not in use. Be careful to check that any high-magnification immersion lens is 'unlocked' before use. If it is an infinity-corrected objective, it will still work, but it certainly will not be parfocal with the other objectives!

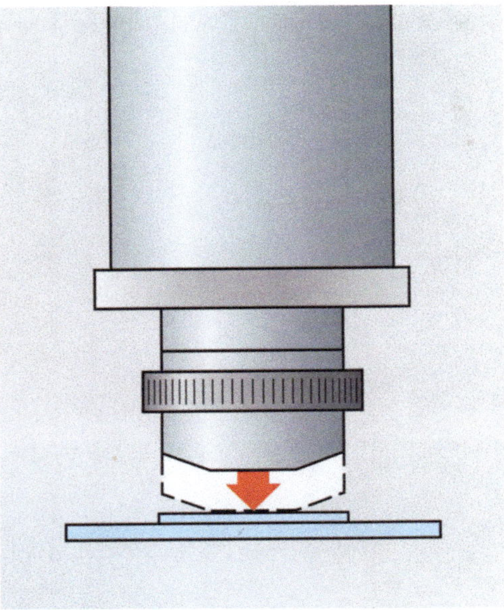

Source: Kapitza 1997. Reproduced with permission from ZEISS Microscopy.

These sixteen steps (often referred to colloquially as 'Köhler illumination') complete adjusting the microscope so that the specimen is properly illuminated. In order to avoid eye strain under conditions of prolonged visual use (e.g. when then image is viewed directly by eye and not projected onto a computer monitor), it is important to complete the following two steps. In any case, it is good practice to follow these steps so that all the objectives remain parfocal with one another. It should then only be necessary to adjust the fine focus slightly when a new objective is selected.

There are two designs of binocular head found on microscopes. The earlier 'Jentsch' design was first introduced by Leitz. It is neat and compact, having a sliding arrangement to alter the distance between the eyepiece tubes to set the inter-ocular (inter-pupillary) distance for each individual user. Adjusting the inter-ocular distance in this way also changes the mechanical tubelength. This necessitates changing the focus, which will introduce spherical aberration unless a compensating adjustment is made on the eyepiece tubes, where the tubelength can be set against a scale on each eyepiece tube marking out the inter-ocular distance that has been set on the slider. Most modern microscopes are now fitted with the 'Siedentopf' pattern of binocular head where the eyepiece tubes rotate in an arc about a fixed point. The tubelength remains constant, which is advantageous, since no spherical aberration is unwittingly introduced, so no compensating adjustment need be made; a constant tubelength is essential for accurate micrometry with the microscope.

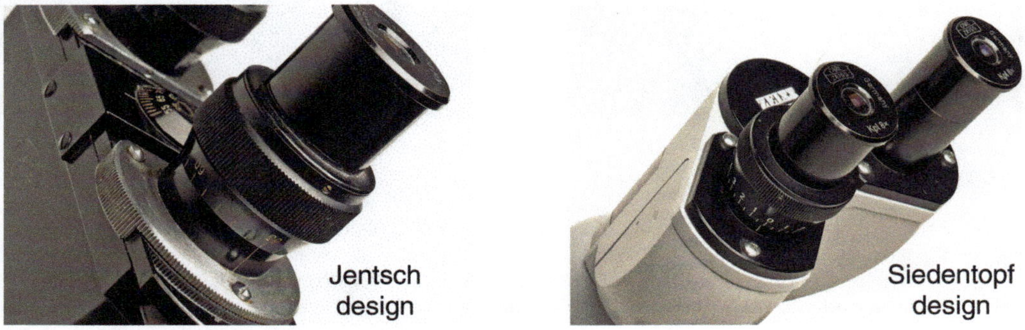

Source: Kapitza 1997. Reproduced with permission from ZEISS Microscopy.

17. If the microscope is fitted with a binocular head, the eyepieces can now be adjusted for comfortable viewing. One or both eyepiece tubes may have adjustable controls capable of adjusting the tubelength of the microscope. If only one eyepiece tube of the binocular has a variable control, set up Köhler illumination with the fixed eyepiece and adjust the variable control for the other eye until the image is in focus while the eyes are relaxed. Try not to close the opposing eye while focusing, nor screw the eyes up, but rather relax them – you should be aiming to 'look through' the microscope to infinity. Fatigued eyes result if the eyepieces are not correctly adjusted for relaxed viewing of an image at infinity – which is how the microscope is designed to be used.
18. If the binocular has two adjustable focusing controls, first focus the microscope using a high-magnification objective (forty times; a 'high-dry' objective – a non-immersion one with high aperture but ideally *without* a correction collar). Change to the ten times (or preferably a lower-magnification objective if it is part of a parfocal set), and focus the eyepiece focusing adjustments separately *without* refocusing the objective.

Reference

Kapitza, H.G. (1997). *Microscopy from the Very Beginning*, 2nd rev. ed. Carl Zeiss, Oberkochen, Jena: Carl Zeiss GmbH.

Box 9.5 Should You Keep Wearing Your Spectacles When Using the Microscope?

Spectacle lenses are either spherical or toric in design. Astigmatism is caused by imperfections or irregularities in the cornea or sometimes in the shape of the lens. This results in unequal refraction, causing distortion of images projected onto the retina. Without correction, a circle will appear elongated or flattened, like a rugby ball. About half of spectacle wearers experience some degree of astigmatism.

On the other hand, long-sightedness (hyperopia) and short-sightedness (myopia) can be corrected with lenses of differing dioptric strength. Since the dioptre adjustment controls of the microscope eyepiece also perform this function, it is possible to use the microscope without wearing your spectacles and (providing that the dioptre range is sufficient) adjust the controls appropriately for comfortable viewing.

However, astigmatism correction requires lenses of toric power – which refract differently in the horizontal and vertical meridians; therefore, if you suffer from astigmatism (which is an unsymmetrical aberration) you should keep your spectacles on when using the microscope: your spectacle lenses provide unique correction, which the microscope eyepiece cannot do.

Test yourself by viewing a simple geometric figure – a square is best – through your spectacles held at arm's length. First hold the glasses horizontally, then turn them through 90° vertically. If the figure appears expanded or compressed by this rotation, you are wearing lenses with a toric, or another non-spherical, power to correct for astigmatism.

High–eye point eyepieces designed for spectacle wearers to use will either have a small icon of a pair of spectacles on the eyepiece next to the field-of-view number or perhaps – on German-made eyepieces – the inscription 'Br' standing for *Brille*, or spectacles.

Eye dominance

Your dominant eye is the one which you unconsciously preferentially use to make alignments. This was more important when using monocular microscopes than it is today, but knowing which is your dominant eye is still helpful. Approximately two thirds of the population is right-eye dominant and one third left-eye dominant. Determine which is your dominant eye by either using the Miles test or the Porta test:

1. The Miles test.
   Extend both arms fully, brings both hands together to create a small opening, then with both eyes open view a distant object through that opening. Close each eye alternately to determine which eye is dominant. The image will shift when seen with the non-dominant eye.
2. The Porta test.
   Extend one arm fully, then with both eyes open align the thumb or index finger with a distant object. Draw the digit back to the face: it will be drawn back to the dominant eye.

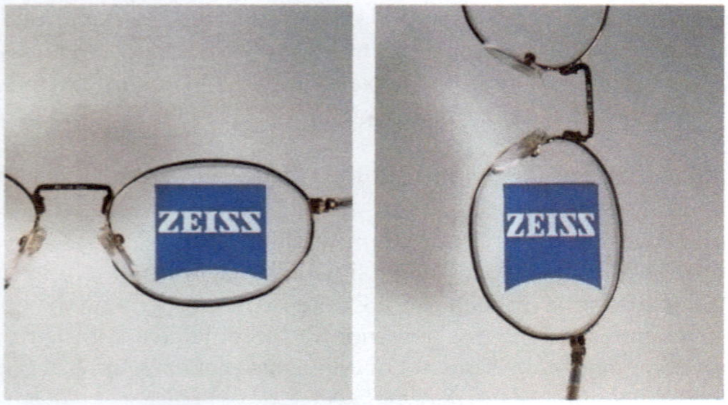

Source: Reproduced with permission from Carl Zeiss Ltd.

Figure 9.25 **Relative positions of the field diaphragm and condenser diaphragm in a reflected-light microscope** *The relative positions of the condenser aperture iris and the illuminated field diaphragm appear to be reversed. See text for details.*

9.12 Köhler Illumination with Reflected Light

With reflected-light illumination, the light path is folded upon itself, and the objective acts as its own condenser. This simplifies matters, but because the field diaphragm cannot be fitted in the condenser/objective (where it would also restrict the imaging aperture), it is located in the next conjugate plane 'upstream', in the normal position of the lamp filament. The filament is then imaged into this plane by an extra lens (Figure 9.20). Because of this, the relative positions of the condenser aperture iris and the illuminated field diaphragm *appear* to be reversed: the aperture diaphragm is nearer the lamp source, and the field diaphragm closer to the condenser/objective (Figure 9.25). With reflected light, or epi-illumination, it is particularly important to pay attention to restricting the area of the illuminated field to avoid stray light – non-image-forming light – which if allowed to intrude into the optical train, will reduce contrast in the image significantly (see Box 9.1 for more on stray light and glare). Further details on setting up Köhler illumination with a reflected-light microscope may be found in Chapter 12, Section 12.4.

9.13 Floaters and Illumination

Muscae volitantes (from the Latin 'flying flies'), or floaters, are cellular inclusions in the vitreous humour (see also Chapter 1, Section 1.3). They may be more prevalent in short-sighted eyes, and, unfortunately, tend to become increasingly annoying as we age. Floaters are seen when the exit pupil is small, and/or is formed by a small source of light, since their shadows are seen more sharply under these conditions. Floaters are therefore most likely to be seen when using objectives and eyepieces of high magnification. The larger the exit pupil of the microscope eyepiece, the less likely are floaters to be observed. Four factors influence the size of the exit pupil of the eyepiece (the older term for which is 'Ramsden disc'), and conspire towards a smaller exit pupil:

- the shorter the focal length of the eyepiece (i.e. the higher the magnification of the eyepiece), the smaller the exit pupil of the microscope will be;
- objectives of small aperture form small exit pupils;
- objectives of short focal length (i.e. high magnification) form small exit pupils; and
- reduction of the illuminating aperture in the condenser if the condenser is focused to low or the condenser diaphragm is not opened sufficiently, such that it does not fill the objective with light, will also give a smaller exit pupil. This is because the condenser diaphragm is in the same set of conjugate planes as the exit pupil, and an image of the front focal plane of the condenser is formed at the exit pupil of the microscope.

The size of the exit pupil is therefore proportional to the degree to which the back aperture of the objective is filled with light, but opening the condenser iris past the 3/4 or 5/8ths-open rule introduces glare (see Box 9.1 and Step 16 of Box 9.4) into the image. Thus, large apertures, low magnifications, both of objective and eyepiece separately, and as full an illuminating cone of light as possible are favoured for the reduction of annoyance caused by 'floaters'.

9.14 Chapter Summary

This chapter explains the different types of light sources available to illuminate the specimen preparation on the stage of the microscope. Until very recently, lamp technology had not progressed very much since the introduction of the incandescent lamp. Now cold light sources and LEDs enable microscopes to be adequately lit with much less wasted power. Besides using efficient lamps, the specimen must be lit with a bright and uniform light that can be fully controlled to allow the full resolving power of the objective to be used. The purpose of Köhler illumination is explained, and a step-by-step procedure is given explaining how to adjust and set up the microscopes perfectly.

1. Tungsten lamps have to be run at high voltage to maintain sufficient brightness and the correct colour temperature. They are inefficient for microscopy and have been replaced by tungsten-halogen lamps, metal-halide lamps or LEDs.
2. The fused-quartz silica glass envelope of tungsten-halogen lamps should not be handled directly with bare hands. Natural oils from the skin of your fingertips can create a potential hotspot and damage the glass envelope.
3. Some designs of tungsten-halogen lamps incorporate an integral reflector with an elliptic paraboloid surface. This is similar in design to a car headlamp.
4. Short-arc mercury lamps, xenon and metal-halide lamps require control transformers. Mercury and xenon lamps require centring; metal-halide lamps are pre-centred.
5. Mercury and metal-halide lamps do not give a constant spectral output but have spectral peaks. These must be considered when selecting fluorophore probes. Xenon lamps exhibit a more constant illumination intensity across the visible spectrum.
6. LEDs tend to have a narrower illumination bandwidth than mercury, xenon or metal-halide lamps, thus minimising heating or toxicity of samples. They also switch on instantaneously, obviating the requirement for slow mechanical shutters.
7. Filters are used to alter or modulate the illumination in the microscope. There are two designs: absorption filters and interference filters.
8. The purpose of Köhler illumination is to provide uniform illumination in the field of view, and fill the aperture of the objective with light to use the full resolving power of the microscope, whilst excluding stray light from reducing contrast in the image.
9. Köhler illumination allows independent control of the illuminating and imaging ray paths, which comprise two different and independent sets of conjugate planes.
10. There are two iris diaphragms in the microscope – one associated with each set of conjugate planes – both must be adjusted properly.
11. The illuminated field diaphragm lies in the field set (or imaging set) of conjugate planes. It is normally found in the proximity of the base of the microscope.
12. The condenser diaphragm, or illuminating aperture diaphragm, lies in the aperture set (or illumination set) of conjugate planes. It is located within, or just below, the condenser.
13. The use of a lamp collector lens to provide uniform illumination from a small light source, together with two interleaved conjugate planes, gives rise to a precisely defined protocol for proper adjustment of the microscope. This Köhler illumination is described here (and should be learned!).
14. Because the objective acts as its own condenser in reflected-light microscopy, the relative positions of the illuminated field diaphragm and condenser diaphragm appear to be reversed.
15. Floaters are cellular inclusions in the vitreous humour. Directions are given to minimise their appearance whilst viewing the image by eye.

Key Reading

Evennett, P.J. (1983). 'Köhler Illumination: A simple interpretation'. *Proceedings of the Royal Microscopical Society 28*/4: 189–192.
- An extremely clear, concise and very helpful explanation of Köhler illumination. I give this to all my students. It should be required reading for all serious microscopists.

Wessels, J.T. et al. (2012). 'Light-emitting diodes in modern microscopy—from David to Goliath?' *Cytometry A 81*/3: 188–197.
- Although a little dated from 2012, this paper is still a good introduction to LED technology for light microscopes.

References

Bancroft, J.D. and Gamble, M. (2008). *Theory and Practice of Histological Techniques*, 6th ed. Churchill Livingstone, Elsevier.

Barron, A.L.E. (1965). Illumination. In: *Using the Microscope*, Chapter 5, 3rd ed. London: Chapman & Hall, pp. 70–90.

Bradbury, S. and Bracegirdle, B. (1998). Practical use of the microscope. In: *Introduction to Light Microscopy*, Chapter 7, 2nd ed. Microscopy Handbook 42. Oxford: Bios Scientific Publishers, pp. 99–118.

Bradbury, S., Evennett, P.J., Haselmann, H., and Piller, H. (1989). *RMS Dictionary of Light Microscopy*. RMS Handbook No. 15. Oxford Science Publication.

Evennett, P.J. (1995). 'Light microscopy'. In: Wootton, R., Springall, D.R., Polak, J.M. (eds.), *Image Analysis in Histology: Conventional and Confocal Microscopy*. Cambridge University Press.

Evennett, P.J. et al. (2009). 'Symmetrical ray diagrams of the optical pathways in light microscopes'. *Quekett Journal of Microscopy 41*/2: 109–118.
Freeman, M.H. and Hull, C.C. (2003). *Optics*, 11th ed. Butterworth-Heinemann, Elsevier.
Hammond, C. (1998). 'Symmetrical representation of the geometrical optics of the light microscope'. *Journal of Microscopy 192*/1: 63–68.
Hammond, C. (2006). 'Symmetrical ray diagrams of the optical pathways in light microscopes'. *Microscopy and Analysis*, September 2006.
Kapitza, H.G. (1997). *Microscopy from the Very Beginning*, 2nd rev. ed. Carl Zeiss Publication G 40-100-e.
Murphy, D.B. and Davidson, M.W. (2013). *Fundamentals of Light Microscopy and Electronic Imaging*, 2nd ed. New Jersey: Wiley Blackwell. Wiley.
Robertson, J.M. et al. (2009). 'Light-emitting diode flashlights as effective and inexpensive light sources for fluorescence microscopy'. *Journal of Microscopy 236*/1: 1–4.
Sartory, C. (2008). 'LED lighting for microscopes'. *Quekett Journal of Microscopy 40*/8: 701–711.
Sartory, C. (2009). 'LED lighting for microscopes: discrete illuminators and lighting for low-power instruments'. *Quekett Journal of Microscopy 41*/2: 145–154.
Sato, T. and Murthy, V.N. (2012). 'Light-emitting diodes for biological microscopy'. *Cold Spring Harbor Protocols* 2012/12.

Further Reading

Bosse, J.B. et al. (2015). 'Open LED illuminator: A simple and inexpensive LED illuminator for fast multicolor particle tracking in neurons'. *PLoS One 10*/11: e0143547.
Boudreau, C. et al. (2016). 'Excitation light dose engineering to reduce photo-bleaching and photo-toxicity'. *Science Report* 6: 30892. doi: 10.1038/srep30892.
Cole, R.W. and Turner, J.N. (2008). 'Light-emitting diodes are better illumination sources for biological microscopy than conventional sources'. *Microscopy and Microanalysis 14*/3: 243–250.
Dawson, T.L. (2010). 'Development of efficient and durable sources of white light'. *Colouration Technology 126*/1: 1–10.
Köhler, A. (1893). *'Ein neues Beleuchtungsverfahren fiir mikrophotographische Zwecke'. Zeitschrift für wissenschaftliche Mikroskopie und für Mikroskopische Technik* 10/4: 433–440—A new system of illumination for photomicrographic purposes. Translated and reprinted in: *Proceedings of the Royal Microscopical Society 28*/4: 181–186.
Proceedings of the Royal Microscopical Society 28/4: 180–201 Köhler Illumination Centenary commemorative issue.
Uhl, R. (2012). 'Arc lamps and monochromators for fluorescence microscopy'. *Cold Spring Harbor Protocols* 2012. doi: 10.1101/pdb.top071001.

Internet Resources

Microscope light sources: http://zeiss-campus.magnet.fsu.edu/articles/lightsources/index.html.
Introduction to LEDs: http://energy.gov/eere/ssl/led-basics.
LED trans-illuminator: http://users.ox.ac.uk/~atdgroup/technicalnotes/TE300%20LED%20system.pdf.
CoolLED: http://www.coolled.com/ and https://www.coolled.com/news/published-articles/.
Excelitas Technologies: http://www.excelitas.com/Pages/Product/Lighting.aspx and http://www.excelitas.com/Pages/Product/X-Cite.aspx/news-webinars.php.

Osram Datasheets

Specialty lamps: https://bit.ly/2S40jdn.
Low voltage halogen lamps: https://bit.ly/2NS7r9B.
HBO Mercury lamps: https://bit.ly/2yrcR6e and https://bit.ly/2S8Wnbx.
XBO Xenon lamps: http://www.msscientific.de/xenon_lamps.htm.
Metal halide lamps: https://bit.ly/2R1ZWPh.

Notes

1. HBO is a registered trademark of Osram, one of the major specialist lamp manufacturers. H stands for Hg or mercury, B is the symbol for luminance and O denotes passive unforced cooling. Likewise, the X in XBO stands for Xe, or xenon. This is not to be confused with the HLX designation for low-voltage lamps.
2. Some authorities have commented on the confocal list server that the green spectral peak of metal-halide bulbs is still weaker than conventional short-arc lamps and inconsistent from one bulb to another.
3. Holonyak also developed the red-light semiconductor laser diode.
4. Unlike incandescent lamps, which are estimated to waste 95% of electrical energy as heat, LEDs are ten times as efficient, converting 50% of the energy to light.
5. The concept of **F**ull **W**idth **H**alf **M**aximum (FWHM) to define bandwidth is explained in Chapter 15, Section 15.8.
6. The Bertrand lens is swung into the optical path when needed and acts together with the eyepiece to form a low-power microscope that is conjugate with the illuminating, or aperture, set of conjugate planes. It allows easy inspection of the back focal plane of the objective for adjustment of Köhler illumination or for conoscopy in polarised-light and materials microscopy.
7. Some microscopists maintain that source-focused illumination differs in practice from Köhler illumination. Exponents of source-focused or critical illumination argue that this form of illumination gives better definition and resolution in the image. However, it is widely accepted that there is no

difference in resolution of fine detail in the image between each method. Source-focused illumination is generally set up using an external lamp and a substage mirror on the microscope, whilst Köhler illumination on a modern microscope stand utilises an internal lamp built into the stand. Nevertheless, it is possible to set up both illumination methods with either an integral or external light source.

8. The positioning of the light source, the lamp collector lens and the condenser is critical: these components must be in precise relation to one another. The microscope will only operate as designed when the lamp collector lens and condenser are correctly adjusted together. The optical axes of the lens and of the eye must also coincide, and the lamp must lie on this common axis. If a diffuser is fitted (as is the case on some cheaper microscopes) the adjustment needs to be less precise, but true Köhler illumination cannot be achieved. A diffuser can be used only when the light source is very bright for bright-field microscopy. It should certainly not be used when every photon is required, as is the case with fluorescence microscopy.
9. The illuminated field diaphragm (IFD) is arranged to lie just after the lamp collector lens.

10

Diffraction and Image Formation in Microscopy

Outline

Understanding Light Microscopy, First Edition. Jeremy Sanderson.

10.1 Introduction

Understanding diffraction and interference in the light microscope is fundamental to appreciate how the microscope objective forms an image of the object and why that image (however good the microscope and its objective) can only be a *representation* of the object. This information is useful from a practical viewpoint, because a basic understanding of image formation allows us to adjust the condenser and set up the microscope properly in order to get the optimum image of the specimen – the object. The image is formed from the interaction of the specimen with light. This chapter gives a theoretical explanation of why we use Köhler illumination, described in Chapter 9, to illuminate the back focal plane of the objective. Diffraction, the very process by which the image is formed, inevitably produces blurring in the image. It is possible to reduce and manage this blurring – in particular fluorescence blurring from self-luminous particles – by deconvolution. This topic is covered in Chapter 21.

So far in this book we have assumed that light travels in straight lines – an assumption usually expressed as 'the rectilinear propagation of light'. Indeed, Newton based his belief that light was composed of particles on this observation, which fitted in well with contemporary knowledge about light and also his work on mechanics. Although Newton considered the possibility of light acting as waves, he concluded that it was particulate. Because of Newton's stature in the scientific community, his conclusions were accepted *de facto* and not questioned too closely.

Huygens, however, questioned Newton's theories. He postulated that wavelets arise from the motion of the æther; also that individual wavelets from neighbouring æther particles could combine to form a single wave envelope, which then serves as the source for another infinite number of spherical waves. Huygens also realised that spherical waves of very large radius can be approximated (Figure 10.1) by plane waves aligned transverse to the direction of propagation. Huygens' theory of the wave nature of light could adequately explain the reflection of light, just as Newton's geometrical corpuscular rays of light could do. It was superior in explaining refraction of light and the action of lenses. The wave theory of the nature of light explained that the edges of shadows are not perfectly sharp, that light bends around into the area of the geometric shadow not directly illuminated and that light passing from a vacuum or rarer medium into a denser medium will slow down.

The work of Huygens was neglected until Thomas Young (Box 10.1) performed his well-known diffraction and interference experiment (Box 10.2, Figure 10.3 and Figure 10.8) His double-slit experiment was strong evidence for the wave model of light. By the close of the 19th century, there seemed to be no doubt that light propagated as a wave. In 1900, following on from apparent inconsistencies in the behaviour of black-body radiation, Max Planck worked out that light was a wave composed of energy packets of particles. Planck's constant, h, (see Box 2.1) links both the wave-like and particle-like nature of light. Five years later Einstein demonstrated the photoelectric effect (for which he received the Nobel Prize) and postulated that the energy packets, or wave trains, making up light rays were photons.

The diffraction of light sets a limit on the resolving power of any optical instrument using a lens, including our eyes. As we saw in Chapter 7, the limit of resolution, or minimum resolved distance (MRD), is defined as the ability to distinguish closely adjacent points with small angular separations (as in the eye, the camera or the telescope) or which are spatially separated by small distances, as in the microscope.

10.2 Rays, Point Sources and Extended Sources of Light

Light originates from the movement of electrons between energy levels within atoms that are excited by virtue of being at high temperature (e.g. lamp filaments or the sun) or by stimulated emission (e.g. lasers). Each atom emits a photon, for about 10^{-8} seconds in the case of incandescent lamp filaments, or from about 10^{-12} s for pulsed lasers to indefinite times for continuous lasers. A light wave does not therefore propagate infinitely, but the wave trains (photons) are of distinct length and travel like a group of ripples from the light source. The length of each photon wave train depends upon the velocity of the light and period of emission. A lamp filament emits photon wave trains about 3 metres long; pulsed laser photon wave trains are much shorter – about 0.3 mm – while photon wave trains from continuous-wave lasers are very long indeed.

Box 10.1 Thomas Young and the Interference of Light

Thomas Young began young: by the age of two he could read, but didn't care for fiction, and by 13 had taught himself to read most of the book of Genesis in the original Hebrew. Young trained formally in medicine but was a wide-ranging polymath. His biographer calls him 'the last man who knew everything' and, justifiably, compares him to Leonardo. Perhaps it fits Young better than Hooke, to whom the epithet 'English Leonardo' has also been applied. Most likely he is not better-known because, following criticism, he preferred to publish anonymously, and his writing style was notably 'wooden'.

'Phenomenon' Young, as he was known at Cambridge, was born in Milverton, Somerset, on the 13th of June 1773, the eldest of 10 children. His parents were stiff Quakers towards whom he felt no warmth, for he was sent away soon after his birth. In an autobiographical sketch written at the close of his life in the third person, Young commented: 'He may be said to have been born old and to have died young'. His contemporaries at Emmanuel College remembered him as a humourless academic recluse, being unable to tell or understand a joke and poor at communicating his knowledge to others. However, he was modest and of equitable temperament, never boasting or losing his temper.

Source: Print Collector/Contributor. Reproduced with permission from Getty Images.

Although he is best remembered in optics for his double-slit experiment with coherent light, there is some doubt that Young actually carried this out. Although Young drew diagrams and described a protocol, he only referred once to this experiment. Like Einstein, he much preferred 'thought experiments' (*Gedankenexperimente*), but he did construct a water ripple tank in order to convince himself of the mechanism of diffraction and interference. Young reported his classical double-slit experiment in 1803 to the Royal Society, to which he had been elected a member at just 20 years of age, having previously correctly researched the focusing mechanism of the eye. In 2002, the readers of *Physics World* voted the double-slit experiment as the most beautiful in physics.

Interestingly, Young's revelation about light came to him when he was moving house, just as the theory of relativity came to Einstein as he moved house a century later. Perhaps changing abode should be made time-limited and mandatory for physicists. The significance of Young's work on the interference of light is that his wave theory challenged Newton's overbearing dominance of classic physics and, ultimately (with Fraunhofer's work), paved the way for Abbe's diffraction theory of microscopical image formation.

Young is also remembered for his work on the Rosetta Stone, a granodiorite stele inscribed with Egyptian hieroglyphs, Egyptian Demotic script and ancient Greek text. It is from this work that we can now read Egyptian hieroglyphs. When he died, just before his 56th birthday, on the 10th of May 1829, he left a vast foundation of work for others to build upon.

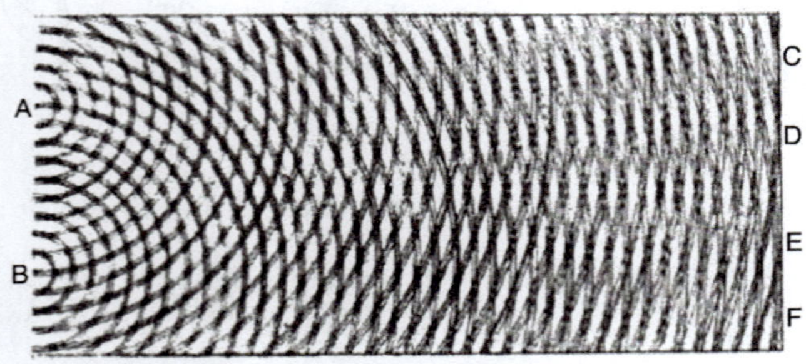

Source: Young 1803.

References

Robinson, A. (2007). *The last man who knew everything*. Oxford: Oneworld Publications.

Young, T. (1804). 'Experiments and Calculations Relative to Physical Optics'. *Proceedings of the Royal Society of London A* 94: 1–16.

A ray of light is a narrow, or fine, idealised beam of parallel light, which undulates as it travels (or propagates), taking the shortest time to travel between two points. A ray indicates the direction of the path taken by the wavefront of a light wave. It is shown as a single line on a diagram, often with an arrow showing the direction of propagation. Where there is no arrow, convention dictates that propagation occurs from left to right across the page. Actual rays are produced when light shines through a small hole, effectively producing a point source of light. Light from a point source has all its photons in phase (and of the same frequency) so is *coherent*. Light which is coherent is only produced by a point source, by a very small part of an extended source[1] or by a laser (Figure 10.2a).

A beam of light consists of many rays; it is produced by a larger hole from a light source such as a lamp or the sun. This we call an extended source (Figure 10.2b). The light from an extended source (or also from two independent sources) is *incoherent*, because the relative phases of the photons change randomly with time. Light from an extended source cannot be focused to a point[2]. Hence you cannot increase the brightness you see in an image illuminated by an extended source merely by using a convex lens. This can be demonstrated by looking through a magnifying glass at a featureless clear or overcast sky (do NOT look directly at the sun). The shadow formed by a point source of light is uniformly dark all over (Figure 10.3b) with sharp edges, evidence for the rectilinear propagation of light. An extended source casts a soft-edged shadow of an object.

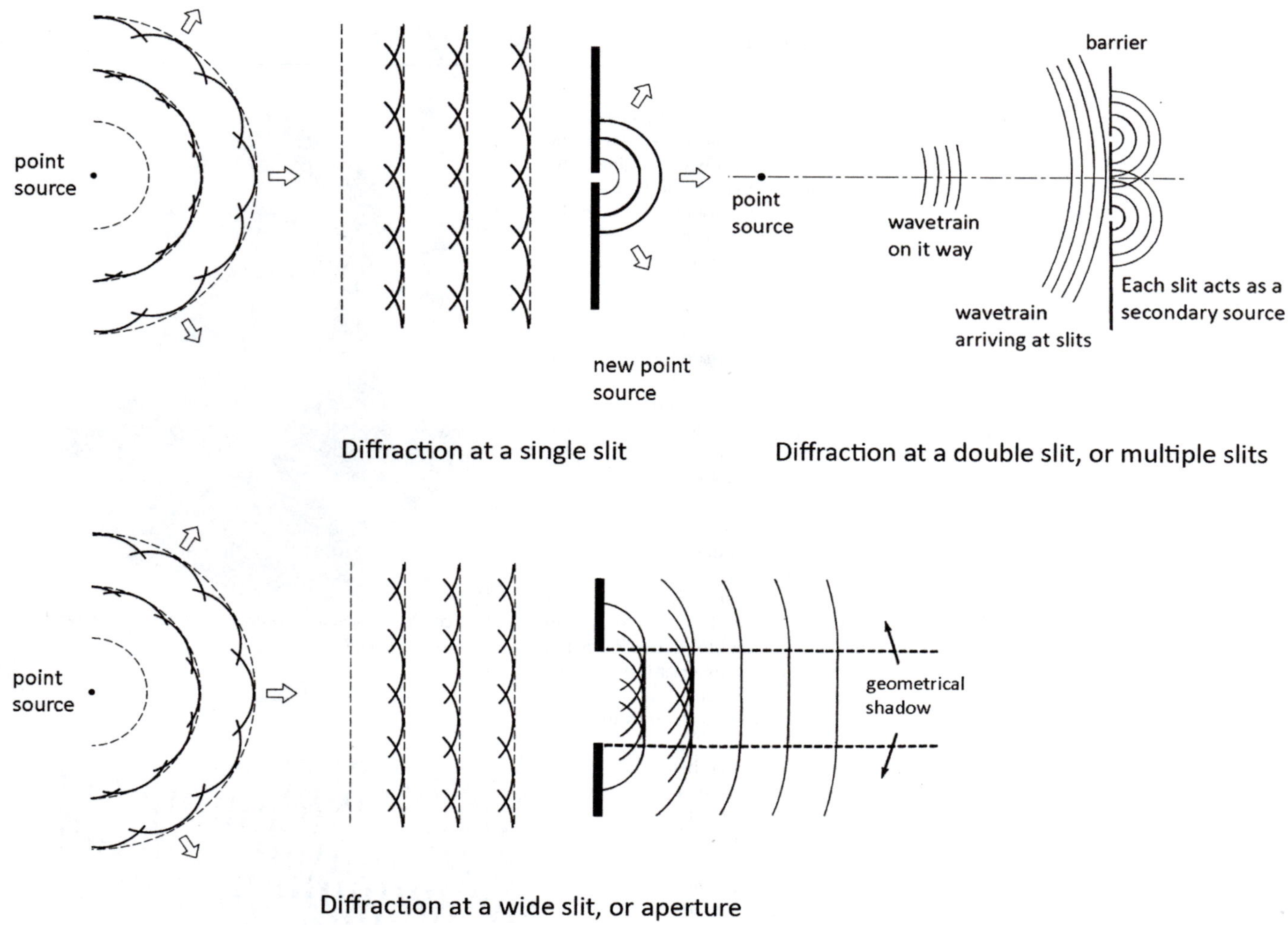

Figure 10.1 **Huygens' approximation of plane waves** *Huygens' principle states that every point on a wavefront acts as a source of new tiny spherical wavelets that move forward with the same speed as the wave. These new sources are also coherent. The plane wave that ultimately is the result, lying tangential to these secondary wavelets, is a small portion of an infinitely large spherical wavefront from a distant point source. Source: Murphy 2001. Reproduced with permission from John Wiley & Sons; Spencer 1982. Reproduced with permission from Cambridge University Press; Robinson and Bradbury 1992. Reproduced with permission from Oxford University Press.*

All ray diagrams, in which rays or beams of light are represented as straight lines, are geometrical expressions of the rectilinear propagation of light. It is an assumption which is perfectly adequate in describing the processes of refraction, reflection and the action of lenses, but it fails on a small scale – a scale corresponding to the wavelength of light itself. Now we need to make use of, and extend, our knowledge of the wave theory of light in order to explain the nature of the scattering processes when light encounters the 'fine detail' in a specimen, the edges of slits or 'pinholes' or the diaphragms of lenses.

10.3 Coherence and the Interference of Waves

A point source produces a spherical wave. Light from infinity (in practice, taken as ≥30 times the focal length of a lens, or several metres for light not passing through a lens) will cause the wavefront of the wave to be sufficiently far from the source so as to form parallel plane waves. For light waves to interfere (whether spherical or parallel), they must originate from the *same* point source and be coherent.

Refer back to Figure 2.2, in Chapter 2. The harmonic motion of a wave can be expressed as a fraction or multiple of radians. One cycle or wavelength corresponds to 2π radians, or 360°. A phase difference occurs when one wave lags behind, or precedes, another in time and space (see also Figures 2.5 and 2.8). Light is said to be in phase when the peaks and troughs of two or more rays or waves of light are in a *constant* relationship to one another over space and time. They may be coincident or not, but the defining relationship of light rays that are in phase is the periodic constancy of one light wave with respect to another. Usually each light wave has an identical wavelength to the others. The clearest interference patterns occur when the two waves have equal or nearly equal amplitudes.

Interference occurs when two waves that are in phase (and have the same polarisation state) superimpose to form a new wave of greater or lesser amplitude. This is called the principle of superposition of waves[3]. The displacement of each wave above and below the

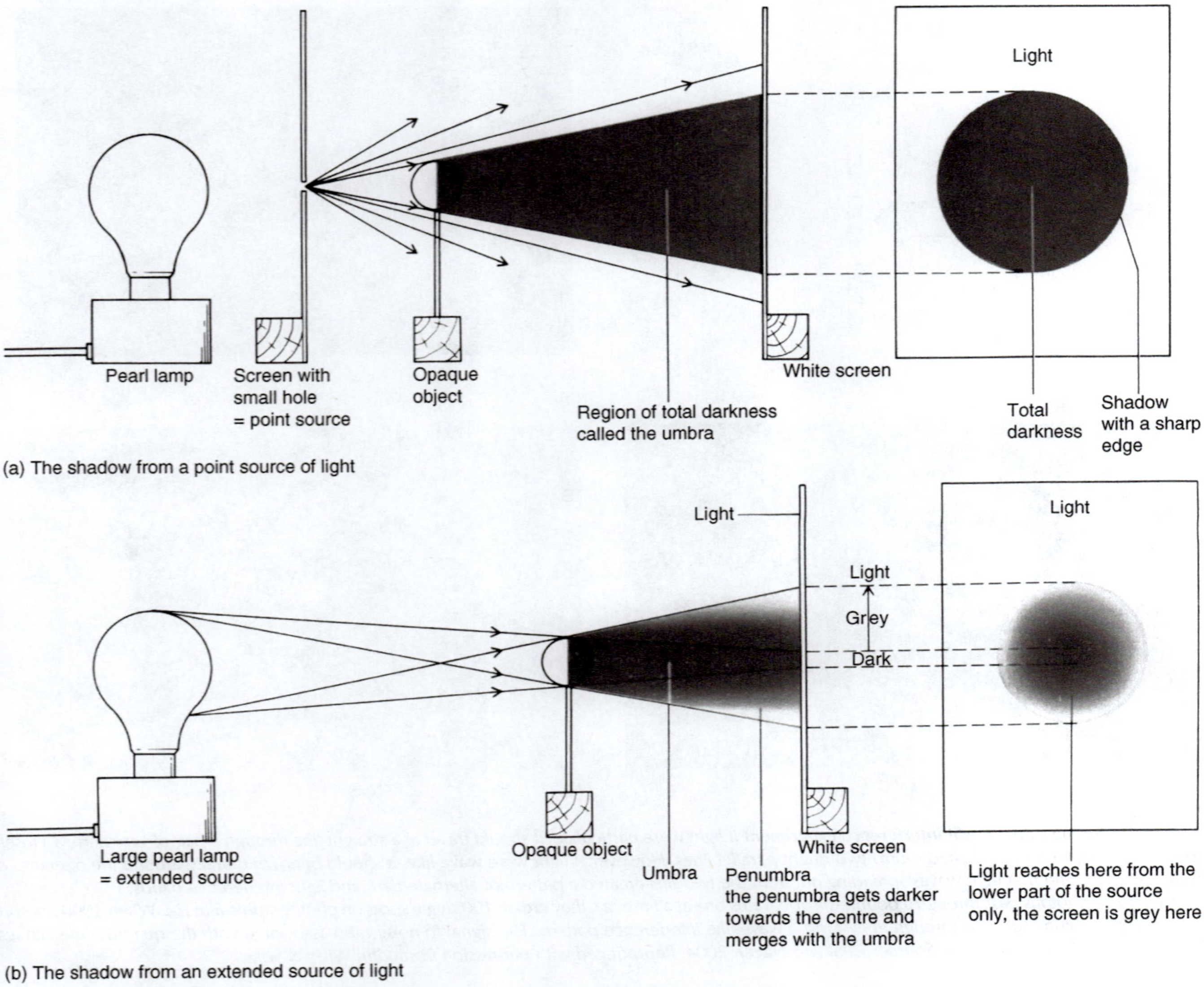

Figure 10.2 ***Point sources and extended sources*** *The differences between these two light sources are shown here and explained in the text. Light from a point source is in phase and so is coherent. Light from an extended source is incoherent. Source: Avison 1989. Reproduced with permission from Oxford University Press.*

direction of propagation (Figure 10.4) is added at each point. The superposition of coherent waves forms a spatial interference pattern that maps how the phase difference between the two waves varies. At its extreme, interference gives rise to a resultant that is completely constructive or else is completely destructive (i.e. the resultant is a flat line along the direction of propagation – with light waves this results in darkness; with sound waves, in silence)[4].

Constructive interference occurs when the phase difference between different coherent waves is a multiple of 2π. In this case the amplitudes of the two waves will add together to produce a larger single wave. The amplitude of the resultant wave is greater than that of the original individual waves. Destructive interference occurs when the difference is an odd multiple of π, and the amplitudes of the two waves cancel out. If the difference between the phases is intermediate between π and 2π, then the magnitude of the displacement of the summed waves lies between the minimum and maximum values.

When two waves that are out of phase by $\lambda/2$ destructively interfere, it is easy to assume that they are annihilated. This is not true: the first law of thermodynamics is concerned with the conservation of energy and matter. Indeed, all of the diffracted photon energy can be recovered at the position of the diffraction pattern. The energy of the photons is redistributed in directions that permit constructive interference and is not lost.

Interference will only occur between two coherent waves (i.e. having the same frequency and being in phase). Incoherent waves do not interfere. We know this instinctively from everyday experience: two light bulbs in the same room produce incoherent light whose waves are out of phase. They are different and independent sources irrespective of whether they are linked together on the same electrical circuit or not. The result of switching on both light bulbs, either together or one followed by the other, is more light. If the sources *were* coherent, an interference pattern of light-and-dark patches would occur in the room, and this is clearly not so.

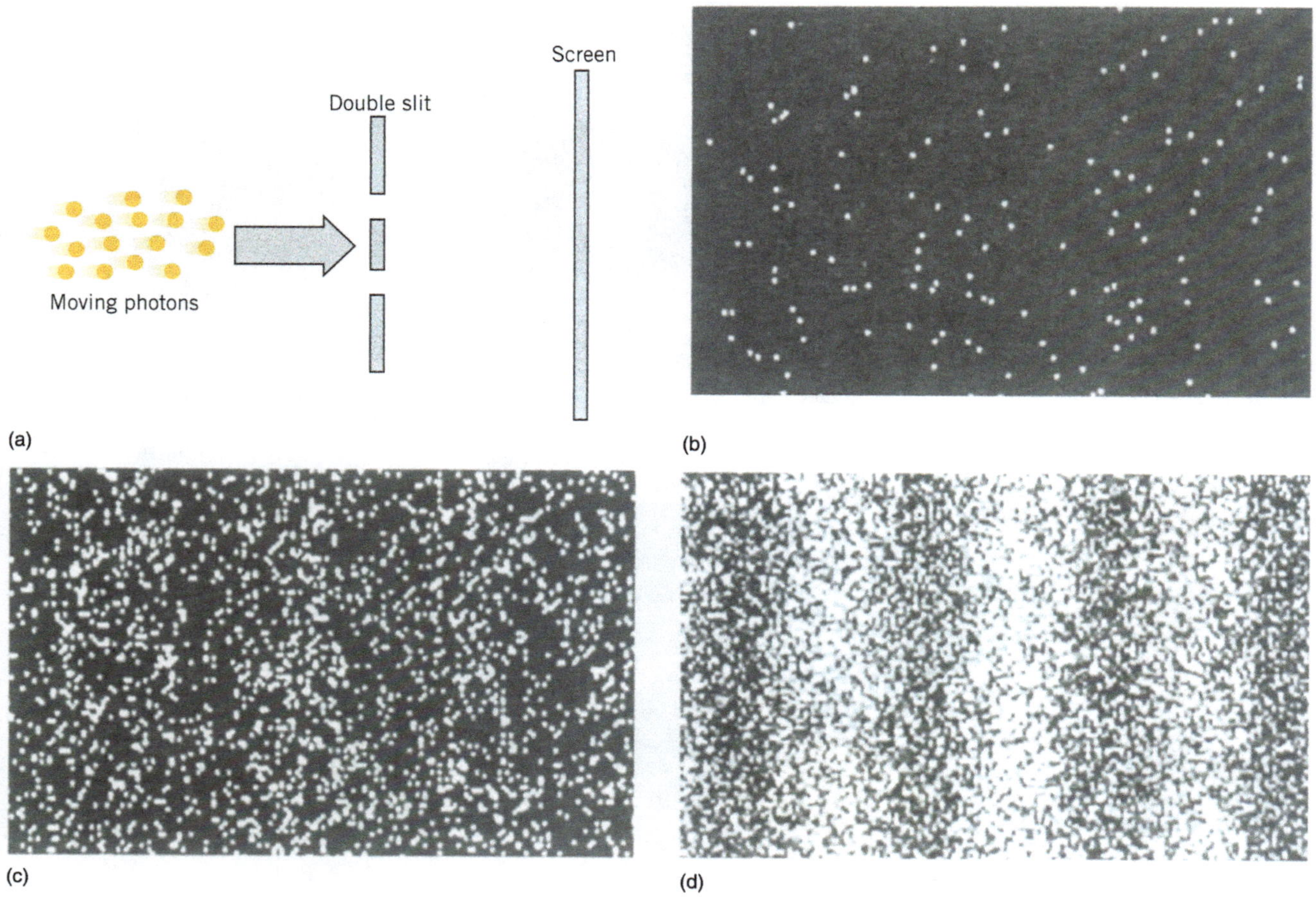

Figure 10.3 **Young's double-slit interference experiment** *If light were particulate, it should travel in a straight line through a pair of very narrow closely apposed slits to give (in a darkened room) two bright parallel lines. However, if light were wave-like, it should bend, or diffract, around the openings of the slits. Young found that light waves spreading out from the two slits created a pattern of alternate dark and light interference bands.*

When 100 photons pass through a double-slit apparatus one at a time (a), they create 100 single spots on photographic film (b). When 3000 photons (c) and 7000 photons (d) pass through, they form a wave-like interference pattern. The signal-to-noise ratio associated with the quantum mechanical photon is due to Poisson statistics. Source: Trefil and Hazen 2004. Reproduced with permission from John Wiley & Sons.

10.4 Diffraction

When waves pass through a small opening, they bend, or spread out, at the edges into the geometrical 'shadow'. This is called diffraction. It is shown in Figure 10.5 using water waves as an example. The extent of the bending of the wave depends upon both the relative size of the wavelength and the size of the opening. If the opening, slit or gap is large in comparison with the wavelength, the waves only spread out a little. As the size of the gap decreases, for the same wavelength, the diffraction of the wave increases. When the size of the gap is less than the wavelength of the incoming plane waves, the gap acts as a point source for the secondary waves (see Figure 10.1), and these are perfectly spherical. As the wavelength increases, then for the same-sized gap, the diffraction of the wave also increases.

Many of the diffraction and interference phenomena we shall encounter are not easily observed because of the small angles of scattering that are involved. Where the wavelength of light (or X-rays or electrons) is small in comparison with the size of the spacings between the gaps, or edges, of the diffracting objects, then the diffraction angles are also correspondingly small. Thus, if you hold a finger in the sun and cast a shadow with it, then either ray or wave optics can be used to predict how the shadow is formed, but the diffraction that actually occurs is so small that we cannot see it. However, if you switch on a light, hold your hand in front of it and look through the narrow slit formed by closing two fingers together, it is possible to see a series of dark lines parallel to the fingers. This is a diffraction pattern formed by interference.

When a plane wave passes through a very small gap, say a slit[5], a new wavefront will form from secondary wave sources according to Huygens' principle (Figure 10.1). One photon wave train is shown here, but the geometrical conditions that apply to this single photon apply to all photons in the ray of light. Unless the wavefront continues to move straight forward, phase differences will occur between the wavelets making up the new wavefront. This will lead to constructive and destructive interference. Huygens' principle also explains why diffraction at the edge of a large gap is less significant than at a small opening. If the gap is large, more secondary wavelets remain in phase, forming a plane wavefront. It is only near the edges of the opening that the phase differences become large enough to cause interference.

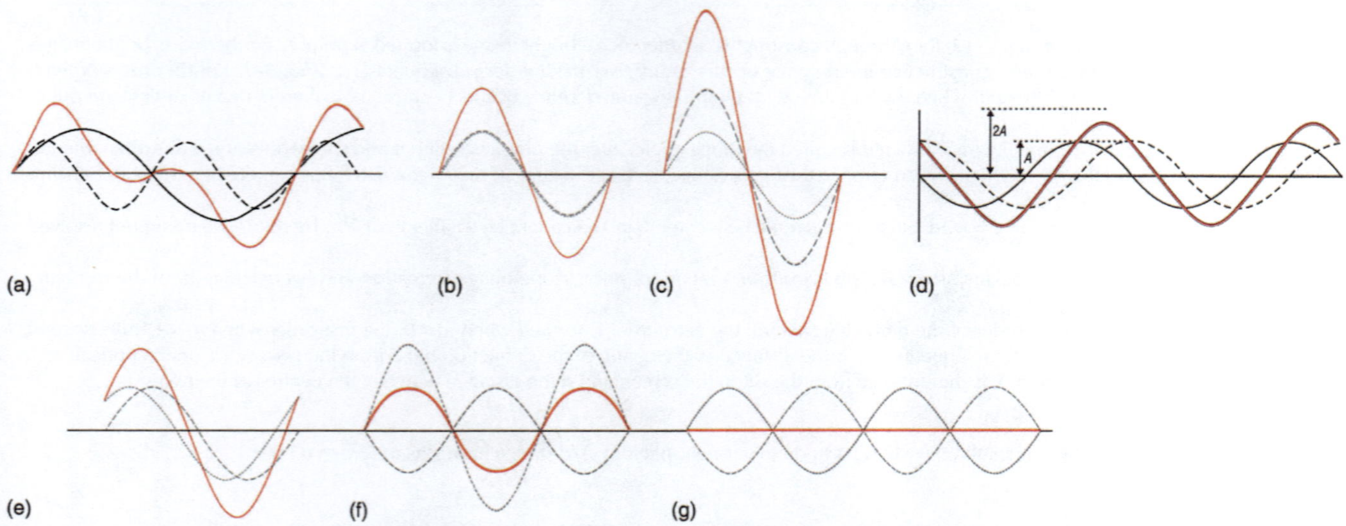

Figure 10.4 **Superposition of waves** *Interference of two identical sinusoidal waves. Two coincident waves can interfere if their electric vectors vibrate in the same plane at their point of intersection. Here, two waves are shown that vibrate in the vertical plane of the page, with the propagation direction running left to right across the page. Both waves have the same wavelength but vary in amplitude. The amplitude of the resultant is the arithmetic sum of the original component waves. Constructive interference occurs for two waves having the same phase. Destructive interference for two waves completely out of phase by λ/2 (if they are of the same amplitude). For phase shifts from λ to λ/2, the amplitude of the resultant lies between fully constructive and destructive interference. Source: (a,d) Pedrotti 2004. Reproduced with permission from SPIE. (b,c,e–g) Courtesy of Mr S. Thomas.*

The diffraction pattern from a single slit that is wider than the wavelength of the incident plane wavefront will form an interference pattern that shows a central maximum[6] with much fainter bands on either side (Figure 10.6). This pattern can be explained by assuming that the slit behaves as though it has a large number of point sources spaced evenly across its width. Both a monochrome and coloured pattern are shown in Figure 10.7. In profile the intensity distribution of this pattern is a Gaussian curve. We shall return to this particular intensity distribution when considering light diffracting around a circular aperture, which then interferes at the (primary) image plane.

Box 10.2 Diffraction by Double Slits

Figure 10.10 shows the arrangement of a modern set-up for Young's double-slit experiment. The first single slit, S, creates a 'point source' from the extended source of the light bulb. For practical purposes, the slits must be about 0.1 mm. If they are wider, no clear diffraction pattern would be observed; if they are narrower, the light intensity will be too low. The two slits (S_1 & S_2) are separated by about 0.25 mm; much further apart and the fringes would be too close together to be distinguishable. The screen is about a metre away from the slits, or approximately 4000 times the slit separation. This distance is labelled L. All points *equidistant* from both slits produce interference maxima with *no path difference* – this is the zero-order.

A close-up of the two slits, S_1 & S_2, is shown in the figure in this Box; the figure is not to scale. If the size of the slits, the width between the two slits, d, and the separation of the interference maxima and minima of the diffraction pattern, y, are all very small compared to the distance, L, from the slits to the fringes of the diffraction pattern projected onto the screen, we find the length of the dotted line ≈L. Then: $d = \lambda L/y$.

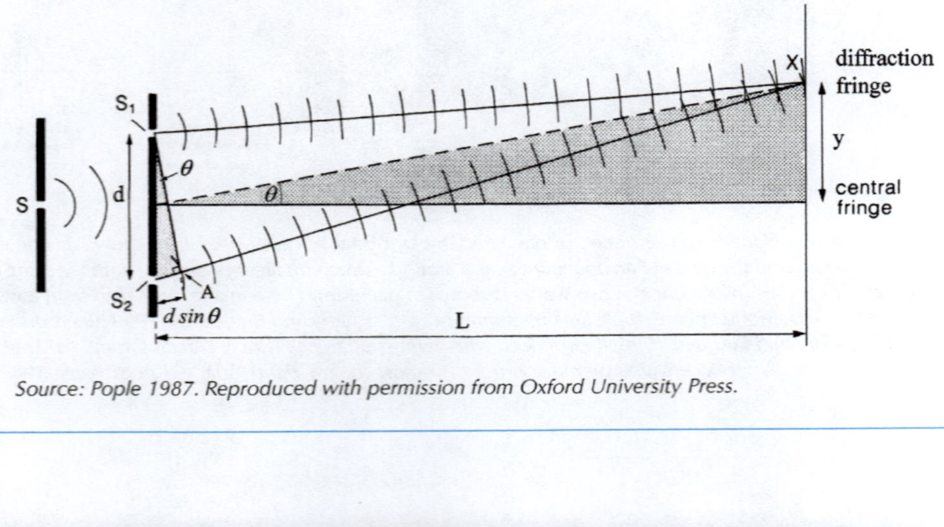

Source: Pople 1987. Reproduced with permission from Oxford University Press.

If the path difference S_2-A is equal to 1 λ, then through constructive interference a bright image is formed at point X. Furthermore, bright images are formed whenever $y = m\lambda L/d$ and constructive interference occurs for to give maxima for subsequent (1st, 2nd, 3rd,...m-th) orders of light. Dark bands (the absence of light) occur when $y = (m \pm 1/2)\ \lambda L/d$. Destructive interference occurs because the coherent beams of light are out of phase by $\lambda/2$.

The angle, θ, through which the light is diffracted is called the Bragg angle, after the physicists, father and son, who studied this phenomenon together. By inspection of the geometrical construction in the figure above, it can be seen that: $\sin\theta = m\lambda/d$. It follows, therefore, that $\lambda = d.\sin\theta/m$ and $d = m\lambda/\sin\theta$.

The first minimum is given by: $\sin\theta = \lambda/2d$. Since $\sin\theta$ cannot be greater than 1, d cannot be smaller than $\lambda/2$. Therefore the minimum resolved distance (MRD), $d \geq \lambda/2\ \sin\theta$.

The expression $d \geq \lambda/2\ \sin\theta$ reoccurs as $d = \lambda/2\ \sin\theta$ in Abbe's theory of microscope image formation and his calculation of the resolving power of an objective.

The integer, m, determines the order of the diffracted spectra: the zero order is formed when $m = 0$; the first order when $m = \pm 1$; the second order when $m = \pm 2$, and so on. In these equations, y is the distance of the centre of the diffraction band from the zero-order on the optical axis, λ is the wavelength of illumination, L is the distance from the slit to the screen and d the distance between the centres of the two slits.

Rearranging this equation, $\lambda = dy/L$,

$$\text{wavelength } (\lambda) = \text{slit separation (d)} \cdot \text{fringe spacing (y)/distance from slits to screen (L)}$$

or $y = L\lambda/d$:

$$\text{fringe spacing (y)} = \text{distance from slits to screen (L)} \cdot \text{Wavelength } (\lambda)\text{/slit separation (d)}$$

Typically, a centre-centre fringe spacing of 2.2 mm (2.2×10^{-3} m), is produced by green light of 550 nm for a slit-to-screen distance of one metre. The separation of the fringes is determined by $L\lambda/d$; thus, the size of the fringe pattern increases with wavelength. Young used this double-slit construction to measure the wavelength of candle-light. He gave a figure of 1/43,636 of an inch, which corresponds to 582 nm. This is accurate for yellow light (see Table 2.2 and Figure 10.10).

Reference

Pople, S. (1987). *Explaining Physics GCSE Edition*, 2nd ed. Oxford: Oxford University Press.

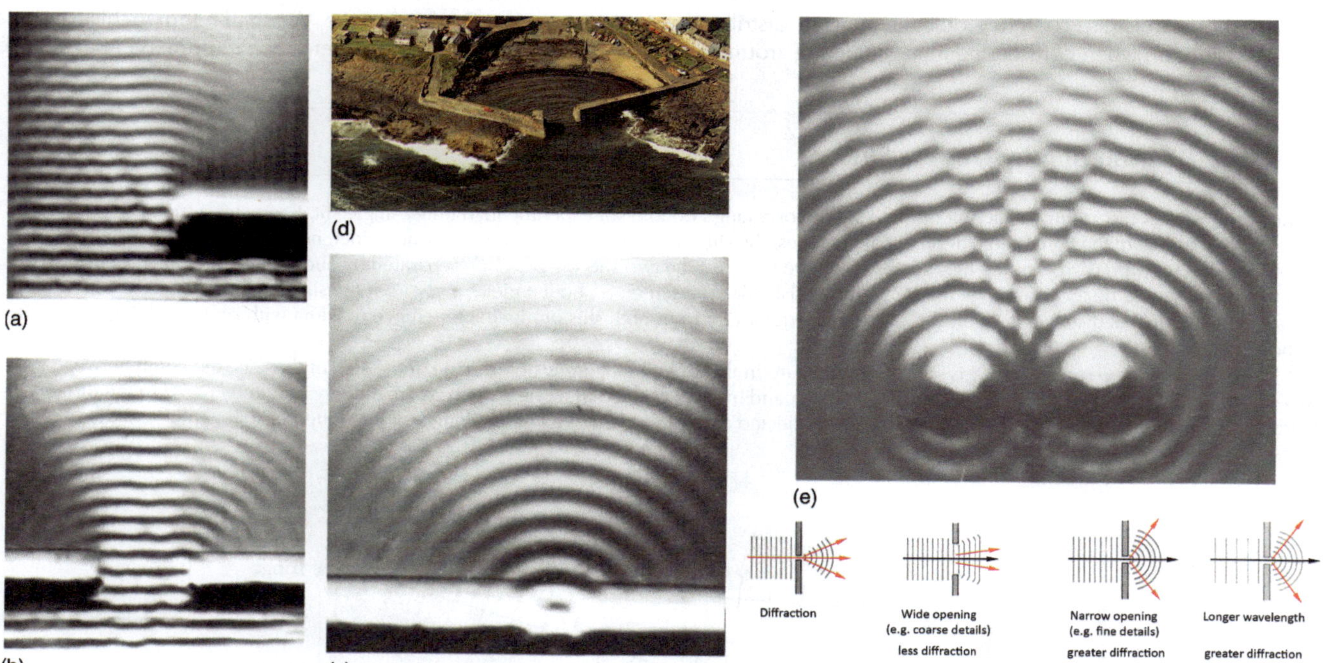

Figure 10.5 Diffraction of water waves *Plane waves are bent, or diffracted, by an obstacle – effectively here shown as one edge of a slit – into the shadow area (a). Waves are diffracted around the edges of an opening (b) and around a much smaller opening (c) with the same effect shown at Craster Harbour on the Northumbrian coast (d). The interference of two waves is shown in panel (e). These images are taken from a water ripple tank used to teach the principles of diffraction. Sources: Monochrome ripple tank photographs (a) Bradbury and Bracegirdle 1998. Reproduced with permission from Taylor & Francis Group. (b,c,e) Bradbury and Evennett 1996. Reproduced with permission from Taylor & Francis Group. (d) England 1995. Reproduced with permission from Hodder Education; Reproduced with permission from Airfotos. (below right) Adapted from Pople 1987 with permission from Oxford University Press.*

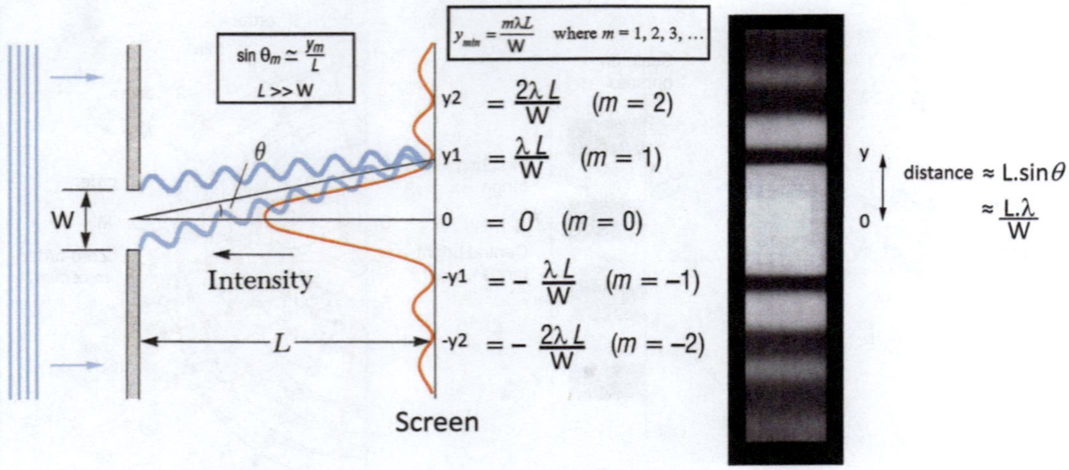

Figure 10.6 Diffraction and interference at a single slit *Where the slit width is less than the wavelength of light, according to Huygens' principle, each slit radiates spherical light waves since it acts as a secondary source. As the slit width increases, the gap W can be regarded as comprising many Huygens' point sources, so diffraction occurs from every part of the wavefront passing through gap W. When the path difference between the light wave diffracting at the top of the slit and the wave diffracting at the bottom is $\frac{1}{2}\lambda$, they destructively interfere to form the first dark minimum at y_1. Source: Serway and Faughn 1989. Reproduced with permission from Cengage Learning; Cagnet et al. 1962. Reproduced with permission from Springer.*

Suppose instead of having only a single narrow slit we have two narrow slits separated by a small distance. Microscope samples can be thought of as being composed of many very small features that admit, or pass, light. In order to model this, it is convenient to see what happens when light passes through two slits set very close together. Figure 10.8 shows a photon wave train from a point source moving as a ray towards two slits. This is the well-known double-slit experiment[7] referred to earlier (Figure 10.3, Box 10.2) performed first by Thomas Young in 1803 and finally published in 1807. An initial single slit (or circular hole), whose aperture is less than the wavelength of light, forms a point source. Two slits (or circular holes) placed perpendicular to the optical axis of the wavefront will then act as secondary sources, producing coherent light that will interfere.

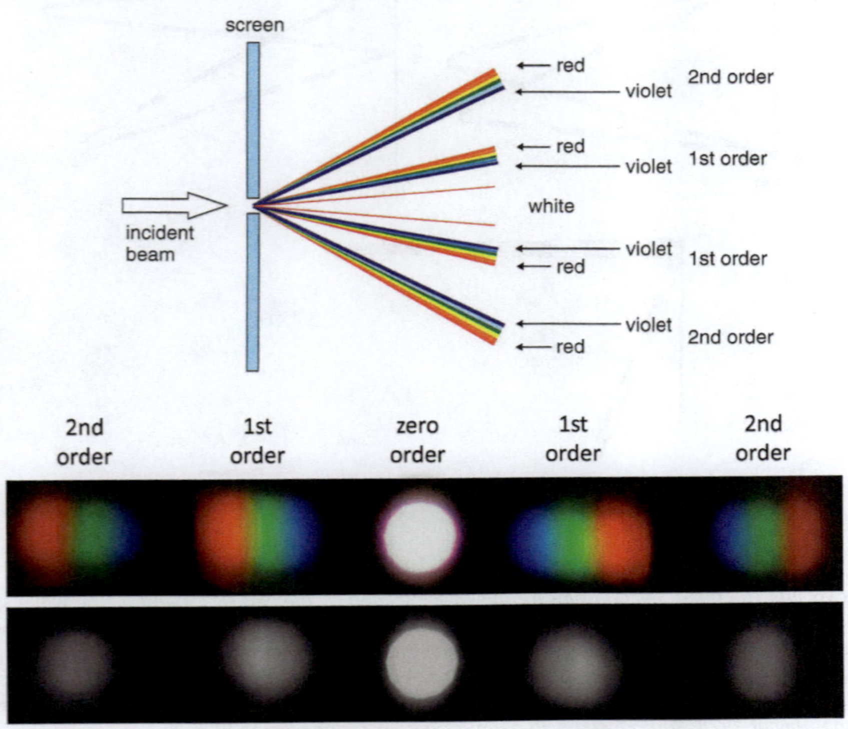

Figure 10.7 Diffraction by monochrome and white light *White light diffracted by a slit is decomposed into spectral colours. Red light is always diffracted to a greater extent than blue light, which is why greater resolution is possible with blue or violet illumination. A monochrome diffraction pattern, which can be achieved by simple filtration, is shown for comparison. Source: Tilley 1996. Reproduced with permission from John Wiley & Sons.*

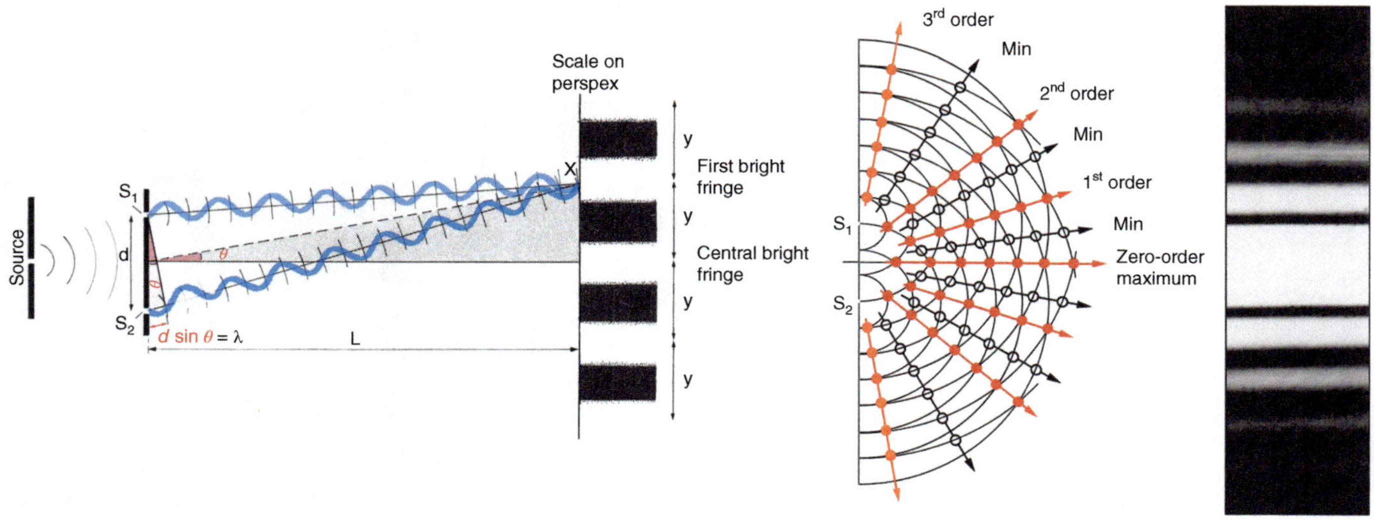

Figure 10.8 **Diffraction and interference at double slits** *This is an illustration of Young's double-slit experiment, which demonstrates the interference of coherent light waves. Slit S_1 and slit S_2 are both illuminated by a single slit so that the light wave that passes through each slit is coherent, having come from a single point source. The left-hand panel shows how light from each slit leaves in phase with light from the other slit. Notice the similarity to Figure 10.6, except that the width of the single slit, W, is replaced by the distance, d, between the two slits. Constructive and destructive interference of these waves occurs to form the diffraction pattern shown, on the right. The middle figure shows the overlapping of the propagating coherent wavefronts and how they interfere to form the diffraction pattern. An antinode is a point where constructive interference occurs (orange circles), whereas a node (open circles) is a point where destructive interference occurs. Antinode and node lines, respectively, join these points. The antinode lines show the diffraction orders that occur. Source: (left) Franklin et al. 2010. Reproduced with permission from John Wiley & Sons. (centre) Franklin et al. 2010. Reproduced with permission from John Wiley & Sons. Adapted from Serway and Faughn 1989. (right) Ohanian 1994. Reproduced with permission from W. W. Norton.*

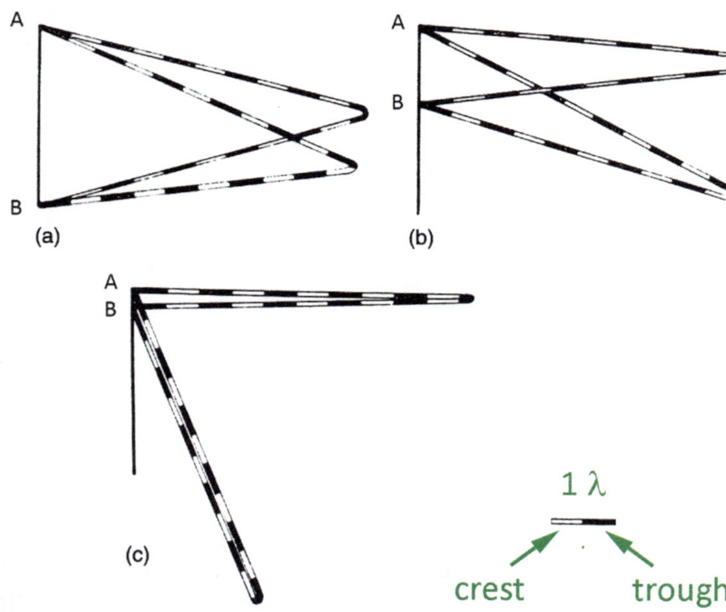

Figure 10.9 **Diffraction at two slits** *The phase relationships for the waves scatted by the two slits A and B. The white stripes indicate the crests and the black stripes the troughs of each wave. In (a) slits A and B are a long way apart, so the phase difference is relatively large, so the orders of light are close together. Where the slits lie closer together (b,c) the light ray must travel a greater distance for a corresponding change in relative phase to permit interference. If the distance between the two point sources of the waves (i.e. A-B) is decreased, the distance between the consecutive node, or antinode, lines increases, as shown. Conversely, where the frequency increases, the wavelength decreases, and the distance between two adjacent node lines, or antinode lines, is smaller. Source: Taylor 1978. Reproduced with permission from Taylor & Francis Group.*

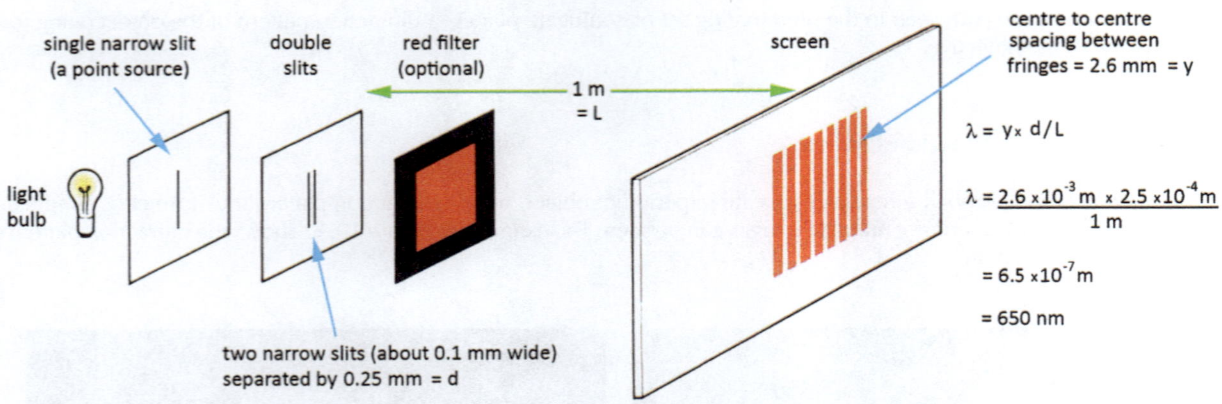

Figure 10.10 **Diffraction at two slits to calculate wavelength** *This illustration, taken from a physics school textbook, shows how the wavelength of light may be calculated from the diffraction fringe widths projected onto a screen. Source: England 1995. Reproduced with permission from Hodder Education.*

Because it is coherent, the wavefront arrives at the same time at the two slits, which will act as secondary point sources according to Huygens' principle, as before with the single slit. As the waves diffract outwards, they meet and interfere. The resulting light-and-dark pattern formed on a screen set perpendicular to the ray is the diffraction pattern. The mathematical relationship between slit spacing, slit width and diffraction angle is explained in Box 10.2. Now we see a new phenomenon: the 'overlapping' of the spherical waves from each slit has given rise to a pattern of broad but distinctive beams – and the narrower the spacing of the slits, the greater is the angle between these beams (Figure 10.9). This is an important point to grasp, because this principle directly affects the number of higher-order diffracted beams an objective is able to collect and resolve. Knowing the spacing between two closely apposed slits, or indeed the spacing in a diffraction grating, the distance to a screen and the spacing between the image of the fringes, the wavelength of the illuminating light may be calculated. This is demonstrated in Figure 10.10.

The phenomenon of interference can be extended to scattering from many parallel slits spaced equal distances apart. Such an object is known as a *diffraction grating*. It is usually an artificially produced series of very closely apposed holes or slits that transmit light or else is a series of very closely apposed lines or ridges that reflect light. A very finely detailed object viewed with the microscope behaves in the same way. The objective collects the diffracted orders of light to form the image.

Figure 10.11 shows how a converging plano-convex lens can be used to supply a parallel wavefront from a relatively near point source of light. The screen that shows the diffraction pattern should normally be placed a long distance away[8] compared to the spacing width between the slits. This may not always be convenient. The same figure also shows how a second converging lens placed after the diffracting object permits a screen to be placed at the rear focal plane to show the diffraction pattern, considerably nearer to the diffracting object than would otherwise be the case. Refer back to the diagram of Köhler illumination in Figure 9.19, and satisfy yourself

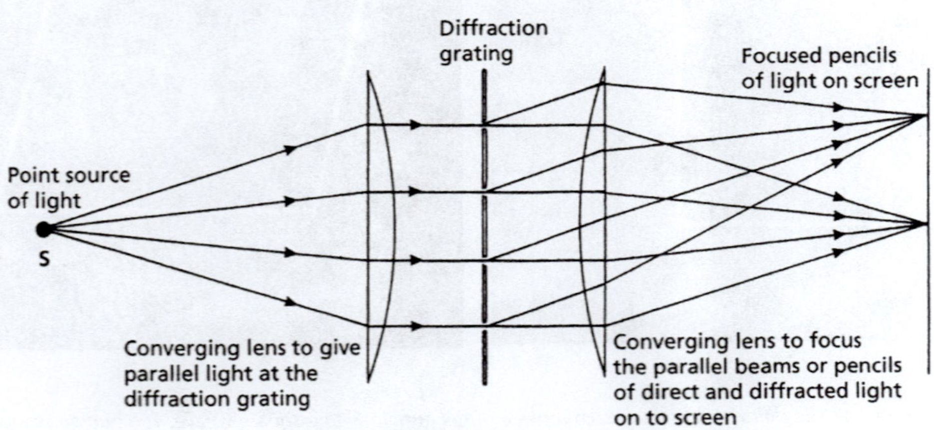

Figure 10.11 **Diffraction with lenses** *The path length between the double slits and the screen in the previous figure has to be as large as possible for accurate measurement of the fringe widths. Using converging lenses of appropriate focal length can shorten the path length as desired. The ray diagram here is very similar to the light path taken in the illuminating, or aperture, set of conjugate planes, with the left-hand converging lens performing the function of the condenser, the diffraction grating acting as the specimen and the right-hand converging lens acting as the objective focusing the diffraction pattern at the back focal plane of the objective. Source: Hammond 2009. Reproduced with permission from Oxford University Press.*

that this is identical to the ray path seen in the illuminating set of conjugate planes, a diffraction pattern of the object being formed in the back focal plane of the objective.

10.5 The Geometry of Diffraction Patterns

By looking at a point source through a regularly repeating (periodic) object, whose diffraction pattern thus-formed is simple and thus easy to study and understand, a series of multiple images can be seen. The example in Figure 10.12 shows the diffraction pattern formed from a piece of nylon net curtain.

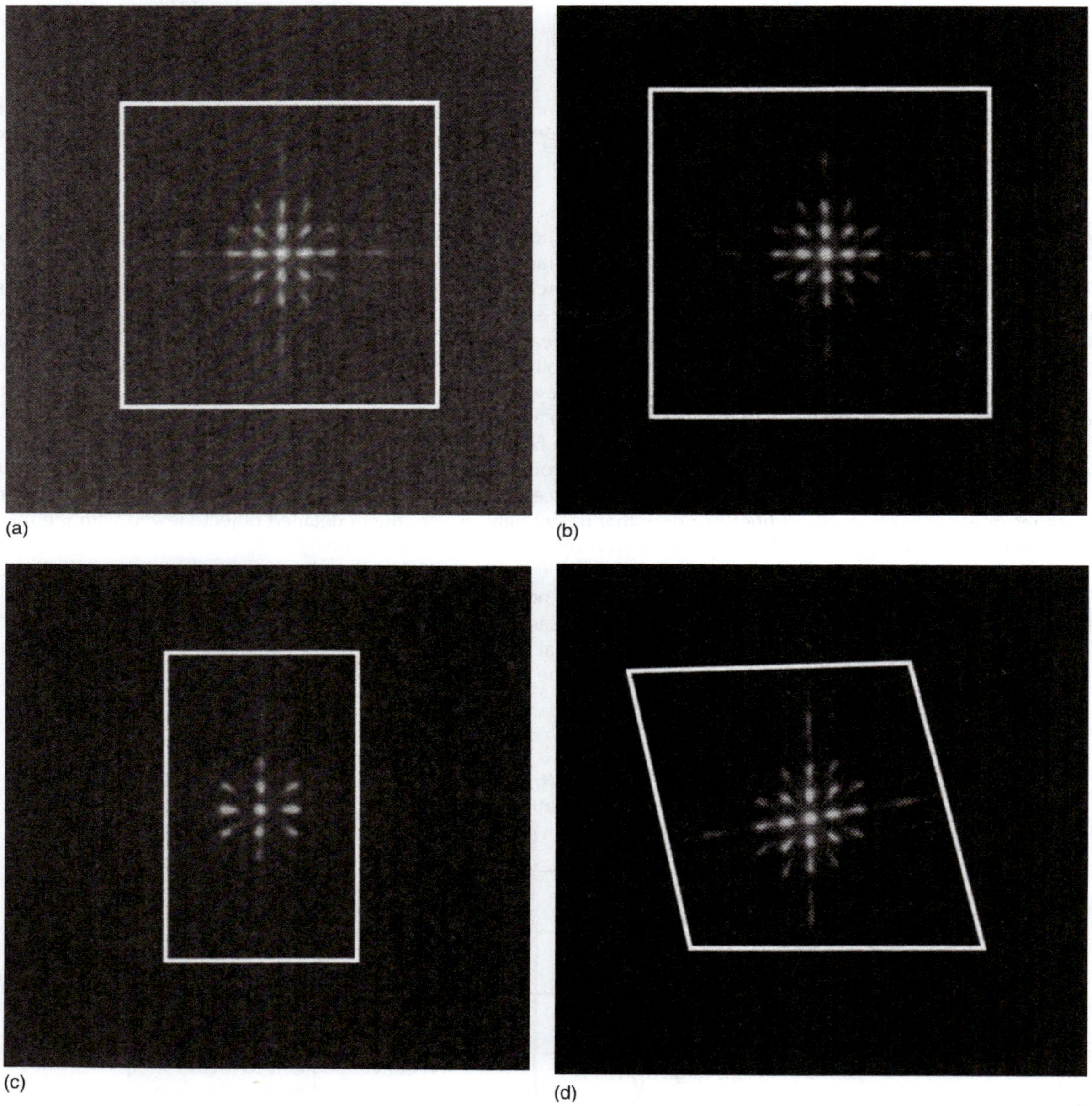

Figure 10.12 **Diffraction patterns** *The diffraction pattern from a piece of net curtain with a square weave. The outline of the net is indicated by the white frame. The identical square weave (a) gives rise to two rows of strong spots with fainter spots also part of the diffraction pattern. If the net is brought closer to the eye (b) there is no change in the size of the diffraction pattern. If the piece of curtain is rotated about a vertical axis (c), thus reducing the effective spacing between the vertical lines of the weave, this results in the spacing between the diffraction spots lying in the horizontal direction becoming larger. Distorting the fabric by inducing a diagonal shear (d) such that the lines are no longer at 90° in a square relationship to one another results in a diffraction pattern sheared orthogonally in the opposite sense – the rows of strong diffraction spots remain perpendicular to the lines of the net. Source: Hammond 2009. Reproduced with permission from Oxford University Press.*

Exercise 10.1 Diffraction Patterns

Take a piece of net curtain (the finer, the better) and observe, say, a street lamp or a small torch bulb – in both cases these are effectively distant point sources.

Figure 10.12 shows the central spot (the image of the source) surrounded by a square pattern of spots. These *diffraction spots* arise from the interference between the waves scattered by the crossed slits or the weave of the net – the horizontal row of spots from the vertical slits in the weave and the vertical row of spots from the horizontal slits in the weave.

Now, say, rotate the net about a vertical axis such that the effective spacing of the vertical slits in the object, with respect to the incoming light, is reduced. The spacing of the horizontal row of spots of the diffraction pattern accordingly increases. The experiments may also be performed using a hand-held laser pointer (a point source of light) and projecting the pattern on to a screen – NOT of course by looking at the laser light directly.

- Firstly, the same diffraction pattern is seen whether one or both eyes are used to view it.
- Secondly (provided that the light source is at infinity and falls parallel on the object), the size of the diffraction pattern is *independent* of the position of the object. The size of the pattern is constant regardless of whether the net is held close to or far away from the eye, indicating that it is formed by parallel light entering the eye from a distant point source.
- Thirdly, the repeat spacing of the diffraction pattern, D, is inversely proportional to the spacing of the regular repeat, d, in the object: $D \approx 1/d$, so the finer the spacing in the object, the more widely spaced the detail in the diffraction pattern, and vice versa.
- Fourthly, the diffraction pattern lies in the same direction as the repeat of the periodic structure in the object and perpendicular to the features in the image.
- When viewed through colour filters[9], it is seen that light of longer wavelengths is diffracted through larger angles than short-wavelength light.

What you have observed with a piece of net curtain can be taken further using microscope specimens, such as diatom frustules (see Section 10.12) with fine periodic details that help to understand the role of diffraction and interference in image formation.

- A proportion of incident light is not diffracted and is transmitted as undeviated rays. These form the central zero-order spot in the diffraction pattern and go on to evenly illuminate the entire image plane.
- Each diffracting feature in the specimen acts as an independent source of diffracted rays (see Figure 10.11) that radiate outwards as Huygens wavefronts towards the objective.
- At least two different orders of light must interfere to form the image. The image formed in the primary image plane is due to the interaction of (usually) the undeviated zero-order with one or more diffracted rays. As a special case if the zero-order ray is excluded, image formation is solely between diffracted rays, and consequently a dark-field image is formed.
- If a specimen does not diffract light, or the objective is of insufficient aperture to capture the diffracted rays, no image will be formed in the primary image plane.

The microscope objective forms a diffraction pattern of the object in the back focal plane of the objective[10]. The primary image is formed by the *recombination and interference* of light that was separated by diffraction when it interacted with the specimen or object. The greater the number of diffracted beams that recombine (usually with the zero-order ray) to form the image, the more faithful a representation that image will be of the original specimen (see Figure 21.7 and Figure 26.11). The diffraction of light determines the resolving power limit of optical instruments, particularly in microscopes and telescopes. The reciprocal relationship between an object and its diffraction pattern is a Fourier transform. The diffraction pattern in the back focal plane, and the image in the next set of conjugate planes, at the primary plane, are inverse Fourier transforms of each other (Figure 10.13; and see Chapter 21 for further details). The Fourier transform of the diffraction pattern is the image formed by interference of the diffracted rays in the primary image plane of the

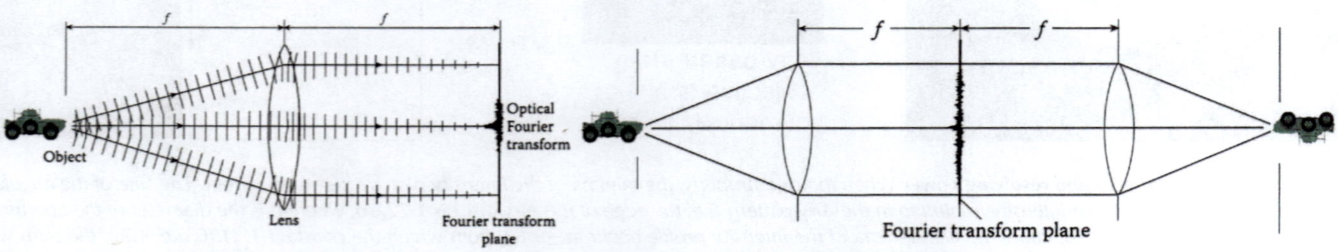

Figure 10.13 ***Diffraction patterns and the Fourier transform*** *The images formed at the imaging and illuminating conjugate planes are situated, respectively, in real space and Fourier space (also called reciprocal, frequency or spatial frequency space, in just the same way as wavelength is reciprocally related to the frequency of light). The diffraction plane is visible as a circular image bounded by the cut-off frequency (f_0 or $\nu_0 = 2NA/\lambda = 1/(\lambda).(f/\#)$). The central zero-order diffraction spot is surrounded by features of increasing spatial frequency until the cut-off point. Source: Saxby 2011. Reproduced with permission from Taylor & Francis.*

microscope. As we shall see later in this chapter, the final image can be artificially changed by modification of the diffraction pattern without altering the properties or structure of the object in any way.

Because the diffraction pattern is the direct result of the interaction of electromagnetic radiation with the specimen, then analysis of the diffraction pattern permits the structure of the object to be calculated, even if it is initially unknown or too small to resolve. Crystal lattices and periodic fibrous structures can be worked out by examination of their X-ray diffraction patterns, the most famous example being the elucidation of the structure of the DNA molecule by Watson & Crick. The helical structure of the molecule was revealed by a fuzzy 'X' seen on the diffraction pattern – the famous 'Photo 51' taken by Rosalind Franklin in Maurice Wilkins' laboratory.

10.6 Airy Patterns

When using a microscope, we use objectives with circular apertures instead of slits. The case of the diffraction grating can be extended to model how a diffraction pattern is formed when the rays of light diffracted by the object are brought to a focus by the objective in the back focal plane of the objective.

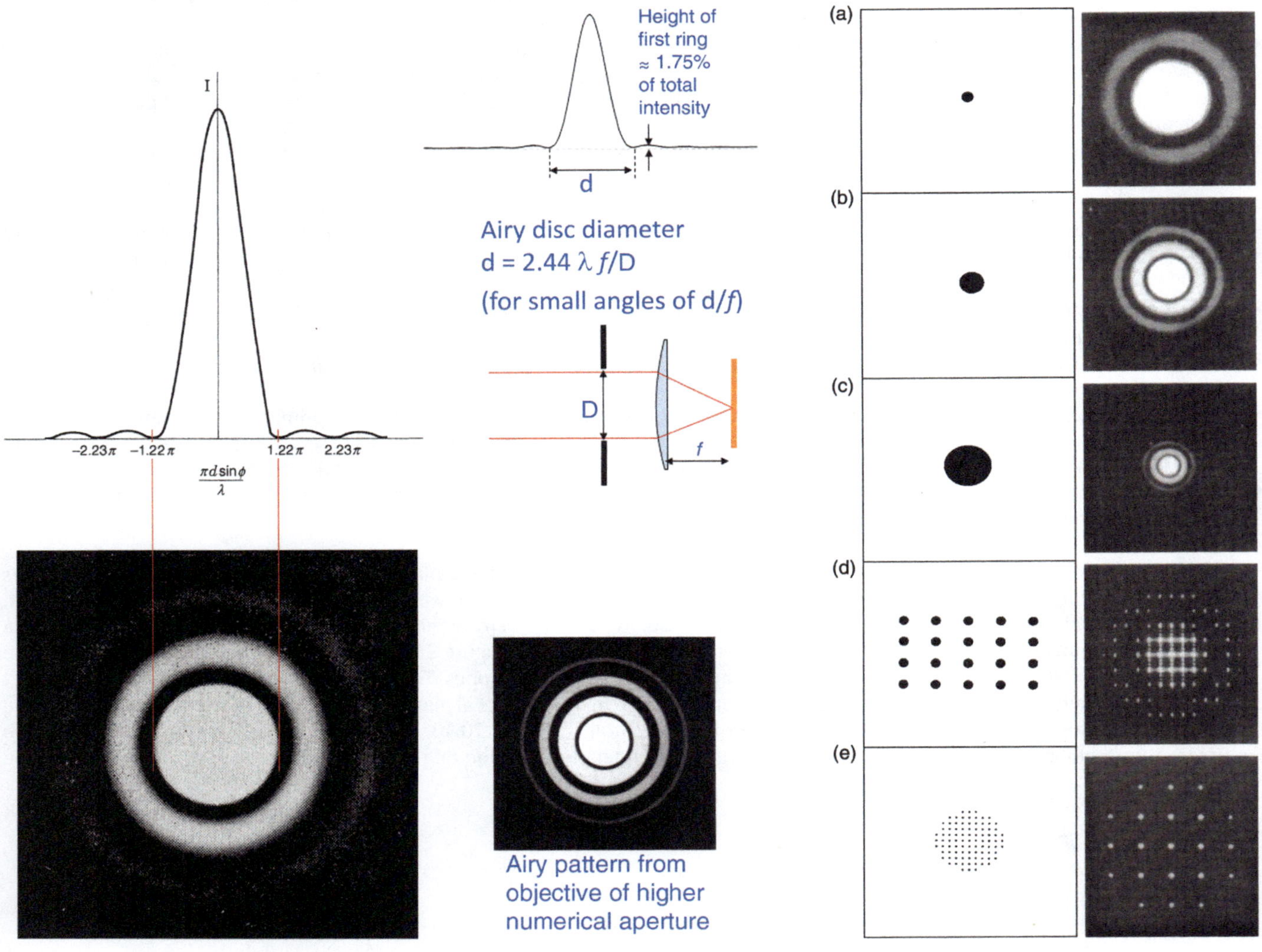

Figure 10.14 **Airy patterns and resolving power** *When focused properly, the minima of the intensity plot are zero throughout. The sine of the angular distance (in radians) to the first intensity minimum in the Airy pattern (i.e. the edge of the Airy disc) is 1.22λ/d, where d is the diameter of the aperture. From expansion of the Bessel function, the first minima of the intensity profile occur at ±3.83, from which the constant 1.2196 (i.e. 1.22, the term we are familiar with) is derived in the expression for the diameter of the Airy disc.*

When a point source is viewed through a restricted aperture in the dark, the Airy disc expands and so covers more cone cells, allowing the Airy disc to be seen, when normally it is not. The intensity profile is very much flatter, but the central maximum is not so high (i.e. the Airy pattern is dimmer). In panels (a) through (e), a series of circular diffracting apertures are shown (left) as black circles on a white background with the corresponding diffraction pattern shown on the right. Source: Murphy 2001. Reproduced with permission from John Wiley & Sons; Adapted from Pawley 2006 with permission from Springer; Hammond 2009. Reproduced with permission from Oxford University Press.

The distribution of light diffracting through a circular aperture was first worked out by Sir George Biddell Airy, Astronomer Royal to Queen Victoria, who using telescopes studied the distribution of light in the images of stars. As we have seen in Chapter 2, Section 2.10, geometrical optics shows that the parallel light from a distant object (e.g. a star) is focused by the objective lens of a telescope at the principal focal plane as a diffraction-limited *point*. However, owing to the scattering of light by the aperture of the lens, the image is not a point, but a disc – the Airy disc – surrounded by fainter rings, called the *Airy pattern* (Figure 10.14a). The Airy patterns corresponding to small, intermediate and large objective lenses are shown in Figure 10.14b. Notice the disc (or image of the star) is much smaller (or sharper) for the lens of large aperture. Exactly the same pattern is formed by a microscope objective.

To understand how a lens forms, or convolves, an image from the object, it is convenient to study a point as an object. In this respect a distant star viewed through a telescope or a sub-resolution fluorescent bead viewed with a microscope are optically identical. Both images are diffraction-limited point spread functions. Sub-resolution particles, which appear as diffraction discs, are only visible because they possess sufficient contrast to be detected against the background. This is the case, for example, with self-luminous fluorescently labelled particles and single molecules. The apparent diameter of the structure remains constant at the size of the Airy disc, but the *true* diameter of the particle or molecule cannot be determined.

A point source will always form an Airy pattern. Its size will decrease with decreasing wavelength of illuminating radiation and with increasing numerical aperture of the objective but will always remain a disc with associated rings and never become a point. For a 60x or 63x objective of NA 1.4 illuminated with blue laser light of 405 nm, the diffraction-limited Airy disc at the primary image plane is 10–11 μm in diameter depending on the exact magnification of the objective. Table 10.1 gives values for other commonly used objectives. An image formed from an extended object other than a single minute point is composed of a series of overlapping diffraction-limited Airy patterns.

The central diffraction spot (the Airy disc) contains ≈ 84% of the light energy from the point source. The first ring (from the first-order diffracted ray) contains a relative intensity of 1.75% of the light, the second ring 0.42%; the third to 0.16%, the fourth to 0.08% and so on. Most of the light forming the Airy pattern therefore falls into the Airy disc, but a much smaller quantity is still seen as faint first- and higher-order diffraction rings. The Gaussian-shaped intensity profile (Figure 10.14) of the Airy pattern can be described mathematically by a complex equation called the Bessel function of the first kind[11]. Airy derived this mathematical description of the diffraction pattern of stars by summing the wavelets diffracting through the aperture of the telescope objective according to Huygens' principle. In three dimensions, the intensity distribution of the Airy pattern is called the point spread function, or PSF.

The point spread function has a complex shape, which at first may be confusing to visualise. In the plane of focus, the PSF is simply the two-dimensional Airy pattern – a bright circular Airy disc in the centre surrounded by annular rings of decreasing intensity. The Gaussian plot associated with the Airy pattern is an intensity plot of the PSF in the focal plane. In theory, the three-dimensional shape of the point spread function should be spherical, but because of the inevitable poorer resolving power axially in the z-plane, the 3D PSF (see Figures 21.2, 21.9 and 26.4) with its associated side lobes assumes the shape of the neck of an hour glass; the concentric rings spread out away from the central plane as a complex set of conical structures. These side lobes are greatly reduced in the confocal microscope, where the 3D PSF is shaped more like a rugby football (see Figure 26.4).

The diameter of the central Airy disc gives a measure of the diffraction occurring at the aperture of the objective. Its size is governed by the wavelength of the illumination and the numerical aperture of the objective. In the microscope the radius of the Airy diffraction disc is given by $r = 1.22\lambda/2\text{NA}$, or $0.61\lambda/\text{NA}$. This distance, r, is sometimes called the Rayleigh distance. Both *f*-number and numerical aperture are measures of the effective aperture angle, light gathering capacity and resolving power of the instrument (Table 10.2), whether camera, telescope or microscope. The mathematical derivation of this expression is given in Box 10.3.

Table 10.1 *Diffraction-Limited Airy Disc Sizes*

Selected Objective	Optical Index (NA/Mag ratio)	Lateral resolving power x,y (μm) @ 488 nm [$d_{x,y} = 0.61\lambda/\text{NA}$]	Size of the Airy disc at the primary image plane (μm) [$d_{x,y} \times \text{Mag}_{obj}$]
6.3x/NA 0.16 dry	25.4	1.86	11.7
10x/NA 0.25 dry	25	1.19	11.9
10x/NA 0.30 dry	30	0.99	9.9
20x/NA 0.5 dry	25	0.60	12.0
20x/NA 0.75 dry	37.5	0.40	8.0
20x/NA 1.0 water	50	0.30	6.0
25x/NA 0.8 multi	32	0.37	9.3
30x/NA 1.05 silicone	35	0.28	8.4
40x/NA 1.30 oil	32.5	0.23	9.2
40x/NA 1.40 oil	35	0.21	8.4
60x/NA 1.22 water	20.3	0.24	14.4
60x/NA 1.30 silicone	21.7	0.23	13.8
60x/NA 1.40 oil	23.3	0.21	12.6
100x/NA 1.30 oil	13	0.23	23.0
100x/NA 1.40 oil	14	0.21	21.0

See also the Airy disc calculator (iscope). URL: http://iscopecalc.com.

Table 10.2 *Relationship between Numerical Aperture and f-No.*

f/0.5 ≈ NA 1	f/3.2 = NA 0.15	f/10 = NA 0.050
f/0.7 ≈ NA 0.7	f/3.5 = NA 0.14	**f/11.3** = NA 0.044
f/1.0 ≈ NA 0.5	**f/4.0** = NA 0.12	f/12.5 = NA 0.040
f/1.2 ≈ NA 0.4	f/4.5 = NA 0.11	f/14 = NA 0.036
f/1.4 ≈ NA 0.3	f/4.8 = NA 0.10	**f/16 = NA 0.031**
f/1.7 = NA 0.29	f/5.0 = NA 0.10	f/18 = NA 0.028
f/1.9 = NA 0.26	**f/5.6** = NA 0.09	f/20 = NA 0.025
f/2.0 = NA 0.25	f/6.3 = NA 0.08	**f/22** = NA 0.023
f/2.4 = NA 0.24	f/6.7 = NA 0.07	f/25 = NA 0.020
f/2.5 = NA 0.20	**f/8.0** = NA 0.06	**f/32** = NA 0.016
f/2.8 = NA 0.18	f/9.5 = NA 0.053	f/36 = NA 0.014

The *f*-number of any lens is the focal length divided by the diameter of the lens (*f*/D). The inverse of this (D/*f*) is the relative aperture of the lens. The maximum theoretical value for the *f*-number is *f*/0.5, although this is never reached in practice.

Since the amount of light admitted by a lens is proportional to the area of the lens, or aperture, then the exposure given by the lens (i.e. 'speed' of the lens) is proportional to the *square* of the *f*-No., which is why the series progresses in steps of $\sqrt{2}$, allowing the exposure to be doubled, or halved, easily. The standard *f*-No. scale (shown in bold) progresses up to *f*/128. This table shows equivalent NA values up to *f*/36. One of the lowest-magnification compound microscope objectives (e.g. 1.25x) has a numerical aperture of 0.03; a long focal length (e.g. 100 mm) photomacrographic lens has a numerical aperture of 0.015 at approximately *f*/36.

The numerical aperture and *f*-number are related: for small NAs, up to about 0.3, the numerical aperture is equal to half the reciprocal of the *f*-No. NA = 1/[2.*f*-No.], and *f*-No. = 1/[2.NA]. This is because, for small angles, the sine and tangent functions are approximately equal and therefore interchangeable. The formula for NA = n.sin α, where α is the half-angle accepting the most marginal ray admitted by the aperture of the objective. For these values, the refractive index term, n, in the expression n.sin α for NA equals 1; therefore, *f*-No. = 1/[2.NA] and NA = 1/[2.*f*-No.]. Since the *f*-No. equals f/D, then the half angle of the angular aperture of the camera lens (α) is cotangent $\alpha/2$, whereas for dry microscope objectives NA = sin α.

See Chapter 6, Section 6.3, for an explanation of the Abbe sine condition, and why the numerical aperture of the microscope is governed by the sine function, as determined by the most marginal ray accepted by the aperture of the objective rather than the tangent, as determined by the ratio of the focal length (i.e. on the optical axis) to the diameter of the camera lens.

Box 10.3 Rayleigh's Resolution Criterion

The diameter of the Airy disc equals 2.44λ(*f*/d), where *f*/d is the focal ratio or *f*-number; *f* is equal to the focal length of the camera or telescope lens or objective, and d is the diameter of the aperture of the lens or objective.

As the angular separation between two stars decreases, their Airy discs overlap. The limit of resolution occurs at the point where the overall intensity profile of the two Airy patterns does not show a discernible 'dip' or decrease between the central maxima of the Airy discs. According to the criterion proposed by Rayleigh (which works well in practice) this point occurs when the first minimum, or dark ring, of one Airy pattern coincides with the zero-order, or centre of the bright Airy disc of the second pattern. The so-called 'zero-order minimum angle for interference' α_{zom} (see Hammond 2009, page 179 and 183), which is the angle where the first order minimum occurs, at which total destructive interference of the zero-order peak intensity occurs. This limit of resolution = α_{zom} = sinα_{zom} = 1.22λ/d

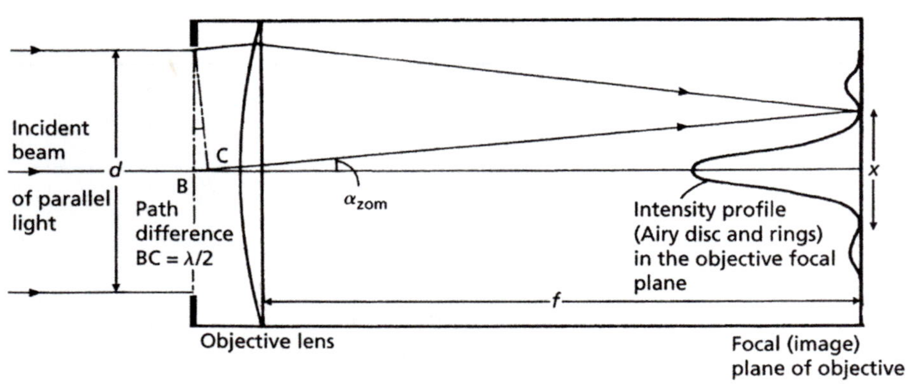

Source: Hammond 2009. Reproduced with permission from Oxford University Press.

In the diagram above, the radius of the Airy disc (x/2) is given as *f*.sinα_{zom}. Substituting the limit of resolution equation above for sinα_{zom}, gives x = 1.22λ(*f*/d), or (1.22λ.*f*)/d.

The limit of resolution in the microscope can be evaluated in the same way. Just as in the telescope, the minimum resolved distance is given when the angular separation between two closely apposed point sources is: α_{zom} = 1.22λ/d.

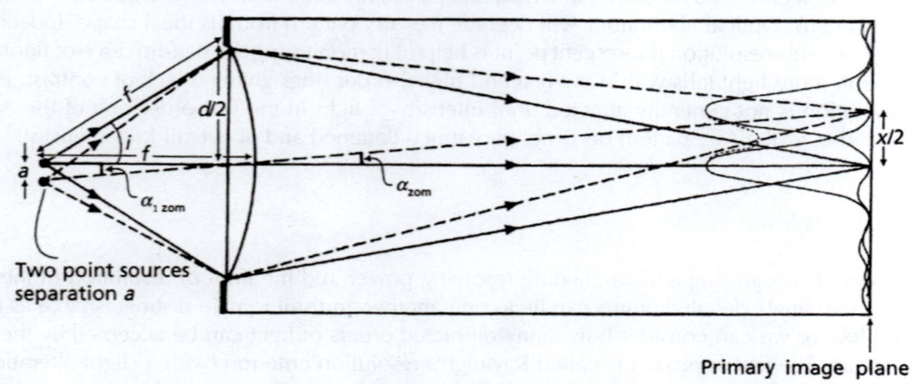

Source: Hammond 2009. Reproduced with permission from Oxford University Press.

For simplicity, the tube-lens has been omitted from the figure above, and the rays refracted from the rear of the objective lens at the convex surface are shown as if the microscope were of finite tubelength (i.e. as if the object lay between *f* and 2*f*). Assuming the objective *is* infinity-corrected, and the focal point emanates from the centre of the convex surface of the plan-convex objective lens, then the focal length is also equivalent to the hypotenuse, as shown in the figure. The angle α_{zom} is that subtended between the two closely apposed points. If the medium between these points and the front of the lens is a medium other than air, then angle α_{1zom} (which is equivalent in air) is smaller than angle α_{zom}, the ratio $\alpha_{zom}/\alpha_{1zom}$ being the refractive index of the immersion medium.

From the figure above, it can be seen that $\sin i = d/2f$ and angle α_{zom} a/*f*. From this result, $\alpha_{zom} = 1.22\lambda/d = a/f$ and $a = 1.22\ \lambda/(f/d)$. By substituting $(f/d) = 2 \sin i$ (when a is very small, $\sin i \approx$ half-angle of acceptance of the objective and thus $(f/d) = 2 \sin i$), we obtain: $a = 1.22\lambda/2 \sin i = 0.61\ \lambda/\sin i$

Both *f*-number and numerical aperture are measures of the effective aperture angle, light gathering capacity and resolving power of the instrument (Table 10.2) whether camera, telescope or microscope. Since, $f/d = 1/2\text{NA}$[i], NA can be substituted above for sin *i*.

The refractive index used will either be $n = 1$ for dry objective lenses or $n > 1$ for immersion objectives. We can extend this result from the case for a dry objective to an immersion objective, to include the effect of the immersion medium which lies between the object and the objective lens. From Snell's law the refractive index, n, is the ratio of the wavelength of light on the object side (λ) to the wavelength on the image side (by convention denoted λ'). By substituting $n = \lambda/\lambda'$ into $0.61\lambda/\sin i$, we get: $a = 0.61\lambda'/\sin i = 0.61\ \lambda/n \sin i$. Since the distance a = the minimum resolved distance in the microscope (d), and $n \sin i = \text{NA}$, we get the Rayleigh criterion for the limit of resolution expressed mathematically as: $d = 0.61\lambda/\text{NA}$.

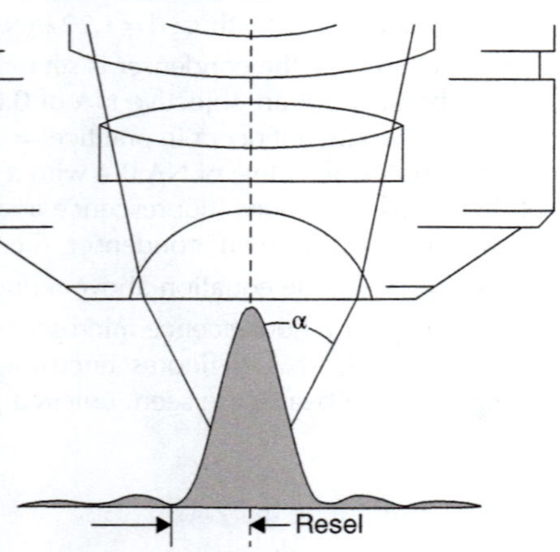

Source: Webb 1996. Reproduced with permission from Institute of Physics.

The diameter of the Airy disc at the focal (i.e. primary image) plane can be calculated by multiplying the value derived using the formula $2.44\lambda/2\text{NA}$ by the initial magnification factor of the objective plus any tube-lens factor.

The resel: The x/2 distance is sometimes referred to a 'resel' – a resolution element, just as a 'pixel' is a picture element. It is the radius of the Airy disc. The sampling effect can be estimated by relating the size of a pixel to the diameter of the Airy disc (see Stelzer, 1998, 2005). Four pixels per Airy pattern are required, along the main axis of two objects, so eight samples per Airy pattern will give a contrast 'dip' of 26.4%. This figure is taken from Figure 4 in Webb (1996). The resel is a useful concept when calculating the magnification required for optimum Nyquist sampling; see Figure 31.9 in Chapter 31 and Section 31.7.

References

Hammond, C. (2009). 'The diffraction of Light'. In: *The Basics of Crystallography and Diffraction*, 3rd ed., Chapter 7. Oxford: OUP, Oxford Science Publications.

Stelzer, E.H.K. (1998). 'Contrast, resolution, pixilation, dynamic range and signal-to-noise ratio: fundamental limits to resolution in fluorescence light microscopy'. *Journal of Microscopy* 189/1: 15–24.

Stelzer, E.H.K. (2005). 'Practical limits to resolution in Fluorescence light microscopy'. In: Yuste, R. and Konnerth, A. (eds.), *Imaging in Neuroscience and Development: a laboratory manual*, Chapter 96. New York: Cold Spring Harbor Laboratory Press, pp. 767–773.

Webb, R.H. (1996). 'Confocal Optical Microscopy'. *Reports on Progress in Physics* 59/3: 427–471.

[i] When calculating the aperture determined by the *f*-number, being the ratio of focal length to lens diameter, the tangent of the half-angle is used. When calculating the numerical aperture, which relates to rays diffracted into the objective aperture, the sine of the half-angle is used. In the comparison of *f*-number and numerical aperture, for large magnifications $f/D = 1/2\text{NA}$, as explained in Loveland, R.P. (1970). *Photomicrography: a comprehensive treatise*, Volume 1, Chapter 5. New York: John Wiley, p. 221.

If an objective is sufficiently well corrected for coma, spherical and chromatic aberration within a quarter of a wavelength, a point will be imaged as an ideal Airy pattern. Optical aberrations will degrade the Airy pattern from its ideal shape. Indeed, the alteration in the point spread function image of a sub-resolution fluorescent point is helpful in monitoring the performance of fluorescence microscopes. When aberrations are present, more light falls within the first and higher-order rings giving a lack of contrast, particularly in areas of fine detail. The overall resolution is not generally affected until intensity of light in the first-order ring of the Airy pattern equals the intensity of the central disc. That is, the Airy pattern becomes very much flattened and of overall low intensity.

10.7 Resolution in the Light Microscope

There are two principal ways of recognising and calculating resolving power and the limit of resolution in the microscope. We can either consider how closely two finely detailed points can lie to one another and still remain distinguishable as discrete entities rather than a single unresolved 'blob' or we can consider how many diffracted orders of light can be accepted by the objective to form the diffraction pattern of an image. The first approach is called Rayleigh's resolution criterion (with a slight alteration proposed by C. M. Sparrow), whilst the second is called Abbe's resolution criterion.

Both Rayleigh's and Sparrow's criteria for resolution have been called 'soft' limits in that they take into account the influence of *contrast* (see Stelzer, 1998) upon the actual minimum resolved distance, and the inter-relationship of contrast and resolving power. These two limits are non-absolute and can also be changed to some extent by altering which light rays contribute to the image (apodization). Abbe's resolution *limit* states exactly (Figure 10.19) which information about the object is transferred by the microscope objective into the image.

For a given magnification – as the angular aperture of the illumination is increased, or else if an objective is used that has a higher optical index (see Chapter 31, Section 31.7) – the diameter of the Airy pattern diffraction spots in the primary image plane become smaller, so the details of the object are better resolved in the image. Equally, objectives of larger numerical aperture that are properly illuminated by correctly adjusted Köhler illumination allow more-widely diffracted rays to contribute to the image, permitting resolution of details that would otherwise not be resolved.

In bright-field microscopy, where the numerical aperture of the condenser is set slightly smaller than that of the objective, the denominator is rewritten to be the sum of both the objective and condenser numerical aperture, whereas the constant 0.61 in the numerator is replaced by 1.22, thus: $d = 1.22\lambda/(NA_{objective} + NA_{condenser})$. The value of $(NA_{objective} + NA_{condenser})$ should only be used when the numerical aperture of the condenser is set only slightly smaller than the NA of the objective; otherwise, a ludicrous result is given. For example, the mean of an objective NA of 0.04 (i.e. objective of 1.25x magnification) used together with a condenser of NA 1.3 (this combination would not occur in practice) = mean NA 0.67; this combination would theoretically give better resolution than a long working distance 20x objective of NA 0.4 with a matched LWD condenser of NA 0.55 = mean NA 0.48!

For self-luminous specimens (fluorescence and dark-field microscopy), the resolving power is given by the formula: $d = 0.61\lambda/NA$. The objective acts as its own condenser (i.e. $NA_{condenser} = NA_{objective}$, self-evidently) or as in dark-ground microscopy (where $NA_{condenser} > NA_{objective}$), the equation above reduces to: $d = 1.22\lambda/2.NA \Rightarrow d = 0.61\lambda/NA$.

Strictly speaking, for a fluorescence microscope, the minimum resolved distance is dependent upon the *emission* wavelength of the fluorophore: $d = 0.61\lambda_{em}/NA$. In fluorescence microscopy, the background should be dark because the object is self-luminous, and only higher-order diffracted beams are seen, unless a poor signal-to-noise ratio is present in the image from high background signal in the specimen.

10.7.1 Resolution Along the z-Axis

The same laws that define the minimum resolved distance laterally in the x- and y-axis also apply axially in the z-axis, since the point spread function is three dimensional and also exists in the z-axis, albeit broader and more spread out (Figure 10.15) than in the x- and y-axis.

A lens can only collect part of the fan of diffracted rays that emanates from an object. Only if the object was placed at the centre of a spherical lens would all the diffracted rays be collected to form an image, and this is impossible. Resolution in z, the axial, direction, is therefore approximately three to four times worse than in x-y, the lateral directions. The depth of field (Box 7.5) is defined as the thickness of the optical section along the z-axis within which objects in the specimen plane are in acceptably sharp focus. The larger the numerical aperture of the objective, the smaller the depth of field will be. The same holds true for cameras: short focal length (fast) lenses (< *f*/4) have a shallow depth of field, whereas lenses with lower aperture (>*f*/16) have greater depth of field.

There are many equations for the depth of field, which determines axial resolution, but for diffraction-limited optics, assuming no spherical or chromatic aberration, the minimum distance resolved in the z-axis by an objective can be given as $d_z = 2.n.\lambda/NA^2$, where *n* is the refractive index of the immersion medium between the objective and sample. In practice, three main factors determine axial elongation in the z-axis.

- The first is the information collected in z is less then laterally, so that x-y resolution is about 3–4 times better than in z.
- The second is seen in much larger structures (>10 μm) due to the mismatch of refractive indexes, between the immersion medium of the objective and the RI of the sample, between the sample mounting medium and the sample itself or from a combination of all of these items. For further information, see Section 18.8.
- The third factor is uncorrected spherical aberration from high-dry objective lenses.

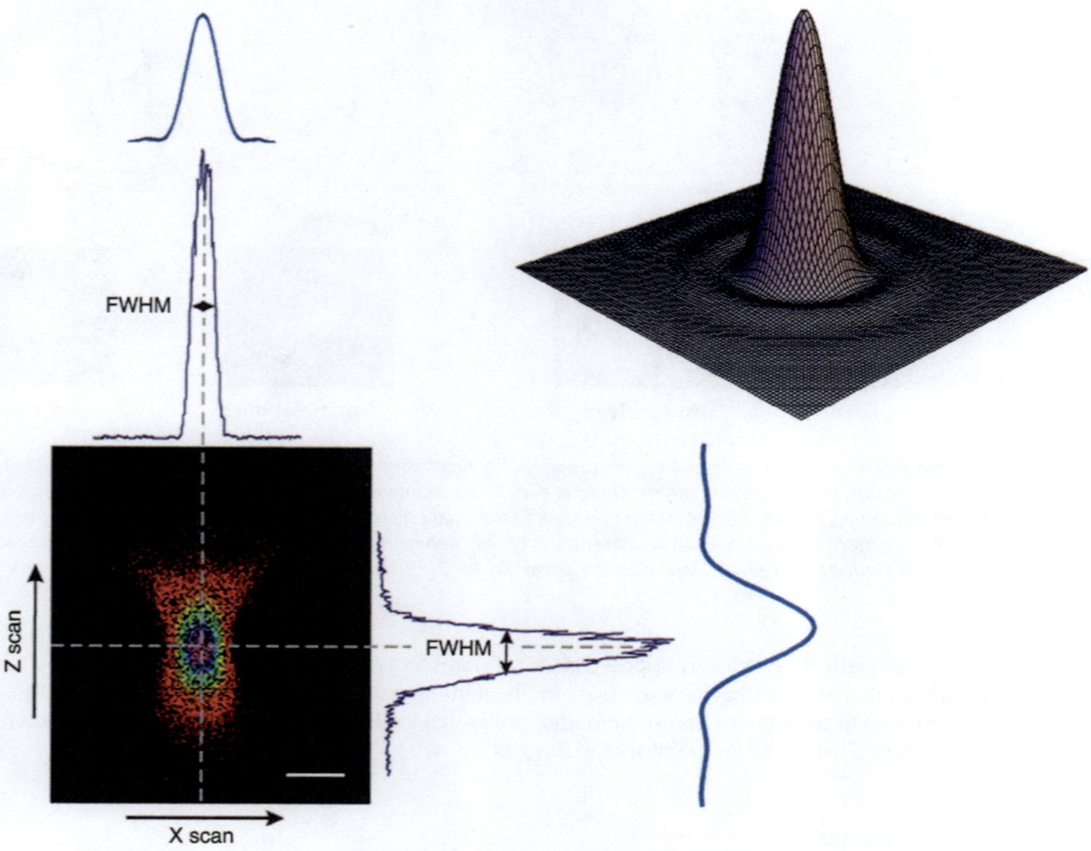

Figure 10.15 **Difference of axial and lateral intensity profiles** *Resolving power is poorer in the axial z direction, and this figure shows the 1D intensity profile along the x-, y- and z-axes. It is taken from a Z-stack of a sub-resolution fluorescent bead. The Full Width Half Maximum value, derived from the intensity plot, can be used to determine the resolving power of the microscope. A 3D representation of the two-dimensional intensity plot of the Airy pattern in the x-, y- and z-axes is shown in the upper right-hand part of this figure. The scale bar represents 300 nm. Source: Hibbs et al 2006. Reproduced with permission from Springer; Adapted from Kubitscheck 2013 with permission from John Wiley & Sons.*

10.8 Rayleigh's Resolution Criterion

The limit to resolution, proposed by John Strutt, Lord Rayleigh, occurs when the first minimum, or dark ring, of one Airy pattern coincides with the zero-order, or centre, of the bright Airy disc of the second pattern. This is illustrated in Figure 10.16, which shows the two-dimensional intensity distribution from a densitometer. At the centre, the trough of the densitometer trace falls by ≈ 26% to ≈ 84% overall of the peak intensity value. This is *just* sufficient contrast such that the two points may still be seen as separate entities. Rayleigh's resolution criterion states that two adjacent objects are just resolved when the diffraction spot (the Airy disc) of one object coincides with the first diffraction minimum of the other, in the image plane.

According to Rayleigh's criterion, the light rays from each of these two closely apposed points is incoherent. Whilst this is certainly true for stars viewed with astronomical telescopes, it does not hold true in microscopy, except for self-luminous points in fluorescence microscopy. Otherwise, the illumination in the microscope is broadly coherent. Box 10.3 shows a ray diagram for two closely apposed points forming an image by diffraction in the microscope.

The Rayleigh criterion assumes a perfectly corrected microscope objective. In practice both spherical and chromatic aberration, as well as stray light, noise and sample preparation all serve to increase the value of the minimum distance resolved by the microscope objective. Theoretically with a wavelength of 405 nm and an NA of 1.4, with oil as the immersion medium, the lateral resolving power is 176 nm, and the axial resolving power is 627 nm. In reality, the lateral resolving power is around 250 nm, and the axial resolving power – at best – is ≈1 μm.

10.8.1 Sparrow's Resolution Criterion

Carroll Mason Sparrow proposed a resolution criterion better suited to astronomy (and digital image recording) than Rayleigh's criterion. He pointed out that Rayleigh's well-used criterion was originally proposed for comparing instruments with one another rather than as a measure of resolving power. Sparrow suggested that two points can *just* be resolved when the central minimum between them was zero. While Rayleigh's criterion rather tacitly assumes incoherence, Sparrow's criterion can be applied to coherent sources and is a more

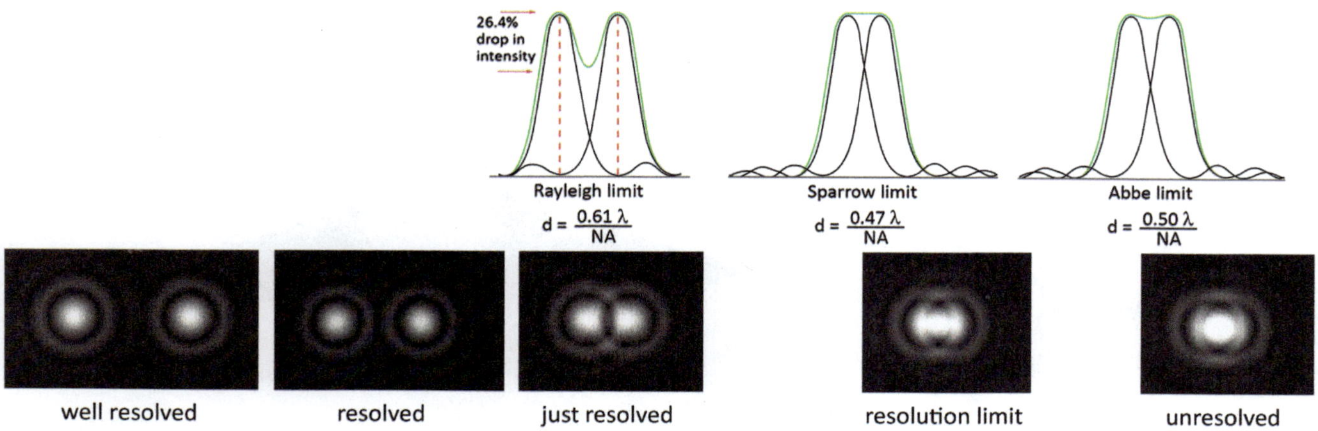

Figure 10.16 **Resolving power criteria** *The well-known Rayleigh limit occurs when the first minimum, or dark ring, of one Airy pattern coincides with the zero-order, or centre, of the bright Airy disc of the second pattern. There is a 26% dip in intensity between the two maxima. The Sparrow limit is sometimes applied in the case of bright fluorescent beads. The Abbe limit equates to the smallest spatial frequency in the object that can be transferred into, and discriminated in, the image. Notice there is still a small dip in intensity discernible between the two maxima in the case of the Abbe limit, whilst the intensity profile of the Sparrow limit is flat-topped. Source: (top) diagram drawn by Mr S. Thomas. (bottom) Franklin et al. 2010. Reproduced with permission from John Wiley & Sons.*

realistic measure of resolution of two equally bright closely apposed stars or fluorescent particles. According to Sparrow's criterion, the constant in the numerator is 0.47 rather than 0.61 so that the expression for the minimum resolved distance now becomes: $d = 0.47\lambda/NA$. However, Sparrow's criterion is sensitive to the signal-to-noise ratio and, unlike Rayleigh's criterion, cannot be readily determined by measuring the full-width half-maximum (FWHM) of the point spread function.

10.9 Ernst Abbe and Image Formation

From the middle of the 18th century it has been possible to manufacture telescopes on rational scientific principles using ray diagrams. Following the development of achromatic microscope objective lenses from the 1830s, it was thought that making lenses for microscope objectives would merely be an extension of these principles in miniature. However, Abbe[12] found that a mathematically constructed objective of low aperture (to reduce spherical and chromatic aberration) performed *worse* than one of larger (i.e. wider) angular aperture constructed by empirical trial and error. Although the work of Young and Fresnel had been known for half a century, it was not until the mid-19th century that microscopists became aware of the role of diffraction in microscopical image formation. By 1869 Abbe had worked out that the so-called 'dark space' surrounding the central illuminating beam of the microscope was not truly dark but held the diffracted light (see Figure 7.7), which carried information about the fine detail of the specimen.

Abbe's theory of image formation has three essential requisites:

- diffraction of light by interaction with the specimen;
- collection of these diffracted rays by the objective; and
- image formation by interference of these diffracted rays and the non-diffracted zero-order ray at the primary image plane of the microscope.

Diffraction by specimens whose dimensions are comparable to the wavelength of the illumination gives rise to orders of light diffracted through very large angles. Abbe realised that it is not valid to assume that microscope image formation follows the same (geometric) laws as the imaging of large objects by a telescope or a photographic camera. He concluded that, upon interacting with the specimen, light was diffracted from the incident axial illuminating beam into the entire aperture of the objective and a diffraction pattern of the object formed (Figure 10.17) in the back focal plane of the objective.

Abbe related resolving power (the minimum resolved distance, MRD) to wavelength, λ, the refractive index, n, of the medium between objective and object/sample and the semi-angle, α, of the angular aperture of the objective by the expression: $d = \lambda/2n.\sin\alpha$. He showed that for image resolution, the zero-order and at least the first order have to enter the objective lens. If the image is unresolved, that is *not* to say that the image is not formed. It *is* formed but sufficient fine detail is *not* seen. See, for example, Figure 10.21.

Since only one diffracted order beam needs to enter the objective together with the zero-order, one way to achieve greater resolution is to tilt the illumination axis (Figure 10.18) so that the zero-order background illuminating ray enters the objective at the extreme margin on one side, and one of the first-order rays will enter the objective at its extreme margin on the other side. This is the basis of oblique illumination (see also Figure 11.8). The price paid for achieving maximum resolution in the image is to sacrifice some directional information from the missing first-order beam.

Recall that $NA = n.\sin\alpha$. The resolving power given by $d = \lambda/NA$ occurs *only* for the special case of illumination of the object by parallel rays from a narrow cone of *coherent* light. If oblique light is used to illuminate the sample, then the objective can accept the

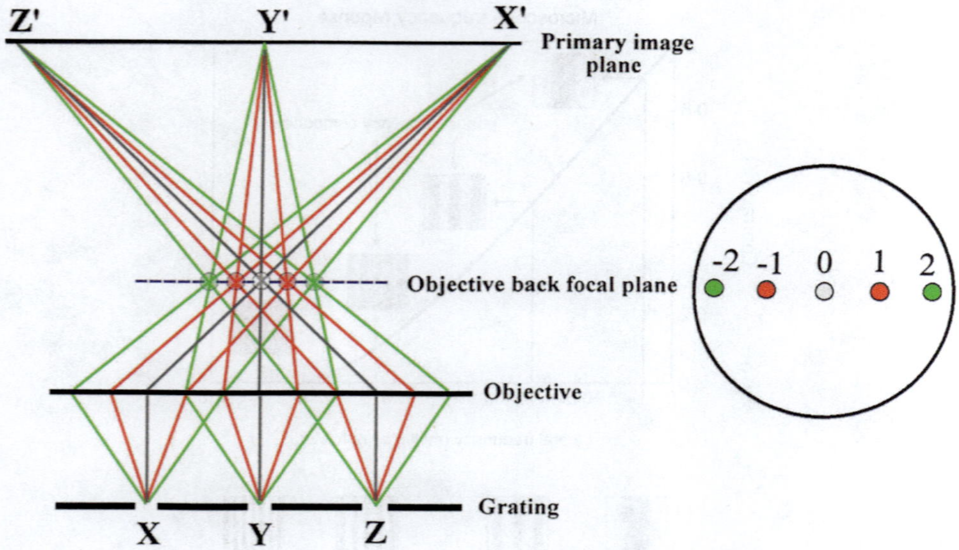

Figure 10.17 ***Image formation in the microscope*** *Image formation in the light microscope. The direct and diffracted beams (zero and higher orders) enter the aperture of the objective and are brought to a focus at the back focal plane of the objective, where a diffraction pattern – the Fourier transform of the image (i.e. in frequency space) – is formed. The rays continue onwards to form an image when the direct beam and diffracted orders recombine and interfere to form the primary image at the primary image plane. Compare this to Figure 10.11. Source: Adapted from Bradbury and Bracegirdle 1998 with permission from Taylor & Francis Group and adapted from Bradbury 1984 with permission from Oxford University Press.*

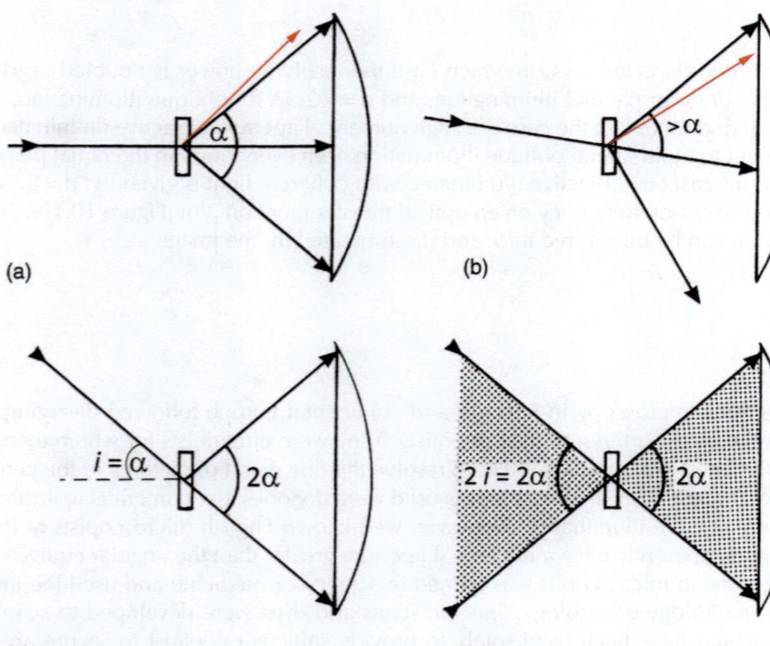

Figure 10.18 ***Increasing resolving power by oblique and conical illumination*** *Panel (a) shows the direct and two first-order beams entering the aperture of the objective and contributing to the image. The incident light is parallel (i.e. no condenser or else the condenser is closed down). Any higher diffracted orders (red arrow) are lost. If the illumination is tilted, a new first-order beam of greater diffraction angle can now enter the objective. This illumination obliquity can be extended to the limiting case (c) where the illuminating beam is inclined at an angle, i, equivalent to the half angle of aperture, α, such that the direct or zero-order just enters at one edge of the objective and one first-order beam, now at twice the angle, 2α, just passes into the objective at the opposite edge. The practical arrangement is shown in (d) with the condenser diaphragm opened to project a convergent cone of light onto the specimen (in practice a little less than 2i) to admit direct and diffracted light into the objective over a whole range of angles. Source: Adapted from Hammond 2009 with permission from Oxford University Press.*

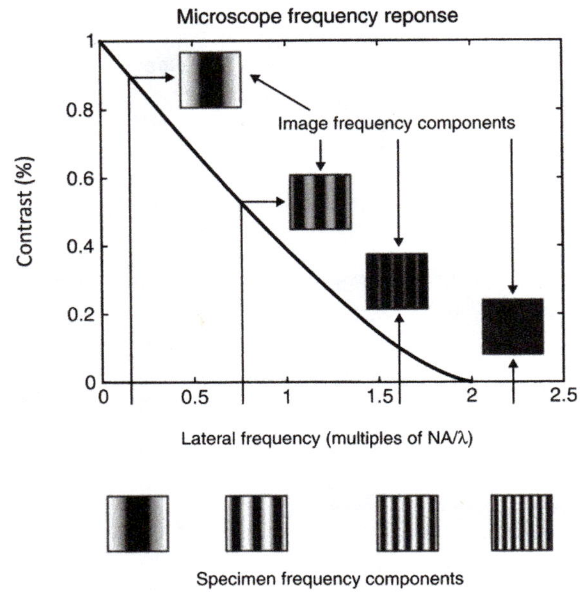

Figure 10.19 **Optical transfer and the Abbe frequency limit** *An optical transfer function (OTF) plot illustrates graphically how any imaging system transfers resolvable object information into the image. The Abbe limit of lateral resolving power occurs at $d = 0.5\lambda/NA$, as shown in the graph at the cut-off point on the abscissa, or x-axis. The higher the frequency of image information carried by the diffracted rays, the more it is attenuated. This lateral band limit means that no objective passes frequencies higher than $2NA/\lambda$ (the axial band limit $= 2NA^2/n\lambda$). Source: Conchello 2006. Reproduced with permission from Scion Publishing Ltd. For further information on the OTF, see Chapter 21, Section 21.3.*

zero-order plus *one* of the first-order diffracted rays, in which case the resolving power is doubled to: $d = \lambda/2NA$. For those readers interested in the proof of why $d = \lambda/NA$ for parallel illumination, and $d = \lambda/2NA$ for oblique illumination, an illustrated explanation is given on pages 185–188 in Hammond (2009). In the case of a high numerical aperture objective illuminated by a solid full cone of light from a condenser, this is equivalent to multi-lateral oblique illumination from every point in the radial plane. In this case, the resolving power of a correctly illuminated microscope objective illuminated with coherent light is given by: $d = 0.5\lambda/NA$. The term is called the Abbe limit and is represented by the cut-off frequency on an optical transfer function plot (Figure 10.19). The Abbe limit is the smallest spatial frequency in the object that can be transferred into, and discriminated in, the image.

10.10 Apparent Paradox

In the latter half of the 19th century, microscopy in England and continental Europe followed diverging paths. The wealthy English 'gentlemen amateurs' were, for the most part, not trained scientists. They were empiricists for whom theory counted for very little, or not at all. As long as they had the latest equipment and could resolve the fine detail of diatoms in the central field of view, they were content. Generally the objectives that the English microscopists used were dry ones, the numerical aperture not exceeding NA 0.8–0.9. The fashion was to use extremely oblique illumination, and even well-known English microscopists of the day could not appreciate why an immersion objective could apparently have a numerical aperture greater than the angular equivalent of 180° in air.

Conversely, the continental European microscopist was trained in science or medicine and used the microscope as a tool to study histological samples relevant to pathology or cytology. Specific stains and dyes were developed to stain tissues and, prior to 1870, a narrow cone of illumination would have been used solely to provide sufficient contrast to secure an image of *unstained* tissues, regardless of resolving power or absolute image quality. Abbe developed his ex-centring substage diaphragm for this particular purpose before he had entirely elucidated the diffraction theory of image formation.

For many years after Abbe published and presented his theory of image formation he was widely misunderstood, by the English microscopists in particular. This is because when the microscope is adjusted correctly, the condenser is used so that most of the aperture of the objective (at the back focal plane) is filled with light. This is equivalent to using oblique illumination from all directions at once. The English microscopists were well aware of the practical value of using a full cone of illumination, for this had been determined empirically. However, when a full cone of illumination is used, equivalent to omni-directional oblique illumination, the diffraction maxima in the back focal plane of the objective overlap and can no longer be distinguished separately. Therefore, having provided a rational explanation for image formation, Abbe *deliberately* used a narrow cone of illumination specifically to demonstrate the diffraction pattern in the back focal plane of the objective by making it visible. In so doing, many were confused, and therefore Abbe's theory took many years and much argument – some of it very heated – before it gained complete acceptance.

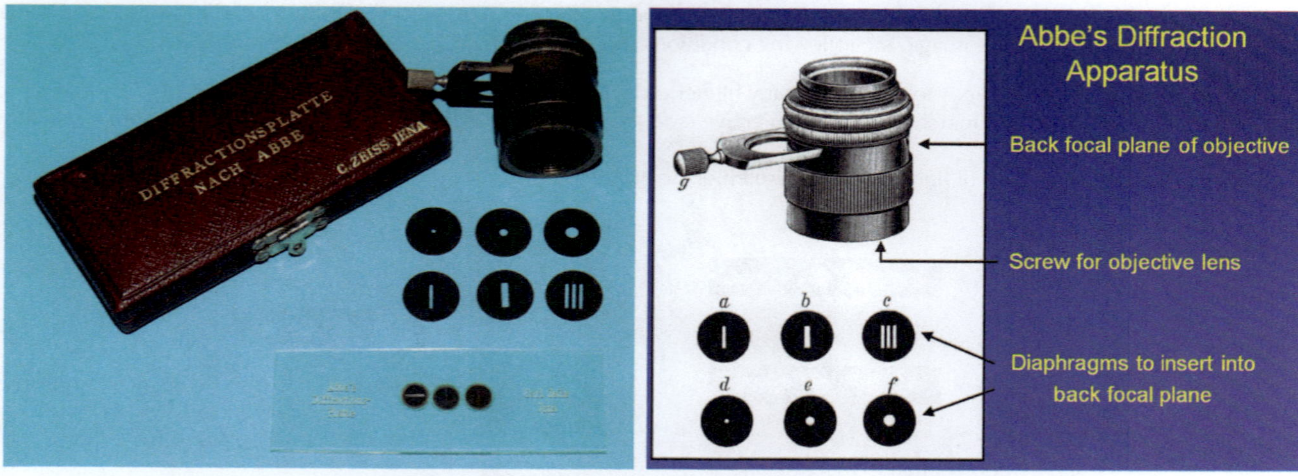

Figure 10.20 **Abbe Diffractionsplatte** *Source: Dr P. Evennett. Reproduced with permission from Dr Evennett.*

10.10.1 Abbe's Diffraction Experiments

In order to convince sceptics of the existence of diffracted light in the back focal plane of the objective and how it contributes to the image formed in the primary image plane of the microscope, Abbe designed and manufactured his 'Diffractionsplatte' (Figure 10.20), a special slide containing fine lines and patterns of dots on an opaque silver background. Clearly, all objects form diffraction patterns but periodic objects possessing a repeating structure (e.g. Abbe's diffraction plate or the frustules of a diatom) form a *regular* diffraction pattern, which is easy to observe and understand.

This diffraction plate was used in conjunction with a holder fitted between the objective lens and microscope stand that allowed diaphragms to be inserted into the back focal plane of the objective. In this way, the diffraction pattern could be selectively altered, and some parts of it prevented from contributing to the image. Using this equipment, Abbe devised a series of elegant and spectacular experiments (Bradbury, 1971; Evennett, 1996, Figure 10.21) to demonstrate and explain image formation in the microscope. Although Abbe wrote very good English, a translation of his work was heavily edited and first published in an obscure journal. This further hindered the acceptance of Abbe's theory together with the English microscopists' insistence on the use of full-cone illumination without understanding the *specific* use of narrow-cone illumination *solely* to show the diffracted orders of light.

Zeiss have not made their diffraction demonstration apparatus for many years. Fortunately Peter Evennett Hon FRMS has devised (Evennett, 1996) a very ingenious demonstration microscope with two CCD cameras imaging the separate ray paths of both independent sets of conjugate planes simultaneously. It is therefore possible to show the diffraction pattern at the back focal plane of the objective and the effect of any alteration of this pattern upon the image at the primary image plane of the microscope. This instrument has been used for many years as a didactic tool on the Royal Microscopical Society's courses to teach the fundamental principles of microscopy. Peter's experiments are now available on YouTube, as five clips, with a total running time of 42:28 minutes, recorded at the MPI-CBG, Dresden. See the URL links at the end of the chapter for these videos.

10.11 Optimising the Image in the Microscope

To achieve maximum resolving power in bright-field microscopy, the numerical aperture of the condenser should theoretically be equivalent to that of the objective. In good practice this is never the case, and the aperture of the condenser is closed *slightly* in order that non-image-forming stray light does not reduce the contrast and hence impede the effective resolution of fine detail in the image. Coherence, or partial coherence, is required for interference of rays of light and image formation. Nearly all 'incoherent' extended sources exhibit some degree of coherence, with rays vibrating in phase with one another, albeit over only short distances.

Closing the condenser diaphragm, thereby reducing the condenser aperture, serves two purposes:

- it reduces, or eliminates, stray non-image-forming light appearing in the image plane; and
- it gives greater contrast in the image by increasing the coherence of the illuminating light.

By selecting a smaller portion of the illuminating source, the phase relationships among diffracted rays become more coherent, or in step with one another. Accordingly, constructive interference of these coherent rays results in overall higher-amplitude differences, thus increasing contrast. With reduced condenser angular aperture, the diffraction patterns become larger. Lines and edges in the object are seen as thicker in the image. The larger diffraction patterns cover a greater number of retinal photoreceptors, thus *initially* improving visibility. Overdoing this process leads to reduced spatial and axial resolution; the thick, dark diffraction edges become objectionable, as does dirt seen in the image above and below the plane of the specimen, due to the increased depth of field. Proper adjustment of the condenser directly affects both spatial and axial resolution in the microscope. For good resolving power in the microscope and resolution of all the fine detail in the image, the following conditions should be met:

- the objective should have a high NA to accept as many higher orders of diffracted light as possible;
- the NA of the condenser should match that of the objective (and then be closed *slightly* to maximise contrast, as explained immediately above); and
- the shortest possible wavelength of light should be used (i.e. violet or blue[13]).

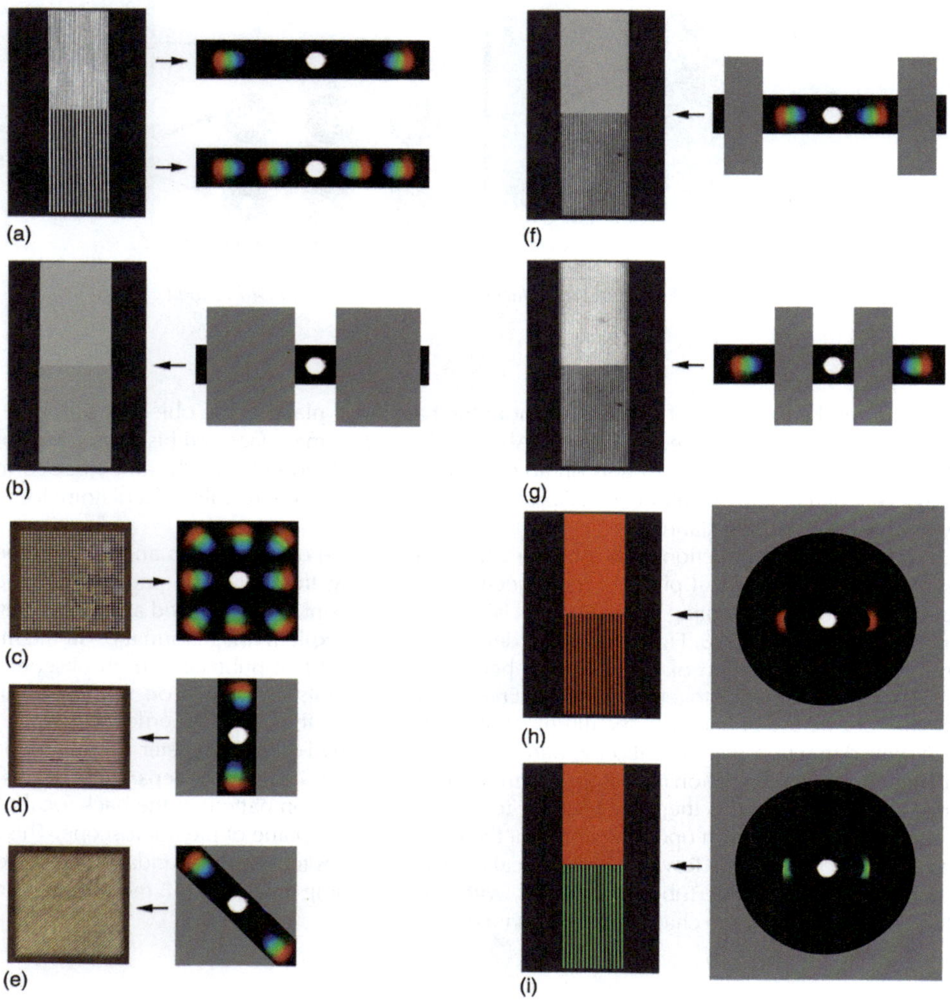

Figure 10.21 **Diffraction experiments** *The Abbe Diffraction Apparatus contains a diffraction plate ruling with two set of periodic lines (a). One set is 8 μm wide (top), the other 16 μm (bottom). The second order of the diffraction pattern of the 'coarse' 16 μm rulings exactly coincides with the first order of the 'fine' 8 μm diffraction pattern, demonstrating the reciprocal relationship between the line spacing and the spacing of the diffracted beams. In (b) these diffracted beams have been occluded – restricting the aperture of the objective – leaving only the zero order illuminating beam, thus demonstrating that diffracted beams must interfere with one another and/or with the zero-order beam to form an image. The zero-order beam is formed by the illumination independently of the object. Panel (c) shows a series of fine circular holes in an opaque film. Because this object has a repeating pattern in two directions, the diffraction pattern similarly repeats. In panels (d) and (e) the diffraction pattern from this square array of holes has been restricted, showing the diffraction pattern arrayed in the same direction as the periodic repeat in the object, but lying at 90° to the orientation formed in the image. In panel (f) the second set of diffracted beams seen in the back focal plane of the objective has been occluded. Since this set corresponds to the first order of the fine 8 μm grating, these lines are not resolved, but the 16 μm lines of the coarser grating are resolved because this first order diffraction pattern can still pass through to form an image. In (g) the first order of the diffraction pattern of the coarse 16 μm ruling is screened, whilst allowing the first order of the fine 8 μm ruling and the second order of the coarse ruling to both pass, giving rise to spurious resolution of detail in the 16 μm coarse grating apparently equal to the fine 8 μm ruling. The appearance of the image depends upon the diffraction pattern from which it is formed. In panel (h) the aperture of the objective has been restricted such that only the spectra of the first set of diffracted beams are seen in the back focal plane. Viewed with a red filter, only the coarse 16 μm grating is resolved from the red portion of the first order diffracted beam. In panel (i) only the coarse 16 μm ruling is seen in the field of view; the 8 μm fine grating has been removed. The aperture of the objective has been further restricted such that only the blue and green spectra of the first diffracted order is seen. Red and green filters have been placed side-by-side in the optical path, each filter across half the field of view. The lines are resolved under shorter wavelength green illumination, but not red. Abbe pointed out that full-cone illumination comprised multiple diffraction orders, all overlapping, so that the apparent so-called 'dark space' between orders – seen here with narrow cone illumination – is no longer visible under normal operation of the microscope. Source: Adapted from Dr P. Evennett with permission from Dr Evennett.*

When the condenser is partially closed, the working numerical aperture for calculating resolution can be calculated from the value of ($NA_{objective} + NA_{condenser}$). As explained in Section 10.7, if the condenser aperture is closed to a very small NA value, this mean term returns an unreasonable result. The influence of the condenser was investigated by Hopkins & Barham, whose rather complex paper was explained by John Baker (Baker, 1952). Although closing the condenser aperture iris control has a significant effect on resolution of fine detail in the image, the influence of the condenser is less than expected. For an objective of NA 1.4 used with a condenser of NA 1.3, if the condenser is closed to NA 0.7, the resolution drops by only 14.5%. Nevertheless, care should be taken to set the condenser correctly for optimum visibility and resolving power.

10.12 Test Objects for the Microscope

Diatoms are unicellular heterokont algae of the kingdom Protista, characterised by their intricate and beautiful silica (hydrated silicon dioxide) shell, or frustule, which is composed of two asymmetric thecae, or valves, held together by silica girdle bands. The role of the frustule is to confer buoyancy and protection.

For the microscopist the architecture of striae (composed of rows of puncta) and puncta seen on the frustules are valuable. Diatoms have long been used as test objects (Box 10.4 and Figure 10.22) for empirically determining the resolving power of the light microscope. The nomogram in Figure 10.23 shows this. Table 10.3 shows the dimensions of the test diatoms – highlighting their natural variability, which makes them unsuitable for *quantifying* resolving power. Nevertheless, individual diatom test slides, or other test objects, are useful for comparing different microscope objectives. The approximate numerical aperture range for which a diatom can be resolved at various wavelengths of light is given both in Figure 10.22 and Table 10.3.

Box 10.4 Test Objects for the Microscope

The very high quality of microscope objective manufacture would suggest that there is no need for test objects, and anything that we use can only be subjective. This is true, but there is still value in using tests for the microscope to check our alignment and use of the instrument. During the 19th century, test objects were very important in causing instrument manufacturers to make continual improvements. They literally pushed forward the technical frontier of microscope development.

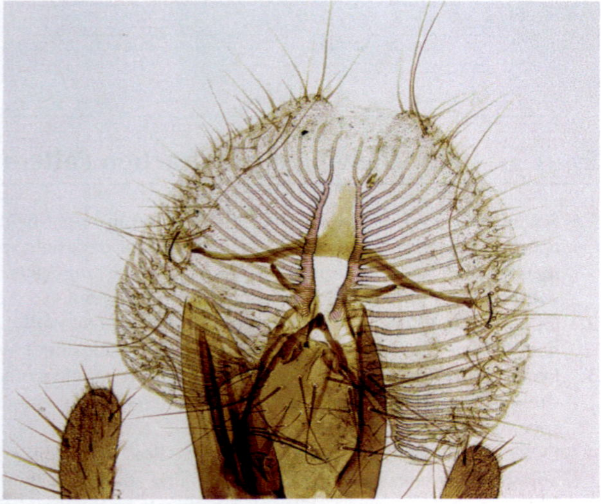

Specimens for measuring the performance of objectives and operator were introduced circa 1824, about the time that improved achromatic objectives became available. The first such 'proofs' were described by Charles Goring. These included the scales of the Large White butterfly, mouse hair, the podura scale of the springtail (*Lepidocyrtus curvicollis*) and the proboscis of the blow fly, illustrated here. All these were natural objects that were easily obtained, displaying fine repeatable details. Goring must have been quite a character, and the purpose of test objects is made clear in the following quotation:

> 'nothing exasperates me more than to hear some half-taught booby bragging of the goodness of a microscope because it shows objects so wonderfully magnified, when, in all probability, it will be found completely inert and ineffective when called upon to show them with a reasonable working power...'

Tiny globules of mercury had initially been used a 'star test' of optical quality; these were replaced in due course by star tests made by depositing a silver or aluminium film onto a measured coverslip, and this is still a useful test if you have access to an EM vacuum sputter coater. Instructions for making a test without the use of such equipment is given in Section 6.11. The star test, Fletcher (1988), shows a diffraction-limited Airy disc, and can reveal the presence of spherical and other aberrations, upon focusing through the image.

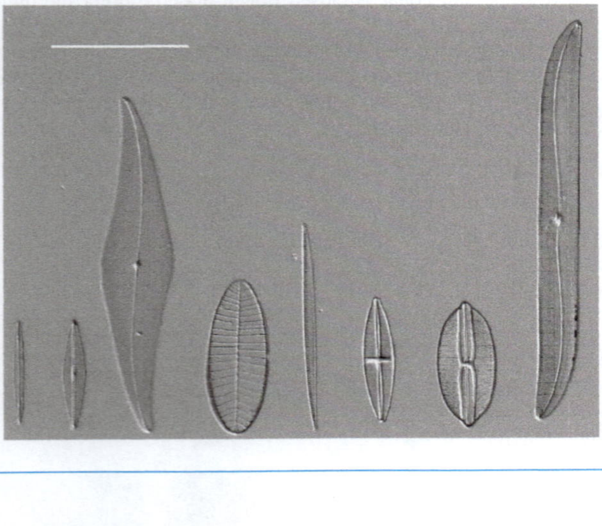

From the middle of the 19th century diatoms became widely used as test objects, and diatom test slides shown here are still sold and used today (see Appendix 5) as subjective tests for microscope performance (see Table 10.3). *Pleurosigma angulatum* and *Amphipleura pellucia* were especially popular. They remain good as tests for medium- and high-magnification objectives; see Sanderson (1990). If you can resolve the striae – let alone the punctae – of *Amphipleura pellucida*, you are doing well.

However the German mechanic Friedrich Nobert realised that despite repeatability of fine detail, organic variation meant that natural test objects have only limited applicability. In the 1840s he set out to rule exceedingly fine lines with a circular dividing engine that used a very finely-cleaved diamond tip to scribe the lines onto a coverslip. Once the noted microscopists of the day, such as the American surgeon J. J. Woodward and the English gentleman Edward Nelson, had resolved one of Nobert's bands, he ruled another, finer, test. Nobert's finest test plate was made in 1873, just before the theoretical work of Abbe on resolving power and image formation was published in English. It featured a final 20th band spacing of 0.11 μm, which cannot be resolved by any light microscope, and so caused great controversy and debate. It was not until an example was examined by electron microscopy that the rulings were conclusively shown to be present. After Nobert, the production of diamond-ruled gratings was continued by Henry Grayson in Australia. The modern equivalent are rulings produced by nanolithography (Oldenbourg et al., 1996).

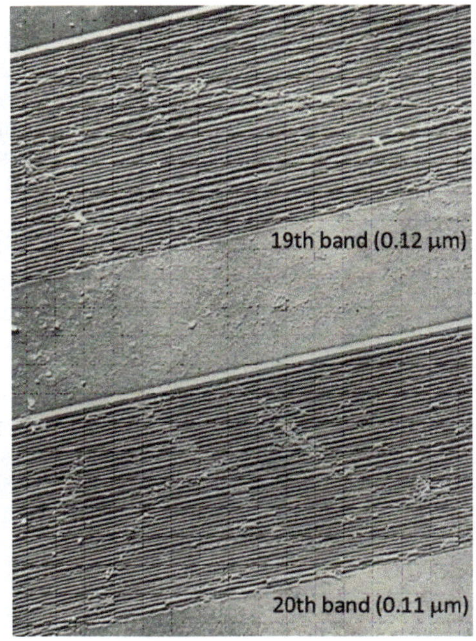

Source: Turner and Bradbury 1966. Reproduced with permission from John Wiley & Sons.

References

Fletcher, J.R. (1988). 'The star test for microscope optics'. *Microscopy* 36/2: 154–159.

Oldenbourg, R. et al. (1996). 'Standard test targets for high resolution light microscopy'. In: *Nanofabrication and Biosystems: Integrating Material Science, Engineering and Biology*, Chapter 8. Cambridge University Press, pp: 123–138.

Sanderson, J.B. (1990). 'The measurement of three light microscope test diatoms by scanning electron microscopy.' *Proceedings of the Royal Microscopical Society* 25/3: 195–203.

Turner, G.L'E. and Bradbury, S. (1966). An electron microscopical examination of Nobert's finest test-plate of twenty bands. *Journal of the Royal Microscopical Society* 85/4: 435–447.

Exercise 10.2 Viewing the Diffraction Pattern and Image of a Diatom

1. Set up the microscope for Köhler illumination in bright-field mode, using a well-stained specimen on a slide, with a variable numerical aperture objective (i.e. an objective designed for dark-field with a variable numerical aperture adjustment collar). Ideally this will be a 40x magnification objective with a numerical aperture ranging from a minimum of NA 0.5 to a maximum of NA 0.95 or 1. The objective supplied with the Zeiss diffraction demonstration apparatus is a 40x Apo/NA 0.5–1.0 oil-immersion objective.
2. Open the aperture adjustment collar on the objective fully. Ensure that, although the objective is operating at full numerical aperture, a narrow cone of coherent illumination is used, by closing down the condenser diaphragm to give a very high contrast image.
3. Exchange the stained slide for an unstained preparation of strewn diatoms. (These can be bought from Klaus Kemp; see Appendix 5). A strew of *Pleurosigma angulatum* is ideal. The reduced condenser numerical aperture should permit focussing of this otherwise faintly visible specimen.
4. Determine which is your dominant eye (see Box 9.5). Remove the eyepiece from the other binocular tube. If available, insert a 'phase-contrast adjustment' telescope, and focus the limiting aperture at the back focal plane of the objective sharply.
5. Close down the aperture collar on the objective. The bright-field illumination intensity may have to be adjusted, but you should be able to see (with *Pleurosigma angulatum*) a hexagonal diffraction pattern (Figure 10.22).
6. Observe the image (through the remaining eyepiece) simultaneously with the back focal plane of the objective. Notice that as the objective aperture collar is closed down, at some point the diffraction pattern is no longer seen, and resolution of the detail in the image is lost.
7. At a point just before the first-order blue diffracted ray is excluded in the back focal plane, the contrast of the image appears very faint as the cut-off for resolution of the image approaches. With a wide-open aperture, a greater number of higher-order diffracted rays contribute to the image, and contrast is better, along with the re-appearance of the resolved fine detail. This exercise demonstrates that contrast and resolving power are interlinked.

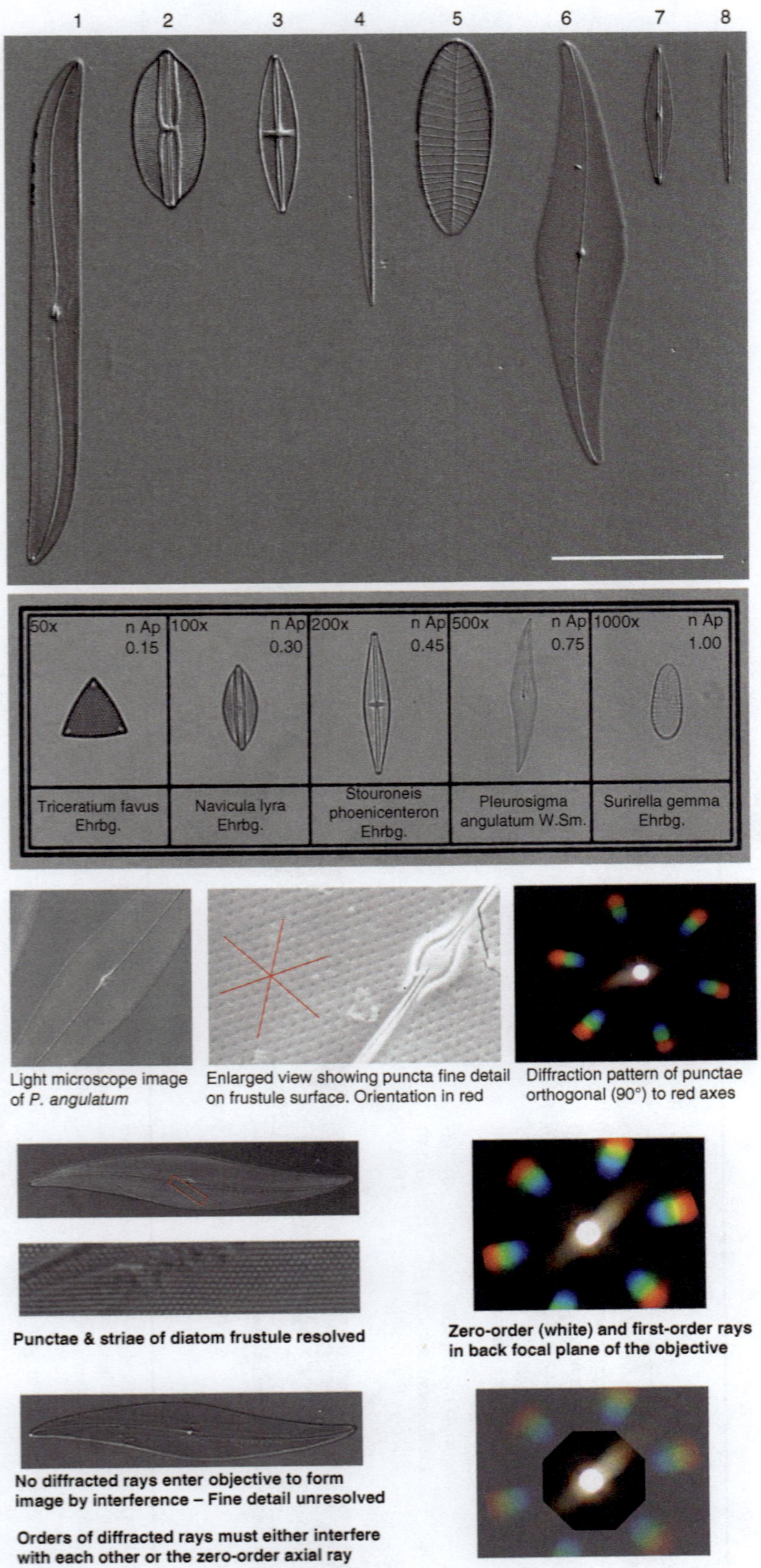

Figure 10.22 ***The diatom test object*** *These numbered diatoms are from a test plate of eight forms, explained in Table 10.3. The scale bar is 100 mm. An older Wild-Heerbrugg diatom slide is shown underneath. The diatom* Pleurosigma angulatum *(No. 6) is shown together with its characteristic hexagonal diffraction pattern, together with resolved and unresolved images. Source: Dr P. Evennett. Reproduced with permission from Dr Evennett.*

Table 10.3 *Typical Values for Test-Plate Diatoms*

No. in Figure 10.22	Species	Length µm	Breadth µm	Striae (per 10 µm)	µm/striae	Puncta (per 10 µm)	µm/puncta	Minimum NA to resolve striae
2.	*Lyrella (Navicula) lyra*	60–160	28–40	8–10	1.25–1	too coarse	–	NA 0.3
1.	*Gyrosigma balticum*	280–310	35–42	12–14	0.83–0.71	12–14	0.83–0.71	NA 0.5
3.	*Stauroneis phoenocenteron*	70–360	16–53	12–20	0.83–0.5	15–30	0.67–0.33	NA 0.7
6.	*Pleurosigma angulatum*	150–360	36–60	16–22	0.63–0.45	16–22	0.63–0.45	NA 0.75
4.	*Nitzschia sigma*	60–285	8–15	22–23	0.45–0.43	32–38	0.31–0.26	NA 0.8
5.	*Petrodictyon (Surirella) gemma*	70–140	30–55	18–28	0.56–0.36	25–30	0.40–0.33	NA 0.95
7.	*Frustulia rhomboides*	40–100	10–30	37–42	0.27–0.29	40–42	0.25–0.24	NA 1.2
8.	*Amphipleura pellucida*	80–140	7–9	37–40	0.27–0.25	53–55	0.19–0.18	NA 1.30
	Nitzschia firthii (singalensis)	180–380	9–11	38–46	0.26–0.22	too fine	–	NA 1.40

Resolved detail

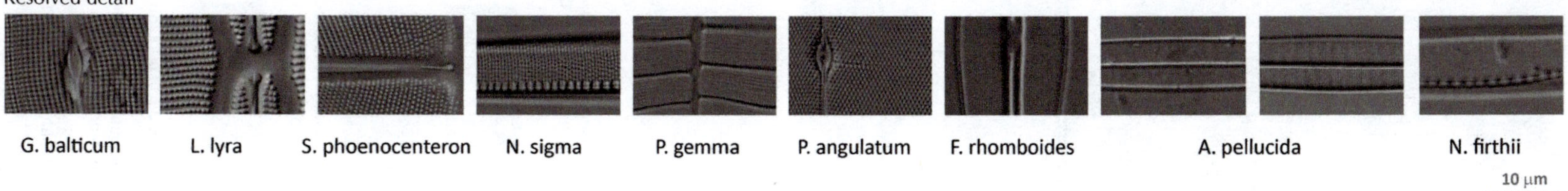

The finest diatoms (e.g. *A. pellucida* and *N. firthii*) will require particularly clean objectives, good microscope alignment, a green filter (to minimise aberrations) and oblique illumination to secure resolved detail with sufficient contrast. If using polarised light, even the striae of *A. pellucida* will just be resolved with an NA 1.2 water-immersion or NA 1.4 oil-immersion objective. With care it is possible to resolve the striae of *A. pellucida* into puncta. Whilst the same diatom preparation is useful for comparing different microscopes, the natural variation inherent in diatom preparations precludes their use for quantitative measurement of resolving power or numerical aperture of the objective. There can be about 15% variation between frustules taken from a single population. However, individual diatoms on a personal test plate are useful for performing comparative evaluations between objectives. The images above were taken with a 100x/NA 1.30 oil-immersion semi-apochromat objective using DIC.

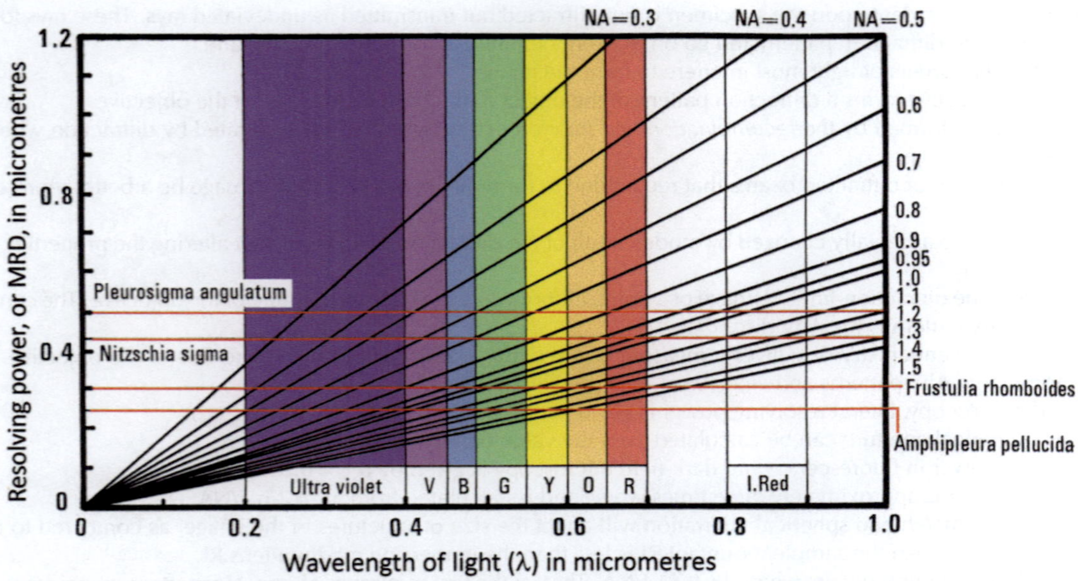

Figure 10.23 ***Diatom resolving power nomogram*** *The nomogram does not take into account individual diatom variability but is included here to show that resolving power achieved may in practice not equate to theoretical resolving power. Read off the actual resolving power on the ordinate (y-axis) from the resolved diatom features on the abscissa (x-axis). MRD stands for minimum resolved distance, an alternative expression for the limit of resolving power. Source: Bingley and Firth 1975. Reproduced with permission from Design Council.*

10.13 Chapter Summary

In this chapter you will have learnt how diffraction and the interference of light form an image in the microscope and how the greater the numerical aperture of the objective, more fine detail in the object is resolved in the image. The next chapter explains how to generate contrast in the image and how contrast and resolving power are inherently interrelated – as the Stelzer (1998) Key reading paper for Chapter 11 explains.

1. Diffraction, the very process by which the image is formed when light interacts with the object, inevitably sets a limit to image fidelity.
2. If the opening, slit or gap in the object is large in comparison with the wavelength, the waves only spread out a little. As the size of the gap decreases, for the same wavelength the diffraction of the wave increases. As the wavelength increases, then for the same-sized gap the diffraction of the wave also increases.
3. The diffraction of light sets a limit on the resolving power of any optical instrument using a lens.
4. Young's double-slit experiment demonstrates the interference of coherent waves of light.
5. Planck's constant, *h*, links both the wave-like and particle-like nature of light.
6. Wave trains (photons) are of distinct length and travel like a group of ripples from the light source.
7. Light from a point source has all its photons in phase (and of the same frequency) so is *coherent*. A point source produces a diffracted spherical wave.
8. Huygens showed that spherical waves of very large radius can be approximated by plane waves aligned transverse to the direction of propagation.
9. Coherent light can only be produced by a point source, a very small part of an extended source or a laser.
10. Light is in phase when the peaks and troughs of two or more rays or waves of light are in a *constant* relationship to one another over space and time.
11. Light from an extended source is incoherent, because the relative phases of the photons change randomly with time.
12. Interference occurs when two coherent waves that are in phase superimpose to form a new wave of greater or lesser amplitude. This phenomenon is called the superposition of waves.
13. Separate, incoherent waves do not interfere.
14. The repeat spacing of the diffraction pattern is inversely proportional to the spacing of the regular repeat in the object.
15. The diffraction pattern lies in the same direction as the repeat of the periodic structure in the object and perpendicular to the features in the image.
16. Longer wavelength light is diffracted through larger angles than short wavelength light.
17. Objects are still resolved when illuminated by shorter wavelength light instead of longer wavelengths. Through a fixed opening just excluding red light, green light will still resolve detail.
18. Resolution in the image is dependent upon (a) the numerical aperture of the objective and (b) the wavelength of the illuminating light. Blue light will resolve more details than red light. Electrons will resolve better than photons.

19. A proportion of light incident upon the specimen is not diffracted but transmitted as undeviated rays. These rays form the central zero-order spot in the diffraction pattern and go on to evenly illuminate the entire image plane.
20. At least two different orders of light must interfere to form the image.
21. The microscope objective forms a diffraction pattern of the object in the back focal plane of the objective.
22. The primary image is formed by the *recombination and interference* of light that was separated by diffraction when it interacted with the specimen or object.
23. The greater the number of diffracted beams that recombine to form the image, so will the image be a better representation of the original specimen.
24. The final image can be artificially changed by modification of the diffraction pattern without altering the properties or structure of the object in any way.
25. The Airy pattern is the diffraction-limited image of a point, as formed by the circular aperture of the objective. The central diffraction spot within the Airy pattern is the Airy disc.
26. The diameter of the central Airy disc gives a measure of the diffraction occurring at the aperture of the objective. Low aperture objectives form large Airy patterns and vice versa.
27. In bright-field microscopy, lateral resolving power is given by: $d = 1.22\lambda/(NA_{objective} + NA_{condenser})$.
28. The working numerical aperture can be calculated from the value of $(NA_{objective} + NA_{condenser})$.
29. Lateral resolving power in fluorescence and dark-field microscopy is given by $d_{x,y} = 0.61\lambda/NA$.
30. Axial resolving power is approximately three times worse and is calculated from: $d_z = 2.n.\lambda/NA^2$.
31. Refractive index mismatch and spherical aberration will affect the size of structures in the image, as compared to their true size. Images are elongated when the sample/mountant RI is less than the immersion medium/lens RI.
32. Rayleigh's limit to resolution occurs when $d = 0.61\lambda/NA$. That is: the first minimum of one Airy pattern coincides with the central peak of the bright Airy disc of the second pattern; incoherence of point sources is assumed.
33. Sparrow's resolution criterion defines resolution when $d = 0.47\lambda/NA$ – when the central minimum between the two Airy pattern intensity distributions is zero.
34. The Abbe limit of resolving power occurs when $d = 0.5\lambda/NA$. It defines the smallest spatial frequency in the object that can be transferred into, and discriminated in, the image.

Key Reading

Evennett, P.J. (1996). 'The Ernst Abbe lecture: Ernst Abbe and the development of the modern microscope'. *Proceedings of the Royal Microscopical Society* 31/4: 283–292.

- An excellent demonstration of the principle of image formation in the microscope using the diffraction equipment that Abbe had made to show this. This lecture should be read in conjunction with the online YouTube videos whose URL links are given below.

References

Avison, J.H. (1989). *The World of Physics*, 2nd ed. Thos. Nelson, Cheltenham.

Baker, J.R. (1952). 'Remarks on the effect of the aperture of the condenser on resolution by the microscope'. *Quarterly Journal of Microscopical Science 93*/4: 375–377.

Bingley, M.S. and Firth, R.I. (1975). 'Tests for a microscope'. *Engineering 215*/1: 33–35.

Bradbury, S. (1971). 'The use of the Abbe diffraction plate in the teaching of microscopy'. *Proceedings of the Royal Microscopical Society 6*/5: 181–189.

Bradbury, S. (1984). 'Some basic concepts'. In: *An introduction to the Optical Microscope*, 1st ed, Chapter 3. Handbook No. 1. Oxford University Press.

Bradbury, S. and Bracegirdle, B. (1998). *Introduction to Light Microscopy*, 2nd ed. Handbook No. 42. Oxford: Bios, Taylor & Francis, Wiley.

Bradbury, S. and Evennett, P.J. (1996). *Contrast Techniques in Light Microscopy*, Handbook No. 34. Oxford: Bios, Taylor & Francis, Wiley.

Cagnet, M., Francon, M., and Thierr, J.C. (1962). *Atlas of Optical Phenomena*. Berlin: Springer-Verlag.

Conchello, J.-A. (2006). An overview of three-dimensional and four-dimensional microscopy by computational deconvolution. In: *Cell Imaging*. David Stephens; Methods Express Series, Chapter 9. Banbury: Scion Publishing, pp. 177–199.

England, N. (1995). *Physics Matters*, 2nd ed. Basingstoke: Hodder Education.

Franklin, K., Muir, P., Scott, T., Wilcocks, L., and Yates, P. (2010). *Introduction to Biological Physics for the Health and Life Sciences*. Chichester: Wiley.

Hammond, C. (2009). 'The diffraction of light'. In: *The Basics of Crystallography and Diffraction*, 3rd ed, Chapter 7. Oxford: Oxford Science Publications, pp. 165–191.

Hibbs, A.R., MacDonald, G., and Garsha, K. (2006). 'Practical confocal microscopy'. In: Pawley, J.B. (ed.), *Handbook of Biological Confocal Microscopy*, 3rd ed, Chapter 36. New York: Springer.

Kubitscheck, U. (2013). *Fluorescence Microscopy: From Principles to Biological Applications*. Wiley Blackwell.

Murphy, D. (2001). *Fundamentals of Light Microscopy and Digital Imaging*, 1st ed. Wiley-Liss.

Ohanian, H.C. (1994). *Principles of Physics*. New York, London: WW Norton & Co.

Pawley, J.B. (2006). *Handbook of Biological Confocal Microscopy*, 3rd ed. New York: Springer.

Pedrotti, L.S. (2004). 'Basic geometrical optics'. In the Fundamentals of Photonics series, SPIE. https://spie.org/Documents/Publications/00%20STEP%20Module%2004.pdf.

Pople, S. (1987). *Explaining Physics GCSE Edition*, 2nd ed. Oxford: Oxford University Press.

Robinson, P.C. and Bradbury, S. (1992). *Qualitative Polarised-Light Microscopy*. Microscopy Handbook No. 9. Oxford: Oxford Science Publications.

Saxby, G. (2011). *The Science of Imaging Diffraction*, 2nd ed. CRC Press – Taylor & Francis imprint.

Serway, R.A. and Faughn, J.S. (1989). *College Physics*, 2nd ed. Orlando, Florida: Saunders College Publishing, Holt Reinhart and Winston.
Spencer, M. (1982). *Fundamentals of Light Microscopy*, IUPAB Biophysics Series. Cambridge University Press.
Taylor, C.A. (1978). *Images*. London: Wykeham Publications Ltd – Taylor & Francis imprint.
Tilley, R.J.D. (1996). *Colour*, 2nd ed. Wiley.
Trefil, J. and Hazen, R.M. (2004). *Physics Matters*. John Wiley.

Further Reading

Abbe, E. (1873). *'Beiträge zur Theorie des Mikroskops und der mikroskopischen Wahrnehmung'* [Contributions to the Theory of the Microscope and of Microscopic Perception]. *Archiv für Mikroskopische Anatomie 9*/1: 413–468.
Andriesse, C.D. (2005). Miedema, S. (transl.) *Huygens: The man behind the principle*. Cambridge University Press.
Den Dekker, A.J. and van den Bos, A. (1997). 'Resolution: a survey'. *Journal of the Optical Society of America 14*/3: 547–557.
Galbraith, W. and Sanderson, R.J. (1980). 'The energy distribution about the image of a point'. *Microscopica Acta 83*/5: 395–402.
Gerth, K. and Wimmer, W. (2005). *Ernst Abbe: scientist, entrepreneur and social reformer*. Jena: Bussert & Stadeler.
Günther, N. (1951). 'Ernst Abbe: Creator of the Zeiss Foundation', translated by Bower, D. I. (2016).
Hopkins, H.H. and Barham, P.M. (1950). 'The influence of the condenser on microscope resolution'. *Proceedings of the Physical Society 63*/10: 737–744.
Kasprowicz, R. et al. (2017). Characterising live cell behaviour: Traditional label-free and quantitative phase imaging approaches. *The International Journal of Biochemistry & Cell Biology 84*: 89–95.
Lacey, A.J. (1968). Some illustrative material for students of microscopy'. *Proceedings of the Royal Microscopical Society 3*/4: 176–185.
Lauterbach, M.A. (2012). 'Finding, defining and breaking the diffraction barrier in microscopy—a historical perspective'. *Optical Nanoscopy 1*: 8 doi: 10.1186/2192-2853-1-8.
Mollon, J.D. (2002). 'The origins of the concept of interference'. *Philosophical Transactions of the Royal Society A 360*: 807–819.
Smith, W.J. (2000). *Modern Optical Engineering*, 3rd ed. New York: McGraw-Hill.
Walker, D. (2008). 'Diatom *Pleurosigma angulatum*—a versatile "demonstration" subject'. http://www.microscopy-uk.org.uk/mag//artsep08/dw-pa-test.html.
Zeiss Innovation 15 (2005). *Issue in memory of Ernst Abbe*. Different parts online, e.g.: http://www.science.marshall.edu/dneff/nobel_etc/earnst%20abbe.pdf.

Internet Resources

James Gregory and diffraction: https://books.google.com.sg/books?id=0h45L_66bcYC&pg=PA254&dq=feather+ovals&ie=ISO-8859-1&output=html&redir_esc=y.
Diffraction and interference phenomena can be seen online in animated form at many websites, for example:
  Wave diffraction: https://www.youtube.com/watch?v=ipl18m5AxqI.
  Water ripple tank: https://www.youtube.com/watch?v=J_xd9hUZ2AY and https://www.youtube.com/watch?v=-8a61G8Hvi0.
  Fraunhofer single slit diffraction: http://hyperphysics.phy-astr.gsu.edu/hbase/phyopt/sinslit.html.
  Interference at two or more slits: http://hyperphysics.phy-astr.gsu.edu/hbase/phyopt/mulslidi.html#c2.
Young's diffraction at two slits: https://www.youtube.com/watch?v=CN-wjj_phVA
Pedrotti, L.S. (2005). *Basic Physical Optics*, Module 1.4 SPIE Fundamentals of Photonics Series. http://spie.org/publications/fundamentals-of-photonics-modules (series) and http://spie.org/Documents/Publications/00%20STEP%20Module%2004.pdf (*Basic Physical Optics*).
Airy disc calculator (iscope): http://iscopecalc.com (see Menu for About details).
Diatom preparations: http://www.diatoms.co.uk/.

Online YouTube Demonstration of the Theory of Microscope Image Formation

Part 1 http://www.youtube.com/watch?v=fAuo7NIS97U	9:37 mins
Part 2 http://www.youtube.com/watch?v=rPAJ2Vs9A_I	9:39 mins
Part 3 http://www.youtube.com/watch?v=CbPoi9bie1I	9:39 mins
Part 4 http://www.youtube.com/watch?v=wZ369WKiJdQ	9:44 mins
Part 5 http://www.youtube.com/watch?v=t6RO7fusHYk	3:49 mins

Notes

1. A very small pinhole is the means by which Isaac Newton and Thomas Young produced coherent light sources for their respective experiments to demonstrate dispersion and interference of light.
2. When focusing beams of light from the sun, this extended source effectively acts as a point source, because the light arrives from infinity as *parallel* waves and so *can* be focused to a point.
3. The superposition principle can be defined as follows: 'When two or more waves move simultaneously through a region of space, each wave proceeds independently as if the other were not present. The resulting wave displacement at any point and time is the vector sum of the displacements of the individual waves'. Quote taken from: Pedrotti, L.S. (2005). *Basic Physical Optics*, Module 1.4. See the Internet link under Internet Resources.
4. This is how noise-cancelling headphones and ANR headsets, used by pilots and armoured vehicle drivers, work. A similar phenomenon is that of standing waves—the superposition of two waves of the same amplitude and frequency but opposite directions of propagation. A standing wave (or stationary wave) can occur when a rope fixed at one end to a wall is vibrated by an observer. A wave pulse sent down the rope is reflected from the fixed point and interferes with incident waves from the source. The standing wave pattern is so-called because specific points, the nodes and antinodes, appear not to move along the rope but to be standing still.
5. Very narrow linear objects (e.g. a hair or a fine wire) or small circular punctuate objects will form the same diffraction patterns as those formed by a slit or hole of the same dimensions. Where the aperture (whether a hole or a slit) is wider and will act as the source of several secondary wavelets, this modifies the intensities but does not change the positions of the diffracted spots.
6. The diffraction image is a pattern that is composed of alternating dark regions (minima) and bright areas (maxima). The maxima are counted as orders starting from the brightest central maximum, which represents the non-diffracted illuminating light.
7. Although the classic experiment is universally referred to as the 'double-slit experiment', in fact Young initially did the reverse: he used a thin card to divide a *single* narrow beam into two coherent beams, each of which diffracted around the edge of the card towards the central axis and also towards the other beam.
8. Diffraction may be classified as either Fresnel or Fraunhofer diffraction. It is Fraunhofer diffraction that applies to imaging in the microscope and the telescope. The incoming light is parallel and is focused to a diffraction pattern by the objective and the image plane is a large (i.e. optically infinite) distance from the diffracting object. Sometimes Fraunhofer diffraction is termed 'far-field' diffraction while Fresnel diffraction is referred to as 'near-field'. Fresnel diffraction tends to apply to diffraction at opaque edges or round small objects, such as a pin head, where the diffraction pattern is very close to the object.
9. The spectral colours that we see in light scattered from the narrowly spaced rows of 'blips' in a CD-ROM also arise from diffraction and interference. The wavelengths of these diffracted spectral colours was accurately calculated by Thomas Young.
10. A diffraction image is also seen at every conjugate plane of the microscope in the aperture set of conjugate planes (e.g. also at the exit pupil, or Ramsden disc, of the microscope), not only the back focal plane of the objective.
11. The Bessel function of the first kind is denoted in textbooks of optics by the term J_1. You can find the derivation in Hecht. (2002). *Optics* 4th ed. Pearson, ISBN 978-080-538566-3 in Chapter 10, section 10.2.5 on page 467 onwards. An abbreviated form in shown in Wayne (2009). *Light & Video Microscopy*. Elsevier, ISBN 978-0-12-374234-6 in Chapter 3, pages 48–49. The constant 0.61 in the equation for the Rayleigh distance, r, is 0.6098 to four decimal places.
12. Ernst Abbe was a brilliant physicist and optician employed by Carl Zeiss. His work is absolutely fundamental to light microscopy (see Box 7.2). Some texts erroneously print his surname with an acute accent on the terminal e, thus: Abbé. I suggest you shun these.
13. Because of the reduced sensitivity of the eye to blue light and also the reduced transmittivity of glass to UV and near-UV light, it is practicable to use blue-green or green filtered light. This gives the best compromise between contrast and resolution of the image. Also, all classes of microscope objectives are best corrected for aberrations in the green region of the spectrum.

11

Contrast Generation and Enhancement

Outline

Understanding Light Microscopy, First Edition. Jeremy Sanderson.

11.1 Introduction

In Chapter 4 we noted that *resolving power, contrast* and *magnification* are the three most important issues to consider in light microscopy. Of these, resolving power and contrast are linked: sufficient contrast is required to resolve detail – there is no point in having the best resolving power possible if the features in the image cannot be made to stand out from the background. Secondly, as the limit of resolving power is approached, it is harder to distinguish finely resolved details, as contrast in the image falls off. Equally, under the right conditions of high contrast (normally using dark-field or fluorescence microscopy) structures can be seen below the resolution limit of the light microscope yet not be resolved per se. Their true size cannot be determined: they are merely represented as points of constant size – the point spread function (PSF) of the microscope (see, for example, Figures 10.15, 21.1 and 21.10).

Both the resolving power of the objective (through the NA value) and the magnification are determined and fixed by the manufacturer in the design of the objective. However, the manner in which light interacts with the specimen – and contrast is generated, or enhanced – is very much the preserve of you, the microscopist. It is worth remembering that some contrast enhancement methods, such as dark-field and its close relative Rheinberg illumination, were devised by non-professional microscopists and very good work continues to be done by such 'amateur microscopists' today. (For example, have a look at the Micscape website, listed under Internet Resources at the end of the chapter.) Indeed, how you introduce contrast into your image is really only limited by your own ingenuity. It is wise, though, first to adjust the microscope and set up Köhler illumination for all forms of contrast enhancement and not just for bright-field examination.

11.2 Perceiving Contrast

Our eyes can only sense changes in the energy of the light wave; that is, we can only see changes in the amplitude (intensity) or a difference in the wavelength (colour) of light. We are unable to see changes in the phase or polarisation vector of light. This applies not only to our eyes but also to digital or video camera sensors and analogue film or TV cameras: all are responsive only to the intensity and/or colour of light. In order to enhance contrast in the image, any phase or polarisation change must first be converted to one of either intensity and/or colour in order to usefully contribute to the image and for us to see it.

Contrast, therefore, describes the degree to which the object is separated from its background arising from local changes in colour or intensity in the image, in order for the necessary detail to be sufficiently resolved and made visible. In bright light, the difference between the object and its background may be as little as 2% for the eye to just discriminate detail in the image (e.g. the figure in Box 11.1), but under poor lighting conditions, or where the resolving power or magnification are insufficient, this figure may rise to 5% or more. Figure 2.15 shows this lack of contrast related to refractive index. The plastic tubing and glass bead are immersed in water and glycerol, respectively. Where the refractive index of the surrounding fluid more closely matches the glass bead and plastic tubing, there is less contrast, and they become less visible.

Box 11.1 Measuring Contrast, Bit Depth and Signal-to-Noise Ratios

A simple definition of contrast, $C = [I_o - I_b] / [I_b]$, where I_b = average intensity of the background, and I_o = intensity of the object. This may also be written $\Delta I/I$. The contrast ratio refers to the ratio of intensities between the most intense (brightest) and least intense (darkest) elements of a scene, an image, or a display. A threshold limit (the Weber fraction) for the eye is approximately 2%. That is, most people can *just* see features in an image distinctly from the background when the difference between them is 2%.

The highest contrast image, which is easiest to see, is a pair or series of black bars on a white background. However, the figure shows a sine wave (which is more realistic, rather than square-topped bars) of increasingly higher frequency and also at different percentage contrast values. The contrast transfer function (CTF) is a graphical display of the ability of the optical system to transmit information from the object into the image, and is a plot of how percentage contrast varies with spatial frequency.

The distance at which contrast becomes zero is the cut-off frequency, measured in lines/mm, or cycles/mm.

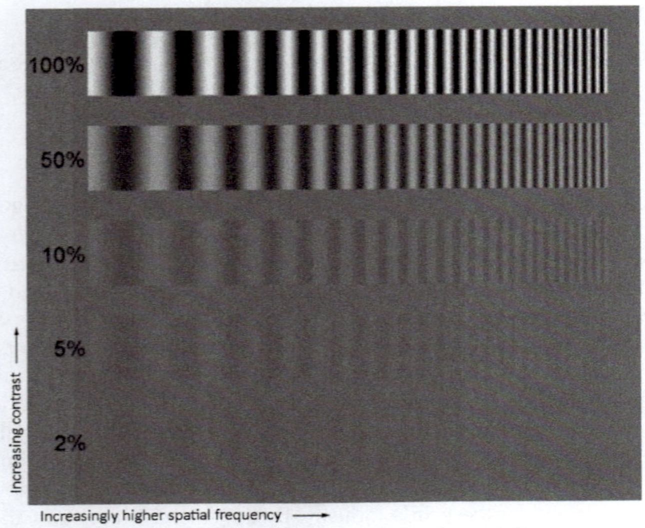

Source: N. Koren. Reproduced with permission from N. Koren.

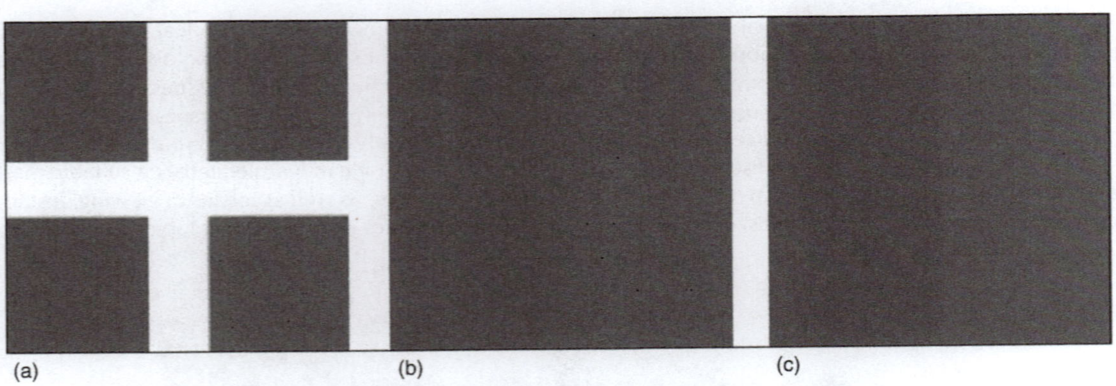

Comparison of grey regions with a 5% brightness difference. Panel (a) separated; the darker pair are on the left. Panel (b) adjacent, but with a gradual tonal charge. Panel (c) adjacent with an abrupt boundary. See http://www.drjohnruss.com/downloads/seeing.pdf, page 8
Source: Gatenby 1919.

The number of binary data bits used to record the intensity (tonal range) and colour of each pixel is described by the bit depth. The human eye can see around 50–60 different grey levels and can only discriminate them with an abrupt boundary. In the figure in (a), the four squares exhibit a 5% difference in tone, which is only seen at a boundary (c); the gradual change in the centre (b) is less obvious. Therefore to achieve the illusion of continuous tone requires 256 levels of brightness from black (0) to white (255). For digital images this dynamic range must be encoded in binary (which computers operate in). Thus the bit depth of the eye is approximately 2^6, because a just perceptible difference of 2% ⇒ a dynamic range of 50, and $\log_2 50 = 5.644$, so our eyes are approximately 6-bit devices. Since $256 = 2^8$, a dynamic range of 256 grey levels is referred to as 8-bit depth. Similarly 12-bit depth encodes 4096 grey levels, and 16-bit grey levels 65 536. A tricolour (red, green, blue, or RGB) image is encoded by 3x 8-bit, often called 24-bit, which can theoretically encode 16 777 216 colours.

However, no imaging system is perfect, and noise from various sources, which obscures the signal, is always present. The dynamic range achievable is also affected by noise. The signal-to-noise ratio (SNR or S/N) is used to quantify how much a signal has been corrupted by noise. SNR and dynamic range are closely related; dynamic range measures the difference between the maximum and the minimum discernible signal, which for most purposes is the noise level. SNR measures the ratio between an arbitrary signal level (not necessarily the most powerful signal possible) and noise. Measuring signal-to-noise ratios requires the selection of a representative or *reference* signal.

$$\mathrm{SNR_{dB}} = 10\log_{10}\left(\frac{A_{\mathrm{signal}}}{A_{\mathrm{noise}}}\right)^2 = 20\log_{10}\left(\frac{A_{\mathrm{signal}}}{A_{\mathrm{noise}}}\right).$$

Because many signals have a very wide dynamic range, SNRs are sometimes expressed using the decibel scale (decibel = ten times the logarithm to base 10; one decibel being a tenth of a bell). As an *amplitude* ratio, this is expressed as 20 $\log_{10}$ (S/N), and thus decibels are conventionally expressed as units of 20 $\log_{10}$ with video equipment or in acoustics for sound pressure. Therefore, a camera signal-to-noise ratio of 60 dB represents a power ratio of 1000:1 between signal power and noise power, not 1 000 000:1. Each binary order of magnitude increases by 20 $\log_{10}(2^n) \approx 6.02.n$. Thus 8-bit = 48 dB, and 16-bit = 96 dB. Light intensity is measured as a power, not amplitude ratio, and therefore photographers use the $\log_2$ (or *f*-stop) scale to express the light-gathering power of cameras, instead of the 10 $\log_{10}$ decibel scale.

Reference

Stelzer, E.H.K. (1998). 'Contrast, resolution, pixelation, dynamic range and signal-to-noise ratio: fundamental limits to resolution in fluorescence light microscopy'. *Journal of Microscopy 189/1*: 15–24.

A high degree of contrast can be seen in the natural world (Figure 11.1). The brightly coloured blue plumage of the male Splendid Fairywren is a clear breeding signal for mates. This is an example of sexual dimorphism, for at other times the male, like the female, assumes a grey-brown colour to blend in with its natural surroundings. Such a lack of contrast (i.e. camouflage) is widespread in nature. However, in both materials and biological microscopy, an inherent lack of contrast in the specimen can present a problem for successfully viewing fine detail. Stratagems for successfully observing opaque and often poorly reflecting samples by reflected light is covered in the next chapter, while this chapter discusses means of contrast enhancement principally by transmitted light.

In its native state, or excised from the body, most biological tissue is opaque or semi-translucent because it scatters light. One means of making the tissue visible is by staining the object, thereby colouring it. Light across the visible spectrum is then differentially absorbed and transmitted, or otherwise reflected off the surface of the object, by the component cells and tissue structures so making the specimen visible. It is only comparatively recently that the importance of limiting the exposure of live cells and tissues to light has been realised (see Chapter 28, Section 28.6). Hitherto, biologists addressed the lack of contrast in virtually invisible tissue by staining it with dyes first developed for the textile industry. By the early 20th century, the instinctive action of a biologist was to kill and fix any living tissue as an essential preliminary to staining sections in order to study the structure and – hopefully, with luck – deduce the function of the organ or organism.

Staining is still important for contrast enhancement – histologists employ a wide range of both empirical and rational histochemical stains in order to selectively impart colour to biopsy or post-mortem tissues for diagnosis. However, biologists now use methods such as phase contrast or differential interference contrast (DIC) to enhance the contrast of living cells and tissues without external alteration by fixatives or stains. Where it is necessary to use stains to selectively and sensitively mark cells, tissues and organelles of interest, fluorescence is widely used. Not only are images of properly prepared fluorescently stained specimens (Figure 11.2) of very high contrast, but the fluorescent stain can often be used in such a low concentration that it can be tolerated as a suitable marker by living cells and tissues. The action of so-called classical (i.e. non-fluorescent) stains and dyes, as well as means of viewing unstained cells and tissues, are covered in this chapter. Using fluorescent markers to stain cells and tissues is an important topic in itself and so is covered in Chapter 15 and onwards.

Figure 11.1 **Contrast and colour in the natural world** *Source: By butupa (IMGP2140) Reproduced under the terms of the Creative Commons Attribution licence CC BY 2.0 (http://creativecommons.org/licenses/by/2.0), via Flickr.*

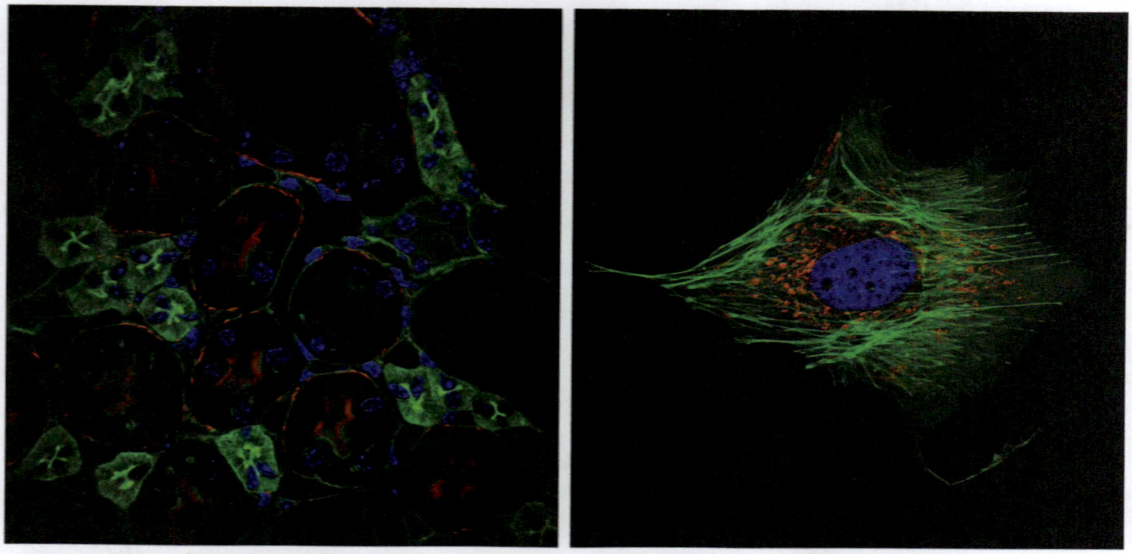

Figure 11.2 **Fluorescently stained specimens** *A section of mouse kidney (left-hand panel) stained with DAPI nuclear stain, Alexa Fluor 488 wheat germ agglutinin targeting the glomeruli and epithelial cells in the proximal tubules. Alexa Fluor 568 phalloidin marks actin in the glomeruli and the brush border of the tubules. The bovine pulmonary artery endothelial cells (right-hand panel) are stained with DAPI, Alexa 488 phalloidin for the filamentous actin and MitoTracker® Red CMXRos for mitochondria. Both prepared slides are sold by Thermo Fisher Scientific. At the time of publication, the kidney section (slide No. 3) has the catalogue number F24630, and the BPAE cells (slide No. 1) the catalogue number F36924. Both slides are useful for teaching microscopy and demonstrate the usefulness of high-contrast fluorescent staining.*

11.3 How Light Interacts with the Specimen

There are six ways in which light can interact with the specimen. These may occur individually or in conjunction with one another. They are:

- absorption, transmission and reflection;
- scattering and diffraction;
- refraction;
- polarisation;
- phase change; and
- fluorescence or phosphorescence.

11.3.1 Absorption, Transmission and Reflection of Light

When light falls upon an object, some of it will be reflected, some absorbed and the remainder will be transmitted. In the daily world in which we live, we are used to seeing objects because they reflect light towards the eye, and the nature of the reflection depends upon the surface. The surface may be matt or else shiny and highly reflective. Other objects may absorb light, yet others will be transparent: these transparent objects transmit the majority of light and so reflect very little back into our eyes. The final appearance and texture of the object will depend upon the relative proportions of reflection, absorption and transmission of light that occur. This is also true of microscopical specimens; whichever mechanism predominates tends to define the technique required to observe the specimen.

Thus a stained histological section will absorb light differentially and so produce contrast. It will transmit light, too, but if the stain was not present, it would only cause an alteration in the phase of the light and so be colourless. In this respect stained tissue sections (Figure 11.3) are entirely analogous to the stained glass of a church window. Objects that primarily absorb light to produce contrast are called *amplitude objects*; they affect the intensity of the light passing through them. If the reduction in the amplitude, or intensity, of the light occurs across the entire visible spectrum, the overall effect is a reduction in tone or brightness (e.g. as in a photographic negative). If the amplitude reduction affects only certain wavelengths, then a colour change in the transmitted light is seen.

11.3.2 Scattering and Diffraction

When light encounters the edge of an object, a certain proportion is scattered. This can either be in an ordered fashion according to the wave-like properties of light, which we know as *diffraction*, or it can be irregular scattering of light, which makes fingerprints, or smears, on glass visible at certain angles, whereas from head-on the fingerprint is not seen. Diffraction and scattering should not be

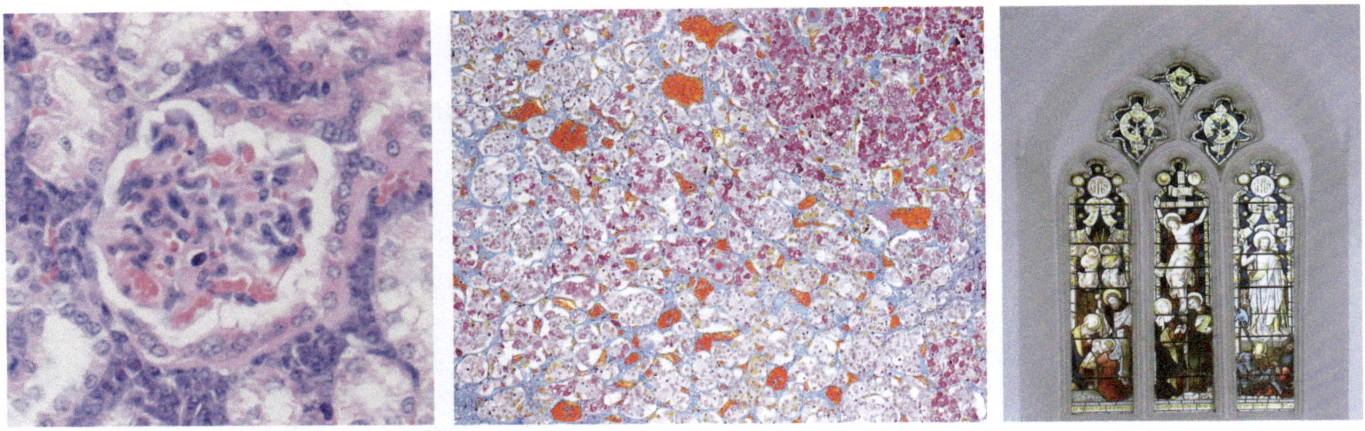

Figure 11.3 **Stained tissue sections** *Both the tissue sections and the stained glass differentially absorb transmitted white light to give colour. The left-hand panel is of an H&E stained kidney glomerulus, the middle panel a section of anterior pituitary stained by Slidders' OFG special stain. The right-hand panel is from All Saints' Church, Chilton, Oxfordshire. Source: Reproduced with permission from the Rector of All Saints Chilton.*

confused with refraction, which causes a bending in the light ray from a change in density of the medium that light travels through. The phenomena of scattering and diffraction are responsible for contrast enhancement at the *edges* of specimens, as well as the formation of images by dark-field microscopy.

11.3.3 Refraction

When a transparent object is immersed in a medium of equal refractive index, it becomes – for all intents and purposes – invisible (see Figure 2.15). This is because light is no longer refracted at the surfaces of the object and immersion medium but passes from one to another undeviated. The refractive index of all materials varies with the wavelength of the light shone through them. This is known as *dispersion* (Chapter 2, Section 2.11) and forms the basis of a form of contrast enhancement known as 'dispersion staining', which is explained in Section 11.10.

11.3.4 Polarisation of Light

Transparent materials may either be crystalline or non-crystalline. Non-crystalline or amorphous materials (e.g. glass and liquids) are isotropic. As such, the speed of light passing through these objects is retarded to the same extent regardless of the direction of propagation. Some crystals with a cubic structure (e.g. sodium chloride or diamond) are also isotropic. All other transparent crystals have structures that are packed differently in one or two dimensions with respect to the second and third or the third dimension, respectively, and are therefore anisotropic. (Some biological structures with orderly repeating 'building blocks', also exhibit anisotropy; see Section 13.6.1). Plane-polarised light vibrating in a certain direction is 'split', or resolved, into two component rays, each travelling at different velocities through the anisotropic material. When these recombine, interference will lead to colour contrast enhancement (see Figures 14.10 and 14.15) according to the nature and thickness of the material and the resulting optical path difference between the rays. The specialised topic of polarised-light microscopy is covered in Chapters 13 and 14.

11.3.5 Phase Change

A completely transparent object immersed in a transparent medium of different refractive index will not affect the *amplitude* of the light[1] passing through it. Equally, if the object is isotropic, it will not affect the plane of polarisation of the incident light. However, the different refractive index of the object will affect the speed of propagation of light: the light will slow down when passing into a transparent object of greater density and higher refractive index. Similarly, the reverse occurs when the light exits the object. These are called phase changes (Figure 11.4) with respect to the light that passes through the medium alone. This alteration in the phase relationships between light waves occurs because of the differing refractive indexes of the object and the medium in which it is immersed. The same effect occurs when light encounters transparent components, and inclusions, of differing refractive index within cells. These transparent entity, made up of components of different refractive index, are called phase objects. They have no (or only minimal) visibility until transformed by optical techniques, or direct treatment, into amplitude objects.

Two optical means of making phase objects visible, which are of interest to us here, employ phase contrast or interference contrast. In both methods, an artificial phase shift between two coherent light rays is introduced. When these rays are combined afterwards at the primary image plane, phase objects are transformed into amplitude objects, which can then be seen. The mechanisms

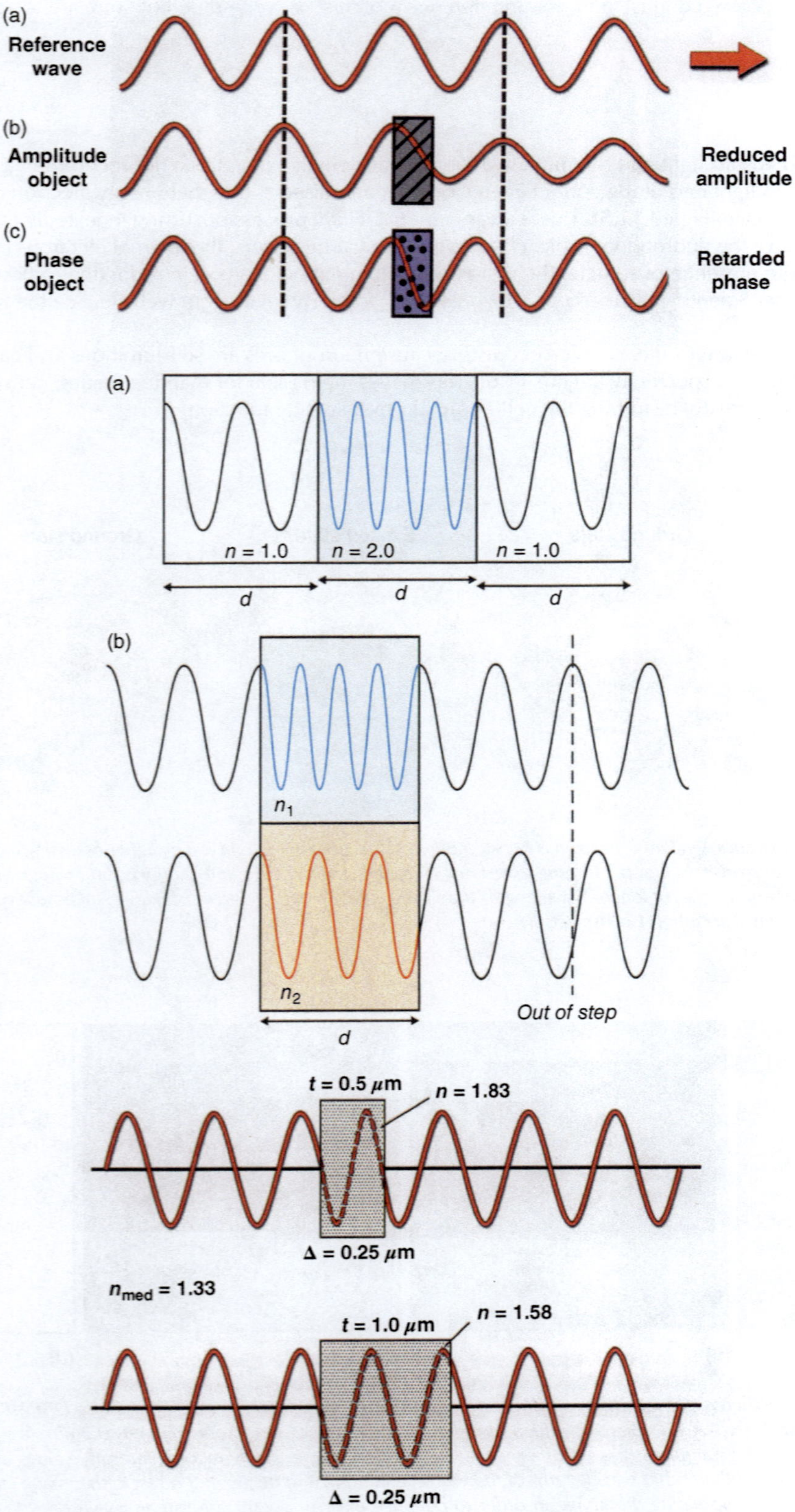

Figure 11.4 **Comparison of phase change and amplitude change** *A so-called amplitude object will reduce the intensity of light passing through it but not necessarily the phase, whereas the opposite is true for a phase object. If the overall optical path difference (OPD) – the product of thickness and refractive index – differs between two objects, the emergent rays will be out of phase with respect to each other. If the overall OPD in each case is the same, the phase relationship between the two rays will be preserved. Source: (top, bottom) Murphy et al. 2013. Reproduced with permission from John Wiley & Sons. (middle) Tilley 1996. Reproduced with permission from John Wiley & Sons.*

by which this is done for phase contrast and for interference contrast is very different and are discussed in Sections 11.12 and 11.14.

11.3.6 Fluorescence

Any molecule that is able to absorb light and re-emit it at a longer wavelength can act as a fluorochrome[2]. Electrons within an excited fluorophore, when exposed to light rays of the correct energy quanta, are raised to a higher energy electron orbital, before returning to their resting orbital or ground state (Figure 11.5). This is a very rapid, cyclical process occurring repeatedly every 10^{-15} to 10^{-9} seconds until light absorption ceases, or the fluorophore is bleached. When bleaching occurs, the internal electrons can no longer be promoted to a higher orbital despite the presence of sufficiently high-energy illumination. A good introduction to fluorescence microscopy can be found on the Thermo Fisher Scientific (formerly Life Technologies; formerly Invitrogen) website. See the list of Internet Resources at the end of the chapter.

Fluorescence is of great value to life-science microscopists because fluorophores are self-luminous and can, under ideal conditions, show great contrast, sensitivity and specificity (Figure 11.6). This makes them ideal for marking tissues, cells and cell components (see also Figure 15.5) often without greatly disturbing the cell's normal physiological function.

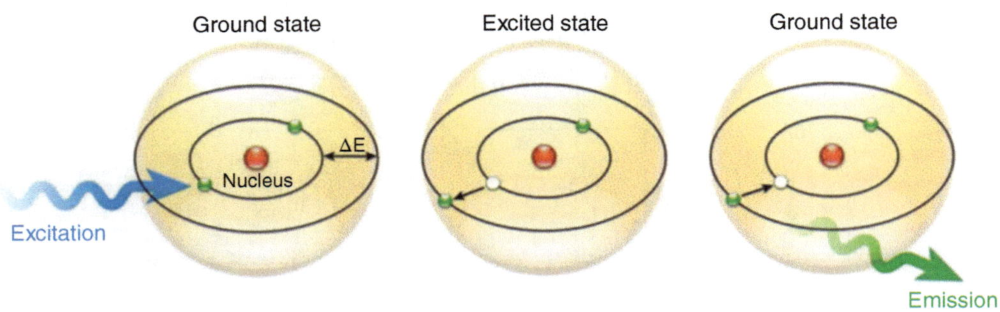

Figure 11.5 **How fluorescence occurs** *A valence electron is boosted briefly to a higher energy orbital by light possessing a quantum of energy equivalent to ΔE, the distance between the two orbitals. Upon returning to rest in the ground state orbital some energy is lost through vibrational relaxation, resulting in emission of a photon of lower energy (i.e. of longer wavelength) than the excitation light. Source: Ishikawa-Ankerhold et al. 2012. Reproduced under the terms of the Creative Commons Attribution License, CC BY 3.0.*

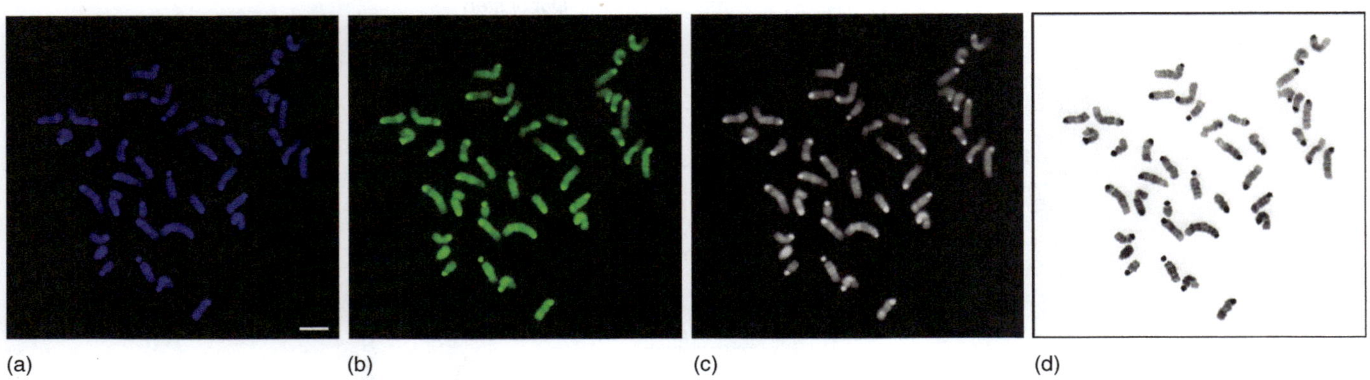

Figure 11.6 **Example of the specificity and sensitivity of fluorescence staining** *These fluorescence images of a DAPI-stained male mouse karyotype were taken with a plan-achromat 100x/NA 1.3 objective, which has a lateral resolving power of approximately 0.2 μm. The blue pseudo-colour (a) helps inform us of the fluorophore used for staining. It has no other value, for the CCD camera only records intensity values as tonal grey levels. Since our eyes are much less responsive to blue than other wavelengths of the visible spectrum, and because contrast in the image falls off as the limit of resolving power is approached, it is worth manipulating the image in order to perceive the fine detail, rather than retain pseudo-colour in images because 'it looks right'. In (b) the karyotype has been pseudo-coloured green, to which our eyes are most sensitive, making it easier to perceive the 'dots' of the centromeres of each chromosome. Removing all colour (c) to achieve the highest possible contrast helps in this respect, yet the mass of back background can still obscure fine detail. Inverting the image (d) and applying an edge-sharpening filter results in pseudo G-banding, which allows the individual chromosomes to be identified and karyotyped. Global image manipulation is usually ethically acceptable, provided full details are published. Scale bar = 2 μm. Source: Courtesy of Dr C. Beechey.*

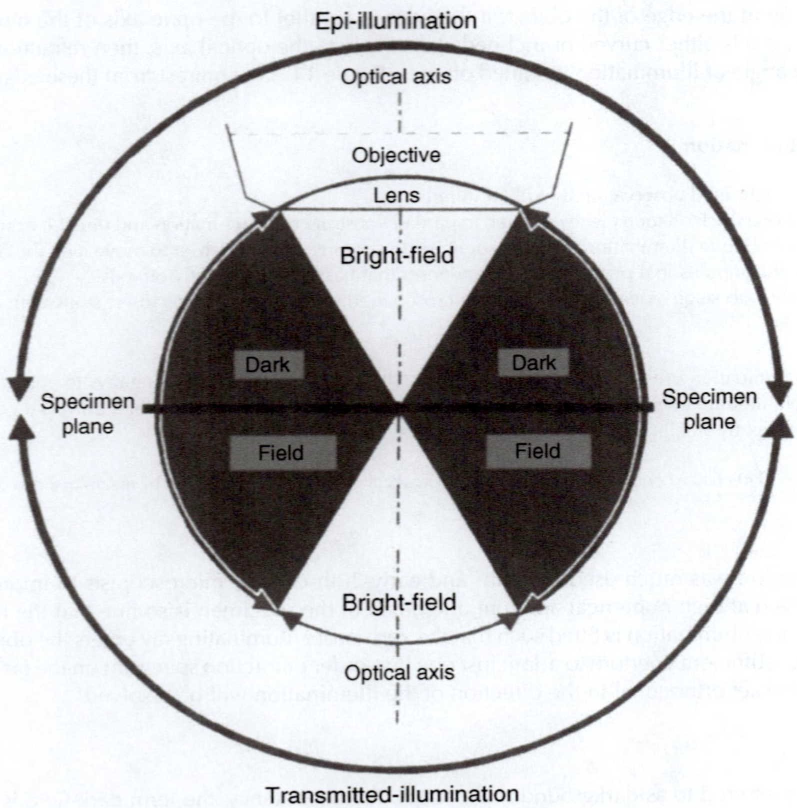

Figure 11.7 **The inter-relationship between bright-field, oblique and dark-field contrast methods** *Light falling onto the sample from above the specimen plane (epi-illumination) is usually employed for opaque samples, the image being formed by reflection off the surface of the specimen. Light incident on the sample from below (transmitted light) forms an image by transmission through the necessarily thin specimen. If the light falls within the shaded area outside the acceptance angle of the objective, a dark-field image is formed. If the illumination is tilted obliquely, either from one azimuth, or as an off-axis cone of light, an oblique contrast image intermediate between bright-field and dark-field is formed. Source: Sanderson 2000. Reproduced with permission from Quekett Microscopical Club. Adapted from Bradbury and Evennett 1996.*

11.4 Bright-Field, Oblique and Dark-Field Microscopy

A transparent specimen is normally illuminated from the opposite side to the objective, regardless of whether the microscope stand is an upright or inverted configuration. Light transmitted along the optical axis of the microscope through the condenser to enter the objective is known as bright-field illumination. Without a specimen in place, this gives a bright, featureless and evenly illuminated disc of light as a background in the field of view. Sometimes the term diascopic is used for transmitted-light illumination (Greek: *dia* – through), particularly by Nikon. Opaque specimens are necessarily illuminated from the same side as the objective. The equivalent in epi-illumination (Greek: *epi* – over) of a featureless empty glass slide is a similarly featureless specularly reflecting surface polished to a mirror-like finish.

Light shone obliquely onto the specimen, either from above or below, may either be incident along a single axis (azimuthal) or else may be directed from all around (annular) with respect to the optical axis. Providing that the angle of illumination to the optical axis is not too great, light will still enter the objective directly. Azimuthal oblique illumination gives rise to asymmetrical shading in the image.

Where the illumination does not enter the objective directly, dark-field illumination occurs, and the image is formed only from light diffracted by the specimen that enters the objective. The difference between these three methods of illumination is shown schematically in Figure 11.7. Correct centration and height adjustment of the condenser is crucial for most forms of contrast enhancement technique, particularly dark-field and phase-contrast microscopy. As mentioned in the Introduction, it is a good policy first to adjust the microscope and set up Köhler illumination for all forms of contrast enhancement and not just for bright-field examination. In reflected-light microscopy (see Chapter 12), the objective acts as its own condenser, so Köhler illumination is simpler to set up for epi-illumination than transmitted-light illumination.

11.4.1 Oblique Illumination

If transparent objects are illuminated with a full axial cone of light, they will lack contrast. However, if their refractive index differs appreciably from the mountant (which may be air), and if the object also has an appreciable thickness, it is possible that some contrast will be generated by refraction at the edge of the object. If the edge is parallel to the optic axis of the microscope, or nearly so, then very little contrast will result. If it is either curved or inclined at an angle to the optical axis, then refraction will cause some degree of contrast enhancement. If the angle of illumination is shifted off-axis (Figure 11.8a), contrast from these edges may be enhanced further.

Protocol 11.1 Oblique Illumination

1. Set up the microscope for bright-field observation by Köhler illumination.
2. Use the edge of your finger or a thick opaque piece of paper, to partially obstruct the illumination and direct it from one side into the condenser.
3. It may also possible produce oblique illumination to some extent by de-centring[i] the condenser to move it off the optical axis of the microscope. Another means is to rotate the annulus in a phase-contrast condenser that has been set slightly off-axis.
4. If a filter tray is built into the sub-stage assembly below the microscope, the best method is to insert stops with a crescent cut-out to provide oblique illumination.

Annular (or circular) oblique illumination uses a patch stop but one that allows peripheral illuminating rays to enter the objective instead of being totally excluded, as in dark-field illumination. If a suitable patch stop is not available in the carousel of a universal condenser, the same effect can sometimes be produced with a slightly smaller mismatched phase-contrast annulus.

[i] Many microscopes built before the 1940's had a condenser built with an aperture diaphragm that could be offset by means of a rack and pinion control.

Azimuthal oblique illumination was much used by 19th- and early-20th-century microscopists to improve resolution in the image (Figure 11.8b). Sometimes, even at high numerical aperture, the detail in the specimen is so fine that the first-order diffraction spectra will not enter the objective. If the illumination is tilted such that the zero-order illuminating ray enters the objective on one side, (Chapter 10, Section 10.9) there may be sufficient aperture to admit just *one* first-order diffraction spectrum on the far side of the objective. Under these circumstances, fine detail set orthogonal to the direction of the illumination will be resolved.

11.4.2 Dark-Field Microscopy

Dark-field microscopy is also referred to as darkground microscopy. For consistency, the term dark-field is used throughout this book. Because of inherently high contrast, and because very weak diffracted rays are captured over the full aperture of the objective, dark-field microscopy (and also fluorescence microscopy) allows us to detect structures far below the resolution limit of the light microscope, although we cannot truly discriminate their size (which will be diffraction-limited). The theoretical aspects are discussed by Françon (1961). Dark-field microscopy is particularly suited to the examination of minute refractile structures, which scatter light well. These include the silica of the frustules (i.e. shells) of diatoms (Figure 11.9), bacteria, aquatic organisms, small inclusions in cells and polymer materials. Dark-field illumination has been known since the beginning of microscopical science. Leeuwenhoek almost certainly knew of the method[3], and Christiaan Huygens wrote in 1678 in his journal:

> 'I look at these animals not directly against the light but on turning the microscope a little which makes them appear on a black ground. One can best discover by this means the smallest animals living and can also distinguish best the parts of larger ones ...'

All that is required for dark-field microscopy is to exclude the zero-order (or direct) illuminating light from entering the acceptance angle of the objective. The background will then be dark, and the image will *appear* self-luminous, shining with high contrast against the background. Unlike fluorescence microscopy, where similar levels of contrast enhancement can occur, the object is not self-luminous but diffracts the image-forming light into the objective. This is akin to dust particles suspended in the air made visible by a shaft of direct sunlight into an otherwise shaded room. Although dark-field is an extension of oblique illumination, and dark-field contrast can also be produced by single beam laterally illuminating at an angle beyond the acceptance angle of the objective lens, the illuminating rays are normally symmetrical about the optical axis rather than occupying a single azimuth. The light from a dark-field condenser forms a hollow cone.

For dry objectives, up to an objective NA of about 0.6, the simplest means of producing dark-field illumination is to use an opaque circle of material, centred with respect to the optical axis of the microscope, to block the direct illumination. This is called a patch stop; see Figure 11.10a. Often the universal condenser has a dedicated patch stop in its carousel, along with the usual bright-field position with its condenser aperture iris, phase annuli and (possibly) Wollaston prisms. The dark-field patch stop is often labelled 'D' for dark-field or the German 'dunkelfeld'.

Ideally, this patch stop should be placed at the front focal plane of the condenser, but the condenser illuminating aperture iris (condenser diaphragm) is normally located in this position. Only in an Abbe illuminator is the first (or front) focal plane beneath the lower element and so as such is easily accessible for a patch stop. However, there is room for latitude: it isn't crucial if the patch stop is not placed in the ideal position. Sometimes, in older condensers, the substage filter tray where the patch stop is placed is far removed from the front focal plane of the condenser. In this case, care should be taken that the patch stop is sufficiently large to ensure a good dark-field effect is obtained. If no dedicated patch stop is present, the annulus for the 100x phase-contrast objective (often denoted with 'Ph3') may be suitable as a dark-field patch stop for lower magnification objectives (i.e. 10x, 16x or 20x LWD-objectives with low numerical apertures). This annulus for phase contrast *is* placed in the correct conjugate plane and also has the means to be independently centred to the optical axis of the microscope.

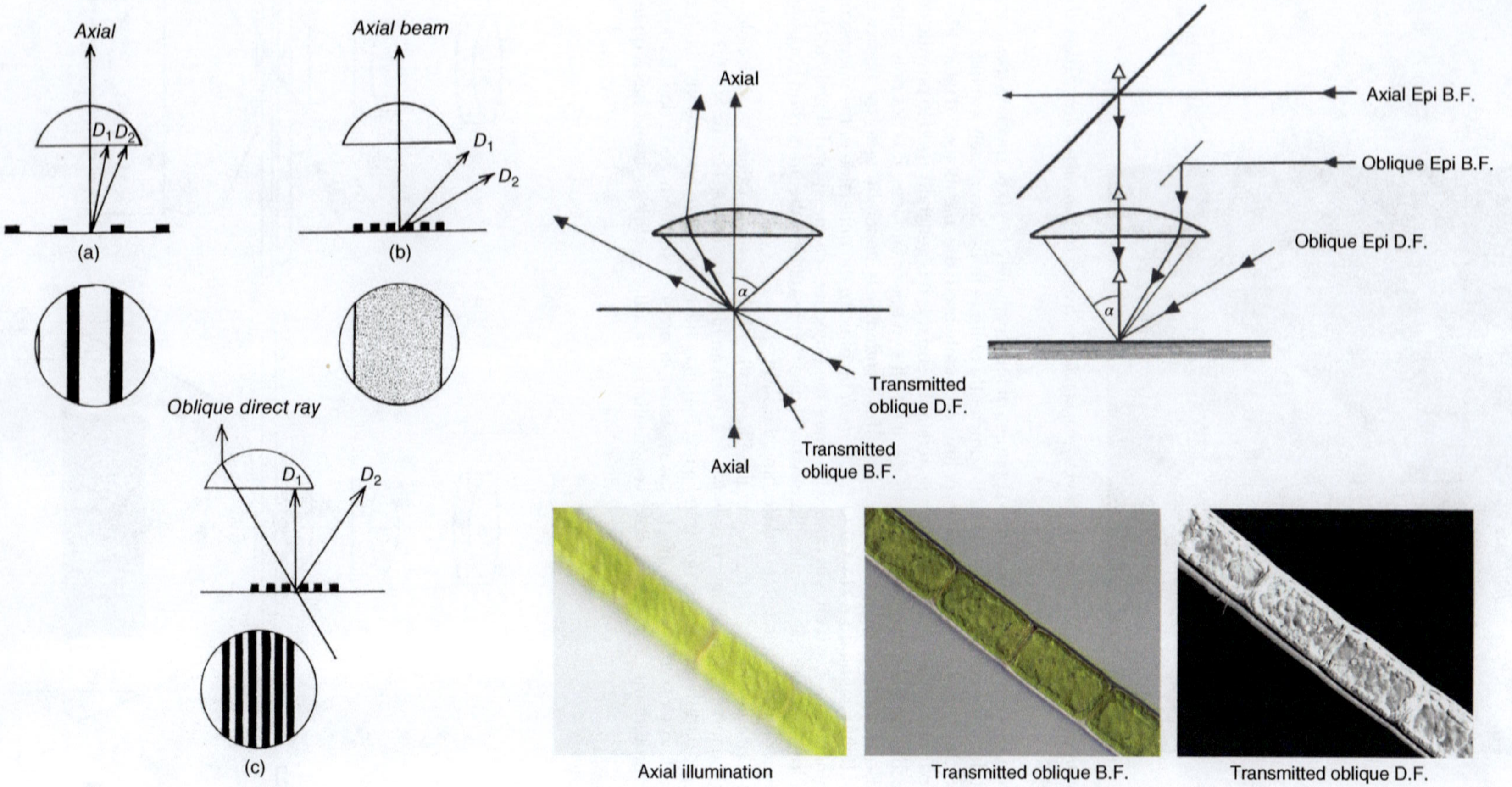

Figure 11.8 **Oblique illumination** *Oblique illumination can be used (a) to increase the resolving power of the microscope, using a single sideband of diffracted rays to form the image whilst the zero-order rays enter the acceptance angle of the objective on the opposite side. Alternatively, if the edges of structures are inclined to the axis of illumination (by tilting the illumination) as in (b), contrast at these edges is enhanced before dark-field illumination occurs. Source: (left) Bradbury 1967. Reproduced with permission from Elsevier. (top right) Bradbury and Evennett 1996. Reproduced with permission from Taylor & Francis Group.*

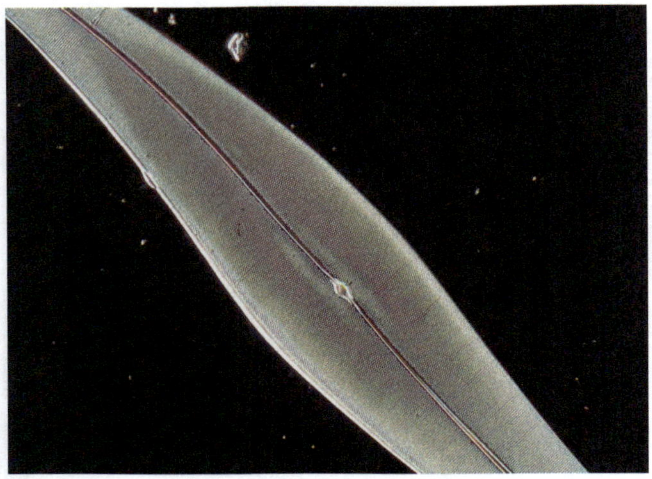

Figure 11.9 **Dark-field microscopy** *An example of the test diatom Pleurosigma angulatum imaged by transmitted-light dark-field illumination.*

For use with dry and immersion objectives above numerical apertures of approximately NA 0.6, specially designed condensers have to be used. This is to ensure that the cone of illuminating light is sufficiently oblique to be completely excluded from the acceptance angle of high-NA objectives. Two designs of dark-field condenser are commonly used: the 'paraboloid' (Figure 11.10b) and 'cardioid' (Figure 11.10c) condensers. The paraboloid condenser has a single reflecting surface; it is generally used to provide conical illumination of NA 0.7–1.0 for dry objectives. The cardioid condenser reflects the light off two mirror surfaces; it is not only aplanatic and thus free from spherical aberration but also free from chromatic aberration. This condenser is generally used for higher numerical aperture objectives and supplies a cone of light ranging from NA 1.0 to 1.4. It *must* be oiled to the underside of the microscope slide to prevent unwanted refraction of the illuminating rays, which will otherwise enter the objective – a messy and slightly tricky business. You can use water for convenience, but a lower numerical aperture illumination cone will result, and the immersion coupling may well dry out over prolonged use – it is then difficult to replace without disturbing the slide.

Careful inspection of Figure 11.10 will show that the NA of the illuminating light has two values. A dark-field condenser may be cited as NA 1.4, but if the NA of the inner beam of light is only NA 1.2, then an objective of NA 1.25 or NA 1.3 will admit some light. This is why objectives designed for dark-field microscopy (see Figure 7.4) may have a variable aperture iris[4] built into the back focal plane. This can then be adjusted and closed down as much as is required in order to secure a good dark background field of view for high contrast. Note that this aperture iris (although it looks very similar) is of a different design, and has a quite different purpose,

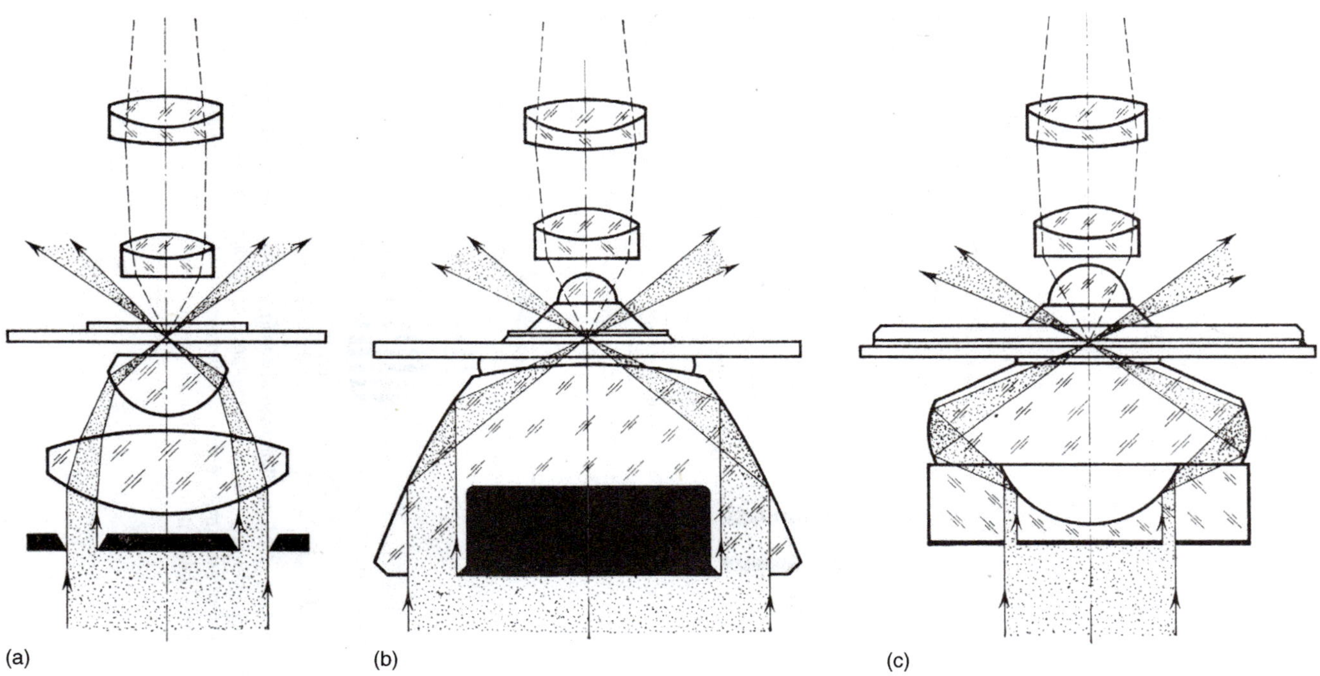

Figure 11.10 ***The design of dark-field condensers*** *Source: Pluta 1988.*

from the correction collar fitted to objectives of high numerical aperture to control spherical aberration (see Chapter 6, Section 6.3 and Chapter 7, Section 7.8).

Protocol 11.2 Dark-Field Illumination

1. Set up the microscope, and establish proper Köhler illumination.
2. Replace any absorption specimen with a transparent one; temporarily close the condenser diaphragm, if needs be, to visualize this.
3. If the condenser has a patch stop, swing this into position, and view a dark ground with an objective up to NA 0.6. This may require centring in the same way as phase annuli are centred. If you have a high-power dark-field condenser, go to step 5.
   (If the condenser does not have a built-in patch stop, follow the instructions below for making a patch stop.)
4. Put the patch stop into the substage filter tray, replace the condenser and proceed.
5. If you have a dry or oil-immersion paraboloid or cardioid condenser, without disturbing the specimen, lower the bright-field condenser and replace it with the dark-field condenser. If it is to be oiled, carefully place a drop of oil on the top surface of the condenser before racking it into position.
6. Look for contact of the oil with the underside of the slide. Very fine control of the condenser height will show when a good dark-field effect with high contrast has been obtained.
7. Even though the previous bright-field condenser was centred, it may still be that the dark-field condenser will require slight centring adjustments. This isn't necessarily easy; any alteration should be made delicately, looking for intensity changes across the field of view. Aim to achieve a uniformly dark ground.
8. When you remove the slide, remember that there is oil on its under surface before you put it down, or store it away!

To make a patch stop:

1. Set up the microscope, and establish proper Köhler illumination.
2. Remove one eyepiece. Look down the body-tube of the microscope, open the condenser aperture diaphragm until it just leaves the field of view of the back aperture (back focal plane) of the selected objective.
3. Without disturbing the aperture diaphragm, remove the substage condenser.
4. Turn the condenser over, and use a vernier caliper to measure the diameter of the aperture diaphragm's opening blades. Add 10% to this value, and make this the diameter of the patch stop for that objective.

Ensure that the correct thickness of slide is used for these dark-field condensers, particularly if they are used with water or oil as immersion fluids. The slide should be a standard thickness of 1 mm. Very much older slides may be too thick, in which case, the illuminated field diaphragm cannot then be focused properly. If, on the other hand, the slide is too thin, it will be difficult to maintain the oil-immersion coupling between the top of the condenser and underside of the slide when the condenser is correctly focused, particularly if the slide is moved much. To overcome these problems, some dark-field condensers (e.g. older ones from Beck and Cooke, Troughton & Simms, CTS) have a variable focus control.

Cleanliness is paramount in dark-field microscopy. Any dirt or other unwanted particles remaining in the preparation (e.g. within the mountant) or adhering to the surfaces of the slide or coverslip, are very likely to scatter light into the objective. This will reduce the contrast and/or obscure the image. Dark-field illumination, like phase contrast (see Section 11.12) does not work well with strongly coloured specimens. Also, like the halo effect in phase contrast, the scattering of light at the edges of refractile specimens can give rise to brightly lit boundaries leading to discontinuities in the image and difficulty in discriminating or accurately measuring edges.

When using bright-field epi-illumination microscopy to observe opaque specimens, the objective acts as its own condenser. Therefore, dark-field microscopy with epi-illumination requires a co-axial illumination system to provide light around the outside of the objective. Dark-field microscopy is used much more widely for studying opaque specimens by reflected light than for transparent specimens by transmitted light. For further details, refer to Chapter 12, Section 12.9.

11.5 Our Perception of Colour

An object's colour depends on how it absorbs, reflects or transmits differing frequencies of light. Our trichromatic colour vision arises from the three different light-sensitive cone cells[5] in our retina. These are receptive to different wavelengths across the visible spectrum, allowing us to see about 10 million different colours. How we perceive colour is a psychophysical phenomenon – we differ from one another in our ability to interpret colour[6]. The average person can identify millions of hues, yet it is difficult to describe individual colours precisely. Words often fail us when it comes to defining nuances of tone with accuracy. Light blue, sky blue, grey-blue: in our everyday lives, we make do with simple terms. It is only those who want, or require, to differentiate more precisely who opt for more unusual descriptions such as cobalt blue, Oxford blue or ultramarine. Nevertheless, all colours may be specified by three parameters: hue, saturation (also called chroma) and 'brightness' (the visual perception of luminance). Hue corresponds to the spectral wavelength; saturation, to the proportion of white light mixed with each hue – to give pastel shades. Brightness describes the luminous intensity (i.e. the number of photons reaching the eye). This is the HSL descriptive colour model.

Colour constancy refers to our ability, as part of our visual perception, to perceive colours of objects independently of the spectrum of the illuminant and to see consistency in the overall colour. An apple which is red in bright sunlight will also be perceived as red

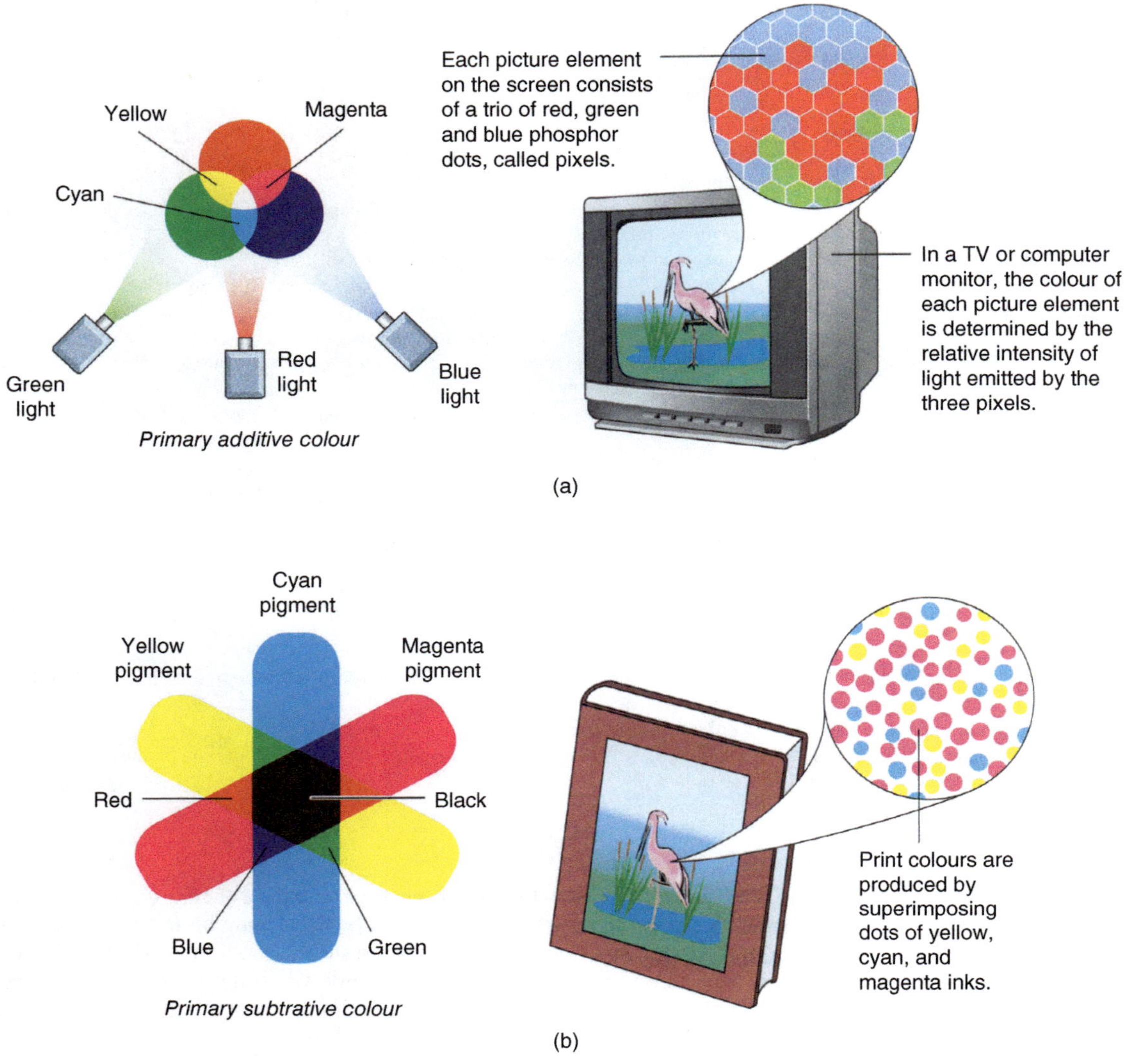

Figure 11.11 **Colour formation by additive and subtractive means** *The colours of many objects: images seen on computer monitors and digital device screens, as well as the printed page, are produced by combining the three primary colours. The three primary additive colours of light – red, green and blue – are shown here forming digital images, whereas the three subtractive primary colours – magenta, yellow and cyan – are used in printing. Both the secondary subtractive, or absorption, colours and the primary, or additive, colours of the spectrum can be arranged sequentially together as a 'colour circle'. Colours which are complementary to one another lie opposite each other on the colour circle. Source: Trefil and Hazen 2004. Reproduced with permission from John Wiley & Sons.*

whether it is partially in the shade or else seen under candlelight in the evening. Equally, the page of this book will appear white under different lighting conditions.

Additive colour mixing[7] (Figure 11.11) occurs when two, or more, beams of differently coloured light combine. White light can be formed by adding all three primary colours: red, green and blue light. These three primary colours of light when added together in various proportions give rise to a range, or gamut, of colours.

The three spectral ranges associated with the three cones of the eye define the tristimulus values of any colour that we perceive. A full plot of all visible colours is represented by a three-dimensional figure, although it is much more convenient to represent the colour gamut as a two-dimensional plot, or chromaticity diagram. Since the three values are normalised, only two (hue and saturation) are required to define any colour. The tristimulus value for luminance is not plotted in the 2D colour space: it comes out of the page in the z-axis in a 3D model. The gamut – or reproducible colour space – of all visible chromaticities is the tongue-shaped CIE colour space (Figure 11.12) of human colour perception represented in two dimensions. The outer curved boundary of the plot is the spectral (or monochromatic) locus, with wavelengths shown in nanometres. The right diagonal edge (called *the purple line*) represents the extra-spectral crimson, magenta and purple hues that are mixtures of red and violet light. These colours, although they are on the border of the gamut, have no counterpart in monochromatic light. Less saturated colours appear in the interior of the figure with white at the centre. The gamut of most monitors is shown by the sRGB triangle drawn in grey. The CIE chromaticity diagram is useful to ensure that colour imaging recording and display devices (e.g. computer monitors) match the gamut of human vision as closely as is possible.

Wavelength/nm	Colour	Wavelength/nm	Colour absorbed	Complementary colour
380–420	Violet	400–435	Violet	Yellow–green
421–445	Indigo	435–480	Blue[a]	Yellow[b]
446–485	Blue[a]	480–490	Blue–green	Orange
486–495	Cyan[b]	490–500	Green–blue	Red
496–570	Green[a]	500–560	Green[a]	Magenta[b]
		560–580	Yellow–green	Violet
571–590	Yellow[b]	580–595	Yellow	Blue
591–620	Orange	595–605	Orange	Blue–green
621–710	Red[a]	605–700	Red[a]	Cyan[b]

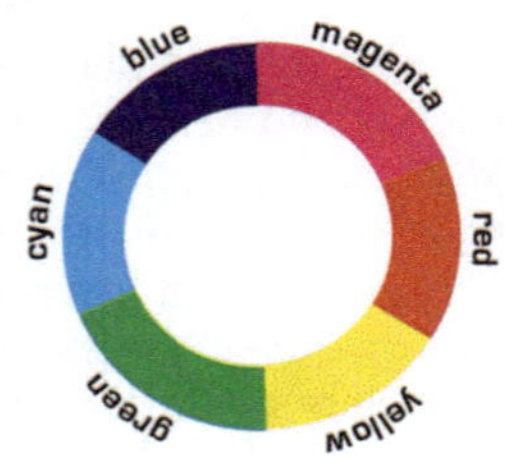

[a] Additive primary colours.
[b] Subtractive primary colours.

Magenta is an extra-spectral colour, not found in the visible spectrum. It is perceived as an equal mixture of red and blue in the absence of green, which is how it is formed on a computer monitor. It is therefore not seen in the two lists of spectral colours, but only as a complementary colour. It occupies the purple line on the CIE 1931 Colour space, Figure 11.12. Magenta joins the red and blue sectors on the colour wheel, but in reality these colours lie at opposite ends of the spectrum.

Filter	Darkens	Lightens	For maximum contrast
Red	Blue & green (cyan)	Red	For red stains use a cyan or green filter
Orange-red	Blue	Red & orange	For orange-brown stains use a blue filter
Yellow	Blue & violet	Yellow	For yellow stains use a blue filter
Green	Blue & red (magenta)	Green	For green stains use a magenta filter
Blue	Red & orange	Blue	For blue stains use a orange-red filter
Violet	Yellow	Violet	For violet or purple stains use a green filter
Magenta	Green	Magenta	For magenta stains use a green filter

Figure 11.11 (Continued)

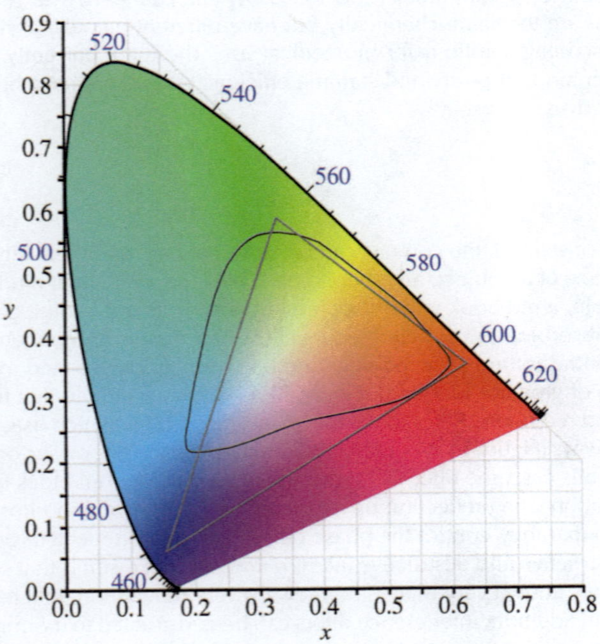

Figure 11.12 **The CIE colour space plot** *The CIE 1931 Yxy colour space chromaticity diagram. The x-y coordinates map the chromaticity values, hue and saturation. Refer to the text for an explanation. Mixing red (700 nm) and blue-green (492 nm) in the proportions 39% to 61%, respectively, will give white light. Complementary colours lie opposite a given colour on a straight line through white in the centre. The gamut of normal human vision covers the entire CIE diagram whilst the gamut of an RGB monitor is contained within the grey triangular region. Adobe RGB, a slightly larger gamut, is preferred when the intended output is high-quality inkjet prints, since it encompasses more of the curved black locus denoting the CMYK gamut of printing inks. Chromaticity coordinates are a precise means of specifying colour. They do not convey a visual perception, which is unique to each person. Source: BenRG adapted from PAR, via Wikimedia Commons.*

11.6 Contrast Using Colour: Staining and Filters

Ideally, all biological specimens for microscopy would be thin and possess strong natural colour but with clearly recognisable differences in regions of interest. Such specimens could then merely be mounted and viewed without further ado. A few naturally coloured objects do exist (e.g. some plant specimens), but often the colour is very intense and masks internal detail. These internal components can be bleached and re-stained in a controlled manner. However, most living tissues are virtually transparent, and so discerning fine detail is difficult. Nevertheless, refractive index differences between cell walls, membranes and internal organelles allowed microscopists, before the widespread uses of stains and dyes (from about the 1860s), to elucidate most plant and some animal cell structures. Some authors distinguish between the terms 'dyeing' and 'staining'[8], but for simplicity the two will be used synonymously here.

The traditional route of the histologist is to section, or cut, fixed and embedded tissues into slices between 2 and 8 μm, mount these onto a glass slide as a transparent mechanical support. The thin section of tissue is then differentially coloured, and a coverslip applied to protect it. This allows an examination of the microanatomy of the tissue. The stains normally employed in histology are empirical or 'irrational'; that is, the exact chemical mechanism is not known, but the method is a tried-and-tested recipe. They are just as valid as rational ones, where the mechanism of staining is well understood.

Many dyes attach to ionic radicals within tissue. For example, eosin, an acidic xanthene dye, is used to stain the cytoplasm of tissues various shades of red and pink. To differentiate the nuclei precisely and well, the dark-blue stain haematin (commonly called haematoxylin) is used. In this case, haematin is combined with a solution of metal ions (a mordant solution) to increase the staining potency. The result is the classical haematoxylin and eosin 'H&E' stain widely used for pathological diagnoses (Figure 11.3). For the most effective result, the contrasting acidic and basic stain pairs should be in highly contrasting complementary colours, and the nuclei, or smaller cellular features, should be in a saturated colour (e.g. blue or red at the opposite ends of the visible spectrum) so that they stand out, and the cytoplasm is stained an unsaturated colour (e.g. orange or yellow) as the background[9]. Although eosin is not an unsaturated colour, it is effective because it is applied as a very pale tint.

Histological stains also work by other means, such as hydrogen- and covalent-bonds, as well as solvent-solvent interactions. A very good review of the mechanism of empirical staining, as well as histochemical methods, is given by Kiernan (2006). The scientific use of dyes as histological stains comprises only a small proportion of the dyeing industry, with most dyes being used for textiles. The variability of dyes affects the histological and biological market to a far greater extent than the clothing industry. Dyes are highly variable for many reasons. They are not pure substances; a particular manufacturing process may vary from time to time, for dye-making is generally a complex stepwise process. Fashion and economic trends also play a part, for example, the principal use of nuclear fast red (C.I. 60760) is to colour wall paper: once that shade of pink is out of fashion, production ceases until the pendulum of public taste swings back.

The names of individual dyes are often chosen irrationally, are inconsistent with related compounds and above all are confusing[10]. The Society of Dyers and Colourists publish a colour index (C. I.) to classify each different dye. These C. I. numerals are 5-digit numbers assigned uniquely to identify dyes that are the same chemically but have different names, or vice versa. Furthermore, the Biological Stain Commission certifies dyes and staining solutions for biomedical use. There are currently around 60 certified dye powders; by using these for staining you can be certain that purity and staining efficiency standards have been met and thus compensate for any batch-to-batch variability of particular dyes.

11.7 Colour Filters

Whether a tissue is naturally coloured or stained, the contrast can be further enhanced or, equally, decreased with colour filters. When we view a colour reflected off the surface of an object, all other colours from the white light illumination are absorbed, and the colour seen is reflected to our eye. For example, a red book will reflect red but absorb the remaining colours of the spectrum. A purple book will reflect both blue and red, while absorbing the middle green, yellow and orange component of the spectrum. With a red-stained specimen, features will appear less contrasty through a red filter and be more contrasty – and much darker – when viewed through a cyan filter or a filter with a proportion of green or blue in it. Figure 11.13 shows a sample that has been selectively stained with blue and red stains viewed through blue and red filters. Refer to the table in Figure 11.11, which lists the complementary colours.

A filter works in one of two different ways (Figure 11.14). Either the filter (made of glass, plastic or gelatin) is stained throughout its mass with a particular colour – e.g. blue. In this case, the filter will transmit the blue colour but block red and other colours of the spectrum. The modern method is to achieve filtration using reflection by interference. A series of thin films is coated onto one surface of a glass substrate. These layers are designed so that they control the phase differences of the reflected- and transmitted-light waves, which are either intensified or attenuated by constructive and destructive interference. The net result is that selected wavelengths are reflected and all others, transmitted. Interference filters absorb very little light; they are very much more efficient than simple absorption filters dyed throughout the mass of their substrate. In addition, interference filters can be constructed to (normally) reflect shorter wavelength of light and allow longer wavelengths to be transmitted. Such dichromatic beam splitters are, functionally, at the heart of the epi-fluorescence microscope.

A green filter can be used to improve the definition of finely resolved detail (Figure 11.15) particularly when using achromat objectives. Since spherical aberration is minimised for green light of about 550 nm wavelength, using monochromatic illumination will eliminate any chromatic aberration and also minimise spherical aberration. This should be considered if recording monochrome images. Also, the eye is most sensitive for green light; therefore, a green filter may help detect and locate isolated fine detail in the image. For phase contrast, the $\frac{1}{4}\lambda$ wavelength shift has to be for an agreed value, and since λ is taken to be 550 nm, phase-contrast equipment should include a green filter for optimum image definition.

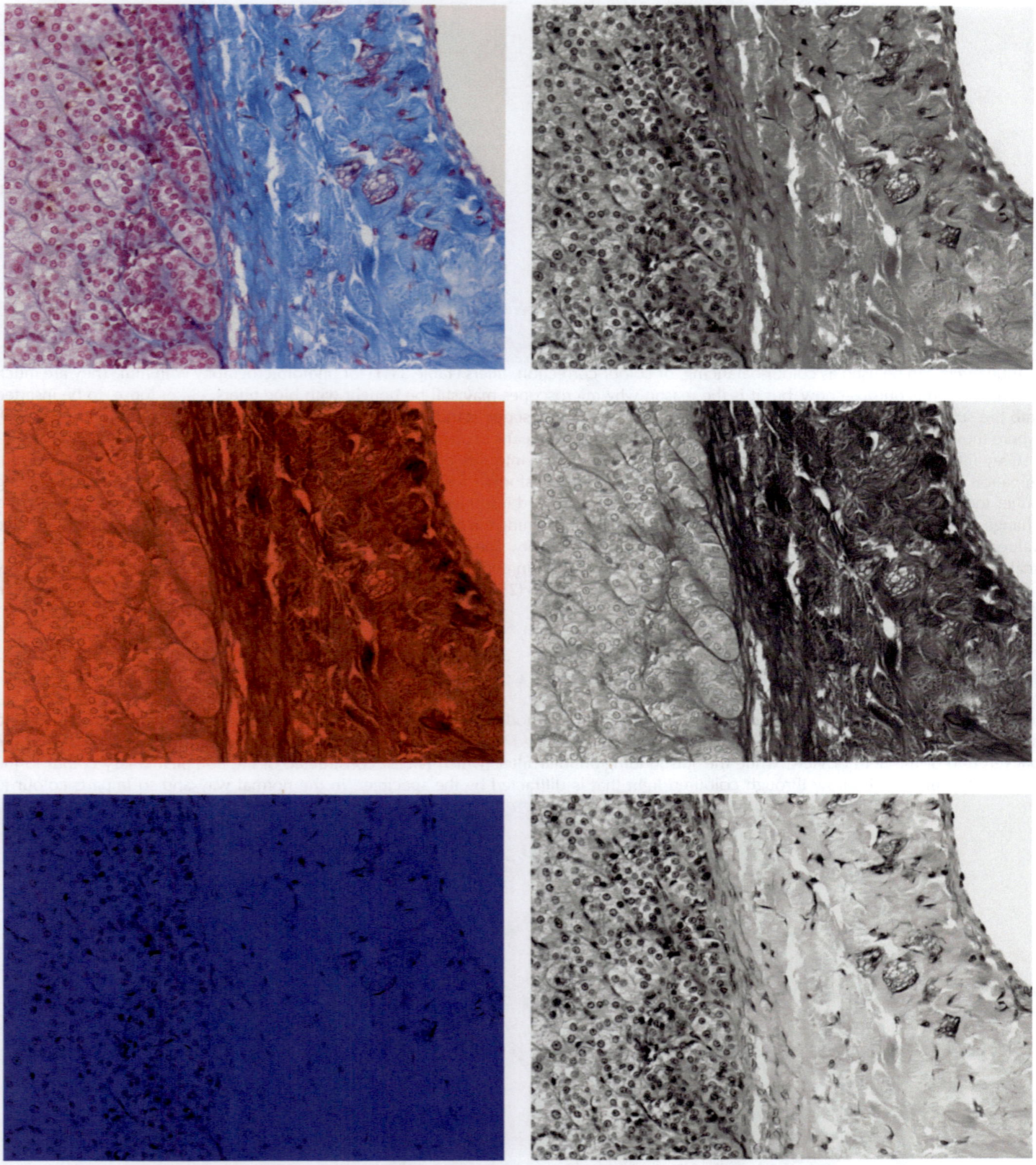

Figure 11.13 **Contrast enhancement of coloured specimens using filters** *A specimen of adrenal gland stained by Heidenhain's azan method. Nuclei are stained magenta-red and the tissue matrix, blue. The monochrome equivalent image is shown to the right of the coloured image. Each filter enhances the contrast of the complementary colour, whilst diminishing the contrast of the similarly-coloured stain.*

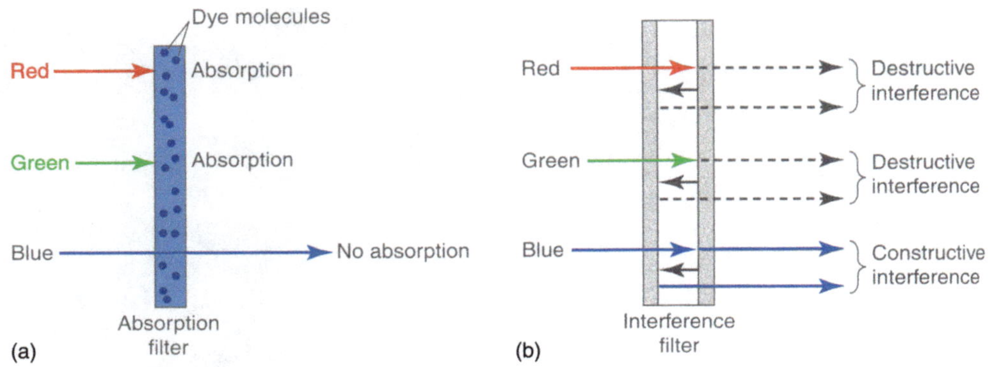

Figure 11.14 **How filters work** *Source: Adapted from Chandler and Robertson 2009 with permission from D. Chandler.*

Blue filters can be used as colour-balancing, or colour-correction, filters (Table 31.1) for photomicrography with film. They are much less used for this purpose now, but it is the reason why microscopes may still be bought with blue filters, either loose (to be inserted into the substage filter tray) or built into the bottom of the microscope stand (sometimes alongside green and neutral-density filters), where they can be inserted into the illumination path, often by means of a spring-loaded pushrod.

Grey filters are used as neutral-density filters, to cut down the intensity of light evenly throughout the visible spectrum. If the illumination intensity is altered by adjusting the rheostat on the transformer for the lamp, the colour temperature will change, becoming either too blue upon increasing the illumination or too orange-red if turning it down. To avoid this, neutral-density filters are used to change the intensity of illumination while simultaneously maintaining the correct colour temperature. These neutral-density filters have values expressed as percentage transmission and/or optical density. The optical density (OD) is a logarithmic, rather than arithmetic, scale. This has the advantage that OD values of two filters used in combination are added together (see Table 31.2). Generally, only five different filters are required: one OD 0.3 (50%), two OD 0.6 (25%) and one OD 0.9 (13%) to perform any required reduction in the intensity of illumination.

11.8 Rheinberg Illumination and Variable Asymmetrical Contrast

In 1896, Julius Rheinberg devised the contrast-enhancement method named after him by incorporating colour filters into the dark-field apparatus on his microscope. If the central patch stop is replaced by, ideally, a very dark filter (e.g. dark blue) this will colour the background of the field of view, while the surrounding annular filter (which is best constructed of the complementary colour to the central stop filter) will allow through coloured light that is diffracted by the specimen in the normal way and so imparts colour to

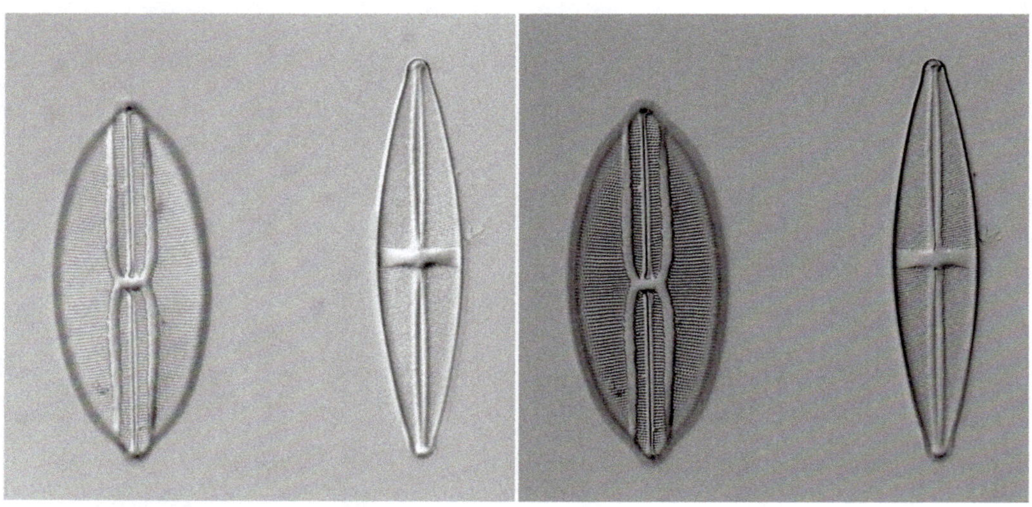

Figure 11.15 ***Using a green filter to enhance resolved detail*** *The left-hand panel shows an image of the two diatoms* Navicula lyra *and* Stauroneis phoenocenteron *(see Table 10.3) acquired in white light by bright-field microscopy with a 40x/NA 0.75 plan-Neofluar semi-apochromat objective without using a filter. The right-hand panel shows the benefit of using a green filter to enhance fine detail.*

the image. Although the title of Rheinberg's paper stated this method was an 'addition to microscopical research', it wasn't adopted seriously by biomedical investigators. This was perhaps because, firstly, the method requires a different patch stop to be constructed for different numerical aperture objectives and can only be used with dry objectives up to about NA 0.6. Beyond this aperture it is necessary to use a paraboloid or cardioid condenser, with which Rheinberg illumination is unworkable.

Protocol 11.3 Making Rheinberg Illumination Patch-Stop Filters

1. Set up the microscope, and establish proper Köhler illumination.
2. Remove one eyepiece. Look down the body-tube of the microscope, open the condenser aperture diaphragm until it just leaves the field of view of the back aperture (back focal plane) of the selected objective.
3. Without disturbing the aperture diaphragm, remove the substage condenser.
4. Turn the condenser over, and use a vernier caliper to measure the diameter of the aperture diaphragm's opening blades. Add 10% to this value, and make this the diameter of the disc stop for that objective.
5. Replace the condenser, and repeat entire process for any further objectives.
6. To find the full diameter of a filter, measure the diameter of the substage filter tray – this may be a standard 32 mm diameter.
7. Create a square image file using photo-editing software (e.g. Adobe Photoshop, Illustrator, or Macromedia Freehand) with dimensions equal to the total diameter of the filter. Using a black 1-pixel line, draw a circle that exactly fits, centred, inside that square.
8. Draw another circle with a black 1-pixel line whose diameter equals the diameter of the disc stop from Step 4 above. Save the image as a template for the measured objective.
9. The template is used to cut circles and sectors from polyester lighting filters. A compass cutter is useful to cut out these segments.

To learn more, see Rheinberg filter construction, Ian Walker Micscape-uk. http://www.microscopy-uk.org.uk/mag/indexmag.html?http://www.microscopy-uk.org.uk/mag/artsep05/iwcolourmagic.html.

Nowadays if you wish to use Rheinberg illumination for contrast enhancement, you have to make your own patch stops. A very useful URL link is given in the Internet Resources at the end of the chapter. Three-colour Rheinberg illumination is particularly effective for objects that show striking periodicity, such as diatoms and textiles, including silk, where the warp appears in one colour and the weft in another (see Box 11.2). Variable dark-field or Rheinberg illumination can be made by constructing the central patch stop from polarising material. Using a rotary analyser (or a rotatable polarising filter over the eyepiece) the intensity of the illumination can be varied continuously. An alternative method of contrast enhancement is variable asymmetrical contrast (VAC contrast; Wilkinson, 2011) based on dark-field and oblique illumination combined, as devised by Arthur Strange.

Box 11.2 Rheinberg and 'Spikeberg' Illumination

Combining a coloured filter of two complementary colours underneath a dark-field patch stop will result in coloured dark ground, named after Julius Rheinberg, who published an account of his technique, following an experiment with two differently coloured filters, which he had inserted in the substage filter holder. He called his method 'multiple or differential color illumination'. The central stop should be made of the darker, more-saturated colour and is surrounded by an annulus made of the complementary hue on the colour wheel.

Because the diffracted ray forming the image is perpendicular to the feature that gives rise to it, Rheinberg constructed quadrant sector stops, which are useful when the specimen has a definite orientation, such as the medullary rays (pith) as distinct from the fascicular cambium in the longitudinal wood section illustrated, or the warp and weft (woof) of fabric.

Prior to the discovery and development of phase contrast, Zeiss made a universal Rheinberg system called the 'Mikropolychromar' in 1933. This apparatus replaced the usual condenser iris assembly, and employed a hollow glass tube over which filters and opaque stops could be placed. A series of irises delimited the radii of the central stop and outer annulus, to give complete control overall.

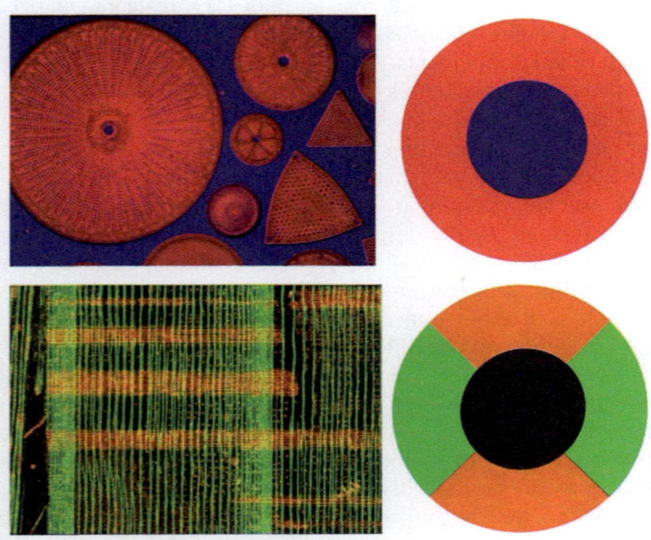

Source: Bradbury and Bracegirdle 1998.

One well-known contemporary microscopist, and winner of many photomicrographic competitions with his beautiful pictures, is MI (Spike) Walker. By using Cokin *Variocolor* bi-coloured polarisers together with phase-contrast illumination, he has created 'Spikeberg' illumination, which is very similar in appearance to Rheinberg's coloured dark-field. The P170 red-green combination Variocolor filter is the best. In this case, the colour element is due to polarisation, one colour orientated N-S passing through the phase annulus with the linear polarising film (and forming the zero-order background illumination), and the complementary colour passing through the remainder of the transparent annulus to give colour to the diffracted rays forming the image.

The success of the method is due to the production in the early 1970s of the annular plate by Zeiss (Oberkochen) in their VZ series of condensers (46 52 78) for phase contrast combined with transmitted-light fluorescence. Admittedly, this is a niche technique, and the condensers have long since ceased to be made, but the results are stunning; the red image on a green background gives the best contrast. The example shown is crystals of Vitamin C.

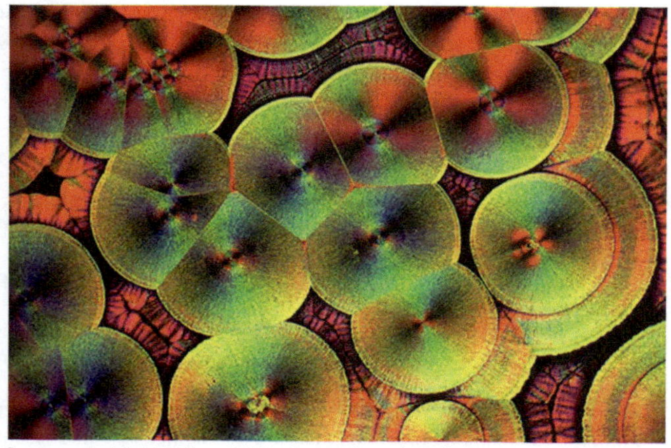

Source: © M I Walker 2005. Reproduced with permission from M. I. Walker.

11.9 Contrast by Exploiting Refractive Index Differences

Where two substances meet at a common interface, refraction can cause relief contrast. The greater the difference in refractive index between the two substances, the more pronounced the relief effect, which is caused by refraction at those parts of the interface that are inclined and not exactly at right angles to the direction of propagation of the light. From our perspective looking down on the optical axis, the refracted light is diverted from its original propagation direction and is reduced in intensity viewed along the vertical axis. Specimens with a refractive index higher than the surrounding medium show a positive relief; those with a lower refractive index show negative relief.

Most of the remaining contrast-enhancement techniques described in this chapter depend upon a difference in refractive index between the specimen and its mountant (which may include air) in order to generate contrast. Dispersion staining exploits chromatic difference of refraction at the boundaries of particles to give coloured contrast. Phase-contrast microscopy exploits sudden refractive differences at the edges of specimen boundaries ('edges'), and both Hoffman modulation contrast (HMC) and differential interference contrast (DIC) exploit gradual differences at the edges of the specimen boundary ('wedges') to produce a visible image from a transparent object.

The optician Horace Dall (Dall, 1985) introduced the 'visibility index' (VI) to quantify refractive index differences. It is quoted as the modulus (i.e. the positive difference without regard to sign) between the refractive index (RI) of the sample and that of the mounting medium. For example, the silica of diatom frustules has an RI of 1.46; water has a RI of 1.33; oil & Canada balsam, an RI of 1.52; Naphrax (a mountant for diatoms), an RI of 1.73; realgar (a classical mountant for diatoms) has a variable RI, from 2.18 upwards, and sometimes as high as 2.61 – Dall gives the RI as 2.45. The visibility indexes for silica are therefore: 0.13, 0.06, 0.27 and $\approx$ 1.0 respectively for each of these mountants. Air-mounted diatom preparations (VI = 0.46) are therefore superior to some mounted preparations and exhibit better contrast, unless the mountant has a high refractive index. However, in realgar refraction of light at the edges of the object produces a very thick delineation in the image. This means that whilst the contrast is inherently good, the edge effect arising from refraction inevitably obscures visibility of fine detail in this region.

The Becke line is a halo that forms around the periphery of a particle or fibre where there is a refractive index mismatch between the particle and the mounting medium that it is immersed in. This line can also be seen surrounding cells in phase-contrast microscopy, where it is more usually described as the phase halo. It is due to refraction and diffraction of the background illumination. When illuminated with axial (i.e. almost parallel) light from a condenser with a closed-down aperture diaphragm, grains, fibres and particles will either diverge or converge the light (Figure 11.16) at their edges, by refraction, according to whether their refractive index is lower or higher than that of the medium that surrounds them.

The Becke line moves with adjustment of the plane of focus. It is sometimes helpful to use monochromatic illumination to avoid any effects of dispersion staining (see Section 11.10): either a yellow 589 nm filter for the sodium D lines or a green 550 nm filter are recommended. However, the most important thing is to procure axial (i.e. virtually parallel) coherent illumination by considerably under-filling the numerical aperture of the objective. A quick and efficient way to do this is to focus a good image with a 40x/0.65 NA objective and then flip out the top lens of the condenser. The best and most reliable Becke lines are found with medium magnification objectives of NA between 0.5 and 0.7. Thus use a 20x, 25x or 40x objective. Another way to secure axial illumination is to focus the particle with a low magnification 10x objective, and then turn in a 20x or 40x objective without re-adjusting the condenser aperture iris. The smaller the refractive index difference in the preparation between particle and mountant, the smaller the illuminating aperture must be in order to be able to visualise the Becke line for useful analysis.

If the Becke line moves from the particle into the liquid upon increasing the distance between objective and stage (i.e. above the plane of best focus on an upright microscope), then the liquid has the higher medium ('focus high: goes high'). If you focus below best

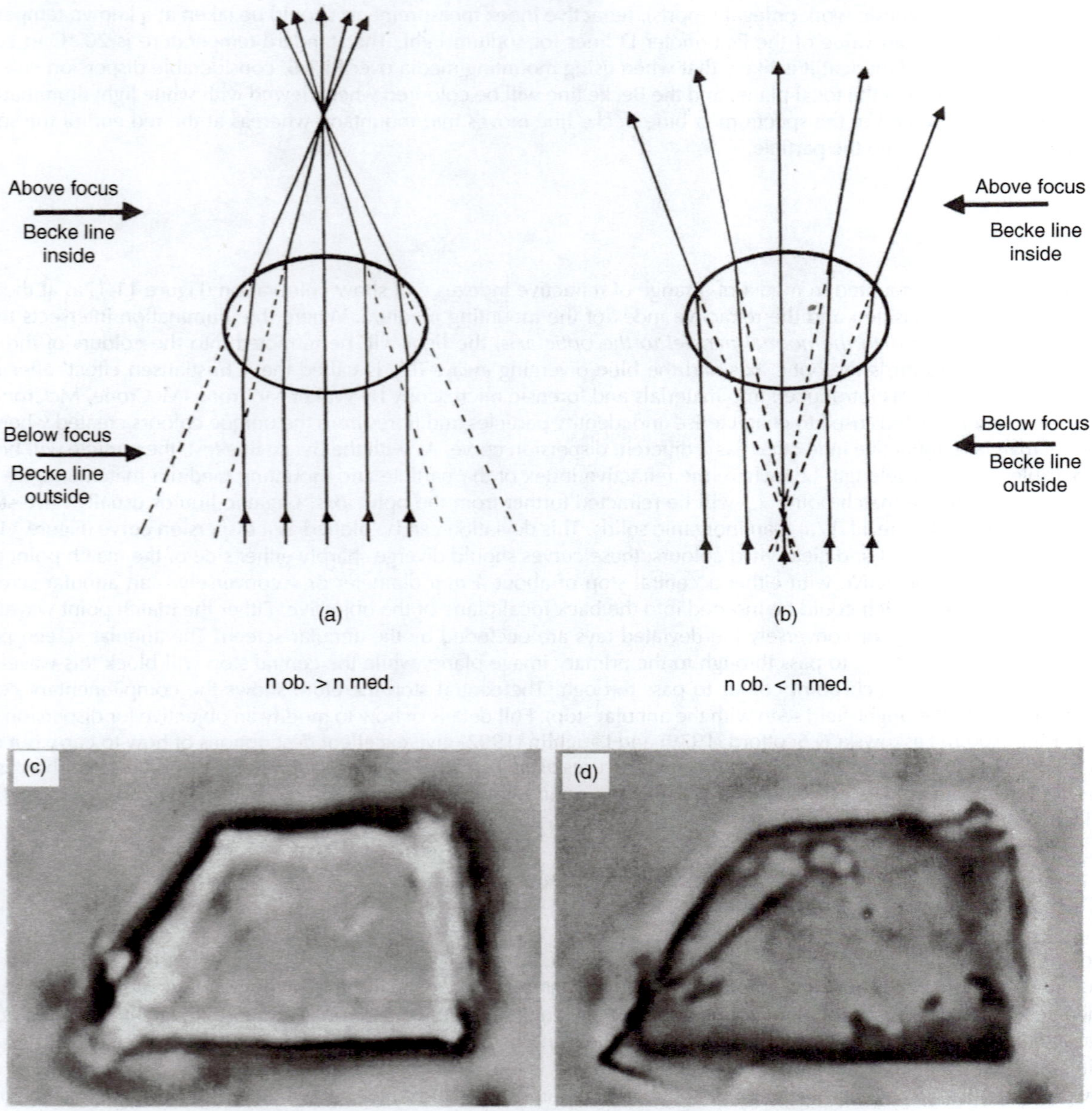

Figure 11.16 ***How the Becke line is formed*** *Named after the Austrian mineralogist and petrologist Friedrich Becke, this method provides a very useful means of estimating the refractive index of grains, fibres or particles of unknown materials as an aid to eventual identification. In (a) the transparent object has a higher refractive index than the surrounding mounting medium. Upon raising the objective above the plane of sharp focus, the bright Becke line appears inside the specimen. Where (b) the specimen now has a refractive index lower than the mountant, the situation is reversed and so on raising the objective above focus, the Becke line, still bright, appears to surround the object. The lower two panels show this occurring with a salt crystal mounted in oil. The effect is obvious in (c), but the outer Becke line is harder to discern in (d); it is not the more obvious dark edge of the crystal but can be seen surrounding this edge. Source: Bradbury and Evennett 1996. Reproduced with permission from Taylor & Francis Group.*

focus, the opposite effect occurs ('focus low: goes low'), and the Becke line moves into the substance with the lower refractive index: in our example, the particle. To match and determine the refractive index of the particle the medium must be substituted with another liquid of higher or lower RI, as appropriate, until (a) the Becke line can no longer be discerned or (b) the particle or grain disappears (or virtually disappears) within the mountant.

The most commonly used oils for refractometry are Cargille oils. They are supplied in small (1 oz. or 15 mL bottles, with refractive index steps of 0.01, 0.002 or 0.004 RI). If you have a deconvolution microscope in the imaging facility, this ought to have a suitable series of Cargille oils; otherwise, Hartshorne & Stuart (1970) give a list of suitable fluids on page 265 of *Crystals and the Polarising Microscope*. Eventually, you should have a single oil that matches the RI of the particle or the two oils either side of the requisite refractive index. A small volume of equal quantities of these two oils can be mixed on a slide to give the exact intermediate refractive-index oil for use. Remember that temperature affects the refractive index, particularly of liquids. Therefore, for accurate measurements to have any meaning (e.g. for analysis, forensic work or legal reports), refractive index measurements should be taken at a known temperature and usually at the 589.29 nm mean value of the Fraunhofer D lines for sodium light. The standard temperature is 20 °C in Europe and 25 °C in the United States of America. It is likely that when using mounting media over RI 1.6, considerable dispersion will be shown upon focusing above and below the focal plane, and the Becke line will be coloured when viewed with white light illumination. Upon raising the focus at the blue end of the spectrum, a blue Becke line moves into mountant, whereas at the red end of the spectrum, a red-orange Becke line moves into the particle.

11.10 Dispersion Staining

Small, transparent particles mounted in media of a range of refractive indexes will show colouration (Figure 11.17a) at their borders, depending upon their composition and the refractive index of the mounting medium. Where the illumination intersects the edge of the crystal at an angle *other than the normal parallel to the optic axis*, the light will be refracted into the colours of the spectrum, with the red converging towards the optic axis and the blue diverging away. This is called the 'Christiansen effect' after its Danish discoverer, and the method was introduced into materials and forensic microscopy by Walter McCrone (McCrone, McCrone & Delly, 1979). Dispersion staining is thus used to characterise and identify particles and fibres from the unique colours created when mounted in a medium of *matching* refractive index that has a different dispersion curve. As with the Becke line test, the particle will be invisible, or almost invisible, at the wavelength (λ_o) where the refractive index of the particle and mounting medium match and are the same. Wavelengths further from the match point, λ_o, will be refracted further from the optic axis. Organic liquids usually have steeper and more pronounced curves (Figure 11.17a) than inorganic solids. This deviation can be plotted as a dispersion curve (Figure 11.17b). For a good dispersion staining effect and clear vivid colours, these curves should diverge sharply either side of the match point at λ_o.

McCrone developed an objective with either a central stop of about 4 mm diameter or – conversely – an annular screen with a hole about 2 mm diameter, which could be inserted into the back focal plane of the objective. Either the match point wavelength, λ_o, is occluded by the central stop or conversely the deviated rays are occluded by the annular screen. The annular screen permits the wavelength at the match point, λ_o, to pass through to the primary image plane, while the central stop will block this wavelength and allow white light minus the match point colour to pass through. The central stop therefore shows the complementary colour on a dark-field (as opposed to the bright-field seen with the annular stop). Full details of how to modify an objective for dispersion staining is described by King (2009). Laskowski & Scotford (1979) and Laughlin (1992) give excellent descriptions of how to carry out dispersion staining. It is important to remember that dispersion staining uses *axial* instead of full-cone illumination and thus 'resolves' colour and not fine detail in the image. Overall resolution of the image is poor, but the colour contrast can be used diagnostically, particularly in the identification of various types of asbestos.

11.11 The Importance of Conjugate Planes

In Chapter 9 the importance of the imaging and illuminating sets of conjugate planes when properly adjusting the microscope to form a good and scientifically robust image was explained and emphasised. As will be explained below, in some forms of contrast enhancement (e.g. phase contrast, Hoffman modulation contrast, and differential interference contrast), the light is modulated by a stop or a special opening (e.g. phase annulus) or prism (DIC) at the front focal plane of the condenser. The subsequent conjugate plane in which an image of the front focal plane of the condenser is formed is the back focal plane of the objective. Any appropriate amplitude filter, phase plate or prism placed here will complete the light manipulation and so form the required contrast enhancement.

11.12 Phase Contrast

Until the widespread use of phase-contrast microscopy from the 1950s onwards, dark-field or oblique illumination microscopy were the only methods of visualising thin unstained cells, tissues or other similar specimens. These samples absorb light to a negligible degree, and/or exhibit very few differences in refractive index between their components and the medium in which they reside or are placed upon, and so are virtually invisible. It is possible to use a crude form of phase contrast to show an amplitude change that we can see by closing down the condenser iris diaphragm and defocusing a little, but the resolving power of the microscope is then seriously compromised and significantly reduced.

As we saw in Chapter 10, all specimens diffract light, and the information about their structure is carried by these diffracted rays. The direct, zero-order, beam passes through without deviation or change in phase. The zero-order and diffracted beams are brought to a focus by the objective at the primary image plane, where they interfere to form the image. When light passes through a thin section stained with dye, the diffracted light will be retarded in phase by one half wavelength ($\frac{1}{2}\lambda$).

The overall difference in intensity between the *resultant wave* – which is the product of the interaction of the direct (zero-order) beam and higher-order diffracted beams – and the direct beam is sufficient for our eyes to perceive this difference as a change in amplitude or brightness, even though the diffracted rays are still in phase with the zero-order direct beam. This situation is shown in the upper ray diagram on the right-hand side in Figure 11.19a. However, the resultant wave arising from the interaction between the direct zero-order

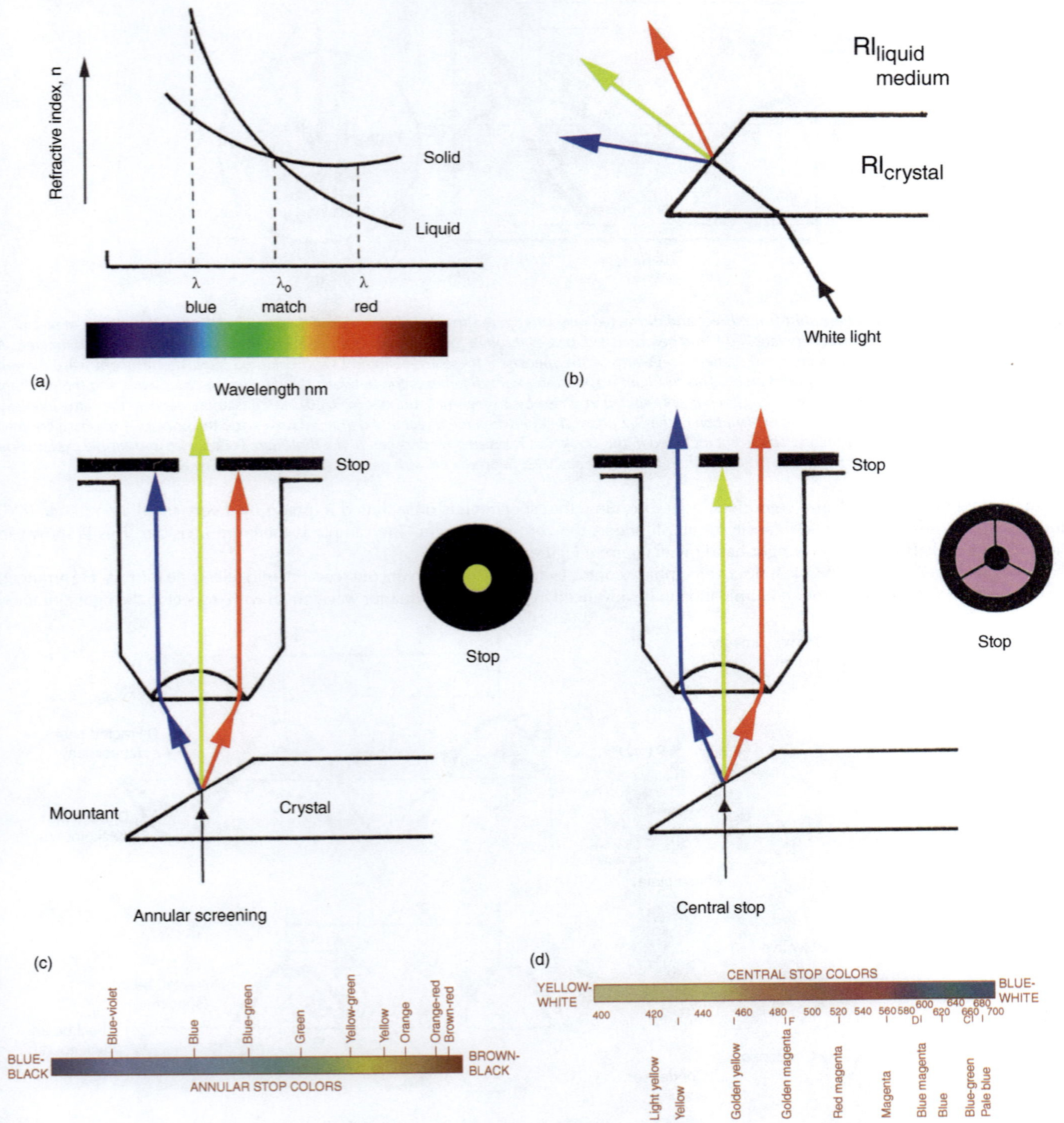

Figure 11.17 **Dispersion staining** *To produce dispersion staining colours, both particles and immersion fluids must have different dispersion curves (a) that intersect sharply in the visible-light region. A Cargille matching series of liquids of differing RI values is normally used for this purpose, whose dispersion is greater than the particle. (b) When illuminated with white light, red wavelengths bend towards the crystal and optical axis, and the blue diverges away towards the liquid medium. A blue Becke line will therefore move outwards and a red Becke line inwards, upon raising the objective and focusing upwards. Figures (c) and (d) show how dispersion colours are formed with annular and central stop dispersion objectives. Source: Adapted from Bradbury and Evennett 1996; Laughlin 1992. Reproduced with permission from John Wiley & Sons.*

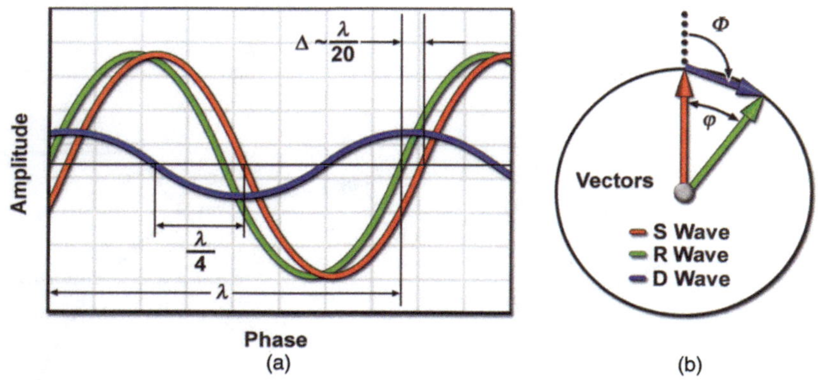

Figure 11.18 **Formation of phase shift from direct and diffracted rays** *This figure shows the behaviour of light passing through a transparent so-called 'phase' object. A zero-order wave, here labelled S and coloured red, passes through the specimen but does not interact with it and is not diffracted. A small proportion of waves are diffracted and scattered regularly by the specimen; these are denoted D and coloured blue. The resultant wave formed at the primary image plane, R, is coloured green. This resultant wave arising from the interaction between the direct zero-order wave and the resultant wave in a colourless, non-absorbing, thin specimen is only shifted in phase by a very small amount (≈ λ/20), and it is approximately the same intensity or amplitude as the zero-order beam. An explanation of the ¼λ phase shift between the direct and diffracted waves and the principles of phase contrast is elegantly explained by vector theory. This is not included in this book, but interested readers can find a thorough, yet concise and simple explanation in the Appendix to Bradbury & Evennett (1996). Source: Murphy et al. 2013. Reproduced with permission from John Wiley & Sons.*

beam and the diffracted rays in a colourless, non-absorbing, thin specimen is only shifted in phase by a very small amount (≈ λ/20) and it is approximately the *same* intensity or amplitude as the zero-order beam. The sample is therefore invisible. This is shown in Figure 11.18 and also in the lower right-hand ray diagram in Figure 11.19a.

Frederick (Frits) Zernicke (Box 11.3) discovered phase contrast whilst he was carrying out research into telescope mirrors. He arranged for the direct zero-order background illumination to be advanced in phase by one quarter wavelength with respect to the light diffracted

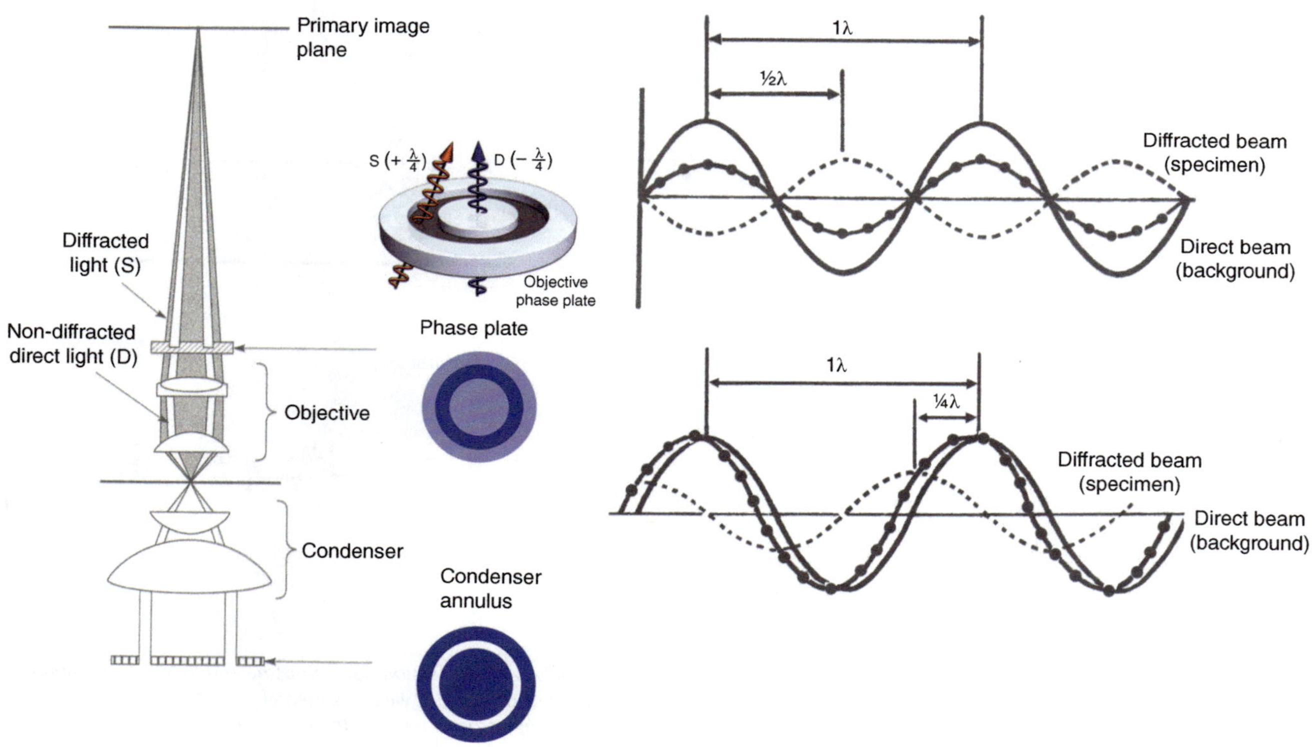

Figure 11.19 **Ray diagram in a phase-contrast microscope** *The direct zero-order light is constrained by an annular circular slit and retarded in phase with respect to the diffracted rays by passing through a specially constructed phase plate. The annulus lies at the front focal plane of the condenser and the phase plate at the back focal plane of the objective. An amplitude image is formed at the primary image plane when the zero-order light and phase-shifted diffracted rays are recombined and interfere to form the primary image. Source: Murphy 2001. Reproduced with permission from John Wiley & Sons; Sanderson 2002. Reproduced with permission from John Wiley & Sons; Bradbury and Evennett 1996. Reproduced with permission from Taylor & Francis Group; Murphy et al. 2013. Reproduced with permission from John Wiley & Sons.*

by the specimen. Since the specimen also introduces a phase shift of difference $\frac{1}{4}\lambda$ anyway, the *overall* phase shift at the primary image plane, when the direct beam and diffracted beams are recombined is now one half wavelength, and so the phase specimen will now be seen as if it were an amplitude (or absorbing) specimen.

Box 11.3 Frits Zernike and Phase-Contrast Development

Source: Bettmann / Contributor. Reproduced with permission from Getty Images.

Prior to the discovery of imaging by phase contrast, dark-field illumination was the only way to make the image of a transparent specimen visible. This is because the eye discriminates only the energy of light falling on the retina and thus can only detect changes in the wavelength and intensity. Any change in the phase of the light cannot be seen.

In order to see the phase variations, it is necessary to combine the light passing through the sample with a reference beam so that interference between the two beams coverts the phase difference into an amplitude difference, thus revealing the phase structure of the sample in the image. A no-phase-change condition gives rise to completely destructive interference.

This was first realised by Frits Zernike, a theoretical physicist at the University of Amsterdam, whilst studying spectral lines using diffraction gratings, which he worked on from about 1930. Zernike discovered that the so-called *Rowland ghost lines*, which occur on either side of a grating line that have their phase shifted from that of the primary spectral line by 90°. These ghosts are caused by slight imperfections associated with ruling the gratings during manufacture. When Zernike focused his viewing telescope onto the gratings, something surprising happened – the striped ghost lines disappeared.

Through a series of experiments, Zernike came to understand that the direct light and the diffracted light were actually in different phases. When the telescope was focused on the gratings, the resulting image, created through interference, was rendered invisible because the eye cannot pick up phase differences. During these studies he appreciated that (a) it is necessary to interfere with a reference beam to maximise the contrast, and (b) it is necessary to introduce a phase shift to this reference so that the no-phase-change condition gives rise to completely destructive interference. Zernike recalled the work of Lord Rayleigh, who described in 1900 an operation to make shallow etchings in glass with very dilute hydrofluoric acid. Using this approach, Zernike created 'phase strips', in which the direct light hit a thin etching on a glass surface while the diffracted light passed through the glass. The result was to shift the direct light in phase by 90° with respect to the diffracted rays.

It was at a medical physics Congress at Wageningen in 1933 that Zernike first described his phase-contrast technique in microscopy. He had previously been to the Zeiss works at Jena, where he was received with disdain. One long-standing senior manager said:

> 'If this had any practical value, we would ourselves have invented it long ago!'

The trouble was, since about 1890 when Abbe turned more to managing the firm, his staff of scientific collaborators and acolytes formed the impression that everything worth trying had already been done. It wasn't until the early 1940s that the Zeiss R&D department with Kurt Michel then working for Köhler would build a prototype phase-contrast microscope and take a beautiful cine film of spermatogenesis in the grasshopper *Psophus stridulus*. Otherwise, it is just possible that this method of contrast might have been known as 'Bratuscheck contrast', for in 1892 Karl Bratuscheck (1865–1913) described the phenomena of phase retardation with an attenuating carbon film. Indeed, Bratuscheck was hired by Abbe in 1892; it is believed that only his poor health prevented him from developing phase contrast.

In order for the direct zero-order light to be artificially retarded in phase with respect to the diffracted rays, the background illumination must be controlled. This is done by constraining the illumination to shine through an annulus placed at the front focal plane of the condenser. The annulus provides a hollow cone of illumination, with the specimen placed at its apex. The undiffracted zero-order light then enters the objective as an inverted hollow cone, and an image of the phase annulus is formed in the back focal plane of the objective as a ring of light (see the diagram of the light path in the microscope shown on the left in Figure 11.19). This annulus is made equivalent in diameter to nearly the full numerical aperture of the objective. Special objectives, fitted with phase plates, are required for phase contrast. The diameter of the etched 'trench' as a ring in the phase plate is the same size as the projected image of the phase annulus. By this means, all the zero-order background illumination can be controllably retarded in phase with respect to the diffracted rays that enter the objective and pass through the rest of the back focal plane of the objective. If the annulus is not properly centred with respect to the fixed phase plate in the back focal plane of the objective (Figure 11.20), then negligible or reduced contrast will be the result.

Provided that the diffracted and undiffracted beams are out of phase by $\frac{1}{2}\lambda$ overall, it does not matter whether the diffracted beams are advanced or retarded by $\frac{1}{4}\lambda$ with respect to the direct, zero-order, beam. Two forms of phase-contrast microscopy are therefore possible: if the path length of the diffracted beams is longer than the direct beam, the diffracted beams being advanced by $\frac{1}{4}\lambda$, this is called positive phase contrast and is the most commonly used design. The zero-order beam travels through a shorter overall path length, because the phase ring is etched as a trench on the carrier plate situated in the back focal plane of the objective and so is retarded in phase. For negative phase contrast, the reverse is true: the extra path length being provided by an annular *ridge* instead of a trench (see Figure 11.22). In positive phase contrast, structures with refractive indexes higher than their surroundings appear darker in the image against a lighter background. Cell organelles having a lower refractive index than the surrounding cytoplasm appear bright against a grey background. It is with positive phase contrast that we recognise a cell with a dark nucleus, dark membrane and dark lysosomal

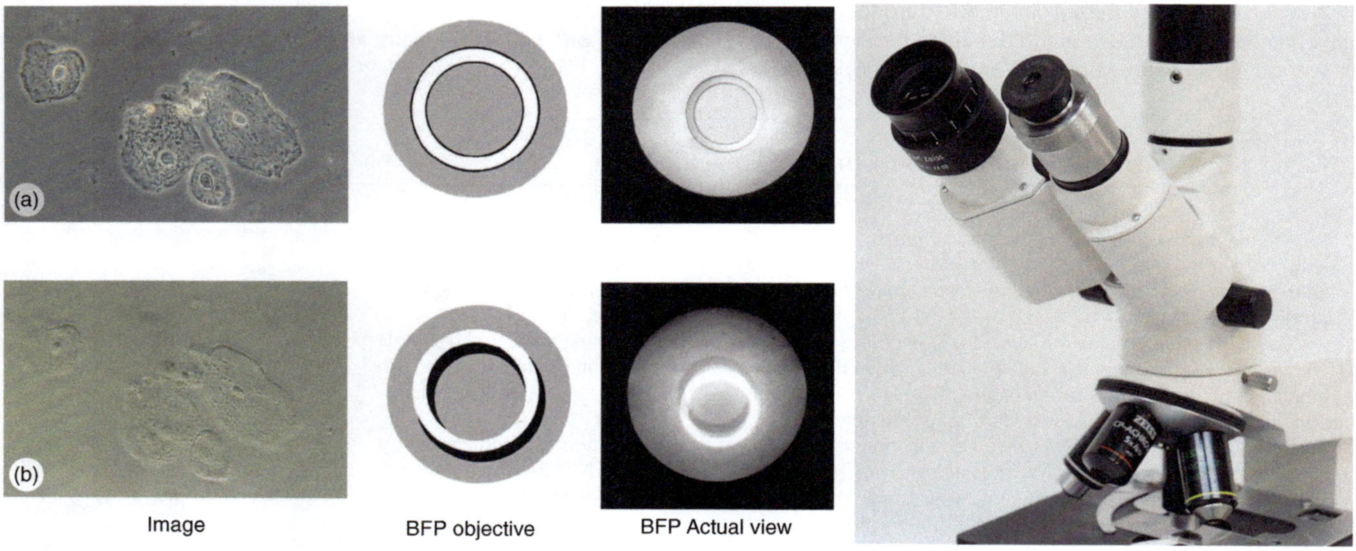

Figure 11.20 **Aligning the phase annulus to the phase plate** *The phase annulus in the condenser illuminating the specimen must be aligned with the fixed phase plate located in the back focal plane of the objective. This is done either by temporarily removing an eyepiece; by using a phase telescope to magnify the image of the back focal plane of the objective, as shown here in the right-hand eyepiece tube, or by retaining the eyepiece and using a Bertrand lens inserted into the optical path.*

and cellular inclusions against the lighter background of the cytoplasm and the neutral grey of the background illumination. Negative phase contrast, which nowadays is seldom encountered, shows a reversed image to that seen with positive phase contrast.

Some people refer to 'high refractive index' or 'low refractive index' objects in a phase-contrast image; this is technically incorrect and only strictly true if the objects being compared have the *same thickness*, for it is the optical path difference, the product of refractive index and thickness, that gives rise to the contrast enhancement. Thus a thin object with high refractive index and a thick object with low refractive index will appear the same intensity, or shade of grey, in the image if their optical path differences are the same.

Because the intensity of the diffracted light forming the image is much less than the zero-order illumination forming the background, the intensity of the direct, zero-order, beam must be reduced in order that glare does not reduce the visibility of the phase-contrast image in the primary image plane. It is only because the otherwise invisible trench of the phase ring is coated with a layer of light-absorbing material working as a neutral-density filter that we can see it as a dark grey ring in the back focal plane of the objective.

Since the image of the phase annulus situated in the condenser is projected into the back focal plane of the objective by the objective, each individual objective requires an annulus to be matched to the size of phase ring etched into its back focal plane. Low magnification objectives have large diameter phase annuli (normally inscribed 'Ph1' for 'phase 1' and suitable for 5x or 10x objectives); intermediate magnification objectives have Ph2 annuli (e.g. 20x and 40x objectives) and the 60x or 100x objectives have the smallest diameter annuli, generally inscribed 'Ph3'. It is convenient to house these various annuli in the carousel of one universal condenser along with a dark-field patch stop and Wollaston prisms for DIC, and other contrast-enhancement methods, as well as an empty space fitted with an illuminating aperture (condenser) diaphragm to be used for bright-field microscopy.

In order for the image of the phase annulus to be aligned accurately with its corresponding phase ring in the objective, centring adjustments (usually recessed at the back of the condenser carousel, see Figure 11.21) are provided. It is important to remember that these individual phase annulus adjustments are quite separate from the centring screws or adjustment provided for ensuring that the entire condenser is aligned whilst setting for Köhler illumination. Once centred on-axis, it is usually not necessary for the phase annuli to be adjusted further. The image of the illuminated field diaphragm that is centred whilst adjusting the microscope and setting up Köhler illumination is in the imaging – or field – set of conjugate planes (Chapter 9, Section 9.10). The phase ring and image of the phase annulus are both seen in the illuminating – or aperture – set of conjugate planes. To see the illuminating set of conjugate planes, an eyepiece is removed to observe the back focal plane of the objective down the body tube of the microscope.

Because any image seen at the back focal plane of the objective is very small, particularly for higher magnification objectives, it is helpful to use a magnification aid to accurately centre the image of the phase annulus onto the phase ring of the objective. This can either be done using an auxiliary telescope, otherwise often referred to as a 'phase telescope', or it can be achieved with a built-in 'Bertrand lens' system. When using the phase telescope, an eyepiece is removed and the telescope substituted in its place (Figure 11.20). The Bertrand lens – if fitted – is used with the eyepiece remaining in place and thus is slightly quicker and more convenient to use. The Bertrand lens projects the illuminating set of conjugate planes into the focal plane of the eyepiece, allowing a magnified image of both phase annulus and phase ring to be seen together. The Bertrand lens is often incorporated as a lens system into a rotatable auxiliary magnifier situated on the binocular head of the microscope. Together with the eyepiece, the Bertrand lens forms a low-power microscope. Both the phase telescope and the Bertrand lens are fitted with an independent means of focusing, because objectives of different magnification have different focal lengths, and their back focal planes lie at different levels within the microscope.

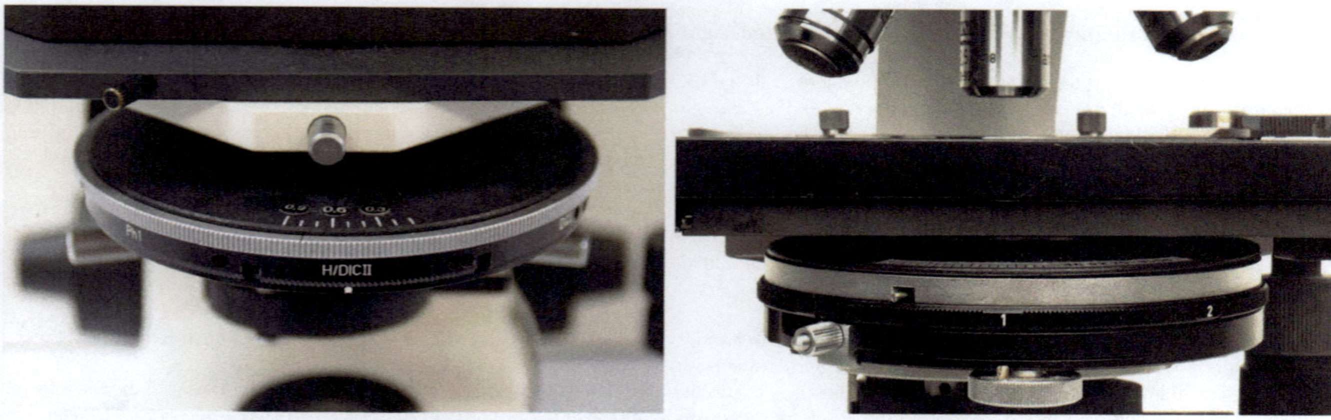

Figure 11.21 **Centring controls for the phase annulus** *On all but entry-level microscopes, the phase annulus can be centred with respect to the phase plate independently of the condenser alignment required to set up Köhler illumination. On a modern condenser, shown in the left-hand panel, centring is performed via inset controls manipulated with a hexagonal Allen key. On the right, an older condenser with fixed lateral controls: a captive lever moving through an arc and a knurled rotating thumbscrew is shown. Source: Courtesy of Mr K. Glover.*

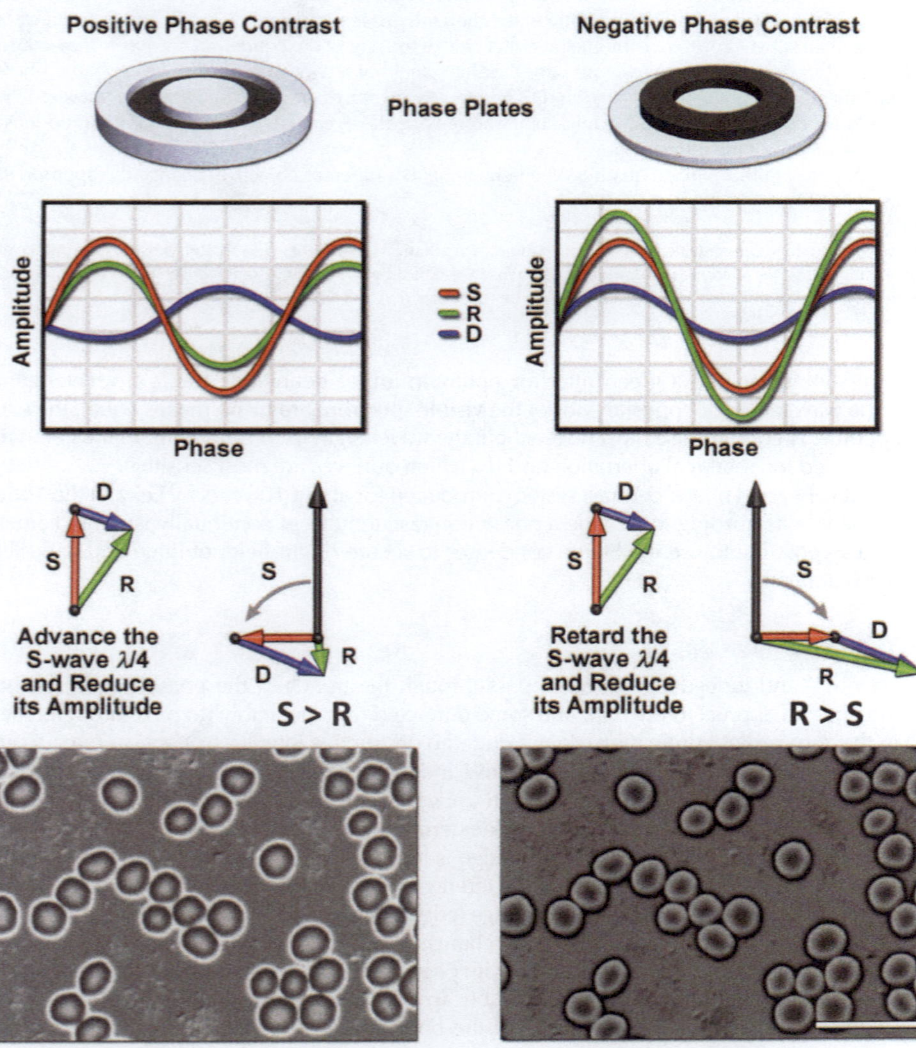

Figure 11.22 **Comparison of positive and negative phase contrast** *Erythrocytes shown by phase contrast. The phase plate advances the zero-order, or surround wave, S, relative to the diffracted wave, D. The amplitude of the resultant wave, R is lower (positive) or higher (negative phase contrast) than the zero-order wave, causing the high-refractive index object to look darker or lighter, respectively, against the background. Scale bar = 20 μm. Source: Murphy et al. 2013. Reproduced with permission from John Wiley & Sons.*

Protocol 11.4 Examination of a Specimen by Transmitted-Light Phase Contrast

1. Prepare a phase contrast test slide. To prepare the cheek cells, gently scrape the inside of your mouth with the tip of a plastic pipette or clean finger nail and spread the cells and saliva on the centre of a clean No. 1.5 coverslip. Invert the coverslip quickly onto a clean slide, and allow the fluid to be drawn throughout the area of the coverslip by capillary attraction. Generally it is not necessary to press down to spread the cell preparation into a thin layer. Seal the edges of the coverslip with a thin layer of nail polish. This will keep the unfixed preparation for an hour or so and prevent the coverslip from falling off due to drying out or being pulled off when oil is used.[i]
2. Using the lowest magnification phase-contrast objective, set up the microscope for bright-field observation by Köhler illumination using a well-stained 'amplitude' specimen. Do this using the bright-field position in the universal condenser carousel. Correct condenser *height* is important with phase contrast as with DIC, so ensure that the illuminated field diaphragm is imaged sharply and that the condenser is centred with the normal condenser-centring screws.
3. Exchange the stained slide for an unstained phase specimen.
4. Remove an eyepiece, and insert a phase telescope if one is to hand, *or* rotate the Bertrand lens into the optical path of the microscope if this is fitted (there is no need to remove the eyepiece if the Bertrand lens is used).
5. Use the focusing adjustment on either the phase telescope or the Bertrand lens, and secure a sharp image of the phase ring in the back focal plane of the objective. This is easiest to do if the condenser carousel is still set at the bright-field position, as in Step 1. Do not try to alter the focal position of the specimen with the microscope focusing controls. These should be left alone.
6. Move the appropriate phase illuminating annulus into position, which may be marked on the side of the condenser by a fiducial mark. The size of the annulus should correspond with the inscription on the objective. Provided the condenser is set at the correct height and is centred on axis (Step 2), the annulus should appear the same size as the phase ring.
7. Still using the phase telescope or Bertrand lens, centre the image of the phase annulus precisely onto the phase ring. Remember, the centring controls for this adjustment of each individual phase annulus are quite separate from those used to centre the entire condenser for Köhler illumination. The thickness of the annulus should be slightly wider than the phase ring, and it will, of course, be much brighter.
8. These centring controls are often set at 120° to one another at either side of the rear of the condenser carousel. They may use separate hexagonal Allen keys (sometime housed in slots on the microscope stand), or have their own captive screws or levers. If the phase ring and annulus are slightly misaligned (rotate the turret slightly), the background light intensity increases, and the quality of the phase-contrast image falls.
9. Once the centring operation is complete, remove the telescope and replace the eyepiece, or remove the Bertrand lens from the optical path. Focus the image with the fine focus of the microscope.
10. Insert a green filter, if desired, to optimise image quality. When changing to a different objective, change the appropriate illuminating annulus also, and centre this, as needed.

[i] Nail varnish, which has been allowed to dry thoroughly (leave it at least 10 minutes to allow all the solvent to evaporate), is better than either vacuum grease or petroleum jelly for sealing temporary preparations, for it is all too easy to overlook how close the objective is to the edge of the coverslip when search for a specimen or the correct focal plane. Cleaning grease or jelly off a dry objective is difficult and not to be recommended.

Phase-contrast equipment should include a green filter for optimum image definition. The $\frac{1}{4}\lambda$ wavelength shift has to be for an agreed value; in any case, the wavelength at opposite ends of the visible spectrum are by no means equal, the wavelength for red being approximately twice that of blue. A compromise must be reached: the wavelength used is 550 nm, in the green region of the spectrum, for which objectives are corrected for spherical aberration and for which our eyes are most sensitive.

The Heine condenser was a very good phase-contrast system introduced for about 10 years by Leitz in the 1950s based on a cardioid condenser unit sliding vertically in its housing to provide a phase-contrast annulus of continually varying diameter to match the phase plate of any objective. It is also possible to use the Heine condenser to secure bright-field, oblique and dark-field illumination and to move from one to the other seamlessly.

11.12.1 Interpreting the Phase-Contrast Image

Most diffracted beams will not – and indeed should not – pass through the trench of the phase ring in the back focal plane of the objective. However, this ring has an appreciable width, and some diffracted rays inevitably do pass through. They remain out of phase only by $\frac{1}{4}\lambda$ with respect to the zero-order illumination, do not undergo destructive interference and so remain bright. This effect causes the bright haloes that surround each dark feature of high refractive index (Figure 11.23). These edge features are also lined internally with a darker halo, but this is more difficult to see than the well-known bright halo. The converse situation occurs for features in the specimen of lower refractive index than the background: a dark external and a bright internal halo. Since some of the diffracted rays are attenuated when passing through the phase plate, a sharp edge is not formed around features in the image. Phase-contrast haloes are especially noticeable around large objects formed by diffracted rays of low spatial frequency.

The phase-contrast effect is maximal at regions of sudden change (edges) in optical path difference (the product of refractive index and geometrical thickness) and is not so evident at areas where the change in optical path difference (OPD) is less pronounced (wedges). This is known as 'shading-off' and is another artifact associated with phase contrast. Shading-off, otherwise known as the zone-of-action effect, occurs with large extended phase objects (Figure 11.24) because central zones of uniform refractive index and refractile edge zones both diffract light differently. Light from central zones of the object is mostly diffracted through small angles so that it passes through the phase plate and so is not fully retarded in phase and contrast enhancement does not occur. The full phase-contrast effect is only seen at the edges of refractile structures. Qualitatively the shading-off effect is not necessarily undesirable, because individual refractile particles near the cell centre appear darker against a lighter background that would otherwise be equally darker and so obscure them.

The effect of both the halo and the shading-off effect can be minimised by using apodized phase plates in the back focal plane of the objective. This newer design of phase plate has two concentric neutral-density filters lying either side of the trench of the phase plate.

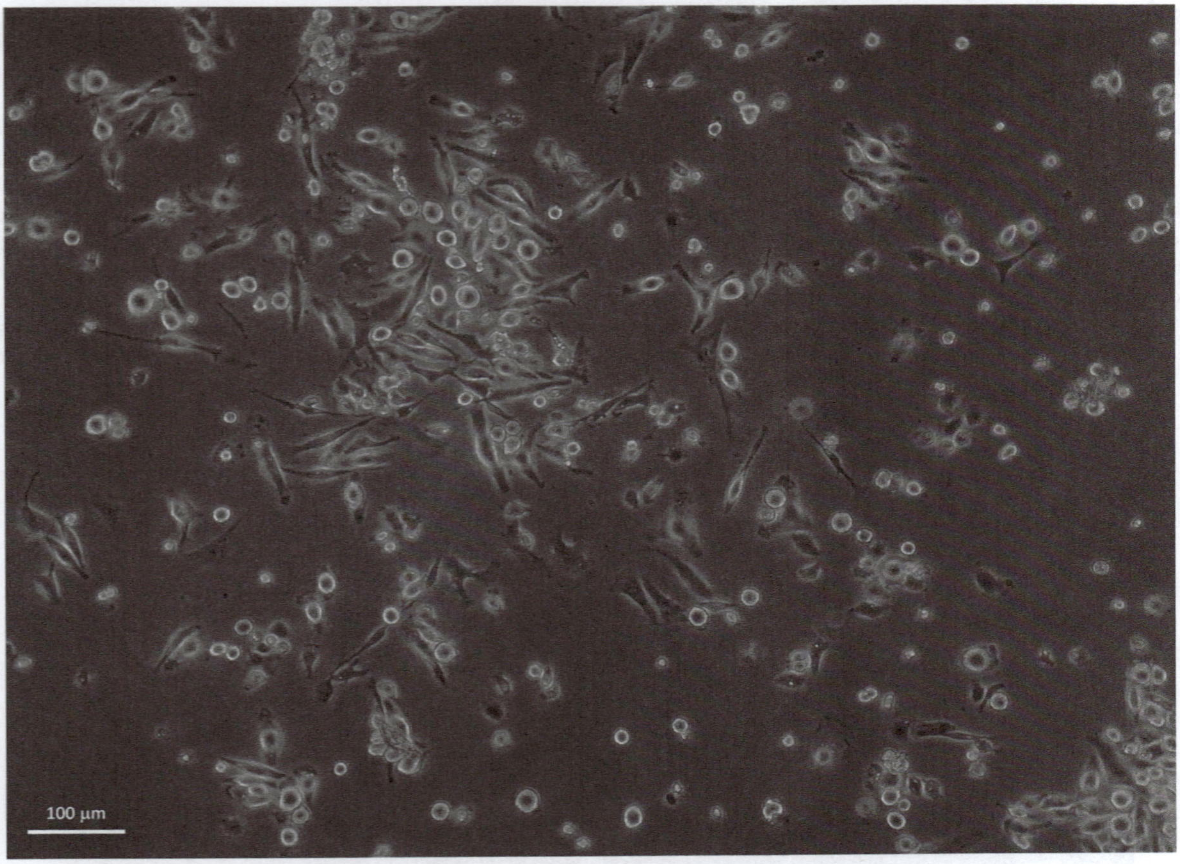

Figure 11.23 **The halo effect in phase contrast** *Macrophages imaged with positive phase contrast using a 10x objective. A noticeable halo surrounds the highly refractile rounded cells. The halo is less evident around the flatter adherent cells, which refract to a lesser extent.*

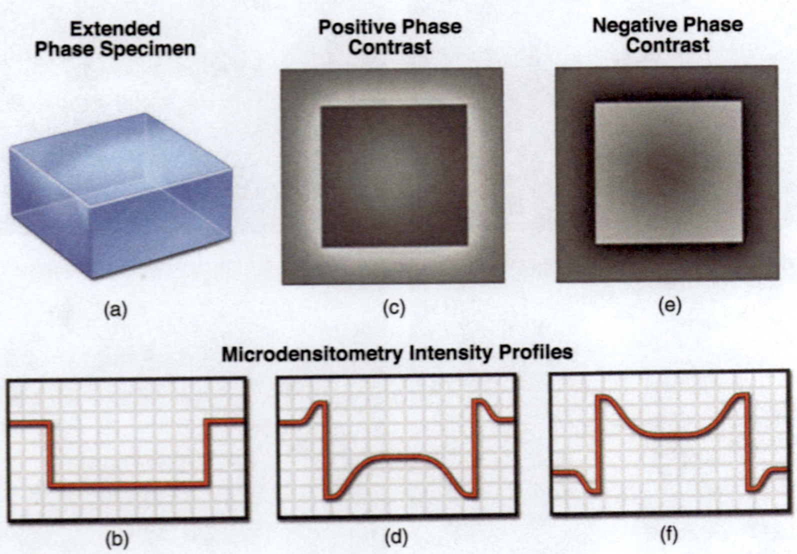

Figure 11.24 **The 'shading off' effect** *The artifact occurs because for a large, thick specimen of uniform refractive index, the centre portion and the refractile edge zones both diffract light differently. If the specimen is large enough, the intensity of the central portion will tend towards that of the background. Figure (a) shows a large phase specimen, (b) the intensity profile of this specimen, whilst (c) and (e) show the images formed by positive and negative phase contrast, with (d) and (f) showing the corresponding intensity profiles. Source: Murphy et al. 2013. Reproduced with permission from John Wiley & Sons.*

On top of the classical phase plate is a thin evaporated-metal coating, reducing transmission of the direct, zero-order, light by about 75% to be of approximately the same intensity and amplitude as the rays diffracted by the specimen. The extra neutral-density filters in an apodized phase plate transmit 50% of the light and reduce the intensity of light diffracted from the specimen at small angles that passes through the phase plate.

Phase contrast is better suited to structures with large and sudden changes in optical path difference: an 'edge' rather than with a 'wedge'. For specimens with features showing gradually changing optical path differences, use differential interference contrast (DIC, Section 11.14). As with interference contrast, remember that the intensities seen in the image are wholly representative of optical path difference (the product of refractive index and geometrical thickness of the specimen) and thus may not necessarily represent the actual height and depth of structural features in the object. For this reason, phase contrast is not ideally suited to making accurate measurements of, or from, the image, since it can be difficult to precisely determine the exact position of the edge of a feature in the specimen.

Phase contrast has been developed for reflected light, but this method is not now widely used. For further details see Chapter 12, Section 12.10. Phase contrast is not suited to coloured samples (Figure 11.25) since the stain introduces unwanted amplitude changes when the zero-order light is already modulated to suit the intensity of the diffracted rays forming the image. Phase contrast with stained

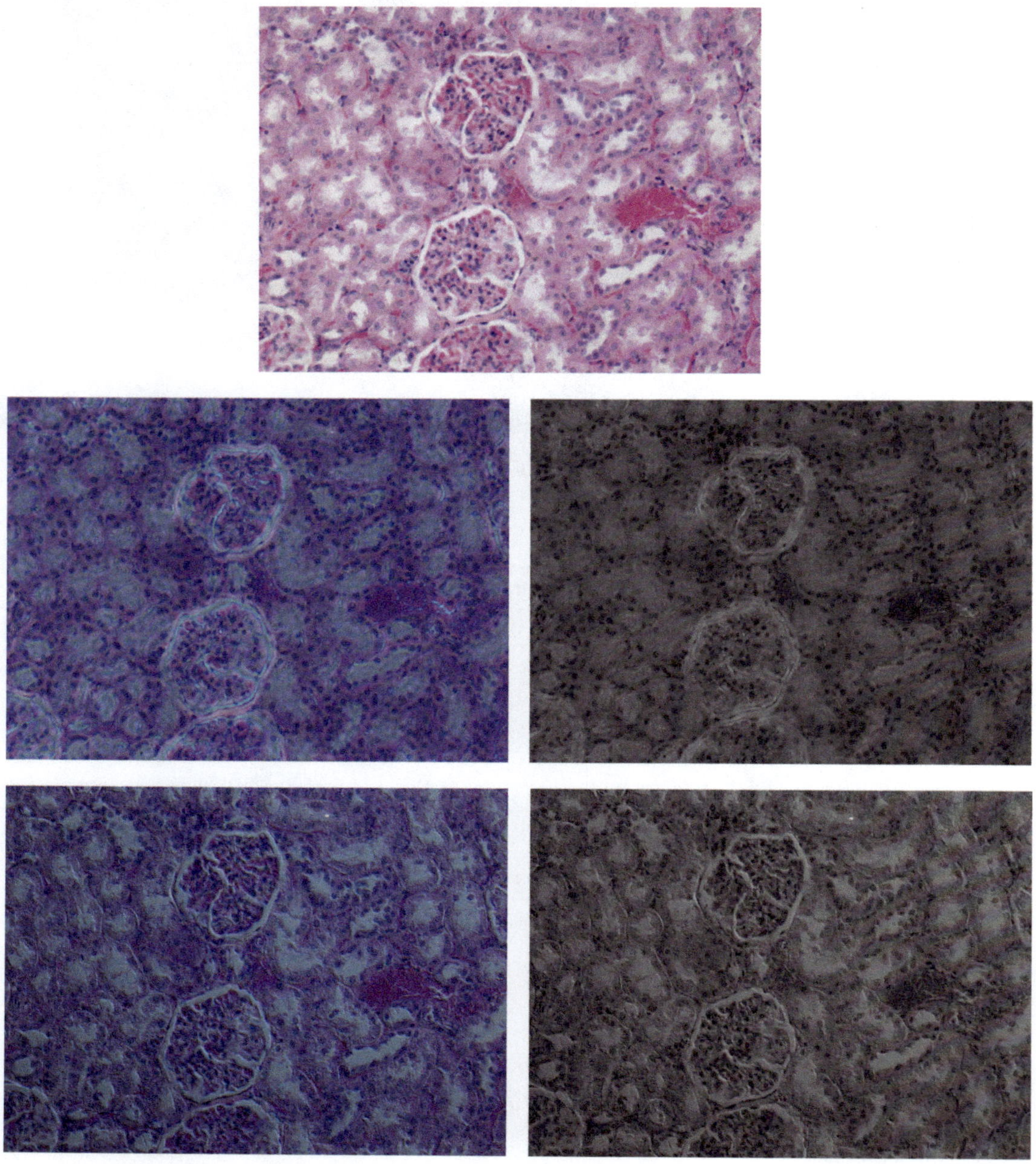

Figure 11.25 **Colour with phase contrast and interference contrast** *Comparison between phase contrast and DIC imaging of coloured specimens. For explanation, see the text. The top image is an image of a stained kidney section taken in bright-field. The middle row shows the same specimen imaged by phase contrast and the lower two panels by differential interference contrast.*

specimens, and tissue sections, will give rise to a 'muddy' and muted image without good contrast or any crisp detail. Ideally, phase contrast is suited to thin phase specimens, whereas if the object is too thick, contrast reversal can be seen. Conversely, interference contrast gives good images with coloured or thick phase specimens.

11.13 Hoffman Modulation Contrast

Hoffman modulation and similar relief contrast techniques are sophisticated methods of oblique illumination. The modulator in HMC is built asymmetrically into the back focal plane of the objective (Figure 11.26) and has three regions that differ in transmittance. Most of the area is clear, transmitting 100%, but a narrow strip transmits 15% of the light, and the very edge of the modulator transmits only 1%. On the opposite side of the optical axis of the microscope, a slit in the front focal plane of the condenser is arranged so that light falls obliquely on the modulator in the objective. Any phase structures in the specimen that are gradient- or prism-shaped will refract light into the objective modulator and thus form a shadow-cast image (Figure 11.26). In Hoffman modulation contrast, phase gradients are changed to gradients of light intensity and so form a 3D pseudo-relief image, similar to oblique illumination and DIC. Phase gradients in one direction (generally the positive phase gradients) are always rendered light, and those in the opposite direction (negative gradients) show up dark in the image.

The slit in the condenser is partially covered with polarising material so that its effective width can be controlled with a polariser situated below the condenser. When the vibration direction of the polariser is 'crossed' with respect to the slit in the condenser, the latter will appear narrow, and when the vibration direction of the polariser is parallel to the slit in the condenser, that slit will be wider. An image of a narrow slit will only fall on the grey portions of the modulator in the objective and so generate a high-contrast modulation image. When the slit width is greater, its image will not only fall on the grey portions of the modulator but also on the clear area. This will reduce the contrast and give much better images of thick objects that have large differences in refractive index and relatively long optical path lengths between one part of the object and another. A narrow condenser slit is better suited to thin, flat specimens such as buccal squamous epithelial cells from the inside of the cheek. It is possible (Hammond, 2007) to replace the three-zone Hoffman modulator in the condenser by a knife-edge diaphragm and perform a variant of Schlieren microscopy. Gilbert Hartley, an eminent English microscopist, was unenthusiastic about HMC (Hartley, 1977) remarking it offered nothing new over classical oblique illumination or DIC. Indeed, Hoffman modulation contrast remains more widespread in the USA than in the UK or Europe.

Protocol 11.5 Examination of a Specimen by Hoffman Modulation Contrast

1. Remove the Hoffman slit diaphragm from the condenser (this may simply rotate out of the optical axis of the microscope), and remove the substage polariser.
2. Set up the microscope, and establish proper Köhler illumination.
3. Remove an eyepiece, and insert a phase telescope if one is to hand, *or* rotate the Bertrand lens into the optical path of the microscope if this is fitted (there is no need to remove the eyepiece if the Bertrand lens is used).
4. Use the focusing adjustment on either the phase telescope or the Bertrand lens, and secure a sharp image of the Hoffman modulator in the back focal plane of the objective.
5. Move the slit diaphragm into position in the condenser, and adjust it so that the image of the slit falls onto the grey region of the modulator plate. There are several designs of adjustment: the simplest is a planetary movement within a gripping holder; otherwise, screws set at 120° operating against a spring may be provided. Do not disturb the centration of the condenser set for Köhler illumination.
6. If the condenser slit has a built-in region of polarising material, insert the polariser between the condenser and illuminated field diaphragm and rotate it until it is 'crossed' with respect to the strip in the condenser, and the narrowest possible illuminating slit is obtained.
7. Remove the phase telescope and replace the eyepiece or remove the Bertrand lens.
8. Rotate the polariser to adjust the image for optimum contrast.

The major microscope companies have produced and marketed similar relief contrast methods of contrast enhancement, for these are ideally suited to examining unstained cells in tissue culture facilities. Leica have a system called Integrated Modulation Contrast (IMC), which uses a modulator in the primary image plane and thus does not require modified objectives. Nikon Advanced Modulation Contrast (NAMC), Olympus Relief Contrast (RC) and Zeiss Varel contrast all use one form of modulator or another in the back focal plane of the objective. Jörg Piper (Piper, 2007) has developed relief phase contrast, which has the advantage that it uses existing phase-contrast objectives and so does not require modification with a special modulator in the back focal plane of the objective. The phase annulus is replaced by a smaller transparent sector, which can be crescent-shaped, or a small round aperture, and this provides the necessary oblique illumination.

11.14 Interference Contrast

Most biologists are accustomed only to using differential interference contrast (DIC) and may be unaware that any other interference contrast methods exist. Francis Smith was one of the first to patent the interference contrast microscope in the 1930s, nearly 20 years before Georges Nomarski developed his DIC system. Essentially, the principles are the same: the illumination is split into two coherent

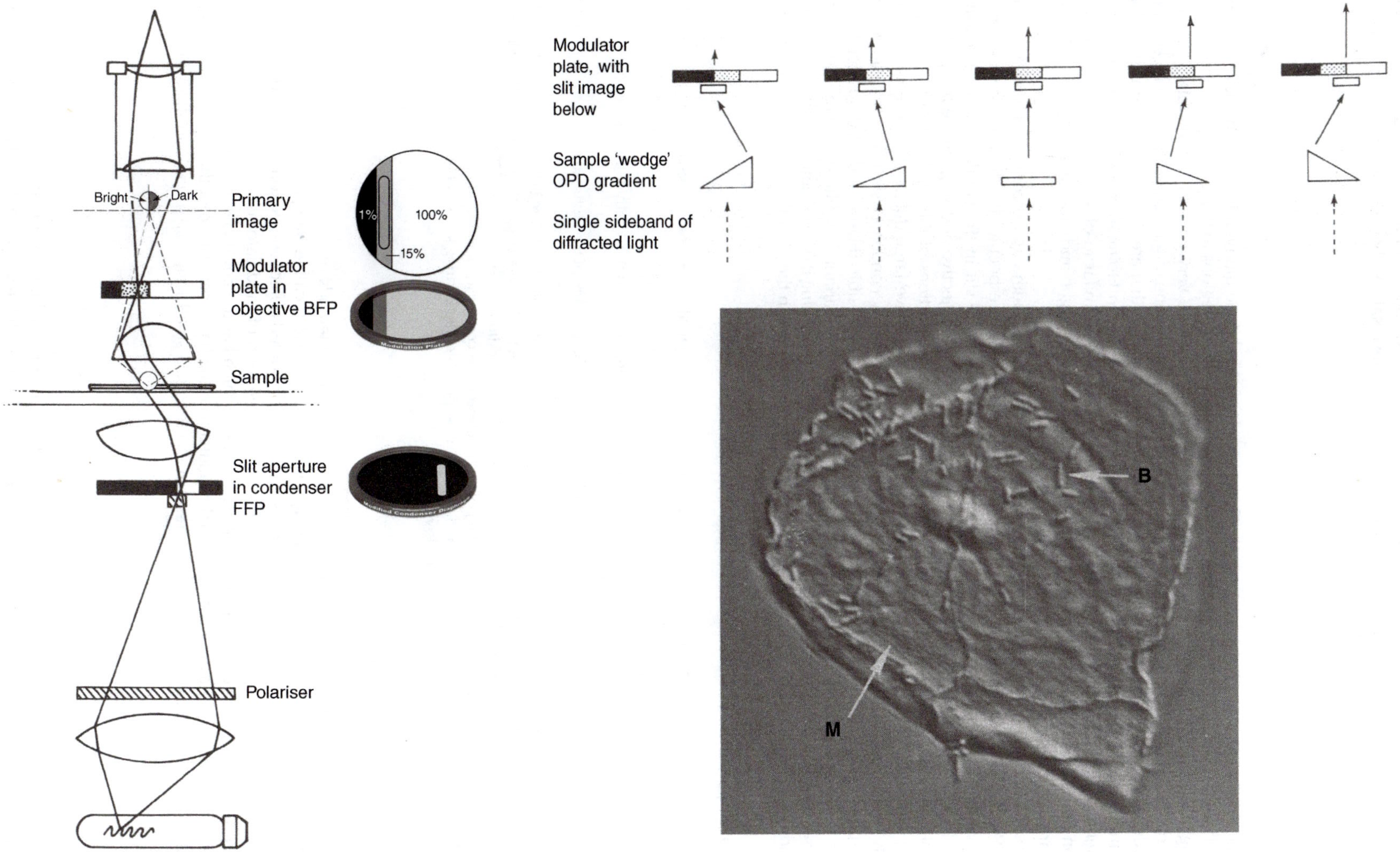

Figure 11.26 **Hoffman modulation contrast microscopy** *The left-hand panel shows the ray diagram of a microscope adapted for Hoffman modulation contrast. The polarising material partly covering the condenser slit aperture enables the lower polariser to be used to virtually increase or decrease the slit width by rotation. An image of the slit can be seen superimposed upon the central 15% transmission neutral-density material in the modulator. The upper right-hand panel shows how the modulator produces intensity variations within a range of optical path differences in the specimen. The length of the vertical arrow after the modulator indicates the relative intensity of the image-forming light. A squamous epithelial cheek cell is shown for comparison with other contrast modes shown in Figure 11.29. Adherent bacteria (B) and membrane ruffles (M) can be discerned.*

The advantage of Hoffman modulation contrast over differential interference contrast is that, unlike DIC, HMC uses no beam-splitting prisms. Furthermore, HMC can be used with birefringent specimens, or cells growing on tissue culture plastic, which are not amenable to DIC. The condenser slit must be aligned with the grey areas of the modulator plate to control the zero-order beam.

Source: (left) Adapted from James and Tanke 1991 with permission from Springer and Professor H. J. Tanke. (middle) Murphy et al. 2013. Reproduced with permission from John Wiley & Sons. (right) Hoffman 1977. Reproduced with permission from John Wiley & Sons.

beams, which are then altered in phase with respect to one another after interaction with the specimen. The image is formed when the two beams interfere upon recombination at the primary image plane.

The differential interference contrast microscope converts changes in optical path difference occurring in an invisible 'phase specimen' into visible amplitude changes. The two coherent rays, which interfere to form the image, are produced by a Wollaston prism beam splitter. This prism is made from two wedges of birefringent (doubly refracting) quartz or calcite cut such that their optical axes (see Chapter 13, Section 13.7) are at right angles to one another. A beam of plane-polarised light will be split into two perpendicularly polarised rays, separated by a small angle (Figure 11.27; see also Figure 13.5 and Figure 14.9). The beam-splitting Wollaston prism is situated at the front focal plane of the condenser. This prism is set at 45° in relation to the polariser. The design of the Wollaston prism ensures that the pair of perpendicularly polarised rays is sheared *laterally* by a small amount less than the resolving power of the objective (i.e. between 0.2 and 2 μm) and so passes through closely adjacent areas of the specimen without forming a double image. After traversing the specimen and undergoing an *axial* phase shift relative to one another, the two rays are brought to a focus at the back focal plane of the objective, where a second Wollaston prism recombines the two individual rays back into one beam of light.

If no specimen is in position on the microscope stage, the coherent rays have the same optical path lengths, and on recombination the o-ray and e-ray[11] have electric vectors vibrating in the same direction as the polariser. The recombined beam is therefore blocked by the analyser. If the two rays encounter and pass through a phase gradient, they will become differentially shifted in phase. Since the two rays originate from a common point in the filament, they are coherent and so will interfere. After passing through the second Wollaston

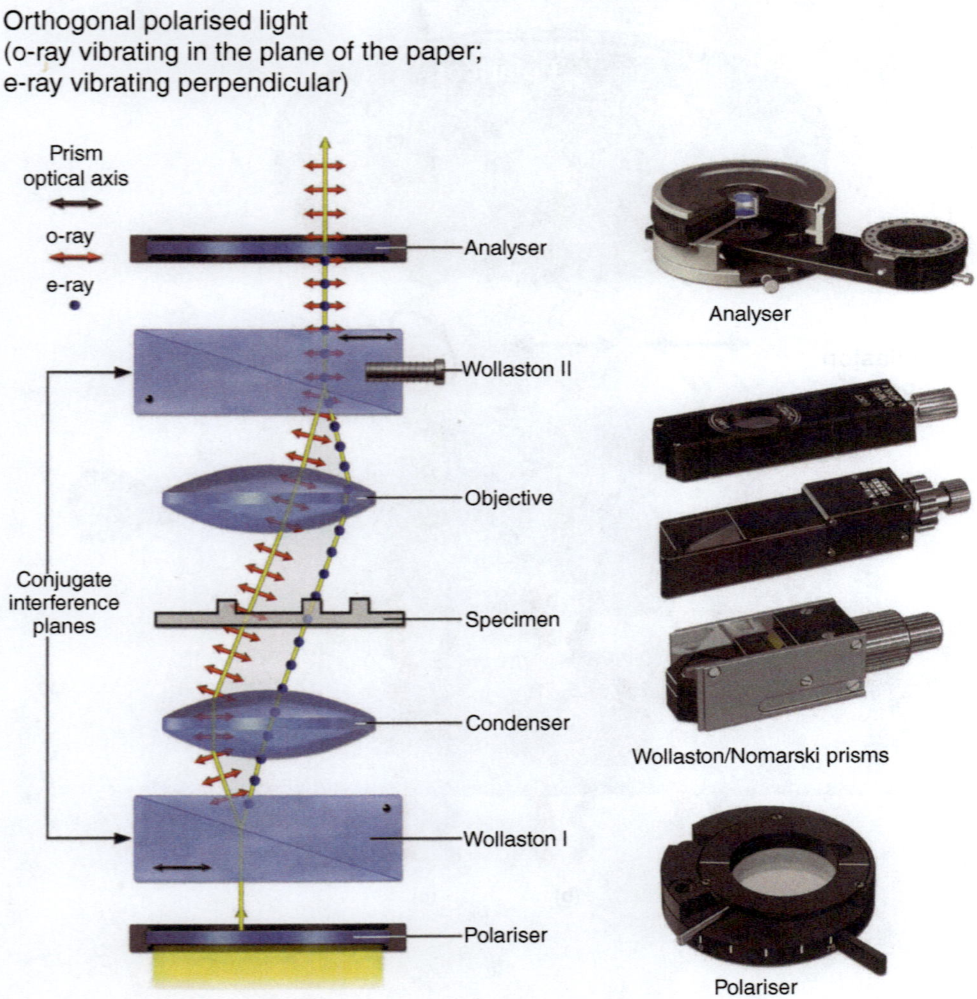

Figure 11.27 **Ray diagram showing how the image is formed in a differential interference contrast microscope** *A beam of linearly polarised light is split by the condenser Wollaston prism into two coherent (o- and e-) rays orthogonal to one another. The two parallel rays, separated by a distance smaller than the resolving power of the objective, follow separate paths between the condenser and the objective. In the absence of a phase sample introducing an optical path difference, these rays are recombined by the upper Wollaston/Nomarski prism, giving linearly polarised light vibrating in the plane of the polariser that is consequently blocked by the analyser. If the sample introduces an optical path difference, one ray is advanced with respect to the other thus forming elliptically polarised light, which can pass in part through the analyser, giving rise to amplitude contrast of differing intensities in the image. Source: Murphy et al. 2013. Reproduced with permission from John Wiley & Sons.*

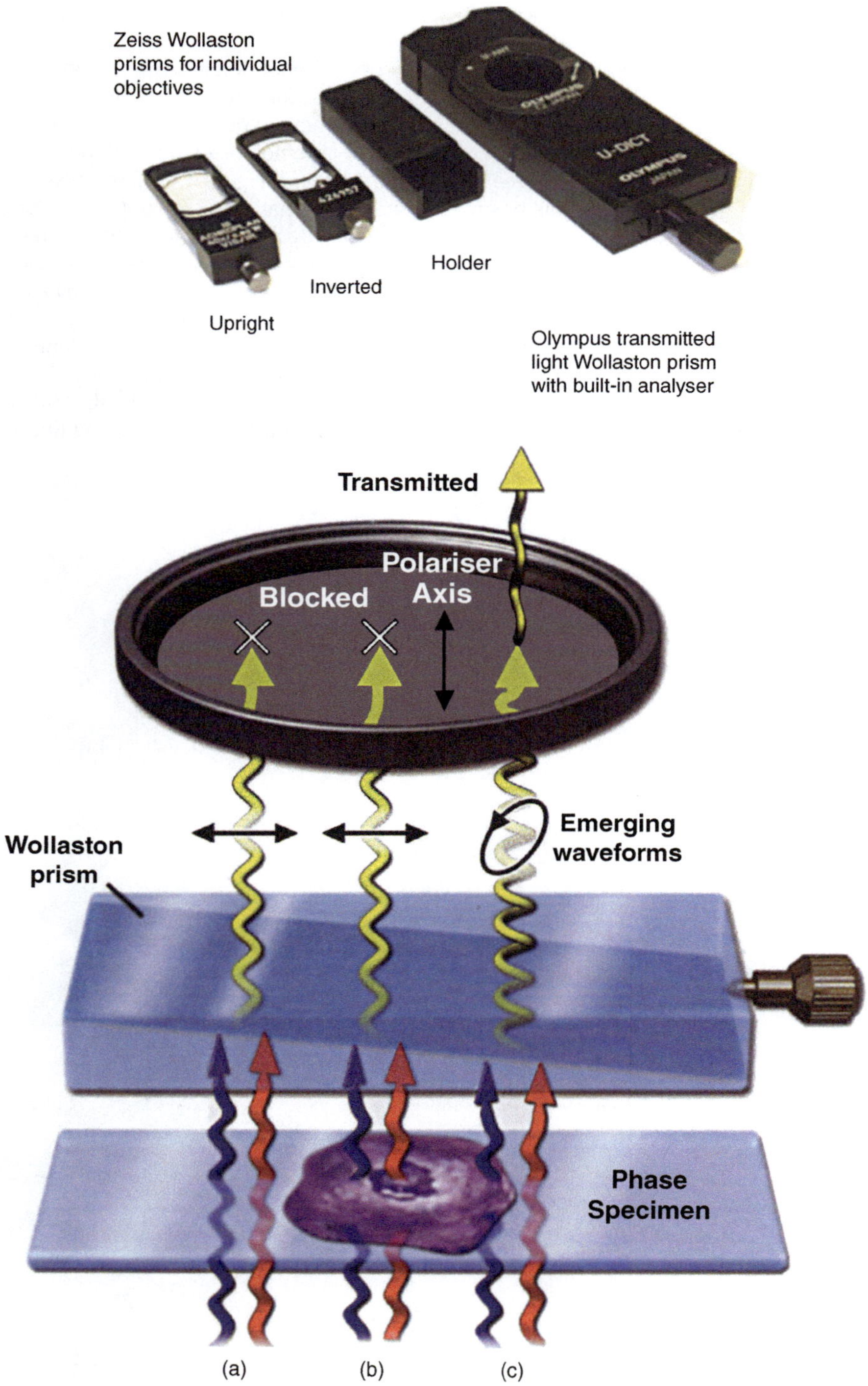

Figure 11.27 (Continued)

prism, the two rays are still polarised at 90° to one another within the beam. This beam emerges from this second beam-recombining Wollaston prism as elliptically polarised light. The elliptical components of these two rays, which are vibrating in the permitted plane of vibration of the analyser, pass through the analyser and go on to interfere and form an image at the primary image plane. By sliding the second Wollaston prism to alter the thickness of each wedge presented to the two coherent rays, the optical path difference between the two beams – and hence the phase difference between them – can be altered. This will alter the contrast of the image. If, however, a DIC adjustable Nomarski prism of the incorrect numerical aperture is used, which is not matched for the objective in use, image contrast will be impaired (Figure 11.28).

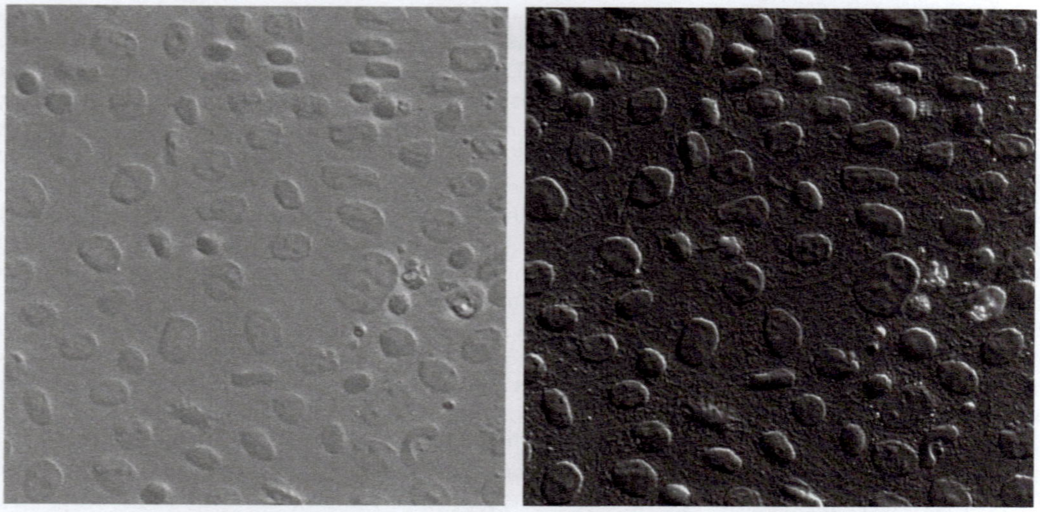

Figure 11.28 ***Consequence of using the wrong DIC Wollaston/Nomarski sliding prism*** *The left-hand panel shows the washed-out reduced contrast that results, compared to the right-hand panel where the correct Nomarski prism is used.*

Box 11.4 Why is Polarised Light Used with DIC?

For interference contrast to work, crossed polars are required. This is true also for Differential Interference Contrast. Why, therefore, is an analyser needed in the microscope together with the polariser? Secondly, why is the polariser on its own not sufficient? Thirdly, why do the Wollaston prism beam splitter and the Wollaston/Nomarski prism beam recombiner both have to be set at 45° to the crossed polars?

A more interesting question is why is a polariser is needed in the first place, since the Wollaston prism will split non-polarised light into the two orthogonally polarised beams – the o-ray and the e-ray, and it is these two coherent rays that interfere with each other to give rise to an amplitude difference and hence contrast.

If you remove the polariser, the image becomes much brighter and very washed out but doesn't disappear altogether. This is because the beam-splitting Wollaston prism (made of calcite) will split the light into two orthogonally polarised beams. The non-polarised light significantly degrades the visibility: thus the polariser acts as a selective spatial filter to cut out the non-image-forming light. On the other hand, if you rotate the polariser through 45 degrees, there should be no contrast visible in the image because all the light will now go down one of the two beam paths, for the o-ray or e-ray.

The crossed polars (polariser and analyser together) act as spatial filters to isolate the light so that coherent ray bundles can interfere, and incoherent rays cannot merely reduce contrast in the image.

In the absence of a specimen, the coherent o-ray and e-ray of each wave subtend the same optical path length. The upper objective beam-combining Wollaston/Nomarski prism recombines the two rays, generating a linearly polarised wave, whose permitted vibration direction is in the same plane as the polariser. The analyser blocks this light, so the background is black (or at least very dark grey).

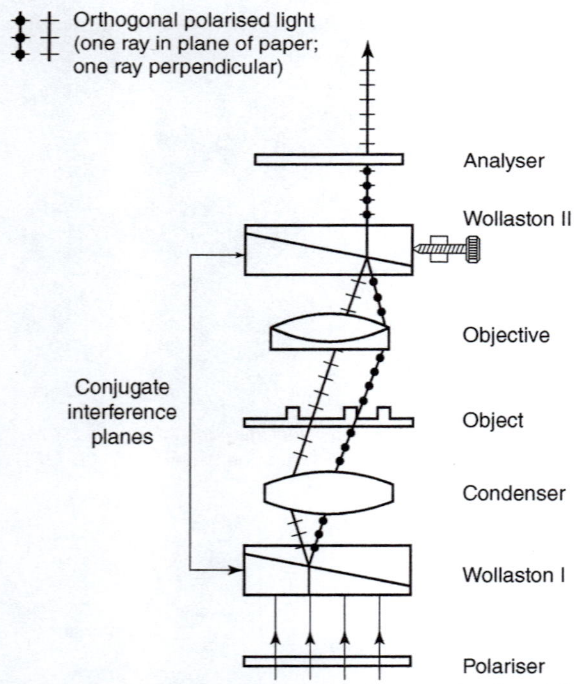

Source: Murphy 2001. Reproduced with permission from John Wiley & Sons.

With a specimen in place, the o-ray and e-ray pair undergo a phase shift relative to one another. On passing through the beam recombining Wollaston/Nomarski prism, the wave becomes elliptically polarised light, because of the phase difference between the o-rays and the e-rays. The analyser is required simply to allow the two orthogonally polarised beams (o-rays and e-rays) to interfere. A component of each beam gets through the analyser, and since these components are now oscillating in the same plane again, they will interfere, and form an amplitude image.

Reference

Murphy, D.B. (2001). *Fundamentals of Light Microscopy and Electronic Imaging,* 1st ed. New York: Wiley-Liss.

The advantage of DIC is that the separation of the two interfering beams occurs below the resolving power of the objective so that a double image is not seen, and it is not apparent to the observer or detector that the image is made up of separate image components from each beam. Each point in the image is made up from light from each separate beam, and the appearance of that point depends on the phase shift between the two beams and their respective optical path difference due to their interaction with the object. This contrast-enhancing method is described as differential because the appearance of the image depends upon the *rate of change* (differential as in mathematics) of the optical path length (which depends upon both the thickness and refractive index of the object at that point) and not to the OPD directly. Thus Figure 11.29 shows the graph of a curve describing the optical path difference across the specimen (e.g. across the diameter of an oocyte). In Figure 11.29a, the graph describes the actual optical path difference across the object. Figure 11.29b is the first derivative of that curve: the first change (increase) in OPD translates into the first 'hillock' in the curve; the second inverted 'hillock' represents the decrease in OPD. We see the intensity change, shown in Figure 11.29b, as light and dark shadows in regions of OPD change in the image.

Because the full numerical aperture of the condenser is used (a phase annulus will restrict the effective numerical aperture very slightly), differential interference contrast is capable of optical sectioning of thick samples. This technique allows a single focal plane to be recorded in sharp focus from a thick phase specimen without details from features lying outside of the immediate focal plane, or stray light, compromising image quality as can happen with phase contrast. Figure 11.30 shows images of a squamous epithelial buccal cell in various contrast modes to compare differences between bright-field, dark-field, phase and differential interference microscopy.

The shadowing effect will confer a 'pseudo-3D' appearance upon the image. This is why students often assume that thin cells (Figure 11.31) have the physical appearance and topology of a fried egg, whereas the image is merely a map of optical path differences,

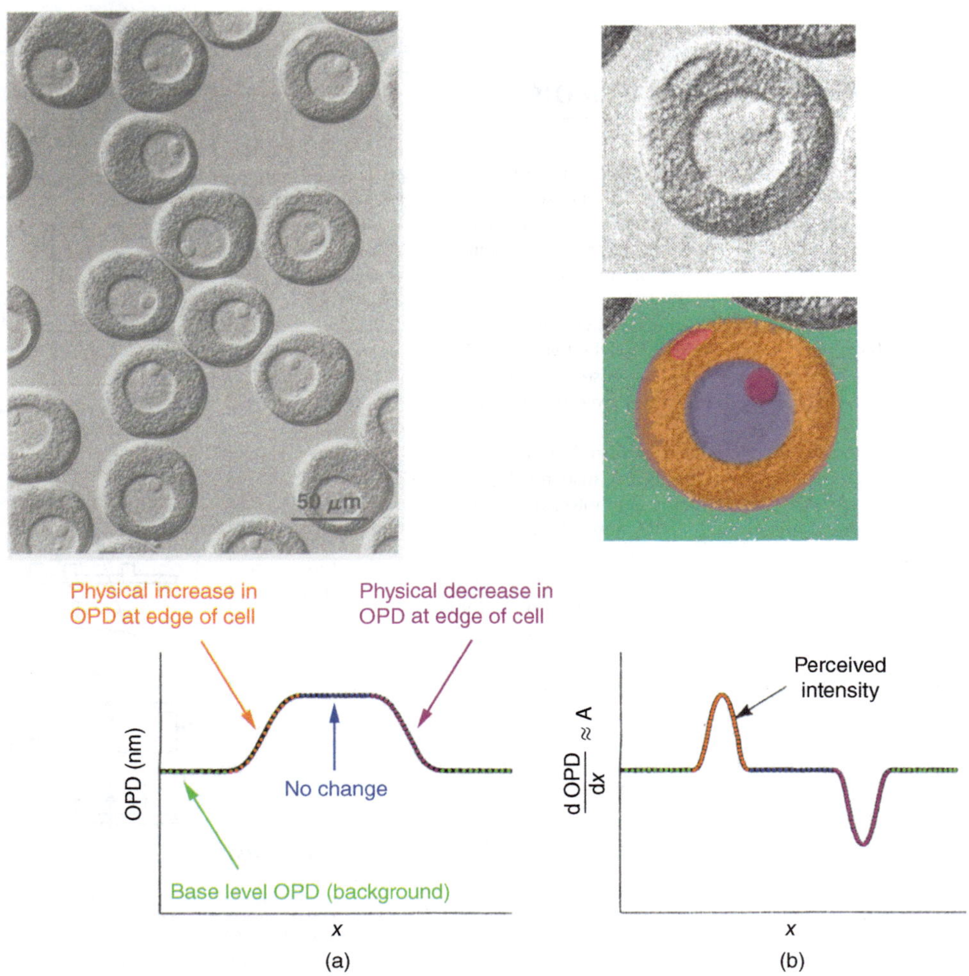

Figure 11.29 ***How a differential interference contrast image is formed*** *Primary oocytes of the surf clam* Spissula solidissima *give a well-defined DIC image. The relief seen is due to optical path differences (a) but it is impossible to distinguish whether the phase gradient in the object is due to specimen thickness or differences in refractive index between the specimen and its surroundings, or both. In DIC microscopy, the amplitude intensity seen as relief in the image corresponds to the* derivative *of the optical path difference (b) rather than the OPD, hence the name. Contrast is a function of the rate of change of optical path across the specimen, as seen in graph (b) corresponding to the pseudo-coloured image. Source: Adapted from Murphy 2001 with permission from John Wiley & Sons.*

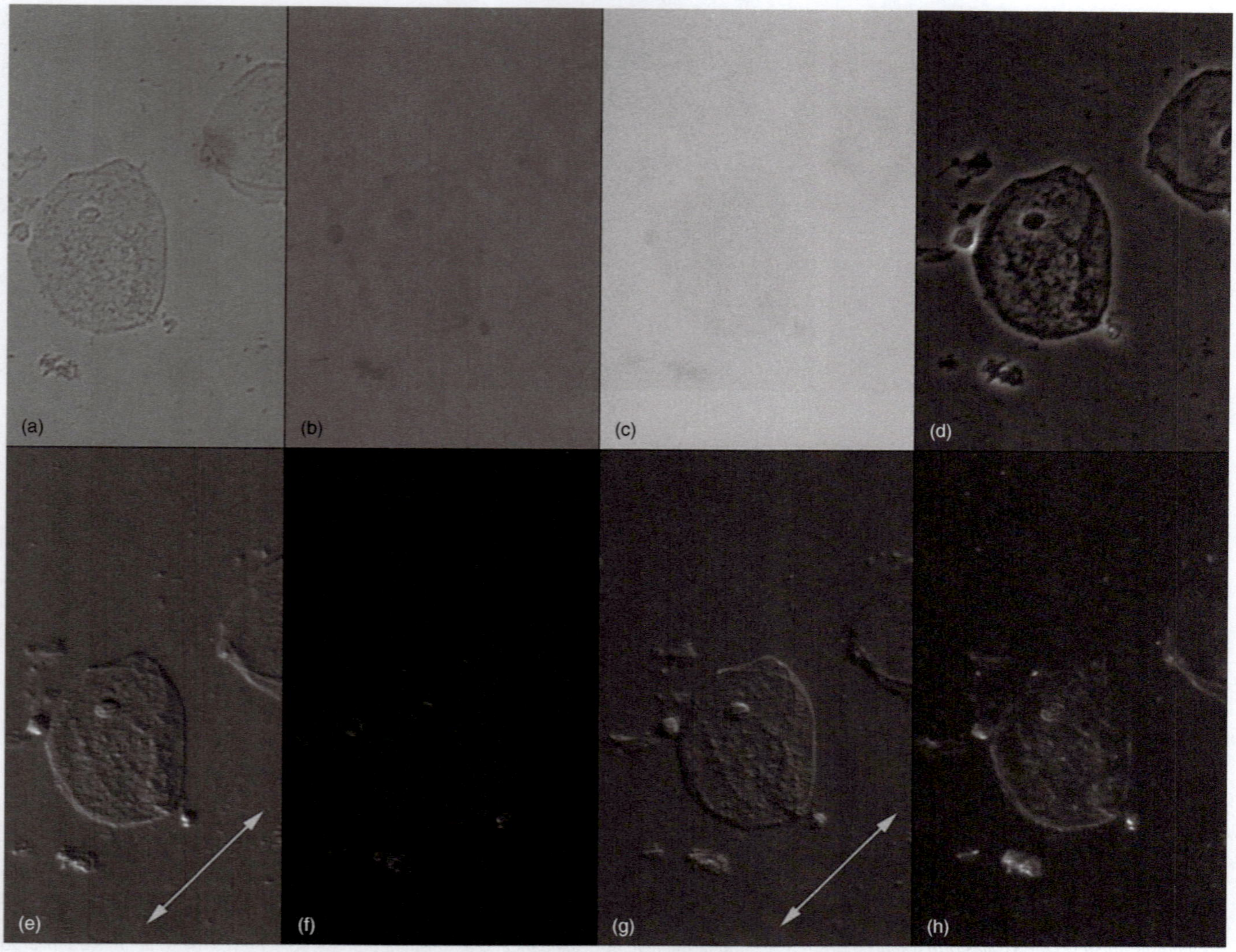

Figure 11.30 **Squamous epithelial cheek cell by different contrast modes** *Comparison of squamous epithelial cheek cell imaged by bright-field (a) – (c), phase contrast (d), DIC (e) – (g) and dark-field microscopy (h). Compare these images with HMC in Figure 11.25. With bright-field mode, panel (a) shows the condenser iris closed too far, to give a reduction in the overall numerical aperture of the microscope – compared to the objective NA – resulting in increased contrast at the expense of resolution of fine detail. Panel (b) shows the condenser iris correctly set with respect to the numerical aperture of the objective. Panel (c) shows the effect on the image of opening the condenser iris too wide: contrast is significantly reduced. In phase contrast (d) the cells are surrounded by an artifactual halo. In two of the three DIC images (e and g), the shear direction is shown, the cells in DIC being illuminated as if by incident light with a wavefront orthogonal to the shear direction. The features with bright edges have a higher refractive index than the background, or surrounding features, in the epithelial cell. The null position where DIC contrast is minimal, with an appearance akin to dark-field, is shown in (f). Here the phase retardation between the o-ray and the e-ray wavefronts is zero, or nearly so.*

and the nucleus appears raised because of its greater refractive index compared to the surrounding cytoplasm, rather than thickness alone. For opaque objects seen with reflected-light DIC, the image can be (unlike transmitted-light DIC) taken as a true 3D topological representation of the surface *providing* the shear setting of the Wollaston prism has been set correctly. If it has *not*, the raised and lowered parts of the image will be a reverse representation of the topology of the opaque surface of the object.

As mentioned in Section 11.12.1, contrast enhancement in phase contrast is due to 'edges' whereas in differential interference contrast it is due to 'wedges', i.e. the steepness – or rate of change – of the optical path difference gradient. The steeper the wedge, the greater the contrast and the darker the image will be at that particular point. The shear introduced by the Wollaston prisms in DIC will reinforce fine detail set orthogonally at 90° to the shear direction (Figure 11.32). A linear periodic structure (e.g. striae on a diatom frustule) may well disappear if aligned parallel to the shear direction but will show maximum contrast if set at 90° to the shear direction. For this reason, it is useful to have a rotating stage fitted to an interference microscope in order to quickly and easily examine all possible appearances of the image. To avoid this 'azimuth effect', circular-polariser filters – instead of plane-polarising filters – are sometimes used.

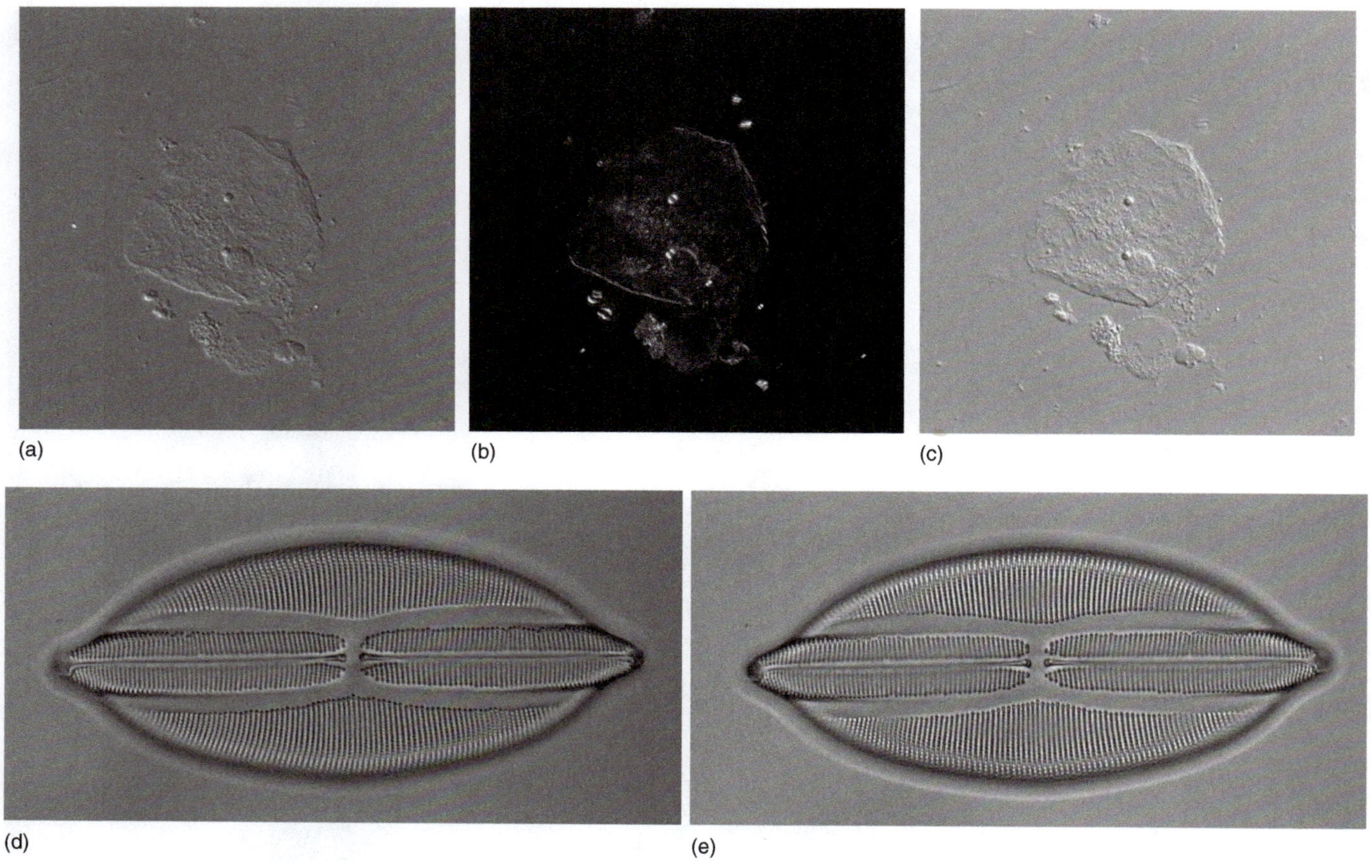

Figure 11.31 **Positive and negative bias in interference microscopy** *Differing bias retardation shown with a squamous epithelial cheek cell and the diatom* Navicula lyra.

The degree of the 'three-dimensional' appearance depends upon the amount of bias retardation between the o-rays and the e-rays introduced by laterally shearing the upper Wollaston/Nomarski prism. Bias retardation adjustment can be made either by advancing or retracting a positioning screw at the end of the upper Nomarski prism or, if the prism is fixed in conjunction with a $\lambda/4$ de Sénarmont plate, by rotating the lower polar. The amount of displacement is small, usually $< \lambda/10$.

When the shear on the Wollaston/Nomarski prism is set so that the field appears dark and an unstained phase object gives rise to an image similar to that in dark-field, it is because there is no phase shift between the o-ray and e-ray, so the electric vectors are vibrating, upon recombination, in the same direction as the polariser. The recombined ray is therefore blocked by the analyser. This is also shown in Figure 11.29 (f).

In practice, a prism setting giving total extinction is not used. Rather, a partial phase displacement between the o-rays and the e-rays is introduced. This is called bias retardation. These coherent waves, with a phase difference between them, are recombined by the Nomarski prism to form an elliptical emergent wave (see Figure 11.27) that partially passes through the analyser to cause the background to look uniformly grey with the image of the object to show dark shadows, bright highlights and intermediate values of grey intensities, according to the phase gradients and optical path differences in the object. Shifting the position of the prism setting in the opposite direction by means of the screw controlling the amount of lateral shear reverses the intensity and apparent depth of the features.

Although the DIC microscope is largely a qualitative instrument, a compensator can be used to manipulate the amount of bias retardation between the o-rays and the e-rays more precisely with greater control. A full-wave plate can be used to colour the specimen by introducing a spectrum of interference colours against a red first-order background. The colours at the edges of the object can be compared with the background using a Michel-Lévy chart to estimate the magnitude of the optical path difference.

Because differential interference contrast microscopy uses polarised light to produce the beam pairs, it is not a good technique to use with birefringent specimens and/or carrier substrates like plastic Petri dishes. Extruded and heat-formed plastic materials generally exhibit strain birefringence (Chapter 13, Box 13.3) and so depolarise the light, giving poor, unclear images of low contrast thus rendering DIC useless. Hoffman modulation contrast produces a similar 3D pseudo-relief but is not limited in this way to isotropic – non-birefringent – specimens and materials.

Differential interference contrast requires two Wollaston prisms; one to be placed at the front focal plane of the condenser and the other at the back focal plane of the objective, in the illuminating set of conjugate planes (see Box 11.4). From a mechanical point of view, it is reasonably easy to construct a Wollaston prism at the front focal plane of the condenser, for these are quite large, normally about 30 mm in diameter. Different prisms are used, according to the numerical aperture and magnification of the objective in use. The prisms in the condenser are usually housed in a carousel immediately beneath the optical components of the condenser along

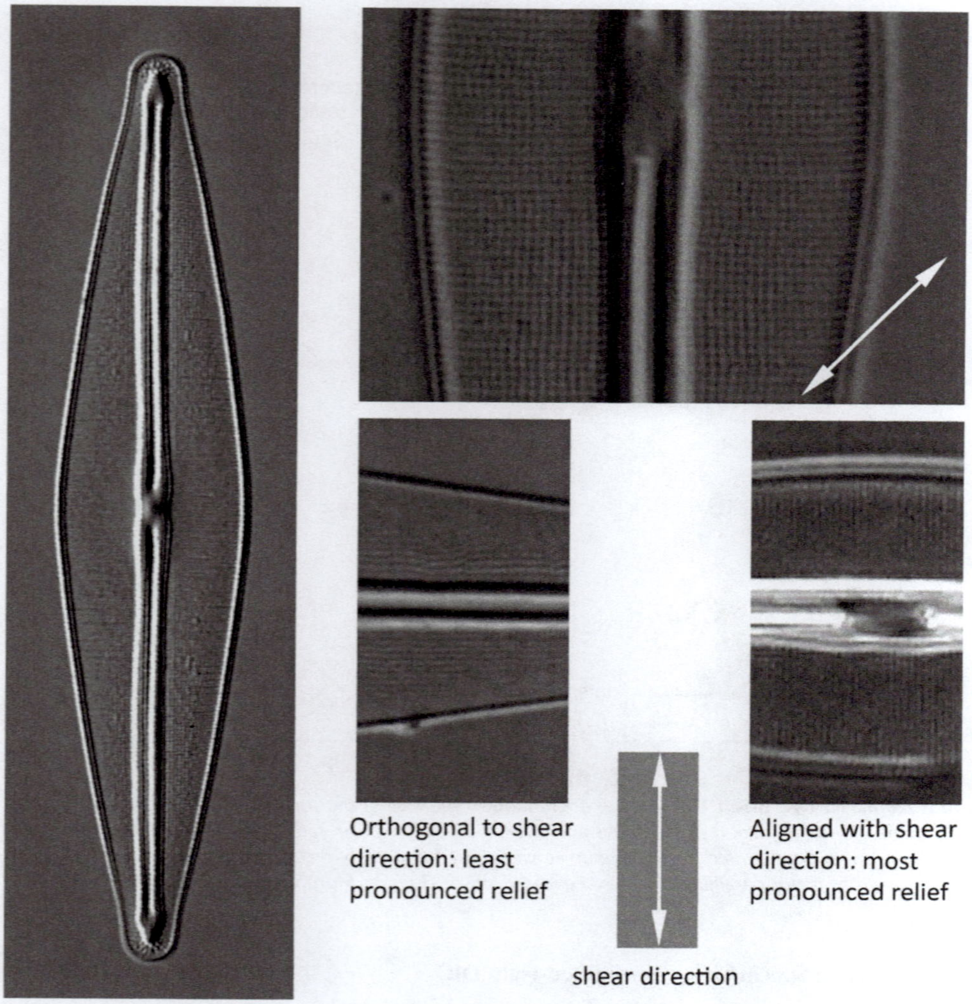

Figure 11.32 **The azimuth effect in DIC microscopy** *The only way of changing the shear axis relative to the specimen is to rotate the specimen, since the optical components used for DIC microscopy are fixed in the microscope. The diatom* Frustulia rhomboides, *which possesses punctae arranged in a square grid pattern (upper panel), shows how punctae, which are resolved with the shear at 45° to the specimen, may present as linear striae when the diatom is either aligned with or set orthogonally to the shear direction by rotation of the stage. In most circumstances, the differing appearance is of no consequence, except when the specimen exhibits periodic features.*

with phase-contrast annuli and dark-field patch stops. Because of the smaller dimensions of the objective, and particularly because the focal length of high-magnification objectives is very small, the back focal plane of most objectives is normally situated between lens elements deep within the barrel of the objective. It is therefore not at all easy to place an adjustable Wollaston prism into the objective at this position.

Georges Nomarski contrived to modify how the wedges of the Wollaston prism are cut so that the prism can be placed some distance outside the back focal plane of the objective (Figure 11.33) and the crossover of the two perpendicular rays still be *situated* at the back focal plane of the objective. The upper wedge of a Nomarski prism is identical to the conventional Wollaston prism, but the lower wedge is modified by cutting the quartz crystal to give an oblique optical axis with respect to the flat surface of the prism. Thus the interference plane where the two polarised beams cross over and diverge is at a point *outside* the prism, rather than at the cemented boundary, as in a conventional Wollaston prism. In practice, because it gives the designers more latitude, Nomarski-modified Wollaston prisms are usually built into the condenser also.

Different manufacturers have slightly different designs of DIC, particularly in the way in which the beam-combining Wollaston/ Nomarski prism is incorporated into the microscope. In some instruments a single Nomarski DIC prism is matched for all objectives and is situated above the nosepiece of the microscope. There is normally a screw to move the prism and to provide contrast. In other instruments each objective has its own individual Nomarski design of prism incorporated into a slider, again with a screw mechanism to provide the shear (see Figure 11.27). Those prisms designed for use with individual objectives are either housed in a spacer between the objective and the nosepiece or within a slot above the objective, which is integral with the nosepiece. With inverted microscopes it is usually necessary to insert the DIC prism upside down, but please do first check the manufacturer's instructions.

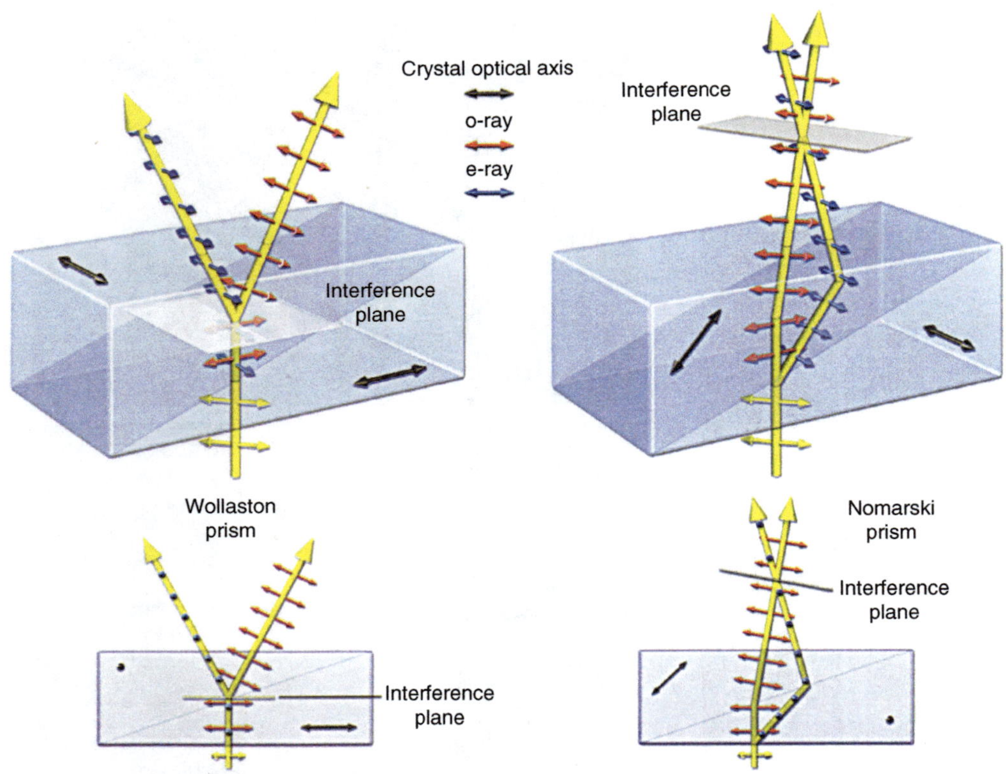

Figure 11.33 **Design of a Nomarski DIC prism** *The oblique orientation of the axis of the upper quartz wedge of a Wollaston prism results in the projection of the apparent plane of recombination of the o-rays and e-rays outside the prism rather than at the interface of the wedges. This allows the Nomarski-modified Wollaston prism to be mounted outside the objective whilst the plane of recombination lies at the back focal plane of the objective. Note that the plane of the wavefront is not parallel to the surface of the prism. Source: Murphy et al. 2013. Reproduced with permission from John Wiley & Sons.*

Protocol 11.6 Examination of a Specimen by Transmitted-Light DIC

1. Prepare a test slide, as given in Step 1 of Protocol 11.4 for phase contrast.
2. Remove any polariser, analyser, full-wave compensator plate and both Wollaston prisms from the light path.
3. Using the lowest magnification DIC objective, set up the microscope for bright-field observation by Köhler illumination using a well-stained 'amplitude' specimen. Do this using the bright-field position in the universal condenser carousel. Correct condenser *height* is important with phase contrast as with DIC, so ensure that the illuminated field diaphragm is imaged sharply and that the condenser is centred with the normal condenser-centring screws.
4. Insert or swing in the polariser and analyser into the light path of the microscope, and ensure that they are crossed (the view through the microscope should be maximally dark. It may help to view the crossed polars at the objective back focal plane). Conventionally, the lower polariser should be set East-West and the analyser North-South.
5. Remove the stained slide.
6. Insert the correct lower (beam-splitting) Wollaston prism in the condenser for the objective in use into place by rotating the condenser carousel into the appropriate position, which may be marked on the side of the condenser by a fiducial mark. The prism should already be oriented so that it lies at 45° between both crossed polars.
7. Insert the correct Wollaston/Nomarski prism above the objective or, if it is built into the objective, rotate the appropriate objective into position.
8. If the objective Wollaston/Nomarski prism has an axial control, move this up and down to produce an evenly illuminated field of view of a uniform shade of grey.
9. Remove an eyepiece, and adjust the lateral shear on the upper Wollaston/Nomarski prism (usually by adjusting a screw control) so that a black zero-order fringe is seen diagonally across the centre of the back focal plane of the objective. Replace the eyepiece.
10. Exchange the stained slide for an unstained phase specimen, suitable for DIC.
11. Adjust the objective Wollaston/Nomarski prism lateral control by means of the screw or lever attached to it, and observe the range of contrast possible when altering the bias retardation in this way. Select the optimum contrast for the specimen for your purposes. If the objective prism is fixed and does not have this adjustment, rotate the polariser away slightly from its crossed position.
12. Where colour is desired, insert a 1λ full-wave red 'sensitive tint' compensating plate into the slot provided in the microscope. This is usually between the objective Wollaston/Nomarski prism and the analyser, but the compensating plate may be set anywhere between the two polars. Again, select the optimum contrast in the image for your purposes.

A protocol for setting up and operating DIC for reflected-light is given in Chapter 12, Section 12.12.

When the two DIC prisms are first aligned, and the zero-order fringe is centred in the back focal plane of the objective, there is zero path difference between the two beams, and the field is maximally dark. As the beam-combining prism is adjusted laterally in either direction, the optical path difference will, if great enough, show the background field as Newton's interference colours (Chapter 14, Section 14.10) from grey, through white, yellows, blue and the higher-order colours. Generally, thin specimens will show best contrast whilst still against a grey background. It is possible to introduce a significant optical path difference with a full-wave red sensitive tint compensating plate or a compensating quartz wedge, in which case the background can be coloured deep purple, or another colour, as desired.

When DIC and fluorescence are imaged simultaneously, the fluorescence signal can sometimes be impaired by the DIC prisms and the polarising filters in the light path. One means of circumventing this called 'infrared scanning gradient contrast' (SGC) imaging has been devised by Hans-Ulrich Dodt (Dodt and Zieglgänsberger, 1990). It is good for observing cells in thick tissues which scatter light (e.g. patch-clamp work on thick brain slices; phase contrast does not work on these thick slices and DIC components can attenuate the infrared light used to produce the image) and as a relief contrast method is similar to Hoffman modulation contrast. A sector in the condenser combined with a diffusing screen illuminates the field of view with oblique illumination. Any thick phase object will refract light; the contrast can be altered by changing the position of the diffusing screen axially and/or rotating the quarter-circle sector. If the distance between the sector and diffuser is increased, there is less shadow cast over the field of view, and contrast is decreased. Rotating the sector will alter the angle of oblique illumination with respect to the orientation of specimens within the field of view. Another means is to manipulate bright-field images with phase modulating software and layer these digitally on top of a fluorescence image (see Cody et al., 2005).

With the PlasDIC modification introduced by Zeiss there is no need for a polariser in front of the condenser. Instead it is situated behind the objective within a unique PlasDIC prism. This means that cells in plastic tissue-culture dishes can be examined, and there is no requirement for strain-free objectives for high-quality DIC. The light is made coherent by passing through a slit in the condenser, as is the case for other relief contrast methods (Section 11.13). The condenser slit is carried on a slider, to enable easy switching from PlasDIC to other forms of contrast enhancement.

11.15 Post-Acquisition Contrast Enhancement in the Image

Once an image has been collected, it is possible to further manipulate contrast. Classically this is done when recording onto 35 mm roll film or 5″ × 4″ sheet film by choosing (a) the appropriate film stock, (b) the developer and (c) the printing paper. The dynamic range of an image, whether taken with film, by video or CCD camera, describes the ratio between the maximum and minimum signal. It is, therefore, a measure of contrast in the image and, when applied to grey levels, a measure of brightness or tonal range. Both digital and video microscopy allow an image to be captured with much greater dynamic range than the eye can perceive; this improves the signal-to noise-ratio (SNR) and enables features in the image, which would otherwise be lost or hidden, to be 'lifted' above the noise 'floor'.

Robert Day Allen and Shinya Inoué both applied analogue video recording to collecting microscope images. These methods are known as VE-DIC and VE-POL, for use with DIC and polarised light respectively. The 'VE' stands for 'video-enhanced'; sometimes 'A' for Allen is added as a prefix (AVEC-DIC). The ability of analogue video cameras to capture image frames in real time is an advantage over digital acquisition. Both video and digital recording also allow for easy and immediate contrast enhancement, especially by manipulating the histogram of intensity values with the many and varied imaging-editing and manipulation programmes for both raster and vector-based images, such as Adobe Photoshop, Adobe Illustrator, Image J/Fiji, or GIMP, to name a few (for a comparison of raster and vector graphics editors, see the URL link under Internet Resources at the end of the chapter). Most raster-based image editors will allow local contrast enhancement via an unsharp mask function, which can remove haze but without sharpening edges unduly. Image editing involves scientific work in the area of ethics, a topic of considerable interest and discussion to journal editors and reviewers, which is discussed in Chapter 31 on applied digital imaging.

11.16 Chapter Summary

At the end of this chapter you should understand how light interacts with the specimen and how these interactions may be used to generate and enhance contrast to make the image visible. You should appreciate that contrast and resolving power are related; that as the limit of resolution is approached, contrast in the image tends to zero, and that without sufficient visibility, resolution of fine detail – the prime purpose of the light microscope – cannot be achieved.

From the remainder of the chapter, you should have an awareness of the main forms of contrast enhancement, given that our eyes and cameras will only respond to changes in wavelength and brightness across the field of view of the image.

1. Resolving power, contrast and magnification are the three essentials. Contrast is under the control of the microscopist.
2. Our eyes can only see changes in the intensity or a difference in the colour of light. Contrast enhancement must effect a change to the amplitude and/or the wavelength of the image-forming light.
3. Most biological tissue lacks contrast. It must either be stained or contrast introduced optically.
4. Light interacts with matter to form an image in six ways: by absorption, transmission and reflection; by scattering and diffraction; by refraction; by polarisation; by a phase change or by fluorescence.

5. Bright-field, oblique and dark-field microscopy are all related and depend upon the illumination angle, whether it falls inside the objective aperture or is excluded.
6. Colours may be specified by hue, saturation (chroma) and 'brightness' (luminance).
7. Additive colour mixing occurs when two, or more, beams of differently coloured light combine.
8. White light can be formed by adding all three primary colours: red, green and blue light.
9. Subtractive colour mixing occurs by the absorption of light by a filter or coloured substance (e.g. a painted surface).
10. Mixing green and blue light gives the complementary colour cyan. Mixing red and blue gives magenta, and mixing red and green gives yellow.
11. Filters work by absorption or interference to selectively isolate chosen wavelengths from the spectrum.
12. Where two substances meet at a common interface, refraction can cause relief contrast.
13. The Becke line is a halo that forms around the periphery of a particle or fibre where a refractive index mismatch exists between the particle and mounting medium.
14. The Becke line moves with adjustment of the plane of focus, dependent upon RI of the sample and mounting medium.
15. The Christiansen effect occurs when a crystal mounted in a liquid of different refractive index disperses oblique white light at its edges to appear fringed with colour.
16. For a stained (amplitude) specimen, light will be retarded in phase by one half wavelength.
17. For an unstained sample the phase difference is only one quarter wavelength.
18. A phase plate artificially introduces an extra quarter wavelength to make otherwise invisible 'phase' specimens visible.
19. Positive phase contrast occurs when the optical path length of the diffracted beams is longer than the direct beam, the diffracted beams being advanced by $\frac{1}{4}\lambda$.
20. With positive phase contrast, structures with refractive indexes higher than their surroundings appear darker in the image against a lighter background.
21. Phase contrast does not work well with stained specimens.
22. Phase-contrast equipment should include a green filter for optimum image definition.
23. The phase contrast effect is maximal at regions of sudden change in optical path difference.
24. A halo is formed when diffracted rays inevitably pass through the phase ring in the back focal plane of the objective. They remain only $1/4\lambda$ out of phase with respect to the zero-order illumination, do not undergo destructive interference and so remain bright.
25. Hoffman modulation contrast uses oblique illumination to change phase gradients into gradients of light intensity and so form a 3D pseudo-relief image.
26. Interference microscopy splits the illumination into two coherent rays, which are altered in phase with respect to one another. After interaction with the specimen, an amplitude image is produced when the two rays interfere upon recombination at the primary image plane.
27. In DIC, the two coherent rays are sheared *laterally* by a small amount less than the resolving power of the objective. A double image is therefore not formed.
28. In DIC, the appearance of the image depends upon the rate of change of the optical path length – 'wedge' contrast as opposed to the 'edge' contrast in phase-contrast microscopy.
29. DIC does not work well with plastic substrates, which generally exhibit strain birefringence.
30. Nomarski modified the wedges of the Wollaston prism so that the apparent plane of recombination of the two coherent beams occurred outside the prism exactly in the back focal plane of the objective.
31. Image contrast can be manipulated post-acquisition. This is easy, using a histogram to alter the dynamic range and gamma function of the image.

Exercise 11.1 Viewing the polarisation cross under crossed polars

1. With the microscope set up for bright-field microscopy and adjusted for Köhler illumination, insert a polariser and analyser into the optical train.
2. View the back focal plane of the objective by removing an eyepiece, using a phase-contrast centring telescope or a Bertrand lens. A dark-field (a) should be the result. Shown here, it is slightly off-centre, for a lighter region on the right-hand side can be seen, as indicated by the white arrow.
3. When centred, there may be four lighter areas (b) at the edges of the dark-field. These are formed from the extinction cross (see Chapter 14 Protocol 14.1).
4. Insert the objective Wollaston/Nomarski prism between two crossed polars. A bright-field will be seen in the back focal plane of the objective with a dark zero-order fringe running diagonally (c) from northwest to southeast across the field.

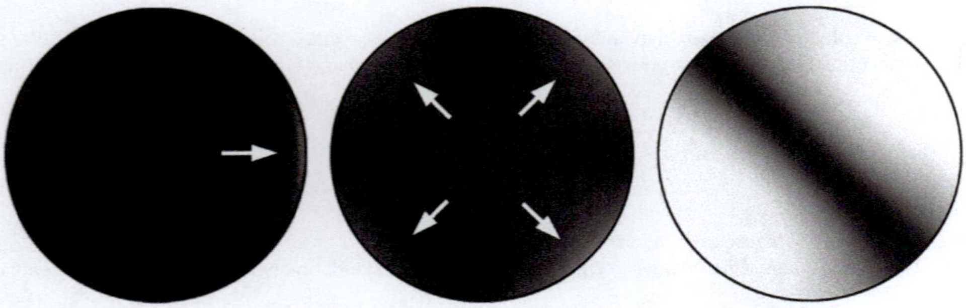

(a) (b) (c)

Source: Murphy et al. 2013. Reproduced with permission from John Wiley & Sons.

5. The position of this fringe can be displaced laterally by turning the objective prism screw, which introduces shear into the image.

Exercise 11.2 Contrast formation in differential interference contrast

1. Set up Differential Interference Contrast as in Protocol 11.6.
2. Remove the polariser from the light path, and the image becomes brighter and washed out but doesn't disappear altogether.
3. Now rotate the polariser through 45 degrees, parallel with the axis of the Wollaston prism. No contrast visible in the image because all the light will now go down one of the two coherent ray paths (see Box 11.4).

Exercise 11.3 Fringe orders of a Nomarski DIC prism

1. Place a Nomarski DIC prism between two crossed polars on a light box.
2. Rotate the prism until a central dark zero-order interference fringe is obtained.
   This will be found to be at 45° to the crossed polars.
3. The zero-order fringe results from interference between the o-rays and the e-rays that have equal path lengths in the middle of the prism.
4. If white light is used, higher-order fringes either side of zero are coloured.
   Low–NA prisms have just one fringe; high-NA prisms have several fringes.
5. Notice that the fringes appear to float above the prism. This is because the interference plane (cross-over point) on a Nomarski-modified Wollaston prism is located outside the prism.
6. As the prism is rotated, four positions – each parallel with the vibration direction of first one then the other polar – give extinction. The retardation between the o-rays and the e-rays introduced by one wedge comprising the Nomarski prism is exactly reversed by the other wedge so that light emerges plane parallel vibrating in the same plane as the polarised light and so is blocked by the upper polar.

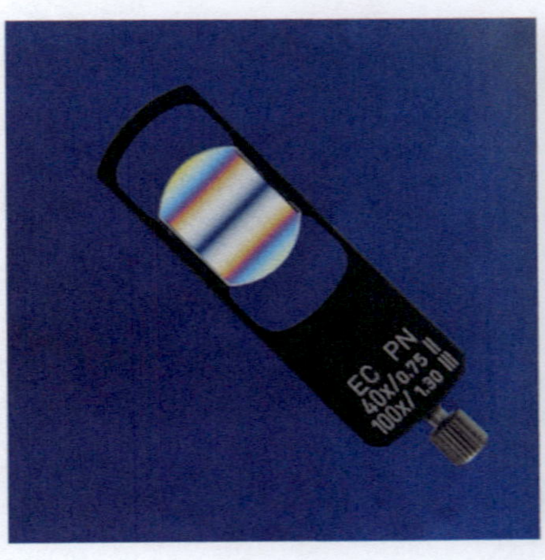

Key Reading

Stelzer, E.H.K. (1998). 'Contrast, resolution, pixelation, dynamic range, and signal-to-noise ratio'. *Journal of Microscopy* 189/1: 15–24.

- A clear and readable explanation of how resolving power and contrast are interrelated and how pixel size, dynamic range and signal-to-noise ratio affect image quality.

References

Bradbury, S. (1967). *The Evolution of the Microscope*. Oxford: Pergamon Press.
Bradbury, S. and Evennett, P.J. (1996). *Contrast Techniques in Light Microscopy*. RMS Handbook No. 34. Oxford: Bios Scientific Publishers (currently Taylor & Francis).
Dall, H.E. (1985). 'Mounting method for high resolution optical microscopy'. *Microscopy 35*/4: 280–284.
Dodt, H.U. and Zieglgänsberger, W. (1990). Visualizing unstained neurons in living brain slices by infrared DIC-videomicroscopy'. *Brain Research 537*(1–2): 333–336.
Chandler, D.E. and Roberson, R.W. (2009). *Bioimaging*. Jones & Bartlet Publisher.
Cody, S.H. et al. (2005). 'A simple method allowing DIC imaging in conjunction with confocal microscopy'. *Journal of Microscopy 217*/3: 265–274.
Françon, M. (1961). *Progress in Microscopy*, International Series of Monographs in Pure and Applied Biology, vol. 9. Oxford: Pergamon Press, p. 58.
Gatenby, J.B. (1919). *Proceedings of the Royal Microscopical Society*. *June 39*/2: 97–114, illustration on page 104.
Hammond, C. (2007). 'A modified modulation contrast technique for transmitted and reflected light microscopy'. *Quekett Journal of Microscopy 40*/6: 495–505.
Hartshorne, N.H. and Stuart, A. (1970). *Crystals and the Polarising Microscope*, 4th ed. London: Arnold, p. 265.
Hartley, W.G. (1977). 'A consideration of the Hoffman modulation contrast system'. *Microscopy 33*/3: 147–152.
Hoffman, R. (1977). 'The modulation contrast microscope: principles and performance'. *Journal of Microscopy 110*/3: 205–222.
Ishikawa-Ankerhold, H.C., Ankerhold, R., and Drummen, G.P. (2012). Advanced fluorescence microscopy techniques – FRAP, FLIP, FLAP, FRET and FLIM. *Molecules 17*/4, 4047–4132. doi: 10.3390/molecules17044047.
James, J. and Tanke, H.J. (1991). *Biomedical Light Microscopy*. Dordrecht: Kluwer Academic Publishers.
Kiernan, J.A. (2008). *Histological & Histochemical Methods: Theory & Practice*, 4th ed. Oxford: Scion Publishing.
Kiernan, J.A. (2006). 'Dyes and other colorants in microtechnique and biomedical research'. *Colouration Technology 122*/1: 1–21.
King, M. (2009). 'Make your own central-stop dispersion staining objective'. *The Microscope 57*/3: 123–125.
Landini, G. and Perryer, D.G. (2009). 'Digital enhancement of Haematoxylin and Eosin stained histological images for red-green colour blind observers'. *Journal of Microscopy 234*/3: 293–301. See also: *Nature Methods* (2011) *8*/11: 891–892.
Laskowski, T.E. and Scotford, D.M. (1979). 'Measurement of refractive index in thin section using dispersion staining and oil immersion techniques'. *American Mineralogist 64*/4: 440–445.
Laughlin, G.J. (1992). 'Microscopy in the identification of asbestos'. *Proceedings Royal Microscopical Society 27*/1: 9–15.
McCrone, W.C., McCrone, L.B., and Delly, J.G. (1979). *Polarised Light Microscopy*. Ann Arbor, MI: Ann Arbor Science Publishers.
Murphy, D.B. (2001). *Fundamentals of Light Microscopy and Electronic Imaging*, 1st ed. New York: Wiley-Liss.
Murphy, D.B. and Davidson, M.W. (2013). *Fundamentals of Light Microscopy and Electronic Imaging*, 2nd ed. New Jersey: Wiley-Blackwell, JohnWiley Inc. USA.
Pluta, M. (1988). *Advanced Light Microscopy*, vol. 2. Amsterdam/ Oxford: Elsevier.
Sanderson, J.B. (1994). *Biological Microtechnique*. RMS Handbook No. 28. Oxford: Bios Scientific Publishers (currently Taylor & Francis).
Sanderson, J.B. (2000). 'The theory of contrast control in the microscope'. *Quekett Journal of Microscopy 38*/7: 617–627.
Sanderson, J.B. (2002). 'Phase Contrast Microscopy'. *Encyclopedia of Life Sciences*, secondary article. Macmillan Publishers Ltd, Nature Publishing Group.
Tilley, R.J.D. (1996). *Colour*, 2nd ed. Wiley.
Trefil, J. and Hazen, R.M. (2004). *Physics Matters*. Hoboken, NJ: John Wiley & Sons Inc.
Wilkinson, B. (2011). 'Arthur Strange's Variable Amplitude Contrast Method - VAC 2'. *Balsam Post 91*: 12–21.

Further Reading

Allen, R.D. et al. (1981). 'Video-enhanced contrast polarization (AVEC-POL) microscopy'. *Cell Motility 1*: 275–289.
Allen, R.D. et al. (1981). 'Video-enhanced contrast differential interference contrast (AVEC-DIC) microscopy'. *Cell Motility 1*: 291–302.
Bracegirdle, B. (1978). *A history of microtechnique*. London: Heinemann.
Dunn, G.A. (1998). 'Transmitted-light Interference microscopy: a technique born before its time'. *Proceedings RMS 33*/3: 189–196.
Franchi, L. (2008). 'Oblique illumination revisited'. *Quekett Journal of Microscopy 40*/8: 675–684.
Hoffman, R. and Gross, L. (1975). 'The Modulation Contrast microscope'. *Nature 254*: 586–588.
Piper, J. (2007). 'Relief phase contrast: a new technique for phase-contrast light microscopy'. *European Microscopy and Analysis 21*/4: 9–12.
Piper, T. and Piper, J. (2014). 'Variable multimodal light microscopy with interference contrast and phase contrast; dark or bright field'. *Journal of Microscopy 255*/1: 30–41.
Ross, K.F.A. (1988). 'Phase Contrast and Interference Microscopy'. *Microscopy 36*/2: 97–123.
Strange, A. (1987). 'Variable Asymmetrical Contrast Microscopy'. *The Microscope 35*: 279–289.
Strange, A. (1989). 'New Developments in Variable Asymmetrical Contrast Microscopy'. *The Microscope 37*: 355–376.
Triantaphillidou, S. (2011). *Introduction to colour science*. In: Allen, E. and Triantaphillidou, S. (eds.), *The Manual of Photography*, 10th ed. Oxford: Focal Press, pp. 77–102.

Weiss, D.G. et al. (1999). *Video microscopy*. In: Lacey, A.J. (ed.), *Light Microscopy in Biology: a practical approach*, 2nd ed. Oxford: IRL Press, pp. 73–149.
Zernicke, F. (1953). 'How I discovered phase contrast' – Nobel Lecture, December 1953.

Internet Resources

Introduction to fluorescence: http://www.thermofisher.com/uk/en/home/support/tutorials.html and https://www.thermofisher.com/sg/en/home/references/molecular-probes-the-handbook/introduction-to-fluorescence-techniques.html.
Oblique illumination: http://www.microscopy-uk.org.uk/mag/indexmag.html?http://www.microscopy-uk.org.uk/mag/artdec08/dw-obliquetip.html and http://www.microscopy-uk.org.uk/mag/indexmag.html?http://www.microscopy-uk.org.uk/mag/artfeb00/pjoblique.html.
Circular oblique illumination: http://www.microscopy-uk.org.uk/mag/indexmag.html?http://www.microscopy-uk.org.uk/mag/artdec02/pjcol.html.
Diatoms & Contrast, René van Wezel Micscape-uk: http://www.microscopy-uk.org.uk/mag/indexmag.html?http://www.microscopy-uk.org.uk/mag/artapr09/rvw-contrast.html.
Colour management: http://www.normankoren.com/color_management_2.html}Working.
Raster versus graphics editors: https://en.wikipedia.org/wiki/Graphics_software.
Contrast enhancement with an unsharp mask: https://luminous-landscape.com/contrast-enhancement/.
Micscape: http://www.microscopy-uk.org.uk/mag/indexmag.html → Article Library/Issue Archive.

Notes

1. There may be some scattering of light at the edges of the object, but the amplitude of these scattered rays will be so low in comparison to the much brighter light passing directly through that any edge contrast is likely to be invisible.
2. For an explanation of the difference between the term 'fluorochrome' and 'fluorophore', see Chapter 15, Section 15.1.
3. The quotation from Leeuwenhoek's letter to the Royal Society, dated 22nd January 1675 is cited on page 30 in the excellent handbook on contrast by Bradbury & Evennett (1996) published by Bios for the Royal Microscopical Society.
4. Older microscopes sometimes used funnel stops to reduce the numerical aperture of the objective in order to secure good dark-field. These were dropped into the objective from the rear. They had a fixed NA but could not be adjusted and so are now obsolete. Objectives designed for dark-field with a variable aperture iris are now used instead.
5. Red sensitive (L) cones respond to light from 500 to 700 nm with a peak at around 560 nm; green (M) cones to 450–630 nm with a peak value around 530 nm and blue (S) cones to 400–500 nm with a peak around 420 nm.
6. The subject of colour blindness is discussed in Chapter 31, Section 31.13.
7. The mixing of paints, inks or dyes is known as subtractive colour mixing, which involves the absorption of light by a filter or coloured (e.g. painted) surface. The three subtractive primary colours are magenta, yellow and cyan (sometimes referred to as 'red', yellow & 'blue'). These are the complementary colours of the additive primary colours, red, green and blue. In practice, mixing black from cyan, magenta and yellow does not result in a fully saturated black, hence printers use a separate black ink – the CMYK system.
8. For an explanation of the different terms, as well as a discussion of the different mechanisms of histological staining, see: *Staining and Dyeing*, chapter 6, pp 159–190 in: Sanderson, J. B. (1994). *Biological Microtechnique*. Oxford: Bios. A more complete account may be found in Kiernan, J. A. (2008). *Histological & Histochemical Methods: Theory & Practice* 4th ed. Oxford: Scion Publishing. The best account of the history of dyeing is Chapter 4 in Bracegirdle (1978).
9. Little research has been carried out on the efficacy of the H&E stain for colour-blind people. One useful study is that by Landini & Perryer (2011). For further details, see the reference list in Chapter 31 on applied digital imaging.
10. Congo Red, CI 22120, the first artificial direct dye (not requiring a mordant) was first synthesised in 1883. It is also logically called Direct Red 28, but the more common name was adopted by AGFA since it was a catchy name in Germany at the time of the 1884 Berlin West Africa Conference, an important event instigated by Bismarck to resolve the control of trade in the Congo River basin. It was an effective marketing ploy, and other Congo dyes followed suit.
11. When non-polarised light is decomposed by anisotropic material into two polarised rays vibrating orthogonally to one another, one of these rays is termed the *ordinary ray* (since it obeys Snell's law of refraction) and the other the *extraordinary ray* (since it does not obey Snell's law). This phenomenon is explained in Chapter 13, Section 13.2.

12

Reflected-Light Microscopy

Outline

Understanding Light Microscopy, First Edition. Jeremy Sanderson.

12.1 Introduction

Whereas we view the world around us almost wholly by reflected light, the majority of bright-field light microscopy is carried out by transmitted light.

In this chapter we are concerned only with the imaging and contrast of flat specimens with little or no surface relief – the equivalent of the thin sections from transmitted-light microscopy. Thus we are restricted to the study of polished and etched metal surfaces (metallography) or of minerals, rocks (petrography) and ceramics (ceramography) – in general, materials microscopy. The use of reflected light, or epi-illumination (from the Greek: epi meaning 'above') in biological science is largely confined to fluorescence microscopy. Sometimes the term 'episcopic' is used to refer to reflected-light microscopy – as opposed to 'diascopic' for transmitted-light microscopy. Nikon tends to adopt this nomenclature. Reflected-light microscopy is sometimes called 'incident-light' microscopy, but this is incorrect, for all light is incident upon a specimen, whether it falls on it from above or below.

The grain size and microstructure of metals, minerals and ceramics is much smaller than the standard 30 μm thickness of thin sections observed by transmitted light. The overlap of detail within the thickness, or extent, of these so-called thin sections obscures most microstructure. A significant advantage of reflected-light microscopy over the examination of thin sections by transmitted light is that microstructural detail at the surface can be examined without obstruction caused by the superimposition of out-of-focus image detail.

Naturally occurring specimens generally exhibit surface relief in excess of the depth of field of objectives with numerical apertures much above 0.15 (i.e. whose depth of observable surface relief ≈50 μm or less). The light scattering that occurs from reflected-light (i.e. wide-angle) illumination is generally such that relatively little light enters the objective lens. Such specimens are best observed with stereomicroscopes custom designed for the examination of thick opaque samples with a relatively large field of view by illumination from above the specimen. See Chapter 4, Section 4.6, for information about stereomicroscopes.

12.2 Reflectance of Light

When light falls on glass or water – or indeed any transparent material – it is not only refracted as described in Chapter 2, Section 2.9 but is also reflected (Figure 12.1; see also Figure 2.12). It is a matter of common experience that the reflectivity (or reflectance) of a glass or water surface varies with angle, and a window pane or a sheet of unruffled water observed at small glancing angles (i.e. large angles of incidence), has a mirror-like appearance. The reflectance off a surface is always less than the radiant flux of incident illumination. The relative proportions of light reflected and refracted at the surface of a transparent material (see Box 12.1) is a matter of some complexity because the reflection and refraction vary not only with the angle of incidence but also with the refractive index of the material; additionally the state of polarisation of the reflected and refracted rays must also be taken into account. This particular phenomenon of polarisation is discussed further in Chapter 13, Section 13.2 onwards.

Box 12.1 The Mathematics of Reflectance

An interface is the planar boundary occurring at the junction of two media of different refractive indexes. When light incident upon the interface moves from one medium to another, both reflection and refraction of light may occur, as predicted by Fresnel's equations and Snell's law, respectively. The reflectance values, R, for light incident *normal* to the interface can be expressed by an extension of Fresnel's equations[i], which involves not only n (a refractive index can also be specified for opaque materials) but also an absorption term k (the absorption coefficient or extinction index, k, is zero for transparent materials). The exact relationship is given by Fresnel's equations, which for normal incidence is:

$$R = \frac{(n - \mathrm{N})^2 + k^2}{(n + \mathrm{N})^2 + k^2} \quad \text{which in air reduces to:} \quad R = \frac{(n - 1)^2 + k^2}{(n + 1)^2 + k^2}$$

where R – the reflectance – is the ratio of the intensities of the incident and reflected rays; n is the refractive index of the substance and N is the refractive index of the medium (e.g. immersion oil, water or air).

In the case of strongly absorbing materials, e.g. metals, the absorption coefficient k dominates, and R approaches unity – i.e. metals are strong reflectors, approaching 100% for aluminium and silver. For these metals R is constant over the whole spectral range, and hence aluminium and silver appear white. For copper and – to a lesser extent – gold, k (and hence R) increases with wavelength; these metals therefore appear red/yellow. Conversely, for iron, k (and hence R) decreases slightly with wavelength, which explains the bluish colour of polished steel.

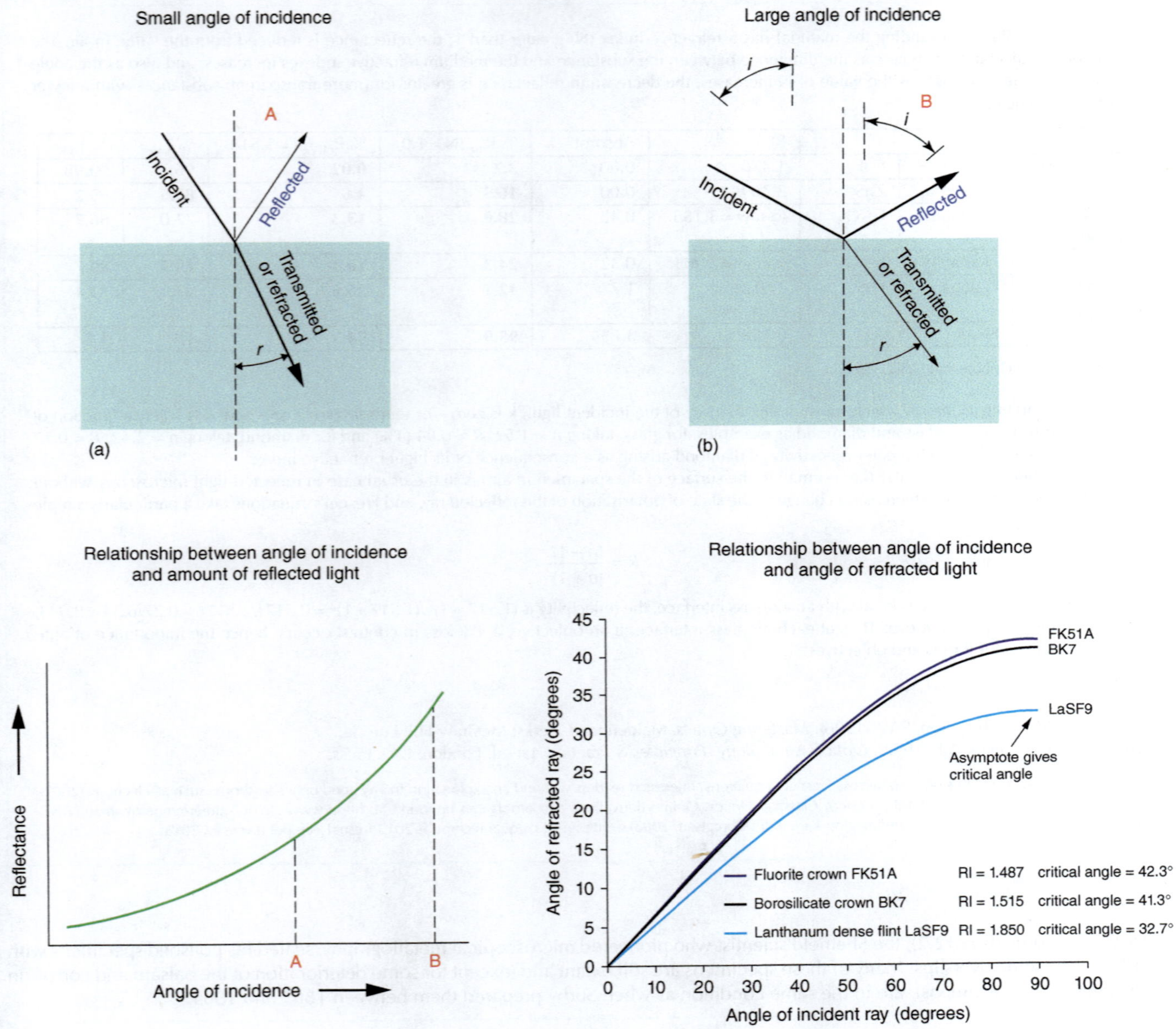

Figure 12.1 ***Ray diagram of reflection of light off a glass surface*** *When light is incident from air upon a transparent slab of glass sufficiently thick such that the lower surface may be ignored, some of it will be reflected and some transmitted at all angles of incidence. The greater the angle of incidence, the greater the amount of reflected light, as shown in the upper two figures and the graph below. The angle of refraction, r, is related to the angle of incidence, i, such that sin i/sin r = RI of the glass (here, 1.52).*

Light may either undergo specular reflection when incident upon a smooth surface, in which case the angle at which the ray of light is reflected (r) is equal to the angle of incidence (i), or it is reflected diffusely from a rough surface. If the material is transparent, a certain proportion of light is also and undergoes refraction through the material. See also Figure 2.11 and Figure 2.12, in Chapter 2.

Source: Adapted from Nesse 2009 with permission from Oxford University Press.

When the medium surrounding the material has a refractive index (N) greater than 1, the reflectance is reduced from the value in air. The magnitude of reflectance increases as the difference between the substance and the medium refractive indexes increases, and also as the angle of incidence increases. Also, as the value of N increases, the decrease in reflectance is greater for (more transparent) substances with a lower absorption coefficient.

	Substance		RI, n	Absorpt^n, k	% R_{air} (N = 1.0)	% R_{oil} (N = 1.518)	% T_{air}	% T_{oil}
Transparent	Fluorite	CaF_2	1.476	0.00	**3.7**	**0.02**	96.3	99.98
	Sphalerite	ZnS	2.369	0.00	**16.5**	**4.8**	83.5	95.2
Weakly absorbing	Hematite	Fe_2O_3	(o-ray) = 3.185	0.42	**28.0**	**13.3**	72.0	86.7
	Hematite	Fe_2O_3	(e-ray) = 2.905	0.32	**24.3**	**10.3**	75.7	89.7
Strongly absorbing	Galena	PbS	3.912	1.7	**42.1**	**26.6**	57.9	73.4
	Silver	Ag	0.150	3.65	**95.9**	**94.3**	4.1	5.7

Source: Data from Gribble and Hall (1992).

With transparent specimens, which transmit the majority of the incident light, k is zero – or tends to zero – and now R is solely a function of refractive index. Consider glass and diamond as examples: for glass, taking $n = 1.52$, $R \approx 0.04$ (4%) and for diamond, taking $n = 2.42$, $R = 0.17$ (17%). Here we see the much greater reflectivity of diamond arising as a consequence of its higher refractive index.

For light incident perpendicular (i.e. normal) to the surface of the specimen in air (as in the usual case in reflected-light microscopy without grossly oblique illumination) there is no change to the state of polarisation of the reflected ray, and Fresnel's equations take a particularly simple form, viz:

$$R = \frac{(n-1)^2}{(n+1)^2}$$

For a ray incident along the normal with an air/glass interface, the reflectivity is $(1.517 - 1)^2/(1.517 + 1)^2 = 0.517^2/2.517^2 = 0.27/6.34 = 0.043$, which is the figure quoted above. Thus at each air-glass interface in an objective, a 4% loss in contrast occurs; hence the importance of anti-reflection coatings for lenses and objectives.

References

Blackwell, J.M. and Thornton, S.W. (1996). *Mastering Optics*. Maidenhead, Berks: McGraw-Hill Europe.
Gribble, C.D. and Hall, A.J. (1992). *Optical Mineralogy: Principles & Practice*, 1st ed. London: UCL Press.

[i] It is beyond the scope of this book to discuss Fresnel's equations. Interested readers will find an explanation in any good optics textbook, such as Hecht, E. (2002). *Optics*, 4th ed. or Brooker, G. (2003). *Modern Classical Optics*. Online derivations and proofs can be found at: http://www.uio.no/studier/emner/matnat/ifi/INF-GEO4310/h09/undervisningsmateriale/imaging-kap2.pdf (Albregtsen, 2008) or http://iqst.ca/quantech/pubs/2013/fresnel-eoe.pdf (Lvovsky, 2013).

12.3 Coverslip Requirement

Henry Clifton Sorby (Box 12.2), the Sheffield scientist who pioneered microscopical metallography, sealed his polished specimens with Canada balsam and coverslips. Many of these specimens are still extant and (except for some deterioration of the balsam and corrosion at the edges of the specimens), are in the same condition as when Sorby prepared them between 1863 and 1865.

Box 12.2 Henry Clifton Sorby

Henry Clifton Sorby was born on 10 May 1826 at Woodbourne, Yorkshire, into a wealthy family who had been successful cutlers in Sheffield for many years. Sorby did not enter the family business, as might have been expected, for he wished to be a scientist. In 1847 at the age of 21, Sorby's father died, and he had no need to earn his own living. Instead, he financed his own laboratory, workshop and yacht to pursue his wide-ranging research interests.

Sorby is principally remembered for his work in earth and material sciences. He founded the disciplines of petrology for examining rock thin sections and, later, microscopic metallurgy. Sorby was attentive to detail and methodical, placing great emphasis upon accurate observation and measurements. From his methods of 'physical analysis' came the practice of using stereology in petrology, as well as an innovative method of refractometry. Often, having established a new discipline, he would turn to other interests leaving other people to follow in his footsteps. From 1863, his petrological research led to work on meteors, which in turn led to studies of iron and steel. These like rocks, assume a granular structure; in later life, Sorby commented:

> 'In those early days, if a railway accident had occurred and I had suggested that the company should take up a rail and have it examined with the microscope, I should have been looked upon as a fit man to send to an asylum. But that is what is now being done…'

Source: Unknown artist, c. 1890. Reproduced with permission from the University of Sheffield.

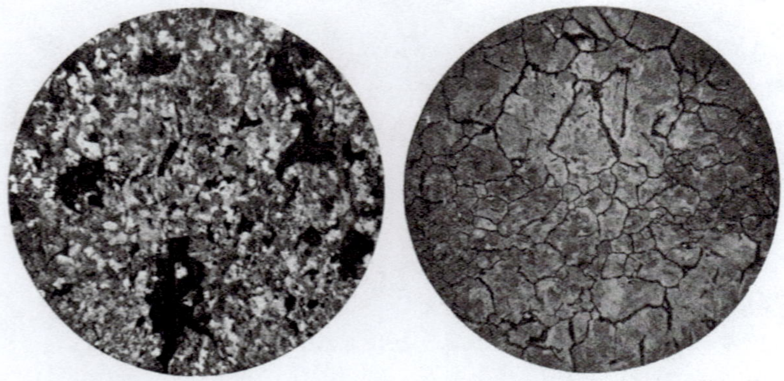

Source: Dr C. Hammond. Reproduced with permission from Dr Hammond.

Sorby was unusual in that his main vehicle for communicating his research was the Sheffield Literary and Philosophical Society, which he joined at the age of 20, remaining an active member for the next 62 years. He reported in the journals of well-recognised learned societies to a much lesser extent. He never went to university, never married, and was elected as a Fellow of the Royal Society at just 31 for successfully determining the phenomenon of the cleavage of sedimentary slate. In 1849 Sorby began the work for which he is perhaps best known: the preparation of thin sections of rocks, about 0.001″ thick, for microscopic study.

Sorby chose 30 µm as a standard, judging the correct thickness by the pale yellow-to-white interference colour of quartz grains. This is still the method used today, and it was Sorby who reintroduced the quartz wedge as a compensator for determining the vibration directions of birefringent crystals. In the 1860s Sorby devised an eyepiece spectroscope for detecting blood and pigment traces. However, he was led into error. Whilst studying jargon ($ZrSiO_4$) from Ceylon, he thought he had discovered a new element 'jargonium' and was strongly encouraged to publish this finding, only to realise soon after that it was the spectrum lines of uranium he had seen. This was his only major setback and damaged the credibility of his blood tests for forensic use.

After the death of his beloved mother in 1878, Sorby bought himself the 75-ton yacht *Glimpse* as a floating laboratory, which he used during the summer months to study meteorology, botany, geology and marine biology. Typically, he became an expert on yacht design and construction before making his purchase. Sorby devised and perfected a technique to mount and preserve large invertebrates as projection slides, so he could display and communicate his findings easily to large audiences. The specimens are held between two pieces of glass $3^1/_4$″ square, preserved so as to be translucent yet retaining remarkable colour. The technique involves rendering the specimens almost completely flat without distorting the shape: a very difficult thing to do. The exact method has, unfortunately, been lost, but his collection is still kept in Sheffield; a small part can be seen in the foyer of the eastern entrance of the Alfred Denny building of the university.

Henry Sorby was elected 18th president of the Royal Microscopical Society from 1875 to 1877. He was a founding member of the Iron and Steel Institute and was also president of the Geological Society. During the course of a scientific career spanning more than 60 years, Sorby wrote over 200 papers on every variety of topic that caught his interest. Henry Sorby worked hard to found a university in his home city, this being done in 1905. He died, aged 82, on the 9th of March 1908, leaving a considerable sum to Sheffield University to endow a chair in geology. It is regrettable that the Geology Department finally closed in 2001, and Sorby's legacy no longer exists at Sheffield.

References

Edyvean, R.G.J. and Hammond, C. (1997). The metallurgical work of Henry Clifton Sorby and an annotated catalogue of his extant metallurgical specimens. *Journal of the Historical Metallurgy Society 31/2*: 54–85.

Edyvean, R.G.J. (1988). Henry Clifton Sorby (1826–1908): studies in marine biology – the algal lantern slides. *Archives of Natural History 15/1*: 35–44.

Hadfield, R. (1920). 'The Great Work of Sorby'. *Transactions of the Faraday Society 16*: 114–118.

Higham, N. (1963). *A Very Scientific Gentleman*. Oxford: Pergamon.

Hammond, C. (1989). 'The contribution of Henry Clifton Sorby to the study of reflected light microscopy of iron and steel'. *Historical Metallurgy: The Journal of the Historical Metallurgy Society 23/1*: 1–8.

Hunter, R.A. (2015). *Sorby's Legacy: Geology at the University of Sheffield* 2nd Edition. Publisher RA Hunter, Sheffield.

The Sorby slides at http://www.shef.ac.uk/aps/about/sorby (accessed November 2014)

Today coverslips are almost never used when preparing opaque specimens for viewing with reflected light. There are a few exceptions, such as employing reflected light to study diatoms prepared for examination by transmitted light. Using homogeneous immersion helps to manage the effects of glare on covered specimens. For further information see also Box 12.3, and for practical details see Strange (1994).

For reflected-light materials microscopy, objectives must be used either (a) for which coverslip correction is not important (in general objectives of NA 0.25 or lower) or (b) which are designed as coverslip correction zero (i.e. a coverslip is not used). In materials science the no-coverslip objective may be found as a dry or immersion design, but in biological science microscopy, a water-immersion dipping objective (for viewing living samples held in water) requires no coverslip (see Figure 12.2 for examples of these objectives). In the first case where the use of a coverslip is immaterial, the objective has a '-' (dash) inscribed on the barrel instead of '0.17'. In the second case, an objective designed for use without a coverslip has a '0' (zero) inscribed on the barrel rather than '0.17'.

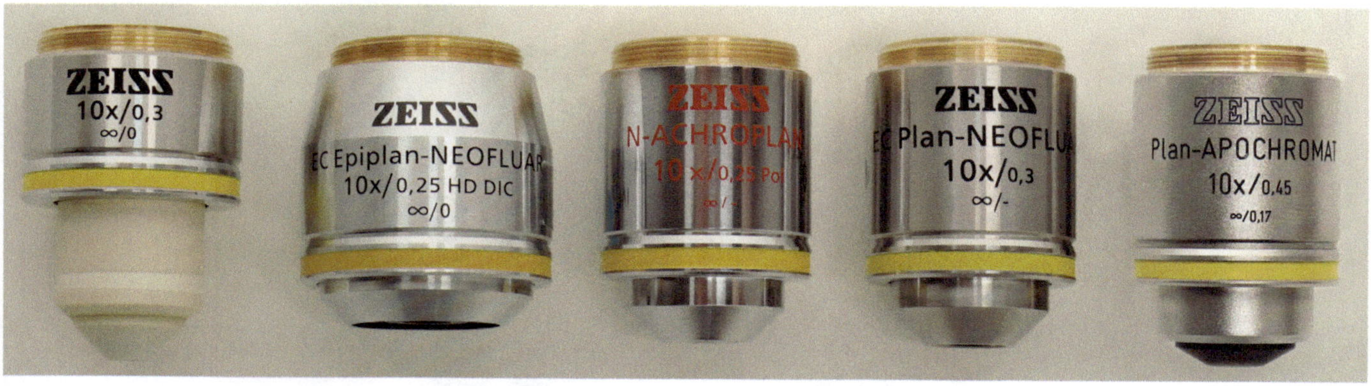

Figure 12.2 **Markings on the barrel for transmitted-light and reflected-light objectives** *Metallurgical objectives do not require correction for use with coverslips and are designated '0' as shown. Dipping objectives, which use water as an immersion medium, and are used for studying aquatic specimens or for physiology and in-vivo work, also do not require a coverslip. Those objectives that do require a coverslip of correct thickness is employed to complete the preparation have the designation '0.17' inscribed on the barrel, or the tolerance range of coverslip thickness, where an adjustable collar for correcting residual spherical aberration is fitted. Where the numerical aperture of the objective is sufficiently low that it does not matter whether a coverslip is used or not, the barrel of the objective is inscribed with a dash. Source: Courtesy of Mr K. Glover.*

12.4 The Optical Pathways in Reflected-Light Microscopy

When working with low-magnification objectives with a long working distance, a microscope configured for transmitted-light bright-field illumination can be used with light shone obliquely from one side onto the specimen. The disadvantage of such directional illumination is that shadows are formed with irregularly shaped specimens: even illumination of the entire field of view is difficult to achieve. Older microscope objectives were sometimes fitted with Lieberkühn illuminators. These allowed for symmetrical reflected-light epi-illumination to be secured with transmitted light. Figure 12.3 shows these means of producing reflected-light illumination

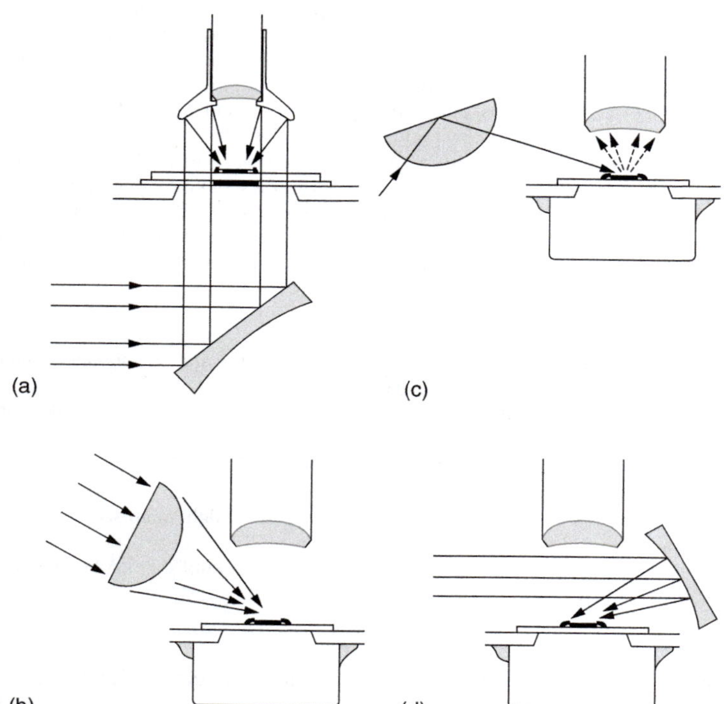

Figure 12.3 **Illumination methods for low-power objectives** *With some ingenuity, various lenses and mirrors can be used to direct light onto the surface of a specimen. The arrangement in (a) shows an older Lieberkühn reflector in use with a transmitted-light source. In Panels (b), (c) and (d) a light trap is shown underneath the sample, attached to the stage by Plasticine® or Blu-Tack®. This can be made from a bottle top or small black plastic container.*

A Mangin mirror (a negative meniscus lens with a reflective surface on the rear side of the glass forming a curved mirror that reflects light without spherical aberration) can be used (d) for low-power reflected light illumination. Source: Sanderson 2002. Reproduced with permission from The Quekett Microscopical Club.

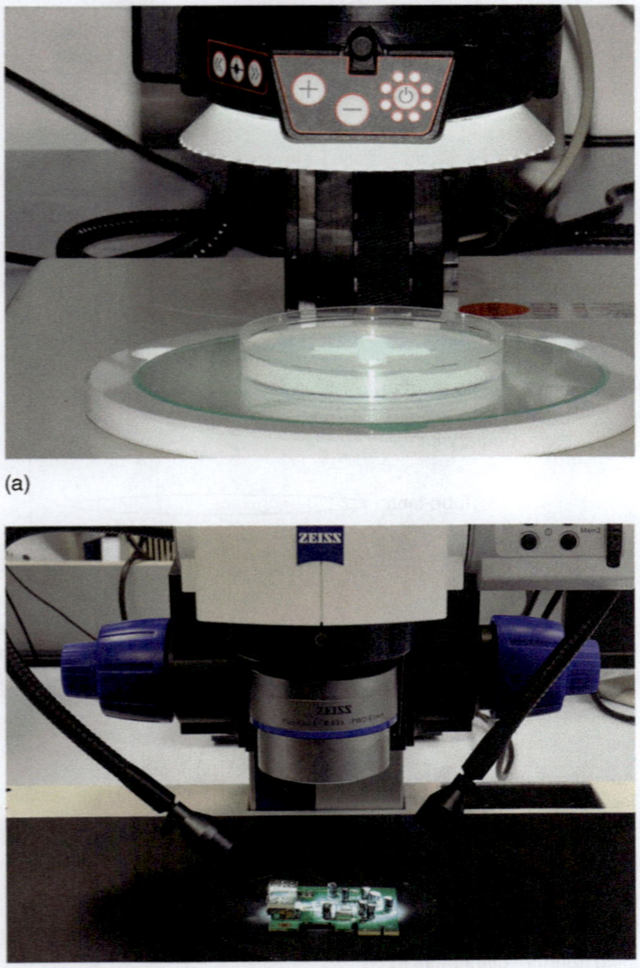

(a)

(b)

Figure 12.4 **Annular and oblique illumination for stereomicroscopes** *Ring illuminators and oblique illuminators can be used, respectively, to minimise or emphasise relief in the surface features and detail of the specimen, which may be shown as shadow or contrast enhancement in the image. Modern ring illuminators can be wired to light up particular sectors or the entire circle of LED illuminators. A ring illuminator is shown in the upper panel (a) whilst a cold light fibre optic swan-neck arrangement is shown in panel (b).*

for low-magnification objectives. The modern equivalent of the Lieberkühn illuminator (Figure 12.3a) is the annular LED illuminator (Figure 12.4a), which can be arranged to fit around the objective. Directional (or nearly symmetrical) reflected-light illumination can also be achieved for stereomicroscopes and compound microscopes with low-magnification objectives (Figure 12.4b) by using one (or more) swan-neck fibre optic illuminators.

For medium- and high-power magnification objectives, epi-illuminators are used to reflect light onto the surface of the specimen. The key difference between transmitted-light microscopy and reflected-light microscopy is that for the latter, the objective lens now has two functions: it acts as a condenser lens to focus light onto the specimen and also as an imaging lens to collect the reflected light.

Because *the objective acts as its own condenser*, and the ray path is consequently folded upon itself, the ray path of the reflected-light microscope is much simpler (Figure 12.5) than that of the transmitted-light microscope. In the case of reflected light, the ray diagram for transmitted light may be considered to be folded twice over: firstly folded in the plane of the object (so that the illuminating and imaging optical components coincide) and secondly folded in a plane at 45° to the optical axis in the parallel-light region between the objective lens and the tube-lens. The latter folding corresponds to the position of a semi-transparent beam splitter, which admits light from the illuminating system at a right angle to the major optical axis and which, in practice, is situated just above the nosepiece.

Since in bright-field reflected-light microscopy the condenser lens is also the objective lens, then once set, the illuminated field diaphragm (IFD) need not be changed with a change of objective lens (Figure 12.6) because each objective projects an appropriately sized image of the diaphragm onto the object. The IFD should be set at the edge of the field of view; stray light results if it is set wider than this value – a particular problem in reflected-light microscopy. The setting of the condenser illuminating aperture diaphragm should always be checked by observing its image in the objective back focal plane and adjusted to about 4/5ths – or 2/3rds where contrast and

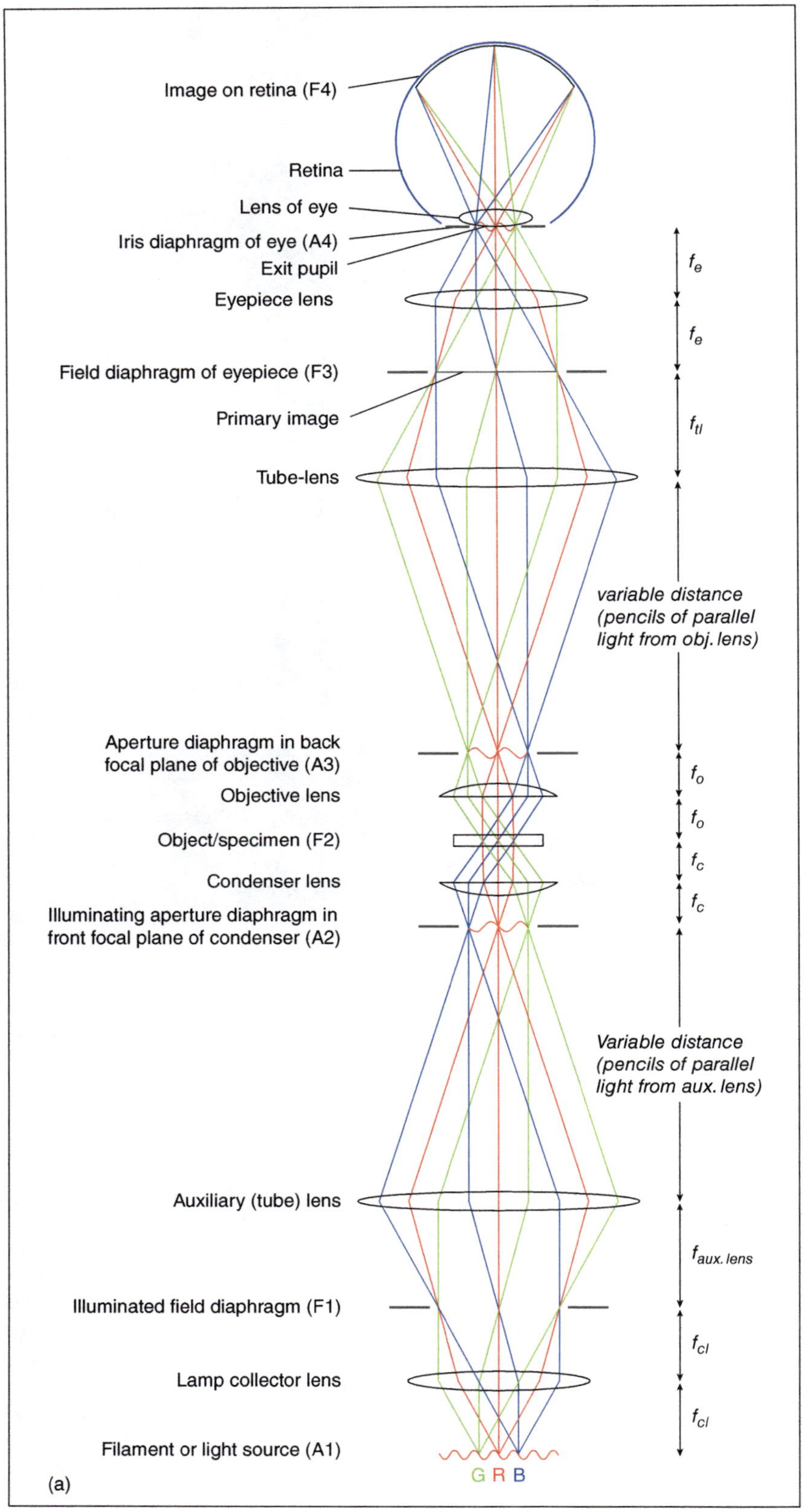

Figure 12.5 **The ray path of the reflected-light microscope** *Showing the difference between transmitted-light (a) and reflected-light ray paths (b,c) and how the reflected-light ray path is folded over about the reflector unit so that the objective becomes its own condenser. The relative positions of the illuminated field diaphragm and the condenser diaphragm are interchanged in the transmitted-light and reflected-light microscope (refer also to Figure 12.7). Source: (a,c) Hammond and Heath 2006. Reproduced with permission from John Wiley & Sons. (b) Dr P. Evennett. Reproduced with permission from Dr Evennett. (d,e) Bradbury et al. 1989. Reproduced with permission from Oxford University Press.*

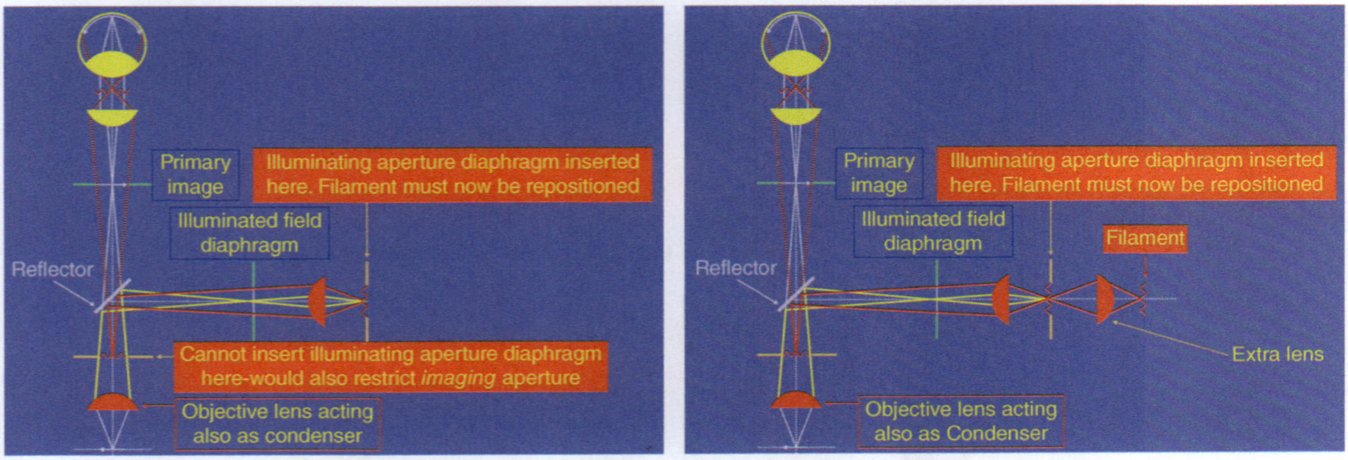

(b)

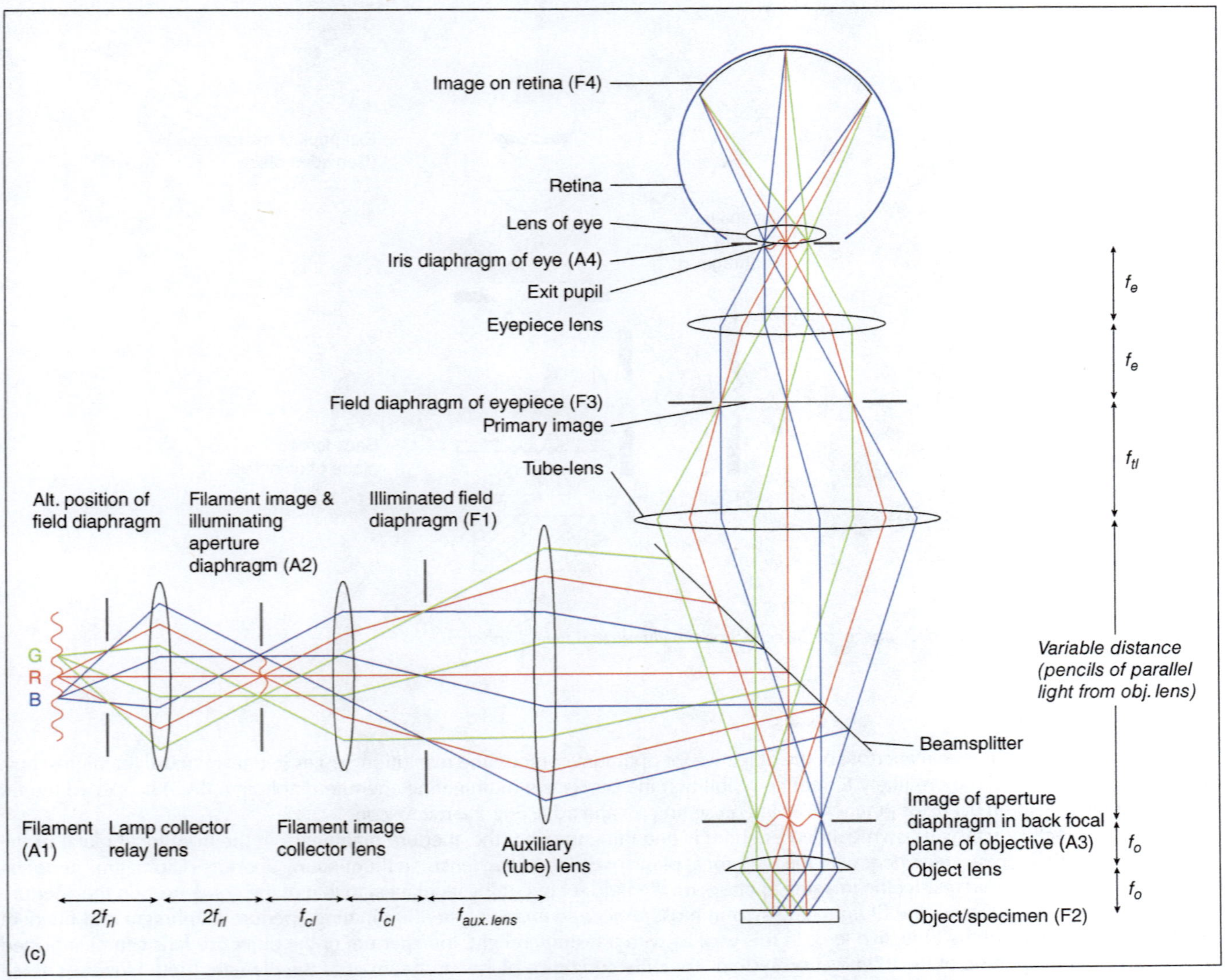

(c)

Figure 12.5 (*Continued*)

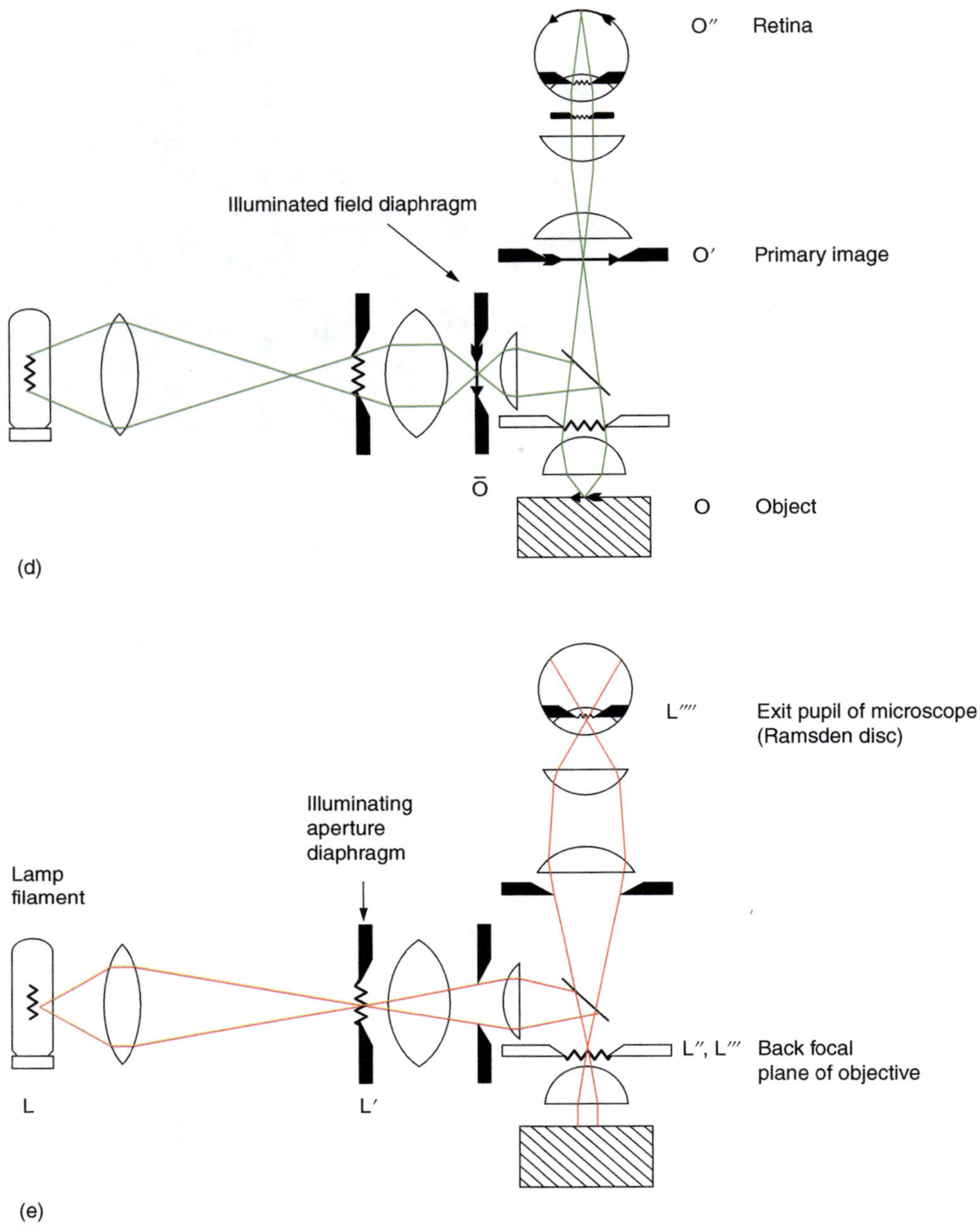

Figure 12.5 (Continued) $\overline{O}$ is the 'back image' of the object on the illuminated field diaphragm.

visibility is poor – of the numerical aperture of the objective for optimum contrast and resolution, just as in transmitted-light microscopy. Stray light (Box 12.3) is even more likely to reduce visibility if the condenser illuminating aperture diaphragm (IAD) is opened too far (refer to Chapter 9, Section 9.11, for practical advice on setting up and adjusting the microscope).

Because the objective acts as its own condenser, there is one thing missing: the aperture diaphragm in the front focal plane of the condenser lens (which is now *coincident* with the back focal plane of the objective lens). An illuminating aperture diaphragm cannot be placed here because it would restrict the imaging aperture (in the field set of conjugate planes) to that of the condenser (in the aperture set of conjugate planes). Instead, the illuminating system must *project an image* of the illuminating aperture diaphragm and filament into the back focal plane of the objective lens. In this way, as with transmitted light, the aperture of the objective lens can be adjusted independently of the aperture of the illumination system. The only conjugate plane 'upstream' is at the filament itself. However, since this is inside a lamp, it is usually inaccessible. The illuminating system therefore incorporates a further lens upstream, generally called a relay lens, which projects an image of the filament into this position. Such a relay lens is shown in the illuminating part of the system in Figure 12.5b.

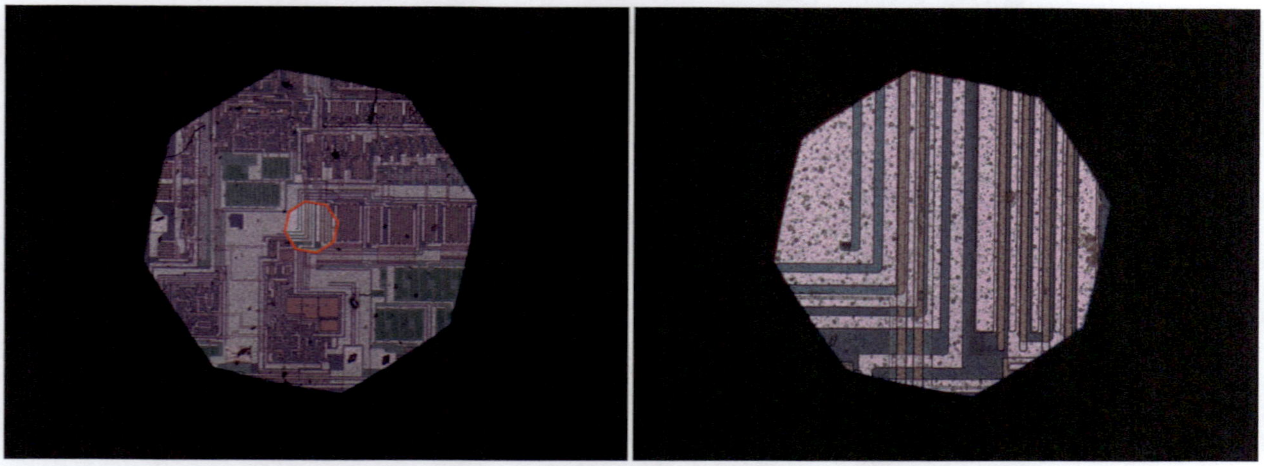

Figure 12.6 ***The constancy of the IFD when changing objectives in the reflected-light microscope*** *The illuminated field diaphragm does not have to be changed in reflected-light microscopy when changing objectives. The figure on the left has been taken with a 2.5x objective, the figure on the right with a 20x objective. The field of view of the 20x objective is outlined in red.*

Careful examination of Figure 12.5b explains a commonly heard query: 'Why are the positions of the field and aperture diaphragms reversed in reflected light?' The answer is: 'Because the aperture diaphragm has leapfrogged one conjugate plane upstream'. The position of the illuminated field diaphragm and the condenser illuminating aperture diaphragm are interchanged (Figure 12.7) compared to a transmitted light epi-illuminator. Equally, the field diaphragm could in fact leapfrog to a position just in front of the filament (i.e. the focal length of the relay lens). Which position is chosen is a matter of individual microscope design (indeed, sometimes both are used), but the geometrical principles of all the designs are simple and the same.

12.5 Types of Reflector

The simplest type of reflector is a flat glass plate – often a coverslip was used in early designs – set at 45° to the optical axis (Figure 12.8a). This is called a Gauss illuminator. However, it is not very efficient: even with a partially mirrored surface on the glass to aid reflection of the incident illumination downwards onto the specimen, only about 20–25% of the initial illumination reaches the eyepiece(s). Furthermore any linearly polarised illumination is markedly rotated as it reflects off the plate or slip of glass.

Two other designs (Figures 12.8b and 12.8c), therefore, have additionally been used in reflected-light microscopy. The first is the totally reflecting prism (or mirror) where light is reflected down half the aperture of the objective, being reflected back behind the prism up the other half of the objective aperture to the eyepieces. This has an improved efficiency of 50%, but only half the numerical aperture of the objective is used, and if the prism is above the back focal plane of the objective, it causes a shadow. The incident illumination is oblique giving rise to some aberrations but is also an advantageous source of contrast, just as in transmitted-light microscopy. The

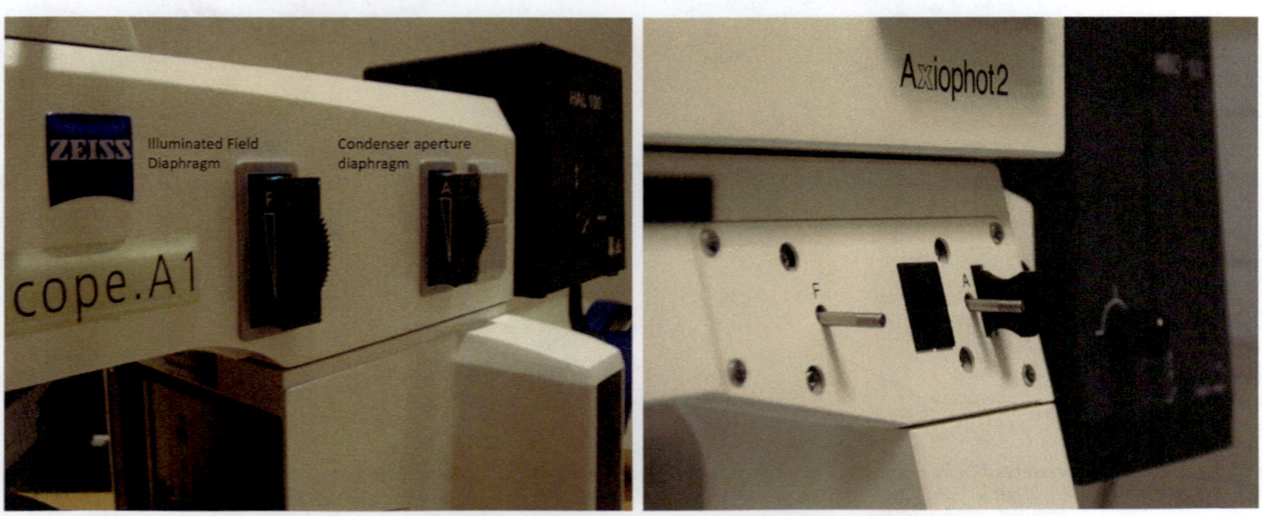

Figure 12.7 ***Position of the field diaphragm and condenser diaphragm in reflected-light microscopy*** *See the text for a description and further details.*

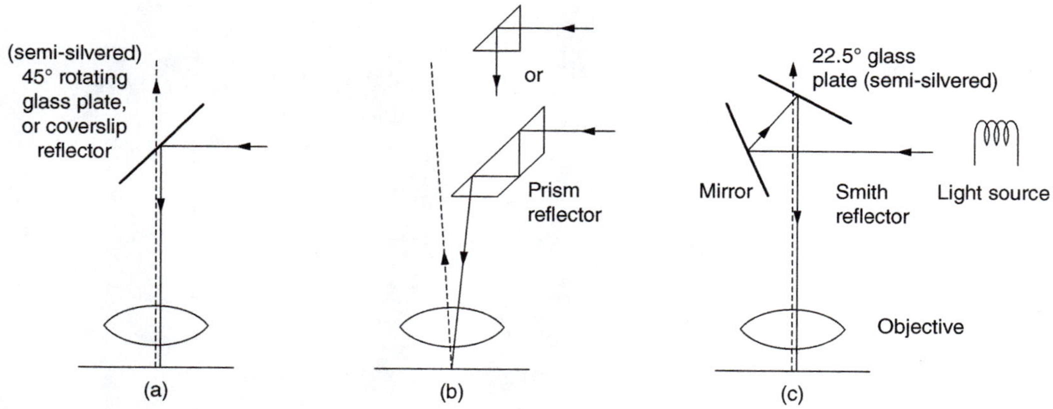

Figure 12.8 **Gauss-type coverslip reflector, prism illuminator, Smith-type illuminator** *See the text for a description and further details.*

second (preferred) design is the Smith illuminator (Figure 12.8c), which considerably cuts down unwanted reflections from the surface of the glass plate, which is inclined at 22.5° to the optical axis. The epi-illumination is first shone onto a mirror set at 67.5°; this reflects the light onto the plate glass reflector at 22.5°, and thus the epi-illumination is turned through 90° onto the specimen. Besides reducing unwanted reflections, the Smith illuminator offers much better extinction between crossed polars[1] than the simple plate-glass reflector design.

12.6 Practical Use of the Reflected-Light Microscope

A microscope set up for reflected-light illumination may have a transmitted light path associated with it. Figure 12.9 shows two microscope stands designed for using reflected-light illumination alone. The older model, represented by the Vickers M73, can still be found.

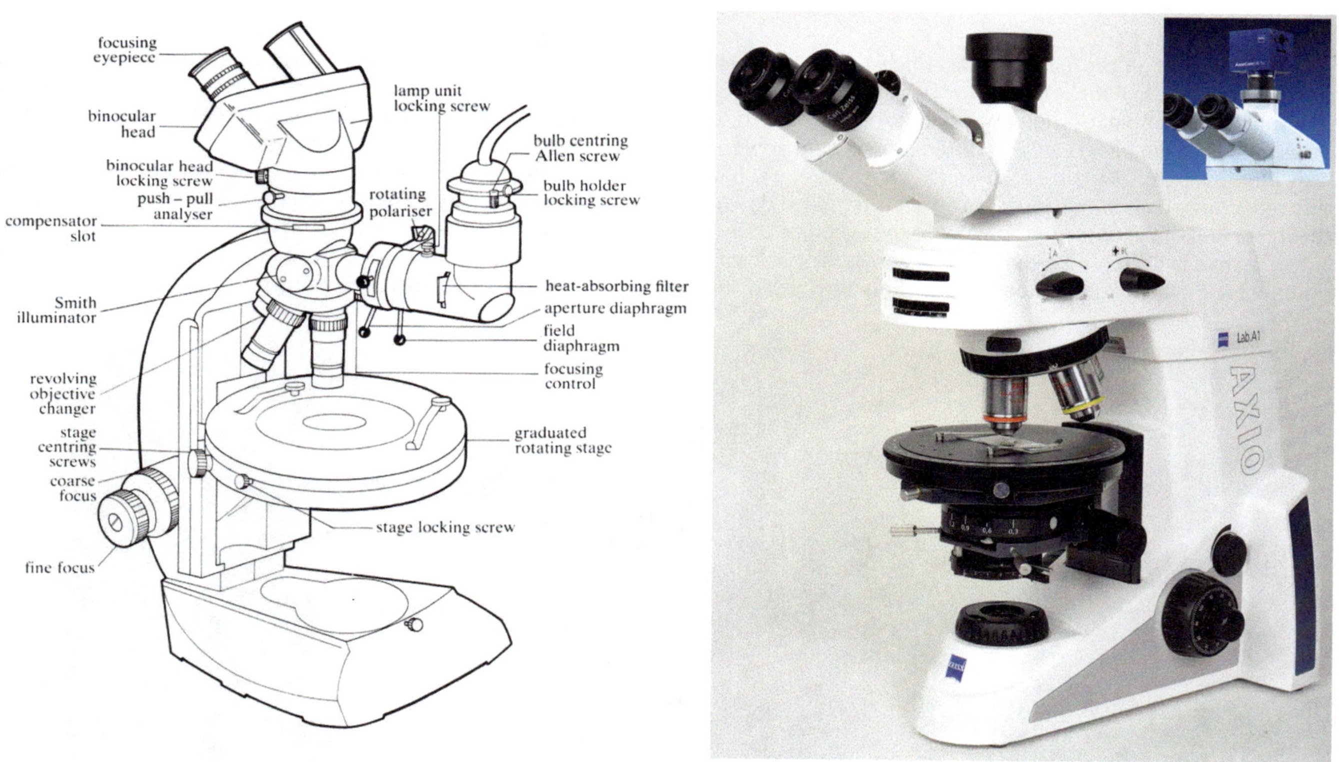

Figure 12.9 **Vickers M73 reflected-light microscope and Zeiss Axio Lab A1 reflected-light stand** *Two different designs of reflected-light microscope. The older Vickers design, on the left, is still used today. This model does not also have transmitted-light illumination, although the cover in the base could be removed to incorporate such a unit. The modern Zeiss instrument on the right shows both reflected and transmitted light incorporated into a single stand. Source: (left) Gribble and Hall 1992. (right) Courtesy of Mr K. Glover.*

The newer model shown here is the Zeiss Axio Lab A1. The advantage of this, and equivalent stands from other manufacturers, is that the modular construction means that the configuration can be altered or upgraded to suit a particular requirement. Remember that the controls for the illuminated field diaphragm and the illuminating aperture iris are reversed in configuration in the illuminator for reflected-light, compared to transmitted light illuminators, as explained above.

12.7 Contrast Mechanisms in Reflected-Light Microscopy

All the contrast mechanisms discussed for transmitted-light microscopy (Chapter 11) have their analogues in reflected light and hence a less detailed treatment is given here. Students who are used to viewing transparent sections by transmitted light may, at first, be confused by the appearance of sections or opaque samples viewed by reflected light. As when starting examination with transmitted light, always use a medium – or low-magnification objective first. For the very lowest magnifications, reflected-light illumination is especially suited to stereomicroscopy. For further details of the illuminators designed for this work, please refer to the section on stereomicroscopes in Chapter 4, Section 4.6.

12.7.1 Increasing Contrast in Epi-Microscopy

Glare, the result of stray light (see Box 9.1 and Box 12.3), is seen as a low-intensity haze of light over the entire image, which diminishes contrast. The reflectivity (for normal incidence) of light at an air-glass interface[2] is 0.04 (4%). If this occurs at several interfaces at separate lens elements within the objective, a significant proportion of stray light may be generated. Unlike strongly reflecting samples, where the contrast is inherently high, this stray light generated at air-glass interfaces causes a problem when examining weakly reflecting samples because the stray light constitutes a significant proportion of the light passing through the microscope and so can swamp the true signal. This is why anti-reflection coatings (Chapter 7, Section 7.12) are so useful – and why objectives must be carefully cleaned to avoid scratching them. Stray light may also originate as a result of extreme peripheral rays refracted by the front hemispherical lens element (especially with high magnification dry objectives) hitting the inside of the lens mount. This is the main reason for judicious control of the illuminating aperture diaphragm (condenser diaphragm) iris, so as not to use too large a cone of light for the numerical aperture of the objective in use. Careful adjustment also of the illuminated field diaphragm (IFD) in the base of the microscope, so as not to illuminate too large an area of the specimen, additionally helps to prevent stray light from reducing image contrast.

One simple way to improve the contrast in bright-field imaging is to close down the IFD to well within the field of view. The contrast should improve, although only the central part of the field will be illuminated, because glare through stray light reflecting off the tube of the microscope and being refracted into the image from the edges of the internal lens elements of the objective will be much reduced

Table 12.1 *Appearances of Thin Sections and Polished Surfaces*

Feature	Transmitted light	Reflected light
Transparent phases (minerals, glass, liquid)	Appear white to coloured because they transmit most of the illumination.	Appear dark grey, because they reflect a small proportion of incident light. Bright patches may be seen within transparent mineral areas where they reflect from under the polished surface.
Absorbing phases (opaque structure and ore minerals)	Appear black, may be coloured, particularly red to brown, since these longer wavelengths are absorbed less than blue.	Appear grey to white, since they reflect more of the incident light. There may be faint colour tints.
Grain boundaries and micro-fractures	Thin black lines, since light is scattered and refracted at these boundaries.	Grain boundaries may show as dark or light lines. Small holes may have dark and indistinct margins.
Holes and fractures	Appear white where the rock is missing.	Appear black, although some light may be reflected from crystal faces in the holes.
Relief	Minerals with $n < 1.54$ (RI of mounting resin) show negative relief. Those with $n > 1.54$ show positive relief.	Polishing relief may cause dark or light lines along grain boundaries because of differing hardness.
Artifacts	Bubbles appear as small, round, highly refractile dark areas, sometimes with a clear central area. They may move. Grit may appear as black dust or blue-black angular grains accumulated in fractures or at the edge of the section.	Patches of moisture or oil may be seen as dark or iridescent patches. Tarnish is seen as a coloured hue. Specks of dust appear black and may accumulate in fractures and scratches. Scratches appear as long straight or curving lines. They may refract light or cause colour effects.

Source: Data from Gribble and Hall 1992.

Reference

Gribble, C.D. and Hall, A.J. (1992). *Optical Mineralogy: Principles & Practice*, 1st ed. London: UCL Press, Taylor & Francis e-library 2004, 2009.

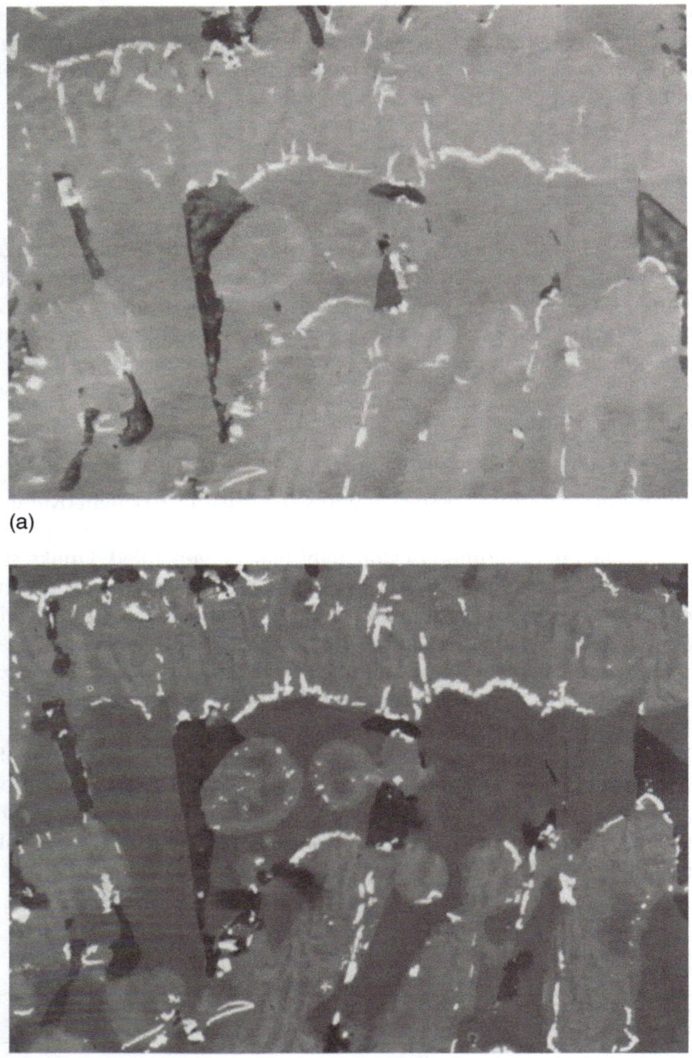

(a)

(b)

Figure 12.10 **The use of homogeneous (oil) immersion to improve image contrast** *Upper panel shows slag rich in phosphorus imaged with a dry 40x/NA 0.65 objective. The lower panel shows the same field of view acquired with an oil-immersion 40x/NA 0.85 objective. The lowering of reflectance by using an immersion medium for weakly absorbing structures (see Box 12.1) with low absorption coefficients improves contrast. Source: Reproduced with permission from ZEISS Microscopy.*

or eliminated. When using a camera, the film or CCD detector may only record the central area of the entire field of view as seen in the eyepieces, so it makes sense in this case to close down the IFD to just outside the limit of the *camera frame*, even though it may be seen over most of the field of view by eye.

Using an immersion medium of homogeneous refractive index with that of the sample, to cut down specular reflections at the top surface of the uncovered preparation, is another means for improving contrast (compare Figure 12.10a and Figure 12.10b) for highly reflective (e.g. metal) samples, because surface reflection is reduced, reflections from sub-surface features can be seen. However, light reflected from sub-surface features in ceramics can cause glare. Although it is convenient to use dry objectives to change between non-immersed specimens, homogeneous immersion not only improves contrast for certain *highly reflective* samples but concomitantly reveals much more fine detail due to the higher resolving power of the objective. Very occasionally a very high refractive index immersion medium, such as methylene iodide, may be used instead of oil. However, this technique requires special objectives and is seldom employed. Overall, homogeneous (i.e. oil) immersion is little used in reflected-light microscopy because much greater contrast can generally be obtained with dark-field and/or polarised-light microscopy using dry objectives.

The reflectivity of light at an isotropic surface decreases as the refractive index of the immersion medium better matches the refractive index of the material of which the surface is composed. Using bright-field immersion microscopy is therefore no good for examining

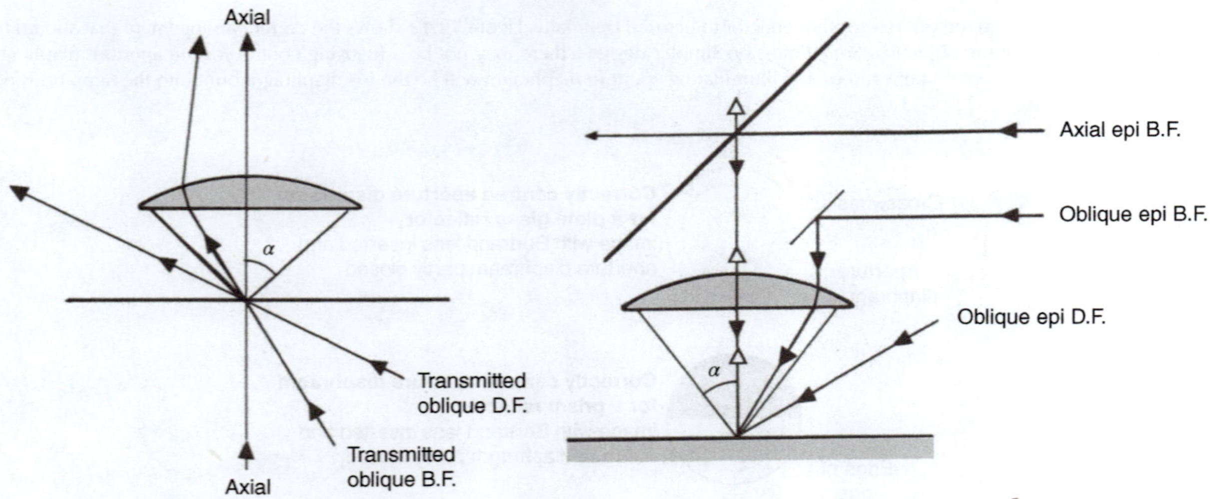

Figure 12.11 **Comparison between reflected-light bright-field compared to transmitted-light bright-field ray paths** *See the text for a description and further details. Compare this figure with Figure 11.8b, showing the same information in a different format. Source: Bradbury and Evennett 1996. Reproduced with permission from Taylor & Francis Group.*

weakly reflecting materials (e.g. most ceramic, coal or other rock samples). The best contrast comes from using dark-field (*all* the stray surface light reflected is eliminated) or polarised light using crossed polars. For materials microscopy it is much more common to use dark-field or epi-polarised light to give enhanced contrast and better visibility of fine detail in the image. Two other stratagems are to use reflection-contrast or 'Antiflex' microscopy (Section 12.13). The value of such low magnification (10x or 40x) Antiflex objectives in the study of materials of low reflectivity (e.g. coal) deserves to be more widely known.

12.8 Bright-Field Reflected-Light Microscopy

When an objective in reflected-light or epi-illumination mode (Figure 12.11), directs a cone of incident light down onto a perfectly flat, equally reflecting featureless surface, the surface acts as a mirror or specularly-reflecting surface. All the light is reflected back into the objective. This is the equivalent, in transmitted-light microscopy, of no specimen being present.

Protocol 12.1 Reflected-Light Bright-Field Microscopy

Required

- A specimen with well-defined surface features (e.g. a piece of old CD observed from the non-printed side; miniature electronic circuit component or polished, etched, metal).
- An epi-illuminator with adjustable illuminated field diaphragm

Procedure

1. Place the specimen slide onto the stage of the microscope.
2. Select a low-magnification objective (e.g. 10x). Look sideways on and raise the stage until the specimen lies within, or above, the position of focus for the objective. Lower the stage and bring the specimen into focus.
3. Open the IFD to *just* outside the field of view. This is more critical than with transmitted light microscopy, in order to minimise unwanted reflections and glare in the image.

   Adjust the intensity of the illumination to be even with neutral-density filters, or use the transformer (which may be built into the stand) rheostat control.

   An intense light source is usually required for viewing with crossed polars. Use neutral-density filters to maintain the correct colour temperature (Chapter 31, Section 31.4) where this is important.
4. Remove an eyepiece, or introduce a Bertrand lens into the light path, or use a centring-telescope ('phase' telescope). Observe the image of the illuminating aperture diaphragm in the back focal plane of the objective, and adjust this to about 80% of the diameter of the back focal plane of the objective to secure optimum contrast.

The lamp needs to be centred on-axis for the reflector to be used optimally. Figure 12.12 shows the correct arrangement that should be seen at the back focal plane of the objective. Sometimes, on simpler designs, there may not be a focusing control for the aperture diaphragm. On older microscopes with external light sources, the illuminating aperture diaphragm will be the iris diaphragm built into the lamp housing.

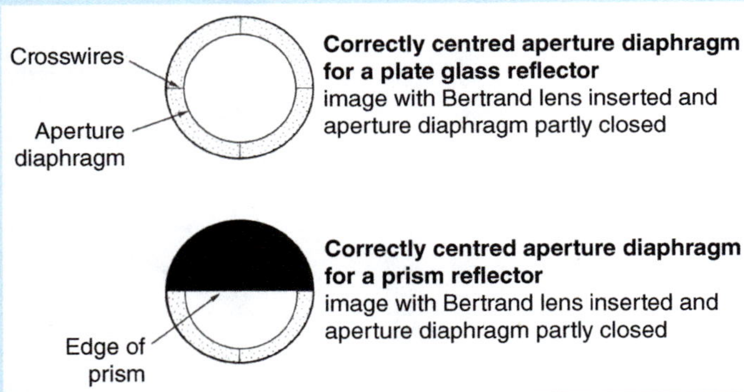

Figure 12.12 ***Adjustment of the aperture diaphragm with different reflector designs*** *Refer to the protocol on adjusting the microscope for bright-field reflected-light microscopy. Source: Gribble and Hall 1992.*

5. Set the correct inter-ocular spacing on the eyepieces; on a Jentzsch-pattern binocular head, this is done by moving the eye-tubes closer together or farther apart in a linear fashion. The desired inter-ocular (inter-pupillary) distance is set on each of the eyepiece tubes to correct for the altered optical tubelength. On a Siedentopf-pattern head, the inter-ocular distance is changed by rotating the eyepiece tubes about a central axis so that the optical tubelength is not altered. The eyepiece with the cross-hairs is usually the fixed eyepiece with no dioptre control. Focus the cross-hairs sharply.

With a typical etched metal or ceramic surface the situation is different. The incident light is now reflected or scattered at different angles as a result of surface relief, or is reflected by different amounts according to the varying reflectivity of the phases present. The light/dark contrast is determined by the extent to which the light, reflected or scattered at different angles, falls outside the objective or by the variation in reflectivity over the surface. Figure 12.13 shows examples of lapis lazuli and colour film illuminated with reflected and transmitted light under conditions of bright-field and polarised-light illumination, respectively. Prior to the Industrial Revolution and the artificial manufacture of aniline dyestuffs from coal tar waste, the only way to achieve a deep rich blue was by using ultramarine from lapis lazuli, applied in pure form as a glaze (Figure 12.14). Yet, as Figure 12.13 shows, under transmitted light, lapis lazuli is quite unremarkable. In the same way, thin histological sections coloured with absorption stains and intended to be imaged with transmitted light, exhibit low contrast and poor visibility (Figure 12.15) under reflected light. The contrast method chosen should depend upon the nature of the specimen.

Microstructural inclusions with differing reflectance from the surrounding mass (e.g. pearlite zones in ferrous materials) can easily be distinguished using a dry objective with reflected-light microscopy. Ceramics, however, present a challenge with reflected light bright-field illumination. The semi-transparent surface of ceramic materials allows some light to penetrate below the surface, where it scattered back by internal structural features, disturbing the image of the surface. This scattered stray light (of generally weak intensity) may reduce contrast of the image of the surface. It is particularly noticeable with high magnification dry objectives.

Box 12.3 Stray Light and Glare – Reflected Light

As the condenser numerical aperture is increased, the potential for stray light to reduce contrast also increases. When ultimate resolving power is not required, close the condenser illuminating aperture diaphragm slightly. Also close down both the illuminated field diaphragm to illuminate only that part of the field of view under observation. This is important when observing opaque specimens by reflected light, where visibility can be poor due to low contrast in the image. Stray light can arise from being scattered or reflected by the object without contributing to image formation. It may also arise from dust or other structures in the light path, either being luminescent themselves or reflecting light into the field of view.

Glare is slightly different, although the terms 'glare' and 'stray light' are often used synonymously; put simply, glare is the result of stray light. Glare is the increased flux of light either disturbing the proper function of the eye (as with headlights 'blinding' a driver) within the field of view or is caused by very high luminance (and reduced visibility) across the whole field of view. Lenticular glare is caused by reflection of light from the glass-air interfaces of the lens elements making up the objective. Glare is nowadays very much reduced because of high-quality anti-reflection coatings on the lenses. This is why objectives must be cleaned carefully: a scratched front lens will invariably have a damaged anti-reflection coating, which will admit stray light into the image.

In the reflected-light microscope, the careful use of a polariser can help reduce glare, just as polarisers in sunglasses block the sunlight specularly reflected off water or road surfaces. In this case, the reflected sunlight is mostly polarised horizontally, parallel to the reflecting surface, so the glare can be blocked by polarisers in the sunglasses arranged with their permitted vibration direction set vertically. Not all stray light is eliminated, for some light that is reflected diffusely will be reflected or scattered several times, and so be unpolarised.

The use of homogeneous (e.g. oil) immersion will also help reduce glare (see Figure 12.10). Stray light reflected off the surfaces of the coverslip and glass slide is eliminated or, at least, reduced. It is possible that some glare can arise from reflections from the eyepiece, spectacles or the external room illumination. If you see glare on a camera image for which you cannot identify the cause, there is always the possibility that it is entering the microscope via the eyepieces of a trinocular photo-tube from the room lights (see Figure 12.14, reflected-light dark-field, top-left panel). Some microscopes dedicated to photomicrography have been designed with extra-wide body-tubes, to minimise the reflections reducing contrast in the image. Some research microscopes are made with black velvet lining the inside of the body and photo-tube to minimise glare.

Sources of stray light

Bright-field microscopy

Excessive condenser aperture
Uncoated or dirty lens surfaces
Reflected-light from body-tube inner surfaces
Reflected-light from glass slide/coverslip surface

Dark-field microscopy

Smears or dirt on the preparation
Excessive objective aperture
Uncoated or dirty lens surfaces

Polarised-light and interference microscopy

Poor exitinction of crossed polars
Defects, holes, hot-spots, in polarising filters
Strain birefringence in lenses
Dirt or surface imperfections on lenses
Sub-maximal compensation

Fluorescence

Autofluorescence
Poor staining: high background
Poor antibody: non-specific staining
Excessive field aperture

Source: Adapted from Table 1, page 80, Chapter 3, Lacey 1999, adapted from Lacey 1989.

References

Lacey, A.J. (ed.) (1989). *Light Microscopy in Biology: A Practical Approach*, 1st ed, Chapter 8. Oxford: IRL Press. p. 225.

Lacey, A.J. (ed.) (1999). *Light Microscopy in Biology: A Practical Approach*, 2nd ed, Chapter 3. Oxford: IRL Press. p. 80.

Martin, L.V. (1975). 'Some aspects of vertical illumination'. *Microscopy (Jour. Quek. Microsc. Club) 32*: 489–502.

McClure (1963). 'Glare control in the superficial illumination of mounted objects by means of the vertical illuminator'. *Jour. Quekett Microscopical Club 29/5*: 105–110 and *Jour. Quek. Microsc. Club 29/6*: 147–153.

Strange, A. (1994). 'Visibility and resolution of covered specimens with incident-light illumination'. *Journal of Biological Photography 62/3*: 91–102.

12.9 Dark-Field Reflected-Light Microscopy

The principal object of dark-field (sometimes referred to as darkground) imaging is to prevent the specularly-reflected illumination – which carries no information about the structural detail – from entering the objective. As in transmitted-light microscopy, only the diffracted information-carrying light contributes to the image. Reflected-light dark-field illumination requires the use of specially designed (catadioptric or dioptric) objectives in which an annular ring of light passes around the perimeter of the objective proper and is projected on to the specimen by a system of curved mirrors and/or prisms at such an angle that the specularly-reflected light falls outside the aperture of the objective lens. Figures 12.16a–c show, in simplified form, the ray paths and the construction of such objectives, and Figures 12.16d–e, the front face of a 10x and 20x HD objective[3] and a high-power 100x dry objective for reflected-light microscopy. The annulus of light is provided by a stop in the illuminating system: at the same time, the aperture diaphragm must be opened fully to allow light to pass around the perimeter of the stop.

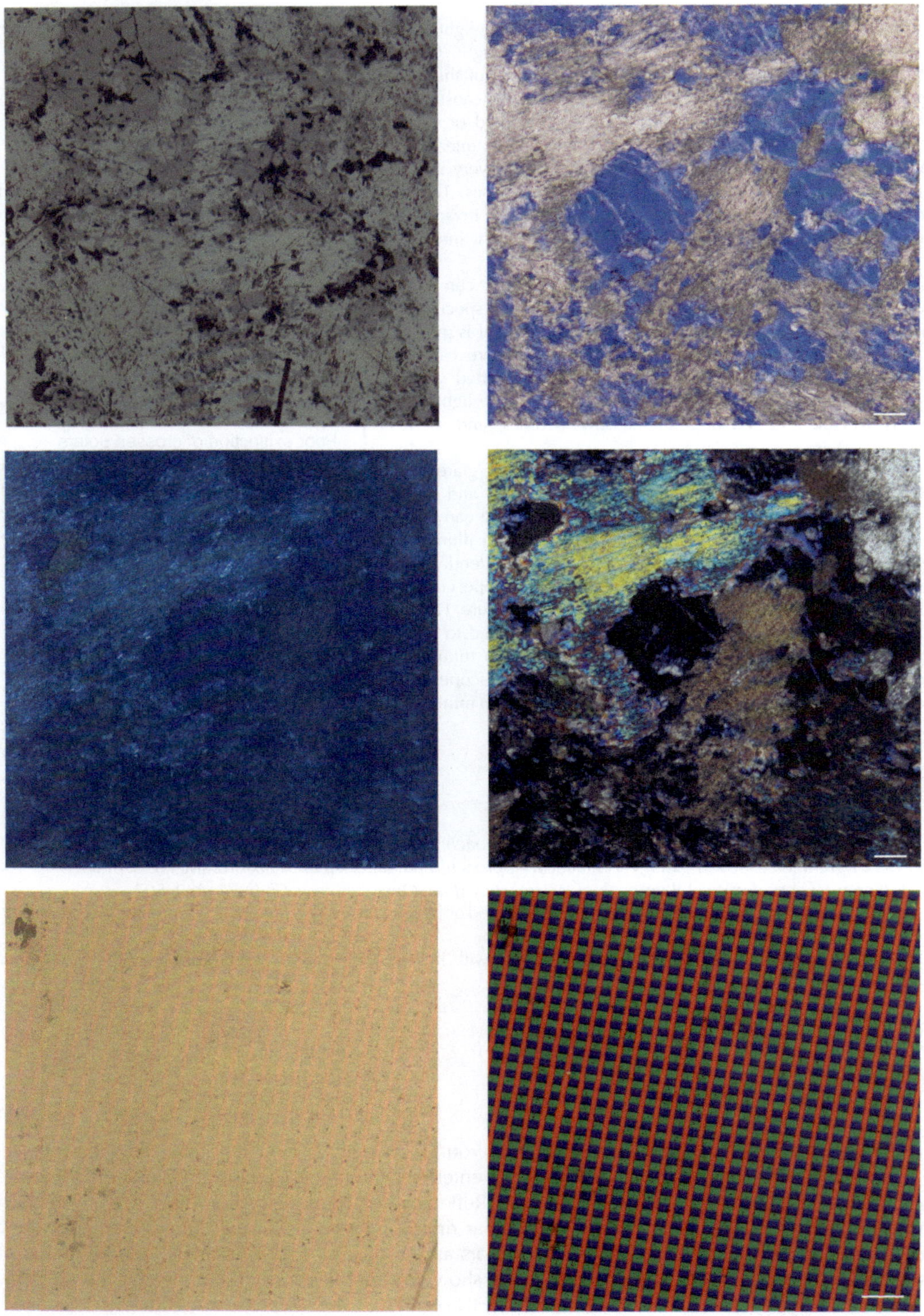

Figure 12.13 **Comparison between reflected-light and transmitted-light images** *Top row: reflected-light bright-field (left) and transmitted-light bright-field (right) of a 30 μm thin section of lapis lazuli. Middle row: same lapis lazuli thin section under crossed polars by reflected light (left) and transmitted light (right). Compare these images taken by bright-field and polarised light with the equivalent images by dark-field and reflected light of an opaque slice of lapis lazuli, shown in Figure 12.14.*

The bottom panels of this figure show the réseau of developed unexposed Dufaycolour film by reflected-light (left) and transmitted-light (right) bright-field microscopy. EC-Epi-plan 5x objective. Scale bar 100 μm.

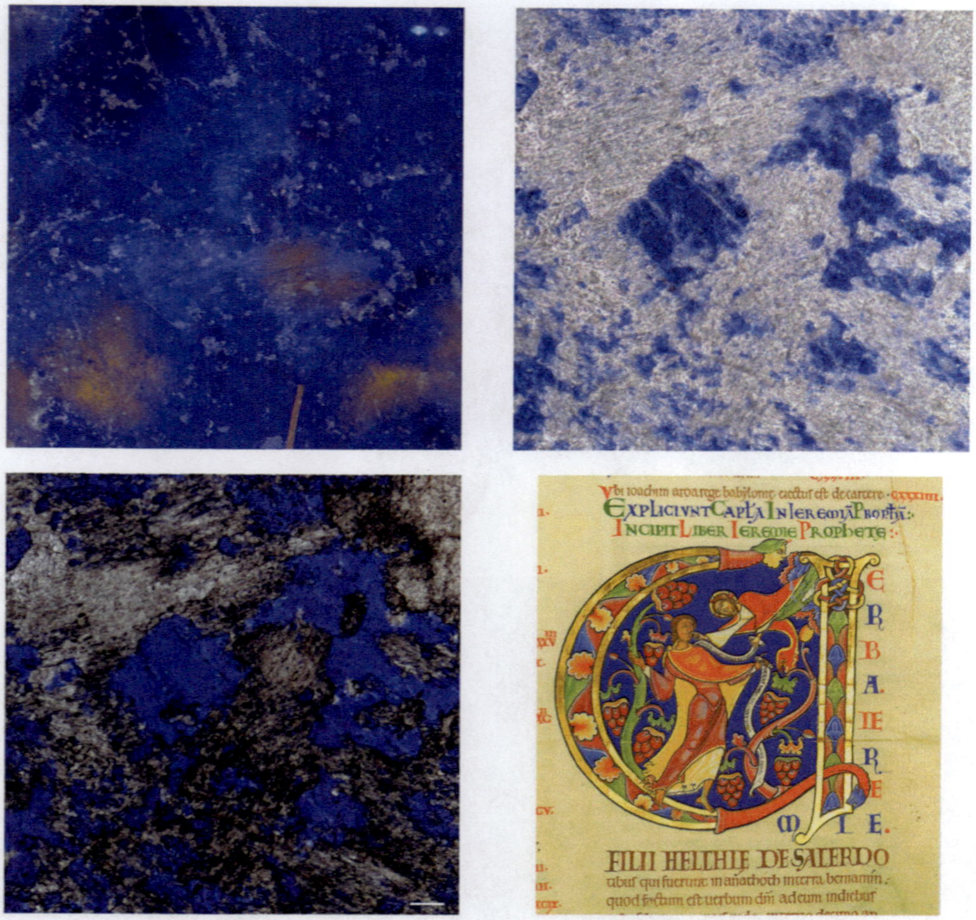

Figure 12.14 **Lapis lazuli by reflected light** *Top row: reflected-light dark-field (left) of an opaque chip of lapis lazuli and (right) transmitted-light dark-field of the same thin section as in Figure 12.13.*

Bottom row: Reflected-light bright-field (left) of the opaque chip of lapis lazuli seen top left.

For comparison (right) the use of lapis lazuli as a medieval pigment. Natural blue pigments occur rarely in nature; prior to the Industrial Revolution, the only way to achieve a deep rich blue was by using ultramarine from lapis lazuli. Dürer complained of the prohibitive cost and difficulty of preparing the pigment. This example is taken from the Winchester Bible and dates from the 12th century. Source: Reproduced with permission from Winchester Cathedral.

During preparation of these figures, one image was inadvertently acquired with the trinocular photo-tube set for visual observation as well as photography. Light has reached the sensor faceplate of the CCD camera via the eyepieces, resulting in a bright image of the inset ceiling lights in the top right-hand corner of the reflected-light dark-field image.

Protocol 12.2 Reflected-Light Dark-Field Microscopy

Required

- A specimen with well-defined surface features (e.g. a piece of old CD mounted on a slide and viewed from the non-printed side, or an opaque polished/etched metal sample).
- Specialised HD objective with co-axial annular mirror system (as shown in Figure 12.16).

Procedure

1. Focus on a specimen suited for bright-field, and adjust the microscope, if needed, for reflected-light Köhler illumination. Remove this specimen.
2. Open fully any iluminated field diaphragm (IFD) fitted to the epi-illuminator.
3. Place the specimen slide for dark-field onto the stage of the microscope.
4. Swing in, or fit into place, the reflected-light dark-field objective.
5. Look sideways on and raise the stage until the specimen lies within the position of focus for the objective. Lower the stage and bring the specimen into focus.

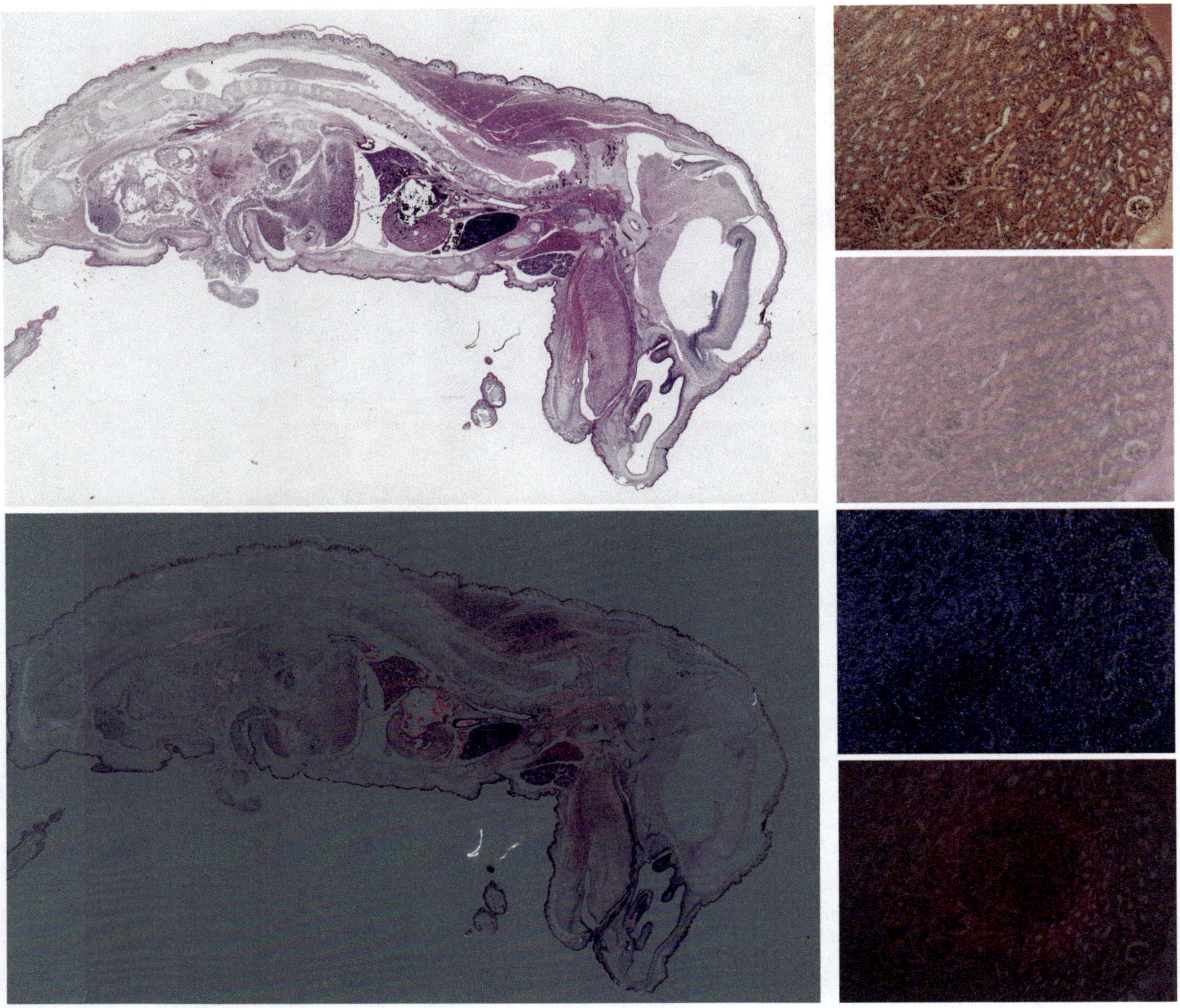

Figure 12.15 ***Stained thin sections viewed by reflected light*** *Stained histological sections cut thinly for examination by transmitted light do not give good images when illuminated by reflected light. Contrast in the image is low and a washed-out appearance results. Use each illumination method as appropriate to the sample type, its thickness, its transparency and surface relief. A mouse thin section is shown in transmitted-light bright-field (upper left) and reflected-light bright-field (lower left). A kidney section stained with Verhoeff's haematoxylin and van Gieson stain on the right; from top to bottom: transmitted-light bright-field; reflected-light bright-field; transmitted-light dark-field; reflected-light dark-field.*

In metallic specimens only surface detail that scatters light within the aperture of the objective lens is imaged, and Figure 12.17 shows the enhanced contrast that can be achieved as compared with that under bright-field illumination. Dark-field microscopy accentuates surface irregularities, such as grain boundaries, cracks and pits in the surface of the specimen. However, in semi-translucent ceramics, sub-surface detail – which would otherwise be swamped by specular reflection from the surface – can also be detected. Dark-field illumination is much more widely used in reflected-light examination of materials than in transmitted-light microscopy of biological specimens.

12.10 Phase Contrast with Reflected Light

Both phase contrast and its close relation, modulation contrast, can in principle be applied to reflected-light microscopy. Unlike the situation in transmitted light (Chapter 11, Section 11.12), the reflected-light phase-contrast microscope is a compromised design

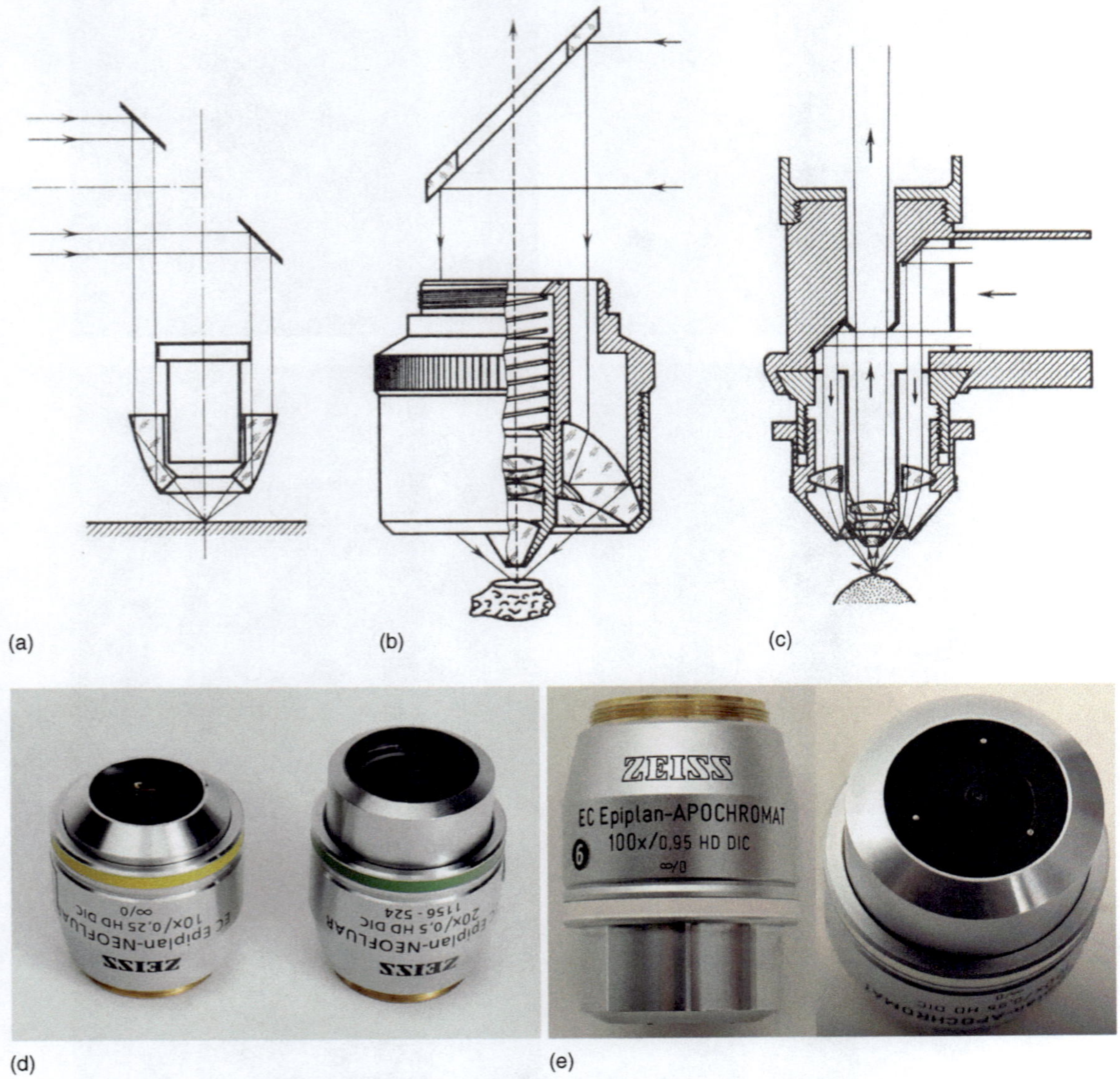

Figure 12.16 **Ray path and appearance of reflected-light dark-ground objectives** *(a) Catoptric condenser – can also be formed from a mirrored surface (b) catadioptric condenser (c) dioptric condenser employing a lens. In each case, the objective inside the reflector 'skirt' is a regular objective, which can also be used for bright-field imaging with transmitted light. (d) Examples of 10x and 20x bright-field / dark-field (HD) objectives showing the design of the outer reflector skirt (e) the 100x Epipolan HD objective, showing three lugs designed to protect the objective and prevent contact of the front lens element with the surface of the specimen. This is necessary because of the short working distance and very shallow depth of field of a high numerical aperture objective of this magnifying power. Source: (a–c) Pluta 1988. (d,e) Courtesy of Mr K. Glover.*

(as explained below). It is nowadays very little used, because reflected-light Nomarski differential interference contrast (DIC) has proved to be much more versatile.

In the same way that the illuminating aperture diaphragm of the condenser cannot be situated within the objective lens (where it would restrict the imaging rays), an image of the phase contrast illuminating annulus is projected into the back focal plane of the objective. Similarly, the phase plate is not built into the back focal plane of the objective because the annulus of incident light is partially absorbed and partially reflected at the phase ring (i.e. before it meets the specimen), giving rise to glare and lack of contrast. Also, any phase plate situated at the back focal plane of the objective would reduce the intensity of the illuminating rays.

A solution to the problem is to place the phase plate *above* the 45° illuminator such that only the light reflected from the specimen passes through it (Figure 12.18a). This means that one common phase plate can be used for all objectives, and one set of objectives can be used for both phase-contrast and bright-field microscopy. However, this position is well above the back focal plane of all except the lowest power objectives, and thus an adjustable 'Telan' negative lens system is required to project an image of the phase plate

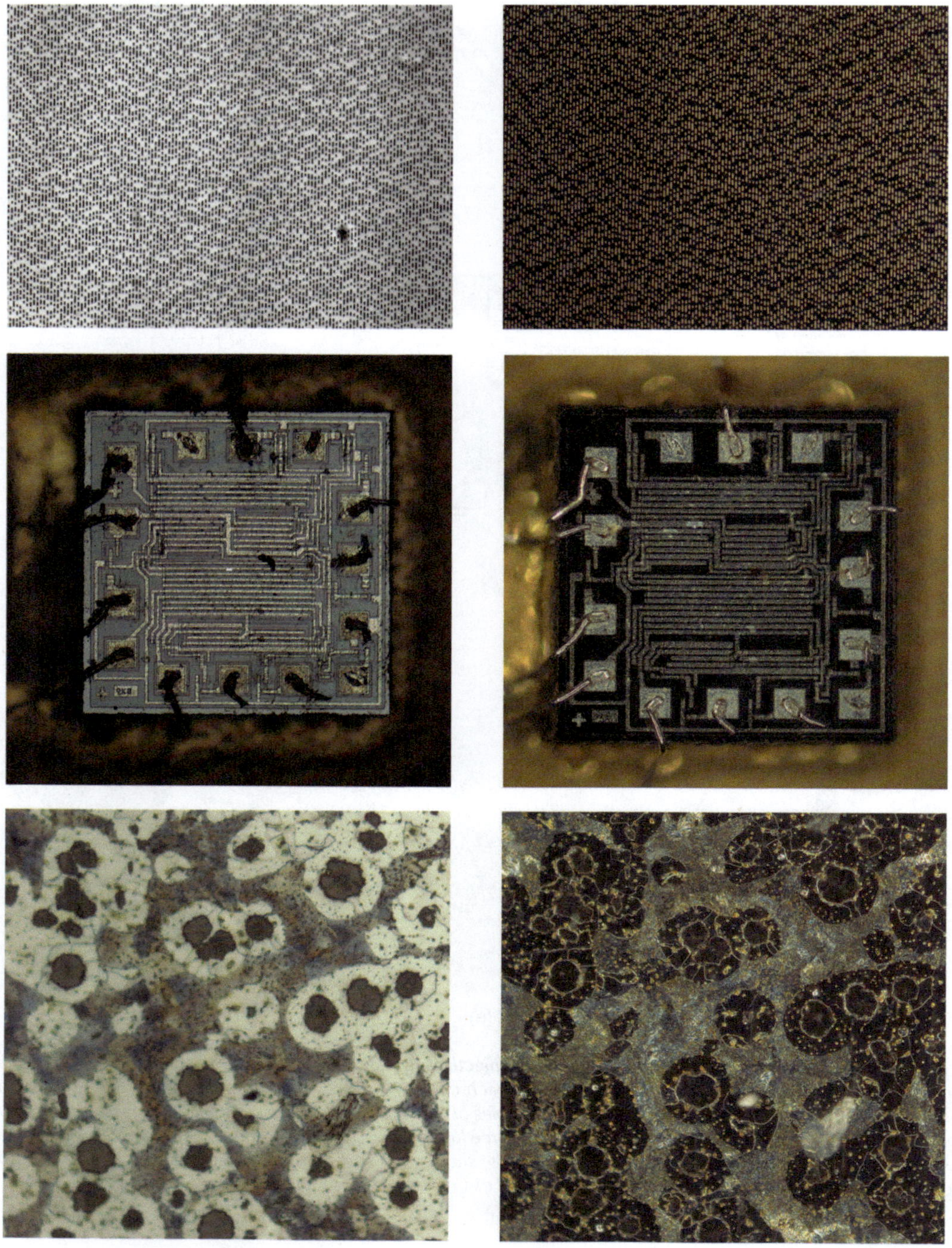

Figure 12.17 **Comparison of reflected-light bright-field compared to reflected-light dark-field images** *Three samples (the surface of a compact disc, top row; an integrated circuit, middle row and the polished surface of cast iron containing ferrite, bottom row) are shown with the bright-field image in the left-hand column and the corresponding dark-field image in the right-hand column. The nodules of ferrite are revealed as white in the bright-field image and dark in the dark-field image; graphite is seen as dark nodules in both images, whilst pearlite, also, is seen in both images as grey within the matrix of the iron.*

for a range of numerical apertures. This represents something of an optical compromise. Reflected-light phase-contrast microscopy equipment is no longer offered by any manufacturer.

The contrast, it should be noted, arises from variations in the optical path difference (OPD) of light reflected from the specimen surface in an analogous way to transmitted light, but rather than acquiring the phase shift from the edges of the specimen, as in transmitted light, the phase differences in reflected light are caused by the *relief* on the surface of the specimen. Features that are physically higher

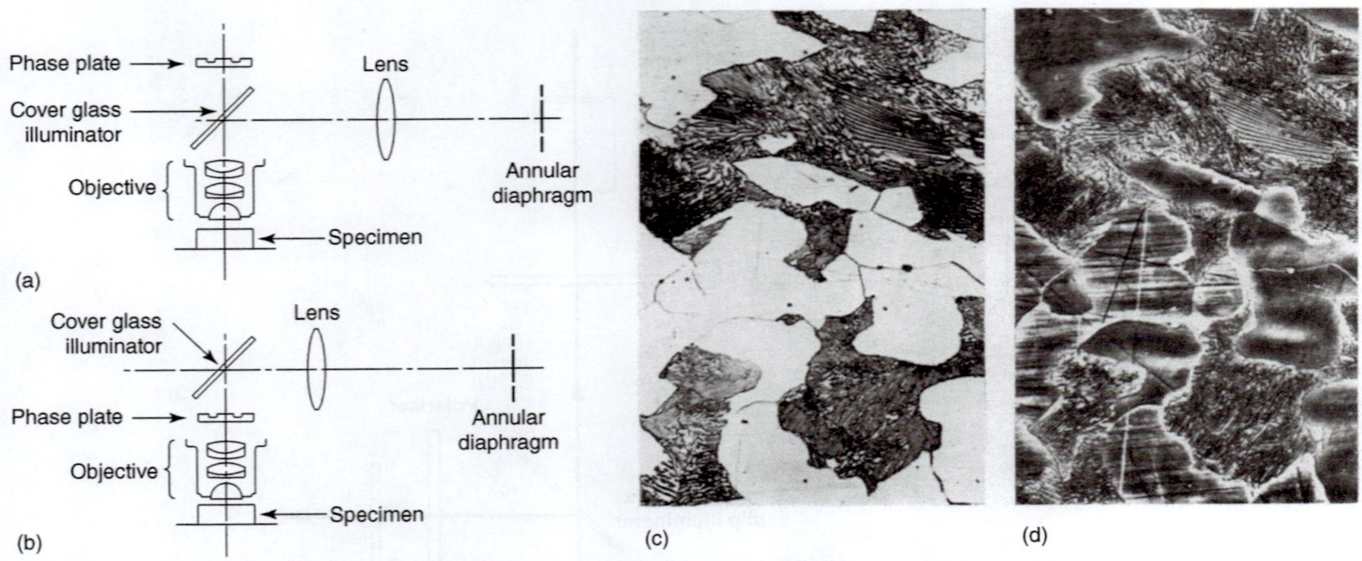

Figure 12.18 ***Reflected-light phase contrast*** *The phase plate may be situated either above or below the reflector unit (left). An image of polished perlite and ferrite is shown (centre: bright-field; right: phase contrast). Phase contrast does not add greatly to the metallographic information, save that artifacts arising from inadequate specimen preparation are revealed. Source: (a,b) Hammond 2007. Reproduced with permission from The Quekett Microscopical Club. (c,d) Haynes 1984. Reproduced with permission from Springer.*

than their surroundings appear darker in positive phase contrast, and vice versa. Figure 12.18d shows an example of a phase contrast image.

12.11 Polarisation Contrast in Reflected Light

Polarised light is useful for examining opaque specimens of ceramics, ores and minerals. By convention the polarising filter in the illumination train is orientated east-west (E-W) on the microscope, and the analyser above the objectives, orientated north-south (N-S). The reasons for this are explained in the next chapter (Chapter 14, Section 14.3). Some older microscopes had the lower polar set N-S and the analyser E-W. The correct orientation can usually be seen where the index mark on the polar is set at zero. If the field of view is dark but not entirely black (i.e. at extinction), check that the lower polar is not misadjusted, nor the analyser loose, broken or rotated in its mount.

Some metals, chiefly titanium, zirconium, zinc and tin, have optically anisotropic inclusions. These will rotate polarised light out of its original vibration direction. Through crossed polars, these inclusions are revealed in different colours due to different extinction conditions. It is necessary to look closely for these changes: they are generally only obvious if anisotropy is pronounced. Polarised light is also useful for studying ceramics. The light scattered off structures beneath the semi-transparent glazed surface can be minimised by careful orientation of the sample using a rotating stage.

12.12 Nomarski Differential Interference Contrast

As with transmitted-light microscopy, reflected-light DIC utilises crossed polars but with the addition of a single Wollaston prism in the optical path. The image of this prism is projected into the back focal plane of the objective. Because the objective acts as its own condenser in reflected-light microscopy, the pathway for operating DIC in reflected light is simpler, requiring fewer components than transmitted-light microscopy (Figure 12.19). Besides the simpler setting up and adjusting for Köhler illumination in reflected-light mode, the fact that the objective acts also as its own condenser is the reason why only a *single* Nomarski/Wollaston prism is required: the one prism performs the roles of both beam splitting and beam recombining of the image-forming rays. The prism is usually tilted slightly from the horizontal plane in order to (a) minimise unwanted reflections and (b) to make the plane of the cross-over of the split rays horizontal[4] in the objective back focal plane.

Some manufacturers incorporate a separate beam-splitting prism into each objective – the effective point of the beam divergence of the prism being matched to the back focal plane of objective. In this case, the Nomarski prism is strictly of the original Wollaston design. Other manufacturers use a single Nomarski prism for all objectives. In this case an additional (vertical) adjustment is needed

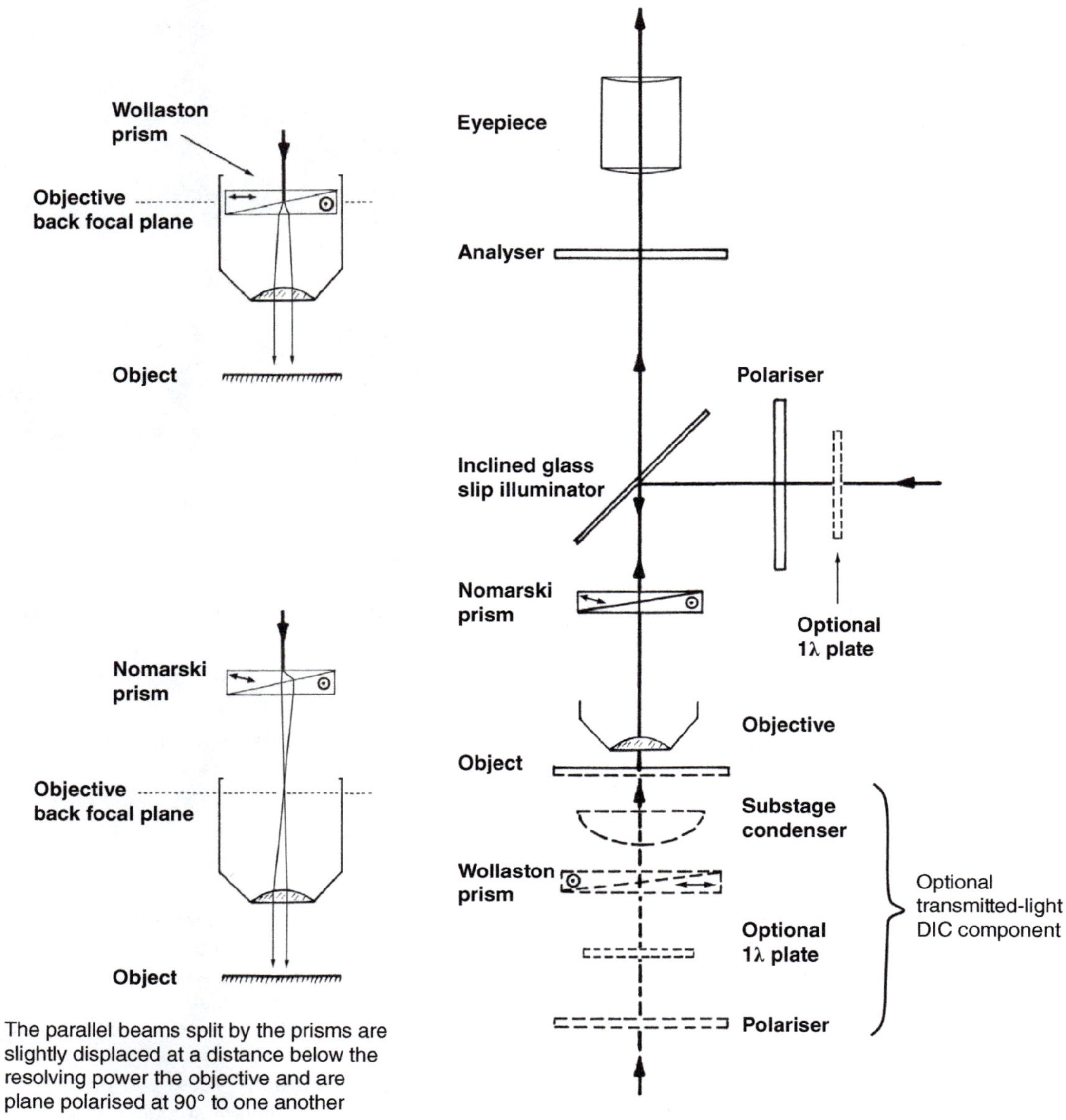

Figure 12.19 **Ray path for reflected-light DIC** *Recall from Figure 11.33, that the Nomarski prism is a modified Wollaston prism to allow the prism physically to be placed outside the objective, and not at the back focal plane which is (more often than not) buried within the objective body. Epi-DIC can either use individual Nomarski prisms above each objective or use a single slider for all objectives, with a vertical adjustment to focus the image of the prism into the back focal plane of the selected objective. This latter single-slider design is often employed for C-DIC, where circular polarisers are used to avoid the azimuth effect arising from linearly polarised-light illumination. Source: Adapted from Haynes 1984 with permission from Springer; Dr C. Hammond. Reproduced with permission from Dr Hammond.*

to match the effective point of beam divergence to the back focal plane of each objective because of the differing focal length of each objective. The correct setting is obtained when the brightness or colour is uniform over the whole field of view.

The contrast (Figure 12.20) in reflected-light DIC arises from the optical path differences (OPDs) between the split ray components reflected from points of different heights across the specimen surface (analogous, in transmitted light, to the split rays passing through regions of different refractive index – Chapter 11, Section 11.14). In general the height differences, and hence OPDs, are small, leading to low-order 'grey-white' interference colours. It is a great temptation to shift the Nomarski prism laterally such that the OPD between the split ray components is increased – resulting in a beautiful range of interference colours. However, it is the grey-white contrast that most clearly reveals surface detail. In general, DIC is not an appropriate technique for partially translucent materials since the varying optical path differences arising from reflections at different points below the specimen surface can give rise to a confused image. C-DIC, using circular polarisers, allows the DIC prism to be rotated without any loss of contrast. This section only deals with reflected-light DIC; for information on the less common interference contrast techniques, see Hammond (2001) and Piller (1977).

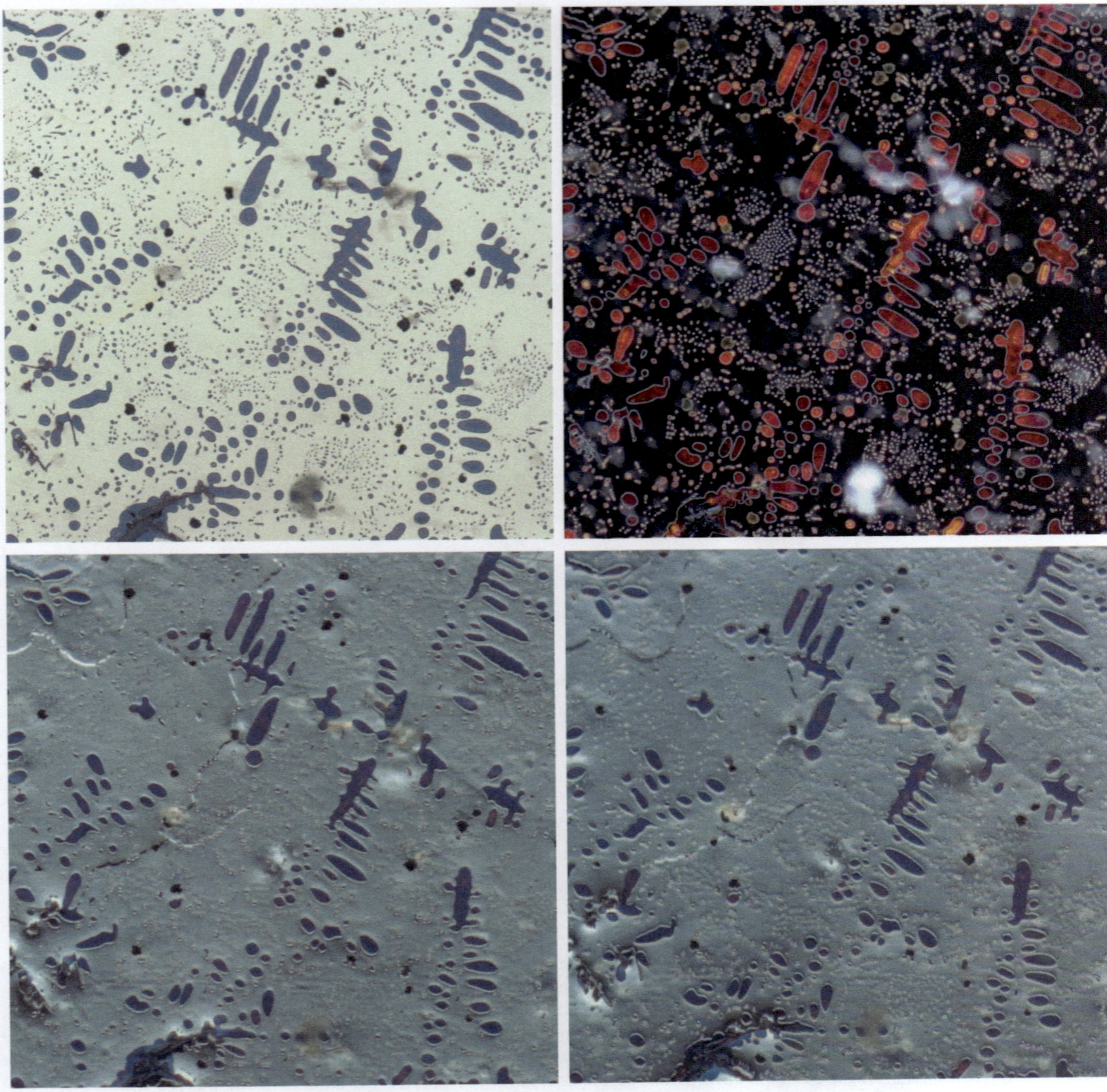

Figure 12.20 Image of opaque copper sample by reflected-light DIC *A specimen of copper with copper oxide inclusions is shown. The top-left panel shows the bright-field image; the top-right panel, the reflected-light dark-field image. The lower two panels shown two DIC images with different shear settings of the Nomarski prism. Dark-field illumination shows the true colour of the sample. The reflected-light DIC images (unlike transmitted-light DIC) show the three-dimensional structure of the surface of the specimen, provided the shear on the prism is set correctly.*

In the bright-field image, flecks of surface dirt or contamination, coloured grey/brown, can be seen. In the dark-field image these are seen as bright, highly reflective, out-of-focus areas in the image. They can be eliminated from the image to a certain extent by the optical sectioning properties of differential interference contrast.

Protocol 12.3 Reflected-Light Differential Interference Contrast (DIC)

Required

- Rotatable polariser in the illuminating path (should be set east-west)
- Strain-free objective(s)
- Nomarski prism with its axis set at 45° between the crossed polariser and analyser
- Analyser above the objectives (should be set north-south)
- One-lambda full-wave 'sensitive tint' plate (optional)

Procedure

1. With the polariser, analyser and Nomarski prism removed from the light path, set up the microscope for Köhler illumination in bright-field mode. (If the Nomarski prism is fitted as an integral part of the objective, and no objective without a prism is available, select the lowest magnification DIC objective possible).
2. Insert the polariser and analyser and ensure that they are crossed for maximum extinction so that the field of view is as dark as possible.
3. Insert the Nomarski prism.
4. If the Nomarski prism has an axial adjustment, use this to produce a field of view of a uniform shade of colour or greyness.
5. Adjust the sideways shear of the prism by the screw on the slider to optimise the contrast in the image (a uniform shade of grey is best). Where the prism is fixed, slightly rotate the polariser away from its crossed position to achieve the same effect.
6. Insert a full-wave compensating plate (optional), and repeat step 5.

12.13 Reflection-Contrast and 'Antiflex' Microscopy

An ingenious approach to reducing unwanted reflections in epi-illumination microscopes – particularly those employed in ore petrography where the inherent contrast of the sample is very low – is to use the 'Antiflex' method developed by Zeiss. This contrast-enhancing equipment is made for both stereo- and compound microscopes.

Linearly polarised light is directed onto the specimen from above and, as it passes through the objective (which acts as its own condenser), it travels through a rotatable quarter-wave plate specially mounted on the front of the objective (turned to achieve maximum brightness). An analyser crossed with respect to the polariser will block any unwanted reflections off the surface of the specimen (Figure 12.21). However, the image-forming rays properly reflected off the surface of the specimen have passed through the quarter-wave plate *twice*, and so their polarisation plane is rotated through 90° and so will pass unimpeded through the analyser. The image-forming light is circularly polarised (Chapter 13, Section 13.9) by the $1/4\lambda$ plate but is linearly polarised once again upon its return through the $1/4\lambda$ plate. This method is principally intended for use with low-NA oil-immersion objectives, which are generally of low magnifying power. The quarter-wave plate on the front of the objective is made to be rotatable to achieve optimum contrast and visibility. Antiflex microscopy is also useful for eliminating the 'hot-spot' that may occur – without Antiflex illumination – when using low-magnification reflected-light objectives. This so-called hot-spot shows up as a prominent central reflection and is usually pinkish in colour.

Reflection-contrast microscopy (RCM) was devised by J. S. 'Bas' Ploem in conjunction with Leitz. RCM is very closely related to the Antiflex method but was developed principally for high-NA immersion objectives for examining reflective tissue markers, such as immuno-silver, immuno-gold and 3,3′-diaminobenzidine (DAB) histochemical substrates in living cells. Both the Antiflex and RCM methods are suited to very weakly-reflecting samples of <10% reflectance. For untreated living cells to be seen, these must lie close to the surface of a reflecting substrate, such as a glass coverslip. For highest contrast in the image, both methods should use a green interference filter of either 550 or 546 nm, this being the wavelength for which the $1/4\lambda$ plate is constructed. In reflection contrast microscopy, a central stop is usually fitted in the illumination pathway to provide annular illumination. This further reduces stray light or glare arising from reflections off the glass-air interface of the bottom of the microscope slide or tissue-culture vessel, improving contrast in the image. Reflection contrast microscopy can be an alternative to, or used with, TIRF microscopy (see Chapter 23) for studying surface details of biological specimens, but Leica no longer support the RCM method in the infinity optics range. Further details of these methods are found in Piller (1977), Sanderson (1998) and Ploem et al. (1999).

12.14 Reflected-Light Fluorescence Microscopy

Epi-fluorescence microscopy is very widely used in biomedical science, and as such is the subject of Chapter 15 and onwards. Fluorescence microscopy is used in materials science studies to study wood specimens, as well as organic maceral inclusions within coals, which often autofluoresce red or orange. Fluorescence microscopy is very useful for revealing artifactual cracking and general porosity of the sample, as well as microstructures and natural grain boundaries (see Figure 29.3) within epoxy-embedded concrete and petrology specimens. The epoxy resin embedding medium is fluorescent, usually green/yellow. Another area in which fluorescence microscopy is used for opaque specimens is in the semi-conductor industry in the manufacture of printed circuit boards. Photo-resist is activated by UV light and must subsequently be removed. Since photo-resists are autofluorescent, reflected-light fluorescence microscopy can be used in quality control.

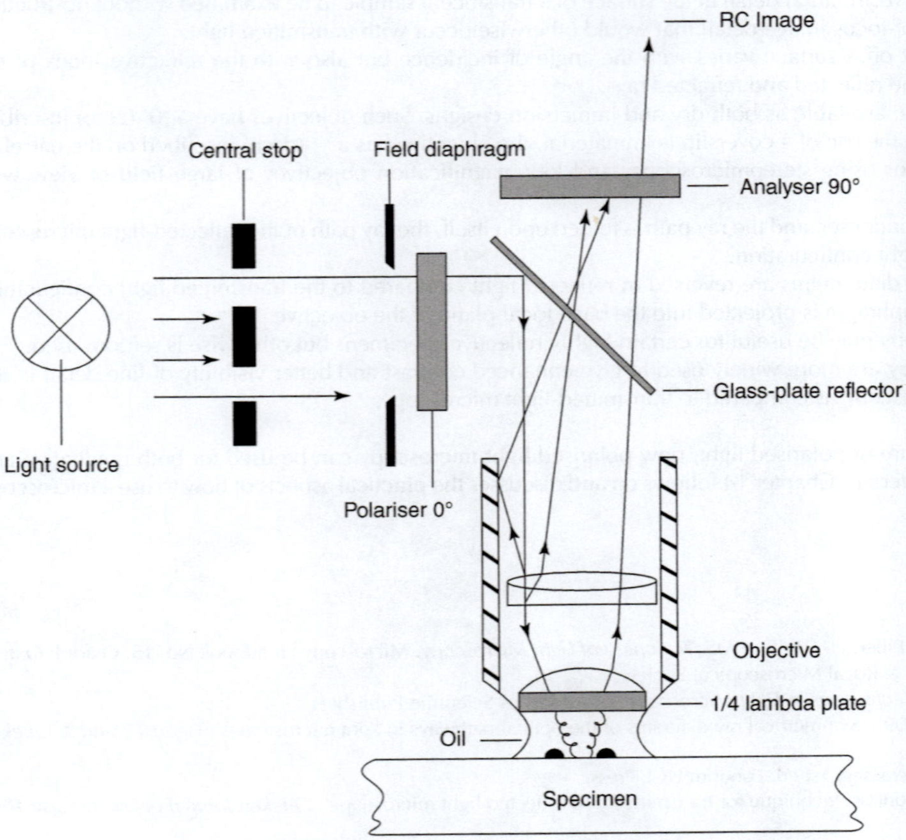

Figure 12.21 **Ray diagram of the 'Antiflex' and RCM method for low-contrast specimens** *Plane-polarised light is reflected down onto the specimen, and is circularly polarised by the 1/4 λ plate fixed to the front lens of the objective. The signal reflected off the specimen is further rotated by the 1/4 λ plate, becoming plane-polarised again and is thus is passed by the analyser. Stray light, from the objective lens surfaces, and within the microscope, is blocked by the analyser. Source: Sanderson in Cellis 1998. Reproduced with permission from Elsevier.*

The pair of images on the right are of spherulites in a polished iron sample, with (top panel) and without (bottom panel) Antiflex contrast enhancement applied. The rotatable and adjustable 1/4 λ Antiflex cap permits internal detail of the spherulties to be seen in colour and is useful for eliminating or reducing stray light that occurs particularly with low-magnification objectives. Source: Reproduced with permission from ZEISS Microscopy.

12.15 Samples for Reflected-Light Microscopy

Details of sample preparation are given in Chapter 29; however, check the following points before starting work. Polished samples should be level. If required, use a levelling device (Figure 29.13) to press the resin block holding the sample into a small ball of Plasticine on a microscope slide. Machined and polished blocks can usually be placed directly upon the stage.

A sample that is level should appear uniformly illuminated. To test this, focus on the sample, and close the aperture diaphragm. The illumination seen in the back focal plane of the objective should not wobble as the stage is rotated. Any preparation method carries the risk of introducing artifacts. Ideally, there should be no deformation (e.g. surface damage) or introduction of foreign particles from grinding, lapping and polishing. Nor should there be false relief: scratching or pull-outs where microstructural elements are lost from the polished specimen or thin section.

12.16 Chapter Summary

At the end of this chapter you should understand how the microscope can be used with reflected light to visualise the surfaces of opaque materials. You should be aware of how the illumination systems between transmitted light illumination and reflected-light illumination differ: principally that in the latter case the objective acts as its own condenser. This ensures a simpler illumination system to set up and use.

1. We view the world around us almost wholly by reflected light, the majority of bright-field light microscopy is carried out by transmitted light.

2. Reflected-light microscopy enables microstructural detail at the surface of a translucent sample to be examined without obstruction caused by the superimposition of out-of-focus image detail that would otherwise occur with transmitted light.
3. The reflectivity (or reflectance) of light off a surface varies with the angle of incidence but also with the refractive index of the material and the polarisation state of the reflected and refracted rays.
4. Objectives not requiring a coverslip are available as both dry and immersion designs. Such objectives have a '0' (zero) inscribed on the barrel rather than '0.17'. Where the use of a coverslip is immaterial, the objective has a '-' (dash) inscribed on the barrel.
5. Annular LED illuminators are useful for using stereomicroscopes and low-magnification objectives of large field of view with reflected-light illumination.
6. Because the objective acts as its own condenser, and the ray path is folded upon itself, the ray path of the reflected-light microscope is much simpler than the transmitted light configuration.
7. The positions of the field and aperture diaphragms are reversed in reflected light compared to the transmitted light configuration because an image of the condenser diaphragm is projected into the back focal plane of the objective.
8. Oil-immersion reflected-light microscopy may be useful for certain highly reflective specimens but otherwise is seldom used.
9. Dark-field and epi-polarised microscopy are more widely used to give enhanced contrast and better visibility of fine detail in the image when using reflected-light microscopy as compared to transmitted-light microscopy.

Chapter 13 now tells you about the nature of polarised light; how polarised light microscopy can be used for both qualitative and quantitative studies of doubly refracting materials. Chapter 14 follows on and discusses the practical aspects of how to use a microscope configured for work with polarised light.

References

Bradbury, S., Evennett, P.J., Haselmann, H., and Piller, H. (1989). *RMS Dictionary of Light Microscopy*. Microscopy Handbook No. 15. Oxford: Oxford Science Publications, Oxford University Press & Royal Microscopical Society.

Bradbury, S. and Evennett, P.J. (1996). *Contrast Techniques in Light Microscopy*. Oxford: Bios Scientific Publishers.

Evennett, P.J., Hammond, C., and Heath, J.P. (2009). 'Symmetrical ray diagrams of the optical pathways in light microscopes', Figures 2 and 3. *Quekett Journal of Microscopy 41*/2: 109–118.

Gribble, C.D. and Hall, A.J. (1992). *Optical Mineralogy*, 1st ed. London: UCL Press.

Hammond, C. (2007). 'A modified modulation contrast technique for transmitted and reflected light microscopy'. *Quekett Journal of Microscopy 40*/6: 495–505.

Hammond, C. and Heath, J. (2006). 'Symmetrical ray diagrams of the optical pathways in light microscopes', Figures 2, 3 and 4. *Microscopy & Analysis 115*, September 2006, pages 5–8. http://www.microscopy-analysis.com/sites/default/files/magazine_pdfs/mag%20708_2006_Sept_Hammond_1.pdf.

Haynes, R. (1984). *Optical Microscopy of Materials*. Blackie, Glasgow: International Textbook Company.

Nesse, W.D. (2009). *Introduction to Optical Mineralogy*. Oxford University Press.

Piller, H. (1977). *Microscope Photometry*. Berlin: Springer-Verlag.

Ploem, J.S., Prins, F.A., and Cornelese-ten Velde, I. (1999). Reflection-contrast microscopy. In: A.J. Lacey (ed.), *Light Microscopy in Biology: A Practical Approach*, 2nd ed. Oxford: OUP, pp. 275–310.

Pluta, M. (1988). *Advanced Light Microscopy* Vol. 2. Amsterdam/ Oxford: Elsevier.

Sanderson, J.B. (1998). Contrast enhancement techniques for light microscopy. In: J. Cellis (ed.), *Cell Biology: A Laboratory Handbook*, 2nd ed, vol 3. San Diego: Academic Press, pp. 15–33.

Sanderson, J.B. (2002). 'Practical control of contrast in the microscope'. *Quekett Journal of Microscopy 39*/3: 275–288.

Strange, A. (1994). 'Visibility and resolution of covered specimens with incident-light illumination'. *Journal of Biological Photography 62*/3: 91–102.

Further Reading

Edyvean, R.G.J. and Hammond, C. (1997). 'The metallurgical work of Henry Clifton Sorby and an annotated catalogue of his extant metallurgical specimens'. *Journal of the Historical Metallurgy Society 31*/2: 54–85.

Galopin, R.E. and Henry, N.F.M. (1972). *Microscopic Study of Opaque Minerals*. Cambridge: Heffer & Sons.

Hammond, C. (2001). Exploiting light microscopy contrast techniques to the full. In: A. Strang and J. Cawley (eds.), *Quantitative Microscopy of High Temperature Materials*. Leeds: Maney Publishing, pp. 69–78.

Roberts, E.W. and Robinson, P.C. (1985). 'Light microscopy of ceramics'. *Journal of Microscopy 140*/2: 137–158.

Notes

1. A polar is any device that produces plane-polarised light (or selects it from natural light) or which permits the passage and emergence of light restricted to vibrating in one plane only.
2. Reflectivity at an air-water interface is approximately 2%.
3. This type of objective, which can be used either for bright-field or alternatively dark-field imaging sometimes carries the notation 'HD', which refers to 'Hell, Dunkel' the German for 'Bright, Dark'.
4. In Chapter 11, refer to Figure 11.33 where it is shown that the wavefront of the recombined e- and o-rays is not parallel to the surface of the Nomarski prism.

13

Polarised-Light Microscopy: Part 1 – Theory

Outline

Understanding Light Microscopy, First Edition. Jeremy Sanderson.

13.1 Introduction

In order to understand how the polarised-light microscope functions and to use it to best advantage, it is necessary to understand something about the propagation of light rays as waves, to understand coherence, phase relationships and how light interacts with the specimen (which is often crystalline in nature). It is also helpful to appreciate how the interference of rays of light upon recombination at the primary image plane forms the image. In the same way that absorption stains exploit the chemical properties of the specimen, polarised light interacting with anisotropic specimens exploits the physical properties of the specimen to generate colour contrast enhancement. This is useful analytically in a non-destructive manner.

The wave nature of light has been explained in Chapter 2. Light rays propagate with electric and magnetic vector components orthogonal to one another (Figure 13.1). Many people visualise light as a sine wave, but it is also useful to consider the appearance of

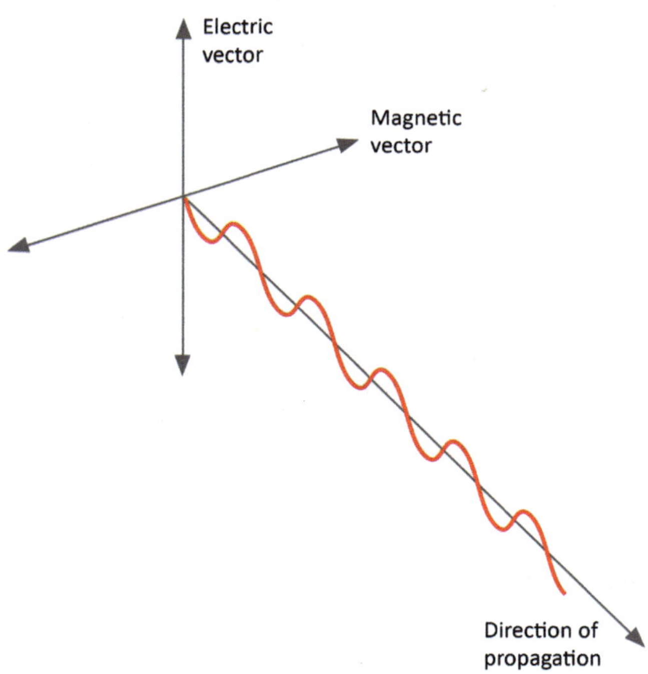

Figure 13.1 **Ray of light** *The direction of propagation of a ray of light is represented by a straight line. The magnetic field oscillates perpendicular to the electric field and at the same frequency. The electric and magnetic vectors, which express both the amplitude and vibration directions, are orthogonal to the direction of propagation and also to one another. See also Figure 2.4. Source: Courtesy of Mr S. Thomas.*

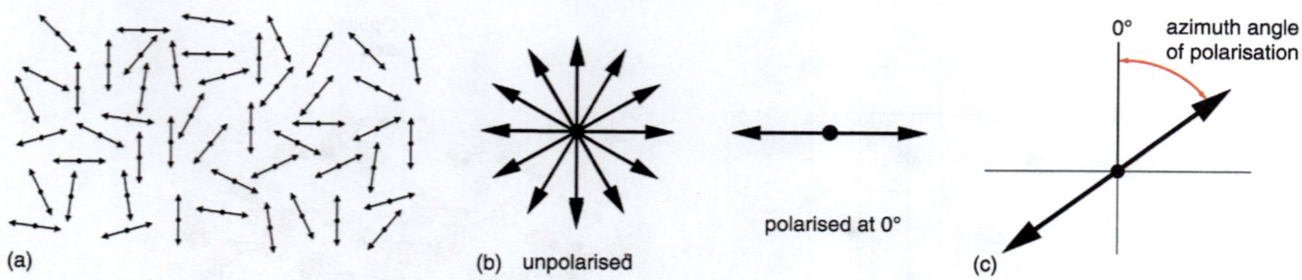

Figure 13.2 **Unpolarised and polarised light** *In Figure 13.2a (left) the vibration directions of several rays in a beam of light whilst looking head-on to the wavefront, are shown. The propagation direction is coming out of the plane of the paper; the vibration directions of each ray lie within the plane of the paper. With unpolarised light, the plane of vibration of the individual E-vectors of each ray in the wave train may be at any angle. In Figure 13.2b (middle) a single beam of unpolarised light is shown, vibrating randomly in all directions as it propagates. In Figure 13.2b (right) the electric vectors of the light rays are all restricted to vibrating in the same plane; the light is, therefore, polarised. In Figure 13.2c, the azimuth of the vibration direction of polarised light is described relative to a fixed reference plane at 0°.*

light rays from 'end-on', looking down the axis of propagation. When studying polarised light microscopy it is helpful to describe light rays as vectors in order to predict or understand their behaviour.

The phenomenon of polarisation is caused by a property of all electromagnetic waves: namely that they can oscillate with more than one orientation, and particularly that their electric and magnetic field directions are perpendicular to their direction of propagation. Rays of white light normally contain waves of different frequencies, of all possible phase relationships, vibrating in all possible planes (Figure 13.2a) at right angles to the direction of propagation. Several rays of light make up a beam of light (Figure 13.2b). In polarised light beams, the usual random vibration directions which occur in natural (i.e. non-polarised) light are now restricted to one vibration direction alone (Figure 13.2c). The phenomenon of polarisation is strong evidence that the oscillation of the wave is transverse to the direction in which the ray, or beam, of light travels.

As mentioned previously, the human eye generally[1] has no ability to distinguish polarised light from unpolarised light: we require a device – usually a filter, called a *polar, polarising filter* or *polariser* – to see evidence of polarisation. Some examples are shown in Figure 13.3. Polaroid® is the term reserved for the commercial product used as a polarising filter. When a polarising filter is used to determine the plane of vibration of the light, or interrogate the rays emanating from a sample on the microscope stage, the same filter is called an *analyser*. Linear polarisers have a unique transmission axis marked on them (usually as a double-head arrow) that indicates the plane of polarisation of the light rays. A method on how to do this is given in the first practical Exercise in Chapter 14, using a light source whose polarisation transmission axis is known. Further details of polarising filters, how to convert an ordinary microscope into a polarised-light microscope and descriptions of the components found on polarising microscopes will be found in the next chapter, from Section 14.2 onwards. The remainder of this chapter explains the essential theoretical principles of polarised light microscopy. For still further details, see the specialist introductory text written by Robinson & Bradbury (1992) or the references and online resources given at the end of this chapter.

Figure 13.3 **Examples of polarising filters** *Several different examples of polarising filters. The left-hand filter is a photographic polar that can simply be placed over the field diaphragm. The square is a sample of HN22 sheet available commercially, with the polarisation axis of transmission direction drawn on with a fine-tipped marker pen. The two right-hand filters are standard 32 mm diameter polars available from microscope manufacturers. Note that one of these circular Polaroid filters shows a central yellow mark. This is heat damage; the dichroic filter has absorbed electromagnetic radiation as heat, as well as light, and so has suffered damage.*

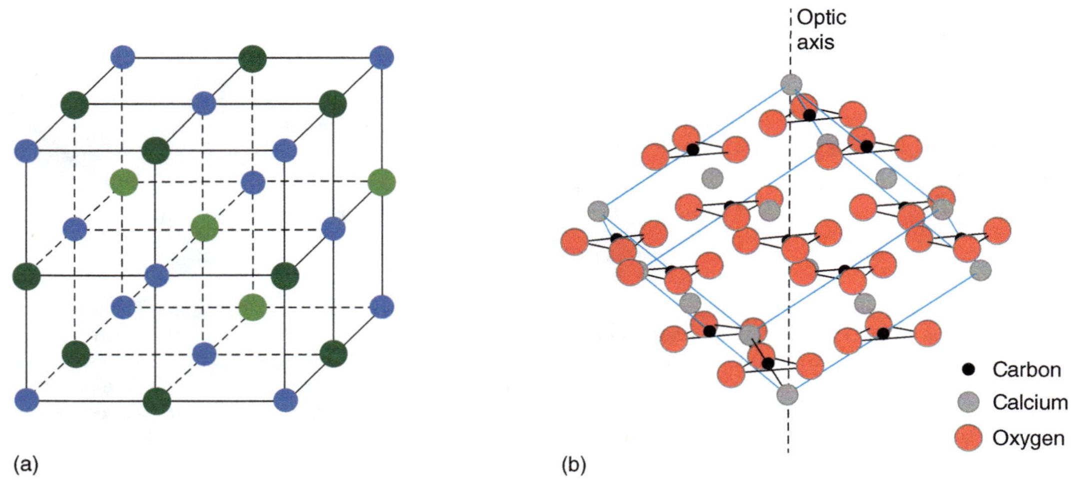

Figure 13.4 ***Isotropic crystal structure*** *Crystal lattice of an isotropic cubic crystal. Sodium chloride is shown here as an example. Since the molecular structure of an isotropic material is equal in all three dimensions, the refractive index (RI) of light passing through the material is the same, regardless of the direction of propagation of the light. Source: Gao 1999. Reproduced with permission from John Wiley & Sons.*

Anisotropic crystal structure *Crystal lattice of an anisotropic cubic crystal. Calcite is shown here as an example. The molecular or crystalline structure of an anisotropic material is more densely packed in one orientation than another. Hence it follows intuitively that the refractive index (RI) of plane-polarised light should be different for propagation through the material in different directions because the E-vector vibration encounters different densities of electrons in the material. Anisotropic crystals can possess two or three different refractive indexes. Source: Adapted from Hecht 2001 with permission from Pearson Education Inc.*

13.2 Isotropic and Anisotropic Materials

We saw in Chapter 2 how light can interact with matter in one of six ways: It may pass straight through matter and be partially absorbed. Light may also be reflected or refracted by matter. Light may excite matter to cause the emission of fluorescence. Light may be scattered, diffracted (i.e. scattered in a regular fashion) or undergo a phase change as a result of passing through matter. Finally, light may be polarised by reflecting off, or passing through, matter. Those materials that have the ability to polarise light are called anisotropic materials, and those whose structure has no effect are isotropic.

In an isotropic material:

- light velocity is uniform with respect to the vibration direction of light;
- Snell's law applies generally, since it has one refractive index;
- non-polarised light remains non-polarised when transmitted through it – it is not split into mutually orthogonal polarised rays;
- plane-polarised light remains polarised in its original vibration direction – isotropic substances do not change the vibration direction of polarised light passing through them; and
- only the refractive index and absorption colour, if any, can be used for identification since isotropic substances show *only* the effects of refraction and/or dispersion of light.

Isotropic materials are gases, liquids and non-crystalline solids. The chemical bonds in these materials are randomly oriented, so that light rays in all possible planes of polarisation are transmitted with equal velocity. If air was not isotropic, it would be impossible to carry out polarised light microscopy! Cubic crystals or minerals are also isotropic. In this case the atomic or molecular distribution within the crystal is equal in every azimuth, so light propagates through cubic crystals with equal velocity, irrespective of its plane of polarisation. The common example cited of an isotropic cubic crystal (Figure 13.4a) is sodium chloride, NaCl.

In an anisotropic material:

- light is transmitted at differing velocities according to the vibration direction of the beam;
- there is a range of refractive indexes according to the direction that the light travels through the material and the polarisation direction. For each wavelength, this range of RI values lies between two extreme values of refractive index;
- there are two or three refractive indexes, depending on its structure – uniaxial materials have two refractive indexes, corresponding to these extreme values, and biaxial materials also have a third refractive index, intermediate between the other two;
- non-polarised light is polarised when transmitted through an anisotropic material, split into mutually orthogonally-polarised rays;
- plane-polarised light may also split into mutually orthogonal polarised rays; and
- light remains polarised when exiting the material.

Transparent anisotropic materials (Figure 13.4b) decompose non-polarised light into two rays polarised orthogonally to one another. One of these rays is termed the *ordinary ray* (or *o-ray*; since it obeys Snell's law of refraction) and the other the *extraordinary ray* (or *e-ray*; since it does not obey Snell's law). For example, when light is incident at 0° to the surface of the transparent anisotropic material, the o-ray passes through undeviated, but the e-ray is refracted slightly[2], but this angle of refraction changes according to the angle of the incident beam of light. The velocity of the o-ray is uniform: the RI value (n_o, sometimes called n_{omega}; or n_ω) does not change. The velocity of the other ray, called the extraordinary ray or *e-ray* (RI = n_e; $n_{epsilon}$ or n_ε) has a velocity – and hence refractive index – that varies between a maximum value (n_e) and n_o depending upon the direction of travel through the anisotropic material. For example, the o-ray in calcite has a uniform value (n_o) of 1.6584; the refractive index of the e-ray (n_e') varies between 1.6584 and 1.4864. This is why the e-ray does not obey Snell's law. The propagation directions of the o-ray and e-ray are called privileged directions.

Uniaxial anisotropic materials exhibit isotropy *only* when light propagates along the optic[3] axis. Biaxial materials, as the name suggests, have two optic axes. They behave as isotropic substances when light travels along either of these two axes. Crystals of calcite (Figure 13.4b), a uniaxial material, possesses a high birefringence value. Calcite is therefore ideal for explaining and demonstrating double refraction. Optical anisotropy and double refraction (colloquially known as 'birefringence') are discussed further in Sections 13.4–13.6.

Interference of polarised light beams only occurs if these polarised beams are coherent *and* if they have components vibrating in the same plane[4], namely that their electric vectors are parallel to one another. For further explanation of coherence and interference of light, see Chapter 10, Section 10.3. The use of polarised light microscopy for examining different classes of crystals is not covered here. The interested reader is referred to Wood (1977) and Hammond (1992) for introductory texts on crystallography. For an advanced textbook on the subject, see Hammond (2009). Further information on the use of polarised light microscopy in the analysis, identification and characterisation of minerals can be found in Nesse (2009). There are also various online resources, which are given under Internet Resources at the end of the chapter.

13.3 Formation of Polarised Light

When you drop a stone into a pond, the waves 'diffract' radially away from the point of origin. The actual water droplets travel up in crests and down in troughs (as a wavefront transverse to the direction of propagation), while the wave propagates along the surface of the water. For example, a cork floating on a water wave moves up and down axially while the wave propagates laterally. Because they are *restricted* in one plane, water waves are polarised 'up and down' axially. Sound waves, which (in air[5]) are longitudinal waves, normally cannot be polarised, whereas light, which is a transverse wave, can be polarised.

When light travels along an axis, it possesses an electric field vector[6], E, orthogonal, or perpendicular, to the optical axis, which is the direction of propagation, or travel (Figure 13.1 and see Chapter 2, Section 2.3). For light from an incandescent source, the plane of vibration of the electric field vectors (also called E-vectors) may be at any angle (Figure 13.2a). The photons emanate as rays in *all* directions perpendicular to the optical axis: the light is said to be *unpolarised*. However, if the electric field vectors of all the waves vibrate in the same plane, the light is restricted to vibrating in *one* plane perpendicular to the optical axis; it is then *linearly polarised* (Figure 13.2b). Polarisation is essentially a filtering process. Box 13.1 explains the irradiance (i.e. intensity, see endnote 1 to Chapter 1) of light passed by both the polariser and analyser. The phase of the light waves is unimportant. The light output of a laser is typically highly polarised, whereas that of an ordinary incandescent light bulb, an LED or the sun is not. If the rays of polarised light cover an extended area, they are *plane parallel polarised*. Since the plane of vibration can occur at any angle, θ, the orientation of the plane of vibration – its azimuth – is described relative to a fixed reference plane at 0° (Figure 13.2c).

Light can be polarised by four means:

- by selective absorption (dichroism);
- by double refraction through transparent anisotropic materials;
- by reflection off a smooth surface; and
- by scattering.

We are primarily concerned with the first two methods: absorption and refraction. The most common method of producing polarised light is to use a dichroic absorption filter, such as Polaroid. These filters polarise light by selective absorption of all vibration directions except one. Alternatively, transparent anisotropic media (Section 13.6) plane-polarise rays of light by double refraction into two mutually perpendicular waves (Figure 13.5). Prior to the introduction of Polaroid sheet in the 1930s, specially-cut Nicol prisms, made of two pieces of calcite, were used to polarise light[7]. Polaroid sheet is mostly used nowadays because it can be produced in large sizes, without restriction on overall numerical aperture.

Besides dichroic absorption and double refraction, there are two further ways of producing polarised light: by reflection and by scattering. Reflected light is partly – or fully – polarised by certain materials except when the incident light is normal (i.e. perpendicular) to the surface. The reflected light is not completely plane-polarised unless the incident angle is such that the angle between the reflected and refracted rays is 90° (see Figure 13.7). Light may also be polarised by molecules that scatter light. These molecules act as dipole radiators, polarising light at 90° to the dipole axis. The scattering of sunlight off gas molecules and very tiny particles ($<1/10\ \lambda$) in the atmosphere is called Rayleigh[8] scattering.

Box 13.1 Malus' Law and Polarisers

A piece of Polaroid sheet is placed on a light box with its permitted vibration direction horizontal. A second sheet is arranged on top so that its vibration direction is parallel with the first sheet. An initial reading, I_0, of the intensity of light passed is taken.

When set at an angle to the first, the vibration direction of this second polar may be described by a vector. Plane-polarised light passing through this second polar has both a vertical and a horizontal vector component, and any vector can be described by its two coordinate axes. Because the light is already polarised by the first polar set horizontally E-W, the vertical coordinate component is zero. The horizontal coordinate component can be calculated by dropping a vertical dashed line onto the horizontal line P-P, representing the vibration direction of the E-W polar. Since the intensity of any wave depends upon the square of its amplitude, the transmitted intensity varies as $\cos^2\theta$. Therefore, the intensity of the light, I, transmitted by this second polar can be calculated from Malus' law, $I = I_0 \cos^2\theta$. Light of variable intensity and polarisation can be produced by rotating and changing the orientation of this second polar.

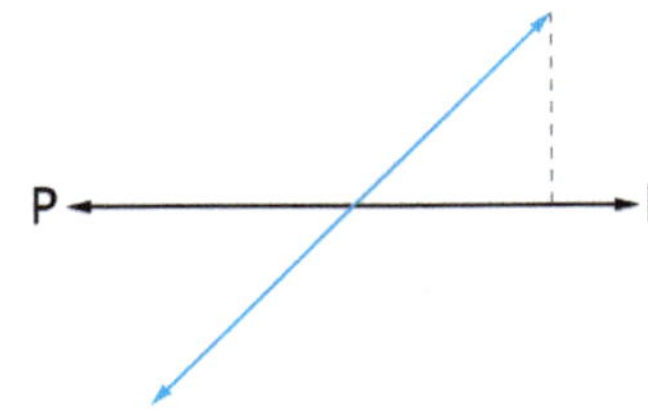

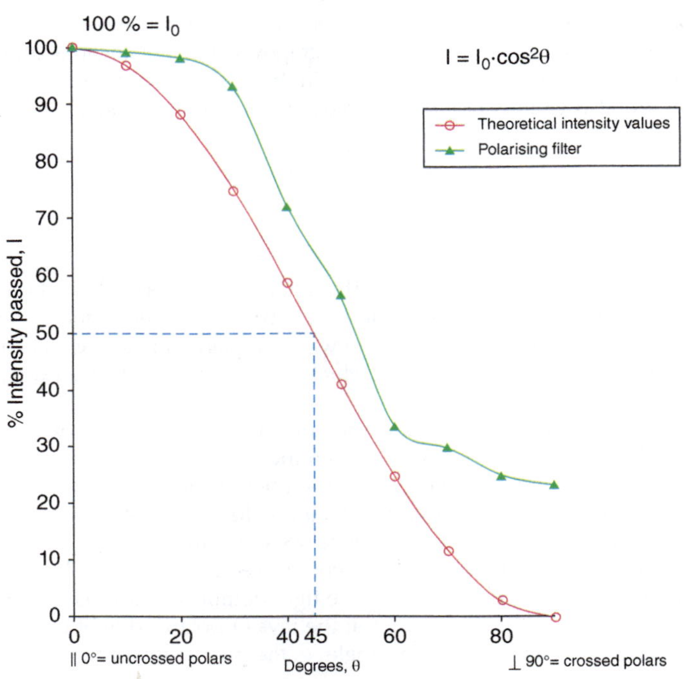

Suppose that this second polar is now gradually turned such that its vibration direction ultimately finishes in the vertical orientation, and light intensity readings are taken at, say, every 10°. The graph shows a plot of light intensity as the second, upper, polar is successively rotated until it is crossed with respect to the first polar. The intensity of light passing through a polariser that behaved perfectly would correspond to the red line, which is a plot of the function $I = I_0 \cos^2\theta$. If the starting intensity of light *before* passing through the first polar is unity, so $I_0 = 1$ (or 100% on the graph), then at 45° $I = 0.5$, since cos 45° = 0.707 and the square of 0.707 is 0.5. Two crossed polars should, in theory, transmit no light. The green line shows a polaroid gel, which is not of the highest optical quality. A high-quality polar made of Polaroid should have a maximum extinction coefficient of about 1:500, or 2×10^{-3}, whilst a high-quality calcite Nicol prism is in the order of 1×10^{-6}.

As a theoretical exercise, insert a third polar at an angle between two crossed polars over a light box. This polar will now transmit some light, seen when looking through all three polars together. This is because the permitted vibration direction of this third polar has both horizontal and vertical vector components. The vertical component is allowed through the uppermost (second) polar, which acts as an analyser. From Malus' law the maximum light intensity is passed when the intermediate, third, polar is set at 45°; $I = 0.5^3$, or 1/8th I_0. When the intermediate polar aligns with either the first or second (crossed) polars, light passed will be blocked and the intensity will fall to zero in each case.

Three states of polarisation exist: linear, circular and elliptical. Linear polarisation, where the vibration direction is restricted to a single plane, is the simplest concept to understand. Circular polarised light (see Section 13.8) is produced by two waves of the same amplitude, out of phase by $^1/_4\,\lambda$, whose electric field vector vibration directions lie at right angles to each other. If the optical path difference (OPD) of the two waves is anything other than $^1/_4\,\lambda$, the result is elliptically polarised light. In this chapter we will restrict ourselves mainly to discussing linear polarised light.

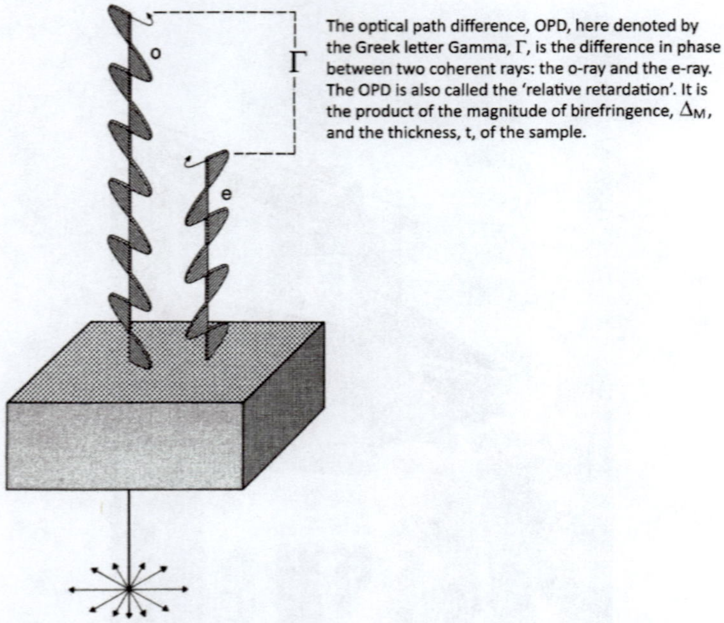

Figure 13.5 **Polarisation of light by double refraction** *The formation of mutually perpendicular waves by anisotropic media. Unpolarised light is incident upon the specimen from below. It splits into two polarised rays vibrating at right angles to one another. NB here these rays are shown separated from one another to show clearly their orthogonal relationship and the phase shift between them due to birefringence and retardation.*

***NB** Light propagating through anisotropic materials does not separate into independent plane-polarised rays that later recombine. This description, as with much in science, is simply a convenient analytical model that explains the observed phenomena best and most closely. Source: Patzelt 1974. Reproduced with permission from Leitz, Leica Microsystems.*

13.4 Formation of Polarised Light by Reflection and by Scattering

When light moves between two media of differing refractive indexes, some of it is refracted according to Snell's law, and some of it is reflected at the boundary. Light reflected by shiny transparent dielectric materials is partly – or fully – polarised, except when the incident light is normal to the surface. A dielectric material is an electric insulator, but one that can be polarised when an electric field is applied. For our purposes it is sufficient to appreciate that highly reflective metal surfaces, being conductors, cannot polarise light in this way. This polarisation by reflection will also happen at the highly curved surfaces of some lens doublets in objectives and can affect and reduce the extinction achieved under polarised light microscopy.

Étienne-Louis Malus first noticed the polarisation of light by reflection (Figure 13.6). He was also the first to use the term 'polarisation'. Malus observed, most likely through a calcite crystal, the reflections of the setting sun off the surface of a window of the Luxembourg Palace across the street. He *expected* to see two images of the sun's orb of *equal* brightness. However, Malus noticed how the intensity of the double images of the sun's orb varied when he rotated the crystal: that at two positions, one image then the other reduced so much in intensity that they appeared to vanish. However, the double image of the window (which also rotated) did not similarly vanish. The sun, or another light source, reflected off glass or a dish of water only produced one image when viewed directly without the anisotropic calcite crystal. The light must therefore be polarised. Malus calculated the relationship between the polarising angle of reflection and the refractive index of water, which led to the formulation of his 'cosine-squared' law (see Box 13.1 for the mathematical explanation). He was unable to extend this to glass, since early glass formulations were not of sufficient quality and contained components of varying refractive indexes. Later, Sir David Brewster used better quality glass to calculate this relationship.

The plane of incidence is perpendicular to the plane of a reflecting surface and contains the direction of propagation of the light ray. Light with its electric field vector (E-vector) parallel to this plane is *p*-polarised. Conversely, light whose electric field vector is perpendicular to the plane of incidence is termed *s*-polarised light (from the German *senkrecht* meaning perpendicular).

Brewster showed (Figure 13.7) that at a particular angle of incidence (the Brewster angle; the 'polarisation' angle) such that when the *reflected* ray is at 90° to the light ray *refracted* through the material, the reflected ray is completely plane-polarised parallel to the reflecting surface[9]. At Brewster's angle, no *p*-polarised light is reflected from the surface; thus all reflected light must be *s*-polarised, with an electric field perpendicular to the plane of incidence and parallel to the plane of the reflecting surface (i.e. interface between the two media). At this angle, the reflected beam is polarised to its maximum extent. The refracted light is polarised with its vibration direction in a plane set at right angles to that of the plane of vibration of the reflected light. Brewster's law states that $n = \tan i$, where n is the refractive index of the material forming the reflective surface. This (see Box 13.2) is simply a special case of Snell's law where $n = \sin i / \sin r$ and where $i + r = 90°$.

Figure 13.6 **Polarisation of light by reflection** *The experiment of Malus. The acute angle, I_B, shown is Brewster's angle. In this case for an air-glass interface. Depending upon the refractive index of the glass, I_B will have a value between 56.08° for crown glass (RI = 1.487) and 61.61° for a dense flint glass (RI = 1.850). Although the pinhole (in the black card) gives a clearer demonstration of the effects of the polarised light, it is not crucial to the performance of the experiment.*

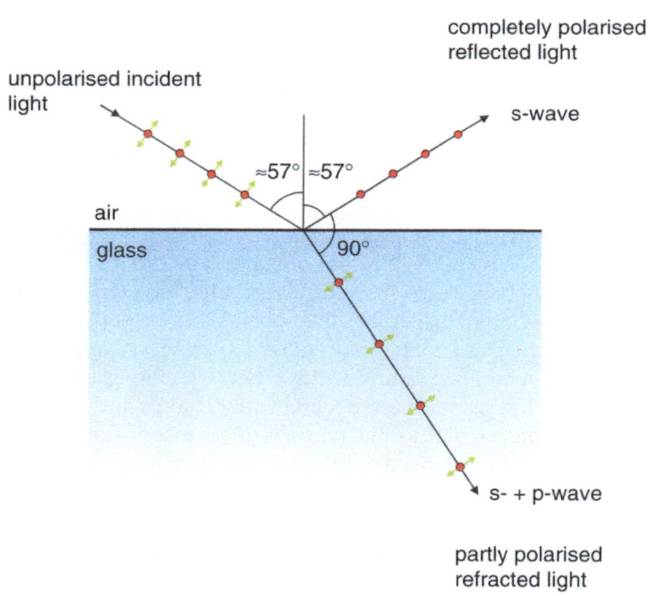

Figure 13.7 **Reflected light and Brewster's angle** *When unpolarised light strikes a transparent isotropic surface at Brewster's angle, the reflected light is always s-polarised. Remember that 's' polarised light will 'skip' off a Brewster boundary, and 'p' polarised light (German:* **p**arallele*) will 'plunge' through. The s-polarised light ray lies perpendicular (German:* **s**enkrecht *– perpendicular) to the plane of incidence and parallel to the reflecting surface. Unpolarised light incident upon a single sheet of high-quality optical glass at the Brewster angle will reflect 100% s-wave polarised light and transmits 42% s-wave and 58% p-wave polarised light. In this case, for glass, the Brewster angle is ≈57° (see Box 13.2 for an explanation of the mathematical reasoning). Source: Tilley 2011. Reproduced with permission from John Wiley & Sons.*

The relationships between the intensities of the plane-polarised reflected and refracted rays (as a proportion of the incident intensity) were first worked out by Augustin Jean Fresnel, and were described by him in a set of equations that bear his name – the Fresnel equations (see Box 12.1). We need note only two important predictions of these equations. First, the reflected light is plane-polarised in a direction parallel to the surface (perpendicular to the plane containing the incident, reflected and refracted rays) and second, the refracted light is *partially* plane-polarised, but orthogonal to the reflected light. The intensity of the reflected beam is weaker than that of the transmitted, or refracted, beam (see Figure 2.12).

Box 13.2 The Mathematics of Polarisation by Reflection

The critical angle of incidence for polarisation is given by: $\tan_c = n_1/n_2$, where n_1 and n_2 are the refractive indexes of the media above (e.g. n_1 = air) and below the reflecting surface (i.e. n_2 = solid or liquid material), respectively. It is simply a special case of Snell's law (**n_1.sin *i* = n_2.sin *r***) when going from a medium of refractive index n_1 into one of refractive index n_2:

$$n_2/n_1 = \sin i/\sin r, \text{ and so } n_2 = n_1.\sin i/\sin r \text{ and since } n_1 = 1 \text{ in air} \Rightarrow n_2 = \sin i/\sin r, \text{ in which } i + r = 90°.$$

Brewster's angle (θ_B, the particular value for *i* at which the reflected and refracted rays are completely plane-polarised) depends upon the refractive index of the glass, according to the formula: $\tan i = n_2/n_1$, where *i* is the angle of incidence at which the refracted ray makes an angle of 90° with the reflected ray. Thus

$$\theta_B = \arctan\left(\frac{n_2}{n_1}\right).$$

The proof is as follows:

$$\text{Since, from above: } i + r = 90°, \text{ then } r = 90° - i \text{ and } \sin r = \sin(90° - i); \text{ therefore, } \sin r = \cos i.$$

From Snell's law, $n_1.\sin i = n_2.\sin r$, and the angle (θ_B) at which all the *p*-polarised light is refracted is given by $n_1.\sin i = n_2.\sin r$, or $n_2.\sin(90° - i)$, and since $n_1 = 1$, then: $\sin i = n_2.\sin(90° - i) = n_2 \cos i$ and $\sin r = \sin i/n_2$, which is equal to $\cos i$; thus, $n_2 = \sin i/\cos i = \tan i$, so $i = \theta_B = \arctan(n_2/n_1)$.

For crown glass of RI = 1.5172, the angle *i*, or Brewster's angle, $= \theta_B$, is $\approx 56.7°$.

When unpolarised light strikes a surface at Brewster's angle, the reflected light is always *s*-polarised. Remember that 's' polarised light will "skip" off a Brewster boundary and 'p' polarised light will "plunge" through. The *s*-polarised light ray lies perpendicular (German: *senkrecht* – perpendicular) to the plane of incidence, and parallel to the reflecting surface.

As an extension of the above argument, when light passes from a medium of higher refractive index (e.g. glass, n_2) to one of lower refractive index (e.g. water or air, n_1), so that $n_1/n_2 < 1$, the emergent ray is bent by refraction towards the interface, and away from the optical axis normal to the interface. As the angle of incidence (*i*) increases, so does the angle of refraction (*r*) away from the normal, such that the refracted ray approaches the interface.

At a certain angle of incidence, the critical angle θ_c, the refracted ray (here *i*) is such that $i = 90°$. The beam of light is now refracted parallel to the surface, does not emerge at the interface, and therefore no light propagates through to the other side. When this occurs, $\sin i = 1 \Rightarrow \theta_c = \arcsin(1/n_2)$. When the angle of incidence, *i*, is equal to or exceeds the critical angle, the incident light is totally internally reflected (see also https://preview.tinyurl.com/zvcjtwo).

Light from a blue sky is polarised because it is light scattered from molecular fluctuations in the upper atmosphere, and the polarisation is strongest at 90° from the sun (Figure 13.8). Light from clouds, or from a hazy sky, is not polarised. Thus if the sun is low in the sky, reflected light is mostly *s*-polarised. Sunglasses use the principle of Brewster's angle to reduce glare from the sun reflecting off horizontal surfaces such as water or a road. This is why the Polaroid in sunglasses is orientated to transmit vertically-polarised light, to cut out the glare of light reflected off these horizontal surfaces.

13.5 Optical Anisotropy and Polarisation of Light by Double Refraction

An optically *isotropic* material has only one refractive index for all directions of the electric field vector: light will propagate through the material at the same velocity irrespective of its direction. However, most crystalline materials are optically *anisotropic* – the structure, or the pattern of atoms or molecules, is different in different directions (Figure 13.4b). Therefore the velocity, and hence refractive index, is dependent upon the polarisation direction of the light.

An anisotropic material, or crystal, is said to be birefringent, or doubly refracting. A very simple description of double refraction would be: one ray goes in, two come out. The phenomenon is properly called *double refraction*. The numerical difference between the refractive indexes is the *birefringence*. However, the term birefringence is commonly used to mean double refraction, as in 'a birefringent material'.

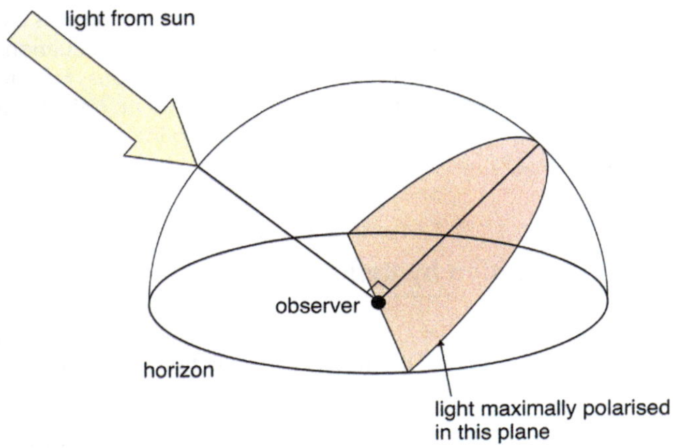

Figure 13.8 **Polarisation of light from the sun by scattering** *Light scattered by small molecules in the air is maximally polarised at 90° to the direction of the incident light, as discovered by Dominique Arago.*

On a clear day, if you look at the blue sky through a polar, you can turn the filter so that the sky goes darker and white clouds stand out clearly. Photographers use polarising filters for this purpose to reduce glare. Light is polarised due to Rayleigh scattering; it is almost unpolarised in the direction of the sun, but 75–85% polarised in a plane set at 90° to the altitude angle of the incident sunlight, with the observer at the centre of the base of the plane.

Scattering is also responsible for the blue colour of clear sky, since the scattering is wavelength dependent: $1/(\lambda^4)$. Blue light is, therefore, scattered better than red and is directed towards earth. Source: Tilley 2011. Reproduced with permission from John Wiley & Sons.

The basis of double refraction is due to the regular atomic or molecular arrangement of materials being different in different directions. For example, in nylon fibres the long molecules are not random but run largely parallel along the fibre: the refractive index for incident light plane-polarised along the length of the molecules is higher than that which is plane-polarised perpendicular to the molecules (the refractive indexes for nylon are approximately 1.575 and 1.526, respectively). Using the polarising microscope to determine these unique refractive indexes in anisotropic materials allows petrologists, geologists and forensic scientists to positively identify unknown rocks and minerals, particulates and fibres[10].

While it is easy to produce polarised light by reflection, the output is weak, unless a stack of plate reflectors is used (Figure 13.9). A better means is to use a doubly refracting material and eliminate one of the rays, either by absorption or total internal reflection. This is the basis of the Nicol prism, used extensively to polarise light before the introduction of Polaroid. A rhombohedron of calcite naturally forms a rhombus parallelogram with an acute angle of 70.9° and an obtuse angle of 109.1° (see Section 13.15). William Nicol reground the end face of the rhombohedron to an angle of 68°. The crystal is then bisected, one half inverted and glued back onto the first half (Figure 13.10), such that the glued interface makes an angle of 90° to the end face of the crystal. Canada Balsam ($n_D = 1.552$) was used originally to glue the two halves together.

When non-polarised light impinges on the crystal, the resulting slow o-ray is incident upon the Canada balsam layer at an angle of about 78° that results in total reflection at the interface between the two halves. The totally internally reflected ray is lost by absorption

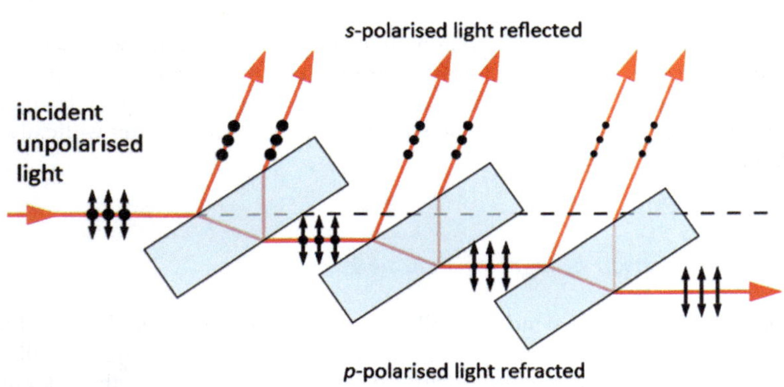

Figure 13.9 **Polarisation by reflection from a stack of glass plates** *A simple linear polariser can be made by tilting a stack of glass plates at Brewster's angle to a beam of light. A stack of half a dozen microscope slides will demonstrate this. At each reflective surface, 16% of the incident beam is reflected as s-polarised light, leaving a greater fraction of p-polarised light to be transmitted. Gas lasers typically use a window tilted at Brewster's angle to allow the beam to leave the laser tube. Source: Adapted from DrBob under the terms of the Creative Commons Attribution Share-Alike licence CC-BY-SA-3.0 (http://creativecommons.org/licenses/by-sa/3.0/), via Wikimedia Commons.*

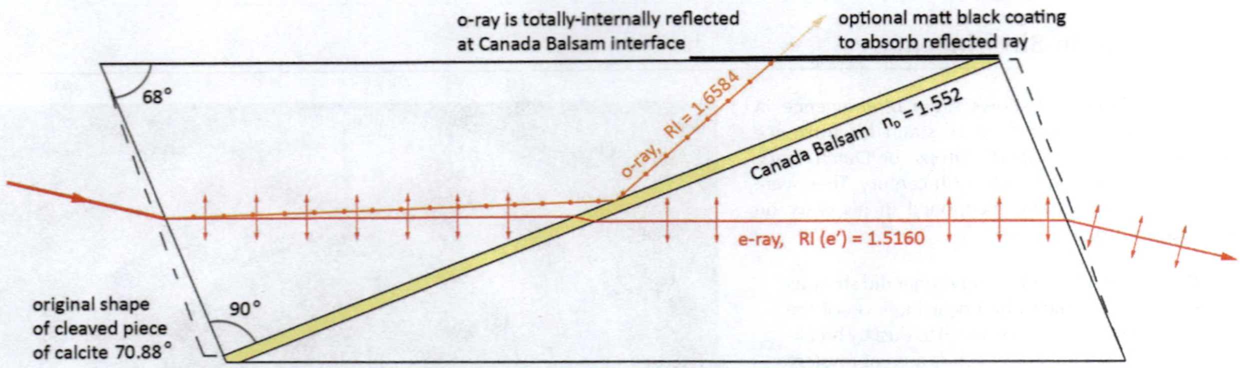

Figure 13.10 **Action of a Nicol prism** *These prisms consist of two right-angled calcite prismatic wedges that are cemented together on their long faces by Canada balsam. The optic axes of the calcite crystals are parallel and aligned perpendicular to the plane of reflection. Anisotropic calcite splits the incident unpolarised light into two mutually orthogonal rays. The ordinary ray (RI = 1.6584) is totally internally reflected (being of higher refractive index than Canada balsam) from the calcite-cement interface, leaving the extraordinary ray to be transmitted as the source of plane-polarised light. A Nicol prism has – at best – only 50% efficiency but until the introduction of Polaroid material, was the only means of producing polarised light. The incident ray is shown at an angle here, but in practice this cannot be more than 14° from the horizontal axis.*

on the matt black sides of the prism, leaving the extraordinary ray as the source of plane-polarised light. The TIR (total internal reflection) angle for the o-ray is arcsin (1.552/1.6584) = 69.36°. This is why the rhombohedron is reground with faces at 68° to the long parallel sides; it also permits the entrance angle of incident light – the field-of-view angle – to be increased. The faster, less refracted e-ray, of lower refractive index, passes through the interface and exits the opposite end of the crystal as a beam of plane-polarised light. It is not totally internally reflected, because the RI of the e-ray (n_D = 1.4864) is less than that of the Canada balsam (note that RI e′ = 1.5160 here). Nicol prisms were common at the later half or the 19th century. Two disadvantages are associated with Nicol prisms. The size of the crystals making up Nicol-type prisms plus the restriction of the total entrance angle of the incident ray to about 28°, for total internal reflection of the o-ray, restricts the available numerical aperture of the microscope. Nowadays, the Glan-Foucault prism is used, employing an air-gap, because Canada balsam absorbs UV light.

13.6 Types of Double Refraction

Anisotropic materials exhibit one of four classes of double refraction, depending upon their molecular structure and gross morphology. (Applying an electric or magnetic field can also cause birefringence, but that is not of concern to us here.)

- Intrinsic birefringence is the largest class and contains all those materials whose birefringence is due to their intrinsic molecular structure (e.g. non-cubic crystals) due to resonance in chemical bonds; e.g. Cellophane.
- Form birefringence arises when an isotropic material, surrounded by another isotropic material, is present as orientated structures such as sheets or rods; for example chitin. Form birefringence occurs widely in biological molecules – see Howie & Brewer (2009) and Howie (2008) for a full explanation of this phenomenon.
- Strain birefringence is due to mechanical stress altering the molecular orientation of the material. Birefringence may be introduced into otherwise isotropic materials during manufacture (e.g. extrusion of plastic products); see Box 13.3.

  It is important, therefore, to select 'strain-free' objectives, with minimal or no residual strain in the lens elements, for polarised light microscopy; otherwise, the strained objective will act as a 'compensator', introducing an unknown degree of birefringence into the image.
- Flow birefringence is the result of the ordered movement or streaming of structures in a liquid, fluid or dispersion (e.g. within solutions of myosin and the pseudopodia of Amoeba).

When a plane-polarised beam of light passes through a lens, it is subject to alteration because of the selective reflection and loss of light at the lens surfaces. This is particularly serious in the four quadrants of the lens where the polarisation direction is at 45° to the lens radius. The original plane-polarisation may be considered to have radial and circumferential components. In these four quadrants, the circumferential component is preferentially reduced by reflection and, as a result, the net polarisation direction of the light transmitted is slightly rotated. Thus even a perfectly isotropic lens shows good extinction between crossed polars only in a cross-shaped area of the lens aperture. With a conventional, non-rectified, polarising microscope the diffraction pattern is distorted and each point in the image becomes anomalously represented by four closely-spaced points in the image. Very weakly anisotropic features are hidden. Rectified optics give an extinction factor of 13 000:1 instead of 600:1.' For further details, see Oldenbourg (1991). Very weak edge birefringence is found at the edges of any specimen, including isotropic materials, where refractive index differences exists at a sharp boundary, and is due to diffraction (see Inoué (1986), Appendix III.10).

Box 13.3 Strain Birefringence

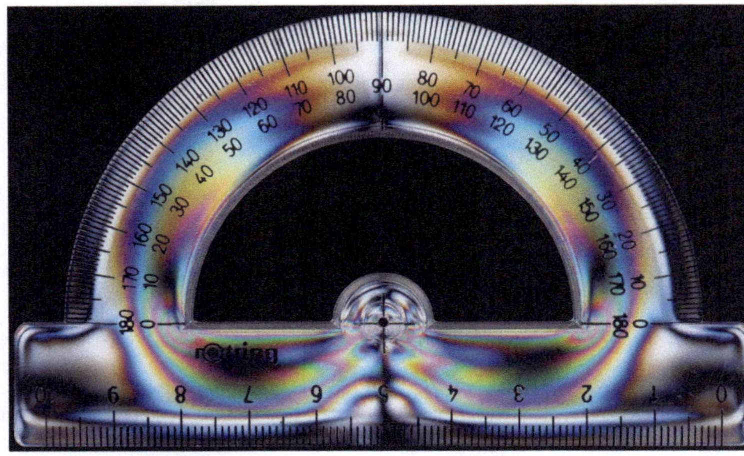

Source: Courtesy of Mr K. Glover.

The protractor illustrated shows strain birefringence in extruded plastic. Another form of strain birefringence can be seen in Prince Rupert's Drops, or Dutch Tears, which were very popular in the 17th century. They were seen by Samuel Pepys and mentioned in his diary on Monday 13th of January 1662:

> 'Mr. Peter [Honywell] who after dinner did show us the experiment (which I had heard talk of) of the chymicall glasses, which break all to dust by breaking off a little small end; which is a great mystery to me.'

Here is the answer to the mystery. The many artistic skills and scientific interests of Prince Rupert made him a colourful character of the Restoration period when his first cousin Charles II returned to England. He converted some of the apartments of Windsor castle into a laboratory, and was made an honorary founder member of the newly established Royal Society in 1664. Rupert had heard of the curiosity named 'Dutch Tears' from his time spent in Europe. He gave them to Charles II, who in turn passed them onto the Royal Society, where their characteristics were investigated and exhibited to an audience. In England the drops are named after Rupert, but Pepys (who had little time for Rupert) makes no mention of him.

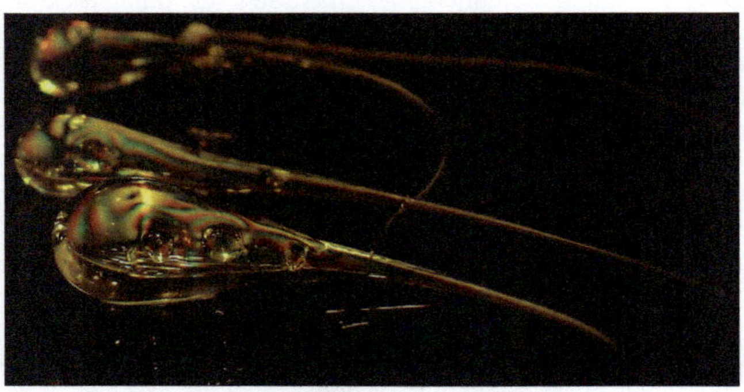
Source: Mg3kc at English Wikipedia, via Wikimedia Commons.

Prince Rupert's Drops are created by dripping molten glass off the end of a rod and letting it fall into a bucket of cold water. When the experiment is successful, the glass cools extremely rapidly into a tadpole-shaped extrusion with a long thin tail. The large volume of water cools the glass on the outside of the drop before the inner bulk. When this mass eventually cools, it contracts the outer shell, setting up compressive stresses on the outside and tensile stresses on the inside held in equilibrium.

The compressed egg-shaped outer shell of the drop can withstand very large forces, such as the blow of a hammer, but the tail is very delicate indeed, and can be broken with the slightest flick of a finger. Because of the very high degree of stress in the entire drop and the stored potential energy, the tail disintegrates explosively when broken – hence the novelty value. It has been shown by very high-speed photography that the crack-front propagates from the tail towards the head at about 1800 metres/s, which is 4000 miles per hour, or approximately Mach 5.4.

Strain birefringence is usually studied (e.g. in glass factories) by the use of a pair of circular polarisers of opposite sense rather than a crossed pair of linear polarisers. This gives an image in which retardations due to birefringence are turned into intensities that are not azimuth-dependent, and therefore is not missed. You can see this loss of information, for example, if you turn a plastic protractor around between two crossed linear polarisers.

13.6.1 Double Refraction in Living Cells

Most biological materials exhibit birefringence but only very weakly and normally only with very poor contrast. For that reason, polarised light microscopy is not as widely employed in biological science as in materials science. Studying birefringence in living cells has mainly concerned work on cell division and motility. The mitotic spindle of a dividing cell is doubly refracting. In the 1960s, Shinya Inoué, a major contributor in the field of biological polarisation microscopy, used rectifying optics[11] to correct for the depolarisation of the light by the condenser and objective lenses (see the reference in Further Reading), which otherwise hid the small magnitude of birefringence.

Birefringence can be observed in regularly ordered macromolecules; for example, the amyloid plaque deposits such as are found in the brains of Alzheimer patients. Modified proteins, such as immunoglobulin light chains, abnormally accumulate between cells and form fibrils. The dye Congo Red is used to identify the presence of amyloid fibrils within plaques in biopsy tissue sections. This dye intercalates between the β-pleated sheets within the amyloid fibrils and causes a pronounced birefringence (Figure 13.11). The colour seen depends upon the thickness of the section cut on the microtome. If it is thinner than 6 μm or so, the colour will be faint red, while those thicker than 12 μm will exhibit a yellow colour. For those sections of ideal thickness of 8–10 μm, the

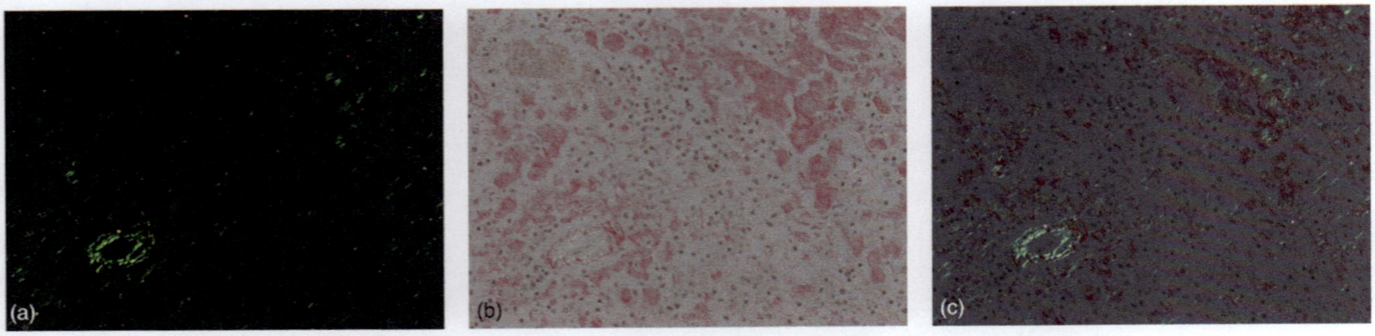

Figure 13.11 **Form birefringence in biology** *Form birefringence of accumulated amyloid plaque deposits in a section of liver stained with Congo Red, classically used to demonstrate amyloid. Panel (a) shows the amyloid coloured 'apple-green' in an 8–10 μm thick section viewed through crossed polars. The same field of view is shown in panel (b) seen with bright-field microscopy – the amyloid is difficult to pick out in the Congo Red section. It is helpful to partially uncross the polars, as in panel (c), to give a useful image showing the distribution of the amyloid fibrils with respect to the overall morphology of the tissue section.*

diagnostic Newton's interference colour of green is seen. For an explanation of the production of Newton's interference colours, see Chapter 14, Section 14.10. Birefringence may also be seen in the anisotropic 'A' band of striated muscle due to the structure of the myosin filaments.

13.7 The Behaviour of Light in Anisotropic Materials

When a ray of polarised light enters a transparent doubly refracting substance, or crystal, *other than along the optic axis*, it is resolved into two mutually perpendicular rays that vibrate in planes orthogonal to one another (Figure 13.5, Figure 13.12a), which do *not* interfere[12]. One of the rays travels at the same velocity in all directions and is called the ordinary ray, or *o-ray*, (RI = n_o) and has a constant velocity through the material irrespective of direction and hence a constant refractive index. This ray obeys Snell's law of refraction. The velocity of the other ray, called the extraordinary ray, or *e-ray*, (RI = n_e) has a velocity – and hence refractive index (n_e') – that varies (Figure 13.12b) between n_e and n_o depending upon the direction of travel through the anisotropic material. The magnitude of birefringence (Δ, Δ_M or Δ_n) is given by the difference between n_e – and n_o. For many materials including calcite, n_e is greater than n_o but for others, such as quartz, it is less (see Section 13.12). The maximum or minimum value for the refractive index of the e-ray (n_e) is called the principal index. It occurs when the e-ray vibrations are parallel to the optic axis. The plane cutting through the anisotropic material that contains both the plane of vibration of the o-ray and also contains the optic axis is called the principal plane of the o-ray. A similar plane can be constructed for the e-ray. The o-ray is polarised with its plane of vibration perpendicular to its principal plane and the e-ray polarised with its plane of vibration lying in its principal plane.

Optical path differences arising from the differing velocities of the o-ray and e-ray are explained further in the next chapter, in Section 13.9 and in Section 14.8. In order for characteristic polarisation colours to be formed by interference, four conditions must be met:

- the initial light must be polarised;
- the light must be split into two differing rays;
- a phase difference must exist between these two rays; and
- the two components of the two rays that are vibrating in the same plane must be recombined (vector analysis helps to visualise this step).

13.8 Using Vectors to Understand Double Refraction

The amplitude and vibration direction of polarised light may be represented by a vector. A vector has two Cartesian components (x,y). The 'side-on' view of a light wave as a sine function can be described as a vector (Figure 13.13a). Equally the same wave, looking 'end-on' down the axis of propagation, as the wave travels out of the plane of the paper perpendicularly, can also be described by a vector (Figure 13.13b) describing both the amplitude and vibration direction of the light.

Vectors are useful in visualising and predicting how the orthogonal mutually perpendicular rays arising from plane-polarised light encountering a doubly refracting material behave and recombine at the analyser of the polarising microscope. Both the o-ray and e-ray can be drawn as vectors. The magnitude of each component can be determined by projecting a line to the respective axis (Figure 13.14).

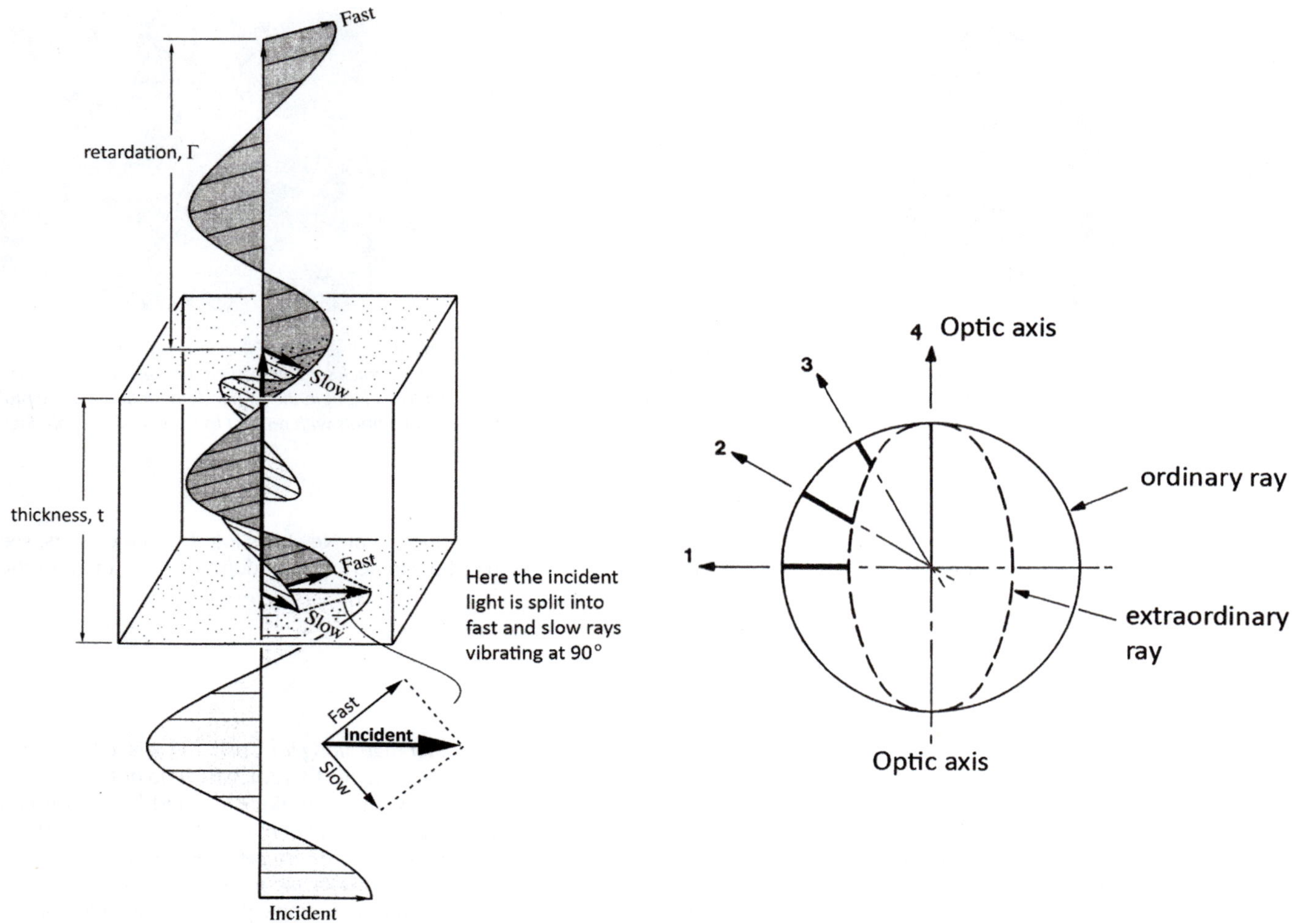

Figure 13.12 **Birefringence** *Calculation of the magnitude of birefringence. The magnitude of birefringence (Δ or Δ_M) is given by $n_e - n_o$. The relative retardation, Γ, is a product of birefringence and the thickness of the anisotropic material.*

The figure on the right shows ray velocity surfaces for a uniaxial positive crystal (e.g. quartz). In this case the RI of the extraordinary ray can vary between the maximum value, n_e, (indicated by ray 1), for which the maximum birefringence value is recorded, through two intermediate values, n_e' (rays 2 & 3). A smaller birefringence value will be recorded for ray 3 than ray 2. When the e-ray coincides with the optic-axis and is equivalent to the velocity of the o-ray (as indicated by ray 4), no birefringence will be shown. Source: Nesse 2009. Reproduced with permission from Oxford University Press; Robinson and Bradbury 1992. Reproduced with permission from Oxford University Press.

This is useful for calculating the amplitude of incident light passed through a polarising filter (for example, the proportion of light passed by the analyser with respect to the orientation of an anisotropic sample between crossed polars).

Supposing the specimen is moved so that the o-ray increasingly aligns parallel to the E-W polariser, the proportion of the o-ray allowed through the analyser gets smaller and tends to zero: this particular vector is now missing. When the specimen is moved so that it is at 90° to the analyser, then the e-ray aligns parallel to the E-W polariser and so cannot get through the analyser.

At 45° the e-ray and o-ray pass the greatest possible components through the analyser and are of equal intensity. The specimen is therefore at its brightest. After passing the analyser, the two component rays are traveling in the same plane – that imposed upon them by the vibration direction of the analyser. These beams will therefore interfere (in the normal way as diffracted rays do) to form an image in the primary image plane.

Hence, in a full rotation (Figure 13.15), the specimen is dark at 0°, bright at 45°, dark at 90°, bright at 135°, dark at 180°, bright at 225°, dark at 270° and bright at 315°, before becoming dark again at 0°. Extinction occurs twice for each vibration direction in each 360° rotation – a total of four times.

It is because of the need to rotate the specimen to achieve maximum brightness and contrast that polarised light microscopes are fitted with a rotating stage, usually graduated to record the exact angles of extinction. Orienting an anisotropic specimen with respect to crossed polars permits the four extinction positions to be determined. This is useful in locating the refractive index of an unknown crystal

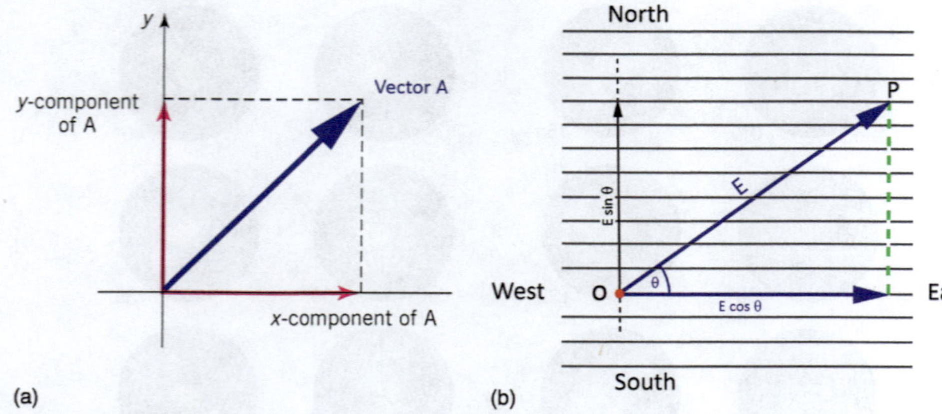

Figure 13.13 ***Vectors*** *A Euclidian vector describes both magnitude and direction. As such, it can represent both the intensity (amplitude) and vibration direction of the light. The Cartesian components (x,y) of the vector describing a beam of polarised light conveniently indicates the magnitude, or intensity, of the light able to pass through crossed polars.*

In Figure 13.13b, the horizontal lines represent the preferred, or privileged, direction of the polariser, set E-W. A light beam (red dot) propagates at right angles to the plane of the paper and emerges at O. The intensity and vibration direction of a particular wave (O-P) is represented by the blue line, E, set at angle θ to the E-W permitted vibration direction. E.cos θ represents the component of the light passed by the polarising filter. This value is obtained by dropping a perpendicular, shown in green, from point P to the axis indicating the permitted vibration direction of the polariser. The intensity is clearly less than the wave before it encounters the filter. Only the E-W component is passed, the N-S component (E.sin θ) is absorbed and thus blocked. Source: Adapted from Robinson and Bradbury 1992 with permission from Oxford University Press.

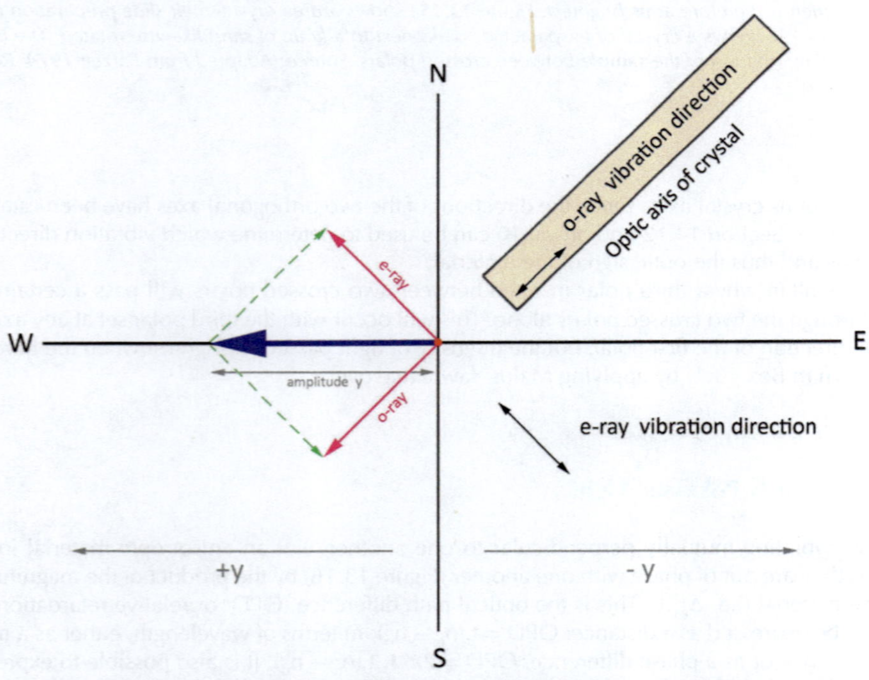

Figure 13.14 ***Using vectors to describe the o-ray and e-ray magnitude*** *A beam of light polarised by the polariser set E-W is emerging from the plane of the paper, as denoted by the red dot. The thick blue vector (of amplitude y) describes the intensity and direction of the beam. An anisotropic crystal inserted between the two polars at 45° will decompose the polarised light beam into two components, the o-ray and the e-ray. In this case, these rays vibrate either perpendicularly or parallel to the optic axis of the crystal, as shown. The rays will have amplitudes determined by the polarising filter. The amplitudes of these rays can be found by dropping perpendiculars from the vector to the permitted vibration direction of the o-ray and e-ray.*

Since the amplitude of the incident beam of light varies from +y to −y (refer back to Figure 2.2 on this point) for one wavelength, so also the amplitudes of the o-ray and the e-ray increase and decrease in the same way. Both the o-ray and the e-ray start with a path difference of zero. However, they travel with different velocities, and so become out of step with one another. Source: Adapted from Robinson and Bradbury 1992 with permission from Oxford University Press.

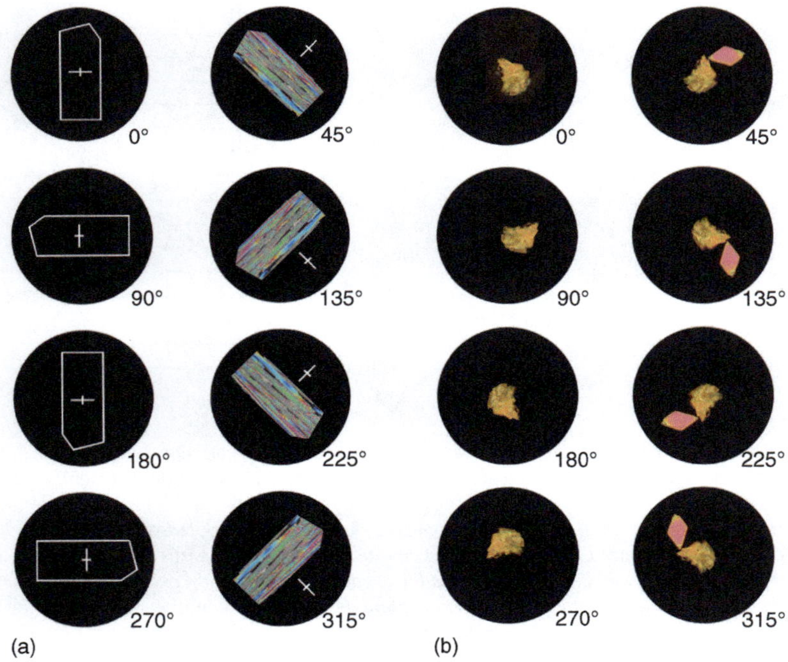

Figure 13.15 The proportion of each of the o-ray and e-ray that are able to pass through the analyser depends upon the orientation of the birefringent sample with respect to the crossed polariser and analyser. At 45° the e-ray and o-ray pass the greatest possible components through the analyser, and are of equal intensity. The specimen is therefore at its brightest. Figure 13.15a shows a urea crystal melt slide preparation rotated successively by 45° between crossed polars. Figure 13.15b shows a crystal of l-asparagine (pink) next to a grain of sand likewise rotated. The intensity of the image of the isotropic sand grain is unaffected by rotation of the sample between crossed polars. Source: Adapted from Patzelt 1974. Reproduced with permission from Leitz, Leica Microsystems.

and determining the location of its crystal axes. Once the directions of the two orthogonal axes have been established, a compensating accessory plate (see Chapter 14, Section 14.12 and onwards) can be used to determine which vibration direction of n_e or n_o is the fast axis and which the slow axis and thus the optic sign of the material.

Vector analysis helps to explain why a third polar inserted between two crossed polars will pass a certain proportion of light that would otherwise not get through the two crossed polars alone. This will occur with the third polar set at any azimuth angle with respect to the East-West vibration direction of the first polar, but the intensity of light passed is brightest when the third polar is set at 45°. This can be determined (as shown in Box 13.1) by applying Malus' law twice over.

13.9 Elliptically and Circularly Polarised Light

When the o-ray and e-ray, vibrating mutually perpendicular to one another, exit an anisotropic material in which the surfaces are parallel to the optical axis, they are out of phase with one another (Figure 13.16) by the product of the magnitude of birefringence (Δ_M) and the thickness (t) of the material (i.e. Δ_M.t). This is the optical path difference (OPD; or relative retardation Γ, capital gamma). The optical path difference can be expressed as a distance: OPD = t.($n_e - n_o$); in terms of wavelength, either as a multiple or a fraction of a wavelength, OPD = t/λ.($n_e - n_o$), or as a phase difference, OPD = $2\pi/\lambda$. t.($n_e - n_o$). It is also possible to express OPD in terms of time elapsed, but since the velocity of light is so fast and the times consequently so short, it is more convenient to express OPD as a phase difference. The combination of these two rays, which remain out of phase and continue vibrating perpendicularly to one another, yields either a circularly polarised wave or an elliptically polarised wave (Figure 13.17).

In unpolarised light the plane of vibration of the electric vector, which is transversely perpendicular to the direction of propagation, may be at any angle. Where the plane of vibration of the electric vector is confined in space to a single plane, the light is then linearly polarised (see Figure 13.1). The transverse electric field wave is accompanied by a magnetic field wave as illustrated. We shall only consider the plane of vibration of the electric vector. If the light is now composed of the superposition of two plane waves of equal amplitude but differing in phase by $\pi/2$ (90°) or by $^1/_4\ \lambda$, it is circularly polarised. The vectors of these two rays can be added together to give a single resultant, which describes a helical path (i.e. of the 'tip' of the electric field vector) along the direction of propagation. By convention, polarisation is defined as seen (suppose for a moment that we could actually see the ray of light and its electric field

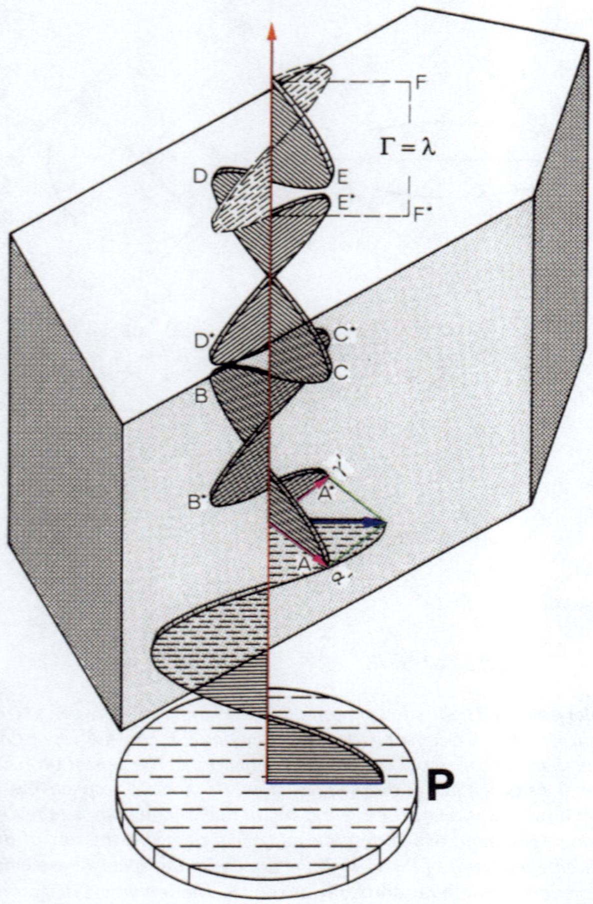

Figure 13.16 How an optical path difference arises between the o-ray and the e-ray. In this case, the relative retardation is shown as 1λ. Source: Adapted from Patzelt 1974. Reproduced with permission from Leitz, Leica Microsystems.

vector propagating) from the receiver looking against the direction of propagation towards the source. The electric field vector makes one complete revolution per wavelength. When this resultant is viewed end-on (i.e. with the direction of propagation coming out of the plane of the paper), the outline of this helical path described by the tip of the electric field vector is a circle. Circular polarisation may be referred to as right-circular polarisation (rotating anticlockwise as seen from the observer's position) or left-circular (rotating clockwise), depending on the direction in which the electric field vector rotates. If circularly polarised light is reflected, the handedness reverses.

From a practical viewpoint, a $^1/_4\,\lambda$ filter (quarter-wave filter) is useful for converting plane-polarised light into circularly polarised light, thus eliminating the extinction effects seen in the image with linearly plane-polarised light. An anisotropic specimen viewed with circular polarised light does not go to extinction four times in 360° (as it will when viewed with plane-polarised light) but shows *as if it were* constantly at the 45° position between crossed polars alone, irrespective of the rotation of the stage and orientation of the sample. Also, when viewing interference figures, the isogyres formed by the polariser and analyser are no longer seen in the conoscopic image.

If the two plane waves are now out of phase by an amount other than 90°, or $^1/_4\,\lambda$, yet nevertheless have a fixed-phase relationship and have equal or differing amplitudes, the resulting electric vector describes an ellipse, and the light is now elliptically polarised. Elliptically polarised waves exhibit left- or right-handed chirality. Since this is what commonly occurs when the o-ray and e-ray exit an anisotropic material, elliptical polarisation is the most general description of polarised light; circular and linearly polarised light can be regarded as special cases of elliptically polarised light.

The elliptical nature of the polarised light is important in determining the polarisation colours formed by birefringence from doubly refracting materials, since the light no longer vibrates only E-W, as defined by the polariser, but contains a component that also vibrates N-S along the axis of the analyser. It is this component allowed through the analyser that gives rise to the polarisation colours that we observe. Circular polarisation does occur in nature (Box 13.4) but only very rarely. Circular polarisation filters[13] are now used to create 3D effects in films. Since we don't have the ability to see circular polarised light, it can be used to create an illusion of 3D

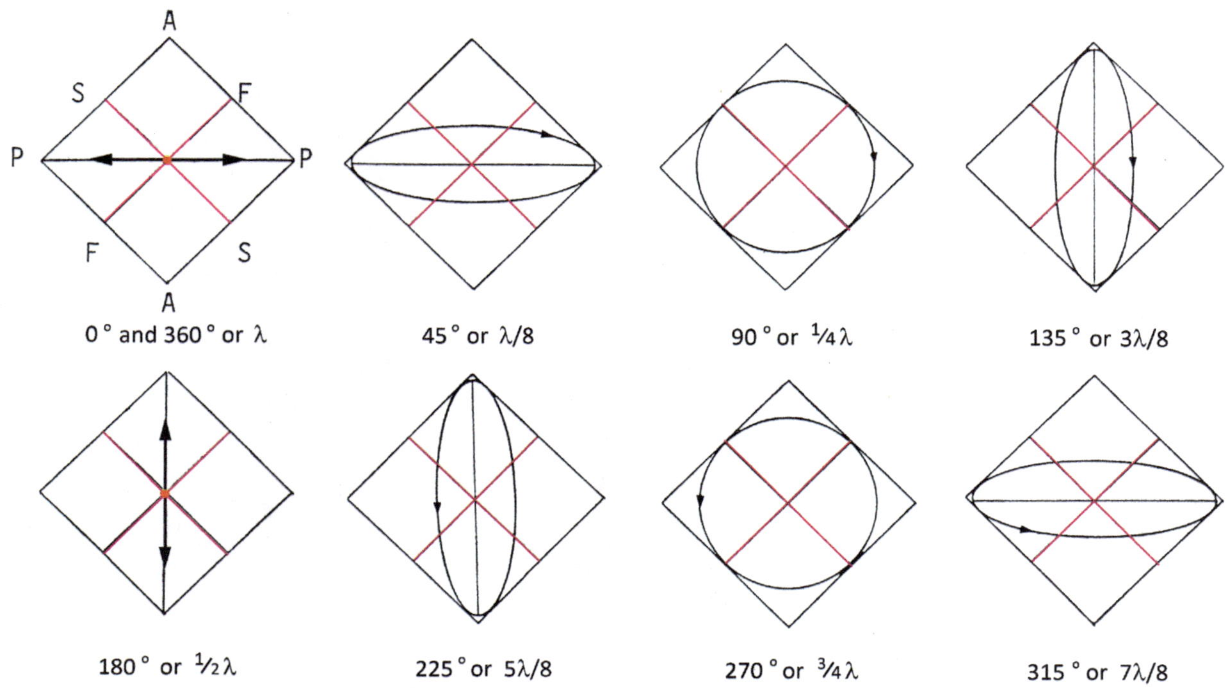

Figure 13.17 **Origin of elliptical and circular polarised light** *This figure shows the relationship between two out-of-phase rays, and the how the resultant forms circularly and elliptically polarised light. The black double-headed arrows along P-P and A-A represent the polariser and analyser respectively. A beam of incident light is emerging out of the plane of the paper, and perpendicular to it. The axis of propagation is shown as a red dot perpendicular to the plane of the paper. The vibration directions of the fast and slow rays, along F-F and S-S respectively, are shown by two the pink lines. Only the plane of vibration of the electric vector of each wave is shown. The electric vector makes one complete revolution per wavelength. The vectors of these two rays can be added together to give a single resultant, which describes a helical path (i.e. the 'tip' of the electric field vector) along the direction of propagation. This first diagram, in the extreme left-hand of the top row, is drawn for the special case where the incident plane-polarised light is set at 45° to the optic axis, c, such that the amplitudes of the o-ray and e-ray are equal. The remaining diagrams show the relationship between these two out-of-phase rays, and the resultant formed for seven different intervals of optical path difference.*

When the phase difference between the two rays is zero, the resultant lies E-W along P-P. As the phase difference increases, the resultant traces out an elliptical path with its major axis parallel to P-P, until at 90° the components of each ray contributing to the resultant are equal, but differing in phase by $\pi/2$ or $1/4\lambda$ and the resultant is a circle. At 180° the resultant is linear, aligned with the analyser N-S. Between 90° and 180° the long axis of the ellipse continues to lie N-S; above 180° the process is repeated, but the 'handedness' of the resultant changes (in this case to anti-clockwise). At 270° or $3/4\lambda$ and the resultant is again a circle, after which the resultant lies E-W. Circular or elliptical polarisation may be referred to as right-handed or left-handed, depending on the direction in which the electric field vector rotates. Source: Adapted from Dr C. Hammond, with permission from Dr Hammond.

stereoscopic perspective without interfering with our vision. Thus the same scene is projected into each eye but from slightly different perspectives.

13.10 Fast and Slow Directions

The vibration direction of the ray of light through any anisotropic material that has the smaller, or lower, refractive index is called the *fast direction*. This is because it has to traverse through less dense material than the ray associated with the greater, or higher, refractive index, which is slowed down more by the greater density of the material in its path. The light waves travelling in the slow direction thus lag behind those in the fast plane of polarisation. *Birefringence is not a fixed value but varies, depending upon the orientation of the doubly refracting object with respect to the polarised-light illumination*. Birefringence is at a minimum if the light propagates along the optic axis, it is maximum when travelling orthogonally to the optic axis and assumes intermediate values for orientations between these axes.

- The wave vibrating in the lower-refractive index direction is the fast wave. Conversely, the wave vibrating in the higher-refractive index direction is the slow wave.
- The velocity of a wave (v) is given by: $v = \lambda\nu$. If the velocity is reduced when the wave passes through an anisotropic material, the wavelength (λ) decreases, since the frequency (ν) remains constant.
- Since the slow wave has a lower velocity, it will have a shorter wavelength while passing through the anisotropic material.

Box 13.4 Circular Polarisation in Nature

The surface of the of the elytra, or wing cases, of the jewel scarab beetle *Chrysina gloriosa* has a microstructure that not only causes iridescence (from the Greek *irides*, meaning rainbow) but also ensures that the reflected light is left-hand circularly polarised. This is very unusual in nature, the one other animal able to detect and exploit circularly polarised light being the purple spot mantis shrimp *Gonodactylus smithii*.

It has been known since 1911 that the light reflected off the back of the scarab beetle is circularly polarised. The beetle, found high in the montane juniper forests of the extreme southwestern United States and northern Mexico, is well-camouflaged (a) in juniper foliage. However, when viewed with right circularly polarising filters (b), the beetles exhibit a stark black contrast. In contrast, *Chrysina woodi*, a close relative (seen here top and left), is not seen so well, and also shows no phototactic discrimination between linear and circularly polarised light. Studies suggest that because this ability is rare, it confers an evolutionary advantage upon *C. gloriosa*, allowing it to communicate within the species while staying hidden from predators.

Source: Warrant 2010. Reproduced with permission from Elsevier.

Source: Reproduced with permission from Professor R. Cauldwell.

On the other hand, *Odontodactylus cultrifer*, the keel tail mantis shrimp, appears to be a carnivorous predator that has taken advantage of both linear and circular polarisation. These shrimps are thought to have very good vision, extending into both the ultraviolet and infrared parts of the electromagnetic spectrum. Nevertheless, since the refractive index of seawater is similar to the bodies of the animals that it eats, they tend to be transparent to a greater or lesser degree. The advantage for the mantis shrimp is that their prey are packed full of polarising sugars, and so they 'light up like Christmas trees'. Like the beetle, it is thought that the shrimps use polarisation cues to communicate with potential mates.

References

Biology: A jewel's true colours (2010). *Nature 464*: 468. doi: 10.1038/464468a.

Chiou, T.-H. et al. (2008). 'Circular polarization vision in a stomatopod crustacean'. *Current Biology 18*/14: 429–434.

Cronin, T.W. et al. (2003). 'Polarization vision and its role in biological signaling'. *Integrative and Comparative Biology 43*/4: 549–558.

Cronin, T.W. et al. (2014). 'Filtering and polychromatic vision in mantis shrimps: themes in visible and ultraviolet vision'. *Philosophical Transactions of the Royal Society A 369*/1636: 20130032.

Kleinlogel, S. and White, A.G. (2008). The secret world of shrimps: polarisation vision at its best. *PLoS One 3*/5: e2190.

Warrant, E.J. (2010). 'Polarisation vision: beetles see circularly polarised light'. *Current Biology 20*/14: R610–R612.

13.11 Pleochroism, Dichroism and Bireflectance

The formation of polarised light by double refraction includes two special cases of this phenomenon: pleochroism and dichroism. Pleochroism is the selective spectral absorption of the wavelength and/or intensity of *transmitted* light, depending upon the vibration direction. It is only exhibited by coloured minerals. For uniaxial anisotropic materials, where there are two refractive indexes (n_o and n_e) and two permitted vibration directions, this pleochroic behaviour is referred to as dichroism. Biaxial media have three permitted vibration directions: n_α, n_β and n_γ, so that three colours[14] (trichroism) will be seen. These colours are maximal when the light is polarised parallel to the crystallographic axes, a, b or c, respectively. Pleochroism has a diagnostic value in crystal and mineral identification; in addition, gems are sometimes cut and set specifically to display pleochroic colours.

The intensity and colour of the pleochroism shown is also dependent upon how the crystal, or doubly refracting material, is oriented with respect to the optic axis of the crystal. Where the crystal lies on the stage of the microscope perpendicular to the optic axis of the crystal, the extraordinary ray will act as an ordinary ray and will not be absorbed, or undergo a change in intensity or change in colour. The degree of pleochroism or dichroism will also depend on the thickness of the specimen and will be weaker and less pronounced with thinner material. In this case, the phenomenon will be observed more readily if the polariser is turned instead of the specimen, which remains static. If the polariser is oscillated $\pm 10°$ either side of the 'crossed' position, then the weak effect of pleochroism will be seen.

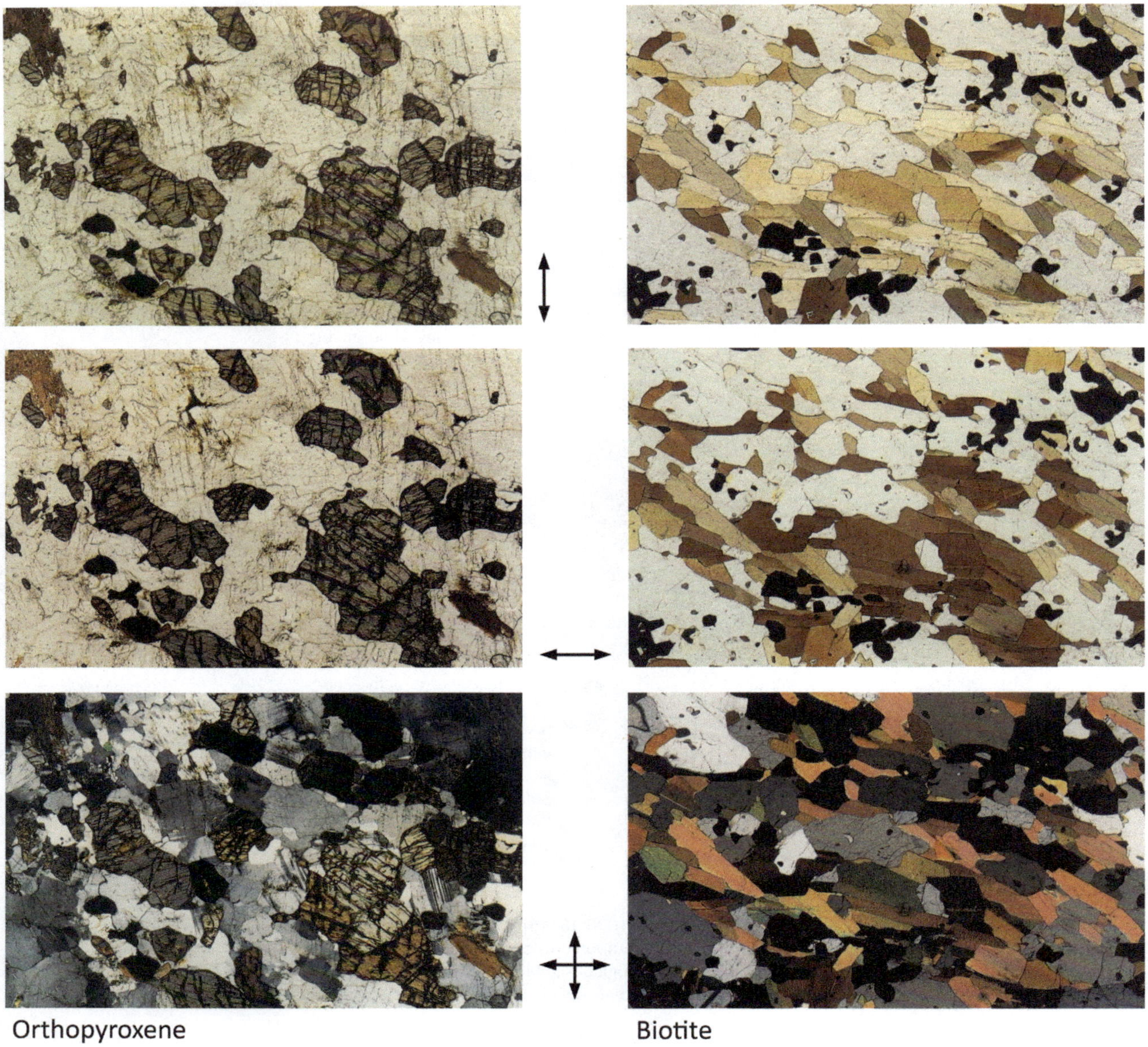

Figure 13.18 **Pleochroism** *The double-headed arrows show the orientation of the polar (without the analyser) in the top and the middle pair of panels. The bottom pair of panels shows, for comparison, the specimen under crossed polars (i.e. with the analyser reinserted).*

The left-hand column shows orthopyroxene, which exhibits a pale pink to green pleochroism. This mineral has a low birefringence value, and so only first-order interference colours are seen between crossed polars.

The right-hand column shows biotite, a mineral that cleaves easily and has a high birefringence, so pleochroism is obvious. The strongest absorption is seen when the cleavage plane of the mineral grain is parallel to the polariser. Brown mica, or biotite, is biaxial and therefore shows trichroism. Further details on the phenomena of pleochroism and the crystallographic basis for these changes in colour and intensity can be found in Robinson & Bradbury (1992), Section 5.3. Higher-order interference colours are seen in biotite as compared to orthopyroxene, because of the larger birefringence value, but biotite exhibits strong absorption colours, which can mask the interference colours.

The trigonal mineral tourmaline provides a good example of dichroism. Other examples (Figure 13.18) are orthopyroxene and biotite. The e-ray vibration direction lies along the length of these prismatic crystals, parallel to the optic axis, while the o-ray vibration is perpendicular, along the narrow width. Most wavelengths are absorbed for vibration directions normal to the optic axis and weakly in the vibration direction parallel to the c-axis. Tourmaline exhibits blue, green or brown colours, depending upon the thickness of the crystal. The colour is darkest and most intense when the long axis of the crystal lies perpendicular to the vibration direction of the polar and is palest when lying parallel. This is a good way of checking, or setting, the correct orientation of the polariser and hence also the analyser (see Figure 14.6). Brown mica, or biotite, which occurs in all granite (thin sections) is biaxial and therefore shows trichroism. Further details on the phenomena of pleochroism, and the crystallographic basis for these changes in colour and intensity, can be found in Robinson & Bradbury (1992), Section 5.3.

Table 13.1 *Refractive Indexes of Some Uniaxial Birefringent Crystals, Measured at $\lambda_0 = 589.3$ nm*

Crystal	n_o	n_e
Tourmaline	1.669	1.638
Calcite	1.6584	1.4864
Quartz	1.5443	1.5534
Sodium nitrate	1.5854	1.3369
Ice	1.309	1.313
Rutile (TiO_2)	2.616	2.903

Source: Hecht 2001. Reproduced with permission from Pearson Education Inc.

Reference

Hecht, E. (2001). Polarization. In: *Optics*, 4th ed., Chapter 8. Boston: Addison-Wesley.

The white light emitted by a computer monitor can be used in a similar manner to a light box to view pleochroism, since an LCD screen emits plane-polarised light. If the sample, or slide, is rotated in front of the screen, any pleochroism or dichroism can be determined. Where there is a pleochroic change in intensity and/or colour of reflecting samples when the stage is rotated while illuminated by plane-polarised light, this special case of pleochroism is called bireflectance. This provides a helpful way to distinguish isotropic minerals such as magnetite from anisotropic ones (e.g. hematite), which otherwise look similar in reflected light.

13.12 Uniaxial and Biaxial Materials

There are two classes of doubly refracting materials: *uniaxial* and *biaxial*, which have one optic axis and two optic axes, respectively. Uniaxial minerals all comprise crystals with trigonal, tetragonal and hexagonal (3-, 4- or 6-fold, respectively) fundamental symmetry. Uniaxial crystals have two refractive indexes and one optic axis (see Figure 13.20). For a uniaxial doubly refracting material, the modulus of the maximum difference between the refractive indexes (n_e, n_o) of the two rays is called the birefringence of the material (Table 13.1). Thus $\Delta n = n_e - n_o$. The birefringence may be positive or negative (Figure 13.19) and is the result of subtracting the refractive index associated with the ordinary ray, n_o, from that of the extraordinary ray, n_e. Where $n_e > n_o$, then the crystal is 'optically positive' (i.e. if $[n_e - n_o]$ is a positive value, the crystal is positive). Quartz is an example of an optically positive material. Where $n_e < n_o$, (i.e. if $[n_e - n_o]$ is a negative value), then the crystal has a negative sign of double refraction and is 'optically negative'. Calcite is an optically negative crystalline mineral.

Suppose you were to take an anisotropic uniaxial crystal and cut several thin sections from it, each at a different, random plane through the crystal. If you then measured the refractive indexes in both the fast and slow directions, then for a positive uniaxial material (e.g. quartz) you would find that n_{fast} ($n_o = 1.5443$) is always the same value, and n_{slow} varies between n_{fast} and some maximum RI value (for quartz $n_e = 1.5534$). The birefringence of quartz is $n_e - n_o = 1.5534 - 1.5443 = +0.0091$. Therefore n_{fast} is the ordinary ray (because it stays the same; n_o) and n_{slow} is the extraordinary ray (n_e'). At the maximum RI value for the e-ray, it is labeled n_e (with no

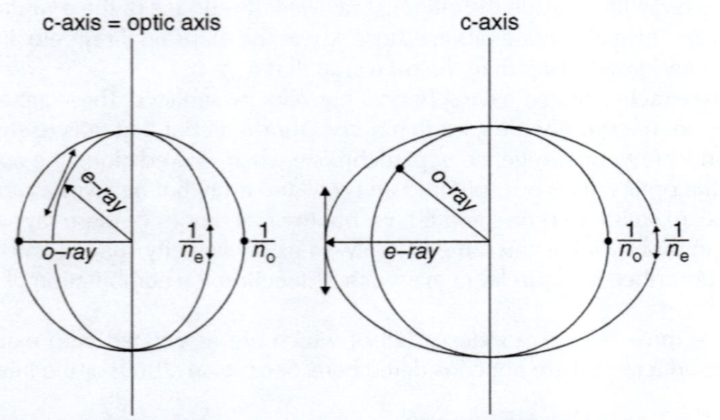

Figure 13.19 **Ray velocity surface** *Differing velocity of the e-ray through anisotropic materials. When the propagation direction of the extraordinary ray is such that it travels along the optic axis of the material, the e-ray will behave exactly like the o-ray, and obey Snell's law. For a uniaxial negative material (e.g. calcite) the e-ray travels faster than the o-ray and $n_e < n_o$, and the ellipsoidal ray velocity surface (RVS) lies outside the spherical RVS. For a uniaxial positive material (e.g. quartz) the converse holds true. Source: Adapted from Robinson and Bradbury 1992 with permission from Oxford University Press.*

prime notation). Conversely, for a negative uniaxial material (e.g. calcite), n_{slow} is always the same value, and n_{fast} varies between n_{slow} and some minimum RI value.

Most crystals and minerals are biaxial. Biaxial crystals have orthorhombic, monoclinic and triclinic symmetry. They have three refractive indexes and two optic axes. Two of these refractive indexes vary between two extremes, but there is also a unique intermediate refractive index. The lowest refractive index (fastest propagating ray) is denoted by n_X, (also sometimes denoted by n_α), the refractive index of the intermediate ray by n_Y (sometimes written n_β) and the highest refractive index (of the slowest ray) by n_Z (sometimes designated n_γ). The birefringence in the case of biaxial crystals is the modulus of the difference between the highest and the lowest refractive indexes ($n_Z - n_X$). The intermediate refractive index, n_Y, may lie closer to n_X or n_Z. If n_Y lies closer to n_X, the crystal is optically positive. Conversely, if n_Y has a value nearer to n_Z, the crystal is optically negative. Further discussion of biaxial crystal geometry is beyond the scope of this book, and interested readers are referred to Wood (1977), Hammond (1992, 2009) or Dyer, Gunter & Tasa (2008) for further detail and discussion.

13.13 Ray Velocity Surfaces and the Optic Axis

The optical properties of an anisotropic material are displayed by a two-dimensional or three-dimensional surface representation called the ray velocity surface (RVS; Figure 13.19). The distance of the surface from the central origin is drawn according to the refractive index and velocity of each ray of light in the radial direction parallel to the electric field vector of vibration of the ray.

The ellipsoid describes the refractive index measured in all directions within the material. The ray velocity surface of the o-ray is a sphere, while the RVS of the e-ray is an ellipsoid of revolution with the rotation axis corresponding to the optic axis of the crystal[15]. The optic axis is the direction of light propagation, or vector, in an optically anisotropic medium where the substance behaves isotropically and the spherical and elliptical RVSs touch at two common points where the cross section of the indicatrix is circular (see Figure 13.19). A cross section through the centre of the ellipsoid, cut perpendicular to the direction of propagation of the light, gives a geometrical construction called the refractive index ellipse for that material. The refractive indexes of the two components of a uniaxial material are inversely proportional to these two radii. This is because the velocity of light is proportional to the inverse of the refractive index multiplied by the velocity in a vacuum, c. The major axis of the ellipse describes the slow wave (with the higher refractive index) and the minor axis, the fast wave (with the lower RI). The wavefronts of the ordinary and extraordinary rays coincide only in the direction of the optic axis.

A section through the ray velocity surface perpendicular to the optic axis is the circular section whose radius is n_o for the o-ray. A principal section through the ray velocity surface that includes the optic axis forms a two-dimensional representation of ray velocity. A random section through the ray velocity surface will produce an ellipse whose axes are n_o for the o-ray, and n_e' for the e-ray, where n_e' assumes a value between n_o and n_e.

This index ellipse also indicates the vibration direction of the electric field vector of the o-ray and e-ray, because the waves vibrate with the electric field vectors along the major and minor axes of the index ellipse, since the rays vibrate with their electric vectors in planes of vibration parallel to the major and minor axes of the ray velocity surface. The vibration direction of the o-ray is perpendicular to the optic axis (coming out of the plane of the paper, as shown by the dot in Figure 13.19). The vibration direction of the e-ray lies in a plane containing the optic axis and the direction of the e-ray, as shown by the double-headed arrows on the ellipses in Figure 13.19.

Two useful mnemonics are given by Robinson & Bradbury (1992) to remember whether a uniaxial anisotropic crystal is positive or negative. Recall that n_o and n_e are sometimes written n_ω and n_ε (see Section 13.2). Remember 'POOF' for POsitive O-rays Fast, and since $n_e > n_o$, then Epsilson Greater than Omega: the EGO is positive! Positive uniaxial crystals have a prolate ellipsoid, where the spherical ray velocity surface of the o-ray lies outside the elliptical ray velocity surface of the e-ray, and the o-rays travel faster than the e-rays (i.e. $n_o < n_e$, see Figure 13.12b). Negative materials are those where the ellipsoidal ray velocity surface lies outside the spherical one (an oblate ellipsoid), and the e-rays travel faster than the o-rays, so that $n_o > n_e$.

Isotropic materials, with a single refractive index, have spherical ray velocity surfaces. These are represented in any particular plane (e.g. of the paper) by a circle. It is not necessarily the case that a crystal or material that fails to show anisotropy is always isotropic. Anisotropic substances will not show either birefringence or pleochroism when viewed along the optic axis, as shown in Figure 13.20. An incident ray that is parallel to the optic axis is not split into an o-ray and e-ray but behaves as an ordinary ray. Incident beams that are perpendicular to the optic axis are split into o-rays and e-rays, but the trajectories of these rays are *coincident* and do not diverge. In this case, the e-ray behaves as an o-ray – it has the same velocity, so its ray velocity surface transverse to the propagation direction of the light along the optic axis is described by a circle. In every other direction the combination of the fast and slow directions of the e-ray describes an ellipsoid.

Biaxial crystalline materials have three refractive indexes, all of which are associated with extraordinary rays. Their ray velocity surfaces describe complex double surfaces and are not considered here. See Nesse (2009) or the Internet Resources given at the end of the chapter for further information.

13.14 The Indicatrix

It is a simple matter to draw both the ray velocity surfaces as a *single* two-dimensional representation on a sheet of paper (Figure 13.21). This is called the *indicatrix*, and for a uniaxial material, it is geometrically similar to the e-ray velocity surface. The indicatrix therefore

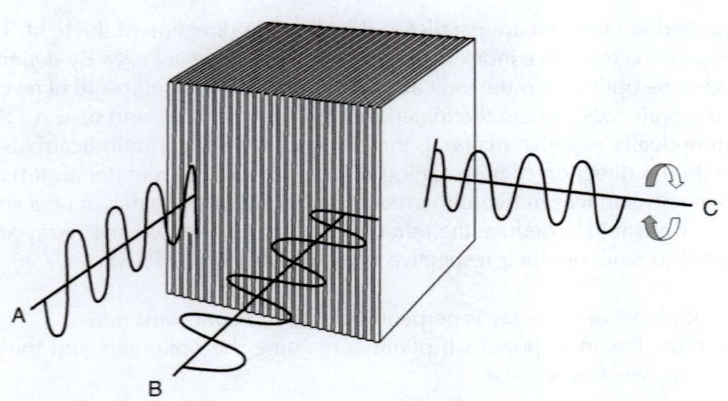

Figure 13.20 **The optic axis, or c-axis** *An imaginary crystal of anisotropic material in which the optically dense molecular groups are arranged in parallel planes. Two polarised rays incident on the crystal from the side will pass through the crystal at different speeds and therefore have different refractive indexes. Ray A passes relatively easily through the crystal, the vibration direction being aligned with the planes of the material. Ray A is therefore the fast ray, with the lower refractive index, while ray B is the slow ray, with the higher refractive index. Ray C is aligned to the optic axis, because it may be polarised in any azimuth (i.e. the plane of polarisation may be rotated about the propagation axis) without encountering different molecular groups in the crystal structure, as do rays A & B. The optic axis of the crystal, is therefore parallel to ray C.*

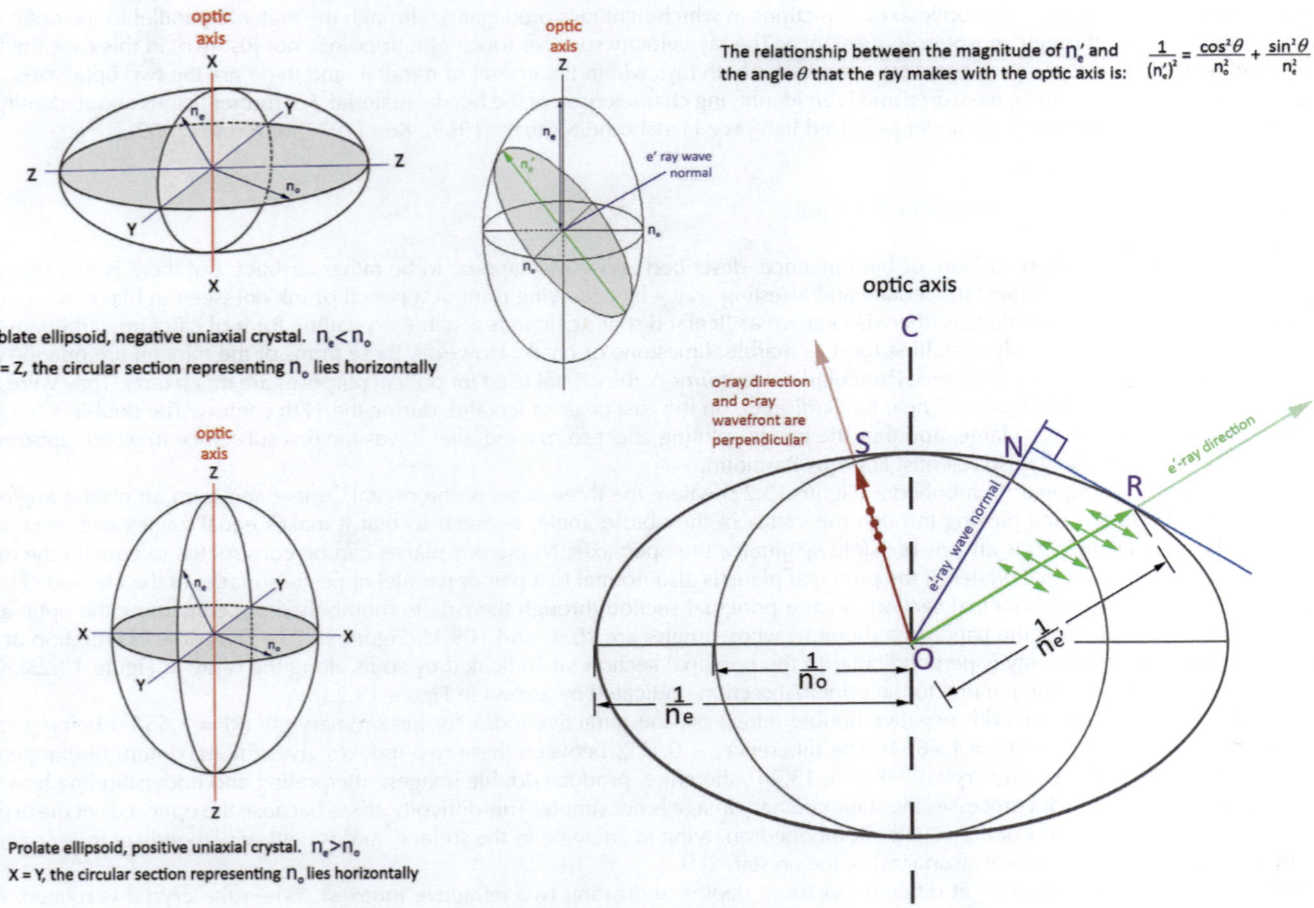

Figure 13.21 **Wavefronts and wave normals** *The relationship of the wavefront and wave normal in the wave velocity surface. The direction of propagation of an arbitrary e-ray (n_e') is shown by the solid green line (O-R). The extraordinary ray waves (in green) always vibrate in a plane, R-O-C, containing the e-ray and the optic axis (i.e. the principal section of the ellipsoid of revolution). The e-waves are shown by the green double-headed arrows, which represent the e-ray wavefront. They vibrate perpendicular to the wave normal and are, therefore, tangential to the elliptical ray velocity surface.*

By constructing a line (shown in light blue) tangential to the e-ray wavefront at point R on the indicatrix, the direction (O-N) wave normal of the e-ray can be deduced. A line dropped perpendicular to the tangent to pass through the central point, O, gives the wave normal.

The ordinary ray always vibrates at right angles to the plane containing the ray and the optic axis, here S-O-X, the term X being the Cartesian coordinate used to define the plane. The wavefront of the o-ray is depicted by closed red circles. The direction of the ray is given by O-S. The wavefront is also tangential to the RVS, but since this is circular, the wavefront is perpendicular to the ray direction.

shows the indexes of refraction plotted as radii that are parallel to the vibration direction of the light. The wavefronts have a shape that is the inverse of the indicatrix because the refractive index is the inverse of the wave velocity. By definition, a principal section through the ray velocity surface that includes the optic axis *is* the indicatrix. The indicatrix is an ellipsoid of revolution whose axes are n_o and n_e with the rotation axis parallel to the optic axis. Where the material is optically positive, and $n_e < n_o$, the indicatrix ellipsoid is prolate. Where $n_e > n_o$, in the case of an optically negative material, the ellipsoid is oblate. The indicatrix is useful as a method of showing how the refractive index varies with the direction of propagation of the *wavefront* in a material. It has the following advantages over the 3D representation of the ray velocity surfaces in two-dimensional space: (i) it can be drawn on a single flat surface (e.g. the page of a book); (ii) from vector analysis it is possible to deduce the refractive indexes of the two rays (e-ray and o-ray) for any given direction of propagation, and (iii) it is possible to work out their respective vibration directions. Thus:

- the vibration direction (principal plane) of the o-ray is perpendicular to the optic axis; and
- the vibration direction of the e-ray lies in a (principal) plane containing the optic axis and the direction of the e-ray, which is tangential to the ellipsoid of the ray velocity surface.

The plane parallel to this tangent (seen here drawn in green) is called the wavefront, which is also parallel to the vibration direction (Figure 31.21) for the o-ray. The *wave normal* is the direction perpendicular to the *tangent* of the wavefront (also called the wave surface). It is not parallel with the ray direction of the e-ray except when (i) either the light propagates along the optic axis (in which case the o-ray and the e-ray have equal velocities) or (ii) when the light is propagating perpendicular to the optic axis, where the differences between the o-ray and the e-ray are maximum. Both the circle of the o-ray and the ellipse of the e-ray touch each other at two points: where they intersect the optic axis (Figure 13.19). In this direction both rays have equal velocities. The main use of the indicatrix is to relate the optical properties of mineral grains to their crystal orientation.

Since biaxial materials have two optic axes, directions in which light rays propagating through the material parallel to the optic axis behave as if they pass through an isotropic substance. The ray velocity surfaces touch at four points, not just two. In this case the four points define the two directions of equal ray velocity for both rays within the crystal or mineral, and these are the two optic axes. The angle between them, 2V, can be measured and is an identifying characteristic of the biaxial material. For further details about identifying biaxial crystals and their behaviour under polarised light, see Hartshorne & Stuart (1969), Kerr (1977) or Nesse (2009).

13.15 Observing Double Refraction in Calcite

The phenomenon of double refraction, or birefringence, described above may appear to be rather abstract. But there is one situation in which it is manifested in a most immediate and arresting way – by observing print or a pencil or ink dot (seen in black) on a piece of paper through a crystal of calcite. Otherwise known as 'Iceland spar', calcite is a stable crystalline form of calcium carbonate that usually occurs in an impure polycrystalline form as marble, limestone or chalk. However, these forms of the mineral are opaque and therefore unsuitable for our requirements. Particularly clear forms of the crystal used for optical purposes are much rarer. They were first discovered at the Helgustaðir Quarry[16] near Reyðarfjörður, on the east coast of Iceland, during the 17th century. The double refraction or birefringence of calcite is so large, and thus the image-splitting effect so marked, that it was the first substance in which anisotropy was noted, in 1669 by the Danish scientist Erasmus Bartholin.

Calcite cleaves into trigonal rhombohedra (Figure 13.22) where the three faces of the crystal[17] meet and form an obtuse angle (of 102°) with each other. A line passing through the vertex of the obtuse angle, oriented so that it makes equal angles with each face (45.4°) and each edge (63.7°) is an axis of 3-fold symmetry, the optic axis. Numerous planes can be constructed to contain the optic axis; these are called principal planes. If the principal plane is also normal to a pair of parallel opposite surfaces of the cleaved calcite, it slices the rhomb across a principal section. Such a principal section through the calcite rhombohedron, containing the optic axis, can be drawn in the plane of the paper as a rhombus whose angles are 70.9° and 109.1° (Figure 13.23). The plane of vibration of the electric field vector of the o-ray is perpendicular to the principal section (as indicated by spots along the o-ray in Figure 13.23). The e-ray has its plane of vibration parallel to the principal section (indicated by arrows in Figure 13.23).

Optically, calcite is uniaxial with negative double refraction, the refractive index for the ordinary ray (RI = 1.6584) being greater than for the extraordinary ray (RI = 1.4864). The difference, – 0.172, between these two indexes gives the maximum birefringence. Objects viewed through a calcite crystal[18] (Figure 13.24), therefore, produce double images. Interpreting and understanding how the extraordinary image of the dot precesses the static ordinary image is not simple. This difficulty arises because the optic axis of the crystal runs through opposite corners of the calcite rhombohedron, lying at an angle to the surface, and is neither perpendicular nor parallel to any of the natural cleavage faces (surfaces) of the crystal.

Two images of the dot are seen at different apparent depths (indicating two refractive indexes). When the crystal is rotated, one image – formed by the extraordinary ray – rotates (or precesses) in the direction of rotation about the other (which is fixed) and formed by the ordinary ray, as shown in Figure 13.25. The moving dot remains constantly between the stationary dot and the obtuse angle of the rhombohedron. A line joining the two images of the dots is parallel to the diagonal that bisects the two obtuse angles of the rhombohedron; this is the optic axis. Looking through the thickness of the rhomb, the dot formed by the o-ray appears higher (because its refractive index is higher/greater) than the moving dot formed by the e-ray. If rhombs of different thickness are compared, it will be seen that the thicker the rhomb, the greater the separation of the two dots.

Now observe the images through a piece of Polaroid (Figure 13.25e & f): on rotation of the Polaroid the images successfully disappear and reappear every 90° rotation, showing clearly that light from each is plane-polarised at 90° and have vibration directions

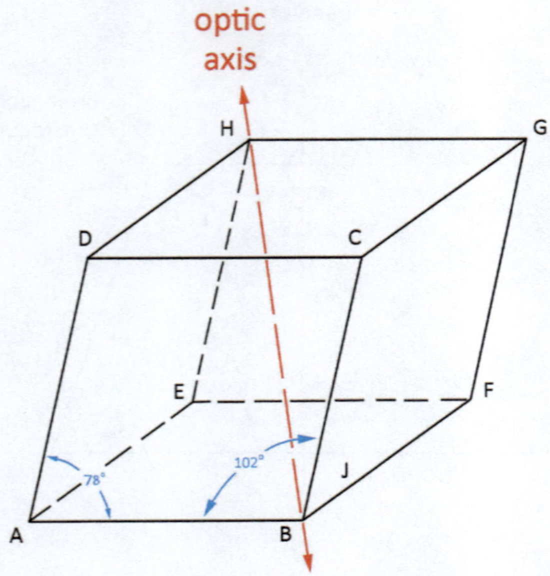

Figure 13.22 **Calcite rhombohedra** *A calcite crystal rhomb, with the principal angles and the position of the c-axis of symmetry about the blunt corners of the rhomb. The crystal cleaves with the sides forming parallelograms of angles 78° and 102°. All the corners except two contain both angles. The two blunt corners contain angles of 102° only, here H and B.*

at right angles to each other – a corroboration of our statement on the splitting of light into two mutually perpendicular plane-polarised components in anisotropic materials, in Sections 13.2 and 13.7. If the polar is orientated at 45°, then both images are seen but at reduced intensity.

Draw a line, instead of a dot as mentioned above, onto a piece of paper. Observe it through the calcite rhombohedron, taking care to look perpendicular to the top face of the crystal. One image of the line will appear to pass straight through the crystal. This is the ordinary ray, behaving according to Snell's law. The other image is offset towards the corner, or vertex, where the three faces of the crystal meet and form an obtuse angle (of 102°) with each other. As the rhombohedron is rotated, the double images merge into a single line. This indicates the optic axis, which bisects the opposite obtuse angles of the rhombohedron.

Christiaan Huygens derived the theoretical basis of the property of double refraction in calcite from a geometric wave construction, extending the construction method he had employed to explain[19] (single) refraction. Huygens deduced that light travelling through doubly refracting materials gives rise to two waves, the ordinary one, which is spherical, and the extraordinary one, which is elliptical. The o-ray wave forms a spheroid of revolution, and the e-ray wave forms an ellipsoid of revolution (Figure 13.26), *both revolving about*

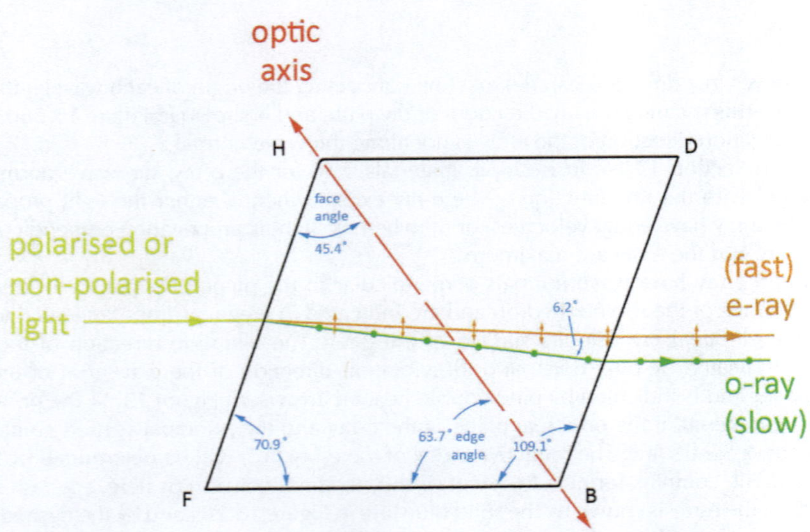

Figure 13.23 **The principal section of calcite** *The behaviour of calcite in splitting non-polarised light into polarised o-ray and e-ray. See the text for details.*

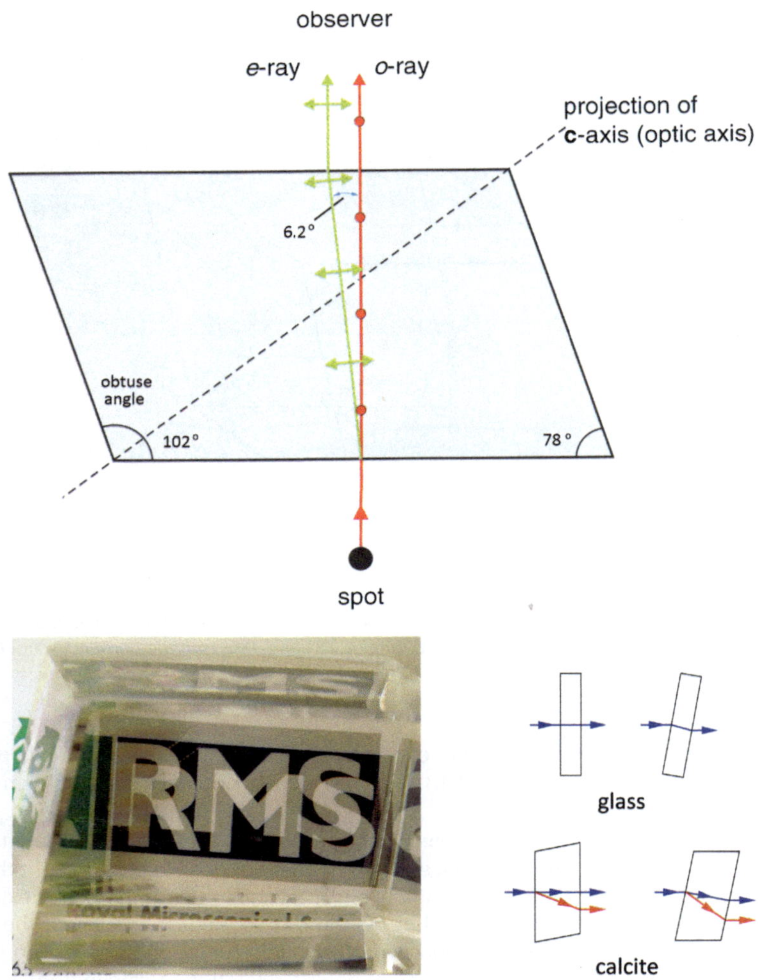

Figure 13.24 **Double refraction with calcite (1)** *Here, a beam of light is shown entering the calcite rhomb from underneath, rather than from the side. The effect is the same: the beam is decomposed into two mutually orthogonal rays. The effect on the RMS logo, seen through a calcite rhomb, is shown. The right-hand panel shows that if the rhomb is tilted, the beam will still decompose into two polarised rays, regardless of the angle of incidence. Source: Tilley 2011. Reproduced with permission from John Wiley & Sons.*

the optic axis. The wave*front* moves in a direction parallel to a line connecting the origin of each wavelet to the point of tangency with the planar envelope of the wave. This is called the ray direction, or ray path, and is shown in Figure 13.26b. The elliptical nature of the e-ray wave means that the propagation direction of the e-ray is not along the wave normal.

Refer back to Figure 13.21, in Section 13.14. In isotropic materials, and for the o-ray, the wave normal travels perpendicular to the wavefront[20]. It is not parallel with the ray direction of the e-ray except when (i) either the light propagates along the optic axis (in which case the o-ray and the e-ray have equal velocities) or (ii) when the light is propagating perpendicular to the optic axis, where the *differences* between the o-ray and the e-ray are maximum.

In Figure 13.26a the o-ray and e-ray have wave normals perpendicular to the plane of paper, in a direction towards the eye. The wavefronts are parallel to the surface of the rhombohedron and are indicated in green, as lines tangential to the spherical ray velocity surface of the o-ray and to the elliptical ray velocity surface of the e-ray. The vibration direction of the wavefront of the o-ray is perpendicular to the page (as indicated by pink dots), and the vibration direction of the wavefront of the e-ray lies (perpendicular to the o-ray) in the plane of paper and is indicated by pink double-headed arrows. In Figure 13.24 the principal section is shown as a rhombus. In a doubly refracting material, if the principal plane of the e-ray and the principal section coincide, then the wave normal (but *not* the e-ray wavefront) obeys Snell's law. The refractive index of the e-ray (n_e') will be determined by the difference between the wave normal and the optic axis. The complex formula to calculate this need not concern us here.

The direction of propagation of the e-ray is shown by the solid blue line in Figure 13.26a and by the dashed blue line in Figure 13.26b, and follows a line from the centre of the dot to the tangent of the extraordinary ellipse. The propagation direction of the e-ray is not parallel with the e-ray wave normal, because the optic axis within the calcite rhombohedron is not parallel or perpendicular to any

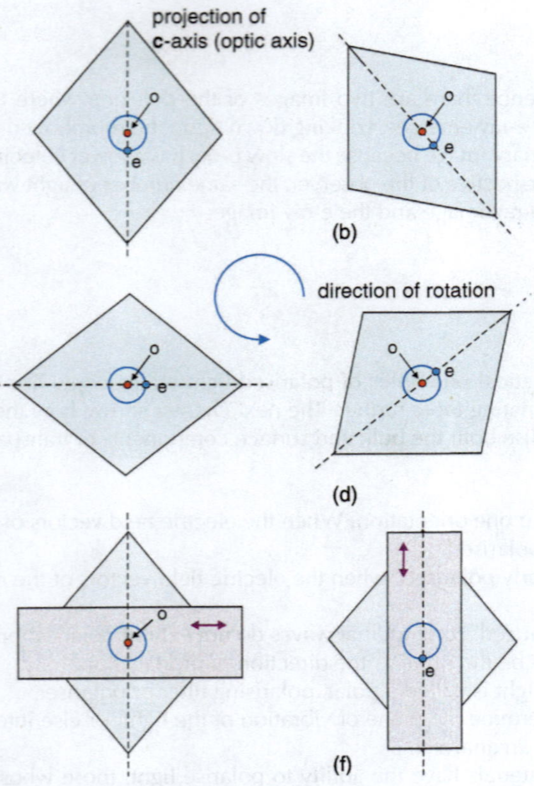

Figure 13.25 ***Double refraction with calcite (2)*** *On rotation of the crystal, one of the images (the 'extraordinary' image, seen in blue) rotates in the direction of rotation, whereas the other (the 'ordinary' image, seen in red) remains fixed (Figure 3.25a–d). This, in fact, is the experiment of Bartholin, which so perplexed Huygens and Newton. When viewed through a polar with its privileged transmission axis indicated by the double-headed arrow, the images of the dot on the paper disappear and reappear every 90° rotation, proving that light is polarised and split into two mutually perpendicular waves by the anisotropic calcite. The transmission axis of the polar placed perpendicular to the optic-axis of the calcite rhomb causes the dot formed by the e-ray to disappear. With the polar set parallel to the c-axis, the o-ray disappears. At the intermediate position of 45°, images of both dots are seen but at reduced intensity. Source: Tilley 2011. Reproduced with permission from John Wiley & Sons.*

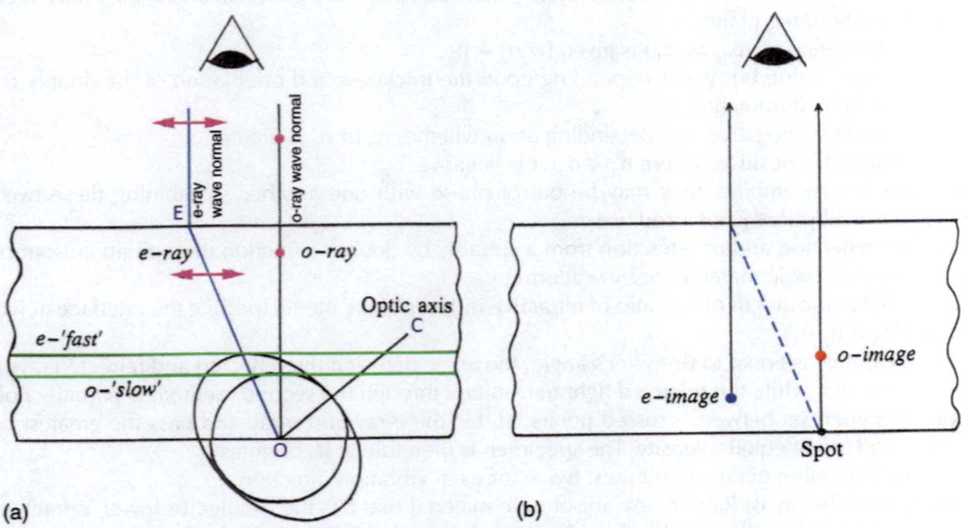

Figure 13.26 ***An explanation for double refraction with calcite*** *A side-on sectional view through a calcite rhomb, resting on a sheet of paper. A spot is drawn on paper at point O, and light rays are assumed to originate from this point. Both the o-ray and e-ray wavefronts (shown in green) are parallel to each other and tangential to their respective ray velocity surfaces. The vibration direction of the e-ray is in the plane of the page (double-headed arrows); the o-ray is perpendicular to the plane of the page and is shown by pink closed circles. Both wave normals are indicated by black lines perpendicular to the sheet of paper, in a direction towards the eye. The 'offset' e-ray path is found by constructing a tangent to the indicatrix parallel to the wavefront. The propagation direction of the e-ray is given by the line from the centre to the tangential point of the wavefront on the RVS ellipse. The e-ray wave normal passes through this point to emerge from the rhomb at the surface at point E.*

Upon careful inspection, the two images of the spot appear to be at different depths. The ordinary image appears to be at a higher plane in the crystal because the o-ray travels slower than the e-ray. If we say that the same number of waves occur between the top surface of the calcite rhomb and the position of the image of the spot, then the o-ray, with a lower velocity and higher refractive index, compresses the same number of waves into a smaller distance (see also the Figure in Box 14.1). Source: Adapted from Robinson and Bradbury 1992 with permission from Oxford University Press.

of the cleavage faces of the crystal. Hence there are two images of the dot: one where the o-ray emerges from the surface of the rhombohedron and the other where the e-ray emerges. Looking down upon the rhombohedron, the ordinary image appears at a higher plane (Figure 13.26b) than the extraordinary image because the slow o-ray has a lower velocity and higher refractive index ($n_o = 1.6584$, compared to $n_e = 1.4864$). From the perspective of the observer, the same number of light waves appear between the top surface of the crystal and the apparent depths of the o-ray image and the e-ray image.

13.16 Chapter Summary

This chapter introduces the basic theoretical principles of polarised-light microscopy. The Further Reading and References are given for those who wish to explore this fascinating topic further. The next chapter shows how the microscope can be used as an analytical instrument with polarised light to visualise both the bulk and surface components of transparent and opaque materials.

1. Waves can oscillate with more than one orientation. When the electric field vectors of all the rays comprising a beam of light are randomly oriented, the light is unpolarised.
2. A wave is plane-polarised, or linearly polarised, when the electric field vectors of the rays are restricted to vibrating in one plane and are all aligned in that plane.
3. Only transverse waves can be polarised. Longitudinal waves do not exhibit polarisation because the vibration direction, in which the medium is displaced, can only be the same as the direction of propagation.
4. A device that produces polarised light is called a polar, polarising filter or polariser.
5. A polarising filter that used to determine the plane of vibration of the light, or else interrogate the rays emanating from a sample on the microscope stage, is called an analyser.
6. Anisotropic (doubly refracting) materials have the ability to polarise light; those whose structure has no effect are isotropic. An optically *isotropic* material has only one refractive index for all directions of the electric field vector. Isotropic materials are gases, liquids and non-crystalline solids.
7. Transparent anisotropic materials decompose non-polarised light into two polarised rays vibrating orthogonally to one another. One of these rays is termed the *ordinary ray* (since it obeys Snell's law of refraction) and the other the *extraordinary ray* (since it does not obey Snell's law).
8. Light rays propagate with electric and magnetic field vector components orthogonal to one another.
9. Uniaxial anisotropic materials exhibit isotropy *only* when light propagates along the optic axis.
10. Interference of polarised light beams only occurs if these polarised beams are coherent *and* if they have electric vectors that are parallel (i.e. vibrating in the same plane).
11. The magnitude of birefringence (Δ, Δ_M or Δ_n) is given by $n_e - n_o$.
12. Birefringence is not a fixed value but varies depending upon the thickness and orientation of the doubly refracting object with respect to the polarised-light illumination.
13. Birefringence has a positive or negative sign depending upon whether n_e or n_o is greater.
14. Where $n_e > n_o$, the material is positive; where $n_e < n_o$, it is negative.
15. When the o-ray and e-ray recombine, they may be out of phase with one another. Combining these two rays yields either a circularly polarised or an elliptically polarised wave.
16. Light can polarised by reflection and/or refraction from a surface, by double refraction through an anisotropic material, through selective absorption by a dichroic material, or by scattering.
17. Brewster's angle is equal to the arctan of the ratio of refractive indexes of the media forming the interface at which light is reflected and/or refracted ($\tan \theta_B = n_2/n_1$).
18. When the angle of incidence is equal to Brewster's angle, the angle between the reflected and refracted rays is 90°. The reflected light is completely polarised, while the refracted light transmitted through the second medium is partially polarised.
19. For an anisotropic specimen set between crossed polars, at 45° the e-ray and o-ray and pass the greatest possible components through the analyser and are of equal intensity. The specimen is therefore at its brightest.
20. In a full 360° rotation, extinction occurs four times: twice for each vibration direction.
21. The vibration direction of the ray of light in any anisotropic material that has the smaller, or lower, refractive index is called the *fast direction*. The light waves travelling in the slow direction thus lag behind those in the fast plane of polarisation.
22. The optical properties of an anisotropic material are displayed by a two-dimensional surface representation called the ray velocity surface.
23. The indicatrix is a principal section through the ray velocity surface that includes the optic axis. It describes a two-dimensional representation of ray velocity.
24. A principal plane contains the optic axis. If the principal plane is also normal to a pair of parallel opposite surfaces of the cleaved calcite, it cuts across a principal section.
25. The plane of vibration of the electric field vector of the o-ray is perpendicular to the principal section. The e-ray has its plane of vibration parallel to the principal section.

26. The wave*front* moves in a direction parallel to a line connecting the origin of each wavelet to the point of tangency with the planar envelope of the wave. This is called the ray direction or ray path.
27. In a doubly refracting material, if the principal plane of the e-ray and the principal section coincide, then the wave normal (but *not* the e-ray wavefront) obeys Snell's law.

Key Reading

Raith, M.M., Raase, P., and Reinhardt, J. (2012). *Guide to Thin Section Microscopy*, 2nd ed. Chantilly, Virginia: Mineralogical Society of America. http://www.minsocam.org/msa/openaccess_publications/Thin_Sctn_Mcrscpy_2_rdcd_eng.pdf.
- A very clear, comprehensive and beautifully illustrated treatise on polarised-light microscopy.

References

Buchwald, J.Z. (2007). 'Huygens' methods for determining optical parameters in birefringence'. *Archive for History of Exact Sciences 61*/1: 67–81.
Dyar, M.D., Gunter, M.E., and Tasa, D. (2008). *Mineralogy and Optical Mineralogy*. Chantilly, Virginia: Mineralogical Society of America.
Gao, H.X. (1999). *Acta Crystallographica Section A 55*: 1014–1025, Figure 1, p. 1014.
Hammond, C. (1992). *Introduction to Crystallography* (revised edition). Microscopy Handbook No. 19. Oxford: Oxford Science Publications.
Hammond, C. (2009). *The Basics of Crystallography and Diffraction*, 3rd ed. Oxford: Oxford Science Publications.
Hartshorne, N.H. and Stuart, A. (1969). *Practical Optical Crystallography*, 2nd ed. London: Edward Arnold.
Hecht, E. (2001). *Optics*, 4th ed. San Francisco: Addison Wesley, Pearson Educational.
Howie, A.J. and Brewer, D.B. (2009). Optical properties of amyloid stained by Congo Red: History and mechanisms. *Micron 40*/3: 285–301.
Howie, A.J. et al. (2008). 'Physical basis of colors seen in Congo Red-stained amyloid in polarized light'. *Laboratory Investigation 88*/3: 232–242. https://www.nature.com/articles/3700714.pdf.
Inoué, S. (1986). *Video Microscopy*, 1st ed. New York: Plenum Press.
Kerr, P.F. (1977). *Optical Mineralogy*, 4th ed. New York: McGraw-Hill.
Nesse, W.D. (2009). *Introduction to Optical Mineralogy*. Oxford: OUP.
Oldenbourg, R. (1991). 'Analysis of edge birefringence'. *Biophysical Journal 60*/3: 629–641.
Patzelt, W.J. (1974). *Polarized-Light Microscopy: Principles, Instruments, Applications*. Leitz Wetzlar, booklet 550–51/Engl.
Robinson, P.C. and Bradbury, S. (1992). *Qualitative Polarised-Light Microscopy*. Microscopy Handbook No. 9. Oxford: Oxford Science Publications.
Tilley, R.J.D. (2011). *Colour and the Optical Properties of Materials*, 2nd ed. Chichester: Wiley.
Wood, E.A. (1977). *Crystals and Light: An Introduction to Optical Crystallography*, 2nd rev. ed. New York: Dover.

Further Reading

Dell, K.R. and Vale, R.D. (2004). 'A tribute to Shinya Inoue and innovation in light microscopy'. *Journal of Cell Biology 165*/1: 21–25.
Gay, P. (1967). *An Introduction to Crystal Optics*. Harlow: Longman.
Huygens, C. (1690). *Treatise on Light*. Project Gutenberg e-book #14725.
Inoué, S. (2008). 'Microtubule dynamics in cell division: exploring living cells with polarized light microscopy'. *Annual Review of Cell and Developmental Biology 24*: 1–28.
Koike-Tani, M. et al. (2013). 'Polarized light microscopy in reproductive and developmental biology'. *Molecular Reproduction & Development* Jul 30. doi: 10.1002/mrd.22221.
McCrone, W.C., McCrone, L.B., and Delly, J.G. (1979). *Polarized Light Microscopy*. Ann Arbor: Ann Arbor Science Publishers.
Pye, J.D. (2001). *Polarised Light in Science & Nature*. Bristol: Institute of Physics.

Internet Resources

These two web pages give some background to polarised light, but are mainly concerned with applied optical mineralogy:
1. Generic reference page of web links: https://serc.carleton.edu/NAGTWorkshops/mineralogy/optical_mineralogy_petrography.html.
2. Professor Stephen Nelson's pages: http://www.tulane.edu/~sanelson/eens211/.

Crystal symmetry: http://www.mindat.org/article.php/2721/Determining+Symmetry+of+Crystals%3A+An+Introduction+
Polarisation of light: Leica technical datasheet: https://www.leica-microsystems.com/products/light-microscopes/details/product/leica-dm750/downloads/.
Circular polarisation animations: https://www.youtube.com/watch?v=fy7EeRSsOsY and https://www.photophysics.com/tutorials/circular-dichroism-cd-spectroscopy/1-understanding-circular-dichroism.
Elliptical polarisation animation: https://www.youtube.com/watch?v=Q0qrU4nprB0 and https://www.youtube.com/watch?v=KZz25bmTWXo.
Edmund Optics web page on circular & elliptically polarised light: https://www.edmundoptics.com/resources/application-notes/optics/introduction-to-polarization/.
SPIE guide on the construction of Nicol polarising prisms: http://spie.org/samples/PM200.pdf.
Delly, J. G. (2008). Selected topics from Essentials of Polarized Light Microscopy. http://www.jysco.com/archives/asbestos/PLM_reading_McCrone.pdf.

Notes

1. Haidinger's brushes is an entoptic phenomenon that may be seen by some people as a very faint yellowish vertical crude 'hourglass' or bow-tie shape with fuzzy ends (hence the 'brush' in the name). Many people can also see two bluish-purple dots at the constriction of the hourglass shape. The effect is due to the anisotropic nature of the visual pigments for colour in the macula.
2. The actual refraction of the e-ray is very slight indeed and is greatly exaggerated in textbooks, in order to illustrate the process. In calcite, which exhibits a very large birefringence value, −0.172, the angle through which the e-ray is refracted (which depends upon the angle of incidence of the unpolarised beam) is about 6° maximum.
3. The term '*optical* axis' refers to the imaginary straight line that passes through the centres of curvature of the lens surfaces in the microscope, whilst '*optic* axis' refers to the direction in a doubly refracting (birefringent) material that is parallel to the direction in which plane-polarised light travel with the *same* speed. The optic axis being the direction in which a ray of light transmitted through the material undergoes no double refraction.
4. Non-polarised light is also reorganised into two mutually perpendicular waves of differing velocities, but we do not see the effect because the non-polarised light is vibrating in all directions transverse to the direction of propagation and is not merely organised selectively to vibrate in one plane, as is plane-polarised light. When more than one wave exists, the phase difference between the two waves then becomes important.
5. In gases, plasma and liquids, sound waves propagate as longitudinal, or compression, waves. In solids, sound can be transmitted as both longitudinal and transverse waves.
6. Do not confuse the term 'e' (meaning extraordinary ray here) with the 'E' term for the electric field vector of light: they are entirely separate.
7. Calcite or quartz prisms, rather than Polaroid sheets, must still be used for high quality imaging (e.g. used to manufacture DIC prisms). They have much higher extinction factors (e.g. 10^6:1) than sheet polars ($\approx 5 \times 10^2 : 1$) because the molecules in synthetic Polaroid are not oriented as perfectly as the ions in a good specimen of crystal. The extinction factor is a measure of the perfection of extinction with crossed polars, expressed as a ratio: $X_{\parallel}/X_{\perp}$ where the X values are the irradiance (intensity) measured between parallel and crossed polars, respectively (see Section 14.4). The greater the ratio, the better the extinction factor. Because calcite is doubly refractive, and does not depend upon pleochroic absorption, it is much less prone to heat damage. A heat-damaged Polaroid filter will characteristically show a central yellow 'scorch' mark (see Figure 13.3). All modern DIC prisms are made with quartz.
8. The two other principal forms of the scattering of light are Mie and Tyndall scattering. For particles whose size is the same or greater than the wavelength of light, Mie scattering predominates. Clouds appear white because of the wavelength-independent Mie scattering of light by suspended water droplets. The beautiful aquamarine colour of the lakes in the Canadian Rockies is due, in part, to Tyndall scattering from the suspended particles forming a colloid in the water.
9. Strictly, this must be the surface of a transparent isotropic material. However, light reflected off any dielectric, or non-electrically conductive insulator, will be polarised. Hecht (*Optics*, 4th ed. 2001) mentions the sheen of a billiard ball, book jacket or a balding head!
10. Birefringence can be measured directly by calculating the difference between the two refractive indexes, but this is less accurate than measuring the specimen thickness and the retardation and calculating the unique birefringence. The degree of retardation is measured using a compensator, and this is explained in Chapter 14, Section 14.12.
11. As with aberration correction, rectified optics use a 1/2-wave compensating plate with an air or glass meniscus to introduce an opposite rotation error into the optical path. The use of a mica Brace-Köhler compensator provides a high-extinction background in order to see clearly weakly polarised features, with very low birefringence [see section III.10 of Appendix 3, pages 493-499 in Inoué (1986) *Video Microscopy*, New York: Plenum Press].
12. When two coherent rays are derived from a common source, they interfere. However, if these rays are polarised at right angles to one another they will not interfere, even if they are brought to a focus in a common plane. Yet if these rays are polarised and *parallel* to one another, they will interfere in the normal way.
13. A right-handed circular polarising filter will not pass left-hand polarised light, and vice-versa.
14. Although pleochroism is a generic term, it is often used synonymously with trichroism specifically to indicate the phenomenon occurring in biaxial media.
15. The electric field vector of the e-ray oscillates (i.e. is polarised) in a plane parallel to the optic axis of the material, whilst the electric field vector of the o-ray is polarised in a plane perpendicular to the optic axis.
16. Before the 20th century Helgustaðir was virtually the only source of optically-pure calcite. The 'quarry' was in fact a cavity within the Tertiary basalt bedrock about 11 m by 5 m in area and approximately 3 m deep. When first discovered, it was filled with a pure crystallised mass containing enormous crystals, some reported to be a metre across. This source is now exhausted and a protected locality. Modern deposits of optical calcite come now from Brazil, Mexico, Africa and Siberia.
17. For our purposes it is sufficient to refer to calcite crystals. Strictly the rhombohedral form of calcite is not a true crystal, but rather a cleavage fragment. Calcite is said to have rhombohedral cleavage, because each of its faces is a rhombus, in which none of the corners are square. When it forms true crystals, calcite takes platy or spiky shapes that give it the common name 'dogtooth spar'.
18. Although the double refraction of some other minerals is greater than that of calcite (the birefringence of calomel is +0.683, of cinnabar is +0.347, of rutile is +0.296 and silver nitrate is −0.251), yet this double refraction is best demonstrated in calcite, since it is a mineral obtainable in reasonably large pieces of perfect transparency.
19. For a detailed explanation of Huygens' method, see Buchwald (2007) and Huygens (1690), *Treatise on Light*, Chapter 5, from page 32 at http://www.gutenberg.org/ebooks/14725.
20. The beam of light travels in a direction parallel to a line connecting the origin of each wave with the tangent point of the planar envelope of the wavefront. This is known as the ray direction. In an anisotropic material, the ray direction of the extraordinary ray is *not* normal to the wavefront. A plane wave will be displaced sideways (by a small amount) as it transverses the crystal. The maximum divergence in calcite between the o-ray and e-ray is 6.2°.

14

Polarised-Light Microscopy: Part 2 – Applied

Outline

14.1 Introduction
- Importance of polarising microscopy
- Specimen types: crushed material, grains & thin sections

14.2 The Polarised-Light Microscope
- Converting a bright-field microscope
- Setting up a transmitted-light polarised light microscope
- Positions of polariser and analyser

14.3 The N-S/E-W Convention for Polars
- Explanation of the convention

14.4 Polaroid Filters
- Polaroid sheet
- The extinction factor

14.5 Objectives and Eyepieces for Polarised-Light Microscopy
- Strain-free objectives
- Inscriptions
- Polarised-light eyepiece designs

14.6 Centring the Microscope
- Centring the stage and/or objectives
- Procedure for centring the stage
- Remarks on centring objectives

14.7 Examining Doubly Refracting Materials with a Single Polar
- 'Twinkling'
- Pleochroism & dichroism explained

14.8 Optical Path Difference
- Phase shift and birefringence explained

14.9 How Optical Path-Length Differences (OPD) Give Rise to Colour
- Optical path differences with Sellotope

14.10 Newton's Colour Scale
- How polarisation colours are formed
- Relative retardation
- Magenta 'sensitive-tint'
- The Michel-Lévy birefringence colour chart

14.11 Polarised-Light Contrast in Reflection
- Newton's rings
- Surface films on metals

14.12 Compensators
- Fixed plate and wedge compensators

14.13 How Compensators Work
- Addition or subtraction of optical path difference

Understanding Light Microscopy, First Edition. Jeremy Sanderson.

14.1 Introduction

The theoretical principles explained in the previous chapter have very practical applications. Polarised-light microscopy (PLM) offers a fast way of identifying and analysing minerals and other anisotropic materials, such as asbestos fibres. Fibrous materials can be identified on the basis of morphology by PLM, which cannot be done by chemical analysis using more modern (e.g. X-ray diffraction) techniques. The polarising light microscope has contributed more to our knowledge of minerals and rocks than any other single instrument. The identity of a mineral can be found relatively quickly by immersion of grains, thereby calculating refractive index changes. It can also be elucidated by the petrological study of thin sections. Microstructural relationships can be viewed directly, or else inferred by calculating the optical class using birefringence and compensation techniques. Furthermore, the polarised-light microscope still plays an important role in quality control of industrial and civic engineering materials. For example, ore minerals can be identified using reflected polarised light. Although some academic institutions have closed geological and petrological teaching courses (due to the higher per capita cost of training students arising from fieldwork trips) the applied use of polarised-light microscopy undoubtedly continues to have a significant impact upon our modern society.

There are several sample types for polarising light microscopy, but three predominate:

- crushed bulk material grains;
- individual crystals or grains; and
- thin sections.

Crushed mineral or crystal grains can be immersed in mounting media of different refractive index. This immersion method employs the Becke line method, explained in Chapter 11, Section 11.9. Usually grain sizes about 100 µm in diameter are examined. These may be obtained by collecting crushed material that will pass through a 100-mesh sieve with holes about of 150 µm diameter, but which will not pass through a 200-mesh sieve with 75 µm holes.

Individual crystals or mineral grains can be observed with a spindle stage. This allows precise measurement of refractive index and study of crystal morphology (see Bowen, 2009, as an example). The spindle stage is a very versatile instrument, allowing retardation and orientation of the optical indicatrix also to be determined. This stage, allowing rotation in two dimensions about a single axis, is a simplified version of the 4-axis universal stage (see Box 29.3). The definitive work on the spindle stage is Bloss (1981). In addition, Mickey Gunter has made excellent teaching and research material available; see the Internet Resources at the end of the chapter for further details. Although very little used nowadays, the universal stage permits minerals and crystals in thin sections to be oriented and their properties analysed. The URL links at the end of the chapter give details of the history and use of the universal stage.

Rocks and other optical mineralogical samples may also be studied as thin sections. The advantage of a thin section over individual grains is that the morphology and inter-relationship of the constituent minerals and crystals can be studied. The standard thickness of a thin section is 30 µm, which assists in calculating retardation and identifying components within the section. The identification of rocks and minerals in thin section using the petrographic microscope is both fascinating and rewarding. Although this subject is beyond the scope of this book, those readers interested in applied petrography are referred to MacKenzie & Adams (1994) and the mineral identification key published by Barker (2010) for further guidance. Raith et al. (2012) have published an excellent online text on *Thin Section Microscopy*; the URL link may be found at the end of the chapter. The best text on mineralogy is *Mineralogy and Optical Mineralogy* by Dyar et al. (2008). Both of these texts are available on the Mineralogical Society of America website (minsocam.org).

It is a relatively simple matter to adapt a microscope designed for bright-field contrast into one also capable of polarised-light microscopy. Custom-designed polarised-light microscopes are usually more sophisticated, with additional features that make it an easy matter to secure good results. The next section discusses the features that should be found on a good polarised-light microscope. The remainder of the chapter discusses how characteristic polarisation[1] colours arise from doubly refracting specimens and how these colours may be analysed with the polarised-light microscope and accessory tools, such as compensators.

14.2 The Polarised-Light Microscope

To convert an ordinary microscope into a polarised-light microscope we need to place two polarising filters, or polars, (usually pieces of Polaroid) into the light path. One is situated before the specimen and one after. These are oriented with their planes of polarisation at 90° relative to one another – called the 'crossed' position. In a polarised-light microscope (Figure 14.1) the first polarising filter, called the polariser, is situated just below the condenser (usually on a swing-out fitting) with its plane of polarisation, or privileged direction, set East-West (E-W) and the second polarising filter, called the analyser, is situated in the body-tube above the objective lens, again preferably in a fitting which allows it to be slid in and out of the light path. On more sophisticated polarising microscopes, the analyser can additionally be rotated. In the crossed position the plane of polarisation of the analyser is set North-South (N-S). There is no effect on the polarisation colour of a transparent anisotropic[2] material if these polars are reversed: each filter may equally take the position of polariser or analyser.

A basic polarising microscope will have only a polariser and analyser. It is helpful to have a rotating stage, because anisotropic materials change their appearance when rotated and viewed between crossed linear polars. In addition, it is useful to have a slot for inserting compensators (see Section 14.12) as well as a Bertrand lens for inspecting a magnified image of the back focal plane of the objective. The Bertrand lens acts together with the eyepiece to form a low-power microscope. It should be focusable since the focal lengths of different objectives will vary. The Bertrand lens shows the conoscopic interference figure (see Section 14.17) of a crystal in the specimen, which can be selected from among other crystals by means of a stop or iris below the condenser which prevents other crystals being illuminated.

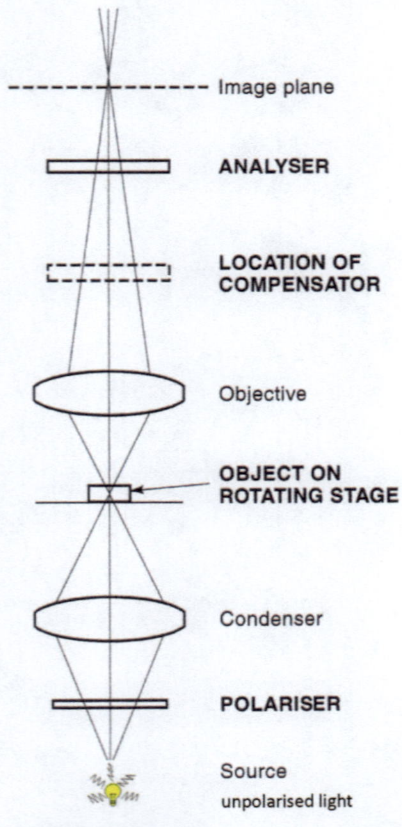

Figure 14.1 **Parts of the polarised light microscope** *Source: Murphy 2001. Reproduced with permission from John Wiley & Sons.*

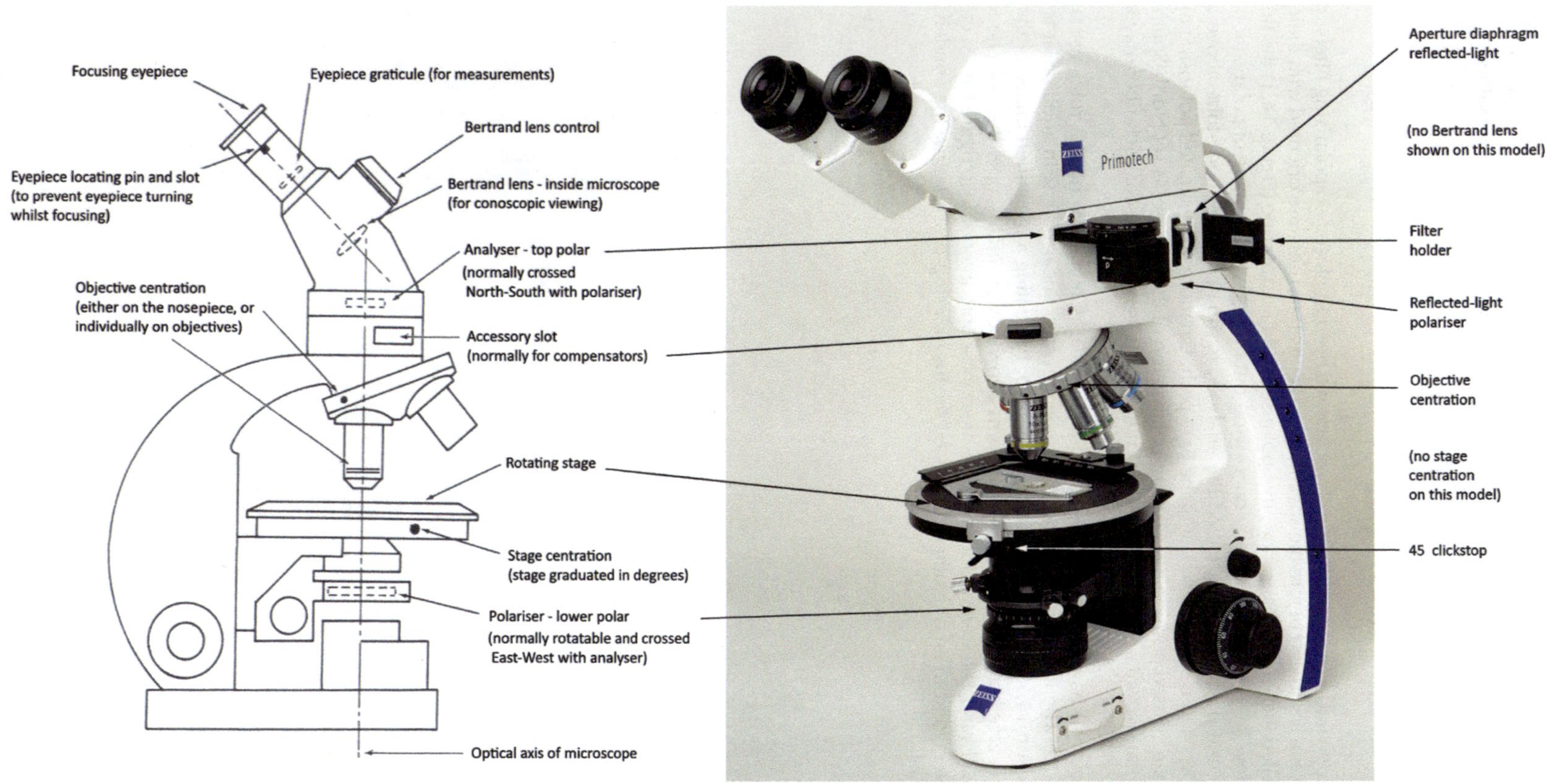

Figure 14.1 *(Continued) Source: (left) Robinson and Bradbury 1992. Reproduced with permission from Oxford University Press. (right) Courtesy of Mr K. Glover.*

Protocol 14.1 Setting Up the Transmitted-Light Polarised-Light Microscope

1. Set up the microscope for transmitted-light bright-field Köhler illumination.
2. Insert a polariser oriented E-W into the light path.
   (This polariser may either be captive and preset (or rotatable) in a swing-in holder, or else a loose filter, to be introduced manually into the light path)

Note: The minerals biotite and tourmaline are strongly dichroic. They can be used to check the E-W alignment of the polarising filter with respect to the eyepiece cross-hairs: Biotite is darkest when aligned parallel to the direction of the lower polar (see Figure 14.5, Section 14.7).

3. Insert an analyser N-S into the light path.
   (The analyser may also have the means to be rotated laterally with respect to the polariser)
4. Either the polariser and/or analyser should be rotatable. Ensure that they are crossed to obtain maximum extinction[3]. This adjustment can only properly be done whilst using the microscope: it is not sufficient to set the polariser and analyser externally.
5. Remove an eyepiece and check that the two polars are fully crossed and extinction gives a maximally dark-field when looking at the back focal plane of the objective.
6. Either use a centring telescope or (if fitted) swing in the Bertrand lens into the optical axis. An extinction ('polarisation') cross is seen at extinction when no object is present.
   This cross, as seen in the back focal plane of the objective, is shown in Figure 14.2. The shape is due to depolarising effects of lenses in the condenser and objective (for an explanation, see Section 13.6). The 'arms' of the cross should be oriented E-W and N-S; in other words, the cross should be 'upright'. If not, the polariser and analyser may need to be adjusted slightly[4]. Close the condenser illuminating aperture iris slightly to improve extinction. Beware that, as when adjusting the microscope for Köhler illumination, closing the condenser aperture diaphragm too much will adversely reduce the resolving power of the microscope.

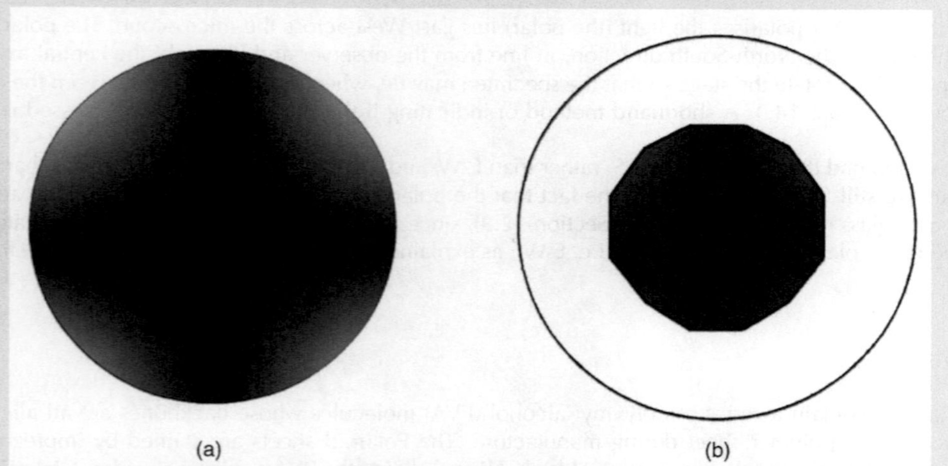

Figure 14.2 **The extinction cross** *Source: Adapted from Murphy 2001 with permission from John Wiley & Sons.*

7. Locate the slot at 45° for inserting a compensation plate.

It is not satisfactory merely to place the analyser over the eyepieces or insert it in any position above the binocular head, because of the complex depolarising properties of the prisms in the viewing head. This leads to a loss of extinction across the field of view. The analyser should be placed between the objective and the eyepieces, usually in a custom slot. The conventional place for a polariser is either on top of (transmitted-light path) or in front of (reflected-light path) the illuminated field diaphragm, or in a filter holder on the substage apparatus just underneath the condenser. If the polariser is placed on top of the illuminated field diaphragm in the base of the microscope, it should be clean, scratch free and of good quality; otherwise, imperfections in, or on, the polar will likely be imaged into the specimen plane. A better place to insert, or hold, the polariser is in the substage filter holder just below the condenser. In this position, the polarising filter is well away from the imaging (field) set of conjugate planes; if anything, it is situated nearer the illuminating (aperture) set. Slight dirt, marks or scratches will not then be projected into the image plane.

14.3 The N-S/E-W Convention for Polars

The polariser and the analyser are crossed at 90° with respect to one another, but in order for this to have any meaning with respect to the specimen, both the polariser and analyser must be set in defined positions. It is accepted by convention (DIN standard) that the

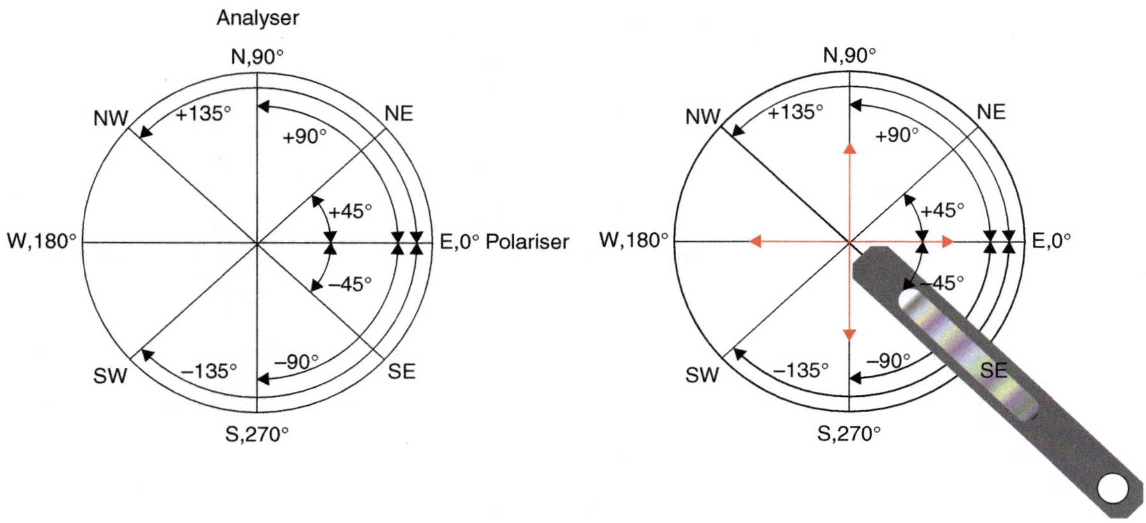

Figure 14.3 **Relationship of polariser, analyser and accessory plates** *The polariser lies East-West whilst the analyser is crossed North-South. These privileged directions are shown in red. Accessory compensator plates are inserted at 45°, as shown with the quartz wedge. Source: Wayne 2009. Reproduced with permission from Elsevier. (Image) Courtesy of Mr K. Glover.*

transmission axis of the filter that polarises the light (the polar) lies East-West across the microscope. The polar that recombines the light (the analyser) then lies in the North-South direction, in line from the observer and 'through' the central axis of the microscope. Ideally, it should be possible to rotate the stage so that the specimen may lie, when required, at 45° between the crossed polars. These relationships are shown in Figure 14.3. A shorthand method of indicating if the polarising filters are crossed is to write $X_{\perp}$, and for non-crossed filters, $X_{\parallel}$.

Earlier microscopes often had the polariser set N-S, rather than E-W, and the analyser likewise E-W rather than N-S. This would not matter (for both polars are still 'crossed') except for the fact that the polarised light would be orientated N-S, and be reflected much less efficiently by a cover-glass reflector (Chapter 12, Section 12.3), since the reflection of polarised light is greatest when its vibration direction is orientated in the plane of the cover glass (i.e. E-W), as explained previously in Section 13.4, on the formation of polarised light by reflection.

14.4 Polaroid Filters

Sheets of Polaroid material contain a series of polyvinyl alcohol (PVA) molecules whose backbones are all aligned in one direction by heating and stretching the polymer sheet during manufacture. The Polaroid sheets are stained by impregnation with dichroic[5] polyiodide micelles (aligned by electrical or magnetic fields) that lie parallel to the PVA carrier molecules. Dichroism means a variation in absorption coefficient with the direction of polarisation of light. The term is widely misapplied to the splitting of light according to wavelength by interference coatings, but this usage is too well-established to change. In this case, dichroism has the latter meaning; polyiodide micelles absorb light that is polarised parallel to the direction of the crystal alignment (i.e. the Polaroid filter goes dark) but transmit light polarised perpendicular to the filter. In good-quality microscopes, there should also be the means to rotate this polar. If you purchase your own polars, use HN32 Polaroid sheet, with a high extinction value. Polaroid can be improved optically by encasing the sheet in glass with a mounting medium of high refractive index. A cheap source of excellent encased anti-reflection coated filters is the photographic camera trade.

Extinction occurs when the magnitude of light transmission through crossed polars drops to a minimum. It describes the condition that occurs when a transparent optically anisotropic, doubly refracting, material appears dark when viewed through these crossed polars. The extinction position refers to the orientation of such an anisotropic material, such that it shows extinction. The extinction factor[6], or extinction ratio, is a measure of the perfectness of extinction between crossed polars, is calculated from the ratio of intensities $X_{\parallel}/X_{\perp}$, where $X_{\parallel}$ is the intensity recorded through parallel polarising filters, and $X_{\perp}$ the intensity through crossed polars. Polars are not perfect blockers of polarised light. The ratio of (unwanted) light passed to the (desired) blocked light is called the extinction ratio and varies from around 500:1 for Polaroid to about 10^6:1 for calcite Nicol polarisers of the modern Glan-Thompson design. A well-designed polarising microscope should have an extinction ratio, or factor, of about 1000:1. Rectified optics gives an extinction factor circa: 13 000:1 instead of 500:1.

14.5 Objectives and Eyepieces for Polarised-Light Microscopy

It is possible to introduce strain into the optical cement at the interface of the glass lens elements of microscope objectives during assembly and alignment in the body of the objective. If this happens, the strained objective will act as a 'compensator' introducing

Figure 14.4 **Objectives for polarised light microscopy** *It is important to select 'strain-free' objectives, with minimal or no residual strain in the lens elements, for polarised light microscopy; otherwise, the strained objective will act as a 'compensator', introducing an unknown degree of birefringence into the image. These are both dry semi-apochromatic objectives. The 40x one on the left is designed for use with a coverslip, the 50x one on the right for uncovered samples. Source: Courtesy of Mr K. Glover.*

an unknown degree of birefringence into the image. All manufacturers offer objectives that have been manually inspected and put to one side during manufacture to be offered as 'strain-free'. For most high-quality microscopy, apochromatic objectives are usually specified. However, these necessarily have a high number of internal lens elements. Not only does this increase the likelihood of any one lens not being strain-free, but these relatively large number of lens elements are also responsible for rotation of the polarisation direction (Section 13.6), which reduces the extinction factor of crossed polars. Polarisation by reflection will also occur at the highly curved surfaces of some lens doublets in objectives. This can affect and reduce the extinction achieved under polarised-light microscopy. For these reasons, achromat or semi-apochromat objectives, and not apochromats, are most often used for polarised-light microscopy.

By international standard, objectives designated for polarised-light microscopy have their inscriptions in red on the barrel (Figure 14.4) and are designated Pol, POL or merely P. The numerical aperture associated with polarised-light objectives of a given magnification is often higher than for biological microscopy because conoscopic viewing of interference figures in the back focal plane of the objective requires high numerical apertures.

Strain-free condensers and eyepieces that have been similarly selected can also be found on high-quality polarised-light microscopes. Even on a microscope configured to the most basic specification for polarised-light microscopy, it is necessary to have a focusing eyepiece fitted with cross-hairs on a graticule in the primary image plane. This helps in adjusting the centration of the objectives and/or stage onto the central optical axis. It also permits interfacial and cleavage angles to be measured in thin sections. Ideally there should be a small slot in the microscope body-tube for a pin or screw on the eyepiece to engage so that the eyepiece is prevented from turning, thus keeping the cross-hairs in registration with the crossed polars.

14.6 Centring the Microscope

Routine alignment of the microscope according to Köhler's principles is an integral part of adjusting the microscope optimally. This is the first step that should be performed (without either the polariser or the analyser inserted into the optical pathway). For most biological and biomedical science studies, this is all that may be required. However, for materials science, particularly those studies involving polarised-light microscopy, the ability to rotate the stage about a central axis common to all the optical components is important. This ability allows you to observe extinction or pleochroism without the sample disappearing out of the field of view. When determining the optical class of crystals and minerals, it is crucial that the sample remains within the field of view when the sample is rotated on the stage.

Polarised-light microscopes should (and often do) have the means to centre the microscope (Figure 14.5) so that the sample can rotate stably around the optical axis. This is usually done *either* by centration of the rotating stage *or* (less commonly) by centring the

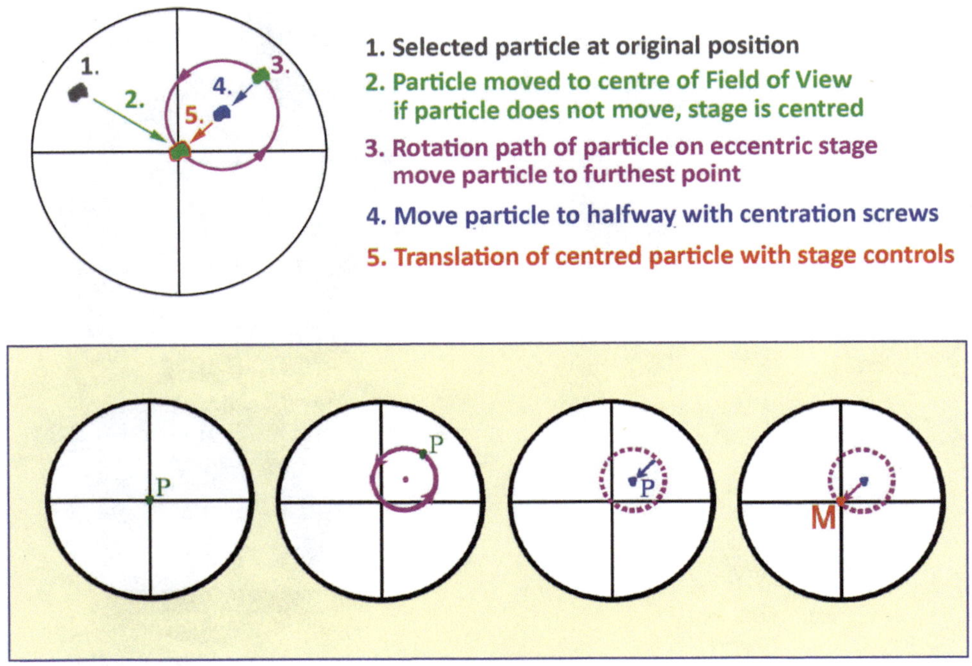

Figure 14.5 ***Centring a polarised light microscope*** *Instructions on how to centre the stage on-axis are given in Protocol 2. Source: Raith et al. 2012. Reproduced with permission from Professor Raith, Dr Raase and Dr Reinhardt.*

objectives on-axis. A rotating stage will either have square or hexagonal lugs set at 120° and bearing on a third, fixed, point, requiring a key to rotate them. Otherwise the centring controls will have a recess for an Allen key. Objectives will either have rotating rings built into the body of the objective or else two small machine screws or grub screws requiring Allen keys on the nosepiece adjacent to the lens mounting hole or on the lens collar of the nosepiece.

The method of making the centre of rotation coincident with the eyepiece cross-hairs is the same, regardless of whether it is performed on the rotating stage or the objectives. Some microscopes may have the means for adjusting for both the stage and the objectives. If your microscope has the means to centre both the stage and the objectives, adjust the stage first, using the highest magnification objective. Proceed as follows:

Protocol 14.2 Centring the Objectives or Rotating Stage

1. Unlock the stage, and identify the centring controls set at 120°. Insert the appropriate key to control their rotation.
2. If necessary, swap the eyepiece for your dominant eye (see Box 9.5) for one with cross-hairs.
3. Focus these cross-hairs, so they are imaged sharply in the primary image plane.
4. Locate a prominent particulate feature in the field of view. Otherwise use a dedicated centring slide for this purpose with a cross-hair marked on it. Focus the particle or cross-hairs of the centring slide. Move the cross-hairs on the particle/slide laterally, to coincide with the eyepiece cross-hairs.
5. Unlock the stage and carefully rotate it. The particle will most likely move, describing a circular off-centre path. If it is centred and rotates whilst remaining stationary, leave well alone, and test the rotation with a second objective.
6. If the particle moves off-centre, rotate the stage until it has moved to the furthest off-centre position. With a centring slide, do the same with the intersecting point of the cross-hairs.
7. Use the stage centring controls to move the particle *half way* along an imaginary line connecting the particle to the intersection of the eyepiece cross-hairs, at the centre of the field of view, on the optical axis. This should be the centre of rotation (see Figure 14.5). Do not move the particle all the way along this line so that it is coincident with the eyepiece cross-hairs at this point.
8. Leave the centring controls alone; now use the lateral controls of the stage to shift the particle back to the centre of the field of view coincident with the eyepiece cross-hairs.
9. If the particle moves so much that it will move out of the field of view with an entire rotation, estimate the centre of rotation, and use the centring controls to move the particle to this point.
10. Repeat the process until the particle remains stationary (or near enough) about the centre of the field of view.

The centration of each objective on the optical axis may well change slightly from one objective to another. If this is the case and the objectives are not parcentral, use the highest magnification objective (i.e. smallest field of view) to perform the most accurate stage centration. Greater precision can be achieved if each of the objectives can be centred independently. Once the stage is centred, proceed with each objective, from highest magnification to the lowest.

14.7 Examining Doubly Refracting Materials with a Single Polar

If the analyser is removed from the optical path of the microscope, two different effects will be noticed when viewing transparent doubly refracting specimens. In the first case, a twinkling effect can be seen when a highly anisotropic sample, such as calcite grains in a rock thin section, is rotated rapidly over a single polariser. This is due to markedly different refractive indexes exhibiting variations in visibility as the crystal grains rotate. Let us assume that the refractive index associated with the fast permitted vibration direction (the lower RI) of the sample is the same as the mounting medium by which the sample is surrounded and possibly infiltrated. When the permitted vibration direction of the sample is parallel with the vibration direction of the polarising filter, it will transmit the plane-polarised light, and the contrast of the sample will be at a minimum. The image of the sample will be invisible or almost so. When the permitted vibration direction of the sample is orientated at 90° to the vibration direction of the polarising filter, then the possible contrast in the image will be at a maximum.

In the second case, pleochroism can be observed where birefringence is associated with the *selective absorption* of transmitted light in one vibration plane. The sample will absorb light, and thus the image will change in intensity and/or colour when the vibration direction of the polarising filter is parallel to the permitted vibration direction of the sample, fast or slow, which is associated with this behaviour (Figure 14.6). There may be more than one tint exhibited if the anisotropic sample is biaxial because biaxial materials have three possible vibration directions. Uniaxial materials have two vibration directions (the o-ray and the e-ray), display dichroism and thus show only two different states.

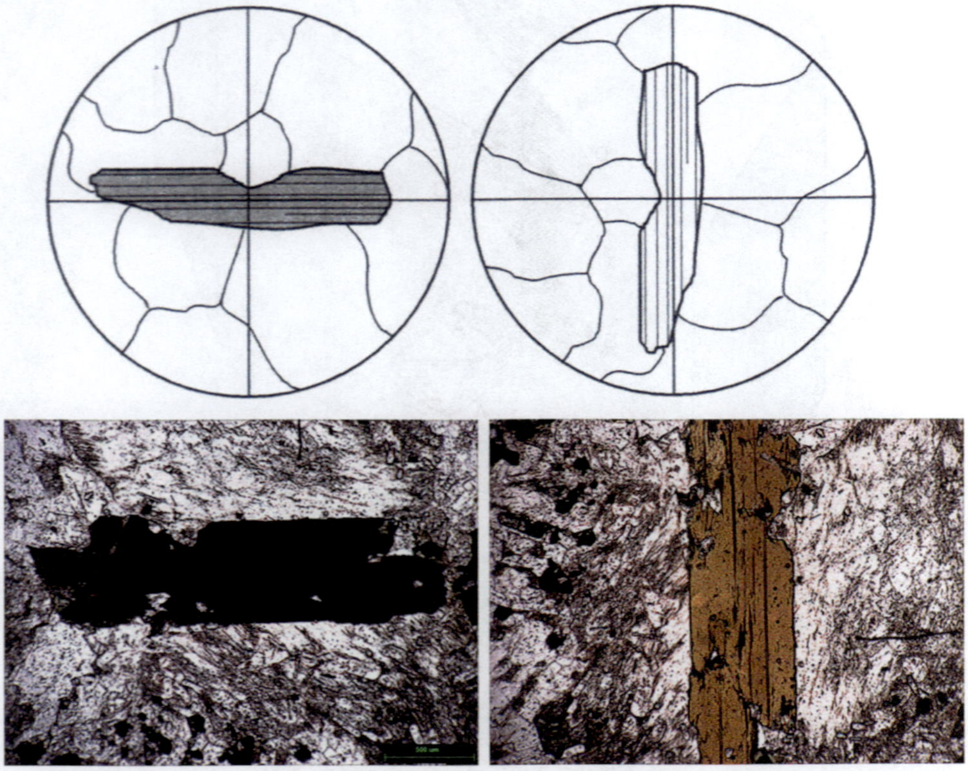

Figure 14.6 **Pleochroic behaviour under one polar** *With the analyser removed, the polariser passes light East-West. A strongly pleochroic grain of biotite (a black mica-like mineral, principally found in igneous and metamorphic rocks, such as granites, diorites, and igneous pegmatites) will be darkest when the grain is oriented with its cleavage aligned East-West, parallel to the polariser. Biotite mica shows perfect basal or pinacoidal cleavage parallel to the base of the crystal, i.e. the {001} plane in the crystal lattice. The c-axis is perpendicular to the plane of the paper. Observed with a 4x objective; field of view approximately 4.5 mm across. Source: Nesse 2009. Reproduced with permission from Oxford University Press.*

Polaroid is an extreme example of dichroism, where the aligned polyiodide micelles absorb light selectively and very strongly in one vibration direction and hardly at all in the other so that the emergent light is plane-polarised. The presence of double refraction does not usually give rise to colour effects per se (e.g. the strongly birefringent block of calcite shown in Figure 13.25 is colourless, because higher-order interference gives rise to white). However, most anisotropic materials when viewed between crossed polars, as grains or in thin section, give rise to a range of polarisation colours.

14.8 Optical Path Difference

Since the two polarised rays produced by double refraction travel through an optically anisotropic material (e.g. a crystal) at equal amplitudes but at different velocities depending upon the unique refractive indexes of each ray, there will be a lag, or shift, in the phase relationship between them when they exit the crystal (Figure 14.7; see also Table 13.1, Figures 13.5 and 13.12). For an explanation of the meaning of 'double refraction' and 'birefringence', see endnote 2 at the end of this chapter. The ray travelling in the slow direction (higher refractive index) lags behind the fast ray. This lag is the optical path-length difference (OPD) or relative retardation (Γ), see Box 14.1.

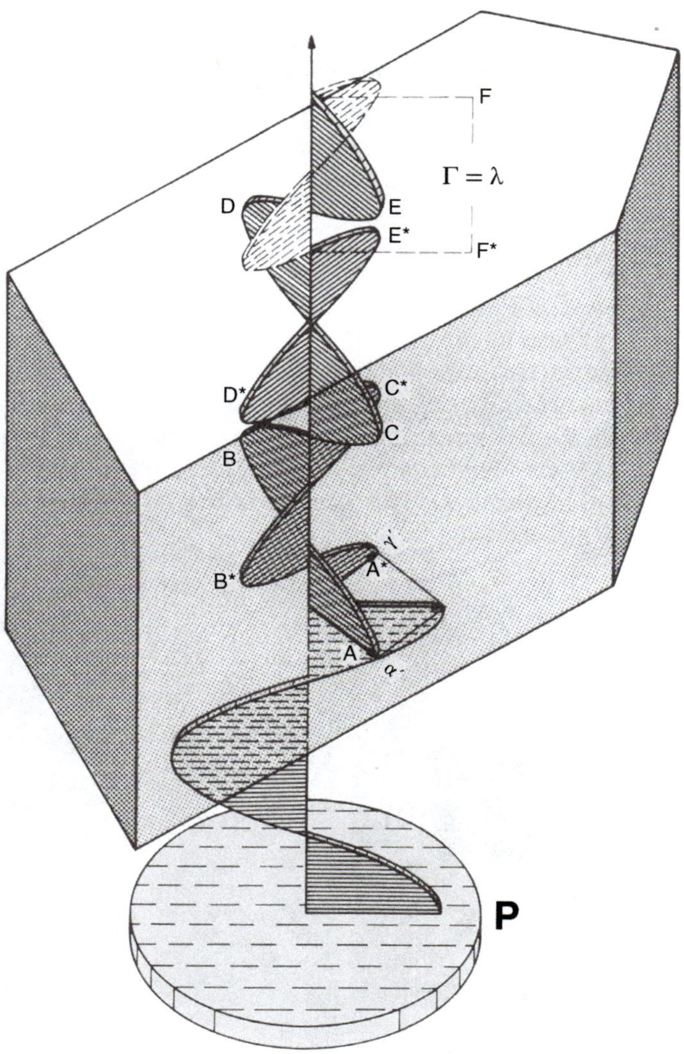

Figure 14.7 **Optical path differences by double refraction** *The incident light is split by the doubly refracting sample into two rays vibrating orthogonally to one another. When the rays leave the crystal, the lag, or retardation, between them is a function of the difference in velocity between the two rays and the thickness of the sample. In this case $\Gamma = \lambda$. Source: Adapted from Patzelt 1974. Reproduced with permission from Leitz, Leica Microsystems.*

Box 14.1 The Mathematics of Optical Path Difference (OPD)

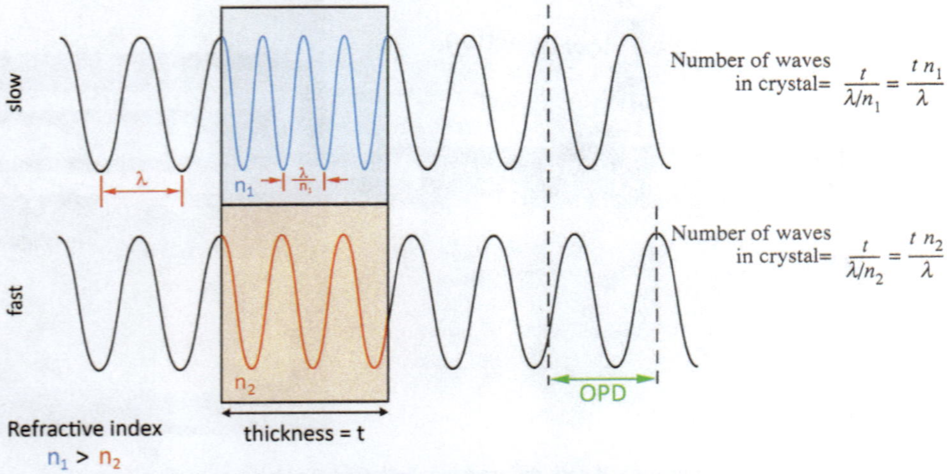

Source: Tilley 2011. Reproduced with permission from John Wiley & Sons.

The illustration shows the general case how an optical path difference arises.

Within two transparent (but not translucent) materials of the same thickness, t, the blue-coloured material has a higher refractive index (n_1) than the orange material (n_2). If $n_1 = 2 \times n_2$, then the blue-coloured material 'contains' twice as many wavelengths as the orange coloured material.

Upon emerging from the materials, the two rays will be out of step by $(|n_1 - n_2|) \times t$, where only the modulus, or absolute value, of the difference between the two refractive indexes is important (in an anisotropic material this modulus is strictly the birefringence). The 'out of step' property is called optical path difference, or OPD for short. Relative retardation is also another term used for optical path difference.

There are a number of ways of expressing optical path difference:

- as the actual distance that one wave is ahead of the other, e.g. OPD = 500 nm, as $t.(n_1 - n_2)$
- in terms of fractions or numbers of wavelengths, e.g. OPD = $1/4\lambda$, $1/2\lambda$ or 2λ, as $t/\lambda.(n_1 - n_2)$
- in terms of phase difference (degrees or radians) e.g. OPD = 90° = $\pi/2$, as $360°/\lambda\ t(n_1 - n_2)$ or $2\pi/\lambda\ t(n_1 - n_2)$

With an anisotropic specimen, retardation is a product of phase shift and specimen thickness and depends upon the difference between the two refractive indexes (i.e. the phase shift between the o-ray and the e-ray) and the thickness of the specimen. The OPD determines how the two rays travelling through the specimen recombine upon emerging and leaving the anisotropic material. From this we can explain how both isotropic thin films (e.g. oil, see Figure 14.15) and so-called 'birefringent' (colloq.) materials form non-spectral interference colours in the former case by reflection, and in the latter by anisotropy, which can be seen with a polarised-light microscope.

In a thin film that can reflect light off both its top and bottom surfaces, these light rays coming from the same source are coherent and can therefore interfere (in this case, the phenomenon is general and not merely confined to polarised light). When the OPD is such that the crests of the waves of one ray of light match the crests of the waves of the other, interference is constructive. Conversely, when crests match troughs, interference is destructive. For a film of thickness t and refractive index n, OPD = $1/2\lambda$ + 2tn.

The $1/2\lambda$ term appears because there is a phase change (i.e. a path change) of the reflection from the top air-film interface of the oil. The term 2tn arises because the wave reflected from the bottom surface of the oil traverses the film twice, and n determines its reduced velocity (change of wavelength) within the film. Hence, constructive interference occurs at OPD = $0\lambda, 1\lambda, 2\lambda, \ldots = 1/2\lambda + 2tn \Rightarrow 2tn = 1/2\lambda, 3/2\lambda, 5/2\lambda \ldots$, and destructive interference occurs at OPD = $1/2\lambda, 3/2\lambda, 5/2\lambda, \ldots = 1/2\lambda + 2tn \Rightarrow 2tn = 0\lambda, 1\lambda, 2\lambda \ldots$

For anisotropic materials (e.g. a strip of Sellotape®, Figure 14.8) the OPD is equal to the product of the birefringence and thickness: $(|X_{\parallel} - X_{\perp}|).t$ where $X_{\parallel}$ and $X_{\perp}$ are the refractive indexes parallel to (higher RI), and perpendicular to (slower RI) the length of the strip of tape respectively, and *t* is the thickness. The resulting polarisation colour is determined by the full white-light spectrum minus the colour of the wavelength that has undergone elimination by destructive interference. This is the basis of Newton's colour scale.

Reference

Tilley, R.J.D. (2011). *Colour and the Optical Properties of Materials*, 2nd ed. Chichester: Wiley. ISBN: 978-0-470-74695-0.

Rutile, which is a uniaxial positive crystal, has one of the highest birefringence values (+0.287) known. Calcite – a component of standard compensators – which is a uniaxial negative crystal, also has a high value (−0.172 = 1.4864 − 1.6584), but quartz[7], a uniaxial positive crystal, has a very low value (+0.0091 = 1.5534 − 1.5443), which makes it useful for precise retardation measurements.

14.9 How Optical Path-Length Differences (OPD) Give Rise to Colour

Consider an anisotropic material of uniform thickness – a strip of Sellotape, or Scotch tape®, is a readily available example. In the case of Sellotape the refractive index for light that is plane-polarised along the length of the strip is greater than the refractive index for light plane-polarised perpendicular to the length of the strip. Therefore the slow direction is oriented along the length of the Sellotape strip

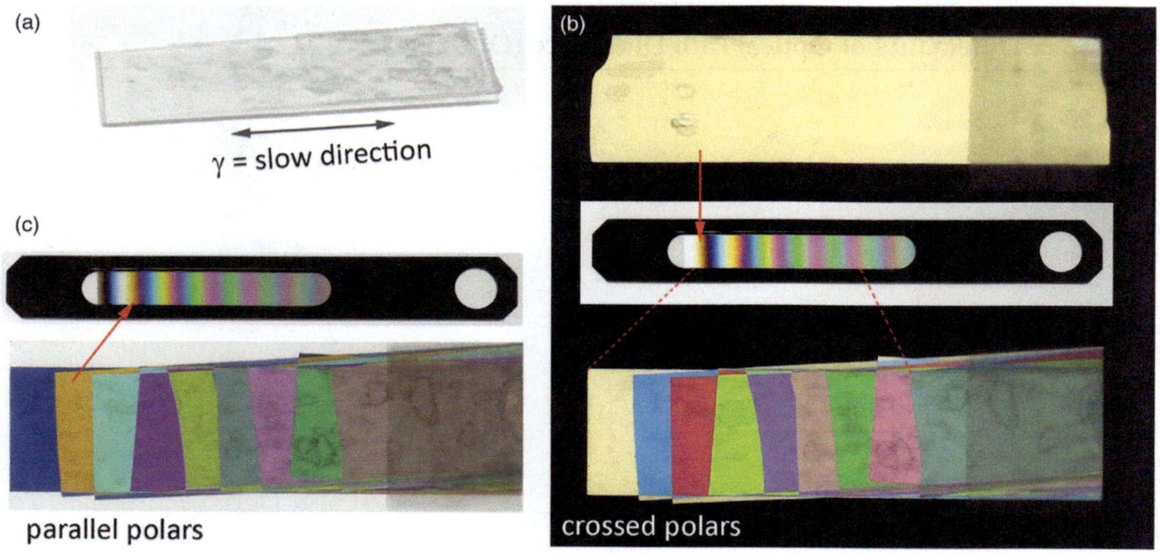

Figure 14.8 **Colour formation by optical path difference in Sellotape (Scotch tape) strips** *A single piece of Sellotape, or Scotch tape, when viewed at 45° between crossed polars on a light box, will show a straw yellow or greenish white colour. The slow direction of the tape is aligned with the long axis of the tape (a) as it comes off the reel. From Exercise 19, the OPD of a single strip of tape is between 220 and 280 nm. The tape acts as a half-wave plate for blue light, which is therefore blocked by the analyser, giving the characteristic straw yellow colour.*

Figure (b) shows several pieces of tape stuck on top of one another, all with the slow axes aligned, to form a wedge. Successive polarisation colours are formed, and these approximate a quartz wedge compensation plate. (The diffuse area at the right-hand end of each slide is the frosted end of this design of microscope slide.)

Figure (c) shows the polarisation colours formed by both the Sellotape wedge and the quartz compensator wedge when viewed between parallel polars (i.e. the analyser is rotated by 90°). These colours are complementary to those seen between crossed polars. Since the analyser is rotated, the o-ray and e-ray, which constructively interfered to produce the yellow colour in the single strip of tape, now destructively interfere to give a darker colour, complementary to straw yellow.

The polarisation colours seen with a quartz wedge out of the microscope are those for reflection. When the analyser is turned to a parallel setting, the monochromatic dark bands exchange places with the light bands, and the polarisation colours are now for Newton's scale in transmitted light. When viewed with the reflected-light box, the polarisation colours shown are those for transmitted light in *the microscope. Since the tape has a retardance, or OPD, of about 250 nm, the second stripe (i.e. two pieces of tape) shows a polarisation colour of 500 nm, just below that of first-order red (546 to 550 nm). Source: (b,c) Courtesy of Mr K. Glover.*

and the fast direction across the width of the strip of tape. Figure 14.8 shows the optical path difference between them as they emerge from the strip: a single strip of tape will be coloured green-grey or straw yellow between crossed polars, depending upon the thickness of the tape.

We are now in a position to understand why the polarisation colours are brightest at the 45° position and fade to extinction once in every 90° of rotation. At 45° the two components have *equal amplitudes* and, as in the case of light reflected equally from the top and bottom surfaces of thin films, the interference effects are maximised; this is constructive interference. Correspondingly, if the Sellotape strip is orientated E-W, then the component plane-polarised perpendicular to the length of the strip is zero; the E-W component travels through the strip 'on its own' and is blocked by the analyser – and similarly with the converse situation if the strip is orientated N-S. Therefore when the permitted vibration directions of the tape are parallel with those of the polars, the Sellotape strip is dark and is in extinction.

14.10 Newton's Colour Scale – Another Manifestation of OPD

This can be seen in the microscope but is best demonstrated on a macro scale with a light box and two large sheets of polarising material. When polarised light passes through a transparent anisotropic material, an optical path difference will arise between the two coherent polarised light rays that are produced. If the polarised white light leaves the doubly refracting material as it entered – along either of the vibration directions – it will be extinguished, to a greater or lesser degree, by the analyser. However, at 45° the polarised white light beam is divided by the specimen into two mutually perpendicular rays orthogonal to one another, and these are out of phase by an amount (the OPD = $t.[n_1 - n_2]$) that is the product of birefringence and the thickness of the material.

These two mutually perpendicular rays have a common origin and can, therefore, interfere. A path difference of (n/2) x λ results for some wavelengths where the complementary colour has a path difference of $n\lambda$. These rays will cancel out destructively to give darkness when they are out of phase by a half wavelength, or any odd-number multiple of half-wavelengths. These two examples are shown in Figure 14.9a and Figure 14.9b.

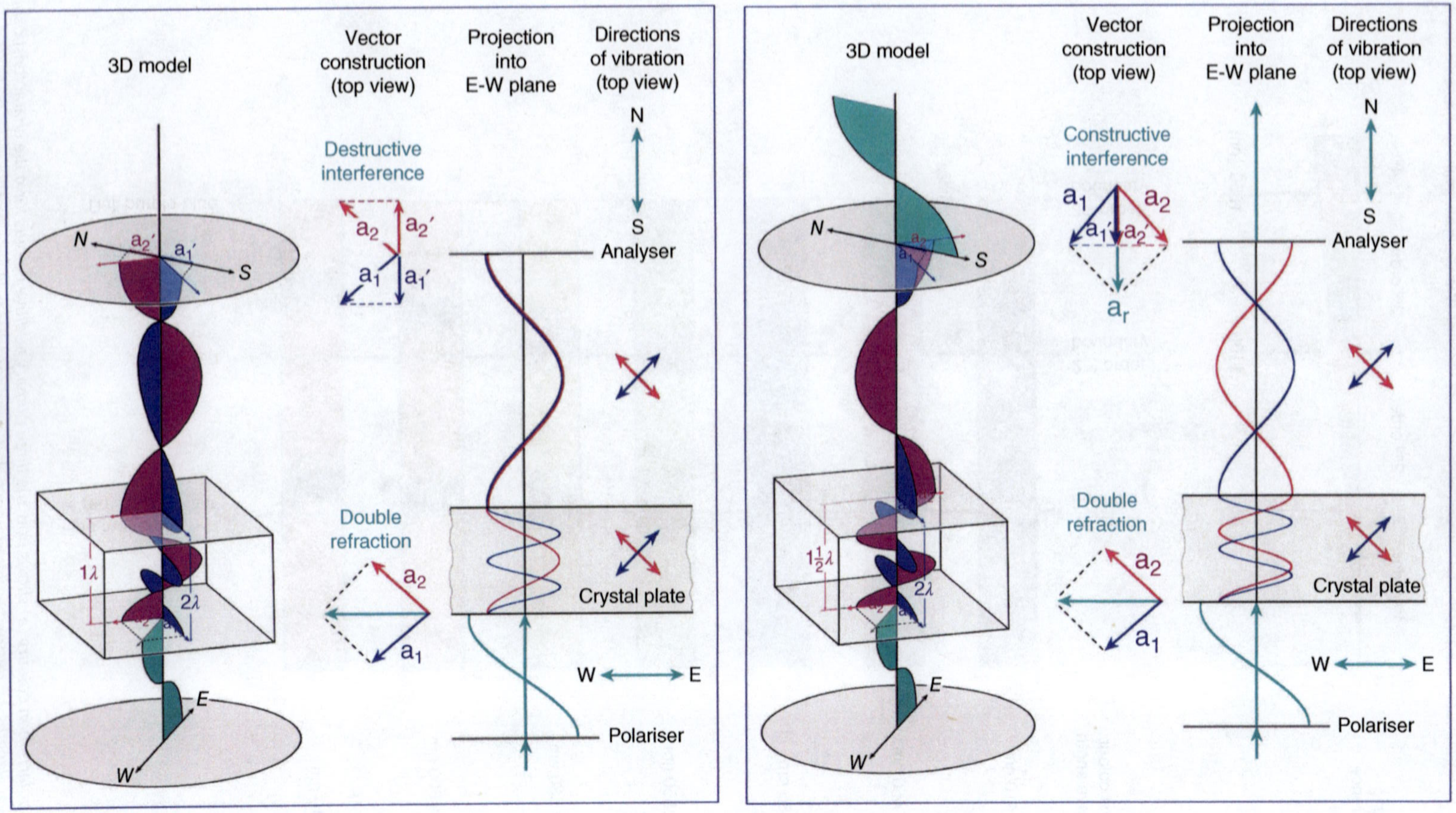

Figure 14.9 **Optical path difference and interference of light** *The left-hand panel shows destructive interference between the o-ray and e-ray (designated a_1 & a_2 in this figure), where the overall retardation is one wavelength or a whole-number multiple of λ (i.e. OPD = $n.\lambda$; n = 0,1,2,3, …). The components of the two rays encountering the analyser are of the same magnitude, but the rays vibrate in opposite directions and so cancel each other out.*

In the right-hand panel, maximum constructive interference is shown. This occurs when the o-ray and e-ray are out of phase by half a wavelength, $\lambda/2$, or if the retardation is an odd number multiple of $\lambda/2$ (i.e. OPD = $[2n + 1].\lambda/2$; n = 0,1,2,3, …). The two rays vibrate in the same direction, so they constructively interfere. The resultant is therefore able to pass through the analyser.

Source: Raith et al. 2012. Reproduced with permission from Professor Raith, Dr Raase and Dr Reinhardt.

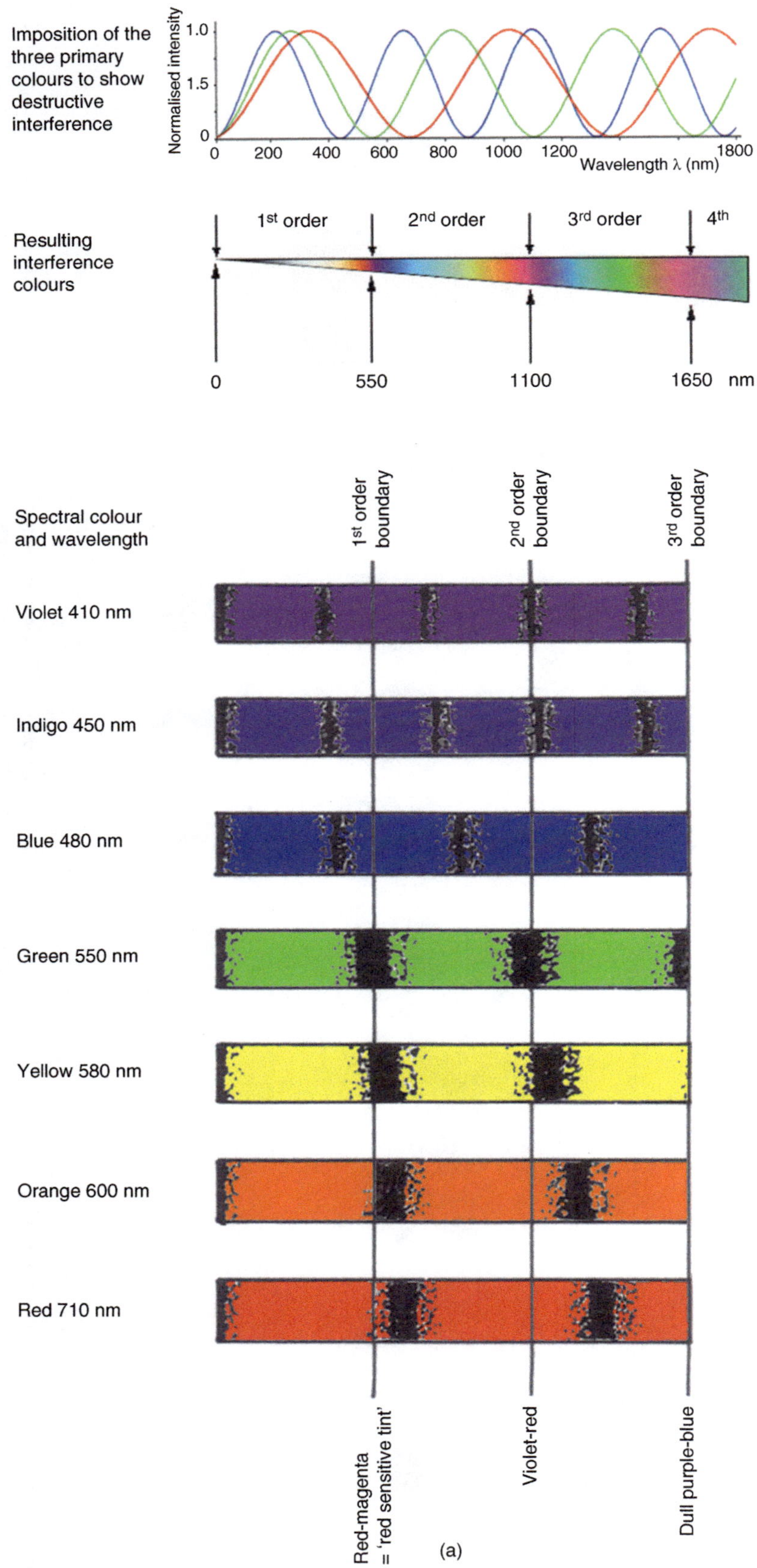

Figure 14.10 **Formation of Newton's polarisation colours** *(a) shows how the three primary additive colours form the characteristic interference colour scale with the elimination of specific wavelengths by destructive interference. If a quartz wedge is illuminated by green light of wavelength 550 nm, destructive interference occurs at optical path differences of $n\lambda$, i.e. at 0 nm, 550 nm, 1100 nm, 1650 nm and so on. This is shown for seven wavelengths across the visible spectrum.*

(b) shows the interference colours arranged in three separate orders, together with the colours extinguished by destructive interference. Compare this figure with Figure 14.11 and Figure 14.20. Source: Adapted from Robinson and Bradbury 1992 with permission from Oxford University Press.

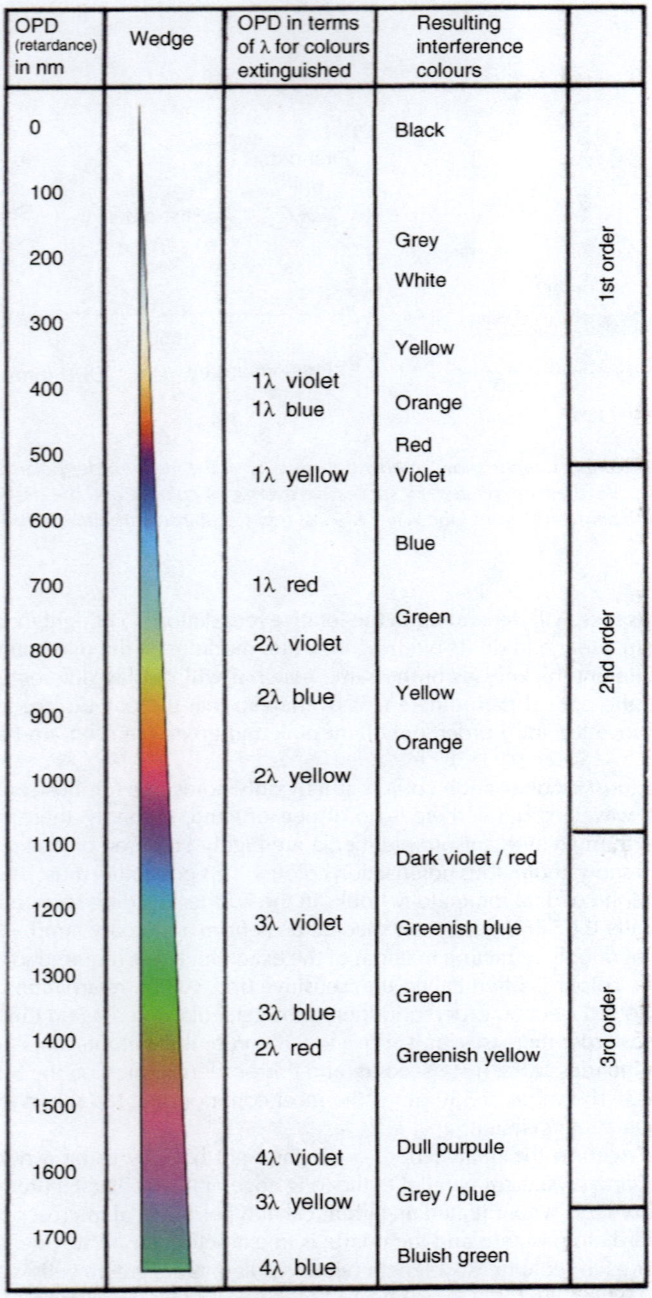

(b)

Figure 14.10 (Continued)

The resulting polarisation colour is determined by the full white-light spectrum minus the colour of the wavelength that has undergone elimination by destructive interference. This is the basis of Newton's colour scale. How these colours are formed is shown in Figure 14.10. Newton divided the sequence of polarisation colours into what he called *orders* (not to be confused with the diffraction orders), each interference colour order terminating with the distinctive red/magenta sensitive tint colour. It should be noted that the colours we see are *not* the spectral colours of the rainbow but rather mixtures of them. For example, the very characteristic and beautiful deep-red colour is a mixture of violet-blue and red light – the colours at either end of the spectrum with the intermediate yellow-green colours left out.

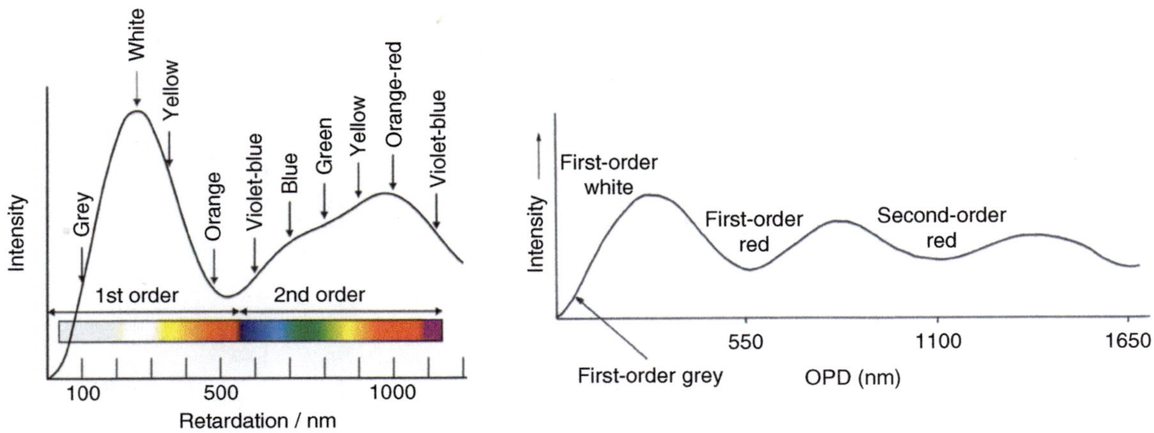

Figure 14.11 **Higher-order polarisation colours** *Graphs showing how the intensity of the higher-order polarisation colours approximates to white. As the interference colours go up Newton's scale, they are 'diluted' by successive overlap of colours, and the destructive interference effect becomes less. Source: (left) Tilley 2011. Reproduced with permission from John Wiley & Sons. (right) Robinson and Bradbury 1992. Reproduced with permission from Oxford University Press.*

Exactly which polarisation colour is seen will depend upon the relative retardation (Γ) of light through the material, and this depends upon both (i) the thickness (t) of the material and (ii) its birefringence (the modulus of the maximum difference between the refractive indexes: n_e, n_o, or n_ε, n_ω) so that different thicknesses of the same material will display different polarisation colours. As the orders increase, their intensity falls off and the optical path differences overlap so that the colours are 'diluted towards white light' and so become very faint (Figure 14.11). Above the third order, only faint pink and green are seen, and above the 10th order, only 'higher-order white' (see also Figure 14.20).

The formation of characteristic *diagnostic* polarisation colours strictly only holds true for those anisotropic materials in which, firstly, the birefringence does not vary with wavelength (i.e. there is no dispersion) and, secondly, there is no selective absorption of light at different wavelengths across the spectrum. Where anisotropic media are highly coloured or else possess high dispersion (e.g. jadeite, tourmaline and plagioclase), they will show anomalous polarisation colours. This possibility must be considered when using polarisation colours diagnostically. Refer to any of the optical mineralogy books in the Further Reading section for further details.

Green light at about 550 nm (actually the 546.1 nm green spectral peak from a mercury lamp) is used as a reference wavelength. A 1λ first-order red plate is a thin slice of doubly refracting material of the exact thickness required to extinguish green light and thus give the polarisation colour magenta. This colour is often called the 'sensitive tint'. As the retardation increases, the colour series repeats every 550 nm, going from magenta to red (second order) and then from magenta to red again (3rd order). The colours become more pastel in hue as the polarisation colour order increases, and above the 4th order the 'colours' appear white. The length 550 nm, and its multiples, therefore comprise the boundaries of the first-, second- and third-orders printed on the Michel-Lévy Interference color chart, whose use is described in Appendix 3. The value of 550 nm is the most common, but the sensitive tint can range from 530–560 nm, depending upon the individual manufacturer's design.

These polarisation colours can be seen in the microscope, or on the light box, by using a quartz wedge with the e-ray (= slow direction for positive quartz) of the quartz crystal cut parallel to the long edge of the wedge (Figure 14.12). This device was invented by H. C. Sorby, the Sheffield scientist (Box 12.2) who initiated metallurgical and petrological microscopy. At the thin end of the wedge, the optical path difference (OPD), or retardation, is zero and the quartz is in extinction for all wavelengths. As the thickness of the wedge increases, the OPD increases, and for each separate wavelength (violet, indigo, blue, green, yellow, orange, red) of the spectrum, there is a series of dark extinction bands at which the OPD is exactly 1λ, 2λ, 3λ and so on (see Figure 14.10).

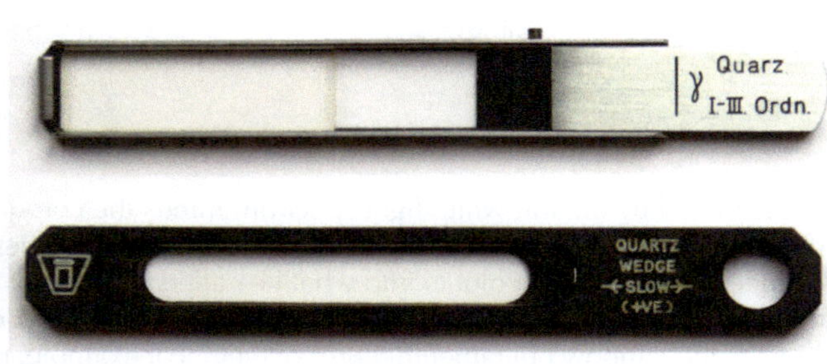

Figure 14.12 **The quartz wedge compensator** *The design of a six-order quartz wedge. Quartz is a uniaxial positive material. The RI of the slow e-ray is higher than that of the o-ray. The slow (+ve) direction is usually cut parallel with the long dimension, but this is not always the case. Source: Courtesy of Mr K. Glover.*

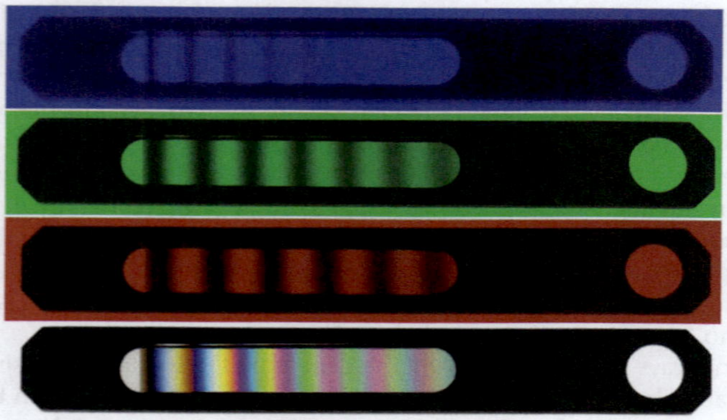

Figure 14.13 **Interference by monochromatic light** *Interference colours seen through a quartz wedge. If a coloured filter is placed over the analyser, the extinction bands on the quartz wedge can be seen clearly for the wavelength of that particular colour. Source: Courtesy of Mr K. Glover. Refer back to Figure 14.10.*

When a quartz wedge is viewed at 45° under crossed polars under *monochromatic* green light, it will show alternating wide light (i.e. green) and narrow dark bands corresponding to OPDs of 0λ (dark), $^1/_2\lambda$ (light), 1λ (dark), $3/2\lambda$ (light), 2λ (dark), $5/2\lambda$ (light) and so on. With a 546 nm green filter (see Figure 14.13), the centres of the dark bands of destructive interference will be at: 546, 1092 and 1638 nm.

We can recognise the polarisation colours that occur in our specimen and then relate them to the sequence of colours in the Michel-Lévy chart (Appendix 3) from which we can read off the optical path difference and retardation. Knowing the thickness of the specimen, we can then determine the birefringence, from which we may be able to identify the specimen. In petrology or optical mineralogy, this developed into a fine art: thin sections of rocks and minerals are ground down to a standard thickness of 30 μm, and by noting the colour and using the radiating scales on the chart, the mineral may be directly identified. A convincing demonstration that the Michel-Lévy chart is derived from the study of real petrological samples and is not merely a helpful academic concept may be seen in Figure 14.14. This figure, taken from Figure 18 on page 23 of MacKenzie & Adams (1994), shows the Michel-Lévy Interference colour chart prepared directly from a quartz wedge, ground from approximately 0 to 160 μm at the top of the 3rd order.

14.11 Polarised-Light Contrast in Reflection

Interference colours arise from the interference of coherent rays of light reflected from both the upper and lower surfaces of transparent thin films. They were first described by Isaac Newton in his *Opticks* when pressing together two uneven plates of glass with a very narrow air gap (i.e. a film of air) between them. Newton also determined, with remarkable accuracy, the film thicknesses (both of air and water) to which each of the interference colours correspond (see Hammond, 2001). So the phenomenon is general and is not confined to the particular situation of polarised light.

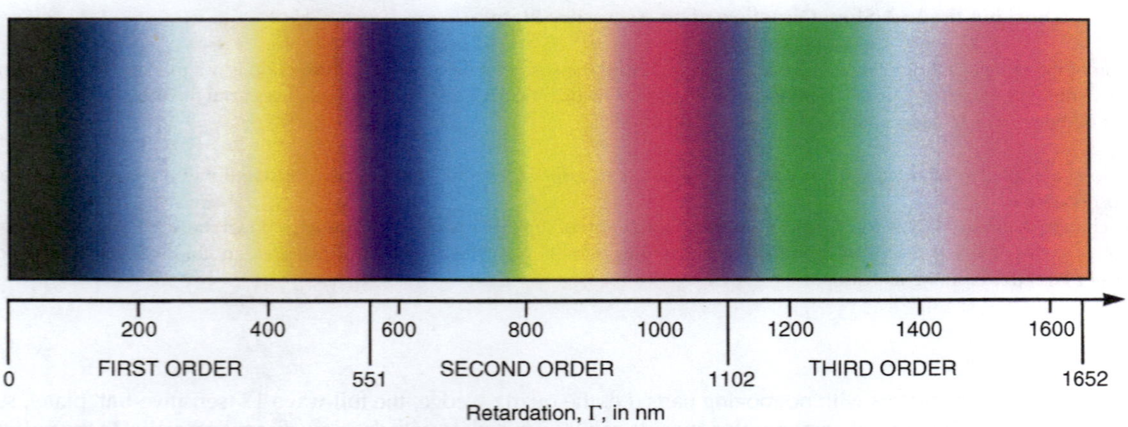

Figure 14.14 **Birefringence and relative retardation** *This illustration shows a quartz crystal ground as a wedge imaged between crossed polars and, as such, demonstrates the Michel-Lévy polarisation colour chart (compare with Figure 14.13, Figure 14.19, and the chart in Appendix 3). Retardation values are given in nm; the thickness of the quartz wedge at the right-hand end is approximately 1.60 μm thick. Source: MacKenzie and Adams 1994. Reproduced with permission from Taylor & Francis Group.*

Newton's colours are also formed by the interference of coherent rays of light being reflected off the upper and lower surfaces of a thin film of oil on the surface of a puddle of water or soap bubbles in air (Figure 14.15) or, as Newton observed them, reflected off the very closely apposed upper and lower surfaces of glass separated by a *very* thin film of air. In this case, the thickness of the film is less than, or of the same dimension as, the wavelength of light. When this occurs, the two rays will interfere constructively and destructively, depending upon the distance between them that is introduced by the thickness of the film. Note also how the colours change as we view an oil film from different angles or as the soap bubble films change their thickness before they burst.

Electron microscopists use the thin film interference phenomenon to calculate the thickness of ultrathin sections cut onto a supporting water surface. Light reflected off the upper air–section interface interferes with light reflected at the lower section–water interface to produce interference colours. The phase difference between the two rays is dependent on the thickness of the ultra-thin section, and this determines the interference colour produced. Sections less than 30 nm are dark grey to grey, sections between 50 and 80 nm are grey to silver and sections from 80 to 100 nm are silver to gold in colour.

Contrast may arise in the following ways from polarised light incident upon the surface of the specimen: (i) as a result of the optical anisotropy of the material or a surface film, (ii) from scattering at inclined surfaces such as grain boundary grooves or facets brought about by etching, or (iii) from scattering at subsurface boundaries in semi-translucent materials, e.g. some ceramics and materials.

With metals it is frequently the optical properties of the surface film, rather than the metal itself, that determines the contrast. For example, zinc (hexagonal anisotropic) shows vivid orientation contrast whereas in another hexagonal metal, such as titanium, the expected orientation contrast is weak or entirely absent – an effect that is almost certainly due to the presence of the tenacious oxide film. On the other hand, heavily etched isotropic metals (brasses, cupro-nickels, nickel, aluminium) can give rise to orientation contrast, an effect that arises at the inclined facets of geometrical etch pits, which are themselves characteristic of the crystal orientation. Finally, in semi-translucent specimens reflection at subsurface boundaries also gives rise to contrast in much the same way as for dark-field illumination.

14.12 Compensators

Having determined, with the aid of crossed polars (Section 13.8 and Figure 13.15 or section 14.7 and Figure 14.6, for pleochroic materials) the directions of the orthogonal anisotropic axes, a compensating retardation plate is now used to determine which of n_e or n_o is the fast axis and which the slow axis. The compensator is designed to be inserted at 45° (usually orientated NW-SE on modern microscopes) into the appropriate slot of the microscope tube between the crossed polars. On older microscopes, the slot dimensions were 10 mm × 3 mm, but manufacturers now usually adopt the DIN standard size of 20 mm × 6 mm.

All compensators consist of thin plates or a wedge of anisotropic material (quartz, selenite, calcite or mica) cut *parallel* to their optic axes. The thickness of the material is such that the compensating plate will introduce a uniform degree of specific retardation of $^1/_4\lambda$, $^1/_2\lambda$, 1λ or several full wavelengths of optical path-length difference across the entire field of view so that, between crossed polars, the specimen appears brighter or darker depending on whether the specimen OPD is added or subtracted, respectively, from the *fixed* OPD of the compensator.

These crystal plates and wedges are mounted in sliders with one of their vibration directions (fast or slow) parallel to their long edge of the slider (and marked as such) so that they can be inserted into the light path, above the specimen and between crossed polars in a known and controlled manner, generally at 45° to the vibration directions of both crossed polars.

Protocol 14.3 Checking the Fast/Slow Direction of an Accessory Plate

The orientation of the compensation plate can be checked quickly (in the case that the slow or fast direction is not marked on the mount) against a birefringent material with characteristic morphology whose fast or slow direction is known. Urea crystallises from the melt as long, thin crystals, the slow direction of the crystals being parallel to the length of the crystal.

1. With the long axis of a selected crystal lying NW-SE on the microscope stage, the compensator is inserted into the slot for accessory plates in the microscope tube.
2. Note the change in colour of Newton's scale. If there is subtraction of the OPD, and the colour goes down the scale through orange to yellow or pale brown, then the fast direction of the compensator lies parallel with the long axis of the urea crystal, and the slow direction of the compensator lies NE-SW along its width.

There are both fixed compensators with no moving parts (i.e. the quartz wedge, the full-wave 1λ 'sensitive-tint' plate), such as those shown in Figure 14.16. A de Sénarmont compensator consists of a $^1/_4$-wave plate with the slow direction parallel to the polariser azimuth and a rotating graduated analyser. There are also compensators with moving parts, such as the Berek, the Ehringhaus, the Soleil-Babinet and the Brace-Köhler. Further details of these compensators are given under Internet Resources. They are not described in this book. Please refer to Hartshorne & Stuart (1970), Bloss (1981) or the operating manuals obtainable from the microscope manufacturer, for full details of their use.

Figure 14.15 **Formation of interference colours** *Light is reflected off the surface of the oil at the air-oil interface (left) but also passes through the film to reflect off the oil-water interface below. Because the light is from the same source and coherent, the two rays will interfere. The phase shift between the rays determines the colours formed by interference. The right-hand panel shows Newton's rings formed from a thin 'film' of air between the glass surfaces of two microscope slides.*

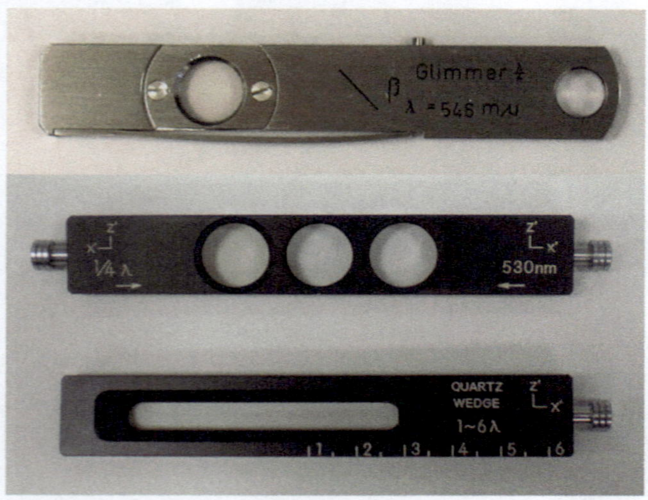

Figure 14.16 **Compensator accessory plates** *At the top is a quarter-wave de Sénarmont compensator with a flat mica plate that retards light by 546/4 or 136.5 nm. The β indicates the intermediate refractive index (refer to Section 13.12). The middle compensator combines a quarter-wave plate on the left with a 1λ sensitive-tint plate on the right. The central hole is blank to permit viewing without using the compensators. The x′ and z′ indicate the fast and slow vibration directions, respectively.*

14.13 How Compensators Work

The anisotropic mineral or crystal should be aligned on the microscope stage with one of its privileged directions oriented at 45° to the vibration directions of both crossed polars. A compensator is inserted into the accessory slot in the body-tube of the microscope between the specimen and the analyser.

If constructive interference occurs, then the slow ray of the sample coincides with the slow ray of the compensator (and hence the two fast rays are also coincident); these slow rays are thus vibrating in the same direction. The OPDs, or relative retardations, from the sample and the compensator will therefore *add* together when they recombine at the analyser. The resulting polarisation colour will be higher (go *up* Newton's scale) to blue or even green. Conversely, if the similar vibration directions between sample and compensator are at 90° to each other so that the fast vibration direction is aligned parallel with the slow vibration direction, then the OPDs *subtract*, and the colours go *down* Newton's scale, first to orange, then to yellow. Figure 14.17a and Figure 14.17b show how the o-ray and e-ray vibration direction vectors produce the OPD, which is added or subtracted from the fixed OPD of the compensator (accessory plate). See Exercise 14.18 to clarify and support this explanation.

Since the orientation of the slow or fast direction of the compensator is normally engraved on the mount, we can also determine the optic sign of the birefringence – i.e. which is the slow and which is the fast direction with respect to some characteristic feature in the specimen – for example, cleavage traces in minerals or the long direction of fibrous materials. Compensators are, therefore, very useful for accurately determining either very low or very high values of retardation in order to calculate the birefringence and so make an identification of the mineral, crystal or other substance being studied.

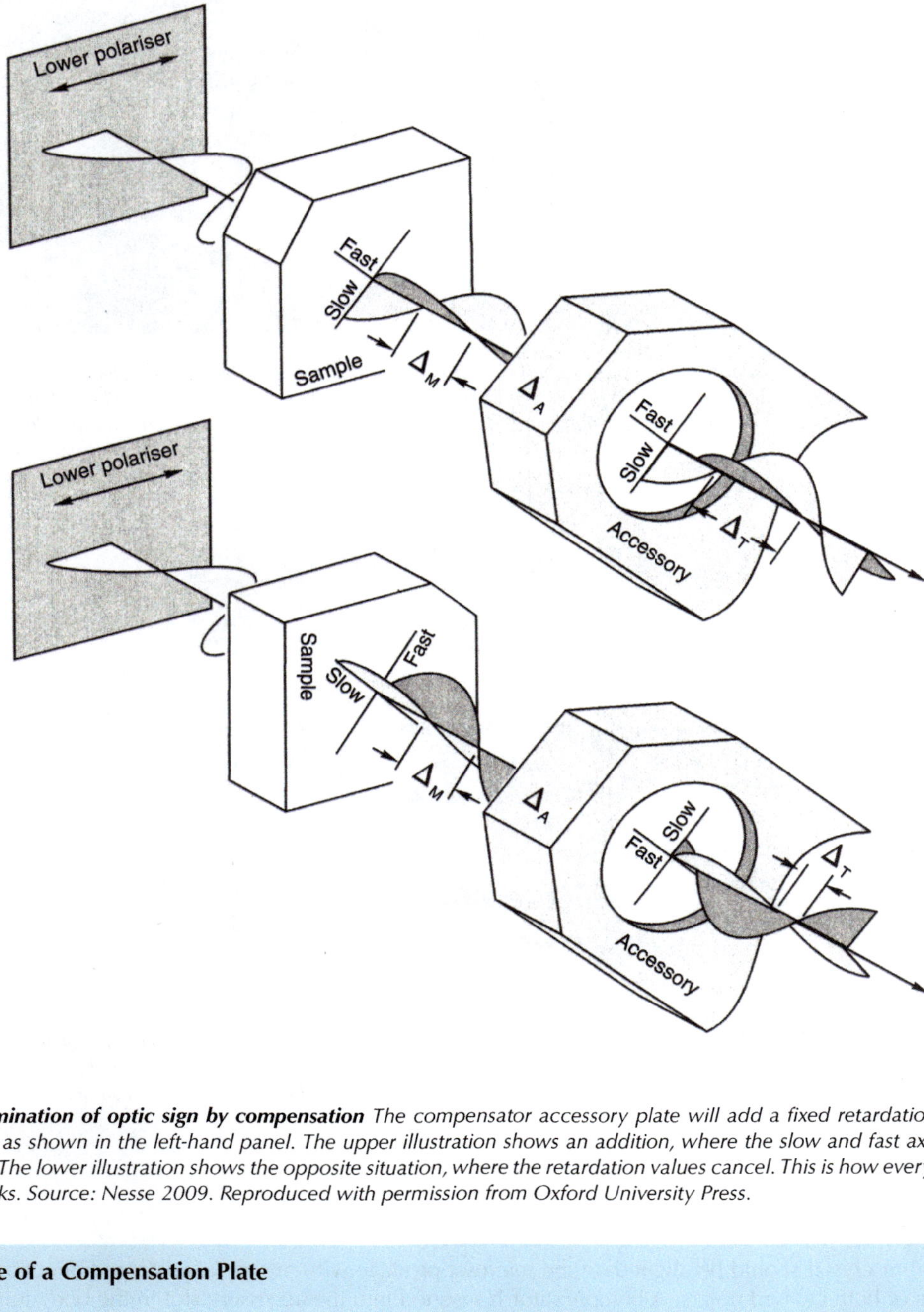

Figure 14.17a ***Determination of optic sign by compensation*** *The compensator accessory plate will add a fixed retardation value uniformly across the entire field of view, as shown in the left-hand panel. The upper illustration shows an addition, where the slow and fast axes of the sample and the compensator coincide. The lower illustration shows the opposite situation, where the retardation values cancel. This is how every retardation and variable compensator plate works. Source: Nesse 2009. Reproduced with permission from Oxford University Press.*

Protocol 14.4 Use of a Compensation Plate

1. Bring the crystal or mineral into the extinction position (i.e. so that either the fast or slow vibration direction is aligned with either the polariser or analyser). Preferably do this by rotating the stage. Check the alignment with the eyepiece cross-hairs.
2. Rotate the sample and set it at the 45° position, for maximum brightness, so that the vibration direction that lay N-S is now orientated NE-SW. As confirmation, the grain should be brightly coloured.
3. Note the interference colour and locate it on the Michel-Lévy chart. Record the magnitude of OPD (Γ, Δ_M), and record the retardation that corresponds to that polarisation colour.
4. Insert the compensation plate. By convention, the compensator slot will usually be aligned NW-SE. Note the fast and slow directions of the compensator. Usually the slow direction (on a 1λ 'sensitive-tint' plate) will be aligned NE-SW.
5. Note whether the colour goes 'up' (to blue) or 'down' (to yellow) on Newton's scale.
   (The colour may not always subtract to yellow, but may assume other hues.)
6. Calculate the new retardation value, adding or subtracting the fixed compensator value. (Where the subtraction value is negative, either use a compensator giving a smaller retardation value or use the modulus, or absolute value)
7. Rotate the sample a further 90° to 135°. Here the opposite addition or subtraction should occur. Use this as a check of the value calculated in step 5.
8. Use the Michel-Lévy chart (Appendix 3) to calculate the birefringence as an aid to identifying the sample.

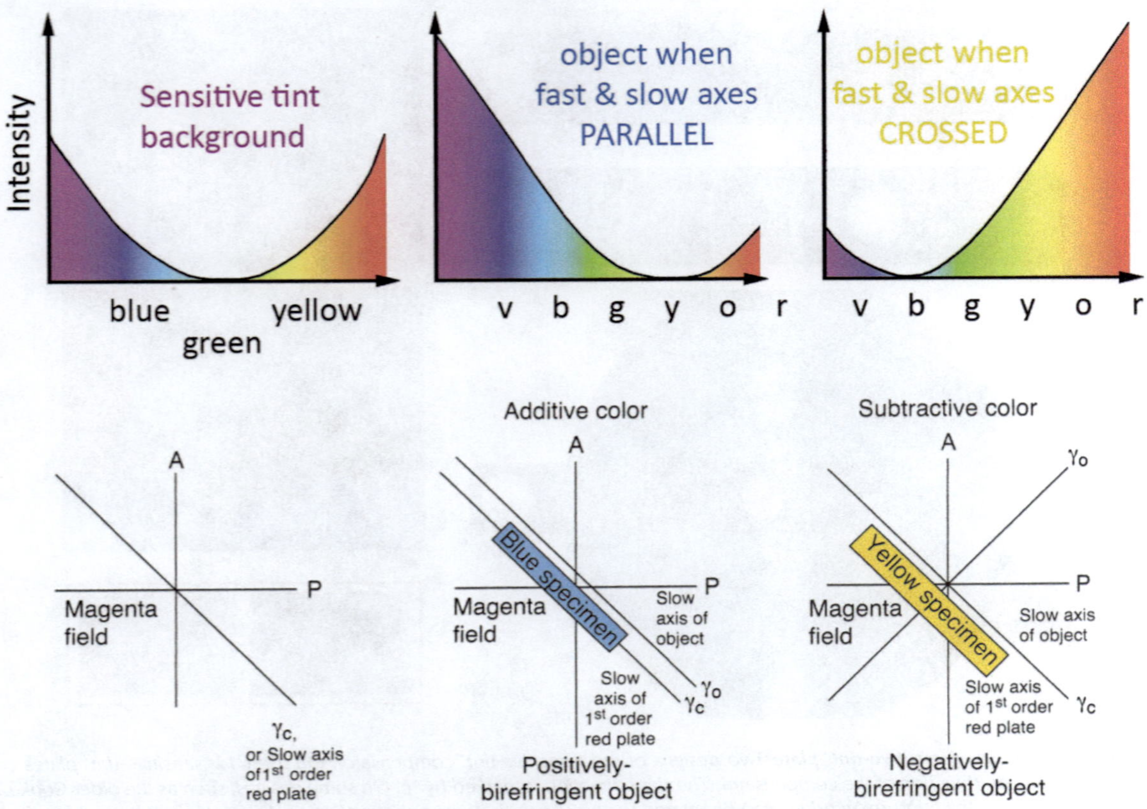

Figure 14.17b **Compensation with a 1λ plate** *The most commonly used retardation plate is the 1λ red 'sensitive-tint' plate. How this works is shown above. The 1λ plate extinguishes green light by introducing a phase difference equal to 546 nm. When a doubly refracting object whose fast and slow axes are aligned with the 1λ retardation plate is placed on the stage, the retardation values add. Yellow hues are lost by destructive interference from the spectrum to give a blue colour – going up Newton's scale. When the axes do not coincide, blue hues are lost, giving a subtractive yellow colour – going down the polarisation colour scale. Source: (top) Adapted from Spencer 1982 with permission from Cambridge University Press. (bottom) Wayne 2009. Reproduced with permission from Elsevier.*

14.14 The 1λ Full-Wave 'Sensitive-Tint' Plate

The 1λ, whole-wave, 'Rot 1' or 'sensitive-tint' plate may be marked 1λ, Rot 1, Δ = 530–560 nm (depending upon the manufacturer) or a variation of the name gypsum (e.g. gypsum plate, Gips, Gyps) or quartz (e.g. quartz sensitive tint) on the frame of the compensator. It is cut from a piece of quartz or gypsum, usually with the fast axis parallel to its length, which is cut along its optic axis (Figure 14.18) so that the o-rays and e-rays are out of phase by 546 nm, i.e. by exactly one wavelength for green light. This means that the slow direction (marked 'slow' or 'γ') runs NE-SW, perpendicular to the slot and at right angles also to the long axis of the 1λ plate. A phase difference (optical path difference) of one lambda for green wavelengths of light are linearly polarised in the same direction as the polariser and thus are blocked by the analyser.

Strictly speaking, the full-wave 1λ accessory plate is a retardation plate rather than a variable compensator (like the quartz wedge). When an isotropic specimen is viewed between crossed polars with the 1λ 'sensitive-tint' plate inserted at 45°, no ellipticity is introduced in the electric vector of the polarised light because the emergent o-rays and e-rays are in phase and combine to form a plane-polarised beam. The analyser therefore extinguishes the green light and allows all others to pass, thus creating the (white minus green, i.e. magenta) 'sensitive tint' colour. This divides the first and second orders of Newton's scale.

The red sensitive tint plate is useful for determining the optical sign and the fast or slow direction of crystals and other anisotropic materials that show low-order interference colours. The crystal either raises the tint to blue or diminishes it to red (and this can happen very quickly at the boundary of the first order, hence the term 'sensitive tint') according to whether the positive direction of the crystal reinforces or compensates for that of the compensator. For very weakly birefringent crystals, which may be obscured against the purple background of the compensator, it is useful if the compensator is rotated slightly off the 45° towards extinction, so as to give a weaker and less intense background with concomitantly more intense yellow and blue interference colours.

The 1λ whole-wave plate can also be used to introduce colour contrast to polarised light images (see Exercise 14.17). This is not only for aesthetic value, but also because colour may usefully be used to enhance contrast in low reflectivity metallurgical and ceramic images.

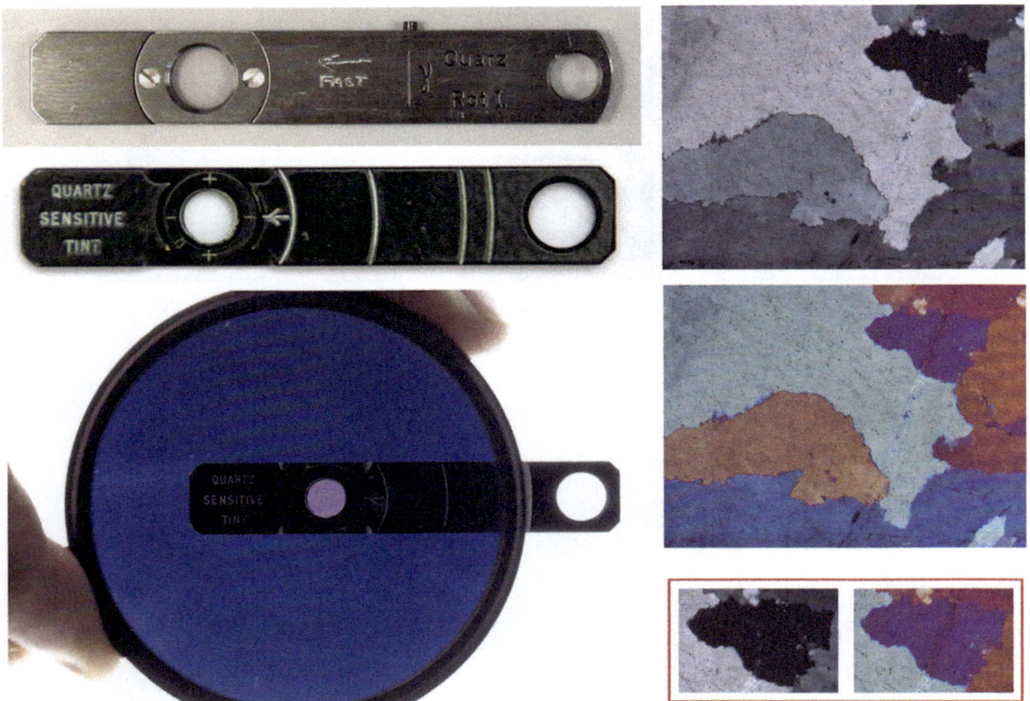

Figure 14.18 **The 1λ full-wave 'sensitive-tint' plate** *Two designs of red 'sensitive-tint' compensator. On fixed 1λ 'sensitive-tint' plates (top), the fast direction is parallel to the long direction of the compensator. The slow direction is marked by 'γ'. On some designs, such as the older Cooke Troughton & Sims accessory plate (bottom), the gypsum 'window' can be rotated to assist analysis. If the anisotropic sample is only weakly doubly refracting, it helps to offset the compensator to give a weaker background against which to view the polarisation colours.*

The + on the CTS compensator indicates the slow direction (higher refractive index). The violet tint of the background will not change unless the compensator is also rotated 45° parallel to the privileged direction of the polarising filters.

The right-hand panels show a section of quartz crystals in various orientations in a thin section. The top panel shows the quartz section between crossed polars only; the middle panel with the addition of a 1λ retardation plate. Accordingly, where the orientation of the individual quartz crystal is aligned with the polariser and analyser, the retardation value is virtually zero, and black results. Because these quartz crystals are oriented differently with respect to the crossed polars, they will exhibit different optical path difference values from the maximum possible value to black. The bottom inset panel shows such a crystal; with the retardation plate, the crystal assumes the 'sensitive tint' (i.e. $\approx$ 0 nm OPD + 546 nm).

Source: Courtesy of Mr K. Glover.

14.15 The de Sénarmont $^1/_4\lambda$–Wave Plate

This compensator is useful for measuring small retardations accurately, particularly those <300 nm in the first-order grey region of Newton's scale. It is made from mica sheet sufficient to retard yellow or green light. Mica is a negative biaxial crystal, with a birefringence value of about +0.04. The mica sheet is cleaved so that the compensation directions will correspond to n_Y and n_Z, often denoted β and γ, respectively, and these directions are marked on the compensator (Figure 14.16). Sometimes the word 'glimmer' is marked on the $^1/_4\lambda$-wave plate, this being the German for mica. The de Sénarmont compensator uses monochromatic light and a rotatable graduated analyser; it is cheaper than a moving-plate Berek compensator and will do the same job. The high RI slow direction is usually orientated parallel to the polariser direction (not at 45° as with the other compensators). For a detailed description of how to use a de Sénarmont compensator, see Delly (2003). The URL is given under Internet Resources at the end of the chapter.

Protocol 14.5 Using the $^1/_4\lambda$-Wave Plate

1. Move the sample until it lies under the cross-hairs. Swing out the top lens of the condenser (to give narrow cone axial illumination). Insert the analyser, and rotate the stage until extinction of the sample is noted when it 'disappears'.
2. Rotate the stage 45° to bring the sample to maximum brightness. Insert a 1λ full-wave plate and use this to orientate the sample's high RI, slow direction, NE-SW. Take care with this step.
3. Remove the first-order red compensator, and insert the de Sénarmont 1/4-wave plate. Place a green narrow-band interference filter into the light path. Rotate the sample until it becomes maximally black; the background will lighten and assume a green colour.
4. Read off the analyser rotation (θ) in degrees. The specified wavelength (λ) is given by the monochromatic light, and the retardation can be calculated from: $r = (\theta/180°)\lambda$. Since $\lambda/180°$ = constant, multiply the constant (e.g. 3.033 nm/degree for a 546 nm green filter) by the analyser reading to get the retardation.

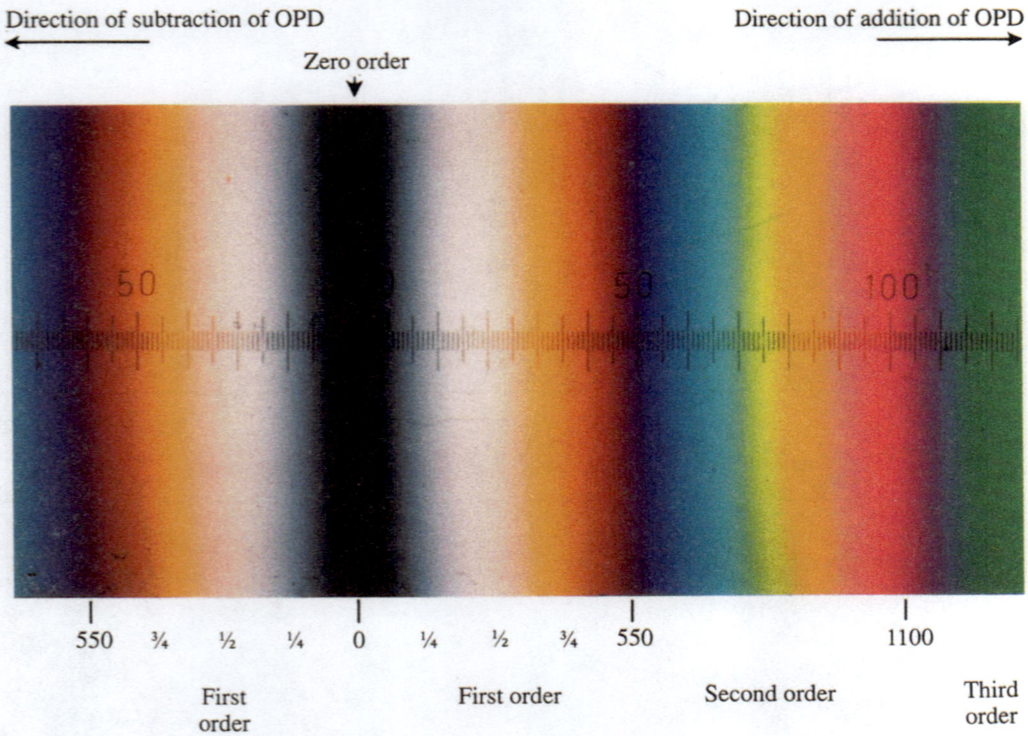

Figure 14.19 **Continuation Michel-Lévy polarisation colour chart** *The colour chart has been superimposed upon a scale, calibrated to represent optical path difference in nanometres. This chart differs from the Michel-Lévy chart reproduced in Appendix 3, which has the gradient-birefringence lines added. This chart does not show gradient-birefringence lines, only optical path differences. Each small division on the scale represents 10 nm. Source: Robinson and Bradbury 1992. Reproduced with permission from Oxford University Press.*

Note: With the $1/4\lambda$ plate it is possible to subtract more than a quarter order, which will give an overall higher-order polarisation colour. Essentially, this 'step down' past the zero OPD black is not possible unless the Michel-Lévy chart has been printed as a continuation. An example of such a chart is given in Figure 14.19, and is taken from the rear cover of Robinson & Bradbury (1992).

14.16 The Quartz Wedge

A quartz wedge (Figure 14.12 and Figure 14.13) works in the same way as the quarter and full-wave plates but has the advantage that it covers several orders (usually three) of Newton's scale. It is therefore a variable compensator. Since compensation only occurs when unlike vibrations in the crystal and the wedge are parallel, it is a guide to the optic sign of the crystal or mineral. The quartz wedge can also be used to exactly determine the true colour value of higher-order interference colours of Newton's scale, which are much paler and washed out than the more saturated lower orders.

In quartz the positive (slow) direction is along the crystal axis. This slow direction is usually set parallel to the short side of the wedge, and marked as such (e.g. γ, z'). The fast direction then lies parallel to the long dimension of the wedge. However, some quartz wedges may have their slow and fast directions reversed. Figure 14.12 shows two such wedges; the Vickers wedge has its slow direction parallel to the long dimension of the wedge.

It may help, while using the wedge, to restrict the illuminating beam to a narrow pencil of light by closing down the condenser aperture iris: we are not concerned with resolution here. Closing down the condenser will increase contrast in the image. The coloured fringes seen near the edge of the crystal grain then move outwards as the wedge is inserted, with still higher orders of colours appearing at the thickest part of the grain (i.e. retardations are added: slow-on-slow or fast-on-fast).

If the crystal grain is turned through 90° on the rotating stage, it will lie in the 'compensating' position. When the wedge is inserted once again, the reverse effect will occur, and the colour fringes will move inwards into the grain, with the orders of the colours diminishing. A black compensation band reaches the thickest part of the grain if the wedge is long enough. At this point the optical path difference due to the grain is equal but opposite in sign to that of the quartz wedge and can be determined. Retardations cancel, or subtract (slow-on-fast or fast-on-slow).

The polarisation colours seen with a quartz wedge out of the microscope are those for reflection. When the analyser is turned to a parallel setting, the monochromatic dark bands in the image exchange places with the light bands, and the polarisation colours are now for Newton's scale in transmitted light. This method is little used but is sometimes convenient for identifying the rather similar tints that occur in the higher orders of Newton's scale by noting the 'parallel' colour (Figure 14.20), which may be easier to identify.

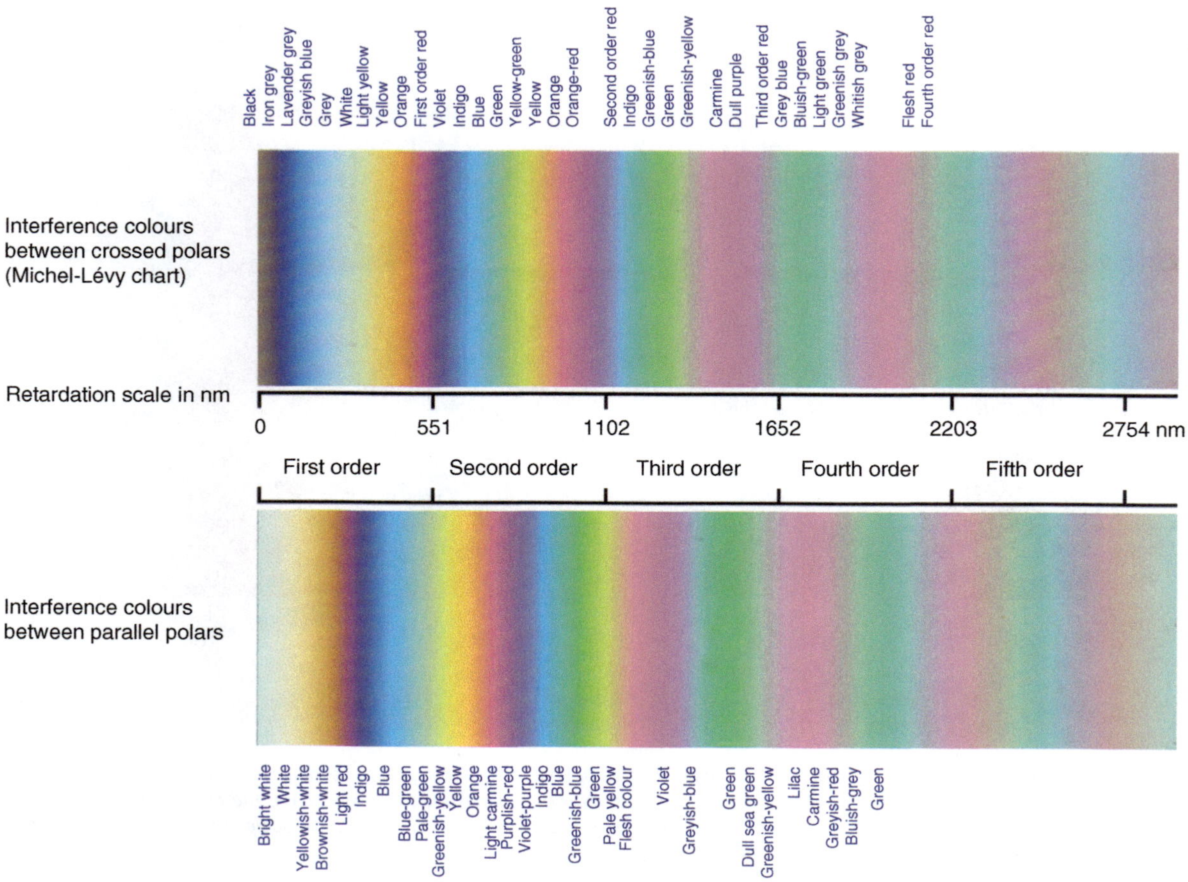

Figure 14.20 **Newton's interference colours through parallel polars** *A unique Michel-Lévy chart in which the interference colours through parallel polars are shown alongside those seen through crossed polars. See also Figure 27 in Delly (2003) and Figure 4.13 in Houck & Siegel (2010). Higher-order tints as seen between parallel polars are sometimes easier to identify (compare this figure with Figure 14.8, where reflection colours from a Sellotape wedge are shown). The list of colours in this figure are taken from those published in Figure 7.19 on page 279 of Hartshorne and Stuart (1970), which were themselves selectively taken from the table of Newton's colour scale according to Quincke (1866). The entire table of Qunicke's colours are reproduced together with a Michel-Lévy interference colour chart between crossed polars, as is usually the case, in Appendix 3. Source: Colours as listed in Hartshorne and Stuart 1970.*

The disadvantage of the fixed quartz wedge is apparent when trying to take a retardation reading at around zero, which coincides with the thin end of the wedge. In the quartz wedge an accurate zero measurement is often not possible because of loss of quartz when grinding the plate during manufacture. The Babinet compensator overcomes this problem, having twin, opposed quartz wedges with mutually perpendicular crystallographic axes. Some $^1/_4\lambda$ and 1λ compensating plates have their birefringent plates mounted on a wheel rotatable only in the plane of the compensator. This allows the retardation to be determined without rotating the analyser and to determine the optic sign quickly and easily.

14.17 Conoscopic Microscopy

This is the method by which convergent light is used to view an image of the interference figure of the crystal in the back focal plane of the objective in order to determine the optic sign of the crystal, as a means of identifying it. Interference figures are not images of the anisotropic material but an interference pattern formed in the illuminating set of conjugate planes, rather than the imaging set. When the birefringence of a material is low, the interference colours tend to cluster around high-order white, because the section is relatively thick. What is needed is to show the interference colours for many angles at once: conoscopy thus provides an angular map of the light through the crystal as viewed from above.

A high-magnification, and hence high-NA, objective is used (e.g. 40x or higher). All doubly refracting materials produce characteristic interference figures, but the resulting image depends upon the *orientation* of the section. The most informative are those interference figures where the points of emergence of the optic axes (the melatope) can be seen. The top lens, or auxiliary lens, of the condenser provides the strongly convergent light for conoscopic observation that passes through the specimen.

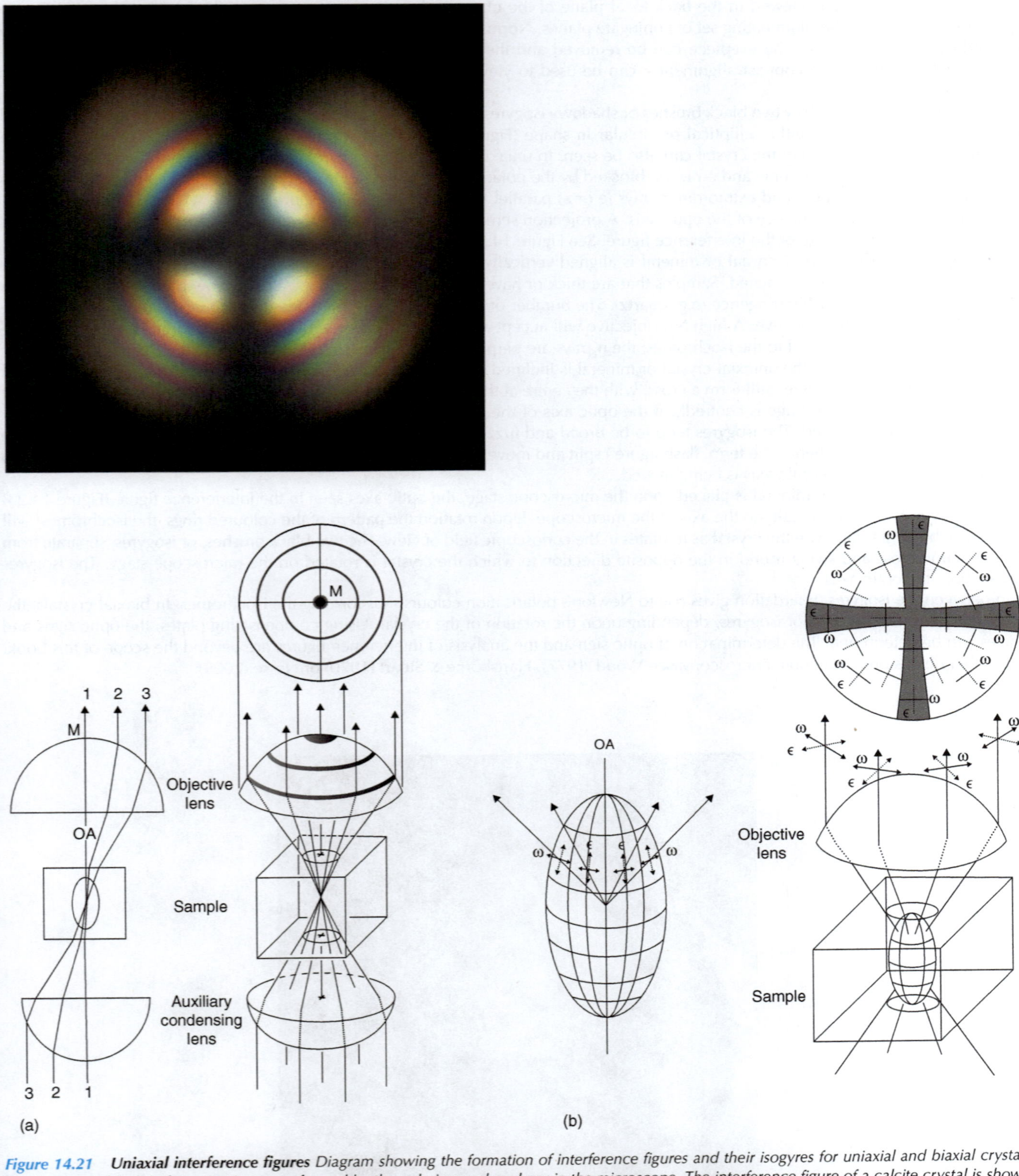

Figure 14.21 ***Uniaxial interference figures*** *Diagram showing the formation of interference figures and their isogyres for uniaxial and biaxial crystals. The isogyre in black is the extinction cross formed by the polariser and analyser in the microscope. The interference figure of a calcite crystal is shown. M shows the emergence of the melatope. Source: Nesse 2009. Reproduced with permission from Oxford University Press. (Image) J. D. Chapman. Reproduced with permission from Mr Chapman.*

The conoscopic image is viewed in the back focal plane of the objective by using an extra lens in the imaging set of conjugate planes in order to view the illuminating set of conjugate planes. Normally this is done by inserting a Bertrand lens between the objective and eyepiece; otherwise, the eyepiece can be removed and the interference figure viewed directly, or a centring telescope – such as that used for phase contrast alignment – can be used to view a magnified image of the interference figure by conoscopic microscopy.

The interference figures show two black brushes or shadowy isogyres in the form of an extinction cross, or two hyperbolae. Coloured rings[8] (isochromes) that are either elliptical or circular in shape (Figure 14.21a) and whose colours and spacing depend upon the birefringence and orientation of the crystal can also be seen. In uniaxial crystals, the isogyres are formed along the optic axes, where the vibration directions of the o-ray and e-ray are blocked by the polariser and analyser. Ordinary rays (o or ω) vibrate parallel to lines analogous to lines of latitude, and extraordinary rays (e or ε) parallel to lines of longitude (Figure 14.21b). Where the isogyres cross, the melatope marks the emergence of the optic axis. A projection screen, sometimes referred to as a Quirke hemisphere, can be used to observe a magnified image of the interference figure. See Figure 14.22 for more details.

If the optic axis of a uniaxial crystal or mineral is aligned vertically to the axis of the microscope, the interference figure will *not* move or change as the stage is rotated. Samples that are thick or have high birefringence (e.g. calcite) show more isochromes than a mineral or crystal with low birefringence (e.g. quartz). The number of isochromes seen in the interference figure also depends on the numerical aperture of the objective. A high NA objective will accept a greater angular cone of light, and so more isochromes will be seen. The n_o rays are tangential to the isochromes; the n_e rays are perpendicular (Figure 14.21b).

If, rather, the optic axis of the uniaxial crystal or mineral is inclined (say $\leq$30°) to the vertical axis of the microscope, the figure will rotate, and although the isogyres still form a cross, with the centre at the optic axis, the melatope will swing in an arc (hopefully about the centre of the field if the stage is centred!). If the optic axis of the specimen is orientated with its optic axis horizontal, or nearly so, a flash figure is produced. The isogyres tend to be broad and fuzzy, rather than well defined, and upon rotation of the stage, the isogyres will very quickly (hence the term 'flash figure') split and move out of the field of view in opposite quadrants, corresponding to the quadrants in which the optic axis is being moved.

When a biaxial crystal or mineral is placed upon the microscope stage, the optic axes seen in the interference figure (Figure 14.23) are normally not aligned vertically to the axis of the microscope. Upon rotation the pattern of the coloured rings (the isochromes) will not change but will rotate with the crystal as it rotates in the conoscopic field of view. The two black brushes, or isogyres, separate from the extinction cross and swing round in the opposite direction to which the crystal is rotated on the microscope stage. The isogyres rotate about the optic axes.

In between the isogyres, retardation gives rise to Newton's polarisation colours thus forming the isochromes. In biaxial crystals, the cross separates into *two* sets of isogyres, depending upon the rotation of the crystal. Using compensating plates, the optic signs and angles can be determined. The determination of optic sign and the analysis of interference figures are beyond the scope of this book. For further details on conoscopic microscopy, see Wood (1977), Hartshorne & Stuart (1970) or Nesse (2009).

Figure 14.22 ***Viewing interference figures*** *A hemispherical screen, named a Quirke hemisphere after its originator, can be used to inspect an interference figure. In this example a chord has been cut from the top of a pearl light bulb and glued to a Polaroid sheet. It is also possible to construct a hemisphere from a table tennis ball. Source: D. Kile. Reproduced with permission from D. Kile.*

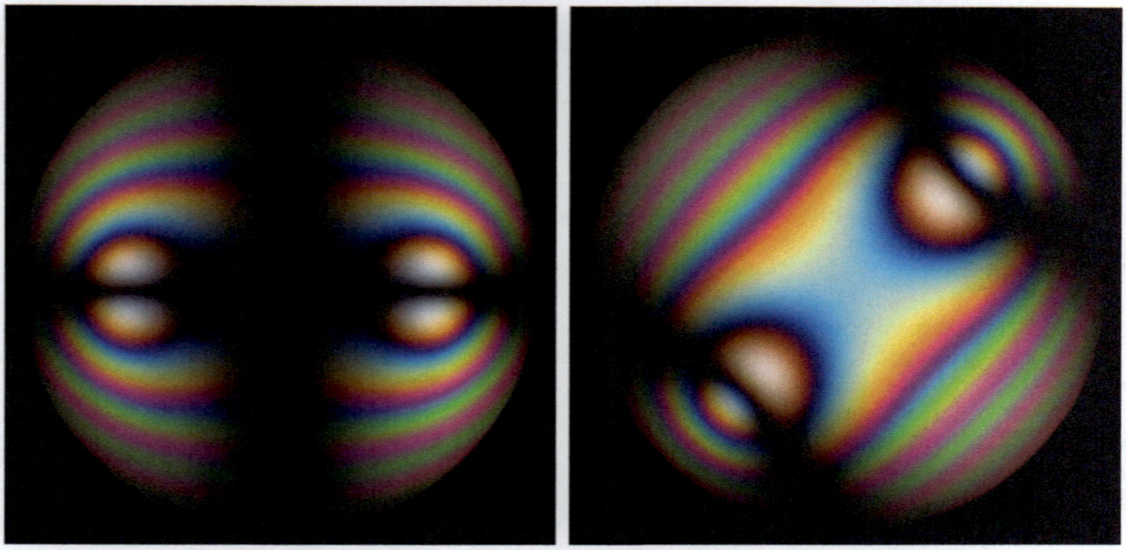

Figure 14.23 **Biaxial interference figure** *The interference figure of a thick section of muscovite viewed conoscopically with linear-polarised light. The melatopes containing the optic axes can be seen left and right and at the bottom left and top right of the figure. The numbers and intensity of the isochromes indicate this is a thick section, but the birefringence of muscovite ($\Delta M = 0.036 - 0.049$) is also relatively high. Therefore, intense second-order colours and well-defined third-order isochromes can be made out. Source: J.D. Chapman. Reproduced with permission from Mr Chapman.*

14.18 Chapter Summary

At the end of Chapter 13 and this chapter, you should understand how the microscope can be used as an analytical instrument with polarised light to visualise both the bulk and surface components of transparent and opaque materials. You should have a basic understanding of how the interaction of polarised light with doubly refracting specimens permits the use of qualitative and quantitative procedures in order to identify unknown crystals, minerals, powders and fibres and to understand and categorise their physical properties.

1. Three methods of preparing samples for polarised-light microscopy predominate. Crushed mineral or crystal grains can either be immersed in mounting media of different refractive index or examined using a spindle stage. Thin sections permit the morphology and inter-relationship of the constituent minerals and crystals to be studied.
2. To convert an ordinary microscope into a polarised-light microscope, two polarising filters, or polars, are the minimum requirement.
3. The polariser and the analyser are crossed at 90° and oriented E-W/N-S with respect to one another.
4. Rotation of the stage allows for observation of extinction or pleochroism.
5. Two polarised rays produced by double refraction travel through an optically anisotropic material at equal amplitudes but at different velocities. There will be a lag, or shift in phase, depending upon the unique refractive indexes of each ray when they exit the material. This lag is the optical path-length difference (OPD), or relative retardation (Γ).
6. The *overall* retardation depends upon the two factors: the difference between the two refractive indexes (i.e. the phase shift between the o-ray and the e-ray: the birefringence, ΔM) and the thickness of the specimen. The OPD, or relative retardation, is a product of these two values.
7. The o-ray and e-ray are mutually perpendicular; they also have a common origin, are therefore coherent and can therefore interfere.
8. Specific polarisation colours are determined by the full white-light spectrum *minus* the colour of the wavelength that has undergone elimination by destructive interference.
9. Polarisation colours are brightest when the anisotropic material is oriented at 45° between crossed polars, because each ray has equal amplitudes.
10. Interference colours arise from the interference of coherent rays of light reflected from both the upper and lower surfaces of transparent thin films.
11. Compensators introduce a uniform degree of specific retardation so that the specimen appears brighter or darker depending on whether the specimen OPD is added or subtracted, respectively, from the *fixed* OPD of the compensator.

12. If constructive interference occurs, the slow ray of the sample coincides with the slow ray of the compensator and the two fast rays are also coincident. The OPDs from the sample and the compensator therefore *add* together when they recombine at the analyser. The resulting polarisation colour will go *up* Newton's scale.
13. If the similar vibration directions between sample and compensator are set at 90° to each other, the fast vibration direction is aligned parallel with the slow vibration direction. Then the OPDs *subtract*, and the colours go *down* Newton's scale.
14. An isotropic specimen viewed between crossed polars with the 1λ 'sensitive-tint' plate inserted at 45° produces no ellipticity in the electric vector of the polarised illumination. This is because the emergent o-rays and e-rays formed from the anisotropic quartz are *in phase* and combine to form a plane-polarised beam. The analyser, therefore, extinguishes green light and allows all others to pass, thus creating the (white-minus-green, i.e. magenta) 'sensitive purple' colour.
15. The red sensitive tint divides the first and second orders of Newton's scale.
16. The 1λ 'sensitive-tint' plate is useful for determining the optical sign and the fast or slow direction of anisotropic materials that show low-order interference colours.
17. Conoscopic microscopy is used to view an image of the interference figure of the anisotropic material in the back focal plane of the objective, to determine the optic sign.
18. Conoscopy provides an angular map of light through a crystal as viewed from above.
19. Interference figures show black isogyres as an extinction cross or two hyperbolae. Coloured rings (isochromes) are either elliptical or circular in shape.
20. Isogyres are formed along the optic axes. Where the isogyres cross, the melatope marks the emergence of the optic axis.

Exercise 14.1 Formation of polarised light by filtration

Take two pieces of Polaroid polarising filters and hold them above a photographic light box, other artificial light source or up towards bright blue sky, and rotate them with respect to one another. Notice that at one point of rotation the field of view will darken. This is the point of *extinction*.

Rotate the pair of crossed polar together as one unit over the light-box, and notice that the intensity of the extinction position does not change.

Notice that with only a single polarising filter above the light source, no change in intensity occurs upon rotating fully through 360°. The light source provides unpolarised light. Also notice that the light seen through the single polar is less intense than when viewed without a polarising filter. The theoretical maximum transmittance of a polar is 50%.

Exercise 14.2 Extinction positions of polariser with analyser

Place a polar, with its transmission axis in the East-West direction, on top of the Illuminated field diaphragm (if it is held here, the polar must be very clean, scratch free and of good quality, since it will be imaged directly into the specimen plane) or else into the substage filter holder on the microscope. Ensure that this polariser can be freely rotated.

Insert a second polar into a designated slot or position between the microscope nosepiece and eyepiece(s). Turn this polar so that its position is aligned North-South, as the analyser filter. (There should also be another slot orientated at 45° to the analyser slot: ignore this, it is for compensation plates.) The field of view should be completely dark, or nearly so, since the filters should be 'crossed' with respect to one another. Without turning the analyser, look down the microscope without any specimen on the stage, and slowly rotate the polarising filter.

Notice that as the polariser is rotated through 360°, the light is extinguished – or at best considerably darkened – twice through one complete revolution of the polariser.

Remove an eyepiece and inspect the back focal plane. A dark polarisation cross (Figure 14.2) should be seen.

Exercise 14.3 Polarisation by reflection: checking the transmission direction of a polar

Stack a dozen glass microscope slides together. Hold the stack of slides with the long (3″/76 mm) edge of the stack facing at an angle of about 60° from the vertical towards the sun or a bright light source.

Look indirectly at the sunlight reflected off the mirror surface through a single polar. Rotate this and notice that it goes slightly darker at one particular orientation. This will be its transmission direction, which will be parallel with the long edge of the stack of slides. In this position the polar is 'crossed' with respect to the horizontal plane of reflected polarised sunlight.

Depending upon the number of slides used or the reflectivity of the surface, only a slight darkening will be seen rather than anything like full extinction (because at a maximum of about 15% the intensity of the reflected light is relatively weak), so you may have to rotate the polar several times to see the effect. Looking at unpolarised light through a polar will, obviously, reveal no intensity change.

Exercise 14.4 Polarised light from an LCD monitor

Create a uniformly blank white screen by opening a blank word processing document on an LCD flat-screen monitor or a laptop or netbook screen. These screens are ideal because they are back-illuminated.

Hold a polar in front of the screen, and rotate it. You should see two points of extinction in each 360° rotation. The light may be vertically or horizontally polarised, or else polarised at 45°, depending upon the design of the LCD flat screen.

Exercise 14.5 Polarisation by reflection: Brewster's angle

Draw a straight base line on a wad of white scrap paper. Arrange a mirror to stand vertically on the line (a hinged compact mirror is suitable). Stick a needle into the paper about 1 cm away from the mirror and 1/3rd in from the edge. Illuminate the needle with a light source, and view the reflected image of the needle from the other side to the light source. Rotate the polar until the background appears at its dimmest. Now alter the azimuth angle at which you view the reflected image of the needle in the mirror. At a certain angle the reflection off the surface of the needle will appear darkest (this change may only be very slight).

Take a pencil and put a mark on the paper directly behind the image of the needle. Remove the needle; join the pencil mark and the pricked point made by the needle, and extrapolate this line to intersect the base line. Construct an orthogonal line to the base line, and measure the azimuth angle. Brewster's angle for glass of RI 1.515 = 56.6°. Within the limits of accuracy for this simple experiment, you should get a value within ±10°.

Exercise 14.6 Polarisation by scattering

Take a clear-sided, unpatterned glass or beaker; fill it with water and add 2–3 drops of milk, or dissolve a *tiny* amount of soap into the water. The solution should have a very faint grey tint, which assumes a very pale blue colour when a light is shone through it (the particles scatter blue light better than longer wavelengths). The colloid particles make the beam of light visible as it passes through the solution. Ensure that the beam of light shines horizontally through the solution. Look through a polar at the beam, and rotate the polar.

Notice that a variation in intensity (indicating the light is polarised) is most noticeable when looking perpendicular to the light beam rather than head-on through the glass. Also, when the vibration direction of the polar is set North-South, it passes more light than when rotated so that the permitted vibration direction is East-West. This shows that the light scattered at 90° from the direction of propagation of the beam is polarised and that this light is polarised with a vertical vibration direction.

Exercise 14.7 Strain birefringence

Using a light box, or bright daylight, arrange a plastic-extruded ruler, set square or protractor between crossed polars, and observe the strain birefringence and the change in colour distribution and intensity – due to extinction – as the top polar is rotated.

Exercise 14.8 Birefringence

Repeat Exercise 14.2, but with a birefringent crystal (such as calcite or mica if you have access to these) on the microscope stage. It is preferable if the crystal can be turned on a rotating circular stage (for angles may be more precisely determined this way), but the same effect can be seen by rotating the polariser, as before.

As the crystal is rotated through 360° the light is extinguished *four times* through one complete revolution of the stage. Extinction occurs each time when the planes of polarisation of the crystal become parallel to the transmission axes of first one then the other polar. As the crystal moves away from extinction, it becomes brightest and most colourful at 45° to the transmission axes of the polariser and analyser. You should therefore see four positions of minimum intensity and another four of maximum intensity through each complete 360° rotation.

Notice that this demonstration requires two polarising filters. If one is omitted, the birefringent specimen shows no intensity change or colour contrast, unless it is dichroic. If you do not have any strongly-birefringent crystals, try single grains of fine sand (quartz).

It is important to realise that the extinction described will only occur with plane-polarised light as the illuminant. Circularly polarised light does not produce this effect.

Exercise 14.9 Fast & slow directions

From a piece of hardwood, cut a block (roughly cut is sufficient) across the grain and another (cut to exactly the same dimensions and volume) parallel with the grain, in the proportions shown in the figure. (Source: Inoué 1986. Reproduced with permission from Springer.) A piece 3 cm^2 on the small face and 9 cm long is sufficient.

Tap the wood on the face that is transverse to the long dimension with a hammer or a piece of heavy cutlery. The sound wave will travel faster where the wood is parallel to the grain rather than across it, and will thus produce a slightly higher sound, the frequency being higher and the energy less impeded.

Across the grain the velocity of the sound wave is slower, and so the tone is lower. Here we are exploiting the property of a sound wave to propagate transversely through a solid. The example given by Inoué for hard oak cites a frequency of 4500 Hz parallel to the grain compared to 2500 Hz across the grain (hardwood is denser and thus gives a clearer result than softwood).

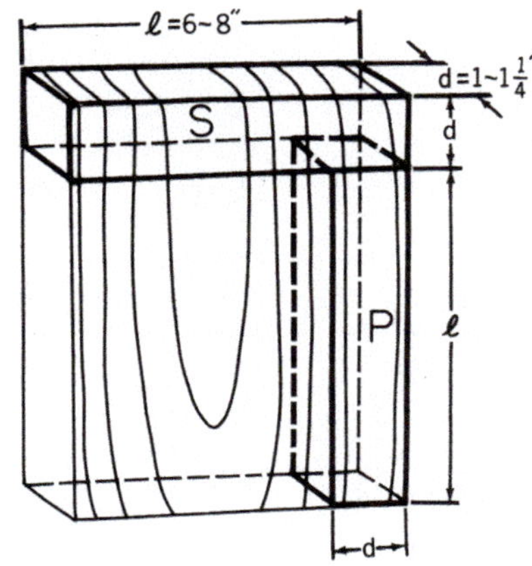

Exercise 14.10 The optic axis

Take a glass slide, and paint a streak of glue along the long dimension. Before it dries, shake some coarse sand along it so that the quartz crystals within the sand mixture adhere to the slide randomly.

Examine the crystals with a dry 40x or 63x objective. Notice that some grains appear not to be birefringent. These are grains whose optic axes lie perpendicular to the plane of the slide.

Exercise 14.11 Double refraction with calcite (1)

Draw a single spot on a piece of white paper. Take a piece of Perspex or a piece of glass, and place it over the spot. Notice that only one image of the spot is seen with this isotropic material. Rotate the Perspex or glass, and notice that the image of the spot does not move or change.

Now take a cleavage rhomb of calcite, and place it over the spot. Notice that two images of the spot are formed. As the calcite rhomb is rotated, one of the spots rotates (precesses) around the other. The two planes of polarisation are fixed with respect to the optic-axis emerging perpendicular to the plane of the paper. With care you should also be able to see that the images of the two spots seen through the calcite rhomb appear to be at different levels. The image formed from the ordinary ray (the spot that does not move) appears to be slightly higher than the image formed from the extraordinary ray (which moves).

Now look at the spot through a polar, preferably one whose preferred vibration direction is marked. Align the polar East-West. Experiment with rotating the calcite rhomb: the spots will appear to rotate. At a certain position, only one image of the spot will be seen. When the image disappears, the permitted vibration direction of the polar is orthogonal to the vibration direction of the light forming the image. Now rotate the polar North-South; only the other image of the spot is now seen. With the polar set at 45° both images of the spot (both at reduced intensity) will once again be seen.

Repeat this exercise, but keep the calcite rhomb static, and rotate a single polarising filter above it. The two spots appear to remain stationary, but disappear and reappear in turn, once in every 90° rotation. At one position both spots can be seen, again both at reduced intensity. This is evidence that the planes of polarisation are fixed orthogonally with respect to one another.

This is also evidence that the ordinary and extraordinary rays are mutually perpendicular to one another, and cannot interfere. This exercise also shows (with the polar set in the 45° position) that part of *both* horizontal and vertical vector components can pass through the polar to form the two images.

Exercise 14.12 Double refraction with calcite (2)

Draw a straight line on a piece of white paper. Take a crystal of calcite, and place it over the line. Rotate the crystal until the double image of the line becomes a single line. This will happen twice in every revolution and will mark the position of the optic axis through the crystal, where it assumes isotropic behaviour.

Exercise 14.13 Interference colours

Press hard together two microscope slides (take great care that they do not break!). A better result may be obtained by pressing together two larger plates of glass, such as the glass of 'clip-type' photograph display frames. This may require several attempts before it is successful.

Observe the intense hues of the thin film interference colours of Newton's rings. Each particular wavelength will occur along the spectrum at discrete intervals. The violet wavelength is every 410 nm, indigo: 450 nm, blue: 480 nm, green: 550 nm, yellow: 580 nm, orange: 600 nm and red every 710 nm. The bands for violet are therefore the most closely spaced and those for red, most widely spaced. For monochromatic light, it is easy to see the dark bands caused by destructive interference in this way.

Exercise 14.14 Double refraction and retardation (1)

Take a clear piece of Scotch tape or Sellotape (those types with a diffuse matt appearance are not suitable), and stick it onto a microscope slide with the long edges of the tape and slide parallel.

View the tape under crossed polars. The tape will appear brightest when it is set at the midway position of 45° between the two crossed polars.

Where the intensity seen in the tape is at a minimum (extinction), the permitted vibration direction of the tape is parallel to and aligned with either the polar or analyser. Determine that one vibration direction lies parallel to the long edge of the tape and one lies transverse, across the width of the tape.

If both polariser and analyser are now 'uncrossed' and set with their permitted vibration directions parallel to one another, the tape will now appear *darkest* at the midway position of 45° between the two polars. This is because each of the fast and slow polarisation vibration directions from the tape is blocked in turn by the analyser.

NB. Some cheap adhesive tape will tear obliquely off the reel. This is because the tape has been cut at an angle during manufacture, and the polymer chains are not aligned parallel to the length of the strip of tape. In this case, extinction of the tape will be set at an angle to the vibration directions of polars set E-W and N-S.

Exercise 14.15 Double refraction and retardation (2)

Take the slide and tape used in the previous experiment. Stick 7 or 8 further pieces of Scotch tape or Sellotape over the first, again with the long edges of the pieces of tape and slide parallel. Ensure that each successive piece of tape leaves 2–3 mm of the tape immediately underneath uncovered, so that a wedge of layers of tape is built up on the microscope slide.

View the adhesive tape wedge under crossed polars. Various interference colours should be seen in distinct steps (see Figure 14.8), in the same sequence of colours as in the Michel-Lévy chart (Appendix 3). The fifth or sixth piece of tape should approximate the 'first-order red' colour. These colours are brightest when the tape wedge is in the 45° position and fade to extinction when it is in the N-S or E-W position.

Exercise 14.16 Pleochroism

Examine a thin section of pleochroic material. A cleaved sheet of biotite ('black mica') is easiest to use. Tourmaline is a suitable alternative. Withdraw the analyser from the light path of the microscope, and examine the biotite or tourmaline under plane-polarised light from the polariser alone. Biotite, for example, will change from pale yellow to dark brown twice in one complete 360° rotation.

Exercise 14.17 Polarisation colours

Plants photosynthesise glucose from carbon dioxide and water, which is converted to starch that is stored as grains. These starch grains are composed of both amylose and amylopectin chains layered around a nucleation centre called the hilum. The short semi-crystalline amylopectin chains show positive form birefringence.

Smear the freshly cut face of a potato onto a slide to deposit grains of starch on its surface. Mount in a very small drop of water, cover and examine between crossed polars. A very small drop of water is necessary to prevent the starch grains moving in the field of view. Each grain will show white patches as quadrants between a distinctive extinction cross. The melatope, which indicates the optic axis, lies at the centre of the 'Maltese' cross. It appears offset to one end of the grain at the hilum.

Insert a full-wave 'sensitive tint' compensator plate into the accessory slot on the microscope. The background will go magenta, whilst opposite quadrants show either addition to give a blue hue or subtraction to give a yellow hue. This effect is consistent for each grain when the stage is rotated and is due to the circularly symmetrical regular arrangement of amylopectin chains making up the starch molecules.

This effect of addition and subtraction of Newton's colours using a compensation plate is also vividly seen with starch grains, in which the molecules have a radial configuration. With crossed polars the grains exhibit a 'Maltese cross' with dark E-W and N-S 'arms', or isogyres, and grey-white interference colours in the quadrants in between. With a 1 λ plate, the quadrants are alternately blue and yellow, and the arms assume the red sensitive tint colour.

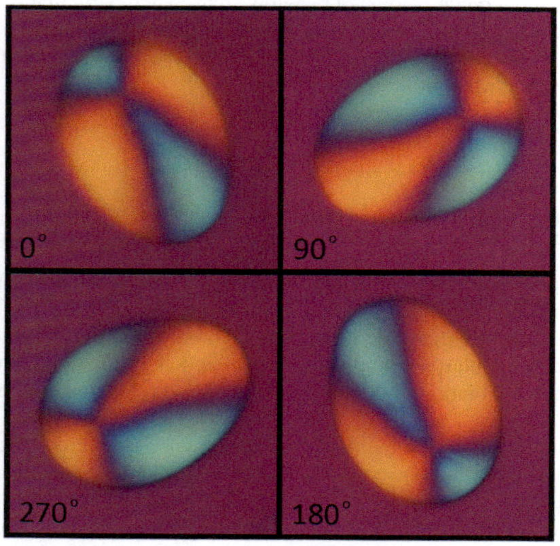

Exercise 14.18 How compensator plates work

Take a sheet of cellophane from a microscope slide box or CD wrapping. Examine this between crossed polars on a light box. As in Exercise 14.8, rotate the sheet between the crossed polars and note the four extinction positions. Between these positions, at 45°, there are four positions where the amplitude of the o-ray and e-ray are equal to one another and at maximal brightness. Mark these positions with a right-angled '+' using a marker pen. Cut the sheet into a smaller square so that the arms of the '+' are parallel to the sides of this square.

Fold over one corner at 45° so that the top layer of cellophane is now orientated at 90° to the bottom layer, and re-examine the folded square of cellophane at 45° between the crossed polars. The top sheet should remain dark because the phase displacement (OPD) of the top sheet cancels out that of the bottom sheet. The top sheet is acting as a retardation plate in this case.

Now unfold the corner and refold the square of cellophane sheet into a rectangle so that the opposite edges are parallel to one another. The top layer is orientated parallel (and not at 90° as before) to the bottom sheet. Placed at 45° between crossed polars, the identical vibrations (slow-slow or fast-fast: we have not determined which) should add and thus a blue or purple colour going up Newton's scale should be seen.

If the polars are set now uncrossed and set parallel, the colour of the cellophane rectangle should go down Newton's scale and be yellow or almost colourless, since blue can now get through the analyser and is not blocked to give the complementary colour yellow.

To determine the fast and slow vibration direction of the cellophane sheet, take two small strips of Sellotape, which has its slow direction parallel to the length of the tape and fast direction parallel to the cut edge of the tape. The Sellotape is being used as a compensator here.

Stick each piece of tape at 90° to each other on the folded rectangle of cellophane, with each piece of tape being stuck parallel to the edges of the cellophane.

Place the folded sheet at 45° between crossed polars so that the polarisation colours of the tape can both be seen. One piece should be pale blue and the other pale yellow against the brighter yellow of the folded cellophane sheet. The slow axis of the cellophane will be parallel to the slow axis of the tape, seen here as blue. This exercise can also be found in Murphy and Davidson (2013).

Exercise 14.19 Determining fast and slow directions

Take the Sellotape wedge made in Exercise 14.15, and set it at 45° between crossed polars on a light box. View this through the analyser with a 1 λ 'sensitive-tint' compensator held over the tape. The compensator should be held between the tape and the analyser.

If the colour observed is higher on Newton's scale (e.g. blue), then the slow direction of the wedge coincides with the slow direction of the Sellotape. If the colour is lower (e.g. yellow), then the fast direction of the Sellotape coincides with the long dimension of the wedge.

Beléndez et al. (2010) have measured the thickness of Sellotape, or Scotch tape, to be 42.5 μm, with a birefringence (ΔM) of 0.0077. Edwards & Langley (1981) cite a tape thickness of 33.5 μm. These values should give OPDs between 220 and 280 nm on the Michel-Lévy chart, depending on the tape used.

Exercise 14.20 Observing interference figures without a microscope

Either hold or set in a clamp two polars such that a sheet of mica can be viewed between them, with a pearl (diffuse) light bulb (or, better, a light box) behind as the illuminant. At a relatively long distance of a metre – with virtually parallel (orthoscopic) illumination – the mica sheet will be seen showing a homogenous polarisation colour, depending upon its thickness.

Bring the mica sandwiched between the two polars very close to your eye. This approximates conoscopic illumination. The angle of the illuminating rays should now be highly divergent, and the angular aperture increases. A biaxial interference figure can be seen. It should be possible to identify dark bands (isogyres) and the curved coloured bands in between these isogyres (isochromes).

Cleave a much thinner sheet of mica off the original material with a fine scalpel. This time the polarisation colour under orthoscopic illumination should go down Newton's scale. Because the sheet is now much thinner, the isogyres will be much less pronounced, or even absent. At a particular orientation (possibly with the edges of the square mica sheet parallel to the crossed polarisers) a clear extinction cross can be seen against a pale grey background. As the mica sheet is rotated to 45° between crossed polars, so the arms of the cross will draw apart as two hyperbolae.

Repeat the exercise with a thicker section of quartz or a small piece of calcite. Why should these both appear white when mica shows distinct polarisation colours? All three anisotropic minerals show polarisation colours between crossed polars, but in the latter cases the thickness of the quartz or calcite is such that only a 'higher-order white' (i.e. colourless) is seen. Mica cleaves easily into sheets less than 200 μm thick and therefore is thin enough to display lower-order polarisation colours.

As an alternative to viewing interference figures close-up by eye, a simple Abbe condenser can be used to project a magnified image of an interference figure onto a screen. A polar is inserted underneath the condenser, with the crystal sample on a slide at the focal point of the condenser. The second polar is held, or placed, above the sample and the image projected onto a screen. Further details may be found in Buckley (1939).

Chapter 15 now tells you about fluorescence and how the use of fluorescent markers allows the detection of multiple cellular components. Fluorophores can be attached to target molecules by specific functional groups. Alternatively, fluorescent proteins can be expressed exogenously or as fusion protein for reporting cell structure and function. Fluorescence microscopy has become a ubiquitous and indispensable tool in biological microscopy.

Key Reading

Barker, A. (2014). *A key for identification of rock-forming minerals in thin section*. Leiden: CRC Press.
- A practical identification key using the principles explained in this chapter

References

Beléndez, A. et al. (2010). *European Journal of Physics 31*: 551–561.
Bloss, F.D. (1981). *The Spindle Stage: Principles and Practice*. Cambridge: CUP.
Bowen, A.M. (2009). 'Optical crystallography of silver sulfadiazine'. *The Microscope 57*/1: 11–18.
Buckley, H.E. (1939). 'Demonstration of Optical Interference', Figures. *Nature 143*/3628: 801–802.
Delly, J.G. (2003). 'The Michel-Lévy Interference Color Chart – Microscopy's Magical Color Key'.
Dyar, M.D., Gunter, M.E., and Tasa, D. (2008). *Mineralogy and Optical Mineralogy*. Chantilly, Virginia: Mineralogical Society of America.
Edwards, S.J. and Langley, A.J. (1981). *Leonardo 14*/3: 187–190.
Hammond, C. (2001). 'A note on Newton's measurements of film thicknesses and their corresponding (interference) colours'. *Notes Rec. Roy. Soc. Lond. 55*/3: 391–394.
Houck, M.M. and Siegel, J.A. (2010). Figure 4.13, page 91, in: *Fundamentals of Forensic Science*, 2nd ed. Burlington, MA: Academic Press.
Hartshorne, N.H. and Stuart, A. (1970). *Crystals and the polarising microscope*, 4th ed. London: Edward Arnold.
Humphries, D.W. (1992). *The Preparation of Thin Sections of Rocks, Minerals, and Ceramics*. Microscopy Handbook No. 24. Oxford: Oxford Science Publications.
Kile, D.E. (2003). *The Petrographic Microscope*. Supplement to *The Mineralogical Record*. Chantilly, Virginia: Mineralogical Society of America.
MacKenzie, W.S. and Adams, A.E. (1994). *A Colour Atlas of Rocks and Minerals in Thin Section*. London: Manson Publishing.
Murphy, D.B. (2001). *Fundamentals of Light Microscopy and Electronic Imaging*, 1st ed. New York: Wiley-Liss.
Murphy, D.B. and Davidson, M.W. (2013). Polarization microscopy. In: *Fundamentals of Light Microscopy & Electronic Imaging*, Chapter 9. Hoboken, New Jersey: Wiley-Blackwell, pp. 162–164.
Nesse, W.D. (2009). *Introduction to Optical Mineralogy*. Oxford: OUP.
Patzelt, W.J. (1974). *Polarized-Light Microscopy: Principles, Instruments, Applications*. Leitz Wetzlar, booklet 550-51/Engl.
Raith, M.M., Raase, P., and Reinhardt, J. (2012). *Guide to Thin Section Microscopy*, 2nd ed. Chantilly, Virginia: Mineralogical Society of America.
Robinson, P.C. and Bradbury, S. (1992). *Qualitative Polarised-Light Microscopy*. Microscopy Handbook No. 9. Oxford: Oxford Science Publications.
Inoué, S. (1986). Figure III-1, Appendix III, Introduction to biological polarization microscopy. In: *Video Microscopy*. New York: Plenum Press, p. 478.
Quincke, G. (1866). Optische Experimental-Untersuchungen. *Annalen Physik Chemie 79*/10: 177–218.
Spencer, M. (1982). *Fundamentals of Light Microscopy*. IUPAB Biophysics Series, Vol. 6. Cambridge University Press.
Tilley, R.J.D. (2011). *Colour and the Optical Properties of Materials*, 2nd ed. Chichester: Wiley.

Vincent, H.C.G. (1955). 'Projector for interference figures and for direct measurement of 2V'. *Minerological Magazine 30*/229: 666–671.
Wayne, R. (2009). *Light and Video Microscopy*, 1st ed. Kidlington, Oxford: Elsevier.
Wood, E.A. (1977). *Crystals and Light: An Introduction to Optical Crystallography*, 2nd ed. New York: Dover.

Further Reading

Delly, J.G. (1998). 'Parallel polarizers'. *The Microscope 46*/4: 207–214.
Gribble, C.D. and Hall, A.J. (1992). *Optical Mineralogy: Principles and Practice*. London: UCL Press.
Nesse, W.D. (2009). *Introduction to Optical Mineralogy*. Oxford OUP.
Perkins, D. and Henke, K.R. (2004). *Minerals in Thin Section*, 2nd ed. New Jersey: Pearson Education.

Internet Resources

Polarised microscope manuals: http://earth2geologists.net/Microscopes/index.htm (James G McHone's excellent website).
Generic reference page of web links: https://serc.carleton.edu/NAGTWorkshops/mineralogy/optical_mineralogy_petrography.html and https://www.mccrone.com/mm/article_pix/070718_ustage/ustage.pdf.
Mineral Society of America: http://www.minsocam.org/.
Delly, J. G. (2008). Selected topics from *Essentials of Polarized Light Microscopy*. http://www.jysco.com/archives/asbestos/PLM_reading_McCrone.pdf.
Michel-Levy chart: https://www.mccrone.com/mm/the-michel-levy-interference-color-chart-microscopys-magical-color-key/ and https://www.sciencedirect.com/science/article/pii/B9780124072305500030.
Using the Spindle stage:

1. http://www.minsocam.org/msa/Monographs/Mngrph_03/MG003_309-318.pdf
2. http://www.science.smith.edu/geosciences/min_jb/ExcaliburInstr.pdf
3. http://www.minsocam.org/ammin/AM44/AM44_1272.pdf
4. http://www.minsocam.org/ammin/AM63/AM63_433.pdf.

The Universal stage:

1. https://www.geochemsoc.org/publications/geochemicalnews/gn140jul09/theuniversalstage/
2. https://www.geochemsoc.org/files/3513/4436/8119/GN140kile.pdf.

Mineral identification table: http://funnel.sfsu.edu/courses/geol426/handouts/mintable.pdf.
Compensators:

1. https://www.olympus-lifescience.com/en/microscope-resource/
2. John Delly (2003) on de Sénarmont use. https://www.mccrone.com/mm/senarmont-compensation-how-to-accurately-measuresmall-relative-retardations-0-1/.

Conoscopy:

1. http://jm-derochette.be/conoscopy.htm
2. http://www.minsocam.org/ammin/AM23/AM23_595.pdf.

Notes

1. Colours produced from the interference of light rays passing through doubly anisotropic 'birefringent' materials should be called polarisation colours and not interference colours. When two coherent rays that are vibrating in the same plane interfere, they can do so constructively or destructively to produce a new wave of different amplitude. The two mutually perpendicular rays arising from a wave of light passing through an anisotropic material have a common origin (i.e. are coherent) but cannot interfere, although components of both can interfere at the analyser. The resulting polarisation colour is determined by the full white-light spectrum *minus* the colour of the wavelength that has undergone elimination by destructive interference between the two rays, as explained in Section 14.10. Interference can also occur when a coherent beam of light is reflected off both top surface and bottom surface of a sufficiently thin, transparent film, for example, an oil film on a puddle of water (see Section 14.11). This is an example of interference occurring in the absence of diffraction or anisotropy.
2. The term 'doubly refracting' is best used to describe optical anisotropy. Birefringence refers to the modulus of the maximum numerical difference in refractive index (i.e. ($|n_1 - n_2|$)) arising from double refraction in anisotropic materials. Since the term birefringence is used colloquially to mean 'double refraction', it is used in this way in this chapter.
3. Crossed polars does not automatically guarantee the polars are set E-W and N-S. Generally the polariser and analyser have click-stop detents to ensure true alignment East-West and North-South.
4. Since the polarising microscope is primarily an analytical instrument, perfect alignment with respect to an azimuth at 0° is necessary for determining angles or if birefringence measurements are made with compensators. Rectified optics (see endnote 11, Chapter 13), if used, will virtually eliminate the extinction cross.
5. In gemmology, calcite is used as a dichroscope: if a material is dichroic the two images are coloured differently.
6. The extinction factor is properly measured with an anisotropic object between the crossed polars. It is expressed as a ratio: $X_{\parallel}/X_{\perp}$, where the X values are intensities measured of the entire unit of polar and anisotropic object between parallel and crossed polars, respectively. The extinction factor is higher for low numerical aperture than objectives of high NA. This reduced extinction factor is due to the rotation of the azimuth of polarisation when polarised light strikes the surface of the lens components in high-NA objectives at large angles of incidence. Such depolarisation is manifested by the 'extinction cross', a pattern seen as a dark cross in the back focal plane when the aperture of the objective is fully illuminated, the polars are crossed, but no anisotropic object lies between these. Until the introduction of rectified optical components by Shinya Inoué, limiting the NA of

the microscope was the only way to achieve reasonably high extinction factors necessary to observe the very small birefringence shown by certain organelles in living cells.

7. Because of its high birefringence value, the minimum thickness that calcite has to be is about 1 mm to resolve two spots distinctly. A sample of quartz would have to be 19 times as thick as calcite to do the same.
8. The colours of the isochromes at any particular point in the interference figure are determined by the optical path difference produced by light travelling in that direction. The number of rings depends on the thickness of the crystal as well as the birefringence. More isochromes may be seen in a thick crystal or material exhibiting high birefringence.

15
Fluorescence Microscopy

Outline

Understanding Light Microscopy, First Edition. Jeremy Sanderson.

15.1 Introduction

Photoluminescence describes the absorption and re-radiation, or emission, of photons. Luminescence is the production of light at low temperature by cold-body radiation, whereas incandescence is the production of light from a heated body (e.g. the sun or a heated filament).

Fluorescence is part of the family of luminescence. Luminescence can be divided into several sub-categories. For example, chemiluminesence is the product of a chemical reaction (e.g. in 'glow-sticks' employing diphenyl oxalate), while cathodoluminescence (e.g. in electron microscopes, cathode ray tubes and old TV tubes) occurs when a beam of electrons impacts upon a luminescent material, such as phosphor, to emit visible light. Cathodoluminescence is also used in material science to identify crystal defects and trace elements in the mineral or material bulk.

Bioluminescence is the production and emission of light by living organisms (e.g. fireflies, glow worms, jellyfish). Most bioluminescence occurs in marine animals, which emit in the blue and green part of the spectrum, since these wavelengths transmit through seawater most easily. Terrestrial bioluminescence is less common but exhibits a wider spectrum of coloured light. In firefly larvae, the bioluminescence serves as a warning to predators, whilst at the adult stage it serves as a mating signal. The enzyme catalyst luciferase acts upon luciferin in the presence of magnesium ions, ATP, and oxygen to produce light. This reaction has been adopted as a non-fluorescent marker in biology (Greer 3rd & Szalay, 2002; Badr & Tannous, 2011; Waidmann et al., 2011).

The main advantage of bioluminescence imaging lies in its ability to detect very low levels of gene expression due to its virtually background-free light emission. Luciferase has a short half-life; it matures quickly and penetrates tissues easily. Since luciferase bioluminescence is a chemiluminescent reaction and does not require excitation light from an external source, there is no autofluorescence. Because the signal intensity is very low, far less than fluorescence, both EM-CCD camera(s) and the microscope must be used in a *totally* dark environment, and excluding all extraneous photons can be a practical problem. In the coelenterate medusa *Aequorea victoria*, luciferin (in this case coelenterazine) is combined with luciferase and oxygen to form the photo-protein aequorin. The reaction is triggered by calcium ions. Green fluorescent protein (GFP) converts the blue signal to green emitted light. GFP and other important (auto)-fluorescent molecules are endogenous expression proteins; they are discussed further in Chapter 16.

Fluorescence, which we are interested in, refers to the luminescent property of some atoms and molecules to absorb light of one colour (i.e. at a particular wavelength) from an external energy source and subsequently to re-emit light of another colour at a longer wavelength after a very brief interval, which is of the order of 10^{-9}–10^{-12} seconds (Box 15.1). This process is red-shifted due to the inevitable loss of energy associated with the process.

Box 15.1 The Fluorescence Process

The Jabłoński diagram (strictly known as a Jabłoński-Perrin diagram) is named after the Polish scientist Professor Alexander Jabłoński. Within each energy level, there are various vibrational spin states.

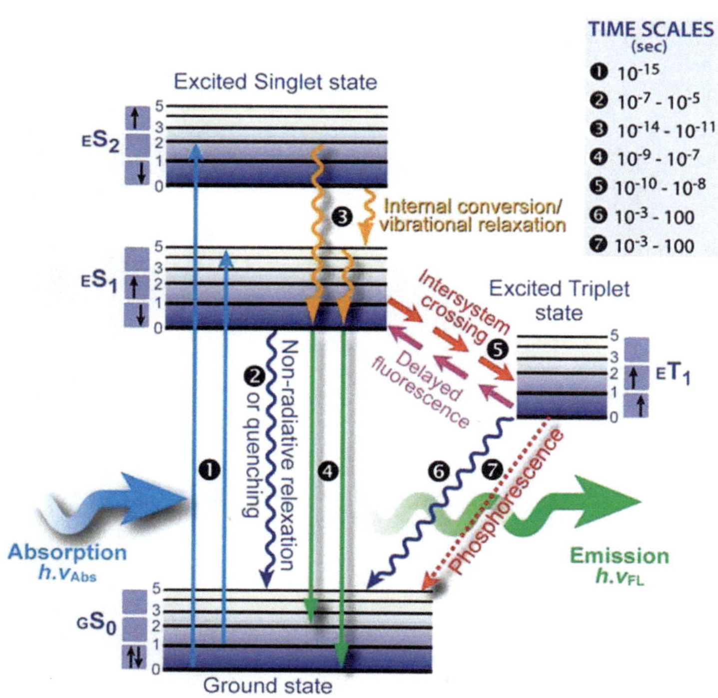

Source: Ishikawa-Ankerhold et al. 2012. Reproduced under the terms of the Creative Commons Attribution License, CC BY 3.0.

When a photon is absorbed, the molecule usually is not merely transferred into an excited electronic state, but also acquires some vibrational energy. Photon absorption is instantaneous because it involves only the rearrangement of inertia-free electrons. The vastly much heavier nuclei vibrate much later. Excitation is therefore represented, according to the Franck-Condon principle, by a vertical arrow in the Jabłoński diagram.

The Morse diagram below shows that there is no displacement of the much heavier nuclei within the femtosecond absorption timeframe. Since the shape of the upper and lower potential energy curves of the ground state and excited state are identical, or at least very similar, in shape, the emission spectrum is therefore a mirror image of the absorption spectrum, but displaced towards a lower energy wavelength (the *Rotverschiebung* or red-shift) more usually called the Stokes shift.

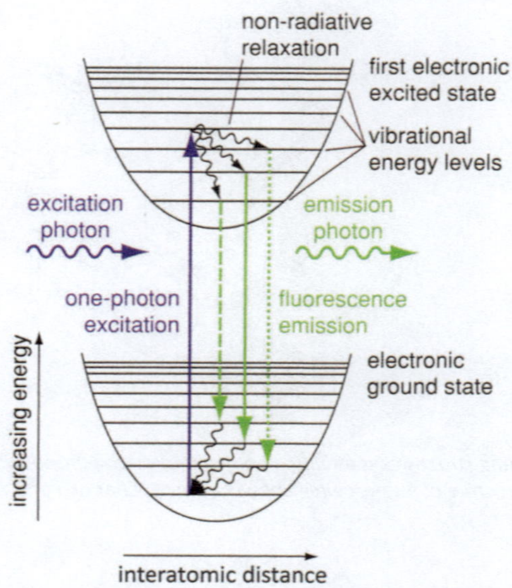

	Time (s)	
Absorption of photon	$<10^{-15}$	
Vibrational relaxation	$10^{-14}-10^{-11}$	
Non-radiative relaxation (quenching)	$10^{-7}-10^{-5}$	
Fluorescence lifetime, τ_f	$10^{-10}-10^{-7}$ →	Fluorescence
Intersystem crossing	$10^{-10}-10^{-8}$	
Internal conversion	$10^{-11}-10^{-9}$	
Lifetime via triplet state T_1	$10^{-4}-10^{-2}$ →	Phosphorescence

Source: Watkins and St. Croix 2013. Reproduced with permission from John Wiley & Sons.

References

Ishikawa-Ankerhold, H.C., Ankerhold, R., and Drummen, G.P. (2012). 'Advanced Fluorescence Microscopy Techniques – FRAP, FLIP, FLAP, FRET and FLIM'. *Molecules 17*/4: 4047–4132. doi: 10.3390/molecules17044047. http://www.mdpi.com/1420-3049/17/4/4047.

Watkins, S.C. and St. Croix, C.M. (eds.) (2013). *Current Protocols Select Imaging & Microscopy*. New Jersey, Hoboken: Wiley.

Phosphorescence is due to the *delayed* emission of photons from an excited state. Due to the nature of quantum mechanics, the absorbed radiation may be re-emitted at a lower intensity for up to several hours. The dials and hands of navigational instruments are often coated with phosphorescent materials. Optical brighteners, such as those found in washing powders, are phosphorescent chemicals. When the UV component in daylight falls on recently washed clothes, atoms of these optical brighteners, left behind by the detergents, become excited and give off a blueish light. More recently, phosphorescent compounds have been incorporated into LEDs to produce a range of colours across the spectrum.

The terms 'fluorochrome' and 'fluorophore' are often used interchangeably. Fluorophore refers to a chemical group that confers fluorescence upon a molecule. Similarly, a chromophore is the part of a molecule responsible for its colour. In this book, although very strictly incorrect, the term fluorophore is used to refer to a fluorescent compound, whether as a stain or dye, an adsorbed inorganic ion or compound, as a tracer or indicator (in fluids or tissues) or conjugated to a non-fluorescent molecule or protein. This terminology is used because 'fluorophore' is much more widely used than the correct word 'fluorochrome'.

15.2 How Fluorescence Occurs

Fluorescent compounds typically have delocalised electron clouds associated with the aromatic ring structures present in the fluorescent molecule. That is, they possess alternating single and double bonds arranged in an aromatic ring (Figure 15.1). The conjugated[1] (π) bonds distribute outer orbital electrons over the surface of the fluorochrome.

A photon of energy ($h\nu_{EX}$) from an external source such as an incandescent lamp or a laser is absorbed by the fluorophore. Alterations in the electronic and vibrational state of the fluorescent molecule occur, promoting electrons to higher energy levels. For example, a valence electron is pushed from the ground state (S_0) to a higher orbital within the fluorophore, creating an excited electronic singlet state (S_1).

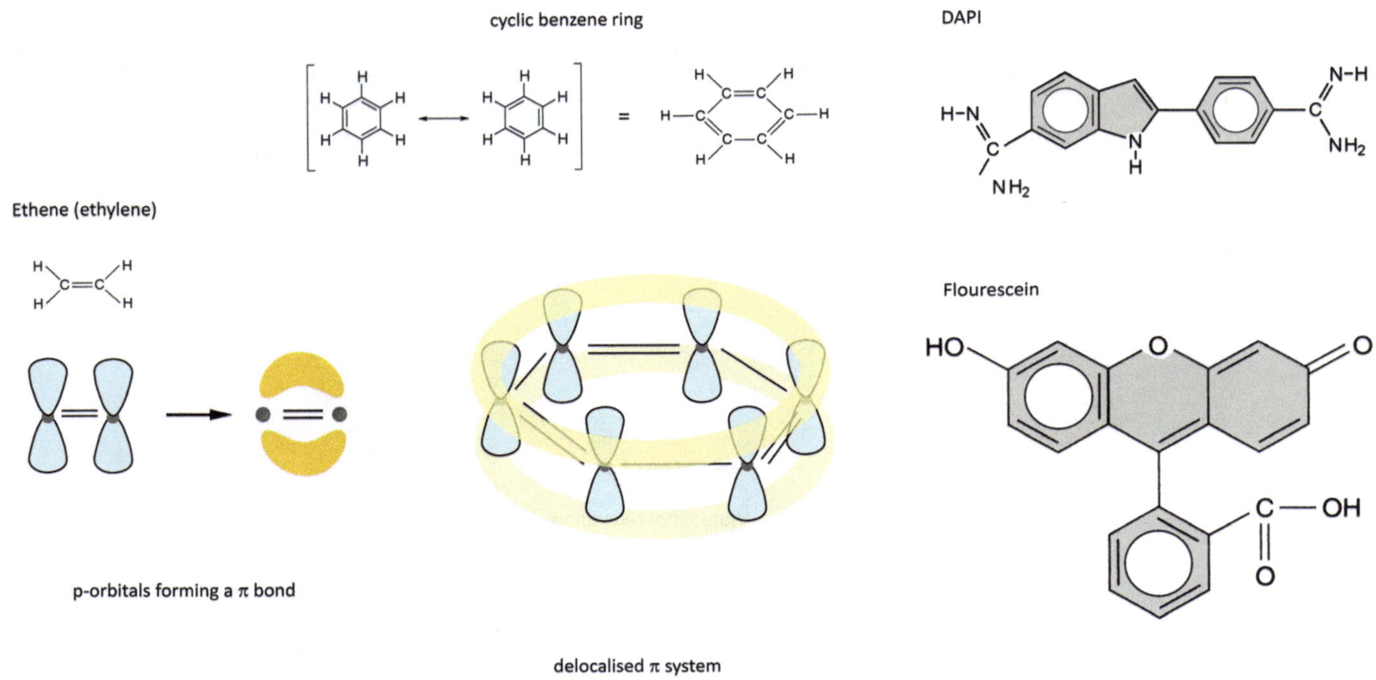

Figure 15.1 ***Fluorescence aromatic ring structure*** *Stains that absorb visible light have ring structures containing conjugated double bonds. In general, the greater the number of rings, the further red-shifted is the fluorophore. The molecular structure of fluorescein is shown. Source: Chandler and Robertson 2009. Reproduced with permission from D. Chandler.*

The structure of the outermost electron orbitals determines the efficiency and the spectral behaviour of the fluorophore. From Chapter 2 we saw that photons are discrete quanta, or 'wave packets', of energy. Absorption of photons occurs at those wavelengths whose quantum energy is equivalent to the difference in energy between the ground electronic orbital state and the higher excitation orbital.

Sometimes the energy from the absorbed photon excites an electron to move not just from its resting orbital (ground state) to a singlet state, S_1, but to a higher one, S_n. Thus a *range* of wavelengths can excite a fluorophore, giving rise to the absorption spectrum (Figure 15.2). The time taken for absorption of the photon is approximately 1 femtosecond (10^{-15} sec).

The excited electron state exists for a finite time. A photon of energy $h\nu_{EM}$ is emitted, returning the electron to its ground state, S_0. The time interval taken for decay from the excited electronic singlet state (S_1) to the ground state (S_0) is known as the fluorescence lifetime. This period lasts from 1×10^{-9} to 1×10^{-6} seconds, a much slower process than photon absorption. The fluorescence lifetime (τ_f) is merely a related way of measuring and describing the characteristics of a fluorophore: by its lifetime rather than by its absorption and emission spectral properties. FLIM, fluorescence lifetime imaging microscopy, has the advantage that it is independent of fluorophore concentration and thus a much more accurate indicator of the environment of the molecule and whether it is bound to, or in close proximity to, another molecule.

Apart from fluorescence emission, there are several other possible fates available to the excited electron. From S_2 the electron may return to S_1 by vibrational relaxation without emitting a photon (typically lasting from 1×10^{-14} to 1×10^{-11} seconds). Some electrons also return to the vibrational ground state, S_0, without emitting a photon, the energy being dissipated as heat. A more common possibility is that the energy of S_1 is partially dissipated, yielding a relaxed singlet excited state (S_1') from which fluorescence emission originates as the electron returns to the ground state, S_0. Other processes such as collisional quenching, fluorescence resonance energy transfer (e.g. FRET) and intersystem crossing (eventually leading to slower decay which is seen as phosphorescence) may also depopulate S_1.

Due to energy dissipation during the excited-state lifetime, the energy of the emitted photon ($h\nu_{EM}$) is lower than that of the excitatory, or excitation, photon ($h\nu_{EX}$). Because wavelength varies inversely to the radiative energy of the photon, fluorescence emission is *always* at a longer wavelength (i.e. of lower energy) than that of the illuminating radiation used to excite the fluorophore. The difference in energy or wavelength, represented by $h\nu_{EX} - h\nu_{EM}$, is called the Stokes shift. The Stokes shift is fundamental to the sensitivity of fluorescence techniques because it allows emission photons to be detected against a low background, isolated from excitation photons.

The entire fluorescence process is cyclical. Until it is irreversibly destroyed in the excited state (an important phenomenon known as photobleaching, Figure 15.3), the same fluorophore can be repeatedly excited and detected, usually for 10 000–100 000 cycles before photobleaching. The fact that a single fluorophore can generate many thousands of detectable photons is fundamental to the high sensitivity of fluorescence detection techniques.

Because the energy levels of the ground state orbital and excited state orbitals are spread slightly and there are several inter-orbital vibrational states, it is possible to excite the fluorophore with photons of a range of different energies (see the Jabłoński diagram in Box 15.1). This gives rise to the unique spectral excitation curve characteristic of the fluorophore. The spectral curve represents

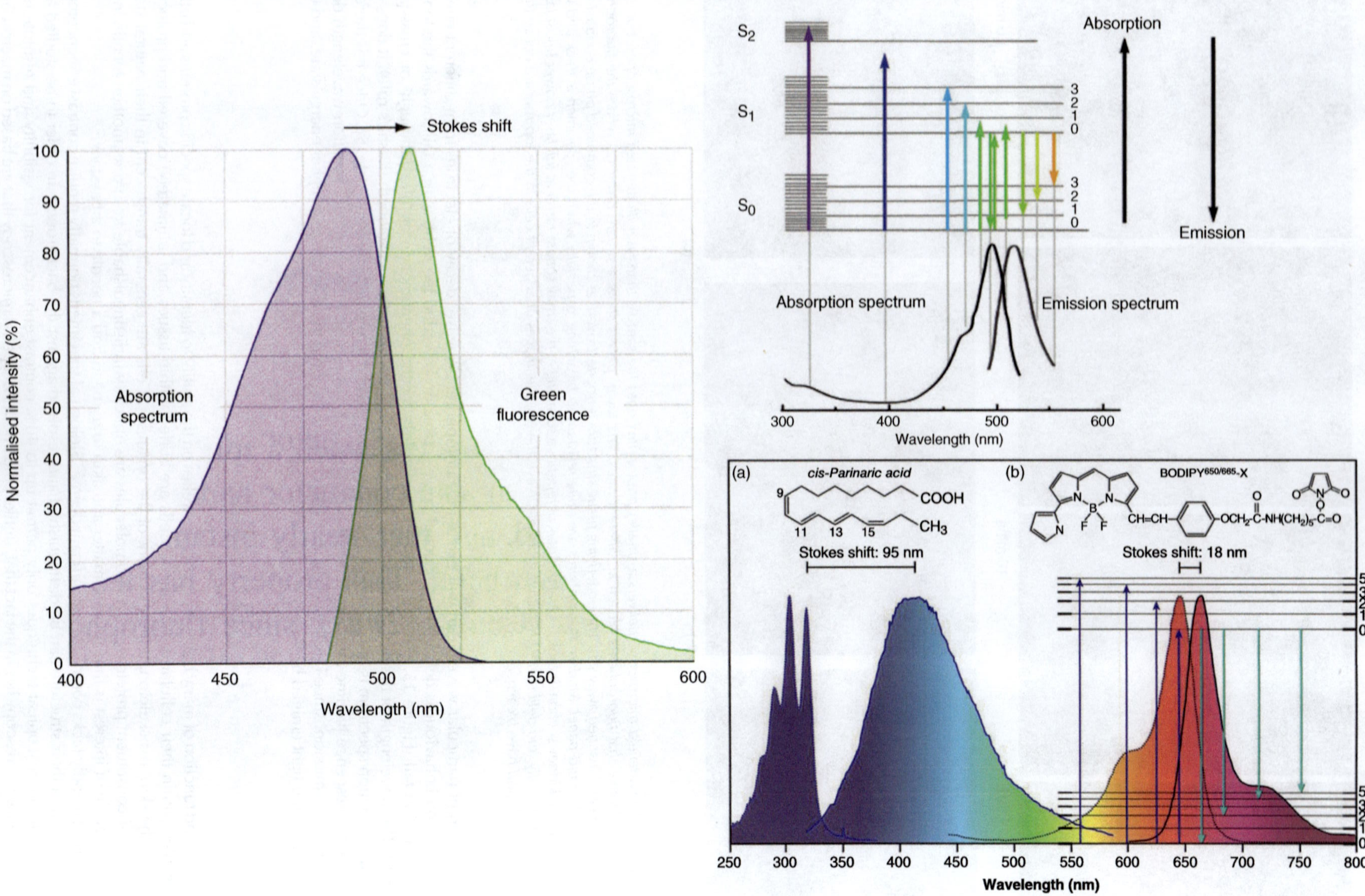

Figure 15.2 ***Fluorophore spectral curves*** *A typical absorption and emission spectrum is shown. Note the tail of the emission spectrum on the right-hand of the figure compared to the relatively steep 'cut-on' of the absorption, or excitation, spectrum on the left. Compare this figure with Figure 15.11, which shows how filter sets are matched to the spectral characteristics of the fluorophore, and to Figure 22.4, which shows the two-photon spectrum of eGFP. Source: (left) Watkins and St. Croix 2013. Reproduced with permission from John Wiley & Sons. (top right) Lichtman and Conchello 2005. Reproduced wilth permission from Macmillan Publishers Ltd. (bottom right) Ishikawa-Ankerhold et al. 2012. Reproduced under the terms of the Creative Commons Attribution License, CC BY 3.0.*

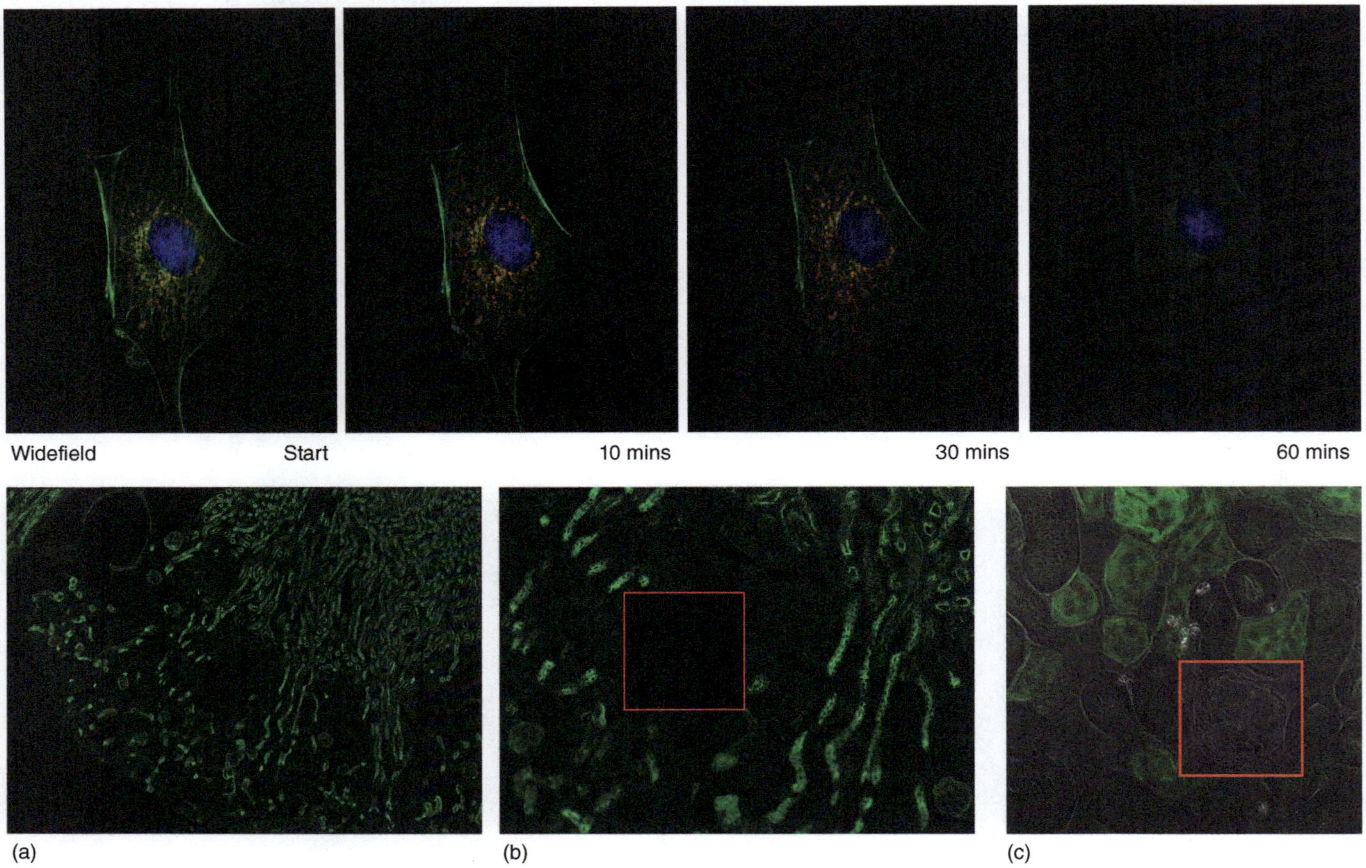

Figure 15.3 **Photobleaching** *The widefield microscope delivers a smaller photon flux onto the sample than laser-illuminated microscopes, but even so, photobleaching still occurs. Notice that fluorophores bleach at different rates. The lower panels show a square raster (a) where the fluorophore labelling the kidney section has been bleached away (b). The underlying tissue structure is not destroyed, as shown in the combined phase contrast and fluorescence image (c) of the right-hand panel. Modern fluorophores are quite resistant to bleaching, especially when the specimen is thick and well stained like this 16 μm thick mouse kidney section. It took several minutes of repetitive scanning of the small square raster at full laser power (most lasers are used at <2% power) to successfully and deliberately bleach the sample. Nevertheless, the images of adherent cells in the top row are not so thick, have not been stained so intensely and have bleached much faster.*

the probability that a photon of particular wavelength will excite the fluorophore. Likewise, decay of electrons from different energy levels means that the emission behaviour is defined by a spectral curve – usually skewed to the right, with a tail towards the longer wavelengths. It is this emission tail (Figure 15.4) that is primarily responsible for the phenomenon of 'bleed-through' or cross-talk, discussed in Chapter 20. The wavelength of the emitted photon is independent of the excitation wavelength (Kasha's rule). It does not matter from where in the excitation spectrum is the wavelength of the electron excited to the higher energy state, S_1. Once it is there, it will decay with the same fluorescence lifetime and the same *emission spectrum*, or profile. However, the excitation wavelength does affect the apparent *intensity* of emission, since it is less likely, in all probability, to excite the electron to a higher energy state, and this is reflected in the intensity of the light emitted by the fluorophore (see Figure 20.6).

15.3 The Importance of Fluorescence in Biological Research

Understanding the nature and function of many biological processes relies on the ability to identify and locate specific proteins and other important macromolecules within their cellular environment. Cells are buried within tissues and (ultimately) covered and protected from the environment by the skin, exoskeleton, cuticle or bark of the organism. Cells are generally translucent in their natural state, and the immense number of constituent proteins and macromolecules are optically indistinguishable from one another. Identifying a single protein or other molecule of interest is thus no easy matter and akin to trying to find a needle in a haystack.

Untreated or living tissue generally lacks contrast to a significant degree. Biological specimens normally require contrast enhancement so that the fine detail inherent in their structure can be distinguished and resolved against the background, in order to be studied and understood. Most bright-field microscopical techniques only permit up to 40% contrast enhancement for visibility and resolution of detail. The value of fluorescence microscopy lies in the fact that, unlike other modes of optical microscopy that are based on macroscopic

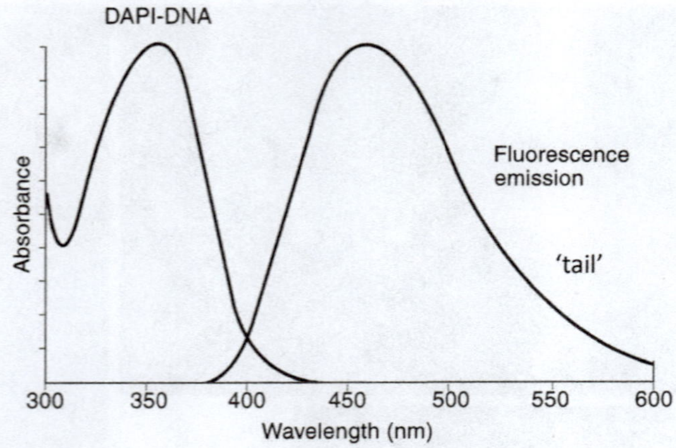

Figure 15.4 ***Fluorophore emission tail*** *The tail of the emission spectrum of DAPI is very large compared to other fluorophores. It should not be used too intensely as a stain, and fluorophores used with it should be selected carefully to avoid, or at least minimise, bleed-through. Source: Adapted from Murphy 2001 with permission from John Wiley & Sons.*

specimen features, such as phase gradients and birefringence, fluorescence microscopy is capable of imaging with very high contrast and visibility: the distribution of a single molecular species can be accurately determined based solely on the properties of bright fluorescence emission.

By the use of very high-quality filters, the illuminating excitation light can be efficiently separated from the weaker signal emitted from the sample, so only the latter contributes to the image. Conversely, bright-field absorption techniques exhibit much lower contrast because the amount of light absorbed that forms the image is – particularly for small objects – only marginally different from the illuminated background. The magnitude of the Stokes shift determines how easily the fluorophore absorption spectrum may be separated from the emitted signal or from the emission of a second fluorophore within a multiply-labelled sample. The greater the Stokes shift, the easier it is to separate the weaker emitted signal from the excitation illumination (see Figure 15.2, Figure 20.3 and Figure 20.4).

It is therefore possible to mark organelles, proteins and molecules with great specificity, sensitivity and with a very high degree of contrast (see Figure 15.5). Another reason for the widespread use of fluorescent markers in biological research is due to their versatility at being incorporated into living tissues and organisms. Fluorescent markers can be applied as fluorescent dyes, exogenous markers and endogenous genetically derived protein tags. Furthermore the specificity and sensitivity of these antibody-conjugated probes and genetically engineered fluorescent protein constructs allows multiple labelling and the precise location of intracellular components, which can be monitored over time. Fluorescence labelling can also help understand protein diffusion coefficients, transport characteristics and interactions with other biomolecules. In addition, the dramatic response in fluorescence to changes in localised environmental variables enables the investigation of pH, viscosity, refractive index, ionic concentrations, membrane potential and solvent polarity in living cells and tissues.

Concomitant with the development of fluorescent dyes, inorganic probes and natural fluorescent proteins has been the development of high-quality CCD cameras, sCMOS cameras and PMT detectors. For these reasons, fluorescence microscopy has come to dominate biological and medical research. In short, the organism becomes its own 'test tube'. Fluorescence microscopy is used much less by materials scientists (e.g. demonstrating grain boundaries). The advantages of fluorescence for life scientists as a contrast-enhancing labelling technique are:

- Self-luminous fluorophores – exhibit very high contrast (clean labelling experiments with no background or non-specific staining!)
- Very sensitive markers – good detection: can mark single molecules at concentrations that can be tolerated physiologically
- Very specific markers – very good discrimination of individual components
- Multiple labelling possible
- Versatile – antibody markers and endogenous fluorescent proteins available
- Protein kinetics can be analysed – controlled photobleaching used to advantage
- Analysis of interactions between molecules – using FRET and ratiometric markers
- Economical – relatively cheap, also able to label for high-throughput studies
- Very good electronic detectors able to collect weak fluorescent signals

All of these clearly outweigh the few disadvantages:

- blurring – self-luminous signal apparent outside depth of field through entire sample
- bleaching – finite lifetime of electrons cycling between orbitals
- bleed-through – cross-talk and bleed-through from spectral overlap of fluorophores

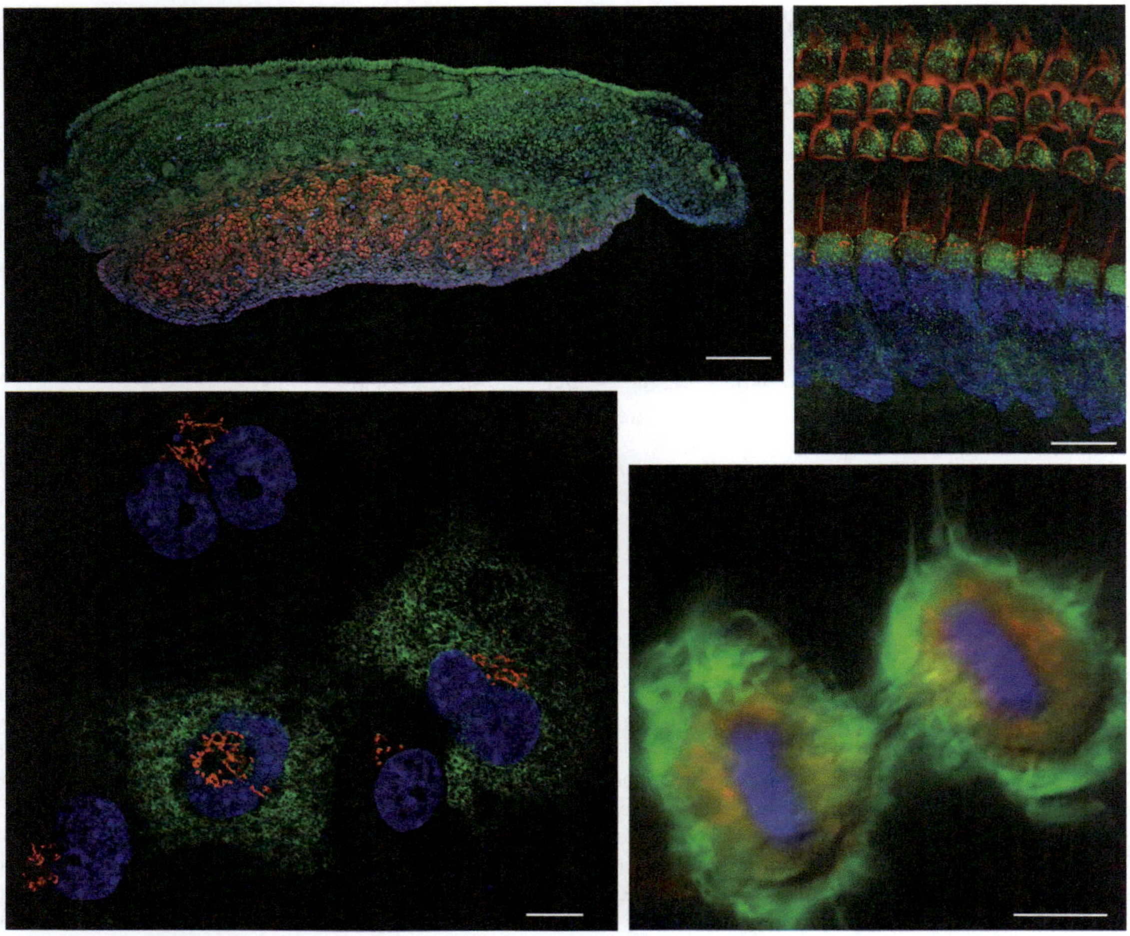

Figure 15.5 ***The specificity and sensitivity of fluorescence*** *Whether in individual cells or thicker tissues, fluorophores can be used to mark structures very specifically, and with sensitivity that often allows very low fluorophore concentrations that can be tolerated over time by living cells and organ explants. The upper left-hand panel shows mouse gonad tissue (Source: Courtesy of Dr Nick Warr); the upper right-hand panel exquisitely stained hair cells of the cochlea (Source: Courtesy of Dr Mike Bowl) and the lower panels various organelles in different cells. The lower right-hand panel shows fluorescence and differential interference contrast combined. Scale bars all 10 μm, except for the gonad tissue (100 μm).*

It is the management and circumvention of fluorescent blurring – the prime disadvantage – that has led to the various strategies of 'optical sectioning' that are explained and discussed in Chapter 17 and also in Chapter 27.

15.4 The Discovery of Fluorescence

For such a wide-ranging technique, fluorescence microscopy has had a very slow start. The first significant observation was by Sir John Herschel in 1845, that quinine sulphate fluoresces blue when illuminated with UV light (from the sun). Sir George Gabriel Stokes studied the phenomenon of fluorescence intensively and coined the term 'fluorescence'[2] in 1852, after the mineral fluorite (fluorspar), which exhibits this behaviour. Stokes (Box 15.2) carried out two important experiments investigating the nature of fluorescence.

In 1904 August Köhler, together with Moritz von Rohr, designed a microscope that operated in the ultraviolet region of the electromagnetic spectrum using lenses made with fused quartz. The aim of the work was to improve the resolving power of the microscope. They found that using UV light from a cadmium arc lamp for illumination did indeed improve resolution but that the cell membranes of woody plant tissues appeared blue. This was later found to be autofluorescence (see Chapter 16, Section 16.7). Although for these resolution experiments, the UV-induced autofluorescence was an undesirable effect, Köhler stated that 'maybe this disturbing luminescence phenomenon might be of use, somehow'. Informally, Köhler referred to this new instrument as his 'Bergkristallmikroskop'. He also noted that some objects emitted light of longer wavelength upon illumination with UV, but it was the physicist Oskar Heimstädt who, in 1911, used this observation as the basis for the construction of the first successful fluorescence microscope.

Box 15.2 Sir George Gabriel Stokes

Source: Smithsonian Institution, via Flickr.

Source: Bricksnite. Reproduced under the terms of the Creative Commons Attribution licence, CC-BY 3.0 (http://creativecommons.org/licenses/by/3.0/), via Wikimedia Commons.

Together with Maxwell and Kelvin, his contemporaries, George Stokes contributed to the studies in mathematical physics at Cambridge in the 19th century. Born in Skreen, northwest Ireland, on the 13th of August 1819, Stokes went up to Pembroke College, Cambridge, from where he graduated as Senior Wrangler and Smith's prizeman four years later. He was thus awarded a fellowship at the college. Stokes was also elected Lucasian professor, was member of Parliament for the university and was created a baronet in 1889. From the time of his first publications in 1842, George Stokes worked prodigiously and deftly in a wide range of fields. Mathematics and experimental physics went hand in hand for him, and the bulk of his work centred on waves, light and fluid motion.

Upon his marriage to Susanna in 1857, Stokes had to give up his fellowship, until a change of college statutes twelve years later. The young couple took up residence in Lensfield Cottage, where Stokes built a laboratory and did much of his best work. He was to remain here until his death on the 1st of February 1903. It was Stokes who demonstrated that the plane of polarisation is perpendicular to the direction of vibration of light and explained why the sky was blue and how clouds are formed. He also researched the absorption of light by colour filters and from that work investigated fluorescence. In 1852 Stokes published the first of his two famous papers on the change in 'refrangibility' (refraction) of light.

Stokes noticed that ultraviolet light was absorbed by several substances (including fluorite, uranium glass and quinine sulphate) and then re-emitted as longer-wavelength, lower-energy, visible light. Stokes transmitted the sun's rays through a blue filter onto a solution of quinine sulphate. The longer-wavelength emission passed through a glass of white wine acting as a pale yellow filter. However, if the filters were switched, no UV radiation reached the quinine solution, and no fluorescence emission was observed.

In a second experiment, using prisms, Stokes formed a visible spectrum of the sun's rays. A test-tube filled with a dilute solution of quinine sulphate was placed just outside the longer-wavelength red end of the spectrum and gradually moved towards the shorter-wavelength blue end. No fluorescence was seen until the test-tube 'arrived nearly at the violet extremity', whereupon:

> 'a ghost-like gleam of pale blue light shot right across the tube. On continuing to move the tube, the blue light at first increased in intensity and afterwards gradually died away. It did not, however, cease to appear until the tube had been moved far beyond the violet extremity of the spectrum. It was certainly a curious sight to see the tube instantaneously lighted up when plunged into the invisible rays: it was literally *darkness visible*. Altogether the phenomenon has something of an unearthly appearance.'

Stokes initially called fluorescence *dispersive reflexion*. Later he wrote:

> 'I confess I do not like this term [dispersive reflexion]. I am almost inclined to coin a word, and call the appearance *fluorescence*, from fluor-spar, as the analogous term *opalescence* is derived from a mineral.'

Reference

Stokes, G.G. (1852). 'On the Change of Refrangibility of Light'. *Philosophical Transactions of the Royal Society of London*, Vol. 142: 463–562 and Part II (1853) on Vol. 143: 385–396.

Heimstädt used dark-field illumination (Figure 15.6). This ensured that (with a transmitted-light configuration) no excitation light would enter the objective lens and so provided high contrast for the fluorescence signal. However, the reliance upon autofluorescence and the requirement for dark-field illumination limited the initial applications of the fluorescence microscope. Later, in the 1930s, several workers found that acridine orange was a potent vital fluorescent stain, which obviated the need to rely on autofluorescence. This was an improvement, but it still meant fluorescence was limited to using non-specific, empirical stains. In the next decade Albert Coons used fluorescent dyes conjugated to antibodies to stain target organelles and molecules in a specific fashion, and immunofluorescence became widely used. More details can be found in Rost (1995), Rusk (2009), Masters (2010) and Renz (2013), listed in the Further Reading section at the end of this chapter.

15.5 Fluorescence Yield

The fluorescence quantum[3] yield (Φ) is the ratio of the number of fluorescence photons emitted to the number of photons absorbed. It is essentially a measure of how bright the emission of the fluorophore is. Generally, the greater the number of conjugated double

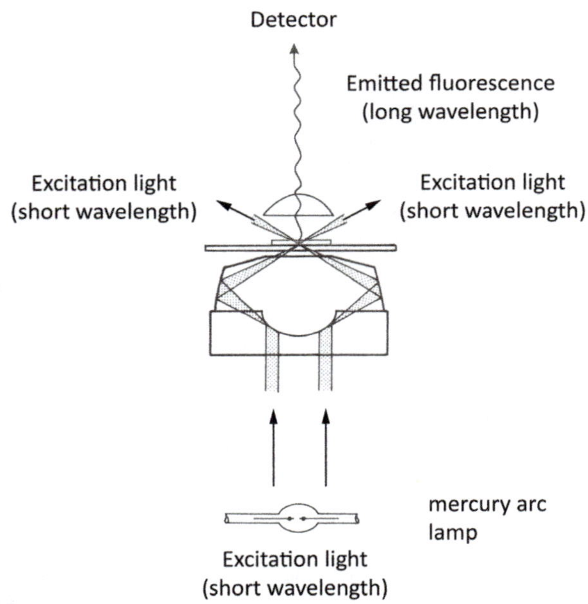

Figure 15.6 ***Transmitted-light fluorescence diascopic light path*** *Ray path of the original diascopic transmitted-light fluorescence using a dark-field condenser to prevent the excitation light entering the objective, rather than the universally employed modern method of epi-illumination fluorescence where the objective acts as its own condenser thus giving a brighter fluorescence signal. Source: Adapted Ploem and Tanke 1987 and Herman 1998.*

π-bonds in the molecule, the lower the absorption energy requirement (the 'redder' the fluorophore) and the higher its quantum yield. The local environment of the fluorophore also determines its quantum yield. For soluble fluorophores the quantum yield can be as high as 90%, but for protein-bound fluorophores, it is generally between 30 and 70%.

The *molar extinction coefficient* (ε_{max}) of the fluorophore describes the potential for a fluorophore to absorb light at a reference wavelength (which is usually the maximum absorption wavelength) and to emit a useful signal. ε_{max} is cited in units of M^{-1} cm^{-1}; the brightness of a fluorophore is the product of quantum yield and extinction coefficient, $\Phi \times \varepsilon_{max}$. For examples of these parameters, see Table 16.2 and for fluorescent proteins, Table 16.3. Clearly, high values for $\Phi \times \varepsilon_{max}$ indicate a 'good' fluorophore, useful for labelling very small populations of target molecules or for time-lapse imaging with low light intensities. At low concentrations of fluorophore (or a dilute solution in vitro) the intensity, or brightness, of the fluorescence is linearly related to concentration. However, at high concentrations (>10–100 μM) of fluorophore, more of the incident radiation is excited without corresponding fluorescence emission. In other words, the fluorophore is quenched. Where concentrations of the fluorophore are potentially high, it may be advantageous to illuminate the fluorophore at a wavelength away from the maximum excitation wavelength, in order to improve fluorescence emission at high fluorophore concentrations.

Quenching and blinking – just like photobleaching – lead to a loss of emitted signal. The former two processes are reversible, whereas photobleaching is not. Quenching occurs when excited-state electrons lose energy and relax to the ground state without emitting a photon. Various elements and compounds can act as quenching agents, including oxidising agents, halogens tissue amines, heavy metals and other organic molecules. The size of the Stokes shift also has a bearing on how easily a fluorophore may be quenched. Those fluorophores that have a small Stokes shift often self-quench, particularly when several molecules of the same fluorophore are bound to a single macromolecule. Fluorescein is particularly sensitive to self-quenching. Conversely, fluorophores that display a large Stokes shift, such as FM 4-64 or Coumarin dyes, do not self-quench. These dyes are typically more polar, thus are more soluble and so less prone to aggregation. Quenching can also be caused by non-radiative transfer of energy from one fluorophore to another that is in very close proximity – Förster resonance energy transfer. Blinking in fluorescent nanocrystals and quantum dots is well known. It was predicted as a quantum mechanical event by Niels Bohr and is thought to involve transition between a singlet and dark triplet state. The phenomenon does not usually cause problems, except for stochastic location super-resolution methods, such as single-molecule PALM and STORM microscopy.

Most photobleaching occurs when the excited electrons undergo intersystem crossing, entering a long-lived triplet state. Absorption of a further photon of light prevents normal cycling between ground and excited states, causing irreversible photobleaching. Use an antifade reagent in the mountant of your preparation to retard bleaching. Antifade reagents are antioxidant, reducing agents that scavenge free oxygen radicals from the mountant, or act as triplet state quenchers, depopulating excited states other than S_1. They also retard the ability of the fluorophore to act as an electron acceptor and be bleached. Antifade formulations do not inhibit the fading of fluorescent proteins, but the β-barrel of the GFP protects the fluorophore to some extent.

Bleaching is directly related to photon flux whether antifading reagents are used or not. Therefore, avoid fluorescence saturation by controlling the laser power and keeping it at a low level. Use the minimum of laser power consistent with a good signal-to-noise ratio. Bleaching is proportional to the laser intensity *and* also the frame rate. Nipkow scanners can use a lower laser intensity per frame but may bleach with a higher frame rate (high frame rates being their raison d'être). This is also the reason why very fast line scanners, such as the 7-Live or LiveScan SFC designs, have the potential to bleach very quickly and unexpectedly, unless exposure per frame and overall frame rates are considered whilst first planning the experiment.

Just as road safety campaigners urge other drivers to 'think bike', microscopists should '**think bleaching**'. Most people will protect their samples by covering slide trays with silver foil and keeping them out of light in the fridge or a cold room when not in use, but some will then needlessly overexpose the sample at the microscope. If possible, use lower-intensity white light to focus your sample, with bright-field, phase contrast or DIC microscopy. Once you have used fluorescence to focus your sample, *do not forget* to close the shutter on the epi-fluorescence microscope! It is situated between the lamp house and the nosepiece of the microscope, or else may be situated on the fluorescence lamp unit itself. Closing this shutter after visual inspection, in order to look after your fluorescent specimen, should be habitually automatic.

15.6 Choosing a Fluorophore

Despite a strong illuminant, the numbers of photons in the emitted fluorescent signal may also be low for the simple reason that the number of fluorophores labelling the structure of interest are very few in the first place. If (in a living cell) a fluorescent dye has a physiological concentration of 1 μM inside, say, a labelled small mitochondrion, which may be just resolvable at 200 nm wide and 400 nm in length, then there may be only 30 dye molecules inside the mitochondrion. The volume sampled by a high numerical aperture objective may only be 20–50 attolitres (1 attolitre = 10^{-18} litres). Only a fraction of the light emitted from the fluorescent dye molecules in this volume will be collected (some estimates put this as low as 10%) by the objective and formed into the image. This is a fairly extreme example, but it does occur in live-cell imaging and shows why implementing strategies to conserve and collect photons without excessive photobleaching or phototoxic damage to the cell is important. Thus, the following parameters must be considered when selecting a fluorophore:

- spectral absorption and emission curves (maximise absorption coefficient);
- Stokes shift (large Stokes shifts mean good contrast, may mean low quantum yield);
- quantum yield, Φ (maximise quantum yield to maximise brightness);
- molar extinction coefficient, ϵ (maximise extinction coefficient to maximise brightness);
- brightness (but lower brightness and higher photostability can give better SNR for time lapse);
- photobleaching resistance (lower brightness and higher photostability can give better SNR, or signal-to-noise ratio, for time-lapse);
- maximum concentration tolerated within the cell;
- the brightest and most photostable fluorophores should be reserved for the least abundant cellular targets;
- in general, dyes with green and red fluorescence emission tend to be brighter than those emitting in the blue and far-red portions of the spectrum;
- the localised environment significantly influences fluorophore absorption and emission spectral profiles, so the spectra published by manufacturers of isolated fluorophores in solution can differ markedly from those observed under actual experimental conditions;
- the shape of the absorption and emission spectral curves will indicate whether excitation cross-talk or emission bleed-through is likely when used with other fluorophores; and
- use wide bandwidth excitation and emission filters to maximise the signal, unless there is significant bleed-through.

The Internet provides several useful resources for selecting fluorophores that are appropriate both for your experiment and available microscope equipment. The most comprehensive of these resources, tailored to the needs of light microscopists, are given in the Internet Resources at the end of the chapter, together with information from filter and microscope manufacturers to help you select the right filter sets for your application.

15.7 Fluorescence Microscope Design

The first fluorescence microscopes were based on the transmitted-light (diascopic) configuration (Figure 15.6), with the illuminator, condenser, specimen, objective and eyepiece laid out in order along the optical axis. This design relied upon a dark-field condenser and appropriate filters to illuminate the specimen yet prevent the stronger illuminating light from reaching the detector. Without this dark-field condenser, the transmitted-light configuration would deleteriously reduce the contrast of the emitted fluorescent signal.

The diascopic configuration has several disadvantages: both the condenser and objective must be perfectly aligned on the optical axis with respect to one another, and it is difficult to completely suppress the illuminating light, which excites the fluorophore. This leads to a reduction in contrast and may also present a danger to the eye. It is not generally possible to combine fluorescence microscopy with phase-contrast or differential interference contrast (DIC) microscopy (these are useful for imaging the morphology of the cells or tissue in order to provide a reference map for the localisation of the fluorescent signal), since the condenser is normally a dark-field one. Most important, the brightness, or intensity (photon flux), of the emitted fluorescence signal in the transmitted-light configuration is proportional only to the square of the numerical aperture divided by the square of the magnification (Intensity $\propto NA_{obj}^2$/Total mag^2).

The efficiency of fluorescence contrast microscopy is very much improved by use of fluorescence *epi-illumination* (reflected-light) rather than transmitted-light illumination. The first epi-illumination microscope was designed by Philipp Ellinger and August Hirt in 1929 (Ellinger, 1940). Using a reflected-light configuration (Figure 15.7) rather than the earlier transmitted-light pathway, allows the much

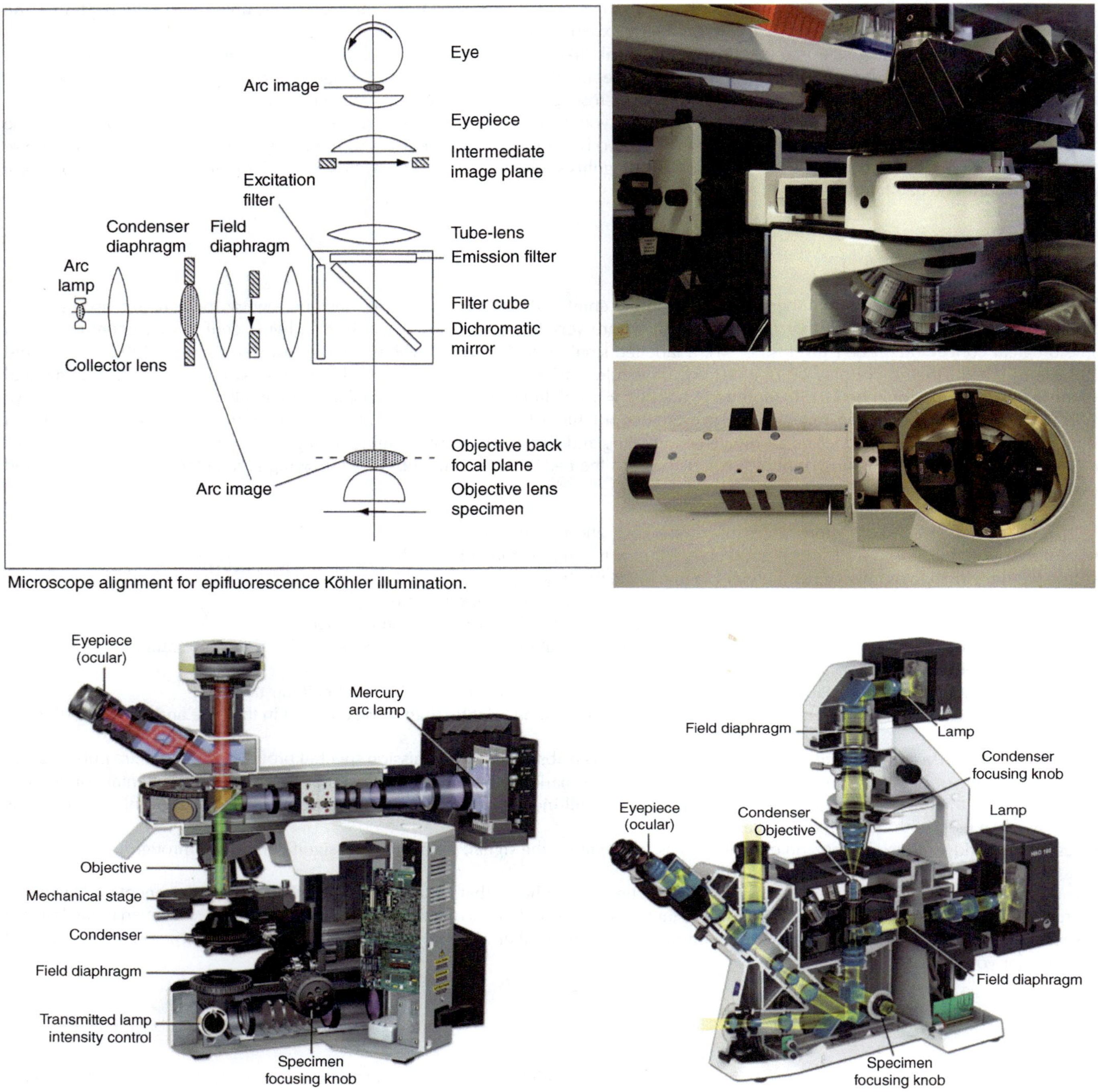

Figure 15.7 **Reflected-light configuration** *Epi-illuminator carousel holding a series of fluorescence filter sets, according the design of 'Bas' Ploem (upper left). The upper right-hand illustration shows the automated carousel unit of an Olympus BX-61 upright fluorescence microscope and the lower right-hand, the automated carousel of an Olympus IX-81 inverted fluorescence microscope.*

Diagram of the components used in a wide field epi-fluorescence microscope. The arrangement and ray path are similar as for a reflected-light microscope set up for bright-field illumination, except that the arc lamp is replaced by a tungsten halogen or LED lamp, and the fluorescence filter cube is replaced by a half-silvered mirror arrangement.

Source: (top left) Watkins and St. Croix 2013. Reproduced with permission from John Wiley & Sons. (top right) Sanderson 2008. Reproduced with permission from The Quekett Microscopical Club. (bottom) Murphy and Davidson 2013. Reproduced with permission from John Wiley & Sons.

less intense emitted fluorescence signal to be effectively separated from the bright and intense high-energy illumination. It is a Dutch investigator, J.S. (Bas) Ploem, who is credited with the introduction of the first practical epi-illuminator for fluorescence microscopy. His design, first marketed by Leitz in 1967 as the Ploemopak, is virtually unchanged today over forty years later and is in use worldwide.

In the epi-illumination configuration, light is directed from the side and reflected via a special two-way mirror (dichroic mirror[4], dichromatic beam-splitter) through the objective onto the sample. The dichromatic beam splitter is fixed at a 45° angle to the incoming excitation light. The mirror reflects up to 95% of this shorter-wavelength excitation light but transmits longer-wavelength light emitted by the fluorophore. The emitted signal passes through the dichromatic beam splitter and is detected by the eye or camera. The fluorescence epi-illuminator contains a slider or carousel in the incident light path housing a number of dichromatic beam splitters for different wavelengths of excitation light. Ultimately, Ploem developed the epi-illuminator to contain both the excitation and barrier filters along with the dichromatic beam splitter in one self-contained filter cube or 'filter block'. Four filter cubes can be rotated, in turn, into the optical axis of the epi-illuminator. This convenience enables specimens stained with more than one fluorophore to be examined quickly and easily.

This design very neatly solves several problems associated with the earlier diascopic configuration: the illuminating light is not on-axis with the eye, so any stray illumination is far less likely to be harmful; the objective acts as its own condenser so that alignment difficulties are obviated. There is no dark-field stop to attenuate much of the illuminating light, and the condenser may now be used to provide a phase-contrast or DIC image. Because the objective performs both illumination and detection roles, the brightness of the fluorescent signal[5] is proportional to the fourth power of the numerical aperture of the objective (Intensity $\propto NA^4/Mag^2$). The fluorescent image is much brighter with epi-illumination. This is crucial when the signal from living samples may be weak in the first place.

Although there is usually no means of changing the magnification of the eyepieces and/or the relay optics to the camera, you should keep the overall magnification low in order to ensure maximum signal intensity and brightness. Thus for viewing, use the lowest magnification eyepieces possible (x5 or x8 are preferable to the ubiquitous x10 eyepieces). However, keeping the objective magnification low – in order to keep the overall magnification low – is counterproductive. Since, with series of objectives, numerical aperture normally increases as magnification rises, the quadratic NA term outweighs the square of the magnification term. It pays, therefore, to use the highest numerical aperture objective possible, regardless of magnification, to secure the brightest fluorescence signal. Thus a 40x/NA 0.65 objective transmits an image approximately 5 times less bright than the 40x/NA 1.0 objective, a 40x/NA 1.3 objective gives an image 3 times as bright as the 40x/NA 1.0 objective and a 40x/NA 1.4 objective, four times as bright. This quadratic relationship holds true regardless of the magnification or numerical aperture of the objective and is illustrated graphically in Figure 15.8 for a series of 20x objectives of differing numerical aperture.

Good epi-fluorescence microscopy demands high numerical aperture objectives, good filters and very high quality dichromatic beam splitters. Zeiss introduced the Neofluar objective in 1953, using artificial optical-grade calcium fluoride. The semi-apochromatic plan-neofluar design remains the workhorse objective irrespective of the manufacturer, because it contains fewer glass elements than apochromat objectives. That said, most fluorescent images are collected using apochromats either because they are fitted on the microscope by default or because of concerns regarding chromatic aberration, but it is worth remembering that semi-apochromat objectives are likely to collect a more intense signal.

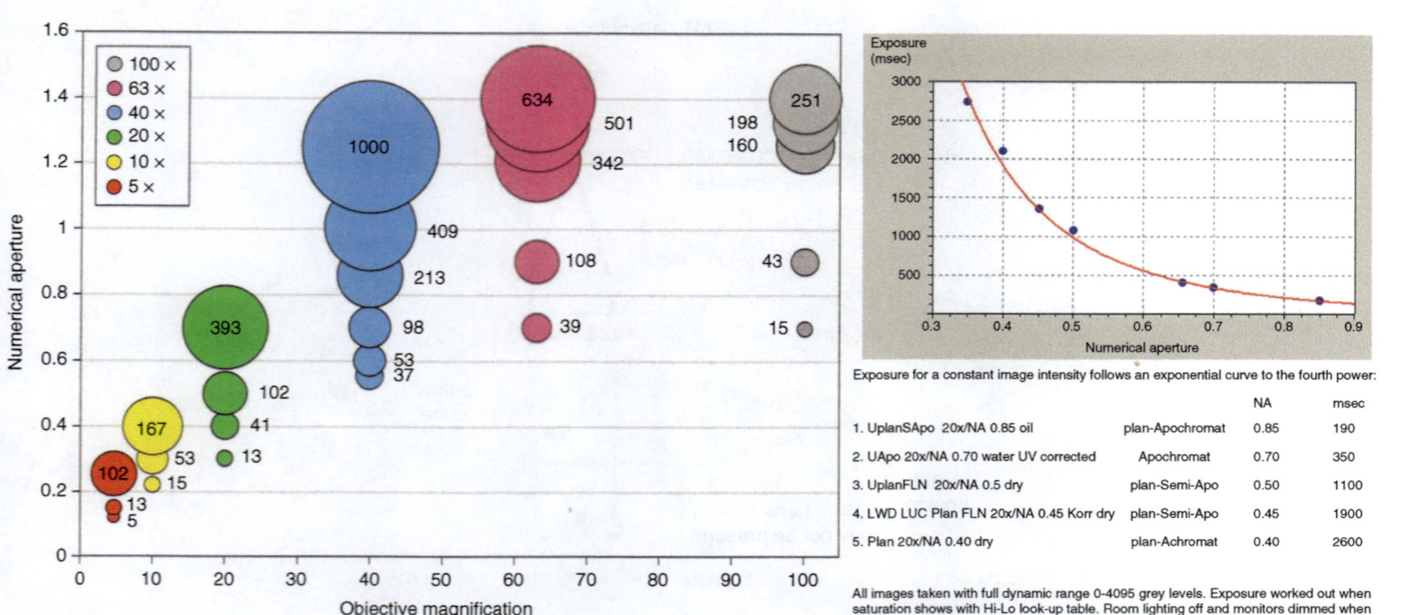

Exposure for a constant image intensity follows an exponential curve to the fourth power:

		NA	msec
1. UplanSApo 20x/NA 0.85 oil	plan-Apochromat	0.85	190
2. UApo 20x/NA 0.70 water UV corrected	Apochromat	0.70	350
3. UplanFLN 20x/NA 0.5 dry	plan-Semi-Apo	0.50	1100
4. LWD LUC Plan FLN 20x/NA 0.45 Korr dry	plan-Semi-Apo	0.45	1900
5. Plan 20x/NA 0.40 dry	plan-Achromat	0.40	2600

All images taken with full dynamic range 0-4095 grey levels. Exposure worked out when saturation shows with Hi-Lo look-up table. Room lighting off and monitors dimmed when making the exposure.

Figure 15.8 ***Fluorescence exposure graph – quadratic relationship*** *The left-hand panel shows the relative light-gathering capacity of objectives depending upon magnification and numerical aperture. Source: Watkins and St. Croix 2013. Reproduced with permission from John Wiley & Sons. The graph in the right-hand panel shows the quadratic relationship between exposure and numerical aperture.*

In theory, the dichromatic beam splitter should act as an excitation filter (to reflect the illuminating light sideways down onto the specimen) and as a barrier filter (blocking the illumination on its return path on-axis to the eye), but in practice separate excitation filters are used in conjunction with the dichromatic beam splitter whose sole purpose is to separate the high-energy (short-wavelength) illumination from the low-energy (longer-wavelength) emitted fluorescence signal. Each filter cube, therefore, contains three components: two filters (excitation and emission filters) and a dichromatic beam splitter.

Filter sets are essential for widefield fluorescence microscopy, because a certain spectral bandwidth must be isolated from the broadband illuminating spectrum of the mercury, metal-halide, xenon or LED lamp. The excitation spectrum must then be prevented from reaching the detector by means of the dichromatic mirror and emission filter, which together allow only the emitted signal to reach the eye or camera. When using lasers for illumination, the situation is simplified: because they are highly monochromatic sources of light, no excitation filter is required. Only dichromatic beam-splitting mirrors are required to separate the excitation wavelength from the emission signal, together with an emission filter to ensure that the correct signal bandwidth is transmitted and to block any excitation light from reaching the detector.

15.8 Fluorescence Filter Sets

High-quality precision-made filters, with very precisely defined reflection and transmission wavelengths, are crucial to good fluorescence microscopy. These were first developed in the 1940s; see Table 15.1 for the websites of modern filter manufacturers. The dichromatic beam splitter (Figure 15.9; Figure 15.13) is a plate of optically flat glass coated on one side with very thin dielectric metal coatings evaporated evenly onto the surface of the glass. Modern fluorescence excitation and emission filters are also made with thin dielectric films covering a glass substrate. The nature of these coatings determines which wavelengths of light are reflected onto the sample and which are permitted to transmit back to the eye or camera. They work by destructive interference, (Figure 11.14) allowing certain wavelengths to be transmitted through and others to be reflected off the coating. To cut down back-reflections and improve the signal-to-noise ratio, the back of the filter set may have a hole cut in it (Figure 15.10) and/or the emission filter may be offset. These filter sets are supplied by specialist companies (Table 15.1), as well as being available from the major microscope manufacturers directly. Two Internet Resources available to help you select the right filter sets for your application are given at the end of the chapter. Figure 15.11 shows spectral curves for the excitation filter, the dichromatic beam splitter and the emission filter suitable for GFP overlaid on the absorption and emission spectra shown in Figure 15.2.

The lists of filters and their associated spectral curves may seem complicated to understand, but they are straightforward once you understand how to read the filter specification. The excitation filter must allow light above, between or sometimes below certain wavelengths to pass through to excite the specimen. The dichromatic beam splitter will (usually) reflect light below a stated wavelength onto the sample and transmit emitted signal above that wavelength from the sample to the detector. Likewise the emission filter will allow light above a certain wavelength to pass to the eye or the detector and block unwanted shorter wavelengths (e.g. residual excitation light). Besides the wavelength given, filters also have certain letters to denote their function. The letters 'LP' stand for *long pass*, the

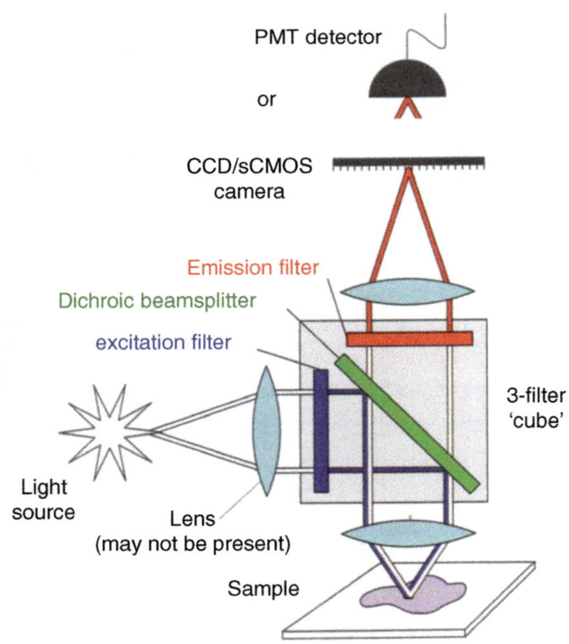

Figure 15.9 **Fluorescence filter set** *Filter and dichromatic beam splitters are used to separate the excitation light from the emitted signal, hence keeping the signal-to-noise ratio, and thus the contrast, high. The collimating lens in front of the excitation filter may not always be present. Source: Watkins and St. Croix 2013. Reproduced with permission from John Wiley & Sons.*

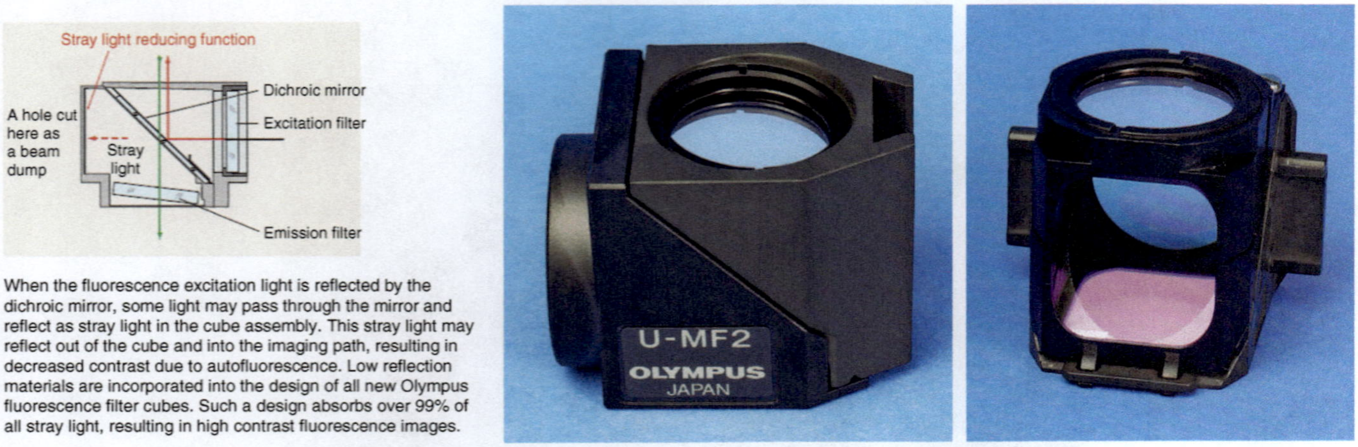

Figure 15.10 ***Filter set with hole in it*** *The design of modern filter sets incorporates advanced technology in anti-reflection coatings to ensure a good signal-to-noise ratio of the final signal but also reliance on simple 'older technology' solutions, such as cutting a hole in the rear face of the filter block to let out stray light and/or tilting the emission filter to prevent back-reflections reducing the overall signal-to-noise ratio. Source: (left) Reproduced with permission from Olympus Europa SE & Co. KG. (right) Courtesy of Mr K. Glover.*

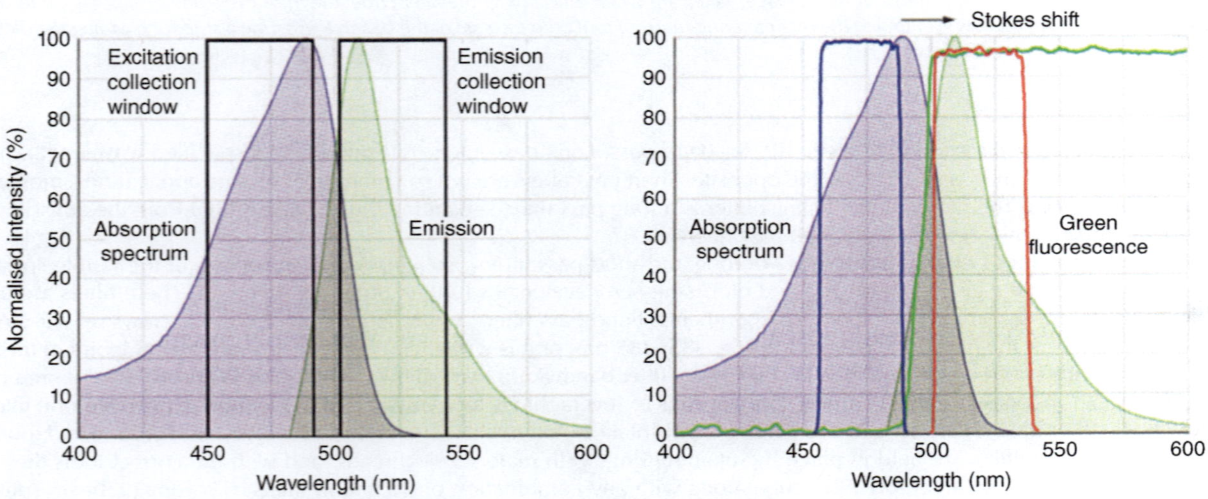

Figure 15.11 ***Spectral curves of a typical filter set*** *Knowing the spectral absorption and emission characteristics of a fluorophore allows laser lines and filter sets to be matched so that a good signal-to-noise ratio is achieved. All microscope and filter manufacturers offer information on the spectral characteristics of both fluorophores and filter sets. These may be found under Internet Resources at the end of the chapter. Source: Adapted from Watkins and St. Croix 2013 with permission from John Wiley & Sons.*

Table 15.1 *Fluorescence Filter-Set Manufacturers*

1. **Chroma**:	https://www.chroma.com
Catalogue PDF:	https://www.chroma.com/sites/default/files/chroma-catalog-en.pdf
Information:	https://www.chroma.com/knowledge-resources
2. **Horiba** (Glen Spectra):	http://www.horiba.com/uk/scientific/glen-spectra
Catalogue on-line:	http://www.horiba.com/uk/scientific/products/optical-filters/
Information:	http://www.horiba.com/uk/scientific/products/optical-filters/product-line/fluorescence-filters/choose-fluorescence-filters/
3. **Omega Optical**:	http://www.omegafilters.com
Catalogue on-line:	http://www.omegafilters.com/products/fluorescence-sets.html
4. **Schott**:	http://www.schott.com/uk/english/index.html
Catalogue on-line:	https://www.schott.com/advanced_optics/english/products/optical-components/optical-filters/interference-filters/index.html
5. **Semrock**:	http://www.semrock.com
Catalogue on-line:	https://www.semrock.com/product-families.aspx
Information:	https://www.semrock.com/technical-faq.aspx
Filter set selection – Microscopy company websites:	
Leica:	https://www.leica-microsystems.com/
Nikon:	https://www.microscopyu.com/techniques/fluorescence/nikon-fluorescence-filter-sets
Olympus:	https://www.olympus-lifescience.com/en/microscope-resource/
Zeiss:	https://www.micro-shop.zeiss.com/?s=55711561de9533&l=en&p=us&f=f

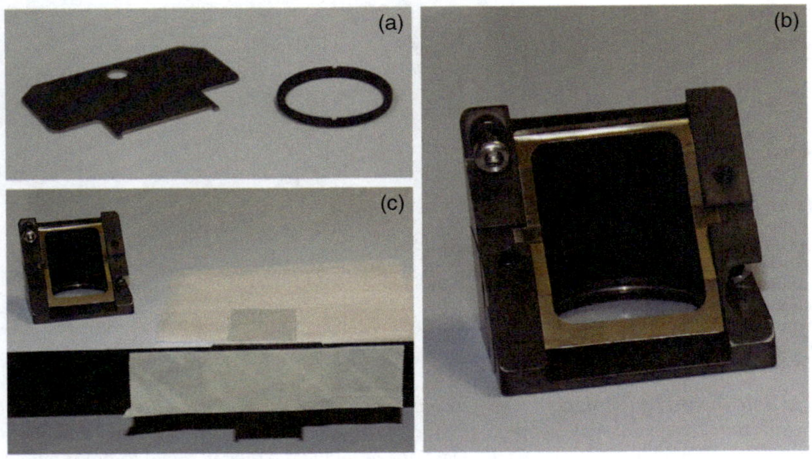

Figure 15.12 **Filter set, retaining rings and diagram** *(a) Tool for removing filter-retaining rings from the filter cube or block. (b) The sprung plate assembly, which holds the dichromatic mirror firmly in place. Take care when dismantling the filter cube; the mirror is prone to slipping off its mounting and onto the floor unless care is taken! (c) taping the retaining ring-removal tool onto the side of the bench to provide a stop for cleaning the dichromatic mirror. Source: Courtesy of Mr K. Glover.*

letters 'SP' or 'KP' for *short pass* and the letters 'BP' for *band pass*. Long pass filters transmit above a specified wavelength, blocking shorter wavelengths whilst short pass filters do the opposite. Short pass filters are not so common in routine applications, most are long pass or band pass designs. The dichromatic beam splitter is a long pass filter, separating the excitation light from the emission signal, while the excitation and emission filters are band pass filters.

A band pass filter combines both characteristics of long and short pass filters, isolating a certain portion of the visible spectrum. It allows light above a certain wavelength to pass and blocks higher wavelengths further along the spectrum. These filters are useful to prevent bleed-through of signal in dual-labelling experiments. Band pass filters are designated in one of two ways: either the cut-on and cut-off wavelengths of the band are given in full (e.g. 335–385 nm) or it is written in abbreviated form: the midpoint of the band is given, e.g. 360, then the width of the band (i.e. FWHM of the filter bandwidth) that will pass light. Thus the band 335–385 nm can also be written 360/50. It is possible to combine the characteristics of several filters for viewing multi-labelled samples into one filter set.

The filter cubes, filter blocks or filter sets are generally left intact in the microscope and not removed or dismantled. However, as Figure 15.12a shows, the filters are held in place by retaining rings with male screw threads, and with the correct tools they can be removed and incorporated into an empty filter cube along with any combination of excitation filters, dichromatic beam splitter and emission filter. This is useful if you have to make up custom filters for, say, specific fluorescent proteins. The dichromatic beam splitter is usually held in place by a thin sprung metal plate (Figure 15.12b), which is screwed or clipped into place, and this should be handled with care.

When removing the circular filters and dichromatic beam splitter, make especial note of the correct orientation of the filter and mirror into the filter cube. This is important for correctly reassembling the filter block afterwards and is a point very easily overlooked. The excitation and emission filters usually have an arrow or 'V' pointer (Figure 15.13) to indicate which surface should face inwards towards the dichromatic beam splitter when inserted into the filter cube or you can add your own with a fine marker pen, as shown. The dichromatic mirror itself usually has a very small 'V' notch inscribed on the edge of the glass plate. Otherwise, when the reflective side is uppermost, the logo, corner chamfer or etched line is in either the far-left corner or the near-right corner. If these distinguishing marks are not present, examine the two surfaces of the dichromatic mirror very carefully: the surface with the dielectric dichromatic coating can usually be distinguished from the reverse surface by its degree of reflectivity. The surface coated with the interference coating is very slightly sharper; when viewed through the thickness of the glass substrate, reflections are slightly softer. You can see a double reflection of a bright object, and the thickness of the filter at the far edge is apparent with the reflective interference coating held underneath, away from the eye. The coated surface of the dichromatic beam splitter should face towards the excitation filter.

Since the dielectric interference coating vacuum-deposited onto the glass filters and mirror is very thin and delicate, it can be very easily marked or scratched if great care is not taken when handling them. This is particularly true if they are older and have 'soft-coating' interference films deposited on their surface, rather than the modern 'hard coatings'. You should insist on using modern 'hard' coatings: they pass up to 30% more light than the traditional 'soft-coated' filters. Always put filters and mirror down onto 2–3 sheets of Whatman lens cleaning tissues, and handle them with clean, dry hands. If you wear gloves, ensure that they are not a powdered variety. To clean the filters, use analytical-grade ethanol and clean cotton buds. Clean from the centre outwards in a spiral, and let the residual ethanol evaporate. Further information is given on the Semrock website (see Internet Resources at the end of the chapter). An alternative method for cleaning both circular filters and the dichromatic mirror is to tape a spare filter cube key (Figure 15.12c) onto the side of the laboratory bench so that it protrudes slightly to form a retainer. Place several sheets of Whatman lens cleaning tissue up against the edge of the key, and lay the filter or dichromatic mirror on these. Take a fresh single sheet of Whatman lens cleaning tissue and lay it over the filter or mirror. Apply no more than 1–2 drops of analytical-grade ethanol to the tissue, and draw it across the surface of the

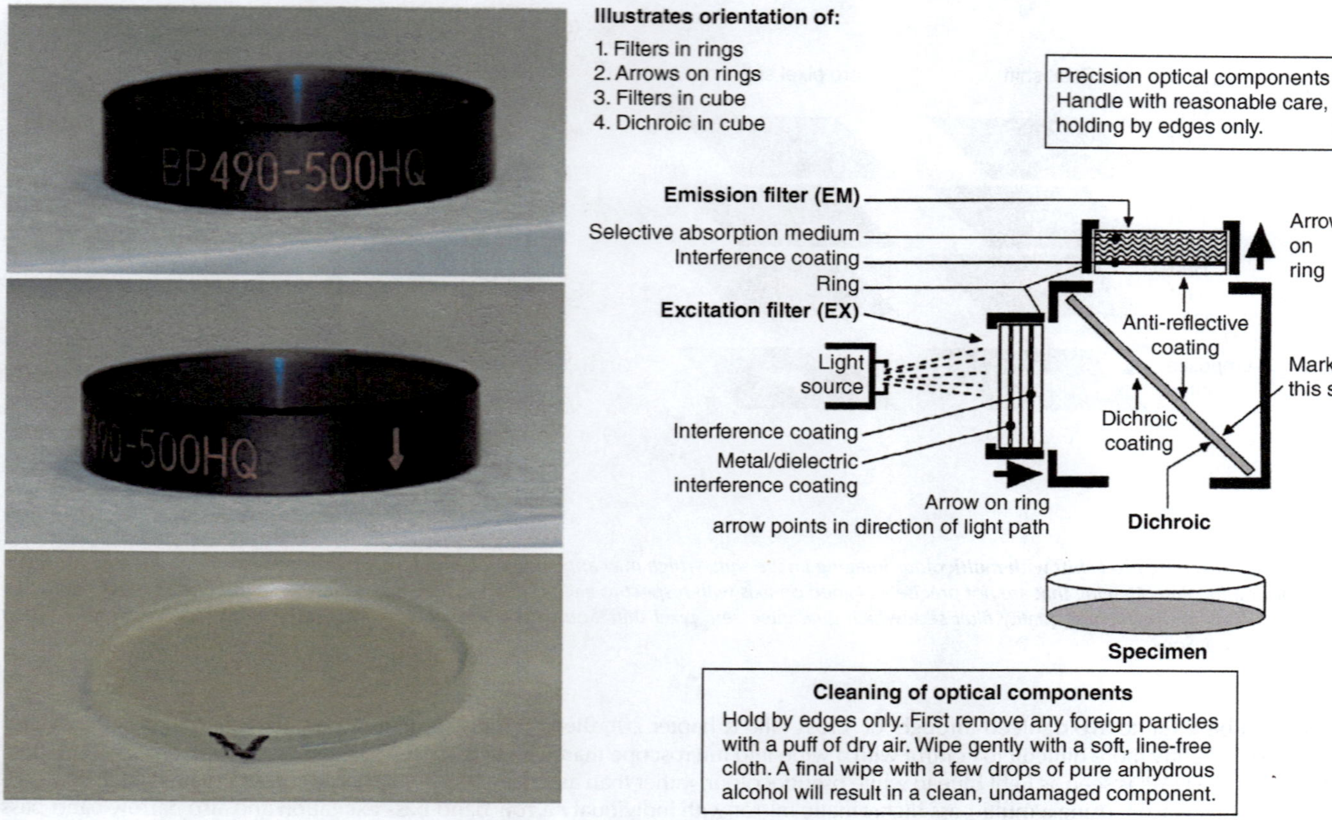

Figure 15.13 **Orientation of filters** *Filter components. The two upper panels show the marking on a narrow bandwidth excitation filter 495/10 for eGFP or a similar fluorophore. The arrow shows the orientation of the filter into the filter cube. The lower panel shows a self-made marking on an otherwise unlabelled emission filter. Not all marks both face towards the dichromatic mirror. The instructions on the right are taken from the Omega Optical website. In their case, the arrow on the emission filter points towards the eye or camera and not inwards towards the dichromatic mirror. Take extreme care to retain information on how each filter is oriented when you dismantle the filter cube. Obvious advice, but it is very easy to forget or lose sight of the orientation. Source: (left) Courtesy of Mr K. Glover. (right) Reproduced with permission from Omega Optical, Inc.*

filter or mirror in one even sweep. The surface should be entirely clean once any remaining ethanol solvent has evaporated. Repeat the process with a fresh sheet of lens cleaning tissue if needed.

Because filters are built up of layers of thin-film interference coatings sandwiched between glass discs as carrier substrates, it is possible for a small wedge angle in between layers to creep in during manufacture. The consequence of this is poor image registration, or 'pixel shift'. When a multicolour image with, say, blue, green and red channels imaged by three different filter sets is recorded onto a CCD faceplate, then any deviation of the image by one or more of the filter sets will cause a pixel shift (Figure 15.14) that has severe consequences for image clarity and any quantitative studies (e.g. colocalisation). Manufacturers now design 'zero pixel-shift' filters. The permitted tolerance is based on an infinity-corrected microscope with a 200 mm focal length tube-lens, using a CCD camera with a pixel spacing of 6.7 µm, which is a commonly encountered arrangement. Most dichromatic mirrors, because they have a single-sided coating and a thinner substrate, tend to be the primary contributors of pixel shift error.

The certain way to avoid pixel shift error and consequent image misregistration is to use a multi-band filter set. A multi-band-pass dichromatic beam splitter (Figure 15.15a), excitation filter and emission filters are used. Neither the excitation filter, dichromatic beam splitter nor the emission filters physically move, because they transmit all the channels required. Not only do they not suffer from pixel shift, they (obviously) also acquire multi-channel images much faster. However, this arrangement generally suffers from reduced brightness. More seriously, it is particularly prone to cross-talk or bleed-through, of the fluorescent signal between the different channels. If a multi-band-pass dichromatic beam splitter is used, check that it does not reflect portions of the emitted signal(s) that must be passed because the multiple excitation bands overlap the spectral regions of these signals.

To improve matters, an alternative arrangement (Figure 15.15b), sometimes referred to as 'Pinkel filters' after Daniel Pinkel, who first developed them, keeps the multi-pass dichromatic beam-splitting mirror and multi-pass emission filters static (to reduce the possibility of pixel shift) but moves individual excitation filters. The advantages of using the Pinkel filter sets are as follows: the exposure of each colour can be individually optimised, and the imaging is performed one channel at a time, eliminating the need for a colour camera. Since no Bayer colour mask (Chapter 30, Section 30.9) is required, this ensures a higher pixel resolving power in the recorded image. The method is also reasonably fast, since only the excitation filters move, although not as fast as using a fixed multi-band filter set.

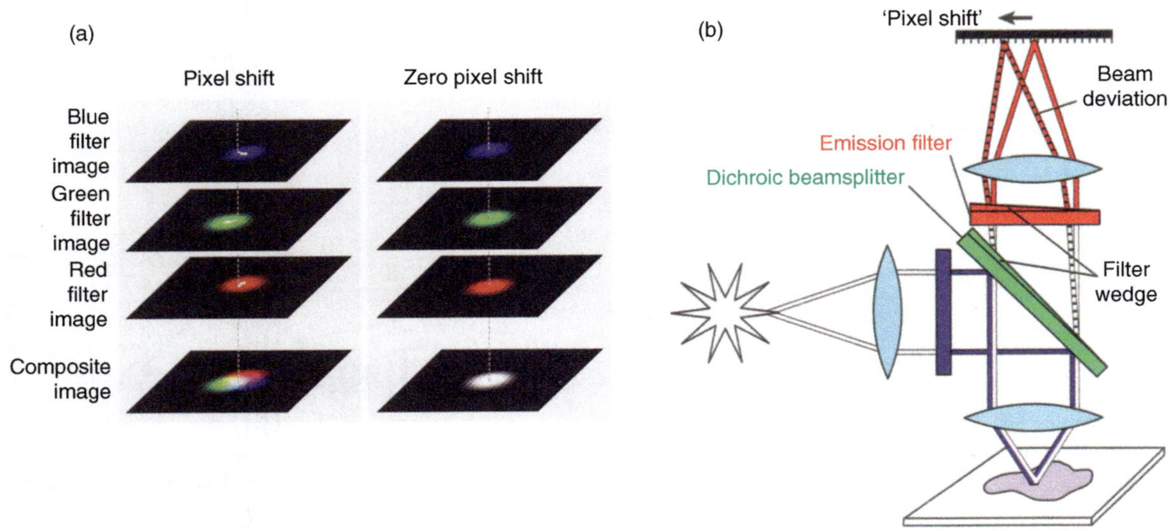

Figure 15.14 ***Avoiding pixel shift with multicolour imaging*** *Image shift, which may appear like gross chromatic aberration, will occur if separate filters are used for multicolour imaging that are not precisely aligned on-axis with respect to each other or where individual filter components in the filter cube are wedged. Look for the highest quality filter sets, which guarantee zero pixel shift. Source: Watkins and St. Croix 2013. Reproduced with permission from John Wiley & Sons.*

Care must be taken to avoid 'bleed-through' or 'cross-talk' (Chapter 20). Bleed-through from DAPI (or any other fluorescent dye for that matter) is slightly more difficult to control with a widefield microscope than with very narrow-bandwidth laser excitation lines on a confocal. Where DAPI is seen as light blue to white pseudocolour, rather than as a deep blue, you should suspect some degree of bleed-through. For that reason, using a multi-pass dichromatic mirror with individual *narrow* band-pass excitation and also narrow band-pass emission filters will ensure minimum bleed-through occurs. This type of filter arrangement (Figure 15.15c) is sometimes referred to as a 'Sedat' set after John Sedat. Pixel registration must be corrected for, but cross-talk is much more effectively managed. Since DAPI has a very broad emission spectral profile, it is very susceptible to bleed-through. Although they allow fast image acquisition, these multi-band filters have two main disadvantages, which must be borne in mind. First they have *very* narrow band-widths to avoid cross-talk and bleed-through. Secondly, the interference coating transmitting one fluorophore may well reflect the fluorescence emission from another: they have a lower signal throughput than using filter sets for individual fluorophores, acquiring images in sequential fashion.

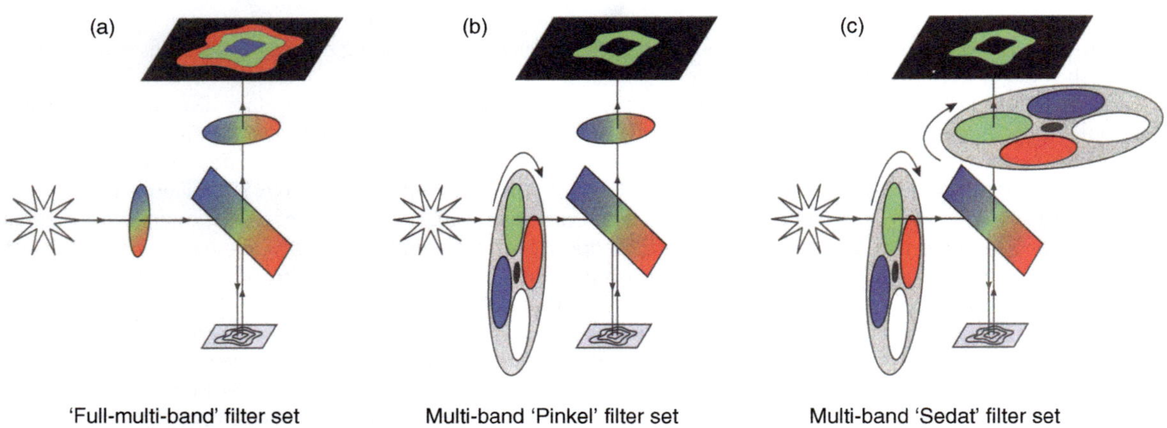

Figure 15.15 ***Multi-pass and Pinkel filter-set diagram*** *The spectral profile of a multi-band beam splitter is shown (left), similar to that in Figure 15.14. This can be used either with a fixed multi-band excitation filter or a fixed multi-band dichroic mirror and fixed multi-band emission filter (middle) with individual excitation filters in a 'Pinkel' filter set arrangement where individual narrow band excitation filters are used to cut down cross-talk.*

A further arrangement in the 'Sedat' design (right) is to have moving emission filters as well as moving excitation filters, whilst keeping the dichromatic beam splitter – the component most prone to pixel shift – static. This is slower than the arrangements shown in (a), but correctly selected narrow bandwidth excitation and emission filters can reduce cross-talk significantly. If a dichromatic beam splitter is subsequently used to project two or more channels onto a single CCD faceplate for simultaneous multicolour imaging, it is best to use software tools and calibration with multicolour beads to correct for pixel shift. This has the advantage that it will also at the same time correct for the fixed chromatic aberration arising from the objective. Source: Watkins and St. Croix 2013. Reproduced with permission from John Wiley & Sons.

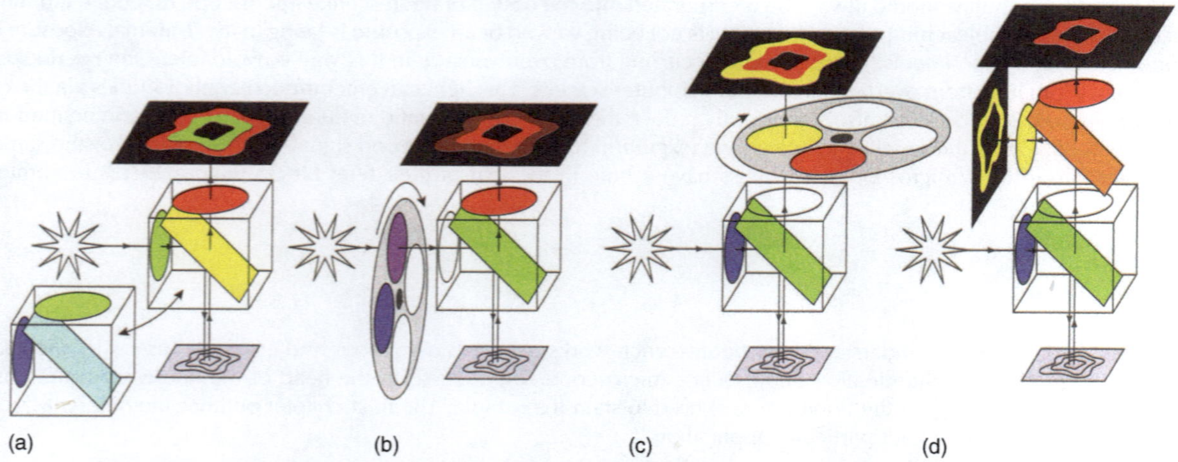

Figure 15.16 **Different possible arrangements of excitation and emission filters** *The previous figure showed arrangements for multicolour imaging; here, various filter configurations for single-colour fluorescence imaging are shown. The usual method (a) is to rotate individual filter cubes into position, but this is slow. Higher temporal resolution can be achieved with excitation and emission filter wheels (b, c). The arrangement in (d) shows an additional dichroic beam splitter in use for FRET imaging – to separate the donor and acceptor signals – whilst they are simultaneously collected on a widefield microscope. Source: Watkins and St. Croix 2013. Reproduced with permission from John Wiley & Sons.*

There are other means of improving temporal resolution without having to use narrow bandwidths, and Figure 15.16 shows these. Excitation and/or emission filters may be held in a filter wheel, rather than separate filter cube. The filter wheel may be rotated into position faster than a heavier filter cube.

Box 15.3 A Defective Filter Set

Sometimes, an image artifact may be unrelated to the specimen. In this case, the optical surfaces of the microscope, objectives, eyepieces, condensers, DIC sliders, camera faceplate – to name a few – must be checked for cleanliness. It is also important to include a check of any fluorescence filter-cubes used.

In the example below, the central haze intermittently visible with a single filter set was initially a puzzling phenomenon. Tilting the illumination arm of the inverted microscope helped locate this fault within the illumination pathway. The haze was only visible with a long exposure (10 s) and served to obscure fine detail captured at very low light intensities. It turned out that a metal-halide light source left on overnight had burned through the thin interference coating on one of the filter surfaces within a filter cube. This used to be a common phenomenon when organic dyes were used to coat filters (excess heat also commonly degrades polarising filters) but is far less commonly seen with modern hard coatings formed by ion-beam sputtering and magnetron deposition.

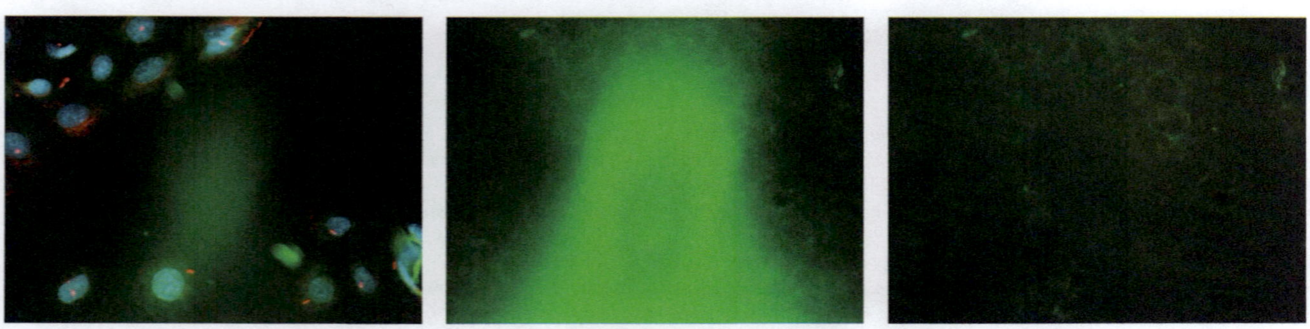

Original image of cells (left) showing infrared haze passed by the defective green filter. The exposure times for both the weak green and punctate red signals were 10 seconds each. The nuclear DAPI signal was 50 msec. The red filter cube is unaffected. The exposure using the DAPI filter cube is too short to tell if this filter set is defective. The central image shows the same green filter set used to image a similar weak green signal, with a longer exposure giving significant haze in the image. The right-hand image shows the same specimen imaged in the green channel with an identical perfect filter set for the same long exposure time.

A manual or automatic shutter should always be incorporated into the design of fluorescence microscope to reduce illuminating the sample and thus cause photobleaching when the sample is not being viewed or an exposure is being made. Minimal exposure is crucial when performing live-cell experiments. Photo-toxicity occurring from overexposure in this way is avoidable. One overlooked source of noise is ambient room light from overhead lights and computer screens. This light can enter trinocular photo-tubes via the eyepieces (e.g. Chapter 12, Figure 12.14) and can also significantly reduce the signal-to-noise ratio of fluorescence images. Some manufacturers' software (e.g. Zeiss ZEN) runs the monitor with a grey background to help maintain good signal-to-noise ratios. For the same reason, Nikon use a light trap in their microscopes and Zeiss have a hole in the rear of their filter blocks (Figure 15.10) to eliminate stray light.

15.9 Chapter Summary

This chapter should enable you to understand how fluorescence works, why it is ubiquitous and extremely useful in microscopy and how to best use the modern epi-illumination fluorescence microscope. The filter set is the heart of the fluorescence microscope, of which there are many, depending on the fluorophore(s) used to stain the sample. The next chapter outlines the different fluorophores available and which to choose for your particular application.

1. Photoluminescence is the absorption and emission of photons. Luminescence is the production of light by cold-body radiation.
2. Fluorescence is the absorption of light at one wavelength and emission at a longer wavelength.
3. Fluorescence eventually ceases from interruption of the external photon source or by photobleaching of the fluorophore.
4. A Jabłoński diagram illustrates the electronic states of a fluorophore and the transitions that may occur between states – the physical process of fluorescence.
5. The magnitude of the Stokes shift determines how easily the fluorophore absorption spectrum may be separated from the emitted signal.
6. Fluorescence allows organelles, cells and tissues to be labelled with great specificity, sensitivity and with high contrast.
7. Fluorescence quantum yield is the ratio of the number of fluorescence photons emitted to the number of photons absorbed.
8. The molar extinction coefficient describes the absorption potential of the fluorophore.
9. Fluorophore brightness is the product of quantum yield and the extinction coefficient.
10. The localised environment – in particular the pH – can significantly influence fluorophore absorption and emission spectral profiles.
11. Modern fluorescence microscopes are built as epi-illumination stands. The objective acts as its own condenser.
12. Fluorescence intensity in the epi-fluorescence microscope is proportional to NA^4/Mag^2.
13. Modern filter sets incorporate three components: an excitation filter, a dichromatic beam splitter and an emission filter.

Key Reading

Brown, C.M. (2007. 'Fluorescence microscopy – avoiding the pitfalls'. *Journal of Cell Science 120*/10: 1703–1705.
- A very concise summary of the essential points you should know for fluorescence image acquisition.

Ishikawa-Ankerhold, H.C. et al. (2012. 'Advanced Fluorescence Microscopy Techniques - FRAP, FLIP, FLAP, FRET and FLIM'. *Molecules 17*/4: 4047–4132.
- An excellent paper giving an all-round introduction to fluorescence microscopy. I give out this paper to all the students I teach.

References

Badr, C.E. and Tannous, B.A. (2011). 'Bioluminescence imaging: progress and applications'. *Trends in Biotechnology 29*/12: 624–633.

Chandler, D.E. and Roberson, R.W. (2009). *Bioimaging Current Concepts in Light and Electron Microscopy*. Sudbury, MA: Jones and Bartlett Publishers.

Ellinger, P. (1940). 'Fluorescence microscopy in biology'. *Biological Reviews 15*/3: 323–350.

Greer 3rd, L.F. and Szalay, A.A. (2002). 'Imaging of light emission from the expression of luciferases in living cells and organisms: a review'. *Luminescence 17*/1: 43–74.

Herman, B. (1998). *Fluorescence Microscopy*, 2nd ed. No. 40. Bios/Taylor & Francis/Wiley.

Lichtman, J.W. and Conchello, J-A. (2005). 'Fluorescence microscopy'. *Nature Methods 2*/12: 910–919.

Murphy, D.B. (2001). Fluorescence microscopy. In: *Fundamentals of Light Microscopy & Electronic Imaging*, Chapter 11. Wiley-Liss.

Murphy, B.D. and Davidson, M.W. (2013). *Fundamentals of Light Microscopy and Electronic Imaging*, 2nd ed. Wiley Blackwell.

Ploem, J.S. and Tanke, H.J. (1987). The fluorescence microscope. In: *Introduction to Fluorescence Microscopy*, Chapter 2, 1st ed. RMS Handbook Series No. 10. OUP/Bios/Taylor & Francis/Wiley.

Sanderson, J.B. (2008). 'The role of the centralised light microscope light microscope imaging facility'. *The Quekett Journal of Microscopy 40*/8: 715–739.

Waidmann, M.S. et al. (2011). 'Bacterial luciferase reporters: The Swiss army knife of molecular biology'. *Bioengineered Bugs 2*/1: 8–16.

Watkins, S.C. and St. Croix, C.M. (eds.) (2013). *Current Protocols Select Imaging & Microscopy*. Hoboken, NJ: Wiley.

Further Reading

Campbell, A.K. (2003). 'Save those molecules! Molecular biodiversity and life'. *Journal of Applied Ecology 40*/2: 193–203.
Davis, I. and Parton, R. M. (2006). 'Fluorescent reagents for live cell imaging and their introduction into cells'. *Cold Spring Harbor Protocols* 6. pii: pdb.ip20.
Diaspro, A. et al. (2006). Photobleaching. In: J.B. Pawley (2006). *Handbook of Biological Confocal Microscopy*, Chapter 19, 3rd ed. New York: Springer, pp. 690–702.
Dickson, R.M. et al. (1997). 'On/off blinking and switching behavior of single molecules of green fluorescent protein'. *Nature 388*/6640: 355–358.
Johnson, S. and Rabinovitch, P. (2012). 'Ex vivo imaging of excised tissue using vital dyes and confocal microscopy'. *Current Protocols in Cytometry* Unit 9.39.
Marrison, J. et al. (2013). Ptychography – a label free, high-contrast imaging technique for live cells using quantitative phase information. *Scientific Reports 3*: 2369. doi: 10.1038/srep02369.
Masters, B. (2010). 'The development of fluorescence microscopy'. *Encylopedia of Life Sciences*. Wiley.
Mayer, A. and Neuenhofer, S. (1994). 'Luminescent labels – more than just an alternative to radioisotopes?' *Angewandte Chemie 33*/10: 1044–1072.
Ozawa, T. et al. (2012). 'Advances in fluorescence and bioluminescence imaging'. *Analytical Chemistry 85*/2: 590–609.
Petty, H.R. (2007). 'Fluorescence microscopy: established and emerging methods, experimental strategies, and applications in immunology'. *Microscopy Research and Technique 70*/8: 687–709.
Rathbun, C.M. and Prescher, J.A. (2017). 'Bioluminescent Probes for Imaging Biology beyond the Culture Dish'. *Biochemistry 56*/39: 5178–5184.
Renz, M. (2013). 'Fluorescence Microscopy – a historical and technical perspective'. *Cytometry 83A*/9: 767–779.
Rost, F.W.D. (1995). 'The history of fluorescence microscopy.' In: *Fluorescence microscopy*, Chapter 18, vol. 2. Cambridge University, pp. 183–195.
Rusk, R. (2009, 1911, 1929, 1967). 'First fluorescence microscope, First epifluorescence microscope, The dichroic mirror'. *Nature Milestones in Light Microscopy* No. 4.
Sanderson, M.J. et al. (2014). 'Fluorescence microscopy'. *Cold Spring Harbor Protocols*. doi: 10.1101/pdb.top071795.
Cold Spring Harbor Laboratory Press (http://www.cshlpress.com) publish the following five books, which have practical in-depth chapters by authors who are specialists in their field. The topics in these books cover all forms of microscopy, fluorescence microscopy in particular:
Goldman, R.D., Swedlow, J.R., and Spector, D.L. (eds.) (2010). *Live Cell Imaging: A Laboratory Manual*, 2nd ed.
Helmchen, F. and Konnerth, A. (eds.) (2011). *Imaging in Neuroscience: A Laboratory Manual.*
Sharpe, J. and Wong, R. (eds.) (2011). *Imaging in Developmental Biology: A Laboratory Manual.*
Spector, D.L., and Goldman, R.D. (eds.) (2006). *Basic Methods in Microscopy, Protocols and Concepts from Cells: A Laboratory Manual.*
Yuste, R. (ed.) (2011). *Imaging: A Laboratory Manual.*
Wiley (http://eu.wiley.com) publish two similar volumes:
Kubitscheck, U. (ed.) (2013). *Fluorescence Microscopy: From Principles to Biological Applications*.
Watkins, S. and St. Croix, C. (eds.) (2014). *Current Protocols Select: Methods and Applications in Microscopy and Imaging*.

Internet Resources

Fundamentals of fluorescence tutorials: http://www.thermofisher.com/sg/en/home/references/molecular-probes-the-handbook/introduction-to-fluorescence-techniques.html.
Calculating fluorescence quantum yields: http://www.horiba.com/fileadmin/uploads/Scientific/Documents/Fluorescence/quantumyieldstrad.pdf.
Semrock, cleaning filters: https://www.semrock.com/cleaning-optical-filters.aspx.
Fluorescence spectral databases URLs: https://www.thermofisher.com/sg/en/home/life-science/cell-analysis/labeling-chemistry/fluorescence-spectraviewer.html and https://www.micro-shop.zeiss.com/index.php?l=en&p=de&f=f.
Leica Fluoscout: https://www.leica-microsystems.com/ and https://www.leica-microsystems.com/science-lab/online-tool-to-determine-the-optimal-combination-of-fluorescence-filter-cubes-fluorophores-and-light-sources/.
Fluorophores table: http://www.geomcnamara.com/data.

Notes

1. Alternating single and double bonds form a conjugated system with connected p-orbitals and delocalised electrons, as shown in the figure. This generally lowers the overall energy of the molecule and increases stability.
2. In a paper read on 26th June 1853, Stokes includes a long footnote to his experiments on page 387 explaining his choice of the term 'fluorescence': Stokes G. G. (1853). 'On the Change of Refrangibility of Light – No. II'. *Phil. Trans. Roy. Soc. 143*: 385–96. See: http://www.jstor.org/stable/108570.
3. Fluorophores that absorb light strongly have ε values between 20 000 and 100 000. Quantum yield is sometimes incorrectly referred to as quantum efficiency. This latter metric, QE, refers to the efficiency by which photons are converted to photo-electrons. It is a measure of, say, a CCD's electrical sensitivity to light.
4. Beam-splitting mirrors, having thin interference coatings, that reflect and transmit specific wavelengths of light may variously be referred to as dichromatic or dichroic mirrors. Here, the term 'dichromatic' is used because dichroism is a property of pleochroism (Chapter 14), which describes the ability of a material to differentially absorb light, depending on its azimuth of polarisation.
5. When the image is not diffraction limited, and the source of fluorescence is extensive, the total light flux (signal) will depend on NA^4, and the brightness on NA^4/Mag^2. If the fluorescence is collected over the entire field of view (such as labelled cells distributed in a monolayer viewed by widefield microscopy), both signal and brightness will depend on NA^4/Mag^2 because the number of illuminated cells in the field of view decreases by the factor $1/Mag^2$.

16

Fluorophores and Fluorescent Proteins

Outline

16.1 Introduction

Fluorophores can be used in several ways to mark or target tissues, cells and cell organelles. They are mainly used conjugated, or covalently bonded, to an antibody or to a reporter molecule. Fluorophores can also be linked to enzyme substrates (Wysocki and Davis, 2011) or else incorporated into fluid tracers (Wouterloud et al., 2014) or applied to tissues as stains. A recent development, optogenetics (Pastrana, 2011; Gautier et al., 2014) uses light and caged ligands to control cell function. There are – literally – thousands of different fluorophores to choose from. The best way to select the most appropriate fluorophore for your particular application is to talk to co-workers or collaborators in your field and/or go online to the websites of the major supply companies and then consider the

Understanding Light Microscopy, First Edition. Jeremy Sanderson.

Table 16.1 Selected Fluorophore Suppliers

The following dye families are trademark groups:

Fluorophore dye type	Manufacturer/distributor	Company web page
Abberior dyes	Abberior	http://www.abberior.com/
Alexa Fluor dyes	Thermo Fisher Scientific	https://www.thermofisher.com/uk/en/home.html
APC, APCXL, RPE, BPE	Phyco-Biotech	http://phyco-biotech.com/en
Atto & Tracy dyes	Sigma Aldrich	https://www.sigmaaldrich.com/
Bodipy	Thermo Fisher Scientific	https://www.thermofisher.com/uk/en/home.html
Brilliant Violet	Biolegend	https://www.biolegend.com/
Cellular Lights, Organelle Lights	Thermo Fisher Scientific	https://www.thermofisher.com/uk/en/home/life-science/cell-analysis/cell-structure/celllight-ready-to-use-fluorescent-protein-based-reagents.html
CF dye	Biotium	https://biotium.com/
Chromis	Cyanagen	http://www.cyanagen.com/product
Chromeo dyes	Active Motif	https://www.activemotif.com/
Cy dyes	Thermo Fisher, Lumiprobe	https://www.lumiprobe.com/
DyLight Fluor	Pierce	UK = https://www.thermofisher.com/uk/en/home/life-science/protein-biology/protein-labeling-crosslinking/protein-labeling/fluorescent-protein-labeling.html & https://www.fishersci.com/content/dam/fishersci/en_US/documents/programs/scientific/brochures-and-catalogs/guides/thermo-scientific-pierce-fluorescent-products-guide.pdf
DY MegaStokes dyes	Dyomics	https://dyomics.com/
FluoProbes	Interchim	http://www.interchim.com/
Quasar & Cal Fluor dyes	Biosearch Technologies	https://www.biosearchtech.com/
Seta, SeTau & Square dyes	SETA BioMedicals	https://www.setabiomedicals.com/
Sulfo Cy dyes	Cyandye	https://www.cyandye.com/
SureLight Dyes	Columbia Biosciences	http://stores.columbiabiosciences.com/

Selected fluorophore and antibody supply companies:
AbD Serotec
Abcam
Biostatus
Jackson
Pierce Antibodies
Sigma Aldrich
Stratech
Thermo Fisher Scientific (formerly Invitrogen, then Life Technologies)
The antibody resource: https://www.antibodyresource.com/onlinecompiz.html.
NB: this list cannot be exhaustive

properties of those fluorophores you have short-listed. To help you get started, a selection of these companies is given in Table 16.1. Fluorophores are generally catalogued and described according to their absorption and fluorescence properties. This list shows the other characteristics to take into account when choosing a suitable fluorophore:

- spectral absorption and emission curves (maximise absorption coefficient);
- Stokes shift (large Stokes shifts mean good contrast; may mean low quantum yield);
- quantum yield, Φ (maximise quantum yield to maximise brightness);
- molar extinction coefficient, ε (maximise extinction coefficient to minimise brightness);
- brightness (but lower brightness and higher photostability can give better SNR for time lapse);
- photobleaching resistance;
- maximum concentration tolerated within the cell;
- saturation of the fluorophore;
- the brightest and most photostable fluorophores should be reserved for the least abundant cellular targets;
- dyes with green and red fluorescence emission generally tend to be brighter than those emitting in the blue and far-red portions of the spectrum;
- the localised environment (pH, oxidising and reducing agents) significantly influences fluorophore absorption and emission spectral profiles, so the spectra published by manufacturers of isolated fluorophores in solution can differ markedly from those observed under actual experimental conditions;
- the shape of the absorption and emission spectral curves will indicate whether excitation cross-talk or emission bleed-through is likely when used with other fluorophores; and
- use wide bandwidth excitation and emission filters, to maximise signal, unless there is significant bleed-through.

When a single fluorophore is used to label tissues, cells or specific organelles, generally only the brightness, photostability and whether the cells or tissue will autofluoresce need be considered. When two or more fluorophores are required to mark different

Table 16.2 *Characteristics of Some Common Organic Fluorophores and Fluorescent Proteins*

Fluorophore	Absorption max λ (nm)	Emission max λ (nm)	Stokes shift (nm)	Quantum yield ratio (Φ)	Extinction coefficient ε $M^{-1}cm^{-1}$	Brightness Φ × ε	Photostability $t_{1/2}$ (s) WFM
DAPI	344	450	106	0.04 (+DNA = 0.92)	21 000	840 (+DNA = 19 320)	25
Hoechst 33258	345	507	162	0.83	46 000	39 560	~100
To-Pro 3	642	661	19	0.11	102 000	11 220	~10
DraQ 5	598	697	99	0.004	21 000	84	high
Alexa 488	495	519	24	0.92	71 000	65 320	~200
Cy 2	489	506	17	0.12	150 000	18 000	-
ATTO 488	501	523	22	0.8	90 000	72 000	-
FM 4-64	515, 558	640, 734	125, 176	0.6	46 000	27 600	-
Alexa 568	578	603	25	0.69	91 300	63 000	430
Cy 3	550	570	20	0.15	150 000	22 500	230
Texas Red	589	615	26	0.93	139 000	129 270	-
Alexa 633	632	647	15	0.05 (estimated)	147 000	7400	-
Alexa 647	650	668	18	0.33	239 000	78 870	220
Cy 5	649	670	21	0.28	250 000	70 000	120
typical quantum dot	continuous	narrow	size dependent	~0.70	100 000–1 300 000	~400 000	very high
eGFP	488	507	19	0.6	56 000	34	174
mTagBFP	399	456	57	0.63	52 000	33	34
mVenus	515	528	13	0.57	92 000	53	15
mKO2	551	565	11	0.62	63 800	40	9
tdTomato	554	581	27	0.69	138 000 (69 000 × 2)	95	70
mCherry	587	610	23	0.22	72 000	16	111
mKate 2	588	633	45	0.4	63 000	25	69
mPlum	590	649	59	0.10	41 000	4	53

This table is by no means comprehensive. Its purpose is to show how both fluorophores and fluorescent proteins differ between one another in their essential characteristics of Stokes shift, brightness and photostability, and how they compare to one another. Refer also to Table 16.3. When DAPI is DNA-bound, the abs/em figures are 358 and 461 nm respectively. The local environment can alter abs/em values. For example, DNA-bound Hoechst 33258 has abs/em values of 352/455; dissolved in dimethyl formamide, the values are 354/486 and dissolved in water the values are 345/507 (data from *Fluorophores.org*). Hoechst is less toxic to living cells than DAPI. Photostability values are always higher for confocal than widefield (WFM).

Sources: (a) Table 1, Dempsey, G.T. et al. (2011). *Nature Methods 8*/12: 1027; (b) https://en.wikipedia.org/wiki/Fluorophore; (c) Poster: https://www.nature.com/nrm/posters/fluorescent/index.html; (d) I love GFP: https://sites.google.com/site/ilovegfp/Home/fps; (e) Table 1 & Supp. Table 1, Shaner, N. et al. (2005). *Nature Methods 2*/12: 905–909. http://www.tsienlab.ucsd.edu/Publications/Shaner%202005%20Nature%20Methods%20-%20Choosing%20fluorescent%20proteins.pdf; also Shaner, N.C. et al. (2007). *Jour. Cell Science 120*/24: 4247–4260. http://jcs.biologists.org/content/joces/120/24/4247.full.pdf; (f) Resch-Genger, U. et al. (2008). *Nature Methods 5*/9: 763–775. http://invibe.net/biblio_database_dyva/woda/data/att/304b.file.07088.pdf.

structures, then additionally the potential for bleed-through or cross-talk (see Chapter 20) must also be taken into account. The Stokes shift values, excitation/emission maxima and spectral curves are useful when selecting appropriate fluorophore combinations. Table 16.2 shows the characteristics of some commonly used man-made fluorophores.

The earliest and possibly best-known fluorophores are the xanthene dyes fluorescein iso-thiocyanate (FITC) and tetramethyl rhodamine iso-thiocyanate (TRITC). Although still widely used, FITC and TRITC, as well as the biaminated acridine derivative acridine orange, are essentially first-generation fluorophores. Modern antibody-conjugated probes or quantum dots should be selected, since these are brighter, more photostable and generally less sensitive to local changes in pH than traditional dyes.

Historically FITC and TRITC (rhodamine) were selected as fluorescent dyes because they matched the spectral characteristics of the mercury short-arc lamp. FITC, the iso-thiocyanate salt of fluorescein, is a bright fluorophore, and suitable for the mercury lamp, which emits relatively weakly in the blue portion of the spectrum, whereas TRITC is a much less bright fluorophore and is matched to the strong emission of the mercury lamp around 546 nm. Thus, used together, these two fluorophores exhibit roughly the same intensity when observed down the eyepieces of the microscope using a 50 or 100 watt mercury lamp. The brightness will change when the same fluorophore in the same field of view is either (a) observed with a xenon lamp or is (b) illuminated with a laser for confocal microscopy, because the spectral characteristics of these illumination sources are different.

16.2 Chemical Fluorophores

Most fluorophores are small organic molecules comprising between 20 and 100 atoms, whereas natural fluorophores are generally around 250 to 1000 times larger (Figure 16.1 and Figure 16.2). Fluorescent molecules typically possess aromatic ring structures with several π bonds, which, in turn, have delocalised electron clouds associated with them (see Section 15.2) that are distributed throughout the ring structure rather than being associated with a single atom or covalent bond. In compounds with alternating single and double

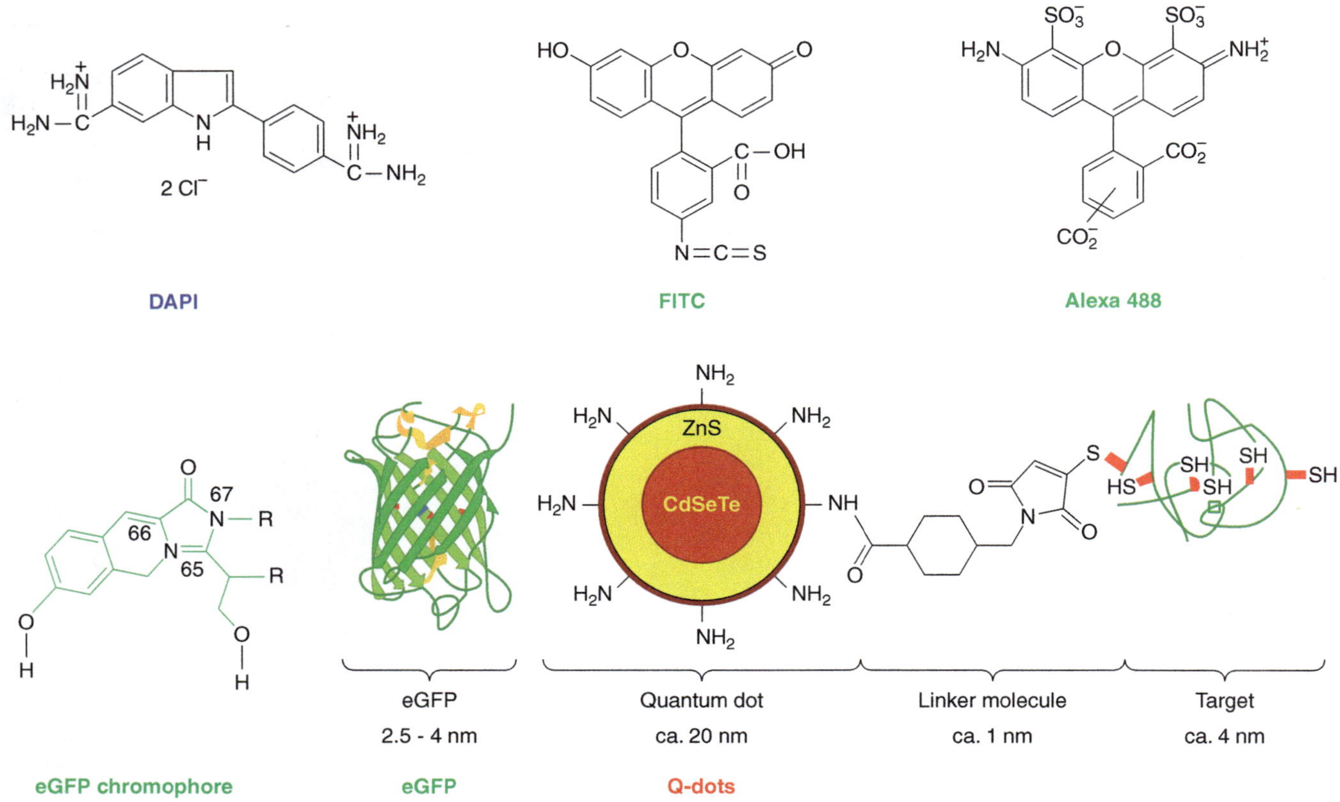

Figure 16.1 ***The size of fluorophores*** *Source: Adapted from Thermofisher and Zeiss.*

bonds, this averages out the overall energy across the molecule and increases stability. Benzene, the simplest aromatic hydrocarbon (an arene) is excited at 254 nm and emits at 300 nm. Other fluorophores are derivatives of similar ring structures: oxazine, naphthalene, coumarin, cyanine, arylmethine, pyrene and tetrapyrrole, for example. Additionally, various functional groups (e.g. amino, carboxyl or thiol groups) can alter the properties (e.g. solubility) of the fluorophore, and can also be used to conjugate proteins to the fluorescent molecule.

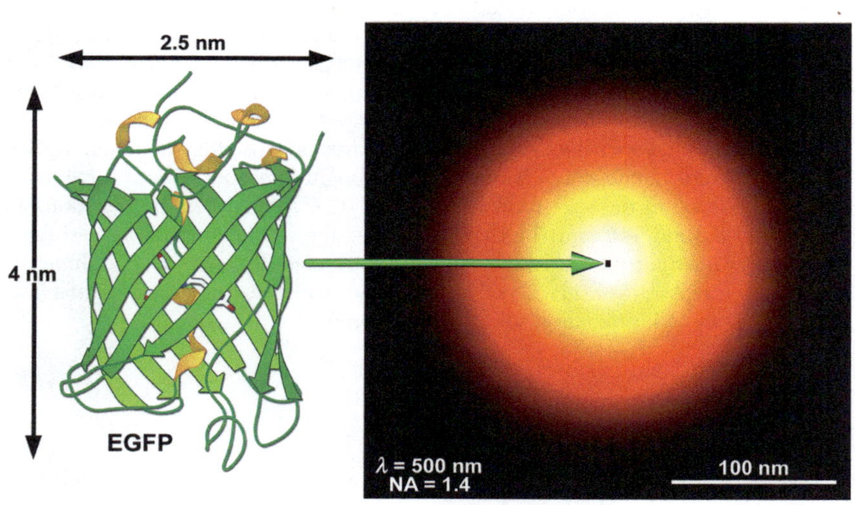

Figure 16.2 ***A comparison of the physical size of eGFP to the size of its image in the fluorescence microscope*** *Source: Murphy and Davidson 2013. Reproduced with permission from John Wiley & Sons.*

As increasing numbers of aromatic rings are added to the molecule, the fluorescence emission tends to shift further towards the red end of the spectrum. Although fluorescent proteins are increasingly widely used, small molecular organic compounds remain important, both as fluorophore labels and activatable sensors. They do not require prior work to introduce them into the genome of the organism being studied. They are relatively inexpensive, easier to handle and use than fluorescent proteins, amenable to chemical modification and generally have higher brightness and signal-to-noise ratios.

16.3 Natural Fluorophores – GFP

The emergence of genetically encoded endogenous fluorescent proteins from cnidarians in the 1990s meant that virtually any non-fluorescent protein of interest can now be made to fluoresce by the introduction of the fluorescent protein codon (Figure 16.3) into its own gene sequence. Green fluorescent protein was the first to be isolated, characterised and used as an endogenous biological marker. Fluorescent proteins are very stable, being largely unaffected by fixation or temperature changes (see Becker et al., 2012; Joosen et al., 2014). They are coded for by a single gene, so they are easy to transfect into the host organism, require no post-transcriptional modification or exogenous co-factors to work and are directly visible with any fluorescence microscope. Fluorescent proteins are normally minimally invasive natural proteins that can be used at physiological concentrations in the host cell or organism, allowing us to follow the localisation of the protein or the dynamics of labelled cell populations. In his Nobel Laureate lecture, the late Roger Tsien remarked:

> 'One of the most common requests I get is to explain why GFP and related FPs are significant, preferably in lay terms. My usual answer has been that their genetically encoded fluorescent colors can make many biochemical processes directly visible inside living cells and organisms. Using standard molecular biological tricks to connect host cell genes to FP genes, we can watch when and where those host cell genes get switched on and off, when the protein products are born, where they travel, with what other proteins do they interact, and how long they survive. Even when we cannot directly tag the signals of interest, e.g. Ca^{2+}, cAMP, protease or kinase activities, we can often engineer FP-based indicators to image those signals directly.'

The green fluorescent protein (GFP) was discovered in 1961 in the hydromedusa *Aequorea victoria* by Osamu Shimomura (Box 16.1) after he was first asked to investigate bioluminescence in a marine seed shrimp. A great deal of sheer hard work was involved: Shimomura and his supervisor, Frank Johnson, with their families collected over 850 000 *Aequorea* specimens from Friday Harbor, near Seattle, in the course of two decades of research. About 50 000 jellyfish were needed to produce a milligram of aequorin (an apoaequorin-coelenterazine luminophore complex).

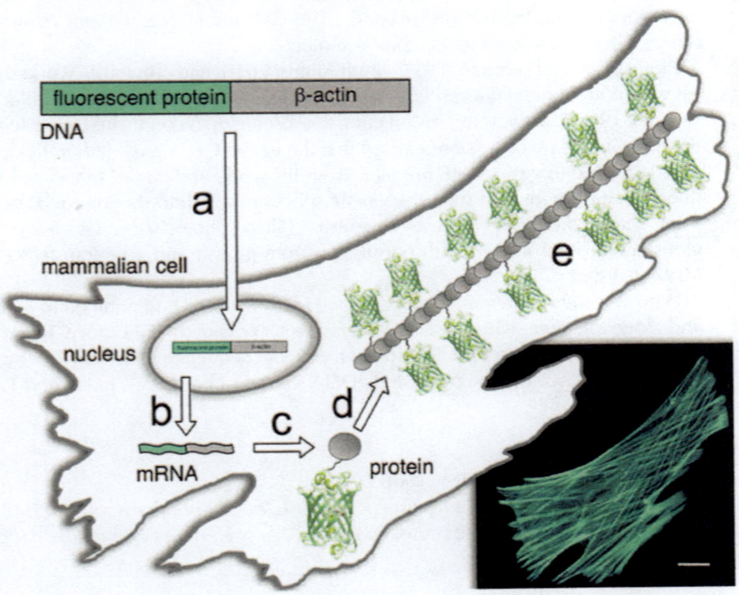

Figure 16.3 **How a fluorescent protein is manufactured** *A schematic representation of how a monomeric fluorescent protein is employed for imaging of β-actin. A mammalian cell is transfected (a) with a cDNA chimera composed of a fusion of the genes encoding the fluorescent protein and β-actin. The gene is transcribed (b) to produce mRNA that is then translated (c) to form the chimeric protein. The trafficking and localisation (d) of the protein is dictated by the protein of interest and the fluorescent protein, ideally, does not interfere. In the case of β-actin, the chimera is incorporated into actin filaments (e) along with the endogenous protein. Shown in the inset is a fluorescence image of a grey fox lung fibroblast (FoLu) cell that has been transfected with mTFP1-β-actin (Ai et al., 2008). Scale bar represents 10 microns. (The figure legend has been taken directly from Figure 9 on this web page: http://www.scholarpedia.org/article/Fluorescent_proteins. Location: http://www.scholarpedia.org/article/File:Fluorescentprotein_localization.png.) Source: Campbell 2008. Reproduced with permission from Professor R.E. Campbell.*

Box 16.1 GFP and the road to Gold

The jellyfish *Aequorea victoria* [*Murbach & Shearer, 1902*] is found off the coast of North America, and as a defence mechanism emits bioluminescence from the photocytic cells around the margin of its umbrella. When the Victorian naturalist PH Gosse (who developed and popularised the aquarium) tapped a glass jar containing the jellyfish it quickly lit up with a ghostly greenish-blue light. This is the story of the men who won the Nobel Prize for work on GFP.

Source: The Asahi Shimbun/ Contributor. Reproduced with permission from Getty Images.

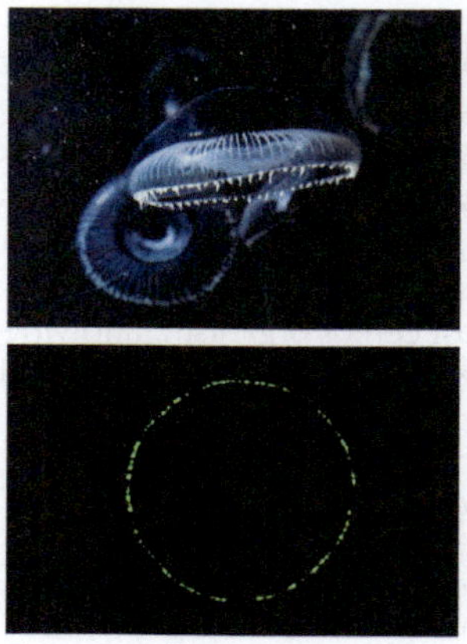

Source: Shimomura, 2005. Reproduced with permission from John Wiley & Sons.

Aequorin is a calcium ion-activated photoprotein which in conjunction with the luciferin coelenterazine produces blue light of around 470 nm. This blue light is taken up by green fluorescent protein (GFP) and re-emitted as green light at around 504 nm. The transfer of energy from aequorin to GFP is non-radiative by FRET, so no blue photons are seen. The gene for GFP encodes its own fluorophore, which is what makes it so useful. What is the story behind this protein that has revolutionised modern biological science?

Osamu Shimomura (27th August 1928 – 19th October 2018) certainly didn't have an easy start, and only dogged determination and a brilliant mind changed his fortune for the better. Caught up in wartime Nagasaki his schooling was severely disrupted, yet in post-war Japan he managed to isolate coelenterazine from *Cypridina* (*vargula hilgendorfii*), a purification which had defeated an American group at Harvard. As a result Shimomura was invited to America to work on the bioluminescent reaction of *Aequora*. There he worked out the bioluminescent production of blue light when aequorin was activated (at will by the jellyfish) using calcium ions in seawater. The protein giving off green light was crystallised in 1974 and was named – rather pragmatically and uninspiringly – Green Fluorescent Protein. The fluorophore was determined in 1979.

For the next 15 years little happened concerning GFP until Martin Chalfie, a molecular biologist heard about it during a lunchtime seminar and wondered if it could be expressed in another organism: *Caenorhabditis elegans* a thin, transparent nematode worm amenable to study by light microscopy, and whose cell fate was being worked out, for which a fluorescent marker would make a superb tool. Chalfie was looking for a dynamic marker that could be introduced non-invasively into living organisms, to supersede the existing static antibody and enzyme-substrate staining protocols then in current use. His ignorance of fluorescent proteins kept him unaware of the difficulties most people – including Shimomura – considered there would be to get GFP to fluoresce in isolation outside of its parent organism. Eventually, in 1993 Chalfie succeeded in introducing the cDNA of GFP heterologously into, first the bacterium *E. coli*, then into *C. elegans*. The GFP fluoresced without requiring any co-factors, enzymes or other proteins. It was an ideal reporter molecule.

Roger Tsien (1st February 1952 – 24th August 2016) had previously worked on calcium signalling. His interest was in identifying fluorescent markers to dissect and analyse the 'cellular anthropology' of higher organisms. By 1992 Tsien was micro-injecting fluorescent dyes conjugated to protein into cells. With the announcement of Chalfie's work, Tsien realised that the native GFP signal – 'dim, fickle and spectrally impure' – could be engineered to glow much brighter. From his work, and that of others, we now have a complete palette of fluorescent proteins that span the visible spectrum and into the infra-red. In 2008, Shimomura, Chalfie and Tsien shared the Nobel Prize in Chemistry. (These photos show Glo-fish, or fluorescent zebrafish. Source: glofish.com. Reproduced with permission from glofish and GFP-transfected mice: photograph courtesy of Mr M. Gardiner).

Who was the fourth man? Someone had identified and had gone on to isolate the cDNA sequence of GFP and clone the gene, which was crucial to the success of the GFP story. That man was Doug Prasher. After his work on GFP and before publication of the cDNA sequence, Prasher's funding dried up and he eventually left science. Only two people requested cDNA samples from him: Chalfie and Tsien.

References

Chalfie, M. (2009). 'GFP: Lighting up life'. *PNAS 106*/25: 10073–10080.

Shimomura, O. (2005). 'The discovery of aequorin and green fluorescent protein'. *Journal of Microscopy 217*/1: 3–15.

Tsien, R.Y. (2009). 'Constructing and exploiting the fluorescent protein paintbox'. *Angewandte Chemie International Edition 48*/31: 5612–5626.

Fortunately – although GFP requires Ca^{2+} ions and aequorin as co-factors to fluoresce naturally in *A. victoria*, when expressed exogenously, it only requires activation by UV light in the presence of oxygen. GFP is a small 27 kDa molecule; it is relatively inert, making it a very powerful marker of dynamic cellular processes. GFP is an extremely stable protein of 238 amino acids. It is membrane-permeant, can mark any protein as a fusion protein and generally does not obstruct the functionality of the molecule to which it is bound. GFP and other fluorescent proteins are not toxic to cells in the same way as synthetic fluorophores and dyes, because the fluorophore is buried within the β-barrel of the GFP structure (Figure 16.2; Figure 16.4). GFP can also be utilised as a reporter (Enterina et al., 2015) and a biosensor (Mishin et al., 2015) as well as a fluorescent tag. If GFP is linked to the gene promoter, then it will be

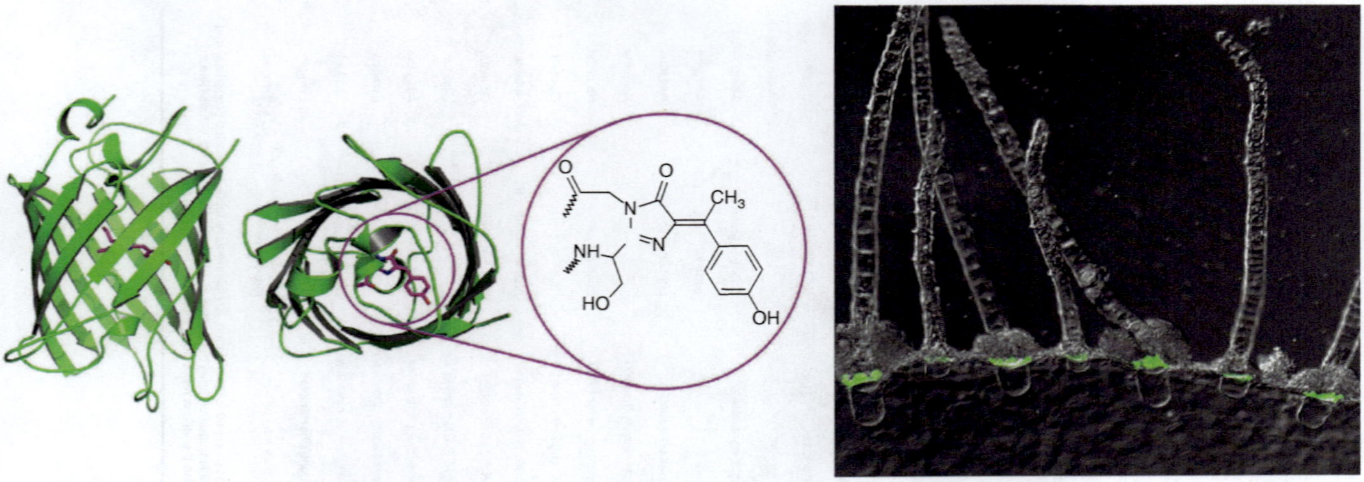

Figure 16.4 **Green fluorescent protein** *The molecular structure of GFP is shown in the left-hand panel; the fluorophore is shown in the centre panel of the figure. The GFP chromophore is 4-[p-hydroxybenzylidene]-5-imidazolidinone.*

Whilst jellyfish from which GFP was isolated were collected from Friday Harbor in the Puget Sound of Washington state in the United States, the right-hand panel shows the medusa of the jellyfish Obelia lucifera *collected on a field trip at Plymouth Marina in the United Kingdom.*

Some texts claim that the blue sheen that may be seen in pictures of the Aequorea victoria *jellyfish is bioluminescence or fluorescence: it is not. This blue colour arises from reflected light when using a flash to photograph the jellyfish. The true bioluminescent reaction is green, as shown here and in the figure in Box 16.1.*

Source: (left) Stepanenko et al. 2013. Reproduced with permission from Elsevier; Reproduced with permission from B. Moeyaert. (right) Courtesy of Professor A.K. Campbell.

expressed in place of the protein, enabling the location and time of synthesis of the protein to be established. GFP is heritable, and it is possible to create whole organisms that will fluoresce.

The GFP gene is introduced (transfected) into cells as a circular plasmid using calcium phosphate transfection, vector-based systems or a recombinant virus (Snapp, 2005; Ai et al., 2014). BacMam technology (Boyce and Bucher, 1996; Khan, 2013) employs double-stranded DNA insect viruses (baculovirus) as vehicles efficiently to deliver and express genes in mammalian cells. Because of its insect origins, the virus can penetrate into the cell but cannot replicate. These fusion proteins give highly specific labelling for cell organelles, and are available from Thermo Fisher Scientific.

The wild-type GFP (wt-GFP) has two excitation peaks, a major one at 395 nm and a minor one at 475 nm with extinction coefficients of 25 000 and 9500 M^{-1} cm^{-1}, respectively[1]. Excitation of the wild-type GFP at 395 nm leads to rapid quenching of the fluorescence with an increase in the 475 nm excitation band. The commonly-used mutant form, eGFP, has an excitation peak at 490 nm with the emission at 508 nm. This is not only a brighter fluorophore, with an extinction coefficient of 55 000 M^{-1} cm^{-1}, but is conveniently and easily detected with standard 'green' fluorescence filter sets[2]. The brightest version at the time of publication is mNeonGreen (Shaner et al., 2013; Table 16.3, Figure 16.5), but mWasabi developed from mTFP1 and mClover (Lam et al., 2012) are also very bright monomeric green fluorescent proteins, similar to tdTomato or the monomeric TagRFP, the brightest red fluorescent protein (Shaner, 2008). If the protein being studied is transient, brightness may trump everything else: with extra brightness it is possible to reduce the exposure time – a vital concern for live-cell work – and still pick up the signal.

Although GFP continues to be used worldwide and is ubiquitously useful, it does have certain disadvantages. Its size can perturb protein folding and trafficking; it also has limited brightness and photostability, which is a problem for single-molecule imaging and super-resolution studies (see Costantini and Snapp, 2013). Although fluorescent proteins are generally more photostable than synthetic fluorophores, the photostability of the chosen fluorescent protein is an important cornerstone of any successful imaging experiment, and photostability is crucial for long-term time-lapse work, for gathering weak fluorescent signals and for quantitative measurements, including FRET techniques. When examining GFP-expressing tissues, it is worth remembering that GFP is soluble in water and buffers and is quenched by the solvents used to embed tissues into paraffin wax (e.g. see Malkani and Schmid, 2011). Animals are best perfusion-fixed, the tissues immersed in 30% sucrose at 4°C and frozen in OCT compound for cutting as fixed frozen sections. Post-fix in 1% EM-grade paraformaldehyde or 1% periodate-lysine-paraformaldehyde (PLP) for 10 minutes avoiding any organic media, which will quench the fluorescent signal. Block with 20 mM *fresh* sodium borohydride; 50 mM ammonium chloride or 300 mM glycine buffer (see Section 16.7). To boost any loss of signal, many workers apply an anti-GFP (e.g. chicken) polyclonal antibody.

16.4 Other Fluorescent Proteins

It is the modification of the wild-type GFP by random or directed molecular mutagenesis into enhanced (eGFP) and variant forms, principally by Roger Tsien's research group (e.g. Tsien, 2009; Tsien, 2005; Subach et al., 2011), that has made GFP so useful. The focus

Table 16.3 Selected fluorescent proteins

Protein	Excitation maximum (nm)	Emission maximum (nm)	Stokes shift (nm)	Suggested filter set	Suggested laser line (nm)	**Absorption coefficient ($M^{-1}cm^{-1}$)**	Fluorescence Quantum Yield	Brightness (cf. eGFP)	pKa	Photostability	$t_{1/2}$ (s)	Cranfill (2016) $t_{1/2}$ @80 μW	Oligomeric state	Maturation $t_{1/2}$ (hr)	Notes	Suggested choices	Commercially available from:
eBFP2	383	448	65	DAPI/BFP	351	32000	0.56	53%	5.3	OK	55	15.3 ± 0.3	weak dimer	0.4	v. bright, v.v. photostable		Life Technologies
mTagBFP2	399	454	55	DAPI/BFP	405	50600	0.64	120%	2.7	OK	57	6.2 ± 0.1	monomer	0.2	bright, pH insensitive	popular blue FP brightest blue FP	Addgene
mTurquoise2	434	474	40	CFP	440, 458	30000	0.93	85%	3.1	good	90	71.7 ± 1.3	monomer	<0.2	bright	highest QY, good for FRET compare with mClover3	Addgene
mCerulean 3 (CFP)	433	475	42	CFP	440, 458	40000	0.87	105%	4.7	very good	1100	76.8 ± 1.4	monomer	<0.2	very photostable FP good, bright, can aggregate	popular cyan FP	Addgene
Tag CFP	458	480	22	CFP	458	37000	0.57	63%	4.7	OK	55	—	monomer	0.4	pH insensitive, matures fast		Evrogen
mTeal (TFP1)	462	492	30	CFP	458	64000	0.82	162%	4.3	very good	132	72.3 ± 1.4	monomer	0.4	bright, stable, pH insensitive	brightest cyan FP	Allele Biotech
emGFP (emerald)	487	509	22	FITC/GFP	488	57500	0.68	116%	6.0	good	100	79.4 ± 0.7	weak dimer	0.2	pH insensitive, photostable	fast initial bleacher	Thermo Fisher
eGFP (enhanced)	490	508	18	FITC/GFP	488	55000	0.60	100%	6.1	very good	150	179.2 ± 2.1	weak dimer	0.2	sensitive to pH	gold standard FP consider meGFP	Clontech
mAG1(Azami-Green)	492	505	13	FITC/GFP	488	55500	0.74	121%	5.8	OK	50	—	monomer	<0.2	bright		MBL-International
mNeon Green	506	518	11	FITC/GFP	488	116000	0.80	276%	5.7	very good	158	197.2 ± 2.8	monomer	<0.2	v.v. bright, photostable	brightest green FP	Allele Biotech
mClover3	506	517	12	FITC/GFP	488	109000	0.78	258%	6.5	OK	80	53.2 ± 0.2	monomer	0.5	v.v. bright	good for FRET	Addgene
Venus	515	528	13	YFP	514	92200	0.57	159%	6.0	poor	15	—	monomer	<0.1	bright, not photostable	popular yellow FP	Addgene
mCitrine (YFP)	516	529	13	YFP	514	77000	0.76	174%	5.7	OK	49	15.7 ± 0.1	monomer	0.4	better pH stability than eYFP		Addgene
YPet	517	530	13	YFP	514	104000	0.77	238%	5.6	poor	49	30.8 ± 0.4	weak dimer	0.3	very bright, good for FRET	brightest yellow FP	No outlet
mOrange 2	549	565	16	TRITC/Cy 3	543	58000	0.60	104%	6.5	very good	122	353.6 ± 5.9	monomer	4.5	use mOrange2 for time-lapse		Clontech
mKO2 (Kusabira-Orange)	551	565	14	TRITC/Cy 3	543	63800	0.62	118%	5.5	poor	40	51.2 ± 0.3	monomer	1.2	less stable than mCherry	popular orange FP	MBL-International
tdTomato	554	581	27	TRITC/Tx Red	561	138000	0.69	283%	4.7	good	70	31.8 ± 0.3	tandem dimer	1.0	very bright, photostable	brightest red FP	Clontech
Tag RFP	555	584	29	TRITC/Tx Red	561	100000	0.48	142%	<4.0	poor	48	29.0 ± 1.4	monomer	1.7	use tagRFP-T for time-lapse	tagRFP-T most photostable monomer FP	Evrogen
mRuby 3	558	592	34	Texas Red	561	128000	0.45	175%	4.8	very good	176	—	monomer	2.1	pH insensitive, photostable	good for FRET	Addgene
FusionRed	580	608	28	Texas Red	594	94500	0.19	53%	4.6	very good	150	7.5 ± 0.0	monomer	2.2	v. low cytotoxicity, photostable		Evrogen
mCherry	587	610	23	Texas Red	594	72000	0.22	50%	<4.5	very good	111	318.9 ± 7.6	monomer	0.6	pH insensitive, photostable	very popular far red FP can aggreagate, thus consider mKate/FusionRed	Clontech
tdKatushka2	588	633	45	Tx Red/Cy 5	594	132500	0.37	146%	5.4	very good	144	—	tandem dimer	<0.3	very bright, photostable		Evrogen
mKate2	588	633	45	Tx Red/Cy 5	594	62500	0.40	76%	5.4	good	69	51.6 ± 0.6	monomer	0.8	10x brighter than mPlum	brightest long wavelength monomeric FP	Evrogen

A tandem-dimer contains *two* copies of the fluorescent protein bound by a short linker and is therefore effectively a monomer. As such, tandem dimers are both very bright and photostable. Since this Table was prepared for publication, Supplementary Table 1 (spectral properties) and Supplementary Table 2 (photobleaching properties) have been published by Cranfill et al. (2016). *Nature Methods* May 30. doi: 10.1038/nmeth.3891. Their photobleaching values, the result of much work by the lead author, have been cited in addition to the literature values here. An ultra-stable monomeric GFP (muGFP) has been developed which can withstand tissue clearing with CLARITY (Scott et al. (2018). *Scientific Reports* 8: 667 Jan 12. doi:10.1038/s41598-017-18045-y. See also the values reported in Table 1 of Thorn (2017).

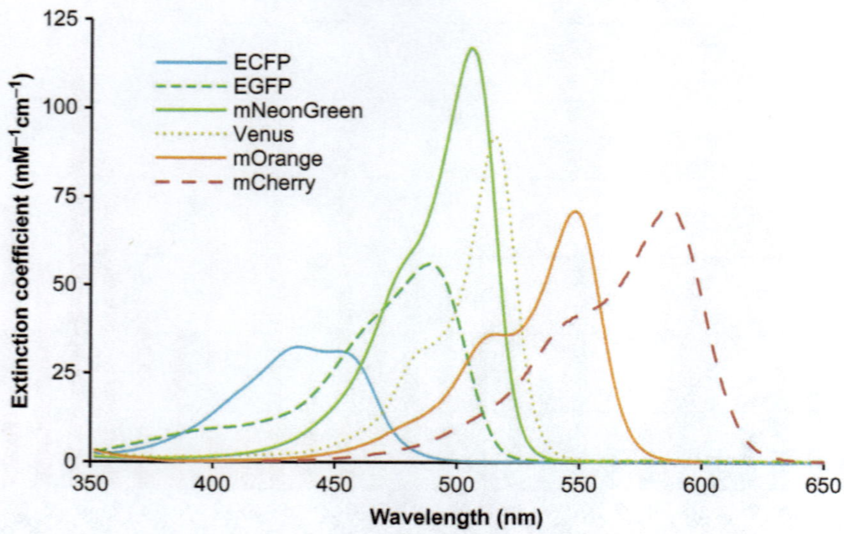

Figure 16.5 **Absorbance spectra of some fluorescent proteins** *The brightness of a fluorescent protein depends on how well it absorbs light and how well this energy is emitted as a useable fluorescent signal. This is determined by the extinction coefficient and the quantum yield, brightness being the product of these two factors. The absorbance curves are scaled to their peak extinction coefficients. mNeon and mCherry are the brightest with eGFP and mOrange third equal. The values (taken from Waters and Wittmann, 2014) are: mNeonGreen 100%, mCherry 95%, eGFP 73%, mOrange 72%, Venus 66% and eCFP 45%. See Shaner (2013) for further details. Source: Waters and Wittman 2014. Reproduced with permission from Elsevier.*

was to characterise brighter forms of GFP as well as isolate other fluorescent proteins capable of emitting different colours for use in multiple-labelling experiments. Generally, for fluorescent proteins with similar molar extinction coefficients, the one with the higher quantum yield will be brighter, and these tend to be clustered around the centre of the visible spectrum (Figure 16.6). eGFP remains one of the most useful mutants of wt-GFP in the fluorescent protein palette. Nevertheless, fluorescent proteins of lower brightness but higher photostability can result in better signal-to-noise ratio, especially for long time-lapse experiments. Therefore, brightness should not be the only parameter considered while choosing a fluorescent protein. It can sometimes be a challenge to achieve an emission signal of maximum potential brightness from some fluorescent proteins, because the filter sets available are not optimised for imaging the protein in question. It is only the largest imaging facilities that will have an inventory of filters to make up custom filter sets. Most users will be restricted to using those supplied with their microscope.

Typical GFP-like chromophores cannot be red-shifted by mutagenesis beyond 540 nm (yellow), and yellow fluorescent protein (YFP) remains the most red-shifted GFP mutant. Red fluorescent proteins were first discovered when Mikhail Matz, in the Lukyanov group, found some GFP-like proteins in *Discosoma* coral from a hobbyist's aquarium in a Moscow apartment in 1999 (Pieribone and Gruber, 2005). Before Lukyanov, no one had tried looking for GFP-like proteins in corals because corals do not exhibit bioluminescence. A whole palette of other probes have since made their appearance with fluorescent proteins now available in every colour of the spectrum. Significantly, the fluorescent protein variants formed by mutagenesis are sometimes less bright than the original GFP or DsRed from which they have been formed. The first of these red-shifted mutant fluorescent proteins derived from mRFP1 (monomeric DsRed fluorescent protein, which is more useful than DsRed alone) was nicknamed mCherry because of its purplish hue under reflected light and deep red fluorescent glow. Subsequent mutants were also named after fruits (see Day and Davidson, 2009, and Shaner et al., 2008).

Red-shifted fluorescent proteins have greater applicability than GFP and its homologues for four reasons: firstly, autofluorescence of animal tissues is much less pronounced at the red end of the spectrum, giving a much better signal-to-noise ratio (SNR) than the short wavelength end of the spectrum; secondly, red illumination is less harmful to living cells and tissues; thirdly, red light scatters much less when passing through thick cells and tissues, meaning that clear images may be acquired deeper into the specimen than with short-wavelength illumination and fourthly, very recently developed fluorescent proteins are excited in the short wavelengths but emit in the red or far-red, making them ideal for dual labelling without significant bleed-through and also for multi-photon microscopy, e.g. mT-Sapphire, mAmetrine and mKeima[3]. These are known as large Stokes shift (LSS) variants (Table 20.1; Yang, 2013; Shcherbakova et al., 2012a; Piatkevich et al., 2010a, 2010b). However, bright far-red fluorescence[4] is not a natural property of fluorescent proteins; the brightest are green fluorescent proteins, suggesting an approximate Gaussian-like histogram of fluorescent protein brightness across the spectrum with the brightest (Figure 16.6) being green or yellow. Hence future development of red-shifted fluorescent proteins most likely will be limited.

GFP can donate electrons to vitamins (e.g. riboflavin) present in culture media, so pH and culture medium composition can significantly affect the photostability of GFP (see Bogdanov et al., 2009). Fluorophores often produce reactive oxygen species; GFP is no exception, and depending on the fusion copy number, some cells react to transfection and fluorescence observation better than others. Finally, all GFP-like proteins have a tendency to oligomerise. Even *Aequorea victoria* GFP, which is considered to be naturally monomeric, will form dimers at high concentrations corresponding to physiological conditions in jellyfish. Most fluorescent

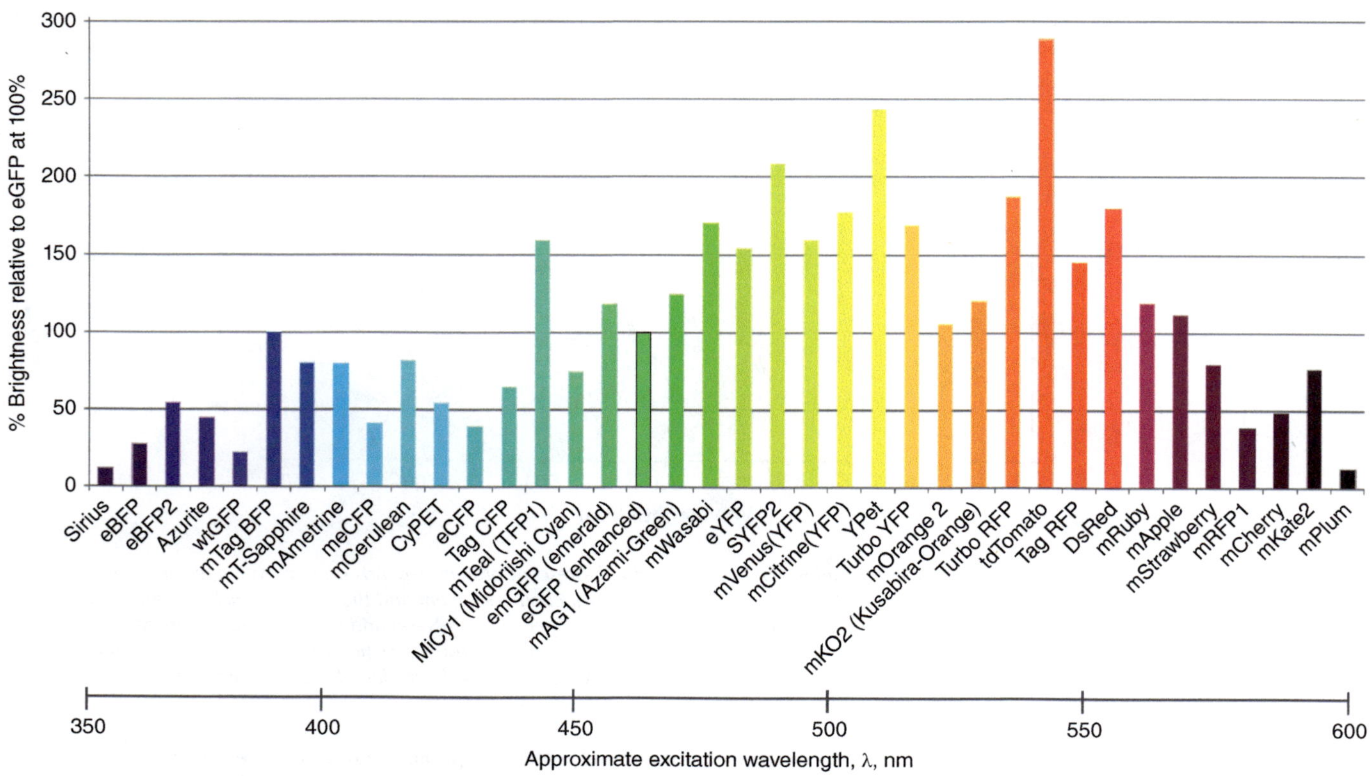

Figure 16.6 ***The palette of fluorescent proteins.***

proteins from non-bioluminescent *Anthozoa* form very stable homo-tetramers even at very low (nanomolar) concentrations. However, for labelling proteins, a fluorescent protein must be monomeric[5]. Finally, fluorescent proteins can take several hours to fully mature[6] into a fluorescent moiety after initial expression in cells.

Once synthesised, fluorescent proteins must fold into shape and oxidise to form the conjugated bonds of the fluorophore. It is this oxidation step that is the longest (Miyawaki et al., 2003). For practical use, the fluorescent protein maturation rate should be reasonably fast to obtain a bright and stable fluorescent signal. The temperature of the marine environment of the jellyfish is 4–5°C, very different from the laboratory where fast folding and maturation are required. Red-shifted fluorescent proteins derived from *Anthozoa* generally express much better at ambient temperature and at 37°C, most likely because they live in warmer habitats. Most fluorescent proteins have a maturation half-time from approximately 40 min to 1–2 h, which is sufficient to label cells, organelles and proteins of interest and to perform various quantitative experiments. However, for some time-limited applications, such as early detection of promoter activation, labelling transient proteins with a short lifetime or monitoring single translational events, very fast maturing fluorescent protein tags are needed. Table 16.3 shows a range of fluorescent proteins, and these are only a selection of the dozens available. Choosing the most suitable fluorescent protein for your experiment can be a real dilemma. Often the choice is constrained due to the microscope and filter set/laser lines available or else because the DNA plasmid was a gift from another laboratory. Ideally, the best fluorescent proteins are the brightest, most photostable, fastest-folding, least phototoxic monomeric candidate available.

Several photoactive fluorescent proteins (FPs) have also been discovered, and these fall into three distinct classes:

- photoactivatable FPs;
- photoconvertible FPs; and
- photoswitchable FPs.

Photoactivatable proteins change from a non-fluorescent state to a fluorescent one (Adam, 2014). For example, when activated by near-UV light around 411 nm (in practice with a 405 nm diode laser), photoactivatable GFP (PA-GFP) will increase its emission in the green by 100-fold. This is useful for analysing the mobility of specific protein pools in the cell. A photoconvertible FP changes from one already coloured fluorescent state to another spectral class. Examples of photoconvertible FPs are EosFP, *Kaede* (from the Japanese for maple leaf) and Dendra2, which are all irreversibly switched from a green to red fluorescent state. These FPs are used for protein mobility pulse-chase studies (Nemet et al., 2015) and as internal controls of protein mobility but also for tracking entire cells in their fates, shape and migration (Patterson, 2011; Wu et al., 2011; Tiwari and Nagai, 2013; Uno et al., 2015). A fluorescent timer changes colour over time, allowing temporal and spatial analysis of cellular fates. Kusabira Green Orange (mK-GO) is one such example. The photoswitchable FPs: *Dronpa* (Japanese for appear and disappear) and so-called 'kindling proteins' such as KFP-Red, which can be reversibly switched on and off at will from a non-fluorescent base state to a stable fluorescent form (Zhou and Lin, 2013). These can be used for studying fast protein dynamics at multiple time-points, as well as super-resolution localisation microscopy. These so-called 'optical highlighter' photoswitchable fluorescent proteins are also useful for calculating the velocity of proteins in living cells and tracking them directly in a spatially and temporally controlled manner, which cannot be done with merely irreversibly photo-activated proteins. Bimolecular fluorescence complementation (BiFC) uses half segments of a fluorescent protein (either GFP or YFP/Venus), for example in FRET studies. Fragments of a fluorescent protein can be genetically dissected into two non-fluorescent fragments, each fragment being joined to a separate protein. When these are brought together in close proximity in the cell, they join to form a complete fluorescent protein.

Why have fluorescent proteins not taken over from organic fluorophores and antibody staining? Firstly, any marker attached to a protein or epitope can alter the protein function, either by steric hindrance or disrupting the targeting sequences (Snapp, 2009). All fusion proteins require a functional assay to ensure that the fluorescent protein localises correctly in the cell and behaves in just the same manner as the natural non-fusion protein does. Any fusion protein localisation has to be independently verified with antibody studies. Immunoblots and immuno-labelling are required to check that the tagged protein is similar in size to, and is expressed at similar levels to, the natural protein.

The literature on fluorescent proteins is extensive and increasing rapidly. Selected references are given under Further Reading at the end of this chapter, but you are advised to consult the latest literature, for which PubMed is an excellent start.

16.5 Quantum Dots

Organic fluorophores, whether chemically synthesised dyes or fluorescent proteins, suffer from two limitations: firstly, their complex structures lead to rapid photobleaching; secondly, the relatively narrow absorption spectra of fluorescent dyes usually require individual laser lines or filters for excitation, and their broad emission spectra overlap. Hence neither types of organic fluorophores – chemical dyes or fluorescent proteins – are optimised for multicolour imaging (see Resch-Genger et al., 2008).

Man-made fluorescent probes (quantum dots), which are extremely resistant to photobleaching, have been developed in recent years (Massey et al., 2015; Montalti et al., 2015; Vu et al., 2015; Wegner and Hildebrandt, 2015; Geske-Moritz and Moritz, 2013; Lidke et al., 2007). Whereas a fluorescent dye will bleach in minutes (e.g. fluorescein bleaches entirely in 20 minutes from constant illumination with a 50 mwatt laser), quantum dots will continue to fluoresce for over 12 hours. Quantum dots exhibit large extinction coefficients (1×10^5 to $> 2 \times 10^6$ M^{-1} cm^{-1}) and quantum yields of around 80%, making them very bright fluorophores (e.g. circa 8×10^4–1.6×10^6 M^{-1} cm^{-1}). These extinction values compare to circa 2.5×10^4–1×10^5 M^{-1} cm^{-1} for fluorescent dyes, and 5×10^3–1×10^5 M^{-1} cm^{-1} for fluorescent proteins. Fluorescent semiconductor nanocrystals have a core made up of a few hundred to a few thousand atoms of a semiconductor material (often cadmium mixed with selenium or tellurium) with a zinc-sulphide outer shell. The entire 'dot' (or rod) ranges from 2 to 10 nm in diameter or length. These nanocrystals have the advantage of being excited by a broad spectrum of blue-shifted excitation light yet possess very narrow (FWHM 25–40 nm) emission spectra (Figure 16.7), which can be finely tuned by the size of their shell. A range of different-sized quantum dots will therefore cover the entire spectrum: the smaller dots fluoresce blue and the larger ones, red.

The broad excitation spectrum commonly shared by quantum dots with different emission spectral profiles means that quantum dots can be excited by a single laser line (e.g. four different Q-dots whose emission encompasses the visible spectrum can be excited by the single 488 nm laser line on the confocal) or one excitation filter on a widefield microscope and the emission collected into different PMT channels (confocal) or by different filters (widefield). Quantum dots also have much larger Stokes shifts than classical organic dyes, which significantly aids discrimination of the emitted signal. The narrow emission bandwidth means that quantum dots also have much less emission spectral overlap, which makes simultaneous collection of multicolour fluorescent signals much easier than with organic fluorophores. They are particularly useful as neurone tracers and for tracking dynamic events within cells. The broad excitation spectrum and adaptability for multicolour experiments makes quantum dots particularly well-suited for multi-photon applications, where switching of the pulsed laser wavelength is relatively time-consuming. A single optimised wavelength between 700 and 1040 nm can be chosen to excite multiple quantum dots simultaneously. Furthermore the large absorption cross section of quantum dots enables less photobleaching at greater depths, making quantum dots the candidate of choice for labelling thick tissue samples. The large physical sizes of the quantum dots also favours multi-photon microscopy, since they are more likely to absorb the pulsed excitation light than chemical dye fluorophores. This, in turn, increases the SNR and tissue viability, since less average power can be used for collecting multi-photon images.

Despite being made of cadmium and selenium, these nanocrystals are not toxic to either cells or organisms (Wang, 2013) and can be used to monitor cell fates throughout development. Quantum dots conjugated to antibodies are larger than fluorophores conjugated to antibodies, and it can therefore be more difficult to introduce them into the cell, even when the cell membrane is permeabilised with weak detergents. Nevertheless, quantum dots are brighter, much more photostable and offer greater sophistication for multiple-labelling than organic fluorophores. They have been used for correlative LM-EM studies (Giepmans, 2005). Quantum dots exhibit 'blinking' – the on-off switching of the emitted signal due to charge-trapping during excitation. This is generally only a problem when using quantum dots for tracking single molecules, and attempts have recently been made to prevent this. There is some evidence to suggest that the vitamin E analogue Trolox will eliminate blinking. Clearly, the on-off effect of the phenomenon of blinking cancels out when moieties are stained with populations of large numbers of quantum dots. The chief disadvantage associated with quantum dots is their relative size

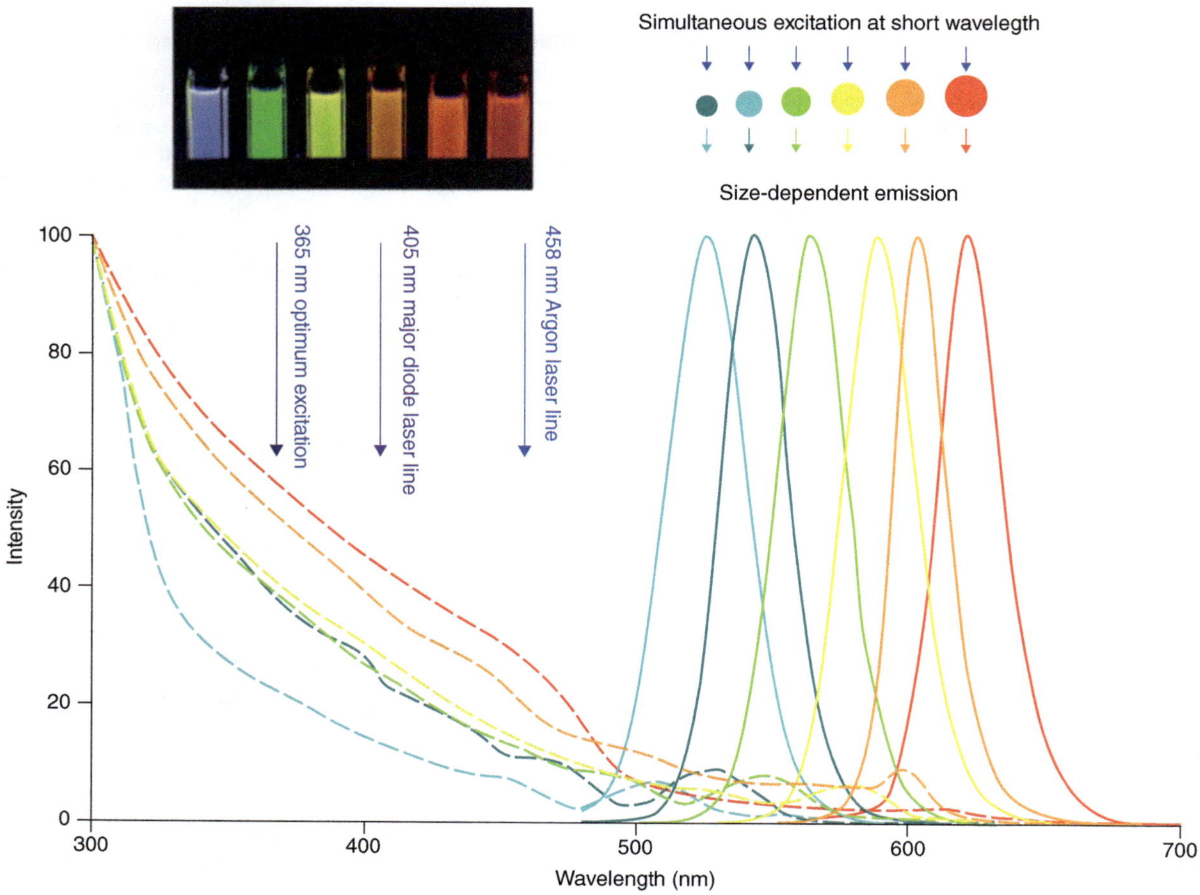

Figure 16.7 **Excitation and emission spectra of quantum dots** *The broad excitation and narrow emission spectra are shown. The broad excitation spectra allow effective excitation by a choice of laser lines (as shown) or filter sets, which can be combined with those exciting other fluorophores. The emission spectra can be tuned according to the size of the nanodot outer shell, and the narrow bandwidth allows effective separation from other fluorophores. Source: Lin et al. 2007. Reproduced under the terms of the Creative Commons Attribution Licence, CC-BY 2.0; Xing 2008. Reproduced with permission from IOS Press; Kubitscheck 2013. Reproduced with permission from John Wiley & Sons; Lidke et al. 2007. Reproduced with permission from John Wiley & Sons.*

and the fact that the external zinc-sulphide shell must be treated, generally with hydrophilic or amphipathic coats (often a polyethylene glycol PEG coating), to make the entry of quantum dots into cells possible, without aggregation or selective trafficking of quantum dots, to allow cell penetration and labelling.

The physical properties of quantum dot nanocrystals make them largely incompatible with common mounting media, such as ProLong Gold, PVA-formulations and other mounting media designed primarily for use with organic dyes. Thermo Fisher Scientific, a major distributor of quantum dot fluorescent probes, offer Qmount. This is a specialised mountant that preserves the fluorescence signal of Q-dot nanocrystals with little to no quenching of the initial intensity of the signal. It is not recommended for use with Alexa Fluor dyes or fluorescent proteins.

16.6 Covalent Labelling Strategies

The choice of labelling to attach a fluorophore to the cell or tissue depends on the biology underlying the research hypothesis. If heterologous fluorescent protein fusions disturb the function of the target protein or the organism is anaerobic (GFP requires to be oxygenated in order to mature and become fluorescent), then alternative labelling methods (Jing and Cornish, 2011) must be used. Other options are to consider a tetracysteine (TC) bi-arsenical labelling system for intracellular proteins (which can give results with a low background) or a biotin ligase system for proteins located on the cell surface (see Corrêa, 2014; Hori and Kikuchi, 2013; Dean and Palmer, 2014; Yan and Bruchez, 2015). The tetracysteine peptide tag is very small (as small as six amino acids), which is an advantage. However, the TC-tag requires a reducing environment (like the cytosol) to express fluorescence, suffers from non-specific labelling and requires incubation with dithiols to minimise this non-specific binding. Since the cysteines must be in reduced form, this method will not work in the endoplasmic reticulum or on the cell surface without further treatment with reducing agents.

The tetracysteine tag causes the bi-arsenides FlAsH and ReAsH to fluoresce when it binds to these molecules; the TC-bi-arsenical labelling system is available from Thermo Fisher Scientific. The SNAP-tag and CLIP-tag and Halo tag technologies all rely upon using a polypeptide tag fused to a carrier fusion protein. In the case of SNAP-tag and CLIP-tag this carrier protein is a mutant of the DNA repair protein O^6-alkylguanine-DNA alkyltransferase (AGT). While these covalent labelling methods do not require time-consuming cloning and evaluation of fluorescent protein expression, the signal intensity is lower than most fluorescent proteins and photo-bleaches relatively quickly (Schneider and Hackenberger, 2017). This method of labelling is available from New England Biolabs. The Halo tag uses the halogene dehalogenase fusion protein and a chloroalkane linker and is applicable to all eukaryotic cells. This system is available from Promega.

16.7 Autofluorescence

Many biological specimens, particularly plant specimens, exhibit significant autofluorescence, which depends upon the method of fixation used and the wavelength at which the sample is illuminated. Generally the shorter, higher energy wavelengths give off more autofluorescence, which can be particularly troublesome from natural fluorophores (e.g. in chlorophyll and polyphenols in plants; pyridinic [NADPH] and flavin coenzymes, also lysosomes, elastin and collagen in animal cells; see Table 16.4). Autofluorescence 'floods' the image obscuring other coexistent fluorescent signals in the same tissue. Unfortunately, autofluorescence cannot be avoided merely by choosing a fluorescent probe with narrow excitation (e.g. quantum dots) and emission spectra outside the spectrum of the autofluorescent molecules, because the spectra of autofluorescent compounds are much broader than those of most fluorescent probes. The most straightforward way to avoid autofluorescence in the image is to use long wavelength fluorophores (e.g. Tynan et al., 2012) in the red, far-red or even infrared part of the spectrum. However, using far-red and infrared probes will mean that samples cannot be checked visually before acquiring digital images.

Formalin-induced fluorescence (FIF, Figure 16.8) occurs when formalin forms aromatic rings by cross-linking proteins. FIF is excited from 350 to 450 nm, and emits between 450 and 600 nm, with a peak around 500 nm. Glutaraldehyde fixation will also cause severe autofluorescence (absorption approximately: 540/25 nm; Emission: 560/40 nm). The presence of autofluorescence can be used to advantage to diagnose certain melanomas and reveal biogenic amines, but otherwise it is unwanted and reduces the overall signal-to-noise ratio.

Where they are not removed by prior fixation and processing, autofluorescent compounds can significantly reduce the signal-to-noise ratio of the experimental fluorescent probe. First, close down the illuminated field diaphragm as much as possible to exclude stray light. In live-cell preparations, avoid using phenol red as a pH indicator in the media and, if possible, avoid serum concentrations >20%. Chemical treatment (e.g. ammonium chloride at 50–100 mM in buffer for three 5-minute washes; 300 mM glycine or 20 mM *fresh* sodium borohydride [10 mg/mL in buffer for aldehyde-induced fluorescence] or 0.3% Sudan Black in 70% EtOH for 20 mins at room temperature) will reduce, or eliminate autofluorescence (Oliveira et al., 2010; Baschong et al., 2001; Eldred et al., 1983) but it can also reduce the true fluorescent signal, especially if the signal is weak. Preparing sodium borohydride solution on ice and performing all subsequent incubations on ice has also been recommended. Apply this solution immediately (while fizzing) to cells or tissue sections. Wash well (minimum 15 minutes) in several changes of 0.1 M phosphate-buffered saline (PBS); proceed with labelling. For glutaraldehyde-fixed cell monolayers, incubate in the sodium borohydride solution for 4 minutes. Replace with fresh sodium borohydride solution for another 4 minutes. For paraformaldehyde-fixed paraffin embedded 7 μm sections, incubate 3 times,

Table 16.4 Spectral Values of Naturally Autofluorescent Compounds

Compound	Excitation λ (nm)	Emission λ (nm)
Tryptophan	280	300–350
Vitamin C	350	430
Folic acid	250–300	450
NAD(P)H	330–380	440–470
Glycation adducts	370	450
Vitamin D	390	470
Collagen and elastin	330–420	400–510
Formaldehyde-induced fluorescence	360–490	420–620
Cytokeratin	280–320	500–530
Lignin	350	400–700
Melanin	420–550	470–650
Retinol and caretinoids, vitamin A	370–380	490–510
Polyphenol	440–450	470–600
Lipofuschins	420–480	500–690
Riboflavin	450–490	550
Protoporphyrin IX	390–420	630–670
Chlorophyll	465, 665	670, 730

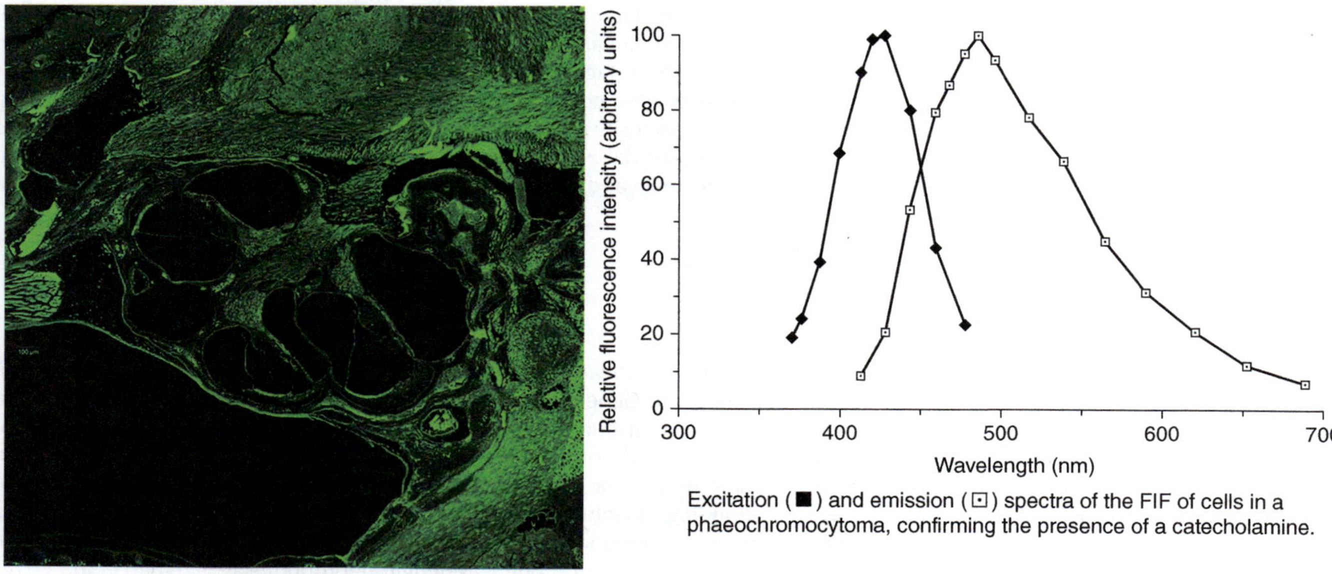

Figure 16.8 **Formalin-induced autofluorescence** *Shorter, higher-energy, wavelengths give off more autofluorescence; the result is a flooding of the image with non-specific signal. It is easy to mistake FIF or formalin-induced fluorescence (seen here) for a true positive signal where none exists. The right-hand panel shows the excitation and emission spectra of FIF autofluorescence. Source: Cochlea section courtesy of Dr Andy Parker. (graph) Rost 1991. Reproduced with permission from Cambridge University Press.*

10 minutes each, in sodium borohydride solution. Sodium borohydride is a strong reducing agent and will likely work where glycine or ammonium chloride do not; it is flammable on contact with water and harmful by ingestion, inhalation. Take adequate precautions. Other stratagems for reducing autofluorescence include staining with trypan blue or pontamine sky blue 5BX at very low dilutions ($\approx$ 0.01%) to red-shift the emission of the autofluorescence by resonant energy transfer. Another approach is to stain the tissues or cells with very dilute osmium tetroxide (0.2% for 5 min, followed by 12 hrs or so in running water). The heavy metal ions effectively suppress the ability of aromatic organic molecules to fluoresce. Pang et al. (2013) advocate using a customised filter set to selectively remove autofluorescence, but this method has not yet gained wide acceptance over chemical methods.

In order to gauge the magnitude of autofluorescence present in your sample, take an image of the non-autofluorescent sample first to establish a threshold for the autofluorescent component, then take images of single-labelled samples in all the fluorescence channels in use in order to characterise the autofluorescence and determine the extent of any bleed-through. If the fluorophore of interest has a narrow excitation, particularly if it is of medium or long wavelength on the visible spectrum, it is often possible to subtract the signal containing autofluorescence, taken at short wavelength, where only autofluorescence is present, from an image containing both true fluorescence signal plus autofluorescence. This can easily be done with free image analysis software such as ImageJ. Thus, if we have a green-emitting fluorophore excited at 480 nm and emitting at 515 nm, and we suspect the presence of shorter-wavelength fluorescence at 350 nm, then if we have an unstained control preparation, proceed as follows:

Elimination of Autofluorescence – Method 1

1. Take an image of the fluorescence excited at 480 nm and collected at 515 nm. This is the specific fluorescence image (SF). If possible, repeat the process, and average the intensities of several images to reduce the noise.
2. Take an image of the fluorescence excited at 350 nm collected at 515 nm. If possible, average over a series of images to improve SNR, as before. This is the autofluorescence image (AF).
3. Repeat steps (1) and (2) with an unstained control image. From the pixel-by-pixel image intensities, calculate the ratio of intensities: $\text{Ratio}_{480/350}$.
4. The intensity of autofluorescence can be corrected for by multiplying the intensities of the autofluorescence image, AF, by the ratio derived from the unstained image:

$$\text{Corrected}_{AF} = (\text{Ratio}_{480/350} \times \text{AF})$$

5. To find the true fluorescence, subtract the corrected autofluorescence from the specific fluorescence:

$$\text{True fluorescence} = \text{SF} - \text{Corrected}_{AF}$$

To use this approach, you have to be confident, through the use of controls, that *only* autofluorescence is present in the images from the higher energy, shorter-wavelength region of the spectrum. If the fluorophore you are interested in has a broad excitation profile, this approach is less likely to be successful. If the bandwidth of the fluorescence excitation is too broad and if there are no control slides available, a slightly different approach is to subtract a background image at two different wavelengths and subtract the autofluorescence image multiplied by the ratio between the two wavelengths, as follows:

Elimination of Autofluorescence – Method 2

1. Take a specific fluorescence image at 480 nm excitation. From this image subtract a background image taken at 480 nm without a sample in place, merely through a blank area of the slide. If possible, average the result over a series of images.
2. Repeat the process in the autofluorescence region of the spectrum, in this case at 350 nm. If possible, take an average intensity value, as before, to improve the SNR.
3. Calculate the ratio of intensities of the two images.
4. True fluorescence = $(SF - SF_{background}) - (AF - AF_{background}) \times Ratio_{SF/AF}$

A similar image-processing approach is to employ spectral unmixing, either on a confocal or widefield microscope (Chapter 20, Section 20.4). Using a widefield microscope with filter sets exhibiting broader bandwidths than confocal lasers, this method relies up on the fact that the autofluorescence emission spectrum is very broad and can easily be resolved, or distinguished. Either the autofluorescence can be treated as a separate fluorophore and its spectrum subtracted from the distribution of the true fluorescent marker or the true fluorescent markers can be spectrally unmixed, with the remaining signal assigned to a 'residual channel' and considered autofluorescence. It is worth trying both methods and considering the common spatial and intensity distribution of both maps.

Autofluorescence is not always entirely responsible for low signal-to-noise ratios in the image. In many cases, background fluorescence due to the labelling procedure is a greater source of problems than autofluorescence *per se*. It also pays to check the alignment of the florescence epi-illuminator and to properly adjust the illuminated field diaphragm and the aperture diaphragm (if fitted), to exclude stray light and improve the signal-to-noise ratio as much as possible.

16.8 Chapter Summary

This chapter introduces the important fluorophores and fluorescent proteins commonly used in biological and medical research. There are many different fluorophores to choose from; this chapter is intended to help narrow the choice and outline the ways in which tissues, cells organelles and individual molecules can be marked with fluorescent labels.

1. When a single fluorophore is used, generally only the brightness, photostability and whether the cells or tissue will autofluoresce need be considered.
2. Cross-talk and bleed-through will need to be taken into account when using more than one marker, whether it be an organic fluorophore, a fluorescent protein or nanocrystal.
3. Modern fluorophores are brighter and more photostable than FITC and TRITC used initially.
4. The greater the number of aromatic rings in the fluorophore molecule, fluorescence emission tends to shift further towards the red end of the spectrum.
5. Fluorescent proteins are coded for by a single gene. They are easy to transfect into the host organism requiring no post-transcriptional modification or exogenous co-factors to work.
6. GFP is stable and membrane-permeant. It is not toxic to cells unless overexpressed. GFP can be used as a reporter as well as a fluorescent tag.
7. A fluorescent protein must be monomeric to label proteins and may take several minutes or hours to fold and mature into an effective fluorescent molecule.
8. Red-shifted fluorescent proteins are generally named after fruits. They are derived from *Anthozoa* species, corals and sea anemones.
9. Photoactivatable fluorescent proteins can be used to track protein dynamics and are used in localisation super-resolution microscopy.
10. Quantum dots are extremely bright and resistant to photobleaching with broad excitation spectra and narrow emission spectra: several can be excited by single laser lines.
11. Self-labelling fusion proteins, such as the SNAP-, CLIP-, or the Halo tag methods are an alternative to fluorescent proteins.
12. Autofluorescence arises from natural fluorophores or aldehyde fixatives.
13. The autofluorescent signal can be chemically quenched, avoided altogether by using red-shifted fluorophores or else subtracted by image manipulation or spectral unmixing.

Key Reading

Cranfill, P.J. et al. (2016). 'Quantitative assessment of fluorescent proteins'. *Nature Methods 13/7*: 557–562.
- The best and most comprehensive review at the time of writing.

Kremers, G.-J. et al. (2011). 'Fluorescent proteins at a glance'. *Journal of Cell Science 124/2*: 157–160.
- An excellent summary with a useful poster. A little dated, but still relevant.

Schneider, A.F.L. and Hackenberger, C.P.R. (2017). 'Fluorescent labelling in living cells'. *Current Opinion in Biotechnology 48*: 61–68.

References

Adam, V. (2014). 'Phototransformable fluorescent proteins: which one for which application?' *Histochemistry and Cell Biology 142/1*: 19–41.

Ai, H-W. et al. (2014). 'Engineering and characterizing monomeric fluorescent proteins for live-cell imaging applications'. *Nature Protocols 9/4*: 910–928.

Baschong, W. et al. (2001). 'Control of Autofluorescence of Archival Formaldehyde-fixed, Paraffin-embedded Tissue in Confocal Laser Scanning Microscopy (CLSM)'. *Journal of Histochemistry & Cytochemistry 49/12*: 1565–1571.

Becker, K. et al. (2012). Chemical clearing and dehydration of GFP expressing mouse brains. *PLoS One 7/3*: e33916.

Bogdanov, A.M. et al. (2009). 'Cell culture medium affects GFP photostability: a solution'. *Nature Methods 6/12*: 859–860.

Boyce, F.M. and Bucher, N.L. (1996). 'Baculovirus-mediated gene transfer into mammalian cells'. *PNAS 93/6*: 2348–2352.

Campbell, R.E. (2008). 'How a fluorescent protein is manufactured'. *Scholarpedia 3/7*: 5410.

Chudakov, D.M. et al. (2005). 'Fluorescent proteins as a toolkit for *in vivo* imaging'. *Trends in Biotechnology 23/12*: 605–613.

Corrêa, I.R. Jnr. (2014). 'Live-cell reporters for fluorescence imaging'. *Current Opinion in Chemical Biology 20*: 36–45.

Costantini, L.M. and Snapp, E.L. (2013). 'Fluorescent proteins in cellular organelles: serious pitfalls and some solutions'. *DNA and Cell Biology 32/11*: 622–627.

Day, R.N. and Davidson, M.W. (2009). 'The fluorescent protein palette: tools for cellular imaging'. *Chemical Society Reviews 38/10*: 2887–2921. Part of the GFP special issue of this journal.

Dean, K.M. and Palmer, A.E. (2014). 'Advances in fluorescence labelling strategies for dynamic cellular imaging'. *Nature Chemical Biology 10/7*: 512–523.

Eldred, W.D. et al. (1983). 'Comparison of fixation and penetration enhancement techniques for use in ultrastructural immunocytochemistry'. *Journal of Histochemistry and Cytochemistry 31*: 285–292.

Enterina, J.R. et al. (2015). 'Emerging fluorescent protein technologies'. *Current Opinion in Chemical Biology 27*: 10–17.

Gautier, A. et al. (2014). 'How to control proteins with light in living systems'. *Nature Chemical Biology 10/7*: 533–541.

Geske-Moritz, M. and Moritz, M. (2013). 'Quantum dots as versatile probes in medical sciences: synthesis, modification and properties'. *Materials Science and Engineering: C 33/3*: 1008–1021.

Giepmans, B.N.G. (2005). 'Correlated light and electron microscopic imaging of multiple endogenous proteins using quantum dots'. *Nature Methods 2/10*: 743–749.

Hori, Y. and Kikuchi, K. (2013). 'Protein labeling with fluorogenic probes for no-wash live-cell imaging of proteins'. *Current Opinion in Chemical Biology 17/4*: 644–650.

Jing, C. and Cornish, V.W. (2011). 'Chemical tags for labeling proteins inside living cells'. *Accounts of Chemical Research 44/9*: 784–792.

Joosen, L. et al. (2014). 'Effect of fixation procedures on the fluorescent lifetimes of Aequorea victoria derived fluorescent proteins'. *Journal of Microscopy 256/3*: 166–176.

Khan, K.H. (2013). 'Gene Expression in Mammalian Cells and its Applications'. *Advanced Pharmaceutical Bulletin 3/2*: 257–263.

Kubitscheck, U. (2013). *Fluorescence Microscopy: From Principles to Biological Applications*. Wiley-VCH.

Lam, A.J. et al. (2012). 'Improving FRET dynamic range with bright green and red fluorescent proteins'. *Nature Methods 9/10*: 1005–1012.

Lidke, D.S. et al. (2007). 'In vivo imaging using quantum dot-conjugated probes'. *Current Protocols Cell Biology* Chapter 25: Unit 25.1. doi: 10.3390/molecules17010796.

Lin, S. et al. (2007). 'Quantum dot imaging for embryonic stem cells'. *BMC Biotechnology 7*: 67. doi: 10.1186/1472-6750-7-67.

Malkani, N. and Schmid, J.A. (2011). 'Some secrets of fluorescent proteins: distinct bleaching in various mounting fluids and photoactivation of cyan fluorescent proteins at YFP-excitation'. *PLoS One 6/4*: e18586.

Massey, M. et al. (2015). 'Mind your P's and Q's: the coming of age of semiconductor polymer dots and semiconductor quantum dots in biological applications'. *Current Opinion in Biotechnology 34*: 30–40.

Mishin, A.S. et al. (2015). 'Novel uses of fluorescent proteins'. *Current Opinion in Chemical Biology 27*: 1–9 [Molecular Imaging special issue].

Miyawaki, A. et al. (2003). 'Mechanisms of protein fluorophore formation and engineering'. *Current Opinion in Chemical Biology 7/5*: 557–562.

Moeyaert, B. (2010). Structural insight into fluorescent protein properties by rational design and characterization of photoconvertible, photoswitchable and phototoxic fluorescent proteins. Master's thesis. The Katholieke Universiteit Leuven (in English: Catholic University of Leuven), abbreviated KU Leuven.

Montalti, M. et al. (2015). 'Nanodiamonds and silicon quantum dots: ultrastable and biocompatible luminescent nanoprobes for long-term bioimaging'. *Chemical Society Reviews 44/14*: 4853–4921.

Murphy, D.B. and Davidson, M.W. (2013). *Fundamentals of Light Microscopy and Electronic Imaging*, 2nd Ed. New Jersey: Wiley Blackwell.

Nemet, I. et al. (2015). 'Applications of phototransformable fluorescent proteins for tracking the dynamics of cellular components'. *Photochemical and Photobiological Sciences 14/10*: 1787–1806.

Oliveira, V.C. et al. (2010). 'Sudan Black B treatment reduces autofluorescence and improves resolution of in situ hybridization specific fluorescent signals of brain sections'. *Histology and Histopathology 25/8*: 1017–1024.

Pang, Z. et al. (2013). 'Autofluorescence removal using a customized filter set'. *Microscopy Research and Technique 76/10*: 1007–1015.

Pastrana, E. (2011). 'Optogenetics: controlling cell function with light'. *Nature Methods 8/1*: 24–25 (A Primer).

Patterson, G.H. (2011). 'Highlights of the optical highlighter fluorescent proteins'. *Journal of Microscopy 243*/1: 1–7.

Piatkevich, K.D. et al. (2010a). 'Monomeric red fluorescent proteins with a large Stokes shift'. *PNAS 107*/12: 5369–5374.

Piatkevich, K.D. et al. (2010b). 'Red Fluorescent proteins and their properties'. *Russian Chemical Reviews 79*/3: 243–258.

Pieribone, V. and Gruber, D.F. (2005) *Aglow in the dark: the revolutionary science of biofluorescence*. Belknap Press. Cambridge (Massachusetts): Harvard University Press.

Resch-Genger, U. et al. (2008). 'Quantum dots versus organic dyes as fluorescent labels'. *Nature Methods 5*/9: 763–775.

Rost, F.W.D. (1991). *Quantitative Fluorescence Microscopy*. Cambridge University Press.

Shaner, N.C. et al. (2008). Improving the photostability of bright monomeric orange and red fluorescent proteins. *Nature Methods 5*/6: 545–551.

Shaner, N.C. et al. (2013). 'A bright monomeric green fluorescent protein derived from *Branchiostoma lanceolatum*'. *Nature Methods 10*/5: 407–409.

Shcherbakova, D.M. et al. (2012a). 'An orange fluorescent protein with a large Stokes shift for single-excitation multicolor FCCS and FRET imaging'. *Journal of the American Chemical Society 134*/18: 7913–7923.

Snapp, E.L. (2005). 'Design and use of fluorescent fusion proteins in cell biology'. *Current Protocols in Cell Biology Supplement* Chapter 21: Unit 21.4.

Snapp, E.L. (2009). 'Fluorescent proteins: a cell biologist's user guide'. *Trends in Cell Biology 19*/11: 649–655.

Stepanenko, O.V. et al. (2013). Beta-barrel scaffold of fluorescent proteins: folding, stability and role in chromophore formation. *International Review of Cell and Molecular Biology 302*: 221–278.

Subach, F.V. et al. (2011). 'Directed molecular evolution to design advanced red fluorescent proteins'. *Nature Methods 8*/12: 1019–1026.

Thorn, K. (2017) 'Genetically encoded fluorescent tags'. *Mol Biol Cell*. 28/7: 848–857.

Tiwari, D.K. and Nagai, T. (2013). 'Smart fluorescent proteins: innovation for barrier-free superresolution imaging in living cells'. *Development, Growth & Differentiation 55*/4: 491–507.

Tsien, R.Y. (2005). 'Building and breeding molecules to spy on cells and tumors'. *FEBS Letters 579*/4: 927–932.

Tynan, C.J. et al. (2012). 'Multicolour single molecule imaging in cells with near infra-red dyes'. *PLoS One 7*/4: e36265.

Uno, S-N. et al. (2015). 'A guide to use photocontrollable fluorescent proteins and synthetic smart fluorophores for nanoscopy'. *Microscopy 64*/4: 263–277.

Vu, T.Q. et al. (2015). 'Quantum dots for quantitative imaging: from single molecules to tissue'. *Cell and Tissue Research 360*/1: 71–86.

Wang, Y. et al. (2013). 'Functionalized quantum dots for biosensing and bioimaging and concerns on toxicity'. *ACS Applied Materials & Interfaces 5*/8: 2786–2799 [Quantum dots special issue].

Waters, J.C. and Wittman, T. (2014). *Quantitative Imaging in Cell Biology Methods in Cell Biology*, vol. 123. Elsevier Inc. http://dx.doi.org/10.1016/B978-0-12-420138-5.00006-9.

Wegner, K.D. and Hildebrandt, N. (2015). 'Quantum dots: bright and versatile *in vitro* and *in vivo* fluorescence imaging biosensors'. *Chemical Society Reviews 44*/14: 4792–4834.

Wouterloud, F.G. et al. (2014). 'A fourth generation of neuroanatomical tracing techniques: exploiting the offspring of genetic engineering'. *Journal of Neuroscience Methods 235*: 331–348.

Wu, B. et al. (2011). 'Modern fluorescent proteins and imaging technologies to study gene expression, nuclear localization and dynamics'. *Current Opinion in Cell Biology 23*/3: 310–317.

Wysocki, L.M. and Davis, L.D. (2011). 'Advances in the chemistry of small molecule fluorescent probes'. *Current Opinion in Chemical Biology 15*/6: 752–759.

Xing, Y. et al. (2008). Quantum dot bioconjugates for in vitro diagnostics & in vivo imaging. *Cancer Biomark. 4*/6: 307–319.

Yan, Q. and Bruchez, M.P. (2015). 'Advances in chemical labelling of proteins in living cells'. *Cell and Tissue Research 360*/1: 179–194.

Yang, J. et al. (2013). 'mBeRFP, an Improved Large Stokes Shift Red Fluorescent Protein'. *PLoS One 8*/6: e64849.

Zhou, X.X. and Lin, M. (2013). 'Photoswitchable fluorescent proteins: ten years of colorful chemistry and exciting applications'. *Current Opinion in Chemical Biology 17*: 682–690.

Further Reading

Bajar, B.T. et al. (2016). 'Improving brightness and photostability of green and red fluorescent proteins for live cell imaging and FRET reporting'. *Scientific Reports 6*: 20889.

Baker, M. (2010). 'Nanotechnology imaging probes: smaller and more stable'. *Nature Methods 7*/12: 957–962.

Billinton, N. and Knight, A.W. (2001). Seeing the wood through the trees: A review of techniques for distinguishing green fluorescent protein from endogenous autofluorescence'. *Analytical Biochemistry 291*: 175–197.

Bindels, D.S. et al. (2014). 'Optimization of fluorescent proteins'. In: Engelborghs, Y. and Visser, A.J.W.G. (eds.), *Methods in Molecular Biology*, Chapter 16, vol. 1076. New York: Humana Press, pp. 371–417.

Bogdanov, A.M. et al. (2012). 'Anti-fading media for live cell GFP imaging'. *PLoS One 7*/12: e53004.

Bruchez, M.P. (2015). 'Dark Dyes–Bright Complexes: Fluorogenic Protein Labeling'. *Current Opinion in Chemical Biology 27*: 18–23.

Campbell, R.E. and Davidson, M.W. (2010). 'Fluorescent reporter proteins'. In: Gambhir, S.S. and Yaghoubi, S.S. (eds.), *Molecular Imaging with Reporter Genes*, Chapter 1. Cambridge University Press, pp. 3–40.

Chalfie, M. (2009). 'GFP: lighting up life (Nobel Lecture)'. *Angewandte Chemie International Edition 48*/31: 5603–5611.

Chudakov, D.M. et al. (2010). 'Fluorescent Proteins and Their Applications in Imaging Living Cells and Tissues'. *Physiological Reviews 90*/3: 1103–1163.

Costantini, L.M. et al. (2015). 'A palette of fluorescent proteins optimized for diverse cellular environments'. *Nature Communications 6*: 7670. doi: 10.1038/ncomms8670.

Crivat, G. and Taraska, J.W. (2012). 'Imaging proteins inside cells with fluorescent tags'. *Trends in Biotechnology 30*/1: 8–16.

Day, R.N. and Davidson, M.W. (2014). *The Fluorescent Protein Revolution*. Boca Raton, FL: CRC Press.

Eason, M.G. et al. (2017). 'Structure-guided rational design od red fluorescent proteins: towards designer genetically-encoded fluorophores'. *Current Opinion in Structural Biology 45*: 91–99.

Gjetting, S.K. et al. (2013). 'Perspectives for using genetically encoded fluorescent biosensors in plants'. *Frontiers in Plant Science 4*: 234. doi: 10.3389/fpls.2013.00234.

Gossa, S. et al. (2014). 'Development of an immunologically tolerated combination of fluorescent proteins for in vivo two-photon imaging'. *Scientific Reports 4*: 6664.

Greer III, L.F. and Szalay, A.A. (2002). 'Imaging of light emission from the expression of luciferases in living cells and organisms: a review'. *Luminescence 17*/1: 43–74.

Hoffman, R.M. (2012). 'Live cell imaging in live animals with fluorescent proteins'. In: *Methods in Enzymology*, Chapter 11, vol. 506. Elsevier, pp. 197–224.

Hong, G. et al. (2017). Near infra-red fluorophores for biomedical imaging. *Nature Biomed. Engineering 1*: 0010. doi: 10.1038/s41551-016-0010.

Horisawa, K. (2014). 'Specific and quantitative labeling of biomolecules using click chemistry'. *Frontiers in Physiology 5*: 457.

Hughes, L.D. et al. (2014). 'Choose Your Label Wisely: Water-Soluble Fluorophores Often Interact with Lipid Bilayers'. *PLoS One 9*/2: e87649.

Johnson, E. et al. (2015). 'Correlative in-resin super-resolution and electron microscopy using standard fluorescent proteins'. *Scientific Reports 5*: 9583.

Jones, A.M. et al. (2012). 'A never ending race for new and improved fluorescent proteins'. *BMC Biology 10*: 39. doi: 10.1186/1741-7007-10-39.

Lin, M.Z. (2011). 'Beyond the rainbow: new fluorescent proteins brighten the infrared scene'. *Nature Methods 8*/9: 726–728.

Lin, M.Z. et al. (2010). 'Fluorescent proteins illuminate cell biology'. Poster from *Nature Reviews Mol. Cell Biol.* with supplementary information here.

Long, M.J.C. et al. (2016). 'On-Demand Targeting: Investigating Biology with Proximity-Directed Chemistry'. *Jour. Am. Chem. Soc. 138*/11: 3610–3622.

McNamara, G. and Boswell, C.A. (2007). 'A Thousand Proteins of Light: 15 Years of Advances in Fluorescent Proteins'. *Modern Research and Educational Topics in Microscopy*. Méndez-Vilas, A. and Diaz, J. vol. 1: 287–296 Formatex.

Nienhaus, K. and Nienhaus, G.U. (2014). 'Fluorescent proteins for live-cell imaging with super-resolution'. *Chem. Soc. Reviews 43*/4: 1088–1106.

Ozawa, T. et al. (2013). 'Advances in fluorescence and bioluminescence imaging'. *Analytical Chemistry 85*/2: 590–609.

Progatzky, F. (2013). 'From seeing to believing: labelling strategies for *in vivo* cell-tracking experiments'. *Interface Focus 3*/3: 20130001.

Rodriguez, E.A. et al. (2017). 'The Growing and Glowing Toolbox of Fluorescent and Photoactive Proteins'. *Trends in Biochemical Sciences 42*/2: 111–129.

Sample, V. et al. (2014). 'Genetically encoded molecular probes to visualize and perturb signaling dynamics in living biological systems'. *Jour. Cell Science 127*/6: 1151–1160.

Schultz, C. (2009). 'Fluorescent Revelations'. *Chemistry & Biology 16*/2: 107–111.

Shaner, N.C. et al. (2007). 'Advances in fluorescent protein technology'. *Jour. Cell Science 120*/24: 4247–4260.

Shcherbakova, D.M. et al. (2012b). 'Red Fluorescent Proteins: advanced imaging applications and future design'. *Angew. Chemie 51*/43: 10724–10738.

Shcherbo, D. et al. (2009). 'Far-red fluorescent tags for protein imaging in living tissues'. *Biochem. Jour. 418*/3: 567–74.

Shimomura, O. (2005). 'The discovery of aequorin and green fluorescent protein'. *Journal of Microscopy 217*/1: 3–15.

Shimomura, O. (2009). 'Discovery of green fluorescent protein (GFP) (Nobel Lecture)'. *Angew. Chem. Int. Ed. 48*/31: 5590–5602.

Steiert, F. et al. (2018) 'Photophysical behavior of mNeonGreen, an evolutionarily distant green fluorescent protein'. *Biophys. Jour. 114*/10: 2419–2431.

Tsien, R.Y. (2009). 'Constructing and exploiting the fluorescent protein paintbox (Nobel Lecture)' *Angew. Chem. Int. Ed. 48*/31: 5612–5626.

Viswanathan, S. et al. (2015). 'High-performance probes for light and electron microscopy'. *Nature Methods 12*/6: 568–76.

Wang, S. et al. (2014). 'Characterization and development of photoactivatable fluorescent proteins for single-molecule–based superresolution imaging'. *PNAS 111*/23: 8452–8457.

Wang, Y. et al. (2008). 'Fluorescence proteins, live-cell imaging, and mechanobiology: seeing is believing'. *Annu. Rev. Biomed. Eng. 10*: 1–38.

Yan, L. (2016). 'Fluorescent nanoparticles based on AIE fluorogens for bioimaging'. *Nanoscale 8*/5: 2471–2487.

Yang, Z. (2013). 'Development of a plastic embedding method for large-volume and fluorescent-protein-expressing tissues'. *PLoS One 8*/4: e60877.

Internet Resources

Key fluorescent protein URLs: http://nic.ucsf.edu/FPvisualization/ http://nic.ucsf.edu/FPvisualization/PSFP.html.

These two resources grew out of: http://nic.ucsf.edu/dokuwiki/doku.php?id=fluorescent_proteins.

Fluorescent proteins: http://www.scholarpedia.org/article/Fluorescent_proteins.

George McNamara fluorophores table: http://www.geomcnamara.com/data.

Zeiss fluorescent protein poster: http://applications.zeiss.com/C125792900358A3F/0/71E93C9C25B25E5EC1257B4800370B67/$FILE/EN_41_050_007_Poster_Imaging-life-proteins.pdf.

Lin et al. Fluorescent poster: https://www.nature.com/nrm/posters/fluorescent/index.html and https://www.nature.com/nrm/posters/fluorescent/fluorescent-s1.pdf.

Zeiss Campus: http://zeiss-campus.magnet.fsu.edu/articles/probes.

Bionumbers database for maturation values: http://bionumbers.hms.harvard.edu/browse.aspx.

Sigma-Aldrich technical data on quantum dots: https://www.sigmaaldrich.com/content/dam/sigma-aldrich/materials-science/material-matters/material_matters_v2n1.pdf.

Nanoparticles overview: http://pubs.rsc.org/en/content/articlehtml/2015/cs/c4cs00392f or http://pubs.rsc.org/en/content/articlehtml/2015/cs/c4cs00392f.

Extinction coefficients: https://tools.thermofisher.com/content/sfs/brochures/TR0006-Extinction-coefficients.pdf.

Notes

1. For the single-photon emission spectrum of eGFP, see Figure 22.4.
2. The wild-type (S65) GFP variant can best be seen with standard filter sets if the excitation filter from a standard 'DAPI' filter set (e.g. LP 365 or 350/50) is exchanged in place of the excitation filter from a 'green' filter set (e.g. 470/40), with the same dichromatic mirror (BS-FT 495) and emission filter (525/50) remaining in the 'green' filter set.
3. mT-Sapphire has a Stokes shift of 112 nm, with an excitation maximum at 399 nm and emitting at 511 nm; mAmetrine has a Stokes shift of 120 nm, having an excitation maximum of 406 nm and emitting at 526 nm; mKeima has a Stokes shift of 180 nm from an excitation maximum of 440 nm emitting at 620 nm.
4. tdTomato (Table 16.3) is the brightest red-shifted fluorescent protein, comparable with mNeonGreen, mClover and YPet. These four fluorescent proteins also mature fast or, in the case of YPet, relatively fast. mNeonGreen is derived from the lancelet (Cephalochordata) rather than jellyfish (Cnidaria), so shares only 26% genome sequence homology. Antibody affinity tools for GFP are unlikely to be effective, see https://resources.chromotek.com/blog/mneongreen-vs-gfp.
5. Using a monomeric fusion protein minimises artifacts such as aggregation, non-specific protein localisation or secondary protein-protein interactions that often occur when fluorescent fusion proteins are expressed at levels exceeding the physiological concentration of the endogenous protein being studied. Artificially derived monomeric fluorescent proteins usually exhibit loss of brightness, longer maturation times and reduced photostability.
6. Besides the figure for maturation times given in Table 16.3, several values can be found on the Bionumbers database. See http://bionumbers.hms.harvard.edu/browse.aspx.

17

Optical Sectioning and Confocal Microscopy

Outline

Understanding Light Microscopy, First Edition. Jeremy Sanderson.

17.1 Introduction

There are many advantages to using fluorescence microscopy, which is why it is very widely used. Fluorophores are self-luminous, so they exhibit very high contrast; they are very sensitive; only small concentrations are required for successful labelling; they are very specific, discriminating individual proteins, organelles and components of cells very well. Fluorophores can be tolerated by living cells in low concentrations and can be produced endogenously within the cell as fluorescent proteins. Fluorophores also lend themselves to tracking protein diffusion kinetics (e.g. FRAP) and also measuring interactions with other biomolecules through FRET and ratio analysis. Nevertheless, to take maximum advantage of this powerful technique, there are three disadvantages that must be successfully managed:

- blurring;
- bleaching; and
- bleed-through.

Unless the sample is very thin, blurring is arguably the most serious disadvantage that must be addressed. Bleaching is – to a large extent – inevitable but can be managed through careful selection of fluorophores and restricting their exposure to light in a controlled manner. Bleed-through will occur only if the sample is stained with, or inherently possesses, more than one fluorophore. Whether the management of bleaching or bleed-through is more important really depends upon the nature of the specimen and the fluorescent markers used.

This chapter discusses the stratagems and approaches to imaging that are available to reduce blurring in the final image or are required to collect blur-free 2D and 3D image data sets. In order to acquire images that are in focus, some compromises must be made. There are several designs of optical sectioning microscope, and which one is best for your purposes will depend upon the nature of the specimen, most often its thickness, how labile it is and whether it is fixed or living. This chapter explains how the point-scanning confocal and other optical sectioning microscopes work. More information and guidance is also given later in Chapter 27 regarding core imaging facilities and on which microscope is the best choice to use.

The single-beam point-scanning confocal is the most widespread and popular of all the optical sectioning fluorescence microscopes, and several designs are available on the market. At the time of writing, typical microscopes are the Zeiss LSM 810 and the LSM 780, the Olympus FluoView FV-3000, the Nikon C2+ and A1R+/A1+ and the Leica SP8. Both Nikon and Leica also produce confocal macroscopes (the Nikon AZ-C1 and the Leica TCS LSI) for low-magnification examination of thick samples and entire organisms in vivo. Macroscopic optical sectioning can be performed on a widefield macroscope – the Zeiss Axio Zoom V16 macroscope is designed for use with the ApoTome structured illumination optical sectioning device.

Manufacturers patent their unique design features so that although the overall concept and design is identical, instruments from different manufacturers differ in small but sometimes significant respects. Of course, the sales staff of each instrument manufacturer will try to convince you that their particular microscope is best suited for your applications! Take advice from co-workers, read the Confocal Listserver[1] also post questions on it, and balance the information and opposing views that you may receive from different sales staff. As a starting point, refer to Appendix 1 if you intend to purchase a high-end research microscope.

17.2 Blurring and Optical Sectioning

Because, firstly, fluorescent molecules are self-luminous and are present throughout the mass of the stained cell or tissue and because, secondly, the depth of field (the distance between the upper and lower planes of the in-focus region along the optical or z-axis) of a high-quality microscope objective is generally much smaller than the thickness of the cell or tissue, that part of the image which is in focus will be degraded by light that is emitted or scattered by the tissue outside the narrow plane of focus (Figure 17.1). The image of a 'thick' specimen will only be wholly in focus throughout if the thickness of the specimen is less than the depth of field for the objective used. Usually, this is not the case.

If a typical cell or tissue section is, say, 10 μm thick and the depth of field of the microscope objective is 1 μm, then 90% of the image seen will be blurred because most of the light is contributed by regions that are not exactly in focus. The contribution of the blurred background light – from the out-of-focus regions – is superimposed over the weaker (less-intense) in-focus image. This will reduce the signal-to-noise (SNR) ratio, and thus the contrast, of the image. This out-of-focus haze will also reduce the resolution of detail in the image.

We can view the entire field of view at once, as we do in bright-field, DIC or phase-contrast microscopy and hence also see the out-of-focus information axially in z, but this does not help with depth discrimination of fluorescently stained cells and tissues, unless they are both thin and also flat. This situation is very restrictive. When cells divide (even those that spend much of the cell cycle in a flattened state) they tend to compress and round up during cell division. For those who wished to study such a fundamental process, this posed a problem, hence the practical development of the laser-scanning confocal microscope. Blurring can be prevented or eliminated

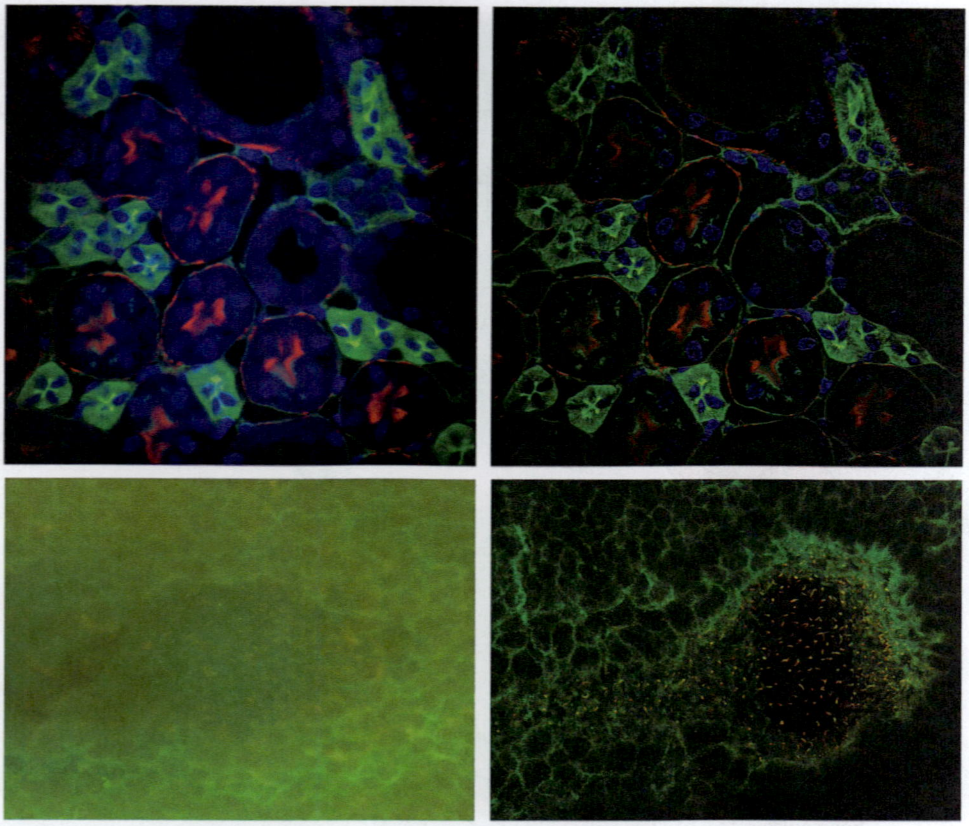

Figure 17.1 **Why a confocal microscope is used** *Showing blurring of signal in two different images in widefield microscopy and the result of optical sectioning. In the widefield images (left-hand panels) the in-focus part of the image is degraded by light emitted or scattered by the tissue outside the narrow plane of focus. The specimen in the top row is a 16 μm thick triple-stained cryostat section of mouse kidney, which is available (F24630) from Thermo Fisher Scientific and is widely used for confocal microscopy training. The bottom row shows double-stained cilia in the node of an 8.5 dpc mouse embryo. The thickness of the whole mount is about 20 μm, it scatters light very effectively and is therefore ideally suited to optical sectioning. Compare both of these images to Figure 18.5, which shows the respective maximum intensity projection as 2D representations of the 3D Z-stack data sets. Mouse images courtesy of Dr J. Keynton.*

by generating an 'optical section' through the sample, whereby only the image information that lies entirely within the objective's depth of field is displayed. Those parts of the sample that are out of focus are excluded from the image. Consequently the image has increased sharpness, increased contrast and increased resolution in the axial direction. As Inoué (2006) points out, the confocal can either increase or decrease the effective depth of field without loss of resolution.

Besides the confocal microscope – the most commonly used 'optical sectioner' – there have also been other approaches to optical sectioning[2] in order to 'clean up' the fluorescent image. These are listed in Table 17.1. In the point-scanning laser illumination confocal microscope, a single pinhole is used as a spatial filter to isolate the in-focus optical section from the out-of-focus portion of the image. The size of the pinhole determines how much of the Airy disc reaches the detector. In the multiple beam-scanning Nipkow spinning disc confocal microscope, a disc perforated with a series of pinholes of fixed diameter performs the same function. Optical sectioning can also be performed with a widefield microscope using the Optigrid or ApoTome (see Section 17.8). Parallel series of grid lines superimposed over the image act as the required spatial filter.

The ideal fluorescent sample prepared for confocal microscopy should be very well stained, to give a good signal with corresponding low noise. In this respect it is wise to use freshly prepared antibodies conjugated to modern 'second and third generation' fluorophores (e.g. Alexa,Cy cyanine and Atto or DyLight dyes) rather than earlier ones such as fluorescein (FITC) or rhodamine (TRITC).

17.3 Confocal Microscopy

The *confocal* microscope is so-called because the illumination pinhole, the plane of focus within the specimen and the detector pinhole are all situated at *con*jugate *focal* planes. Every laser-scanning confocal (a generic design is shown in Figure 17.2) has the following features in common:

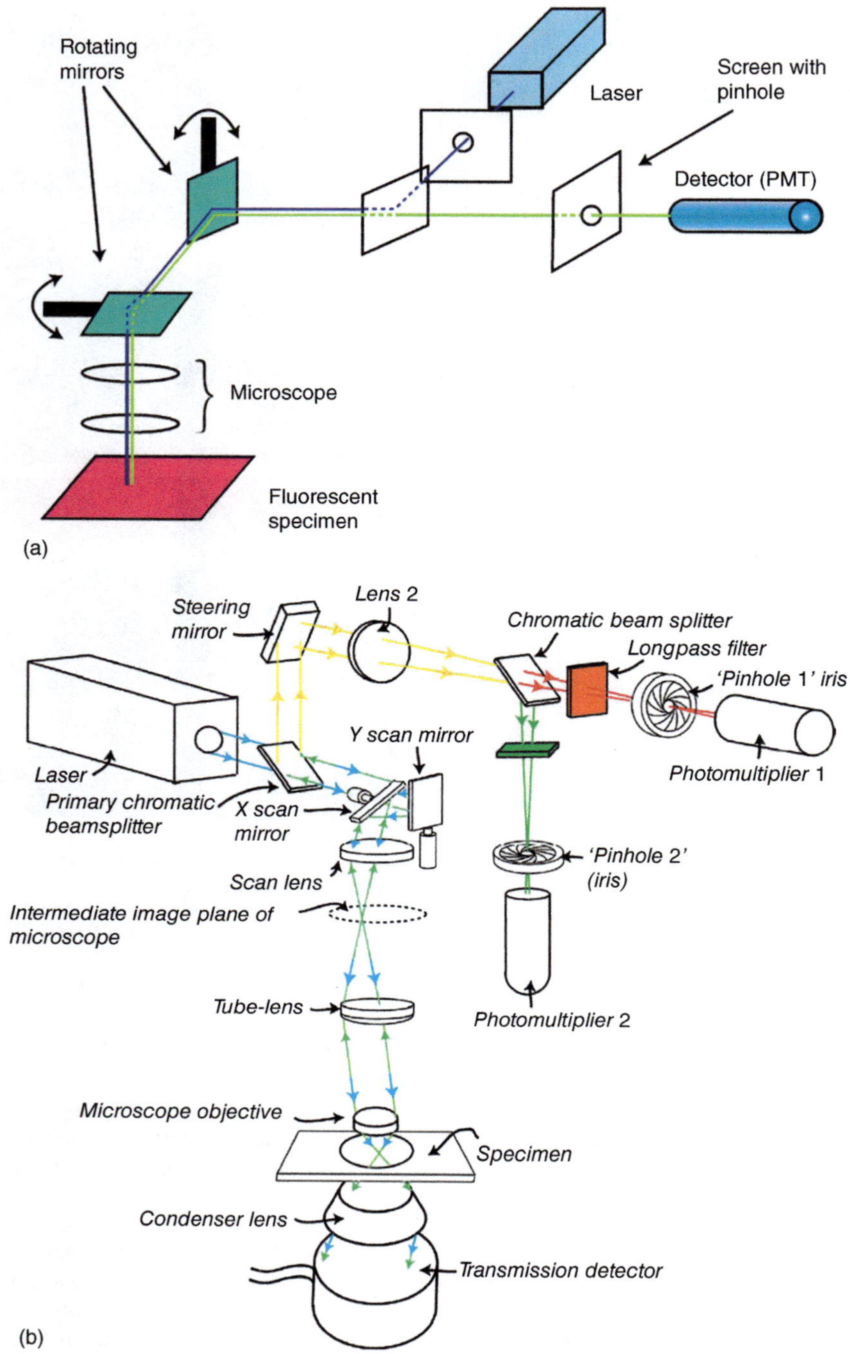

Figure 17.2 **Generic design of single-beam point-scanning confocal** *Features of the confocal microscope. The essential components are shown in (a) and are: the laser for illumination, the illumination pinhole in front of the laser, the galvanometer mirrors and the confocal pinhole in front of the detector. The lower panel (b) shows how the actual components are arranged within a confocal point-scanning microscope. Ideally an iris diaphragm comprises the pinhole, but it may in some designs be as simple as two sliding triangular cut-outs forming a diamond shape, which approximates a 'pinhole'. Source: (a) Semwogere and Weeks 2005. Reproduced with permission from Taylor & Francis. (b) Amos et al. 2012. Reproduced with permission from Elsevier.*

- one or more gas or diode laser(s) providing illumination of specific wavelengths;
- fluorescence filters and a dichromatic beam-splitting mirror as a filter set to direct the illumination onto the sample and direct the emitted fluorescence of specific bandwidth towards the detector;
- a mirror and galvanometer-based raster scanning mechanism;
- one, or more, pinholes; and
- photomultiplier tube (PMT) detectors for each channel.

Table 17.1 *Optical Sectioning Methods*

Microscopy type	Illumination	Detector/camera	When to use	Advantages	Disadvantages
Point-scanning confocal	Laser	PMT	**With thick tissues**	Variable & precise optical sectioning	Lasers photo-bleach. PMT & circuitry noisier than widefield.
Multi-beam Nipkow spinning disc confocal	Mostly laser (can be white light)	EM-CCD or CCD	**For acquisition speed**	For photo-sensitive living samples. High temporal resolution.	Fixed pinhole: fixed optical section thickness. No optical zoom.
Structured illumination	White light	CCD	**For photo-sensitive samples**	Better than confocal at lower mags.	Ronchi grid can burn into image.
Widefield deconvolution	White light	CCD or sCMOS	**For thin fluorescent cells;** for photo-sensitive samples	Good SNR, less noise than confocals.	Requires post-processing. Sensitive to spherical aberration.
Multi-photon (two-photon)	Pulsed laser	PMT	**For very thick tissues**	For very thick, living tissues.	Expensive. Requires dark for NDD detector.
TIRF	Laser	EM-CCD	**For imaging cell membrane**	Very thin optical section 100–200 nm.	Can only image in the evanescent field.

The merits of each type of optical sectioning microscope are shown in further detail in Tables 27.1 and 27.2 in chapter 27.

For those who prefer a more objective approach as to which optical sectioning microscope to use, John Murray, who has written an excellent chapter (Murray, 2011) on imaging fluorescent specimens, suggests the Haziness index, H. The metric H is the ratio of background to signal in a widefield image. The intensity of out-of-focus background is divided by the intensity of signal from a very small in-focus object. For practical purposes, with $H \leq 20$, widefield with deconvolution is appropriate. Where H lies in the range $20 < H < 200$, spinning disc confocals are best, and where $200 < H < 1000$, point-scanning confocals are the microscope of choice. For values of H above 1000, a multi-photon microscope should be used.

Ultimately, scientists wish to carry out 'in toto' imaging; that is, acquiring images and tracking every single cell movement and division that forms a tissue or organ. This approach is powerful for understanding how cell lineage, shape changes and movements control the morphogenesis of a tissue (Megason, 2010). A suitable objective is a water- or multi-immersion objective with a numerical aperture >1.0, yet with sufficient working distance (200–600 μm) to reach and image through the entire cell layer. Zeiss makes two 40x objectives that meet these criteria – the 40x C-Apochromat NA 1.2 (WD = 290 μm) and the 40x LD C-Apochromat NA 1.1 (WD = 600 μm). Zeiss also manufacture the 25x/NA 0.8 multi-immersion objective, which, with a fully open pinhole, can image a monolayer of adherent cells in an optical slice thicker than the cell layer. Zeiss, Leica and Olympus have all manufactured 20x objectives with NAs of 0.95 to 1.05 and long working distances for multi-photon and cleared tissue imaging.

References

Megason, S.G. (2010). 'In toto imaging of embryogenesis with confocal time-lapse microscopy'. *Methods in Molecular Biology 546*: 317–332.

Murray, J.M. (2011). 'Methods for imaging thick specimens: confocal microscopy, deconvolution, and structured illumination'. *Cold Spring Harbor Protocols 2011*/12: 1399–1437.

Presley, J.F. (2006). 'Measurement of protein motion by photobleaching'. In: Stephens, D. (ed.), *Cell Imaging*, Chapter 6, Protocol 2, note c. Bloxham: Scion Publishing, p. 135.

By restricting the illumination to a pinhole rather than illuminating the entire field of view at once, it becomes possible to build up a blur-free image in three dimensions (Figure 17.3). Point-by-point illumination of a small volume of the sample and corresponding synchronous (identical small volume) point-by-point detection provides good spatial discrimination in all three axes. The PMT does not 'see' the entire image as a CCD detector does but very rapidly produces a voltage equivalent to the photon intensity emitted at each sampling point in the object, which is then digitised to form a corresponding pixel in the image.

The entire image is built up point-by-point in each frame, as the diffraction-limited spot of the illuminating beam rasters across the frame, like reading words in lines along a page of script in a book. Each frame corresponds to a single plane in the z-axis. This may be equivalent to, or thinner than, the depth of field of the objective, but sharp discrimination of detail is provided in the z-axis, as well as laterally in the x and y axes. To continue the analogy, each frame can be regarded as a separate page, distinct from those on either side. By using a stepper motor to drive either the stage or the nosepiece axially, through z, a three-dimensional image, which is in-focus throughout, can be acquired of the specimen.

With only a single point illuminated (rather than the entire field of view as in 'widefield' mode) the illumination intensity rapidly falls off above and below the plane of focus as the beam converges and diverges to a point (Figure 17.4). This reduces excitation of fluorescence of these objects situated out of the focal plane, improving depth discrimination. This provides the necessary selective axial illumination along the z-axis, thus giving a sharp, blur-free result. The confocal microscope differs from widefield fluorescence microscopy in that the microscope configuration for widefield (both bright-field and fluorescence) microscopy is designed to be used with Köhler illumination where the excitation illumination is maximally out-of-focus at the specimen plane (i.e. the lamp filament is not imaged at the specimen plane but at a different conjugate plane: the entire purpose of Köhler illumination). In the confocal microscope the illumination *is* focused onto the specimen plane as a diffraction-limited point.

The Airy disc, named after Sir George Biddell Airy, who studied diffraction patterns and image formation in stars, describes the *central* intensity maximum of the two-dimensional pattern of the point spread function, the PSF. The Airy pattern is the PSF projected onto a flat surface[3]. An example is shown in Figure 17.5. In confocal microscopy the Airy unit is used to set the diameter of the pinhole and the thickness of the optical slice relative to the Airy disc. When the Airy unit (AU) is set equal to 1, only the central intensity maximum of the PSF reaches the PMT; this is optimum.

Since the confocal microscope depends upon a pinhole to exclude out-of-focus light, lasers are used because they have sufficient collimation and power to provide the necessary intensity, after passing through the illumination pinhole, to induce sufficient emission of signal[4] from fluorescent markers, which must then pass through the confocal detection pinhole in front of the detector. Lasers produce very intense, coherent, highly directional light beams of very narrow wavelength (usually of 1 nm bandwidth) that are ideal for illumination in a microscope where 90% or more of the emitted signal may (intentionally, due to the pinhole) not reach the detector. However, the use of lasers as illuminants all too easily causes photobleaching of fluorophores and photo-toxicity of biological tissue,

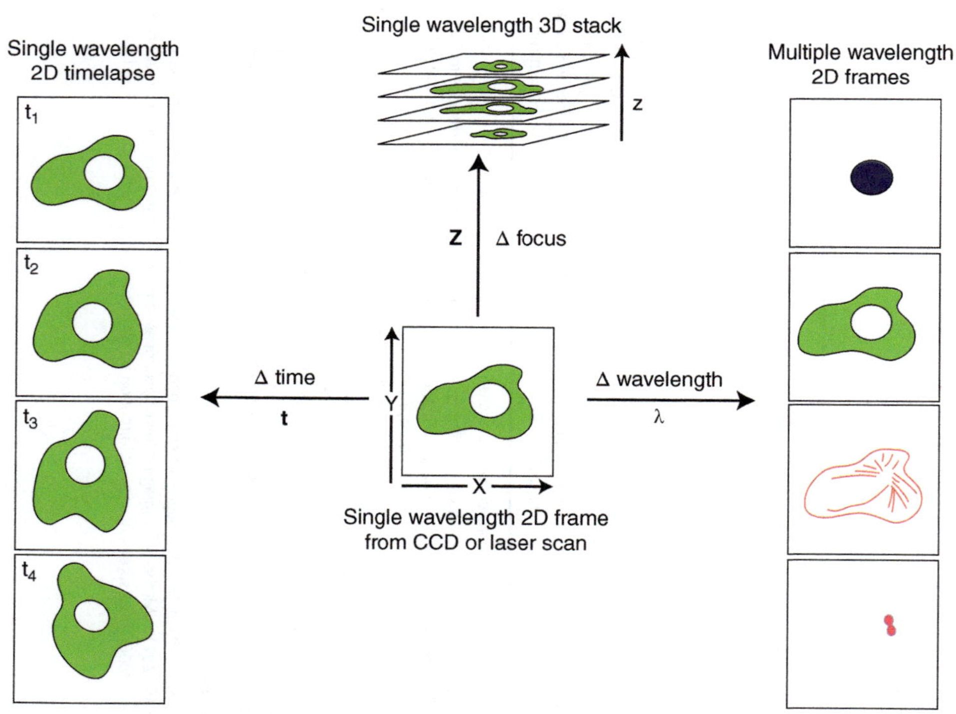

Figure 17.3a **Image acquisition by confocal** *Figure 17.3a summarises the image collection possibilities with the confocal microscope. Source: Andrews et al. 2002. Reproduced with permission from John Wiley & Sons.*

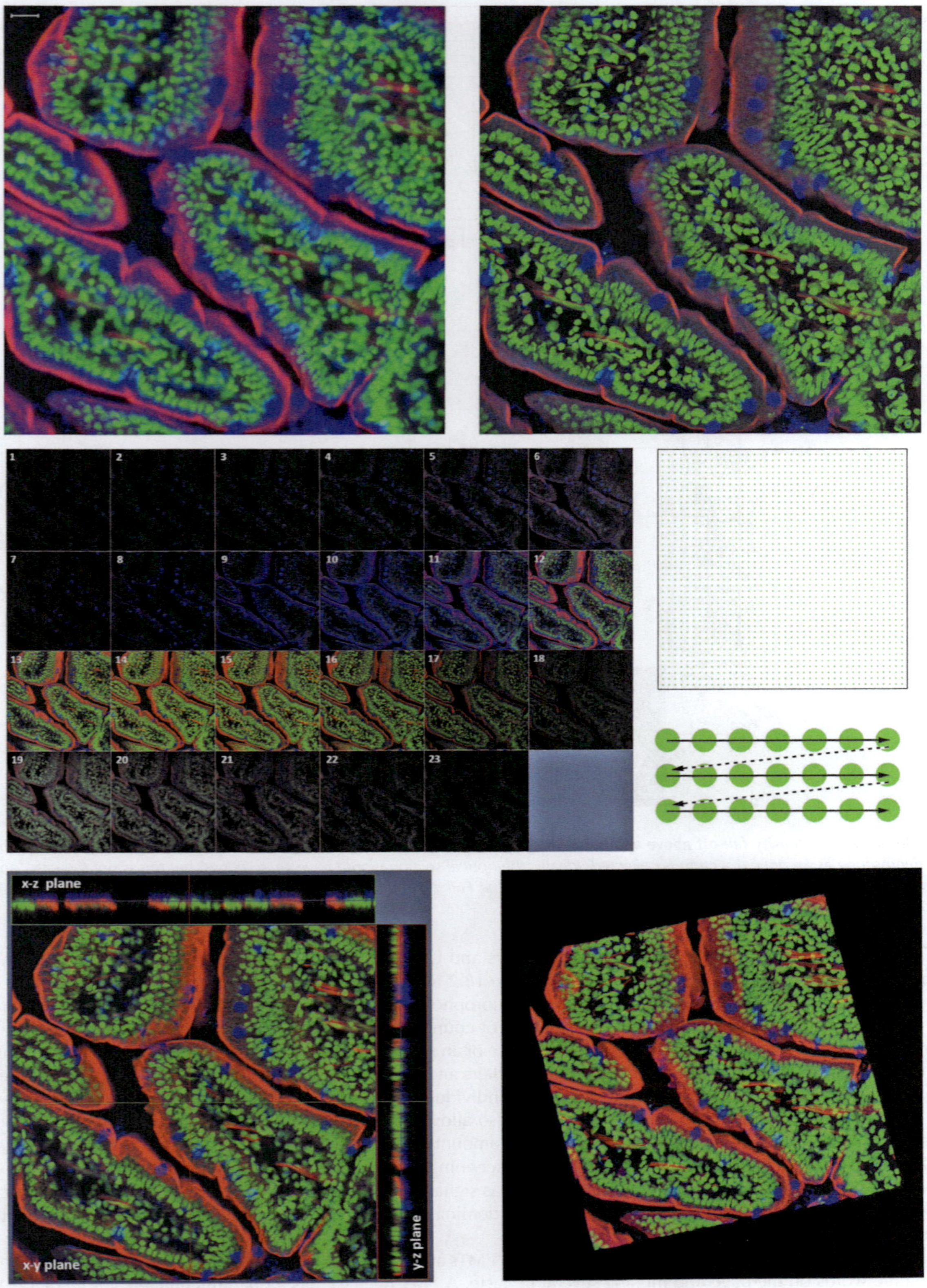

Figure 17.3b ***Pinhole illumination and detection*** *Figure 17.3b shows a 16 μm thick section of triple-stained mouse intestine (Thermo Fisher Scientific F24631) acquired as a widefield image (top left) with consequent blurring, followed by (top right) a single optical section. A series of these optical sections are formed as the laser is rastered in outwards and flyback mode across the field of view, effecting point illumination and point detection. It is possible to collect a second image on the flyback, which increases the collection speed, but the sample must not move, and the scanning mirrors must absolutely be in phase. The Z-stack can be presented as a gallery, orthogonal projection or 3D projected movie.*

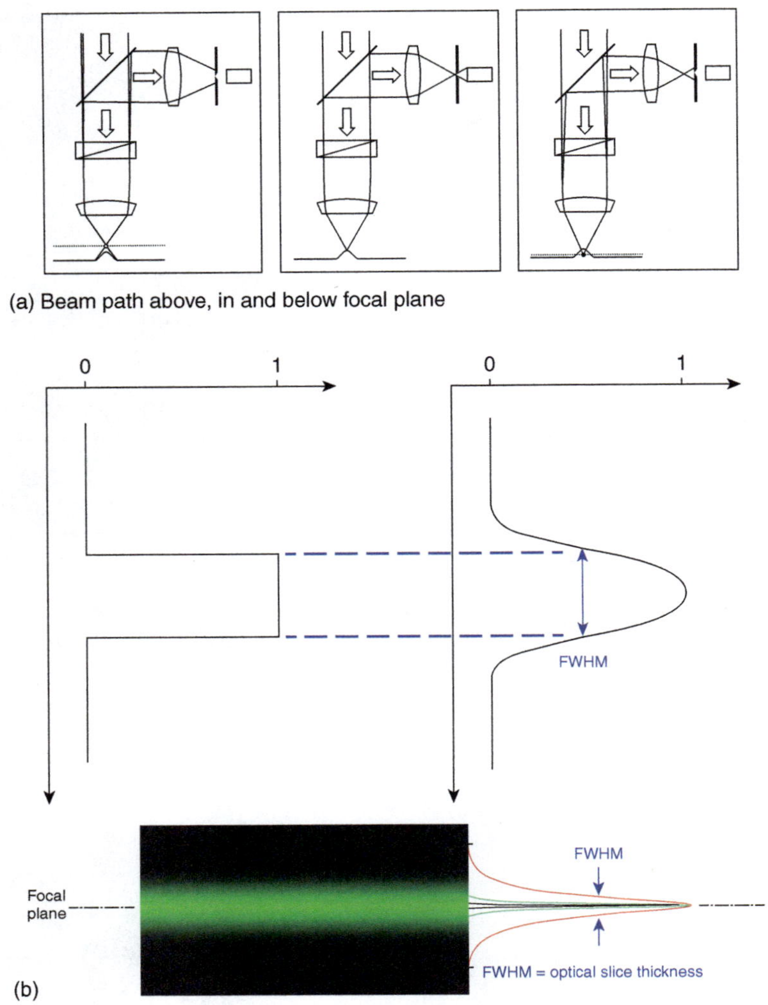

Figure 17.4 ***Illumination intensity fall-off above and below focus*** *Ray diagram showing the intensity fall-off above and below the plane of focus in the confocal microscope as an optical section is collected. Ideally, the optical section would have sharp edges, but this is not the case, and the optical section forms a horizontal Gaussian distribution of fluorescence with the Full Width Half Maximum, a measure of the optical slice set in the software. Source: Adapted from Isenberg 1997 with permission from Springer.*

in particular of living cells. For advice on the safe use of lasers and UK law (Artificial Optical Radiation at Work Regulations 2010) regarding the operation of lasers and intense lamps, see Section 18.2, in the next chapter.

It is essential to be able to adjust the laser intensity so that fluorophores are not saturated and bleached. Lasers are generally used at fractions of a percent of their maximum output power unless, of course, they are intentionally used to bleach the sample. The lasers can be switched very rapidly and adjusted in intensity by the use of an Acousto-Optical Tuneable Filter (AOTF) (Figure 17.6) or neutral-density filters. AOTFs have an advantage over neutral-density filters and are now used exclusively for this purpose, because they allow the intensity of individual laser lines in multi-line lasers to be individually adjusted. AOTFs also permit blanking of the laser beam on the flyback during the raster scan, to reduce exposure. They also allow specific regions of interest to be scanned in the field of view. Furthermore, neutral-density filters tend to permit a significant amount of bleed-through.

Photomultiplier tubes (PMTs) are used as detectors in point-scanning confocal microscopes because these enhance the weak signal and can also very rapidly collect single points of light emitted as signal, whereas using a CCD, suited to widefield microscopy, would be far too slow. PMT technology (Figure 17.7) is dated but still dominates microscopy as the detector of choice at low light levels with a fast response time, despite introducing multiplicative noise.

However, speed in detection comes at the price of efficiency. PMTs acquire the diffraction-limited spot signal at each pixel rather like speed-reading the words on the page and missing several in haste. A PMT detector has a quantum efficiency (QE) of 12–15% at best and so only detects 1 in every 7 or 8 photons of fluorescent signal. By comparison (see Figure 17.8) a charge-coupled device (CCD) camera that is used on a widefield microscope has a QE of 60–70%. Nonetheless, CCDs have lower gain, higher background noise and cannot read the point signal as rapidly as a PMT, because they read out their signal once the entire frame is collected. Signal gains of up to 10^7 are possible with PMTs; however, they amplify noise in the signal at high gain settings. Therefore **to secure the best possible image, the controls for laser strength, gain and pinhole size must all be adjusted and balanced relative to one another**.

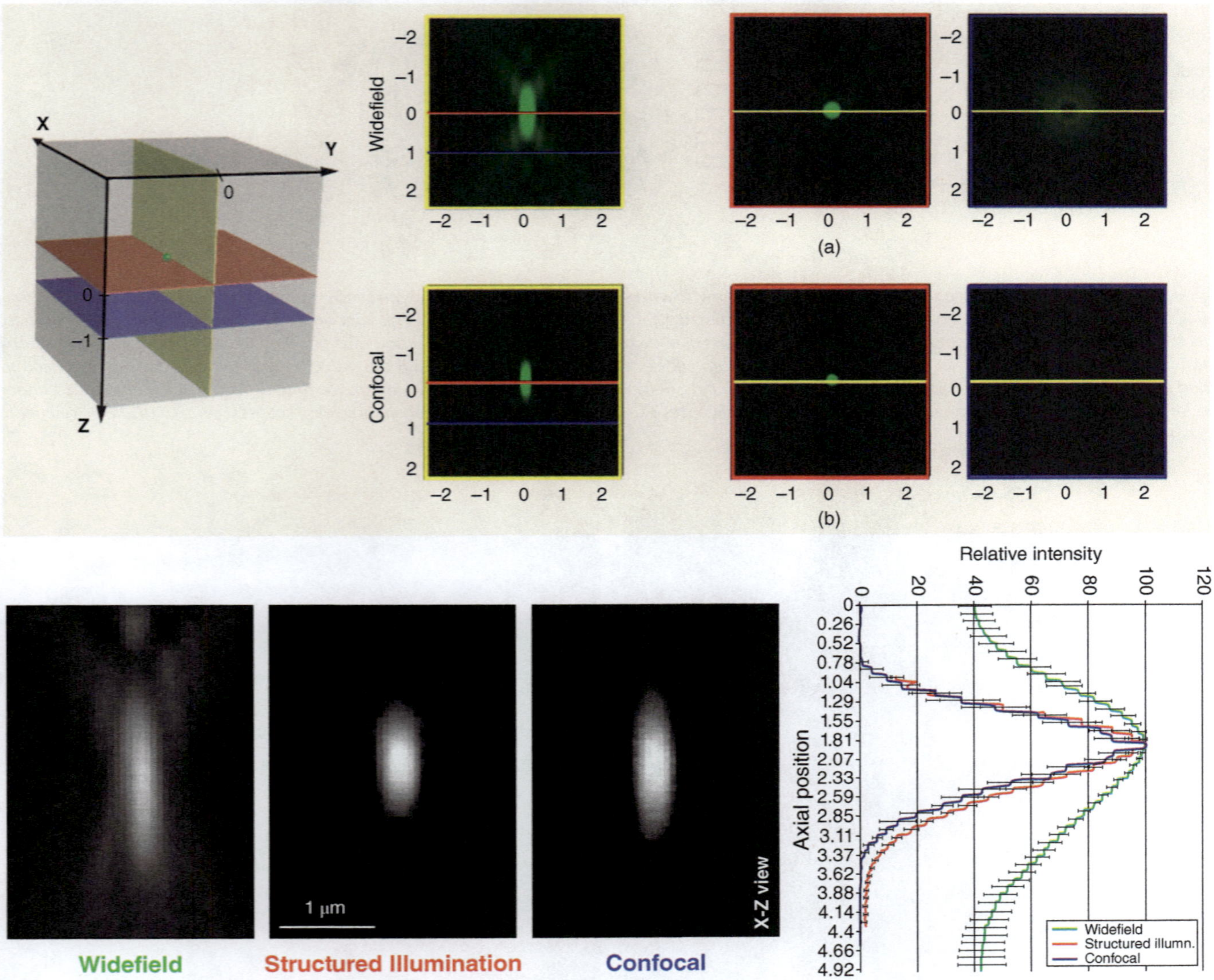

Figure 17.5 ***Confocal PSF*** *Comparison of the relative size and shape of the PSF between the widefield, the point-scanning laser confocal microscope and the structured illumination optical sectioning microscope. Source: (top) Vonesch et al. 2006, Reproduced with permission from IEEE and Dr C. Vonesch. (bottom) Puche 2005.*

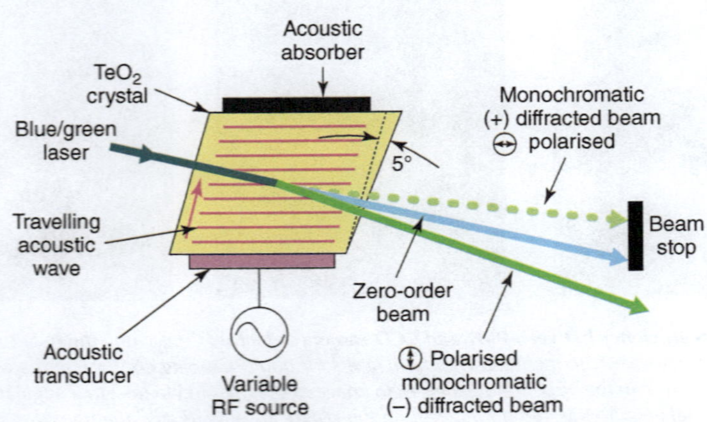

Figure 17.6 ***How the Acousto-Optical Tuneable Filter works*** *The AOTF is a solid state device with no moving parts. Radio signals propagate an acoustic wave through the anisotropic crystal causing it to act as a diffraction grating with a bandpass characteristic, enabling very rapid wavelength tuning. Source: Rietdorf and Stelzer 2006. Reproduced with permission from Springer.*

Figure 17.7 **The design of the photomultiplier tube, and how it works** *The PMT consists of a photo-sensitive surface, the photocathode, electron-multiplying dynodes and a collection anode all within an evacuated glass envelope. Light enters the input window and is absorbed by the photocathode, which emits electrons. These electrons are accelerated to the first dynode by an applied voltage with a potential sufficient to produce secondary electrons upon hitting the dynode. These secondary electrons are in turn accelerated to the next dynode, and the process repeats itself. The resulting electron cloud is collected at the anode to form an amplified signal. Source: (left) Adapted from Harp, via Wikimedia Commons. (right) Poll. Reproduced under the terms of the Creative Commons Attribution Share-Alike licence, CC-BY-SA 3.0 (http://creativecommons.org/licenses/by-sa/3.0/), via Wikimedia Commons.*

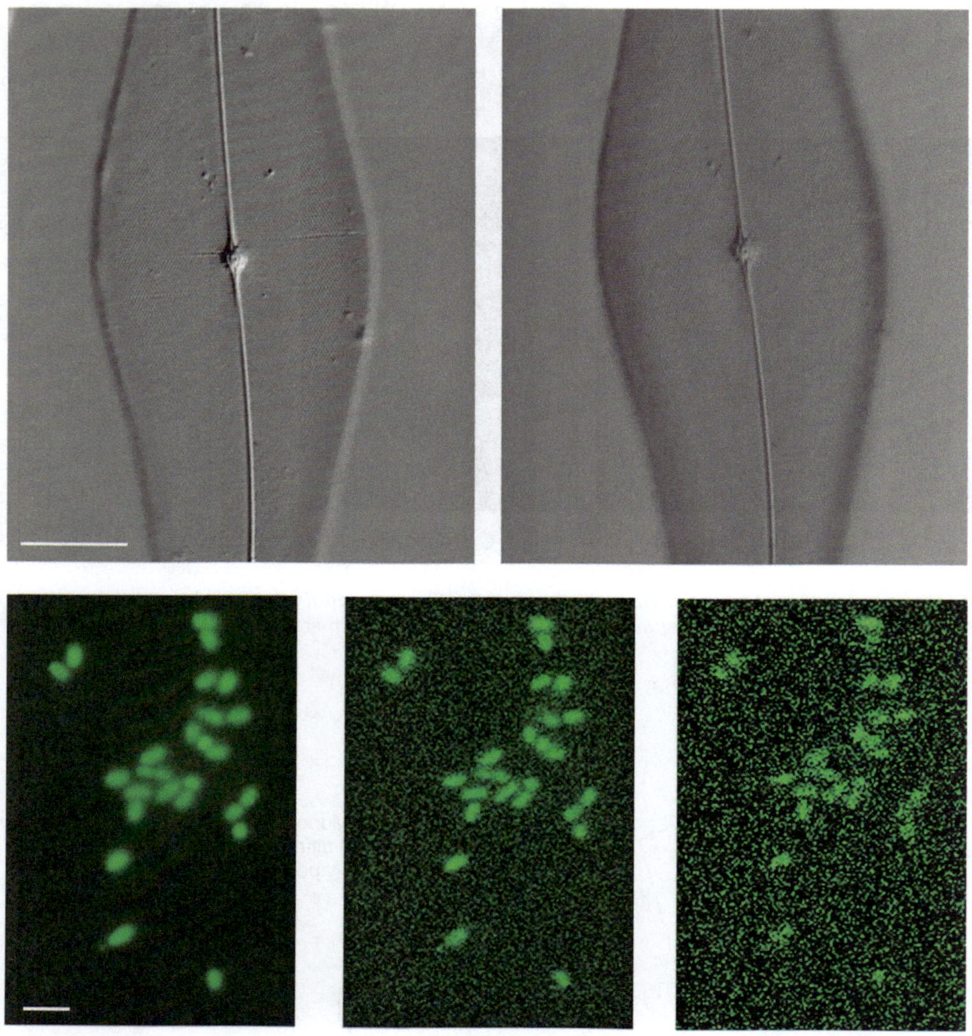

Figure 17.8 **Comparing the quantum efficiency between PMT and CCD camera detectors** *Comparing the quantum efficiency between the PMT and CCD camera detectors. A live-cell time-lapse microscope fitted with both lasers for point-scanning confocal and a widefield CCD camera with 6.45 μm pixels has been used to acquire these images. In the upper two panels is an image of the diatom* Pleurosigma angulatum. *On the left is a widefield image and on the right a non-confocal image taken with the 488 nm laser using the confocal software and the transmission PMT. In the lower three panels is an image of GFP-expressing NTHi* Haemophilus influenzae *bacteria. The widefield image is on the left; the single-scan confocal image on the right. The middle image is a confocal image taken with 16 average scans – the best possible. Although improved, it is still noisier than the widefield image. The scale bar for the* Pleurosigma *images is 20 μm and for the NTHi* H. influenzae, *2 μm.*

The requirement to illuminate living samples with minimum radiant intensity or else to detect weakly emitting samples and still image both scenarios with good signal-to-noise ratios means that improved detectors are always being developed. The first significant development was the avalanche photodioide, or APD. It is basically a silicon-doped PMT designed for photon-counting; the APD has an increased quantum efficiency (about 45%) particularly at the red end of the spectrum, as well as an increased amplification range. Their spectral response range is typically within 200–1100 nm. APDs deliver a higher signal-to-noise ratio than regular photodiodes and also have fast time response, low dark current and high sensitivity. Recently very sensitive gallium arsenide phosphide (GaAsP) and related hybrid PMT (HyD) detectors have entered the marketplace. These combine the advantages of PMTs and silicon-based photodiodes and are now offered by all major manufacturers.

17.4 Development of the Confocal Microscope

The evolution of the confocal microscope has a long history. A 'critical mass' for effective development required the development of fluorophores resistant to photobleaching, high-quality and sufficiently economical lasers, fast computers with enough memory to control the laser scan and the optical expertise to build these instruments. The first system was developed from the flying spot microscope (Amos et al., 2012) by Marvin Minsky[5] in 1957, and was a stage-scanning system, but it gained little interest because stage-scanning confocals possessed inherent disadvantages, not least their slow scan speed. It was not until images were published from the first practical single-beam laser-scanning confocal microscope that interest in the confocal microscope really took off. This design (which kept the specimen stationary and utilised raster-scanning of the illuminating beam) was developed (see Amos and White 2003) by William 'Brad' Amos, John White, Mick Fordham and Richard Durbin at the Medical Research Council laboratory in Cambridge in 1986. It followed earlier crucial development work by Brakenhoff in the Netherlands and by Wilson & Sheppard in Oxford. Stage-scanning confocal microscopes are better suited to non-living, inert, materials specimens. This is because stage movement can perturb living cells, particularly at high speeds. Also, crucially, for work requiring immersion objectives for high-resolution studies (i.e. most life science applications), stage-scanning instruments would disrupt the oil or water connection between objective and microscope preparation. Beam-scanning confocals present more technical challenges to overcome optical aberrations but are faster and better suited to living cells.

The generic features of a typical laser-scanning confocal microscope are shown in Figure 17.2. Fluorescent light emitted by the specimen travels by the reverse beam path through the microscope. Unlike the raster scanning pattern of excitation light passing over the specimen, the fluorescence emission remains in a steady position at the pinhole but fluctuates with respect to intensity over time as the illumination spot traverses the specimen producing variations in signal. Because in the exceedingly brief time required for fluorescence excitation and emission, the scanning mirrors have (for all intents and purposes) not moved, the beam of emitted light is stationary and in register with its point of origin. This return of the emitted signal through the confocal beam path to the pinhole detector is referred to as 'descanning' of the beam. Without the descanning of the emitted signal by the galvanometer mirrors on the return path, the signal would only pass through the pinhole at one position of the raster scan. By descanning the signal in the beam-scanning design of confocal, only one pinhole is required. If the emitted beam of light were not descanned, the emitted signal would move relative to the pinhole. This would necessitate a separate pinhole for each position of the laser during the raster scan, which would be impractical.

The size of the adjustable pinhole in front of the PMT detector governs how much in-focus signal (and hence the thickness of the optical section, or 'slice') is recorded in the image. The thickness of the optical section depends upon the diameter of the detection pinhole. It is less than, or equal to, the depth of field – the focal layer. As the pinhole is opened, a thicker optical section is collected. However, the limited depth of field of high numerical aperture objectives also defines the maximum thickness of the in-focus optical section. See also Protocol 18.1 in Section 18.4 and endnote 2 of Chapter 18.

The first designs of confocal microscope were equipped with individual pinholes in front of each PMT in order to compensate for poorer detection in the red part of the spectrum. There is also a good theoretical reason to have individual pinholes: the PSF alters with wavelength and so the ability to set different pinhole diameters at the UV-blue, the central green-yellow and the far-red part of the spectrum means that a constant thickness for the optical slice can be maintained throughout. The disadvantage of having multiple pinholes means that there is 'more to go wrong'. It is easier for the pinholes to become misaligned and so collect poor optical sections. Most pinhole adjustment protocols run on automated macros in software, but it is possible to make manual adjustments. A brief description of how to do this is given in Chapter 27, Section 27.6. The Zeiss LSM 510 has multiple pinholes in its design, but the design of modern point-scanning confocals, including the LSM 710 and its successor, has only one pinhole. This can be kept constant, or (better) adjusted between scans for the particular wavelength associated with each channel.

Altering the size of the pinhole will proportionally alter the thickness of the optical slice that is collected. The optimum diameter is related to the numerical aperture of the objective and wavelength of the illuminating radiation.

- The larger the pinhole, the thicker – less confocal – is the optical section.
- The higher the NA of the objective, the thinner the optical section.
- The shorter the wavelength of the illuminating radiation, the thinner the optical section.
- A pinhole set at 1 AU collects only the central intensity maximum (the Airy disc) of the PSF.
- The maximum size of the pinhole at which the system is still confocal is $\approx$1.2 AU.

The graph in Table 18.2 shows graphically how altering the pinhole diameter will determine the thickness of the optical section. These optical sections through the cell or tissue (i.e. frames; 'pages') can be collected non-invasively. It is not necessary to physically slice up the specimen to reconstruct it in three dimensions. Because light is used to image each optical section, these will be in perfect register, so they can be reconstructed into a three-dimensional model with ease. A computer is used to control the sequential scanning of the sample, building up the image point-by-point for display onto a monitor.

Because the image is collected from an (analogue) microscope and encoded as a digital signal, the final displayed image can be contrast-enhanced, and image collection can be seamlessly integrated with electronic triggers (e.g. in a physiological experiment) or else a time-lapse movie taken of living cells. The confocal microscope is therefore an integrated electronic microscope system, with sophisticated electronic circuitry, under computer control, synchronising the raster sweep of the point of the laser illumination across the sample in the field of view of the microscope.

17.5 Advantages, Disadvantages and Resolving Power of the Confocal Microscope

Confocal microscope technology has been very widely adopted, and considerable numbers of new instruments continue to be sold worldwide. Confocal microscopes clearly offer considerable advantages over widefield fluorescence microscopy:

- First and foremost, the confocal microscope is a very efficient optical sectioner, providing very clear image slices from thick cells and tissues where finely stained detail would otherwise be lost within the noise arising from the out-of-focus blur.
- Secondly, because the lasers used exhibit very narrow wavelengths, the confocal microscope is ideally suited to imaging multiply-labelled specimens with minimal bleed-through.
- Thirdly, the illumination is rastered across the specimen under computer control. This allows the single-beam laser scanning confocal microscope to acquire images from precisely defined specific areas, or regions of interest, of the specimen.
- Fourthly, computer control of the laser illumination permits controlled bleaching in protein mobility studies employing FRAP and similar techniques.

From a strictly theoretical point of view and under perfect conditions, the confocal microscope offers *at best* slightly better resolving power than the widefield fluorescence microscope. This is because the ray path is focused *twice* by the objective: first the excitation light is focused to a point, then the emitted signal is detected point-by-point: **point illumination** and **point detection**. The total point spread function (PSF) is therefore the *product* of both the excitation PSF and the emission PSF[6]. The resolution is therefore equal to the *square* of the intensity of the PSF. However, the position of the minima of the two Airy discs remains unaltered (so, in a strict sense, the resolution according to the Rayleigh criterion does not change), but the full width at half maximum (FWHM) of the PSF is improved by a factor of $\sqrt{2}$, or 1.41.

Thus the classical equation $d_{x,y} = 0.61\lambda/NA$ (multiplied by $1/\sqrt{2} \approx 0.71$) becomes $d_{x,y} \approx 0.43\lambda/NA$, with the focal spot being 60% of the diameter of that in widefield microscopy. In the axial direction, the PSF of the confocal microscope is also slimmer, and thus smaller, than for the widefield fluorescence microscope, and $d_z = 2\lambda n/NA^2$ becomes $d_z = 1.4\lambda n/NA^2$. Confocal optics do not improve the Rayleigh resolution criterion at all but confers a 1.41x ($\sqrt{2}$) resolving power advantage over the Sparrow criterion as applied to widefield microscopes in both lateral and axial directions.

However, to utilise this advantage in practice, the pinhole must be closed to an extent (< 0.5 AU) where most of the emitted signal is rejected; this is very rarely done. In practice, the primary advantage of the confocal microscope is optical sectioning rather than making possible an increase in resolving power.

There are several different formulae cited in the literature to calculate resolving power in the confocal microscope, and this has led to confusion about which is the 'correct' formula. The formulae turn out to differ according to whether a large or small pinhole or a large or small objective aperture is under consideration. The formulae may also differ depending upon whether the Full Width Half Maximum (FWHM) of the Gaussian distribution of the PSF intensity is measured experimentally using a fluorescent point or a fluorescent sheet. Naredi-Rainer et al. (2013), Amos et al. (2012) and Wilhelm et al. (2010) cite for lateral and axial resolution, respectively:

$$\text{for a pinhole} > 1\ \text{AU: } d_{x,y} \approx 0.51\lambda_{ex}/NA \quad \text{and} \quad d_z = 0.88\lambda_{ex}/[n - \sqrt{(n^2 - NA^2)}]$$

Besides the advantages listed above in the bullet-point list, the single-point laser-scanning confocal microscope has some inherent disadvantages:

- confocal microscopes are relatively noisy compared to widefield microscopes, exhibiting a lower signal-to-noise ratio;
- the laser illumination can easily photo-bleach the sample to a much greater extent than widefield fluorescence microscopes or other designs of optical-sectioning microscopes; and
- confocal microscopes are relatively slow at acquiring full-frame images.

Widefield microscope images are much less noisy for three reasons. Firstly, the illumination is delivered in parallel to all regions of the sample simultaneously, not pixel by pixel, thus eliminating pixel-to-pixel differences across the field of view. Secondly, the CCD has approximately a four-fold higher detection quantum efficiency than the photomultiplier tubes of the laser-scanning confocal microscope. This increases the number of detected photons by four, thus reducing the Poisson shot noise by a factor of two for the same total illumination (the available photon flux emitted for detection is limited in living samples by photobleaching and photo-toxicity). Thirdly, the CCD read-out electronics add less noise to the signal than does the laser-scanning confocal microscope (LSCM) electronic circuitry.

There are three principal sources of noise in the laser-scanning confocal: (a) the gas lasers themselves, (b) the PMT detector electronics, and (c) the fibre optic coupling of laser to microscope. The 2–5% power deviation of LSCMs acts on *each* individual pixel, whereas fluctuations in arc lamps illuminating the widefield microscope act *equally on all* the pixels, or sampling points that form the image. The noise variation can obscure the true signal, especially in weakly stained samples. This was investigated, and is further explained in an important paper by Swedlow et al. (2002); see also Figure 25 on page 275 in Murray (2010) which unfortunately cannot be reproduced here for copyright reasons.

A typical 2D frame scan representing a single optical slice within the thickness of tissue may consist of 512 lines each comprised of 512 points, that is: 262,144 points overall, which are represented as digitised pixels. In order to scan this 2D frame in 1–2 seconds, the laser must dwell over each point – and have the emitted signal detected – in under 4 microseconds. As explained previously in this chapter, a photomultiplier tube is used to enhance the weak signal and read it very rapidly. However, speed in detection comes at the price of efficiency: a PMT has a QE of 12–15%. Furthermore, the shorter the 'dwell time' of the laser over each part of the specimen, the greater the noise.

Because the PMT detector is quantum-inefficient, the resulting image may – unless the sample is well stained – also be noisy unless the dwell time is increased and/or successive images are taken and summed to average[7] out such noise (see Figure 18.12). Both stratagems carry the risk of extra photobleaching. An image taken with the confocal microscope may well suffer from photobleaching (see Figure 15.3) unless care is taken to operate the laser judiciously at the minimum power necessary. For these reasons, the confocal is best suited to collecting images from relatively thick (20–80 μm) well-stained cells and tissue slices: these will survive the effects of photobleaching better than thin samples. It is very easy to forget this minimum power requirement: if the eyepieces of the confocal are not adjusted to be parfocal with the PMT detector, it is possible to secure an initial in-focus image at the start of a microscopy session with the microscope in widefield visual mode, as usual, then switch to digital confocal acquisition only to find very little or no evidence of fluorescence on the monitor. Many users are tempted to open the pinhole, or worse, turn up the laser, at this point to get a smattering of signal to help them refocus an image on the monitor. This will inevitably lead to premature bleaching of the sample.

Single-beam laser scanning confocal microscopes have a limited acquisition speed for two reasons. Firstly, the lasers are used at a high intensity very near the saturation point of most fluorophores, as explained in Section 18.5. A high illumination intensity is required because the pixel dwell time is very rapid – each point in the raster scan being sampled in microseconds. Secondly, the galvanometer mirrors are driven very hard indeed, particularly the X-galvanometer. This is mechanically challenging, and it is not possible to collect full frames at video frame rates of 30 fps. Even so, to scan a 512 × 512 frame in a second or so, each point must be sampled in under 4 μs, and the galvanometer mirrors must oscillate at over 15 000 rotations per second. If they are driven faster, they are heat-distorted[8].

Despite the fact that the PMT is a very rapid detector, the laser beam must still be scanned over the entire sample within the field of view point-by-point. This is done by galvanometer mirrors, which are relatively slow in operation. Since a full frame can take 1–2 seconds to be collected, the point-scanning laser confocal using a single laser beam is not fast enough to capture dynamic events at anything much more than 5–10 frames per second, even when strategies such as pixel binning (using larger pixels at the expense of resolution) and reducing the boundaries of the scanned area are employed to speed up image acquisition. The answer has been either to use a CCD camera on a widefield microscope and subsequently employ mathematical deconvolution techniques or use a multiple laser beam spinning disc confocal scanning microscope to collect the entire field of view almost instantaneously with a CCD camera instead of a PMT detector.

17.6 The Spinning Disc Microscope

In 1884, Paul Nipkow patented a solution to scan a 2D object sequentially, so that the signal could be transmitted via a selenium photocell as a one-dimensional serial signal through a wire and the 2D image reconstructed in a remote location – an utterly revolutionary concept at the time. The so-called 'Nipkow disc' is a rotating circular disc with a helical array of pinholes and is a simple mechanical method to scan a light beam over the photocell. The intensity of the signal detected by the selenium cell was amplified and used to power a remote lamp placed in front of a second Nipkow disc. The image is reconstructed point-by-point by this second disc spinning in phase with the first, with each illuminating pinhole corresponding synchronously with a conjugate imaging pinhole (hence tandem-scanning of both illumination and detection). This was the basis upon which John Logie Baird devised the first mechanical television set.

The fundamental difference between the Nipkow disc and the first tandem-scanning microscope developed in 1965 by Mojmír Petráň was that the latter had a spinning disc containing thousands of pinholes (part of such a disc is shown in Figure 17.9), which were illuminated *in parallel*. Petráň's instrument was, in effect, thousands of confocal optical systems operating simultaneously. Each beam formed, by an individual pinhole, is confocal with the same aperture serving as the excitation pinhole as well as the emission pinhole (upon the return of the emitted signal)[9]. The colloquial name 'Nipkow' spinning disc microscope is therefore – strictly speaking – a misnomer.

This early spinning disc microscope worked well for reflection imaging of opaque materials (e.g. in dentistry); it was also very fast. Also, the image built up in real-time could be viewed by eye or recorded by camera directly. However, the pinholes in this design only occupied 1% of the area of the disc: the pinholes have to be spaced at least some distance apart in order to prevent cross-talk between adjacent pinholes. Petráň's microscope could not be used for studying fluorescently labelled samples.

Engineers at the Yokogawa Institute Corporation solved the poor light budget of the spinning disc by incorporating a second disc (Figure 17.9) rotating in parallel fashion above the first disc, carrying a series of microlenses that focus the illuminating laser light onto the pinholes of the first disc. The emitted fluorescent signal returns through the first disc and is directed via a dichromatic beam splitter

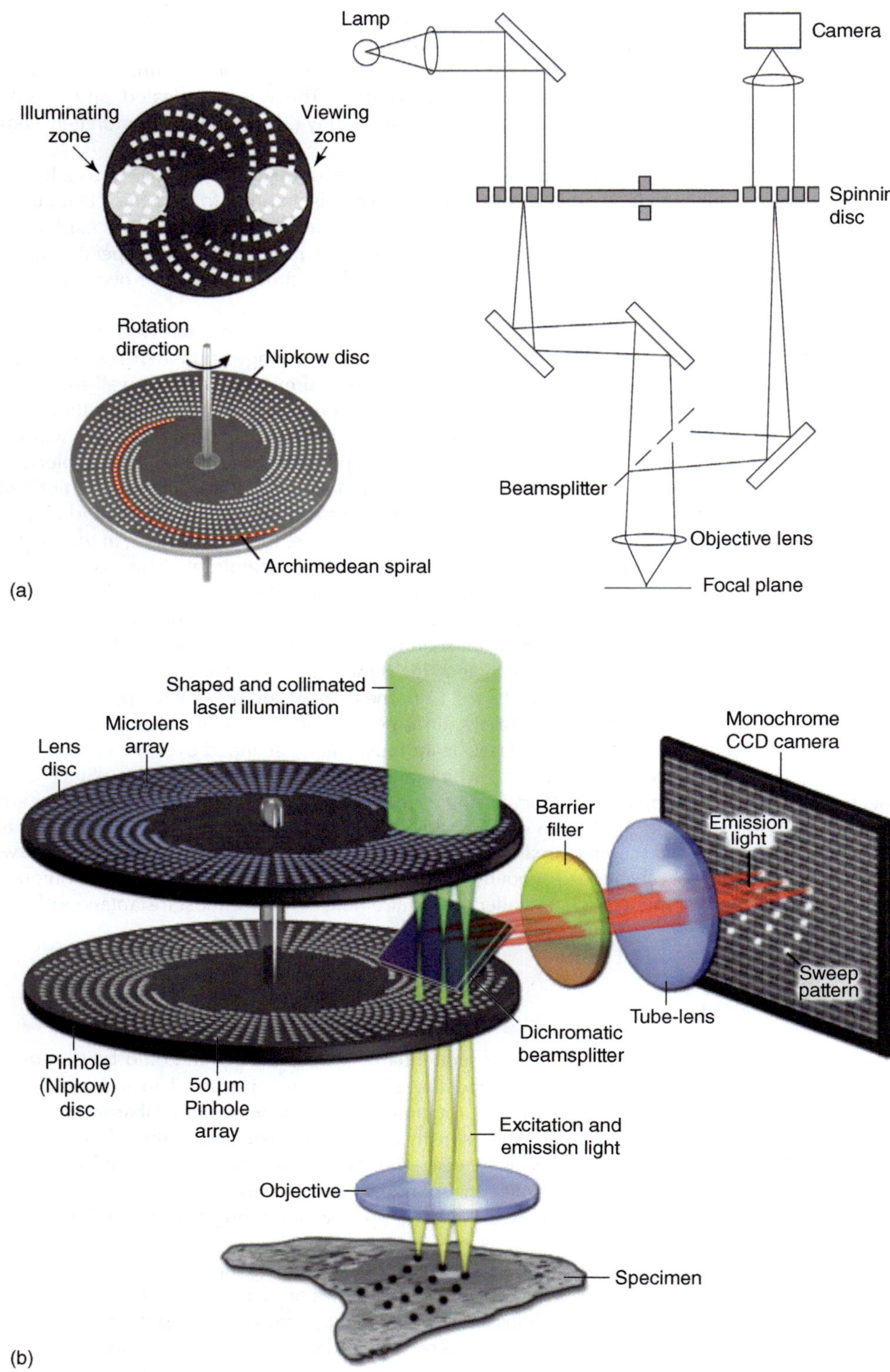

Figure 17.9 **The Spinning disc microscope design, how it works** *Panel (a) shows the Nipkow disc and the Petráň design of the original design of spinning disc microscope. Panel (b) shows the modern tandem-scanning design incorporating the Yokogawa double disc arrangement with the microlenses that significantly improve light throughput. Source: (a) Nipkow disc, reproduced with permission from Carl Zeiss Ltd; Dean 1998. Reproduced with permission from John Wiley & Sons. (b) Murphy and Davidson 2013. Reproduced with permission from John Wiley & Sons.*

to a very sensitive EM-CCD camera (which allows very low exposures of weak[10] signals). Since the image is observed in real time, it can be detected with a higher-QE (quantum-efficiency) sCMOS, CCD or an EM-CCD camera instead of a lower-QE PMT. The frame rate is essentially limited to the camera read-out speed.

Very fast z-sectioning is possible through the use of a high-quality piezo-actuator attached to the stage or objective lens. The spinning disc confocal therefore has a much higher temporal resolution than the single-beam laser-scanning confocal microscope. A high-quality three colour Z-stack, which could take 70 minutes to collect on the point-scanning confocal, is likely to take 70 seconds on the spinning disc – sixty times faster. There are several commercial spinning disc microscopes built around the Yokogawa CSU-X1 unit. These include the Perkin Elmer UltraVIEW VoX, the Zeiss Cell Observer SD and the Andor Revolution XD. Yokogawa have also developed the CSU-W1 with a slower frame rate, but a 4x larger field of view.

The multi-beam spinning disc confocal microscope therefore offers the following advantages over the single-beam LSCM:

- (EM) CCD and sCMOS cameras have a much higher quantum efficiency (65–95%) over PMTs (approximately 12–15%) or avalanche photodiodes (approximately 40%). The signal is collected more efficiently, leading to
- shorter exposure times;
- higher frame rates (for equivalent SNRs);
- more efficient fluorescence detection (lower intensity individual beam over longer exposure); and
- less photobleaching.

The thickness of the optical section in a single-beam laser-scanning confocal is controlled by altering the size of the pinhole. However, in the spinning disc microscope, the size of the pinholes are fixed and are calculated to give confocality only for one objective, usually a 100x/NA 1.4 objective, yielding an optical slice of 0.8 μm. The spinning disc microscope generally has less axial resolution than the point-scanning confocal. It is also not possible to zoom in and out, as can easily be done with a single-beam point-scanning confocal microscope. This means that the optical resolution cannot be matched to the Nyquist sampling requirements of the object and thus can only be optimised for a single magnification.

Although the total light flux may be similar to a point scanner, the spinning disc confocal has a much lower light intensity *per individual scanning spot*, which reduces photobleaching and cellular photo-toxicity. Furthermore, the pixel dwell time of the spinning disc microscope is much higher than a single-beam point-scanning confocal. These two factors, together with the higher quantum efficiency of the EM-CCD detector, cause less photobleaching and the ability to image cells longer over time without saturating the fluorophore. However, bleaching is proportional to both the laser intensity and the frame rate. It is worth remembering that although spinning disc scanners generally use lower laser intensities, they can bleach just as much as a point-scanning confocal *if* images are collected with too high a frame rate.

The multi-beam spinning disc microscope does have a few disadvantages when compared to the single-beam laser-scanning confocal:

- the size of the pinholes are fixed and cannot be altered to match the NA of the objective in use or the wavelength of the excitation illumination in use;
- lack of scan zoom to match the sampling rate to the Nyquist criterion;
- only able to use very simple ROI overlays, unlike the single-beam confocal;
- the field of view is non-uniformly illuminated; and
- pinhole cross-talk, which gives rise to background haze and reduced SNR, as well as reduced axial resolving power.

Just as with the single-beam confocal, the image acquisition rate can be increased by reducing the lateral resolving power and/or reducing the image area by using a region of interest or CCD subarray readout to capture a smaller field of view. Using the zoom function and pinhole diameter to change the size of the voxels[11] imaged is very easy in the single-beam confocal, since the zoom and pinhole functions are very flexible continuously variable controls. Likewise, selecting a reduced field of view in the single-beam confocal by means of drawing a region of interest to restrict the sample area that is rastered by the laser(s) to form the image is also very easy to manage. The spinning disc confocal, however, has no zoom function, and the size of the pinholes is fixed, so adjusting the frame rate by changing the image size can only be done stepwise by changing the objective magnification (and thus resolving power) or by increasing the image pixel size by binning the CCD camera. Sub-array readout is still possible, and this can also increase the frame rate. It is also more difficult to apply a complex region of interest (e.g. for bleaching) using the spinning disc microscope, whereas this is very easily and quickly done with the single-beam point-scanning confocal.

The spinning disc confocal is ideally suited to imaging single-labelled samples stained with just one fluorophore. It is not possible to switch lasers pinhole by pinhole, which would be the equivalent of line-scanning on the single-beam confocal; the stratagem of line-by-line switching using the LSCM single-beam confocal avoids mismatches in register between channels, which could otherwise happen when collecting images frame by frame. Sequential imaging – to avoid bleed-through – whilst using the spinning disc microscope has to be done by switching lasers and filter sets between frames. This will reduce the frame rate to a certain extent, because although switching lasers with an AOTF is fast (2–5 msec), changing filters is an order of magnitude slower ($\approx$ 20 msec) depending on the angular travel (position of filters within the filter wheel) and velocity of the filter wheel.

One way of imaging two multi-channel fields of view simultaneously is to use a beam splitter. There are three options on the market. The first is the DualView and QuadView from Photometrics or Olympus; the second is the Optosplit III from Cairn Research or Andor,

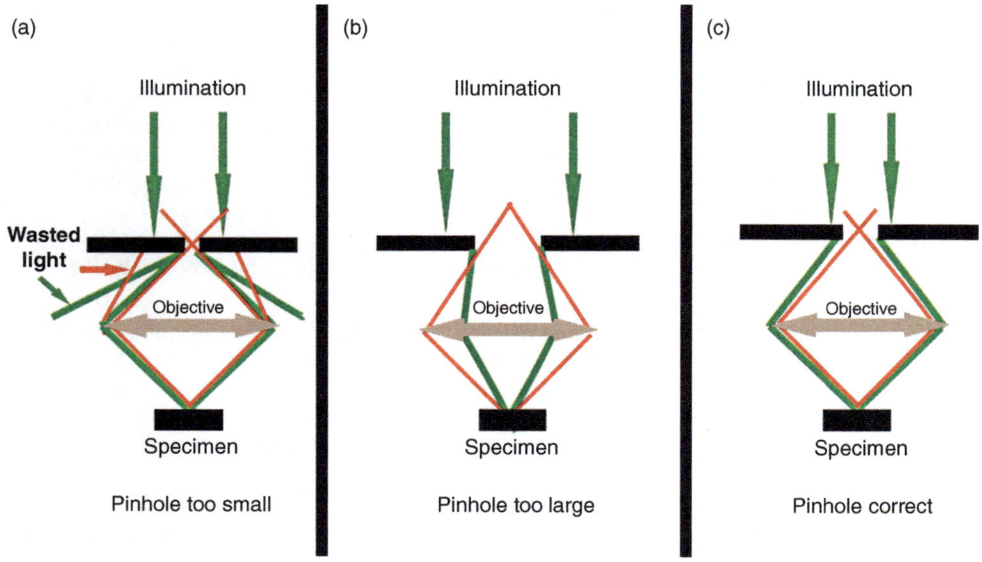

Figure 17.10 ***The Spinning disc microscope, cross-talk*** *Where the size of the pinhole is too small (a) light is wasted, and some higher-order diffracted light may not be collected. Where the pinhole is too large (b) the diffracted light may not adequately fill the back focal plane of the objective. Also, bleed-through will occur as signal returns to the detector through the wrong pinhole. Where the pinhole is of the correct size, all the light diffracted by the specimen is collected, and the proper Nyquist sampling occurs without bleed-through. Source: Fuseler 2011. Reproduced with permission from Springer.*

and the third the W-view from Hamamatsu. An ImageJ plugin is available to display and analyse split fields of view on a single monitor. Beam splitters such as these are also useful for displaying the two different polarisation states (crossed, uncrossed) on a single monitor simultaneously or for FRET and calcium ratio-imaging applications.

Because some light reflected, or emitted, from the sample will reach the detector via pinholes other than the one intended, a small amount of cross-talk (Figure 17.10) will also occur with the spinning disc microscope. Cross-talk will increase as the thickness of the sample increases, say above 50 µm. Cross-talk from one pinhole to adjacent pinholes will also occur if there is a significant degree of refractive index mismatch. Thus the Olympus silicone-immersion objectives or a similar multi-immersion objective (see Figure 7.4), are particularly suited for live-cell imaging of watery samples on the spinning disc microscope.

The camera of choice for the spinning disc confocal has been the EM-CCD, since these designs allow efficient collection of low-light level signals encountered in live-cell imaging, thus enabling fast frame-acquisition rate. Recently the performance of sCMOS cameras has improved to such an extent that, unless the emitted fluorescent signal intensity is 50 photo-electrons or less, the sCMOS is a better detector than the EM-CCD (Oreopoulos et al., 2014). In the summer of 2016, Andor released their Dragonfly microscopy platform, a multi-point confocal, with two selectable fixed pinhole sizes of 25 and 40 µm designed for high speed and high SNR with up to 200 fps, taking advantage of the latest Zyla sCMOS cameras.

An alternative to the Yokogawa spinning design, popular because of the cost benefit, is the aperture correlation microscope sold by Zeiss as the VivaTome II (Figure 17.11), by Andor as the DSD (differential spinning disc) and Aurox – the developer of the technology – as Clarity. Instead of a series of pinholes in the spinning disc, a grid pattern etched into both sides of the disc is used. It is the same structured illumination technology (Section 17.8) as the ApoTome and Optigrid, which are alternatives for the single-beam scanning confocal. As with the ApoTome and Optigrid, the VivaTome, DSD and Clarity use a white-light source. This is cheaper than lasers and provides for specific wavelength illumination, depending on the filter sets used. The double-sided spinning disc collects a transmitted- and reflected-emission image. Computer subtraction of these images differentiates between the in-focus and out-of-focus light from the fluorescence signal to give a confocal optical section. These aperture correlation microscopes can collect signals from a single fluorophore imaged at full frame at between 20–30 frames a second (e.g. at video rate). On some newer models it is also possible to increase the frame rate by binning and/or using a region of interest (ROI) to collect an image from a sub-array of the detector.

It is also possible to use white light for illuminating the spinning disc, using metal-halide, mercury or (best) xenon arc lamps, which latter have a relatively even profile across the visible spectrum. The advantage of white light systems is that the illumination can be better tuned to a greater number of fluorophores than merely with single-line laser illumination, although the introduction of several feasible diode laser sources to cover the near-UV, visible and infrared spectrum makes this less of an advantage than was formerly the case. The chief advantage of white-light spinning disc systems is cost: they can be added very cheaply and easily to an existing widefield fluorescence microscope. The Olympus DSU (disc scanning unit), Andor DSD (differential spinning disc), Clarity^{HS} the Crisel CrEST X-light and the BD Carv II are examples of white-light confocal designs. Single-beam laser scanning confocal microscopes and parallel-beam spinning disc confocal microscopes are complementary technologies, not alternatives to one another.

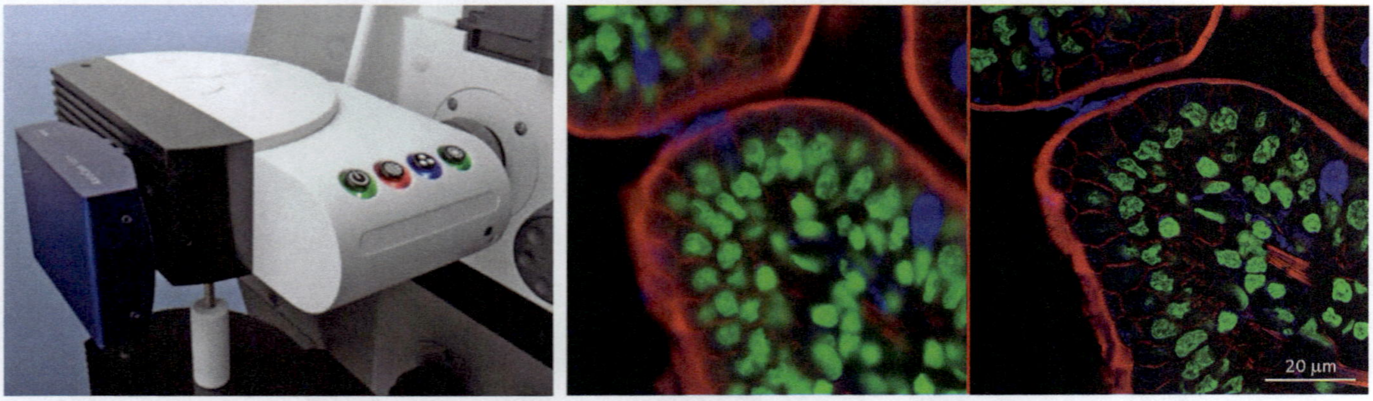

Figure 17.11 **The VivaTome** *The Zeiss VivaTome attached to the side port of a widefield microscope, together with an example of an optical section collected from a triple-stained mouse intestine preparation. Source: Reproduced with permission from Carl Zeiss Ltd.*

17.7 Line-Scanners and Array-Scanning Confocals

Programmable array and line-scanning microscopes are designs that attempt to increase the frame rate, and thus the temporal resolution, of the laser-scanning confocal microscope. Instead of using a diffraction-limited spot formed by a pinhole[12], these microscopes use either slit apertures, and sacrifice some confocality in so doing, or else use spatial grids or resonant scanning devices. In a single-beam point-scanning confocal, the X-galvanometer mirror ('X-galvo') works faster, and is exposed to greater mechanical stress, than the Y-galvo. This is because it must raster the point of the laser beam across the entire line scan, whereas the Y-galvo merely has to 'drop' the beam to the next line. Since the X-galvo is more prone to wear and eventual failure, if we do away with it and replace the mirror by a slit, we can scan each line in one go and merely use the single remaining mirror to raster the entire line across the frame. Since the slit used to illuminate the sample is aligned in the x-direction, some out-of-focus light is also detected, so confocality is partially lost in the x-direction but retained in the y-direction.

Slit scanners, such as the 7-Live from Zeiss or the Nikon LiveScan SFC, use a slit of light to sweep the field of view and so build up an image much faster than a single-beam point-scanning confocal. The Zeiss system uses beam-shaping lenses to create a 1-pixel-wide scanning beam that is detected and read out by a linear-array CCD line detector with 512 elements, which naturally has a very fast read-out time. The Nikon swept-field design is a hybrid model, combining both slits and pinholes, which can be exchanged by the user to suit the application. The scan head is made by Prairie Technologies, and the image is detected with a regular full-frame CCD. Maximum frames rates of over 1000 fps are possible with both models, provided that a sub-array ROI is scanned (also with binning on the full-frame CCD). Slit scanners may be inferior in optical sectioning to a point-scanning confocal, especially when a large fluorescence volume is scanned. They are as expensive, or more so, than point scanners, and they tend to bleach very quickly.

Instead of a disc with pinholes, another possible design is to use a rectangular microlens array where the illuminating beam is scanned across the sample with a single galvanometer mirror, and the emitted signal descanned using the reverse side of the same mirror. The Visitech VT-infinity is based on this design. A further solution, again from Visitech, is the VT-eye. This design uses an acousto-optical deflector (AOD) and since this is a solid-state device, it does not suffer from wear like a galvanometer mirror does. The AOD is wavelength dependent, functioning in a similar manner to an AOTF by passing a sound wave through a crystal in order to isolate a particular wavelength, so the longer-wavelength beam cannot be descanned through the return optics (as in a single-beam point-scanning confocal). Hence, point-scanning excitation and slit-scanning detection are used, resulting in less confocality than for a single-beam point-scanning system. AOD microscopes have not made point-scanners obsolete because of the greater technical challenges of maintaining stability at high speeds and high data transfer rates.

An alternative method of increasing temporal resolution is to retain the single-beam point-scanning design and to increase the rate at which the galvanometer mirrors are driven. Unlike the spinning disc, swept-field or line-scanning confocals, resonant scanning single-beam point-scanning confocal microscopes are able to alter magnification without changing objectives by retaining the versatile confocal zoom functionality. The repetitive, shorter exposures usually lead to brighter images, albeit with slightly worse signal-to-noise ratio but with much less bleaching per scan. The speed of the resonant scanner is fixed, usually at 8000 Hz, and it can be difficult to define regions of interest, except along horizontal lines. Leica, Nikon and now Olympus market resonant scanner units for their point-scanning confocal microscopes.

17.8 The Structured Illumination Approach to Optical Sectioning

Another means of being able to optically section the sample, to provide clear blur-free images, is implemented by using structured illumination microscopy (SIM). Currently there are three designs on the market capable of doing this: the original Optigrid from Qioptiq,

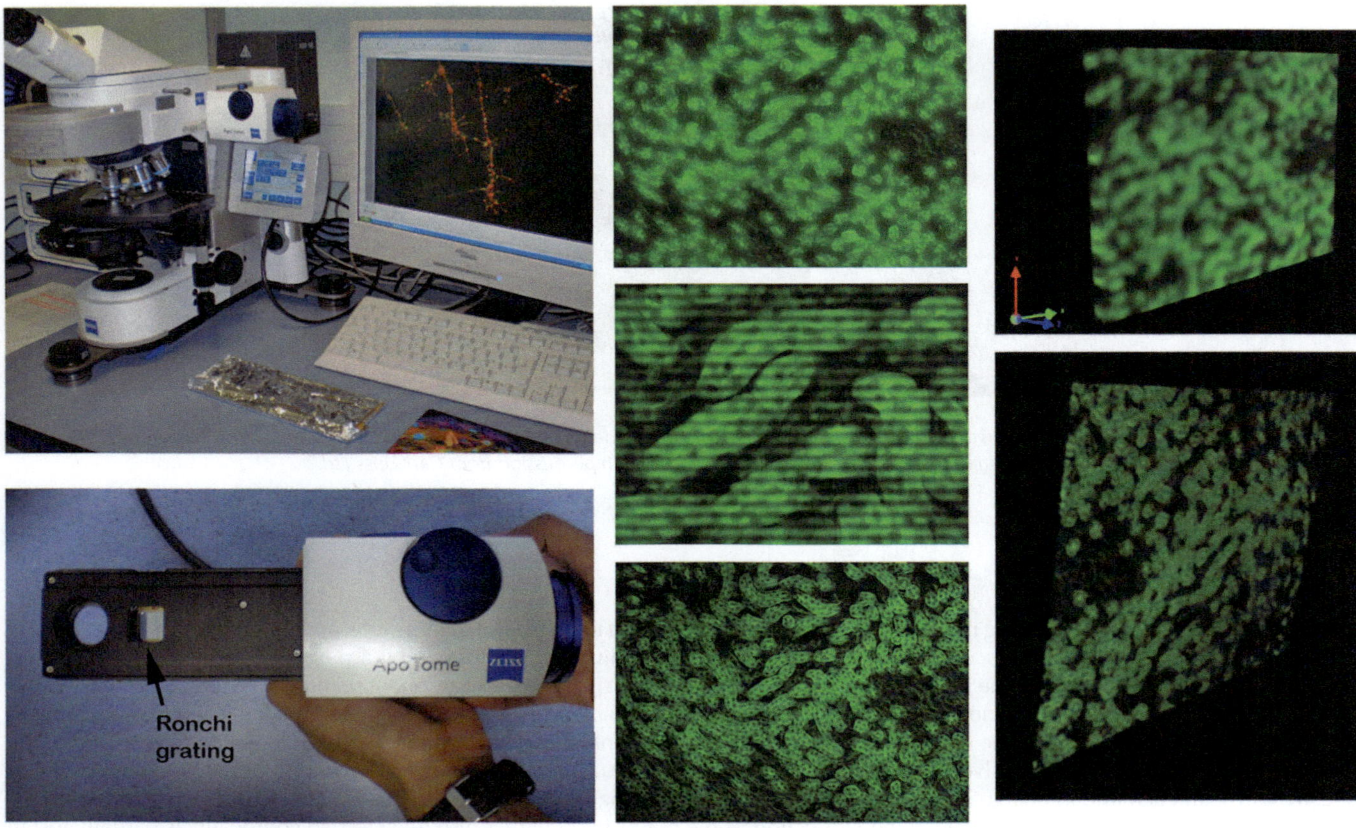

Figure 17.12 **The ApoTome structured illumination optical sectioner** *See the text for explanation of how this and similar structured illumination optical sectioning devices work. Source: Sanderson 2008. Reproduced with permission from The Quekett Microscopical Club.*

the Angstrom sold by Quorum and the Apotome 2 by Zeiss (Figure 17.12). All three structured illumination systems can be retrospectively fitted to modern widefield fluorescence microscopes. The ApoTome design fits only the manufacturer's own brand instruments, but the Angstrom and the Optigrid can be fitted to a range of microscope stands from all main manufacturers.

All designs work in the same way: they superimpose an image of a Ronchi ruling (placed in the excitation light path at the position of the epi-illumination field aperture) onto the sample. A Ronchi ruling is one composed of very high contrast black and clear rulings. When the grid is operating, these rulings can be seen moving up and down as stripes superimposed upon the image in the field of view of the microscope. Three images are acquired by moving the grid laterally (here within the plane of the paper) in three steps, each displaced by 1/3 of the ruling's period. These three grid movements produce one optical section. An algorithm recombines these three images to extinguish any portion of the images that contains unfocused light, from outside the depth of field of the objective (the out-of-focus signal is pretty much constant in all three images and subtracts out), while eliminating the 'grid-image' from the final resultant image. At the plane of focus, the differences between the illumination intensities are weak, whereas the differences in illumination intensity are strongest when the sample is out of focus.

Since it is necessary to collect three images at each z-position for each fluorochrome, this implies that the specimen must be stable and must not move during these three acquisitions. Therefore this technique is not very suitable for live cell imaging, where movement inevitably occurs. Where temporal resolution is required, the aperture correlation microscope (Section 17.6), using a spinning disc etched with a grid pattern, is the structured illumination equivalent of the spinning disc microsope.

These structural illumination microscopes have the same optical sectioning ability as the laser-scanning confocal microscope, but the algorithms won't easily handle *very* thick specimens (>30 μm) and can be prone to stripe artifacts. Nevertheless they can be easier to set up to collect an image and are cheaper to run, since they are fitted to a widefield fluorescence microscope and so use a mercury lamp, rather than a laser, as the fluorescence illuminator. For this reason also, the sample generally suffers less bleaching than would otherwise occur with laser illumination. In addition, the image formed by structured illumination in a widefield microscope is detected with a CCD camera, rather than a PMT with lower quantum efficiency. This makes the structured illumination method ideal for capturing images from very weak or sensitive and photo-labile samples. The structured illumination approach to optical sectioning is particularly suitable when using lower magnification objectives. It is easier to get an optical section from a large field of view at lower magnifications (e.g. less than 20x) with the structured illumination approach than by using point-scanning confocal microscopy.

This method is fundamentally the same as super-resolution SIM (Chapter 26, Section 26.6), but in the latter case, a much finer Ronchi grating is used.

17.9 The Airyscan Principle

The 'iron triangle' of imaging is discussed in Chapter 28 with particular regard to live-cell imaging. It is not possible simultaneously to image at high speed with a good signal-to-noise ratio and with good resolution. You can only have one of these three conditions (speed, SNR, resolving power) at the expense of the other two. With a smaller pinhole set below 1 AU, resolving power increases but primarily at the expense of SNR. Alternatively, SNR can be increased but primarily at the expense of acquisition speed. Fast scanning with the single-beam point-scanning confocal is possible at the expense of SNR, resolving power and the size of the image collected.

One approach to using a small pinhole for resolving power advantage yet simultaneously keeping the SNR high is to use a very sensitive multi-element area detector at the pinhole to collect all the light from an Airy disc simultaneously. This works because a displaced pinhole produces an image about equal to the resolving power of a pinhole aligned to the optical axis but of lower intensity. A high-sensitivity multi-channel 32-element gallium arsenide phosphide (GaAsP) detector constructed rather like an insect's compound eye is able to detect these offset signals and sum the signal shifted back on-axis. In this way much less signal is rejected by the 0.2 AU pinhole and may contribute usefully to image formation. Deconvolution of the image data ensures that resolving power is improved in all directions by a factor of 1.7x to give a lateral resolving power of 140 nm and an axial resolving power of 400 nm. This idea is not new, having been proposed by Pawley in 1996 and recently by Roth et al., 2013 and De Luca et al., 2013. Zeiss have very recently (Weisshart, 2014; Huff et al., 2015) made this idea commercially possible.

17.10 The Mesolens

Although they have been available for many years, the confocal microscope – like its widefield counterpart – can only image a large field of view (mm) at low resolution *or* a small field of view (μm) at high resolution. Some firms (see comments in the Introduction to this chapter) have introduced confocal macroscopes, which can image larger fields of view. However, at low magnification, where the numerical aperture of most objectives is low – and the depth of field large – it is impossible at low magnification to acquire good optical sections that are not excessively thick, because of poor axial resolution in depth.

Currently the only way around this is to use a high-NA objective and tile successive fields to provide a larger image overall. The problem with this approach is that very often the illumination is not perfectly even across the field of view, and so the edges of the tiles are very obvious. Furthermore, this method is time consuming, which can also lead to photobleaching.

A new giant lens is being developed by Amos, McConnell and Dempster, which has an unprecedented numerical aperture of 0.47 at a magnification of 4x, with a working distance of over 3 mm and a field size of 6 mm by 6 mm (McConnell et al., 2016). It is called the Mesolens because it lies between a conventional microscope objective and a close-up photographic lens. For example, the Mesolens can show an entire mouse embryo at 0.6 μm lateral resolution while observing cells and tissues at around 3 μm optical section thickness without losing a sense of depth. The Mesolens has a 3–5 times improvement in lateral resolving power and an improvement of 8–10 times in axial resolving power over a conventional 4x objective.

Such a 4x microscope objective has an NA of about 0.1, with a lateral resolution ($d_{x,y} = 0.61\lambda/NA$) of 3.4 μm and an axial resolution ($d_z = 2n\lambda/NA^2$) of about 100 μm. This means that it would not even be able to resolve detail within a 3 μm optical section. By comparison, a conventional objective of NA 0.47 will normally have a magnification of 20x or more, with a field of view of approximately 1 mm × 0.7 mm, rather than 36 mm^2. This 20x objective will resolve 0.7 μm laterally and about 5 μm axially. The Mesolens is capable of gathering approximately a hundred times[13] the amount of information in a focal series and can capture such a Z-stack in an hour that would take a day stitching and tiling with a 60x/NA 1.4 objective routinely used on a confocal microscope. Further information is available on the Mesolens website.

17.11 Concluding Remarks

At the conclusion of this chapter, it is worth reviewing the technical innovations that in the last forty years have made the laser-scanning confocal the predominant microscope platform for viewing fluorescent specimens, installed worldwide in most universities, hospitals and research institutes. In my view, each of these developments has been key to the evolution of the modern instruments now available as turn-key systems.

In the 1980s incorporating the **pinhole** to make a scanning microscope confocal was the key initial development (see Section 17.4, endnote 5 in Chapter 17 and Amos and White, 2003). The next significant development in the early 1990s was the **AOTF**, or Acousto-Optical Tuneable Filter, which replaced neutral-density filters for individually modulating the intensity of the lasers. The AOTF for the first time allowed blanking of the laser beam on the flyback during the raster scan. This helped significantly to reduce exposure of the specimen to light. The AOTF was first integrated into commercial instruments by Leica and became an essential requirement when tendering for a new instrument. The next development, around 2001, which became a convenient and essential tool to have, was **ROI**

scanning whereby the laser(s) could be controlled to raster within both defined and freehand regions rather than across the entire field of view. Not only did this allow for scanning only the area of interest, thus reducing exposure, but also saved significant time during image acquisition. The key advantage of ROI scanning, however, was to make FRAP a mainstream technique available to all. Just a year later Zeiss introduced their META detector, which permitted **spectral unmixing**, adding another dimension to image data sets. This advance allowed not only spectrally similar fluorophores and fluorescent proteins to be differentiated within the image for the first time but also brought FRET into mainstream imaging. By 2010 Zeiss again led the way by introducing high-sensitivity gallium arsenide phosphide (**GaAsP**) **detectors**, followed shortly after by similar PMT/silicon-based photodiode hybrid detectors. The latest significant development to impact the laser-scanning confocal market is the **Airyscan** introduced in autumn 2014. By incorporating a 32-element GaAsP detector into the confocal pinhole optical plane, it is now possible to secure the resolving power advantage of a reduced pinhole yet still maintain very high signal-to-noise levels. The only alternative is to acquire an image and deconvolve it – as the Airyscan also does – but using ten times the exposure. Like the previous technical advances, this newest technology significantly reduces the exposure of delicate specimens to light, whilst allowing lateral resolving power similar to super-resolution techniques.

17.12 Chapter Summary

This chapter explains how to take advantage of the specificity, sensitivity and potentially very high contrast of fluorescence imaging using confocal microscopes. The next chapter explains how to operate the most commonly used of these confocal microscopes, the point-scanning confocal.

1. There are many advantages to using fluorescence: specificity, sensitivity, high contrast.
2. The three disadvantages are blurring, bleaching and bleed-through.
3. Unless the sample is very thin, blurring is the most serious disadvantage. It occurs because fluorescent samples are self-luminous.
4. To restrict blurring, axially in the z-direction, confocal point-scanning is widely used.
5. The entire image is built up point-by-point in each frame, as the diffraction-limited spot of the illuminating beam rasters across the frame, like reading words in lines along a page of script in a book. Each frame corresponds to a plane in the z-axis.
6. The confocal microscope depends upon a pinhole to exclude out-of-focus light.
7. Altering the size of the pinhole will proportionally alter the thickness of the optical slice that is collected.
8. Photomultiplier tubes are used as detectors in point-scanning confocal microscopes. They very rapidly collect, and enhance, the single points of fluorescent light emitted as signal.
9. The confocal microscope is suited to imaging multiply-labelled specimens with minimal bleed-through. The narrow bandwidth laser illumination can be switched fast and efficiently.
10. Confocal microscopes are relatively noisy and slow at collecting an image compared to widefield microscopes.
11. The spinning disc microscope, containing thousands of pinholes, which are illuminated in parallel, is much faster.
12. The spinning disc confocal has a much lower light intensity per individual scanning spot, which reduces bleaching.
13. Cross-talk between pinholes in the spinning disc microscope occurs with refractive index mismatch and increases in proportion to the thickness of the sample.
14. Programmable array, slit or line-scanning microscopes and resonant scanning microscopes are alternative confocal designs that increase temporal resolution over the single point-scanning confocal microscope.
15. Structured illumination microscopes superimpose an image of a grating on the specimen. Computer algorithms subtract the out-of-focus light to give a confocal optical section. Aperture correlation microscopes are able to image dynamic processes at about video rate.
16. The Airyscan compound GaAsP detector improves resolving power and SNR.

Key Reading

Murray, J.M. (2011). 'Methods for imaging thick specimens: confocal microscopy, deconvolution, and structured illumination'. *Cold Spring Harbor Protocols 2011*/12: 1399–1437. doi: 10.1101/pdb.top066936.

- An excellent review on stratagems to avoid image blurring when collecting data sets from thick specimens. This paper is one I recommend to my students.

Chhetri, R.K. and Keller, P.J. (2016). 'Imaging far and wide'. *eLife 5*: e21072. Commentary on the Mesolens and acquiring large volume datasets.

- With several good references, this paper gives a useful insight into the future of image acquisition by light microscopy of very large 3D data sets.

References

Amos, W.B., McConnell, G., and Wilson, T. (2012). 'Confocal laser scanning microscopy'. In: Schwille, P. (ed.), *Comprehensive Biophysics*, vol. 2. *Biophysical Techniques for Characterization of Cells*, Chapter 2.2. Amsterdam: Elsevier, pp. 3–23.

Amos, W.B. and White, J.G. (2003). 'How the Confocal Laser Scanning Microscope entered biological research'. *Biology of the Cell 95*/6: 335–342.

Andrews, P.D. et al. (2002). 'To 5D and beyond: quantitative fluorescence microscopy in the postgenomic era'. *Traffic 3*: 29–36. Wiley.
De Luca, G.M.R. et al. (2013). 'Re-scan confocal microscopy: Scanning twice for better resolution'. *Biomedical Optics Express 4*: 2644–2656.
Dean, P.N. (1998). 'Confocal Microscopy: principles and practices'. *Current protocols in Cytometry* Unit 2.8: 2.8.1–2.8.12.
Fusler, J. et al. (2011). 'Types of confocal instruments: basic principles and advantages and disadvantages'. In: Price, R.L. and Gray, J.W. (eds.), *Basic Confocal Microscopy*. Springer.
Geyer, S.H. et al. (2009). 'Visualizing vertebrate embryos with episcopic 3D imaging techniques'. *Scientific World Journal 9*: 1423–1437.
Huff, J. et al. (2015). 'Application Note: The Airyscan detector from ZEISS: confocal imaging with improved signal-to-noise ratio and super-resolution'. *Nature Methods 12*: i-ii.
Inoué, S. (2006). 'Foundations of confocal scanned imaging in light microscopy'. In: Pawley, J. (ed.), *Handbook of Biological Confocal Microscopy*, Chapter 1, 3rd ed. New York: Springer, pp. 1–19.
Isenberg, G. (1997). *Modern Optics, Electronics and High Precision Techniques in Cell Biology*. Berlin: Springer.
McConnell, G. et al. (2016). 'A novel optical microscope for imaging large embryos and tissue volumes with sub-cellular resolution throughout'. *eLife 5*: e18659.
Minsky, M. (1988). 'Memoir on inventing the confocal scanning microscope'. *Scanning 10*: 128–138.
Murphy, D.B. and Davidson, M.W. (2013). *Fundamentals of Light Microscopy and Electronic Imaging*, 2nd ed. New Jersey: Wiley Blackwell.
Murray, JM. (2010). 'Confocal microscopy, deconvolution, and structured illumination methods'. In: Goldman et al. (ed.), *Live Cell Imaging*, 2nd ed. Chapter 14. Cold Spring Harbor, NY, US: CSHL Press, pp. 239–279. URL: http://cshprotocols.cshlp.org/content/2011/12/pdb.top066936.full.pdf.
Naredi-Rainer, N. et al. (2013). 'Confocal Microscopy'. In: Kubitscheck, U. (ed.), *Fluorescence Microscopy from Principles to Biological Applications*, Chapter 5. Weinheim: Wiley-VCH, pp. 175–213.
Oreopoulos, J. et al. (2014). 'Spinning-disk confocal microscopy: present technology and future trends'. In: *Quantitative Imaging in Cell Biology: Methods in Cell Biology*, Chapter 9, vol. 123. Elsevier, pp. 153–175.
Puche, A.C. (2005). 'Resolution performance in structured illumination microscopy: Comparison of Optigrid to traditional confocal microscopes'. Independent Resolution Testing, Thales Optem Inc.
Rietdorf, J. and Stelzer, E. (2006). 'Special Optical Elements'. In: Pawley, J.B. (ed.), *Handbook of Biological Confocal Microscopy*, 3rd ed. New York: Springer.
Roth, S. et al. (2013). 'Optical photon reassignment microscopy (OPRA)'. *Optical Nanoscopy 2/5*: 1–6.
Sanderson, J.B. (2008). 'The role of the centralized light microscope imaging facility in modern biology'. *Quekett Journal of Microscopy 40/8*: 715–739.
Semwogerere, D. and Weeks, E.R. (2005). *Encylopedia of Biomaterials and Biomedical Engineering*. Taylor & Francis. doi: 10.1081/E-EBBE-120024153.
Swedlow, J.R. et al. (2002). 'Measuring tubulin content in Toxoplasma gondii: a comparison of laser-scanning confocal and wide-field fluorescence microscopy'. *PNAS 99/4*: 2014–2019.
Vonesch, C. et al. (2006). 'The colored revolution of bioimaging'. *IEEE Signal Processing 20*, May 2006.
Weisshart, K. (2014). 'The basic principle of Airyscanning'. Zeiss White paper.

Further Reading

Conchello, J.A. and Lichtman, J.W. (2005). 'Optical sectioning microscopy'. *Nature Methods 2/12*: 920–931.
Heintzmann, R. (2006). 'Structured illumination methods'. In: Pawley, J.B. (ed.), *Handbook of Biological Confocal Microscopy*, Chapter 13, 3rd ed. New York: Springer, pp. 265–279.
Korobchevskaya, K. et al. (2017). Exploring the potential of Airyscan microscopy for live cell imaging. *Photonics 4*: 41. doi: 10.3390/photonics4030041.
Křížek, P. and Hagen, G.M. (2012). 'Current optical sectioning systems in florescence microscopy'. In: Méndes-Vilas, A. (ed.), *Current Contributions to Advances in Science and Technology*, Chapter 5, vol. 2. *Formatex Microscopy series No. 5*, pp. 826–832.
Lam, F. et al. (2017). Super-resolution for everybody: An image processing workflow to obtain high-resolution images with a standard confocal microscope. *Methods 115*: 17–27.
Langhorst, M.F. et al. (2009). 'Structure brings clarity: Structured illumination microscopy in cell biology'. *Biotechnology Journal 4/6*: 858–865.
Wilhelm, S. et al. (2010). 'Confocal laser scanning microscopy principles'. *ZEISS Jena*. For the latest edition, see below: bit.ly/confocal-principles.
Yang, W. and Yuste, R. (2017). In vivo imaging of neural activity. *Nature Methods 14/4*: 349–359.

Internet Resources

Confocal Principles (Zeiss publication): bit.ly/confocal-principles.
Getting the best from PMTs: http://users.ox.ac.uk/~atdgroup/technicalnotes/Getting%20the%20best%20out%20of%20photomultiplier%20detectors.pdf.
HyD PMT detectors: http://www.formatex.info/microscopy5/book/818-825.pdf.
Spinning disc microscopy: http://cshprotocols.cshlp.org/content/2010/11/pdb.top88.full and http://zeiss-campus.magnet.fsu.edu/articles/spinningdisk/introduction.html.
Resonant scanning microscopy: https://www.microscopyu.com/techniques/confocal/resonant-scanning-in-laser-confocal-microscopy.
Aperture correlation microscopy: http://zeiss-campus.magnet.fsu.edu/articles/opticalsectioning/aperturecorrelation/introduction.html.
High-resolution episcopic imaging (Indigo Scientific): http://www.indigo-scientific.co.uk/Optical-HREM-High-Resolution-Episcopic-Microscopy.
ImageJ split fields of view display plugin: https://imagej.nih.gov/ij/plugins/cairn-splitter.html.
The Mesolens: http://www.strathclydemesolab.com and https://www2.mrc-lmb.cam.ac.uk/group-leaders/emeritus/brad-amos/.

Notes

1. There are several thousand active professional microscopists who read, and contribute to, the Confocal Listserver. Amongst the inevitable chatterboxes, which any forum attracts, a few contributors offer consistently good advice: experience will tell you who these are.
2. Prior to optical sectioning of fluorescent samples, developmental biologists studying embryogenesis resorted to reconstructing analogue 2D sections into 3D models. This was useful and informative but cumbersome, time consuming and prone to distortion. Digital imaging has assisted the re-emergence of this approach called high-resolution episcopic imaging (HREM). Tissues are embedded in plastic resin containing the fluorescent dye eosin and successively sliced or milled. Successive digital images of the block surface are taken in register to create a 3D data set. See Geyer et al. (2009) for further details. This equipment is available through Indigo Scientific, whose details are given in the Internet Resources, at the end of the chapter, and in the list of suppliers in Appendix 5.
3. The point spread function has a complex shape, which at first may be difficult to visualise. In the plane of focus, the PSF is simply the two-dimensional Airy pattern – a bright circular Airy disc in the centre surrounded by a pattern of annular rings of decreasing intensity. The Gaussian plot associated with the Airy pattern is an intensity plot of the PSF in the focal plane. In three dimensions the central part of the PSF assumes the shape of a rugby football, and the concentric rings of the larger Airy pattern spread out away from the central plane as a complex set of conical structures (see Figure 17.5, and also Figure 10.15, Figure 21.2 and Figure 21.9). These side-lobes are greatly reduced in the confocal microscope and virtually non-existent in the multi-photon microscope. Each point in the image is replaced by its corresponding PSF, the intensity of each PSF being directly proportional to that of the point from which it arose.
4. Many people assume that a laser is mandatory as an illuminant for a confocal microscope. It is not. The single-beam scanning confocal fluorescence microscope in the transmitted-light configuration normally does require a laser for the reasons explained above, but Professor Tony Wilson has demonstrated a reflected-light confocal microscope, similar in design to Petráň's tandem-scanning microscope and the aperture correlation microscope, which captured an optical section by the illumination of a single candle. The author has witnessed this demonstration in the old lecture theatre of the Department of Human Anatomy, Oxford.
5. In a delightfully personal memoir, Marvin Minsky emphasises the problem of blur due to the scattering of light and describes how he first developed the principle of the confocal microscope into a working prototype. Carbon arc lamps were used before the advent of the laser; see Minsky (1988). For a fascinating report by Amos and White on how they developed the confocal microscope into a practical instrument for biological research, see Amos and White (2003). These references are given at the end of this chapter.
6. The central maximum intensity (corresponding to the Airy disc) of the PSF is reinforced, while the lower intensity maxima from the surrounding rings, the 'side-lobes' of the Airy pattern, are attenuated. Since these side-lobes give rise to much of the out-of-focus light, the confocal microscope is therefore a good optical sectioner.
7. In a line scan with averaging set to 8, if – due to Poisson noise variation – there is a single noisy pixel value of 30, and remaining pixels are valued at 0 (to make the calculation easy and obvious) in the remaining seven scans, then the average value assigned to that pixel will be ≈ 4. The noise is therefore integrated out of the image. Some people prefer line scanning to the more usually employed frame scanning, because if the noisy pixel occurs in a later frame, the average function is computed from that frame onwards, not the entire eight frames.
8. Zeiss have recently increased scan speed to a notable extent by cooling both the galvanometer mirrors and the electronic circuitry (a significant source of image noise in the single-beam laser-scanning confocal) in the LSM 800 series.
9. The earliest spinning disc designs, developed by Petráň and Hadravsky, were called 'tandem scanners' because they used opposing pinholes on two matched discs. This design has been superseded by the Yokogawa Nipkow disc design with microlenses incorporated, because it is a much more light-efficient design. Modern spinning discs have an array of 20 000 very fine (30–50 μm diameter) pinholes.
10. If the camera exposure is very short (or the rotational velocity of the disc falls, which is unlikely), streaking will occur in the image. If the alignment of the pinholes is not perfect or the timing synchronisation between the camera and the spinning disc is offset, then streaking and scan lines may degrade the image.
11. Just as a pixel (the *picture element*) is the smallest irreducible part of a digital image which represents the sampling intensity of the analogue photon stream emitted from the sample, so a voxel is a 3D volume element, which includes the step size between optical sections along the z-axis. Most voxels are anisotropic voxels because of the relatively poorer resolution along the z-axis, but a few image analysis packages require voxels having equal dimensions. These are called isotropic voxels; effectively the sample is significantly over-sampled along the z-axis to create these isotropic voxels.
12. Note this is the fixed illumination pinhole in front of the laser, not the variable aperture emission pinhole in front of the detector that controls the thickness of the optical section and the signal collected.
13. The Optical Index (see Chapter 31, Section 31.7) of a 60x/NA 1.4 plan-apochromat objective lens is about 23. Some of the best-designed objectives for confocal microscopy have an NA of 1, and an OI of 50. By comparison, the Mesolens has an OI of 117. It packs an unprecedented amount of resolved detail into the primary image plane.

18

Operating the Confocal Microscope

Outline

Understanding Light Microscopy, First Edition. Jeremy Sanderson.

Key Reading
References
Further Reading
Internet Resources

18.1 Introduction

This chapter gives guidance on how to view a specimen with a confocal microscope, set it up to collect in-focus optical sections and to acquire Z-stacks. Like driving a car, once the basic principles are understood and have been learnt, it is possible to obtain images of sufficient quality for publication. This can usually be done regardless of the design of the particular confocal microscope you have access to: it is mostly learning where specific functions are located in the software. Before using a confocal, or any other laser-illuminated microscope, it is extremely important that you are aware of the potential hazards posed by lasers.

Every confocal microscope will have the means to be used as a widefield fluorescence microscope independently of the laser illumination so that the specimen may be set up on the microscope, brought into focus and a suitable field of view chosen. Visual observation of the specimen, using the fluorescence microscope in non-confocal mode, is done using an LED, metal-halide, short-arc mercury or xenon light source. This may be obvious but is worth stating nevertheless: you *absolutely cannot* use laser light to view the specimen by eye down the microscope.

18.2 Laser Safety and Interlocks

A laser cannot be used for visual observation because our eyes are adapted to view objects in the distance. For all intents and purposes, distance viewing over several metres or more represents 'infinity' compared to the focal length of the eye, which is ≈17 mm. Any light from a source placed at infinity will impinge as parallel Huygens wavelets upon a lens; therefore, light from infinity enters the eye as parallel rays (see Figure 2.19 and Figure 8.10). All parallel rays taken up by any lens – including the eye – will always come to a focus in the back focal plane of that lens.

> At the back of each of our eyes our curved retina is situated one focal length away from the lens of the eye. Laser light is intense, highly collimated light. Therefore, if laser light enters the eye, it *will* be brought to a focus at the retina. The eye will suffer immediate and irreparable damage.

Lasers are required for confocal microscopy because they are extremely powerful light sources, which ensures that enough illumination reaches the sample to give sufficient emitted fluorescent signal to pass through the pinhole in front of the detector. The power of the focal spot is extremely high: a typical 20 mW diode laser will deliver 2 mW/cm^2 irradiance into a focused spot 1 mm^2. This is easily enough to cause retinal damage.

Most lasers used for microscopy are rated Class 3B or Class 4. That means that the normal automatic blink and aversion response, which acts over 0.25 seconds is too slow to prevent irreversible damage to the eye. All commercial microscopes sold with lasers incorporated must, by law, have suitable laser interlock systems fitted. This is why it is not possible to go directly from viewing mode to scanning mode. Electronic and/or mechanical interlocks have to be switched in order to go from one mode to another. The precise details depend on the individual microscope/software design.

In the United Kingdom, employers have a duty under the Management of Health and Safety at Work Regulations 1999 to assess the risks to which their workers are exposed and to introduce control measures where appropriate. These duties extend to the protection of workers from the adverse effects of exposure to non-ionising radiation (NIR). Specific requirements in relation to artificial optical radiation have been introduced as a result of a European Directive passed in 2006, which was implemented in the UK by the Control of Artificial Optical Radiation at Work Regulations 2010[1]. These regulations exist to protect people working with lasers and illumination sources from experiencing adverse health effects to the eyes (e.g. corneal or retinal damage, cataracts and blindness) and skin (e.g. reddening/burning, blistering and skin cancer).

The majority of artificial light sources are safe, e.g. light-emitting diodes (LEDs), enclosed tungsten-halogen lights and any Class 1 laser product such as laser printers and bar code scanners. These do not present a risk of harm to health, and there is no need to do anything further. Modern confocal microscopes must be safe by engineering design, and thus most are designated as Class 1 laser products when used in accordance with the manufacturer's instructions. Nevertheless, the presence of Class 3B or Class 4 lasers within, or associated with, the confocal microscope must be declared and appropriate risk assessments drawn up. The meaning 'artificial optical radiation' includes Class 3B and Class 4 lasers (including ultraviolet, visible and infrared lasers), in addition to a wide variety of non-laser optical radiation sources. These non-laser radiation sources in microscopy include high-pressure arc lamps emitting ultraviolet radiation to reveal fluorescent dyes. Clearly, common sense and awareness supplement health and safety legislation; you are advised to check national legislation, supporting regulations and local rules for your geographical area and workplace.

Laser protection supervisors or laser safety officers must be appointed when Class 3B or Class 4 lasers are used in the workplace. Light microscopy imaging facility staff often assume this role. They report to the local laser safety adviser, usually a specialist with overall responsibility for health and safety who can be consulted on complex laser safety management issues. Laser protection supervisors are responsible for the safe operation and maintenance of equipment having lasers.

These duties include:

- maintaining a register of all laser equipment (except for Class 1 lasers) in the workplace;
- consulting with the laser safety adviser in arranging designated areas for the safe use of Class 3B and Class 4 lasers;
- ensuring that effective safety policies and local rules are adhered to and any breaches or accidents reported; and
- ensuring that all registered laser workers receive appropriate instruction and training.

18.3 Lasers for Confocal Microscopy

Inert noble-gas lasers are historically the most commonly used laser illumination sources. These include argon ion and krypton ion, both in single-gas lasers, and helium-neon (He-Ne) mixed-gas lasers. Their advantage is that they emit more than one line, but they are bulky and have relatively short lifetimes. There is now greater emphasis to use diode lasers, which offer longer lifetimes and perform better with regard to fluctuations in their emission intensity, which has been a problem in the past with older designs of lasers. Table 18.1 lists the possible lasers available for optical sectioning microscopy.

Supercontinuum white-light lasers (see Borlinghaus and Kuschel, 2012), launched via a crystal photonic fibre, have recently been introduced to the microscopy market by Leica. Instead of being constrained by the discrete wavelengths of conventional gas-ion and semiconductor diode lasers, it is now possible to select an excitation line from 470 nm to 670 nm in 1 nm increments, allowing fluorophores to be excited at peak quantum efficiency with less power. The patented acousto-optical beam splitter (AOBS) used by Leica to select excitation laser lines in the confocal microscope is particularly suited for work with a white-light laser. Supercontinuum lasers are very powerful and so are rated as Class 4 lasers.

Whether the single-beam raster of the point-scanning confocal or multiple-beam scanning approach of the spinning disc is used, lasers are normally used for illuminating the sample. Increasing the laser power is a simple way of increasing the emitted signal, but this has two disadvantages. Firstly, the risk of fluorescence saturation is higher. Secondly, the fluorophore will bleach more rapidly. Single-beam point-scanning confocals operate with laser intensity levels close to the fluorophore saturation level. Normally, the emitted signal is proportional to the concentration of fluorescent dye in the sample, but very high fluorophore concentrations can be toxic and can also quench the emitted signal.

18.4 Operation of the Confocal Microscope

If the fluorescence is weak or difficult to see with the microscope in visible mode, consider whether the point-scanning confocal is a suitable method. Otherwise use bright-field, phase-contrast or DIC to focus an image of the specimen. The ideal fluorescent sample prepared for confocal microscopy should be very well stained, to give a good signal with corresponding low noise. In this respect it is wise to use freshly prepared antibodies conjugated to modern 'second and third generation' fluorophores (e.g. Alexa, Cy cyanine, ATTO and DyLight dyes) rather than earlier ones such as fluorescein (FITC) or rhodamine (TRITC).

The confocal microscope should first be used in widefield mode to locate an appropriate field of view and the cell(s) or tissue(s) of choice. Once the sample is in focus, the microscope is operated remotely in confocal mode from the laser-scanning software coupled into the instrument. It should not be necessary to switch the microscope back into 'visible fluorescence' mode unless the focal plane is lost whilst setting up for image acquisition or a particular specimen feature needs to be checked.

Table 18.1 Laser Lines for Fluorophore Excitation

351 & 364 nm	405 nm	458 nm	477 nm	488 nm
Alexa Fluor 350	Alexa Fluor 405	Acridine Orange + RNA	Acridine Yellow	Acridine Orange + DNA
DAPI + DNA	ATTO-425	Alexa Fluor 430	Cy 2	Alexa Fluor 488
Dylight 350	DAPI + DNA	ATTO-465	DiO	ATTO-488
eBFP-2	Dylight 405	Cerulean	(this line is useful to supplement bleaching by the 488 nm laser)	BODIPY-FL
Fura-2	eBFP-2	CyPET		Cy 2
Hoechst 33258	Formaldehyde-induced fluorescence (FIF)	eCFP		DiO
Hoechst 33342	Hoechst 33258	tagCFP		Dylight 488
	Hoechst 33342	mTFP-1		FITC
	Lucifer Yellow			FM 4-64
	tagBFP			eGFP
	T-Sapphire			SYTOX green
				mTFP-1
				tagGFP-2
				T-Sapphire
				mWasabi
				YOYO-1
				YOPRO-1

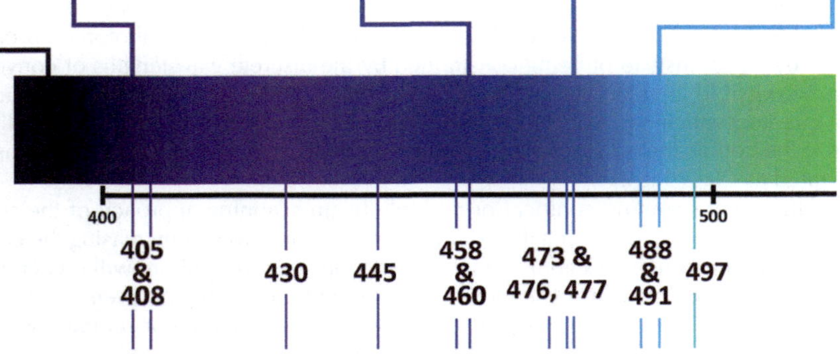

major laser lines **405** **458** **488**

Laser types

DPSS		**405**	**408**	**430**	**445**	**460**	**473**	**488**	**491**
He-Ne									
Argon						**458**	**477**	**488**	**497**
Argon-UV	**351 364**								
Krypton									

100 W Hg lamp

254, 299, 313, 334, 365 405 436

Individual He-Ne and diode lasers give single lines, but they are grouped together in their laser types: He-Ne or diode. The wavelengths of the spectral peaks of the 100 watt short-arc mercury lamp, a common widefield illumination source, are given for comparison.

Table 18.1

514 nm	543 nm	561 nm	594 nm	633 nm
Alexa Fluor 514	Alexa Fluor 546	Alexa Fluor 546	Alexa Fluor 568	Alexa Fluor 633
ATTO-520	BODIPY TMR-X	Alexa Fluor 555	ATTO-594	Alexa Fluor 647
mCitrine	Cy 3	Alexa Fluor 568	BODIPY TR-X	Alexa Fluor 660
Eosin	mKO-2	ATTO-565	mCherry	ATTO-633
Rhodamine 123	mOrange	mCherry	Cy 3.5	ATT0-635
TOTO-1	Propidium Iodide	Cy 3	Dylight 594	ATT0-647
turboYFP	TRITC	Cy 3.5	mKate-2	BODIPY TR
Venus		DiI	mNeptune	Dylight 633
eYFP-2		DsRed-2 (RFP)	mPlum	Dylight 649
YPET		DyLight 550	mRaspberry	DRAQ-5
tagYFP		Phycoerythrin	Texas Red	Cy 5
		Rhodamine		TOPRO-3
		mRuby		TOTO-3
		turboRFP		
		mStrawberry		
		mTangerine		
		td Tomato		
		TRITC		

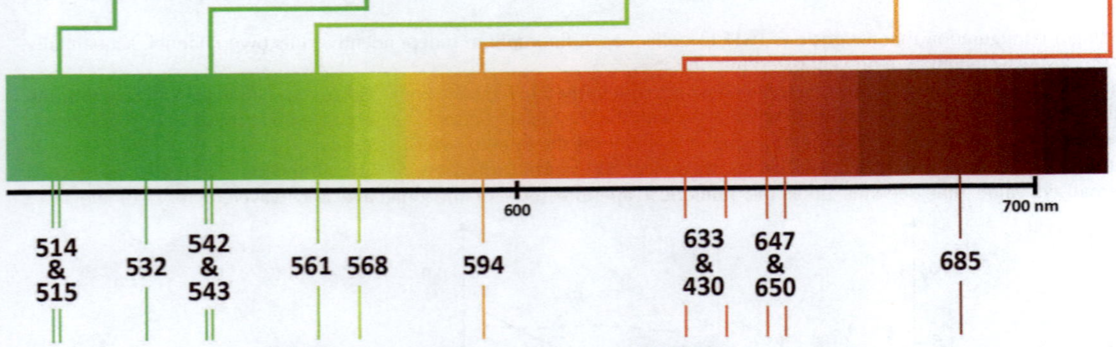

major laser lines	514				561				633				
Laser types													
DPSS	515	532	542		561			594	635	640	650	685	λ (nm)
He-Ne			543					594	633				
	514												
Krypton						568					647		
100 W Hg lamp				546			579						

Protocol 18.1 Operating the Confocal Microscope

Every point-scanning confocal microscope is operated in broadly the same fashion, differing only in minor points imposed by software design. For your particular confocal microscope, consult the manufacturer's specific operating instructions. These should have come in printed form or on a DVD with the microscope but can also be found online, using suitable terms put into a good search engine.

1. Boot up the software and switch on the microscope. It is a good idea – and essential for any form of quantitative analysis of fluorescence – to switch on the lasers at least 30 minutes beforehand in order for them to warm up sufficiently so as to put out a consistent illumination flux.
2. Switch on the microscope, and using the visible fluorescence light source, focus an image whilst the microscope is in widefield mode. Pay attention to using the correct type of objective and corresponding immersion medium so that a focused image can be secured free of spherical aberration.

   Specimens that may still be seen by widefield visible fluorescence microscopy may not be in exact focus when switching to confocal operation. That depends on (a) the eyepiece dioptre controls being set properly so that the plane of focus imaged by the objective is the same as that seen by the eye (which is capable of accommodation) and (b) the setting of the pinhole – indeed the raison d'être of the confocal – can make it difficult to secure an image, unless the specimen is reasonably in focus in the first place! Take care not to bleach the sample whilst searching for the plane of focus.
3. If you intend to overlay the fluorescence image onto a DIC image collected with the non-confocal transmission detector PMT, check at this stage (with the tungsten-halogen illumination) that the microscope is properly set up to collect a DIC image.
4. From a Smart Setup or Experiment Configuration dialogue box, select the required fluorophores, appropriate laser lines and filter combinations. Activate the acousto-optical tuneable filter (AOTF) and select the correct primary dichromatic mirror for directing the emitted signal to the PMTs, or select the acousto-optical beam splitter (AOBS). The AOTF and AOBS essentially do the same job, but an AOBS (Leica) replaces the primary dichroic.

   Select the laser power from each respective beam line to go through the AOTF/AOBS. This is usually kept to a minimum to prevent fluorophore saturation and photobleaching. Most lasers are operated at a fraction of a percent of their peak value. Avoid fluorescence saturation, and consequent bleaching, by controlling laser power carefully.
5. Select which channels are required. This may involve either setting the variable bandwidth on a slider associated with each PMT or selecting the appropriate secondary fixed filter set(s).
6. You can get a 'working image', which allows you to optimise the confocal microscope controls by activating the 'Best Focus' or 'Find' button, or similar software control. This will perform a scan, adjusting the PMT gain and focus automatically to secure an image. I prefer to focus manually beforehand and set the PMT gain to about halfway manually with the slider control to minimise potential bleaching from repeated scans by the software.
7. Bear in mind that selecting a configuration that assigns one PMT to collect each fluorophore independently in its own channel, sequentially from the longest to the shortest wavelength, will prevent bleed-through.
8. Once the correct beam path and/or filter sets for directing the emitted signal into the PMT(s) have been set up, the sample is in focus and some degree of PMT gain has been set, a single sample image may be taken with the 'Fast Scan' button or at a scan speed setting around 400 hertz. The raster scan at this speed is sufficient to allow accurate focusing in real time on the monitor whilst scanning.
9. At this point, the pinhole(s) in front of the PMT must be set to collect the correct thickness of optical section. Normally this would be 1 Airy Unit, a normalised value that depends upon the numerical aperture (NA) of the objective and wavelength (λ) of the laser

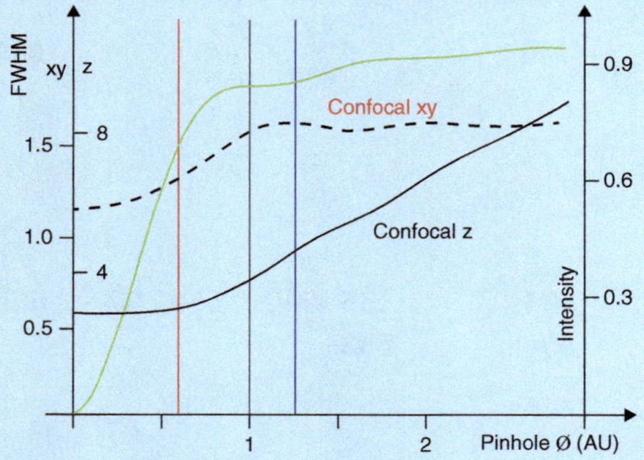

Figure 18.1 ***Pinhole graph*** *Optical performance of a confocal microscope as a function of the diameter of the pinhole (x-axis) in multiples of the diameter of the Airy disc. Green curve: Integrated intensity in the focal plane relative to an infinite pinhole (y-axis on the right). About 90% of the focal intensity is collected if the pinhole diameter is 1 AU (grey vertical), which equates to collecting the central Airy disc. At a diameter of 1.25 AU (blue vertical line) for example, the increase is merely 2%. The optical sectioning deteriorates above 1.25 AU. If the pinhole is opened above 1.35 AU, the increase in brightness is due to collecting unwanted extra-focal light – which was intended to be removed by investing in a confocal microscope!*

Dotted black line: lateral resolution (in arbitrary optical units, y-axis on the left). The lateral resolution is increased substantially when closing the pinhole to 0.6 AU (red vertical) without sacrificing too much signal. The limit is reached for a completely closed pinhole, where the theory predicts an improvement of some 30%. Solid black line: optical sectioning performance as full width half maximum of the intensity profile (y-axis on the left). Source: Adapted from Leica 2016 with permission from Leica and from Macmillan Publishers Ltd. Black curves adapted from Wilson 1990.

illumination equivalent to the diameter of the diffraction-limited Airy disc. The radius of the Airy disc (Chapter 10, Section 10.6) is equivalent to 0.61 λ/NA, and 1 AU is equal to the diameter of the Airy disc (1.22 λ/NA) in micrometres.

When the Airy unit (AU) is set equal to 1, only the central intensity maximum of the PSF reaches the PMT; this is optimum[2]. However, an optimal signal-to-noise ratio, and optimum resolution, is established when the diameter of the pinhole is not 1 AU, but 0.75 AU (75% of the diameter of the Airy disc). However, at settings below 1 AU, the intensity of the signal collected falls off significantly, and so it is normally not worth setting the pinhole at less than 1 AU unless the sample is intensely stained and very bright (see Figure 18.1).

For a given AU value, shorter-wavelength illumination will result in a thinner optical section than for longer wavelengths. This is shown in Table 18.2.

Table 18.2 Optical Section Dependence on Wavelength

The exact optical section thickness will differ slightly, even for the same objective and excitation wavelength used on different microscopes. This will depend upon the design of the microscope stand and the version of the software used. The values in the table below are taken from the software operating the two confocal microscopes in the author's imaging facility, the LSM 710 NLO on the upright Axio-Examiner Z1 and the LSM 700 on the inverted Axio-Observer Z1 microscope stand.

The values in the table are the optical section thicknesses when the pinhole is set at 1 Airy unit for each wavelength. The values given in parentheses are the optical section thicknesses for a pinhole set at 1.2 AU. At this pinhole setting, confocality is maintained, but it is possible to use the same gain setting on the PMT with up to 60% less laser power (see endnote 2 at the end of the chapter).

The optical section thickness values for the objectives are shown in the top graph for the upright LSM 710 on the Axio-Examiner Z1. The lower graph shows the optical section thickness for a series of objectives, dependent upon the pinhole setting in Airy units for the inverted LSM700 on the Axio-Observer Z1.

Objective	Excitation wavelength	Fluorophore colour	Optical section at 1 AU (µm)			
			Upright LSM 710 on Axio-Examiner Z1		Inverted LSM 710 on Axio-Observer Z1	
10x/NA 0.45 dry	λ = 405 nm	near UV-blue	4.9	(5.6)	5.3	(6.3)
	λ = 488 nm	green	5.5	(6.4)	6.4	(7.6)
	λ = 633 nm	far red	6.7	(7.7)	8.4	(9.8)
20x/NA 0.8 dry	λ = 405 nm	near UV-blue	1.5	(1.7)	1.5	(1.8)
	λ = 488 nm	green	1.6	(1.9)	1.8	(2.2)
	λ = 633 nm	far red	2.0	(2.3)	2.4	(2.7)
40x/NA 1.2 water	λ = 405 nm	near UV-blue	0.8	(1.0)	0.7	(0.8)
	λ = 488 nm	green	0.9	(1.1)	0.8	(1.0)
	λ = 633 nm	far red	1.2	(1.4)	1.1	(1.3)
63x/NA 1.4 oil	λ = 405 nm	near UV-blue	0.6	(0.7)	0.6	(0.7)
	λ = 488 nm	green	0.7	(0.8)	0.7	(0.8)
	λ = 633 nm	far red	0.9	(1.1)	0.9	(1.1)

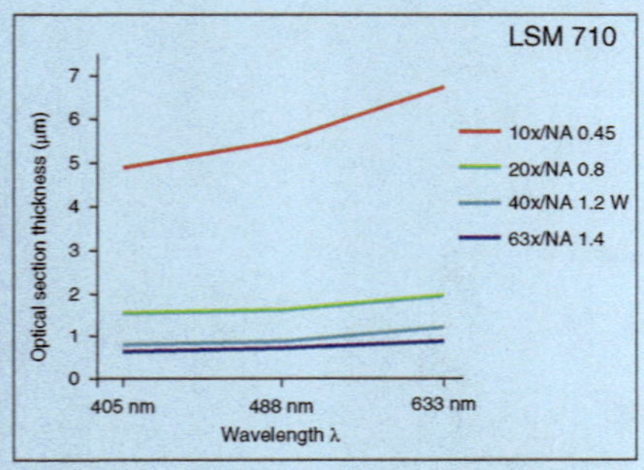

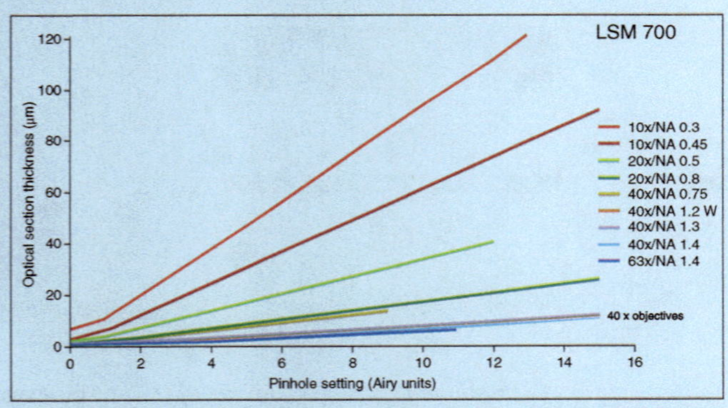

If the confocal microscope has more than one pinhole, these will have to be set at different Airy unit values to obtain the *same* optical section thickness. The same is true when using instruments having only one pinhole: it must be set differently for different channels to maintain a constant optical section when collecting the image. Generally, the pinhole diameter has to be reduced at the longer wavelengths. There may be a 'Match pinhole' or similar macro button to perform this operation. I do it manually, to ensure that this critical step has been carried out.

Clearly, if one fluorophore is much weaker in intensity or photo-bleaches more than another, it makes sense where possible, to assign the larger pinhole to the PMT collecting this weaker or more photo-labile fluorophore and adjust the others accordingly.

10. Select one PMT only (if more than one channel is being used to collect signal from a dual- or multi-labelled sample), and switch any other PMT off. Using the continual 'Fast Scan' or 'Live', scan the sample continuously. Work quickly (to avoid unnecessary bleaching) but efficiently, to adjust first the black level via the 'Offset' control on the PMT and then the signal level via the 'Gain' control. Refer to Box 18.1 for further guidance and explanation on this.

Essentially, you should optimise the offset and gain such that the PMT collects all the signal over its 8-bit depth (256 grey levels) or 12-bit depth (4096 grey levels). There should not be any oversaturation and a minimum of undersaturation (Figure 18.2). If the gain on the microscope is set too high, it is possible to see 'streaking' to one side of the image as the PMT is oversaturated with signal (Figure 18.3). The presence of undersaturated pixels in the image from too-low an offset setting is not quite as serious as oversaturation, but gross undersaturation leads to loss of information in exactly the same way. Very occasionally this can be used to advantage (Figure 18.4) where a sample has been overstained (in this case with DAPI that has leached out of the nucleus into the surrounding cytoplasm).

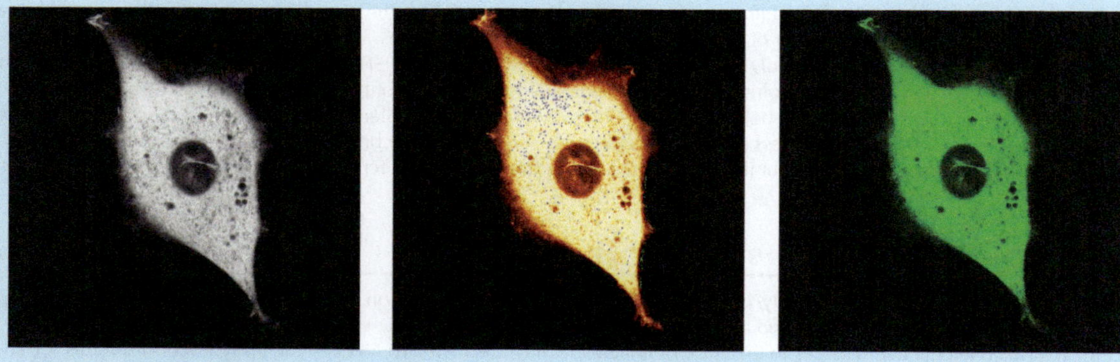

Figure 18.2 **Use of a look-up table for balancing signal dynamic range** *See the text and Box 18.1 for details of how to correctly set the signal dynamic range. The left-hand panel shows the greyscale image that the electronic detector collects; the middle panel shows the look-up table applied to the image to set the correct gain and offset values; the right-hand panel shows the pseudo-colour reapplied once the correct dynamic range has been set.*

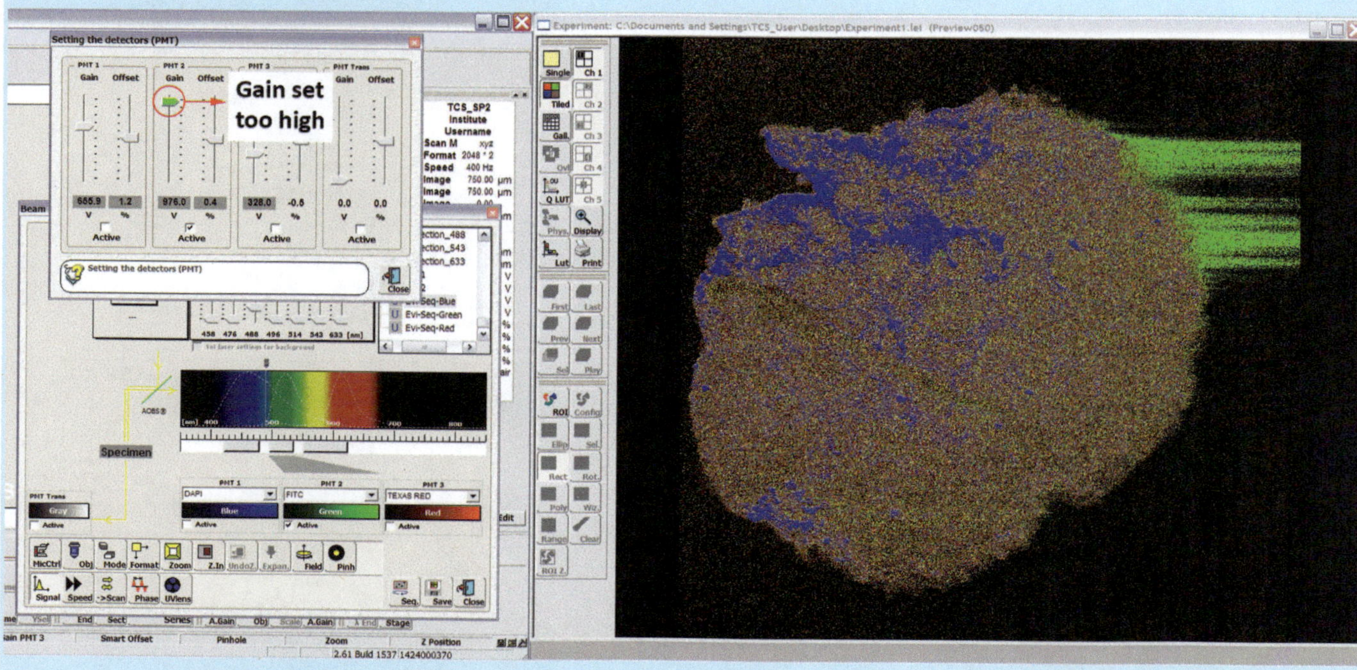

Figure 18.3 **Oversaturation of signal into the PMT** *This streaking is typical of a gain setting that is far too high. Source: Created using Leica Application Suite Advanced Fluorescence.*

11. If you cannot secure an image, do not be tempted to turn up the gain of the PMT to maximum. This probably won't ruin the PMT, but it will certainly bleach the sample, even if you can't see it. Rather, return to non-confocal mode, and locate the correct focal plane by viewing the sample down the microscope. See Box 18.2 for further advice on what to do if you cannot see an image.
12. Because it is easy to zoom using the point-scanning confocal, it is also very easy to bleach the specimen. In visible epi-fluorescence mode it is not possible to see an image that is both large enough to see the finest resolved details and bright enough to view by eye. In the confocal this *is* possible by zooming but only for a brief moment, until bleaching occurs. Therefore, be aware that this can happen in an instant. The tactics concerning zooming are discussed further in Section 18.7.
13. With all channels set with correct PMT offset and gain levels, a single image can be collected. Normally frame-by-frame collection is used, but if you are collecting moving objects, collect your image line by line to minimise any blur in the captured image due to movement of the specimen (it will move far less between lines than between each frame, i.e. 1/512th or 1/1024th of the distance if imaged frame by frame). Another reason to use line-by-line scanning is if you wish to see each fluorophore of a multicolour experiment at once.
14. Image quality can be improved by either collecting a series of scans (averaging) to improve the signal-to-noise ratio or by reducing the scan speed so that the laser dwells on each sampling point for longer, allowing more emitted signal to be collected. Both stratagems can be used together, and both should be used with care to guard against photobleaching. Tactics concerning improving image quality and signal-to-noise ratio (SNR) are discussed further in Section 18.6, (see Figure 18.12).

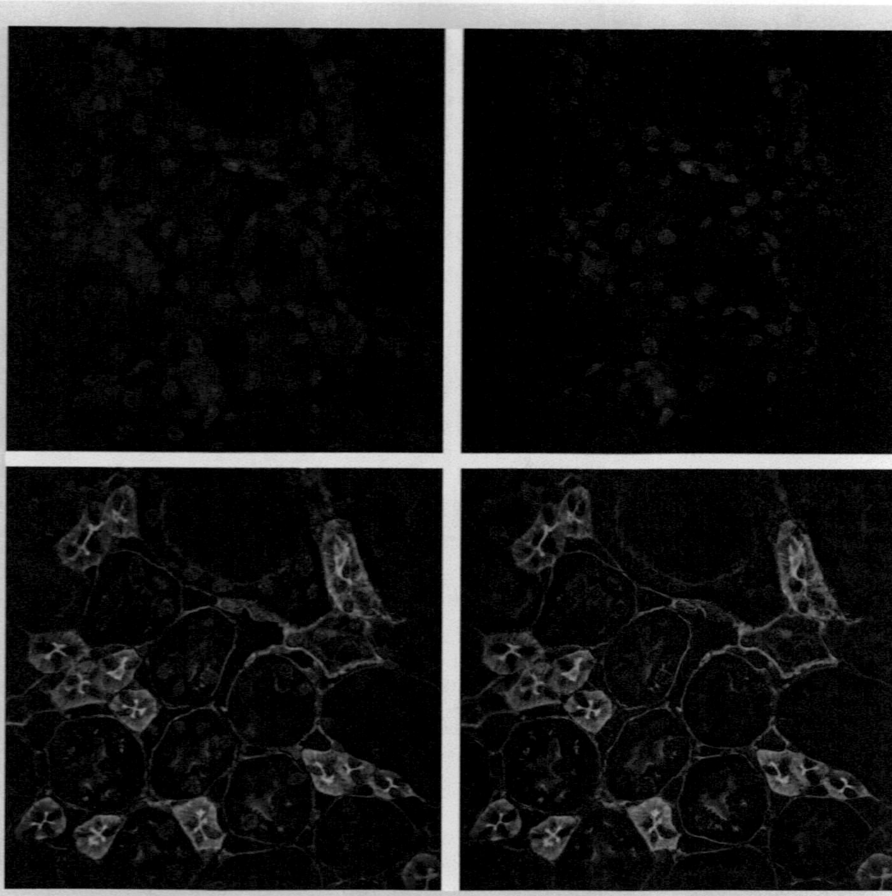

Figure 18.4 **Undersaturated signal level in the image** *Occasionally, by setting the black level via the offset control lower than normal, extraneous, non-specific signal may be deliberately excluded from the image. The DAPI has leached into the cytoplasm (because it was applied to the preparation dissolved into the mounting medium, rather than as a separate step), and by adjusting the black-level offset very low, only the nuclei are seen. This is helpful if the nuclei are only acting as morphological markers and are not the subject of experimental analysis.*

15. To collect a Z-stack, open the Z-stack dialogue box, then manually focus through the sample and mark the first and last points in the stack. In most software, it is possible to perform a rapid X-Z scan through the sample to help set these limits. If the limits of the Z-stack are set outside the surface of the specimen, you risk scanning and collecting empty optical sections. This wastes time and needlessly bleaches the sample.

    Sometimes the cells or features that require collecting as a Z-stack do not fill the entire field of view or are repeated throughout. In this case, it is more efficient in terms of sample stability/bleaching and time saved to collect a region of interest within the field of view. In some software packages, the Z-stack limits have to be set before an ROI is applied from a separate Regions menu.
16. You should also use the Nyquist calculation macro provided with the software. This calculates the sampling Z-slice (sometimes referred to as the Z-step size) as 1/2 the thickness of the optical section determined by the numerical aperture of the objective, illumination wavelength and pinhole diameter. For better results, without information loss between Z-slices, the optical section should be divided by 2.3 or 3 to get the best Z-slice setting. For further information, see Section 18.7 and Box 18.4).

    Because the axial resolution in the z-direction is poorer than laterally in x and y (Chapter 7, Section 7.4) it is not always necessary to sample the Z-stack with cubic voxels. Setting the Z-slice to the same dimension as for correct lateral Nyquist sampling in x, y will oversample axially in z.

    It is irritating that most proprietary confocal software packages do not tell you how long a Z-stack will take to collect *before* you start scanning. As explained above, it can take many minutes to collect a Z-stack, and it may be necessary to compromise on some parameters (e.g. Nyquist sampling in z by adjusting the thickness of the Z-slice) in order not to photo-bleach unduly.
17. When a very large Z-stack is being collected deep into tissue, it may be necessary either to alter the laser power, gain or offset values to maintain signal intensity when focusing through tissue that scatters both the illumination and the emitted signal. Most software has an 'Auto Z Brightness' function, or dialogue box, to perform these alterations. Similarly, there is usually also a means of correcting (to some extent) for refractive index differences where non-homogeneous immersion conditions (e.g. between the objective and the specimen) exist.
18. The Z-stack data set contains a series of in-focus optical sections. As such it can be 'collapsed' into a 2D data set for publication. There are several algorithms that will do this, but the most common is to render the image as a maximum intensity projection (sometimes referred to as an MIP). An example is shown in Figure 18.5. The brightest pixel is used for each location, regardless of which focal plane in the Z-stack it originated from. The MIP is useful where structural elements span several optical sections. These cell or tissue components (e.g. neurons) can then be seen in their entirety in 2D. The disadvantage of the maximum intensity projection is, of course, loss of spatial information along the z-axis. As such, it is not suitable to make an MIP from a data set for colocalisation studies.

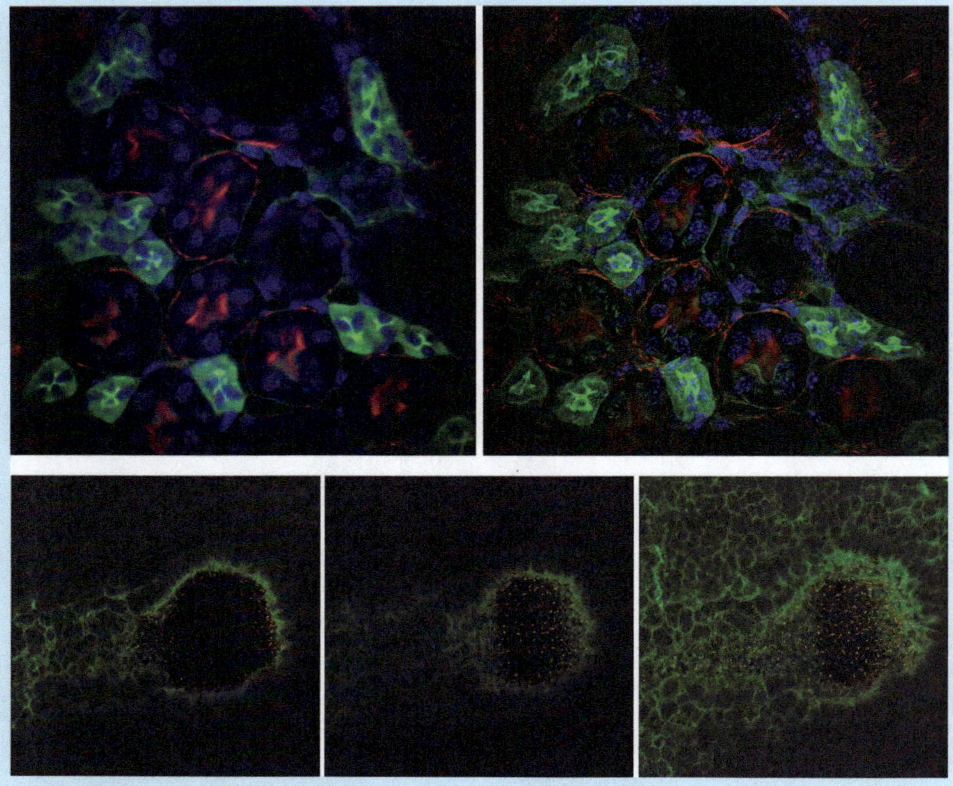

Figure 18.5 **Maximum intensity projection** *A maximum intensity projection allows the in-focus detail captured in each optical section of a three-dimensional Z-stack to be displayed simultaneously – and published – in two dimensions. It also allows, in the images of cilia in the mouse embryo node, to see the distribution of the entire cilia population in depth. Two separate optical sections (lower left and middle panels) are shown together with the MIP (right). Compare this figure with Figure 17.1. Mouse images courtesy of Dr J. Keynton.*

19. The confocal microscope can be used to collect a transmission image using phase contrast or DIC with laser illumination. This provides a useful means to overlay the fluorescent image over a greyscale image (Figure 18.6) containing morphological information about the cell or tissue. The contrast enhancement technique should be set up in the usual way using visible bright-field illumination before switching to using a laser. Since there is no need for a pinhole in front of the transmission PMT, the image is non-confocal and limited only by the depth of field of the objective. Generally only a very low laser intensity need be used: about 1–2%. If the captured image is kept darker than usual, any superimposition of a fluorescence image will not oversaturate areas within the combined image.

(a) (b)

Figure 18.6 **Combined fluorescence and phase or DIC image taken by confocal** *Combining these two methods of imaging helps put a fluorescence signal into context amongst the morphological features of the specimen, shown by phase contrast (a) or differential interference contrast (b). The underlying morphological features may not be immediately apparent in the fluorescence image, although nuclear fluorescent staining patterns tend to reveal similar tissue morphology in the image. Compare this figure to Figure 15.3, used to illustrate fluorescence beaching, which can also be seen to a lesser extent in this figure.*

20. Lastly, users sometime say that their images 'lose intensity' when viewed on another monitor. Most modern confocal microscopes are supplied with high-quality monitors featuring high dynamic ranges, which show images brightly. Different monitors will display luminance values differently. Ideally monitors should be properly calibrated, and specialist webpages (e.g. Dry Creek Photo and Norman Koren) exist to help do this (see the Internet Resources at the end of the chapter). One way to get around poor display of luminosity values is to increase the brightness and gamma of an image on the microscope monitor before exporting it as a *.tif file, which will not have the meta-data header of the original proprietary file format (this original raw data should always be saved prior to export).

The generic Protocol above is detailed and necessarily long as a result; refer to the instruction manual of your own microscope in conjunction with Box 18.3 for advice on acquiring a good image. Also refer to the summary checklist in Table 18.3 to ensure you have done everything to collect a good image.

Box 18.1 Adjusting the PMT Signal Gain

A photomultiplier tube (PMT) photo-multiplies the weak signal emitted by the fluorophore so that it can be detected. Photomultiplier tubes are extremely sensitive and very rapid detectors of light and can multiply the fluorescent signal by a factor as much as 10^8. Because they are fast, speed in detection by a PMT comes at the price of efficiency: a PMT has a QE of 12–15% at best and so only detects 1 in every 7 or 8 photons of fluorescent signal.

As the voltage gain applied to the PMT increases, so the detected signal displayed on the monitor gets brighter until it saturates, contrast is lost, and no more detail can be discriminated against the background.

There are two rules that apply to image collection with the confocal microscope above all others:

1) **Do not bleach your sample with the laser – keep the power to a minimum.**
Otherwise, there is no signal to detect in the first place!
2) **Do not oversaturate the emitted signal into the display.**
Otherwise, no detail can be seen in the image, and 3D reconstruction is impossible.

A photomultiplier will only see shades, or tones, of grey, not colour. The colour applied to the display is a false pseudo-colour to help us recognise each fluorophore. The photomultipliers can operate in either 8-bit or 12-bit depth. The bit depth defines the number of grey levels discriminated by the PMT: 8-bit depth = 2^8 = 256 grey levels from zero (black) to 255 (white). 12-bit = 2^{12} = 4096 grey levels. CCD cameras are usually 12-bit; by comparison, colour images are 24-bit depth (3 × 2^8; one 8-bit depth per red, green and blue channel).

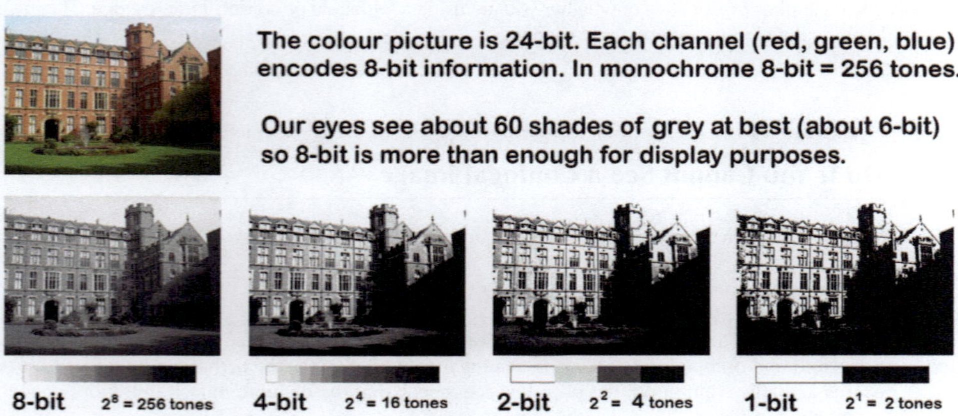

Source: Sanderson. Reproduced with permission from University of Sheffield.

Computers have on-off switches to encode information. They therefore store data in binary (base 2) form, unlike humans who have ten digits and so count in base 10. Hence the bit depth for encoding increasing tones, or shades, of grey, depends upon the exponential power of 2.

For most applications, use 8-bit detection unless you are collecting quantitative data from which you must extract numerical measurements, in which case 12-bit depth is better.

Our eyes only see information either as colour (change in wavelength) or intensity (change in amplitude of the wave). We are better at discriminating colour than tones of grey, and our eyes will therefore only see the central part of the 8-bit grey scale because we can only see about 60 different tones of grey. For the mathematical explanation, refer to Box 30.1 and endnote 5 in Chapter 30.

In order to contrast-stretch the emitted signal from the sample into the PMT, and not over- or under-saturate it, we must use a false-colour look-up table (LUT) to pseudo-colour the display (typically the top 2% and the bottom 2%) so that we can accurately set the voltage gain on the emitted signal so that it fills the entire PMT range without under-saturation and (much more important) without over-saturation.

A. **Set the black level of the PMT first.** This is controlled by the amplifier offset. (It is usually a more sensitive control than the detector, or HV, gain.)
B. Once the amplifier offset has been used to set the background at just pure black, increase the detector gain (you are increasing the voltage of the PMT by doing this) until the image *just* shows 1–2 pixels saturated.

Here is an example taken from the Zeiss LSM 710 confocal. In both the Zeiss and Olympus confocals, the LUT sets under-saturated pixels as blue and over-saturated pixels as red. In the Nikon confocal you can set the pixels to a colour of your choice, which is really good. In the Leica SP-series confocals, the under-saturated pixels are set as green and the over-saturated pixels as blue. You can usually get away with more under-saturated blue pixels than shown here (the ideal), but too many over-saturated red pixels will mean a loss of contrast, loss of fine detail and the inability to collect good 3D reconstructions.

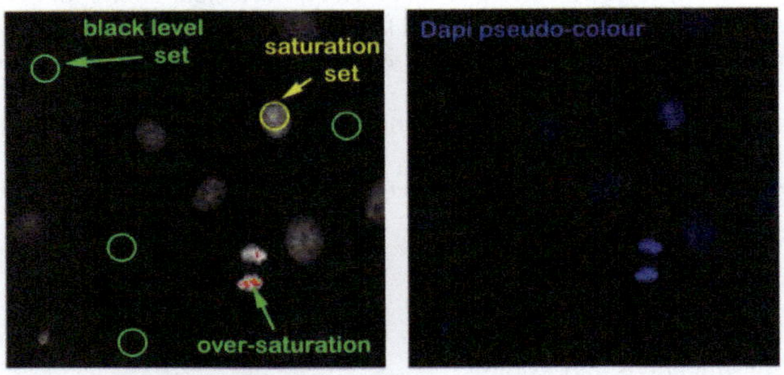

Black level (blue pixels circled in green) and saturation (red, circled in yellow) set. So as to get contrast in the Interphase nuclei, the denser M nucleus is over-exposed.

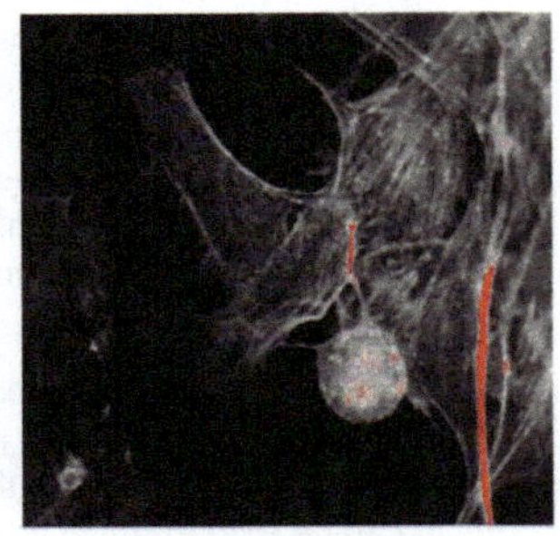

Over-saturation (red). No detail seen in this area. Black level not set to proper limit (no blue pixels).

Our eyes perceive colour with differing sensitivity, and we can see green better than red or blue. We pseudo-colour the PMT signal in order to relate quickly and efficiently to the fluorophores we are detecting. If it is important to pick up fine detail in the image, consider using white as your pseudo-colour to give the highest contrast against the black background. In addition consider inverting the image so that the background is white. See Figure 11.6. This advice applies irrespective of whether you are using a widefield or confocal microscope. The image of the karyotype in Figure 11.6 was collected on a widefield fluorescence microscope.

Box 18.2 What to Do If You Cannot See a Confocal Image

1. Stop scanning – so that you don't bleach the sample (do not panic and turn up the laser illumination intensity in order to try and search for the specimen).
2. Check that the sample is actually in focus by viewing through the microscope.
3. Check if the sample has been stained properly.
4. Check the laser interlocks are switched to scanning from observation mode.
5. Check if the stand of an inverted microscope is tilted back, activating the interlocks (i.e. check that a shutter is not blocking the beam).
6. Check that the laser beam is actually exiting from the objective (i.e. check that the dichroic mirror and fluorescence filter set is not the wrong combination blocking the beam).
7. Check that the correct laser and filter sets for imaging your fluorophore(s) are in place.
8. Check that the channel has not been switched off in the display software (easily overlooked).
9. Check that sufficient gain has been applied to form an image on the monitor.
10. Open the pinhole to allow sufficient signal through to the detector.
11. Check that no very high zoom value, or a misplaced ROI, is preventing the image being seen.
12. Check that the right dichroic filters are inserted in the light path.
13. Check that the laser beam is not being attenuated by DIC optical components.
14. Check that the offset value is not set abnormally low and so is attenuating the signal.
15. Check that the sample is not too thick or opaque that the laser cannot reach the focal plane, or any emitted signal can't reach the pinhole due to excessive loss through scattering.

Box 18.3 Acquiring a Good Confocal Image

Confocal imaging requires a trade-off between:

- resolution (image quality)
- scan time (acquisition speed)
- specimen bleaching (excitation light intensity)
- signal-to-noise ratio

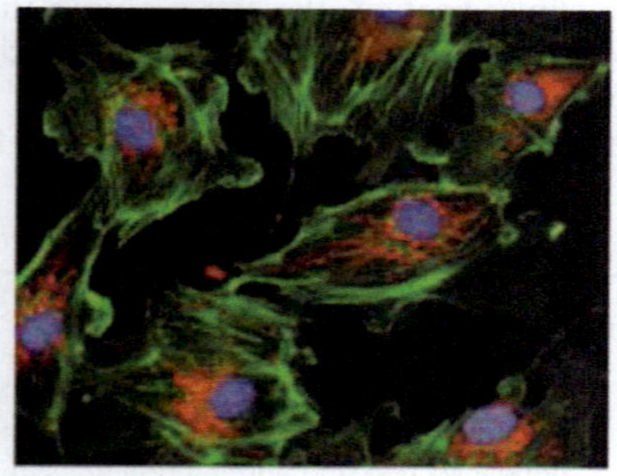

A. Don't waste photons
B. Use all the optical performance you've paid for

1. **Have a well-prepared and well-stained specimen**
For non-living tissues a key to good morphology is proper fixation. Be aware that aldehyde fixation can give rise to formaldehyde-induced (auto) fluorescence: FIF. For confocal microscopy a sample with a greater stain intensity than for a section examined by widefield microscopy helps generate a strong signal, requiring low gain settings, and a good SNR.
2. **Set the channels you require**
Check the spectral signatures of the fluorophore combinations for bleed-through. Determine whether to use simultaneous scanning (for speed) or sequential scanning (minimal bleed-through). Use line-by-line scanning to see multiple channels at once with sequential imaging 'on the fly'. Set up parameters for Z-series, time series, λ-series as required.
3. **Set the pinhole diameter for optical section thickness 'confocality' and signal intensity**
Check the noise in the image with respect to the gain setting on the PMT. This will guide you as to whether you can keep the pinhole(s) at the '1 Airy unit' setting; use this unless the signal is too weak and the signal-to-noise ratio (SNR) is poor. In this case, you will have no option other than to collect a thicker optical section for a brighter signal.
4. **Check laser saturation and bleaching**
Optimise the laser intensity. Use the highest level that won't photo-bleach unduly. Too high a laser power will bleach the fluorophore and photo-damage living cells. At fluorophore saturation, the emission signal no longer increases in proportion to the intensity of the excitation light. Any increase in laser intensity only serves to bleach the fluorophore, and the signal disappears. Using the zoom control will very rapidly bleach the sample.
5. **Optimise the PMT tube amplifier circuitry**
Set the black-level and gain settings, in that order: black-level (offset) first. It is best for the signal not to be 'clipped' by the dynamic range of the PMT. Use the false-colour look-up table (LUT) for this: our eyes will not see the full dynamic range seen by an electronic detector. For an 8-bit detector (a standard default), the lower limit will be 0 (black) and the upper limit 255 (white). Because of Poisson statistics, set the brightest pixel to be $\approx$240, rather than 255.
6. **Decide whether averaging is required to increase SNR**
Optimise the SNR by combining several scans into one image (averaging or Kalman averaging) and/or reducing the pixel dwell time of the laser to scan at a lower rate and collect more photons per pixel. You must offset both these gains against the increased fluorophore saturation and photobleaching; generally, averaging bleaches less than implementing a slower scan speed.
7. **Check that Z-stacks conform to the Nyquist criterion**
Most Z-stacks are collected with non-cubic voxels. To use cubic voxels grossly over-samples in z. Use the confocal software macro, or manually divide the optical section thickness by 2.3, to Nyquist-sample the Z-stack sufficiently. As a rule of thumb, the pixel size in z should be about three times that in x,y. Check that each channel is sampled with the same optical section thickness. This will require opening the pinhole above 1 AU for shorter-wavelength channels and closing it for longer-wavelength channels.
8. **Adjust for any depth attenuation**
The software should allow you to set the laser power to increase with depth when collecting a very thick Z-stack, to keep a constant SNR as the sample bleaches. Collect the stack from the top (coverslip side) first, to minimise bleaching.
9. **Check the zoom is set for Nyquist sampling**
As the zoom function is increased, so the lateral pixel size becomes smaller. For each objective there will be an optimum zoom setting to conform to 2.3x Nyquist for best sampling. The pixel size of PMT must be $\leq d/2.3$. So a 63x NA 1.4 needs a 90 nm pixel size. On the Leica TCS SP5, this requires a zoom of 5.2x at 512 × 512 and a zoom of 2.8x at 1024 × 1024; on the Zeiss LSM 710 at 512 × 512, a zoom of 28–31x and at 1024 × 1024 a zoom of 14–15x. You will need a higher zoom for lower-magnification objectives, because the Airy disc is magnified less by the objective onto the PMT (see Table 10.1).

When imaging living cells, be prepared to open the pinhole, rather than keep it at the 1 AU default without question. Sacrifice spatial resolution to acquire a signal at suitably low illumination levels: use a 128 × 128 or 256 × 256 raster, rather than the default 512 × 512. Do not zoom: on some microscopes it is possible to go to zoom 0.5x, which is preferable. Consider under-sampling according to Nyquist's sampling criterion, both laterally in x,y in the focal plane and also axially in z if taking a Z-stack. Also, use a high PMT gain setting, as well as a high digital gain. All these strategies will allow you to keep the laser power, and overall irradiance of the sample, low.

To collect a useful and informative confocal image, you must learn to operate the microscope with several parameters in mind. Confocal imaging requires a compromise, or trade-off, between resolution, scan time and specimen bleaching, or to put it another way: between image quality, acquisition speed and excitation light intensity. Broadly speaking, when instrument settings are adjusted, there is a choice to be made between specimen damage and image quality. The major points to consider are summarised in Box 18.3. Generally, those actions that lead to an improved signal-to-noise ratio tend to damage the specimen through bleaching, photo-toxicity or sometimes heating. The aim is always to adjust the image acquisition parameters in the confocal microscope software so that the image is a faithful representation of the specimen. A good image should fill the 8-bit or 12-bit dynamic range available. The SNR should be good, with minimal noise present, but this should not be achieved at the expense of undue bleaching or harming living cells and tissues. Finally the image should exhibit all the necessary resolved detail, with good contrast and brightness.

Table 18.3 Basic Checklist of Point-Scanning Confocal Parameters

1.	Sample preparation is paramount: well-stained and either well-fixed or else healthy cells and tissues.
2.	Set the channels you need to image, bearing in mind the wavelength of each fluorophore and whether bleed-through or cross-talk may arise.
3.	Set the pinhole – usually to (or near to) 1 Airy unit. Since the size of the Airy unit is wavelength dependent, with multicolour preparations each channel should be the same optical slice. Adjust the pinhole in each channel accordingly. A large pinhole size will create a thicker optical section and admit a brighter signal to the detector.
4.	Use the minimum laser power concomitant with adequate signal-to-noise ratio (SNR). Too much power will excite *all* the fluorophores within the focal volume, leading to excited-state saturation. This causes irreversible photobleaching.
5.	Set the PMT amplifier offset (black level) so as not to exclude signal from the recorded dynamic range (see Box 18.1).
6.	Set the PMT gain to ensure all the recorded signal fills the dynamic range of the detector whilst keeping the SNR high and the shot noise level low.
7.	Perform frame averaging to improve the SNR. If the signal is very weak, increase the laser dwell time as well as performing averaging to improve the SNR.
8.	Balance dwell time against photobleaching versus better SNR. Consider averaging before reducing the dwell time, although both can be used together to improve SNR.
9.	Check that the Nyquist sampling rate matches the resolving power of the objective. In the lateral (x,y) directions, a scan of 1024 × 1024 is more often used instead of 512 × 512.
10.	Check that the Nyquist sampling rate (especially when collecting Z-stacks) matches the resolving power of the objective. The optical section should be sampled *at least* twice. If you must undersample below the Nyquist limit, open the pinhole to better match the pixel size.
11.	Ensure any zoom factor satisfies Nyquist's sampling criterion for the lateral pixel size. As the zoom increases, pixel size gets smaller. Lower-magnification objectives, of lower NA, require a higher zoom setting than high-magnification objectives. Reduce the laser power when zooming to prevent photobleaching of a smaller area (see Box 18.4).

The higher the resolution, the more time required for the scan and the longer the fluorophore is exposed to the laser. Improving one of these variables usually requires sacrificing quality on one or both of the other variables. Just as this 'iron triangle' holds true for live-cell imaging (Chapter 28, Section 28.6), so it also holds true for confocal microscopy. For example, higher-quality images can be captured if slower acquisition speeds are used, but the likelihood of bleaching is also higher. Acquisition speeds can be faster if brighter excitation light intensities are used, but noisier images, and possibly bleached samples, may result. Bleaching can be reduced if scanning is fast and resolution is reduced, but image quality and signal-to-noise ratios may suffer. Opening the pinhole lets more emitted signal through to the detector (Figure 18.7), permitting shorter scan times, or less photobleaching during a time course, but axial resolution is adversely affected, and so on.

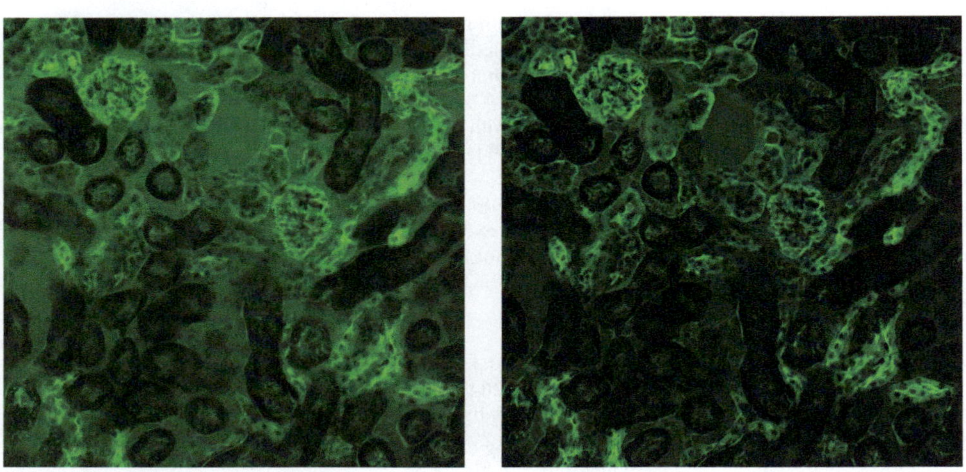

Figure 18.7 **Pinhole setting and the optical section** *With the pinhole set wide open to maximum, the thickest possible optical section (left) is collected. Although not as thick as a corresponding widefield image because of the presence of the pinhole, the blurring is clearly evident compared to an optical section collected (right) with the pinhole set at the optimum 1 Airy unit. Source: Adapted from Isenberg 1997 with permission from Springer.*

All confocal microscope software has a 'reuse' facility in one form or another. Generally, the image or the experiment where it resides is selected and the instrument settings reapplied in order to collect the new image under the same conditions: laser intensity, PMT settings and so on. The reuse button takes the settings from the metadata stored in the header of the image file. If some of the microscope components are not motorised, you will have to ensure that the software 'knows' that they have been selected if no encoder is built into the microscope. For example, you may have to manually turn the nosepiece to bring the correct objective into position if this is the component that is not motorised (e.g. on an upright confocal scanning microscope) after activating the reuse button in software.

It is possible to waste both time and expense if samples are stained with fluorophores for which there is no appropriate laser on the instrument. The absorption spectrum of DAPI, for example, ranges from about 260 to 420 nm; it is best excited at 358 nm. This is not usually a problem with widefield microscopes, for the illumination of a metal-halide lamp is usually sufficiently broad to accommodate most fluorophores, and if the appropriate filter set is not present, it can be purchased relatively cheaply. However, lasers possess a very narrow bandwidth, just 1 nm, compared to the tens of nanometres bandwidth of a widefield excitation filter. Furthermore, they are extremely expensive to add to an instrument configuration; so inappropriately stained samples have to be discarded and the experiment started afresh. With DAPI, optimum excitation is with water-cooled gas UV lasers giving lines at 351 and 361 nm. These are very expensive to run and very rarely encountered nowadays. They have been replaced with blue-violet diode lasers with a laser line at 405 nm. If this laser is not fitted to the confocal microscope, DAPI cannot be imaged, and an alternative nuclear dye (e.g. a far-red dye such as Draq-5) will have to be selected. In any case, 405 nm lies right towards the right-hand part of the DAPI absorption spectrum, so the 405 nm laser line only excites DAPI with 9% efficiency. This means that the sample must be well stained. A semi-apochromat (which contains fewer glass elements) will usually acquire a better image than a plan-apochromat objective.

There are several stratagems for testing confocal microscopes (e.g. Zucker, 2014, and Hng and Dormann, 2013) to ensure that they are working optimally and collecting high-quality images. These tests are also applicable to widefield fluorescence microscopes, and the subject of testing is very extensive. A few essential tests are described in Chapter 27, Section 27.6.

18.5 Fluorescence Intensity

Although the illumination in the typical laser scanning confocal system appears to be extremely bright, the *average* illumination at any given point on the specimen is relatively moderate, due to the fact that many points are scanned per second, and the highly sensitive photon detectors (which do not require spatial discrimination because they scan diffraction-limited points) instead respond very quickly with a high level of sensitivity to a continuous flux of varying light intensity. The majority of emitted photons are not detected by the photomultiplier tube detectors, because they perforce must operate very fast, and so PMTs have a correspondingly poor quantum efficiency – around 12–15% at best as compared to 60–80% for CCD or sCMOS detectors.

Because confocal microscopes are used with high numerical aperture lenses (to collect a bright signal and which have a small depth of field), they have an inherent speed limitation in fluorescence because the pixel dwell time[3] on each point cannot be too fast. This arises because of a limitation in the amount of light that can be obtained from the small volume of fluorophore sampled by a high numerical aperture objective. This volume is about 20–50 attolitres and yields about 3–4000 photons from about 20 fluorophore molecules (assuming a dye concentration of 1 μM) contained within the focal volume of the scanned beam (see also Box 18.5). If a fluorophore is used to dye a structure at a concentration of 1 μM, within a volume of 40 attolitres, there may only be 10–30 molecules of dye present. No fluorophore is excited 100% efficiently; the quantum yield of Alexa Fluor 488, for example, is 0.92. Assuming that all the fluorophore in the sampling volume is excited, each dye molecule will give off about 180 photons in about 1 μsec (the approximate time taken to sample one point in a 1024^2 frame raster collected over one second) – the time taken for the signal from a single voxel in the specimen to be collected. If 20 dye molecules are present, then each voxel will yield about 3600 photons. These photons will be emitted in *all* directions. The estimate of how many photons are detected via the objective and the detector may be as low as 10% of the emitted light. Hence samples must be well stained and deliver very bright fluorescence emission.

Due to the optical sectioning achieved by the confocal approach, the fluorescence in a single optical section within a thick specimen is less intense compared to the entire thickness of the sample viewed with a conventional widefield epi-fluorescence microscope. A sample viewed by confocal may require longer exposure times – and thus is at greater risk of photobleaching. Hence samples prepared for confocal microscopy normally require increased staining times or stain concentrations, and if evaluated in the conventional microscope, may appear to be over-stained. The advantage of having well-stained, intensely labelled samples means that not only can exposure times be shortened, thus reducing photobleaching, but the gain setting on the PMT can be set lower than would otherwise be the case. This reduces noise in the image and, in turn, reduces the need to average scans together – also reducing the risk of photobleaching. However, do not over-stain too intensely, since fluorophore dyes will self-quench if too high a concentration forces fluorophore molecules too close to one another.

Images from a somewhat slightly deeper part of a specimen can be obtained by using dyes that are excited at longer wavelengths (such as Cy5), as opposed to those that require shorter wavelength excitation (e.g. Alexa 405). The use of longer-wavelength illumination will, however, slightly reduce the maximum resolution that can be achieved in comparison to images acquired at shorter wavelengths. This stratagem works because longer-wavelength red light is scattered less by cells and tissues, and is employed in multi-photon microscopy (Chapter 22, Section 22.2). It is a good stratagem to consider, particularly if the specimen need not be viewed at the very limit of resolution necessitating short-wavelength light. Not only will the emitted signal be scattered less, autofluorescence in the specimen is very much reduced. This is particularly important is samples have been fixed with aldehydes, such as formalin and paraformaldehyde, which can give rise to formaldehyde-induced fluorescence if this is not quenched during the specimen preparation stage.

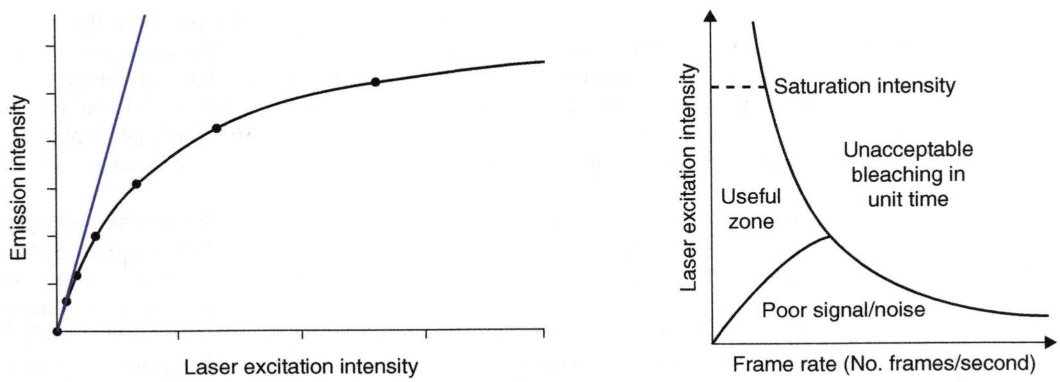

Figure 18.8 **Fluorescence saturation** *The left-hand panel shows how the intensity of the emission signal falls off with increasing laser power. In the widefield microscope the intensity of the signal is linearly proportional to the radiant intensity from the lamp. In the confocal microscope, it is all too easy to deliver too much power to the specimen. In this case, the fluorophore saturates: the signal is no longer linear with excitation intensity. Start with lower laser power values, and increase to the minimum required.*

The right-hand panel shows the same information plotted in a different manner. Laser excitation intensity is plotted against frame rate. The area to the right of the 'fatal hyperbola' causes unacceptable bleaching. Absolute units are not given because they vary according to the nature of the specimen. The curve is hyperbolic because the amount of bleaching is the product of the laser intensity and the frame rate. If laser intensity is increased to provide better SNR with increasing frame rate, take care the sample does not bleach. Source: Amos 2000, Figure 14. Reproduced with permission from Oxford University Press.

The speed of the point-scanning confocal microscope is fundamentally limited by fluorescence saturation, as illustrated schematically in Figure 18.8. When this occurs, all the fluorophores illuminated by the laser are excited to a higher-energy state – even by a short pixel dwell time and brief exposure to the beam – because the laser is so intense. Thus the number of photons emitted by the fluorophore falls off, because as soon as the fluorophore returns to the ground state, S_0, it is re-excited. This is ground-state depletion (this dark-state *is* useful in super-resolution microscopy but *not* in confocal imaging). The emission signal no longer increases in proportion to the intensity of the excitation light. Any increase in laser intensity only serves to bleach the fluorophore, and the signal disappears. Thus to go faster and collect a higher frame rate, it is better to either use a resonance scanner or spinning disc confocal, rather than to attempt to increase the signal reaching the PMT by increasing the laser power, which already operates near fluorescence saturation anyway. This is the fundamental limit to temporal resolution and why frames rates are so low in the single-beam point-scanning confocal. It is a mistake to assume that increasing the laser intensity will increase the signal intensity. In fact, only a fraction of the available laser power is generally needed to maximally excite fluorophores in confocal microscopy. Any superfluous laser light will lead to quenching of the fluorophores, more rapid photobleaching and heating of the cells and tissues (disrupting both morphology and – if living – physiological function). You should be aware, also, that using the zoom control will very rapidly bleach the sample. Conventional mercury lamps generally do not saturate the fluorescent signal.

18.6 Signal-to-Noise Trade-Offs

The number of photons emitted as signal from a fluorophore is finite. This is because fluorophores are, ultimately, destroyed by free radicals generated as side products of the cyclical process of fluorescence. A typical fluorophore emits 3–4 × 10^4 photons before photobleaching irreversibly. Of these only a maximum of approximately 20% will be emitted in the right direction to be captured by the angular aperture of the objective. Further losses occur in transit through the optical components in the microscope, to just a few percent: about 800 photons per fluorophore (see Section 18.5). Although only a few tens of photons are required for detection – about 500 photons (Amos et al., 2012) are required to resolve a closely apposed pair of fluorophores according to the Rayleigh criterion – because of the limitations imposed by various sources of noise on confocal imaging, it is unlikely that single molecules (which can be detected with suitable widefield fluorescence microscopes) will ever be *resolved* with a confocal microscope.

When using the confocal microscope for live-cell imaging, cells are prone to the effects of both photobleaching and photo-toxicity. Therefore, use the minimum possible laser illumination; be prepared to open the pinhole rather than keep it at the 1 AU default without question and be prepared also to sacrifice spatial resolution for acquiring a signal at suitably low illumination levels. Thus use a 128 × 128 or 256 × 256 raster, rather than the default 512 × 512 or a higher raster. Do not use a zoom above 1.0x, because this will bleach significantly; on some microscopes it is possible to zoom out to 0.7x or 0.5x, which is preferable. Consider under-sampling according to Nyquist's sampling criterion, both laterally in x,y in the focal plane, and also axially in z, if taking a Z-stack. Also, use a high-gain setting; all these strategies will allow you to keep the laser irradiation, and overall irradiance of the sample, low.

If line scanning is selected rather than frame scanning, then any living or moving samples will be captured with less blurring, since each line is captured in, say 1/512th the time of each frame. Normally the scan is only made on the outward sweep, with the beam flying back without imaging on the return. To speed up the rate of image capture by a factor of 2x, which may be an important consideration

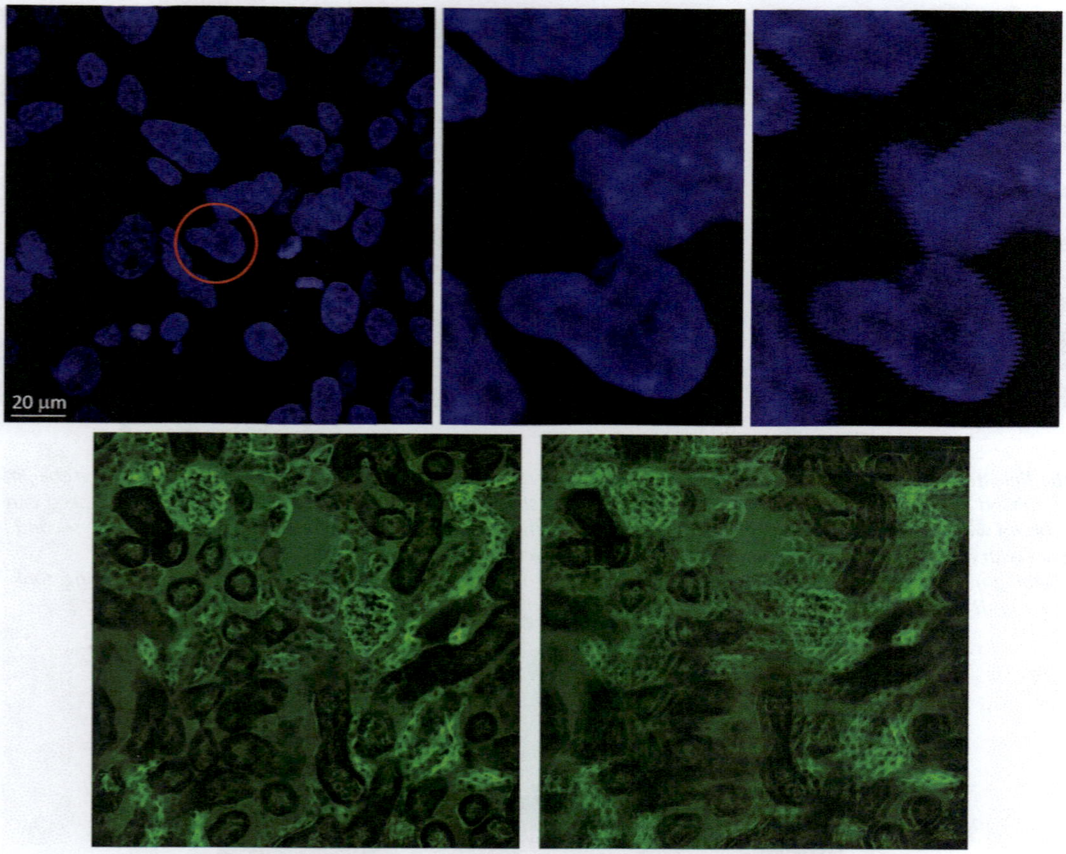

Figure 18.9 **Bidirectional scanning and phase shift distortion** *Bidirectional scanning is useful when imaging living specimens or those that are moving. However, the phase of the forward and reverse scan must be equal. If they are not, image shift will lead to blurring. This is especially noticeable if the X-galvanometer is not synchronised. The central figure of the upper row of three figures of nuclei shows a 2% shift in the X-galvo scans – present but hardly noticeable. This central figure also shows a 50% shift in the Y-galvo, which likewise gives an apparently normal image. The right-hand figure shows a 20% lag in phase between the forward and reverse scan. The lower figures of the kidney section likewise show a 2% phase lag versus a 20% lag. The phase of the raster scan can be adjusted and corrected with sliders in the software. For comparison the effects of vibration on the image can be seen in Figure 18.11.*

for photobleaching and living samples, bi-directional scanning can be engaged to scan and collect an image on the return sweep also. This will halve the time taken to collect an image but may introduce some phase shift distortion (Figure 18.9) that must be corrected (which is why it is seldom used except for living, moving samples or those with potential to move).

It is also possible to scan fewer lines in the raster (Figure 18.10) using the 'Line Step' function, or equivalent control. Usually the value is left at 1 – the default figure – so that every line in the raster is scanned. If '2' is input into this feature, then every other line will be scanned with a consequent time saving. As higher values are used, and the lines scanned become fewer, so the image will become blurred. Again, when required, this is a choice based on compromise: balancing speed of image acquisition versus image quality and the intended use for that image. Both phase shift distortion and blurring from omitting lines from the raster scan show effects very similar to vibration (Figure 18.11) showing in the image. All these effects – besides that of a dirty objective (e.g. oil on the front surface of a dry objective) – must be borne in mind when diagnosing the cause of image unsharpness.

Because the X-galvanometer mirror – which sweeps the laser beam horizontally – operates faster than the Y-galvanometer mirror, there is a speed advantage in setting an ROI for image capture or bleaching in a landscape 'letter-box' format, with the long axis of the rectangle horizontal (e.g. refer to Figure 24.2 in Chapter 24 on FRAP/FRET), rather than in the portrait format with the long axis vertical. You should consider setting an ROI whenever possible for image collection, particularly when collecting Z-stacks, so as to reduce both bleaching and the duration of scanning.

During normal operation of beam-scanning confocal microscopes, the beam dwells on each pixel for 1–20 μsec. For example, if the default 512 × 512 raster is made over a field of view which is collected in 1 second, each of the 262 144 individual points must be sampled in 3.8 μsec; for a 1024 × 1024 frame, the dwell time is 0.95 μsec. On average, in 4 μsec the laser intensity fluctuates by 2–3%, but fluctuations up to 10–15% are observed with shorter sub-microsecond dwell times. This introduces significant noise, which may drown out weakly fluorescent structures; it is not due to poor resolving power.

When a single integrated image is taken from a sequence of images averaged over a much longer time (msec to sec), the amplitude of the fluctuation is small, in the range of 0.5–1%. However, on this microsecond and sub-microsecond timescale, fluctuation of the laser

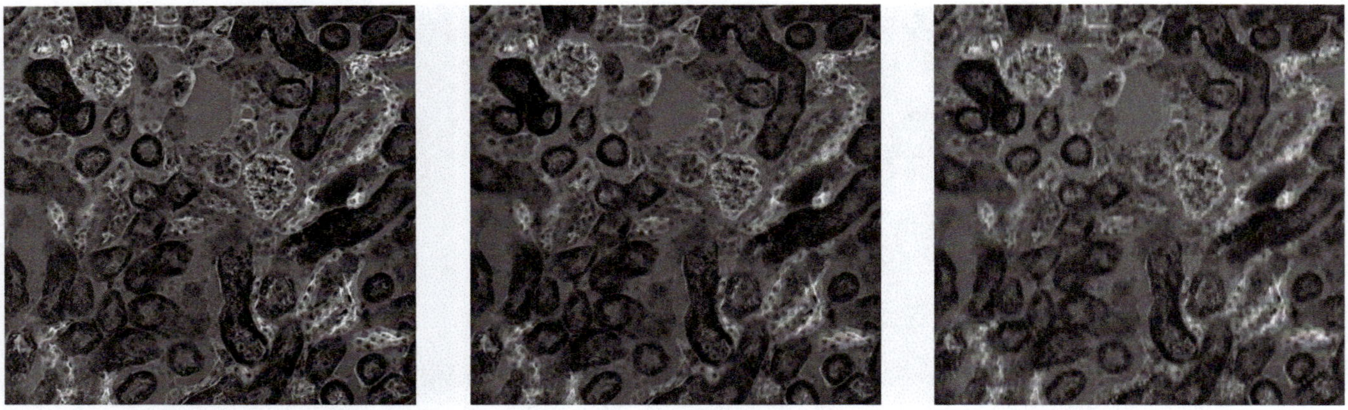

(a) (b) (c)

Figure 18.10 **The line step function** *The line scan control is, in general, rarely used. Its overuse results in blurred and unusable images. However, if every other line is 'missed out', and the Y-galvo drops every second line to start the new raster, an acceptable advantage in speed can be achieved. It is worth using this control with bidirectional scanning. The figure to the left shows a normal scan, the central panel with every second line recorded and the right-hand figure with every fourth line recorded. Here the effect is noticeable.*

The full image, with a maximum of 16 average scans and a line scan of 1 (i.e. no lines missed) and unidirectional scanning, took 30.86 seconds to acquire. With every fourth line recorded, the scan took 7.71 seconds.

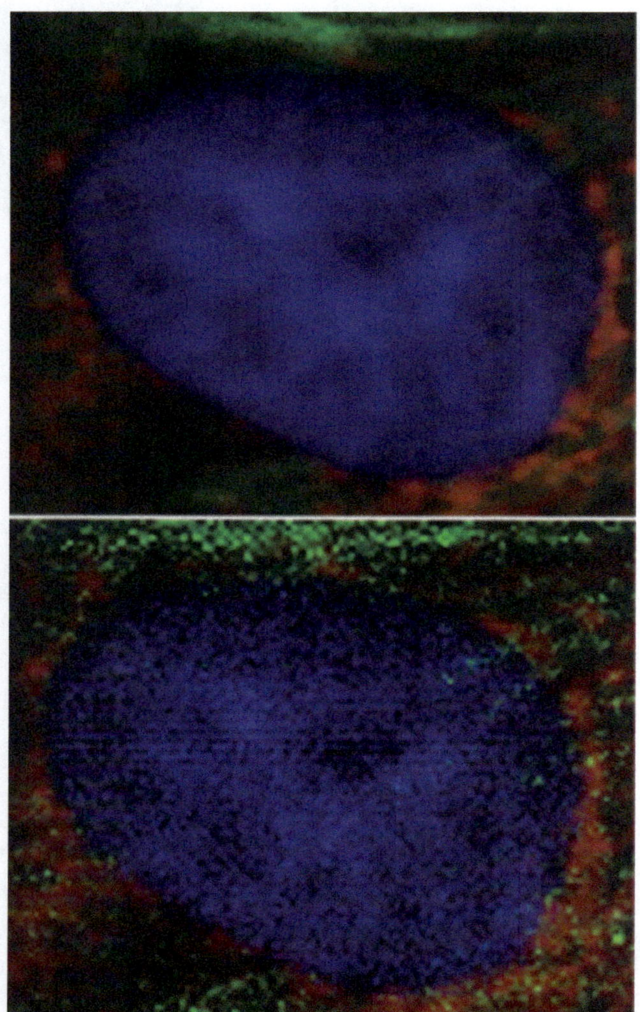

Figure 18.11 **The effect of vibration on the image**

illumination intensity within the fibre optic coupling of laser to microscope and noise introduced by the electronic circuitry controlling the confocal microscope may become significant.

Because each sampling point in the image is only scanned for a very short time and because the quantum efficiency of the PMT detector is low, only a small number of photons of emitted signal are collected from each pixel. Poisson-distributed shot noise[4] dominates, and this noise, as well as the low detector QE, contributes to the graininess in the image. Increasing the laser power will only serve to saturate the fluorophore and ultimately bleach the signal. The answer is to collect several scans, sum the values and from this sequence of scans produce a single integrated image. The intensity value at each pixel, therefore, is based on the arithmetic mean calculated from the individual pixel values of each scanned image. This is known colloquially as 'averaging'; the overall effect is to reduce the noise of the signal, increasing the signal-to-noise ratio. Please note: these averaging algorithms do not actually *remove* the noise but collect more photons to contribute to each sampling point that form pixels in the image. Averaging preserves those persistent pixel values that constitute specific signal while those fluctuated values that are mostly noise are divided away. The overall statistical variations (see Table 18.4) between each of the many thousands of pixels that make up the image is reduced.

With line scanning the averaging is performed upon each line, with repeated imaging and then averaging of the result before the laser drops to the next line. Line averaging is slightly faster than frame averaging and is good for live-cell work: because of the much poorer temporal resolution in acquiring an image, during averaging any cell movement is far more obvious if it occurs between successive frames than between lines. The crucial disadvantage of line scanning is that you cannot tell if two, four or more frames will improve the image. To take advantage of line-scan averaging, you must first try frame scanning.

With a simple averaging algorithm, the SNR improves in direct proportion to the square root of the number of frames that are averaged. The improvement is $1^1/_2$ times to fourfold, depending upon the number of scans. The effect is asymptotic: you achieve a greater noise reduction from using a few scans, and increasing the number of scan does not give a correspondingly linear improvement. Therefore aim to use between 4 and 8 average scans, although the software between manufacturers differs slightly. Another commonly used related algorithm is the Kalman average filter. Provided that the signal is sufficiently strong and the number of photons per pixel is >20 (it normally is), then Kalman averaging will achieve the same result with fewer scans, typically less than five. With Kalman averaging, unlike the simple averaging algorithm, the more-recently collected frames (which are assumed to have less noise) contribute significantly more to the final image than those frames collected earlier in the averaging sequence.

Ways to increase the signal-to-noise ratio:

- open the pinhole;
- increase the laser power;
- decrease the scan rate to increase the dwell time per pixel;
- increase the zoom factor;
- apply averaging by collecting multiple images;
- increase the PMT gain and apply averaging;
- decrease the PMT offset slightly; and
- ensure the correct filter sets are selected and/or the AOTF is optimally adjusted.

Frame *accumulation*, or summing, also takes a specified number of scan and adds up the pixel values, using the sum of the intensities at each pixel in the final image, without calculating a mean value. Scanning using the summing mode doubles the image brightness after two scans (from 1–2), then increases that by a half on the next (from 2–3) and by a third on the next (from 3–4) and so on. As such, this means of scanning multiple images is used to intensify a dim image without increasing the PMT gain or laser power unnecessarily. Although the brightness of the image increases by a set proportion with each scan, random noise will occur in a different place at each scan, so it should not cause an increase in brightness to the same extent as the (static) signal. Noise will remain and may still detract from high-image quality. This multiple-scanning procedure to increase a signal is not often used.

Some of these steps (e.g. increasing the laser power, decreasing the scan rate, increasing the zoom factor and increasing the PMT gain, as well as excessive averaging) will serve to bleach the sample and so should be used judiciously, particularly with photo-labile and/or living samples. Figure 18.12 shows the effect on the signal-to-noise ratio of applying averaging and/or reducing the scan speed by increasing the dwell time of the raster scan[5].

18.7 Nyquist Sampling

The default raster scan is 512 × 512 pixels on most confocal microscopes. Usually this is insufficient to give adequate resolution, and most routine scanning will be done at 1024 × 1024 pixels per frame in the x,y directions. However, do bear in mind that with increased image resolution from smaller pixels, there will be a concomitant increase in noise and a reduction in the SNR value, because fewer photons are collected from a smaller area, and photon intensities naturally fluctuate. This is shown in Figure 18.13. If the dwell time is increased or averaging applied, or the intensity of the laser is increased (all stratagems to collect more photons) then the amount of photobleaching and/or photo-toxicity (i.e. specimen damage) per pixel will increase.

Table 18.4 *Poisson Statistics Underlie Intrinsic Shot Noise*

Photons collected	0	5	15	20	200	400	600	800	1000	2000	4000	6000	8000	10 000
Square root	0	2.2	3.9	4.5	14.1	20	24.5	28.3	31.6	44.7	63.2	77.5	89.4	100
Shot noise values (range)	0	3–7	11–19	15–25	186–214	380–420	575–625	772–828	968–1032	1955–2045	3937–4063	5923–6078	7911–8090	9900–10 100
No. of possible noise photons in the range	0	5	9	11	29	41	51	57	65	91	127	156	180	201
Noise as percentage of signal	0	44	26	22.5	7.1	5	4.1	3.5	3.2	2.2	1.6	1.3	1.1	1

As quantum particles, photon wave packets are discrete entities. The camera can never collect fractions of photons. The number of photons falling upon each individual point in the detector is sampled to form a pixel of the intensity of light at that point in the image. The Poisson distribution describes the probability of a number of independent events occurring in a fixed time – in this case photons falling upon each point or light-sensitive element in the detector. Shot noise is therefore expressed as: **Shot noise** = $\sqrt{\textbf{signal}}$

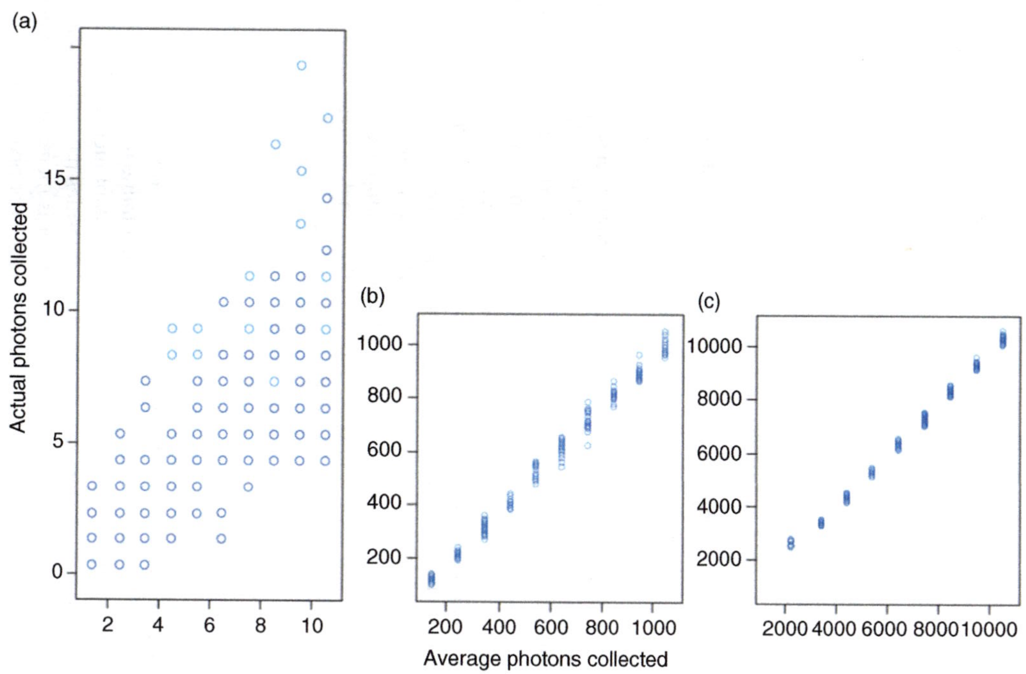

If the emitted light signal is so small such that the potential variation in intensity at an individual pixel is large (this may be demonstrated in practice by a significant variation in intensity between one pixel and its neighbour where the intensity of the two adjacent pixels should be the same), the image will be noisy due to shot – or photon – noise, and the image will appear grainy. The table shows how increasing the number of photons detected will reduce the stochastic fluctuation. Despite the increase in the number of photons that represent noise in the image, this range of values represents only a small fraction of the overall signal. This is shown in the accompanying figure for the three signal ranges shown in the table. (Source: Hammamatsu. Reproduced with permission from Hammamatsu.)

(a) At low light levels, the variance introduced by photon shot noise contributes to a large proportion of the signal. (b,c) At higher light ranges, the variance introduced by photon shot noise is a smaller percentage of the overall signal

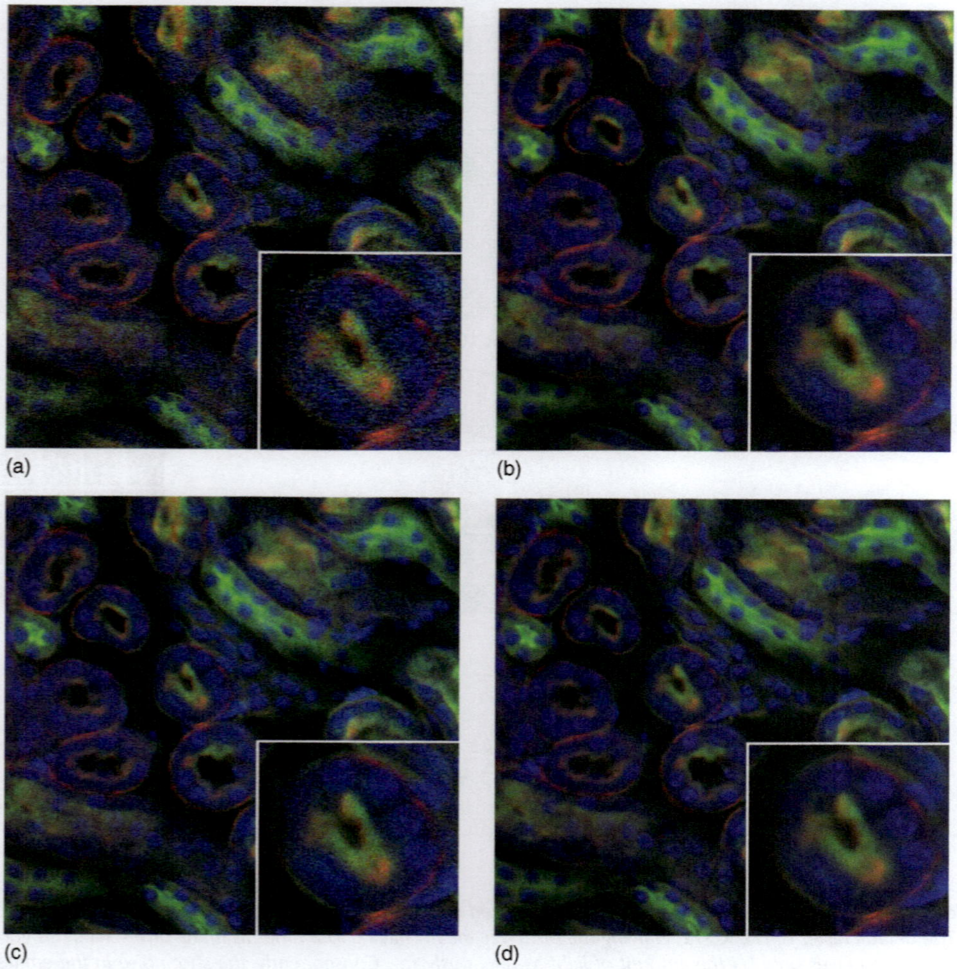

Figure 18.12 **Average scanning to improve SNR** *All images were taken at 512 × 512 pixels per frame with a Zeiss 40x/NA 0.75 EC plan-Neofluar semi-apochromat objective. The specimen was a 16 μm thick triple-stained kidney cryosection, as sold by Thermo Fisher Scientific (F24630). The pinhole was set at 1 AU: optical slice = 1.6 μm at 488 nm wavelength. The laser value was set at 2% output for acquiring all four images.*

Figure (a) *(top left panel) single frame collected in 906 milliseconds. Pixel dwell time = 1.61 μs, laser speed rating = 9 (fast). Mean averaging value set to 1 (no averaging). The image is noisy, as seen in the insert showing the proximal tubule in the centre of the image, magnified for emphasis.*

Figure (b) *(top right panel) single frame collected in 7.3 seconds. Pixel dwell time = 1.61 μs, laser speed rating = 9 (fast). Mean averaging value set to 8 (out of a possible maximum value of 16). Notice that the SNR has improved in this integrated image of the average of 8 scans, as compared to Figure (a).*

Figure (c) *(bottom left panel) single frame collected in 7.2 seconds. Pixel dwell time = 12.8 μs, laser speed rating = 5 (slow). Mean averaging value set to 1 (no averaging). The SNR improvement is due solely to an increase in the pixel dwell time; the laser excites more fluorophore per unit time.*

Figure (d) *(bottom right panel) single frame collected in 63 seconds. Pixel dwell time = 12.8 μs, laser speed rating = 5 (slow). Mean averaging value set to 8. There is a risk of photobleaching the sample due to the time taken to collect this image, but the SNR improvement over Figure (a) is the best of all. Generally applying averaging, rather than reducing speed and increasing the dwell time, gives a better SNR with less bleaching.*

This sample is well stained, the laser power is kept low to try to avoid fluorescence saturation: it is possible in this case to combine an increased dwell time and frame averaging together to improve SNR significantly. Where the time available for image acquisition is a limiting factor or there is a risk of photobleaching or photo-toxicity of the specimen, then applying some average scanning at a high laser speed and the shortest possible dwell time (Figure (b)) is recommended.

For most applications, mean average scanning gives the best results. Kalman scanning can be applied together with an increase in pixel dwell time (i.e. reduction in scan speed) to collect more signal. However, slower scans will bleach to a greater extent. Combining multiple repeat scans to average out the noise (i.e. as in (b)) will also bleach the sample, but it is generally the least inconvenient option and bleaches to a lesser extent than increasing the dwell time.

Ideally, the sample should be adequately sampled according to Nyquist's criterion (Chapter 31, Section 31.6), particularly if three dimensional Z-stacks are being collected (for an online 'Nyquist calculator', see the Internet Resources given at the end of the chapter). Some software automatically assigns a pixel value based on the resolving power of the objective and the Nyquist sampling requirement.

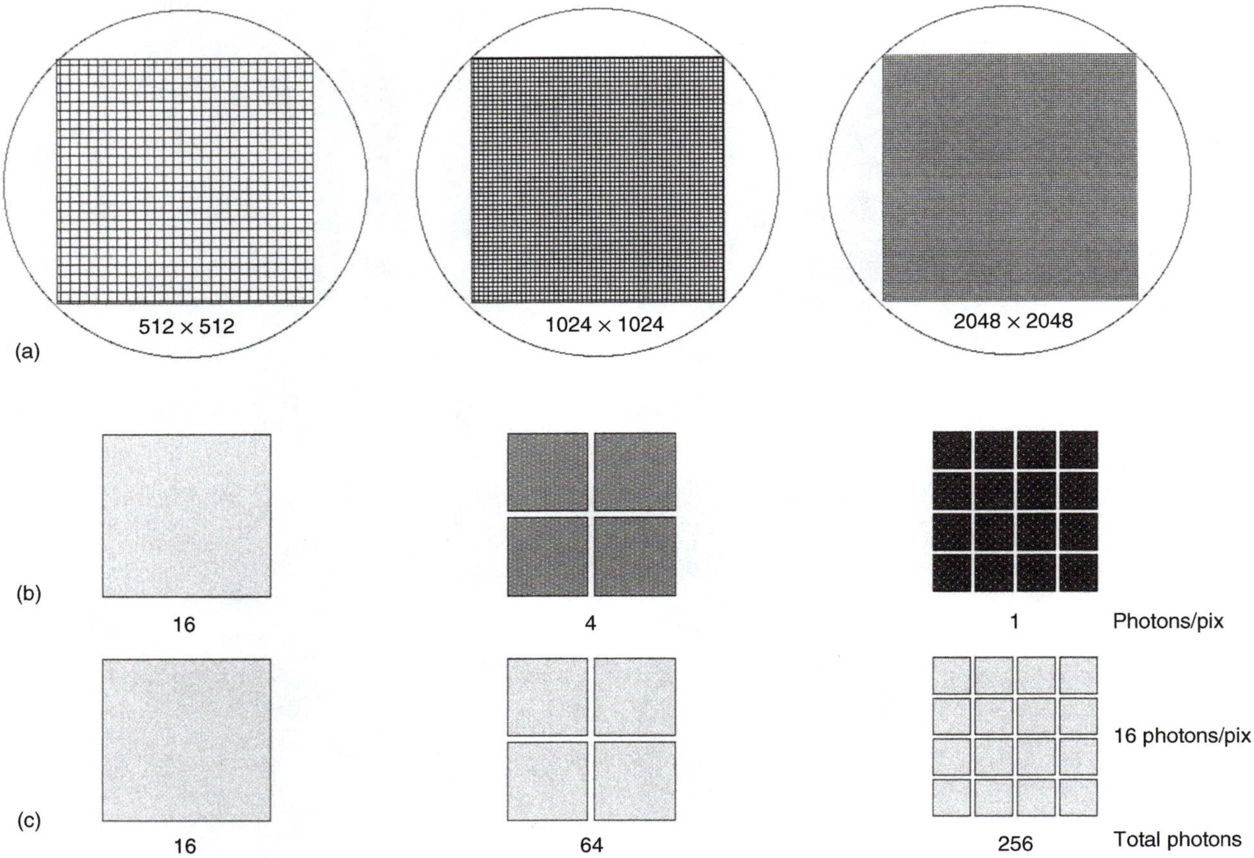

Figure 18.13 ***The relationship between pixel size and SNR*** *There is a fixed relationship between the raster scan size, pixel size and the radiant energy density, or 'light dose'. The default raster is 512 × 512. Halving the pixel size means having four times as many pixels, which is good for spatial resolution, but each pixel is small and so receives only 1/4 the amount of light as previously (a). Consequently, the shot noise in the emitted signal that forms the image will increase (b). If the laser beam scans more slowly or the laser power is increased, to ensure the same amount of signal is collected from each pixel (c) the photobleaching and/or photo-toxicity will increase. Source: Adapted from Pawley 2006 with permission from Springer.*

All manufacturers include a macro, or macros (one for x,y, laterally, e.g. an optimum button, and a separate button for z) in their software (Figure 18.14) to calculate the correct sampling distance, particularly in the axial (z) dimension. In most cases this will give the minimum sampling distance of 2x (i.e. Nyquist's criterion), whereas it might be better to sample at 2.3x or 3x Nyquist. However, in many cases adherence even to the 2x Nyquist limit is over-optimistic, simply because of the relatively slow scan speed of the point-scanning confocal microscope. It may take an hour or more to faithfully sample a triple-stained sample and average at moderate or slow dwell times in order to keep the SNR high. Almost inevitably, the sample will be photo-bleached by the time the Z-stack is completed, although most software will allow you to automatically adjust the gain on the PMT, or the laser power, in order to compensate for the reduced signal as the laser scans through the thickness of the sample, bleaching as it goes. Often a compromise has to be made, and the specimen will be under-sampled in z whilst collecting a Z-stack. In such cases, a Z-slice spacing of 1 μm should be considered.

There are good theoretical reasons why confocal data sets should be deconvolved, and these are explained in Chapter 21 (Section 21.8). However, in practice, confocal data tends not to be deconvolved because it has already been 'sampled' with the pinhole to produce an optical section. Jim Pawley (2006) is firm on this point:

> 'Poisson noise [shot noise; photon noise] can introduce features into the data that appear to be about 1 pixel in size when the smallest Nyquist-sampled feature should be greater than 5 pixels across. These [spurious] noise features can and should be eliminated by deconvolving the data, preferably in 3D.'

If the pinhole is opened much wider than 1 AU, then it becomes more imperative to sample adequately according to Nyquist's criterion (e.g. up to four times Nyquist) and to deconvolve the data set after it has been collected. This is because, effectively, you are sampling a larger voxel.

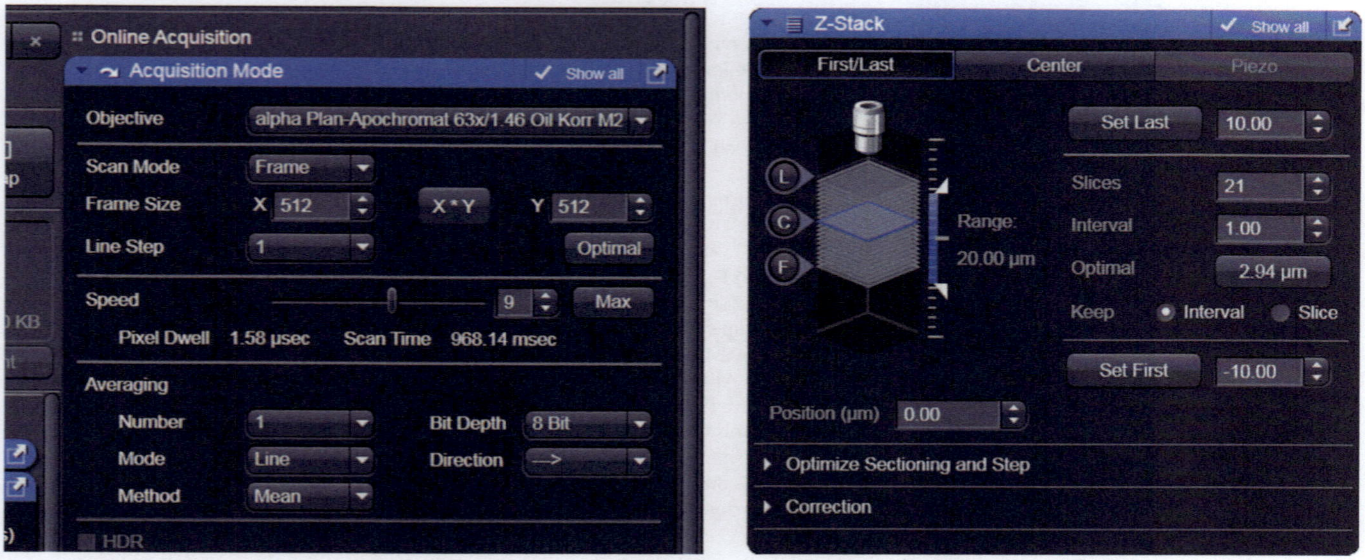

Figure 18.14 ***Implementing Nyquist's criterion*** *The Optimal macro buttons on the Zeiss Zen software, as used in the author's imaging facility, to calculate the sampling requirement for Nyquist. As is the case with most software, this only calculates Nyquist/2, the minimum requirement. Other values must be calculated and entered manually. Where conditions dictate compromise and a degree of undersampling, using a value higher than half Nyquist is immaterial. Source: Reproduced with permission from Carl Zeiss Ltd.*

Because a specific pixel size is required in the image to satisfy Nyquist's sampling criterion in the lateral (x,y) direction according to the NA of the objective used and the wavelength of the illuminating laser, there will be a specific zoom factor to give the particular pixel size required in the image. Lower magnification objectives often have a higher Optical Index[6] than high-magnification objectives and so may require a higher zoom setting to adequately sample the detail resolved in the object and transfer this into the image (see Box 18.4 for the requisite mathematical calculations).

Box 18.4 Using the Correct Zoom Factor

A specific pixel size is required in the image to satisfy Nyquist's sampling criterion. The software controlling the operation of the microscope will display the pixel size being used.

Objective	Minimum resolved detail (MRD, d)	Pixel size required at Nyquist = MRD/2	Pixel size required at Nyquist = MRD/2.3	Pixel size required at Nyquist = MRD/3	Pixel size at scan size 1.0 and raster 1024 × 1024	Zoom factor required for Nyquist = MRD/2	Zoom factor required for Nyquist = MRD/2.3	Zoom factor required for Nyquist = MRD/3
5x/NA 0.16	1.56 µm	0.78 µm	0.68 µm	0.52 µm	1.25 µm	1.6x	1.8x	2.4x
10x/NA 0.30	0.83 µm	0.42 µm	0.36 µm	0.28 µm	0.63 µm	1.5x	1.8x	2.3x
10x/NA 0.45	0.55 µm	0.28 µm	0.24 µm	0.18 µm	0.63 µm	2.3x	2.6x	3.5x
20x/NA 0.80	0.31 µm	0.16 µm	0.13 µm	0.10 µm	0.31 µm	1.9x	2.4x	3.1x
20x/NA 1.0	0.25 µm	0.13 µm	0.11 µm	83 nm	0.31 µm	2.4x	2.8x	3.7x
40x/NA 0.75	0.33 µm	0.17 µm	0.14 µm	0.11 µm	0.16 µm	0.9x	1.1x	1.5x
40x/NA 1.3	0.19 µm	95 nm	83 nm	63 nm	0.16 µm	1.7x	1.9x	2.5x
40x/NA 1.4	0.18 µm	90 nm	78 nm	60 nm	0.16 µm	1.8x	2.1x	2.7x
63x/NA 1.4	0.18 µm	90 nm	78 nm	60 nm	0.10 µm	1.8x	1.3x	1.7x

NB. The pixel size, which is calculated by the software at scan zoom 1.0, will change according to the design of microscope and also for the size of the raster scan. For example, the values given in the table above are the pixel sizes for zoom factor = 1.0 with a raster scan of 1024 × 1024 on the Zeiss LSM 700 in the author's laboratory.

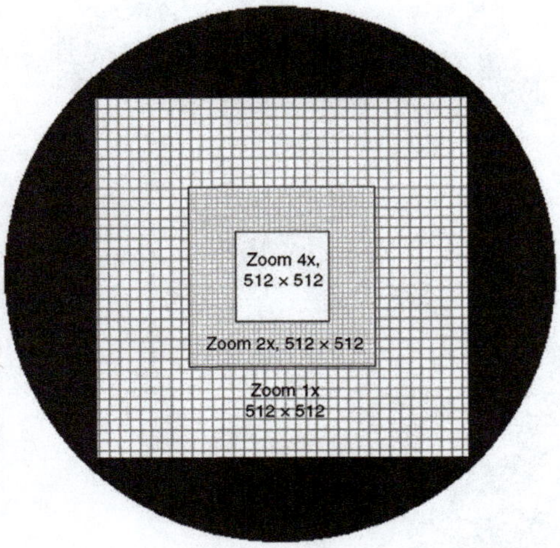

Source: Pawley 2006. Reproduced with permission from Springer.

As the zoom factor is increased, the area sampled decreases, and the pixels that constitute the image, which arise from the sampling area in the object, become smaller.

The zoom magnification should be adjusted such that the pixel size is about 1/2 or 1/3 the Abbe resolution limit. This limit, the minimum resolved distance (MRD), will depend upon the excitation wavelength used and the numerical aperture of the objective.

The formula used to calculate MRD is: $d = 0.51 \times \lambda_{ex}/NA$ – as shown in Section 17.5. The Nyquist sampling frequency and MRD are calculated assuming excitation at 488 nm. Compare the values in the table above to Table 1, page 1936 in: Cole, R. W., et al. (2011), where a Nyquist sampling frequency of MRD/3 has been used.

This value of MRD/3 arises because sampling MRD/2 is all very well if features in the specimen inhabited the Cartesian coordinates x and y. However, features in the specimen extend in two dimensions in all directions. The pixel size calculated at MRD/2 will be 1.41 times too big to properly sample features running along a line oriented at 45° midway between the x- and y-directions, as shown in the figure below. However, other pixels in the square raster array will also sample this line. Therefore a pixel only has to be made 0.71 times smaller (i.e. 1.41/2) and so a pixel size of (2/0.71 = 2.8x) ≈ MRD/3 is sufficiently small to adequately sample the specimen and reconstruct the digital image.

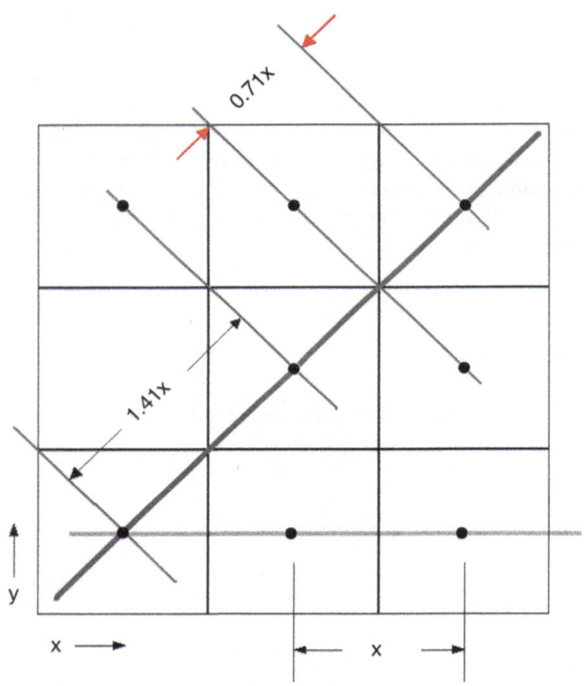

Source: Pawley 2006. Reproduced with permission from Springer.

If this argument were extended to the 3D data sets, the diagonal of a cube is 1.73x longer than its side. However, axial resolution in z is always three times worse than laterally in x,y, so this factor may be ignored. Confocal microscopes have a macro (e.g. see Figure 18.14) to set the Nyquist frequency at MRD/2, to sample half the minimum resolvable element (sometimes called a resel; see Box 10.3). Applying Nyquist's sampling frequency at half the MRD is usually acceptable unless you wish to see the finest details that the objective is capable of resolving, in which case apply a Nyquist frequency of MRD/3.

The pixel dimension is determined by the scan size of the frame and the number of sampling points along one dimension of the raster scan. For example, a 10x/NA 0.45 objective with a resolving power of 0.55 μm and a Nyquist sampling requirement for Nyquist/2.3 requires a pixel size of 0.55/2.3 = 0.24 μm. If the raster scan being used is 1024 × 1024, and the scan size of the frame is 639.5 μm, then the pixel size on the raster scan is 639.5/1024 = 0.63 μm. The length of raster scan in micrometres (here, 639.5 μm) can be read from the software. Therefore, the zoom factor for this plan-apo 10x/NA 0.45 objective is: 0.63/0.24 = 2.6x. If a higher zoom factor is used, the pixels will be smaller than required, and the specimen will be oversampled. If the zoom is less than 2.6x, under-sampling occurs.

The effect of over- and under-sampling can be seen in the panel below:

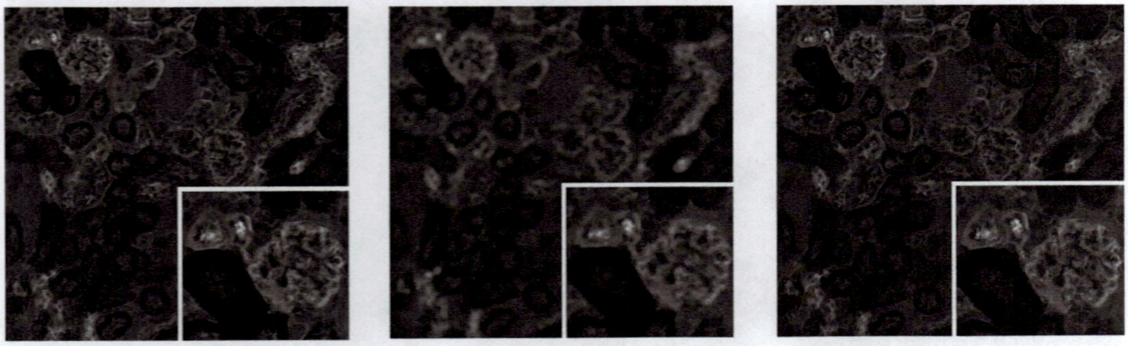

The left-hand panel shows an image of kidney sampled at the minimum acceptable frequency of Nyquist/2. The middle panel shows loss of resolved detail and addition of spurious detail from under-sampling, and the right-hand panel shows bleaching and loss of signal from oversampling.

The table above can be extended; various zoom factors for each objective when using different raster scan sizes can be constructed. The table below shows zoom factors for four objectives and three different raster scans for a Nyquist criterion using a sampling element of 1/3rd the MRD (i.e. Nyquist by 3). Such a table can be left by the microscope for easy reference.

Objective	Minimum resolved detail (MRD, d)	Pixel size required at Nyquist/2.3	Pixel size at scan size 1.0 and raster 512 × 512	Pixel size at scan size 1.0 and raster 1024 × 1024	Pixel size at scan size 1.0 and raster 2048 × 2048	Zoom factor required for raster scan 512 × 512	Zoom factor required for raster scan 1024 × 1024	Zoom factor required for raster scan 2048 × 2048
10x NA 0.30	0.83 μm	0.36 μm	1.25 μm	0.63 μm	0.31 μm	3.5 x	1.8 x	0.9 x
20x NA 0.80	0.31 μm	0.13 μm	0.63 μm	0.31 μm	0.16 μm	4.8 x	2.4 x	1.2 x
40x NA 1.3	0.19 μm	83 mm	0.31 μm	0.16 μm	80 nm	3.7 x	1.9 x	1.0 x
63x NA 1.4	0.18 μm	78 mm	0.20 μm	0.10 μm	50 nm	2.6 x	1.3 x	0.6 x

The image collected by a CCD camera on a widefield (or a spinning disc) microscope cannot be zoomed in the same was as for a point-scanning confocal. The means for calculating the correct objective, intermediate magnification factor and camera binning factor is explained in Chapter 31, Section 31.7.

References

Cole, R.W. et al. (2011). 'Measuring and interpreting point spread functions to determine confocal microscope resolution and ensure quality control'. *Nature Protocols* 6/12: 1929–1941.

Pawley, J.B. (2006). 'Points, pixels and gray levels: digitizing image data'. In: Pawley, J.B. (ed.), *Handbook of Biological Confocal Microscopy*, Chapter 4, 3rd ed. New York: Springer, pp. 59–79.

If you plan to zoom into an image, it may be helpful to first take a screen shot (PrtScn/Print Screen on PCs; Command-Shift-4 on Macs). This can be loaded (Figure 18.15) into Windows' standard image editor, Microsoft Paint or saved as a PNG file in Macs. Although it is worth taking a snapshot of representative images with scale bars, for reproduction and publication later, it is possible to calculate the magnification associated with the image from the metadata, which can be displayed in software and printed out, usually as a text file.

18.8 Effect of Refractive Index Mismatch

Although manufacturers design objectives such that spherical aberration is minimised, it is possible for these corrections to be upset by the microscopist and thus reintroduce spherical aberration into the image. The major causes of reintroduced spherical aberration (Figure 18.16) being inappropriate coverslip thickness and/or refractive index mismatch. Most people assume that when imaging cells by fluorescence, any refractive index changes are very small or negligible. This is certainly not the case. Variations in refractive index associated with the nucleus and other cell organelles are the basis of contrast formation in phase-contrast and DIC images: the degree of contrast enhancement being proportional to the differences in refractive index between specimen features and the background. Also, spherical aberration can affect image quality more than most people realise. Focusing into thick samples (even well within the working

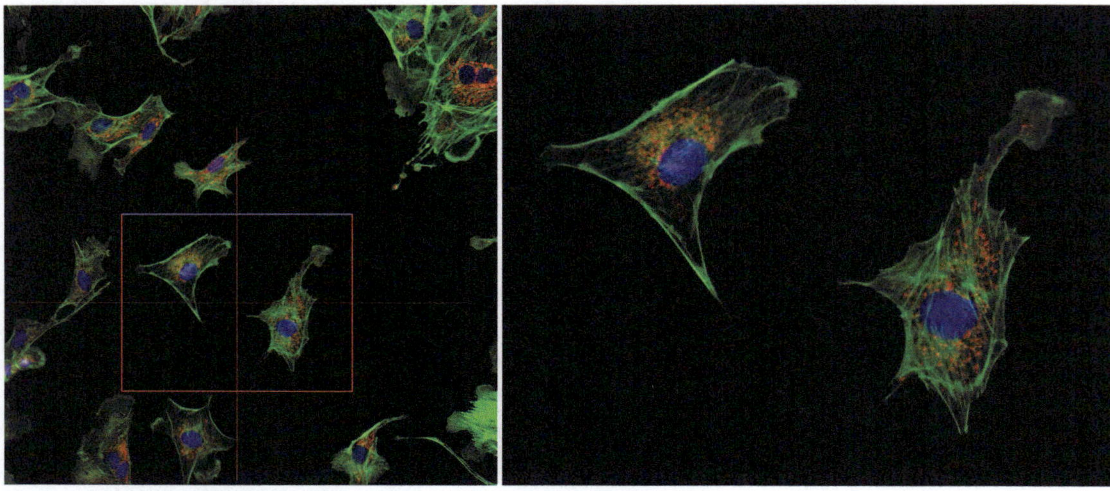

Figure 18.15 **Recording a zoom setting** *A screen-grab is a useful way of capturing a record of the intended zoom in relation to the larger field of view just before the zoom is made. Source: Reproduced with permission from Carl Zeiss Ltd.*

distance of a high numerical aperture objective) will result in loss of resolution, intensity and contrast in the image. To quote Jim Pawley again:

> 'The traditional description of diffraction-limited image formation in the microscope, as laid out by Ernst Abbe in the 1870s, has been accepted as dogma for so long that it is not uncommon for important conditions associated with its attainment to be overlooked. It is tolerably satisfied when viewing thin cells, or tissue sections, mounted on the underside of the coverslip – not on the slide – and embedded in mounting media having a refractive index equivalent to that of the coverslip: e.g. homogeneous immersion. In practice the Abbe limit is rarely reached, because of photon noise, bleaching, specimen movement and vibration.
>
> It is necessary to use water-immersion, or multi-immersion, objectives to image aqueous specimens with water between objective and coverslip, *or* to use an oil of higher-than-normal refractive index to compensate for the spherical aberration introduced by using an oil objective on an aqueous specimen. In practice, using a high RI oil in this way only works well within a narrow depth range below the coverslip.

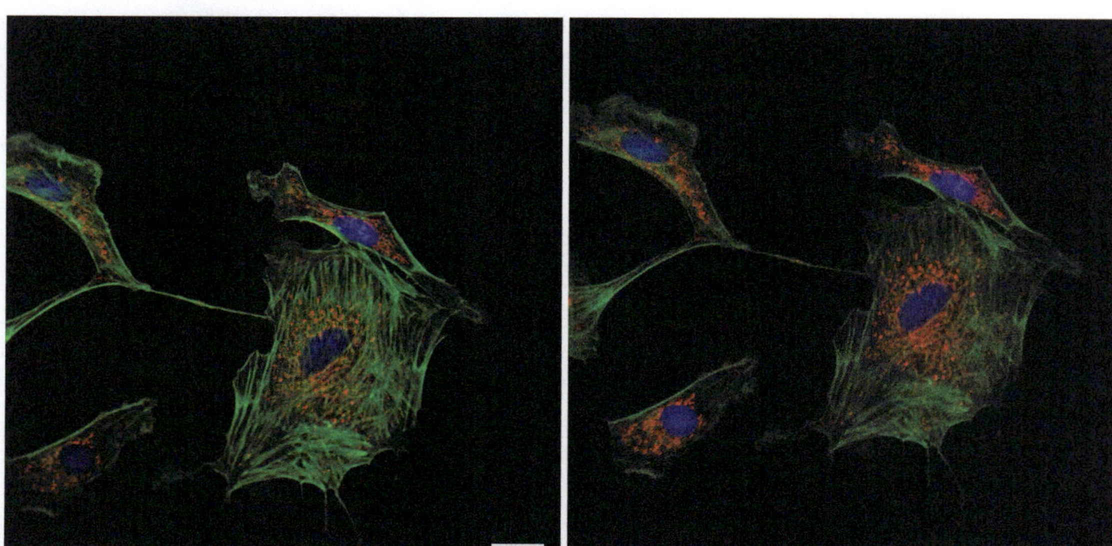

Figure 18.16 **Effect of refractive index mismatch** *The image in the left-hand panel (a) was taken with a 40x/NA 1.2 water-immersion objective with a correction collar. The collar was set for the correct coverslip thickness to give a sharp image and Immersol W (Zeiss), immersion oil corrected for the refractive index of water, was used to acquire an aberration-free image. The immersion medium was then changed for immersion oil of RI 1.518 to take the image in panel (b). This image could just be focused (sometimes this is impossible), but the PMT gain settings had to be increased in all channels because of the significant loss of signal intensity. Notice also a loss of resolving power.*

> Refractive index anomalies associated with cell organelles can displace the location of optical sections by several nanometers or even micrometers. Confocal microscopy is inextricably linked to the concept of the optical section, which occurs because light originating from out-of-focus planes is excluded by the pinhole in front of the PMT. **This idea is so firmly embedded in the psyche of modern biological microscopy that few individuals stop to question whether the 'plane-of-focus' is an actual geometrical plane, resembling a mechanical section, or merely the surface described by the array of points at which the rastered laser beam happens to reach best focus.** In fact, this plane will be close to a geometrical plane *only* if the specimen is optically homogeneous with constant, or near constant, refractive index. Few living cells are optically homogeneous, so we must generally accustom ourselves to compromise when viewing reconstructed samples in three dimensions.'

Besides introducing spherical aberration, the occurrence of refractive index mismatch means that the perceived depth of a structure (or the position of the focal plane) will differ from the true depth. The consequence of this is that 3D reconstructions will be distorted axially in the z-direction. Any measurements in z must be corrected for. It is helpful to know the refractive indexes of the media in the microscope preparation. A series of refractive index values are listed in Table 28.1 in the chapter on biological specimen preparation.

Since the refractive index (n_1) of the immersion medium is often greater than that of the mounting medium (n_2), the actual focal position (AFP) is closer to the coverslip than the nominal, or perceived, focal position (NFP). The position of the object will *appear* to be further away (see Figure 18.17) from the coverslip than it actually is: the image formed is larger than the actual size of the object and is elongated axially in the z-direction (Figure18.18). Likewise, the thicker the coverslip with a departure from the ideal of 170 μm, the more elongated the image becomes. Conversely, when the sample medium is denser than the objective immersion medium, objects appear smaller than they are (i.e. collapsed). Egner and Hell (2006) state that if a relatively thick specimen is mounted in a watery medium and observed with an oil-immersion objective, the focal shift (i.e. NFP – AFP) is 13λ at a depth of 100λ beneath the coverslip.

A related factor to consider when making 3D reconstructions is how far below the surface of the coverslip useful data can be obtained. The primary – and most obvious – limitation is the mechanical one imposed by the fixed working distance of the objective. Most users fail to appreciate – or perhaps remember – the optical factors that limit the depth to which a Z-stack, or a single plane at some depth within the specimen, may usefully be acquired. Firstly, the excitation illuminating beam is either scattered or absorbed as it passes through the upper layers of the specimen (longer wavelengths scatter less, hence the use of multi-photon microscopy). Secondly, the emitted signal is likewise scattered or absorbed between the focal plane and the detector (which is one reason why non-descanned detectors are used in multi-photon microscopy: to collect this emitted signal, which is otherwise lost). Thirdly, any optical inhomogeneity (i.e. refractive index mismatch) causing a departure from the ideal of homogeneous immersion will tend to defocus both

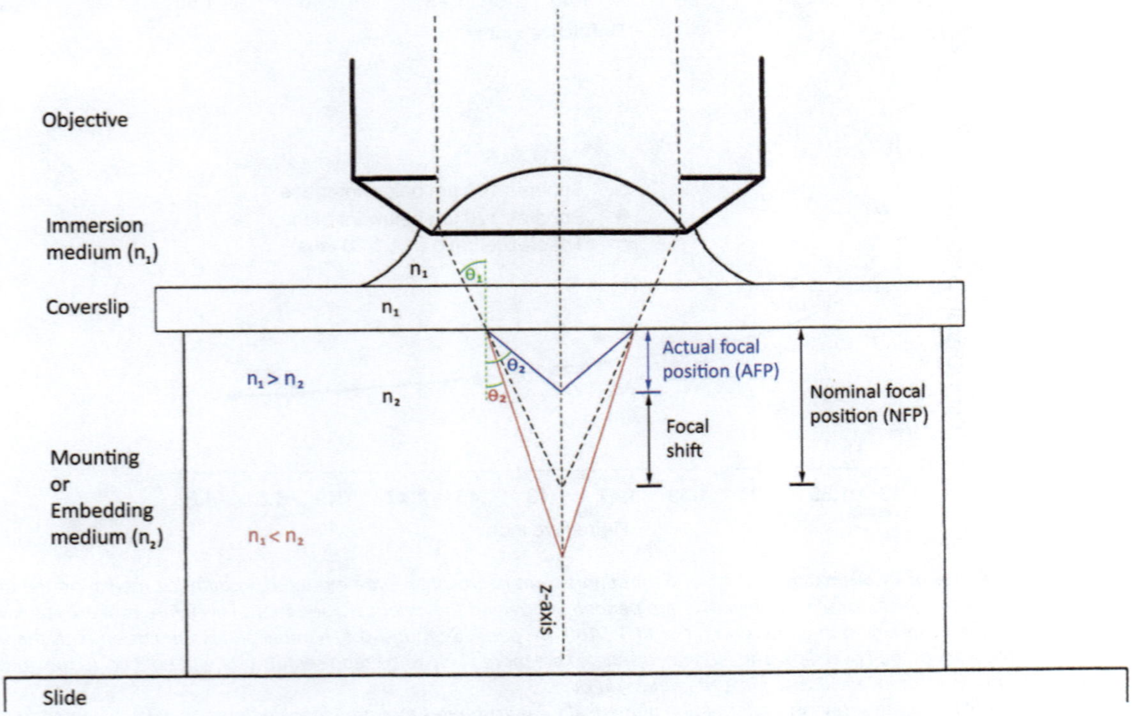

Figure 18.17 **Refractive index mismatch** *Where the refractive indexes n_1 and n_2 are perfectly matched, homogeneous immersion occurs. Where $n_1 > n_2$, for example when using an oil-immersion objective to study a specimen mounted in water, the angle of refraction (θ_2) is greater than the angle of incidence (θ_1), the actual focus position (AFP) is shifted higher within the mounting medium (see also Figure 2.14, Chapter 2, showing refraction) and the object will appear compressed. The correction factor for paraxial rays at an oil-water interface is 0.88; for the marginal rays it is 0.37. Therefore the apparent depth, or length, of an object in the image will be between $(0.88)^{-1}$ and $(0.37)^{-1}$ the actual depth, that is, overestimated between 1.1x and 2.7x. The full calculations may be found in Visser and Oud (1992 and 1994). If the reverse case occurs, and $n_1 < n_2$, such as using a water-immersion objective to image a specimen immersed in a medium with a greater refractive index than water, then the apparent depth will be deeper, and the object will appear elongated. Source: Adapted from Egner and Hell 2006 with permission from Springer.*

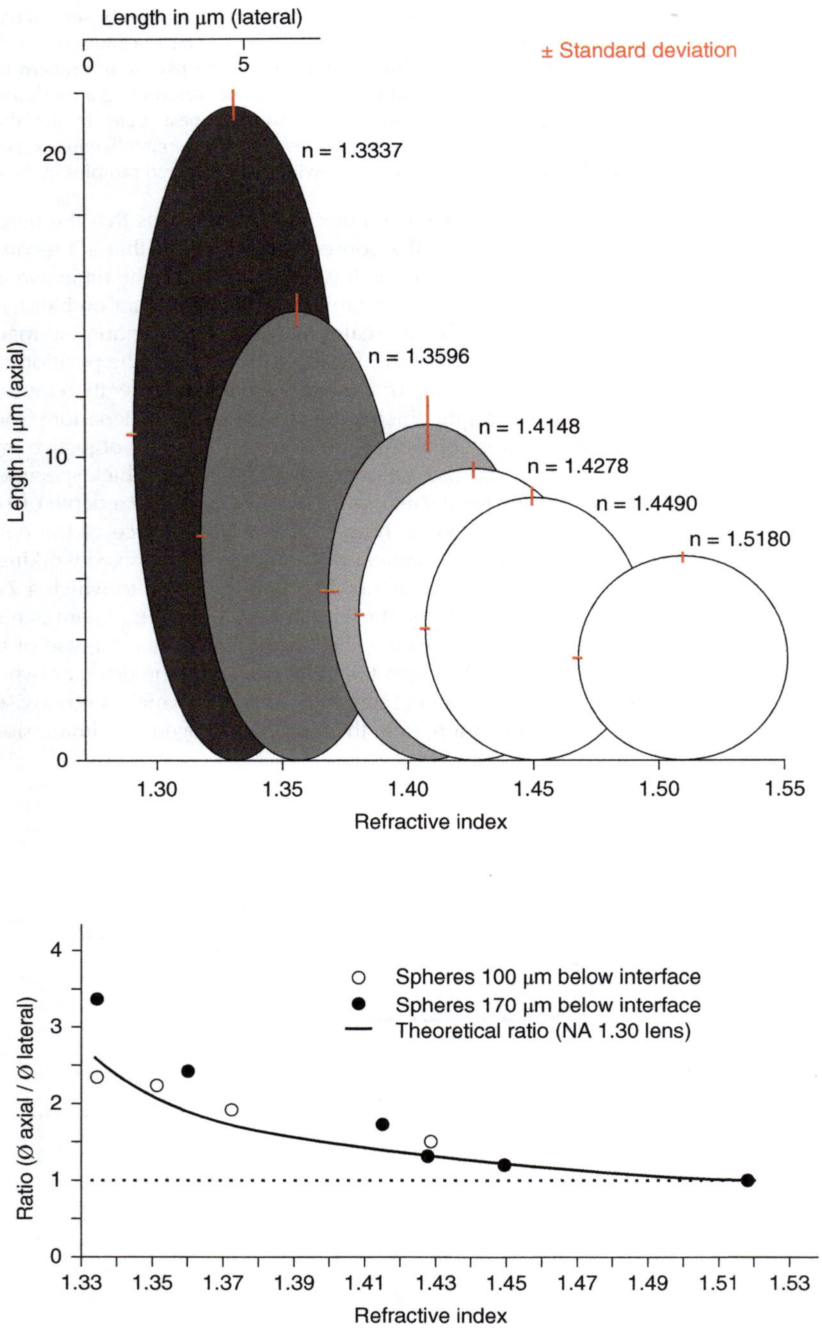

Figure 18.18 ***Effect on the image of RI mismatch*** *A series of 6 mm fluorescent microspheres are mounted, in differing media, on the underside of a measured coverslip exactly 170 mm thick, for which objectives are perfectly corrected for spherical aberration. This figure is of a Z-stack viewed in the x-z plane. Only those microspheres mounted in immersion oil of RI 1.5180 are perfectly round (i.e. homogeneous immersion with the immersion oil used for the objective and the glass of the coverslip). The elongated spheres are more likely to be representative of what occurs to the images of watery biological objects mounted in media of lower RI than glass or immersion oil.*

The graph in the lower part of the figure shows this information numerically. Microspheres mounted in water have an axial diameter up to 3.4x that of the real (lateral) diameter. Source: Visser and Oud 1994. Reproduced with permission from John Wiley & Sons.

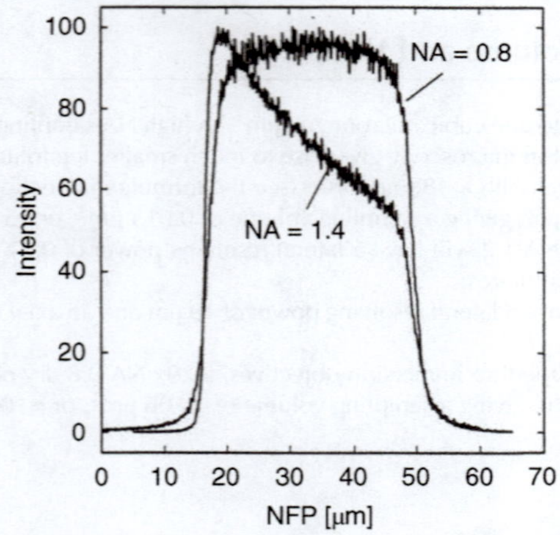

Figure 18.19 ***The edge effect on the image of RI mismatch*** *These curves show signal intensity plotted against nominal focus position for an objective of numerical aperture of 0.8 and one of NA 1.4. The gradual loss in resolution and intensity, shown by the decrease in the NA 1.4 curve, are due the effects of high NA, rather than quenching or bleaching of the fluorophore. Source: Egner and Hell 2006. Reproduced with permission from Springer.*

the excitation illumination and emitted signal. This defocus effectively spreads the size of the point spread function (PSF) both laterally and especially axially[7]. The defocus concomitantly increases the size of the focused spot at the detector pinhole (a loss of confocality) so that a proportion of the signal that should pass through the pinhole to form an optical section does not do so. If you are imaging details at the limit of the resolving power of the objective, this defocus will cause a loss of spatial resolution. The degree to which these three optical factors affect the quality of the image depends upon the nature and optical properties of the specimen. Pawley (2006) notes that an immersion objective of numerical aperture less than NA 0.85 may help with thick specimens or samples mounted in a deep preparation (which may not be adjacent to the underside of the coverslip), because the defocusing effects are less serious under these circumstances than with higher numerical aperture objectives. This is illustrated graphically in Figure 18.19.

In a very good chapter on aberrations caused by refractive index mismatch (Egner and Hell, 2006), the following advice is given:

- Always use the appropriate objective.

  Large NA objectives with high resolving power are designed to be used with thin specimens mounted on the underside of a measured coverslip 170 μm thick. Objectives with correction collars are used to minimise spherical aberration.

- Thick specimens should only be observed in matched systems.

  Homogeneous immersion is the ideal state: use a water-immersion objective for aqueous samples and multi-immersion or a glycerol immersion objective for living samples or those mounted in a glycerol-based mountant.

- Avoid situations where the sample is thin but far away from the coverslip.

  This frequently happens when tissue sections are attached to a slide or cells are attached to or grown upon the slide (or a filter) rather than a coverslip. An unknown thickness of transparent mounting medium together with the thickness of the coverslip used in the preparation acts optically as a much thicker coverslip.

- If you cannot match the RI of the immersion medium to the mounting medium, try to get it as close as possible.

  You can achieve this by using glycerol and water or adding dextrose to the mounting medium. This helps when imaging at depth.

- For unmatched systems, increase the confocal detection pinhole as a function of depth.

- If you have to image really deep into samples, use a lower NA immersion objective.

- Spherical aberration, not absorption, is the major loss of signal intensity with focus depth.

- Living specimens are optimally viewed with a water-immersion dipping objective and no coverslip.

Where the confocal microscope is to be used for quantitative studies, particular attention should be paid to fluorophore bleaching, ground-state depletion (which keeps fluorophores in a dark state) and fluorophore saturation. If these states occur, then fluorescence emission will not increase linearly with illumination intensity or the fluorophores will be bleached. The subject of quantitative fluorescence microscopy is discussed further in Chapter 27, and several references are given. Two recent chapters specifically dealing with quantitative confocal microscopy are Murray (2013) and Jonkman et al. (2014). Besseling et al. (2015) have published a method to calibrate the axial distances in the confocal as a function of refractive index, for use in quantitative studies.

Box 18.5 Confocal Focal Spot Volume and Voxels

One femtolitre, or 1000 attolitres, is equivalent to one cubic micrometre (μm^3). A high NA sampling volume in confocal or widefield fluorescence microscopy will be 15–25 attolitres. Multi-photon microscopy gives rise to much smaller femtolitre volumes.

A 63x/NA 1.4 objective used on the confocal with a 488 nm laser (see the formulae in Section 17.5) will have a lateral resolving power of 0.18 μm and an axial resolving power of 0.46 μm, giving a sampling volume of 0.015 μm^3, or 15 attolitres.

A 40x or 63x water-immersion objective of NA 1.2 will have a lateral resolving power of 0.21 μm and an axial resolving power of 0.57 μm, giving a sampling volume of 0.025 μm^3 or 25 attolitres.

A 40x/NA 1.3 oil immersion objective will have a lateral resolving power of 19 μm and an axial resolving power of 0.58 μm, giving a sampling volume of 0.021 μm^3, or 21 attolitres.

For comparison with these high-numerical aperture immersion objectives, a 20x/NA 0.8 dry objective will have a lateral resolving power of 0.31 μm and an axial resolving power of 1.1 μm, giving a sampling volume of 0.106 μm^3, or ≈100 attolitres.

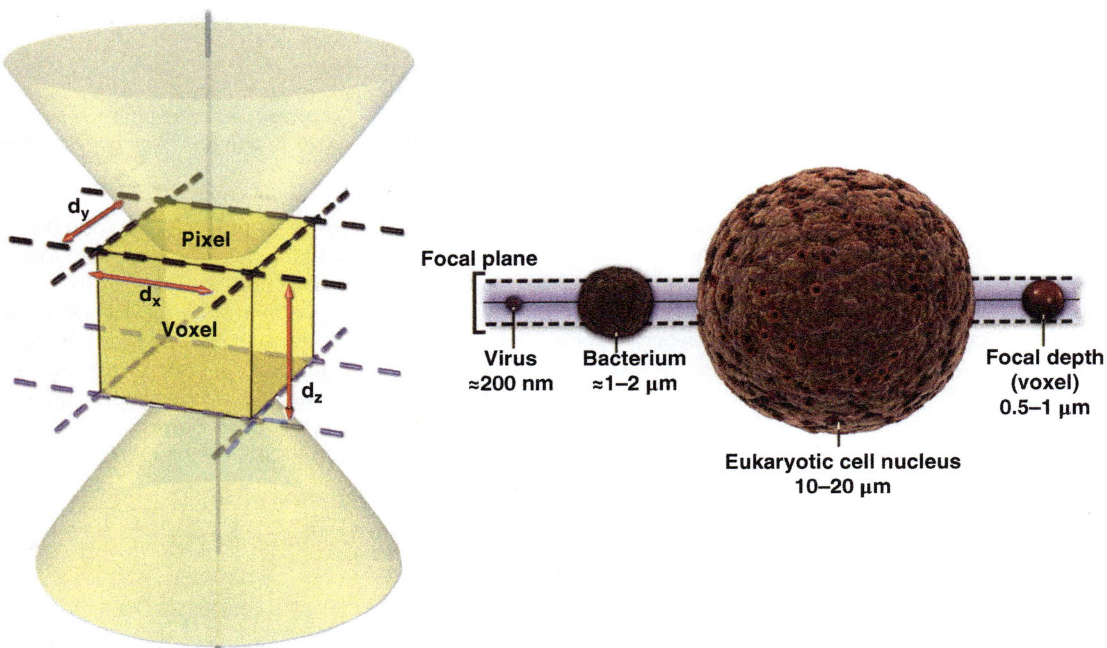

Source: Adapted from Hazlewood et al. 2007 with permission from Springer and adapted from Murphy and Davidson 2013 with permission from John Wiley & Sons.

Just as a pixel (picture element) is the smallest irreducible part of a digital image and represents the photon intensity value from sampling the signal emitted from a particular point in the object, so a voxel is a volume element, the three-dimensional equivalent.

Since the axial resolution limit is larger – by about three times – than the limit for lateral resolution, a voxel is generally *not* an equal-sided cube and will have a z dimension that can be several times greater than the x and y dimension (this is entirely different from, and irrespective of, the influence of elongation in the axial direction arising from refractive index mismatch).

A few image analysis packages require voxels having equal dimensions. These are called isotropic voxels; effectively the sample is significantly over-sampled along the z-axis to create these isotropic voxels.

References

Hazelwood, K.L. et al. (2007). Figure 1.9, Title: The voxel concept, page 22, 'Entering the portal'. In: Shorte, S. and Frischknecht, F. (eds.), *Imaging Cellular and Molecular Biological Functions*, Chapter 1. Berlin Heidelberg: Springer-Verlag, pp. 3–43.

Murphy, D.B. and Davidson, M.W. (2013). Confocal Laser Scanning Microscopy. In: *Fundamentals of Light Microscopy and Electronic Imaging*, Chapter 13, 2nd ed. New Jersey: Wiley Blackwell, p. 279.

Roux, P. et al. (2004). 'Focusing light on infection in four dimensions'. *Cellular Microbiology* 6/4: 333–343.

18.9 Objectives for Optical Sectioning

The best objective to use is one with the highest possible numerical aperture and longest working distance. Use the manufacturers' literature and Internet databases, and also see Table 7.2 and Table 7.5 to find this information. The high NA of the objective gives a shallow depth of field and high image brightness, while the working distance dictates how deep into the tissue the objective can image to collect a three-dimensional Z-stack. If you are viewing living cells with watery cytoplasm, aqueous specimens or those prepared in a mountant with a refractive index close to that of water (rather than oil) and the specimen is thick (≥40 μm, or so), then a water-immersion objective of NA 1.2 with a correction collar to adjust for spherical aberration arising from refractive index mismatch is the best choice (see Figure 27.1a). There is no advantage in using an objective with a numerical aperture greater than 1.3 NA if the specimen is aqueous or mounted in water, since the refractive index of water will limit the half-angle of the most oblique ray taken up by the objective (see Chapter 7, Section 7.4).

Multi-immersion objectives (Figure 7.4) are also useful, since they give ideal image brightness at medium magnification and also have a correction collar to adjust for any refractive index mismatch. The advantage of a 20x/NA 0.7 or 25x/NA 0.8 multi-immersion objective is that with a fully open pinhole, these objectives give an optical section between 20 and 30 μm, greater than the thickness of most monolayer tissue-cultured cells (see the notes in Table 17.1). In-focus fluorescence is effectively collected from throughout the depth of the cell, and changes in focus caused by stage heating and drift, vibrations or settling of the tissue culture preparation in the dish are minimised. Also, an area of interest can be selected with minimal photobleaching for further study by confocal microscopy.

18.10 Should a 100x Objective Be Used on a Point-Scanning Confocal?

Unlike the widefield fluorescence microscope, the single-point scanning laser confocal microscope has a zoom control. It is therefore possible to magnify the final image at will. This is why very few 100x (NA 1.3 or NA 1.4) objectives are sold for use on a confocal microscope, when their resolving power is equivalent or slightly less to the 40x (NA 1.3) or 60x (1.3 or 1.4) objective, and furthermore they do not capture such a bright image (Intensity $\propto NA^4/Mag^2$).

In a widefield fluorescence microscope or spinning disc confocal microscope, where there is no inherent continuous zoom function (we are not considering a magnification changer here) a 100x objective *is* useful because a smaller field of view can be obtained, as required. Using a magnification changer in conjunction with a 60x objective will inevitably lead to a less bright image, of possibly poorer optical quality and certainly a lower signal-to-noise ratio in the image because of the extra glass in the magnification changer. In any case, with a magnification changer a discrete magnification increase of 1.5x or 2x will only be possible instead of the exact 1.67x (100/60) required. In a widefield microscope, given two objectives of the same numerical aperture, the objective with the larger magnification will exhibit a lower image brightness and poorer SNR, because the available light is spread over a larger area in proportion to the square of the magnification.

In point-scanning confocal microscopes, the light-gathering efficiency is only proportional to the fourth power of the NA (Intensity $\propto NA^4$) and is unaffected by the overall magnification. This is because the continuous zoom function and pinhole adjustment allows the magnification of a 60x/NA 1.4 objective to match that of a 100x/NA 1.4 objective. The 60x/NA 1.4 objective at zoom 1x (for the same illumination wavelength) will give a brighter image than the 100x/NA 1.4 objective at zoom 1x. Each objective will capture the same level of resolved detail; the 100x objective simply magnifies this detail more in the image plane.

To achieve the same magnification as the 100x objective, a digital, or electronic, zoom of 1.67x applied when using the 60x objective will give the same image size, albeit at the expense of empty magnification and some photobleaching, because the same laser power is concentrated into a smaller area. Adjusting the pinhole to 1.67x for the 100x objective will give the same signal-to-noise ratio as the 60x objective. The drawback, in this case, is that a thicker optical section is collected.

Thus, a 100x objective is useful for a widefield microscope or spinning disc microscope (particularly if the size of the pinholes of the Nipkow disc are optimised for the depth of field of a 100x objective), but is not required for a point-scanning confocal microscope with a continuous zoom function and, strictly, should not be used on one.

The same arguments extend to using a 40x/NA 1.4 oil objective for all confocal scanning instead of a 63x/NA 1.4 or 100x/NA 1.3 objective. Whilst the image may be two and a half times brighter, this advantage is counterbalanced by the need to employ a much higher zoom to satisfy the Nyquist criterion (most of the time you will be under-sampling when using this 40x/NA 1.4 objective, which has a very high Optical Index). Any advantage in signal brightness is also counterbalanced by the reduced working distance of the 40x objective. However, the key advantage of using this 40x/NA 1.4 objective is in live-cell imaging (where fulfilling the Nyquist criterion is unimportant) and for cells – grown onto a measured coverslip – these can be imaged with much less exposure due to the high light transmission of this particular objective.

Some manufacturers, notably Leica, have fitted high NA objectives with variable aperture collars on them (e.g. 40x plan-Apo NA 0.75–1.25) onto confocal microscopes. Whilst there is nothing wrong with using these objectives for optical sectioning by confocal microscopy or other means, care must be taken to ensure that the collar is set fully open in order to give the highest resolving power and, more importantly, the highest light transmission and the shallowest depth of field. It is easy to overlook this crucial point at the start of an imaging session. You can check the collar setting (when the objective is off the microscope) while holding the objective and looking through the barrel from the rear. Generally, the collar is fully open when it is turned anti-clockwise. A strip of Parafilm or plumber's Teflon® tape stretched slightly and wound around the objective collar and barrel under tension (Figure 18.20) will keep the variable numerical aperture collar held open on the microscope, obviating any risk of it being closed down inadvertently.

Figure 18.20 **An objective with a variable aperture diaphragm used for confocal** *A variable aperture objective can be used on the confocal, but please be aware that if the collar is inadvertently moved – or set wrongly from the outset – and the diaphragm within the objective is closed, then poor results will ensue. To prevent this from happening, this objective has had Parafilm wrapped around the barrel to stop the diaphragm from being closed down to the low numerical aperture setting. Source: Courtesy of Mr K. Glover.*

18.11 Reflection Confocal Scanning

A seldom-used but valuable technique is to operate the confocal microscope in reflection (or 'back-scattered') mode (Figure 18.21) in order to image opaque specimens, or those samples stained with markers that reflect light. Both the point-scanning confocal and the spinning disc can be used in reflection mode, the irradiating laser being scattered off the sample into the PMT. The correct filter and dichromatic mirror must be selected, or the bandwidth in front of the PMT (if a prism or grating is used to direct the light towards the PMT) adjusted, to detect the reflected light. Sometimes a partial 80/20 neutral-density primary dichromatic mirror, reflecting 20% of the light down onto the sample and allowing 80% to be reflected back to the PMT for all wavelengths in the spectrum, is useful for this technique. Since any surface that gives a change in refractive index in the preparation will backscatter light (e.g. the coverslip), this can

Figure 18.21 **Imaging opaque specimens with the confocal microscope** *An old farthing scanned in reflection confocal mode with multiple tile scans (left). These tiles have been stitched together (middle) into a single image. A photograph (right) is shown for comparison. Source: Courtesy of Mr K. Glover.*

Figure 18.22 **The appearance of the laser in reflection confocal scanning** *Concentric Newton's interference rings seen in the image from the laser entering the PMT detector directly.*

interfere with the final image. Fortunately, the reflected light image is confocal, and so it is possible to use the pinhole to exclude the coverslip reflection from entering the PMT and contributing to the image.

Whenever the laser is allowed (most often inadvertently) to enter the PMT directly, concentric interference rings will be seen (Figure 18.22) overlaying the image. Very often a reflection spot will be seen when using the confocal in reflection or backscattered mode. The simplest way to remove unwanted reflections is to place a polarising filter in the beam path to cut out most of the polarised laser light causing the reflections. The image-forming light scattered off the sample is randomly polarised and so passes through the filter. Since the laser beam is linearly polarised, there is no need for an analyser in the beam path between scan head and specimen, where it would reduce the epi-fluorescence signal. This enables a (non-confocal) DIC image to be collected alongside the fluorescence one and both merged into a single image.

Most optical sectioning using confocal microscopes is done by biologists. However, reflection confocal microscopy is very useful for examining the surfaces of features non-destructively, avoiding direct contact with the surface of the specimen. White-light confocal microscopes have been developed for this purpose. Because it is not necessary to scan fast with living tissues that photo-bleach, it is possible to use the optically simpler stage-scanning design for assessing surface profile and roughness in materials science. For further details, see Hovis and Heuer, 2010.

18.12 Chapter Summary

This chapter should help you understand how to use a point-scanning confocal microscope to collect high-quality images containing scientifically valid data. This point bears repeating: an image is a data set of photon intensity. It is not merely a pretty picture – your image may indeed have artistic merit, command respect and win photographic competitions, but it is first and foremost a scientific data set. To collect an appropriate image to support the scientific hypotheses underlying the experiment, various parameters (e.g. pinhole size, PMT gain, signal intensity, optical section thickness, raster speed and dwell time, SNR) have to be prioritised, and set rationally, for operating a confocal microscope involves some degree of compromise to balance conflicting parameters. The conditions required to collect one image, Z-stack or set of images may be quite different from those supporting a separate experiment, and these considerations must be taken into account when operating the confocal microscope.

1. A laser cannot be used for direct visual observation.
2. Most lasers used for microscopy are Class 3B or Class 4 and are fitted with interlocks.
3. Single-beam point-scanning confocals operate with laser intensity levels close to the fluorophore saturation level.
4. A protocol for acquiring images with the single-beam point-scanning confocal microscope is given.
5. The speed of the single-beam point-scanning confocal microscope is fundamentally limited by fluorescence saturation but also by the speed at which the scanning mirrors operate.
6. Confocal microscopes have an inherent speed limitation because the pixel dwell time on each point cannot be too fast.

7. Confocal imaging may require compromising between conflicting parameters, depending on the nature and purpose of the images being collected.
8. PMT detectors have a poor quantum efficiency compared to a CCD camera.
9. Samples for confocal microscopy normally require increased staining times or stain concentrations.
10. The confocal microscope is an inherently noisier instrument than a widefield fluorescence microscope. Performing an average scan will help improve the signal-to-noise ratio in the image.
11. The sample should be adequately sampled according to Nyquist's criterion.
12. Confocal data sets should be deconvolved but often are not because the confocal pinhole is considered to have 'sampled' the image. Without the benefit of deconvolution, image artifacts may still remain.
13. The effect of refractive index mismatch is often overlooked. This, and the use of the wrong thickness of coverslip or erroneous specimen preparation, can inadvertently reintroduce spherical aberration into the image.
14. Refractive index mismatch means that the perceived depth of a structure (or the position of the focal plane) will differ from the true depth.
15. Refractive index mismatch will also tend to defocus both the excitation illumination and emitted signal, reducing signal intensity and – by spreading the PSF – also reduce resolution.
16. High NA water-immersion and multi-immersion objectives are useful in managing refractive index mismatch. Otherwise use an immersion objective below NA 0.85, if this is available.
17. High NA 100x objectives are not required for confocal microscopy.
18. Reflection confocal microscopy is useful for imaging opaque specimens or samples stained with markers that reflect light.

Key Reading

Egner, A. and Hell, S.W. (2006). 'Aberrations in confocal and multi-photon fluorescence microscopy induced by refractive index mismatch'. In: Pawley, J.B. (ed.), *Handbook of Biological Confocal Microscopy*, Chapter 20, 3rd ed. New York: Springer, pp. 404–413.

- A very concise yet thorough explanation of the effect of refractive index mismatch

Pawley, J.B. (2006). *Handbook of Biological Confocal Microscopy*, 3rd ed. New York: Springer.

- This massive tome, first published in 1989, is the pre-eminent reference work in the field and is sometimes referred to as 'the blue bible' on the confocal listserv and elsewhere. It is not light reading, but it does cover every aspect of the theory and practice of biological confocal microscopy that you are likely to encounter.

References

Amos, W.B. (2000). 'Instruments for fluorescence imaging', Figure 14, page 96. In: Allan, V.J. (ed.), *Protein Localization by Fluorescence Microscopy–A Practical Approach*, Chapter 4. IRL Practical Approach series. Oxford University Press, pp. 67–108.

Amos, W.B., McConnell, G., and Wilson, T. (2012). 'Comprehensive Biophysics'. *Confocal Microscopy*, vol. 2, Chapter 2.2. Waltham, MA: Academic Press, pp. 3–23.

Besseling, T.H. et al. (2015). 'Methods to calibrate and scale axial distances in confocal microscopy as a function of refractive index'. *Jour. Microscopy 257*/2: 142–150.

Borlinghaus, R.T. and Kappel, C. (2016). 'HyVolution – the smart path to confocal super-resolution'. *Nature Methods* 13.

Borlinghaus, R.T. and Kuschel, L.R. (2012). 'The white confocal – spectral gaps closed'. In: Méndes-Vilas, A. (ed.), *Current Contributions to Advances in Science and Technology*, Chapter 3, vol. 2. *Formatex Microscopy series No. 5*, pp. 809–817.

Egner, A. and Hell, S.W. (2006). 'Aberrations in confocal and multi-photon fluorescence microscopy induced by refractive index mismatch'. In: Pawley, J.B. (ed.), *Handbook of Biological Confocal Microscopy*, 3rd ed. New York: Springer.

Hng, K. and Dormann, D. (2013). 'Confocal check – a software tool for the automated monitoring of confocal microscope performance'. *PLoS One 8*/11: e79879. doi: 10.1371/journal.pone.0079879.

Hovis, D.B. and Heuer, A.H. (2010). 'The use of laser scanning confocal microscopy (LSCM) in materials science'. *Jour. Microscopy 240*/3: 173–180.

Isenberg, G. (1997). *Modern Optics, Electronics and High precision Techniques in Cell Biology*. Berlin: Springer.

Jonkman, J. et al. (2014). 'Quantitative confocal microscopy: beyond a pretty picture'. In: Waters, J.C. and Wittmann, T. (eds.), *Quantitative Imaging in Cell Biology Methods in Cell Biology*, Chapter 7, vol. 123. Elsevier Inc., pp. 113–134.

Murray, J.M. (2013). 'Practical aspects of quantitative confocal microscopy'. In: Sluder, G. and Wolf, D.E. (eds.), *Digital Microscopy*, Chapter 18, 4th ed. *Methods in Cell Biology* vol. 114. Elsevier Inc., pp. 427–440.

Visser, T.D. and Oud, J.L. (1994). 'Volume measurements in three-dimensional microscopy'. *Scanning 16*: 198–200.

Visser, T.D. et al. (1992). 'Refractive index and axial distance measurements in 3-D microscopy'. *Optik 90*/1: 17–19.

Zucker, R.M. (2014). 'Evaluating confocal microscopy system performance'. In: Paddock, S.W. (ed.), *Confocal Microscopy*, Chapter 17, 2nd ed. *Methods in Molecular Biology* vol. 1075. New York: Human Press, pp. 321–374.

Further Reading

Cody, S.H. et al. (2005). 'A simple method allowing DIC imaging in conjunction with confocal microscopy'. *Jour. Microscopy 217*/3: 265–274.

Kam, Z. et al. (1997). 'Three-dimensional microscopy in thick biological samples: a fresh approach for adjusting focus and correcting spherical aberration'. *Bioimaging 5*: 40–49.

Internet Resources

The Control of Artificial Optical Radiation at Work Regulations 2010: http://www.legislation.gov.uk/uksi/2010/1140/made.
Non-ionising radiation guides: http://www.ohsbok.org.au/wp-content/uploads/2013/12/25-Hazard-Non-ionising-radiation.pdf and guide for employers: http://www.hse.gov.uk/radiation/nonionising/employers-aor.pdf.
Radiation protection supervisors: http://www.hse.gov.uk/pubns/irp6.pdf and safe working: http://www.hse.gov.uk/pubns/indg334.pdf.
An online Nyquist calculator can be found at: http://www.svi.nl/NyquistCalculator.
Monitor adjustment: Dry Creek Photo: https://www.drycreekphoto.com/Learn/Calibration/monitor_sensitivity.html.
Norman Koren: http://www.normankoren.com/makingfineprints1A.html and the homepage: http://www.normankoren.com.

Notes

1. For details of Control of Artificial Optical Radiation at Work Regulations 2010 online, see the URL links given under Internet Resources at the end of this chapter. The latest British Standards Institute laser safety standard is BS EN 60825-1 (31/08/14).
2. Where the sample is photosensitive, the pinhole may be opened to 1.25 AU without appreciable loss of image clarity. Yet to give the same signal intensity in the image, only 60% of the laser power needed when imaging at 1 AU is required (see Figure 7.3, page 117 in Jonkman, J. et al., 2014). The reference is given at the end of this chapter.
3. The dwell time is the period over which the fluorescence intensity in the specimen is sampled at a particular point during the raster scan to be combined into a single pixel value.
4. Shot noise, or photon noise, is intrinsic noise arising from the uncertainty associated with counting quantum-mechanical events. Quantum particles only occur as discrete entities, and so the camera or PMT detector can never collect fractions of photons. The noise distribution is modelled by a Poisson distribution such that the noise component is equal to the square root of the signal intensity. Like diffraction, at low light intensity levels (i.e. low irradiance), shot noise is a rigid physical limitation to image quality. See Chapter 30, Section 30.10.
5. The difference between SNR and signal-to-background ratio (SBR) is explained in Chapter 30, in endnote 10. A confocal microscope reduces SBR, rejects out-of-focus and scattered light by means of the pinhole, thus giving a high contrast image with low background. It may still be noisy; indeed, point-scanning laser confocal microscopes are inherently noisy, as discussed in Section 18.6.
6. The Optical Index is the ratio of the objective's numerical aperture to its magnifying power. See Chapter 31, Section 31.7 for further details.
7. The broadening of the PSF in the axial direction gets progressively worse as the focusing depth increases.

19

Light-Sheet Microscopy

Outline

Understanding Light Microscopy, First Edition. Jeremy Sanderson.

19.1 Introduction

Light-sheet microscopy allows biologists to image tissues very fast in three dimensions with very little photobleaching. In this design of microscope, optical sectioning is performed by a sheet of laser light used to illuminate the specimen orthogonally from side-on rather than through the objective, as is the case with widefield and confocal fluorescence microscopes (Figure 19.1). By rotating the sample

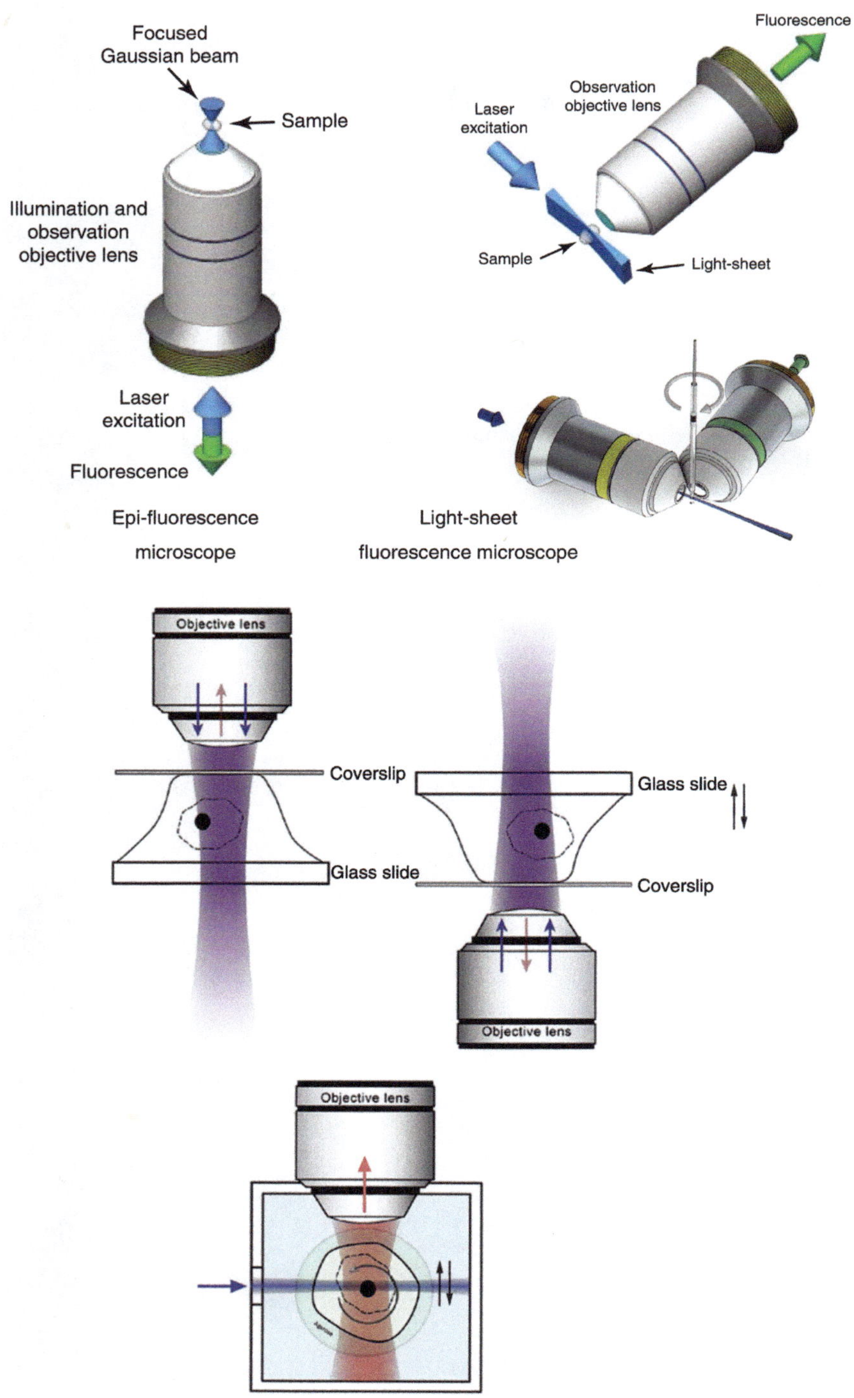

Figure 19.1 **The light-sheet microscope** *In a conventional widefield fluorescence or confocal microscope, the objective acts as its own condenser with both illuminating light and emitted signal travelling on-axis through the objective. The light-sheet microscope is arranged with the illuminating objective and detection objective orthogonally to one another. This allows for optical sectioning with a very thin light sheet of illumination. Source: Capoulade et al. 2014. Reproduced with permission from John Wiley & Sons; Weber and Huisken 2012. Reproduced with permission from Macmillan Publishers Ltd; Reproduced with permission from Carl Zeiss Ltd.*

and also by illuminating the sample from multiple angles, a 3D tomographic data set can be built up of cells, tissues, embryos and entire small organisms. Tomographic imaging (e.g. magnetic resonance imaging, MRI, computed tomography, CT or μCT and optical coherence tomography, OCT) is widely used as a clinical diagnostic tool. The sample is normally rotated and imaged along multiple angles, these different views being merged into a single 3D data set. These tomographic methods were developed to image macroscopic samples in three dimensions. However, the large volume capability comes at the cost of limited resolving power, which is generally not less than 10–20 μm. Optical coherence tomography has a resolving power limit of 2 μm, but it suffers from a very limited depth penetration into tissues. Light-sheet microscopy allows functional imaging of labelled proteins at high resolution, with a large field of view, minimal photo-toxicity at very high frame rates.

A form of light-sheet microscopy was used to image colloids in the 1900s. Since its reintroduction in 2004 (Huisken et al., 2004) most light-sheet platforms were self-built (there is a vibrant self-build community supported by OpenSPIM). The first commercial design, introduced in 2009, a macroscope, was manufactured by LaVision Biotec, followed in 2012 with a light-sheet microscope made by Zeiss, the Lightsheet Z1. The third commercial platform at the time of writing is an inverted light-sheet configuration manufactured by Applied Scientific Instrumentation. This instrument has the orthogonally set illumination and detection pathways tilted at 45° to the horizontal axis of the microscope stage.

In a conventional widefield fluorescence or confocal microscope, the objective acts as its own condenser with both illuminating light and emitted signal travelling on-axis through the objective. The confocal pinhole rejects out-of-focus light, but this can only be efficiently achieved (particularly in large or thick samples) when imaging cells and tissues that do not scatter light. Images from thick, scattering tissues show a significant drop in brightness with increasing image depth. Moreover, the negative effects associated with confocal imaging scale with the number of focal planes collected in a Z-stack data set. In contrast to one-photon confocal or two-photon microscopy, light-sheet microscopy can acquire data sets from much thicker tissues. Light-sheet combines the speed of spinning disc confocal microscopy with the advantage of point-scanning confocal microscopy to give optical sectioning and reduced out-of-focus blur. Like widefield microscopy, it delivers very much reduced photobleaching and reduced sample photo-toxicity.

In a significant commentary (Scherf and Huisken, 2015), the authors suggest a paradigm shift from traditional microscope design to construct smart and gentle microscopes that are built around the sample. Classical designs of upright and inverted light microscopes traditionally accommodate a sample embedded into a slide preparation. In fluorescence microscopy, the exposure of the sample to photons and the number of photons emitted as signal are critically important (Box 19.1). The light-sheet microscope exposes the specimen

Box 19.1 Healthy Exposure Values for Living Cells

There is no precise answer as to how much light a living biological specimen should be exposed to on the microscope. What modern microscopists now realise is that the tolerable irradiation intensity is very much less than was once thought. Most live-cell time-lapse experiments are run under conditions with exposures that are far too high, particularly since the presence of conjugated fluorophores also has a negative impact.

A study on yeast (Carlton et al., 2010) showed that irradiation intensity[i] had to be 1/100th the usual value [$\approx 5 \times 10^{-5}$ W/μm^2] in order that normal cell growth is not arrested. That is, 500 nW/μm^2/sec [0.5 μJ/μm^2] – an extremely low value.

Swedlow (2010) in a commentary on this work gives the exposure from a mercury lamp on a widefield fluorescent lamp as 100 μW/cm^2. Stelzer (2015) related the radiant flux of the microscope to that of the sun. The solar constant is 1.4 kW/m^2 at the equator and around 1 kW/m^2 in central Europe. In microscopic terms, this is about 1 nW/μm^2 or 100 mW/cm^2. The maximal radiant exposure should be around 0.5 mJ/μm^2 (even this is 1000x the exposure value given by Carlton) and infer that cells and small model embryos should not be exposed to more than a few millijoules and a few hundred millijoules, respectively. For a cell of 100 μm diameter, a 'single sun' exposure of 10 minutes is a total of approximately 5 mJ, whilst for a small embryo, the same 10 minute exposure is approximately 400 mJ.

Very few microscopes can operate at these low levels, but the light sheet–based fluorescence microscope is one, whereas confocal and even widefield fluorescence microscopes usually operate in a 'multiple-suns' regime, with consequences for imaging living cells and tissues for any length of time.

That said, the editorial on the subject of photo-toxicity in *Nature Methods* comments: 'Even light-sheet microscopy can cause severe phototoxic effects in spite of delivering far less light than other techniques during time-lapse three-dimensional imaging, not least because users push application boundaries.'

References

Carlton, P.M. et al. (2010). 'Fast live simultaneous multiwavelength four-dimensional optical microscopy'. *PNAS 107*/37: 16016–16022.

Editorial (2013). 'Artifacts of light'. *Nature Methods 10*/12: 1135.

Stelzer, E.H.K. (2015). 'Light-sheet fluorescence microscopy for quantitative biology', Box1. *Nature Methods 12*/1: 23–26.

Strack, R. (2015). 'Death by super-resolution imaging'. *Nature Methods 12*/12: 1111.

Swedlow, J.R. (2010). 'Advanced hardware and software tools for fast multidimensional imaging of living cells'. *PNAS 107*/37: 16005–16006.

[i] Radiant flux [W] is the quantity of energy of light emitted per second in all directions. Photon flux is the number of photons emitted, transferred or received per unit area per unit time. Irradiance (energy flux) is energy received by a surface per unit area [W/m^2].

to between one to four orders of magnitude less radiative energy than widefield or confocal fluorescence microscopes. The light-sheet technique is proving to be of great interest, in particular to developmental biologists, because it allows high-resolution studies of entire small organisms and embryos to be undertaken by tomography; for example, see Table 1 in Strobl and Stelzer, 2014. Multicellular life is both complex and dynamic, and light-sheet microscopy is ideal for non-invasively examining growing tissues over time. For example, with light-sheet microscopy it is possible to record a beating fish heart in three dimensions with sufficient temporal resolution to do so without blurring (Trivedi et al., 2015), to image developing mouse embryos (Ichikawa et al., 2013; Udan et al., 2014), *Tribolium* insect embryos (Strobl and Stelzer, 2014), or plants (Ovečka et al., 2015; Berthet and Maizel, 2016).

Light-sheet microscopy is also generically referred to as SPIM: selective plane illumination microscopy. The microscope originally developed by Huisken at EMBL (Huisken, 2004) used a cylindrical lens to shape the illuminating beam to form a static light sheet (Figure 19.2), and thus the term SPIM was adopted. However, a better light sheet can be formed by using laser scanning. This is incoherent, and so is less affected by specimen-induced scattering of light. Unfortunately, in a rapidly developing field, there are many acronyms used to describe different modalities of light-sheet microscopy. For a chart of these, see Figure 8.2 in Colombelli and Lorenzo (2014). Key acronyms are listed in Table 19.1.

19.2 Historical Development of the Light-Sheet Microscope

The principle of light-sheet microscopy is not new. At the beginning of the 20th century several researchers were trying to increase the resolving power of the light microscope. Siedentopf and Zsigmondy employed dark-field illumination to study colloidal gold and called their microscope an ultramicroscope. This work was reported at the beginning of the 20th century (Siedentopf and Zsigmondy, 1903) and is often taken as the starting point of light-sheet microscopy. The short working distance of the high numerical aperture objectives used meant it proved impossible to orientate one objective orthogonally to the other (sometimes still a problem today), and so the front lenses and mounts of both objectives were partially cut away at a 45° angle and polished flat. This early orthogonal illumination microscope has been recreated, described and illustrated by Mappes et al. (2012) in a centenary anniversary paper. Fortunately, for most applications, no high NA objective is required to form a light sheet on the illumination side. A standard 10x/0.3 objective will generate a light sheet of 2–4 μm.

The first modern use of light-sheet microscopy was by Voie et al. in 1993 employing a macroscope, when orthogonal illumination was used to make tomographic recordings of cleared[1] guinea pig cochleae (Santi, 2011). They called their method orthogonal-plane fluorescence optical sectioning (OPFOS) but, despite being published in the *Journal of Microscopy*, the technique was not widely adopted, possibly because of the 'wire model' image analysis methods, rather than 3D reconstruction, used. Also in the 1990s, the Stelzer lab was investigating the use of confocal theta microscopy (Lindek et al., 1997) to improve axial resolving power. Both Voie's OPFOS instrument and Stelzer's confocal theta microscope were the forerunners of the selective plane illumination (SPIM) design of light-sheet microscope. These two instruments between them anticipated (Pampaloni et al., 2015) several features that ensured the success of light-sheet microscopy: laser illumination, illumination and detection paths set orthogonally to each other, water-dipping objectives for both illumination and detection, multi-view tomographic data sets by independent image acquisition from several sides and mounting and rotation of the specimen within a capillary tube.

The Ultramicroscope II (which is in fact a macroscope, see Chapter 4, Section 4.6) sold by LaVision Biotec was developed from a design developed by the Dodt group in Vienna (Dodt et al., 2007) and has its origins in the design built by Voie. The macroscope is built around an upright configuration, catering for sample sizes from a few micrometres up to one centimetre thick, which are immersed in buffer or clearing solution. Image acquisition is with a 20x/NA 0.5 objective, and the sample is moved vertically through a horizontal light sheet, which can illuminate the sample from one or two directions, as required. The Zeiss lightsheet Z1 is based on the horizontal setup introduced by Husken and Stelzer. The magnification offered for the detection objective is from 5x to 63x, all immersion objectives, with sequential acquisition of multiple stacks of optical sections from different rotation angles. Applied Scientific Instrumentation offer a single-sided and dual-sided illumination inverted SPIM configuration. Although the objective choice is limited, because they have to be placed very close together, magnification is from 10x to 40x. The advantage is greater versatility for sample preparation, accommodating glass slides and Petri dishes. The dual illumination diSPIM (Wu et al., 2011, 2013; Chhetri et al., 2015) microscope offers z-axis resolution isotropic with lateral resolution. It is possible to home-build this configuration (see Kumar et al., 2014) and there now exists a vibrant and experienced home-build community OpenSPIM (see Section 19.6).

The most recent advances in light sheet have been twofold: in improving the design of the illumination sheet and in managing very large data sets. Incoherent (i.e. non-diffracting) Bessel beams now achieve better isotropic resolution over larger fields of view; this approach is particularly suited to two-photon light-sheet microscopes and has been combined (Planchon et al., 2011) with structured illumination microscopy. As an extension of this, the lattice light sheet, also from the Betzig lab (Chen et al., 2014) offers two key advantages of much less photon flux that the sample is exposed to – better than even the Gaussian beam light sheet, for which low photobleaching, for low photo-toxicity is a major advantage, and much faster image acquisition speeds than its predecessors.

The latest development at the time of writing (Guan et al., 2015) is the compact plane illumination plugin (PIP) device. The base component is 3D printed; the device measures 15 cm × 5 cm × 5 cm and can be mounted onto the stage of a conventional inverted microscope. A similar device has been prosed by Cutrale and Gratton (2014), who also review the advantages of inclined selective plane microscopy.

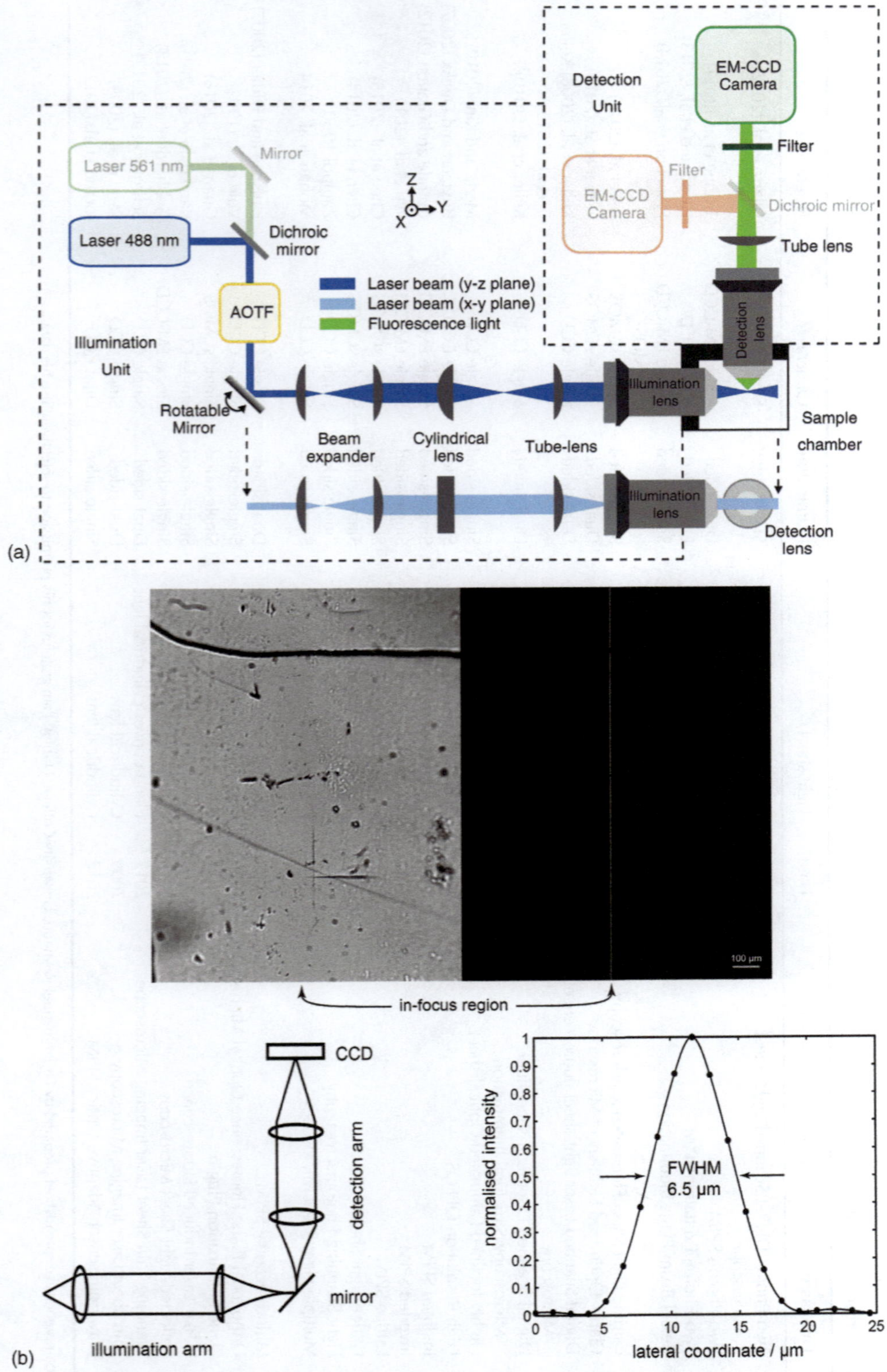

Figure 19.2 **Architecture of the light-sheet microscope** *Panel (a) a beam expander, cylindrical lens and a tube-lens together form the Gaussian light sheet. The essential components are shown in bold, with the optional second red 561 nm laser and second sCMOS camera in lighter colours. The light sheet is shown in plan view and, below this, in side elevation also. Panel (b) is taken from Figure S5 in Huisken et al. (2004) showing the sectioning performance of the light-sheet microscope. Under white light illumination (left-hand) a mirror surface shows a large depth of focus and lack of sectioning when imaged with a 5x/NA 0.25 objective. The darker (right-hand) field of view shows the good sectioning ability and small depth of focus of the light-sheet microscope, marked in each case by the arrows. The lower part of this panel shows the basic configuration of the light-sheet microscope, and the full width half maximum (FWHM) achieved. Source: (a) Capoulade et al. 2014. Reproduced with permission from John Wiley & Sons. (b) Huisken et al. 2004. Reproduced with permission from AAAS.*

Table 19.1 *Light-Sheet Microscopy Acronyms*

Acronym	Full name	Date	Light sheet	Illumination	Camera	Reference
2P-DSLM (2nd harmonics)	Two-Photon Digital Scanned Light-Sheet Microscope	2012	Tube-lens	Single-sided	Single CCD	Olarte et al. (2012)
2P-SPIM	Two-Photon SPIM	2011	Tube-lens	Dual-sided	Single EM-CCD	Truong et al. (2011)
2PE-SPIM	Two-Photon Excitation SPIM	2013	Cylindrical lens	Dual-sided	Single CCD	Lavagnino et al. (2013)
BBPI	Bessel Beam Plane Illumination	2011	Two-photon annulus lens	Single-sided	Single EM-CCD or sCMOS	Planchon et al. (2014)
c-LSFM	cardiac Light Sheet Fluorescence Microscopy	2016	Cylindrical lens	Single-sided	Single sCMOS	Fei et al. (2016)
COLM	Clarity Optimized Light-sheet Microscopy	2014	Cylindrical lens	Dual-sided	Single sCMOS	Tomer et al. (2014)
DSLM	Digital Scanned Laser light-sheet fluorescence Microscopy	2008	F-theta lens, scanning lens	Dual-sided	Single CCD	Keller et al. (2008) Keller et al. (2011)
DSLM-SI	Digital Scanned Laser light-sheet fluor. Microscopy-Structural Illumination	2010	F-theta lens, scanning lens	Dual-sided	Single CCD	Keller et al. (2010)
HILO	Highly Inclined Laminated Optical sheet	2010	Scanned laser line	Single-sided	Single CCD	Mertz and Kim (2010)
HR-OPFOS	High Resolution OPFOS	2007	Cylindrical lens	Single-sided	Single CCD	Buytaert and Dirckx (2007)
iSPIM	Inclined SPIM	2012	Cylindrical lens	Single-sided	Single EM-CCD	Cutrale and Gratton (2012)
iSPIM	Inverted SPIM	2011	Cylindrical lens	Single-sided	Single EM-CCD	Wu et al. (2011)
Lattice SPIM	Lattice SPIM	2015	2 cylindrical lenses	Single-sided	Single sCMOS	Chen et al. (2015)
LLS	Lattice Light Sheet	2014	2 cylindrical lenses	Single-sided	Single sCMOS	Chen et al. (2014)
LSP	Light Scanning Photomacrography	1960	White light	Single-sided	Single CCD	Zampol 1960
MLSM	Multiple Light Sheet Microscopy	2014	Cylindrical lens and horizontal filter	Single-sided	Single CCD	Mohan et al. 2014
mSPIM	Multidirectional SPIM	2007	Pivoting cylindrical lens	Dual-sided	Single CCD	Huisken and Didier (2007)
OPFOS	Orthogonal Plane Fluorescence Optical Microscopy	1993	Cylindrical lens	Single-sided	Single CCD	Voie et al. (1993)
PIP	Plane Illumination Plugin	2016	Cylindrical lens	Single-sided	Single sCMOS	Guan et al. (2016)
PLIF	Planar Laser Induced Fluorescence	2009	Powell lens	Single-sided	Single CCD	Karaköylü et al. (2016)
RLSM	Reflected-Light Sheet Microscopy	2013	Cylindrical lens	Single-sided	Single EM-CCD	Gebhardt et al. (2013)
sTSLIM	Scanning Thin Sheet Laser Imaging Microscope	2012	Plano-convex cylindrical lens	Dual-sided	Single CCD	Schröter et al. (2012)
TSLIM	Thin Sheet Laser Imaging Microscopy	2009	Cylindrical lens	Dual-sided	Single CCD	Santi et al. (2009)
WAO-SPIM	Wavefront sensor Adaptive Optics SPIM	2012	Cylindrical lens	Single-sided	Dual CCD	Jorand et al. (2012)

This table is taken from a comprehensive spreadsheet prepared by Dr Emmanuel Reynaud, University College, Dublin. I am grateful for his permission to selectively use his data.

19.3 The Rationale of Light-Sheet Microscopy

Light-sheet microscopes use laser-generated external illumination from side-on to excite only the fluorophores at the focal plane of the objective (Figure 19.2) thus creating an optical section. Very thin light sheets (≈1 μm) are used to create thin optical sections through smaller samples (e.g. 20–100 μm), whereas a thicker light sheet (5–10 μm) has to be used for larger samples (1–5 mm) for the illumination to remain even across the field of view (see Box 19.2 for details of the light-sheet parameters). Because the illumination optical path is independent of the detection path, it is not constrained by the working distance or the numerical aperture of the illuminating objective and exposes only a very narrow volume of tissue about the focal plane to light, greatly reducing photobleaching. The Gaussian beam light sheet of the SPIM setup is formed by first expanding a laser beam with a collimating beam expander lens, then flattening this in the axial direction by using a cylindrical lens to form a lateral (i.e. in the x-y plane) sheet of light. The sample is then translated in the z-axis. The sample can be rotated and manipulated in the light sheet beam so that a high-resolution 3D image can be built up by combining images from several angles by optical tomography, a technique normally associated with electron microscopy and the CT scanner (which uses X-rays), but which is being adopted in light microscopy also.

Box 19.2 Light-Sheet Parameters

When a light sheet is formed by a slit placed in front of a cylindrical lens, the laser line remains unaltered along the y-axis, but is formed into a hyperbolic profile in the x-z plane by the cylindrical lens:

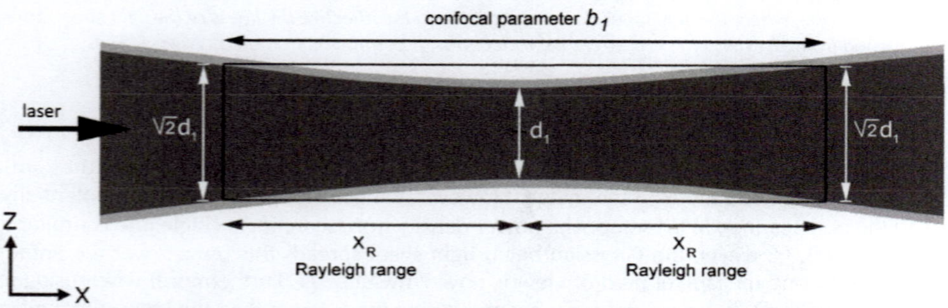

Source: Buyaert et al. 2012. Reproduced under the terms of the Creative Commons Attribution Share-Alike licence, CC-BY-SA 3.

The thickness of the light sheet in the z-axis increases either side of the point of focus, d_1, where the beam has minimal thickness. The Rayleigh range, X_R, is the distance either side of d_1 where the hyperbolic profile thickens to $\sqrt{2}.d_1$.

The confocal parameter, b_1, is a measure of the depth of field for which the thickness of the light sheet can be used for optical sectioning where the depth of field $\leq \sqrt{2}.d_1$. b_1 also defines the field of view into which the sample should ideally fit.

$$b_1 = 2X_R = \pi d_1^2/2\lambda \approx d_1/\mathrm{NA} \text{ and } d_1 \approx 2\lambda/\pi.\mathrm{NA}, \text{ so } \sqrt{2}.d_1 \approx 1/\sqrt{b_1},$$

where NA is the NA of the cylindrical lens and λ is the laser wavelength forming the light sheet.

References

Adams, M.W. et al. (2015). 'Light sheet fluorescence microscopy (LSFM)'. *Current Protocols in Cytometry* 12.37.1–12.37.15. doi: 10.1002/0471142956.cy1237s71.

Becker, K. et al. (2013). 'Ultramicroscopy: light-sheet-based microscopy for imaging centimetre-sized objects with micrometer resolution'. *Cold Spring Harbor Protocols* 8: 704–713.

Buytaert, J.A.N. et al. (2012). 'The OPFOS microscopy family: high-resolution optical sectioning of biomedical specimens'. *Anatomy Research International* 2012, Article ID 206238. doi: 10.1155/2012/206238.

Another means of illuminating the sample is to scan a laser sheet across the sample (see Figure 19.3). A moving mirror rapidly scans a micrometre-thin spherical focus of laser light into a plane. This is called digitally scanned laser light sheet microscopy (DSLM; Keller et al., 2008). DSLM has the advantage of more even illumination intensity (95% compared to 3% for a cylindrical lens) more efficiently

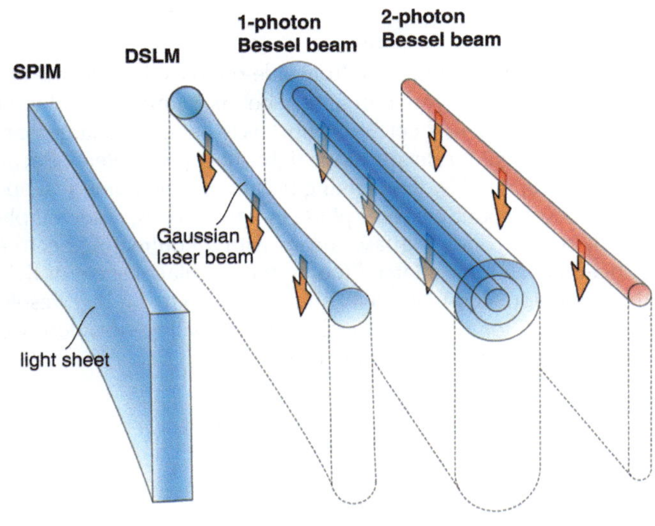

Figure 19.3 **Design of the illuminating light sheet** *In the simplest SPIM configuration of a light-sheet microscope, the light sheet is formed by a static Gaussian beam (SPIM). In the digital scanning (DSLM) mode, this Gaussian beam is scanned through the sample to provide more even illumination. Both SPIM and DSLM light sheets suffer from uneven thickness, and so the Bessel beam was developed. Although the core of the Bessel beam is suitably thin and long, allowing greater fields of view to be evenly illuminated, the side lobes do thicken the beam, which is why the Bessel beam is best suited for a two-photon light-sheet microscope, where the side lobes do not contribute to the effective thickness of the light sheet. Source: Weber and Huisken 2011. Reproduced with permission from Elsevier; Hulsken 2012. Reproduced with permission from John Wiley & Sons.*

yet with high SNR and very low excitation energy values. Image quality is improved because slit apertures are not used to form the illuminating sheet, resulting in fewer aberrations in the image. However, the scanned laser beam can potentially cause much greater local photobleaching than the non-scanned SPIM setup. The power density from scanning a single line is similar to that of the spinning disc confocal: about 1 μwatt/pixel. However, the Gaussian beam light sheet spreads this power over the entire field of view, so the power density for a 1024 × 1024 one megapixel field of view is now 1 nwatt/pixel. Furthermore, when using a lower-magnification detection objective, such as a 20x, this will have a field of view about nine times larger than the highest magnification objective used, the 63x objective. This in turn means a potential reduction in power density across the field of view of one order of magnitude than when a small field of view is magnified at the largest possible magnification. The trade-off of the Gaussian beam design of light sheet is that thicker light sheets are required to cover larger fields of view, which degrades the axial resolution, one of the driving forces for developing digitally scanned laser light sheet microscopy. The Bessel beam is thinner than the Gaussian beam, allowing imaging over a greater field of view with a better depth of field, but deconvolving the data sets is more difficult than with Gaussian beams. Where the sample is thin, the Gaussian beam configuration is advantageous; for larger samples, the Bessel beam configuration is likely to be advantageous, but the decision will depend on deconvolution requirements and post-processing of the data sets.

Light-sheet microscopy has tremendous potential to completely image sensitive living specimens (albeit only up to 2–3 mm in size), so that whole organs in, say, an intact living embryo, may be studied, allowing developmental biologists to move away from merely looking at single cells on a slide[2]. This has important consequences for organogenesis: the formation and development of organs and gene or protein expression can be tracked over days. It is possible to record a beating fish heart in three dimensions with sufficient temporal resolution to do so without blurring. Because one objective is used for illumination and another (higher NA objective) for detection, the depth of penetration of the beam is good, being equivalent to a confocal microscope, yet this system operates in widefield mode with a CCD or sCMOS camera.

For example, the ability to image embryos over a long time with minimal photobleaching and photo-toxicity is more important than absolute resolving power per se. Although light-sheet microscopy does not enhance the lateral resolving power and thus is not a super-resolution technique, axial resolution is improved down to 1.7 μm (Table 19.2). Purely in terms of resolving power, the confocal microscope outperforms the Gaussian beam SPIM when using detection objectives with NAs greater than 0.8 (Engelbrecht and Stelzer, 2006). The theoretical lateral resolution limit of a light-sheet microscope is given by the classical formulae (see Section 7.4) although the pixel pitch of the sCMOS camera (0.5–3 μm) imposes the resolving power limit, especially for a large field of view. The axial resolving power is determined by the thickness of the illuminating sheet of light (2–6 μm) when using long working distance objectives with a large depth of field.

Spherical aberrations arising from refractive index mismatch (e.g. Hufnagel and Pepperkok, 2016) can be minimised by using immersion objectives to view the sample mounted in ≈ 1% agarose, or Matrigel, suspended in an aqueous fluid within an especially designed chamber. Typical preparation methods are discussed in the next section and shown in Figure 19.6. The sample is often contained in a FEP (fluorinated ethylene propylene) capillary tube (Kaufmann et al., 2012) of equivalent refractive index to water (RI = 1.338). For further details, including guidance on specimen preparation, see Huisken and Stainier (2009).

The advantages of light-sheet microscopy are:

- good optical sectioning – approx. equivalent to confocal for objectives < 0.8 NA; sectioning on the excitation side: low NA objectives have good working distance; z-resolution not important;
- less blurring of thick samples than widefield;
- excitation intensity much less than for confocal microscopes;
- sCMOS cameras possess a very high dynamic range and large field of view;
- can capture very fast frame rates over the large field of view;
- high dynamic range and high frame both lead to reduced exposure to light; and (hence)
- very much reduced photo-damage and thus photo-toxicity[3].

The disadvantages (aside from expense) are few:

- sample preparation can be challenging and thus limit the organisms studied;
- the illumination can cause stripes in the image (illumination from different angles helps); and
- optical sectioning limited by the excitation sheet thickness (from 10 to 1 μm).

Table 19.2 Resolving Power Comparison

Detection objective	Technique	FWHM $_{lateral}$ (μm)	FWHM $_{axial}$ (μm)	PSF volume (attoL)
10x/NA 0.3 W	Widefield FM	1.00	15.17	7943.0
	Confocal	0.68	10.38	2513.1
	Two-photon	1.22	18.57	14 472.0
	SPIM	**1.00**	**5.73**	**3000.2**
40x/NA 0.8 W	Widefield FM	0.37	1.94	139.1
	Confocal	0.25	1.33	43.5
	Two-photon	0.45	2.38	252.3
	SPIM	**0.37**	**1.65**	**118.3**
100x/NA 1.0 W	Widefield FM	0.29	1.15	50.6
	Confocal	0.20	0.79	16.5
	Two-photon	0.36	1.40	95.0
	SPIM	**0.29**	**0.99**	**43.6**
100x NA 1.2 W	Widefield FM	0.24	0.69	20.8
	Confocal	0.16	0.47	6.3
	Two-photon	0.29	0.84	37.0
	SPIM	**0.24**	**0.65**	**19.6**

Source: Data from Table 2, page 1479, Engelbrecht and Stelzer 2006.
The calculated FWHM and PSF values are for PSFs derived from numerical simulation. These artificial values differed from PSF values, experimentally acquired from 0.1 μm diam. Orange fluorescent microspheres, by a maximum of 4% in the lateral and 8% in the axial direction (see Table 1 in Engelbrecht and Stelzer). The biggest discrepancy was for the 40x/NA 0.8 objective, whereas the 10x and 100x objectives differed between the simulated and experimental PSF values by no more than 0.6–2%. Similar discrepancy values, not exceeding 7%, were found between the PSF volumes, in attolitres.

Reference

Engelbrecht, C.J. and Stelzer, H.K. (2006). 'Resolution enhancement in a light-sheet based microscope (SPIM)'. *Optics Letters* 31/10: 1477–1479.

The light sheet can be both absorbed and scattered by the sample; to some extent this is inevitable. The result of scattering is seen as striped lines and shadows along the direction of illumination. To avoid this, the sample can be imaged sequentially (and with pivotal scanning on the Lightsheet Z1) from both sides of the light sheet using two objectives (Figure 19.4) and the two Z-stacks digitally fused into a single data set. Also see the commentary by Hufnagel and Pepperkok (2016). Because the sample is imaged in a chamber filled with water or aqueous medium, it can be kept alive for time-lapse work by heating and the supply of warmed 5% CO_2 (Figure 19.5). Since the sample is commonly mounted in an agarose cylinder and introduced into the imaging chamber from the top, it can be rotated about the vertical axis, which allows a tomographic 3D data set (of living organisms) to be constructed.

19.4 Sample Handling

For answering questions regarding the fundamental mechanisms of life, model organisms and tissues may be too complex, and so cells are studied. However, isolated cells grown in suspension or as adherent monolayers do not correspond to the in vivo environment of highly organised 3D populations of heterogeneous cells and an extracellular matrix tightly packed together (e.g. see Yamada and Clark, 2002). Cells naturally organise into tissues having a common functional purpose comprising differing cells within an extracellular matrix, so organotypic explant cultures maintain the original tissue architecture very well. Whist cell monolayers are undoubtedly easier

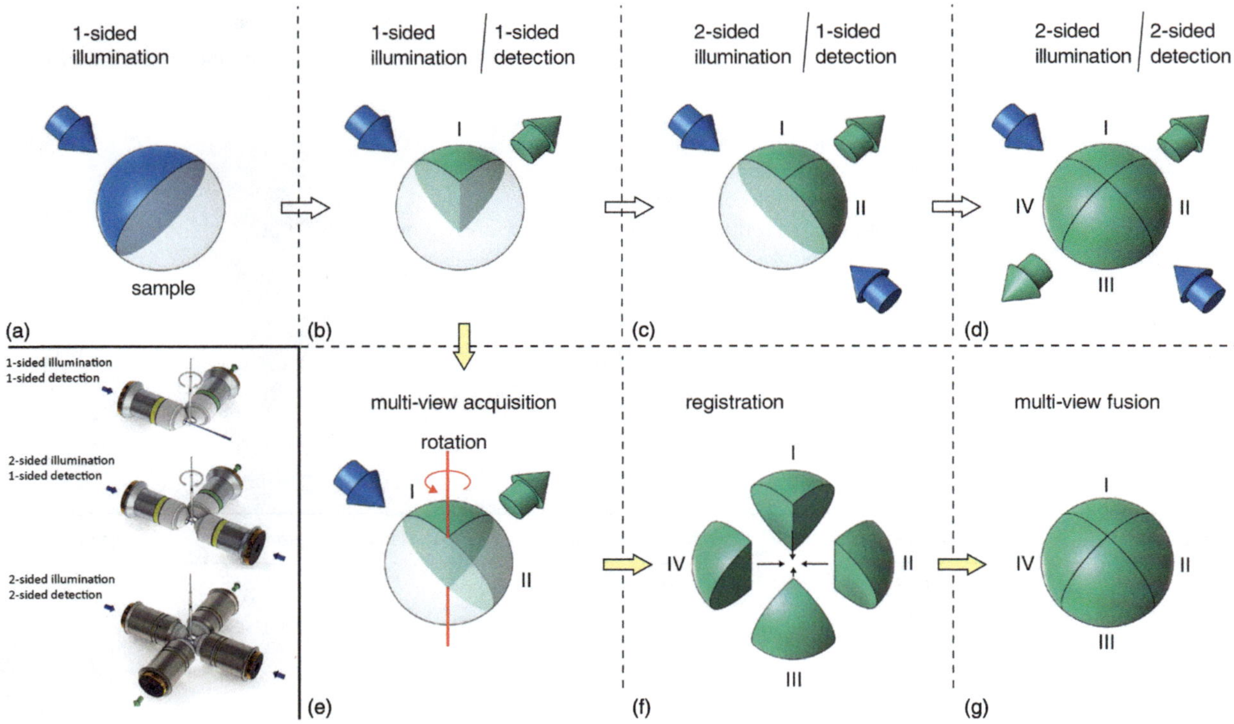

Figure 19.4 ***Multi-view illumination and detection*** *In a scattering sample, the illumination will penetrate both unevenly and only part way through the sample, on the side facing the illumination (a). Likewise, the emitted signal can only be acquired on the side facing the detection objective (b). By adding a second illuminating objective, quadrant II can be imaged (c), whilst adding a second detection objective, the final two quadrants, III and IV, can also be imaged. These architectures are shown together in the lower left-hand panel of the figure. An alternative means of collecting information from all four quadrants is to rotate the sample (e). These views need to be registered to sub-pixel accuracy (f) and fused (g) to form a 3D image of the entire data set as a multi-view reconstruction. Source: Weber and Huisken 2011. Reproduced with permission from Elsevier; Weber and Huisken 2012. Reproduced with permission from Macmillan Publishers Ltd.*

to maintain in culture, they are known to lose their in vivo tissue-related properties (De Vos et al., 2014). Histotypic cultures of cell spheroids are easier to culture and image by time-lapse microscopy than organotypic explants (Figure 19.6). As such, cell spheroid samples can be easily prepared and imaged with light-sheet microscopy.

In their concise overview paper, Reynaud et al. (2015) classify five methods (Figure 19.7) of sample preparation: embedding, clipped, enclosed, flat and flow-through, which are described in detail in the excellent white paper available as a PDF download from Zeiss. Stability is the key to sample preparation, since living samples potentially must be imaged continually over many days. Please bear in mind irradiance values; see the endnote in Chapter 22 on multi-photon microscopy (Section 22.2, endnote 4) and in Chapter 28 discussing live-cell imaging (Section 28.6, endnote 9). The gold standard for experimental fidelity is to succeed in getting the sample – whether cell culture, tissue explant, embryo or entire organism – to survive and successfully undergo further cell division or to produce viable offspring.

In order to rotate the sample in the chamber, most samples have rotational symmetry about their long axis; unlike a typical slide preparation, consideration must be given to the orientation of the specimen with respect to both the illumination and detection axes of the light-sheet microscope. The light sheet needs to pass through the region to be imaged, whilst ideally avoiding highly refractile or absorbent areas of tissue. The embryos or organisms are mounted into low-melting point agarose within a glass capillary or a plastic FEP tube. Once set, excess agarose at the end of the cast cylinder is removed, and the entire embedded specimen in its cylinder is extruded from the tube into the fluid-filled sample chamber. The 1.5% agarose matches nearly enough the refractive index of the medium in the chamber, which lies between that of water and that of the proteins of biological tissue. With care, the glass capillaries are reusable and are easy to handle.

It is important to include fluorescent beads (1:10 000 dilution of 0.5 μm diameter microspheres) dispersed within the agarose. These are recorded along with the image data and provide fiducial markers for later reconstruction of the 3D data set. Individual cells and cell spheroids are ideally mounted within the space formed in a hollow agarose or Matrigel cylinder enclosed in transparent foil. FEP tubes (Kaufmann et al., 2012) are useful for long-term time-lapse experiments, but care and forward planning is advised, since developing organisms will inevitably grow and change shape. In this case the FEP tube should be filled with liquid or low-viscosity media to allow for expansion (see Udan et al., 2014, for details). The tube can be closed with a plug of 3% low–melting point agarose. Besides a tube for holding samples, these can be suspended from a hook or glued (e.g. with Vetbond) to a rod. Figure 19.7 f–h, taken from Strobl, Schmitz, and Stelzer (2015), shows the egg of an insect *Tribolium* embryo mounted onto an agarose hemisphere, using liquid agarose as glue. An alternative is to 3D print (Jeandupeux, E., et al., 2015) a small chamber (Desmaison et al., 2013) from resin to hold a casting of hydrogel to hold the sample.

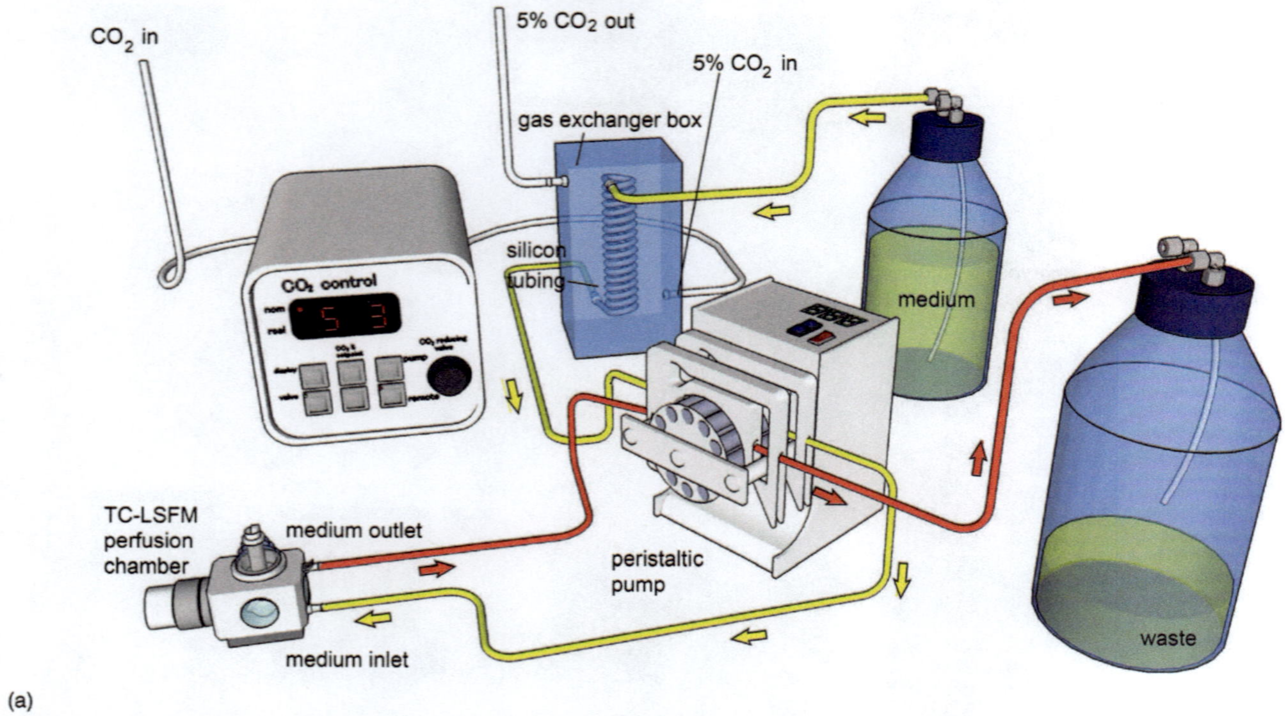

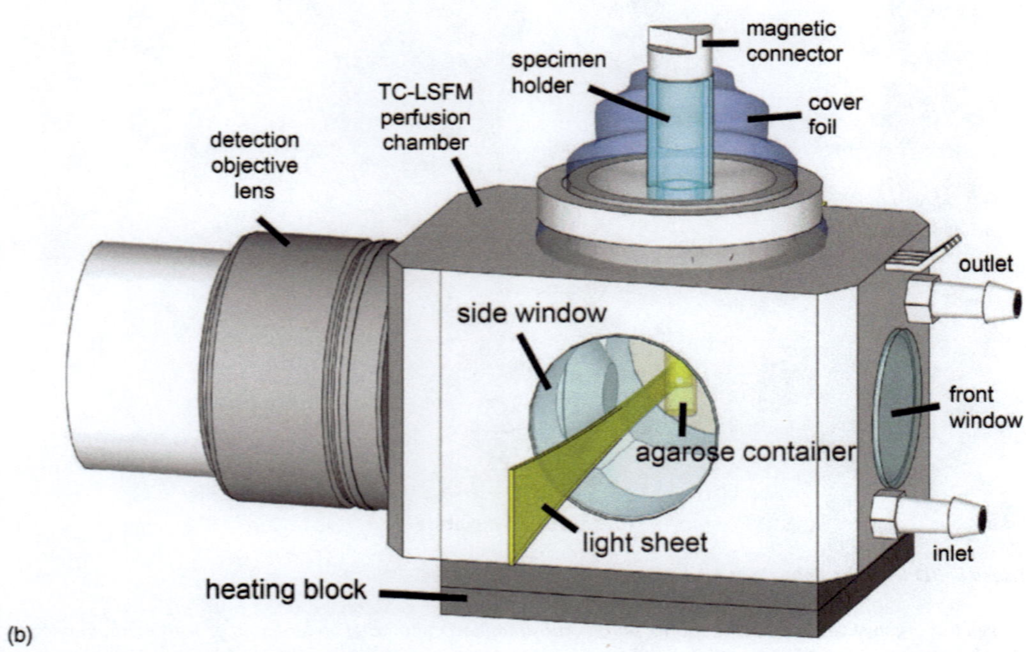

Figure 19.5 ***Live imaging with the light-sheet microscope*** *The low photobleaching and low photo-toxicity (see Box 19.1) means that the light-sheet microscope is ideally suited to long-term time-lapse experiments of living tissues. The fluid in the imaging chamber can be renewed and kept replenished using a peristaltic pump. The gas-permeable silicon tubing (about 4 m long) is tightly wound and placed within a gas-tight chamber and the gas mixture regulated by the CO_2 controller to saturate the perfusion fluid. The fluid is shown leaving the supply bottle in yellow and returning to the waste bottle in red. For further details, see Pampaloni, F., et al. (2014). Source: Pampaloni et al. 2014. Reproduced with permission from The Royal Society of Chemistry.*

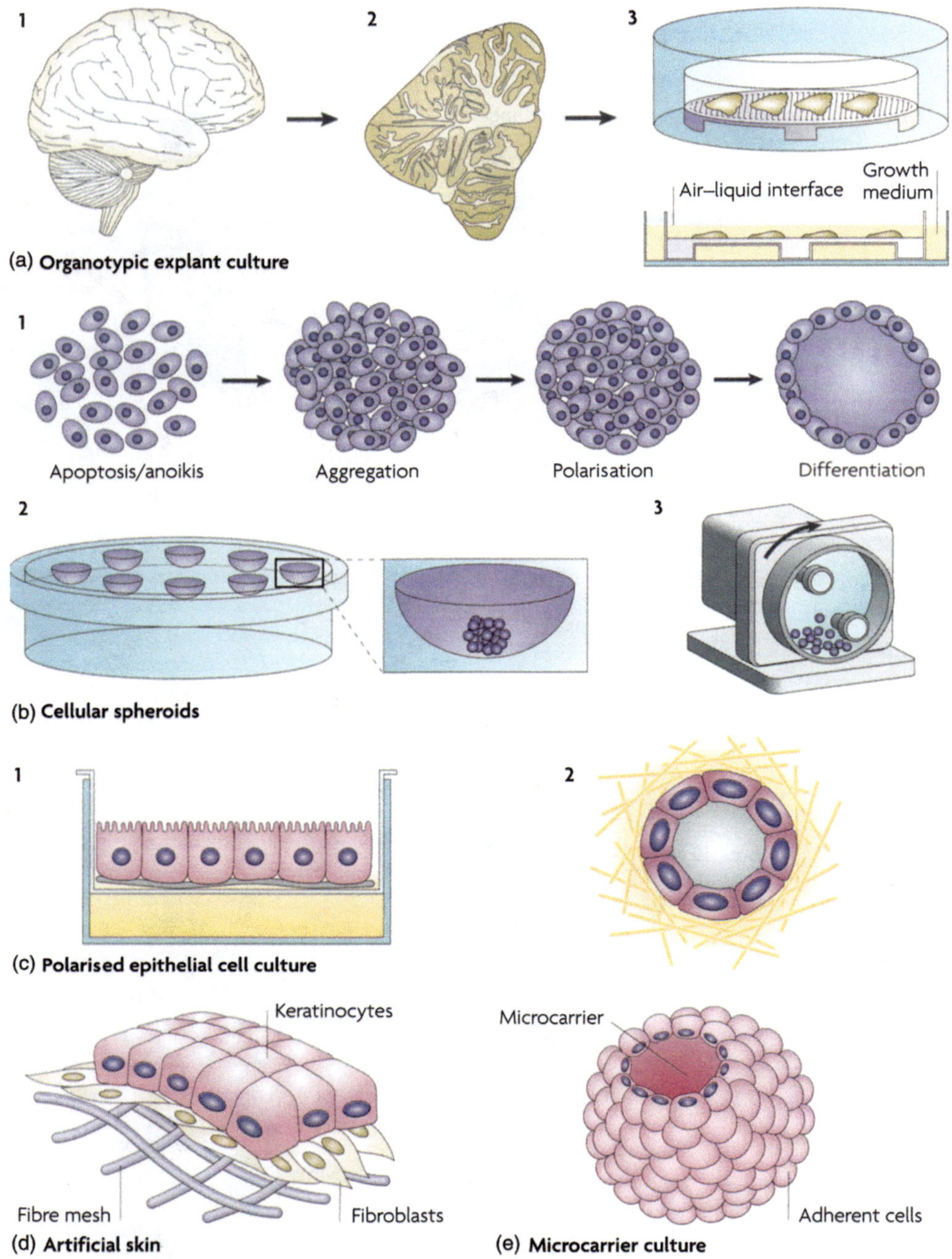

Figure 19.6 ***Studying cells and tissues in 3D with the light-sheet microscope***

(a) Organotypic explant culture. Dissected organ slices are placed upon a porous substrate and cultured at an air-liquid growth medium interface.
(b) Cellular spheroids. Endothelial cells cluster and form a differentiated hollow monolayer spheroid. These are obtained by culture in a hanging drop or in a rotating-wall culture vessel.
(c) (1) Polarised epithelial cell culture. Cells grown at the air-medium interface on porous membrane supports form polarised monolayers. (2) Three-dimensional growth of MDCK cells in hydrated collagen gel yields spherical monolayered cysts.
(d) Artificial skin. Cultured primary fibroblasts seeded onto a biodegradable fibre mesh; keratinocytes placed several weeks after onto this biomesh form an epidermal layer.
(e) Micro-carrier materials. Dextran, gelatine, glycosaminoglycan and porous polymer beads are used as a support for adherent anchorage-dependent mammalian cell lines.

Source: Text adapted from Figure 1, Pampaloni et al. 2007. Image reproduced with permission from Macmillan Publishers Ltd.

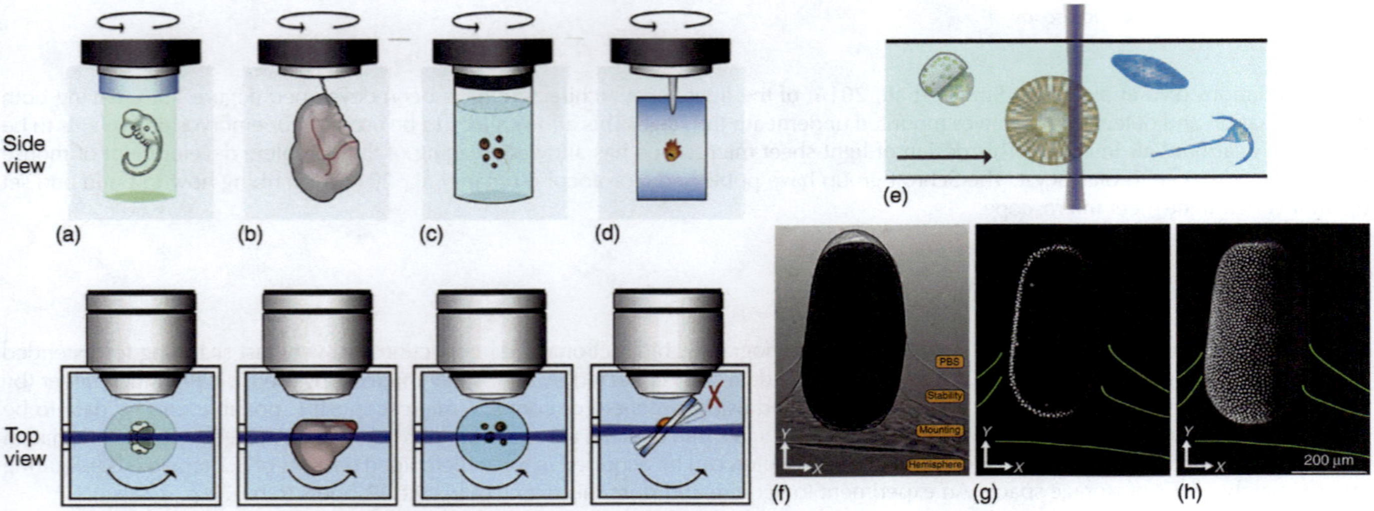

Figure 19.7 ***Specimen handling in the light-sheet microscope*** *The light-sheet microscope is most often used with a tissue explant or small embryo mounted in agarose within a glass capillary or a plastic FEP tube (a), but it can also be suspended on a hook (b), held within a polymer foil tube (c), attached to a coverslip (d) or allowed to flow through a capillary (e). Stability is key for long-term imaging of large data sets, and this will dictate how the sample is mounted in the microscope. Comprehensive details are given in the Zeiss Lightsheet Z1 white paper – see the Internet Resources at the end of the chapter. In panel (f) a bastoderm tribolium embryo is glued with liquid agarose into a previously set hemisphere and surrounded with buffer. Panel (h) shows the embryo at the conclusion of a 50 hour time-lapse. For further details, see Strobl and Stelzer (2014). Source: (a–d) Reproduced with permission from Carl Zeiss Ltd. (e) Reynaud et al. 2015. Reproduced with permission from Macmillan Publishers Ltd. (f–h) Strobl et al. 2015. Reproduced with permission from Macmillan Publishers Ltd.*

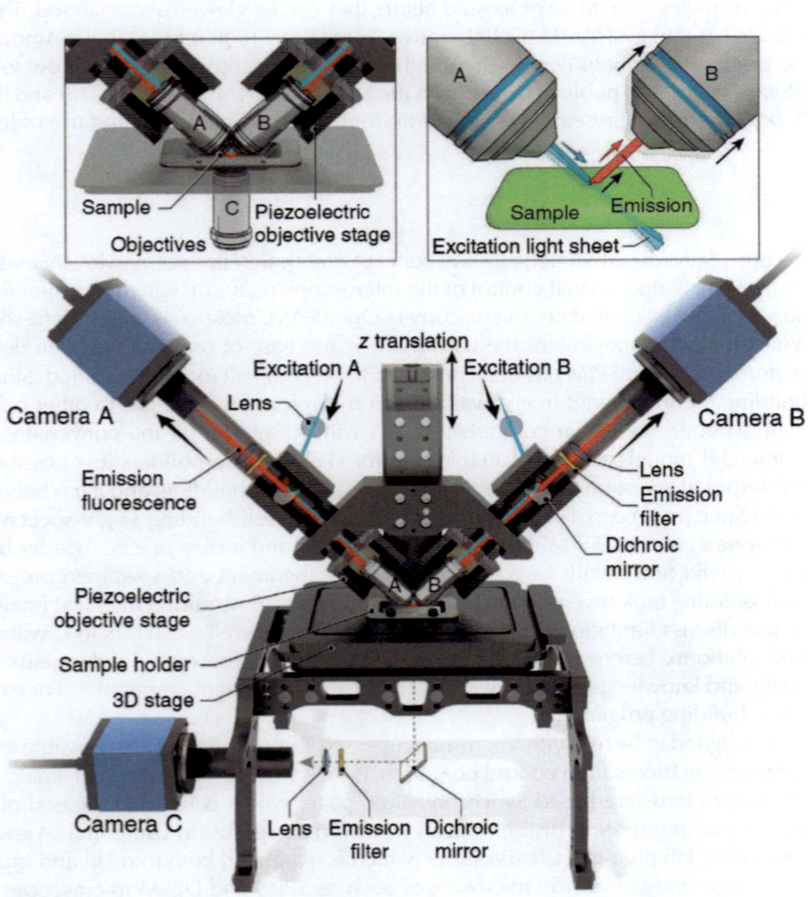

Figure 19.8 ***The inverse light-sheet microscope*** *This inverted configuration, using two 40x/NA 0.8 water-immersion custom-designed objectives (A,B), allows greater versatility for sample preparation, accommodating glass slides and Petri dishes and for samples to be grown at an interface. See Wu et al. (2013) and Chhetri et al. (2015) for further details. Source: Wu et al. 2013. Reproduced with permission from Macmillan Publishers Ltd.*

Two variations (Wu et al., 2013; Strnad et al., 2016) of the light-sheet architecture have been developed (Figure 19.8) having both the illumination and detection objectives mounted underneath the stage. This allows slides to be used, or for embryos or explants to be held at an agar/fluid/air interface. This design of light-sheet microscope has allowed imaging of the complete development of mouse embryos from zygote to blastocyst. The Schroff group have published a protocol (Kumar et al., 2014) describing how to build and set up an inverted light-sheet microscope.

19.5 Data Handling

Because of the relatively large sample size, the use of tomography, bidirectional data collection and very fast scanning for extended periods with fast high-resolution sCMOS cameras, the 3D data sets[4] collected by light-sheet microscopy are typically much larger (by about three orders of magnitude) than image files acquired using widefield or confocal microscopy. It is possible for raw data to be generated at rates anywhere from 300 MB/s to 1 GB/s. To put this into context, refer to Müllenbroich et al. (2015). They calculate that an adult mouse brain has a volume of 0.5 cm^3, an image of which can be acquired in a parallelepiped data set of 8.2 teravoxels, requiring approximately 15 TB of storage space. An experiment to acquire and store this data would take 11 hours to run.

Some cameras stream data to an internal storage for read-out after acquisition; others require a frame grabber to stream data off to separate storage. The overall file sizes can be reduced by defining a region of interest and only collecting data from this part of the sample (see Reynaud et al., 2015). Light microscopy facility managers reiterate to their users that images are not pretty pictures but photon intensity data sets, and as such should be acquired and stored as uncompressed *.tif or *.bmp image files rather than a compressed file, such as *.jpg, since this leads to data loss. However, when collecting and manipulating gigabyte and terabyte image files, image compression may be inevitable (Tomer et al., 2012). One company that can help and advise in this respect is Acquifer, based in Karlsruhe; see the Internet Resources for contact details.

Unless some thought is given to how data is to be streamed off the microscope computer, how it is to be displayed, analysed and effectively stored, the image curation part of the experiment will all too easily become a bottleneck. Multi-view tomographic images acquired from two or more different angles need to be processed before they can be viewed and analysed. There are Fiji/ImageJ2 plugins, such as SiMView (Tomer et al., 2012) and CATMAID (Collaborative Annotation Toolkit for Massive Amounts of Image Data, Saalfeld et al., 2009) to facilitate these processes. To help new users of light-sheet microscopy understand how to apply these computational processes, two very good protocols have been published, one from the Keller group (Amat et al., 2015) and the other from the Tomancak group (Schmied et al., 2014), both pioneering researchers at the forefront of the development and use of light-sheet microscopy.

19.6 Self-Built Microscopes

Most people self-build microscopy platforms either because the concept and design has not previously existed or because they require radically different architecture and/or full operational control of the microscopy platform, which a commercial model cannot offer[5]. An exciting off-shoot from the development of light-sheet microscopy is OpenSPIM. Most customised light-sheet microscopes cost in the region of £250k, or more. Whilst it is worth paying for the convenience and ease of use that has been designed into the commercial instruments, these may be unaffordable. OpenSPIM has been designed to be compact (originally called 'SPIM in a suitcase') and simple with instructions for home-building the instrument freely available. In this respect it is similar to other open-source projects, such as Raspberry Pi, which allows you to build your own computer. Just as with Raspberry Pi, the convenience and software capabilities and sophistication of the commercial models are not available, but for skilled home-builders it is possible to build a working SPIM microscope. The advent and widespread adoption of 3D printing has facilitated self-building, and it has been known for a demonstration light-sheet microscope to be fabricated from Lego. For comprehensive details on self-building a light-sheet microscope, see Greger et al. (2007), Pitrone et al. (2013), Pitrone et al. (2015), Müllenbroich et al. (2015) and a very practical paper by Girstmair et al. (2016). A parts list, assembly instructions, a wiki and details of online tutorials on alignment and specimen preparation are all given in this paper. Girstmair reports that self-building took seven months from ordering parts to acquiring their first image. However, some steps are time consuming. The authors also discuss limitations of the OpenSPIM project as well as factors to consider (e.g. trying your research samples on an existing self-built platform) before embarking upon building the microscope. An alternative to completely self-building from scratch is to buy in the help and knowledge of specialist microscope development companies. For example, Cairn Research (see Appendix 5) offer a collaborative building programme.

The OpenSPIM microscope is designed to be run with Micro-manager (see Table 27.6). It is also possible to increase image acquisition speeds by incorporating a hardware synchronisation control box, such as the ESio TTL controller (Girstmair, 2016), but other controllers can be found on the Micro-manager Hardware-based Synchronization page, which is listed at the end of the chapter under Internet Resources. Image acquisition, data set registration (using fiducial beads), data set fusion using multi-view deconvolution and image processing can all be performed using Fiji plug-ins. ClearVolume, which is integrated both into Fiji and ImageJ2, is a real-time live 3D visualisation library designed for high-end volumetric microscopes such as SPIM and DLSM microscopes. With ClearVolume (Royer et al., 2015; Internet Resources) you can see Z-stacks in real time, giving an intuitive and comprehensive view of your data sets. Another helpful Fiji/ImageJ plug-in is BigDataViewer (Pietzsch et al., 2015; Internet Resources) which can handle large 3D data sets, either on a local machine or a remote server.

It is easy to be seduced by the economic argument for self-building, but the old adage 'time is money' is never truer than when self-building anything – microscopes included! A high-end self-built light-sheet microscope equivalent in performance to a commercial model can easily cost in the region of £300 000, which is not far off what can be purchased from a manufacturer off the shelf, with fully

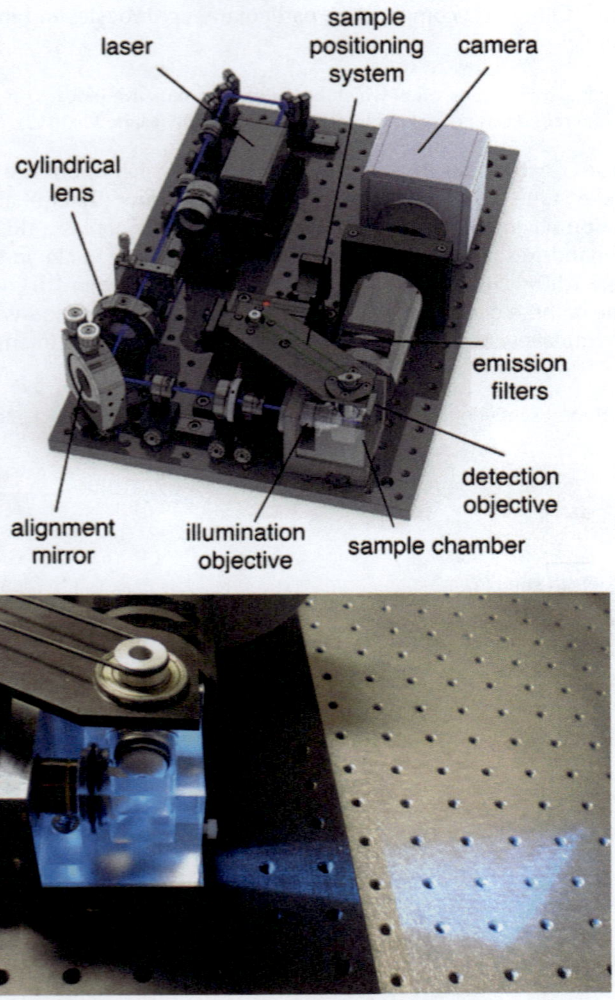

Figure 19.9 **Light sheet and chamber from the OpenSPIM microscope** *The OpenSPIM Gaussian beam light-sheet microscope built on a breadboard. The upper panel shows the individual components. The lower panel shows a 488 nm laser light sheet illuminating the sample chamber. From a microscope built at the Max Planck Institute of Molecular Cell Biology and Genetics, Dresden. Source: (top) Pitrone et al. 2015. Reproduced with permission from John Wiley & Sons.*

functional software as well as training and applications support. In the past, figures as low as £10 000 have been cited for the overall cost of a self-built microscope, but this will allow you to build an entry-level system, which will take considerable time that may have to be found in the evening and at weekends.

19.7 Clearing Tissues

The translucent 'milky' appearance of most tissues is caused by the elastic scattering of light[6]. Traditionally, immersing tissues into particular solvents to render them transparent has long been used in histology and specimen preparation. Traditional histological clearing media have included xylene, acetone, chloroform, methyl salicylate or essential oils such as cedar wood oil. These organic solvents are also sometimes referred to as 'transition media', for they permit tissues to be dehydrated and transferred into wax for microtomy, followed by transition to a hydrated state for staining with aqueous dyes, then taken back again through the reverse procedure through to a clearing medium, then a mounting medium whose refractive indexes are equivalent – or very similar – to that of the glass slide and coverslip between which the stained section is sandwiched.

The purpose of clearing media when used with the light-sheet microscope is similar. Opaque tissues need to be rendered either translucent (letting some light pass through the specimen) or rendered transparent (letting all the light pass through the structure of the sample). Complete transparency is difficult and often time consuming to achieve, but it is helpful to render tissues translucent to a greater or lesser degree. Tissues that are not cleared either absorb and block the illuminating light or scatter it. This scattering lowers image contrast, causes poor resolution of fine detail and limits the penetration depth of the illumination through the tissue. Lipids both absorb and scatter light whilst proteins and nucleic acids are intrinsically transparent, so most clearing solutions are used that dissolve lipids, ideally leaving proteins in situ[7]. Other cell components, particularly erythrocytes in blood (being lenticular in shape), scatter light strongly. In the words of Eric Betzig:

> 'Cells contain water, proteins, carbohydrates and lipids, all of which have different refractive indexes. In these samples, light rays are scrambled, like the view through a windshield when rain drops collect before the wipers are on.' [Marx, V. (2013). *Nature Methods* 10/12: 1163]

Apart from perfusion, which removes erythrocytes from tissues, one way in which to obviate the effects of these scatterers is to use a solvent to match their refractive index. One of the first to do this was Robert Barer (Barer et al., 1955), who cleared a suspension of cells by immersion in a protein solution having the same refractive index as the cytoplasm (RI = 1.35–1.38) of the cells. Nuclei and cytoplasmic organelles have refractive indexes between 1.38 and 1.41, with the RI of nuclei in the range 1.39–1.40. Lipids have an RI of 1.48, whilst proteins lie in the range 1.50–1.58. The RI of dry erythrocytes lies between 1.61 and 1.66, whilst the RI of human whole blood is in the range 1.36–1.40. Some of these clearing solutions will simultaneously decolourise tissues. For example, amino-alcohols release haem from erythrocytes. Unfortunately, no clearing techniques exist for living specimens.

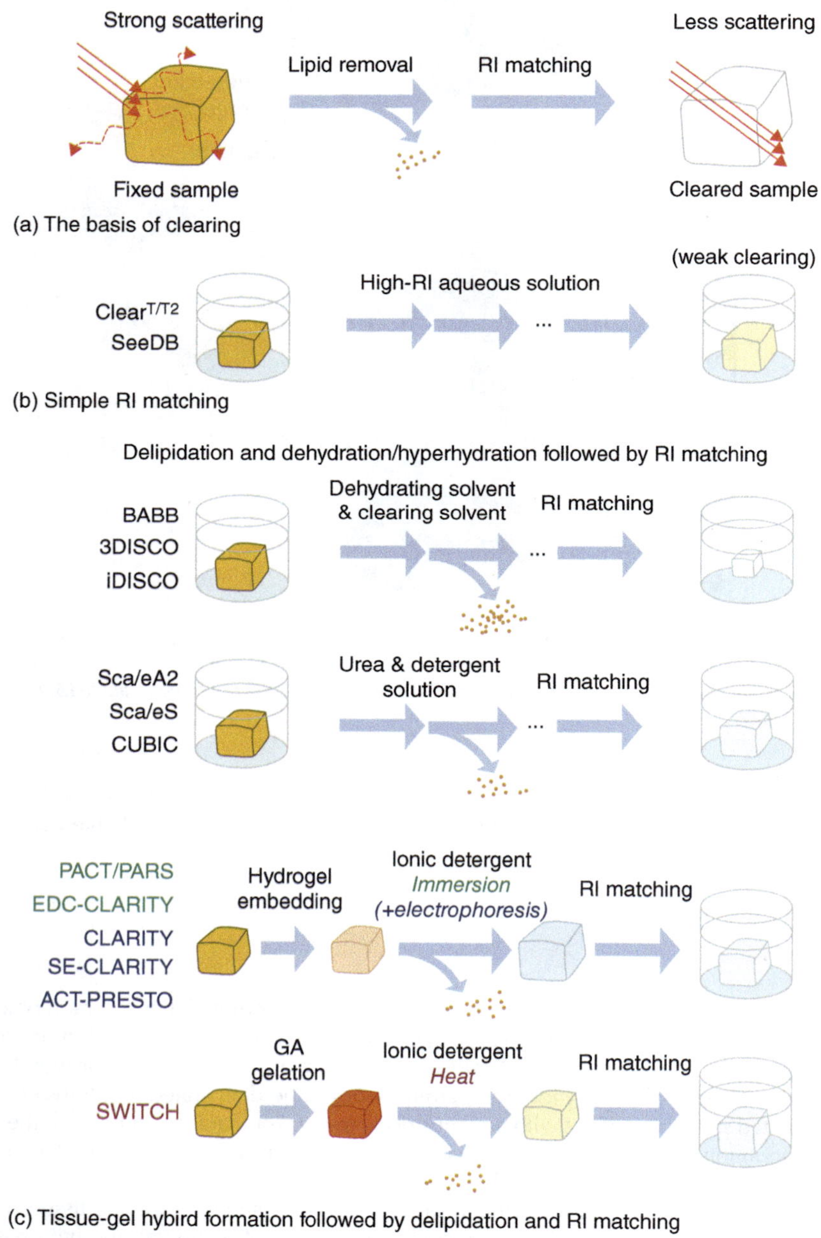

Figure 19.10 ***Methods for clearing tissues*** *Light scattering, which causes translucency and opacity of tissues, can be reduced or in some case obviated, firstly by refractive index matching and secondly by removal of blood and lipids. See the text and Table 19.3 for details. Source: Seo et al. 2016. Reproduced with permission from The Korean Society for Molecular and Cellular Biology.*

Clearing type	Final refr. index (RI)	Main agent	Clearing time	Detergent used	Lipid extracted	Alteration to tissue morphology	Fluorophore stability	Immuno-staining compatibility	Year	Reference
Solvent-based										
Spalteholtz	1.56	methyl salicylate benzyl benzoate	Weeks/months	None	Yes	Shrinkage	No	Incompatible	1911	Azaripour et al. (2016)
BABB	1.55 (1.53–1.57)	benzyl alcohol benzyl benzoate	2–5 days	None	Yes	Shrinkage	Some, limited	Limited	2007	Appleton et al. (2009) Becker et al. (2008)
3DISCO	1.56	benzyl ether	1 day	None	Yes	Some shrinkage	Some, transient	Limited	2012	Becker et al. (2012) Ertürk et al. (2012)
iDISCO	1.56	DCM/DBE/BABB	2–5 days	None	Yes	Shrinkage	Some	Compatible	2014	Renier et al. (2014)
Immersion										
Sucrose	1.44	sucrose	1 day	Triton X-100	No	Shrinkage	Yes, stable	Compatible	2009	Tsai et al. (2009)
Hoyer's solution	1.43	chloral hydrate	2–5 days	None	No	Shrinkage	No	Incompatible	1882	Lersten (1967)
FocusClear	1.45–1.47	diatrizoic acid	1–2 days	Tween 20	Yes	None known	Yes, stable	Compatible	2002	Chiang et al. (2002)
Clear T	1.44	formamide	1–2 days	None	No, retained	None known	Some	Incompatible	2012	Kuwajima et al. (2013)
Clear T2	1.44	formamide/PEG	1–2 days	None	No, retained	None known	Yes, stable	Limited	2013	Kuwajima et al. (2013)
ClearSee	1.41	urea/sodium deoxycholate	Days/weeks	None	Yes	None known	Yes, stable	Compatible	2015	Kurihara et al. (2015)
See DB	1.49	Fructose α-Thioglycerol	Several days	None	No, retained	None known	Yes, stable	Very limited	2013	Ke et al. (2013) Ke & Imai (2014)
TDE	1.52 (1.33–1.52)	2,2′-thiodiethanol	Days/weeks	8% SDS (optional)	Yes	None known	Yes, stable	Compatible	2009	Appleton et al. (2009) Aoyagi et al. (2015)
FRUIT	1.48	fructose/urea α-Thioglycerol	Several days	None	No, retained	Minimal expansion	Yes, stable	Incompatible	2015	Hou et al. (2015)
Hyperhydration										
Sca*l*e	1.38	urea	Weeks/months	Triton X-100	Yes	Significant swelling	Yes, stable	Compatible	2011	Hama et al. (2011) Warner et al. (2014)
Sca*l*e S	1.44	urea/sorbitol	Weeks/months	Triton X-100	Yes	Some swelling	Yes, stable	Compatible	2015	Hama et al. (2015)
Sca*l*e A2	1.36	urea/glycerol	Weeks/months	Triton X-100	Yes	Some swelling	Yes, stable	Limited	2011	Hama et al. (2011)
Sca*l*e U2	1.36	urea/glycerol	Months	Triton X-100	Yes	None known	Yes, stable	Not known	2011	Hama et al. (2011)
CUBIC	1.38/1.48	urea/sucrose/TKED	Days/weeks	Triton X-100	Yes	Transient Swelling	Some	Compatible	2014	Susaki et al. (2014)
Whole body CUBIC	1.38	urea/perfusion	Several days	Triton X-100	Yes	Transient Swelling	Yes, stable	Compatible	2014	Tainaka et al. (2014)
Hydrogel embedding										
CLARITY	1.45	SDS/FocusClear glycerol	Days/weeks	8% SDS	Yes	Transient swelling	Yes, stable	Limited	2013	Chung et al. (2013) Epp et al. (2015)
PEA-CLARITY	1.43–1.45	SDS/boric acid	Weeks/months	4% SDS	Yes	Some swelling	Yes, stable	Limited	2015	Palmer et al. (2015)
PACT	1.38–1.48	SDS/Histodenz	Days/weeks	8% SDS	Yes	Some swelling	Some	Compatible	2014	Yang, B et al. (2014)
PARS	1.38–1.48	SDS/Histodenz	Several days	8% SDS	Yes	None known	Yes, stable	Compatible	2014	Treweek et al. (2015)
SWITCH	1.47	SDS	Days/weeks	6% SDS	Yes	Large swelling	Yes, stable	Compatible	2015	Murray et al. (2015)
ACT-PRESTO	1.43–1.48	SDS/boric acid	1–2 days	4% SDS	Yes	Transient swelling	Yes, stable	Compatible	2016	Lee et al. (2016)

References

Azaripour, A. et al. (2016). *Progress in Histochemistry and Cytochemistry*. pii: S0079–6336(16)30004-3 (in press).

Costantini, I. et al. (2015). *Scientific Reports 5*: 9808.

Hama, H. et al. (2015). *Nature Neuroscience 18*/10: 1518–1529.

Marx, V. (2014). *Nature Methods 11*/12: 1209–1214.

Marx, V. (2016). *Nature Methods 13*/3: 205–209.

Richardson, D.S. and Lichtman, J.W. (2015). *Cell 162*/2: 246–257.

Seo, J. et al. (2016). *Molecules and Cells 39*/6: 439–446.

Abbreviations: ACT-PRESTO: Active Clarity Technique-Pressure Related Efficient and Stable; Transfer of macromolecules into Organs; BABB: Benzyl Alcohol-Benzyl Benzoate; CLARITY: Clear Lipid-exchanged Acrylamide-hybridized Rigid Imaging/Immunostaining/*In situ* hybridization-compatible Tissue-hYdrogel; CUBIC: Clear, Unobstructed Brain/Body Imaging Cocktails and Computational analysis; DCM: dichloromethane; DBE: dibenzyl Ether; 3DISCO: 3D Imaging of Solvent-Cleared Organs; iDISCO: Immunolabeling-enabled 3D Imaging of Solvent-Cleared Organs; PACT: Passive CLARITY Technique; PARS: Perfusion-assisted Agent Release *in Situ*; SE-CLARITY: Stochastic Electrotransport CLARITY; SPED: SPherical aberration assisted Extended Depth-of-field; SWITCH: System-Wide control of Interaction Time and kinetics of Chemicals; THF: tetrahydrofuran.

Clearing methods fall into three broad groups:

- Dehydration, removal of lipids and replacing water in the sample with media closer to the RI of lipids and other cell constituents. However, water molecules necessary for fluorescence are removed. These are the solvent-based examples, shown in Table 19.3.
- Water-based clearing agents containing urea or formamide. Passive immersion to remove lipids is followed by hydration to lower RI of remaining tissue components. Examples are FocusClear and similar solutions, such as RIMS, TDE, FRUIT (a combination of SeeDB and Scal*e*). The immersion and hyperhydration examples are in this group.
- Fixation and cross-linking with a hydrogel mesh to limit protein loss, followed by lipid removal with detergents. This method can be passive by simple immersion or active, involving electrophoresis, heating or pressure.

It is very easy to quench fluorescence in acidic pH environments or if solvents are used for clearing. For example, fluorescence fairly rapidly fades in the classical clearing agent BABB in just a few hours. While BABB clears very rapidly and thoroughly, it is immiscible in water, so dehydration is necessary prior to clearing, yet most fluorescent proteins require an aqueous environment. Most new formulae for clearing media (Table 19.3) attempt to retain endogenous fluorescence and immunological capabilities. Becker et al. (2012) report that dehydration with tetrahydrofuran rather than ethanol, and clearing with dibenzyl ether (phenylmethoxymethylbenzene) tends to preserve GFP signals. Protocols are given in Ertürk et al. (2012; 2014). 3DISCO was developed in place of BABB to enable clearing of lipid-rich tissues. It is extremely rapid, taking only 1 day to clear. iDISCO (Renier et al., 2014) is a modification of the 3DISCO procedure favouring antibody labelling.

Since organic solvents are very good at clearing; a further modification, uDISCO, was recently published (Pan et al., 2016). Many clearing solutions shrink tissues, and whilst considerable effort has been made to stop this effect, Pan et al., from the Ertürk group, have

Table 19.4 Some Objectives Suitable for Use with Clearing Methods

Objective	Numerical Aperture (NA)	Manufacturer	Working distance	Refractive index range	Example clearing media
LVMI-Fluor 4x/NA 0.30	0.30	LaVision BioTec	6.00 mm	1.33–1.56	CLARITY, glycerol, Scaleview-A2, SeeDB
LVMI-Fluor 20x/NA 0.45	0.45	LaVision BioTec	5.50 mm	1.45–1.57	CLARITY, glycerol, Scaleview-A2, SeeDB
HC PL I-R APO 20x/0.75 W	0.75	Leica	0.60 mm	1.330	Water, PBS
HCX APO L 20x/0.95 IMM	0.95	Leica	0.63 mm	1.563	BABB
HC FLUOTAR L 25 × /1.00 IMM (n_e = 1.457) motCORR VIS-IR	1.00	Leica	6.00 mm	1.457	CLARITY
HC PL APO 63x/1.30 GLYC CORR CS2	1.30	Leica	0.28 mm	1.447–1.455	Glycerol, SeeDB
CFI 75 Apo LWD 25x/1.1 W MP	1.1	Nikon	2.00 mm	1.330	Water, PBS
10 × UIS2- XL Plan N 10x SVMP	0.6	Olympus	8.00 mm	1.33–1.52	CLARITY, glycerol, Scaleview-A2, SeeDB
25 × UIS2- XLSL Plan N 25x SVMP	0.95	Olympus	8.00 mm	1.33–1.40	CLARITY, glycerol, Scaleview-A2, SeeDB
25 × UIS2-XL Plan N 25x SVMP	1.0	Olympus	4.00 mm	1.33–1.40	CLARITY, glycerol, Scaleview-A2, SeeDB
25x UIS2 -XLSL Plan N 25x Glyc MP	1.0	Olympus	8.00 mm	1.41–1.52	Glycerol, CLARITY, SeeDB
EC Plan-NEOFLUAR 5 × /0.16	0.16	Zeiss	5.60 mm	1.45	CLARITY, CUBIC
LSFM Clearing 20 × /1.0	1.0	Zeiss	5.60 mm	1.450 ± 0.03	CLARITY, CUBIC
LD Plan-Apochromat 20 × /1.0	1.0	Zeiss	5.60 mm	1.380 ± 0.03	Scal*e*
LD LCI Plan-Apochromat 25x/0.8 Imm Corr DIC M27	0.8	Zeiss	0.57 mm	1.330–1.518	CLARITY, glycerol, Scaleview-A2, SeeDB

Source: Adapted from Table 1, Marx 2014 with permission from Macmillan Publishers.

Range of objectives for use with clearing fluids from Olympus

Objective for LaVision BioTec GmbH

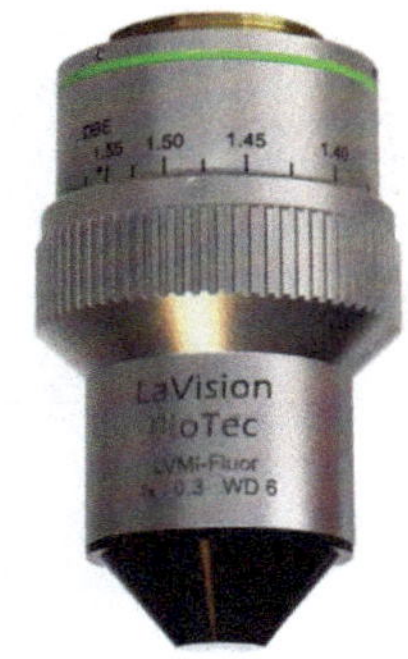

Source: (left) Olympus Europa SE & Co. KG. (right) LaVision BioTec GmbH.

Reference

Marx, V. (2014). 'Microscopy: seeing through tissue'. *Nature Methods 11*/12: 1209–1214 (published online 25 November 2014).

maximised the shrinkage phenomenon (claimed to be up to 65% in volume) turning it to advantage to enable entire organisms to fit into the light-sheet microscope. uDISCO employs tert-butyl alcohol as a dehydration fluid to replace water in the tissues, dichloromethane to remove lipids and diphenyl ether (RI = 1.58) as a solvent mixed with BABB (RI = 1.55 to 1.56) to preserve the fluorescent signal. It is claimed that uDISCO not only clears to a greater degree of transparency than similar clearing solutions, preserves both fluorescent antibody staining and endogenous fluorescent protein signal, but bony tissues are also cleared, without the prior use of EDTA, through simple refractive index matching.

Sca*l*e and its variants (Hama et al., 2011; 2015) employ urea as a hydrophilic clearing agent. Glycerol is added to reduce fragility of tissues as a result of protein loss. SeeDB (Ke et al., 2013; Ke and Imai, 2014) was developed to address the issue of sample fragility with Sca*l*e immersion, as was ClearT2 (Yu et al., 2018), with polyethylene glycol added to stabilise proteins. Most of these hydrophilic methods have been developed for brain tissue but are less successful in tissues with extracellular matrix, for which solvent-based fluids are better suited. CUBIC (Susaki et al., 2014; 2015; Matryba et al., 2018) is a modification of the Sca*l*e procedure, based on urea, triton X-100 and ethylenedinitrilotetraethanol (TKED), which passively removes lipids.

For really clear transparent tissues, dissolving lipids alone is not enough, for they provide structural support, and membranes provide compartmentation and integrity. CLARITY was developed to address this issue. Tissues are embedded in a polyacrylamide hydrogel mesh, which preserves the morphology and spatial integrity of the proteins; the lipids are removed by detergents speeded by electrophoresis to preserve both fluorescence and immunological epitopes. Although widely used, CLARITY requires the use of FocusClear, which is extremely expensive[8], and uses customised electrophoresis equipment. Several groups have worked on optimising the CLARITY technique. At the time of writing, see for example, Lee et al. (2014), Poguzhelskaya et al. (2014), Epp et al. (2015), Magliaro et al. (2016), Tomer et al. (2014) and for ACT-PRESTO, the latest variant using pressure (Lee et al., 2016). Alternative methods PACT (passive CLARITY) and CUBIC as simpler clearing protocols have also been developed (Woo, J. et al., 2016; Gomez-Gaviro et al., 2017; and Chen et al., 2017). These take longer, but the hydrogel is not altered in its structural integrity by electrophoresis or heating. Since the CUBIC reagents are much cheaper than FocusClear or RIMS (Refractive Index Matching Solution; Yang, B., et al., 2014; Marx, 2014) they may be more widely used, particularly for clearing large specimens.

Clearing plant tissues has also benefitted from the recent development of various clearing fluids. Classically chloral hydrate, glycerol and methyl salicylate have been used. Due to its use as a sedative, chloral hydrate can be replaced by Viskol (Villani, 2013). ClearSee is a variant of Sca*l*e developed (Kurihara et al., 2015) for clearing plant tissues, and PEA-CLARITY (Palmer et al., 2015) uses enzyme degradation for clearing plant tissues (also see Lersten, 1967; Warner et al., 2014). For a review of imaging plants with light-sheet microscopy, see Berthet and Maizel (2016).

Observing samples in the 1–10 mm range is referred to as mesoscopy. The light-sheet microscope successfully occupies this range. The Mesolens (Section 17.10; McConnell et al., 2016) usefully permits very large 3D data sets of very large fields of view to be collected very rapidly. Because the Mesolens has a magnification of 4x, it is able to collect very large fields of view, which are otherwise too large to acquire using either a conventional fluorescence microscope or indeed a light-sheet microscope without excessive stitching of multiple fields of view. With slightly lower resolving power yet an unprecedented optical index, it complements, rather than competes with, light-sheet technology.

19.8 Chapter Summary

1. Light-sheet microscopy allows biologists to image tissues very fast in three dimensions with very little photobleaching.
2. By rotating the sample and also by illuminating it from multiple angles, a 3D tomographic data set can be acquired.
3. In both widefield and confocal microscopes, the objective acts as its own condenser. With the light-sheet microscope, the sample is illuminated orthogonally to the axis of detection, using a separate objective. Optical sectioning is set by the thickness of the light sheet.
4. The light-sheet microscope exposes the specimen to between one to four orders of magnitude less radiative energy than widefield or confocal fluorescence microscopes.
5. The Gaussian beam light sheet of the SPIM microscope can be held static or digitally scanned. Alternatively, a thinner Bessel beam can be used to form the illuminating light sheet.
6. The pixel pitch of the camera imposes the lateral resolving power limit, whilst the thickness of the light sheet sets the axial limit of resolving power.
7. sCMOS cameras can capture very fast frame rates over the large field of view. This and the very low photobleaching makes light-sheet microscopy ideal for long-term time-lapse experiments.
8. Light-sheet microscopes can generate very large terabyte-sized data files.
9. Tissues have to be cleared for image acquisition. Apart from immersion in high–refractive index solutions to match tissue to the RI of proteins and glass, water-based clearing methods, such as hyperhydration and using hydrogel meshes, are used to preserve fluorescence signal and immuno-staining capabilities.

Key Reading

Feuchtinger, A. et al. (2016). 'Deep tissue imaging: a review from a preclinical cancer research perspective'. *Histochemistry and Cell Biology 146*/6: 781–806.

Höckendorf, B. et al. (2014). 'Making biology transparent'. *Nature Biotechnology 32*/11: 1104–1105.

Scherf, N. and Huisken, J. (2015). 'The smart and gentle microscope'. *Nature Biotechnology 33*/8: 815–818.

- An excellent argument for designing microscopes to suit specimen requirements

Reynaud, E.F. et al. (2015). 'Guide to light-sheet microscopy for adventurous biologists'. *Nature Methods 12*/1: 30–34.

- A concise review of the fast-moving field of light-sheet microscopy

Richardson, D.S. and Lichtman, J.W. (2015). 'Clarifying tissue clearing'. *Cell 162*/2: 246–257.

- The best review paper on the fast-developing field of tissue clearing for microscopy

Yu, T. et al. (2018). 'Optical clearing for multiscale biological tissues'. *Journal Biophotonics 11*: e201700187.

References

Amat, F. et al. (2015). 'Efficient processing and analysis of large-scale light-sheet microscopy data'. *Nature Protocols 10*/11: 1679–1696.

Becker, K. et al. (2012). 'Chemical clearing and dehydration of GFP expressing mouse brains'. *PLoS One 7*/3: e33916.

Berthet, B. and Maizel, A. (2016). 'Light sheet microscopy and live imaging of plants'. *Journal of Microscopy 263*/2: 158–164.

Capoulade, J., Reynaud, E., and Wachsmuth, M. (2014). *Imaging Marine Life: Macrophotography and Microscopy Approaches for Marine Biology*. Wiley-VCH.

Chen, B.-C. et al. (2014). 'Lattice light-sheet microscopy: Imaging molecules to embryos at high spatiotemporal resolution'. *Science 346*/6208: 1257998–1257998.

Chen, L. et al. (2017). 'UbasM: An effective balanced optical clearing method for intact biomedical imaging'. *Scientific Reports 7*: 12218.

Chung, K. et al. (2013). 'Structural and molecular interrogation of intact biological systems'. *Nature 497*/7449: 332–337.

Colombelli, J. and Lorenzo, C. (2014). Light Sheet Fluorescence Microscopy Applications for Multicellular Systems. In: Cornea, A. and Michael Conn, P. (eds.), *Fluorescence Microscopy, Super Resolution and Other Techniques*, Chapter 8. Academic Press, pp. 109–120. ISBN 9780124095137.

Cutrale, F. and Gratton, E. (2014). 'Inclined selective plane illumination microscopy (iSPIM) adaptor for conventional microscopes'. *Microscopy Research and Technique 75*/11: 1461–1466.

De Vos, W.H. et al. (2014). 'Invited review article: Advanced light microscopy for biological space research'. *Review of Scientific Instruments 85*/10: 101101.

Desmaison, A. et al. (2013). 'A versatile sample holder for single plane illumination microscopy'. *Journal of Microscopy 251*/2: 128–132.

Dodt, H.U. et al. (2007). 'Ultramicroscopy: three-dimensional visualization of neuronal networks in the whole mouse brain'. *Nature Methods 4*/4: 331–336.

Engelbrecht, C.J. and Stelzer, H.K. (2006). 'Resolution enhancement in a light-sheet based microscope (SPIM)'. *Optics Letters 31*/10: 1477–1479.

Epp, J.R. et al. (2015). 'Optimization of CLARITY for clearing whole-brain and other intact organs'. *eNeuro 2*/3. pii: ENEURO.0022-15.2015.

Ertürk, A. et al. (2012). 'Three-dimensional imaging of solvent-cleared organs using 3DISCO'. *Nature Protocols 7*/11: 1983–1995.

Ertürk, A. et al. (2014). 'Imaging cleared intact biological systems at a cellular level by 3DISCO'. *Journal of Visualized Experiments 89*: 51382. doi: 10.3791/51382.

Girstmair, J. et al. (2016). 'Light-sheet microscopy for everyone? Experience of building an OpenSPIM to study flatworm development'. *BMC Developmental Biology 16*: 22.

Gómez-Gaviro, M.V. et al. (2017). 'Optimized CUBIC protocol for three-dimensional imaging of chicken embryos at single-cell resolution'. *Development 144*/11: 2092–2097.

Greger, K. et al. (2007). 'Basic building units and properties of a fluorescence single plane illumination microscope'. *Review of Scientific Instruments 78*/2: 023705.

Guan, Z. et al. (2015). 'Compact plane illumination plugin device to enable light sheet fluorescence imaging of multi-cellular organisms on an inverted widefield microscope'. *Biomedical Optics Express 7*/1: 194–208.

Hama, H. et al. (2011). 'Sca*l*e: a chemical approach for fluorescence imaging and reconstruction of transparent mouse brain'. *Nature Neuroscience 14*/11: 1481–1488.

Hama, H. et al. (2015). 'Sca*l*eS: an optical clearing palette for biological imaging'. *Nature Neuroscience 18*/10: 1518–1529.

Hufnagel, L. and Pepperkok, R. (2016). 'Let there be light – at the right place'. *Nature Biotechnology 34*/12: 1246–1247.

Huisken, J. (2012). 'Slicing embryos gently with light sheets'. *Bioessays 34*/5: 406–411.

Huisken, J. and Stainier, D.Y. (2009). 'Selective plane illumination microscopy techniques in developmental biology'. *Development 136*/12: 1963–1975.

Huisken, J. et al. (2004). 'Optical sectioning deep inside live embryos by selective plane illumination microscopy'. *Science 305*/5686: 1007–1009.

Ichikawa, T. et al. (2013). 'Live imaging of whole mouse embryos during gastrulation: migration analyses of epiblast and mesodermal cells'. *PLoS One 8*/7: e64506.

Jeandupeux, E. et al. (2015). '3D print customized sample holders for live light sheet microscopy'. *Biochemical and Biophysical Research Communications 463*/4: 1141–1143.

Kaufmann, A. et al. (2012). 'Multilayer mounting enables long-term imaging of zebrafish development in a light sheet microscope'. *Development 139*/17: 3242–3247.

Ke, M.T. et al. (2013). 'SeeDB: a simple and morphology-preserving optical clearing agent for neuronal circuit reconstruction'. *Nature Neuroscience 16*/8: 1154–1161.

Ke, M.T. and Imai, T. (2014). 'Optical clearing of fixed brain samples using SeeDB'. *Current Protocols in Neuroscience 66*: Unit 2.22.1–2.22.19.

Keller, P.J. et al. (2008). 'Reconstruction of zebrafish early embryonic development by scanned light sheet microscopy'. *Science 322*/5904: 1065–1069.

Kumar, A. et al. (2014). Dual-view plane illumination microscopy for rapid and spatially isotropic imaging. *Nature Protocols 9*/11: 2555–2573. doi: 10.1038/nprot.2014.172.

Kurihara, D. et al. (2015). 'ClearSee: a rapid optical clearing reagent for whole-plant fluorescence imaging'. *Development 142*/23: 4168–4179.

Marx, V. (2013). Is super-resolution microscopy right for you? *Nature Methods 10*/12: 1157–1163.

Lee, H. et al. (2014). 'Improved application of the electrophoretic tissue clearing technology, CLARITY, to intact solid organs including brain, liver, kidney, lung, and intestine'. *BMC Developmental Biology 14*: 48. doi: 10.1186/s12861-014-0048-3.

Lee, E. et al. (2016). 'ACT-PRESTO: Rapid and consistent tissue clearing and labeling method for 3-dimensional (3D) imaging'. *Scientific Reports 6*: 18631.

Lee, E. et al. (2016). 'See-through technology for biological tissue: 3-dimensional visualization of macromolecules'. *International Neurourology Journal 20*/Suppl 1: S15–S22.

Lersten, N.R. (1967). 'An annotated bibliography of botanical clearing methods'. *Iowa State Journal of Science 41*: 481–486.

Lindek, S. et al. (1997). 'Single-lens theta microscopy – a new implementation of confocal theta microscopy'. *Journal of Microscopy 188*/3: 280–284.

Magliaro, C. et al. (2016). 'Clarifying CLARITY: quantitative optimization of the diffusion based delipidation protocol for genetically labelled tissue'. *Frontiers Neuroscience 10*: 179.

Mappes, T. et al. (2012). 'The invention of immersion ultramicroscopy in 1912 – the birth of nanotechnology?' *Angewandte Chemie International Edition 51*/45: 11208–11212.

Marx, V. (2014). 'Microscopy: seeing through tissue'. *Nature Methods 11*/12: 1209–1214.

Marx, V. (2016). 'Optimizing probes to image cleared tissue'. *Nature Methods 13*/3: 205–209.

Matryba, P. et al. (2018). 'Optimized perfusion-based CUBIC protocol for the efficient whole-body clearing and imaging of rat organs'. *Journal Biophotonics 2018*: e201700248.

McConnell, G. et al. (2016). 'A novel optical microscope for imaging large embryos and tissue volumes with sub-cellular resolution throughout'. *eLife 5*: e18659.

Müllenbroich, M.C. et al. (2015). 'Comprehensive optical and data management infrastructure for high-throughput light-sheet microscopy of whole mouse brains'. *Neurophotonics 2*/4: 041404.

Ovečka, M. et al. (2015). 'Preparation of plants for developmental and cellular imaging by light-sheet microscopy'. *Nature Protocols 10*/8: 1234–1247.

Palmer, W.M. et al. (2015). 'PEA-CLARITY: 3D molecular imaging of whole plant organs'. *Scientific Reports 5*: 13492. doi: 10.1038/srep13492.

Pampaloni, F. et al. (2007). 'The third dimension bridges the gap between cell culture and live tissue'. *Nature Reviews Molecular Cell Biology 8*/10: 839–845.

Pampaloni, F. et al. (2014). 'Tissue-culture light sheet fluorescence microscopy (TC-LSFM) allows long-term imaging of three dimensional cell cultures under controlled conditions'. *Integrative Biology 6*: 988–998.

Pampaloni, F. et al. (2015). 'Light sheet-based fluorescence microscopy (LSFM) for the quantitative imaging of cells and tissues'. *Cell Tissue Research 360*/1: 129–141.

Pan, C. et al. (2016). 'Shrinkage-mediated imaging of entire organs and organisms using uDISCO'. *Nature Methods 13*/10: 859–867.

Pietzsch, T. et al. (2015). 'BigDataViewer: visualization and processing for large image data sets'. *Nature Methods 12*/6: 481–483.

Pitrone, P. et al. (2013). 'OpenSPIM: an open-access light-sheet microscopy platform'. *Nature Methods 10*/7: 598–599.

Pitrone, P.G. et al. (2015). 'OpenSPIM: A do-it-yourself open access light sheet fluorescence microscope'. *Microscopy and Analysis 29*/16: 7–11.

Planchon, T.A. et al. (2011). 'Rapid three-dimensional isotropic imaging of living cells using Bessel beam plane illumination'. *Nature Methods 8*/5: 417–423.

Poguzhelskaya, E. et al. (2014). 'Simplified method to perform CLARITY imaging'. *Molecular Neurodegeneration 9*: 19.

Preibisch, S. et al. (2009). 'Globally optimal stitching of tiled 3D microscopic image acquisitions'. *Bioinformatics 25*/11: 1463–1465.

Preibisch, S. et al. (2010). 'Software for bead-based registration of selective plane illumination microscopy data'. *Nature Methods 7*/6: 418–419.

Preibisch, S. et al. (2014). 'Efficient Bayesian-based multiview deconvolution'. *Nature Methods 11*/6: 645–648.

Renier, N. et al. (2014). 'iDISCO: a simple, rapid method to immunolabel large tissue samples for volume imaging'. *Cell 159*/4: 896–910.

Reynaud, E.F. et al. (2015). 'Guide to light-sheet microscopy for adventurous biologists'. *Nature Methods 12*/1: 30–34.

Royer, L.A. et al. (2015). 'ClearVolume: open-source live 3D visualization for light-sheet microscopy'. *Nature Methods 12*/6: 480–481.

Saalfeld, S. et al. (2009). 'CATMAID: collaborative annotation toolkit for massive amounts of image data'. *Bioinformatics 25*/15: 1984–1986.

Santi, P.A. (2011). 'Light sheet fluorescence microscopy: a review'. *Journal of Histochemistry and Cytochemistry 59*/2: 129–138.

Scherf, N. and Huisken, J. (2015). 'The smart and gentle microscope'. *Nature Biotechnology 33*/8: 815–818.

Schmied, C. et al. (2014). 'Open-source solutions for SPIMage processing'. In: Waters, J.C. and Wittman, T. (eds.), *Methods in Cell Biology, Quantitative Imaging in Cell Biology*, Chapter 27, vol *123*. Cambridge, Mass: Academic Press, pp. 505–529.

Seo, J. et al. (2016). 'Clearing and labeling techniques for large-scale biological tissues'. *Molecules and Cells 39*/6: 439–446.

Strnad, P. et al. (2016). 'Inverted light-sheet microscope for imaging mouse pre-implantation development'. *Nature Methods 13*/2: 139–142.

Strobl, F. and Stelzer, E.H.K. (2014). 'Non-invasive long-term fluorescence live imaging of *Tribolium castaneum* embryos'. *Development 141*/11: 2331–2338.

Strobl, F., Schmitz, A. and Stelzer, E.H.K. (2015). 'Live imaging of *Tribolium castaneum* embryonic development using light-sheet–based fluorescence microscopy'. *Nature Protocols 10*: 1486–1507. doi: 10.1038/nprot.2015.093 (published online 03 September 2015).

Stuurman, N. et al. (2007). 'μManager: Open Source software for light microscope imaging'. *Microscopy Today 15*/3: 42–43.

Susaki, E.A. et al. (2014). 'Whole-brain imaging with single-cell resolution using chemical cocktails and computational analysis'. *Cell 157*/3: 726–739.

Susaki, E.A. et al. (2015). 'Advanced CUBIC protocols for whole-brain and whole-body clearing and imaging'. *Nature Protocols 10*/11: 1709–1727.

Tinling, S.P. et al. (2004). 'Microwave exposure increases bone demineralization rate independent of temperature'. *Journal of Microscopy 215*/3: 230–235.

Tomer, R. et al. (2012). Quantitative high-speed imaging of entire developing embryos with simultaneous multiview light-sheet microscopy. *Nature Methods 9*/7: 755–763. doi: 10.1038/nmeth.2062.

Tomer, R. et al. (2014). 'Advanced CLARITY for rapid and high-resolution imaging of intact tissues'. *Nature Protocols 9*/7: 1682–1697.

Trivedi, V. et al. (2015). 'Dynamic structure and protein expression of the live embryonic heart captured by 2-photon light sheet microscopy and retrospective registration'. *Biomedical Optics Express 6*/6: 2056–2066.

Udan, R.S. et al. (2014). 'Quantitative imaging of cell dynamics in mouse embryos using light-sheet microscopy'. *Development 141*/22: 4406–4414.

Voie, A.H. et al. (1993). 'Orthogonal-plane fluorescence optical sectioning: three-dimensional imaging of macroscopic biological specimens'. *Journal of Microscopy 170*/3: 229–236.

Warner, C.A. et al. (2014). 'An optical clearing technique for plant tissues allowing deep imaging and compatible with fluorescence microscopy'. *Plant Physiology 166*/4: 1684–1687.

Weber, M. and Huisken, J. (2011). 'Light Sheet microscopy for real-time developmental biology'. *Current Opinion in Genetics & Development 21*/5: 566–572.

Weber, M. and Huisken, J. (2012). 'Omnidirectional microscopy'. *News & Views, Nature Methods 9*/7: 656–657.

Woo, J. et al. (2016). 'Optimization of the optical transparency of rodent tissues by modified PACT-based passive clearing'. *Experimental and Molecular Medicine 48*/12: e274.

Wu, Y. et al. (2011). 'Inverted selective plane illumination microscopy (iSPIM) enables coupled cell identity lineaging and neurodevelopmental imaging in *Caenorhabditis elegans*'. *PNAS 108*/43: 17708–17713.
Wu, Y. et al. (2013). 'Spatially isotropic four-dimensional imaging with dual-view plane illumination microscopy'. *Nature Biotechnology 31*/11: 1032–1038.
Yamada, K.M. and Clark, K. (2002). 'Survival in three dimensions'. *Nature 419*/6909: 790–791.
Yang, B. et al. (2014). 'Single-cell phenotyping within transparent intact tissue throughout whole-body clearing'. *Cell 158*/4: 945–958.
Yu, T. et al. (2017). 'UbasM: An effective balanced optical clearing method for intact biomedical imaging'. *Scientific Reports 7*: 12218.
Yu, T. et al. (2018). 'RTF: a rapid and versatile tissue optical clearing method'. *Scientific Reports 8*: 1964.

Further Reading

Adams, M.W. et al. (2015). 'Light Sheet Fluorescence Microscopy (LSFM)'. *Current Protocols in Cytometry* 12.37.1–12.37.15. doi: 10.1002/0471142956.cy1237s71.
Aoyagi, Y. et al. (2015). 'A rapid optical clearing protocol using 2,2′-thiodiethanol for microscopic observation of fixed mouse brain'. *PLoS One 10*/1: e0116280.
Azaripour, A. et al. (2016). 'A survey of clearing techniques for 3D imaging of tissues with special reference to connective tissue'. *Progress in Histochemistry and Cytochemistry*. pii: S0079-6336(16)30004-3.
Bassi, A. et al. (2015). 'Optical tomography complements light sheet microscopy for *in toto* imaging of zebrafish development'. *Development 142*/5: 1016–1020.
Becker, K. et al. (2008). 'Ultramicroscopy: 3D reconstruction of large microscopical specimens'. *Journal of Biophotonics 1*/1: 36–42.
Buytaert, J. et al. (2013). 'MicroCT versus sTSLIM 3D imaging of the mouse cochlea'. *Journal of Histochemistry and Cytochemistry 61*/5: 382–395.
Buytaert, J. et al. (2012). 'The OPFOS microscopy family: high-resolution optical sectioning of biomedical specimens'. *Anatomy Research International* 2012, Article ID 206238. doi: 10.1155/2012/206238.
Chhetri, R.K. et al. (2015). 'Whole-animal functional and developmental imaging with isotropic spatial resolution'. *Nature Methods 12*/12: 1171–1178.
Chmielewski, A.K. et al. (2015). 'Fast imaging of live organisms with sculpted light sheets'. *Scientific Reports 5*: 9385.
Costantini, I. et al. (2015). 'A versatile clearing agent for multi-modal brain imaging'. *Scientific Reports 5*: 9808.
Ebeling, C.G. and Jorgensen, E.M. (2013). 'Two views on light sheets'. *Nature Biotechnology 31*/11: 992–993.
Glaser, A.K. et al. (2016). 'Assessing the imaging performance of light sheet microscopies in highly scattering tissues'. *Biomedical Optics Express 7*/2: 454–466.
Gualda, E. et al. (2014). 'Going "open" with mesoscopy: a new dimension on multi-view imaging'. *Protoplasma 251*/2: 363–372.
Hou, B. et al. (2015). 'Scalable and DiI-compatible optical clearance of the mammalian brain'. *Frontiers in Neuroanatomy 9*: 19. doi: 10.3389/fnana.2015.00019.
Jensen, K.H.R. and Berg, R.W. (2016). 'CLARITY-compatible lipophilic dyes for electrode marking and neuronal tracing'. *Scientific Reports* (in press, June 2016).
Jorand, R. et al. (2012). 'Deep and clear optical imaging of Thick Inhomogeneous Samples'. *PLoS One 7*/4: e35795.
Kuwajima, T. et al. (2013). '*Clear^T*: a detergent- and solvent-free clearing method for neuronal and non-neuronal tissue'. *Development 140*/6: 1364–1368.
Lavagnino, Z. (2016). '4D (x-y-z-t) imaging of thick biological samples by means of Two-Photon inverted Selective Plane Illumination Microscopy (2PE-iSPIM)'. *Science Reports 6*: 23923.
Lux, A. et al. (2005). 'An improved method for clearing and staining free-hand sections and whole-mount samples'. *Annals Botany 96*/6: 989–996.
Mayer, J. et al. (2014). 'OPTiSPIM: integrating optical projection tomography in light sheet microscopy extends specimen characterization to nonfluorescent contrasts'. *Optics Letters 39*/4: 1053–1056.
McGloin, D. and Dholakia, K. (2005). 'Bessel beams: diffraction in a new light'. *Contemporary Physics 46*/1: 15–28.
De Medeiros, G. et al. (2016). 'Light-sheet imaging of mammalian development'. *Seminars in Cell and Developmental Biology 55*: 148–155.
Mertz, J. (2011). 'Optical sectioning microscopy with planar or structured illumination'. *Nature Methods 8*/10: 811–819.
Morawetz, J.J. (2013). 'A clearing protocol for whole tissues: An example using haustoria of *Orobanchaceae*'. *Applications in Plant Sciences 1*/1. pii: apps.1200361.
Murray, E. et al. (2015). 'Simple, scalable proteomic imaging for high-dimensional profiling of intact systems'. *Cell 163*/6: 1500–1514.
Pampaloni, F. et al. (2014). 'Tissue-culture light sheet fluorescence microscopy (TC-LSFM) allows long-term imaging of three-dimensional cell cultures under controlled conditions'. *Integrative Biology 6*/10: 988–998.
Pitrone, P.G. et al. (2015). 'OpenSPIM: A do-it-yourself open access light sheet fluorescence microscope'. *Microscopy & Analysis* 7–11.
Poguzhelskaya, E. et al. (2014). 'Simplified method to perform CLARITY imaging'. *Molecular Neurodegeneration 9*: 19.
Power, R.M. and Huisken, J. (2017). 'A guide to light-sheet fluorescence microscopy for multiscale imaging'. *Nature Methods 14*/4: 360–373.
Santi, P.A. et al. (2009). 'Thin-sheet laser imaging microscopy for optical sectioning of thick tissues'. *Biotechniques 46*/4: 287–294 (contains component list for building).
Sharpe, J. et al. (2002). 'Optical Projection Tomography as a tool for 3D microscopy and gene expression studies'. *Science 296*/5567: 541–545.
Siedentopf, H.F.W. and Zsigmondy, R.A. (1903). 'Uber Sichtbarmachung und Grössenbestimmung ultramikroskopischer Teilchen, mit besonderer Anwendung auf Goldrubingläser'. *Ann. Phys.* (Leipzig) *10*/Ser. 4: 1–39.
Silvestri, L. et al. (2013). 'Micron-scale resolution optical tomography of entire mouse brains with confocal light sheet microscopy'. *Journal of Visualized Experiments 80*: e50696. doi: 10.3791/50696.
Silvestri, L. et al. (2016). Clearing of fixed tissue: a perspective from a microscopist's perspective. *Journal of Biomedical Optics 21*/8: 081205.
Tainaka, K. et al. (2014). 'Whole-body imaging with single cell resolution by tissue decolorization'. *Cell 159*/4: 911–924.
Treweek, J.B. et al. (2015). 'Whole-body tissue stabilization and selective extractions via tissue-hydrogel hybrids for high-resolution intact circuit mapping and phenotyping'. *Nature Protocols 10*/11: 1860–1896.

Villani, T.S. (2013). 'An improved clearing and mounting solution to replace chloral hydrate in microscopic applications'. *Applications in Plant Sciences 1/5*: apps.1300016.
Wong, M.D. et al. (2013). 'Design and implementation of a custom built optical projection tomography system'. *PLoS One 8/9*: e73491. doi: 10.1371/journal.pone.0073491.
Yang, Z. et al. (2014). 'A compact Airy beam light sheet microscope with a tilted cylindrical lens Biomedical'. *Optics Express 5*/10: 3434–3442.

Internet Resources

Open SPIM Software in Fiji: http://fiji.sc/SPIM_Registration.
OpenSPIM wiki: http://openspim.org/Welcome_to_the_OpenSPIM_Wiki.
Fiji: https://fiji.sc/ and http://imagej.net/2012-06-29_-_Fiji_papers_at_Nature_Methods.
Micromanager: https://micro-manager.org/. http://valelab.ucsf.edu/publications/2007stuurmanmicros.pdf. http://www.andor.com/pdfs/manuals/Andor_MicroManager_Guide_For_Neo_and_Zyla.pdf.
Micro-manager Hardware-based Synchronization: https://micro-manager.org/wiki/Hardware-based_Synchronization_in_Micro-Manager.
SPIM Bead registration: http://imagej.net/SPIM_Bead_Registration.
ImageJ stitching plugin: http://fly.mpi-cbg.de/~preibisch/software.html.
ImageJ data handling http://imagej.net/SPIM_Registration_on_cluster_(deprecated).
Multi-view fusion: http://imagej.net/Multi-View_Fusion.
Multi-view reconstruction: http://imagej.net/Multiview-Reconstruction.
Big Data viewer: http://imagej.net/BigDataViewer and https://lmb.informatik.uni-freiburg.de/resources/opensource/imagej_plugins/hdf5.html.
Clear Volume: http://imagej.net/ClearVolume and https://clearvolume.github.io/.
Knime: http://www.knime.org/ and http://tech.knime.org/book/clearvolume.
Multi-view deconvolution: http://imagej.net/Multi-View_Deconvolution.
Acquifer (data pipelines and handling): https://www.acquifer.de/contact.
CATMAID: http://fly.mpi-cbg.de/~saalfeld/catmaid/ and (old page) http://catmaid.readthedocs.io/en/s/.
General clearing article (Leica): https://www.leica-microsystems.com/science-lab/clarity/clearing-procedures-for-deep-tissue-imaging/.
CLARITY: http://wiki.claritytechniques.org/index.php/CLARITY_Technique.
CLARITY protocol: http://wiki.claritytechniques.org/index.php/ISH.
Optimised CLARITY: http://capture-clarity.org/optimized-clarity/. http://capture-clarity.org/ and http://clarityresourcecenter.org/.
CUBIC protocol: http://cubic.riken.jp.
SeeDB resources: https://sites.google.com/site/seedbresources/.
iDISCO: http://idisco.info/ with a list of validated antibodies.
Ultramicroscope website: https://www.lavisionbiotec.com/products/UltraMicroscope/scanner-options.html.
SWITCH method: http://www.chunglabresources.com/sw1/.
2, 2′-thiodiethanol TDE https://arxiv.org/ftp/arxiv/papers/1504/1504.03855.pdf.
FocusClear: http://www.celexplorer.com/product_list.asp?MainType=107&BRDarea=1.
Easy Index: http://www.microsystem.se/lifecanvas.
RapiClear, Sunjin lab: http://www.sunjinlab.com and http://www.sunjinlab.com/modules.php?name=introduction&op=product and http://www.sunjinlab.com/modules.php?name=introduction&op=faq.

Notes

1. Samples are mostly opaque so must be 'cleared' and made transparent or translucent to varying degrees to allow the laser light to pass through. This is covered in Section 19.7. Voie used Werner Spalteholz's fluid (RI ≈ 1.56), which consists of a 5:3 mixture of methyl salicylate (RI = 1.54) and benzyl benzoate (RI = 1.57) with a refractive index close to that of protein (RI = 1.50–1.58). Spalteholz's fluid is insoluble in water but soluble in ethanol. Although specimens can be stained with inorganic fluorophores and immersed in Spalteholz's fluid, it will bleach endogenous GFP expression. Spalteholz was a pioneer in clearing tissues, writing: 'the transparency of tissues depends first of all on the refraction index of permeating liquid'. The calcium atoms in bone scatter light strongly, so it must be decalcified with EDTA. This lengthy process can be speeded up (Tinling et al., 2004) through the use of thermostatically controlled microwaves, which avoid adversely heating the sample.
2. OPT, or optical projection tomography microscopy builds a 3D data set, using bright-field and/or fluorescence illumination, by acquiring different images from different directions and reconstructing the 3D data set using a back-projection algorithm. In this sense, it is a tomographic technique which involves rotating the sample, just as with light-sheet microscopy. OPT and light-sheet are fundamentally different, and therefore incompatible, because OPT uses very low numerical aperture objective with a very large depth of field, whereas light-sheet benefits from a shallow depth of field. OPT is better suited to imaging larger samples that would otherwise scatter the SPIM light sheet. However, SPIM gives better resolution in the image and captures the data set much faster than with OPT. Micro-CT (micro computed tomography) is very similar to OPT but illuminates the sample using X-rays. Buytaert et al. (2013) have made a comparison of micro-CT and scanning thin light-sheet microscopy (Santi et al., 2009), which they call sTSLIM. OCT is an interferometric technique that takes advantage of the scattering of light by sub-surface tissues to form an image; it is effectively 'optical ultrasound'.
3. About 15% fluorescence loss after 27 000 frames. This was confirmed during product development of the Zeiss Lightsheet Z1: 10 fps for 45 minutes = 27 000 frames. This represents up to 100 TB of data; pers. comm. Dr Chris Power.
4. OpenSPIM colloquially refers to individual 3D Z-stack data sets taken through a 2D plane, each stack taken from multiple angles, as 'views'. Hence, the names of the following ImageJ plugins: SimView, Multi-view fusion and Multi-view deconvolution.
5. It is estimated that fewer than 1% of point-scanning confocals and about 15% of spinning disc microscopes are home-built. The effective advertising of the OpenSPIM wiki may lead one to think that most people self-build light-sheet microscopes, but in fact the percentage is thought to be 10–20%.

Multi-photon microscopes, estimated at 30%, are most often self-built because these are a niche technique where flexibility and specific sample requirements are paramount. Some companies, such as Scientifica, therefore offer modular components to assist such projects.

6. A very thorough treatment of why tissue scatters light can be found in Richardson and Lichtman (2015). To summarise: haemoglobin, myoglobin, melanin primarily absorb visible light, whilst lipids primarily scatter light. Energy absorbed is usually insufficient to promote an electron to a new orbital, as occurs in fluorescence. Rather, the energy causes electronic vibration and with a slight phase delay, scattering in all directions. This effect occurring multiple times is responsible for slowing down the propagation of light (RI change) with the heterogenic scattering between different regions of tissues causing the milky translucent appearance. Clearing techniques, therefore, work by equilibrating – or matching – refractive indexes through the tissue sample to reduce these inhomogeneities in light scattering.
7. In tissues that have a high percentage of extracellular matrix, such as collagen and elastin, removing lipid alone is insufficient to render optical transparency. Also, the high ECM density generally hinders penetration of antibodies. For nuclear stains, the To-Pro dyes give the best results with cleared tissues.
8. FocusClear (RI = 1.46) is a commercial clearing fluid that contains DMSO, diatrizoate acid, EDTA, glucamine, β-nicotinamide adenine dinucleotide phosphate and a derivative of polyoxyalkalene as an emulsifier and detergent, and is mentioned in papers and protocols on CLARITY. Because of its expense, people tend now not to use it, preferring to use RIMS (Yang, B *et al.*, 2014, Supp 4; Marx, 2014) or RapiClear from the Sunjin lab in Taiwan. There are also alternatives such as Easy-Index (LifeCanvas Technologies), X-Clarity Mounting solution (Logos Biosystems).

20

Bleed-Through and Spectral Unmixing

Outline

20.1 Introduction

If you have only one target that you need to investigate, a single fluorophore marker is all that is required. Cross-talk and bleed-through will therefore not normally[1] be present to cause problems. However, if you wish to distinguish and study two or more different proteins or cellular moieties, using multiple fluorescent antibody labelling and/or endogenous fluorescent protein reporters, erroneous signal crossover though the 'wrong' filter or into the 'wrong' detector must be taken into account and corrected or else allowed for.

Understanding Light Microscopy, First Edition. Jeremy Sanderson.

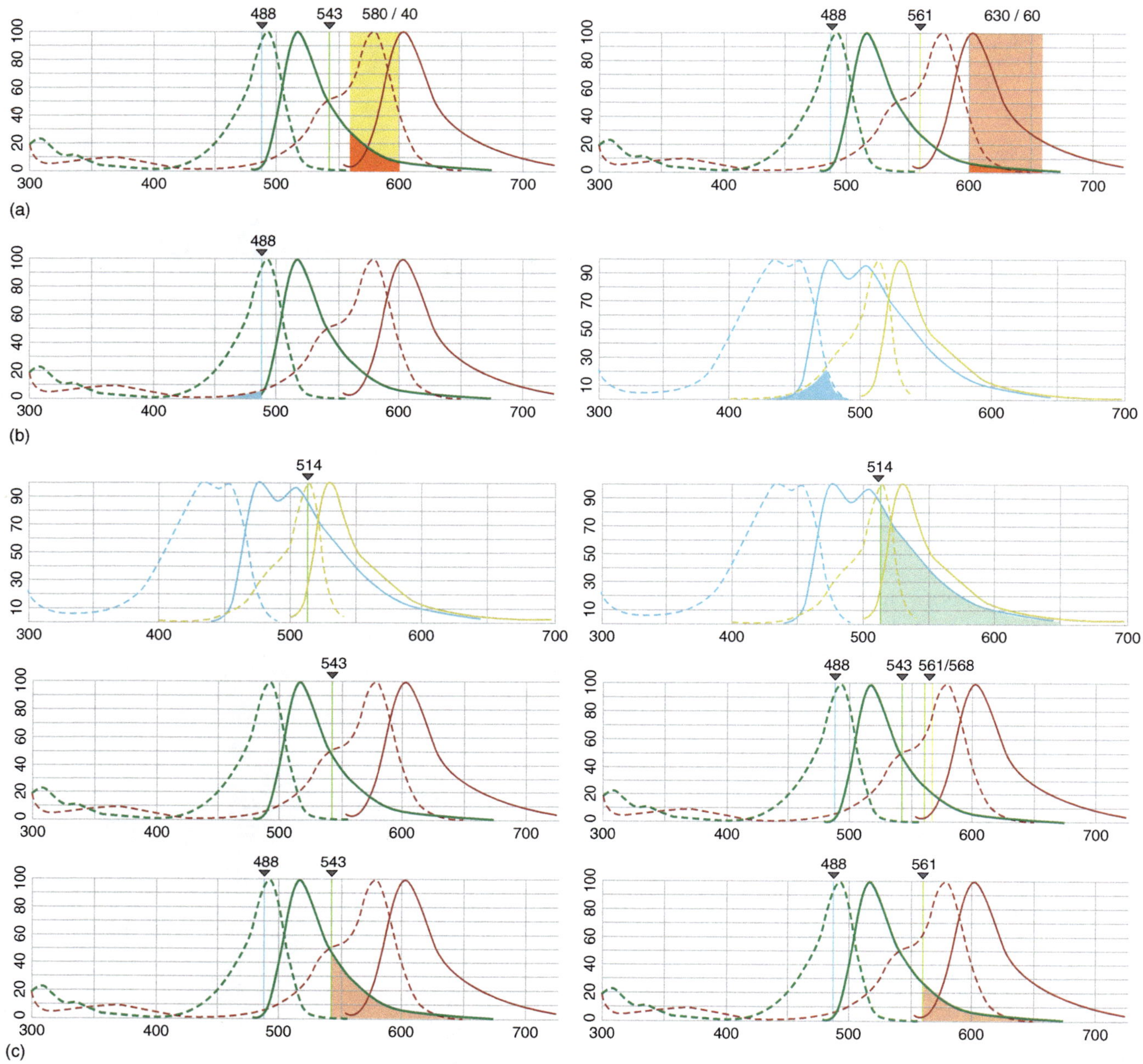

Figure 20.1 **Categories of bleed-through** *Three categories of bleed-through. (a) Cross emission. Shifting the excitation to the right (i.e. using 568 or 561 nm laser excitation instead of 543 nm) will reduce bleed-through and in this case increases the excitation of the second fluorophore. Thus a 630/60 bandwidth emission filter will collect the second fluorophore emission better than the 580/40 option. (b) Cross-excitation. The dyes (left) exhibit less cross-excitation (blue region) than the CFP-YFP fluorophores. (c) FRET. Excitation of donor emission by acceptor excitation. Bleed-through from excitation of the emission spectrum of the green fluorophore by the laser line intended for excitation of the red fluorophore. By selecting a different laser line, bleed-through can be reduced. In all these graphs, and subsequent ones, the excitation spectra are shown with a dashed line, and the emission spectra with a solid line.*

When more than one fluorophore is used to stain cells or tissues, the excitation and emission spectra of each fluorophore may overlap and allow cross-talk or bleed-through. There are three categories of bleed-through, each of which is illustrated in Figure 20.1:

1. Overlap between the emission spectra of two different fluorophores: cross emission.
   This is relatively common.
2. Overlap between the excitation spectra of two different fluorophores: cross excitation.
   This is much less common.
3. Interaction between emission from one fluorophore and excitation of another fluorophore.
   This is the situation that most commonly occurs (e.g. FRET) but does not normally pose a problem when routinely imaging two or more fluorophores.

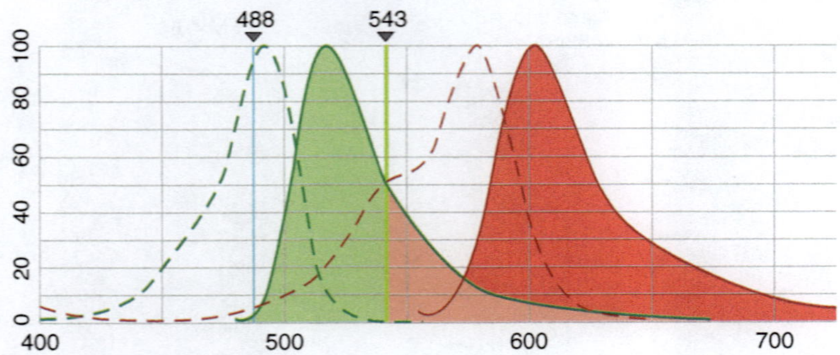

Figure 20.2 **Typical overlapping spectra for bleed-through** *Alexa 488 and Cy3 spectra. Bleed-through on excitation of Alexa 488 emission signal (green spectral curve) by 543 nm laser intended to excite Cy3. The overlap area between the two emission and excitation spectra, which leads to bleed-through, is shown in orange.*

Where two fluorophores lie physically very close together, the energy from photons emitted by the first, short-wavelength fluorophore (called the donor) is sufficient to be absorbed by – and excite – the second fluorophore (called the acceptor). This is called FRET, or Förster resonance energy transfer. It is an example of where cross-talk is used to advantage; see Chapter 24.

In describing these spectral overlap artifacts, the terms 'bleed-through', 'crossover' and 'cross-talk' are often used interchangeably. In this book, the term 'cross-talk' is used for cross-excitation and 'bleed-through' for cross-emission. Cross-talk and bleed-through occur because the excitation and emission spectra of fluorophores partly overlap along the spectrum, and these individual spectra are excited by different filter sets or by different laser lines (Figure 20.2). Properly, cross-talk refers to the (cross-) excitation of one fluorophore by the illumination intended to excite another quite different fluorophore in the experiment. Bleed-through occurs when emission from a short wavelength fluorophore overlaps the excitation filter, or laser line, illuminating subsequent longer-wavelength fluorophores, and the re-excited fluorescence is collected erroneously by the detector assigned to the second fluorophore, rather than the detector for the first.

Most fluorophores have very broad bandwidths, and both the excitation and the emission spectra (Figure 20.3 and also Figure 15.2) are skewed asymmetrically. This skewed part of the spectral curve is often colloquially referred to as the 'tail' of the fluorophore. The fluorescence emitted is distributed over a wide spectral range because of the chemical properties of the fluorophore. If these spectra overlap, cross-talk and/or bleed-through will result unless the spectral profiles of each the individual fluorophores are well separated (e.g. Alexa 488 with FM4-64, Figure 20.4). In particular circumstances (e.g. FRET), overlap of the donor emission and acceptor spectra *should* be maximised, but in routine fluorescence studies, this must be avoided. Unfortunately, most fluorophores only have typical Stokes shift[2] values between 20 and 30 nm. This means that it is difficult to adequately separate the fainter emission signal from the much more intense absorption, or excitation, signal. Long-Stokes-shift (LSS) fluorophores (Table 20.1) have been developed, which help minimise bleed-through, whilst at the same time allowing up to four or five fluorescent dyes to be imaged simultaneously.

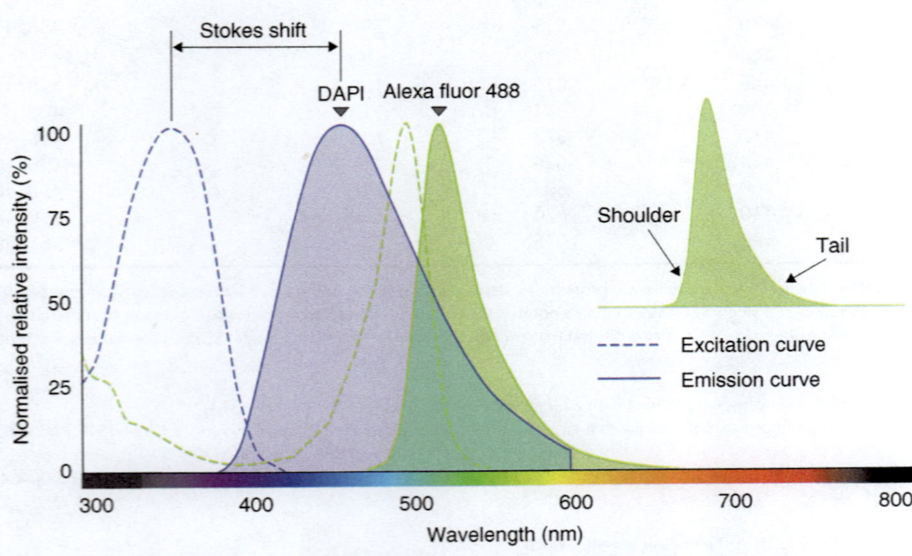

Figure 20.3 **Characteristic skew on typical spectra** *The skewed emission tail towards the low-energy end of the spectrum occurs because the most probable electronic energy transitions are the lowest few vibrational states of S_0. The width and shape of the excitation spectrum is determined by the S_1 vibrational levels, whilst the characteristics of the emission spectrum are determined by the S_0 vibrational levels. See also Box 15.1 and Figure 15.2.*

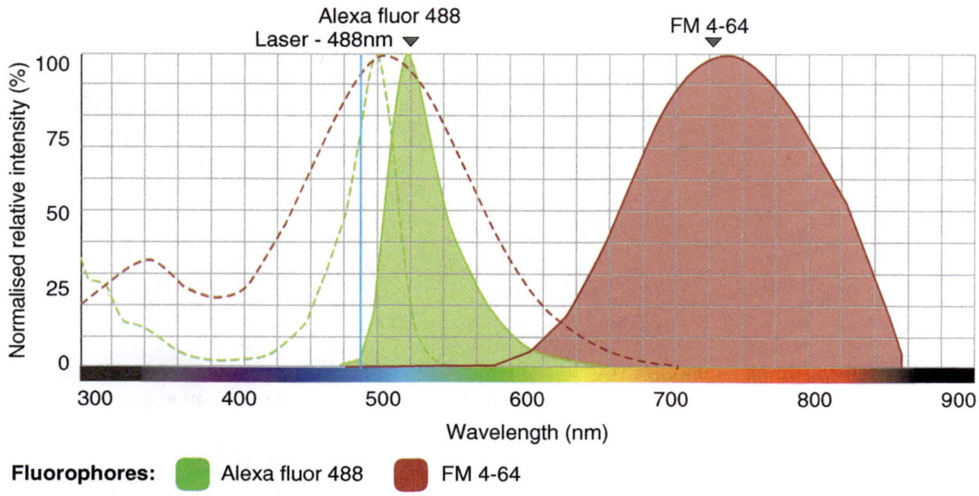

Figure 20.4 ***Alexa 488 with FM4-64*** *The benefit of a large Stokes shift allows the emission signal to be well separated from the excitation of the fluorophore. It also minimises bleed-through and allows more than one fluorophore to be excited with a single laser line or filter set. In this example, the 488 nm laser line can excite both GFP and FM 4-64 simultaneously with minimal bleed-through. This does away with the need to collect each signal sequentially. Simultaneous collection can be used, which is simpler to set up and is faster.*

Table 20.1 *Long Stokes Shift (LSS) Fluorophores*

Dye, Fluorescent protein	Ex_{max}	Em_{max}	Stokes shift (nm)	Suitable excitation lasers (nm)
Atto 430 LS	433	547	114	458, 460, 488
Atto 490 LS	496	661	165	488
Chromeo 494 active motif	494	628	134	488
DyLight 485-LS	485	559	74	488
DyLight 510-LS	509	590	81	488, 514
DyLight 515-LS	519	648	129	488, 514, 532
DyLight 521-LS	526	666	140	488, 514, 532
Epicocconone (deep purple)	390, 490	610	220, 120	375, 405, 488, 432
FM 1-43	479	598	119	488
FM 4-64	515	740	225	488, 514
PerCP-Cy5.5	482	695	213	488
PF-395-LSS	396	572	176	375, 405
PF-488-LSS	485	560	75	458, 460, 488
PF-510-LSS	509	590	81	488, 514
STAR 440SXP	436	515	79	458, 460, 488
mAmetrine	406	526	120	405
mBeRFP	446	615	169	458, 488
mT-Sapphire	399	511	112	405
LSS-mOrange	437	572	135	458, 488
LSS-mKate 1	463	624	161	458, 488
LSS-mKate 2	460	605	145	458, 488
mKeima Red	440	620	180	458, 488
TagRFP675 (from mKate)	598	675	77	561, 594, 607

The LSS dyes share a common excitation range, while their emission properties extend from the green to the far-red wavelengths. They can be used for recording two emission colours (i.e. signals) from excitation at a single wavelength that is optimal for both dyes. Long Stokes shift organic fluorophores and LSS fluorescent proteins lend themselves to multi-labelling and FRET experiments as well as two-photon microscopy. A representative selection of LSS dyes and fluorescent proteins are given in this table.

References

Piatkevich, K.D. et al. (2010). 'Monomeric red fluorescent proteins with a large Stokes shift'. *PNAS 107*/12: 5369–5374.

Shcherbakova, D.M. et al. (2012). 'An orange fluorescent protein with a large Stokes shift for single-excitation multicolor FCCS and FRET imaging'. *Journal of the American Chemical Society 134*/18: 7913–7923.

Yang, J. et al. (2013). 'mBeRFP, an improved large Stokes shift red fluorescent protein'. *PLoS One 8*/6: e64849. doi: 10.1371/journal.pone.0064849.

URL Links

Kurt's Microscopy Blog: http://nic.ucsf.edu/blog/2013/09/long-stokes-shift-dyes/.

Chroma Spectra Viewer: https://www.chroma.com/spectra-viewer?fluorochromes=10501.

Thermo Fisher Scientific Spectra Viewer: http://www.thermofisher.com/sg/en/home/life-science/cell-analysis/labeling-chemistry/fluorescence-spectraviewer.html.

Cross-talk is usually less troublesome than emission bleed-through. This is because, although the excitation spectrum does exhibit a slight skew towards shorter wavelengths (i.e. on the left) before the peak, this part of the spectral curve is usually very steep. Absorption values rise rapidly from a minimum value to the peak excitation value. The 'tail' of the spectral curve occurs towards the longer wavelengths. The intensity of the signal emitted by the fluorophore is determined by two factors: (a) the spectral range detected by the emission filter or PMT bandwidth and (b) the efficiency of fluorophore excitation. Therefore, even if a fluorophore has a significant 'tail' (i.e. on the right, after the peak value) of its emission spectrum, which may potentially give rise to bleed-through, cross-talk may be avoided by choosing an appropriate excitation wavelength for the second fluorophore. This is illustrated in Figure 20.5.

The possibility particularly of bleed-through must be considered, taken into account and managed when performing colocalisation studies (Chapter 25). If this is not done, then colocalisation may too easily be inferred, or at least overestimated, when it does not, in fact, occur. Even where colocalisation does occur to some extent, its reported value will be inflated by any signal bleed-through that does occur.

Where the broad excitation and emission spectra of two fluorophores overlap too much, it may be difficult to distinguish, or resolve, the two signals separately. If two emission signals pass through the bandwidth of the same filter or the signals cannot be distinguished as different hues by eye, then they will be perceived as one. Nevertheless, each of the fluorophores will still possess individual spectral profiles, and can therefore be computationally unmixed using special algorithms. The results can then be displayed with sufficiently different pseudo-colour overlays that the eye can easily distinguish between them. This is the technique of spectral unmixing, which is covered later in the chapter.

20.2 Managing Bleed-Through

The amount of cross-talk and bleed-through depends on the concentration of fluorophore, and either (a) the filters used in the widefield microscope or (b) the lasers and photomultiplier bandwidth or emission filter settings used in the confocal microscope. Bleed-through can only be reduced by using optimised fluorescence filter sets (Figure 20.5) and/or photomultiplier detector slit widths but is never completely eliminated. This is due to the fact that fluorescent dyes have very broad absorption and emission spectra that exhibit a significant degree of overlap.

Excitation (i.e. absorption) spectra are generally slightly skewed towards shorter wavelengths whereas emission spectra are skewed to a greater extent towards longer wavelengths, with 'tails' extending into the red part of the spectrum that may cover regions of tens to hundreds of nanometres. Thus, emission from a green fluorophore can often be detected through red emission filters, but a red dye is only seldom imaged through a green emission filter. For this reason, multicolour fluorescence imaging should be conducted with collection of the reddest (longest-wavelength peak emission) dye imaged first, using (where necessary) excitation wavelengths that are only minimally absorbed by the skewed spectral tails of the bluer dyes.

The fluorescence emission spectral profiles of common fluorophores differ significantly with regards to bandwidth, peak emission wavelength, symmetry, and number of maxima. In many cases, the most common cause of bleed-through is unequal fluorophore staining intensity. This should be corrected by altering the staining protocol and trying to achieve emission signals balanced in intensity rather than adjusting exposure times in the microscope or performing extensive image rehabilitation with post-acquisition processing.

The effect of the discrete spectral peaks present in short-arc mercury lamps or metal-halide lamps is often forgotten and not taken into account when deciding what concentration of fluorescent stain to use in an experiment. For example, at equal concentrations, rhodamine is excited more effectively than fluorescein in widefield fluorescence (by a factor of 10) due to the bright 546-nanometre emission line in the mercury arc spectrum.

In multiply-labelled specimens, if the degree of labelling and the intensity of fluorescence emission from the fluorophores is not equally balanced, brighter signals can overwhelm and penetrate the barrier filters of channels reserved for weaker, lower-intensity fluorophores or those stained sufficiently but marking fewer, less abundant targets. For example, bleed-through is particularly pronounced where the signal from a blue-shifted (lower-, shorter-wavelength) fluorophore (e.g. DAPI) is much stronger than the emission signal from the other red-shifted (higher-, longer-wavelength) fluorophore(s)[3]. Also, any attempt to see the weaker fluorophore better on the monitor by amplifying the signal directly either (a) through the microscope and camera controls or PMT gain controls or (b) by altering the gamma on the digital display will also amplify the bleed-through signal. This is one reason why control slides should be used.

During imaging, the apparent fluorophore signal intensity can be adjusted through the use of neutral-density filters in widefield microscopes or by altering the gain or laser power for the individual detection channels in confocal microscopy. However, merely balancing the concentration of fluorophore, selecting spectrally well-separated fluorophores and choosing saturated colours for target of low abundance during the specimen preparation stage *before* image acquisition will alleviate many problems when adjusting the microscope software and imaging and will, in general, lead to superior images. Therefore, optimise the initial concentrations of fluorescent probes during specimen preparation rather than attempt to alter variables whilst acquiring images afterwards.

A fluorophore is not only excited by the wavelength at its peak absorption – it is merely excited most efficiently at this point. A fluorophore gives off photons of the same emission spectral wavelengths, regardless of the excitation locus on the absorption spectrum. At excitation values away from the peak value, the intensity of the emission spectrum will be proportionally reduced (Figure 20.6). This can be used to advantage when devising stratagems to manage bleed-through.

For example, the 488 nm spectral line of an argon-ion laser will excite fluorescein[4] with 77% efficiency of the peak value (at 498 nm) and will also excite rhodamine 6G, although to a lesser degree. However, these two fluorophores would represent a very poor combination, because although rhodamine 6G is only excited by the 488 nm laser at 46% of its peak value (at 525 nm), this 46% value will still cause significant bleed-through, since it represents nearly 2/3^{rds} of the intensity of maximum excitation of fluorescein

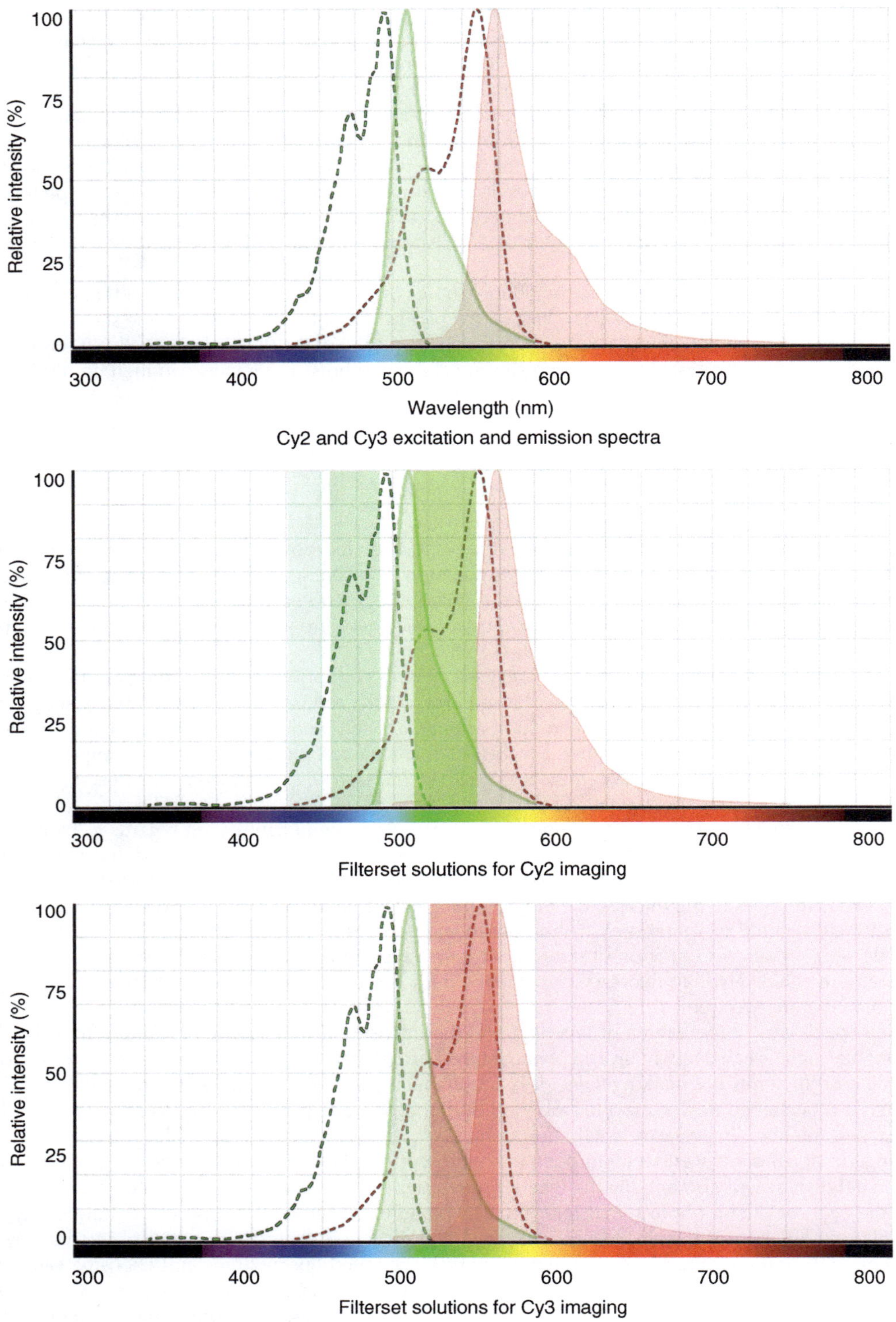

Figure 20.5 **Minimising cross-talk and bleed-through by utilising a tail: filter set selections** *If well-separated fluorophore pairs, or those with long Stokes shifts, cannot be chosen, the only option on a widefield microscope is to use carefully selected narrow bandwidth filter combinations. In this example, Cy2 can be excited with a 469/35 filter, but this will allow some Cy3 to be excited also. The narrower the bandwidth of the band-pass filter, the better it can separate fluorochromes with close emission spectra. A blue-shifted 434/18 filter exciting from 425 to 443 nm will excite Cy2 to a much lesser extent, but the cross-talk is very much reduced. However, the narrower the filter, the less intense is the signal collected. This factor must be considered when designing the experiment and choosing appropriate fluorophores. With only minimal Cy3 excitation and consequently very little Cy3 emission (the graphs in this figure show normalised values) a 525/30 filter set can be used to collect Cy2 emission.*

Because Cy3 is the longest-wavelength fluorophore, there are more choices to excite and collect the emission from this fluorophore. The 535/40 filter set may not excite Cy3 at its peak value, but it is just upstream of Cy2, which is what matters. A long-pass LP 590 filter set collects all the Cy3 emitted signal.

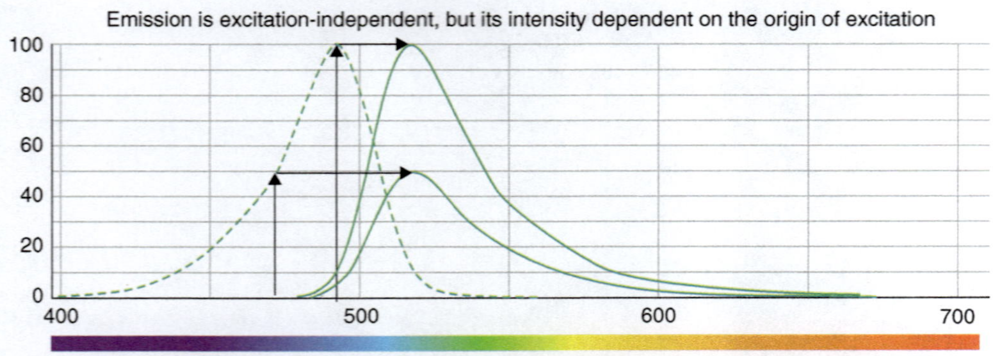

Figure 20.6 **Fluorophore excitation** *When a fluorophore is excited at a wavelength away from the peak (i.e. optimal) value, the same emission spectral curve is produced, but the intensity differs according to the quantum efficiency of the excitation photon.*

(Figure 20.7a). A better choice to use with fluorescein would be Rhodamine red, where excitation by the 488 nm laser is only 6%. If the concentration of fluorescein label is carefully increased (i.e. so that it does not create significant background staining), it may then be possible to switch to exciting fluorescein with the 458 nm laser (at 28% of peak intensity). Using the 458 nm laser line ensures that Rhodamine red is only excited at 1%. At this level, the cross-talk is negligible and can be managed by adjusting the PMT detector gain with single-labelled controls.

The example in Figure 20.7b shows the excitation profile of Cy3. The ideal laser to excite the fluorophore is the 543 nm He-Ne laser at around 75% efficiency, but this laser is only 1 mwatt, whereas the 514 nm line from the multi-line argon ion laser is 25 mwatts, even though it only excites Cy3 at 50% efficiency. The advantage of using the 514 instead of the 543 nm line is that slightly greater bandwidth is obtained for collecting the emission signal (which is independent of the excitation wavelength used). For this reason, using the 561 nm laser to excite Cy3, even though it is a 25 mwatt laser and excites the dye at 85% efficiency, is not the first choice. It is too far red-shifted to be much use, because the emission signal collected is of much narrower bandwidth than if a 514 nm laser line was used to excite Cy3.

The principle of optimising illumination and detection also applies to widefield microscopes (Figure 20.7c). For example, the fluorescent protein mOrange 2 is viewed with a standard TRITC filter or Cy3 filter (e.g. emission = 605/70). Whilst the excitation filter of the TRITC/Cy3 filter set is suitable (excitation = 545/30), collecting the emission signal on the tail for mOrange 2 (>590 nm) means that only half the emission signal is collected unless a custom filter is used that also collects emitted signal from 560 nm upwards instead of merely over 590 nm. This means now exciting the fluorescent protein at 515–530 nm and 50% efficiency, instead of 530–560 and 100% efficiency, but the emission signal is optimised.

The major disadvantage of using filter sets to detect fluorescence emission is that *all* photons collected by a filter are treated in the same manner, regardless of source. Even when probe concentrations are carefully controlled with regards to quantum yield and target staining, the level of signal cross-talk – or more commonly bleed-through – generally remains between 10 and 15% unless the emission spectral maxima are separated by 100 to 150 nanometres or more. If it is the case that the different emission spectra lie very close to each other and overlap to a significant extent, consider using spectral unmixing (Section 20.4) to differentiate between fluorophores. This unmixing technique is sometimes referred to as emission fingerprinting. Spectral unmixing can be performed with widefield as well as with confocal microscopes; protocols are given at the end of this chapter.

Sequential collection, or 'multi-tracking', is a means whereby the laser lines on a confocal can be used one after the other, either frame by frame or line by line, to excite individual fluorophores to control emission bleed-through (Figure 20.8). The use of lasers as very narrow bandwidth single nanometre wavelength excitation sources (rather than broader bandwidth fixed filters on a widefield microscope) is one advantage of the confocal microscope. The other advantage is that the lasers are under precise electronic control and can be switched on and off very rapidly. Line-scan switching, rather than sequential collection frame by frame achieves the same result but is much faster. Where simultaneous scanning is to be preferred for its speed advantage, consider sequential scanning line by line[5]. Some people maintain that there is a possibility of galvanometer error with sequential collection, giving rise to spatially misregistered features in images. If this is likely to be a concern, collect the data simultaneously and correct for bleed-through as appropriate with controls.

20.3 Controls for Cross-Talk and Bleed-Through

A quick assessment to control bleed-through can be accomplished by exciting only one fluorophore at a time, while monitoring emission settings of all the other fluorophores (e.g. on the confocals, exciting with only one laser at a time and look at all emission channels). Ideally, the 'wrong' fluorophore channels should appear dark (no signal, no cross-talk). For absolute rigor, you should prepare control samples. The bleed-through controls are necessary to determine the amount of signal gain possible in each channel without initiating bleed-through into adjacent channels.

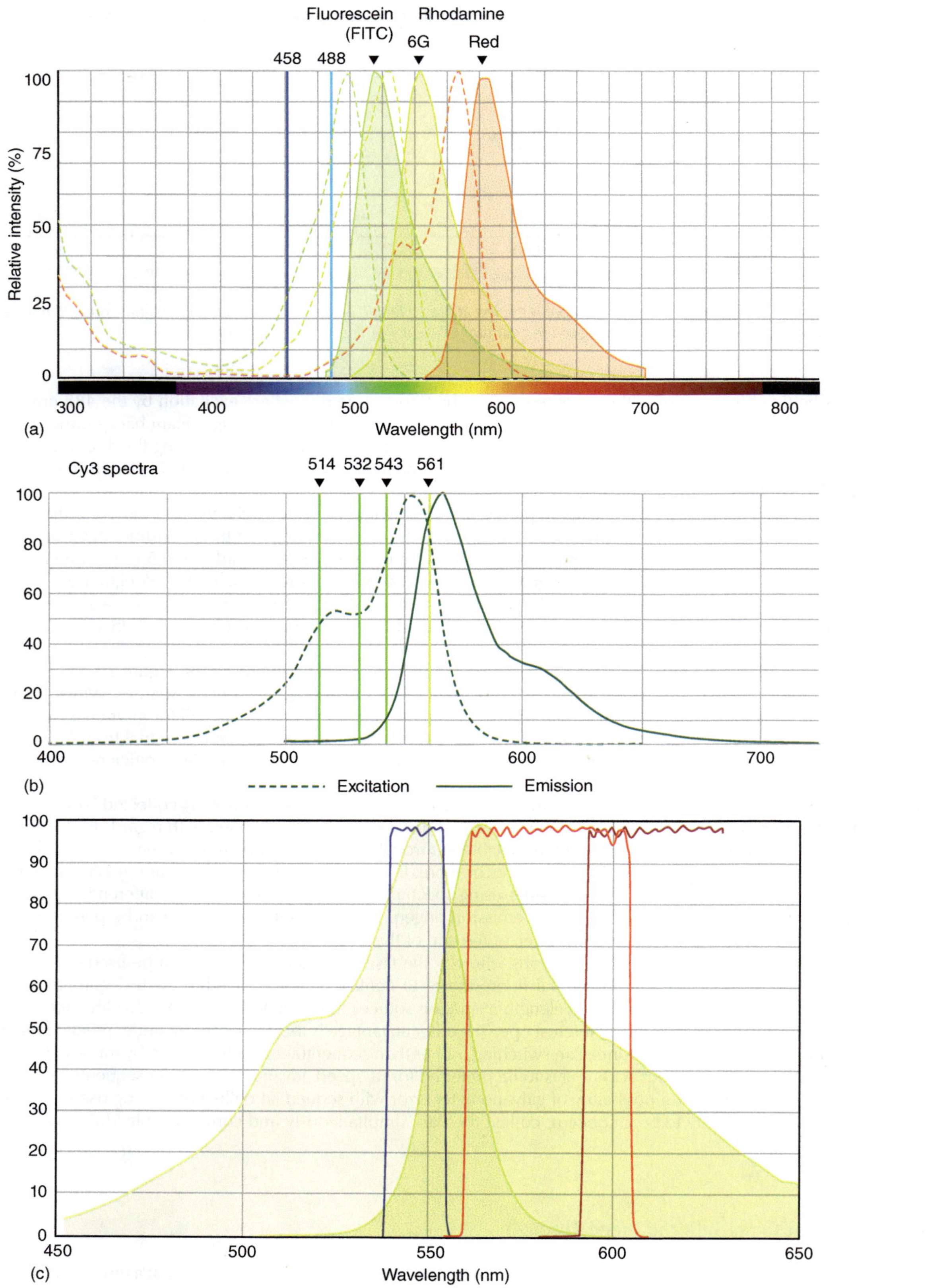

Figure 20.7 **Selection of laser lines for optimal excitation** *Using the 561 nm laser line is better, since there is less potential for bleed-through from the Alexa 488. Likewise, the 561 nm line is a better choice than the weaker 543 nm laser line, because it is separated further from blue-shifted fluorophores. However, the relatively broad Cy3 excitation spectrum allows the potential selection of one of several different laser lines to image this fluorophore.*

At least five controls are required to adjust the collection settings for each channel:

(a) all fluorophores together, and (if more than two) in paired combination;
(b) separate single-labelled (bleed-through) controls for each fluorophore; and
(c) an unstained slide (or one without secondary antibodies) for each channel for background determination.

Use the background control (without added fluorophore) to set the gain limit of signal amplification – determined for each channel – before background signal or autofluorescence becomes a problem. The background control should be examined independently with each laser and detection channel to set the limits of signal gain and offset that should be employed for the final imaging series. All channels that will be used to image a multiply-labelled specimen must be subjected to an independent background correction because the level of autofluorescence in each channel can vary substantially.

Use the single-labelled controls to assess, in turn, the degree of cross-excitation and emission bleed-through. Cross-talk, or cross-excitation, *cannot* be prevented via sequential collection, unlike emission bleed-through. Adjust the gain, photomultiplier voltage or laser power for the individual detection channels in confocal microscopy or use neutral-density filters in widefield fluorescence. Once the collection settings have been optimised, go back to the mixed slide set and collect an image. The only reliable way to manage significant cross-excitation is by spectral unmixing.

20.4 Spectral Imaging

We humans have trichromatic vision (Chapter 1, Figure 1.12) and see three primary colours at different sensitivities across the visible spectrum. Our brains are able to perceive about 10 million different hues of colour from the neural signals triggered by the three primary colour visual pigments. Because the brain 'constructs' different colour hues from varying signals of these three pigments, it is possible for people to disagree about the exact colour of an object. Since most of us have similar sensitivities, we can generally agree on primary colours. If we only had one or two visual pigments, we would not be able to 'unmix' the primary colours across the visible spectrum in the way that we do[6]. Hence, spectral unmixing, a technique first used in remote sensing from satellites, is useful to discriminate between different hues that fall below the resolution limit of colour perception by our eyes. Mixing of hues also occurs when using fluorescent filter sets and camera detectors: if two emission signals pass through the bandwidth of the same filter, they cannot be perceived as different yet can be resolved, or differentiated, by spectral unmixing.

In conventional multicolour imaging, separate fluorophores can be used to target specific areas of the specimen. The goal is to choose each fluorescent marker with sufficient separation between the emission spectra so that the signal arising from each fluorophore is collected into its own individual detection channel. Many fluorophores present collection problems because their broad excitation and emission spectra overlap too much. Accordingly it can be very hard to assign a structure in the image to a specific fluorescent label. It can also be time consuming to try to overcome this by trying to reduce bleed-through using sequential scanning, particularly if taking an extended Z-stack. Band pass filters are appropriate if there is only a small amount of spectral overlap of excitation and emission spectra. Even if filter combinations are available that eliminate overlap by allowing collection of emission signals from different parts of the spectrum, it is often at the expense of a severe reduction in light intensity due to the very narrow collection bandwidths necessary to achieve realistic dye separation. Usually large amounts of signal must be discarded to ensure reliable separation, and a compromise must be made between using narrow excitation and emission filters that efficiently keep signals separate yet at the same time maintaining sufficient bandwidth to collect enough signal.

One solution is to take a fluorescence emission spectrum whilst exciting simultaneously with one or more wavelengths. Normally fluorescence emission throughout the entire spectrum is confined to collection at one image point or image plane. In this case the reverse occurs, and a spectral scan is performed at several sampling points, as a number of discrete intensity measurements (Figure 20.9) over several different wavelengths across the spectrum. This may be as few as two intensity measurements or as many as several tens of measurements, each into its own detection channel, with the proviso that there must be *at least* as many detection channels as individual fluorophores.

So, for example, a spectral stack may have a series of measurements across 10 nm bandwidths spanning the visible spectrum from 380 to 720 nm, represented as 32 measurements[7]. These 32 measurements may require more than one pass to scan the full spectral profile, perhaps as many as four separate passes. Alternatively, a series of eight of the minimum bandwidths of 10 nm each could be binned (the same principle as when binning in a rectangular CCD array) to form one sampling element 85 nm wide. In this case, the full spectrum of 340 nm from 380 to 720 nm could be sampled in one pass.

Since each fluorophore has a unique spectral profile, this signature can provide a means of reliably discriminating each fluorophore from another (including separation of autofluorescence as an independent 'hue') in the specimen. This process of spectral unmixing is also sometimes referred to as spectral deconvolution. The relative contribution of each fluorophore to individual bands (of predetermined sampling bandwidth) along the spectrum is determined. One image is produced of the entire specimen for each band of the detector. The result is called a 'lambda (λ) stick' or 'lambda stack' or 'lambda scan'.

Although used mainly on confocal microscopes, with their ability to switch rapidly between laser excitation wavelengths, a lambda scan can also be collected on a widefield microscope. At its simplest, a lambda stack from a widefield microscope can be collected by using a series of filters, say from violet, deep blue, cyan, green and so on through the spectrum to far red. Successive images taken with a monochrome camera, or other detector, will result in a lambda scan across the image plane. This approach demands the use of many filters and is time consuming, requiring each successive filter to be rotated into the optical path. There is the risk of image shift

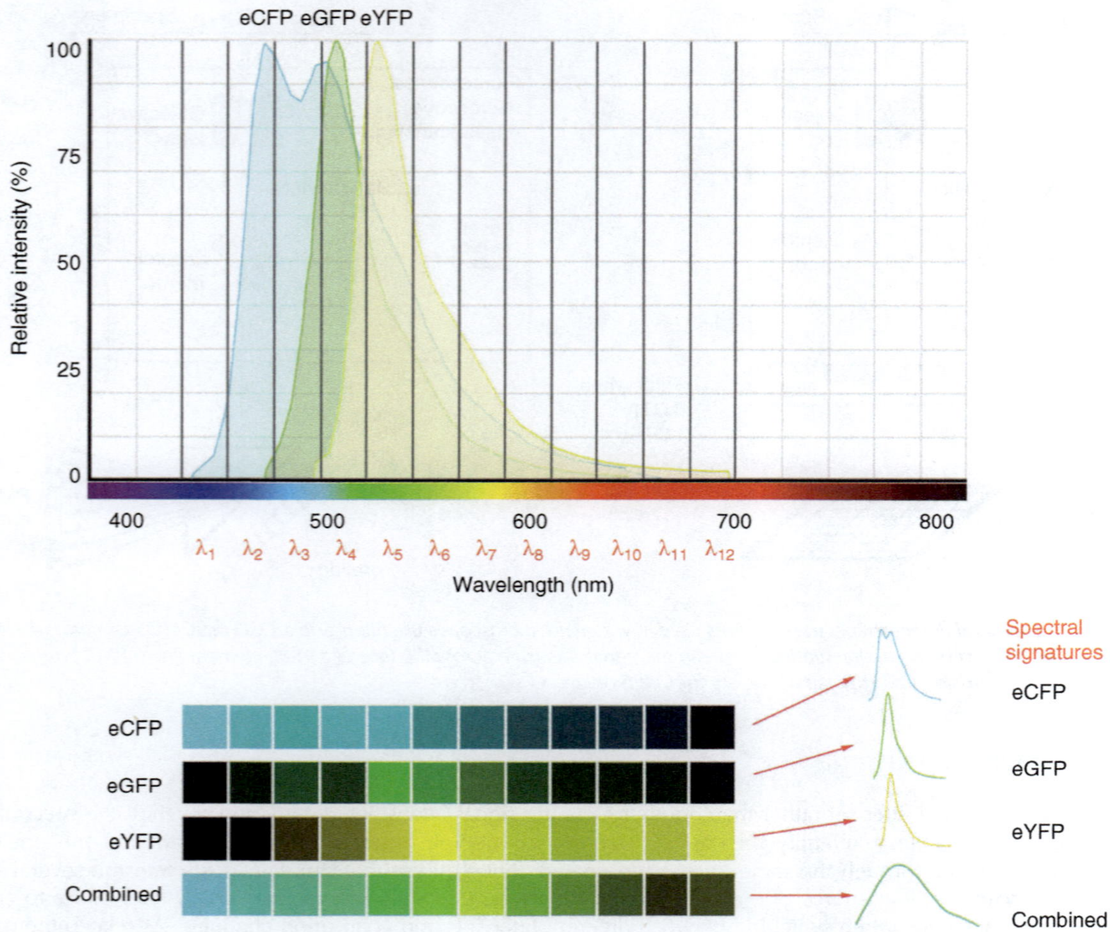

Figure 20.9 ***Collecting a lambda stack, taking spectral intensity measurements*** *A series of detection channels of, say, 10 nm, is applied as a wavelength scan across the fluorescence emission of each fluorophore to create a unique spectral signature. Our eyes, and the digital detector, see the combined result, but the spectral signature can be collected from control slides or single pixels containing only each fluorophore. These reference spectra are used to unmix each individual signal. Source: Reproduced with permission from Carl Zeiss Ltd.*

and vibration artifacts in the final lambda scan of the image. A faster, more reliable, means of collecting a lambda scan on a widefield microscope is to use a liquid crystal tuneable filter[8] (LCTF) – related technology to the AOTF or acousto-optical tuneable filter – which can be changed rapidly without mechanical vibration or risk of pixel shift in the image. The rapid switching technology incorporated into the point-scanning confocal design offers the best solution to spectral unmixing.

In a confocal microscope, a lambda scan is collected by means of a prism or a grating (Figure 20.10) through which the emission signal is directed. The signal is either collected with 32 mini-PMT detectors arranged linearly or by switching rapidly (using an AOTF or AOBS system) into a single PMT detector. The Zeiss system uses a diffraction grating and linear PMT array (Quasar and the earlier Meta-detector) with 3 nm and 10 nm step sizes. The Nikon A1Rsi and Olympus FV3000 confocal microscopes also have spectral detection units employing a diffraction grating (the Nikon has a 2.5 nm minimum step size; the Olympus, 2 nm) whilst the Leica SP-8 uses an AOBS and prism with a single PMT detector having 2 nm resolution. Most systems collect across the spectrum from 380 to 720 nm. That said, a lambda stack acquisition with a spectral resolution of 5 nm or less is really only useful when examining specimens that are labelled with very bright fluorophores. In most cases, a spectral resolution of 10 nm or more is sufficient.

In the example (Figure 20.11) a linear spectral unmixing algorithm uses a set of linear equations (see Zimmermann, 2005, and Zimmermann et al., 2014, for mathematical details) to relate the concentration of each dye in the spectral stack to the amount of light measured at each pixel sampled in the image. For these equations to work, the number of spectral detection channels must be at least equal to the number of fluorophores in the sample. Also you either have to provide the software with a known reference spectrum of each dye from appropriate control slides, or you have to generate a spectral signature from an area of the image (as internal control) that contains *only one* of the fluorophores used to label the specimen. This may be as small as a single pixel across the depth of the entire lambda stack, but it must be purely one fluorophore, so as to generate an accurate spectrum of that dye. The software can then clearly separate the contribution of the individual fluorophores to the overall signal. Obviously, independent, rather than internal, controls are preferable, but these may not be available.

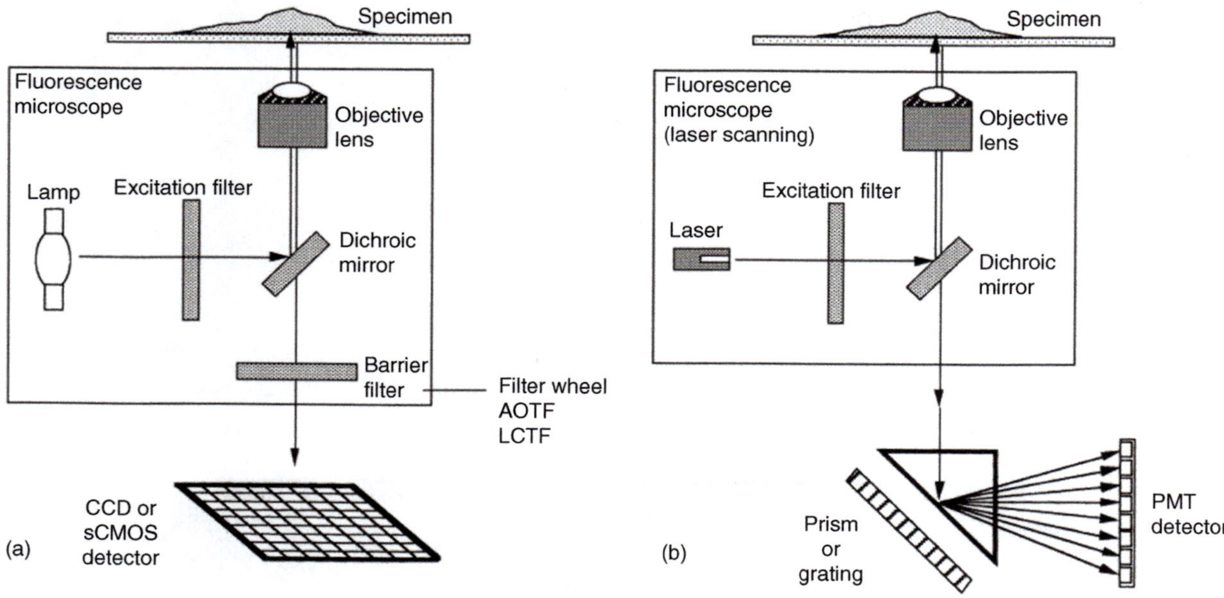

Figure 20.10 ***Types of spectral detector*** *Spectral detectors used in widefield microscopes use filters and a CCD or sCMOS camera, whereas a confocal spectral detector utilises a prism or diffraction grating to spread the signal spectrum across the face of a multi-element linear PMT. Source: Adapted from Hiraoka et al. 2002 with permission from the Japan Society for Cell Biology.*

20.5 Sources of Experimental Error

Individual dust particles on one filter set rather than another can alter signal intensities slightly and so 'shift' the spectral signature – leading to inaccuracies. If possible, to quantify and control this error, use multiply-labelled fluorescent beads and only one filter setting, to ensure images are collected through the same filter. Multi-coated (Pinkel) filters that simultaneously transmit several wavelengths are useful for this purpose. Another source of error is misaligned images. Use shift-free, or 'zero shift', filter cubes to ensure image registration to sub-pixel accuracy when switching between different filter sets and acquisition channels. Also keep the parameters of the experimental setup as close as possible to those used for acquiring the reference spectra. Be aware that any 'undetermined' effects, such as FRET bleed-through, are properly accounted for; otherwise, linear spectral unmixing will lead to false results. Filter sets are relatively cheap and with careful selection can be used to avoid bleed-through and cross-talk whilst achieving maximum sensitivity and signal collection. Be prepared to change the filter sets that come with the microscope: they can be changed easily for relatively little cost; all too often this point is overlooked.

If image deconvolution is performed before spectral unmixing, it is imperative that signal intensities in the output image should match those of the images acquired before deconvolution. Since spectral unmixing is a pixel-based method, it is also prone to errors in spectra generated by shot noise or else by detector read-out noise. Any signal not originating from the fluorophores that are being spectrally unmixed must be removed from the background. This is usually straightforward in the case of fluorescence where the background is (ideally) sufficiently black. If it is not, the background (e.g. if it originates as autofluorescence) can be assigned its own channel and be duly unmixed from the data. You need a good signal-to-noise ratio to begin with. Spectral unmixing does not generate signal out of nothing, it merely 'pushes' the signal into the correct channel representing the fluorophore in question in the image. It is also important that the image and/or the background are not noisy. Where noise is present (e.g. low signal; high background or high detector read-out noise) it is better to set up spectral detection with fewer channels of large bandwidth, rather than use narrow channels. If too much noise is present, the spectral algorithms will tend not to give a precise result. For accurate results from confocal spectrophotometers, you may wish to check that these are accurately calibrated. For practical guidance on how to check and correct this, see Zucker et al. (2007) and also Cole et al. (2013).

20.6 Protocols for Spectral Unmixing

Any suitable pairs of spectrally similar dyes of fluorescent proteins are suitable to use as a test sample (e.g. FITC/Cy2 /Alexa Fluor 488/Alexa Fluor 514, or GFP/ YFP). Alternatively, Thermo Fisher Scientific (old Life Technologies) sell custom-made test samples on slides (F36905 to F36908, individual green, orange, red or far-red pairs of dyes) with 6 μm diameter spectrally overlapping beads, or the Focal Check slide #2 (F36913) can be used. Each slide has beads mixtures of dye pairs as well as individually stained control bead populations.

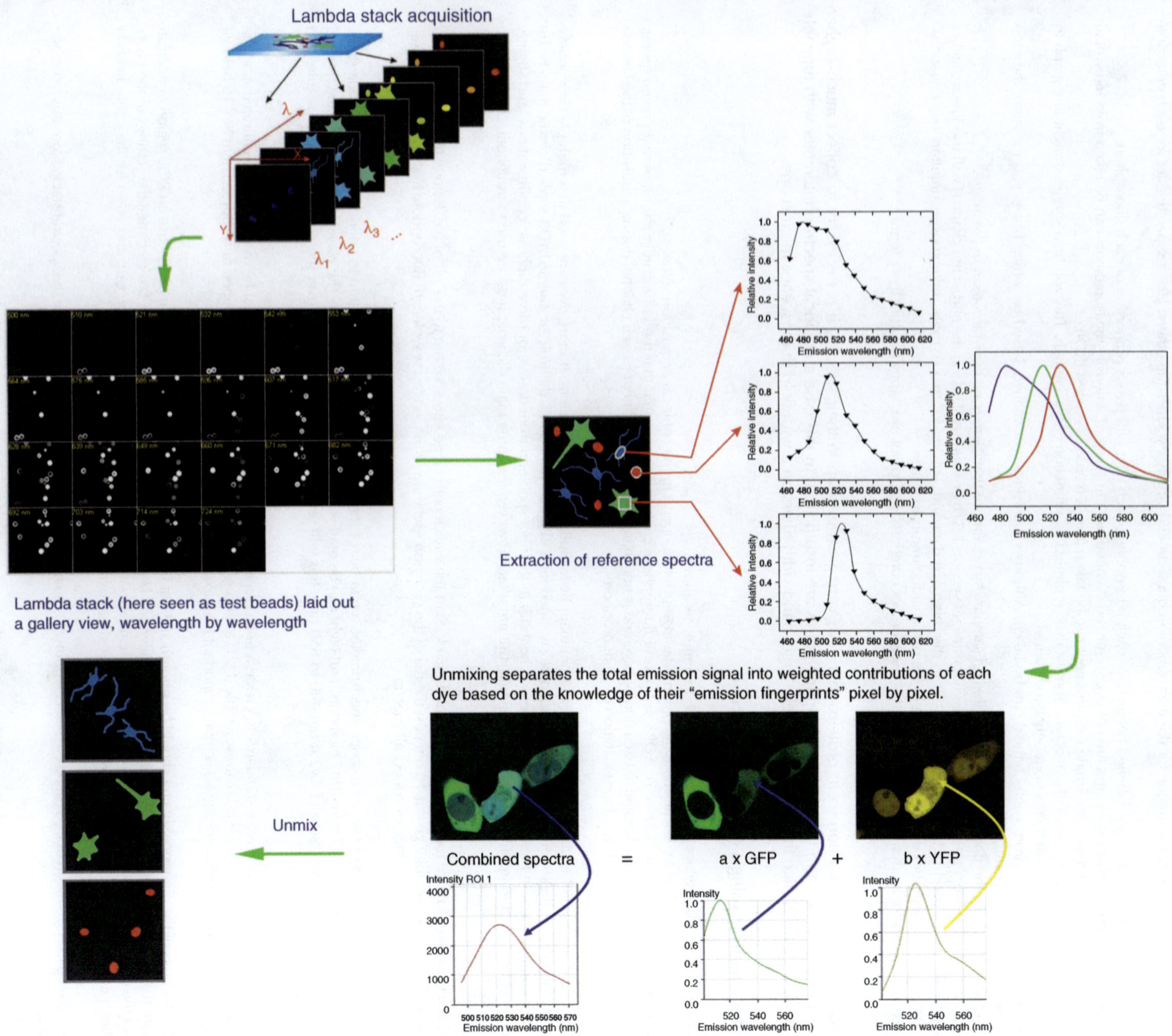

Figure 20.11 **Spectral unmixing** *Unmixing separates the total emission signal into weighted contributions of each dye, based on the knowledge of their emission 'fingerprints', pixel by pixel. Source: Reproduced with permission from ZEISS Microscopy.*

A. Spectral unmixing using a widefield microscope

1. Check the spectral behaviour of the fluorescent dyes that you wish to unmix. For example, use the Thermo Fisher Scientific spectra viewer or a similar Java application (see URLs at the end of the chapter).
2. Select a filter bandwidth that is unique for the fluorophore in question and largely excludes the others. Use a separate filter for each detection channel.
3. Take reference images with single-labelled controls for each fluorophore. Each filter will pass a proportion of each fluorophore.
4. It is important to ensure that the images are not over-saturated ('whited-out') in any part. Intensity measurements can be taken from individual pixels in each reference image, or a single measurement can be taken from a region of interest (ROI).
5. The reference intensity measurements (averages from ROIs) are plotted as a function of wavelength. This will be a simple straight line graph in the minimum case of measurements from only two channels.
6. A series of measurements from each channel can be taken. Although intensities will change, based on which pixels you measure, the *ratio* between the measured points will be the same.
7. Collect an image from the double- or multiply-labelled sample. Again, ensure that there is no over-saturation in the image.
8. A spectral unmixing programme (e.g. in freeware Fiji or ImageJ version 1.37b or later; see URLs at the end of the chapter) will use linear algebra to interrogate every pixel in the image, compare it to the reference data and, solving for two or more unknowns in simultaneous equations, evaluate the final intensities of each fluorophore in the pixels of the unmixed image.
9. Collect the images for analysis in *.tif format so that they are not compressed. Do not use *.jpeg or other lossy format.

For those fluorophores or fluorescent proteins that have very great overlap in their spectra, it is necessary to collect much–higher resolution spectral stacks, and two or three sampling points are insufficient. In this case a confocal spectrophotometer with tuneable bandwidths is a better choice than fixed, relatively large, bandwidth filters on a widefield fluorescence microscope.

B. Spectral unmixing using a confocal microscope

1. Check the spectral behaviour of the dye you wish to unmix, as above.
2. Select the spectral unmixing, linear unmixing or lambda fingerprinting option in the confocal acquisition software.
3. Select the appropriate main dichroic or AOBS settings for fluorophore capture. It is essential that the *same* configuration of main beam splitter filter is used for the channel allocated to each fluorophore and is not changed between collecting each channel during acquisition of the lambda stacks. Normally, one laser is used for excitation of all fluorophores.

Thus, if Cy2 or Alexa 488 is used alone, a single 488 nm main beam splitter filter is sufficient. If more dyes are used in a multiple fluorophore experiment, then the single 488 nm filter may require to be swapped for a 488/561 nm double dichroic or 488/561/633 triple dichroic main beam splitter filter configuration. In these latter cases, there will be a dip in the filter spectrum from double or triple dichroic configurations, and slightly less signal will be collected from these than from a single 488 nm filter. However, the same main dichroic must be used throughout, despite some signal loss for the Cy2 signal.

4. Define a wavelength range for collecting the lambda stack. Start the wavelength range either 10 nm to the left of (i.e. below) the wavelength of the laser line and (if possible) go out 10 nm past the tail of the emission spectrum of the longest wavelength fluorophore that you are acquiring. Use the same wavelength range to collect all spectra.

Depending upon the design of the microscope spectrometer, a number of sequential passes may be required to collect the entire lambda stack. Modern spectral detectors have 32 mini-PMT elements, and eight can be read out at one time. If, for example, the wavelength range for the spectral stack is 170 nm to cover the emission of Cy2 (from 480 to 650 nm), this will require 16 PMT elements to cover the range, which will require two passes by this spectral detector.

5. Image the single-stained reference standards to create a unique spectral signature. These images must not be over- or under-saturated. Use the proprietary range indicator (e.g. Hi-Lo) lookup table (LUT) to ensure this. The pinhole can normally be opened fully (but see item No. 10 below): spectral resolution (i.e. sufficient signal) is more important than axial or spatial resolution.

Ideally, the plots of the reference spectra should be smooth and not jagged. Collect the spectra from several pixels (use an ROI, region of interest) and ensure the signal is bright but not over-saturated. Normally one laser is used, but if using more than one laser, change the settings of the lasers to control the intensity of emitted signal into two different channels, since the spectral detector only uses *one* gain and offset for the entire lambda stick. Alter the digital gain as needed to optimise the signal without over-saturation.

6. Use a pixel picker or region of interest box to select a point or area from which to create the spectrum from the spectral stack. Store this reference spectrum for unmixing later.

If no independent controls exist, create a spectrum on the fly. Use an area that uniquely contains each dye to be unmixed to create a spectral signature that can be used for unmixing. It is possible to use different slides and areas to capture different spectra, provided the filter configurations remain the same. Where no region can be found that is unique to one spectrum and devoid of signal in the other channels, the algorithm selects an area with the lowest signal intensity in the other channels instead and considers that area to be a pure fluorophore. This technique is known as automatic component extraction (**ACE**) and automatically unmixes fluorescence signals to remove excitation cross-talk.

7. Image the test image. The gain values can be altered from those of the reference images, and this will usually be required. However, the filter configuration, wavelength range and the laser used must remain the same as those used to collect the reference spectra.

   If needs be, use the digital gain to roughly match the intensities of the signals collected.

8. Load the test image into the unmixing software. Recall and load the reference images.
9. Use the unmixing software to unmix the result from the test image.
10. To eliminate autofluorescence from the test image, treat it as an extra channel, take a reference spectral signature collected on the fly and unmix it. Then delete this from the unmixed result. If the signal from this residual channel gives rise to a poor result, try reducing the pinhole.
11. Most software packages will allow you to combine images taken in channel mode from those taken by spectral unmixing, simply by adding images together in the processing mode of the software.

20.7 Chapter Summary

Control of bleed-through, and to a lesser extent, cross-talk, is essential when acquiring images from multiply-labelled fluorescent specimens so that the signal emitted from each channel is collected only by the detector associated with that channel. It is important to take care to ensure minimal bleed-through, particularly when performing colocalisation analysis. Proper control of signal transfer – both bleed-through and cross-talk – is usually managed by switching laser lines on confocal microscopes or the careful selection of filter sets on widefield instruments. Well-designed controls are necessary to manage bleed-through for quantitative analysis and are essential for preventing cross-talk. Where it is not possible effectively to manage bleed-through or cross-talk, then spectral unmixing allows two different signals to be separated one from another. Emission fingerprinting involves acquiring reference spectral signatures to separate fluorophores otherwise indistinguishable in hue using linear unmixing of the emitted signal. It can be used to eliminate background and/or autofluorescence as well as fluorophores excited by the same laser line.

1. Bleed-through – that is, overlap between the emission spectra of two fluorophores – is a relatively common situation.
2. Cross-talk – interaction via overlap between the excitation spectra of two different fluorophores – is much less common.
3. Overlap between emission from one fluorophore and excitation of a second, longer-wavelength fluorophore is the commonest situation.
4. FRET, or Förster resonance energy transfer, is an example of where cross-talk is used to advantage.
5. Most fluorophores have very broad bandwidths; both the excitation and the emission spectra are skewed asymmetrically. There is usually a longer 'right-hand' emission tail and only a shorter, steeper excitation slope before the peak absorption value.
6. The intensity of the emitted fluorescence signal is determined by two factors: (a) the spectral range detected by the emission filter or PMT bandwidth, and (b) the efficiency of fluorophore excitation.
7. A fluorophore gives off photons of the same emission spectral wavelengths, regardless of the excitation locus on the absorption spectrum.
8. Where the fluorophore is excited away from the peak absorption value, the intensity of the emitted fluorescence signal is correspondingly reduced.
9. In multiply-labelled specimens, if the labelling from the fluorophores is not equally balanced, brighter signals can overwhelm weaker ones, and thus bleed-through.
10. Discrete spectral peaks present in short-arc mercury lamps and metal-halide lamps can excite fluorophores differently and cause bleed-through from differing signal intensities.
11. Sequential collection, or 'multi-tracking', is used on a laser scanning confocal microscope to excite individual fluorophores, one after the other, to control emission bleed-through.
12. Cross-talk, or cross-excitation, cannot be prevented via sequential collection, unlike emission bleed-through.
13. Use control samples to adjust the collection settings for each channel.
14. Spectral unmixing is useful to discriminate between different fluorophores whose colour hues appear the same to our eyes. The spectral profile is called a lambda stack.
15. The spectral unmixing algorithm requires a known pure spectrum of each dye from appropriate control slides or a spectral signature from an area of the image (an internal control) that contains the single fluorophore used to label the specimen.
16. Spectral unmixing is a useful means of eliminating autofluorescence, whose spectrum can be assigned to its own channel in the image and subtracted.

Key Reading

Garini, Y., Young, I.T., and McNamara, G. (2006). 'Spectral imaging: principles and applications'. *Cytometry A 69A/8*: 735–747.

- A good overview of spectral imaging, of how spectral data sets are collected and analysed

References

Cole, R.W. et al. (2013). 'International test results for objective lens quality, resolution, spectral accuracy and spectral separation for confocal laser scanning microscopes'. *Microscopy & Microanalysis 19/6*: 1653–1668.

Hiraoka, Y. et al. (2002). 'Multispectral imaging fluourescence microscopy for living cells'. *Cell Structure and Function 27*, 367–374.

Zimmermann, T. et al. (2014). 'Clearing up the signal: spectral imaging and linear unmixing in fluorescence microscopy'. In: Paddock, S.W. (ed.), *Confocal Microscopy, Methods in Molecular Biology*, Chapter 5, 2nd ed., vol. 1075. Springer Protocols. New York, Heidelberg: Humana Press, pp. 129–148.

Zimmermann, T. (2005). 'Spectral imaging and linear unmixing in light microscopy'. *Advances in Biochemical Engineering/Biotechnology 95*: 245–265.

Zucker, R.M. et al. (2007). 'Reliability of confocal microscopy spectral imaging systems: use of multispectral beads'. *Cytometry A 71/3*: 174–189.

Further Reading

Favreau, P. et al. (2014). 'Thin-film tunable filters for hyperspectral fluorescence microscopy'. *Journal of Biomedical Optics 19/1*: 011017.

Kraus, B., Ziegler, M., and Wolff, H. (2007). 'Linear fluorescence unmixing in cell biological research'. *Modern Research and Educational Topics in Microscopy 2*: 863–872.

Lerner, J.M. and Zucker, R.M. (2004). 'Calibration and validation of confocal spectral imaging systems'. *Cytometry A 62/1*: 8–34.

Lerner, J.M. et al. (2010). 'Approaches to spectral imaging hardware'. *Current Protocols in Cytometry* 12.20.1–12.20.40 July 2010.

Mylle, E. et al. (2013). 'Emission spectra profiling of fluorescent proteins in living plant cells'. *Plant Methods 9/10*.

Neher, R. and Neher, E. (2004). 'Optimizing imaging parameters for the separation of multiple labels in a fluorescence image'. *Journal of Microscopy 213/1*: 46–62.

Internet Resources

Chroma Spectra Viewer: https://www.chroma.com/spectra-viewer?fluorochromes=10501.

Thermo Fisher Scientific Spectra Viewer: http://www.thermofisher.com/sg/en/home/life-science/cell-analysis/labeling-chemistry/fluorescence-spectraviewer.html.

Promokine Long Stokes shift dyes: https://www.promokine.info/products/fluorescent-labeling/promofluor-dyes/promofluor-lss-dyes/.

Optimum Fluorophore combinations: https://www.chromocyte.com/educate/Beginners-Guide-Training-Resources/Which-Fluorochrome-Combination.

Channel optimizer for ImageJ: https://www.mh-hannover.de/cellneurophys/poissonNMF/TIS/TestImagingSettings.html.

Spectral unmixing theory: https://newenglandcytometry.files.wordpress.com/2011/06/spectral-unmixing-an-imaging-perspective-bartek-rajwa.pdf.

Zeiss tutorials on Spectral Unmixing: http://zeiss-campus.magnet.fsu.edu/articles/spectralimaging/introduction.html and http://zeiss-campus.magnet.fsu.edu/tutorials/spectralimaging/linearunmixing/indexflash.html.

Spectral unmixing for Zeiss Zen software: http://mudayja.capeville.wfunet.wfu.edu/microscopy/wp-content/uploads/2012/03/spectralimaging-protocol-2012-3-5.pdf.

Spectral unmixing algorithms for ImageJ/Fiji: https://imagej.nih.gov/ij/plugins/spectral-unmixing.html

Spectral Unmixing test slides

Thermo Fisher Scientific – (F36905 to F36908): http://tools.thermofisher.com/content/sfs/manuals/mp07234.pdf.

Focal Check slide #2 (F36913): http://tools.thermofisher.com/content/sfs/manuals/mp36909.pdf.

Also from Spherotech: http://www.spherotech.com/flu_par_sli.htm, and Bangs labs: https://www.bangslabs.com/products/fluorescent-microspheres.

Notes

1. This statement assumes that autofluorescence (usually present at the shorter wavelengths of the spectrum) does not pose a problem. Autofluorescence may be collected into a separate channel and unmixed from the image. This technique is dealt with in Section 20.4.
2. The Stokes shift is the difference (usually given in nanometres) between the peak excitation and the peak emission wavelengths of a particular fluorophore. Most illustrations of the excitation and emission spectra show these spectra as normalised plots; it is easy to forget that the emission signal is mostly significant weaker than the absorption spectrum.
3. DAPI and Hoechst 33342 have relatively long emission spectra that are likely to bleed through into the fluorescent protein (FP) channel. This problem is made worse if the DAPI/Hoechst signal is very strong and the target protein relatively weak. In this case, consider using a far-red dye, such as Draq5, RedDot-1, Draq7 or RedDot-2, to stain the nuclei.
4. Fluorescein-conjugated compounds should *not* be used for labelling, since there are much brighter, more photostable labels available nowadays. Fluorescein is only cited here as an example.
5. Scanning a selected smaller region than full-frame, particularly a 'letter-box' format with the x-dimension longer than the y-scan, will also help reduce the time taken for line-by-line or frame-by-frame sequential scanning. This stratagem is used in FRAP imaging.
6. This lack of unmixing spectral hues is one way to regard colour-blindness, see Chapter 31, Section 31.13.
7. This is what happens in a commercial spectral detector. The range 380–720 spans 340 nm, so 34 discrete measurements each of 10 nm bandwidth are expected, and 340 nm/32 = 10.625 nm. In fact, this 0.625 residual is taken up as 'blind spots' in sampling the 340 nm spectrum, because the linear 32-array PMT has walls built in between each element of the PMT array.
8. The LCTF may only transmit a low-intensity value even at peak transmission due to the polarising elements used in its design.

21
Deconvolution

Outline

21.1 Introduction

From Chapter 10, we have seen that the image formed is never a wholly true representation of the object, even for a perfect optical system (in which we assume that the influence of spherical aberration, chromatic aberration, stray light, optical noise, sample preparation artifacts, temperature changes and vibration are all zero) because of the effect of *diffraction* on the image. As the size of the object approaches the wavelength of the illuminating radiation, the effect of diffraction becomes particularly significant in determining the quality of the image that we see.

When any image is formed, or convolved, from an object, it suffers degradation. This may not seem apparent at first: surely the image of the pencil that you see is a faithful representation of the object lying on the desk in front of you. Well, 'yes', but it is a matter of degree. Your eye is a low-numerical aperture imaging device, and that pencil is much larger than the wavelength of the light that illuminates it. Any lens used to form an image has a finite aperture. The limit to the aperture of the lens – and hence its resolving power – means that the most widely diffracted rays transferring high-frequency information, which encodes the finest detail in the object, are lost. With a low-numerical aperture objective, a greater number of widely diffracted rays are lost than with a high-numerical aperture objective.

Because of the effects of diffraction, a sub-resolution particle in the object (i.e. a point object) is imaged as a point spread function, or PSF. The wave nature of light gives rise from a point in the object plane to a 3D point spread function, including out-of-focus diffracted light above and below the focal plane in the image. If a lens behaved perfectly and points in the object were not smeared into a PSF, then a defocused image of fluorescent beads would appear black in the field of view (Figure 21.1, top Z-stack), but this is not the case.

The spot formed from a single sub-resolution point of light will be a bright point at the focal plane, possibly showing some faint diffraction rings, and of a size matching the resolving power of the objective. Either side of the focal plane, the spot is transformed into a disc that becomes both larger and dimmer in intensity. Ideally, the defocused image should be the same either side of the focal plane, but again this is usually not the case (Figure 21.1, bottom Z-stack) because of the presence of aberrations. In three dimensions this image is seen as an hour-glass shape (Figure 21.2) and describes the PSF characteristic of that particular microscope.

Furthermore, in fluorescence microscopy, the image is also separately degraded by out-of-focus blur from self-luminous fluorophores. This blur contributes to the image from above and below the plane of focus. When viewing cells or tissues by fluorescence microscopy, significant amounts of intensity are recorded at many microns distance from the actual focal position of the fluorophores. Only a small amount of light recorded in the image actually emanates from the focal plane where the features in the object are seen sharply in focus. This blurring degrades the contrast, dynamic range and resolution of the image.

Because any image can be reduced to its sub-resolution components, image restoration and sharpening by deconvolution is a useful technique in microscopy borrowed from astronomy. It was first applied to fluorescence studies in biological science by David Agard in 1983. The DeltaVision was the first dedicated commercial deconvolution microscope system and entered the marketplace about ten years later in the early 1990s.

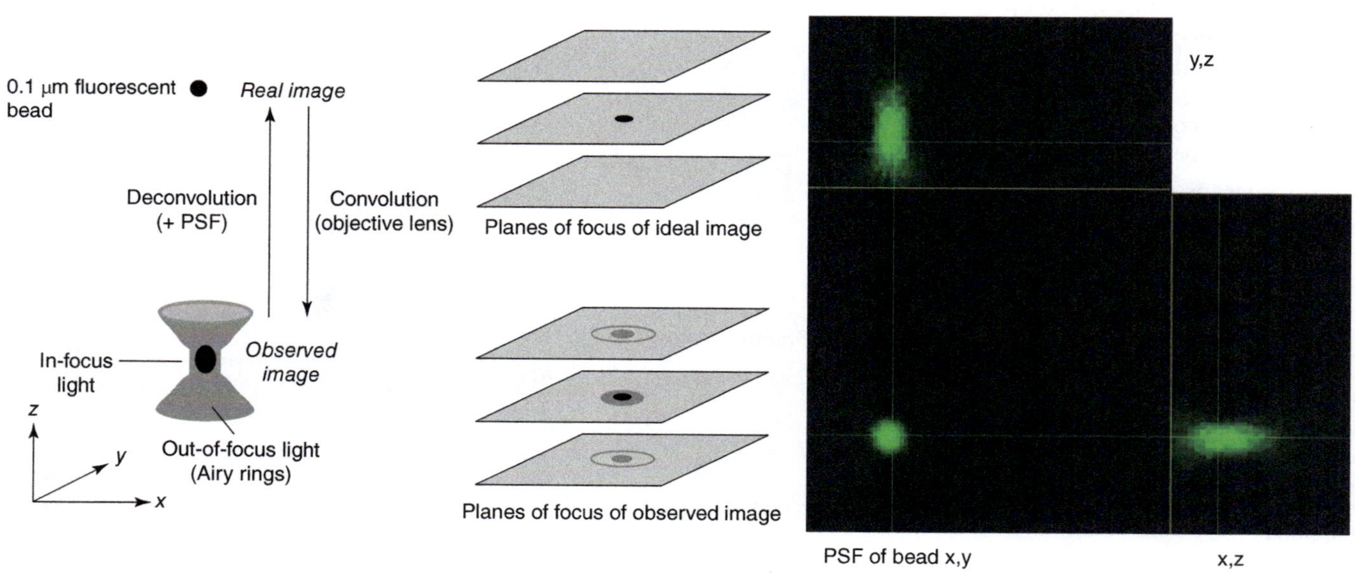

Figure 21.1 ***The point spread function*** *In a perfect world, the image point would equate exactly to the object point (middle, top). It would not be smeared laterally within the focal plane (left) or above and below into adjacent planes (middle, bottom). To the right is an actual PSF, showing the hourglass shape illustrated on the left of the figure. A good widefield PSF should form a symmetrical hourglass shape; any asymmetry above and below the focal plane is usually due to significant spherical aberration, which has a profound effect on axial resolution. Source: Davis 1999. Reproduced with permission from Oxford University Press.*

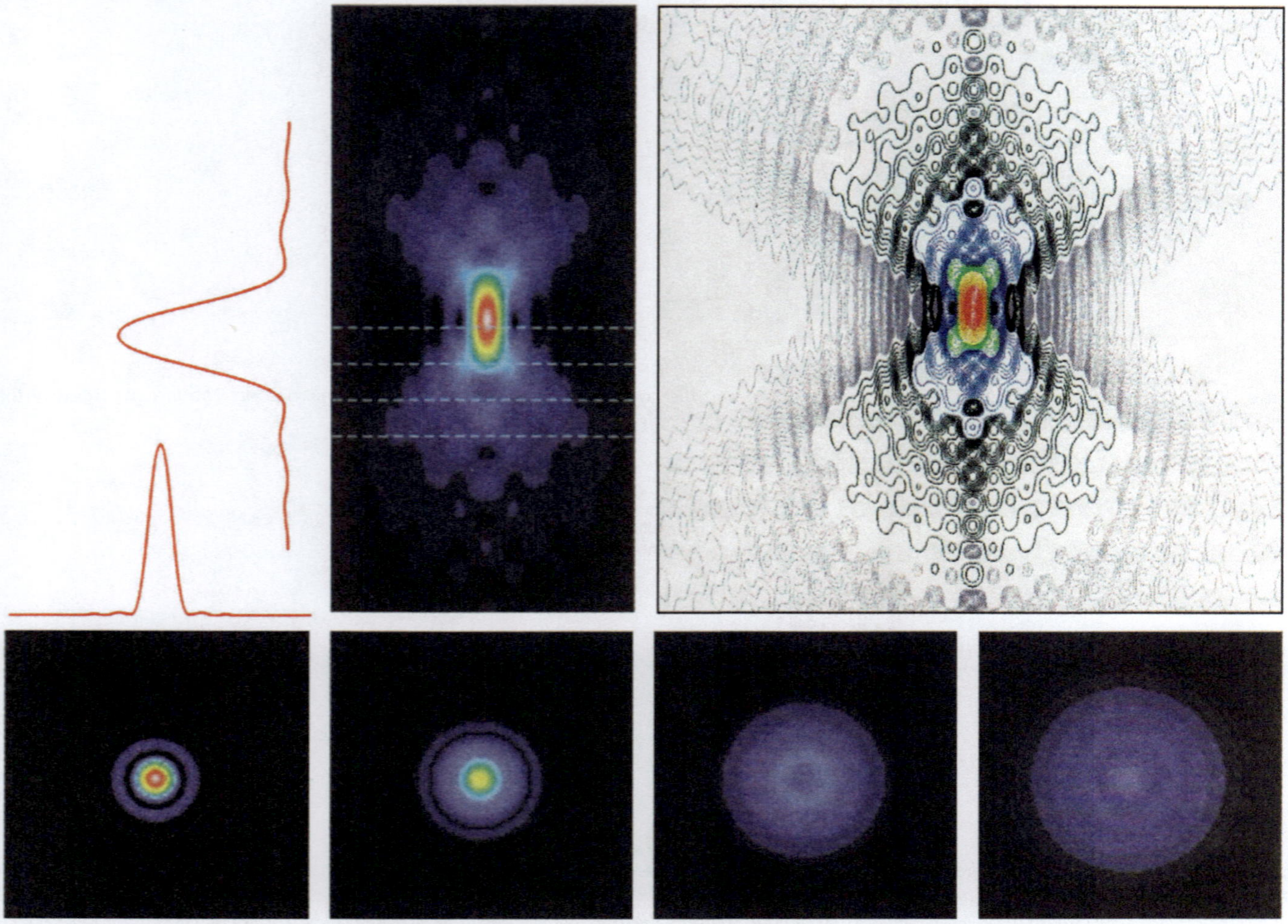

Figure 21.2 ***The point spread function*** *The point spread function is composed of a diffraction pattern. The lateral Airy pattern intensity distribution for one-half of the PSF is shown in the row of panels in the lower half of the figure. The intensity distributions for both lateral and axial components of the PSF are shown on the left, with the axial PSF shown in the upper centre and upper right of the figure. Source: Sibarita 2005. Reproduced with permission from Springer.*

21.2 The Process of Convolution

This process of image formation is termed 'convolution' (described by the symbol ⊗ or *, Figure 21.3). Let the object be represented by a single point (Figure 21.4) referred to earlier, as a single integer within a matrix of zero intensity values. These values are individually multiplied with the convolution mask – a blurring function that mathematically describes the way in which that particular lens or microscope system operates – to form the image. This convolution process is shown in Figure 21.5 and Figure 21.6. For simplicity only the two lateral dimensions (x,y) are shown, but convolution simultaneously occurs axially in the z-direction also. The result of the convolution mask acting upon a single sub-resolution point is the PSF.

Mathematically, convolution involves replacing each point in the specimen during the process of image formation to form a (blurred) point in the image. The final image is the sum of all these points. It is assumed that the imaging process is linear and shift invariant. This generally holds true for both bright-field and fluorescence microscopy but not phase contrast or differential interference contrast. Linearity means the final image is the sum of the images of its parts; shift invariant means that the point spread function is the same over the entire field of view. Phase contrast and DIC images are not linear because their contrast depends upon phase differences and phase gradient differences, respectively, within the object.

$$\text{image} = \text{object} \otimes \text{PSF}$$

Figure 21.3 ***The convolution function symbol.***

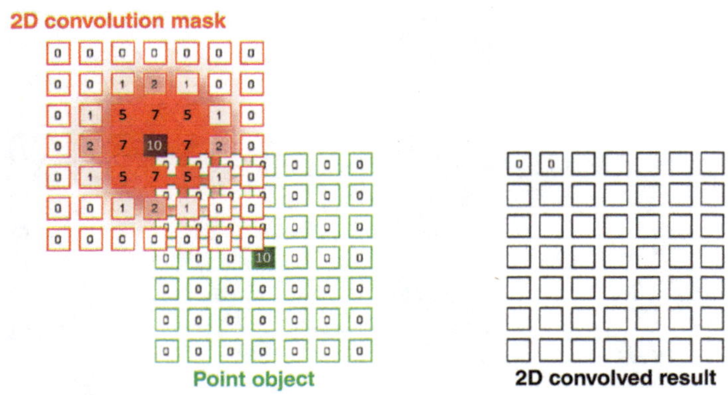

Figure 21.4 **Convolution upon a single point** *A point object and the convolution mask 'signature' of the microscope. The first two steps of the convolution operation. Source: Holmes et al. 2006. Reproduced with permission from Springer.*

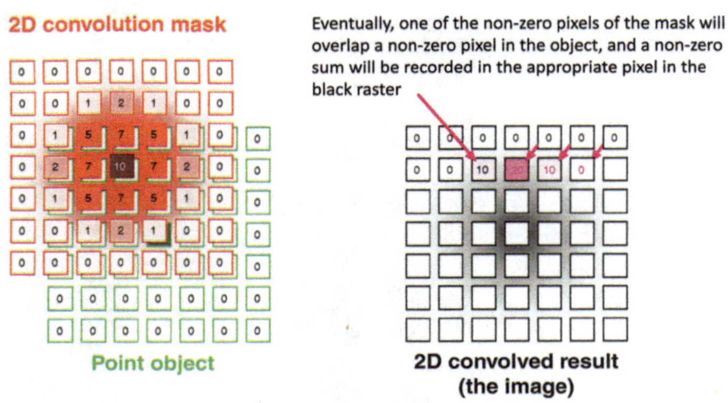

Figure 21.5 **Image formation with the convolution 'mask'** *The photon intensity values at each sampling point in the object are multiplied by the point spread function to form the image. The purple numbers (small arrows) will be recorded as the mask continues to step to the right. Convolving the point object by the convolution mask replicates the mask and in so doing forms an image called the PSF. Source: Holmes et al. 2006. Reproduced with permission from Springer.*

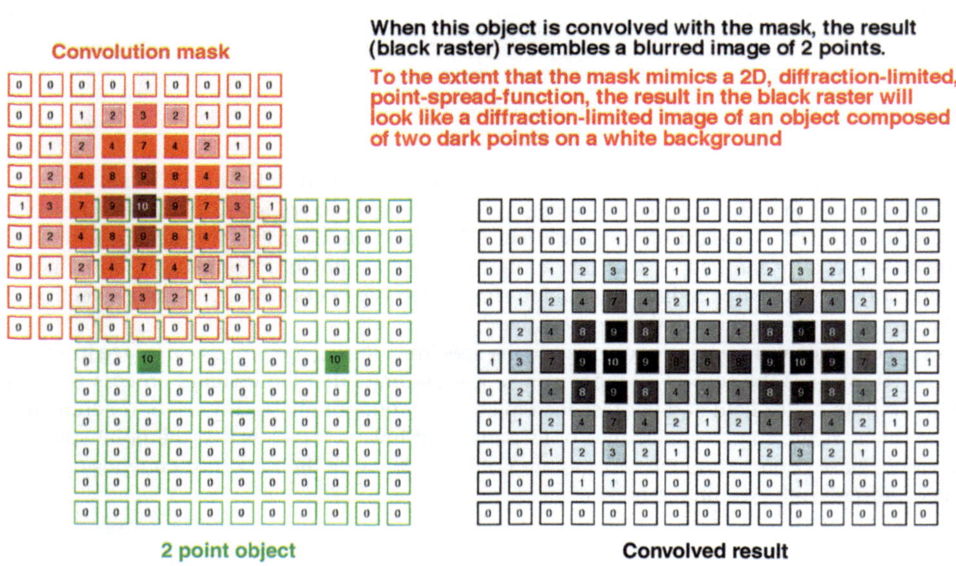

Figure 21.6 **Image formation of two points by convolution** *The resulting image of two closely apposed points formed from a two-point object. Here each point object is separated by five zero-intensity sampling points as shown in green. Since the pixel size in the image is close to half the Rayleigh limit of resolution (i.e. obeying Nyquist sampling), the image of the two points will be separated by sufficient distance to form two discernible features in the image. Source: Holmes et al. 2006. Reproduced with permission from Springer.*

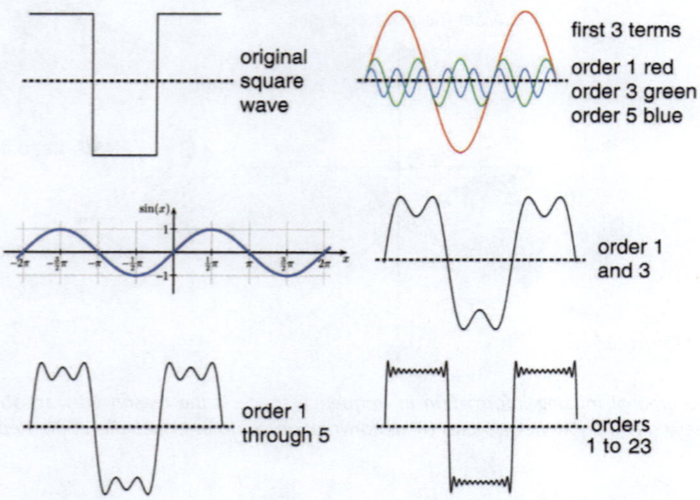

Figure 21.7 **Fourier transform description of a square wave 'object'** *A periodic object, such as the square wave, can be represented by a number of sine waves, each a harmonic of the frequency of that structure. As more components (e.g. higher-order diffracted rays forming the image) are added to the sum, the resulting image looks more like the original object. Source: Pawley 2006. Reproduced with permission from Springer.*

This direct convolution operation may appear complex, and indeed it is very computationally time consuming. However the mathematical operations described above in 'real space' can be simplified by computation in the frequency domain, or 'Fourier space' using Fourier transforms.

21.3 Fourier Transforms and the Missing Cone

Fourier transforms describe images mathematically in terms of their sine and cosine components (Figure 21.7) and so describe images as a continuous series of wave functions (i.e. their periodic frequency components). As an analogy, it is akin to using the basic components of chords and notes that make up music to describe a particular composition. For example, in Chapter 10 we noted that the diffraction pattern formed at the back focal plane of the objective is a Fourier transform (FT) of the object situated at the front focal plane. The Fourier transform of the PSF is called the optical transfer function[1], or OTF. The optical transfer function is so-called because it characterises how the microscope transfers spatial frequency information from the object into the image (Chapter 10, Section 10.19). Mathematically, the convolution operation reduces to a simple multiplication function, and the relationship given in Figure 21.3 becomes:

$$FT_{\text{microscope image}} = FT_{\text{object}} \times FT_{\text{PSF}} \text{ (or OTF)}$$

To convolve, all that is necessary is:

(a) to calculate the 3D Fourier transform of the object;
(b) calculate the 3D Fourier transform of the PSF (= OTF);
(c) multiply these two transforms together; and
(d) calculate the inverse Fourier transform of the product.

Knowing the PSF (or OTF), deconvolution can easily be performed by mathematical algorithms acting upon the intensity values that comprise the image to become a more faithful representation of the original object.

Because any lens is imperfect and collects only a proportion of the light emanating in all directions from the point of origin, the 'missing cone' (Figure 21.8) describes those low-frequency axial components lost as a result of diffraction and the limiting numerical aperture of the objective lens (Biggs, 2010) above and below the plane of focus[2]. This is why a widefield microscope is less well-suited to imaging sudden changes in intensity in the axial direction, such as when imaging thick samples. The PSF signatures of both confocal and multi-photon microscopes are ovoid rather than hour-glass shaped, because the PSF is the square of the product of both the excitation PSF and the emission PSF (Figure 21.9). These microscopes do not exhibit 'missing cones' like the widefield microscope, because in Fourier space such a multiplication function becomes a convolution, and convolution of one toroid by another tends to fill this central conical void. For further information on Fourier optics, refer to Parton and Davis (2006) and Appendix 3 in Saxby (2011).

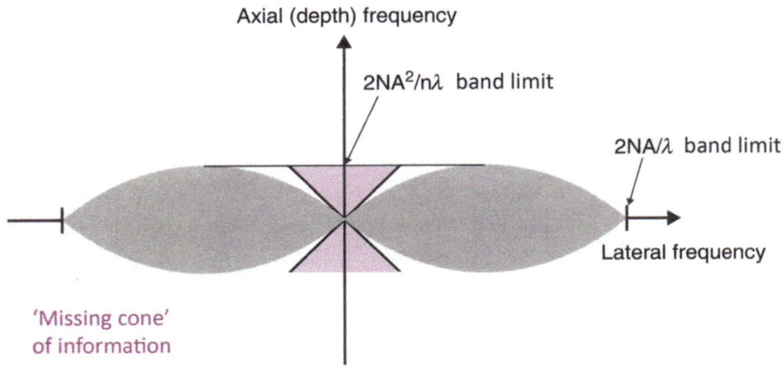

Figure 21.8 ***The 'missing cone'*** *The area of missing information in frequency space is the reason why an object point is elongated axially in the image. It is not possible to recover these missing low frequencies by deconvolution. Source: Conchello 2006. Reproduced with permission from Scion Publishing Ltd.*

21.4 Deconvolution

Deconvolution is the reverse process of convolution: (a) divide the 3D Fourier transform of the object by the 3D Fourier transform of the PSF (i.e. divide by the OTF), and (b) calculate the inverse Fourier transform of the result. Theoretically, deconvolving with a linear inverse filter should clean up the observed image to form the real image, which is a better approximation of the original object. However, this simplistic approach does not take account of noise inherent in image formation, and this noise component is amplified by a simple linear inverse filter. Furthermore, there is the problem with lost information, as represented by the missing cone. Therefore deconvolution algorithms of varying complexity have been devised to manage this missing image information and – as far as possible – 'clean up' the noise component of the image.

The best deconvolution algorithms attempt to correctly 'reassign' each photon of out-of-focus light from its position within each optical section of the object to the point of origin, as seen within the image. While the term 'reassignment' is consistent with the effect of their use, obviously these algorithms cannot *physically* replace photons to their source. However, iterative restorative deconvolution will shift and concentrate the majority of the out-of-focus fluorescence back to the focal plane in the image corresponding to the origin from which it emanated in the object.

In order to perform meaningful deconvolution, the correct algorithm must be selected for the task or application in hand *and* the correct point spread function, whether calculated or experimentally acquired, must also be used. Cannell, McMorland and Soeller (2006) report that measured PSFs are usually 20% larger than calculated PSFs and nearly always are non-symmetrical due to the presence of spherical aberration. Calculated PSFs are smaller and exhibit perfect axial symmetry. Without the most appropriate algorithm or correct PSF, there is little or no benefit to post-processing the image, and the old computer adage holds true: 'Garbage in, garbage out'. Deconvolution will not make bad data good; it will only make good data better. The key to successful deconvolution is acquiring an accurate PSF.

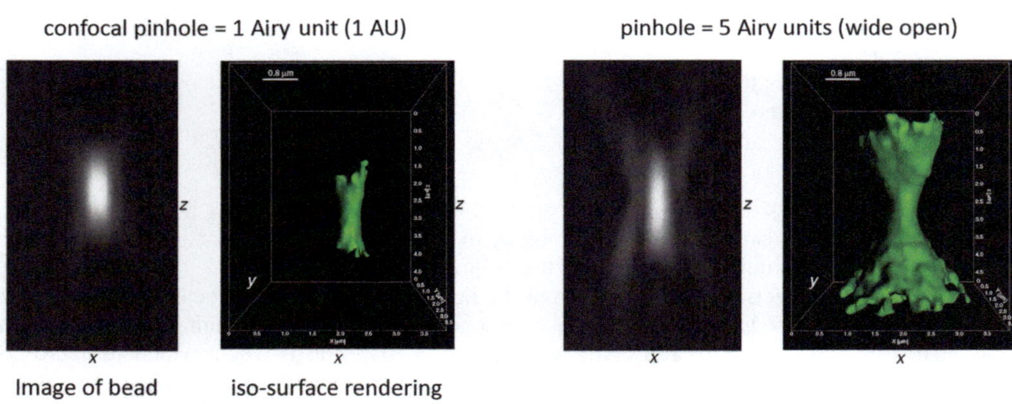

Figure 21.9 ***The ovoid PSF of the confocal microscope*** *With the pinhole set at 1 Airy unit, the central intensity distribution of the confocal PSF looks rather like a rugby ball. The iso-surface rendering more clearly shows the out-of-focus diffracted light above and below the focal plane. As the pinhole is opened on the confocal microscope, as expected, the ovoid PSF assumes a shape much like the widefield microscope, and an hourglass shape becomes apparent (see Cole et al., 2011). Source: Adapted from Cole et al. 2011 with permission from Macmillan Publishers Ltd.*

21.5 The PSF (Point Spread Function)

A microscope objective can be characterised by its PSF (Figure 21.1 and Figure 21.2), a 2D or 3D description of how a single point of light (in practice a sub-resolution fluorescent bead) in the specimen is formed into an image (or, to put it another way, how the structural information inherent in the object is *transferred* into the image by the lens). The point spread function is the *signature* of the microscope[3]. Since the PSF describes how an objective behaves, a good objective should exhibit a symmetrical PSF, with the size and number of the rings increasing, and the intensity of the rings decreasing as you move away from the focal plane.

The size – and to some extent the shape – of the PSF of an objective depends upon its numerical aperture (NA), its design (whether [plan]Apo, semi-apo or achromat). The illuminating wavelength and the contrast mode (bright-field, phase, DIC, fluorescence) also determine the PSF. Examples of a widefield PSF, a confocal and a multi-photon PSF are shown in Figure 21.10 for comparison. The three-dimensional sum of all the PSFs constitutes the image. In most (e.g. bright-field, phase or DIC) images, any trace of the Airy ring diffraction pattern arising from individual points of out-of-focus light are obscured due to the summation of so many PSFs. Nevertheless, these PSFs exist; it is simply easier to see PSFs with fluorescence because of the high contrast and dark background.

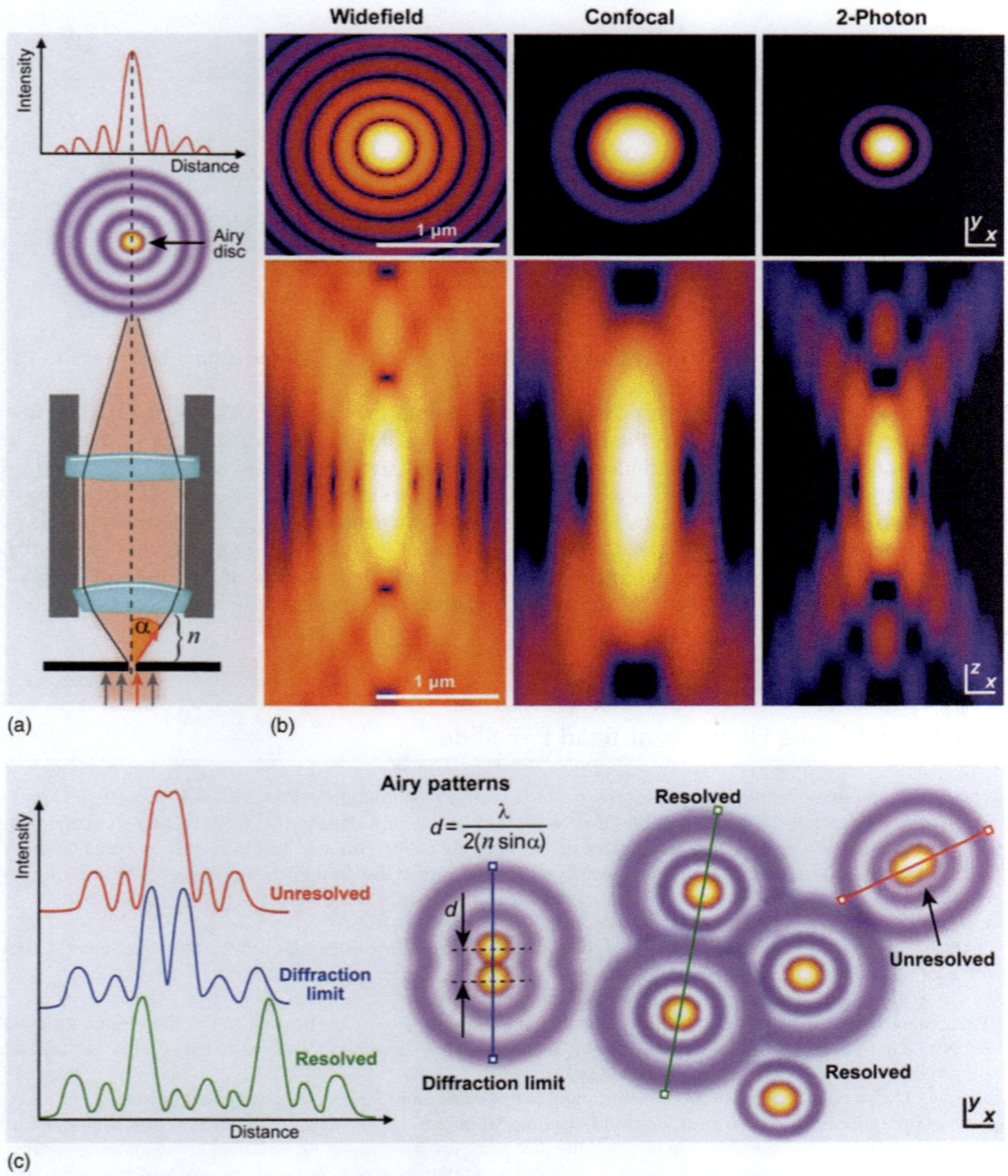

Figure 21.10 **Comparisons of differing point spread functions** *The half-angle of the sides of the cone of the widefield PSF is determined by the numerical aperture of the objective. The confocal 3D PSF is ovoid, shaped like a 'rugby ball', and that of the multi-photon is smaller still due to the nonlinear nature of the excitation volume of the multi-photon microscope. Source: Diaspro et al. 2006. Reproduced under the terms of the Creative Commons Attribution Licence, CC-BY 2.0.*

The quality of the deconvolved and restored image is directly dependent a priori upon knowing very accurately the imaging characteristics of the objective used to form the image of the specimen, in other words, by using an accurate PSF model for deconvolution[4].

The ideal solution is to:

(a) use a measured PSF derived from the Z-stack of a fluorescent sub-resolution bead; once collected, the data set can be used repeatedly for the same objective;
(b) calculate an estimate of the PSF from the wavelength of the illumination and the numerical aperture of the objective used, together with the pertinent refractive indexes used – this is quicker and easier method than deriving a measured PSF; and
(c) estimate the PSF directly from the observed data set, as is done with blind deconvolution (see Section 21.9).

21.6 Obtaining the PSF

Calculating a PSF requires knowledge of the following parameters:

- the numerical aperture of the objective;
- the wavelength of the emitted fluorescence signal;
- the working distance of the objective;
- the x,y,z dimensions of the PSF;
- the pixel size of the image;
- the refractive index of the mounting medium;
- the refractive index of the immersion medium;
- the refractive index of the sample;
- the distance of the sample from the underside of the coverslip; and
- the temperature.

Good deconvolution software will allow you to enter these values to calculate a PSF. Some online URL links to PSF calculators are given in the Internet Resources at the end of the chapter.

Measuring a PSF is a much more accurate way to perform deconvolution than using a calculated PSF. This is achieved using a sub-resolution fluorescent bead that is smaller than the resolution limit so that it has no structure per se that can be resolved but is rather a complete representation of the point spread function. The optimum size of the bead can be determined using the formula: bead diameter $= 0.4\lambda/\text{NA}$. For a given resolving power and NA of the objective to be tested, use a bead that is approximately 2/3rd in diameter of the minimum resolving power. So, for a NA 1.4 objective imaging a bead emitting green fluorescent light ($\lambda =$ circa 500 nm; resolving power $= 0.22$ µm), the diameter of microsphere should be around 0.14 µm, so a 0.15 µm diameter bead will be fine. Confocal microscopes may require the use of slightly larger, brighter beads because of the attenuation of the fluorescent signal by the confocal pinhole, which should normally be set at 1 Airy unit (1 AU) for PSF collection. In any case, the size of the bead used should still be of a size that is sub-resolution.

Box 21.1 Making and Using Fluorescent Bead PSF Slide

1. Shake or vortex bead tube immediately before use. These beads settle to the bottom and tend to clump together. Dilute the stock bead (0.1 and 0.5 µm diameter) mixture 1:500 in ethanol. Use Thermo Fisher Scientific TetraSpeck™ T7284 or a similarly manufactured bead preparation. The Davis reference, below, also recommends a red intensity calibration, e.g. Thermo Fisher Scientific InSpeck™ I7224.
2. Pipette a 1–5 mL drop into the centre of a high-tolerance coverslip, and spread the drop out using a pipette tip. Also put some beads onto the surface of the slide.
3. Dry the coverslip and the slide on a 50°C hotplate for at least 20 minutes (but also see Notes, below).
4. Mount the coverslip, bead side down, in mountant of the same refractive index as to be used for the experiment. Use about 10 µL of mountant with a 22 × 22 mm square coverslip.
5. Seal the coverslip to the slide with colourless nail varnish. Store in the dark at 4°C in the fridge.
6. To collect a PSF Z-stack, focus on the coverslip. Chose a bead that is as bright as possible but that has no discernible structure. It will help to calculate the theoretical resolving power of the objective to locate a suitably sized bead. Make sure the bead is well separated from others so that signal from other beads does not degrade the Z-stack.
7. With a widefield CCD camera, ensure that any pixel binning is turned off, and for the point-scanning confocal, ensure that pinhole size and Nyquist sampling are optimised. Whether widefield or confocal microscope is used, it is critical to ensure that the brightness of the image of the bead does not exceed the dynamic range of the detector. Any over-saturation will lead to meaningless results. Focus through the bead to ensure that complete Airy rings are seen at different focal planes. If they are not, this may be due to the presence of an air bubble (see figure, panel e).
8. Capture a PSF Z-stack. Use the microscope orthogonal viewer software to run through the Z-stack. If the Airy rings spread out asymmetrically, then spherical aberration is present and should be corrected.

9. Find the exact focal plane where the intensity of the bead image is maximal, and make a note of this. Focus through the Z-stack, and make a note of the z-positions either side of centre where the intensity is 50% of the maximum. The diameter of the image of the bead at this point is the full width half maximum (FWHM) and gives a measure of the lateral resolving power of the microscope.
10. The two z-positions you have noted either side of exact focus will give the axial FWHM, the resolving power of the microscope in Z. This will never be less than 1.7 times the axial resolving power and will usually be greater (see Davis, 1999, and Figure 21.12).
11. Where the Airy rings seen in the Z-stack of the PSF are not symmetrical, it may be necessary either to adjust the correction collar on the objective, if fitted, or to try different mixtures of oil (e.g. Cargile oils) to adjust the overall refractive index of the immersion medium. If the Airy rings spread away from the lens, increase the RI of the immersion oil, and vice versa (see figure, panel d).

Once found, the immersion oil with the best-matching RI will give an intense bead with – crucially – Airy rings of much reduced intensity. Both the z-resolving power and the signal-to-noise ratio should both be significantly improved.

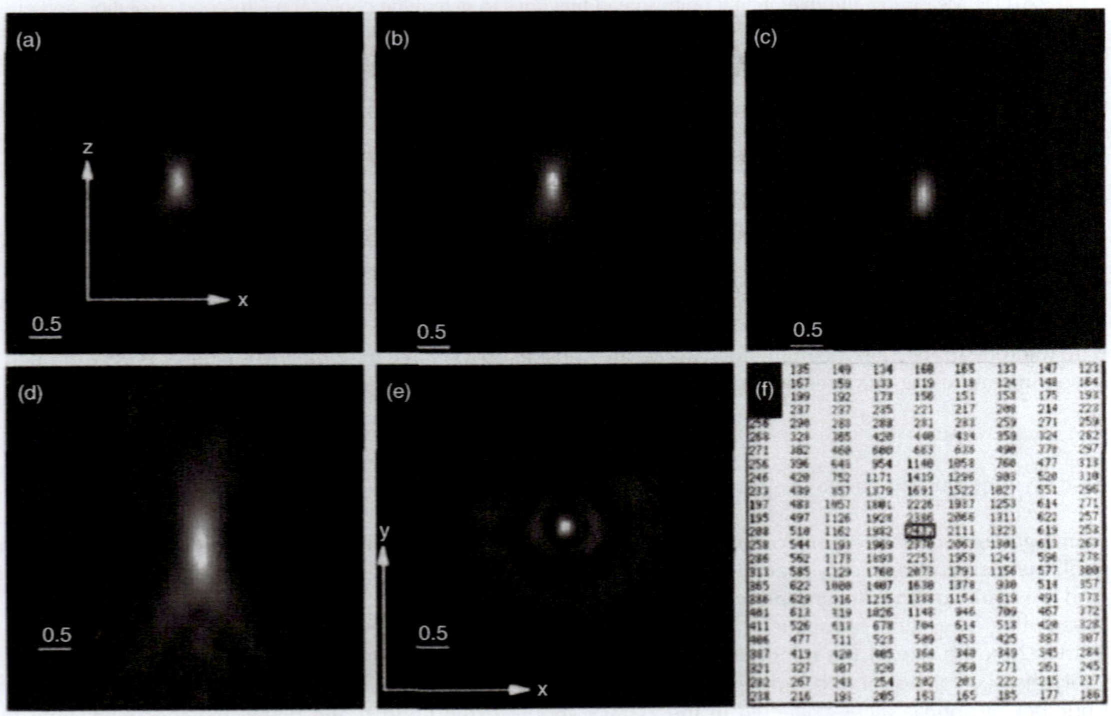

Panels (a)–(d) show the PSF of a 0.1 μm fluorescent bead side on (x-z view). Panel (a) is the PSF of a bead mounted just beneath the surface of a coverslip of correct thickness and RI 1.151 immersion oil. It shows no visible spherical aberration and some out-of-focus light. Panel (b) illustrates the PSF of a bead located 15 μm below the surface of the coverslip, showing slight spherical aberration and more out-of-focus information. Panel (c) shows the deconvolved result of the correctly imaged bead in Panel (a). Panel (d) shows a bead, again located 15 mm below the surface of the coverslip, but this time imaged using immersion oil of slightly reduced refractive index (RI = 1.514). This shows that refractive index matching is important. Panel (e) shows Airy rings seen laterally, with an incomplete segment due to the presence of an air bubble in the immersion oil. Panel (f) shows the intensity of light collected at each pixel, which is plotted in Figure 21.12b. The scale bar represents 0.5 μm. (Source: Davis 1999. Reproduced with permission from Oxford University Press.)

Notes

Mounting is critical to the successful use of these beads. The beads must be immobile, for Brownian motion, even in a viscous fluid renders them unusable, and accurate imaging impossible.

These PSFs are collected at the coverslip surface. This is the ideal and suitable for deconvolving images of cells grown on coverslips. Where the sample as appreciable thickness and is imaged by confocal, it is preferable to collect a PSF at approximately the same axial distance in Z.

I am grateful to Colin Gray at the University of Sheffield for the following suggestion: Take a slice of silicone rubber with a small hole punched in it. Dampen the rubber, or if required apply a smear of vacuum grease to the surface and stick it to a measured coverslip of 0.17 mm, or a high tolerance coverslip. Cool the coverslip and rubber mould on a metal block held in an ice bucket. Make up the bead suspension in 15 low melting point agarose and pour this carefully into the mould. Once set, cover the top with another measured coverslip (C. Gray, personal communication).

Cole et al. (2011) recommend a using a 1×10^6 dilution of beads with 20 minutes sonication to break up clusters. They recommend cleaning slide and coverslip with 70% ethanol, and drying the preparation without heat for 1–2 hours or overnight. Some authorities say that heat-drying of the beads will cause shrinkage.

Paul Goodwin (2007) remarks that premixed mounting media designed to gel or polymerise to form a hardened preparation without the necessity to ring or varnish the edge of the coverslip to ensure immobility, can lead to problems acquiring good images. Any medium that is not thoroughly mixed to form a homogenous mixture may contain particles which act as miniature microlenses and severely degrade the signal to noise ration and amplify non-specific background in the image. Polyvinyl alcohol (PVA) which is often added to gel the preparation, is a particular culprit and premixed formulations that are non-miscible with water can also lead to problems when coverslipping wet samples (as is the norm). Tiny bubbles of water form with the mounting media and again degrade the image due to refractive index mismatching.

References

Cole, R.W. et al. (2011). 'Measuring and interpreting point spread functions to determine confocal microscope resolution and ensure quality control'. *Nature Protocols* 6/12: 1929–1941.

Davis, I. (1999). 'Visualizing fluorescence in *Drosophila*'. In: Allan, V.J. (ed.), *Protein Localization by Fluorescence Microscopy: A Practical Approach*, Chapter 6. OUP, pp. 133–162.

Goodwin, P.C. (2007). 'Evaluating optical aberration using fluorescent microspheres: methods, analysis, and corrective actions'. In: Sluder, G. and Wolf, D.E. (eds.), *Digital Microscopy, Methods in Cell Biology*, Chapter 18, 3rd ed., vol. 81. San Diego: Academic Press, Elsevier, pp. 397–413.

The most important aspect of preparing a reliable bead slide is to ensure that the beads are dried down sufficiently long (either very gently on a hotplate, or for 1–2 hrs at room temperature) so that, once the slide has been prepared, the microspheres attach firmly to the coverslip and do not move by Brownian motion during Z-stack collection. Suitable beads may be obtained from Thermo Scientific (e.g. G50, the 0.05 μm diameter bead), or from Thermo Fisher Scientific (e.g. PS-Speck Microscope Point Source Kit, P7220). The InSpeck intensity standard microbeads (I7224) recommended above are considered too large by some investigators. A very through and detailed reference on preparing and using a fluorescent sub-resolution test slide to collect a PSF has been published by Cole et al. (2011).

21.7 Practical Aspects of Image Collection

A modern research-grade fluorescence microscope is required with a highly accurate and repeatable Z-focusing mechanism, under computer control, that can ensure evenly spaced Z-slices, together with a cooled digital CCD (or sCMOS) camera optimised for both low read-out and low dark current background noise.

Correct data sampling is essential for accurate deconvolution. Ideally, the minimum Nyquist factor should be 2.3 times, so that the pixels are at least 1/2.3x the size of the minimum resolvable detail (i.e. highest frequency, see Chapter 31, Section 31.7 for further details). Preferably the pixel size should be arranged such that four pixels sample the minimum detail resolved by the microscope.

Ideally, the number of optical slices collected in the Z-stack should extend above and below the specimen, until the defocused features are blurred to a uniform haze. In practice, this may entail collecting a large number of Z-slices. If this is excessive, leading to concern about photobleaching or time constraints, consider collecting a Z-stack that is half the apparent specimen thickness, and take slices amounting to one-quarter the thickness of the sample above and below the middle plane of focus. This midplane can usually be identified because it exhibits the densest and most intensely stained tissue elements. Thus, for a sample 10 μm thick, collect additional slices 3–4 μm above and below the middle of the sample.

Optical sections comprising the Z-stacks taken for deconvolution should *not* be oversaturated, since saturated regions in the image cannot be accurately restored. This point is crucial and cannot be overemphasised. The exposure should be minimised while keeping sufficient intensity of emitted signal to overcome inherent Poisson noise in the image (this is usually not a problem unless the specimen fluoresces very weakly). A high signal-to-noise ratio in the image data set is good for deconvolution.

All deconvolution methods assume that the sample is evenly illuminated both in time and space. To correct for uneven spatial illumination of the illumination source as well as of the CCD detector, images should ideally be flat-field corrected before deconvolution. Flat-field correction helps even out the illuminated field of view, especially if high resolving power is retained and 'binning' (i.e. averaging signal and thus intensity across several pixels) is not used. The DeltaVision microscope addresses this problem by using an optical fibre to even out the illumination and a xenon lamp, which has a more even spectral distribution of its illumination than a mercury lamp (see Figure 9.2), is generally much more stable and exhibits much less temporal flicker. If alternating bright-dark stripes are present in successive deconvolved Z-slices (best seen in x-z view), this is a classic flicker artifact. Even without the presence of lamp flicker, stripes can arise when too few slices are collected and insufficient out-of-focus light captured. Such truncated raw data sets do not provide enough out-of-focus light for some deconvolution algorithms.

The influence of spherical aberration in degrading the image and its associated PSF is well known (see, for example Nasse and Woehl, 2010; Sibarita, 2005 or Model et al., 2011). The image can also be degraded by poor choices of mounting medium, particularly those containing DAPI as a nuclear stain or contain premixed polyvinyl alcohol. If these mountants are not mixed thoroughly after thawing from frozen, RI differences, which scatter light, can be introduced. The influence of chromatic aberration is often disregarded or not taken

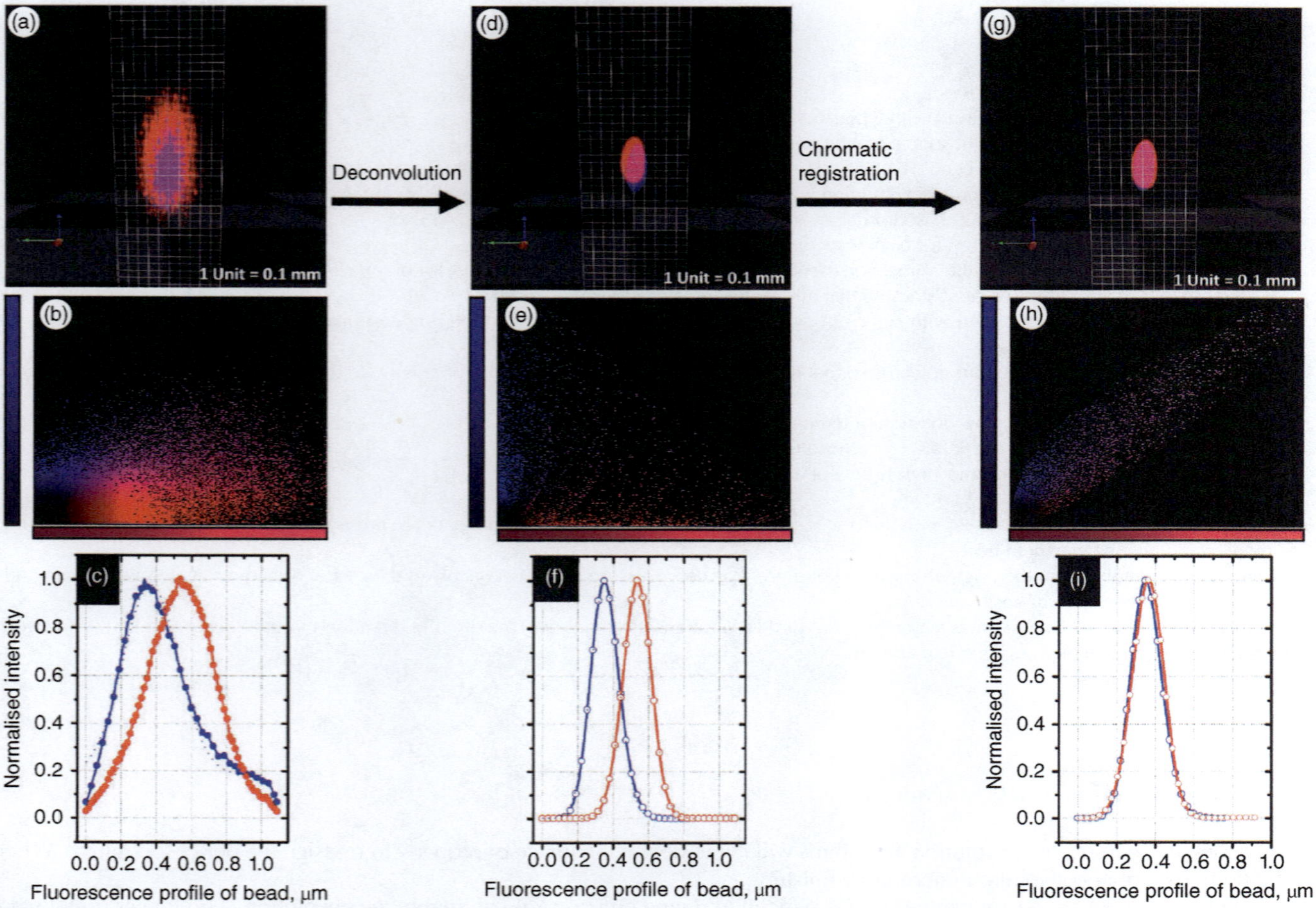

Figure 21.11 ***Correcting for chromatic aberration*** *Panels (a), (d), and (g) show three images of a fluorescent bead as a point source. Panels (b), (e), and (h) show the corresponding colocalisation scatterplots, and panels (c), (f), and (i) show the each channel as an intensity profile of the bead. The left-hand panels, (a), (b), and (c) show an undeconvolved bead with no chromatic correction applied to the blue and red emission channels. Panels (d), (e), and (f) show the deconvolved bead still exhibiting a chromatic shift. Panels (g), (h), and (i) show the same deconvolved bead with the appropriate image shift applied to ensure the emission signal from each channel corresponds exactly. The scatterplots best show (compare panels (b) and (h)) the effect of applying the necessary deconvolution and registration. Source: Abraham et al. 2010. Reproduced with permission from Elsevier.*

into account either because workers may be unaware of its influence, or because people assume – or hope – that chromatic aberration is sufficiently minimised through the use of apochromat objectives. Using a multicolour fluorescent bead to collect registration images is worthwhile. Figure 21.11, taken from Figure 3 in Abraham et al. (2010), shows both a lateral and axial colour correction shift applied to deconvolved beads, permitting accurate colocalisation analysis.

Image restoration by computational deconvolution is not often successful with live specimens. It takes time to collect a Z-stack and even slowly-moving cells or organelles may well move while you acquire the stack. Furthermore, many living samples have poor signal-to-noise ratios, which can interfere with successful deconvolution. When collecting any Z-stack, acquiring the entire Z-stack in one sweep before changing wavelength to a different fluorophore is faster than collecting multiple wavelengths at each focal plane of the Z-stack before moving up to the next plane of focus. In other words, Z then channels is faster and better suited when imaging living samples. Collecting channels then Z is slower but may give better image registration if the filter set design and camera arrangement is not pixel-perfectly aligned.

Aggressive deconvolution and contrast enhancement can give rise to small artifactual spots in the image. A second artifact that may be encountered is evident as a darker region surrounding the brighter sources of light in the specimen. If structures in the image show marked asymmetry, this can often be traced back to an asymmetric PSF often due to refractive index mismatch in the imaging setup. If noise in the original data set of the observed image is now seen as fine spots or as structural elements in the deconvolved real image, then too much processing has been performed. Use fewer iterations and a conservative setting for the algorithm rather than an aggressive one, if that is possible. If the contrast drops and very dim features are lost, this may be due to excessive processing but can equally be due to the loss of Poisson noise by properly applied deconvolution. Finally, do note that deconvolution can reveal flaws in the specimen that may not be so obvious in the raw image. Fluctuations in illumination intensity and inadequate sample preparation may be much more obvious after deconvolution.

Practical Steps for Deconvolution

1. Preferably use apochromat objectives.
2. Check that the illumination is uniform with a plastic fluorescence slide.
3. Use a stable emission intensity light source (e.g. xenon).
4. Use multi-spectral sub-resolution beads to check registration between wavelengths, for correction to lateral and axial chromatic aberration.
5. Calibrate the lateral pixel resolution from emission wavelength, objective NA and any magnification factor.
6. Set the axial sampling distance in Z, based on Nyquist's sampling theorem.
7. Use the same pixel and slice settings for each wavelength.
8. Check that the refractive index of the immersion medium matches that of the preparation. A list of refractive indexes is given in Table 28.1.
9. Set the top and bottom limits of the Z-stack so that all features of interest are defocused.
10. Check that the data set is collected with sufficient dynamic range (e.g. 12-bit) and that the brightest pixel is ≈90% of the dynamic range to avoid oversaturation at any point.
11. Collect the Z-stack from the bottom upwards so that the focussing mechanism works against gravity for the best accuracy and repeatability in sampling each focal plane.
12. Minimise any stray room light, any vibration or temperature and humidity variation.
13. Collect the Z-stack and check the data set for oversaturation (there should be none).
14. Check for illumination flicker or unevenness between slices. Correct this as needed.
15. Deconvolve the data set.
16. Check that the deconvolved data set with the original Z-stack that features in the original data are present in the restored data set. Maximum intensity projections may help here.
17. Check that any noise from the original image is not present as fine structure in the deconvolved data set. If it is, run the algorithm again and reduce the number of iterations.
18. Check for any residual haze, which may mean that further iterations are required. Haze in one direction through the Z-stack may, however, indicate the presence of gross spherical aberration.

21.8 Deconvolution of Thick Specimens

If the specimen is too thick, deconvolution algorithms will fail to deblur the image or properly to reassign the out-of-focus light. Where does the usefulness of deconvolution microscopy finish?

The exact thickness where deconvolution fails is difficult to define but, as a rule of thumb, deconvolution is no longer useful when the image contrast between the signal and the background noise falls to <1% as seen through the entire thickness of the cell or tissue (Swedlow et al., 2002). In general, deconvolve where the specimen is 'thin' (i.e. <10 times the depth of field of the objective used). Specimens having a moderate degree of thickness (2–15 μm) will produce dramatically improved images with confocal, and/or deconvolution techniques. The thickest specimens (20 μm and thicker) will suffer from a tremendous amount of residual extraneous light in the image, originating in out-of-focus regions and are probably best imaged using confocal or multi-photon techniques.

Deconvolution is also applicable to images collected by confocal and multi-photon microscopy, since they involve convolution to form the image and in practice rarely exclude all out-of-focus information. For most practical purposes, confocal data is not deconvolved, because the pinhole excludes out-of-focus light, and applying deconvolution as a 'post-processing' technique is time consuming.

Quite apart from the fact that laser scanning confocals are more inefficient at collecting light than widefield microscopes because they are inherently noisier, confocal microscopy is a light *rejection* strategy (the light excluded by the pinhole can never be recovered) whereas the widefield microscope collects the *entire* fluorescent signal and then applies mathematical deconvolution algorithms to 'deblur' the image, or reassign the out-of-focus signal. While excluded photons cannot be restored, the confocal pinhole can be opened to allow more light through to the detector, and this image can be deconvolved. Use four times Nyquist sampling if this is the case (refer to Table 21.1). Axial resolution, in particular, is improved by deconvolution (see Figure 21.12).

Some people consider that the pinhole has already 'sampled' the data, rendering post-acquisition deconvolution less important than with widefield microscopy. However, there are very sound reasons why data sets from confocal and other optical sectioning microscopes should be deconvolved.

Firstly, if properly applied, deconvolution of 3D data sets acquired with a laser-scanning confocal will confer a measure of resolution improvement and contrast enhancement[5]. Noisy signals, particularly those of low intensity, contain a lot of high-frequency Poisson, or shot, noise. This noise can be erroneously sampled and digitised into non-existent single-pixel 'features' that could never have been resolved by the microscope. Deconvolving confocal data removes these noise artifacts. If the contrast drops, the gamma display of the monitor or the contrast in the printed output can always be increased. Boutet de Monvel et al. (2001) demonstrate the use of wavelet denoising algorithms for confocal microscopy to help prevent these artifacts.

Secondly, deconvolution averages the signal over the many voxels in the image needed to sample the signal from a single point in the object sampled at Nyquist frequency. This is a very effective way of reducing the overall noise and boosting the SNR. To achieve the same result with the confocal microscope would require Kalman averaging of the signal anywhere between 80 to 130 times. This is clearly not feasible and would – in any case – bleach and damage the specimen.

Table 21.1 Table of Z-Stack Spacing Values to Satisfy the Nyquist Sampling Criterion for Deconvolution

Selected objective	NA	Refractive index, RI = $n_{D\,23}$	Lateral resolving power x,y (µm) @ 488 nm [d = 0.61 λ/NA]	Optimum X,Y-sampling step (µm) to sample for Nyquist criterion @ Nyquist = 2x calculated from lateral resolving power = d/2	Optimum X,Y-sampling step (µm) to sample for Nyquist criterion @ Nyquist = 2.3x calculated from lateral resolving power = d/2.3	Optimum X,Y-sampling step (µm) to sample @ 488 nm for deconvolution calculated for Nyquist = 2x from [d = 0.25λ/NA]	Optimum X,Y-sampling step (µm) to sample @ 488 nm for deconvolution calculated for Nyquist = 2.3x from [d = 0.22λ/NA]	Depth of field Z (µm) @488 nm [d = 2nλ/(NA)²]	Optimum Z-sampling step (µm) to sample for Nyquist criterion @ Nyquist = 2x calc from d/2	Optimum Z-sampling step (µm) to sample for Nyquist criterion @ Nyquist = 2.3x calc from d/2.3	Optimum Z-sampling step (µm) to sample @ 488 nm for deconvolution $d_z = 0.5\lambda/(RI - \sqrt{[RI^2 - NA^2]})$ for Nyquist = 2x	Optimum Z-sampling step (µm) to sample @ 488 nm for deconvolution $d_z = 0.43\lambda/(RI - \sqrt{[RI^2 - NA^2]})$ for Nyquist = 2.3x	**Maximum Z-step** (µm) @ 1.5x calculated (mean of values calculated) for $Nyq_2/Nyq_{2.3}$
6.3x/NA 0.16 dry	0.16	1.0	1.86	0.93	**0.80**	0.76	**0.67**	38.13	19.07	**16.6**	18.8	**16.3**	26.5/**24.5**
10x/NA 0.25 dry	0.25	1.0	1.19	0.60	**0.52**	0.49	**0.43**	15.62	7.81	**6.8**	7.6	**6.6**	10.8/**10.0**
10x/NA 0.30 dry	0.30	1.0	0.99	0.50	**0.43**	0.41	**0.36**	10.84	5.42	**4.7**	5.3	**4.5**	7.5/**6.9**
20x/NA 0.5 dry	0.5	1.0	0.60	0.30	**0.26**	0.24	**0.21**	3.90	1.95	**1.7**	1.8	**1.6**	2.6/**2.5**
20x/NA 0.75 dry	0.75	1.0	0.40	0.20	**0.17**	0.16	**0.14**	1.74	0.87	**0.76**	0.72	**0.62**	1.11/**1.04**
		1.332 = water						2.03	1.02	**0.88**	0.91	**0.77**	1.34/**1.24**
25x/NA 0.8 multi	0.8	1.47 = glycerol	0.37	0.19	**0.16**	0.15	**0.13**	2.24	1.12	**0.97**	1.03	**0.89**	1.50/**1.40**
		1.518 = oil						2.31	1.16	**1.00**	1.07	**0.92**	1.55/**1.44**
30x/NA 1.05 silicone	1.05	1.406 = silicone	0.28	0.14	**0.12**	0.12	**0.10**	1.24	0.62	**0.54**	0.52	**0.46**	0.79/**0.75**
40x/NA 1.30 oil	1.30	1.518 = oil	0.23	0.12	**0.10**	0.09	**0.08**	0.88	0.44	**0.38**	0.33	**0.29**	0.53/**0.50**
40x/NA 1.40 oil	1.40	1.518	0.21	0.11	**0.09**	0.09	**0.08**	0.76	0.38	**0.33**	0.26	**0.23**	0.44/**0.42**
60x/NA 1.22 W	1.22	1.332	0.24	0.12	**0.10**	0.10	**0.09**	0.87	0.44	**0.38**	0.31	**0.26**	0.51/**0.48**
60x/NA silicone	1.30	1.406	0.23	0.12	**0.10**	0.09	**0.08**	0.81	0.41	**0.35**	0.28	**0.24**	0.47/**0.44**
60x/NA 1.40 oil	1.40	1.518	0.21	0.11	**0.09**	0.09	**0.08**	0.76	0.38	**0.33**	0.26	**0.23**	0.44/**0.42**
100x/NA 1.30 oil	1.30	1.518	0.23	0.12	**0.10**	0.09	**0.08**	0.88	0.44	**0.38**	0.33	**0.29**	0.53/**0.50**
100x/NA 1.40 oil	1.40	1.518	0.21	0.11	**0.09**	0.09	**0.08**	0.76	0.38	**0.33**	0.26	**0.23**	0.44/**0.42**

$d_{x,y} = 0.25\lambda/NA$ $d_z = 0.5\lambda/(RI - \sqrt{[RI^2 - NA^2]})$ for Nyquist = 2x, and constants 0.22 and 0.43 for Nyquist = 2.3x.

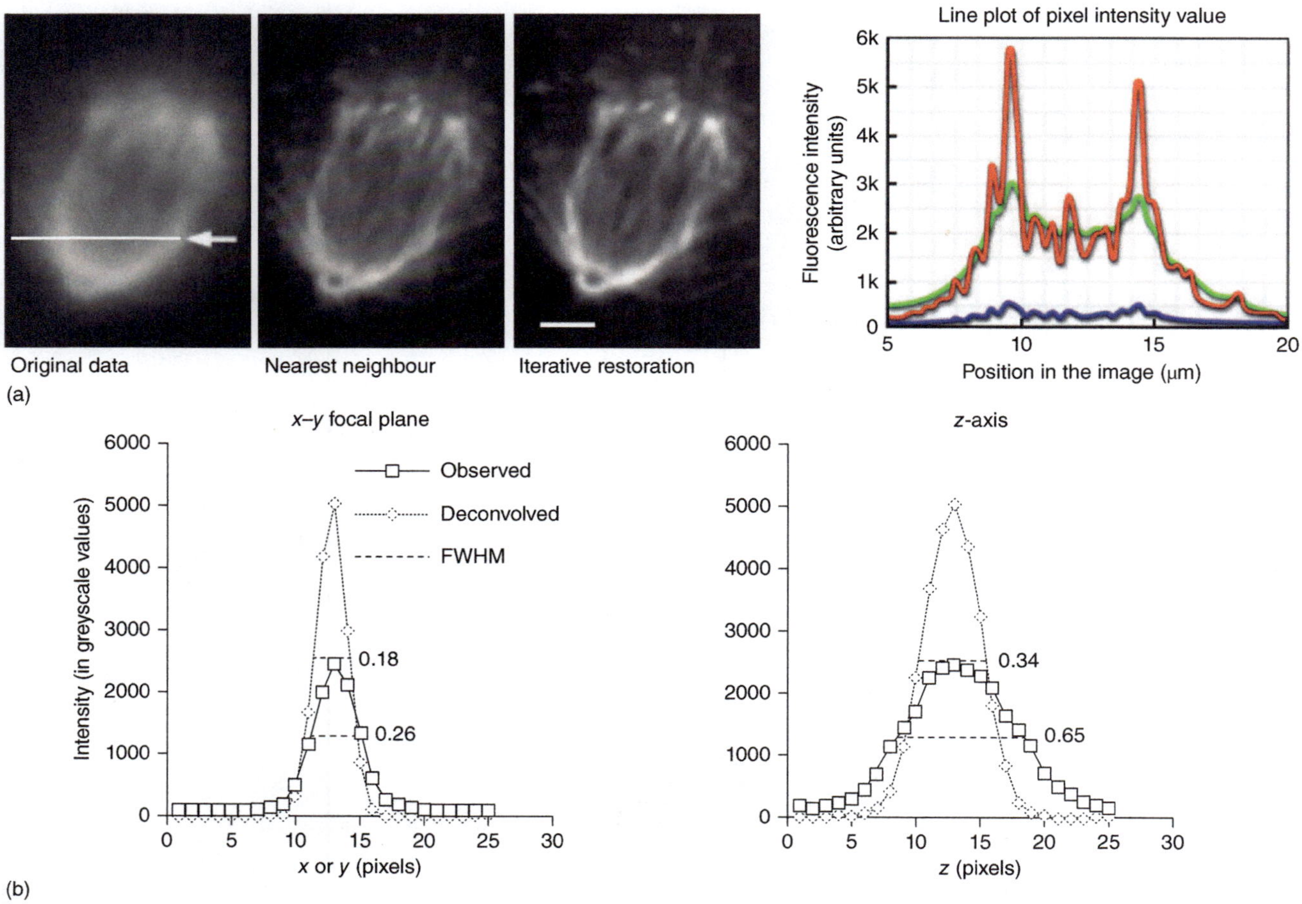

Figure 21.12 **Improving resolution by deconvolution** *(a) The left-hand original data panel shows a single optical section from a cell stained to reveal tubulin. The middle panel shows deblurring with a nearest-neighbour algorithm and the right-hand panel, the effect of constrained iterative restoration. The arrow indicates where the line intensity profiles have been taken in each image. The intensity profile of the original data is shown in green, the deblurred image data by the blue line and the iteratively restored data in red. Deblurring, which is quick, causes a significant loss of intensity across the entire image, whereas restorative deconvolution properly applied improves the SNR in the image and hence the resolution of fine detail. Source: Wallace et al. 2001. Reproduced with permission from J.R. Swedlow, PhD.*

(b) Graphs of the intensity values of a fluorescent bead before (observed values) and after deconvolution. The left-hand graph shows the intensity values plotted against the lateral pixel dimensions of the bead image. The right-hand graph shows the intensity values axially along the z-axis. The solid line shows the observed intensity values and the dotted line, the deconvolved result. The horizontal lines give the full width half maximum (FWHM) values in each case. The lateral x-y FWHM is reduced from 0.26 to 0.18 after deconvolution, a 1.4x improvement. The corresponding improvement for the axial resolution is 1.9x. Deconvolution improves the lateral/axial resolution ratio from as much 2.5x and never less than 1.7x the lateral resolving power. Source: Davis 1999. Reproduced with permission from Oxford University Press.

A third reason to deconvolve data collected by the confocal microscope concerns properly displaying an analogue signal reconstructed from the digitised data. It is often forgotten that the Nyquist theorem stipulates that the bandwidth of the (output) display device must be equal to that of the (input) digitising device. The input device is the microscope and detector together. The frequencies that the microscope objective can handle are limited, and a point object should be sampled by 4–5 pixels (Chapter 18, Section 18.7 and Chapter 31, Section 31.6 and Section 31.7). The bandwidth of the output monitor could, in principle, be matched to that of the microscope by reducing its resolution 4–5 times. Clearly, this is not feasible. Furthermore, reducing the monitor resolution acts only in two dimensions, whereas deconvolution can be done in 3D. Deconvolution is the correct way to implement this (often forgotten) condition of Nyquist reconstruction.

21.9 Which Deconvolution Algorithm to Choose?

There are two main types of deconvolution algorithms: deblurring and restorative (Figure 21.13). Deblurring algorithms are nearest-neighbour based; they use an estimate of the image blurring function to subtract blurred fluorescence from an image. The degree of haze arising in a given optical section from the neighbouring sections on either side is estimated and duly removed by computation

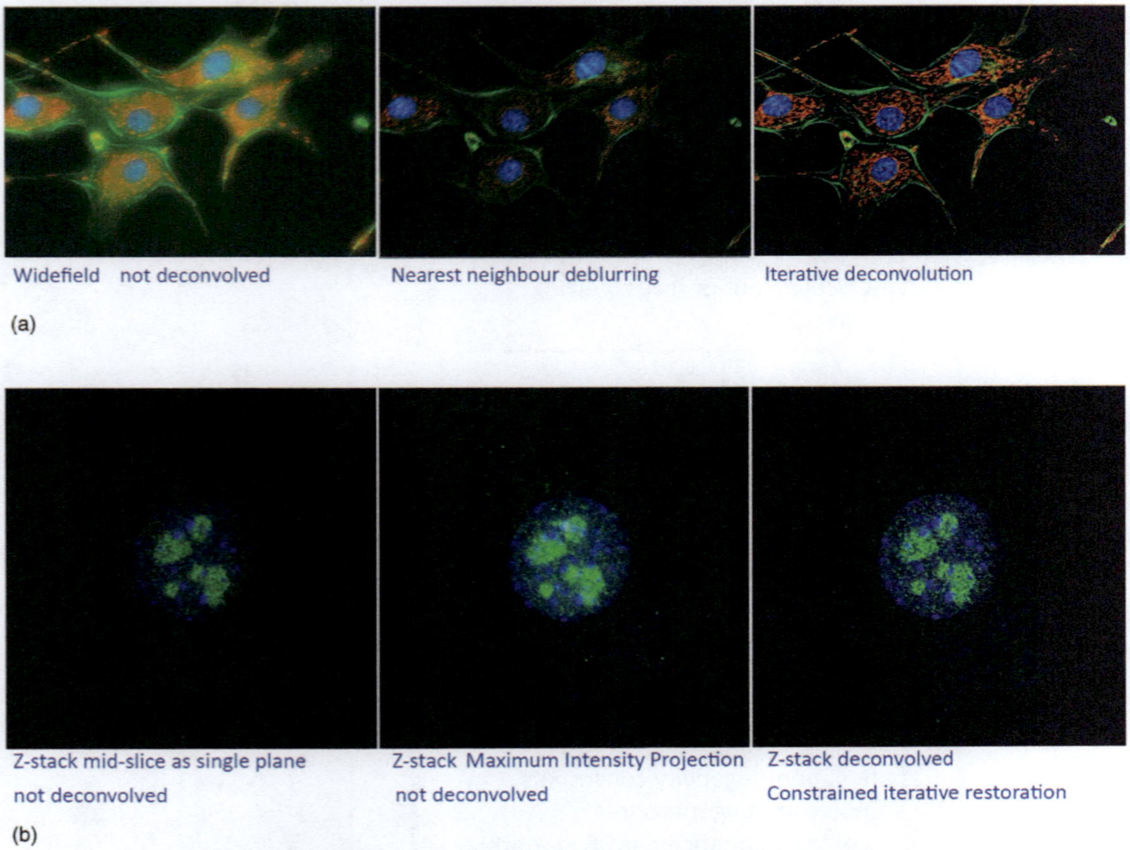

Figure 21.13 **Examples of deconvolved images** *(a) Examples of a thin (circa 5 μm) cell monolayer of bovine endothelial cells stained for the nucleus, cytoskeleton and mitochondria (FluoCells® prepared slide #1, Thermo Fisher, F36924). Compare these images with those in Figure 21.12a.*

(b) The left-hand panel shows a single mid-slice in a Z-stack with 40 optical sections 0.1 μm apart. The centre panel shows a maximum intensity projection (MIP) of the entire Z-stack. The right-hand panel shows this same MIP restored by constrained iterative deconvolution.

(hence deblurring algorithms are sometimes called 'digital confocal' software). These deblurring methods sacrifice some signal (the SNR falls) in return for an improvement in contrast within the image. Deblurring methods do not process image noise well and usually introduce artifacts into the deconvolved image when applied to noisy data sets.

Nearest-neighbour deblurring methods use a theoretically calculated PSF rather than a measured one and are best for thin samples. They work best with a Z-stack of three slices at a time: the blur in the centre slice is approximated from the blur inherent in the outer two slices and is subtracted to leave the in-focus information. Where a Z-stack is greater than three slices, this process is repeated throughout the whole 3D stack. Images deconvolved by deblurring cannot be used for quantitative measurement of intensities or calculation of ratios between fluorophores (e.g. as in calcium imaging). If the objective you have used is of low NA (i.e. is <NA 1) and/or the Z-stack is undersampled in Z, with relatively large spaces between each slice, then nearest-neighbour algorithms are a good choice.

The no-neighbours method is similar to the nearest-neighbour method but only considers a single slice at a time. It has the advantage of speed and is similar to the un-sharp masking algorithm used in photo-editing and image-manipulation software. Ideally, deconvolution requires a 3D stack in order to estimate or quantify the out-of-focus signal. It is not possible to indicate what part of a single 2D image contains out-of-focus information as distinct from true signal, which limits the usefulness of the no-neighbours algorithm. This algorithm tends to work better with punctate images or those of filaments (i.e. high-frequency images) with image details near the limit of resolution.

Restoration algorithms are more accurate and can be used for quantitative purposes because they keep the relative intensity relationships seen in the data forming the original image (although you should always quantify or extract numerical data or analysis from the raw non-deconvolved data). Restoration algorithms treat the entire three-dimensional stack as a single entity, and 'reassign', or shift, by computation, the out-of-focus information from its recorded location back to its original location in the sample, *as seen in the image*. These latter algorithms treat the out-of-focus haze that is generated by the mass of fluorophore stain as useful signal, rather than throwing it away. Restoration algorithms are also particularly useful when examining small (and hence, dim) samples or those photo-sensitive samples affected by bleaching where the signal intensity may well be low and the SNR poor.

Blind deconvolution (more properly referred to as adaptive blind deconvolution) is an alternative restorative method that extracts the PSF *directly* from the image data. This may seem counter-intuitive, since the algorithm is trying to compute a solution for both the image and the PSF. The PSF is not known beforehand, hence the term blind; the PSF is adaptively estimated directly from the data set,

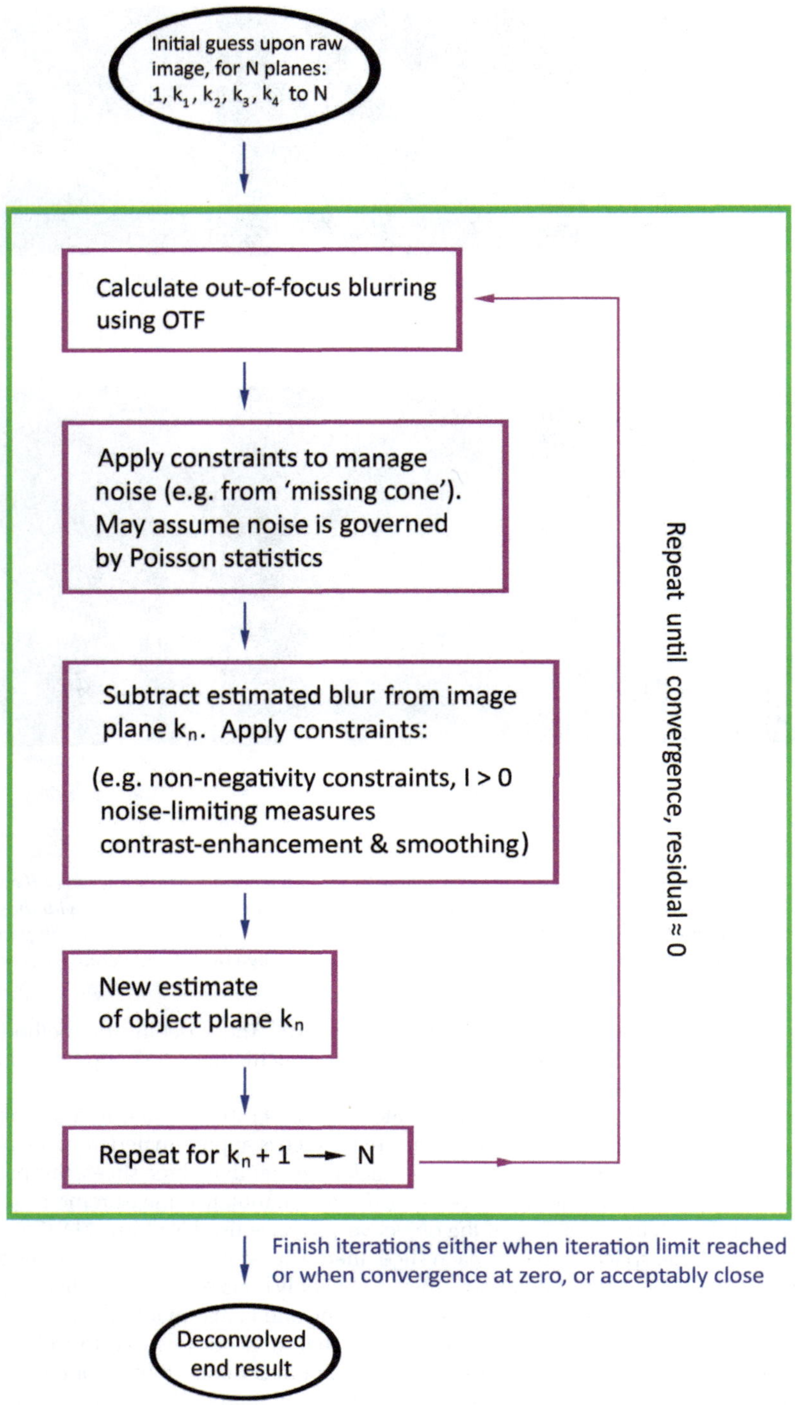

Figure 21.14 **Iteration** *A generalised scheme showing how a constrained iterative deconvolution algorithm works. The optical transfer function is applied to the image, any constraints applied to manage noise amplification and the result compared to the original. Subtraction is performed to remove blurring. This 'guess' is compared to the original image and the process repeated (either by automatic or by user-managed iteration) until the result converges to a sufficiently acceptable approximation of the original object. Source: Biggs 2010. Reproduced with permission from John Wiley & Sons.*

usually employing a calculated PSF as a starting point. The reasoning for using blind deconvolution being that the best PSF that fits the actual data is obtained. Blind deconvolution is best suited to confocal microscopy using variable pinhole settings or where a degree of spherical aberration is known to contribute to the image.

Most restoration algorithms are iterative. With iterative deconvolution, the ideal 'model' image is compared with the results of computation (Figure 21.14) successively allowing for noise and signal integration. The iterations are stopped when the normalised residual (i.e. computational comparison of the deconvolved result to the original image) falls to around zero. This is to prevent over-iteration, which can lead to unwanted noise amplification. Iterative deconvolution is good for colocalisation analysis to reduce false colocalisation from residual noise.

21.10 Chapter Summary

Any image is only a representation of the object that it is formed from. Deconvolution is a mathematical stratagem to reduce blurring and out-of-focus fluorescence signal in the image. It is crucial for good deconvolution to calculate, or accurately measure, the point spread function formed by the objective.

Deconvolution is also applicable to images collected by confocal and multi-photon microscopy, although deconvolution algorithms will fail with very thick specimens. The algorithms available fall into two broad groups: they either deblur the image or they attempt to restore the signal – to 'reassign' it to the point of origin, as seen within the image.

1. Convolution is the process whereby an image is formed. It can be characterised mathematically.
2. Any object can be regarded as a series of sub-resolution points. The PSF is the *signature* of the microscope. It describes what happens to each point source of light after it passes through the imaging system. The PSF convolution mask applied to each object point forms the complete image.
3. Deconvolution is the mathematical reverse process of convolution, applying the PSF on the image to give a more faithful representation of the original object.
4. Fourier transforms describe images mathematically in terms of their sine and cosine components. They describe images as a continuous series of wave functions (i.e. their frequency components).
5. Properly applied, deconvolution gives improved resolution of fine detail, as well as better contrast and signal-to-noise ratio in the image.
6. A miscalculated, or poorly measured, PSF will not improve a thick and/or blurred image. Deconvolution will not make bad data good; it will only make good data better.
7. Deblurring algorithms are fast to implement and merely qualitative. These nearest-neighbour algorithms generally do not process image noise well. They may introduce artifacts into the deconvolved image when applied to noisy data sets.
8. No-neighbours deblurring algorithms only consider a single 2D focal plane.
9. Restoration algorithms are more accurate. They treat the out-of-focus haze generated by the mass of fluorophore stain as useful signal and so attempt to 'reassign' it to the point of origin, as seen within the image.
10. Restoration algorithms, which are usually iterative, preserve the *relative* signal intensities in the image, and so permit quantitative analysis of the restored image.

Key Reading

Shaw, P. (1994). 'Deconvolution in 3-D optical microscopy'. *Histochemical Journal 26*/9: 687–694.
- A dated review but a very clear explanation from an experienced practitioner in the field

Cole, R.W. et al. (2011). 'Measuring and interpreting point spread functions to determine confocal microscope resolution and ensure quality control'. *Nature Protocols 6*/12: 1929–1941.
- An excellent protocol paper explaining all you need to know about preparing and using fluorescent bead preparations

References

Abraham, T. et al. (2010). 'Deconvolution and chromatic aberration correction in quantifying colocalization of a transcription factor in three-dimensional cellular space'. *Micron 41*/6: 633–640.

Biggs, D.S.C. (2010). '3D deconvolution microscopy'. *Current Protocols in Cytometry*. Chapter 52, Unit 12.19. Wiley Interscience.

Boutet de Monvel, J. et al. (2001). 'Image restoration for confocal microscopy: improving the limits of deconvolution with reference to the visualization of the mammalian hearing organ'. *Biophysical Journal 80*/5: 2455–2470. See also Von Tiedemann, M. et al. (2006). *Microscopy Research and Technique 69*/1: 10–20.

Cannell, M.B., McMorland, A., and Soeller, C. (2006). 'Image enhancement by deconvolution'. In: Pawley, J.B. (ed.), *Handbook of Biological Confocal Microscopy*, Chapter 25, 3rd ed. New York: Springer, pp. 488–500.

Cole, R.W. et al. (2011). 'Measuring and interpreting point spread functions to determine confocal microscope resolution and ensure quality control'. *Nature Protocols 6*/12: 1929–1941.

Conchello, J.-A. (2006). 'An overview of three-dimensional and four-dimensional microscopy by computational deconvolution'. In: David, S. (ed.), *Cell Imaging: Methods Express Series*, Chapter 9. Banbury: Scion Publishing. pp. 177–199.

Davis, I. (1999). 'Visualizing fluorescence in Drosophila'. In: Allan, V.J. (ed.), *Protein Localization by Fluorescence Microscopy: A Practical Approach*. OUP, pp. 133–162.

Diaspro, A., Bianchini, P., Vicidomini, G., Faretta, M., Ramoino, P., and Usai, C. (2006). 'Multi-photon excitation microscopy'. *BioMedical Engineering OnLine 5*: 36.

Holmes, T.J. et al. (2006). 'Blind deconvolution'. In: Pawley, J.B. (ed.). *Handbook of Biological Confocal Microscopy*, 3rd ed. New York: Springer, pp. 468–487.
Model, M.A. et al. (2011). '3D deconvolution of spherically aberrated images using commercial software'. *Journal of Microscopy 241*/1: 94–100.
Nasse, M.J. and Woehl, J.C. (2010). 'Realistic modelling of the illumination point spread function in confocal scanning optical microscopy'. *Journal of the Optical Society of America A 27*/2: 295–302.
Parton, R.M. and Davis, I. (2006). 'Lifting the fog: Image restoration by deconvolution'. In: Celis, J.L. (ed.), *Cell Biology: A Laboratory Handbook*, Chapter 19, 3rd ed. San Diego: Academic Press, pp. 187–200.
Pawley, J.B. (2006). *Handbook of Biological Confocal Microscopy*, 3rd ed. New York: Springer.
Saxby, G. (2011). 'The Fourier model for image formation'. In: *The Science of Imaging*, Appendix 3, 2nd ed. Boca Raton: Taylor & Francis, pp. 311–318.
Sibarita, J.-B. (2005). 'Deconvolution microscopy'. In: Rietdorf, J. (ed.), *Microscopy Techniques*, Advances in Biochemical Engineering/Biotechnology, Chapter 8, vol. 95. Berlin: Springer-Verlag, pp. 201–243.
Swedlow, J.R. et al. (2002). 'Measuring tubulin content in *Toxoplasma gondii*: A comparison of laser-scanning confocal and wide-field fluorescence microscopy'. *PNAS 99*/4: 2014–2019.
Wallace, W. et al. (2001). 'A workingperson's guide to deconvolution in light microscopy'. *Biotechniques 31*: 1076–1097.

Further Reading

Davis, I. (1999). 'Visualizing fluorescence in *Drosophila*'. In: Allan, V.J. (ed.), *Protein Localization by Fluorescence Microscopy: A Practical Approach*, Chapter 6. OUP, pp. 133–162.
Goodwin, P.C. (2014). 'Quantitative deconvolution microscopy'. In: Waters, J.C. and Wittman, T. (eds.), *Methods in Cell Biology*, Chapter 10, vol. 123. Elsevier BV, pp. 177–192.
Heintzmann, R. (2006). 'Band-limit and appropriate sampling in microscopy'. In: Celis, J.E. (ed.), *Cell Biology: A Laboratory Handbook*, Chapter 3, vol. 3, 3rd ed. Elsevier Academic Press, pp. 29–36.
Heintzmann, R. and Sheppard, C.J. (2007). 'The sampling limit in fluorescence microscopy'. *Micron 38*/2: 145–149.
Hiraoka, Y., Sedat, J.W., and Agard, D.A. (1990). 'Determination of three-dimensional imaging properties of a light microscope system'. *Biophysical Journal 57*/2: 325–333.
Mizutani, R. et al. (2016). 'A method for estimating spatial resolution of real image in the Fourier domain'. *Journal of Microscopy 261*/1: 57–66.
Pawley, J. and Holmes, T.J. (2006). 'Convolution: an introduction for those who don't remember second-year calculus', Appendix to Chapter 24, 'Blind Deconvolution'. In: Pawley, J.B. (ed.), *Handbook of Biological Confocal Microscopy*, 3rd ed. New York: Springer, pp. 485–487.
Preibisch, S. et al. (2014). 'Efficient Bayesian-based multi-view deconvolution'. *Nature Methods 11*/6: 645–648.
Schmid, B. and Huisken, J. (2015). 'Real-time multi-view deconvolution'. *Bioinformatics 31*/20: 3398–3400.
Sun, Y. et al. (2009). 'An open-source deconvolution software package for 3-D quantitative fluorescence microscopy imaging'. *Journal of Microscopy 236*/3: 180–193.
Tadrous, P.J. (2010). 'A method of PSF generation for 3D brightfield deconvolution'. *Journal of Microscopy 237*/2: 192–199.
Wallace, W. et al. (2001). 'A workingperson's guide to deconvolution in light microscopy'. *Biotechniques 31*/5: 1076–1097 (1076-8, 1080, 1082 passim).

Internet Resources

Deconvolution software links: http://www.deconvolve.net/DNLinks.html. ImageJ plugins are at:

1. EPFL deconvolution page: http://bigwww.epfl.ch/deconvolution/.
2. Piotr Wendykier's page: https://sites.google.com/site/piotrwendykier/software/deconvolution/paralleliterativedeconvolution.
3. Iterative deconvolution: http://imagej.net/Parallel_Iterative_Deconvolution.
4. The PSF: https://www.ibiology.org/talks/point-spread-function/.
5. The OTF: https://www.telescope-optics.net/mtf.htm#The_OTF.
6. The MTF: https://www.edmundoptics.eu/resources/application-notes/optics/introduction-to-modulation-transfer-function/.
7. The Fourier transform: https://www.ibiology.org/talks/fourier-transform/.

PSF Calculators

Good deconvolution software will allow you to enter values to calculate a PSF, but if you need free software, then consider:

1. PSF_Lab http://en.wikipedia.org/wiki/PSF_Lab and http://people.uwm.edu/woehl/.
2. Optinav: http://www.optinav.info/Diffraction-PSF-3D.htm.
3. Iscope calculator: http://iscopecalc.com.
4. PSF generator: http://bigwww.epfl.ch/algorithms/psfgenerator.

Notes

1. The optical transfer function is a combination of the modular transfer function (MTF) and the phase transfer function (PTF). The MTF shows how well an imaging system performs (in our case the microscope objective) transferring contrast at a particular resolution from the object to the image. Normally a periodic black-white bar grating object is used, and the magnitude response of the objective to sinusoids of different frequencies is plotted. The MTF therefore incorporates the imaging system's handling of both resolving power and contrast. The PTF describes the aberrations inherent in the objective. The MTF of an ideal non-aberrated objective can be calculated from diffraction equations. Strictly, the Fourier transform of the PSF is equivalent to the OTF, providing that the phase component of the optical transfer function is negligible.
2. The lateral band limit is $2NA/\lambda$; the axial band limit is $2NA^2/n\lambda$ and the angle of the missing cone is arcsin(NA/n).
3. For example, quite apart from its value in establishing resolving power or use for deconvolution, analysis of the Z-stack of the PSF of a sub-resolution fluorescent bead allows rejection of poorly manufactured or damaged objectives.
4. Ideally we would inject fluorescent beads as an internal control into specimens, but this is not practical. Instead we use fluorescent beads mounted in the same refractive index medium as the sample dried down onto the underside of a measured coverslip of 0.17 mm thickness.
5. It may appear counter-intuitive, but deconvolution properly applied should cause at first an apparent drop in contrast, with possibly an 'unsharp' appearance to the image. These 'sharp' features that have been removed are noise artifacts; the gamma display can be raised to match the output characteristics of the monitor.

22

Multi-Photon Microscopy

Outline

22.1 Introduction

The structural components of cells and tissues can strongly scatter light in the visible region. Two-photon microscopy was devised in 1990 as a method of imaging deeper into tissues, with much less scattering of illumination and emitted signal, than single-photon laser scanning confocal microscopy was capable of. Generally an improvement in depth imaging along the z-axis over the single-photon LSCM is possible by about four to six times. The technique was developed by the neurologist Winfried Denk and is also called multi-photon or nonlinear microscopy.

Neural tissue is relatively transparent to infrared (IR) light. Many fewer excitation IR photons – roughly half compared to visible light – are scattered for every 50–100 μm of tissue thickness. Because IR photons are also absorbed to a lesser degree than

---

Understanding Light Microscopy, First Edition. Jeremy Sanderson.

shorter-wavelength light, IR light penetrates further into tissue and excites fluorophores more efficiently. Red and infrared light also produces less autofluorescence than blue or green illumination and, importantly, is very much less harmful to living cells and tissues than shorter-wavelength, higher-energy radiation. It has been possible to image hamster blastocysts continuously over 24 hours and then re-implant them to grow to full term (e.g. see White et al., 2001). Similarly prepared two-cell embryos viewed with a confocal using lower-wavelength illumination (514/532/568 nm) were unable to develop to the blastocyst stage. Long-wavelength red and IR light exhibits almost no autofluorescence. For this reason, it is better suited to long-term time-lapse studies especially with thick living tissues such as developing embryos (see Hardin, 2006; Truong et al., 2011; Supatto et al., 2011) and explant tissues (Zinselmeyer et al., 2009; Herz et al., 2012).

Safety Warning:

The British Standard sets out seven classes of laser; these are Class 1, Class 1M, Class 2, Class 2M, Class 3R, Class 3B and Class 4. The pulsed lasers used for multi-photon microscopy are Class 4.

Class 4 lasers have output power >500 mW and are capable of causing both immediate severe skin injury and irreparable eye injury. Damage to the eye is instantaneous. Safety precautions *must* be planned and enacted beforehand, for last-minute avoidance is *not* a possibility. The green pump laser produces about 10 watts of power, but the infrared laser produces about 8 watts of *invisible* radiation!

These Class 4 lasers will also present a fire hazard if sufficiently high output powers are used. The beam of the tunable infrared laser – which is normally invisible – should be enclosed and exit only at the objective.

The enclosures should *only* be removed by qualified service personnel, and mirror test samples should not be used. It is advisable to remove rings and other specularly-reflecting items, to prevent unintended exposure to the laser beam.

For further advice, refer to PHE in the UK: https://bit.ly/2CxrZ1N, or OSHA in the USA: https://bit.ly/2EswaAT

The first commercial multi-photon microscopes were built by Bio-Rad (like the laser-scanning confocal, beforehand) in 1996, ten years after the introduction of the LSCM. The principles of nonlinear microscopy were first elucidated by the Nobel Prize–winning physicist Maria Göppert-Mayer in 1931, but it was not until the development of lasers that the effect could be observed experimentally.

The multi-photon microscope is good for producing well-defined optical sections through tissues but, just like the single-photon confocal microscope, it can photo-damage and photo-bleach the specimen unless care is taken to select the correct laser power and exposure to light. The absolute maximum depth penetration of the laser-scanning confocal into a sample is limited to about 100 µm, depending upon the refractive properties of the specimen – with 50 µm being a more realistic depth limit. For sections ≈ 100 µm, the absolute maximum section thickness in practical terms means that less than 1% of the emitted signal actually contributes to the image.

Conversely, the multi-photon microscope can produce clear images from thicker tissues. Some tissues are virtually transparent to far-red wavelengths. In very exceptional cases, the laser can penetrate (e.g. brain) to produce images up to 1000 µm into tissue depending upon laser power and the nature of the sample, although clear imaging through depths of 200–600 µm is more common. One reason that the multi-photon microscope can image deeper into tissues than the single-photon confocal microscope is because more photons penetrate the tissue. Thus a higher proportion of photons that are able to excite the fluorophore actually reach the focal spot in multi-photon microscopy. This advantage increases as the focal plane moves deeper into the tissue. The second reason that the multi-photon microscope will image deeper into tissues is because IR photons are scattered much less by the sample.

Despite the increase in popularity and the development of well-designed turnkey systems, both two-photon microscopy and TIRF microscopy still tend to be niche techniques. Often this is because of the high cost of dedicated multi-photon microscopes. However, since the initial patent expired in 2009, firms such as Scientifica and Olympus now offer modular microscopy platforms, effectively reducing start-up costs. Multi-photon and TIRF systems are used on a regular basis to a much lesser extent than confocal or widefield fluorescence microscopy, yet they complete the palette of methods available for optical sectioning (Table 27.1) and for generating three-dimensional images of cells and tissues.

22.2 Two-Photon and Multi-Photon Microscopy

Multi-photon microscopy normally uses two photons (Figure 22.1) to work although, theoretically, there is no limit to the number photons that can be used. In practice it is two, occasionally sometimes three (Ware, 2014; Rodriguez et al., 2018), long-wavelength photons that are used to excite fluorescence dyes respectively in the visible (two-photon) and near-UV (three-photon) regions of the electromagnetic spectrum. For example, two red IR-wavelength photons (e.g. excitation at 960 nm for fluorophore emission at 515 nm can excite Alexa Fluor 488, whereas three photons of IR-wavelengths – 1050 nm – can be used to excite UV dyes[1] at 350 nm). In this chapter, multi-photon microscopy refers to both two- and three-photon microscopy.

A wavelength of 1500 nm (see Norris et al., 2012), achieved using an optical parametric oscillator (OPO; Mojzisova and Vermot, 2011) for three-photon microscopy gives even better multi-photon results. The focal volume will be smaller than that for two-photon microscopy, and short pulses are essential, since signal intensity is proportional to the inverse square of the pulse length. Because of the quantum-mechanical properties of the photon and the physics of fluorescence absorption, the photon density required for three-photon microscopy is only ten times (rather than 10^6 times) greater than that required for two-photon microscopy; this makes three-photon microscopy possible. However, the higher-order nonlinear relationship of three photon microscopy results in weaker emission signals being generated, and so it is not as often used as two-photon microscopy. With care, it is possible to select fluorophores so that

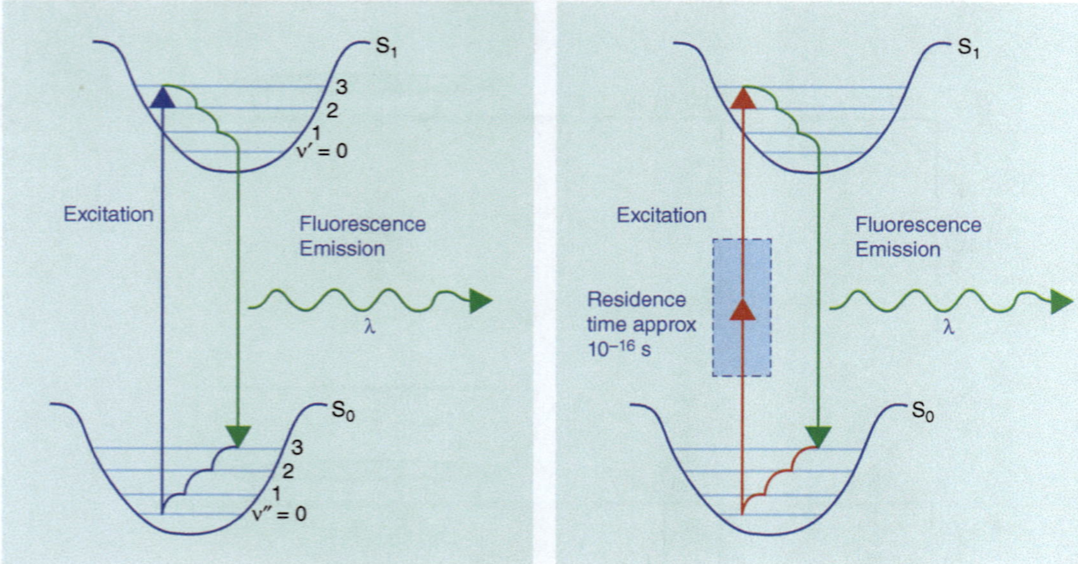

Figure 22.1 **Two-photon absorption of photons** *The Franck-Condon energy level diagram shows the vibrational levels (v) within the ground state (S_0) or first excited singlet state (S_1). This diagram is similar to the Jabłoński diagram, but inter-nuclear separation – rather than multiple spin states – is shown on the horizontal axis. As in a Jabłoński diagram, straight lines show the absorption or emission of a photon between electronic energy levels, whilst wavy lines denote non-radiative molecular internal conversion or non-radiative relaxation. Note that two long-wavelength IR photons, of comparably lower energy, are effectively absorbed together in a fraction of a femtosecond for multi-photon excitation to occur. Source: Gadella et al. 2003.*

multiple-labelled samples can be imaged by combination of two- and three-photon excitation, using a *single* infrared laser excitation wavelength. Since three-photon excitation extends the useful imaging into the deep UV (i.e. 720 nm used to excite a fluorophore absorbing at 240 nm), this means that fluorophores that are otherwise difficult to excite (e.g. hydroxycoumarin-based dyes) with conventional non-quartz optics can be imaged.

Multi-photon excitation fluorescence depends upon the near-simultaneous, within ≈ 0.5 femtosecond (5×10^{-16} sec), absorption of two, or three, long-wavelength photons each of which contain half (or one third) the energy required to excite the fluorophore. The time interval is so short, that the absorption of two or more photon occurs as a single quantum event. The success of excitation is therefore probabilistic: photons must be sufficiently crowded at the focal point to deliver two (or more) red-light photons (e.g. 960 nm) at the same time, and their combined energies will act as a single blue photon (e.g. 480 nm) and thus excite the GFP-tag or green fluorophore. Clearly, very high photon fluxes are required precisely at the focal point to generate nonlinear events. It is because multi-photon excitation depends upon simultaneous absorption of the energy of the photons that the resulting fluorescence emission depends upon the *square* of the excitation intensity. For example, to double the absorption by two-photon methods, the intensity of excitation illumination must be quadrupled. It is this quadratic relationship that makes possible the unique advantages of multi-photon microscopy.

Away from the focal point, the photon intensity falls off quadratically (to the fourth power of the distance) rather than linearly, and so fluorophores and tissue elements outside the focal spot do not absorb photons (see Figure 22.2 and Figure 22.3). With a pulsed laser, the fluorescence intensity at the focal spot depends on the average squared intensity rather than the squared average intensity, hence the alternative name of *nonlinear optics* for multi-photon microscopy.

Whilst there are no differences in the emission spectra of fluorophores in single-photon and multi-photon imaging, the excitation (absorption) spectral profile of most fluorophores is considerably different between single- and multi-photon imaging (e.g. GFP Figure 22.4, Table 22.1), and the overall spectrum tends to be blue-shifted. The combination of two or more photons relaxes the quantum mechanical rules, allowing higher excitation states (S_2–S_4) to be reached, resulting in broader multi-photon excitation spectra. The emission profile of the fluorophore is the same, because fluorescence emission occurs from the same singlet state as for single-photon microscopy.

Because two-photon excitation is a quantum mechanical event, light absorption in this case cannot be described in the same way as single-photon excitation, where light absorption can be described by Beer's law and molar extinction coefficients. Rather, absorption is related to a probability of molecular excitation as described by a two-photon *cross section*, with the GM unit. This unit is named after Maria Göppert-Mayer, where 1 GM = 10^{-50}cm^4. s.photon^{-1}. Molecules with large cross sections make good labels for two-photon microscopy; most lie between 1 and 300 GM. GFP has a value of 10 GM, whereas quantum dots have values in the region 10^4–10^5 GM. An excellent explanation can be found in Xu and Zipfel (2015). Refer to Drobizhev et al. (2011) and Speiss et al. (2005) for excitation properties of fluorescent proteins.

A Titanium-Sapphire (Ti:S) pulsed femtosecond laser[2] beam delivers a very high number (8×10^7 pulses sec^{-1}) of short pulses[3] (pulse length = 10^{-13} sec; one tenth of one millionth of one millionth of a second; ≈ 100 femtosecond pulse duration) focused into a very tight focal volume (Figure 22.2), which is diffraction-limited (<1 μm laterally) and from 1 to 0.1 femtolitre, depending on the NA of

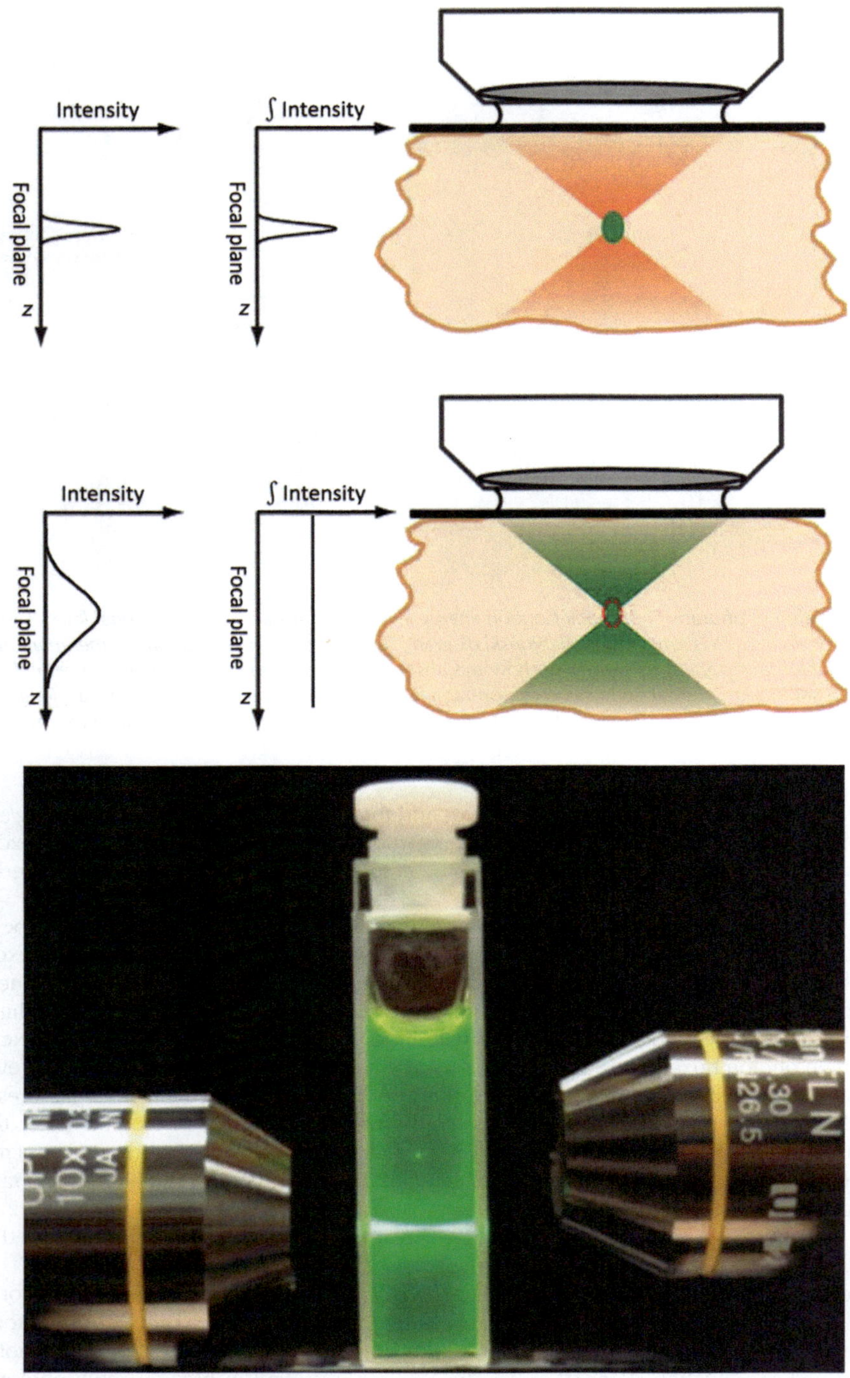

Figure 22.2 **Light absorption differences between single-photon and multi-photon microscopy** *In multi-photon microscopy, fluorophores and tissue elements outside the focal spot do not absorb photons. Light intensity falls off quadratically because the probability of excitation is proportional to the square of the laser intensity – i.e. is nonlinear – so absorption only occurs within a very small focal spot where the photon flux is highest. The graphs show fluorescence intensity at each focus plane plotted on a graph. The integrated intensity shows that equal intensity is seen at all planes in 1-photon microscopy, with no depth discrimination, whereas in the 2-photon case, both local and integrated intensity decrease rapidly away from the focal plane. The photograph shows two identical objectives, the one on the left producing one-photon (i.e. confocal) illumination in the lower part of the cuvette. The objective on the right, in the upper part of the cuvette, produces the very tight two-photon focal volume. Source: (top) Rocheleau and Piston 2013. Reproduced with permission from John Wiley & Sons. (bottom) Ustione and Piston 2011. Reproduced with permission from John Wiley & Sons.*

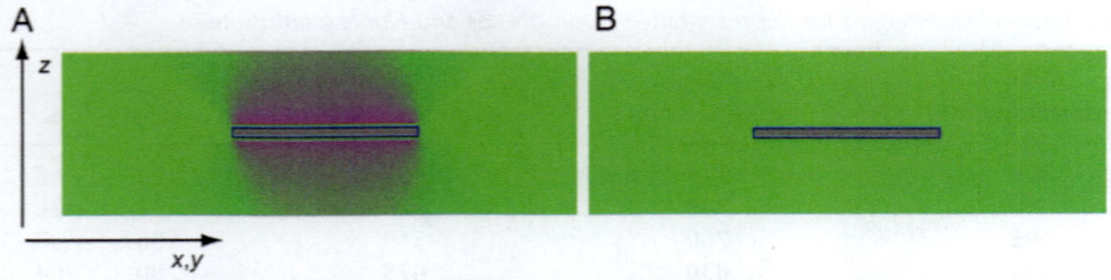

Figure 22.3 **Bleach volume differences between single-photon and multi-photon microscopy** *A single-photon confocal microscope (left) and the same microscope operating in two-photon mode (right) were used to photo-bleach a region of interest (ROI) in a plastic block impregnated with fluorescent dye (a thick tissue section can also be used for this demonstration). A Z-stack taken subsequently through the bleached area shows, in the case of the single-photon confocal microscope, the two cones meeting at the focal plane as a large bleached area, whereas two-photon excitation bleached a sharply defined line at the focal plane. Source: Adapted from Ustione and Piston 2011 with permission from John Wiley & Sons.*

the objective (i.e. NA 0.8–1.2, respectively; Figure 22.5). The pulse frequency of 80 MHz means that the duration between the pulses is about 12 nanoseconds (10^{-8} s). This focusing of pulses crowds the photons in both time and space by a factor of 10^5 over that of a single-photon confocal. Without a Ti:S pulsed laser, even a high NA objective would focus insufficient light into the focal spot, such that less than one photon of emitted signal per pixel would be collected. Any signal this weak would be difficult, if not impossible, to differentiate from the background Poisson-distributed shot noise. The 10^5 factor increases the number of photons per pixel to about 100 (which would have a standard deviation of ±10 photons as shot noise).

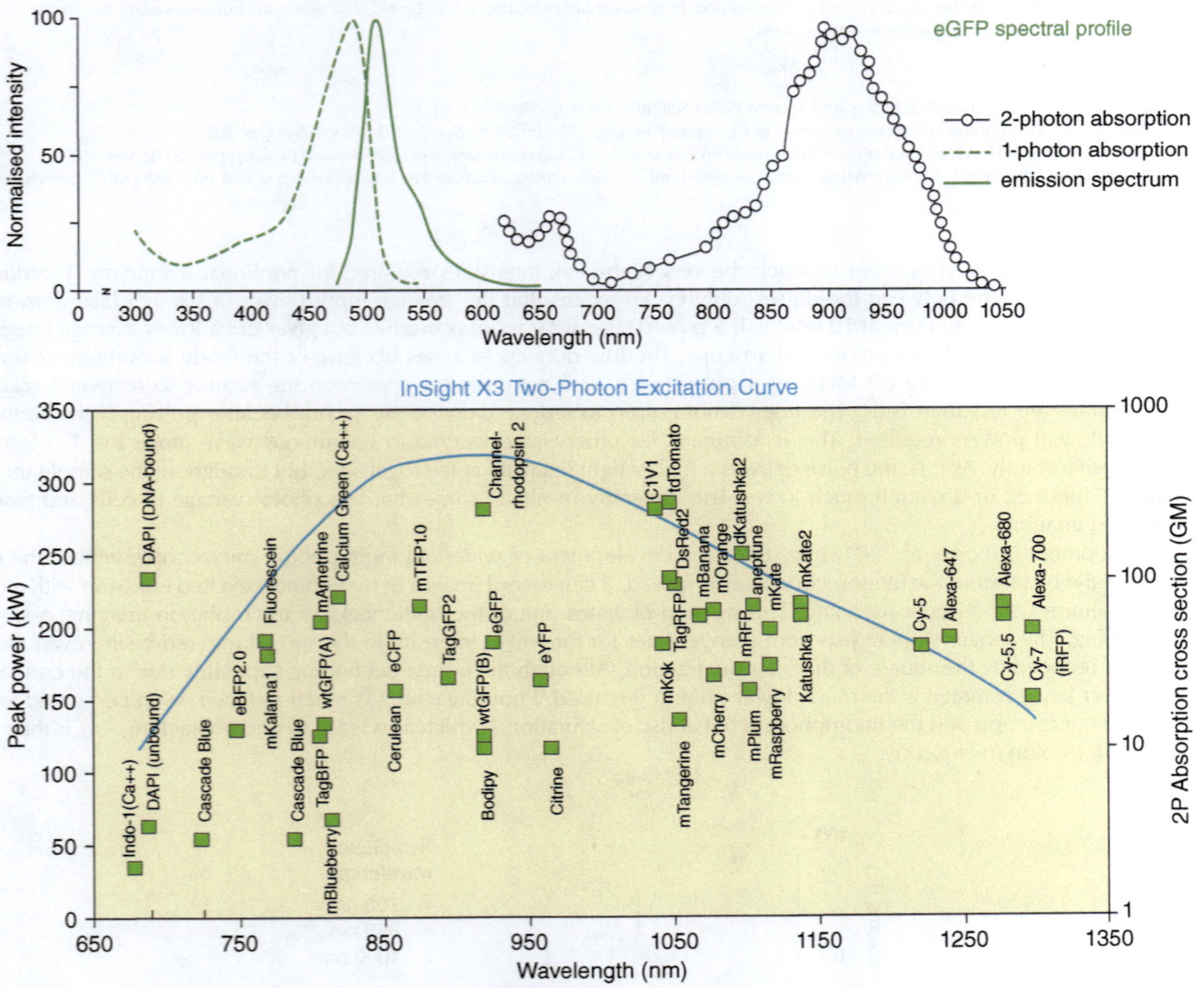

Figure 22.4 **Spectral profile of eGFP** *Both fluorescent proteins and intrinsic fluorophores exhibit differing absorption spectra for single-photon and multi-photon absorption but identical emission spectra because the route for photon relaxation to the ground state is the same for both single- and multi-photon modes. The power curve for the ytterbium InSight X3 femtosecond laser is shown in blue. Source: Adapted from Cornell; Drobizhev et al. 2009. Reproduced with permission from American Chemical Society; Reproduced with permission from Newport Spectra-Physics Ltd.*

Table 22.1 *Absorption and Emission Wavelengths for Representative Fluorophores and Fluorescent Proteins*

Fluorophore	λ 1P excitation (nm)	λ 2P peak* excitation (nm)	λ emission (nm)	GM values		Photostability (s)
DAPI	358	**700**	461	102	7	50
FITC	490	**780**	525	32	26	30
Alexa 488	495	**760**	519	96	37	65
Rhodamine B	544	**830**	625	280	204	35
Alexa 568	578	**760**	603	173	11	45
Alexa 594	590	**780**	617	100	3	65
Cy5	649	**800**	669	40	20	60
Alexa 647	650	**790**	665	44	20	80
eCFP	433	**860**	475	23	319	64
eGFP	484	**940**	507	148	181	34
eYFP	514	**960**	527	116	238	60
mCitrine	516	**970**	529	75	318	49
tdTomato	558	**1050**	581	278	200	64
mCherry	587	**1080**	610	27	6	68
mKate2	586	**1140**	630	72	30	5
mPlum	590	**1100**	649	22	3	53
mNeptune	600	**1100**	650	70	12	160

*In certain cases, the peak can be quite broad, and a range of values about the maximum excitation value is cited. The exact value depends upon the chemical diluent or cellular environment in which the absorption value was measured.

References

Zeiss Multi-photon dye sheet: https://bit.ly/2LMN8Kq, and Thermo Fisher Scientific list: https://bit.ly/2UFIEtT.

Drobizhev, M. et al. (2011). 'Two-photon absorption properties of fluorescent proteins'. *Nature Methods 8*/5: 393–399. Table 1, p. 397.

Mojzisova, H. and Vermot, J. (2011). 'When multiphoton microscopy sees near infrared'. *Current Opinion in Genetics and Development 21*/5: 549–557.

Cornell cross-section values. http://www.drbio.cornell.edu/cross_sections.html see also http://daniel.fiole.free.fr/twophotondyes and from the LMU, Munich https://bit.ly/2QQgKME.

The Ti:Sapphire laser is pulsed in order to reach the very high-peak intensities required for nonlinear excitation. The duration of the pulses divided by the time between the pulses (duty cycle) means that the average input power of the Ti:S laser into the tissue is <10 milliwatts at the objective. Even at 10 mW, a Ti:S pulsed laser is far more powerful[4] but gives out a lower average intensity than the continuous wave lasers used for confocal microscopy. The low-duty cycle arises because of the mode locking mechanisms and the repetition rate, which is usually 80 MHz (or a pulse every 12.5 ns) gives the specimen the chance to recover because most fluorescence lifetimes are less than 8 ns. The laser is not pulsed to reduce damage *per se* but because pulsing is the simplest way to generate the kilowatt powers required. This is fortunate, for otherwise operating in continuous wave mode the Ti:S laser would vaporise the sample instantly. As it is, the pulse delivers a higher light intensity at the focal spot, but throughout the sample the *average* photon intensity (Figure 22.6) is overall much lower. This generally results in somewhat less photo-damage of cells and tissues than with single-photon imaging.

A recent development (Amor et al., 2016) has been the development of widefield multi-photon microscopy without the need for beam scanning. If the beam from the femtosecond laser is spread, it can record images of two-photon-excited emission with an sCMOS camera from the entire field without scanning. This method obviates one of the bottlenecks of multi-photon imaging: a slow frame rate due to scanning. The longer camera exposure compensates for the lower intensity in the spread infrared beam, even though the excitation level is reduced by the square of the intensity fraction. Although the optical sectioning capability due to the confined focal spot is lost, another key advantage is the much lower photon flux used. Photobleaching is much reduced, even compared to (single-photon) widefield microscopy, and the fluorophore is not at risk of saturation – which also leads to photobleaching – as is the case with confocal and multi-photon microscopy.

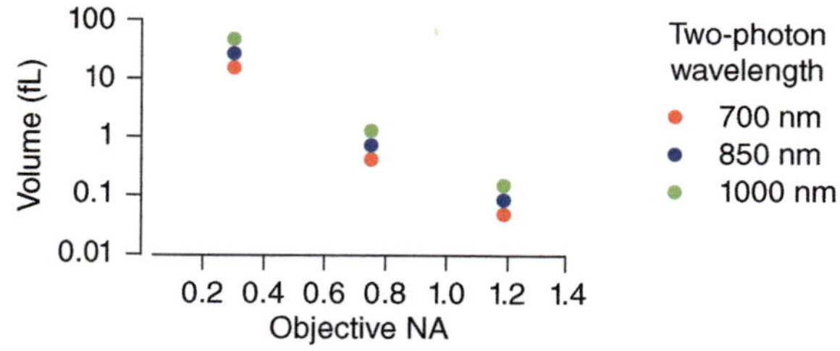

Figure 22.5 ***Volume of the focal spot*** *The volume of the focal spot is approximately one femtolitre, but the exact volume depends upon the illuminating wavelength and the numerical aperture of the objective. Source: Rubart 2004. Reproduced with permission from Wolters Kluwer Health.*

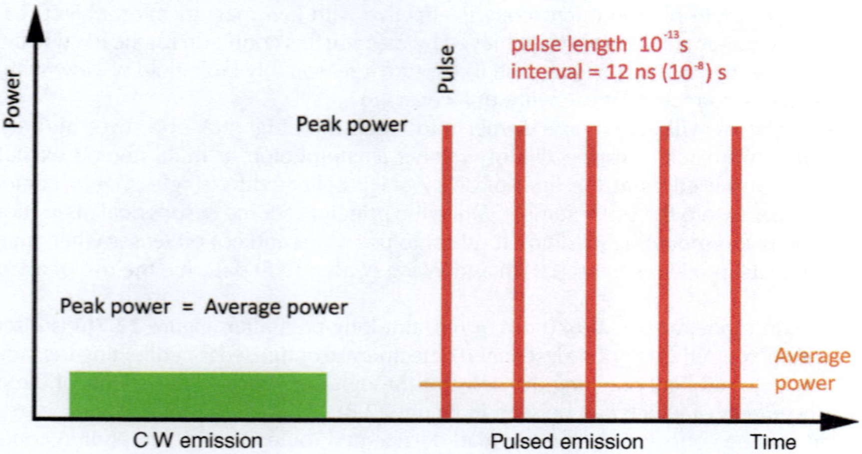

Figure 22.6 ***Comparison of the power output of pulsed and continuous wave lasers*** *The Ti:S laser is off for more than 99% of its cycle. The pulse it generates delivers a beam of higher light intensity at the focal spot, but throughout the sample the average photo-intensity is lower overall than the continuous wave lasers used for confocal microscopy. It is often overlooked, or forgotten, that the photon irradiance incident upon the sample can be very high indeed. Peak confocal power* $\approx 2.5 \times 10^5$ *W/cm²*, *and average power* ≈ 0.25 *W/cm² (assuming a 1024 × 1024 raster). Peak multi-photon power* $\approx 5 \times 10^{20}$ *W/cm² and average power is* $\approx 2.5 \times 10^4$ *W/cm². For comparison, the mean solar irradiance value is 0.13 W/cm².*

Early designs of the multi-photon microscope required the Ti:S pulsed laser to be finely tuned (selectable from 690 to 1040 nm) and mode-locked, using an oscilloscope and a lot of patience, in order to deliver a correctly pulsed excitation beam at the correct wavelength. Nowadays tuning the laser across its output wavelengths to suit different fluorophores is under computer control and thus is very much easier to operate. This ease of operation, with no longer any necessity to manually retune the laser for exciting several fluorophores, makes multiple-labelling experiments much more straightforward than was formerly the case. Suitable lasers are either the Newport Spectra-Physics Mai Tai HP/Deep See, the Insight X3 or the Coherent Chameleon Discovery models. Ti:S and ytterbium lasers operate most efficiently in the 800 nm region of the spectrum. Their output (IR power) falls off as the laser is tuned towards the far-IR region of the spectrum: this is normal. The upper limit of the tunable pulsed lasers is around 1100 nm. To image with illumination above this wavelength, an optical parametric oscillator (OPO) must be used and this can be teamed with a number of different laser types, including TI:S, bulk-based and fiber lasers. Finally, the very recent development of variable repetition rate fiber lasers provides the possibility for even more efficient multi-photon excitation and even deeper imaging (e.g. KMLabs Y-Fi NOPA and OPA). For further details on lasers for multi-photon microscopy, see Wise (2013) and Charan et al. (2018).

In confocal microscopy the excitation photons are absorbed throughout the sample, generally within the V-shaped cones above and below the focal spot (Figure 22.2). In the multi-photon microscope there is no out-of-focus absorption as the beam passes through the sample to the focal point. Approximately 80% of the excitation illumination is absorbed within this tightly defined focal volume. This leads to reduced background and therefore a better SNR and higher contrast than (single-photon) confocal microscopy. The exact dimensions of this volume are dependent upon objective numerical aperture, but for a typical large NA fluorescence objective used at near-infrared wavelengths, this focal volume is defined by an ellipsoid having a lateral dimension of ≈ 0.3 μm in diameter and an axial length of ≈ 1 μm. Like TIRFM, multi-photon microscopy achieves sharper images by restricting the *excitation* volume. A confocal microscope, on the other hand, achieves a sharp image by restriction of the fluorescence emission collection volume[5] using a pinhole in front of the detector.

The resolving power of multi-photon microscopy with its long-wavelength excitation illumination would be expected to be worse than single-photon fluorescence microscopy. However, because the process is nonlinear, the intensity of the PSF is also squared, which reduces its volume to a certain extent. The factors that determine resolving power in multi-photon microscopy are complex, but for objectives of NA 1.4, NA 1.0 and NA 0.8, those factors that determine resolving power are given in Box 22.1.

Because the multi-photon microscope excites only a very small volume of fluorophore ($\approx$ 1 femtolitre ≈ 1 μm³), there is no chance of two long-wavelength excitation photons scattering precisely to the focal point at the same time. The multi-photon microscope is not sensitive to light scattering. Compared to a confocal microscope, photon detection is much more effective since even scattered photons contribute to the usable signal. No pinhole is required: the small focal volume determines the optical sectioning capability. A non-descanned detector can therefore be used to collect all the emitted photons, and since there is no pinhole to attenuate the emitted signal, the non-descanned detector (NDD) collects *all* the light returning from the sample[6].

These two reasons (small excitation volume and insensitivity to scattering events) mean that the multi-photon microscope can generally image about six times deeper into tissue compared to single-photon confocal mode, although this depends very much upon the physiological and anatomical nature of the tissue. By reducing the pulse frequency and simultaneously increasing the average power, it is possible to image up to 1 mm in depth, but poor SNR and lack of contrast, as in the confocal, likewise eventually become limiting factors. Collagen (which induces second harmonic signals), myelin and blood-rich tissues or micro-vasculature all degrade the focus and depth penetration of the laser.

For best results, a high numerical aperture (ideally an immersion) objective, of low overall magnification (i.e. high brightness) is required with the emission detector (Figure 22.7a) placed as close as possible to the back focal plane of the objective. Unlike single-photon confocal microscopy, multi-photon microscopy is effective with low magnification objectives of reasonably moderate numerical aperture, provided that wide-angle detection is achieved by meeting this condition for the ideal location of the NDD detector. Employing a low-magnification objective is useful if you wish to capture a reasonably large field of view with a long working distance dipping objective (e.g. 10x), which is corrected for use without a coverslip.

Lower numerical aperture objectives will also image deeper into tissue than high NA objectives and are less prone to spherical aberration due to refractive index mismatch – irrespective of whether single-photon or multi-photon excitation is used. You should switch to multi-photon mode when the edges at the limit of any Z-stack collected by single-photon confocal microscopy become unclear and you still wish to go deeper into the tissue sample. Since the principal source of spherical aberration is from refractive index differences between the immersion and mounting medium, it is best to use water and not oil lenses when imaging aqueous specimens more than 15 μm beneath the coverslip. Heintzmann (2010) and Wang et al. (2015) describe the use of adaptive optics to correct for spherical aberration.

Whilst scanning in multi-photon mode, a special 670 nm or 690 nm long-pass filter (Figure 22.7b) is inserted in the beam path to direct the emitted signal to the detector. Although non-descanned detectors are optimised for collecting the emitted multi-photon signal, they are *very* susceptible to any external light reaching the detector through the optical components of the microscope (e.g. through the objective and eyepieces, the effects of which can be seen in Figure 22.8).

The multi-photon signal can also be collected via the 'normal' de-scanned route (as in single-photon confocal mode) into the fully enclosed 'de-scanned' PMTs if the pinhole(s) in front of the PMTs is opened fully. This is recommended if light pollution is a severe problem. Directing the multi-photon emission signal through the internal light path into the de-scanned PMT detector also allows the signal to be spectrally unmixed, either by stepwise collection or with a lambda stack spectral detector. Collecting such a lambda stack can be useful for precisely determining the emission peak of an SHG signal (second harmonic generation, Section 22.5).

Besides collecting the fluorescent signal emitted from the sample in the usual way by reflected-light non-descanned detectors (NDDs), it is also possible to use transmission NDDs to collect the proportion of light that is scattered and transmitted *through* the sample. With thick tissues or whole organisms, the emitted signal scattered in the forward direction may be very small, and so – clearly – the reflected-light detection architecture dominates in multi-photon design. However, if a tissue section is, say, 400 μm thick, just as much signal will be scattered upwards (i.e. forwards) through 200 μm as downwards (backwards) through the remaining 200 μm, since signal emanates in all directions from the focal point. Very often transmitted-light NDDs are omitted for both cost and convenience, but where the signal is weak, they can – if fitted – make a significant difference to the quality of the image that can be collected. These transmitted-light fluorescence NDDs are also very useful for detecting second harmonic generation. Using transmission NDDs demands a high numerical aperture (e.g. NA 1.4) oil-immersion condenser to maximise signal collection, which can be tricky and slightly messy to use.

Box 22.1 Resolving Power in Multi-Photon Microscopy

In theory, the resolution of the multi-photon microscope would appear to be a factor of 2–3 times worse than the single-photon confocal microscope, because of the longer excitation wavelength, which broadens the PSF. However, the smaller focal volume and higher S/N ratio generated by the multi-photon microscope confer an advantage in terms of improved resolving power.

Because these opposing factors of excitation wavelength, focal volume and SNR tend to cancel each other out, the resolving power of the two-photon microscope is similar to the confocal (resolving power in the single-photon confocal microscope only improves significantly when the pinhole is closed to about 0.6 Airy units or less).

From inspection of the equations below, it is apparent that if an object cannot be resolved in the confocal microscope, especially with a small pinhole, then the same feature will not be resolved with a multi-photon microscope. The advantage of using the multi-photon instrument lies in improvement of imaging deep into tissue and a better signal-to-noise ratio.

Theoretically, the lateral resolution of the multi-photon microscope is:

$d_{x,y} \approx 0.7\lambda/\text{NA}$ for lateral resolution and $d_z \approx 2.3\lambda n/\text{NA}^2$ for resolution in the axial direction,

compared to the values for the confocal microscope:

$d_{x,y} \approx 0.43\lambda/\text{NA}$ for lateral resolution and $d_z \approx 1.4\lambda n/\text{NA}^2$ axially.

Objective & RI	wavelength (nm)	Lateral resolution (μm)	Axial resolution (μm)
NA = 0.8	λ = 488 nm (single photon)	0.26	1.07
n = 1.00 (air)	λ = 900 nm (**multi-photon**)	**0.79**	**3.23**
NA = 1.0	λ = 488 nm	0.21	0.68
n = 1.00 (air)	λ = 900 nm	**0.63**	**2.07**
NA = 1.0	λ = 488 nm	0.21	0.91
n = 1.33 (water)	λ = 900 nm	**0.63**	**2.76**
NA = 1.4	λ = 488 nm	0.15	0.53
n = 1.52 (oil)	λ = 900 nm	**0.45**	**1.61**

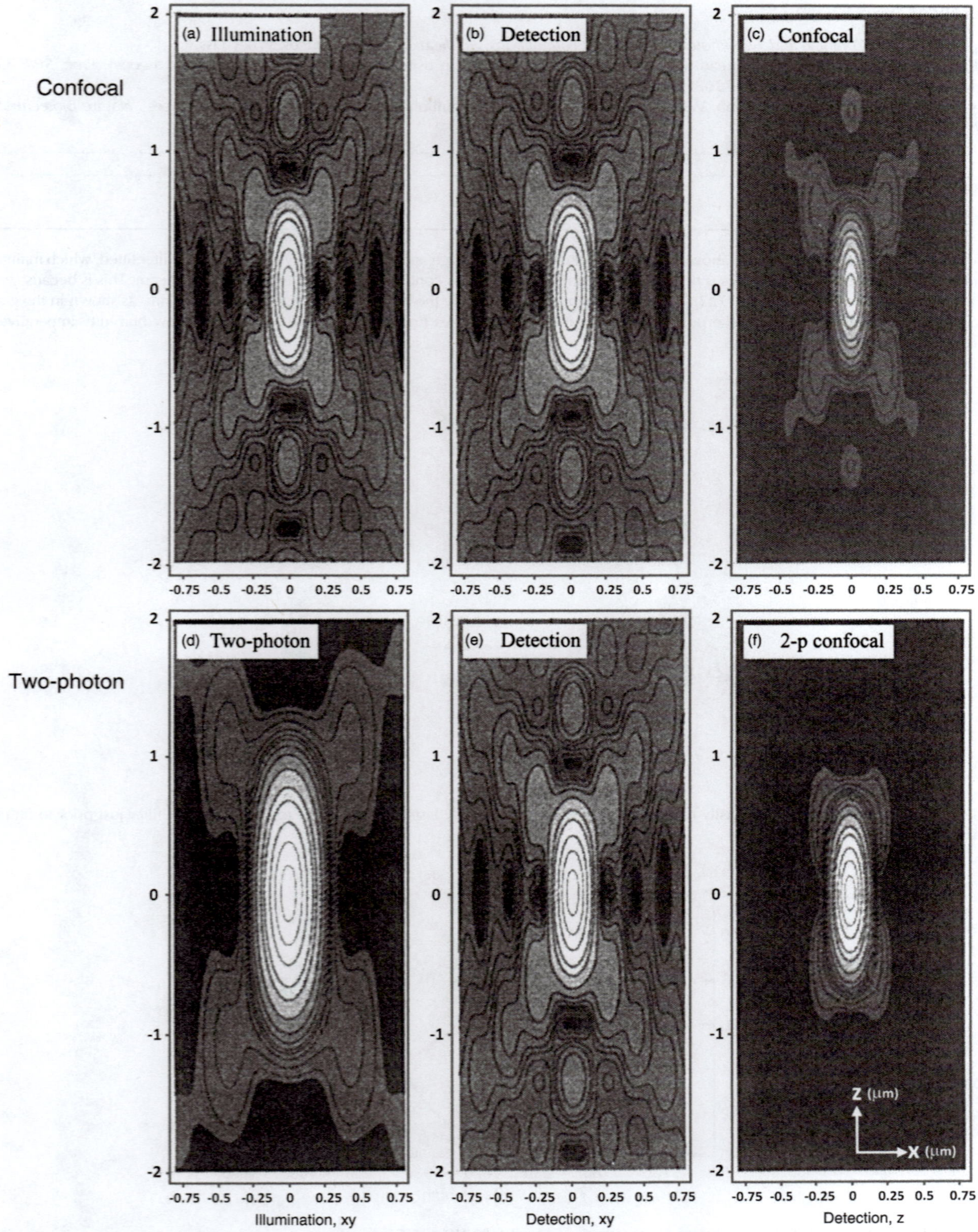

Source: Figure 2 from page 105, Diaspro (2002). Reproduced with permission from John Wiley & Sons. The values in the table above are derived from the resolving power calculations given in the text and cited here. Further equations to calculate the FWHM of the PSF of sub-resolution fluorescent beads can be found in Guldbrand et al. (2009).

References

Diaspro, A. (2002). *Confocal and Two-Photon Microscopy: Foundations, Applications and Advances*. Wiley-Liss.
Guldbrand, S. et al. (2009). 'Point spread function measured in human skin using two-photon fluorescence microscopy'. *Proc. SPIE* 7367, Advanced Microscopy Techniques, 73671R. doi: 10.1117/12.831542.
Zipfel, W.R., Willimas, R.M., and Webb, W.W. (2003). 'Nonlinear magic: multiphoton microscopy in the biosciences'. *Nature Biotechnology* *21*/11: 1369–1377.

Box 22.2 Purge Filters

All far-red pulsed lasers, which operate above 700 nm, irrespective of their design and manufacture, will have a purge filter fitted, which maintains the internal humidity below 10%. If this is not done, it becomes impossible to continuously mode-lock the IR laser beam. This is because water effectively absorbs far-red light around 722, 760 and 823 nm, and particularly the infrared wavelengths at 895–980 nm, as shown in the graph. Concomitant with water absorption, the pulsed laser output power decreases from a peak of 800 nm, making low humidity imperative for successful multi-photon microscopy.

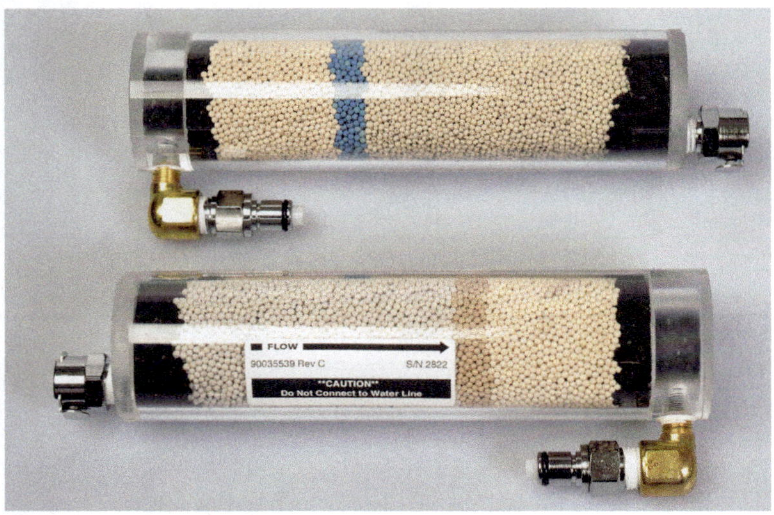

The purge filter is designed to be easily replaced by the user and normally lasts for a year. The top shows a new filter just prior to fitting to replace the exhausted unit below.

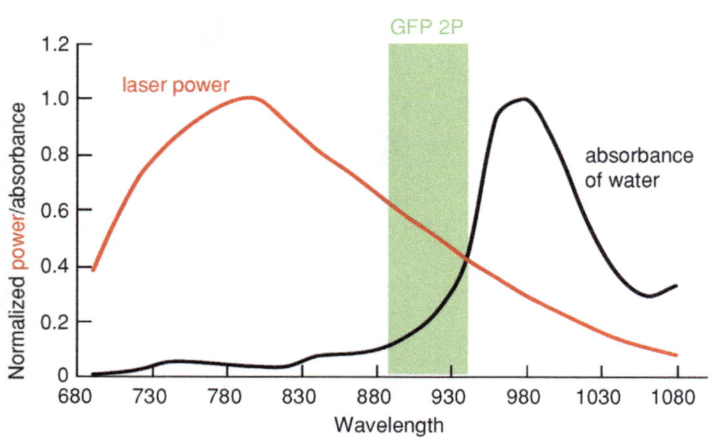

Source: Padmanabhan et al. 2010. Reproduced with permission from John Wiley & Sons.

Reference

Padmanabhan, K., Andrews, S.E., and Fitzpatrick, J.A. (2010). Multi-photon imaging. *Current Protocols in Cytometry*. Chapter 2: Unit2.9. doi: 10.1002/0471142956.cy0209s54.

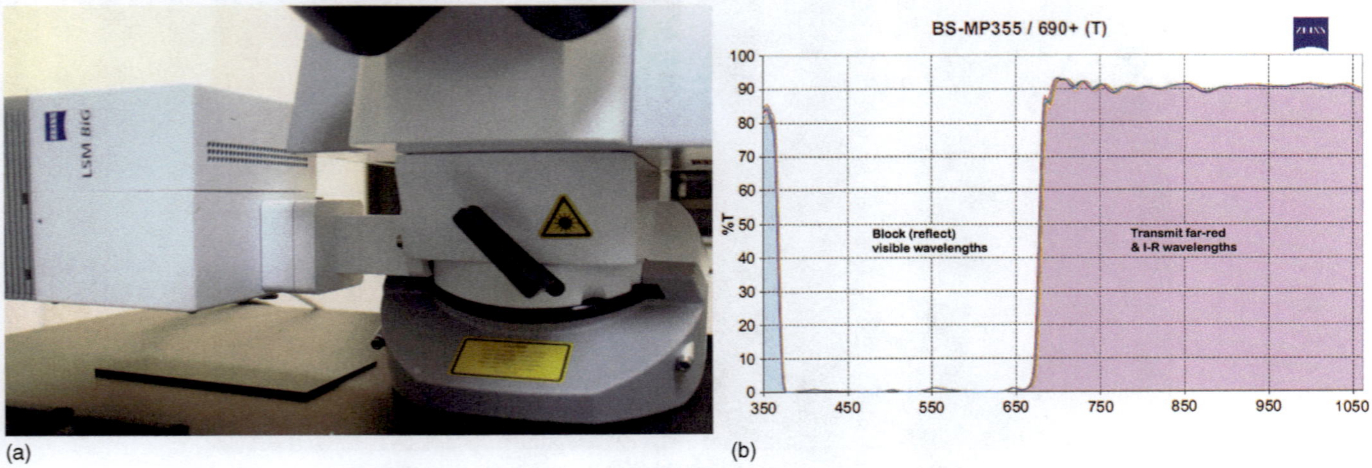

Figure 22.7 **Location of the non-descanned detector and long-pass detection filter** *(a) Ideally, the NDD should be located as close to the back focal plane of the objective as possible. (b) Spectral characteristics of the long-pass filter used to collect the multi-photon signal. Source: Reproduced with permission from Carl Zeiss Ltd.*

The spectra of fluorophores is different for nonlinear microscopy than single-photon excitation; the excitation spectra are generally much broader (Table 22.1), meaning that a single excitation wavelength selected by tuning the Ti:S laser can simultaneously excite multiple fluorophores, each with distinct emission profiles. Multi-photon microscopy also confers the advantage that fluorophores can be used at higher concentrations than for single-photon widefield or confocal microscopy because self-absorbance of the fluorophore is significantly reduced where the absorbance of the dye in the red or IR regions of the spectrum is low. Multi-photon microscopy also comes into its own when UV or blue dyes must be used for multi-labelling experiments. A tuneable multi-photon laser gives more choice for optimally exciting UV dyes than the few laser lines available to single-photon microscopy.

Just as with single-photon confocal microscopy and selective plane illumination (light-sheet) microscopy, it can be advantageous to optically clear tissue prior to acquiring images. Benzyl alcohol (Clendenon et al., 2011), benzyl alcohol with benzyl benzoate (BABB; see Becker et al., 2012), methyl salicylate (McDonald et al., 2008) and mineral oil with glycerol (Wilson et al., 2012) have all been used to clear thick tissues to reduce scattering. A review may be found online by Thorn et al.; the URL is given at the end of the chapter. Following the introduction of the Sca*l*e clearing method (Hama et al., 2011), the Clarity method (Lee et al., 2014) and subsequent protocols (see Chapter 19, Section 19.7; for an excellent review see Richardson and Lichtman, 2015). Specialist objectives designed to have long working distances for imaging cleared tissues have been made by the major microscope manufacturers (see Marx, V., 2014). This has opened the potential for thick-tissue imaging, particularly in neurology, and a host of different clearing protocols have recently been developed.

22.3 Equipment for Multi-Photon Microscopy

The specification of pulsed laser, the configuration of the microscope stand, the choice of detectors and the objectives used are the major considerations in specifying equipment for multi-photon microscopy. It is still possible to build a dedicated multi-photon microscope; for example, see Majewska et al. (2000), Nikolenko and Yuste (2013) and Rosenegger et al. (2014), but dedicated modular turnkey systems using open-source software are available commercially.

Because of their broad multi-photon excitation spectral profiles, most fluorophores can be excited at around 800 nm. However, since red-shifted fluorescent proteins have been developed for live-cell imaging, there is a requirement for excitation wavelengths in excess of 1000 nm. Nonetheless, a Ti:S laser gives much less output power at the limits of its tuning range. Currently, the wavelength range (690–1040 nm) of the standard Ti:S laser can be extended by using an optical parametric oscillator (OPO) unit to around 1600 nm, although ytterbium-doped lasers are being developed, which provide greater power at longer wavelengths.

Most multi-photon microscopes are additionally fitted with what is called a 'pre-chirp' unit. As the laser pulse travels along the fibre optic that transmits the beam, through the objective and also as it penetrates into tissue, the pulse will be dispersed and undergo chromatic aberration (Figure 22.9). The degree of spreading of each laser pulse is inversely proportional to the length of the pulse: thus shorter (femtosecond) pulses are spread more than longer (picosecond) pulses. This is because the longer-wavelength 'red' portion of the pulse travels faster than the shorter 'blue' component. In effect, the 100 fs pulses are broadened to 3–400 fs, rendering the advantage of femtosecond lasers over their picoseconds predecessors negligible and a consequent reduction in excitation efficiency. This pulse-spreading aberration is known as group velocity dispersion (GVD) with the signal being chirped[7]. Chirping is an undesirable effect, reducing the efficiency of the Ti:S laser. To counteract this GVD effect, a negative ('pre-chirp') pulse compression is applied, by means of prisms or diffraction gratings, to ensure that the Ti:S laser pulse is of the correct duration and wavelength and that the duration of the pulse entering the sample is the same as when it was generated and left the laser cavity. The negative pre-chirp unit forces the longer-wavelength 'red' portion of the laser pulse to travel a greater distance than the 'blue' part of the pulse, so that the

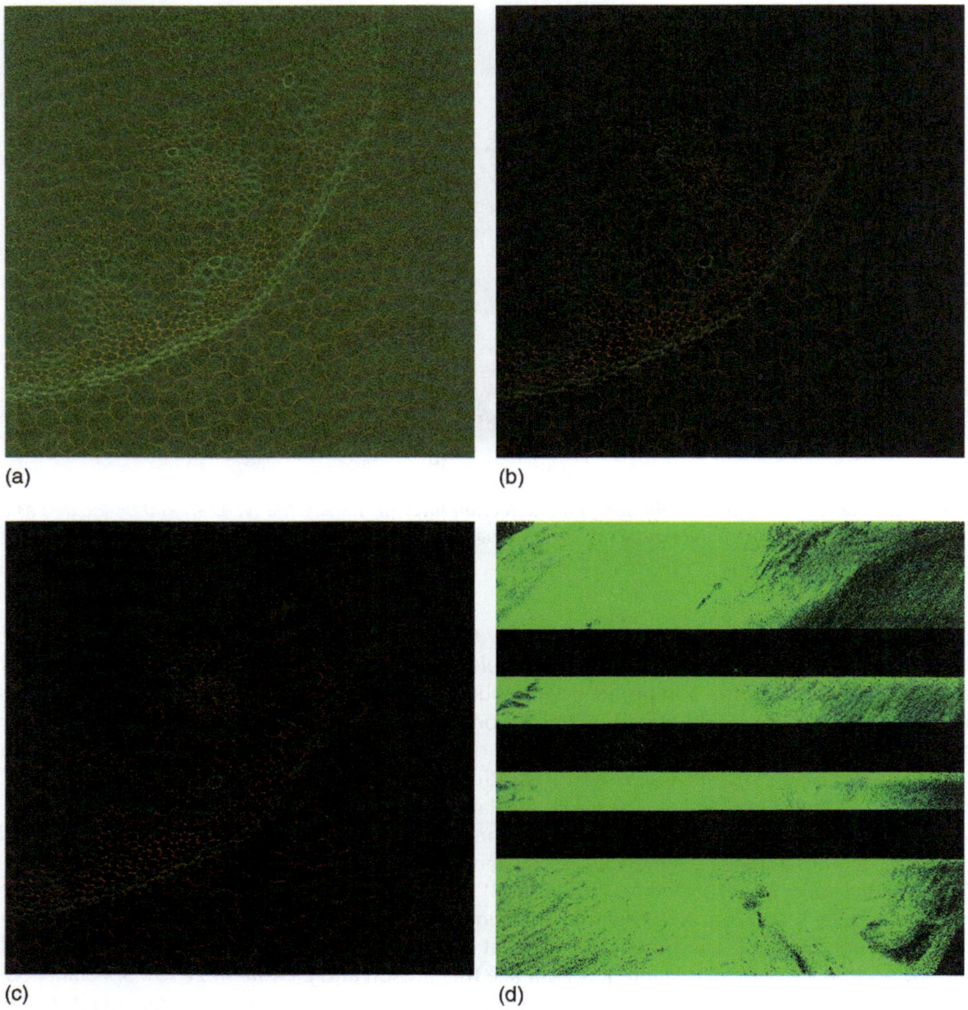

Figure 22.8 **Sensitivity of the non-descanned detector** *The GaAsP NDDs are incredibly sensitive. This is good for obtaining maximum signal but makes them extremely susceptible to stray light. They must be used in complete darkness: the imaging area of the microscope must either be enclosed in a blacked-out incubator or an old style photographer's cloth draped over the instrument when scanning. In (a) the room lights and computer monitor were left on whilst scanning. In (b) the light from the computer screen still contributes to background noise, particularly in this weakly stained* Convalaria *sample, which has been selected to illustrate this point. In (c) the microscope stand has been draped with a cloth, cutting out all potential stray light and glare. In (d) the signal intensity is so high (rather than the external room illumination) that the automatic shutter has activated to protect the GaAsP detector.*

shorter wavelength 'blue' part of the pulse is in front of the 'red' portion in the preconditioned beam. The longer wavelength part of the pulse then closes up as the pulse travel through the optical components and the tissue to the focal point.

The pulsed laser should have a software control box that, ideally, should also be integrated into the microscope manufacturer's software to allow control of the GVD prism or grating to give the best SNR for each particular objective used. This involves calibrating the GVD compensation via the laser control box to get the best SNR for each objective fitted to the microscope. Once this is done for three or more wavelengths for each objective, the software will make the appropriate compensation once it knows the objective in use.

Whether you chose an upright or inverted microscope for a multi-photon set-up very much depends upon the possible range of cells, tissues and whole organisms that may conceivably be studied during the lifetime of the instrument. Whether confocal or multi-photon, these instruments are more often sold as inverted microscopes, because an inverted design allows for better examination of living cells and tissues bathed in fluid. Using an inverted design also means that cells and tissues can be examined without recourse to elaborate means of tissue immobilisation in buffer or other fluid. If a dipping objective is used from above on an upright microscope, any focusing and stage movement can set up currents in the fluid, which can displace the specimen, unless it is adherent to the dish or slide or securely attached. The advantage of using an upright microscope is greater working distance for thick samples or whole organisms – as in intra-vital microscopy.

There are two potential ways to secure long working distances in order to image deep into tissues on an inverted microscope. Dipping objectives offer longer working distances than coverslip-corrected objectives, but they cannot easily be used on an inverted

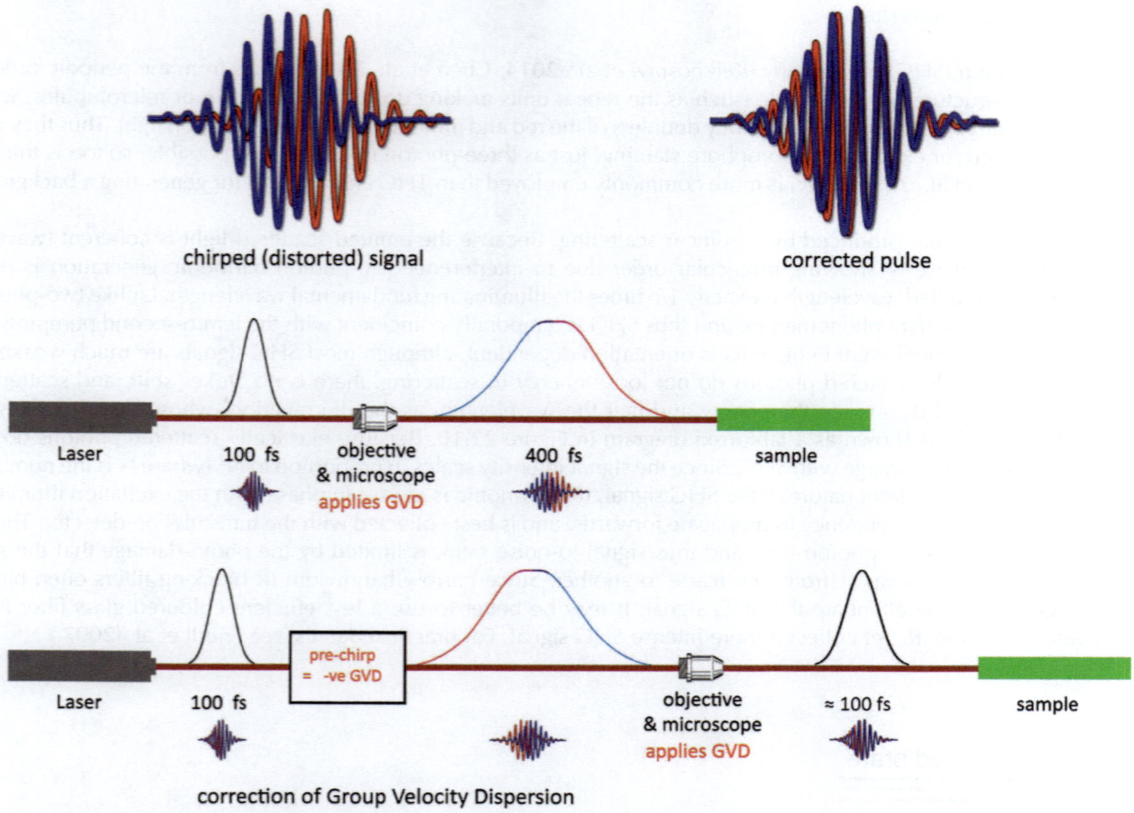

Figure 22.9 ***GVD and signal chirp correction*** *Each pulse is a small wave packet, and so the effect of red light having a lower refractive index in its passage through the microscope optics than blue light causes group velocity dispersion. The red wavelengths travel faster, and so the pulse is broadened. The chirped signal, which suffers a change in frequency, is corrected by using prisms to apply negative 'pre-chirp' pulse compression. This correction can either be performed directly through a control box attached to the laser or through the microscope software. Source: Murphy et al. 2013. Reproduced with permission from John Wiley & Sons.*

design (unless an objective inverter is used, such as that sold by LSM Tech[8]). However, it is possible to use a dipping objective inverted if you use ultrasound gel as the 'immersion' medium to couple the objective to the sample, but it must be a colourless variety, which is not always easy to locate and purchase. In any case, the non-coverslip-corrected dipping objectives will only work in an inverted configuration effectively through the dish/slide at low to medium numerical apertures. Another way to achieve long working distances is to use long working-distance tissue culture objectives designed specifically for inverted microscopes, but the correction collars present on the higher magnification lenses are usually awkward to adjust. Alternatively, LaVision Biotec manufacture a Doubleheader stand, which incorporates both an upright and an inverted microscope stand coupled to a single laser system.

This chapter does not give advice on maintaining alignment of the multi-photon pulsed laser. For information on this topic, see Cannell et al. (2006). For those users with turnkey systems under maintenance contract, the service engineers for the pulsed laser and for the microscope will normally work together to achieve a well-aligned instrument.

22.4 Fast Frame Multi-Photon Microscopy

LaVision Biotec offer the TRIM Scope, a commercial multi-beam multi-photon microscope. Like the single-beam design, it requires no pinholes on the detection side. This multi-beam multi-photon microscope operates very much like a multiple-beam confocal microscope, enables collection with a high frame rate and collects its signal with a CCD camera rather than a PMT. Multi-photon microscopy combined with scanned light-sheet (i.e. selective plane illumination) microscopy offers fast frame rates coupled with low rates of photobleaching. Light-sheet multi-photon microscopy is particularly suited to studying embryonic development and living tissues (Tomer et al., 2012); for a good review see Supatto et al. (2011). Just as with single-photon confocal microscopy, a multiple-beam approach confers a speed advantage. The performance characteristics of single beam versus multi-beam multi-photon microscopes have been investigated by Niesner et al. (2008) and Andresen et al. (2012). The reader is referred to these references for more details on this specialist topic.

22.5 Second Harmonic Generation

Second harmonic generation (SHG) microscopy (Keikhosravi et al., 2014; Chen et al., 2012) results from the periodic ordered nature of non-centrosymmetric structures in the sample (such as the repeat units making up collagen, myosin or microtubules, which lack a generalised mirror symmetry) functioning as frequency doublers of the red and infrared photons of excitation light. Thus they can be seen unstained, without the need for exogenous fluorophore staining. Just as three-photon microscopy is possible, so too is third harmonic generation (THG; Weigelin et al., 2016). SHG is more commonly employed than THG and is useful for generating a background image of the tissue.

SHG images (Figure 22.11) are produced by nonlinear scattering. Because the emitted scattered light is coherent (waves vibrating in the same plane and in phase) – showing molecular order due to interference – n-photon harmonic generation is essentially a scattering process, and the emitted wavelength is exactly 1/n times the illuminating fundamental wavelength. Unlike two-photon excited fluorescence, SHG is not a resonant phenomenon, and thus SHG is temporally coincident with the femto-second pumping beam. Like polarisation microscopy with incoherent light, SHG is orientation dependent, although most SHG signals are much weaker than two-photon fluorescence. Elastically scattered photons do not loose energy in scattering: there is no Stokes shift, and scattered photons have twice the frequency (and thus twice the energy and half the wavelength) as the illumination, whose wavelength is called the fundamental wavelength. This is shown as a Jabłoński diagram in Figure 22.10. Because elastically scattered photons do not absorb energy, there is much less photo-damage with SHG. Since the signal intensity scales in proportion to N^2 (where N is the number of active SHG molecules) and also the coherent nature of the SHG signal, the harmonic is always in phase with the excitation illumination. This means the SHG signal has a strong tendency to propagate forwards, and is best collected with the transmission detector. The maximum illumination intensity as defined by photon flux, and thus signal-to-noise ratio, is limited by the photo-damage that the specimen is capable of withstanding, and this varies from one tissue to another. Since narrow bandwidth IR blocking filters often have multiple interference coatings, these may attenuate the SHG signal. It may be better to use a less efficient coloured glass filter to block the fundamental wavelength in the IR, yet collect a more intense SHG signal. For practical details, see Friedl et al. (2007.)

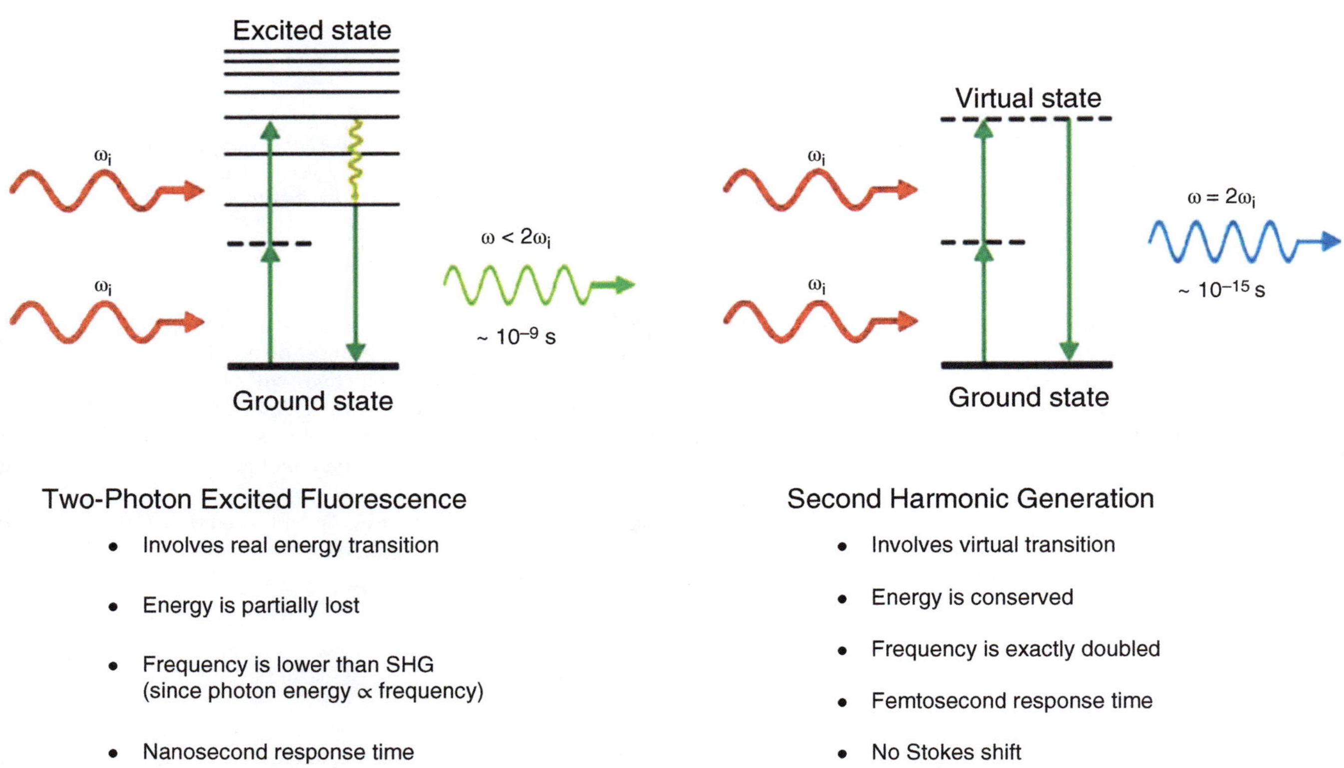

Figure 22.10 ***Jabłoński diagram showing how SHG occurs*** *Comparison of two-photon excited fluorescence versus second harmonic generation. Intense light shone on materials that do not have an inversion mirror symmetry is polarised, re-emitting light at the input frequency ϖ, and also 2ϖ, the SHG frequency. Both two-photon microscopy and SHG are nonlinear processes. They depend on the simultaneous input of two photons from pulsed lasers; the efficiency of both processes depends upon the square of the excitation intensity. The intensity at the focal plane also falls off quadratically. However, SHG is coherent involving virtual energy transitions and does not suffer from dye saturation, bleaching or photo-toxicity, problems associated with both single- and multi-photon processes. Source: Pantazis et al 2010. Reproduced with permission from National Academy of Sciences.*

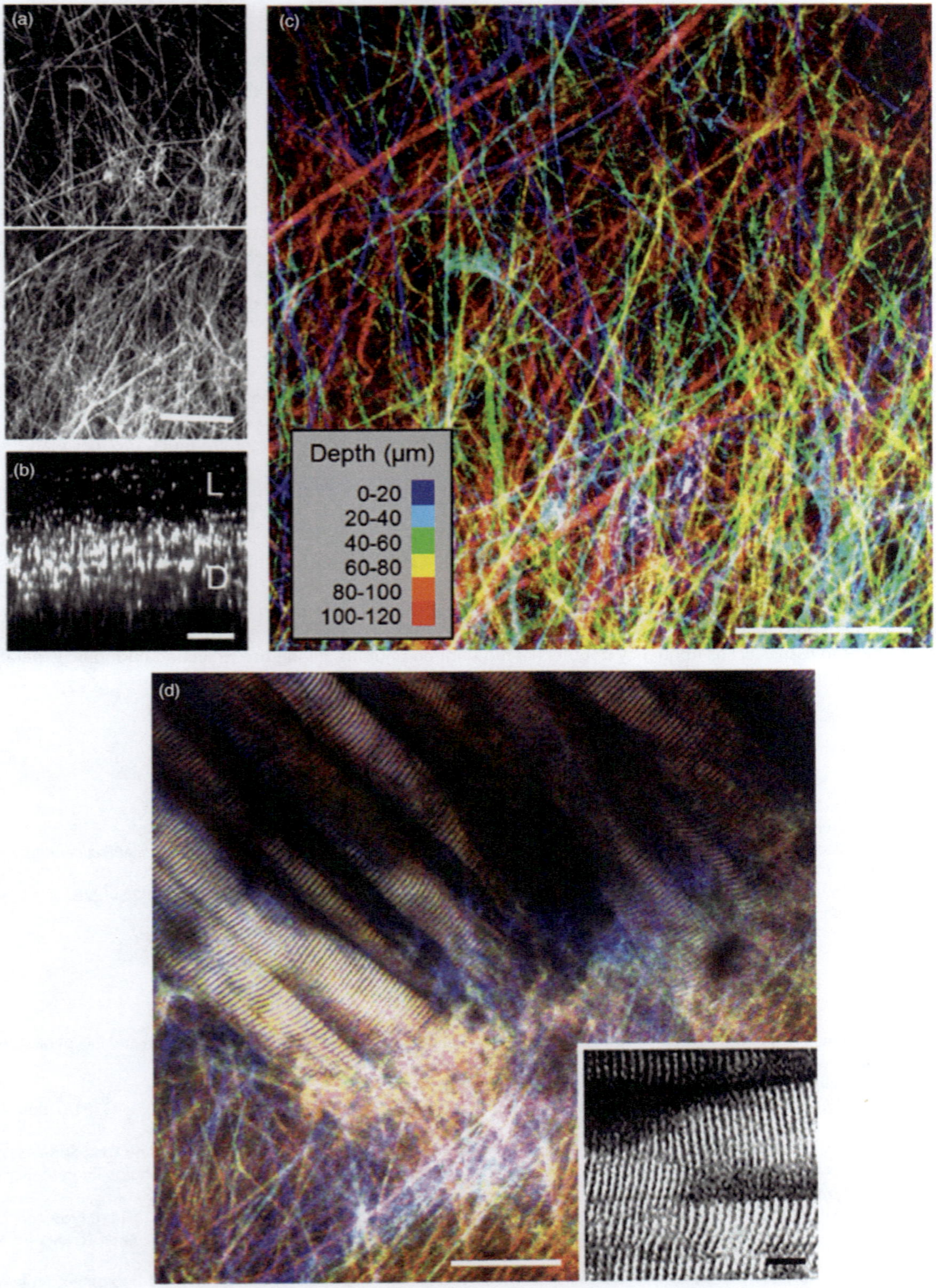

Figure 22.11 **Second harmonic generation imaging** *Intra-vital SHG of individual collagen and striated muscle fibres in mouse dermis at 800 nm. Images were collected with a 20x/NA 0.75 water-immersion objective. SHG was detected at 400 ±20 nm. (a) shows two images taken laterally in the x,y direction, (b) shows an x-z section of loose and dense dermal tissues, whilst panel (c) shows this information as a false-colour depth-coded image of penetration into the sample. (d) shows SHG detection of striated muscle and collagen-rich fascia. The scale bars are 50 μm (a–c); 30 μm D and 10 μm (d inset). Source: Friedl et al. 2007. Reproduced with permission from John Wiley & Sons.*

22.6 Chapter Summary

This chapter explains how multi-photon microscopes work and the advantages they confer over confocal microscopy. At the end of this chapter you should be aware of when to use multi-photon microscopy, of the potential dangers associated with pulsed lasers and the factors to consider in order to acquire good images.

1. Two-photon microscopy enables deep imaging into tissues with less signal scattering than with single-photon confocal microscopy. An improvement of 4–6 times is possible.
2. The absolute maximum depth penetration of the confocal into a sample is limited to about 100 μm.
3. Infrared photons are both absorbed and scattered less than visible light, so they penetrate further into tissue.
4. Red and infrared light also produce less auto-fluorescence and is less harmful to living cells than shorter-wavelength, higher-energy green and blue radiation.
5. Multi-photon excitation fluorescence depends upon the near-simultaneous absorption of two, or three, long-wavelength photons to excite the fluorophore.
6. Outside the focal point, photon intensity falls off quadratically.
7. Photobleaching and photo-damage are confined to a smaller excitation volume.
8. There are no differences in the emission spectra of fluorophores between single-photon and multi-photon imaging, but the excitation (absorption) spectral profiles differ markedly.
9. A single laser can excite multiple fluorophores because multi-photon absorption spectra are much broader than the equivalent single-photon spectra.
10. Ti:S and ytterbium lasers operate most efficiently in the 800 nm region of the spectrum. The tuning range is from around 690 to 1040 nm. The wavelength can be extended with an optical parametric oscillator up to 1300 nm.
11. Efficient signal detection can be achieved using both scattered and non-scattered photons with non-descanned detectors.
12. Point-scanning still limits acquisition speed, and careless use of the pulsed laser can still inflict photo-damage and -bleaching of the sample.
13. The laser pulse suffers chromatic group velocity dispersion. GVD is corrected by applying negative 'pre-chirp' pulse compression to the pulse to maintain signal quality.
14. Periodic structures in tissues act as frequency doublers of red/infrared photons, giving rise to second harmonic generation images. These tissues can be seen unstained.

Key Reading

Helmchen, F. and Denk, W. (2005). 'Deep tissue two-photon microscopy'. *Nature Methods 2*/12: 932–940.
- A concise and thorough review of the principles of multi-photon microscopy

Miller, D.R. et al. (2017). Deep tissue imaging with multiphoton fluorescence microscopy. *Current Opinion in Biomedical Engineering 4*: 32–39.
- A recent review of the principles of, and advances in, multi-photon microscopy

Ustione, A. and Piston, D.W. (2011). 'A simple introduction to multiphoton microscopy'. *Journal of Microscopy 243*/3: 221–226.
- A short concise introduction to multi-photon microscopy

References

Amor, A. et al (2016). Widefield two-photon excitation without scanning: live cell microscopy with high time resolution and low photo-bleaching. *PLoS One 11*/1: e0147115.

Andresen, V. et al. (2012). 'High-resolution intravital microscopy'. *PLoS One 7*/12: e50915.

Becker, K. et al. (2012). 'Chemical clearing and dehydration of GFP expressing mouse brains'. *PLoS One 7*/3: e33916. doi: 10.1371/journal.pone.0033916.

Cannell, M.B. et al. (2006). Appendix 1. In: Pawley, J.B. (ed.), *Handbook of Biological Confocal Microscopy*, 3rd ed. New York: Springer, pp. 900–905.

Charan, K., Li, B., Wang, M., Lin, C.P., and Xu, C. (2018). 'Fiber-based tunable repetition rate source for deep tissue two-photon fluorescence microscopy'. *Biomedical Optics Express 9*/5: 2304–2311.

Chen, X. et al. (2012). 'Second harmonic generation microscopy for quantitative analysis of collagen fibrillar structure'. *Nature Protocols 7*/4: 654–669.

Clendenon, S.G. et al. (2011). 'Deep tissue fluorescent imaging in scattering specimens using confocal microscopy'. *Microscopy and Microanalysis 17*/4: 614–617.

Drobizhev, M. et al. (2009). Absolute two-photon absorption spectra and two-photon brightness of orange and red fluorescent proteins. *Journal of Physical Chemistry B 113*/4: 855–859. doi: 10.1021/jp8087379.

Drobizhev, M. et al. (2011). 'Two-photon absorption properties of fluorescent proteins'. *Nature Methods 8*/5: 393–399.

Friedl, P. et al. (2007). 'Biological second and third harmonic generation microscopy'. *Current Protocols in Cell Biology*. Chapter 4: Unit 4.15.

Gadella, B.M., Haeften, T.W., van Bavel, K., and Valentijn, J.A. (2003). 'Multi-photon excitation microscopy for advanced biomedical imaging'. *Veterinary Sciences Tomorrow* 2003.

Hama, H. et al. (2011). '*Scale*: a chemical approach for fluorescence imaging and reconstruction of transparent mouse brain'. *Nature Neuroscience 14*/11: 1481–1488.

Hardin, J. (2006). 'Confocal and multi-photon imaging of living embryos'. In: Pawley, J.B. (ed.), *Handbook of Biological Confocal Microscopy*, Chapter 43, 3rd ed. New York: Springer, pp. 746–768.
Heintzmann, R. (2010). 'Correcting distorted optics: back to the basics'. *Nature Methods 7*/2: 108–110.
Herz, J. et al. (2012). 'Two-photon imaging of microbial immunity in living tissues'. *Microscopy and Microanalysis 18*/4: 730–741.
Keikhosravi, A. et al. (2014). 'Second-harmonic generation imaging of cancer'. In: Waters, J.C. and Wittman, T. (eds.), 'Quantitative Imaging in Cell Biology', Chapter 28, vol. 123. *Methods Cell Biology*. Amsterdam: Elsevier BV, pp. 531–546.
Lee, H. et al. (2014). 'Improved application of the electrophoretic tissue clearing technology, CLARITY, to intact solid organs including brain, pancreas, liver, kidney, lung, and intestine'. *BMC Developmental Biology 14*: 781–787.
Majewska, A. et al. (2000). 'A custom-made two-photon microscope and deconvolution system'. *Pflügers Archiv: European Journal of Physiology 441*/(2–3): 398–408.
Marx, V. (2014). 'Microscopy: seeing through tissue'. *Nature Methods 11*/12: 1209–1214.
McDonald, G.H. et al. (2008). 'Three-dimensional imaging of the intact mouse cochlea by fluorescent laser scanning confocal microscopy'. *Hearing Research 243*/(1–2): 1–10.
Mojzisova, H. and Vermot, J. (2011). 'When multiphoton microscopy sees near infrared'. *Current Opinion in Genetics & Development 21*/5: 549–557.
Murphy, D.B. and Davidson, M.W. (2013). *Fundamentals of Light Microscopy and Electronic Imaging*, 2nd ed. Wiley Blackwell.
Niesner, R.A. et al. (2008). 'Intravital two-photon microscopy: focus on speed and time resolved imaging modalities'. *Immunological Reviews 221*: 7–25.
Nikolenko, V. and Yuste, R. (2013). 'How to build a two-photon microscope with a confocal scan head'. *Cold Spring Harbor Protocols*. doi: 10.1101/pdb.ip075135.
Norris, G. et al. (2012). 'A promising new wavelength region for three-photon fluorescence microscopy of live cells'. *Journal of Microscopy 246*/3: 266–273.
Pantazis, P. et al. (2010). 'Second harmonic generating (SHG) nanoprobes for in vivo imaging'. *PNAS 107*/33: 14535–14540.
Richardson, D.S. and Lichtman, J.W. (2015). 'Clarifying tissue clearing'. *Cell 162*/2: 246–257.
Rocheleau, J.V. and Piston, D.W. (2003). 'Two-photon excitation microscopy for the study of living cells and tissues'. *Current Protocols in Cell Biology 20*: 4.11:4.11.1–4.11.15.
Rodriguez, C. et al. (2018). 'Three-photon fluorescence microscopy with an axially elongated Bessel focus'. *BioRxiv*. doi: 10.1101/240960.
Rosenegger, D.G. et al. (2014). 'A high performance, cost-effective, open-source microscope for scanning two-photon microscopy that is modular and readily adaptable'. *PLoS One 9*/10: e110475.
Rubart, M. (2004). 'Two-photon microscopy of cells and tissue'. *Circulation Research 95*/12: 1154–1166.
Speiss, E. et al. (2005). 'Two-photon excitation and emission of the green fluorescent protein variants ECFP, EGFP and EYFP'. *Journal of Microscopy 217*/3: 200–204.
Supatto, W. et al. (2011). 'Advances in multiphoton microscopy for imaging embryos'. *Current Opinion in Genetics & Development 21*/5: 538–548.
Tomer, R. et al. (2012). 'Quantitative high-speed imaging of entire developing embryos with simultaneous multiview light-sheet microscopy'. *Nature Methods 9*/7: 755–763.
Truong, T.V. et al. (2011). 'Deep and fast live imaging with two-photon scanned light-sheet microscopy'. *Nature Methods 8*/9: 757–760.
Ustione, A. and Piston, D.W. (2011). 'A simple introduction to multiphoton microscopy'. *Journal of Microscopy 243*/3: 221–226.
Wang, K. et al. (2015). 'Direct wavefront sensing for high-resolution *in vivo* imaging in scattering tissue'. *Nature Communication 15*/6: 7276.
Ware, L.A. (2014). 'Three photons are better than two'. *Biophotonics 57*/5: 237–239.
Weigelin, B. et al. (2016). 'Third harmonic generation microscopy of cells and tissue organization'. *Journal of Cell Science 129*/2: 245–255.
White, J.G. et al. (2001). 'Applying multiphoton imaging to the study of membrane dynamics in living cells'. *Traffic 2*/11: 775–780.
Wilson, J.W. et al. (2012). 'Optical clearing of archive-compatible paraffin embedded tissue for multiphoton microscopy'. *Biomedical Optics Express 3*/11: 2752–2760.
Wise, F. (2013). 'Lasers for nonlinear microscopy'. *Cold Spring Harbor Protocols*. doi: 10.1101/pdb.top073551.
Xu, C. and Zipfel, W.R. (2015). 'Multiphoton excitation of fluorescent probes'. *Cold Spring Harbor Protocols*. doi: 10.1101/pdb.top086116.
Zinselmeyer, B.H. et al. (2009). 'Two-photon microscopy and multidimensional analysis of cell dynamics'. In: *Methods in Enzymology*, Chapter 16, vol. 461, pp. 349–378.

Further Reading

Appleton, P.L. et al. (2009). 'Preparation of wholemount mouse intestine for high-resolution three-dimensional imaging using two-photon microscopy'. *Journal of Microscopy 234*/2: 196–204.
Benninger, R.K.P. and Piston, D.W. (2013). 'Two-photon excitation microscopy for the study of living cells and tissues'. *Current Protocols in Cell Biology*. Chapter 4: Unit 4.11.1–24.
Bestvater, F. et al. (2002). 'Two-photon fluorescence absorption and emission spectra of dyes relevant for cell imaging'. *Journal of Microscopy 208*/2: 108–115.
Bixler, J.N. (2014). 'Assessment of tissue heating under tunable near-infrared radiation'. *Journal of Biomedical Optics 19*/7: 070501.
Dempsey, W.P. et al. (2012). 'SHG nanoprobes: advancing harmonic imaging in biology'. *Bioessays 34*/5: 351–360.
Horton, N.G. et al. (2013). '*In vivo* three-photon microscopy of subcortical structures within an intact mouse brain'. *Nature Photonics 7*/3. doi: 10.1038/nphoton.2012.336.
König, K. (2006). 'Cell damage during multi-photon microscopy'. In: Pawley, J.B. (ed.), *Handbook of Biological Confocal Microscopy*, Chapter 18, 3rd ed. New York: Springer, pp. 680–689.
Masedunskas, A. et al. (2012). 'Intravital microscopy: a practical guide on imaging intracellular structures in live animals'. *BioArchitecture 2*/5: 143–157.
Molina, R.S. et al. (2017). 'Blue-shifted green fluorescent protein homologues are brighter than enhanced green fluorescent protein under two-photon excitation'. *Journal of Physical Chemistry Letters 8*/12: 2548–2554.
Mütze, J. et al. (2012). 'Excitation spectra and brightness optimization of two-photon excited probes'. *Biophysical Journal 102*/4: 934–944.
Ntziachristos, V. (2010). 'Going deeper than microscopy: the optical imaging frontier in biology'. *Nature Methods 7*/8: 603–614.

Ouzounov, D.G. et al. (2017). In vivo three-photon imaging of activity of GCaMP6-labeled neurons deep in intact mouse brain. *Nature Methods 14*/4: 388–390.
Reeve, J.E. et al. (2010). Dyes for biological second harmonic generation imaging. *Physical Chemistry Chemical Physics 12*/41: 13484–13498.
Rowlands, C.J. et al. (2017). Wide-field three-photon excitation in biological samples. *Light: Science & Applications 6*: e16255. doi: 10.1038/lsa.2016.255.
Salomonnson, E. et al. (2012). 'Cell-based and in vivo spectral analysis of fluorescent proteins for multiphoton microscopy'. *Journal of Biomedical Optics 17*/9: 96001.
Sofroniew, N.J. et al. (2016). 'A large field of view two-photon mesoscope with subcellular resolution for in vivo imaging'. *eLife 14*: 5. pii: e14472.
Tillo, S. et al. (2010). 'A new approach to dual-color two-photon microscopy with fluorescent proteins'. *Biotechnology 10*: 6.
Wang, B.G. et al. (2010). 'Two-photon microscopy of deep intravital tissues and its merits in clinical research'. *Journal of Microscopy 238*/1: 1–20.
Yang, X. et al. (2012). 'Two-photon microscopy of the mouse cochlea *in situ* for cellular diagnosis'. *Journal of Biomedical Optics 18*/3: 031104. doi: 10.1117/1.JBO.18.3.031104.
Zipfel, W.R., Williams, R.M., and Web, W.W. (2003). 'Nonlinear magic: multiphoton microscopy in the biosciences'. *Nature Biotechnology 21*/11: 1369–1377.

Internet Resources

Zeiss Multi-photon dye sheet: https://bit.ly/2LMN8Kq.
Thermo Fisher Scientific list of multi-photon probes: https://www.thermofisher.com/sg/en/home/references/molecular-probes-the-handbook/technical-notes-and-product-highlights/fluorescent-probes-for-two-photon-microscopy.html#.
Objective inverter (LSM Tech): http://www.lsmtech.com/product_objective_inverter.html.
LaVision Biotec Doubleheader: https://bit.ly/2LAzMUM https://www.lavisionbiotec.com/products/doubleheader.html
Thorn et al., Clearing and Mounting media: http://nic.ucsf.edu/dokuwiki/lib/exe/fetch.php?media=mounting_media_tests.pdf.
Multi-photon microscope users' group: https://b.gatech.edu/2OflsiX.
The Cornell website for 2-P spectra: https://zipfellab.bme.cornell.edu/cross_sections.html.
Daniel Fiole's spectral database: http://daniel.fiole.free.fr/twophotondyes.
Ludwig-Maximilians-Universität resource: https://bit.ly/2QQgKME.
SVI Nyquist calculator: http://www.svi.nl/NyquistCalculator and http://www.svi.nl/NyquistRate.
For calculating the ideal Nyquist sampling rate for multi-photon data you must select the confocal mode on the SVI Nyquist calculator and then set the number of excitation photons on '2'.

Notes

1. Because of the very large spectral cross section of most fluorophores, it is more usual to image DAPI by two-photon excitation. This is because DAPI is used as a counterstain in conjunction with longer-wavelength dyes. A downloadable PDF of suitable multi-photon dyes and appropriate excitation wavelengths is produced by Carl Zeiss; similar information is on a Thermo Fisher Scientific web page. The URLs are given at the end of the chapter.
2. The Ti:S designation refers to the lasing medium, a crystal of sapphire doped with titanium ions. The Ti:S laser is pumped with a green laser in the region from 514 to 532 nm. See Wise (2013) for further details of lasers for multi-photon microscopy.
3. The femtosecond pulse may be very short (10^{-15} s), but compared to the timescale of the absorption event of a typical fluorophore (10^{-18} s), a 100 fsec pulse is 1000x longer.
4. The peak intensity of light emitted by a femtosecond laser is 10^{14} W/m^2, whereas the average intensity is 10^6 W/m^2. The equivalent intensity of light (solar irradiance) on earth from the surface of the sun is 10^4 W/m^2; the peak power of a 1 mW continuous laser is 10^8 W/m^2, and its average power is 10^2 W/m^2. The pulse repetition rate of 80 MHz, though well-suited for biological imaging, was not chosen with that application in mind. The first demonstration of a mode-locked Ti:Sapphire laser had a repetition rate of close to 80 MHz because that's the frequency at which the AOM mode-locking device operated. Others repeating this work had no cause to change the frequency, and when it was discovered to be at a frequency useful for biological imaging, it stuck.
5. In single-photon confocal microscopy, excitation photons that are scattered in the sample will cause background fluorescent haze, particularly as the laser power is increased to try and image deeper into the tissue sample. In single-photon LSCM, the emitted photon can also be scattered as it returns through the sample. The scattered photon will not go through the pinhole, and similarly unintended photons *will* by scattering erroneously enter the detector (this is analogous the formation of the halo in phase contrast microscopy) so that as one penetrates deeper into tissue, the resulting background haze obscures any resolved detail, contrast is reduced and visibility lost.
6. In single-photon confocal microscopy, the emitted signal is passed back along the same route (i.e. via the scanning mirrors; descanned detection) as the excitation pathway to block out-of-focus emission – this is unnecessary in multi-photon microscopy. It is possible to use a confocal PMT to collect the two-photon signal, but the pinhole and scatter from the surfaces of the mirrors reduce signal intensity.
7. A chirp is a signal in which the frequency increases ('up-chirp') or decreases ('down-chirp') with time.
8. This inverter arm device allows conversion of an inverted system into an upright one, but at the inevitable cost of the loss of some excitation and emission light. This situation can be redeemed to a certain extent by using high quantum efficiency GaAsP (gallium-arsenide-phosphide) photocathode detectors. These GaAsP detectors are at least twice as efficient (up to 40%) than regular PMT designs, and unlike PMTs are sensitive into the far red (ca. 700 nm) region of the spectrum.

23

Total Internal Reflection Fluorescence Microscopy

Outline

23.1 Introduction

Like multi-photon microscopy covered in the previous chapter, total internal reflection microscopy (TIRF or sometimes TIRFM) is a niche technique designed for specific applications. TIRF is characterised by superb axial optical sectioning along the z-axis with concomitant high signal-to-background (SBR) ratios. A TIRF microscope captures a full field rather than scanning the field of view point by point. Its popularity has increased recently and turnkey systems are now available with high-quality, high-NA, objectives that make TIRF microscopy much easier to perform than when it was first introduced in the early 1980s by Daniel Axelrod.

Safety Warning:

Some designs of TIRF microscope, especially if 'home-built', may have the laser illuminating beam exposed during routine operation. Also, the design may entail 'walking' the laser illumination from an exposed position (usually vertically along the optical axis out of the objective of an inverted microscope) directed onto the ceiling then down horizontally and thus through eye level into a totally internally reflected configuration for use.

Understanding Light Microscopy, First Edition. Jeremy Sanderson.

Great care must be taken, particularly when the laser beam is at eye level, not to allow laser light to enter your eyes. You should be aware of, and adhere to, the local laser safety rules of your institution and use the appropriate personal protective equipment.

23.2 The Phyiscal Principles of TIRF Microscopy

Light hitting a dielectric[1] surface is partially reflected and refracted. Snell's law (Chapter 2, Section 2.9, and Box 2.3) governs the direction of propagated light, whereas the intensities of the transmitted and reflected components are determined by Fresnel's coefficients (Chapter 12, Box 12.1) for transmission and reflection, respectively.

In TIRF microscopy, this dielectric interface occurs where the glass coverslip and the watery medium of the cell or sample are adjacent to one another. The lower refractive index of the cell, or liquid medium, is denoted as n_1 and the higher, denser, RI of the glass as n_2. (This convention is not strictly applied. In some books and papers, the notations may be reversed: n_1 applying to the higher RI of the glass substrate and n_2 to the lower RI of the cell constituents.) At the critical angle, θ_c, measured relative to the optical axis normal to the surface, the value of θ is such that $\theta_{incident} = 90°$, the beam of light is refracted parallel to the surface and therefore no light propagates through to the other side.

If the angle of incidence, θ_i, is less than θ_c, as is usual during routine epifluorescence microscopy, the vast majority of the excitation light passes through the coverslip-sample interface into the sample (Figure 23.1a) leading to widefield illumination. When θ_i exceeds the critical angle, the incident light is totally internally reflected off the interface and back through the thickness of the coverslip (Figure 23.1b, see also Chapter 2, Section 2.10, and Figure 2.16 and Figure 2.17).

When total internal reflection occurs, a standing wave propagates normal to the interface (because light interactions are quantum-mechanical events) through the medium of lower refractive index and into the sample. This creates an evanescent field, which is due to 'near-field' effects as predicted by Maxwell's equations[2]. This evanescent wave has the same frequency, wavelength and polarity as the incident illumination. The intensity of this wave decreases exponentially from the interface. In practice it extends between

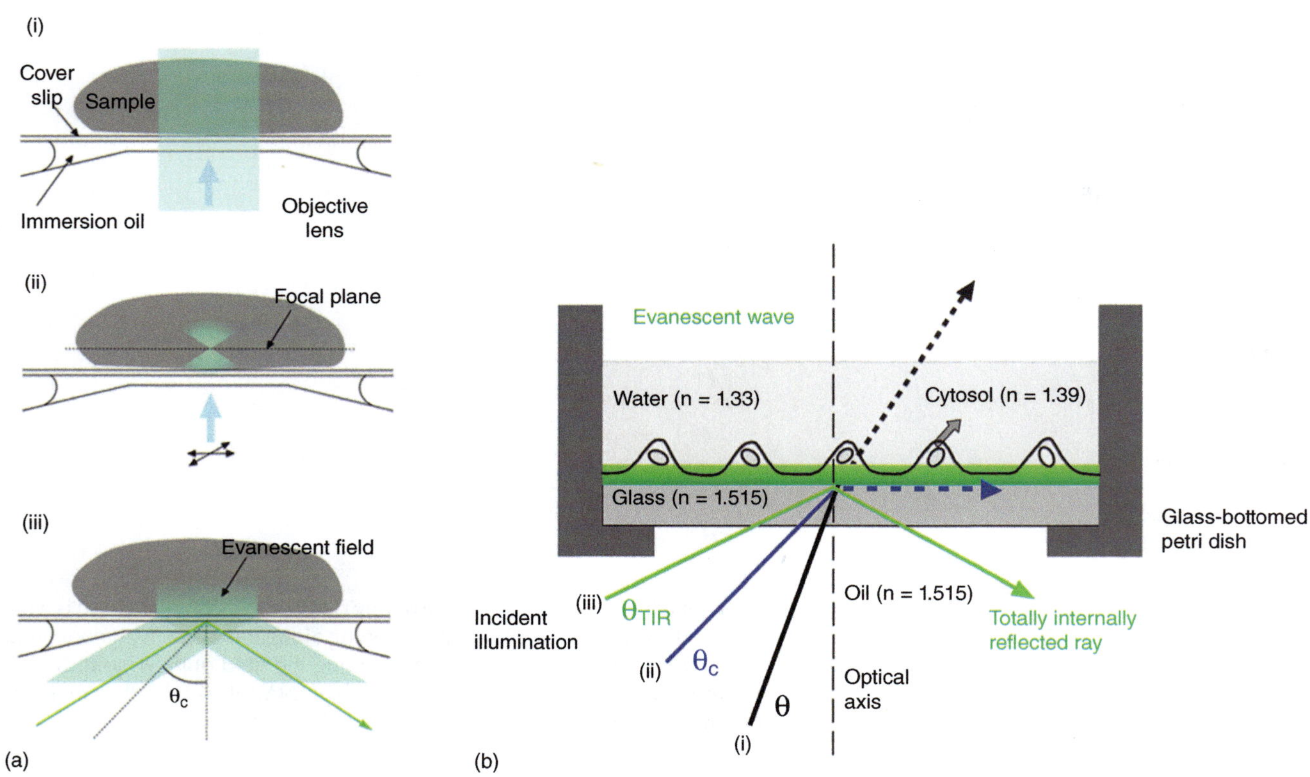

Figure 23.1 **Principles of TIRF microscopy** *(a) In routine fluorescent imaging, (i) the beam of illumination is usually incident at 90° to the preparation, lighting up the entire sample. In laser-scanning confocal microscopy, (ii) the illumination is constrained to the focal plane by means of a pinhole to optically section the sample. Total internal reflection microscopy (iii) generates a very thin evanescent field (not drawn to scale!), which decays exponentially within a few hundred nanometres (<400nm) of the coverslip-sample interface, to optically section the sample. Source: Vizcay-Barrena et al. 2011. Reproduced with permission from Oxford University Press.*

(b) Where the angle of the incident illumination, θ (shown by the solid black line), is less than the critical angle, θ_c (shown by the solid blue line), most of the excitation light passes into the sample. When θ exceeds the critical angle, the incident light is totally internally off the interface (since $\theta_{TIR} > \theta_c$) and back through the thickness of the coverslip (n_2). A thin evanescent wave is formed, which is used to image the sample (n_1). The NA of the objective used for TIRF microscopy must be greater than n_1. Source: Adapted from Jaiswal and Simon 2003 with permission from John Wiley & Sons.

approximately 60 and 200 nm beyond the upper surface of the coverslip into the sample. TIRF only enhances the axial resolution to below the resolution limit: the lateral limit of resolving power remains the same as for widefield microscopy.

Since the evanescent field is so thin, it provides superb optical sectioning and a very high signal-to-background ratio. Any signal or autofluorescence arising from cell components several micrometres away from the surface is simply neither illuminated nor detected. Because of the extreme thinness of the TIRF optical section and the fact that moving particles (i.e. within living cells) can move in and out of it rapidly, interpreting and analysing TIRF images is not as straightforward as other modes of fluorescence microscopy. TIRF is particularly useful when studying cell motility and attachment, vesicle-recycling events or other dynamic processes that occur at the surface of, or just within, the cell membrane.

TIRF has been little used for botanical study, generally because of the thick nature of individual cells and the range of differing refractive indexes found in plants. For a useful review see Vizcay-Barrena et al. (2011). However, it is possible, particularly with prism-based TIRF configurations (see below) to alter the illumination angle of the laser to change between TIRF and oblique fluorescence microscopy. This method is also called variable-angle epifluorescence microscopy (VAEM), with a specific version called HILO (highly inclined laminated optical sheet) microscopy (Tokunaga et al., 2008). With VAEM the narrow band of illumination generated will optically section the sample very close to the coverslip, but this illumination is not an evanescent field, yet VAEM will image structures lying deeper within the sample than TIRF microscopy can achieve.

Clearly, TIRF is not suitable for stuying non-adherent cell types. It may be necessary to pre-treat the coverslips with poly-L-lysine (Chapter 28, Section 28.5) to ensure adherence of the cells. TIRF can be combined with bleaching techniques, such as FRAP and FRET. Although a widefield technique, a TIRF attachment can be added to a spinning disc confocal to be used when needed, rather than having two microscope platforms.

An important source of background is fluorescence from sources other than the target molecules, which will reduce the signal-to-background ratio (Vizcay-Barrena et al., 2011). Some glass used for microscope slides and coverslips is inherently autofluorescent due to impurities within the glass, and this can pose a significant problem, particularly with prism designs of TIRF microscope. Most modern immersion oil used for fluorescence microscopy is of a low-fluorescence formulation and does not generally cause problems. However, if any medium (e.g. nitrocellulose) is used to immobilise and anchor cells or single molecules onto the surface of the coverslip, this should be checked for autofluorescence.

23.3 Designs of TIRF Microscope

The evanescent wave can be generated through three different arrangements in the microscope (Figure 23.2). In Figures 23.2a, 23.2b and 23.2c, a prism is used to reflect the wave through the sample from either the top or the bottom of the sample. Prisms allow the

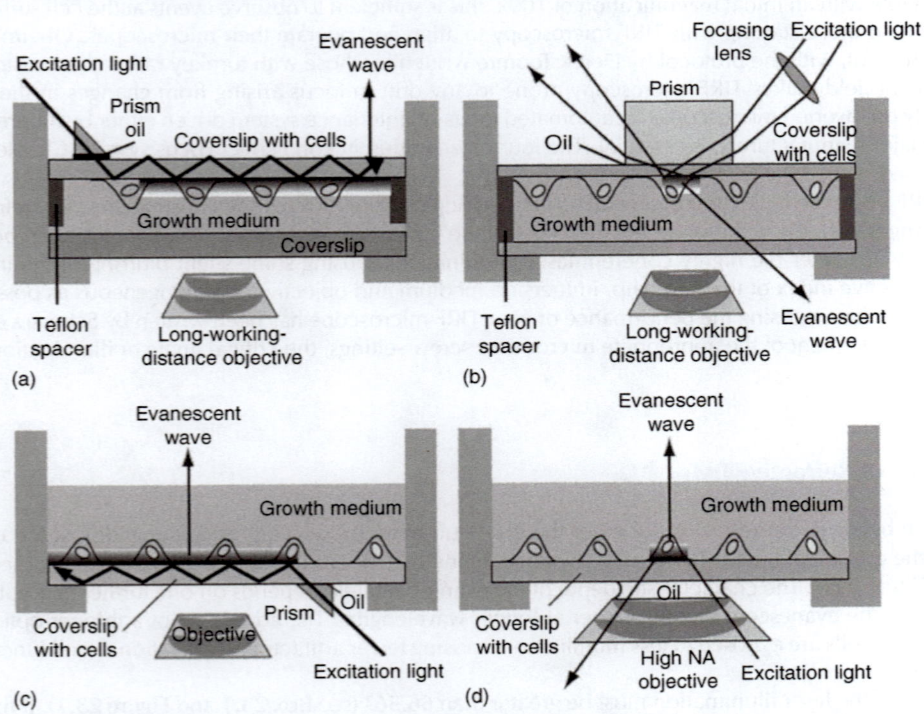

Figure 23.2 **Designs of TIRF microscope** *This figure shows four possible arrangements for achieving total internal reflection fluorescence. The prism designs are relatively cheap and give a high SNR and a cleaner evanescent field, with less scattered light than the objective-based design. Prism systems also allow for 'routine' objectives to be used, which are not of extremely high numerical aperture. However, the prism designs are challenging to set up and align optimally. Microscopes employing specially designed high-NA objectives are routinely used for TIRF. They are easy to align and use and can be switched very easily to widefield fluorescence mode, to allow comparison images to be acquired. Source: Jaiswal and Simon 2003. Reproduced with permission from John Wiley & Sons.*

incidence angle of the laser beam to be varied widely and give a better signal-to-noise ratio than when using a high–numerical aperture objective, but they are technically much more challenging to set up and can restrict sample accessibility. On an inverted microscope, the laser beam is exposed, so great care must be taken (and health and safety regulations adhered to) when using this method. With the top prism design (Figures 23.2a and 23.2b), a separate objective (often a long working distance design) is used to view the evanescent field from the opposite side. With a prism used from the underside (Figure 23.2c), a higher NA objective, usually a 'high-dry' objective, is used, because any oil- or water-immersion coupling will destroy the TIRF effect. Such an objective will require very careful setting of the correction collar to minimise spherical aberration.

In the last arrangement shown in Figure 23.2d, a single very high–numerical aperture objective is used. This latter configuration is nowadays the configuration of choice, because it is easier to align and switch between TIRF and standard fluorescence. Some microscopes have the TIRF unit as a separate slider, which can be inserted as required. Using an objective rather than a prism also allows a stage environmental chamber and any physiology equipment (such as a patch clamp electrode) to be used above the stage on an inverted microscope. The numerical aperture of the objective *must* be greater than the refractive index of the sample (NA > n_1), and so objectives of NA 1.4 (better: NA 1.45, 1.49 or 1.65; Table 23.1) and greater are normally employed for TIRF microscopy. It is easier to align a microscope for TIRF imaging with higher NA objectives. An inverted microscope is normally used: the illumination enters the sample from the outer edge, on one side of the front lens of the objective, at such an oblique angle that it is totally internally reflected off the underside of the coverslip. The angle of incidence is generally controlled by hand with a micrometer gauge or automatically through software.

23.4 Practical Considerations

Although an NA 1.4 objective is theoretically suitable for TIRF, only ≈2% of the peripheral aperture can be used to achieve total internal reflection of the laser beam. The critical angle, θ_c, is only just surpassed (≈2.8° beyond θ_c), whereas with an NA 1.49 objective ≈9% of the outer periphery of the aperture can be used, and with an objective of 1.65, the figure is ≈20%. The NA 1.45 and NA 1.49 objectives can be used with standard coverslips (n_D = 1.515) and immersion oil (n_D = 1.518), but the NA 1.65 objectives require expensive fused silica 'quartz' coverslips (RI = 1.788) and volatile, toxic, immersion oil (RI = 1.78, which passes blue light poorly). Both coverslips and oil are available from Cargille or Olympus (e.g. Ohara LAH64 coverslips of 0.150 mm, or the Schott glass SF11 coverslips, again 0.150 mm thick). For multicolour experiments use the Ohara glass coverslips, which are less autofluorescent and exhibit less chromatic error.

The high-NA 1.65 objective is easier to align than one of lower NA, gives a brighter and more evenly illuminated evanescent field. Although it is only available with an initial magnification of 100x, this is sufficient to observe events at the cell surface. There are several protocols published to help those starting with TIRF microscopy to align and operate their microscopes. Oheim (2001b) and Toomre (2012) are particularly helpful, with the protocol by Derek Toomre written for those with turnkey rather than custom-built systems. The thinness of the evanescent field makes TIRF microscopy prone to any drift in focus arising from changes in the local environmental temperature. Particularly on inverted microscopes an automated focus-maintenance system driven either by infrared laser or white light, available from all the major manufacturers, is essential. The fourth generation Nikon Perfect Focus system (Chapter 28, Section 28.6) is particularly recommended.

Several designs of TIRF objective have correction collars fitted to adjust for refractive index mismatches, any refractive index changes due to temperature changes and any variation in the thickness of the coverslip. Any refractive index mismatch between the specimen and surrounding media will scatter the highly coherent laser illumination, causing some slight blurring of the image. It is important, therefore to keep the refractive index of the coverslip, immersion medium and objective as homogeneous as possible.

An excellent chapter on characterising the performance of your TIRF microscope has been written by Schwarz et al. (2011). It guides you through practical measurement of the appropriate micrometer screw settings, the critical angle of illumination and the evanescent field depth.

23.5 The Importance of Refractive Index

The larger the difference between the refractive index of the glass substrate (n_2) and the refractive index of the cell/medium (n_1), the thinner is the depth of the evanescent field. The greater the angle of incident illumination, the thinner the evanescent field, also. This is shown in the graph in Figure 23.3. The characteristic depth of the evanescent field depends on one further factor: the wavelength of the illumination. The depth of the evanescent field is greater at longer wavelengths. The advantage of a thinner optical section[3] from the evanescent field means that cells are exposed to less illumination, causing fewer artifacts and alteration to cell function and morphology associated with photo-toxicity.

The angle of incidence of the laser illumination must be greater than 66.56° (see Box 23.1 and Figure 23.1). If fixed or permeabilised cells are used, then the effective refractive index of the medium is reduced to RI 1.33, and the critical angle of incidence to achieve total internal reflection is reduced to 61.42°. When preparing fixed cell samples, do not use a mounting medium that hardens or contains glycerol but an aqueous one, such as phosphate-buffered saline. Using a higher RI mountant, which approaches the refractive index of the glass substrate at the interface, demands a higher angle of incident illumination to overcome the critical angle and so will eventually render TIRF impossible to perform.

Table 23.1 Characteristics of TIRF Objectives

Representative objective	NA	Working distance (mm)	Immersion medium	Coverslip thickness	n_1 of cell cytosol	n_2 of coverslip	$\sin\theta$ (NA/n_2)	Objective half-angle, α	Critical angle, $\theta_c = \sin^{-1}(n_1/n_2)$	Peripheral angle available for TIRF	% difference of objective BFP for TIRF, based on $n_1 = 1.33$ (fixed cells in water or medium)	Typical minimum depth of evanescent field (based on $n_1 \approx 1.37$)
 Mag = 150x	1.45	0.08	Standard oil	0.13–0.19 Correction collar	1.37	1.515	0.9571	73.16°	64.73°	8.4°	9%	≈100 nm
 Mag = 60x or 100x	1.49	0.1	Standard oil	0.13–0.19 Correction collar	1.37	1.515	0.9835	79.58°	64.73°	14.9°	12%	≈70 nm
 Mag = 100x	1.70	0.08	Proprietary high RI oil	0.15 Correction collar	1.37	1.788	0.9508	71.95°	50.02°	17.3°	24%	≈40 nm

Note: A representative objective typical for each NA class is shown. Olympus specialise in a range of high-NA TIRF objectives, so they have been used to illustrate this table. (Image source: Olympus Europa SE & Co. KG.) This is for illustrative purposes only, and it is not intended to confer product endorsement to the detriment of other manufacturers' objectives. Nikon have 60x and 100x NA 1.49 TIRF objectives; Leica have a 100x/NA 1.46 objective and Zeiss manufacture 60x and 100x TIRF objectives of NA 1.45 and NA 1.46. For details of these, see the appropriate manufacturers' websites.

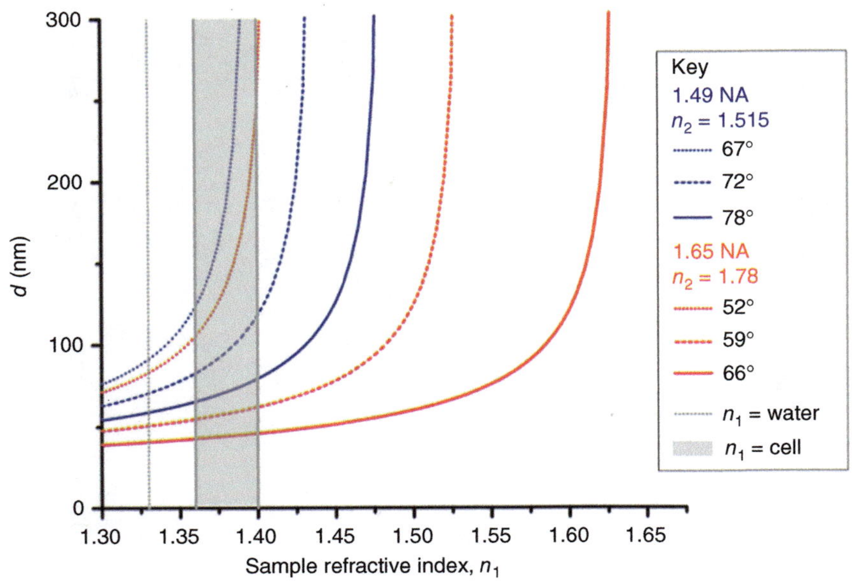

Figure 23.3 ***TIRF evanescent field depth*** *Graph showing typical evanescent field depths for an objective of NA 1.49 (blue lines) and NA 1.65 (red lines). Incidence angles were chosen at 1.5° above the critical angle (dotted lines), the midpoint of all possible TIRF angles (dashed lines), and 1.5° less than the maximum possible incidence angle (solid lines). As the angle of incidence is increased, the sensitivity of* d *to the sample refractive index decreases in the range for cells. The NA 1.65 objective has a larger range with lower sensitivity to sample refractive index than the NA 1.49 objective. (Text taken directly from Figure 3, page 3624 in Mattheyses et al., 2010.* Jour. Cell Science *123, 3621–3628.) Source: Johnson et al. 2012. Reproduced with permission from John Wiley & Sons.*

Box 23.1 The mathematics of TIRF microscopy

1. From Snell's law: sin (i) /sin (r) = n_1/n_2, and in this case sin θ/ sin (r) = n_1/n_2.
   The critical angle, θ_c = arcsin (n_1/n_2), which can also be written as $\theta_c = \sin^{-1}(n_1/n_2)$.
   Total internal reflection occurs when $n_2 > n_1$, so that: $[n_1/n_2]/\sin\theta \geq 1$, then angle r must exceed 90°, and light cannot propagate through the interface into the medium, since $\theta > \theta_c$, where $n_1/n_2 < 1$. In the case of TIRF microscopy, n_1 refers to the sample refractive index and n_2 to the RI of the coverslip.
2. The penetration depth of the evanescent field[i] depends upon three factors: the wavelength of the illuminating radiation, λ; the angle of incidence, θ_i, and the ratio of the refractive indexes n_1 and n_2. The equation describing the thickness of the evanescent field and hence the optical section generated by TIRF microscopy is: depth = $(\lambda/4\pi).\ \sqrt{[n_2^2 \sin^2\theta - n_1^2]}$,
   which is sometimes written: depth = $(\lambda/4\pi).\ \left(NA_i^2 - n_1^2\right)^{-1/2}$, or $d = \lambda(4\pi\ n_2).\ ([\sin^2\theta/\sin^2\theta_c - 1])^{-1/2}$,
   where λ = wavelength of the illumination; $NA_i = n_2 .\ \sin\theta$; θ = angle of incidence; θ_c = critical angle.
   The larger the difference between NA_i and n_2, the smaller is the value d for the penetration depth of the evanescent field.
3. The intensity of fluorescence at any distance, z, from the interface is given by: $I_z = I_0 e^{-z/d}$
   where I_z = intensity at point z, I_0 = intensity of fluorescence at the interface and d = depth of the evanescent field.
4. The refractive index (RI) of the cellular medium is around 1.33 for water but can lie anywhere between 1.332 and 1.334. The RI of the cell cytosol is around 1.36–1.40. In the following calculations, a value of 1.39 is assumed for the living cell cytosol (n_1) and a value of 1.515 for the glass coverslip (n_2). To achieve total internal reflection, the angle of incidence must therefore be greater than arcsin(1.39/1.515), or >66.56°. To be useful for TIRF, the NA of the objective must be greater than n_1.

[i] A macro plugin is available for ImageJ to calculate the evanescent field penetration depth: http://rsbweb.nih.gov/ij/plugins/tirf/index.html. For proof of the theory of TIRF, see (a) Axelrod, D., Burghardt, T.P., and Thompson, N.L. (1984). 'Total Internal Reflection Fluorescence'. *Ann. Rev. Biophys. Bioeng. 13*: 247–268, (b) Gingell, D., Heavens, O.S., and Mellor, J.S. (1987). 'General electromagnetic theory of total internal reflection fluorescence: the quantitative basis for mapping cell-substratum topography'. *Jour. Cell Sci. 87*/5: 677–693.

23.6 Multicolour TIRF

TIRF captures a full field rather than scanning the field of view point by point. Normally an EM-CCD camera is used for high sensitivity in low light, but this is not mandatory, and a CCD camera can be used, depending upon the sample and intensity of the emitted signal. An image splitter, such as those from Cairn Research, Hamamatsu and Optical Insights, is useful for imaging two or more fluorophores with a multicolour TIRF configuration. As usual with image-splitting devices, you must ensure that you achieve perfect image registration when overlaying two spectrally distinct channels. Use multicolour beads (e.g. 0.1 μm TetraSpeck beads (T7279) from Thermo Fisher Scientific) for this.

It is important to be able to change the incident angle of the illumination to control the penetration depth for multicolour TIRF microscopy with different fluorophores. Blue light has a smaller critical angle than red light because the refractive index of blue light is greater than the refractive index of red. As the refractive index increases, the critical angle decreases. Performing multicolour TIRF experiments is not trivial, because as the wavelength increases, the penetration of the evanescent field also increases, and the longer the wavelength, a greater incidence angle is required to maintain the same depth of evanescent field. An ImageJ plug-in for calculating penetration depth is given in the Internet Resources at the end of the chapter. Fish (2009) has written a good practical guide to the principles governing multicolour TIRF. For further details, particularly with the practical aspects of troubleshooting TIRF systems – which is beyond the scope of this book – you are referred to Toomre (2010), Johnson, Jaiswal and Simon (2012), Mattheyses et al. (2010) and the two review papers by Oheim (2001a and 2001b).

23.7 Chapter Summary

In this short chapter, the principles of TIRF microscopy are explained. The principal advantage of TIRF is the very thin optical sectioning capability. The flip side reveals its prime restriction: the technique only reveals fluorescent events at the interface between the cell membrane and supporting substrate.

1. The evanescent wave extends no more than 200 nm beyond the upper surface of the coverslip into the sample.
2. TIRF microscopy can be achieved using a prism to totally internally reflect the laser illumination. It is more common to use a very high-NA apochromat objective to achieve the same purpose.
3. TIRF microscopy is very prone to any drift in focus.
4. The efficiency of TIRF microscopy is dependent upon the refractive index difference between the sample and substrate forming the interface.
5. Multicolour TIRF is possible; longer wavelength illumination gives a greater depth of evanescent field but requires a more oblique inicidence angle than for shorter-wavelength illumination.

Key reading

Martin-Fernandez, M.L., Tynan, C.J. and Webb, S.E.D. (2013). 'A "pocket guide" to total internal reflection fluorescence'. *Journal of Microscopy 252*/1: 16–22.

- A clear concise explanation of TIRF microscopy published in the RMS pocket guide series

References

Jaiswal, J.K. and Simon, S.M. (2003). 'Total internal reflection fluorescence microscopy for high-resolution imaging of cell-surface events. *Current Protocols in Cell Biology*. Chapter 4: Unit 4.12.1–4.12.15.

Johnson, D.S., Jaiswal, J.K., and Simon, S.M. (2012). 'Total internal reflection fluorescence (TIRF) microscopy illuminator for improved imaging of cell surface events'. *Current Protocols in Cytometry*. Chapter 61: Unit 12.29.

Mattheyses, A.L. et al. (2010). 'Imaging with total internal reflection fluorescence microscopy for the cell biologist'. *Journal of Cell Science 123*/21: 3621–3628.

Oheim, M. (2001a). 'Imaging transmitter release. I. Peeking at the steps preceding membrane fusion'. *Lasers in Medical Science 16*/3: 149–158.

Oheim, M. (2001b). 'Imaging transmitter release. II. A practical guide to evanescent-wave imaging'. *Lasers in Medical Science 16*/3: 159–170.

Schwarz, J.P., König, I., and Anderson, K.I. (2011). 'Characterizing system performance in total internal reflection fluorescence microscopy'. In: Wells, C.M. and Parsons, M. (eds.), *Cell Migration: Developmental Methods and Protocols*, Methods in Molecular Biology, Chapter 25, vol. 769. New York: Humana Press, pp. 373–386.

Tokunaga, M., Imamoto, N., and Sakata-Sogawa, K. (2008). 'Highly inclined thin illumination enables clear single-molecule imaging in cells'. *Nature Methods 5*/2: 159–161.

Toomre, D. (2010). 'Cellular imaging using total internal reflection microscopy'. In: Goldman, R.D., Swedlow, J.R., and Spector, D.L. (eds.), *Live Cell Imaging: A Laboratory Manual*, Chapter 36, 2nd ed. New York: Cold Spring Harbor Laboratory Press, pp. 675–696.

Toomre, D. (2012). 'Alignment and calibration of total internal reflection fluorescence microscopy systems'. *Cold Spring Harbor Protocols 2012*/4: 504–509. doi: 10.1101/pdb.prot068668.

Vizcay-Barrena, G. et al. (2011). 'Subcellular and single-molecule imaging of plant fluorescent proteins using total internal reflection fluorescence microscopy (TIRFM)'. *Journal of Experimental Botany 62*/15: 5419–5428.

Further Reading

Fish, K.N. (2009). 'Total internal reflection fluorescence (TIRF) microscopy'. *Current Protocols in Cytometry*. Chapter 12: Unit 12.18.

Simon, S.M. (2009). 'Partial Internal reflections on total internal reflection fluorescent microscopy'. *Trends in Cell Biology 19*/11: 661–668.

Steyer, J.A. and Almers, W. (2001). 'A real-time view of life within 100nm of the plasma membrane'. *Nature Reviews Molecular Cell Biology 2*/4: 268–275.

Toomre, D. and Manstein, D.J. (2001). 'Lighting up the cell surface with evanescent wave microscopy'. *Trends Cell Biology 11*/7: 298–303.

Toomre, D. (2012). 'Cellular imaging using total internal reflection fluorescence microscopy: theory and instrumentation'. *Cold Spring Harbor Protocols 2012*/4: 414–424. doi: 10.1101/pdb.top068650.

Internet Resources

TIRF microscopy – Fluorescence Foundation: http://www.fluorescence-foundation.org/lectures%5Cgenova2009%5Clecture8.pdf.
Penetration depth calculator: https://imagej.nih.gov/ij/plugins/tirf/index.html.
Objective parameter calculator (for critical angles for TIRF): https://imagej.nih.gov/ij/plugins/calc-objective/index.html.

Notes

1. A dielectric is a non-conductor, or poor conductor, of electric current. For practical applications, most dielectric materials are solid insulating materials such as ceramics, mica and glass, but a vacuum is also an efficient dielectric as is any dry gas, including air.
2. The mathematical theory is given in the references within Box 23.1.
3. A wedge design testplate of fluorescent beads lying on an incline can be used to measure the penetration depth of the evanescent field. For further details, see: Figure 5, page 49 in: Fiolka, R., et al. (2008). 'Even illumination in total internal reflection fluorescence microscopy using laser light'. *Microsc. Res. Tech. 71*/1: 45–50. See also Fish, K. N. (2009) and Schwarz (2011) cited in the Further Reading section, for other practical methods of measuring the depth of the evanescent field.

24

FRAP and FRET

Outline

24.1 Introduction
- Photobleaching
- Definitions

24.2 FRAP – Fluorescence Recovery After Photobleaching
- Setting FRAP parameters
- FRAP protocol
- A FRAP test

24.3 FLIP and FLAP
- Fluorescence loss in photobleaching
- iFRAP – inverse FRAP
- Fluorescence localisation after photobleaching
- Photo-activation
- Fluorescence correlation spectroscopy

24.4 FRET – Förster Resonance Energy Transfer
- FRET efficiency
- R_0 – the Förster radius
- Requirements for FRET & FRET pairs
- Methods for measuring FRET
- SBT – Spectral bleed-through

24.5 FLIM – Flourescence Lifetime Imaging MIcroscopy
- How fluorescence lifetime microscopy works

24.6 Chapter Summary
- 13 points

Key Reading
References
Further Reading
Internet Resources

24.1 Introduction

When teaching fluorescence imaging, it is generally instilled into new microscopists that bleaching is detrimental and should be avoided as far as possible. However, when used in a controlled and known manner, selective bleaching of fluorophores can be very useful indeed for revealing the rate and extent of movement, by diffusion or transport, of fluorescently labelled proteins and other labelled molecules. FRAP and its corollary, FLIP, are the most commonly used methods. FRET is used to detect molecular associations and interactions, although it is a more complex method for which to prepare specimens and to manage image acquisition. It will therefore only be described briefly here.

Understanding Light Microscopy, First Edition. Jeremy Sanderson.

Definitions

FRAP
Fluorescence Recovery After Photobleaching. A fluorescent label (e.g. GFP) is attached to the molecules of interest. Photobleaching is performed through short intensive laser illumination of selected regions that results in a localised destruction of fluorophores. After the bleach pulse, a recovery of fluorescence within the bleached region can be observed if mobile photo-bleached molecules exchange with non-bleached molecules from a neighbouring area. The kinetics of the fluorescence recovery measured within the bleached region provides a quantitative measure of the mobility of the labelled molecules.

FLIP
Fluorescence Loss In Photobleaching. Labelling is performed as in FRAP experiments. Repetitive bleaching of the same region of interest leads to a successive depletion of fluorescence in continuous cellular structures. During bleaching, the loss of fluorescence in one or more regions, outside of the bleach region, is monitored. The kinetics of the fluorescence loss in these unbleached regions correlates with the mobility of the labelled molecules.

iFRAP
inverse Fluorescence Recovery After Photobleaching; iFRAP, is a variation of FLIP. Labelling is performed as in FRAP experiments. The entire cell is bleached except for a small region of interest. After bleaching, there is a redistribution of mobile fluorescent molecules with photo-bleached molecules causing a decrease in fluorescence intensity within the bleach-protected region. In contrast to FRAP experiments, iFRAP enables direct tracking of fluorescent molecules that originated from a structure of interest.

FLAP
Fluorescence Localisation After Photobleaching. Two spectrally distinct fluorescent labels are required to conduct FLAP experiments of which one will provide a 'reference signal' whereas the other will be photo-bleached. At the beginning of a FLAP experiment, the detectable emission ratio of both labels is identical throughout the cell. One of the two labels is selectively bleached in a defined region. This results in local change of the, say, red-to-green ratio. After redistribution, the ratio of the two labels in any region of the cell can be measured as an indicator for the presence of molecules originating from the bleach area.

24.2 FRAP – Fluorescence Recovery After Photobleaching

Essentially a FRAP experiment is a time series with controlled bleaching. It is the simplest photobleaching kinetic technique to set up and perform. An initial series of unbleached (pre-bleach) images are collected for a short time of a few seconds to give a baseline reference of fluorescence intensity. A defined region of interest (ROI) is then illuminated with high-intensity light causing permanent and irreversible photobleaching. The diffusion of unbleached fluorescent molecules back into the bleach area ROI is recorded and tracked over time (Figure 24.1). The bleach is a single event of continuous duration, albeit of several iterations. Laser-scanning microscopes, both point-scanners and spinning-disc designs, are generally used to perform FRAP experiments. This is because lasers can be switched on and off very rapidly. They can be electronically rastered to provide a tightly defined region of interest – if necessary, of sophisticated shape – and the intensity of the laser can be controlled virtually instantaneously through the AOBS or AOTF on the confocal microscope. Measurement of the diffusing molecules allows an assessment of the kinetics of the mobility of the tagged protein or other molecule.

The mobility of diffusing fluorescent molecules into the bleach ROI depends upon the nature of the cell, its local environment, the presence and types of membrane present and whether the fluorescent target is wholly mobile or partially bound to membrane or organelles within the cell. See Mueller et al. (2010) and Day et al. (2013) for reviews and discussions of different kinetic models, and Blumenthal et al. (2015) for a suitable algorithm – using ImageJ – to calculate diffusion coefficients. It is also important to take into account any photo-switching of the target fluorophore back from a dark state. Do not assume that all the bleached fluorescent molecules remain bleached (Mueller et al., 2012; Yang et al., 2010) or that adjacent regions are not affected by scattered high-intensity light (Dobrucki et al., 2007). Fluorescence recovery depends upon two factors. Firstly, the movement of free fluorescent signal back into the bleached region, which depends upon Brownian motion, diffusion and active transport mechanisms. Secondly, the slower recovery of the bleached structure as the bleached molecules recover or else are replaced with new fluorescence molecules. Full recovery of fluorescence within the bleached region may well not occur, allowing a mobile and non-mobile fraction of fluorescent molecules to be identified (graph, Box 24.1). This data is used to gain better insight into the properties and interactions of molecules, cellular components and organelles within the cell. As in other areas of fluorescence microscopy, the introduction of fluorescent proteins has allowed FRAP analysis to be applied to the study of a wide range of subjects in cell biology.

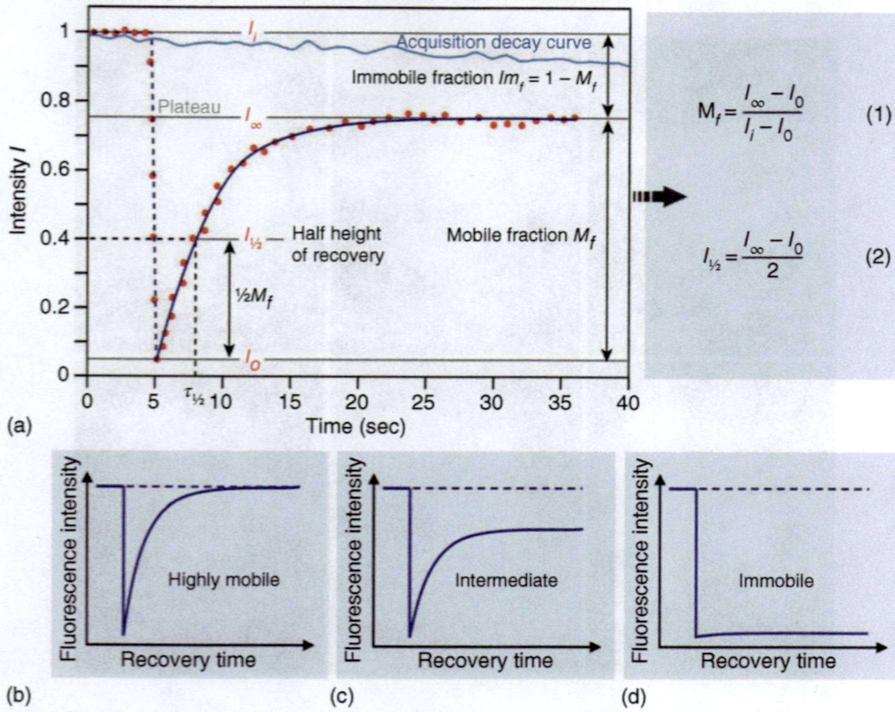

Figure 24.1 **A typical FRAP curve** *Anatomy of a typical FRAP curve. (a) From the initial (pre-bleach) fluorescence intensity (I_i), the signal drops to a particular low value (I_0) as the high-intensity laser beam bleaches fluorophores in the ROI. Over time the signal recovers from the post-bleach intensity (I_0) to a maximal plateau value (I_∞). From this plot and equations 1 and 2, the mobile fraction (M_f), immobile fraction (IM_f), $I_{1/2}$ and corresponding time ($\tau_{1/2}$ – the time for the exchange of half the mobile fraction between bleached and unbleached areas) can be calculated (light blue line: decay curve is the reference photobleaching curve to correct for fluorescence loss during data acquisition). The information from the recovery curve (from I_0 to I_∞) can be used to determine the diffusion constant and the binding dynamics of fluorescently labelled proteins. Based on different recovery profiles, the protein mobility can be classified as (b) highly mobile with virtually no immobile fraction, (c) intermediate mobile with an immobile fraction, or (d) immobile. Source: Ishikawa-Ankerhold et al. 2012. Reproduced under the terms of the Creative Commons Attribution License, CC BY 3.0.*

Box 24.1 FRAP Curves and Data-Fitting

After the time course of the FRAP experiment, a plot of fluorescence intensity against time will usually reveal mobile and immobile fractions. The mobile fraction can be determined by comparing the immediate post-bleach intensity (F_0) with the asymptote of the recovery curve (F_∞).

Calculate the mobile fraction (M_f) and the immobile fraction (I_f) by the following equations:

$M_f = (F_\infty - F_0)/(F_i - F_0)$, where F_∞ is the fluorescence intensity after full recovery, F_i is the fluorescence intensity before photobleaching, and F_0 the fluorescence intensity immediately post-bleaching. The immobile fraction, $I_f = 1 - M_f$.

To allow for photobleaching during the time of the experiment, the prebleach intensity is calculated from the following equation (Snapp et al. 2003). The prebleach intensity is defined as 100% fluorescence intensity:

$$\%M_f = 100 \times \frac{F_{\text{precell}} - F_{\text{background}}}{F_{\infty\text{cell}} - F_{\text{background}}} \times \frac{(F_{\infty\text{cell}} - F_{\text{background}}) - (F_0 - F_{\text{background}})}{(F_{\text{pre}} - F_{\text{background}}) - (F_0 - F_{\text{background}})}$$

The half-life, $t_{1/2}$, is only relevant to each individual microscope and set of experimental conditions. It can be estimated from a graphical plot, or calculated from: $F(t) = [F_0 + t/t_{1/2})]/[1 + (t/t_{1/2})]$ where F(t) is fluorescence intensity at timepoint *t*, and the other terms are as defined above. The function $\ln(F_\infty - F_t) = -kt + b$ gives a straight-line plot with the slope −k, a constant, proportional to $t_{1/2}$ and the term b, at the y-axis intercept, related to F_0. Further calculations for the value of $t_{1/2}$ can be found in Snapp et al. (2003), and further details regarding the calculation given above in Feder et al. (1996).

If the sample is sufficiently thin, the diffusion rate, D, can be calculated from a two-dimensional diffusion analysis model (Ellenberg et al., 1997):

$$D = w^2\gamma/4t_{1/2} \text{ and also fitting the recovery curve to: } I(t) = F_\infty[1 - (w^2/w^2 + 4\pi Dt)^{-1/2}],$$

where I(t) = intensity as a function of time (with t_{zero} being the midpoint of the bleach), w^2 = the radius of the bleached area, γ is a correction factor for unavoidable bleaching, $t_{1/2}$ = the half-life of recovery, which is the time taken to recover half of its initial integrated intensity, and F_∞ = the fluorescence intensity corresponding to the asymptote of the mobile fraction. The diffusion rate is given in units of $\mu m^2/s$. Presley (2006) gives a γ value of 0.88, from the original paper by Axelrod et al (1976), which applies only to a *circular* bleach spot. Typical values of D of membrane-bound proteins in cells range from 0.2 to 0.5 $\mu m^2 s^{-1}$; freely diffusing molecules may have D values ten times greater.

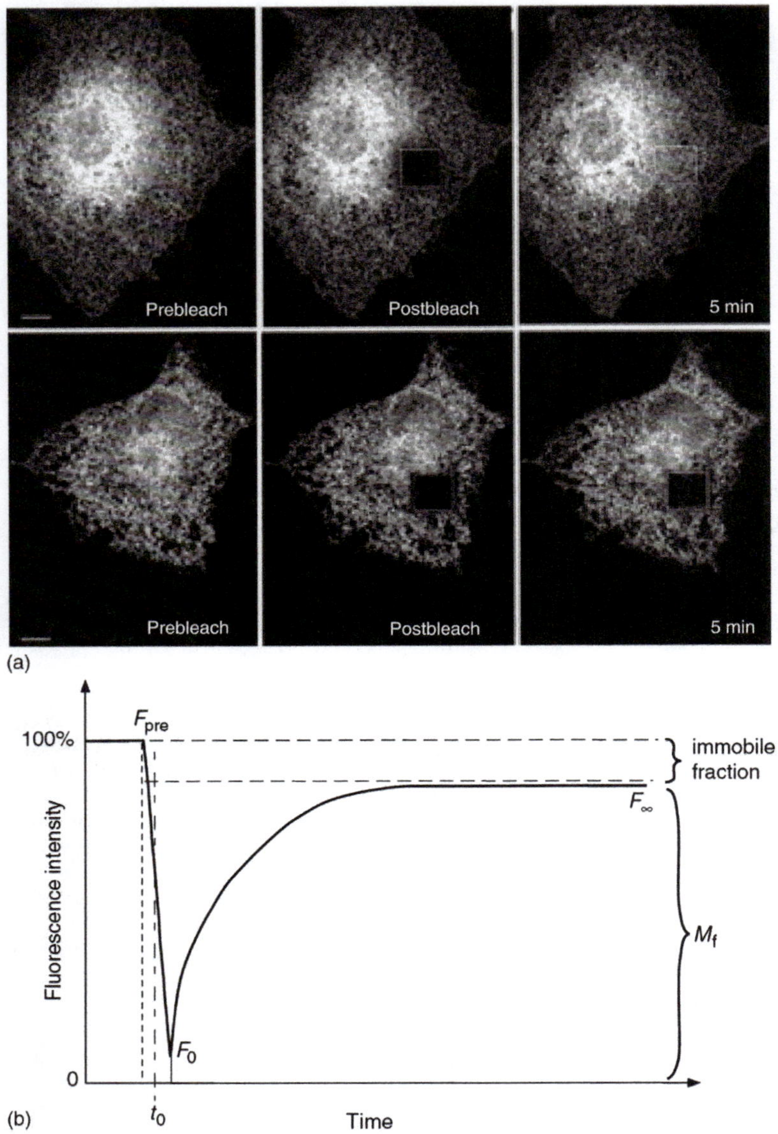

The calculation of $t_{1/2}$ requires, firstly, conversion of the time for recovery to plateau, with t_0 taken as the half-time of the bleach duration, as shown and, secondly, conversion of the bleach recovery data to fractional fluorescence. For details, see Snapp et al. (2003). (Figure reproduced with permission from John Wiley & Sons.)

References

Axelrod, D. et al. (1976). 'Mobility measurement by analysis of fluorescence photobleaching recovery kinetics'. *Biophysical Journal 16*/9: 1055–1069.

Blumenthal, D. et al. (2015). 'Universal approach to FRAP analysis of arbitrary bleaching patterns'. *Scientific Reports 5*: 11655.

Day, C.A. et al. (2013). 'Analysis of protein and lipid dynamics using confocal fluorescence recovery after photobleaching (FRAP)'. In: Watkins, S.C. and St. Croix, C.M. (eds.), *Current Protocols Select Imaging & Microscopy*. Hoboken, New Jersey: Wiley.

Ellenberg, J. et al. (1997). 'Nuclear membrane dynamics and reassembly in living cells: targeting of an inner nuclear membrane protein in interphase and mitosis'. *The Journal of Cell Biology 138*/6: 1193–1206.

Feder, T.J., Brust-Mascher, I., Slattery, J.P., Baird, B., and Webb, W.W. (1996). Constrained diffusion or immobile fraction on cell surfaces: A new interpretation. *Biophysical Journal 70*/6: 2767–2773.

Lippincott-Schwartz, J. et al. (2001). 'Studying protein dynamics in living cells'. *Nature Reviews Molecular Cell Biology 2*: 445–456.

Presley, J.F. (2006). 'Measurement of protein motion by photobleaching'. In: Stephens, D. (ed.), *Cell Imaging*, Chapter 6. Bloxham: Scion Publishing Ltd., pp. 119–142.

Snapp, E.L. et al. (2003). 'Measuring protein mobility by photobleaching GFP chimeras in living cells'. *Current Protocols in Cell Biology*. Chapter 21: Unit 21.1.

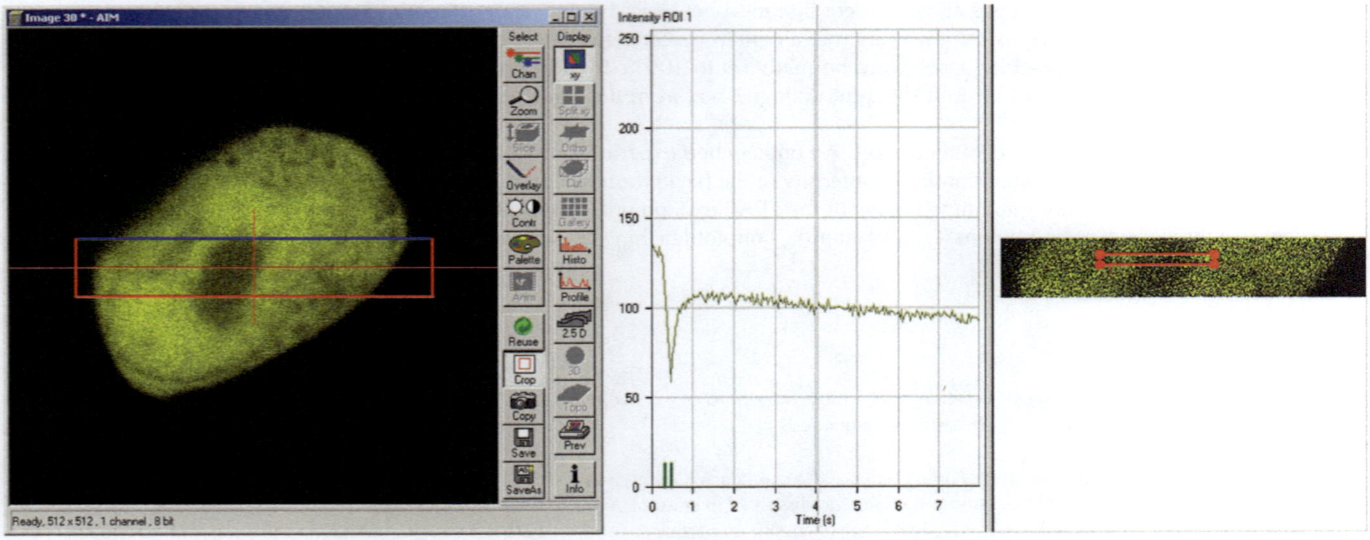

Figure 24.2 **Optimal letter-box bleaching shape for FRAP** *Author's photomicrograph. Source: Reproduced with permission from Carl Zeiss Ltd.*

For a FRAP experiment to have any meaning, it is important that the sample is not photo-bleached during the pre-bleach or recovery phase, or if it is, a good idea of the bleaching behaviour is known so that loss of intensity prior to the controlled bleaching event can be subsequently corrected for in the data. It is also important that the emitted signal into the PMT or CCD is not saturated, because fluorescence intensity is plotted over time to generate a FRAP curve. Normally a compromise, or trade-off, has to be sought between spatial and temporal resolution in a FRAP experiment. If we image slowly, we may get a good image showing morphological detail of the cell, showing precisely where diffusion of molecules back into the bleach region is occurring. However, the kinetic data may well lead to erroneous conclusions, because the recovery phase, which started immediately upon cessation of bleaching, with diffusion back into the bleach ROI, was not recorded fast enough because of poor temporal resolution. In order to record each time point sufficiently fast, it is usually necessary to sacrifice overall image quality. For quantitative FRAP particularly, temporal resolution is far more important than image quality (spatial resolution), so that the recovery kinetics can be accurately plotted and calculated.

In relation to this, it is better to use a letter-box shape for the ROI (Figure 24.2). This is because in a point-scanning confocal the X-galvanometer operates faster than the Y-galvanometer. Thus both bleaching and imaging occur faster with this shape rather than a circular or square ROI. Because of the fundamentally different design, the shape of the ROI has no bearing on bleaching or image acquisition when performing a FRAP experiment on a spinning disc microscope.

Protocol 24.1 FRAP Protocol

1. Pre-warm the lasers for at least 10 minutes to avoid significant intensity fluctuations during the experiment.
2. Set up the imaging parameters and acquire a fluorescent image in the normal way. Use a 512 × 512 raster and 8-bit depth for the dynamic range. Use a weak laser to image, and a short pulse with very high laser power to bleach to prevent diffusion back into the ROI during bleaching.
3. Open the pinhole to maximum. Spatial resolution is not important here; ensure that the image is neither undersaturated nor oversaturated.
4. Zoom in on the cell. At high zoom factors, the dwell time is increased, increasing the bleaching effect.
5. Draw a region of interest to form a bleaching area. This should be about 10–20 μm^2 and preferably a spot, or letter-box shape for good temporal resolution. A letter box is better where the concentration of the target molecule, and hence the signal level, is low.
6. If possible, draw a second ROI in the background of the cell to form an internal control. This ROI can be used for FLIP analysis. Otherwise, image a separate cell in the field of view. This control enables an allowance to be made for the photobleaching rate, P_R.
7. Draw a third ROI in the image outside the cell to monitor the background level.
8. Set a number of time-lapse pre-bleach image acquisitions, without bleaching (normally 5–10). This is suitable where the acquisition frame rate is less than or equal to one per second. If the frame rate is faster, collect 30–50 frames before acquisition.
9. Set the bleaching lasers and laser power for the first ROI bleach spot (e.g. 477 & 488 nm at 100%), including a number of bleach iterations to destroy the fluorescence in the bleach spot, but not to overbleach the entire cell. This will require trial bleaches and is generally of relatively short duration.
10. Check that the time interval will sufficiently resolve the rate of recovery. Recheck the focus.
11. Set a number of post-bleach image acquisitions sufficient to reach a recovery plateau in order to accurately determine $t_{1/2}$, the half-life of recovery. Generally this will require 40–50 data points. If the frame rate is high, up to 200 images may need to be collected.
12. Initiate the time course: pre-bleach, bleach, recovery to plateau.
13. Check that the cell has not moved: the ROI should be fixed during the time course.

Collecting several pre-bleach images allows for equilibration between the dark states (i.e. triplet state) and fluorescent states of the population of fluorophore molecules. At any one time, a certain percentage of the fluorescent label is in the dark state where it does not fluoresce. Generally, the bleaching laser(s) are normally set to 100%. In most confocal software packages, it is usual to have the controls of the bleaching lasers set up in a different dialogue box from the imaging lasers within the tracks and channels dialogue windows.

There must be at least two experimental controls. An unbleached experimental region (i.e. a cell not undergoing FRAP) as well as a background region (to show any alteration in the intensity of the background) should be acquired with the imaging lasers to provide a measure of photobleaching during the time course of the FRAP experiment. Calculate the photobleaching rate (P_R) by comparing the fluorescence of the control region before (F_0) and after (F_∞) photobleaching. $P_R = F_\infty/F_0$.

A FRAP test

The following test[1] is quick and simple to set up, and although it does not involve living cells, it does show how the principles of FRAP work very well indeed, allowing suitable recovery curves to be generated.

1. Cut up a sheet of Whatman 105 lens cleaning tissue (GE Healthcare, Chicago) into small pieces about 0.5 cm square.
2. Dye the paper briefly in a 0.025% solution of fluorecein dissolved in ethanol. You can substitute turmeric spice powder for fluorescein: make up a saturated 0.1% solution in ethanol and spin out the excess spice particles. Turmeric will absorb light broadly in the same range as fluorescein and can be excited with either the 488 or 514 nm laser.
3. Allow the paper to dry thoroughly, and mount the pieces of paper in DPX or other hardening mountant that does *not* contain antifade reagent.

In order to get an effective FRAP curve, you need to have some idea of the speed of diffusion, whether there is a fast or slow component. You also must normalise the data to take account of unavoidable general photobleaching that occurs during the time course of the experiment. Calculate this photobleaching by monitoring cells or part of the cell that is not photo-bleached. Using a separate cell from the one being photo-bleached is best, since any FLIP effect can then be ruled out.

Essentially, the rate of recovery of the fluorophore labelling the protein or macromolecule of interest is compared to the FRAP recovery and straightforward diffusion rate of an inert non-binding fluorescently labelled molecule. The mobile fraction will (ideally) recover to a plateau with the rate indicating the degree of local binding. A fast recovery will indicate less binding than a slow recovery of signal within the bleached area. Not only will a fast recovery tend to indicate diffusion of free, unbound molecules, but bleaching with different spot sizes will tend to support this conclusion also, since diffusion is normally very rapid irrespective of bleach spot size.

GFP alone (molecular mass = 27 kDa) will generally diffuse back into the bleached region in ≈1 second with a 2 μm bleach spot; larger molecules will diffuse more slowly, between about 1.5 and 3.5 seconds. GFP-bound molecules will recover more slowly, and a molecular sieving effect may retard the progress of these larger molecules. Other factors may also be responsible: the local environment may be viscous, or the protein may be interacting transiently with a fixed substrate; it may be forming immobile aggregates. Also, be aware that the photo-toxic effects from bleaching the cell with laser light will lead to the formation of free radicals that may well impair the function of the cell during the fluorescence recovery phase and alter the kinetics of the process. Many potential biological factors may influence the behaviour of the labelled molecule. Hence the curve fitting and interpretation of the data is usually complicated (see Box 24.1). A review of the different FRAP kinetic models for analysing the mobility of molecules bound to cellular components, which is beyond the scope of this book, can be found in Mueller et al. (2010), Presley (2006), Yang et al. (2010) and Bancaud et al. (2010). Also see Blumenthal et al. (2015) for a suitable algorithm to calculate diffusion coefficients.

24.3 FLIP and FLAP

The counterpart of FRAP, Fluorescence Loss In Photobleaching (FLIP) generates an inverse curve. FLIP is a complementary technique whereby an ROI in the cell is repeatedly bleached (in FRAP the ROI is bleached in *one* session, albeit one of several iterations) and the reduction in fluorescence throughout the entire cell monitored as the entire fluorescent complement moves *into* the bleach region and is bleached away. FLIP is useful for studying contiguous structures, such as membranes, or membrane organelles (e.g. the Golgi, ER or nucleus). FLIP follows the path of fluorophores away from the repeatedly bleached area. As the bleach spot erodes successive incoming molecules, eventually the entire organelle decreases in intensity, or is bleached away entirely. The method can also be useful to track diffusion of molecules across membranes into cellular compartments.

In Figure 24.3, the dotted curve shows the expected result from a FLIP experiment run in association with a FRAP experiment. Inverse FRAP, or iFRAP, is a variation of FLIP. The entire cell is repeatedly bleached, with the *exception* of the region of interest. During bleaching and after it stops, the loss of fluorescence in the unbleached ROI is monitored, whereas FLIP monitors the depletion of pools of fluorescence external to a continually bleached ROI. In an iFRAP experiment, the rate of diffusion of fluorescent molecules *out* of the ROI is measured. The advantage of both FLIP and iFRAP is that data is acquired from areas of the cell that are not bleached directly.

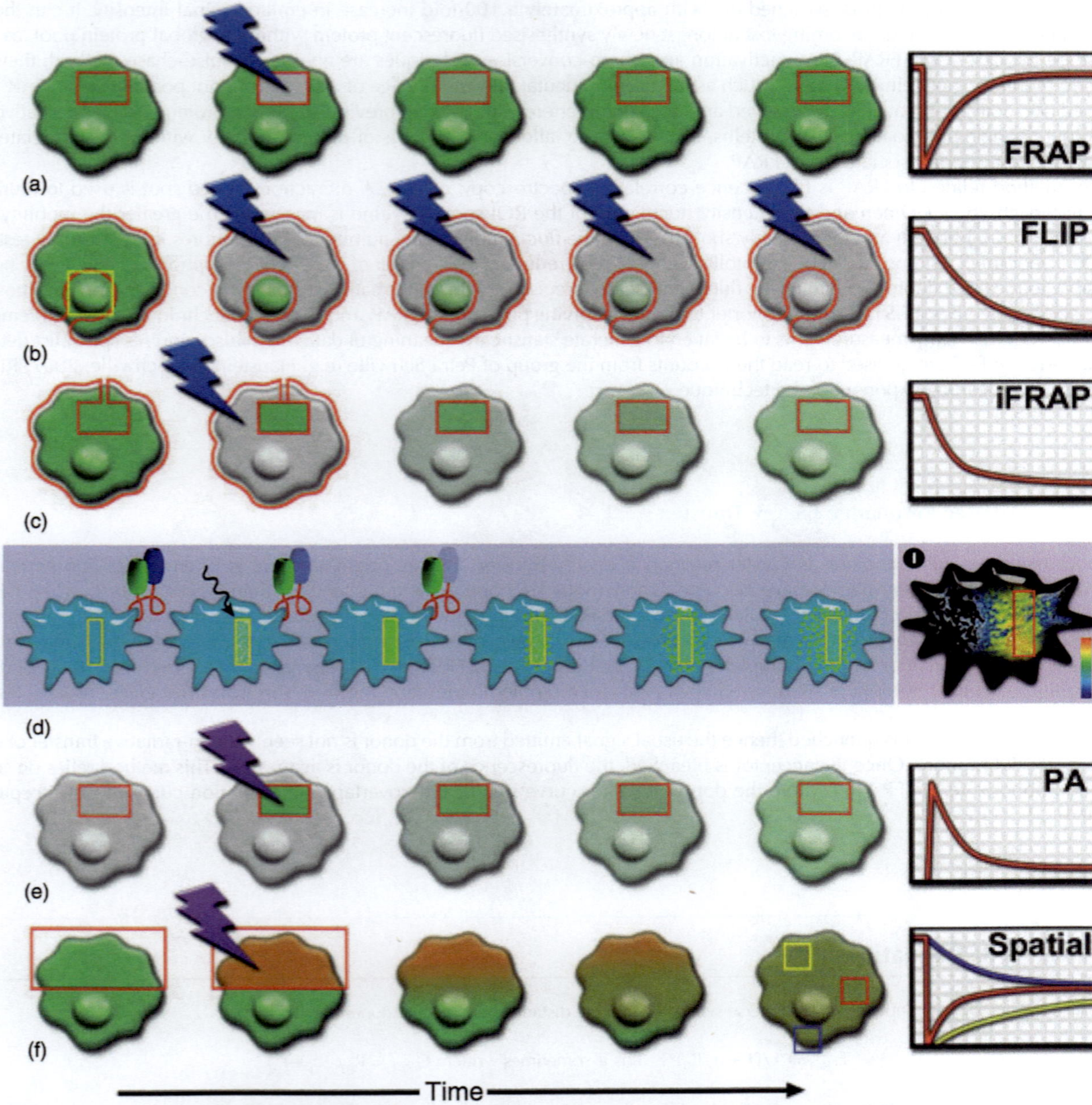

Figure 24.3 ***Various bleaching stratagems*** *Techniques used in FRAP and related dynamic imaging methods. The blue and purple lightning bolts represent high-intensity laser excitation at 488 and 405 nm, respectively. The series of graphs illustrate typical recovery kinetics. (a) shows FRAP; (b) FLIP; (c) iFRAP; (d) FLAP; (e) photo-activation and (f) photo-conversion. In (f) the three separate coloured curves relate to the respective coloured ROIs. Source: (a–c, e–f) Murphy and Davidson 2013. Reproduced with permission from John Wiley & Sons. (d) Ishikawa-Ankerhold et al. 2012. Reproduced under the terms of the Creative Commons Attribution License, CC BY 3.0.*

A similar method, fluorescence localisation after photobleaching, FLAP (Dunn et al., 2002) enables molecules dual-labelled with different fluorophores, or differently labelled molecules, which colocalise absolutely, to be tracked dynamically. If one fluorophore is bleached, subtraction from a reference image containing both fluorophores will enable both the bleached population of molecules as well as the unbleached population to be localised. FLAP is therefore valuable in tracing the turnover and relocation of molecules during steady-state dynamics that would otherwise appear as a static population.

Photoactivatable GFP can be used (Patterson and Lippincott-Schwartz, 2002) – as can a photo-convertible fluorescent protein such as Kaede, Dendra2 and mEos2 (Sattarzadeh et al., 2015) – as a complementary approach to FRAP. PA-GFP has an excitation profile similar to wtGFP; it can thus be activated by a 405 or 408 nm blue laser and the emitted signal subsequently visualised with a 488 nm laser. The PA-fluorescent protein is 'switched on' with approximately a 100-fold increase in emitted signal intensity. It can therefore be individually tracked without becoming lost amongst newly synthesised fluorescent protein within the global protein pool, as would otherwise be the case using FRAP. Photo-activation and photo-conversion techniques are an optical pulse-chase method; they allow longer-term monitoring of cellular changes, such as the developmental fate and lineage of a single cell, not possible with FRAP. Newly synthesised non-activated protein is unobserved and does not interfere with tracking previously formed protein pools. Alternatively, the use of colour-changing photoactivatable proteins, such as Kaede, allows the analysis of protein mobility without the associated high risk of damage to the cell associated with FRAP.

Another method related to FRAP is fluorescence correlation spectroscopy, or FCS. A diffraction-limited spot is used to illuminate, rather than bleach, the specimen and the intensity fluctuation of the ROI spot over time is measured. The greater the mobility of the molecules, the more frequent and increasingly short-lived are the fluctuations of the number of fluorophores in the ROI. By restricting the ROI to a very small volume (about one femtolitre) and thereby reducing the number of recorded fluorophores to a very low number, it is possible to measure fluorophore intensity fluctuations very precisely, giving information on both the concentration and the size of the fluorescent molecules. FCS has a much shorter temporal resolving power than FRAP, and so the two techniques are complementary. FCS not only requires many measurements to be taken to generate statistically meaningful data, but it also requires specialist detectors. Those interested in FCS are advised to read the accounts from the group of Petra Schwille (e.g. Haustein and Schwille, 2007; Ries and Schwille, 2012), a leading exponent of the technique.

24.4 FRET – Förster Resonance Energy Transfer

The correct name for this technique is Förster resonance energy transfer after its originator[2] but is often called *fluorescence* resonance energy transfer. These names have become synonymous by popular default, so it is often worth using both in database searches. FRET is used to detect inter-molecular (and to a lesser extent intra-molecular) interactions between a pair of fluorophores – usually independently attached to two macromolecules, or else to two opposing arms of a single molecule which may fold (this is known as bimolecular fluorescence complementation, or BiFC). When FRET occurs, energy is transferred from the donor fluorophore to the acceptor fluorophore by non-radiative resonance dipole-dipole coupling over very short distances, usually less than 10 nm apart.

In a steady state, the donor is quenched (hence the usual signal emitted from the donor is not seen) by non-radiative transfer of energy to the acceptor fluorophore. Once the acceptor is bleached, the fluorescence of the donor is again seen. This method relies on specific pairs of fluorophores (e.g. CFP-YFP) where the donor emission curve significantly overlaps the excitation curve of the acceptor (see Figure 20.1c).

Box 24.2 FRET Equations

The FRET efficiency, E_{FRET}, depends on the inverse sixth power of the distance between the donor and acceptor.

$$E_{FRET} = 1/[1 + (r/R_0)^6] \quad \text{this is sometimes written: } E_{FRET} = R_0^6/(R_0^6 + r^6),$$

where r is the distance between the two molecules, and R_0 – the Förster radius – the distance when the FRET efficiency between the FRET pair is 50%. (i.e. 50% of excited donors are deactivated by FRET, Figure 24.4). Since the FRET efficiency, E_{FRET}, can be calculated by experiment, the distance, r, between the two fluorescently labelled molecules of interest can be calculated from the rearranged expression:

$$r = R_0\{(1/E_{FRET}) - 1\}^{1/6}$$

From these R_0 can be calculated: $R_0\ (\text{nm}) = [8.79 \times 10^{-22}.\ n^{-4}.\ QY_D.\ \kappa^2.\ J(\lambda)]^{1/6}$,
where n is the refractive index (generally assumed to be 1.4 for protein), QY_D the quantum yield of the donor and κ^2 the dipole orientation factor between the fluorophore pair (range 0 → 4 but assumed to be 2/3). The term $J(\lambda)$ is the overlap integral, the region of overlap between the two spectra: $J(\lambda) = \int F_D(\lambda).\ \varepsilon_A(\lambda).\ \lambda^4 d\lambda / \int F_D(\lambda) d\lambda$

In practice, the R_0 value is taken from the literature (see Table 24.1).

Reference

Pietraszewska-Bogiel, A. and Gadella, T.W.J. (2010). 'FRET microscopy: from principle to routine technology in cell biology'. *Journal of Microscopy 241/2*: 111–118.

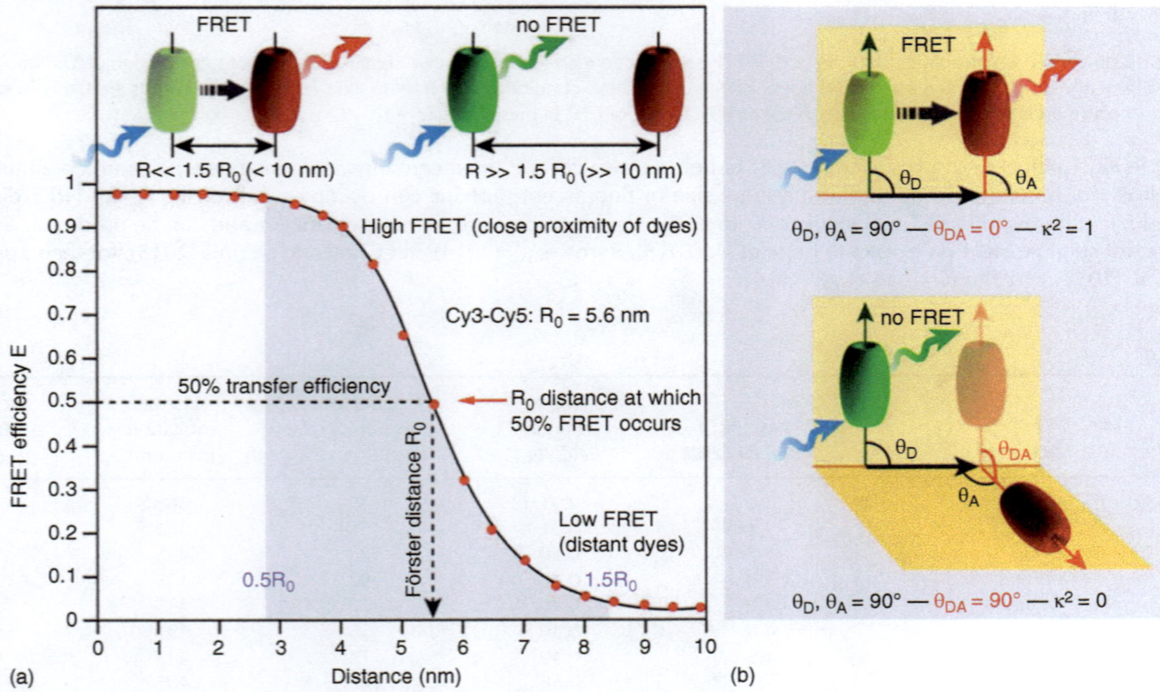

Figure 24.4 ***FRET efficiency R_0 curve*** *Panel (a) The inverse sixth power dependence of the curve of the plot of E_{FRET} versus r means that for distances less than R_0 the FRET efficiency is close to maximal, whereas for distances greater than R_0 the efficiency falls off. Panel (b) FRET is also dependent upon the mutual molecular orientation of the dipoles of the molecular pair. The emission dipole of the donor and the acceptor absorption dipole of the donor must be parallel, rather than perpendicular, to one another. Source: Ishikawa-Ankerhold et al. 2012. Reproduced under the terms of the Creative Commons Attribution License, CC BY 3.0.*

Because of the sixth-power term in the equation, FRET efficiency is very sensitive to distance, but only near to the R_0 value. The magnitude of R_0 is dependent on the donor quantum yield, fluorophore dipole orientation[3] and the spectral overlap between the donor emission spectrum and the acceptor excitation spectrum of the fluorophore pair. This distance R_0 can be calculated for the particular pair of fluorescent molecules selected and used for FRET. Selected R_0 values are given in Table 24.1; for example, R_0 for CyPET-YPET = 5.2 nm. In Figure 24.4, the Cy3-Cy5 organic dyes pair is used as an example. From Förster's original work, Lubert Stryer and Richard Haugland suggested that FRET could be used as a 'spectroscopic ruler' in the 1–6 nm range to reveal the close proximity between labelled proteins and macromolecules and to help decide whether they may be functionally colocalised[4] or not. Because of the difficulty in quantifying fluorophore dipole orientation and the degree of spectral overlap, most FRET studies are used to qualitatively localise molecular interactions within cells, rather than derive quantitative data.

For FRET to occur successfully, the emission spectrum of the shorter-wavelength donor must significantly overlap the excitation (absorbance) spectrum of the longer-wavelength acceptor fluorophore (Figure 24.5). If this is the case, excitation of the donor alone results in the emission of a photon by the acceptor at the acceptor fluorophore's usual emission wavelength. The donor emission signal is quenched, and the acceptor emission signal sensitised, or increased: hence the name sensitised emission for one of the FRET methods described below. The greater the degree of overlap between the donor emission spectrum and the acceptor excitation spectrum, the higher the value of R_0, and thus the greater the chance of FRET being detected.

The stoichiometry of the two fluorophores or fluorescent proteins should, ideally, be 1:1 and the brightness of the pair should also match. A mismatch in these respects can lead to bleed-through from the brighter fluorophore saturating the FRET channel detector, whilst the signal from the dimmer fluorophore is lost. The classical, well-known FRET pair is CFP as donor and YFP as acceptor, although this combination has generally been replaced by brighter variants such as CyPET and YPET or Cerulean and Venus fluorescent proteins, which have been specifically designed for FRET (Ai et al., 2006). eCFP ($\varepsilon = 32\,500$; QY = 0.4; brightness = 13.0; photostability $t_{1/2} = 64$ s in PBS-glycerol $t_{1/2} = 78$ s) has fourfold less brightness than its common yellow acceptor partner, EYFP (Ex 83 400; QY = 0.61; brightness = 50.9. photostability $t_{1/2} = 60$ s in PBS-glycerol $t_{1/2} = 21$ s), which has a relatively high quantum yield. In contrast to eCFP, mCerulean3 has Ex 40 000; QY = 0.48; brightness = 34.8. photostability $t_{1/2} = 1100$ s. CFPs can photo-activate at YFP excitation wavelengths. Malkani and Schmid (2011) show that while CFP and YFP bleached at about the same rate in live cells, CFP bleaches significantly faster in commercial mounting fluids but is more photostable in PBS-glycerol mountant. The opposite is observed for YFP, and they conclude:

> 'FRET microscopy based on bleaching of YFP in a CFP/YFP-sample can lead to false-positive signals due to photo-conversion of YFP into a CFP-like fluorophore and due to an increase of CFP-fluorescence under these conditions. Contrary to that, false-negative results are likely in samples embedded in commercial mounting fluids due to the fast bleaching of CFP in these media.'

Other possible FRET pairs are shown in Table 24.1, but please also consult the recent literature since new pairs are continually being developed. It is worth bearing in mind that the large size of fluorescent proteins can occupy much of the useful FRET distance of 1–10 nm, reducing the overall FRET efficiency to ≈ 40–50%. Quantum dots are, of course, smaller and can be used both as a donor and acceptor: for suitable FRET pairs refer to Bajar et al. (2016), Byrne et al. (2014) and Chou and Dennis (2015); for their application, see Algar et al. (2014) and Shamirian et al. (2015).

Table 24.1 FRET Pairs

Donor	Acceptor	R_0 (nm) [FRET efficiency = 50%]	Donor quantum yield (Φ_D)	Acceptor extinction coeff ε (mM^{-1} cm^{-1})	Donor excitation max (nm)	Acceptor emission max (nm)
eBFP	eCFP	3.8	0.31	33	383	477
eBFP	eGFP	4.1	0.31	55	383	508
eCFP	eYFP	4.7	0.40	83	435	535
eCFP	mVenus	5.0	0.40	92	435	528
eCFP	mCitrine	4.8	0.40	77	435	529
eCFP	eGFP	4.8	0.40	55	435	508
eCFP	mCherry	4.5	0.40	72	435	610
mAmetrine	tdTomato	5.3	0.58	138	406	581
mCerulean	eYFP	5.4	0.62	83	433	535
mCerulean	mVenus	5.3	0.62	92	433	528
cyPet	YPet	5.2	0.51	104	435	530
eGFP	mCherry	5.3	0.60	72	490	610
eGFP	Alexa Fluor 546	5.1	0.60	112	490	573
eYFP	mCherry	5.7	0.61	72	514	610
eGFP	eYFP	5.7	0.60	83	490	535
mTFP1	mCitrine	5.7	0.82	77	462	529
mTFP1	mOrange	5.7	0.82	71	462	562
TagGFP	TagRFP	5.7	0.59	100	482	584
MiCy	mKO	5.3	0.90	52	472	559
tSapphire	TagRFP	5.7	0.60	100	399	580
mOrange2	mCherry	6.3	0.60	72	549	610
mCitrine	mKate2	5.8	0.76	63	516	633
Clover	mRuby2	6.3	0.76	113	486	605
mClover 3	mRuby3	6.5	0.78	128	506	592
Clover	mCherry	5.8	0.76	72	486	610
mVenus	mCherry	5.8	0.57	72	515	610
mVenus	tdTomato	5.9	0.57	138	515	581
mNeon Green	mRuby3	6.5	0.80	128	506	592
Alexa Fluor 350	Alexa Fluor 488	5.0	0.55	73	346	519
Alexa Fluor 488	Alexa Fluor 546	6.4	0.92	112	495	573
Alexa Fluor 488	Alexa Fluor 555	6.3	0.92	155	495	565
Alexa Fluor 488	Alexa Fluor 568	6.2	0.92	88	495	603
Alexa Fluor 488	Alexa Fluor 594	5.6	0.92	92	495	617
Alexa Fluor 488	Alexa Fluor 647	5.1	0.92	270	495	668
Alexa Fluor 546	Alexa Fluor 568	6.8	0.79	88	556	603
Alexa Fluor 546	Alexa Fluor 594	6.9	0.79	92	556	617
Alexa Fluor 546	Alexa Fluor 647	6.5	0.79	270	556	668
Alexa Fluor 555	Alexa Fluor 594	4.7	0.10	92	555	617
Alexa Fluor 555	Alexa Fluor 647	5.1	0.10	270	555	668
Alexa Fluor 568	Alexa Fluor 647	6.9	0.69	270	578	668
FITC	Cy5		0.92	136	495	667
Rhodamine 6G	Malachite green	6.1	0.95	Non-fluorescent acceptor	526	Non-fluorescent acceptor
mKO	mCherry	6.4	0.60	72	548	610
yPet	mRuby2	6.3	0.77	113	517	605
mTurquoise2	mVenus	5.8	0.93	92	434	528

Table 24.1 (Continued)

Donor	Acceptor	R_0 (nm) [FRET efficiency = 50%]	Donor quantum yield (Φ_D)	Acceptor extinction coeff ε (mM^{-1} cm^{-1})	Donor excitation max (nm)	Acceptor emission max (nm)
mTurquoise2	mRuby2	5.6	0.93	113	434	605
mTurquoise2	YPet	5.7	0.93	104	434	530
mTFP1	mRuby2	5.6	0.82	113	462	605
YPet	TagRFP	6.0	0.77	100	517	584
Cy3	Cy5	5.4	0.15	136	554	667

A larger R_0 value indicates a greater likelihood of FRET occurring. A donor with a high quantum yield, an acceptor with a large extinction coefficient and a FRET pair with a larger spectral overlap area than the minimum 30%, all lead to large R_0 values and the increased probability of FRET.

Shcherbo et al. (2009) state that the classical CFP/YFP FRET pair cannot be completely spectrally separated; cyan autofluorescence and photo-toxicity further complicate the use of this classic FRET pair. Engineered cyan proteins, such as mTurquiose, mCerulean or tSapphire, have longer Stokes shifts, higher extinction values and quantum yields; therefore, less energy is required to excite these fluorophores, which are less prone to photobleaching.

The Alexa dyes have been cited (https://bit.ly/2uXyee3) since they are listed in Byrne et al. (2014), but the equivalent DyLight and Atto dyes can equally well be substituted. See the current Atto dye catalogue and for the Dylight dyes, the Thermo Fisher Scientific website. The R_0 values for the Alexa Fluor FRET pairs given in Byrne et al. (2014) sometimes differ from the values cited elsewhere.

eBFP has a low photostability and quantum yield and is rarely used as a donor in FRET pairing. For FRET pair behaviour, see https://bit.ly/2PEnd8T.

References

Arai, Y. and Nagai, T. (2013). 'Extensive use of FRET in biological imaging'. *Microscopy 62/4*: 419–428.

Bajar, B.T. et al. (2016). 'A guide to fluorescent protein FRET pairs'. *Sensors 16/9*. pii: E1488.

Byrne, A.G. et al. (2014). 'Data'. In: Medintz, I. and Hildebrandt, N. (eds.), *FRET - Förster Resonance Energy Transfer: From Theory to Applications*, Chapter 14. Weinheim: Wiley-VCH, pp. 657–755.

Day, R.N. and Davidson, M.W. (2009). 'The fluorescent protein palette: tools for cellular imaging'. *Chemical Society Reviews 38/10*: 2887–2921.

Ishikawa-Ankerhold, H.C. et al. (2012). 'Advanced fluorescence microscopy techniques - FRAP, FLIP, FLAP, FRET and FLIM'. *Molecules 17/4*: 4047–4132.

Kremers, G.-J. and Goedhart, J. (2009). 'Visible fluorescent proteins for FRET'. In: Gadella, T.W.J. (ed.), *Laboratory Techniques in Biochemistry and Molecular Biology*, vol. 33, *FRET and FLIM Techniques*. Oxford: Elsevier, pp. 171–223.

Lam, A.J. et al. (2012). 'Improving FRET dynamic range with bright green and red fluorescent proteins'. *Nature Methods 9/10*: 1005–1012.

Müller, S.M. et al. (2013). 'Quantification of Förster resonance energy transfer by monitoring sensitized emission in living plant cells'. *Frontiers Plant Science 4*: 413.

Panchuk-Voloshina, N. et al. (1999). 'Alexa dyes, a series of new fluorescent dyes that yield exceptionally bright, photostable conjugates'. *Journal of Histochemistry and Cytochemistry 47/9*: 1179–1188.

Scott, B.L. and Hoppe, A.D. (2015). 'Optimizing fluorescent protein trios for 3-way FRET imaging of protein interactions in living cells'. *Scientific Reports 5*: 10270.

Shcherbo, D. et al. (2009). 'Practical and reliable FRET/FLIM pair of fluorescent proteins'. *BMC Biotechnology 9*: 24.

The three common methods for measuring FRET are:

- sensitised emission FRET;
- a variant of sensitised emission: spectral FRET; and
- acceptor photobleaching FRET (also called donor dequenching).

There are other variations of FRET methods applicable to microscopy, but only these three are described here. Acceptor photobleaching FRET is the simplest method, but is not used for quantitative studies because it does not correct for spectral bleed-through (SBT).

In sensitised emission studies (see Figure 24.6), only the donor molecule is excited and the emitted FRET signal measured only from the acceptor. Theoretically only two images need be collected: one of the donor alone excited at its excitation wavelength and one of the donor and acceptor together excited at the same wavelength. However, some of the signal recorded in the detector assigned to the acceptor channel will be from direct excitation of the acceptor by light used to excite the donor – that is, excitation of acceptor molecules by the donor excitation wavelength, or cross-excitation. This leads to an artificially reduced FRET value and must be corrected for.

Another contamination of the true FRET signal will arise from bleed-through of fluorescence emission from the donor fluorophore being detected in the channel assigned to the acceptor emission signal. This leads to an artificially inflated FRET value and likewise must be corrected for, thus making it impossible to determine with a single measurement exactly what fraction of fluorescence in the FRET channel arises only from the acceptor. The influence of both donor spectral bleed-through and acceptor spectral bleed-through must be taken into account when measuring FRET efficiency.

An ideal FRET pair for sensitised emission should have extensive (> 30%) overlap of the donor emission and acceptor absorbance/excitation spectra, but – to eliminate acceptor cross excitation – there should ideally be no overlap of the donor excitation spectra with the acceptor excitation spectrum. For most pairs this cross-excitation value is very low indeed, but YFP as an acceptor is particularly prone to acceptor excitation overlap with the CFP donor excitation spectrum. Also – to eliminate donor spectral bleed-through – there would ideally be no overlap of donor emission and acceptor emission spectra. Unfortunately, no FRET pair is ideal; donor bleed-through inevitably occurs (Figure 24.5), and controls must be run to manage the inevitable cross-talk.

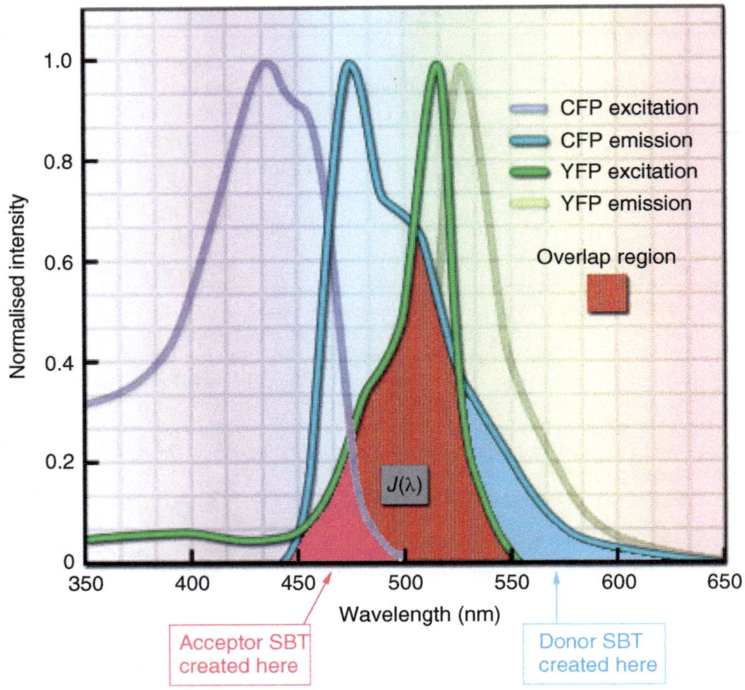

Figure 24.5 **Requirement of donor/acceptor spectral overlap for FRET** *FRET occurs when the donor emission spectrum overlaps the acceptor absorption spectrum by at least 30% of the acceptor excitation spectrum. Source: Murphy and Davidson 2013. Reproduced with permission from John Wiley & Sons.*

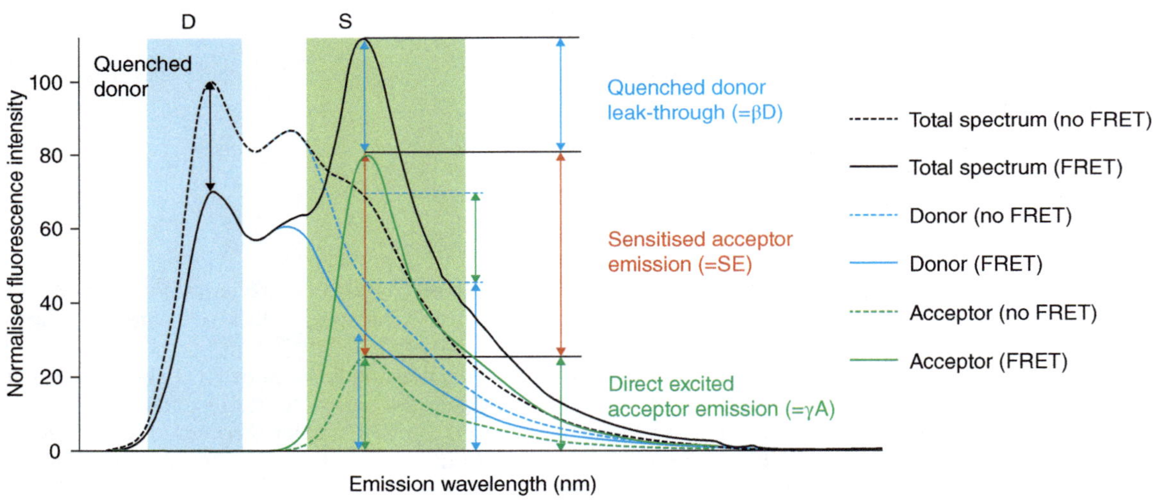

Figure 24.6 **Intensity-based and spectral FRET** *Fluorescence emission spectra of donor (cyan line) and acceptor (green line) in absence (dashed line) and presence (solid line) of FRET are shown. Note the quenched donor and sensitised acceptor emission as a result of energy transfer. For intensity-based FRET, small spectral regions (blue and green shaded areas) are selected for detection (e.g. by using bandpass emission filters). In the absence of FRET, the detected intensity in the sensitised emission image (S) contains direct acceptor excitation and donor emission contribution (bleed-through) (green and blue dashed arrows, respectively). In case of FRET, the S image contains direct acceptor excitation, donor bleed-through and sensitised emission (green, blue and red solid arrows, respectively). Here the donor emission filter to acquire the D image was chosen to avoid acceptor emission bleed-through. The calculation of net sensitised emission and E requires careful calibration with reference samples and determination of several correction factors (β, γ and G in the formula, see Jalink and van Rheenen 2009). Source: Pietraszewska-Bogiel and Gadella 2010. Reproduced with permission from John Wiley & Sons.*

The FRET signal is always contaminated by (a) bleed-through of the donor emission signal into the acceptor channel and (b) cross-talk excitation of the acceptor fluorophore by the wavelength used for donor excitation. Both of these signals constitute spectral bleed-through (SBT) into the acceptor channel.

Protocol 24.2 Sensitised Emission FRET Protocol

To quantify the true FRET signal, nine images[5] are required:

1. Cells expressing donor fluorophore only:
   (a) measure donor signal in the donor channel ($donor_{ex}/donor_{em}$) – positive control
   (b) measure acceptor signal ($acceptor_{ex}/acceptor_{em}$) – negative control
   (c) measure FRET channel ($donor_{ex}/acceptor_{em}$) – donor SBT control

   Any signal from the donor in the FRET channel is cross-talk of the donor into the FRET channel; it will occur to some degree. This is measured as a ratio of signal in the FRET channel to the donor signal (i.e. (c)/(a) above), and is subtracted from the FRET measured in the experimental cells. This ratio of (c)/(a) is called 'A'. This inevitable spectral bleed-through is generally in the region of 50–70%.

2. Cells expressing acceptor fluorophore only:
   (d) measure donor signal ($donor_{ex}/donor_{em}$) – negative control
   (e) measure acceptor signal ($acceptor_{ex}/acceptor_{em}$) – positive control
   (f) measure FRET channel ($donor_{ex}/acceptor_{em}$) – acceptor SBT control

   Bleed-through of the acceptor signal into the FRET channel is measured as a ratio of signal in the FRET channel to the acceptor signal (i.e. (f)/(e) above) and is subtracted from the FRET measured in the experimental cells. This ratio of (f)/(e) is called 'B'. This inevitable spectral bleed-through is generally in the region of 10–20%.

3. The actual FRET sample, having both labels:
   For the sensitised emission experiment, the following images are captured from cells expressing both fluorophores:
   (g) measure donor signal ($donor_{ex}/donor_{em}$)
   (h) measure acceptor signal ($acceptor_{ex}/acceptor_{em}$)
   (i) measure FRET channel ($donor_{ex}/acceptor_{em}$)

   The degree of FRET, E_{FRET} = FRET signal in (i) – {[B × (h) – [A × (g)]}

Negative controls are crucial because there is significant bleed-through of both donor and acceptor fluorophores into the FRET channel. Single-labelled controls are also required to subtract the bleed-through contribution of each fluorophore from the FRET channel (Broussard 2013). The low levels of fluorescent signal arising from live-cell microscopy and the significant degree of inevitable bleed-through present between donor excitation and acceptor emission mean that the sensitised emission method is much less robust in live-cell microscopy than the acceptor bleaching method upon fixed cells. Ideally you should have approximately the same brightness between donor and acceptor, because you have to use the same exposure between the measurements above. PixFRET, an ImageJ plug-in to calculate sensitised emission FRET, has been published by Feige et al. (2005).

With spectral FRET, after excitation of the donor alone, the emission spectrum of donor emission and acceptor emission (together in the FRET channel) is collected, as in sensitised emission FRET. However, instead of acquiring data through two individual channels as single intensity values, the entire emission spectrum containing both donor and acceptor fluorescence is collected upon excitation of the donor. The donor spectral bleed-through contribution is defined by reference to *individual* donor emission and acceptor emission reference spectra, and so the emission signal from the donor can be removed from the FRET signal by linear unmixing. What cannot be removed is the acceptor spectral bleed-through component from direct excitation of the acceptor, because this is identical to the signal in the FRET channel and cannot be unmixed. Acceptor spectral bleed-through is normally defined by comparison to the control expressing the acceptor only: (e) in the list above. The advantage of spectral FRET lies in the fact that both the quenched donor signal and the sensitised acceptor emission can be measured together simultaneously.

The acceptor bleaching technique was originally developed for small molecular weight fluorophores. In the acceptor bleaching method, the fluorescence intensity of the donor fluorophore is measured: the acceptor fluorophore is bleached and the intensity of the donor measured again. The difference between these two measurements allows FRET to be calculated.

The FRET efficiency, E_{FRET}, is calculated as:

$$E_{FRET} = 1 - [F_{donor\ quenched}/F_{donor}] \text{ sometimes also written: } E_{FRET} = 1 - (I_{DA}/I_D) = (I_D - I_{DA})/I_D$$

or $E_{FRET} = 1 -$ [donor emission before acceptor photo-bleach/donor emission after acceptor photo-bleach]

where $F_{donor\ quenched}$ is the fluorescence intensity of the donor (I_{DA}) in the presence of the acceptor, and F_{donor} the fluorescence intensity of the donor (I_D) after photobleaching the acceptor.

In the presence of FRET, the donor signal is enhanced after bleaching of the acceptor, whereas in the absence of FRET, the donor fluorescence does not increase. Ensure that the acceptor bleach filter (in widefield) or the laser line (in confocal) does not also bleach the donor fluorophore; otherwise, FRET will be underestimated. Monitor a control ROI of the acceptor that is not photo-bleached in order to assess any level of non-FRET bleaching during the course of the experiment; the acceptor should be photo-bleached down to about 10% of its initial intensity. By bleaching the acceptor fluorophore, only donor emission is measured, and thus the issue of bleed-through from the donor emission channel into the acceptor excitation channel is avoided, because it cannot occur. Nevertheless, contamination of donor emission channel by acceptor emission can occur.

The acceptor bleaching method can be used as a verification of the sensitised emission method. Use an exposure long enough to completely bleach the acceptor, using the excitation filter or laser for the acceptor, and take an image. Then switch to using the excitation filter or laser for the donor and the emission filter for the donor. Take an image. Subtract the ($donor_{ex}/donor_{em}$) image from (g), above, from the ($donor_{ex}/donor_{em}$) image of the bleached sample that you have just taken. The brightest areas in the subtraction image are the areas where FRET occurred, because they are no longer quenched.

It is much simpler to perform acceptor-bleaching FRET than sensitised emission FRET. An ImageJ plug-in to calculate FRET using the acceptor photobleaching method has been published in a clear and concise account by Roszik et al. (2008). The advantage of the acceptor photobleaching method is that each cell acts as its own control: the same specimen is used to determine the quenched and dequenched donor signals. Clearly, it works best with fluorescent pairs where the donor is photo-stable and the acceptor photo-labile. Because of the long exposures required to perform bleaching with the consequent release of reactive products, acceptor photobleaching is best suited to fixed samples.

Protocol 24.3 Acceptor–Photobleaching FRET Protocol

1. Cells expressing acceptor fluorophore only:

   measure acceptor signal ($acceptor_{ex}/acceptor_{em}$)
   (check that you are not measuring obvious bleed-through)

2. The actual FRET sample, having both labels:
   (a) Collect a pre-bleach image using settings for the donor fluorophore only
   (b) Collect a pre-bleach image using settings for the acceptor fluorophore only (used to check the presence of acceptor labelling)
   (c) Using the donor settings and bleach the acceptor completely
   (d) Collect background images for both donor and acceptor.

$$FRET_{Efficiency} = \frac{(Donor_{Post} - BkGrd_{Post}) - (Donor_{Pre} - BkGrd_{Pre})}{(Donor_{Post} - BkGrd_{Post})} \times 100$$

$Donor_{Post}$ = Donor emission after photo-bleach
$BkGrd_{Post}$ = Background after photo-bleach
$Donor_{Pre}$ = Donor emission prior to photo-bleach
$BkGrd_{Pre}$ = BkGrd prior to photo-bleach

Photobleaching the sample is invariably photo-toxic to living cells, and an accurate FRET measurement is often negated by recovery of the acceptor fluorophore before the second donor measurement in the absence of acceptor (in other words an 'unwanted' FRAP experiment). This effect reduces the magnitude of the second donor measurement and the degree of true FRET. Therefore, most FRET experiments on living cells are best conducted by the sensitised emission method, which requires several controls, or by FLIM (fluorescence lifetime imaging microscopy) in which the donor lifetime decreases in the presence of FRET. Since only the donor signal is monitored, spectral bleed-through, probe concentration and stoichiometry are irrelevant. Although turn-key systems are now available, FLIM requires sophisticated time-gated and photon-counting measuring equipment and so is not covered here. For further details, see the references given at the end of this chapter.

FRET can be carried out using widefield, confocal or two-photon microscopes. On a widefield microscope, three filter sets (donor excitation/donor emission; acceptor excitation/acceptor emission and the FRET channel: donor excitation/acceptor emission) are used to collect the required images. Narrow bandpass excitation and emission filters (see Figure 24.7) help reduce the inevitable spectral bleed-through. Neutral-density filters are required on widefield microscopes to modulate the excitation intensity using a single dichromatic beam splitter together with separate excitation and emission filters housed in filter wheels, to acquire *both* donor and acceptor images, which is important in order to eliminate pixel shift. This is because the FRET image is derived from pixel-by-pixel calculation. One drawback of the using a widefield system for FRET is that signal above and below the focal plane will degrade the FRET signal, unless a wide area (e.g. an entire cell) is analysed. Confocal microscopes, particularly those capable of spectral analysis, are better suited to acquiring FRET images.

Finally, a point easily overlooked: do not use any anti-fade reagent in the mounting medium of a sample used for FRET studies.

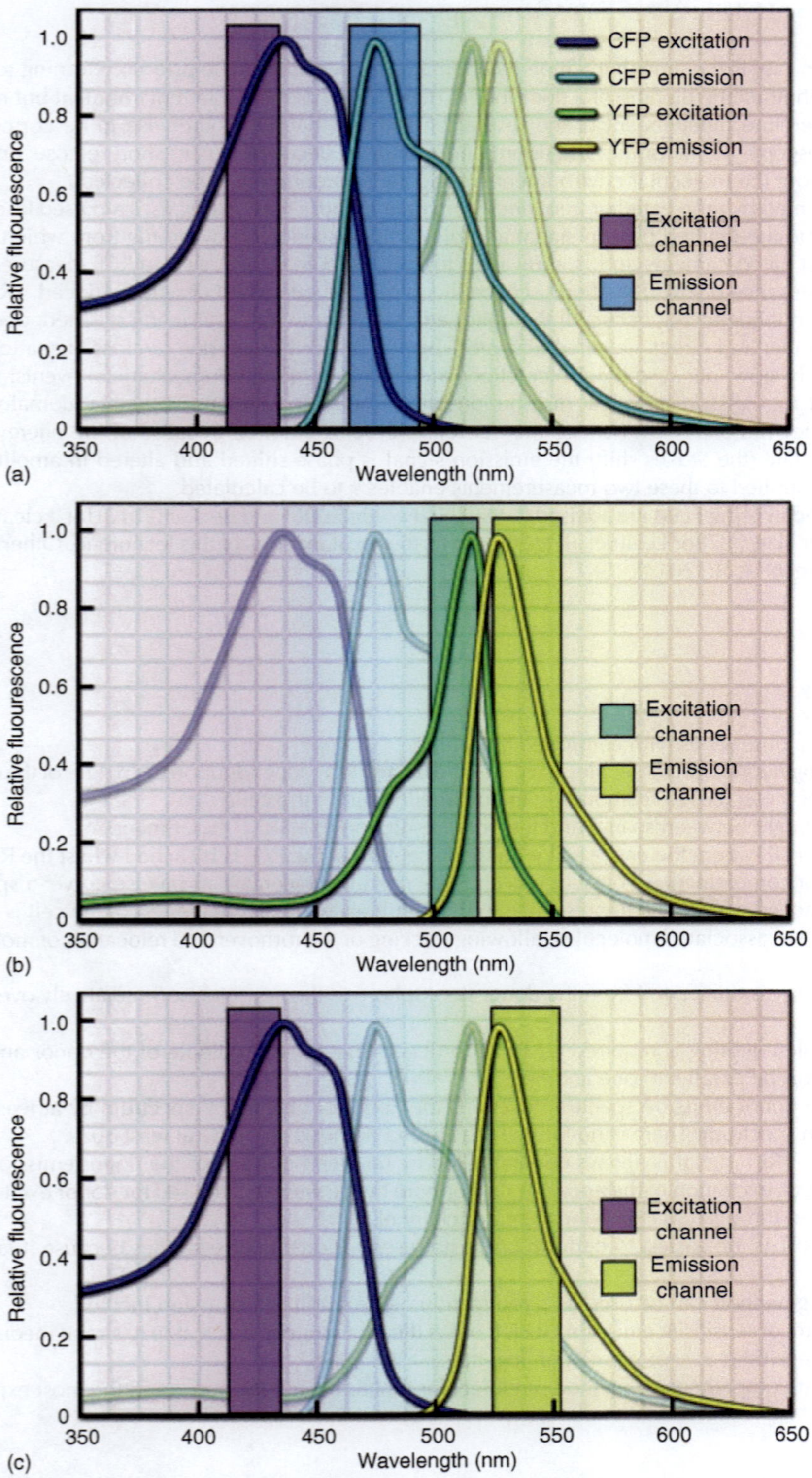

Figure 24.7 ***Widefield FRET with bandpass filters*** *Three filter cubes are used to acquire FRET images using a widefield microscope. Panel (a) the cube for donor excitation and donor emission gives a specific donor fluorescence signal but will also excite, and therefore pass, some acceptor signal as cross-excitation by the donor emission filter. In some cases, the donor emission overlaps the spectral region used for detecting acceptor emission. Panel (b) the cube for acceptor excitation and acceptor emission gives a specific acceptor fluorescence signal, which is usually free from bleed-through and is acceptor-specific. Panel (c) the FRET filter cube has bandpass filters for the donor excitation and acceptor emission. The FRET signal contains unwanted donor spectral bleed-through passed by the acceptor emission filter. Source: Murphy and Davidson 2013. Reproduced with permission from John Wiley & Sons.*

24.5 FLIM – Fluorescence Lifetime Imaging Microscopy

The fluorescence lifetime (τ, tau) is the time that a molecule remains in an excited state prior to returning to the ground state with the consequent emission of a photon. The lifetime of a fluorophore depends on its molecular environment but not upon its concentration. Fluorescence lifetime is therefore independent of the several parameters that have otherwise to be corrected for when performing intensity-based FRET microscopy. By measuring the fluorescence lifetime decay over very short (picosecond to nanosecond) periods with a pulsed laser, data about the molecular environment around the fluorophore can be collected.

FLIM can either be performed by time-gated or frequency domain methods. Time-gating uses a pulsed laser together with a synchronised detector to accumulate the emitted photon signal and plot a fluorescence decay profile from which τ can be calculated. This technique requires very fast gated cameras and is equivalent to 'microscopical MRI'. FLIM-FRET measurements are made upon the donor fluorophore. In the presence of FRET, the fluorescence lifetime of the quenched donor is reduced. FRET efficiency is calculated from the extent to which donor quenching occurs. If the donor emission bandwidth is properly selected, donor spectral bleed-through (SBT) into the acceptor channel and directly-excited acceptor spectral bleed-through due to cross excitation should not occur. Time-gated FLIM-FRET is relatively slow, because light interactions are stochastic quantum-mechanical events: sufficient photons have to be counted to take care of the Poisson statistical distribution of fluorophore emission. Frequency domain FLIM uses a light source modulated by high-frequency radio waves to excite the fluorophore being studied. Because of the energy loss on relaxation of the fluorophore to the ground state (the Stokes shift) the emission signal is phase-shifted and altered in amplitude in comparison to the excitation. Fourier analysis applied to these two measurements enables τ to be calculated.

Readers who are interested in using FLIM are advised to read Hallworth (2006) and Becker (2012) for a clearly presented introduction, and Gadella (2009), Festy et al. (2007) or Levitt et al. (2009) for further explanations of this technique. Other references are Llères et al. (2007a and 2007b) and Ebrecht et al. (2014).

24.6 Chapter Summary

1. A FRAP experiment is a time series with controlled bleaching.
2. The mobility of diffusing fluorescent molecules back into the bleach ROI depends upon the nature of the cell, its local environment, the size and binding of molecules to membranes and organelles within the cell.
3. A trade-off has to be sought between spatial and temporal resolution in a FRAP experiment.
4. iFRAP is a variation of FLIP where the entire cell with the exception of the ROI is bleached whilst the ROI is monitored, giving an inverse curve. FLIP is a complementary bleaching technique that also generates an inverse curve: a specific region of the cell is repetitively photo-bleached and loss of fluorescence in the non-bleached parts of the rest of the cell is measured.
5. FLAP bleaches one of two associated molecules, allowing tracking of the turnover and relocation of molecules during steady-state dynamics.
6. FRET occurs when energy is transferred from the donor fluorophore to the acceptor non-radiatively over very short distances, less than 10 nm apart.
7. Molecular dipole-dipole coupling is required for FRET to occur. The emission dipole of the donor and the acceptor absorption dipole of the donor must be parallel to one another.
8. FRET occurs when the donor emission spectrum overlaps the acceptor absorption spectrum by at least 30% or more by area of the acceptor excitation spectrum. There is no FRET if the spectra do not overlap by at least 30%.
9. As a consequence, the FRET signal is always contaminated by (a) bleed-through of the donor emission signal into the acceptor channel and (b) cross-talk excitation of the acceptor fluorophore by the wavelength used for donor excitation. Both of these signals constitute spectral bleed-through (SBT) into the acceptor channel.
10. Acceptor-bleaching FRET is the easiest to perform but is not used for quantitative studies because it fails to account for spectral bleed-through.
11. The acceptor bleaching method can be used as a verification of the sensitised emission method.
12. Spectral FRET is a variant of sensitised emission FRET. It offers the advantage of being able to simultaneously measure the quenched donor signal and the sensitised acceptor emission together.
13. FLIM is the surest, most comprehensive method of molecular functional association – and the most expensive – available.

Key Reading

Sprague, B.L. and McNally, J.G. (2005). 'FRAP analysis of binding: proper and fitting'. *Trends in Cell Biology 15*/2: 84–91. See also the follow-up paper: Mueller et al. (2010). *Current Opinion in Cell Biology 22*/3: 403–411.

- A comprehensive explanation of FRAP curve analysis

Pietraszewska-Bogiel, A. and Gadella, T.W.J. (2010). 'FRET microscopy: from principle to routine technology in cell biology'. *Journal of Microscopy 241*/2: 111–118.

- FRET explained by one of the leading scientists in the field

References

Ai, H.-W. et al. (2006). 'Directed evolution of a monomeric, bright and photostable version of *Clavularia* cyan fluorescent protein: structural characterization and applications in fluorescence imaging'. *Biochemical Journal 400*/3: 531–540.

Algar, W.L. et al. (2014). 'Semiconductor quantum dots and FRET'. In: Medintz, I. and Hildebrandt, N. (eds.), *FRET – Förster Resonance Energy Transfer: From Theory to Applications*, Chapter 12. Weinheim: Wiley-VCH, pp. 475–605.

Bajar, B.T. et al. (2016). 'A guide to fluorescent protein FRET pairs'. *Sensors 16*/9. pii: E1488.

Bancaud, A. et al. (2010). 'Fluorescence perturbation techniques to study mobility and molecular dynamics of proteins in live cells'.In: Goldman, R.D., Swedlow, J.R., and Spector, D.L. (eds.), *Live Cell Imaging: A Laboratory Manual*, Chapter 5, 2nd ed. pp. 67–93.

Becker, W. (2012). 'Fluorescence lifetime imaging – techniques and applications'. *Journal of Microscopy 247*/2: 119–136.

Blumenthal, D. et al. (2015). 'Universal approach to FRAP analysis of arbitrary bleaching patterns'. *Scientific Reports 5*: 11655.

Broussard, J.A. et al. (2013). 'Fluorescence resonance energy transfer microscopy as demonstrated by measuring the activation of the serine/threonine kinase Akt'. *Nature Protocols 8*/2: 265–281.

Byrne, A.G. et al. (2014). 'Data'. In: Medintz, I. and Hildebrandt, N. (eds.), *FRET – Förster Resonance Energy Transfer: From Theory to Applications*, Chapter 14. Weinheim: Wiley-VCH, pp. 657–755.

Cardullo, R.A. (2007). Chapter 23. In: Sluder, G. and Wolf, D.E. (eds.), *Digital Microscopy*, 3rd ed. *Methods in Cell Biology series*, vol. 81, pp. 479–494.

Chou, K.F. and Dennis, A.M. (2015). 'Förster resonance energy transfer between quantum dot donors and quantum dot acceptors'. *Sensors 15*/6: 13288–132325.

Day, C.A. et al. (2013). 'Analysis of protein and lipid dynamics using confocal fluorescence recovery after photobleaching (FRAP)'. In: Watkins, S.C. and St. Croix, C.M. (eds.), *Current Protocols Select Imaging & Microscopy*. Hoboken, New Jersey: Wiley.

Dobrucki, J.W. et al. (2007). 'Scattering of exciting light by live cells in fluorescence confocal imaging: phototoxic effects and relevance for FRAP studies'. *Biophysical Journal 93*/5: 1778–1786.

Dunn, G.A. et al. (2002). 'Fluorescence localization after photobleaching (FLAP): a new method for studying protein dynamics in living cells'. *Journal of Microscopy 205*/1: 109–112.

Ebrecht, R. et al. (2014). 'Fluorescence lifetime imaging microscopy in the medical sciences'. *Protoplasma 251*/2: 293–305.

Festy, F. et al. (2007). 'Imaging proteins in vivo using fluorescence lifetime microscopy'. *Molecular BioSystems 3*: 381–391.

Feige, J.N. et al. (2005). 'PixFRET, an ImageJ plug-in for FRET calculation that can accommodate variations in spectral bleed-throughs'. *Microscopy Research and Technique 68*/1: 51–58.

Gadella, T.W.J. (ed.) (2009). *Laboratory Techniques in Biochemistry and Molecular Biology*, vol. 33, *FRET and FLIM Techniques*. Oxford: Elsevier.

Hallworth, R. et al. (2006). 'Studying inner ear protein-protein interactions using FRET and FLIM'. *Brain Research 1091*/1: 122–131.

Haustein, E. and Schwille, P. (2007). 'Fluorescence correlation spectroscopy: novel variations of an established technique'. *Annual Review of Biophysics and Biomolecular Structure 36*: 151–169.

Ishikawa-Ankerhold, H.C., Ankerhold, R., and Drummen, G.P. (2012). 'Advanced fluorescence microscopy techniques – FRAP, FLIP, FLAP, FRET and FLIM'. *Molecules 17*/4: 4047–4132. doi: 10.3390/molecules17044047.

Levitt, J.A. et al. (2009). 'Fluorescence lifetime and polarization-resolved imaging in cell biology'. *Current Opinion in Biotechnology 20*/1: 28–36.

Llères, D., James, J., Swift, S., Norman, D.G., and Lamond, A.I. (2007a). 'Quantitative analysis of chromatin compaction in living cells using FLIM-FRET'. *Journal of Cell Biology 187*/4: 481–496.

Llères, D., Swift, S., and Lamond, A.I. (2007b). 'Detecting protein-protein interactions in vivo with FRET using multiphoton fluorescence lifetime imaging microscopy (FLIM)'. *Current Protocols in Cytometry* Chapter 12: Unit 12.10.

Malkani, N. and Schmid, J.A. (2011). 'Some secrets of fluorescent proteins: distinct bleaching in various mounting fluids and photoactivation of cyan fluorescent proteins at YFP-excitation'. *PLoS One 6*/4: e18586.

Mueller, F. et al. (2010). 'FRAP and kinetic modelling in the analysis of nuclear protein dynamics: what do we really know?' *Current Opinion in Cell Biology 22*/3: 403–411.

Mueller, F. et al. (2012). 'Minimizing the impact of photoswitching of fluorescent proteins on FRAP analysis'. *Biophysical Journal 102*/7: 1656–1665.

Murphy, D.B. and Davidson, M.W. (2013). *Fundamentals of Light Microscopy and Electronic Imaging*, 2nd ed. New Jersey: Wiley Blackwell.

Patterson, G.H. and Lippincott-Schwartz, J. (2002). 'A photoactivatable GFP for selective photolabeling of proteins and cells'. *Science 297*/5588: 1873–1877.

Pietraszewska-Bogiel, A. and Gadella, T.W.J. (2010). FRET microscopy: from principle to routine technology in cell biology. *Journal of Microscopy 241*/2: 111–118. doi: 10.1111/j.1365-2818.2010.03437.x.

Presley, J.F. (2006). 'Measurement of protein motion by photobleaching'. In: Stephens, D. (ed.), *Cell Imaging*, Chapter 6. Bloxham: Scion Publishing Ltd., pp. 119–142.

Ries, J. and Schwille, P. (2012). 'Fluorescence correlation spectroscopy'. *Bioessays 34*/5: 361–368.

Roszik, J. et al. (2008). 'AccPbFRET: an ImageJ plugin for semi-automatic, fully corrected analysis of acceptor photobleaching FRET images'. *BMC Bioinformatics 9*: 346.

Sattarzadeh, A. et al. (2015). 'Green to red photoconversion of GFP for protein tracking *in vivo*'. *Scientific Reports 5*: 11771.

Shamirian, A. et al. (2015). 'QD-based FRET probes at a glance'. *Sensors 15*/6: 13028–13051.

Stepensky, D. (2007). 'FRETcalc plugin for calculation of FRET in non-continuous intracellular compartments'. *Biochemical and Biophysical Research Communications 359*/3: 752–758.

Yang, J. et al. (2010). 'An improved strip FRAP method for estimating diffusion coefficients: correcting for the degree of photobleaching'. *Journal of Microscopy 238*/3: 240–253.

Further Reading

Axelrod, D., Koppel, D.E., Schlessinger, J., Elson, E., and Webb, W.W. (1976). 'Mobility measurement by analysis of fluorescence photobleaching recovery kinetics'. *Biophysical Journal 16*/9: 1055–1069.

Bajar, B.T. et al. (2016). 'Improving brightness and photostability of green and red fluorescent proteins for live cell imaging and FRET reporting'. *Scientific Reports 6*: 20889.

Berney, C. and Danuser, G. (2003). 'FRET or no FRET: a quantitative comparison'. *Biophysical Journal 84*/6: 3992–4010.

Clegg, R.M. (2012). 'Fluorescence and FRET: theoretical concepts 101'. In: Dyson, H.J. (ed.), *Comprehensive Biophysics*, Chapter 24, vol. 1. Amsterdam: Elsevier B.V, pp. 592–617.

Day, R.N. and Davidson, M.W. (2012). 'Fluorescent proteins for FRET microscopy: Monitoring protein interactions in living cells'. *Bioessays 34*/5: 341–350.

Grecco, H.E. and Verveer, P.J. (2010). 'FRET in cell biology: still shining in the age of super-resolution'. *ChemPhysChem 12*/3: 484–490.

Hoi, H. et al. (2014). 'FRET with fluorescent proteins'. In: Medintz, I. and Hildebrandt, N. (eds.), *FRET – Förster Resonance Energy Transfer: From Theory to Applications*, Chapter 11. Weinheim: Wiley-VCH, pp. 433–473.

Jares-Erijman, E.A. and Jovin, T.M. (2003). 'FRET imaging'. *Nature Biotechnology 21*/11: 1387–1395.

Lorén, N. et al. (2015). 'Fluorescence recovery after photobleaching in material and life sciences: putting theory into practice'. *Quarterly Reviews of Biophysics 48*/3: 323–387.

Medintz, I. and Hildebrandt, N. (eds.), (2014). *FRET – Förster Resonance Energy Transfer: From Theory to Applications*. Weinheim: Wiley-VCH. (Especially see Chapter 5: 'How to apply FRET: from experimental design to data analysis'; Chapter 11: 'FRET with fluorescent proteins'; Chapter 12: 'Semiconductor quantum dots and FRET' and Chapter 14: 'Data'.)

Miyawaki, A. (2011). 'Development of probes for cellular functions using fluorescent proteins and fluorescence resonance energy transfer'. *Annual Review of Biochemistry 80*: 357–373.

Padilla-Parra, S. and Tramier, M. (2012). 'FRET microscopy in the living cell: different approaches, strengths and weaknesses'. *Bioessays 34*/5: 369–376.

Phair, R.D. and Misteli, T. (2001). 'Kinetic modeling approaches to in vivo imaging'. *Nature Reviews Molecular Cell Biology 2*/12: 898–907.

Rodighiero, S. et al. (2008). 'Fixation, mounting and sealing with nail polish of cell specimens lead to incorrect FRET measurements using acceptor photobleaching'. *Cellular Physiology and Biochemistry 21*/(5–6): 489–498.

Roy, R. et al. (2008). 'A practical guide to single-molecule FRET'. *Nature Methods 5*/6: 507–516.

Tramier, M. et al. (2006). 'Sensitivity of CFP/YFP and GFP/mCherry pairs to donor photobleaching on FRET determination by fluorescence lifetime imaging microscopy in living cells'. *Microscopy Research and Technique 69*/11: 933–939.

Trembecka, D.O. (2010). 'Conditions for using FRAP as a quantitative technique – influence of the bleaching protocol'. *Cytometry 77A*/4: 366–370.

Internet Resources

ImageJ SimFRAP curve fitting: https://imagej.nih.gov/ij/plugins/sim-frap/index.html.

General FRET scripts: http://karri.anu.edu.au/corry/software.html#fretty.

FRET pair behaviour: https://www.microscopyu.com/tutorials/f%C3%B6rster-fluorescence-resonance-energy-transfer-with-fluorescent-proteins.

Roszik, J. et al. (2008). 'AccPbFRET: an ImageJ plugin for semi-automatic, fully corrected analysis of acceptor photobleaching FRET images'. *BMC Bioinformatics 9*: 346. http://biophys.med.unideb.hu/accpbfret/.

Alternative acceptor photobleaching plugin:

Stepensky, D. (2007). 'FRETcalc plugin for calculation of FRET in non-continuous intracellular compartments'. *Biochemical and Biophysical Research Communications 359*/3: 752–758. https://imagej.nih.gov/ij/plugins/fret/fret-calc.html.

Feige, J.N. et al. (2005). 'PixFRET, an ImageJ plug-in for FRET calculation that can accommodate variations in spectral bleed-throughs'. *Microscopy Research and Technique 68*/1: 51–58. http://www.med.unc.edu/microscopy/files/courses/spring-2013-lm/7-molecular-interactions-from-afar/pixfret-userguide and http://www.unil.ch/cig/home/menuinst/research/research-groups/prof-desvergne.html#tabs-ad1640159a2e492d8b8dd070d9b95ae6.

Notes

1. I am grateful to Trevor Booth, at Newcastle University's Bio-Imaging Facility, for details of this test specimen.
2. The technique is named after Theodor Förster, who published the first paper in 1946. In a second paper two years later, Förster postulated that the rate of energy transfer, the FRET efficiency, E_{FRET}, depends on the inverse sixth power of the distance between the donor and acceptor fluorophore molecules.
3. The calculation of R_0 involves complex equations, which are not covered here. For further information, see: Pietraszewska-Bogiel and Gadella (2010) or Cardullo (2007). The emission dipole of the donor and the acceptor absorption dipole of the donor must be parallel, rather than perpendicular, to one another. An analogy is the position of a radio antenna, which for best results must point towards the transmitting station.
4. For FRET to occur, the fluorescently labelled pair of molecules must lie within 10 nm to one another.
5. Since these images are to be used for numerical analysis, it is best to collect the images over a 12-bit, rather than 8-bit, dynamic range.

25

Colocalisation

Outline

25.1 Introduction

Colocalisation is an essential part of image analysis. It is a topic that only recently has a clear consensus of a 'correct' way to do this, and many different approaches have been suggested. Furthermore, misunderstandings and errors have been published, even in well-respected review articles. Using Pearson's correlation coefficient as a statistical test to measure relationships between two fluorophores in an image was first introduced into microscopy in 1992. Whilst this was good, the confusion surrounding the topic of colocalisation is not helped by the fact that the original paper attempted to measure two variables (correlation and overlap) with a single metric. This misunderstanding of the true nature of Pearson's correlation coefficient has been, in part, responsible for the introduction of up to ten different metrics (e.g. PCC, MOC, k_1, k_2, M_1, M_2, m_1, m_2, PDM and ICQ). In most cases, no more than three (PCC, M_1 & M_2) are required to accurately and precisely describe colocalisation.

This chapter does not attempt to describe the entire field of colocalisation. Object-based correlation analysis is not covered in depth, since pixel-based intensity analysis is much more common. For those interested, some references to object-based colocalisation are given at the end of the chapter. Various software packages use different approaches through the use of different algorithms. The simplicity with which these turn-key software packages can be used and the ease with which different parameter settings can be adjusted uncritically to tweak colocalisation values means that similar image data sets can give widely differing results unless some thought is given to the analysis. This chapter refers to general principles and points to watch out for. Each of the major microscope companies includes colocalisation modules within their image acquisition software. In addition, there exist dedicated commercial image analysis software packages as well as the freeware JACoP 2 and Coloc 2 supported by Fiji and ImageJ.

25.2 Colocalisation Defined

Colocalisation refers to the observation of the spatial overlap between two different fluorescent labels within the sampling pixels of an image. Generally only two labels are analysed in each image, and these are commonly pseudo-coloured red and green. The purpose is to see if the two differently labelled target proteins or molecules are *located* in the same area of the cell or very near to one another – colocalisation cannot prove interaction. We are evaluating correlation in terms of *intensity* at each 2D sampling pixel, or 3D sampling voxel.

The definition of colocalisation can be split into two different phenomena: *co-occurrence* which refers to the presence of the emission signal of two (possibly unrelated) fluorophores recorded in the same pixel, and *correlation*, a much more significant statistical relationship (measured by Pearson's correlation coefficient) between the two fluorophores, suggestive – but not indicative – of a biological interaction (Adler et al., 2008). A statistical relationship suggests co-distribution of the two fluorescent probes in proportion to one another, both within and between structures.

There are several caveats associated with the mere 'yellow pixels equals red-green image merge' approach to determine whether co-occurrence is present. These are shown in Box 25.1. However, with due care in controlling the Nyquist sampling, intensity, display gamma, resolving power of the objective used and region of the cell that is imaged, it is possible to use this approach to draw conclusions about co-occurrence between two fluorophores, and hence the proteins that these fluorophores are targeted to. Controlling all these image acquisition parameters takes time and effort, whereas easier image collection demands a more complex statistical analysis using correlation.

Correlation and linear regression analysis are not the same. Correlation is almost always used when measuring both variables together, to calculate the probability of a relationship between *pairs* of each variable. It does not matter which of the two variables is named 'A' and which is named 'B': the same correlation coefficient is computed if the variables are swapped over. The degree, the strength, otherwise referred to as the *quality*, of the correlation is given by Pearson's correlation coefficient (see Section 25.5).

Box 25.1 Merged 'Red-Green' Image Caveats

There is much published work suggesting that correlation and colocalisation can be inferred from inspecting a merge of red and green pixels, shown as yellow pixels in the image. This is not possible. Co-occurrence refers to the presence of the emission signal of two (possibly unrelated) fluorophores recorded in the same pixel. With care, conclusions about co-occurrence may be drawn from a merged image of red and green pixels. The following caveats apply:

- First, the pixel *intensities* from both channels determine the merge colour.
- Second, the gamma of the display – which can be set quite arbitrarily – will influence the yellow merge colour.
- Third, whether you use a CCD on a widefield microscope or PMT on a confocal, the camera has a spectral sensitivity curve. Two signals of different wavelength that actually have the same intensity will appear to be of differing brightness.
- Fourth, our eyes are prone to optical illusions and are very poor at determining colour brightness or intensity. In the figure below, the grey squares of the Adelson illusion (left) are displayed on a computer monitor at the same intensity and hue, whilst the 'blue' and 'green' colours seen in the spiral are the same colour, so there are three colours, not four. See: https://en.wikipedia.org/wiki/Checker_shadow_illusion for an explanation of the Adelson illusion and http://imagej.net/Colocalization_Analysis and http://imagej.net/Colocalization_-_hardware_setup_and_image_acquisition from the Fiji and ImageJ analysis webpages for further information on the spiral illusion.
- Fifth, each grey value in the image, which encodes intensity, is independent of the display.
- Sixth, in a two-dimensional image, the overlap may suggest correlation, which may, in fact, be co-occurrence because the fluorophores do not overlap axially in the third (unseen) dimension.
- Seventh, if you take this crude approach, it is quite likely that you have not considered Nyquist's sampling theorem and have not sampled the specimen properly. In this case, false red-green overlap will lead to a much higher 'colocalisation' result than is actually the case.

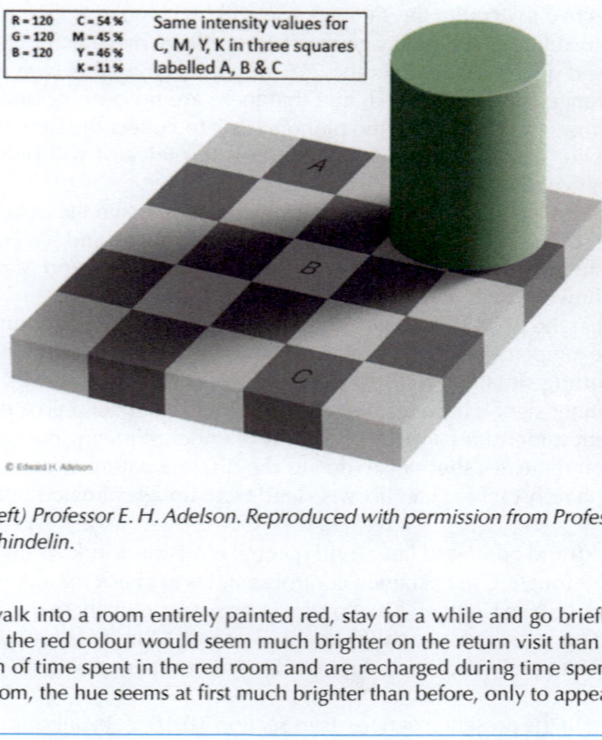

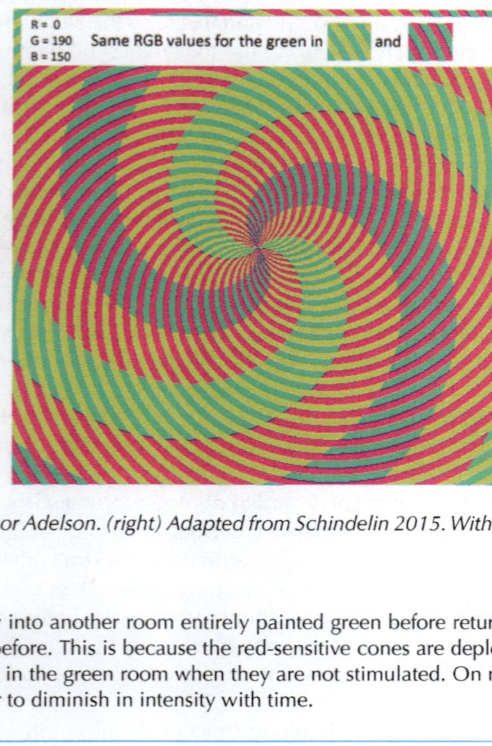

Source: (left) Professor E. H. Adelson. Reproduced with permission from Professor Adelson. (right) Adapted from Schindelin 2015. With permission from J. Schindelin.

If we walk into a room entirely painted red, stay for a while and go briefly into another room entirely painted green before returning to the red room, the red colour would seem much brighter on the return visit than before. This is because the red-sensitive cones are depleted during any length of time spent in the red room and are recharged during time spent in the green room when they are not stimulated. On returning to the red room, the hue seems at first much brighter than before, only to appear to diminish in intensity with time.

Linear regression analysis is appropriate to use in experimental analysis when one independent variable is manipulated to determine the effect of alteration upon the other dependent variable. Linear regression quantifies goodness of fit with the coefficient of determination, r^2, sometimes shown in uppercase as R^2. The higher the value of the coefficient of determination, the more accurately the variance of the dependent variable is explained by variance of the independent variable. If the same data is put into correlation calculations (which is not appropriate), the square of Pearson's *r*, from correlation analysis, will equal r^2 from regression. In the case of regression analysis it *does* matter which variable is termed the independent variable (x-axis) and which the dependent variable (y-axis). The best-fit line that most accurately predicts A from B is not the same as the line that predicts B from A (however both those lines have the same value for r^2). Since – in an image – *two signals* of varying intensity are mapped to the set of pixels that constitute that image and these signals are *collected together*, correlation is therefore the appropriate method of analysis.

Proteins and macromolecules are of much smaller size (ca. 1–50 nm) than the resolving power of a conventional fluorescence microscope. It may be possible, due to high contrast, to discriminate two (punctuate) fluorescent proteins well separated by a distance greater than the resolving power of the microscope. However, the minimum size of the visible signal will always be set by the point spread function of the microscope. This means that two adjacent proteins will always be non-resolvable, regardless of their colour or the pixel size of the detector. Super-resolution techniques may improve matters slightly and improve the chances of discriminating two colocalised proteins that are correlated, but any conclusion of functional colocalisation must be underpinned by other means, such as FRET/FLIM or electron microscopy (<10 nm resolution), single molecule imaging (e.g. Koyama-Honda et al., 2005; Malkusch et al., 2011) or biochemical techniques, such as co-immunoprecipitation (Masters, 2004). Other approaches for studying the behaviour of protein complexes are given in Dwane and Kiely (2011).

25.3 Collecting Images, Locating your Protein in the Cell

Good sample preparation (see Chapter 28) is crucial for accurate results. Colocalisation analysis is sensitive to artifacts; therefore, specimens prepared for analysis must be of the highest quality and care taken with setting up the microscope for image acquisition. Any antibodies used should be of high quality, be as specific as possible (use Fab fragments for better specificity where endogenous immunoglobulins may be present) and not cross-react. For further information, see the DAKO Handbook, 6th Edition (2013). Some anti-fade media can increase background fluorescence. Too high a background signal level will overestimate colocalisation: it buries low-intensity signal and obscures image detail.

The microscope should also be properly aligned, and ideally multiply-stained fluorescent beads (e.g. Tetraspeck or similar, see Chapter 27) should be used to ensure the two fluorescent channels are in absolute registration with respect to one another, and that any shift to compensate for chromatic aberration (see Figure 21.11) is applied. Failure to carry out this crucial step will lead to false results, most likely underestimation of colocalisation. See Abraham et al. (2010) for a well-written example of how to ensure accurate channel registration. Use apochromat objectives, because these are best corrected for chromatic aberration. They will cause the minimum amount of misregistration between channels. Use the same objective to acquire the samples you will be comparing.

Although not necessary, it helps to use the greatest bit depth possible: spread the image data across a large dynamic range. 12 bit and 16 bit depth are better for collection information that will be used quantitatively than the 256 grey levels available with the dynamic range of the detector set to 8 bit. Make sure that the dynamic range is properly filled, and that there are no over- or under-saturated pixels clipping the photon intensity data set. On a confocal microscope, ensure that the pinhole is set to collect the same optical slice thickness (this will probably require different Airy Unit settings for each channel). If the section is too thick, it will blur the image. Blurred images will overestimate colocalisation, giving false-positive results.

If possible, deconvolve your images. Deconvolution of both widefield and confocal data sets will help reassign the out-of-focus blur to its origin, reducing the likelihood of a false-positive result (refer to the Abraham et al., 2010, paper cited above and see Figure 21.11). Deconvolution also suppresses background and sharpens high-frequency fine detail of low intensity (Landmann and Marbert, 2004) reducing the possibility of false colocalisation. Deconvolution, however, will not entirely remove background signal.

Proper Nyquist sampling should be adhered to. This means that the pixel size should be at least one-half, preferably one-quarter of the minimum resolvable distance of the microscope, in both the x,y plane laterally as well as along the z-axis, axially. Undersampled images, with larger pixels or voxels (the more commonly occurring situation) will overestimate the degree of colocalisation. Avoid oversampling with pixels that are too small; this results in a dimmer signal, because the signal is spread out over a greater number of pixels. Images which are oversampled will give weaker fluorescence intensities and the possibility of underestimating the colocalisation by the presence of a large number of low intensity pixel pairs, and greater shot noise (due to the discrete nature of the signal) in the data set. Oversampling can also lead to photobleaching. Gross photobleaching can likewise lead to an underestimated result, because signal is lost relative to the remaining background.

Use optimised excitation filters and emission filters (in widefield) and optimised lasers and spectral emission windows (in confocal) to maximise the emission signal while avoiding cross-talk or bleed-through. Use unstained control samples to check for autofluorescence and singly-stained samples to check for cross-talk and bleed-through (see Chapter 20). Acquire images sequentially to minimise bleed-through. At least five controls are required to adjust the collection settings for each fluorophore channel (see Section 20.3). Use the single-labelled controls to assess the degree of cross-excitation and emission bleed-through. Use the unstained slide(s) to check the background signal.

Dark noise is random and should not affect colocalisation – but in the presence of noise (see Section 30.10) colocalisation is underestimated. If noise significantly affects the correlation value, then the source of the noise in the sample or microscope should be addressed. Normally, both (i.e. red and green) channels should be collected at the same time and under the same conditions, so inherent noise should not unduly affect the result. Poisson noise can be minimised by extending the image acquisition time, by sacrificing spatial resolution for a good signal-to-noise ratio and by binning, or summing replicate images by averaging. A stratagem for correcting for excessive noise is given later in the chapter in Box 25.4.

25.4 Red-Green Analysis for Co-occurrence

There is no such thing as a green dye or a red dye, since they have broad emission spectra, not a single wavelength corresponding to a certain 'colour'. You can change the look-up tables (LUTs) of the images so that the emission signal in one channel is coloured

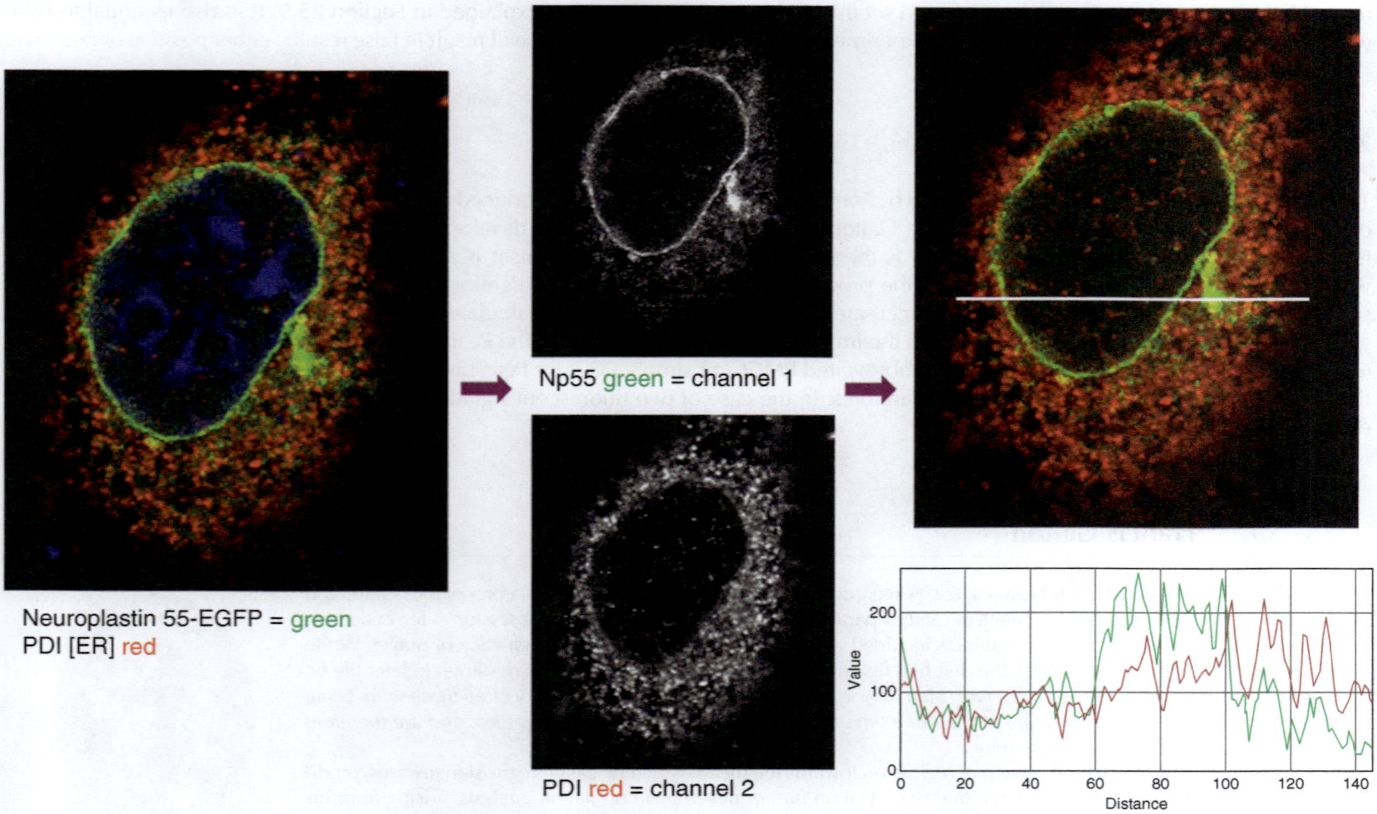

Figure 25.1 ***Analysing red-green co-occurrence*** *Using the RGB line profiler plug-in from ImageJ/Fiji to analyse co-occurrence. See the text for further details.*

'green' and the other is 'magenta'. Clearly, the colours used are always false. These false colours are only useful (to us, as apes with trichromatic vision) to tell which individual channel relates to the fluorophores in use. The electronic detectors we use (CCD, PMT) only see photons: colour *per se* is irrelevant. The spectral range of the signal collected is determined by the fluorescence emission filters that we use in front of the detector.

The advantage of a simple red-green analysis is that the interpretation is very simple. It can work to indicate co-occurrence (but not correlation) if you are careful. However, this simplistic approach to establishing colocalisation is very prone to misinterpretation and error. It is also highly susceptible to differences in pixel intensities between the two fluorophore channels within the image. In the RGB additive colour model, combining the primary colours red and green results in a yellow hue. The brightness of this combination colour is interpreted as a degree of colocalisation. You *cannot* assume correlation merely because your digital image shows yellow pixels where a red pseudo-coloured image and green pseudo-coloured apparently overlap (see Box 25.1). If you are careful, you can assume the presence of co-occurrence, but it must not be confused with correlation. This point cannot be overemphasised, because this mistake is most often made[1] when first attempting to assess colocalisation through correlation analysis.

Unequal red-green signal intensities will result in false-negative, or underestimated, results, since one signal can 'swamp' and occlude the other. Poor human judgement and/or analysis of 3D data sets collapsed into a two-dimensional image (see Figure 2 in Scriven et al., 2008) can lead to a false-positive result. There are other reasons for false results, the principal one arising from bleed-through between fluorophores. The phenomena of excitation cross-talk and emission bleed-through is discussed in Chapter 20. Both will lead to false-positive results.

Co-occurrence can be verified qualitatively by drawing a line profile (Figure 25.1) across the area of interest within the image. This can be done relatively easily with freeware image analysis packages, such as ImageJ or Fiji using the RGB Profiler plug-in. If one stain is very diffuse and the other is discretely localised, it may appear that co-occurrence occurs, but this may not, in fact, be the case. A line profile will help show co-occurrence or exclusion of signals. Another approach when confronted with diffuse signal present in all – or almost all – pixels, is to calculate the product difference from the mean (PDM) values (Section 25.10) and test whether each fluorophore signal varies in synchrony or not.

The only unambiguous way to determine whether colocalisation occurs is to perform a statistical analysis with a scatterplot on a properly sampled image, calculate a thresholded Pearson's correlation coefficient (tPCC) *and relate this data back to what you know of the underlying biological phenomena*. How to set thresholds on the scatterplot is explained in Section 25.9. It is also essential to have well-prepared and well-stained control and experimental specimens, since artifacts will result in false results: either positive or negative.

25.5 Measuring Colocalisation – Pearson's Correlation Coefficient

The correlation coefficient was first developed by Francis Galton (Box 25.2) and introduced in its modern form by Karl Pearson, a student of Galton's. Although the imaginative ideas of Galton laid the foundation, Pearson developed the rigorous mathematics of correlation. It was Pearson who remarked that "statistics is the grammar of science". This truism is all too easily forgotten by those who say "I want to do colocalisation" and assume that the procedure is both routine and straightforward. The arithmetic is straightforward, but setting the thresholds precisely to enable an accurate correlation coefficient to be calculated is the difficult part. In truth, this requires a good knowledge of the scientific basis behind the images that you have collected. The Pearson product-moment correlation coefficient (denoted[2] by the symbol r for a sample, and abbreviated PMCC, or simply PCC, for Pearson's correlation coefficient) is a measure of the linear correlation between two non-ranked variables. In the case of two fluorescent signals within an image, these variables are often denoted R (for red) and G (for green).

Box 25.2 Francis Galton

Sir Francis Galton (born 16th February 1822) is regarded as founder of biometry. He was a precocious child, and throughout his life had a prolific intellect, with a particular interest – bordering upon obsession – for collecting measurements and analysing data. Subjects included measuring female beauty and the efficacy of prayer. While still a young man, the death of his father left him financially independent but emotionally destitute. In later life he was, from his own experience, to observe: '*Men who leave their mark on the world are very often those who, being gifted and full of nervous power, are at the same time haunted and driven by a dominant idea, and are therefore within a measurable distance of insanity*'.

Galton elucidated the natural pattern of regression towards the mean (which he called regression towards 'mediocrity') in successive generations. From that work Galton further developed correlation analysis, arising from his studies concerning heredity. He was a self-taught naturalist, anthropologist, astronomer, meteorologist and psychologist. Charles Darwin was his much better known half-cousin: they corresponded after publication of *On the Origin of Species*. In the last part of his life, Galton applied statistics to the measurement of human traits.

Source: Galton.org.

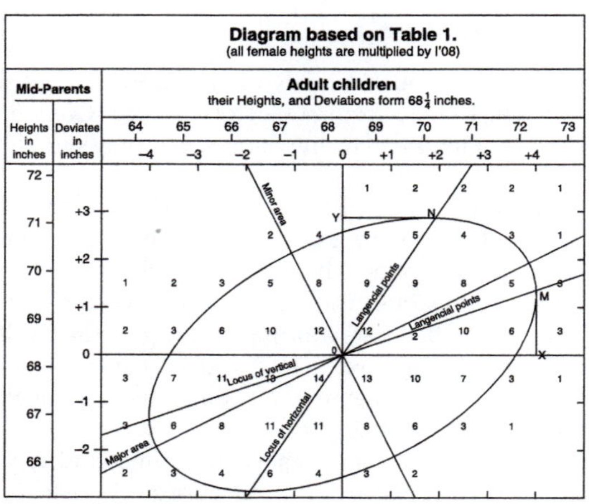

Source: Galton 1875.

About 1875, Galton used sweet pea plants as a model to study how strongly the characteristics of one generation manifested in the next. Intensely practical, he enlisted the help of nine of his friends to grow up and harvest the next generation of seeds, then return them for analysis. A slope of 45° would indicate that the size of the daughter pea would equal that of the mother pea. No slope would indicate that the daughter pea diameter was the same, regardless of the size of mother pea. Galton realised that the median weights of daughter seeds from a particular size of mother seed approximately described a straight line with a modern r value of 0.33, indicating a regression to the mean. Later, Galton set up a biometric laboratory (now part of University College, London) and extended this work to measuring humans, including in January 1884 publishing 'The weights of British noblemen during the last three generations' (*Nature 29*: 266–268).

Galton had built a machine that he called the quincunx. A working example can be seen at https://www.youtube.com/watch?v=AUSKTk9ENzg. It was designed to illustrate the central limit theorem: how a number of independent events could lead to a normal distribution. Pearson became Galton's protégé sometime in the last decade of the 19th century and published his work on the correlation coefficient in 1896. Later Pearson took over the research laboratory founded by Galton and assumed the first chair of the new department of applied statistics (set up with a bequest from Galton) upon his death on 17th January 1911.

References

Pearson, K. (1896). 'Mathematical contributions to the theory of evolution. III. Regression, heredity and panmixia'. *Philosophical Transactions of the Royal Society A 187*: 253–318.

Rodgers, J.L. and Nicewander, W.A. (1988). 'Thirteen ways to look at the correlation coefficient'. *The American Statistician 42*: 59–66.

The value of r lies between +1 and −1. The Pearson correlation is +1 in the case of a perfectly correlated relationship; 0 indicates no linear relationship, or complete independence; and −1 a perfectly anti-correlated relationship. Note that an anti-correlation (e.g. where protein A expression suppresses protein B) is not the same as no correlation (e.g. nuclear and mitochondrial staining). Pearson's correlation coefficient was introduced into microscopy (Box 25.3) by Eric Manders in 1992. Upon inspection, this equation is less formidable than at first sight. For each pixel pair, the product is positive when both red and green intensities are either above or below their respective means and negative when one intensity value is above and the other value below their respective mean values.

Box 25.3 Pearson's Correlation Coefficient

The formula for the Pearson's product-moment correlation coefficient (PCC, or r) is:

$$r = \frac{\sum (R_i - R_{av}) \cdot (G_i - G_{av})}{\sqrt{\sum (R_i - R_{av})^2 \cdot \sum (G_i - G_{av})^2}}$$

The numerator is the sum of the product of the intensities of the two fluorophores (red and green) minus their mean values, taken pixel by pixel over the entire data set. The sum in the numerator is maximised when pairs of both red and green fluorophores have similar departures from the mean (i.e. high:high and low:low pairs), particularly when this difference from the mean is large. Conversely, a combination of high with low pixel pair intensities reduces the sum of their products.

The denominator calculates the maximum possible product if the red and green data sets of pixel intensities are perfectly correlated. The denominator limits the range of the Pearson's correlation co-efficient from −1, through 0 to +1.

Pearson's correlation coefficient

R_i = intensity (pixel) value in red channel (x)
R_{av} = mean intensity in red channel (y)
G_i = intensity (pixel) value in green channel
G_{av} = mean intensity in green channel

Since the PCC is based on differences from the mean, intensity pairs near the mean have far less weight than outlying values near the extremes of the distribution. Hence, the consequence of including background pixels (i.e. in effect calculating a global PCC value) will be to over-estimate the true PCC.

The Manders' overlap coefficient (MOC) differs from the PCC as shown. The two measures are similar. The MOC uses absolute intensities, the PCC the difference from the mean, a small but crucial difference. The MOC is blind to low intensity values (see also endnote 5).

$$MOC = \frac{\sum (R_i \cdot G_i)}{\sqrt{\sum (R_i)^2 \cdot \sum (G_i)^2}}$$

Manders introduced the MOC because 'a major advantage' is that it is not sensitive to differences in signal intensities, implying that the PCC was sensitive, besides 'the negative values of the correlation coefficient being difficult to interpret' statement, which is also untrue providing that the thresholded PCC is used. It is recommended that the MOC is not used (see Adler and Parmryd, 2010; Barlow et al., 2010).

Manders' colocalisation coefficients M_1 and M_2 describe the proportions of pixels of non-background intensities (i.e. those above the threshold) in one channel that are also found with intensities above background and the threshold in the second channel. To be classified as 'co-localised', a pixel from the first channel must be paired with a non-background pixel intensity from the second channel.

The M_1 colocalisation coefficient is defined as the ratio of the 'summed intensities of pixels from channel 1 for which the intensity in channel 2 is above zero (or the threshold)' to the 'total intensity in channel 1', and M_2 is defined conversely for the other channel. Therefore, M_1 (or M_2) is a good indicator of the proportion of one signal coincident with the signal from the other channel.

where $i_1 > T_1$ and $i_2 > T_2$

$$M_1 = \frac{\sum_i Ch1_{i,coloc}}{\sum_i Ch1_i} \qquad M_2 = \frac{\sum_i Ch2_{i,coloc}}{\sum_i Ch2_i}$$

$$Ch1_{i,coloc} = Ch1_i \text{ if } Ch2_i > \text{Threshold}$$

When two fluorophores do not co-occur (e.g. a cytoplasmic and nuclear signal) the PCC may report a negative value. This is why it must be cited (using properly set thresholds) in conjunction with M_1 and M_2 colocalisation coefficients.

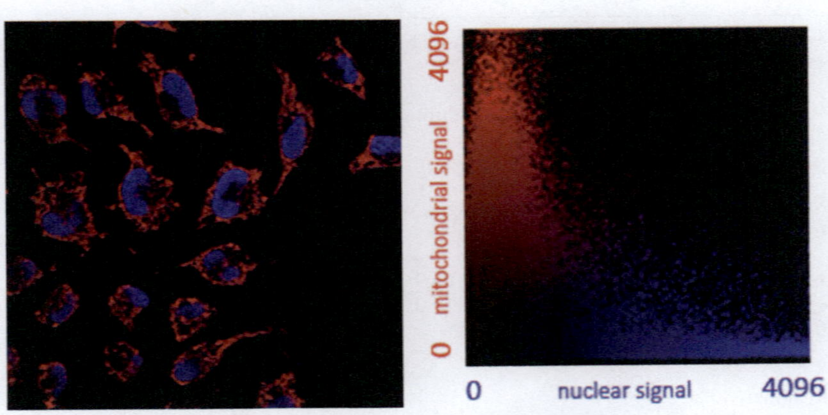

Global Pearson's CC = + 0.13
Thresholded Pearson's CC = -0.231
Manders' Overlap Coeff. = + 0.145
M_1 (blue, nuclei) = 0.263
M_2 (red, mitochondria) = 0.08

Because Pearson's correlation coefficient is a statistical measure, whenever a sample has a restricted range of scores, the correlation will potentially be reduced, and the associated standard deviation will be large; it will be a less reliable metric than one obtained from a large population of values. In most cases this point is immaterial and does not usually matter with most image data sets, because the pixel array in a CCD or PMT is normally sufficiently high to give a large number of pixel pairs (i.e. scores). However, where a small region of interest (ROI) is set, the number of pixel pairs will be reduced. In this case, consider collecting a larger image with a greater number of pixel values. For example, on the confocal microscope, collect a 1024×1024 size image rather than a default 512×512 image. The standard deviations associated with a noisy data set will also be larger than for a 'clean' image with a high signal-to-noise ratio.

Two other conditions must, as far as possible, be met for the correlation coefficient to be accurate. Firstly, the two fluorescent signals must be bivariate normally-distributed data[3]. Secondly, the two signals (i.e. variables) must also represent a single linear distribution, rather than exhibit a quadratic relationship. Since two fluorescent signals are being analysed, there is never a problem with the data not being bivariate, but biological relationships are not always conveniently normal or linear in their distribution. For Pearson's correlation coefficient to report a meaningful relationship between the two variables, there must be no outlying values. Even a small departure from a truly linear relationship can reduce (i.e. underestimate) the value of Pearson's correlation coefficient. Other factors influencing PCC are summarised in Table 25.1.

If there are multiple distributions of differing intensities within the image, each must be analysed separately. This situation often occurs with images showing cells transfected with endogenous fluorescent protein. The expression levels may well be different in different cells and show up as distinct and different distributions in the scatterplot. A zero, or near-zero, value for Pearson's correlation coefficient does not necessarily mean that there is no correlation. A quadratic relationship – rather than a linear one – will produce this zero result, because PCC significantly underestimates nonlinear relationships. This point was made clear by Anscombe (1973) with his quartet data[4]. Where a nonlinear distribution is known, or suspected, Spearman's rank correlation coefficient, a non-parametric test, can be substituted for PCC. See Adler and Parmryd (2010) and French et al. (2008) for details on applying Spearman's rank correlation coefficient.

Linear transformations have no effect on Pearson's correlation coefficient: it is independent[5] of gain and offset, provided that the two entire data sets of values remain within the saturation range of the PMT or CCD camera. Equally, Pearson's correlation coefficient is not sensitive to a *global* change in the intensity of the background pixels. It is only affected by variability in *both* the signal and the background. However, proper calculation of the PCC assumes that the image is noise free. This is almost never the case. With noise in excess of about 10 dB (i.e. an increase of 10^1 times), the PCC will be under-reported, or underestimated. One approach to correcting for noise in the image is to apply a replicate correction factor (Adler et al., 2008; Bergholm et al., 2010; Daly et al., 2012). Providing that the images from each channel are replicated (i.e. two 'red' and two 'green' images are acquired) a noise correction factor can be applied to the PCC. This procedure is shown in Box 25.4.

> Pearson's correlation coefficient cannot be used to describe the *amount* of colocalisation. Its sole purpose, when calculated accurately, is to describe the *quality* of a single linear relationship between two variables. The PCC has no means of describing the extent of overlapping of the two fluorophores. The Manders' colocalisation coefficients, M_1 and M_2 (see below), are used to describe the distribution, or degree (i.e. extent) of overlap of the two signals.

Table 25.1 *The causes of over- and under estimation of PCC values*

Factors causing under-estimation of PCC	Factors causing over-estimation of PCC
High background containing non-specific signal; not perfectly black or very low intensity	Too many low-intensity background pixels will give an over-estimation of colocalisation
Noise, fewer pixels remain which sample true signal	Cross-talk or bleed-through
Chromatic shift; non registration of channels	Blurred images
Over-sampling, weaker fluorescence intensities, the presence of a large number of low intensity pixel pairs, and greater shot noise in the data set	Under-sampled images, with larger pixels or voxels (the more commonly occurring situation)
Unequal stoichiometry of red-green signal intensities outside the dynamic range of the detector	Costes thresholding method reports over-estimate with high staining density, or asymmetric staining distribution.
Smoothing data using 'de-noising' filters	
PCC significantly under-estimates non-linear relationships (shown by Anscombe quartets)	

Box 25.4 Adler and Parmryd's RBNCC Noise Correction

In the presence of noise, colocalisation is under-estimated. If noise is introduced into the image greater than about 10 times the background (i.e. 10 dB), the correlation coefficient is significantly under-reported. The replicate-based noise-corrected correlation (right-hand figure) is a method employing four images – a pair from each channel – to calculate a correction factor to apply to Pearson's correlation coefficient. The method is explained in Daly et al. (2012); the formal proof given in Bergholm et al. (2010).

1. Acquire two images of each fluorophore.
2. Calculate the Pearson's correlation coefficient, **r**, between the replicate pairs from each channel. PCC $\mathbf{r}_{AA}$ is calculated from image A1 and A2, and $\mathbf{r}_{BB}$ likewise from B1 and B2.
3. Calculate the correction factor: $C_{AB} = 1/\sqrt{r_{AA}.\, r_{BB}}$
4. Calculate each PCC value for all four permutations: A1-B1, A1-B2, A2-B1 and A2-B2.
5. Calculate the arithmetic mean, R_{AB}, of all four inter-fluorophore correlation coefficients.
6. Use the correction factor to adjust the mean correlation coefficient: $R_{AB} = r_{AB}/\sqrt{r_{AA}.\, r_{BB}}$

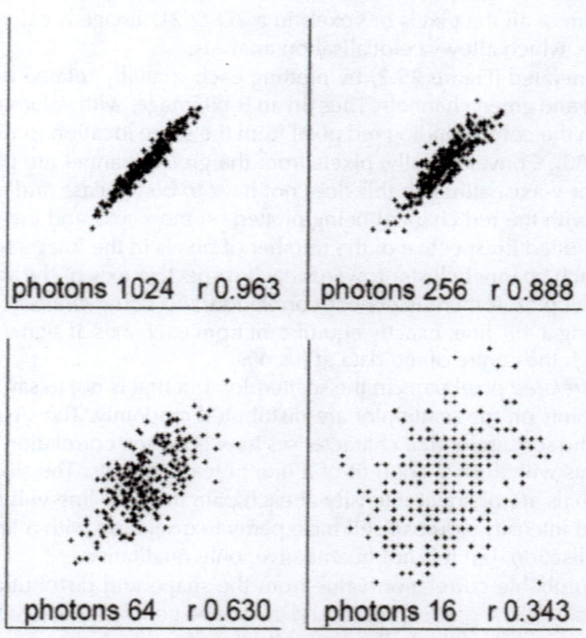

These scatterplots illustrate the dependence of measured colocalisation on noise. Poisson noise, arising from the number of photons in the data, degrades otherwise perfectly correlated data. As the numbers of photons drop, the dispersion of the data increases, and the PCC value falls. Source: Figure 2, page 8, Adler et al. 2008. Reproduced with permission from John Wiley & Sons.

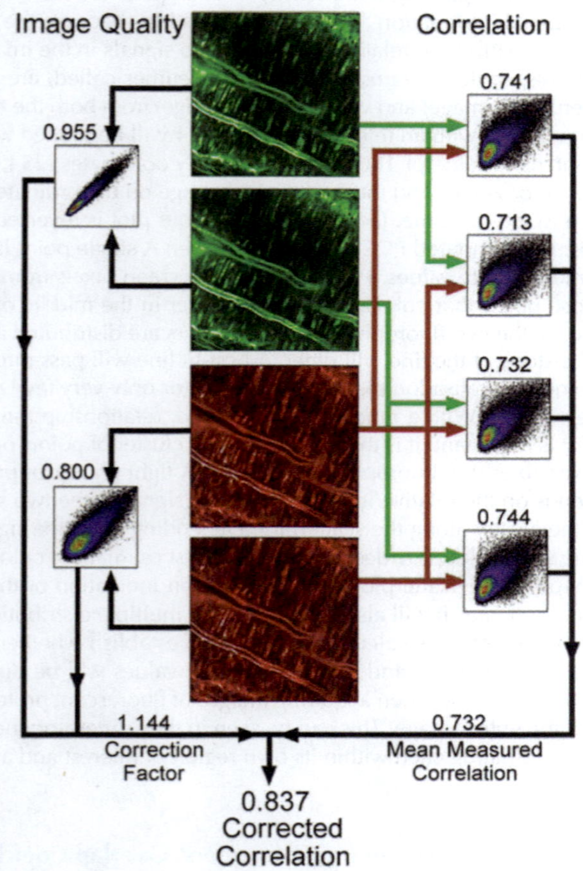

Quality			Between Fluorophore					Final
A1-A2 r_{AA}	B1-B2 r_{BB}	Correction C_{AB}	A1-B1	A1-B2	A2-B1	A2-B2	Mean A-B r_{AB}	Corrected R_{AB}
0.955	0.8	**1.144**	0.741	0.713	0.732	0.744	**0.732**	**0.837**

In the example shown, taken from Figure 4, Daly et al. (2012; reproduced with permission from Springer), the image quality of the green fluorophore (A) is of higher quality than that of the red fluorophore, B, which is noisy. When thresholds are applied to the image, the noise-correction factor is smaller than when applied to the image globally. This underlines the importance of thresholding, to exclude background and non-signal pixels from the correlation calculation.

References

Alder, J. et al. (2008). 'Noise and colocalization in fluorescence microscopy: solving a problem'. *Microscopy & Analysis*. Wiley.

Bergholm, F., Adler, J., and Parmryd, I. (2010). 'Analysis of bias in the apparent correlation coefficient between image pairs corrupted by severe noise'. *Journal of Mathematical Imaging and Vision 37*: 204–219.

Daly, C.J., Parmryd, I., and McGrath, J.C. (2012). 'Visualisation and analysis of vascular receptors using confocal laser scanning microscopy and fluorescent ligands'. In: *Methods in Molecular Biology*, Chapter 5, vol. 897, Receptor Binding Techniques. New York: Springer, pp. 95–107.

When two fluorophores that do not co-occur are tested for correlation (e.g. a nuclear stain with a localised cytoplasmic or a mitochondrial stain) with Pearson's correlation coefficient, a negative PCC value may be reported. It is important to distinguish between a truly negative correlation, where a high-intensity signal in one channel correlates with a low-intensity signal in the other, from an artifactual pseudo-correlation where the signals are not located in the same place in the specimen or the image. The cases are distinct; it is only relevant to measure correlation using pixels that contain signal from both fluorophores. Adler and Parmryd (2014) advise that when the fluorophores are in different places, to simply report the lack of co-occurrence and omit these uninformative negative correlations. To avoid this error, thresholding should be carried out prior to the calculation of Pearson's correlation coefficient to exclude pixel pairs – such as background – which do not contain both signals[6]. In addition, biologically distinct regions should be analysed separately. How to set a threshold on the data set for colocalisation analysis is explained in Section 25.9.

25.6 The Purpose of Scatterplots

A scatterplot summarises a relationship between two variables by plotting the value of each variable for each data point in a pair. For images this means plotting the intensity of each pixel for the first channel against the intensity of the equivalent pixel in the second channel. Spatial information about the image is lost (because the sum of all the pixels or voxels in a 2D or 3D image is calculated) in favour of the distribution relationship of the two signals in the image, which allows colocalisation analysis.

Scatterplots, or fluorograms as they are sometimes called, are generated (Figure 25.2) by plotting each spatially related pixel (in a two-dimensional image) and voxel (in a 3D image) from both the red and green channels. Thus (in an 8-bit image, with values from 0 to 255) a green pixel with an intensity value of 254 will be plotted with the corresponding red pixel from the same location in the image, with an intensity value of 180 at the Cartesian coordinates (254, 180). Conventionally, pixels from the green channel are plotted on the abscissa, or x-axis, and those of the red image on the ordinate, or y-axis, although this does not have to be the case and makes no difference to the outcome. (Sometimes the colour plot is reversed, with the red channel being plotted on the x-axis and the green on the y-axis; the computed PCC value is the same.) A single point is plotted irrespective of the number of pixels in the image with these exact same intensity values. Pure red and pure green pixels from each channel cluster towards each respective axis of the scatterplot while those pixels that colocalise tend to cluster in the middle of the plot. If there is perfect correlation *and equal intensity* between the signals of the two fluorophores, all pixel pairs are distributed along a 45° line, exactly equidistant from each axis. If signal intensity differs, the slope of the line will differ. A best-fit line will pass through the centre of the data at ($\bar{X}$, $\bar{Y}$).

Where no colocalisation exists, there are no (or only very few) *correlated* pixel pairs in the scatterplot, but that is not to say there are only a few pixels. With a random, uncorrelated, relationship the points on the scatterplot are distributed randomly. The *shape* of the scatterplot is important: it is the tightness of the cluster of points on the scatterplot that characterises how good the correlation between the two variables (i.e. fluorophore signals) is. A tight cluster of points will allow a good fit of a line of least squares. The slope of the line depends on the relative intensities of each signal. If the two signals are of equal intensity at each pair, then the line will lie at 45° through the cluster along the scatterplot. Depending upon the signal intensity, you can still have perfect correlation with a line of any slope. The scatterplot provides a good visual first estimate of colocalisation, but it is not quantitative, only qualitative.

The shape of the scatterplot not only gives an indication of the probable correlation value from the shape and distribution of the population of points, it will also reveal whether multiple distributions occur. A scatter plot should always be generated before Pearson's correlation coefficient is calculated in order to establish whether there exists a single linear distribution, for which the use of PCC is appropriate. Both global and thresholded PCC values will be diminished if there are mixed-intensity distributions within the image. These commonly arise when acquiring images of fluorescent protein chimeras that show very different expression levels across the field of view of the entire image. This can be seen in the scatterplot; the solution is to calculate different correlation values for each separate distribution, isolating each within its own region of interest and analysing them separately.

25.7 Measuring Overlap Using Manders' Overlap Coefficient

Manders (and others) have taken the view that negative (anti-correlated) relationships are difficult to interpret. Manders thus simplified Pearson's correlation coefficient (Manders' overlap coefficient, MOC[7], also called Manders' colocalisation coefficient, MCC) by removing the average term and subtracting it from the individual grey-value intensities of each red and green channel. Manders' formula is shown with Pearson's correlation coefficient in Box 25.3.

Manders' overlap coefficient is mentioned here for completeness because several software packages include it. Adler and Parmryd (2010) have examined both PCC and MOC and also Li's ICQ for establishing correlation. They conclude that both PCC and MOC are independent of differences in labelling intensities between the pair of fluorophores, are independent of photobleaching and independent of the gain setting on the photomultiplier detectors. Because it lacks mean values, Manders' overlap coefficient is less affected per se by high background values, but it is affected by high-intensity pairs of values, which makes it a poor reporter of co-occurrence. To describe any inequality between the two signals being analysed, Manders introduced the split overlap coefficients[8] – otherwise known as colocalisation coefficients – k_1, k_2 and M_1, M_2. The coefficients k_1 and k_2 are intended to show co-occurrence but have no additional usefulness to M_1 and M_2. Manders overlap coefficient uses global values (background and signal values) when calculating correlation. This can lead to a positive bias to the calculation of the correlation coefficient. This is not merely due to the different formula used to calculate MOC: Pearson's correlation coefficient also gives a similar positive value when applied *globally* to the entire image without

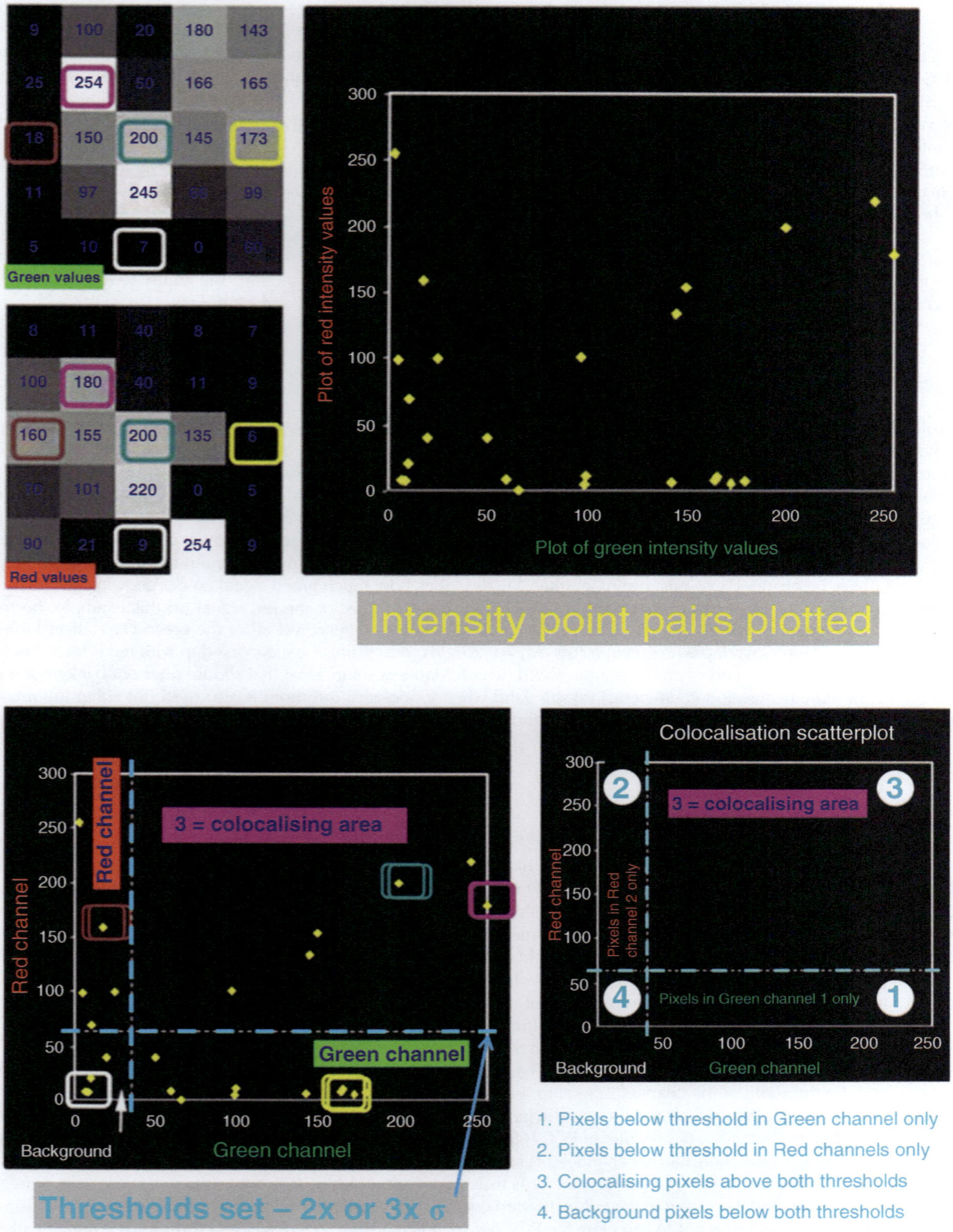

Figure 25.2 ***The scatterplot*** *How a scatterplot is constructed and used to derive tPCC values. See the text for further details. The slope of the least-squares line is determined by the intensities of the signals in each channel, relative to one another. Pearson's correlation coefficient is sensitive to outlying values, which detract from a linear relationship between the paired pixel intensity values. This is shown, for example, in the 'Anscombe quartets' data.*

setting a threshold to isolate only the fluorescent signals. MOC is also affected by changes in the intensity of background pixels and of the offset value, which Pearson's correlation coefficient is not[9]. MOC increases with a positive offset increasing the intensity of background pixels. Both Adler and Parmryd (2010) and Barlow et al. (2010) recommend not using MOC to report colocalisation.

Since PCC is based upon differences from the mean, intensity pairs near the mean influence the result less than those at the extremes of the intensity range. Thus PCC is affected by background (i.e. many low-intensity uncorrelated pixels), reporting a higher positive correlation value (a false-positive result) where many background pixels are included in the calculation because of the global effect of these low-intensity background pixels upon reducing the mean. Including these background pixels, which are devoid of true signal, yields *mean* values for each channel that are lower than most of the pixel values where non-background signal is present. Eventually the mean becomes so small, that global PCC becomes indistinguishable from Manders' overlap coefficient. Therefore, in the extreme situation where an anti-correlated relationship exists between a relatively small number of pixels in an image having many background pixels, almost *all* the pixel pairs will contribute positively to the final PCC metric, despite the fact that actual signal intensities within the image are anti-correlated! Manders' overlap coefficient cannot accommodate anti-correlated (i.e. inverse) relationships. This is the reason why the MOC was derived in the first place, as explained at the beginning of this section. **The ability to discriminate between positive and negative correlations gives thresholded PCC a validity that both global PCC and MOC lack.**

25.8 Manders' Colocalisation Coefficients M_1 and M_2

Manders' colocalisation coefficients M_1 and M_2 are normalised against total pixel intensity, so they help to avoid issues relating to pixel intensity. They relate *overlap stoichiometry* of each channel to the correlation value. M_1 and M_2 apply to the (red and green) *signals* within the image being analysed for colocalisation. In order to be classified as 'colocalised', a pixel must be paired with a pixel of non-background intensity from the second channel. For this reason, Manders' colocalisation coefficients M_1 and M_2 are useful and should be quoted along with the Pearson's correlation coefficient.

These colocalisation coefficients relate to the entire image. Sometimes the values are written in lower case: m_1 and m_2. The specific metrics, m_1 and m_2, refer to a region of interest (ROI) drawn upon the image. If green pixels are represented by channel 1 (x-axis, abscissa) and red pixels by channel 2 (y-axis, ordinate), then M_1 describes how much green signal co-occurs with red and vice versa for M_2. In the examples of synthetic data sets shown in Figure 25.3 (row 5), most of the red signal (in this example the red signal is in channel 1 and the green signal in channel 2) does not colocalise with the green, yet all of the green colocalises with the red; therefore M_1, describing the red overlap with green, is low ($M_1 = 0.25$). M_2, describing the green overlap with red is total, i.e. $M_2 = 1.0$. The colocalisation coefficients M_1 and M_2 require a threshold (which Manders set to zero) that should filter out background pixels of zero, but crucially these colocalisation coefficients are still valid when a non-zero threshold is also used (set either automatically or empirically by the user) to filter out background pixels. M_1 and M_2 are sensitive to poor threshold correction.

25.9 Thresholding

All image analysis packages will provide some means of thresholding the background, to exclude it from calculation of the correlation coefficient. Most use the method of Costes et al. (2004), which attempts to provide an unambiguous objective method of setting a threshold. When manually setting thresholds, it is difficult not to be subjective: a visual estimation will lead to inconsistency and inaccuracy. For a thresholding method to address this, see Adler and Parmryd (2014) who recommend setting the threshold at twice the local standard deviation above the local mean background intensity, or employing the method of Phansalkar (Phansalkar et al., 2011). Barlow (Barlow et al., 2010) recommends setting the threshold at three times the standard deviation value above the mean background intensity. The purpose of the Costes' algorithm is to set thresholds in an objective manner in order to calculate Manders' colocalisation coefficients, M_1 and M_2, and use these to determine colocalisation overlap. Rather than performing data analysis to calculate and cite Pearson's correlation coefficient, the PCC is used as a tool ultimately to generate M_1 and M_2 values.

This is how the Costes' algorithm works: each pixel pair in the images is plotted on the scatterplot according to its intensity. A least-squares orthogonal regression line is then fitted to the data (Figure 25.4) and plotted on the scatterplot. The threshold of channel 1 is set to its maximum, and the threshold of the channel 2 is set to the value predicted by the regression line (whilst the threshold of the first channel is set to the maximum). Pearson's correlation coefficient (PCC) is calculated for all pixel pairs for which the intensity of at least one channel is fainter than its threshold. If the value of PCC is found to be greater than zero, the value of the threshold for channel 1 is decremented, and the value for the threshold for channel 2 is set to the new value predicted by the regression line. The test is repeated iteratively until a value for the thresholds of channel 1 and 2 is found beneath which PCC is not positive, i.e. it is zero or negative. At this point, the threshold is set, and the values of the colocalisation coefficients M_1 and M_2 calculated to indicate colocalisation overlap. The purpose of the Costes' algorithm is not to calculate the PCC per se, it is merely a tool used to set thresholds automatically in order to generate Manders' colocalisation coefficients M_1 and M_2.

The Costes thresholding method is effective for images with a high signal-to-noise ratio, but for images with low signal, high noise or high intensity values in the background pixels, the threshold is set too low. This means that labelled structures are not discriminated from the background intensity values. Conversely, where many background pixels of uniformly low intensity are included, the high threshold (as set by Costes) results in the values of the colocalisation coefficients M_1 and M_2 being increasingly underestimated and inaccurate. Similarly, the algorithm struggles to set a realistic threshold when a large number of very densely labelled structures (i.e. those with

	Synthetic data sets	PCC	**tPCC**	M_1	M_2
		Applies to images		Applies to channels	
1	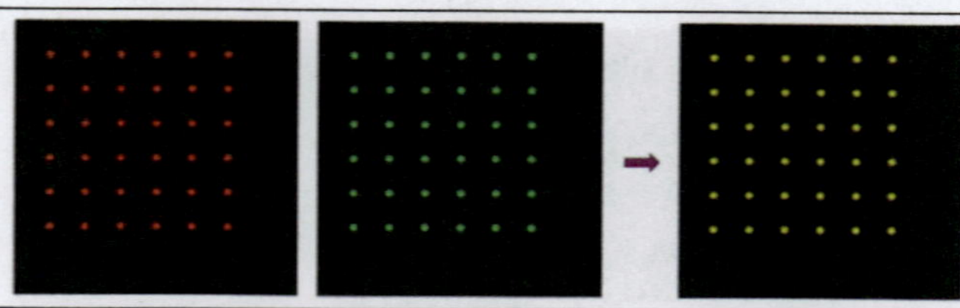	1.00	1.00	1.00	1.00
2	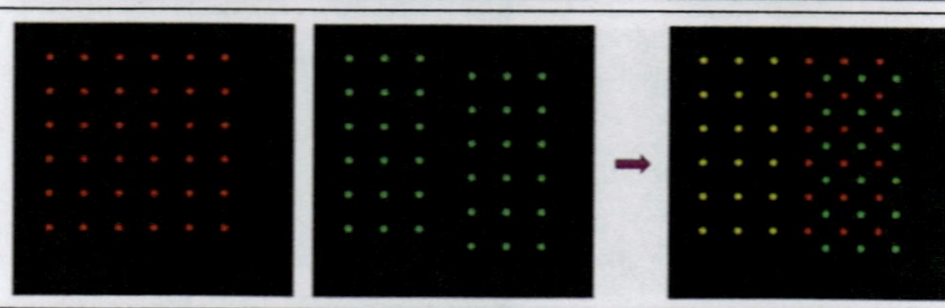	0.483	1.00	0.50	0.50
3	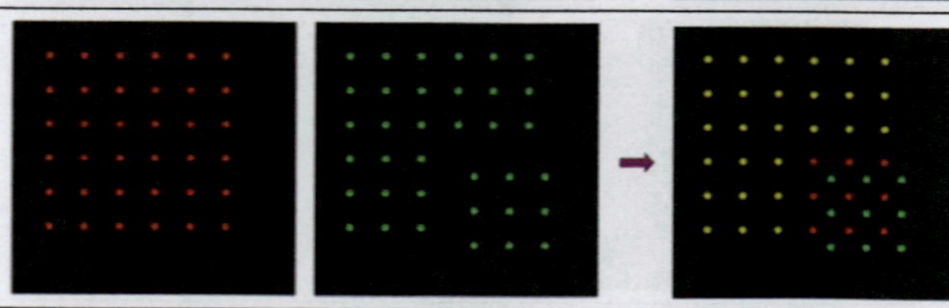	0.74	1.00	0.75	0.75
4	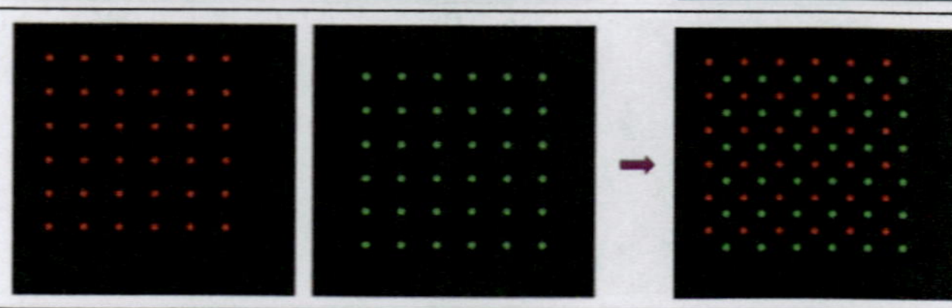	-0.03	0	0	0
5	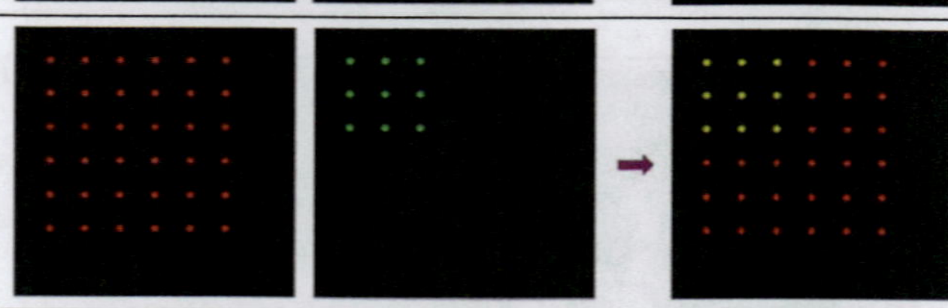	0.49	1.00	0.25	1.0
6	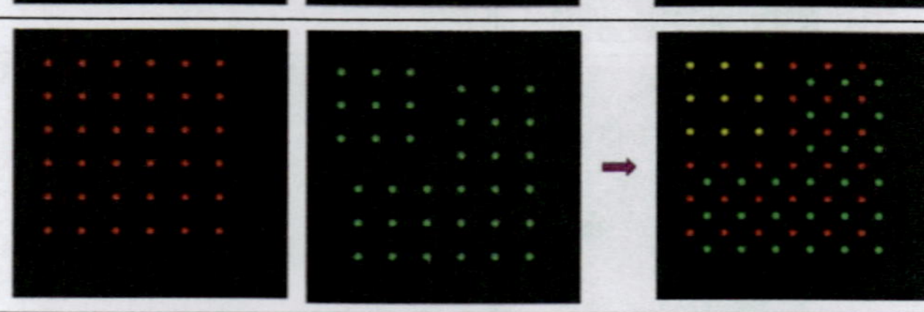	0.23	1.00	0.25	0.25
7	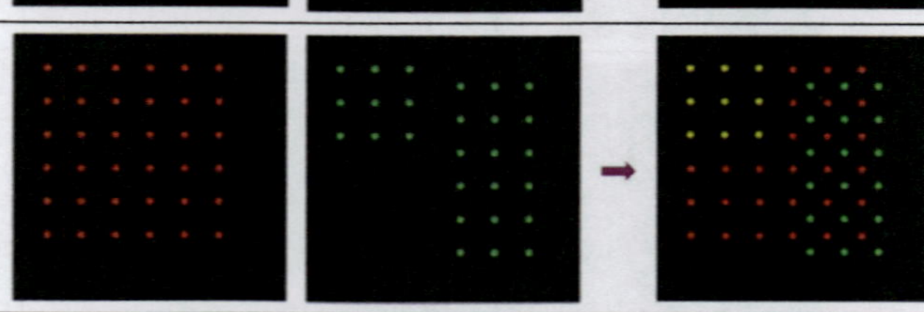	0.27	1.00	0.25	0.33
8	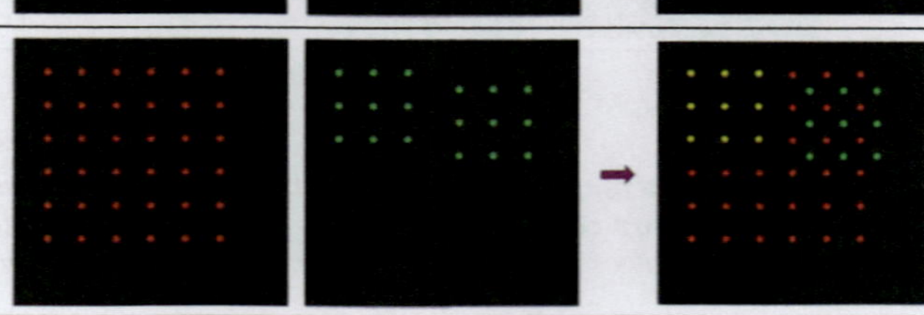	0.34	1.00	0.25	0.50

Figure 25.3 ***How Manders' colocalisation coefficients work in the context of tPCC*** *Manders used synthetic data to introduce the concept of overlap coefficients. Unfortunately, the original illustrations, in Fig. 1, page 377 & Fig. 2, page 380 of Manders et al. (1993) were printed in monochrome (which has, inevitably, lead to further confusion). Colour examples of these synthetic data sets can be found in Table 1, page 712, Barlow et al. (2010). Source: Barlow et al. 2010. Reproduced with permission from Cambridge University Press and A. Barlow.*

	Synthetic data sets	PCC	tPCC	M_1	M_2
		Applies to images		Applies to channels	
9	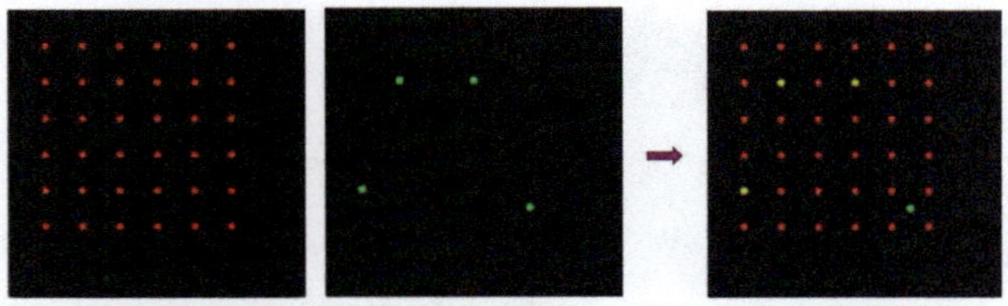	0.24	1.00	0.08	0.75
10	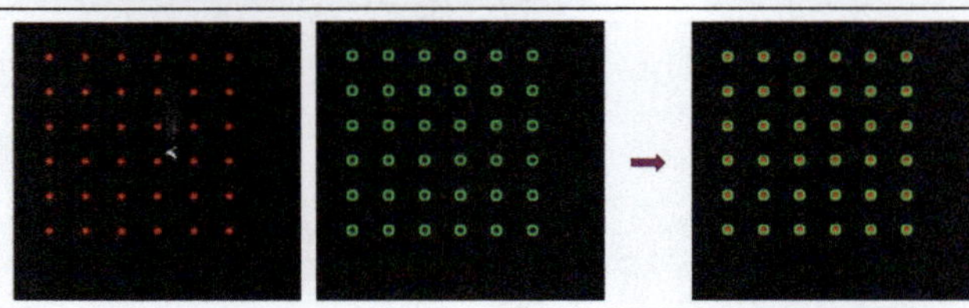	0.495	-1.00	0.98	1.0
11	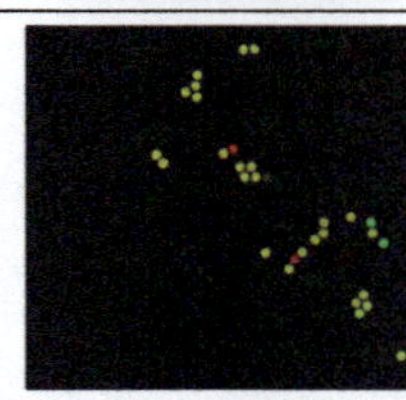	0.915	0.96	0.92	0.92
12	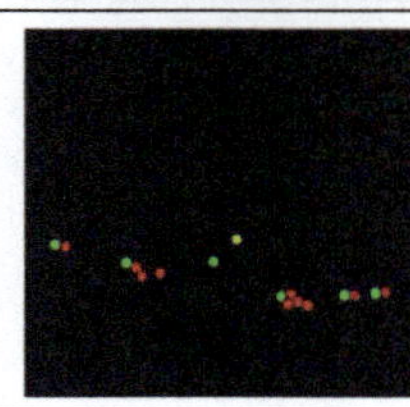	0.086	0.945	0.086	0.134
	Effect of noise				
13	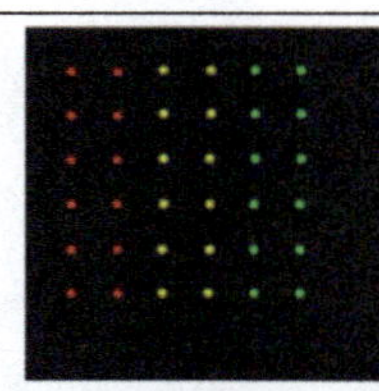	0.446	0.50	0.50	0.50
14	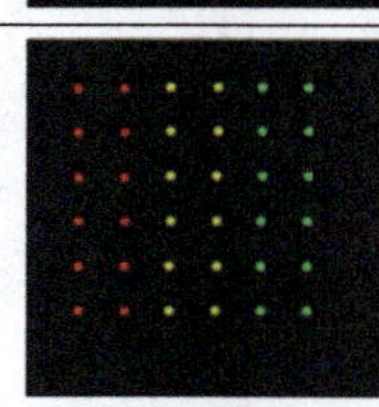	0.442	0.50	0.26	0.49
15	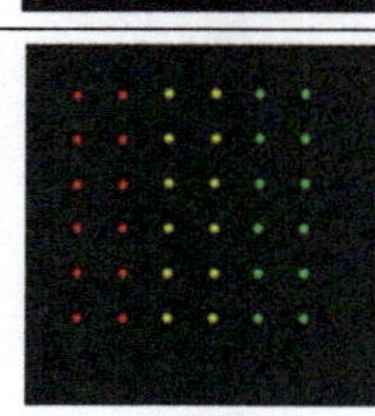	0.398	0.50	0.18	0.49

Figure 25.3 *(Continued)*

Notes:

In row 9, The TPCC is 1.0 because where the red and green signal are greater intensity than both the thresholds they overlap and do so perfectly. For the same reason, in row 12, the tPCC is 0.95 rather than 0.09. The tPCC result must be cited in the context of the colocalisation (overlap) coefficients M_1 and M_2. The tPCC applies only 13% of the green and 8% of the red signal intensities in the image. The tPCC is similar in row 11, but applies to a greater proportion of intensities.

In row 10, the size of the red circle and green ring are the same size: they are perfectly overlapped throughout their diameter, hence M_1 and $M_2 = 1.0$ but the intensity of the green ring is much lower on the inside than outside, and for the red circle, the opposite is the case, leading to true negative correlation, hence a tPCC of -1.0.

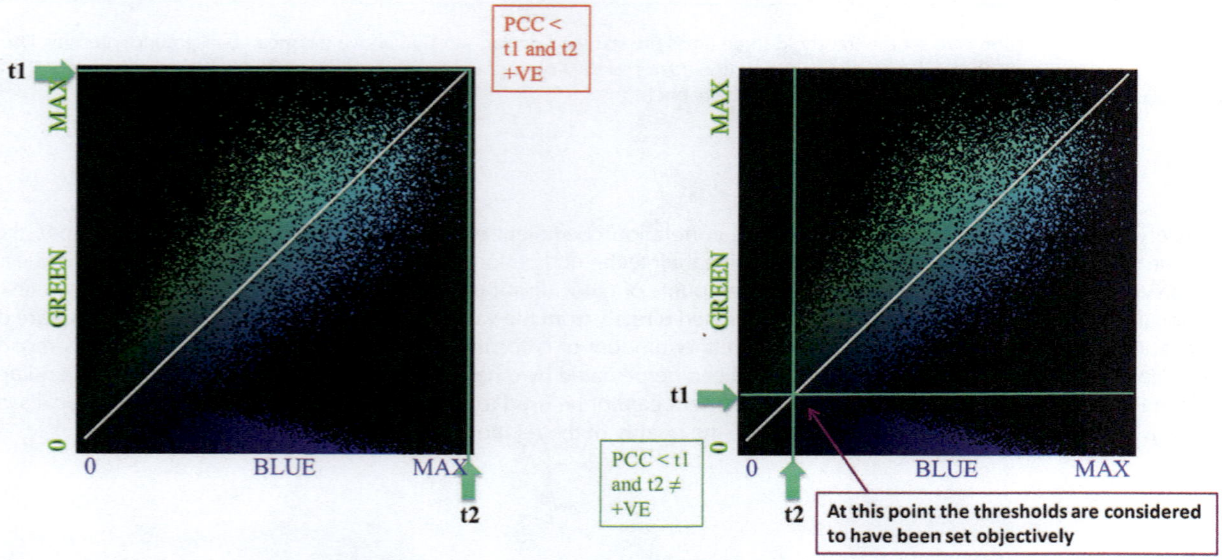

Figure 25.4 ***Thresholding setting by the Costes' algorithm*** *See the text for an explanation of how the Costes' algorithm works. Source: Barlow et al. 2010. Reproduced with permission from Cambridge University Press and A. Barlow.*

intensities substantially above background) are present in the image. Costes' method also has difficulty setting a meaningful threshold when there is a significant departure from 1:1 labelling stoichiometry, with large differences in the number of structures labelled with each fluorophore. Finally, The Costes' algorithm fails completely to set a proper threshold for fluorophores that exhibit no correlation and also ignores anti-correlated pixel pairs, which give rise to valid negative values for Pearson's correlation coefficient.

Costes et al. recognised that their algorithm set thresholds that excluded both colocalised pixel and background pixels and suggest that trying to exclude low intensity background pixels is too difficult. They fail to make the distinction between pixels containing signal and low-intensity background pixels. This means that the final thresholds set by their widely used algorithm are determined not by the quality of the relationship between colocalised signal pixels but by the proportion of non-background signal pixels in one channel that are colocalised with low-intensity background pixels from the second channel, which are given equal weight with the positively correlated signal pixel intensities. As the proportion of non-background pixels that are colocalised with background pixels increases, the Costes' algorithm sets higher thresholds, thus underestimating the colocalisation coefficients, M_1 and M_2.

Van Steensel's approach shifts the pixels of one channel relative to the other channel, which is kept static, and plots the PCC as a function of successive pixel shift. A bell-shaped curve is the result when positively correlated colocalisation is present, peaking at zero pixel shift. If the maximum peak of the bell curve is shifted from the zero pixel shift point, then partial colocalisation occurs. A flat line results when there is no colocalisation and a dip or inverted curve, when negative correlation occurs. This analysis method (van Steensel et al., 1996) is useful for small particles and is helpful for detecting low values of Pearson's correlation coefficient. The combined displacement algorithm (CDA) method (Ramirez et al., 2010) is an extension of van Steensel's pixel shift analysis. It was developed to reduce the risk of false-positive colocalisation results arising from the Costes' and others' image block-scrambling methods as a statistical significance test. The original CDA algorithm plug-in for ImageJ/Fiji developed by Ramirez has been modified by Alex Herbert to include support for 16-bit and 3D images. The URL for this plug-in can be found at the end of the chapter under Internet Resources.

Barlow et al. (2010) have investigated these thresholding methods and recommend setting the threshold at a value equal to three times the standard deviation of the mean value of the background pixels. This approach requires fewer manual interventions; it is more rigorous than Costes and embraces negatively correlated data. A thresholded PCC (tPCC) value is calculated only for those pixels above the threshold in *both* channels. If global background pixels are included, artificially low mean values are calculated for PCC, which results in a disproportionately large number of pixels above the mean and disproportionately few below the mean. Positive differences from the mean become very large whilst negative differences become smaller. Since intensity values far from the mean carry greater weight than those which lie near the mean, there is a bias towards positive PCC values, which makes negative values reporting anti-correlation less likely, leading to a false-positive result for Pearson's correlation coefficient.

Thresholded PCC applies to the *signals*, rather than the entire image (see Barlow et al., 2010). Thresholding also allows weak positive correlations to be detected, which would otherwise be 'diluted out' by global PCC measurements. This is important, and accurate thresholding is crucial to meaningful colocalisation analysis. Regions of interest (ROIs) are similar to thresholds. They do not cut off intensities below a minimum value for analysis but instead are used to identify areas of biological relevance. By isolating a single cell, for example, a single distribution can be analysed. The scenario whereby multiple distributions contribute to the analysis can be avoided.

The best measure of correlation is to set the threshold three times the standard deviation value above the mean background intensity. Use the thresholded tPCC value to report the degree of correlation (the *quality* of the colocalisation), and qualify it with coefficients M_1 and M_2, which report co-occurrence (the *quantity* of colocalisation in each channel).

Unfortunately Costes et al. not only define Pearson's correlation coefficient erroneously at the beginning of their paper, they also suggest that the colocalisation coefficients calculated using their technique performed better than PCC as descriptors of the true amount of overlap in synthetic images, "Before evaluating the amounts of colocalisation M_1 and M_2 in a region of the image, we first check that true colocalisation in present. This cannot be determined directly from the value of *r* since intermediate positive values are difficult to interpret" and "We see that *r* evaluates inaccurately either amount of colocalisation [in red and green channels] in comparison to our current algorithm". These statements – which have been perpetuated by others – represent a fundamental misunderstanding of the purpose of the Pearson correlation coefficient. The coefficient *cannot* be used to describe the quantity, or extent, of colocalisation; its sole purpose, when calculated accurately, is to describe the quality of the relationship between variables.

25.10 Analysing Product Differences from the Mean (PDM)

If two fluorophores colocalise, the pair of intensities at each pixel (the pixel pair) will *both* tend to lie above or lie below the mean intensity values. A scatterplot of each pixel pair reduces spatial information in the image to reveal the intensity distribution and any pattern of correlation. Plotting the data as product differences from the mean (PDM) qualitatively shows the trend of colocalisation, contra-localisation (i.e. negative correlation) or independent distribution. A normalised scatterplot of both dye intensities is still constructed, but the product differences from the mean are plotted for each pixel pair within each channel. If the intensities in two images vary in synchrony (i.e. they are *dependent*), they will vary around their respective mean image intensities together. The shape and skewness of the scatterplot reveals the degree of correlation between the two fluorophores.

The Intensity Correlation Quotient (ICQ, see Box 25.5 and Khanna et al. 2006) is a quantitative means of describing this relationship. The ICQ tests for covariance – whether when the intensity of one stain increases, the other varies in synchrony. ICQ does not test for colocalisation overlap: in the absence of covariance between two signals that overlap to a large extent, a high ICQ value is *not* generated. The ICQ simply expresses the number of global matching pixel pairs when both are either above or below their mean, as a fraction of the total number of pixels, and then subtracts 0.5 from the result. The subtraction ensures that negative correlations have a negative quotient, within a −0.5 to +0.5 scale, in a similar fashion to the PCC. It is possible to double the ICQ result after subtraction to fit a −1 to +1 range, as for Pearson's correlation coefficient. Because the intensity correlation quotient is still dependent upon mean values, any background subtraction and thresholding should still be carried out, as described earlier in the chapter. The ICQ is available as an ImageJ/Fiji plug-in. This is sometimes referred to as Li's intensity correlation quotient method from the paper by Li et al. (2004), which first described this method of intensity correlation analysis.

The ICQ method is particularly useful when one fluorophore is distributed across most, or all, of the cell and is present in almost every pixel or voxel (Figure 25.5). In this case, the second fluorophore, with a much smaller distribution, will tend to be identified as colocalised even though its distribution is totally random with respect to the first. This can occur, for example, when labelling a cytosolic protein or a widely distributed protein, such as actin. The ICQ will not report these fluorophores as associated because they do not vary in synchrony. For the ICQ to be meaningful you *must* limit the analysis to a region of interest (ROI) that is representative. Thus, including cells and the spaces round the cells in the ROI will give you an artificial positive ICQ value – you are then comparing cell space with background and the ICQ will always be positive. ICA analysis is only useful if you look at a single structure. It cannot be performed upon multiple cells at the same time – that comparison will almost certainly report a strongly positive value. The more clearly defined the region selected, the more meaningful the analysis.

PDM values can be plotted as an overlay of the original image by some commercial software packages (e.g. Volocity), or using Gorlewicz's ImageJ plug-in Colocalization Colormap (see Internet Resources, at the end of the chapter). The colour map, developed by Frédéric Jaskolski et al. (2005) employs Li's PDM method but calculates correlation between individual pixel pairs rather than between global images. By this means, spatial (i.e. image) information is retained.

In Volocity, pixels (and voxels) whose intensity for both channels are either both above or both below the mean channel values show positive correlation and are coloured yellow in the positive PDM overlay, whilst pixels whose intensity for either channel are below the mean for that channel and are above the mean in the other channel show negative correlation and are included in the negative PDM overlay and coloured purple. The Jaskolski algorithm normalises the product difference from the mean (called the nMDP), so that the nMDP index lies in the range −1 to +1, representing non-correlated and correlated pixel pairs respectively. The absolute value of each pixel in the colour map depends upon the deviation from the mean of the pixel intensity of both channels, thus showing the relative relationship between these channel intensities at each pixel. The nMDP colour map method is only slightly affected by background noise in the image.

Box 25.5 PDM Analysis and ICQ

The PDM value is calculated by the co-variance (dye A intensity – mean dye A intensity) × (dye B intensity – mean dye B intensity, or put mathematically PDM = $(G_i - G_{av}) \cdot (R_i - R_{av})$, where G_i and R_i represent the green and red intensities of an individual pixel, and G_{av} and R_{av} represent the average pixel intensity for each green and red image. With random staining, the overall difference of pixel intensities from the mean (dye A intensity – mean dye A intensity) for each individual channel, is zero ($\Sigma_i (G_i - G_{av}) = 0$). Thus the sum of the product of the two equalities ($\Sigma_{n\ pixels} (G_i - G_{av}) (R_i - R_{av})$ also tends to equal zero.

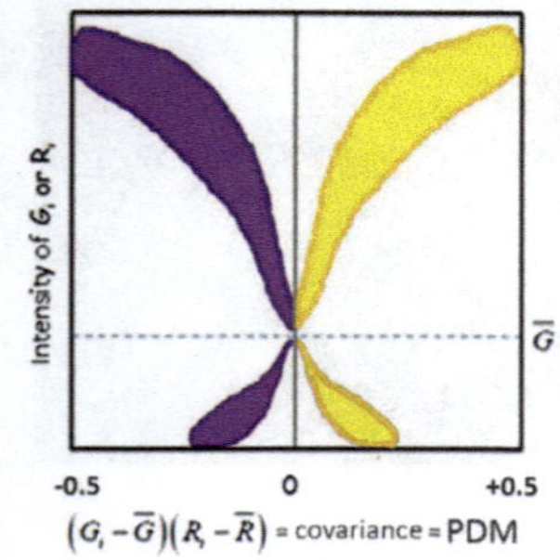

When two signals vary dependently, as predicted when two proteins are part of the same molecular complex, the mean deviation product trends toward a positive value, $\Sigma_{n\ pixels}(G_i - G_{av})(R_i - R_{av}) > 0$. When two signals vary independently, the mean deviation products tends toward a negative value, $\Sigma_{n\ pixels} (G_i - G_{av}) (R_i - R_{av}) < 0$. The intensity correlation quotient (ICQ) was devised to account for the covariance of mean deviation products in a defined intracellular area. The ICQ is the ratio of the number of positive $(G_i - G_{av}) (R_i - R_{av})$ values to the total number of pixel pairs. ICQ is given as follows:

$$ICQ = \frac{\sum |(G_i - G_{av}) \cdot (R_i - R_{av})|}{N_{TOTAL}} - 0.5$$

where $\sum |(G_i - G_{av}) \cdot (R_i - R_{av})|$ is the number of positive values for PDM, and N_{TOTAL} is the number of pixels that do not have a value of zero in each channel.

By subtracting 0.5 from the resulting ratio, the ICQ is finally normalised to a scale of −0.5 for inverse correlation, 0.0 for randomness, and +0.5 for complete positive correlation.

Intensity correlation analysis is illustrated by two graphs, showing these normalised intensities (0–1) as a function of covariance, the product of both channels (−0.5 to +0.5). The x-axis is the co-variance, and the y-axis the normalised intensity values of the channel being analysed. Distributions that skew to the right reflect dependent staining patterns, although adopting a 'C' shape, where the two pixel staining intensity values vary in synchrony. Distributions that are symmetrical about the 0 value on the axis indicate random staining, while those that skew to the left reflect staining patterns where the pixel staining intensity values vary inversely with respect to one another. The spread of the cloud is dependent on the intensity distribution of the channel as a function of *both* channel intensities. Note that a high staining overlap in the absence of covariance will not generate a high ICQ value. The disadvantage of the ICQ statistic compared to the Pearson correlation coefficient is that pixels with intensity values marginally above or below the mean carry the same weight as those with more extreme intensities. With the PCC, extreme values carry a greater weight than pixel pair values nearer the mean.

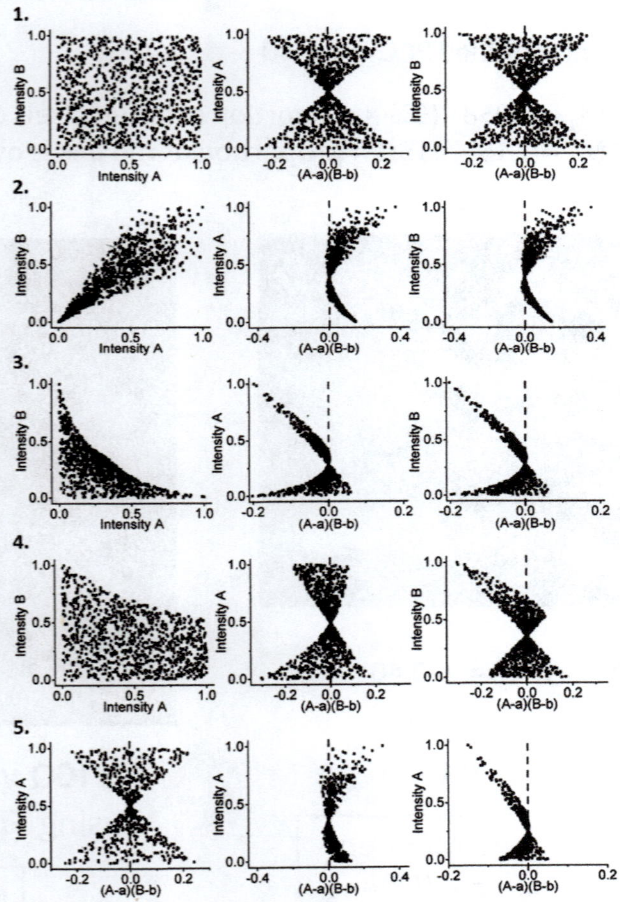

Intensity correlation analysis, simulated models. In each of rows 1 through to 4 the left Panel shows a scatterplot of 1000 paired A_i versus B_i staining intensities (normalised to 1.0). The centre and right Panels show plots of individual A_i and B_i staining intensities against their respective calculated $(A_i - a)(B_i - b)$ values.

1. Plot of random *A* and random *B* intensity values. ICQ = 0.023.
2. Intensity plot simulating dependent staining. ICQ = +0.39.
3. Intensity plot simulating non-correlated staining. ICQ = -0.269.
4. Random staining distribution in channel *A*, but with segregated staining in channel *B*. ICQ = −0.12.
5. Effect of thresholding channel *A* values by deleting the lowest 1/3rd of the data set in the centre Panels of Figures 1–3, above. The *A* and *B* data sets were normalised, and the $(A - a)(B - b)$ and ICQ values recalculated. ICQ values: left, random distributions = +0.003; centre, dependent distribution ICQ = +0.250; right, non-correlated distribution ICQ = −0.287. The same key distribution features of the non-thresholded plots are retained after thresholding.

Source: Adapted from Figure 2, page 4073, Li et al. 2004.

References

Brown, M. et al. (2010). 'Quantitative distribution and colocalization of non-muscle myosin light chain kinase isoforms and cortactin in human lung endothelium'. *Microvascular Research 80*/1: 75–88.

Khanna, R. et al. (2006). 'N type Ca^{2+} channels and RIM scaffold protein covary at the presynaptic transmitter release face but are components of independent protein complexes'. *Neuroscience 140*/4: 1201–1208.

Li, Q. et al. (2004). 'A syntaxin 1, $G\alpha_o$, and N-type calcium channel complex at a presynaptic nerve terminal: analysis by quantitative immunocolocalization'. *Journal of Neuroscience 24*/16: 4070–4081.

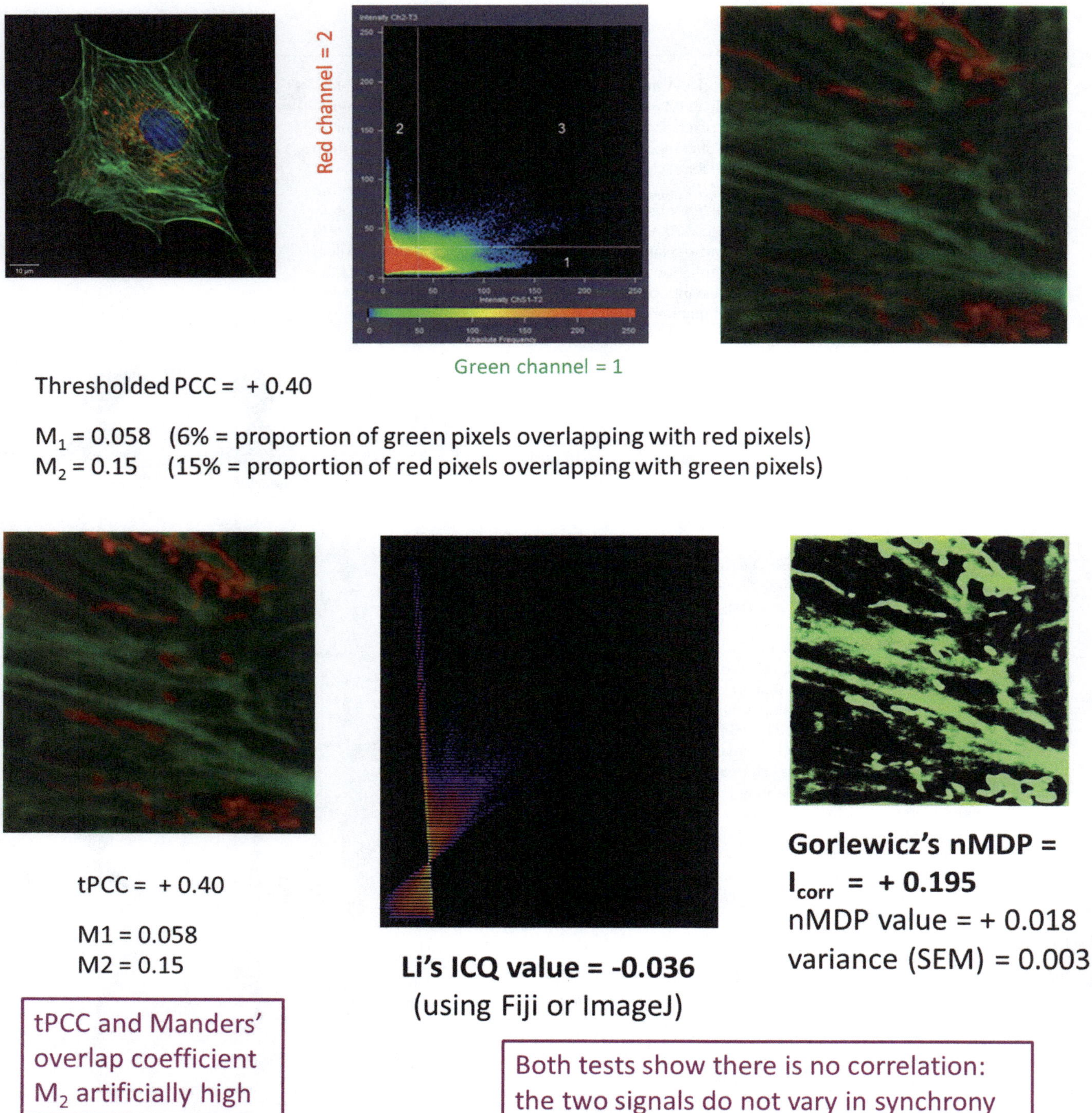

Figure 25.5 **Testing unequal fluorophore signal distributions** *Using Li's intensity correlation quotient and Gorlewicz's colocalisation colour map with ICA synchrony analysis.*

25.11 Analysis Choices and Software

There is disagreement between authors and reviewers (e.g. Manders, Costes, Bolte, Dunn, Barlow and Adler) on which metric to use to describe colocalisation and which is the best range of values to use. Some argue – like Manders – that only positive values (0 to 1) should be used for correlation to describe colocalisation in microscopy. Others, e.g. Costes and Zinchuk, suggest that only extreme values for correlation allow conclusions about colocalisation to be drawn, and that midrange values (around ±0.5) are inconclusive. Yet others (e.g. Barlow, Dunn and Adler) argue that the full range of possible correlation values (from −1 to +1) are valid and are supported by natural examples, not merely synthetic data.

Despite the confusion and disagreement, several software programmes use algorithms to generate both Pearson's correlation coefficient and the Manders' overlap coefficient. Automatic thresholding, as developed by Costes, is also widely used as an objective means to set thresholds, despite certain flaws. The author prefers to use Volocity (although this programme is no longer actively supported by Perkin-Elmer) because the software developer behind this commercial programme has thought rigorously and deeply about the application of correlation analysis to fluorescent images.

25.12 Statistical Analysis

The scientific method involves formulating a hypothesis from reasoning and conjecture, deriving logical predictions and testing these by experiment. The cornerstone of the scientific method is hypothesis testing. This is done in a seemingly upside-down fashion by testing the opposite, or 'null', hypothesis, H_0, rather like playing the devil's advocate. The null hypothesis states generally that no relationship exists between two measured phenomena, or that there is no difference between groups. It is generally assumed to be true until evidence indicates otherwise. To test the null hypothesis, statistical analysis is used, with the assumption that the null hypothesis is true. Note that you cannot accept the null hypothesis, but only find evidence against it. You can either reject it or fail to reject it.

The statistician Ronald Fisher advocated the significance testing approach developing the p-value as an informal means to judge whether the evidence was significant and worthy of more critical analysis. If this probability (i.e. p-value) is low, then the null hypothesis is rejected. Herein lies a further twist to the complexity of correlation and colocalisation analysis. Fisher's academic rivals, Jerzy Neyman and Egon Pearson (the son of Karl Pearson) advocated a different approach. They compared the null hypothesis with an alternative, opposing, hypothesis, H_1 and rejected the null hypothesis, or not, on the basis of the data, with certain error[10] rates (defined by a variable parametric limit, α, also called a significance level) set prior to the statistical analysis. Both sides polemicised each other. Neyman and Pearson ignored the p-value whilst others created a hybrid analysis system that 'crammed Fisher's easy-to-calculate p-value into [the framework] of Neyman and Pearson's reassuringly rigorous rule-based system. This is when a p-value of 0.05 became enshrined as "statistically significant"' (Nuzzo, 2014) rather than merely a guide towards drawing conclusions.

In a colocalisation analysis experiment, the null hypothesis is usually taken to be that there is no correlation: namely that the true correlation coefficient ρ (rho: see endnote 2) is equal to zero. It is usual to test H_0 at a nominal significance level of 0.05 using the two-tailed Student's *t*-test. The 0.05 p-value measures not the probability of the null-hypothesis being true, but – more correctly – the probability of obtaining at least as extreme a value for the statistic under consideration if the null hypothesis is true. In other words, if the experiment is repeated many times over, the null hypothesis is true 95% of the time. If the Student's *t*-test gives a p-value less than 0.05, the null hypothesis is rejected. Simply reporting a p-value by itself is insufficient[11], since by itself, it provides no information on the strength or size of the effect and only indirectly on the power of the statistical test. The confidence level, PCC value and degree of freedom should also be given.

A one-tailed Student's *t*-test is used when you have a prior hypothesis as to the sign (+ or −) of the correlation. A two-tailed test is used if there is no statement in the null hypothesis regarding the sign of the correlation value – this is the more commonly occurring situation. The correct degree of freedom for Student's *t*-test is $n-2$, where n is equal to the number of pixel pairs. Locate the degree of freedom in the Pearson critical value tables (see Figure 25.6 and the Internet Resources at the end of the chapter), and use the $(n-2)$ row of values. If the value of r exceeds the critical value at the 0.05 significance level, there is significant linear correlation. The null hypothesis is rejected. If the null hypothesis were true, such a high value of r would occur by chance once or fewer than one in twenty times: a probability of less than 5%. Figure 25.6 shows the route to performing this statistical analysis and Pearson's correlation coefficient cited to show the degree of freedom and associated p-value.

Unless the data has been taken from a very small region of interest that has also been thresholded, the number of data points (i.e. pixel pairs) will be very large. In this case, the critical value against which you compare the Pearson metric falls to a very low value for very large degrees of freedom. This means that a very low value for Pearson's correlation coefficient may turn out to be statistically significant (leading to a rejection of the null hypothesis) but entirely insignificant in terms of being unimportant. Thus, red and green pixel pairs with no real association will appear to be correlated with each other. This phenomenon can also occur with autocorrelation, whereby the value of one pixel is likely to be similar to that of its neighbours. This is why smoothing data with a low-pass kernel 'de-noising' filter is a bad approach. Deconvolution is a better stratagem. Not only will smoothing data underestimate the correlation coefficient, but it will give false-positive significant tests from artificially low p-values. However, this statistical test assumes that the intensity values constituting each pixel pair from which Pearson's r is calculated are independent, and this may not be the case.

PEARSON'S CORRELATION COEFFICIENT *r* (Critical Values)

	Level of Significance for a One-Tailed Test										
	.05	.025	.01	.005	.0005		.05	.025	.01	.005	.0005
	Level of Significance for a Two-Tailed Test										
df = (N-2)	.10	.05	.02	.01	.001	*df* = (N-2)	.10	.05	.02	.01	.001
1	0.988	0.997	0.9995	0.9999	0.99999	21	0.352	0.413	0.482	0.526	0.640
2	0.900	0.950	0.980	0.990	0.999	22	0.344	0.404	0.472	0.515	0.629
3	0.805	0.878	0.934	0.959	0.991	23	0.337	0.396	0.462	0.505	0.618
4	0.729	0.811	0.882	0.971	0.974	24	0.330	0.388	0.453	0.496	0.607
5	0.669	0.755	0.833	0.875	0.951	25	0.323	0.381	0.445	0.487	0.597
6	0.621	0.707	0.789	0.834	0.928	26	0.317	0.374	0.437	0.479	0.588
7	0.582	0.666	0.750	0.798	0.898	27	0.311	0.367	0.430	0.471	0.579
8	0.549	0.632	0.715	0.765	0.872	28	0.306	0.361	0.423	0.463	0.570
9	0.521	0.602	0.685	0.735	0.847	29	0.301	0.355	0.416	0.456	0.562
10	0.497	0.576	0.658	0.708	0.823	30	0.296	0.349	0.409	0.449	0.554
11	0.476	0.553	0.634	0.684	0.801	40	0.257	0.304	0.358	0.393	0.490
12	0.457	0.532	0.612	0.661	0.780	60	0.211	0.250	0.295	0.325	0.408
13	0.441	0.514	0.592	0.641	0.760	120	0.150	0.178	0.210	0.232	0.294
14	0.426	0.497	0.574	0.623	0.742	∞	0.073	0.087	0.103	0.114	0.146
15	0.412	0.482	0.558	0.606	0.725						
16	0.400	0.468	0.542	0.590	0.708						
17	0.389	0.456	0.529	0.575	0.693						
18	0.378	0.444	0.515	0.561	0.679						
19	0.369	0.433	0.503	0.549	0.665						
20	0.360	0.423	0.492	0.537	0.652						

1. Specify H_0, the α significance level and whether the *t*-test is one-tailed or two-tailed. A one-tailed test is used when you have a prior hypothesis as to the sign (+ve or –ve) of the correlation. Use a two-tailed test if no hypothesis regarding correlation.
2. From the sample size N, calculate the degrees of freedom (= N-2).
3. Locate the DoF value in the table and for the level of significance determine the critical region for which you reject H_0 if your statistic is a larger value.
4. Calculate Pearson's r, and the *t* value.

Example

If correlation is expected, then the null hypothesis, H_0, is that $\rho = 0$. Say N = 18 and $\alpha = 0.05$, for a two-tailed test the critical value is 0.444.

Calculate the statistic *t* from the formula. If this calculation is below the critical value, H_0 is not rejected. For *t* values above 0.444 we assume correlation.

the variable

$$t = r\sqrt{\frac{n-2}{1-r^2}}$$

has a Student's *t* distribution in the null case

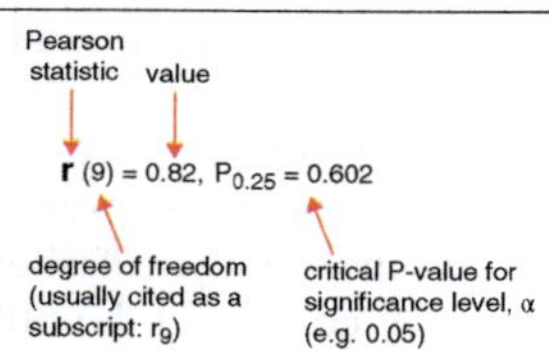

Figure 25.6 **Testing correlation** *Testing the statistical significance of the tPCC value. Further details are given in the text.*

25.13 Object-Based Colocalisation

Identifying, and then quantifying, the distribution of objects in the image to see whether colocalisation exists between them or not usually requires some post-acquisition image analysis. This involves segmenting the image to partition it into regions, normally by applying a threshold. Background can be subtracted from the image using a rolling-ball filter; background signal is not as influential upon object-based colocalisation as it is with intensity-based colocalisation analysis. Object-based colocalisation clearly lends itself well to the analysis of discrete punctate targets, such as endocytosis or nuclear speckles.

One method is to draw a vector through discrete structures whose size is close to, or larger than, the resolving power of the microscope. The intensities of both fluorophores, on the y-axis, should be plotted against the vector length on the x-axis. If the full width of the overlap region at half maximum intensity exceeds the minimum resolved distance of the microscope, the two objects can be considered to colocalise. Another method is to plot the centroids of objects and investigate whether the centroid of objects in one channel overlap object areas in the other (see the flow chart in Figure 25.7). Refer to Silver and Stryker (2000), Lachmanovich et al. (2003), Zhang et al. (2008), Fletcher et al. (2010), Helmuth et al. (2010), and Pageon et al. (2016) for further details. It is possible to carry out object-based colocalisation using ImageJ/Fiji with the JACoP plug-in, as well as the OBCOL plug-in and the DiAna plugin developed by Gilles et al. (2017). Links to both of these are given under the Internet Resources.

25.14 Chapter Summary

Always restrict the analysis of colocalisation to the biological structure you are interested in measuring. The best method of analysing colocalisation is only to include pixels that contain signal and to exclude background intensities in the image. It is recommended that the threshold for each channel is set at a value equivalent to the mean intensity of the background plus three times the standard deviation value (assuming the noise is normally distributed). The *thresholded* Pearson's correlation coefficient (tPCC) value is used to report the degree of correlation (the quality of the colocalisation). This *r* value is qualified with Manders' overlap coefficients, M_1 and M_2, which report co-occurrence (the quantity, or extent, of colocalisation in each channel). Taken with a scatterplot (and possibly a colocalisation

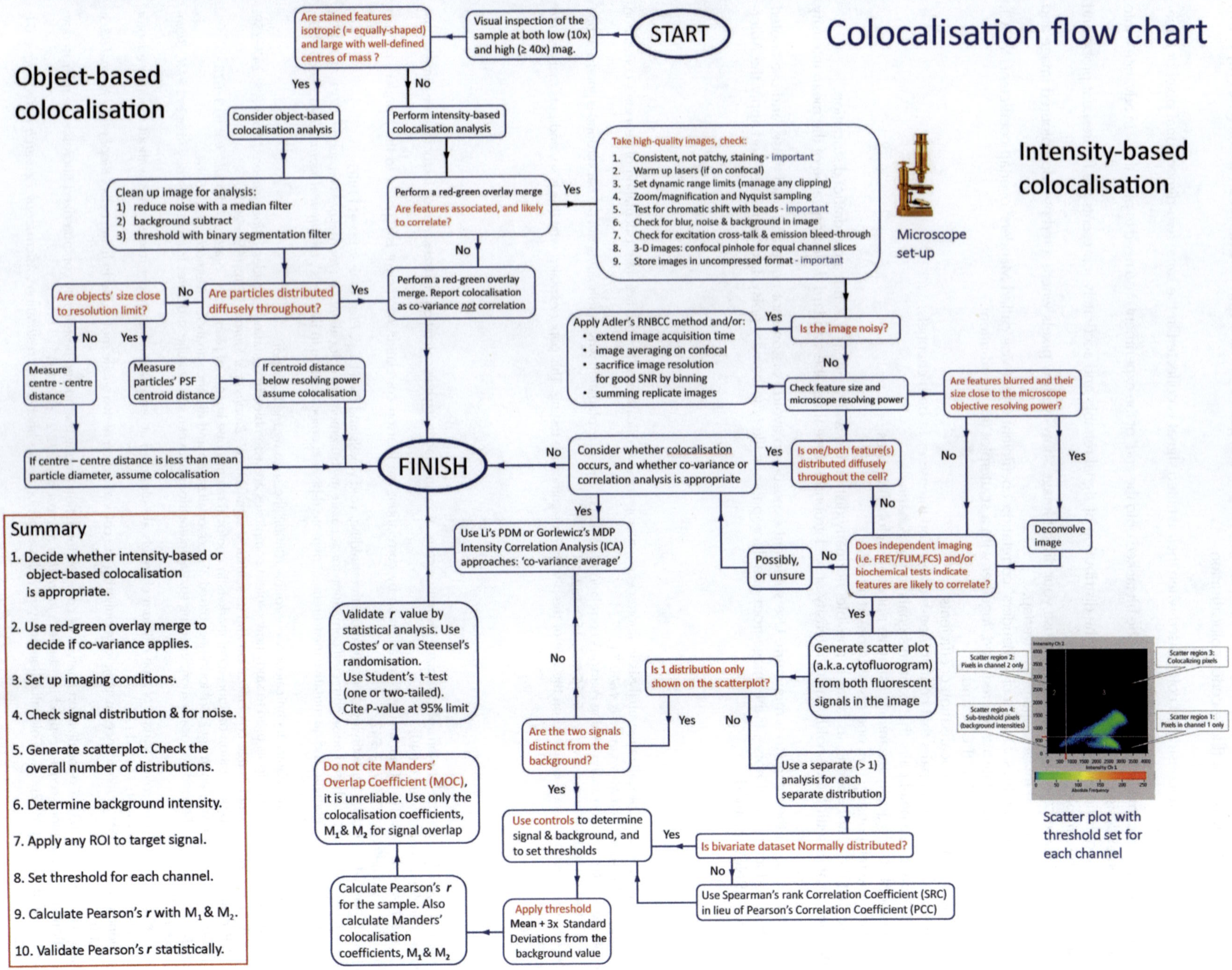

Figure 25.7 **Flow diagram** *Decision tree for colocalisation analysis*

colour map to report on distribution(s) within the data), thresholded PCC together with the split colocalisation coefficients, M_1 & M_2, give the best possible analysis of fluorescence colocalisation.

1. Two fluorochromes are considered colocalised when their emitted signal is collected in the *same* two-dimensional pixel or three-dimensional voxel.
2. Colocalisation, like imaging, should *not* be 'the bit tacked onto the end' of the experiment. You need to plan this – in advance – into the experimental design.
3. Good image acquisition is required so that the distribution of signal reliably and accurately represents the fluorescent probe and the target in the sample.
4. You *cannot* assume correlation merely because your digital image shows yellow pixels where a red pseudo-coloured image and green pseudo-coloured image apparently overlap.
5. Ignore meaningless metrics. Do not cite Manders' correlation coefficient (otherwise called Manders' overlap coefficient, MOC). Only three descriptive statistics are required, together with a scatterplot (for distribution):
   - thresholded Pearson's *r* (tPCC); and
   - Manders' M_1 & M_2 colocalisation coefficients.
6. Analyse only those pixel pairs that contain non-background intensities in both channels.
7. Analyse only those pixel pairs that contain signals in both channels.
8. Analyse only biologically meaningful signals, not entire (global) images.
9. Analyse one relationship (i.e. one distribution) at a time.
10. Ensure that the distribution is, as far as is possible, normally distributed with a linear (i.e. not a quadratic) distribution.
11. Be consistent in setting thresholds: mean intensity of background + three times the standard deviation value of the mean intensity of the background.
12. Keep the noise in the image to a minimum. Use sufficient exposure to avoid Poisson shot noise and read noise, and use cooled CCDs to reduce thermal noise. If the image from one fluorophore is likely to be *very* noisy, take replicates and apply the Adler-Parmryd RBNCC method.

Key Reading

Barlow, A.L. et al. (2010). 'Colocalization analysis in fluorescence micrographs: verification of a more accurate calculation of Pearson's correlation coefficient'. *Microscopy and Microanalysis 16*/6: 710–724.
- A critique of colocalisation analysis using Pearson's correlation coefficient and an explanation why thresholding out the background is good practice

Nuzzo, R. (2014). 'Statistical errors'. *Nature 506*/7487: 150–152.
- Too often too much reliance is placed on the role of the p-value in significance testing. This paper explains why this is so, and puts matters into perspective.

References

Abraham, T. et al. (2010). 'Deconvolution and chromatic aberration corrections in quantifying colocalization of a transcription factor in three-dimensional cellular space'. *Micron 41*/6: 633–640.

Adler, J. and Parmryd, I. (2010). 'Quantifying colocalization by correlation: the Pearson correlation coefficient is superior to the Mander's overlap coefficient'. *Cytometry A 77*/8: 733–742.

Adler, J. and Parmryd, I. (2014). 'Quantifying colocalization: thresholding, void voxels and the H_{coef}'. *PLoS One 9*/11: e111983.

Adler, J. et al. (2008). 'Replicate-based noise corrected correlation for accurate measurements of colocalization'. *Journal of Microscopy 230*/1: 121–133.

Costes, S.V. et al. (2004). 'Automatic and quantitative measurement of protein-protein colocalisation in live cells'. *Biophysical Journal 86*/6: 3993–4003.

DAKO Handbook, 6th ed. (2013).
URL: https://www.agilent.com/cs/library/technicaloverviews/public/08002_ihc_staining_methods.pdf.

Dwane, S. and Kiely, P.A. (2011). 'Tools used to study how protein complexes are assembled in signaling cascades'. *Bioengineered Bugs 2*/5: 247–259.

Fletcher, P.A. et al. (2010). 'Multi-image colocalization and its statistical significance'. *Biophysical Journal 99*/6: 1996–2005.

French, A.P. et al. (2008). 'Colocalization of fluorescent markers in confocal microscope images of plant cells'. *Nature Protocols 3*/4: 619–628.

Gilles, J.-F. et al. (2017). 'DiAna, an imageJ tool for object-based 3D co-localization and distance analysis'. *Methods 115*: 55–64.

Helmuth, J.A. et al. (2010). 'Beyond co-localization: inferring spatial interactions between sub-cellular structures from microscopy images'. *BMC Bioinformatics 11*: 372.

Khanna, R. et al. (2006). 'Long splice variant N type calcium channels are clustered at presynaptic transmitter release sites without modular adaptor proteins'. *Neuroscience 138*/4: 1115–1125.

Koyama-Honda et al. (2005). 'Fluorescence imaging for monitoring the colocalization of two single molecules in living cells'. *Biophysical Journal 88*/3: 2126–2136.

Lachmanovich, E. et al. (2003). 'Co-localization analysis of complex formation among membrane proteins by computerized fluorescence microscopy: application to immunofluorescence co-patching studies'. *Journal of Microscopy 212*/2: 122–131.

Landmann, L. and Marbert, P. (2004). 'Colocalization analysis yields superior results after image restoration'. *Microscopy Research & Technique 64*: 103–112.

Li, Q. et al. (2004). 'A syntaxin 1, Gα_o, and N-Type calcium channel complex at a presynaptic nerve terminal: analysis by quantitative immunocolocalization'. *Journal of Neuroscience 24*/16: 4070–4081.
Malkusch, S. et al. (2011). 'Coordinate-based colocalization analysis of single-molecule localization microscopy data'. *Histochemistry and Cell Biology 137*/1: 1–10.
Manders, E.M.M. et al. (1992). 'Dynamics of three-dimensional replication patterns during the S-phase, analysed by double labelling of DNA and confocal microscopy'. *Journal of Cell Science 103*/3: 857–862.
Manders, E.M.M. et al. (1993). 'Measurement of co-localization of objects in dual-colour confocal images'. *Journal of Microscopy 169*/3: 375–382.
Masters, S.C. (2004). 'Co-immunoprecipitation from transfected cells'. In: Fu, H. (ed.), *Protein-Protein Interactions. Methods in Molecular Biology*, Chapter 23, vol. *261*. Totowa, NJ: Humana Press, pp. 337–348.
Pageon, S.V. et al. (2016). Clus-DoC: a combined cluster detection and colocalization analysis for single-molecule localization microscopy data. *Molecular Biology of the Cell 27*/22: 3627–3636.
Phansalkar, N. et al. (2011). 'Adaptive local thresholding for detection of nuclei in diversely stained cytology images'. International Conference on Communications and Signal Processing, pp. 218–220. https://bit.ly/2LhnSMQ.
Ramirez, O. et al. (2010). 'Confined displacement algorithm determines true and random colocalization in fluorescence microscopy'. *Journal of Microscopy 239*/3: 173–183.
Scriven, D.R.L. et al. (2008). 'Image acquisition for colocalization using optical microscopy'. *American Journal of Physiology-Cell Physiology 294*/5: C1119–C1122.
Silver, M.A. and Stryker, M.P. (2000). 'A method for measuring colocalization of presynaptic markers with anatomically labelled axons using double label immunofluorescence and confocal microscopy'. *Journal of Neuroscience Methods 94*/2: 205–215.
van Steensel, B. et al. (1996). Partial colocalization of glucocorticoid and mineralocorticoid receptors in discrete compartments in nuclei of rat hippocampus neurons. *Journal of Cell Science 109*/Pt 4: 787–792.
Zhang, B. et al. (2008). 'Statistical colocalization in biological imaging with false discovery control'. 5th IEEE international symposium on biomedical imaging, ISBI: From nano to macro, pp. 1327–1330. doi: 10.1109/ISBI.2008.4541249.

Further Reading

Adler, J. et al. (2008). 'Noise and Colocalization in fluorescence microscopy: solving a problem'. *Microscopy & Analysis*, September 2008: 7–10.
Jaskolski, F. et al. (2005). 'An automated method to quantify and visualize colocalized fluorescent signals'. *Journal of Neuroscience Methods 146*/1: 42–49.
Pike, J.A. et al. (2017). 'Quantifying receptor trafficking and colocalization with confocal microscopy. *Methods 115*: 42–54.
Reitan, N.K. et al. (2012). 'Quantitative 3-D colocalization analysis as a tool to study the intracellular trafficking and dissociation of pDNA-chitosan polyplexes'. *Journal of Biomedical Optics 17*/2: 026015.
Trinkle-Mulcahy, L. and Swift, S. (2007). 'Caught on camera with another protein – just good friends, or something more?' *Infocus* (8): 48–57.
Wang, Y. et al. (2012). 'A robust co-localisation measurement utilising z-stack image intensity similarities for biological studies'. *PLoS One 7*/2: e30632. doi: 10.1371/journal.pone.0030632.
Woodcroft, B.J. et al. (2009). 'Automated organelle-based colocalization in whole-cell imaging'. *Cytometry A 75*/11: 941–950. See also OBCOL link, below.

Statistics Texts and References

Altman, N. and Krzywinski, M. (2017). 'P values and the search for significance'. *Nature Methods 14*/1: 3–4.
Anscombe, F.J. (1973). 'Graphs in statistical analysis'. *The American Statistician 27*: 17–21. See also: https://en.wikipedia.org/wiki/Anscombe's_quartet.
Dahiru, T. (2008). 'P-value, a true test of statistical significance? A cautionary note'. *Annals of Ibadan Postgraduate Medicine 6*/1: 21–26.
Drummond, G. (2013). 'Statistics: a journey that needs a guide'. In: Langton, P.D. (ed.), *Essential Guide to Reading Biomedical Papers: Recognising and Interpreting Best Practice*, Chapter 3. Chichester: Wiley, pp. 17–26.
Drummond, G.B. and Vowler, S.L. (2012). 'Categorized or continuous? Strength of an association – and linear regression'. *Microcirculation 19*/4: 373–376.
Dytham, C. (2010). *Choosing and Using Statistics: a Biologist's Guide*, 3rd ed. Oxford: Wiley-Blackwell. (This book has a key to help select statistical tests.)
Greenland, S. et al. (2016). 'Statistical tests, P values, confidence intervals, and power: a guide to misinterpretations'. *European Journal of Epidemiology 31*: 337–350.
Halsey, L.G. et al. (2015). 'The fickle P value generates irreproducible results'. *Nature Methods 12*/3: 179–185.
Lew, M.J. (2012). 'Bad statistical practice in pharmacology (and other basic biomedical disciplines): you probably don't know P'. *British Journal of Pharmacology 166*: 1559–1567 (the p-value explained).
McKillup, S. (2011) *Statistics Explained*, 2nd ed. Cambridge: CUP.
Moore, J.H., Thornton, T.A., and Ritchie, M.D. (2003). 'Basic Statistics'. *Current Protocols in Human Genetics Curr. Protoc. Hum. Genet.* Appendix 3M, Supplement 37. doi: 10.1002/0471142905.hga03ms37.
Morton, J.P. (2009). 'Reviewing scientific manuscripts: how much statistical knowledge should a reviewer really know?' *Advances in Physiology Education 33*: 7–9.
Mudge, J.F. et al. (2012). 'Optimizing a for better statistical decisions: A case study involving the pace of life syndrome hypothesis'. *Bioessays 34*/12: 1045–1049.
Nuzzo, R. (2014). 'Statistical errors'. *Nature 506*/7487: 150–152.

Internet Resources

1. General statistics web pages:
   (a) http://www.sportsci.org/resource/stats/contents.html.
   (b) https://www.statsdirect.co.uk/help/basics/p_values.htm.
2. Statistical power: https://www.nature.com/articles/nmeth.2738.pdf.
3. Pearson critical value tables:
   (a) http://www.statisticssolutions.com/table-of-critical-values-pearson-correlation/.
   (b) http://users.sussex.ac.uk/~grahamh/RM1web/Pearsonstable.pdf.
   (c) http://capone.mtsu.edu/dkfuller/tables/correlationtable.pdf.
4. Anscombe's quartets:
   (a) https://en.wikipedia.org/wiki/Anscombe's_quartet.
   (b) https://www.nature.com/articles/nmeth.1829.
5. Sample size calculator:
   (a) http://www.surveysystem.com/sscalc.htm.
   (b) http://www.nss.gov.au/nss/home.nsf/pages/Sample+size+calculator.
6. ImageJ & Fiji plugins:
   (a) ImageJ: http://rsbweb.nih.gov/ij/download.html.
   (b) Fiji: http://imagej.net/Fiji.
   (c) RGB Line profiler: https://imagej.nih.gov/ij/plugins/rgb-profiler.html.
   (d) Coloc 2 manual: http://fiji.sc/Colocalization_Analysis.
   (e) JACoP v2.0 download: http://imagejdocu.tudor.lu/doku.php?id=plugin:analysis:jacop_2.0:just_another_colocalization_plugin:start.
   (f) JACoP v2.0 manual: http://imagejdocu.tudor.lu/lib/exe/fetch.php?media=plugin:analysis:jacop_2.0:just_another_colocalization_plugin:jacop_ijconf2008.pdf.
   (g) Li & Stanley's ICQ download: http://wwwfacilities.uhnresearch.ca/wcif/software/Plugins/ICA.html.
   (h) Li & Stanley's ICQ manual: http://wwwfacilities.uhnresearch.ca/wcif/PDF/ICA.pdf.
   (i) Colocalization Colormap: https://sites.google.com/site/colocalizationcolormap/home.
   (j) Confined Displacement Algorithm: http://imagejdocu.tudor.lu/doku.php?id=plugin:analysis:confined_displacement_algorithm_determines_true_and_random_colocalization_:start.
   (k) EPFL Denoising algorithm: http://bigwww.epfl.ch/algorithms/denoise/.
   (l) OBCOL (Organelle based colocalisation): http://obcol.imb.uq.edu.au/index.html. (m) Colocalisation Analyser – University of Sussex: http://www.sussex.ac.uk/gdsc/intranet/microscopy/omero/scripts/colocalisation.
   (m) Colocalisation Analyser – University of Sussex: http://www.sussex.ac.uk/gdsc/intranet/microscopy/omero/scripts/colocalisation.
   (n) University of Sussex Colocalisation Manual: http://www.sussex.ac.uk/gdsc/intranet/pdfs/Colocalisation.pdf.
   (o) DiAna ImageJ plugin: http://imagejdocu.tudor.lu/doku.php?id=plugin:analysis:distance_analysis_diana_2d_3d_:start.

Notes

1. In a survey of three high-impact journals (Barlow et al., 2010, supplementary table 1) approximately 50% of articles calculated colocalisation by means of a red-green overlay and only 3% used a scatterplot or any form of mathematical analysis.
2. In keeping with mathematical convention, the Roman letter r is used when the PCC is applied to a sample, whilst the Greek letter ρ denotes PCC of an entire population. The mathematical definition of the term 'moment' recognises that PCC analyses the shape distribution of the two data sets being analysed for correlation (the product-moment being the mean of a set of products). See: Pearson, K. (1896). 'Mathematical Contributions to the Theory of Evolution. III. Regression, Heredity, and Panmixia'. Phil. Trans. Roy. Soc. Lond. A 187: 253–318.
3. A normal distribution is assumed because the distribution of r for a series of random samples taken from a population where there is no correlation between two variables and $\rho = 0$, is normally distributed with a mean of zero.
4. A PCC of zero does not necessarily mean the two variables are independent. Variables equally spaced in the x-axis about zero and related by $y = x^2$ have a correlation coefficient of zero.
5. Because Pearson's correlation coefficient subtracts the mean intensity from each pixel intensity value, it is independent of signal level (or gain) and also independent of offset. This is extremely useful and is *not* the case with Manders' overlap coefficient (MOC), which increases with a positive offset level to both channels and decreases with an offset to one channel. Neither PCC nor MOC are susceptible to multiplicative gain changes applied to intensity values, except at very low gains when low signals are compressed when using an 8-bit dynamic range. For quantitative work, always use 12-bit or higher dynamic ranges.
6. It is particularly important to carefully consider cause and effect when calculating correlation values and performing colocalisation analysis. Consider three analogies: (1) Lice make you healthy. On islands in the South Pacific, healthy people have lice, sick people do not. Is there a correlation? Yet it is not that lice make you healthy, rather being sick with a fever drives off lice. (2) Clerics and spirits. There is a good relationship, shown by a nice tight scatterplot, between the salaries of Baptist ministers in Oxford and the price of rum in Havana. Do the ministers support the rum trade, or do they profit from it? Worldwide inflation affects salaries and the price of rum equally and independently. (3) Medics and patients colocalise pretty well in hospitals where their respective concentrations are high, do not colocalise in general, and may even contra-localise when richer and poorer quarters of a city are considered. Therefore, consider the numbers produced by analysis and statistical testing in the context of the biology and available facts.
7. The Roman letter R is sometimes used to denote the Manders' overlap coefficient, to distinguish it from the PCC. The similar nomenclature of MOC to the colocalisation overlap coefficients, M_1 & M_2, which *are* useful, and the differing use of each description in the literature further confuse the field of colocalisation and correlation analysis. In this book the M_1 & M_2 coefficients – which should be cited with the thresholded PCC – are

called Manders' colocalisation coefficients. The values m_1 and m_2, where the colocalisation coefficients M_1, M_2 apply to regions of interest, have also been suggested and are useful in this specific context.

8. The split overlap coefficients k_1 and k_2 are intended to show co-occurrence but depend upon absolute intensity (since they depend on the sum of the products of the intensities of two channels), requiring the absolute numbers of fluorophore being sampled in each pixel, and are sensitive to variation in pixel intensity in the opposite channel. They are moderately useful to show whether PMT detector gain or CCD exposure has been set too low in one channel or whether one channel has bleached more than the other, thus affecting the *k* value for that channel. The coefficients k_1 and k_2 are rarely used and are not considered further here. Refer to Manders et al. (1992) and Adler and Parmryd (2010) for further information.
9. The colocalisation coefficients M_1 and M_2 can also be immune to offset, providing the correct thresholds are set.
10. A type I error (a false positive) leads to a rejection of a null hypothesis when it is, in fact, true. The significance level, α, is the probability of a type I error (e.g. a type I error occurring 5% of the time). A type II error (a false negative) leads to the failure to reject a null hypothesis that is false.
11. The p-value has no precision and is not a dichotomous variable. It is a poor measure of the strength of evidence against the null hypothesis, for unless the power of the test is very high (>90%), the p-value may show wide sample-to-sample variability and does not reliably indicate the strength of evidence against the null hypothesis.

26

Super-Resolution Fluorescence Microscopy

Outline

Understanding Light Microscopy, First Edition. Jeremy Sanderson.

26.1 Introduction

Ever since Ernst Abbe elucidated and published the principles of image formation in 1873, it has been assumed that the wave-like nature of light imposed a fundamental limit on resolving power, that these physical laws are immutable. Even with,

- perfect objectives,
- minimal aberrations,
- precise optimal alignment,
- evenly illuminated field of view,
- homogeneous refractive index,
- good signal-to-noise ratio and
- large numerical apertures,

it is impossible to resolve detail whose size is smaller than approximately half the wavelength of light. The physical limit is set primarily by the wavelength of the illuminating radiation (strictly speaking, for a fluorescence microscope, the minimum resolved distance is dependent upon the *emission* wavelength of the fluorophore: $d = 0.61\lambda_{em}/NA$; see Section 10.7) and by the NA of the objective. Lateral resolution is therefore limited to $\lambda/2.n.\sin\alpha$, or $\lambda/2.NA$, in the focal plane, and $2\lambda/n.\sin^2\alpha$, or $2\lambda/NA^2$, axially. The best oil-immersion objectives therefore give a point spread function (PSF) having radial symmetry laterally in the x,y plane and extended in z, whose dimensions are about 200 nm laterally by about 600 nm along the optical axis. (In practice, the bandwidth of the optical microscope is set by the cut-off point, k_0, of the optical transfer function, OTF; see Figure 10.19). The OTF shows graphically the efficiency at which the microscope transfers spatial frequency information from the object into the image. The Abbe limit ($d = 0.5\lambda/NA$) occurs at the spatial cut-off frequency (where OTF = zero and the cut-off $k_0 = 2.NA/\lambda_{em}$; see Section 10.9) The smaller the PSF, the more remote the outer boundary of the cut-off point as the abscissa on the x-axis.

The point spread function has a complex shape. In the plane of focus, the PSF is simply the two-dimensional Airy pattern – a bright circular Airy disc in the centre surrounded by annular rings of decreasing intensity. The Gaussian plot associated with the Airy pattern is an intensity plot of the PSF in the focal plane. In theory, the three-dimensional shape of the point spread function should be spherical, but because of the inevitable poorer resolving power axially, in the z-plane, the 3D PSF (e.g. Figure 21.1 and Figure 21.9) assumes the shape of a rugby football; the concentric rings spread out away from the central plane as a complex set of conical structures. These side lobes are greatly reduced in the confocal microscope and virtually non-existent in the multi-photon microscope. Each sub-resolution point in the image is replaced by the PSF, the intensity of each PSF being directly proportional to that of the point from which it arose. In confocal microscopy, in which the image is built up by point illumination and point detection, the PSF is applied twice: once by the illuminating pinhole and again by the detecting pinhole. The confocal PSF is therefore the original 'widefield' PSF formed by the objective, squared. The central maximum intensity (corresponding to the Airy disc) is reinforced while the lower intensity maxima from the surrounding rings, the 'side lobes', are attenuated. Since these side lobes give rise to much of the out-of-focus light, the confocal microscope is therefore a good optical sectioner.

The dynamic components that control living cells occur at the level of macromolecular complexes, in the size range of circa 10 nm to a few hundred nanometres, tantalisingly beyond the limit set by the effects of diffraction and the size of the PSF. Electron microscopy can cover this size range from a few nanometres upto and beyond the LM resolution limit of approximately two hundred nanometres, but EM techniques are inherently artifactual, far more invasive than fluorescence microscopy and exclude the possibility of examining living cells at high resolution in their unperturbed state. Furthermore, fluorescence microscopy offers a wide range of staining and targeting methods, including those for observing living cells and tissues, which are not available at the EM level. Even with fixed cells, the range of versatile markers to label proteins in a specific manner offered by light microscopy is greater than for electron microscopy. Thus super-resolution light microscopy techniques are at the forefront of the quest ultimately to monitor macromolecular dynamic processes without affecting the viability of the cell. The possibility of actually exploiting the advantages of light microscopy to resolve details once only thought possible with electron microscopy first arose in 1976 when Hirschfeld used fluorescence microscopy to image single molecules, followed by Betzig (Betzig and Chichester, 1993), who used near-field scanning optical microscopy to acquire images of single fluorophores. Near-field scanning microscopy is, however, both complex and – obviously – limited in the depth to which it can image. The practical application of super-resolution methods lies in far-field microscopy using dioptric lenses, therefore necessitating the circumvention of the Abbe limit by exploiting fluorescence saturation states or by precise localisation. These methods (Table 26.1) are described in this chapter. Super-resolution microscopy only applies to fluorescence microscopy, where self-luminous point spread functions can be clearly defined and manipulated. This is not possible in bright-field microscopy, where the PSFs overlap and are indistinguishable.

Super-resolution microscopy has started to become a mainstream imaging technique in these first two decades of the 21st century but is still far from being routinely adopted in the way that optical sectioning microscopy was very soon after the introduction of the first confocal microscopes. This is due to four reasons: firstly, turn-key super-resolution microscopes are still very expensive, and the alternative of self-building a super-resolution platform requires a great deal of time, together with skills in engineering and physical optics. Secondly, specimen preparation demands more time and care than is required for routine fluorescence microscopy. Thirdly, image processing is required post-acquisition, particularly for localisation methods. Fourthly, super-resolution microscopes demand much more time than a routine confocal or widefield fluorescence microscope to maintain at their operational peak. Whilst it is ideal to routinely run a performance quality control audit for the confocal microscopes in an imaging facility, it is possible to go a fair period of time between tests: this is not so for super-resolution microscopes, which demand close attention.

Since this is an introductory text to light microscopy, this chapter covers the basic principles of super-resolution microscopy but recognises that essential operational details can only be learned within an experienced imaging facility of a research group where

Table 26.1 The Development of 'Super-Resolution' Microscopy Methods

Method	Acronym	Year	Circumvents Abbe limit	Lateral resolution (x,y)	Axial resolution (z)	Realistic imaging depth	Acquisition time/image	Advantages	Disadvantages	References
4pi microscopy using confocal (I^5M for widefield)	4pi/I^5M	1992	No	220–240 nm	80–140 nm	80–200 µm	msec	Forerunner to STED microscopy (Forerunner to SIM microscopy)	Both are fluorescence-enhanced microscopy techniques	Hell and Stelzer (1992) Gustafsson et al. (1999)
Near-field scanning optical microscopy	NFSOM	1982	Yes	30–100 nm	10 nm	50–100 nm	msec	Forerunner to localisation microscopy		Betzig et al. (1986) Lereu et al. (2012)
Total Internal Reflection Fluorescence Microscopy	TIRFM	1981	Yes*	180–250 nm	100 nm	100 nm	msec	v. good optical sectioning	Limited depth close to coverslip	Axelrod (2001) Martin-Fernandez et al. (2013)
Stimulated Emission Depletion microscopy	STED	1994	Yes	20–60 nm	500–700 nm recently c 100 nm	>20 µm	msec–min	No image processing, fast	Limited dyes, high photobleaching, small FoV	Hell and Wichmann (1994) Eggeling and Hell (2015)
Reversible Saturable Optical Fluorescence Transitions	RESOLFT/ GSDIM	1994	Yes	—	—	—	msec–min	Much lower light depletion than STED	High SNR required, post-acquisition computation	Böhm et al. (2016) Hell and Kroug (1995)
Structured Illumination Microscopy fluorescence-enhanced microscopy technique	SIM SSIM	1997	No	100–130 nm	150–340 nm	10–20 µm	sec–min	Best method for live-cell imaging	Slow, needs many images Requires stable fluorophores	Gustafsson et al. (2008) Rego and Shao (2015)
Photoactivated Localisation Microscopy	PALM	2006	Yes	20–30 nm	30–80 nm	10 µm	min	Simple to do, can use WF microscope	Limited to PA-fluorophores, long acquisition & processing	Betzig et al. (2006) Temprine et al. (2015)
Fluorescence Photoactivation Localisation Microscopy	FPALM	2006	Yes	20–50 nm	20–80 nm	10 µm	min–hr	Simple to do, can use WF microscope	Require bright dyes, slow, long acquisition & processing	Hess et al. (2006) Gould et al. (2009)
Interferometric Photoactivation & Localisation Microscopy	iPALM	2009	Yes	10–20 nm	5–10 nm	0.5 µm	min–hr	Ultra-high resolution in 3D	Slow, limited penetration depth	Shtengel et al. (2009) Shtengel et al. (2014)
Stochastic Optical Reconstruction Microscopy	STORM	2006	Yes	20–50 nm	20–80 nm	10 µm	min–hr	v. high resolution wide range of dyes	slow, artifacts long acquisition & processing	Rust et al. (2006) Dempsey (2013)
Direct Stochastic Optical Reconstruction Microscopy	dSTORM	2008	Yes	20–50 nm	20–80 nm	10 µm	min–hr	Simple, cheap, can be used on tissues	Poor Mab labelling causes poor localisation accuracy	Heilemann (2008) van de Linde et al. (2011)
Classical microscopy										
Widefield microscopy (bright-field & fluorescence)	WFM	19th C	No	250 nm	>500 nm	0–10 µm	msec–min	Low cost, easy to use	Out-of-focus blur, low contrast	Sanderson et al. (2014) Lichtman and Conchello (2005)
Point-scanning confocal	LSFM	1986	No	180–250 nm	>500 nm	5–100 µm	sec–min	Good optical sectioning	Slow, poor SNR photobleaching	Conchello and Litchmann (2005) Amos et al. (2012)
Spinning-disc confocal	SDCM	1967	No	250 nm	>500 nm	5–50 µm	msec–min	Good SNR; live-cell Very fast frame rate, sensitive	No low mag. Imaging, pinhole cross-talk, no ROI	Stehbens et al. (2012) Hosny et al. (2013) Gräf et al. (2005)

(*continued*)

Table 26.1 (Continued)

Method	Acronym	Year	Circumvents Abbe limit	Lateral resolution (x,y)	Axial resolution (z)	Realistic imaging depth	Acquisition time/image	Advantages	Disadvantages	References
Multi-photon	2MP	1990	No*	300–500 nm	400–600 nm	50–1000 μm	sec–min	Penetration depth, SHG	Photo-damage, poor SNR	Ustione and Piston (2011) Helmchen and Denk (2005)
Light-sheet microscopy	SPIM DSLM (Table 19.1)	1993	No	250 nm	700–900 nm or light sheet thickness	>1000 μm	msec–sec	Very fast, very little photo-damage, good 3D data sets	Z resolution limited by light-sheet thickness, sample mounting complex	Mappes (2012) Huisken et al. (2004) Reynaud et al. (2015)

Notes: The Yes* and No* are qualified responses. Yes, TIRF in theory can circumvent the Abbe limit for a diffraction-limited dioptric lens system, because the evanescent field is a near-field technique. However, the improvement in resolution only lies along the z-axis. No, multi-photon microscopy (MPM) does not offer a practical advantage in resolving power because of the long wavelengths used. In theory the nonlinear process (that is, a nonlinear dependence of emission signal on local excitation) is not diffraction limited. In MPM the emitted signal is dependent upon the square of the excitation intensity process. With fluorescent methods, an increase in resolving power is only possible through nonlinear methods. It is not possible to employ oblique illumination, as with bright-field microscopy, because the emitted signal from a particular fluorophore molecule is not coherent with – and so cannot interfere with – the excitation illumination or indeed with light emitted from any other fluorescent molecule in the population.

References

Amos, W.B., McConnell, G., and Wilson, T. (2012). 'Comprehensive Biophysics'. In: *Confocal Microscopy*, Chapter 2.2, vol. 2. Academic Press, pp. 3–23.
Axelrod, D. (2001). 'Total internal reflection fluorescence microscopy in cell biology'. *Traffic* 2/11: 764–774.
Betzig, E. et al. (1986). 'Near-field scanning optical microscopy (NSOM)'. *Biophysical Journal 49*/1: 269–279.
Betzig, E. et al. (2006). 'Imaging intracellular fluorescent proteins at nanometer resolution'. *Science 313*/5793: 1642–1645.
Böhm, U. et al. (2016). '4Pi-RESOLFT nanoscopy'. *Nature Communications 7*: 10504. doi: 10.1038/ncomms10504.
Conchello, J.A. and Lichtman, J.W. (2005). 'Optical sectioning microscopy'. *Nature Methods 2*/12: 920–931.
Dempsey, G.T. (2013). 'A user's guide to localization-based super-resolution fluorescence imaging'. *Methods in Cell Biology 114*: 561–592.
Eggeling, C. and Hell, S.W. (2015). 'STED Fluorescence nanoscopy'. In: Tinnefeld, P., Eggelin, C., and Hell, S.W. (eds.), *Far-Field Optical Nanoscopy*, Chapter 1. Springer-Verlag, pp. 3–26.
Gould, T.J. (2009). 'Imaging biological structures with fluorescence photoactivation localization microscopy'. *Nature Protocols 4*/3: 291–308.
Gräf, R. et al. (2005). 'Live cell spinning disc microscopy'. *Advances in Biochemical Engineering/Biotechnology 95*: 57–75.
Gustafsson, M.G. et al. (1999). 'I^{5}M 3D widefield light microscopy with better than 100 nm axial resolution'. *Journal of Microscopy 195*/1: 10–16.
Gustafsson, M.G. et al. (2008). 'Three-dimensional resolution doubling in wide-field fluorescence microscopy by structured illumination'. *Biophysical Journal 94*/12: 4957–4970.
Heilemann, M. (2008). 'Subdiffraction-resolution fluorescence imaging with conventional fluorescent probes'. *Angewandte Chemie International Edition 47*/33: 6172–6176.
Heintzmann, R. and Ficz, G. (2006). 'Breaking the resolution limit in light microscopy'. *Briefings in Functional Genomics & Proteomics 5*/4: 289–301.
Hell, S.W. and Kroug, M. (1995). 'Ground-state depletion fluorescence microscopy, a concept for breaking the diffraction resolution limit'. *Applied Physics B 60*/5: 495–497.
Hell, S. and Stelzer, E.H.K. (1992). Fundamental improvement of resolution with a 4Pi-confocal fluorescence microscope using two-photon excitation. *Optics Communications 93*/5–6, 277–282.
Hell, S.W. and Wichmann, J. (1994). 'Breaking the diffraction resolution limit by stimulated emission: stimulated-emission-depletion fluorescence microscopy'. *Optics Letters 19*/11: 780–782.
Helmchen, F. and Denk, W. (2005). 'Deep tissue two-photon microscopy'. *Nature Methods 2*/12: 932–940.
Hess, S.T. et al. (2006). 'Ultra-high resolution imaging by fluorescence photoactivation localization microscopy'. *Biophysical Journal 91*/11: 4258–4272.
Hosny, N.A. et al. (2013). 'Super-resolution imaging strategies for cell biologists using a spinning disk microscope'. *PLoS One 8*/10: e74604.
Huisken, J. et al. (2004). 'Optical sectioning deep inside live embryos by selective plane illumination microscopy'. *Science 305*/5686: 1007–1009.
Lereu, A.L. et al. (2012). 'Near field optical microscopy: a brief review'. *International Journal of Nanoscience 9*/(3–7): 488–501.
Lichtman, J.W. and Conchello, J.-A. (2005). 'Fluorescence microscopy'. *Nature Methods 2*/12: 910–919.
Mappes, T. et al. (2012). 'The invention of immersion ultramicroscopy in 1912 – the birth of nanotechnology?' *Angewandte Chemie International Edition 51*/45: 11208–11212.
Martin-Fernandez, M.L. et al. (2013). 'A "pocket guide" to total internal reflection fluorescence'. *Journal of Microscopy 252*/1: 16–22.
Rego, E.H. and Shao, L. (2015). 'Practical structured illumination microscopy'. *Methods in Molecular Biology 1251*: 175–192.
Reynaud, E.F. et al. (2015). 'Guide to light-sheet microscopy for adventurous biologists'. *Nature Methods 12*/1: 30–34.
Rust, M.J. et al. (2006). 'Sub-diffraction-limit imaging by stochastic optical reconstruction microscopy (STORM)'. *Nature Methods 3*: 793–795.
Sanderson, M.J. et al. (2014). 'Fluorescence microscopy'. *Cold Spring Harbor Protocols*. doi: 10.1101/pdb.top071795.
Shtengel, G. et al. (2014). 'Imaging cellular ultrastructure by PALM, iPALM, and correlative iPALM-EM'. *Methods in Cell Biology 123*: 273–294.
Shtengel, G. et al. (2009). 'Interferometric fluorescent super-resolution microscopy resolves 3D cellular ultrastructure'. *PNAS 106*/9: 3125–3130.
Stehbens, S. et al. (2012). 'Imaging intracellular protein dynamics by spinning disk confocal microscopy'. *Methods in Enzymology 504*: 293–313.
Temprine, K. et al. (2015). 'Three-dimensional photoactivated localization microscopy with genetically expressed probes'. *Methods in Molecular Biology 1251*: 231–261.
Ustione, A. and Piston, D.W. (2011). 'A simple introduction to multiphoton microscopy'. *Journal of Microscopy 243*/3: 221–226.
van de Linde, S. et al. (2011). 'Direct stochastic optical reconstruction microscopy with standard fluorescent probes'. *Nature Protocols 6*/7: 991–1009.

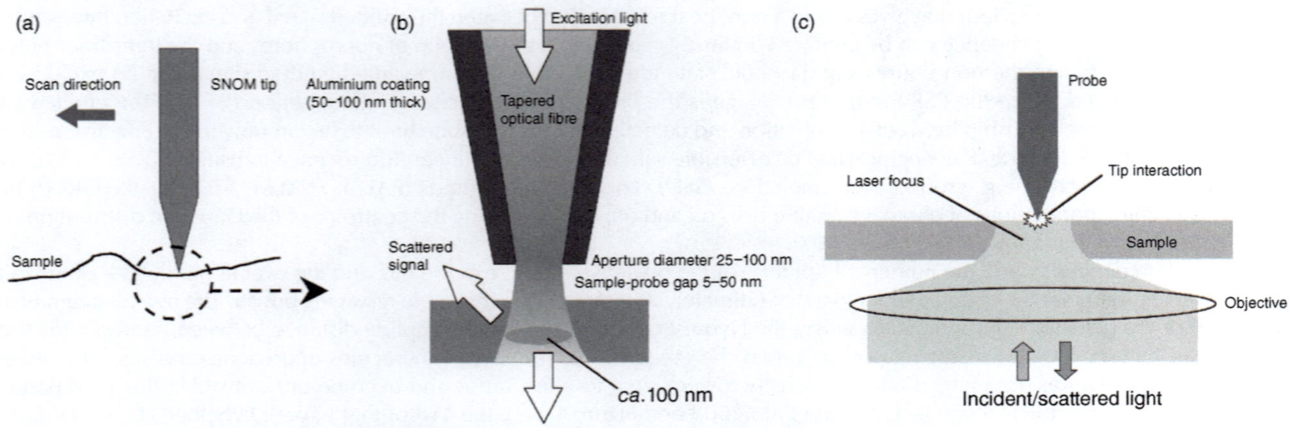

Figure 26.1 **Near-field scanning microscopy** *An optical fibre tip (SNOM: scanning near-field optical microscopy) pulled from quartz and coated with aluminium (a) is scanned over the specimen very near to the surface, within a fraction of a wavelength of visible light. In (b) a feedback system precisely controls the physical separation of the probe and specimen. With the aperture of the fibre at 100 nm diameter, only 1 photon in 10 000 will pass through; with an aperture of 50 nm, one photon in 10^8 passes through. The input power cannot arbitrarily be increased to compensate for poor SNR because the coating will melt, so sometimes a metal tip alone (c) is used. If a sharp metal tip with a diameter around 10 nm is placed in the focus of a laser beam, an effect called local field enhancement will give an increased signal in the electric field. The most recent commercial SNOM instruments combine the scanning force techniques with the optical detection capabilities of conventional optical microscopy (c) to overlay the near-field signal onto a far-field image. Source: Richards 2003. Reproduced with permission from the Royal Society.*

super-resolution microscopes are a central tool or are else being developed. A recent book (Birk, 2017) and several excellent reviews are available for further information. For example, see Turkowyd et al. (2016), Cremer and Birk (2016), Cox (2015), Eggeling et al. (2015); Han et al. (2013), Ball et al. (2012), Patterson et al. (2010), Schermelleh et al. (2010) and Hell (2009).

26.2 Circumventing Abbe's Limit

When faced with unsurmountable odds, the most effective military tactic is to evade or bypass the obstacle. Although the diffraction limit postulated by Abbe has not been broken as such, it has been circumvented[1]. To get around the Abbe resolution limit, it is necessary to exploit properties of how the illumination interacts with the sample – the very basis of diffraction. This was first done by employing various forms of near-field microscopy. In far-field microscopy, there are, broadly, two means of circumventing the Abbe limit: either (a) by switching fluorophores from one state to another and stochastically plotting their precise position or (b) by PSF engineering using nonlinear fluorescence excitation and patterned illumination.

Just as the stature and influence of Newton's particulate theory overshadowed the work of Huygens on the wave nature of light (Chapter 10), Abbe's theory of image formation in the microscope 'directed too much attention on waves and the passband…dismissing any hope for better resolution with conventional objective lenses.…Therefore, in the quest for nanoscale optical resolution, it seemed natural to trade off lenses and free wave propagation for mechanical confinement [of light] by nanosized tips, which is the philosophy of near-field optics' (Hell, 2009, page 29).

Near-field microscopy (NSOM, SNOM) does away with focusing objective lenses altogether and instead exploits the properties of evanescent waves, placing the light source and detector very close to the sample (Figure 26.1), whereby the resolving power of the microscope is determined by the sub-wavelength aperture of the scanning tip in front of the sample and/or the size of the detector. This bypasses the diffraction limit in all three directions, giving a lateral resolving power between 30 and 100 nm and a much smaller axial resolving power of 5–10 nm. Near-field methods are intrinsically limited to deriving surface information. Total internal reflection fluorescence (TIRF) is a form of near-field microscopy that has been widely adopted and is discussed in Chapter 23. TIRF does not improve the lateral resolution, only the axial resolution. For a review of other methods of near-field optical microscopy, see de Lange et al. (2001) and Richards (2003).

Far-field microscopy uses an aperture-limiting dioptric objective lens to focus light onto the sample. The classical Abbe limit can be circumvented in far-field microscopy because this long-held limitation only holds true if three assumptions are made (Gustafsson, 1999). First, that light is collected by a single objective; second, that the excitation light is uniformly incident upon the sample and third, that fluorescence occurs through normal linear absorption followed by emission of a single photon such that there is a linear relationship between the excitation intensity and the intensity of the emitted signal. If these assumptions are ignored, or circumvented, it is possible to extend the resolving power of the light microscope by:

- gathering light over a different, preferably larger, set of angles around the sample;
- using nonlinear fluorescence excitation;
- switching between two stable fluorophore transition states; and
- using patterned excitation illumination.

Since fluorophores exhibit behaviour that gives rise to a nonlinear relationship between the emitted signal and excitation intensity[2], and as long as such nonlinear dependence can be confined (in time or space) to a small group of fluorophores and the transitions between the fluorescing 'on' state and the non-fluorescent dark 'off' state are detectable, then unwanted emitted signal can be excluded and super-resolution achieved. Since the PSF forming the image is the *product* of the excitation illumination PSF and the emitted signal PSF, breaking this linear relationship between illumination and detection allows the Abbe limit to be circumvented. Effective nonlinear super-resolution methods include 'PSF engineering' or reversible saturable optically linear fluorescence transitions (RESOLFT) through switching fluorescence states (e.g. ground state depletion, GSD). Localisation methods (PALM, FPALM, STORM, dSTORM[3]) entail bleaching or switching populations of photoactivatable dyes on and off, thus localising the centroids of the Gaussian distribution of the PSF with high precision (for a review see Hell, 2007).

Whilst resolution depends upon the numerical aperture of the objective, the size of the PSF and the overall bandwidth of the microscope, it is easy to forget that there are other factors that ultimately determine the achievable resolving power. The overall magnification and pixel size of the detector is important. To satisfy the Nyquist criterion, the mean sampling distance between fluorophores should be smaller than half the desired resolving power. When the mean distance between molecules approaches the size of features in the specimen, these features are imaged at progressively lower signal-to-noise ratios and become unresolvable. This is of particular importance in localisation microscopy (see, Shroff et al., 2008 and Figure 1 on page 418 of that paper). Whether 2D or 3D data sets are collected is also important: if 10^4 molecules/μm^2 are required for a 2D image, this will rise to 10^6 molecules/μm^2 for a 3D data set. Another trade-off with localisation microscopy becomes sample photostability versus the time required to collect a high-quality image with sufficient SNR. For such stochastic localisation methods, the number of fluorophores per pixel, F, will vary by $\sqrt{F}$, introducing noise either at low-labelling densities or if a detector with small pixel sizes is used to sample the specimen in forming the image.

The term 'super-resolution' has become part of the microscopy lexicon, yet it is disliked by many, for it is imprecise. Stefan Hell, who with Eric Betzig and William Moerner was jointly awarded the 2014 Nobel Prize in Chemistry for 'the development of super-resolved fluorescence microscopy' prefers to call this subject of nanoscopy 'diffraction-unlimited fluorescence microscopy'.

26.3 Increasing the Size of the Collection Aperture

Because the microscope objective collects diffracted light from only one side of the specimen to form the image, a significant proportion of information is lost. If the specimen were situated at the centre of a spherical lens, then all the diffracted light could theoretically be collected and used to form the image. In this case, the axial resolution would equal the lateral resolution in the focal plane, and the PSF would be isotropic – an ideal situation. Clearly this ideal is not possible, but by increasing the overall collection angle, the axial resolution is improved to be – for all intents and purposes – isotropic with the lateral resolving power.

Although not strictly super-resolution techniques, in that they do not circumvent the Abbe limit, some methods use more than one objective (gathering light from larger aperture angles) and interference to increase the resolving power in z, along the optical axis, for this is 2–3 times worse than the lateral resolving power of the microscope. 4-pi microscopy, I^nM microscopy and confocal theta microscopy are examples of this approach – they only push resolving power to the Abbe diffraction limit. The other important technique that has been widely adopted, which improves the resolving power to the maximum that it is possible to achieve without circumventing the Abbe limit is SIM, or structured illumination microscopy, discussed in Section 26.6. 4-pi, I^5M and SIM all depend on a linear relationship between the excitation intensity and the intensity of the emitted signal and so cannot circumvent the Abbe limit due to nonlinear fluorophore excitation.

Prior to the development of STED microscopy, Hell introduced the 4-pi microscope[4], manufactured by Leica Microsystems. In the 4–pi system, two high–numerical aperture objectives are set orthogonally to one another, to synthesise a larger collection angle with two lenses. The illumination for both lenses is precisely aligned and coherent so that the two wavefronts of the excitation light constructively interfere with each other to produce a very small focal spot. This approach demands both a very well-aligned laser and well-aligned microscope system with matched objectives. Another related approach to 4-pi microscopy is confocal theta microscopy, which was the forerunner of digital light-sheet microscopy. All these techniques are very sensitive to any localised or transient misalignment due to temperature shifts or air currents but improve axial resolving power fivefold to around 100 nm. I^5M is a related technique to 4-pi microscopy; it uses incoherent widefield illumination with post-acquisition deconvolution.

Standing wave fluorescence microscopy, embodied in the two-photon 4-pi microscope and the widefield I^5M microscope, enabled researchers to see up to the maximum theoretical resolution (Egner and Hell, 2005), but – because of secondary intensity maxima (side lobes) in the PSF – the samples were restricted to thin cell layers between two coverslips; the alignment of the microscope was also challenging. Stefan Hell and Mats Gustafsson, the pioneers of 4-pi and I^5M, went on to develop STED and SIM microscopy, respectively.

26.4 RESOLFT and STED Microscopy

The first practical implementation of super-resolution imaging in far-field microscopy with lenses was by STED, or stimulated emission depletion, in 1994. In his fascinating Nobel lecture (Hell, 2015), Professor Hell describes how he first looked at quantum mechanics and molecular states to try and overcome the diffraction barrier. The key lay in controlling how the transition states of a fluorophore switched from an 'on' (fluorescing) to an 'off' (dark) state. This could be achieved either by RESOLFT or emission depletion.

RESOLFT is an acronym that stands for reversible saturable optical fluorescence transitions. It is a means of reversibly switching a fluorophore between an 'on' state and a dark, 'off', state in order to confine a particular state to a sub-diffraction region. STED microscopy differs from RESOLFT in that it uses nonlinear excitation saturation (Figure 26.3) to switch fluorophores into the dark S_0, ground, state by emission depletion. The usual application of the RESOLFT concept not involving localisation microscopy is ground state depletion (GSD), which uses a population of metastable triplet states, in a microscope also manufactured by Leica.

In conventional fluorescence microscopy (Figure 26.3; see also Section 15.2) delocalised electrons are transferred from a ground state, S_0, to an excited state, S_1 or higher. They emit fluorescence as they fall back to the ground state. In this case, the ground state, S_0, is the dark state. GSD switches most of the fluorescent molecules from S_1 to a dark triplet state T_1 – in practice the fluorophores light up and 'blink' in the specimen. This dark triplet state occurs naturally with only very low probability but can be artificially populated with electrons by repetitively exciting the fluorophore to the first excited singlet state S_1 by nonlinear stimulated emission to increase the probability of a non-radiative intersystem crossing from S_1 to T_1. Fortunately, this de-excitation protects the fluorophore from immediate photobleaching. The only caveat is that the process must be reversible, that the fluorophore is able to recover once the depletion laser is switched off. The other means of switching is, of course, employed in localisation microscopy with the bright and dark states of a reversibly photoswitchable fluorophore[5]. Those optical highlight fluorescent proteins (e.g. Eos, Kaede, Dendra2, and PA-GFP) that are permanently converted from one state to another are not suitable for RESOLFT techniques because they are not reversible.

The STED microscope is configured (Figure 26.3b) so that a far-red-shifted STED pulse shaped like a doughnut selectively inhibits the outer side lobes of the PSF of the focal spot rather like an 'optical pencil sharpener' to sharpen and slim down the imaging PSF. This is an example of 'PSF engineering' and provides increased localisation accuracy of fluorophores in the image. The lateral diameter of the region from which the fluorophores are excited is reduced from around 200 nm to 50 nm (because the STED PSF is very thin and elongated; see Figure 26.4. Attempts have been made to improve the axial resolution [isoSTED; Schmidt et al., 2008] using opposing objectives). Making the region of excitation smaller in this way is not the same as closing the confocal pinhole. The size of the pinhole only determines what signal is detected; the size of the illuminated region remains the same. With STED the size of the illuminated region is made smaller: the excited fluorescent molecules in the side lobes of the imaging spot are actively pumped down – literally absorption driven backwards by photons – out of their excited S_1 state by 'stimulated emission' of the STED pulse.

The depletion wavelength should lie outside the excitation range of the target fluorophore; otherwise, unwanted excitation will occur in the depletion area of the PSF. Also, the red STED laser should not be absorbed by the excited S_1 or triplet state; otherwise, photo-bleaching occurs. STED and RESOLFT techniques (Figure 26.5) require three illumination steps: one for activation, one for depletion and the final one for detection of the remaining active fluorophores. There are limits to the wavelength of the excitation STED beam that can be used: it is difficult to perform multicolour labelling because only very few fluorophores have non-overlapping absorption profiles and are depleted by the same STED wavelength. STED has high rates of bleaching and photo-toxicity, but it can be performed at video rate and is not dependent upon intensive image post-processing, as are the other super-resolution methods.

STED is a relatively mature technology, so suitable fluorophores have been well characterised, but for it to work well, the saturation intensity and photostability of the fluorophore is important (e.g. see Wurm et al., 2012), limiting this method to a few dyes, quantum dots and fluorescent proteins. This technique works best with bright fluorophores of high quantum efficiency that have a large Stokes shift so that minimal absorption and excitation by the STED-pulse laser occurs, such as the long-wavelength ATTO dyes, which are able to withstand repeated cycles of excitation and depletion. Photobleaching and photo-toxicity can be a problem. Although live-cell imaging is possible, STED is best-suited to well-stained fixed samples. There are three principal advantages to using STED for super-resolution studies: (a) it is fast; (b) what you see is what you get: there is no extensive post-processing required, and (c) STED can image relatively deep (ca. 20 mm) into cells, and the lateral resolving power can be tuned to give limiting values down to 20 nm in fixed samples, 30–50 nm in living cells, and around 60 nm in 3D data sets.

The improvement in resolving power is given by the formula: $\Delta_{min} \approx \lambda/2.n.\sin\alpha\ (\sqrt{[1 + I_{max}/I_{sat}]})$ where λ is the wavelength of the depletion laser, n is the refractive index of the sample and sin α the aperture of the objective, so the objective NA forming the STED depletion 'doughnut' can replace the n.sin α term. I_{max} is the peak intensity of the depletion laser and I_{sat} the saturation intensity of the fluorophore at which 50% of the fluorescent signal is depleted (see Donnert et al., 2006, and Hell, 2015). This equation is an extension of the Abbe equation $d = \lambda/2NA$; for oblique illumination, see Section 10.9.

In order to satisfy the saturated depletion equation, the term I_{max}/I_{sat} must be >1, so there is at least one photon that drives the molecule down to the depleted ground state. The probability of transition between one fluorophore state to the other depends exponentially upon the applied intensity of light, which must be inversely proportional to the concentration of fluorophore. The saturation intensity, I_{sat}, scales inversely with the lifetimes of the two states: the longer the lifetime, the smaller is I_{sat}, and the diffraction barrier can be circumvented at much lower light levels. Applying a switching intensity $I_{max} > 5.I_{sat}$ makes the switching probability greater than 99% certain. Provided that the intensity of the depletion laser is much greater than the threshold I_{sat}, the molecules in the outer part of the PSF will remain 'off'. The diameter of the central region where the fluorophores are 'on' depends upon the ratio of (I_{max}/I_{sat}). Δ_{min} scales inversely with the square root of the ratio (I_{max}/I_{sat}); the greater this ratio, the smaller the resolving power.

The size of the central fluorescing region can be made smaller by increasing I_{max}, the peak intensity of the depletion laser, or by using a fluorophore with a long excited state lifetime, which means a low saturation intensity, I_{sat}. Crucially, the longer the fluorescence lifetime, the smaller is the intensity threshold, I_{sat}. Hence, the diffraction barrier can now be broken at much lower light levels. For STED I_{sat} is measured in megawatts; STED typically uses a high I_{sat} value ($\approx 10^7$ W/cm^2), requiring an intense[6] depletion laser greater than 10^9 W/cm^2.

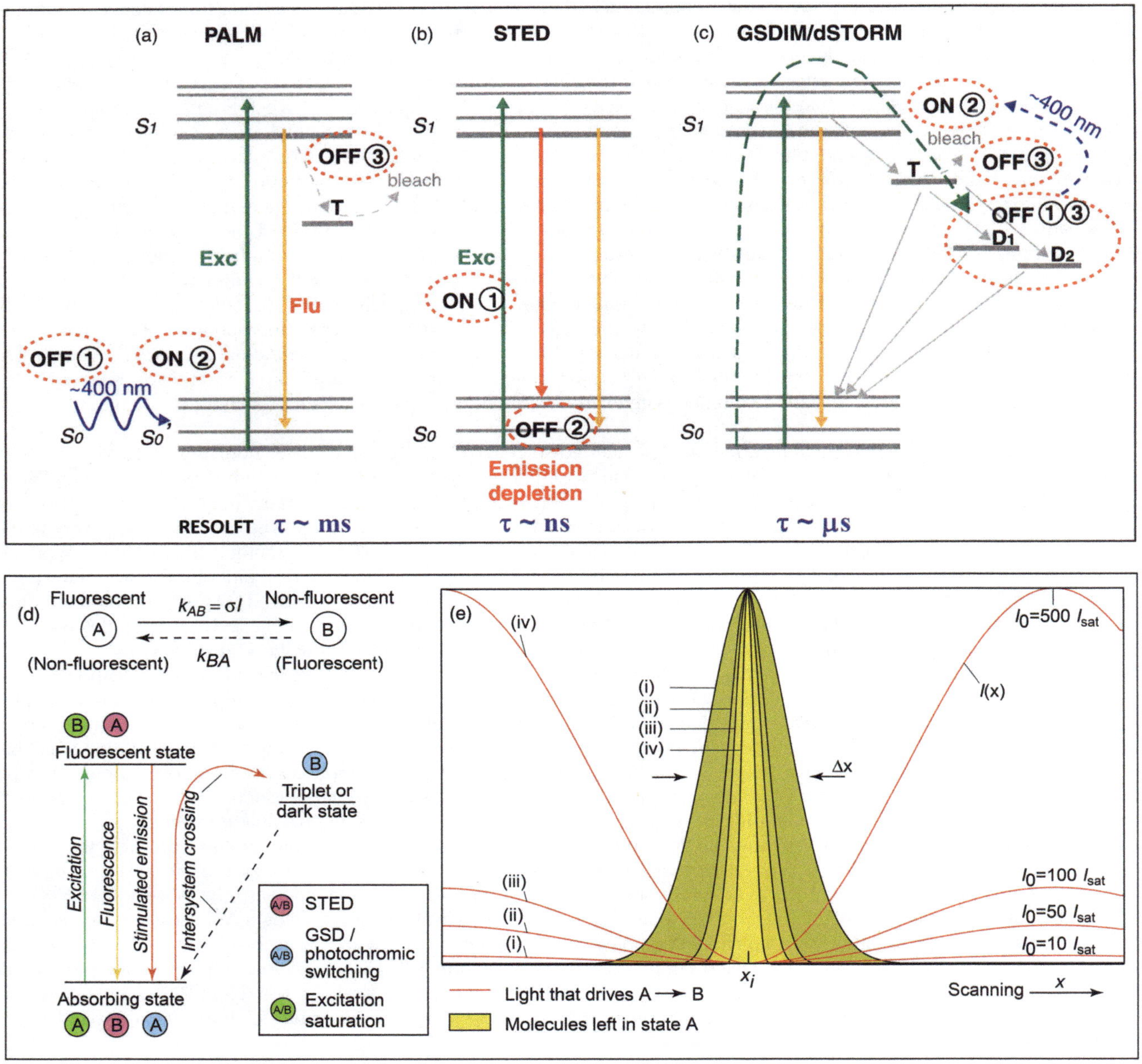

Figure 26.2 **Jabłoński diagrams of conventional, RESOLFT and STED transitions** *These diagrams show the 'ON/OFF' electronic transition states of fluorescent dyes that is crucial for super-resolution microscopy. At any particular moment, only a very few molecules are fluorescing: most reside in the 'OFF' state. (a) PALM – these photoactivatable/photoswitchable molecules absorb light (here, 400 nm blue) going from 'OFF' (step 1) to 'ON' (step 2) to fluoresce normally (yellow). Ideally, this stochastic step is irreversible; eventually, irreversible bleaching (step 3) does occur. (b) STED – fluorescent molecules are excited 'ON' followed by spontaneous emission (yellow) and a return to the ground state, S_0, in the usual way. Intense light (supplied by the depletion laser) is absorbed at the red-shifted end of the excitation spectrum of the fluorescent dye with durations shorter than the S_1 lifetime. Spontaneous emission can be supressed (i.e. depleted) back to S_0 with emission of a photon signal (2). Only those photons in the zero intensity centre of the depletion 'doughnut' fluoresce normally give rise to a smaller PSF. (c) GSDIM/dSTORM – fluorophores may return from the excited S_1 state via a long-lived triplet state (grey), which can be converted to a long-lived metastable dark state (D1, D2) by oxidation or reduction. This process is cyclical. The triplet state can be reactivated by blue light (1). After many cycles, the dyes eventually photo-bleach, or are trapped in a triplet state (3) both of which turn the dye 'OFF'. Source: Xu et al. 2011. Reproduced with permission from Elsevier.*

(d) A fluorescent label resides in two states (A/B; 'OFF/ON') at a rate K_{AB} that is proportional to the light intensity applied (the reverse k_{BA} returns the molecule to its initial state). The Jabłoński diagram is a combination of those shown in (b) and (c) in the upper panel. (e) shows the intensity profile of molecules left 'OFF' in state A. A standing wave of light is shown in red with various peak intensities and a common zero intensity node at x_i, the position of the scanned spot. Increasing the intensity of light up to 500 times saturation intensity, I_{SAT}, records a 'negative image' of 'ON' molecules in state B, which can be recovered by mathematical post-processing of the scanned image. Resolution is now defined by the ratio I_0/I_{SAT}, as explained in the text. Source: Hell et al. 2004. Reproduced with permission from Elsevier.

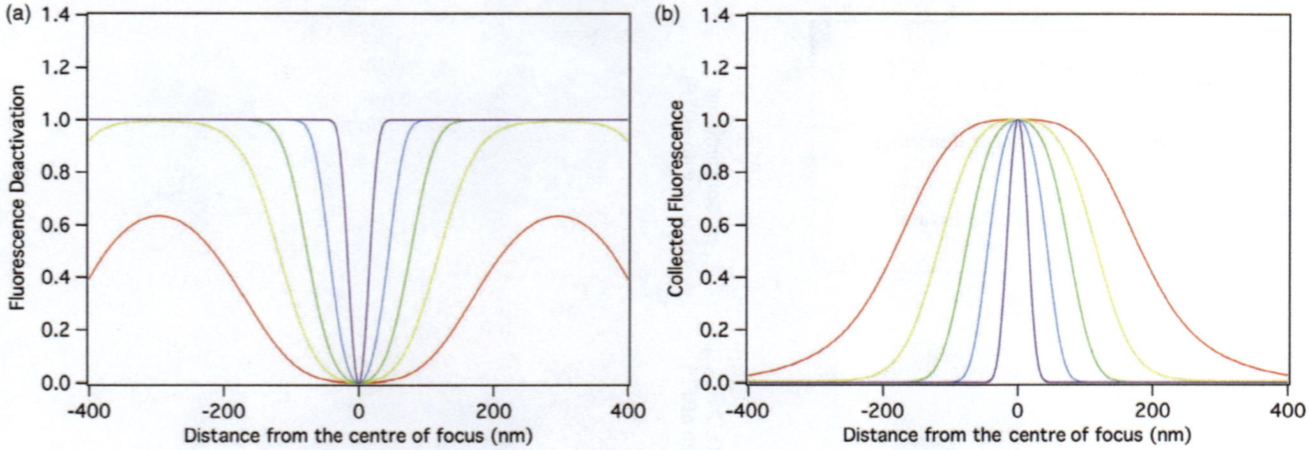

Figure 26.3 **The effect of nonlinear emission saturation** *(a) The diffraction-limited excitation intensity is shown by the red curve, and it does not change. As the excitation intensity is increased, the emission saturation changes, giving rise to distributions with sharp edges and strongly confined non-emissive regions (compare this graph with Figure 26.12b and the distribution of the red curves in the middle panel of this figure). The fluorescence emission is effectively confined to a spot with sub-diffraction dimensions, as shown in (b), which is shown as the yellow-filled Gaussian distributions in the middle panel, above. These are two ways of illustrating the same phenomenon of nonlinear fluorophore saturation. Source: Dedecker et al. 2008. Reproduced with permission from Elsevier.*

GSD with its longer-lasting triplet state, I_{sat}, requires only kilowatts of laser power. For switching times of a few milliseconds, I_{sat} is a few watts/cm^2, and a photoswitchable fluorescent protein requires only 600 W/cm^2 to initiate switching, so fewer photons are required to establish the on-off state difference required to separate sub-diffraction features. Where N is the number of fluorophore molecules within an area of diameter $\lambda/2.NA$, using a photon flux of intensity $\ll I_{sat}/N$ ensures the 'on'-state fluorophore molecules are far enough apart to be detected individually. Higher intensities risk recording multiple molecules as a single switching event, both in time and space, whereas very low photon intensities slow down the recording and may undersample according to the Nyquist criterion.

The first STED microscope used a pulsed laser for both excitation and depletion because it delivered a very high photon flux for stimulating the fluorophore within a short (1–5 ns) excited lifetime. This calls for precise timing of the depletion pulse: to be longer than the vibrational relaxation time (>50 ps) but shorter than the fluorescence lifetime (< 250 ps) of the fluorophore.

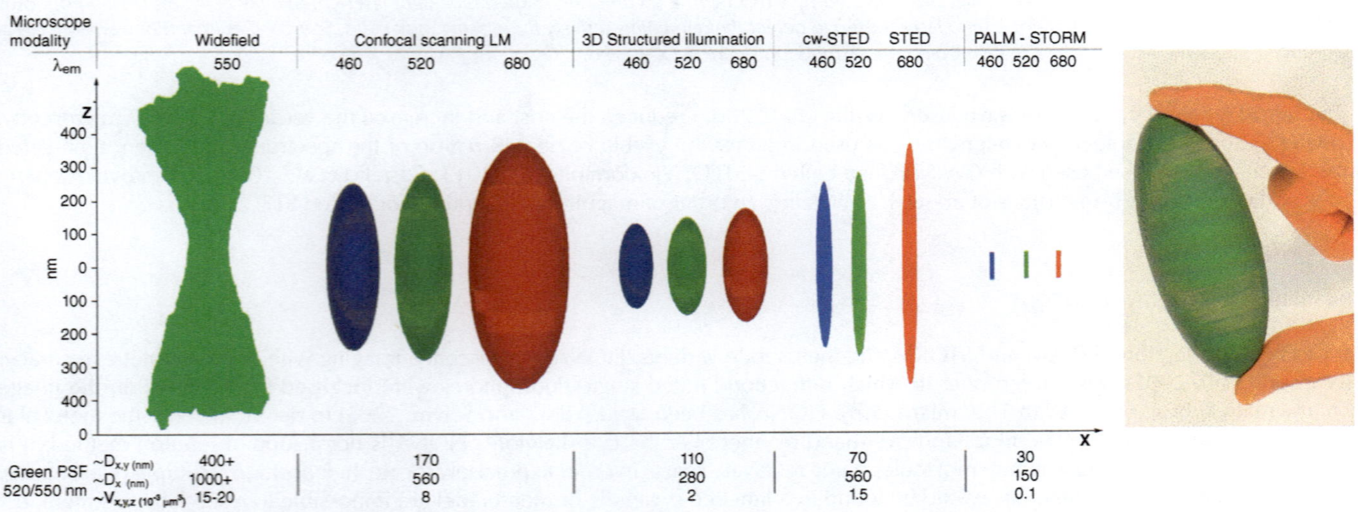

Figure 26.4 **PSFs of the various super-resolution techniques** *A schematic 3D representation of the focal volumes of super-resolution methods compared to widefield and confocal laser-scanning microscopy (CSLM). Deconvolution can further improve (i.e. reduce) these volumes. The PALM/STORM volumes are estimates based on the localisation precision in conjunction with the z-range of the evanescent field of the TIRF microscope used. These volumes approximate attainable resolution, but actual resolving power depends upon fluorophore orientation, refractive index differences, optical aberrations, camera and image sampling quality and the noise present. The dimensions and focal volume of the widefield PSF shown for comparison depends upon the objective used and can be as much as twice the size depicted here (for example, see Figure 21.10). The 3D printed PSFs depicted in this figure, which are excellent teaching models, are a generous gift from Dr Ian Dobbie, Micron Advanced Bioimaging Unit, Oxford. Source: Adapted from Schermelleh et al. 2010. Courtesy of Dr I. Dobbie, Micron Oxford, Advanced Bioimaging Unit.*

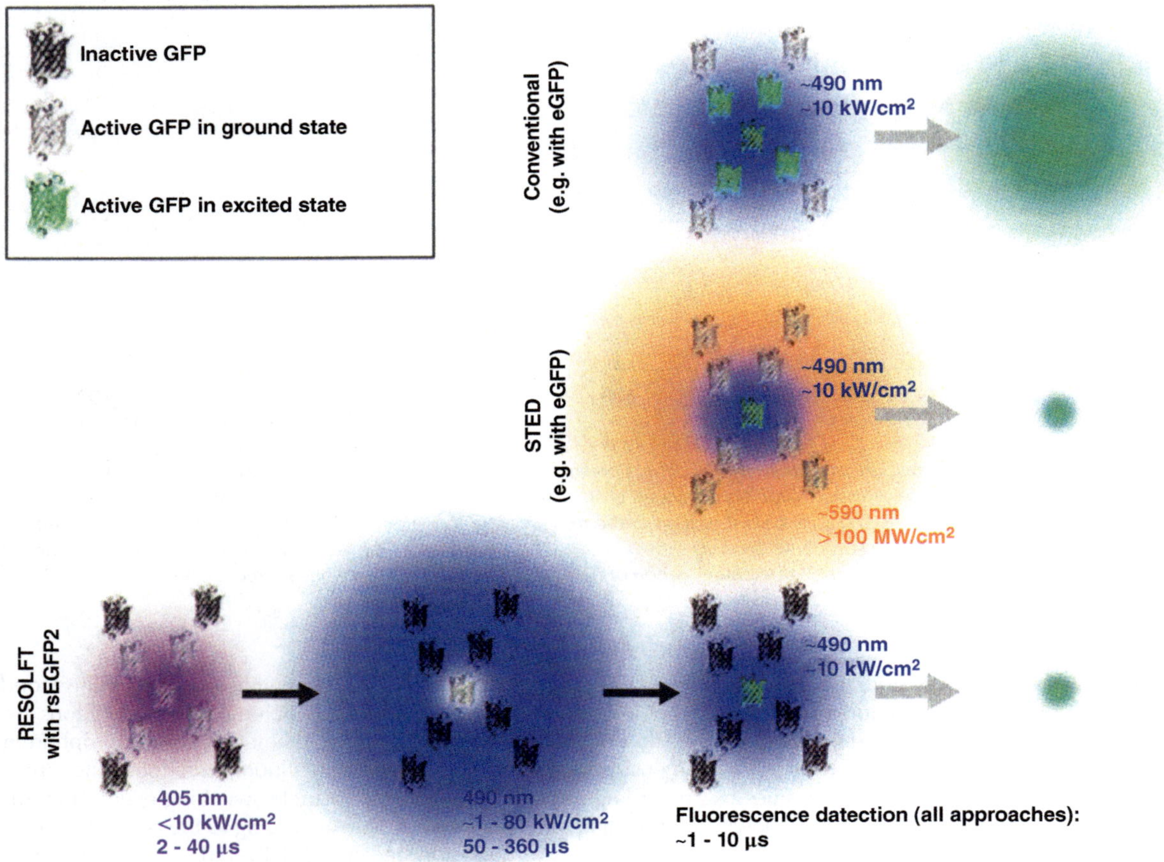

Figure 26.5 ***The STED microscope-1*** *The top figure shows the usual fluorescence excitation of GFP to produce a diffraction-limit focal spot. The middle figure shows the action of the STED depletion laser. In the lower figure in this panel, the RESOLFT principle is shown with the reversibly switchable molecule rsEGFP2. Non-fluorescing molecules are activated by violet light; a second ring-shaped blue-light laser continually activates the fluorophores until they switch back to the dark state. A third laser is used to detect the remaining actively fluorescing molecules. Source: Gould and Bewersdorf 2012. Reproduced under the terms of the Creative Commons Attribution Licence, CC-BY-4.0.*

The introduction of continuous-wave lasers (Willig et al., 2007) reduced the cost and increased the versatility of STED microscopy, allowing almost any depletion wavelength to be used in either the visible or near-IR region of the spectrum. Since then, time-gated pulsed excitation has been used with CW-STED (so-called g-STED; Vicidomini et al., 2013; Castello et al., 2016) to improve resolving power and operate at I_{max} intensities of about 10 MW/cm^2, an order of magnitude lower than for pulsed STED.

26.5 Localisation Microscopy

Prior to the introduction of PALM and STORM and their subtle variants, FIONA (fluorescence imaging with one nanometre accuracy; Thompson et al., 2002) was a forerunner in which diffraction-limited single fluorophores were localised from the PSF in the image by mathematical processing. With TIRF microscopy, FIONA has been used (Yildiz and Selvin, 2005) to determine how the molecular motors myosin and kinesin move in a hand-over-hand manner over the cytoskeleton[7]. FIONA is not a super-resolution method; it is reliant upon the density of fluorophore molecules being relatively sparse in order to precisely locate their position. Multiple fluorophores in close proximity to one another (as would be found in a labelled organelle or membrane) are impossible to resolve by this method.

Whilst spectroscopic methods could, theoretically, be used to separate individual molecules and distinguish them apart, it is not possible to achieve in practice. The breakthrough came when it was realised that time, acting through the 'on/off' fluorescence transition states, photoswitchable dye pairs (such as Cy3-Cy5) or photoactivatable fluorescent proteins (e.g. PA-GFP; PA-mCherry, Figure 26.7) or photoswitchable fluorescent proteins (rsEGFP2; Dreiklang) exploiting a *cis-trans* isomerisation as a switching mechanism, could be used to separate and precisely locate the position of different fluorophore molecules (e.g. see Shcherbakova et al., 2014; Zhou and Lin, 2013).

Descriptions of PALM, STORM and FPALM were all published in 2006 by different groups (see Table 26.1). Eric Betzig et al. described PALM in September (epub 10th August); Rust et al. on STORM in October (epub 9th August) and Samuel Hess et al. in December (epub 15th Sept) for FPALM. These methods rely upon switching on and off single molecules (Figure 26.8) and plotting the centroid of each PSF

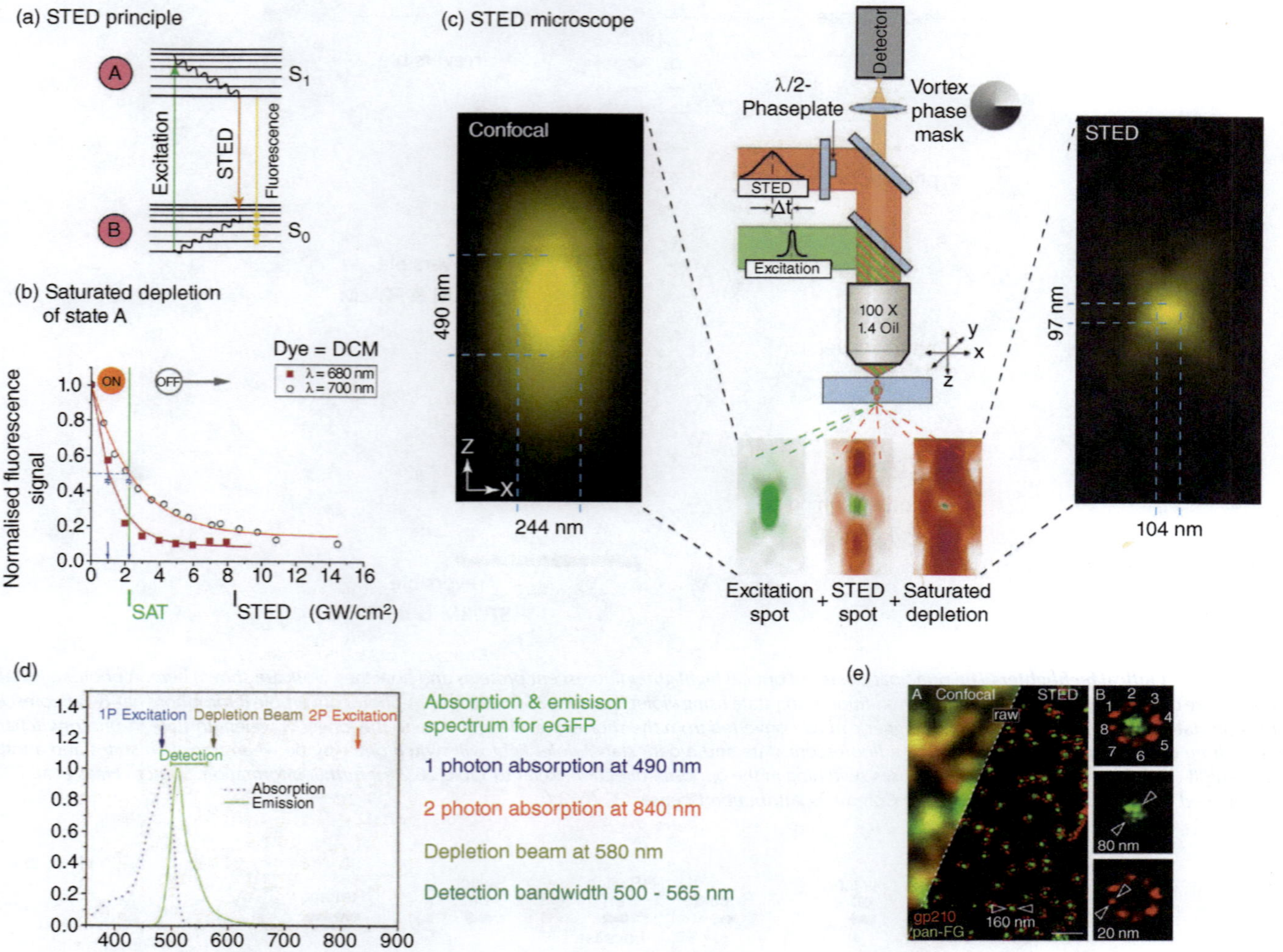

Figure 26.6 ***The STED microscope-2*** *The STED Jabłoński diagram in the original figure from which this figure is taken, is shown in (a). This is identical to the Jabłoński diagrams shown in Figure 26.2. The graph in figure (b) shows how the fluorescent molecule DCM in the excited state A is depleted with increasing probability above saturation intensity. Provided that the STED beam does not initiate significant two-photon absorption, the intensity of the fluorescence signal decreases exponentially with increasing STED power. The saturation intensity value at which stimulated emission depletion occurs is taken at the intensity, I_{SAT}, where a mean 50% depletion occurs. This is shown by the blue arrow for each wavelength. (c) shows the setup of the STED microscope, how a vortex phase plate forms the 'doughnut depletion beam' and the resulting shape of the PSF along the x,z coordinates. The STED laser requires zero intensity at its centre, rising to non-zero intensity at the periphery. The central zero intensity is formed by passing the laser beam through a phase plate. (d) shows how the wavelength of the depletion beam, and the bandwidth of the emitted signal is determined according to the spectral behaviour of the fluorophore, in this case eGFP. The depletion beam has to be re-shifted sufficiently to allow sufficient signal to be emitted but not shifted so far that two-photon absorption occurs. (e) shows STED and confocal images compared. The octameric nuclear pore glycoprotein-210 (gp210) is labelled with a red fluorophore, Abberior STAR635P; the Pan-FG protein central subunit is immuno-labelled with Abberior STAR580 and pseudocoloured green. Compare this STED image with the dSTORM image in Figure 26.8c. The scale bar in Panel A is 500 nm. Source: (a,c) Hell et al. 2004. Reproduced with permission from Elsevier. (b) Adapted from Scheul 2011 with permission from OSA. (d) Leung et al. 2011. Reproduced with permission from SAGE. (e) Göttfert et al. 2013. Reproduced with permission from Elsevier.*

with greater precision than the size of the PSF itself. Each fluorophore is diffraction-limited, but provided sufficient signal photons are collected, its central intensity can be localised with very high precision indeed, thus circumventing the Abbe limit. In theory, detecting a single photon is enough for localising the position of a molecule; in practice several photons are required, because of signal-to-noise factors. The super-resolved image, which contains high lateral resolution, is effectively built up like a pointillist painting by merging all the thousands of individual images of single molecule positions into one. This is a stochastic method, hence the names stochastic optical reconstruction microscopy (STORM) and photoactivation localisation microscopy (PALM).

The original STORM method used the orange and far red-emitting dye pair Cy3-Cy5 for photoswitching. Cy5 normally fluoresces when illuminated with a 633 or 647 nm laser, but illumination at these wavelengths can also drive the Cy5 fluorophore back into a dark state. Subsequent exposure to green light (e.g. a 532 nm diode laser) converts Cy5 from the dark state back to a fluorescent state, but this recovery depends on the close proximity of a Cy3 fluorophore molecule, which acts as an activator molecule, and converts

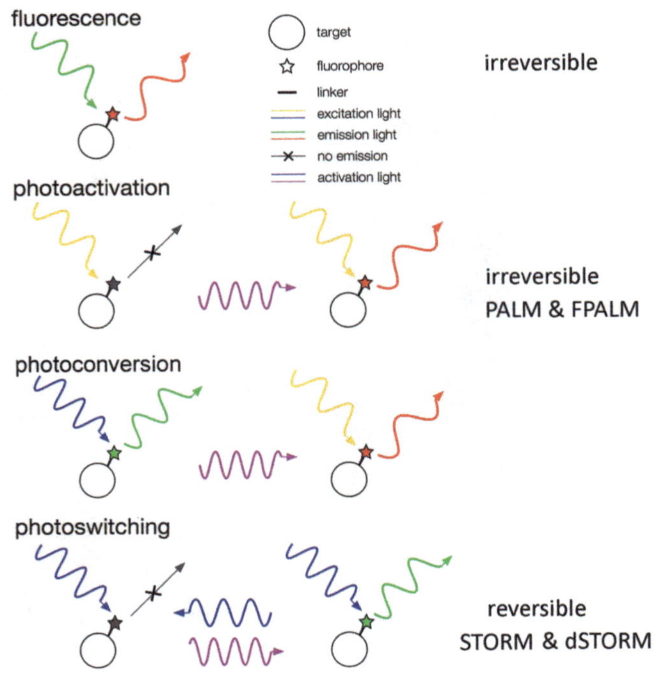

Figure 26.7 **Optical highlighters** *The principal classes of optical highlighter fluorescent proteins and how they work are shown here. A photoactivatable fluorophore can be switched on from a dark non-fluorescing state using violet light ($\lambda = 350$–420 nm). A photoconvertible fluorophore has two fluorescent emission states and using an external light trigger can be converted from the shorter-wavelength state to the longer-wavelength state. A photoswitchable fluorophore can reversibly change between a fluorescent state and a dark state; violet light will switch one way between transition states and another wavelength (here blue, $\lambda = 473$–488 nm) causes switching in the opposite direction. Refer to Table 26.2 for further information. Source: Haas et al. 2014. Reproduced under the terms of the Creative Commons Attribution License, CC BY 3.0.*

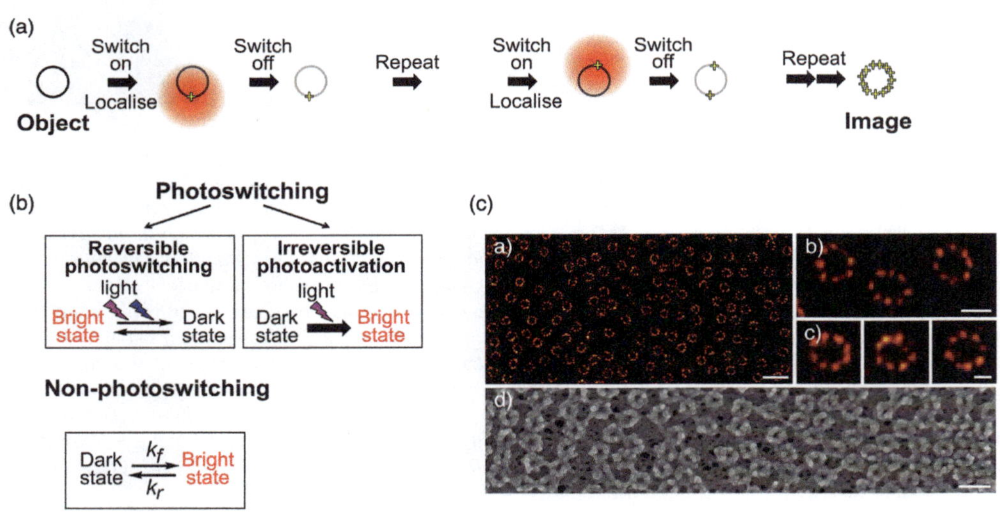

Figure 26.8 ***How localisation microscopy works*** *(a) An object labelled with switchable fluorophores is shown as a black circle. The illumination is switched on, and a single molecule, seen here as a diffraction-limited Airy pattern in red, the centre of which, in yellow, is precisely located. This fluorophore is switched off, and another illuminated, localised and plotted. This process is repeated in turn until an image of the structure of interest with very high structural resolution is assembled by combining the data from each round of activation. During localisation the sample must not move significantly during the acquisition time, nor should the fluorescently labelled molecules diffuse, in order to allow confident determination of the point spread functions resulting from single molecules. Panel (b) shows the possible transition states of different photoswitchable fluorophores. The non-photoswitching fluorophores, such as the regular dyes – Alexa Fluor and ATTO dyes for example – can be reduced to a long-lived dark state by reduction in the presence of electron donors (e.g. thiols). This is the basis of dSTORM: direct STORM. Panel (c) shows the nuclear pore glycol-protein gp210 immuno-labelled with a secondary mAB conjugated to Alexa Fluor 647. The confocal image panel, which is part a) of panel (c), is shown for comparison with the dSTORM images in parts b) and c) showing the octameric gp210 structure in nuclear pore complexes. Part d) of panel (c) shows a negatively stained electron micrograph. Within panel (c) the scale bars are 500 nm for part a); 200 nm for b) and the electron micrograph d) and 100 nm for c). Source: (a,b) Dempsey et al. 2012. Reproduced with permission from Macmillan Publishers Ltd. (c) Adapted from Linde et al. 2013. Reproduced with permission from Elsevier and A. Löschberger.*

Cy5 (the 'reporter') back to a fluorescent form that can be re-excited with far-red illumination. Cy5 can therefore be switched between fluorescent 'on' and dark 'off' transition states in a controlled and reversible manner using different wavelengths of light. Other dye pairs work similarly: Alexa Fluor 405 is another activator molecule, working with Alexa Fluor 488, Alexa 555 and Alexa 647. Cy2 with Cy5 is another pair. STORM works best in the presence of reducing agents, which is why some thiol-based dyes are used for this technique. It has to some extent been replaced by dSTORM (van de Linde et al., 2013; Chozinski et al., 2014; Endesfelder and Heilemann, 2015) in which single organic fluorescent dyes (Alexa Fluor 647 switches well, and is commonly used) can be switched on and off. The exact position of individual molecules can be plotted and a single super-resolution image generated from thousands of individual images.

In the period since PALM, STORM and FPALM were first introduced, a great deal of consideration has been given to precisely and accurately localising fluorophores in localisation microscopy. The reader is advised to start with a *Nature Methods* review by Deschout et al. (2014) and a very practical *Scientific Reports* paper by Whelan and Bell (2015). There is also Shivanandan et al. (2014) and Fox-Roberts et al. (2014) for further practical details to be considered in a localisation experiment. Metcalf et al. (2014) have published details of three test samples: fluorescent dextran, actin filaments and fiducial microspheres. A selection of software algorithms and ImageJ/Fiji plug-ins have been given under Internet Resources at the end of the chapter. See Baddeley et al. (2010) and Fox-Roberts et al. (2017) for guidance on data sampling and visualisation. Additionally, Sage et al. (2015) have evaluated over 30 software packages for single molecule localisation; this paper is recommended reading for anyone setting up a localisation microscopy platform. Table 26.2 gives further references to suitable fluorophores for super-resolution imaging.

The mean distance between labelled molecules (F) must be at least twice as fine as the desired resolution (i.e. 2/d, the limitation imposed by adherence to the Nyquist sampling criterion); that is, there must be at least two fluorophores situated within the sampling area required for adequate Nyquist sampling. The Poisson noise varies as $\sqrt{F}$, introducing noise at lower labelling densities or smaller sampling pixels at higher resolution. The density of the photoswitching molecules determines the resolution. Most of the fluorophores are kept in the dark 'off' state, with an average of less than one molecule per resolvable diffraction-limited volume being in the excited 'on' state. The precision of localisation (for further details, see Box 1 in Deschout et al., 2014, and Adam, 2014) is proportional to the inverse square root of the number of photons detected: $1/\sqrt{N}$; thus, the spatial resolving power, d, is improved by a factor of $1/\sqrt{N}$. $\lambda/2.NA$, where N = total no. photons detected[8].

Table 26.2 *Markers for Super-Resolution Imaging*

The following references, ranging from 2011 to 2016, discuss the properties and applications of organic fluorophores and endogenous fluorescent proteins. Some markers, such as mMaple (McEvoy et al., 2012) can be used for more than one technique, in this case FPALM/STORM and SIM.

Adam, V. (2014). 'Phototransformable fluorescent proteins: which one for which application?' *Histochemistry and Cell Biology 142*/1: 19–41.

Chen, Z. et al. (2013). 'Chemical tags: inspiration for advanced imaging techniques'. *Current Opinion in Chemical Biology 17*/4: 637–643.

Chozinski, T.J. et al. (2014). 'Twinkle, twinkle little star: photoswitchable fluorophores for super-resolution imaging'. *FEBBS Letters 588*/19: 3603–3612.

Dempsey, G.T. et al. (2012). 'Evaluation of fluorophores for optimal performance in localization-based super-resolution imaging'. *Nature Methods 8*/12: 1027–1036.

Deschout, H. et al. (2014). 'Precisely and accurately localizing single emitters in fluorescence microscopy'. *Nature Methods 11*/3: 253–266; Box 4, p. 262.

Haas, B.L. et al. (2014). 'Imaging live cells at the nanometer-scale with mingle-molecule microscopy: obstacles and achievements in experiment optimization for microbiology'. *Molecules 19*/8: 12116–12149.

Jing, C. and Cornish, V.W. (2011). 'Chemical tags for labeling proteins inside living cells'. *Accounts of Chemical Research 44*/9: 784–792.

Jones, S.A. et al. (2011). 'Fast three-dimensional super-resolution imaging of live cells'. *Nature Methods 8*/6: 499–508.

Kasuboski, J.M. et al. (2012). 'Super-resolution microscopy: A comparative treatment'. *Current Protocols in Cytometry* Unit 2.17.1.

Lukinavičius, G. and Johnsson, K. (2011). 'Switchable fluorophores for protein labelling in living cells'. *Current Opinion in Chemical Biology 15*/6: 768–774.

McEvoy, A.L. et al. (2012). 'mMaple: a photoconvertible fluorescent protein for use in multiple imaging modalities'. *PLoS One 7*/12: e51314.

Müller, T. et al. (2012). 'STED microscopy and its applications: new insights into cellular processes on the nanoscale'. *ChemPhysChem 13*/8: 1986–2000; Table 1, p. 1992.

Nienhaus, K. and Nienhaus, G.U. (2014). 'Fluorescent proteins for live-cell imaging with super-resolution'. *Chemical Society Reviews 43*/4: 1088–1106.

Patterson, G.H. (2011). 'Highlights of the optical highlighter fluorescent proteins'. *Journal of Microscopy 243*/1: 1–7.

Requejo-Isidro, J. et al. (2013). 'Fluorescence nanoscopy. Method & applications'. *Journal of Chemical Biology 6*/3: 97–120.

Shcherbakova, D.M. et al. (2014). 'Photocontrollable fluorescent proteins for superresolution imaging'. *Annual Review of Biophysics 43*: 303–329.

Shim, S.H. et al. (2012). 'Super-resolution fluorescence imaging of organelles in live cells with photoswitchable membrane probes'. *PNAS 109*/35: 13978–13983.

Shtengel, G. et al. (2014). 'Imaging cellular ultrastructure by PALM, iPALM and correlative iPALM-EM', Chapter 15. *Methods in Cell Biology 123*: 273–294.

Tiwari, D.K. and Nagai, T. (2013). 'Smart fluorescent proteins: innovation for barrier-free superresolution imaging in living cells'. *Development, Growth & Differentiation 55*/4: 491–507.

Uno, S-N. et al. (2015). 'A guide to use photocontrollable fluorescent proteins and synthetic smart fluorophores for nanoscopy'. *Microscopy 64*/4: 263–277.

van de Linde, S. et al. (2011). 'Direct stochastic optical reconstruction microscopy with standard fluorescent probes'. *Nature Protocols 6*/7: 991–1009.

Yang, Z. et al. (2016). 'Super-resolution fluorescent materials: an insight into design and bioimaging applications'. *Chemical Society Reviews 45*/17: 4651–4667.

Zeiss campus: http://zeiss-campus.magnet.fsu.edu/articles/superresolution/palm/introduction.html (accessed 7 June 2018).

Zhang, Y. et al. (2015). 'Photoswitchable fluorescent nanoparticles and their emerging applications'. *Nanoscale 7*/46: 19342–19357.

Zhou, X.X. and Lin, M. (2013). 'Photoswitchable fluorescent proteins: ten years of colorful chemistry and exciting applications'. *Current Opinion in Chemical Biology 17*: 682–690.

Notes: The 'on-off' contrast is the brightness of the fluorophore in the 'on' state divided by that of the 'off' state; a high ratio is better, especially for small, or weakly labelled, areas in the image. The greater the number of switching cycles, the better the fluorophore for STED and RESOLFT techniques. A high number of switching cycles averages out stochastic errors in PALM/STORM, but for these methods the number of detected photons per switching cycle is more important, because uncertainty of the position of the PSF depends directly upon the inverse square root of the number of photons. The switching rate determines the speed with which images may be recorded. The duty cycle, the time the fluorophore spends in the 'on' state, limits the number of fluorophores that may be localised in a diffraction-limited region. This limit on fluorophore density has a direct bearing on the Nyquist sampling, since the resolution achievable is equal to twice the average distance between two fluorophores.

The problem lies in that whilst a sparse distribution (such that separation between fluorophores is greater than the Abbe diffraction limit) favours accurate localisation, to enable accurate reconstruction of the overall structure of the object, the object must be sampled with a frequency higher than 2/d. Too low a density leads to loss of features through under-sampling; too high a density satisfies the Nyquist criterion, but multiple points are not distinguished as separate and resolved and thus are seen as one. A solution to this apparently irreconcilable situation was proposed by Betzig. A PA-GFP molecule (Patterson and Lippincott-Schwartz, 2010) was used that could first be activated by irradiation at 413 nm to subsequently fluorescence when irradiated at 488 nm; after intense 488 nm radiation, the PA-GFP is irreversibly photo-bleached. This molecule allowed sequential sampling of sparse subsets within a high-density labelled fluorophore.

The data from these super-resolution techniques has to be interpreted with care (Endesfelder and Heilemann, 2014). Both primary data acquisition and post-processing are prone to artifacts (see Whelan and Bell, 2015, for practical advice on specimen preparation). Ideally, each fluorescing molecule would be switched on and off and counted only once. Due to misfolding and incomplete maturation of fluorescent proteins or limited photoconversion between transient 'on' and 'off' states, only a fraction of fluorophores may be active, so undercounting can occur from a limited detection efficiency. A figure of 53–60% has been reported for the relatively bright mEos2 (Annibale et al., 2012; Durisic et al., 2014); many other fluorescent proteins perform worse. Other factors that can lead to undercounting are imperfect sample fixation and permeabilisation or clustering of the fluorescent maker. A clustered appearance is the most commonly observed artifact in super-resolution images and is often the result of poor sample fixation and preparation. Effective photoswitching requires a buffer containing redox-active chemicals to eliminate free radicals, for unlike the β-barrel structure that affords a measure of protection to fluorescent proteins, organic and synthetic molecules are exposed to their immediate environment. Nearby natural electron donors such as tryptophan or guanidine can quench fluorescence, leading to the switching 'off' of the fluorophore before due time with consequent undercounting of the true signal. On the other hand, it is possible to overcount the true signal if fluorophores photo-blink rather than switching between an 'on' and 'off' state cleanly and precisely. The fluorophore enters a dark state and is counted again later on when it re-emits a signal, to be counted multiple times by a localisation algorithm that does not correct for this phenomenon. When detecting single molecules, it is possible for intrinsic noise to appear the same as a single-molecule signal, again leading to overcounting, especially when the background is high and the fluorophore is dim.

Rather like the challenges facing the early Victorian photographers with their slow film emulsions, which required subjects to sit utterly still for minutes on end, sometimes with the aid of a head brace, PALM and STORM localisation microscopy is very susceptible to any sample movement during the long acquisition times required to take the thousands of raw images needed. Because they are bright, organic fluorophores can be switched very quickly, with the best frame rates being 1–2 seconds when sufficiently strong illumination is used. Improvements have been made in reducing the overall acquisition time (e.g. Jones et al., 2011; Lakadamyali, 2013; Winter and Shroff, 2014), but these methods are only really able to capture relatively slow processes occurring in living cells.

26.6 Structured Illumination Microscopy (SIM)

While STED microscopy was being developed by Stefan Hell's research group, Heintzmann and Gustafsson were independently developing structured illumination microscopy (SIM). A moveable grating is used to diffract coherent collimated light into a grid pattern (Figure 26.9), which, projected onto the sample by the objective, interacts with the periodic features of the object to produce a sinusoidal pattern of moiré fringes by interference. In the resulting moiré pattern[9], higher spatial frequencies are shifted to lower spatial frequencies below the diffraction limit and so can be imaged. In effect, the numerical aperture of the objective is increased to about NA 2.7. Since both the patterned illumination mask and the sample itself are diffraction-limited, SIM is only capable of a maximum of doubling the spatial resolution in the image (i.e. halving the resolving power of the microscope).

Because when any image is formed, information about the image is set orthogonal to the axis of the diffraction pattern, the moiré fringes are rotated in three different orientations and phase-shifted three times to collect nine individual data sets, which are recovered by Fourier transformations to give an amplitude function and a phase function. Structured illumination microscopy is usually performed in three dimensions (3D-SIM). In this case, a phase grating is used to project three coherent beams into the back focal plane of the objective to form a three-dimensional pattern: the Talbot effect. In 3D-SIM, each rotation of the patterned illumination grid is phase-shifted five times (Figure 26.10) to collect 15 raw images per axial plane. Doing this allows the 'missing-cone' of information along the z-axis (Chapter 21, section 21.3) to be filled in, with a consequent improvement in axial resolution.

Fourier transformations allow complex image information such as phase and amplitude (brightness) to be processed as simpler oscillating sine wave components. These frequency components are analogous to using individual notes, and their related scales, to describe more complex chords and – ultimately – a piece of music. Applying a Fourier transform to the PSF yields the optical transfer function (OTF) of the microscope. The OTF has two components: the phase transfer function, PTF, and the modulation transfer function, MTF, the amplitude of which serves as a measure of how well contrast is transferred by the microscope from the object into the image. The amplitude image – the reciprocal transform – corresponds to the diffraction image seen at the back focal plane of the objective. Spatial frequencies below the cut-off point ($k_0 = 2NA/\lambda_{em}$) are said to fall within the microscope passband. This is shown in Figure 26.11. An inverse Fourier transform of both amplitude and phase images generates a real image showing the extra resolved detail. The mathematical details can be found in Hayashi (2016), Allen et al. (2014) and Heintzmann and Cremer (1999).

SIM employs an illumination grating in exactly the same way that the ApoTome structured illumination optical sectioning device uses a Ronchi grating, but in this latter case the grating is too coarse to achieve any significant improvement in resolving power, and it uses incoherent illumination from an arc or metal-halide lamp. It is crucial (Wicker and Heintzmann, 2014) to use incoherent illumination, and not coherent or scattered illumination, for SIM.

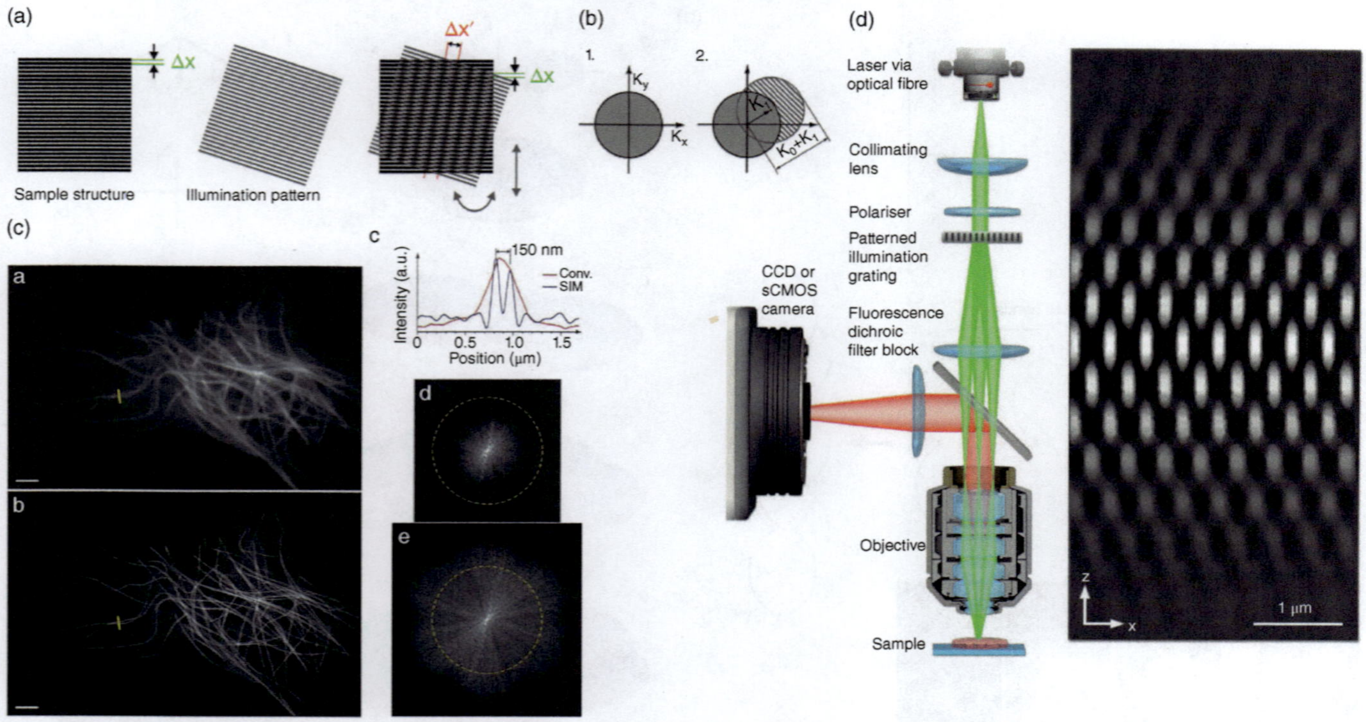

Figure 26.9 ***How a moiré pattern is formed in structured illumination microscopy*** *If an unknown sample structure (a) is multiplied by a known regular illumination pattern, a beat pattern (moiré fringes) will appear. To easily demonstrate how SIM works, the 'sample' in (a) is a periodic structure with a width between lines of ΔX. Overlaying a structured illumination pattern forms moiré fringes with a larger dimension, ΔX'. The unresolved structure of the sample can be deduced from the fringes. During image acquisition, the illumination pattern is both shifted and rotated to collect a series of individual images that comprise the final computationally restored image. Panel (b) shows the set of spatial frequencies in Fourier space seen in a conventional, non-super-resolution fluorescence microscope with the cut-off limit of* $2NA/\lambda$ *of radius* k_0. *Patterned excitation light of frequency* k_1 *gives rise (hatched circle) to new information encoded in the resulting moiré fringes. Those fringes will be observable in the microscope if* $|\mathbf{k}-\mathbf{k}_1| < k_0$, *that is, if* $\mathbf{k}$ *lies within a circle of radius* k_0 *around* $\mathbf{k}_1$. *Since* $\mathbf{k}_1$ *cannot be made larger than* $2NA/\lambda_{ex} \approx k_0$, *the new resolution limit* $k_0 + k_1$ *can be at most* $\approx 2k_0$. *Panel (c) shows a conventional image of cytoskeleton microtubules compared to the SIM image and the corresponding data in Fourier space for each image. The cut-off for a conventional microscope is shown as the yellow dashed circle, equivalent to 6* μm^{-1}. *The sample information, seen as a 'bright starburst' within this circle, indicates that effective resolving power for most high NA objectives is often lower than predicted by theory. Panel (d) shows how a structured illumination pattern is formed in 3D SIM. Source: (a) Heilemann 2010. Reproduced with permission from Elsevier. (b) Gustafsson 2005. Reproduced with permission from National Academy of Sciences. (c) Kner et al. 2009. Reproduced with permission from Macmillan Publishers Ltd. (d) Allen et al. 2014. Reproduced with permission from John Wiley & Sons.*

Since there is no reliance upon saturated emission photophysics, practically all modern fluorophores can be used with structured illumination microscopy. Also, several fluorophores can be simultaneously imaged in multiply-labelled samples. Because the illumination intensity for SIM is much lower than that required for STED and it requires fewer raw images, taking less time to generate a super-resolved image as with localisation methods, structured illumination microscopy is the best suited of all these methods to live-cell imaging. Normally SIM is limited only by the readout time of the camera to about 100 ms per image. To improve the temporal resolution for live-cell imaging, Hirvonen et al. (2009), Shao et al. (2011) and Song et al. (2016) have used a liquid crystal light modulator and fast-readout rolling shutter cameras to rapidly acquire the raw images plane by plane through the 3D data set. SIM is sensitive to aberrations, spherical aberration in particular, and so demands careful specimen preparation, microscope calibration and refractive index matching to minimise image artifacts. If this is not done, the PSF generated between experiments will vary, leading to errors.

Nonlinear illumination has been used with SIM, applying the RESOLFT approach of saturable fluorescence emission, to give saturated structured illumination microscopy (SSIM; Heintzmann et al., 2002; Gustafsson, 2005). The result is that the sinusoidal peaks of the SIM excitation pattern can be clipped flat (Figure 26.12), adding higher-order spatial frequencies to the excitation pattern. This saturable SIM excitation pattern can be used to detect sub-resolution detail (Li et al., 2015). Like STED and PALM/STORM methods, the spatial resolution of SSIM is limited only by the level of fluorescence saturation that can be achieved in practice, but the samples have to be well stained, bright and photostable, since SSIM uses a much higher photon flux and collects twelve times as many raw widefield images as SIM.

26.7 Future Directions

Most of the figures given for super-resolution resolving power (Table 26.1) are ultimate records established by the originators and/or expert practitioners and not necessarily reproducible in routine laboratories, even as these microscopes become developed as

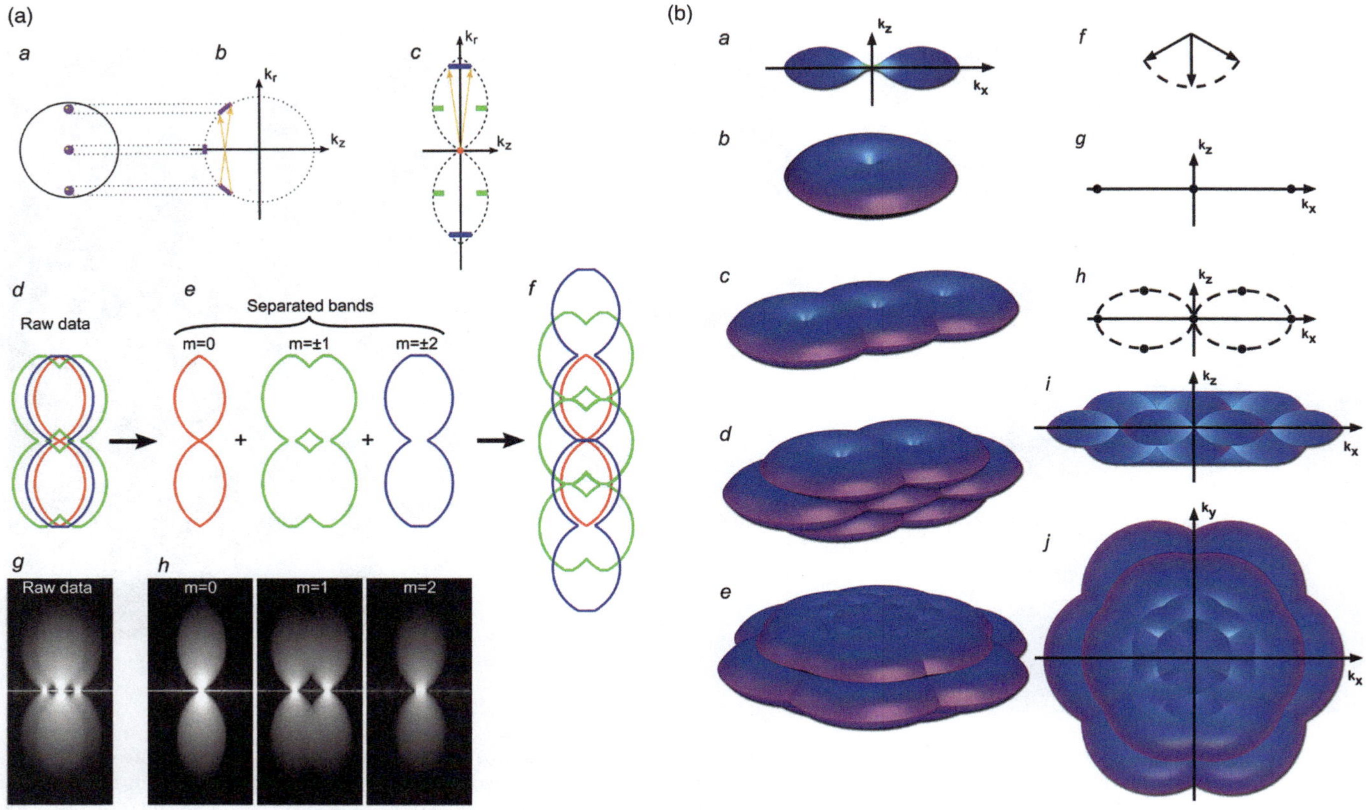

Figure 26.10 **Rotation and phase-shifting of the SIM grid to fill the 'missing cone'** *Left-hand panel: Part* (a) *shows the three illumination beams, seen in green in Figure 26.9d, and here in (a) as three purple dots in Fourier space in the back focal plane of the objective. Each point within each beam is mutually coherent only with the corresponding point of each of the other beams that originates from the same point of the light source; the yellow points illustrate one such coherent point triplet. (b) Spatial frequency components of the light amplitude. For an objective lens that satisfies the sine condition, the pupil plane maps directly onto the lateral spatial frequency* (dotted lines *from a to b). The difference vectors between mutually coherent amplitude components (*amber arrows *in b) will therefore have the same lateral extent, but will differ axially, for different source points. (c) Resulting spatial frequency components of the illumination intensity; the $m = 0, \pm 1$ and ± 2 components are shown in red, green and blue, respectively. The nonzero-m components, which get a contribution from each coherent point pair (*amber arrows, *corresponding to those in (b), are axially broadened into line segments but remain delta-function-like laterally. The raw data (d and g) is a sum of five information bands, corresponding to the five lateral illumination-intensity frequencies. Once five images are observed with different pattern phases, the bands can be separated (e and h; only the $m = 0$ [red], 1 [green] and 2 [*blue*] bands are shown), and computationally moved back to their correct lateral position (f). Each band has been broadened by convolution with the corresponding frequency component of the illumination intensity (corresponding colours); the order 1 component, for example, consists of two parts (e,* green*). The observed raw data (g) and separated bands (h) for a point-source object are in excellent agreement with the theoretical expectations. Part* (b) *shows how combining the raw data images collected from rotating and phase-shifting the illumination pattern fills the 'missing cone' of information in Fourier space, shown in (a), whose OTF represented in three dimensions is shown in (b) as a toroidal shape. (c) and (d) show the extended OTF with three and seven harmonics, respectively, whilst (e) shows how rotation of the phase-shifted pattern from (d) extends the OTF in all directions. The extended OTF in (e) is shown laterally in (i) and axially in (j). (f) shows the wave vectors representing the three illumination beams, while (g) shows the three harmonics created by a two-beam illumination pattern and (h) the seven harmonics from a three-beam illumination pattern. Source: (left) Gustafsson et al. 2008. Reproduced with permission from Elsevier. (right) Allen et al. 2014. Reproduced with permission from John Wiley & Sons.*

commercial turn-key systems. Also, as the focal volume of these far-field methods decreases in order to give the ideal resolving power to observe dynamic macromolecular events, the signal-to-noise ratio becomes progressively worse. A smaller focal spot inevitably means fewer fluorophores per voxel, such that a twofold increase in resolving power means an eightfold (2^3) decrease in brightness and a corresponding increase in the Poisson-distributed shot noise of the system. Any increase in excitation intensity to compensate for this runs the risk of increased fluorophore photobleaching and photo-toxicity of the cell (Strack, 2015). Most super-resolution techniques are slow, because of the need to satisfy Nyquist's sampling criterion; these factors also present challenges to increasing temporal resolution. Once again, we are faced with the constraints of the 'iron triangle' (Chapter 28, Section 28.6).

Your choice of super-resolution technique will probably depend upon which instrument you have in your institution or is available to use elsewhere. If you are considering buying a super-resolution microscope or are in the fortunate position of being able to choose which technique to use, four recent reviews (Demmerle et al., 2015; Bachmann et al., 2015; Tam and Merino, 2015, and Wegel et al., 2016) may be helpful in making a decision. Holm et al. (2014) describe how to put together a cost-effective home-built localisation microscope at 1/10th of the price of a commercial system. Before starting a super-resolution experiment, optimise the cell growth and

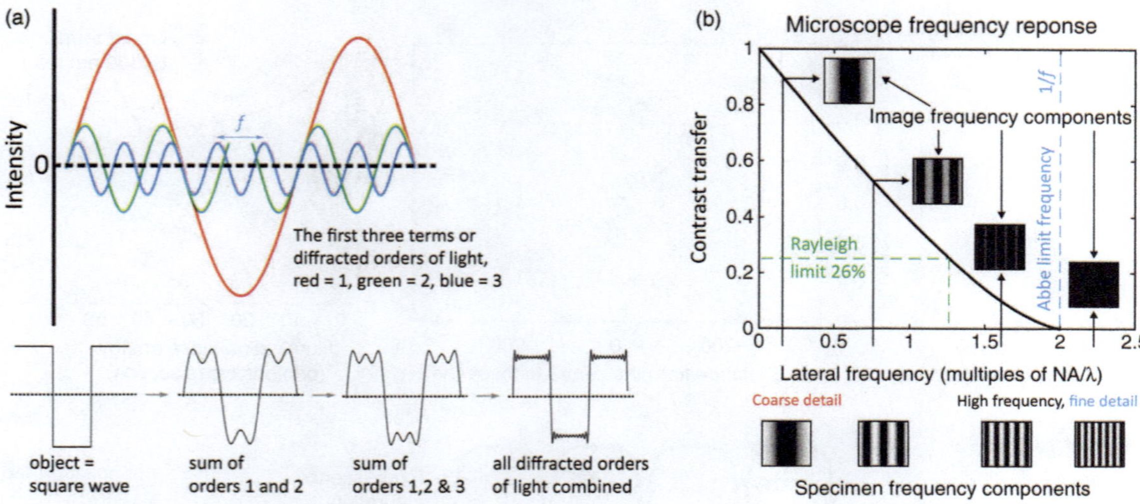

Figure 26.11 **Fourier space and the Abbe limit to resolving power** *Every object, whether it is periodic or not, can be described as the sum of a series of sinusoidal curves, or waves, of differing spatial frequencies (a) each of which is harmonic of the frequency of the object. As more components (i.e. greater numbers of higher-order diffracted rays) are added, the result (i.e. the image) increasingly resembles the original object. The process of convolution to form an image can be better understood by illustrating the convolution operation in Fourier space. This is shown in (b) as the Fourier transform of the PSF which is the optical transfer function, or OTF. The Abbe limit ($d = 0.5\lambda/NA$) determines the smallest periodic structure (i.e. of high contrast) that can be discriminated in the image. See also Chapter 21, section 21.3. Source: (a) Pawley 2006. Reproduced with permission from Springer. (b) Conchello 2006. Reproduced with permission from Scion Publishing Ltd.*

density, and first check these on a widefield microscope to understand and control any factors that may be sub-optimal or derail the experiment. It is advisable first to know how your chosen fluorophore(s) and fluorescent protein(s) work in your laboratory with existing equipment to allow you to plan the design of your super-resolution experiment. This is something that cannot be predicted from the literature. Crucially, do some initial experiments to work out the optimum activation density of your fluorescent markers. If too many probes are activated at once, the centre of localisation may not represent a single molecule but an average of several. This may be hard to identify, for the image may look like a good super-resolution localisation image but in fact be of very poor quality.

Where will the advances in the next few years be? Most researchers believe that better performing fluorophores – specifically engineered for particular applications (just as the fluorescent protein palette has been engineered in a targeted fashion) – will be the principal advance for diffraction-unlimited microscopy. Hell (2009) suggests that spectral switching of fluorophores, with interstate transitions differing in excitation and emission wavelengths, will speed up localisation methods because fewer molecules will be in the 'off' state. The real challenge so far has been to apply these super-resolution methods to imaging live cells (e.g. Cox, 2015; Taraska, 2015, Jones et al., 2011). STED requires a very high photon flux of the depletion laser, whilst the blue or UV wavelength used, almost continuously, for switching in PALM is also deleterious to living cells. A means of combining the advantages of localisation microscopy and STED to significantly reduce photo-toxicity called MINFLUX has been developed by the Hell research group (see Strack, 2017). The speed of these techniques also presents a challenge in collecting multiple frames for live-cell time lapses. In this respect, SIM offers the greatest advantage for live-cell super-resolution microscopy even though it does not break the Abbe diffraction limit. The biggest challenges are improving signal-to-noise ratios due to the Poisson (shot) noise arising from low-intensity signals. Nonlinear SSIM will overcome the diffraction barrier but at the expense of much greater illumination intensities. The other major challenge with SIM is to improve both the temporal resolution (due to sCMOS cameras and spatial light modulators, which is now in the region of several hundred frames per second) and write improved data reconstruction algorithms.

Most super-resolution methods exhibit a common drawback: with respect to the total number of emitted photons, they are less efficient than standard widefield microscopy. STED, for example, shaves the PSF rather like a pencil sharpener, and localisation methods are limited in terms of both SNR and temporal resolution by the numbers of photons that can be collected. Classically, the *étendue* – the product of area and solid angle – can never decrease in an optical system. Roth et al. (2016) by photon reassignment (not to be confused with deconvolution) in a method called 'superconcentration' are able to break the classical light concentration limit and increase the detected signal. Another approach is to couple non-fluorescent dyes to an apoprotein activator (Bruchez, 2015), which switches the fluorophore to a very bright 'on' state when the two molecules are joined. It took 30 years from the initial concept of optical sectioning by confocal microscopy until the introduction of a workable point-scanning system. Therefore super-resolution microscopy, coupled with advances in digital 'de-noising' algorithms and superconcentration, is likely to provide a breakthrough in our visualisation and understanding of cells at the molecular level but could take time to fully implement.

At the time of writing, the latest approach to resolving details below the Abbe limit is by expansion microscopy (Chang et al., 2017; Chen et al., 2015; Gao et al., 2017; Karagiannis and Boyden, 2018; commentary: Geertsema and Ewers, 2016). The concept, like most elegant ideas, is simple: after labelling the tissue with fluorescent antibodies, the cells are soaked in a polymerising polyelectrolyte gel, sodium polyacrylate. They are then fixed with a cross-linking reagent, such as glutaraldehyde or methacrylic acid succinimidylester, to hold the fluorophore markers in position within the matrix-infused cells. Once the fluorescent markers are fixed in position and the gel

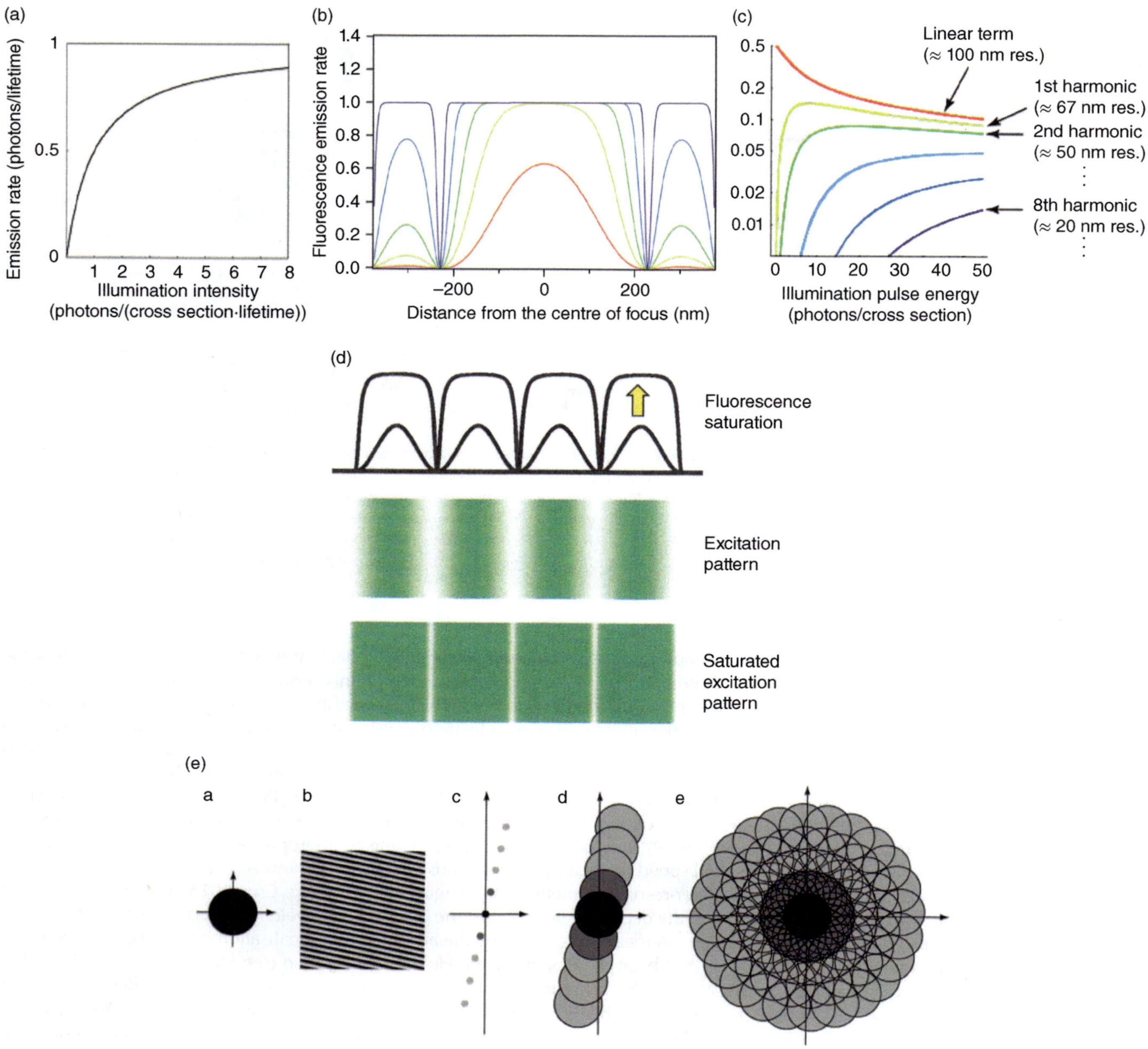

Figure 26.12 **Clipping of the SIM sinusoidal frequencies in nonlinear SSIM** *Where the emission intensity is nonlinear with respect to the excitation illumination intensity, as shown in graph (a). The cross section of the nonlinear emission along a the focal plane of a microscope is shown for different excitation illumination intensities, represented by differently coloured curves in (b). The diffraction-limited excitation intensity is shown by the red curve, and it does not change. As the excitation intensity is increased, the emission saturation changes, giving rise to distributions with sharp edges and strongly confined non-emissive regions. Panel (c) shows each emission pattern as harmonics in Fourier space. As the intensity reaches saturation, more harmonics contribute to image formation of increasing resolving power. The practical application can be seen in (d) where the sinusoidal pattern produced in SIM has alternating peak and zero intensities. Strong nonlinear illumination saturates the fluorescence emission at the peaks but not the fluorophores at zero intensity, leading to narrowly defined high-frequency stripes in the saturated excitation patterned illumination. This allows features well below the diffraction limit to be resolved. Panel (e) and part (a) shows the Fourier frequency space of a conventional microscope in black; (b) shows a SIM sinusoidal illumination pattern and (c) the frequency components of a conventional microscope (black), of linear-structured illumination (dark grey) and the three lowest harmonics of nonlinear-structured illumination (light grey). These translate into the observable regions in Fourier space represented by the circles in (d) and (e) the filling of the 'missing cone' by combining raw images from rotation and phase shifting of the illumination pattern in the usual way. Source: (top) Reproduced with permission from National Academy of Sciences; Dedecker et al. 2008. Reproduced with permission from Elsevier. (middle) Huang et al. 2009. Reproduced with permission from Annual Reviews. (bottom) Reproduced with permission from National Academy of Sciences.*

polymerised, the cells are digested with protease, then dialysed in water to cause even and isotropic expansion by about 4–5 times in volume. It is claimed there is no distortion at the gross anatomical level. The lipids, proteins and most tissue constituents are largely removed by this process, so the cells are cleared. Rehydration of the remaining tissue and the subsequent expansion is the key to this process, allowing previously sub-resolution features to be observed with routine confocal microscopy.

26.8 Chapter Summary

This chapter explains the manner in which the properties of fluorophores have been exploited to circumvent the resolving power limit imposed by dioptric lenses and diffraction in far-field microscopy. Whilst these techniques hold great promise for enabling biologists to elucidate the structure and function of biologically active proteins and macromolecules, the field is in its relative infancy. Super-resolution methods are generally expensive to implement and require high photon fluxes and post-processing of data to construct a super-resolved image. These techniques are unlikely yet to supplant classical fluorescence and bright-field microscopy with other contrast methods but rather complement these long-established methods.

1. It is impossible to resolve detail whose size is smaller than half the wavelength of light.
2. Lateral resolution is limited to $\lambda/2.NA$ in the focal plane and $2\lambda/NA^2$ axially.
3. The Abbe limit ($d = 0.5\lambda/NA$) occurs at the spatial cut-off frequency where OTF = zero at $2.NA/\lambda_{em}$.
4. There are two means of circumventing the Abbe limit: either (a) by switching fluorophores from one state to another and stochastically plotting their precise position or (b) by PSF engineering using nonlinear fluorescence excitation and patterned illumination.
5. Using two objectives to gather light from larger aperture angles and employing standing wave fluorescence microscopy increases the resolving power in z, along the optical axis.
6. RESOLFT (reversible saturable optical fluorescence transitions) reversibly switches a fluorophore between an 'on' state and a dark 'off' state, to image a sub-diffraction region of signal.
7. STED and RESOLFT techniques require three illumination steps: one for activation, one for depletion and the final one for detection of the remaining active fluorophores.
8. A STED depletion laser inhibits the outer side lobes of the PSF of the focal spot – shaving the PSF.
9. The size of the central fluorescing region can be made smaller by increasing the peak intensity of the depletion laser, or by using a fluorophore with a long excited state for low saturation intensity.
10. The super-resolved localisation image is built up like a pointillist painting by merging all the thousands of individual images of single molecule positions into one.
11. The localisation precision is proportional to $1/\sqrt{N}$, where N is the number of photons detected.
12. The density of photo-switching molecules determines the resolution, according to the Nyquist criterion. Too low a density leads to loss of features through under-sampling; with a high density, multiple points are not resolved separately but as one.
13. In SIM, a patterned grid of light interacts with the specimen to form moiré fringes. Finely structured details are shifted to a lower spatial frequency, below the diffraction limit, and so can be seen.
14. An inverse Fourier transform of both amplitude and phase images generates a real SIM image.
15. Nonlinear SSIM combines SIM and RESOLFT to circumvent the Abbe diffraction barrier.

Key Reading

Cox, S. (2015). 'Super-resolution imaging in living cells'. *Developmental Biology 401*/1: 175–181.
- A concise overview of super-resolution methods

Turkowyd, B. et al. (2016). 'From single molecules to life: microscopy at the nanoscale'. *Analytical and Bioanalytical Chemistry 408*/25: 6885–6911.
- Another excellent review, which I give to my students

Vandenberg, W. et al. (2015). 'Diffraction-unlimited imaging: from pretty pictures to hard numbers'. *Cell and Tissue Research 360*/1: 151–178.
- A reminder that careful post-processing is required when analysing super-resolution images

Vangindertael, J. et al. (2018). 'An introduction to optical super-resolution microscopy for the adventurous biologist'. *Methods Applied Fluorescence 6*/2: 022003.
- Also an excellent review, which I give to my students

Nobel Lectures

Betzig, E. (2015). 'Single molecules, cells and super-resolution optics' (Nobel Lecture). *Angewandte Chemie International Edition 54*/28: 8034–8053.
Hell, S.W. (2015). 'Nanoscopy with focused light' (Nobel Lecture). *Angewandte Chemie International Edition 54*/28: 8054–8066.

References

Allen, J.R. et al. (2014). 'Structured illumination microscopy for superresolution'. *ChemPhysChem 15*/4: 566–576.
Annibale, P. et al. (2012). 'Identification of the factors affecting co-localization precision for quantitative multicolor localization microscopy'. *Optical Nanoscopy 1*: 9. doi: 10.1186/2192-2853-1-9.

Babcock, H. et al. (2012). 'A high-density 3D localization algorithm for stochastic optical reconstruction microscopy'. *Optical Nanoscopy 1*: 6. doi: 10.1186/2192-2853-1-6.

Bachman, M. et al. (2015). 'Practical limitations of superesolution imaging due to conventional sample preparation revealed by a direct comparison of CLSM, SIM and dSTORM'. *Journal of Microscopy 262*/3: 306–315.

Baddeley, D. et al. (2010). 'Vizualization of localization microscopy data'. *Microscopy and Microanalysis 16*/1: 64–72.

Birk, U.J. (2017). *Super-Resolution Microscopy: A Practical Guide*. Wiley-VCH GmbH.

Brede, N. and Lakadamyali, M. (2012). 'GraspJ: an open source, real-time analysis package for super-resolution imaging'. *Optical Nanoscopy 20121*: 11. doi: 10.1186/2192-2853-1-11.

Bruchez, M.P. (2015). 'Dark dyes–bright complexes: fluorogenic protein labeling'. *Current Opinion in Chemical Biology 27*: 18–23.

Castello, M. et al. (2016). 'Gated-sted microscopy with subnanosecond pulsed fiber laser for reducing photobleaching'. *Microscopy Research and Technique 79*/9: 785–791.

Chang, J.-B. et al. (2017). 'Iterative expansion microscopy'. *Nature Methods 14*/6: 593–599.

Chen, F. et al. (2015). 'Expansion microscopy'. *Science 347*/6221: 543–548.

Chozinski, T.J. et al. (2014). 'Twinkle, twinkle little star: photoswitchable fluorophores for super-resolution imaging'. *FEBS Letters 588*/19: 3603–3612.

Chozinski, T.J. et al. (2016). 'Expansion microscopy with conventional antibodies and fluorescent proteins'. *Nature Methods 13*/6: 485–488.

Conchello, J.A. (2006). 'An overview of three-dimensional and four-dimensional microscopy by computational deconvolution'. In: Stephens, D. (ed.), *Cell Imaging. Methods Express Series*. Banbury: Scion Publishing.

Cox, S. et al. (2011). 'Bayesian localization microscopy reveals nanoscale podosome dynamics'. *Nature Methods 9*/2: 195–200.

Cremer, C. and Birk, U. (2016). 'Perspectives in super-resolved fluorescence microscopy: what comes next?' *Frontiers of Physics 4*: 11. doi: 10.3389/fphy.2016.00011.

de Lange, F. et al. (2001). 'Cell biology beyond the diffraction limit: near-field scanning optical microscopy'. *Journal of Cell Science 114*/23: 4153–4160.

Dedecker, P. et al. (2008). 'Diffraction-unlimited optical microscopy'. *Materials Today* (special issue) 11, Supplement: 12–21.

Demmerle, J. et al. (2015). 'Assessing resolution in super-resolution imaging'. *Methods 88*: 3–10. doi: 10.1016/j.ymeth.2015.07.001.

Dempsey, G.T. et al. (2012). 'Evaluation of fluorophores for optimal performance in localization-based super-resolution imaging'. *Nature Methods 8*/12: 1027–1036.

Deschout, H. et al. (2014). 'Precisely and accurately localizing single emitters in fluorescence microscopy'. *Nature Methods 11*/3: 253–266.

Donnert, G. et al. (2005). 'Macromolecular-scale resolution in biological fluorescence microscopy'. *PNAS 103*/31: 11440–11445.

Durisic, N. et al. (2014). 'Quantitative super-resolution microscopy: pitfalls and strategies for image analysis'. *Current Opinion in Chemical Biology 20*: 22–28.

Eggeling, C. et al. (2015). 'Lens-based fluorescence nanoscopy'. *Quarterly Reviews of Biophysics 48*/2: 178–243.

Egner, A. and Hell, S.W. (2005). 'Fluorescence microscopy with super-resolved optical sections'. *Trends in Cell Biology 15*/4: 207–215.

Endesfelder, U. and Heilemann, M. (2014). 'Art and artifacts in single-molecule localization microscopy: beyond attractive images'. *Nature Methods 11*/3: 235–238.

Endesfelder, U. and Heilemann, M. (2015). 'Direct stochastic optical reconstruction microscopy (dSTORM)'. *Methods in Molecular Biology 1251*: 263–276.

Fox-Roberts, P. et al. (2014). 'Fixed pattern noise in localization microscopy'. *ChemPhysChem 15*/4: 677–686.

Fox-Roberts, P. et al. (2017). 'Local dimensionality determines imaging speed in localization microscopy'. *Nature Communications 8*: 13558. doi: 10.1038/ncomms13558.

Gao, R. et al. (2017). 'Q&A: expansion microscopy'. *BMC Biology 15*: 50. doi: 10.1186/s12915-017-0393-3.

Göttfert, F. et al. (2013). 'Coaligned dual-channel STED nanoscopy and molecular diffusion analysis at 20 nm resolution'. *Biophysical Journal 105*/1: L01–L03. doi: 10.1016/j.bpj.2013.05.029.

Gould, T.J. and Bewersdorf, J. (2012). 'Nanoscopy at low light intensities shows its potential'. *eLife 1*: e00475.

Gustafsson, M.G. (2005). 'Nonlinear structured-illumination microscopy: wide-field fluorescence imaging with theoretically unlimited resolution'. *PNAS 102*/37: 13081–13086.

Gustafsson, M.G. et al. (2008). 'Three-dimensional resolution doubling in wide-field fluorescence microscopy by structured illumination'. *Biophysical Journal 94*/12: 4957–4970.

Haas, B.L. et al. (2014). 'Imaging live cells at the nanometer-scale with mingle-molecule microscopy: obstacles and achievements in experiment optimization for microbiology'. *Molecules 19*/8: 12116–12149.

Han, R. et al. (2013). 'Recent advances in super-resolution fluorescence imaging and its applications in biology'. *Journal of Genetics and Genomics 40*/12: 583–595.

Hayashi, S. (2016). 'Resolution doubling using confocal microscopy via analogy with structured illumination microscopy'. *Japanese Journal of Applied Physics 55*: 082501. doi: 10.7567/JJAP.55.082501.

Heilemann, M. (2010). 'Fluorescence microscopy beyond the diffraction limit'. *Journal of Biotechnology 149*/4: 243–251.

Heintzmann, R. et al. (2002). 'Saturated patterned excitation microscopy – a concept for optical resolution improvement'. *Journal of the Optical Society of America A 19*/8: 1599–1609.

Heintzmann, R. and Cremer, C.G. (1999). 'Laterally modulated excitation microscopy: improvement of resolution by using a diffraction grating'. *Proceedings of SPIE 3568*: 185–195. doi: 10.1117/12.336833.

Hell, S.W. (2009). 'Microscopy and its focal switch'. *Nature Methods 6*/1: 24–32.

Hell, S.W. (2007). 'Far-field optical nanoscopy'. *Science 316*/5828: 1153–1158.

Hell, S.W. et al. (2004). 'Concepts for nanoscale resolution in fluorescence microscopy'. *Current Opinion in Neurology 14*/5: 599–609.

Hess, S.T. et al. (2006). 'Ultra-high resolution imaging by fluorescence photoactivation localization microscopy'. *Biophysical Journal 91*/11: 4258–4272.

Huang, B. et al. (2009). 'Super-resolution fluorescence microscopy'. *Annual Review of Biochemistry 78*: 993–1016.

Huang, B. (2010). 'Super-resolution optical microscopy: multiple choices'. *Current Opinion in Chemical Biology 14*/1: 10–14.

Jones, S.A. et al. (2011). 'Fast three-dimensional super-resolution imaging of live cells'. *Nature Methods 8*/6: 499–508.

Karagiannis, E.D. and Boyden, E.S. (2018). 'Expansion microscopy: development and neuroscience applications'. *Current Opinion in Neurology 50*: 56–63.

Kner, P. et al. (2009). 'Super-resolution video microscopy of live cells by structured illumination'. *Nature Methods 6*/5: 339–342.

Lakadamyali, M. (2014). 'Super-resolution microscopy: going live and going fast'. *ChemPhysChem 15*/4: 630–636.
Leung, B.O. et al. (2011). 'Review of super-resolution fluorescence microscopy for biology'. *Applied Spectroscopy 65*/9: 967–980.
Li, D. et al. (2015). 'Extended-resolution structured illumination imaging of endocytic and cytoskeletal dynamics'. *Science 349*/6251: aab3500. doi: 10.1126/science.aab3500.
Metcalf, D.J. et al. (2013). 'Test samples for optimizing STORM super-resolution microscopy'. *Journal of Visualized Experiments 79*. doi: 10.3791/50579.
Munck, S. et al. (2012). 'Sub-diffraction imaging on standard microscopes through photobleaching microscopy with non-linear processing'. *Journal of Cell Science 125*/9: 2257–2266.
Nienhaus, K. and Nienhaus, G.U. (2014). 'Fluorescent proteins for live-cell imaging with super-resolution'. *Chemical Society Reviews 43*/4: 1088–1106.
Patterson, G. et al. (2010). 'Supperresolution imaging using single molecule localization'. *Annual Review of Physical Chemistry 61*: 345–367.
Pawley, J.B. (2006). Points, pixels and gray levels: digitizing image data. In: *Handbook of Biological Confocal Microscopy*, Chapter 4, 3rd ed. New York: Springer, pp. 59–79.
Richards, D. (2003). 'Near-field microscopy: throwing light on the nanoworld'. *Philosophical Transactions of the Royal Society A 361*: 2843–2857.
Roth, S. et al. (2016). 'Superconcentration of light: circumventing the classical limit to achievable irradiance'. *Optical Letters 41*/9: 2109–2112.
Sage, D. et al. (2015). 'Quantitative evaluation of software packages for single-molecule localization microscopy'. *Nature Methods 12*/8: 717–724.
Schermelleh, L. et al. (2010). 'A guide to super-resolution fluorescence microscopy'. *Journal of Cell Biology 190*/2: 165–175.
Scheul, T. et al. (2011). 'Two-photon excitation and stimulated emission depletion by a single wavelength'. *Optics Express 19*/19: 18036–18048.
Schmidt, R. et al. (2008). 'Spherical nanosized focal spot unravels the interior of cells'. *Nature Methods 5*/6: 539–544.
Shao, L. et al. (2011). 'Super-resolution 3D microscopy of live whole cells using structured illumination'. *Nature Methods 8*/12: 1044–1046.
Shcherbakova, D.M. et al. (2014). 'Photocontrollable fluorescent proteins for superresolution imaging'. *Annual Review of Biophysics 43*: 303–329.
Sheppard, C. (2007). 'Fundamentals of superesolution'. *Micron 38*/2: 165–169.
Shivanandan, A. et al. (2014). 'Challenges in quantitative single molecule localization microscopy'. *FEBS Letters 588*/19: 3595–3602.
Shroff, H. et al. (2008). 'Live-cell photoactivated localization microscopy of nanoscale adhesion dynamics'. *Nature Methods 5*/5: 417–423.
Small, A. and Stahlheber, S. (2014). 'Fluorophore localization algorithms for super-resolution microscopy'. *Nature Methods 11*/3: 267–279.
Song, L. et al. (2016). 'Fast structured illumination microscopy using rolling shutter cameras'. *Measurement Science and Technology 27*/5: 055401. doi: 10.1088/0957-0233/27/5/055401.
Stephens, D. (ed.) (2006). *Cell Imaging. Methods Express Series*. Banbury: Scion Publishing.
Strack, R. (2015). 'Death by super-resolution imaging'. *Nature Methods 12*/12: 1111.
Strack, R. (2017). 'Taking nanoscopy to the limit'. *Nature Methods 14*/3: 221.
Tam, J. and Merino, D. (2015). 'Stochastic optical reconstruction microscopy (STORM) in comparison with stimulated emission depletion (STED) and other imaging methods'. *Journal of Neurochemistry 135*/4: 643–658.
Taraska, J.W. (2015). 'SIMply better resolution in live cells'. *Trends in Cell Biology 25*/11: 636–638.
Thompson, R.E. et al. (2002). 'Precise nanometer localization analysis for individual fluorescent probes'. *Biophysical Journal 82*/5: 2775–2783.
van de Linde, S. et al. (2013). 'Investigating cellular structures at the nanoscale with organic fluorophores'. *Chemical Biology 20*/1: 8–18.
Vicidomini, G. et al. (2013). 'STED nanoscopy with time-gated detection: theoretical and experimental aspects'. *PLoS One 8*/1: e54421.
Wegel, E. et al. (2016). 'Imaging cellular structures in super-resolution with SIM, STED and localization microscopy: a practical comparison'. *Scientific Reports 6*: 27290.
Whelan, D.R. and Bell, T.D. (2015). 'Image artifacts in single molecule localization microscopy: why optimization of sample preparation protocols matters'. *Scientific Reports 5*: 7924.
Wicker, K. and Heintzmann, R. (2014). 'Resolving a misconception about structured illumination'. *Nature Photonics 8*: 342–344. doi: 10.1038/nphoton.2014.88.
Willig, K.I. et al. (2007). 'STED microscopy with continuous wave beams'. *Nature Methods 4*/11: 915–918.
Winter, P.W. and Shroff, H. (2014). 'Faster fluorescence microscopy: advances in high speed biological imaging'. *Current Opinion in Chemical Biology 20*: 46–53.
Wurm, C.A. et al. (2010). 'Sample preparation for STED microscopy'. *Methods in Molecular Biology 591*: 185–199.
Wurm, C.A. et al. (2012). 'Novel red fluorophores with superior performance in STED microscopy'. *Optical Nanoscopy 1*: 7. doi: 10.1186/2192-2853-1-7.
Xu, Y., Melia, T.J., and Toomre, D.K. (2011). 'Using light to see and control membrane traffic'. *Current Opinion in Chemical Biology 15*/6: 822–830.
Yildiz, A. and Selvin, P.R. (2005). 'Fluorescence imaging with one nanometer accuracy: application to molecular motors'. *Accounts of Chemical Research 38*/7: 574–582.
Zhou, X.X. and Lin, M. (2013). 'Photoswitchable fluorescent proteins: ten years of colorful chemistry and exciting applications'. *Current Opinion in Chemical Biology 17*: 682–690.

Further Reading

Abbe, E. (1873). 'Beiträge zur Theorie des Mikroskops und der mikroskopischen Wahrnehmung' ['Contributions to the Theory of the Microscope and of Microscopic Perception']. *Archiv für Mikroskopische Anatomie 9*/1: 413–468.
Adam, V. (2014). 'Phototransformable fluorescent proteins: which one for which application?' *Histochemistry and Cell Biology 142*/1: 19–41.
Aspelmeier, T. et al. (2015). 'Modern statistical challenges in high-resolution fluorescence microscopy'. *Annual Review of Statistics and Its Application 2*: 163–202.
Baddeley, D. et al. (2011). '4D super-resolution microscopy with conventional fluorophores and single wavelength excitation in optically thick cells and tissues'. *PLoS One 6*/5: e20645.
Ball, G. et al. (2012). 'A cell biologist's guide to high resolution imaging'. *Methods Enzymology 504*: 29–55.
Bon, P. et al. (2015). 'Three-dimensional nanometre localization of nanoparticles to enhance super-resolution microscopy'. *Nature Communications 6*: 7764.

Coltharp, C. et al. (2014). 'Quantitative analysis of single-molecule superresolution images'. *Current Opinion in Structural Biology 28*: 112–121.

Cordes, T. et al. (2011). 'Mechanisms and advancement of antifading agents for fluorescence microscopy and single-molecule spectroscopy'. *Physical Chemistry Chemical Physics 13*/14: 6699–6709.

Engel, U. (2014). 'Structured Illumination superresolution imaging of the cytoskeleton'. In: Waters, J. and Wittmann, T. (eds.), *Methods in Cell Biology*, Chapter 17, vol. 123, *Quantitative Imaging in Cell Biology*. Amsterdam: Academic Press, Elsevier, pp. 315–333.

Fernández-Suárez, M. and Ting, A.Y. (2008). 'Fluorescent probes for super-resolution imaging in living cells'. *Nature Reviews Molecular Cell Biology 9*/12: 929–943.

Griffiths, N. et al. (2016). 'Photo-convertible fluoreecent proteins as tools for fresh insights on subcellular interactions in plants'. *Journal of Microscopy 263*/2: 148–157.

Hell, S.W. (2003). 'Towards fluorescence nanoscopy'. *Nature Biotechnology 21*/11: 1347–1355.

Henriques, R. et al. (2011). 'PALM and STORM: unlocking live-cell super-resolution'. *Biopolymers 95*/5: 322–331.

Holm, T. et al. (2014). 'A blueprint for cost-efficient localization microscopy'. *ChemPhysChem 15*/4: 651–654.

Hosny, N.A. et al. (2013). 'Super-resolution imaging strategies for cell biologists using a spinning disk microscope'. *PLoS One 8*/10: e74604.

Hoyer, P. et al. (2016). 'Breaking the diffraction limit of light-sheet fluorescence microscopy by RESOLFT'. *PNAS 113*/13: 3442–3446.

Kaufmann, R. et al. (2014). 'Fluorescence cryo-microscopy: current challenges and prospects'. *Current Opinion in Chemical Biology 20*: 86–91.

Klauss, A. et al. (2015). 'Upgrade of a scanning confocal microscope to a single-beam path STED microscope'. *PLoS One 10*/6: e0130717.

Komis, G. et al. (2015). 'Super-resolution microscopy in plant cell imaging'. *Trends in Plant Science 20*/12: 834–843.

Lauterbach, M.A. (2012). 'Finding, defining and breaking the diffraction barrier in microscopy – a historical perspective'. *Optical Nanoscopy 1*: 8. doi: 10.1186/2192-2853-1-8.

Lidke, D.S. and Lidke, K.A. (2012). 'Advances in high-resolution imaging – techniques for three-dimensional imaging of cellular structures'. *Journal of Cell Science 125*/11: 2571–2580.

McBride, D. et al. (2013). 'A low cost and versatile STED superesolution fluorescent microscope'. *Modern Instrumentation 2*/3: 41–48.

Millis, B.A. et al. (2013). 'Superesolution imaging with standard fluorescent probes'. *Current Protocols in Cell Biology* Unit 21.8. doi: 10.1002/0471143030.cb2108s60.

Nelson, A.J. and Hess, S.T. (2014). 'Localization microscopy: mapping cellular dynamics with single molecules'. *Journal of Microscopy 254*/1: 1–8.

Patterson, G.H. (2011). 'Highlights of the optical highlighter fluorescent proteins'. *Journal of Microscopy 243*/1: 1–7.

Sauer, M. (2013). 'Localization microscopy coming of age: from concepts to biological impact'. *Journal of Cell Science 126*/16: 3505–3513.

Snapp, E.L. (2009). 'Fluorescent proteins: a cell biologist's user guide'. *Trends in Cell Biology 19*/11: 649–655.

Stemmer, A. et al. (2008). 'Widefield fluorescence microscopy with extended resolution'. *Histochemistry and Cell Biology 130*/5: 807–817.

Truckenbrodt, S. et al. (2017). x10 Expansion microscopy enables 25 nm resolution on conventional microscopes. *bioRxiv* preprint August 2017. doi: 10.1101/172130.

Vicidomini, G. et al. (2018). 'STED super-resolved microscopy'. *Nature Methods 15*/3: 173–182.

Single Molecules References

Single molecule techniques are valuable because they allow study of scarce molecules that would otherwise be hidden in the cellular milieu in an overview study; also because they permit study of molecular kinetics in vivo that previously could only be performed as biochemical analyses in vitro with bulk purified components. The 2014 Nobel Laureate William Moerner's address has been included in the this Key reading section, for it gives a very personal touch and shows how this work depended upon collaboration with other scientists. The co-Laureate addresses of Betzig and Hell are cited above for the same reason.

Claridge, S.A. et al. (2011). 'Electrons, photons, and force: quantitative single-molecule measurements from physics to biology'. *ACS Nano 5*/2: 693–729.

Coelho, M. et al. (2013). 'Single-molecule imaging in vivo: the dancing building blocks of the cell'. *Integrative Biology 5*/5: 748–758.

Forties, R.A. and Wang, M.D. (2014). 'Discovering the power of single molecules'. *Cell 157*/1: 4–7.

Gust, A. et al. (2014). 'A starting point for fluorescence-based single-molecule measurements in biomolecular research'. *Molecules 19*/10: 15824–15865.

Ha, T. (2014). 'Single-molecule methods leap ahead'. *Nature Methods 11*/10: 1015–1018.

Klein, T., Proppert, S., and Sauer, M. (2014). 'Eight years of single molecule localization microscopy'. *Histochemistry and Cell Biology 141*/6: 561–575.

Malkusch, S. and Heilemann, M. (2016). 'Extracting quantitative information from single-molecule super-resolution imaging data with LAMA – LocAlization Microscopy Analyser'. *Scientific Reports 6*: 34486.

Moerner, W.E. (2014). 'Single-molecule spectroscopy, imaging and photocontrol: foundations for super-resolution microscopy' (Nobel Lecture). *Angewandte Chemie International Edition 54*/28: 8067–8093.

Persson, F., Barkefors, I., and Elf, J. (2013). 'Single molecule methods with applications in living cells'. *Current Opinion Biotechnology 24*/4: 737–744.

Walter, N.G. et al. (2008). 'Do-it-yourself guide: how to use the modern single-molecule toolkit'. *Nature Methods 5*/6: 475–489. Note: This June 2008 issue of *Nature Methods* contains a special Focus section on single molecule imaging methods.

Zeng, Z. and Xi, P. (2016). 'Advances in three-dimensional super-resolution nanoscopy'. *Microscopy Research and Technique*. doi: 10.1002/jemt.22719 (Epub ahead of print). PMID: 27412634. doi: 10.1002/jemt.22719.

Correlative Microscopy References

Correlative microscopy has become a subject of increasing importance, and offers the potential to gain much more information about the biological specimen, its structure and function than light or electron microscopy are able to reveal individually. These references have been selected to give an introduction to this technically demanding, yet fast growing subject.

Bell, K. et al. (2013). Correlative imaging of fluorescent proteins in resin-embedded plant material. *Plant Physiology 161*/4: 1595–1603.

de Boer, P. et al. (2015). Correlated light and electron microscopy: ultrastructure lights up! *Nature Methods 12*/6: 503–513.

Caplan, J. et al. (2011). The Power of Correlative Microscopy: Multi-modal, Multi-scale, Multi-dimensional. *Current Opinion in Structural Biology 21*/5: 686–693.

Collinson, L.M. et al. (2017). Correlating 3D light to 3D electron microscopy for systems biology. *Current Opinion in Structural Biology 3*: 49-55.

Fonta, C.L. and Humbel, B.M. (2015). Correlative microscopy. *Archives of Biochemistry and Biophysics 581*: 98–110.
Giepmans, B.N.G. (2008). Bridging fluorescence microscopy and electron microscopy. *Histochemistry and Cell Biology 130*/2: 211–217.
Hodgson, L. (2014). Development of a quantitative Correlative Light Electron Microscopy technique to study GLUT4 trafficking. *Protoplasma 251*/2: 403–416.
Jahn, K.A. et al. (2012). Correlative microscopy: providing new understanding in the biomedical and plant sciences. *Micron 43*/5: 565–582.
Kellner, M. et al. (2016). A combined method for correlative 3D imaging of biological samples from macro to nano scale. *Scientific Reports 6*: 35606.
Kopek, B.G. et al. (2017). Diverse protocols for correlative super-resolution fluorescence imaging and electron microscopy of chemically fixed samples. *Nature Protocols 12*/5: 916–946.
Kukulski, W. et al. (2011). Correlated fluorescence and 3D electron microscopy with high sensitivity and spatial precision. *Journal of Cell Biology 192*/1: 111–119.
Legant, W.R. et al. (2016). High-density three-dimensional localization microscopy across large volumes. *Nature Methods 13*/4: 359–365.
Mateos, J.M. (2016). Topographic contrast of ultrathin cryo-sections for correlative super-resolution light and electron microscopy. *Scientific Reports 6*: 34062.
Mueller-Reichert, T. and Verkade, P. (eds.) (2012). *Correlative Light and Electron Microscopy I. Methods in Cell Biology*, vol. 111. Academic Press.
Mueller-Reichert, T. and Verkade, P. (eds.) (2014). *Correlative Light and Electron Microscopy II. Methods in Cell Biology*, vol. 124. Academic Press.
Mueller-Reichert, T. and Verkade, P. (eds.) (2017). *Correlative Light and Electron Microscopy III. Methods in Cell Biology*, vol. 140. Academic Press.
Perkovic, M. et al. (2014). Correlative Light- and Electron Microscopy with chemical tags. *Journal of Structural Biology 186*/2: 205–213.
Robinson, J.M. et al. (2001). Correlative fluorescence and electron microscopy on ultrathin cryosections: Bridging the resolution gap. *Journal of Histochemistry & Cytochemistry 49*/7: 803–808.
Russell, M.R.G. et al. (2017). 3D correlative light and electron microscopy of cultured cells using serial blockface scanning electron microscopy. *Journal of Cell Science 130*/1: 278–291.
Takizawa, T. et al. (2015). FluoroNanoGold: an important probe for correlative microscopy. *Journal of Chemical Biology 8*/4: 129–142.
Timmermans, F.J. and Otto, C. (2015). Contributed review: Review of integrated correlative light and electron microscopy. *Review of Scientific Instruments 86*/1: 011501.
Tuijtel, M.W. et al. (2017). Inducing fluorescence of uranyl acetate as a dual-purpose contrast agent for correlative light-electron microscopy with nanometre precision. *Scientific Reports 7*/1: 10442.
Viswanathan, S. et al. (2015). High performance probes for light and electron microscopy. *Nature Methods 12*/6: 586–576.
Wiley Essential Knowledge introduction: https://www.essentialknowledgebriefings.com/downloads/correlative-light-electron-microscopy?edd-flush=1 and http://www.magersci.com/brochures/leica/CryoCLEM_ebook.pdf.
Wolff, G. et al. (2017). Towards correlative super-resolution fluorescence and electron cryo-microscopy. *Biology of the Cell 108*/9: 245–258.

Internet Resources

Fluorescent dyes for STED: https://nanobiophotonics.mpibpc.mpg.de/dyes/
Abberior Instruments: http://www.abberior.com.
Sample preparation for GSD (Leica): https://www.leica-microsystems.com/fileadmin/academy/2013/Leica_SR_GSD-Protocol_Guide_v2_0.pdf.
Sample preparataion for STORM (Nikon): http://www.mvi-inc.com/wp-content/uploads/N-STORM+Protocol.pdf.
Localisation software list: http://bigwww.epfl.ch/palm/software.
3D STORM localisation algorithm: http://zhuang.harvard.edu/software.html.
Cell Profiler, for image segmentation: www.cellprofiler.org.
Soliman, K. (2015). 'CellProfiler: Novel automated image segmentation procedure for super-resolution microscopy'. *Biological Procedures Online 17*: 11.
Expansion microscopy: http://expansionmicroscopy.org.
3B Bayesian analysis ImageJ plugin: http://www.coxphysics.com/3b/ and https://www.edwardrosten.com/work/rosten_2013_plugin.pdf.
Rosten, E. et al. (2013). 'ImageJ plug-in for Bayesian analysis of blinking and bleaching'. *Nature Methods 10*: 97–98. See the supplementary information to this paper for the plugin, source code and manual.
GraspJ: https://code.google.com/archive/p/graspj/downloads.
Brede, N. and Lakadamyali, M. (2012). 'GraspJ: an open source, real-time analysis package for super-resolution imaging'. *Optical Nanoscopy 20121*: 11. doi: 10.1186/2192-2853-1-11.
QuickPalm: http://imagej.net/QuickPALM.
Henriques, R. et al. (2010). 'QuickPALM: 3D real-time photoactivation nanoscopy image processing in ImageJ'. *Nature Methods 7*/5: 339–340.
Palm3d: https://code.google.com/archive/p/palm3d/.
York, A.G. et al. (2011). 'Confined activation and subdiffractive localization enables whole-cell PALM with genetically expressed probes'. *Nature Methods 8*/4: 327–333.
rapidSTORM: https://github.com/stevewolter/rapidSTORM.
Wolter, S. et al. (2012). 'rapidSTORM: accurate, fast open-source software for localization microscopy'. *Nature Methods* 9/11: 1040–1041.
fairSIM: www.fairsim.org.
Müller, M. et al. (2015). 'Open-source image reconstruction of super-resolution structured illumination microscopy data in ImageJ'. *Nature Communications 7*: 10980.
SIMcheck: https://github.com/MicronOxford/SIMcheck.
Ball, G. et al. (2015). 'SIMcheck: a toolbox for successful super-resolution structured illumination microscopy'. *Scientific Reports 5*: 15915.
See also: Sage, D. et al. (2015). 'Quantitative evaluation of software packages for single-molecule localization microscopy'. *Nature Methods 12*/8: 717–724.
TestSTORM: http://titan.physx.u-szeged.hu/~adoptim/?page_id=183.
Novák, T. et al. (2017). 'TestSTORM: Versatile simulator software for multimodal super-resolution localization fluorescence microscopy'. *Scientific Reports 7*/1: 951.

ThunderSTORM: https://zitmen.github.io/thunderstorm/.
Ovesný, M. et al. (2014). 'ThunderSTORM: a comprehensive ImageJ plug-in for PALM and STORM data analysis and super-resolution imaging'. *Bioinformatics 30*/16: 2389–2390.
NanoJ-SQUIRREL: https://bitbucket.org/rhenriqueslab/nanoj-squirrel/wiki/Home.
Culley, S. et al. (2017). 'Quantitative mapping and minimization of super-resolution optical imaging artifacts'. *Nature Methods*. doi: 10.1038/nmeth.4605.
Software for single molecule localisation microscopy (SMLM): http://bigwww.epfl.ch/smlm/software/.

Notes

1. For those wishing to understand the mathematical theory behind circumventing the Abbe diffraction limit, the paper by Sheppard (2007) is a useful starting point. In 1927, about 50 years after Abbe's work, Werner Heisenberg formulated his uncertainty principle, which states that the location and momentum of a quantum particle are interdependent, and only one of these parameters may be precisely known. Abbe's principles of image formation and Heisenberg's uncertainty principle are related mathematically via Fourier analysis. In a sense, the principle of localisation microscopy is to determine the location of a single molecule with greater precision than the width of its PSF in the image.
2. In theory, super-resolution can be achieved by multi-photon microscopy since the absorption of light by the fluorophore is proportional to the square of the excitation light intensity and therefore nonlinear, but the super-resolution advantage of the potentially small focal volume achieved by multi-photon microscopy is negated by the longer-wavelength infrared excitation illumination used.
3. The human fascination with acronyms has, in this field of diffraction-unlimited fluorescence microscopy – just as with light-sheet microscopy – spawned an entire series of acronyms. In this chapter we discuss the major ones: STED, RESOLFT, PALM, STORM and SIM, but there are others. For example, see Table 1, page 12119 in Haas et al. (2014) with original references.
4. The 4-pi microscope is so-named because if the object were at the centre of a perfectly spherical lens, all the diffracted rays would be collected and the resolving power laterally in x,y and axially in the z Cartesian coordinate would be isotropic, since the lens aperture would extend over 4π steradians of solid angle. Two objectives set orthogonally at right angles to one another are the best possible attempt to collect as many diffracted rays as possible and increase axial resolving power.
5. STORM and STED rely either on switching or depleting fluorophores, respectively; both methods have been compared: e.g. see Tam and Merino (2015).
6. Although the pulsed STED laser operates at very high intensity and is two orders of magnitude more intense than a widefield or confocal microscope, the STED laser is still three orders of magnitude lower than the Ti:S pulsed laser used for multi-photon microscopy.
7. A similar method to these early approaches has been published (Cox et al., 2011) utilising high-speed imaging with a widefield microscope rather than specialised and expensive super-resolution instruments using light at intensities than damage live cells and tissues. The images are post-processed by Bayesian statistics to achieve a lateral resolving power of 50 nm and 4 seconds temporal resolution. This method is sometimes referred to as 3B-Bayesian analysis of blinking and bleaching. A similar approach (Munck et al., 2012) called PiMP (photobleaching microscopy with nonlinear processing) acts on a large set of molecules, rather than the small subset photo-bleached or -switched in PALM and STORM.
8. The minimum labelling density required is given by $(2/a)^2$, where a feature of size **a** is to be sampled and resolved in the image. Thus, while a labelling density of 100 labelled molecules per μm^2 satisfies the Nyquist sampling criterion, for diffraction-limited resolution at the 200 nm limit, 10 000 labelled molecules per μm^2 are required to resolve 20 nm features. For successful sampling in 3D, a minimum molecular labelling density of 1×10^6 labelled molecules per μm^3 is required. As the molecular density decreases, the signal-to-noise ratio also decreases, and features are no longer resolvable when the mean molecular separation approaches the size of the features of interest. Since more time is required to localise large numbers of molecules, there is an inevitable trade-off between spatial resolution and temporal resolution.
9. A moiré pattern is an interference pattern created when two semitransparent objects, each with a repetitive, periodic pattern that exceeds the resolving power of the sensor, are placed over another and offset or rotated. The resulting moiré pattern is a beat pattern that contains high-frequency information about the structure of the original object. A slight motion of the object creates large-scale changes in the moiré pattern, which, in effect, 'magnify' high-frequency information: the spacing between moiré fringes being greater than the periodic information in the object. The term 'moiré' is named after a silk fabric that exhibits a fine rippled appearance like a water surface.

27

Choosing a Microscope Platform and Core Imaging Facilities

Outline

Understanding Light Microscopy, First Edition. Jeremy Sanderson.

27.1 Introduction

We have seen in Chapter 17 how the requirement for optical sectioning is determined by the self-luminous nature of fluorophores. Even when viewing relatively thin fluorescent specimens, less than 5 μm thick, 80–90% of the light may be coming from out-of-focus regions. This is because the depth of field (see Chapter 7, Section 7.11) of a high numerical aperture objective is less than 1 μm.

'Confocal versus widefield' is a topic of much debate in the microscopy community. Ultimately, the answer to the question *"Which approach is best?"* depends on your particular sample and requirements. Sometimes the only way is to try all approaches available to you and see which one works best to give the data required in order to answer the hypothesis underlying the imaging experiment.

This chapter also discusses aspects of experimental design, working with staff in a core imaging facility and the various excellent software programmes that have been developed to help maintain research microscopes, to keep them working optimally and to record a quality control audit trail.

27.2 Which Optical Sectioning Method to Use?

The thickness of your sample primarily determines which microscope to use. First examine your sample with a widefield fluorescence microscope (WFM). Very thick tissues and organ slices greater than 100 μm thick that produce blurred images can only usefully be imaged with a multi-photon microscope. Generally an improvement in depth imaging along the z-axis over the single-photon laser-scanning confocal microscope (LSCM) by about four to six times is possible.

Use the point-scanning or spinning disc confocal for tissue samples or slices from 10 μm to about 100 μm thick. These samples will still appear indistinct and virtually featureless when examined with a widefield microscope. However, there will be regions of just sufficient contrast to allow you to set approximate focus and see some level of detail. Confocal microscopes detect less fluorescence at each focal plane compared to widefield microscopes because of the pinhole(s) present. Samples prepared for confocal analysis may require increased staining times or the use of increased stain concentrations to ensure an adequate signal-to-noise ratio in the image.

Those samples under 10 μm whose images are not obscured by fluorescence above and below the plane of focus are generally thin enough to set the focal point easily. These are best imaged with widefield fluorescence microscopy, particularly if they are no more than 5 μm thick. Further sharpening of fine detail in the image can be achieved by deconvolving the image after acquisition. In theory, confocal methods offer a slight advantage in resolving power and an increase in contrast, but in practice deconvolution of thin widefield samples gives superior images.

Total internal reflection fluorescence, or TIRF, microscopy, is capable of capturing an image of a very thin optical section adjacent to the surface of a cell and the substrate upon which it is supported. TIRF is a near-field optical microscopy technique; it generates an optical section around 100 nm thick compared to a confocal optical section of around 0.2–1.0 μm or thicker. TIRF microscopy is discussed in Chapter 23.

There are some broad pointers why one might choose confocal or widefield with deconvolution, and these are presented in Box 27.1 and Table 27.1, together with the advantages and disadvantages of each system, in Table 27.2.

Box 27.1 Optical Sectioning Microscope Choices

First, determine how thick your sample is. As a rule of thumb, if it is:

1. greater than 100 μm thick, choose a multi-photon microscope;
2. from 10 μm to 100 μm thick, choose a confocal microscope; and
3. under 10 μm thick, choose a widefield microscope with or without deconvolution.

Then consider:

- Sensitivity of detection, acquisition speed and resolution.
- This is the 'iron triangle' that applies to live-cell imaging.
- How long do you want to image for – will the specimen remain alive?
- How is the specimen mounted – whether solid mountant or fluid; how long will it last?
- How bright is the sample; how robust the fluorophores – at what rate will they photo-bleach?
- Are there different wavelengths to sample? Wavelength switching is done easily with a laser-illuminated confocal microscope. Filter wheels on widefield systems can take time to switch; otherwise, use LEDs on widefield microscopes with digital shutters/switching circuits and Pinkel filter blocks.
- Optical sections of photosensitive samples should, if possible, be imaged using a structured illumination microscope (e.g. Optigrid, Apotome, Angstrom), moving samples with the aperture correlation equivalent (e.g. Clarity). These designs do not use lasers.
- Light-sheet (selective plane illumination; SPIM) microscopes achieve 3D data sets with minimal photobleaching.
- Image time-lapses of very sensitive samples should be made with an EM-CCD or sCMOS camera on a widefield microscope set up for live-cell work or a fast camera arranged to expose the sample to light only for image acquisition.
- Image rapidly moving, photosensitive samples with a Nipkow spinning disc confocal equipped with an EM-CCD or sCMOS camera.

Table 27.1 *Optical Sectioning – Which Microscope to Use*

Microscopy type	Illumination	Detector/ camera	When to use	Advantages	Disadvantages	When not to use
Point-scanning confocal	Laser Peak intensity/μm^2 = 1 mW. Peak signal 20–100 photons/pixel	PMT	**With thick tissues** For control of precise optical section thickness Region of interest scanning, ideal for spectral unmixing	For control of precise optical section thickness. Generating 3D images and intensity projections. Laser switching good for control of bleed-through	Lasers photo-bleach. PMT and circuitry noisier than widefield. PMT QE = 5–15%. Requires well-stained samples. Slower scan than spinning disc.	For photosensitive samples or weakly stained or very thin samples. For rapidly moving specimens
Nipkow spinning disc confocal	Mostly laser (can also be white light, e.g. Carv) Peak intensity/μm^2 = 1 μW. Peak signal ≈ 5000 photons/pixel	EM-CCD sCMOS	**For acquisition speed** and very high frame rates of living cells and tissues and/or photosensitive tissues, over time. Slit-scanners good for multi-channel collection	For photosensitive living samples. High temporal resolution. Excellent for time-lapse. Also collects Z-stacks very fast: thus minimal photobleaching EM-CCD QE = 60–90%	Fixed pinhole: fixed optical section thickness. Can get bleed-through between pinholes. No optical zoom, like point-scanner	When very low SNR is likely (Spinning Disc can be noisy). When spectral unmixing is important. When optical section thickness is important
Structured illumination (Ronchi grid and aperture correlation types; correlate respectively to point-scanner and Nipkow disc above)	White light e.g. ApoTome (static Z-stack) or VivaTome, Clarity (moving objects)	CCD	**For photosensitive** samples that would be bleached by laser. For routine optical sections at low magnifications	Fast to use, cheaper than LSCM, can add to existing WF stand. Better than confocal at lower objective magnifications	Ronchi grid can burn stripes into image. Can be difficult to focus sharply. Poor with periodic structures and very high-magnification objective	When you need control of optical section thickness. Cannot do ROI selection or photo-activation. Requires good fluorescence signal
Widefield deconvolution	White light Peak intensity/μm^2 = 10 W. Peak signal 20–100 photons/pixel	CCD/sCMOS	**For thin fluorescent cells** For small, photosensitive samples that would be bleached by lasers. Good with low-NA objectives	For capturing all the fluorescent signal (not rejected by pinhole). Good SNR. High sensitivity. Less noise than confocals CCD QE = 50–80%	Requires post-processing. Not best suited to moving/living tissues. Sensitive to spherical aberration. No zooming	For tissues greater than 20 μm thick, or so. Less well suited to time-lapse. Needs lots of computer processing power

(continued)

Table 27.1 (Continued)

Microscopy type	Illumination	Detector/ camera	When to use	Advantages	Disadvantages	When not to use
Multi-photon (two-photon)	Pulsed laser Peak intensity/μm^2 = 10 nW. Peak signal 20–300 photons/pixel	PMT (mainly NDD)	**For very thick tissues** For imaging living tissues with low photo-toxicity over time	Will image sharply very deep into thick tissues without scattering. Good for living tissues.	Expensive, can cause high localised bleaching unless care is taken. Longer IR λ means lower resolution. Same PMT restriction as confocal: PMT QE = 5–15%	For very thin fluorescent cells and tissue slices. External light sources must be excluded from the detector
Total Internal Reflection Fluorescence Microscopy (TIRF)	Laser (peak intensity and signal as for point-scanning confocal)	EM-CCD sCMOS	**For imaging cell membrane** also rapid dynamic events at cell-coverslip interface	Very thin optical section 100–200 nm. High frame rates possible	Can only image fluorescence in evanescent field. Can bleach sample	For samples thicker than 200 nm, not adhering to glass-fluid interface
Light-sheet or Selective Plane Illumination Microscopy (SPIM)	Laser Peak intensity/μm^2 = 1 nW (SPIM) and 1mW (digital LSM). Peak signal 500–30 000 photons/pixel	CCD/sCMOS	**For thick tissues**, embryos and small organisms For imaging living tissues in 3D with low photo-toxicity over a very long time	For photosensitive living samples. High temporal resolution. Also collects Z-stacks very fast: thus minimal photobleaching	Sample preparation can be complex. No commercial instrument with dry lenses. 3D depth limited by working distance of lens	For very large samples, may have to use optical projection tomography (OPT)

References

Centonze, V. and Pawley, J.B. (2006). Tutorial on Practical Confocal Microscopy and use of the confocal test specimen, 'Which 3D Method is best?'. In: Pawley, J.B. (ed.), *Handbook of Biological Confocal Microscopy*, Chapter 35, 3rd ed. New York, Frankfurt: Springer, pp. 644–648.

Sanderson, J. (2010). Light Microscopy Imaging Facilities. In: *Encyclopedia of Life Sciences (ELS)*. Chichester: Wiley. doi: 10.1002/9780470015902.a0022189. Table 4, p. 6.

Peak intensity and peak signal data from Shaw, P.J. (2006). Comparison of widefield/deconvolution and confocal microscopy for three-dimensional imaging. In: *Handbook of Biological Confocal Microscopy*, Table 23.1, Chapter 23, 3rd ed, pp. 453–467.

Table 27.2 *Advantages-Disadvantages of Each Type of Microscopy*

1. **Laser Scanning Confocal Microscopy**

Advantages
- Non-invasive optical sectioning – sharply defined slices
- Reduction of background – reduced blurring of the image from light scattering
- Smaller PSF – better lateral and axial resolving power than widefield microscopy
- Coherent laser illumination reduces chromatic aberration for co-localisation studies
- Can use deconvolution algorithms to sharpen images – PSF usually better predicted
- Improved signal-to-noise (S/N) ratio compared to widefield
- Laser switching – better elimination of cross-talk and bleed-through, which is easily eliminated
- Laser switching line by line instead of frame by frame for living, moving samples
- Region of interest scan very easy – reduce scan times for sensitive, living samples
- Magnification zoom can be adjusted electronically – PMT not limited by the matrix of a photo-detecting CCD array
- Greater control of regions of interest for bleaching – FRAP, FRET
- Can easily capture 2–4 fluorophores simultaneously
- Consider fluorescent protein overlap – LSCM good at spectral separation

Disadvantages
- Requires regular evaluation to maintain optimum performance
- Light scattering/refractive index mismatches limits depth penetration to 100–200 μm
- Real-time collection difficult, unless slow-scan methods used – need to wait to collect a low-noise signal
- Large volume of specimen still illuminated
- Non-image forming light passing through pinhole generates noise
- PMT noisier than CCD, lower quantum efficiency (QE) – majority of photons not detected
- Laser power, noise and fibre optic coupler (change polarisation state) leads to artifactual pixel-pixel fluctuations
- Laser results in very high light intensity at the specimen plane
- Laser bleaching of sample at higher rate than with spinning disc confocal
- LSCM scanning with galvanometers is slow: typical confocal image takes 10 s to collect; widefield takes 300 ms to collect
- The maximum speed attainable with LSCM is about 10 frames per second, and that compromises image quality
- Samples must be well stained

2. **Widefield microscopy (deconvolution)**

Advantages
- Bright fluorescence signal is not attenuated by confocal pinhole
- High acquisition speed over entire field by CCD with high quantum efficiency
- Broad-band mercury arc lamp excites more fluorophores over the visible spectrum
- Less photobleaching, photo-toxicity and specimen damage – widefield microscopy (WFM) is good for tissue cultured cells
- WFM can visualise weak fluorescent structures, where contrast is low (2–3%) and which would otherwise be drowned out by the noise inherent in LSCM.
- The intrinsic noise level is lower in WFM than in LSCM
- Useful increase in S/N and contrast when speed and sensitivity requirements preclude collection of Z-stack

Disadvantages
- High non-specific background labelling is common
- Out-of-focus blurring degrades image and reduces visibility of fine detail – PSF (point spread function) less accurately modelled in thick tissues.
- WFM not ideally suited for imaging thick tissues
- WFM less good than confocal for real-time imaging of ratiometric probes (e.g. Ca^{2+})
- Deconvolution to sharpen images requires accurate knowledge of measured or theoretical PSF.
- Inconsistency in a single optical section can affect the deconvolution of the entire Z-stack
- Image restoration by computational deconvolution is often not so successful with live specimens. Any movements that occur in the specimen while a stack of optical sections are being recorded will abrogate successful deconvolution
- Sensitive to spherical aberrations
- No zooming function
- Region of interest function, if present at all, is much less sophisticated than LSCM
- Images from faint sources that have poor signal-to-noise ratios, such as those typically obtained from living specimens
- Refractive index mismatches can lead to PSF being less accurately modelled

3. **Nipkow Spinning Disc confocal microscopy**

Advantages
- Much higher frame rate possible than with LSCM
- Image collection is routinely at 10–15 frames per second. Can exceed video rate for small frames, if the fluorophores are bright enough and/or the CCD camera is sensitive enough
- Spinning disc – ideal for real-time processes
- High acquisition speed over entire field by CCD with high quantum efficiency
- Less photobleaching, photo-toxicity and specimen damage than with LSCM
- Light efficient CCD – the high QE of the CCD means it is more efficient detecting emitted signal
- Good for long-term time-lapses

(continued)

Table 27.2 (Continued)

Nipkow Spinning Disc confocal microscopy

Disadvantages

- Can get streaking on the image, disc synchronisation problem
- Must synchronise camera and Nipkow disc – not always easy
- Some cross-talk between adjacent pinholes/microlenses
- Non-modular: Can only be matched to one objective magnification, usually 100x. Thus less optical sectioning versatility than LSCM
- No zooming function
- Region of interest function, if present at all, is much less sophisticated than LSCM
- Can only bin pixels to increase signal gain, and thus will lower image resolution
- Nipkow spinning disc systems most commonly sample multiple colours concurrently (and not simultaneously), thus increasing the time needed to acquire images and making multi-D acquisition vulnerable to artifacts
- The centre of the field is always brighter than the edges

4. **Structured Illumination – Optigrid/Angstrom/Apotome**

Advantages

- Used on a widefield microscope; therefore, all the advantages of WFM over LSCM
- Simpler to use than a LSCM and faster overall when repeatedly imaging through a set of slides
- Gives better confocal sections when using lower-magnification objectives for larger fields of view
- Faster acquisition than a LSCM – will do about 9 frames per second at full resolution and up to video rate (30 fps) with binning and cropping
- Better than the LSCM for imaging photosensitive samples (photobleaching leads to a low signal-to-noise ratio)
- With the xenon metal-halide light source, the Optigrid-Apotome can image many more fluorophores and is limited only by the filter blocks and not the availability of the lasers
- Cheaper to use than LSCM of Nipkow spinning disc (no laser sources)
- If collecting a single fluorophore, it is better than LSCM

Disadvantages

- Optigrid cannot do ROI photobleaching and photo-activation
- Cannot change the thickness of the optical section. Depth of field is fixed by objective used
- Needs care in setting the illuminated field diaphragm for a good S/N ratio
- The confocal is good for excluding out-of-focus fluorescence from strongly scattering samples (like fat tissue in zebrafish embryos); structured illumination is not good for this. Neither is it good for thick largely homogeneous structures (e.g. bone), or for periodic samples, which exhibit very strong and regular structures

5. **Two-photon microscopy**

Advantages

- Low photobleaching outside focal volume
- Increased depth penetration – multi-photon is the best choice for thick tissue imaging
- Longer wavelengths are scattered less
- Any photon scattered from PSF used for imaging, irrespective of its route back to PMT; hence multi-photon microscope can view deep into tissue by about a factor of 4x – makes maximum use of light
- Non-descanned detectors at pinhole position; no pinhole in emission path – very sensitive
- Simpler optical configuration than confocal: signal does not have to be descanned and focused onto a pinhole
- Increased signal-to-noise ratio – no confocal aperture to restrict the signal
- Excitation only at focal point, smaller excitation volume, smaller PSF
- Decreased photo-toxicity, due to illumination by longer-wavelength light
- Simultaneous excitation of a variety of dyes
- Can do long-duration live-cell imaging

Disadvantages

- Very high costs – intensive service and running costs
- Theoretically lower resolution than confocal: uses longer-wavelength illumination
- Can get more bleaching at focal plane than LSCM or widefield, because of laser power
- Thermal damage to chromophores that absorb in the infrared

6. **TIRF microscopy**

Advantages

- Extremely thin optical section
- Very high S/N ratio with low background
- Low photobleaching and photo-damage (illumination only by evanescent field)
- Fast image acquisition compared to confocal methods: good for living adherent cells

Disadvantages

- Optical section only at cell-substrate interface: limited applications
- Method critically dependent upon optical quality of coverslip
- Objective method for TIRF requires a very high-NA objective, sometimes with specialised immersion medium
- Prism method for TIRF gives an evanescent field with less scattered light but can be difficult to set up and align
- Difficult to use with a flow chamber

Table 27.2 (Continued)

7. **Light-sheet/SPIM microscopy**

Advantages

- Very low photobleaching rate
- Very high frame rate for acquisition
- Ideally suited to long-term time-lapse
- sCMOS camera has a high quantum efficiency
- Very high S/N ratio with low background

Disadvantages

- Sample preparation can be complex
- 3D depth limited by working distance of lens
- At the time of publication, still an evolving technology

Factors to consider when choosing a microscope

This table and Table 27.1 give the advantages and disadvantages of the principal methods of fluorescence microscopy. Several factors govern the optimum choice of instrument, but the thickness of the sample is the most decisive. Which filter sets and which fluorophores/fluorescent proteins are available must also be taken into account. If the sample is fixed, it is better to use longer-wavelength fluorophores if the sample has been aldehyde-fixed, to avoid autofluorescence. For living cells and tissues, 'light budgets' and exposure must be balanced against acquisition speed and resolution. It is also better to use longer-wavelength fluorophores for live-cell imaging; see Chapter 28, Section 28.6. For both fixed and living cells/tissues, the length of exposure or length of time-lapse (for photobleaching) should also be considered.

Both two-photon microscopy and TIRF microscopy are niche techniques, used much less than confocal or widefield fluorescence microscopy, yet they complete the palette of methods available for optical sectioning and generating three-dimensional images of cells and tissues and so are included in the tables here.

For those who prefer a more objective approach, Murray (2010), who has written an excellent chapter on imaging fluorescent specimens, suggests the Haziness index, H. The metric H is the ratio of background to signal in a widefield image. The intensity of out-of-focus background is divided by the intensity of signal from a very small in-focus object. For practical purposes, with $H \leq 20$, widefield with deconvolution is appropriate. Where H lies in the range $20 < H < 200$, spinning disc confocals are best, and where $200 < H < 1000$, point-scanning confocals are the microscope of choice. For values of H above 1000, a multi-photon microscope should be used.

Before rejecting widefield fluorescence microscopy in favour of optical sectioning, bear in mind that a greater number of photons are elicited and detected with widefield than the other techniques. To collect an image in the same time, the laser used for confocal must be more intense than the LED, metal-halide or mercury lamp used for widefield microscopy. In order to operate closer to optimum fluorophore saturation, one must either take more time to collect an image by confocal means, or settle for detecting fewer photons per pixel than with widefield microscopy. This is further explained in chapter 35 of Pawley (2006) by Centonze and Pawley and by Murray et al. (2007).

27.3 Which Objective to Use?

Match your objective, as best you can, to the refractive index (RI) of the sample and the mounting medium. Immersion objectives with high numerical apertures (NA > 1.3) are designed for use with homogeneous immersion. That is, the objective front lens element, the immersion medium between objective and coverslip, the coverslip itself and the mounting medium permeating the specimen should all possess exactly the same refractive index. That way, light rays will pass from the specimen into the objective without being refracted or lost by total internal reflection. In practice, this is only achieved with only a few types of preparation. A well-prepared histological thin section comes close to this ideal.

If a living or freshly killed, unfixed sample is mounted in an aqueous medium, this prerequisite of homogeneous immersion is rendered void. Three undesirable optical effects occur: (a) the most widely diffracted rays (which transfer the finest detail into the image) are lost by total internal reflection at the water-coverslip interface, (b) the incidence angle of the rays that *do* enter the objective is reduced by refraction at the water-glass interface (also leading to the loss of high-frequency image information), and (c) the sine condition (Chapter 6, Section 6.3), which corrects for spherical aberration and coma, no longer holds true. Unless the object is situated directly against the under-surface of a coverslip of the correct (0.170 ± 0.005 mm) thickness, the image will suffer from lack of sharpness and also brightness: it will be less intense.

For example, live zebrafish mounted in agar should be imaged with a glycerol-immersion or water-immersion or multi-immersion objective rather than an oil-immersion objective. A silicone-immersion objective can also be used, but these are extremely expensive. These water- or specialised immersion objectives should be used because the refractive indexes of the agar and body constituents of the zebrafish are closer to the refractive index of water than of oil.

Where the sample is located at the underside of the coverslip, an oil-immersion objective will often secure a brighter image (Figure 27.1a), despite the object being mounted in water. However, a water-immersion objective will function more or less optimally, even when the object is located at a distance from the surface of the coverslip (Figure 27.1b). Further guidance on the selection of objectives

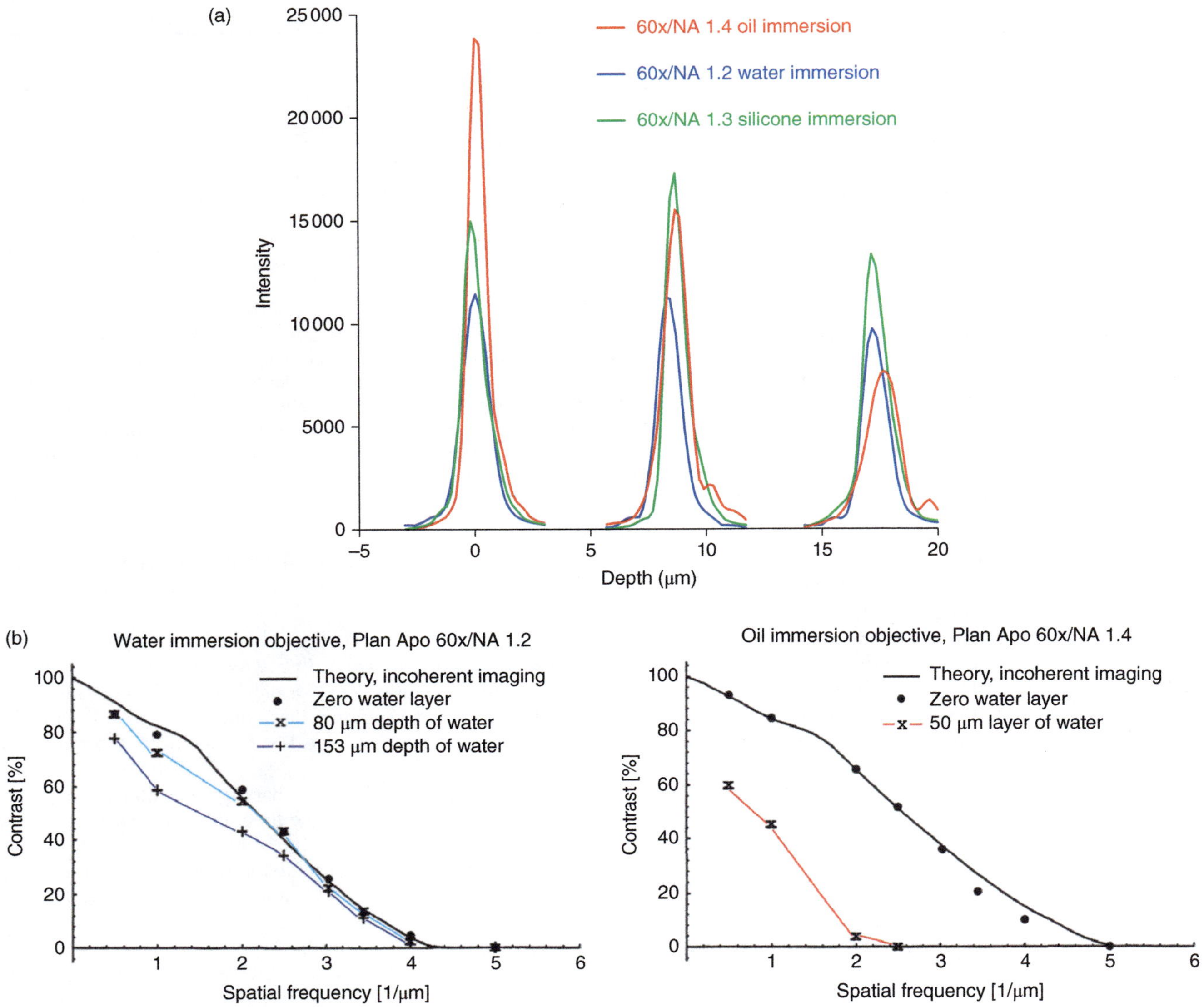

Figure 27.1 **Objective lens performance** *(a) The brightness of the image and the depth into the sample to which an objective can secure a sharply focused image is shown in the graph above. Three-dimensional Z-stacks of fluorescent beads immersed in water were acquired using an oil-immersion objective lens, a water-immersion objective lens and a silicone-immersion objective lens. The graph shows the intensity of the image of an individual bead in each focal plane of the 3D stacks for three beads, each located at different distances from the underside of a #1.5 coverslip of 0.17 mm thickness. At depth 0 μm, the bead is adherent to the surface of the coverslip; the other two depths are approximately 8.5 μm and 17 μm beneath the coverslip.*

For objects located immediately adjacent to the coverslip, in the correct position optically, the oil-immersion objective gives a brighter image and sharper axial responses than the other objectives. With increasing depth, the performance of the oil-immersion objective is increasingly degraded. The other two objectives are less sensitive to spherical aberration arising from the change in depth because they are more closely matched to the refractive indexes of the mountant (water) and of the beads. Since living tissues are likely to be watery and not immediately adjacent to the underside of the coverslip, it pays to use a water-immersion, silicone-immersion, glycerol-immersion or multi-immersion objective.

(b) These contrast transfer function (CTF) graphs show that water-immersion objectives perform better when used with aqueous mountants, even when the object is mounted at some distance from the underside of the coverslip. The theoretical resolving power (for an aberration-free objective with an illumination wavelength of 546 nm) for incoherent imaging, in both cases, is shown by the solid black line.

In the left-hand graph, the water-immersion objective performs near to the theoretical limit with both an 80 μm and 153 μm depth of water between the object and coverslip. By contrast, the performance of the oil-immersion objective is very significantly reduced with only a 50 μm layer of water.

can be found in Chapter 7, Table 7.7). Information specifically regarding objectives suitable for live-cell and time-lapse imaging can be found in Chapter 28 (section 28.6).

A flow chart, such as that made for the Light Microscopy Facility at the MRC Mammalian Genetics Unit (Figure 27.2; see also Table 1 and Figure 4 in Sanderson, 2010) can be constructed for the microscopes in your institution or microscope facility. It is useful to help novice users make a decision about which instrument to use, and/or to initiate a discussion with light microscopy facility staff.

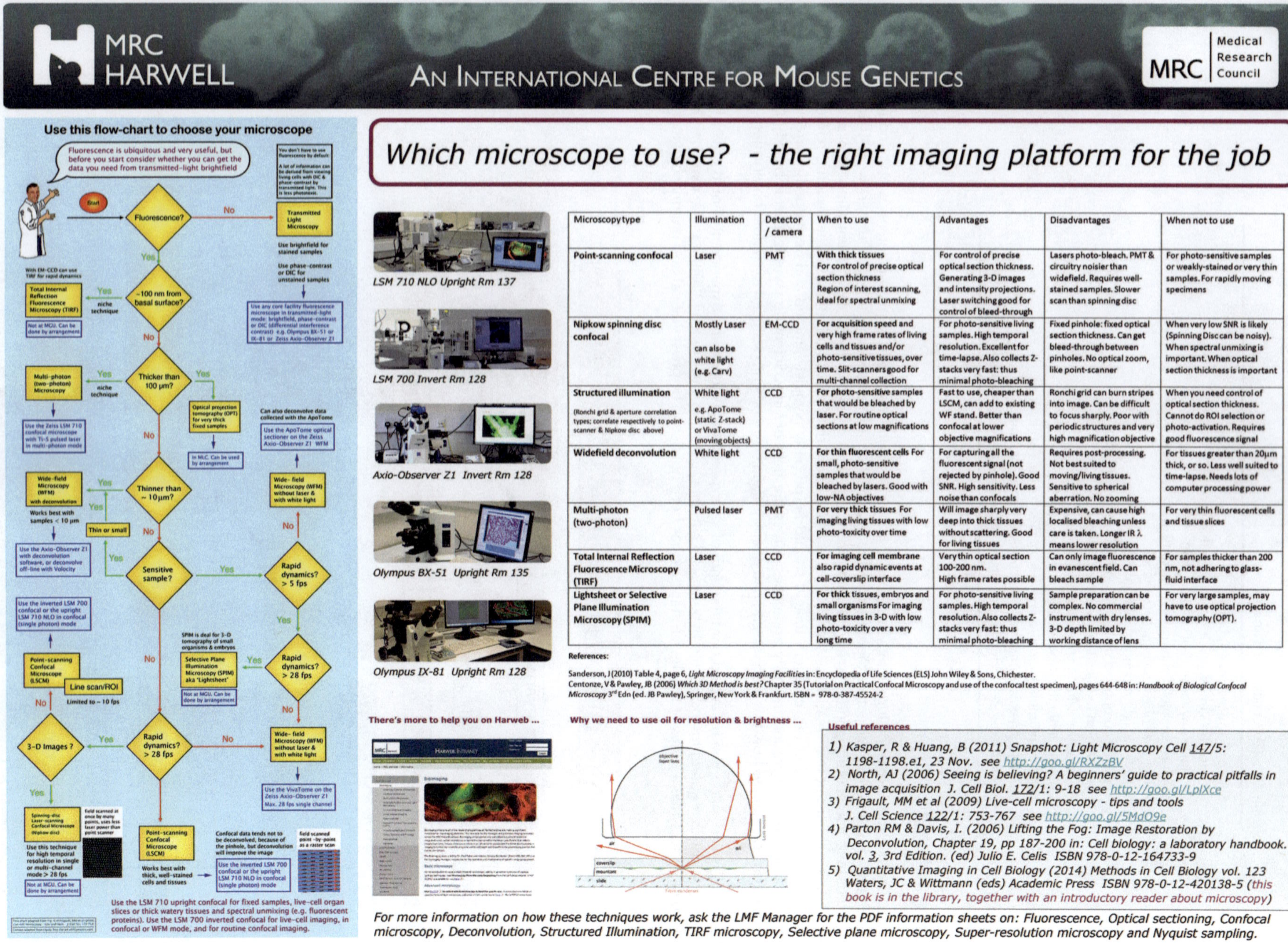

Figure 27.2 **Image facility flow chart** *This flow chart illustrates the decision-making process when selecting a fluorescence microscope for optical sectioning of cells and tissues. It presents basic information to the user of the light microscope imaging facility, from which an informed discussion can take place with LMF staff. Source: Reproduced with permission from University of Sheffield, Department of Biomedical Science; Adapted from Sanderson 2010 with permission from John Wiley & Sons. Adapted from Frigault 2009; Reproduced with permission from MRC Harwell; Data from Sanderson 2010; from Centonze and Pawley, in Pawley 2006.*

Table 27.3 Collecting Good Images

1. Set up the microscope properly;
2. Prepare the sample carefully; remove fluorescent media;
3. Eliminate any movement or focus drift;
4. Minimise photobleaching, and use antifade reagents;
5. Select the correct objective for the job;
6. Select the right mounting medium;
7. Choose fluorophores that give good SNR with your filter sets;
8. Pay attention to 'bleed-through' or cross-talk;
9. Avoid spherical and chromatic aberrations;
10. Collect within the dynamic range of the detector;
11. Set the minimum exposure and field of view of the specimen;
12. Fulfil the Nyquist sampling limit; consider deconvolution; and
13. Monitor your cells after the experiment for healthy appearance and behaviour.

27.4 Planning an Imaging Experiment

Scientific images are not just 'pretty pictures'; they are spatial and/or temporal data sets of photon intensity. You should therefore expend sufficient time and effort in collecting the best possible images. The adage 'do work to save [further] work [later]' applies, and Table 27.3 will help you acquire good images. Without proper planning of your imaging experiment, time spent acquiring and analysing your images is wasted (e.g. read Pearson, 2007; the *Nature* Editorial, 2007; and also the Key Reading reference North, 2006, in the references at the end of this chapter). Images should be properly Nyquist-sampled and be properly corrected for non-uniform illumination and/or background intensity anomalies across the field of view. Scientific investigation is based upon testable hypotheses, and Figure 27.3 shows how this is an integral part of any imaging experiment, designed to explore observations and answer questions.

Statistical analysis is very often 'the bit tacked onto the end' of an experiment, whereas the experiment should rather be designed around the correct statistics required to answer the hypothesis. Dytham (2010) and van Emden (2008) are two extremely good books on statistics for biologists, while a series of articles on statistics has been published by physiological and pharmacological societies in

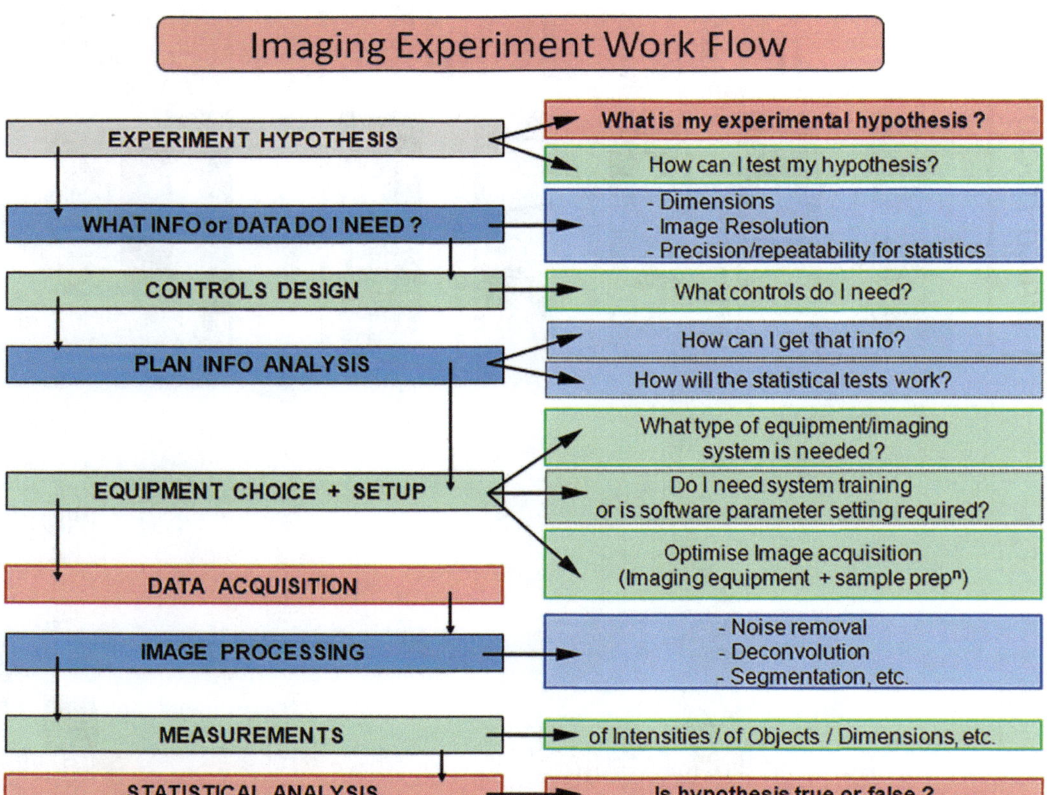

Figure 27.3 **Imaging experiment flow chart** *A flow chart for planning, initiating and performing an imaging experiment. (Original ppt slide courtesy of Dr Michael Doube, MPI-CBG, Dresden & Royal Veterinary College, London.) Source: Dr M. Doube. Reproduced with permission from Dr Doube.*

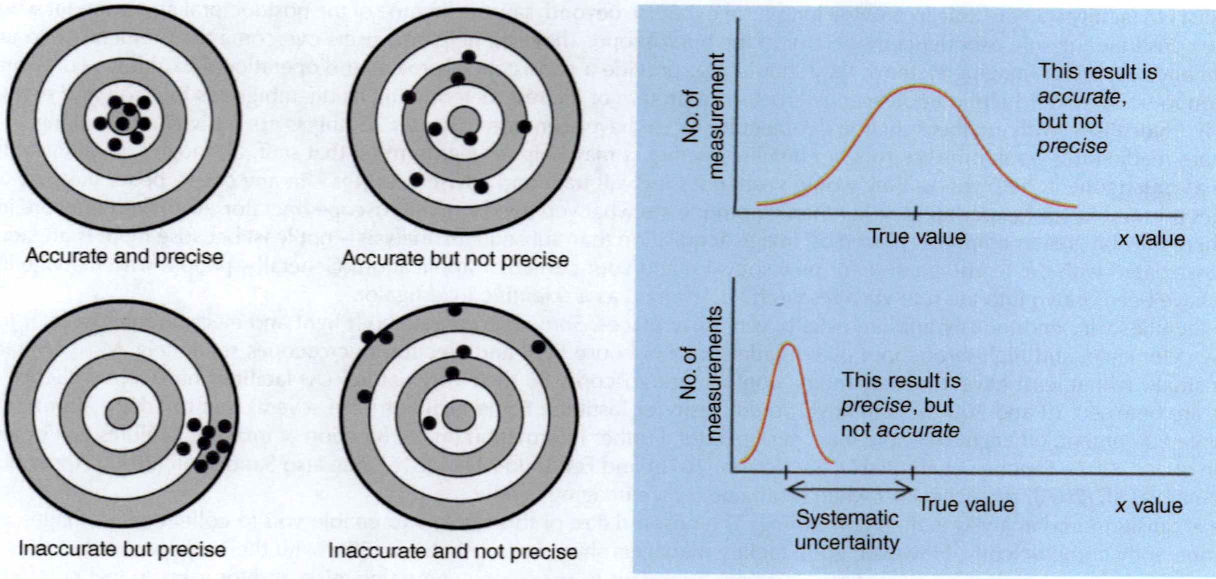

Figure 27.4 **Accuracy and precision** *Since images represent scientific data that will later be subjected to statistical analysis, it is important to collect sufficient data so that the conclusions drawn are meaningful. Both accuracy and precision need to be considered when designing an experiment and collecting data. Source: (left) Meah and Kebede-Westhead 2012. Reproduced with permission from John Wiley & Sons. (right) Adams and Allday 2000. Reproduced with permission from Oxford University Press.*

the UK as well as in *Nature Methods*[1]. Aitken et al. (2009) is helpful for those who wish to recap on basic mathematics as applied to scientific investigation. For a concise biostatistics introduction, see the *Cell* paper by Dukes and Sullivan (2007). The URL link is given at the end of the chapter. Also highly recommended is the paper on common statistical pitfalls by Sullivan et al. (2016).

Quantitative microscopy is a complete subject in its own right and will not be covered here. In volume 123 of the *Methods in Cell Biology* series (Elsevier), there are three good chapters on quantitative microscopy. Waters and Wittman (2014) introduce the essential concepts of measuring fluorescence quantitatively; Jonkman et al. (2014) apply these principles to confocal microscopy, and Verdaasdonk et al. (2014) and Coffman and Wu (2012) discuss how to determine absolute protein numbers using fluorescence microscopy.

When collecting data, it is worth keeping in mind the difference between accuracy and precision (Figure 27.4) and the need to collect sufficient images for data extraction for the statistical analysis to be relevant. The following references are essential reading for anyone serious about extracting qualitative or quantitative data from their images. These are Waters (2009), Waters and Swedlow (2007) and Kedziora et al. (2011). A worked example is given in Mutch et al. (2011) and a further example in Terasaki (2006). For information on the subject of laboratory measurements, accuracy and precision, see Meah and Kebede-Westhead (2012). A useful paper with guidance on selecting sample sizes for analysis is Silcocks (1983). Stereology is used to elicit quantitative data regarding 3D volumes from 2D sections. The sample should be randomly sampled without bias – not always easy to do if the numbers of well-stained representative structures in a field of view are limited, for example. The definitive text is RMS Handbook No. 41 *Unbiased Stereology* (1998) by Howard and Reed (2nd Edn (2004) published by Garland Science). See also Browne et al. (1995) and Weibel et al. (1966). The importance of stereology is underlined by a recent special 2015 issue of *Cell & Tissue Research* (von Bartheld and Wouters, editorial 2015; Geuna and Herrera-Rincon, 2015).

The increasing ease of acquiring digital data coupled with the ability to capture many fields of view from many samples – 'high-throughput' analysis, bioimage informatics, or bioinformatics – has become a powerful tool for biologists and geneticists. Bioinformatics serves two purposes: it allows imaging data to be stored and subsequently retrieved with molecular biological data from other sources, and it permits analysis of data holistically across previously disparate disciplines. Ultimately, the aim within biomedical research is to offer therapeutic solutions in combating disease. A themed issue on bioinformatics has been published in *Nature Methods* for July 2012; in particular refer to Eliceiri et al. (2012).

27.5 Light Microscope (LM) Imaging Facilities

Modern fluorescence microscopes are both very expensive and complex, because viewing anything but the thinnest fluorescent cells and tissues requires optical sectioning to discriminate fine detail within the sample. These instruments are not cheap, being well beyond the means of all but the very best-funded research groups. Therefore, institutions and university departments pool their resources to fund and operate these instruments within centralised facilities. Despite using core facilities to spread the financial load, many still operate at a net loss. The price of hourly charges is a regular topic on listservers and microscopy forums; these may appear excessive even when subsidised heavily by the parent institution.

Specialist LM facility staff are able to provide long-term expertise beyond, say, the lifetime of the postdoctoral appointment who might otherwise constitute the sole experienced operator of the microscope. They can help new users overcome their reluctance to use very expensive and unfamiliar imaging systems; they should also provide a consistent approach and operational workflow. For example, it is common to be asked for help in fluorescently labelling a protein of interest, determining its unambiguous location and establishing whether it 'colocalises' with another functional molecule. As we have seen from Chapter 25, this is not a trivial undertaking.

If you are reading this book prior to using an imaging facility, it may help to bear in mind that staff, although knowledgeable, may not have as much time to help you as they would wish, but they will train and advise you; it is – in any case – better that *you* acquire the images relevant to your experiment rather than communicate what you need to a microscope operator. Also, within a core imaging facility there may be greater emphasis placed on image acquisition than subsequent analysis – not least because there is an 'activation energy' associated with the learning curve for new software and your particular application. Generally, people who manage imaging facilities have been drawn into the role via a research background as a scientific investigator.

Image facilities vary enormously and are usually very busy places. Some incorporate both light and electron microscopes together with flow cytometers and high-throughput plate readers; others house light and electron microscopes separately. Most LM facilities, however small, will at least have a laser-scanning confocal microscope. By their very nature, LM facilities have many clients: typical numbers are between 50 and 300 users in any one university or institute. Some facilities have several staff to advise about, maintain and operate equipment; others have a sole manager/operator. Further information on the function of imaging facilities can be found in Anderson et al. (2007), Sánchez et al. (2011), Sanderson (2010) and Ferrando-May (2016). See also Sanderson (2017) Appendix 4.4.1 in McNamara et al. (2017) for a checklist when setting up a core imaging facility.

Image acquisition and analysis is time consuming. The raison d'être of this book is to enable you to collect good images unaided both competently and efficiently. However, good facility managers should not only be familiar with their equipment, but they should also be approachable towards their client base. It is very important to encourage communication and for users to feel comfortable in asking questions, in seeking advice and obtaining support. The entire purpose of the image facility is summarised in this quotation from *Nature*:

> 'To correctly capture images using a modern microscope, researchers must have a good grasp of optics, an awareness of the microscope's complexity...such skills can take months or even years to master [which] owing to inexperience or the rush to publish, are all too often squeezed into hours or days. Inept microscopy, and subsequent analysis, can easily generate results that are misleading or wrong and it is estimated that between 5–10% of microscopy images do not match what is reported in the text.'

Thus microscope imaging facility staff have a key role to play as specialist advisers and educators so that worthwhile and scientifically meaningful data is derived from the images acquired on core facility microscopes. They also have a role to play ensuring that sophisticated equipment is maintained correctly.

27.6 Performance Testing for Quality Control

When a microscope system is installed, the vendor company's engineers will run performance tests. Clearly, these will vary from one company to another and will need to be repeated, since instrument performance invariably changes over time. It is always sensible to arrange for service maintenance contracts, particularly where lasers and complex electronics are part of the hardware of the microscope system.

Many facility managers run their own routine quality control checks. These have two purposes: to reassure users that the microscopes are operating optimally and to help the service engineer when they must be called in, either to diagnose an emergent fault or to perform a routine service. When a result is not as anticipated by the user, it helps to be able to rule instrument variation and performance out of the equation. Regard regular testing as a service history applied by facility staff to microscopes to ensure that instruments in a centralised core facility have an audit trail to show that they are working optimally and that they can be used to collect meaningful data.

Testing the performance of fluorescence microscopes is a comprehensive subject in its own right. A very brief treatment is given here. More extensive discussions of testing instruments for quality control can be found in Chapter 36 of Pawley (2006) by Hibbs et al. and in the excellent papers and book chapters by Bob Zucker (e.g. Zucker, 2005) and Robert Stack et al. (2011) as well as the *PLoS* paper by Hng and Dormann (2013), Eason et al. (2014), Theer et al. (2014) and Deagle et al. (2017). The five tests given below will serve as a good starting point to develop your own routine to suit the microscopes that you use.

As a first indication, take benchmark shots immediately after the microscope has had a thorough overhaul and preventative service by the company service engineer. Acquire and store a 2D image of a nanolithography resolution test target[2] (Figure 27.5). If this is not possible (and for most users it will not be), acquire a typical image of your sample that displays the finest detail that you need to resolve. Also take 3D Z-stacks of sub-resolution fluorescent polystyrene microsphere beads. Do this for each objective and laser wavelength or filter set combination, and store the images in a separate folder for each objective.

For each image, note the offset settings that give zero intensity in the image, and also note the gain setting corresponding to peak intensity *just* below saturation of the detector. Also record laser current, AOTF settings, detector or filter set bandwidth, beam splitter, dwell time, pixel size and any averaging or integration time. Fortunately, much of this information is collected in the metadata header of the test image and can be reused at will, when required. It is then an easy matter to compare these benchmark images with those collected subsequently as a broad indicator of instrument performance. Zeiss manufacture a special calibration objective that, with custom software routines, makes confocal pinhole alignment and instrument calibration straightforward.

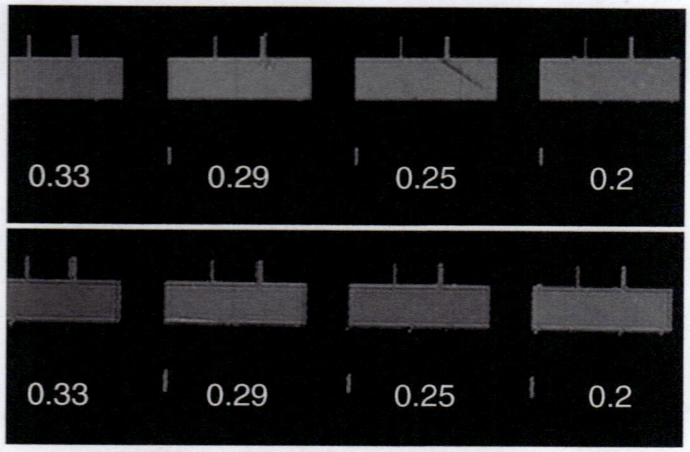

Figure 27.5 **Image of a nano-lithography resolution test plate** *A lithographically-etched resolution test plate. The results of imaging two objectives, each from one of the 'big four' microscope manufacturers, is shown. Both objectives have an initial magnification of about 60x and equivalent numerical apertures (NA = 1.4 in each case). The objective that acquired the upper image is of slightly better quality and resolving power. The figures give the spacing period of the high contrast Ronchi rulings in micrometres.*

27.6.1 Determining Lateral Resolving Power

Use sub-resolution beads, and refer to Cole et al. (2011). Precise instructions are given in this excellent paper and in Chapter 21, Box 21.1 so are not repeated here. Dry the beads down on a measured coverslip that must be 0.17 mm thick. These can be selected from a box of Zeiss narrow tolerance 18 mm × 18 mm No. 1½ coverslips (Chapter 6, Figure 6.18, Part Number 474030-9000) or the 22 mm × 22 mm size (Part Number 474030-9020). These can also be bought from Marienfeld GmbH or CellPath, UK. Use a PVA-type hardening mountant[3] (e.g. Mowiol or Prolong Gold) – to prevent Brownian motion. A Z-stack within Nyquist sampling limits must be collected. For details of how to prepare and collect a Z-stack from a sub-resolution fluorescence bead, see Chapter 21 on deconvolution (Sections 21.6 and 21.7).

The plot of intensity against focus gives a Gaussian – or bell-shaped – curve, of which the width at half-peak intensity (full width half maximum, FWHM) is the resolving power. PSF measurements will vary from one objective lens to another but should be within 30–40% of the theoretical resolution (for these values, see Table 7.5). Cole et al. (2011, 2013) report standard deviations from the mean of 4–6% for the x- and y-axis FWHM values and ≈12% for the z-axis FWHM for PSF measurements of different microspheres imaged within the same session. The presence of spherical aberration, which will decrease the lateral resolving power of the objective, will be apparent on scrolling through the Z-stack. Spherical aberration causes the PSF to be non-symmetrical.

Wollaston prisms, polarising filters and other optical components for differential interference contrast (DIC) must be removed from the light path before performing this test. Any prism shears the illuminating light into two waves, below the resolving power of the objective. This will give a double image of the sub-resolution fluorescent beads, thus distorting the PSF. Dirty objective lenses and dirt elsewhere in the optical path will similarly distort the PSF, as will misaligned confocal pinholes. These are the most common reasons for poor PSF data, which will result in gross underestimation of the resolving power of the microscope and its objectives.

27.6.2 Determining Axial Resolving Power

The axial resolving power can be determined using a sub-resolution bead Z-stack, as above, and the data processed with the MetroloJ or PSFj plug-in or similar, for Fiji or ImageJ. However, taking an intensity profile through the image of a silver 'mirror' film is the classical way of determining axial resolving power. From a theoretical point of view, imaging a PSF is preferable because the emitted fluorescence signal is incoherent. For a further explanation of why this is so, see Cole et al. (2011). The Z-stack of the PSF can be analysed with the programme PSFj (Theer et al., 2014). For further details, refer to the link to the PSFj online manual given under Internet Resources.

To test axial resolution with a mirror slide, a narrow tolerance No. 1½ coverslip (Zeiss, details above) is taken and covered with an opaque evaporated silver film. The coverslip is then mounted face down onto a slide with a hard-setting mountant such as DPX. This film will have slight imperfections in it, which can be used to focus on the front surface of the film – on the underside of the coverslip. A Z-series, well within Nyquist sampling limits for the objective numerical aperture, is collected and the intensity measured at each plane collected. Use a 488 nm laser to illuminate the bead, with a moderate zoom of 10x, or so. Lower zoom factors will decrease the accuracy; higher zooms may influence the tests due to scanning instability.

The plot of intensity against focus gives a Gaussian distribution of intensity, of which the full width half maximum (FWHM) is the axial resolution (see Figure 10.15; Section 17.5 and see also Figure 21.12). It is important that (a) the image is not oversaturated, in which case the true height of the intensity peak cannot be displayed, and (b) that the image is not undersaturated. The black-level offset must be set correctly to ensure this so that measurements are taken from a true baseline of intensity; a false offset will give erroneous results. The line section through x-z (Figure 27.6) should generate a symmetrical distribution, without any 'shoulder' to one side or the other.

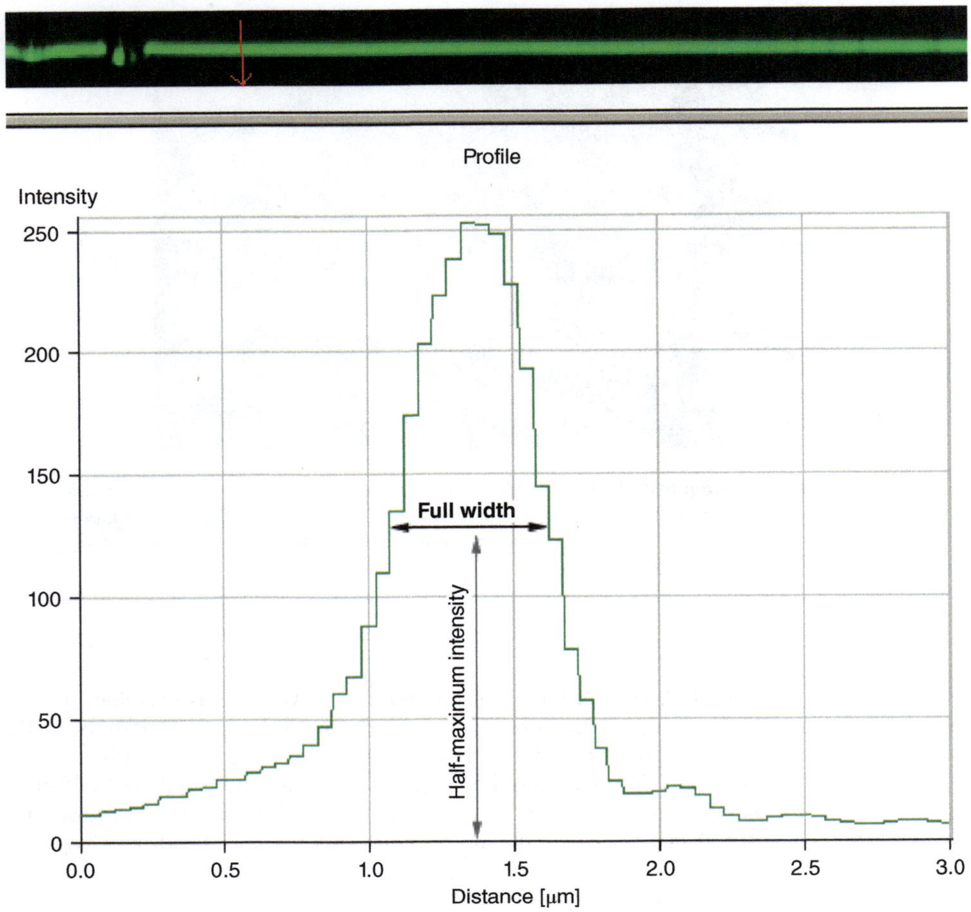

Figure 27.6 **Screengrab of mirror slide analysis** *An x-z section through a silvered mirror, shown here in profile pseudo-coloured green. The intensity plot below has been taken at the point of the red arrow, avoiding the imperfection in the mirror surface. The full width half maximum value is calculated as shown. Here the FWHM corresponds to 550 nm; this is on high side. One manufacturer cites ≤350 nm for optimum performance of a 63x/NA 1.4 plan-apochromat oil-immersion objective.*

27.6.3 Chromatic Aberration Accuracy

Although transmittivity of near U-V signals is poorer for apochromatic objectives these are, obviously, best-suited to multi-channel fluorescence studies. For testing chromatic aberration, use multi-Speck beads with different colours. Lateral aberration should be low, but there is always some lateral and longitudinal chromatic aberration in any objective.

Use multicoloured fluorescent beads (Table 27.4) to check for residual chromatic aberration and for colocalisation accuracy. Make a slide preparation of these in the same manner as given above. Compare the profiles of individual beads at different wavelengths, both in the xy-plane and the xz-plane. Even the best corrected objectives will exhibit some chromatic aberration (Figure 27.7), often as much as 0.2 μm between blue and red wavelengths at the opposite ends of the visible spectrum. Knowing that the triple colours of the beads are exactly coincident allows a correction shift to be applied to the image post-acquisition. These multicoloured bead slides are thus essential for correcting chromatic shift (see Figure 21.11) for accurate colocalisation studies. Some software (e.g. DeltaVision SoftWorx) will allow you to shift the image to overlay each channel. The shift required, as measured by the beads, can then be applied to the image. You can also use a plug-in for ImageJ to correct for chromatic aberration. See the Internet Resources at the end of the chapter.

When working with fluorescent microsphere beads, especially at moderate or high zoom factors, it is important to have the microscope mounted upon an adequate anti-vibration table or platform. Details of the various designs available are given in Chapter 4, Section 4.4. The effect of vibration in the image can be seen in Box 4.1 and in Figure 18.11.

27.6.4 Concentric Field Illumination Check

Use a fluorescent plastic slide (e.g. from Chroma, part no. 92001; see Internet Resources), which is easier and more convenient than making up a fluorescent 'sea' solution. A well-aligned objective, with an even field of illumination, does not always mean that other objectives on the nosepiece are similarly aligned. Before running this check, ensure that the pinhole on your confocal microscope is correctly aligned (test 5, below).

Table 27.4 *Multicoloured Microsphere Bead Characteristics*

The TetraSpeck (T7284, T7279) beads from Thermo Fisher Scientific have wavelengths peaks as follows:

1.	360 ex	430 em	Blue	DAPI filter	405 nm laser < 5%
2.	505 ex	515 em	Green	FITC/GFP filter	488 nm laser 60%
3.	560 ex	580 em	Orange	TRITC/Rhod filter	561 nm laser 50%
4.	660 ex	680 em	Far Red	Cy 5 filter	633 nm laser 60%

Suggested filter sets for widefield microscopy are given above, together with optimum laser lines and power intensities
TetraSpeck beads: http://www.thermofisher.com/order/catalog/product/T7279.
Sampler kit: http://www.thermofisher.com/order/catalog/product/T7284.
These two products are the bead suspensions. If you purchase the beads already made up on slides (e.g. T14792, F24633 and F36909), these are now mounted onto a No. 1½ thickness grade coverslip.
Alternative Suppliers:
Bang Labs: http://www.bangslabs.com/products/fluorescent-microspheres.
Polysciences: http://www.polysciences.com/default/catalog-products/microspheres-particles.
Spherotech: http://spherotech.com/multi_flu.htm.
Information: http://spherotech.com/fluorescent%20particles%20catalog%202010-2011%20rev%20a.pdf.
Mounting is critical to the successful use of these beads. The beads must be immobile, for Brownian motion, even in a viscous fluid renders them un-useable, and accurate imaging impossible.
They should lie in a single plane, but not be allowed to dry too quickly onto a coverslip surface, otherwise they will shrink. On a coverslip cleaned with abrasive domestic cream cleaner, a very dilute suspension of beads in distilled water is applied and allowed to dry overnight. The beads become firmly attached, and remain when mounted in Gelvatol medium. Gelvatol is a PVA formulation which sets hard. This is the recipe for Gelvatol; for the very similar Mowiol, see Protocol 28.4 in Section 28.5.
Add 21g Poly Vinyl Alcohol (Sigma P8136) to 42 mL glycerol. Add 52 mL distiiled water, to which a crystal of sodium azide has been added as a preservative. Add 106 mL of 0.2 Molar TRIS buffer pH 8.5. Stir, and warm gently, if necessary to dissolve. Centrifuge at 5000 g for 15 minutes to clarify and remove undissolved components. Aliquot into Eppendorf tubes at store at 4°C. If 2% w/v DABCO (diazo-bis-cyclo-actane) is added as an antifade compound, this results in slower setting & hardening.

Reference

Cole, R.W., Jinadasa, T. and Brown, C.M. (2011). 'Measuring and interpreting point spread functions to determine confocal microscope resolution and ensure quality control'. *Nature Protocols* 6/12: 1929–1941.

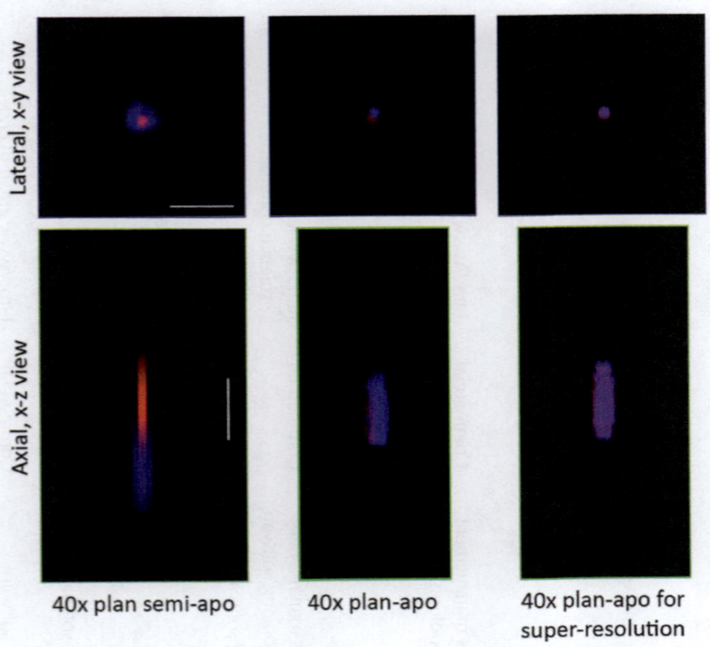

Figure 27.7 **The effect of chromatic aberration** *Three 40x objectives from the same manufacturer show very different degrees of axial chromatic aberration along the optical axis. As intended, the fluorite plan semi-apochromat objective is less well corrected, whilst the two plan-apochromat objectives are highly corrected. Those objectives which have high Strehl ratios (>90%) are specially selected for super-resolution microscopy and work with the Airyscan.*

The top three panels show a single image taken of the focal plane through the centre of the bead. The lower three panels show the corresponding 'Z-stack' series of images of TetraSpeck fluorescent microspheres are taken with both red and blue wavelengths at each focal plane. The scale bars represent 5 μm.

There is virtually no longitudinal chromatic aberration (chromatic difference of magnification), even in the objective showing severe axial chromatic aberration, because the diameter and length of the image of the fluorescent microsphere 'bead' is virtually equal between each wavelength. For the apochromat objective shown in the middle, the distance between centres of the blue and red profiles in both the x-y and x-z panels is 114 nm. This is the image shift that must be applied post-acquisition for any quantitative microscopy or colocalisation studies.

Table 27.5 *Freeware and Commercial Facility-Management Software*

Software	Function	Type	Web page	Manual/Support	Trial demo	Contact
Book-it Lab	Facilities management and booking software	Commercial	http://bookit-lab.com	Manual – no Support – unknown	via web page	contact@prog4biz.com
Bumblebee	Facilities management and booking software	Freeware	http://bumblebeeman.sourceforge.net	Manual – yes http://bumblebeeman.sourceforge.net/documentation/ Support – no	N/A	Limited
Brownbear Software	Booking software	Commercial	http://www.brownbearsw.com	Manual – yes https://www.brownbearsw.com/calcium/docs/ and https://www.brownbearsw.com/ical/icalpage.html Support – unknown	via web page	http://www.brownbearsw.com/contactus.html
Calpendo	Facilities management and booking software	Commercial	http://www.calpendo.com	Manual – yes http://www.exprodo.com/docs/ and videos: http://www.exprodo.com/video-tutorials/ Support – unknown	via web page	info@exprodo.com
Core Vision in Science	Facilities management and booking software	Commercial	http://www.corevision-science.com/marie	Manual – no Support – yes	via web page	marie@corevision-science.com
Faces	Booking software	Freeware	http://faces.ccrc.uga.edu	Manual – yes http://homer.chem.sc.edu/faces.pdf and https://faces.ccrc.uga.edu/ccrcfaces/faq0.php Support – no	N/A	Limited
Google Calendar	Booking software	Freeware	http://www.google.com/calendar	Manual – no Support – no	N/A	None
Great Hill Software	Booking software	Commercial	http://www.greathill.com	Manual – yes http://www.greathill.com/support_documentation.php Support – unknown	via web page	info@greathill.com
Idea Elan	Facilities management and booking software	Commercial	http://www.ideaelan.com	Manual – no Support – yes	via web page	support@ideaelan.com

(continued)

Table 27.5 (*Continued*)

Software	Function	Type	Web page	Manual/Support	Trial demo	Contact
iLab	Facilities management and booking software	Commercial	http://www.ilabsolutions.com	Manual – no Support – unknown	via web page	info@ilabsolutions.com
Kerio Connect		Commercial	http://www.kerio.com	Manual – yes http://www.kerio.com/support Support – yes	via web page	
Quartzy	Facilities management and booking software	Freeware	http://www.quartzy.com	Manual – yes, videos: https://support.quartzy.com/hc/en-us/articles/202777638 Support – yes https://support.quartzy.com/hc/en-us	Via webinar https://www.quartzy.com/webinar	support@quartzy.com
Scionics	Facilities management and booking software	Commercial	http://www.scionics.com	Manual – no Support – no	FAMS 1.3 no longer supported	products@scionics.com PyRAT now only available
Stratocore/PPMS	Facilities management and booking software	Commercial	http://stratocore.com	Manual – no Support – unknown	via web page	info@stratocore.com

The booking systems listed are those regularly discussed at the UK national microscopy facility managers' meetings. The commercial facility management software options usually provide more than a booking system, offering integrated billing and the ability to generate usage statistics and reports. As such, these are more expensive solutions but are designed to save administrative time. Some suggest that up to a fifth of a manager's administrative time can be saved with a well-designed facility management system. The small print: the information in this table is given in good faith for information only. None of the products is endorsed by the author or by John Wiley & Sons Inc. The commercial software may be trialled before purchase.

Reference

List of things to think about on setting up an Imaging Facility. URL: http://www.microscopist.co.uk/imaging-facilities-useful-links/.

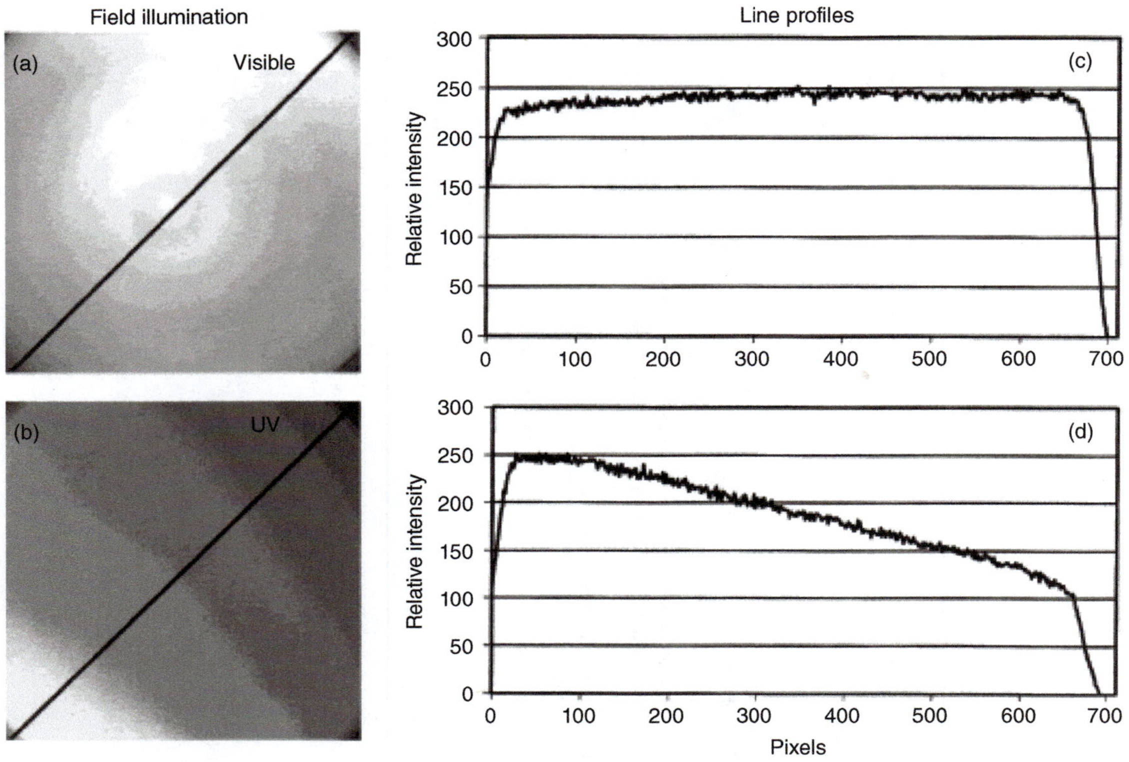

Figure 27.8 **Screengrab of field illumination intensity analysis** *An intensity plot taken across the diagonal of the field of view. The full bit-depth of greyscale values has been sub-divided with software into 'contour' isolines. The line profile shows where (e.g. UV) illumination intensity falls to unacceptable levels. In this case, the 'invisible' fibre optic cable carrying the UV laser signal will require realignment. Source: Zucker 2006. Reproduced with permission from John Wiley & Sons.*

Take a single image, draw a diagonal region of interest (ROI) line from one corner to the opposite corner (Figure 27.8) and plot the intensity values along this ROI. The values at each corner should, ideally, not vary more than 10% from the central intensity value. A complete protocol for analysing the intensity data using ImageJ can be found in Model and Blank (2008). Two other papers are Sawyer et al. (2017) and the very practical paper by Brown et al. (2015) again with Richard Cole as author. A protocol for calibrating image intensity using a fluorescent 'sea' of flourescein mixed into PVA as a test slide is given in Kedziora et al. (2011), and a similar algorithm has been published by Kask et al. (2016). PSFj can also be used to analyse a microscope's performance across a large field of view.

Dirt, for example, on the surface of the front lens of the objective can disrupt the field illumination result. The 405 UV diode laser tends to have an elliptical profile and is prone to misalignment. This may easily be detected, since equally intensely-stained nuclei will show a variation in image intensity across the field of view, particularly as one chosen nucleus is traversed from one side, or corner, to another.

It is important to measure the field illumination at a specific depth, since the intensity distribution will change as one goes deeper into the slide. The surface of the plastic slide is the brightest point, emitting the most intense fluorescence. Therefore mount a measured coverslip onto the surface of the fluorescent slide, and image the surface of the plastic slide through this. Mounting a coverslip onto the slide will also serve to protect the surface of the slide from scratches.

For judging the extent of plan correction of field curvature, Z-stacks of beads from the centre of the field of view and the periphery should be acquired. The focus in z can be determined and compared, as can the exact position of the focal plane laterally in x-y, to determine any focal shift due to lack of plan correction and field curvature.

27.6.5 Pinhole Check

This test is only applicable to the point-scanning laser confocal microscope, but a well-aligned pinhole is crucial to ensure optimum image registration. A pinhole that is out of alignment will have an adverse effect upon the illumination intensity across the field of view. Most manufacturers will have an automatic software routine to align the pinhole and associated collimators of the fibre optics for both the visible and UV lasers. It is not recommended that the user aligns the collimators – these are best left to a service engineer.

Automatic software routines are both accurate and reproducible, but pinhole alignment can be performed manually, as follows. With a 10x objective in place, a silvered mirror slide is imaged in reflection mode with the 80/20 primary beam splitter. The intensity is adjusted so that the image is just saturated, and the pinhole controls are adjusted to give optimum brightness in the centre of the field of view.

Table 27.6 Open-Source Software and Freeware

Software	Function	Web page	Manual	Reference
Bioformats	Data reader	https://docs.openmicroscopy.org/bio-formats/5.8.2/about/index.html	No	
Micro-Manager μ–Manager	Microscope control and acquisition	https://micro-manager.org/ Device support: https://micro-manager.org/wiki/Device%20Support	Yes https://micro-manager.org/wiki/Overview%20of%20the%20documentation http://valelab.ucsf.edu/publications/2010EdelsteinCurrProt.pdf	Edelstein, A.D. et al. (2014). *Jour. Biol. Methods 1*/2: e10. Edelstein, A.D. et al. (2010). *Curr. Protoc. Mol. Biol*: Unit 14.20.
Micropilot	Microscope control and acquisition	https://www.embl.de/services/core_facilities/almf/services/high-throughput-microscopy/microscope-automation/micropilot/index.html	No	Conrad, C. et al. (2011). *Nature Methods 8*/3: 246–249.
SciScan	Two-photon microscope control and acquisition	http://sciscan.scientifica.uk.com/download	Yes http://sciscan.scientifica.uk.com/knowledge-base/	Labview: Ogren, P.J. (1996). *Jour of Chem Ed 73*/12: 1115–1116.
VisBio	2D, 3D and 4D data visualisation	http://loci.wisc.edu/software/visbio	No	Rueden, C. et al. (2004). *Traffic 5*/6: 411–417.
Big Data Viewer	Large data set visualisation and image analysis/processing	http://imagej.net/BigDataViewer	Yes https://arxiv.org/pdf/1412.0488.pdf	Pietzsch, T. et al. (2015). *Nature Methods 12*/6: 481–483.
3D Slicer	3D data visualisation	http://www.slicer.org/	Yes https://www.slicer.org/wiki/Documentation/4.8/Training	Farhat, N. (2014). *Tutorial: Preparing Data for 3-D Printing using 3D Slicer.*
FluoRender	3D data visualisation	http://www.sci.utah.edu/software/fluorender.html	Yes http://www.sci.utah.edu/software/fluorender.html	Wan, Y. et al. (2012). IEEE Pac. Vis. Symp. 2012: 201–208.
cellPACK	3D data visualisation	http://www.autopack.org/	No	Johnson, G.T. et al. (2015). *Nature Methods 12*/1: 85–91.
COSMOS	Deconvolved 3D data visualisation	http://cirl.memphis.edu/cosmos.php (replaces XCOSM)	Yes http://www.memphis.edu/cirl/	Preza, C. and Conchello, J.-A. (2004). *Jour. Opt. Soc. Am. A 21*/9: 1593–1601.
ImageJ	Data visualisation and image analysis	http://rsb.info.nih.gov/ij/ http://en.wikipedia.org/wiki/ImageJ	Yes https://imagej.nih.gov/ij/docs/index.html	Schneider, C.A. et al. (2012). *Nature Methods 9*/7: 671–675.
Fiji	Data visualisation and image analysis	http://fiji.sc/Fiji	Yes http://imagej.net/Introduction	Schindelin, J. et al. (2012). *Nature Methods 9*/7: 676–682.
Icy	Image analysis, tracking and database	http://icy.bioimageanalysis.org/	Yes http://icy.bioimageanalysis.org/doc/icy-introduction.pdf	de Chaumont, F. et al. (2012). *Nature Methods 9*/7: 690–696.

(continued)

Table 27.6 (*Continued*)

Software	Function	Web page	Manual	Reference
Vaa3D/V3D	Data visualisation and image analysis	www.vaa3d.org/	Yes http://home.penglab.com/papersall/docpdf/2014_NPROT_vaa3d.pdf	Peng, H. et al. (2014). *Nature Protocols 9*/1: 193–208. Peng, H. et al. (2010). *Nature Biotechnol. 28*/4: 348–353.
Blender	3D data creation, visualisation and rendering	https://www.blender.org/	Yes https://www.blender.org/support/	Asadulina, A. et al. (2015). *BMC Bioinformatics 16*: 229.
Clear Volume	Data visualisation, image analysis and volume rendering	http://imagej.net/ClearVolume and https://clearvolume.github.io/	No	Royer, L.A. et al. (2015). *Nature Methods 12*/6: 480–481.
VolumeJ	Data visualisation and volume rendering	http://bij.isi.uu.nl/	Yes http://bij.isi.uu.nl/vj_manual.htm	Abràmoff, M.D. et al. (2004). *Biophotonics International 11*/7: 36–42.
Voxx	Data visualisation and volume rendering	http://www.indiana.edu/~voxx/	No	Clendenon, J.L. et al. (2002). *Am. Jour. Physiol. Cell Physiol. 282*/1: C213–C218.
Cell Explorer	3D image analysis	http://goo.gl/LMgpfj	No	Liu, X. et al. (2009). *Cell 139*/3: 623–633.
Go Figure	Data visualisation, image analysis and segmentation	http://gofigure2.org/	Yes http://gofigure2.org/index.php/videos.html	Megason, S.G. and Fraser, S.G. (2003). *Mech. Development 120*/11: 1407–1420.
GraspJ	Super-resolution localisation data analysis	https://code.google.com/archive/p/graspj/downloads	Yes https://static-content.springer.com/esm/art%3A10.1186%2F2192-2853-1-11/MediaObjects/13689_2012_9_MOESM1_ESM.pdf	Brede, N. and Lakadamyali, M. (2012). *Optical Nanoscopy 1*: 11.
QuickPalm	Super-resolution localisation data analysis	http://imagej.net/QuickPALM	Yes https://code.google.com/archive/p/quickpalm/wikis	Henriques, R. et al. (2010). *Nature Methods 7*/5: 339–340.
Palm3d	Super-resolution localisation data analysis	https://code.google.com/archive/p/palm3d/	Yes https://code.google.com/archive/p/palm3d/wikis/Tutorial.wiki	York, A.G. et al. (2011). *Nature Methods 8*/4: 327–333.
Cell Profiler	Image analysis	http://cellprofiler.org/	Yes http://cellprofiler.org/manuals/ http://cellprofiler.org/tutorials/	Kamentsky, L. et al. (2011). *Bioinformatics 27*/8: 1179–1180.
Cell Profiler Tracer	Cell tracking	http://cellprofiler.org/tracer	http://cellprofiler.org/tracer/#tutorial	Bray and Carpenter. (2015). *BMC Bioinformatics 16*: 368.

(*continued*)

Table 27.6 (Continued)

Software	Function	Web page	Manual	Reference
Endrov	Image analysis and data processing	www.endrov.net/	Yes http://www.endrov.net/wiki/index.php?title=Users_Guide	Henriksson, J. et al. (2013). *Nature Methods 10*/6: 454–456.
BioimageXD	Image analysis and data processing	http://bioimagexd.net/	Yes http://priede.bf.lu.lv/ftp/pub/TIS/atteelu_analiize/BioImageXD/BioImageXD_GettingStarted.pdf	Kankaanpää, P. et al. (2012). *Nature Methods 9*/7: 683–689.
Cell-ID	Image quantification and segmentation	http://www.embnet.qb.fcen.uba.ar/embnet/Cell-ID-Rcell/Cell-ID-main.htm	Yes http://www.embnet.qb.fcen.uba.ar/embnet/Cell-ID-Rcell/Cell-ID-Segmentation-Algorithm-and-Quantification.htm	Bush, A. et al. (2012). *Curr. Protoc. Mol. Biol.* Unit 14.18.
Cell Organiser	Image classification and segmentation	http://cellorganizer.org/	No	Buck, T.E. et al. (2012). *Bioessays 34*/9: 791–799.
Daime	Image classification and segmentation	http://www.microbial-ecology.net/daime	No	Daims, H. et al. (2006). *Environ. Microbiol. 8*/2: 200–213.
ilastik	Image classification and segmentation	http://ilastik.org/	Yes http://ilastik.org/documentation/index.html	Sommer, C. et al. (2011). *Eighth IEEE (ISBI) Proceedings*: 230–233.
Itk-SNAP	Image classification and segmentation	www.itksnap.org/	Yes http://www.itksnap.org/pmwiki/pmwiki.php?n=Documentation.SNAP3	Yushkevich, P.A. et al. (2006). *Neuroimage 31*/3: 1116–1128.
Cell Tracker	Cell tracking	https://warwick.ac.uk/fac/sci/dcs/people/till_bretschneider/celltracker	Yes https://warwick.ac.uk/fac/sci/dcs/people/till_bretschneider/celltracker/celltracker-manual.pdf	Du, C.-J. et al. (2010). *Cytometry Part A, 77A*: 1137–1147.
ParticleStats	Cell and particle tracking	http://www.ParticleStats.com	Yes http://www.darogan.co.uk/ParticleStats/ParticleStats_UserGuide.pdf	Hamilton, R.S., et al. (2010). *Nucleic Acid Res. 38* (suppl 2): W641–W646.
iTrack4U	Cell and particle tracking	https://sites.google.com/site/itrack4usoftware/home	No	Cordelières, F.P. et al. (2013). *PLoS One 8*/11: e81266.
μ–track 2.0	Cell and particle tracking	https://www.utsouthwestern.edu/labs/danuser/	Yes https://downloads.openmicroscopy.org/u-track/2.1.1/artifacts/u-track-2.1.1.pdf	Jaqaman, K. et al. (2008). *Nature Methods 5*/8: 695–702.
Traxtile	Cell and particle tracking	https://github.com/braunb/traxtile-public	Yes http://goo.gl/StwsDb	Braun, B.S. (2015). *Biotechniques 59*/2: 82–86.

(continued)

Table 27.6 (*Continued*)

Software	Function	Web page	Manual	Reference
TrackMate	Cell and particle tracking	https://imagej.net/TrackMate	Yes http://imagej.net/Getting_started_with_TrackMate	Tinevez, J.-Y. et al. (2017). *Methods 115*: 80–90.
Bisque	Image analysis and database	http://bisque.ece.ucsb.edu/	No	Kvilekval, K. et al. (2010). *Bioinformatics 26*/4: 544–552.
Omero	Image analysis and database	http://www.openmicroscopy.org/	Yes https://docs.openmicroscopy.org/omero/5.4.6/	Allan, C. et al. (2012). *Nature Methods 9*/3: 245–253; Cho, B. H., et al. (2012). *Nature Methods 9*/7: 633–634.
Zegami	Image classification and database	http://zegami.com/	No	Taylor, S. and Kemp, F. (2014). *Isis Innovation* 9th Feb 2016.

Overview papers

Freeman, J. (2015). 'Open source tools for large-scale neuroscience'. *Current Opinion in Neurology 32*: 156–163.
Barry, D.J. et al. (2015). 'Open source software for quantification of cell migration, protrusions, and fluorescence intensities'. *The Journal of Cell Biology 209*/1: 163–180.
Bioimaging software tools. URL: https://bit.ly/2PwYTGb.
Hamilton, N.A. (2012). 'Open source tools for fluorescent imaging'. *Methods Enzymology 504*: 393–417.
Walter, T. et al. (2010). 'Visualization of image data from cells to organisms'. *Nature Methods 7*/3s: S26–S41.

References

Amat, F. et al. (2014). 'Fast, accurate reconstruction of cell lineages from large-scale fluorescence microscopy data'. *Nature Methods 11*/9: 951–958.
Antony, P.M.A. et al. (2013). 'Light microscopy applications in systems biology: opportunities and challenges'. *Cell Communication & Signalling 11*/1: 24. doi: 10.1186/1478-811X-11-24. http://www.biosignaling.com/content/11/1/24. (Tables 1 and 2 in this paper give comprehensive summaries of commercial and open-source software, respectively.)
Beneš, M. and Zitová, B. (2015). 'Performance evaluation of image segmentation algorithms on microscopic image data'. *Journal of Microscopy 257*/1: 65–85.
Carpenter, A.E., Kamentsky, L. and Eliceiri, K.W. (2012). 'A call for bioimaging software useability'. *Nature Methods 9*/7: 666–670.
Chenouard, N. et al. (2014). 'Objective comparison of particle tracking methods'. *Nature Methods 11*/3: 281–289.
Davis, M.A. et al. (2015). 'Semi-automated object tracking methods in biological imaging'. *Current Protocols in Cytometry 71*: 12.38.1–12.38.21.
Eliceiri, K.W. et al. (2012). 'Biological imaging software tools'. *Nature Methods 9*/7: 697–710.
Free deconvolution software: http://www.deconvolve.net/DNLinks.html.
Gustafsdottir, S.M. et al. (2013). 'Multiplex cytological profiling assay to measure diverse cellular states'. *PLoS One 8*/12: e80999.
Huth, J. et al. (2010). 'Significantly improved precision of cell migration analysis in time-lapse video microscopy through use of a fully automated tracking system'. *BMC Cell Biology 11*: 24.
Jaqaman, K. and Danuser, G. (2009). 'Computational image analysis of cellular dynamics: a case based on particle tracking'. *Cold Spring Harbor Protocols 4*/12. doi: 10.1101/pdb.top65.
Leicester College of Medicine, Biological Sciences and Psychology software list: https://www2.le.ac.uk/colleges/medbiopsych/facilities-and-services/cbs/AIF/software-1/other-software.
Ljosa, V. et al. (2009). 'Introduction to the Quantitative Analysis of Two-Dimensional Fluorescence Microscopy Images for Cell-Based Screening'. *PLOS Computational Biology 5*/12: e1000603.
Long, F. et al. (2012). 'Visualization and analysis of 3D microscopic images'. *PLOS Computational Biology 8*/6: e1002519.
Lynch, A.E. et al. (2014). 'Low-cost motility tracking system (LOCOMOTIS) for time-lapse microscopy applications and cell visualisation'. *PLoS One 9*/8: e103547.
Meijering, E. et al. (2012). 'Methods for cell and particle tracking Methods in Enzymology'. In: Conn, P.M. (ed.), *Imaging and spectroscopic analysis of living cells*, vol. 504, Chapter 9. San Diego: Academic Press, pp. 183–200 (refer to Table 1).
Schroeder, T. (2011). 'Long-term single-cell imaging of mammalian stem cells'. *Nature Methods 8*/4 Suppl: S30–S35.
Tracking plugins: http://www.imagescience.org/meijering/software/mtrackj/ and http://fiji.sc/Particle_Tracker.
Visualisation tools: http://bivi.co/visualisations.
Volume rendering plugins: http://bij.isi.uu.nl/.

27.7 Free and Open-Source Imaging Software

ImageJ and Fiji have already been referred to throughout this book as cross-platform (i.e. can be used on both Windows PCs and Apple Mac computers) freeware. The history of the development of ImageJ (Schneider et al., 2012) and an explanation of Fiji (Schindelin et al., 2012) has been published in a themed issue of *Nature Methods* (volume 9, No. 7; July 2012). Icy is a similar programme (de Chaumont et al., 2012) albeit with a smaller community. Of these three free, open-source image display and analysis packages, ImageJ is the most widely used. There are literally hundreds of plug-ins (a small stand-alone software component that extends specific functionality of the ImageJ programme) and macros (a simple programme that automates a set of instructions) available for ImageJ. The developers of Fiji based Fiji on ImageJ source code but bundled a set of approved plug-ins specific for microscopy. Fiji is continually updated. Icy was developed so that research could be easily reproduced by others. Table 27.6 summarises the major freeware software packages available. Several open-source software projects have been used by major research groups to build complete microscope systems. The Open-SPIM project (Pitrone et al., 2013, and Gualda et al., 2014) are further referred to in Chapter 19, 'Light-Sheet Microscopy', also known as selective plane illumination microscopy, or SPIM. A similar project (Wong et al., 2013) gives online details of how to construct an optical projection tomography (OPT) microscope and a low-cost fluorescence imaging system (Nuñez et al., 2017). An open-source project for high-content imaging is that developed by the Vojnovic group (Barber et al., 2013) and for an inverted fluorescence microscope (Vojnovic et al., 2011) at the Gray Institute, Oxford. Whilst these open-source projects can be straightforward, they usually require the support of both very good IT/programming departments and scientific engineering workshops.

27.8 Chapter Summary

At the end of this chapter, you should understand whether to use widefield microscopy for imaging your sample or to choose an optical sectioning method. Suggested guidance is given to plan a successful imaging experiment. You should have some understanding of the role of light microscopy imaging facilities and of the importance of instrument performance checks to provide a quality control audit trail. The next two chapters provide information on biological and materials specimen preparation, whilst the last two chapters of this book discuss the theoretical and practical aspects of image recording.

1. Which imaging method is chosen primarily depends upon the thickness of your sample.
2. Each microscope design is optimised for a specific application; see Table 27.1.
3. The correct objective should be selected to match your imaging application: use oil-immersion objectives for samples mounted on the underside of the coverslip and water-immersion or multi-immersion objectives for aqueous and/or living samples.
4. It is worth spending time planning your imaging experiment and data analysis; both are time consuming.
5. The role of light microscopy facilities is explained.
6. Five simple performance checks are given to maintain fluorescence microscopes in optimum working order.
7. A guide to freeware and open-source software is given.

Key Reading

North, A. (2006). 'Seeing is believing? A beginners' guide to practical pitfalls in image acquisition'. *Journal of Cell Biology 172*/1: 9–18.
- A good, concise review on collecting good and scientifically rigorous images

Ferrando-May, E. (2016). 'Advanced light microscopy core facilities: Balancing service, science and career'. *Microscopy Research and Technique 79*/6: 463–479.
- How to set up and run a core imaging facility from some of the best in the business

References

Adams, S. and Allday J. (2000). *Advanced Physics*. Oxford University Press.

Aitken, M.R.F., Broadhurst, B., and Hladky, S. (2009). *Mathematics for Biological Scientists*. New York: Garland Science.

Anderson, K.I., Sanderson, J., and Peychl, J. (2007). 'Design and function of a light microscopy facility'. In: Shorte, S.L. and Frischknecht, F. (eds.), *Imaging Cellular and Molecular Biological Functions*, Chapter 4. Berlin: Springer, pp. 93–113.

Barber, P.R. et al. (2013). 'The Gray Institute "open" high-content, fluorescence lifetime microscopes'. *Journal of Microscopy 251*/2: 154–167.

Brown, C.M. et al. (2015). 'A quantitative measure of field illumination'. *Journal of Biomolecular Techniques 26*/2: 37–44. doi: 10.7171/jbt.15-2602-001.

Browne, M.A. et al. (1995). 'Principles of Stereology'. In: Wootton, R., Springall, D.R., and Polak, J.M. (eds.), *Image analysis in histology*, Chapter 7. Cambridge: Cambridge University Press, pp. 96–120.

de Chaumont, F. et al. (2012). 'Icy: an open bioimage informatics platform for extended reproducible research'. *Nature Methods 9*/7: 690–696.

Coffman, V.C. and Wu, J.-Q. (2012). 'Counting protein molecules using quantitative fluorescence microscopy'. *Trends in Biochemical Sciences 37*/11: 499–506.

Cole, R.W., Jinadasa, T., and Brown, C.M. (2011). 'Measuring and interpreting point spread functions to determine confocal microscope resolution and ensure quality control'. *Nature Protocols 6*/12: 1929–1941. Supplementary methods: https://media.nature.com/full/nature-assets/nprot/journal/v6/n12/extref/nprot.2011.407-S1.pdf.

Cole, R.W. et al. (2013). 'International Test Results for Objective Lens Quality, Resolution, Spectral Accuracy and Spectral Separation for Confocal Laser Scanning Microscopes (CLSM)'. *Microscopy and Microanalysis 19*/6: 1653–1668.

Deagle, R.C. et al. (2017). 'Reproducibility in light microscopy: Maintenance, standards and SOPs'. *The International Journal of Biochemistry & Cell Biology 89*: 120–124.

Dukes, K.A. and Sullivan, L.M. (2007). 'A review of biostatistics'. In: Zuk, D. (ed.), *Evaluating Techniques in Biochemical Research*. Cambridge, MA: Cell Press, pp. 50–56. For a colour version: http://mcb.med.stu.edu.cn/%E7%94%B5%E5%AD%90%E4%B9%A6%E7%B1%8D/Evaluating%20Techniques%20in%20Biomedical%20Research.pdf (scroll down).

Dytham, C. (2010). *Choosing and Using Statistics: a Biologist's Guide*, 3rd ed. Chichester: Wiley Blackwell.

Eason, B. et al. (2014). 'Microscope maintenance and quality control: a practical guide'. In: Méndez-Vilas, A. (ed.), *Microscopy: Advances in Scientific Research and Education*, vol. 2. Extremadura: Formatex Research Center, pp. 713–724.

Editorial (2007). 'Under the microscope – the use of "black box" techniques carries risks'. *Nature 447*/10: 116.

Eliceiri, K.W. et al. (2012). 'Biological imaging software tools'. *Nature Methods 9*/7: 697–710.

Geuna, S. and Herrera-Rincon, C. (2015). 'Update on stereology for light microscopy'. *Cell and Tissue Research 360*/1: 5–12.

Gualda, E. et al. (2014). 'Going "open" with mesoscopy: a new dimension on multi-view imaging'. *Protoplasma 251*/2: 363–372.

Hibbs, A.R., MacDonald, G., and Garsha, K. (2006). 'Practical confocal microscopy'. In: Pawley, J.B. (ed.), *Handbook of Biological Confocal Microscopy*, 3rd ed, Chapter 36. New York: Springer, pp. 650–671.

Hng, K. and Dormann, D. (2013). 'Confocal check – a software tool for the automated monitoring of confocal microscope performance'. *PLoS One 8*/11: e79879. doi: 10.1371/journal.pone.0079879.

Howard, C.V. and Reed, M.G. (1998). *Unbiased Stereology: three-dimensional measurement in microscopy*. RMS Handbook No. 41 Bios Scientific Publishers.

Jonkman, J. et al. (2014). 'Quantitative confocal microscopy: beyond a pretty picture'. In: Waters, J.C. and Wittmann, T. (eds.), *Quantitative Imaging in Cell Biology*. Methods in Cell Biology, vol. 123, Chapter 7. Elsevier Inc., pp. 113–134.

Kask, P. et al. (2016). 'Flat field correction for high-throughput imaging of fluorescent samples'. *Journal of Microscopy 263*/3: 328–340.

Kedziora, J. et al. (2011). 'Method of calibration of a fluorescence microscope for quantitative studies'. *Journal of Microscopy 244*/1: 101–111.

McNamara, G. et al. (2017). Sanderson, J. (2017) Appendix 4.4.1, in: Microscopy and Image Analysis. *Current Protocols in Human Genetics* 2017 Jul 11;*94*: 4.4.1–4.4.89. doi: 10.1002/cphg.42.

Meah, M. and Kebede-Westhead, E. (2012). 'Measurements and Calculations'. In: *Essential Laboratory Skills for Biosciences*, Chapter 1. Chichester: Wiley, pp. 1–22 (see also: 'Basic concepts of preparing solutions', by Flinn Scientific. https://bit.ly/2wCnfYd.

Model, M.A. and Blank, J.L. (2008). 'Concentrated dyes as a source of two-dimensional fluorescent field for characterization of a confocal microscope'. *Journal of Microscopy 229*/1: 12–16.

Murray, J.M. (2010). 'Confocal Microscopy, Deconvolution, and Structured Illumination Methods'. In: Goldman, R.D., Swedlow, J.R., and Spector, D.L. (eds.), Live Cell Imaging: A Laboratory Manual, 2nd ed, Chapter 8. New York: Cold Spring Harbor Laboratory Press, pp. 139–181.

- The same information in Murray, J.M. (2010) has been reprinted from the 1st Ed. of: Murray, J.M. (2006). 'Confocal Microscopy, Deconvolution, and Structured Illumination Methods'. In: Spector, D.L. and Goldman, R.D. (eds.), *Basic Methods in Microscopy*, Chapter 2. New York: Cold Spring Harbor Laboratory Press, pp. 43–81.

Murray, J.M., Appleton, P.L., Swedlow, J.R., and Waters, J.C. (2007). 'Evaluating performance in three-dimensional fluorescence microscopy'. *Journal of Microscopy 228*/3: 390–405.

Mutch, S.A. et al. (2011). 'Determining the number of specific proteins in cellular compartments by quantitative microscopy'. *Nature Protocols 6*/12: 1953–1968.

Nuñez, I. et al. (2017). 'Low cost and open source multi-fluorescence imaging system for teaching and research in biology and bioengineering'. *PLoS One 12*/11: e0187163. doi: 10.1371/journal.pone.0187163.

Pearson, H. (2007). 'The Good, the Bad and the Ugly'. *Nature 447*/10: 138–140.

Pitrone, P.G. et al. (2013). 'OpenSPIM: an open-access light-sheet microscopy platform'. *Nature Methods 10*/7: 598–599.

Sánchez, C. et al. (2011). 'Setting up and running an advanced light microscopy and imaging facility'. *Current Protocols in Cytometry* 12.22.1–12.22.21. July 2011.

Sanderson, J. (2010). 'Light Microscopy Imaging Facilities'. In: *Encyclopedia of Life Sciences* (ELS). Chichester: John Wiley & Sons.

Sawyer, W.T. et al. (2017). 'Evaluation of illumination system uniformity for wide-field biomedical hyperspectral imaging'. *Journal of Optics 19*/4: 0405301. http://iopscience.iop.org/article/10.1088/2040-8986/aa6176/meta.

Schindelin, J. et al. (2012). 'Fiji: an open-source platform for biological-image analysis'. *Nature Methods 9*/7: 676–682.

Schneider, C.A. et al. (2012). NIH Image to ImageJ: 25 years of image analysis. *Nature Methods 9*/7:671–675.

Silcocks, P.B.S. (1983). Measuring repeatability and validity of histological diagnosis – a brief review with some practical examples. *Journal of Clinical Pathology 36*/11: 1269–1275.

Stack, R.F. et al. (2011). 'Quality Assurance Testing for Modern Optical Imaging Systems'. *Microscopy and Microanalysis 17*/4: 598–606.

Sullivan, L.M. et al. (2016). 'Common statistical pitfalls in basic science research'. *Journal of the American Heart Association 5*: e004142.

Terasaki, M. (2006). 'Fluorescence quantitation in thick specimens, with an application to cyclin B-GFP expression in starfish oocytes'. *Biology of the Cell 98*/4: 245–252.

Theer, P. et al. (2014). 'PSFj: know your fluorescence microscope'. *Nature Methods 11*/10: 981–982.

van Emden, H. (2008). *Statistics for Terrified Biologists*. Chichester: Wiley Blackwell.

Verdaasdonk, J. et al. (2014). 'Determining absolute protein numbers by quantitative fluorescence microscopy'. In: Waters, J.C., and Wittmann, T. (eds.), *Quantitative Imaging in Cell Biology*. Methods in Cell Biology, vol. 123, Chapter 19. Elsevier Inc., pp. 347–365.

Vojnovic, B. et al. (2011). 'An inverted fluorescence microscope assembly'.

von Bartheld, C.S. and Wouters, F.S. (2015). 'Quantitative techniques for imaging cells and tissues'. *Cell and Tissue Research 360*/1: 1–4 (editorial).

Waters, J.C. (2009). 'Accuracy and precision in quantitative fluorescence microscopy'. *Journal of Cell Biology 185*/7: 1135–1148.

Waters, J.C. and Swedlow, J.R. (2007). 'Interpreting fluorescence microscopy images and measurements'. In: Zuk, D. (ed.), *Evaluating Techniques in Biochemical Research*. Cambridge, MA: Cell Press, pp. 37–42.

Waters, J.C. and Wittmann, T. (2014). 'Concepts in quantitative fluorescence microscopy'. In: Waters, J.C., and Wittmann, T. (eds.), *Quantitative Imaging in Cell Biology*. Methods in Cell Biology, vol. 123, Chapter 1. Elsevier Inc., pp. 1–18.

Weibel, E.R. et al. (1966). 'Practical stereological methods for morphometric cytology'. *Journal of Cell Biology 30*/1: 23–38.
Wong, M.D. et al. (2013). 'Design and Implementation of a Custom Built Optical Projection Tomography System'. *PLoS One 8*/9: e73491. doi: 10.1371/journal.pone.0073491.
Zucker, R.M. (2005). 'Evaluation of Confocal Microscopy System Performance'. In: Taatjes, D.J. and Mossman, B.T. (eds.), *Methods in Molecular Biology*, vol. 319, Chapter 5. *Cell Imaging Techniques: Methods and Protocols*. Totowa, NJ: Humana Press, pp. 77–135.
Zucker, R.M. (2006). Cytometry Part A 69A/7: 659–676. International Society for Analytical Cytology. (Published by John Wiley.)

Further Reading

Bernas, T. et al. (2007). 'Precision of light intensity measurement in biological optical microscopy'. *Journal of Microscopy 226*/2: 163–174.
Centonze, V. and Pawley, J.B. (2006). 'Tutorial on practical confocal microscopy and use of the confocal test specimen'. In: Pawley, J.B. (ed.), *Handbook of Biological Confocal Microscopy*, 3rd ed, Chapter 35. New York: Springer, pp. 627–649.
Combs, C.A. and Shroff, H. (2017). 'Fluorescence microscopy: a concise guide to current imaging methods. *Current Protocols in Neuroscience* Unit 2.1.1–2.1.25 April 2017. doi: 10.1002/cpns.29.
Fischer, R.S. et al. (2011). 'Microscopy in 3D: a biologist's toolbox'. *Trends in Cell Biology 21*/12: 682–691.
Garsha, K. (2008). 'Quantitative Fluorescence Microscopy: Considerations and Controls'. In: Resch-Genger, U. (ed.), *Standardization and Quality Assurance in Fluorescence Measurements II*, Chapter 3. Berlin: Springer-Verlag, pp. 55–88.
Shaw, P.J. (2006). 'Comparison of Widefield/Deconvolution and Confocal microscopy for Three-Dimensional Imaging'. In: Pawley, J.B. (ed.), *Handbook of Biological Confocal Microscopy*, 3rd ed, Chapter 23. New York: Springer, pp. 453–467.
Swedlow, J.R. (2012). 'Innovation in biological microscopy: Current status and future directions'. *Bioessays 34*: 333–340.
Wheeler, A. and Henriques, R. (eds.) (2018). *'Standard and Super-Resolution Bioimaging Data Analysis: A Primer'*. Wiley.
Wollmann, T. et al. (2017). 'Workflows for microscopy image analysis and cellular phenotyping'. *Journal of Biotechnology 261*: 70–75.
Yang, W. and Yuste, R. (2017). 'In vivo imaging of neural activity'. *Nature Methods 14*/4: 349–359.

Internet Resources

List of things to think about on setting up an Imaging Facility: http://www.microscopist.co.uk/imaging-facilities-useful-links/
Confocal Microscopy listserver: http://lists.umn.edu/cgi-bin/wa?A0=confocalmicroscopy.
Statistics papers series:

1. https://bpspubs.onlinelibrary.wiley.com/doi/full/10.1111/j.1476-5381.2012.01999.x Drummond and Vowler (2012)
2. http://jcp.bmj.com/content/36/11/1269.full.pdf+html Silcocks (1983)
3. http://mkweb.bcgsc.ca/pointsofsignificance
4. http://mmb.bme.wisc.edu/stuff/GeneralInfo/website/Biostatisticsreview.pdf Dukes and Sullivan (2007).

Bioinformatics:

1. http://www.nature.com/nmeth/focus/bioimageinformatics/index.html
2. http://www.nature.com/nmeth/journal/v9/n7/pdf/nmeth.2084.pdf

ImageJ/Fiji:

1. http://imagej.net/Welcome
2. http://imagej.net/History
3. http://fiji.sc/Fiji

Chromatic aberration correction: http://www.lightmicroscopy.cruk.cam.ac.uk/image-chromatic-aberration-corrector-colour-alignment/.
PSFj plugin for PSF evaluation and microscope performance checks:

1. http://www.knoplab.de/psfj/overview/
2. http://www.knoplab.de/psfj/online-manual/.

Plastic fluorescent 'sea' slides:

1. Chroma: https://www.chroma.com/products/accessories/92001-autofluorescent-plastic-slides
2. Labrigger Open Source laboratory solutions blog: http://labrigger.com/blog/2011/04/14/test-slides/.

Notes

1. See: *Br. Jour. Pharmacol.* (2011) *163*/2: 207 and *Br. Jour. Pharmacol.* (2012) *166*/7: 1977. See also the *Nature Methods* series referred to in the text. An extensive treatment on statistics 'Points of Significance' has also been published by *Nature Methods*, starting from September 2013.
2. See: Oldenbourg, R. et al. (1996). 'Standard test-targets for high-resolution light microscopy', Chapter 8. In: Hoch, H.C., Jelinski, L.W. and Craighead, H.C. (ed.). *Nanofabrication and Biosystems*. Cambridge: CUP, pp. 123–138.
3. If PVA added to glycerol is not evenly dispersed throughout, it will clump (like microscopical tapioca). Each of these clumps will act as microlenses to scatter the light and severely degrade the signal. See Goodwin (2007), page 412, for further explanation.

28
Biological Specimen Preparation

Outline

28.1 Introduction

The guidance given in this chapter is a basic introduction to preparing biological samples and acquiring images of living cells and tissues with the microscope. Most preparation methods are specific for particular cells or tissues. They have been known for a considerable time and have been developed on either an empirical or a rational basis because they work successfully. Particularly with live-cell imaging, only very general considerations can be offered, because each combination of sample, microscope, imaging room (i.e. environment) and operational protocol is unique and can only be worked out in practice, often by trial and error, at the coalface. Indeed, it is entirely possible that living preparations and cell cultures can vary in their response to imaging from one day to the next, depending upon their particular micro-environment or the medium in which they are held. Nevertheless, some general principles have reliably stood the test of time, and these are given here. Living specimens examined by light microscopy are usually individual cells, tissue explants or else small whole organisms, and sufficient attention should be paid to maintaining their natural environment as closely as possible. Most samples, if they are not observed live, are fixed to prevent decay, stained as required and mounted in a transparent medium of suitable refractive index between a glass slide and coverslip. A basic protocol for fluorescence staining of adherent cells using antibodies is given. There are diverse ways to prepare non-fluorescent biological samples for viewing with the microscope. For more detail on this topic, see Sanderson (1994).

28.2 Initial Specimen Preparation

Most organisms and tissues are usually too large to be examined whole under the microscope and thus are examined in part either as individual cells, as groups of cells or as tissue biopsies, having first been excised, killed, fixed and perhaps embedded and sliced into thin sections. Biological tissues contain water, which must be removed as part of the procedure to prevent decay and disorganisation. Once cells or tissues are removed from the organism, decay will set in. This decay is due to the cell or tissues' own proteolytic enzymes as well as bacterial action. This must be arrested and the tissues kept in as natural a state as possible with a minimum of artifacts. It is impossible to entirely avoid artifacts. Fixatives harden tissues, and the subsequent processing steps shrink the volume of the tissues and cells (see Sanderson, 1994, and Dorph-Petersen et al., 2000). The key to successful specimen preparation (Figure 28.1) is to minimise these artifacts as much as possible at each step of the process.

There is no one universal fixative; you must select the most suitable for the cell, organelle or tissue that you wish to study (for a review, see Eltoum et al., 2001). Fixatives act either by chemical cross-linking or else by protein coagulation and precipitation due to solvent extraction[1]. The most common fixative used in light microscopy is aldehyde (a cross-linking, additive fixative), either as the monomer formalin or the polymerised solid paraformaldehyde (PFA). Formalin is stronger and is used in diagnostic pathology laboratories; paraformaldehyde is purer, being converted to the monomer by heating with alkali. Other cross-linking fixatives are glutaraldehyde and osmium tetroxide. Typical precipitating (non-additive) fixatives, which fix by denaturation and coagulation, are alcohols: ethanol and methanol, but also acetone, an organic dehydrating solvent. Acidic fixatives, such as acetic acid and picric acid, also precipitate proteins but do not fix carbohydrates or lipids. In general, precipitating fixatives cause less damage to tissue antigenic epitopes than cross-linking fixatives, but they do wash out lipids, and allow some tissue macromolecules to diffuse away from their origin. Precipitating fixatives also allow better access to antigens and can be quicker to make up and apply. Formalin-fixed tissues require post-fixation treatment (Grizzle, 2009; Otali et al., 2009) to unmask and recover antigenic epitopes for immunocytochemistry, since formalin fixation causes the loss of most immuno-reactivity. Somogyi and Takagi (1982) have combined both aldehyde and picric acid to form a penetrative fixative suited for immunocytochemistry as well as correlative LM-EM studies[2]. Another compound fixative mixture, periodate-lysine-paraformaldehyde (PLP) has also been used (McLean and Nakane, 1974) for immunocytochemistry. Ideally, it is best not to fix at all, and very labile antigens may necessitate using unfixed frozen cells or tissue sections in order to have any chance of being detected.

Animal tissues lack the support that plant cells walls offer. Therefore the water, whose pressure helps maintain cell shape and turgor, must be replaced with a suitable embedding material to support the fixed cells or tissues and permit them to be opened up and studied yet at the same time retain as much as possible of their original shape and morphology. Histological laboratories routinely use paraffin wax to embed tissues. The embedding medium must penetrate the tissue easily yet also support the tissue sufficiently without damage when hardened. The embedding medium must therefore be of approximately equal density and resilience to the biological material so that tissues cut evenly and do not separate under the force of the knife when cutting thin sections. For harder tissues or very thin sections, wax is sometimes replaced by a resin. Yang et al. (2013) report on the efficacy of different resins for fluorescent probes and endogenous reporters. Plant tissues are normally embedded in resin because their tough cell walls have to be softened to slice them open effectively.

Because wax is immiscible with water, the tissues are transferred stepwise through a series of increasingly concentrated solvents (usually ethanol) to dehydrate them. The tissues are then transferred into a transition medium miscible with both the dehydrating solvent and the embedding medium. Xylene and toluene were once widely used as transition media, but they have largely been replaced by safer terpene-based alternatives from essential oils, limonene-based formulations such as Histoclear® or aliphatic hydrocarbon formulations such as Sub-X™. Another important function of the transition medium is to 'clear' the tissues. The refractive indexes of these transition media are similar to tissue, which is consequently rendered transparent before being embedded. Table 28.1 gives a series of refractive index values for reference. This facilitates the study, once they have been embedded and sectioned, of otherwise opaque or translucent tissues[3]. 2-2 thiodiethanol can be used both as a clearing agent (Aoyagi et al., 2015) and is reported being used as a mountant (Staudt et al., 2007) albeit for thick tissues – because of its clearing properties.

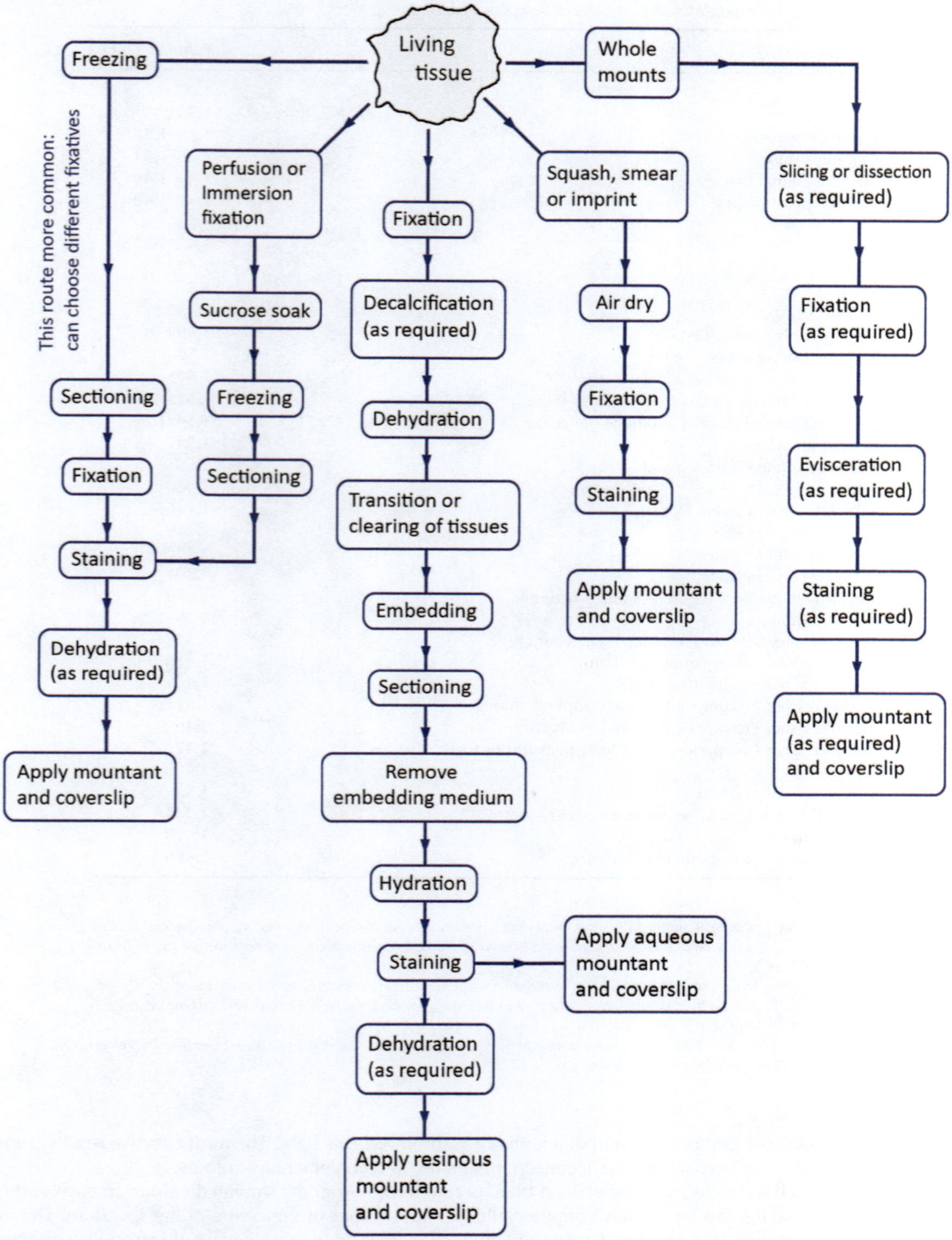

Figure 28.1 ***Tissue processing chart*** *Flow chart showing an overview of tissue-processing steps for making a microscope slide preparation.*

Sections a few micrometres thick of the embedded material are then sliced off with a microtome and mounted onto a microscope slide. These thin sections are generally stained for contrast enhancement before mounting between glass. The transparent appearance of the fixed, embedded and sectioned tissue means that the preparation has a virtually homogeneous refractive index throughout. This, in turn, renders the section a uniform and uninteresting dull grey colour when viewed by transmitted-light bright-field microscopy. To distinguish and differentiate the various components of the cells and structure of the tissue, it is essential to stain the preparation in contrasting colours (see Figure 11.3, Section 11.3). Histology uses stains and dyes[4] developed by the textile industry. Dyes work by

Table 28.1 Selected Refractive Indexes of Samples and Mounting Media

Medium	Refractive index
Air	1.000
Water	1.333
Living cytoplasm	1.35–1.38
Cell membrane	1.46–1.54
General animal tissues	1.43–1.49
General plant tissues	1.41–1.48
Collagen	1.43
Lipids	1.45–1.47
Cartilage	1.49
Bone	1.56
Diatom silica	1.430
Dimethylsulphoxide (DMSO)	1.484
Xylene clearing agent	1.50
Cedar wood oil clearing agent	1.515
2,2-thiodiethanol, TDE	1.515
BABB 1:2 benzyl alcohol to benzyl benzoate	1.54–1.57
Methyl salicylate	1.541
Chloral hydrate clearing agent	1.55
100% Glycerol	1.466
80% Glycerol in water	1.446
50% Glycerol in PBS	1.416
50% Glycerol in water	1.406
40% Glycerol in water	1.391
Vectashield glycerol-based mounting medium	1.458
CyGel mountant	1.36–1.38
Polyvinyl alcohol PVA mountant (Sigma-Alrich)	1.377
Taab Watermount (aqueous, sets hard)	1.4105
Slow-Fade non-setting medium	1.415
Thermo Fisher Prolong Gold (newly-applied, liquid)	1.415
Thermo Fisher Prolong Gold (cured, set hard)	1.46
Thermo Fisher Component E (for Tetraspeck test beads)	1.47
Mowiol	1.490
Polyvinylalcohol (PVA)	1.52–1.55
Imersol W (Zeiss oil for water-immersion objectives)	1.334
Immersion oil	1.518
Silicone immersion medium (Olympus)	1.406

References

Bacallao, R. et al. (2006). 'Guiding principles of specimen preservation for confocal fluorescence microscopy'. In: Pawley, J.B. (ed.), *Handbook of Biological Confocal Microscopy*, 3rd ed, Chapter 18. New York: Springer, pp. 368–380; Table 18.2, p. 377.

Davis, I. (2000). 'Visualizing fluorescence in Drosophila – optimal detection in thick specimens'. In: Allan, V.J. (ed.), *Protein localization by fluorescence microscopy – a practical approach*, Chapter 6. Oxford: Oxford University Press, pp. 133–162; Table 3, p. 149.

Sanderson, J.B. (1994). *Biological Microtechnique*. RMS Handbook No. 28. Oxford: Bios Scientific Publishers, Routledge/Garland Science/Taylor & Francis.

selectively binding to different tissue components and differentially absorbing visible light. The most effective results come from using a basic and acid stain, applied to the preparation in succession, preferably in complementary colours.

Small cellular inclusions, such as nuclei, are best stained blue, purple or red: from the saturated colours at each end of the visible spectrum. Background counterstains can be chosen from the yellow or green hues of the centre of the spectrum. The eye perceives yellow and green as unsaturated colours; if small inclusions and organelles are stained thus, they would effectively disappear against a more highly saturated background. One of the most widely used stains is haematoxylin and eosin (Figure 28.2), commonly referred to as 'H&E'. Haematoxylin is a very versatile stain; a mordant is used to improve its binding fast to tissue, where it is used to stain nuclei blue, blue-purple or black (depending upon the exact stain protocol). Eosin is useful because it stains cytoplasm various shades of red and pink (eosin also fluoresces). The family of eosin xanthene dyes are derived from fluorescein, itself widely used as a fluorescent label. For staining protocols for H&E and other stains, as well as further explanations of the action of dyes and stains, see Chapter 6 in either Sanderson (1994) or Cook (2006) or refer to the 2nd edition of the Dako *Special Stains and H&E* Education Guide.

Whether the histochemical action of the dye is understood rationally or whether its action is known and accepted empirically, there are a wide range of stains available to distinguish and reveal many different tissue components and inclusions. However, fluorescent staining offers very sensitive and specific marking of cells and their internal organelles and is widely used in all branches of biomedical research. The following sections describe how to prepare immuno-stained and dye-stained cells for examination with a fluorescence

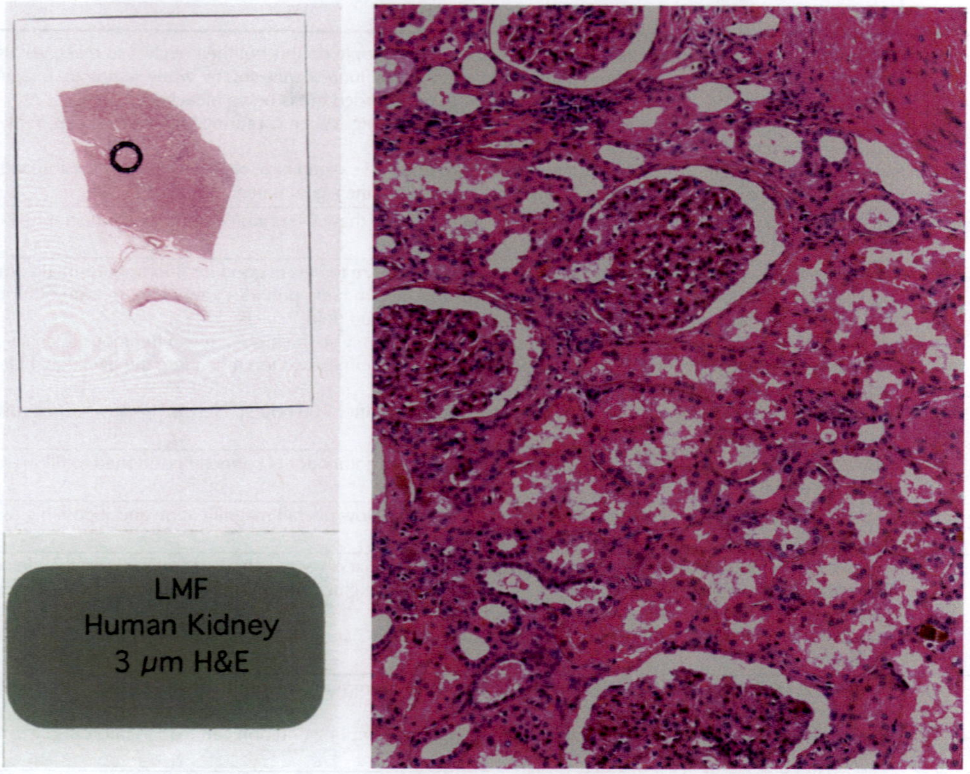

Figure 28.2 ***The H&E stain; marking a location*** *Kidney cortex stained by the classical haematoxylin and eosin stain. The picture of the slide shows an inked relocation mark. The slide marker, which has an objective thread enabling it to be screwed into the nosepiece and so form the relocation mark, has largely been superseded by software position-marking stored in the metadata header of the image file. Further details about relocating areas of interest can be found in Chapter 31, Section 31.10.*

microscope. Histology is a separate topic and apart from this brief description, is not covered here. For further information on the histological preparation of cells and tissues, see Sanderson (1994), Kiernan (2008), Orchard and Nation (2017), Suvarna (2012) or Carson and Cappellano (2015).

Where the solvent and embedding medium are likely to interfere with using antibodies, it is possible to freeze the cells and tissues, either embedded in sucrose or within their own watery constituents, surrounded by a supporting medium – e.g. Tissue-Tek OCT compound[5] – to cut frozen sections. This helps preserve labile epitopes (the antigen determinant) on and within the cells and tissues for immunological study of structure and function. See Figure 28.1 for how to do this. A protocol for cryosectioning tissues has been published by Fischer et al. (2008). For three practical guides on immunocytochemistry, see Burry (2010), Buchwalow and Böcker (2010) and the 6th edition Dako Education Guide (2013). The URL for the Dako guide is given in the Internet Resources, at the end of the chapter.

28.3 Preparing Fluorescently Labelled Samples

Immunohistochemistry is the use of labelled antibodies (conjugated to stains or fluorescent dyes) as specific markers for the localisation of antigenic tissue constituents in situ. Immunocytochemistry refers to this procedure as applied to cells. For the reasons discussed in Chapter 15, fluorescent markers are widely used because they are self-luminous, thus giving very high contrast. They are potentially very specific in their action, labelling individual cell and tissue components accurately with specific discrimination. Fluorophores are versatile, being available to stain many biological components; they are sensitive, able to mark single molecules when required. Because they can successfully be used in very low concentrations, most fluorophores can be tolerated by living cells. Indeed, cells can generate their own endogenous fluorescent proteins – a significant advantage for biomedical research. Finally, fluorescent markers are relatively economical and straightforward to apply to cell and tissue preparations. To take maximum advantage of fluorescent labelling, you should take care to avoid bleaching and fading of the fluorescent signal. Try not to leave too much time between specimen preparation and image acquisition. When not acquiring images, keep the slides covered and store them in a fridge or cold room. It is also good to take steps (see Chapter 16, Section 16.7) to reduce non-specific autofluorescence. Table 28.2 lists some of the possible causes that can give rise to poor image quality and Table 28.3 on steps to take to optimise signal-to-noise ratio in the image.

To maximise the photon signal, it is necessary to retard bleaching by incorporating antifade reagents into the mounting medium. Antifade reagents act by scavenging oxygen radicals from the mountant (Cordes et al., 2011); otherwise, the fluorophore acts as an electron acceptor and is bleached (see Figure 15.3, Section 15.2). Antifades are thus anti-oxidant reducing agents. Most are toxic, but

Table 28.2 What to Do If Your Florescence Image Is Dim and/or of Poor Quality

Bleaching	Keep samples cooled and in the dark when not in use. Many people do this but then neglect to make visible observation and exposure for the minimum time, and often forget to close the fluorescence shutter while setting up imaging parameters (e.g. for digital acquisition). All the while the sample is not being recorded but is being bleached. Use minimum exposure both for observation and image recording. When capturing high frames rates, remember that overall exposure is a product of average intensity *and* frame rate. Use anti-fade reagents in the mounting medium to scavenge free oxygen radicals. Keep oxygen tension and photo-toxicity low by using minimum exposure and intensity. Integrate or pixel bin the signal if needed.
Fluorophore degradation	Store the samples in a refrigerator or freezer. If the latter is used, thaw the sample to ambient temperature before using immersion fluid.
Check the illumination	Mercury lamps have discrete intensity peaks (Figure 9.2). They are therefore good for fluorophores that coincide with these peaks, but they are not good for ratiometric dyes, like Fura-2. A Xenon lamp provides a much more even output across the spectrum. Use a fibre optic or light guide to provide an evenly-illuminated field. Unlike lasers, you cannot saturate fluorophores with an arc lamp. Use the highest mean luminous intensity possible (e.g. 100 watts instead of 50 watts) providing that the image of the arc is not minimised too much into the back focal plane of the objective (which reduces intensity). Check the lifetime of the mercury arc lamp (the intensity drops after 200 hours). Hence, modern sources use 120 watt 2000 hour metal-halide lamps.
Check that the lamp source is centred	Check the lamp centration. Check also that the epi-illuminator collimator is correctly positioned to fill the back focal plane of the objective evenly.
Optimise the objective used	Where possible use an objective with high numerical aperture, low overall magnification and if possible use an immersion, rather than a dry, objective.
Optical path transmittivity	Use objectives with good UV and/or IR transmittivity. Apochromat objectives have more elements than lesser-corrected objectives. Dyes that absorb in the UV, such as DAPI, generally appear brighter when using optimised achromat and semi-apochromat (e.g. Fuor, Fl, Fluotar or Fluar) objectives. Ensure that magnification changers, de Sénarmont sliders for DIC and other non-essential optical components are removed from the beam path.
Check the overall magnification used	Sometimes to get full resolving power and satisfy the Nyquist sampling criterion, you must magnify the widefield image. This can lead to low-intensity images. Zooming on a point-scanning confocal to satisfy the Nyquist sampling criterion can lead to fluorophore saturation and rapid bleaching.
Use modern fluorophores	These are often more expensive but worth the extra money. Use modern, photostable fluorophores of high brightness, high absorption coefficient and quantum efficiency. If necessary, amplify the signal (e.g. by using secondary antibodies by indirect staining).
Fluorophore concentration	Ensure that sufficient fluorophore has penetrated throughout the tissue and that cells or tissues for confocal microscopy are well stained. Stain longer with both primary and secondary antibody, rather than use higher concentrations of secondary antibody, in order to keep the background low and SNR high.
Fluorophore quenching	Fluorophores are susceptible to pH and local ionic concentrations. The presence of ammonium, iodine and metal ions can all affect dye intensity.
Autofluorescence	Use red-shifted fluorophores to minimise autofluorescence. The emitted signal from these tends to be scattered less than blue fluorophores.
Background fluorescence; poor SNR	Set the field diaphragm and aperture diaphragm (if fitted) correctly on the fluorescence epi-illuminator to improve contrast and visibility. Check that image quality is optimised, and spherical aberration is not causing reduced signal-to-noise ratio. Fixatives (e.g. aldehydes) can increase background. Change the tissue/cell fixative protocol. Poor antibody quality, poor storage and/or repeated freeze-thaw cycles can give rise to high background. Use a respected supplier. Poor blocking of endogenous proteins can also increase background and reduce signal. Presence of unblocked false epitopes can lead to non-specific 1° and/or 2° antibody binding. Use pre-adsorbed 2° antibody to reduce or prevent species cross-reaction. Incubate with normal serum obtained from species other than from the species that were the source of the primary antibody. Include sufficient bovine serum albumin or fish skin gelatin in diluent buffer to block exogenous immunoglobulins. Include a weak detergent (e.g. Triton X-100) to permeabilise membranes or uncover epitopes. Suggested concentrations are 0.02% Triton X-100 and 0.5% BSA. Inadequate, scanty, washing technique between staining steps can also lead to high background and poor signal intensity. Poor signal-to-noise ratio may also arise from using too high a concentration of secondary antibody. Both primary and secondary antibodies should first be titrated out to determine the optimum concentrations to use for staining.
Check filter set excitation and emission bandwidth	Unless bleed-through is a problem, use the maximum excitation and emission band width possible.
Check the camera/ detector used	Use a high QE detector; use a cooled CCD to reduce thermal noise. Reduce readout noise by increasing the exposure. Use an sCMOS or EM-CCD for low signal with high frame rates.
Check the pixel size	Use large pixel sizes where possible. Increase the effective pixel size by binning the signal (you must consider signal brightness versus image resolution).
Repeat the staining	Consider this option if all else fails; redesign the experiment.

For further information online, see:

1. General: https://www.thermofisher.com/sg/en/home/life-science/protein-biology/protein-biology-learning-center/protein-biology-resource-library/pierce-protein-methods/overview-immunohistochemistry.html.
2. Troubleshooting: http://www.ihcworld.com/troubleshooting.htm.
3. Background: https://sharedresources.fredhutch.org/training/sources-background-immunohistochemistry and https://www.agilent.com/cs/library/technicaloverviews/public/08002_ihc_staining_methods.pdf.

Table 28.3 *Optimising the Signal-to-Noise Ratio*

To increase signal	To decrease background/noise
Use a high-NA, oil-immersion, minimally corrected objective with the lowest acceptable magnification	Close down the field diaphragm to just outside the CCD chip, or raster scan, field of view. This minimises stray light
Align microscope and fluorescence illuminator for Köhler illumination	Turn off room lights (light enters via microscope eyepieces)
Use a clean coverslip of 170 µm thickness; if in doubt, measure this before sterilisation and use	Mount the sample as close to the underside of the coverslip as possible
Open the confocal pinhole: collect thicker sections	Deconvolve the final image (2D section or stack)
Choose a bright, 'third generation' fluorophore	Close down the condenser diaphragm to the smallest diameter compatible with good resolving power. This minimises stray light
Choose fluorescence filter sets that match fluorophores and minimise bleed-through	Perform good antibody staining with adequate blocking, washing and controls
Use proven antibodies from established and suppliers. Mabs can differ between batches/lots.	Unmask antibody epitopes by using antigen retrieval protocols, where needed
Use antifade inhibitors to retard photobleaching	Select mounting media carefully; some have high background. Depending on formulation, some fluorophores are quenched
On a laser-scanning confocal microscope, use the slowest acceptable scan speed without causing undue photobleaching	On a laser-scanning confocal microscope, use the lowest gain setting(s) possible
Remove DIC Wollaston prism(s) and analyser from fluorescence light path	Do not use coloured nail varnish to seal specimen preparations; use phenol red-free media
Use camera binning to increase signal intensity	Integrate the signal; avoid very short exposures
On a widefield microscope, use a cooled CCD camera with at least 60% quantum efficiency	Use cooled CCD or sCMOS camera with low readout noise

Source: Data from Table 1, page 116, Waters 2007.

To increase signal:

- Choose a bright fluorophore.
- Use a clean 170 µm coverslip (No. 1.5). Measure these before sterilisation and use.
  (You can now get coverslips of 0.17 mm thickness with a narrow distribution from Zeiss and Paul Marienfeld GmbH)
- Mount specimen as close to distal side of coverslip as possible.
- Use high-NA, minimally corrected oil-immersion clean objective with lowest acceptable magnification.
- Chose fluorescence filter sets that match fluorophores and minimise bleed-through.
- Use inhibitors to retard photobleaching.
- Remove DIC Wollaston prism(s) and analyser from light path.
- Use cooled CCD camera with at least 6% quantum efficiency (QE).
- Use camera binning to increase signal intensity.

To decrease background and/or improve signal-to-noise ratio (SNR):

- Turn off room lights (light enters via microscope eyepieces).
- Close down field diaphragm.
- Use phenol red-free media.
- Perform good antibody staining with adequate blocking and controls.
- Do not use coloured nail varnish to seal specimen preparations.
- Use cooled CCD camera with low readout noise.

Summary

- Work out particular pH, CO2, humidity and temperature needs of your cells.
- Leave sufficient time to allow temperature hoods to equilibrate.
- Use phenol red-free and CO2-independent media to maintain pH, where possible.
- Minimise photobleaching and use antifade reagents.
- Immobilise/anaesthetise samples where necessary to avoid undue movement.
- Take measures to avoid evaporation, which can affect exposure through the time-lapse.
- Measure grey scale values in image software to determine exposure.
- Monitor your cells after the experiment for healthy appearance and behaviour.

Reference

Waters, J.C. (2007). 'Live-Cell Fluorescence Imaging'. In: Sluder, G. and Wolf, D.E. (eds.), *Methods in Cell Biology*, vol. 81, *Digital Microscopy*, Chapter 7. Amsterdam: Academic Press, pp. 115–140.

n-propyl gallate can be used in reduced concentration with live cells, as can Trolox (Cordes et al., 2009; CAS No. 53188-07-1; available as an antifade in glycerol mounting media from Vectashield: CB-1000). It has been reported (Tycon et al., 2012) that ascorbic acid can cause DNA damage, so these additives should be used with caution.

Check that the anti-fade that you select is compatible with the fluorophores that you have used. For example, aqueous media containing *n*-propyl gallate (NPG), typically at 1%, or those with 1–2.5% DABCO (1,4-diazabicyclo[2.2.2]octane) antifade (e.g. DABCO in Invitrogen/Thermo Fisher Scientific Prolong and Slowfade products) can be used with cyanine dyes and are good for Alexa dyes but poor when used with fluorescent proteins. Any mounting medium using *p*-phenylene diamine (PPD, e.g. Vectashield) certainly quenches Cy2 and Cy5, causing weakened and diffuse staining. PPD also will tend to quench DNA-binding fluorophores. However, PPD is better for Texas Red, which does not react well to DABCO or NPG, and NPG is best for rhodamine. For those fluorophores not quenched by it, PPD remains one of the most potent antifade reagents, but it will darken and go brown with time. Both NPG and DABCO can also be used in low concentration with living cells, but don't use NPG, DABCO or PPD for quantum dots: use Mowiol or a PVA-based mountant.

If you are using GFP or any other fluorescent protein alone as a single 'stain', then it is best to leave out any antifade, which can quench GFP. Pay attention to the cell culture media used, for there are reports (Bogdanov et al., 2012) that GFP can donate electrons to vitamin (e.g. riboflavin) additives in culture media, which will lead to decreased photostability and consequent quenching of GFP. To look at GFP in tissues, it is best to perfusion fix the animal, soak the tissues in 30% sucrose, embed in OCT and cut frozen sections. Fix in 1–2% EM-grade PFA (do not use PBS buffer as a diluent) or use 1% periodate-lysine-paraformaldehyde. The amount of GFP fluorescence lost to PFA fixation is highly variable, depending on the construct created. Using an anti-GFP antibody label (e.g. anti-GFP chicken polyclonal) is a good way in which to amplify the GFP signal.

28.4 Immunohistochemistry

Good antibody staining can be both very specific and very sensitive, allowing you to detect the protein or molecule in which you are interested with high precision. However, unless care is taken at *each* stage of the staining process, it is possible to end up with very much less than perfect results. The direct method, using just one antibody, has been superseded by the indirect method (Figure 28.3), using a second antibody directed against the first (it is possible to add further steps, but two only is usual). The advantage of the indirect method, with both primary (1°) and secondary (2°) antibodies, is amplification of detection sensitivity.

Check the optimum fixation and permeabilisation conditions for the cells and tissues being studied. Prior to fixation, some workers advocate washing the cells or tissue sections with serum-free media or buffer[6] so that exogenous proteins are not thrown down onto the sample. Titrate out the 1° antibody to work out the best concentration to use that is consistent with strong signal and low background. Determine the best method of blocking non-specific binding; use conjugated 2° antibodies that are bright, photostable, have well-separated emission spectra (to minimise bleed-through), which match the laser lines or filter sets used to excite and detect the fluorescent signal, and which also have minimal cross-reactivity.

Monoclonal antibodies are pure, but the epitope that they recognise may be carried on proteins other than the target or be masked by fixation. When using monoclonal antibodies, most people select mouse[7] as the host species. Several companies now produce rabbit monoclonal antibodies (RMAb), and results suggest that these offer an increased sensitivity with no apparent loss of specificity. Polyclonal antibodies are found in serum containing a mixture of antibodies recognising different epitopes of the same antigen. The signal is likely to be strong, but polyclonal antibodies have to be purified, and the likelihood of non-specific binding (leading to unwanted background signal) is high. The choice of secondary antibody depends upon the species which the 1° antibody was raised in and the potential for cross-reactivity. The 2° antibody is usually an IgG fragment. If working with tissues or cells that have Fc receptors (thymus, spleen, blood, leukocytes, B cells, etc.) choose, where possible, an $F(ab')_2$ fragment for the 1° antibody to eliminate non-specific binding. Some people advocate chicken IgY – a functional equivalent to IgG – as an alternative to rabbit polyclonal antibodies.

Antibodies are expensive; you should apply them to cells and tissues with the smallest volume possible. Ideally, cells should be grown directly onto the coverslip. Tissue sections are ideally mounted onto slides, for it requires larger volumes of fluid to stain free-floating tissue sections. When staining, the sections are held in a humid chamber (see Figure 28.3) to prevent evaporation of reagents and non-specific staining caused by the sections drying out. The free-floating technique is generally reserved for enzyme histochemical methods. Where living cells, small tissue biopsies or tissue sections have to be bathed in fluid, use a multiwell chamber slide with a coverslip bottom ensuring that in between the staining steps the compartments are washed thoroughly but gently so that reagents from the previous step do not remain trapped in the corners of the well. Poor washing technique causes problems with poor or irregular staining and high background.

Antibodies vary in how well they last. They can also vary (Marx, 2013) from one host animal to another and between batches from the same host; this important point is often overlooked. Some antibodies can be stored at 4°C; others must be stored frozen. It is wise to avoid repeated freeze-thaw cycles; therefore, aliquot the antibody (this advice applies to all reagents) upon reconstitution, and store smaller volume aliquots (e.g. 500 or 100 μL) appropriately. Primary antibodies are normally used between 1:100 to 1:10 000 dilution and secondary or polyclonal antibodies between 1:50 and 1:5000. Once a working stock has been made from an aliquot, use it once only; do not store it for reuse. Glycerol is a good stabilisation agent for antibodies, and if they are not to be used on living cells or tissues, 0.02% sodium azide is recommended by some as a fungal inhibitor, although most workers leave it out. Others add bovine serum albumin at 0.1–1% to reduce antibody loss from polymerisation and/or adsorption during storage.

Protocol 28.1 A General Immuno-Staining Protocol

1. prepare sample: grow cells on coverslips, or excise tissue;
2. embed, section and mount tissue onto slides/coverslip; (frozen tissues are sectioned and mounted before fixation);
3. fix cells or tissue cryosections;
4. permeabilise membranes where necessary;

5. block unwanted epitopes with normal serum from the secondary (2°) antibody host;
6. incubate with primary (1°) antibody;
7. rinse 1° antibody reagent free;
8. incubate with 2° antibody;
9. rinse 2° antibody reagent free;
10. counterstain nuclei, or other cell components, as a separate step (optional); and
11. mount and seal the preparation to finish.

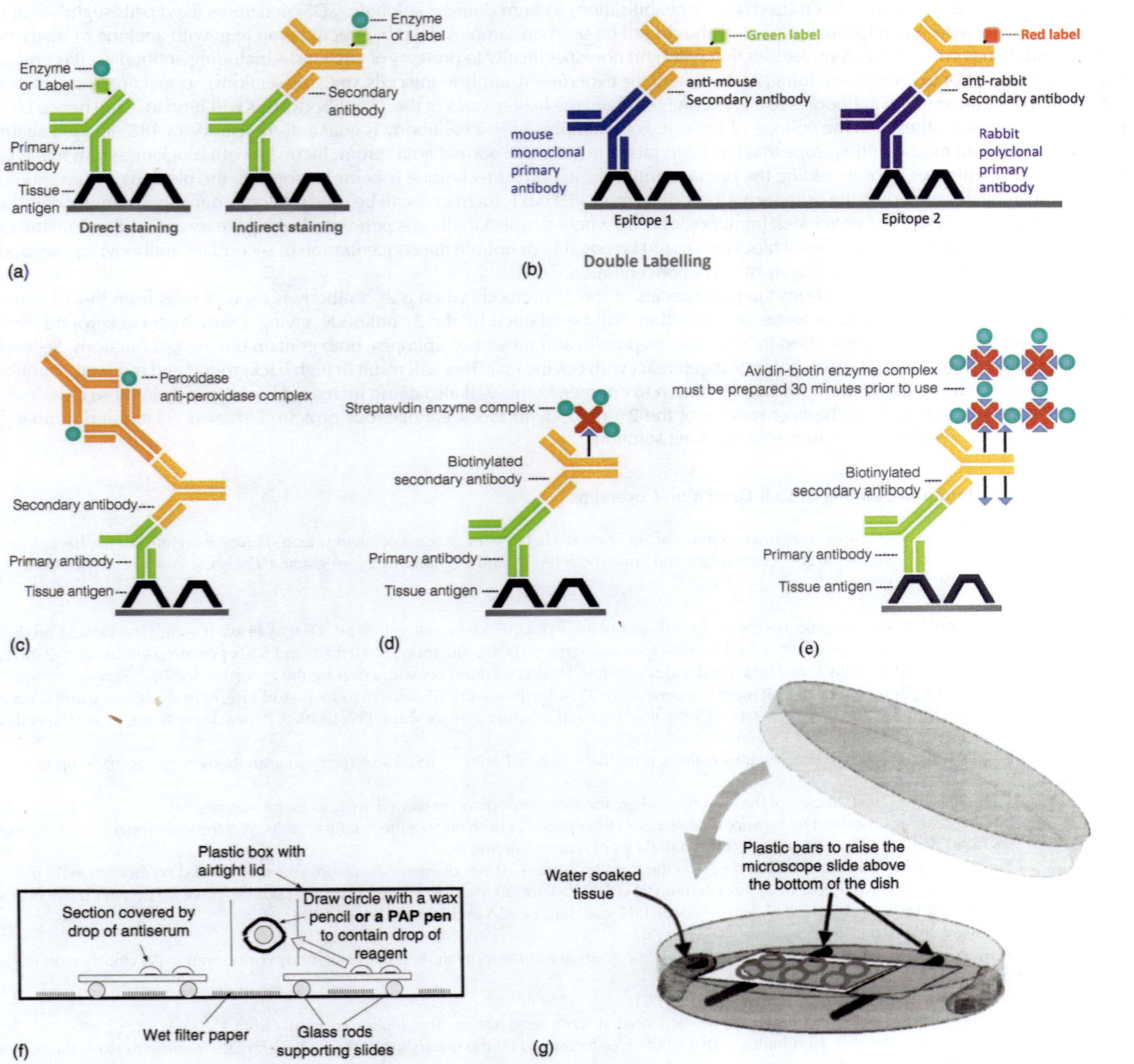

Figure 28.3 ***Immuno-staining and a humidity chamber*** *Top panel. Indirect immuno-staining with conjugated antibody labels. The primary antibody bound to a specific epitope is revealed (a) by a secondary antibody conjugated to a fluorophore or detected (b) by double labelling or (c) by a peroxidase anti-peroxidase (PAP) method. Avidin-Biotin and streptavidin-biotin amplification (d,e) are other means of detecting antigenic epitopes, particularly where there are only low levels of antigen. The avidin-biotin complex is the strongest known non-covalent interaction between a protein and ligand. Streptavidin binds better than biotin alone to conjugated avidin.*

Bottom panel. A humid incubation chamber for antibody staining. The lids of these are usually covered with silver foil to exclude light during long staining steps. Source: Albert et al. 2014. Reproduced with permission from Garland Science.

Permeabilisation is necessary to allow antibodies access across cell membranes and contact with intracellular structures. The detergents saponin, Triton X-100 or Tween 20 are used. Since solvent fixatives wash out lipids from membranes, these permeabilisation detergents are only required with cross-linking (e.g. aldehyde) fixation. Use 0.1% Triton X-100, 0.02% Tween-20 or 0.5% saponin in buffer for 5 or 10 minutes at room temperature. Triton X-100 and Tween 20 are non-ionic, non-denaturing detergents. Permeabilisation can be incorporated into a single step, but any quenching of fixative autofluorescence should be done separately and immediately after fixation. Triton X-100 will also permeabilise the nuclear membrane. Saponin is gentler in its action; it dissolves cholesterol but can leave membranes intact. Saponin should therefore be included in all antibody solutions, all blocking and incubation steps and all washing steps in order to work. In electron microscopy it is well-known that a balance has to be worked out between fixation and permeabilisation (Schnell et al., 2012), but this crucial factor and its effect on the structure and distribution of the target protein in light microscopy is less well known. When used for permeabilisation, sodium dodecyl sulphate (SDS) denatures fixed cells slightly and may be used to reveal masked antigenic epitopes. It should not be used on samples fixed by precipitation (e.g. with acetone or methanol).

Cells and tissues contain macromolecules that will bind non-specifically to proteins of any kind – including antibodies. It is crucial to block these sites prior to immuno-staining. At the start of the experiment, apply to the cells, or tissue sections, normal non-immune serum (serum that does not contain antibodies to a given antigen) from the host species of the 2° antibody. This will bind to – and hence block – protein-binding sites other than the epitope of interest. For example, if the 2° antibody is goat anti-mouse Alexa 488 directed against a primary antibody of mouse anti-epitope that has been raised in goat, use normal goat serum. Incubate with blocking serum made up in buffer for 10–30 minutes prior to adding the primary antibody. If a direct technique is being performed, the blocking serum should be from the host species providing the primary antibody (e.g. mouse). $F(ab')_2$ fragments can be used to block endogenous immunoglobulins to reduce background staining or to block immunoglobulins where double labelling is performed with primary antibodies from the same host species. The concentration for used blocking should be equal to or double the concentration of secondary antibody; e.g. secondary 1:100, blocking 1:50 to 1:100 and – at most – 5% concentration.

Never block with normal serum from the host species of the 1° antibody when a 2° antibody is used. If IgGs from the 1° antibody bind non-specifically to the cells or tissue sections, they will be labelled by the 2° antibody, giving a very high background. Bovine serum albumin (BSA) and dry milk, used to block excess protein and unwanted epitopes, both contain bovine IgG fractions. Secondary antibodies directed against cow, goat, horse or sheep react with bovine IgG. This will result in high background and reduced 2° antibody effectiveness (increasing the 2° antibody concentration to counteract this will also cause increased background). In these cases use 5% (v/v in PBT) normal serum from the host species of the 2° antibody to block endogenous protein. Fish skin gelatin and normal goat serum (NGS) are other alternatives to use as blocking solutions.

Protocol 28.2 **Immuno-Staining of Cell Grown on Coverslips**

Ideally the longer you can stain with lower concentrations of antibody, the better your chances of having clean staining with little, or no, background. If possible, fix at 4°C overnight, block at 4°C overnight and apply both the 1° and 2° antibody overnight at 4°C also. If you wish to complete the staining within one day, proceed as follows:

1. Plate out cells onto 25 mm Ø glass coverslips in a 6-well plate at 1×10^5 cells/well with 2 mL DMEM in each well. (The DMEM medium contains 250x conc. Sodium pyruvate @ 4 mL/L; 10% fetal calf serum; 100x L-glutamine @ 10 mL/L and 500x pen-strep antibiotic @ 2 mL/L.)
2. Incubate the cells at 37°C, with 2 mL phenol-red and serum-free DMEM medium per well covering the coverslip, for 30 minutes.
3. Fix coverslips in 4% paraformaldehyde at room temperature or 37°C by increasing dilutions of fixative. Add 1 mL of freshly thawed and warmed 4% paraformaldehyde for 1 minute. Aspirate off 1 mL fluid and add a further 1 mL of warm PFA fixative. Repeat three times. Leave the cells at room temperature to fix for a further 15 minutes.
4. Wash the cells with 5 washes over 5 minutes with 1x phosphate buffered saline (PBS). Have ready your antibody reagents made up in small volumes.
5. Take a strip of Parafilm®, wet an area of the bench*, and lay the film down onto the wetted area, so that it adheres flat.
6. Arrange rolled pieces of absorbent tissue wipes, or pieces of filter paper, at both ends of the Parafilm® strip. Wet these thoroughly with distilled water. This raises the humidity, and prevents the small drops of reagents drying out.
7. Lay out the first reagent in well spaced out drops on the Parafilm®. For a 20 mm diameter coverslip, you should need no more than 50 µL.
8. Block the coverslips in normal goat or 2° host serum (1:500) in PBT for 30 mins. to 1 hour. (PBT = 0.0002% Triton-X 100 with 0.5% BSA to reduce background in 1x PBS, pH 7.2–7.4. For 100 mL PBT, use 0.5 g of BSA and 20 µL Triton-X 100.)
9. Wash the cells with 5 washes over 5 minutes with 1x PBT.
10. Incubate coverslips in 1° antibody diluted in PBT for 1.5–2 hours at room temperature. Use the antibody concentration previously worked out by titration.
11. Wash the cells with 5 washes over 5 minutes with 1x PBT.
12. Incubate coverslips in 2° antibody diluted in PBT for 1 hour at room temperature. The dilution is usually from 1:200–1:500.
13. Wash the cells three times over 15 minutes with 1x PBS. Use forceps to lift the coverslips in the multiwell plate in order to wash the reverse side free of trapped media/fixative.
14. Stain with DAPI 5–10 minutes at 1–5 µg/mL. The nuclear counterstain is best applied as a separate step. It can be added in the mounting medium, but this tends to continue to penetrate the tissue and so gradually stain non-specifically over time. From the 5 mg/mL 1000x stock solution, use diluted 1:1000 to 1:5000. Add 12 µL stock to 12 mL 1x PBS, with 2 mL per well incubation at room temperature for 15 minutes. (The 10 mg vial should be diluted into 2 mL of dimethyl formamide to make the stock solution. Store this aliquoted stock at ≤ −20°C.) An alternative is to use the ready-made NucBlue™ and NucRed™ stains from Thermo Fisher Scientific.

    An alternative is to use either DRAQ-5 or ToPro-3. With DRAQ-5, dilute stock supplied as 5 mM in water; dilute to use 10 µM (1:500) or 5 µM (1:1000). Stain coverslips on drops in the dark for 5 minutes; miss out step 15, go to step 16. With ToPro-3, stain 1:1000 from 1 mM stock solution in DMSO. Store this aliquoted stock at ≤ −20°C. Dilute the thawed stock to 1 µM working solution in PBS; stain for 15 minutes.

15. Wash the cell three times over 5 minutes with 1x PBS.
16. Remove the coverslips from the multiwell plate with forceps, dip twice in Milli-Q distilled water for a few seconds to rinse (don't drop it and remember which side the cells adhere to!), drain on filter paper and invert the coverslip, cells down, onto a 10 µL drop of mounting medium. This is 1% *n*-propyl gallate in 90% glycerol in 1x PBS (4% or 5% *n*-propyl gallate has been reported to alter to a lowered pH and cause quenching of fluorescent signal).
17. Seal the coverslips with VALAP or colourless nail varnish.

*If staining tissues adhering directly on the slide, use a hydrophobic PAP pen to draw a fluid-retaining ring around the section on the glass (Figure 28.3). The slides can then be contained within a humid box for staining or enclosed in a similar manner to the coverslips, above.

Antibodies from one particular species (e.g. rat) can cross-react with another closely-related species (e.g. mouse) unless they have been adsorbed and purified against that host species (i.e. mouse). For example, if working with mouse tissues or cells, choose a secondary (rat) antibody that has been adsorbed with mouse serum or mouse IgG. This removes the mouse IgG fractions to which the rat antibody may otherwise react endogenously, causing high non-specific background staining. This is an important consideration when selecting 2° antibodies. If closely related IgG fractions are not present, then do not use an adsorbed, purified mouse antibody because such adsorbed antibodies have greatly reduced epitope recognition. Without adsorption the non-adsorbed antibody will give a stronger signal in cases where the epitope is weak or in low abundance. For diagnosing problems in immunohistochemistry, refer to Table 28.4 as a starting point.

Table 28.4 *Troubleshooting Immuno-staining*

(This table gives a very brief checklist. For further information, see the references given below.)

1. No staining of either controls or specimen
   - check that no reagents were omitted and that they were added in the right order;
   - check the right antibodies and isotype (e.g. IgG/IgM) were used for specific tissues (use positive control tissue blocks to check this);
   - check the expiry date of reagents;
   - check antibody titrations and dilutions (consider increasing these);
   - check that the blocking solution does not adsorb or interfere with antibody; and
   - if using fixed wax-embedded tissue, consider omitting this step and using frozen material, fixed by paraformaldehyde perfusion fixation, or fresh tissue, fixed later.
2. Weak staining
   In addition to above,
   - ensure that the staining is consistent between positive control and sample;
   - consider over-fixation or poor antigen retrieval/unmasking; and
   - change the fixation protocol, or omit and use frozen material.
3. Background staining
   - check the antibody titre (consider decreasing this);
   - check the blocking serum/solution used; try different normal non-immune sera;
   - check any cross-reaction with 2° antibody;
   - check cells or tissues have not dried out at any step; and
   - check if any particulates are in the antibodies.

References

Buchwalow, I.B. and Böcker, W. (2010). *Immunohistochemistry: Basics and Methods*. New York: Springer.
Burry, R.W. (2010). *Immunocytochemistry: A Practical Guide for Biomedical Research*. New York: Springer.
Dako Education Guide (2013). *Immunohistochemical Staining Methods*, 6th ed. (PDF).

Protocol 28.3 Non-Immuno-staining Protocol for Flat Adherent Cells

For reliable staining of cells to give three fluorescent labels without using antibodies, lectin stains and vital dyes are useful. In the following protocol fixable mitochondrial dyes are combined with a fluorophore-conjugated lectin and intercalating nuclear counterstain. This preparation is a very useful test sample for teaching purposes. The commercial equivalent is the BPAE slide F36924.

Stains:		**Confocal**	**Widefield (approx. values)**		
	Organelle	**laser line**	**Excite**	**Dichroic**	**Emit**
Mitotracker Orange	mitochondria	543 nm	510–560	570	580–630
Alexa-488 phalloidin	F-actin	488 nm	450–490	500	510–560
DAPI	nuclei	405 nm	340–380	410	420–480
DRAQ-5/TO-PRO-3	nuclei	633 nm	625–655	660	670–720

1. Plate out cells onto 25 mm Ø glass coverslips in a 6-well plate at 1×10^5 cells/well with 2 mL DMEM in each well. (The DMEM medium has 250x conc. sodium pyruvate at 4 mL/L; 10% fetal calf serum; 100x L-glutamine @ 10 mL/L and 500x pen-strep antibiotic at 2 mL/L.)
2. 1 mM stock of fixable Mitotracker™ Orange CMTMRos (Cat. No. M-7510) was diluted 5000x = 200 nMolar (recommended working stock is 100 – 500 nM). Keep the stock solution in the dark, in the freezer at −20°C.

3. 50 mL of *serum-free* DMEM was split into 2 x 25 mL aliquots in Falcon tubes and pre-warmed at 37°C, and 5 μL of 1 mM stock of Mitotracker™ Orange was added to one 25 mL aliquot of media. (Serum-free media is used in order to remove exogenous proteins that would otherwise be thrown down onto the cell by the coagulating action of the PFA fixative.)
4. Incubate the cells at 37°C in serum-free medium containing Mitotracker™ Orange for 40 minutes (recommended incubation time is 15–45 minutes).
5. Aspirate the medium, wash the cells twice in serum-free medium and replace with warmed 1x PBS at 37°C. Repeat this step. (Alternatively, replace the media with PBS by increasing dilution: remove half the volume of DMEM media and replace with an equal volume of PBS. Repeat this 3–4 times until most or all the media has been replaced by PBS.)
6. Fix the cells with pre-warmed *methanol-free* 4% PFA in 0.1 molar sodium cacodylate buffer pH 7.4 for 30 minutes at room temperature in the fume hood.
7. Wash the cell three times over 15 minutes with 1x PBS. Use forceps to lift the coverslips in the multiwell plate in order to wash the reverse side free of trapped media/fixative.
8. Wash twice with 1x phosphate-buffered Triton-X 100 (PBT) at room temperature, over 10 minutes. (PBT = 0.1 % Triton-X 100 with 0.5% BSA to reduce background in 1x PBS, pH 7.2–7.4. For 100 mL PBT, use 0.5 g of BSA and 100 μL Triton-X 100.)
9. Stain with Alexa-488 phalloidin (Cat. No. A12379) for F-actin.

   Methanol fixative will disrupt F-actin. Stain for 1 hour at room temperature. From the stock solution dilute 1:50 and 1:100 into PBT. The vial contents should be dissolved in 1.5 mL Analar methanol, to yield a final concentration of 200 units/mL, which is equivalent to approx. 6.6 μM (Mwt Alexa-488 phalloidin = 1320). The Thermo Fisher Scientific PDF recommends dilution of 1:40 of 5 μL stock into 200 μL PBS and staining for 20 minutes between 4°C and 37°C, with subsequent draining and mounting in Prolong Gold. Phalloidin is not a cell permeant phallotoxin and so is not suitable for living cells.
10. Wash the cells three times over 15 minutes with 1x PBS. Use forceps to lift the coverslips in the multiwell plate in order to wash the reverse side free of trapped media/fixative.
11. Stain with DAPI for 15–30 minutes. From the 5 mg/mL 1000x stock solution, use diluted 1:1000; add 12 μL stock to 12 mL 1x PBS, with 2 mL per well incubation at room temperature for 15 minutes. (The 10 mg vial should be diluted into 2 mL of dimethyl formamide to make the stock solution. Store this aliquoted stock at ≤ −20°C.)

    An alternative is to use either DRAQ-5 or ToPro-3. With DRAQ-5, dilute stock supplied as 5 mM in water; dilute to use 10 μM (1:500) or 5 μM (1:1000). Stain coverslips on drops in the dark for 5 minutes; miss out step 12, go to step 13. With ToPro-3, stain 1:1000 from 1 mM stock solution in DMSO. Store this aliquoted stock at ≤ −20°C. Dilute the thawed stock to 1 μM working solution in PBS; stain for 15 minutes.
12. Wash the cell three times over 5 minutes with 1x PBS.
13. Remove the coverslips from the multiwell plate with forceps, dip twice in Milli-Q distilled water for a few seconds to rinse (don't drop it and remember which side the cells adhere to!), drain on filter paper and invert the coverslip, cells down, onto a 10 μL drop of mounting medium. This is 1% *n*-propyl gallate in 90% glycerol in 1x PBS (4 or 5% *n*-propyl gallate has been reported to alter to a lowered pH and cause quenching of fluorescent signal).
14. Seal the coverslips with VALAP or colourless nail varnish.

Controls are vital to verify your experimental results. Include a known positive control antibody, on separate cells or tissue (e.g. another antibody directed against the same antigenic epitope) to your sample, to show that the fixation and any permeabilisation were successful. Also include two negative controls omitting the 1° antibody (replace it with (a) buffer and (b) IgG to host of the 1° mAb at exactly the same concentration, not just 1:100 but concentration as mg/mL). Check the specificity of the 1° antibody by comparing it to a general IgG fraction (monoclonal) or pre-immune serum (polyclonals) and (if possible) against a cell type not expressing the antigen. Use an 'incorrect' 2° antibody to check for species cross-reactivity. Test for autofluorescence by omitting both the 1° antibody and 2° antibody together. All image acquisition settings should be established using the real sample (or a positive control), and then the negative control is imaged using identical settings. The 'autoexpose' or 'find' functions in the microscope acquisition software should never be used when imaging negative controls, since the camera or PMT detectors will attempt to compensate for the low signal. For further details regarding immuno-fluorescence controls, see Section 6, page 20, in Allan (2000).

28.5 Finishing the Preparation

The influence of coverslip thickness on spherical aberration in the image is discussed in Chapter 7, Section 7.7. For the best results, use coverslips of thickness 0.17 mm, and culture the cells, or mount the tissue section, directly on the underside of the coverslip. Sometimes cells will not adhere well to the borosilicate glass surface of the coverslip, and so the glass must be cleaned (e.g. either with 70% ethanol or more rigorously with acid) and coated with a high ionic charge substance, such as poly-L-lysine – for a protocol, see Allan (2000).

Wash the coverslip having sections or adherent cells briefly, by dipping rapidly in distilled water and wicking off the drops of water before mounting. This will remove any buffer salts, which will otherwise dry on the top surface of the coverslip (Figure 28.4). The edges of the coverslip of preparations made with non-curing or non-setting mounting medium should be sealed to prevent desiccation, formation of air spaces or bubbles and/or slippage or movement of the coverslip. If the coverslip moves, the shear forces can easily destroy the sample. There are several choices of sealing medium. Clear nail varnish can be used, which is both quick and easy to apply, but some reports suggest that the isopropanol or acetone solvents will quench fluorescent signals. Another choice is Watermount (Taab), which contains water-soluble plant resins and is easy to work with. Use a minimal volume of mountant (e.g. 8 μL for an 18 mm Ø coverslip), and let this spread to the edge of the coverslip by capillary action. Apply sufficient mountant to prevent ingress of air, but not so much that excess seeps out at the edges, with the risk of smearing onto the top of the coverslip. Mowiol® is a hydrolysed polyvinyl alcohol mounting medium that hardens overnight. A protocol for making Gelvatol is given at the bottom of Table 27.3.

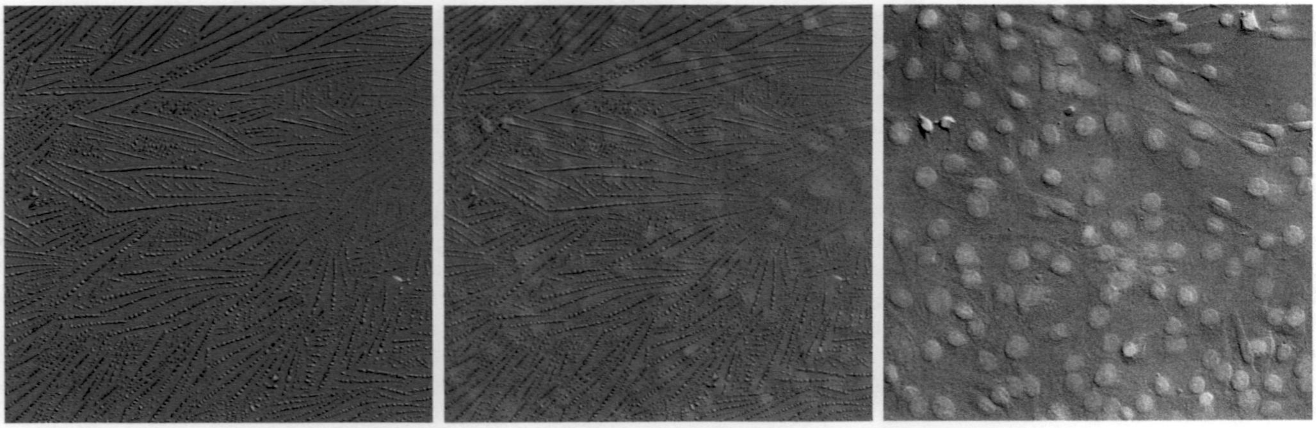

Figure 28.4 **Sample preparation artifacts** *Besides poor washing technique causing high background and non-specific staining in immunohistochemistry, a failure to wash the coverslip before mounting can lead to deposition of buffer salts, which form a characteristic pattern overlaying the image (middle panel). The left-hand panel shows this layer of salts in sharp focus, while the right-hand panel shows a good image secured after cleaning the preparation and its coverslip.*

Protocol 28.4 Protocol for making Mowiol with DABCO antifade

1. Put 6 g of glycerol in a 50 mL plastic centrifuge tube, and add a small stir bar.
2. Add 2.4 g Mowiol 4-88, and stir well to mix. **Mowiol is carcinogenic – use gloves.**
3. While stirring, add 6 mL distilled water, stir again.
4. Add 12 mL 0.2 M Tris (pH 8.5), and leave stirring overnight at room temperature.
5. Add 0.16 g of sodium azide such that the final concentration is approx. 0.02% (optional).
6. Incubate the tube in hot water (50–60°C) for 10 minutes to dissolve the Mowiol.
7. Add 0.2 g DABCO to a final concentration of approx. 0.5% (optional).
8. Centrifuge at 5000 g for 15 minutes to remove any undissolved solids.
9. Store 1 mL aliquots in Eppendorf tubes at −20°C. Use once, then discard.

Sigma-Aldrich sell polyvinyl alcohol (PVA) with DABCO antifade (No. 10981) as a very similar ready-to-use 100 mL solution, which can be aliquoted and stored frozen. Premixed mounting media can – if they are not mixed thoroughly – introduce RI differences and scatter light, particularly those that are commercially supplied and are water-immiscible.

Be very careful *not* to get nail varnish from a freshly sealed preparation onto the front lens of your objectives! It is *very* difficult to remove, especially from dry, non-immersion, objectives with concave front lenses. Always allow sufficient time for the varnish to dry completely before viewing the preparations.

Collecting good images – a summary

1. set up the microscope properly;
2. prepare the sample carefully; remove fluorescent media such as phenol red;
3. eliminate any movement or focus drift;
4. minimise photobleaching, and use antifade reagents;
5. select the correct objective for the job;
6. select the right mounting medium;
7. choose fluorophores that give good SNR with your filter sets;
8. pay attention to bleed-through or cross-talk;
9. where possible, avoid spherical and chromatic aberration;
10. collect within the dynamic range of the detector;
11. set the minimum exposure and field of view of the specimen;
12. fulfil the Nyquist sampling limit; consider deconvolution; and
13. monitor your cells after the experiment for healthy appearance and behaviour.

Table 28.2 gives a list of possible causes of sub-optimal image quality that may arise from poor staining

28.6 Live-Cell Imaging

Time-lapse imaging (for reviews see, Frigault et al., 2009; Coutu and Schroeder, 2013) provides a fuller, more accurate record than mere snapshots (Landecker, 2009; Fink, 2011) and a full record can only be made by acquiring time-lapse data of living cells. The analogy often used is observing or teaching football – a dynamic sport – using stills, or snapshots, alone. However, this is only relevant if the cells and tissues are kept healthy (for dead cells don't dance) and are not subjected to very great photon fluxes.

It is easy to obtain a clear, bright image of fixed, dead cells labelled with exogenous fluorophores. When imaging fixed cells, a maximum signal-to-noise ratio (SNR) may be achieved by either using longer exposure times in widefield systems or in the confocal microscope by using a higher laser power or by increasing the time the laser dwells on each pixel (using slower scan speeds) and/or by averaging the overall signal (summing the signal from several frames). However, light damages living cells. It is much more challenging to keep cells healthy and alive for any length of time under the microscope at the correct pH and temperature – ensuring that temperature changes do not disturb the plane of focus – and then reliably collect the weak endogenous fluorescent signal emitted by the specimen above the background noise generated collectively by both the specimen and the imaging system[8]. Frankly, noisy images are inevitable. If possible use a CCD or sCMOS camera instead of a PMT on a confocal, and pay attention to inherent noise from the detector.

Live-cell and long-term time-lapse imaging is a specialist subject requiring a book in itself, (see Goldman et al., 2010; Watkins and St. Croix, 2014); also much can only be learned by experience for each cell or tissue experimental setup under the environmental conditions in your own laboratory, for each cell, tissue explant or organism is unique. It is not uncommon to require several months work to ensure consistently successful long-term time-lapse imaging in your laboratory. However, it can be done, and acquiring a useful live-cell time-lapse data set is very rewarding.

When performing live-cell imaging, you cannot get away from the *'iron triangle'* and must choose (Figure 28.5) between sensitivity (SNR), resolution and acquisition speed, or overall exposure. Light causes more damage to living samples than was once realised. To maintain healthy cells under the microscope for any significant time, you must adopt strategies that minimise the exposure of the cells and tissues to light, in particular high-energy, short wavelength, light[9]. This is sometimes referred to as the 'photon budget'. Adopting strategies (reduced exposure time, reduced laser power, shorter dwell times, larger pinhole, greater PMT gain or CCD signal binning and Nyquist undersampling) to limit the number of photons illuminating the tissue inevitably means poorer lateral and temporal resolution and reduced SNR. A point often overlooked is to remove DIC prisms, polarisers and magnification changers from the light path if not required for imaging. This will increase the light-collection efficiency of the detector, reducing unwanted exposure of the specimen.

- budget your photons: minimise exposure;
- consider the overall time scale of the live-cell experiment;
- consider the sensitivity, SNR (at the expense of speed of acquisition and overall length or resolution);
- plan the speed (at the expense of resolution or sensitivity – i.e. SNR – and contrast); and
- determine the resolution of detail (at the expense of SNR and speed).

The measure of success of a live-cell time-lapse experiment is that the cells remain healthy, do not shrink, form blebs, or otherwise look unnatural (Figure 28.6). Healthy cells are, quite simply, those that are able to continue to live and divide after exposure to light (the photon insult) at the end of the experiment. A study on yeast (Carlton et al., 2010) showed that the irradiation intensity[10] had to be

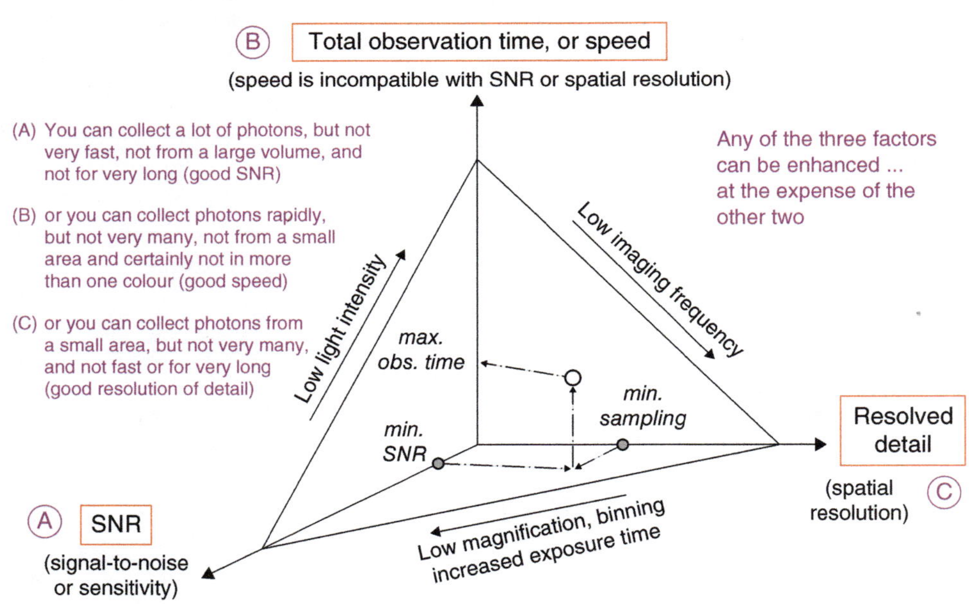

Figure 28.5 **The 'iron triangle'** *The restrictions imposed by the photon budget and the very low tolerance of cells to light means that sensitivity (SNR), resolution and acquisition speed must be balanced and a best compromise decided. It is better to have initial noisy data that can be 'cleaned up' by post-processing than dubious data taken from dead or dying cells. Source: Adapted from Dorn et al. 2008 with permission from Elsevier.*

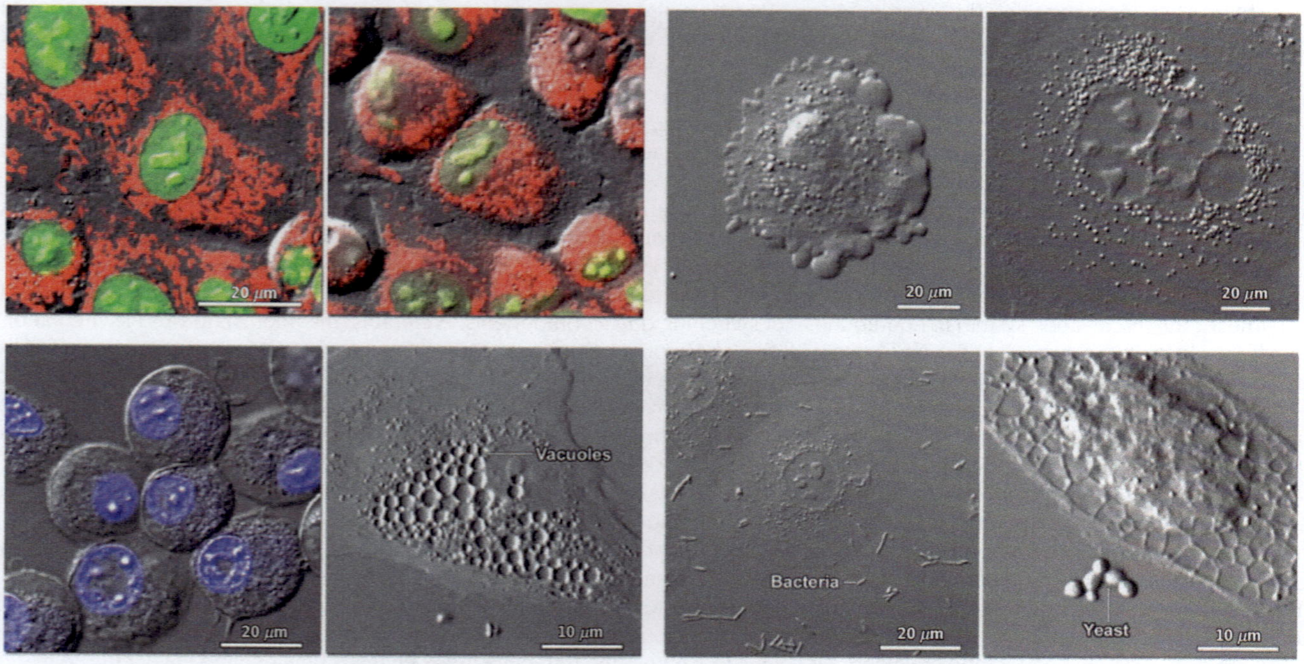

Figure 28.6 **Features of unhealthy cells** *In the four left-hand panels, the effects of photo-toxicity are seen. In the top two illustrations a rabbit kidney cell line (RK-13) expressing eYFP fused to a nuclear location signal has been stained with Mitotracker red (far left top panel) and time-lapse imaged for 2 hours at 15-minute intervals. Fluorescence and DIC modes shown in the image. In the corresponding field, the mitochondria have fragmented, and the cells have rounded up. In the bottom left-hand illustration, HeLa cells stained with Hoechst 33342 for 30 minutes at 10-second intervals with a 405 nm laser in a confocal microscope have become detached from the coverslip and very much rounded up, instead of being healthy, flat, adherent cells. In the corresponding illustration vacuoles have formed after imaging for 8 hours at 30-second intervals with tungsten halogen illumination using DIC. Even with a significantly reduced photon flux in widefield non-fluorescence microscopy than fluorescence widefield or confocal imaging, cells still suffer photo-toxicity and show signs of photo-damage.*

In the right-hand panel of four illustrations, further signs of ill health are shown. Membrane blebbing, a classic sign of cells entering apoptosis or dying from photo-toxicity is shown in the top left-hand illustration. Binucleated cells are shown in the top right-hand illustration; bacterial contamination, especially mycoplasma infection, which results from poor tissue culture technique, is shown in the lower left-hand panel. The lower right-hand panel shows yeast contamination and extensive vacuole formation.

Source: Murphy and Davidson 2013. Reproduced with permission from John Wiley & Sons.

1/100th the usual value [$\approx 5 \times 10^{-5}$ W/μm^2] in order that normal cell growth is not arrested. That is, 500 nW/μm^2/sec – an extremely low value (see also Box 19.1). The authors considered that normal growth and division through five subsequent cell cycles was a reliable indicator that the cells were not damaged or perturbed. Other studies on mammalian cells have shown similar results (Rieder and Cole, 1998; Krawczyk et al., 2008). In these circumstances, the images have to be de-noised (Carlton et al., 2010; Luisier et al., 2010; Yang et al., 2010; Arigovindan, M., et al., 2013) to perceive the image signal above the noise floor. The light-sheet microscope (Chapter 19) is ideal in the respect, exposing the specimen to between 1–4 orders of magnitude less radiative energy than the widefield or confocal fluorescence microscopes. Increasingly, experimenters are becoming aware of the crucial importance of managing the 'photon budget' to limit and ameliorate the time-lapse by controlling the effect of photo-toxicity wherever possible. The *Nature Methods* Editorial *Artifacts of light* is a short and succinct introduction that everyone should read (Anonymous, *Nature Methods 10*/12: 1135). Two recent papers dealing with the assessment and management of photo-toxicity are Icha et al. (2017) and Laissue et al. (2017) besides those given elsewhere in this Section of Chapter 28.

There are three principal ways in which damage may be caused. Firstly, UV light can induce DNA lesions, forming pyrimidine dimers from DNA, and initiate apoptosis and be genotoxic in conjunction with DNA probes such as Hoechst and Draq-5 (Zhao et al., 2009; Ge et al., 2013). Syto17 is to be preferred as a stain. Secondly, visible light is also harmful, causing the photochemical formation of excessive concentrations of reactive oxygen species (ROS, see Halliwell and Aruoma, 1991; Dixit and Cyr, 2003; Khodjakov and Rieder, 2005, and Wang et al., 2013). Image-iT^{TM} live (Thermo Fisher Scientific) can be used to detect the presence of ROS in live cells. Thirdly, photophysical mechanisms, such as localised heating at the point of focus, causes damage, a factor which is often overlooked (König, 2006; Bixler et al., 2014). Where possible, longer-wavelength red light should be used for imaging, since this is much less harmful to living tissues than higher-energy UV or blue wavelengths. Ideally, the illumination exposure cycle can be controlled (Hoebe et al., 2007, 2010; Magidson and Khodjakov, 2013) in software. Controlled light exposure microscopy (CLEM) is fitted to Nikon fluorescence microscopes and adopts a twofold approach to minimising exposure of the sample to light. Firstly, if there is no sample at the pixel detected by the PMT or CCD, the illumination is not turned on. Secondly, if there is a signal of detectable intensity at the pixel, the illumination is switched on but is switched off once a statistically valid signal has been collected from which a final intensity value can be obtained by extrapolation. This overcomes any Poisson noise arising from insufficient signal being collected, but since CLEM analyses the sample pixel by pixel, the overall result keeps exposure of the sample to harmful photons to an absolute minimum. Another very good way to reduce overall illumination intensity is to use LEDs (Cole and Turner, 2008; Cole, 2014) rather than lasers, mercury arc-lamps or metal-halide illuminators. LEDs require no mechanical shutters[11] (which reduces vibration) and they can be pulsed (Nishigaki, T., et al., 2006) which reduces the total photon flux. It is estimated that pulsed LEDs achieve the same image quality using less than 25% exposure time of a shuttered lamp. An excellent and very practical paper (Bosse et al., 2015) describes a protocol to build and operate a pulsed LED illuminator using open-source software.

Besides cell health, an equally important parameter that must be controlled is focus drift. To overcome this problem of focus shift-over-time, the environment must have a constant temperature for at least 4–8 hrs before starting a long time-lapse experiment. Focus drift mainly arises as a result of thermal influence on the geometry of objective, nosepiece, stand and stage. Open windows and air conditioners (Adler and Pagakis, 2003) also cause variation in the temperature of the immediate environment, as does waiting for too brief a time for the microscope system to equilibrate after switching on all components. Preferably, the microscope and its incubator (or 'hotbox') should be left on overnight to equilibrate with the laboratory environment and achieve a constant temperature. Some facilities keep their live-cell microscopes on and heated all the time. One of the leading firms with years of experience of making bespoke microscope incubators is Solent Scientific. Their address and contact details are given in Appendix 5. Alternatively, many facilities construct their own incubators; instructions for two designs, one from the Gray Institute at the University of Oxford, and the other from the Stowers Institute, are given under the Internet Resources, at the end of the chapter.

Changes to the focal plane can also arise from flexing of the coverslip in the culture dish, from either thermal changes or fluid pressure caused by perfusion/adding experimental reagents, external vibration (see Chapter 4) and specimen movement (e.g. cells rounding up as they pass through mitosis). Both software and hardware solutions exist to compensate for changes in focal position during time lapse. Software algorithms search for maximum contrast and resolved detail within several images in a Z-stack to set the point of best focus, but this takes time and is only suitable for time lapses with long intervals (in the order of minutes) between each exposure.

All four microscope manufacturers manufacture a hardware focus compensation system (Figure 28.7) for their instruments. These measure the distance between the objective and the coverslip – where the refractive index changes between the glass–culture medium interface or in the case of a dry objective, the lower coverslip surface–air interface. An offset is then calculated so that the position of focus is corrected to lie within the specimen adherent to the coverslip. The first of these compensation systems, with patented features that make it the fastest on the market, is the Perfect Focus System (PFS) produced by Nikon, now in its fourth version at the time of writing. This uses an 870 nm LED in the infrared range and an internal linear (i.e. fast) CMOS detector, to detect the focal point, so it does not intrude on wavelengths used for observation, allowing simultaneous observation and focus maintenance when switching to image acquisition. Multiple Cartesian coordinates (x,y,z) can be recorded to enable collection of long-term multi-dimensional

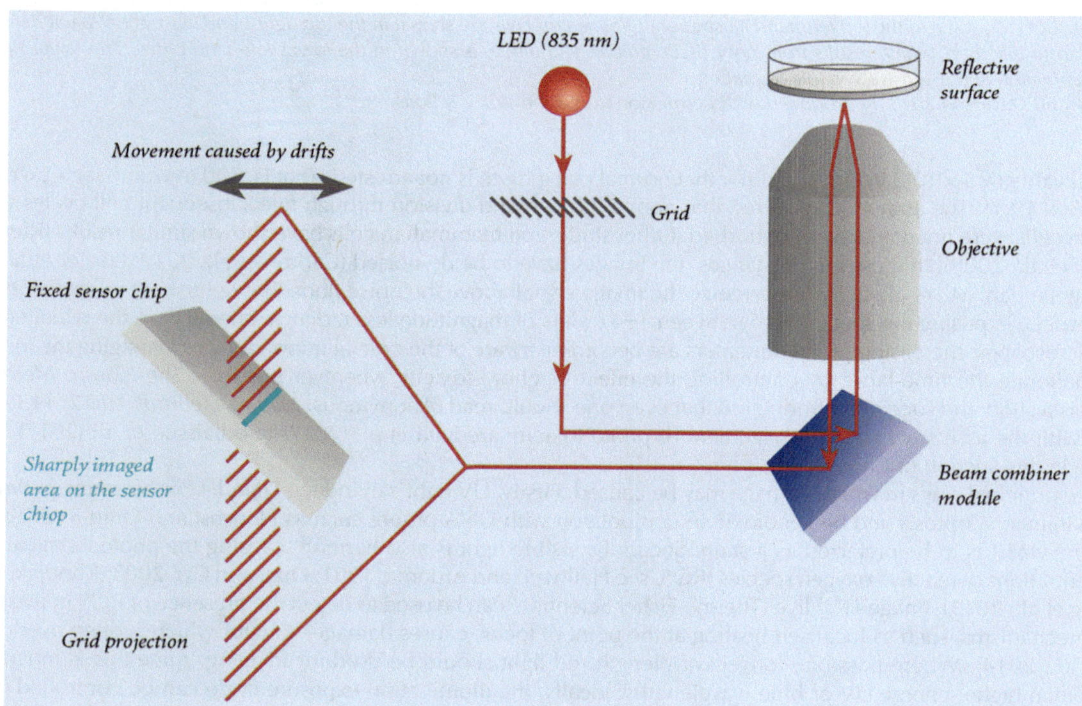

Figure 28.7 **Automatic focus compensation** *A focus compensation system projects an image of a grid using a far red 830–870 nm LED. This image is reflected off the coverslip-media interface onto a CMOS sensor fixed at a slant, which can very sensitively detect changes in position and thus counteract any drift. Source: Reproduced with permission from Carl Zeiss Ltd.*

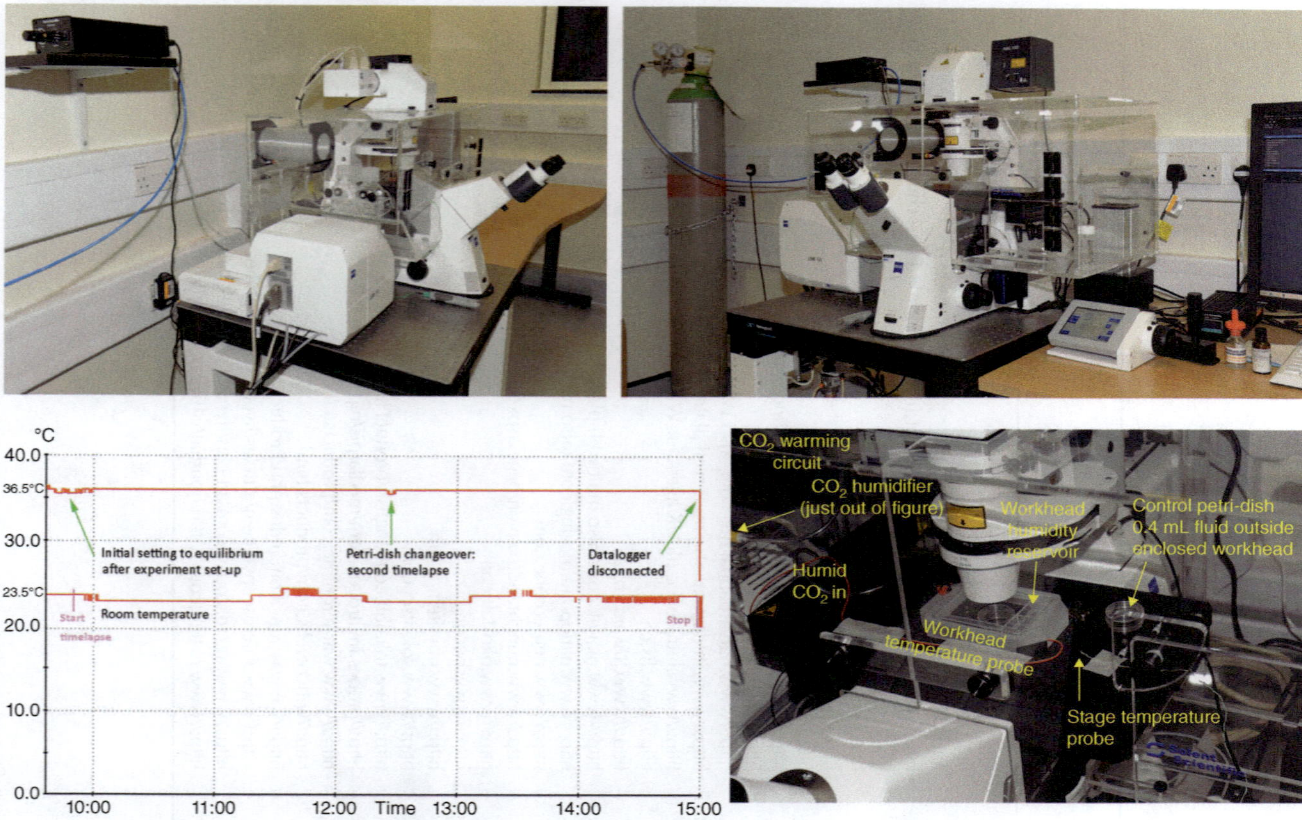

Figure 28.8 ***A typical live-cell time-lapse apparatus*** *The microscope incubator contains a gas warmer, two humidifying bottles and a chamber (here called the workhead) to control the micro-environment at the focal plane on the microscope stage. The supply of 5% CO_2 in air, the gas mixer and flow regulator, and the incubator heater are external to the microscope and outside the picture. Notice that the focus controls remain external to the incubator for ease of use without disturbing the heated environment. Source: Courtesy of Mr K. Glover.*

time-lapse experiments. Leica call their system Adaptive Focus Control; Olympus have Zero Drift Control, and Zeiss refer to their system as Definite Focus. A few researchers have installed non-commercial autofocus feedback mechanisms; refer to Shen et al. (2006) and Smirnov et al. (2017) for further details.

Besides keeping the temperature surrounding the microscope constant, it is essential for long-term time-lapse studies also to control the environment immediately at the dish holding the cells or tissue explant at the plane of focus. Most cells require humidified 5% CO_2 in 95% air to be bubbled through the culture medium to survive. This is usually about the rate of 1 bubble per second, which can be seen as the gas bubbles through the humidity bottles (Figure 28.8) but must be optimised by trial and error. The CO_2 buffers the media and maintains physiological pH. Two alternatives are to use 25 mM HEPES buffer at pH 7.4, which is CO_2 independent, but for no longer[12] than 8 hours, or Leibovitz L-15 medium. For timescales longer than 8 hours, exogenous humid CO_2 should be bubbled through the media. Stevenson et al. (2008) propose a modified Carrel flask instead of glass-bottomed Petri dishes (Table 28.5) to maintain a CO_2-laden atmosphere in which to both grow and visualise cell cultures. Most cells will withstand deliberate oxygen depletion to prevent the formation of ROS, because most tissues exist at a partial pressure of oxygen (pO_2) of 40 mm Hg, which is 26% of the pO_2 in air.

All media used for image acquisition and live-cell work should be 'phenol-red free'; otherwise, the phenol red pH indicator will fluorescence and interfere with the true emitted signal. If the correct pH is not maintained, the cells will suffer within a few minutes, particularly in alkaline media. Although rarely considered, another factor to take into account is the osmotic pressure of the media. Mammalian cells should be kept within a range of 260–330 mOsm/kg solute; the human blood plasma value is 290 mOsm. Figure 28.8 shows the chamber controlling this micro-environment within the incubator. This particular chamber is manufactured by Solent Scientific, but alternatives can be bought from PeCon GmbH, Oko-labs, Tokai-Hit, Ibidi, Chamlide and Bioptechs. URL links to these companies' websites are given at the end of the chapter. Details for home-building three designs of environmental chambers have been published by Heidemann et al. (2003), Kulesa and Kasemeier-Kulesa (2007), and Cole (2014).

Many advanced long-term live-cell imaging platforms have some means (Figure 28.9) of heating the objective[13]. Okolabs also provides an objective water-cooling jacket which can be useful for imaging at lower temperatures to stabilise the heat efflux from a warm sample to the colder microscope. Another firm, Cherry Biotech also manufacture a stage-top heater/cooler, although at present this equipment will only fit standard 3″ × 1″ slide preparations (e.g. for microbiology) rather than glass-bottomed Petri dishes, it is remarkably accurate.

Table 28.5 *Solutions for Live-Cell Imaging Dishes, Slides, Chambers and Heaters*

Company	Petri-dish coverslip bottom	Chamber slide	Perfusion chamber	Incubator and/or heated stage insert	Website URL link
20-20 Technology	✓	✓	✓	✓	http://www.20-20tech.com
ALA Scientific			✓	✓	http://www.alascience.com
AutoMate Scientific			✓		http://www.autom8.com
BD Falcon		✓			http://www.bdbiosciences.com/external_files/dl/doc/mkt_lit/brochures/B00F022v1.pdf
Bellco glass	✓		✓		http://www.bellcoglass.com
Bioptechs	✓		✓	✓	http://www.bioptechs.com
					http://www.bioptechs.com/Products/Dishes/Dishes.html
BioscienceTools	✓	✓	✓	✓	http://www.biosciencetools.com
C & L Instruments			✓		http://www.fluorescence.com/products/index.htm#v-cell
Grace Bio-Labs		✓	✓		http://www.gracebio.com
Harvard Apparatus		✓	✓	✓	http://www.harvardapparatus.com (also Warner Instruments)
Integrated Biodiagnostics (ibidi)	✓	✓	✓	✓	http://www.ibidi.com (also muti-well plates & microfluidics)
Intracel	✓		✓	✓	http://www.intracel.co.uk UK agent for Willcowells dishes
Life Imaging Services			✓	✓	http://www.lis.ch
Linkam Scientific Instruments			✓	✓	http://www.linkam.co.uk
MatTek Corporation	✓				http://www.mattek.com (also multiwell plates)
Nanopoint Inc.		✓			http://www.nanopointimaging.com (also microfluidics)
Neuroprobe		✓	✓		http://www.neuroprobe.com
Nunc		✓			http://www.nuncbrand.com/en/page.aspx?ID=229 (also multiwell plates)
Oko Lab			✓	✓	http://www.oko-lab.com
PeCon			✓	✓	http://www.pecon.biz
Physitemp Instruments				✓	http://www.physitemp.com
Solent Scientific				✓	http://www.solentsci.com
Tokai Hit			✓	✓	http://www.tokaihit.com/english/product%20guidance/index.html
Thermo Fisher (Invitrogen)		✓			http://probes.invitrogen.com/media/pis/mp18120.pdf
Warner Instruments			✓	✓	http://www.warneronline.com/products.cfm
Willco wells	✓				http://willcowells.com/products.html
World Precision Instruments	✓		✓		http://www.wpiinc.com/index.php/Products.html
Further resources					http://www.microscopyu.com/articles/livecellimaging/livecellmaintenance.html
					http://www.microscopyu.com/articles/livecellimaging/chamberresources.html
					http://www.nature.com/nature/journal/v466/n7310/pdf/4661141a.pdf

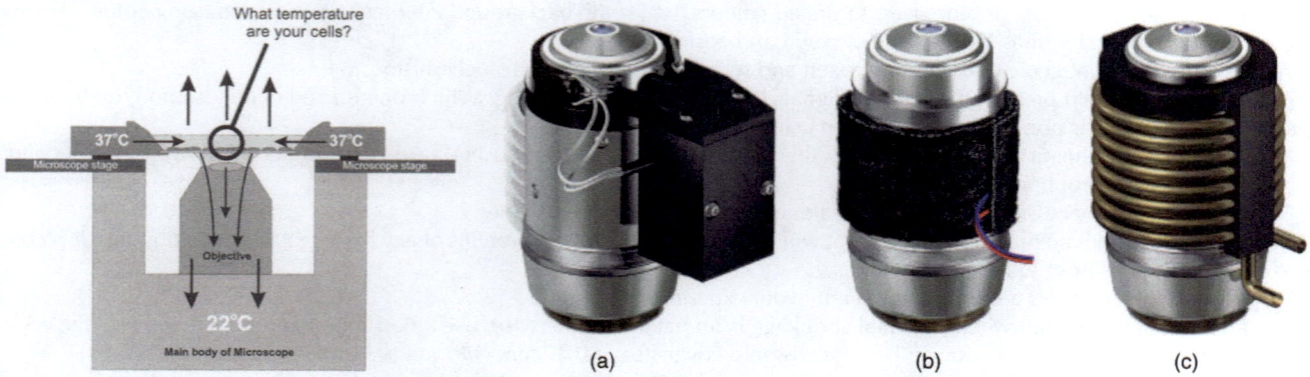

Figure 28.9 **Designs of objective heaters** *A heated chamber on the microscope stage may be able to control the temperature of the micro-environment very accurately, but if the objective acts as a heat sink, this cannot be known precisely and may very well be much lower than the optimum temperature hoped for. The most efficient objective heater is the design shown in (a) which has a closed feedback loop that only heats the top of the objective near the front lens element. The heating blanket design (b) and water jacket (c) are less efficient, with heat also being radiated away to the area beneath the specimen instead of being concentrated upon the objective. Source: (left) Hibbs 2004. Reproduced with permission from Springer. (right) Murphy and Davidson 2013. Reproduced with permission from John Wiley & Sons.*

Further details may be found in Velve-Casquillas et al. (2010) and the Internet Resources, at the end of the chapter. Zeiss have developed calorimetrically insulated objectives, called Zeiss i-objectives, for live-cell imaging, in which this heat loss is minimised. Heating the objective not only helps to prevent the microscope acting as a heat-sink but also helps to maintain the actual temperature of the cells or tissues at a temperature at or near the physiological optimum of 37°C (for mammalian cells). Hypothermia at 20°C can inhibit or arrest mitosis (see Figure 4, page 5 in Cole, 2014) in a homeothermic organism, but this temperature may be ideal for poikilothermic *Drosophila* tissues (Kuntz and Eisen, 2014).

Not only can the sample go out of focus because of thermal drift, it can also move out of the plane of focus and also out of the field of view. Sometimes it is necessary to anchor the cells or tissues to the glass or plastic substrate of the dish, or it may be necessary to immobilise the sample. This can be done in several ways (Aufderheide, 2008; Westendorf et al., 2010; Yan et al., 2014): by using a viscous oil layer over the cell layer – this has the additional advantage of excluding oxygen; by overlaying the cells with a thin layer of agarose or by using a microfluidic chamber. Adherent cells can be grown on a coverslip forming the chamber floor treated with a physiologically compatible adhesive (poly-L-lysine, concanavalin A). Tissues or small organisms can be trapped in a small volume of fluid, or a small cell of agarose or self-adhesive reinforcement rings (Michaels and Büntzow, 2010) sandwiched between coverslips (Ettinger and Wittmann, 2014) or within a rotocompressor. Further illustrated details are given in the cited papers.

Most software programmes allow you to mark the location of cells and features of interest and store this information as a separate file within the metadata of an acquired image. The best analogue means of locating a position on a microscope slide, so that this field of view may be revisited, is by using an England Finder, whose use is described in Chapter 31, Section 31.10.

Intravital microscopy is the recording of cellular and tissue function in live animals. As such, it is subject to stringent ethical procedures overseen in the UK by the Home Office, who grant licences for approved projects, and by the Department of Agriculture in the USA. The microscopes have to be used and maintained within an animal research facility. Intravital microscopy is a very demanding research topic to plan and execute, and such is not covered in this book. For good introductions to intravital microscopy, see Beerling et al. (2011), Timpson et al. (2011), Andresen et al. (2012), Masedunskas et al. (2012), Nobis et al. (2013) and Ellenbroek and van Rheenen (2014).

Whether performing time-lapse microscopy over a few hours or a few days, the final data set will be in the order of tens or hundreds of gigabytes, or more. An experiment may well require more than one time-lapse data set for meaningful analysis. Data of this magnitude has to be processed and analysed by protocols that have been validated beforehand and built into the experimental design. Cohen (2014) discusses how to extract meaning from biological imaging data in a paper of that name.

28.7 Chapter Summary

At the end of this chapter you should be able to prepare samples for histology, immunohistochemistry and fluorescence microscopy. You should be aware of the factors to take into consideration in order to acquire images of living cells as a time-lapse recording.

1. Tissues require fixation to prevent decay, to preserve morphology and where possible also preserve antigenic epitopes. Fixation also hardens tissues sufficiently to withstand subsequent processing and sectioning for specimen preparation.
2. There is not one universally suitable fixative. Fixatives may broadly be divided into cross-linking and coagulant (precipitating) types. In histology, 10% buffered formalin is routinely used, but tissues generally require post-treatment for immunohistochemistry.
3. Precipitating fixatives generally allow better access to antigens. Cross-linking fixatives will normally require post-fixation antigenic epitope unmasking.

4. Nuclei are stained with red or blue dyes, saturated colours, whilst the background counterstain is a contrasting colour. The most common histological stain is H&E – haematoxylin and eosin.
5. Tissues with labile antigenic epitopes are frozen and sectioned for immunohistochemistry.
6. Antifade reagents help preserve fluorescent signals by free oxygen scavenging. Which one is used depends upon the fluorescent stain used. Antifade is normally avoided when using fluorescent proteins.
7. Most immunohistochemistry is performed by the indirect staining method, using a primary antibody followed by a secondary conjugated to a fluorophore.
8. Permeabilisation is necessary to allow antibodies access across cell membranes.
9. Cells and tissues contain macromolecules that will bind non-specifically to proteins of any kind – including antibodies. It is crucial to block these sites prior to immuno-staining.
10. Properly planned controls are essential to verify your experimental results.
11. Use a hardening mountant when potential shrinkage is immaterial. Otherwise, use a non-setting mountant, and seal the edge of the coverslip. For best results, grow adherent cells onto coverslips of 0.17 mm (170 μm ± 5 μm) thickness.
12. When live-cell imaging, give due regard to the 'iron triangle'. Balance acquisition speed against SNR and resolution. Cells require minimal light exposure to remain healthy.
13. Warm up the microscope before a time-lapse experiment to manage focus drift. Pay attention to the immediate environment at the microscope stage and focus plane as well as the incubator and room environment.
14. Check the light exposure levels, the temperature, humidity, pH and oxygen content. For some cells and tissues, also check the osmolality.

Key Reading

Eltoum, I. et al. (2001). 'Introduction to the theory and practice of fixation of tissues'. *Journal of Histotechnology 24*/3: 173–190.
- Selection of the proper fixative is essential and can dictate the success or otherwise of subsequent experimental work. This paper is one of the best reviews.

Vermot, J. et al. (2008). 'Fast fluorescence microscopy for imaging the dynamics of embryonic development'. *HFSP Journal 2*/3: 143–155.
- This paper tells you how to make every photon count and achieve the best temporal resolution with the different microscopy platforms available.

References

Adler, J. and Pagakis, S.N. (2003). 'Reducing image distortions due to temperature-related microscope stage drift'. *Journal of Microscopy 210*/2: 131–137.

Alberts, A., Johnson, A., and Lewis J. (eds.) (2014). *Molecular Biology of the Cell*, 6th ed. Garland Science, an imprint of Taylor & Francis.

Allan, V.J. (2000). 'Basic Immunofluorescence'. In: Allan, V.J. (ed.), *Protein Localization by Fluorescence Microscopy – a practical approach*, Chapter 1. Oxford: OUP, pp. 1–26.

Andresen, V. et al. (2012). 'High-resolution intravital microscopy'. *PLoS One 7*/12: e50915.

Anonymous. Artifacts of Light. *Nature Methods 10*/12: 1135.

Aoyagi, Y. et al. (2015). 'A rapid optical clearing protocol using 2,2′-thiodiethanol for microscopic observation of fixed mouse brain'. *PLoS One 10*/1: e0116280.

Arigovindan, M. et al. (2013). 'High-resolution restoration of 3D structures from widefield images with extreme low signal-to-noise ratio'. *PNAS 110*/43: 17344–17349.

Aufderheide, K.J. (2008). 'An overview of techniques for immobilizing and viewing living cells'. *Micron 39*/2: 71–76.

Beerling, E. et al. (2011). 'Intravital microscopy: new insights into metastasis of tumors'. *Journal of Cell Science 124*/3: 299–310.

Bixler, J.N. et al. (2014). 'Assessment of tissue heating under tunable near-infrared radiation'. *Journal of Biomedical Optics 19*/7: 070501.

Bogdanov, A.M. et al. (2012). 'Anti-fading media for live cell GFP imaging'. *PLoS One 7*/12: e53004.

Bosse, J.B. et al. (2015). 'Open LED Illuminator: a simple and inexpensive LED illuminator for fast multicolour particle tracking in neurons'. *PLoS One 10*/11: e0143547.

Buchwalow, I.B. and Böcker, W. (2010). *Immuno-histochemistry: basics and methods*. New York: Springer.

Burry, R.W. (2010). *Immunocytochemistry: a practical guide for biomedical research*. New York: Springer.

Carlton, P.M. et al. (2010). 'Fast live simultaneous multiwavelength four-dimensional optical microscopy'. *PNAS 107*/37: 16016–16022.

Carson, F.L. and Cappellano, C. (2015). *Histotechnology: a self-instructional text*, 4th ed. Chicago: American Society for Clinical Pathology.

Cohen, A.R. (2014). 'Extracting meaning from biological imaging data'. *Molecular Biology of the Cell 25*/22: 3470–3473.

Cole, R. (2014). 'Live-cell imaging: the cell's perspective'. *Cell Adhesion & Migration 8*/5: 452–459.

Cole, R.W. and Turner, J.N. (2008). 'Light-emitting diodes are better illumination sources for biological microscopy than conventional sources'. *Microscopy and Microanalysis 14*/3: 243–250.

Cook, J. (2006). *Cellular Pathology*, 2nd Rev. Ed. Banbury: Scion Publishing Ltd.

Cordes, T. et al. (2011). 'Mechanisms and advancement of antifading agents for fluorescence microscopy and single-molecule spectroscopy'. *Physical Chemistry Chemical Physics 13*/14: 6699–6709.

Cordes, T. et al. (2009). 'On the mechanism of Trolox as antiblinking and antibleaching reagent'. *Journal of the American Chemical Society 131*/14: 5018–5019.

Coutu, D.L. and Schroeder, T. (2013). 'Probing cellular processes by long-term live imaging – historic problems and current solutions'. *Journal of Cell Science 126*/17: 3805–3815.

Dako Education Guide. *Immunohistochemical Staining Methods*, 6th ed. (PDF).

Dako Education Guide. *Demasking of Antigens*, 2nd ed. (PDF).

Dako Education Guide. *Special Stains and H&E*. (PDF).

Dixit, R. and Cyr, R. (2003). 'Cell damage and reactive oxygen species production induced by fluorescence microscopy: effect on mitosis and guidelines for non-invasive fluorescence microscopy'. *The Plant Journal 36*/2: 280–290.

Dorn, J.F. et al. (2008). 'Computational Processing and Analysis of Dynamic Fluorescence Image Data'. In: *Fluorescent Proteins*, 2nd ed, Chapter 22. *Methods in Cell Biology*, vol. 85. Elsevier, pp. 497–538.

Dorph-Petersen, K.-A. et al. (2000). 'Tissue shrinkage and unbiased stereological estimation of particle number and size'. *Journal of Microscopy 204*/3: 232–246.

Ellenbroek, S.I. and van Rheenen, J. (2014). 'Imaging hallmarks of cancer in living mice'. *Nature Reviews Cancer 14*/6: 406–418.

Ettinger, A. and Wittmann, T. (2014). 'Fluorescence live cell imaging'. In: *Methods in Cell Biology*, Chapter 5, vol. 123, pp. 77–94.

Fink, R. (2011). 'A single frame: imaging live cells twenty-five years ago'. *Genesis 49*/7: 484–487.

Fischer, A.H. et al. (2006). 'Preparation of cells and tissues for fluorescence microscopy'. In: Spector, D.L. and Goldman, R.D. (eds.), *Basic Methods in Microscopy*, Chapter 4. New York: Cold Spring Harbor Laboratory Press, pp. 105–123.

Fischer, A.H. et al. (2008). 'Cryosectioning Tissues'. *Cold Spring Harbor Protocols 3*/8. doi: 10.1101/pdb.prot4991.

Frigault, M.M. et al. (2009). 'Live-cell microscopy – tips and tools'. *Journal of Cell Science 122*: 753–767.

Ge, J. et al. (2013). 'Standard fluorescent imaging of live cells is highly genotoxic'. *Cytometry A 83*/6: 552–560.

Goldman, R.D. et al. (eds.) (2010). *Live Cell Imaging: a laboratory manual*, 2nd ed. New York: Cold Spring Harbor Laboratory Press.

Grizzle, W.E. (2009). 'Models of Fixation and Tissue Processing'. *Biotechnic & Histochemistry 84*/5: 185–193.

Halliwell, B. and Aruoma, O.I. (1991). 'DNA damage by oxygen-derived species. Its mechanism and measurement in mammalian systems'. *FEBS Letters 281*/(1–2): 9–19.

Heidemann, S.R. et al. (2003). 'Open-dish incubator for live cell imaging with an inverted microscope'. *Biotechniques 35*/4: 708–716.

Hibbs, A.R. (2004). *Confocal Microscopy for Biologists*. New York: Kluwer Academic/Plenum Publishers.

Hoebe, R.A. et al. (2007). 'Controlled light-exposure microscopy reduces photobleaching and phototoxicity in fluorescence live-cell imaging'. *Nature Biotechnology 25*/2: 249–253.

Hoebe, R.A. et al. (2010). 'Noise effects and filtering in controlled light exposure microscopy'. *Journal of Microscopy 240*/3: 197–206.

Icha, J. et al. (2017). 'Phototoxicity in live fluorescence microscopy, and how to avoid it'. *Bioessays 39*/8. doi: 10.1002/bies.201700003.

Khodjakov, A. and Rieder, C.L. (2005). 'Imaging the division process in living tissue culture cells'. *Methods 38*/1: 2–16.

Kiernan, J.A. (2008). *Histological and Histochemical Methods*, 4th ed. Banbury: Scion Publishing Ltd.; New York: CSHL Press.

Krawczyk, P.M. et al. (2008). 'Analysis of the mobility of DNA double-strand break-containing chromosome domains in living mammalian cells'. *Methods in Molecular Biology 463*: 309–320.

Kulesa, P.M. and Kasemeier-Kulesa, J.C. (2007). 'Construction of a Heated Incubation Chamber around a Microscope Stage for Time-Lapse Imaging'. *Cold Spring Harbor Protocols* pdb.prot4792.

Kuntz, S.G. and Eisen, M.B. (2014). '*Drosophila* Embryogenesis Scales Uniformly across Temperature in Developmentally Diverse Species'. *PLoS Genetics 10*/4: e1004293.

Laissue, P.P. et al. (2017). 'Assessing phototoxicity in live fluorescence imaging'. *Nature Methods 14*/7: 657–661.

Landecker, H. (2009). 'Seeing things: from microcinematography to live cell imaging'. *Nature Methods 6*/10: 707–709.

Lepe-Zuniga, J.L. et al. (1987). 'Toxicity of light-exposed Hepes media'. *Journal of Immunological Methods 103*/1: 145.

Luisier, F. et al. (2010). 'Fast interscale wavelet denoising of Poisson-corrupted images'. *Signal Processing 90*/2: 415–427.

Magidson, V. and Khodjakov, A. (2013). 'Circumventing photodamage in live-cell microscopy'. In: *Methods in Cell Biology*, Chapter 23, vol. 114: 545–560.

Marx, V. (2013). 'Finding the right antibody for the job'. *Nature Methods 10*/8: 703–707.

Masedunskas, A. et al. (2012). 'Intravital microscopy: a practical guide on imaging intracellular structures in live animals'. *Bioarchitecture 2*/5: 143–157.

McLean, I.W. and Nakane, P.K. (1974). 'Periodate-lysine-paraformaldehyde fixative a new fixative for immunoelectron microscopy'. *Journal of Histochemistry and Cytochemistry 22*/12: 1077–1083.

Michaels, J. and Büntzow, M. (2010). 'Assessment of Congo red as a fluorescence marker for the exoskeleton of small crustaceans and the cuticle of polychaetes'. *Journal of Microscopy 238*/2: 95–101.

Murphy, D.B. and Davidson, M.W. (2013). *Fundamentals of Light Microscopy and Digital Imaging*, 2nd ed. Wiley Blackwell.

Nishigaki, T. et al. (2006). 'Stroboscopic illumination using light-emitting diodes reduces phototoxicity in fluorescence cell imaging'. *Biotechniques 41*/2: 191–197.

Nobis, M. et al. (2013). 'Advanced intravital subcellular imaging reveals vital three-dimensional signalling events driving cancer cell behaviour and drug responses in live tissue'. *FEBS Journal 280*/21: 5177–5197.

Orchard, G. and Nation, B. (eds.) (2017). *Histopathology*, 2nd ed. Oxford: Oxford University Press.

Otali, D., et al. (2009). 'Combined effects of formalin fixation and tissue processing on immunorecognition'. *Biotechnic & Histochemistry 84*/5: 223–247.

Rieder, C.L. and Cole, R. (1998). 'Perfusion chambers for high-resolution microscopic studies of vertebrate cell monolayers: some considerations and a design'. *Methods in Cell Biology 56*: 253–275.

Sanderson, J.B. (1994). *Biological Microtechnique*. RMS Handbook No. 28. Oxford: Bios Scientific Publishers (currently published by Taylor & Francis).

Schnell, U. et al. (2012). 'Immunolabeling artifacts and the need for live-cell imaging'. *Nature Methods 9*/2: 152–158.

Shen, F. et al. (2006). 'Digital Autofocus Methods for Automated Microscopy'. *Methods in Enzymology 414*: 620–632.

Smirnov, M.S. et al. (2017). 'Automated Remote Focusing, Drift Correction, and Photostimulation to Evaluate Structural Plasticity in Dendritic Spines'. *PLoS One 12*/1: e0170586.

Somogyi, P. and Takagi, H. (1982). 'A note on the use of picric acid-paraformaldehyde-glutaraldehyde fixative for correlated light and electron microscopic immunocytochemistry'. *Neuroscience 7*/7: 1779–1783.

Staudt, T. et al. (2007). '2,2′-thiodiethanol: A new water soluble mounting medium for high resolution optical microscopy'. *Microscopy Research and Technique 70*/1: 1–9.

Stevenson, D.J. et al. (2008). 'Long-term cell culture on a microscope stage: the Carrel flask revisited'. *Microscopy and Analysis* Jan 2008, 9–11. https://bit.ly/2rMWZYi.
Suvarna, K.S. et al. (2012). *Bancroft's Theory and Practice of Histological Techniques*, 7th ed. Oxford: Churchill-Livingstone, Elsevier.
Timpson, P. et al. (2011). 'Imaging molecular dynamics in vivo – from cell biology to animal models'. *Journal of Cell Science 124*/17: 2877–2890.
Tycon, M.A. et al. (2012). 'Quantification of dye-mediated photodamage during single-molecule DNA imaging'. *Analytical Biochemistry 426*/1: 13–21.
Velve-Casquillas, G. et al. (2010). 'A fast microfluidic temperature control device for studying microtubule dynamics in fission yeast'. *Methods in Cell Biology 97*: 185–201.
Wang, X. et al. (2013). 'Imaging ROS signaling in cells and animals'. *Journal of Molecular Medicine* (Berlin) *91*/8: 917–927.
Watkins, S.C. and St. Croix, C.M. (2014). 'Building a live-cell microscope: what you need and how to do it'. In: Watkins, S. and St. Croix, C. (eds.), *Current Protocols Select: Methods and Applications in Microscopy and Imaging*. New Jersey: Wiley.
Westendorf, C. et al. (2010). 'Live-cell flattening – traditional and novel approaches'. *PMC Biophysics 3*/1: 9. doi: 10.1186/1757-5036-3-9.
Yan, Y. et al. (2014). 'A Microfluidic-Enabled Mechanical Microcompressor for the Immobilization of Live Single- and Multi-Cellular Specimens'. *Microscopy and Microanalysis 20*/1: 141–151.
Yang, L. et al. (2010). 'An adaptive non-local means filter for denoising live-cell images and improving particle detection'. *Journal of Structural Biology 172*/3: 233–243.
Yang, Z. et al (2013). 'Development of a plastic embedding method for large-volume and fluorescent-protein-expressing tissues'. *PLoS One 8*/4: e60877.
Zhao, H. et al. (2009). 'Induction of DNA damage response by the supravital probes of nucleic acids'. *Cytometry A 75*/6: 510–519.

Further Reading

Bacallao, R. et al. (2006). 'Guiding principles of specimen preservation for confocal fluorescence microscopy'. In: Pawley, J.B. (ed.), *Handbook of Biological Confocal Microscopy*, 3rd ed, Chapter 18. New York: Springer, pp. 368–380.
Daddysman, M.K. et al. (2014). 'Photoinduced damage resulting from fluorescence imaging of live cells'. In: *Photoswitching Proteins: Methods and Protocols, Methods in Molecular Biology*, Chapter 1, vol. 1148, pp. 1–17. doi: 10.1007/978-1-4939-0470-9_1.
Dailey, M.E. et al. (2011). 'Maintaining live cells and tissue slices in the imaging setup'. In: Yuste, R. and Konnerth, A. (eds.), *Imaging in Neuroscience & Development: a laboratory manual*, Chapter 1. New York: Cold Spring Harbor Laboratory Press, pp. 1–8.
Dailey, M.E. et al. (2006). 'Confocal microscopy of living cells'. In: Pawley, J.B. (ed.), *Handbook of Biological Confocal Microscopy*, 3rd ed, Chapter 19. New York: Springer, pp. 381–403.
Davis, I. and Parton, R.M. (2006). 'Fluorescent reagents for live cell imaging and their introduction into cells'. *Cold Spring Harbor Protocols* 6. pii: pdb.ip20.
Dean, K.M. and Palmer, A.E. (2014). 'Advances in fluorescence labelling strategies for dynamic cellular imaging'. *Nature Chemical Biology 10*/7: 512–523.
Diaspro, A. et al. (2006). 'Photobleaching'. In: Pawley, J.B. (ed.), *Handbook of Biological Confocal Microscopy*, 3rd ed, Chapter 19. New York: Springer, pp. 690–702.
Domozych, D.S. (2012). 'The quest for four-dimensional imaging in plant cell biology: it's just a matter of time'. *Annals of Botany 110*: 461–474.
Hardin, J. (2006). 'Confocal and multi-photon imaging of living embryos'. In: Pawley, J.B. (ed.), *Handbook of Biological Confocal Microscopy*, 3rd ed, Chapter 43. New York: Springer, pp. 746–768.
Jaqaman, K. et al. (2008). 'Robust single-particle tracking in live-cell time-lapse sequences'. *Nature Methods 5*/8: 695–702.
Johnson, S. et al. (2013). 'Assessment of Cell Viability Current Protocols Cytometry'. Chapter 9, Unit 9.2. doi: 10.1002/0471142956.cy0902s64.
Jones, E.A. et al. (2002). 'Dynamic in vivo imaging of postimplantation mammalian embryos using whole embryo culture'. *Genesis 34*/4: 228–235.
König, K. (2006). 'Cell damage during multi-photon microscopy'. In: Pawley, J.B. (ed.), *Handbook of Biological Confocal Microscopy*, 3rd ed., Chapter 18. New York: Springer, pp. 680–689.
Mubaid, F. and Brown, C.M. (2017). 'Less is more: Longer exposure Times with Low Light Intensity is Less Photo-Toxic'. *Microscopy Today 25*/6: 26–35.
Nketia, T.A. et al. (2017). 'Analysis of live cell images: Methods, tools and opportunities'. *Methods 115*: 65–79.
Nowotschin, S. (2014). 'Live imaging mouse embryonic development: seeing is believing and revealing'. *Methods in Molecular Biology 1092*: 405–420.
Ohn, J. et al. (2011). 'High-speed multicolor microscopy of repeating dynamic processes'. *Genesis 49*/7: 514–521.
Richter, K.N. et al. (2017). 'Glyoxal as an alternative fixative to formaldehyde in immunostaining and super-resolution microscopy'. *The EMBO Journal 37*/1: 139–159.
Satiat-Jeunemaitre, B. and Hawes, C. (2001). 'Immunocytochemistry for light microscopy'. In: Satiat-Jeunemaitre, B. and Hawes, C. (eds.), *Plant Cell Biology*, 2nd ed, Chapter 10. IRL Practical Approach series No. 250. Oxford: OUP, pp. 207–233.
Schneckenburger, H. et al. (2011). 'Light exposure and cell viability in fluorescence microscopy'. *Journal of Microscopy 245*/3: 311–318.
Shaw, S.L. (2006). 'Imaging the live plant cell'. *The Plant Journal 45*: 573–598.
Sivakumar, S. et al. (2014). 'Live cell fluorescence imaging for phenotypic analysis of mitosis'. *Methods in Molecular Biology 1170*: 549–562.
Swedlow, J.R. (2010). 'Advanced hardware and software tools for fast multidimensional imaging of living cells'. *PNAS 107*/37: 16005–16006.
Terasaki, M. et al. (1998). 'Fluorescent staining of subcellular organelles: ER, Golgi complex, and mitochondria'. *Current Protocols in Cell Biology* Unit 4.4.1. doi: 10.1002/0471143030.cb0404s00.
Tinevez, J.-Y. et al. (2012). 'A quantitative method for measuring phototoxicity of a live cell imaging microscope'. *Methods in Enzymology 506*: 291–309.
Waters, J.C. (2007). 'Live-Cell Fluorescence Imaging'. In: Sluder, G. and Wolf, D.E. (eds.), *Digital Microscopy*, 3rd ed, Chapter 7. Methods in Cell Biology, vol. 81. Amsterdam: Academic Press, pp. 115–140.
Wiedenmann, J. et al. (2009). 'Fluorescent proteins for live cell imaging: opportunities, limitations and challenges'. *IUBMB Life 61*/11: 1029–1042.
Wilson, R.S. et al. (2016). 'Automated single particle detection and tracking for large microscopy datasets'. *Royal Society Open Science 3*/5: 160225.
Wolf, F. and Geley, S. (2006). 'A simple and stable auto focusing protocol for long multidimensional live cell microscopy'. *Journal of Microscopy 22*/1: 72–77.

Correlative Microscopy

Correlative microscopy has become a subject of increasing importance and offers the potential to gain much more information about the biological specimen, its structure and function than light or electron microscopy are able to reveal individually. These references have been selected to give an introduction to this technically demanding yet fast-growing subject.

Mueller-Reichert, T. and Verkade, P. (eds.) (2012). *Correlative Light and Electron Microscopy I*. Methods in Cell Biology, vol. 111. Academic Press.
Mueller-Reichert, T. and Verkade, P. (eds.) (2014). *Correlative Light and Electron Microscopy II*. Methods in Cell Biology, vol. 124. Academic Press.
Bell, K. et al. (2013). 'Correlative imaging of fluorescent proteins in resin-embedded plant material'. *Plant Physiology 161*/4: 1595–1603.
Caplan, J. et al. (2011). 'The Power of Correlative Microscopy: Multi-modal, Multi-scale, Multi-dimensional'. *Current Opinion in Structural Biology 21*/5: 686–693.
Cortese, K. et al. (2009). 'Advanced Correlative Light/Electron Microscopy: Current Methods and New Developments Using Tokuyasu Cryosections'. *Journal of Histochemistry and Cytochemistry 57*/12: 1103–1112.
de Boer, P. et al. (2015). 'Correlated light and electron microscopy: ultrastructure lights up!' *Nature Methods 12*/6: 503–513.
Jahn, K.A. et al. (2012). 'Correlative microscopy: providing new understanding in the biomedical and plant sciences'. *Micron 43*/5: 565–582.
Johnson, E. et al. (2015). 'Correlative in-resin super-resolution and electron microscopy using standard fluorescent proteins'.
Mateos, J. M. (2016). 'Topographic contrast of ultrathin cryo-sections for correlative super-resolution light and electron microscopy'. *Scientific Reports* 6: 34062.
Takizawa, T. et al. (2015). 'FluoroNanoGold: an important probe for correlative microscopy'. *Journal of Chemical Biology 8*/4: 129–142.
Timmermans, F.J. and Otto, C. (2015). 'Contributed review: Review of integrated correlative light and electron microscopy'. *Review of Scientific Instruments 86*/1: 011501.
Viswanathan, S. et al. (2015). 'High performance probes for light and electron microscopy'. *Nature Methods 12*/6: 586–576.
Wiley Essential Knowledge introduction: http://www.magersci.com/brochures/leica/CryoCLEM_ebook.pdf.

Internet Resources

Fixation (an on-line series by Leica): https://www.leicabiosystems.com/pathologyleaders/fixation-and-fixatives-1-the-process-of-fixation-and-the-nature-of-fixatives/#c22446.
Aldehyde fixation summary (Kiernan): http://publish.uwo.ca/~jkiernan/formglut.htm.
Specimen preparation (as as on-line series by Leica): https://www.leicabiosystems.com/pathologyleaders/an-introduction-to-specimen-preparation/.
Cryosectioning (CSHL Protocols): http://cshprotocols.cshlp.org/content/2008/8/pdb.prot4991.full.pdf+html.
Clearing & Mounting media: http://nic.ucsf.edu/dokuwiki/lib/exe/fetch.php?media=mounting_media_tests.pdf.
Intravital microscopy: http://users.ox.ac.uk/~atdgroup/technicalnotes/Notes%20on%20intravital%20microscopy.pdf.
Bioscience Tools: http://www.biosciencetools.com/index.html.
Indigo Scientific: http://www.indigo-scientific.co.uk.
Harvard Apparatus/Warner Instruments: https://www.warneronline.com/corporate.cfm.
Bioptechs: http://www.bioptechs.com/.
Cherry Biotech: http://www.cherrybiotech.com.
LCI/Chamlide: http://www.chamlide.com/.
Linkam: http://www.linkam.co.uk/.
Cairn Research: https://www.cairn-research.co.uk.
Cool-LED: http://www.coolled.com.
Controlled Light Exposure (CLEM): https://www.imaging-git.com/science/light-microscopy/controlled-light-exposure-microscopy-clem.
Further information on culture chambers: https://www.microscopyu.com/applications/live-cell-imaging/live-cell-imaging-culture-chambers.
Nikon MicroscopyU, Maintaining live-cells on the microscope stage: https://www.microscopyu.com/applications/live-cell-imaging/maintaining-live-cells-on-the-microscope-stage.
Photo-toxicity: https://www.nature.com/articles/nmeth.2760 and http://blogs.nature.com/methagora/2013/11/is-phototoxicity-compromising-experimental-results.html.
Valap original reference: McGee-Russell, S. M., and Allen, R. D. (1971). 'Reversible stabilization of labile microtubules in the reticulopodial network of Allogromia'. *Advances in Cell and Molecular Biology 1*: 153–184. It is a 1:1:1 mixture of vaseline, lanolin and 56°C melting-point paraffin wax. I substitute 1/3rd of the paraffin wax with beeswax, which makes the valap stickier and more adhesive onto the glass surface of the slide.
Immunology protocol notes: https://www.agilent.com/en/products/immunohistochemistry.
Jackson website: https://www.jacksonimmuno.com/technical/products and https://www.jacksonimmuno.com/technical/products/groups and https://www.jacksonimmuno.com/technical/products/groups/fab.

Cleaning Slides

Fischer, A.H. et al. (2008). 'Preparation of Slides and Coverslips for Microscopy'. *Cold Spring Harbor Protocols*. doi: 10.1101/pdb.prot4988.
[This information is also in Fischer, A.H. et al. (2006). 'Preparation of cells and tissues for fluorescence microscopy'. In: Spector, D.L. and Goldman, R.D. (eds.), *Basic Methods in Microscopy*, Chapter 4. New York: ColdSpring Harbor Laboratory Press, pp. 105–123.]

Dako Education Guides

Dako Education Guide *Immunohistochemical Staining Methods*, 6th ed. https://www.agilent.com/cs/library/technicaloverviews/public/08002_ihc_staining_methods.pdf
Troubleshooting causes of background staining: https://bit.ly/2NG75n1 and https://bit.ly/2NEhzDj and https://bit.ly/2PyBrbR
Choosing the correct secondary antibody: https://bit.ly/2NIAcGb and https://bit.ly/2NFNEuE and https://bit.ly/2NaWtj1
Demasking antigens and epitope retrieval: https://bit.ly/2Pqo0KT
Dako Education Guide. *Special Stains and H&E* (PDF). https://www.agilent.com/en/dako-pathology-education-guides

Gray Institute links

http://users.ox.ac.uk/~atdgroup/technicalnotes/.
Getting the best from PMTs: http://users.ox.ac.uk/~atdgroup/technicalnotes/Getting%20the%20best%20out%20of%20photomultiplier%20detectors.pdf.
Tissue autofluorescence: http://users.ox.ac.uk/~atdgroup/technicalnotes/Tissue%20autofluorescence_watch%20out%20what's%20on%20the%20menu.pdf.
Making an incubator enclosure: http://users.ox.ac.uk/~atdgroup/technicalnotes/TE300%20incubator%20enclosure.pdf.
Air flow temperature regulator for incubator enclosures: http://users.ox.ac.uk/~atdgroup/technicalnotes/A%20simple%20air%20flow%20temperature%20regulator%20for%20small%20enclosures.pdf.
LED trans-illuminator: http://users.ox.ac.uk/~atdgroup/technicalnotes/TE300%20LED%20system.pdf.
Stowers incubator design in: Kulesa, P.M. and Kasemeier-Kulesa, J.C. (2007). *Cold Spring Harbor Protocols*. doi: 10.1101/pdb.prot4792.
Oxford incubator design: http://users.ox.ac.uk/~atdgroup/technicalnotes/TE300%20incubator%20enclosure.pdf.

Notes

1. Coagulant fixatives used for light microscopy preserve microanatomical detail, organ components and cells in spatial relation to one another. Proteins forming the ultrastructure of cells and tissues clump into a spongy mass, rather like the white of an egg on cooking (where heat is used as a fixative). On the other hand, cross-linking fixatives, which are primarily non-coagulant or cytological fixatives, convert the cytoplasm of the cell into an insoluble gel. Penetration is poorer, so much smaller tissue blocks are fixed by immersion, but cell organelles and cellular ultrastructure are well preserved. However, non-coagulant fixatives used alone give poor tissue consistency for embedding.
2. Correlative LM-EM specimen preparation has texts dedicated to it as a subject in its own right. Ten references, books, reviews and papers are given at the end of this chapter.
3. Clearing large tissues for the acquisition of 3D tomographic data sets is a rapidly developing technology and is covered further in Chapter 19 on selective plane illumination microscopy. See section 19.7.
4. Strictly, the term 'dyeing' refers to tissue taking up only the dye molecule from the staining solution, leaving the solvent, whereas 'staining' implies that both dye and solvent are taken up together. The chromophore contains the coloured part of the dye and is usually an aromatic resonance hybrid structure.
5. OCT is an acronym meaning optimal cutting temperature. OCT compound is a water-soluble chemically inert mixture of glycol, resin and polyvinyl alcohol.
6. Many workers advocate buffer with permeabilisation detergent: PBT (PBS + 0.1% Triton plus 0.5% BSA).
7. Clearly for mouse tissue another host is preferably required. If using mouse antibodies on mouse tissues is unavoidable, endogenous immune-globulins may be blocked with a monovalent Fab fragment of unconjugated anti-mouse IgG (H+L).
8. The primary noise source is Poisson (shot) noise. Variations in illumination intensity are also common, both in widefield and particularly confocal laser scanning microscopes. The efficiency of the CCD or sCMOS camera will be determined by its dark current and readout noise. For further details on the effects of noise, see Chapter 17, Section 17.5 and Chapter 30, Section 30.10.
9. It is often overlooked, or forgotten, that the photon irradiance incident upon the sample can be very high indeed. Peak (instantaneous) and mean solar irradiance values are both 0.13 W/cm^2. Peak power in the confocal is of the order of megawatts and in the multi-photon, multi-gigawatts, simply because of the small moving spot of a confocal continuous wave laser or a multi-photon pulsed laser. Peak confocal power = 2.5×10^5 W/cm^2, and average power = 0.25 W/cm^2 (assuming a 1024×1024 raster). Peak multi-photon power = 5×10^{20} W/cm^2 and average power is 2.5×10^4 W/cm^2.
10. Radiant flux is the quantity of the energy of the light emitted per second in all directions. Photon flux is the number of photons per area per time. Irradiation intensity (irradiance, energy flux) is energy received per unit area [W/m^2]. Many authors do not consider the effects of photo-toxicity upon the experimental results they report and publish. See the Editorial in *Nature Methods* (2013, vol. 10/12, page 1135) for a pertinent commentary.
11. Mechanical shutters are also slow. Changing filters housed in individual filter cubes is often the bottleneck to achieving high temporal resolution. Where possible, use fast shutters and filter switching by motorised filter wheel holding lightweight filters. Monochromators and LEDs provide much faster illumination switching solutions. See Figure 15.16. Confocal scanning mirrors are also relatively slow and laser illumination intense. Although it is possible to perform live-cell imaging using a confocal microscope, a specially designed widefield fluorescence microscope is very much to be preferred.
12. There are reports that HEPES buffer exposed to UV light can be cytotoxic (Lepe-Zuniga et al. 1987). On the first day that HEPES buffer is used, add 95 mg glucose and 95 mg albumin to each 50 mL of buffer. HEPES will also reduce growth rates since bicarbonate ions are essential for cell growth, this HEPES can only be a temporary solution. Some cell lines do not respond well to Leibovitz L-15, which should be tested on cell cultures beforehand.
13. Heat has a negative impact on the lifetime of complex objectives (delamination of optical cement). Some imaging facilities which maintain dedicated instruments continually kept heated for live-cell work, the cost of annual replacement of objectives is factored into the operating costs. All objectives used at temperatures different from 20°C will introduce slight spherical aberration into the image. This is critical if structures are very minute with lots of background signal, and why correction collars have a red fiducial mark for setting correctly at elevated temperatures (see Figure 7.4).

29
Materials Specimen Preparation

Outline

Understanding Light Microscopy, First Edition. Jeremy Sanderson.

29.1 Introduction

Petrographic examination of geomaterials for the construction industry is of global importance. Since the 19th century, the examination of the surfaces of polished samples and of thin sections has played a vital role in the safe use of many differing types of construction materials. The examination and testing of these building materials for quality control – to understand their physical limits and potential for deterioration – are carried out according to international standards. Some of these can be found in Chapter 1 of Ingham (2010), on the websites of the International Organisation for Standardisation (ISO) and the American Society for Testing and Materials (ASTM). Links to these websites are given under Internet Resources, at the end of the chapter.

Opaque materials specimens are normally prepared either as thin sections or as polished samples (Figure 29.1). Thin sections are ground from bulk material, but they can also be prepared as a peel from a prepared flat surface. Rocks that are loose or friable should be impregnated with resin before cutting into blocks; otherwise, a grain mount (see Section 29.7) can be made. Both thin sections and grain mounts can be examined by both reflected and – if the grains are small enough – by transmitted light. Polished samples, however, are examined solely by reflected light, but the surface must be perfectly perpendicular to the optical axis and polished to a high-quality finish.

Historically, polished blocks and thin sections were ground by hand. This was the way that Henry Sorby (Box 12.2) made his first sections, and they can still be usefully studied today. Preparing thin sections by hand is more an art than science and requires considerable experience, but it is the most economical way to make thin sections, especially if only a few samples are being prepared. The alternative is to contract out specimen preparation to specialist firms. MetPrep and Pelcon are two companies known to the author, and their web pages URL links are given at the end of the chapter. Otherwise, ask colleagues or the application specialists in the companies supplying equipment and consumables. It is suggested that you not merely rely on an Internet search engine to throw up an unknown contractor. The contact details of selected equipment supply companies are given in Appendix 5.

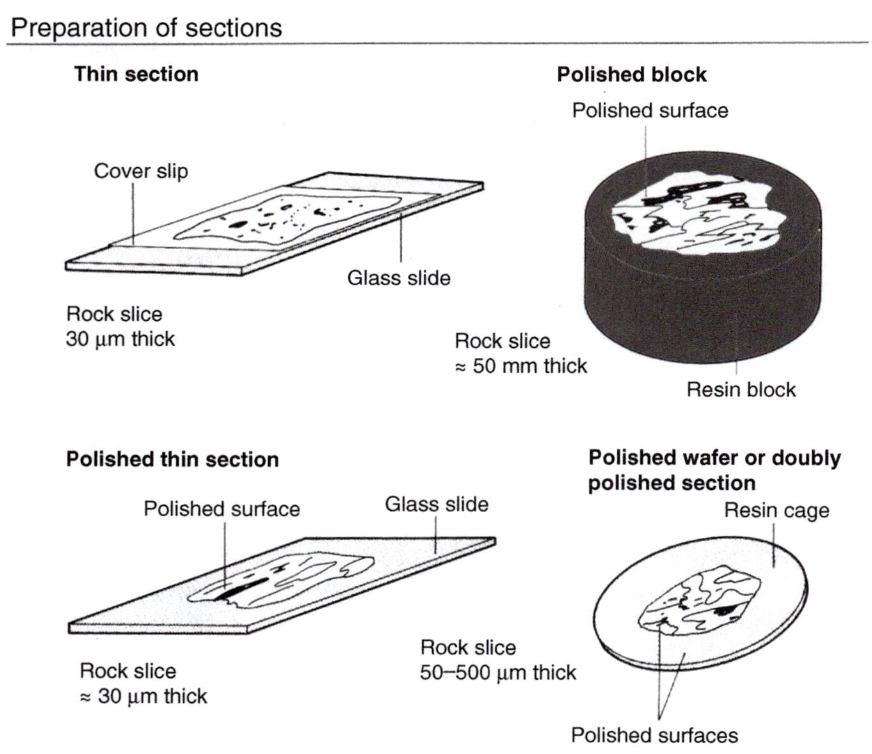

Figure 29.1 ***Types of preparation*** *Polished blocks are simple to prepare and are robust, since only one surface has to be polished.*

A thin section allows identification by reflected light of various opaque metallic minerals present in most rocks. Depending upon the material and the analysis requirements, the section is either coverslipped after grinding or polished further and left exposed.

A doubly polished thin section mounted on a slide provides a much sharper image under the microscope. The greater transparency permits crystalline structures within the material (e.g. zoning phenomena especially in low-reflectance minerals) to be studied in greater detail and at higher resolution.

The doubly polished unmounted thick section is not as common as slide-mounted thin sections or polished samples but is of intermediate thickness. A wafer of this nature, polished on both sides and between 150 and 200 µm thick is ideal for studying microfossils and fluid inclusions, particularly for porous, friable or water-sensitive material. Source: Gribble and Hall 1992.

The entire process of preparing opaque blocks and thin sections has been automated; machines not only take care of the laborious effort but are faster and produce polished surfaces and thin sections that are consistently of higher quality than hand-prepared samples. This is the key to consistent and economical specimen preparation. For those who wish to make thin sections by hand, the best reference is Humphries (1992).

The size of structures studied in materials science, both by transmitted light and reflected light, rarely approach the Abbe limit of resolving power of the microscope. The structures observed are generally at a scale of 1 μm or greater (Box 29.1). This is unlike the study of biological specimens, where the size of cells organelles is at, or beyond, the limit of resolution. In petrological or ceramic thin sections of standard 30 μm thickness, overlapping fine detail within the thickness of the sample makes observation of structures close to the Abbe limit rather difficult. In transmitted-light materials microscopy there is no equivalent to the 1–2 μm 'thin' sections used in high-resolution biological transmitted-light microscopy. In reflected-light microscopy, grain and crystal sizes typically occur within a wide size range from 5 to 500 μm. Because small grains (e.g. 5–10 μm in size) will overlap throughout the thickness of a standard 30 μm 'thin' section and be out of focus (e.g. ceramics with their smaller grain sizes), it is now common practice with modern precision equipment to prepare thin sections 20 μm thick.

Box 29.1 The Krumbein φ Scale for Particle Sizes

1 to 0	0.5–1 mm	coarse sand
2 to 1	0.25–0.5 mm	medium sand
3 to 2	125–250 μm	fine sand
4 to 3	63–125 μm	very fine sand
5 to 4	31–63 μm	coarse silt
6 to 5	15.6–31 μm	medium silt
7 to 6	7.8–15.6 μm	fine silt
8 to 7	3.9–7.8 μm	very fine silt
14 to 8	0.06–7.8 μm	colloid/clay

Sources: (table from http://en.wikipedia.org/wiki/Particle_size_%28grain_size%29) Williams et al. 2007; (photograph) PaleoMal. Reproduced under the terms of the Creative Commons Attribution Share-Alike licence, CC-BY-SA 4.0 (http://creativecommons.org/licenses/by-sa/4.0/), via Wikimedia Commons.

The successful resolution of particles within crystal grains (rather than simply their detection) is usually hampered by the necessity of etching, which makes it difficult to distinguish the nature and size of the particle itself from its surrounding etched area. Similarly, in the identification of particulate or fibrous specimens of industrial or forensic importance (e.g. asbestos) it is the *optical* and *morphological* characteristics, and not ultimate resolving power, which are of prime importance. If you have to go beyond the 1 μm size limit, consider using TEM or SEM in parallel with light microscopy. Modern quantitative microstructure analysis uses scanning electron microscopy with electron backscatter diffraction analysis (EBSD) to examine the crystallographic orientation of rocks, minerals and metals.

Whether making a thin section for transmitted-light observation, with two perfectly flat and parallel surfaces, or a polished specimen for reflected-light examination with a single high-quality surface, there are two considerations that govern the entire process of specimen preparation.

- Firstly, the prepared surface should be a true representation of the material being studied. This might, at first, be self-evident, but each step of cutting, grinding, lapping and polishing to produce a flat surface introduces damage. The purpose of each subsequent step is to remove the damage created by the previous one until the surface is polished and perpendicularly flat (to the optical axis) within acceptable limits. For example, lapping removes the damage and relief caused at the grinding stage, and polishing removes the damage caused by lapping. If the experimental analysis demands an active oxide step, this chemical-mechanical step removes the last traces of abrasive damage caused by polishing.
- Secondly, it is important to realise that during each stage of cutting, grinding, lapping and polishing, it is possible both to remove material from the sample as well as introduce adventitious material into the preparation. Both these processes are artifactual and should be avoided – or at least minimised – where possible. You should aim to only remove as much material as required. It may be necessary only to lap and polish with a 3 μm diamond paste rather than continue to the surface finish produced with a 0.25 mm diamond paste if a macro-assessment of the surface features are all that it required.

In this chapter is included general guidance for preparing a polished opaque surface or a ground thin section (Section 29.3 onwards) and a rock peel (see Section 29.8) to illustrate the principle and the steps involved in preparing material for microscopical examination. This chapter is not intended to be a comprehensive guide or a more detailed treatment of the subject; please refer to the selected references given at the end of the chapter and the forthcoming *The Preparation of Geomaterials for Microscopical Study: A Laboratory Manual* by Green and Wells in the Wiley RMS Handbook series. Also, because the nature of the material (as well as, perhaps the equipment available) determines the exact preparation protocol followed, there will be differences between laboratories, even for the same material, according to the experience and preferences of the technical staff making the preparation.

Typical preparation steps of samples for materials microscopy:

1. cutting the sample from bulk material;
2. vacuum-embed as required: to fill cavities/protect porous nature;
3. precision-sectioning of the embedded material down to a slice;
4. first face lapping of the slice;
5. grind/polish the slide so that it is absolutely parallel[1]; frost the slide (optional);
6. mount the slice onto a prepared slide (frosted for opaque samples); glue to slide – hot or cold (UV) bonding with a bonding jig;
7. cut sample down to between 500 and 200 μm with a Petrothin saw, or similar;
8. grind & lap out saw damage;
9. polish one face of the slice;
10. re-glue permanently to a slide (for thin section preparation);
11. grind, lap and final polish of material to 30 μm; and
12. cover with coverslip as required.

29.2 Fidelity of Specimen Preparation

When killing and fixing cells for examination or removing a tissue sample from a living organism, it is all too easy for artifacts to arise in the preparation. It is a misconception that this is only true for biological material. The act of *mechanical* specimen preparation can also very easily introduce artifacts into specimens prepared for materials microscopy. Potential artifacts can be minimised by systematic and reproducible preparation methods. This requires consumables of high standard and uniform quality. The cheapest consumable preparation materials do not always give the most economical result. In order for the finished sample to reveal the true structure of the material, any artifactual damage should be of a size below the resolution limit of the objective used to examine the sample. The nearly perfect condition, with only superficial damage remaining, is commonly referred to as the true structure.

This true structure is defined by what should *not* be present: Ideally, there should be no deformation, no surface scratches, no loss or pull-out of material and no introduction of foreign material (such as grinding and polishing abrasives or lubricating fluids[2]). There should also be no structural change of the material of the sample or surface smearing. Care should particularly be taken to ensure that when the sample is cut, it does not overheat which may give rise to artifactual cracks or changes in the microstructure.

There should be no adventitious material introduced into the preparation. Extraneous material (most often seen as a fine, loose aggregate) may be introduced from the cutting saw from silicon carbide (carborundum) or diamond grit from the slurry used for lapping (Figure 29.2) or grinding material held in the nap of the polishing cloth. The simplest way to avoid this happening is to fill any pores and cracks that can be easily identified in the sample with coloured or fluorescent low-viscosity epoxy mountant (Figure 29.3). This avoids, for example, mistaking silicon carbide and the finest 0.25 μm diamond paste (when coloured white) for titanium oxide minerals. When examining the section or polished sample for adventitious material, use a dry 10x or 20x objective: with an oil objective, it is difficult to determine whether suspected extraneous material is part of the section or polished sample or not. Finally there should be no surface relief in the polished surface of the finished preparation.

29.3 Sampling and Cutting the Bulk Material

Historically the initial chipping, breaking or sawing and slicing of a representative sample from the bulk material was not considered an important step. However, unless this first step is carried out carefully, it can introduce the most damage of all. This is particularly true when handling brittle, friable or poorly consolidated materials. Ultimately, the final information and data collected from examination and analysis of the microscope preparation will only be as good as the quality of the sampling procedure.

Initial examination of the bulk material should be done by eye. A stereomicroscope can sometimes be used to identify a region for taking a sample for further preparation and examination using a compound microscope. Fragments broken from the bulk are generally not suitable for microscopical examination. It is difficult to distinguish and differentiate between defects inherent within the sample and those arising from damage caused when sampling. It may not always be possible to drill out a core sample, but where possible, these are preferred because the edge is regular and easily recognisable within the field of view of the microscope. For aggregate products, such as concrete, the core size should be at least three times the maximum size of the particles comprising the aggregate material.

(a) 600 grit slurry

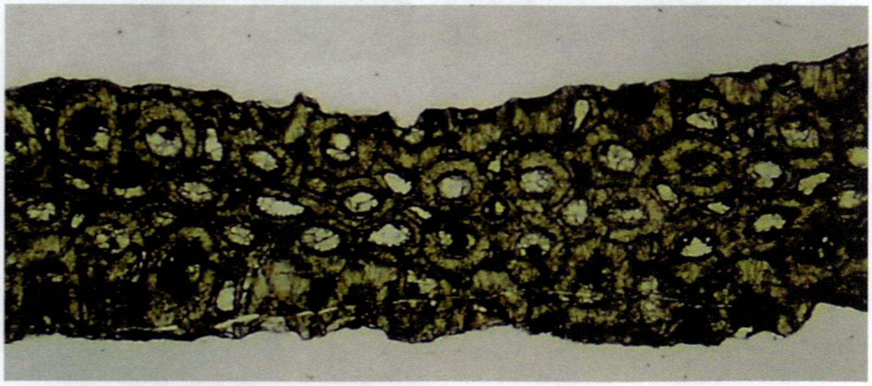

(b) 600 grit SiC paper

Figure 29.2 **Artifacts from slurry** *Adventitious material introduced into a thin section of fossilised bryozoan during lapping with diamond or carborundum slurry (top). It is better to use fixed diamond rollers, or abrasive carborundum papers, for grinding rather than traditional loose abrasives (bottom). Here the sample has been ground to a thickness of ~100 μm to avoid crystal shatter, then successively lapped with a series of graded papers mounted on rotating wheels with running water to remove the ground material and to lubricate the paper. A finishing grade of paper may range from 600–2500 grit size, depending on what the section will be used for and/or how fine any structure is that must be examined. On some occasions the paper may need to be as fine as 4000 grit. (Figure courtesy of Chris Jones, NHM.) Source: The Natural History Museum, London. Reproduced with permission from The Trustees of the Natural History Museum, London.*

Sawing and slicing of samples is nowadays done by machine (Figure 29.4) rather than by hand, which is more consistent and leads to higher-quality samples with better end results. It is preferable to cut the material into as thin a slice as possible in order to reduce the amount of grinding and lapping required. All the major manufacturers publish guides – both in print and online – which advise on the most suitable cut-off wheels for their machines. Table 29.1 gives a typical classification of different types of wheel available. Modern cut-off machines are electronically controlled (e.g. Struers Accutom-5) with a list of wheels stored in an internal database, making it easy and convenient to control both the speed and force used to cut particular materials.

A billet is a sliced rectangle of rock, or other material, used to make a thin section. You should aim to end up with a series of billets cut from the initial slab of material, each of a different orientation. Specimens taken as samples from the bulk rock should be sliced in a number of different orientations. This increases the chances of sampling the several distinct structural fabrics of the rock. From a practical viewpoint, many rocks and minerals are structurally anisotropic and so will polish very well in one orientation and less well in another. This is particularly the case for hard minerals (such as magnetite, haematite and pyrite), those which cleave well (such as micas and amphiboles and feldspar) and those with a sheet-like structure (such as molybdenite and graphite).

It is not always necessary to use a diamond saw, which is hard but not sharp. It pays to use a relatively cheap silicon carbide saw for a sharp, accurate cut, which – size for size – leaves less damage at the cutting stage than diamond. Use the diamond saw when cutting smaller-sized billets. Again, the saw type is determined by the material; as a rule of thumb, you should use a cutting or grinding material that is twice as hard as the sample; see Box 29.2 for an explanation of how material hardness is classified. Ceramics will need a diamond saw but aluminium does not, and a silicon carbide saw will suffice. An annular cup saw or wire saw will provide a gentler cut than a circular rock trimming saw.

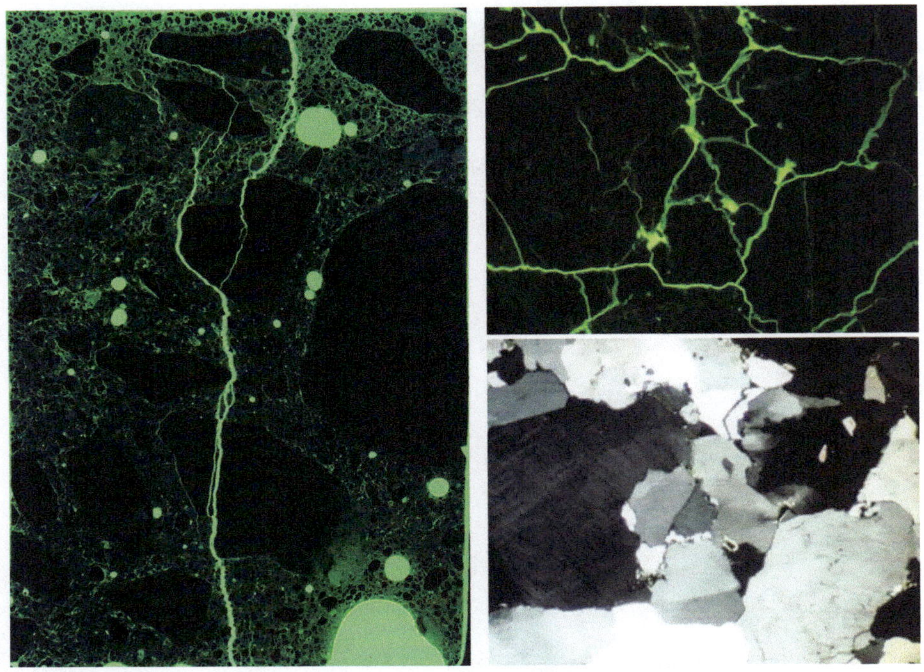

Figure 29.3 **Low viscosity mountant** *Rock thin section impregnated with fluorescent dye within the epoxy mounting medium (e.g. Greenland airfield runway surface frost damage). Air bubbles act as expansion chamber for guarding against frost damage in concrete. The cracks and micro-porosity within the material can be seen clearly. The samples on the right are similarly prepared; the lower panel being taken under crossed polars with transmitted light for comparison with the upper right-hand panel. Source: Reproduced with permission from PELCON.*

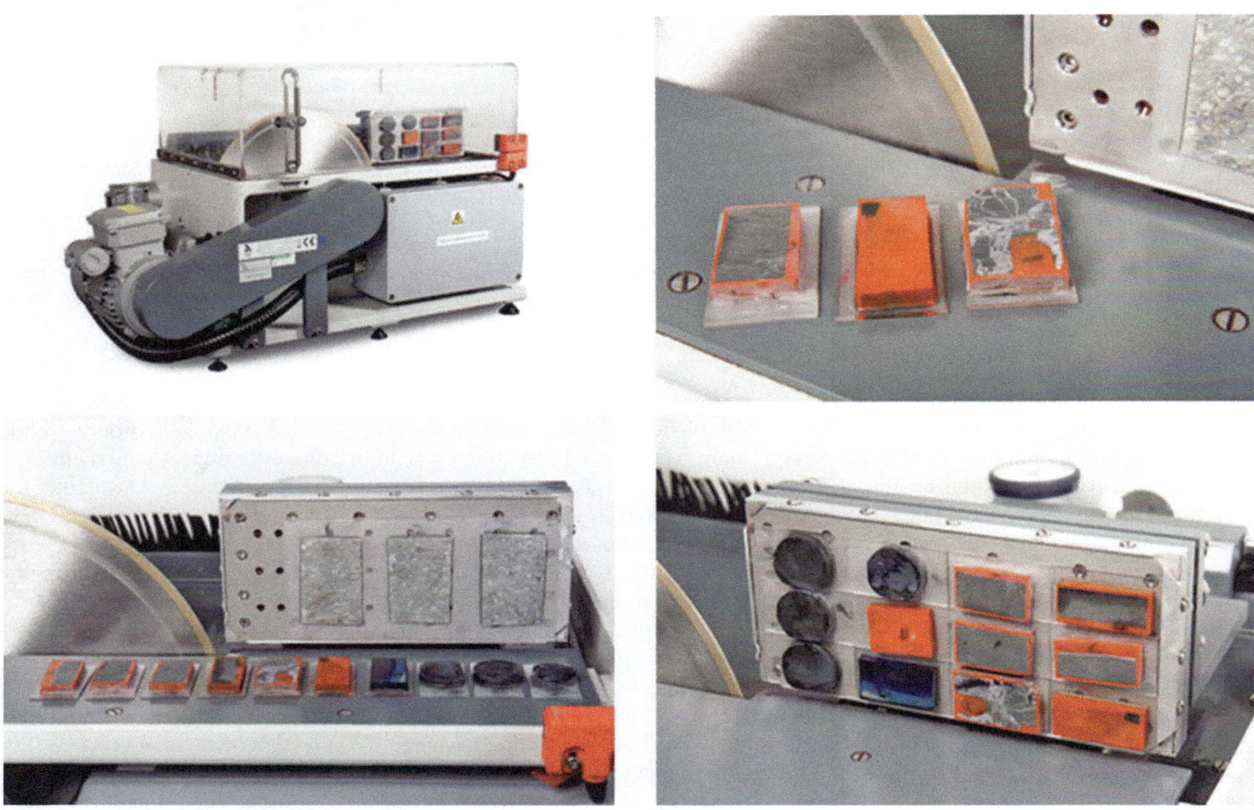

Figure 29.4 **Sawing and slicing samples** *The Logitech GTS1 thin section cut-off and trimming saw, with examples of sliced preparations. Source: Logitech. Reproduced with permission from Logitech; Reproduced with permission from PELCON.*

Table 29.1 *General Purpose Cut-off and Cutting Wheels*

Abrasive	Type	Materials	Hardness	
			Rockwell	Vickers
Aluminium oxide	8A80NBK	Low carbon, mild or case-hardened steels	20–36	100–350
Aluminium oxide	8A80MBK	Treated steels, stainless steels, refractory steels, alloyed cast iron	30–64	300–600
Aluminium oxide	8A80KBK	Nickel-based super alloys, carburised steels, treated steels with fragile coatings, hardened high-carbon steels	30–64	300–600
Silicon carbide	BC80NBK	Non-ferrous materials, aluminium and alloys, copper and alloys, white, grey and spheroidal cast iron	20–36	80–350
Silicon carbide	GC80JBK	Titanium and alloys		80–970
High concentration Metal bond diamond	–	Ferrous materials, medium/hard ceramics	>70	>1000
Low concentration Metal bond diamond	–	Ceramics, glass, geological samples	>70	>1000
Continuous rim diamond	–	Petrographic samples	>70	>1000

This table is taken as an example from the Kemet materialographic catalogue of equipment and consumables for sample preparation. Other manufacturers publish similar guides and tables – both as hard copy and online – to help select the most appropriate consumables for their preparation equipment.

Box 29.2 Hardness Testing

Since preparation of thin sections and opaque reflective samples involves grinding, lapping and polishing surfaces flat, it is important to know how hard your material is because this is a measure of how materials resist abrasion. Hardness is related to the strength of a material, but a range of different tests have been independently developed, and so hardness is not a well-defined property. This box compares the main scales in use today.

The Mohs scale is defined by how well a substance will resist scratching by another substance. It is used for classifying minerals. The Vickers hardness scale was developed as an easier test to apply than the Brinell hardness test. Both use an indenter to deform the test material under load. The Vickers unit of hardness (HV) is known as the Vickers Pyramid Number or Diamond Pyramid Hardness (DPH). Because a diamond indenter is used, the advantage of the Vickers test is that one scale covers the entire hardness range, so it is widely used to measure metals. Vickers indentation values are independent of the force used, but below 200 grammes force, caution must be used comparing results.

Mohs scale	Mohs test mineral		Knoop value	Vickers value	Rockwell C value	Brinell W value
1	Talc		–	–	–	–
2	Gypsum		(30)	(50–60)	–	(55)
3	Calcite	Coin	140	125	–	130
4	Fluorite		160	145	–	145
5	Apatite	Tooth	430	420	44	396
6	Feldspar	Most glass	560	530	52	500
7	Quartz	Steel file	820	800	63	–
8	Topaz		1350	1300	(70)	–
9	Corundum		2100	2150	–	–
9+	Silicon carbide		2480	2550	–	–
10	Diamond		7000	10 000	–	–

NB these values are approximate. Refer to the conversion charts and protocols given in the References for further details.
The Brinell test employs a hardened steel ball, usually 3/8th inches in diameter. The Rockwell scales are used in the USA, whilst Vickers, Knoop and Brinell scales are used in both the UK and USA. The advantage of the double load Rockwell method is its ability to display hardness values directly without the calculations required for the Vickers or other methods. When used, the parameters should be defined for each test. As the table shows, the values of each scale differ widely (the loads applied are not shown in the table: it is for general comparison only). Different comparative scales are published for different materials.

References

General introduction: Wikipedia https://en.wikipedia.org/wiki/Indentation_hardness and https://www.struers.com/Knowledge/Hardness-testing and http://www.materials.co.uk/microhard.htm.

Hardness conversion tables: http://www.engineershandbook.com/Tables/hardness.htm.

ASTM E140-12be1 provides tables to convert from one scale to another: Standard Hardness Conversion Tables for Metals Relationship Among Brinell Hardness, Vickers Hardness, Rockwell Hardness, Superficial Hardness, Knoop Hardness, Scleroscope Hardness, and Leeb Hardness.

International Standards, Knoop: ASTM E92, ISO 4545; Vickers: ASTM E384, ISO 6507; Rockwell: ASTM E18, E110, ISO 6508; Brinell: ASTM E10-12, E110, ISO 6506.

A photographic record of the location and orientation of field samples should be made, supplemented by written notes and sketch drawings. Samples should be individually bagged and labelled. Wrap those that are liable to be affected by carbonation from the carbon dioxide in the air within cling film for transportation back to the laboratory.

Specimens for either transmitted-light analysis as a thin section or for examination as an opaque sample by reflected light should: (a) be in the size range 10–40 mm (Figure 29.5), (b) leave at least 2 mm free from the edge of the slide to allow bonding for the surface dimension – small enough to fit onto a standard 46 × 27 mm slide[3] – and (c) for examination by reflected light, be between 2 and 10 mm thick. Ideally a polished sample should not be thicker than 10 mm to enable the sample to fit onto the stage of the microscope within the focus range. Nevertheless, the specimen should be a good representative sample of the bulk from which it is cut or taken. When it is taken, note down the particular orientation of the specimen within the bulk material.

29.4 Impregnation and Mounting of Samples

Some sedimentary samples can be very friable and break up easily, particularly if they are poorly lithified and/or contain gross organic inclusions. These samples need to be vacuum impregnated with either an epoxy or acrylic resin (Table 29.2) to give support during both cutting and grinding. First the sample should be dried at no more than 30–40°C for 12–16 hours[4] to a constant mass before impregnation with a very low-viscosity resin, preferably under vacuum. The vacuum should be broken and restored occasionally. This will force the impregnating medium into the small pores of the rock. This process is usually complete after a few minutes. A wide choice of casting

Figure 29.5 **Slides for petrographic thin sections** *Example of slide sizes for petrographic thin sections. These may well not be a standard size. This simple, effective jig is available from Pelcon. The silicon carbide paper is to smooth the edges of the glass slide. Source: Reproduced with permission from PELCON.*

Table 29.2 Resin embedding media

Selection Guide for Hot mounting

Resin	ClaroFast	ConduFast	DuroFast	LevoFast	PolyFast	MultiFast	Pre-Mount

Material	Acrylic resin	Acrylic resin with iron filler	Epoxy with mineral filler	Melamine with mineral and glass filler	Bakelite with wood filler	Bakelite with wood filler	Bakelite with wood filler
Specific properties	Clear transparent	Electrically conductive	Very low shrinkage. Very low removal rate.	Very low shrinkage. High removal rate.	Very low shrinkage. Medium removal rate.	Medium shrinkage. Medium removal rate.	In pellet form for easy handling
Recommended use	Clear mounts. Porous specimens.	Electrolytic polishing.	Excellent edge-retention for hard materials.	Excellent edge-retention for soft to medium hard materials.	Fast mounting process. SEM examination.	Routine examination of soft to medium hard materials. Color coding.	Serial mounting of uncomplicated specimens.

Selection Guide for Cold mounting

	ACRYLICS					EPOXIES			
Material	**VersoCit-2**	**ClaroCit**	**DuroCit**	**LevoCit**	**ViaFix**	**CaldoFix-2**	**SpeciFix-40**	**SpeciFix-20**	**EpoFix**

Curing time	10 min.	20 min.	20 min.	20 min.	20 min.	1 ½ hour in oven at 75 °C / 149 °F	3 ½ hours in oven at 50 °C / 122 °F	8 hours	Approx. 12 hours
Shrinkage From 1-4 (1 is best)	****	***	*	**	***	*	*	*	*
Application	**For routine examination** • Routine examination of soft to medium hard materials	**For extraordinarily clear mounts** • For universal use • Target preparation	**Fast curing and no shrinkage** • For medium hard and hard ferrous metals and other hard materials – including ceramics, carbides etc. • For specimens where protection of layers is important e.g. coated specimens • Excellent edge-retention and planeness	**Good edge-retention and planeness** • For non-ferrous metals and soft ferrous metals • Low shrinkage • Low peak temperature	**For vias and microvias** • Excellent for filling of microvias. ** Affected by alcohol. When using diamond products or lubricants containing alcohol, the surface will be affected and the structure of the polymer beads will appear.*	**For all-round vacuum impregnation** • Short curing time • Low viscosity • Relatively hard after curing	**Extremely good adhesion** • Relative fast curing time • Very clear colourless mounts • Cures in oven or Drybox	**For vacuum impregnation of small specimens** • Ideal for small specimens • Very good adhesion • Very low curing temperature	**For vacuum impregnation - low viscosity** • Can be used on all types of specimens • Extremely low curing temperature – Very good for heat sensitive specimens • Superior penetration of cracks and pores • Excellent adhesion
Mixing ratio weight recommended	Liquid: 2 parts Powder: 3 parts	Liquid: 6 parts Powder: 10 parts	Liquid: 9 parts Powder: 25 parts	Liquid: 1 part Powder: 2 parts	Liquid: 9 parts Powder: 11 parts	Resin: 25 parts Hardener: 7 parts	Resin: 5 parts Curing Agent: 2 parts	Resin: 7 parts Curing Agent: 1 part	Resin: 25 parts Hardener: 3 parts
Mixing ratio volume	Liquid: 1 part Powder: 2 parts	Liquid: 2 parts Powder: 5 parts	Liquid: 1 part Powder: 2 parts	Liquid: 1 part Powder: 2 parts	Liquid: 1 part Powder: 2 parts	Resin: 31 parts Hardener: 10 parts	Resin: 10.5 parts Curing Agent: 5 part	Resin: 26 parts Curing Agent: 5 part	Resin: 15 parts Hardener: 2 parts
Mixing time	30 s	1 ½ min.	1 ½ min.	45 s	30 s	5 min.	3 min.	3 min.	2 min.
Potlife	3 min.	1 ½ min.	1 ½ min.	1 ½ min.	2 min.	> 60 min.	> 60 min.	60 min.	30 min.
Colour	Dull yellowish, partly transparent	Colourless, clear (extremely clear when cured under pressure)	Light green	Off-white	Colourless, clear (extremely clear when cured under pressure). Otherwise semi-transparent.	Clear, transparent	Clear, transparent Refractive index: ND = 1.573	Clear, transparent Refractive index: ND = 1.573	Clear, transparent Refractive index: ND = 1.578
Can be coloured with EpoDye		X			X	X	X	X	X
Can be coloured with AcryDye	X	X	X	X	X				
Peak temperature	100 °C / 212 °F	90 °C / 194 °F	80 °C / 176 °F	75 °C / 167 °F	115 °C / 239 °F	170 °C / 338 °F	100 °C / 212 °F	60 °C / 140 °F	40 °C / 104 °F
Hardness	82 Shore D	85 Shore D	92 Shore D	84 Shore D	83 Shore D	85 Shore D	82 Shore D	84 Shore D	78 Shore D

(30 mm dia. mount without specimen at 21 °C / 70 °F)

Table 29.2 shows a typical guide to the different hot-mounting and cold-mounting options available as epoxy and acrylic resins. (Taken from page 14, Struers, 2014. 'Materialographic Consumables Catalogue'. Reproduced with permission from Struers APS.)

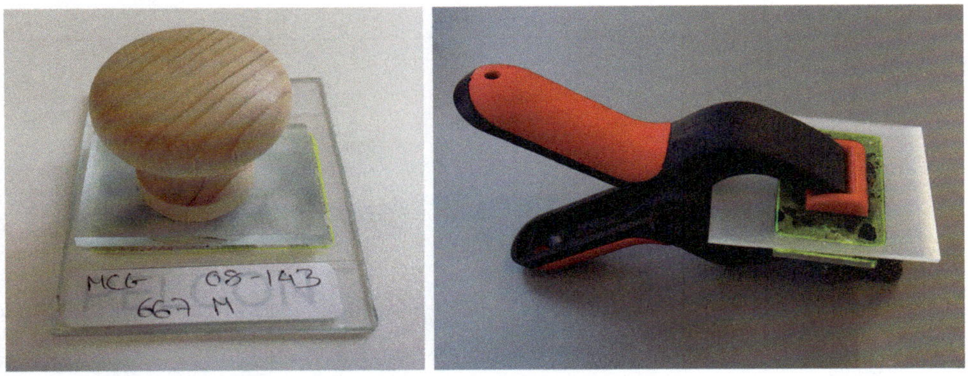

Figure 29.6 **Support clips** *Examples of support clips used when casting opaque samples for reflected-light examination. Provided they deliver sufficient pressure, either by gravity or sprung compressive force, these solutions need not involve great expense. The priority is that they are quick, simple and efficient in use. Source: Reproduced with permission from PELCON.*

epoxy or acrylic resins is available to choose from. Epoxy resins that are cured under heat and pressure are the cheapest options but require an initial outlay for a curing oven. These hot-mounting resins when cured are extremely hard, are acid and alkali resistant (which is an advantage for etching) and exhibit low shrinkage. Cold casting acrylic resins are generally more expensive to use but require no initial capital cost outlay for heat-curing equipment and are better suited to delicate samples that cannot withstand excessive heat and pressure[5]. The samples are cast into either polypropylene or silicone moulds and can be coloured with added dye, as required. After impregnation is complete, remove the impregnation mould and billet from the vacuum, and allow it to cure overnight.

Besides holding friable and poorly consolidated material together, casting specimens into an epoxy or acrylic resin block helps to protect the edge of the sample for examination, where this may be important in metal or ceramic specimens. Compression helps minimise shrinkage, but cool the sample to room temperature before removing it from the press. Unfused casting material remaining in the mould is usually caused by insufficient temperature or pressure, whilst bulging or soft mounts are caused by too short a curing time. Uncured mounts can also occur where the casting powder has not been kept under airtight conditions and has taken up atmospheric moisture.

Support clips (Figure 29.6) are used to prevent the specimen from sinking into the mould and to keep the surface for analysis presented at the top of the sample, since heat-cured resins will initially thin as they are heated before curing fully. Use a plastic support clip when the sample is to be etched, as these will not interfere with the etching process, as a metal support clip might. It is helpful to use a fluorescent dye mixed with the impregnation medium, for easy identification (Figure 29.3) of the filled pores or cracks propagating through the bulk material. The dye is generally either fluorescent yellow or blue. This helps show up porosity of the sample, natural grain boundaries and any artifactual cracking. Fluorescent low-viscosity epoxy resin can be obtained from Pelcon, whose contact details, along with other main suppliers, can be found at the end of the chapter, under Internet Resources.

A specimen label is normally inserted into the resin. Some laser printers use ink that can produce labels whose text does not run or smear when the resin is cast. This can be an advantage over handwritten labels in pencil. Buehler recommend their TransOptic powder added in two layers with a label in between at the base of the mould before filling with the specimen and casting material. Slides used for petrology must be perfectly flat. These can be bought as such (see the suppliers in the list at the end of the chapter) or ground accurately to the required thickness. Once the slide has been ground perfectly parallel, it can either be frosted for mounting opaque samples or else polished for use with thin sections. A frosted slide, with a surface roughened for a depth of 15–20 μm, will help bond the mounted sample to the slide. Specimens are usually bonded to the slide with a pressure of 1 kg/cm^2. This can be done with a spring-loaded bonding jig (Figure 29.7), to apply pressure evenly throughout the slice. Aim for a clear, bubble-free bond with minimal thickness (<1 μm) of glue holding the slice to the slide. Good bonding requires both the sample and the glass slide to be well cleaned, degreased and dried.

Slices or billets of rock can be glued onto the prepared glass slide with either a permanent fixative or, if the slice is to be dismounted in order to bond the prepared surface to a slide later, a thermolabile fixative that holds the slice temporarily is used. There are various permanent bonding glues available: Petrobond, Locktite 358, EpoSpeed and Dymax 211 all require UV light to cure. The latter has a bright blue colour while uncured and cures to a colourless state. This is useful, since any adjustment may be made, or a mistake rectified, before curing takes place. EpoSpeed S is a thermally soluble resin, also UV cured, which becomes fluid when heated above 45°C. This allows preparation of one face of the slice, so that the chip or slice can be removed from the glass substrate when heated.

29.5 Grinding and Lapping Thin Sections

Grinding (Figure 29.8) is the relatively rapid removal of material from the sample, to reduce it to a suitable thickness. Lapping is the removal of further material in a slower fashion to produce a smooth, flat unpolished surface. Both procedures together thereby produce a first-finish flat surface. Grinding is fast (200–1000 rpm) but not so precise, whereas lapping is slower (<80 rpm) but more precise.

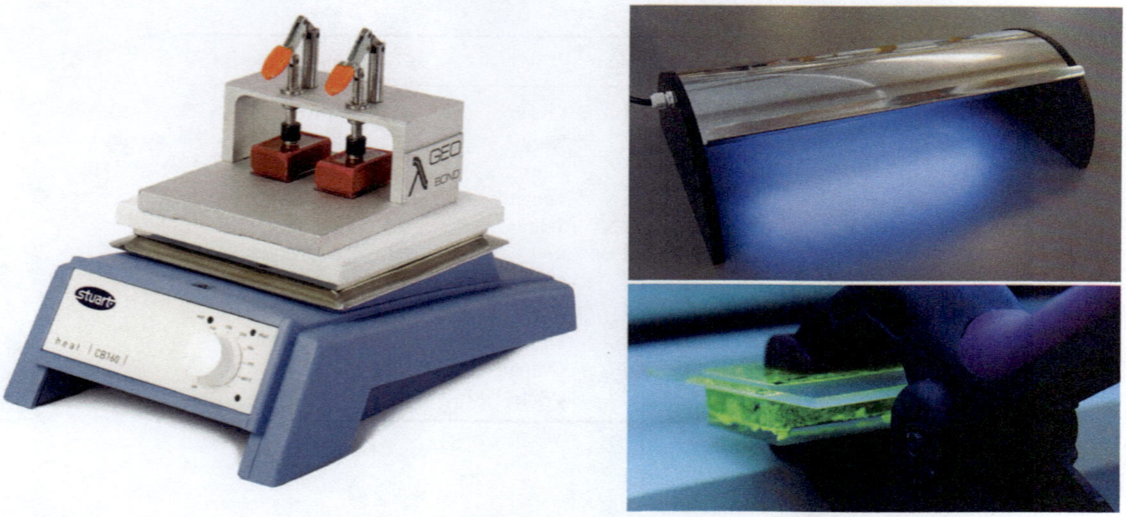

Figure 29.7 **Spring-loaded bonding jigs and cramps** *Devices to hold slices securely onto microscope slides. Either thermo-setting adhesives (left, Logitech CH30-CB160 bonding jig and hotplate) can be used, or else cold-setting UV curing adhesives, depending upon the sample types and the protocols used. The UV curing system is that supplied by Pelcon. Source: Logitech. Reproduced with permission from Logitech.*

Therefore, first cut the sample as thinly as possible: down to 0.3–0.5 mm, or so, and grind down to about 100 µm. Secondly, lap the ground surface to a thickness of about 30 µm. Diamond paste can be bought in various grit sizes; these are coloured for easy selection when grinding, lapping and polishing. Table 29.3 shows the common sizes and corresponding colour code.

Synthetic polycrystalline diamond gives higher and more even removal rates as well as a better final finish than natural industrial-grade mono-crystalline diamond. Three diamond-embedded rollers are used to grind the samples, which are typically held in place under vacuum. A thin layer of vacuum grease is applied around each slide to protect the bearings on the heads of the diamond grit rollers. The coarse roller 1 (64 µm grit size) removes material from the rough-cut slice from about 500 µm down to 200 µm. Roller 2 (46 µm) grinds down to 100 µm; roller 3 (16 µm) down to 40–30 µm. The last 10 µm is then polished. The diamond-embedded rollers will last for approximately 150 thin sections, although this is very dependent upon the nature of the sample. Each roller will cost approximately £150–200 to be repainted with diamond. This can be done by DK Holdings in Staplehurst, Kent.

With grinding, the abrasive is fixed in contact with the sample (even if it is sprayed on and held in a cloth), whereas when lapping, the slurry is free to roll around, causing a micro-dimpled surface, rather than one with micro-scratches. Normally, loose silicon carbide abrasive powders, with grit sizes of 600 (P1200) or 1000 (P2000) are used. Traditional lapping can give very fine control over the thickness and finish of the preparation. Kemet grade their lapping material by weight, rather than sieving it. The material is put into

Figure 29.8 **Grinding sections** *A classic grinding plate and the modern Pelcon automatic thin section machine with diamond rollers. Source: Reproduced with permission from PELCON.*

Table 29.3 Diamond Grit Sizes

Grit size (μm)	Colour		Corresponding mesh size
0.10		Charcoal	140 000
0.25		Grey	25 000
1.0		Blue	14 000
3.0		Green	8000
6.0		Yellow	3000
8.0		Red	1800
14		Brown	1200
25		Mahogany	600
45		Purple	300
60		Orange	230
90		White	170

an air tunnel and drawn off at particular heights. This prevents long thin particles going through the sieve mesh, only to act as larger abrasive particles (which will introduce scratches) once present in the lapping lubricant. To prevent scores and deformation when they do occur, the lapping cloth can be changed for one made of more resilient material and/or the force applied to the rotating plate can be increased by 10%, or so, and the process repeated as required.

Usually the lapping plate (Figure 29.9) is cast iron but can be changed to suit the sample. Copper plates can be used for softer materials (for samples of 5–8 on the Mohs scale) whereas composite plates (7–9 Mohs) and cast iron plates (8–10 Mohs) are in general use. The cast iron lapping platen has very fine ridges milled into its top surface. These surface ridges are to hold the diamond spray lubricant and to act as a reservoir for very fine 'pull-outs' of material which can then be removed at the conclusion of lapping. Prepare the lapping platen by scraping it with a blade (e.g. Olfa LBB-10 Excel Black, 18mm Snap Off Blades or Stanley knife blades in a strip) lubricated with vegetable oil. This procedure is akin to removing debris from a runway that would otherwise damage the aeroplane – you don't want scratches put into the polished surface.

A calibrated reflective quartz optical flat is laid on top on the recently lapped workpiece and viewed under a lamp giving monochromatic light – usually a sodium lamp. A series of Newton's interference rings should be seen. If these rings are perfectly straight, then the lapping plate (whose surface profile is a mirror image of the workpiece) is perfectly flat. If the rings are concentric and move towards slight finger pressure applied to one edge of the optical flat, then the workpiece is convex, and the lapping plate is concave. The converse also holds true: Newton's rings moving away from the point of pressure indicate a convex lapping plate, which must be measured and reground. The flatness of the lapping plate can be checked with a flatness gauge. These can be obtained from Kemet and other major suppliers.

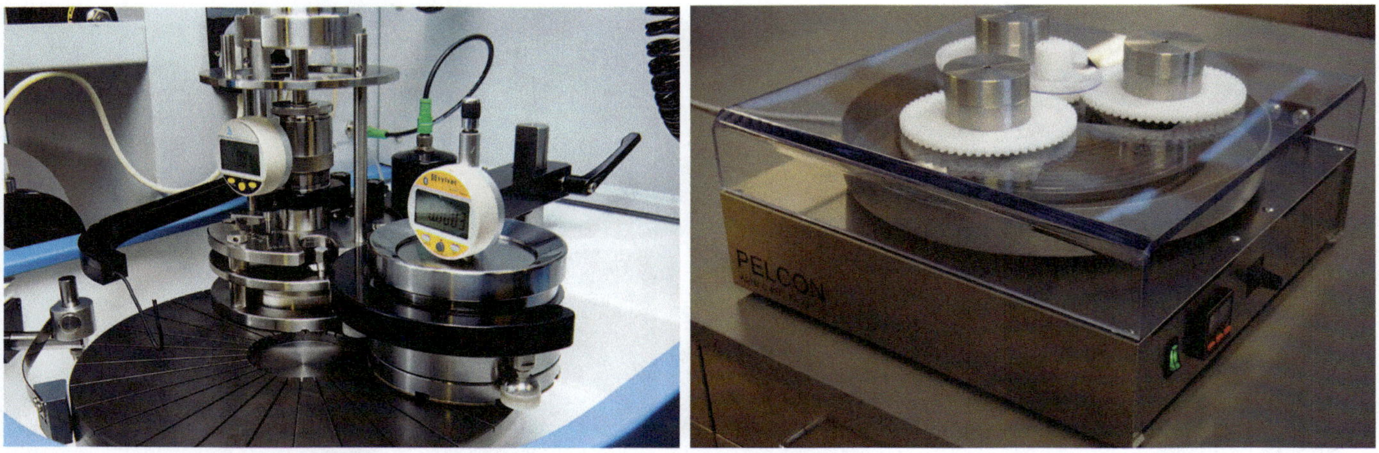

Figure 29.9 ***Lapping plates*** *The Logitech PM6 and the Pelcon lapping and polishing machine. Source: (left) Logitech. Reproduced with permission from Logitech. (right) Reproduced with permission from PELCON.*

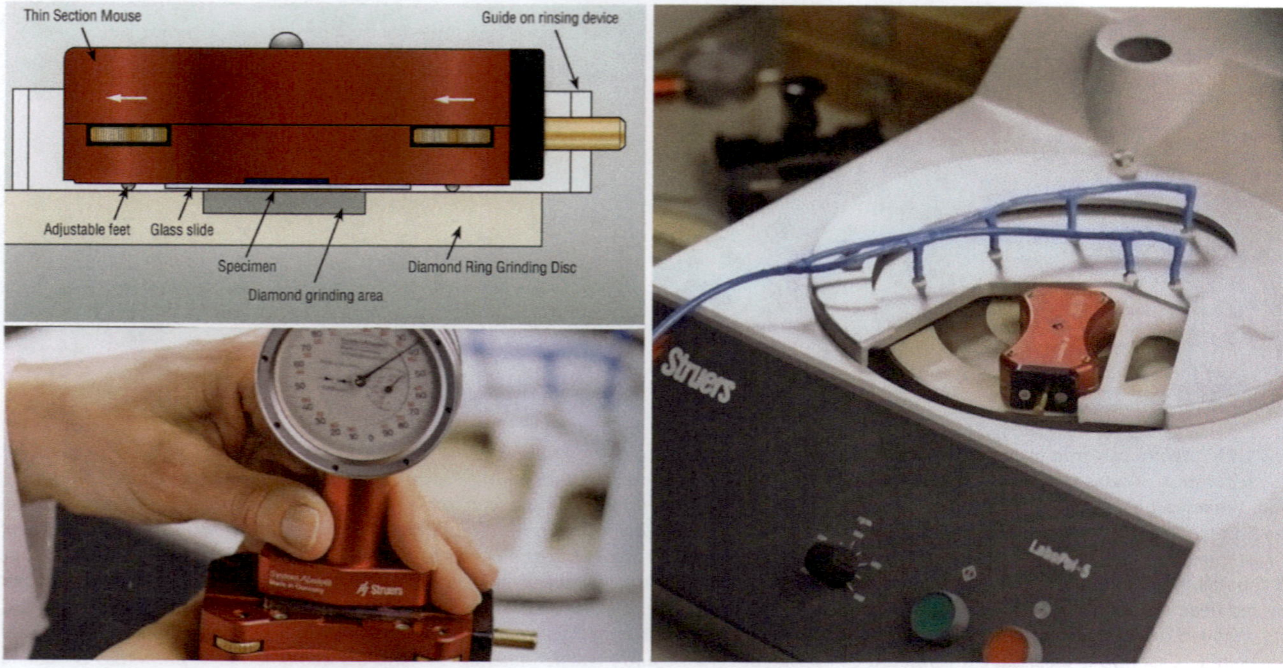

Figure 29.10 ***Struers System Abele*** *The Struers System Abele specimen preparation equipment with 'intelligent ball' system. The thin section 'mouse' feet can be set to the required thickness and the sample ground to the requisite thickness. It is then lapped and polished on the portable variable rotating LaboPol-5. Source: Reproduced with permission from Struers APS.*

Modern grinding/lapping machines have an 'intelligent ball' as part of the foot (see Figure 29.10 for one example) in the sample-holding system, which stops the lapping process once the section is of the correct thickness. If an automatic system such as this is not available, the thickness of 30 μm can be calculated from careful examination of the slice under the microscope. The presence of silica showing a first-order yellow interference colour and of quartz showing a grey first-order colour will show that the slice is 30 μm thick. Instructions of how to use a Michel-Lévy chart to do this are given in Appendix 3.

The sample *must* be washed when transferring from one stage to the next. Depending on the sample, it can be ultrasonically cleaned for a short period. This is not suited to friable samples, but it does ensure that all oil lubricant and abrasive particles are removed before bonding the first face to the microscope slide.

29.6 Polishing Samples

The object of polishing is to obtain a mirror-like surface, free of unevenness in relief (height) and without scratches (see Figure 29.11). Polishing should only be very brief to take off surface roughness. A few minutes is all that should be necessary, and no more than 20 minutes. Final polishing is traditionally done with a cloth-covered lapping plate impregnated with fine abrasive and oil lubricant. Work the abrasive thoroughly into the lap with lubricant before polishing. Some workers do not use a cloth: their reasoning being that cloths introduce very fine surface relief. The specimen should be rotated periodically, until all previous scratches are eliminated. Typically, for best results, 6 μm, 1 μm and 0.25 μm diamond in suspension is used for polishing: the time, lubricant, lap speed and pressure being dependent upon the material and according to experience. The sample must be weighed down, usually with between 300 and 800 g. Care must be taken not to use too heavy a weight, or the polish sample will burnish – the result of over-polishing.

Inspect the ground surface under reflected light (Figure 29.12) to check that it is flat. If the surface is evenly reflective, then this is a good sign. Otherwise, if the surface is matt, it may need further polishing (although some rocks, such as biotite, can still appear matt when polished).

For 'ultrathin' preparations that have been ground and lapped very thinly to about 10 μm (to allow the internal crystal grain structure to be seen clearly), conventional polishing is too harsh, and vibration or chemical-mechanical polishing must be used. Chemical-mechanical polishing uses a neoprene chemical-resistant cloth and creates a reaction layer that is dissolved away; therefore, nothing is added. Besides being suitable for ultrathin samples, it is good for removing damage from 1 μm scratches. The chemical-mechanical process requires care in storing samples, which can dry out and crystallise when exposed to air. Some minerals take a poor polish, due

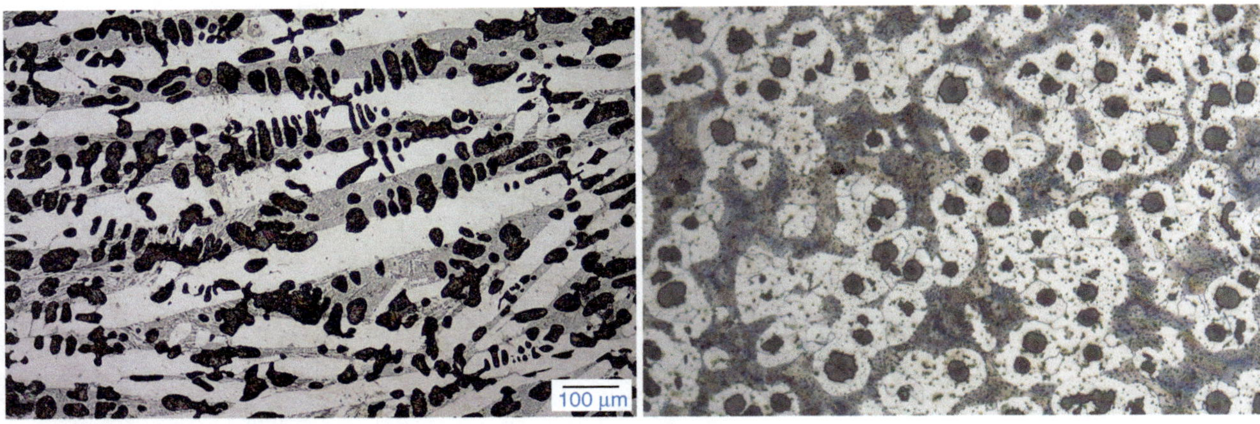

Figure 29.11 **White cast iron example specimen** *Example of polished opaque preparation.*

This preparation of white cast iron illustrates the features that can be detected and analysed from a polished opaque sample taken from the bulk material. White cast iron (left) is so-called because of its low carbon content, present as iron carbide (cementite), which has an orthorhombic crystal structure. Cementite is a hard, brittle material, essentially a ceramic in its pure form. It forms directly from the melt and dominates the microstructure of white cast iron. Very little graphite is present; the composition of cementite allows fracture cracks to propagate through the material. White cast iron is thus too brittle for structural use but has a high compressive strength and excellent wear resistance, and it retains its hardness for limited periods even up to a red heat. It is therefore suitable for use where hard-wearing surfaces are required, such as bearings and shell liners. The properties of cast iron may be changed by adding various alloying elements, or alloyants. Adding silicon forces carbon out of the molten iron solution to form graphite rather than cementite. Cast iron is essentially a cermet: a composite of ceramic and metallic materials. An unusual combination of properties is obtained in ductile cast iron (right) because the graphite occurs as spheroids rather than as individual flakes as in grey iron, due to the addition of small amounts of magnesium. Ductile iron has high impact and fatigue resistance, due to its spherical graphite inclusions.

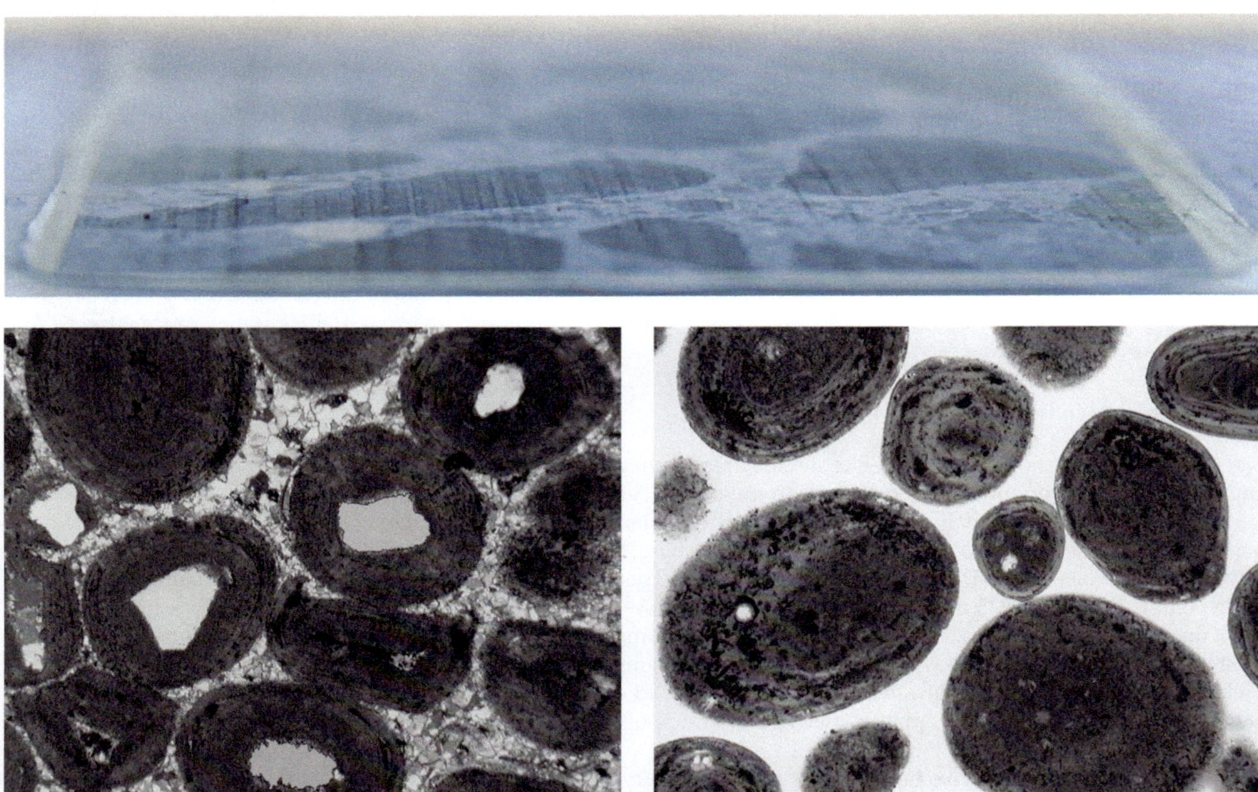

Figure 29.12 **Inspecting the ground surface under reflected-light, Marks to polish out** *A ground thin section held at a shallow visual angle to the light (upper panel) to reveal fine lines, of 1–2 µm relief, which must be polished out for a high-quality finished thin section. These Oolitic limestone preparations show how modern diamond grinding methods (right-hand panel) leave the centres of the ooids intact. Source: Reproduced with permission from PELCON.*

Figure 29.13 ***Sample press*** *This spring-loaded press presses resin samples into a bed of Plasticine® or Blu-tack® onto a 3″x1″ black aluminium microscope slide. Source: Sample press; kindly loaned by Mr T. Cooper, courtesy of TAAB Laboratories Equipment Ltd; photograph courtesy of Mr K. Glover.*

to their hardness or structural anisotropy. Other minerals pluck out because of their perfect cleavages; the triangular pits in galena are a classical example of this. Some metals are too soft to take a good polish and also smear, notably native gold and silver. You are advised to finish polishing very soft and ductile metals, such as gold, on the surface of a brown paper bag[6].

Opaque samples that have been cast into a resin or acrylic block and have had one face polished flat are examined held on a slide so that the slide holder on the stage of the microscope can translate the sample laterally. Figure 29.13 shows a suitable press (available from TAAB Laboratories) and how Plasticine® or specialist putty can be used to attach the sample to a slide. Black matt anodised slides do not reflect unwanted illumination, which would otherwise reduce contrast in the image.

Most highly polished specimens that are left uncovered with a glass coverslip are usually hygroscopic to some extent. These water sensitive samples should be stored in a desiccator to prevent damage from atmospheric moisture. After examination, uncovered polished preparations should be cleaned of any immersion oil and stored in dry, air-tight containers or a desiccator in order to avoid tarnishing, etching or incipient alteration. Many ore minerals will tarnish in a few days, but this can sometimes be removed from the surface using an organic solvent, such as ethanol. Finally, whether grinding, lapping or polishing, *always remember to put the slide the right way onto the machine*! We all make this mistake at least once!

29.7 Preparing a Grain Mount

Grain mounts are used to investigate unconsolidated, loose or very friable material. They can also be used to investigate individual crystal and mineral components of rocks (e.g. with a spindle stage). If the grains do not occur naturally from unconsolidated material, the bulk rock must be crushed and sieved. For later analysis with a spindle stage (see Box 29.3), sieve the material through a 100-mesh sieve (150 μm holes), such that the particles are larger than the 75 μm holes of a 200-mesh sieve, the next size down.

The material should be cleaned of any heavy liquids used in separation, then washed and dried before mounting. The grains are sprinkled either directly onto a slide or into the base of a mould. In the latter case, resin is carefully poured over and allowed to set. Preparations of grains and small fragments are suitable for unconsolidated material, but any textural and spatial information is lost, so grain mounts should be prepared together with sections of the original rock – if possible.

Box 29.3 Stages for Petrography

The Spindle Stage

In a perfect world, birefringent crystals would be orientated perfectly to produce ideal interference figures so that we could determine whether the crystal was uniaxial or biaxial, their optical sign and measurement of 2V – the acute angle between the optic axes. With good orientation of the crystal it is also possible to perform a Becke line test to determine the two refractive indexes of a uniaxial mineral and the three refractive indexes of a biaxial one.

In order to carry out these tests, the optic axis must be vertical, aligned perpendicular to the stage of the microscope. The spindle stage allows you to rotate a mineral or crystal grain about an axis that is perpendicular to the rotation axis of your microscope stage. The crystal, which is temporarily glued to the end of the spindle, can be orientated as desired and immersed in fluids of different refractive index for refractive index matching by the Becke line test.

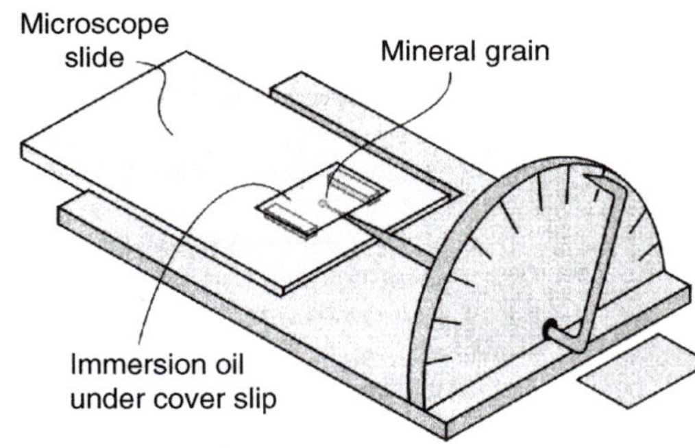

Source: Nesse 2009. Reproduced with permission from Oxford University Press.

> As Bloss remarks, 'Anisotropic crystals become amazingly tractable in optical crystallography if they can be rotated about an axis that is parallel to the stage of a polarising microscope.'

The advantages of the spindle stage are as follows: it is easy to use, and a version can be cheaply home-made. Any direction of the crystal can be rotated into the plane of the microscope stage. This allows all three refractive indexes to be determined without correction for angular rotation.

The more versatile cousin of the spindle stage was the universal stage, a 4-axis or 5-axis goniometer stage for the petrographic microscope, now no longer made. It is useful for the assessment of crystal optics (i.e., principal vibration directions, designated α or x, β or y, γ or z) relative to the crystallographic axes (designated a, b, c) of rock thin sections, rather than just minerals.

Source: Kile 2003.

References

Bloss, F.D. (1980). *The Spindle Stage: principles and practice*. Cambridge: CUP.

Nesse, W.D. (2009). The petrographic microscope. In: *Introduction to Optical Minerology*, Chapter 2. OUP, Oxford: New York.

Kile, D.E. (2003). The Petrographic Microscope, supplement to The Mineralogical Record.

The Universal Stage. https://www.geochemsoc.org/publications/geochemicalnews/gn140jul09/theuniversalstage/.

29.8 Preparing a Rock-Section Peel

Applying a thin coat of transparent plastic onto the surface of a rock is a much quicker way to examine the surface components than grinding a thin section. A very thin layer of surface material is glued to the liquid plastic. When this hardens into a transparent layer and is pulled away, the surface of the rock is also removed. This technique is particularly useful in the study of carbonate-containing rocks after etching with acid. The peel shows some features better than a thin section: grain boundaries are more distinct, small grains are more readily resolved and a photographic image usually has sharper definition. See Wilson and Palmer (1989) and Holmes and Lopez (1986) for further descriptive details and applications to paleontological plant tissues.

A rock sample[7] is ground and polished on one surface, with 400, then 800 grade carborundum grit on a glass lapping plate. For very fine detail, finish with a polishing step using either 1 µm Aloxite or diamond abrasive. The face of the sample is then carefully washed and dried to ensure no lapping abrasive remains. An aqueous 2% solution of hydrochloric acid is used for a few seconds to etch the surface of the rock[8]. Greater control can be achieved by using a stronger (20%) solution of the milder aqueous acetic acid for up to 5 minutes. Dip the sample face up in the solution, and cover it well. A rock dipped face up rather than face down will not trap carbon dioxide bubbles. If these remain, they can lead to rather unselective etching. A second reason for immersing the sample face up, is that a rock placed thus in the etchant can, with care, be examined using a stereomicroscope to determine the optimum endpoint of etching. After etching, wash the rock face thoroughly and dry the face carefully. The rock is then set almost horizontally and leaning backwards slightly, on a bed of Plasticine®, Blu-Tack® or modelling clay within a suitable tray. Figure 29.14 shows how this is done.

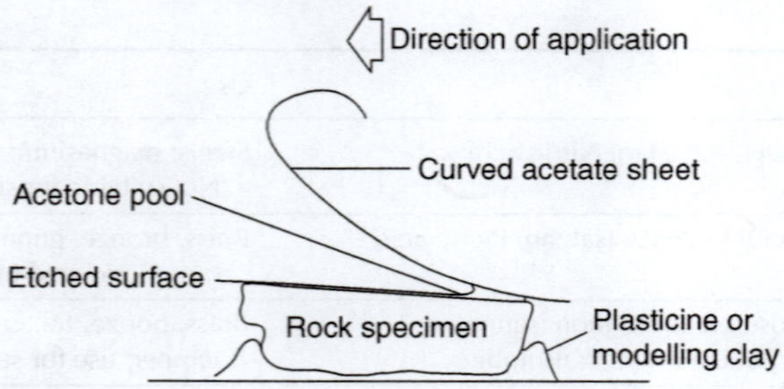

Figure 29.14 **Making an acetate peel** *Illustration showing the steps involved in making an acetate peel. Source: Tucker 1988. Reproduced with permission from John Wiley & Sons.*

The surface of the rock is then flooded with acetone (or other suitable solvent[9]), and a piece of acetate sheet 0.1 mm thick, such as those formerly used for overhead projectors, laid over the rock face. The size of the sheet should be about 1 cm greater all round than the face of the rock, and initially it should float slightly on the pool of liquid. As the solvent evaporates, the soft, partially dissolved film settles down and can afterwards be stripped off as a 'peel'. This takes time: peel the acetate sheet off about 60–90 minutes after it is applied to the rock surface. Do not attempt to do this too early. Try to strip off the peel in one continuous movement. If there are trapped air bubbles or blemishes, the procedure can be repeated. Mount the finished peel onto a glass slide with clear matt adhesive tape. The peel can be examined with an objective requiring no cover glass or mounted dry under a coverslip. A slightly different approach is to take a self-adhesive label, cut a 10 mm (1/2″) square hole with a scalpel and ruler, and peel off the backing paper. The label can then be stuck face down onto the polished rock surface. Cellulose nitrate 'dope' such as that used for model aircraft is then painted over the window and onto the surface of the paper label. Again, allow the lacquer to dry thoroughly, and mount the 'peel' as before. This method circumvents the use of large amounts of solvent.

The slice used for the peel can be mounted and prepared into a thin section to complement the peel, if desired. It is possible to stain both rock thin sections and peels. This is not described here. See Humphries (1992), Miller (1988) or Füsun et al. (2005) for details of these staining techniques; the references are given at the end of the chapter.

29.9 Etching Samples

For many materials (e.g. rocks, ceramics) once the sample is polished, the preparation is finished. Identification, characterisation and study in the sample of the various constituent phases are revealed by differences in colour, bireflectance and surface relief. Etching can be used to accentuate these features and improve contrast.

Etching is the selective attack of grain and interface boundaries or of the phases within polished metal or alloy samples. Before using any etchant, check the sample for polishing artifacts, and record its appearance. Look out, also, for the presence of coarse inclusions or porosity, which may be unduly affected by the etching process. Etchants are usually aqueous salts, acids or alkalis or else are alcoholic. They are normally *very corrosive* and should not be handled without experience or supervision, and all local health and safety procedures must be followed. Here is a typical procedure:

1. Flood the base of a shallow, acid-resistant container (e.g. a shallow glass baking dish) with the etching fluid. Because the acid should be replaced frequently, a shallow bath is recommended. Do not use a glass container when using hydrofluoric acid.
2. Either support the slice on two glass slides, plastic bar cut to the same dimensions or hold the rock slice by hand. Do not allow the polished surface to touch the base of the container.
3. Move the slice around for a few seconds to ensure good contact with the etchant.
4. Rinse the slice, partially dry it and use a microscope to determine if the etching should be repeated.
5. When finished, rinse the surface with distilled water, and drain the slice on a paper towel. Dry thoroughly in a slightly warm oven for a few minutes.

There are a wide range of etchants, and only experience can tell which are better suited to the particular sample under consideration. Some of the more commonly used etchants are listed in Table 29.4; for further comprehensive lists, see Petzow (1999). Fine etching can be partially combined with the final stage of polishing: the etchant is added to the polishing compound, and a final etch is given after polishing. Contrast between phases in the polished surface can be developed by anodizing oxides of metal specimens or deposited by vacuum evaporation or reactive-ion sputtering. Etching of rocks for removal of carbonate is relatively simple, as described using hydrochloric acid for rock peels, above.

Table 29.4 Etchants

Name	Constituents	Use
Nital	Ethanol (EtOH) + 2% (aq) Nitric acid	Steels; magnesium; a general etchant; use for up to 1 minute. Not suitable for stainless steels or aluminium.
Picral	EtOH or MeOH + 2–3% (sat. aq) Picric acid	Brass, bronze, gunmetal; up to 2 minutes. Do not allow to dry out: picric acid in its dry form is an explosion hazard.
Klemm's reagent	Sodium thiosulphate solution (saturated, aq) + 1% potassium metabisulphite	Brass, bonze, tin, cast iron, ferrite, martensite, zinc and steel temper; use for seconds up to 3 minutes.
Aquaregia	Conc. HNO_3 + Conc. HCl (mix fresh, ratio 1:3)	Titanium, platinum and corrosion resistant metals; use for up to 1 minute.
Hydrochloric acid	5–10% (aq) HCl	For carbonate minerals; from a few seconds up to 1–2 minutes at room temperature.
Alcoholic acid	1% HCl in ethanol	Tin or tin-alloys; zinc or zinc-alloys.
Carapella's reagent	5% $FeCl_3$ + 5–30 mL HCl + 70-95 mL EtOH	Nickel or copper and their alloys; also steel: for grain contrast.
Hydrofluoric acid	0.5–2% (aq) Hydrofluoric acid	Aluminium or titanium and their alloys; selective revealing of fossils in silicate rocks; cast iron; ceramics; oxides and glass.
Keller's etch	2.5% HNO_3 + 1.5% HCl + 1% HF (aq)	Aluminium and titanium alloys; 10–30 seconds.
(ortho)Phosphoric acid	37% (aq) H_3PO_4	Dental enamel; use for 1–2 minutes.

Safety note

All mineral acids are dangerous. Wear full protective clothing. Always use the appropriate eye goggles and gloves (double-glove with neoprene gloves) and use a mask where necessary. Work within a fume hood and follow the risk assessments for your institution. This is a legal requirement. Take particular care using Hydrofluoric acid (HF): it dissolves oxides and glass and is extremely poisonous. As an example, see this PDF from the safety office of the Department of Chemistry and Chemical Biology at Harvard: http://chemistry.harvard.edu/files/chemistry/files/safe_use_of_hf_0.pdf.

References

Haynes, R. (1984). *Optical Microscopy of Materials*. Glasgow: International Textbook Company, Blackie group. Table 9.1, p. 102.
Petzow, G. (1999). *Metallographic Etching*, 2nd ed. Ohio: ASM International.
Walker, P. and Tarn, W.H. (eds.) (1991). *Handbook of Metal Etchants*. Boca Raton: CRC Press.
Etchant databases: http://www.metallography.com/etching/table.htm and http://www.steeldata.info/etch/html/overview.html.

29.10 Staining Samples

Histological stains are used widely to differentiate cells and tissues in biological science. The purpose of staining rock and mineral samples is the same: to differentiate components that look very similar and are otherwise undistinguishable. However, the repertoire of stains used in materials science is much smaller. They are principally used to differentiate carbonate materials in gypsum and limestone or feldspars, as well as some clays and ceramics. For example, Alizarin S stains calcite; anhydrite, which is found in association with gypsum and sometimes calcite, does not stain: it must be identified by a process of elimination. Table 29.5 gives a list of some of the more commonly used stains. For further information, consult the references given in the table.

It is useful to have an old polarised-light microscope set up in the polishing laboratory for examining thin sections easily. If the grinding or polishing steps are not sufficiently automated to stop at the requisite thickness, it will be necessary to check quartz or other mineral inclusions in the preparation to judge when to stop removing material and when to perform the final lapping and polishing of the section.

29.11 Chapter Summary

At the end of this chapter, you should be aware of the principles involved in preparing samples for materials science microscopy. Chapters 30 and 31, the last two in the book, now tell you about the use and selection of cameras and digital detectors to acquire and record the image that you have learnt to prepare.

1. Opaque materials specimens are normally prepared either as thin sections or as polished samples.
2. Thin sections are ground from bulk material, but they can also be prepared as a peel from a prepared flat surface.
3. The structures observed in thin section are generally at a scale of 1 μm or greater. In reflected-light microscopy, grain and crystal sizes typically occur within a wide size range from 5 to 500 μm.
4. Each step of cutting, grinding, lapping and polishing to produce a flat surface introduces damage. The purpose of each subsequent step is to remove the damage created previously.
5. It is possible both to remove material from the sample as well as introduce adventitious material into the preparation.

Table 29.5 *Example Stains*

Name	CI number	Constituents	Directions	Use
Haines' method: Alizarin Red S	58005	Etch 3–5 secs in 5% (aq) HCl 0.2% Alizarin Red S + 0.2% (aq) HCl	Stain for 1–2 minutes	For identifying carbonates: calcite pink/red; brucite purple; aragonite pink/red; dolomite and siderite unstained
Dickson & Evamy's method: Alizarin Red S plus Potassium ferricyanide	58005	0.2% Alizarin Red S + 2% $C_6N_6FeK_3$ (potassium ferricyanide) in 1.5% (aq) HCl	After 15 second etch in dilute HCl, stain for 30–40 seconds. Make up fresh.	For staining limestones. Calcite stains pink-red/brown; dolomite is colourless; ferroan calcite mauve/blue; ferroan dolomite pale blue
Tucker's method	42000 22120 52015	0.1% (aq) Malachite green congo red or methylene blue, or in 50% (aq) EtOH	Stain for 20–30 minutes. Use the dye that contrasts best by eye with the bulk sample	For staining clays
Houghton's method	–	48% Hydrofluoric acid etch Sodium cobaltinitrite Barium chloride (fixer) Potassium rhodizonate	Etch with HF *vapour* for 30 seconds. Immerse in sodium hexanitritocobaltate (III). Wash and blot dry. Immerse in 2% barium chloride. Wash. Drop on 0.05% potassium rhodizonate.	For plagioclase and alkali feldspars. Leave potassium rhodizonate until plagioclase stains pink. Alkali feldspars stain yellow. Quartz is unstained.
Freidman's method	58005	0.2g Alizarin Red S added to 25 mL methanol + 50 mL 5% (aq) sodium hydroxide	Stain for a few minutes.	Gypsum stains purple; anhydrite and calcite are unstained; dolomite stains very pale purple

References

Haynes, R. (1984). *Optical Microscopy of Materials*. Glasgow: International Textbook Company, Blackie group. Table 9.1, p. 102.
Ingham, J. (2010). *Geomaterials Under the Microscope*. Oxford: Elsevier. Appendix B, pp. 176–177.
Department of Earth Sciences, University of Cambridge. https://www.esc.cam.ac.uk/resources/facilities/laboratories/geochemistry-labs/carbonate-staining.
Preparation of solutions. https://www.struers.com/en-GB/Knowledge/Etching#etching-about and https://www.kemet.co.uk/blog/metallography/etching-in-metallography.

6. Poorly consolidated samples are vacuum impregnated before cutting. Fluorescent dye helps identify filled pores or cracks.
7. Grinding is the rapid removal of material from the sample, to reduce it to a suitable thickness. Lapping is the removal of further material in a slower fashion to produce a smooth, flat unpolished surface.
8. The sample *must* be washed when transferring from one stage to the next.
9. Polishing gives a mirror-like surface, free of unevenness in relief without scratches.
10. Grain mounts are used to investigate unconsolidated, loose or very friable material.
11. Acetate peels grain boundaries and are useful for etched carbonate-containing rocks.
12. Etching and staining further reveal crystal grain boundaries and morphology.

Key Reading

Barker, A.J. (2014). *A Key for Identification of Rock-Forming Minerals in Thin Section*. Boca Raton, FL: CRC.
- A well-designed and illustrated identification key with clear concise explanations of crystal optics and behaviour under crossed polars

MacKenzie, W.S. and Adams, A.E. (1994). *A Colour Atlas of Rocks and Minerals in Thin Section*. Boca Raton, FL: CRC Press.
- This well-known atlas complements Dr Barker's key; both are essential guides for petrology students.

References

Füsun, A. et al. (2005). 'Acetate peel technique: a rapid way of preparing sequential surface replicas of dental hard tissues for microscopic examination'. *Archives of Oral Biology 50*/10: 837–842.
Gribble, C.D. and Hall, A.J. (1992). Introduction to the microscopic study of minerals. In: *Optical Mineralogy, Principles and Practice*, Chapter 1. London: UCL Press. p. 33.
Holmes, J. and Lopez, J. (1986). 'The disappearing peel technique: an improved method for studying permineralized plant tissues'. *Palaeontology 29*/4: 787–808.
Humphries, D.W. (1992). *The Preparation of Thin Sections of Rocks, Minerals, and Ceramics*. RMS Handbook No. 24. Oxford: Oxford Science Publications.

Miller, J. (1988). 'Microscopical Techniques: 1. Slices, slides, stains and peels'. In: Tucker, M. (ed.), *Techniques in Sedimentology*, Chapter 4. London: Blackwell Scientific Publishers, pp. 85–107.
Petzow, G. (1999). *Metallographic etching*, 2nd ed. ASM International.
Wilson, M.A. and Palmer, T.J. (1989). 'Preparation of acetate peels'. Special publication No. 4. *In*: Feldmann, R.M., Chapman, R.E., and Hannibal, J.T. (ed.), *Paleotechniques*. The Paleontological Society Special Publication *4*, pp. 142–145.

Further Reading

Allman, M. and Lawrence D.F. (1972). *Geological Laboratory Techniques*. London: Blandford Press.
Green, O.R. (2001). *A Manual of Practical Laboratory and Field Techniques in Palaeobiology*. Dordrecht: Kluwer Academic.
Haynes, R. (1984). 'Specimen Preparation'. In: *Optical Microscopy of Materials*, Chapter 9. Glasgow: International Textbook Company, pp. 94–110.
Ingham, J. (2010). *Geomaterials Under the Microscope: A Colour Guide*. Boca Raton, FL: CRC Press.
Ivarsson, M. (2006). 'Advantages of doubly polished thin sections for the study of microfossils in volcanic rock'. *Geochemical Transactions 7*: 5. doi: 10.1186/1467-4866-7-5.

Internet Resources

American Society for Testing and Materials (ASTM): https://www.astm.org/Standard/standards-and-publications.html.
International Organization for Standardization (ISO): https://www.iso.org/home.html.
Hardness testing (see also Box 29.2): http://www.materials.co.uk/microhard.htm.
Measuring lapping plate flatness: https://www.kemet.co.uk/blog/lapping/how-to-measure-flatness-technical-article.
Specimen preparation: http://www.thin.stir.ac.uk.

Contracting specimen preparation

MetPrep: http://www.metprep.co.uk.
Pelcon: http://pelcon.dk/.

Company contacts

Buehler (USA): http://www.buehler.com.
DK Holdings (diamond rollers): http://www.dk-holdings.co.uk/flex/flexintr.html.
Intertronics (optical bonding glues and UV cured media): http://www.intertronics.co.uk.
Kemet: http://www.kemet.co.uk.
Logitech Ltd: https://logitech.uk.com/.
MetPrep: http://www.metprep.co.uk.
Pelcon: http://pelcon.dk/.
Struers: http://www.struers.com.
Making acetate peels: http://woostergeologists.scotblogs.wooster.edu/2015/05/01/woosters-fossil-of-the-week-how-to-make-brilliant-acetate-peels-with-a-jurassic-coral-example/.

Notes

1. The International Standard provides for each side of the slide to be parallel to within five minutes of arc. This can be improved on a lapping and polishing machine.
2. Potentially harmful solvents such as trichloroethylene have been used in the past to clean samples. These solvents are now being replaced by biodegradable detergents for removing contamination such as lapping and polishing residue.
3. The other sizes of microscope slide that tend to be used are 51 mm square, 76 mm × 39 mm, 76 mm × 52 mm or 110 mm × 76 mm.
4. Concrete must not be heated above 50°C, for example, and must be ground or lapped with an oil-based rather than a water-based lubricant. Some expert practitioners advocate not drying above 25°C ambient temperature.
5. Cold-casting can be a misnomer, because the exothermic nature of the curing reaction can generate considerable heat (especially while acrylic resins are curing), and you should be aware of this for temperature-sensitive samples. Consult the supplier for further information if in doubt on this.
6. I am grateful to Dr Chris Jones, at the Natural History Museum, UK, for this information.
7. For small samples (e.g. teeth) which are difficult to hold steady on the lapping wheel, embed these first in epoxy resin.
8. If a dental peel is being taken, use 37% (aq) orthophosphoric acid for 1–2 minutes.
9. Acetate sheet is also soluble in methyl acetate, 4-Hydroxy-4-methyl-pentan-2-one and 1,1,1-trichloroethane.

30

Recording the Image: Part 1 – Theory

Outline

30.1 Introduction
- Digital imaging: terminology
- The decline of analogue film

30.2 Advantages of Digital Imaging
- Digital imaging advantages listed
- The transient nature of digital storage

30.3 How digital Images Are Formed and Encoded
- Binary notation
- Vector and raster images
- Pixels

30.4 Overview of Electronic Camera Operation
- Photo-sites of CCD & sCMOS chips
- Photon conversion to photo-electrons

30.5 Digital Image Resolution
- The different meanings of 'resolution'
- Pixel resolution of the camera sensor
- Resolution of the monitor or printer

30.6 Bit Depth and Dynamic Range
- The binary basis of bit depth notation
- Grey levels and visibility

30.7 Black Pixels and Pixel Pathology
- Hot pixels, stuck pixels and dead pixels

30.8 Types of Electronic Detector
- Video recording
- CCD cameras
- Full-well capacity, how CCD cameras work
- How EM-CCD cameras work
- How sCMOS cameras work
- Global and rolling shutters

30.9 Encoding Colour in Digital Images
- RGB and the Bayer mask
- Co-site sampling and microscanning
- Demosaicing

30.10 Digital Camera Noise
- Random signal fluctuations
- Periodic noise and Fourier transform removal
- Poisson (shot) noise
- Read noise

Understanding Light Microscopy, First Edition. Jeremy Sanderson.

30.1 Introduction

A digital imaging system comprises a microscope and CCD/sCMOS camera, PMT, GaAsP or APD detector together with a computer, display monitor and imaging software. Digital, or electronic, imaging has transformed both amateur and professional picture taking in microscopy as well as every other field of image acquisition. Photomicrography is the recording by photography of an image formed by a microscope. As explained in Box 30.1, it should not be confused with microphotography. For the first 150 years of photomicrography, a photosensitive silver halide emulsion spread onto glass or film was used to record and reproduce the image. Continual improvements in computer and electronic camera technology have rendered analogue film recording obsolete. Digital imaging is faster, cheaper and much more accurate than classical photography. This chapter covers what digital images are, how they are recorded in the microscope, the architecture of digital cameras and the various forms of noise they may suffer from. These charge-coupled device (CCD) and complementary metal oxide semiconductor (CMOS/sCMOS) electronic detectors employ very similar technologies, each with different advantages. This chapter concludes with a section on how to choose the best electronic sensor for your particular application. The next chapter gives practical advice on coupling a microscope and camera together as part of an imaging system, as well as practical guidance on imaging collection and display or output.

Photomultiplier tubes (PMTs), gallium arsenide phosphide (GaAsP) detectors and avalanche photodiodes (APDs) are special cases of digital detector. They are used to read the photon signal very rapidly from a very small area (effectively a diffraction-limited point) of the microscope field of view and convert this to an electronic signal for digitisation. Although PMTs and related sensors can be raster-scanned to produce a digital image as the end result, just like a CCD or sCMOS camera, they are associated with point-scanning confocal microscopes and so are discussed in Chapter 17 rather than here.

Silver halide film is gradually being withdrawn from the market by Kodak and other manufacturers. It is hardly used nowadays and is not discussed further in this book. For information on the exposure, developing and printing of classical film emulsions, see *The Manual of Photography*, 10th edition (Allen and Triantaphillidou, 2009 or *The Science of Imaging*, 2nd edition (Saxby, 2011). The best practical texts on using a microscope to capture analogue photomicrographs with film are still Bracegirdle and Bradbury (1995) and Delly (1988). The clearest and most concise explanation of photomicrography is that by Evennett (1989). Although that particular chapter concerns the use of film, there is much on image recording that does not date and is still relevant today. For a comprehensive treatise on photomacrography, see Bracegirdle (1995). All these references, together with a history of photomicrography, which has many useful citations (Bracegirdle, 2010), are given in the Further Reading section, at the end of the chapter.

30.2 Advantages of Digital Imaging

Digital images can be copied (cloned) exactly without loss; they can also be easily altered, transported and stored. Optical duplication of film and paper-based images always results in a sequential loss of information from one step to the next, be that from film to print or from print to copy. Digital image acquisition is much easier and less complex than photographic recording and development. Using dedicated software, images can be enhanced and changed very easily, whereas film requires intensive darkroom work to achieve even part of the same result. Within seconds of creation, digital images can be sent worldwide via the Internet. They can also be published in journals online without the need for printing as hard copy. It is possible to store digital images in easily searchable databases, which is not the case for negatives or prints. Digital images are also potentially easier to retrieve than hardcopy media, which can be misplaced! Digital images can be viewed very soon after acquisition, making it possible to evaluate the image in a way not previously available with analogue film. Very often it is possible to retake a better image while the specimen is still available under the microscope. This ability immediately to reacquire the image together with the potential to change the exposure conditions and manipulate it are powerful strengths of digital imaging.

Box 30.1 Photomicrography and Microphotography

Source: Shadbolt 1857.

These two terms are sometimes confused; it is not helped by the fact that the German translation of photomicrography is *mikrophotographie*. The distinction between photography, photomicrography and microphotography, as well as close-up photography, macro-photography and photo-macrography lies first in the lens system used to form the image, and next in the means of generating the image.

The reproduction ratio is the ratio of the actual size of an object to the size of the image on the sensor. **Photomicrography** requires a compound microscope (i.e. an objective used with an eyepiece) and is the correct term for recording an image seen down the eyepiece, or on the monitor, with reproduction ratios from 10:1 to an upper limit around 2000:1 (where the image of the object on the sensor is 2000x its actual size). Beyond that is the province of electron micrography.

Microphotography, on the other hand, involves the production of minute photographs: it uses reversed objectives to form the image by reduction, with reproduction ratios from 1:10 to 1:200, or less. It is a reduction method, requiring a microscope to magnify and thus observe the image. In preparing a microphotograph, the objective used to reduce the image onto the film is carried in the sub-stage of the microscope, and the microscope itself (with objective and eyepiece as usual) is merely used to ensure that the image of the object being photographed is accurately focused in the film plane. In German the equivalent term is *mikrokopie*.

John Benjamin Dancer (1812–1887) was the inventor of microphotography in 1839, using the Daguerreotype method and a first reduction of 1:160. The most complete account of Dancer's life and of his microphotographs has been written by Bracegirdle & McCormick (1993). No illustration of Dancer's microphotographic apparatus survives, but that of George Shadbolt, does.

Microphotography was dismissed by the *Dictionary of Photography* in 1858 as "of little or no practical utility" and "somewhat childish and trivial", yet its potential was noted by Sir David Brewster, a prominent physicist and Principal of Edinburgh University, who was enthusiastic about Dancer's work. He predicted that microphotographs might one day be used to transmit messages secretly during wartime. A few years later, during the siege of Paris in the Franco-Prussian war of 1870, Brewster was proved correct when Dagron, a French microscopist, having escaped by balloon, used carrier pigeons to send reduced size messages into the beleaguered city.

Shown here is a microphotograph of a well-known illustration of Antoni van Leeuwenhoek, and the microscope slide from which it was taken. The impact of microphotography on the modern world is all around us, for the circuits of microchips are photographically reduced 1:200 onto photo-resist on a glass carrier prior to forming silicon circuits.

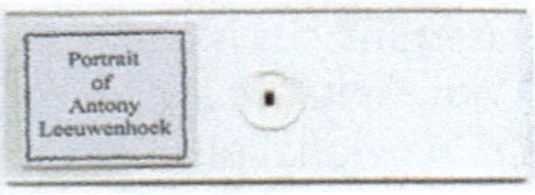

This image is taken from the Frontispiece to: *The Select works of Antony van Leeuwenhoek concerning his microscopical discoveries in many of the works of nature*, by Samuel Hoole. London, Whittingham and Horlis, 1816. Picture painted by Verkilje and engraved by Anker Smith, ARA. (Source: Hoole, 1816.) The actual size of the microphotograph mounted on the 3″ × 1″ slide is 1.28 × 0.92 mm. The film used was Kodak SO-343, of very high resolving power for microfiche recording (with an unexposed grain size of circa 40 nm), which is no longer made.

References

Bracegirdle, B. and McCormick, J.B. (1993). *The Microscopic Photographs of J.B. Dancer*. Chicago: Science Heritage Ltd.

Groom, N.G. (1986). 'Micrography – the making of microscopic size photographs'. *Microscopy 35*/6: 445–450.

Shadbolt, G. (1857). 'On the mode of producing extremely minute photographs for microscopic examination'. *Journal of the Photographic Society of London 4*: 78–81.

Stevens, G.W.W. (1968). *Microphotography: Photography and Photo-Fabrication at Extreme Resolution*, 2nd ed. NY: Wiley.

Generally, less photographic expertise is required to acquire digital images, 'point-and-shoot' pocket-sized digital cameras being an obvious example. To facilitate digital photomicrography, the major manufacturers continually update and streamline their software into (hopefully intuitive) workflows that are both quick and easy to use. Well-designed intuitive software is very powerful and takes care of much of the arduous and time-consuming aspects of setting up a microscope and camera to acquire images. For this reason, it is important to remember, and not to forget, that the quality of the final image – whether digital or analogue – is largely dependent upon the care taken to set up the microscope correctly. As stated throughout this book, a microscope can record an image however poorly it is set up. Both film and digital imaging systems can reveal imperfections in the image (e.g. the colour temperature or unevenness in the background illumination) that are not immediately visible when looking through the microscope eyepiece. This is because the eye is unconsciously tolerant of such imperfections that are not selectively filtered out by artificial recording systems.

Nevertheless, for all its convenience, digital imaging is relatively new technology, and archiving remains an issue. Under the correct storage conditions, film has a known life expectancy greater than 150 years. The means of storing and reading digital media has changed so rapidly, together with rapid changes in successive versions of computer operating systems, that images created only a few years ago and stored on obsolete discs or media often can no longer be read. The longevity of images stored on local or remote servers (i.e. the cloud) remains to be seen, but the challenges of handling the volume of data now being generated are immense.

30.3 How Digital Images Are Formed and Encoded

We humans count in decimal, or denary, notation, otherwise called base 10, because we have eight fingers and two opposable thumbs: ten digits. However, it would be more logical to count in dozenal or base 12, because 12 is divisible by 2, 3, 4 and 6. Even so, we are slow at arithmetical calculation and prone to error: it is much quicker and more accurate to use computers. These machines use base 2, called binary, because they are comprised of switches which are either 'on' or 'off'. Each 'on/off' state is represented by a 'one' or a 'zero' respectively; this is the smallest, irreducible, piece of digital data, called a bit (a contraction of *binary digit*). A bit is the smallest unit of data: it is either on or off, black or white, a 'one' or a 'zero'. Eight bits are arranged into one byte of information. A single byte encodes 256 units of information (2^8 is equal to 8 successive bits, or $2 \times 2 \times 2 \times 2 \times 2 \times 2 \times 2 \times 2$) and can be expressed in hexadecimal (base 16) using two characters.

Numbers that are short in denary are represented by quite long strings of figures in binary (Table 30.1). The power of digital imaging lies in the way in which computers can process a sequence of numbers extremely fast (several million operations a second). It becomes, therefore, largely irrelevant that a simple instruction requires many 1s or 0s as a string of numbers to represent a computer command, text or pictorial image component.

Digital images are encoded in one of two ways: (a) as vector images or (b) as raster images. Vector images are comprised of geometric points, lines and curves encoded mathematically and typically are produced with illustration software. Vector images are widely used for lettering and shapes in graphic design; they do not lend themselves to encoding complex patterns such as encountered in microscope images. An advantage of vector images is that they can be scaled to any size and detail. The quality of a vector image is limited only by the resolution of the display device, and because the information is stored algorithmically, they only take up a fraction of the storage space required for a raster image.

Raster images (from the Latin *rastrum* meaning 'rake') are composed of small geometrically square non-overlapping Cartesian subunits called pixels (pixel = a portmanteau of *picture element*). These pixels are the smallest addressable points in a raster image and are encoded as bytes of information. The address of a pixel uniquely encodes (a) its coordinates, (b) its intensity and, where applicable, (c) its colour. When a raster image is continuously enlarged, these pixels eventually become visible. Therefore, when collecting a pixelated raster image, it is important to consider how it will be displayed and to collect sufficient signal at sufficiently high resolution so that pixelation does not occur, which leaves images looking 'blocky' and unsharp. It is relatively easy to convert from vector images to raster images for display purposes. It is much more difficult to transfer an image format from a raster to a vector image, and data compression of raster images is ineffective for the complexity of biological images unless they are preprocessed to mask the unwanted areas.

With digital images taken through the microscope, we are concerned with raster images (Figure 30.1) composed of pixels. While each individual pixel generally has a 1:1 aspect ratio (i.e. it is square), this is not an absolute requirement. Nevertheless, square pixels ensure minimal distortion when displayed, as well as maximum compatibility with image processing algorithms. Having square pixels as the smallest addressable sampling point does not preclude the image from having a different aspect ratio overall. The aspect ratio is found by dividing the image width by the height, and is determined by the dimensions of the CCD faceplate onto which the image falls. A 1388 × 1040 CCD sensor array will form a rectangular image, of square pixels, with an aspect ratio of 1.33:1.

The number of pixels in an image file is normally determined at the time of acquisition by the number of photo-sites in the electronic detector, whether it be a CCD, sCMOS or PMT. It is sometimes possible to select the desired pixel resolution when capturing the image, but most users leave the image size as set by the hardware limitation of the detector. Thereafter the number of pixels can only be changed by using image editing software (e.g. Adobe Photoshop) to downsample by deleting pixels or upsample inserting new pixels by interpolation. It is worth remembering that a scientific image is not just a pretty picture but is in fact a scientific data set of photon intensity, from which measurements can be derived to support or reject hypotheses. When upsampling, these 'new' pixels can never

Table 30.1 *Binary and Denary Equivalents*

Table 30.1 shows the equivalent notation in denary and binary up to 10 and selected integers up to 256. Two switches can code for up to 0 to 3 (00 to 11), three switches for 0 to 7 (000 to 111) and so on, up to eight switches for 0 to 255 (eight bit depth).

Denary	0	1	2	3	4	5	6	7	8	9	10	11	12	16	19	20	21	22	23	24	25
Binary	0	1	10	11	100	101	110	111	1000	1001	1010	1011	1100	10000	10011	10100	10101	10110	10111	11000	11001
Power		2^0	2^1		2^2				2^3					2^4							

29	30	32	63	64	100	127	128	253	254	255	256	4095	4096
11101	11110	100000	111111	1000000	1100100	1111111	10000000	11111101	11111110	11111111	100000000	111111111111	1000000000000
Power		2^5		2^6			2^7			Fills 8 places	2^8	Fills 12 places	$2^{12} = 13$ places

Figure 30.1 ***Formation of a raster digital image*** *The Bayer mask overlying the CCD/sCMOS faceplate generates an RGB colour image. An image intensity value is addressed at each sampling point in a raster fashion to form a pixel value. The raster pattern is highlighted as a pattern of green dots, for emphasis. Each intensity value is written in binary. If the image is enlarged, the underlying square foundation of the pixelated images becomes evident. Source: Stemmer Imaging 2016; Gillman & Soame. Reproduced with permission from Gillman & Soame photographers and H. Sanderson.*

replace the original data, and downsampling can introduce moiré artifacts. Some thought should therefore be given at the time of image acquisition as to how the image is to be displayed and the size required, so that any alteration may, as far as possible, be avoided.

30.4 Overview of Electronic Camera Operation

An analogue image is produced when light is focused onto a specimen. This image contains a wide spectrum of intensities and colours as a blended continuous tone of shades and hues. How colour images are captured and digitised is discussed later in Section 30.9. For the moment, it is easier to consider only an image formed using a monochrome (greyscale) detector. It is transformed into a series of brightness, or intensity, values by two processes: sampling and quantisation.

The heart of any digital camera is the CCD or CMOS chip. These devices consist of a series of light-sensitive silicon-based photodiodes, coupled to an electron-storage well to form photo-sites (Figure 30.2), which convert the accumulated photons from an analogue image into an electronic charge. Each photo-site is embedded in a square 'well' on the faceplate of the CCD or CMOS sensor, whose dimensions are anything from 2.5 to 25 μm. These photo-sites are often referred to as the 'pixels' of the camera, because each encodes information that is displayed as one pixel in the digital image.

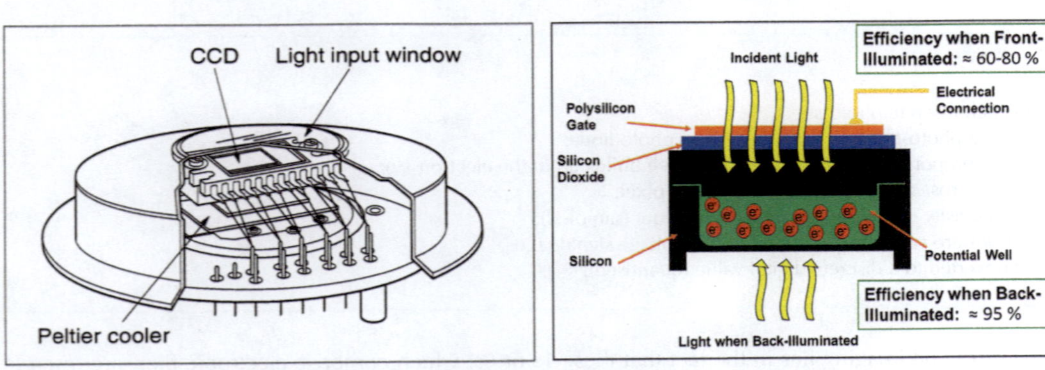

Figure 30.2 ***Photon capture and conversion to an electronic signal*** *The CCD/sCMOS faceplate is hermetically sealed within the body of the camera (left) allowing it to be cooled without misting occurring from condensation. The sensor is cooled by a Peltier device, which creates a heat flow at the junction between two dissimilar semiconductors. Cooling reduces thermal, or dark, noise. The right-hand panel shows how photons incident upon the MOS (metal-oxide semiconductor) capacitor – which consists of a metal electrode (or gate), an insulating film of silicon dioxide, and a silicon substrate and forms the heart of the CCD – accumulates a photo-electron charge in the potential well capacitor, which is then read out. The signal is proportional to the photon intensity falling on each photo-site. Source: (left) Hamamatsu 2009. Reproduced with permission from Hamamatsu Photonics K. K. (right) Sabharwal 2011. Reproduced with permission from John Wiley & Sons.*

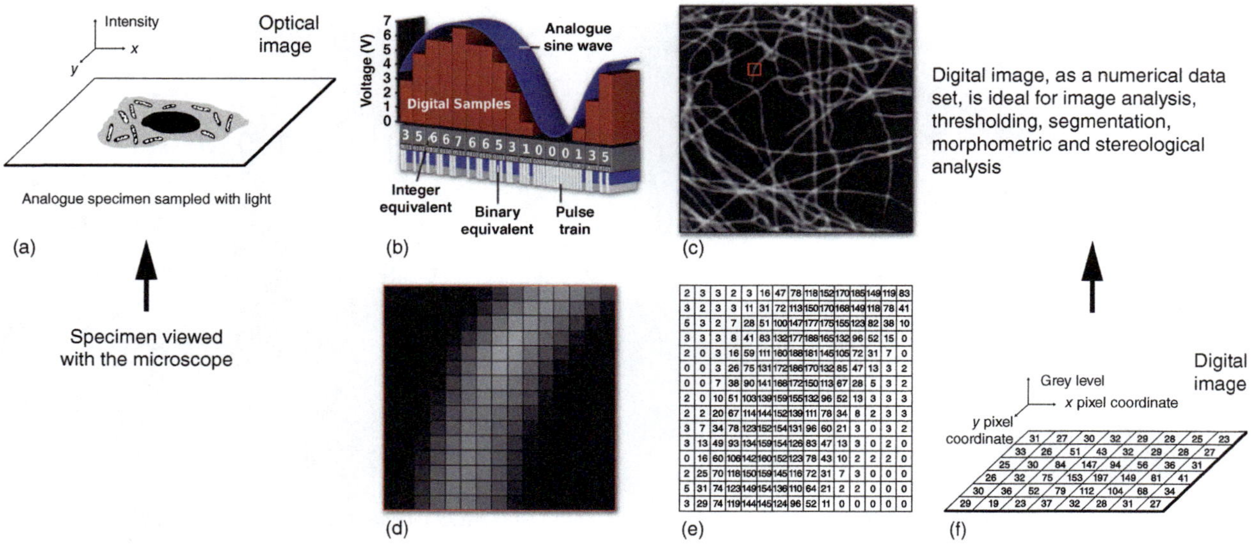

Figure 30.3 ***Digital image formation*** *A specimen interacting with light (a) forms an image optically. The stream of photons captured by the objective generates an analogue sine wave (b) whose amplitude is proportional to the intensity of the incident light. This continuous tone image comprises various shades and hues blended together. It is divided into individual brightness values by sampling and quantisation. The pulse train ensures regular sampling (c) and (d) with the integers converted to binary values to form a raster image (e) and (f). The images shown in (c) and (d) are of a-tubulin fibres. The raster pattern (e) showing how different intensity values are encoded as grey levels. Source: Shotton 1995. Reproduced with permission from Springer; Murphy and Davidson 2013. Reproduced with permission from John Wiley & Sons. Hazlewood et al. 2007. Reproduced with permission from Springer.*

The size of the pixels on the CCD/sCMOS faceplate determines the smallest sampling area applied to the analogue image. A sensor built with large pixels can capture more light and will have higher (i.e. better) signal-to-noise ratios. They are best suited to acquiring images under low-light conditions, such as live-cell fluorescence microscopy. Having smaller pixels enables better image resolution at the expense of signal-to-noise and dynamic range.

A photon of light falling on the photodiode of the electronic sensor releases a photo-electron from the silicon of the photodiode. This photo-electron is held in the potential well of the photodiode (the photodiode acts as a capacitor). The associated electronics move the charge across the sensor via successive wells to the read-out amplifier (Figure 30.6). Pixel by pixel, the read-out amplifier converts the photo-electrons (which directly represent the intensity of the photon signal) into a continuous analogue voltage signal. These continuous voltage values are then quantised into discrete integers by the analogue-to-digital (A/D or ADC) converter, depending upon the dynamic range of the detector. This process is illustrated in Figure 30.3 and explained in more detail for each type of camera in Section 30.8. The sensor array can only detect brightness. The brightness of the original photon signal is directly represented as an intensity greyscale (see Figure 30.1). The greyscale value of each pixel is related to (but not equal to) the number of photons emitted (in fluorescence microscopy) or transmitted through, or reflected off (in bright-field microscopy) the area of the specimen that corresponds to the pixel.

CCD operation

- light falls onto each discrete photo-site;
- a photon releases a photo-electron from the silicon photodiode;
- a potential charge proportional to light intensity (pixel) builds up in the electron-storage well;
- the charge moves across the CCD faceplate pixel by pixel;
- row by row the intensity of the photon signal is read out (sampled);
- the photo-electrons are converted to a continuous voltage signal; and
- the voltage is converted to a discrete digital value (quantised).

The Achilles heel of digital imaging lies in the fact that these 1s or 0s, which comprise electronic files, are *discrete* whole number *integers*. This is all very well where a text file is concerned, but any continuous data, such as a stream of photons or frames in a movie, that form an image must be adequately sampled in order to faithfully represent the original continuous analogue signal. This is why Nyquist's sampling criterion (Chapter 31; Pawley 2006a) is important and must be adhered to for meaningful, scientifically rigorous, results. The process of digitising a continuous-tone analogue signal to a discrete digital signal always results in a loss of spatial resolution (defined by the pixel size), and a tonal loss of the signal (recorded by the bit depth of the digitised image) because the number of grey levels is limited. For the most part, these losses are extremely small and not noticeable, but they do occur when a continuous analogue signal is digitised.

Please remember that, although a raster image is displayed as a series of (normally square) pixels containing brightness (the visual perception of luminance) and colour (chrominance) information that correlate with the photo-sites[1] of the CCD electronic detector, pixels are *sampling elements* of the continuous analogue data of the original image. To be exact, each pixel value represents a statistical estimate of the intensity of the photon flux recorded as a constant value of luminance from the area sampled in a particular region of the specimen. Pixels are not *just* little squares!

30.5 Digital Image Resolution

The term resolution can have different meaning in different contexts. When applied to the microscope, we are referring to the resolving power of the objective to transfer fine detail from the object into the image. This is determined by the numerical aperture of the objective and the wavelength of the illumination, as explained in Chapter 7, Section 7.4, and Chapter 10, Section 10.7.

There are several factors that ultimately determine the detail both recorded and displayed in a digital image:

- the magnification and resolving power of the microscope objective;
- the wavelength of the illumination;
- the size of the pixels in the electronic detector;
- the number of pixels in the array of the CCD or sCMOS faceplate;
- any zoom factor or binning applied;
- the Nyquist sampling frequency used; and
- the resolution of the output device, whether printer or monitor.

Resolution is often used to denote pixel density or pixel count in digital imaging, even though international standards specify that it should not be so used. To reiterate: a pixel is not a little square *per se*, although it is displayed as a raster square of constant value luminance.

Resolution

- optical resolution – of the microscope: determines fine detail seen in image;
- digital image resolution – size and number of pixels in the image: determines image enlargement;
- Bit-depth resolution – encoding tonal values in the image;
- Screen resolution – pixel pitch of the monitor: determines clarity of display;
- display/output resolution – either as ppi (pixels per inch) on monitor or dpi in printer: determines image size.

The overall number of pixels in an image is derived by multiplying the number of pixels formed horizontally in the CCD by those formed vertically. The spatial resolution of a digital image is determined by the number of pixels available on the CCD faceplate to sample the analogue image and whether the signal from groups of these pixels has been collected as one sample. This is called binning (Figure 30.4; see also Section 31.5). Binning increases the signal intensity but decreases the spatial resolution. The number of pixels used to record the image is fixed by the camera sensor. This can only be changed on high-quality scientific-grade cameras, where sub-array image capture is possible. Quite separately, the output resolution of the digital image can also be changed by upsampling (by interpolation) or downsampling, to change the overall size of the image. The other measure of resolution of the digital image is the bit-depth, or grey level, resolution, which digitally encodes the tonal values in the image.

A large CCD/sCMOS sensor with a large number of pixels has the capacity to form a very large image file. Such an image can be displayed or printed greatly enlarged without pixelation or loss of detail. Higher pixel density within a given sensor size means better

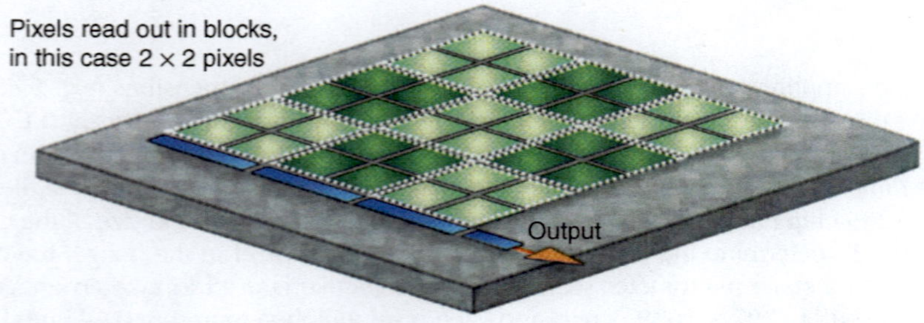

Figure 30.4 ***Binning*** *Groups of pixels, here 4 pixels in a 2 × 2 binning raster, can be read out as a single sampling point. Binning improves the signal-to-noise ratio, due to reduced read noise contributions. Binning also permits increased frame rates, albeit at the expense of spatial resolution. Compare these tonal scales with the image shown at different bit depths in Figure 31.6, where the effect of binning may more readily be seen. Source: Stemmer Imaging 2016.*

resolving power, but smaller pixels mean a smaller full-well capacity with less dynamic range, increased Poisson (shot) noise and lower (worse) signal-to-noise ratio. Having a large number of megapixels is not the full story. A more modest CCD, with larger pixels, can often produce much higher-quality images because of reduced noise. This is especially important in low-light applications.

The optimum sampling frequency (the Nyquist sampling frequency) depends upon the microscope, the resolution or resolving power of the objective and the size of the CCD sensor. However, the image is more likely to be under-sampled by a camera sensor with large pixels (see Section 31.6 on Nyquist sampling, in the following chapter). The 6.45 μm pixel size of the widely used Sony ICX285 CCD chip is sufficiently small to adequately sample the image formed by a 40x/NA 1.3 and a 60x/NA 1.4 objective.

The resolution of the image may be linked to the output device and given in pixels per inch (ppi) in the case of a monitor or the equivalent dots per inch (dpi) for a printer. The terms 'ppi' and 'dpi' are often used interchangeably. Dots per inch refers to the printing density that an inkjet or laser printer can place on the paper. A printer must put down multiple tiny dots of ink to simulate the colours of each pixel in a digital photo. For example, when printing a digital photo at 360 pixels per inch, the printer may place picolitre drops of each ink in up to eight distinct locations to produce a single pixel with a 70 μm^2 area. The term 'pixels per inch' *only* has meaning when a digital photo is printed or displayed. For printers, a 4″ × 5″ image always remains 4″ × 5″. Increase the resolution of the image, and you'll be able to display a sharper image or a greater dynamic range, but the physical size remains the same. For display, the size of the image is dependent on the resolution of your display device. The higher the resolution of your display device, the smaller your image appears. The resolution of the image itself doesn't matter to your monitor.

The computer display monitor also uses pixels (inspect them with an inverted eyepiece as a magnifier held very close to the surface of the screen), but these should not be confused with the pixels that make up the digital image *displayed* on the monitor. A computer monitor will display images at 72 or 96 ppi or higher. The display resolution of the computer monitor depends upon the memory available on the graphics card in your computer, and the upper limit is ultimately set by the maximum resolution of the monitor. If the monitor is older and comprises relatively few pixels (e.g. 640 × 480), images will lack clarity and be very large, necessitating scrolling to view the entire image. At a 1:1 display resolution, the image pixels are the same size as the screen display pixels. When an image is displayed at a smaller overall size, the image pixels are not made smaller: some of them are simply not displayed on the screen. If the image is enlarged, the individual image pixels will eventually be seen, and pixelation occurs. This gives rise to a blurred image[2], the digital equivalent of 'empty magnification'. If the aspect ratio of the physical display and the selected resolution are different and the pixels in the image do not map directly to those of the monitor, the monitor will resample the image, and thus 'stretch' or 'squash' the image to fill the entire display. This can result in the image being aliased. To view images at higher resolution, more pixels must be added – this is resampling. Images that are 'downsampled' lose information irretrievably. For an aesthetically pleasing result, an interpolation filter is required in both upsampling and downsampling; such a low-pass filter implements anti-aliasing. Note that upsampling requires an interpolation filter after increasing the data rate, and downsampling requires filtration before decimation.

From the above it should be realised that the pixels comprising a raster image (say, 1024 × 1024 pixels) on the screen are not the same as the image sample points commonly referred to as pixels. The sampling points have a discrete size determined by the size of the pixels on the detector, whereas the pixels displayed on screen have no absolute or defined size. Their size is determined by the resolution of the monitor or screen and any zoom factor used to display the image. The overall clarity of an image[3] is determined by three factors: screen resolution in pixels per inch (i.e. the pixel pitch of the monitor); the display resolution, or the number of pixels used (e.g. 1024 × 768) on the monitor – a greater number of pixels leads to a less pixelated display – and the number of pixels within the image (e.g. 1024 × 1024), which determines how much the image can be enlarged.

At 72 ppi and greater, the eye perceives a continuous tone image. However, if a digital image of this resolution were to be printed, it could only be done at a very small size. Most images for printing must be acquired at 300 dpi or higher. For more information on displaying digital images correctly, see Section 31.9 in the next chapter.

> The final size of the image is determined by the number of pixels and how that image is displayed or printed out. The output size of an image must be considered *before* acquisition. It must be sampled sufficiently and also be large enough to display without pixelation.

While the pixels that form the sampling area of the CCD faceplate have absolute dimensions (e.g. 6.45 or 13 μm), the pixels that constitute the digital image are dimensionless until they are calibrated. This is for two reasons: firstly, a CCD detector employing, say, the Sony ICX285 chip with a fixed number of pixels both horizontally (i.e. 1388) and vertically (i.e. 1040) can form images at different magnifications. Secondly, the dimensions of the pixels in the image depend upon the size at which the file is displayed.

In the first case, the Sony ICX285 chip has 1388 × 1040 6.45 μm pixels. This means that the size of the CCD faceplate is 8.95 mm × 6.71 mm, with 1 443 520 pixels. To determine the distance represented by each pixel in the image, the digital imaging system must be calibrated from the outset with a stage micrometer scale. If a 10x objective is used to take an image, the field of view will be 836.31 μm × 664.09 μm, comprised of 1297 × 1039 pixels, and each pixel will then represent 0.645 μm linearly, whereas with a 40x objective, the pixel dimension will represent 0.16 μm. If the digital file, with 1388 × 1040 pixels is displayed on a monitor at 72 ppi, then the entire image will cover 48.97 cm × 36.69 cm, and only part of the image will be displayed, regardless of the magnification of the objective used or the size of the features in the image. If, on the other hand, the digital file is printed at 300 dpi, then the size will be 11.75 cm × 8.81 cm, and the entire image will fit easily onto a sheet of A5 paper.

30.6 Bit Depth and Dynamic Range

It is simpler and quicker for machines to count in groups of eight (octal notation) and convert this to binary. This gives rise to a *byte* as an eight-bit number, sufficient to encode one character of alphanumeric data. Thus, computer memory is cited in bytes by which the size of digital files and computer storage capacity is measured, for example kilobytes (1024 bytes), megabytes, gigabytes, terabytes and so on. A single bit translates to black or white at each pixel in the image. If two bits are associated with each pixel, then four (2^2) different combinations are possible: 00, 01, 10 and 11. These will code for black, white and two shades of grey per pixel (see Figure 30.5). Eight bits are grouped as a byte in computer memory; therefore, each pixel of a monochrome image has one single byte associated with it, which is 8-bit depth. When distinguishing luminance ('brightness'; grey levels) in a digital photodetector, it makes sense to use the -bit notation with its finer divisions. Brightness resolution by the digital camera is the number of grey levels quantised through the A/D converter. The total number of grey levels generated by the A/D converter on the camera is equal to $2^{(\#\text{bits})}$. For monochrome imaging, 8-bit depth, $2^8 = 256$ grey levels are used, with 0 representing pure black and 255 representing pure white. There are 254 shades of grey in between the black and white limits.

The number of different shades of grey that humans can see ranges approximately from 50 to 60, so for display purposes and for publication, images captured in 8-bit depth are perfectly adequate. Because the human eye can (under bright conditions) *just* detect a 1–2% increase in brightness, its dynamic range is about 67 (the maximum signal divided by the minimum detectable increase). This makes the eye (roughly) a 6-bit device; see endnote 5, at the end of the chapter. The dynamic range describes the maximum number of tonal values, or contrast, that the recording medium can register and record. Dynamic range is determined by the black level: the darkest tone, or shade, where contrast can no longer be discerned. The black level is in turn determined by image noise, so the dynamic range is calculated (see also Box 30.2) by dividing the full-well capacity by the read noise and dark noise. Modern cameras characteristically have very low read noise values (typically 1–6 electrons), so the dynamic range can potentially be very large: 60–80 000 grey levels. Cameras having larger pixel sizes can convert a greater number of photons to photo-electrons and so have a greater dynamic range[4] than those with smaller pixels. Knowing the dynamic range is useful for determining the bit depth required to accommodate the signal. The bit depth is different from the dynamic range: It is the number of bits per channel used to quantise the continuous signal, or how many 'pieces' the camera can use to divide up the dynamic range.

Photodiodes exhibit a much more linear response to light than classic film emulsion, which is a distinct benefit for quantitative scientific imaging – always a difficult subject to pursue correctly. However, camera sensors lack the 'cushioning' effect of the toe and top slope of the H & D characteristic curve for film, meaning that highlights and shadow tend not to be recorded distinctly in the image as well as with monochrome film. This digital compression of dynamic range at the upper and lower limits is more apparent with cheaper point-and-shoot digital cameras.

Our eyes see 8-bit images as continuous tone images, as they would an analogue image recorded and printed by classical photography. For extracting quantitative data, it is a different matter, and high-quality scientific digital cameras provide pixel intensity data ranging from 8-bit through to 12-bit (4096 grey levels) or 16-bit depth (65 536 grey levels), much higher than the 8-bit depth found in consumer cameras. Neither a computer screen nor a printer is capable of displaying even the 256 grey levels of an 8-bit image nor can our eyes see this entire range of intensity[5], never mind images of greater bit depth. It is worth capturing data at higher than 8-bit depth when the image is to be used for 'number crunching' and scientific analysis in support of experimental hypotheses.

To display images encoded by large bit depths, the number of grey levels must be 'downsampled'; otherwise, high bit-depth images will be displayed dark on screen, because (say) the first 64 grey levels near to black in a 12-bit image correspond to the first 4 grey levels

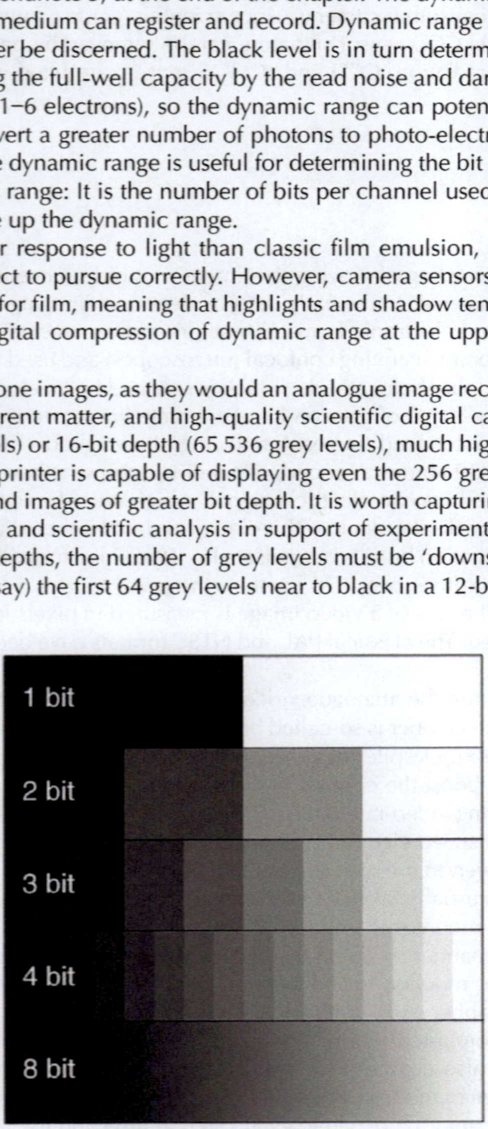

Figure 30.5 ***Bit depth*** *Our eyes perceive 8-bit depth as a smooth transition of grey levels from black to white, since our eyes are limiting to perceiving 50–60 grey levels. At lower bit-depth values, discrete steps in graduation can be seen. At 1-bit depth, the black-white transition is obvious. See also the figure in Box 11.1 and Figure 31.6 showing how dithering may help blend grey values in printing. Source: Lambert and Waters 2014. Reproduced with permission from Elsevier.*

in an 8-bit image, but these 64 dark levels from the 12-bit image take up 25% of the display of the 8-bit monitor (see also Figure 31.18). When a digital image must be displayed, it must be converted back to an analogue voltage signal in order to be viewed on a CRT monitor or one accepting a VGA (video graphics array) signal. This is done by a digital-to-analogue converter (DAC) housed within the computer or the monitor. Modern flat panel liquid crystal display (LCD) monitors with DVI (digital visual interface) or HDMI (high definition multimedia interface) connections operate digitally, and no signal conversion is required.

30.7 Black Pixels and Pixel Pathology

Around the periphery of the sensor there are pixels covered by a metallic shield. These are dummy pixels whose signal comprises the dark current (see Section 30.10 on noise), the dark current being spurious electron accumulation of 'thermal' electrons in the CCD or sCMOS charge wells. The average dark current is subtracted from each reading to give a black reference level for the image. Despite very high manufacturing standards, chips with upwards of one million individual pixels will inevitably have a certain proportion of defective pixels. So-called 'hot' pixels are those that have a higher rate of dark current leakage than their neighbours. They may appear white or, if a Bayer mask is fitted, may be individually coloured. These hot pixels are more noticeable under low-light imaging conditions and/or when using long exposures. Stuck pixels only register a maximum value regardless of exposure; they are usually seen as white or coloured dots in the image. Dead pixels register a minimum value and are either black or may be coloured from the Bayer filter. These errant pixels are more often seen on consumer cameras, since scientific-grade cameras are inevitably made to much higher and tighter quality control standards. The unique CCD sensor signature can be removed by performing a flat-field correction (see Section 31.5). Since silicon is sensitive to infrared radiation, CCD and sCMOS sensors are fitted with IR cut-off filters to minimise noise and colour contamination of the signal.

30.8 Types of Electronic Detector

30.8.1 Video Recording

The versatility of digital imaging permits recording of moving specimens in rapid succession, which is always more realistic and representative than static snapshots of living material. In the early days of electronic imaging, video was the only means of encoding the microscope image into an easily manipulated electronic format. These first video cameras were developed from photomultiplier tubes (which are still used to record images from point-scanning confocal microscopes) and used in microscopy from the 1970s. An analogue video camera projects the optical image onto a photocathode, which is scanned by an electron beam. The analogue image is scanned in horizontal lines (either 625 or 525) every 1/60th of a second and alternate lines scanned in successive frames so that the complete image is scanned twice as two successive interlaced frames. Interlacing allows good visual quality within the limitations of narrow bandwidth. Video is broadcast at 24, 25 or 30 frames a second, depending upon whether you are in Europe, North America or Japan.

The analogue photon signal is transformed into an electrical signal, which encodes light intensity and positional information as a negative voltage synchronisation pulse for the display device. This electrical signal is then digitised. The signal is referred to as *composite* because it not only carries luminance information but a 'sync pulse' also. PAL video format is specified as *576i50*, where *576* indicates the vertical line resolution (with extra blank lines over 576 for the sync pulse to give a total of 625), *i* indicates interlacing and *50* indicates 50 half-frame fields per second. The size of a video image is measured in pixels for digital video or horizontal scan lines and vertical lines of resolution for analogue video. The classical PAL and NTSC formats have been, or are being, replaced by the fully digital ATSC and DVB-T equivalents, respectively.

A frame-grabber board is required to digitise the analogue video signal, and the length of the analogue signal thus digitised defines the size of the digital pixel output. The frame grabber is so-called because early versions only had enough memory to acquire, or 'grab' a single analogue video frame for digitisation. Despite these sequential steps, video recording made capturing images of living cells straightforward. It is also easy to artificially boost the contrast in the image and subtract noise, dirt or uneven lighting in the image by making a 'background subtraction'. Capturing video-rate data at 25 or 30 frames per second, which was once relatively easy with off-the-shelf video cameras, now requires high speed electronic cameras and vast digital bandwidth. The minimum frame rate to achieve the illusion of a moving image is about fifteen frames per second, although this differs for rods and cones and also with the intensity level. The human flicker fusion threshold is usually taken as 16 Hz, or 16 frames per second.

The standard video camera fitting is a C-mount (see endnote 1 in Chapter 31), which allows it to be fitted easily onto a trinocular photo-port on a microscope, in the same manner as a film or digital camera. The minimum equipment required is a video camera, video recorder and a TV or dedicated video monitor, linked with coaxial cables to carry the electrical signal to the digitiser. Where a computer is fitted with an internal frame grabber card and has sufficient storage space and memory (RAM) or else the ability to stream data directly to the hard disc storage, this computer system can replace the video recorder. Besides allowing easy recording of moving specimens in real time, video microscopy also enables low-contrast very small differences in intensity in the image to be amplified and detected. An excellent practical account of modern video microscopy has been written by Weiss et al. (1999) and Hinchcliffe and Sluder (2007). Video devices are ideal for capturing dynamic events in real time and for adjusting image contrast, but the technology is limited by the 525 or 625 horizontal line resolution and difficulties of integration with printers, video recorders and display monitors that this presents. The dynamic range of an analogue video camera is between 70:1 to 100:1, or 6–7 bit depth, about the same as the eye. This is less than film (≈10-bit depth) or CCD/sCMOS sensors (4000:1 to 65 500:1, or 12–16 bit-depth). These CCD and sCMOS solid-state sensors are now comprehensively replacing the older video technology.

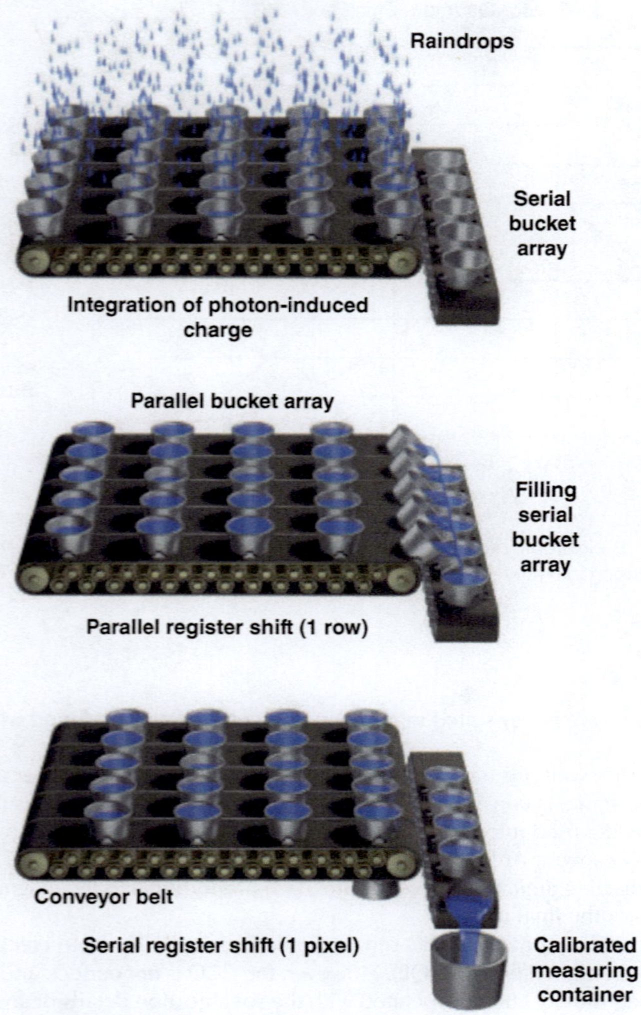

Figure 30.6 **CCD operation** *At its simplest, the manner in which the charge accumulated in each photo-site of the CCD array is read out can be visualised as an array of buckets (the individual photo-sites or pixels) collecting rainwater (photons). Each bucket in the array is exposed for the same time to the rain. The buckets fill up with varying amounts of charge. The CCD is then read one bucket at a time. This process is initiated by pouring water into the adjacent empty column. The buckets in this column transfer their charge down to a final pixel where the electronics of the camera read out this particular pixel (i.e. the computer measuring the bucket) and turn it into a number that can be understood and stored by a computer. Source: Murphy and Davidson 2013. Reproduced with permission from John Wiley & Sons; Hazlewood et al. 2007. Reproduced with permission from Springer.*

30.8.2 CCD Cameras

Modern CCD cameras are all based upon silicon semiconductors for their operation. A voltage applied across a p-n silicon semiconductor junction forms semiconductor-diode transistors. These act as switches. A metal oxide semiconductor (MOS) capacitor, which can store an electronic charge, is the heart of the CCD device: light falling on the silicon semiconducting layer transfers electrons into the p-type silicon substrate, where the electrons remain as surface charges. The electronic charge in each well accumulates in *proportion*[6] to the incident photons. At the end of the exposure, the charge is passed sequentially as a 'charge packet' along each row of the CDD into a shift register, where it is converted to a voltage signal (Figure 30.6). The term 'charge-coupling' refers to the way in which the electrons are transferred from one capacitor forming each pixel of the CCD to the next. An applied 'clocking voltage' moves the charges sequentially from one well to the next, like a 'bucket brigade' transferring water, to be recorded sequentially at the end of the line, or CCD faceplate.

If the charge in each well is not moved along, or the incident photon signal is too bright anyway, the capacity (full-well capacity) of the pixel is exceeded, it overfills and the resulting saturation is seen as *blooming*. The phenomenon of blooming causes vertical streaks of light to appear in the final images, as additional electrons to spill over into adjacent wells in the read-out chain. Moving photo-electrons is a simple matter on a linear CCD, such as found in a modern flatbed scanner, but moving charges on a two-dimensional pixel array of a CCD faceplate is slightly more complex. Anti-blooming gates tend not to be incorporated into scientific-grade CCDs, where the exposure can be more precisely controlled, because it reduces the quantum efficiency (QE) of the camera at the red end of

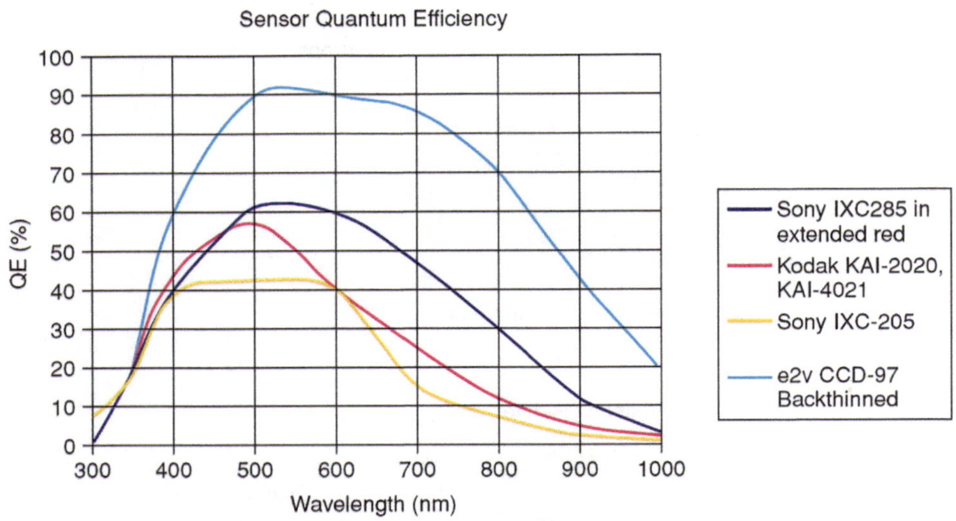

Figure 30.7 ***Quantum efficiency values*** *As electronic sensors improve, QE values continually increase. Some modern CCD and sCMOS can have QE values in excess of 90%, with greater extension into the red end of the spectrum. Source: Watkins and St. Croix 2013. Reproduced with permission from John Wiley & Sons.*

the spectrum. Anti-blooming stratagems are better suited to removing specular reflections found when using consumer-grade cameras in everyday situations.

The photo-electron packets from each well are transferred row by row to a serial shift register at the top of the chip, which sends them in sequence to an amplifier. The signal is converted to a continuous analogue voltage that is proportional to the intensity of light falling upon each well. This voltage is digitised into binary code by analogue-to-digital conversion circuitry (ADC), which is built into the electronic architecture of the CCD camera. An analogue-to-digital converter (abbreviated ADC, A/D, ADU or A to D) converts the continuous voltage signal to a set of discrete digital numbers (Figure 30.3) proportional to the magnitude of the voltage. This allows the computer to store, display, copy or alter the final image.

In an ideal world every 100 photons falling onto the CCD sensor would generate 100 photo-electrons, and 100 units of signal would be detected. This is defined as 100% quantum efficiency (QE). However, the CCD is not perfect, and its relative efficiency is wavelength dependent, as shown by the spectral sensitivity curve associated with the specification details of any high-quality CCD camera. Digital sensors have typical QE values (Figure 30.7) of around 45–60%, which is much higher than film, which captures only about 2% of light incident on the emulsion. Generally, peak sensitivity is found in the green region of the spectrum and falls off in the red and particularly in the blue regions of the spectrum. Coating the sensor with UV-sensitive phosphor can improve QE into the near UV part of the spectrum. Reflection and absorption of light by the overlying silicon photodiodes of front-illuminated CCDs give rise to low QE in the blue region. A list of factors affecting QE can be found on page 927 of Pawley (2006b).

Whilst the charge is moved and read out, the photo-sensitive area must be masked off from further incident light. The simplest means of doing this is with an electromechanical shutter covering a full-frame CCD architecture (Figure 30.8a). However, this is not common and is inherently slow. A quicker way is to have a second CCD, covered with a metal shield, arranged in tandem with the photon-collection CCD. This is called a frame-transfer CCD (Figure 30.8b). This second storage array collects the charge that is read out, obviating the need for any shuttering device. The shift of the entire charge from the imaging array to the storage array takes from 200 μsec to 2 msec; once the charge has been shifted across and readout, the imaging array can capture the next exposure. In addition to this, modern solid-state LED and laser light sources can be modulated in the micro-second domain, and most scientific cameras provide the timing signal to 'shutter' the light source during read-out or frame transfer. This is especially important in live-cell imaging, where exposure of the sample to photons should be minimised.

A frame-transfer CCD is an expensive solution, because CCD wafers are costly. It can also cause vertical smearing of the image due to the long read-out time (however modulating the light source as described above prevents this). A cheaper way of achieving the same end (Figure 30.8c) is to use the interline CCD architecture. Most CCDs utilise this interline design[7]. This configuration combines two parallel registers into a single array (at the expense of light collection efficiency). One vertical (odd-numbered) column of pixels is light sensitive, whilst the adjacent (even-numbered) column is light insensitive and contains read-out circuitry. Transfer to the read-out rows only requires pixels to shift one row, so transfer speeds are very high and do not require a mechanical shutter; they can therefore produce a continuous stream of images at relatively high frame rates. To address the poor fill factor and consequent low effective quantum efficiency, microlenses are added to the CCD surface to direct light away from the opaque regions and on the photo-sensitive area. Scientific interline cameras with well-designed microlens arrays have traditionally been the workhorse detectors of fluorescence imaging systems, although for high-speed applications they are now increasingly being replaced with scientific CMOS cameras.

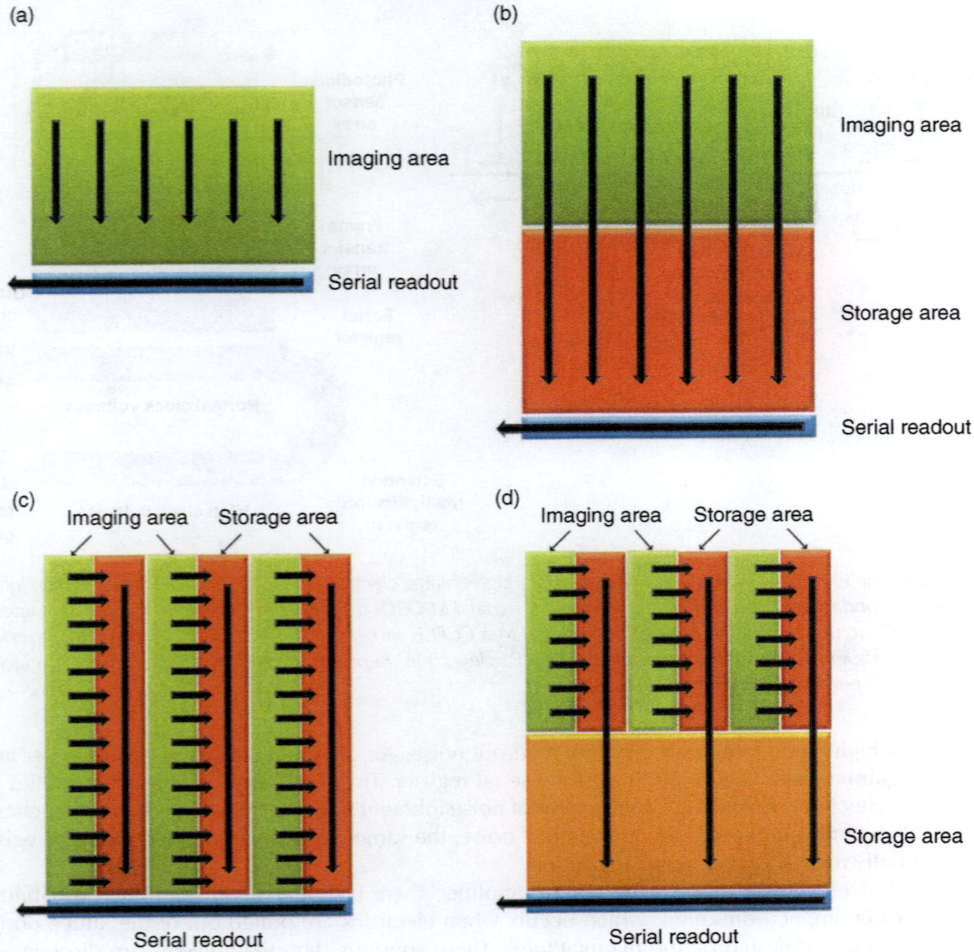

Figure 30.8 ***CCD designs*** *This figure shows (a) a full frame readout, usually with a mechanical shutter to avoid charge spillage and smearing in the image. This design is normally adopted by entry-level DSLRs rather than scientific-grade CCDs; (b) frame transfer CCDs overcome the requirement for a mechanical shutter by providing a storage area on the chip which is shielded from light. The transfer of the image can be achieved very quickly, with a higher frame rate than the full frame readout mechanically shuttered CCD. This design is expensive but captures a great deal of light. The interline CCD design is shown in (c); the strips between the columns increase the rate of data transfer and frame rate. However, the fill factor of the sensor and hence the quantum efficiency is reduced, so microlenses are added above the sensor array to compensate for this and increase photon capture. In (d) a hybrid full-frame and interline design is shown. Source: Allen and Triantaphillidou 2011. Reproduced with permission from Taylor & Francis Group.*

The number of pixels in a CCD chip can be calculated by dividing each dimension of the chip by the pixel size, which ranges from around 3 to 20 μm. Typical CCD chip faceplates can measure up to 13 mm or more. For example, the Sony ICX285 chip is an interline-transfer architecture with 6.45 μm square pixels (i.e. the pixel pitch is 6.45 μm). The pixel dimensions are 1388 × 1040 effective pixels so that the size of the CCD chip is 8.95 × 6.71 mm. However, the fill factor is the ratio of light sensitive area of a pixel versus total area, since part of an image sensor pixel is always used for transistors, electrodes or registers. This value can range from 30 to 50%, and with microlenses over the photo-sites, up to 90%. The saturation charge (i.e. the maximum number of electrons that each photo-site can collect), or full-well capacity, is approximately 1000 times the pixel size in square micrometres. Thus, for the ICX285 chip, the area = $8.95 \times 6.71\ mm^2 \times 0.32$ (fill factor) ⇒ full-well capacity = $60.055\ mm^2 \times 0.32 \times 1000 = 19\,217$. The full-well capacity specified by Sony is 16 000 electrons.

30.8.3 EM-CCD Cameras

It is possible to link an image-intensifying camera to a CCD camera to detect very low-light signals and to count single-photons. However ICCDs have a relatively low quantum efficiency, relatively low resolving power, poor dynamic range and are susceptible to damage by too great a photon flux. They have been rendered virtually obsolete by EM-CCD technology.

CCD cameras have slow read-out rates in order to keep the read-out noise as low as possible. Also the amount of noise in the image rises proportionally with the size of the charge register that amplifies the voltage before digitisation. This means that high frame rates to record living samples at low-light levels is not possible. This is the purpose of an EM-CCD; it essentially does one job, and does

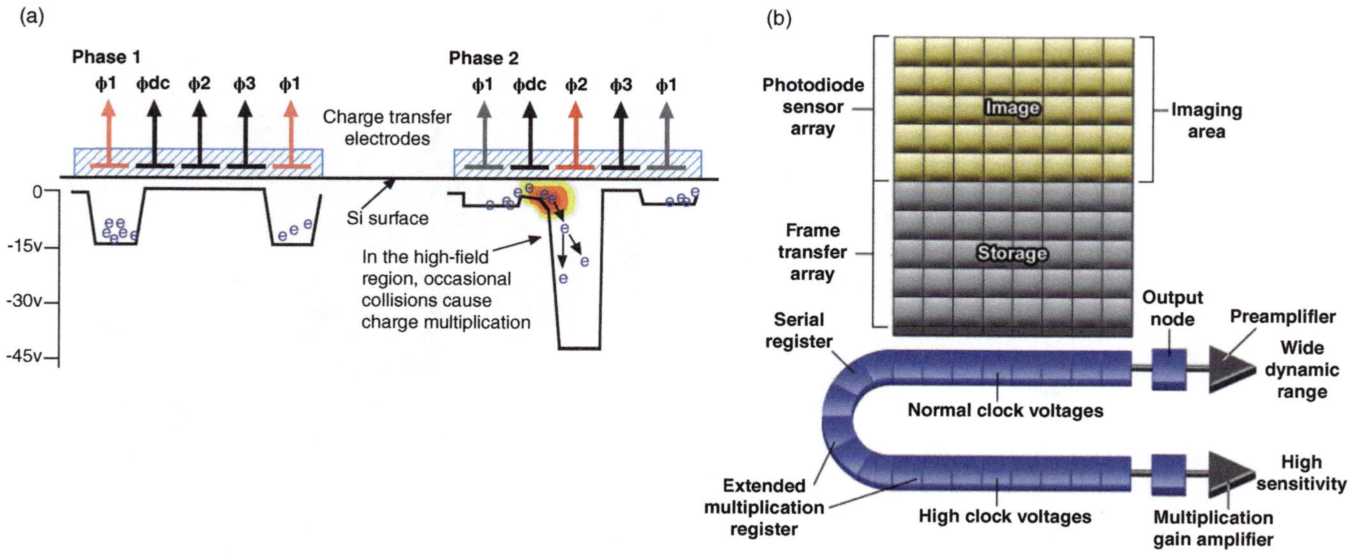

Figure 30.9 ***EM-CCD designs*** *The EM-CCD has, as the name suggests, a high-voltage electron-multiplying serial register, which increases the probability of each electron producing a secondary electron and so multiplying the signal. EM-CCDs are therefore very good for low-light applications. The process is shown in (a) while the overall architecture of the EM-CCD compared to a CCD is shown in (b) because the multiplication process occurs hundreds of times, it only has to be 1% efficient to be successful. Source: (left) Pawley 2006. Reproduced with permission from Springer. (right) Murphy and Davidson 2013. Reproduced with permission from John Wiley & Sons.*

it well: reading out signal at high frame rates with very low read-out noise. An EM-CCD utilises a frame-transfer architecture and has an extra multiplication (or gain) register (Figure 30.9) after the serial register. This multiplication register amplifies the signal *before* it enters the charge register by effectively exploiting a minor form of noise inherent in CCDs called clock-induced-charge (CIC). Although CIC on the sensor is usually an undesirable additional source of noise, the same effect is exploited in the gain register of EM-CCDs to provide low-noise gain, usually referred to as impact ionisation.

When a charge is shifted pixel to pixel towards the output amplifier, there is a very small but finite probability that charges can knock off additional charges by impact ionisation, which occurs when electrons are pulled out of the lattice during charge transfer. It is controlled by careful electronic design by the manufacturer. These spurious 'hot-pixel' events are discrete and can be seen as anomalous bright pixels distributed randomly and changing position randomly from frame to frame. The effect is present in all CCDs but is generally managed by good design and is so low that it is masked by the read noise or dark noise components and seen only when binning a very large number of pixels. In EM-CCD design, impact ionisation is exploited to multiply the signal above the read noise of the camera. Although the probability of secondary electron generation is low (≈1%), the amplification of the signal above the noise floor is considerable. The camera sensor and gain register *must* be cooled and the temperature tightly regulated, to minimise dark noise and achieve a stable electron-multiplication gain.

For low light signals, as the electron multiplying charge is increased, so the dynamic range of the camera increases, because the multiplication register has a higher full-well capacity than the regular CCD. Once the multiplication register is full, the dynamic range reaches a limiting plateau. Back-thinning the sensor and illuminating from the reverse side allows photons to reach the photoactive silicon more directly and can afford cameras with quantum efficiencies >90%. This inherent sensitivity combined with high dynamic range and very low read-out noise makes back-illuminated cooled EM-CCDs the camera of choice for single-photon counting experiments[8]. Electron multiplication was conceived to improve the signal-to-noise ratio during low-light imaging, where the read noise dominates, typically at high frame rates. It is important to consider that there is a statistical noise associated with the EM process equal to the square root of two. If the demands of the experiment allow extended integration times to capture more photons, rather than requiring temporal resolution, it may be better to use a conventional deep-cooled CCD, which will usually deliver better results. Fortunately, many back-illuminated EM-CCD cameras also include a conventional read-out amplifier, and since these have relatively deep wells (giving a high shot-noise limit), the same camera can be used for imaging across a range of time and light domains.

EM-CCDs do age, and their ability to multiply the incoming signal falls off. This is known as gain-ageing and is thought to arise from electronic charge becoming permanently embedded in the insulating material between the photo-sites. It is also possible to damage the EM-CCD chip by applying too high a gain for too long, so they should be used with care. Avoid over-saturation, and switch off or minimise the gain applied when it is not needed.

30.8.4 CMOS Cameras

Both CCD and CMOS were developed in the late 1960s and early 1970s and use very similar technology. They both convert photons into electrical charge and process this charge into a digitised electronic signal. In 1992, Eric Fossum developed the active pixel sensor with each photo-site containing a photodetector together with an active amplifier. The individual amplifiers help reduce both noise and

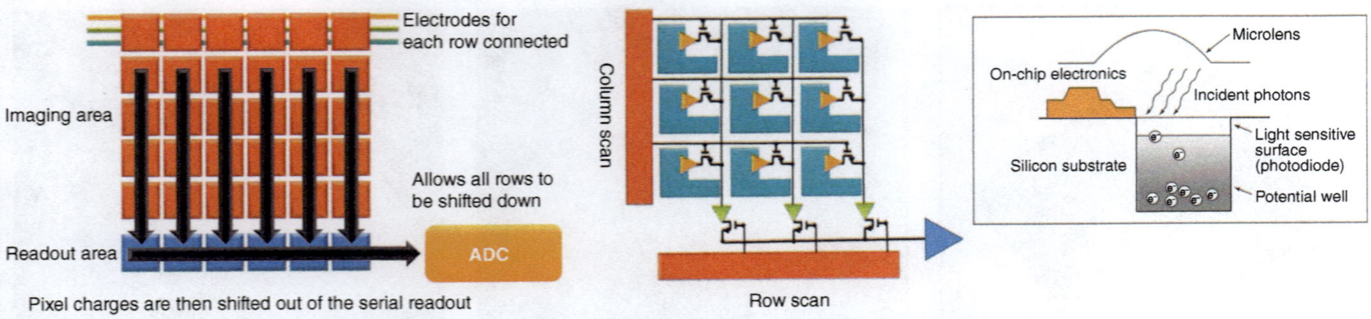

Figure 30.10 ***CMOS design*** *The CCD (left) is more complex to manufacture and requires more timing circuits than sCMOS designs (right), which use less power, can offer larger fields of view and can read out the data faster with lower overall read noise. The sCMOS has 'active pixels' – that is, amplifiers and readout transistors associated with each pixel (shown here in orange). Source: Allen and Triantaphillidou 2011. Reproduced with permission from Taylor & Francis Group; Adapted from Allen and Triantaphillidou 2011 with permission from Taylor & Francis Group; Bernas 2005. Reproduced with permission from John Wiley & Sons.*

distortion but can introduce a fixed pattern noise, which is seen as speckle. CCDs continue to dominate the scientific market because, until very recently, they gave far superior images than the CMOS detector.

In a CCD sensor the charge from each photo-site is transferred through a single serial register to be converted to voltage and sent off-chip as an analogue signal (Figure 30.10). All of the photo-sites can be devoted to light capture, and so uniformity – a key factor in image quality – is high. In a CMOS sensor, each photo-site has its own charge-to-voltage conversion and digitisation circuits so that the chip sends out a digital signal directly. This increases the design complexity and reduces the photo-sensitive area available for light capture. This also means that, until recently, CMOS chips suffered from more non-uniform pixels, more noise, more locked (maximum output) pixels and lower overall sensitivity. With each photo-site performing its own photon/electron conversion, uniformity is lower, but the chip can be built to require less off-chip circuitry for basic operation. CMOS pixels, being independent, do not suffer from column streaking due to blooming. Recent improvements in wafer fabrication technology now mean that CMOS sensors can be manufactured to give greater uniformity between pixels. Alongside the general improvements in CMOS technology for consumer and machine vision markets, there have been specific enhancements to make so-called scientific CMOS, or sCMOS, cameras for the biosciences and other demanding applications.

One of the dominant sources of noise in CMOS sensors is associated with resetting each pixel after reading the photon signal. The sCMOS architecture incorporates an additional transistor per pixel so that the reset noise can be read and offset automatically using on-chip correlated double sampling (CDS). CDS combined with advanced microlenses to improve fill factor with field programmable gate array on-chip processing to further remove fixed pattern noise and deep wells to give higher dynamic range, has resulted in cameras that outperform CCD detectors in most cases. sCMOS sensors also tend to be large and operate at clock speeds at least an order of magnitude higher than CCD or EM-CCD cameras. Typical sCMOS sensors have 6.5 μm pixels and resolution of 2048 × 2048, giving an overall sensor size of 13.3 × 13.3 mm with capture rates at 100 frames per second (fps). This fully uses the field of view of a microscope camera port (typically 18 mm diameter) and also pushes modern PCs and hard discs to their limits in terms of data capture and subsequent processing. Using standard 16-bit digitisation, each image occupies 8 MB, and hence the camera produces 800 MB per second. This is beyond the USB 3.0 rate of 335 MB/second and hence requires specialist, multi-tap (parallel), Cameralink cards. Of course the camera is capable of building data sets at a rate of nearly 1 GB per second, which requires RAM or a solid-state hard disc in order to keep up with the transfer speed. When considering making best use of the speed and field of view of sCMOS cameras, it is important to consider the computing power with much more care than when using CCDs.

The ability to start and stop an exposure is called shuttering (even if a real photographic-type mechanical shutter is not used). sCMOS cameras use either global shuttering, rather like a camera taking a snapshot, or a rolling shutter, which takes an exposure and reads out the data in staggered fashion, one row after another. Global shuttering is noisy and slower than a rolling shutter, but rolling shutters can cause distortion, particularly when imaging fast-moving objects. The solution to this is to synchronise the light source with the camera, and commercial sCMOS cameras provide the appropriate timing signals to modulate a solid-state light source to be 'on' only during the virtual global shutter period when all pixels are active.

Back-illuminated EM-CCDs still have an advantage over sCMOS detectors in low-light situations with higher quantum efficiency and a read-out noise of <1 photon per pixel, as opposed to 2 or 3 photo-electrons for sCMOS cameras. However this advantage is under threat as a new generation of back-illuminated sCMOS cameras becomes available with very high quantum efficiency. Currently the sCMOS sensor format is larger, which gives it the ability to acquire a whole cell into a larger field of view (see Figure 30.11). At least three leading camera companies – Andor (Neo and Zyla sCMOS), Hamamatsu (Orca Flash 2.8 and Orca Flash 4.0) and Roper Scientific (Prime and Prime 95b back-illuminated) – produce scientific-grade CMOS cameras. With low noise, high frame rates, small pixels and high dynamic range, sCMOS cameras are the choice for the future for most applications. In 2015, Sony announced that it would stop production of CCD sensors from 2017 with availability ceasing in 2026, but they have since changed their minds. Nevertheless the future of traditional CCD cameras looks rather bleak: they will probably be dominated in the marketplace by sCMOS detectors.

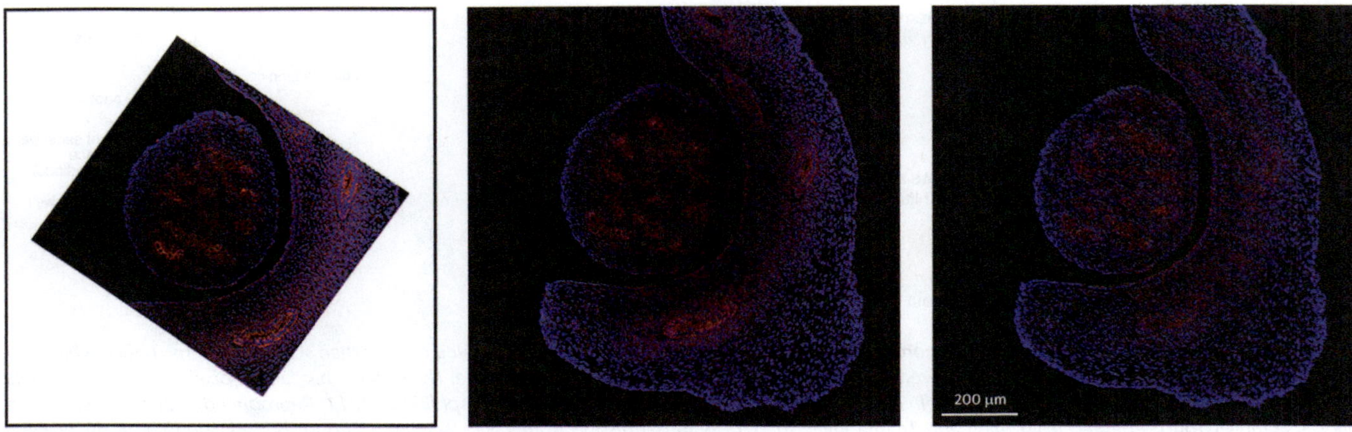

CCD Sony ICX285 AL = 6.45 μm pixels | **sCMOS** Hamamatsu FL-280 = 3.63 μm pixels | **Confocal PMT** effective pixel sixe 3.03 μm

Figure 30.11 **CCD versus sCMOS camera designs compared** *A comparison of the size of the field of view of a CCD, sCMOS and PMT-generated image from the same specimen. With smaller pixels, the sCMOS camera has a larger field of view than the CCD, one which can be matched by the confocal PMT with an appropriate zoom factor. sCMOS cameras also consume less power, offer faster frame rates, and have an increased dynamic range.*

30.9 Encoding Colour in Digital Images

The pixels in a CCD or sCMOS sensor are responsive only to the intensity of the photons and not their wavelength. High quality and high-resolution cameras (e.g. as used for fluorescence imaging) are therefore monochrome; the displayed image being pseudo-coloured post-acquisition.

A colour image is created by using the additive colour RGB (red, green, blue) model. A separate monochrome image (typically 8-bit) encoding intensity for each colour is combined into the final colour image. The number of shades of grey or colours (which affect contrast) associated with each pixel in the image is termed the 'bit depth' of that image. Colour images use one byte per primary colour; hence, there are 256 different shades each of red, green and blue. This is known as True Colour or 24-bit depth, which allows 256^3, or $2^{24} = 16\,777\,216$, or 16.7 million colours. True Colour 32-bit is the same as 24-bit colour, with an extra byte reserved for the alpha channel (which encodes transparency and other information for the image) to improve the colour display.

There are two principal ways to record a colour image. Either three (red, green and blue) filters are used separately and the three separate images combined into one, or a single one image is recorded through a Bayer mask (Figure 30.12) designed to match the sensitivity of the eye. High-quality cameras use the former method, to avoid any loss of spatial resolution. A high-resolution image is captured either by using three separate CCD chips, each with a red, green or blue filter in front of it, or else by rotating a filter wheel holding the separate RGB filters in front of a single chip. Mid-range CCD cameras designed for colour imaging use a Bayer matrix of coloured filters overlying each well, or 'pixel', on the CCD sensor. This mimics the distribution of cones within the retina of the eye. The Bayer mask has twice as many green-filtered sensors in order to closely approximate the spectral response of the human eye. Using a Bayer mask to generate a colour image works well for uniform-coloured areas in the image but results in the loss of accurate spatial information at high resolution. Colour subsampling will result in aliasing, sometimes seen as a coloured moiré pattern or false colour artifacts at edges in the final image. Cheaper digital cameras include a low-pass optical anti-aliasing filter to correct this effect.

To overcome this colour defect, high-quality cameras use either a technique known as *co-site sampling* or another called *microscanning*. A Bayer mask is still built into the camera, but several image sub-frames are acquired sequentially. This is faster, cheaper and more sensitive than using three monochrome CCD cameras, each with its own coloured filter. With co-site sampling, after the acquisition of each sub-frame, a piezoelectric crystal mechanism moves the sensor by precisely the distance of one pixel. With microscanning the sensor is moved by a fraction ($^1/_2$, $^1/_3$, $^1/_4$ or $^1/_6$) of a pixel. Another frame is taken through the Bayer mask. Thus each point in the analogue image is sampled by a red, green and blue pixel at least once. At the end of the acquisition routine, all the individual sub-frames are combined, with the RBG colour information sampled at each pixel location contributing equally to the final image. Four (2 × 2), nine (3 × 3), sixteen (4 × 4) or thirty-six (6 x 6) sub-frames can be microscanned, collected and combined for improved colour reproduction in the final image. Co-site sampling can be implemented at high speed, allowing for reasonable frame rates. Microsampling is not suited to capturing images of moving specimens.

The latest technique to encode colour in an image takes advantage of the different penetration of silicon by different wavelengths of light to create a liquid crystal RGB tunable filter. This can either be an integral part of the camera or added later (e.g. Varispec or MicroColor). This allows separate RGB images to be captured at full resolution, with a monochrome camera and without the loss incurred using a Bayer mask. It takes about 1 msec to change between each colour channel when taking an image. Ideally, the filter should be separate from the camera and able to slide out of the light path for fluorescence, polarised light microscopy and low-light imaging. Since each pixel is filtered to record the luminance from only one of three colours (R,G,B) the data from each of these pixels cannot fully determine the true additive colour at the position of the sampling pixel on its own. A full-colour image is derived

Figure 30.12 **The Bayer mask** *An electronic sensor will only register the intensity of light, not differences in wavelength, so the Bayer mask is a means of introducing colour to digital images. Named after the Kodak engineer Bryce E. Bayer, the Bayer mask has twice as many green filtered sensors in order to closely approximate the spectral response of the human eye. Source: Allen and Triantaphillidou 2011. Reproduced with permission from Taylor & Francis Group.*

using a *demosaicing* algorithm to interpolate the information from neighbouring pixels (Figure 30.13) to determine the missing colour information and generate a complete set of red, green, and blue values for each point in the image.

RGB is a *device-dependent* colour model: different devices detect or reproduce a given RGB value differently. This is because the phosphors that make up the display device colour elements vary from manufacturer to manufacturer (or even in the same device design over time). For critical printing of images, some kind of colour management of the display monitor is necessary. This is because whilst computer monitors use the RGB additive model, printing presses uses the CMYK subtractive colour model[9]. Light from the display monitor stimulates our cones directly, whilst a printed ink CMYK image reflects coloured light into our eyes. The amount of data required to encode even a simple colour image is vast, and image files of upwards of five to fifty megabytes (MB) are not unheard

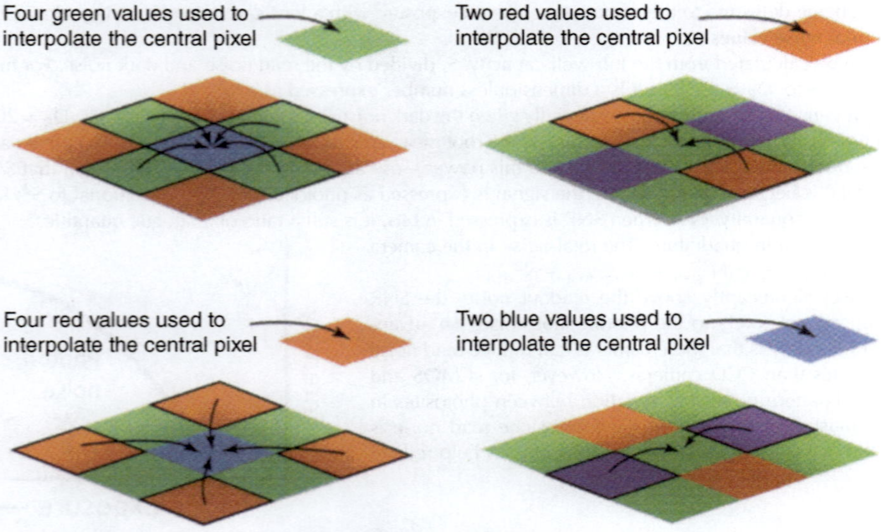

Figure 30.13 **Demosaicing** *Besides co-site sampling and micro-scanning, using a demosaicing algorithm to interpolate the information from neighbouring pixels is another means of generating a complete set of red, green and blue values for each point in the image. Source: Stemmer Imaging 2016.*

of. An extremely fast and memory-powerful computer with plenty of RAM (read-write random-access memory) is required to handle such images.

For scientific imaging and in particular fluorescence measurements, the colour images, as seen by the user, are likely to comprise monochrome images with either coloured look-up tables (LUTs) to express different features in the same image or pseudocolour images to visualise large dynamic ranges. These are both display tools and the data are likely to be 12 or 16-bit (monochrome) intensity values.

30.10 Digital Camera Noise

Noise in a digital image causes variance in the pixel greyscale intensity values and reduces contrast in the image. Noise is not merely background[10]. It arises from random fluctuations in the electronic signal (Box 30.2). Photo-electrons and signal-independent electrons are both collected by the camera but cannot be distinguished from one another. There are several sources of noise in an electronic camera, all of which must be managed and, where possible, reduced to a minimum. The signal-to-noise ratio (SNR) describes the proportion of signal electrons to noise electrons, and we want this ratio to be as high as possible. Noise may be either random (stochastic, normally-distributed or Poisson-distributed) or else periodic.

Box 30.2 The Mathematics of Noise

1. **Photon (shot) noise**
   Photon, or shot, noise is approximately described by a Poisson distribution with a standard deviation equal to the square root of the pixel value. Shot noise = $N_{shot} = \sqrt{\text{signal}}$ (electrons or photons). It cannot be eliminated, and becomes important when the signal is very small or limited. Reduce shot noise (i.e. increase the photon flux) by increasing the exposure time and/or by binning.
2. **Read noise**
   Read noise is the price paid for imprecise reading out of the signal from each photosite by the camera. Read noise is measured in ±RMS (root mean square) electrons of noise. The root mean square, or quadratic mean, is a measure of the magnitude of a varying quantity. It is a better measure of deviation than the standard deviation, where the variates sampled are both positive and negative. The RMS is the square root of the mean of all the squares of the values taken over a sampling interval. It is not synonymous with standard deviation.
   If the dimmest pixel of your image has registered 900 electrons, the shot noise will be 30 electrons. Adding a further 30 electrons of read noise to the background will make little difference: a total of 60 electrons, which will be insignificant when added to the greater Poisson noise of the actual signal forming the image. The total noise from these 60 electrons will be $\sqrt{60} \approx 8$ electrons. If the photon flux is reduced (e.g. for live-cell imaging) and the background signal is now only 50 electrons, the shot noise is now $\approx \pm 7$ electrons. Adding ±30 electrons of read noise will be unacceptable if you have to take measurements from the dimmest areas of your image.
3. **Dark noise**
   Dark noise is dependent upon the signal integration time (τ) as well as the temperature of the CCD chip, so $N_{dark} = \sqrt{(C + \tau)}$ where C is the rate of spontaneous thermal electron creation (the dark current), and τ = integration time. Like photon noise, dark noise follows a Poisson relationship to the dark current of the detector. With each 10°C fall in temperature, the dark current is reduced by half. This is why most sensor chips are (Peltier) cooled. Overall camera noise, including read noise, can be seen by collecting a so-called 'dark image': one collected with no light being delivered to the camera (i.e. take an exposure with a lens cap on and/or using a low exposure).
4. **Dynamic range and total noise values**
   The dynamic range can be calculated from the full-well capacity, S, divided by the read noise and dark noise. For most situations, the dark noise is zero, this simplifies to: $D_R = S/N_{read}$. It is a dimensionless number expressed as a ratio.
   Another way of expressing the dynamic range (especially when the dark noise is cooled to virtually zero) is: $D_R = 20 \log_{10} f/n_{RMS}$ decibels, where f is the number of electrons in the full-well, and n_{RMS} the root mean square value of the read noise (to convert a dB value to a dynamic range ratio, divide the dB number by 20. Then raise 10 to this power – use an online calculator). The reason that S/N dB = 20 $\log_{10}$ (S/N) rather than 10 $\log_{10}$ (S/N) is because the power – if the signal is expressed as photocurrent – is proportional to S^2; it is a root-power (field) quantity rather than a power quantity. Even when SNR is expressed in bits, it is still a ratio of analogue quantities.
   Uncorrelated noise is added in quadrature. The total noise in the camera can be estimated by: $\text{Noise}_{total} = \sqrt{(N^2_{shot} + N^2_{readout} + N^2_{dark})}$.
   As the signal increases significantly above the readout noise, the SNR becomes directly proportional solely to the photon shot noise. An advantage of sCMOS camera circuitry is that these cameras can reduce read noise at much faster frame rates than CCD cameras. However, for sCMOS and EM-CCD cameras, fixed-pattern noise (the variation between photosites in collection efficiency) must also be taken into account since read noise is not constant across the detector chip. Flat-field correction can help reduce fixed-pattern noise.

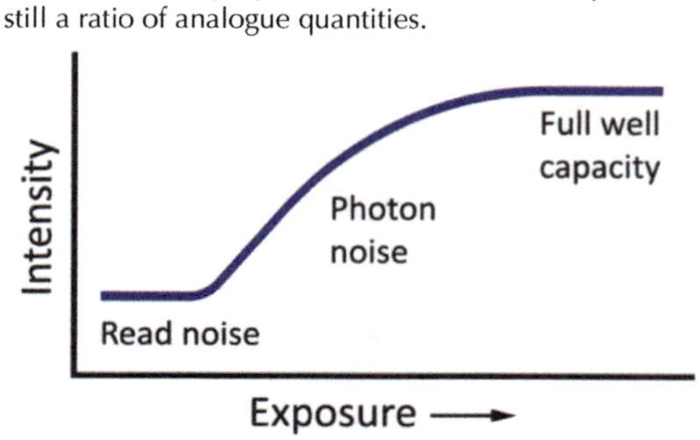

5. **Signal-to-noise ratio**
   The SNR can be calculated from: $SNR = P.Q_e.\tau / \sqrt{[P.Q_e.\tau + N_D.\tau + N_R^2]}$, where P is the incident photon flux (photons/pixel/second); Q_e is the quantum efficiency and τ is the exposure or integration time (seconds). N_D is the dark noise value (electrons/pixel/second; e^-/p/s) and

N_R the read noise (electrons RMS/pixel). Quantum efficiency is a measure of how many photons are converted to electrons by the sensor. Notice that RMS noise signals are added as the square root of the sum of the squares of each component.

Do not confuse noise with background. Noise causes variance, or imprecision, in measuring photon intensity values than are sampled to form the pixels in the digital image whereas background signal adds to the true signal that is of interest. Background can be subtracted from the signal, and various stratagems employed to reduce it (e.g. using blocking techniques in immuno-histological staining) but noise is not constant, so cannot merely be subtracted from the image. Noise can be reduced or frame-averaged out (a common tactic with scanning microscopes) but it must be recognised and dealt with appropriately.

References

Waters, J.C. (2009). 'Accuracy and precision in quantitative fluorescence microscopy'. *The Journal of Cell Biology 185/7*: 1135–1148.

Hamamatsu guide: http://camera.hamamatsu.com/jp/en/technical_guides/about_noise/index.html and https://www.hamamatsu.com/us/en/community/optical_sensors/articles/signal-to-noise_ratio_in_microscope_cameras/index.html.

Photometrics guide: https://www.photometrics.com/resources/technotes/pdfs/snr.pdf.

Andor Guide: http://www.andor.com/LearningPdf.aspx?id=148 and http://www.andor.com/learning-academy/ccd-sensitivity-and-noise-what-is-noise-and-what-does-it-matter.

Some images contain periodic noise that impedes analysis; this can easily be removed by editing a two-dimensional Fourier transform of the image. The freeware image analysis software ImageJ has plug-ins to do this. When you perform a forward fast Fourier transform (FFT), you view the power spectrum of the image, which reveals the location of periodic noise that can then be suppressed or removed using the editing tools. Once you have edited the frequency information, perform an inverse Fourier transform to create an image free of periodic noise. Random noise is most familiarly seen as 'snow' in an image. It may also be called 'salt-and-pepper' noise. This type of noise may occur from errors introduced by the A/D converter. It can be background subtracted. We have to consider three types of random noise:

- shot noise – signal dependent;
- read noise – fixed; exposure dependent;
- dark charge – temperature dependent.

Shot noise is also called Poisson noise or photon noise. It describes the statistical variation in the signal that occurs naturally due to the quantum nature of light (Figure 30.14). Each pixel is a statistical estimate of the photon flux at that point associated with the area of the specimen. Since each pixel represents a sampled measurement of a Poisson process, the probability that the pixel value represents the photon flux at that point is approximately described by a normal (Gaussian) distribution with a standard deviation equal to the square root of the pixel value. Any quantitative measurement of the photon flux must take account of this variation in signal magnitude as well as the noise of the system. This is why the signal-to-noise ratio is used as a measure of image quality.

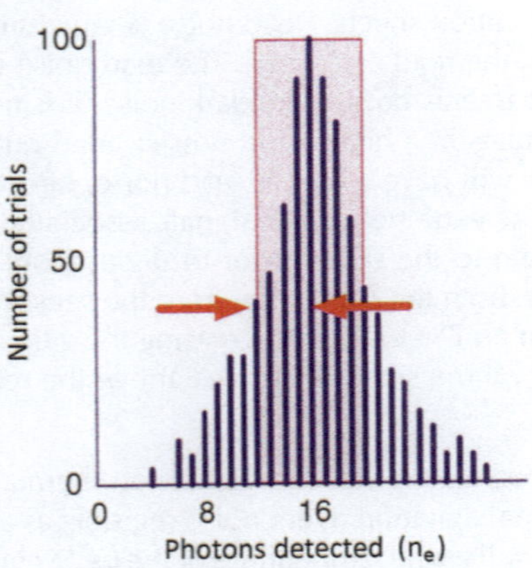

Figure 30.14 ***Poisson shot noise*** *Detecting photons is a quantum-mechanical event. There is an intrinsic uncertainty in the number actually detected on any given trial. This uncertainty is manifested as Poisson, or shot, noise, which is equal to the square root of the number of photons detected. In this example, counting 16 photons is really like counting 16 ± 4 photons. The only way to reduce the uncertainty regarding the photon intensity sampled at this particular point is to increase the irradiant photon flux. When there are few photons, the signal level will exhibit greater variation and greater noise. Therefore, use larger sampling pixels in low light (e.g. live imaging) situations. Here, 700 identical exposures and a histogram plotted of the photoelectrons detected (assuming no read noise and a detector with a QE of 100%). 63% of exposures gave a value for n_e between 12 and 20 detected photons, indicated by the pink shaded box. The half-width of this distribution (4) as delimited by the red arrows, indicates the root mean square (RMS) shot noise value of this measurement. Source: Pawley 2006. Reproduced with permission from Springer.*

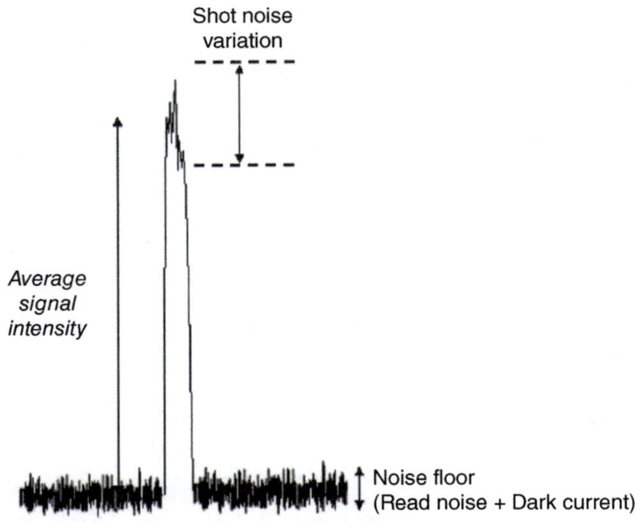

Figure 30.15 **The relative contributions of noise to the image** *Source: Reproduced with permission from Carl Zeiss Ltd.*

As the exposure and/or light intensity increases, shot noise decreases; hence, longer exposures from large pixels exhibit less noise. That is why shot noise can be reduced by binning adjacent pixels (Section 30.5). Because it is inherent and occurs naturally, shot noise *cannot* be removed by taking a flat-field correction.

The remaining sources of noise variation – principally read noise and dark noise – contribute to the overall system noise. The true signal has to be greater than the 'noise floor' (Figure 30.15) imposed by the read noise and dark noise together. In practical terms, the noise floor is the minimum grey level that can be detected by the CCD. Because shot noise is inherent and cannot be reduced by camera design but only by increased photon flux, it sets a minimum achievable noise floor. Otherwise, in the presence of high irradiance, the noise floor is set by the read noise and dark noise together. When the dark noise is minimised by cooling, the read noise alone sets the noise floor, and the camera is said to be read noise–limited. A high quantum efficiency results in a better signal/shot noise ratio and a lower shot noise value. Because shot noise is inherent, it is consequently desirable to operate a camera under conditions that are limited only by read noise, which then defines the camera detection limit, with other noise components being reduced to relative insignificance.

Read noise, or read-out noise, arises from the charge amplifier responsible for measuring the charge packet associated with each pixel and converting the analogue signal (i.e. the electron voltage from each photo-site) into a digital signal. Think of it as a 'toll' charge that has to be paid for accessing and reading the charge stored on the CCD. In most cases, the read noise imposes the actual CCD or sCMOS detection limit because short exposures, high illumination and a cooled chip make the dark noise negligible.

Read noise is Gaussian distributed so is measured in ±RMS (root mean square) electrons of noise, which is equivalent to the standard deviation and should be cited on all good specification sheets. Read noise is large under fast read-out conditions (required for higher frame rates) and increases with the square root of the read-out speed. The read noise can be reduced simply by reading out the signal in a slower fashion, and this is what most CCD cameras do. Unlike dark noise, it is not temperature dependent. A choice of read-out speeds is offered by most cameras: faster frame rates at a higher, but noisier, read-out speed, or a better SNR at slower speeds. For a fixed frame rate, a CCD chip with smaller pixels will have a higher read noise, because it must read out faster. The higher the noise floor on a detector, the less able it is to read out the extremely weak signals associated with low-light imaging. EM-CCDs are designed to effectively 'reduce' read noise by applying gain to the signal prior to digitisation. Of course this gain is also applied to the dark current, so EM-CCDs must be cooled to minimise both the dark noise from the sensor and the thermal effects of the impact ionisation process. The read noise can be further reduced in an EM-CCD by increasing the gain on the signal. Under low light levels, read noise exceeds photon noise, and becomes significant. As the signal increases above the read-out noise, shot noise predominates, and the SNR is directly proportional to the shot noise.

Dark noise, or dark charge, arises from spurious electron accumulation of 'thermal' electrons in the wells or pixels of the sensor. These are freed from the silicon substrate by thermal agitation. Dark noise registers as an output current without an input. Dark noise is dependent upon the signal integration time (τ) as well as the temperature of the CCD chip. Dark noise will increase with longer exposure times. It can be virtually eliminated (0.001 electron/pixel/sec, or e/p/s) by cooling the chip, usually by Peltier cooling to −20°C (because even 1 e/p/s is now considered a major noise source). Very deep cooling beyond −20°C is used for high-end EM-CCD cameras and when extended exposures are required for luminescence studies. Provided that the camera is sufficiently well cooled, the dark noise can be kept below the read noise and so is not regarded as a significant source of noise.

If the CCD is not cooled or else is poorly cooled, the average value of the dark charge can be subtracted from each pixel by subtracting a 'dark image' as part of a flat-field correction. Performing a flat-field correction can reduce the average dark noise value despite the high Poisson noise associated with a reading of only a very few photons/pixels. Because of pixel inhomogeneity, a very few pixels will collect more charge than others. It is then necessary to collect a featureless 'white' flat field that will also remove any irregularity in the background illumination that is otherwise correctly detected by the CCD. Thus taking a flat-field image takes care of both of these concerns (see Chapter 31, Section 31.5).

To give an idea of the effect of dark current noise, at room temperature, thousands of electrons per second of dark noise are generated. With each 10°C fall in temperature, the dark current is reduced by half, until below −60°C, dark noise is as low as 1 electron per pixel per second. The sensor faceplate is sealed in a vacuum to prevent condensation build-up when operating below −20°C. If the vacuum is compromised, it is possible to have ice crystals obscuring the image, and the CCD must then be sent away for repair.

As the signal increases significantly above the read noise, the SNR becomes directly proportional solely to the photon shot noise. To achieve a good SNR, long exposures are better, since the signal approaches the full-well value, thus achieving a good dynamic range in the image. A higher quantum efficiency value means that fewer photons are required to overcome the noise of the camera. Don't make the mistake of looking at a weak signal full of Poisson noise and blame this on the camera. Judge noise on a black image taken with the CCD sensor capped off. Provided this is not due to read noise (use a reasonable exposure time) or dark current (cool the CCD), assume that you can reduce noise only by using more signal per 'pixel' (brighter light or longer exposure).

CCD cameras give their optimum performance when capturing images at approximately 1 frame per second (fps). This is because read-out noise increases with shorter exposures. Most cameras have two read-out speeds. The Sony ICX285 chip has a read-out speed of 10 MHz giving ≈7 frames/sec and 20 MHz giving ≈14 frames/sec. This 'image time' includes charge integration or exposure time, charge transfer, digitisation time, signal to computer time and display time. Exposure time is determined by the strength of the signal, charge transfer and digitisation time by the CCD architecture, signal to computer time by the PCI (or Firewire) image capture board and the display time by the graphics card.

A PMT records single counts from a point very rapidly. When properly operated, it never records a negative value. It is thus very easy, as on an analogue video camera, to set the offset to correspond accurately with a zero-light signal. However, because the electronic read-out noise of a CCD is an RMS (root mean square) function determined by Poisson statistics, with both positive and negative values possible, there will be some pixels that measure below the zero-light level and so clip, or truncate, the signal. To avoid this and to ensure that all the photon data is encoded, scientific-grade CCDs are set so that the zero-light signal is registered at a few tens of units above zero by the analogue-to-digital encoding unit. This compulsory offset makes it more difficult, or impossible, to apply an offset and gain normalisation in the same way as for PMT operation.

Fixed pattern noise occurs because some pixels collect more charge than others. This is not due to 'hot' pixels, which are defective. Fixed pattern noise is normal. It is due to slight differences in the electrical properties of each photo-site or due to differences arising from the shift register and A/D converter. These inhomogeneities can be normalised by taking an image of a featureless illuminated 'white' field without any object feature in the field of view. This is known as background subtraction or flat-field correction and also usefully serves to remove small out-of-focus features, such as specks of dirt or dust on the slide or coverslip of the preparation.

It is not often realised that the microscope is, in fact, a 'photon-loss' machine. A high-quality apochromat objective (100x NA 1.3) of high numerical aperture will collect approximately 37 photons within its angular aperture of 134° from a fluorescent molecule emitting 100 photons. Approximately 5% of these photons will be lost within the objective, reflected off the sides or from the glass-air interfaces of the lens elements. This leaves 35 photons. A further 20% will be lost passing through the filter sets, leaving 28 photons. Another 10% might be lost at the surfaces of any DIC components, the mirror in the trinocular tube head, or the projection lens linking the camera to the microscope, leaving 25 photons. At this point, shot noise means that the signal might be as low as 20 photons regardless of any other noise factors. Now factor in the QE of the sensor, and the number of detected photons may range from 4 to 12% of the original signal emitted.

If you divide the full well capacity of your camera by the bit depth, here 16-bit (e.g. 30 000/65 536), the number generated is the conversion factor from grey levels to photo-electrons = 0.458. Identify a pixel or region of interest comprising a number of pixels, and measure the intensity in grey levels. Subtract from this the value of the camera offset, and multiply the result by the conversion factor (e.g. from a starting value of – say – 1000, subtract 40 grey levels of offset; then multiply 960 grey levels × 0.458 electrons/grey level to get the number of photo-electrons = 440 e-. The electrons can be converted to photons by dividing this result by the QE of the camera: e.g. 440/0.82 = 537. This is an approximate calculation of how many photons per pixel were acquired from your sample.

30.11 Choosing a CCD or sCMOS Camera

How well the signal is captured is the primary criteria for the choice of digital camera. The overall sensitivity of the CCD camera is determined by the pixel size, the read noise floor (assuming the shot noise and dark noise are negligible) and the quantum efficiency of the camera. The QE tells you how many photons are *not* counted. The fill factor is the fraction of the sensor surface than actually collects photons. This can be very high for frame transfer CCDs but is less for the popular interline CCDs simply because half the surface of the chip is used to transfer the charge rapidly. Microlens arrays can increase the QE of interline CCDs to over 70%, but this is rarely as much as with a frame-transfer CCD design. Cooled cameras also have better SNR, and full-frame transfer CCDs have a better SNR than interline CCDs for a given pixel size. Back-illuminated cameras have a better QE and better SNR than front-illuminated ones. This is because the path length for the photo-electron is shorter, and it is not scattered by the transparent polysilicon charge-transfer electrodes on the front face of the chip. They are, however, more expensive.

The read noise is the next important criteria to consider. A high read noise tends to reduce the dynamic range and swamp low-intensity signals. A camera with a high read noise can still have a high dynamic range if it has large pixels and a large full-well capacity. But a camera with a small dynamic range and smaller pixels can have less read noise per pixel (see the examples given in Table 30.2).

Table 30.2 *The Effect of Read Noise*

	12-bit camera	14-bit camera
Sensor size	13.3 × 13.3 mm	8.19 × 8.19 mm
Sensor diagonal	18.8 mm	11.6 mm
Flange distance	17.5 mm	25.5 mm
Effective No. pixels	2048 × 2048	512 × 512
Photo-site (pixel) size	6.5 μm × 6.5 μm	16 μm × 16 μm
Readout noise [RMS]	0.9 e-	8 e-
Read noise level	±1.8 e-	±16 e-
Full well capacity	30 000 e-	340 000 e-
Dynamic range	33 000:1	42 500:1
Dark current	0.14 RT 0.04 (−10°C)	0.0005 e-/pixel/s (−80°C)
Max QE	82%	96%
Clock rate	216 MHz	0.6875 MHz
Binning	2 × 2, 3 × 3, 4 × 4, 8 × 8	2 × 2 and 4 × 4
Frame rate	53 f/s	70.4 f/s
Price	£ 9500	£ 14 000

These cameras have been selected from the range offered by two digital camera market leaders. They are both designed for specific purposes; no implication is intended that one is better than the other. The bit-depth does not necessarily indicate a better camera: look at the entire specification and the purpose for which the camera was designed. To be detected, the signal in the 14-bit camera has to be approximately nine times greater than with the 12-bit camera.

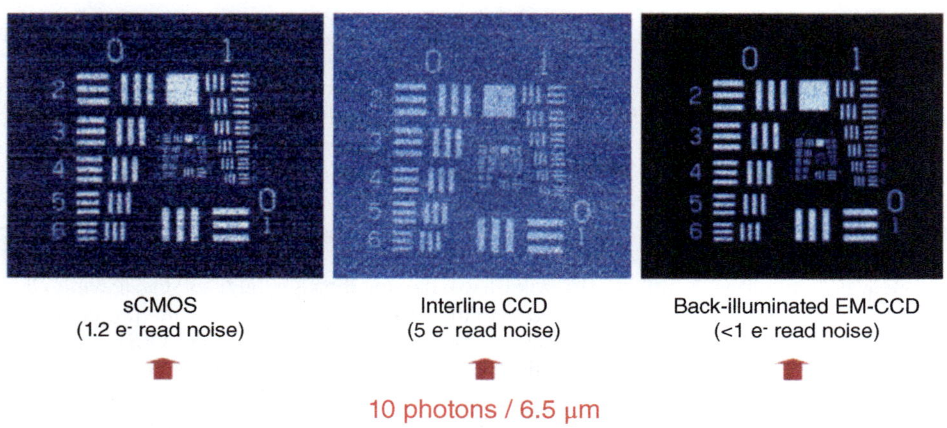

This image (Source: Andor/Oxford Instruments. Reproduced with permission from Andor) shows the effect of read noise on a test image. Where the photon flux is very low, the read noise, rather than the shot noise, becomes limiting and influences the quality of image. Here 10 photons fall upon each 6.5 μm photo-site for these three sensors.

Factors to take account of when choosing a camera:

Quantum efficiency (QE %); photo-site fill factor or full well capacity (FWC, e-), which influences dynamic range; readout noise (N_r e- rms) also influences dynamic range; dark current (e-/p/sec); bit depth (8, 10, 12, 14 or 16 bits); dynamic range (ratio or dB) important for maintaining image contrast; pixel size (μm x μm) influences resolution in the image and Nyquist sampling of the specimen; pixel number and camera chip faceplate array size (mm x mm) determines the proportion of the field of view acquired; binning factor determines signal intensity, detail resolved and potential acquisition frame rate; frame rate and subarray collection, determines temporal resolution in the image data set; shutter stability and vibration influences accuracy of acquiring images of moving specimens; cooling determines the influence of dark noise.

In this case, the CCD with smaller pixels may successfully display a weaker signal better than a high dynamic range camera with a higher noise floor. Overall, high read-out noise can be tolerated: a very low read noise is only essential when using the camera in low-light situations to view very dim specimens. That is, when the read noise exceeds the statistical noise (variation) of the photon signal in the dimmest pixel.

If the light level, or photon flux is extremely dim, in the range of 10^3 to 10^5 photons/second/cm^2, then an EM-CCD is the likely choice, but do also consider the new back-thinned sCMOS cameras. EM-CCD cameras are often preferred for spinning disc confocals and for TIRF microscopy where their relatively large pixel size satisfies the Nyquist sampling criterion and where high frame rates are required. EM-CCD cameras go fast and 'lift' the signal above the read noise floor of the CCD. The ability of EM-CCDs to multiply weak signals above the detector's read noise, by applying gain, effectively eliminates read noise at any speed. EM-CCDs are ideal for low-light, single-molecule detection and long-term live-cell fluorescence microscopy. To get the best sensitivity from an EM-CCD, it requires effective thermoelectric Peltier cooling. When a CCD sensor is not Peltier-cooled to reduce dark charge, a process called 'multipinned phasing' is used to reduce the dark noise by a factor of 1000x, making it negligible in exposures up to 50–60 seconds. This feature should be sought in uncooled CCDs. A low read-out noise is useful, but it is only critical if you want to record dim images from weakly stained specimens under low-light conditions. Further details on camera types to consider and strategies to employ for low-light imaging are given in a comprehensive chapter by Moomaw (2007); a comparison between EM-CCD and sCMOS cameras for live-cell imaging is published in Beier and Ibey (2014).

If the light level, or photon flux, is greater than 10^5 photons/second/cm^2, then integration with a regular sCMOS or CCD camera is advisable. Longer integration times from longer exposures will give a brighter image. If the image is moving, then an EM-CCD camera may be necessary to capture the image at a sufficiently high frame rate to prevent blurring from movement during acquisition. sCMOS cameras give the advantage of higher frame rates, lower read noise, greater dynamic range and larger field of view, whereas most cooled scientific CCD cameras allow hardware binning and will give quieter images of static samples.

Consider also the size of chip that you require and the pixel size so as to choose the best trade-off between resolution and field of view. Do not be fooled by 'mega-pixels', which is a common marketing ploy in the consumer camera market (for more on this see Section 31.11). For a given sensor size, the higher the 'mega-pixel' number, the smaller the pixel size on the chip, the less sensitive the sensor will be and the more prone to noise variations it will be because it has a smaller capacity to capture photons and hold electrons, respectively. sCMOS detectors used to be the poor relation of electronic sensors, unsuitable for scientific-grade cameras. They can now offer advantages over CCDs in what were once mutually exclusive factors:

- extremely low noise;
- rapid frame rates;
- wide dynamic range;
- high resolution; and
- large field of view.

sCMOS cameras now outperform interline CCD cameras for most dynamic biological applications.

In putting all these factors together, read the camera specification sheets that all reputable manufacturers offer, for they will give you the data to make comparisons, as Table 30.2 shows, although for live-cell dynamic measurements it is best – if possible – to test an EM-CCD, sCMOS and interline CCD on the same sample. For example, see the table on the Cairn Research website: *'Which camera to go with my microscope objective?'* The URL link is given at the end of the chapter. However, do establish with the company representative whether the figures given represent an average case scenario or guaranteed values. There will always be a certain very small amount

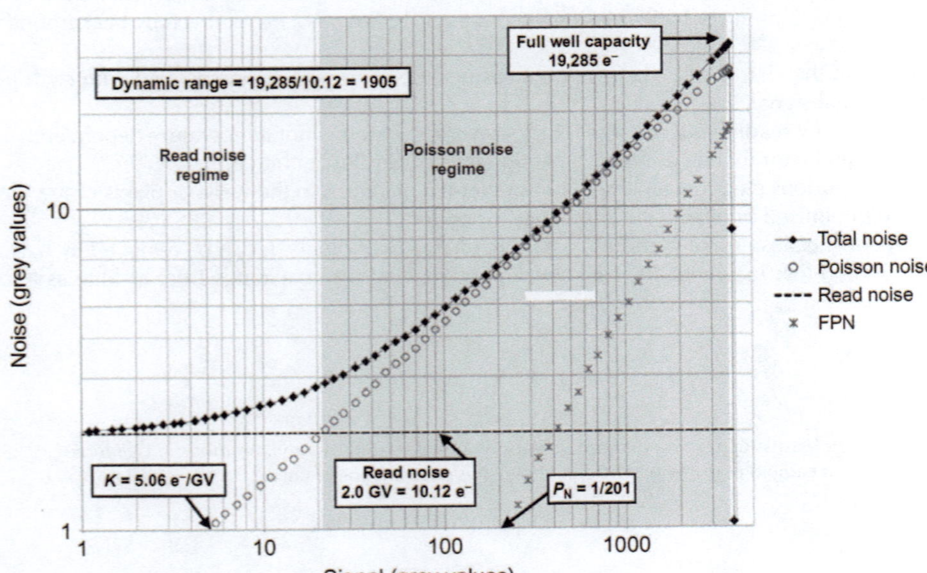

Figure 30.16 **Photon transfer curve** *A representative photon transfer curve generated with a 12-bit interline CCD. The total noise is composed of: read noise, Poisson noise and fixed-pattern noise (the dark noise is negligible due to cooling). On the log-plot, read noise is constant and has a value of zero; Poisson noise scales with the square root of the signal intensity and has a slope of 0.5; fixed-pattern noise is proportional to the signal and has a slope of 1. The three noise regimes are defined by the dominant noise source over a given range of signal intensity. In this camera, FPN is low enough that the full-well capacity (pixel saturation) was reached before fixed-pattern noise is significant. GV = grey value; K = electron conversion factor; P_N = FPN quality factor. See Lambert & Waters (2014) for further details. Source: Lambert and Waters 2014. Reproduced with permission from Elsevier.*

of variation between different cameras of the same model; such is the nature of production line goods. Different firmware/software and camera drivers can also lead to differences in performance between the same camera being used on one manufacturer's microscope than another imaging platform. It is best therefore to test your own camera on your microscope. This can be done by generating a photon transfer curve, an analysis method developed by NASA. A photon transfer curve (Figure 30.16) provides an empirical measure of all the critical camera performance characteristics: read noise, dark noise, full well capacity and dynamic range. An easy-to-follow protocol for creating a photon transfer curve can be found in an excellent book chapter by Lambert and Waters (2014) under Key Reading.

30.12 Chapter Summary

This chapter discusses how digital images are formed from an analogue stream of photons, how electronic sensors such as CCD and sCMOS cameras can convert the analogue signal into an electronic image. Factors such as dynamic range, bit depth and electronic noise all have to be taken into account when choosing an appropriate camera or manipulating image acquisition parameters.

1. Digital images can be cloned exactly without loss and easily altered, transported and stored.
2. Digitally archived material can potentially quickly date.
3. Digital images, encoded in binary, may be raster- or vector-based images.
4. Raster images are composed as pixels, a sample of photon intensity given off by or associated with the object at that point.
5. A photon converted to an electron is quantised into discrete integers by the A/D converter to record signal intensity in the image.
6. Resolution may apply to optical resolution of the microscope, to digital image resolution – the size and number of image pixels – to screen resolution, which determines display clarity, or to output resolution, which governs display or print size.
7. Bit depth determines the number of grey levels in the image.
8. Humans can see roughly 60 grey levels; 8-bit equates to 256 grey levels.
9. CCD cameras may be either frame-transfer or interline designs. The latter are being replaced by sCMOS technology. EM-CCD cameras are good for applications recording very low light levels.
10. The quantum efficiency of a camera measures its efficiency at converting photons to photo-electrons.
11. Additive colour RGB is encoded as 24-bit true colour.
12. A Bayer mask is used to convert intensity grey levels to colour and to match the sensitivity of the eye, at the expense of spatial resolution.
13. Microscanning and pixel interpolation by a demosaicing algorithm can preserve spatial resolution.
14. Noise in a digital image causes variance in the pixel greyscale intensity values and reduces contrast in the image. It is not merely background.
15. Periodic noise can be removed by Fourier transform analysis.
16. Poisson, or shot, noise is natural statistical variation in the signal due to the quantum-mechanical nature of light.
17. Shot noise can be reduced by binning or by increasing the photon flux. It cannot be removed by background subtraction (flat-field correction).
18. Read-out noise arises from the charge amplifier, which measures the charge packet associated with each pixel and converts the analogue input into a digital signal.
19. Read noise can be reduced by reading out the signal in a slower fashion. It is not temperature dependent.
20. Read noise is the important factor that determines dynamic range, regardless of bit depth.
21. Dark noise is caused by spurious electron accumulation of thermal electrons in the wells or pixels of the sensor. It is temperature dependent and can be minimised or effectively eliminated by cooling.
22. Fixed pattern noise occurs because some pixels collect more charge than others. It can be removed by background subtraction.
23. The microscope is a photon-loss machine. The number of photons detected may range from as little as 4 to 12% of the original signal emitted.

Key Reading

Lambert, T.J. and Waters, J.C. (2014). 'Assessing camera performance for quantitative microscopy'. *Methods in Cell Biology 123*: 35–53.
- This chapter explains how to test camera performance by calculating the photon transfer curve.

References

Allen, E. and Triantaphillidou, S. (2009). *The Manual of Photography*, 10th ed. Routledge, Abingdon: Focal Press.

Beier, H.T. and Ibey B.L. (2014). 'Experimental comparison of the high-speed imaging performance of an EM-CCD and sCMOS camera in a dynamic live-cell imaging test case'. *PLoS One* 9/1: e84614.

Bernas, T. (2005). 'Basics of digital microscopy'. *Current Protocols in Cytometry*. Chapter 12: Unit 12.2.

Evennett, P.J. (1989). 'Image recording'. In: Lacey, A.J. (ed.), *Light Microscopy in Biology*, 1st ed, Chapter 3. Oxford: Oxford University Press, pp. 61–102.

Hamamatsu Photonics UK (2009). EM-CCD Technical Note. p. 4, Figure 7. http://www.tecnicaenlaboratorios.com/HAMAMATSU/Catalogos/e_imagemtec.pdf.
Hazlewood, K.L. et al. (2007). 'Entering the Portal'. In: Shorte, S.L. and Frischnecht, F. (eds.), *Imaging Cellular and Molecular biological Functions*, Chapter 1. Berlin: Springer-Verlag.
Hinchcliffe, E.H. and Sluder, G. (2007). 'Working with classic video'. In: Sluder, G. and Wolf, D.E. (eds.), *Digital Microscopy*, 3rd ed, Chapter 3. Methods in Cell Biology, vol. 81. Amsterdam: Academic Press, Elsevier, pp. 141–157.
Moomaw, B. (2007). 'Camera technologies for low light imaging: overview and relative advantages'. In: Sluder, G. and Wolf, D.E. (eds.), *Digital Microscopy*, 3rd ed, Chapter 13. Methods in Cell Biology, vol. 81. Amsterdam: Academic Press, Elsevier, pp. 251–283.
Mortensen, K.I. and Flyvbjerg, H. (2016). '"Calibration on the spot": how to calibrate an EM-CCD camera from its images'. *Scientific Reports* 6: 28680.
Murphy, D.B. and Davidson, M.W. (2013). *Fundamentals of Light Microscopy and Digital Imaging*, 2nd ed. Wiley Blackwell.
Pawley, J.B. (2006a). 'Points, pixels, and gray levels: digitizing image data'. In: Pawley, J.B. (ed.), *Handbook of Confocal Biological Microscopy*, 3rd ed, Chapter 4. New York: Springer, pp. 59–79.
Pawley, J.B. (2006b). 'More than you ever really wanted to know about charge-coupled devices'. In: Pawley, J.B. (ed.), *Handbook of Confocal Biological Microscopy*, 3rd ed, Appendix 3. New York: Springer, pp. 918–931.
Sabharwal, Y. (2011). 'Digital camera technologies for scientific bioimaging. Part 1: the sensors'. *Microscopy & Analysis* Image analysis supplement, May 2011. Wiley.
Saxby, G. (2011). *The Science of Imaging*, 2nd ed. Boca Raton, FL: CRC Press.
Shotton, D.M. (1995). 'Electronic light microscopy: Present capabilities and future prospects'. *Histochemistry and Cell Biology 104*/2: 97–137.
Watkins, S.C. and St. Croix, C.M. (2013). *Current Protocols Select Imaging & Microscopy*. New Jersey, Hoboken: Wiley.
Weiss, D.G. et al. (1999). 'Video microscopy'. In: Lacey, A.J. (ed.), *Light Microscopy in Biology: A Practical Approach*, Chapter 8. Oxford: IRL Press, pp. 221–278.

Further Reading

Bernas, T. (2005). 'Basics of digital microscopy'. *Current Protocols in Cytometry*. Chapter 12: Unit 12.2.
Bernas, T. et al. (2007). 'Precision of light intensity measurement in biological optical microscopy'. *Journal of Microscopy 226*/2: 163–174.
Hazelwood, K.L. et al. (2007). 'Entering the Portal: Understanding the digital image recorded through a microscope'. In: Shorte, S.L. and Frischknecht, F. (eds.), *Imaging Cellular & Molecular Biological Functions*, Chapter 1. Berlin: Springer, pp. 3–43.
Pang, Z. et al. (2012). 'Dark pixel intensity determination and its applications in normalizing different exposure time and autofluorescence removal'. *Journal of Microscopy 246*/1: 1–10.

For those interested in the use of classical film emulsions:
Bracegirdle, B. (1995). *Modern Photomacrography*. RMS Microscopy Handbook No. 31. Oxford: Bios Scientific Publishers.
Bracegirdle, B. (2010). *A history of photography with the light microscope*. Quekett Microscopical Club.
Bracegirdle, B. and Bradbury, S. (1995). *Modern Photomicrography*. RMS Microscopy Handbook No. 33, Oxford: Bios Scientific Publishers.
Delly, J.G. (1988). 'Photography through the microscope P-2'. Rochester, NY: Eastman Kodak.
Evennett, P.J. (1989). 'Image recording'. In: Lacey, A.J. (ed.), *Light Microscopy in Biology*, 1st ed, Chapter 3. Oxford: Oxford University Press, pp. 61–102.
Jacobson, R.E., Ray, S.F., Attridge, G.G., and Axford, N.R. (2000). *The Manual of Photography*, 9th ed. Oxford: Focal Press.
Kodak Basic Photographic Sensitometry Workbook. (2006).
Saxby, G. (2002). *The Science of Imaging*, 1st ed. Bristol: Institute of Physics Publishing.

Internet Resources

Cairn Research: https://www.cairn-research.co.uk/support/technical-notes/.
Andor: http://www.andor.com/learning-academy.
Andor Cameras for Microscopy: http://www.andor.com/cameras-for-microscopy/.
Andor sCMOS & CCD: https://bit.ly/2MrRGVV.
Andor Rolling and global shutters: https://www.oxinst.com/learning/view/article/rolling-and-global-shutter.
Andor EM-CCD: http://www.emccd.com/.
Andor Noise & Sensitivity: http://www.andor.com/cameras/ixon-emccd-camera-series.
Hamamatsu: http://hamamatsu.magnet.fsu.edu/.
Hamamatsu CCD noise: http://hamamatsu.magnet.fsu.edu/articles/ccdsnr.html.
Hamamatsu sCMOS noise: http://camera.hamamatsu.com/jp/en/index.html.
QImaging EM-CCD: https://www.qimaging.com/resources/pdfs/emccd_technote.pdf.
Roper CCD primer: https://www.roperscientific.de/encyclopedia.html.
Roper CCD architecture: https://www.roperscientific.de/architectures.html.
Dead pixel test: http://www.photo-freeware.net/download/action/show/Item/dead-pixel-test/.
Signal-to-noise calculator: https://www.photometrics.com/resources/imaging-tools/signal-to-noise-calculator.php.
Read noise calculator: https://www.photometrics.com/resources/imaging-tools/read-noise-calculator.php.
Calculate CCD read-out noise (ImageJ plugin): https://imagej.nih.gov/ij/plugins/ccd-noise/index.html.

Notes

1. Within the CCD and sCMOS architectures, the photo-site is the light-sensitive well that collects, or samples, the photons to form a pixel: a record of the light intensity at that point within the raster image. No further distinction is made in this book between photo-sites and pixels; for clarity the term 'pixel' is used interchangeably for photo-site, according to context.
2. Sometimes the display software will smooth a pixelated image by interpolation. It is still over-magnified, and no extra information, or resolved detail, is contained within the enlarged image.
3. For professional printing a digital image must have a minimum of 300 pixels per inch; domestic photographs require 200 ppi.
4. High quality DSLRs have around 10–12 f-stops of dynamic range compared to about 7–8 f-stops for reversal slide film and to 6–8 stops for an entry-level point-and-shoot digital camera. As far as dynamic range is concerned, DSLRs are superior to all reversal (transparency or slide) films, very close to the best colour negatives and not quite as good as the best monochrome negatives.
5. A false-colour look-up table (LUT) is required successfully to set the limits of the dynamic range for an 8-bit or 12-bit camera or PMT. Provided you set the limits correctly, all the contrast in the image will be recorded. It is impossible to do this without using a false colour look-up table, which assigns all those pixels that are zero (black) and below to a single pseudocolour (e.g. blue) and all those pixels at 255 (white) and above to another pseudocolour (e.g. red). This is necessary because our eyes are only 5 to 6-bit devices, perceiving 50–60 grey levels. (The reason why the eye is only a 6-bit device is because the dynamic range is the signal level at saturation divided by the dark noise of the detector. SAT/N_{dark} = dynamic range = maximum signal/minimum detectable increase ≈2% for human eye, so dynamic range $\approx \log_2 50 = 5.644 \approx 6$, and $2^6 = 64$.
6. This proportion is linear even at long or short exposures, usually <1%, which is one advantage over using film. Linearity in a CCD camera only falls off as the full-well capacity is reached. Silver halide films exhibit a distinct shoulder and toe at each end of the sensitometric characteristic curve for the emulsion. See the references under Further Reading and the URL links on classic film emulsions, at the end of the chapter, for information on sensitometry and the H&D characteristic curve.
7. See: https://www.roperscientific.de/architectures.html for an online demonstration of the operation of different CCD designs.
8. For a protocol on how to calibrate an EM-CCD for photon counting, see Mortensen and Flyvbjerg (2016).
9. CMYK refers to the four inks used in colour printing: **c**yan, **m**agenta, **y**ellow, and **k**ey black. The 'k' refers to key in key black, and not to 'k' of black. Colour half-toning permits a full continuous range of colours, which can be produced from seven basic colours.
10. The signal-to-noise ratio (SNR) should not be confused with, or erroneously equated to, the signal-to-background ratio (SBR), which is a measure of contrast in the image. SNR describes the variability in photon intensity at a single sampling point, or pixel. Confocal and TIRF microscopes generate images with high SBR compared to a widefield microscope.

31

Recording the Image: Part 2 – Applied

Outline

Understanding Light Microscopy, First Edition. Jeremy Sanderson.

31.1 Introduction

The purpose of this chapter is to help you couple a digital camera to a microscope, to acquire correctly sampled images and to display them properly, for subsequent analysis, to show resolved details present in the specimen or to print them as a permanent record. Most digital images or photographs will never be printed out but rather are displayed on a monitor (including those submitted to journals, which are included in the online supplementary information). The question becomes rather: 'How will I display my digital image?' This topic is discussed in Section 31.9.

Although very little used by professional scientists, drawing was the first means of reproducing the microscope image. In skilled hands, drawing still has the advantage of bringing together information into a single illustration that direct recording by a camera cannot do. Elements of differing contrast enhancement, methods from different magnifications, fields of view and planes of focus can be selectively combined to direct the eye to important features. Equally, non-essential information can be omitted to emphasise a particular feature or make a point. Suitable camera lucidas or their modern equivalents, drawing tubes, for projecting the image onto a sheet of paper can still be obtained second-hand. Even without these, an eyepiece fitted with a net graticule, which comprises a square grid pattern, can be used as a drawing guide. Further details on drawing techniques can be found in Bracegirdle and Bradbury (1995).

Photomacrography is a specialist field allied to photomicrography. Special lenses take the place of the objective, and the image is taken directly, without secondary magnification employing an eyepiece. Photomacrography covers the range 1:1 to 50:1. The reproduction ratio is the ratio of the actual size of an object to the size of the image on the electronic sensor. Below reproduction ratios of 1:1, ordinary photographic equipment can be used for close-up work (e.g. 1:1 to 1:10), for general photography utilises ratios from 1:1 to 1:infinity. The reproduction ratio indicates the size of the image relative to the object, viz. reproduction ratio = image size/object size. Thus a ratio of 20:1 means that the image is 20 times the real-life size of the object. With reproduction ratios greater than about 50:1, the nominal limit of photomacrography, a compound microscope is used to record photomicrographs. The upper limit of approximately 2000:1 for photomicrography is set by the useful magnification associated with the limit of resolving power for the light microscope.

The larger the field of view, the more challenging it is to secure an evenly lit field. There are other difficulties too: the depth of field is often limited; vibration can affect the specimen or equipment, and thus be recorded into the image. At low magnifications, focusing accurately can be a problem, since the depth of focus is large, and the eye may accommodate on the image above or below the actual focal plane. A comprehensive book on the subject, Bracegirdle and Bradbury (1995), has been supplemented by two very practical papers: Bracegirdle (2010) on digital photomacrography and Bracegirdle (2012) on suitable specimen supports. Both these papers are fully illustrated in colour. Those interested in photomacrography should consult these references as well as Cremona (2014) and Cremona (2018); the subject is not covered further in this book.

Well-designed modern commercial digital imaging systems make capturing a good digital image both quick and simple. If so, why is there this chapter dealing with the practical aspects of acquiring digital images? Firstly, it is easy to over-rely upon an automated system, with the result that poor images are taken. It is still necessary to set up the microscope to illuminate the specimen properly according to the principles of Köhler illumination and form a real image on the camera sensor. Secondly, it is usually necessary to subtract any unevenness in background illumination from the image and to set the white balance correctly for proper colour rendition. Thirdly, the analogue photon signal has to be sampled correctly so that the resolving power of the camera matches that of the microscope. Nyquist's sampling criterion should be observed; otherwise, all the finest detail resolved by the objective will not be recorded.

> The image must be projected onto the detector surface such that the minimum distance resolved at the object is magnified to cover *at least* twice the size of the photo-sites of the electronic sensor or rods and cones of the retina.

Fourthly, scale bars should be added to images, and consideration must be given to how the image is to be displayed. All these aspects of image recording are covered here.

31.2 Producing a Real Image onto the Electronic Camera Sensor

The image-forming rays leaving the microscope at the exit pupil are parallel: when they enter the relaxed eye, they are focused by the lens of the eye to form an image on the retina. The rays in the aperture set of conjugate planes (Chapter 9, Section 9.10) come to a focus at the exit pupil (or Ramsden disc) of the microscope, and it is here that the lens of the eye is placed to see the image (Figure 31.1). We see a virtual image, which appears to represent a magnified object situated at infinity. The eyepiece thus works in conjunction with the eye to produce a real image on the retina. A real image is one that can be formed on a surface, such as a screen, retina, film or CCD/sCMOS/PMT sensor.

When a microscope is to be used with a camera for photomicrography, it must be adapted in one of several ways (Figure 31.2) so that it forms a real image onto the digital camera sensor. If the position of the primary image is lowered relative to the position of the eyepiece (Figure 31.2b) by refocusing the microscope or if the specimen plane is lowered, the rays exiting the eyepiece will no longer be parallel but will converge to form a real image. This method is simple and requires no extra equipment, but it is not ideal. The conjugate distances between, firstly, the specimen and the objective lens and, secondly, the conjugate distance between the objective lens and primary image are now longer and shorter, respectively. Any alteration in the proper conjugate distances will introduce noticeable spherical aberration (Figure 31.3a) into the image. This simple method of forming an image may be adequate when using equipment for large-format photography (which nowadays is rarely the case) because the small amount of refocusing represents a very small change in length compared to the overall long distance between eyepiece and film. It may also be possible to use this method with objectives of small NA because they are less sensitive to this effect.

The opposite action is to raise the eyepiece slightly (Figure 31.2c) to raise the primary image to eye lens conjugate distance, while leaving the focus properly adjusted for visual use. This method has the advantage of ensuring that the conjugate distance between the specimen and objective lens remains undisturbed. The eyepiece can be held in place by a piece of adhesive tape or a specially made ring. A similar way to produce a real image is merely to move the top eye lens element of the eyepiece, by refocusing it (Figure 31.2d), leaving the body of the eyepiece and the lower field lens *in situ*. This is a focusable projective lens. Such lenses are sometimes erroneously referred to as 'photographic eyepieces' or 'projection eyepieces'. They cannot, however, be used with the eye. The photographic projection lens need not be focusable. Some are designed without focusing means for specific microscopes and cameras. Both types of projective lens perform the same function: converging the image-forming rays to form a real image.

The best way in which to form a real image for photomicrography is to mimic the lens of the eye and use a converging lens (Figure 31.2e) above a 'normal' eyepiece intended for visual observation. Any lens on the camera is removed; the camera body, whether digital or analogue, fits above this relay lens. A beam-splitting cube can be incorporated between the regular eyepiece and the camera relay lens (Figure 31.2f) to allow a focusing telescope (e.g. a small 3 × 12 monocular telescope focused at infinity) that is conjugate with the film plane to be used. This is necessary with a sealed camera back using analogue film, but it is also useful for focusing a sharper image than relying upon the screen on the digital camera.

31.3 Coupling a Camera to the Microscope

A camera can be attached to the microscope either to record the primary image directly or by use of an eyepiece or projection lens to record a secondary image. The simplest way in which a digital sensor can be connected to a microscope is to use a webcam. These have a pixel density of around 1.3 megapixels, and a 4 mm square CMOS sensor size. These can be arranged to pick up the primary image directly if the front lens can be unscrewed, but the image is generally only suitable for display on a computer monitor and shouldn't be printed out at anything greater than standard postcard 6″ × 4″ format. This type of camera is not considered further here.

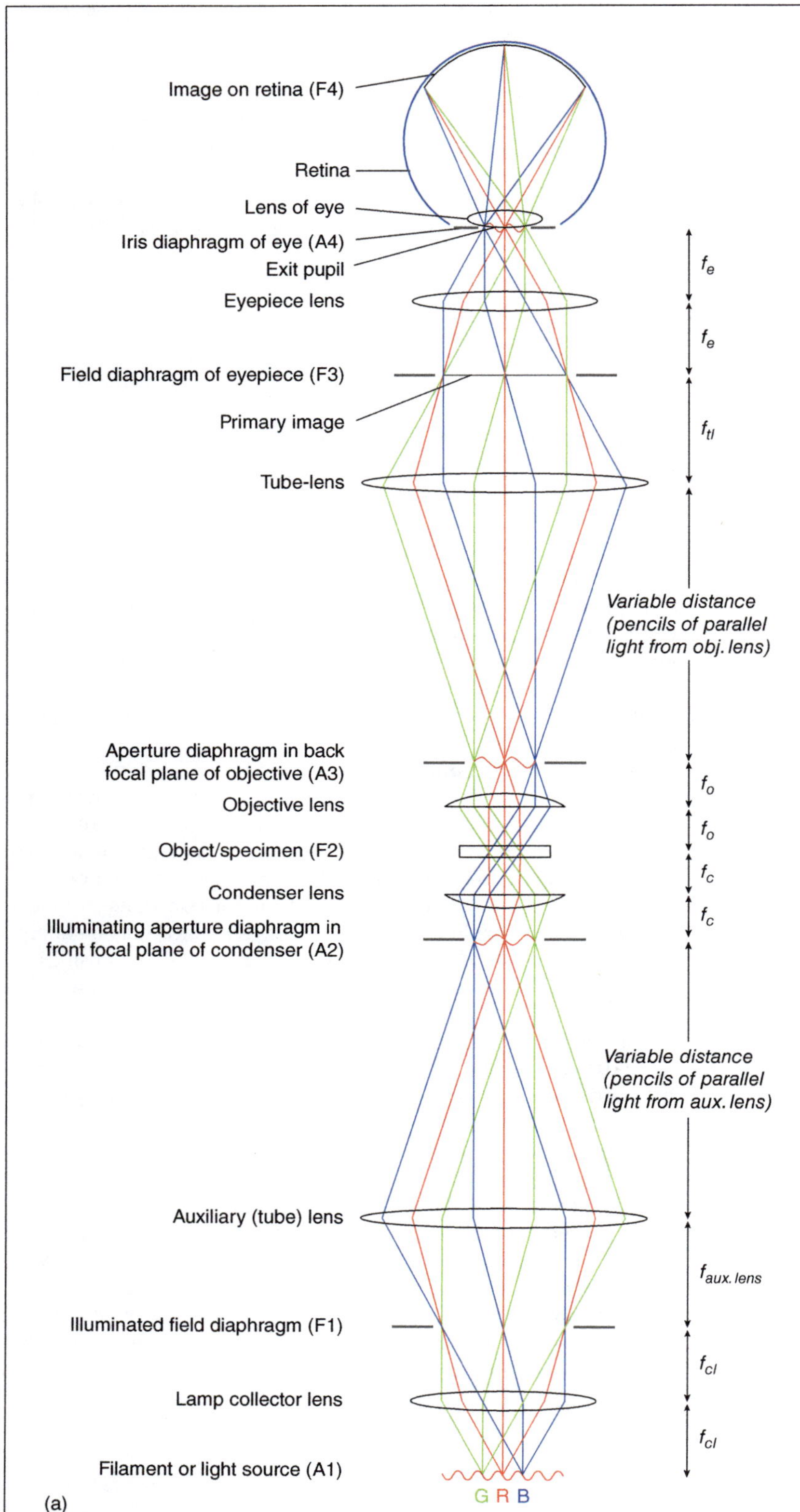

Figure 31.1 **Ray diagrams for visual and photographic observation** *The ray diagram for Köhler illumination as shown in Figure 9.19. The field and aperture sets of conjugate planes are labelled F1–F4 and A1–A4, respectively. The focal lengths (f) of the tube-lens, the eyepiece and the projection lens are shown. The eyepiece works in conjunction with the eye to produce a real image on the retina, whereas the projection lens alone produces an equivalent real image on the detector plane. Source: Hammond 2006. Reproduced with permission from John Wiley & Sons.*

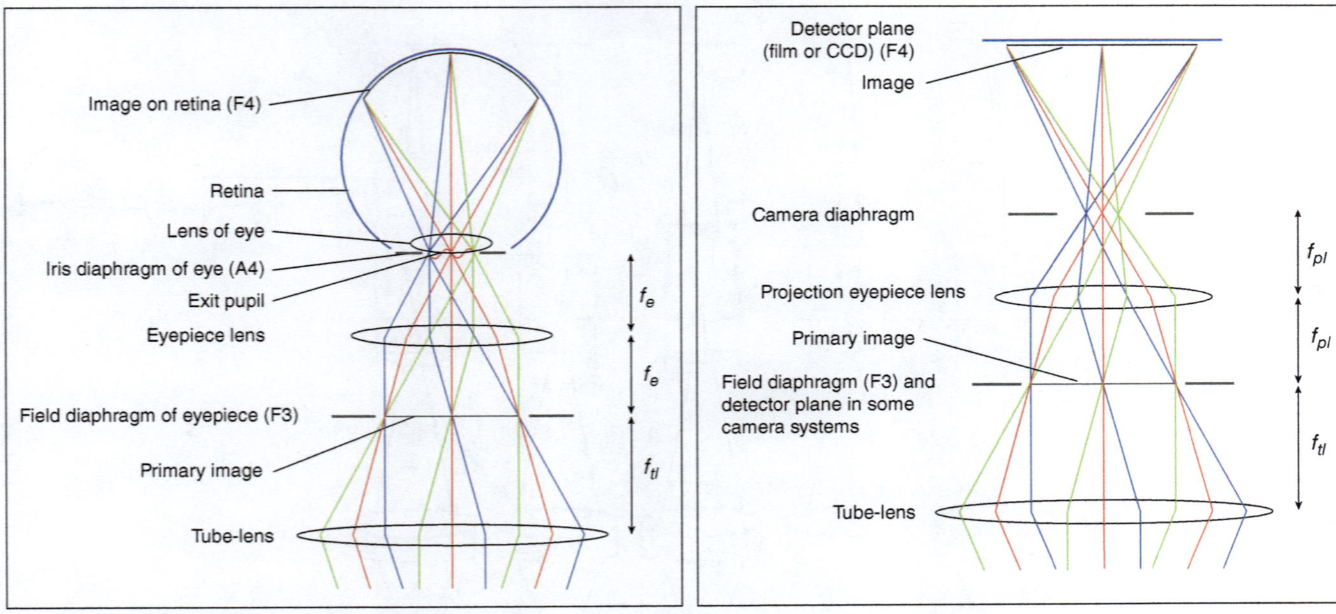

Figure 31.1 (Continued)

Try not to use a point-and-shoot consumer camera with a retractable lens. These cameras do not have a screw thread or other means to attach the camera easily to the microscope or eyepiece. Although the camera can be positioned in the plane of the exit pupil of the microscope to focus an image, this can be awkward to do without an independent support, such as a retort stand. Photos taken in this way will exhibit poor quality because the optical alignment is not correct. If the camera is of the design that moves the lens whilst zooming, this makes the exercise doubly awkward. Digital SLR cameras with removable lenses are widely available, are of better quality and are a much better choice for photomicrography.

Compact digital cameras have a smaller sensor size, and are generally designed with a fixed zoom lens built in. With a fixed lens camera, the camera lens takes the place of the eye, so a conventional eyepiece can be used if there is a female filter screw thread to screw a suitable adaptor into. Use a 'high-eye-point' eyepiece to give sufficient clearance for the camera lens. Some of these cameras, such as the Nikon Coolpix 950 and 995 models, could take a Leitz Periplan eyepiece with an equivalent male thread, mimicking the eye-eyepiece combination perfectly. These compact cameras will only work provided that image zooming function is digital and not optical.

A scientific-grade digital camera is connected to a microscope either by a C-mount[1], or by non-standardised adapters via a dedicated trinocular tube-head, photo-tube, or side-port. Check the optical correction of the coupler when using older microscopes. For example, Nikon's C-mount adaptor is designed without compensation, since the primary image is fully corrected. For older microscopes, this will produce fringing at the edge of the image. Consumer digital SLR cameras usually have removable front lenses; they can be connected to microscopes with a T-2 adaptor or Nikon bayonet F-mount. These provide a universal M42 42 mm female thread for attaching the camera body to suitable coupling units. These can be found from various firms via the Internet, and a selection is listed at the end of this chapter.

You will have to take account of the register distance (Figure 31.4) to ensure parfocality between the camera body and the eyepieces of the microscope. This value is different for different camera systems (for lists of these, see the Internet Resources, at the end of the chapter). Seating errors of as little as 0.01 mm will manifest themselves critically on the imaging plane: the camera will not be in sharp conjugate focus with the eyepieces, and the microscope will not be parfocal across its range of objectives, particularly when using high-magnification objectives having a high NA and consequently very small depth of focus.

The imperial units for camera sensor sizes (1/2″, 1/3″ and 2/3″) are anachronisms from earlier times coupling TV and cine cameras to microscopes via the C-mount. For example, the 1/2″ dimension refers not to the actual size of the sensor but to the outer diameter of the camera tube. The actual sensor was 8 mm across its diagonal. The consequence is that a 1/2″ chip requires a 0.5x reducer to fit most of the primary image onto the faceplate of the CCD sensor. With a 1x microscope coupler (i.e. without any optical elements), the diagonal of the digital image will be 8/25 mm, or a third of the field of view covering the image seen down the eyepieces. Using a 0.5x coupler will ensure that more of the visual field of view is captured digitally, but it will also lower the sampling resolution of the specimen in the final image because the primary image is reduced in size onto the same size pixels in the camera sensor.

Unless provision is made to lock the SLR viewfinder mirror out of the way, its movement upon releasing the shutter will cause unacceptable vibration and camera shake. Particularly at higher magnifications, camera vibration will be evident. To allow this to settle down, exposures will have to be over 3 seconds. A live-view function on the camera allows a real-time image to be displayed on the camera LCD screen or (with appropriate software) on the computer monitor. It is also useful to have remote control of the camera shutter

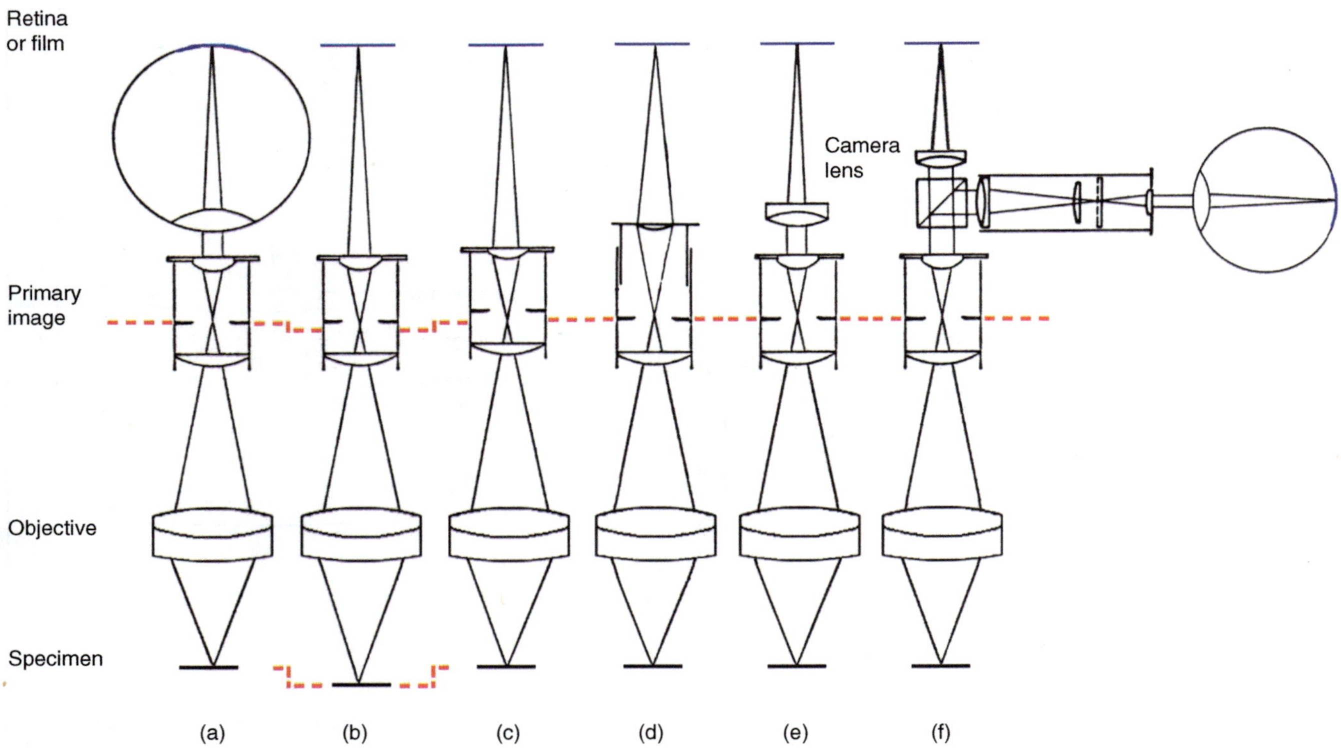

Figure 31.2 **Methods of forming a real image**

(a) *How the eye forms an image with the microscope in normal use. The rays exiting the eyepiece are parallel and thus do not form a real image. The cornea and lens of the eye focus the rays on the retina to form a virtual image.*
(b) *The microscope is refocused to increase the specimen-to-objective distance so that the primary image is lowered and falls below the proper location at the focal plane of the eyepiece. The rays leaving the eyepiece now converge to form a real image, and film or a CCD faceplate can be arranged to be placed at this focal plane, but spherical aberration is no longer properly corrected, and the performance of the objective will be impaired. The image will be unsharp to a greater or lesser degree because of the spherical aberration.*
(c) *The microscope is not refocused, so the two conjugates – specimen and primary image – are in their correct positions with respect to one another, but instead the eyepiece is raised to increase the primary image-to-eye lens distance and so converge the rays exiting the eyepiece to form a real image. This method is preferred to (b) because the specimen-to-objective distance is not disturbed, so spherical aberration is still properly corrected. However, objectives of differing focal lengths do not remain parfocal with one another.*
(d) *A special projection lens or 'photographic eyepiece' sometimes fitted with a means of varying the primary image-to-eye lens distance and designed to form a real image at the film plane utilising the normal settings of the microscope is employed.*
(e) *A normal eyepiece is used together with a special relay lens to converge the light rays onto the film plane. This is optically identical to using the eye in (a) for visual observation.*
(f) *As above in (e) but with an extra relay lens to provide a visual image conjugate with the film plane to allow focusing of the film or CCD camera.*

Source: Adapted from Bracegridle and Bradbury 1995 with permission from Taylor & Francis Group and adapted from Lacey 1989 with permission from Oxford University Press.

to prevent or minimise vibrations, and this can usually be done via a USB connection to the computer. For vibration-free images at high magnification, look for electronic shuttering that circumvents using a mechanical mirror (such as Canon's Silent Live View, designed to prevent shutter noise scaring wildlife).

Some trinocular photo-heads have 'all or nothing' 100% splits to the eyepieces or camera. Other designs permit either an 80-20% or 50-50% split between both. With these latter designs, it is possible for light from overhead room lighting to enter the eyepieces (see Figure 12.14 as an example) and so be recorded by the film or CCD camera. With a digital camera, it is easy to rectify this, since the lights will show up as an unnatural regular pattern on the monitor often towards the top edge of the screen. The image will disappear when the photo-head is switched to 100% camera option, cutting out the eyepieces.

It is helpful to have a suitable camera driver file for your camera. This can often be downloaded from the Internet from the manufacturer's website. Control of the camera from the computer is advantageous because it offers preview of the image, control of ISO, colour temperature and exposure control, as well as triggering the camera shutter (to reduce vibration) and convenient image capture without the need to transfer from the camera card. Breeze Systems, Cam4you and digiCamControl all offer related software; details are given at the end of the chapter.

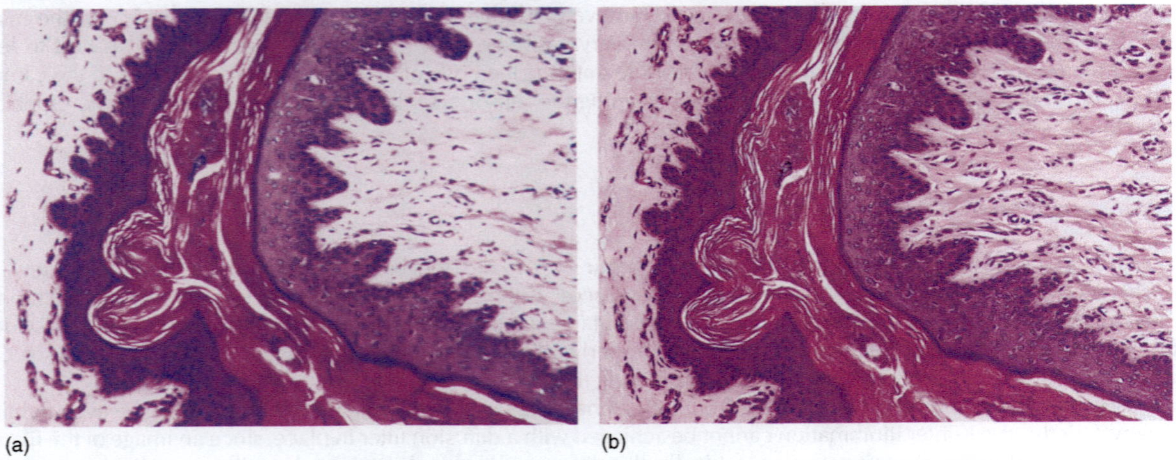

Figure 31.3 Spherical aberration in the recorded image *Left-hand panel (a): Leitz Dialux – trinocular head, 2/3″ chip CCD with 0.5x:1/2 Zoll reduction lens. 25:1 NA 0.5 plan-Fluotar semi-apo. objective. 0.5 ND filter – manual adjustment of white balance on Leica QWin image capture software. Microscope stage adjusted downwards within the correct focus of the objective to focus camera as sharply as possible on the monitor. This is situation (b) shown in Figure 31.2. Microscope set correctly for Köhler illumination, spherical aberration present. Right-hand panel (b): Olympus BX-51 – Trinocular head, Olympus DP25 2/3″ chip CCD with U-TV 0.63 reduction lens; UPlanFIN 20x NA 0.5 semi-apo. objective. Automatic adjustment of white balance on Olympus Cell D software. No adjustment of microscope stage required for correct focus.*

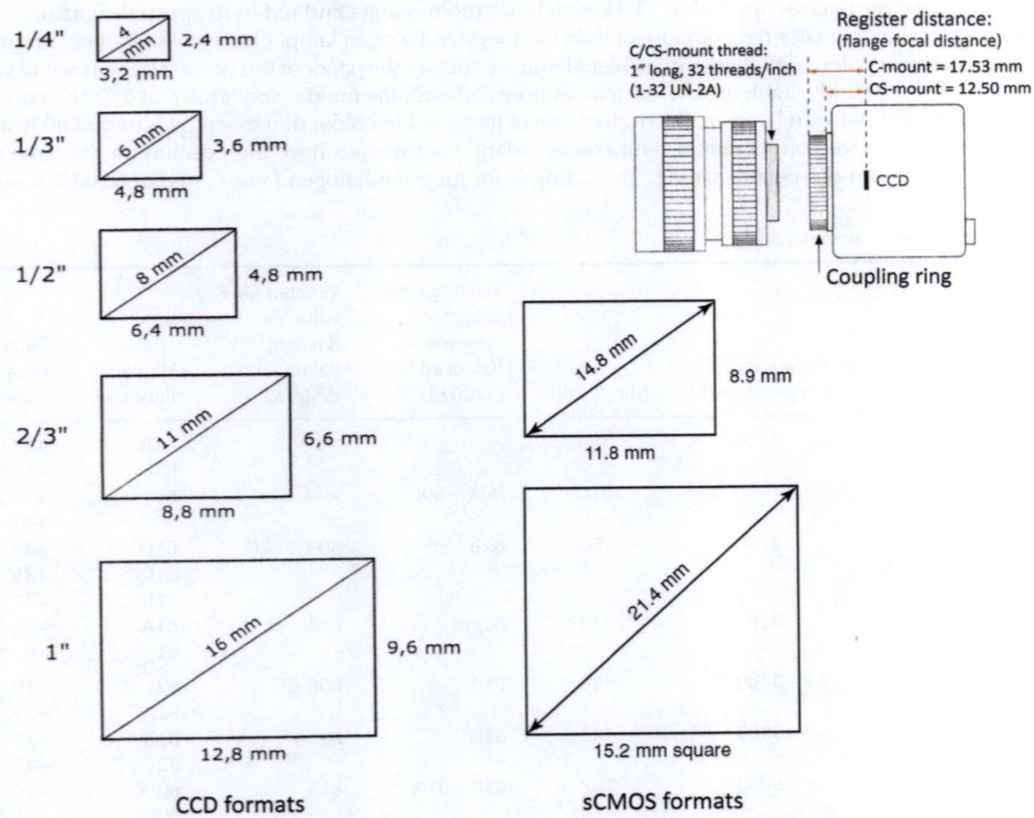

Figure 31.4 Register distance and the C-mount *The register distance is the length between the lens mounting flange (the metal ring on the camera and the rear of the lens) and the image pane. It is also called the flange focal distance. C-mount couplings have a male thread, which mates with a female thread on the camera; the register distance is 17.526 millimetres. The CS-mount is identical to the C-mount but has a smaller register distance of 12.50 mm to accommodate smaller (< 1/2″) chip formats.*

The diameter of the object field of view at the primary image plane is: FN/Mag_{obj} × Q, where FN is the Field of View number, Mag_{obj} is the objective magnification, and Q the magnification of any camera coupler, tube-lens or intermediate magnification. Thus Q will be 0.6x with a 0.6x reduction coupler. A useful rule of thumb is to convert the chip size fraction to a decimal to choose the correct adaptor for coupling. For a 1/2″ chip use a 0.5x adaptor; for a 2/3″ chip use a 0.6x adaptor. Source: Goodwin 2007. Reproduced with permission from Elsevier.

Finally, please be aware that, once fitted, it is best not to remove or change over the CCD or sCMOS camera on the microscope. While these SLR cameras or dedicated scientific cameras can very easily be removed from the microscope, it is best to leave them mounted on the microscope port to prevent dust and debris from entering the body of the microscope or adhering to the camera image sensor window (an artifact promoted by static electricity). Cleaning the sensor faceplate window is a thankless task, especially when the goal is to completely remove every particle of dust.

31.4 Illuminating the Specimen

Film and digital camera sensors are much more sensitive than our eyes to differences in intensity across the field of view. Photomicrography, whether by film or digital means, cannot 'make allowances' as the eye naturally does, for variations in illumination intensity across the field of view. This may be because of a poorly centred filament, poorly centred fluorescence light source or a condenser that is off-centre. It is therefore important that the microscope is properly adjusted for Köhler illumination (Figure 31.1 and see Chapter 9, Section 9.11) at the outset to avoid wasted time and disappointing results.

Some microscopes have a diffusion filter of finely ground or etched glass in their illumination train to help give even lighting across the field of view. Strictly, true Köhler illumination cannot be achieved with a diffusion filter in place, since an image of the filament is no longer present in the aperture set of conjugate planes. Ideally, this diffusion filter can be removed at will (generally the control to do this is found in the base of the microscope stand), although it may sometimes be fixed in place. A diffusion filter is primarily of importance when using low-magnification, low NA, objectives with a large depth of field for photomicrography. With these objectives but without a diffusion filter, it is possible to record an out-of-focus image of the filament in the field set of conjugate planes even when the microscope is properly adjusted. The diffusion filter prevents this, as well as any specks of dirt or uneven illumination showing in the final image. It may be necessary to defocus the condenser slightly, after setting it correctly for Köhler illumination, to avoid including the granular structure of the diffusion filter in the image.

What we perceive as 'white light' is not actually white. The midday sun has a greater proportion of blue light than the filament of a tungsten-halogen lamp used for microscopy (Table 31.1), which has more orange and red in its spectral signature. A domestic light bulb is slightly 'cooler' still, with a greater red component than the tungsten-halogen lamp. Our eye – or, rather, our brain – is adept at compensating unconsciously for colour differences in ambient light: we still see the paper of this printed page as white whether we see it by daylight, electric light or even candle light. In absolute temperature (kelvins), the midday sun is rated at 5500 K, a domestic light bulb at around 2800 K and a tungsten-halogen lamp at 3200 K. Because of the possible colour differences, it is important to adjust the supply voltage of the lamp carefully to give correct colour temperature. Many microscopes have the position on the lamp rheostat control marked for photography, as an 'on-off' default setting. This rating is for tungsten-halogen lamps rated at 3200 K. Clearly, if the lamp

Table 31.1 'White' Light and Colour Temperature

Illuminant	Approximate temperature (K)	Mired value	Wratten filter value for tungsten-balanced film (3200 K)	Wratten filter value for daylight-balanced film (5500 K)	Wratten filter code	Filter mired value	Filter colour
Match light	1700	588	–	–	85A	+141	
					85B	+131	
Candle light	1930	518	80A + 80C	–	85	+112	
					85C	+81	
Tungsten filament lamp @ 6 volts	2780	360	82B	80A + 82B	81D	+42	
Domestic tungsten light bulb ≈60 W					81C	+35	
					81B	+27	
Tungsten halogen lamp	**3200**	313	None	80A	81A	+18	
Tungsten filament lamp @ 9 volts					81	+9	
Photoflood lamp	3400	294	81A	80B	82	−10	
					82A	−21	
Daylight (average, midday)	**5500**	182	85B	None	82B	−32	
					82C	−45	
Electronic flash	6000	167	85B + 81A	81A	80D	−56	
					80C	−81	
Daylight (overcast)	8000	125	85A + 81D	81D	80B	−112	
					80A	−131	
CRT screen	9300	108	–	85C			

Although film is rarely used, these filter combinations are still useful because filters can still be used to change the colour temperature of the illumination regardless of whether a film camera or electronic sensor is used to record the image. Because colour correction is not a linear process, the reciprocal of the temperature in kelvins (K) is assigned to each illumination source. The term mired is short for micro-reciprocal degree, and is 1×10^6/K. Subtract the mired value of the light source from that of the film. If the answer is positive, you should use a red colour-correction filter; if negative, use a blue colour-correction filter. Thus to increase the colour temperature, use a blue filter; to decrease the colour temperature, use a red filter. Example: to use daylight film with a tungsten-halogen lamp, mired value of film = 182; mired value of light source = 313. 182 − 313 = −131, so use an 80A red filter. Filter series 80 and 85 are called colour-correction filters, and series 81 and 82 are called light-balancing filters by Kodak. Where no filter combinations are suggested, this is because the exposure factor and filter density is too excessive for the film to record the colour accurately.

Table 31.2 Neutral-Density Filter Values

Colour	Optical density (OD) value	Percentage transmission	Change in exposure relative to OD = 0.3	*f*-stop reduction (stops)
Black/opaque	ND 4.0	0.01%	x13.3	13 + 1/3
	ND 3.0	0.1%	x10	10
	ND 2.0	1%	x6.7	6 + 2/3
	ND 1.0	10%	x3.3	3 + 1/3
	ND 0.9	13%	x3	3
	ND 0.8	16%	x2.6	2 + 2/3
	ND 0.7	20%	x2.3	2 + 1/3
	ND 0.6	25%	x2	2
	ND 0.5	32%	x1.4	1 + 2/3
	ND 0.4	40%	x1.3	1 + 1/3
	ND 0.3	50%	x1	1
	ND 0.2	63%	x0.67	2/3
	ND 0.1	80%	x0.33	1/3
	ND 0.05	90%	x0.14	1/6
Clear	–	100%	–	0

Neutral-density filters have values expressed as percentage transmission and/or optical density. The optical density (OD) is a logarithmic, rather than arithmetic, scale. This has the advantage that OD values of two filters used in combination are added together. Generally, only five different filters are required: one OD 0.3 (50%), two OD 0.6 (25%) and one OD 0.9 (13%) to perform any required reduction in the intensity of illumination.

The OD value of a filter is calculated as the intensity of the source in the numerator (I_o) divided by the intensity of the attenuated light after passing through the filter (I_T). Therefore, $OD = \log I_0/I_T$, or $\log_{10}(1/T)$, or $T = 10^{-OD}$. Each 0.3 increment cuts the light intensity by half ($\approx \log_{10} 2$). The OD value is defined as the $\log_{10}$ (to base 10) of the opacitance (or degree of absorption) of a filter. The opacitance is the reciprocal of (and inversely proportional to) the transmittance.

As an example, a filter with a transmittance of 50%, or 0.5 times the original light intensity has an opacitance of 2. The log base 10 of 2 is approximately 0.3, which is the nominal density value. Thus two neutral-density filters of OD = 0.3 have a combined density value of 0.6. However, their combined transmittance would be found by multiplying $0.5 \times 0.5 = 0.25$, or 25%, of the original light intensity.

intensity is too high for comfortable viewing, you must use neutral-density filters to adjust the intensity of the illumination, rather than turning down the rheostat. Neutral-density filters have values expressed as percentage transmission and/or optical density. The optical density (OD) is a logarithmic, rather than arithmetic, scale. This has the advantage that OD values (see Table 31.2) of two filters used in combination are added together. Generally, only five different filters are required: one OD 0.3 (50%), two OD 0.6 (25%) and one OD 0.9 (13%) to perform any required reduction in the intensity of illumination. White LEDs are increasingly being used as an alternative to incandescent sources. These can be varied in intensity without noticeable colour shifts, preventing the need for neutral-density filters, however they typically produce 'colder' light, which has a significant effect on the colour balance.

31.5 Flat-Field Correction

It is a very quick and simple procedure, using modern digital image acquisition software, to subtract a background image of the shading (otherwise known as a flat-field correction) caused by the illumination and so ensure an evenly illuminated field (see Figure 31.5). This

Figure 31.5 ***Flat field correction*** *An original image (a) and the inhomogeneous background (b) have been processed as details in the text to give a flat field image (c) with a homogeneous background.*

is done by recording an image with the microscope properly adjusted and with the sample removed or else with the slide and subject *very much* out of focus so that only the shading due to the illumination is recorded. There is usually a button in the exposure setup dialogue box that will do a background subtraction automatically. If not, this can be done according to the protocols given here:

Exercise 31.1 Performing a Flat-Field Correction in Bright-Field Microscopy

1. In ImageJ or Fiji open the original image (image 1) and a blank flat-field image (image 2).
2. Highlight image 2 and select all (Ctrl + A). A yellow ROI should appear around the boundary.
3. Measure the average value (Ctrl + M), and note this down as constant, k1.
4. Open the Calculator Plus function (Process/Calculator Plus or Analyse/Tools/Calc. Plus).
5. Insert image 1 into i1, image 2 into i2 and set k1 to the measured value. Leave k2 = 0.
6. Check the tick-box: Create new window.
7. Use the brightness-contrast control (Image/Adjust/Brightness-Contrast) to make final adjustments.

You can also use the Image Calculator function. See which gives the best result for your images. This is found under Process/Image Calculator.

8. Divide the flat-field image into the original image.
9. Tick both check-boxes; it is important to tick '32-bit (float) result'.
10. Use the brightness-contrast control (Image/Adjust/Brightness-Contrast) to make final adjustments.

For a fluorescence image, take a uniform background of the same spectral characteristics as your fluorophore (Kask et al., 2016). This can either be from a uniformly fluorescent slide or a film made from your fluorescent-conjugated protein or a fluorescent dextran solution. If you allow a drop to sit for 30 minutes on a coverslip and then gently wash off the excess and invert the coverslip on a slide, sufficient fluorescent protein will stick to the glass to form a thin and even fluorescent film.

Exercise 31.2 Performing a Fluorescence Flat-Field Correction

1. Find a homogeneous area of the calibration slide and defocus slightly.
2. Take an image of the homogenous field of view.
3. Take a background image, using the same slide, in a region of no fluorescence.
4. Subtract the background image from the homogenous image to give a background-corrected homogeneous fluorescence image. This can be done in ImageJ or Fiji with the command Process/Image calculator, as above
5. Take an image of your sample.
6. Likewise take a background image in a region of no fluorescence, and background correct the fluorescence image of your sample, as you did with the flat-field calibration image.
7. Divide the corrected image (6) by the background-corrected image (4) to perform the flat-field correction, using the Image Calculator function.

The methods described above for correcting uneven illumination are prospective methods, relying on a reference image collected simultaneously. Retrospective methods rely solely upon the image itself, and for that reason are regarded as a less reliable means of flat-field correction. Besides the smoothing method given below, a retrospective method called CIDRE (corrected intensity distributions using regularised energy minimisation; Smith et al., 2015) claims to achieve correction quality similar to that of prospective methods.

Another way to achieve even illumination is to use a low-pass Gaussian kernel filter to smooth out the differences in illumination. This has the advantage of working with the original image without capturing a flat-field image, so it can be performed retrospectively. This may be necessary for cleaning up the background when the microscope used to acquire the image or the specimen are no longer available. However, this procedure will alter the original data set, so keep a copy of the original file, and be prepared to declare the result as a manipulated image.

Exercise 31.3 Smoothing the Image Background

1. Convert the colour or monochrome image to an 8-bit greyscale image.
2. Use a Gaussian blur kernel filter in the freeware Gimp or commercial Adobe Photoshop (these instructions are for Adobe Photoshop). The Gaussian Blur command is found in: Filter/Blur/Gaussian Blur.
   Set the radius at 120 to start; some trial and error is required to get the best result.
   Smooth the image until it is featureless but corresponds to the illumination pattern.
3. Invert the smoothed image, and place it as a layer on top of the original. (either Ctrl + I or Colours/Invert)
4. Blend it using the 'hard light' function. (In the Layers window change Normal to Hard Light.)
5. Step 4 will over-correct, so set the opacity of the filter layer to approximately 50%.
6. Apply a contrast expansion, if desired, to correct for the contraction in dynamic range. (Go to Image/Adjustments/Levels.)

Binning is a technique for combining adjacent pixels (typically squares) and treating the accumulated charge as one sampling point to create a single pixel in the image. Many scientific CCD cameras allow hardware binning, which pools photo-electrons from adjacent pixels prior to readout, with consequent reduction in read noise, resolution and digitisation time. Alternatively, binning can be performed retrospectively in software, which doesn't offer the read-noise and speed advantages, but does preserve the full spatial resolution of the raw image. Hardware binning is widely used in real-time low-light imaging to increase the sensitivity of the camera, allowing a shorter exposure time with reduced bandwidth and hence a faster readout rate. Binning also reduces the shot noise in the image, increasing the signal-to-noise ratio. In fact, if you are *not* concerned about utilising the ultimate resolving power of the objective, binning is a really good way to increase the SNR of the image. The other advantage of binning is that it improves the effective pixel well depth and hence the dynamic range of the camera. The trade-off is loss of spatial resolution (Figure 31.6), and images that are successively 'binned' are smaller and thus show a pixelated appearance when displayed or printed at the same size as the original un-binned image. Binning is a good stratagem to consider for live-cell imaging.

There are currently three practical methods of acquiring very dim images. One is to amplify the image using an image intensifier or EM-CCD. The second is to accumulate (integrate) the image over a longer period of time. The third is to bin the image and sacrifice resolving power. Typical integration times can range from a few milliseconds to several seconds. The times are influenced by such

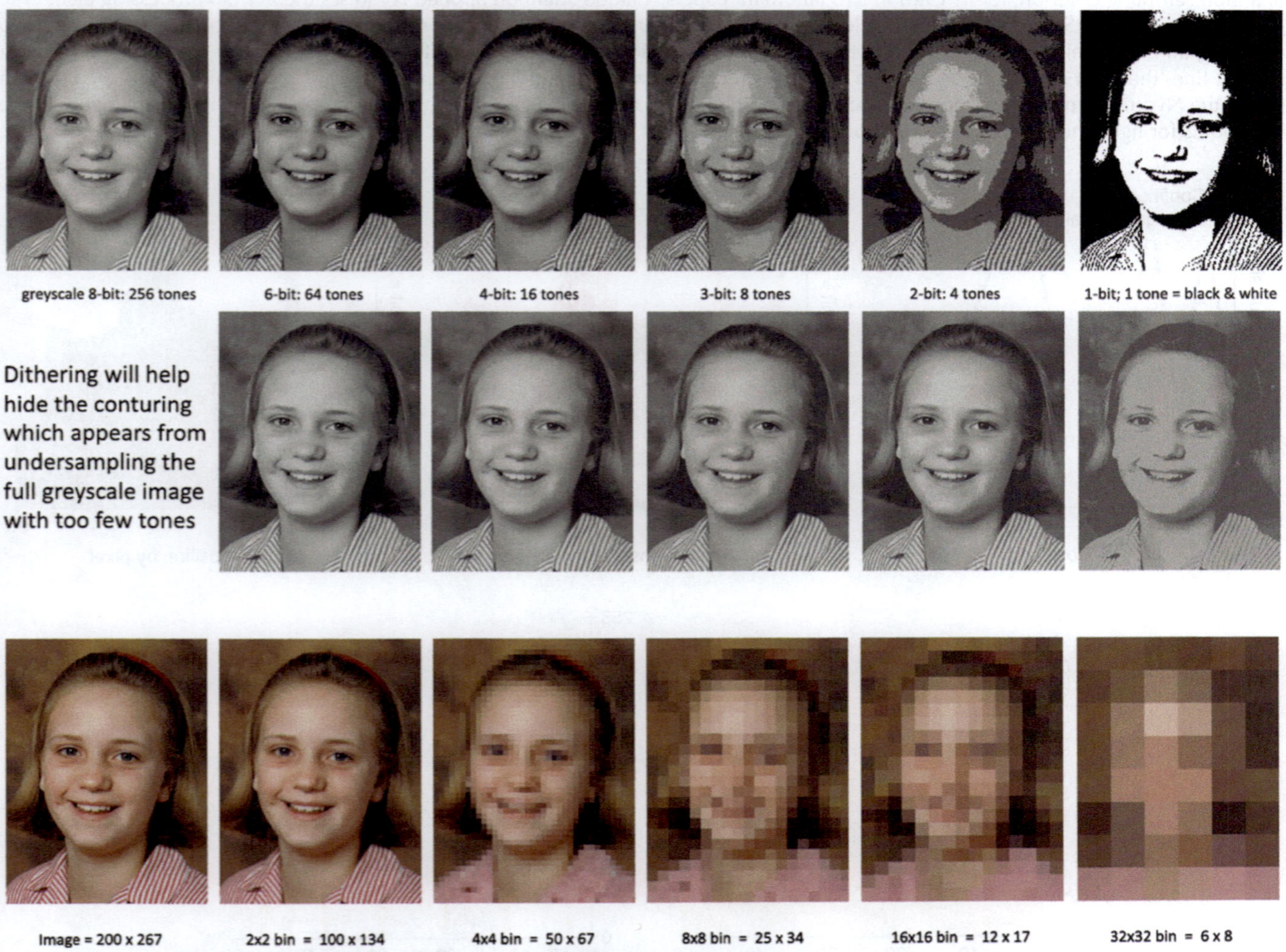

Figure 31.6 ***Binning and bit depth*** *The effect of reducing bit depth and the separate stratagem of binning are shown together in this figure. Compare these images of different bit depths with the tonal scales shown in Figure 30.5. Printers use dithering to converting a greyscale image to binary or a lower bit depth, such that the density of black dots in the new image approximates original bit depth. Similarly, dithering can also create the illusion of colour depth in images when printing with a limited colour palette. Half-toning is similar, using dots, varying either in size or in spacing to simulating continuous tone images. In any analogue-to-digital conversion, dithering is thus an intentionally applied noise to randomise quantisation error arising from undersampling.*

The advantages of binning are brighter images, shorter exposures, faster readout and higher frame rates and a reduction in read noise and Poisson noise, giving a higher SNR and increased dynamic range. The disadvantage, of course, is loss of spatial resolution and the risk of Nyquist undersampling. It is not always easy to see a loss of resolution in biological images, but the effect of binning can be recognised very clearly in these images of the young girl's face. Pull back from the page or screen, and the effect of binning/pixellation is reduced. Source: Reproduced with permission from H. Sanderson.

factors as the susceptibility of the particular dye to photobleaching, the quality of the microscope (especially the NA of the objective) and the sensitivity of the camera. Movements of more than several pixels during integration will cause noticeable blurring of the image.

31.6 Nyquist Sampling

Whenever a continuous wave signal is collected from an analogue system (in this case the microscope) to be presented as a digital output, it is sampled from a continuous stream of data to form a discretely sampled data set (see Figure 31.7). The photon flux incident upon the microscope field of view is sampled by the photo-sites of the electronic sensor, and the light intensity that falls on each photo-site is stored as a number corresponding to an individual pixel in the image (see Figure 30.3). Each pixel is therefore a sampling point of the original signal.

In 1928 the Swedish electrical engineer Harry Nyquist published his paper on transmission of Morse code information pulses via telegraph lines. He specified the Nyquist criterion: the sampling frequency must be greater than twice the bandwidth of the input signal to reconstruct the original input from the sampled data. In microscopy this means that the smallest resolvable detail must be sampled at least twice, be it laterally or, in an optical slice, axially. The sampling criterion should properly be known as the Nyquist-Shannon sampling criterion, for in 1948 in two landmark papers, Claude Shannon applied Nyquist's theorem to processing electronic information. For this, Shannon has been called by some 'the father of the digital age', for the rules of the sampling criterion apply to digitising all communication channels, whether that be transmission of eBay auction site bids or microscopy of cells. Apart from a straight line, the most regular curve that can be reconstructed from a population of sampled points is a sinusoid. This is the basis on which the Nyquist sampling criterion[2] works: creating a digital sinusoid curve which faithfully represents the original analogue signal. This works for light and sound, which both propagate as sinusoidal waves.

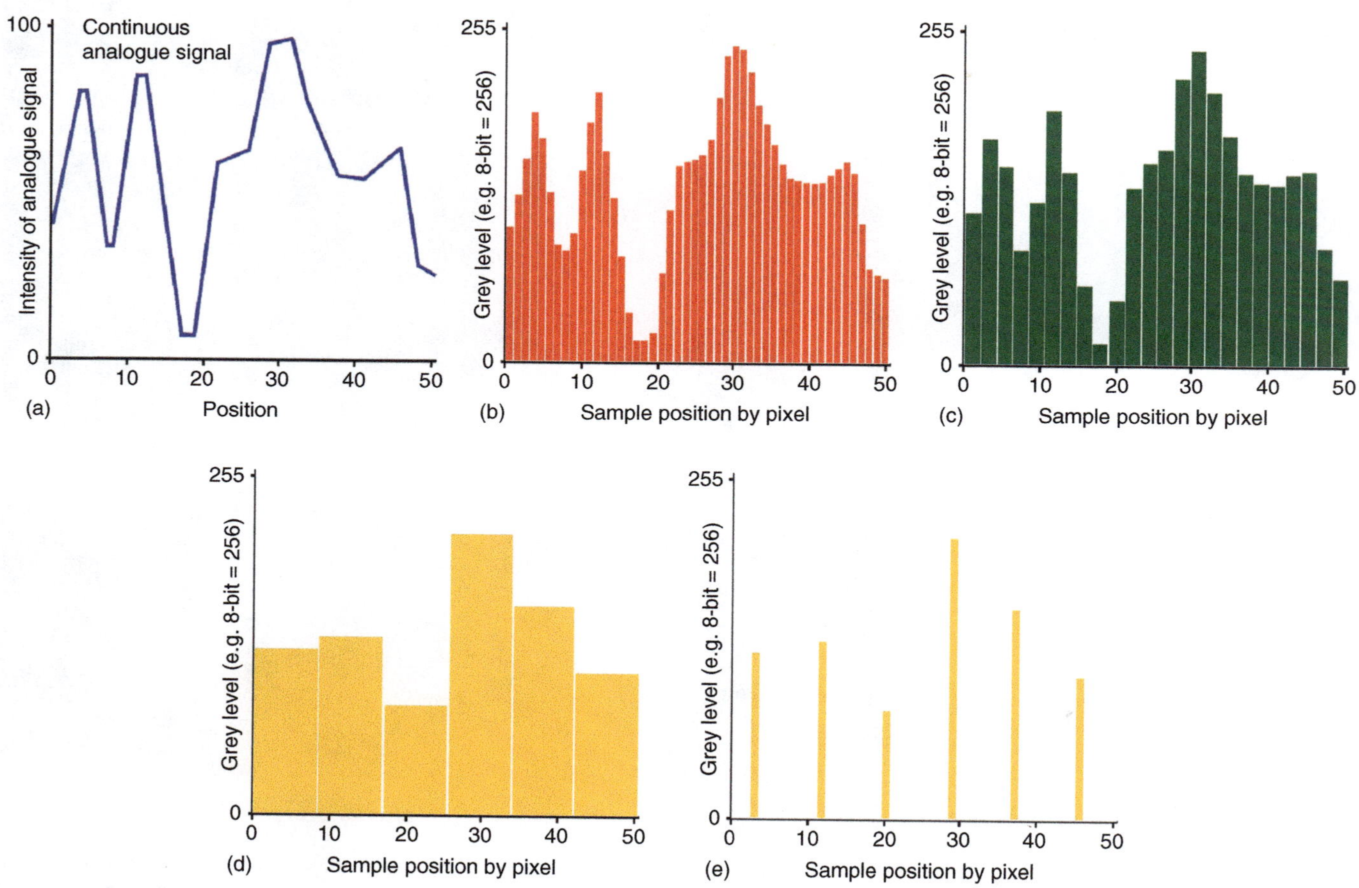

Figure 31.7 **Nyquist sampling-1** *An analogue signal (a) is continuous in some form: in tone, hue, temporally. When digitised, it must be converted to discrete integers with the concomitant risk of losing data or else introducing spurious features by undersampling, or bleaching by over-sampling. In (b) the signal has been oversampled; although the resolved fine details are recorded faithfully, and the position of sub-resolution features can be pinpointed with accuracy, oversampling will lead to over-exposure, bleaching, loss of contrast, reduced dynamic range and possibly increased noise in the image. There is little justification for over-sampling above the Nyquist limit. The green histogram (c) shows the result of sampling correctly according to Nyquist's criterion: a faithful representation of the analogue signal. In (d) gross under-sampling causes a loss of high spatial frequency data while simultaneously introducing spurious lower frequency data, which doesn't actually exist. The graph in (d) has been redrawn as panel (e) to show the effect more clearly. Source: Adapted from Hazlewood et al. 2007 with permission from Springer.*

An aliased signal provides a poor representation of the analogue signal. When aliasing happens, the original analogue signal cannot be correctly reconstructed by the digital data set (Box 31.1). Continuous signals of differing frequency become indistinguishable from one another (aliases of one another) when sampled. Aliasing can be seen in images as moiré fringing displayed as a false period in images with regular features, induced by sampling at less than two times the repetitive pattern frequency. In audio, aliasing introduces spurious beats into the original signal. Aliasing also occurs where the display device has a periodic structure, such as a CCD or TV monitor with geometric pixels or raster lines. Under-sampling leads to progressively coarser images, and pixel 'blockiness' becomes evident with a corresponding loss of detail and, ultimately, contrast.

Box 31.1 Aliasing

Photons transmitted through, reflected off or emitted from a specimen comprise a continuous analogue signal. When these photons are transformed into a digital electronic signal, they can only be represented as discrete intensities and/or colours within pixels. This means that the analogue signal must be sampled, and the digital signal is only a representation. How good that representation is will depend on whether or not you have taken Nyquist's sampling theorem into account. This is explained in the text.

In order to record the image adequately, we must know the resolving power of the microscope, and sample the image so that (a) no high-frequency fine detail is lost nor (b) is spurious information encoded into the digital image by the sampling process. Both these events will occur through under-sampling, which leads to aliasing. The most well-known example of aliasing is the jagged edges – especially along diagonal lines – seen in under-sampled and pixelated images. Ultimately, contrast will be lost as well as resolved detail.

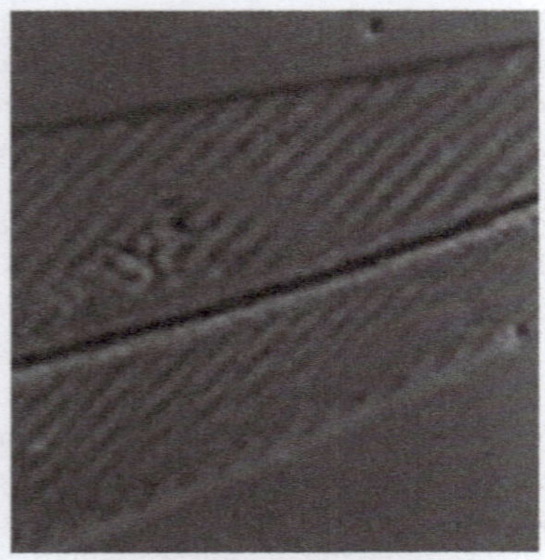

An under-sampled aliased image

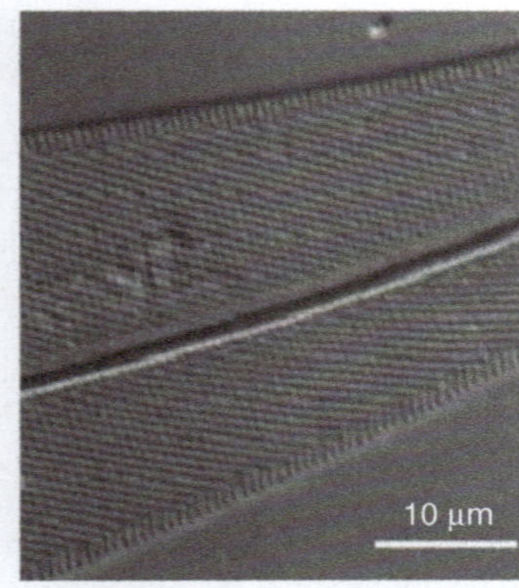

Correct sampling showing all the detail resolved by the objective

Original size, pixel size = 35 µm

4x zoom, pixel size = 141 µm

Correct display size of the 2x zoom image, pixel size = 70 µm at half-size to avoid aliasing & pixellation

8x zoom, pixel size = 282 µm

32x zoom, pixel size = 1.10 mm

Correct display size of the 4x zoom image, pixel size = 141 µm, at quarter-size

An aliased signal provides a poor representation of the analogue signal. When aliasing happens, the original analogue signal cannot be correctly reconstructed by the digital data set.

Continuous signals of differing frequency become indistinguishable from one another (aliases of one another) when sampled.

Aliasing can be seen as moiré fringing displayed as a false period in images with regular features. It also occurs where the display device has a periodic structure, such as CCD or TV monitor with geometric pixels or raster lines.

Since Nyquist's sampling criterion also holds good for temporal sampling as well as sampling in the x,y,z spatial dimensions, this is the reason why aliasing can be seen in old films, which are under-sampled when they are digitised or displayed on TV. A well-known example is the appearance of spoked carriage wheels turning differently, or in reverse, behind a galloping horse. The number of frames per second by which the film is sampled does not match the speed of the wheel, or else the rotating wheel is illuminated by flickering light of a different frequency to the true rotation.

Since under-sampling causes loss of true information and addition of false information into the digital image, why not routinely over-sample? Over-sampling necessitates smaller pixels. If the pixels sampling the image are too small, the signal transferred will be dimmer, of lower dynamic range and the contrast of small features will be lost in the increased shot noise of the system. Also, over-sampling will cause increased bleaching and photo-toxicity of the specimen. However, sampling by the extra pixels does not theoretically contribute to the spatial resolution but may help improve the accuracy of feature measurements taken from a digital image. There may be no choice but to under-sample, especially in live-cell imaging, where the prime consideration is to collect a meaningful signal regardless of ultimate resolution of detail (which may not be possible to see because of noise and/or aberrations inherent in the image). Under-sampling at spatial frequencies below the Nyquist criterion – using large pixels (e.g. by 'binning' the signal) will lose resolving power but will give a brighter signal and will mean less exposure, which in turn leads to less photobleaching and photo-toxicity.

The *minimum* sampling required by Nyquist's theorem is two pixels per resolvable unit seen in the microscope (see Table 31.3). This is the basis of most macros used by confocal microscopes and is generally inadequate. Where structures lie perfectly orthogonal to the x,y direction of the sampling pixel raster on the CCD faceplate, then Nyquist's sampling criterion of a pixel size of half the minimum resolved distance holds good. However, biological structures are very rarely laid down geometrically in the image! Because diagonal edges must be reproduced by square pixels oriented North-South and East-West along their dimensions and to adequately capture details lying along these diagonals, Nyquist's requirement[3] becomes 2.8, and there is good argument for using 3, 4 or 5 pixels per minimum resolved unit. See Figure 31.8 for a further explanation. Providing the sample is not bleached or damaged, over-sampling will help when samples exhibit only low contrast against the background, but it will never compensate for a lack of resolution of fine detail in the original image.

Because the axial resolution of the microscope is approximately three times greater (or worse) than the lateral dimensions, it is important to consider how you will sample in the axial direction when collecting Z-stacks. Ideally the smallest resolvable feature in the image in the axial direction should also be sampled by 4–5 pixels across its width. You therefore need to know the resolving power of the microscope objective that you are using. Over-sampling in Z may be desirable in this case because interpolation during image alignment can degrade resolution. If fading is a problem, reducing the lateral resolution in order to sample adequately in Z is better than under-sampling axially. Over-sampling is also required if you intend to use deconvolution (see Chapter 21) to increase resolution of fine detail; the final deconvolved image must also meet the Nyquist sampling criterion. Values of 3–8 pixels per resolvable element have been cited, but Nyquist sampling with 4 pixels per resolvable element is a good starting point.

Since Nyquist's sampling criterion also holds good for temporal sampling as well as sampling in the x,y,z spatial dimensions, aliasing can be seen in old films, which are under-sampled when they are digitised or displayed on TV. A well-known example is the appearance of spoked carriage wheels turning differently, or revolving in reverse, behind a galloping horse. The number of frames per second which samples the film does not match the speed of the wheel, or the rotating wheel is illuminated by flickering light of a different frequency to the true rotation. The human eye will only see about 24 images per second; hence the standard 30 images per second will blend into continual movement. You must, therefore, also consider the frame rate at which you sample your live-cell movies, and ideally you should not sample at less than 24–30 frames per second. If a movie is played back at fewer than 15 frames a second, we don't perceive it as continuous.

Strictly you should never evaluate *raw* confocal or widefield data. It should always be deconvolved first[4]. Low-intensity signals contain a lot of high-frequency Poisson, or shot, noise, which can be erroneously sampled and digitised into non-existent single-pixel 'features' that could never have been resolved by the microscope. The effect of deconvolving confocal data – that is, averaging the signal over the number of voxels (say $30 \approx 3 \times 3 \times 3$) required to sample the signal at the frequency demanded by the Nyquist criterion – is the same as Kalman average scanning by 30 times to improve the SNR. The difference, of course, is that deconvolution doesn't take so long or damage the specimen as much. However, when confocal data is deconvolved, the contrast drops, and the image may not look as sharp. This is not a retrograde step: these sharp features are the noise artifacts referred to above, which require removal from the image. The contrast can always be raised, using the display histogram, to match the display characteristics of the monitor. That said, confocal data tends not to be deconvolved in practice because it has already been 'sampled' with the pinhole to produce an optical section.

31.7 Matching the Microscope Magnification to the CCD Pixel Size

The practical application of Nyquist's sampling criterion means that you should select a pixel size in the final image that is at least half and preferably 1/3, 1/4 or sometimes even smaller than the size of the smallest resolvable detail required (which may or may not equate to the minimum distance capable of being resolved by the microscope objective). When acquisition speed and photobleaching are not constraints that need to be taken into account, it may be useful to over-sample (i.e. project the Airy disc over 6–8 pixels rather than the minimum 4 pixels). This will allow for images that can then withstand some degree of cropping and enlargement, as well as lead to better acquisition of low-contrast features in the image.

With a point-scanning confocal, it is easy to use the zoom control to adjust the raster scan to create smaller pixels by reducing the sampling interval of the raster sweep of the point illumination (see Table 31.3). For example, the 40x/NA 1.4 objective has a resolving limit of 0.177 μm with 488 nm illumination. The calculated pixel size in the image must be 89 nm^2 or smaller for correct Nyquist sampling. The pixel value determined by the LSM 700 software is 90 nm^2 for Nyquist/2 and for 71 nm^2 Nyquist/3. The zoom factor for Nyquist/2 sampling with 90 nm^2 pixels is ≥1.7x for a 1024×1024 raster, while a zoom factor ≥2.1x applies to a raster of 1024×1024

Table 31.3 *Typical Zoom Factors Required for Nyquist Sampling: Confocal and Magnification: Widefield*

Selected Objective	Optical index (NA/Mag ratio)	Refractive index, n	Lateral resolving power x,y (µm) @ 488 nm [$d_{x,y} = 0.61\ \lambda/NA$]	Minimum pixel size x,y (µm) on point-scanning confocal PMT @ Nyquist/2 [$=d_{x,y}/2$]	Raster scan size when minimum pixel size selected	Typical value for optimum zoom setting on a point-scanning confocal with 512 x 512 raster @ 488 nm illumination	Typical value for optimum zoom setting on a point-scanning confocal with 1024×1024 raster @ 488 nm illumination	Depth of Field Z (µm) @ 488 nm [$d_z = 2n\lambda/(NA)^2$]	Calculated Z-sampling step (µm) to sample on a point-scanning confocal @ Nyquist/2 [$=d_z/2$]	Actual Z-sampling step (µm) to sample on an LSM 700 point-scanning confocal @ Nyquist/2 [$=d_z/2$]	Overall magnification for a widefield microscope using a camera with 6.45 µm² pixels @ Nyquist/2	Extra magnification changer required to Mag_{obj}
10x/NA 0.30 dry	30	1.0 = air	0.99	0.43	1348×1348	≥2.6 x	≥1.4 x	10.84	5.42	11.09	≥13 x	≥1.3 x
10x/NA 0.45 dry	45	1.0 = air	0.66	0.32	2020×2020	≥3.9 x	≥2.0 x	4.82	2.41	7.29	≥20 x	≥2.0 x
20x/NA 0.8 dry	40	1.0 = air	0.37	0.16	1020×1020	≥4.0 x	≥2.0 x	1.53	0.76	2.05	≥35 x	≥1.8 x
25x/NA 0.8 multi-imm	32	1.33 = water	0.37	0.16	816×816	≥3.2 x	≥1.6 x	2.03	1.02	2.36	≥35 x	≥1.4 x
		1.47 = glycerol						2.24	1.12			
		1.518 = oil						2.31	1.16			
40x/NA 1.30 oil	32.5	1.518 = oil	0.23	0.10	1660×1660	≥3.1 x	≥1.5 x	0.88	0.44	0.96	≥56 x	≥1.4 x
40x/NA 1.40 oil	35	1.518 = oil	0.21	0.09	1788×1788	≥3.5 x	≥1.7 x	0.76	0.38	0.89	≥61 x	≥1.6 x
63x/NA 1.22 water	19.4	1.33 = water	0.24	0.10	972×972	≥2.0 x	≥1.0 x	0.87	0.44	059	≥54 x	≥1.2 x
63x/NA 1.40 oil	22.2	1.518 = oil	0.21	0.09	596×596	≥2.2 x	≥1.1 x	0.76	0.38	0.57	≥61 x	≥1.2 x
100x/NA 1.30 oil	13	1.518 = oil	0.23	0.10	288×288	≥1.2 x	≥0.6 x	0.88	0.44	0.47	≥56 x	N

The software and electronic control in hardware of the galvanometer mirrors means that zoom factors can differ between point-scanning confocal microscopes offered by different manufacturers. These values have been generated using the Zeiss LSM 700 confocal in the author's core imaging facility. Zoom factors for a standard 512×512 frame size are given. For larger pixels (e.g. 256×256) increase the size required, in proportion. For smaller pixels (e.g. 2048×2048) decrease the zoom size. Note that the 100x/NA 1.4 objective requires a zoom of 0.6x for a raster scan of 1024×1024 pixels. This is equivalent to using a 60x/NA 1.4 objective; there is no benefit to having a 100x objective installed on a single-point-scanning confocal, as explained in Chapter 18, Section 18.10.

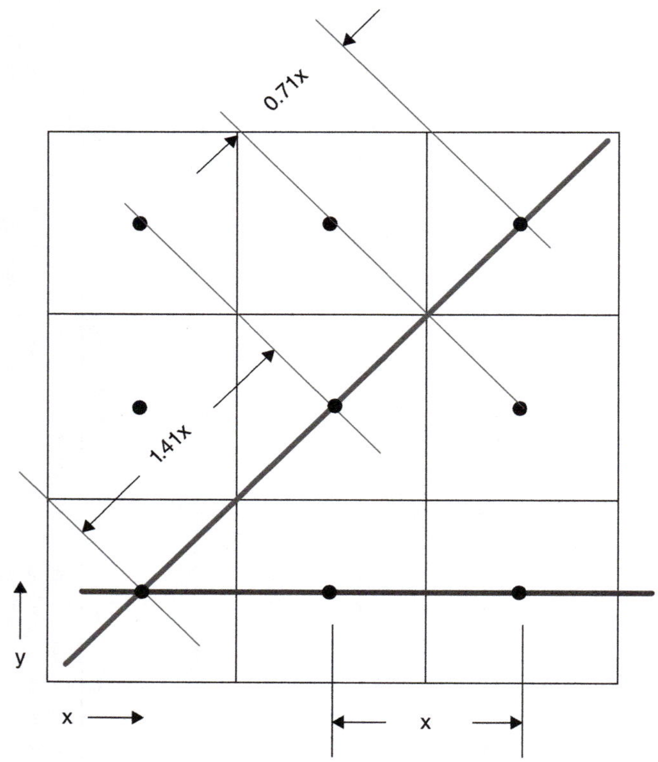

Figure 31.8 **Nyquist sampling-2** *The 3 × 3 array of squares represents a small raster and the central dots the sampling points within each square. Along the diagonal the distance between each sampling point is 1.41x. A common assumption is that 1.41x is too big to sample a structure (say a fibre) that extends diagonally at 45° across the field of view. However, this is generally not too far apart to sample properly, as might be thought, because structure extends laterally in two dimensions across neighbouring squares and the sampling points of the adjacent diagonal rows of pixels actually sample at 0.71x of the x- or y-raster pitch. However, to be certain of adequate sampling some workers use Nyquist/(2 × 1.41), where 2 × 1.41 = 2.82 ≈ 3, so use Nyquist/3 for sampling. Source: Pawley 2006. Reproduced with permission from Springer.*

for a sampling criterion of Nyquist/3. Different values may occur between different instruments, where the zoom function may differ slightly. The LSM 710 requires a zoom factor ≥2.2x to satisfy Nyquist/2 and a zoom ≥2.8x for Nyquist/3 when applied to a raster of 1024 × 1024 pixels (see also Section 18.7 and Box 18.4).

With a spinning disc confocal or widefield microscope, the situation is slightly more complex since the overall magnification must be adjusted to the photo-site (pixel) size of the CCD to ensure that all the resolved detail is sampled adequately. This is done as follows:

1. The camera manufacturer's specification sheet will give you the pixel pitch (P, the size of the pixels) of your camera. This usually ranges from 4 to 16 μm.
2. Multiply this value P by any binning factor. For 2 × 2 binning, multiply P by 2.
3. Calculate the minimum resolved distance, $d = 0.61\lambda_{em}/NA$.
4. Note the magnification of the objective used. If any intermediate magnification changer is used, multiply the Mag_{obj} by this factor to give overall magnification, M.
5. The overall magnification required is M = 2.P/d. Where a sampling frequency of Nyquist/3 is used instead of the minimum Nyquist/2, the formula becomes: M = 3.P/d, and so on.

Since the pixel pitch of the Sony ICX285 CCD camera is 6.45 μm, using a wavelength of illumination (λ) of 0.488 mm with no binning, the magnification required (for Nyquist/2) for the 63x/NA 1.4 objective is $M = 2 \times 6.45/0.213 \geq 60.6$x.

In this case, the 63x/NA 1.4 objective has tolerable magnification so that the image is spread over sufficient sensor 'pixels' to ensure good sampling (Figure 31.9). If, however, a more stringent sampling of N/3 is applied, the magnification required is ≥90.8x. In this case, a 1.6x magnifier will be required with the 60x objective, or it will have to be changed for a 100x objective.

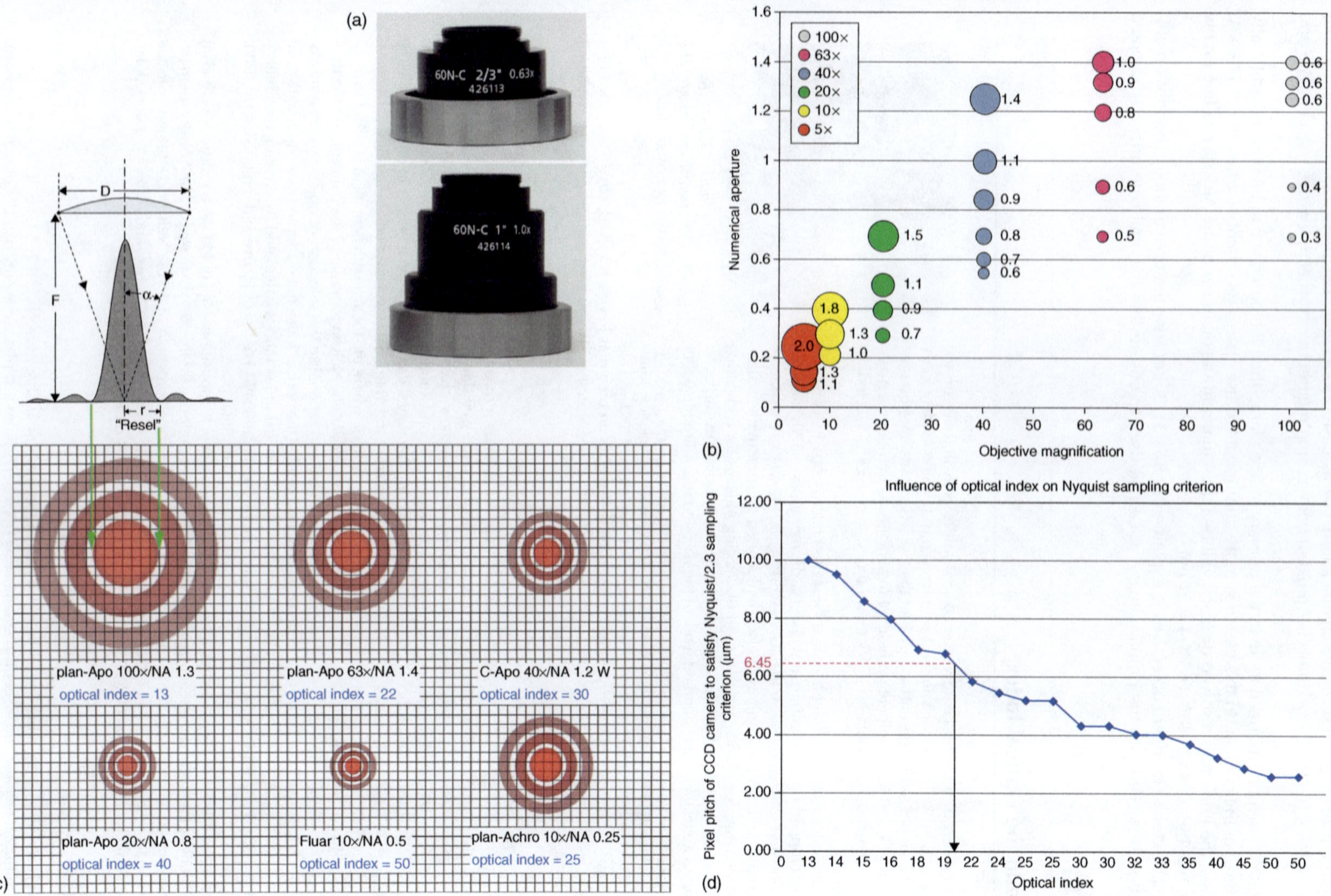

Figure 31.9 **Nyquist sampling-3** *The picture (a) in the top-mid panel is a C-mount coupler incorporating a 0.63x magnification reduction lens (compared to a 1x coupler, a bare tube with no lens), which is required to reduce the size of the Airy disc at the sensor plane when over-sampling would otherwise occur. The reverse situation, use of a magnifying lens incorporated into the C-mount coupler, or elsewhere in the imaging train, is required to prevent under-sampling. The top right-hand panel (b) shows the optimal C-mount magnification factors required for objectives of differing magnification and numerical aperture when used with a CCD camera having 6–7 μm square pixels, assuming Nyquist's sampling criterion N/2.3, or 2.3 pixels per resel (i.e. minimum resolved distance, elsewhere referred to as 'd' in this book). The lower-left panel (c) shows a raster array of 6.45 μm square pixels with examples of Airy patterns from three commonly used objectives superimposed upon the CCD faceplate.*

The Airy disc radius (i.e. the resel distance) of the diffraction spot for a 40x/NA 1.2 objective, with excitation illumination of 488 nm, is: $40 \times (0.61 \times 0.488)/1.2 = 9.92$ μm.

This is under-sampled by the 6.45 μm square camera pixels ($9.92/6.45 = 1.5$x) so a 1.8x coupler ($1.3/2.3 = 1.78$) is required for adequate sampling at a criterion of N/2.3 times. Put another way, for a pixel pitch of 6.45 μm, the objective selected must have an optical index of **≤20**, *for the image to be adequately sampled, as shown in (d) in the lower-right panel. With high-magnification objectives (e.g. 100x/NA 1.3) having a low optical index and correspondingly large Airy disc (23 μm), it is possible to bin 2 × 2 (pixel pitch now 12.9 μm) and get away with only just under-sampling at N/2. Source: (b) Watkins and St. Croix 2013. Reproduced with permission from John Wiley & Sons. Top diagram in (c) Adapted from Webb and Dorey 1995.*

Usually camera choice is limited, and it is therefore possible only for the objective/magnifier to be changed to achieve the correct overall magnification to sample the specimen correctly. However, the question can be posed the other way, namely: what maximum size of pixel is required to sample correctly given a fixed magnification? In this case, the formula becomes: $P = M.d/2$, and in our example, the pixel of the CCD must be: $\leq [63 \times 0.213]/2$ µm ≤ 6.7 µm, but with a sampling frequency of N/3, the size must be ≤ 4.5 µm, a more stringent requirement.

If the CCD faceplate of the digital camera is arranged to lie in the primary image plane, then the contrast can be boosted artificially by control of the offset or black level of the camera. In this case it is more appropriate to use Sparrow's criterion, where $d = 0.47\lambda/NA$. For the same 63x/NA 1.4 objective, the minimum resolved distance will be 0.164 µm, and the diameter of the Airy disc falling on the CCD in the primary image plane will be: $0.94 \times 0.488 \times 63/1.4 \approx 21$ µm.

The concept of using the optical index to compare objectives is explained in Box 31.2. Because the optical index describes mathematically how closely information is packed into the primary image, it can be used to ensure that the analogue image is sampled adequately by a digital camera.

Box 31.2 EM Nelson and the Optical Index

Source: Bradbury 2000. Reproduced with permission from The Quekett Microscopical Club.

One of the greatest gentleman amateurs of the 19th century was Edward Milles Nelson, born on the 24th of January 1851 to a doctor and his wife. Nelson's interest in microscopy became evident early on, for when he was about 11 or 12 he begged his parents to buy him the 'Society of Arts' microscope. This was a simple stand, the winning design in a competition held by the Society to introduce a mass-market microscope because of its tremendous popularity in Victorian times. It cost £5 (£420 today); young Edward's request was refused.

Nelson was to be more fortunate in later life, recalling later that he started microscopy in 1866. He began work as a surveyor, having dropped out of Cambridge without a degree. In the 1870s Nelson's parents died and, like Sorby, he inherited sufficiently never to have to work again. He was free to pursue his real loves of microscopy, horology and, later, astronomy. Nelson had a good grasp of mathematics, designed his own lenses and escapements for clocks, and was fond of devising indexes for comparing equipment. In microscopy he was largely self-taught but went on to contribute anonymously to Carpenter & Dallinger's *The Microscope and its Revelations*, the standard work of the day, and was a prolific writer to the *English Mechanic* as well as the journals of the two microscopical societies.

At one time Nelson had 17 different clocks in his bedroom, and devised an 'efficiency number' to compare them. He did the same for objectives too, and called it the 'optical index'. He was a superb microscopist, arguably the best-ever of the 19th century. It was Nelson who is on record as having resolved, circa 1885, the finest 11th line of Nobert's final 20-band test-plate of 0.21 µm spacing. There was a 'rubicon' between the 11th and remaining bands that could not be resolved with the light microscope.

Mag.	NA	OI		Mag.	NA	OI	
2,5	0.08	**32**	achro	40	0.65	**16**	achro
6,3	0.16	**25**	achro	40	0.80	**20**	semi-apo
10	0.22	**22**	achro	40	0.95	**24**	apo
10	0.33	**33**	apo	40	1.00	**25**	apo, oil
10	0.45	**45**	apo, oil-imm	40	1.22	**31**	apo, water
16	0.35	**22**	achro	40	1.40	**35**	apo, oil
20	0.40	**20**	achro	63	0.95	**15**	apo
20	0.75	**35**	apo	63	1.40	**22**	apo, oil
25	0.45	**18**	achro	100	1.25	**12**	achro, oil
25	0.80	**32**	semi-apo, oil	100	1.40	**14**	apo, oil

Nelson's optical index is a means of comparing the efficiency of different objectives, in the same way that NA can be compared directly. The OI of an objective describes how closely information is packed into the primary image. The larger the optical index, the better the objective. It has a modern use when coupling digital cameras to a microscope. In photography the focus-to-aperture ratio, the *f*/No., defines the speed and quality of the lens. In microscope objectives, where the focal length is small, it is easier to measure magnification accurately. Nelson formulated the concept and the term optical index to try and rationalise the selection of objectives on a microscope, and he intended the optical index to be adopted as a comparative aid for objectives. The optical index ratio is calculated by dividing the numerical aperture by the magnification of the objective and multiplying by 1000, to clear it of several decimal places.

Nelson was a lone voice in support of Abbe's theory of image formation, and he faced much opposition. He could take constructive criticism well, but when he considered his honour impeached, he reacted strongly. Having worked tirelessly for the Royal Microscopical Society for many years, and having served as President, he fell out with some Fellows, severing all connection sometime in 1914. Nelson continued with the Quekett Microscopical Club who, to its credit, had continued to publish his views on Abbe even when both Nelson and Abbe were unpopular. In later life he was elected an Honorary member of the club. Edward Nelson died in Bristol on July the 20th 1938.

References

Bradbury, S. (2000). 'EM Nelson (1851–1938), a notable English microscopist'. *Quekett Journal of Microscopy 38/7*: 547–572.

Turner, GL'E, and Bradbury, S. (1966). 'An electron microscopical examination of Nobert's finest test-plate of twenty bands'. *Journal of the Royal Microscopical Society 85/4*: 435–447.

Table 31.4 *The number of square CCD camera elements needed to give full resolution with a field of view equivalent to that of 35 mm colour reversal film.*

NA/M_{obj}	O.I	Wavelength (nm)		
		400	500	650
0.0325	32.5	4550 × 3033	3640 × 2427	2800 × 1867
0.0300	30	4200 × 2800	3360 × 2240	2585 × 1723
0.0220	22	3080 × 2054	2464 × 1623	1896 × 1264
0.0140	14	1960 × 1307	1568 × 1046	1207 × 804

Source: Entwistle 1998. Reproduced with permission from John Wiley & Sons.
The dimensions of a frame of 35 mm film are 36 mm along the horizontal axis and 24 mm along the vertical axis. The higher the optical index, the more information resolved in the primary image plane, and therefore the greater the number of photo-sites ('pixels') required to adequately sample the image according to Nyquist's criterion (see Box 31.1).

Reference

Entwistle, A. (1998). A comparison between the use of a high-resolution CCD camera and 35 mm film for obtaining coloured micrographs. *Journal of Microscopy 192/2*: 81–89.

Entwistle (1998) cites a different, but equally valid, formula for calculating the number of pixels, N, required in the sensor array as: $N \geq (4.A/M_{path}) \times (1/\lambda) \times (NA/Mag_{obj})$, where A is the width of the sensor array and M_{path} the magnification factor of any magnification changer. This last term (NA/Mag_{obj}) is the optical index ratio. As the optical index increases, the sampling requirements become more stringent, with a greater number of sampling pixels required and/or a greater magnification required at the image plane (see Table 31.4).

For low-magnification objectives (which have optical indexes >20) use the 1.6x magnification changer (see Table 31.3) to increase the overall magnification so that the minimum distance resolved by the objective is sufficiently sampled by the size of the CCD detector pixel. In other instances, e.g. for high numerical aperture objectives (such as 40x/NA 1.4, OI = 35), the pixel size of the image will also be insufficient to provide adequate sampling. The 40x/1.4 objective has the same resolving power as the 63x/NA 1.4 objective (OI = 22.2) but projects an image 60% the size of that projected by the 63x objective. In this case, it may be also be useful to use a 1.6x magnification changer with the 40x objective.

31.8 Scale Bars

For any meaningful evaluation of your images and photomicrographs, include a scale bar to indicate the size of the subject. This is necessary because although the eye is very good at colour and pattern recognition, it is poor at estimating – let alone making an absolute measurement – size, length, area or numbers. Optical illusions depend upon this weakness in order to work. We learn to estimate size from experience of our life-size three-dimensional environment.

To measure an object in real life, we use an absolute scale, e.g. a piece of string knotted at regular intervals or, for greater accuracy, a ruler. By using a recognised absolute scale (such as a ruler divided into centimetres or micrometres), other people can place confidence in our findings and assertions. In microscopy, unless we are very familiar with the two-dimensional image presented to us, we have no idea of scale. The problem is compounded when studying a small field of view at high magnification with few or no reference marks that we can recognise. Is the image taken with a 10x or a 100x objective, has it been further enlarged? We most often cannot tell and so need to include an absolute scale.

We cannot easily place a ruler on top of the specimen on the microscope stage, so we have to transfer a scale into the image in order to measure and define it. To record the size of an image and/or the width of a field of view, acquire an image of a micrometer scale bar (a miniature ruler, Figure 31.10) substituted for the specimen on the microscope stage at the same magnification. The image of the scale bar can be enlarged to the same degree when making the digital scan or final print, or it may be used to calculate the overall magnification of the displayed image. Many digital image analysis programmes, *once calibrated*, include an automatic scale bar function in their software. It is wise to check this calibration and not assume that it has already been done. The calibration figure in micrometres per pixel (μm/pixel) can be input into the image analysis software. Some packages (e.g. Amira, Imaris, Volocity) will automatically read the metadata from supported proprietary file formats. However *.tif files do not save metadata, and so the appropriate calibration figures will have to be entered by hand.

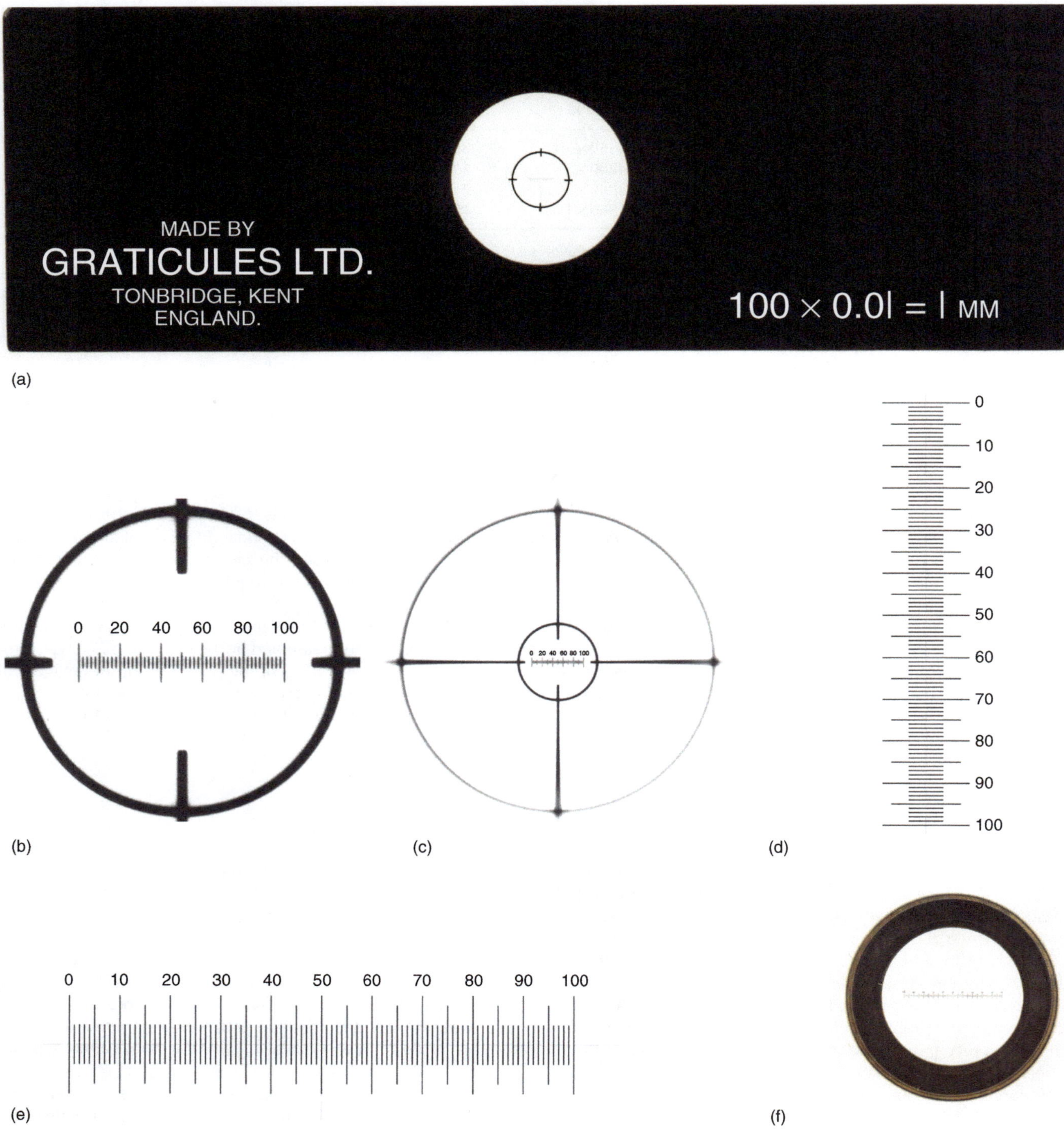

Figure 31.10 **Scale bar** *The micrometer scale – an absolute scale shown in (a) – is used to calibrate either an arbitrary graticule scale (d–f) in the eyepiece, or calibrate the image pixel dimensions in software. The arbitrary scale in (f) is of a similar design to that shown in (e). Both are calibrated with a stage micrometer scale. The illustrations in (b) and (c) show the micrometer scale close up. (b) is a ruling of 1 mm divided into one hundred 10-μm divisions, while (c) is a 0.1 mm (100 μm) micrometre scale divided into 50 divisions of 2 μm, for high-magnification objectives.*

Calibration Using a Scale Bar

1. Insert a focusing eyepiece containing a graticule into one of the microscope's viewing tubes.
2. Focus the graticule scale sharply using the focusing adjustment of the eyepiece.
3. Place the stage micrometer scale on the stage, and focus it sharply, using the microscope focus control.
4. The images of the stage micrometer scale and of the graticule scale in the eyepiece should be superimposed (Figure 31.11), and both should be in sharp focus. Rotate the eyepiece, if needs be, so that the two scales are lined up and overlapping.

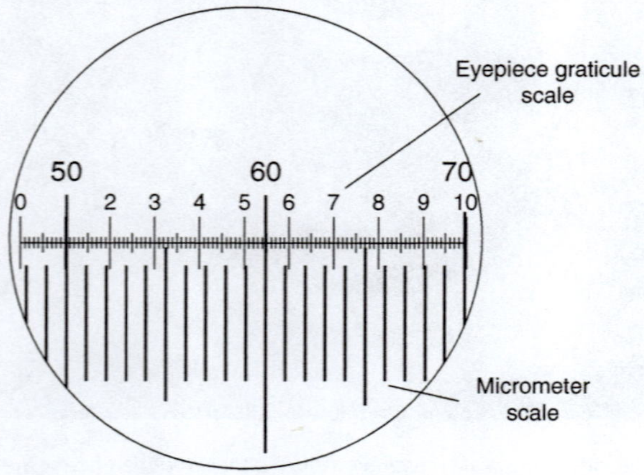

Figure 31.11 **Calibrating the eyepiece graticule** *Because the specimen plane and the primary image plane are both in the field (imaging) set of conjugate planes, an image of the eyepiece graticule is seen superimposed upon an image of the micrometer scale bar, as shown, and can therefore be calibrated for any desired objective.*

5. Record the true length of the full number of divisions of the eyepiece graticule for each of the objective lenses of the microscope. (There will usually be 10 or 100 divisions, depending upon the design of the engraved markings on the graticule.) This record needs to be performed only once for each objective magnification, assuming that no other factors in the imaging system are changed.
6. Draw a graph (e.g. Figure 31.12) relating graticule units to true dimensions.

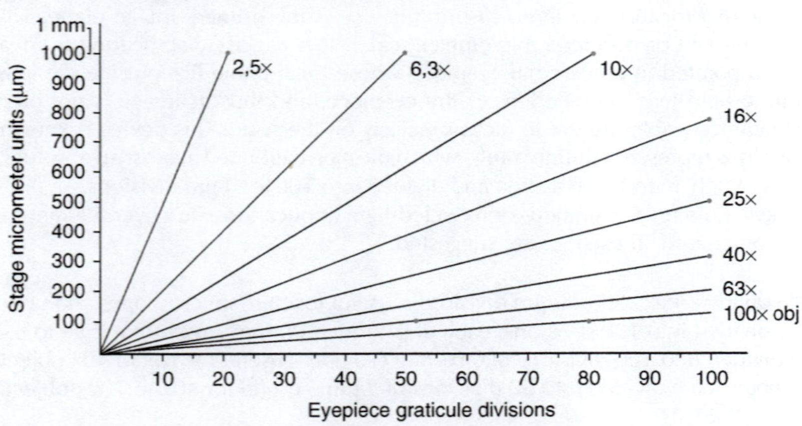

Figure 31.12 **Scale bar calibration graph** *A graph showing the calibration of the eyepiece micrometer graticule for several different objectives. This allows easy determination of the length of an object without the need for arithmetic calculation. Source: Adapted from Bradbury 1991 with permission from Oxford University Press.*

This method has the advantage of only requiring one measurement to plot a straight line from zero and being able to accommodate any auxiliary magnification incorporated between objective and eyepiece.

Calibrating Images or Other Image-Analysis Software

1. Take an image of a stage micrometer scale at the required magnification. The image *must* be in sharp focus, but it is not important whether the background illumination is entirely even.
2. Determine the number of pixels along each length of the image. If the image is full frame, this will correspond to the specified pixel dimensions of the CCD or sCMOS sensor. Otherwise, you need to calculate the pixel dimensions of any sub-array sampled image.
3. Calculate the (μm/pixel) value, and enter this into the software.

Figure 31.13 **Filar micrometer scale** *Since the measuring scale is engraved around the circumference of the drum, this is equivalent to a much extended linear scale. There is no need to estimate fractions of graduation intervals as with the eyepiece micrometer scale. That said, digital measuring in software – which has all but replaced these analogue methods – is both fast and accurate provided the software has been calibrated properly at the outset when first installed. The knurled ring on the eye lens, which can be seen just below the K8x inscription, allows the lines of the graticule scale to be sharply focused by different observers. Source: Courtesy of Mr K. Glover.*

Always include a scale bar within the image, rather than quote a measurement or magnification in the accompanying figure legend. If the journal compositor resizes the image to fit the page, the information will be incorrect, and it is easy to miss any alteration when checking proofs. If the length of the scale bar is not labelled in the image, remember to refer to it in the figure legend.

The commonest and most practical location for the measuring device is the primary image plane. This usually takes the form of a *graticule* (sometimes called a reticule or eyepiece micrometer scale). It is a glass disc bearing a linear scale, generally of 10 or 100 divisions. It is normally incorporated in an external eyepiece whose focal plane lies outside the lenses of the eyepiece (so that the graticule lies outside both the field lens and eye lens of the eyepiece, and thus its image is not prone to distortion) and which incorporates a focusing adjustment to enable the eye to focus precisely on the scale. This device is known as a *micrometer eyepiece*. The graticule or micrometer divisions represent arbitrary units, which must be calibrated against an absolute scale, the *stage micrometer scale*, usually one millimetre accurately marked on a slide and divided into 100 × 10-μm divisions.

Stage micrometer scales are available for transmitted- or reflected-light applications. To cover the range of objectives used, different stage micrometer scales can be purchased. These four are suggested:

1. A 20 mm micrometer scale divided into 2000 10-μm divisions – useful for stereomicroscopes and up to 5x objectives.
2. A 10 mm micrometer scale divided into 100 divisions, each of 0.1 mm (100 μm) – useful for 2.5x to 6.3x objectives.
3. A 1 mm micrometer scale divided into 100 divisions of 0.01 mm (10 μm) – useful for 10x to 40x objectives.
4. A 0.1 mm (100 μm) micrometer scale divided into 50 divisions of 2 μm – useful for 40x to 100x objectives.

A more accurate method of taking linear measurements is by using a filar micrometer eyepiece (Figure 31.13). In this case the graticule is arranged so that turning a graduated drum on the side of the eyepiece moves a fiducial line either side of a central fixed datum line. The graduations of the drum are marked in arbitrary units, and these are calibrated in the same way as a micrometer eyepiece.

When recording images, whether on film or digitally, it is not necessary to use an eyepiece graticule: the stage micrometer scale can be recorded directly as a separate exposure, at each magnification. The micrometer scale can then be printed or displayed in the same way as the images of the specimen and thus provides a convenient 'mini-ruler' with which to measure image features or from which to draw a scale bar. This is easily done with an image-editing programme such as Adobe Photoshop, Gimp or Pixlr.

31.9 Resolution and Display or Print Size

Table 31.5 shows the number of pixels required in the image in order to print out a sharp image. As a rule of thumb, you should have a minimum of 300 pixels per inch, so an 8″ × 10″ print requires [8 × 300 × 10 × 300 = 7 200 000 pixels, or 7.2 megapixels]. Printed images are normally sent to an inkjet printer at between 150 and 360 pixels per inch (ppi) resolution. The human eye can, at very best, resolve about 0.8 arcminutes. This equates to 0.0023″ (59 μm) at the least distance of distinct vision of 254 mm (10″). One arcminute

Table 31.5 *Pixel Dimensions and Print Sizes*

Print size (inches)	Metric size equivalent (mm)	Pixel dimensions @ 300 ppi	Approx. file size (MB) at 300 ppi	Number of megapixels for given print quality			View 300 ppi @ 2 feet: MP	View 300 ppi @ 4 feet: MP	View 300 ppi @ 6 feet: MP
				300 pixels/inch	240 pixels/inch	600 pixels/inch			
2.5″ × 3″ Wallet	64 × 76	750 × 900	2	0.68	0.43	2.70	0.34	0.17	0.08
4″ × 6″ Print	102 × 152	1200 × 1800	6	2.16	1.38	8.64	1.08	0.54	0.27
5″ × 7″ Print	127 × 178	1500 × 2100	9	3.15	2.02	12.60	1.58	0.79	0.39
A5 5.8″ × 8.3″	148 × 210	1740 × 2490	13	4.33	2.77	17.33	2.17	1.08	0.54
8″ × 10″ Print	203 × 254	2400 × 3000	21	7.20	4.61	28.80	3.60	1.80	0.90
A4 8.3″ × 11.7″	210 × 297	2490 × 3510	25	8.74	5.59	34.96	4.37	2.18	1.09
11″ × 14″	279 × 356	3300 × 4200	40	13.86	8.87	55.44	6.93	3.47	1.73
A3 11.7″ × 16.5″	297 × 420	3510 × 4950	50	17.37	11.12	69.50	8.69	4.34	2.17
16″ × 20″	406 × 508	4800 × 6000	83	28.80	18.43	115.2	14.4	7.20	3.60
A2 16.5″ × 23.4″	420 × 594	4950 × 7020	100	34.75	22.24	139.0	17.4	8.69	4.34
20″ × 30″	508 × 762	6000 × 9000	155	54.00	34.56	216.0	27.0	13.5	6.75

corresponds to 0.0029″ (74 μm), which is the accepted value for the resolving power of a very healthy human eye (see Chapter 3, Section 3.11, Box 3.4). The inverse of 0.0029″ gives 345 'dots' per inch. This is why 300 dpi is set as an optimum print density (85 μm spacing), with 350 dpi (≈ 73 μm spacing) being the maximum a typical human can comfortably resolve in print.

Under domestic (800 lumen/60 watt) lighting at about 1 metre, it is impossible to distinguish between a print made at 300 dpi and one of 600 dpi. Desktop printers are now sold that print at 1200 dpi. This is because in strong daylight, or under very bright artificial light, a few people can tell the difference between 600 and 1200 dpi if the print quality is good. If the maximum resolving power of the eye is 345 dpi, and Nyquist's criterion states that 2–3 pixels (or cones) should cover each resolved dot, then the limit that the eye will discriminate is 800 dpi at 1 arcminute and at very best 1000 dpi at 0.8 arcminute – hence the use of 1200 dpi inkjet and laser printers.

You can relax the print quality to 240 ppi at 8″ × 10″ or A4 size and generally get away with this. If you want to be able to crop pictures before printing a certain size, then you need a greater pixel density (i.e. more megapixels) because cropping is essentially a zooming function. For critically sharp prints, raise the pixel density to print out at 600 pixels per inch. Taking the estimated megapixel equivalent of film into account, a 35 mm monochrome film negative will produce a sharp 12-inch by 18-inch print when scanned at a resolution of 300 dots per inch (dpi) by photo printing lab equipment. However, this does not take the viewing distance into account.

If you sit too close to your television, the picture appears to lose quality. This is because TV displays are designed for certain viewing distances. The same is true for prints; large ones are meant to be viewed from a few feet away, not inches. This means that an image with fewer pixels than optimum can still be printed at the same size, without reduction, and with satisfactory results. An 8″ × 10″ is the smallest standard size print designed to be viewed from a distance: at least 2 feet away. This means that while 7.2 megapixels of information is optimally required to record a 300 dpi 8″ × 10″, a longer viewing distance lowers the amount of information needed for a visually good print. As a rule of thumb, divide the pixels needed by the viewing distance (in feet) to calculate how few pixels you can get away with. When given freedom of choice, viewers of a large print tend to position themselves at a distance that is approximately equal to a diagonal of the print[5]. Hence, the viewer of a 24″ × 30″ print naturally drifts toward a position about 38–40 inches from the print's surface.

31.10 Relocating the Image

In conjunction with automated motorised stages, most commercial image acquisition programmes allow you to record slide coordinates so that you can return to the same position later. This is extremely useful, yet very difficult to do blind without the assistance of such software or the presence of a *very* notable feature, such as a large blood vessel. One way around without a dedicated location marking function in software is to mark the slide yourself.

There used to be sold ink markers that could screw into the objective nosepiece (Figure 31.14), but it is simpler to do this with a very fine-tipped indelible marker pen on the rear of the slide. Once the area is located, swing a low magnification (e.g. 5x or 10x) objective into position and (if using fluorescence) illuminate the slide temporarily by bright-field microscopy. Draw carefully around the circle of light seen on the slide. If you are using an upright microscope, remove the slide from the stage, invert it and copy the circle on the reverse of the slide. This is because any oil, when the specimen is viewed using oil-immersion, can dissolve the ring drawn on the obverse of the coverslip.

An alternative is to use an England Finder. These are still manufactured by Pyser-SGI Ltd; it is a 3″ × 1″ (76 mm × 26 mm) slide (Figure 31.15) marked with an alphanumeric reference pattern. This pattern consists of 75 columns of squares numbered 1–75, and 25 rows (A–Z, omitting I). Each square is ≈1 mm^2 and is divided into 4 numbered quadrants, with the square number and letter in the centre.

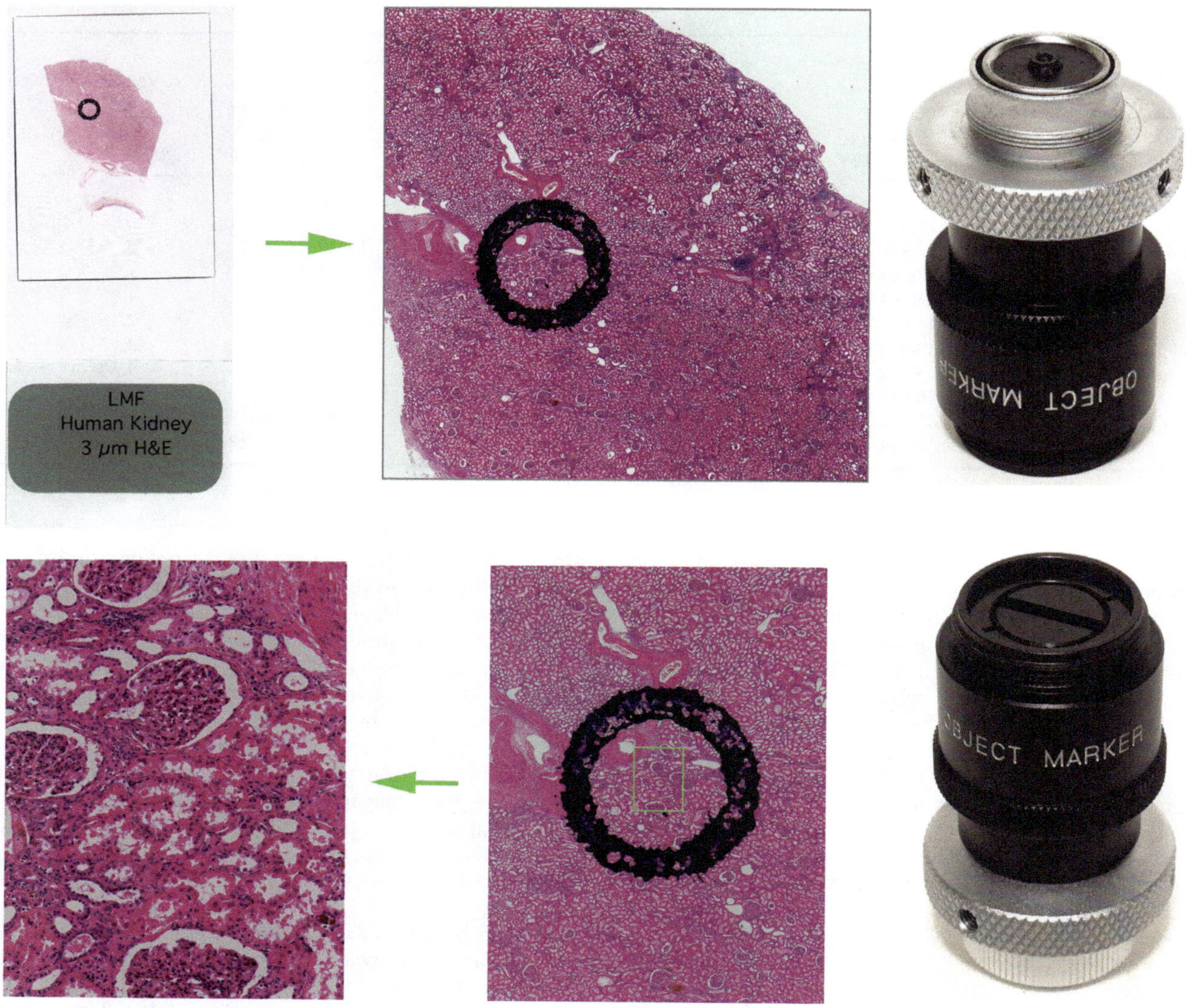

Figure 31.14 **Relocating a field of view** *Here a slide marker, which has an objective thread to screw into the nosepiece and a tiny circular pad that can be inked, has been used to ring the field of view of interest. Although these items of equipment are no longer common, they may still be found in laboratories. It is also useful to record the location of any marked feature, either in software or by using an England Finder (see Figure 31.15). Source: Courtesy of Mr K. Glover.*

Three sides are marked with double locating arrows; these edges are guaranteed to be accurately ground so that the slide can be placed on the stage in a consistent manner. Use these edges to place the Finder on the stage, even if the image appears inverted. The orientation of the specimen slide on the microscope is noted. This is important. Once an interesting feature is located, the England Finder is substituted for the specimen slide, and the reference is seen in the field of view recorded.

Coverslips marked with alphanumeric grids are available for locating adherent cell populations grown in culture. Cellattice^{TM} is a plastic substrate No.1 thickness 18 mm^2 coverslip ruled into 100-μm squares with 25-μm divisions per alphanumeric square. Bellco Glass manufacture a similar No.2 thickness 18 mm^2 glass coverslip with 60-mm squares. Eppendorf make the CELLocate, which is a 3 mm diameter area ruled within a No. $1^1/_2$ (0.17 mm) 12 mm diameter glass coverslip, with two grid sizes: 175 μm for locating cell groups and 55 μm for locating individual cells.

31.11 How Important Are Megapixels?

Consumer camera buyers are encouraged to go for ever more 'megapixels': it is a classic marketing ploy. As an analogy, salespeople will try to convince you that megapixels are equivalent to horsepower. However, just as with extra horsepower in cars, which you can only drive to a maximum speed limit, having extra megapixels is overkill. In fact, as the number of pixels increases, the noisier will the image become because the noise is dependent on the quality of the diode that forms the photo-site (read noise) and its temperature (dark

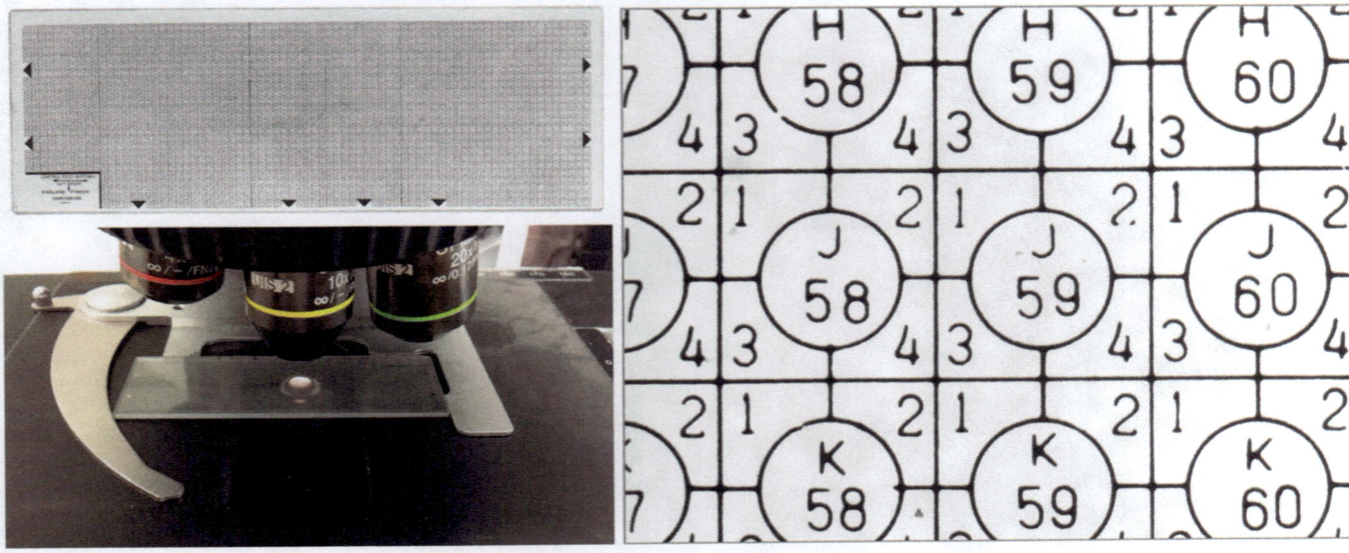

Figure 31.15 **The England Finder** *Using the England Finder to record the position of a field of view for later relocation. The slide is marked with a grid pattern of letters and numbers, with two sides of the slide marked for insertion into the sprung slide holder on the microscope stage. Each single-lettered quadrant is 100 μm square.*

noise). Smaller pixels have a smaller capacity to capture photons and hold electrons. Generally, larger pixels collect more photons and thus produce images with less shot noise. A larger chip size with larger pixels may have poorer resolution, but it will have the advantage of a greater full-well size and so have a larger dynamic range and higher signal-to-noise ratio. Small-chip consumer cameras struggle to produce 8-bit depth, whereas a large chip can easily capture 12 bits or more. There is a trade-off between resolving power and noise. For *consumer* cameras, with a pixel size of 3 μm, 6 MP is the optimum number of megapixels. More expensive and scientific-grade cameras have chips with fewer dead (zero output) or stuck (maximum output) pixels. However, lighting of the subject and the resolving power of the objective play a greater part in determining the overall quality of the image.

It is possible to get full-frame (i.e. 36 mm × 24 mm sensor size) digital SLR cameras (e.g. Nikon D800) up to 36 megapixels, with a pixel pitch of 4.89 μm. The Canon 5D series has a pixel pitch between 8.2 μm (12.8 MP) and 6.25 μm (22.3 MP), depending upon the version. Cameras with the smaller APS-C (23.6 mm × 15.7 mm) have pixel pitches from 3.85 μm (Nikon D3200, 24.2 MP) to 4.8 μm (Pentax K-01, 16.3 MP) or 4.94 μm (Nikon D3100, 14.2 MP). However, many designers consider that 12 megapixels is sufficient for most needs, and to go above this value requires smaller pixels, compromising light capture and dynamic range.

31.12 Gamma

Gamma is the function relating digital intensity to display brightness. If this is nonlinear (often an advantage), proportionally more of the grey levels of the display are devoted to correctly displaying the low-signal regions of the image. Sometimes the range of intensities present in a specimen demands a large (e.g. 12-bit) dynamic range, since 8-bit depth cannot display all the features without either over-saturation and loss of high-intensity details or else loss of darker, low-intensity, features[6]. This situation occurs when 12-, 14- or 16-bit images must be printed out or displayed on an 8-bit computer monitor. Trying to linearly compress 12-bit depth into 8-bit depth (neutral-contrast stretching) is only successful if part of the 12-bit depth has been used; where most of the 4096 grey levels contain data, this approach achieves nothing and loses information. Likewise, merely adjusting the brightness and contrast may introduce saturation and certainly discards information. Some form of nonlinear contrast adjustment is required, and this is why gamma adjustment (Figure 31.16) is important: it does not over-saturate the data, or inappropriately skew it, but preserves the ranking of the intensities of the data. Applying gamma values <1 stretches out changes in intensity occurring at the lower end of the greyscale values, and compresses those at the top end. Conversely, applying gamma values >1 does the reverse.

Even when an adjustment has been made for displaying image information in 8-bit depth, the working of the cathode-ray tube or electronic display device must also be taken into account. Unlike a CCD camera, the brightness of the display screen is not proportional to the input voltage applied. It follows a 'power law' (a number raised to a power). The numerical value of that power is what we call the gamma of the monitor or system. Thus the gamma of the display device describes the relationship between the brightness of a pixel as it appears on the screen and the numerical value of that pixel. Typical standard gamma values are 1.8 (Macintosh standard) and 2.2 (Windows standard). However, gamma 2.2 is rather dark, and if the gamma of your monitor is not set correctly, you won't see colours (hues) or tones (grey levels) as others do or be able to print out correctly. URL links are given in the Internet Resources at the end of the chapter to help you adjust your monitor. Whilst appropriate gamma correction of the monitor is useful, if your camera gives an option for gamma control, then it is usually best to default to a setting of 1 so that the raw data exhibits linear gain with incident photons.

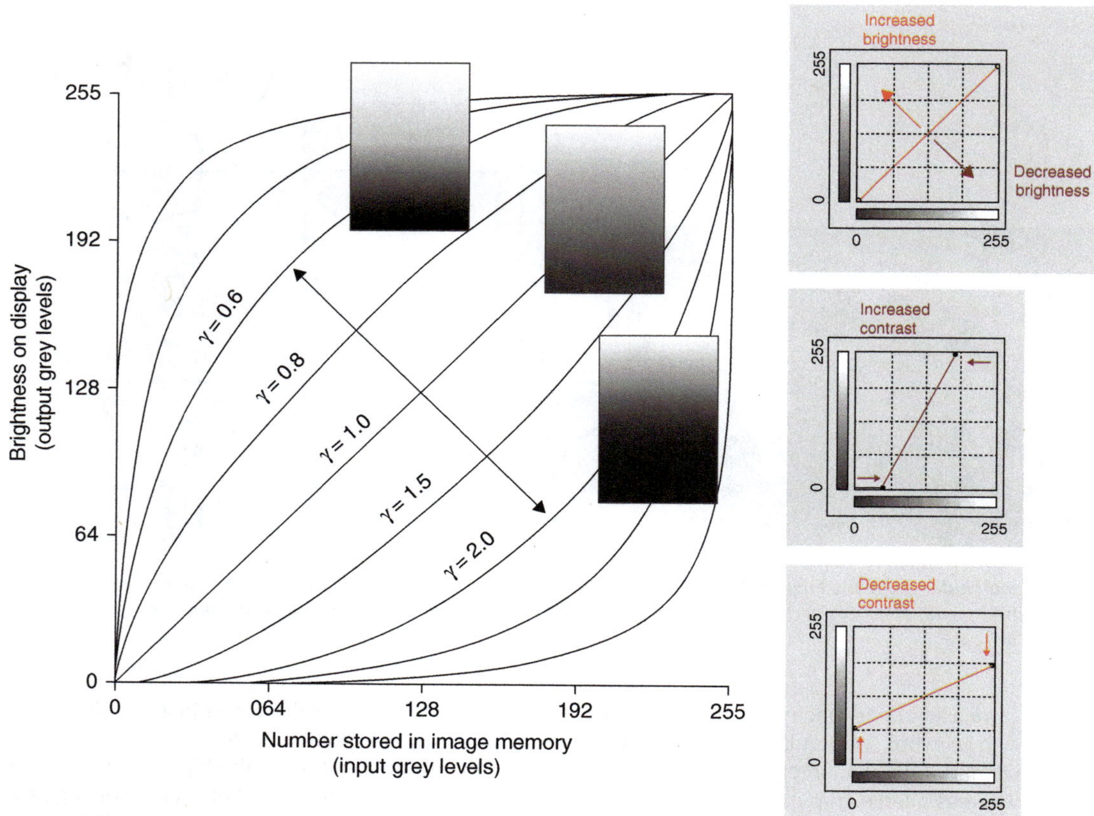

Figure 31.16 **Gamma adjustment of the display** *Where $\gamma = 1$, the relationship between the data input values and the display output values is linear. Where $\gamma > 1$, high numbered grey level values (approaching saturated white) are selectively spread over a greater number of grey levels in the display. Conversely, if $\gamma < 1$, darker areas of the image are enhanced by being spread over a greater number of grey values. Gamma can also describe the relationship between photons detected and the output intensities of our eye, of film and electronic sensors. CCD and sCMOS chips have relatively linear input-output, with $\gamma \approx 1$. We have a $\gamma = 0.7$, since our visual system can discern faint objects in low light areas of a scene whilst preventing bright areas from glare in our eyes. Provided the horizontal and vertical axes are equally scaled, the gamma of a photographic film is described by the slope of the linear portion of the characteristic H&D curve. The best gamma values for viewing digital images is a value between 1.0 and 1.2. Source: Pawley 2006. Reproduced with permission from Springer.*

31.13 Colour Management

If you recall watching a bank of television sets in an electronics shop, you'll probably remember that while they move in perfect synchrony, they display the same image in slightly different hues. This is because the colour profile of each TV display is slightly different from its neighbour, regardless of whether it is made by the same manufacturer or not. The phosphors on the screen wear at different rates, and the contrast, or gamma, controls may be designed and adjusted differently. This holds true for all other colour output devices, whether they are designed to produce additive RGB or subtractive CMYK colour, because they have different colour gamuts, or ranges of reproducible colour (also referred to as 'colour space'). RGB and CMYK values represent control signals, or recipes for colour, rather than specific hues, and each device responds differently to the control signals.

RGB and CMYK are often called *device-specific* or *device-dependent* colour models, precisely because they really only produce predictable results for a given device. Colour management software (a colour engine) matches the unique RBG/CMYK device-dependent colour profile of each device (printer, scanner, monitor) and compares it to an internationally agreed standard, namely the CIELAB 1976 colour space model, rather like a dual-language dictionary. This International Color Consortium colour profile is a standard that represents colour in terms of device-independent human perception, allowing profiles to be matched between devices. For example, if you want to translate between German and Ubuntu, but only have English-German and English-Ubuntu dictionaries, English takes the place of the device-independent international standard. Similarly, you need one profile to describe the colour and two profiles (source and destination) to match colour across two devices. Without colour management, only a trial-and-error approach will eventually produce a reasonable colour match.

Between 5–8% of men and 0.5% of women are colour-blind, or more correctly colour-deficient[7]: most can see colour but perceive it differently as a narrowed colour spectrum[8]. There are several different classes of colour blindness, but they can be broadly divided into protanopia (missing/malfunctioning red L-cone) and deuteranopia (missing/malfunctioning the green M-cone), both of which have difficulty discriminating red and green hues. Oranges, pale reds, browns and yellows all appear as yellowish hues, distinguished only by their intensity. A very few people suffer from tritanopia (missing/malfunctioning blue S-cone) and confuse blue with green and yellow with violet. Protanopes have a truncated perception of the red end of the spectrum, which appears dark, and the central blue-green part appears dark grey. Protanopes see longer-wavelength hues less brightly than other people. Deuteranopes also see green hues as shades of grey but equally magenta, the complementary colour of green, as tending to grey, also. Both protanopes and deuteranopes register blue and yellow hues well, and these tend to dominate their spectrum.

One consequence of colour-blindness is that the commonly used red channel and green channel for deriving scatterplots and calculating possible colocalisation becomes confusing, since both the red and green channels are seen as yellow and/or grey. Also, a double-labelled fluorescence image with a green and a red fluorophore will present as overall yellow. It is better under these circumstances either to use magenta or vermillion for the red channel and to change the green channel to blue-green or turquoise, or else to display them separately as monochrome greyscale images. For successful colour management, it is recommended to:

- avoid using all the colours of the rainbow together;
- use yellow with care; contrast it with blues;
- use differential intensity to reinforce colour shades;
- avoid yellow-green colours;
- avoid using green completely;
- avoid using red and green together;
- avoid pastel colours; and
- use a few bright, high-contrast, primary colours (white, black, blue, yellow).

An excellent one-page explanation of colour-blindness and how to prepare suitable illustrations and figures accessible to those with colour-deficient vision has been published by Wong in the *Nature Methods* 'Points of view data visualization' series (Wong, 2011).

31.14 Image Stacking

Using conventional objectives, it is only possible to get a large depth of field with a low magnification objective having a low numerical aperture. Otherwise, a series of optical sections must be taken or a Z-stack deconvolved and in each case a projection (e.g. maximum-intensity) image formed. However, these two approaches are only suitable for fluorescent sections.

To create sharp high-resolution images of large focal depth in bright-field microscopy, specialist software is required to stack a series of individual widefield images together. Combine ZM and Helicon Focus are recommended; the former is freeware and the latter shareware. URL links are given at the end of the chapter. These programmes, together with those for high dynamic range rendering have been evaluated by Piper (2010). The two very practical books by Julian Cremona (Cremona 2014; 2018) are also highly recommended.

31.15 Image Analysis

There are several image manipulation and image analysis software packages available for preparing your images for publication and/or extracting measurements and other data from them. Some of these freeware programmes have been mentioned in Chapter 27, and listed in Table 27.6. It is important to remember, particularly when considering ethics (see Section 31.16), that image manipulation programmes, such as Adobe Photoshop, Gimp or Pixelmator, are designed primarily for the artistic and graphics markets. Therefore they should be used sparingly on scientific images and in a well-defined, controlled manner with an accompanying audit trail. Some aspects of image manipulation may be essential, such as altering gamma values to display, or show, features within the image.

Images recorded with the microscope are scientific data sets, as previously stated in Section 27.4, on planning imaging experiments and repeated in the previous chapter, in Section 30.3. Image analysis techniques are used to extract meaningful information from these temporal and/or spatial data sets of photon intensity. Pixels in the digital image represent sampled photon intensity measurements. It is relatively easy to calibrate the linear distance – and hence area – of each pixel by using a stage micrometer scale to calibrate the digital imaging system. Details on how to do this are given in Section 31.9, on how to produce scale bars in the image. It is more difficult to calibrate intensity values absolutely, because very often calibration standards are not available or there are too many variables (e.g. sample, illumination or lamp intensity, exposure time, camera gain settings) to control. Often, the best that can be achieved is a comparison of *relative* intensity values either of differing areas within the same image or of images collected sequentially (to minimise inter-image variation). Some guidance on calibrating images to be able to derive photon intensity values, and the measurement errors involved, is given in Wolf et al. (2007). This is an essential chapter to read for anyone considering quantitative analysis of microscope images, and the details will not be covered here. See also Lambert and Waters (2014) cited in Chapter 30.

A separate aspect of image analysis is automatic counting and studies of shape(s) within images. Segmentation is the process of dividing the image into multiple parts in order to analyse particular features or to count components. Feature extraction by segmentation involves thresholding or edge detection to identify regions of interest in the image. Thresholding is best applied to greyscale images. For an explanation of all these operations, see Roeder et al. (2012). Morphometry is the quantification of a structure in two dimensions, whilst stereology is a technique for obtaining data on a number of identifiable objects in a three-dimensional structure by sampling in two dimensions. Because they involve sampling, most morphometric and stereological methods produce estimates rather than exact answers. As with any statistical analysis, repeating the counting process will confer a greater degree of accuracy on the final result.

Those interested in applying morphometry or stereology to analyse their images should consult Aherne and Dunnill (1982) and Howard and Read (2010).

31.16 Saving Images and Data

As an LM facility manager, I am often surprised at how cavalier some people are at saving the experimental data that they have laboured so long to acquire. Many people leave their images on the local computer of the microscope, hoping to return to it when needed for writing-up. Be aware that a shared light microscopy facility will normally have a policy of clearing out hard drives on microscope computers to keep them running optimally, and so you may find your data has been deleted when you need it again! The best policy is to save multiple copies and at least one copy onto a server that is backed-up regularly.

It is also best to have a logical naming routine – indeed some shared facilities insist on this – to help you relocate your images and relate them back to their associated experiment after a period of time has elapsed. Besides a good laboratory notebook, in which all details should be recorded *at the time of image acquisition*, you should have a recognised format to name your images. Some programmes (such as that used with the Leica confocal SP series) will append the channel number to each image.

Digital images can be manipulated with great ease. The filters used to do this are computer algorithms that perform mathematical operations upon the numbers that represent the intensity values of the image. Contrast alteration and stretching is one of the most commonly used operations. Ideally, images should be collected within the dynamic greyscale range available with the offset and gain properly set. Because our eyes perceive a smaller tonal range than the digital sensor (PMT/sCMOS/CCD) can capture, we use a false-colour look-up table (LUT) to set the detection limits appropriately.

The 'standard' file format for continuous tone grey or colour images is TIFF (tagged image file format, *.tif); 8-bit TIFF files display images in 256 grey levels. 24-bit RGB TIFF files display images in colour, using a red, green and blue channel (RGB) with 256 intensity levels for each primary colour. 16-bit TIFF files contain 65 536 grey levels. Digital images generated during microscopy also often contain 12-bit (4096 grey levels). These are generally saved in 16-bit format (with much of the intensity range of the 16-bit file not being used).

Use the histogram function to ensure that the tonal range (often denoted 'levels') of both colour and greyscale images fits within the dynamic range of the display monitor (Figure 31.17). Stretching an image that has only 150 grey levels over the full 256 tonal values in the display will improve the contrast of the image. Any bit depth between 2-bit to 8-bit will occupy the 256 greyscale range as a direct mapping function. Hence if you have a 2-bit image only able to fill the very lowest grey levels of the image, it won't show on a 256 greyscale monitor without remapping (sometimes called 'clipping') the image. Similarly, a 12-bit image displayed on an 8-bit monitor usually shows only black. This is not because image information has been lost; rather, the first 256 levels on a 4096 (12-bit), 16 384 (14-bit) or 65 536 (16-bit) grey level image will only encode black, and only these first 256 levels are displayed (Figure 31.18).

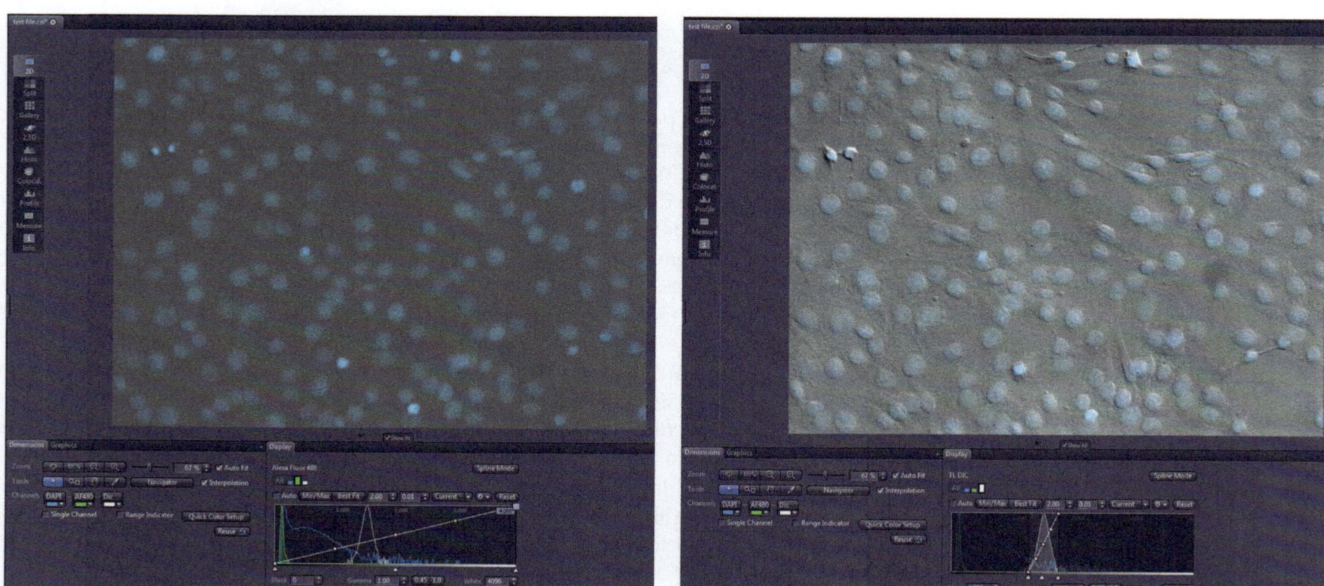

(a) (b)

Figure 31.17 **Digital contrast enhancement in the image/histogram adjustment of the display** *This example of an image of unstained cell monolayer taken with DIC (a) shows very little inherent contrast. Using the histogram and gamma functions, the contrast in the image is enhanced (b) whilst keeping the original data set unchanged. This operation is very easy and straightforward to perform in modern image acquisition software. Note: if the image is then saved in format (e.g. *.tif) that differs from the proprietary acquisition format (e.g. *.czi/*.lif/*.nd2/*.vsi), the original metadata associated with the file may be lost. It is then impossible to return and re-alter the contrast in the image.*

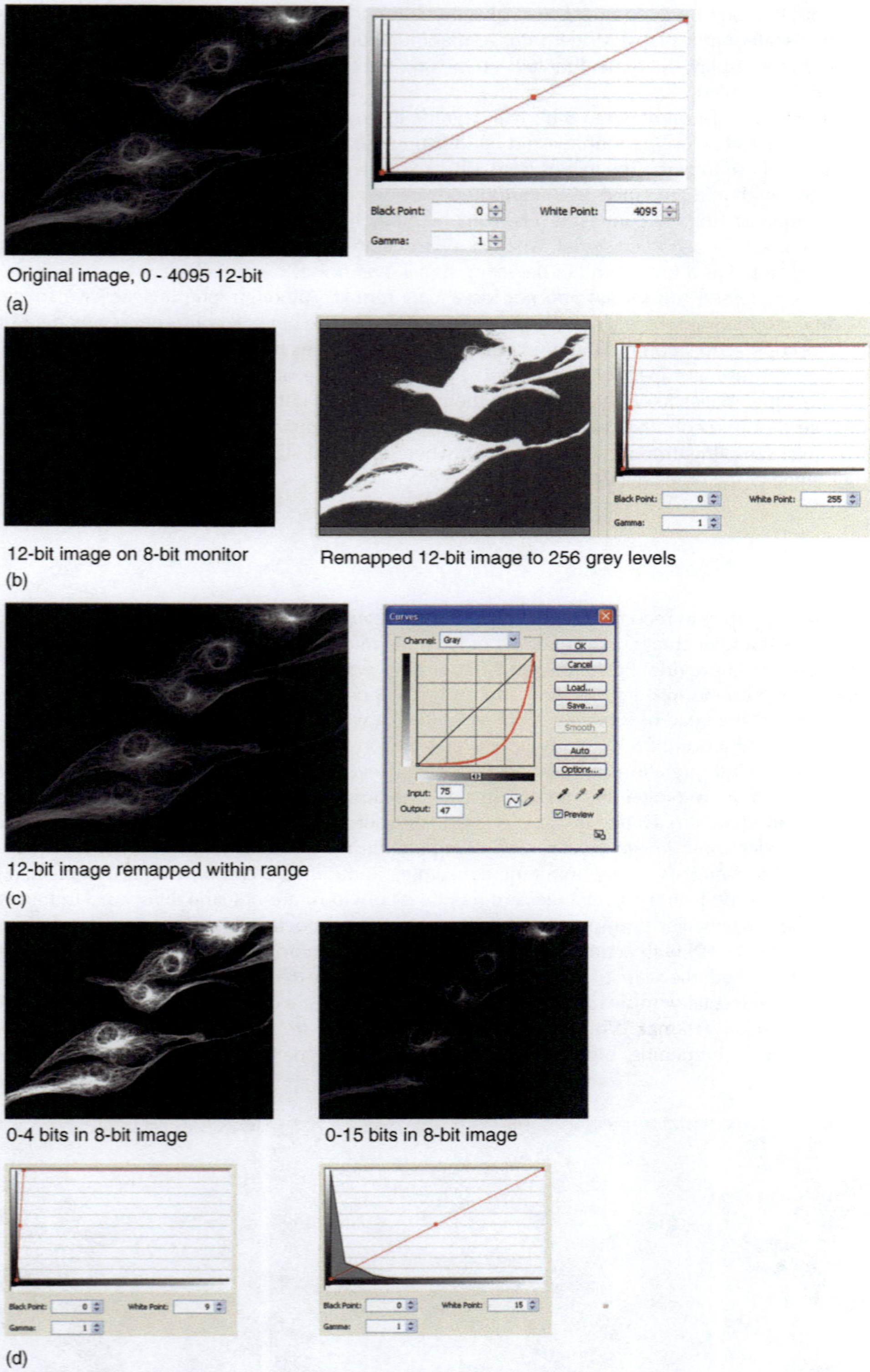

Figure 31.18 ***Bit depth and image display*** *A 12-bit image of actin tubules from the Thermo-Fisher fluorescence test slide of BPAE cells (F36924) was captured with a 12-bit camera using Zen Blue software. Panel (a) shows this 12-bit image displayed in 4095 grey levels with the corresponding histogram. Panel (b) shows the same 12-bit image displayed on an 8-bit monitor: only the first 256 grey levels are displayed, hence the black image. Mapping this 12-bit image directly results in over-saturated display of the upper grey levels present in the image. Panel (c) shows the 12-bit image properly displayed by alteration of the gamma function, rather than by direct display, using the curve gamma function in Adobe Photoshop. Panel (d) shows the effect of stretching too few grey levels over the full 8-bit 256 display: the individual grey levels become very obvious. Above 8-bit and up to 16-bit, the look-up table now has to give much broader representations of the range of incoming bit depth for every increment in the 256 grey scales that it maps.*

Above 8-bit and up to 16-bit, the look-up table now has to give much broader representations of the range of incoming bit depth for every increment in 256 grey scale that it maps. Modern image manipulation software (e.g. Adobe Photoshop) will allow conversion from 12-bit and 16-bit to 8-bit for display, but remember to keep the original data and convert on a copy of the image, for this operation cannot be undone once the file is saved.

Compressing an image from 12 or 14 bit down to 8 bit will degrade the signal-to-noise ratio. This is because 50 steps from 3950 to 4000 may appear as only 5 to 10 steps on the monitor and be indistinguishable from background noise. One answer is to threshold all values below 3900, and set these to zero. The values from 3901 to 4095 can be rescaled so that these 195 values are represented by the full 256 values of 8-bit depth and the range 3950 to 4000 covers 67 grey levels. However, if the feature only covers, say, 5 grey levels in the original image, then scaling the contrast levels won't help much in raising it out of the background noise. Because of the nature of Poisson (shot) noise upon the recorded signal, whether a particular intensity value is signal or noise is a question of statistics. An SNR of 2.7:1 has been suggested as a lower limit of detecting signal over noise.

Capture all images in uncompressed *.tif format and not lossy *.jpg format. Although jpeg images save space, they change pixel resolution and intensity. Any information thus lost may be imperceptible to the eye but will compromise any measurement and analysis performed later. Most microscope acquisition programmes will capture *.tif images by default. When compressing an image, either to display on an 8-bit monitor or to use as *.jpeg for convenience, always keep the original *.tif file. In my imaging facility, we record images having a scale bar as *.jpg images to easily distinguish them from the *.tif data sets.

A properly sampled image should record the detail in the specimen without apparent loss for the human eye. In microscopy, this is achieved by using the Nyquist sampling criterion (Section 31.6), where the pixel size should be equal to or less than half the size of the smallest resolvable distance of the objective used.

31.17 Digital Imaging Ethics

Before the advent of photomicrography to record images, reproducing drawings was inexact and utterly dependent upon the draughtsman and engraver having a clear and accurate understanding of what the microscopist had observed. Fortunately, Robert Hooke, who was one of the first microscopists to provide illustrations of what he saw, was a very good draughtsman. This was unusual, however, and reproducing drawings was prone to misinterpretation and error. The possibility also existed of fitting data to preconceived ideas, such as drawing a homunculus in the head of spermatozoon, as well as of wilful misrepresentation.

With the advent of photography around 1840, many hoped that no longer using a draughtsman or engraver as an intermediary would remove all possibility of misinterpretation and misunderstanding. Nevertheless, from the earliest days of photography it has been possible to superimpose two or more photographs into one image or to airbrush out unwanted details. Nowadays it is very rare that microscopical images are captured on film. Digital imaging is faster, cheaper and more sensitive than using silver halide. The versatility of digital recording has made electronic image capture and manipulation much easier to perform. Techniques that once were the preserve of a few highly skilled technicians, being both time consuming and difficult, can now be accomplished by almost everyone in a 'digital darkroom'. The downside is that any changes are much less obvious, are not straightforward to trace and audit and there is plenty of opportunity to alter images. For example – a very simple one – a mitochondrial fluorescent probe in red can be made to appear to be 'colocalised' (Figure 31.19) with actin fibres stained with a green fluorophore, by shifting the red pixels to yellow.

Some pictures are obviously falsified: the satire of political cartoonists being an extreme example. The danger lies where any alteration is subtle and intended to enhance results or mislead the reader. Scientific images are considered as data by most publishers and should be available for examination in the raw format. Whilst it is simple – and tempting – to alter digital images, most of the software designed for this purpose has filters (such as sharpening, blurring and cloning tools) that change the digitally encoded data in the image. This

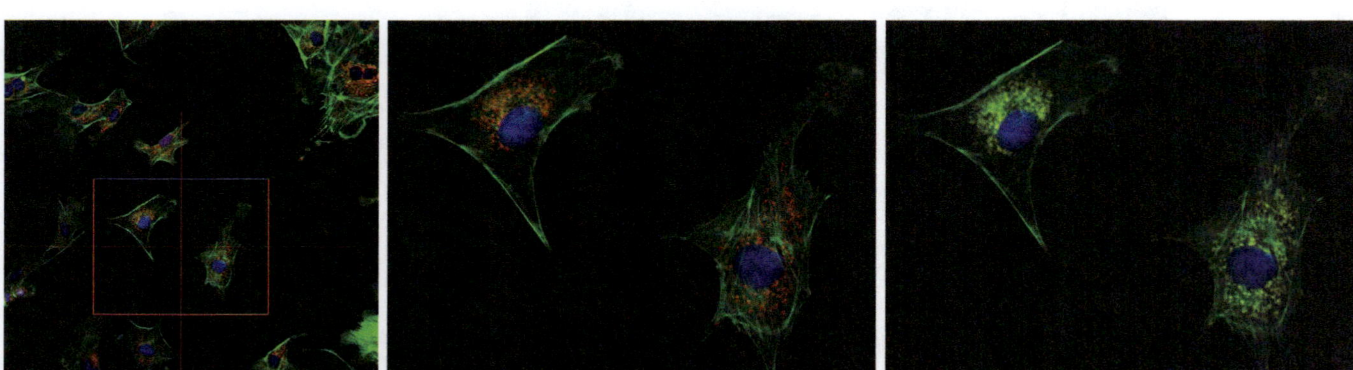

(a) (b) (c)

Figure 31.19 ***Image manipulation*** *From the overview image in (a) showing bovine pulmonary arterial endothelial (BPAE) cells stained green for filamentous actin and red for mitochondria, two have been selected (b) for image manipulation (c) to give the impression that the mitochondria are colocalised with the actin, whereas this is evidently not the case.*

is fine for holiday snaps, but these software filters are not designed for use with scientific images. The fact that the description of the experiment in the Methods section of most papers is sparse in the extreme and gives little indication of exactly how the images were collected only serves to complicate matters. In any case, skilled editorial staff can spot alterations; scientific journals now publish detailed guidelines to limit digital manipulation of images[9]. The list of acceptable 'do's' and 'don'ts' varies from one journal or authority to another, but most are agreed that:

- any digital manipulation should be made on copies of the original raw data, and the raw data should be kept for inspection and verification;
- adjustments of brightness, contrast or colour balance are acceptable if applied *globally* to the whole image. Also allowed is applying a false-colour look-up table (LUT) with appropriate explanation;
- using software convolution filters and alteration of all or part of the image by changing gamma, threshold manipulation or truncating the bit depth is normally not allowed;
- changing the size and resolution of a digital image can add spurious detail through aliasing artifacts and moiré patterns arising from under-sampling. Setting the displayed or printed image size to be greater than that of the original ('upsampling') causes the software to insert pixels between existing ones, thus creating artifacts. This should be avoided;
- reducing the size and pixel resolution of an image also creates new pixels and throws away data. In this case the new pixel is an average and not the artifactual creation of data that was not originally present. This is generally allowed, as is cropping an image;
- cloning an image or the background to 'clean' it up is not allowed. Also not allowed is juxtaposing separate fields of view to appear as one. In this case thin lines should separate each field;
- digital images that will be compared to one another should be acquired under identical conditions;
- any quantitation should be performed only from the original raw image data; and
- a copy should be kept of a history file, and all manipulations should be declared in the figure legends. Ideally, all image acquisition parameters, including the microscope, camera, bit depth, exposure time and analysis software with version number, should be stated.

Rules are made for 'the blind obedience of fools and guidance of the wise'. The best advice on whether to implement any image manipulation has been given by Entwistle (2005) and is worth citing in full:

> First, when you are about to change the appearance of some image data you have to ask yourself, "if I do this could I, even inadvertently, mislead an informed observer?" If the answer to that question is 'yes', then carrying out that procedure is inappropriate.
>
> Secondly, providing that the answer to the first question is 'no', you have to ask yourself, "if I do this will I give the observer some extra information?" If the answer to this question is 'yes', then that procedure ought to be carried out, and a [full] record made of what was done.

Image processing is primarily the means by which metric and numerical data is extracted from images. For an introductory review, see Sysko and Davis (2010). It should not be regarded as a fall-back to clean up or improve poor image capture. Therefore, use it sparingly and in a planned manner. In 2010 the *Journal of Cell Biology* launched a data viewer (Linkert et al., 2010) for readers to view the original data files supporting scientific papers. This has been built upon the Open Microscopy Environment (OME; Allan et al., 2012; Burel et al., 2015), an approach to streamlining multi-dimensional microscopy image data acquired from a variety of platforms with different proprietary file formats.

Finally, do include a full account of how you record your images, what equipment is used, the objectives used and their numerical apertures, the sensor (CCD, sCMOS or PMT) used to capture the image, the microscope settings as well as the reagents and fluorescent probes used. Also include a description of how you displayed and interpreted your data. An example of scientifically-useful digital image manipulation is shown in Figure 11.6. If these details are included in any supplementary data additional to the online or hard-copy published paper, ensure you state this. In short, give enough information so that others can repeat your experiments and verify your results. All too often all the information anyone is given amounts to: 'the images were collected with a confocal microscope'.

31.18 Chapter Summary

At the end of this chapter you should have an understanding of how an analogue image is collected and converted into digital format and how to optimise this process to give a high-quality image. You should be able to connect an electronic camera to a microscope, sample the analogue image appropriately, add a scale bar to your final result and display or print your image for others to evaluate.

Chapter 1 opened with the statement 'Ah, I see!' showing how dependent we are on vision. Our sense of vision is also the primary way in which scientists acquire information from their experiments. However, we are programmed to compare images, and since our visual goal is recognition, it can be very inefficient using an image as a scientific descriptor, especially of unfamiliar situations and objects. We are prone to optical illusions, so our perception of an image may therefore have very little in common with a direct photographic recording. Herein lies the power of political cartoons: we can easily recognise parallels with the merest of guidance, filling in the 'gaps' with our minds. The mental impression of a scene is transitory; perhaps that's why image and imagine have the same root. Because our visual system evolved to compare new information to previously recognised events and scenes, we are disadvantaged in direct quantitative evaluation of images without extracting and reducing the image information to numerical data. Context is important for evaluating scientific image data: this is particularly true when trying to reconstruct 3D models from 2D data sets. As Russ (see Further Reading) remarks: 'nothing in our evolutionary experience has prepared us to handle section images, we are much better at tracking movement'.

The purpose of this book is primarily to enable you to set up and use a light microscope to best advantage practically, but it is also worth remembering that whether you view the image directly or use a camera to record the image from the microscope, human vision and the way we evaluate pictorial information is the ultimate arbiter of our use of the light microscope.

1. The larger the field of view, the more challenging it is to secure an evenly lit field.
2. Image-forming rays leaving the microscope at the exit pupil are parallel. They are focused by the lens of the relaxed eye to form an image on the retina.
3. The best way in which to form a real image for photomicrography is to mimic the lens of the eye and use a converging lens. This is usually referred to as a Telan or tube-lens and is an intrinsic component of modern microscopes used to form the primary image plane where the camera sensor should be located.
4. Digital camera sensors are more sensitive than our eyes to differences in intensity across the field of view.
5. Flat-field correction involves subtracting a background image of shading from uneven illumination. It also removes local inhomogeneities and specks from the image field.
6. Binning combines the signal from adjacent pixels, increases the sensitivity of the camera and improves dynamic range and the signal-to-noise ratio. The trade-off is loss of spatial resolution.
7. The *minimum* sampling required by Nyquist's sampling criterion is two pixels per resolvable unit.
8. Data should ideally always be deconvolved after acquisition.
9. Use a scale bar in your images because the eye is poor at estimating length.
10. An image should have a minimum of 300 pixels/inch for printing as hard copy.
11. Gamma preserves the ranking of the intensities of the data, relating digital intensity to display brightness.
12. For colour-deficient individuals, use magenta or vermillion for the red channel and blue-green or turquoise for the green channel pseudocolour.
13. Morphometry quantifies a 2D structure; stereology derives 3D data by sampling in two dimensions.
14. Save images in uncompressed *.tif format. Other formats (e.g. jpeg) compress data.
15. Scientific journals publish detailed guidelines to limit digital manipulation of images.

Key Reading

Pawley, J.B. (2006). 'Points, Pixels and Gray Levels: Digitizing Image Data'. In: Pawley, J.B. (ed.), *Handbook of Biology Confocal Microscopy*, Chapter 4. New York: Springer, pp. 59–79.

- The information and advice given in this chapter is often forgotten. It is an excellent explanation of how to digitise and display image data.

References

Aherne, W.A. and Dunnill, M.S. (1982). *Morphometry*. London: Edward Arnold.

Allan, C. et al. (2012). 'The Open Microscopy Environment Remote Objects (OMERO) software platform provides a server-based system for managing and analyzing microscopy images and non-image data'. *Nature Methods 9*/3: 245–253.

Bracegirdle, B. (2010). 'D.I.Y. digital photomacrography'. *Quekett Journal of Microscopy 41*/3: 183–191.

Bracegirdle, B. (2012). 'Specimen support for photomacrography'. *Quekett Journal of Microscopy 41*/7: 575–584.

Bracegirdle, B. and Bradbury, S. (1995). *Modern PhotoMICROgraphy*. Royal Microscopical Society Handbook 33. Oxford: Bios Scientific.

Bradbury, S. (1991). *Basic Measurement Techniques for Light Microscopy*. Royal Microscopical Society Handbook No. 23, Oxford Science Publications. Oxford: OUP.

Burel, J.M. et al. (2015). Publishing and sharing multi-dimensional image data with OMERO. *Mammalian Genome 26*/9–10: 441–447.

Cremona, J. (2014). *Extreme Close-Up Photography and Focus Stacking*. Marlborough: The Crowood Press Ltd.

Cremona, J. (2018). *Beyond Extreme Close-Up Photography*. Marlborough: The Crowood Press Ltd.

Entwistle, A. (1998). A comparison between the use of a high-resolution CCD camera and 35 mm film for obtaining coloured micrographs. *Journal of Microscopy 192*/2: 81–89.

Entwistle, A. (2005). 'Digital images in science: Fair or fraud?' *The Biochemist 27*/5: 17–22.

Goodwin, P.C. (2007). 'Evaluating optical aberration using fluorescent microspheres: methods, analysis, and corrective actions'. In: Sluder, G. and Wolf, D.E. (eds.), *Digital Microscopy*, 3rd ed, Chapter 18. Amsterdam, Boston, Oxford: Academic Press, Elsevier.

Hammond, C. (2006). 'Symmetrical ray diagrams of the optical pathways in light microscopes'. *Microscopy and Analysis* September 2006. Wiley.

Hazelwood, K.L. et al. (2007). 'Entering the portal: understanding the digital image recorded through a microscope'. In: Shorte, S.L. and Frischknecht, F. (eds.), *Imaging Cellular and Molecular Biological Functions*, Chapter 1. Berlin: Springer, pp. 3–43.

Howard, C.V. and Read, M.G. (2010). *Unbiased Stereology*, 2nd Rev. Ed. QTP Publications.

Kask, P. et al. (2016). 'Flat field correction for high-throughput imaging of fluorescent samples'. *Journal of Microscopy 263*/3: 328–340.

Lambert, T.J. and Waters, J.C. (2014). 'Assessing camera performance for quantitative microscopy'. *Methods in Cell Biology 123*: 35–53.

Landini, G. and Perryer, D.G. (2009). 'Digital enhancement of haematoxylin and eosin stained histological images for red-green colour blind observers'. *Journal of Microscopy 234*/3: 293–301. See also: *Nature Methods* (2011) *8*/11: 891–892.

Linkert, M. et al. (2010). 'Metadata matters: access to image data in the real world'. *Journal of Cell Biology 189*/5: 777–782.

Piper, J. (2010). 'Software-based stacking techniques to enhance depth of field and dynamic range in digital photomicrography'. In: Hewitson, T.D. and Darby, I.A. (eds.), *Histology Protocols Methods in Molecular Biology*, vol. 611, Chapter 16. New York: Humana Press, pp. 193–210.

Roeder, A.H.K. et al. (2012). 'A computational image analysis glossary for biologists'. *Development 139*/17: 3071–3080.

Smith, K. et al. (2015). 'CIDRE: an illumination correction method for optical microscopy'. *Nature Methods 12*/5: 404–406.

Sysko, L.R. and Davis, M.A. (2010). 'From image to data using common image processing techniques'. *Current Protocols in Cytometry*. Chapter 12: Unit 12.21.

Webb, R.H. and Dorey, C.K. (1995). 'The pixelated image'. In: Pawley, J.B. (ed.), *Handbook of Biological Confocal Microscopy*, 1st ed, Chapter 4. New York Springer, pp. 41–51.

Wolf, D.E., Samarsekera, C. and Swedlow, J.R. (2007). 'Quantitative analysis of digital microscope images'. In: Sluder, G. and Wolf, D.E. (eds.), *Digital Microscopy*, 3rd ed, Chapter 17. Amsterdam: Academic Press, pp. 365–396.

Wong, B. (2011). 'Color blindness'. *Nature Methods 8*/6: 441. See also: Wong, B. (2011). 'Points of view: Avoiding color'. *Nature Methods 8*/7: 525.

Further Reading

Bradbury, S. (1991). *Basic Measurement Techniques for Light Microscopy*. Royal Microscopical Society Handbook No. 23, Oxford Science Publications. Oxford: OUP.

Cromey, D.W. (2013). 'Digital images are data: and should be treated as such'. In: Taatjes, D.J. and Roth, J. (ed.), *Methods in Molecular Biology*, Chapter 1, vol. 931. *Cell Imaging Techniques*. New York: Humana Press, pp. 1–27.

Eliceiri, K.W. et al. (2012). 'Biological Imaging software tools'. *Nature Methods 9*/7: 697–710.

Greaves, P.M. (2010). 'Digital camera and microscope'. *Quekett Journal of Microscopy 41*/4: 279–291.

Heintzmann, R. (2006). 'Band-limit and appropriate sampling in microscopy'. In: Celis, J.E. (ed.), *Cell Biology: A Laboratory Handbook*, vol. 3, 3rd ed, Chapter 3. Elsevier Academic Press, pp. 29–36.

Hinchcliffe, E. and Sluder, G. (2007). 'Working with classic video'. In: Sluder, G. and Wolf, D.E. (eds.), *Digital Microscopy*, 3rd ed, Chapter 7. Amsterdam: Academic Press, pp. 141–157.

Model, M. (2014). 'Intensity calibration and flat-field correction for fluorescence microscopes'. *Current Protocols in Cytometry 68*: 10.14.1–10.

Rossner, M. and Yamada, K.M. (2004). 'What's in a picture? The temptation of image manipulation'. *Journal of Cell Biology 166*/1: 11–5.

Russ, John C. (2004). 'Seeing the Scientific Image'.
Part 1: *Proceedings of the Royal Microscopical Society* June *39*/2: 97–114.
Part 2: *Proceedings of the Royal Microscopical Society* Sept *39*/3: 179–193.
Part 3: *Proceedings of the Royal Microscopical Society* Dec *39*/4: 267–281.
http://www.drjohnruss.com/downloads/seeing.pdf.

Web, R.H. and Dorey, C.K. (1990). 'The pixelated image'. In: Pawley, J.B. (ed.), *Handbook of Biology Confocal Microscopy*, 1st ed, Chapter 4. New York: Springer, pp. 41–51.

Weiss, D.G., Maile, W., Wick, R.A., and Steffen, W. (1999). Video Microscopy. In: Lacey, A.J. (ed.), *Light Microscopy in Biology*, 2nd ed, Chapter 3. Oxford: Oxford University Press, pp. 73–149.

Internet Resources

Camera coupler contacts

1. Eric Best Scientific Couplers: https://www.bestscientificcouplers.com/.
2. SRB-Griturn: http://www.srb-griturnengineering.co.uk/.
3. Indigo Scientific: http://www.indigo-scientific.co.uk/microscope-to-camera-couplers.html
4. Zeiss: https://www.micro-shop.zeiss.com (register for password).
5. Richard Kinch: http://www.truetex.com/micad.htm.

Register distance

List of values: http://m43photo.blogspot.com/2010/04/register-distance.html.
List of values: http://www.graphics.cornell.edu/~westin/misc/mounts-by-register.html.
List of values: https://en.wikipedia.org/wiki/Lens_mount.

Eyepiece adapters

Alan Wood website: http://www.alanwood.net/photography/olympus/eyepiece-adapters.html.

Remote capture software

1. Canon: http://cpn.canon-europe.com/content/education/infobank/image_download/remote_capture.do.
2. Cam4you: http://alkenius.no-ip.org/Cam4you/.
3. Breeze Systems: http://www.breezesys.com/products.htm.

Ruled coverslips

Cellattice: http://www.nexcelom.com/Products/Cellattice.html.
Scalebars and graticules Pyser-SGI Ltd: https://www.pyseroptics.com/graticules/stage-micrometers-calibration-scales-grids/.
Graticule (reticule) diameters for common microscopes https://www.pyseroptics.com/graticules/technical-guidance-notes/reticle-diameters-for-common-microscopes/.

England Finder

Pyser-SGI Ltd: Fircroft Way, Edenbridge, Kent, United Kingdom. TN8 6HA.
www.pyser-sgi.com, or any good microscopy equipment supplier (e.g. TAAB).

Colour-blind display

1. http://vischeck.com.
2. http://www.colorvisiontesting.com/what-colorblind-people-see.htm.
3. http://jfly.iam.u-tokyo.ac.jp/color/index.html.
4. https://www.adobe.com/accessibility/products/photoshop.html.
5. *Nature Methods* Points of view data visualization series.
   http://blogs.nature.com/methagora/2013/07/data-visualization-points-of-view.html.

Gamma adjustment

1. http://www.photoscientia.co.uk/Gamma.htm.
2. http://w4zt.com/screen/.
3. http://www.normankoren.com/makingfineprints1A.html.

Image-stacking software

1. http://chdk.wikia.com/wiki/Stacking_Software.
2. https://en.wikipedia.org/wiki/Focus_stacking.

Combine ZM – freeware: http://combinezm.en.lo4d.com.
Picolay – freeware: http://www.picolay.icbm.de.
Helicon Focus – shareware: http://www.heliconsoft.com/.
Tim Niemann webpage: https://www.epaperpress.com/psphoto/stacks.html.
ImageJ plugin: http://bigwww.epfl.ch/demo/edf/.

High dynamic range: https://en.wikipedia.org/wiki/High-dynamic-range_imaging.

Printer calibration

1. https://www.color-management-guide.com/printer-calibration.html.
2. http://asktimgrey.com/2016/08/23/printer-calibration-frequency/.

Monitor calibration

1. https://www.epaperpress.com/monitorcal/index.html.
2. http://www.lagom.nl/lcd-test/?.
3. https://www.imaging-resource.com/ARTS/MONCAL/CALIBRATE.HTM.
4. https://www.cnet.com/how-to/how-to-calibrate-your-monitor/.
5. http://www.cambridgeincolour.com/tutorials/monitor-calibration.htm.

Colour management:

Norman Koren webpage: http://www.normankoren.com/color_management_2.html#Working.
Raster versus graphics editors: https://en.wikipedia.org/wiki/Graphics_software.

Nyquist sampling calculators

1. SVI = http://www.svi.nl/NyquistCalculator.
2. Iscope = http://iscopecalc.com.

Notes

1. A C-mount has a male thread of 1″ diameter, 32 threads per inch and a shoulder designed to be 0.69″ (17.526 mm) from the surface of the CCD/sCMOS detector. This screws into a female thread on the camera. The C-mount has survived from the days when cine cameras, then video cameras, were coupled to microscopes. Similarly, the term 'ENG-bayonet' derives from the acronym for **E**lectronic **N**ews **G**athering, a term from the world of TV reporters who use these cameras. A trend towards smaller camera chips and lenses of shorter focal length has resulted in so-called CS cameras. These should be attached to a C-mount with a spacer of 5 mm to give the correct C-mount-to-chip distance. An alternative mount (e.g. Nikon F-mount) should be used if filter wheels placed in front of the camera need more clearance than 17.5 mm.
2. See the Internet Resources in this chapter for an online Nyquist sampling calculator.
3. In Chapter 4, 'The Pixelated Image' in the first edition (1990) of the *Handbook of Biological Confocal Microscopy*, a Nyquist sampling limit of 2.3 is advocated by Webb and Dorey on the basis that sampled information must be reconstructed through a filter, such as the eye, half-tone printing or the A/D converter which limits f_H (where f_H is 1/resolving power, the optical bandpass of the microscope. f_0 is the cut-off frequency; it is often equivalent to f_H, but may differ) contributing to the image. Webb and Dorey argue that Nyquist/3 is unnecessary over-sampling. The radius of the Airy disc is sometimes known as a resolving element, or resel (Figure 31.9).
4. Applying Nyquist's sampling criterion is not only about recording data, it is also about treating it properly before it is 'reconstructed'. See Pawley (2006) and Webb and Dorey (1990) for a detailed explanation of the theory behind why you should deconvolve all data.
5. Strictly speaking, this relationship applies only to prints that have a 4:5 aspect ratio. For consistency, use the formula 1.28H, where H is the long dimension of the print. In accordance with this expression, the normal viewing distance for a 90″ × 144″ print becomes 185 inches.
6. Another approach is to use HDR (high dynamic range) software to combine images, in much the same way as image stacking for depth of focus; see: https://www.techsupportalert.com/best-free-high-dynamic-range-hdr-software.htm.
7. John Dalton first wrote about colour blindness in 1794. The condition is sometimes called 'Daltonism' in his honour. Dalton requested that his eyes be examined after his death, in an attempt to discover the cause of his colour blindness. DNA analysis by the PCR in 1995 showed that he suffered from protanopia. Aside of blue and purple, he could only recognise yellow, confusing scarlet with green, and pink with blue.
8. There is every reason to alter images and digital slideshows to accommodate colour-blind people, and very good software exists to do this, for example Daltonize on the Visicheck website. For further details, see the Internet Resources given in this chapter, together with Section 11.5 in Chapter 11 and Landini, G. and Perryer, D.G. (2009).
9. Guidelines, for example, for three leading journals can be found at http://jcb.rupress.org/fig-vid-guidelines, https://www.cell.com/figureguidelines, and https://www.nature.com/authors/policies/image.html.

Appendix 1
Buying, and Tendering for, a Light Microscope

Purchasing a high-end research microscope of any sort involves complex decision making. Even experienced microscopists will not have at their fingertips all the relevant details – this is where you must establish a relationship with and, to some extent, rely upon the sales people and company applications specialists. It is very possible that you will be more familiar with the staff and company profile of one manufacturer than another. Hard though it may be, it is wise not to let this detract from the merits of other instruments on offer. Some facility managers will go for one particular brand regardless, and they are well known in the community of microscopists. There is nothing inherently wrong with this; however, you will serve your purposes best if you can keep an open mind. Buying a new microscope and its associated instrumentation is all about selecting on merit alone the microscope and the company who supplies and/or manufactures it.

Reputable companies will offer a demonstration using your samples. Ideally they should come to you and do this in-house. If you are taken as a guest to their demonstration centre, it is good to have a fixed list of tests that can be replicated on other companies' microscopes, or in your own institution. Most companies will be prepared to leave their equipment for you to try in your laboratory. Clearly, this is an expensive undertaking for them, with no guaranteed prospect of a sale. In practice it is very difficult to fully evaluate any microscope and associated software in such a short time window. The applications specialist will be familiar with the microscope and software and be able to effortlessly produce good results. It may be a very different matter for you or your colleagues and a very time-consuming learning curve. Instrument foibles, and your preferences, only become apparent much later. Nevertheless, think about which test specimens to use (see also the references at the end of this Appendix). Have a list of specific questions that require answering, and be prepared to weigh their importance for tender scoring later. Be specific, because general questions have very little value and can always be answered in a positive light by any good salesman.

- Will your samples give a fully representative picture of the research you want to do?
- What is the lifetime of the equipment? Will it become obsolete; are there sufficient spares available?
- How adaptable is the equipment? Is it future-proof for upgrading to meet future requirements? This might include fitting live-cell imaging equipment, or upgrading a confocal to a multi-photon role. Or perhaps as mundane as a Windows operating system upgrade forcing the purchase of new hardware and/or firmware and software.
- Does the microscope use a Linux-based real-time controller? This helps maintain an accurate clock in time-lapse experiments and allows multiple processing operations to be performed in real time on a Windows OS.
- Is the software easy to learn and use? Is it sufficiently flexible for the job in hand?
- How well does the hardware and software workflow operate; is it efficient, or over-complicated?
- Is the software GUI well laid out? Can *you* easily teach colleagues and students how to use it later?
- How frequently is the software upgraded, and will this be a significant extra cost? How many off-line licences can you get – are these free or do they cost a lot? Can the image files be opened and worked on in ImageJ/Fiji?
- How well does the software run on Windows or Mac platforms? Can your IT department maintain the software? Do you need to write macros?
- How good is the applications and service support? What is the cost of service support out-of-warranty?
- Who provides service applications support? (If you are a lone operator of an imaging service provider, trust me: you *will* rely upon your company specialists.) Can they visit you easily, and are they sufficiently experienced?
- How quickly do the company service engineers respond? Is this verified independently by other users?
- Is the company reputable? Is it stable? And a very different issue: are key staff likely to remain?

Because the sales people are under pressure to show a return for their time and the company investment in a demonstration, be aware that time to carry out a full testing regime may be limited. You will generally only hear about the positive attributes of the microscope and associated instrumentation, unless some issue is well known, glaringly obvious, or the company representative is brutally honest – and we rarely are all the time. For the same reason, you will usually hear about negative aspects and weaknesses from competitors' sales

Understanding Light Microscopy, First Edition. Jeremy Sanderson.

people. It is then a question of balancing and weighing all the information that you receive, for the most accurate overall picture – your end aim – truly is the sum of many small things.

The Confocal Listserver is a good source to pose questions and receive advice, but you have little means of weighing the merits of what you might be told, save that several respondents may be in agreement, since there is no way to peer-review online material. Usually, any recommendation, whether from a local colleague or solicited online, reduces a complex purchasing decision to an artificially simplistic answer. Whilst it is worth having recommendations, especially from colleagues known to you, do bear in mind that once people have purchased very expensive equipment, they generally have a vested interest, even unconsciously, of defending their purchase as the correct one. It is easy to lose sight of the fact that their purchasing criteria will probably be different from your own. Your circumstances, the work you do, the people you support (if you run an imaging service facility) are unique to your organisation.

Company salesmen will always prefer to know the budget that you have to spend. Whether you tell them or not is up to you. There is a school of thought that you shouldn't tell them, and you keep your trump cards to yourself. My policy has always been to establish a relationship with the companies and to be open and up-front, but it is up to you what you chose to do. Your aim is to achieve a successful win-win outcome. It is worth remembering that the world of professional microscopists is a small one: what goes around comes around. Like any other relationship, *trust* and friendship with the company applications specialists and service engineers *will* pay dividends. Once you've bought your microscope on its merits, you have to work with the company staff: play the futures game and win their respect. The view that company representatives are merely hawkers is an outdated and short-sighted view, but surprisingly it does persist. All good salespeople can accept why they did not win a sale. What they dislike intensely are customers who are not straight with them. You can't buy from everyone, and it's never nice to give bad news to good people, but this cannot be avoided. Don't hide behind your procurement staff: tell them yourself and be honest; they can read between the lines just as well as you can.

When buying very expensive equipment, you will have to put the purchase out to tender if the amount exceeds £30 000. In both Europe and the USA this is regulated by law in order to give vendors a fair and level playing field. As a guide, an EC tender is required for amounts over £156 000 and in the United States the equivalent is $100 000.

Some of the things to think about concerning the microscope and associated hardware are:

- Think about the performance specifications that you want to include in the tender. Can these be met?
- Do you have sufficient space, electrical power, compressed air (anti-vibration tables) or gas cylinders (A-V tables; and live-cell requirements)? Can you fit a third-party environmental chamber to the microscope?
- Do you have a budget for third-party image analysis software? Is it compatible with the image files?
- Do you need to incorporate third-party cameras or detectors? Will the microscope software support these, or will a software upgrade be required, and who will pay for this?
- What objectives are offered on the microscope?
- What contrast-enhancement options are there?
- How many fluorescence filter blocks can be held in the microscope? Can these be added or exchanged easily by the user (e.g. to take advantage of new fluorophores/fluorescent proteins)?
- What lasers might you need? Can these be upgraded or replaced easily – is it expensive to do so?
- How much automation do you require? Do you need a motorised stage in x, y, and z, or a fast z-step motor?
- Is the focus control stable over time? (absolutely essential for live-cell time-lapse work)
- Can the stage holder accommodate all the sample holders that you might need (e.g. for live-cell work)?
- Will you have money later for add-ons (e.g. time-lapse incubators), or can they be bought at a discount now with the rest of the equipment in one package?
- How good are the digital detectors? Are PMTs noisy because of the company's unique circuitry?
- Are there sufficient USB ports? Even unused plugged-in USB peripherals use bandwidth.
- Can you fit, or is there supplied, a separate graphics card for camera(s). This will help not to slow down the PC.

The funds available for the microscope purchase may be less than the total available to you if you have to make alterations to the building or the room in which you intend to install the new microscope. A good sales support person will help you check that all the equipment can indeed be delivered and installed. It is awkward and potentially very embarrassing if, for example, the lift access or a crucial doorway on the route in is found to be too small. Check that there will be adequate vibration isolation for the microscope: high-end research instruments invariably require some form of anti-vibration table or stand. It is very important to check the air-conditioning and the stability of temperature and humidity in the room into which the new microscope will go. Any large or harsh changes will cause the focus of the microscope to drift. Do not position the microscope directly under the air-conditioning vent, and fit these (and possibly the door sills) with dust filters. Try to keep the room air-conditioning under your local control rather than being administered remotely by the estates department of your organisation. If performing fluorescence microscopy or if using cameras, check that the room lighting can be controlled, so that images with an optimum signal-to-noise ratio can be collected without additional glare from the overhead room lighting. Is there sufficient space to have a functional work-flow around the microscope? Are health & safety regulations complied with as far as loose electrical cables are concerned? Do you need to factor in the cost of extra power sockets and/or data networking points? Bear in mind that modern microscopes have the potential to generate many large data sets. It is quite likely that you may need your own dedicated terabyte server to store the data generated, and the cost of buying and running this needs to be considered. Is there an easy means to stream data off a network, or will users be forced to use USB sticks?

The final decision is made on the sum of small things. The major microscope manufacturers' product(s) that you are considering are all fairly similar, and there is a good chance that each will do 90% of the tasks that you have specified. This makes the final decision both easy and difficult. What you are aiming to purchase is the most cost-effective and versatile microscope concomitant with superb image quality and service support. Some new technological development or unique requirement of your research work may make this straightforward; otherwise, you really will have to thoroughly sift through all the data and answers to your questions that you have collected. Remember that an instrument may be written down by your organisation over 8 to 12 years; the evolution of new technology generally occurs every 5 to 7 years. Are you considering 'old' technology and will your purchase soon become obsolete, or will it upgrade successfully to meet future needs? How soon will you be able to apply for a new grant to raise money for the replacement? Upgrading firmware and hardware forced upon you by a change in computer operating systems can be very expensive, and are often unavoidable.

Before making a judgement on price alone, draw up a shortlist of suitable microscopes on their merits, listing all the pros and cons, as well as convenience and ergonomic features that make the microscope functional and attractive to you and your potential users. Remember to include the applications and service support that you expect, and how well the companies will deliver on this post-sales support. If, when you have narrowed down your choice of microscope, the one that you prefer is more expensive than another, negotiate with the representative, or get the purchasing officer of your organisation (if one exists) to do so. Factor in the intangible requirements, such as training: your expectations may not wholly align with what the company intends to deliver and should be negotiated and clearly understood. Once you have bought an instrument – believe me – the relationship changes: you are dependent upon the company, and you need to be confident that they will support you adequately.

Further Reading

Centonze, V. and Pawley, J.B. (2006). 'Tutorial on practical confocal microscopy and use of the confocal test specimen'. In: Pawley, J.B. (ed.), *Handbook of Biological Confocal Microscopy*, 3rd ed, Chapter 35. New York: Springer, pp. 627–649.

Murray, J.M. et al. (2007). 'Evaluating performance in three-dimensional fluorescence microscopy'. *Journal of Microscopy 228*/3: 390–405.

McNamara, G. et al. (2017). 'Microscopy and image analysis'. In: Sanderson, J. (ed.), *Current Protocols in Human Genetics*, Appendix 4.4.1, *94*: 4.4.1–4.4.89. doi: 10.1002/cphg.42.

Zucker, R.M. and Chua, M. (2010). 'Evaluation and purchase of confocal microscopes: numerous factors to consider'. *Current Protocols in Cytometry* 2.16.1–2.16.12, October 2010. doi: 10.1002/0471142956.cy0216s54.

Appendix 2
Troubleshooting Poor Image Quality

The flow chart is designed to help you set up a bright-field light microscope and adjust it for proper Köhler illumination, as described in Chapter 9, Section 9.11. The lists labeled A–G, printed below and referred to within the flow chart, identify and suggest the most likely causes of poor image quality and the issues which may need addressing.

List A **Checking an obstruction in the light path**

List B **The light coverage of the illuminated field is inadequate**

List C **The image is blurred, hazy and indistinct**

List D **Specimen is of low contrast**

List E **Does the image drift or move?**

List F **Causes of poor resolution of the image**

List G **Setting the eyepiece dioptre controls**

List A Checking an obstruction in the light path

The key to locating an obstruction in the light path is to move or rotate optical components, filters or accessories, in turn, that may be placed in either the imaging or illuminating sets of conjugate planes.

1. A grossly-misaligned condenser may occlude the light path even when the diaphragm is fully open. Check the condenser is properly located in the condenser substage, is set at the correct height and is centred on-axis. On an older stand, also check the mirror.
2. Check the objective is properly clicked into its detent position.
3. Check for filter holders being half pushed into position – these show as crescents in the field of view. Sometimes a blanking plate has been pushed part way in.
4. Check that all filters are out of the light path and that the polar and analyer filters are 'uncrossed'.
5. Check that any phase-contrast condenser annulus or dark-field stop (if fitted) has not been left in position in the condenser carousel, or left mis-aligned.
6. Check that any photo-tube push rod is located correctly – allowing light to the eyepieces – and not half in position. Likewise, ensure any software-driven shutter is open.
7. Some microscopes have a 'dark slide' to prevent fluorescence light reflecting off the top lens of the condenser. If no transmitted-light is seen, check that this dark slide is not in position.

List B The light coverage of the illuminated field is inadequate

1. Is the condenser properly seated in the substage, focused on the specimen and centred onto the optical axis? Where a diffuser is supplied, this may help correct an unevenly-illuminated field of view.
2. If the illuminated field is too small, either flip-out, or remove, the condenser top lens and use an objective less than 10x magnification. Or, with the top lens remaining in, try a higher magnification objective to give a smaller field of view.
3. Where dark spots are seen with uneven illumination, this is most likely due to the condenser height being wrongly set, the condenser diaphragm being wrongly set, or the top-lens being filled out.
4. Refer to List C and check for any dirt on objectives, eyepieces, condenser and surface of the illuminated field diaphragm.

Understanding Light Microscopy, First Edition. Jeremy Sanderson.

List C The image is blurred, hazy and indistinct

1. Is the objective dirty? This is the most common cause, and the first to consider. Is there unwanted oil on dry objectives, or old oil dried onto immersion objectives?
2. Are there unsharp spots in the image? Check for dirt on the lens surfaces or CCD camera faceplate.
3. Is the slide upside down on the stage? Is the specimen too thick (i.e. thicker than the depth of field of the objective)?
4. Is the specimen poorly prepared and mounted, the preparation 'cloudy' or not cleared properly? Are there air bubbles in the mounting medium and/or immersion oil? Has mounting medium or old oil been left on top on the coverslip? Is the coverslip clean?
5. Has a coverslip of the correct thickness been used? Is the wrong refractive index immersion medium being used?
6. Is the objective fitted with a correction collar that is not properly set?
7. Does image sharpness change with focus at the edge of the field of view? Use a plan-objective.
8. Is the condenser top-lens inadvertently flipped out of the optical path: poor resolution?

List D Specimen is of low contrast

1. Are either the illuminated field- or condenser-diaphragms opened too far? Is the light too bright?
2. Is the specimen intrinsically of low contrast anyway?
3. Is the layer of mountant too thick, or has the wrong mountant been used?
4. Has the wrong colour filter (if any) been used? Use a complementary colour filter.
5. Are the objectives, eyepieces and any other components in the optical train clean?

List E Does the image drift or move?

1. If the mountant is liquid, is the specimen or coverslip moving? Sample not secure within the preparation, or moves because it is living.
2. If the coverslip moves, do not use immersion objectives: shear forces will damage the sample. Either allow the mountant to harden, remove excess fluid, or let the specimen settle.
3. A drifting cloud, or a line forming the edge of a bubble, preventing clear focusing of the image means an air bubble is present in the oil (or water) immersion medium. There also may be insufficient immersion medium.
4. A focus drift when altering the fine focus may indicate the surface of the coverslip is stuck to the objective with oil. Replenish the oil. Secure the coverslip on the sample preparation.
5. Is the objective securely screwed into the nosepiece and centred on the optical axis?
6. If the above do not apply, there may be a possible fault with the focus mechanism.
7. If the image drifts out of focus immediately, the microscope body or stage may sink under its own weight. Get a specialist engineer to check and relubricate the rackwork or reset the electronic focusing mechanism.
8. If the focus drift is over time, it is likely to be thermal drift if using a heated stage incubator or heated incubation chamber for time-lapse live imaging. Let the system settle – see Section 28.6.

List F Causes of poor resolution of the image

1. Is the numerical aperture of the objective too low to resolve detail? Can a 'rotten' image be seen?
2. Is the eyepiece magnification too low to present all resolved detail to the eye?
3. Is the condenser NA set too low, and so the back focal plane of the objective is not fully filled with light? This will also occur if the condenser is set too low at the wrong height to start. Is the condenser top-lens inadvertently flipped out of the optical path?
4. Is the objective fitted with a collar to adjust the numerical aperture? Is this closed or set too low?
5. Are the objectives, eyepieces, condenser and other optical components all clean?

List G Setting the eyepiece dioptre controls

1. Set the microscope up using a contrasty, fine-detailed specimen.
2. Set the interocular distance on the binocular head (if fitted) for comfortable use.
3. If the microscope has only one adjustable eyepiece, ensure the fixed eyepiece is in the dominant eye tube.
4. Focus precisely using the fine focus adjustment.
5. Change to the lowest magnification objective but *do not* readjust microscope focus control.
6. Focus image for each eye by using the individual adjustment for each eyepiece.
7. For a stereomicroscope, the procedure differs slightly. Use a contrasty, fine-detailed specimen, raise the magnification to maximum using the magnification or zoom step changer.
8. Focus precisely using the microscope's main focus control.
9. Lower the magnification to its minimum value but *do not* readjust the microscope focus control.
10. Focus image for each eye by using the individual dioptre adjustment control for each eyepiece.
11. Repeat steps 1–4; the stereomicroscope should maintain focus throughout its magnification range.

Appendix 3

The Michel-Lévy Interference Colour Chart

The Michel-Lévy chart shown can also be obtained from your local Zeiss representative or agent. A PDF reproduction can be found online in Science Direct at https://www.sciencedirect.com/science/article/pii/B9780124072305500030.

The Michel-Lévy chart is used for quantitative analysis[1] of the colours shown by birefringent samples, usually thin sections. It relates birefringence (Δn), the numerical difference between the two refractive indexes (n_1, n_2), thickness (t) and relative retardation or optical path difference (Γ) for transparent and colourless or lightly coloured samples. Figure 14.10, Section 14.10, shows the derivation of the Michel-Lévy chart from Newton's polarisation colours as a result of destructive interference of individual spectra throughout the visible spectrum.

Since retardation increases linearly with both thickness and birefringence, minerals have specific and predictable colours in polarised light between crossed polars. This chart can help identify minerals because the birefringence values are known for each one.

Providing that two of the three variables from birefringence, thickness or optical path difference are known, the third can be calculated. These are the three equations to use:

retardation (OPD)

$$\Gamma = t \cdot (n_2 - n_1) \quad \Gamma = \Delta n \cdot t \cdot 1000$$

birefringence

$$(n_2 - n_1) = \frac{\Gamma}{t} \quad \Delta n = \frac{\Gamma}{t \cdot 1000}$$

thickness

$$t = \frac{\Gamma}{\Delta n} \quad t = \frac{\Gamma}{\Delta n \cdot 1000}$$

Since birefringence is commonly expressed in nanometres and thickness is measured in micrometres, use the 1000 term to achieve equality of units (right-hand formula in each pair). Use the formula $\Delta n = (n_2 - n_1)$ to calculate birefringence. When using the chart, it is important to remember these points apply:

- Isotropic and opaque minerals or metals cannot be identified.
- Rotate the stage in order to get the maximum birefringence under crossed polars.
- Each mineral, depending on the orientation that it is cut in thin section, may not exhibit the maximum birefringence. It is important to try and find multiple samples in order to get the best value of birefringence.
- Maximum brightness occurs at 45° between crossed polars.
- Uniaxial minerals can look isotropic (i.e. always extinct) if the mineral is cut perpendicular to the optic axis.
- The particular interference colour seen therefore depends upon birefringence, thickness *and* the orientation of the mineral with respect to the optic axis.
- The standard thickness for petrological thin sections is 30 µm.
- Use a compensator to accurately determine the order of the colour of thick specimens of high birefringence. The most sensitive area of the chart is the first-order red, where small changes in retardation cause a large colour shift.

A recent paper (Sørensen, 2013) describes a redesigned Michel-Lévy chart to take account of changes in colour reproduction and printing, computer display colour gamuts, as well as a more precise match to the actual observations under modern illumination in the optical microscope.

Understanding Light Microscopy, First Edition. Jeremy Sanderson.

Worked examples

1. The birefringence of quartz is given as 0.009. This gives the entry point on the chart along the top abscissa or right-hand ordinate. The colour recorded on the Michel-Lévy chart is pale straw. Where the diagonal from 0.009 reaches the correct colour, read off the thickness: 30 μm.

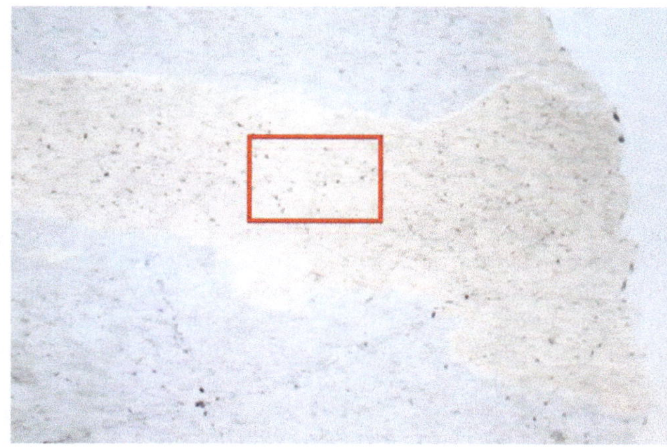

Where the section is thicker, extrapolate upwards. For a quartz colour of second-order green, the thickness of quartz will be ≈ 80 μm; or calculate the thickness from the formula t = 747/(0.009 × 1000) = 83. Thickness = 83 μm.

2. An unknown synthetic fibre examined under crossed polars shows a maximum interference colour of deep blue.

   The fibre has been measured in cross section to be 11 μm thick. From the deep-blue colour at the centre of the fibre, count the number of reds seen between the centre and both edges. There is one, so the blue is second order. Enter the chart on the abscissa at blue = 640 nm. Intercept the diagonal at 11 μm, follow out to the right-hand ordinate and read off the birefringence at 0.060. This equates to nylon, a common carpet fibre.

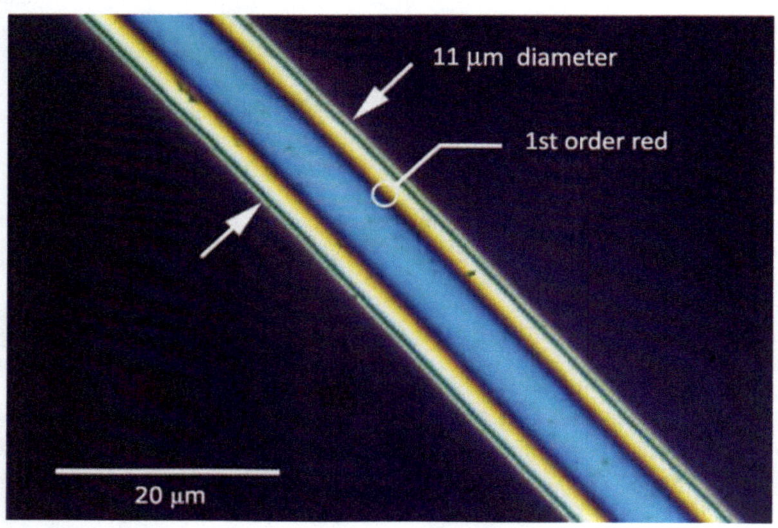

The table below is Newton's colour scale according to Quincke. These colours are found listed in one form or another on most Michel-Lévy interference colour charts. The interference colours seen between both crossed and parallel polars are listed. Both these sets of colours are shown in Figure 14.20, Section 14.16, for the higher-order tints as seen between parallel polars are sometimes easier to identify.

No.	Retardation (nm)	Interference colour between CROSSED polars	Interference colour between PARALLEL polars	Order
1	0	Black	Bright white	First order
2	40	Iron-grey	White	
3	97	Lavender-grey	Yellowish white	
4	158	Greyish blue	Brownish white	
5	213	Clearer grey	Brownish yellow	
6	234	Greenish white	Brown	
7	259	Almost pure white	Light red	
8	267	Yellowish white	Carmine-red	
9	275	Pale straw-yellow	Dark reddish brown	
10	281	Straw-yellow	Deep violet	
11	306	Light yellow	Indigo	
12	332	Bright yellow	Blue	
13	430	Brownish yellow	Grey-blue	
14	505	Reddish orange	Bluish green	
15	536	Red	Pale green	
16	**551**	Deep red	Yellowish green	
17	565	Purple	Lighter green	Second order
18	575	Violet	Greenish yellow	
19	589	Indigo	Golden yellow	
20	664	Sky-blue	Orange	
21	728	Greenish blue	Brownish orange	
22	747	Green	Light carmine-red	
23	826	Lighter green	Purplish red	
24	843	Yellowish green	Violet-purple	
25	866	Greenish yellow	Violet	
26	910	Pure yellow	Indigo	
27	948	Orange	Dark blue	
28	998	Bright orange-red	Greenish blue	
29	**1102**	Dark violet-red	Green	
30	1128	Light bluish violet	Yellowish green	Third order
31	1151	Indigo	Impure yellow	
32	1258	Greenish blue	Flesh-coloured	
33	1334	Sea-green	Brownish red	
34	1376	Brilliant green	Violet	
35	1426	Greenish yellow	Greyish blue	
36	1495	Flesh-coloured	Sea-green	
37	1534	Carmine-red	Green	
38	**1652**	Dull purple	Dull sea-green	
39	1662	Violet-grey	Yellowish green	Fourth order
40	1682	Greyish blue	Greenish yellow	
41	1711	Dull sea-green	Yellowish grey	
42	1744	Bluish green	Lilac	
43	1811	Light green	Carmine	
44	1927	Light greenish grey	Greyish red	
45	**2203**	Whitish grey	Bluish grey	

Source: Quincke, 1868.

References

Delly, J.G. (2003). 'The Michel-Lévy Interference Color Chart – Microscopy's Magical Color Key'. https://bit.ly/2JbLwtf.

Linge Johnsen, S.A. et al. (2018). 'Accurate representation of interference colours (Michel-Lévy chart): from rendering to image colour correction'. *Jour. Microsc. 269*/3: 321–337.

Quincke, G. (1868). 'Ueber Newton'sche Farbenringe und totale Reflexion des Lichtes bei Metallen'. *Pogg. Ann.* 1868. CXXIX, p.177.

Sørensen, B.E. (2013). 'A revised Michel-Lévy interference colour chart based on first-principles calculations'. *European Journal of Mineralogy 25*/1: 5–10.

Note

1. The original form of the Michel-Lévy chart, from which the copies are printed, was hand-painted for Zeiss in 1961–1962. This was done using a tungsten-halogen lamp rated at 3,200 K, and therefore – strictly – to derive proper colour hues from the chart, only this type of lamp ran at this voltage should be used for analysis. However, this is not done in practice.

Appendix 4

Cleaning and Maintenance of the Light Microscope

Poor image quality can usually be attributed to objectives whose front lenses are dirty, contaminated with oil or that have fingerprints on them. If you have hazy images lacking in crisp definition or simply cannot focus sharply, inspect the objectives, eyepieces and condenser front lens, and be prepared to clean them.

All microscopists have their own method of cleaning which they swear by. It is paramount not to scratch the delicate anti-reflection coating that covers the front surface of the lenses of both objectives and eyepieces. Poor cleaning practices can easily make things much worse. Forming a microscopic grit from dirt and cleaning solvents can ruin both glass surfaces and anti-reflection coatings; caustic solvents can also ruin these coatings and dissolve the cement holding lenses in place. Therefore, cleaning should be done sparingly and with care.

It is also important to remember that *immersion oil is not inert*. It can dry down to form a coat on lenses, which can then be difficult to remove. Oil is especially prone to seep by gravity into the internal parts of objectives on inverted microscopes if they are left unattended without cleaning. Immersion oil is the single most serious contaminant in microscopy – yet it is a necessary evil.

High-magnification objectives have short working distances. This is true for both dry and oil-immersion objectives. A 20x/NA 0.8 dry objective will have a working distance of 0.55 mm; moreover, the front lens is concave unlike the flat surface of an oil-immersion objective. It is very prone to collecting immersion oil off the surface of the coverslip. This can happen inadvertently as an oil objective is rotated into position (particularly with a motorised nosepiece). Oil on the front lens of a dry objective is the commonest reason why suddenly you cannot focus such a lens.

Some inverted microscopes have removable slide carriers that are held by spring(s) in a machined recess in the stage. Often the front of the objective passes very close to the underside of the slide carrier as it is rotated in or out of the optical axis. It is thus common to find that oil accumulates on the underside of this carrier, which must periodically be wiped clean with ordinary tissues.

You are advised to confine your cleaning to the front lenses of objectives and condensers; eyepieces, the external glass covering of the illuminated field diaphragm, the CCD camera faceplate (with extreme care!) and the surface of the stand. It is best to get a service engineer to clean internal components requiring disassembly, since attempting to clean these can lead to loss of collimation (for example) of the binocular head or scratching of filters.

Use only specialist lens-cleaning tissues for cleaning any optical surface. Disposable boxed tissues (e.g. 'Kleenex', 'KimWipes' or similar) contain wood pulp and – in some cases – silica particles, and so are exactly analogous to using abrasive sandpaper in the toilet or bathroom for personal hygiene. They *will* scratch and ruin delicate anti-reflection coatings. Zeiss specifically recommend Whatman Lens Cleaning Tissue 105, now made by GE Healthcare Lifesciences. As the advertising blurb says, 'Lens Cleaning Tissues are very soft, high-absorbent, strong, chemically pure, and free from silicones and other additives. Only 0.035 to 0.040 mm thick, they ensure safe removal of surface moisture and grease without leaving fibres'.

A systematic approach will aid cleaning with the minimum risk of ruining objectives:

- locate dirt;
- inspect the surface; and
- clean the lens surface.

Wearing gloves when cleaning is down to personal preference. Some people take the view that gloves prevent contamination from the hands; equally, they can protect the hands from solvents. Ensure that you select a brand of glove that is talc- or powder-free.

Locating dirt

Partially rotate each component (e.g. eyepieces) in turn. Partly swing the objective off-axis, or carefully part-rotate it on its thread on-axis. Likewise, a CCD camera can be rotated on its thread. Check the condenser by moving it up and down with the substage control and moving the top lens out of the field of view. Slide DIC prisms and filters in and out of the light path. Dirt will follow the direction of movement. The only exception to this is that dirt on a CCD camera faceplate will remain static on the monitor.

Understanding Light Microscopy, First Edition. Jeremy Sanderson.

Dirt is commonly found on the upper (eye) lenses of the eyepiece(s), the coverslip, the objective (particularly if inverted), the condenser top lens (particularly if on an upright stand), the underside of the condenser and the glass window of the illuminated field diaphragm in the base of the microscope. It is also possible – but less likely – that dirt can get onto the faceplate of any CCD camera and the lens surfaces of any camera adapter optics (particularly if these components are regularly interchanged).

A blurred image may not always be due to dirt seen in the field of view; consult the fault-finding chart in Appendix 2 to check whether the wrong thickness of coverslip or the wrong immersion oil is introducing spherical aberration and causing poor image quality.

Inspecting the surfaces of objective lenses

Source: Courtesy of S. Laber.

1. Unscrew the objective (use both hands: they don't bounce), and inspect it either with a stereomicroscope (better) or by using an inverted eyepiece (convenient).

   If using the stereomicroscope, screw the lens into the cap of its pot holder, to prevent damage to the rear lens elements, which may protrude beyond the objective screw-thread. You may find that some stereomicroscopes have a limited working distance, and it is not possible to view the top lens of the objective.
2. An alternative method of inspection is to hold the objective up to a strong light and an evenly lit surface, and view any oil or dirt on the top surface of the front lens through the barrel of the objective. This works best for lower-magnification objectives with relatively large-diameter rear lens elements.

Cleaning Procedure

1. Begin by blowing off any dust with a dry 'air duster'. I prefer to use a hand-operated bellows ball rather than a canister with propellant, since if the can is inverted, an invisible film of aerosol propellant can coat the objective.

   Resist the tendency to blow particles off with your breath. The saliva mist, together with the pre-existing dirt, and any particles from your mouth will form an abrasive micro-slurry that will cause more harm than good.
2. Never dry wipe an objective, since there is even a risk that dry lens tissues may cause scratching. Always use them dampened with a suitable solvent. However, it is possible to use an ordinary disposable tissue to wipe excess oil off the objective mount *surrounding* the front lens, provided that you take care not to touch the lens element itself with the tissue. This leaves much less oil in place to be cleaned off at the next step.
3. Clean objectives by placing them face upwards screwed into the lid of their storage pot, and draw across a single lens tissue wetted with solvent. There is no need to press onto the surface of the lens when doing this – you are not window-cleaning.

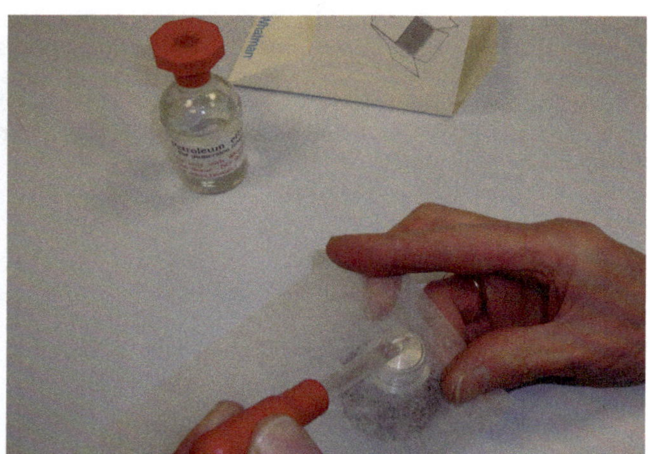

   Repeat a couple of times with fresh tissue, and let the solvent clean off the oil. With a bit of practice, it is possible to use two or three separate *areas* of one piece of lens tissue to do this. However, it is best to use one piece of lens tissue only once.
4. Clean oil-contaminated dry objectives with a clean cotton bud; soak the tip in solvent, and wipe out the excess oil. Repeat with an unused, freshly wetted cotton bud.

   Take care to use real cotton buds, not synthetic ones made of polyester. This latter can dissolve in solvents used for cleaning and end up smearing the objective with a 'glue' that will be difficult to remove.

5. Clean eyepieces with a cotton bud or cotton wound onto an applicator stick. Clean from the centre outwards and not in a zig-zag motion across the eye lens surface. This latter action is likely to sweep the dirt around to another location on the lens. Neither the lower field lens nor the internal surfaces of the eyepiece should require cleaning, and the eyepiece should not be dismantled.
6. Filters (e.g. in fluorescence filter cubes) should not need cleaning, since they are housed within the body of the epi-illuminator. However, if they do need cleaning, clean the outer surface of the circular excitation and emission filters as for eyepieces, bearing in mind that the anti-reflection coatings will be much more delicate than those of the eye lens of the eyepiece.

   If you do need to clean the dichromatic mirror, tape a thin piece of card (or the flat metal key used to unscrew the filters from the fluorescence filter block) edge onto the bench, so that it protrudes a few millimetres and provides a 'stop' for holding the dichromatic mirror (this is shown in Figure 15.12). Lay down a few sheets of lens tissue, place the mirror on these up against the stop and clean the surface by dragging a fresh piece of lens tissue, wetted with ethanol, across the top surface of the mirror.

   Differential interference contrast (DIC) sliders should only very rarely need cleaning, since they are normally located within the microscope stand for much of their lifetime. If they do need cleaning, do so with ethanol or a mild surfactant, such as Tenside solution. Do not use organic solvents, such as acetone.
7. With microscopes fitted with CCD cameras, any dirt on the CCD faceplate is difficult to remove. It is highly likely that in removing one dirt particle, another will be introduced. Therefore, approach cleaning with extreme patience and care, or otherwise call in an accredited service engineer. Only attempt to clean those CCD cameras whose faceplate sensor is protected by a filter or glass window. In some cameras, the sensor surface is directly exposed within the camera body and is highly likely to attract dust and other debris. This should be left to a trained engineer, in order to avoid direct mechanical and static charge damage to the CCD faceplate.

   If the dirt cannot be blown away, wind fresh cotton wool onto a long applicator stick, then wind on a strip of lens tissue, and use this. Wet the lens tissue and remove the dirt. Take great care to limit the pressure put on the window surface. Wind a fresh piece of lens tissue onto the applicator stick for each cleaning attempt.
8. The best way to clean non-optical parts of the microscope is to use a soft, slightly water-dampened cloth. If you do use a solvent, test it first on a tiny part of the rear surface of the stand! Do not use an organic non-polar solvent such as xylene, chloroform or acetone, because they tend to attack paint and some may attack the metal surface. Xylene is carcinogenic and since it is oily, will leave a greasy residue on the surface of the glass once it has evaporated.

Sometimes surfaces may be contaminated with sticky organic compounds. These will usually clean very well with a cloth dampened with any alcohol. Cover your microscope when it is not in use. A microscope that remains unused for any length of time can accumulate dust and debris from the air.

Fungal contamination will occur in tropical climates. This is seen as cloudiness in the image and on the objective lens. Fungal hyphae grow on the lens surfaces and also penetrate between the lens elements. It is impossible to get rid of them, and contaminated objectives and optical components can only be thrown away. Covering the microscope can exacerbate fungal contamination, and in this case the microscope should be kept in a dry, air-conditioned room and/or kept under an infrared lamp or at low temperatures during periods of high humidity.

Microscope parts that are lubricated with grease can become stiff with age, and trying to move them will cause mechanical damage. Lubricants by Nye (Nyogel 795A; Reference 3) are recommended. Only lubricate moving parts of the stand; *never* lubricate optical components. Never apply an oil- or grease-based lubricant to a surface lubricated with graphite, or vice versa. Doing so will very effectively seize up the working part. In this case n-hexane should be used to dissolve the solidified grease. If metal surfaces become jammed, then a very small amount of a light penetrating oil, such as WD-40, can be painted on and allowed to seep into the joint.

Cleaning solvents

Methanol, ethanol and propan-2-ol are polar solvents and are generally safe for lenses, providing they are used sparingly. I use 85% petroleum ether as a solvent (this is an *n*-hexane, a light petroleum fraction) with 15% propan-2-ol, as recommended in the Zeiss publication '*The Clean Microscope*'. Some people do not recommend *n*-hexane for heavy use or frequent use, preferring instead filtered methanol or methylated spirits. This was the practice of two of the best Zeiss service engineers I ever met. Otherwise, use absolute ethanol-ether (30–70%), or propan-2-ol (isopropyl alcohol). Do not use the 'rubbing alcohol' form of propan-2-ol; this contains too much water. I find that any solution with ether is volatile, and its smell can cause headaches. Since ether is also highly flammable, I avoid it.

Ordinary glass slides can be cleaned very effectively by the green version of 'Mr Muscle' Window & Glass cleaner, which does not contain ammonia but uses vinegar. You are advised not to use it for objectives or other optical components – although some people do. Do not use proprietary window cleaning fluids (e.g. 'Sparkle', 'Windex'), since these contain ammonia. Older lenses mounted with Canada balsam will dissolve in most aromatic solvents, and these should be avoided. Instead try using a high vapour solvent (e.g. ether) sparingly on a cotton bud. Some optical components are surrounded by black anti-reflective lacquer. These can be damaged by organic non-polar solvents such as acetone, which can also damage plastic parts and the rubber eyepiece cups.

Cleaning summary

- Remove oil before it dries onto, or seeps into, the objective.
- Use specialist lens-cleaning tissues – do not use general wipes.
- Use each tissue on a one-use basis. Then discard.
- Use a hexane-based solvent. Non-polar solvents can do damage.
- Take care not to scratch delicate optical surfaces and coatings.

References

Zeiss. *'The Clean Microscope'*. https://bit.ly/2KEbdS8. Also at: https://bit.ly/2KEHJUj and http://www.microscopist.co.uk/general-techniques/objective-care/
Zeiss campus on cleaning. http://zeiss-campus.magnet.fsu.edu/articles/basics/care.html.
Advice on microscope lubricants (Ian Walker, Microscopy UK, 2004). https://bit.ly/2OnOwE9.
Micro-Tools (Europe). http://www.micro-tools.com.

Appendix 5
Selected Suppliers

It is difficult to give a comprehensive list of suppliers, for people and places/addresses and circumstances change. These are companies that I have used in the UK, which may help you get started. In three decades as a microscopist, I have known most of these people for many years. I make no apologies for the personal touches to the recommendations, for they are friends.

Microscope lamp suppliers

There are many good suppliers of microscope illuminators; I have routinely used the following three firms in the UK and have never been disappointed:

Alan Todd
Microscope Services Ltd
22 Ashford Close
Woodstock
Oxfordshire, OX20 1FF
Tel: + 44 (0)1993 810000 or (0)1993 810081
E-mail: info@microscopeservices.co.uk.
Web: http://www.microscopeservices.co.uk.
Contact: http://www.microscopeservices.co.uk/contact.

Alan has been in the microscope business for what seems like forever. He is hugely knowledgeable and works with a consortium of similar businesses in the UK. Alan, of course does more than merely supply lamps. He is a man of the utmost integrity. I've bought teaching microscopes, stereomicroscopes and microscope accessories from him in the past.

TLS UK Ltd
922–923 Yeovil Road
Slough Trading Estate
Slough, SL1 4NH
Tel: + 44 (0)1753 576888
E-mail: sales@tls.uk.com.
Web: http://www.tlsuk.com.
Contact: http://tls.uk.com/contact/contact.html.

Alan Tilley
Atlantic Imaging Ltd
Tel: 07730 437925
E-mail: alan@atlantic-imaging.co.uk.
Web: www.atlantic-imaging.co.uk.

Understanding Light Microscopy, First Edition. Jeremy Sanderson.

Medlamps
Millhouse Business Centre
Station Road
Castle Donington
Derby DE74 2NJ
Tel: 01332 813921
E-mail: info@medlamps.co.uk
Web: https://www.medlamps.co.uk
Contact: https://www.medlamps.co.uk/contacts

I have not personally used this firm, but they have a clear and straightforward website.

Independent service agents

The major microscope manufacturers support their equipment post-sales with service engineers trained to maintain, fault-find and repair microscopes. You are advised to take out service contracts to cover the lifetime of the instruments, particularly with high-end laser-illuminated research microscopes, such as confocal, spinning disc and multi-photon microscopes. However, for service, maintenance and support of widefield microscopes, an independent agent may well provide a cost-effective solution. Most independent firms employ service engineers who have previously worked for the major microscope manufacturers. In the UK, I suggest the following:

Kevin Champion
K-Tec Microscope Services Ltd
The Business & Technology Centre
Bessemer Drive
Stevenage
Hertfordshire, SG1 2DX
Tel: 01438 315262
E-mail: info@ktecmicroscopes.co.uk.
Web: http://ktecmicroscopes.co.uk.

Paul Wheatley
Lightscope Services Ltd
5 Priory Lane, Pillerton Priors
Warwick, CV35 0PR
Tel: 01789 740419
E-mail: paul@lightscopeservices.co.uk.
Web: www.lightscopeservices.co.uk.

Paul and Kevin are both extremely competent engineers with many years' experience of servicing microscopes to draw upon. I rate their work highly and recommend them equally.

Microscope automation, cameras & Illumination systems

Alan Tilley
Atlantic Imaging Ltd
Tel: 07730 437925
E-mail: alan@atlantic-imaging.co.uk.
Web: www.atlantic-imaging.co.uk.

Alan works with a consortium of similar businesses in the UK. Like Alan, most of the people running these businesses have worked for established microscopy companies and have used their extensive and long-standing expertise to run successful independent companies. Go to Atlantic Imaging for everything from high-end camera equipment, automated components, bleaching laser engines, software and servicing requirements. I've known Alan for years: the microscope community would be poorer without people like him. He supports the image analysis software at my place of work, and I've bought several microscopes from him.

Indigo Scientific Ltd
Suite G
15 Royston Road
Baldock
Herts
SG7 6NW
Tel: + 44 (0)1462 633500
E-mail: sales@indigo-scientific.co.uk.
Web: http://www.indigo-scientific.co.uk.

Chris set this company up after many years working for one of the big four microscope companies. Very highly regarded, this is another company managed by first-class people. Apart from microscopes and camera systems, Indigo are agents for OKO incubation and cooling systems, which I have used successfully in previous core imaging facilities. They also make and sell globally the only commercially available Optical HREM system (see Chapter 17, Section 17.2 and endnote 2).

Fluorescence microscopes, illumination & detection hardware, filters and custom solutions

Cairn Research
Graveney Road
Faversham
Kent, ME13 8UP
Tel: + 44 (0)1795 590140
E-mail: sales@cairn-research.co.uk.
E-mail: tech@cairn-research.co.uk.
Web: https://www.cairn-research.co.uk.

Cairn design and manufacture an extensive range of illuminators and image-splitting devices and are UK distributors for the excellent Chroma filter range. They also specialise in configuring and supporting complete widefield and confocal fluorescence microscopes and are happy to take on custom design, modifications and upgrades of existing frames. I have known the founder, Martin Thomas, for many years and have huge respect for him, his ethics, and his company.

Camera connectors

Dr Eric Best
Best Scientific Couplers
6, Spencers Orchard
Wroughton
Swindon
Wiltshire, SN4 9NT
Tel: + 44 (0) 1793 812477
Mob: 07850449800
E-mail: eric@bestscientificcouplers.com.
Web: https://www.bestscientificcouplers.com.

Best Scientific offer a specialist camera coupler business but also operate a microscopy and industrial vision business (http://www.bestscientificweb.com).

Microscope incubation chambers

Solent Scientific Limited
4 Portsmouth Enterprise Centre
Quartremaine Road
Portsmouth PO3 5QT
United Kingdom
Tel (dialling from UK): 0870 774 7140
Tel (dialling from USA): 011 44 870 774 7140
E-mail: sales@solentsci.com.
Web: http://www.solentsci.com.
FAQ page: https://www.solentsci.com/incubator_faqs.htm.

Orry Dugdale was in the 'hot-box' business from the start. He's very capably supported by Derrick Atkins. Solent make bespoke equipment tailored to your microscope. I have nothing but praise for this company and the people; they are very efficient.

Image capture and analysis software

Most big manufacturers and suppliers usually tie you into using proprietary software. Indeed the 'feel' of the software is a significant choice when choosing microscopes and supporting capital equipment. In this regard, 'small is beautiful' very much applies. Digital Scientific offer a flexible and bespoke service with their software package, with utter integrity and value for money. Like Atlantic Imaging Ltd – who has exactly the same integrity and business ethos – they offer a telephone support service with the ability to take remote control of your software: a very useful advantage indeed. Both firms operate in the UK and internationally.

DSUK Limited
The Commercial Centre
6 Green End
Cambridge CB23 7DY
Tel: + 44 (0)1223 911215
E-mail: info@dsuk.biz.
Web: http://www.dsuk.biz.

Alan Tilley
Atlantic Imaging Ltd
Tel: 07730 437925
E-mail: alan@atlantic-imaging.co.uk.
Web: www.atlantic-imaging.co.uk.

Stage Micrometre & Graticule suppliers

TAAB Equipment
3 Minerva House
Calleva Park
Aldermaston
Berkshire, RG7 8NA
Tel: +44 (0)118 981 7775
E-mail: sales@taab.co.uk.
Web: http://www.taab.co.uk.

TAAB offer much more than merely measuring devices. They are also excellent for general microscopy supplies and electron microscopy goods. I use them frequently and have never been let down. TAAB may be a small firm, but the unique, fast and flexible service they offer far outweighs their larger competitors. They are second to none in their business because their office manager Alison has all the facts at her fingertips and misses nothing. The manufacturer of precision measuring equipment and accessories in the UK is Pyser-SGI. Anything I need from their range, I can get through TAAB.

Suppliers of arranged diatom test-plates

Klaus Kemp is world-renowned for his arranged diatom test-slides. It is a unique business underpinned by years of experience. Klaus supplies worldwide.

Klaus Kemp
Microlife Services
Blautannen, Wickham Way
East Brent, Somerset
TA9 4JB, England
Tel: +44 (0)1278 760411
E-mail: klaus.kemp@talktalk.net.
Web: http://www.diatoms.co.uk.

Appendix 6
Historical Background

The precise origins and originators of the light microscope are disputed and lost in history; simple glass lenses and water-filled globes have been used to assist the examination of small objects for more than two millennia, and images seen in natural water drops can hardly have escaped notice before then. Later in the Middle Ages, the Arabic scholar and polymath Alhazen described the use of magnifying lenses (besides much else) in his *Kitab al-Manazir* or *Book of Optics*, around 1000 AD.

The microscope and the telescope are related optically, yet until people experimented with combinations of different lenses, no great effort would have been given to systematically exploring the unknown and hidden world beyond the limits of the eye. Hartley (1993) postulates that simple magnifying glasses were known since time immemorial and had no need of description, whereas maritime and military necessity drove forward the invention of the telescope in the 16th century. The telescope was *invented*; the microscope only *discovered* later. There was clearly profit to be made in telescope production, yet no immediate benefit (aside of natural curiosity) from developing the microscope.

The origin of the telescope is unclear, but it is certain that it predates the microscope. In 1608 a patent dispute between the opticians Johannes Lippershey, Jacob Metius and Zacharias Janssen of Middleburg (in modern-day Holland) is recorded. This telescope patent was denied because other spectacle-makers had made similar claims; the design was considered too simple to keep secret, and the authorities in Holland recognised the military advantages of the telescope. Galileo Galilei, working at the University of Padua, heard of the 'Dutch spyglass' and constructed his own telescope of nine times magnification about the second half of 1609. In Venice he had access to exceptionally pure and clear Murano glass, which could be ground into high-quality lenses.

There are two designs of refracting telescope, the first being the Dutch or Galilean model, which uses a convex objective lens and concave eyepiece lens. This telescope produces an upright image but suffers from a small field of view, a relatively low magnification (maximum of about 30 times) and generally poor image quality. For this reason it is sometimes referred to as the terrestrial telescope. The second design is the astronomical Keplerian refracting telescope with a convex eyepiece. This produces an inverted image (which does not matter in astronomy) but could be made of greater overall magnification with a wider field of view and brighter image. At some point the microscope was discovered when somebody looked through the 'wrong' end of the Keplerian telescope, for the one is merely an inversion of the other. Until the separate coining of the terms 'telescope' and 'microscope' by the end of the first quarter of the 17th century, the words *perspicillum, tubus opticus* and *occhiale* were interchangeable neologisms. Since both instruments produced their image by dioptric magnification through refractive enlargement of the visual angle, this can only have added to the confusion about the exact date of their origin.

The first popular and comprehensive book on microscopes and what they could reveal was Robert Hooke's *Micrographia* (literally: *Small Drawings*), dated 1665 and released in October 1664. Hooke was not sufficiently confident to use the word 'microscope' coined forty years earlier, referring instead to 'magnifying glasses' in the subtitle of his book. Significantly, Hooke called the compartments *cells* that he could see in a thin slice of cork. In *Observ. XVIII* [No. 18] *'Of the Schematisme or Texture of Cork, and of the Cells and Pores of some other such frothy Bodies'*, he likened them to the orderly arrangements of small rooms inhabited by hermit monks within a religious community. It is a good analogy: individual cells are the basic functional unit of life within the entire organism.

It seems that Antoni van Leeuwenhoek, a draper living in Holland (Bracegirdle, 1986) and contemporary of Hooke's, inspired by reading *Micrographia*, was not deterred by the 'troublesome' difficulties using high-power single magnifying glasses *'offensive to my eye because of their smallness, and the nearness of the Object'* as encountered by Hooke. Leeuwenhoek would have been familiar with using a magnifying glass for inspecting cloth. He studied objects that had never before been seen, including bacteria, protozoa and human sperm, even the explosion of gunpowder under the microscope – nearly blinding himself in the process! Leeuwenhoek's microscope consisted of little more than a simple metal frame to hold the lens and the specimen. This demonstrates convincingly that the only truly essential part of a microscope is that first tiny lens just after the specimen; all of the other parts simply add to the convenience of use. It is this objective lens that is all-important in collecting information about the object, and any information that does not enter the objective lens will not be seen. Those of Leeuwenhoek's microscopes that survive have been intensively investigated and shown to have magnifications up to several hundred times, able to resolve between 1 and 2 micrometres.

Compound microscopes, with multiple lenses, were inferior to the single lens (or simple) microscope due to the cumulative effects of the aberrations present in each lens. The more lenses that were added to the microscope, the worse it performed because each lens

Understanding Light Microscopy, First Edition. Jeremy Sanderson.

successively magnified the aberration present. Thus the resolving power of the simple microscope was approximately five times better than the compound microscope. The simple microscope tended to be used for higher-power studies by transmitted light, whereas the compound microscope, close relative of the telescope, was largely used to investigate solid objects by reflected light at relatively low magnification.

Throughout the 18th century, after the deaths of both Hooke and Leeuwenhoek, the microscope degenerated largely into a beautifully constructed plaything, primarily for the cabinets of English gentlemen of leisure. The apotheosis of such instruments is George Adams' silver microscope made for King George III (which may be seen in the Museum of the History of Science, Oxford). Hence, microscopy came to be despised and distrusted by scientists of the 18th and early 19th centuries because of the erroneous interpretations that could arise all too easily from the poor and defective images. It was not until the 1820s that significant advances were made so that the compound microscope could surpass the simple microscope in convenience, versatility and performance.

Various enterprising individuals tried constructing tiny achromatic doublets for the microscope. In order to minimise spherical aberration in the objectives of compound microscopes, it was usual practice to insert a restricting aperture between the lenses. This was none too effective but did, however, cause serious loss of both resolving power and brightness. A wine merchant named Joseph Jackson Lister (Bracegirdle, 1987) made it his hobby to study the optics of the microscope and successfully combined two plano-convex achromatic doublets in an aplanatic arrangement to minimise spherical aberration. Lister's work (Lister, 1830) revolutionised microscope design, since progress was made not only in minimising chromatic and spherical aberrations but also by increasing the useful aperture of the objectives. The improved microscopes allowed for advances to be made in cell biology and medicine.

It was thought that the angular aperture of an objective was important, and a situation akin to that of the 18th century arose: objectives were made with far too much magnifying power (empty magnification) in relation to their aperture. However, objective manufacture was entirely empirical, no one knew the true basis of image formation, and the uncertainty threatened to put people like Carl Zeiss out of business. Zeiss turned for help to an academically successful 26-year-old named Ernst Carl Abbe (Volkmann, 1966; Gerth and Wimmer, 2005; Bower, 2016), who started work with the firm in 1866. Abbe abandoned all preconceived ideas about objective design and set to work to establish the entire process of design and manufacture upon a firm basis of theoretical calculations: to rationalise and standardise the process for absolute consistency. He was successful, and from the early 1870s all the lenses made in Jena were made to pre-calculated computations instead of relying (as elsewhere) solely upon the skill of individual craftsmen. Abbe elucidated the diffraction theory of image formation and the *'Abbe sine condition'*, which he published in December 1873 (Abbe, 1873). This single technical advance was to have a greater impact upon microscopy than the work of Lister 40 years previously.

From the early 1870s, Abbe devoted his energy to improving achromatic objectives, which still exhibited residual colour fringes and some spherical aberration. This necessitated finding new types of optical glass, and Abbe learnt that high-quality fluorite could be obtained from the Oltscherenalp in Switzerland. He used this because the exceptionally clear crystal from the area offered the required dispersive properties to minimise residual errors in the objective. Again Abbe was successful, and the light microscope reached the zenith of optical development in 1886, which would last for the next century, with the production of highly corrected apochromat lenses (Haselmann, 1986).

The closing years of the 19th century saw the introduction of a reliable method of adjusting the microscope to exploit the full aperture of the objective – Köhler illumination (Köhler, 1893; Haselmann, 1993). Improvements during the 20th century broadly centred on photographic image recording and developing contrast-enhancement methods. The development of electron microscopy in the second half of the 20th century led many to believe that the light microscope had only a limited future. However, realisation of the power of fluorescence microscopy, together with the advent of the laser, powerful computers and electronic detectors for digital imaging, has led to the renaissance of the light microscope in the late 20th century (Lippincott-Schwartz, 2011). The newest advances in physics and optics are helping to close the gap between the limit of resolving power of the light microscope and electron microscopy, recognised by the award of the 2014 Nobel Prize in Chemistry to Eric Betzig, Stefan Hell, and William Moerner for surpassing the limitations of the light microscope (van Noorden, 2014; Möckl et al., 2014). Scientists are beginning to use the renewed power and versatility of light microscopy to explore, *in-situ* and in real time, how proteins and other cellular components function and interact within their natural environment.

References

Abbe, E. (1873). 'Beiträge zur Theorie des Mikroskops and der mikroskopischen Wahrnehmung'. *Archiv für Mikroskopische Anatomie* 9: 413–468.

Bower, D.I. (2016). 'Ernst Abbe: creator of the Zeiss Foundation'. Transl. of 1951 volume by N. Günther.

Bracegirdle, B. (1986). 'Famous microscopists: Antoni van Leeuwenhoek, 1632–1723'. *Proceedings of the Royal Microscopical Society 21*/6: 367–373.

Bracegirdle, B. (1987). 'Famous microscopists: Joseph Jackson Lister, 1786–1869'. *Proceedings of the Royal Microscopical Society 22*/5: 273–297.

Gerth, K. and Wimmer, W. (2005). *Ernst Abbe (1840-1905): scientist, entrepreneur, social reformer*. Jena: Bussert & Stadeler.

Hartley, G.W. (1993). *The Light Microscope: its use & development*, Chapter 1. Oxford: Senecio, pp. 1–6.

Haselmann, H. (1986). 'One Hundredth Anniversary of Abbe's Apochromats'. *Proceedings of the Royal Microscopical Society 21*/4: 197–201.

Haselmann, H. (1993). 'One Hundredth Anniversary of Köhler illumination'. *Proceedings of the Royal Microscopical Society 28*/4: 186–188.

Köhler, A. (1893). 'Gedanken zu einem neuen Beleuchtungsverfahren für mikrophotographische Zwecke'. *Zeitschrift für wissenschaftliche Mikroskopie 10*: 433–440. A reprint translated into English of Köhler's original paper: *A new system of illumination for photomicrographic purposes* may be found in: *Proceedings of the Royal Microscopical Society* (1993) 28/4: 181–186.

Lippincott-Schwartz, J. (2011). 'Bridging structure and process in developmental biology through new imaging processes'. *Developmental Cell 21*/1: 5–10.

Lister, J.J. (1830). 'On some properties in achromatic object-glasses applicable to the improvement of the microscope'. *Philosophical Transactions of the Royal Society 120*: 187–200.
Möckl, L. et al. (2014). Super-resolved fluorescence microscopy: Nobel Prize in Chemistry 2014 for Eric Betzig, Stefan Hell, and William E. Moerner. *Angewandte Chemie International Edition 53*/51: 13972–13977.
van Noorden, R. (2014). Insider view of cells scoops Nobel. *Nature 514*/7522: 286.
Volkmann, H. (1966). 'Ernst Abbe and his work'. *Applied Optics 5*/11: 1720–1731.

Further Reading

Amos, B. (2000). 'Lessons from the history of Light Microscopy'. *Nature Cell Biology 2*/8: E151–E152.
Bracegirdle, B. (1978). 'The performance of seventeenth- and eighteenth-century microscopes'. *Medical History 22*/2: 187–195.
Bracegirdle, B. (ed.). *Beads of Glass: Leeuwenhoek and the early microscope*. Catalogue of an exhibition in the Museum Boerhaave Nov 1982–May 1983 and in the Science Museum May–Oct 1983.
Bradbury, S. (1967). 'The quality of the image produced by the compound microscope: 1700–1840'. *Proceedings of the Royal Microscopical Society 2*/1: 151–173.
Bradbury, S. (1996). 'The reception of Abbe's Theory in England'. *Proceedings of the Royal Microscopical Society 31*/4: 293–299.
Bradbury, S. (2001). 'Ernst Abbe's theories and his letters to English Microscopists'. *Quekett Journal of Microscopy 39*/2: 139–161.
Dobell, C. (1960). *'Antony van Leeuwenhoek and his "Little Animals"'*. New York: Dover. Unabridged reprint of Dobell, C. (1932). London: Constable.
Fripp, H.E. trans. (1874–1875). 'Abbe, E. A contribution to the theory of the microscope, and the nature of microscopic vision'. *Proceedings of the Bristol Naturalists' Society* **I** pt. II: 200–261.
Fripp, H.E. (1876–1877). 'On Professor Abbe's experimental demonstration of diffraction spectra in the microscope'. *Proceedings of the Bristol Naturalists' Society* **II** pt. I: 24–38.
Journal of Microscopy, volume 155 part 3, September 1989
(In celebration of the 150th anniversary of the Royal Microscopical Society). Including:
Bennett, J.A. (1989). 'Social history of the microscope', pp. 267–280.
Bradbury, S. (1989). 'Landmarks in biological light microscopy', pp. 281–305.
Bracegirdle, B. (1989). 'Development of preparative techniques', pp. 307–318.
Hartley, W.G. (1989). 'Triumphs and conflicts in microscopy', pp. 319–326.
Piston, D.W. (2009). 'The impact of technology on light microscopy'. *Nature Milestones* S23–S24.
Robertson, L.A. (2015). 'van Leeuwenhoek microscopes – where are they now?' *FEMS Microbiology Letters 362*/9: 2015 fnv056.
Turner, G.L'E. (1990). '1590 and all that'. *Proceedings of the Royal Microscopical Society 25*/6: 423–424.

Appendix 7
Timeline of Key Events

c 3000 BC	The Assyrian quartz rock crystal 'Nimrud' lens found (by Austen Layard, 1850) in Iraq. Sir David Brewster suggested that it was used as a magnifying lens, for intricate work, or as a burning lens
c 800 BC	Egyptian hieroglyphic inscriptions depict simple meniscal lenses
c 450 BC	First record of using a pinhole to form an image by the Chinese philosopher Mo-Tsu
424 BC	Aristophanes refers to 'burning spheres' – a glass sphere filled with water to focus sunlight
c 300 BC	Euclid defines the law of reflection and suggests that light travels in straight lines
c 200 BC	Vaisheshika Indian school of philosophy suggests light rays are a high velocity stream of fiery atoms
c 60 AD	Hero of Alexandria studies reflection of light, and formulates the 'shortest path length' principle
c 63 AD	Seneca reports that small, dim writing is seen clearly enlarged through an Aristophanes lens
180	Ting Huan constructs the first zoetrope for stop-frame animation
525	Anicus Boethius beheaded in Rome for using 'magic' to attempt to determine the speed of light
c 900	Viking explorers use Icelandic calcite 'sunstone' for navigational purposes
c 984	Abu Said al-Ala Ibn Sahl writing *Fī al-'āla al-muḥriqa* (On the Burning Instruments), states law of refraction
1021	The Arabic scholar Ibn al-Haytham (Alhazan) studies lenses and publishes *Kitab al-Manazir* (*Book of Optics*). A comprehensive treatment, describes the camera obscura, deduced the linearity of light
1267	Roger Bacon writes in *Opus Majus* about magnification with convex lenses to aid reading
c 1300	Spectacles "newly invented" mentioned in a Florentine manuscript
1304	Theodoric of Freiburg explains, and verifies experimentally, the geometric formation of rainbows
1502	Hematein dye from the logwood tree *Haematoxylon campechianum* discovered by the Spanish in Campeche, Mexico
1513/4	Leonardo da Vinci grinds single lenses to observe and draw the moon in *Codex Atlanticus*
1565	Nicolás Monardes describes the blue fluorescence opalescence from the alkaline water infusion of the wood of *Eysenhardtia polystachya*, a Mexican tree used to treat kidney ailments
1589	Giambattista della Porta reports in *Magia Naturalis* using single lenses to correct visual defects
1590	The date often given for the discovery of the microscope. Evidence does not support this; Zacharias Janssen, son of spectacle-maker Hans Martens, would have been two years old.
1602	Thomas Harriot describes (Snell's) law of refraction – rediscovered by Snellius in 1621
1604	Johannes Kepler explains how convex and concave lenses work in *Astronomiae Pars Optica*
c 1608	Discovery of the refracting telescope with convex objective and concave eyepiece
1609	Thomas Harriot, the English astronomer, sketched the moon through a telescope magnifying 5x
1609	Galileo Galilei builds a refracting telescope magnifying 9x, based on Lippershey's 'Dutch spyglass'
1611	Kepler publishes a better refracting telescope design with two convex lenses in *Dioptrice*; he also includes a scheme for a microscope – an inverted Keplerian telescope
1611	Giovanni Demisiani coins the word 'telescope' at Galileo's induction to the Lincei Academy
1611	Johannes Kepler describes total internal reflection
1617	Mizar, the first binary star is seen by telescope probably by Benedetto Castelli, friend of Galileo
1619	Cornelius Drebbel designs a shorter form of Keplerian telescope as a microscope
1621	Willebrord Snellius (or Snel) describes the ratio between the sines of the angles of incidence and refraction, now known as Snell's law.
1624	Galileo sends a Drebbel microscope to his patron, Federico Cesi, an apiarist
1625	Isaack Beekman of Middelburg records Janssen's microscope in his diary
1625	Giovanni Faber, a contemporary of Galileo, coined the word 'microscope' – the new occhiale
1637	René Descartes describes the law of refraction in terms of sines; explains the formation of rainbows

Understanding Light Microscopy, First Edition. Jeremy Sanderson.

1647	Bonaventura Cavalieri describes the lensmaker's formula: the relationship between the radius of curvature for the surface of a thin lens and its focal length
1657	Pierre de Fermat's 'principle of least time' describes the passage of light rays and proves Snell's law
c 1660	Francesco Maria Grimaldi coins the term 'diffraction'. Work published posthumously in 1665
1661	Henry Power writes his poem *'In Comendation of ye Microscope'* referring to Galileo
1664	Robert Boyle investigates luminescence, influenced by pH, in Mexican 'kidney wood' of Monardes
1664	Publication by Henry Power's *Experimental Philosophy*, the first English work on microscopy
1665	Robert Hooke publishes *Micrographia* and, inter alia, describes 'cells' in a cork slice
1666	Sir Isaac Newton discovers that white light is decomposed into the spectrum by a prism
1669	Erasmus Bartholinus reports double refraction in Iceland spar – calcite
1672	Newton publishes on light and colour in the *Philosophical Transactions* of the Royal Society
1673	Antoni van Leeuwenhoek corresponds with the Royal Society regarding his simple microscope
1674	Christoph Adolph Balduin discovers calcium nitrate and its deep red phosphorescent properties
1676	van Leeuwenhoek first describes living bacteria from stagnant water
1676	Ole Rømer measures the speed of light as from eclipses of Jupiter's moon Io: 215,000 km/s
1678	Christiaan Huygens publishes his wave theory of light – the Huygens principle. He also discovers the polarisation of light passing through calcite and uses Rømer's data to recalculate the speed of light
1678	Robert Hooke suggests the immersion principle for increased 'clearness and brightness'
1690	Huygens publishes *Traité de la Lumière*, with his wave theory of light
1704	Newton publishes in *Opticks* his view of the corpuscular theory of light
1714	van Leeuwenhoek stains muscle tissues with saffron
1718	James Jurin uses pre-measured snippets of hair in a preparation to measure the size of erythrocytes
1722	van Leeuwenhoek uses a sand grain to indicate the size of his "little animalcules"
1732	Benjamin Martin invents his micrometer for measuring distances and sizes in the microscope image
1733	Chester Moore Hall invents the achromatic doublet lens for the telescope
1738	Johann Lieberkühn develops a concave highly polished speculum for illuminating opaque samples
1742	Publication of the first edition of *The Microscope Made Easy*, by Henry Baker
1746	Abraham Trembley reports on the fission of the algal diatom *Synedra*: division from existing cells
1750	John Cuff builds a low-power dissecting microscope and makes a silver wire lattice micrometer
1758	John Dollond patents and manufactures achromatic lens combinations for telescopes
1770	First description of a microtome by John Hill, in the investigation of timber
1770	Hill uses the natural dyestuff cochineal, or carmine, to stain wood sections
1773	Diatoms, later to become widely used test objects, are first described by Otto Müller
1783	Jean Rome de L'Isle lays the foundation of crystallography, recognising constant interfacial angles
1784	Abbé René-Just Haüy discovers cleavage planes. He describes the mathematics of crystal structure
1785	The American David Rittenhouse makes the first diffraction grating, from hairs, of 100 lines per inch
1793	Thomas Young describes the accommodation of the eye by curvature of the lens
1795	Joachim Graf von Sternberg first observes fractures in wrought iron with the microscope
1795	Abraham Ypelaar uses venetian terpentine for the first resin-mounted preparations
1800	Sir Frederick William Herschel discovers the infrared part of the electromagnetic spectrum
1801	Johann Wilhelm Ritter notices the effect of ultraviolet radiation on the blackening of silver
1801	Thomas Young discovers interference of light and challenges Newton's corpuscular theory
1801	Augustin-Jean Fresnel describes, by equation, partial refraction and reflection at an interface
1802	Sir Humphrey Davy creates the first incandescent light with a carbon arc lamp
1806	William Hyde Wollaston devises the prism named after him to produce plane-polarised light, using his patented *camera lucida*
1806	First recorded attempt by van Deijl to sell achromatic microscope lenses
1807	Young devises his double-slit experiment to demonstrate the interference of coherent light rays
1808	Etienne Louis Malus observes plane polarisation of light by reflection; coins the term 'polarisation'
1809	Dominique François Jean Arago discovers polarisation of light from a clear blue sky
1811	Arago shows that two superimposed birefringent minerals sections exhibit a reduction in interference colour. The same year he also demonstrates rotary polarisation in quartz crystals
1812	Friedrich Mohs devises his hardness scale
1812	Sir David Brewster formalises the polarisation of incident reflected light – the Brewster angle
1813	Jean-Baptiste Biot suggests a method to work out unknown vibration directions in mineral crystals
1813	Brewster dips the objective – a fluid front lens – to correct chromatic aberration
1814	Joseph von Fraunhofer, using his spectroscope, discovers the 574 dark lines in the solar spectrum
1815	Brewster explains Malus' observation, and describes the relationship between the refractive index of a reflective substance and the angle at which light striking the substance will be polarised
1816	Fresnel demonstrates that two rays polarised at 90° to each other never interfere

1817	Fresnel devises his 'Fresnel rhomb' and describes circular polarisation
1818	Brewster first describes the optical characters of mineral crystals as 'positive' and 'negative'
1819	Arago demonstrates the Poisson bright spot to prove Fresnel's theory of the wave nature of light
1821	Fraunhofer builds his diffraction grating composed of 260 very closely spaced parallel wires
1821	Fresnel publishes his finding that light waves are transverse to the direction of propagation
1823	Fresnel installs his lens, developed for lighthouses, in the Cordouan lighthouse on the Gironde
1824	Dr Charles Goring introduces test objects for the microscope – feathers and insect scales
1825	François-Vincent Raspail proposes 'omnis cellula e cellula' – from microscopical studies of plants
1826	Joseph-Nicéphore Niépce produces the first photograph
1826	Joseph Jackson Lister designs and commissions an achromatic microscope from William Tulley
1828	William Nicol invents the 'Nicol prism' – a polarising prism made of calcite
1830	Improved achromatic objectives described by Lister, in a paper published by the Royal Society
1830	Brewster devises the glass plate reflecting polariser, for filling large fields of view with polarised light; he also suggests use of a quartz or gypsum crystal as a first-order compensator of retardance
1830	First use of Canada balsam reported as a resinous mountant for microscopical preparations
1830	Haematoxylin first used as a histological dye, but without a mordant
1831	Robert Brown names the nucleus 'an opaque areola' from the Latin for 'little nut'
1832	Fresnel deduces his laws describing the intensity and polarisation of reflected and refracted light
1832	Joseph Plateau develops the phenakistoscope to create an illusion of motion from static pictures
1833	Camille Nachet builds the first polarising microscope to use crossed Nicol prism polars
1833	Jacobson first suggests use of chromic acid as a hardening and fixing agent for microscopy
1833	Brewster saw red fluorescence from alcoholic chlorophyll solution; he thought it was dispersion
1834	Pleochroism demonstrated in the microscope by William Henry Fox Talbot with a Nicol prism. This is the first record of the use of a basic polarising microscope of the design we know today
1835	Robert Remak discovers that the origin of cells is by the division of pre-existing cells
1835	Sir George Biddell Airy publishes his work on the point spread function diffracted by a circular aperture
1836	Revd. Joseph Reade describes dark-field illumination
1837	Andrew Ross invents the correction collar for 'high-dry' objectives to correct spherical errors
1838	Christian Gottfried Ehrenberg feeds protozoa with carmine as a vital dye
1838	Sir Charles Wheatstone describes stereoscopic vision and develops the stereoscope
1838–39	Matthias Jakob Schleiden (1838) and Theodor Schwann (1839) propose the 'cell theory' for plants and animals
1839	Louis Daguerre & Henry Fox Talbot independently refine photographic recording onto plates and paper
1839	John Benjamin Dancer exploits photography to record microphotographs at a reduction of 1:160
1839	Foundation of the Royal Microscopical Society, as the Microscopical Society of London
1840	Standardised microscope slide dimensions (3″ × 1″ and 3″ × 1$^1/_2$″) proposed by the Microscopical Society of London
1840	Alfred Donné exhibits photomicrograph of *Trichomonas vaginalis* to Academy of Sciences, Paris
1840	Henri de Sénarmont's method for using a $^1/_4\,\lambda$ compensator plate to determine retardation colours
1841	First published report of diatoms being used as test objects – *Gyrosigma hippocampus*
1842	Jean-Daniel Colladon reports trapping light in a water jet for fountain displays
1844	Giovanni Battista Amici uses a calcite rhomb over the eyepiece as an analyser
1844	Wilhelm von Haidinger describes the visual perception of polarised light known as Haidinger's brush
1845	Friedrich Nobert manufactures his first ruled test-plate to replace variable natural specimens
1845	Fluorescence phenomenon 'epipolic dispersion' discovered in quinnine sulphate by John Herschel
1846	Michael Faraday suggests that electricity, magnetism and light may be related
1846	Carl Zeiss opens his optical workshop in Jena
1847	The origin of stereology by the French geologist Auguste Delesse in studying rock sections
1849	Henry Clifton Sorby pioneers 'microscopic petrography' viewing rock thin sections
1849	Carl Kellner opens his optical workshop in Wetzlar – later to become Leitz, subsequently Leica
1849	A Hippolyte L. Fitzeau uses the first terrestrial method to measure the speed of light: 313,300 km/s
1850	Jean Bernard Léon Foucault measures the speed of light through water as less than air: 225,563 km/s
1850	Francis Wenham devises a paraboloid dark-field illuminator
1851	Alfonso Corti uses carmine, upon the inner ear, to first stain nuclei
1851	Henri de Sénarmont uses a $^1/_2\lambda$ compensator plate to elucidate all three optic axes of crystals
1851	Sorby produces first account of the mineral composition of a rock, differentiating between agate (quartz) and calcereous spar (calcite)
1852	Sir George Gabriel Stokes coins the term 'fluorescence'
1853	Amici invents the water-immersion objective, giving a brighter image
1853	Delves & Highley publish technical papers on photomacrography
1854	John Tyndall popularises Colladon's work, demonstrating the forerunner of the fibre optic cable
1855	Francis Wenham manufactures a microscope with an asymmetric binocular head

1855	James Glaisher publishes on snow crystals viewed with the microscope. His later work helps secure the Royal Charter of the Microscopical Society in 1866, which adopts a snowflake as its emblem
1856	William Perkin discovers aniline dye – named mauveine – used ultimately for histological staining
1857	The standardised Whitworth screw thread (the 'RMS thread') for connecting objectives to the nosepiece is introduced by the Microscopical Society of London.
1858	Joseph von Gerlach popularises the use of carmine as a histological stain for nuclei on brain slices
1858	Rudolf Virchow popularises Raspail's 'all cells arise from cells' statement – without attribution!
1860	Gustav Kirchhoff found that incandescent tourmaline exhibits polarised light due to dichroism
1862	Jean Foucault finally determines the first accurate value for the speed of light: 299,796 km/s
1863	Sorby uses the microscope to discover the micro-crystalline structure of steel at Sheffield
1864	N. Chrzonszczewsky uses carmine as the first vital dye on the kidney
1865	Beck constructs the vertical illuminator, in favour of the Lieberkühn reflector
1865	James Clerk Maxwell concludes that light is a form of electromagnetic radiation
1865	F. Böhmer uses haematoxylin with alum mordant as a histological nuclear stain
1866	Royal Charter awarded to the Royal Microscopical Society (RMS)
1866	Ernst Carl Abbe taken on as a technical adviser by Carl Zeiss
1869	Abbe develops his two-lens substage illuminator and adjustable iris for oblique illumination
1870	Abbe calculates his 'sine condition' and elucidates the diffraction theory of image formation
1870	G. W. Royston-Piggott shows a water-immersion objective has a larger angular aperture than a dry one
1870	Franco-Prussian war: Rene Dagron uses microphotography to send messages into besieged Paris
1871	Wenham disputes the efficacy of homogeneous immersion – the 'Aperture Wars'
1871	Adolf von Bäyer synthesises the first fluorescent dye, fluorescein
1871	John William Strutt, 3rd Baron Rayleigh, explains why the sky looks blue due to particle scattering
1872	The first colour photograph taken by Louis Ducos du Hauron, using the subtractive method
1872	Zeiss manufactures objectives according to Abbe's theory of image formation
1873	Ernst Abbe publishes his diffraction theory of image formation
1873	Robert Tolles produces a 1/10" objective NA 1.25 for homogeneous immersion using soft balsam
1874	Abbe produces his test plate for evaluating the quality of objectives
1874	Heinrich Caro synthesises Eosin, named not after the Greek for 'dawn' but rather the nickname of a girl he admires
1875	Bausch & Lomb start manufacturing microscopes in the USA
1876	First recorded use of surgical loupes by Edwin Theodor Sämisch
1876	Wissowsky proposes eosin as a counterstain to haematoxylin – introduction of the H&E stain
1876	Abbe produces his diffraction kit to demonstrate the theory of microscope image formation
1877	Zeiss manufacture the first useful oil-immersion objectives according to Abbe's design
1877	Abbe produces his apertometer to measure numerical aperture
1877	Koch reports on culturing live anthrax bacillus in a microscope stage chamber
1878	The 'Aperture Wars' debate settled, after investigation, in favour of immersion and Tolles et al.
1878	Ernst Abbe elected as Honorary Fellow of the Royal Microscopical Society
1878	Émile Bertrand introduces the convergent lens bearing his name for viewing the objective back focal plane
1879	Thomas Edison perfects, patents and mass-produces the incandescent electric light
1881	Eadweard James Muybridge and Étienne-Jules Marey collaborate on chronophotography, the forerunner of cinematography. Muybridge developed the zoöpraxiscope to project motion pictures
1881	Silvanus Thompson improves the Nicol prism with Paul Glan to become the Glan-Thompson prism
1882	Carl Zeiss and Otto Schott open a glass-working laboratory and factory
1882	Robert Koch described the tubercle bacillus using an oil-immersion objective of high aperture
1882	Walther Flemming coins the term 'mitosis' to describe chromosome pairing in cell division
1882	Étienne-Jules Marey develops the *Fusil Photographique* motion-picture camera
1884	Paul Nipkow develops his picture transmission system using a perforated spiral scanning disc
c 1885	Edward Nelson resolves the finest band on Nobert's test plate – an average spacing of 0.21 μm
1885	Horace Darwin markets the 'Cambridge Rocker' microtome
1886	Apochromatic objectives described by Abbe. Carl Zeiss produce the first high-quality microscope
1886	Sorby examines the minute pearlite structure of steel at high magnification
1886	The principle of anti-reflective coatings first noticed by John Strutt, Lord Rayleigh
1887	Michelson & Morley disprove the existence of æther and show the speed of light is constant
1888	Auguste Michel-Lévy publishes, with Lacroix, his interference colour chart
1888	Heinrich Hertz verifies experimentally that electromagnetic radio waves travel at the speed of light
1889	2.5 mm focal length 100:1 Zeiss monobromonaphthalene apochromatic objective of NA 1.60 developed
1890	Ferdinand Hurter & Vero Driffield introduce the characteristic curve in photography
1891	Klein's rotation apparatus for observing rock fragments manufactured by R Fuess

1892	Karl Bratuscheck describes the phenomena of phase retardation with an attenuating carbon film
1892	Evgraf Stepanovich Fedorov devises the 4-axis universal rotary stage to characterise the optical character of crystals, calculate the optical angle 2 V and to determine the crystal indicatrix
1893	August Köhler devises his method of illumination
1893	Rudolf Winkel devises semi-apochromatic (so-called 'fluorite') objectives, using fluorspar lenses
1893	Formalin used as a fixative for tissues (anthrax-infected mouse) first reported by Ferdinand Blum
1893	Elucidation and introduction of the Becke line test by Friedrich Johann Karl Becke
1893	J.A. Sirks develops the first transmitted-light interference microscope, naming it as such
1893	Edward Milles Nelson devises the 'Optical Index' to compare objective performance one to another
1896	Lord Rayleigh formulates his criterion for image resolution
1896	Julius Rheinberg describes his coloured dark-field contrast method
1896	Horatio Greenough develops the first stereomicroscope with Ernst Abbe
1897	Sir Joseph John Thompson discovers the existence of electrons, which are related to photons
1897	R. Meyer coins the term 'fluorophore' to describe chemical groups associated with fluorescence
1899	Henry Joseph Grayson reports (*Jour RMS*) his ruling of a 120 000 lines per inch diffraction grating
1899	The RMS proposes standard eyepiece diameters and substage fittings for microscopes
1900	Max Planck formulates his quantum theory relating wavelength and energy via his constant, h
1900	Alfred Ewing and Walter Rosenhain publish their work on the crystalline structure of metals
1902	Miers develops Klein's rotation apparatus into the spindle stage, manufactured by J. Swift & Son
1902	Siedentopf & Zsigmondy develop the forerunner of the light-sheet microscope to study colloids
1904	Ultraviolet fluorescence in the microscope first noted by August Köhler and Moritz von Rohr
1905	Einstein uses Planck's quantum theory to demonstrate light exists as both waves and photons
1907	Julius Ries makes the first time-lapse film of the development of sea urchin eggs
1907	Lumière Autochrome mosaic screen plate colour process, using potato starch grains, is first marketed
1907	Ross Granville Harrison creates the first batch of tissue-cultured (nerve) cells in lymph fluid
1907	Einstein first uses the term 'c' to denote the speed of light in a vacuum
1911	Oskar Heimstädt constructs the first successful fluorescence microscope
1911	August Köhler introduces parfocal objectives
1913	Max Berek, working for Leitz, introduces his tilting calcite variable wave plate compensator
1913	Leitz introduce the Jentsch pattern symmetrical sliding binocular head
1916	Carroll Mason Sparrow proposes his criterion for image resolution in astronomy
1917	Establishment of Nikon Corporation as a microscope and optical component manufacturer
1919	Formation of Takachiho Works Co. Ltd., which would later become the Olympus Corporation
1920	F. Weigert discovers that fluorescent light emitted from dye solutions is polarised
1921	Carl Nylén uses his monocular compound surgical microscope for otitis media surgery
1922	R.L. Smith and G. E. Sandland devise the Vickers indentation improved hardness test
1924	Louis-Victor de Broglie proposes that electrons have wave-like properties
1924	Reichert produce of the first infinity-corrected objective for a metallographic microscope
1925	Leitz introduce the first 35 mm camera for photomicrography
1926	Gilbert Lewis coins the word 'photon' to express a quantum of light
1926	Zeiss introduce the Siedentopf binocular head, which maintains constant tubelength
1926	Enrique Gaviola successfully measures the 2 nsec fluorescence lifetime of aqueous rhodamine
1927	Max Berek formulates his equation for the depth of field of objectives
1928	Edward Synge conceives the principle of near-field imaging
1928	Harry Nyquist publishes his paper on transmission of Morse code information pulses via telegraph lines
1929	Edwin Herbert Land develops and patents Polaroid – the first synthetic polarising filter sheet
1930	Heinrich Lamm, a medical student, develops a prototype optical fibre to carry an image
1931	Ernst Ruska and colleagues build the first transmission electron microscope
1931	Two-photon excitation predicted by Maria Göppert-Meyer
1932	Dufaycolour reseau-based colour additive film marketed by Spicers
1933	The resolving power of the early electron microscope surpasses the light microscope
1934	Frederik (Frits) Zernicke publishes a description of his phase-contrast technique
1935	Max Haitinger introduces secondary fluorescent staining and coins the word 'fluorochrome'
1935	Alexander Jabłoński publishes his energy level diagram to explain the phenomenon of fluorescence
1936	First multi-stage photomultiplier tube (PMT) developed at RCA by Vladimir Zworykin et al.
1936	Agfa introduce their 3-layer subtractive colour transparency film with integral colour couplers
1936	Interference-type anti-reflection coatings for lenses first developed by Adolf Smakula at Zeiss
1938	'Plan-' (flat-field) correction invented by Dr Hans Boegehold at Zeiss
1938	Manfred von Ardenne builds the first scanning electron microscope

1939	First commercial transmission electron microscope produced by Siemens-Halske
1939	Walter Geffcken creates the first interference filters using a dielectric coating
1941	Kurt Michel produces the first phase-contrast live-cell film of spermatogenesis in grasshoppers
1942	Fluorescein IsoThioCyanate (FITC) fluorophore conjugated to mouse antibody by Albert Coons et al.
1943	Harold William Chalkley introduces point-counting as a stereological method
1943	Building of ENIAC – the first electronic computer
1948	Claude Shannon 'father of the digital age' publishes two papers applying Nyquist's sampling criterion to processing electronic information
1951	TV tubes coupled to microscopes to give increased contrast and spectral sensitivity
1952	Francis Smith patents his interference microscope, which uses Wollaston prism beam splitters
1953	Georges Nomarski invents and patents his differential interference contrast (DIC) method
1953	Leitz market phase contrast employing the cardioid variable Heine condenser
1953	Standard colour code of objectives was introduced by Dr Kurt Michel at Zeiss
1954	Indirect (two-layer) fluorescent immunocytochemistry demonstrated by Weller and Coons
1954	Horst Piller devises the 'Antiflex' reflection contrast method for low reflectance (i.e. ore) samples
1955	Marvin Minsky patents the moving-stage confocal microscope
1956	Charles Ginsberg et al. invent the video camera, the first machine to record both image & sound
1956	Lawrence Curtiss develops the first glass-clad optical fibre as a successful light guide
1957	The American Optical Company introduces the first Common Main Objective stereomicroscope
1959	RE Wilcox improves Miers' rotating apparatus as the Spindle stage
1959	Gordon Gould coins the term 'laser': light amplification by stimulated emission of radiation
1960	Theodore Maiman builds the first ruby laser
1960	First commercial integrated circuit silicon transistor produced by Fairchild Electronics
1962	First semi-conductor diode lasers are invented at General Electric
1962	Osamu Shimomura et al. report Green Fluorescence Protein in bioluminescent *Aequorea* jellyfish
1963	Lukosz & Marchand propose structured illumination to increase resolving power, forerunner of SIM
1964	Adam Curtis uses interference reflection microscopy to study cell-substrate interactions in living cells
1965	First commercial scanning electron microscope produced by the Cambridge Instrument Company
1965	First spinning disc confocal microscope produced by Milan Hadravský, Mojmír Petráň & David Egger
1965	James Russell invents the compact disc player – ultimately for data and multimedia storage
1966	Walter Lukosz achieves near-field imaging at three times Abbe diffraction limit, $\lambda/2$NA
1967	Johan (Bas) Ploem develops the dichromatic beam splitter and epi-fluorescence microscope
1968	First commercial CMOS integrated circuit developed by Albert Medwin et al. at RCA
1969	First working charge coupled device (CCD) camera produced by Willard Boyle and George E. Smith at Bell Labs
1969	Harris & Wallace publish their design for the first Acousto-Optic Tuneable Filter (AOTF)
1970	Maurer, Keck, and Schultz of Corning Glass develop the first successful glass fibre optic cable
1972	Ash & Nicholls break the Abbe diffraction limit with near-field imaging of 3 cm microwaves
1972	Calculation by K.M. Evenson et al. at the US National Bureau of Standards using laser interferometry of the most accurate measurement of the speed of light: $299{,}792{,}456.2 \pm 1.1$ m/s
1973	Fairchild Electronics produce the first 100×100 pixel uncooled commercial CCD chip
1974	Bähr et al. use fluorescence recovery after photobleaching (FRAP) to analysis of protein motility
1975	Robert Hoffman introduces the modulation contrast named after him
1975	César Milstein & Georges Köhler report the production of monoclonal antibodies
1975	Steve Sasson, at Kodak, develops the first digital camera based upon the Fairchild CCD
1976	Fernandez and Berlin use Förster resonance energy transfer (FRET) to study protein interactions
1977	The first mass-produced personal computer, the Apple II, introduced onto the market
1978	Colin Sheppard & Tony Wilson develop the moving-stage confocal microscope
1978	First multi-immersion objective (water, glycerol, oil) designed by Carl Zeiss
1981	Daniel Axelrod devises Total Internal Reflection Fluorescence (TIRF) microscopy
1981	Video-enhanced contrast (VEC) digital light microscopy developed separately by Inoué and Allen
1982	First commercial stage-scanning confocal microscope S-100 produced by Oxford Optoelectronics
1983	David Agard and John Sedat use deconvolution methods to view thick polytene nuclei sharply
1983	The metre is redefined against the speed of light: path travelled *in vacuo* as exactly 1/299,792,458 s
1984	Lewis et al. & Pohl et al. both develop near-field scanning optical microscopy using evanescent waves
1986	Horio & Hotani use dark-field microscopy to dynamically track individual microtubule assembly
1986	Development of the moving beam point-scanning confocal microscope by John White & Brad Amos
1988	First commercial point-scanning confocal microscope produced by BioRad, the Lasersharp MRC-500
1989	Cyanine fluorescent dyes introduced
1990	Multi-photon excitation microscopy invented by Denk, Strickler and Webb

1991	Tim Berners Lee uses hypertext to put up the first web page at CERN
1992	Eric Fossum invented the CMOS active pixel sensor (APS) technology at the NASA Jet Propulsion Lab
1993	Voie et al. use light-sheet microscopy (OPFOS) to investigate cleared cochleae
1993	Leica first use an Acousto-Optical Tuneable Filter (AOTF) on a confocal microscope
1994	Martin Chalfie et al. introduce Green Fluorescent Protein (GFP) as a fluorescent marker
1994	STED microscopy, the first technique to break the diffraction limit, introduced by Stefan Hell
1996	Tony Wilson et al. develop aperture correlation and structured illumination using a Ronchi grating
1997	Molecular Probes introduce Alexa fluorescent dyes
1998	Leica introduce the spectrally tuneable detector on the first SP range of confocal microscopes
1999	Matz et al. isolate non-bioluminescent red fluorescent proteins from *Anthozoa* coral
1999	Lene V. Hau et al. slow light to 17 m/s in an atomic sodium Bose-Einstein condensate
1999	Heintzmann & Cremer demonstrate resolution improvement by structured illumination (SIM)
2000	Mats Gustafsson publishes SIM method with Moiré gratings in the *Journal of Microscopy*
2000	ATTO fluorescent dyes introduced
2001	First commercial EM-CCD camera made available by Andor
2001	Zeiss introduce spectral unmixing META detector onto laser-scanning confocal microscopes
2002	Leica introduce the acousto-optic main beam splitter (AOBS) on confocals to replace static filters
2002	Quantum dots coated in micelles for use as *in-vivo* fluorescent markers
2002	Non-linear structured illumination, to circumvent diffraction barrier, described by Heintzmann et al.
2002	Photoactivatable GFP, later permits development of high-resolution methods STORM & PALM
2004	Stelzer et al. develop Selective Plane Illumination Microscopy (SPIM) for 3D *in-vivo* imaging
2004	Bio-Rad first introduce GaAsP detectors for confocal microscopes
2005	Dyomics introduce DyLight conjugated dyes
2006	Breaking the diffraction limit by stochastic methods – STORM, PALM, FPALM
2006	Interchim introduce FluoProbes dyes
2008	Nobel Prize for Chemistry shared by Chalfie, Shimomura and Tsien for work on GFP
2009	Commercial introduction of Hybrid APD detectors for research microscopes
2010	Scientific-grade CMOS cameras introduced by Hamamatsu, Andor, Fairchild and PCO
2011	Hama et al. develop the Scale clearing method
2011	Abberior dyes introduced for STED microscopy
2013	Deisseroth et al. publish the Clarity tissue-clearing method
2014	Nobel Prize for Chemistry shared by Betzig, Hell and Moerner for work on super-resolution
2014	Zeiss introduce the Airyscan for improved speed, resolution and signal-to-noise ratio
2015	Chen et al. publish the concept of expansion microscopy for super-resolution
2016	Roth et al. achieve super-concentration of étendue for circumventing the diffraction barrier
2017	Nobel prize for Chemistry shared by Dubochet, Frank and Henderson for work on cryo-electron microscopy
2018	Nobel prize for Physics awarded to Ashkin, Mourou and Strickland for optical tweezers & short optical pulsing

Index

All entries are given as page numbers. Definitions of words and concepts are listed as separate entries. The seven Appendices and an abridged list of the important tables, all the boxes and all the exercises are listed at the front of the book. Major entries for each subject are highlighted in bold where multiple pages are cited. Where endnotes at the end of each chapter are referred to, these are written in the format 610n.6. This refers to endnote 6 on page 610; where several endnotes on a single page are referred to, these are given in the form 287nn.6,9,10. A subject entry may be explained under a single Section heading within the relevant chapter. These sections are not referred to directly, but are cited as single page numbers for the beginning of the Section. The chapter summary pages are highlighted in bold font. Most definitions of terms occur in the Glossary and Definitions in the front matter of the book, for which lowercase roman numerals are used here to cite these entries.

Understanding Light Microscopy, First Edition. Jeremy Sanderson.

Poor Image Quality – fault-finding chart

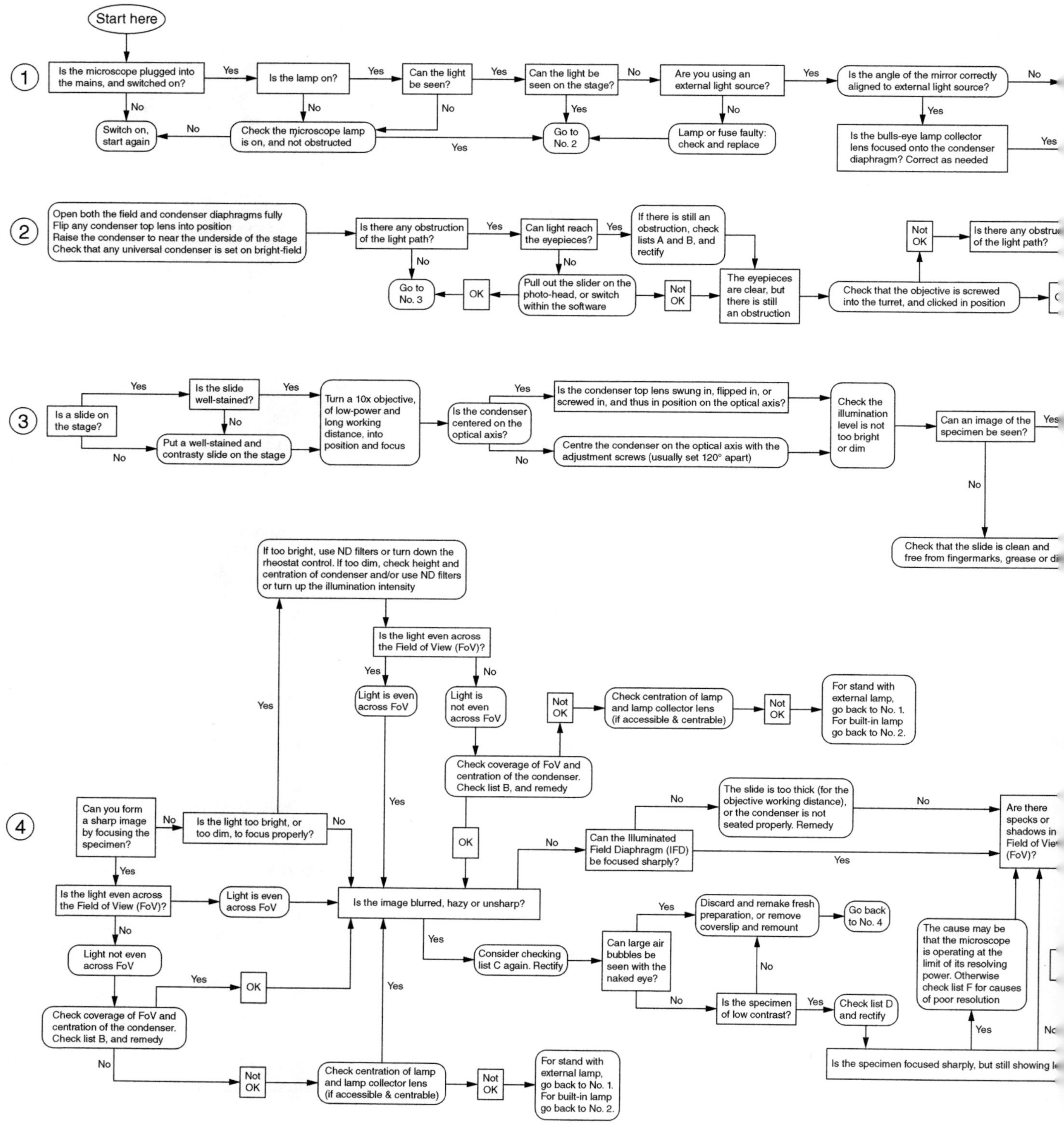

Checklist

Step ① – check the light source and illumination train: use list A.

Step ② – open field and condenser diaphragms: use lists A and B; check illumination and for any obstructions.

Step ③ – obtain an image: use lists C, D and E to check image quality.

Step ④ – further check on image quality: also use lists F and G to obtain a good quality image.

Finish – the microscope is now properly set up and adjusted correctly for Köhler illumination.

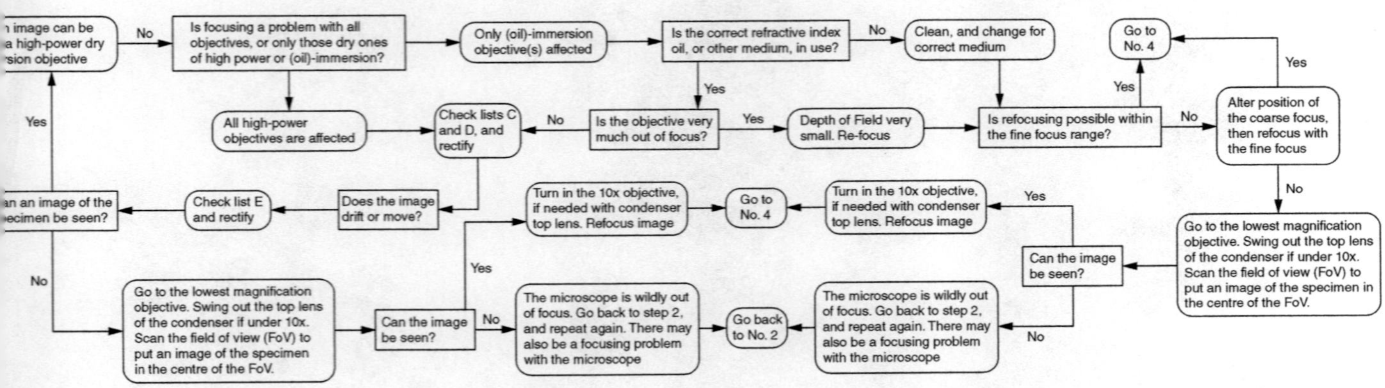

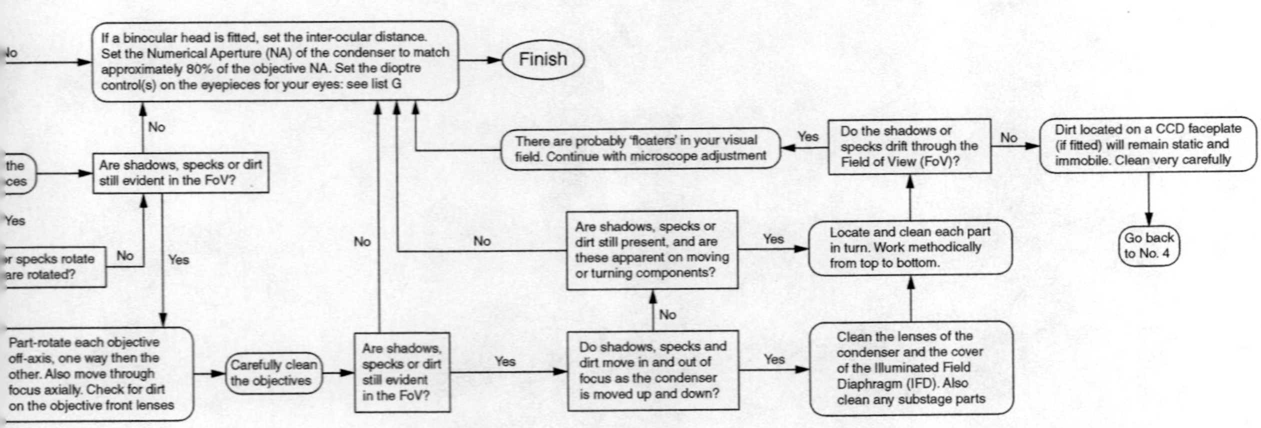

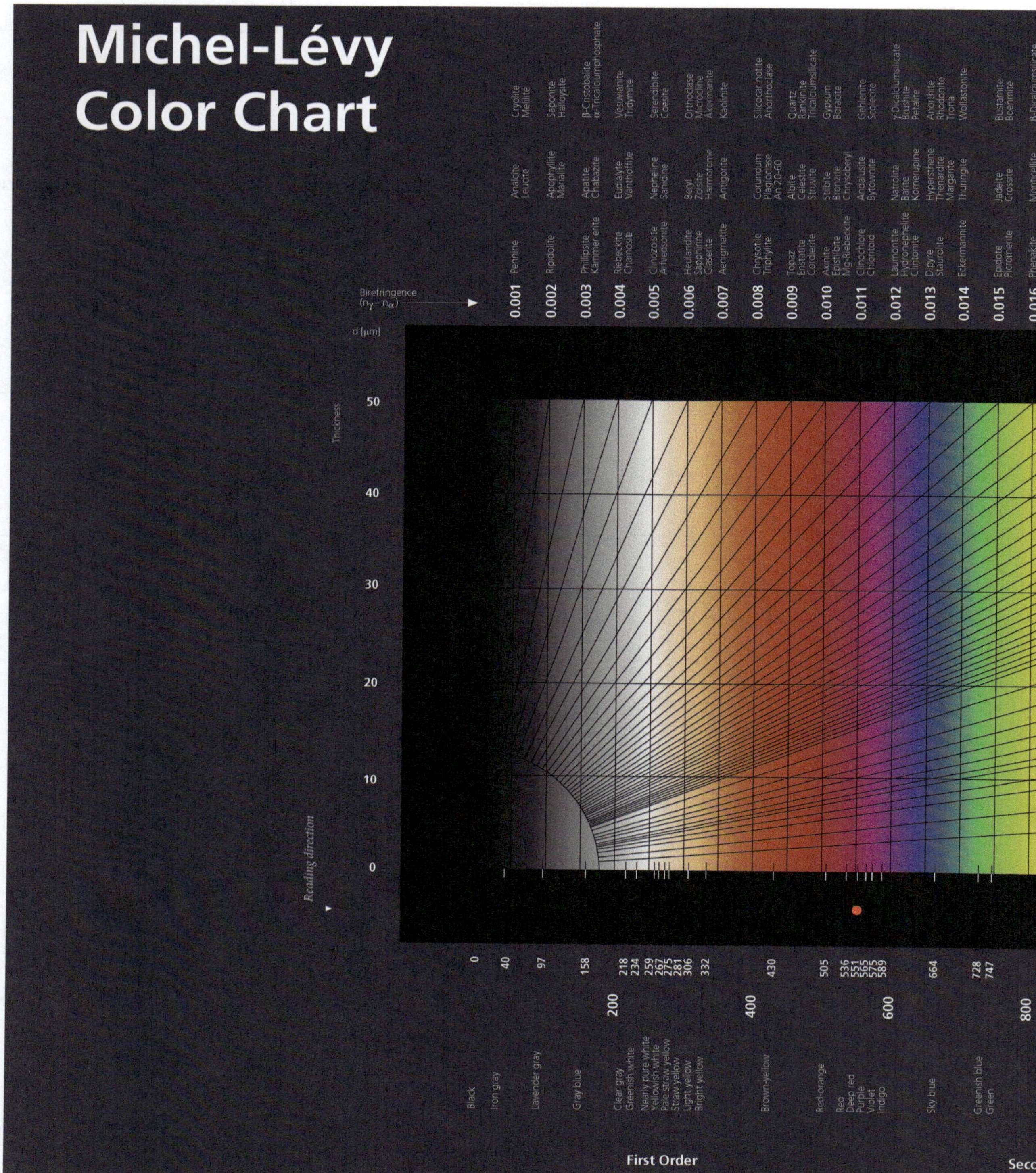

Carl Zeiss Microscopy GmbH
microscopy@zeiss.com
www.zeiss.com/microscopy

Source: Reproduced with permission from Carl Zeiss Ltd.